Abridged Table of Atomic and Molar Masses of the Elements*

ELEMENT	SYMBOL	ATOMIC MASS(u) MOLAR MASS(g mol⁻¹)	ELEMENT	SYMBOL	ATOMIC MASS(u) MOLAR MASS(g mol⁻¹)
Aluminum	Al	26.98	Lithium	Li	6.941
Argon	Ar	39.95	Magnesium	Mg	24.30
Arsenic	As	74.92	Manganese	Mn	54.94
Barium	Ba	137.3	Mercury	Hg	200.6
Beryllium	Be	9.012	Neon	Ne	20.18
Boron	B	10.81	Nickel	Ni	58.69
Bromine	Br	79.90	Nitrogen	N	14.01
Calcium	Ca	40.08	Oxygen	O	16.00
Carbon	C	12.01	Phosphorus	P	30.97
Cesium	Cs	132.9	Potassium	K	39.10
Chlorine	Cl	35.45	Rubidium	Rb	85.47
Chromium	Cr	52.00	Silicon	Si	28.09
Cobalt	Co	58.93	Silver	Ag	107.9
Copper	Cu	63.55	Sodium	Na	22.99
Fluorine	F	19.00	Strontium	Sr	87.62
Gold	Au	197.0	Sulfur	S	32.07
Helium	He	4.003	Tin	Sn	118.7
Hydrogen	H	1.008	Titanium	Ti	47.88
Iodine	I	126.9	Tungsten	W	183.8
Iron	Fe	55.85	Vanadium	V	50.94
Krypton	Kr	83.80	Xenon	Xe	131.3
Lead	Pb	207.2	Zinc	Zn	65.39

*The values in this table are to four significant figures.

A complete table is given in Appendix B.

CHEMISTRY

CHEMISTRY

SECOND EDITION

Ronald J. Gillespie
McMASTER UNIVERSITY

David A. Humphreys
McMASTER UNIVERSITY

N. Colin Baird
UNIVERSITY OF WESTERN ONTARIO

Edward A. Robinson
UNIVERSITY OF TORONTO

ALLYN AND BACON
Boston
London
Sydney
Toronto

To

Madge, Vivienne, and Jennifer

for their encouragement, support, and understanding

Library of Congress Cataloging-in-Publication Data

Chemistry.

Includes index.
1. Chemistry. I. Gillespie, Ronald J. (Ronald
James)
QD31.2.C39 1988   540   88-7920
ISBN 0-205-11795-3

Printed in the United States of America

10 9 8 7 6 5 4 3 2 1   93 92 91 90 89

Editorial-Production Administrator: Elaine Ober
Composition Buyer: Linda Cox
Manufacturing Buyer: Joe Sita
Cover Coordinator: Linda Dickinson
Cover Designer: Lynda Fishbourne
Interior Designer: Deborah Schneck
Experiment Photographer: Tom Bochsler, Photography
   Limited

CREDITS

All photographs by Tom Bochsler except the following:
CHAPTER 1: Opening photo, page 1: Tom Algire/TOM STACK
AND ASSOCIATES. Figure 1.1, page 9: Courtesy of Texas
Instruments. Figure 1.3, page 9: (left) Grant Heilman Photography; (right) John Gerlach/TOM STACK AND ASSOCIATES.
Figure 1.5, page 11: Grant Heilman Photography/Runk/Schoen-
berger. Figure 1.6, page 13: Courtesy of General Electric Re-
search and Development Center. Page 19: Bettman Archive.
CHAPTER 2: Opening photo, page 49: Sims/Boynton Photog-
raphy. Figure 2.1, page 51, Box 2.2, page 55, and Figure 2.4,
page 63: E.F. Smith Memorial Collection, Center for History of
Chemistry, University of Pennsylvania.   CHAPTER 3: Opening photo, page 99 and Box 3.4, page 121: Gerald Corsi/TOM
STACK AND ASSOCIATES. Figure 3.2, page 103: E.F. Smith
Memorial Collection, Center for History of Chemistry, University
of Pennsylvania.   CHAPTER 4: Opening photo, page 151:
*(Credits continue on page vi, which constitutes a continu-
ation of the copyright page.)*

Brief Contents

Contents

Experiments

To The Instructor

AIMS AND OBJECTIVES

We have written this book not only to help beginning students acquire a basic knowledge and understanding of chemistry, but also to help you make your course as stimulating, interesting, and relevant as possible. But why yet another general chemistry text when there are already several books that deal with the "principles of chemistry" in a competent manner? Because *this* book not only gives a thorough treatment of the principles but also brings descriptive chemistry back into the course in a new and exciting way—a way that makes it easier for students to learn, more interesting for instructors to teach, and that adds interest and significance to the discussion of principles.

DESCRIPTIVE CHEMISTRY

There has been much discussion in recent years concerning the role of descriptive chemistry in the general chemistry course and rather general agreement that students should acquire some basic knowledge of descriptive chemistry, that is, of the properties and reactions of the elements and their compounds. It is unfortunate that general chemistry has been artificially divided into "principles of chemistry" and "descriptive chemistry". These are not two separate subjects, but rather two aspects of chemistry that are intimately related and interdependent. Not only have these two aspects of chemistry been artificially separated but, as increasing emphasis has been given to the principles and theories, descriptive chemistry has been increasingly neglected. This artificial separation between theory and descriptive chemistry has never been made in the introductory organic chemistry course, where principles and theories have never replaced the description of organic reactions. Why, then, should principles be separated from the description of properties and reactions in the general chemistry course?

Descriptive chemistry has acquired an undeserved reputation for being dull and boring. And yet the search for new compounds, new structures, and new properties continues to inspire the efforts of large numbers of chemists. We have tried through our text to transmit to students the fascination of the infinite variety of properties and reactions, in other words the "descriptive chemistry", that inspires many researchers.

AN INTEGRATED APPROACH

The general chemistry course should be based on properties and reactions. The realization that students have in the past acquired little, if any, knowledge of even the simplest descriptive chemistry has led to a demand for its return to the general chemistry curriculum. But it is not satisfactory simply to *add* descriptive chemistry to the end of an already overloaded principles course. It should be integrated into the course and the artificial separation into principles and descriptive chemistry abandoned. This text provides a fully integrated approach.

Chemistry, Second Edition, begins with some simple descriptive chemistry such as the composition of the atmosphere and the formation of oxides, which will be familiar to students from previous science courses or from everyday experience. We use this descriptive chemistry to introduce some basic principles, and then use these principles to discuss more reaction chemistry. In discussing principles we use, as far as possible, only those facts of descriptive chemistry that have already been introduced. We often return to these same facts in several different contexts, reinforcing the student's knowledge of these facts and underscoring their importance.

Some chapters have descriptive titles because they emphasize the reactions and properties of one or more elements. These chapters are not concerned only with descriptive chemistry; rather the descriptive material is used to introduce theories and principles. Other chapters have titles that indicate that they are primarily concerned with principles, such as equilibrium, thermodynamics, and reaction rates. These chapters use illustrative reactions taken from the descriptive material already introduced.

A qualitative discussion of acid–base, oxidation–reduction, and precipitation reactions is given early (Chapter 5) and these important reaction types are referred to frequently in the subsequent chapters. The student is thus thoroughly prepared for the quantitative treatment of these reaction types in Chapters 14, 15, and 17.

We continually make the student aware that theories and principles are developed to explain observations and that it is important to have a knowledge of facts as well as of theories. Descriptive chemistry is made easier to understand and remember because it is presented in small doses together with the appropriate theory. At the same time the discussion of theories and principles is made less abstract because it is associated with the discussion and demonstration of properties and reactions. This integration of descriptive chemistry with theory facilitates the efficient presentation of both, enabling the text to deal with all the theories and principles normally covered in a first-year course, while including somewhat more reaction chemistry than is usually possible.

This approach, coupled with lecture demonstrations and laboratory work, provides the instructor with a proven method for introducing students to descriptive chemistry in a manner that is interesting to teach and easy for students to learn. In our experience, and in the experience of many who used the first edition, students respond favorably to this type of course as they begin to appreciate that chemistry is concerned with substances and their reactions and properties and not only with solving principles-based problems.

CHANGES FROM THE FIRST EDITION

We have changed the order of certain chapters to reflect the constructive criticisms that we have received from many enthusiastic users. In particular we have moved forward the quantum mechanical treatment of chemical bonding (Chapter 7) and the discussion of thermodynamics (Chapter 6) that constituted the last two chapters of the first edition to those positions where most instructors will use them. We have also moved forward the discussion of thermochemistry together with the inorganic part of the chemistry of carbon (Chapter 6). This has the advantage of introducing more quantitative work in the first half of the course. At the same time the discussion of hydrocarbons has been combined with the chemistry of other organic compounds in Chapter 23.

In addition to increasing the number of worked examples we have followed many of the worked examples with a straightforward exercise of the same type which will enable students to test their understanding. Answers to these exercises are provided. Some of the end-of-chapter problems have been replaced and many new ones have been added, so that the total number is now approximately 1500. The new problems include many relating to reaction chemistry that will help students to increase their knowledge and understanding of reaction chemistry, improve their ability to recognize the main reaction types, and make reasonable, if not always completely correct, predictions of the products of reactions between common substances.

LECTURE DEMONSTRATIONS

A special feature of the first edition that we have retained and expanded is the use of illustrations of demonstration experiments—sequences of two to four color photos illustrating reactions and other properties of substances. Descriptive chemistry can only be fully appreciated by direct observation of reactions in the laboratory and in lecture demonstrations. Detailed instructions for performing the lecture demonstrations are given in the *Instructor's Manual*. The photos of the demonstrations in the text bring reality to the written word and stimulate student interest. Although we feel that the importance of lecture demonstrations cannot be overemphasized, we appreciate that not all instructors have the facilities or the assistance needed to perform these demonstrations. So we have made a *videotape* of all the demonstration experiments which can be shown on monitors or on a large screen in the classroom. A live demonstration, or a videotape of it, adds interest to the lecture, focuses the interest of the students, and stimulates questions and discussions. Each experiment on the tape takes only two to three minutes so one or two will easily fit into a lecture. In some cases the videotape can show a close-up view that is superior to the live demonstration and it is often valuable to show the tape as well as do the live demonstration. The videotape is available free to instructors who adopt the text for their class.

SUPPLEMENTS

The ancillary items for Chemistry have been specially designed to complement the text.

Study Guide: Provides a summary review, review questions, learning objectives, problem-solving strategies with worked examples, and a self-test for each text chapter.

Test Bank: Provides over 1500 multiple-choice questions. Available in printed or disk format.

Videotape: Shows every demonstration experiment in full color with voice commentary.

Solutions Manual: Provides detailed solutions to all end-of-chapter problems.

Laboratory Manual: Offers an appropriate collection of laboratory experiments to complement the text material. An Instructor's Manual for the lab is also available.

Instructor's Manual: Provides instructions and cautionary notes on performing the demonstration experiments, commentary notes on each chapter, and an overview of topic organization.

Transparencies: Full-color transparencies of many of the illustrations and figures are available.

We hope that you will enjoy teaching from this text and that your students will find it interesting, easy to read, and easy to learn from. We would very much appreciate your comments; they will greatly assist us to develop and improve our text. Only by continued discussion among all those concerned with the general chemistry curriculum can we ensure that it reflects modern chemistry while continuing to meet the needs of students who will be pursuing their studies in a variety of fields. We have worked hard to make our text as error free as possible but no doubt some errors have remained undetected. Please draw our attention to any that you may discover.

ACKNOWLEDGEMENTS

The following chemistry teachers reviewed the second edition manuscript at various stages in its development or participated in a focus session prior to revision. We are grateful to them for their criticisms and suggestions.

John M. Bellama, *University of Maryland*
Michael Brayditch, *U.S. Air Force Academy*
Arthur Breyer, *Beaver College*
Dewey Carpenter, *Louisiana State University*
Wilbert Hutton, *Iowa State University*
Robert Kerber, *SUNY, Stony Brook*
Jerry Mills, *Texas Tech University*
Joseph Morse, *Utah State University*
Lyle Peter, *Seattle Pacific University*
Allan Pinhas, *University of Cincinnati*

It is a pleasure to acknowledge the enthusiastic support and assistance given to us by the staff at Allyn and Bacon. We especially wish to thank Jim Smith, whose continued support and enthusiasm for our project has enabled us to carry it through to completion; Elaine Ober, who led us so competently through the intricacies and deadlines of production; and Judy Fiske, Production Director.

We also owe a debt of gratitude to Tom Bochsler, who took the outstanding color photos that illustrate the demonstration experiments—his patience and fortitude under sometimes trying circumstances were greatly appreciated.

To The Student

WHY STUDY CHEMISTRY?

Students take the first chemistry course for many reasons; some because they have already discovered the fascination of chemistry and some, perhaps reluctantly, because it is a requirement for their chosen field of study. Whatever your goal, we have written this book for you. Our primary aim is to help you succeed in the course, and to provide you with the background knowledge and understanding of chemistry that is required in many other science courses. We also hope that we will develop your interest in chemistry, and that you will come to understand how important a knowledge of chemistry is in many fields of science and in everyday life. No one can afford to be ignorant about chemistry because so many contemporary problems in fields as diverse as environmental pollution, alternative energy sources, health, drugs, medicine and nutrition that we hear about almost daily in the media, require some knowledge of chemistry for their understanding and solution.

WHAT IS CHEMISTRY?

The material world consists of an enormous variety of substances. Chemistry is concerned with the nature of these substances and with the changes that they undergo when they interact with each other. We call these changes chemical reactions. Whether we are concerned with living systems, as we are in biology and medicine; with materials such as iron, steel, and concrete, as we are in engineering; or with the manufacture of integrated circuits on silicon chips for computers, we are dealing with substances and therefore with chemistry. In today's technological age the applications of chemistry are becoming ever more important and diverse.

The most important idea in chemistry is that matter consists of enormous numbers of incredibly small atoms. Elements have only one type of atom but most substances are compounds that contain two or more different types of atom. The remarkable transformations that accompany chemical reactions—such as the burning of gasoline in air in an automobile engine to give carbon dioxide and water or the conversion of carbon dioxide and water into the complex molecules from which plants are composed in the process of photosynthesis—are simply a consequence of the reshuffling of the atoms of the reacting substances to form new combinations. Making new materials by combining atoms in new ways—whether they are for manufacturing integrated circuits for computers, building new space vehicles, or for making new types of batteries, new fuels, or new drugs—is one of the primary activities of chemists.

Chemistry is not just a collection of a large number of experimental observations—facts—about substances and their reactions. Chemists are constantly striving to understand the facts better in terms of general principles and theories. So the facts about substances and their reactions have been organized into a coherent body of knowledge by the principles and theories that have been developed over the years.

HOW TO STUDY AND UNDERSTAND CHEMISTRY

Chemistry may at first appear to be a difficult subject because it deals with so many different substances and the reactions between them. But in fact the vast majority of the facts about substances can be understood in terms of a rather small number of relatively simple basic concepts. It is important right from the beginning to start thinking as a chemist does, and always to relate the properties of substances that we observe to the atoms of which they are composed, how these atoms are combined, and how they are rearranged in reactions. Keep in mind too that although there are very many substances there are many fewer different types of substances and although there are many different reactions there are many fewer *types* of reactions. So you should learn the characteristic properties of different *types* of substances—metals, nonmetals, ionic salts etc, and you should learn to recognise the important types of reactions—acid–base, oxidation–reduction etc.

HOW TO USE THIS BOOK

First read each chapter or each section fairly quickly to get an overall impression. Then reread the chapter or the section more slowly and carefully making sure that you understand everything. Highlight the text or make notes in the margin and/or in a separate notebook as you go along. Study the *worked examples* and then try the *exercises* which usually follow. The *answers to the exercises* are given in the back of the text. If you do not get the correct answer, go back and study the appropriate section of the text and the worked examples again. Try the worked examples without looking at the solution and then try the exercise again. If you still do not get it right, seek help from your professor or instructor or another student but do not give up.

At the end of each chapter we list the important **key terms** that have been introduced in the chapter and we give a definition of each. These key terms constitute the basic vocabulary of chemistry. When you have studied a chapter test yourself by listing the terms and then close the book and write out a definition of each term. Remember to use the **index** to find further references to a given topic or word that may help you to understand it better.

When you have studied a chapter, work some of the **problems** at the end of the chapter. **Answers** to about half the problems are given at the end of the book. If your answer is not correct read the appropriate section again and try the problem again. If you still do not get the correct answer, seek help. Problems will also be assigned by your instructor and answers will be provided and discussed with you.

Because people remember most easily those things they have experienced directly, color photos of approximately 100 **demonstration experiments** of the properties of substances and their reactions are an important feature of this book. Your instructor may perform these demonstrations in class or show them on a **videotape** that has been specially prepared to accompany this book. Watching the demonstrations carefully and studying the photos in the book will help you to remember many important facts about substances and their reactions.

HOW TO SUCCEED IN THIS COURSE

The two secrets to success in this course, as in many other endeavors, are understanding and systematic work. Do not sit passively in the lecture or read the text as you might casually read a magazine, hoping that the material presented will miraculously be recorded in your memory. Studying chemistry is not a spectator sport! If you do not understand what is being presented in the lecture or in the textbook, seek help from your instructor or fellow students. Take some lecture notes but don't try to understand the material as it is presented. Study your notes as soon as possible after the lecture in conjunction with the text. Correct or rewrite your notes as necessary and make a summary in your own words. A certain amount of material must be memorized, but remember that understanding, not memory, is the key to success.

A **Study Guide** is available if you should feel that you need extra help. It includes a summary of each chapter as well as a self-test, strategies for problem solving, and additional problems.

We wish you every success in this course. If you study the text carefully and follow our recommendations for studying you will do well.

If you find any mistakes while you read the book, or have any suggestions for making future editions of the book easier for students to use, or if you have any questions, or you feel that our book has been particularly helpful to you and you would like to recommend it to other students, please write to us at Allyn and Bacon, 160 Gould Street, Needham Heights MA 02194-2310, U.S.A. We will answer every letter that we receive.

CHEMISTRY

Structure
of Matter

1

The world in which we live is one of infinite variety and continual change. We inhabit the earth along with the trees, plants, fish, birds, and all other forms of life. All depend for their existence on the multitude of different materials that compose the solid surface of the earth, the water in the rivers, lakes, and oceans, and the gases that form the atmosphere. Ever since we learned to communicate by speech and by writing, we have attempted to describe the natural world of which we are a part: its form, its variety, its beauty, and its continual changes. Change is, perhaps, the most striking feature of the natural world. Mountains are worn down by the action of water and of the atmosphere on the rocks that compose them. New mountains are formed as a consequence of changes occurring beneath the surface of the earth. Leaves turn red in the fall, iron rusts, and wood burns. We are born, we breathe—inhaling oxygen and exhaling carbon dioxide—we grow, and we die.

Most of our ideas concerning the nature of the material world have developed over the past two hundred years as a result of careful systematic observations and experiments. Such studies of materials and their transformations constitute the modern science of chemistry. As our understanding of the transformations of materials has increased, we have learned to make many changes in naturally occurring substances. We extract iron ore from the ground and convert it to steel; we use steel to build drilling rigs to obtain petroleum from the ground; petroleum in turn is transformed into gasoline, plastics, and drugs. Indeed, during the past hundred years chemists have transformed the material world by creating a multitude of new substances, from synthetic fibers to antibiotics.

To the beginning student chemistry may sometimes seem to be an overwhelming jungle of formulas, equations, and theories. Remember, however, that behind all the formulas and theories lies the real world, the fascinating world of chemistry, in which we study the materials making up the earth and, in fact, the whole universe. All the formulas, equations, and theories that form part of modern chemistry have been invented by chemists to help us understand the real world around us: its multitude of different substances, both living and nonliving, and the extraordinary variety of changes they undergo.

1.1
Composition of Matter

We can describe the universe, and all the changes occurring in it, in terms of two fundamental concepts: **matter** and **energy**. Matter is anything that occupies space and has mass. Water, air, rocks, and petroleum, for example, are matter, but heat and light are not; they are forms of energy. If a sample of matter has the same properties throughout it is said to be **homogeneous**; if it consists of parts with different properties it is said to be **heterogeneous**. The many different kinds of homogeneous matter are called **substances**. Chemists are concerned with determining the composition and structure of substances, with finding out how the properties of substances depend on their composition and structure, and with understanding the changes substances undergo. *The transformation of one or more substances into one or more different substances is called a* **chemical reaction**. When iron rusts, it is transformed into iron oxide. When gasoline burns, it is transformed into carbon dioxide and water. These transformations are chemical reactions. Thus **chemistry** *is the study of the composition, structure, and properties of substances and the transformations (called chemical reactions) by which substances are changed into other substances.*

The scope of chemistry is very broad; in fact, it overlaps with all the other natural sciences. The biologist studies the substances in living organisms; the geologist is concerned with the rocks and minerals that compose the earth; the astrophysicist is interested in the substances in the stars and interstellar space. This overlap between chemistry and other sciences has led to the development of many interdisciplinary fields such as biochemistry (the chemistry of living matter), geochemistry (the chemistry of rocks and minerals), and cosmochemistry (the chemistry of the planets, the stars, and interstellar space). Thus the study of chemistry is important not only for its own sake but also for a full understanding of many other sciences.

ELEMENTS AND COMPOUNDS

Some of the earliest attempts to understand the natural world were made over two thousand years ago by the ancient Greeks, who proposed that everything in the world was composed of four *elements*, or basic substances, which combined to form rocks, plants, clouds, sunlight and all the other components of the universe. They identified these elements as earth, water, fire, and air. Sunlight, for example, appeared to them to be a mixture of fire and air. Ice, they thought, was water plus the hardness of earth.

Although these ideas might seem rather naive and even amusing to us today, they made several valuable contributions to our understanding of the world. One contribution was an awareness that the different states that a substance can take are important: earth is a *solid*; water is a *liquid*; air is a *gas*. Another contribution was their recognition that fire, or *energy*, is important in the changes that substances undergo. For the moment let us focus on the important idea that even the most complex substances are made up of basic components called elements.

The concept of an **element** has changed over the centuries. With the development of the science of chemistry an element came to mean a substance that is not composed of simpler substances. Although, as we shall see shortly, we can today give a more precise definition, it is clear that earth does not qualify as an element; even a superficial look shows that it is composed of many different substances such as sand, clay, rock, and decaying plant material. Nor does water qualify as an element. If we pass an electric current through water, we can separate it into the gases hydrogen and oxygen (Experiment 1.1, p. 4). But nobody has succeeded in breaking down hydrogen or oxygen into other substances by chemical reactions. They therefore are considered to be elements, whereas the substances, such as water, that are formed when two or more elements combine, are known as **compounds**. Today we recognize just over a hundred elements (Table 1.1, p. 5).

It may surprise you to learn that the number of elements known today cannot be stated with complete certainty. In recent years new elements have been made by nuclear, not chemical, reactions, and in some cases only a few atoms have been made. In such cases establishing the identity of the atoms with complete certainty is sometimes difficult. Table 1.1 lists 103 elements that are

Experiment 1.1

Decomposition of Water by an Electric Current

Water can be decomposed into the elements hydrogen and oxygen by passing an electric current through it. Bubbles of hydrogen can be seen rising from the wire at the left and bubbles of oxygen from the wire at the right. The volume of hydrogen produced is twice the volume of oxygen. A few drops of sulfuric acid have been added to the water to increase the electrical conductivity.

known with complete certainty, and there are probably at least 4 more. Others are likely to be synthesized in the future. We will discuss only about 40 of the elements in this book.

ATOMS

What is the difference between elements and compounds? Once again, some of the ancient Greek philosophers provided the germ of the idea that is our present-day answer to this question. Although Plato and Aristotle believed that matter was continuous, Democritus argued that matter was composed of very small indivisible particles. However, not until the beginning of the nineteenth century did the Englishman John Dalton (Box 1.1, p. 6) show that this idea could form the basis for understanding the nature of elements and compounds and the transformations that they undergo. Dalton called the fundamental particles of which matter is composed **atoms**, from the Greek word *atomos*, meaning "indivisible". He also made the following propositions:

· The atoms of any given element are identical.
· The atoms of one element are different from those of another element.
· Atoms of two or more elements may combine in definite ratios to form compounds.
· Atoms remain unchanged in chemical reactions.

We can therefore define an **element** as *a substance that contains only one kind of atom*. A **compound** is *a substance that contains two or more kinds of atoms combined in fixed proportions*. For example, the element oxygen contains only oxygen atoms, and the element hydrogen contains only hydrogen atoms. But the compound water contains both hydrogen and oxygen atoms combined in the ratio of two hydrogen atoms for every one oxygen atom. We have seen that water can be broken down into hydrogen and oxygen; in other words, the hydrogen atoms can be separated from the oxygen atoms. Not surprisingly, we can also combine hydrogen and oxygen to form water (Experiment 1.2, p. 7).

TABLE 1.1
The Elements

Element	Symbol	Element	Symbol	Element	Symbol
Actinium	Ac	Hafnium	Hf	Promethium	Pm
Aluminum	**Al**	**Helium**	He	Protactinium	Pa
Americium	Am	Holmium	Ho	Radium	Ra
Antimony	Sb	**Hydrogen**	**H**	Radon	Rn
Argon	**Ar**	Indium	In	Rhenium	Re
Arsenic	As	**Iodine**	**I**	Rhodium	Rh
Astatine	At	Iridium	Ir	**Rubidium**	**Rb**
Barium	**Ba**	**Iron**	Fe	Ruthenium	Ru
Berkelium	Bk	**Krypton**	Kr	Samarium	Sm
Beryllium	Be	Lanthanum	La	Scandium	Sc
Bismuth	Bi	Lawrencium	Lr	Selenium	Se
Boron	**B**	**Lead**	Pb	**Silicon**	Si
Bromine	**Br**	**Lithium**	Li	**Silver**	Ag
Cadmium	Cd	Lutetium	Lu	**Sodium**	Na
Calcium	**Ca**	**Magnesium**	**Mg**	**Strontium**	Sr
Californium	Cf	**Manganese**	**Mn**	**Sulfur**	S
Carbon	**C**	Mendelevium	Md	Tantalum	Ta
Cerium	Ce	**Mercury**	Hg	Technetium	Tc
Cesium	**Cs**	Molybdenum	Mo	Tellurium	Te
Chlorine	**Cl**	Neodymium	Nd	Terbium	Tb
Chromium	**Cr**	**Neon**	Ne	Thallium	Tl
Cobalt	**Co**	Neptunium	Np	Thorium	Th
Copper	**Cu**	**Nickel**	Ni	Thulium	Tm
Curium	Cm	Niobium	Nb	**Tin**	Sn
Dysprosium	Dy	**Nitrogen**	**N**	**Titanium**	Ti
Einsteinium	Es	Nobelium	No	Tungsten	W
Erbium	Er	Osmium	Os	Uranium	U
Europium	Eu	**Oxygen**	**O**	**Vanadium**	V
Fermium	Fm	Palladium	Pd	**Xenon**	Xe
Fluorine	**F**	**Phosphorus**	**P**	Ytterbium	Yb
Francium	Fr	Platinum	Pt	Yttrium	Y
Gadolinium	Gd	Plutonium	Pu	**Zinc**	Zn
Gallium	Ga	Polonium	Po	Zirconium	Zr
Germanium	Ge	**Potassium**	**K**		
Gold	**Au**	Praseodymium	Pr		

Note: The elements discussed in this book are in bold type.

Both the decomposition of water to hydrogen and oxygen and the combination of hydrogen and oxygen to give water are chemical reactions. (The arrow denotes a chemical reaction.)

$$\text{Hydrogen} + \text{Oxygen} \longrightarrow \text{Water}$$

and

$$\text{Water} \longrightarrow \text{Hydrogen} + \text{Oxygen}$$

*A **chemical reaction** takes place when atoms are rearranged from their original combinations to form new combinations, the atoms themselves remaining unchanged.*

The silvery white metal magnesium is an element. When it is heated in air, it burns with a brilliant white light, combining with the oxygen in the air to form a white powder, which is the compound magnesium oxide (Experiment 1.3).

BOX 1.1

JOHN DALTON (1766–1844)

John Dalton, the son of a poor weaver, was born in Cumberland, England. He first studied at a village school, and he made such rapid progress that by the age of twelve he became the teacher at the school. Seven years later he became a school principal. In 1793 he moved to Manchester, and he remained there for the rest of his life. He first taught mathematics, physics, and chemistry at a college. But finding that his teaching duties interfered with his scientific studies, he resigned from this post and supported himself by teaching mathematics and chemistry to private pupils. He never married, and he con-

tinued to live a very simple and modest life even after he became famous. Dalton's first scientific investigations were in meteorology, and he continued throughout his life to make daily observations on the temperature, barometric pressure, and rainfall. He described the nature of color blindness, of which he was a victim. A devout Quaker, Dalton always wore plain and somber clothes. Friends were therefore surprised when he wore a scarlet academic robe on being presented to King William IV in 1832. To Dalton, however, it appeared a dull gray, and he wore it without concern.

Dalton put forward his atomic theory in 1803. Although he proposed that compounds were formed by the combination of atoms of different elements in small, whole-number ratios, he had no certain method for determining the ratios in which the different atoms combine. He assumed therefore that, when only one compound of two elements A and B was known, it had the simplest possible formula, AB. On the basis of this assumption and from the masses of different elements that were found to combine, he was able to deduce relative atomic masses. He published the first table of such relative atomic masses. His assumptions about formulas of compounds, however, were not always valid. For example, he assumed that the formula of water was HO, which led to some of the atomic masses in his table being incorrect. In fact, not until 1858 did chemists solve the problem of the correct determination of molecular formulas and therefore of atomic masses. Nevertheless, the credit must go to Dalton for first putting the atomic theory on a quantitative basis and for laying the foundation for the rapid development of chemistry that followed.

Magnesium also burns in steam and in carbon dioxide. When magnesium reacts with carbon dioxide, the magnesium atoms combine with the oxygen atoms of carbon dioxide to form magnesium oxide, leaving the element carbon, which consists only of carbon atoms. In this reaction we note a very important feature of chemical reactions: they are always accompanied by energy changes. The burning of magnesium is accompanied by the emission of heat and light.

The many experiments carried out since Dalton put forward his ideas have fully confirmed that matter consists of atoms. Dalton's proposals form the basis of what we now call the **atomic theory**. We are so used to the idea of atoms today that we cannot imagine that people ever had other ideas about the composition of matter. In the present century, however, scientists have shown that, contrary to what Dalton believed, atoms are not indivisible in all circumstances.

Experiment 1.2

Reaction of Hydrogen with Oxygen

Hydrogen burns in the air with a pale blue, almost colorless flame, combining with the oxygen of the air to form water. The water can be observed condensing on the outside of the cold flask. Here the flame is colored yellow by the sodium in the glass tube.

Rather, each atom consists of still smaller particles. However, in chemical reactions atoms are not split up into their constituent particles; they remain unchanged but are rearranged to form new substances. As we shall see later, it is the structure of atoms that determines how two or more elements may combine to form a compound.

 Experiment 1.3

Reactions of Magnesium

Magnesium burning in air. The white powder in the watch glass and the white smoke are magnesium oxide.

Magnesium burns in substances that normally extinguish flames. Here we see it burning in a flask of boiling water. White magnesium oxide is produced in a different reaction.

Magnesium also burns in carbon dioxide, giving white magnesium oxide and carbon. The carbon can be seen as black specks on the sides of the flask.

SYMBOLS AND FORMULAS

To simplify the representation of elements and compounds, chemists have agreed on an international set of symbols. The symbol for a given element nearly always consists of the first letter of its English name, frequently followed by one other letter (see Table 1.1). For some elements an abbreviation of the name in another language, usually Latin, is used. For example, sodium has the symbol Na, from the Latin *natrium*.

A compound is represented by a formula that indicates the elements that it contains and the relative number of atoms of each element. Water is represented by the formula H_2O because it contains two hydrogen atoms for every oxygen atom. *Methane*, a major component of natural gas, is a compound of the elements carbon and hydrogen and contains four hydrogen atoms for every carbon atom; it has the formula CH_4. Common table salt, *sodium chloride*, contains equal numbers of sodium and chlorine atoms and is represented by the formula NaCl. Carbon and oxygen atoms combine in a 1:2 ratio to form the compound carbon dioxide, which makes up a very small part of the earth's atmosphere. *Carbon dioxide* is represented by the formula CO_2. Carbon and oxygen atoms also combine in a 1:1 ratio to form *carbon monoxide*, a poisonous gas that is present in automobile exhaust; it has the formula CO. Thus two elements can form more than one compound. Many compounds contain more than two elements; for example, *sulfuric acid*, H_2SO_4, contains hydrogen, oxygen, and sulfur. Most compounds, however, contain only a few different elements.

SUBSTANCES AND MIXTURES

All the different materials that we recognize around us either are mixtures of two or more substances or are single substances. Every substance is either an element or a compound. Magnesium is an example of a substance that is an element. Water and sodium chloride are examples of substances that are compounds. Seawater, however, is a mixture of water, sodium chloride, and many other substances.

Every substance has a unique set of properties that allows us to distinguish it from all other substances. We recognize two different types of properties: physical properties and chemical properties. *The **physical properties** of a substance are those properties that can be observed and measured without the substance changing into other substances.* These properties include, for example, physical state at room temperature, melting or freezing point, boiling point, color, solubility in water, and electrical conductivity. Some physical properties of water are that it is a liquid at room temperature, that it freezes at 0 °C and boils at 100 °C, that it is colorless, and that it dissolves sugar, salt, and many other substances. *The **chemical properties** of a substance are those properties that relate to its participation in chemical reactions.* Thus, iron has the property of rusting when it is exposed to air and water. Water has the property of decomposing to hydrogen and oxygen when an electric current is passed through it.

A very important property of a substance is that it has a constant **composition**. If it is an element it contains only one kind of atom. If it is a compound every sample contains exactly the same relative numbers of atoms of each kind. Regardless of the source from which it is obtained, water always has two hydrogen atoms for each oxygen atom and can be represented by the formula H_2O. Carbon dioxide always has two oxygen atoms for each carbon atom; sodium chloride always has one sodium atom for each chlorine atom; and magnesium consists only of magnesium atoms. In short we may define a **substance** as *any homogeneous portion of matter with a definite constant composition and a unique set of properties.*

If a given material is a single substance and not a mixture of substances, we often call it a **pure substance**. In nature very few substances occur in a pure form. One substance may be the major component of a given sample, but other substances, which we called *impurities*, are nearly always present. Thus *purity* is a relative term. No substance can be regarded as absolutely pure; no matter how carefully it has been purified, traces of other, contaminating substances always remain. A substance may be described as 99% pure when it contains 1% by mass of impurities and as 99.99% pure when it contains 0.01% of impurities. Such purities are sufficient for many purposes, and such substances are often described as pure.

Sometimes, though, a substance with a still greater purity is required for a particular purpose. For example, the silicon, Si, used in silicon chips has to be 99.999 99% pure (Figure 1.1). And often we need to know not only the purity of a substance but also the nature of the impurities present. The labels on the containers of many substances used in a chemical laboratory often list the impurities and their amounts (Figure 1.2).

HETEROGENEOUS MIXTURES Many common materials such as soil, rocks, concrete, and wood are mixtures. Different parts of these materials have different properties such as color and hardness (Figure 1.3). Such mixtures are said to be **heterogeneous**; both the properties and the composition of the material are nonuniform.

If we mix powdered sulfur with iron filings, the result is a heterogeneous mixture. We can still discern hard, dark grey iron particles and yellow sulfur powder when we examine the mixture. We can separate the iron and the sulfur by taking advantage of their different physical properties. One way to separate them is with a magnet: the iron filings are attracted, while the sulfur is not (Experiment 1.4).

FIGURE 1.1
Ingots of Ultrapure Silicon.

A large ingot such as this is cut into thin wafers and then into tiny chips for making the many kinds of semiconductors used in radios, pocket calculators, and computers.

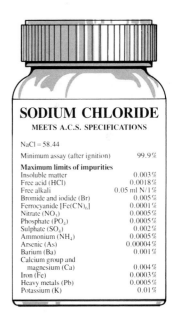

FIGURE 1.2
Label on Bottle of Sodium Chloride Indicating Impurities.

This sodium chloride is described as 99.9% pure. It therefore contains a total of 0.1% of impurities, which are listed on the label. A.C.S. is the abbreviation for the American Chemical Society.

FIGURE 1.3
Heterogeneous Mixtures.

Common materials such as rocks and wood are heterogeneous mixtures. Different parts of the material clearly have different properties, such as hardness and color.

_____ *Experiment 1.4*

Mixtures and Compounds

On the left is a pile of powdered sulfur. On the right is a pile of iron filings. In the middle is a mixture of the iron and sulfur.

The iron filings can be separated from the sulfur by using a magnet.

When the mixture is heated it glows brightly as the iron and the sulfur combine to form the compound iron sulfide.

Iron sulfide is gray-black and brittle, unlike either sulfur or iron. The iron in the compound iron sulfide cannot be separated from the sulfur by means of a magnet.

In general, heterogeneous mixtures can be separated into their components by making use of their different physical properties—that is, by carrying out some **physical change**. *A* **physical change** *is any change in which the composition and amount of each of the substances present does not change.* In a physical change none of the substances present changes into other substances; in other words, no chemical reaction occurs. Boiling and freezing water and dissolving sugar in water are physical changes.

The nonuniform composition of a heterogeneous mixture is not always easily visible. Milk might not appear to be heterogeneous, but under a microscope we can see that it consists of small droplets of fat suspended in a clear liquid (Figure 1.4).

HOMOGENEOUS MIXTURES   A mixture that has uniform properties throughout is a **homogeneous mixture**, or a **solution**. If you stir a spoonful of sugar into a cup of water until all the sugar dissolves, you create a homogeneous mixture, or a solution. The sugar has become completely dispersed in the water so that the mixture is uniform. Any sample of a particular mixture has the same composition and the same physical properties as any other sample—color and sweetness, for example. Homogeneous mixtures, like heterogeneous mixtures, can be separated into their components by physical means. If you heat the mixture of sugar and water, the water evaporates and the sugar is left behind. A solution of salt in water can be separated into its two components in the same way.

Not all solutions are liquids. Pure, dry air is a solution—a homogeneous mixture of gases, principally nitrogen, oxygen, and argon. The gold used in jewelry is not pure but is a solid solution of copper or silver in gold. However, the term solution normally implies a liquid solution, unless it is specifically stated that the solution is a solid solution. Gaseous mixtures are always homogeneous mixtures; they are invariably described as mixtures rather than solutions.

How does a homogeneous mixture differ from a pure substance? The properties of a homogeneous mixture of a given composition do not vary from one part of the mixture to another, but mixtures do not have to have a constant composition or constant properties. Homogeneous mixtures, such as solutions of sugar in water, can have different compositions depending on how much sugar is dissolved in a given amount of water, and they will have correspondingly different sweetnesses. A mixture of CO_2 and CO can have any ratio of oxygen atoms to carbon atoms, from 2:1 to 1:1, depending on the relative amounts of the CO_2 and CO.

In contrast, a compound always has the same composition and properties. Water always has the composition expressed by the formula H_2O, and its color and freezing point are always the same. Moreover, water cannot be separated into its component elements by physical means but only by a chemical reaction that produces two new substances (hydrogen and oxygen). These new substances and the water from which they were obtained have completely different properties. Experiment 1.4 shows the differences between a mixture of iron and sulfur and the compound iron sulfide. Figure 1.5 summarizes the classification of matter and the ways in which one type of matter can be changed into another.

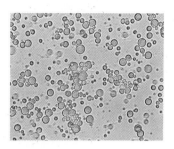

FIGURE 1.4
Milk Under a Microscope.

The heterogeneity of milk is not obvious to the naked eye. But under a microscope we see that it consists of drops of fat suspended in a clear liquid.

═ *Example 1.1* ═══════════════════════════════

CLASSIFICATION OF MATERIALS

Classify the following materials as homogeneous or heterogeneous, substance or solution, compound or element:

(a) Ice   (b) Copper   (c) Blood   (d) Sugar   (e) Wine

Solution

(a) Ice is a homogeneous substance. It is a compound of the elements hydrogen and oxygen.

(b) Copper is a homogeneous substance. It is an element.

(c) Blood is a heterogeneous mixture of red and white blood cells, water, and other substances.

(d) Sugar is a homogeneous substance. It is a compound of the elements carbon, hydrogen, and oxygen.

(e) Wine is a homogeneous solution of ethanol and other substances in water.

FIGURE 1.5
Classification of Matter.

All matter is either homogeneous or heterogeneous. Homogeneous matter may be either a (pure) substance or a homogeneous mixture (solution). A substance may be either an element or a compound.

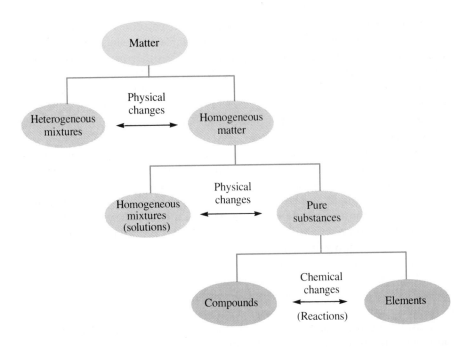

MOLECULES

Not only does a compound have a constant composition, being composed of fixed relative numbers of its component atoms, but these atoms have a definite arrangement with respect to each other. The arrangement of the atoms in a compound or element is known as its **structure**. Because the structure of a substance has a profound effect on its properties, one of the aims of modern chemistry is to relate the properties of a substance to its structure. For instance, the element carbon is known in several different forms, such as diamond and graphite, with very different properties (Figure 1.6). Diamond and graphite both consist of carbon atoms only; their different properties are due to the different ways these atoms are arranged.

The early chemists had no direct method of finding the structure of any given substance, although by studying its properties and reactions they were sometimes able to deduce how its atoms were arranged. Only during the past sixty years have experimental methods been developed that enable chemists to determine the structures of substances directly. Most of these methods are based on the study of how different forms of electromagnetic radiation, such as visible light, infrared radiation, and X rays, interact with matter. Some of these methods are discussed later.

FIGURE 1.6
Diamond and Graphite.

Both substances are forms of the same element, carbon, but they have very different properties. The "lead" in a lead pencil is made from graphite mixed with clay.

Experiments have shown that in many substances atoms are combined in small groups called **molecules**. Water, for example, consists of molecules in which two hydrogen atoms are joined to an oxygen atom. The water molecule has the formula H_2O. This is called a *molecular formula*. It shows the number of atoms of each kind in one molecule of the compound. Carbon monoxide consists of molecules composed of one carbon atom and one oxygen atom; its molecular formula is CO. Carbon dioxide consists of molecules composed of one carbon atom and two oxygen atoms; its molecular formula is CO_2.

Some elements are also composed of molecules. For example, oxygen, nitrogen, and hydrogen each have two atoms in their molecules, and their molecular formulas are O_2, N_2, and H_2, respectively. Some elements have bigger molecules; phosphorus consists of P_4 molecules and sulfur of S_8 molecules. Molecules such as H_2, N_2, O_2, and CO, which are composed of only two atoms, are known as **diatomic molecules**; molecules such as H_2O and CO_2 are called **triatomic molecules**. In general, any molecule containing more than two atoms is called a **polyatomic molecule**.

Not all substances are composed of molecules. For example, the compounds sodium chloride, NaCl, and silicon dioxide, SiO_2, do not have a molecular structure; neither do the elements magnesium and carbon. In magnesium, for example, magnesium atoms are arranged in a regular array or network that extends indefinitely in three dimensions and in which no Mg_2, Mg_3 or similar molecules can be recognized. In silicon dioxide there is a three-dimensional network of silicon and oxygen atoms but there are no SiO_2 molecules. We will discuss the structures of these substances later. For the moment we will restrict our attention to those substances that consist of molecules.

EMPIRICAL AND MOLECULAR FORMULAS

The molecular formula of a substance is not always identical with the *simplest formula* that expresses the relative numbers of atoms of each kind in the substance. Since an element consists of only one kind of atom, the simplest formula for any element is simply the symbol for the element: O for oxygen, N for nitrogen, P for phosphorus, S for sulfur. The molecular formulas of these substances, however, are O_2, N_2, P_4, and S_8.

The composition of *ethane*, which is another component of natural gas in addition to methane, is expressed by the formula CH_3. For each carbon atom in ethane there are three hydrogen atoms. However, ethane consists of C_2H_6 molecules and therefore has the molecular formula C_2H_6. Hydrogen and oxygen combine to give a compound, hydrogen peroxide, which is a colorless liquid like water but has different chemical and physical properties. For example, it has a boiling point of 158 °C. One of its characteristic chemical properties is that it decomposes into oxygen and water in the presence of certain substances such as manganese dioxide and blood, as is illustrated in Experiment 1.5. Hydrogen peroxide is composed of equal numbers of hydrogen and oxygen atoms, and its composition can therefore be represented by the formula HO. However, its molecules each contain two hydrogen atoms and two oxygen atoms, so its molecular formula is H_2O_2.

The simplest formula that correctly expresses the composition of a substance, in terms of whole-number ratios of atoms, is called its **empirical formula**. Thus the empirical formulas of water, hydrogen peroxide, ethane, sulfur, and phosphorus are H_2O, HO, CH_3, S, and P. Since we cannot have a fractional part of an atom (half an atom is no longer an atom), empirical formulas are written with integral numbers of atoms and not fractional numbers. The empirical formulas of water and ethane are written as H_2O and CH_3, not as $HO_{1/2}$ and $C_{1/3}H$.

The **molecular formula** *of a substance tells us how many atoms of each kind there are in one molecule of the substance.* The molecular formulas of water, hydrogen peroxide, ethane, sulfur, and phosphorus are H_2O, H_2O_2, C_2H_6, S_8, and P_4. For some substances, such as H_2O, the empirical and molecular formulas are the same. For others, such as ethane, the molecular formula, C_2H_6,

Experiment 1.5

Properties of Water and Hydrogen Peroxide

Water (left) and hydrogen peroxide (right) are both colorless liquids. They are both compounds of the elements hydrogen and oxygen. When black solid manganese dioxide is added to water no reaction is observed. When manganese dioxide is added to hydrogen peroxide it causes hydrogen peroxide to decompose, producing bubbles of oxygen which carry the manganese dioxide to the surface. The oxygen that is evolved ignites a glowing splint.

When a drop of blood is added to water it slowly mixes but no reaction is observed. When a few drops of blood are added to hydrogen peroxide it decomposes rapidly producing bubbles of oxygen which form a thick foam that fills the beaker and overflows the top.

is a whole-number multiple of the empirical formula, CH_3. Empirical and molecular formulas are compared in Table 1.2. Figure 1.7 illustrates the molecules of hydrogen, oxygen, water, and hydrogen peroxide and summarizes the differences among mixtures, substances, compounds, and elements.

Many molecules, including most of those found in living organisms, are much larger than the simple molecules we have considered so far, and they have more complicated formulas. For example, sucrose (table sugar) has the molecular formula $C_{12}H_{22}O_{11}$; riboflavin (vitamin B_2) has the molecular formula $C_{17}H_{20}N_4O_6$; adenosine triphosphate (ATP) has the molecular formula $C_{10}H_{16}N_5O_{13}P_3$.

STRUCTURE AND SHAPE OF MOLECULES

We can obtain information about the structure of a molecule on several different levels, each level corresponding to a more detailed knowledge (Figure 1.8, p. 16). Once the molecular formula of a substance is known the next question that we can ask is: How are the atoms connected together? To take a very simple example, we know that in the water molecule the atoms are arranged with the oxygen in the middle so we write HOH and not HHO. Many years ago chemists started to draw a line between atoms that they believed were attached to each other and so the water molecule would be written as $H\text{—}O\text{—}H$.

TABLE 1.2
Empirical and Molecular Formulas

Substance	Empirical Formula	Molecular Formula
Water	H_2O	H_2O
Hydrogen peroxide	HO	H_2O_2
Ethane	CH_3	C_2H_6
Sulfur	S	S_8
Phosphorus	P	P_4

This is called a **structural formula** and the lines are called **bonds**. The structural formula of sucrose is shown in Figure 1.9. In later chapters we will be discussing in some detail the nature of bonds and how they hold atoms together. Until about 60 years ago no direct methods for determining the structures of molecules had been developed and chemists had to rely on the reactions of substances to deduce the arrangement of the atoms in their molecules. Clearly,

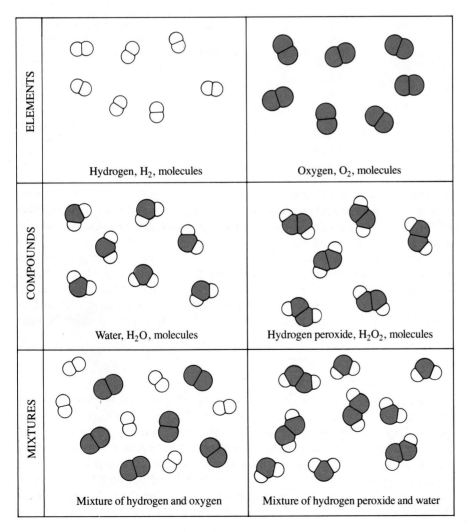

Hydrogen, H_2, molecules

Oxygen, O_2, molecules

Water, H_2O, molecules

Hydrogen peroxide, H_2O_2, molecules

Mixture of hydrogen and oxygen

Mixture of hydrogen peroxide and water

ELEMENTS

COMPOUNDS

MIXTURES

FIGURE 1.7
Mixtures, Substances, Compounds, and Elements.

Hydrogen, oxygen, water, and hydrogen peroxide are all substances. Hydrogen and oxygen are elements; water and hydrogen peroxide are compounds. A mixture of hydrogen peroxide and water is homogeneous; it is a solution. A mixture of hydrogen and oxygen is also homogeneous; it is a (gaseous) solution.

FIGURE 1.8
Formulas and Structures of
Molecules.

Formulas and Structures of Molecules		
Formula or Structure	*Examples*	*Information provided*
Empirical formula	HO CH$_3$	Relative numbers of atoms in molecule
Molecular formula	H$_2$O$_2$ C$_2$H$_6$	Number of atoms of each kind in molecule
Structural formula	H–O–O–H	Connectivity of the atoms
Shape	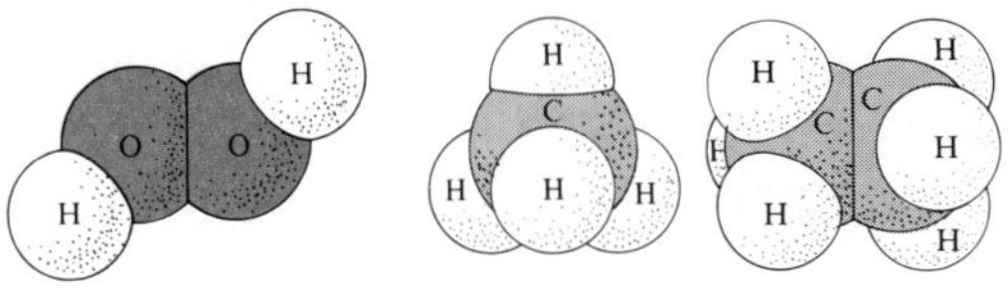	Arrangement of atoms in space
Dimensions	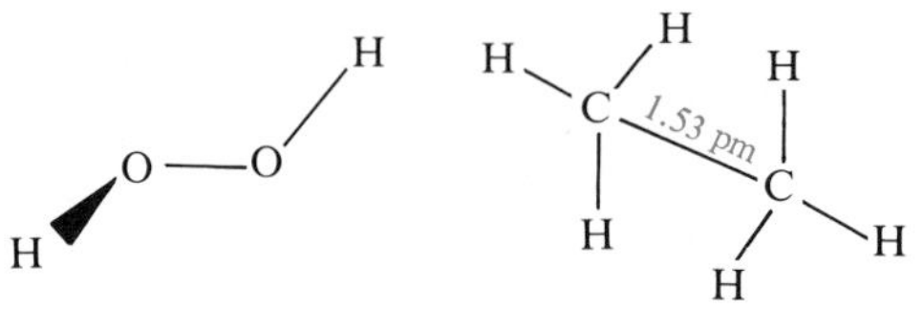	Distances and angles

for a molecule such as sucrose this was quite a feat and involved much pain-staking work as well as considerable intuition, insight and clever arguments. Today, however, we can determine not only how the atoms in a molecule are connected together but how they are arranged in space. In other words, we can find the **three-dimensional structure** or **shape** of the molecule, even for molecules such as proteins which are much more complicated than sucrose. For example,

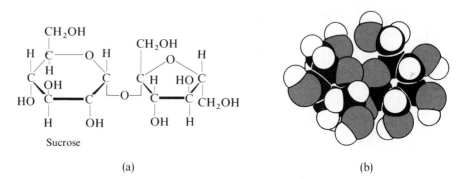

Sucrose

(a)   (b)

FIGURE 1.9
The Structural Formula and Shape of the Sucrose Molecule.

(a) The structural formula shows how the atoms are connected together. (b) This is an illustration of the shape of the molecule: it shows how the atoms are arranged in three dimensions.

we now know that the water molecule has an **angular shape** (Figure 1.10) and that the sucrose molecule has the shape shown in Figure 1.9. With modern experimental methods we can even measure the distances between all the atoms in the molecule and the angles made by the atoms. Thus we know that the distances between the H atoms and the O atom in the water molecule are 0.000 000 000 097 meter and that the three atoms make an angle of 104.5° (Figure 1.10).

The shapes of some other molecules we have encountered so far in this chapter are shown in Figure 1.11. The ammonia molecule has the shape of a *triangular pyramid*, with the nitrogen atom at the apex and the three hydrogen atoms at the corners of the pyramid's equilateral triangular base. The three angles at the apex of the pyramid are each 107.8°. The shape of the methane molecule is a common shape for molecules but is less familiar in everyday experience. It is the *tetrahedron*. Each of the four faces of a tetrahedron is an equilateral triangle, so all the edges have the same length and all the angles between them are 60°, as indicated in Figure 1.12. In the methane molecule the four hydrogen atoms are at the corners of the tetrahedron, and the carbon atom to which they are joined is at the center of the tetrahedron. Another molecule with a tetrahedral shape is the P_4 molecule. In this case there is a phosphorus atom at each corner of the tetrahedron, but there is no atom at the center.

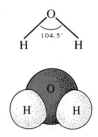

FIGURE 1.10
The Angular Shape of the Water Molecule.

The three atoms make an angle of 104.5°. The distance between the center of each hydrogen atom and the center of the oxygen atom is 0.000 000 000 097 m = 9.7×10^{-11} m.

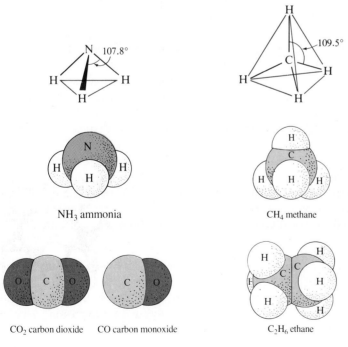

FIGURE 1.11
Shapes of Some Simple Molecules.

Although molecules are very small, the distances between the atoms and the arrangement of the atoms in space can be accurately determined by modern experimental methods.

FIGURE 1.12
The Tetrahedron.

The tetrahedron is a regular solid
with four equivalent equilateral
triangular faces and six equal
edges. The shape of the tetrahedron
is closely related to the shape of
a cube. This relation can be seen
by drawing a diagonal across one
face of the cube and another diagonal at right angles across the
opposite face of the cube as in (a).
Then join the ends of the diagonals, as in (b), to form the tetrahedron. Two views of the methane
molecule are shown in (c) and (d);
two views of the phosphorus molecule are shown in (e) and (f).

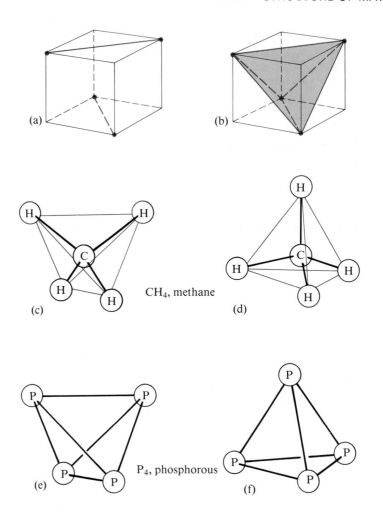

1.2
Units of Measurement

The distances between the atoms in a molecule are extremely small. In the
water molecule the distance between the center of each hydrogen atom and
the center of the oxygen atom is only 0.000 000 000 097 meter. We will also
frequently encounter very large numbers in chemistry. There are, for example,
33 460 000 000 000 000 000 000 molecules in 1.000 gram of water.

SCIENTIFIC NOTATION

To write out these very small or very large numbers in this way is inconvenient.
Scientific, or *exponential*, **notation** is usually used for expressing very small and
very large numbers. Scientific notation uses the form

$$N \times 10^n$$

where N is a number between 1.000 . . . and 9.999 . . . , that is, a number with
one nonzero digit before the decimal point. The number 33 460 000 000 000 000
000 000 is equal to 3.346 multiplied by 10 twenty-two times, that is,

$$3.346 \times 10 \times 10 \times 10 \times 10 \times 10 \times 10 \times 10 \times 10 \times 10 \times 10 \times 10$$
$$\times 10 \times 10 \times 10 \times 10 \times 10 \times 10 \times 10 \times 10 \times 10 \times 10 \times 10$$

In scientific notation this number is written as

$$3.346 \times 10^{22}$$

Thus there are 3.346×10^{22} molecules in 1.000 gram of water.

Similarly, the number 0.000 000 000 097 is equal to 9.7 divided by 10 eleven times, that is,

$$\frac{9.7}{10 \times 10 \times 10 \times 10 \times 10 \times 10 \times 10 \times 10 \times 10 \times 10 \times 10}$$

In scientific notation this number is written as

$$\frac{9.7}{10^{11}} \quad \text{or} \quad 9.7 \times \frac{1}{10^{11}}$$

and $1/10^{11}$ is written as 10^{-11}. Hence

$$0.000\,000\,000\,097 = 9.7 \times 10^{-11}$$

Thus the distance between a hydrogen atom and an oxygen atom in the water molecule is 9.7×10^{-11} meter.

In general, to convert a number to scientific notation, move the decimal point until there is only one nonzero digit in front of it. The number of places that the decimal point is moved gives the exponent of 10. For example,

$$3\,472\,809.0 \rightarrow 3.472\,809\,0 \times 10^6$$
$$\underbrace{\qquad}_{6 \text{ places}}$$

$$0.048\,729 \rightarrow 4.8729 \times 10^{-2}$$
$$\underbrace{\quad}_{2 \text{ places}}$$

In writing numbers and doing calculations, be very careful always to use the correct number of **significant figures**. The number 0.048 729 has five significant figures, and it also has five significant figures when expressed in the form 4.8729×10^{-2}. Significant figures are discussed in Appendix A.

We have described the distance between the hydrogen and oxygen atoms in terms of meters. The meter is just one of many units of measurement. If we had been talking about the distance between two cities, we would have used miles or kilometers. If we were talking about the weight of this book, we probably would use kilograms or pounds and ounces. Unless a number is accompanied by a unit, it tells us nothing about the quantity measured.

Throughout history many units of measurement have been used, but in recent times two sets of units have become widely adopted. In most English-speaking countries length has traditionally been measured in inches, feet, yards, and miles; weight has been measured in ounces, pounds, and tons. In the rest of the world, however, lengths are commonly measured in meters or in decimal multiples or fractions of the meter, such as the kilometer (1000 meters) and the centimeter (0.01 meter). Masses are measured in grams or in decimal multiples or fractions of the gram, such as the kilogram (1000 grams) or the milligram (10^{-3} gram). Weight is very often (incorrectly) expressed in the same units (see Box 1.2).

In ancient Egypt, cards were used as measuring units.

BOX 1.2

MASS AND WEIGHT

The *mass* of an object is the quantity that measures its resistance to a change in its state of rest or motion. For a change in the state of rest or of motion of an object, a force must be applied. An object of small mass such as a pebble needs only a small force to set it in motion if it is at rest or to change its motion if it is already moving. An object of large mass such as a space rocket needs a large force to set it in motion or to change its motion.

Mass can be measured by the force necessary to give an object a given acceleration. On earth we use the force of gravitational attraction of the earth for an object to measure its mass. We call this force the *weight* of the object; it depends on the mass of the object, the mass of the earth, and the distance of the object from the center of the earth.

A given object always has the same mass. But if it were taken to a planet with a mass different from that of the earth, it would be subject to a different gravitational force and would therefore have a different weight. On the surface of the earth, although the mass of the object and that of the earth are constant, the distance of an object from the center of the earth may vary slightly—from the top of a mountain to the bottom of a valley, for example—and therefore its weight will also vary slightly. Because the earth is not a perfect sphere but is slightly flattened at the poles, an object at one of the poles is slightly closer to the center of the earth than is an object at the equator. Thus a person weighing 140.0 lb at the equator would weigh 140.8 lb at the North Pole. So although the mass of an object is a constant, its weight depends on its location. The mass of an object is determined by comparison with a set of standard masses. Since the weights of two objects of equal mass are the same at any one place on the earth's surface, these ob-

The astronaut on the moon has the same mass as on Earth but he weighs much less. Thus he can move easily, despite his bulky clothing and equipment.

jects balance each other when placed on the pans of a balance with arms of equal length.

In SI *mass* is measured in kilograms or convenient multiples of this unit, but *weight*, which is a force, is measured in newtons. In the English system weight is measured in pounds, while mass is measured in slugs, which is a unit you will not encounter in chemistry. Therefore we can quite correctly say that an object with a mass of 1 kg weighs 2.205 lb. But the terms *mass* and *weight* are frequently misused in everyday conversation, so we often speak of a weight of 1 kg rather than a mass of 1 kg. This misuse is unfortunate, but it is unlikely to cause any confusion if we understand the difference between mass and weight and recognize that *weight* is often used when *mass* is meant.

THE METRIC SYSTEM AND SI UNITS

The system of units based on the meter and the gram is known as the **metric system**. This system was first adopted in France after the French Revolution (1789), and its common use quickly spread throughout Europe and beyond. In 1799 Thomas Jefferson tried to persuade the U.S. Congress to adopt the metric system, which he said would eventually become the standard for the world. Today, Great Britain and the United States are almost the only countries still resisting the change from the English to the metric system. In science, however, the metric system was universally adopted, because the decimal relationship between different units of length and mass makes this system very convenient. Table 1.3 includes some metric system–English system equivalents.

TABLE 1.3
Metric System–English System Equivalents

Quantity	Metric and English Units
Length	1 meter = 1.094 yards 2.540 centimeters = 1 inch 1 kilometer = 0.6214 miles
Mass	1 kilogram = 2.205 pounds[a] 453.6 grams = 1 pound[a]
Volume	1 liter = 1.06 quarts (U.S.) 1 cubic foot = 28.32 liters

[a] A pound is, strictly speaking, a unit of force (weight), not mass (see Box 1.2).

Since 1899 international conferences have been held for the purposes of agreeing on systems of units, of providing accurate definitions of units, and of defining any new units that might be required. In 1960 the eleventh International Conference on Weights and Measures proposed some major changes to the metric system and suggested a new name for this modified metric system, the **International System of Units**. This name is abbreviated as **SI**, from the French *Système International d'Unités*. All the major countries in the world have agreed to adopt SI, not only for scientific purposes but also for everyday use.

The seven basic units of SI are given in Table 1.4. For the present we will consider in detail only length and mass. The other basic units will be considered as they are needed.

THE METER AND UNITS OF LENGTH

The meter (m), the basic SI unit of length, is not convenient for expressing the dimensions of the water molecule or for many other purposes. Therefore a series of prefixes that represent various powers of ten have been defined in SI. These prefixes, which are listed in Table 1.5 on page 22, allow us to reduce or enlarge the SI base units to an appropriate size.

The O–H distance in the water molecule, 9.7×10^{-11} m, can then be expressed as 97 picometers (pm) or as 0.097 nanometer (nm). We will normally

TABLE 1.4
The Seven SI Base Units

Quantity Measured	Name of Unit	Symbol of Unit
Length	Meter	m
Mass	Kilogram	kg
Time	Second	s
Electric current	Ampere	A
Temperature	Kelvin	K
Amount of substance	Mole	mol
Luminous intensity	Candela	cd

TABLE 1.5
Prefixes for Decimal Fractions and Multiples of SI Units

Prefix	Symbol for Prefix		Scientific Notation
Exa	E	1 000 000 000 000 000 000	10^{18}
Peta	P	1 000 000 000 000 000	10^{15}
Tera	T	1 000 000 000 000	10^{12}
Giga	G	1 000 000 000	10^{9}
Mega	M	1 000 000	10^{6}
Kilo	k	1 000	10^{3}
Hecto	h	100	10^{2}
Deca	da	10	10^{1}
—	—	1	10^{0}
Deci	d	0.1	10^{-1}
Centi	c	0.01	10^{-2}
Milli	m	0.001	10^{-3}
Micro	μ	0.000 001	10^{-6}
Nano	n	0.000 000 001	10^{-9}
Pico	p	0.000 000 000 001	10^{-12}
Femto	f	0.000 000 000 000 001	10^{-15}
Atto	a	0.000 000 000 000 000 001	10^{-18}

Note: The more commonly used prefixes are printed in bold type.

use the picometer as the unit for molecular dimensions. The N–H distance in the ammonia molecule is 102 pm, and the P–P distance in the phosphorus, P_4, molecule is 221 pm.

Prior to the adoption of SI, chemists commonly used another unit, the *angstrom* (Å), equal to 10^{-10} m, for molecular dimensions. Although such non-SI units are being phased out, you should be familiar with the angstrom because it has been used very widely in the past and will undoubtedly continue to be used for some time in the future. The O–H distance of 97 pm in the water molecule is 0.97 Å. In general, the relationships between the angstrom (Å) and the nanometer (nm) and the picometer (pm) are

$$1 \text{ nm} = 10 \text{ Å} \quad \text{and} \quad 1 \text{ pm} = 0.01 \text{ Å}$$

EXERCISE 1.1

Express each of the following distances in meters using scientific notation: 208 pm, 240 nm, 95 μm, 0.97 Å.

THE KILOGRAM AND UNITS OF MASS

The standard mass of the metric system is a cylinder of corrosion-resistant, platinum–iridium alloy kept at the International Bureau of Weights and Measures in Sèvres, France. Its mass is defined as exactly 1 kilogram (kg): all other masses are determined by comparison with this standard (Figure 1.13). The SI base unit of mass, the kilogram, is unusual because it already contains a pre-

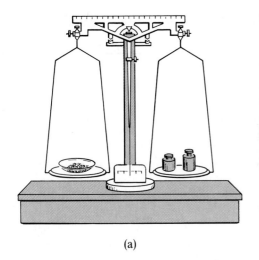

(a)   (b)

FIGURE 1.13
Weighing using a Balance.

(a) A traditional balance had two pans. The object to be weighed was placed on one pan and "weights" (masses) were added to the other pan to balance the mass of the object. (b) A modern electronic laboratory balance has only one pan and a direct digital read-out.

fix. Nevertheless, the standard prefixes are applied to the gram when larger or smaller units are needed. For example, the quantity 10^6 kg (1 million kilograms) is written as 1 Gg (gigagram), not as 1 Mkg (megakilogram). Mass and weight are often confused; the difference between them was explained in Box 1.2.

EXERCISE 1.2

Express each of the following masses in grams using scientific notation: 25 kg, 250 mg, 2.5 μg, 100 Mg.

CONVERSION OF UNITS

On many occasions we must convert from one set of units to another. For instance, we may need to convert one SI unit to another, English units to SI, or vice versa. A very convenient way of doing these conversions, as well as many other types of problems, is the **unit factor method**, which is also called **dimensional analysis**. To use this method we write the units for every quantity used in a calculation, and we carry the units through the calculation, treating them as algebraic quantities.

To illustrate this method, suppose we wish to convert a certain number of minutes (min) into seconds (s). We know the basic relationship:

$$1 \text{ min} = 60 \text{ s}$$

We can therefore write the following equalities:

$$1 = \frac{60 \text{ s}}{1 \text{ min}} \quad \text{and} \quad 1 = \frac{1 \text{ min}}{60 \text{ s}}$$

These quantities are called *unit conversion factors*, or simply **unit factors**, because the overall effect of multiplication by these factors is to multiply by 1. Thus multiplication of a quantity by a unit factor changes the unit in which the quantity is expressed but not its physical meaning. To find the number of seconds in 3.0 min, we write

$$3.0 \text{ min} = (3.0 \text{ min})\left(\frac{60 \text{ s}}{1 \text{ min}}\right) = 180 \text{ s}$$

We cancel the unit minutes, and the result is then the same quantity expressed in seconds.

Suppose we wish to convert a speed of kilometers per hour, for example 50.0 kilometers per hour (50.0 km/h), to meters per second (m/s); we need to use two unit conversion factors. First we convert kilometers per hour to meters per hour using the unit conversion factor (1000 m/1 km):

$$50.0 \text{ km/h} = \left(\frac{50.0 \text{ km}}{1 \text{ h}}\right)\left(\frac{1000 \text{ m}}{1 \text{ km}}\right) = 50\,000 \text{ m/h} = 5.00 \times 10^4 \text{ m/h}$$

Then we convert meters per hour to meters per second using the unit conversion factor (1 h/3600 s):

$$5.00 \times 10^4 \text{ m/h} = \left(\frac{5.00 \times 10^4 \text{ m}}{1 \text{ h}}\right)\left(\frac{1 \text{ h}}{3600 \text{ s}}\right) = 13.9 \text{ m/s}$$

We can shorten this calculation by making both conversions at the same time; that is, by multiplying the original quantity by both unit conversion factors:

$$50.0 \text{ km/h} = \left(\frac{50.0 \text{ km}}{1 \text{ h}}\right)\left(\frac{1000 \text{ m}}{1 \text{ km}}\right)\left(\frac{1 \text{ h}}{3600 \text{ s}}\right) = 13.9 \text{ m/s}$$

═ *Example 1.2* ═══

INTERCONVERSION OF UNITS

The distance between the two hydrogen atoms in the H_2 molecule is 74 pm. Express this length in meters, angstroms, and inches (in.).

Solution

The distance in meters is found as follows:

$$74 \text{ pm} = (74 \text{ pm})\left(\frac{10^{-12} \text{ m}}{1 \text{ pm}}\right) = 7.4 \times 10^{-11} \text{ m}$$

The distance in angstroms is found as follows:

$$74 \text{ pm} = (74 \text{ pm})\left(\frac{10^{-12} \text{ m}}{1 \text{ pm}}\right)\left(\frac{1 \text{ Å}}{10^{-10} \text{ m}}\right) = 74 \times 10^{-2} \text{ Å} = 0.74 \text{ Å}$$

From Table 1.3, 1 in. = 2.54 cm. So the distance in inches is found as follows:

$$74 \text{ pm} = (74 \text{ pm})\left(\frac{10^{-10} \text{ cm}}{1 \text{ pm}}\right)\left(\frac{1 \text{ in.}}{2.54 \text{ cm}}\right) = 2.9 \times 10^{-9} \text{ in.}$$

───

Any number of unit factors may be used as required. In each case they are applied so that the units of the preceding factor cancel. Notice that for every equality such as

$$1 \text{ kg} = 2.205 \text{ lb}$$

there are two unit conversion factors. One is the reciprocal of the other.

$$\frac{1 \text{ kg}}{2.205 \text{ lb}} = \frac{2.205 \text{ lb}}{1 \text{ kg}} = 1$$

As an example of the use of the appropriate unit conversion factor, consider the following problem. If a woman weighs 140 lb, what is her mass in kilograms? We choose the unit factor that cancels the unit pounds:

$$140 \text{ lb} = (140 \text{ lb})\left(\frac{1 \text{ kg}}{2.205 \text{ lb}}\right) = 63.5 \text{ kg}$$

If we had used the other (incorrect) unit factor, we would have obtained

$$140 \text{ lb} = (140 \text{ lb})\left(\frac{2.205 \text{ lb}}{1 \text{ kg}}\right) = 309 \text{ lb}^2/\text{kg}$$

Because we see that the units do not cancel to give the desired units, we know that we have used the wrong unit factor.

> Including the units with every quantity provides a very important check in any calculation. If you do not get the right units for the answer you know that you have made a mistake. You should then go back and carefully check each of the conversion factors.

EXERCISE 1.3

An imported car is advertised as having a gasoline consumption of 6.0 liters per 100 kilometers. Convert this rating to miles per gallon.

VOLUME

Many other units are derived from the seven SI base units. One important derived unit is the unit of volume. If each side of a cube has a length of 1 m, its volume is 1 cubic meter (1 m^3). For many objects this unit of volume is rather large, so the cubic decimeter (dm^3) or, more commonly, the cubic centimeter (cm^3) is used instead.

Because 1 dm = 10^{-1} m, the unit conversion factor is (10^{-1} m/1 dm). Hence

$$1 \text{ dm}^3 = (1 \text{ dm}^3)\left(\frac{10^{-1} \text{ m}}{1 \text{ dm}}\right)^3 = (10^{-1} \text{ m})^3 = 10^{-3} \text{ m}^3$$

Similarly, since 1 cm = 10^{-2} m, the unit conversion factor is (10^{-2} m/1 cm). Hence

$$1 \text{ cm}^3 = (1 \text{ cm}^3)\left(\frac{10^{-2} \text{ m}}{1 \text{ cm}}\right)^3 = 10^{-6} \text{ m}^3$$

Prior to the development of SI, the liter (L) and the milliliter (mL) were used to measure the volume of liquids and solutions. The liter was originally defined as the volume of 1 kg of water at the temperature of its maximum density (3.98 °C), but it has since been redefined as exactly one-thousandth of a cubic meter, that is, 1 dm^3. Hence a milliliter is exactly 1 cm^3. The units liter and cubic decimeter, and milliliter and cubic centimeter, can therefore be used interchangeably. Most laboratory equipment for measuring volumes is graduated in liters and milliliters, and the use of these alternative names will continue for some time. We will use the units liter and milliliter extensively in this book.

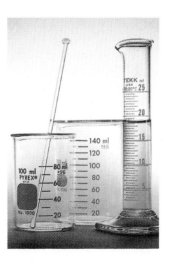

$=$ *Example 1.3* $=$

VOLUME

A cylinder has a diameter of 5.00 cm and contains water to a height of 25.0 cm. Find the volume of water in cubic centimeters, cubic meters, and cubic inches (in.^{3}).

Solution

The volume of a cylinder is given by the expression

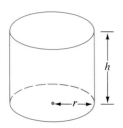

$$\text{Volume} = \pi r^2 h$$

where r is the radius and h the height. For this example the diameter is 5.00 cm, so the radius r is 2.50 cm. Hence

$$\text{Volume of water} = \pi r^2 h = 3.142(2.50 \text{ cm})^2(25.0 \text{ cm}) = 491 \text{ cm}^3$$

Since 1 cm $= 10^{-2}$ m, the unit conversion factor is (10^{-2} m/1 cm). Hence the volume in cubic meters is

$$491 \text{ cm}^3 = (491 \text{ cm}^3)\left(\frac{10^{-2} \text{ m}}{1 \text{ cm}}\right)^3 = 491 \times 10^{-6} \text{ m}^3 = 4.91 \times 10^{-4} \text{ m}^3$$

Since 1 in. $= 2.54$ cm, the unit conversion factor is (1 in./2.54 cm). Hence the volume in cubic inches is

$$491 \text{ cm}^3 = (491 \text{ cm}^3)\left(\frac{1 \text{ in.}}{2.54 \text{ cm}}\right)^3 = 30.0 \text{ in.}^3$$

DENSITY

The common statements that mercury is "heavier" than water or that iron is "heavier" than aluminum are inaccurate—or at least incomplete. What is actually being compared is not mass, which depends on the amount of a substance, but *mass per unit volume*, which is known as **density**.

$$\text{Density} = \frac{\text{Mass}}{\text{Volume}}$$

The mass of 1 cm^3 of water is 1.00 g, while the mass of 1 cm^3 of mercury is 13.6 g at 20 °C. In other words, the density of water is 1.00 g cm^{-3}, and the density of mercury is 13.6 g cm^{-3}. The units for density are commonly written as g/cm^3 or g cm^{-3}. In the first expression the slash denotes "per". In the second expression 1/cm^3 has been written as cm^{-3}, so g/cm^3 becomes g cm^{-3}. Thus units are treated like algebraic quantities; for example, $1/x^2 = x^{-2}$, and similarly $1/\text{cm}^2 = \text{cm}^{-2}$.

Unlike mass and volume, the density of a substance is independent of the amount and the size of the sample. Density can therefore be used as an aid in distinguishing one pure substance from another. Table 1.6 lists the densities of some common substances.

$=$ *Example 1.4* $=$

DENSITY

The density of liquid mercury is 13.6 g cm^{-3}. What is the mass of 1 L of mercury in grams, in kilograms, and in pounds?

TABLE 1.6
Densities of Various Substances at 20 °C

Substance	Physical State[a]	Density (g cm^{-3})
Hydrogen	g	0.000 090[b]
Oxygen	g	0.001 43[b]
Alcohol (ethanol)	l	0.785
Benzene	l	0.880
Water	l	0.998
Magnesium	s	1.74
Salt (sodium chloride)	s	2.16
Aluminum	s	2.70
Iron	s	7.87
Copper	s	8.96
Silver	s	10.5
Lead	s	11.34
Mercury	l	13.6
Gold	s	19.32

[a] The abbreviations are g, gas; l, liquid; s, solid.
[b] At STP (standard temperature and pressure, that is, 0 °C and 1 atmosphere); see Chapter 3.

Solution

We have the relationship $1 \text{ L} = 1000 \text{ cm}^3$, and since density = mass/volume, we can write mass = (volume)(density). Therefore the mass of 1 L of mercury is $(1 \text{ L}) \cdot (13.6 \text{ g cm}^{-3})$, or

$$(1 \text{ L})\left(\frac{1000 \text{ cm}^3}{1 \text{ L}}\right)(13.6 \text{ g cm}^{-3}) = (1000 \text{ cm}^3)(13.6 \text{ g cm}^{-3})$$

$$= 1.36 \times 10^4 \text{ g}$$

The mass in kilograms is

$$1.36 \times 10^4 \text{ g} = (1.36 \times 10^4 \text{ g})\left(\frac{1 \text{ kg}}{1000 \text{ g}}\right) = 13.6 \text{ kg}$$

From Table 1.3 we have the relationship $1 \text{ lb} = 453.6 \text{ g}$, so the mass in pounds is

$$1.36 \times 10^4 \text{ g} = (1.36 \times 10^4 \text{ g})\left(\frac{1 \text{ lb}}{453.6 \text{ g}}\right) = 30.0 \text{ lb}$$

Alternatively, we may use the relationship $1 \text{ kg} = 2.205 \text{ lb}$:

$$13.6 \text{ kg} = (13.6 \text{ kg})\left(\frac{2.205 \text{ lb}}{1 \text{ kg}}\right) = 30.0 \text{ lb}$$

EXERCISE 1.4

A bar of silver with dimensions 2.50 cm × 10.0 cm × 0.50 cm was weighed and found to have a mass of 131 g. What is the density of silver?

1.3
Solids, Liquids, and Gases

One of the most obvious properties of substances is that they can exist as **solids**, **liquids**, or **gases**. We call these the **three states of matter**. When a gas is cooled, it eventually condenses to a liquid and finally freezes to a solid, but it remains the same substance. Water exists in all three forms on the earth's surface. Gaseous water (water vapor) is present in the atmosphere; liquid water is present in rivers, lakes, and oceans; and solid water (ice) is present in snow, in glaciers, and on the surfaces of frozen lakes and oceans.

MACROSCOPIC DESCRIPTION OF THE STATES OF MATTER

A **gas** is distinguished from other states of matter by two characteristic properties: (1) it is a fluid with no definite shape, and (2) it has no definite intrinsic volume but flows and expands to fill any container in which it is placed. If the volume of the container is reduced, the gas is easily compressed to the smaller volume. A **liquid** is also a fluid, but a given amount of liquid has its own definite volume. A liquid flows and takes the shape of a container, but it does not expand to completely fill a container of larger volume. In contrast, a **solid** is not a fluid; it does not flow. Any piece of a solid has a definite size and shape that does not depend on its container. Moreover, this shape can only be changed by exerting considerable forces on the solid. Unlike a gas, solids and liquids are only slightly compressible. A much greater force is needed to compress a liquid or a solid than a gas. A liquid usually has a much greater density than a gas, and a solid usually has a slightly greater density than the corresponding liquid.

These descriptions of solids, liquids, and gases are based on observations with our unaided senses. They are **macroscopic descriptions** of the different states of matter, that is, descriptions in terms of properties such as shape, fluidity, density, and hardness that we can recognize by using our unaided senses.

MICROSCOPIC DESCRIPTION OF THE STATES OF MATTER

One of the main aims of chemistry—and, indeed, of all of the sciences—is to explain the properties of matter in terms of a theory or a model of the fundamental nature of matter. The model that is basic to modern chemistry is the atomic theory that was first clearly enunciated by Dalton, namely, that matter consists of very small particles called atoms. The atomic description of matter is a **microscopic description**, in contrast to the macroscopic description that we obtain with our unaided senses. In chemistry we are constantly striving to obtain a better understanding of the observed properties of matter in terms of the behavior of atoms and molecules.

Dry ice is solid carbon dioxide. It sublimes at $-78\ °C$.

We can use the element bromine as an example of the macroscopic and microscopic descriptions of the three states of matter. At room temperature bromine is a deep red brown liquid with a strong pungent odor. It causes severe burns on the skin. It freezes at $-7.2\ °C$ and boils at $58.8\ °C$. If a small amount of liquid bromine is sealed in a glass tube, we can freeze it to a brown solid by placing the tube in dry ice or an ice–salt mixture. If the tube is gently warmed, the bromine melts to a brown liquid. And if the warming is continued, the tube becomes filled with brown bromine gas (Figure 1.14).

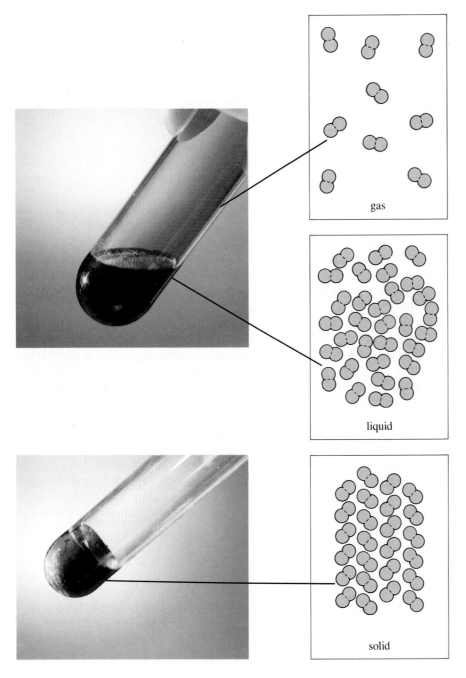

FIGURE 1.14
Solid, Liquid and Gas States of Bromine.

At −15 °C bromine is a dark brown solid. The bromine molecules are packed closely together in a regular arrangement. At room temperature bromine is a dark brown liquid. The molecules are still packed closely together, but their arrangement is less regular, and they are constantly moving around each other. Brown bromine gas is also seen clearly in the tube above the liquid. In the gas bromine molecules are far apart and are moving rapidly so that they fill the whole of the tube. Because the molecules are far apart, the color of the gas is much less intense than the color of the liquid.

gas

liquid

solid

What has happened at the atomic level that would explain these changes? Bromine consists of diatomic molecules, Br_2. The molecules are held together by forces of attraction that act between all molecules. These forces are called **intermolecular forces**; we will discuss them in detail in later chapters. In solid bromine the molecules are packed closely together in a regular manner, as Figure 1.14 illustrates. Although each molecule can move a little, its motion is restricted by the other molecules packed around it, and each individual molecule can do no more than oscillate and vibrate slightly around a fixed mean position.

Because the molecules are packed closely together and cannot be easily displaced, the density of the solid is rather high (4.2 g cm^{-3}), and it is relatively hard and rigid.

As the temperature of solid bromine is increased, the molecules vibrate and oscillate more and more violently. Eventually, they break loose from their fixed positions and move around each other. They remain packed rather closely together but in a random rather than a regular fashion. The solid melts and becomes a liquid, which is fluid because the molecules are now free to move around each other. Because of the increased motion of the molecules and their irregular packing, they take up a little more space than they did in the solid, and the density of the liquid (3.14 g cm^{-3}) is less than the density of the solid (4.2 g cm^{-3}).

As the temperature of the liquid is increased, the molecules move still more rapidly. And some of them move fast enough to fly off the surface of the liquid bromine, becoming bromine vapor, or gas. Eventually, at a temperature of $58.8\ °C$ all the bromine is converted to the gaseous state; in other words, bromine boils at $58.8\ °C$ at normal atmospheric pressure. When a gas is formed from a liquid, it is often called a **vapor**. Thus we often speak of bromine vapor, but there is no difference between a vapor and a gas.

In the gaseous state the bromine molecules are rather far apart and are moving rapidly and randomly through space. Because the molecules no longer stick closely together, the gas expands to fill any container in which it is placed. The density of the gas at the boiling point, 5.9×10^{-3} g cm^{-3}, is much lower than the density of the liquid. One gram of gaseous bromine at normal atmospheric pressure and $58.8\ °C$ occupies a volume that is 500 times greater than the volume of 1 g of liquid bromine under the same conditions. Whereas the molecules in a liquid are packed closely together, the molecules in a gas are far apart. So the total volume occupied by the gas at ordinary pressure is very large compared with the space taken up by the molecules themselves.

You should take particular note that the transformations from solid to liquid to gaseous bromine are *physical changes* and not chemical changes. Bromine consists of the same Br_2 molecules in the solid, liquid, and gaseous states.

= *Example 1.5* =

THE SOLID, LIQUID, AND GASEOUS STATES OF MATTER

Ice has a density of 0.91 g cm^{-3} at $0\ °C$. Liquid water has a density of 1.00 g cm^{-3} at $25\ °C$. Steam has a density of 5.9×10^{-4} g cm^{-3} at $100\ °C$ and 1 atmosphere (atm) pressure. Calculate the volume of 100 g of water (a) as a solid, (b) as a liquid, and (c) as a gas. Also, calculate the ratios of these volumes.

Solution

First, recall that density = mass/volume. Therefore

$$\text{Volume} = \frac{\text{Mass}}{\text{Density}}$$

(a) $$\text{Volume of 100 g ice} = \frac{100 \text{ g}}{0.91 \text{ g cm}^{-3}} = 110 \text{ cm}^3$$

$$= (110 \text{ cm}^3)\left(\frac{1 \text{ L}}{1000 \text{ cm}^3}\right) = 0.11 \text{ L}$$

(b) $\text{Volume of 100 g water} = \dfrac{100 \text{ g}}{1.00 \text{ g cm}^{-3}} = 100 \text{ cm}^3 = 0.10 \text{ L}$

(c) $\text{Volume of 100 g steam} = \dfrac{100 \text{ g}}{5.9 \times 10^{-4} \text{ g cm}^{-3}} = 1.7 \times 10^5 \text{ cm}^3 = 170 \text{ L}$

The ratios of these volumes is as follows:

$$\text{ice:water:steam} = 0.11:0.10:170$$

Dividing by 0.10 gives $1.1:1.0:1700$

Thus steam has a volume that is 1700 times as great as the volume of an equal mass of liquid water. This again emphasizes that the water molecules in steam are much further apart than they are in liquid water. Note also that water is a very unusual liquid in that the density of the liquid is greater than the density of the solid (ice). This unusual property of water is discussed in Chapter 12.

1.4
Solutions

A homogeneous mixture of two or more substances is usually known as a **solution**. Mixtures of gases are homogeneous and they are therefore solutions, but the term is usually employed to describe homogeneous mixtures of two or more liquids or of a liquid and one or more solids. When two substances are mixed to form a solution, one is said to dissolve in the other, or to be soluble in the other. We normally refer to the substance present in largest amount as the **solvent** and the other substances as **solutes**. If the solution consists of a solid in a liquid, however, the solid is always called the solute and the liquid the solvent.

TYPES OF SOLUTIONS

Liquids are probably the most familiar solvents. Many solid substances dissolve in particular liquids. For example, sodium chloride dissolves in water but not in gasoline, whereas paraffin wax dissolves in gasoline but not in water. Liquids may also dissolve in other liquids; for example, ethanol dissolves in water, and lubricating oil dissolves in gasoline. Liquids that are soluble in each other are said to be **miscible**. But not all liquids are soluble in each other. Although ethanol dissolves in water in all proportions, both gasoline and carbon tetrachloride are insoluble in water (Experiment 1.6). Liquids that are insoluble in each other are said to be **immiscible**.

Gases also dissolve in many liquids. A solution of ammonia, NH_3, in water is commonly used as a household cleaner. Oxygen and nitrogen are soluble in water to a small extent. In fact, fish depend on the small amount of dissolved oxygen in rivers, lakes, and oceans. Solutions in water are called **aqueous solutions**.

Gases may also dissolve in solids. For example, hydrogen is soluble in platinum. A solid may also form a solid solution with another solid. Sterling silver is a solution of copper in silver. Solid solutions of metals are generally called *alloys*, although an alloy may also be a heterogeneous mixture of metals.

Experiment 1.6

Immiscible Liquids and Solubility

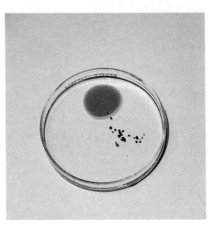

Water and carbon tetrachloride, CCl_4, immiscible. Carbon tetrachloride forms a small pool when added to water. Solid iodine dissolves in the colorless carbon tetrachloride to give a violet solution but it is insoluble in water.

Copper sulfate dissolves in water to give a blue solution but it is insoluble in carbon tetrachloride, which remains colorless.

CONCENTRATION

The term **concentration** is used to describe the amount of solute dissolved in a given quantity of solution. A solution having a relatively small concentration of solute is said to be **dilute**. A solution having a large concentration of solute is said to be **concentrated**.

One way of expressing concentration in a quantitative manner is as the **mass percentage** (mass percent) of the solute, which is the number of grams of solute in 100 g of solution. For example, a solution of 10 g of alcohol (ethanol) in 90 g of water (100 g of solution) is a 10% solution. For a solution containing two components (solute plus solvent),

$$(\text{Mass percent})_{\text{solute}} + (\text{Mass percent})_{\text{solvent}} = 100$$

Beer is typically 4% to 6% alcohol and wine 10% to 14%.

Example 1.6

EXPRESSING CONCENTRATION AS MASS PERCENTAGE

An aqueous solution is prepared by adding 3.42 g of magnesium chloride, $MgCl_2$, and 2.63 g of sodium chloride, NaCl, to 88.20 g of water, H_2O. What are the concentrations, in mass percentages, of (a) NaCl, (b) $MgCl_2$, and (c) H_2O?

Solution

First, we need to know the total mass of the solution:

$$\text{Total mass of solution} = \text{Mass NaCl} + \text{Mass MgCl}_2 + \text{Mass H}_2\text{O}$$
$$= 2.63 \text{ g} + 3.42 \text{ g} + 88.20 \text{ g}$$
$$= 94.25 \text{ g}$$

(a) There are 2.63 g of NaCl in 94.25 g of solution; therefore

$$(2.63 \text{ g})\left(\frac{100 \text{ g}}{94.25 \text{ g}}\right) = 2.79 \text{ g in 100 g solution}$$

The concentration of NaCl is thus 2.79 mass percent.

(b) There are 3.42 g of $MgCl_2$ in 94.25 g of solution; thus

$$(3.42 \text{ g})\left(\frac{100 \text{ g}}{94.25 \text{ g}}\right) = 3.63 \text{ g in 100 g solution}$$

The concentration of $MgCl_2$ is 3.63 mass percent.

(c) There are 88.20 g of H_2O in 94.25 g of solution; therefore

$$(88.20 \text{ g})\left(\frac{100 \text{ g}}{94.25 \text{ g}}\right) = 93.58 \text{ g in 100 g solution}$$

So the concentration of H_2O is 93.6 mass percent.

Finally, as a check, we note that

$$\text{Mass percent NaCl} + \text{Mass percent MgCl}_2 + \text{Mass percent H}_2\text{O}$$
$$= 2.79 + 3.63 + 93.6 = 100.0\%$$

Example 1.7

USING MASS PERCENTAGE

Concentrated aqueous nitric acid has 69.0% by mass of HNO_3 and has a density of 1.41 g cm^{-3}. What volume of this solution contains 14.2 g of HNO_3?

Solution

The 14.2 g of HNO_3 is contained in

$$\left(\frac{100 \text{ g aqueous solution}}{69.0 \text{ g HNO}_3}\right)(14.2 \text{ g HNO}_3) = 20.6 \text{ g concentrated aqueous HNO}_3$$

Using the relationship volume = mass/density, we have

$$\text{Volume} = \frac{20.6 \text{ g}}{1.41 \text{ g cm}^{-3}} = 14.6 \text{ cm}^3$$

EXERCISE 1.5

A 1.00 L volume of battery acid (an aqueous solution of sulfuric acid) contains 368 g of sulfuric acid, H_2SO_4, and has a density of 1.230 g cm^{-3}. What is the mass percentage of sulfuric acid in the solution?

Frequently, we need to give values for the concentrations of substances that are present in solutions in very small amounts. For example, just 2×10^{-5} g of copper per liter of water is believed to be lethal to fish. Such very small concentrations may be more conveniently expressed in *parts per million* (ppm). If there is 2×10^{-5} g of copper per liter of water, there is 2×10^{-5} g per 1000 g of water, since the density of water is 1.00 g cm^{-3}. There is therefore 0.02 g per 1 000 000 g of water, or 0.02 parts per million, that is, 0.02 ppm.

= *Example 1.8* =

EXPRESSING CONCENTRATIONS IN PARTS PER MILLION (ppm)

Seawater contains 0.0064 g of dissolved oxygen, O_2, per liter. The density of seawater is 1.03 g cm^{-3}. What is the concentration of oxygen, in parts per million?

Solution

$$\text{Mass 1 L seawater} = \text{Volume} \times \text{Density}$$

$$= (1 \text{ L})\left(\frac{10^3 \text{ cm}^3}{1 \text{ L}}\right)(1.03 \text{ g cm}^{-3}) = 1.03 \times 10^3 \text{ g}$$

$$\text{Mass } O_2 \text{ in } 10^6 \text{ g seawater} = (6.4 \times 10^{-3} \text{ g } O_2)\left(\frac{1 \times 10^6 \text{ g seawater}}{1.03 \times 10^3 \text{ g seawater}}\right)$$

$$= 6.2 \text{ g } O_2$$

There are 6.2 g of O_2 in 10^6 g of seawater. Therefore, the concentration of O_2 in seawater is 6.2 ppm.

The concentrations of gases present in the atmosphere in small traces are also often expressed in parts per million, but in this case volumes are used rather than masses. For example, 1 ppm of a trace gas is equivalent to dissolving 1 cm^3 of the gas in 1 000 000 cm^3, or 1000 L, of air. The concentration of helium in the atmosphere is 5 ppm. In an industrial city the concentration of sulfur dioxide, SO_2, in the atmosphere may be as high as 5 ppm or more, while in open country at some distance from a large city the concentration of SO_2 may be less than 0.01 ppm.

SOLUBILITY

The extent to which a solid will dissolve in a liquid is limited. If sugar is stirred into water at a given temperature, increasing amounts may be dissolved up to a certain point, after which no more sugar will dissolve. *A solution that contains the maximum amount of solute that can be dissolved at a particular temperature is said to be* **saturated**. The concentration of solute in a saturated solution is called the **solubility** of the solute in that solvent. Solubilities are often expressed in grams of solute per 100 g of solvent, or, alternatively, in grams of solute per liter of solution.

EQUILIBRIUM Once a saturated solution has been prepared at a given temperature, no more solute will dissolve. No matter how much additional solid is added and no matter how long it is left, the concentration of the solute remains unchanged. But this unchanged concentration does not mean that nothing is happening. In fact, solute continues to dissolve, but the amount that dissolves per second is exactly equal to the amount that simultaneously comes out of the solution as solid—that is, by the amount that crystallizes.

In other words, two processes are taking place: the solution process and its reverse, which is called **crystallization**. The solute dissolves and crystallizes at the same rate. In a saturated solution the number of solute molecules going into solution exactly balances, at all times, the number leaving the solution to form the solid. Thus the concentration of solute in a saturated solution and the amount of undissolved solute remain constant at a particular temperature. The saturated solution is said to be in **equilibrium** with the excess solid solute. This situation is an example of a **dynamic equilibrium**, in which two opposing processes occur at the same rate so that there is no overall or net change (Figure 1.15). Using the concept of equilibrium, *we can redefine a saturated solution as one that is in equilibrium with excess undissolved solute*.

SOLUBILITY AND TEMPERATURE The solubility of a substance in a given solvent depends on the temperature. Sugar, for example, is more soluble in hot water than in cold. Although the solubility of many substances in water increases with increasing temperature, as Figure 1.16 illustrates, the solubility of some substances remains nearly constant or even decreases. The solubility in water of many gases such as carbon dioxide and oxygen also decreases with increasing temperature.

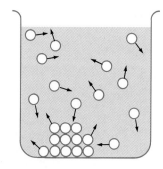

FIGURE 1.15
Dynamic Equilibrium in a Saturated Solution.

Solute molecules leave the crystal surface (dissolve) at the same rate as solute molecules attach themselves to the crystal surface (crystallize).

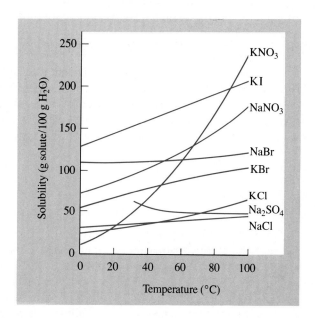

FIGURE 1.16
Temperature Dependence of Solubility.

The effect of temperature on solubility varies from one substance to another. The solubility of potassium nitrate, KNO_3, increases very rapidly with increasing temperature. The solubility of sodium chloride, NaCl, hardly changes, while the solubility of sodium sulfate, Na_2SO_4, decreases with increasing temperature.

1.5
Separation of Mixtures

The separation of mixtures into their component substances is an essential technique both in the laboratory and in industry. Substances rarely occur in nature in the pure state and before they can be used they must often be separated from the other substances that are present (impurities). A chemical reaction often gives several products only one of which is needed, so it must be separated from the other products. Chemists use many procedures for separating mixtures, but three of the most common are crystallization, distillation, and chromatography.

CRYSTALLIZATION

The **crystallization** method of purification depends on the change in solubility of a substance with a change in the temperature. Suppose, for example, that we have a sample containing 80 g of potassium nitrate, KNO_3, and 10 g of potassium chloride, KCl, and we want to obtain a pure sample of potassium nitrate. First, we dissolve the potassium nitrate–potassium chloride mixture in, for example, 100 g of water at a temperature of about 100 °C. If we then allow the solution to cool slowly, it becomes saturated with potassium nitrate at about 50 °C and potassium nitrate starts to crystallize from the solution. If the solution is cooled to 0 °C, 67 g of pure potassium nitrate crystallizes; only 13 g remains in solution because its solubility is 13 g in 100 g of water at 0 °C. Since the solubility of potassium chloride at this temperature is 27 g in 100 g of water, all the 10 g of potassium chloride remains in solution.

   Now we can separate solid potassium nitrate from the solution by the process of **filtration**, which is shown in Figure 1.17. The mixture of saturated solution and crystalline potassium nitrate is poured through a *filter funnel* containing

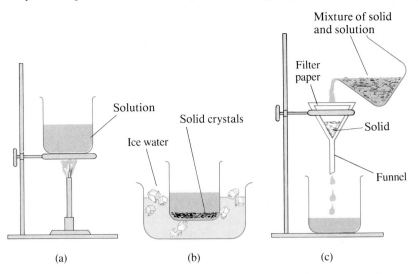

(a)   (b)   (c)

FIGURE 1.17
Crystallization and Filtration.

(a) The solid is dissolved in a minimum amount of hot water. (b) The solution is cooled until some solid crystallizes. (c) The mixture is poured through the filter funnel. The filter paper retains the solid, allowing the solution to pass through.

Filtering a red solid

either a porous paper known as *filter paper* or a *sintered glass disk*. The filter allows the solution to pass through but retains the solid potassium nitrate, which may then be washed free from adhering solution with small amounts of cold water and finally dried.

Crystallization does not separate the potassium nitrate and potassium chloride completely, since some of the potassium nitrate remains in solution with the potassium chloride. But we can, at least, obtain a large proportion of the originally impure potassium nitrate in a pure form. Crystallization is generally a rather efficient process for purification, particularly when the amounts of impurities are not too large and their solubilities are not very small. If one crystallization does not yield sufficiently pure material, it may be recrystallized.

DISTILLATION

Distillation is a convenient method for separating the liquid from a solution of a solid in a liquid or for separating a solution of two or more liquids. The method is based on the differences in the **volatility** of different substances, that is, differences in the ease with which they become gases. For example, pure water can be obtained from seawater in the apparatus shown in Figure 1.18. When seawater is heated, the water eventually boils. But at this temperature sodium chloride and the other solutes are **nonvolatile**; they do not vaporize. The water vapor passes into the condenser, which is cooled by cold water running through its outer jacket. Thus the water vapor is cooled and changed to liquid water. The process is known as **condensation**. The pure water runs out of the condenser into a receiver, where it is collected. The solid that remains in the flask after all the water has been removed contains sodium chloride and all the other substances present in seawater. Water that has been purified by distillation is commonly called *distilled water*. It is used in the laboratory whenever pure water is needed for preparing an aqueous solution.

Distillation can also be used to separate a solution of two or more liquids, although separation is not complete unless the two liquids have very different boiling points. For example, pure water may be obtained by distillation of a mixture of water and glycol, $C_2H_6O_2$ (a major component of antifreeze). The

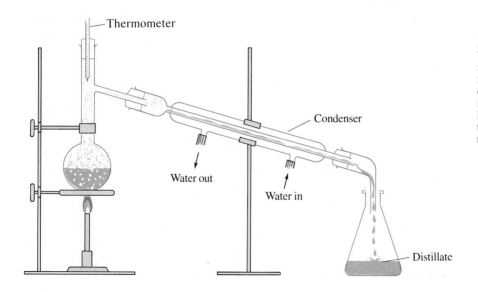

FIGURE 1.18
Distillation Apparatus.

The nonvolatile component of the mixture remains in the distilling flask. The volatile component is converted to vapor, which passes into the water-cooled condenser. Here the vapor is converted back to a liquid which flows into the receiving flask.

FIGURE 1.19
Distillation with Fractionating
Column.

The column is packed with an
inert material, such as glass beads.
Vapor rises up the column and
condenses on the beads. It is re-
vaporized by the hot vapor rising
up the column, and it condenses
again further up in the column,
and so on. The vapor becomes
successively richer in the component of lower boiling point, and
the pure, lower-boiling component
leaves the top of the column and
is reconverted to liquid in the
condenser. The process of condensation and re-evaporation is
equivalent to repeated distillations.

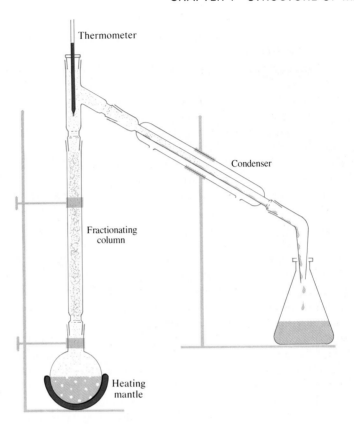

vapor entering the condenser is richer in water vapor than in glycol, which has
a boiling point of 179 °C. So the resulting condensed liquid, which is known as
the **distillate**, contains a higher proportion of water and a lower proportion of
glycol than the original mixture. The liquid remaining in the flask contains a
higher proportion of glycol. Redistillation of the distillate further enriches it in
the lower-boiling-point component, and by repeated redistillations a sample of
pure water can eventually be obtained.

Such a tedious and time-consuming procedure is avoided in practice by use
of a *fractionating column* between the distillation flask and the condenser, as
shown in Figure 1.19. The column is packed with some inert material, such as
glass beads, which does not react with the substances being distilled. The vapor
ascending the column condenses on the glass beads. Then as this liquid trickles
down the column, it is heated by the rising vapor. Some of it again vaporizes,
to give a vapor still richer in the component with the lower boiling point. This
vapor condenses further up in the column, is heated by the rising vapor, and
again is partially vaporized, to give a vapor still richer in the lower-boiling-
point component. The vapor emerging from the top of a column of sufficient
length condenses to a liquid, which is essentially the pure, lower-boiling-point
component. This process is known as **fractional distillation**.

Fractional distillation is important in many industrial processes. For example, petroleum (which is a complex mixture) is separated into gasoline, heating
oil, and other materials by distillation in a fractionating column. Pure oxygen
and pure nitrogen are obtained from air by first liquefying it at a low temperature and then carrying out a fractional distillation.

Experiment 1.7

Paper Chromatography of Ink

A black ink line is drawn at the bottom of a long strip of absorbent paper. The paper is suspended in a mixture of ethanol and water.

As the solvent is drawn up the paper the ink separates into colored bands. Each color corresponds to a compound in the ink.

This shows the progress of the separation with time. The strips were removed from the solvent at different times during the separation.

CHROMATOGRAPHY

The name **chromatography** is based on the Greek word *chrōma*, for "color" since the method was first used for the separation of colored substances found in plants.

There are several different types of chromatography, but they are all based on the same principle. Experiment 1.7 illustrates the use of **paper chromatography**, to separate the different colored substances in ink. The colored substances in the ink have different tendencies to be adsorbed on, that is to stick to, the surface of the paper. The solvent dissolves the colored substances but as the solvent moves up the paper some of the substances move more slowly than the others because they stick to the paper more strongly. In other words, the different substances move up the paper at different rates. Eventually several colored bands corresponding to the various substances in the ink are obtained.

Paper chromatography is useful for identifying substances, but it is not very useful for the separation of significant amounts of the substances in a mixture. For this purpose **liquid-column chromatography** can be used, as illustrated in Figure 1.20 on page 40.

For the separation of gases and volatile liquids **gas–liquid chromatography** is very useful. A long tube is packed with a finely divided and inert solid. The surface of the solid is coated with a liquid that does not react with the substances to be separated (Figure 1.21, p. 40). A stream of an inert gas such as helium, He, or nitrogen, N_2, passes continually through the tube. A mixture of substances to be separated for identification is injected into the stream of inert gas and is vaporized by heating. The vapor is swept down the tube by the inert gas. Those substances that are more soluble in the liquid lag behind, and the less soluble

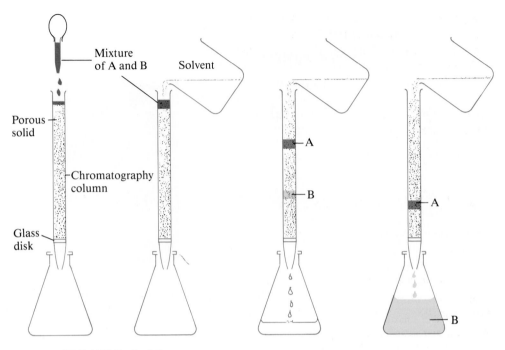

FIGURE 1.20
Chromatographic Column.

The column is packed with an adsorbent material such as aluminum oxide, Al_2O_3.
A solution of the mixture to be separated is poured into the top of the column.
A suitable solvent is then poured slowly through the column. The solvent carries
the different substances in the mixture down the column at different rates so that
they separate into bands, which can be readily identified if they are colored. Each
band can then be separately washed out of the column into the receiving flask.

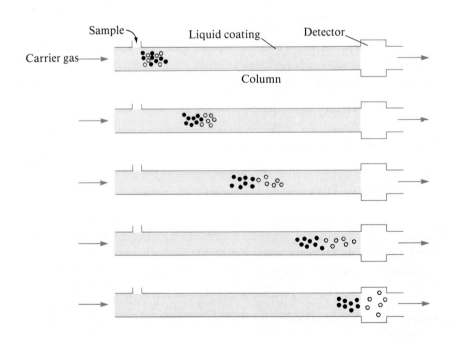

FIGURE 1.21
Schematic Diagram of
Gas–Liquid Chromatography
Apparatus.

components of the mixture move more rapidly down the tube. Eventually, all the components separate from each other, and they emerge one by one from the end of the tube. They can then be separately identified.

The development of the various forms of chromatography in the past twenty-five years has been one of the major advances in experimental chemistry. Separations previously considered impossible are now achieved easily. Perhaps the greatest impact of chromatography has been in biochemistry, where complex mixtures are frequently encountered.

IMPORTANT TERMS

An **aqueous solution** is a solution in water.

An **atom** is the smallest constituent of an element that retains the properties of the element. It consists of a nucleus surrounded by one or more electrons.

A **chemical property** is any property that relates to a substance's participation in a chemical reaction.

A **chemical reaction** is a process by which one or more substances are changed into one or more different substances; a process in which atoms are rearranged to form new combinations.

Chromatography is a physical method, based on the selective adsorption of gases or liquids on the surface of solids or on their relative solubilities in liquids, by which complex mixtures can be separated into their components.

A **compound** is a substance containing two or more kinds of atoms that are present in fixed ratios. In other words a compound always has the same constant chemical composition.

Condensation is a process in which a substance changes from the gaseous form (vapor) to the liquid form.

Crystallization is the formation of crystals from a saturated solution of a substance or from the molten state.

The **density** of a substance is its mass per unit volume:

$$\text{Density} = \frac{\text{Mass}}{\text{Volume}}$$

Diatomic molecules are molecules that contain only two atoms.

Distillation is a method of purifying liquids in which they are changed into a vapor and then condensed back to the liquid form.

Dynamic equilibrium is a situation in which a chemical or physical change and its reverse occur at the same rate so that there is no overall change in the system.

An **element** is a substance all of whose atoms are of the same kind.

A **formula** is a combination of element symbols and numerical subscripts that indicates either the relative numbers of each kind of atom in a substance (*empirical formula*) or the actual numbers of each kind of atom in a molecule (*molecular formula*).

A **heterogeneous mixture** of two or more materials has a nonuniform composition. Different parts of the material have different properties.

A **homogeneous mixture** of two or more substances has a uniform but variable composition and variable chemical and physical properties. A homogeneous mixture is also called a *solution*.

Macroscopic is a term used to describe objects that are large enough to be visible to the naked eye.

Microscopic is a term used to describe very small objects. Literally, it means "visible only under a microscope", but it is applied particularly to objects of atomic or molecular size.

Two liquids are said to be **miscible** when they form a homogeneous mixture or solution. Liquids that do not mix are said to be **immiscible**.

A **molecule** is a combination of atoms that has its own characteristic set of properties.

A **physical change** is any change in which the composition and amounts of each of the substances present do not change. In other words, a physical change is any change in which no chemical reaction occurs.

A **physical property** is any property of a substance that can be observed and measured without the substance changing into other substances. For example, melting point, boiling point, color, and density are physical properties.

A **polyatomic molecule** is a molecule that contains more than two atoms.

A **saturated solution** is a solution that contains the maximum amount of a substance (solute) that can be dissolved in a given amount of solvent at a specified temperature; a solution in which the dissolved solute is in equilibrium with the solid solute.

Solubility is the concentration of a solute in a saturated solution of that solute at a specified temperature.

A **solute** is the minor component of a solution or the component that is solid in the pure state.

Solution. See homogeneous mixture.

A **solvent** is the major component of a solution or the component that is liquid in the pure state.

A **substance** is a homogeneous portion of matter that has a fixed, constant composition and a characteristic set of chemical and physical properties.

A **tetrahedron** is a regular solid with four identical, equilateral triangular faces, six sides of equal length, and four equivalent vertices.

A **vapor** is a gas. The term is usually used for a gas that has been, or is being, formed from a liquid.

A **volatile** liquid is one that readily forms a vapor, that is, a gas.

P R O B L E M S *

Elements, Compounds, Formulas, Mixtures and Solutions

1. Give the chemical symbol for each of the following elements:

   (a) Hydrogen   **(b)** Oxygen   **(c)** Carbon

   (d) Nitrogen   **(e)** Chlorine   **(f)** Magnesium

   (g) Sodium   **(h)** Potassium   **(i)** Sulfur

   (j) Phosphorus

2. Give the chemical symbol for each of the following elements:

   (a) Iron   **(b)** Nickel   **(c)** Mercury

   (d) Silicon   **(e)** Chromium   **(f)** Helium

   (g) Barium   **(h)** Lead   **(i)** Uranium

   (j) Calcium

3. Give the names of the elements with each of the following symbols:

   (a) H   **(b)** He   **(c)** Ne   **(d)** F   **(e)** Mg

   (f) Al   **(g)** P   **(h)** S   **(i)** K   **(j)** Na

4. Give the names of the elements with each of the following symbols:

   (a) Ca   **(b)** Br   **(c)** Fe   **(d)** Mn   **(e)** Cu

   (f) Ag   **(g)** Au   **(h)** Zn   **(i)** As   **(j)** Pt

5. The symbols for the elements iron, copper, lead, and tin are based on their Latin names. Propose two-letter symbols based on the English names that would not duplicate existing symbols.

6. Classify each of the following as an element, a compound, or a mixture:

   (a) Water   **(b)** Iron   **(c)** Beer   **(d)** Sugar

   (e) Wine   **(f)** Steel   **(g)** Sulfur   **(h)** Milk

   (i) Air   **(j)** Magnesium oxide

* Answers to problems numbered in blue appear at the end of the book.

7. **(a)** Classify each of the following as a pure substance or as a mixture.

   (b) Divide the pure substances into elements and compounds, and divide the mixtures into homogeneous or heterogeneous mixtures.

Nitrogen	Gasoline
Salad dressing	Cottage cheese
Iron	Milk
Carbon dioxide	Black coffee
Wet sand	Oxygen
Carbon	Sodium chloride
Iodized table salt	Filtered seawater
Diamond	Distilled water
Nylon	Automobile tires
Vegetable soup	Concrete
Smog	

8. Classify each of the following as a homogeneous or heterogeneous substance:

Soda water	Filtered coffee	Wood	Snow
Soil	"Dry-Ice"	Vinegar	Air

9. Classify each of the following substances as an element, a compound, or a mixture:

   (a) Tin   **(b)** Lemon juice

   (c) Beer   **(d)** Natural gas

   (e) Uranium   **(f)** 3% Aqueous hydrogen peroxide

   (g) Sodium nitrate

   (h) A solution of iodine in tetrachloromethane

   (i) Air   **(j)** Hydrogen peroxide

10. Give at least two physical properties (for example, color, physical state, melting point, density, electrical conductivity) of each of the following:

Water	Sugar	Mercury
Copper	Oxygen	Bromine
Magnesium oxide	Sodium chloride	

11. Give one chemical property (reaction) that is characteristic of each of the following:

Water   Copper
Iron   Magnesium
Hydrogen   Hydrogen peroxide

12. Describe an experiment, or give a simple physical property (not taste or smell), that would help you to distinguish between each of the following:

(a) Two bottles of colorless liquid, one water, and the other a solution of sodium chloride in water

(b) Two samples of white powder, one chalk and the other sodium chloride

13. Write molecular and empirical formulas for molecules of each of the following:

(a) An elemental form of arsenic containing four arsenic atoms

(b) A compound of carbon and hydrogen consisting of molecules containing three carbon atoms and six hydrogen atoms

(c) An oxide of phosphorus consisting of molecules containing four phosphorus atoms and ten oxygen atoms

(d) A compound of xenon and fluorine consisting of molecules containing one xenon atom and four fluorine atoms

14. What is the empirical formula of each of the substances with the following molecular formulas?

(a) N_2O_4 **(b)** N_2O_5 **(c)** P_4

(d) Hg_2Cl_2 **(e)** $AlCl_3$ **(f)** C_3H_8

(g) C_4H_{10} **(h)** $C_6H_{12}O_6$ **(i)** C_6H_{12}

15. What is the empirical formula of each of the substances with the following molecular formulas?

(a) H_2O_2 **(b)** H_2O **(c)** Li_2CO_3

(d) $C_2H_4O_2$ **(e)** S_8 **(f)** C_6H_{14}

(g) B_2H_6 **(h)** O_3

16. Give the name of each of the substances with the following molecular formulas:

(a) H_2O **(b)** H_2O_2 **(c)** CH_4

(d) C_2H_6 **(e)** CO_2 **(f)** CO

(g) $C_{12}H_{22}O_{11}$

17. Give the name of each of the substances with the following molecular formulas:

(a) Mg   **(b)** MgO   **(c)** NaCl

(d) P_4 **(e)** S_8 **(f)** HNO_3

(g) H_2SO_4 **(h)** KNO_3 **(i)** Na_2SO_4

(j) $BaSO_4$

18. Give the molecular formula of each of the following substances:

(a) Hydrogen   **(b)** Oxygen

(c) Nitrogen   **(d)** Water

(e) Hydrogen peroxide   **(f)** Nitric acid

(g) Sulfuric acid   **(h)** Potassium nitrate

19. Give the molecular formula of each of the following substances:

(a) Magnesium oxide   **(b)** Carbon monoxide

(c) Carbon dioxide   **(d)** Bromine

(e) Phosphorus   **(f)** Sulfur

(g) Ammonia   **(h)** Methane

(i) Ethane   **(j)** Sodium chloride

20. Classify each of the following as a diatomic, a triatomic, or a polyatomic molecule:

F_2 NO_2 N_2O C_2H_6 P_4
H_2O_2 P_4O_{10} O_3 HCl CH_4

Molecular Structure

21. The HSH bond angle in the H_2S molecule is very close to 90°. Calculate the H–H distance given that the H–S distance is 134 pm.

22. Draw diagrams to scale showing the bond lengths and shapes of each of the following molecules:

(a) Hydrogen, H_2, H–H = 74 pm; Oxygen, O_2, O–O = 121 pm; Nitrogen, N_2, N–N = 109 pm; Carbon monoxide, CO, C–O = 113 pm

(b) Carbon dioxide, CO_2, linear, C–O = 116 pm

(c) Ozone, O_3, angular, O–O = 128 pm, $\widehat{OOO}$ = 117°

(d) Arsenic, As_4, tetrahedral, As–As = 244 pm

23. What is the shape of each of the following molecules?

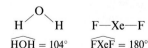

$\widehat{HOH}$ = 104° $\widehat{FXeF}$ = 180°

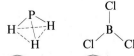

$\widehat{HPH}$ = 92° $\widehat{ClBCl}$ = 120°

*24. By constructing a tetrahedron inside a cube, show by trigonometry that in the methane molecule, where the four hydrogen atoms are at the vertices of the tetrahedron, and the carbon atom is at its center, each of the HCH angles is 109° 28′.

*25. Determination of the molecular structure of water shows that it contains angular H_2O molecules, with identical O–H distances and an HOH bond angle of 104.5°. If the observed H---H distance is 153.3 pm, what is the O—H distance?

26. For a tetrahedron:

   (a) How many faces are there?

   (b) What is the angle formed by any two edges?

   (c) How many such angles are there?

Scientific Notation and Units

27. Which of the following are *exact* numbers?

   (a) There are *12* eggs in a dozen.

   (b) There are *24* hours in a day.

   (c) The newspaper gave the attendance at a football game as *64 000*.

   (d) Peter weighs *165* lb.

   (e) There are *1760* yards in a mile.

   (f) There are *268* students in the chemistry class.

28. Express each of the following numbers in scientific (exponential) notation:

   (a) 12 000   (b) 1 740 312.29

   (c) 0.004 04   (d) −0.049 00

How many significant figures does each number contain?

29. Express each of the following numbers without using scientific notation:

   (a) 3.0×10^2 (b) 1.162×10^5

   (c) 4.8×10^{-3} (d) -6.440×10^{-2}

How many significant figures does each number contain?

30. The number of ways of playing the first ten moves in a game of chess is estimated to be 169 518 829 544 000 000 000 000 000 000. Express this number in scientific notation.

31. Evaluate each of the following expressions, after writing each number in scientific notation:

   (a) (0.000 04)(0.001)(5000)   (b) $\dfrac{0.000\ 12}{0.06}$

   (c) $(200)^3(0.0009)^{1/2}$ (d) $\dfrac{(0.02)(600)(50)}{(0.003)(500)}$

   (e) $(0.000\ 009)^{1/2}(20)^3$

* The asterisk denotes the more difficult problems.

32. What is the value of each of the following expressions? Express each answer in scientific (exponential) notation, giving the correct number of significant figures:

   (a) $\dfrac{(6 \times 10^{-6})(3 \times 10^{14})}{(2 \times 10^3)(1 \times 10^6)}$

   (b) $(6.022 \times 10^{23}) + (7.7 \times 10^{21})$

   (c) $\dfrac{(5000)(0.06)}{0.0003}$

   (d) $\dfrac{(3.6 \times 10^{-5})^{1/2}(0.000\ 12)}{3000}$

   (e) $119.2 + (2.0412 \times 10^2) + (3.734 \times 10^{-4})$

33. What are the basic SI units of (a) length, (b) mass, (c) volume and (d) density?

34. By what power of 10 does each of the following prefixes multiply a basic SI unit?

   (a) kilo   (b) milli   (c) centi

   (d) pico   (e) micro

35. What is the prefix for each of the following multiples of a unit?

   (a) 10^3 (b) 10^{-1} (c) 10^{-2}

   (d) 10^{-6} (e) 10^{-12}

36. Express each of the following in kilograms:

   (a) 5.84×10^{-3} mg   (b) 54.34 Mg

   (c) 0.345 g   (d) 1.673×10^{-21} g

37. What unit factor should be used to make each of the following conversions?

   (a) Inches to feet   (b) Feet to inches

   (c) Miles to kilometers   (d) Kilometers to miles

   (e) Cubic meters to milliliters

   (f) Square miles to square kilometers

   (g) Square centimeters to square meters

38. What factor (or factors) should be used to make each of the following conversions?

   (a) Grams to kilograms

   (b) Milligrams to grams

   (c) Centimeters to meters

   (d) Cubic decimeters to cubic centimeters

   (e) Grams per cubic centimeter to kilograms per cubic meter

39. Using the conversion factors given in Table 1.3, express each of the following in SI units and exponential notation:

   (a) 24 000 miles   **(b)** 150 lb

   (c) 14 lb in.$^{-2}$ **(d)** 60 miles h^{-1}

40. Convert each of the following to SI base units, and give the answers in scientific notation. State the number of significant figures in each case.

   (a) 0.0254 cm   **(b)** 0.30×10^6 g

   (c) 365.0×10^{-5} mm   **(d)** 0.065×10^{-2} pm

   (e) 637.1 mL   **(f)** 13.43×10^4 L

   (g) 0.52×10^{-2} kg L^{-1} **(h)** 63.43×10^3 dm h^{-1}

41. Express each of the following in scientific notation and in SI base units:

   (a) 0.1998×10 Gm s^{-1}

   (b) $(6.022 \times 10^{23})(1.673 \times 10^{-24}$ g)

   (c) 32 150 Å min^{-1} **(d)** 22.41 mL

42. What is the capacity in liters of a 20-gallon tank?

43. Express each of the following in meters:

   (a) 2.998×10^7 km   **(b)** 143 pm

   (c) 0.001 nm   **(d)** 1.54 Å

44. An automobile is advertised as having a gas mileage of 5.0 liters per 100 kilometers. What is the mileage in miles per gallon (U.S.)?

45. The speed limit on many Canadian highways is 100 kilometers per hour. What is this speed limit expressed in miles per hour?

46. A person standing on the equator is moving at a speed of 1039 miles per hour because of the rotation of the earth. What is the circumference of the earth **(a)** in miles and **(b)** in meters?

47. If a runner takes 3 min 50 s to run a 1500 m race, how long would the runner take to run 1 mile, assuming that the same average speed is maintained?

48. If the exchange rate is 1.20 Canadian dollars to the American dollar and if gasoline sells for 46.2 cents per liter in Canada, what is the price in U.S. dollars per gallon?

49. An automobile travels 25.4 miles on a gallon of gasoline. What is the gasoline consumption in liters per 100 km?

50. Convert the following description of the moon to convenient metric units:

The moon revolves around the earth with a period of 27.32 days at a distance of 238 850 miles. The radius of the moon is 1081 miles, and its mass is estimated to be 8.1×10^{19} tons (1 ton = 2000 lb).

51. Convert the following description of the planet Jupiter to SI base units:

Jupiter revolves around the sun with a period of 11.86 years at a distance of approximately 4.84×10^8 miles. The average density of Jupiter is 1.330 g cm^{-3}.

52. Astronomical distances are measured in light-years. (1 light-year is the distance traveled by light in 1 year, at a speed of 3.00×10^8 m s^{-1}.) The distance from the earth to the nearest star, Alpha Centauri, is 4.0 light-years. What is this distance **(a)** in kilometers and **(b)** in miles?

53. The populations of Canada, the United States, Australia, and the United Kingdom in 1980 were, respectively, 23 845 000, 220 090 000, 14 510 000, and 55 819 000.

   (a) Express these populations in exponential notation.

   (b) The areas of the four countries are 9 976 139, 9 519 617, 7 686 849, and 244 013 km^2, respectively. Express these areas in exponential notation, and calculate for each country the average population density per square kilometer.

Volume and Density

54. What factor should be used to convert density in grams per cubic centimeter to the SI unit of density, kilograms per cubic meter?

55. The density of chloroform (trichloromethane), $CHCl_3$, is 1.49 g cm^{-3} at 25 °C.

   (a) What is the volume in cubic centimeters of 75.0 g of chloroform?

   (b) What is the mass of 125 cm^3 of chloroform?

56. The mass of 165.0 mL of ethyl alcohol (ethanol), C_2H_6O, is 129.5 g.

   (a) What is the density of ethanol?

   (b) What mass of ethanol would have a volume of 350 mL?

57. Using the data in Tables 1.3 and 1.6, calculate the volume of each of the following in cubic centimeters:

   (a) A crystal of sodium chloride, NaCl, having a mass of 1.34 mg

   (b) A pool of mercury with a mass of 21.34 g

   (c) 1.00 kg of benzene

   (d) 1.00 quart (U.S.) of water

58. Using the data in Tables 1.3 and 1.6, calculate the mass of each of the following in kilograms:

(a) 1.00 L of concentrated sulfuric acid, H_2SO_4 (density, 1.84 g cm^{-3})

(b) 25.00 mL of ethanol

(c) 1.00 quart (U.S.) of water

(d) A crystal of sodium chloride of volume 1.00 in.^{3}

59. A bar of silver 40 cm long, 25 cm wide, and 15 cm deep has a mass of 157.5 kg. What is the density of silver?

60. Rubbing alcohol (2-propanol) has a density of 6.56 lb gallon^{-1}. What is the density of 2-propanol in grams per cubic centimeter?

61. Propanone (acetone), an important solvent, has a density of 0.785 g cm^{-3}. What is the mass of the contents of a 10 000-gallon tank car filled with acetone?

62. Mercury has a density of 13.594 g cm^{-3} at 25 °C. Mercury is poured into a cylindrical tube with a uniform internal diameter of 8.0 mm until it forms a column of height 78.3 cm. What is the mass of mercury in the tube at 25 °C?

63. The density of air at ordinary atmospheric pressure and 25 °C is 1.19 g L^{-1}. What is the mass, in kilograms, of the air in a room of dimensions 8.5 m × 13.5 m × 2.8 m?

64. One piece of evidence for supposing that the core of the earth is composed largely of iron is the earth's average density. Calculate the average density of the earth from its mass of 6.00×10^{24} kg and its average radius of 6.34×10^3 km. Express the answer in grams per cubic centimeter, and compare it with the actual density of iron.

65. There are 6×10^{28} aluminum atoms in a cube of aluminum having a side of length 1.04 m. What is the volume occupied by one aluminum atom? What is the radius, in picometers, of a sphere of the same volume?

66. (a) What is the mass in grams of a cube of aluminum of side 5.34×10^{-2} m, given that the density of aluminum is 2.7 g cm^{-3}?

(b) What is the mass of a spherical drop of mercury of diameter 5.42 mm, given that the density of mercury is 13.60 g cm^{-3}?

Solutions

67. Explain the difference between a solvent and a solute.

68. When 31.4 g of ammonium chloride, NH_4Cl, is dissolved in 73.8 g of water, the volume of the resulting solution is 82.4 mL. What is the density of the solution?

69. Vinegar is 5% by mass acetic acid in aqueous solution. What mass of acetic acid is contained in 1.00 kg of vinegar?

70. Express the concentration of each of the following as a mass percentage:

(a) Ethanol in a solution containing 5.34 g of ethanol and 121.51 g of water

(b) Sodium chloride in a solution containing 18.12 g of sodium chloride in 250.0 g of aqueous solution

(c) Both hydrochloric acid and sodium chloride in a solution containing 30.1 g of hydrochloric acid and 3.42 g of sodium chloride in 250.0 g of water

(d) Phosphoric acid, H_3PO_4, in 250 mL of an aqueous solution that contains 178 g of phosphoric acid and has a density of 1.35 g mL^{-1}

(e) Water in an aqueous solution of nitric acid, HNO_3, containing 21.35 g of nitric acid and 100.0 mL of water

71. How many grams of sulfuric acid, H_2SO_4, are in 500 mL of concentrated sulfuric acid that is 95% sulfuric acid and 5% water by mass and has a density of 1.84 g mL^{-1}?

72. The density of a concentrated solution of sodium hydroxide, NaOH, is 1.53 g cm^{-3}, and it contains 50% sodium hydroxide and 50% water by mass. Calculate the volume of the solution that contains 20 g of sodium hydroxide.

73. The concentration of oxygen, O_2, dissolved in a sample of seawater is 6.22 ppm (by mass), and the density of seawater is 1.03 g cm^{-3}. What is the mass, and the volume, of this seawater that contains 1.00 g of oxygen?

74. It has been estimated that 1.0 g of seawater contains 4.0 pg of gold. If the total mass of the oceans is 1.6×10^{12} Tg, how many kilograms of gold are present in the oceans?

75. The concentration of bromide ion, Br$^-$, in seawater is 0.06 ppm. What minimum amount of seawater would have to be treated in order to extract 1.00 kg of bromine?

Miscellaneous

76. Give a concise macroscopic description of each of the following:

(a) Solid   (b) Liquid   (c) Gas

77. Briefly explain each of the following terms:

(a) Filtrate   (b) Distillate   (c) Solution

(d) Mixture   (e) Crystallization

78. Suggest possible ways in which to separate each of the following mixtures into its components:

(a) Sugar and water

(b) Water and gasoline

(c) Iron filings and wood sawdust

(d) Sugar and powdered glass

(e) Food coloring in water

79. Suggest suitable methods whereby each of the following mixtures could be separated into its components:

(a) Zinc powder and sulfur

(b) Tetrachloromethane, CCl_4 (b.p. 74 °C) and dichloromethane, CH_2Cl_2 (b.p. 40 °C)

(c) An aqueous solution containing sodium chloride, NaCl, and potassium nitrate, KNO_3

(d) Natural gas (a mixture of gaseous hydrocarbons)

(e) A mixture of iodine and potassium iodide

80. The solubility of potassium sulfate, K_2SO_4, in water is 12 g per 100 g of water at 25 °C. What mass of potassium sulfate is required to give a saturated solution in 30 g of water at 25 °C, and what is the mass percentage of K_2SO_4 in the resulting solution?

81. The solubility of sodium chloride, NaCl, at 25 °C is 36.0 g in 100 g water. If 60.0 g of sodium chloride is dissolved in 150 g of water at 100 °C and the resulting solution is cooled to 25 °C, what is the maximum amount of sodium chloride that can be separated at 25 °C by filtration, and what is the mass percentage of sodium chloride in the filtrate?

Stoichiometry

2

We have said that there are just over a hundred different elements and, therefore, the same number of different kinds of atoms, but we have not yet discussed how atoms differ. To do this we need to consider the structures of atoms. We will see that all atoms are made up of the same basic particles, but they contain different numbers of these particles. A very important property of an atom is its mass. Dalton proposed that all the atoms of a given element have the same mass and that this mass is different from the mass of an atom of another element. We will see that this postulate has had to be slightly modified in the light of experimental evidence.

Finally, we will see how to use atomic masses to determine the formulas of substances and to calculate the amounts of substances taking part in chemical reactions. If we burn 1.00 g of carbon in oxygen we obtain 3.66 g of carbon dioxide. How can we use this information to deduce the formula of carbon dioxide? Ammonia is made by the reaction of nitrogen with hydrogen. The chemist in charge of an ammonia plant needs to know how much hydrogen and how much nitrogen are needed to prepare, say, 1 ton of ammonia. It is the answers to questions such as these that we will study in this chapter. All the quantitative aspects of chemical composition and reactions are referred to collectively as **stoichiometry**, a term that is derived from Greek words meaning "element" and "measure".

2.1
Essentials of Atomic Structure

THE HYDROGEN ATOM: ELECTRONS AND PROTONS

The atom with the smallest mass and the simplest structure is that of hydrogen. The *hydrogen atom* consists of two electrically charged particles. One particle is a small, positively charged entity called the **proton**. The other particle is negatively charged and is called the **electron**. The charge on the electron is 1.6022×10^{-19} coulomb (C), and the charge on the proton is of equal magnitude but opposite sign. The hydrogen atom therefore has no net charge overall. In other words, it is electrically neutral.

The *mass* of the hydrogen atom is concentrated in the proton, which has a mass of 1.6726×10^{-27} kg; the electron has a mass of 9.1096×10^{-31} kg, which is only $\frac{1}{1836}$ of the mass of the proton. The negatively charged electron is attracted by the positively charged proton, and it moves around the proton. It is not, however, to be pictured as moving in a well-defined orbit—such as the earth's orbit around the sun—although this was an early model of the hydrogen atom. It is not possible to describe the motion of the electron around the proton precisely but, as a result of this motion, the electron effectively fills a relatively large spherical space around the very small proton. Our current view of the structure of an atom dates back to 1910, when Ernest Rutherford (Figure 2.1) showed that all atoms consist of a very small, positively charged particle called a **nucleus** surrounded by one or more negatively charged electrons (Box 2.1). In the hydrogen atom the nucleus consists of a single proton. The nuclei of other atoms are more complex, but for all atoms the mass of the nucleus constitutes a very large part of the mass of the atom. Nevertheless the nucleus is indeed tiny compared with the the space occupied by the electrons that surround

FIGURE 2.1
Ernest Rutherford (1871–1937).

Rutherford was born on a farm in New Zealand in 1871, the grandson of Scottish immigrants and the second of twelve children. On graduating from the University of New Zealand, he obtained a scholarship to Cambridge University and began his research in the then new field of radioactivity. At the age of twenty-seven Rutherford was appointed professor of physics at McGill University, Montreal. In 1907 he moved to the University of Manchester, England, and in 1908 he was awarded the Nobel Prize in chemistry. Among his many discoveries were showing that alpha particles are the nuclei of helium atoms and finding the evidence that led to our present model of the atom. In 1917 Rutherford made another discovery of fundamental importance. By bombarding nuclei with subatomic particles, he changed the atoms of one element into atoms of another element, thus achieving the ancient dream of the alchemists. In 1919 he returned to Cambridge as head of the internationally renowned Cavendish Laboratory.

the nucleus. If we imagine an atom magnified to the size of a hot-air balloon about 10 m in diameter, the nucleus would still be a tiny speck of dust only about 0.1 mm in diameter. We obviously cannot represent this realistically in drawings of the atom such as that in Figure 2.2 and we must bear this in mind.

It is often convenient to think of the electron not as a very small, rapidly moving, negatively charged particle but as a cloud of negative charge surrounding the nucleus. This picture of the atom would be obtained if we could take a time exposure photograph. The electron would appear not as a single dot but as a cloud of negative charge. The electron cloud in an isolated hydrogen atom is spherical, but it is not of uniform density. It is most dense near the proton and rapidly diminishes in density as the distance from the proton increases; in other words the electron spends more time near the proton and less time further away from the proton (Figure 2.2).

FORCES The electron is held to the proton in the hydrogen atom because there is an electric force of attraction between them. We are familiar with the

Because the nucleus is so small and contains almost all the mass of the atom, its density is truly enormous—of the order of 10^{13} g cm^{-3}, an inconceivable 10 million tons per cubic centimeter. It is believed that some collapsed stars consist of closely packed nuclei from which all the electrons have been stripped off and that consequently they have densities of this order.

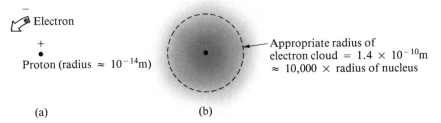

Electron

+
•
Proton (radius ≈ 10^{-14}m)

——Appropriate radius of
electron cloud = 1.4×10^{-10}m
≈ 10,000 × radius of nucleus

(a)   (b)

FIGURE 2.2
The Hydrogen Atom.

(a) The electron is moving at high speed around the nucleus. (b) If we could take a time exposure photograph of a hydrogen atom, its electron would appear, because of its rapid motion, as a cloud of negative charge surrounding the proton. This cloud has no precise boundary, but it is much larger than the very small proton. A radius of 1.4×10^{-10} m would include 90% of the electron cloud. Neutral atoms of other elements have more electrons and protons than hydrogen has, but in these atoms, too, the mass is concentrated in the nucleus, and the electrons account for most of the size of the atom.

BOX 2.1

RUTHERFORD AND THE NUCLEAR ATOM

Our present model of the atom as consisting of a very small nucleus, in which most of the mass of the atom is concentrated, surrounded by one or more electrons, which occupy a very much larger volume, dates from 1910. In that year, in Manchester, England, Ernest Rutherford and two of his collaborators, Hans Geiger and Ernest Marsden, a 20-year-old student, carried out an experiment on the scattering of alpha particles by thin metal foil. Rutherford had earlier found that some radioactive elements—that is, elements that have unstable nuclei—such as radium, decompose by emitting positively charged particles, which he called alpha (α) particles, and electromagnetic radiation, or gamma (γ) rays. He later showed that alpha particles are the nuclei of helium atoms. They are emitted from unstable nuclei at very high speeds ($\approx 10^7$ m s^{-1}). They can travel several centimeters through the air, or approximately 0.1 mm through solids, before they are stopped by collisions with atoms. They can pass right through very thin metal foil, but in so doing, they are deflected from their original path; they are said to be *scattered*. Rutherford observed that when they passed through the thin gold foil most of the alpha particles were only deflected by very small angles. A certain number, however, were scattered through large angles, and some even bounced right back toward the source. This result was very surprising, and Rutherford later remarked: "It was quite the most incredible

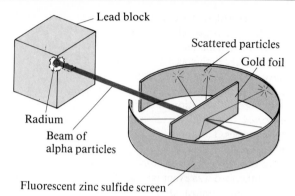

Rutherford's Apparatus for Studying Scattering of Alpha Particles. Alpha particles emitted by radium are blocked by a lead plate, but a small hole in the plate allows a narrow beam of particles to pass through. This beam then passes through a very thin gold foil and strikes a screen coated with zinc sulfide. A momentary flash is observed on the screen whenever it is struck by an alpha particle. (The screen in a television set works on the same principle.)

event. . . . It was almost as surprising as if a gunner fired a shell at a piece of tissue and the shell bounced right back."

At the time, ideas about the structure of atoms were rather vague. The most popular model, proposed by J. J. Thompson in 1898, held that atoms consisted of electrons embedded in a uniform distribution of positively charged matter. Because the alpha particles are posi-

idea of a force from everyday experience. We exert a force when we kick a football or hit a tennis ball. A **force** acting on an object changes its velocity. The magnitude of the force is directly proportional both to the mass being moved and to the rate of change of its velocity, as shown in the following equation. Rate of change of velocity is called **acceleration** and has the units meters per second per second, or meters per second squared (m s^{-2}).

$$\text{Force} = \text{Mass} \times \text{Acceleration} \qquad \text{or} \qquad F = ma$$

A greater force is needed to change the velocity of a heavy object by a certain amount than is needed to change the velocity of a light object by the same amount. For a given mass a greater force is needed to accelerate it to a high velocity than to a low velocity in the same time interval.

The SI unit of force is the **newton** (N). A newton is the force that, acting on a mass of 1 kg, causes an acceleration of 1 m s^{-2}:

$$1 \text{ N} = 1 \text{ kg m s}^{-2}$$

Only three types of forces are known in nature:

1. **Gravitational forces** act between all objects and are proportional to their masses.

tively charged, they were expected to be attracted by the negatively charged electrons and repelled by the positively charged matter. But the electrons were not expected to deflect the high-speed alpha particles significantly; they have a much smaller mass than alpha particles and would simply be pushed out of the path of the alpha particles. A uniform distribution of positively charged matter was expected to deflect the alpha particles through rather small angles. This model of the atom was therefore completely unable to explain the very large deflections that were observed for some alpha particles.

Rutherford was eventually led to the unexpected conclusion that almost all the mass and all the positive charge of an atom are concentrated in a very small particle situated at the center of the atom. He called this particle the nucleus. Since most of the volume occupied by an atom is filled only by a few electrons, most of the alpha particles pass through this space and are only deflected through very small angles. It is only when an alpha particle comes very close to the heavy, positively charged nucleus of the atom that it is deflected through a large angle. The very few particles that make a more or less direct hit on the nucleus are bounced right back in the direction from which they came. Calculations of the number of alpha particles that would be expected to be scattered through any given angle on the basis of Rutherford's model gave excellent agreement with the experimental observations. Moreover, he was able to calculate from his results that the charge on the gold

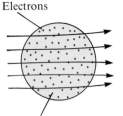

Positively charged matter

Scattering of Alpha Particles according to Thomson's Model of the Atom. The alpha particles are only slightly deflected.

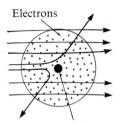

Positively charged nucleus

Scattering of Alpha Particles according to Rutherford's Model of the Atom. A few alpha particles are deflected through large angles.

nucleus, that is, its atomic number, is 100 ± 20, which is in good agreement with the value of 79 that was established later, bearing in mind the simple nature of his experiment and the difficulty of counting the flashes on the screen and measuring their angle. He was also able to calculate that the radius of the nucleus is only approximately $1/100\,000$ of the radius of the gold atom. It is thus not surprising that the vast majority of the alpha particles that pass through an atom do so without suffering any appreciable deflection and only the very few that happen to pass close to the nucleus are scattered through large angles. Rutherford's model of the atom was quickly and universally adopted.

2. **Electric forces** act between charged objects and are proportional to their electric charges. There are also magnetic forces that arise from charges in motion, but they are not fundamentally different from electric forces.

3. **Nuclear forces** act between subatomic particles and hold together the particles (protons and neutrons) of which a nucleus is composed. Physicists recognize two types of nuclear force called strong and weak but the difference between these will not concern us.

In chemistry only electric forces are important. Gravitational forces are so much weaker than electric forces that they can be completely ignored. For example, the electric force of attraction between a proton and an electron is 10^{39} times stronger than the gravitational attraction between them. Nuclear forces are of importance only over extremely small distances, such as the distances between the particles inside the nucleus; they are completely negligible over larger distances.

COULOMB'S LAW The electric force between two charges of magnitude Q_1 and Q_2 is given by **Coulomb's law**:

$$F = k \frac{Q_1 Q_2}{r^2}$$

where r is the distance between the charges and k is a constant called the Coulomb constant. When the two charges are of opposite sign, the force is attractive; in other words, the charges are drawn to each other. When the two charges are of the same sign, the force is repulsive; that is, like charges repel each other.

Physicists believe that quarks and certain other fundamental particles which have very short lives may have charges that are only a fraction of the charge on an electron. These particles are not of any importance in chemistry so we will not be further concerned with them.

THE COULOMB   The unit of charge is the **coulomb** (C). We will not give the formal definition of the coulomb, which is expressed in terms of magnetic forces. For our purposes it is more useful to think of the coulomb in terms of the number of individual elementary charges that add up to a coulomb of charge. All charges, both positive and negative, occur only in multiples of 1.602×10^{-19} C. The electron has a charge of -1.602×10^{-19} C and the proton has a charge of $+1.602 \times 10^{-19}$ C. A total of 6.241×10^{18} electrons have a charge of -1 C and a total of 6.241×10^{18} protons have a charge of $+1$ C. The amount of charge 1.602×10^{-19} C has been given a special name, the **electron charge**, and a special symbol, e:

$$e = 1.60 \times 10^{-19} \text{ C}$$

The charge on an electron was first measured by the American physicist Robert Millikan (Box 2.2).

Charges on particles such as the electron or proton are usually expressed as multiples of e and not in coulombs. Thus an electron is said to have a charge of -1, that is, a charge of $-1e$, or -1.60×10^{-19} C. Similarly, a proton is said to have a charge of $+1$, that is $+1e$, or $+1.60 \times 10^{-19}$ C.

THE HELIUM ATOM: NEUTRONS

An atom of the element *helium* has two electrons surrounding a nucleus with a charge of $+2$; thus the atom is neutral. Like the nuclei of all atoms, the helium nucleus is very small and occupies only a tiny fraction of the total volume of the atom, although it contains most of the mass.

Because the charge on the helium nucleus is twice the charge of the proton, we might expect the helium nucleus to consist of two protons and to have a mass twice the mass of the proton. But it does not. Instead, the mass of the helium nucleus is very nearly *four times* the mass of the proton. Two protons account for the charge of $+2$ but only one-half of the mass, so there must be additional constituents in the helium nucleus. These additional particles are called **neutrons**. A neutron is a particle that has very nearly the same mass as the proton but zero charge.

The detailed structure of the nucleus is still under active investigation by nuclear physicists.

Although the detailed structure of the nucleus is not fully understood, for our purposes we can consider the nuclei of all atoms to be made up of protons, each with a charge of $+1$, and a certain number of neutrons. *The charge on a nucleus is determined by the number of protons that it contains. The mass of a nucleus is determined by the total number of protons and neutrons*, which are collectively called **nucleons**. Thus the nucleus of a helium atom consists of two protons and two neutrons; it has a charge of $+2$ and a mass that is approximately four times the mass of a proton. The nucleus of a fluorine atom consists of nine protons and ten neutrons; it has a charge of $+9$ and a mass that is approximately 19 times the mass of the proton. Except for the nucleus of the hydrogen atom, all nuclei are composed of both protons and neutrons.

BOX 2.2

MILLIKAN'S DETERMINATION OF THE CHARGE OF AN ELECTRON

The charge of the electron was first accurately determined by the American physicist Robert Millikan at the University of Chicago during the period 1909–1913. Millikan allowed a fine spray of very small oil drops to fall between two metal plates, one charged positively and the other negatively. Air between the plates was irradiated with X rays, knocking electrons out of nitrogen and oxygen molecules. Some of these electrons collided with the oil drops and became attached to them, giving them a negative charge.

Millikan observed individual oil drops with a microscope and adjusted the charge on the plates until the force acting on the drop due to the charged plates just balanced the force due to gravity and a drop remained stationary. Knowing the charge on the plates and using Coulomb's law, Millikan was able to calculate the charge on the drop. On repeating the experiment for very many individual drops, he found that the charge on a drop was always -1.60×10^{-19} C or an integral multiple of this charge. In other words, he found that there is a smallest possible amount of

Robert Millikan (1868–1953).

electric charge, indicating that electric charge is not continuous but consists of particles, each having the same charge—the particles that we call *electrons*. The oil drops were picking up one or more electrons and therefore always had the charge of one or more electrons. Millikan's experiments not only established an accurate value for the charge on the electron but also provided the first conclusive proof for the existence of electrons.

ATOMIC NUMBERS

We can summarize the constitution of any nucleus by just two numbers:

1. The number of protons in the nucleus is called the **atomic number**, Z.
2. The total number of protons and neutrons is called the **mass number** or the **nucleon number**. It determines the mass of the nucleus.

The difference between the mass number and the atomic number is equal to the number of neutrons in the nucleus.

By convention, the atomic number is written at the bottom left corner of the symbol of the atom, and the mass number is written at the top left corner. For example, we write $^1_1$H, $^4_2$He, $^7_3$Li, and $^{12}_6$C. The symbol $^{12}_6$C indicates that there is a total of 12 particles (nucleons) in the nucleus of a C atom, 6 of which are protons. Thus there must be $12 - 6 = 6$ neutrons.

Because each atom has a zero charge overall, its nuclear charge, which is $+Z$, must be balanced by the total charge of the surrounding electrons, each of which has a charge of -1. For example, a $^7_3$Li atom, which has $Z = 3$, has a nuclear charge of $+3$ and 3 surrounding electrons, for a total charge of 0. A $^{12}_6$C atom, $Z = 6$, has 6 electrons surrounding the nucleus, and $^{16}_8$O has 8 electrons. Figure 2.3 shows the structure of a few atoms in a diagrammatic way. It is the nuclear charge—that is, the atomic number Z—that differentiates the atoms of one element from the atoms of another. Thus an **element** *may be defined as a substance whose atoms have the same atomic number.*

≡ Example 2.1

ATOMIC NUMBER AND MASS NUMBER

For both $^7_3$Li and $^{32}_{16}$S give (a) the mass number, (b) the number of protons in the nucleus, (c) the number of neutrons in the nucleus, and (d) the number of electrons in the neutral atom.

Solution

For an isotope ^A_ZX, A is the mass number and Z the atomic number, that is, the number of protons. The number of neutrons is the mass number A (the total number of nucleons) minus the number of protons Z, that is $A - Z$, and the number of electrons is equal to the number of protons Z. Thus

	$^7_3$Li	$^{32}_{16}$S
(a) Mass number A	7	32
(b) Number of protons Z	3	16
(c) Number of neutrons $A - Z$	4	16
(d) Number of electrons Z	3	16

EXERCISE 2.1

For each of the following, what is the mass number? How many protons are in the nucleus? How many neutrons? How many electrons are in the neutral atom?

(a) $^1_1$H   (b) $^4_2$He   (c) $^{19}_9$F   (d) $^{118}_{50}$Sn   (e) $^{238}_{92}$U

FIGURE 2.3
Structures of Some Atoms.

These diagrams do not represent the relative sizes of the electron cloud and the nucleus. They are meant to show only the numbers of particles of each kind.

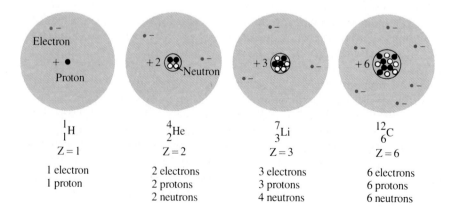

$^1_1$H	$^4_2$He	$^7_3$Li	$^{12}_6$C
$Z = 1$	$Z = 2$	$Z = 3$	$Z = 6$
1 electron	2 electrons	3 electrons	6 electrons
1 proton	2 protons	3 protons	6 protons
	2 neutrons	4 neutrons	6 neutrons

ISOTOPES

One further important aspect of nuclei must be considered. We have said that all the atoms of a given element have nuclei containing the same number of protons and having the same charge. However, the nuclei of all the atoms of a given element do not necessarily contain the same number of neutrons. For example, all oxygen atoms have atomic number $Z = 8$ and therefore have 8 protons in the nucleus and 8 electrons surrounding the nucleus; these features distinguish oxygen atoms from atoms of all the other elements. But three different kinds of oxygen atoms are found in nature: $^{16}_{8}O$ atoms with 8 protons and 8 neutrons in the nucleus, $^{17}_{8}O$ atoms with 8 protons and 9 neutrons in the nucleus, and $^{18}_{8}O$ atoms with 8 protons and 10 neutrons in the nucleus. All oxygen atoms have the same atomic number, $Z = 8$, and their nuclei have the same nuclear charge of $+8$, but they have different masses depending on the number of neutrons in their nuclei. Atoms of an element that have different masses are called **isotopes**.

Many of the elements have two or more naturally occurring isotopes. Even the simplest element, hydrogen, has two stable isotopes, $^{1}_{1}H$ and $^{2}_{1}H$. The nucleus of the lighter isotope is the proton; the nucleus of the heavier isotope contains a proton and a neutron. The mass of the $^{2}_{1}H$ isotope is nearly twice the mass of $^{1}_{1}H$. Isotopes are distinguished by the mass number in the symbol for the isotope, and they are not usually given different names. However, the $^{2}_{1}H$ isotope of hydrogen is an exception. It is called *deuterium* or *heavy hydrogen* and is sometimes given the symbol D.

Some elements have just a single naturally occurring isotope, for example, $^{19}_{9}F$ and $^{31}_{15}P$. But many have two or more isotopes, for example, helium, $^{3}_{2}He$ and $^{4}_{2}He$; lithium, $^{6}_{3}Li$ and $^{7}_{3}Li$; and oxygen, $^{16}_{8}O$, $^{17}_{8}O$, and $^{18}_{8}O$. In addition to naturally occurring isotopes, others can be made artificially by bombarding atoms with nuclear particles, such as protons, neutrons, and helium nuclei (Chapter 25). In these nuclear reactions, nuclei are transformed into different nuclei—that is, atoms are transformed into different atoms. In chemical reactions, however, atoms remain unchanged; they simply form new combinations.

Because the atomic number is implied by the symbol of the element, it is often omitted. The isotopes of oxygen, for example, are often represented simply by the symbols ^{16}O, ^{17}O, and ^{18}O, and they are referred to as oxygen-16, oxygen-17, and oxygen-18.

The isotopes of an element are found not only in the element itself but also in all the compounds of the element. Naturally occurring water, for instance, consists mainly of molecules containing two $^{1}_{1}H$ atoms and an oxygen atom, some molecules containing a $^{1}_{1}H$ atom and a $^{2}_{1}H$ atom, and a very few molecules containing two $^{2}_{1}H$ atoms and an oxygen atom—in other words, H_2O, HDO, and D_2O molecules. Water that consists entirely of $^{2}_{1}H_2O$ (D_2O) molecules is called *deuterium oxide* or, more commonly, *heavy water*. It is used to slow down neutrons in some types of nuclear reactors (Chapter 25).

Substances containing different isotopes of an element have very similar, but not identical, properties. The differences are more marked for ^1H and ^2H because the ratio of their masses is larger than the mass ratios for the isotopes of any other element. Ordinary water freezes at 0 °C and boils at 100 °C, whereas heavy water (^2H_2O) freezes at 3.82 °C and boils at 101.42 °C. Such differences in physical properties, even though they are very small, allow isotopically different molecules to be separated. For example, heavy water can be obtained

There is also an unstable (radioactive) isotope of hydrogen, called tritium, $^{3}_{1}H$ (T). For more on radioactive nuclei see Chapter 25.

by fractional distillation of ordinary water. Heavy water is more commonly obtained by electrolysis of ordinary water. This is a process in which an electric current is passed through water. The electric current decomposes the water into hydrogen, H_2, and oxygen, O_2. But $^1_1H_2O$ molecules are decomposed slightly faster than $^2_1H_2O$ molecules so that, if a large quantity of water is electrolyzed until only a very small amount remains, this small amount consists almost entirely of $^2_1H_2O$ molecules.

2.2
Atomic Mass

The mass of an atom is much too small to be conveniently expressed in kilograms or grams. Because the mass of the proton is 1.6726×10^{-27} kg, a unit of 10^{-27} kg would be a convenient size. However, this unit is not used. In the past, several different units have been used, but today the unit in which atomic masses are measured is defined by international agreement as $\frac{1}{12}$ of the mass of one $^{12}_6C$ atom. This unit is called the **unified atomic mass unit** and is given the symbol **u**. In other words, the mass of one $^{12}_6C$ atom is defined as *exactly* 12 u. The masses of the electron, the neutron, the proton, and a number of lighter atoms on this scale are given in Table 2.1. An important advantage of this scale is that the atomic masses have values that are close to the mass (nucleon) number, the total number of protons and neutrons in the nucleus.

AVERAGE ATOMIC MASS

Because elements occur naturally as mixtures of their isotopes, the mass of a given number of atoms of an element depends on the relative abundance of each isotope in the sample. A hundred atoms of hydrogen, for example, could have a mass between 100.783 and 201.410 u, depending on the relative numbers of ^1H and ^2H atoms in the sample. During the earth's long geological history, however, the isotopes of most elements have become thoroughly mixed with each other. As a result, the isotopic composition of most elements is constant throughout the surface of the earth; only a few elements have an isotopic composition that varies significantly from one sample of a substance containing the element to another sample. Wherever water and other hydrogen-containing compounds are found on the earth's surface they are composed of the same proportions of ^1H and ^2H atoms: 99.985% of the hydrogen atoms are ^1H atoms and 0.015% are ^2H atoms.

If we know the relative abundances—that is, the percentages—and masses of the isotopes of an element, we can calculate the average mass of an atom of the element. The masses and abundances of isotopes can be measured with a *mass spectrometer*, which is described in Box 2.3. Table 2.1 lists the masses and abundances of the isotopes of the elements with atomic numbers from 1 to 19. To calculate the average mass of an atom of an element, we simply multiply the fractional abundance of each isotope by its mass and then add the products. For example, Table 2.1 tells us that in a naturally occurring sample of hydrogen 99.985% of the atoms are ^1H atoms and 0.015% are ^2H atoms. Thus the fractional abundances are ^1H, 0.999 85 and ^2H, 0.000 15. Since a ^1H atom has a

Particle or Isotope	Mass (u)	Percent Natural Abundance	Isotope	Mass (u)	Percent Natural Abundance
Proton	1.007 28		$^{23}_{11}$Na	22.989 77	100
Neutron	1.008 66		$^{24}_{12}$Mg	23.985 04	78.7
Electron	0.000 548		$^{25}_{12}$Mg	24.985 84	10.2
$^{1}_{1}$H	1.007 83	99.985	$^{26}_{12}$Mg	25.986 36	11.1
$^{2}_{1}$H	2.014 10	0.015	$^{27}_{13}$Al	26.981 53	100
$^{3}_{2}$He	3.016 03	0.000 13	$^{28}_{14}$Si	27.976 93	92.18
$^{4}_{2}$He	4.002 60	99.999 87	$^{29}_{14}$Si	28.976 50	4.71
$^{6}_{3}$Li	6.015 12	7.42	$^{30}_{14}$Si	29.973 77	3.12
$^{7}_{3}$Li	7.016 00	92.58	$^{31}_{15}$P	30.973 77	100
$^{9}_{4}$Be	9.012 19	100	$^{32}_{16}$S	31.972 07	95.02
$^{10}_{5}$B	10.012 94	19.77	$^{33}_{16}$S	32.971 46	0.76
$^{11}_{5}$B	11.009 31	80.23	$^{34}_{16}$S	33.967 86	4.22
$^{12}_{6}$C	12.000 00	98.892	$^{36}_{16}$S	35.967 10	0.014
$^{13}_{6}$C	13.003 36	1.108	$^{35}_{17}$Cl	34.968 85	75.76
$^{14}_{7}$N	14.003 07	99.635	$^{37}_{17}$Cl	36.965 90	24.24
$^{15}_{7}$N	15.000 11	0.365	$^{36}_{18}$Ar	35.967 54	0.337
$^{16}_{8}$O	15.994 91	99.759	$^{38}_{18}$Ar	37.962 73	0.063
$^{17}_{8}$O	16.999 13	0.037	$^{40}_{18}$Ar	39.962 38	99.60
$^{18}_{8}$O	17.999 16	0.204	$^{39}_{19}$K	38.963 71	93.22
$^{19}_{9}$F	18.998 40	100	$^{40}_{19}$K	39.964 01	0.012
$^{20}_{10}$Ne	19.992 44	90.92	$^{41}_{19}$K	40.961 83	6.77
$^{21}_{10}$Ne	20.993 85	0.257			
$^{22}_{10}$Ne	21.991 38	8.82			

TABLE 2.1
Mass and Natural Abundance of Some Fundamental Particles and Isotopes

mass of 1.007 83 u and a ^{2}H atom has a mass of 2.014 10 u, the average mass of an atom of hydrogen can be found as follows:

Contribution of ^{1}H atoms	$0.999\,85 \times 1.007\,83$ u $= 1.007\,68$ u
Contribution of ^{2}H atoms	$0.000\,15 \times 2.014\,10$ u $= \underline{0.000\,30\text{ u}}$
Sum of contributions from ^{1}H and ^{2}H atoms	$1.007\,98$ u

Thus the average mass of one hydrogen atom is 1.007 98 u.

═ *Example 2.2* ═

AVERAGE ATOMIC MASS FROM ISOTOPIC ABUNDANCES

Naturally occurring oxygen consists of a mixture of the isotopes ^{16}O, ^{17}O, and ^{18}O, in the relative abundances 99.759%, 0.037%, and 0.204%. Their masses are 15.994 91, 16.999 13, and 17.999 16 u. What is the average mass of an oxygen atom?

Solution

To find the average mass, we calculate the masses contributed by each isotope and then add these amounts. It is convenient first to convert the percent abundances to

BOX 2.3

THE MASS SPECTROMETER: MEASURING THE MASS OF ATOMS AND THE ABUNDANCE OF ISOTOPES

In a mass spectrometer, one or more electrons are removed from atoms to produce positively charged particles known as positive *ions*. A beam of these ions is passed through a magnetic field, which exerts a force on the positively charged ions at right angles to their direction of motion. If the beam contains ions of different masses, the heavier ions are deflected less than the lighter ions, and the beam splits up into separate beams, one for each mass. From the amount by which each beam is deflected, the mass of each of the ions in the beam can be calculated.

In 1913 the British physicist J. J. Thomson was working at the Cavendish Laboratory in Cambridge, England, on methods to determine the masses of individual atoms. For this work he developed an instrument known as the *mass spectrometer*. It is based on the principle that if a force is applied at right angles to the path of a moving object, the force will change the object's direction of motion. A light object will be deflected from its original path more than a heavy object.

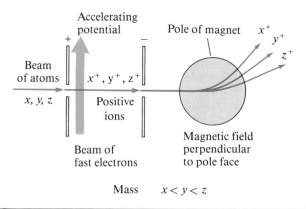

fractional abundances. For example, percent abundance 99.759 = fractional abundance 0.997 59.

Mass Number	Mass		Fractional Abundance		Mass × Abundance
^{16}O	15.994 91 u	×	0.997 59	=	15.9564 u
^{17}O	16.999 13 u	×	0.000 37	=	0.0063 u
^{18}O	17.999 16 u	×	0.002 04	=	0.0367 u
Average mass per O atom					15.9994 u

Example 2.3

CALCULATING ISOTOPIC ABUNDANCES

Chlorine has two isotopes, $^{35}_{17}Cl$ and $^{37}_{17}Cl$, which have masses of 34.968 85 u and 36.965 90 u, respectively. Its average atomic mass is 35.453 u. What are the percent abundances of the two isotopes?

Thomson found, to his surprise, that a sample of pure neon gas gave two deflected beams rather than just one. This discovery showed for the first time that neon contains atoms with two different masses, and was thus the first demonstration of the existence of isotopes.

The simple instrument invented by Thomson has been developed into the modern mass spectrometer, which is capable of measuring the abundance and masses of isotopes with great precision. Almost all the values given in the table of atomic masses have been determined in this way. Today the mass spectrometer is used primarily for the identification of substances and the analysis of mixtures (see Chapter 23).

In the accompanying diagram of a mass spectrometer the substance to be studied is introduced in the form of a gas at A. It passes into the region between the two metal plates B and C, which have a large electric potential between them. High-energy electrons, which are injected at D, collide with the atoms or molecules of the gas. If the gas is a simple monatomic gas such as neon or argon, one or more electrons are knocked out of the atoms to give positively charged ions such as Ne^+, Ar^+, and Ar^{2+}. These ions are attracted toward the negatively charged plate C and are accelerated to high speeds. Some of the ions pass through the slit in the plate to form a narrow beam which then passes between the poles of a powerful magnet. The magnet deflects the beam into a circular path whose curvature depends on the mass of

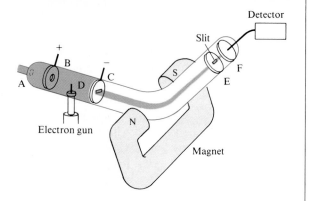

the ions and on their charge. The lighter ions are deflected more than the heavier ions, and the more highly charged ions are deflected more than the ions with a single positive charge.

With a given strength of magnetic field, only ions with a given value of Q/m, the ratio of charge to mass, pass through the slit E to reach the plate F, where they produce an electric current which can be measured. When the magnetic field is varied, ions with different values of Q/m reach the plate F. From the magnitude of the current produced at F by each of the ions of different mass, their relative abundances can be found.

Solution

Let the fractional abundance of $^{35}_{17}Cl$ be x. Then the fractional abundance of $^{37}_{17}Cl$ is $1 - x$. Thus we have

$$(34.968\ 85\ u)x + (36.965\ 90\ u)(1 - x) = 35.453\ u$$

$$(34.969\ u - 36.966\ u)x + 36.966\ u = 35.453\ u$$

$$x = \frac{36.966\ u - 35.453\ u}{36.966\ u - 34.969\ u} = 0.7576$$

Hence the abundance of $^{35}_{17}Cl$ is 75.76% and the abundance of $^{37}_{17}Cl$ is 24.24%.

EXERCISE 2.2

The average atomic mass of gallium is 69.72 u, and naturally occurring gallium consists of two isotopes, $^{69}_{31}Ga$ and $^{71}_{31}Ga$, with atomic masses 68.93 u and 70.92 u, respectively. What is the percent abundance of each isotope?

The average atomic masses of the elements are usually referred to simply as *atomic masses*, although the older and, strictly speaking, incorrect term *atomic weight* is still widely used. Often they are expressed on a relative basis, which avoids the use of the atomic mass unit. The *relative average atomic mass* of an element is defined as the average mass of one of its atoms *relative to* $\frac{1}{12}$ of the mass of a ^{12}C atom. Relative average atomic masses have no units; they are simply numbers. The numerical values of average atomic masses expressed in atomic mass units and as relative atomic masses are the same, and the difference between the two has no great practical significance.

A table of atomic masses appears in Appendix B. Notice that the accuracy with which atomic masses are quoted—that is, the number of significant figures—varies. The more constant the isotopic composition of an element, the greater is the number of significant figures cited in the atomic mass. For example, the atomic mass of sulfur, which has the isotopes ^{32}S, ^{33}S, ^{34}S, and ^{36}S, is given as 32.06 u. Only four significant figures are quoted because the isotopic composition of sulfur, and therefore its average atomic mass, varies slightly, depending on its source. In contrast, the atomic mass of fluorine, which has only one naturally occurring isotope, ^{19}F, is given as 18.998 403 u. For most purposes, including most of the calculations in this book, values of the average atomic masses rounded off to four significant figures are adequate. A table of atomic masses rounded off to four significant figures is given on the inside front cover.

CONSERVATION OF MASS AND ENERGY

We might think that we could calculate the mass of an atom from the masses of its electrons, protons, and neutrons. But we cannot. Consider, for example, a ^4_2He atom. The sum of the masses of its protons, neutrons, and electrons is

$$
\begin{aligned}
2 \text{ protons} &= 2 \times 1.007\,28 \text{ u} = 2.014\,56 \text{ u} \\
2 \text{ neutrons} &= 2 \times 1.008\,66 \text{ u} = 2.017\,32 \text{ u} \\
2 \text{ electrons} &= 2 \times 0.000\,55 \text{ u} = \underline{0.001\,10 \text{ u}} \\
&\phantom{= 2 \times 0.000\,55 \text{ u} = } 4.032\,98 \text{ u}
\end{aligned}
$$

The total is greater than the experimentally determined mass of the helium ^4_2He atom, which is 4.002 60 u. This discrepancy illustrates a rule, not an exception: the mass of any atom is always slightly less than the sum of the masses of the constituent electrons, protons, and neutrons. This difference is called the **mass defect**.

The mass defect arises because a very large amount of energy is released when protons and neutrons combine to form a nucleus. The amount of energy is so large that it has a significant mass equivalent, which is given by the equation

$$E = mc^2$$

where E is energy, m is mass, and c is the velocity of light (3.00×10^8 m s^{-1}). This equation, proposed by Albert Einstein in 1905 and subsequently verified in many experiments, is one of the most important and fundamental equations in all science. It gives the relationship between the two fundamental concepts, mass and energy, that we use to describe the universe.

However, we are not directly concerned with this relationship in most of chemistry, although it is of great importance when we are dealing with *nuclear* reactions (Chapter 25). In nuclear reactions nuclei are split apart or are fused together. Very large energy changes are associated with nuclear reactions, which are accompanied by significant changes in mass. For instance, we have seen that when a helium nucleus is formed from two neutrons and two protons, there is a significant decrease in the total mass. In contrast, *chemical reactions* involve only the rearrangement of atoms. The energy changes associated with these rearrangements of atoms are too small for the accompanying changes in mass to be detectable. For example, when 1000 metric tons (10^6 kg) of gasoline are burned, the energy released is equivalent to only 5×10^{-10} kg, which is much too small to be detected experimentally.

Not surprisingly, therefore, one of the earliest laws of chemistry was the **law of conservation of mass**:

Mass can neither be created nor destroyed.

This law was first formulated more than two hundred years ago by the French chemist Antoine Lavoisier (Figure 2.4). By careful weighing Lavoisier found for many different reactions that there was no detectable difference between the total mass of the substances reacting together (the *reactants*) and that of the substances produced (the *products*). We can therefore state that in any chemical reaction the total mass of the system is conserved. In terms of the atomic theory the **law of conservation of mass** may be restated as follows:

In a chemical reaction atoms are neither created nor destroyed; they are merely rearranged.

FIGURE 2.4
Antoine Laurent Lavoisier (1743–1794).

Lavoisier was the son of a wealthy Parisian lawyer, who hoped that his son would follow him in that profession. Although Lavoisier qualified as a lawyer in 1764 he became interested in science and he devoted much of his life to research in chemistry. He is shown here with his wife, Marie-Anne Pierrette, who assisted him in his experiments. Before he was twenty-five he had lectured to the French Academy on such diverse subjects as the divining rod, hypnotism, and the construction of invalid chairs. He was fascinated by the phenomenon of burning and, rich enough to have the best in apparatus and chemicals, he worked tirelessly to try to understand it. He was the first to clearly recognize that air consists of two gases, one of which supports combustion and the other (nitrogen) which does not. Lavoisier called this gas oxygen and recognized that it was an element. He was one of the first chemists to recognize the importance of accurate quantitative measurements, and his accurate determination of the masses of substances formed in reactions led him to the law of conservation of mass. He was also one of the first to demonstrate clearly the nature of elements and compounds. In a textbook that he wrote he gave a table of elements that included many that we would recognize today. It was a great tragedy that he was arrested and guillotined during the French Revolution.

Similarly, although energy is changed from one form to another during a chemical reaction, the total energy of the reacting system and its surroundings is constant. In other words, the **law of conservation of energy** applies to chemical reactions. It states that

> Energy can neither be created nor destroyed but only changed from one form to another.

Strictly speaking, only the sum of the mass and energy remains constant, but in chemical reactions the mass changes are so small that they can be ignored. We cannot ignore them for processes involving changes in nuclei, however, in which the energy changes are very large and there are significant accompanying mass changes. In such nuclear reactions only the sum of the mass and energy remains constant.

2.3
The Mole

One of the most important uses of the table of atomic masses is to find the masses of the substances that react together in a chemical reaction and the masses of the products formed.

FORMULA MASS AND MOLECULAR MASS

The mass of a molecule is the sum of the masses of its component atoms, that is, the sum of the masses of the atoms in the molecular formula. This mass is called the **molecular mass**. For example, the molecular formula of hydrogen peroxide is H_2O_2. Therefore the molecular mass is $2(1.008 \text{ u}) + 2(16.00 \text{ u}) = 34.02 \text{ u}$. Because of a long-standing tradition many chemists still use the term *molecular weight* rather than molecular mass just as they use the term atomic weight instead of atomic mass.

Many substances do not consist of molecules, so for these substances we cannot calculate a molecular mass. We can instead calculate a **formula mass**, which is simply the sum of the masses of the atoms in the empirical formula. The empirical formula of sodium chloride is $NaCl$. Therefore the formula mass of sodium chloride is $22.99 \text{ u} + 35.45 \text{ u} = 58.44 \text{ u}$. We can also calculate a formula mass for substances which exist as molecules. If the empirical formula differs from the molecular formula, the empirical formula mass will differ from the molecular mass. In the case of hydrogen peroxide, for example, the empirical formula is HO and the empirical formula mass is therefore $1.008 \text{ u} + 16.00 \text{ u} = 17.01 \text{ u}$, which is one-half the molecular mass.

═ *Example 2.4* ═══════════════════════════════════

EMPIRICAL FORMULA MASS AND MOLECULAR MASS

What is the empirical formula mass of ethane? What is its molecular mass?

Solution

The empirical formula of ethane is CH_3 and the molecular formula is C_2H_6. Thus the empirical formula mass is $12.01 \text{ u} + 3(1.008 \text{ u}) = 15.03 \text{ u}$. The molecular mass is $2(12.01 \text{ u}) + 6(1.008 \text{ u}) = 30.07 \text{ u} = 2 \text{ (empirical formula mass)}$.

THE AVOGADRO CONSTANT

In studying reactions in the laboratory, we deal with very large numbers of atoms and molecules rather than with single atoms and molecules. Moreover, we normally measure the masses of substances in grams rather than in atomic mass units. We therefore need a convenient way of translating the information provided by chemical formulas and the table of atomic masses into practical units that apply to convenient amounts of substances.

The formula of CO tells us that one atom of oxygen combines with one atom of carbon to give one molecule of carbon monoxide.

For 1 atom of C and 1 atom of O:

$$C + O \qquad \text{gives } CO$$
$$1 \text{ atom C} + 1 \text{ atom O} \qquad \text{gives } 1 \text{ molecule CO}$$
$$1 \text{ atomic mass C} + 1 \text{ atomic mass O} \longrightarrow 1 \text{ molecular mass CO}$$
$$12.01 \text{ u C} + 16.00 \text{ u O} \longrightarrow 28.01 \text{ u CO}$$

For 100 atoms of C and 100 atoms of O:

$$100 \text{ atoms C} + 100 \text{ atoms O} \longrightarrow 100 \text{ molecules CO}$$
$$100(\text{atomic mass C}) + 100(\text{atomic mass O}) \longrightarrow 100(\text{molecular mass CO})$$
$$1201 \text{ u C} + 1600 \text{ u O} \longrightarrow 2801 \text{ u CO}$$

These numbers of atoms still correspond to extremely small amounts of carbon and oxygen. Normally, in any reaction carried out in the laboratory, we would be dealing with much larger numbers of atoms. We could, for example, consider the reactions of **6.022×10^{23}** atoms of carbon with the same number of oxygen atoms to give carbon monoxide molecules.

For 6.022×10^{23} atoms of C and 6.022×10^{23} atoms of O:

$$6.022 \times 10^{23} \text{ atoms C} + 6.022 \times 10^{23} \text{ atoms O} \longrightarrow 6.022 \times 10^{23} \text{ molecules CO}$$
$$(6.022 \times 10^{23} \times 12.01 \text{ u}) \text{ C} + (6.022 \times 10^{23} \times 16.00 \text{ u}) \text{ O} \longrightarrow (6.022 \times 10^{23} \times 28.01 \text{ u}) \text{ CO}$$

The number 6.022×10^{23} is a very large number. If we had 6.022×10^{23} marbles, they would cover the surface of the earth to a depth of 2 km. Why did we choose this particular very large number? We did so because 6.022×10^{23} u is 1.000 g. Hence

$$6.022 \times 10^{23} \times 12.01 \text{ u} = 12.01 \text{ g}$$
$$6.022 \times 10^{23} \times 16.00 \text{ u} = 16.00 \text{ g}$$
$$6.022 \times 10^{23} \times 28.01 \text{ u} = 28.01 \text{ g}$$

Thus for the reaction

$$C + O \longrightarrow CO$$

12.01 g of carbon react with 16.00 g of oxygen to give 28.01 g of carbon monoxide. In other words, we can write

$$\begin{array}{ccc} C & + & O & \longrightarrow & CO \\ 12.01 \text{ g} & & 16.00 \text{ g} & & 28.01 \text{ g} \end{array}$$

Thus 6.022×10^{23} is the number of atoms in 12.01 g of carbon, or the number of atoms in 16.00 g of oxygen, or the number of molecules in 28.01 g of carbon monoxide. Using the relationship

$$1 \text{ g} = 6.022 \times 10^{23} \text{ u}$$

we can express the atomic mass unit in grams:

$$1 \text{ u} = \frac{1 \text{ g}}{6.022 \times 10^{23}} = 1.661 \times 10^{-24} \text{ g}$$

The number 6.022×10^{23} is called the **Avogadro constant**, after the Italian scientist Amedeo Avogadro (1776–1856); it is given the symbol N_A:

$$N_A = 6.022 \times 10^{23}$$

Such an enormous number is so far outside the range of ordinary experience that you might naturally wonder how its value could possibly be determined. It is certainly not possible to do it simply by counting because even if a computer counted 10 million atoms a second it would take 2 billion years to count all the atoms in 12.01 g of carbon. However, there are several experimental methods by which the value of the Avogadro constant can be determined. Some of these will be discussed in Chapters 11 and 17. The presently accepted value of the Avogadro constant is $6.022\,045 \times 10^{23}$, which gives a value of $1.660\,566 \times 10^{-24}$ g for 1 u. Values of $N_A = 6.022 \times 10^{23}$ and $1 \text{ u} = 1.661 \times 10^{-24}$ g are sufficiently accurate for our purposes in this book.

We may summarize our discussion as follows:

- 1 C atom has a mass of 12.01 u.
- 6.022×10^{23} C atoms have a mass of 12.01 g.
- 1 O atom has a mass of 16.00 u.
- 6.022×10^{23} O atoms have a mass of 16.00 g.
- 1 CO molecule has a mass of 28.01 u.
- 6.022×10^{23} CO molecules have a mass of 28.01 g.

Because of the very special significance of 6.022×10^{23} atoms or molecules, the amount of any substance containing this number of atoms or molecules is called a **mole** (abbreviated mol) of the substance. Thus

$$1 \text{ mol of atoms is } 6.022 \times 10^{23} \text{ atoms}$$

$$1 \text{ mol of molecules is } 6.022 \times 10^{23} \text{ molecules}$$

We see that a mole is simply a counting unit just like a pair and a dozen:

$$1 \text{ pair of shoes is 2 shoes}$$

$$1 \text{ dozen eggs is 12 eggs}$$

The formal SI definition of a mole is

> **The mole is the amount of any substance that contains as many elementary entities as there are atoms in exactly 0.012 kg (12 g) of carbon-12. When the mole is used, the elementary entities must be stated. They may be atoms, molecules, ions, electrons, or other entities.**

One mole of carbon-12 atoms has a mass of 12.00 g, and 1 mol of (naturally occurring) carbon atoms has a mass of 12.01-g. One mole of oxygen atoms has a mass of 16.00 g, and 1 mol of oxygen molecules (O_2) has a mass of 32.00 g.

One mole of atoms has a mass in grams numerically equal to the atomic mass in unified atomic mass units:

Atomic mass C = 12.01 u   1 mol C atoms has a mass of 12.01 g

Atomic mass O = 16.00 u   1 mol O atoms has a mass of 16.00 g

Atomic mass Al = 26.98 u   1 mol Al atoms has a mass of 26.98 g

One mole of molecules has a mass in grams numerically equal to the molecular mass in atomic mass units:

Molecular mass CO = 28.01 u   1 mol CO molecules has a mass of 28.01 g

Molecular mass CH_4 = 16.04 u   1 mol CH_4 molecules has a mass of 16.04 g

When we are dealing with a compound such as sodium chloride, which does not exist in the solid state in the form of molecules, we speak of a mole of empirical formula units:

1 mol NaCl (empirical formula units) has a mass of 22.99 g + 35.45 g = 58.44 g

Figure 2.5 shows 1 mol of the elements mercury (Hg), copper (Cu), and sulfur (S_8), and 1 mol of the compounds water (H_2O) and sugar ($C_{12}H_{22}O_{11}$).

═ Example 2.5 ═

CONVERSION OF GRAMS TO MOLES

How many moles of aluminum atoms are in 50.00 g of aluminum?

Solution

We can summarize the problem as follows:

$$50.00 \text{ g Al} = ? \text{ mol Al atoms}$$

Since the atomic mass of aluminum is 26.98 u, 1 mol of aluminum atoms has a mass of 26.98 g. So we can set up the unit conversion factor

$$\frac{1 \text{ mol Al atoms}}{26.98 \text{ g Al}}$$

which enables us to convert grams to moles. Therefore

$$(50.00 \text{ g Al})\left(\frac{1 \text{ mol Al atoms}}{26.98 \text{ g Al}}\right) = 1.853 \text{ mol Al atoms}$$

FIGURE 2.5
The Mole.

One mole of each of the following substances (clockwise from the bottom): mercury, copper, water, sulfur, and sugar (sucrose).

═ *Example 2.6* ═══

CONVERSION OF GRAMS TO MOLES

How many moles of carbon dioxide molecules are in 10.00 g of carbon dioxide?

Solution

We can summarize the problem as follows:

$$10.00 \text{ g } CO_2 = ? \text{ mol } CO_2 \text{ molecules}$$

The molecular mass of carbon dioxide is 12.01 u + 2(16.00 u) = 44.01 u. So 1 mol of carbon dioxide molecules has a mass of 44.01 g. The unit conversion factor

$$\frac{1 \text{ mol } CO_2 \text{ molecules}}{44.01 \text{ g } CO_2}$$

enables us to convert grams to moles. Therefore

$$(10.00 \text{ g } CO_2)\left(\frac{1 \text{ mol } CO_2 \text{ molecules}}{44.01 \text{ g } CO_2}\right) = 0.2272 \text{ mol } CO_2 \text{ molecules}$$

═ *Example 2.7* ═══

CONVERSION OF MOLES TO GRAMS

What is the mass of 2.500 mol of ethanol, C_2H_5OH?

Solution

The molar mass of ethanol is 2(12.01) + 6(1.008) + 16.00 = 46.07 g. So the mass of ethanol is

$$(2.500 \text{ mol})\left(\frac{46.07 \text{ g}}{1.00 \text{ mol}}\right) = 115.2 \text{ g}$$

EXERCISE 2.3

(a) A silicon chip from a microcomputer has a mass of 4.62 mg. How many moles of silicon are in the silicon chip?

(b) What is the mass of 10.0 mol of aluminum?

(c) How many moles are 15.32 g of sodium chloride?

(d) What is the mass of 3.00 mol of carbon dioxide?

MOLAR MASS

The mass of 1 mol of a substance is called its **molar mass**. The molar mass of hydrogen *atoms* is the mass of 1 mol of hydrogen *atoms*, which is 1.008 g mol^{-1}. The molar mass of hydrogen *molecules* is the mass of 1 mol of hydrogen *molecules*, H_2, which is 2(1.008) = 2.016 g mol^{-1}. The molar mass of water molecules, H_2O, is the mass of 1 mol of water molecules, or 18.02 g mol^{-1}. The molar mass of NaCl is the mass of 1 mol of NaCl empirical formulas or 58.44 g mol^{-1}.

TABLE 2.2
Use of the Terms Molar Mass, Atomic Mass, Molecular Mass and Formula Mass

Atoms	Atomic Mass (u)	Molar Mass (g mol^{-1})	Mass of One Atom (g)
H	1.008	1.008	1.661×10^{-24}
O	16.00	16.00	2.657×10^{-23}
Ne	20.18	20.18	3.351×10^{-23}
Molecules	**Molecular Mass (u)**	**Molar Mass (g mol^{-1})**	**Mass of One Molecule (g)**
H_2	2.016	2.016	3.322×10^{-24}
N_2	28.02	28.02	4.653×10^{-23}
NH_3	17.04	17.04	2.830×10^{-23}
H_2O	18.02	18.02	2.992×10^{-23}
Three-Dimensional Network Solids	**Formula Mass (u)**	**Molar Mass (g mol^{-1})**	**Mass of One Formula Unit (g)**
NaCl	58.44	58.44	9.704×10^{-23}
SiO_2	60.09	60.09	9.984×10^{-23}

When using moles and molar quantities such as molar mass, we must, strictly speaking, specify the nature of the entities being considered (atoms, molecules, empirical formulas, electrons). Nevertheless, we can speak of the molar mass of water without ambiguity, because the fact that water consists of H_2O molecules is well known. We might also speak of the molar mass of oxygen and assume that it is 32.00 g mol^{-1}, since oxygen normally consists of O_2 molecules. We must be careful to specify the entities, however, if the substances concerned are not in their ordinary or familiar states and there could be some doubt about the entity concerned. For example, in the following section we will be making calculations involving moles of oxygen *atoms* so we will use a molar mass of 16.00 g mol^{-1} and not 32.00 g mol^{-1}.

Table 2.2 illustrates the use of the terms molar mass, atomic mass, molecular mass and formula mass.

EXERCISE 2.4

Calculate the molecular mass (weight) and the molar mass of (a) limestone, $CaCO_3$, and (b) saccharin, $C_7H_5NO_3S$, an artificial sweetener.

2.4
Stoichiometric Calculations: Compositions and Formulas

Atomic masses, formulas and the mole are the basic tools that we need to solve problems such as the following: How much carbon is in a kilogram of sugar given that the molecular formula of sugar (sucrose) is $C_{12}H_{22}O_{11}$? If we have prepared a substance that is made up of 56.8% carbon, 28.4% oxygen, 8.28% nitrogen, and 6.56% hydrogen, what is its empirical formula?

PERCENTAGE COMPOSITION

From the formula of a compound we can find out how much of each of the elements is present in a given amount of the compound. Iron is obtained from iron oxide, Fe_2O_3, which is found as the ore hematite. The formula Fe_2O_3 tells us that for every two iron atoms there are three oxygen atoms and also that in 1 mol of Fe_2O_3 there are 2 mol of Fe atoms and 3 mol of O atoms. Since the molar mass of Fe_2O_3 is 2(55.85 g) + 3(16.00 g) = 159.7 g we know that there are 2 × 55.85 g = 111.7 g of iron and 3 × 16.00 g = 48.00 g of oxygen in 159.7 g of Fe_2O_3. In other words we could, in principle at least, obtain 111.7 g of iron from 159.7 g of the ore hematite. It is particularly convenient to express the composition of a compound in terms of the mass percent of each element:

$$\text{Percent mass of element in compound} = \frac{\text{Mass of element in 1 mol compound}}{\text{Molar mass of compound}} \times 100\%$$

Let us calculate the percentage composition of iron oxide, Fe_2O_3:

$$\text{Percent iron} = \frac{\text{Mass of iron in 1 mol } Fe_2O_3}{\text{Molar mass } Fe_2O_3} \times 100\%$$

$$= \frac{111.7 \text{ g}}{159.7 \text{ g}} \times 100\% = 69.94\%$$

We calculate the percentage of oxygen in the same way. One mole of Fe_2O_3 contains 3 mol of O atoms. So

$$\text{Percent oxygen} = \frac{48.00 \text{ g}}{159.7 \text{ g}} \times 100\% = 30.06\%$$

Or, since Fe_2O_3 contains only iron and oxygen, we could have found the percentage of oxygen more simply by subtracting 69.94% from 100%, that is,

$$\text{Percent oxygen} = 100.00\% - 69.94\% = 30.06\%$$

=== *Example 2.8* ===

PERCENTAGE COMPOSITION

What is the percentage composition by mass of (a) ammonia, NH_3, and (b) glucose, $C_6H_{12}O_6$?

Solution

(a) The molar mass of ammonia is 14.01 g + 3(1.008 g) = 17.03 g.

$$\text{Mass percent N in } NH_3 = \frac{14.01 \text{ g}}{17.03 \text{ g}} \times 100\% = 82.27\%$$

$$\text{Mass percent H in } NH_3 = \frac{3.024 \text{ g}}{17.03 \text{ g}} \times 100\% = 17.73\%$$

(b) The molar mass of glucose is 6(12.01 g) + 12(1.008 g) + 6(16.00 g) = 180.2 g.

$$\text{Mass percent C in } C_6H_{12}O_6 = \frac{6 \times 12.01 \text{ g}}{180.2 \text{ g}} \times 100\% = 39.99\%$$

$$\text{Mass percent H in } C_6H_{12}O_6 = \frac{12 \times 1.008 \text{ g}}{180.2 \text{ g}} \times 100\% = 6.71\%$$

$$\text{Mass percent O in } C_6H_{12}O_6 = \frac{6 \times 16.00 \text{ g}}{180.2 \text{ g}} \times 100\% = 53.27\%$$

EXERCISE 2.5

What is the percentage composition by mass of each of the following compounds: (a) methane, CH_4; (b) nitric acid, HNO_3; (c) calcium sulfate, $CaSO_4$?

DETERMINATION OF EMPIRICAL FORMULAS

In the study of any new compound it is important to determine by experiment which elements are present and to find the percentage of each one. The experimental determination of the composition of substances forms an important part of an area of chemistry called *chemical analysis* which is also concerned with the separation and identification of substances in mixtures. The composition of a substance can be determined by a variety of different methods, a few of which we will illustrate in this section. Once we know the percentage composition of a substance, we can determine its empirical formula by reversing the calculations described in the previous section.

For example, water can be found, by analysis, to have the composition 11.19% hydrogen and 88.81% oxygen by mass. From these data we can determine the empirical formula of water. The simplest way to proceed is to assume that we have a 100.00-g sample of water. The percentage composition then tells us that 100.00 g of water contain 11.19 g of hydrogen atoms and 88.81 g of oxygen atoms. From the table of atomic masses we find that 1 mol of hydrogen atoms has a mass of 1.008 g, and 1 mol of oxygen atoms has a mass of 16.00 g. So the unit conversion factors are

$$\left(\frac{1 \text{ mol H atoms}}{1.008 \text{ g H}}\right) \quad \text{and} \quad \left(\frac{1 \text{ mol O atoms}}{16.00 \text{ g O}}\right)$$

Therefore

$$11.19 \text{ g H} = (11.19 \text{ g H})\left(\frac{1 \text{ mol H atoms}}{1.008 \text{ g H}}\right) = 11.10 \text{ mol H atoms}$$

and

$$88.81 \text{ g O} = (88.81 \text{ g O})\left(\frac{1 \text{ mol O atoms}}{16.00 \text{ g O}}\right) = 5.55 \text{ mol O atoms}$$

Thus in water the ratio of moles of hydrogen atoms to moles of oxygen atoms is 11.10:5.55.

Since a mole of atoms always contains the same number of atoms ($N_A = 6.022 \times 10^{23}$), the ratio of moles of atoms in a compound is also the ratio of the numbers of atoms. Hence the ratio of hydrogen atoms to oxygen atoms is 11.10:5.55. We can convert both numbers to integers by dividing each by the smaller of the two numbers, 5.55:

$$\frac{11.10}{5.55} = 2.00 \quad \text{and} \quad \frac{5.55}{5.55} = 1.00$$

Thus the ratio of H atoms to O atoms is 2:1, and the empirical formula of water is therefore H_2O.

⹀ Example 2.9

EMPIRICAL FORMULA FROM PERCENT COMPOSITION

A white compound is formed when phosphorus burns in air. Analysis shows that the compound is composed of 43.7% P and 56.3% O by mass. What is the empirical formula of the compound?

Solution

The simplest procedure is to assume that we have a 100.0-g sample of the compound. According to the percentage composition given, 100.0 g of the compound contain 43.7 g of P atoms and 56.3 g of O atoms. Using the atomic masses of P and O, we can convert these masses to moles by using the appropriate unit conversion factors:

$$43.7 \text{ g P} = (43.7 \text{ g P})\left(\frac{1 \text{ mol P}}{30.97 \text{ g P}}\right) = 1.41 \text{ mol P}$$

$$56.3 \text{ g O} = (56.3 \text{ g O})\left(\frac{1 \text{ mol O}}{16.00 \text{ g O}}\right) = 3.52 \text{ mol O}$$

Thus the ratio of moles of P atoms to moles of O atoms is 1.41:3.52. Hence the ratio of P atoms to O atoms is also 1.41:3.52. We now divide by the smaller number, 1.41, so that one number in the ratio is 1:

$$\text{Ratio of P atoms to O atoms} = \frac{1.41}{1.41} : \frac{3.52}{1.41} = 1.00:2.50$$

This ratio gives us the formula $PO_{2.5}$. To obtain the empirical formula (the simplest formula containing whole numbers of atoms) we must multiply by the smallest factor that will make both numbers integral. In this case the appropriate factor is 2. So, finally, we have

$$\text{Ratio of P atoms to O atoms} = 2.00:5.00$$

The empirical formula is therefore P_2O_5.

Mole ratios do not always work out to exactly whole numbers, both because the numbers used in the calculations are rounded off and because there may be experimental errors in the percentage composition. When the experimental errors are small, there is little doubt about the appropriate whole numbers, as in the following example.

Example 2.10

EMPIRICAL FORMULA FROM PERCENT COMPOSITION

The compound adrenaline is released in the human body in times of stress; it increases the body's metabolic rate. Like many compounds in living systems, it is composed of carbon, hydrogen, oxygen, and nitrogen. It was found by experiment to have the composition 56.8% C, 6.50% H, 28.4% O, and 8.28% N. What is the empirical formula of adrenaline?

Solution

As in Example 2.9, we will assume that we are dealing with 100.0 g of the compound. For convenience we set up this type of calculation in a table.

	Carbon		Hydrogen		Oxygen		Nitrogen
Mass	56.8 g		6.50 g		28.4 g		8.28 g
Moles of atoms	$(56.8 \text{ g})\left(\dfrac{1 \text{ mol}}{12.01 \text{ g}}\right)$		$(6.50 \text{ g})\left(\dfrac{1 \text{ mol}}{1.008 \text{ g}}\right)$		$(28.4 \text{ g})\left(\dfrac{1 \text{ mol}}{16.00 \text{ g}}\right)$		$(8.28 \text{ g})\left(\dfrac{1 \text{ mol}}{14.01 \text{ g}}\right)$
	= 4.73		= 6.45		= 1.78		= 0.591
Mole ratio (atom ratio)	$\dfrac{4.73}{0.591}$	:	$\dfrac{6.45}{0.591}$	:	$\dfrac{1.78}{0.591}$	:	$\dfrac{0.591}{0.591}$
	8.00	:	10.9	:	3.01	:	1.00

Thus the empirical formula of adrenaline is $C_8H_{11}O_3N$.

EXERCISE 2.6

Phosphoric acid has the composition 3.09% H, 31.60% P, and 65.31% O. What is its empirical formula?

There are several important types of experiment by which the composition and empirical formula of a compound may be determined. For example, the compound may be synthesized from its elements, it may be decomposed into its elements, or it may be converted to compounds of known composition. In the following examples we will illustrate some of these methods, and at the same time show how the empirical formula may be determined directly from the analytical results rather than from the percentage composition.

Example 2.11

EMPIRICAL FORMULA BY SYNTHESIS

When the element antimony, Sb, is heated with excess sulfur, a reaction occurs to give a compound containing only antimony and sulfur (Figure 2.6). On further heating, excess sulfur is burnt off, forming gaseous sulfur dioxide, SO_2, and the substance

FIGURE 2.6
Antimony, Sulfur,
and Antimony Sulfide.

Antimony is an element with a
grey metallic appearance. Sulfur is
a yellow, nonmetallic element.
Antimony sulfide is an orange
compound.

left is a pure compound of antimony and sulfur. In one experiment 2.435 g of antimony were used, and the mass of the pure compound of antimony and sulfur was found to be 3.397 g. What is the empirical formula of the antimony–sulfur compound?

Solution

From the law of conservation of mass we have

$$\text{Mass compound} = \text{Mass sulfur} + \text{Mass antimony}$$

$$\text{Mass sulfur} = \text{Mass compound} - \text{Mass antimony}$$

$$= 3.397\ \text{g} - 2.435\ \text{g} = 0.962\ \text{g}$$

Thus we see that 2.435 g of antimony combine with 0.962 g of sulfur to give the antimony–sulfur compound.

Next, we proceed as in previous examples to find the ratio of atoms of each element in the compound.

	Antimony		*Sulfur*
Mass	2.435 g		0.962 g
Moles of atoms	$(2.435\ \text{g})\left(\dfrac{1\ \text{mol}}{121.8\ \text{g}}\right) = 0.0200$		$(0.962\ \text{g})\left(\dfrac{1\ \text{mol}}{32.06\ \text{g}}\right) = 0.0300$
Mole ratio (atom ratio)	$\dfrac{0.0200}{0.0200}$	:	$\dfrac{0.0300}{0.0200}$
	1.00	:	1.50

This result corresponds to the formula $SbS_{1.5}$, but to obtain the empirical formula we must multiply by a suitable factor to obtain whole numbers. In this case the factor 2 is required. Thus the relative numbers of atoms are 2.00 Sb and 3.00 S. The empirical formula is therefore Sb_2S_3.

EXERCISE 2.7

1.234 g of aluminum were burned completely in oxygen to give 2.322 g of a white oxide. What is the empirical formula of the oxide of aluminum?

In Example 2.11 we found the empirical formula of antimony sulfide by synthesizing it from the elements. More commonly, the composition of a substance is obtained not by synthesizing it in this way, but by carrying out a reaction that converts it into products of known formulas and, hence, known compositions. If the masses of these products obtained from a known mass of the original compound are determined, its composition can then be found.

For example, many of the carbon compounds known as organic compounds contain only carbon and hydrogen or only carbon, hydrogen, and oxygen. The composition of these organic compounds is often determined by burning them in an excess of dry oxygen gas, that is, sufficient oxygen to convert all the hydrogen in the compound to water vapor and all the carbon in the compound to carbon dioxide. The gases are then passed through a tube that contains a substance that absorbs the water and through another tube that contains a substance that absorbs the carbon dioxide, as shown in Figure 2.7. These tubes are weighed both before and after the experiment. The increase in the mass of the first tube is the mass of water absorbed, and the increase in mass of the second tube is the mass of carbon dioxide absorbed. From this information we can calculate the amounts of carbon and hydrogen in the original compound and thus determine its empirical formula, as the following example demonstrates.

═ Example 2.12 ═

EMPIRICAL FORMULA BY COMBUSTION OF AN ORGANIC COMPOUND

Ascorbic acid (vitamin C) is known to contain only C, H, and O. A 6.49-mg sample was burned in an apparatus like the one in Figure 2.7. The increased weights of the absorption tubes showed that 9.74 mg of CO_2 and 2.64 mg of H_2O were formed. What is the empirical formula of ascorbic acid?

Solution

We first find the masses of C and H in the sample of ascorbic acid from the masses of CO_2 and H_2O. For this calculation we need the molar masses of CO_2 and H_2O:

$$\text{Molar mass } CO_2 = 12.01 + 2(16.00) = 44.01 \text{ g}$$
$$\text{Molar mass } H_2O = 2(1.008) + 16.00 = 18.02 \text{ g}$$

Then we have

$$\text{Mass C} = (9.74 \times 10^{-3} \text{ g } CO_2)\left(\frac{12.01 \text{ g C}}{44.01 \text{ g } CO_2}\right) = 2.66 \times 10^{-3} \text{ g C}$$

$$\text{Mass H} = (2.64 \times 10^{-3} \text{ g } H_2O)\left(\frac{2.016 \text{ g H}}{18.02 \text{ g } H_2O}\right) = 0.295 \times 10^{-3} \text{ g H}$$

The mass of oxygen in the compound cannot be found from the masses of CO_2 and H_2O because some of this oxygen comes from the O_2 that was used to burn the

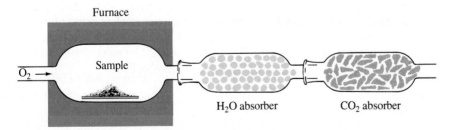

FIGURE 2.1
Combustion Apparatus for Analyzing Organic Compounds.

A weighed sample of an organic compound is burned in a stream of oxygen. The reaction produces gaseous H_2O and CO_2. These gases then pass through a series of tubes. One tube contains a substance that absorbs H_2O; the substance in another tube absorbs CO_2. By comparing the masses of these tubes before and after the reaction, one can determine the masses of hydrogen and of carbon present in the compound that was burned.

compound. But since we know that ascorbic acid contains only C, H, and O, the mass of oxygen can be found by subtraction:

$$\text{Mass O} = 6.49 \text{ mg ascorbic acid} - (2.66 \text{ mg C} + 0.30 \text{ mg H})$$

$$= 3.53 \text{ mg} = 3.53 \times 10^{-3} \text{ g}$$

Now that we have the mass of C, H, and O in a given mass of ascorbic acid, we can proceed to find the empirical formula in the usual way.

	C	H	
Mass	2.66×10^{-3} g	0.295×10^{-3} g	3.53×10^{-3} g
Moles	$(2.66 \times 10^{-3} \text{ g})\left(\dfrac{1 \text{ mol}}{12.01 \text{ g}}\right)$	$(0.295 \times 10^{-3} \text{ g})\left(\dfrac{1 \text{ mol}}{1.008 \text{ g}}\right)$	$(3.53 \times 10^{-3} \text{ g})\left(\dfrac{1 \text{ mol}}{16.00 \text{ g}}\right)$
	$= 2.21 \times 10^{-4}$	$= 2.93 \times 10^{-4}$	$= 2.21 \times 10^{-4}$
Mole ratio (atom ratio)	$\dfrac{2.21 \times 10^{-4}}{2.21 \times 10^{-4}}$	$\dfrac{2.93 \times 10^{-4}}{2.21 \times 10^{-4}}$	$\dfrac{2.21 \times 10^{-4}}{2.21 \times 10^{-4}}$
	1.00	1.33	1.00

Finally, we convert these values to integers. Multiplying by 2 does not give integral values, so we try multiplying by 3, which gives

	3.00	3.99	3.00

Thus the empirical formula of ascorbic acid is $C_3H_4O_3$.

EXERCISE 2.8

Phenol is an important industrial chemical used in the production of plastics. It contains only carbon, hydrogen, and oxygen. When a sample of phenol with mass 0.088 74 g was burned in excess oxygen, 0.2491 g of CO_2 and 0.0510 g of H_2O were obtained. What is the empirical formula of phenol?

If in addition to the percentage composition or the analytical data for a molecular compound we also know its molar mass, we may find its molecular formula as well as its empirical formula, as Example 2.13 shows. Some methods for determining molar mass are described in later chapters.

$=$ *Example 2.13* $=$

MOLECULAR FORMULA FROM COMPOSITION AND MOLAR MASS

Napthalene, which is best known as "moth balls", has the composition 93.75% C and 6.25% H and it has a molar mass of 128 g mol^{-1}. What are the empirical and molecular formulas of naphthalene?

Solution

Assume that we have 100.0 g of naphthalene.

	C	H
Mass	93.75 g	6.25 g
Moles	$\dfrac{93.75 \text{ g}}{12.01 \text{ g mol}^{-1}} = 7.80 \text{ mol}$	$\dfrac{6.25 \text{ g}}{1.008 \text{ g mol}^{-1}} = 6.20 \text{ mol}$
Mole ratio (atom ratio)	$\dfrac{7.80}{6.20}$	$\dfrac{6.20}{6.20}$
	1.25	1.00

Since one of these values is not close to a whole number, we multiply by a suitable factor, which in this case is 4, giving

5.00	4.00

Thus the empirical formula of naphthalene is C_5H_4. The mass of 1 mol of empirical formulas is therefore $5(12.01) + 4(1.008) = 64.08$ g mol^{-1}. The given molar mass of 128 g mol^{-1} is twice this value and so the molecular formula is twice the empirical formula and is $C_{10}H_8$.

EXERCISE 2.9

Acetic acid is the main component of vinegar. Analysis shows that it has the composition 40.0% C, 6.71% H, and 53.29% O. Its molar mass is 60.0 g mol^{-1}. What are the empirical and molecular formulas of acetic acid?

2.5
Stoichiometric Calculations: Reactions

In this section we will see how we can answer a question such as: How much oxygen is needed to burn 1 kg of methane to give carbon dioxide and water? The first step in solving such a problem is to write down the chemical equation for the reaction.

CHEMICAL EQUATIONS

A **chemical equation** is a shorthand description of a reaction that gives the formulas for all the reactants and all the products. For example,

$$\text{Reactants} \qquad \text{Products}$$
$$2H_2 + O_2 \longrightarrow 2H_2O$$
$$C + O_2 \longrightarrow CO_2$$
$$CH_4 + 2O_2 \longrightarrow CO_2 + 2H_2O$$

Most importantly, an equation must be consistent with the law of conservation of mass, which tells us that atoms are neither created nor destroyed in chemical reactions. The total numbers of atoms of each element must be the same in the products and in the reactants; that is, they must be the same on both sides of the equation. Thus two molecules (or four atoms) of hydrogen react with one molecule (or two atoms) or oxygen to give two water molecules, in which there are a total of four hydrogen atoms and two oxygen atoms. Every chemical equation must *balance* in this way.

Balancing equations is not difficult, but practice is required. As an example, consider the equation representing the burning of methane in oxygen to give carbon dioxide and water. We first write down the reactants and products as follows:

$$CH_4 + O_2 \longrightarrow CO_2 + H_2O \qquad \text{(Unbalanced)}$$
$$\underbrace{\hphantom{CH_4 + O_2}}_{\text{Reactants}} \qquad \underbrace{\hphantom{CO_2 + H_2O}}_{\text{Products}}$$

This equation is called the unbalanced equation because it correctly indicates the reactants and products but not the relative numbers of molecules of each. Note that before any equation can be balanced, *all* the reactants and products and their correct formulas must be known. If all this information is not available, then a correct, balanced equation for a reaction cannot be written.

The simplest procedure for balancing an equation is to consider first those elements that appear the least frequently in the equation. In this example hydrogen and carbon appear in only two formulas each, while oxygen appears three times. So we begin by balancing the numbers of carbon and hydrogen atoms. All the carbon in methane, CH_4, must be converted to carbon dioxide, CO_2, because no other product contains carbon. Thus one molecule of CH_4 must give one molecule of CO_2, as already written. Each molecule of CH_4, however, contains four hydrogen atoms; and since all the hydrogen ends up in water molecules, two water molecules, H_2O, must be produced for each methane molecule. Hence we must place a 2 in front of the formula for water, to give

$$CH_4 + O_2 \longrightarrow CO_2 + 2H_2O \qquad \text{(Unbalanced)}$$

Now we can balance the remaining element, oxygen. We note that there are four oxygen atoms on the right-hand side of the equation: two in the CO_2 molecule and two in the two H_2O molecules. We must therefore place a 2 in front of the formula for oxygen, O_2, so that we have four oxygen atoms on each side of the equation:

$$CH_4 + 2O_2 \longrightarrow CO_2 + 2H_2O \qquad \text{(Balanced)}$$

We can now make a final check of the numbers of each kind of atom on both sides of the equation. We find that we have one C atom, four H atoms, and four O atoms on both sides of the equation. The equation is therefore balanced.

═ *Example 2.14* ═══

BALANCING EQUATIONS

Butane, C_4H_{10}, is one of the gaseous components of petroleum. When liquified under pressure, it is sold as bottled gas for use as a fuel. When it is burned in sufficient oxygen, O_2, the only products are carbon dioxide, CO_2, and water, H_2O. Write a balanced equation to describe this reaction.

Solution

First, we write an unbalanced equation showing the correct formulas of all the reactants and products:

$$C_4H_{10} + O_2 \longrightarrow CO_2 + H_2O \qquad \text{(Unbalanced)}$$

Since O atoms appear in three of these formulas, whereas C and H atoms appear in only two, we balance C and H first. There are 4 C atoms in C_4H_{10}, and CO_2 is the only product containing carbon. So 4 CO_2 molecules must be formed:

$$C_4H_{10} + O_2 \longrightarrow 4CO_2 + H_2O \qquad \text{(Unbalanced)}$$

The 10 H atoms in C_4H_{10} must appear in the products as 5 H_2O molecules:

$$C_4H_{10} + O_2 \longrightarrow 4CO_2 + 5H_2O \qquad \text{(Unbalanced)}$$

We now have a total of 13 O atoms on the right-hand side. The equation could be balanced by using a coefficient of $\frac{13}{2}$ in front of O_2 on the left-hand side:

$$C_4H_{10} + \tfrac{13}{2}O_2 \longrightarrow 4CO_2 + 5H_2O$$

We would then have 13 O atoms on each side of the equation. Customarily, however, we do not write fractional coefficients in equations, because they could be interpreted as meaning fractions of a molecule. One-half of an oxygen molecule is an oxygen atom, which has properties quite different from those of an oxygen molecule. Ambiguity is avoided by multiplying both sides of the equation by 2 to obtain the final balanced equation:

$$2C_4H_{10} + 13O_2 \longrightarrow 8CO_2 + 10H_2O$$

A final check should be made by counting the numbers of atoms of each kind on both sides.

$$2C_4H_{10} + 13O_2 \longrightarrow 8CO_2 + 10H_2O$$

	C	H	O		C	H	O
Number of atoms	8	20	26		8	20	26

Although this equation looks complicated, we balanced it rather easily because we used a logical, systematic approach.

Some remarks on incorrect ways of balancing equations might be helpful. The equation for the reaction between hydrogen and oxygen to give water,

$$H_2 + O_2 \longrightarrow H_2O$$

cannot be balanced by writing a 2 after the O in H_2O,

$$H_2 + O_2 \longrightarrow H_2O_2$$

This equation is balanced, but it is *not* the equation for the reaction between hydrogen and oxygen to give *water*. The product has been changed to hydrogen peroxide, and this equation is now for a *different reaction*.

The equation should also not be balanced as

$$H_2 + O_2 \longrightarrow H_2O + O$$

Although this equation is also balanced, another product, namely O atoms, has been introduced, which was not originally specified. In balancing an equation, we can make no change in the nature of either the reactants or the products. Coefficients must be inserted to balance the equation without introducing new products or reactants. The balanced equation for this reaction is

$$2H_2 + O_2 \longrightarrow 2H_2O$$

Finally, we note that equations can only be balanced after all the products are known. Incomplete equations such as

$$H_2 + O_2 \longrightarrow$$

cannot be balanced in an unambiguous way because no product is specified, and therefore we have no way to choosing between the three possibilities discussed above. Do not be tempted to invent products simply to balance an equation. You *must* know all the reactants and products before you can write a balanced equation. The only way that we can be certain what the products of a reaction are is by carrying out appropriate experiments. After completing this course you should know the products of a variety of important reactions. You should also be able to make reasonable predictions about the products of many others by analogy with the reactions that you know. For the moment we are primarily concerned with balancing an equation once the products are known. In the following exercise the products of each of the reactions are given.

EXERCISE 2.10

Write balanced equations for each of the following reactions:

(a) The combustion of ethanol, C_2H_6O, in excess oxygen to give CO_2 and H_2O

(b) The reaction of nitrogen, N_2, and hydrogen, H_2, to give ammonia, NH_3

(c) The reaction of iron with oxygen, O_2, to give the oxide Fe_2O_3

It is important to realize that you are already in a position to make a prediction about the products of one type of reaction. We have seen that when both methane and butane are burned in excess oxygen the products are carbon dioxide and water. Methane and butane are examples of a class of compounds called hydrocarbons. They contain only carbon and hydrogen. Now if you were asked to write the equation for the combustion of the hydrocarbon propane, C_3H_8, in an excess of oxygen and were given no information on the products, you could reasonably assume that the products would be carbon dioxide and water.

EXERCISE 2.11

Write a balanced equation for the combustion of propane in C_3H_8 in an excess of oxygen.

Often in the formula of an element or a compound in an equation, we add a symbol to indicate whether the substance is in the gaseous, liquid or solid state. The designations customarily used are (s) for solid, (l) for liquid, (g) for gas, and (aq) for substances in aqueous solution. Thus the equation

$$CH_4 + 2O_2 \longrightarrow CO_2 + 2H_2O$$

could be more completely written as

$$CH_4(g) + 2O_2(g) \longrightarrow CO_2(g) + 2H_2O(g)$$

Whether we write $H_2O(g)$ or $H_2O(l)$ depends on the temperature at which the products of the reaction are formed. Because combustion reactions occur with the evolution of a considerable amount of heat, the water is normally formed as steam in this reaction, that is, as $H_2O(g)$. That is why we have written $H_2O(g)$. However, if the products were allowed to cool to room temperature we would write $H_2O(l)$. For the reaction of magnesium with carbon dioxide that we saw in Experiment 1.3 we could write the equation

$$2Mg(s) + CO_2(g) \longrightarrow 2MgO(s) + C(s)$$

EXERCISE 2.12

Rewrite the equations in Exercises 2.10 and 2.11 indicating whether the substances involved are in the solid, liquid or gaseous states, at room temperature.

CALCULATION OF THE AMOUNTS OF REACTANTS AND PRODUCTS

From the balanced chemical equation and a knowledge of atomic masses, we can find the masses of the reactants and the products in any reaction. For example, from the atomic masses of hydrogen and oxygen and the equation

$$2H_2(g) + O_2(g) \longrightarrow 2H_2O(g)$$

we know that

$$2 \text{ molecules } H_2 + 1 \text{ molecule } O_2 \longrightarrow 2 \text{ molecules } H_2O$$
$$4.032 \text{ u} \qquad\qquad 32.00 \text{ u} \qquad\qquad 36.03 \text{ u}$$

Normally, we are not concerned with such small numbers of molecules. As we have seen, we may, much more conveniently, consider the reaction of 2 mol of hydrogen ($2 \times 6.022 \times 10^{23}$ molecules) with 1 mol of oxygen (6.022×10^{23} molecules) to give 2 mol of water ($2 \times 6.022 \times 10^{23}$ molecules):

$$2H_2 \qquad + \qquad O_2 \qquad \longrightarrow \qquad 2H_2O$$

2 mol	1 mol	2 mol
$2(6.022 \times 10^{23})$ molecules	6.022×10^{23} molecules	$2(6.022 \times 10^{23})$ molecules
4.032 g	32.00 g	36.03 g

We conclude that 4.032 g of hydrogen react with 32.00 g of oxygen to give 36.03 g of water.

In Experiment 1.3 we saw that magnesium burns in air, combining with the oxygen of the air to form white solid magnesium oxide, MgO. Now we are in a position to consider the quantitative aspects of this reaction. For example, if we have 0.187 g of magnesium, how much oxygen will this combine with and how much magnesium oxide will be formed? The first step in any problem of this type is to write the *balanced* equation for the reaction:

$$2Mg(s) + O_2(g) \longrightarrow 2MgO(s)$$

This equation tells us that 2 mol of Mg react with 1 mol of O_2 to give 2 mol of MgO. So the next step is to convert the mass of Mg to moles:

$$(0.187 \text{ g Mg})\left(\frac{1 \text{ mol Mg}}{24.31 \text{ g Mg}}\right) = 7.69 \times 10^{-3} \text{ mol Mg}$$

Now we can use the equation to set up the conversion factor to convert moles Mg to moles O_2. The required factor is

$$\left(\frac{1 \text{ mol } O_2}{2 \text{ mol Mg}}\right)$$

and so we have

$$(7.69 \times 10^{-3} \text{ mol Mg})\left(\frac{1 \text{ mol } O_2}{2 \text{ mol Mg}}\right) = 3.85 \times 10^{-3} \text{ mol } O_2$$

Finally we convert moles O_2 to mass of O_2:

$$(3.85 \times 10^{-3} \text{ mol } O_2)\left(\frac{32.00 \text{ g } O_2}{1 \text{ mol } O_2}\right) = 0.123 \text{ g } O_2$$

To obtain the mass of magnesium oxide, MgO, formed we follow the same steps:

$$(7.69 \times 10^{-3} \text{ mol Mg})\left(\frac{2 \text{ mol MgO}}{2 \text{ mol Mg}}\right) = 7.70 \times 10^{-3} \text{ mol MgO}$$

$$(7.69 \times 10^{-3} \text{ mol MgO})\left(\frac{40.31 \text{ g MgO}}{1 \text{ mol MgO}}\right) = 0.310 \text{ g MgO}$$

Or more simply, making use of the law of conservation of mass,

$$\text{Mass MgO} = \text{mass Mg} + \text{mass } O_2 = 0.187 \text{ g} + 0.123 \text{ g} = 0.310 \text{ g}$$

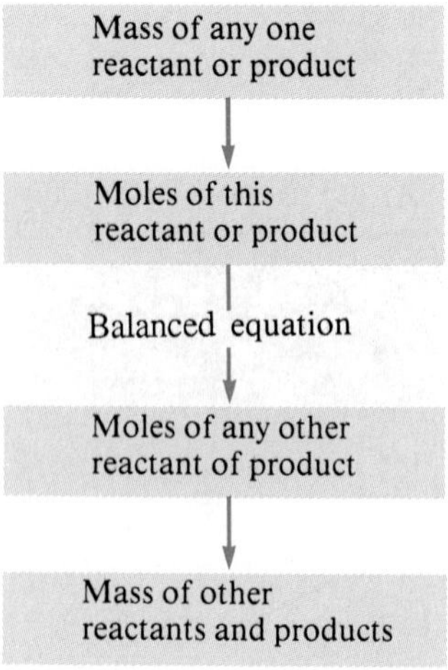

In the above example we made use of four steps and the same four steps are applicable to all problems involving the calculation of the masses of substances taking part in chemical reactions:

1. Write the balanced equation for the reaction.
2. Convert the known mass of one reactant to moles.
3. Use the balanced equation to set up the appropriate conversion factor(s) to find the number of moles of another reactant or product(s).
4. Convert from moles back to grams.

Mass of any one
reactant or product

↓

Moles of this
reactant or product

Balanced equation

↓

Moles of any other
reactant of product

↓

Mass of other
reactants and products

= Example 2.15 =

AMOUNTS OF REACTANTS AND PRODUCTS

When a spark is passed through a mixture of hydrogen, H_2, and chlorine, Cl_2, an explosive reaction occurs in which hydrogen chloride, HCl, is formed. How many grams of chlorine, $Cl_2(g)$, are needed to react completely with 0.245 g of hydrogen, $H_2(g)$, to give hydrogen chloride, HCl(g)? How much HCl(g) is formed?

Solution

First, we write the balanced equation

$$H_2(g) + Cl_2(g) \longrightarrow 2HCl(g)$$

Next we convert the mass of H_2 to moles of H_2:

$$(0.245 \text{ g } H_2)\left(\frac{1 \text{ mol } H_2}{2.016 \text{ g } H_2}\right) = 0.122 \text{ mol}$$

Then we find the number of moles of Cl_2 that react with 0.122 mol of H_2. Since 1 mol of Cl_2 reacts with 1 mol of H_2

$$(0.122 \text{ mol } H_2)\left(\frac{1 \text{ mol } Cl_2}{1 \text{ mol } H_2}\right) = 0.122 \text{ mol } Cl_2$$

Then we convert 0.122 mol of Cl_2 to grams:

$$(0.122 \text{ mol } Cl_2)\left(\frac{70.90 \text{ g } Cl_2}{1 \text{ mol } Cl_2}\right) = 8.65 \text{ g } Cl_2$$

The amount of HCl produced must equal the total amount of H_2 and Cl_2 consumed in the reaction, which is $8.65 + 0.245 = 8.90$ g HCl. Or we could calculate the amount of HCl by making use of the appropriate conversion factors:

$$(0.122 \text{ mol } H_2)\underbrace{\left(\frac{2 \text{ mol } HCl}{1 \text{ mol } H_2}\right)}_{\substack{\text{From the} \\ \text{balanced} \\ \text{equation}}}\underbrace{\left(\frac{36.46 \text{ g } HCl}{1 \text{ mol } HCl}\right)}_{\substack{\text{From the molar} \\ \text{mass of HCl}}} = 8.90 \text{ g } HCl$$

EXERCISE 2.13

What mass of CO_2 and what mass of H_2O are formed when 1.53 g of butane, C_4H_{10}, are burned in an excess of oxygen? See Example 2.14 for the balanced equation.

LIMITING REACTANT

When a hydrocarbon such as butane is burned in an excess of air or oxygen, it is clearly the amount of the hydrocarbon that determines the amounts of CO_2 and H_2O that are formed, and not all the oxygen that is present is used up. It is often the case that substances that are reacting together are not present in exactly the same ratios as in the balanced equation. For example, suppose that we passed a spark through a mixture of 2.5 mol of H_2 and 1 mol of Cl_2. Only 1 mol of H_2 would react according to the equation

$$H_2 + Cl_2 \longrightarrow 2HCl$$
$$1 \text{ mol} \quad 1 \text{ mol} \qquad 2 \text{ mol}$$

so that 1.5 mol of H_2 are left in excess. Thus the amount of Cl_2 determines the amount of HCl that is formed, and so Cl_2 is called the limiting reactant. In general the **limiting reactant** in any reaction *is that reactant among two or more whose amount determines the amount of products that may be formed*, leaving excess amounts of the other reactants unused.

═ *Example 2.16* ══

LIMITING REACTANT

When zinc and sulfur are heated together, they react to form zinc sulfide, according to the equation

$$Zn(s) + S(s) \longrightarrow ZnS(s)$$

Suppose 12.00 g of zinc are heated with 7.50 g of sulfur.

(a) Which is the limiting reactant?

(b) How much ZnS is formed?

(c) How much of one of the reactants remains unreacted?

Solution

(a) We convert the mass of each reactant to moles:

$$(12.00 \text{ g Zn})\left(\frac{1 \text{ mol Zn}}{65.38 \text{ g Zn}}\right) = 0.184 \text{ mol Zn}$$

$$(7.50 \text{ g S})\left(\frac{1 \text{ mol S}}{32.06 \text{ g S}}\right) = 0.234 \text{ mol S}$$

We see from the equation that 1 mol of Zn reacts with 1 mol of S. Therefore there is more S than required; thus Zn is the limiting reactant.

(b) The amount of product formed depends on the amount of the limiting reactant, Zn, and not on the amount of S. The equation shows that 1 mol of Zn gives 1 mol of ZnS. Therefore 0.184 mol of Zn gives 0.184 mol ZnS. The mass of ZnS formed is thus

$$(0.184 \text{ mol ZnS})\left(\frac{97.44 \text{ g ZnS}}{1 \text{ mol ZnS}}\right) = 17.9 \text{ g ZnS}$$

(c) The excess amount of sulfur is $0.234 \text{ mol} - 0.184 \text{ mol} = 0.050 \text{ mol S}$. We convert this result to grams:

$$(0.050 \text{ mol S})\left(\frac{32.06 \text{ g S}}{1 \text{ mol S}}\right) = 1.60 \text{ g excess S}$$

EXERCISE 2.14

When iron oxide is heated with aluminum powder a very vigorous reaction occurs in which molten iron is produced (see Experiment 10.1, page 485) according to the equation

$$Fe_2O_3(s) + 2Al(s) \longrightarrow Al_2O_3(s) + 2Fe(l)$$

A mixture of 30.0 g of aluminum and 100.0 g of Fe_2O_3 is heated.

(a) Which is the limiting reactant?

(b) How much iron will be formed?

(c) How much of one of the reactants remains when the reaction is complete?

In solving any limiting reactant problem we use the same four steps that we use in solving any problem involving the amounts of reactants and products in a reaction, except that we add the step of deciding which is the limiting reactant. So the steps are as follows:

1. Write the balanced equation for the reaction.
2. Convert grams of reactants to moles of reactants.
3. Use the balanced equation to decide which is the limiting reactant.
4. Use the balanced equation to convert moles limiting reactant to moles products.
5. Convert moles products to grams products.

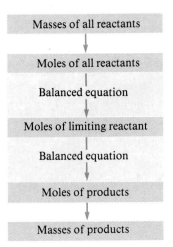

The amount of a product that is formed when the limiting reactant is completely used up is called the *theoretical yield* of that product. The amount of product that we calculated in the previous examples was in each case a theoretical yield. It was the maximum amount of product that could possibly be obtained. In practice in the laboratory, the calculated amount of product, that is, the theoretical yield, is hardly ever obtained. There are several reasons for this. During the handling of solutions and solids small amounts may be lost, other reactions than the desired reaction may use up some of the reactants, and, as we will see later, the main reaction may not go to completion, that is, it may stop before the reactants are used up. It is customary to express the *actual yield* as a percentage by mass of the theoretical yield and to call this the *percent yield*:

$$\text{Percent yield} = \frac{\text{actual yield}}{\text{theoretical yield}} \times 100\%$$

═ *Example 2.17* ═══════════════════════

PERCENT YIELD

Chromium can be made by heating chromium oxide, Cr_2O_3, with aluminum. Aluminum oxide, Al_2O_3, is the other product and the reaction is analogous to the reaction for preparing iron described in Exercise 2.14. When 18.7 g of Cr_2O_3 were heated with excess aluminum, 10.8 g of chromium were isolated from the products. What was the percent yield of chromium?

Solution

We first write the balanced equation which is

$$Cr_2O_3(s) + 2Al(s) \longrightarrow 2Cr(s) + Al_2O_3(s)$$

To find the percent yield of chromium we must first find the theoretical yield. We do this by converting mass Cr_2O_3 to moles Cr_2O_3, to moles Cr, and finally to grams Cr, as follows:

$$(18.7 \text{ g } Cr_2O_3)\left(\frac{1 \text{ mol } Cr_2O_3}{152.0 \text{ g } Cr_2O_3}\right)\left(\frac{2 \text{ mol } Cr}{1 \text{ mol } Cr_2O_3}\right)\left(\frac{52.00 \text{ g } Cr}{1 \text{ mol } Cr}\right) = 12.8 \text{ g } Cr$$

This mass of Cr is the theoretical yield, in other words the maximum amount that could be obtained. To find the percent yield we divide the actual yield by the theoretical yield and multiply by 100%:

$$\text{Percent yield} = \frac{10.8 \text{ g } Cr}{12.8 \text{ g } Cr} \times 100\% = 84.4\%$$

EXERCISE 2.15

Hydrogen is made industrially by heating carbon with steam at 725 °C. The other product is carbon monoxide.

$$C(s) + H_2O(g) \longrightarrow CO(g) + H_2(g)$$

What is the theoretical yield of hydrogen from 10.0 metric tons of carbon and excess steam? If the actual yield in a particular plant was 1.49 tons, what was the percent yield?

2.6
Molar Concentrations

Because so many reactions are carried out in solution, it is convenient to be able to measure out a given amount of a substance by taking a known volume of a solution of known concentration. There are a number of ways of expressing the concentration of a solution. For instance, we previously defined the concentration of a solution in terms of grams of solute in 100 g of solution (mass percent). But often expressing concentrations in terms of moles of solute in a given volume of solution is more convenient. Thus the **molarity** of a solute in solution is defined as the number of moles of solute contained in 1 L (1 dm^3) of solution:

$$\text{Molarity} = \frac{\text{Moles of solute}}{\text{Liters of solution}}$$

The units of molarity are thus moles per liter (mol L^{-1}), or moles per cubic decimeter (mol dm^{-3}). **Molar concentration** is given the symbol M (M = mol L^{-1} or mol dm^{-3}).

═ *Example 2.18* ═══

SOLUTION MOLARITY

A solution is prepared by dissolving 20.36 g of sodium chloride, NaCl, in sufficient distilled water to give 1.000 L of solution. What is the molarity of NaCl in the solution?

Solution

$$\text{Molar mass NaCl} = (22.99 + 35.45) = 58.44 \text{ g}$$

To obtain the number of moles of NaCl, we multiply the mass of NaCl (20.36 g) by the appropriate unit conversion factor:

$$(20.36 \text{ g NaCl})\left(\frac{1 \text{ mol NaCl}}{58.44 \text{ g NaCl}}\right) = 0.3484 \text{ mol NaCl}$$

This is the amount of NaCl dissolved in 1.000 L of solution. Thus

$$\text{Molarity NaCl} = 0.3484 \text{ mol L}^{-1} = 0.3484M$$

— *Example 2.19* ————————————————————————

SOLUTION MOLARITY

Suppose 20.36 g of NaCl are dissolved in sufficient distilled water to give a solution with a volume of 250 mL. What is the molar concentration of NaCl?

Solution

As in Example 2.18, moles of NaCl = 0.3484, but the volume of the solution is now 250 mL. Thus

$$\text{Molarity NaCl} = \left(\frac{0.3484 \text{ mol}}{250 \text{ mL}}\right)\left(\frac{1000 \text{ mL}}{1 \text{ L}}\right) = 1.39 \text{ mol L}^{-1} = 1.39M$$

E X E R C I S E   2 . 1 6

Calculate the molarity of each of the following aqueous solutions:

(a) 1.20 L containing 123 g of sodium carbonate, Na_2CO_3

(b) 0.50 L containing 9.8 g potassium nitrate, KNO_3

(c) 250 mL containing 15.2 g magnesium chloride, $MgCl_2$

Volumes of solutions are measured in the laboratory in several ways. Some of the more common apparatus for measuring the volume of solutions are shown in Figure 2.8. If we measure a given volume of a solution of known

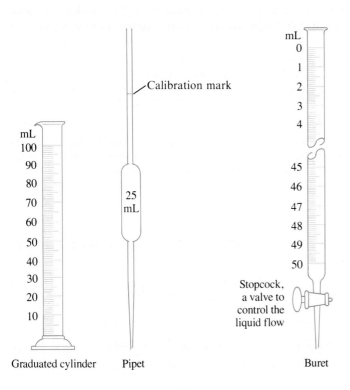

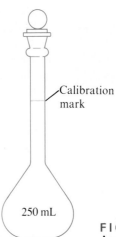

FIGURE 2.8
Apparatus for Measuring Volumes of Solutions.

concentration, we must be able to find how much solute this volume of solution contains. This amount may be calculated as shown in the following example.

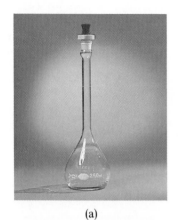

(a)

(b)

(c)

FIGURE 2.9
Using a Volumetric Flask.

(a) An accurately weighed amount of solute is added to the flask.
(b) Distilled water is added and the flask is shaken to dissolve the solute. (c) More distilled water is added until the level of the solution reaches the mark on the neck of the flask.

Example 2.20

SOLUTION MOLARITY

How many moles and how many grams of sodium hydroxide, NaOH, are in 25.0 mL of a 0.500M NaOH solution?

Solution

A 0.500M NaOH solution contains 0.500 mol of NaOH in 1.000 L, or 1000 ml, of solution:

$$0.500M \text{ NaOH} = \frac{0.500 \text{ mol NaOH}}{1000 \text{ mL solution}}$$

We can use this ratio as a conversion factor to convert volume of solution to moles:

$$(25.0 \text{ mL solution})\left(\frac{0.500 \text{ mol NaOH}}{1000 \text{ mL solution}}\right) = 0.0125 \text{ mol NaOH}$$

Now we can convert this result to grams:

$$(0.0125 \text{ mol NaOH})\left(\frac{40.00 \text{ g NaOH}}{1 \text{ mol NaOH}}\right) = 0.500 \text{ g NaOH}$$

Note that molarity is defined as moles of solute in a given volume of *solution*. Thus a 1M solution of sodium hydroxide *cannot* be prepared by adding 1 mol of sodium hydroxide (40.00 g) to 1 L of water. Because of the additional volume occupied by the sodium hydroxide, the total volume of the solution would not be exactly 1.000 L, and therefore the concentration of the solution would not be exactly 1.000M. A solution of known molarity is usually prepared by using a *volumetric flask* (Figure 2.8). This flask has a graduation mark on the neck; when the flask is filled with a solution exactly to this point, the volume of the solution, for example, 250 or 500 mL, is precisely known. The procedure for making up a solution of known concentration by using such a flask is illustrated in Figure 2.9.

To make up a certain volume of a solution of known concentration, we must know what mass of the solute will be needed. The required amount may be calculated as shown in the following example.

Example 2.21

SOLUTION MOLARITY

How many grams of sodium sulfate, Na$_2$SO$_4$, are required to prepare 250 mL of 0.500M Na$_2$SO$_4$ solution?

Solution

This problem can be restated as follows: How many grams of Na$_2$SO$_4$ are in 250 mL of 0.500M Na$_2$SO$_4$ solution? A 0.500M Na$_2$SO$_4$ solution contains 0.500 mol in 1 L of solution. To find the number of moles of Na_2SO_4 in 250 mL, we use the conversion factor 0.500 mol/1000 mL. Therefore 250 mL of solution contain

$$(250 \text{ mL})\left(\frac{0.500 \text{ mol}}{1000 \text{ mL}}\right) = 0.125 \text{ mol}$$

Now we can convert to grams, using the unit conversion factor

$$\frac{142.0 \text{ g } Na_2SO_4}{1 \text{ mol } Na_2SO_4}$$

Therefore

$$(0.125 \text{ mol } Na_2SO_4)\left(\frac{142.0 \text{ g } Na_2SO_4}{1 \text{ mol } Na_2SO_4}\right) = 17.8 \text{ g } Na_2SO_4$$

Or we can do the calculation in one step, as follows:

$$? \text{ g } Na_2SO_4 = (250 \text{ mL})\left(\frac{0.500 \text{ mol } Na_2SO_4}{1000 \text{ mL}}\right)\left(\frac{142.0 \text{ g } Na_2SO_4}{1 \text{ mol } Na_2SO_4}\right) = 17.8 \text{ g } Na_2SO_4$$

EXERCISE 2.17

How many grams of solute would be needed to prepare 500 mL of each of the following aqueous solutions?

(a) 0.100M silver nitrate, $AgNO_3$

(b) 1.00M sodium bromide, $NaBr$

(c) 0.200M barium chloride, $BaCl_2$

Acids such as sulfuric acid are available commercially as concentrated aqueous solutions. These acids must be diluted by adding them to more water in order to prepare the more dilute solution that might be required in the laboratory.

$=$ Example 2.22 $=$

PREPARATION OF A DILUTE ACID SOLUTION

What volume of concentrated, aqueous sulfuric acid, which is 98.0% H_2SO_4 by mass and has a density of 1.84 g cm^{-3}, is required to make 10.0 L of 0.200M H_2SO_4 solution?

Solution

We find the number of moles of H_2SO_4 in 10.0 L of 0.200M H_2SO_4 solution:

$$(10.0 \text{ L})\left(\frac{0.200 \text{ mol } H_2SO_4}{1 \text{ L solution}}\right) = 2.00 \text{ mol } H_2SO_4$$

Thus we require sufficient concentrated acid to give 2.00 mol of H_2SO_4, or

$$(2.00 \text{ mol } H_2SO_4)\left(\frac{98.08 \text{ g } H_2SO_4}{1 \text{ mol } H_2SO_4}\right) = 196 \text{ g } H_2SO_4$$

Since the concentrated (conc) acid is 98.0% by mass, we need

$$(196 \text{ g } H_2SO_4)\left(\frac{100 \text{ g conc acid}}{98.0 \text{ g } H_2SO_4}\right) = 200 \text{ g conc acid}$$

The density of the concentrated acid is $1.84 \text{ g cm}^{-3} = 1.84 \text{ g mL}^{-1}$. So we need

$$(200 \text{ g conc acid})\left(\frac{1 \text{ mL conc acid}}{1.84 \text{ g conc acid}}\right) = 109 \text{ mL conc acid}$$

With experience we can do this problem in one step, as follows:

$$? \text{ g conc acid} = (10.0 \text{ L})\left(\frac{0.200 \text{ mol } H_2SO_4}{1 \text{ L solution}}\right)\left(\frac{98.08 \text{ g } H_2SO_4}{1 \text{ mol } H_2SO_4}\right)$$

$$\times \left(\frac{100 \text{ g conc acid}}{98.0 \text{ g } H_2SO_4}\right)\left(\frac{1 \text{ mL conc acid}}{1.84 \text{ g conc acid}}\right)$$

$$= 109 \text{ mL conc acid}$$

EXERCISE 2.18

Concentrated nitric acid, $HNO_3(aq)$, is a 69% by mass aqueous solution with a density of 1.41 g mL^{-1}. What volume of concentrated nitric acid is needed to prepare 250 mL of $0.100M$ HNO_3?

Many reactions are carried out by using aqueous solutions of known concentration. The amount of a product of a reaction can be found from the volumes of the solutions of the reactants and their concentrations, as shown in the following example.

═ *Example 2.23* ═══

REACTIONS IN SOLUTION

An aqueous solution of sodium sulfate, Na_2SO_4, reacts with an aqueous solution of barium chloride, $BaCl_2$, to give a precipitate of insoluble barium sulfate, $BaSO_4$. Suppose 250 mL of $0.500M$ Na_2SO_4 solution are added to an aqueous solution of 10.00 g of $BaCl_2$. How many moles and how many grams of $BaSO_4$ are obtained?

Solution

First, we must write the balanced equation for the reaction. The unbalanced equation is

$$BaCl_2(aq) + Na_2SO_4(aq) \longrightarrow BaSO_4(s) + NaCl(aq) \qquad \text{(Unbalanced)}$$

The Ba, S, and O are already balanced. Because there are 2 Na and 2 Cl on the left, we need 2 formula units of NaCl on the right. The balanced equation is

$$BaCl_2(aq) + Na_2SO_4(aq) \longrightarrow BaSO_4(s) + 2NaCl(aq)$$

Next, since we know the amounts of both reactants, we must find which is the limiting reactant.

$$\left(\frac{0.500 \text{ mol } Na_2SO_4}{1 \text{ L}}\right)\left(\frac{1 \text{ L}}{1000 \text{ mL}}\right)(250 \text{ mL}) = 0.125 \text{ mol } Na_2SO_4$$

$$(10.00 \text{ g } BaCl_2)\left(\frac{1 \text{ mol } BaCl_2}{208.2 \text{ g}}\right) = 0.04803 \text{ mol } BaCl_2$$

The balanced equation shows us that 1 mol of $BaCl_2$ reacts with 1 mol of Na_2SO_4, so $BaCl_2$ is the limiting reactant. The equation also shows that 1 mol of $BaCl_2$ gives 1 mol of $BaSO_4$, so the mass of $BaSO_4$ formed is:

$$(0.04803 \text{ mol } BaCl_2)\left(\frac{1 \text{ mol } BaSO_4}{1 \text{ mol } BaCl_2}\right)\left(\frac{233.4 \text{ g } BaSO_4}{1 \text{ mol } BaSO_4}\right) = 11.21 \text{ g } BaSO_4$$

E X E R C I S E   2 . 1 9

When an aqueous solution of sodium chloride, NaCl, is mixed with an aqueous solution of silver nitrate, $AgNO_3$, a precipitate of insoluble silver choride, AgCl, is obtained. What mass of silver chloride would be precipitated when 150 mL of a $0.109M$ $AgNO_3$ solution are mixed with an excess of NaCl solution?

IMPORTANT TERMS

The **atomic mass** of an atom is the mass of the atom expressed in atomic mass units.

The **atomic mass unit**, u, is $\frac{1}{12}$ of the mass of a $^{12}_6C$ atom.

The **atomic number**, Z, of an element is the number of protons in the nucleus of an atom of the element; it is equal to the number of electrons surrounding the nucleus in the neutral atom.

The **Avogadro constant** is the number of elementary entities (atoms, molecules, electrons, and so on) in 1 mol of entities; $N_A = 6.022 \times 10^{23} \text{ mol}^{-1}$.

A **chemical reaction** is a process in which atoms are rearranged to form new substances. Atoms are neither created nor destroyed; the total number of each kind of atom is conserved.

Coulomb's law states that the force, F, between two charges, Q_1 and Q_2, is proportional to the product of the charges and inversely proportional to the square of the distance, r, between the charges: $F = k(Q_1Q_2/r^2)$.

An **electron** is a fundamental particle having a charge of -1.60×10^{-19} C and a mass of 0.000 548 u, which is only 1/1836 of the mass of a proton.

An **element** is a substance composed of atoms of only one kind, all of which have the same atomic number Z.

The **empirical formula** of a substance is the simplest formula that expresses the relative numbers of atoms as whole numbers.

The **empirical formula mass** is the mass obtained by adding up the masses of all the atoms in the empirical formula of a substance.

Isotopes are atoms with the same atomic number Z but with different masses.

The **law of conservation of energy** states that energy can neither be created nor destroyed but only changed from one form to another. Energy is conserved in chemical reactions.

The **law of conservation of mass** states that the total mass of the products of a chemical reaction is equal to the total mass of the reactants; atoms are neither created nor destroyed.

A **limiting reactant** is the particular reactant among two or more reactants whose amount determines the amount of product that may be formed.

The **mass defect** is the difference between the mass of an atom and the sum of the masses of its constituent particles (protons, neutrons, and electrons).

The **mass number** (*nucleon number*) is the sum of the number of protons and the number of neutrons in an atom.

Molarity is the concentration of a solute in solution, expressed in moles per liter ($M = mol\ L^{-1}$).

Molar mass is the mass of 1 mol of a substance.

A **mole** is the amount of a substance that contains as many elementary entities (for example, atoms, molecules, empirical formula units) as there are atoms in 0.012 kg (12 g) of carbon-12; 1 mol = 6.022×10^{23} entities.

A **molecular formula** shows the numbers of atoms of each kind in a molecule of a substance. It is always some integral multiple of the empirical formula.

The **molecular mass** is the mass obtained by adding the masses of all the atoms in the molecular formula of a substance.

Nucleon number. See mass number.

The **nucleus** of an atom is the very small positively charged particle at the center of an atom, around which the electrons are moving. The nucleus is composed of protons and neutrons and it makes up almost all the mass of the atom.

A **neutron** is a fundamental particle that has a mass of 1.008 66 u, which is almost the same as that of a proton, but zero charge.

A **proton** is a fundamental particle having a charge of $+1.60 \times 10^{-19}$ C and a mass of 1.007 28 u.

The **percentage composition** of a substance is its composition expressed as the mass percentage of each of its constituent elements.

Stoichiometry is the term used to refer to all the quantitative aspects of chemical composition and chemical reactions.

PROBLEMS *

Atoms, Isotopes, Atomic Mass

1. What are the charges associated with the electron, the proton, and the neutron, and what are their relative masses?

2. What are the chemical symbols and the atomic numbers of each of the following?

(a) Hydrogen   (b) Oxygen   (c) Fluorine

(d) Neon   (e) Magnesium   (f) Phosphorus

(g) Chlorine   (h) Calcium   (i) Zinc

3. Which elements have atoms with nuclei having each of the following numbers of protons?

5   9   32   54   92

4. How many protons are there in the nucleus of an atom of each of the following?

(a) Boron   (b) Nitrogen   (c) Hydrogen

(d) Neon   (e) Chlorine   (f) Oxygen

(g) Sulfur   (h) Potassium   (i) Iron

5. Identify each of the following neutral atoms and give the symbol of the particular isotope:

(a) An atom with 1 proton and 1 neutron

(b) An atom with 8 protons and 9 neutrons

(c) An atom with 6 protons and 8 neutrons

(d) An atom with 19 nucleons and 9 electrons

* Answers to problems numbered in blue appear at the end of the text.

(e) An atom with 32 nucleons and 16 electrons

(f) An atom with 12 neutrons and 12 electrons

(g) An atom with 45 neutrons and 35 electrons

6. Give the symbol for each of the following isotopes:

	Atomic Number	Mass Number
(a)	19	40
(b)	14	30
(c)	18	40
(d)	7	15
(e)	16	32
(f)	11	23
(g)	13	27

7. Complete each of the following symbols and give the numbers of protons, neutrons, and electrons, respectively, in one atom of each:

(a) ^{4}He   (b) ^{27}Al   (c) ^{14}C   (d) ^{31}P

(e) ^{37}Cl   (f) ^{85}Rb   (g) ^{108}Ag   (h) ^{131}Xe

8. What are the numbers of protons, neutrons, and electrons in each of the following neutral atoms?

$^2_1$H $^{19}_9$F $^{40}_{20}$Ca $^{112}_{48}$Cd $^{117}_{50}$Sn $^{131}_{54}$Xe

9. Complete the following table:

Atom symbol	^9_4Be	$^{15}_7N$	$^{18}_8O$	—	—
Mass number	9	—	—	—	23
Atomic number	4	—	—	—	11
Number of protons	4	—	—	6	—
Number of electrons	4	—	—	6	—
Number of neutrons	5	—	—	6	—

10. Complete the following table:

Atom symbol	$^{24}_{12}Mg$	$^{106}_{47}Ag$	$^{137}_{56}Ba$
Mass number	—	—	—
Atomic number	—	—	—
Number of protons	—	—	—
Number of electrons	—	—	—
Number of neutrons	—	—	—

11. Write the symbols for each of the isotopically different molecules of water, given that the isotopes of hydrogen and oxygen are ^1H, ^2H, ^{16}O, ^{17}O, and ^{18}O.

12. Boron has isotopes with masses 10.012 94 u and 11.009 31 u, with respective abundances of 19.77% and 80.23%. What is the average atomic mass of boron?

13. (a) What information is given by the symbol $^{24}_{12}Mg$?

(b) From the data in the accompanying table, calculate the average atomic mass of magnesium:

Mass Number	Abundance (%)	Atomic Mass (u)
24	78.60	23.993
25	10.11	24.994
26	11.29	25.991

14. Explain what is meant by the term *isotope*, and define *mass number* and *atomic number*. Naturally occurring copper has an average atomic mass of 63.55 u, and contains isotopes with masses 62.9298 and 64.9278 u. What is the symbol for each isotope of copper, and its relative abundance?

15. Gallium, Ga, has two isotopes of atomic mass 68.926 u and 70.926 u. Write the symbol for each isotope, given that

* The asterisk denotes the more difficult problems.

the atomic number of gallium is 31. How many protons, and how many neutrons, are present in the nucleus of each isotope? What is the natural abundance of each isotope, if the average atomic mass of gallium is 69.72 u?

16. Uranium has an average atomic mass of 238.03 and consists of ^{235}U, mass 235.044 u, and ^{238}U, mass 238.051 u. The ^{235}U isotope is the fuel used in nuclear power reactors. What is the percentage abundance of ^{235}U in natural uranium?

17. Naturally occurring chlorine has an atomic mass of 35.45 u, and consists of isotopes with mass numbers 35 and 37, respectively. What is the approximate abundance of each isotope? What is the average mass of a Cl_2 molecule?

*18. Bromine atoms and chlorine atoms combine to give BrCl molecules. Bromine chloride is found to consist of molecules with approximate masses 114, 116, and 118 u, and it is known that chlorine has just two isotopes, with mass numbers 35 and 37.

(a) Deduce the possible isotopes of bromine, and write their symbols.

(b) Give the formula of each of the isotopically different BrCl molecules.

The Mole

19. The charge on the electron is $1.602\,19 \times 10^{-19}$ C. What is the charge on 1 mol of electrons in coulombs (C)? What is the charge on 1 mol of protons?

20. An atom of element X has a mass of 3.155×10^{-23} g. What is the atomic mass of X on the scale $^{12}_6C = 12$ u? Write the symbol for X.

21. The artificial sweetener, NutraSweet®, is the compound aspartamine which has the molecular formula $C_{14}H_{18}N_2O_5$. (a) What is the mass of 1.00 mol of aspartamine? (b) How many moles of aspartamine are present in 6.22 g of the substance? (c) What is the mass, in grams, of 0.245 mol of aspartamine? (d) How many molecules are present in 4.28 mg of aspartamine?

22. What factors convert the mass of an atom in atomic units, u, to

(i) grams   (ii) kilograms   (iii) lbs?

23. What is the mass in grams of each of the following?

(a) 7.1×10^{-3} mol of oxygen gas

(b) 5.43 mol of iron

(c) 3.14×10^{-2} mol of phosphorus, P_4

(d) 9.6×10^{-4} mol of sulfur, S_8

(e) 1.263×10^{-2} mol of $CuSO_4 \cdot 5H_2O$

(f) 0.452 mol of magnesium phosphate, $Mg_3(PO_4)_2$

24. How many atoms of mercury are there in exactly 1 g of the metal?

25. **(a)** How many moles of sulfur dioxide, SO_2, are in 0.028 g of SO_2?

(b) What mass of SO_2 contains exactly 3 mol of SO_2?

26. How many moles are there in 10.00 g of each of the following:

(a) Water, H_2O **(b)** Sulfur dioxide, SO_2

(c) Acetic acid, $C_2H_4O_2$ **(d)** Sulfuric acid, H_2SO_4

27. The pain reliever, aspirin, is the compound acetylsalicylic acid, $C_9H_8O_4$.   **(a)** What is the molecular mass of aspirin?   **(b)** How many moles of $C_9H_8O_4$ molecules and how many $C_9H_8O_4$ molecules are present in a 500 mg aspirin tablet?

28. One molecule of a compound has a mass of 2.653×10^{-22} g. What is its molar mass?

29. How many molecules of oxygen, O_2, are there in 2.00 g of O_2? If the O_2 molecules were split into atoms, how many moles of oxygen atoms would be obtained?

30. Magnesium, Mg, has a molar mass of 24.31 g mol^{-1}, and a density of 1.738 g cm^{-3} at 20 °C. Calculate

(a) The average mass of one magnesium atom

(b) The volume of 1 mol of magnesium atoms

(c) The average volume of one magnesium atom

(d) The approximate radius of a spherical magnesium atom, in picometers

31. **(a)** Assume that the human body contains 6×10^{13} body cells, and that the earth's population is 4×10^9 persons. Approximately how many moles of living human body cells are there on the earth?

(b) Assume that the human body mass is 80% water. How many water molecules are there in the body of a person with a mass of 65 kg?

32. Carbon monoxide, CO(g), taken into the lungs reduces the ability of the blood to transport oxygen. It is fatal if its concentration reaches 2.38×10^{-4} g L^{-1}. Calculate the number of CO molecules that must be emitted from an automobile exhaust to produce a fatal concentration of CO in a garage of volume 150 m^3.

33. Using the atomic masses given inside the front cover, calculate the molar mass of each of the following compounds:

H_2O H_2O_2 NaCl $MgBr_2$ CO

CO_2 CH_4 C_2H_6 NH_3 HCl

34. Using the atomic masses given inside the front cover, calculate the molar mass of each of the following compounds:

NH_4Cl $Ca_3(PO_4)_2$ KI

$BaCl_2 \cdot 2H_2O$ PCl_5 C_4H_{10}

35. What is the empirical formula mass, and the molecular mass, of each of the following:

(a) Acetic (ethanoic) acid, $C_2H_4O_2$

(b) Formic (methanoic) acid, H_2CO_2

(c) Cane sugar, $C_{12}H_{22}O_{11}$

(d) Butane, C_4H_{10}

(e) Diborane, B_2H_6

(f) Hydrogen peroxide, H_2O_2

36. What is the molecular mass of benzene, C_6H_6? How many molecules of benzene are in exactly 1 cm^3 of benzene, given that its density is 0.880 g cm^{-3}? How many moles of atoms are obtained if each benzene molecule is completely split into atoms?

Empirical Formulas and Composition

37. What are the mass percentage elemental compositions of each of the following?

(a) Water, H_2O **(b)** Sodium chloride, NaCl

(c) Ethane, C_2H_6 **(d)** Magnesium bromide, $MgBr_2$

(e) Carbon dioxide, CO_2

38. What are the mass percentage elemental compositions of each of the following?

(a) Ammonia, NH_3 **(b)** Chlorine, Cl_2

(c) Sodium hydroxide, NaOH   **(d)** Ethanol, C_2H_6O

(e) Nitrobenzene, $C_6H_5NO_2$

39. Calculate the mass percentage of each element in each of the following compounds:

(a) Ethene, C_2H_4

(b) Ethyne, C_2H_2

(c) Magnesium chloride hexahydrate, $MgCl_2 \cdot 6H_2O$

(d) Iron(II) sulfate heptahydrate, $FeSO_4 \cdot 7H_2O$

(e) Tetrammine copper(II) sulfate dihydrate, $[Cu(NH_3)_4]SO_4 \cdot 2H_2O$

40. What is the mass of oxygen in 3.40 g of nitric acid, HNO_3?

41. Ammonium sulfate, $(NH_4)_2SO_4$, is a common agricultural fertilizer. What is the percentage of nitrogen by mass in this compound? What mass of ammonium sulfate contains 100 g of nitrogen?

42. What is the mass percent of sulfur in antimony sulfide, Sb_2S_3? What mass of sulfur is contained in 28.4 g of this compound? What mass of the compound contains 64.4 g of sulfur?

43. A compound contains 7.00% C and 93.00% Br, by mass. What is its empirical formula and its empirical formula mass?

44. The hydrocarbon anthracene has the composition 94.33% C and 5.67% H, by mass. What is its empirical formula and empirical formula mass?

45. A compound of sulfur and fluorine was analyzed and found to contain 70.3% fluorine by mass. What is the empirical formula of the compound?

46. A compound was analyzed and found to contain 21.7% C, 9.6% O, and 68.7% F, by mass. What is its empirical formula and empirical formula mass?

47. Caffeine is a compound with the composition 49.5% C, 5.2% H, 28.8% N, and 16.6% O, by mass. What is the empirical formula and the empirical formula mass of caffeine?

48. When 3.10 g of a compound containing only carbon, hydrogen, and oxygen were completely burned in oxygen, 4.40 g CO_2 and 2.70 g H_2O were produced. What is the empirical formula of the compound? If its molecular mass is 62.1 u, what is its molecular formula?

49. When 3.62 g of a compound containing carbon, hydrogen, and oxygen were burned completely in air, 5.19 g of CO_2 and 2.83 g of H_2O were produced. What is the empirical formula of the compound?

50. When 0.100 mol of a compound of carbon, hydrogen, and nitrogen was burned completely in oxygen, 26.4 g of CO_2, 6.30 g of H_2O, and 4.60 g of nitrogen dioxide, NO_2, were produced. What is the empirical formula of the compound?

51. Cyclopropane is a compound of carbon and hydrogen that is used as a general anesthetic. When 1.00 g of this substance was burned completely in oxygen, 3.14 g of CO_2 and 1.29 g of H_2O were produced. What is the empirical formula of cyclopropane?

52. A sample of mass 6.20 g of a compound containing only sulfur, hydrogen, and carbon was reacted completely with chlorine and gave 21.9 g of hydrogen chloride, HCl, and 30.8 g of tetrachloromethane, CCl_4. What is the empirical formula of the compound?

53. Vanadium forms oxides with the following compositions:

Oxide	Mass percent V	Mass percent O
A	76.10	23.90
B	67.98	32.02
C	61.42	38.58
D	56.02	43.98

What is the empirical formula of each of these oxides?

54. An oxide of nitrogen contains 3.04 g of nitrogen in a sample of mass 9.99 g. Its molecular mass is determined to be 92 u. What is the empirical formula, and the molecular formula, of the nitrogen oxide?

55. Analysis of nicotine gave 74.0% C, 8.7% H, and 17.3% N, by mass. Its molar mass was determined to be 162 g. What is the empirical formula and the molecular formula of nicotine?

56. On heating in air, 2.862 g of a red copper oxide reacted to give 3.182 g of a black copper oxide. When the latter was heated strongly in hydrogen, it gave 2.542 g of pure copper. What are the empirical formulas of the two copper oxides?

57. The composition of hydrated lithium sulfate may be expressed by the empirical formula $Li_2SO_4 \cdot xH_2O$, where x is an unknown integer. When 3.25 g of the hydrated lithium sulfate were heated, all the water was driven off to give 2.80 g of anhydrous Li_2SO_4. What is the value of x in the empirical formula of hydrated lithium sulfate?

58. When heated, 0.2800 g of blue hydrated copper sulfate, $CuSO_4 \cdot xH_2O$, gave 0.1789 g of colorless anhydrous copper sulfate, $CuSO_4$. What is the empirical formula of hydrated copper sulfate?

59. Polychlorinated biphenyls, PCBs, now known to be dangerous environmental pollutants, are a group of compounds all having the general empirical formula $C_{12}H_mCl_{10-m}$, where m is an integer. What is the value of m, and hence the empirical formula, of the PCB that contains 58.9 mass % chlorine?

Balancing Equations

60. Which of the following equations are unbalanced? Balance those that are unbalanced.

(a) $2SO_2 + H_2O + O_2 \rightarrow 2H_2SO_4$

(b) $CH_3OH + 2O_2 \rightarrow CO_2 + 2H_2O$

(c) $H_2O_2 \rightarrow H_2O + O_2$

(d) $H_2SO_4 + KOH \rightarrow KHSO_4 + H_2O$

(e) $Zn + HCl \rightarrow ZnCl_2 + H_2$

61. Balance each of the following equations:

(a) $Al + HCl \rightarrow AlCl_3 + H_2$

(b) $C_5H_{12} + O_2 \rightarrow CO + H_2O$

(c) $C_3H_8 + H_2O \rightarrow CO + H_2$

(d) $Na_2CO_3 + HCl \rightarrow NaCl + H_2O + CO_2$

(e) $Al + O_2 \rightarrow Al_2O_3$

(f) $Al_2O_3 + H_2SO_4 \rightarrow Al_2(SO_4)_3 + H_2O$

62. Balance each of the following equations:

(a) $S + O_2 \rightarrow SO_3$

(b) $C_2H_2 + O_2 \rightarrow CO + H_2O$

(c) $Na_2CO_3 + Ca(OH)_2 \rightarrow NaOH + CaCO_3$

(d) $Na_2SO_4 + H_2 \rightarrow Na_2S + H_2O$

(e) $Cu_2S + O_2 \rightarrow Cu_2O + SO_2$

(f) $Cu_2O + Cu_2S \rightarrow Cu + SO_2$

63. Balance each of the following equations:

(a) $Na_2SO_4(s) + C(s) \rightarrow Na_2S(s) + CO_2(g)$

(b) $Cl_2(aq) + H_2O(l) \rightarrow HCl(aq) + HOCl(aq)$

(c) $PCl_3(l) + H_2O(l) \rightarrow H_3PO_3(aq) + HCl(aq)$

(d) $NO_2(g) + H_2O(l) \rightarrow HNO_3(aq) + NO(g)$

(e) $P_4O_{10}(s) + H_2O(l) \rightarrow H_3PO_4(aq)$

(f) $Na_2O(s) + H_2O(l) \rightarrow NaOH(aq)$

64. Write balanced equations for the reactions between the following reactants, to give the indicated products:

Reactants	Products
(a) Sulfur, oxygen	Sulfur trioxide
(b) Methane, oxygen	Carbon monoxide, water
(c) Magnesium, steam	Magnesium oxide, hydrogen
(d) Butane, steam	Carbon monoxide, hydrogen

65. Write balanced equations for each of the following reactions:

(a) Elemental phosphorus, $P_4(s)$, with excess oxygen, to give $P_4O_{10}(s)$

(b) Sodium metal with water, to give aqueous sodium hydroxide, NaOH, and hydrogen

(c) The decomposition of ammonium nitrate, $NH_4NO_3(s)$, on heating, to give dinitrogen monoxide, N_2O, and water

(d) The decomposition of lead nitrate, $Pb(NO_3)_2$, on heating, to give lead monoxide, PbO, nitrogen dioxide, NO_2, and oxygen

Stoichiometry: Reactions

66. When magnesium metal is ignited in oxygen, the white oxide MgO(s) is formed. What mass of magnesium reacts completely to give 1.000 g of MgO(s)?

67. Cesium chloride contains 78.94% Cs by mass. What is the empirical formula of cesium chloride? How many grams of cesium metal are required to give 4.34 g of cesium chloride, when its reaction with chlorine gas is complete?

68. Xenon gas and fluorine gas react to give xenon tetrafluoride, XeF_4, a white solid. What mass of fluorine is required to react completely with 2.50 g of xenon, and what mass of XeF_4 would be obtained?

69. Ammonia gas, $NH_3(g)$, and hydrogen chloride gas, HCl(g), react to give the white solid ammonium chloride, $NH_4Cl(s)$. Write the balanced equation for this reaction, and calculate the mass of HCl that reacts completely with 0.200 g of $NH_3(g)$.

70. Phosphorus trichloride, PCl_3, is a colorless liquid made by passing a stream of chlorine gas over phosphorus, and condensing the product in a cooled dry flask. What minimum mass of phosphorus is required to give 100 g of PCl_3?

71. Benzene, C_6H_6, burns completely in oxygen to give carbon dioxide and water. Write the balanced equation for the reaction. What mass of CO_2, and what mass of water, is produced by the complete combustion of 0.434 g of benzene?

72. Phosphoric acid, H_3PO_4, reacts with calcium hydroxide, $Ca(OH)_2$, to give calcium phosphate, $Ca_3(PO_4)_2$, and water. Write the balanced equation for this reaction, and calculate the mass of $Ca(OH)_2$ that reacts completely with 30.0 g H_3PO_4. What mass of $Ca_3(PO_4)_2$ is produced?

73. Write the balanced equation for the reaction of $Cl_2(g)$ with silicon dioxide (silica), $SiO_2(s)$, and carbon, $C(s)$, to give silicon tetrachloride, $SiCl_4(l)$, and carbon monoxide, $CO(g)$. What is the maximum amount of $SiCl_4$ that can be obtained from 15.0 g of SiO_2?

74. A sample of an oxide of barium of unknown empirical formula and of mass 5.53 g gave 5.00 g of pure barium oxide, $BaO(s)$, and oxygen when heated to constant mass. What is the empirical formula of the original oxide of barium?

75. Sulfuric acid is formed when sulfur dioxide and water react with oxygen in the presence of a catalyst:

$$2SO_2(g) + O_2(g) + 2H_2O(l) \longrightarrow 2H_2SO_4(l)$$

What mass of O_2 and what mass of H_2O react with 0.320 g of SO_2 to give what mass of sulfuric acid?

76. A sample of 1.000 kg of impure limestone, containing 74.2 mass % calcium carbonate, $CaCO_3(s)$, and 25.8% of inert impurities, is heated until all the carbonate is decomposed to calcium oxide, $CaO(s)$, and carbon dioxide. What mass of carbon dioxide is produced?

77. When 1.000 g of a mixture containing only the minerals $As_4S_6(s)$ and $As_4S_4(s)$ was burned in excess oxygen, the products were $As_4O_6(s)$ and $SO_2(g)$, and the solid product had a mass of 0.905 g.

   (a) Write the balanced equations for the reactions of As_4S_6 and As_4S_4, respectively, with oxygen.

   (b) Calculate the mass percent of As_4S_4 in the original mixture.

78. Dry boron oxide, $B_2O_3(s)$, reacts on strongly heating with magnesium powder to give a mixture of magnesium oxide, $MgO(s)$, and magnesium boride, $Mg_3B_2(s)$. Write the balanced equation for the reaction. Magnesium boride reacts with dilute aqueous hydrochloric acid, $HCl(aq)$, to give a hydride of boron of formula B_4H_{10}, $MgCl_2$, and H_2. What is the maximum possible yield of B_4H_{10} from 10.00 g B_2O_3?

79. A 1.00-g sample of enriched water (containing both H_2O and D_2O) was electrolyzed, and the $H_2(g)$ and $D_2(g)$ obtained reacted completely with $Cl_2(g)$ to give a mixture of $HCl(g)$ and $DCl(g)$. When this mixture of HCl and DCl was dissolved in 1.000 L of water, a 25.00-mL sample of the aqueous solution reacted with excess aqueous silver nitrate,

$AgNO_3(aq)$, to give 0.3800 g of solid silver chloride, $AgCl(s)$. What was the mass percent of D_2O in the original sample of enriched water?

80. Metals such as magnesium and zinc react with dilute sulfuric acid to give hydrogen, according to the equation

$$M(s) + H_2SO_4(aq) \longrightarrow MSO_4(aq) + H_2(g)$$

When 5.00 g of a mixture of finely divided Mg and Zn were dissolved in dilute sulfuric acid, 0.284 g of hydrogen was collected. What was the composition of the mixture of Mg and Zn expressed as mass percent? If the volume of acid used was 100 mL, what were the final molar concentrations of $MgSO_4$ and $ZnSO_4$?

Limiting Reactant: Percent Yield

81. Sulfuric acid is produced when sulfur dioxide reacts with oxygen and water in the presence of a catalyst:

$$2SO_2(g) + O_2(g) + 2H_2O(l) \longrightarrow 2H_2SO_4(aq)$$

If 5.6 mol of SO_2 react with 4.8 mol of O_2 and a large excess of water, what is the maximum number of moles of H_2SO_4 that can be obtained?

82. Phosphorus trichloride, PCl_3, reacts with chlorine, Cl_2, to give phosphorus pentachloride, PCl_5. What mass of PCl_5 results from the reaction of 5.15 g of PCl_3 with 3.15 g Cl_2?

83. Determine the maximum mass of barium sulfate, $BaSO_4(s)$, that can be obtained by mixing aqueous solutions containing 6.00 g K_2SO_4 and 8.00 g $Ba(NO_3)_2$, respectively.

84. The unbalanced equation for the reaction of aluminum with hydrogen chloride is

$$Al(s) + HCl(g) \longrightarrow AlCl_3(g) + H_2(g)$$

Determine the maximum mass of $AlCl_3$ produced, and the mass of Al or HCl that remains unreacted, when 2.70 g of Al react with 4.00 g HCl.

85. A strip of zinc metal immersed in an aqueous solution of copper chloride, $CuCl_2(aq)$, reacts to give solid copper and zinc chloride, $ZnCl_2(aq)$:

$$Zn(s) + CuCl_2(aq) \longrightarrow ZnCl_2(aq) + Cu(s)$$

If a zinc strip has a mass of 2.00 g, and the solution initially contains 2.00 g $CuCl_2$, what mass of copper is produced?

86. Carbon in the form of graphite was heated strongly with sulfur, and the resulting carbon disulfide, $CS_2(l)$, distilled off and condensed to liquid. If 2.530 g of graphite gave 12.50 g of CS_2, what was the percentage yield?

87. On strongly heating, potassium nitrate, $KNO_3(s)$, decomposes to potassium nitrite, $KNO_2(s)$, and oxygen. When 2.500 g of $KNO_3(s)$ was heated, the resulting solid mixture of $KNO_3(s)$ and $KNO_2(s)$ had a mass of 2.210 g. What mass of oxygen gas was produced, and what was its percentage yield?

88. When 1.000 g of magnesium was dissolved in excess $HCl(aq)$, hydrogen was evolved and a solution containing $MgCl_2(aq)$ was obtained. On evaporation of the solution and cooling, 5.62 g of the salt $MgCl_2 \cdot 6H_2O$ crystallized from the solution. What was the percentage yield of the hydrated salt?

89. What is the maximum amount of barium sulfate, $BaSO_4(s)$, that can be obtained from the reaction of 27.42 mL of $0.112M$ $BaCl_2(aq)$ with 32.30 mL of $0.096M$ $H_2SO_4(aq)$?

Molar Concentrations

90. How many moles of sodium hydroxide are contained in each of the following?

  (a) 1.00 L of $0.0100M$ $NaOH(aq)$

  (b) 250 mL of $0.0100M$ $NaOH(aq)$

  (c) 25.15 mL of $0.0100M$ $NaOH(aq)$

  (d) 25.15 mL of $0.0134M$ $NaOH(aq)$

91. How many moles of sulfuric acid are contained in each of the following?

  (a) 4.00 L of $0.100M$ $H_2SO_4(aq)$

  (b) 125 mL of $0.100M$ $H_2SO_4(aq)$

  (c) 31.46 mL of $0.100M$ $H_2SO_4(aq)$

  (d) 31.46 mL of $0.151M$ $H_2SO_4(aq)$

92. 12.00 g of potassium permanganate, $KMnO_4(s)$, were dissolved in sufficient distilled water give 2.00 L of solution. What was the molarity of potassium permanganate in the solution?

93. (a) What mass of glucose, $C_6H_{12}O_6(s)$, must be dissolved in water to give 0.250 L or a $0.100M$ solution?

  (b) What volume of the resulting solution contains 0.0010 mol of glucose?

94. How many milliliters of concentrated sulfuric acid, which is 98 mass % H_2SO_4 and has a density of 1.842 g mL^{-1}, are required to prepare 500 mL of a $0.175M$ solution of $H_2SO_4(aq)$?

95. What volume of 85 mass % H_3PO_4 (density 1.659 g mL^{-1}) is required to prepare 2.50 L of $1.50M$ $H_3PO_4(aq)$?

*96. How could each of the following be prepared?

  (a) 6.30 L of $0.003M$ $Ba(OH)_2(aq)$, starting with $0.100M$ $Ba(OH)_2(aq)$ solution

  (b) 750 mL of $0.025M$ $Cr_2(SO_4)_3(aq)$, starting with a solution containing 35 mass % $Cr_2(SO_4)_3$, density 1.412 g cm^{-3}

97. Concentrated nitric acid, $HNO_3(aq)$, is 69.0 mass % HNO_3 and has a density of 1.41 g mL^{-1}. What volume of concentrated nitric acid is needed to prepare 250 mL of $0.100M$ nitric acid?

98. What mass of solid silver chloride results from the reaction of excess $HCl(aq)$ with 25.00 mL of $0.068M$ $AgNO_3(aq)$?

99. What volume of $0.250M$ $HCl(aq)$ is required to react completely with 22.6 g of sodium carbonate, $Na_2CO_3(s)$, according to the equation

$$Na_2CO_3(s) + 2HCl(aq) \longrightarrow 2NaCl(aq) + H_2O + CO_2(g)$$

100. What volume of $0.0120M$ $HCl(aq)$ is needed to react completely with 0.1240 g of magnesium metal, to give a solution of $MgCl_2(aq)$?

101. When 0.1573 g of zinc was completely reacted with 25.00 mL of $0.300M$ $HCl(aq)$, the hydrogen gas evolved was burned to give 0.0434 g of water.

  (a) What is the balanced equation for the reaction of zinc with dilute hydrochloric acid?

  (b) What was the concentration of zinc chloride and the concentration of unreacted $HCl(aq)$ in the solution after reaction was complete?

Miscellaneous

*102. **(a)** Balance the following equation:

$$UF_5 + H_2O \longrightarrow UO_2F_2 + UF_4 + HF$$

  (b) What is the maximum mass of UF_4 that could be obtained from 10.00 g of UF_5?

*103. A nuclear reaction that may be usable in the future for power generation involves the fusion of two deuterium nuclei:

$$2{}_1^2H \longrightarrow {}_2^3He + 1 \text{ neutron}$$

Suppose that 1.00 g of deuterium reacts in this way.

  (a) What change in mass accompanies the reaction?

  (b) What is the energy equivalent of this mass loss?

  (c) What mass of water must be processed to extract 1.00 g of deuterium, given that naturally occurring hydrogen contains 0.015 mass % deuterium?

The Atmosphere and the Gas Laws

3

We live on the surface of the earth, immersed in a mixture of gases called the atmosphere. Not only is the atmosphere essential to life, but it also plays a vital role in determining the temperature of the earth and in producing the weather. It is also the source of several elements of great industrial importance. The atmosphere has played an important role in the history of chemistry. Air was the first gas to be studied, and these studies produced some of the first scientific laws.

In this chapter we first examine the nature of the atmosphere and, in particular, the gaseous elements oxygen and nitrogen, which are its two principal components. We also consider another important gaseous element, hydrogen. Then we turn to the general properties of all gases. The physical behavior of all gases is very nearly the same. We will see that, unlike solids and liquids, all gases behave in much the same way with changing conditions, such as pressure and temperature, and we will see why this is.

3.1
The Atmosphere

The **atmosphere** is a mixture of gases held to the earth by gravity. This gaseous envelope is most dense at sea level and thins rapidly with increasing altitude. Almost all the atmosphere (99%) lies within 30 km of the earth's surface. Except for variable amounts of water vapor, this lowest layer of the atmosphere, which we call air, has a constant composition because it is continually mixed up by the winds that blow constantly around the earth.

Pure, dry air consists largely of oxygen and nitrogen; they make up 99% of its volume (Table 3.1). Among the other components are the *noble gases*, helium, neon, argon, krypton, and xenon, of which argon is the most abundant. Carbon dioxide, although it is only present in very small amounts (0.3%), is nevertheless important to life. Plants synthesize from carbon dioxide and water the complex substances they need in order to grow and reproduce in the process called *photosynthesis*. The atmosphere retains its constant composition up to a height of about 80 km. Above this height the atmosphere is no longer mixed up by air currents and it has a composition which changes with increasing height.

TABLE 3.1
Composition of Dry Air

Component	Formula	Percent by Volume
Nitrogen	N_2	78.084
Oxygen	O_2	20.948
Argon	Ar	0.934
Carbon dioxide	CO_2	0.031 4
Neon	Ne	0.001 82
Helium	He	0.000 52
Methane	CH_4	0.000 2
Krypton	Kr	0.000 11
Hydrogen	H_2	0.000 05
Dinitrogen monoxide	N_2O	0.000 05
Xenon	Xe	0.000 008

Over the many millions of years since the earth was formed the lighter atoms and molecules have gradually diffused away from the earth's surface because they are subject to a smaller gravitational attraction. Thus the gases with larger molecular masses are closest to the earth's surface and those with smaller molecular masses are further away. Thus, with increasing height, molecular nitrogen, N_2 (molar mass 28 g), is gradually replaced by atomic oxygen, O (molar mass 16 g), then helium, He (molar mass 4 g), and finally atomic hydrogen, H (molar mass 1 g). There is very little molecular oxygen, O_2, or molecular hydrogen, H_2, in the upper layers of the atmosphere because these molecules are dissociated into atoms by intense ultraviolet radiation and fast-moving electrons and protons coming from the sun. Because of the formation of these very reactive atoms many complex reactions occur in the upper atmosphere. One of these is the formation of a layer containing ozone molecules, O_3, by the reaction

$$O + O_2 \longrightarrow O_3$$

Ozone absorbs ultraviolet radiation and thus prevents most of the intense ultraviolet radiation emanating from the sun from reaching the surface of the earth, thus protecting us and all other living things from its harmful effects (see Chapter 18). There is no definite upper boundary to the atmosphere. At a height of approximately 10 000 km the concentration of hydrogen atoms in the earth's atmosphere becomes equal to their concentration in interplanetary space.

Most of the hydrogen and the helium that might originally have been part of the atmosphere of the earth has been lost to outer space. Considerably larger amounts of hydrogen and helium are found in the atmospheres of the four larger, outer planets of the solar system: Jupiter, Saturn, Uranus, and Neptune. These planets have masses from 17 to 318 times that of the earth, so they exert a much greater gravitational force and have therefore retained more hydrogen and helium in their atmospheres. In contrast, the moon has lost all its atmosphere; because its mass is only $\frac{1}{6}$ of the mass of the earth, it exerts a correspondingly smaller gravitational attraction.

The atmosphere has not always contained the large amount of oxygen that is present today. It is probable that at the time life first appeared on earth, more than 3000 million (3×10^9) years ago, there was little, if any, oxygen in the atmosphere. Early forms of bacteria obtained hydrogen from hydrogen sulfide in order to synthesize the compounds that they needed for growth and reproduction. Later the blue-green algae developed. They contained chlorophyll, the substance that enabled them to carry out photosynthesis and to utilize the most abundant source of hydrogen, namely water, liberating oxygen in the process. All the higher plants that subsequently developed made use of chlorophyll to carry out photosynthesis. The oxygen they produced has accumulated over the millenia to form the oxygen-rich atmosphere that we know today.

Ozone sonde balloon launch at Amundsen-Scott Station, Antarctica. This facility is located at the south pole and is owned by the National Science Foundation.

There is strong evidence that the concentration of ozone in the ozone layer is being reduced by reaction with chlorofluorocarbons (CFCs), which have been widely used in spray cans and refrigerators. If the concentration of ozone in the upper atmosphere continues to decrease it could have serious consequences for life on earth (see Box 18.3).

3.2
The Abundance of the Elements

Although nitrogen is the most abundant element in the atmosphere, oxygen is by far the most abundant element in the whole of the earth's crust. The crust consists of the relatively thin layer of solid rocks called the *lithosphere*, the oceans, which are called the *hydrosphere*, and the *atmosphere* (Figure 3.1).

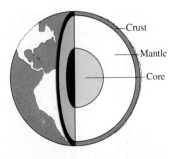

FIGURE 3.1
Structure of the Earth.

We do not have reliable information about the composition of the
deep interior of the earth, but the
core, with a 3500-km (2200-mile)
radius, is believed to be composed
mainly of iron. Surrounding the
core is fluid material called the
mantle. It is 2900 km (1800 miles)
thick and is thought to be composed primarily of silicon, oxygen,
iron, calcium, and magnesium.
The very thin outer, solid layer is
the *crust*. It is 10–50 km (8–26
miles) thick and is composed
of the hydrosphere, lithosphere and
atmosphere.

Helium is formed in the
earth's solid crust by the disintegration of radioactive
elements such as uranium. It
diffuses through fissures and
collects in pockets of oil and
natural gas.

Oxygen, in the form of its many compounds, constitutes almost half the earth's
crust by mass. Indeed, there are more oxygen atoms than the total of all other
kinds of atoms (Table 3.2). In contrast, despite the abundance of molecular
nitrogen, N_2, in the atmosphere, nitrogen constitutes only 0.03% of the earth's
crust (lithosphere, hydrosphere, and atmosphere). This striking difference between oxygen and nitrogen is a reflection of the fact that oxygen is a very reactive element and forms many compounds. It is a major component of almost
all rocks and, of course, water. Nitrogen is a much less reactive element and is
found in far fewer compounds in the lithosphere.

After oxygen the next most abundant element in the earth's crust is silicon;
together, silicon and oxygen compose 75% of the crust. Only 10 elements (O,
Si, Al, Fe, Ca, Na, K, Mg, H, and Ti) together comprise 99.2% of the mass of
the earth's crust. They are all light elements. Fifteen of the 20 most abundant
elements have atomic numbers below 21. This is consistent with the belief that
the elements are built up in stars from protons, neutrons, and electrons, forming
first hydrogen and helium and then other, heavier elements. In fact, hydrogen
and helium are by far the most abundant elements in the universe as a whole,
although they have been very largely lost from the earth's atmosphere. Helium
forms no compounds and is a very rare element on the surface of the earth; it
occurs only in very small amounts in the atmosphere and in association with
oil and natural gas. Hydrogen, however, is relatively common in the form of
its many compounds. Indeed, on an atom percentage basis it is the third most
common element.

The data in Table 3.2 refer to the earth's crust only. They do not take into
account the material of the core, which is believed to be mainly iron, nor the
material of the fluid mantle that surrounds the core, which is believed to contain
mainly iron, calcium, magnesium, silicon, and oxygen.

TABLE 3.2
Abundance of the 20 Most Common Elements in the
Earth's Crust (Atmosphere, Hydrosphere and Lithosphere)

Element	Z	Mass Percent		Atom Percent
Oxygen	8	49.4 ⎫ 75.2		55.1
Silicon	14	25.8 ⎭		16.3
Aluminum	13	7.5		5.0
Iron	26	4.7		1.5
Calcium	20	3.4		1.5
Sodium	11	2.6	99.2	2.0
Potassium	19	2.4		1.1
Magnesium	12	1.9		1.4
Hydrogen	1	0.9		15.4
Titanium	22	0.6		0.2
Chlorine	17	0.2		
Phosphorus	15	0.12		
Manganese	25	0.10		
Carbon	6	0.08		
Sulfur	16	0.06		
Argon	18	0.04		
Nitrogen	7	0.03		
Rubidium	37	0.03		
Strontium	38	0.03		
Fluorine	9	0.03		

The elements that are gases at 25 °C and 1 atm pressure are hydrogen, nitrogen, oxygen, fluorine, chlorine, and the noble gases, helium, neon, argon, krypton, xenon, and radon (Table 3.3). The noble gases are monatomic. The other gaseous elements are diatomic molecules. In the upper atmosphere oxygen is also found in the form of triatomic molecules O_3 and is then called ozone. In the following sections we describe some of the properties and reactions of oxygen, nitrogen, and hydrogen. We consider fluorine and chlorine, with bromine and iodine, in Chapter 5 and we discuss the noble gases in Chapter 20.

TABLE 3.3 The Gaseous Elements	
Hydrogen	H_2
Nitrogen	N_2
Oxygen	O_2, O_3
Fluorine	F_2
Chlorine	Cl_2
Helium	He
Neon	Ne
Argon	Ar
Krypton	Kr
Xenon	Xe
Radon	Rn

3.3
Oxygen and Nitrogen

Oxygen was not recognized as a distinct substance until the late eighteenth century. In fact, before then nobody knew that air was a mixture of gases. Experiments by the English chemist Joseph Priestley (Figure 3.2) and the French chemist Antoine Lavoisier in the latter part of the eighteenth century showed that when a substance burns in air, or when a metal is heated in air, it combines with only part of the air (Box 3.1). Thus they concluded that air must be a mixture of a reactive substance, which was used up in burning and which combined with metals when they were heated, and another less reactive substance. Lavoisier called the reactive component of the air oxygen and he recognized it as an element. We now know that the other less reactive component of air is mainly nitrogen.

OXYGEN: OXIDES AND OXIDATION

A colorless gas at ordinary temperatures, oxygen condenses to a blue liquid at −183 °C and freezes to a pale blue solid at −218 °C. It consists of diatomic

FIGURE 3.2
Joseph Priestley (1733–1804).

The son of a Nonconformist minister, Priestley was born in Yorkshire, England. As a young man he had many interests and he taught himself several languages, including Arabic and Hebrew, as well as philosophy and science. He had radical religious beliefs and he eventually became a Unitarian Minister. In 1766 he met Benjamin Franklin who was on a visit to London to attempt to settle the dispute over taxation between the American colonists and the British government. Franklin had been studying electricity and this meeting inspired Priestley to begin his own research in the field. He was the first to show that graphite is a good conductor of electricity. He then turned to chemistry. As a result of his experiment on the decomposition of mercury oxide by heat he discovered oxygen (Box 3.1). He was the first to collect gases over mercury, and he thus discovered several water-soluble gases, including ammonia, hydrogen chloride, and sulfur dioxide. An outspoken man with very liberal religious and political views, he was sympathetic to the American and French revolutions and was viewed with suspicion by the conservative British majority. In 1791 his house and laboratory were burned down by an angry mob. Priestley managed to escape, went into hiding for a time, and eventually emigrated to the United States. He spent the last ten years of his life in relative seclusion in Pennsylvania.

BOX 3.1

THE DISCOVERY OF OXYGEN

For much of the eighteenth century chemists believed that when a substance was burned in air or when many metals were heated in air a substance called *phlogiston* was released. It was thought that a flame in a closed container was eventually extinguished because the air became saturated with phlogiston. In August 1774 Priestley strongly heated mercury oxide, HgO, by focusing sunlight onto it with a lens. He observed that mercury was formed and that simultaneously a gas was formed. On studying this gas he found that a candle burned much more brightly in it than in ordinary air. Priestley had discovered oxygen but he called the gas *dephlogisticated air*. He believed that a candle burned less brightly in ordinary air because air contained some phlogiston.

In October of that year Priestley visited Lavoisier in Paris and told him about his experiment. Lavoisier immediately began an experiment of his own in which he heated mercury in a sealed air-filled flask in a furnace. Small red specks appeared on the surface of the mercury and the amount of this red substance slowly increased. After 12 days Lavoisier extinguished the furnace and examined the gas left in the flask. He found that it no longer supported combustion. He then weighed the mercury plus red substance and found that its total mass was greater than the mass of the mercury with which he had started. If the phlogiston theory were correct, the mass of the mercury should have decreased as it released phlogiston. Some defenders of the phlogiston theory went so far as to conclude that phlogiston had negative mass

Silver liquid mercury and red solid mercury oxide, HgO.

but Lavoisier correctly concluded that, on heating in air, rather than releasing phlogiston mercury combines with a part of the air and thus its mass increases.

Lavoisier next heated the red solid, which was mercury oxide. He obtained mercury and a gas that supported combustion much better than air, just as Priestley had found. Clearly this gas was one of the components of air. Lavoisier correctly recognized it as an element and called it oxygen. This reaction of mercury with oxygen to give mercury oxide and its decomposition back to the elements we would now represent by the equation

$$2Hg(l) + O_2(g) \rightleftharpoons 2HgO(s)$$

These experiments spelled the end of the phlogiston theory and clearly showed for the first time that air was a mixture containing the element oxygen.

O_2 molecules. Oxygen is obtained on a large scale by liquefying air and then distilling it to separate the components. Very pure oxygen can be obtained by electrolysis, that is, by passing an electric current through water (Experiment 1.1). Small amounts of oxygen can be made in the laboratory by heating potassium chlorate, $KClO_3$, with manganese dioxide, which behaves as a catalyst:

$$2KClO_3(s) \xrightarrow{\text{heat}} 2KCl(s) + 3O_2(g)$$

A **catalyst** is a substance that increases the rate of a reaction without changing the nature of the products. For example, manganese dioxide and blood are catalysts for the decomposition of hydrogen peroxide (Experiment 1.5). We will discuss in Chapter 19 how a catalyst operates.

Oxygen reacts directly with almost all the other elements. The reactions with many elements are quite slow at room temperature; frequently, no reaction at all is apparent. However, when the temperature is raised, the rate of reaction increases. Reactions of the elements with oxygen liberate heat, which is often sufficient to keep the temperature high enough that the reaction continues rapidly

Experiment 3.1

Reactions of Metals and Nonmetals with Oxygen

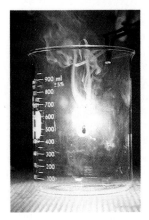   

Burning magnesium burns even more violently in oxygen, forming a white smoke of solid particles of magnesium oxide, MgO.

If steel wool is first heated in a flame it ignites and burns vigorously in oxygen, giving a shower of sparks and forming brown solid iron oxide, Fe_2O_3.

White phosphorus ignites spontaneously in oxygen and burns with a very bright flame, forming a dense white smoke of solid P_4O_{10}.

If sulfur is warmed until it melts, it burns in oxygen with a bright blue flame, producing the pungent smelling gas sulfur dioxide, SO_2.

without the need to supply more heat. The element is said to be undergoing **combustion**, or burning. Sulfur, for example, shows no tendency to react with oxygen at room temperature. However, if sulfur is heated until it melts and is then placed in oxygen, it reacts rapidly, burning with a bright blue flame to form sulfur dioxide, a colorless gas with a pungent odor (Experiment 3.1). Magnesium does not react with oxygen at ordinary temperatures, but when heated it burns rapidly, emitting a brilliant white light.

The products of the reactions of the elements with oxygen are known as **oxides**. Table 3.4 lists some typical reactions in which elements are converted to their oxides.

Only a very few elements do not react directly with oxygen. They are the noble gases, He, Ne, Ar, Kr, Xe, and Rn, and the metals gold, Au, and platinum, Pt. Nevertheless, oxides of some of these elements—for example, XeO_3, XeO_4, PtO_2, and Au_2O_3—can be made from other compounds of these elements.

Many compounds are converted to the oxides of their constituent elements on reaction with oxygen. For example, the compounds of carbon and hydrogen, called **hydrocarbons**, burn in oxygen to give carbon dioxide and water. Propane, C_3H_8, reacts with oxygen according to the equation

$$C_3H_8(g) + 5O_2(g) \longrightarrow 3CO_2(g) + 4H_2O(g)$$

A simple **test for oxygen** is to hold a glowing splint of wood in a tube of the gas suspected to be oxygen. In oxygen the combustion of wood, which is a complex mixture of organic compounds of carbon, hydrogen, and oxygen, is much faster than it is in air, and the wood splint bursts into flame.

TABLE 3.4
Reactions of Some Elements with Oxygen

Element	Reaction with Oxygen	Oxide
Copper, a reddish metal	$2Cu + O_2 \longrightarrow 2CuO$	Copper oxide, a black solid, insoluble in water
Mercury, a silvery liquid metal	$2Hg + O_2 \longrightarrow 2HgO$	Mercury oxide, a red solid, insoluble in water
Magnesium, a silvery metal	$2Mg + O_2 \longrightarrow 2MgO$	Magnesium oxide, a white solid, insoluble in water
Sulfur, a yellow solid	$S + O_2 \longrightarrow SO_2$	Sulfur dioxide, a colorless and pungent-smelling gas, soluble in water
Phosphorus, a red solid	$4P + 5O_2 \longrightarrow P_4O_{10}$	Phosphoric oxide, a white solid, soluble in water
Hydrogen, a colorless gas	$2H_2 + O_2 \longrightarrow 2H_2O$	Water
Carbon (graphite), a black solid	$C + O_2 \longrightarrow CO_2$	Carbon dioxide, a colorless gas, slightly soluble in water

When an element combines with oxygen to form an oxide it is said to have been **oxidized** *and the reaction is called an* **oxidation reaction**. Thus copper is said to be oxidized to copper oxide. When propane burns to form CO_2 and H_2O the carbon in propane is oxidized to CO_2 and the hydrogen to water. We will discuss a more general definition of oxidation in some detail in Chapter 5.

EXOTHERMIC AND ENDOTHERMIC REACTIONS

Burning or combustion—and, indeed, many other oxidation reactions—are accompanied by the liberation of energy in the form of heat. All reactions that occur with the liberation of heat are known as **exothermic reactions**. Experiment 3.2 provides another example of an exothermic reaction. If a reaction occurs with the absorption of heat, it is called an **endothermic reaction**.

Since the combustion of hydrocarbons is associated with the liberation of a considerable amount of heat, hydrocarbons are often used as fuels. Natural gas, which is mainly methane, CH_4, is often used in homes for heating and cooking. Propane, C_3H_8, and butane, C_4H_{10}, which are available compressed in cylinders, are used for the same purposes and also, for example, for camp stoves. Gasoline and diesel fuel, which are mixtures of a large number of hydrocarbons, are burned in internal combustion engines to provide the necessary power for automobiles, airplanes, trains, and ships. The combustion of acetylene, C_2H_2, is the basis of the oxyacetylene torch, which gives a very hot flame which can be used for welding metals.

Oxidation reactions form the basis for life. Animals inhale oxygen from the atmosphere, and this oxygen is absorbed in the lungs by the hemoglobin of the blood and distributed to different parts of the body, where it is used for the oxidation of many different substances. These oxidation reactions are exothermic and they provide the energy that is needed to maintain life.

This Pak-heat pouch contains fine iron power in a porous bag. When the porous bag is removed from the sealed pouch, oxygen from the air comes in contact with the iron powder and oxidizes it to iron oxide in an exothermic reaction, producing heat over a period of an hour or more.

Experiment 3.2

The Exothermic Reaction Between Iron and Sulfur

A mixture of iron and sulfur (see Experiment 1.4) reacts to form iron sulfide, FeS, when it is heated.

If the tube is removed from the flame it continues to glow very brightly as the heat evolved in the reaction maintains the high temperature of the mixture.

REDUCTION

Since most elements combine with oxygen, and since oxygen is very abundant on the surface of the earth, many elements, particularly metals, are found in the form of their oxides (for example, H_2O, Al_2O_3, Fe_2O_3, MnO_2, CuO). Frequently, these oxides are major sources of the elements, which can be obtained by removing the oxygen in a process known as **reduction**. Iron is made industrially by reducing the oxide Fe_2O_3 with carbon monoxide, CO, which combines with the oxygen to form carbon dioxide, CO_2:

$$Fe_2O_3(s) + 3CO(g) \longrightarrow 2Fe(s) + 3CO_2(g)$$

Copper oxide, CuO, can similarly be reduced with carbon and also with hydrogen (see Experiment 3.3):

$$2CuO(s) + C(s) \longrightarrow 2Cu(s) + CO_2(g)$$
$$CuO(s) + H_2(g) \longrightarrow Cu(s) + H_2O(l)$$

When oxygen is removed from an oxide of an element the oxide is said to have been **reduced** *and the reaction is called a* **reduction reaction**. Thus when oxygen is removed from copper oxide by reaction with hydrogen the copper oxide is said to have been reduced to copper. We note that in this reaction hydrogen is, at the same time, oxidized to water. Oxidation is always accompanied by reduction and *vice versa* as we will see in Chapter 5.

Experiment 3.3

Reduction of Copper Oxide with Carbon

A mixture of black copper oxide and black carbon.

When the mixture is heated the copper oxide is reduced to copper and the carbon is oxidized to carbon dioxide. The formation of carbon dioxide is shown by passing the evolved gases through limewater—an aqueous solution of calcium hydroxide, $Ca(OH)_2$. A white precipitate of calcium carbonate is formed.

$$Ca(OH)_2(aq) + CO_2(g) \longrightarrow$$
$$CaCO_3(s) + H_2O(l)$$

When the tube is cooled after the reaction the red-brown copper that has been produced is clearly seen.

NITROGEN

Although nitrogen forms the major part of the atmosphere, it is not a very abundant element on the earth because very little occurs in the form of solid compounds in the lithosphere (Table 3.2). The only important mineral sources of nitrogen are sodium nitrate, $NaNO_3$, and potassium nitrate, KNO_3. Large deposits of these nitrates occur in the arid desert regions of Chile. Nevertheless, nitrogen, together with carbon, oxygen, and hydrogen, is one of the principal elements found in the compounds that constitute living matter.

Elemental nitrogen is a colorless, odorless, tasteless gas, consisting of diatomic N_2 molecules. It boils at $-196\ °C$ and freezes at $-210\ °C$. Like oxygen it can be obtained from the air by liquefaction and distillation (Figure 3.3).

Nitrogen is a relatively unreactive element. It reacts with very few other substances at ordinary temperatures, but it is somewhat more reactive at high temperatures. For example, on strongly heating it reacts with hydrogen to give ammonia,

$$N_2(g) + 3H_2(g) \longrightarrow 2NH_3(g)$$

with oxygen to give nitrogen monoxide,

$$N_2(g) + O_2(g) \longrightarrow 2NO(g)$$

and with a few metals to give nitrides, such as magnesium nitride,

$$3Mg(s) + N_2(g) \longrightarrow Mg_3N_2(s)$$

The reaction between nitrogen and oxygen to produce NO occurs in the atmosphere in lightning discharges during thunderstorms. The nitrogen and oxygen in the air also react in the same way at the high temperature produced in the cylinders of gasoline and diesel engines.

FIGURE 3.3
Liquid Nitrogen

The white mist is a cloud of fine water droplets condensed from the air as it is cooled by the liquid nitrogen.

Since the only widely available source of nitrogen is the nitrogen of the atmosphere, all nitrogen compounds have ultimately to be obtained from this source. The process of converting the nitrogen of the air into useful compounds is known as **nitrogen fixation**. This process is extremely important because of the great need for nitrogen fertilizers. The only practical method of large-scale nitrogen fixation known at the present time is the preparation of ammonia by the reaction of nitrogen with hydrogen. In the industrial process known as the **Haber process**, nitrogen and hydrogen are heated together at about 400 °C at a high pressure, and in the presence of a catalyst, to give ammonia.

Certain bacteria found in the soil and in the root nodules of leguminous plants, such as peas and beans, are able to carry out nitrogen fixation at ordinary temperatures. In other words, they convert the nitrogen of the atmosphere into nitrogen compounds that plants can assimilate.

The NO produced by automobiles has become a serious atmospheric pollutant in an increasing number of cities around the world. Considerable efforts are now being made to reduce the amount of NO in automobile exhaust emissions (see Box 18.1).

3.4
Hydrogen

Hydrogen is the most abundant element in the universe. Interstellar space is very sparsely filled with atoms that are predominantly hydrogen. The stars consist mostly of hydrogen, and the uppermost region of the atmosphere consists mainly of hydrogen atoms. Elsewhere in the earth's crust hydrogen is not present as the free element, but hydrogen compounds are very common. Hydrogen is the third most abundant element on an atom basis, and the ninth most abundant on the basis of mass (Table 3.2). It is found in water, hydrocarbons, proteins, carbohydrates, and almost all the other substances in living organisms.

Hydrogen is a colorless, odorless, tasteless, nonpoisonous gas consisting of H_2 molecules. It combines readily with many other elements. It has a very low boiling point (-252.8 °C) and a very low melting point (-259.1 °C). A simple test for hydrogen is to bring a flame to the open end of a test tube containing the gas suspected to be hydrogen. If the gas is hydrogen, a characteristic "pop" is heard as H_2 and O_2 combine rapidly to form H_2O, causing a small explosion.

PREPARATION OF HYDROGEN

Water is by far the cheapest and most abundant source of hydrogen; all the large-scale methods of making hydrogen are based on the removal of oxygen from water. Oxygen can be removed from water by combining the oxygen with carbon or with metals such as iron to form oxides.

In the production of hydrogen on an industrial scale, steam is passed over coke, an impure form of carbon, at about 1000 °C:

$$C(s) + H_2O(g) \longrightarrow CO(g) + H_2(g)$$

Carbon is oxidized to carbon monoxide, and water is reduced to hydrogen. This mixture of $CO(g)$ and $H_2(g)$ is known as *water gas*, an important industrial fuel.

If pure H_2 is needed, it is separated from the CO by mixing the water gas with steam and passing the mixture over a catalyst at 500 °C, to convert CO to CO_2:

$$\underset{\text{Water gas}}{[CO(g) + H_2(g)]} + H_2O(g) \longrightarrow 2H_2(g) + CO_2(g)$$

A catalyst is needed here because otherwise the reaction would be much too slow to be useful. Because carbon dioxide is much more soluble in water than is hydrogen, it is easily removed from the H_2–CO_2 mixture by passing the mixture, under pressure, into water.

Hydrogen is also made industrially by passing steam over heated iron:

$$3Fe(s) + 4H_2O(g) \longrightarrow Fe_3O_4(s) + 4H_2(g)$$

In this reaction iron is *oxidized* to Fe_3O_4—it adds oxygen—and the steam is *reduced* to hydrogen—it loses oxygen. In the laboratory the reduction of water with a metal can be conveniently demonstrated by the reaction of steam with heated magnesium (Experiment 1.3):

$$Mg(s) + H_2O(g) \longrightarrow MgO(s) + H_2(g)$$

But magnesium is too expensive for this to be used as an industrial method.

Very pure hydrogen is made by the **electrolysis** of water. When an electric current is passed through water, it is decomposed into hydrogen and oxygen, which can be collected separately (Experiment 1.1):

$$2H_2O(l) \xrightarrow{\text{Electric current}} 2H_2(g) + O_2(g)$$

Because electrolysis uses large amounts of electrical energy, this method is a rather expensive way of producing hydrogen. A related process is the manufacture of sodium hydroxide, NaOH, chlorine, and hydrogen by the electrolysis of an aqueous sodium chloride solution (see Chapter 17).

An increasingly important source of hydrogen is methane, CH_4, the major component of natural gas. When methane mixed with steam is passed over a heated nickel catalyst, a mixture of carbon monoxide and hydrogen, called *synthesis gas*, is produced:

$$CH_4(g) + H_2O(g) \longrightarrow CO(g) + 3H_2(g)$$

Synthesis gas is the starting material for the industrial production of a number of important compounds (see Chapter 23).

In the laboratory, small amounts of hydrogen can be made by the reaction of a metal such as zinc with hydrochloric acid (see Hydrogen Chloride below), as shown in Figure 3.4:

$$Zn(s) + 2HCl(aq) \longrightarrow ZnCl_2(aq) + H_2(g)$$

FIGURE 3.4
Preparation of Hydrogen.

This illustrates the apparatus that is commonly used in the laboratory for the preparation of small amounts of hydrogen. Hydrogen has only a very small solubility in water so it can be collected in a jar by displacing water from the jar as shown. This method is useful for collecting samples of other gases, such as oxygen, that are not very soluble in water.

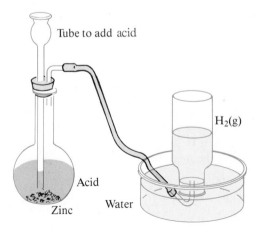

Tube to add acid

$H_2(g)$

Acid

Zinc   Water

But hydrogen is normally available in the laboratory from high-pressure stainless steel cylinders.

COMPOUNDS OF HYDROGEN

Under appropriate conditions hydrogen, H_2, combines with most elements to form compounds. Compounds of hydrogen with metals are known as **hydrides**. Examples include sodium hydride, NaH, calcium hydride, CaH_2, and aluminum hydride, AlH_3. Many common compounds of hydrogen, particularly those with nonmetals, have special names, such as methane, CH_4, ammonia, NH_3, and water, H_2O.

WATER The reaction of hydrogen with oxygen to form water is a very exothermic reaction. A practical use of this reaction is in the oxyhydrogen torch, where the reaction generates temperatures up to 2800 °C, which are useful for welding materials that have high melting points. The combustion of hydrogen and oxygen is also used to fuel rockets (Box 3.2). Mixtures of H_2 and O_2 are explosive, particularly when the $H_2:O_2$ ratio is approximately 2:1.

AMMONIA The reaction of nitrogen with hydrogen to form ammonia, NH_3, occurs much less readily than the reaction between hydrogen and oxygen. We have seen that in the Haber process for the manufacture of ammonia, nitrogen and hydrogen are heated together at a high pressure and temperature in the presence of a catalyst:

$$N_2(g) + 3H_2(g) \longrightarrow 2NH_3(g)$$

Ammonia is a colorless gas (boiling point −33.4 °C) with a unique penetrating odor. It is very soluble in water. Ammonia solutions are widely used as household cleaning agents.

One of the major uses of ammonia is for the manufacture of *fertilizers*. Plants require nitrogen, but they cannot use nitrogen from the atmosphere directly. Instead, they take up nitrogen compounds from the soil. When the same land is used repeatedly for crops, the nitrogen compounds in the soil are depleted. Farmers therefore add nitrogen-containing compounds known as fertilizers. Liquid ammonia stored at a low temperature can be added directly to the soil, but more commonly it is converted to other nitrogen-containing compounds, such as ammonium sulfate, $(NH_4)_2SO_4$, ammonium hydrogen phosphate, $(NH_4)_2HPO_4$, and urea $(NH_2)_2CO$.

METHANE Although methane, CH_4, can be made by the direct reaction of carbon and hydrogen, this is not an important reaction as ample supplies of methane are available from natural gas. A very small amount of methane (0.0002%) is found in the atmosphere. It is formed by the bacterial decomposition of vegetable matter under water, and it is produced by some animals, such as cows, during the digestion of plant material.

Methane is a colorless gas with a very low melting point (−182 °C) and a very low boiling point (−162 °C). It burns readily in air in an exothermic reaction:

$$CH_4(g) + 2O_2(g) \longrightarrow CO_2(g) + 2H_2O(g)$$

Methane is the major component of natural gas. In this form it is widely used as a fuel.

BOX 3.2

HYDROGEN AS A FUEL

Columbia Space Shuttle. *The rocket that launched the shuttle used liquid hydrogen as a fuel. It was stored in a tank 40 m long and 8.4 m in diameter having a capacity of 385 000 gallons. The oxygen for burning the hydrogen was stored in a similar tank containing 143 000 gal of liquid.*

Hydrogen is an important rocket fuel. A primary consideration for a rocket fuel is that its mass be as small as possible for a given amount of energy produced. Hydrogen was used in the *Saturn V* rocket that enabled the first astronauts to land on the moon, and it is the main fuel in the space shuttle rockets. Both the hydrogen and the oxygen needed to burn the hydrogen are carried on the rocket in liquid form.

Hydrogen also has attractive features as a fuel for more general use. But its use at present is hampered by the difficulty of safely storing, transporting, and distributing such a highly flammable, potentially explosive material. In addition, because there are no natural sources of hydrogen, energy from another source must be expended in order to obtain H_2 from hydrogen compounds such as water. Hydrogen is said to be an energy carrier rather than an energy source. Both sunlight and excess electric energy from nuclear reactors are being studied as possible sources of energy for producing hydrogen from water. Thus widespread use of hydrogen as a fuel will be economical only when hydrogen can be produced rather cheaply.

If the problems of safety and economics are resolved, hydrogen may eventually be delivered to homes and industry by pipelines, as natural gas is delivered today. Hydrogen might even be used to fuel vehicles. For this purpose it might be stored as a liquid at very low temperatures or as a solid metal hydride that, when heated, decomposes to hydrogen and metal. Hydrogen has already been used on a trial basis as a fuel for jet airplanes. An important advantage of hydrogen as a fuel for automobiles and airplanes is that the only product of combustion is water, and there is no emission of CO, SO_2, and hydrocarbons, which pollute the atmosphere.

A Prototype Car that Uses Hydrogen as a Fuel. *The hydrogen is stored as an iron titanium hydride from which the hydrogen can be released by heating. The heat from the exhaust can be conveniently used for this purpose.*

HYDROGEN CHLORIDE Hydrogen chloride, HCl, is a colorless gas with a pungent, irritating odor. We have already seen in Example 2.15 that it can be made by the reaction between hydrogen and chlorine either when a spark is passed through a mixture of the two gases or when they are strongly heated:

$$H_2(g) + Cl_2(g) \longrightarrow 2HCl(g)$$

It is very soluble in water and is usually encountered in the laboratory as an aqueous solution called hydrochloric acid (see Chapter 5). Upon contact with gaseous ammonia, hydrogen chloride forms white solid ammonium chloride, NH_4Cl (Figure 3.5):

$$NH_3(g) + HCl(g) \longrightarrow NH_4Cl(s)$$

═ Example 3.1 ═

REACTIONS AND BALANCED EQUATIONS

Write a balanced equation for each of the following reactions:

(a) The reaction of burning magnesium with steam

(b) The combustion of ethane in excess oxygen

(c) Steel wool burning in oxygen

(d) The reaction between atmospheric nitrogen and oxygen during a thunderstorm

Solution

(a) We saw in Experiment 1.3 that magnesium continues to burn in steam, being oxidized to white solid magnesium oxide:

$$Mg(s) + H_2O(g) \longrightarrow MgO(s) + H_2(g)$$

Note that we have seen two other oxidation reactions in which water at high temperature (steam) gives up its oxygen to another substance and is thus converted to hydrogen:

$$C(s) + H_2O(g) \longrightarrow CO(g) + H_2(g)$$

and

$$CH_4(g) + H_2O(g) \longrightarrow CO(g) + 3H_2(g)$$

(b) Ethane is the hydrocarbon C_2H_6. We have seen that hydrocarbons are oxidized by oxygen at high temperature to carbon dioxide and water:

$$2C_2H_6(g) + 7O_2(g) \longrightarrow 4CO_2(g) + 6H_2O(g)$$

(c) We saw in Experiment 3.1 that steel wool burns in oxygen to form the brown solid iron oxide Fe_2O_3:

$$4Fe(s) + 3O_2(g) \longrightarrow 2Fe_2O_3(s)$$

(d) Nitrogen combines directly with oxygen to give nitrogen monoxide only at very high temperatures such as we find in a lightning discharge:

$$N_2(g) + O_2(g) \longrightarrow 2NO(g)$$

FIGURE 3.5
The Reaction between HCl and NH_3.

When a stream of HCl(g) (left) meets a stream of NH_3(g) (right) a cloud of solid ammonium chloride NH_4Cl is formed.

EXERCISE 3.1

Write a balanced equation for each of the following reactions:

(a) The reaction of burning magnesium with carbon dioxide

(b) The reaction of copper oxide, CuO, with hydrogen at high temperature

(c) The reaction of nitrogen with hydrogen at high temperature

(d) The reaction of phosphorus with excess oxygen

═ *Example 3.2* ═══════════════════════════════════

PROPERTIES OF THE ELEMENTS

From the elements that we have so far described in Chapters 1–3 name

 (a) Three that are gases at room temperature

 (b) Two that are liquids at room temperature

 (c) Three that are metals

Solution

 (a) You have your choice of H_2, N_2, O_2, Cl_2, He. The only other elements that are gases at room temperature are F_2, Ne, Ar, Kr, Xe, Rn.

 (b) The only two elements that are liquid at room temperature are Br_2 and Hg.

 (c) The elements that are metals that we have mentioned so far are Mg, Zn, Fe, Hg, Au, Al, Cu. There are many other elements that are metals.

E X E R C I S E   3 . 2

From the elements that we have described so far name

(a) One that is a yellow solid

(b) One that is reddish metal

(c) One that does not combine with oxygen under any conditions

(d) The most abundant element in the earth's crust

(e) The most abundant metal in the earth's crust

3.5
Physical Properties of Gases and the Gas Laws

We have seen that several important elements and compounds are gases under ordinary conditions. The study of gases has been important since the early days of science. Air was the first gas to be studied, and it was studied long before scientists understood that it is a mixture of a number of different elements and compounds. That air is, in fact, a mixture caused no problems, because, with respect to changes in pressure, volume, and temperature, all gases behave approximately in the same way; thus in many ways air behaves like a single substance.

Any given sample of a gas can be described in terms of four fundamental properties: *mass, volume, pressure,* and *temperature*. The investigation of these properties of air led to the discovery of the quantitative relationships between them. The statements of these quantitative relationships constitute some of the earliest scientific laws (see Box 3.3 on page 118).

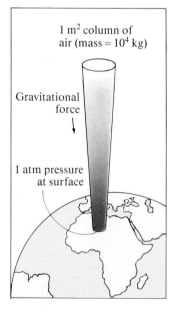

1 m^2 column of air (mass = 10^4 kg)

Gravitational force

1 atm pressure at surface

FIGURE 3.6
Atmospheric Pressure.

A column of the atmosphere 1 m^2 in cross-sectional area and extending to the top of the atmosphere has a mass of approximately 10 kg. It exerts a force $F = mg = (10\ kg)(9.81\ m\ s^{-1}) = 1 \times 10^5$ N where g is the acceleration due to gravity. Thus the pressure P exerted by the atmosphere is $P = F/A = 1 \times 10^5$ N/1 m^2 = 1×10^5 Pa = 1×10^2 kPa.

THE PRESSURE OF THE ATMOSPHERE: PRESSURE UNITS

We live on the surface of the earth at the bottom of the atmosphere. The gravitational attraction of the earth on the gases of the atmosphere causes the atmosphere to exert a force on the earth's surface. Pressure is defined as force per unit area. The atmospheric pressure is the force exerted by the atmosphere on a unit area of the earth's surface (Figure 3.6). The pressure exerted by the atmosphere was first measured by the Italian physicist Evangelista Torricelli (1608–1647) in 1643. He used what we now call a *Torricellian* or *mercury barometer* (Figure 3.7). A long glass tube closed at one end is filled with mercury. The open end is temporarily closed and, with the tube vertical, is inserted under the surface of mercury in an open dish. When the end of the tube is opened mercury runs out into the dish until the surface of the mercury in the tube is at a height of approximately 76 cm above the surface of the mercury in the dish. The pressure exerted by the atmosphere on the surface of the mercury in the dish which is pushing mercury up into the tube is then exactly counterbalanced by the downward pressure due to the gravitational attraction of the earth on the mercury in the tube. The height of the mercury column supported by the atmosphere varies somewhat with the atmospheric conditions (the weather) and it decreases with altitude, but it is normally about 76 cm = 760 mm at sea level. Any liquid could in principle be used in a barometer but mercury, because of its high density (13.6 g cm^{-3}) gives a column of convenient height. Water, which has a much lower density (1.00 g cm^{-3}), would give a column 13.6 times as high (10.3 m)! Because the mercury barometer is widely used, pressures are frequently expressed in terms of the height of a mercury column, that is, in *millimeters of mercury* (mm Hg). One millimeter of mercury is often called a *torr* in honor of Torricelli. A pressure of 760 mm Hg is called a *standard atmosphere* (atm):

$$1 \text{ standard atmosphere} = 1 \text{ atm} = 760 \text{ mm Hg} = 760 \text{ torr}$$

However, since pressure is force per unit area, it should be measured in corresponding units such as pounds per square inch (psi) or in the SI unit of newtons per square meter (N m^{-2}) which is called a *pascal* (Pa):

$$1 \text{ Pa} = 1 \text{ N m}^{-2}$$

The pascal is a very small unit, so *kilopascals* (kPa) are more frequently used. The relationship between kilopascals and the standard atmosphere is

$$1 \text{ atm} = 101.325 \text{ kPa}$$

Although the atmosphere is not an SI unit, it is a convenient unit for many purposes; we will use it often in this book.

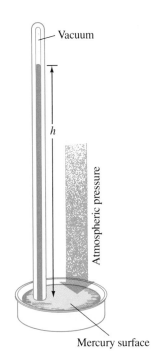

FIGURE 3.7
The Mercury Barometer.

Atmospheric pressure may be measured by the height, h, of the column of mercury. Average atmospheric pressure at sea level supports a column of mercury 760 mm high.

PRESSURE AND VOLUME: BOYLE'S LAW

We are all familiar with the fact that gases are compressible. When the pressure on a certain amount of gas is increased, as in a bicycle tire pump, the volume of the gas decreases; the greater the pressure, the smaller the volume is. In 1660 Irish chemist Robert Boyle (1627–1691) studied the effects of pressure on the volume of air by using the apparatus shown in Figure 3.8. He found that if he doubled the pressure, the volume of the air was halved; if the pressure was

The SI unit of pressure, the pascal (Pa), is named for the French mathematician Blaise Pascal (1623–1662).

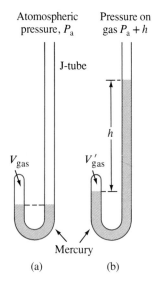

Atomospheric   Pressure on
pressure, P_a gas $P_a + h$

J-tube

h

V_{gas} V'_{gas}

Mercury

(a)   (b)

FIGURE 3.8
Boyle's Apparatus.

Boyle used this apparatus to study the volume and pressure of a gas sample. One end of the J-tube is closed, trapping air at the end of the tube. (a) When the height of the mercury is the same in the open and the closed parts of the tube, the pressure exerted on the gas is equal to the atmospheric pressure. (b) The pressure of the gas is increased by adding mercury to the tube. Then the pressure exerted on the gas is equal to h (the difference in the heights of the two mercury surfaces) plus the atmospheric pressure, and the volume of the gas is smaller.

increased four times, the volume was decreased to one-quarter of its original value. In other words, Boyle found that, in general, the volume, V, of a given mass of air is inversely proportional to its pressure, P, if the temperature, T, is held constant:

$$V \propto \frac{1}{P} \qquad \text{or} \qquad V = \text{Constant} \times \frac{1}{P} \qquad (T \text{ constant})$$

This relationship has been found to hold for *any gas*. It can also be stated in the following form:

Pressure times volume is constant for a given amount of gas at a constant temperature:

$$PV = \text{Constant} \qquad (T \text{ constant})$$

This relationship is known as **Boyle's law.**

To compare the same gas sample at constant temperature under different pressure and volume conditions, we can write Boyle's law conveniently as

$$P_1 V_1 = P_2 V_2 \qquad (T \text{ constant})$$

If the initial pressure and volume of a given quantity of gas are initially P_1 and V_1, and the pressure is changed to P_2, then the new volume, V_2, is given by this relationship. Figure 3.9 shows the relationship between P and V graphically.

Boyle's law expresses quantitatively the important fact that a gas is compressible. The more a gas is compressed, the greater is its pressure. At any altitude the atmosphere is compressed by the mass of gas above it. So the higher the altitude, the less the air is compressed, and the lower its pressure. Thus at 2500 m (8000 ft) in the Rocky Mountains, the pressure is only 0.75 atm, and at 8000 m (26 000 ft) in the Himalayas, the world's highest mountains, the atmospheric pressure is only 0.47 atm.

At high altitudes the amount of oxygen that the body takes in during each breath is considerably decreased. This decreased amount of oxygen makes any exertion very difficult and causes the weakness and headaches known as altitude sickness. Mountain climbers who tackle Everest and other very high peaks must

FIGURE 3.9
Boyle's Law.

Boyle's law gives the relationship between pressure and volume at constant temperature. If the temperature is held constant, the volume of a given amount of gas is inversely proportional to its pressure. (a) A plot of P against V gives a curve (a hyperbola). (b) A plot of $1/P$ against V or $1/V$ against P gives a straight line.

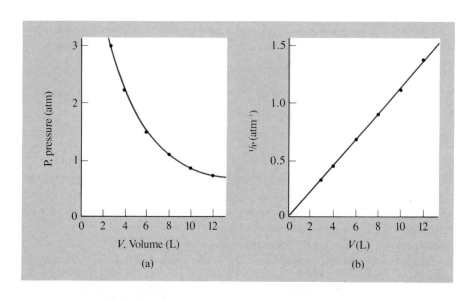

undergo a long period of acclimatization at high altitude to allow the body to adapt to the low oxygen pressure, or they must carry cylinders of oxygen for breathing. For the same reasons jet aircraft, which fly at altitudes up to 10 000 m, must be pressurized and equipped with emergency oxygen in case the pressurization should fail.

≡ *Example 3.3* ══

BOYLE'S LAW

A sample of hydrogen has a volume of 6.20 L at a pressure of 1.05 atm. What is its volume if it is compressed until its pressure is 3.00 atm?

Solution

We will use subscript 1 to denote the original conditions of the hydrogen and subscript 2 to denote the final conditions. Thus we have

$$P_1 = 1.05 \text{ atm} \qquad V_1 = 6.20 \text{ L}$$
$$P_2 = 3.00 \text{ atm} \qquad V_2 = ? \quad \text{(To be found)}$$

We can rearrange Boyle's law,

$$P_1 V_1 = P_2 V_2$$

to obtain an expression for the unknown volume, V_2, as follows:

$$V_2 = \frac{P_1 V_1}{P_2}$$
$$= \frac{1.05 \text{ atm} \times 6.20 \text{ L}}{3.00 \text{ atm}} = 2.17 \text{ L}$$

Thus the sample of hydrogen gas occupies 2.17 L when the pressure is increased to 3.00 atm.

EXERCISE 3.3

What pressure is needed to compress the sample of hydrogen in Example 3.3 to a volume of only 1.00 L?

TEMPERATURE AND VOLUME: CHARLES'S LAW

Studies on the properties of gases continued after Boyle's work, and in 1787 the French scientist Jacques Charles (1746–1823) reported that the volume of a given amount of a gas increased in a linear manner with increasing temperature if the pressure was kept constant. In other words, a plot of the volume against the temperature is a straight line as shown in Figure 3.10. He found that the volume of any sample of a gas increased by 1/273 of its volume at 0 °C for every 1 °C rise in temperature. Conversely, on cooling, the volume of a gas decreased by 1/273 of its volume at 0 °C for every 1 °C decrease in temperature. This leads to the surprising conclusion that its volume would become zero at −273 °C and become *negative* at still lower temperatures. This can be seen by extrapolating the straight-line plots of volume against temperature in Figure

BOX 3.3

THE SCIENTIFIC METHOD: LAWS, HYPOTHESES, AND THEORIES

Boyle's law, Charles's law, and the law of conservation of mass are statements that in each case summarize a large number of observations. The fundamental activity of science is making **observations** of the world around us. If observations are made under carefully controlled conditions, they can be repeated by any other person who has the appropriate equipment. In this way the facts of nature are established. When a large number of observations have been made and a number of facts established, regularities and consistencies in a set of facts may enable one to make a concise statement or give a mathematical equation that summarizes the observed facts. Such a statement is known as a scientific **law**, for example, Boyle's law. Another familiar example is the law of gravity, which summarizes the numerous observations that show that separate masses attract each other.

Once a law has been established, a scientist asks the question: "Why does nature behave in the way that is summarized by the law?" Why, for example, is the volume of a gas inversely proportional to the pressure, as stated by Boyle's law? A tentative answer to such a question is known as a **hypothesis**. If it is to be useful, a hypothesis must suggest new experiments that will either verify or refute the hypothesis. A hypothesis that continually withstands such tests develops into a **theory**. The theory that provides an explanation for Boyle's law is the kinetic molecular theory. A theory is a model of nature that enables us to understand our observations better.

Theories are, however, only tentative. A theory continues to be useful only as long as we fail to find any experimental facts that cannot be accounted for by the theory. But only one fact that the theory cannot explain will cause the theory to be modified or replaced by a new theory. Dalton's atomic theory continues to be very useful today, but it has been modified from its original form in that we no longer believe that an atom is indivisible nor that all the atoms of an element have the same mass.

Observations that have been verified by repeated experiments will never be changed, but the theories invented to explain these observations may well be replaced or at least modified in the future. In this sense the facts are more important than the theories. Thus it is a mistake to believe that if one knows all the laws and theories that are derived from experimental observations one need not know the experimental facts. New theories can only be developed by those who have a wide knowledge of the facts relating to a particular field, particularly those facts that have not been satisfactorily accounted for by existing theories.

The deduction of laws and the development of hypotheses and theories that lead to predictions—which in turn must be tested by experiment, which then lead to new observations—is a continuous and never-ending process. This process is known as the *scientific method*. Although in principle it appears to be a logical process, science does not, in fact, advance in a completely organized and logical manner. Success in scientific research often depends on the ability to observe and interpret the unexpected. The experiment that "does not work" can be the clue to an important discovery. New theories are generally not developed in an entirely logical and planned manner but through a slow and tortuous process in which many incorrect hypotheses may be made before an adequate theory is formulated. Theories may depend as much on flashes of insight and inspiration as on logical argument.

3.10 to zero volume. In every case the straight lines meet the temperature axis at $-273\ °C$. Since we can attach no meaning to a negative volume, we are forced to assume that we cannot obtain a lower temperature than $-273\ °C$, which is therefore called the **absolute zero of temperature**.

THE KELVIN TEMPERATURE SCALE In practice, all gases condense to liquids and solids at temperatures above $-273\ °C$, so no gas can, in fact, be cooled until it has zero volume. But the idea of a lowest possible temperature—that is, an absolute zero of temperature—is exceedingly important. Instead of arbitrarily choosing the melting point of ice as the zero of the temperature scale, as is done on the Celsius scale, we can logically, and also conveniently, choose the absolute zero as the zero of a temperature scale. This choice of zero is the basis of the

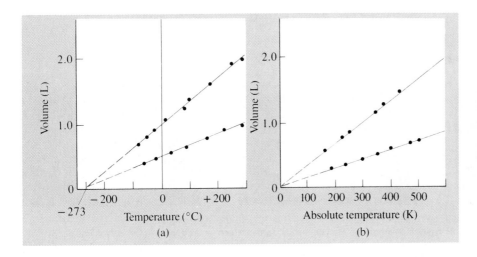

FIGURE 3.10
Temperature and Volume.

(a) For a fixed quantity of gas at constant pressure, a 1 °C change in the temperature alters the volume of the gas by $\frac{1}{273}$ of its volume at 0 °C and the volume extrapolates to zero at −273 °C. The volume of a gas is a linear function of its Celsius temperature, but volume and temperature on the Celsius scale are not directly proportional. (b) The volume of a fixed quantity of gas at constant pressure is directly proportional to the absolute (Kelvin) temperature.

Kelvin temperature scale, first suggested by the British scientist Lord Kelvin (1824–1907). According to accurate measurements, the absolute zero of temperature is −273.15 °C. Thus 0 K = −273.15 °C, and the Kelvin (K) scale is related to the Celsius scale by the expression

$$T = t + 273.15$$

where T is the temperature on the Kelvin scale and t is the temperature on the Celsius scale. In other words, temperatures on the Celsius scale are converted to temperatures on the Kelvin scale simply by adding 273.15. Notice that, by convention, the degree sign (°) is not used when we are expressing temperatures on the Kelvin scale. The unit on the Kelvin scale is the kelvin (K), and a temperature such as 100 K is read as "one hundred kelvins".

CHARLES'S LAW If the temperature is expressed on the Kelvin scale, the volume of a gas is directly proportional to the temperature, which is not true if the temperature is measured on the Celsius scale (Figure 3.10):

$$V \propto T \qquad \text{or} \qquad V = \text{Constant} \times T \qquad (P \text{ constant})$$
$$\frac{V}{T} = \text{Constant}$$

or

$$\frac{V_1}{V_2} = \frac{T_1}{T_2} \qquad (P \text{ constant})$$

where T is the temperature on the Kelvin scale. These expressions summarize **Charles's law**:

At constant pressure the volume of a given mass of gas is proportional to its temperature on the Kelvin scale.

The variation of the volume of a gas with temperature is demonstrated in Experiment 3.4. Hot-air balloons are an interesting application of Charles's law (see Box 3.4).

Experiment 3.4

Liquefaction of Air and Charles's Law

When a balloon is plunged into liquid nitrogen the nitrogen (boiling point −196 °C) and the oxygen (boiling point −183 °C) are both liquefied and the balloon collapses almost completely. In the center we see the collapsed balloon immediately after it has been removed from the liquid nitrogen. The liquid oxygen and nitrogen then rapidly evaporate and, as we see on the right, the balloon slowly expands as its temperature increases and its volume increases in proportion to the absolute temperature (Charles's law).

Example 3.4

CHARLES'S LAW

A sample of nitrogen has a volume of 80.4 mL at 50 °C. What volume will the sample occupy at 0 °C if the pressure remains constant?

Solution

Using subscript 1 for initial conditions and subscript 2 for final conditions, we have

$$t_1 = 50 \text{ °C} \qquad t_2 = 0 \text{ °C}$$
$$V_1 = 80.4 \text{ mL} \qquad V_2 = ?$$

Since the volume–temperature relationship (Charles's law) involves Kelvin scale temperature, we convert t_1 to T_1, rounding off 273.15 to three significant figures:

$$T_1 = 273 + 50 \text{ °C} = 323 \text{ K}$$
$$T_2 = 273 + 0 \text{ °C} = 273 \text{ K}$$

Now we can use Charles's law:

$$\frac{V_1}{V_2} = \frac{T_1}{T_2}$$

Rearranging and substituting the known values gives

$$V_2 = \frac{V_1 T_2}{T_1} = \frac{80.4 \text{ mL} \times 273 \text{ K}}{323 \text{ K}} = 68.0 \text{ mL}$$

EXERCISE 3.4

To what temperature must a sample of nitrogen, with a volume of 900 mL at 25 °C, be cooled in order to reduce its volume to 350 mL?

BOX 3.4

BALLOONS

You may have seen a brightly colored hot-air balloon floating across the sky on a summer day. The sport of hot-air ballooning has revived man's earliest means of getting into the air.

A balloon filled with any gas less dense than air rises in the atmosphere because the mass of the air displaced by the balloon is greater than the mass of the balloon. Since the density of the atmosphere decreases with increasing altitude, the balloon rises until the mass of the balloon and its load equals the mass of the air displaced by the balloon, at which point the balloon floats in the atmosphere. Two French scientists, Jacques Charles and Joseph Gay-Lussac, who made some of the first quantitative studies of the properties of gases, were pioneers in the use of balloons. The first balloon flight was made in France by the Montgolfier brothers in June 1783 using a hot-air balloon. But a few months later in August of the same year Charles filled a balloon with hydrogen which he made by the reaction of about 500 lb of acid and 1000 lb of iron! The balloon remained in the air for 45 min and traveled a distance of about 25 km. In 1803 Joseph Gay-Lussac set a record by ascending to 23 000 ft (7 km). He used balloon ascents to carry out experiments that included studies of the composition of the atmosphere and variations in the earth's magnetic field. Hydrogen-filled balloons have continued to be used for weather observations up to the present day. Because of the very low density of hydrogen they can reach heights of approximately 40 km.

By the 1930s a logical development of the hydrogen balloon, the airship, was providing regular transportation across the Atlantic. Instead of just drifting in the wind, airships were driven by engine-powered propellors and included cabins for passengers. The disastrous fire that destroyed the German airship *Hindenburg* in 1937 marked the end of the airship era, but the possibility of building airships that use helium, which is nonflammable but expensive, has attracted new interest. Although they would be too slow to compete with jets, helium airships might have other uses, such as transporting timber in forestry operations.

STANDARD TEMPERATURE AND PRESSURE (STP)

Since the volume of a gas changes with both pressure and temperature, stating that a certain gas sample has a particular volume is not sufficient; the pressure and the temperature must also be specified in order to characterize it fully. To simplify comparisons the volume of a given sample of gas is normally reported at 0 °C (273.15 K) and 1 atm (101.33 kPa); these conditions are known as **standard temperature and pressure**, abbreviated **STP**.

COMBINED GAS LAW

By combining Boyle's law, which states that at constant temperature $V \propto 1/P$, and Charles's law, which states that at constant pressure $V \propto T$, we obtain a relationship between the volume, temperature, and pressure of a given amount of gas:

$$V \propto \frac{T}{P} \quad \text{or} \quad V = \text{Constant} \times \frac{T}{P}$$

Rearranging this relationship, we find that PV/T = constant for a given mass of gas, or, for two different sets of conditions,

$$\frac{P_1V_1}{T_1} = \frac{P_2V_2}{T_2}$$

This equation is known as the **combined gas law**.

═ *Example 3.5* ═══════════════════════════════════

THE COMBINED GAS LAW

If a sample of oxygen gas has a volume of 425 mL at 70 °C and a pressure of 0.950 atm, what will its volume be at STP?

Solution

We summarize the data given in the statement of the problem as follows:

$$P_1 = 0.950 \text{ atm} \qquad P_2 = 1.00 \text{ atm}$$
$$V_1 = 425 \text{ mL} \qquad V_2 = ? \quad \text{(To be found)}$$
$$T_1 = 343 \text{ K (70 °C)} \qquad T_2 = 273 \text{ K}$$

Because both pressure and temperature are varying, we use the combined gas law,

$$\frac{P_1V_1}{T_1} = \frac{P_2V_2}{T_2}$$

which we can rearrange to obtain an expression for the unknown volume, V_2:

$$V_2 = \frac{P_1V_1T_2}{P_2T_1}$$

Substituting the values given we have

$$V_2 = \frac{0.950 \text{ atm} \times 425 \text{ mL} \times 273 \text{ K}}{1.00 \text{ atm} \times 343 \text{ K}} = 321 \text{ mL}$$

EXERCISE 3.5

A meteorologist fills a weather observation balloon with hydrogen at a pressure of 1.00 atm at 25.0 °C. If the volume of the balloon is 30.0 L when it is on the ground, what will its volume be at an altitude of 15.0 km where the temperature is −50.0 °C and the pressure is 0.100 atm? Assume that the pressure inside the balloon remains equal to the outside pressure.

3.6
Kinetic Molecular Theory of Gases

Because a given mass of a substance occupies much more space as a gas than as a liquid, and because a gas is much more compressible than a liquid, it is reasonable to assume that the molecules in a gas are far apart and that much of a gas consists of empty space. But how can something that is mostly empty space exert a pressure on its surroundings?

We can explain how a gas exerts a pressure and many other properties of a gas by an extension of the atomic theory known as the **kinetic molecular theory**. It assumes that not only do all substances consist of atoms or molecules, but also these atoms and molecules are in a constant state of motion. The kinetic molecular theory for gases is based on the following four assumptions:

1. A gas is composed of molecules that are far apart from each other in comparison with their own dimensions. Most of the volume occupied by a gas is empty space.

2. The gas molecules are in constant random motion. Each molecule continues to move in a straight line, unless it collides with another molecule or with a wall of the container.

3. The molecules exert no force on each other or on the container, except when they collide with each other or with the walls of the container. These collisions are *elastic*; that is, one molecule may gain energy and the other may lose energy, but their total energy remains constant. As a result, the total energy of all the molecules remains constant.

4. The average kinetic energy of the molecules of a gas is proportional to the absolute temperature. The kinetic energy of a molecule is equal to $\frac{1}{2}mv^2$, where m is its mass and v is its speed.

$$K.E. = \frac{1}{2}mv^2$$

EXPLAINING THE GAS LAWS

Let us now see how the kinetic theory enables us to understand the gas laws. According to the kinetic molecular theory, the pressure exerted by a gas on the walls of its container results from the continual bombardment of the walls by the fast-moving molecules. Every time a molecule collides with a wall, it exerts a force on it. Since pressure is force per unit area, the pressure is proportional to the number of collisions per unit area in a given time. Suppose we decrease the volume of a given mass of gas to one-half its original value. There will then be twice as many molecules in any given volume of the gas, and hence they will make twice as many collisions per unit area with the walls of the container. Consequently, the pressure will be doubled (Figure 3.11). Thus the kinetic theory provides a simple explanation of Boyle's law that pressure and volume are inversely proportional.

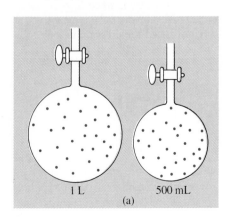

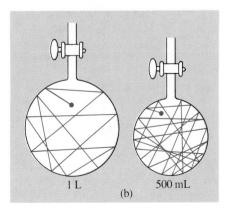

FIGURE 3.11
Volume and Pressure.

(a) If the volume of gas is halved while the temperature and number of molecules are held constant, the molecules are packed closer and the density is doubled. (b) On the average, a molecule now hits the wall twice as often. The total number of impacts with the wall is therefore doubled, which doubles the pressure.

The kinetic theory also explains why the volume of a gas increases with increasing temperature. Suppose that we have a gas confined in a container of variable volume, such as the cylinder with a movable piston shown in Figure 3.12. The atmospheric pressure causes a constant force to act on the piston from above. This force is exactly counterbalanced by that due to the pressure of the gas confined by the piston. A basic assumption of the kinetic theory is that the kinetic energy of the molecules of a gas is proportional to the absolute temperature. If the temperature is increased, the kinetic energy of the molecules increases and so their speeds increase. They then make more frequent collisions with the piston and these collisions have greater energy so they exert a greater force on the piston. Thus the pressure exerted on the piston by the gas increases and is greater than the pressure exerted on the piston by the atmosphere outside. The piston therefore moves outward, increasing the volume of the gas. This increase in volume reduces the number of collisions of the gas molecules with the piston and therefore reduces the pressure. Thus the piston moves outward until the pressure of the gas again equals the atmospheric pressure. The volume of a gas increases as the temperature increases if the pressure is held constant, as is observed experimentally. A more detailed mathematical treatment shows that the volume is directly proportional to the absolute temperature, thus providing an explanation for Charles's law.

KINETIC ENERGY AND TEMPERATURE

The assumption that the average kinetic energy of the molecules of a gas is proportional to the temperature is not so much an assumption as a definition of absolute temperature. It applies not only to gases but also to liquids and solids, even though their molecules are packed more closely together. Suppose that a gas is confined in a container that is initially at a higher temperature than the gas. In the collisions between the slower molecules of the gas and the faster molecules of the container walls, the gas molecules are speeded up and those of the container walls are slowed down. The average kinetic energy of the molecules of the container walls therefore decreases, and the average kinetic energy of the gas molecules increases. Energy flows from the container to the gas; this flow of energy is called **heat**. We say that heat flows from the hot container to the cooler gas. Thus we see that temperature is simply a way of describing the average kinetic energy of the molecules of a substance; it

FIGURE 3.12
Effect of Temperature on Volume.

When the temperature is raised, the molecules move faster and make more collisions and more energetic collisions with the walls of the vessel. The pressure inside the vessel is increased. Since the pressure inside the vessel is greater than the pressure outside, the piston moves outward, increasing the volume and decreasing the pressure in the vessel. The volume continues to increase, reducing the pressure exerted by the molecules in the vessel, until the pressure of the gas is once again equal to the atmospheric pressure. Thus the volume increases with increasing temperature.

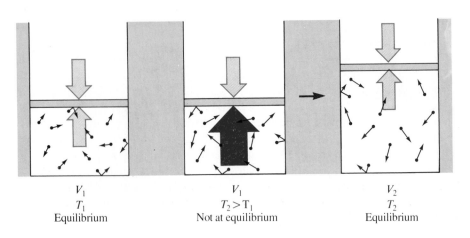

V_1	V_1	V_2
T_1	$T_2 > T_1$	T_2
Equilibrium	Not at equilibrium	Equilibrium

is a property of a substance that is proportional to the kinetic energy of its molecules.

As a gas is cooled the average speed and kinetic energy of its molecules decrease. Thus, the gas will eventually reach a temperature at which the average speed and kinetic energy of its molecules are zero. Since the speed and kinetic energy cannot be decreased further, this temperature must be the lowest possible temperature, that is, the absolute zero, 0 K.

EXPLAINING THE STATES OF MATTER: INTERMOLECULAR FORCES

In fact, no gas can be cooled to 0 K, because all substances become liquids or solids before this temperature is reached. This change of state results from attractive forces that act between all molecules. These forces are called **intermolecular forces**. In a substance that is a gas at ordinary temperature, these forces are relatively weak, and their influence on the motion of the molecules is negligible. But if a gas is cooled to low temperatures, the average speed and kinetic energy of its molecules are greatly reduced. Thus at low temperatures even weak intermolecular forces cause the molecules to stick together to some extent. As a result, they no longer move independently and randomly, as is assumed by the kinetic molecular theory. When the temperature is low enough for a sufficiently large number of molecules to stick together, the gas changes to a liquid.

In a liquid the molecules are packed rather closely, so a liquid has a relatively high density and is not very compressible. But the molecules still have enough kinetic energy to move around each other. When the temperature is lowered still further, the molecules no longer have sufficient energy to be able to jostle past each other. Each one becomes trapped in a hole formed by the surrounding molecules, and it can only rotate and oscillate about a fixed mean position. The molecules then usually take up the regular ordered arrangement that is characteristic of most solids.

The fact that all substances become liquids or solids at sufficiently low temperatures shows that there must be attractive forces between all molecules. Thus the assumption of the kinetic molecular theory of gases that there are no forces between molecules is not correct, but it is valid to a very good approximation for substances that are gases at ordinary temperatures. For substances that are gases even at low temperatures, such as O_2, H_2, N_2, and Ar, the intermolecular forces must be very weak. In contrast, a substance such as sulfur, which consists of S_8 molecules and is a solid at room temperature, melting at 112.8 °C, must have relatively strong intermolecular forces.

Because the volume of the molecules of a gas is very small compared with the space that they occupy, and because the intermolecular forces are almost negligible, the properties of gases are essentially independent of the nature of the molecules. In other words, all gases behave in the same way. They all obey the combined gas law, to a reasonable approximation. In contrast, in a solid and in a liquid the molecules are held close together by intermolecular forces. The properties of solids and liquids therefore depend very much on the sizes and shapes of the molecules of which they are composed and on the strength of the intermolecular forces. Thus the properties of liquids and solids cannot be described, even approximately, by a single equation like the combined gas law.

3.7
The Ideal Gas Law

We have seen that the pressure of a gas is proportional to the number of collisions of gas molecules per unit area of the vessel wall in a given time. If the volume and the temperature of a vessel are constant, the number of collisions per unit area in a given time is proportional to the number of molecules. Hence the pressure is proportional to the number of molecules. Since 1 mol of molecules always contains the same number of molecules (6.022×10^{23}), it follows that *the pressure is proportional to the number of moles of molecules, n,* or

$$P \propto n \quad \text{(at constant } V \text{ and } T) \quad \text{or} \quad \frac{P}{n} = \text{Constant} \quad \text{(at constant } V \text{ and } T)$$

In previous sections we have seen that

$$PV = \text{Constant} \quad \text{(at constant } T)$$

$$\frac{V}{T} = \text{Constant} \quad \text{(at constant } P)$$

Combining these relationships gives us the equation

$$\frac{PV}{nT} = \text{Constant} \quad \text{or} \quad PV = nRT$$

where R is a constant called the **gas constant**. If pressure is expressed in atmospheres, volume in liters, and temperature in kelvins, R is found to have the value 0.0821 atm L mol^{-1} K^{-1}. If the pressure, volume, and temperature are expressed in SI units, the value of R is 8.31 kPa dm^3 mol^{-1} K^{-1}.

This equation is known as the **ideal gas equation**, or the **ideal gas law**. It is obeyed closely, but not exactly, by nearly all gases. Different gases differ slightly in their behavior because the size of the molecules is not completely negligible compared with the distances between them and because there are intermolecular forces. A hypothetical gas in which the molecules have zero volume and in which there are no intermolecular forces would obey the ideal gas equation exactly and is known as an **ideal gas**. For the calculations in this book we may assume that real gases behave like an ideal gas.

If we know any three of the variables P, V, n, and T, which describe the physical state of a gas, we may calculate the fourth by using the ideal gas equation.

—— *Example 3.6* ——————————————

THE IDEAL GAS LAW

A bulb with a volume of 100 cm^3 is filled with hydrogen at a temperature of 300 K and a pressure of 750 mm Hg. How many moles of hydrogen are there in the bulb?

Solution

Either of the two values of the gas constant may be used provided that appropriate units are used for P, V, and T. If we use $R = 0.0821$ atm L mol^{-1} K^{-1}, then we convert the units of P to atmospheres and the units of V to liters:

$$V = (100 \text{ cm}^3)\left(\frac{1 \text{ L}}{1000 \text{ cm}^3}\right) = 0.100 \text{ L}$$

$$P = (750 \text{ mm Hg})\left(\frac{1 \text{ atm}}{760 \text{ mm Hg}}\right) = 0.987 \text{ atm}$$

$$T = 300 \text{ K}$$

Rearranging the ideal gas law, we obtain

$$n = \frac{PV}{RT} = \frac{0.987 \text{ atm} \times 0.100 \text{ L}}{0.0821 \text{ atm L mol}^{-1} \text{ K}^{-1} \times 300 \text{ K}} = 4.01 \times 10^{-3} \text{ mol}$$

The calculation is only slightly different if we use SI units. In this case $R = 8.31 \text{ kPa dm}^3 \text{ mol}^{-1} \text{ K}^{-1}$:

$$V = (100 \text{ cm}^3)\left(\frac{1 \text{ dm}}{10 \text{ cm}}\right)^3 = 0.100 \text{ dm}^3$$

$$P = (750 \text{ mm Hg})\left(\frac{101.3 \text{ kPa}}{760 \text{ mm Hg}}\right) = 100 \text{ kPa}$$

$$T = 300 \text{ K}$$

$$n = \frac{PV}{RT} = \frac{100 \text{ kPa} \times 0.100 \text{ dm}^3}{8.31 \text{ kPa dm}^3 \text{ mol}^{-1} \text{ K}^{-1} \times 300 \text{ K}} = 4.01 \times 10^{-3} \text{ mol}$$

Example 3.7

THE IDEAL GAS LAW

Calculate the pressure exerted by 3.00 g of N_2 gas in a container of volume 2.00 L at a temperature of $-23 \degree C$.

Solution

The volume is 2.00 L. The temperature given must be converted to kelvins:

$$T = 273 + t = 273 + (-23) = 250 \text{ K}$$

The mass of N_2 must be converted to moles of N_2, using its molar mass of 28.02 g:

$$n = (3.00 \text{ g N}_2)\left(\frac{1 \text{ mol N}_2}{28.02 \text{ g N}_2}\right) = 0.107 \text{ mol N}_2$$

Now we can obtain the value for the pressure P from the ideal gas law:

$$P = \frac{nRT}{V} = \frac{0.107 \text{ mol} \times 0.0821 \text{ L atm mol}^{-1} \text{ K}^{-1} \times 250 \text{ K}}{2.00 \text{ L}} = 1.10 \text{ atm}$$

EXERCISE 3.6

Jacques Charles made the first balloon flight in which hydrogen rather than hot air was used. The balloon was filled with approximately 2.5 kg of hydrogen. Assuming that the temperature was 20 °C, and the pressure 755 mm Hg, what was the approximate volume of the balloon? What was its approximate diameter? (Volume of a sphere = $\frac{4}{3}\pi r^3$.)

DETERMINATION OF MOLAR MASS

As shown in Example 3.6, we can use the ideal gas law to find the number of moles of a gas if we can measure V, P, and T. If we also know the mass of the gas, we can calculate its molar mass. The **molar mass**, M, is by definition

$$M = \frac{\text{Mass of gas}}{\text{Number of moles}} = \frac{m}{n}$$

Thus we can deduce the molar mass, M, of a gas if we know P, V and T for a sample with a mass, m. A simple procedure for finding the molar mass is as follows:

1. Weigh an evacuated flask of known volume.
2. Fill it, at a known temperature and pressure, with the gas whose molar mass is to be determined; weigh the flask again.
3. Calculate the increase in the mass of the flask; this increase equals the mass of the gas.

Example 3.8

MOLAR MASS FROM P, V AND T

A flask of 0.300-L volume was weighed after it had been evacuated. It was then filled with a gas of unknown molar mass, at 1.00 atm pressure and a temperature of 300 K. The increase in the mass of the flask was 0.977 g.

(a) What is the molar mass of the gas?

(b) Assuming that the gas is a compound of sulfur and oxygen only, what is the molecular formula of the gas?

Solution

(a) We first find the number of moles of the gas:

$$n = \frac{PV}{RT} = \frac{1.00 \text{ atm} \times 0.300 \text{ L}}{0.0821 \text{ atm L mol}^{-1} \text{ K}^{-1} \times 300 \text{ K}} = 0.0122 \text{ mol}$$

Then we can find the molar mass M:

$$M = \frac{m}{n} = \frac{0.977 \text{ g}}{0.0122 \text{ mol}} = 80.1 \text{ g mol}^{-1}$$

Thus the molar mass is 80.1 g mol^{-1}.

(b) By trial and error we find that the molecular formula could be either SO_3 ($M = 80.06$ g mol^{-1}) or S_2O ($M = 80.12$ g mol^{-1}). Any other combinations of sulfur and oxygen have molar masses quite different from 80 g.

EXERCISE 3.7

Chlorofluorocarbons (CFCs) are gases that are widely used in refrigerators and spray cans. As their name implies they are compounds of carbon, fluorine, and chlorine. An evacuated 500-mL flask was filled with 2.50 g of a chlorofluorocarbon at 22 °C and a pressure of 1.02 atm. (a) What is the molar mass of the chlorofluorocarbon? (b) Assuming that the molecule contains only one carbon atom, what is the molecular formula of this compound?

For a substance that is not a gas at room temperature, we can find the molar mass by a modification of the method just described if the substance has a relatively low boiling point, for example, below 100 °C. A flask containing only air is weighed. A sample of the liquid to be studied is added, and the flask is heated, usually in a bath of boiling water (see Figure 3.13). Heating is continued until all the liquid has evaporated and the large amount of vapor thus formed has completely displaced all the air from the flask. The flask then contains only the vapor of the substance under investigation at 100 °C and atmospheric pressure.

The flask is then closed to the atmosphere and cooled to room temperature. The vapor condenses to a liquid, its mass remaining unchanged. The pressure in the flask is now low, and when we open the flask, air rushes in until the pressure in the flask is again atmospheric.

The flask is then reweighed. The difference between the mass of the flask now and its original mass is the mass of the condensed vapor, which is the mass of the vapor that filled the flask at 100 °C and atmospheric pressure. We thus have all the information required to calculate the molar mass, and if we also know the empirical formula we can find the molecular formula as well, as the following example shows.

= *Example 3.9* =

MOLAR MASS AND MOLECULAR FORMULA FROM *P, V, T,* AND EMPIRICAL FORMULA

Benzene, a colorless liquid, is another example of the large class of compounds containing only carbon and hydrogen that are called hydrocarbons. Benzene has the empirical formula CH. A flask of volume 247.2 cm³ had a mass of 25.201 g when it contained only air. A sample of benzene was added to the flask and heated to 100 °C. The benzene vaporized and drove all the air from the flask. The flask was then cooled to room temperature, opened to the atmosphere, and weighed. It had a mass of 25.817 g. The barometric pressure was 742 mm Hg. Calculate the molar mass and the molecular formula of benzene.

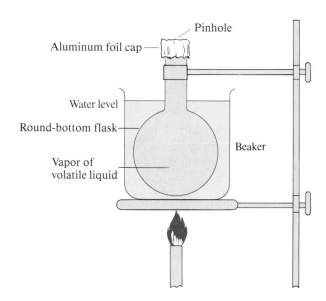

Pinhole

Aluminum foil cap

Water level

Round-bottom flask

Vapor of
volatile liquid

Beaker

FIGURE 3.13
Determination of the Molar Mass of a Volatile Liquid.

A small amount of a volatile liquid is placed in the flask. The flask is heated in a bath of boiling water until all the liquid has evaporated and its vapor has completely filled the flask, driving out all the air that was originally in the flask. From the volume of the flask and the mass of the vapor filling the flask the molar mass of the liquid can be calculated.

Solution

From the values of P, V, and T we can find the number of moles of benzene, n. From the mass of the flask before and after it was filled with benzene vapor we can find the mass of benzene, m. Hence we can find its molar mass $M = m/n$. First we find the number of moles of benzene in the flask. We convert the information given to appropriate units:

$$V = (247.2 \text{ cm}^3)\left(\frac{1 \text{ L}}{1000 \text{ cm}^3}\right) = 0.2472 \text{ L}$$

$$T = (273 + 100) \text{ K} = 373 \text{ K}$$

$$P = (742 \text{ mm Hg})\left(\frac{1.00 \text{ atm}}{760 \text{ mm Hg}}\right) = 0.976 \text{ atm}$$

$$n = \frac{PV}{RT} = \frac{0.976 \text{ atm} \times 0.2472 \text{ L}}{0.0821 \text{ atm L mol}^{-1} \text{ K}^{-1} \times 373 \text{ K}} = 7.88 \times 10^{-3} \text{ mol}$$

Mass of benzene = (Mass of flask + air + condensed vapor) − (Mass of flask + air)

$$= 25.817 \text{ g} - 25.201 \text{ g} = 0.616 \text{ g}$$

Thus the molar mass is

$$M = \frac{m}{n} = \frac{0.616 \text{ g}}{7.88 \times 10^{-3} \text{ mol}} = 78.2 \text{ g mol}^{-1}$$

The empirical formula CH corresponds to a molar mass of $(12.01 + 1.008)$ g mol^{-1} = 13.02 g mol^{-1}. The experimentally determined molar mass is six times larger:

$$\frac{78.2 \text{ g mol}^{-1}}{13.02 \text{ g mol}^{-1}} = 6.01$$

Hence the molecular formula of benzene must be C_6H_6.

EXERCISE 3.8

The molar mass of chloroform was determined by the method described in Example 3.9. It was found that 0.495 g of chloroform occupied a flask of volume 127 cm^3 at 100 °C. The atmospheric pressure was 753 mm Hg. What is the molar mass of chloroform? If the empirical formula of chloroform is CHCl$_3$, what is its molecular formula?

GAS DENSITY

We have seen that the determination of the molar mass of a gaseous substance depends on finding the mass of gas that occupies a known volume at a known P and T. Since density = mass/volume, or $d = m/V$, we can find the molar mass of a gaseous substance if we know its density, pressure, and temperature.

Using the relationship

$$\text{Number of moles} = \frac{\text{Mass}}{\text{Molar mass}} \quad \text{or} \quad n = \frac{m}{M}$$

we may express the ideal gas law, $PV = nRT$, in the form

$$PV = \frac{m}{M}RT \quad \text{or} \quad P = \frac{m}{VM}RT$$

Now since

$$d = \frac{m}{V}$$

we have

$$P = \frac{d}{M} RT \quad \text{or} \quad M = \frac{dRT}{P}$$

Using this equation, we can find the molar mass M of a gaseous substance from its density at a given temperature and pressure. Or in general, we may find any one of the variables P, M, T, and d if we know the other three. Although you could memorize this equation it is not necessary to do so because you can easily derive it from the three important relationships that you already know, namely

$$PV = nRT \quad n = m/M \quad d = m/V$$

Example 3.10

GAS DENSITY AND MOLAR MASS

The density of a gas was found to be 2.06 g L^{-1} at STP. What is its molar mass?

Solution

At STP

$$P = 1.00 \text{ atm} \quad T = 273 \text{ K}$$

Substituting in the equation $M = dRT/P$, we have

$$M = \frac{(2.06 \text{ g L}^{-1})(0.0821 \text{ atm L mol}^{-1} \text{ K}^{-1})(273 \text{ K})}{1.00 \text{ atm}} = 46.2 \text{ g mol}^{-1}$$

EXERCISE 3.9

Ethene is an important product of the petrochemical industry that is used, for example, for the manufacture of polyethylene and poly(vinyl chloride) (PVC). Ethene has a density of 1.117 g L^{-1} at 25 °C and a pressure of 740 mm Hg. What is the molar mass of ethene? The empirical formula of ethene is CH_2. What is its molecular formula?

MOLAR VOLUME OF A GAS: AVOGADRO'S LAW

If we write the ideal gas equation in the form

$$V = \frac{nRT}{P}$$

we see that at the same temperature and pressure a given number of moles of any gas has the same volume. In other words

The volume of a given sample of a gas is proportional to the number of moles in that sample at constant pressure and temperature.

$$V \propto n \quad \text{or} \quad V = \text{Constant} \times n \quad (\text{Constant } P \text{ and } T)$$

BOX 3.5

HOW DID THE EARLY CHEMISTS FIND THE CORRECT FORMULAS FOR SUBSTANCES?

Today we accept without question that the formula of water is H_2O. Modern experimental techniques enable us to measure the properties of individual water molecules and find not only that there are two H atoms and one O atom, but also the distances between the atoms. But a century ago none of these techniques was available and chemists had to use other less direct methods for determining the formulas of molecules. Moreover it was necessary to have the correct formula of a compound before its mass composition could be used to determine an atomic mass. Consequently it is not surprising that in the early days of chemistry there was much confusion concerning the correct formulas of substances and correct atomic masses. Dalton assumed that elements consisted of single atoms such as H and O and that atoms normally combined in a 1:1 ratio. So he assumed, for example, that the formula for water was HO. These assumptions, which we now know to be incorrect, in turn led to an incorrect atomic mass for oxygen. It was not until 1860 that much of this confusion was cleared up by the Italian chemist Cannizzaro who succeeded in convincing other chemists that the idea proposed by Avogadro in 1811 provided a solution to the problem of determining correct formulas and atomic masses.

Avogadro put forward what was then called his *hypothesis* and which he stated in the form

Equal volumes of gases at the same temperature and pressure contain equal numbers of molecules

in order to explain the experimental observations of the French chemist Joseph Gay-Lussac (1778–1850) which can be summarized in the statement

When gases combine at constant temperature and pressure, the volumes of the reactants and of the gaseous products are always in the ratio of small whole numbers.

This statement is known as the *law of combining volumes.* Gay-Lussac found for example that

2 volumes hydrogen + 1 volume oxygen $\longrightarrow$

2 volumes water vapor

1 volume hydrogen + 1 volume chlorine $\longrightarrow$

2 volumes hydrogen chloride

2 volumes carbon monoxide + 1 volume oxygen $\longrightarrow$

2 volumes carbon dioxide

These statements are usually known as **Avogadro's law** (Box 3.5). The volume of 1 mol of an ideal gas at STP is given by

$$V = \frac{nRT}{P} = \frac{1 \text{ mol} \times 0.082\,06 \text{ atm L mol}^{-1} \text{ K}^{-1} \times 273.15 \text{ K}}{1 \text{ atm}} = 22.41 \text{ L}$$

The volume of 1 mol of a substance is called its *molar volume.* Thus the molar volume of an ideal gas is 22.41 L at STP. Most gases obey the ideal gas equation closely, but not exactly, so the molar volume of most gases at STP is approximately, but not exactly, equal to 22.41 L, as we can see in Table 3.5. For most purposes it is a good approximation to take the molar volume of a gas at STP to be 22.4 L.

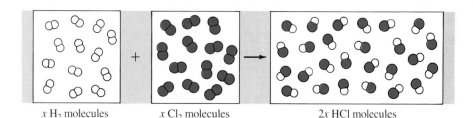

| x H$_2$ molecules | x Cl$_2$ molecules | $2x$ HCl molecules |

These results were difficult to understand in terms of Dalton's assumptions concerning formulas because he wrote, for example,

$$H + Cl \longrightarrow HCl$$

which in terms of Avogadro's hypothesis would imply

$$1 \text{ volume} + 1 \text{ volume} \longrightarrow 1 \text{ volume}$$

whereas Gay-Lussac had found

$$1 \text{ volume} + 1 \text{ volume} \longrightarrow 2 \text{ volumes}$$

This in turn would seem to indicate that hydrogen atoms and chlorine atoms were split in half on forming HCl! Avogadro's brilliant solution to this problem was to postulate, as we now know correctly, that the common gaseous elements such as chlorine, hydrogen, and oxygen had diatomic molecules. Then of course it is easy to understand Gay-Lussac's result as we can write

$$H_2 \quad + \quad Cl_2 \quad \longrightarrow \quad 2HCl$$

| 1 molecule | 1 molecule | 2 molecules |
| 1 volume | 1 volume | 2 volumes |

Similarly we can conclude that the formula of water must be H$_2$O. Gay-Lussac's experiments showed that

$$\text{Hydrogen} + \text{Oxygen} \longrightarrow \text{Water vapor}$$

| 2 volumes | 1 volume | 2 volumes |

According to Avogadro's law,

$$2 \text{ molecules} + 1 \text{ molecule} \longrightarrow 2 \text{ molecules}$$

Then if the formulas of hydrogen and oxygen are H$_2$ and O$_2$, we can see that the formula of water must be H$_2$O:

$$2H_2 + O_2 \longrightarrow 2H_2O$$

However, Dalton was unable to accept that atoms of the same kind might combine to form molecules. So he rejected Avogadro's ideas, and he tried to show that Gay-Lussac's experiments were not accurate. Not until 1860, almost fifty years later, did Avogadro's countryman Stanislao Cannizzaro succeed in convincing chemists of the correctness of Avogadro's ideas. He was able to deduce correct formulas and a consistent and reasonably correct table of atomic masses, which did much to clear up the confusion reigning at the time. This in turn paved the way for another important step forward in chemistry: the formulation of the periodic table by Mendeleev, which we describe in the next chapter.

Using Avogadro's law we can work with volumes of reactants rather than with moles of reactants if we are dealing with gases. For example, hydrogen reacts with chlorine according to the equation

	H$_2$	+	Cl$_2$	$\longrightarrow$	2HCl
	1 mol		1 mol		2 mol
or	22.4 L		22.4 L		44.8 L
In general	1 volume		1 volume		2 volumes
For example	1.0 L		1.0 L		2.0 L

TABLE 3.5
Molar Volumes of Some Gases

Gas	Molar Volume at STP ($L\ mol^{-1}$)
Hydrogen, H_2	22.43
Neon, Ne	22.44
Oxygen, O_2	22.39
Nitrogen, N_2	22.40
Carbon dioxide, CO_2	22.26

≡ *Example 3.11*

REACTIONS AND AVOGADRO'S LAW

The gas propane, C_3H_8, burns in oxygen to give carbon dioxide and water. Write the balanced equation for the reaction. How many liters of carbon dioxide are formed at 25 °C and 1 atm if 5.00 L of propane at 25 °C and 1 atm is burned in oxygen?

Solution

The balanced equation is

$$C_3H_8 + 5O_2 \longrightarrow 3CO_2 + 4H_2O$$

We see from the equation that 1 L of C_3H_8 gives 3 L of CO_2. Hence

$$V_{CO_2} = (5.00\ L\ C_3H_8)\left(\frac{3\ L\ CO_2}{1\ L\ C_3H_8}\right) = 15.0\ L\ CO_2$$

≡ *Example 3.12*

REACTIONS AND AVOGADRO'S LAW

How many liters of carbon dioxide are formed at 1.00 atm and 900 °C if 5.00 L of propane at 10.0 atm and 25 °C is burned in excess air?

Solution

In this case we need to take account of the fact that the carbon dioxide is not formed at the same temperature and pressure as the propane. We begin by finding the volume that the propane would occupy at 900 °C and 1.00 atm pressure, the conditions cited for the formation of carbon dioxide. We convert the temperatures given to kelvins:

$$25\ °C = 273 + 25 = 298\ K \qquad (T_1)$$
$$900\ °C = 273 + 900 = 1173\ K \qquad (T_2)$$

Now we use the combined gas law to find the volume of the propane:

$$V_2 = \frac{V_1 P_1 T_2}{P_2 T_1}$$

$$V(C_3H_8, 900\ °C, 1\ atm) = 5.00\ L \times \frac{10.00\ atm}{1.00\ atm} \times \frac{1173\ K}{298\ K} = 197\ L$$

The equation for the reaction (Example 3.11) indicates that 1 L of C_3H_8 gives 3 L of CO_2. Hence

$$V_{CO_2} = (197 \text{ L } C_3H_8)\left(\frac{3 \text{ L } CO_2}{1 \text{ L } C_3H_8}\right) = 591 \text{ L } CO_2$$

EXERCISE 3.10

At 500 °C, and in the presence of a catalyst, steam reacts with carbon to give a mixture of hydrogen and carbon monoxide. What is the total volume of products that would be obtained from 150 L of steam at 500 °C and 0.95 atm? What would be the volume of the products at STP?

STOICHIOMETRY OF REACTIONS INVOLVING GASES

In many reactions we are concerned with solids and liquids as well as gases. In calculations involving such reactions we work in moles, converting volumes of gaseous reactants to moles and moles of gaseous products to volumes as necessary.

═ Example 3.13 ═

GAS LAWS AND STOICHIOMETRY

What volume of oxygen, O_2, at STP can be obtained by heating 10.00 g of potassium chlorate, $KClO_3$, in the presence of manganese dioxide, MnO_2, as a catalyst?

Solution

The equation for the reaction is

$$2KClO_3 \longrightarrow 2KCl + 3O_2$$
$$\text{2 mol} \qquad\qquad\qquad \text{3 mol}$$

Because 1 mol of an ideal gas occupies 22.4 L at STP, we can also write

$$2KClO_3 \longrightarrow 2KCl + 3O_2$$
$$\text{2 mol} \qquad\qquad\qquad 3 \times 22.4 \text{ L}$$

Therefore

$$V = (10.00 \text{ g } KClO_3)\left(\frac{1 \text{ mol } KClO_3}{122.5 \text{ g } KClO_3}\right)\left(\frac{3 \times 22.4 \text{ L } O_2}{2 \text{ mol } KClO_3}\right) = 2.74 \text{ L } O_2$$

In the above example we were asked to find the volume of oxygen at STP so we were able to use the fact that 1 mol of a gas has a volume of 22.4 L at STP. In the following example the gases are not at STP so we will have to use the ideal gas equation to convert from moles to volumes.

$=$ *Example 3.14* $=$

GAS LAWS AND STOICHIOMETRY

In the production of water gas, which is used as a fuel, coke is heated in steam. What volume of water gas at a temperature of 20 °C and 100 atm pressure can be obtained by heating 1.00 metric ton (10^6 g) of coke in steam at 1000 °C?

Solution

The equation for the reaction is

$$C + H_2O \longrightarrow CO + H_2$$

Thus 1 mol of C gives 1 mol of CO and 1 mol of H_2, that is, a total of 2 mol of gas (water gas). Hence 1.00 metric ton C gives

$$10^6 \text{ g C} \left(\frac{1 \text{ mol C}}{12.01 \text{ g C}}\right)\left(\frac{2 \text{ mol gas}}{1 \text{ mol C}}\right) = 1.67 \times 10^5 \text{ mol gas}$$

Now we can use the ideal gas equation to calculate the volume of the gas:

$$V = \frac{nRT}{P} = \frac{(1.67 \times 10^5 \text{ mol})(0.0821 \text{ atm L mol}^{-1} \text{ K}^{-1})(293 \text{ K})}{100 \text{ atm}}$$

$$= 4.02 \times 10^4 \text{ L water gas}$$

EXERCISE 3.11

What volume of hydrogen at 26 °C and 740 mmHg would be obtained from the reaction of 2.00 g of zinc with excess hydrochloric acid, HCl(aq)?

DALTON'S LAW OF PARTIAL PRESSURES

In a mixture of gases that do not react with each other, each molecule moves independently, just as it would in the absence of molecules of other kinds. Each gas distributes itself uniformly throughout the container. The molecules strike the walls with the same frequency and force—therefore exert the same pressure—as they do when no other gas is present. In other words, the pressure exerted by any one gas in a mixture is the same as it would be if the gas occupied the container by itself. This pressure is called the **partial pressure** of the gas. Furthermore

In a mixture of gases the total pressure exerted is the sum of the pressures that each gas would exert if it were present alone under the same conditions.

Thus the total pressure of a gas mixture is given by

$$P_{\text{total}} = p_1 + p_2 + p_3 + \cdots$$

where p_1, p_2, and so on represent the partial pressures of each gas in the mixture. This law was formulated by John Dalton in 1803 and is known as **Dalton's law of partial pressures**.

We can apply the ideal gas equation to any particular gas in the mixture. Thus for gas 2, for example, $p_2V = n_2RT$, where V is the *total* volume of the gas mixture. And in general, for gas i, we have

$$p_iV = n_iRT$$

We see that the partial pressure of a gas is proportional to the number of molecules of the gas in the mixture, or $p_i \propto n_i$. For example, in the atmosphere 78% of the molecules are nitrogen, 21% are oxygen, and 1% are argon. If the total pressure is 1.00 atm, the nitrogen has a partial pressure of 0.78 atm, the oxygen has a partial pressure of 0.21 atm, and the argon has a partial pressure of 0.01 atm.

― *Example 3.15* ═══════════════════════════════════════

PARTIAL PRESSURES

A 2.00-L flask contains 3.00 g of carbon dioxide, CO_2, and 0.10 g of helium, He, and it has a temperature of 17 °C. What are the partial pressures of CO_2 and He? What is the total pressure exerted by the gas mixture?

Solution

According to Dalton's law, we can treat each gas as if it alone occupied the container. First, we apply the ideal gas law to CO_2 to find its partial pressure. To do so, we need to find n, the number of moles of CO_2:

$$n_{CO_2} = (3.00 \text{ g } CO_2)\left(\frac{1 \text{ mol } CO_2}{44.01 \text{ g } CO_2}\right) = 0.0682 \text{ mol } CO_2$$

Now we use the ideal gas law:

$$p_{CO_2} = \frac{n_{CO_2}RT}{V} = \frac{(0.0682 \text{ mol})(0.0821 \text{ atm L mol}^{-1} \text{ K}^{-1})(290 \text{ K})}{2.00 \text{ L}}$$

$$= 0.812 \text{ atm}$$

Similarly,

$$n_{He} = (0.10 \text{ g He})\left(\frac{1 \text{ mol He}}{4.003 \text{ g He}}\right) = 0.025 \text{ mol He}$$

$$p_{He} = \frac{(0.025 \text{ mol})(0.0821 \text{ atm L mol}^{-1} \text{ K}^{-1})(290 \text{ K})}{2.00 \text{ L}} = 0.30 \text{ atm}$$

Thus the partial pressures of CO_2 and He are 0.81 atm and 0.30 atm.
    The total pressure is given by

$$P_{total} = p_{CO_2} + p_{He} = 0.81 \text{ atm} + 0.30 \text{ atm} = 1.11 \text{ atm}$$

EXERCISE 3.12

A scuba diving tank is filled by pumping in 40.0 L of O_2 and 11.0 L of He at 25 °C and 1.00 atm. If the volume of the tank is 4.8 L, what is the partial pressure of each gas, and the total pressure, at 25 °C?

3.8
Diffusion and Effusion

We see in Experiment 3.5 that if we place a drop of bromine in the bottom of a tube the bromine evaporates forming a brown gas (vapor). The brown bromine gas slowly moves up the tube as the bromine molecules mix with the air molecules. Similarly, we smell substances such as perfume or coffee because molecules of the perfume or coffee mix with other molecules in the air and eventually reach our noses. This mixing of one gas with another is called **diffusion**, and it illustrates that the molecules of a gas are in motion, as assumed by kinetic theory.

A closely related phenomenon is called **effusion**. Effusion is the process by which a gas escapes from a container through a very small opening. Like diffusion, effusion results from the random motions of the molecules of a gas. For example, a toy balloon gradually deflates because the apparently solid rubber skin of the balloon actually contains very small holes through which the gas in the balloon escapes. Because the pressure inside the balloon is greater than the

Experiment 3.5

Diffusion of Bromine Vapor

A few drops of bromine are placed at the bottom of a tube containing air at atmospheric pressure.

After 12 s the bromine molecules have diffused a short distance up the tube. Although the bromine molecules are moving at speeds of several hundred meters per second their constant collisions with nitrogen and oxygen molecules of the air cause them constantly to change direction so that their overall movement up the tube is rather slow.

After 1 min 30 s some of the bromine molecules have now traveled about half way up the visible portion of the tube.

pressure of the atmosphere outside, more molecules pass out through the hole than pass into the balloon from the air outside, so the pressure in the balloon slowly decreases and the balloon deflates.

MOLECULAR SPEEDS

How fast do gases diffuse or effuse? If we open a bottle of ammonia, several seconds pass before the smell reaches our noses even if we are only about a meter away.

The rate of diffusion of a gas will clearly depend on how fast the molecules of the gas are moving. The kinetic theory tells us that if the temperature is constant, the average energy and the average speed of the molecules are also constant. But since the molecules of a gas are moving randomly and colliding with each other, they cannot all have the same speed or the same kinetic energy. When one molecule collides with another, generally one will be speeded up and the other slowed down. Thus the speeds and the energies of individual molecules change constantly, and they vary over a wide range.

The distribution of energies and speeds can be obtained from experimental measurements, as shown in Figure 3.14, or from a detailed treatment of kinetic theory. The distributions of molecular speeds in oxygen at 25 °C and at 1000 °C are shown in Figure 3.15. Notice that many molecules have speeds that are close to the average value, but some have very low speeds and a few have very high speeds.

The average kinetic energy ($\frac{1}{2}mv^2$) of the molecules of a gas depends only on the temperature; the average kinetic energy of all gases is the same at the same temperature. If two gases at the same temperature had *different* average kinetic

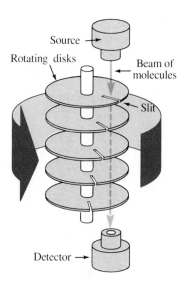

FIGURE 3.14
Determination of Molecular Speeds.

Molecular speeds can be measured by passing a beam of molecules through a velocity selector. This device consists of a series of disks mounted on a common axis. Each disk has a slit, and each slit is displaced by the same angle from the preceding slit. When the disks are rotated, the slits arrive successively in the same angular position. If a molecule moves the distance between the disks in the same time that it takes the next disk to arrive in the same angular position, the molecule will pass through each slit and arrive at the detector. Molecules moving at all other speeds will not arrive at a disk simultaneously with a slit and will therefore not pass through. From the speed of rotation of the disks, the speed of the molecules arriving at the detector can be calculated. Then if the speed of rotation of the disks is varied, the number of molecules arriving at the detector for each different speed can be found. Curves such as those in Figures 3.15 and 3.16 can then be plotted.

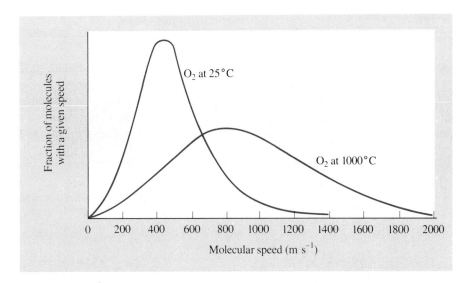

FIGURE 3.15
Molecular Speeds of Oxygen.

Molecules have a wide range of speeds, but a large fraction have speeds close to the average. The distribution of speeds changes if the temperature changes. At 1000 °C the average speed of oxygen molecules is greater than the average speed at 25 °C, and the speeds have a wider distribution.

energies, then when we mixed the gases, collisions between the molecules of the two gases would increase the average kinetic energy of one gas and decrease the average kinetic energy of the other gas. In other words, energy (heat) would flow from one gas to the other. However, we know that no heat flows if two gases are at the same temperature. Hence two gases at the same temperature must have the same average kinetic energy, or

$$\tfrac{1}{2}m_1 v_1^{\,2} = \tfrac{1}{2}m_2 v_2^{\,2}$$

Therefore

$$\frac{v_1}{v_2} = \sqrt{\frac{m_2}{m_1}}$$

The average speed of the molecules of a gas is inversely proportional to the square root of the molecular mass. Molecules with a low molecular mass have higher average speeds than those with a high molecular mass, as Figure 3.16 illustrates.

═ Example 3.16 ═

MOLECULAR SPEEDS

The average speed of oxygen molecules at 25 °C is 450 m s^{-1}. What is the average speed of ammonia molecules at 25 °C?

Solution

We can write

$$\frac{v_{\mathrm{NH_3}}}{v_{\mathrm{O_2}}} = \sqrt{\frac{m_{\mathrm{O_2}}}{m_{\mathrm{NH_3}}}}$$

and so

$$v_{\mathrm{NH_3}} = v_{\mathrm{O_2}}\sqrt{\frac{m_{\mathrm{O_2}}}{m_{\mathrm{NH_3}}}} = (450\ \text{m s}^{-1})\sqrt{\frac{32.00\ \text{u}}{17.03\ \text{u}}} = 617\ \text{m s}^{-1}$$

We have seen in Example 3.16 that the average speed of ammonia molecules at 25 °C is 617 m s^{-1}. Why then does it take several seconds after we open a bottle of ammonia before we detect the smell, even if we are only 1 m away

FIGURE 3.16
Distribution of Molecular Speeds for Some Gases at 25 °C.

Light gases such as hydrogen and helium have higher average speeds and a wider distribution of speeds than heavier gases such as oxygen and nitrogen.

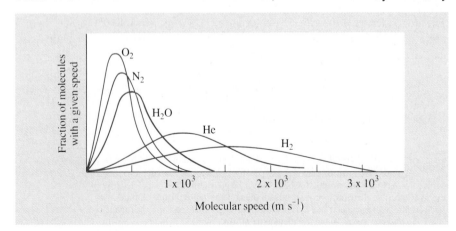

from the bottle? The reason that molecules diffuse rather slowly, even though they are moving at very high speeds, is that they are constantly colliding with each other.

At 1 atm pressure a molecule makes about 10^{10} collisions a second with other molecules. Any one molecule therefore has a very tortuous path, since its direction is changed in every collision. The average distance traveled by a molecule between collisons is known as the *mean free path*. The mean free path of oxygen molecules at 0 °C and 1 atm is only 60 nm, or 6×10^{-9} m. Thus although an oxygen molecule is moving at high speed, the large number of collisions it makes means that it is constantly changing direction, so its resultant motion in any particular direction is slow (Figure 3.17).

FIGURE 3.17
Molecular Speeds and Rates of Diffusion.

In traveling from A to B a molecule makes an enormous number of collisions, so the actual distance that it travels is very much greater than the direct distance between A and B.

GRAHAM'S LAW

Although the rate of effusion or diffusion of a gas is not equal to the speed at which its molecules are moving, we may reasonably assume that the rate of effusion or diffusion is *proportional* to the average speed of its molecules. Hence if r_1 is the rate at which gas 1 effuses from a small hole and if r_2 is the rate at which gas 2 effuses from the same hole, we may write

$$\frac{r_1}{r_2} = \frac{v_1}{v_2} = \sqrt{\frac{m_2}{m_1}} = \sqrt{\frac{M_2}{M_1}}$$

where m_1 and m_2 are the molecular masses and M_1 and M_2 are the molar masses. The expression

$$\frac{r_1}{r_2} = \sqrt{\frac{M_2}{M_1}}$$

is **Graham's law**. This law states the following:

> The rate of effusion of a gas is inversely proportional to the square root of its molar mass.

Hydrogen therefore effuses more rapidly than other gases (Experiment 3.6). Similarly, assuming that the rate of diffusion of a gas into another gas is proportional to the speed of its molecules, we see that Graham's law also applies to diffusion, and so it can also be stated as follows:

> The rate of diffusion of a gas is inversely proportional to the square root of its molar mass.

The dependence of the rate of diffusion on molar mass is demonstrated in Experiment 3.7. The laws of effusion and diffusion were first deduced by the Scottish chemist Thomas Graham (1805–1869) from the results of his experiments.

═ *Example 3.17* ═══════════════════

RATES OF DIFFUSION

(a) What are the relative rates of diffusion of NH_3 and HCl?

(b) If Experiment 3.7 is carried out in a tube 1.00 m long, at what position in the tube do you expect to see the formation of a white cloud of NH_4Cl?

Solution

(a) The molar mass of NH_3 is 17.03 g mol^{-1}, and the molar mass of HCl is

Experiment 3 . 6

Rates of Effusion

 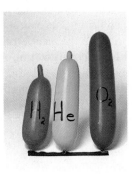 

Three balloons of the same size are filled with hydrogen, helium, and oxygen.

After a few hours the balloons have decreased in size because the gases effuse through tiny holes in the porous rubber skin of the balloon. The hydorgen balloon is maller than the helium balloon which is in turn smaller than the oxygen balloon. Thus, the rates of effusion are $H_2 > He > O_2$.

A large beaker containing hydrogen is inverted over a porous clay cylinder. The hydrogen molecules effuse through the holes in the porous cylinder faster than the oxygen and nitrogen molecules inside effuse out, creating an excess pressure inside the cylinder and the flask to which it is connected. The flask is filled with colored water and the excess pressure forces this water out through the tube on the right.

36.46 g mol^{-1}. From Graham's law we have ("diff" in the subscript stands for "diffusion"):

$$\frac{r_{\text{diff}}(NH_3)}{r_{\text{diff}}(HCl)} = \sqrt{\frac{M_{HCl}}{M_{NH_3}}} = \sqrt{\frac{36.46}{17.03}} = 1.46$$

(b) The NH_3 diffuses 1.46 times more rapidly than HCl. Therefore the distance moved by NH_3 down the tube is 1.46 times the distance moved by HCl in the same time:

$$\frac{d_{NH_3}}{d_{HCl}} = \frac{r_{\text{diff}}(NH_3)}{r_{\text{diff}}(HCl)} = 1.46$$

$$d_{NH_3} = 1.46 d_{HCl}$$

When the NH_4Cl cloud forms, because the tube is 1.00 m long,

$$d_{NH_3} + d_{HCl} = 1.00 \text{ m}$$

so that

$$1.46 d_{HCl} + d_{HCl} = 1.00 \text{ m}$$

or

$$d_{HCl} = \frac{1.00 \text{ m}}{2.46} = 0.41 \text{ m} = 41 \text{ cm}$$

Thus the NH_4Cl cloud forms at 41 cm from the HCl end of the tube, or at 59 cm from the NH_3 end.

Experiment 3.7

Graham's Law of Diffusion

Concentrated aqueous solutions of HCl and NH_3 are injected simultaneously into cotton swabs at opposite ends of the long glass tube, NH_3 on the left and HCl on the right.

Gaseous HCl and NH_3 diffuse slowly down the tube from the opposite ends and after about a minute they meet forming a white cloud of ammonium chloride NH_4Cl. The cloud forms at 23 cm from the left. Thus NH_3 has diffused $23 - 2 = 21$ cm in the time taken for the HCl to diffuse $17 - 2 = 15$ cm. Show that these distances are in accord with Graham's law (see Example 3.17).

EXERCISE 3.13

If Experiment 3.7 was repeated using HI instead of HCl, calculate the position at which a white cloud of ammonium iodide, NH_4I, would be expected to form in a tube 1.00 m in length.

Graham's law provides us with another method for finding the molar mass of a gas. By comparing the rate of diffusion or effusion of a gas of unknown molar mass with that for a gas of known molar mass, we can calculate the unknown molar mass, as shown in the next example.

Example 3.18

RATES OF EFFUSION

A gaseous compound known to contain only carbon, oxygen, and sulfur was found to effuse from a porous container at a rate of 100 mL in 5.00 min. Molecular oxygen effused from the same container under the same conditions at a rate of 137 mL in 5.00 min. What is the molar mass of the gas? What is its molecular formula?

Solution

The molar mass is found as follows ("eff" in the subscript stands for "effusion"):

$$\frac{r_{eff}(O_2)}{r_{eff}(gas)} = \frac{137}{100} = \sqrt{\frac{M_{gas}}{M_{O_2}}} = \sqrt{\frac{M_{gas}}{32.00 \text{ g mol}^{-1}}}$$

$$M_{gas} = \left(\frac{137}{100}\right)^2 (32.00 \text{ g mol}^{-1}) = 60.1 \text{ g mol}^{-1}$$

If the gas contains C, O, and S, it must have the molecular formula COS, for which $M = 60.07$ g mol^{-1}. Any other combination of C, O, and S atoms would have a higher molar mass.

EXERCISE 3.14

It takes 135 s for 1.00 mL of N_2 to effuse from a porous container. If it takes 240 s for 1.00 mL of a gas of unknown molar mass to effuse from the same container under the same conditions, what is the molar mass of the gas?

NONIDEAL GASES

The kinetic molecular theory assumes that the volume of the gas molecules is negligibly small compared with the volume that they occupy and that there are no forces acting between the molecules. On the basis of these assumptions the ideal gas law

$$V_{ideal} = nRT/P$$

can be derived. Real gases do not obey the ideal gas law exactly for two reasons:

1. The volume of the gas molecules may not be negligible compared with the space that they occupy. If this is the case the volume of the gas is greater than the ideal volume.
2. There are intermolecular forces acting between the molecules. These forces tend to pull the molecules together so that the volume of the gas is smaller than the ideal volume.

The finite volume of the molecules becomes more important at high pressures as the volume occupied by the gas decreases. The effect of intermolecular forces becomes more important at low temperatures when the molecules have a lower kinetic energy and are thus less able to overcome the intermolecular forces pulling them together. Thus gases deviate more from the ideal gas law at low temperatures and at high pressures. For most common gases at ordinary temperatures and at pressures of 1 atm or less, deviations from the ideal gas law can be neglected for many purposes.

IMPORTANT TERMS

Avogadro's law states that equal volumes of gases at the same temperature and pressure contain equal numbers of molecules.

A **barometer** is an instrument that measures the pressure of the atmosphere.

Boyle's law states that the volume of a given amount of gas at constant temperature is inversely proportional to the pressure.

Charles's law states that the volume of a given amount of gas at constant pressure is directly proportional to the absolute (Kelvin) temperature.

Combustion is an exothermic reaction that is self-sustaining.

Dalton's law of partial pressures states that in a mixture of gases the total pressure is the sum of the partial pressures of the gases present.

Diffusion is the process by which one gas mixes with another.

Effusion is the process by which a gas escapes from a container through a very small hole.

An **endothermic reaction** is a reaction that absorbs heat.

An **exothermic reaction** is a reaction that liberates heat.

The **gas constant** is the constant R in the ideal gas equation, $PV = nRT$. It has the value 0.0821 atm L mol^{-1} K^{-1} or 8.31 kPa dm^3 mol^{-1} K^{-1}.

Graham's law states that the rate of effusion (or diffusion) of a gas is inversely proportional to the square root of its molecular mass (or molar mass).

A **hydride** is a compound of a metallic element and hydrogen.

Hydrocarbons are compounds of carbon and hydrogen only.

An **ideal gas** is one whose behavior can be predicted by the ideal gas equation.

The **ideal gas equation** relates the pressure, volume, temperature, and number of moles of a gas through the expression $PV = nRT$, where R is the *gas constant*.

Intermolecular forces are the forces of attraction that exist between all molecules.

The **kinetic molecular theory** assumes that the atoms and molecules of which all substances are composed are in constant motion at any temperature above 0 K.

Nitrogen fixation is a process by which atmospheric nitrogen is converted to useful compounds such as ammonia.

An **oxidation reaction** is a reaction in which an element or compound combines with oxygen. (A more general definition is given in Chapter 5).

An **oxide** is a compound of an element and oxygen.

The **partial pressure** of any one gas in a mixture of gases is the pressure that the gas would exert if it occupied the container by itself. The total pressure exerted by a gas mixture is the sum of the partial pressures of all the gases in the mixture.

A **reduction reaction** is a reaction in which oxygen is removed partially or completely from a compound. (See Chapter 5 for a more general definition).

Standard temperature and pressure (STP) are 273.15 K or 0 °C and 101.33 kPa or 1 atm or 760 mm Hg.

PROBLEMS *

The Atmosphere: Oxygen, Nitrogen, and Hydrogen

1. **(a)** What are the three most abundant gases in the atmosphere?

   (b) Name three gaseous constituents of the atmosphere that are chemically reactive.

   (c) Comment on the role played by carbon dioxide in the atmosphere.

2. **(a)** What are the two most abundant elements in the earth's crust?

   (b) What is the most abundant metal in the earth's crust?

   (c) What is the most abundant element in the universe?

* Answers to problems numbered in blue appear at the end of the text.

3. From the elements that we have mentioned in this chapter name **(a)** two that are monatomic gases; **(b)** two that are diatomic gases; **(c)** one that exists as a triatomic gas; **(d)** a nonmetal that does not react with oxygen under any conditions; **(e)** a metal that does not react with oxygen under any conditions; **(f)** an element that can be prepared by passing steam over hot coke.

4. **(a)** Name the four most abundant elements in the earth's crust **(i)** by mass, **(ii)** by atom percentage.

   (b) Classify each of the above elements as a metal or a nonmetal.

5. With the aid of two examples in each case, explain what is meant by the terms

   (a) Oxidation   (b) Reduction

6. Write balanced equations to describe the reactions that occur when

   (a) Sulfur burns in air

   (b) Magnesium burns in air

   (c) Methane burns in air

7. Write balanced equations to describe each of the following:

   (a) The reduction of $Fe_2O_3(s)$ to $Fe(s)$ with $CO(g)$

   (b) The reduction of $Fe_2O_3(s)$ to $Fe(s)$ with $H_2(g)$

   (c) The reduction of $CuO(s)$ to $Cu(s)$ with $CO(g)$

   (d) The oxidation of $Mg(s)$ with $H_2O(g)$

8. Write a balanced equation for the reaction of steam at high temperature with each of the following:

   (a) Iron   (b) Magnesium   (c) Methane

9. Write a balanced equation to describe the reaction of nitrogen at high temperature with each of the following:

   (a) Hydrogen   (b) Oxygen   (c) Magnesium

10. Explain what is meant by the term *catalyst*. Give two examples of reactions in which a catalyst is used.

11. Explain why the carbon monoxide in a mixture of carbon monoxide and hydrogen must be converted to carbon dioxide before it can be easily separated from the hydrogen.

12. (a) Describe a simple test that you could use to confirm that a test tube filled with a colorless gas contained hydrogen.

   (b) Describe another simple test that you could use to confirm that another test tube contained oxygen gas.

   (c) Describe what is observed in (a) and (b) above, and write balanced equations for any reactions that occur.

13. Under the action of lightning, some of the nitrogen and oxygen of air react to form nitrogen monoxide, $NO(g)$. This subsequently oxidizes further and reacts with water to form nitric acid, which forms a dilute solution in rainwater.

   (a) Balance the equation $N_2 + O_2 + H_2O \rightarrow HNO_3$.

   (b) Calculate the maximum amount of nitric acid that can be formed from 0.76 g of nitrogen, 0.24 g of oxygen (the constituents of 1 g of air), and excess water.

14. Write the balanced equations for *three* different reactions by which hydrogen may be produced from water.

15. From the substances discussed in this chapter name and give the formula of each of the following:

   (a) A gas that reacts with $NH_3(g)$ to give a white solid

   (b) A metal that reacts with steam to give hydrogen

   (c) A gas that is used to make fertilizers

   (d) A gas that burns when ignited in air

   (e) A gas that explodes when mixed with oxygen and ignited

16. Explain how ozone is formed in the upper atmosphere. What is the importance of this ozone layer for life on earth?

17. Write a balanced equation for each of the following reactions:

   (a) The reaction of carbon when heated in air to give carbon monoxide

   (b) The reaction of calcium with oxygen to give calcium oxide

   (c) The decomposition of potassium chlorate when heated in the presence of manganese dioxide catalyst

   (d) The excess combustion of propane, $C_3H_8(g)$, in oxygen

   (e) The excess combustion of ethanol, $C_2H_6O(l)$, in oxygen

18. Write a balanced equation for each of the following reactions:

   (a) The reaction of copper oxide with carbon to give copper metal and carbon dioxide

   (b) The decomposition of lead dioxide, $PbO_2(s)$, on heating, to give lead monoxide, $PbO(s)$, and oxygen

   (c) The reaction of magnesium with nitrogen to give magnesium nitride

   (d) The reaction of calcium with hydrogen to give calcium hydride

   (e) The reaction of calcium hydride with water to give calcium hydroxide, $Ca(OH)_2(s)$, and hydrogen

19. Classify each of the following elements as a metal or a nonmetal, and give the physical state (gas, liquid, or solid) at room temperature:

   (a) Magnesium   (b) Nitrogen   (c) Oxygen
   (d) Ozone   (e) Sulfur   (f) Phosphorus
   (g) Copper   (h) Hydrogen   (i) Bromine

Boyle's Law

20. An automobile tire of volume 28 L is filled with air at a pressure of 2.3 atm. What volume does this air occupy at atmospheric pressure (1.00 atm) at the same temperature?

21. A cylinder in a gasoline engine initially has a volume of 0.50 L and contains gases that exert a pressure of 1.00 atm. What is the pressure of the gases when the volume of the cylinder is reduced to 0.20 L, assuming that there is no change in temperature?

22. A meteorological balloon has a volume of 150 L when filled with hydrogen at a pressure of 1.00 atm. To what volume will the balloon expand when it rises in the atmosphere to a height of 2500 m (8000 ft), where the atmospheric pressure is only 0.75 atm, assuming that the temperature remains constant?

23. A sample of hydrogen gas has a pressure of 0.98 atm when it is confined in a bulb of volume 2.00 L. A tap connects the bulb to another bulb that has a volume of 5.00 L and has been completely evacuated. What will be the pressure in the two bulbs after the tap is opened?

Charles's Law

24. A sample of nitrogen has a volume of 400 mL at 100 °C. At what temperature will it have a volume of 200 mL if the pressure does not change?

25. To what temperature must 30.0 L of helium at 25 °C be cooled at 1 atm pressure for its volume to be reduced to 1.00 L at the same pressure?

26. A balloon filled with helium has a volume of 1.60 L at a pressure of 1.00 atm at 25 °C. What will be the volume of the balloon when it is cooled to −196 °C, the temperature of liquid nitrogen, assuming that the pressure remains constant at 1.00 atm?

Combined Gas Law

27. The temperature of a closed vessel containing air at a pressure of 1.02 atm is raised from 20 to 200 °C. What is the pressure inside the vessel if its volume increases by 10% as the temperature is increased from 20 to 200 °C?

28. A sample of helium occupies a volume of 10.0 L at 25 °C and a pressure of 850 mm Hg. What volume will it occupy at STP?

29. What will be the volume of the meteorological balloon described in Problem 22 at a height of 2500 m, if the tem-perature at that altitude is −10 °C and if the balloon was filled at a temperature of 29 °C?

30. A sample of oxygen occupies a volume of 0.840 L at a pressure of 0.450 atm at 37 °C. What will be the pressure when it is cooled to −13 °C and compressed to a volume of 0.150 L?

31. A metal cylinder, which has a safety valve which opens at a pressure of 100 atm, is to be filled with nitrogen gas and heated to 300 °C. What is the maximum pressure to which it can be filled at 25 °C?

32. A gas occupies a volume of 0.500 L at a pressure of 730 mm Hg and 25 °C. What will be its volume when the pressure is increased by 10% and the temperature is lowered by 5 °C?

33. An automobile tire of fixed volume is inflated to a pressure of 2.50 atm at 20 °C. What is the pressure inside the tire at **(a)** 30 °C and **(b)** −10 °C?

Ideal Gas Law

34. Calculate the pressure exerted by 5.29 g of nitrogen dioxide, $NO_2(g)$, in a 5.00-L flask at 30 °C.

35. How many moles of methane, $CH_4(g)$, occupy a volume of 4.00 L at 1.00 atm pressure and 25 °C?

36. What volume is occupied by 0.200 mol of oxygen gas at 20 °C and a pressure of 740 mm Hg?

37. The overall reaction for the metabolic oxidation of glucose is the same as that for the combustion of glucose in air. The equation for the reaction is

$$C_6H_{12}O_6(s) + 6O_2(g) \longrightarrow 6CO_2(g) + 6H_2O(l)$$

What volume of carbon dioxide will be produced at body temperature (37 °C) when 7.20 g of glucose is oxidized?

38. Magnesium metal reacts with aqueous sulfuric acid according to the equation

$$Mg(s) + H_2SO_4(aq) \longrightarrow MgSO_4(aq) + H_2(g)$$

How many grams of magnesium react with excess $H_2SO_4(aq)$ to give 174.1 mL of $H_2(g)$ at 28 °C and a pressure of 1.00 atm?

39. A 1.00-L evacuated vessel is to be filled with carbon dioxide gas at 300 K and a pressure of 500 mm Hg by placing a piece of "dry-ice", $CO_2(s)$, inside the flask and allowing it to vaporize. What mass of dry-ice should be used?

The Ideal Gas Law and Molar Volume

40. What mass of gaseous hydrogen chloride, HCl(g), is needed to exert a pressure of 0.240 atm in a container of volume 250 mL at 37 °C?

41. A NASA scientist proposed that barium peroxide, $BaO_2(s)$, be used on a space capsule in order to supply emergency oxygen. It decomposes on heating according to the equation

$$2BaO_2(s) \longrightarrow 2BaO(s) + O_2(g)$$

   (a) What mass of $BaO_2(s)$ would be needed to supply enough oxygen to fill a 10 000-L space capsule to a pressure of 0.20 atm at 25 °C?

   (b) How long would this oxygen last if the crew operating at 20 °C used 1.00 L min^{-1} in respiration?

42. The atmosphere of Venus is mostly carbon dioxide. At the surface of Venus, the temperature is about 800 °C and the pressure is about 75 atm. In the event that an inhabitant of Venus (if any!) took these values as STP for Venus, what value would the Venusian find for the molar volume of an ideal gas?

43. At STP the mass of 1.00 L of a gas is 1.89 g. What mass of gas will occupy 1.00 L at 200 °C and a pressure of 1.25 atm?

44. As a publicity stunt, a water bed retailer in New Brunswick, Canada, filled a water bed bag with helium and floated it above his store. (Unfortunately it escaped and has not been seen since.) Calculate the mass of helium required to fill a water bed bag at a pressure of 1.03 atm at 23 °C, if its dimensions are 2.00 m × 1.50 m × 0.20 m.

45. The density of a gas at STP is 1.62 g L^{-1}. What is its density at 302 °C and a pressure of 0.950 atm?

46. An anesthetic gas used to relax patients contains oxygen mixed with dinitrogen monoxide, $N_2O(g)$. The mixture has a density of 1.482 g L^{-1} at 25 °C and 0.980 atm. What is the mass percentage of $N_2O(g)$ in the gas mixture?

47. What is the density of the gaseous chlorofluorocarbon CF_2Cl_2 at 1.00 atm and 20 °C?

Determination of Molar Mass

48. A sample of a noble gas had a mass of 0.20 g and exerted a pressure of 0.48 atm in a container of volume 0.26 L at 27 °C. Was the gas helium, neon, argon, krypton, or xenon?

49. Oxygen can exist not only as diatomic molecules, O_2, but also as ozone. What is the molecular formula of ozone, given that at the same pressure and temperature it has a density 1.50 times that of O_2?

50. A gas at a pressure of 740 mm Hg at 20 °C occupied a volume of 1.00 L and had a mass of 1.134 g. What is its molar mass? If the empirical formula of the gas is CH_2, what is its molecular formula?

51. The density of a gaseous chlorofluorocarbon at 23.8 °C and 432 mm Hg is 3.23 g L^{-1}. What is the molar mass of the compound? If it contains only one carbon atom per molecule what is its molecular formula?

52. A gas has a density of 1.275 g L^{-1} at 18 °C and 740 mm Hg pressure.

   (a) What is the molecular mass of the gas?

   (b) How many molecules are in 0.010 mL of the gas under these conditions?

53. A gas has a density of 1.402 g L^{-1} at 20 °C and 740 mm Hg pressure. What is the molar mass of the gas?

54. Several grams of a volatile liquid (boiling point less than 100 °C) were heated at 100 °C in a 350-mL flask until all the air was expelled. It was then cooled to room temperature, allowing air to re-enter the flask. The flask then had a mass 0.750 g greater than the empty (air-filled) flask. If the atmospheric pressure was 0.980 atm, what is the molar mass of the liquid?

55. A volatile liquid was found to have the composition 62.04% carbon, 10.41% hydrogen, and 27.55% oxygen, by mass. At 100 °C and 1.00 atm pressure, 440 mL of the gaseous compound had a mass of 1.673 g.

   (a) What is the molar mass of the compound?

   (b) What is its molecular formula?

56. The molar mass of ozone was determined by weighing an evacuated bulb of known volume, adding some ozone, reweighing the bulb at a known temperature, and measuring the pressure of the ozone at the same temperature. The following data were obtained:

Mass of evacuated bulb	6.208 g
Volume of bulb	1.500 L
Mass of bulb + ozone	7.319 g
Temperature	25.0 °C
Pressure of ozone	287 torr

Calculate the molar mass of ozone, a form of oxygen. What is the molecular formula of ozone?

57. At 25 °C, 2.41 g of a compound containing only boron and hydrogen filled a 2.22-L flask at a pressure of 730 torr. What is the molecular formula of the compound?

Partial Pressures

58. A mixture of 0.200 g of helium and 0.200 g of hydrogen is confined in a vessel of volume 225 mL at 27 °C.

(a) What is the partial pressure of each gas?

(b) What is the total pressure exerted by the gaseous mixture?

59. A 1.00-L flask contains 2.34 g of carbon monoxide and 1.56 g of carbon dioxide at 30 °C. What is the partial pressure of each gas, and the total pressure?

60. One liter each of oxygen, nitrogen, and hydrogen gases, all originally at 1.00 atm pressure, are forced into a single evacuated flask of capacity 2.00 L. What is the resulting pressure if the temperature remains unchanged? What is the partial pressure of each gas in the mixture?

61. On a certain day in Los Angeles, the concentration of nitric oxide, NO(g), in the atmosphere was 0.94 ppm by volume. The atmospheric pressure was 750 mm Hg and the temperature was 30 °C.

(a) What was the partial pressure of NO(g)?

(b) What was the number of NO molecules per cubic meter of air?

62. The partial pressure of oxygen in air at 37 °C is 159 mm Hg. When a human exhales a breath, the partial pressure of oxygen is only 115 mm Hg. How many oxygen molecules per liter of inhaled air are used by the lungs? (Body temperature is 37 °C.)

***63.** Solid copper oxide, CuO(s), reacts with gaseous ammonia, NH_3(g), to give copper metal, gaseous nitrogen, and gaseous water. An evacuated 80.0-L vessel contains 5.00 g of CuO(s). Ammonia gas is introduced into the reaction vessel until the pressure reaches 1.00 atm at 180 °C. What is the partial pressure of each gas in the reaction mixture at 180 °C when the reaction is complete?

64. Is the density of moist air greater than, less than, or the same as the density of dry air at the same temperature and pressure? Explain your answer.

Graham's Law

65. The average speed of oxygen molecules at 25 °C is 450 m s^{-1}. What are the average speeds of each of the following gaseous molecules at 25 °C?

(a) Hydrogen   (b) Chlorine

(c) Carbon monoxide   (d) Water

(e) Carbon dioxide   (f) Hydrogen sulfide

* The asterisk denotes the more difficult problems.

66. Which effuses faster, molecular nitrogen or molecular oxygen? What is the ratio of their rates of effusion?

67. Calculate the ratio of the rates of diffusion of hydrogen gas, H_2, and heavy hydrogen gas, D_2.

68. The original separation of ^{235}U (required for atom bombs) from ^{238}U was achieved by exploiting the slight difference in effusion rates of $^{235}UF_6$ and $^{238}UF_6$. Calculate the ratio of the rates of effusion of these two gases, assuming that the mass of each isotope is exactly equal to its mass number.

69. The average speed of helium atoms is 0.707 miles s^{-1} at 25 °C. What is the average speed of nitrogen molecules at the same temperature?

70. Suppose Experiment 3.7 is carried out with NH_3(g) and HBr(g), which react to give a white cloud of NH_4Br(g). Predict at what point in a tube of length 1.00 m the white cloud will be formed.

71. A gaseous element that exists as a diatomic gas at STP effused through a small hole at 0.324 times the rate of effusion of helium gas under the same conditions. Identify the element.

72. A given volume of oxygen diffuses through a porous plug in 10.0 s, while an equal volume of another gas takes 9.36 s, under the same conditions. On the basis of its calculated molar mass, suggest three possible molecular formulas for this gas. How could the possible gases be distinguished chemically?

73. A 0.492 g sample of a boron hydride gave B_2O_3(s) and 0.617 g of H_2O(l) when it was burned in excess oxygen. A 0.147 g sample of the same boron hydride exerted a pressure of 110 mm Hg in a 250 mL flask at 23 °C. Find the empirical and molecular formulas of the boron hydride.

74. Ethanol, C_2H_5OH, burns in air to give carbon dioxide and water. Calculate the volume of air at 25.0 °C and 770 torr required to burn completely 252 g of ethanol. Air is 21.0% O by volume.

Stoichiometry of Gas Reactions

75. An important reaction in the production of nitrogen fertilizers is the oxidation of ammonia, NH_3, to nitrogen monoxide, NO, at 500 °C:

$$4NH_3(g) + 5O_2(g) \longrightarrow 4NO(g) + 6H_2O(g)$$

How many liters of O_2, measured at 25 °C and 0.896 atm, must be used to produce 500 L of NO(g) at 500 °C and 740 mm Hg, assuming that the reaction goes to completion?

76. How many liters of oxygen are required to burn 1.00 L of **(a)** methane, and **(b)** hexane, C_6H_{14}, to carbon dioxide and water, if initially all the gases are at the same temperature and pressure?

77. Determine the amount of carbon dioxide, in liters at STP, that can be obtained by heating 1.00 kg of calcium carbonate, $CaCO_3(s)$, which decomposes according to the equation

$$CaCO_3(s) \longrightarrow CaO(s) + CO_2(g)$$

78. How many liters of oxygen at 25 °C and 740 mm Hg pressure are required to burn 5.00 g of magnesium completely?

79. Calcium hydride, $CaH_2(s)$, reacts with water to give hydrogen, $H_2(g)$, and calcium hydroxide, $Ca(OH)_2(s)$.

(a) Write the balanced equation for this reaction.

(b) How many grams of $CaH_2(s)$ are needed to prepare 10.0 L of $H_2(g)$ at STP?

80. What mass of potassium chlorate, $KClO_3(s)$, would have to be decomposed to prepare 5.00 L of oxygen at STP?

81. What volume of hydrogen at STP is required to

(a) Reduce 1.00 kg of zinc oxide, $ZnO(s)$, to zinc?

(b) Form 1.00 kg of water on burning in oxygen?

(c) Form 1.00 kg of lithium hydride, $LiH(s)$?

82. Hydrogen peroxide decomposes to give water and oxygen. How many grams of hydrogen peroxide are needed to give 100 L of oxygen at 25 °C and 1 atm?

83. An impure sample of limestone contains calcium carbonate, $CaCO_3(s)$, and inert and involatile impurities. On heating, the reaction

$$CaCO_3(s) \longrightarrow CaO(s) + CO_2(g)$$

occurs. When 1.000 g of the impure limestone was strongly heated to constant mass, 215 mL of gas were evolved at 25 °C and a pressure of 755 torr. What was the mass percentage of $CaCO_3(s)$ in the limestone?

84. Carbon monoxide and oxygen react to give carbon dioxide in an exothermic reaction. If 2.064 mol of $CO(g)$ and 1.032 mol of $O_2(g)$ in a closed flask, and at an initial pressure of 2.30 atm, react completely, what is the final pressure

(a) After reaction when the flask is cooled to 25 °C?

(b) After reaction if the temperature rises to 300 °C?

Miscellaneous

85. The anesthetic cyclopropane is a gas containing only carbon and hydrogen. Suppose 0.550 L of cyclopropane at 120 °C and 0.900 atm reacts with oxygen to give 1.65 L of $CO_2(g)$ and 1.65 L $H_2O(g)$, at the same temperature and pressure. What is the molecular formula of cyclopropane and the balanced equation for its reaction with oxygen?

***86.** Calculate the average volume effectively occupied by each molecule in an ideal gas at 27 °C and 1 atm pressure. Assume the molecules to be spherical and to have a radius of 100 pm, and calculate the actual volume of a molecule. Compare this actual volume with the volume effectively occupied by a molecule in the ideal gas.

87. By what factor does water expand when converted from liquid at 100 °C to vapor at 100 °C at 1 atm pressure, given that the density of liquid water at 100 °C is 0.96 $g \ cm^{-3}$?

***88.** A 50.00-mL bulb has a mass of 67.6259 g when evacuated, and 68.8883 g when filled with xenon gas at 25 °C and 1.00 atm pressure. What is the density of xenon under these conditions? What would be the mass of the same bulb after it is filled with a mixture of composition 35% oxygen and 65% xenon, by volume, at the same temperature and pressure? (The density of oxygen is 1.308 g L^{-1} at 25 °C and 1 atm.)

89. A gaseous hydrocarbon containing 82.7% carbon and 17.3% hydrogen, by mass, has a density of 2.308 g L^{-1} at 30 °C and 750 mm Hg. When the compound is burned in oxygen and converted completely to carbon dioxide and water, how many grams of carbon dioxide can be obtained from 10.00 g of the hydrocarbon?

90. In terms of the kinetic molecular theory of gases, account for each of the following.

(a) A gas exerts a pressure on the walls of any vessel in which it is confined.

(b) The pressure of a given mass of gas increases when its volume is decreased.

(c) Two gases readily mix together, and their total pressure is the sum of the pressures that each gas exerts alone in the same volume at the same temperature.

(d) The absolute zero of temperature is −273.15 °C (0 K).

(e) Real gases obey the ideal gas law to a good approximation, but not exactly.

The Periodic Table and Chemical Bonds

4

Group 1	2											3	4	5	6	7	8
1 H																	He
2 Li	Be				Metals	Nonmetals		Semimetals				B	C	N	O	F	Ne
3 Na	Mg											Al	Si	P	S	Cl	Ar
4 K	Ca	Sc	Ti	V	Cr	Mn	Fe	Co	Ni	Cu	Zn	Ga	Ge	As	Se	Br	Kr
5 Rb	Sr	Y	Zr	Nb	Mo	Tc	Ru	Rh	Pd	Ag	Cd	In	Sn	Sb	Te	I	Xe
6 Cs	Ba	La	Hf	Ta	W	Re	Os	Ir	Pt	Au	Hg	Tl	Pb	Bi	Po	At	Rn
7 Fr	Ra	Ac	104	105	106	107											

Period

The great variety of behavior of the elements and their compounds was as bewildering to early chemists as it must sometimes seem to students at the beginning of a course in chemistry. We have seen that oxygen is a very reactive element that combines readily with almost all the other elements. In contrast, nitrogen is a rather unreactive element that combines directly with far fewer elements than oxygen does. Another difference between elements that became apparent as soon as chemists began to determine the composition of compounds is that the elements have different combining powers. Nitrogen combines with three hydrogen atoms to form ammonia, NH_3; oxygen combines with two hydrogen atoms to form water, H_2O; fluorine combines with only one hydrogen atom to form hydrogen fluoride, HF; and neon does not form compounds with hydrogen—or, indeed, any other element.

In this chapter we will see how the different properties of the elements can be systematized by arranging them in the form of a table called the periodic table. We will use this table extensively throughout this book: it is one of the simplest and most useful tools that we have for understanding the great variety of chemical behavior.

We describe the forces that hold atoms together in molecules as chemical bonds. In this chapter we begin our discussion of two important types of chemical bonds—ionic bonds and covalent bonds—and we will see why it is that the atoms of different elements have different combining powers.

In molecules and crystals the atoms are held together in definite geometrical arrangements. The geometric arrangement of the atoms in a molecule or a crystal of a substance has a profound effect on the properties of the substance and we introduce this important topic in this chapter.

4.1
The Periodic Table

As they increased their knowledge of both elements and compounds, nineteenth-century chemists began to recognize that certain elements, for example, sodium and potassium, have very similar properties, while other elements, such as sodium and chlorine, have very different properties. Attempts to classify the elements in terms of the similarities and differences in their properties culminated in the formulation of the **periodic table** by Dmitri Mendeleev (Box 4.1) in 1869.

Mendeleev listed the elements in order of their atomic masses and found that elements with similar properties recurred in a periodic manner in this list. Because of this regularity in properties, he was able to arrange the elements,

---

BOX 4.1

DIMITRI MENDELEEV (1834–1907) AND THE PERIODIC TABLE

Mendeleev was born in Tobolsk, Siberia, and was the youngest of a family of seventeen. After his father died, his mother, determined that Dmitri should have the best possible education, moved the family to St. Petersburg (now Leningrad). In 1856 Mendeleev obtained a master's degree in chemistry and then taught at the University of St. Petersburg. In 1867 he was appointed professor of inorganic chemistry.

While giving his course of lectures, he felt the need for a new textbook, and he began writing what was to become a famous and widely used textbook, *Principles of Chemistry*. He realized that the order and system that was then becoming apparent in organic chemistry was lacking in inorganic chemistry, and in attempting to remedy this situation, he was led to formulate the periodic table in 1869.

To consider that the properties of the elements were in some way related to their atomic masses was an imaginative idea, since the structures of atoms were unknown at the time. Moreover, to bring certain elements into the correct group from the point of view of their chemical properties, he was bold enough to reverse the order of some pairs of elements and to predict that their atomic masses were incorrect. Some of these predictions were correct, but others were not, because we now know that the fundamental basis of the periodic table is atomic number rather than atomic mass. Also, to keep elements in the correct groups according to their chemical properties, he left gaps in his table, making the prediction that they would be filled by elements that were still undiscovered at that time. From the trends that he observed among the properties of related elements, he predicted the properties of these undiscovered elements.

At first the periodic table attracted little attention. Then, when his predictions concerning undiscovered elements were fulfilled in considerable detail by the successive discoveries of gallium (1874), scandium (1879), and germanium (1885), chemists began to realize that in the periodic table they had a tool of the greatest value. From that time on Mendeleev was recognized as one of the foremost scientists of the day.

Mendeleev was a versatile genius who was interested in many fields of science, both pure and applied. He worked on many problems associated with Russia's natural resources, such as coal, salt, and petroleum, and in

1876 he visited the United States to study the Pennsylvania oil fields. He invented an accurate barometer; he made an ascent in a balloon to study a total eclipse of the sun; and he was interested in the possibility of air travel.

As a professor at the University of St. Petersburg, Mendeleev was unavoidably involved in the political turmoil that affected all nineteenth-century Russia. Although in the middle of the century universities had been left comparatively free to carry out their research and teaching as they saw fit, the government came to suspect that academic freedom encouraged political unrest. Many repressive measures were taken against the universities, and the students reacted with demonstrations and riots. The universities were frequently closed for long periods, and many students were exiled to Siberia. In 1890 Mendeleev resigned from the university because of the government's oppressive treatment of students and the lack of academic freedom. Fortunately, he still had friends at the Czar's court, and he was appointed director of the Bureau of Weights and Measures, where he continued to carry out important research.

Mendeleev was a colorful character as well as a genius. A characteristic always noticed in the portraits of Mendeleev is his enormous head of hair. Biographers say that he only had it cut once a year in the spring and that he would not deviate from this custom even when he had an audience with the Czar.

FIGURE 4.1
The Periodic Table.

The elements are arranged in order of increasing atomic number, which usually (but not always) matches the order of increasing atomic mass. This arrangement places elements with similar properties in the same column. The atomic numbers are noted above the symbols of the elements; average atomic masses appear below each symbol.

listed in order of increasing atomic mass, in the form of a table in which similar elements fell in the same column. A modern form of the table is given in Figure 4.1. In this periodic table the elements are arranged in order of atomic number rather than atomic mass. The order of atomic mass is identical with the order of atomic number, except in a very few cases. The atomic number is the more fundamental property since it determines the number of protons and therefore the number of electrons in an atom, which in turn determines its chemical behavior. The table given in Figure 4.1 does not differ substantially from that proposed by Mendeleev (Figure 4.2) although his table contained only the 60 elements known at that time.

Over the years Mendeleev's original version of the periodic table has been modified in a variety of different ways. Indeed, discussion on how best to formulate the table continues today. It is important to understand that there is no single *correct* form of the periodic table. We can use whichever form is most useful to us. In this book we use a version that is simple and widely used.

The formulation of the periodic table marked the beginning of a new era in chemistry. It led to a much greater understanding of the properties of the elements, and it remains today a most important working tool for the chemist. Learning the general form of the periodic table and the positions of at least the first 20 elements is essential for understanding chemistry.

The periodic table has eight principal vertical columns, called **groups**, and seven main horizontal rows, known as **periods**. The groups are numbered 1 to 8 from left to right, and the periods are numbered 1 to 7 from top to bottom. The elements that fall between groups 2 and 3 are known as **transition elements**. There are ten of these elements in each of periods 4 and 5, which therefore have a total of 18 elements rather than only 8, as in periods 2 and 3. Period 6 between La (lanthanum) and Hf (hafnium) contains an additional 14 elements,

It has recently been proposed that the eight columns of the main groups and the ten columns of transition metals should be labelled 1 to 18 from left to right. The numbers for the main groups then become 1, 2, 13, 14, 15, 16, 17, and 18 and the transition metal groups have the numbers from 3 to 12. Because there has not yet been general agreement to accept this numbering, and because we have little need for special numbers for the transition metal groups, we have not adopted this numbering scheme in this book.

			Ti = 50	Zr = 90	? = 180
			V = 51	Nb = 94	Ta = 182
			Cr = 52	Mo = 96	W = 186
			Mn = 55	Rh = 104,4	Pt = 197,4
			Fe = 56	Ru = 104,4	Ir = 198
		Ni = Co = 59		Pd = 106,6	Os = 199
H = 1			Cu = 63,4	Ag = 108	Hg = 200
	Be = 9,4	Mg = 24	Zn = 65,2	Cd = 112	
	B = 11	Al = 27,4	? = 68	Ur = 116	Au = 197?
	C = 12	Si = 28	? = 70	Sn = 118	
	N = 14	P = 31	As = 75	Sb = 122	Bi = 210?
	O = 16	S = 32	Se = 79,4	Te = 128?	
	F = 19	Cl = 35,5	Br = 80	J = 127	
Li = 7	Na = 23	K = 39	Rb = 85,4	Cs = 133	Tl = 204
		Ca = 40	Sr = 87,6	Ba = 137	Pb = 207
		? = 45	Ce = 92		
		?Er = 56	La = 94		
		?Yt = 60	Di = 95		
		?In = 75,6	Th = 118?		

FIGURE 4.2
Mendeleev's Original Periodic Table Published in 1869

In this original form of the table the periods are the vertical columns and the groups are the horizontal rows. In order to ensure that elements with similar properties came in the same group Mendeleev left several spaces in his table for elements that were not known at the time but which he predicted would be discovered. That several of these predictions turned out to be correct a few years later led to the widespread adoption of the table by chemists.

making a total of 32. These 14 elements are known as the *lanthanides*. There is a similar group of elements in period 7; these elements are called the *actinides*. To give the table a convenient size and shape, the lanthanides and actinides are usually set off separately at the bottom of the table.

Groups 1–8 are usually called the **main groups** to distinguish them from the ten smaller groups of transition metal elements. The latter groups are often numbered as well but, as the numbering of these groups has recently become a controversial issue, and as we do not consider these elements until Chapter 21 and will have little need to use any numbers that might be given to them, we have left them without numbers.

METALS AND NONMETALS

The elements can be broadly classified into metals and nonmetals. The metals are found on the left of the periodic table and the nonmetals on the right, as can be seen in the periodic table at the head of this chapter. **Metals** *have the following characteristic* **physical properties:**

Some familiar metals: aluminum, copper, iron and gold.

- They conduct heat very well.
- They have high electrical conductivities that increase with decreasing temperature.
- They have a high reflectivity and a shiny metallic luster.
- They are malleable and ductile; that is, they can be beaten out into sheets or foil and pulled out into thin wires without breaking.
- With the exception of mercury (melting point $-39\ °C$), they are solids at room temperature.
- They emit electrons when they are exposed to radiation of sufficiently high
- energy or when they are heated. These two effects are known as the *photoelectric effect* and the *thermionic effect*, respectively.

Cesium (melting point 28.5 °C) and gallium (melting point 29.8 °C) melt just above normal room temperature. The very low melting point of mercury could not be predicted from its position in the periodic table. Indeed this is an example of a fact for which there does not seem to be any very satisfactory explanation. Although we can explain a very large number of facts there are still many waiting for future chemists to devise theories or explanations to account for them.

Nonmetallic elements *have the following characteristic* **physical properties:**

- They are poor conductors of heat.
- They are insulators; that is, they are very poor conductors of electricity.
- They do not have a high reflectivity or a shiny metallic appearance.
- They may be gases, liquids, or solids at room temperature.
- In the solid form they are generally brittle and fracture easily under stress, rather than being malleable and ductile.
- They do not exhibit the thermionic or photoelectric effects.

Some common nonmetals: yellow sulfur, dark red phosphorus and yellow-green chlorine.

Not every metal or nonmetal exhibits all the listed properties, and there are other exceptions to these generalizations. For example, the metal chromium is quite brittle, and graphite, a form of the nonmetal carbon, is a fairly good conductor of electricity. Furthermore, the transition from nonmetal to metal is not sharp so that several elements in the vicinity of the borderline have properties that are intermediate between those of a metal and those of a nonmetal. Silicon and arsenic are examples. They are usually called **semimetals**, or **metalloids**. There are only 17 nonmetals but this small number does not represent their relative importance. Most nonmetals are rather common and have many important compounds; quite a few metals are rare and have few important compounds. Of the 20 most abundant elements in the earth's crust, 10 are nonmetals, including the two most abundant elements, oxygen and silicon (see Table 3.2).

EXERCISE 4.1

Classify each of the following elements as a metal, a nonmetal, or a semimetal (metalloid): Cu, O, K, S, Cs, Ga, F, Ge, Ar, B, Mg, P.

FAMILIES OF ELEMENTS

The elements can be further classfied into a number of families whose members all have rather similar properties. These families constitute some of the groups of the periodic table. The resemblances between the physical and chemical properties of the elements in a group are most marked at the sides of the table, that is, in groups 1, 2, 6, 7, and 8. These groups have been given special names.

GROUP 1: THE ALKALI METALS, Li, Na, K, Rb AND Cs The group 1 elements are all metals that have most of the typical physical properties of metals; for example, they are shiny solids and good conductors of heat and electricity. Unlike most metals, however, they are soft enough to be cut with a knife and they have relatively low melting points.

Among their characteristic chemical properties are the following three important reactions.

1. They react with hydrogen to form solid *hydrides* that have the general formula MH, where M represents an alkali metal:

$$2M(s) + H_2(g) \longrightarrow 2MH(s)$$

2. They react with chlorine to form *chlorides*, which are all colorless solids with the formula MCl:

$$2M(s) + Cl_2(g) \longrightarrow 2MCl(s)$$

3. They react vigorously with water, producing hydrogen and a solution of a metal hydroxide (see Experiment 4.1). For example,

$$2Na(s) + 2H_2O(l) \longrightarrow 2NaOH(aq) + H_2(g)$$

GROUP 2: THE ALKALINE EARTH METALS, Be, Mg, Ca, Sr, AND Ba Like the alkali metals, the group 2 elements are all solids with typical metallic properties. Compared with the alkali metals, they are harder, melt at higher temperatures, and are somewhat less reactive. For example, although magnesium reacts with steam (Experiment 1.3), it does not react with cold water, and calcium reacts more slowly than either sodium, or potassium:

$$Ca(s) + 2H_2O(l) \longrightarrow Ca(OH)_2(s) + H_2(g)$$

Experiment 4.1

Reactivity of the Alkali Metals

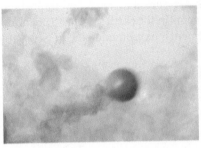

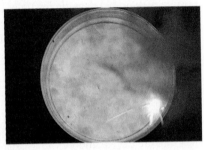

Sodium reacts so vigorously with water that it skates over the surface as hydrogen is evolved. A few drops of the indicator bromothymol blue have been added to the water. The blue trails are due to the sodium hydroxide formed in the reaction, which makes the solution basic as shown by the change in the color of the indicator.

The heat evolved in the reaction melts the sodium, which forms a small shiny metallic bead that floats on the water.

Potassium reacts even more violently, skating very rapidly over the surface leaving blue trails as the indicator changes color. The heat of the reaction ignites the hydrogen. The potassium imparts a characteristic lilac color to the flame. Steam can be seen rising from the water as it is heated by the burning potassium.

The alkaline earth metals form solid hydrides with the general formula MH_2 and solid chlorides with the general formula MCl_2:

$$M(s) + H_2(g) \longrightarrow MH_2(s) \quad \text{and} \quad M(s) + Cl_2(g) \longrightarrow MCl_2(s)$$

GROUP 6: THE CHALCOGENS, O, S, Se, AND Te Except for oxygen, which is a gas, the elements of group 6 are solids. Oxygen and sulfur are typical nonmetals. Selenium and particularly tellurium have some metallic properties; the latter is, in fact, usually classified as a semimetal. The group 6 elements react with hydrogen to form the compounds H_2O, H_2S, H_2Se, and H_2Te, which, with the exception of water, are all gases at room temperature.

GROUP 7: THE HALOGENS, F, Cl, Br, AND I The halogens are all typical nonmetals. Although their physical forms differ—fluorine and chlorine are gases, bromine is a liquid, and iodine is a solid at room temperature—each consists of diatomic molecules: F_2, Cl_2, Br_2, and I_2. The halogens all react with hydrogen to form gaseous compounds, with the formulas HF, HCl, HBr, and HI, all of which are very soluble in water. The halogens all react with metals to give *halides* (see Experiment 4.2). Examples include sodium fluoride, NaF, potassium chloride, KCl, magnesium bromide, $MgBr_2$, and zinc iodide, ZnI_2.

GROUP 8: THE NOBLE GASES, He, Ne, Ar, Kr, Xe, AND Rn These elements are all gases at room temperature and are typical nonmetals. They are all *monatomic*; that is, they consist of single atoms that are not combined to form molecules. They are very inert and, indeed, it was generally believed before 1962, when the first noble gas compound was prepared, that the noble gases formed no compounds. Krypton, xenon, and radon react with fluorine to form fluorides, but they do not react with oxygen or any other elements. No compounds of helium, neon, or argon have yet been made.

═ *Example 4.1* ════════════════════════════════════

REACTIONS OF THE ALKALI AND ALKALINE EARTH METALS

Write balanced equations for the following reactions: potassium with bromine; cesium with water; strontium with iodine; barium with water.

Solution

In order to answer this question we first need to find to which group of the periodic table each element belongs. Then we refer to the general reactions of each group that we described above. Potassium is an alkali metal (group 1). The elements of this group react with the halogens according to the general equation

$$2M + X_2 \longrightarrow 2MX$$

Therefore the equation for the reaction of potassium with bromine is

$$2K(s) + Br_2(l) \longrightarrow 2KBr(s)$$

Similarly, cesium is an alkali metal; therefore the equation is

$$2Cs(s) + 2H_2O(l) \longrightarrow 2CsOH(aq) + H_2(g)$$

Strontium is an alkaline earth metal; therefore the equation is

$$Sr(s) + I_2(s) \longrightarrow SrI_2(s)$$

Barium is an alkaline earth metal; therefore the equation is

$$Ba(s) + 2H_2O(l) \longrightarrow Ba(OH)_2(aq) + H_2(g)$$

 Experiment 4.2

Reaction between Zinc and Iodine

(a) (b) (c)

(a) Crystals of iodine (left) and grey powdered zinc (right) (b) When the iodine and zinc are mixed together they react to form zinc iodide. The reaction is strongly exothermic and sufficient heat is generated to form purple iodine vapor from some of the iodine. (c) The product is zinc iodide, ZnI_2. It is a white solid, which is seen here compared with the two reactants from which it is formed.

VALENCE

A very important property of an element that is related to its position in the periodic table is its combining power for other elements, a property called its **valence**. For example, with very few exceptions, one atom of hydrogen combines with no more than one atom of any other element. Hydrogen is therefore said to have a valence of 1.

An oxygen atom, however, combines with two hydrogen atoms to give the water molecule, H_2O. A nitrogen atom combines with three hydrogen atoms to give the ammonia molecule, NH_3. A carbon atom combines with four hydrogen atoms to give the methane molecule, CH_4. Thus oxygen, nitrogen, and carbon are said to have valences of 2, 3, and 4, respectively.

One example of the usefulness of the periodic table is the fact that

All the elements within a group have the same valence.

For example, all the alkali metals (group 1) have a valence of 1; all the elements in group 2 have a valence of 2. The valences of the elements of periods 2 and 3 are shown in Table 4.1. From these valences we can immediately write the empirical formulas for the compounds of these elements with hydrogen:

Period 2	LiH	BeH_2	BH_3	CH_4	NH_3	H_2O	HF
Period 3	NaH	MgH_2	AlH_3	SiH_4	PH_3	H_2S	HCl

Like hydrogen, fluorine and chlorine have a valence of 1. Hence we may write the empirical formulas of the compounds that they form with the elements of periods 2 and 3 as follows:

Period 2	Fluorides	LiF	BeF_2	BF_3	CF_4	NF_3	OF_2	F_2
	Chlorides	LiCl	$BeCl_2$	BCl_3	CCl_4	NCl_3	OCl_2	FCl
Period 3	Fluorides	NaF	MgF_2	AlF_3	SiF_4	PF_3	SF_2	ClF
	Chlorides	NaCl	$MgCl_2$	$AlCl_3$	$SiCl_4$	PCl_3	SCl_2	Cl_2

Oxygen, with a valence of 2, combines with two atoms of an element with a valence of 1, such as hydrogen, and with one atom of an element with a valence of 2, such as beryllium, Be, to form the compound BeO. In general, in a compound A_yB_z the values of y and z are such that

$$y \times \text{valence of A} = z \times \text{valence of B}$$

On the basis of this rule, the empirical formulas of the oxides of the elements in periods 2 and 3 of the periodic table may be written as follows:

Period 2	Oxides	Li_2O	BeO	B_2O_3	CO_2	N_2O_3	O_2	OF_2
Period 3	Oxides	Na_2O	MgO	Al_2O_3	SiO_2	P_2O_3	SO	Cl_2O

TABLE 4.1
Valences of the Main Group Elements

Group	1	2	3	4	5	6	7	8
Element	H							He
	Li	Be	B	C	N	O	F	Ne
	Na	Mg	Al	Si	P	S	Cl	Ar
Valence	1	2	3	4	3	2	1	0

Some elements, particularly those in periods 3–7 of groups 5–8 have one or more valences in addition to the valence given in Table 4.1. For example, phosphorus has a valence of 5 as well as 3. It forms a compound with the formula PCl_5 as well as one with the formula PCl_3, and it forms an oxide with the empirical formula P_2O_5 and one with the empirical formula P_2O_3. These additional valences are discussed in Chapter 8.

The statement that valence is the combining power of an atom is not a very precise definition. For our present purposes it will suffice to state that

> **The valence of an element is equal to the number of hydrogen atoms or halogen atoms which will combine with one atom of the element.**

When we have studied the arrangements of electrons in atoms and the nature of chemical bonds in the following sections we will be able to give a more general definition.

═ *Example 4.2* ═══════════════════════════════════

PREDICTING FORMULAS OF COMPOUNDS FROM VALENCES

Predict the empirical formulas of the sulfides of C, Si, Na, Mg, and Cl.

Solution

We can write the empirical formula of the sulfide of carbon as C_yS_z. Since carbon is in group 4 its valence is 4; since sulfur is in group 6 its valence is 2. Hence

$$y \times 4 = z \times 2$$

and therefore

$$\frac{y}{z} = \frac{2}{4} = \frac{1}{2}$$

Thus the empirical formula is CS_2. Alternatively since sulfur has the same valence as oxygen the formulas can be predicted by analogy with the formulas of the oxides just given. They are CS_2, SiS_2, Na_2S, MgS, and SCl_2.

E X E R C I S E 4 . 2

Predict the empirical formulas of the bromides and nitrides of the elements Li, Ba, and P, and write balanced equations for the formation of these compounds from the elements.

THE PERIODIC VARIATION OF PROPERTIES

Many of the physical and chemical properties of the elements vary in a periodic manner with atomic number. This variation is not generally as regular as the variation in the valence but nevertheless it is quite evident. For example, Figure 4.3 shows the variation of the molar atomic volume, that is, the volume of one mole of atoms of an element, with the atomic number. This periodic variation in the chemical and physical properties of the elements enables us to make at

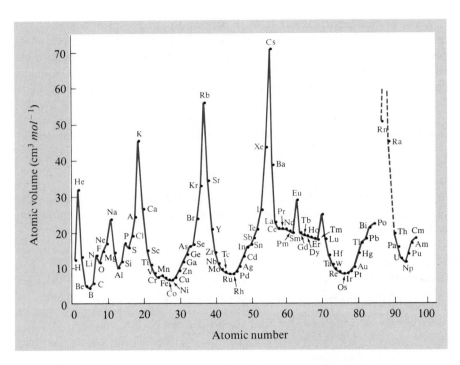

FIGURE 4.3
Atomic Volumes of the Elements.

(Redrawn from Moeller, T. *Inorganic Chemistry*; Wiley: New York, 1952; pp. 130–131.)

least approximate predictions about the properties of an element if we know the properties of its neighbors or of other elements in the same group. Mendeleev was even able to predict the properties of some elements that had not been discovered at that time (Box 4.1). Thus we need to memorize many fewer facts than we would need to if the properties of the elements did not vary in this way. We cannot make an exact prediction of a property such as the melting point, for example, but we can often make an approximate prediction of this and many other physical and chemical properties. Thus the periodic table is a very useful working tool for the chemist and you should try to use it as much as possible in thinking about properties and reactions. Naturally we will make extensive use of the table in this book. But do not rely on it blindly; there are many exceptions, and in the end the only completely reliable way to find out about a given property of an element is to determine it by an appropriate experiment.

4.2
The Shell Model

Why do similar properties occur periodically with increasing atomic number? An obvious answer is that the properties depend on the structure of the atoms of an element and that the structure in turn varies in a regular way with atomic number. Thus to understand the periodic variation in properties, we need to describe the structure of the atom in more detail.

A model of the atom that accounts well for the periodic variation in the properties of the elements is the **shell model**. According to this model, the electrons surrounding a nucleus in an atom are arranged in successive spherical layers called **shells**. The hydrogen atom consists of a nucleus surrounded by a rapidly moving electron that effectively fills a spherical space surrounding the nucleus. The helium atom similarly has two electrons in a similar sphere around the nucleus. The next element, lithium, also has two electrons in a sphere surrounding the nucleus, but the third electron occupies a spherical layer, or shell, that is mostly outside the sphere containing the first two electrons. In the succeeding elements, beryllium, boron, and so on, electrons continue to enter this outer shell until it contains eight electrons. Thus neon ($Z = 10$), which has a total of ten electrons, has two electrons in the inner shell and eight electrons in the second shell.

In sodium ($Z = 11$), the element following neon, the additional electron begins a third shell. Electrons continue to fill this shell up to the element argon ($Z = 18$), which has inner shells containing, respectively, two electrons and eight electrons, with a further eight electrons in the third shell. In the next element, potassium ($Z = 19$), a fourth shell is commenced. The electron arrangements for the first 20 elements are shown in Table 4.2. The successive electron shells are designated by $n = 1$ for the first shell, $n = 2$ for the second shell, and so on. The outer shell of an atom is called its **valence shell**, because it is the electrons in this shell that determine the valence of the atom.

TABLE 4.2
Shell Structure of Atoms of the First 20 Elements

Period	Z	Element	Number of Electrons in each Shell			
			$n = 1$	2	3	4
1	1	H	1			
	2	He	2			
2	3	Li	2	1		
	4	Be	2	2		
	5	B	2	3		
	6	C	2	4		
	7	N	2	5		
	8	O	2	6		
	9	F	2	7		
	10	Ne	2	8		
3	11	Na	2	8	1	
	12	Mg	2	8	2	
	13	Al	2	8	3	
	14	Si	2	8	4	
	15	P	2	8	5	
	16	S	2	8	6	
	17	Cl	2	8	7	
	18	Ar	2	8	8	
4	19	K	2	8	8	1
	20	Ca	2	8	8	2

Note: Color shading indicates the valence shell.

Figure 4.4 shows a simple representation of the shell structures of the first 18 elements. As we will see in Chapter 7, we cannot locate electrons very precisely. Thus the boundaries of each shell are rather fuzzy, and they overlap each other to some extent. Nevertheless, on average, the electrons in the $n = 1$ shell are considerably closer to the nucleus than are those in the $n = 2$ shell, and those in the $n = 3$ shell are still further from the nucleus, and so on.

We see from Table 4.2 and Figure 4.4 that the shell model provides an explanation for the periodic variation in the properties of the elements and for the general form of the periodic table. Each time a new shell commences, we have an element with one electron in its valence shell. These elements are the alkali metals, Li, Na, and K. We have just seen that the alkali metals all have rather similar chemical and physical properties, illustrating the fact that the properties of an element depend on the number of electrons in its valence shell. The elements following the alkali metals all have two electrons in their valence shell; they are the alkaline earth metals Be, Mg, and Ca. The halogens, F and Cl, each have seven electrons in their valence shell.

Elements in the same group of the periodic table have the same number of electrons in their valence shell.

For a main group element (groups 1–8) the number of valence-shell electrons is equal to the group number. The only exception is helium, in group 8, which has only two electrons in its valence shell.

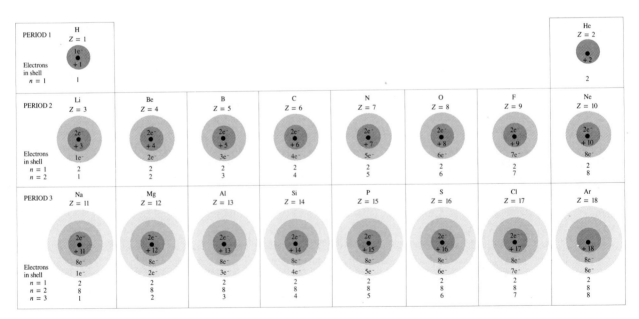

FIGURE 4.4
Shell Model of the First 18 Elements.

These diagrams show how electrons are arranged in shells; they do not indicate the relative size of shells or atoms. Each element within a group has a different number of shells, but they have the same number of electrons in their outermost shell.

IONIZATION ENERGY

Direct experimental evidence for the shell model is provided by ionization energies. *The **ionization energy** of an element is the energy needed to remove an electron from an atom of the element in the gas phase.* Thus the ionization energy of an element M is the energy (ΔE) required for the process

$$M(g) \longrightarrow M^+(g) + e^- \qquad \Delta E = \text{ionization energy}$$

When one or more electrons are removed from a neutral atom, a charged atom called a **positive ion** is formed. Some atoms may also add electrons forming negatively charged atoms known as **negative ions**.

One method of measuring ionization energies, the **electron impact method**, is described in Box 4.2. Atoms are bombarded with fast-moving electrons. If these electrons have sufficient energy, they will, on colliding with an atom, knock out one or more of the atom's electrons. The bombarding electrons can be given any desired energy by accelerating them through an electric potential that can be appropriately adjusted. The ionization of a hydrogen atom by electron bombardment can be summarized as follows:

$$H(g) + e^- \longrightarrow H^+(g) + 2e^-$$

A bombarding electron must have a minimum energy of 2.18×10^{-18} J in order to knock an electron out of a hydrogen atom:

$$H(g) + 2.18 \times 10^{-18} \text{ J} \longrightarrow H^+(g) + e^-$$

or

$$H(g) \longrightarrow H^+(g) + e^- \qquad \Delta E = 2.18 \times 10^{-18} \text{ J atom}^{-1}$$

The joule (J) is the SI unit of energy.
1 joule = 1 kg m^2 s^{-2}
(see Chapter 6).

Thus 2.18×10^{-18} J is the ionization energy of one hydrogen atom. It is more convenient, however, to consider the total energy needed to remove an electron from each of the hydrogen atoms in 1 mol of hydrogen atoms, which is

$$\left(\frac{2.18 \times 10^{-18} \text{ J}}{1 \text{ atom}}\right)\left(\frac{6.022 \times 10^{23} \text{ atoms}}{1 \text{ mol}}\right) = 1.31 \times 10^6 \text{ J mol}^{-1}$$

$$= 1.31 \text{ MJ mol}^{-1}$$

This value is the ionization energy of 1 mol of hydrogen atoms, and we will quote ionization energies in this form.

The ionization energies of the first 20 elements are listed in Table 4.3. They vary in a regular periodic manner with atomic number, as Figure 4.5 illustrates. *Ionization energy generally decreases down any group of the periodic table and increases across any period.* We can understand the variation in the ionization energy by using the shell model in conjunction with Coulomb's law.

Recall from Chapter 2 that according to Coulomb's law the electrostatic force, F, between two charges, Q_1 and Q_2, is proportional to the magnitude of each charge and inversely proportional to the square of the distance, r, between them:

$$F = k\frac{Q_1 Q_2}{r^2}$$

The electrons of hydrogen and helium are both in the first, $n = 1$, shell, and they are therefore at approximately the same distance from the nucleus. But the nuclear charge of helium ($+2$) is greater than that of hydrogen ($+1$). Therefore the force of attraction between the nucleus and the electrons in the helium atom

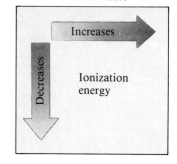

Periodic Table

Increases

Decreases

Ionization energy

BOX 4.2

DETERMINATION OF IONIZATION ENERGIES BY THE ELECTRON IMPACT METHOD

The apparatus used in the electron impact method for determining ionization energies consists of a sealed tube containing a heated metal filament called the *cathode*, which emits electrons and is kept at a negative potential with respect to a wire grid at a positive potential. Another metal plate called the *anode* is kept at a slight negative potential with respect to the grid but positive with respect to the cathode.

The tube is filled with the atoms to be studied in the form of a gas at low pressure. Electrons emitted by the cathode are accelerated toward the grid. Many of the electrons pass through the grid. If they have sufficient energy to overcome the small decelerating potential between the grid and the anode, they reach the anode, causing a current to flow in the external circuit, where it can be measured with a suitable meter.

The energy of the bombarding electrons can be increased by increasing the voltage through which they are accelerated. They collide with atoms on their way down the tube, but until they have enough energy to knock electrons out of any atoms with which they collide, the bombarding electrons simply bounce off with their energy unchanged and continue on their way down the tube to the anode. However, when it has sufficient energy, a bombarding electron will knock an electron out of an atom with which it collides, thereby losing all its own energy. This electron then has too little energy to get from the grid to the anode so the number of electrons reaching the anode falls markedly. Thus the meter registers a correspondingly large decrease in the current.

From the voltage used to accelerate the electrons their energy can be calculated. Hence, from the voltage at which a sudden reduction in the current is noted on the meter, the energy that must be supplied to the bombarding electrons in order to ionize the gaseous atoms can be found. This is the ionization energy.

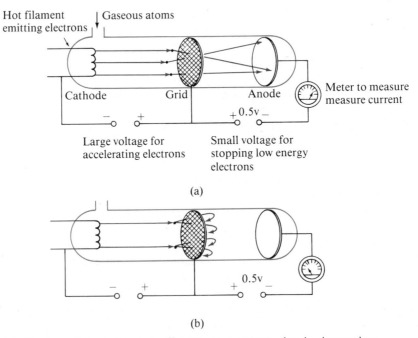

(a)

(b)

(a) The energy of the electrons is insufficient to cause ionization but the electrons have enough energy to pass through the grid and reach the anode where they produce an electric current which is measured by the meter in the circuit. (b) The energy of the bombarding electrons is only slightly greater than needed to cause ionization. So these electrons lose almost all their energy in ionizing the atoms with which they collide. The electrons that are produced by ionization also have very low kinetic energies. Thus almost all the electrons are stopped by the small negative voltage between the grid and the anode and the current registered at the anode is very small.

TABLE 4.3
Ionization Energies of the First 20 Elements

Z	Element	Ionization Energy (MJ mol^{-1})	Shell n	Z	Element	Ionization Energy (MJ mol^{-1})	Shell n
1	H	1.31 ⎱	1	11	Na	0.50 ⎱	
2	He	2.37 ⎰		12	Mg	0.74	
3	Li	0.52		13	Al	0.58	
4	Be	0.90		14	Si	0.79	3
5	B	0.80		15	P	1.01	
6	C	1.09	2	16	S	1.00	
7	N	1.40		17	Cl	1.25	
8	O	1.31		18	Ar	1.52 ⎰	
9	F	1.68		19	K	0.42 ⎱	4
10	Ne	2.08 ⎰		20	Ca	0.59 ⎰	

is greater than the force of attraction between the nucleus and the electron in the hydrogen atom. In other words, removing one of the electrons of the helium atom should take more energy than removing the electron in the hydrogen atom. And as Table 4.3 shows, helium has a higher ionization energy than hydrogen.

We see, however, that for lithium, $Z = 3$, a further increase in the ionization energy with increasing nuclear charge is not observed, and the ionization energy is very much less than that of hydrogen or helium. This observation is consistent with the shell model, which indicates that the third electron occupies a second shell outside the first shell. This electron is at a greater distance from the nucleus than are the two electrons in the inner shell. Since, according to Coulomb's law, the electrostatic force between two charges depends inversely on the square of the distance between them, r^2, the force with which a nucleus attracts an electron drops off rapidly, as the distance between the electron and the nucleus increases. Accordingly, the ionization energy of the outer electron in lithium is much smaller than the ionization energy of either hydrogen or helium.

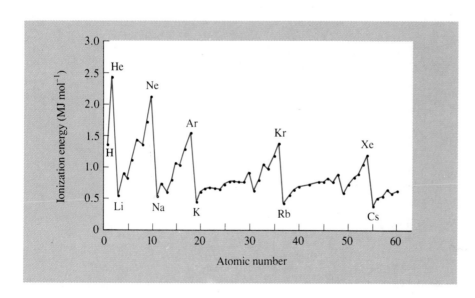

FIGURE 4.5
Ionization Energies.

The ionization energies of the elements vary in a periodic manner with atomic number. Within each period the alkali metal (group 1) has the lowest ionization energy and the noble gas (group 8) has the highest.

Moreover, the electron in the valence shell of lithium is repelled by the two inner electrons. Thus the resultant charge acting on the outer electron is not the nuclear charge of $+3$, but rather the charge of the inner core of the atom, consisting of the nucleus and the two inner-shell electrons, which is $+3 - 2 = +1$. The nucleus plus the completed inner shells of electrons constitute the **core** of an atom. The overall charge on the core is called the **core charge**. *The core charge is equal to the atomic number Z minus the total number of electrons in the inner shells* (see Figure 4.6).

═ *Example 4.3* ═══

CORE CHARGE

Which has a higher core charge, a magnesium atom or an oxygen atom?

Solution

The electron arrangement of magnesium is 2, 8, 2, and it has atomic number $Z = 12$. The core charge is $+12 - (2 + 8) = +2$. The electron arrangement of oxygen is 2, 6, and its atomic number is $Z = 8$. The core charge is $+8 - 2 = +6$. So oxygen has a higher core charge than magnesium.

───

Periodic Table

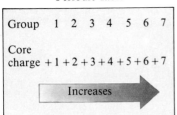

Thus the magnitude of the core charge is simply equal to the number of electrons in the valence shell and therefore to the group number. *The **core charge** increases from 1 to 8 across the main groups of the periodic table and it has the same value for all the elements in a given group.*

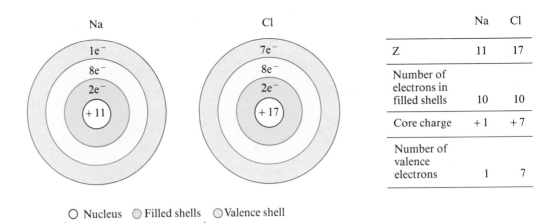

	Na	Cl
Z	11	17
Number of electrons in filled shells	10	10
Core charge	$+1$	$+7$
Number of valence electrons	1	7

○ Nucleus   ◔ Filled shells   ○ Valence shell

Core

FIGURE 4.6
Core Charge.

───

The electrons in the valence shell of chlorine are strongly attracted by a core charge of $+7$. The electron in the valence shell of sodium is only rather weakly attracted by a core charge of $+1$.

EXERCISE 4.3

What is the core charge of each of the following atom: N, F, Be, Ca?

The outer-shell electrons experience the charge of the core rather than the full charge of the nucleus. The inner electrons that surround the nucleus are said to *shield* the nucleus. In fact, because the shells overlap each other and because the electrons in the valence shell repel each other, the resultant charge acting on a valence-shell electron differs somewhat from the core charge. The resultant charge acting on a valence-shell electron is called the **effective nuclear charge**. Since there is no simple way to obtain values for the effective nuclear charge we will use the core charge as a basis for qualitative explanations of the properties of atoms. It is only an approximation, but it is adequate for our purposes.

As the core charge increases across the second period, the ionization energies of the elements also increase, except for two small irregularities that are discussed in Chapter 7. The ionization energy reaches a maximum at neon, at which point the second shell is complete.

The element after neon, sodium, has a nuclear charge of $+11$ and ten electrons in its inner shells, so the core charge is only $+1$. Thus, not surprisingly, the ionization energy of sodium is much lower than that of neon. But it is lower than that of lithium, too, which also has a core charge of $+1$. Because the outermost electron in sodium is in the third shell from the nucleus, it is further from the nucleus than the outer electron in lithium and thus has a lower ionization energy.

As the core charge increases in period 3, from sodium to argon, the ionization energies increase correspondingly. Then for potassium, the first element in period 4, the ionization energy again drops, as we would predict, because potassium has a core charge of $+1$ and an outer electron in the fourth shell from the nucleus. The ionization energy of potassium is therefore even lower than that of sodium.

We may summarize the dependence of the ionization energy of an element on its position in the periodic table by the following two statements:

1. With one or two minor exceptions, the ionization energy increases across each period of the periodic table as the core charge increases. In other words, as we move across a period, an increasing amount of energy is needed to remove an electron from the outermost (valence) shell.

2. The ionization energy decreases down any group of the periodic table because, as the number of shells increases, the distance of the valence-shell electrons from the nucleus increases, correspondingly, while the core charge remains constant.

== *Example 4.4* ===

IONIZATION ENERGIES

Which element in each of the following pairs would be expected to have the higher ionization energy:

(a) Li, Na   (b) Na, Mg   (c) N, F   (d) O, S

Answer this question by reference to the periodic table but not to Table 4.3.

Solution

 (a) Since Li comes above Na in group 1, Li has the higher ionization energy.

 (b) Since Mg follows Na in period 3, Mg has the higher ionization energy.

 (c) Since F comes after N in period 2, F has the higher ionization energy.

 (d) Since O comes above S in group 6, O has the higher ionization energy.

EXERCISE 4.4

Arrange the following atoms in order of increasing ionization energy without reference to Table 4.3: He, O, Be, F.

ELECTRON AFFINITY

We mentioned earlier that an electron may be added to some atoms to form a negative ion. The energy change, ΔE, for this process is called the electron affinity (EA) of the atom:

$$X(g) + e^- \longrightarrow X^-(g) \qquad \Delta E = \text{electron affinity}$$

For example, the electron affinity of the H atom is $-72.8 \text{ kJ mol}^{-1}$. Thus 72.8 kJ is *released* when 1 mol of H atoms combines with 1 mol of electrons to give 1 mol of H^- ions:

$$H(g) + e^- \longrightarrow H^-(g) \qquad \Delta E = -72.8 \text{ kJ mol}^{-1} = \text{EA(H)}$$

When a stable negative ion is formed, energy is released and ΔE is negative. The greater the electron affinity, the more negative is the value of ΔE. Fluorine, for example, has an electron affinity of -322 kJ mol^{-1}:

$$F(g) + e^- \longrightarrow F^-(g) \qquad \Delta E = -322 \text{ kJ mol}^{-1} = \text{EA(F)}$$

Some elements do not form stable negative ions and in these cases the addition of an electron is an endothermic process and ΔE is positive. For example, the electron affinity of neon is $+29 \text{ kJ mol}^{-1}$. This is understandable because the valence shell of neon is completely filled and so it has no tendency to add another electron, whereas fluorine has a vacancy in its valence shell and a high core charge so it attracts an electron strongly to form the F^- ion.

Electron affinity values for the first twenty elements are shown in Figure 4.7. The electron affinity increases across each period from Li to F and from Na to Cl as we expect with increasing core charge, but there are large exceptions to this trend at Be and N in period 2 and at Mg and P in group 3. These exceptions will be discussed in Chapter 8. For the moment the most important points to note are that elements with large ionization energies, such as the halogens, also have large electron affinities, and elements with small ionization energies, such as the alkali metals, also have small electron affinities. Thus a halogen atom has a strong tendency to gain an electron to form a negative ion and it is very difficult to remove an electron to form a positive ion. In contrast, an alkali metal atom has little if any tendency to add an electron to form a negative ion and it is rather easy to remove an electron to form a positive ion.

It can be seen from Figure 4.7 that the change in electron affinity down any group is very small. We might have expected the electron affinity to decrease down any group as the atoms increase in size and the added electron is further

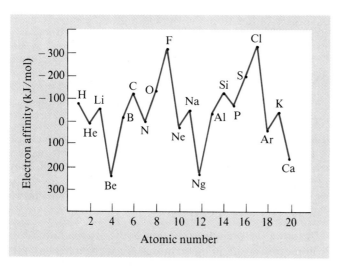

FIGURE 4.7
Electron Affinities of the
First 20 Elements.

from the nucleus and therefore attracted less strongly. But as the atoms get larger the distance between the electrons in the valence shell increases and therefore the electron–electron repulsion decreases and this roughly balances the decreased nucleus–electron attraction.

SIZES OF ATOMS

Another very important property that varies in a regular, periodic manner with atomic number is atomic size. The valence shell of an atom has no precise outer boundary; the electron density of an atom decreases rapidly to a very small value (Figure 2.2), but it does not become exactly zero even at a large distance from the nucleus. Therefore it is impossible to give an exact value for the size of a free atom. However, the distance between the nuclei of any pair of atoms *in a molecule* can be measured rather accurately (see Chapter 11).

The distance between the two nuclei in the H_2 molecule, which is known as the H–H **bond length**, is 74 picometers (pm). If we take one-half this distance, that is, 37 pm, we have a measure of the size of the H atom, which is known as the **atomic**, or **covalent**, **radius** of hydrogen. The distance between the two Cl nuclei in the Cl_2 molecule, the Cl–Cl bond length, is 198 pm. The atomic radius of Cl is one-half of this distance, that is, 99 pm. Other atomic radii can be found in a similar manner.

We cannot obtain values for atomic radii directly from the length of a bond between two dissimilar atoms because we do not know how to divide the distance into two unequal lengths. The length of the C–Cl bond in carbon tetrachloride has been found by experiment to be 176 pm. This distance is the sum of the atomic radii of Cl and C; in other words,

$$r_C + r_{Cl} = \text{C–Cl bond length}$$

If we use the value of 99 pm for the atomic radius of Cl that we calculated from the Cl_2 bond length, we can calculate the atomic radius of C:

$$r_C = \text{C–Cl bond length} - r_{Cl} = 176 \text{ pm} - 99 \text{ pm} = 77 \text{ pm}$$

This calculation assumes that the radius of a Cl atom, or indeed of any atom, is the same in all molecules, and this assumption has been shown in many experiments to be approximately true. The relationships between atomic radii and bond lengths are illustrated in Figure 4.8.

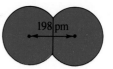

Cl_2
Bond length = 198 pm
rCl = radius of Cl atom
= 99 pm

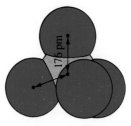

CCl_4
Bond length = 176 pm
$r_C + r_{Cl}$ = 176 pm
r_C = 77 pm

FIGURE 4.8
Atomic (Covalent) Radii and
Bond Lengths.

The length of a bond is the sum
of the atomic (covalent) radii of
the two atoms that are bonded
together.

The relative sizes of the atoms of the main group elements, as indicated by the values of their covalent radii, are shown in Figure 4.9. We note two important trends:

1. *Going down any group of the periodic table the atomic (covalent) radius increases, in other words, the atoms get larger.* This trend is consistent with the fact that an electron shell is added when we pass from one element to the next in a group.

2. *Going across any period the atomic (covalent) radius decreases, in other words, the atoms get smaller.* Although the outer electrons in each of the atoms in the same period are in the same valence shell, they are acted upon by an increasing core charge. This increasing core charge pulls the valence electrons in toward the core, so that the atoms decrease in size from left to right across a period.

Figure 4.10 shows the periodic variation of the atomic radii of the elements with atomic number. The alkali metal atoms are at the maxima of the plot, and the halogen and noble gas (Kr and Xe) atoms are at the minima. No values are given for the radii of He, Ne, and Ar because no molecules containing these atoms are known, and so no bond lengths involving these atoms have been measured.

We will normally use the term covalent radius rather than atomic radius because it emphasizes the fact that the values are for the radius of an atom when it is forming a (covalent) bond in a molecule and are not for the radius of the free atom which cannot be precisely defined.

EXERCISE 4.5

Using the values of the covalent radii given in Figure 4.9 estimate the lengths of the bonds in H_2S, NH_3, PCl_3, and SiF_4.

Periodic Table

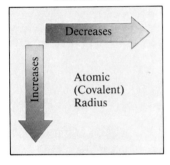

FIGURE 4.9
Covalent Radii.

The sizes of atoms generally increase from top to bottom within a group and decrease from left to right within a period. The radii here are given in picometers. No values are given for He, Ne, and Ar because no compounds of these elements are known.

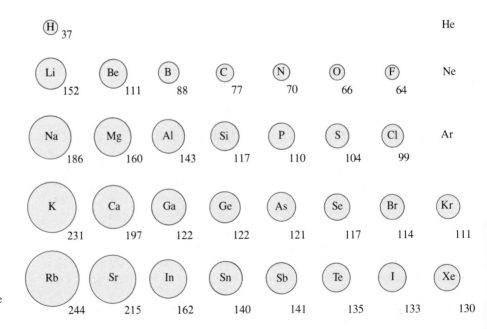

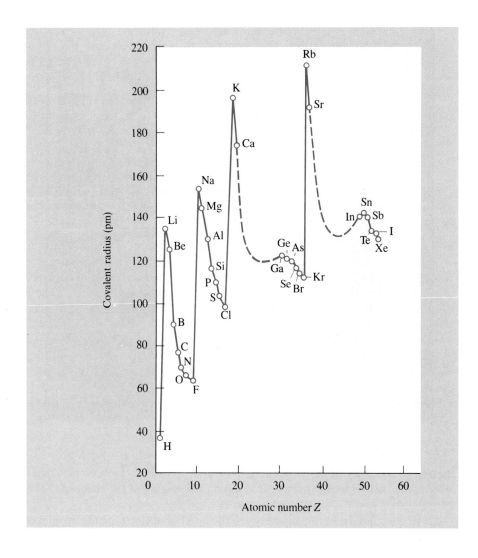

FIGURE 4.10
Periodic Variation of
Covalent Radii.

The largest atoms in any period are
the alkali metals; the smallest
are the noble gases.

4.3
Chemical Bonds and Lewis Structures

We are now in a position to consider the important question of how atoms are
held together, as they are, for example, in compounds such as NaCl and H_2O
and in the molecules of some elements such as H_2 and Cl_2. Whenever atoms are
held strongly together, we say that there is a **chemical bond** between them.
Chemical bonds, then, are strong forces that hold some atoms together.

Because the electrons in the completed inner shells of an atom are held much
more strongly than the electrons in the outer shell, the inner-shell electrons are
not generally involved when atoms combine. In discussing chemical bonds, we
will therefore be concerned only with the valence-shell electrons. The American
chemist Gilbert Lewis (Figure 4.11) introduced very convenient symbols, called

Lewis was born in Massachusetts, but in 1884 his family moved to Lincoln, Nebraska, where he received little formal schooling. He began his college career at the University of Nebraska, but he transferred to Harvard where he obtained his Ph.D. in 1899. After a period of advanced studies in Germany he joined the faculty of the Massachusetts Institute of Technology in 1905. In 1912 he became professor of chemistry at the University of California, Berkeley. Under his guidance the chemistry department at Berkeley gained international recognition.

Lewis's inquisitive, imaginative mind led him to make important contributions in several areas of chemistry. He was the first to propose that atoms could be held together by sharing an electron pair, and he proposed the electron dot (Lewis) structures that we use in this chapter. He made outstanding contributions to thermodynamics (the study of energy changes) and was the coauthor of a textbook that had a profound influence on the teaching of thermodynamics. In addition, he proposed a new definition of acids and bases (which we discuss in Chapter 10) and was the first to prepare and study pure heavy water, ^2H_2O (D_2O).

electron dot symbols, or **Lewis symbols**, in which the electrons in the valence shell of an atom are indicated by dots surrounding the symbol of the element (Table 4.4).

There is a simple relationship between the number of electrons in the valence shell and the common valence of the element.

Groups 1–4   Valence = number of electrons in valence shell

Groups 4–8   Valence = 8 − number of electrons in valence shell

(See Table 4.4.) We can understand this relationship by using the octet rule which we discuss next.

TABLE 4.4
Lewis Symbols and Valences of the Main Group Elements

Group	1	2	3	4	5	6	7	8
Number of valence-shell electrons	1	2	3	4	5	6	7	8
Valence	1	2	3	4	3	2	1	0
Period 2	Li·	·Be·	·B·	·C·	·N·	:O·	:F·	:Ne:
Period 3	Na·	·Mg·	·Al·	·Si·	·P·	:S·	:Cl·	:Ar:

THE OCTET RULE

The valences of neon and argon, both of which have eight electrons in their valence shell, are 0; these elements are not known to form any compounds, and the elements themselves are monatomic. Thus a valence shell containing eight electrons appears to be a specially stable arrangement. This observation led Lewis to suggest the following rule:

In compound formation an atom gains or loses electrons, or shares pairs of electrons, until it has eight electrons in its valence shell.

A valence shell containing eight electrons is called an **octet**, and the above rule is known as the **octet rule**. We will see that it enables us to give a simple explanation for the normal valences of the elements. We will consider first the case where atoms gain or lose electrons to obtain an octet.

IONIC BONDS

Two obvious ways in which an atom can obtain a valence shell of eight electrons are to lose all its outer electrons or to gain additional electrons. For example, sodium reacts with chlorine to form sodium chloride. Sodium has the electron arrangement 2, 8, 1, with only one electron in its valence shell (Tables 4.2 and 4.4). If a sodium atom loses this electron, a positive sodium ion, Na^+, is formed. This ion has the electron arrangement 2, 8, with eight electrons in the $n = 2$ shell, which now becomes the outer, or valence, shell. If chlorine, which has the electron arrangement 2, 8, 7, gains an electron, a negative chloride ion, Cl^-, is formed. This ion has the electron arrangement 2, 8, 8. This is what happens when sodium reacts with chlorine; each sodium atom loses an electron to form a sodium ion, Na^+, and each chlorine atom gains an electron to form a chloride ion, Cl^-. We can summarize this reaction very conveniently in terms of Lewis symbols:

$$Na\cdot + \cdot\ddot{\underset{\cdot\cdot}{Cl}}: \longrightarrow (Na^+)(:\ddot{\underset{\cdot\cdot}{Cl}}:^-)$$

The representation of sodium chloride as $(Na^+)(:\ddot{\underset{\cdot\cdot}{Cl}}:^-)$ is called the **Lewis structure** (or *electron dot structure*) of sodium chloride.

Since opposite charges attract each other, there is an attractive force between the sodium and chloride ions. Thus solid sodium chloride, NaCl, consists of equal numbers of Na^+ ions and Cl^- ions held together by electrostatic attraction (Figure 4.12). The electrostatic attraction between oppositely charged ions is called an **ionic bond**.

We have seen, in Experiment 1.3, that magnesium oxide can be made by burning magnesium in air or oxygen. If magnesium, which has a valence shell with two electrons, loses those two electrons, it becomes the positive ion Mg^{2+}, which has an outer shell with eight electrons. If oxygen, which has six electrons in its valence shell, gains two electrons, it becomes the oxide ion O^{2-}, which has a valence shell with eight electrons. Magnesium oxide consists of Mg^{2+} and

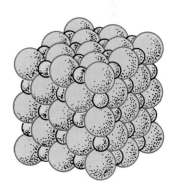

FIGURE 4.12
The Structure of Sodium Chloride.

Solid sodium chloride consists of equal numbers of Na^+ and Cl^- ions held together by electrostatic attraction, that is, by ionic bonds. There are no individual NaCl molecules in this structure.

O^{2-} held together by electrostatic attraction, that is, by ionic bonds. The Lewis structure of magnesium oxide is written as

$$(Mg^{2+})(:\ddot{O}:^{2-})$$

Magnesium also forms the compound magnesium chloride, $MgCl_2$, which occurs, for example, together with sodium chloride, in the sea. This compound consists of magnesium ions, Mg^{2+}, and twice as many chloride ions, Cl^-. It is represented by the Lewis structure

$$(Mg^{2+})(:\ddot{Cl}:^-)_2$$

Compounds such as Na^+Cl^-, $Mg^{2+}O^{2-}$, and $Mg^{2+}(Cl^-)_2$, which are composed of ions held together by electrostatic attraction (ionic bonds), are called **ionic compounds**.

Notice that both sodium and magnesium, which exist as positive ions in NaCl, MgO, and $MgCl_2$, are metals, whereas chlorine and oxygen, which form the negative ions in these compounds, are nonmetals. In general, the ionization energy of a valence electron of a metal is relatively low, so the electron is rather easily removed, leaving a positive ion. Moreover, although the metals have space in their valence shells for more electrons, they have little tendency to attract extra electrons because of their small core charges. We see from Figure 4.7 that they have only small electron affinities. Except for the noble gases, the nonmetallic elements also have space for additional electrons in their valence shells. But, in contrast with the metals, because they have high core charges they strongly attract additional electrons to fill their valence shells. And we can see from Figure 4.7 that in most cases they have high electron affinities.

In general we can state that

> Electrons are transferred from metallic to nonmetallic elements, forming positive metal ions and negative nonmetal ions held together by ionic bonds.

An ionic compound consists of an arrangement of a large number of positive and negative ions, but it has no overall charge, so the total charge on the positive ions is equal to the total charge on the negative ions. Thus in Na^+Cl^-

---

TABLE 4.5
Common Monatomic Ions of the Main Group Elements

			Group			
1	2	3	4	5	6	7
Li^+	Be^{2+}			$:\ddot{N}:^{3-}$	$:\ddot{O}:^{2-}$	$:\ddot{F}:^-$
Na^+	Mg^{2+}	Al^{3+}		$:\ddot{P}:^{3-}$	$:\ddot{S}:^{2-}$	$:\ddot{Cl}:^-$
K^+	Ca^{2+}					$:\ddot{Br}:^-$
Rb^+	Sr^{2+}					$:\ddot{I}:^-$
Cs^+	Ba^{2+}					

Note: All these ions, except Li^+ and Be^{2+}, have eight electrons in their outer shell and have the same electron arrangements as the noble gases Ne to Xe. Both Li^+ and Be^{2+} have only two electrons in the $n = 1$ shell, as in the noble gas He.

and $Mg^{2+}O^{2-}$ there are equal numbers of positive and negative ions, but in $Mg^{2+}(Cl^-)_2$ there are twice as many negative Cl^- ions as positive Mg^{2+} ions. The formula of an ionic compound such as sodium chloride may be written as Na^+Cl^-, but it is often written simply as NaCl, in which case we have to remember that it consists of Na^+ and Cl^- ions. Because ionic compounds are normally solids consisting of a regular arrangement of ions of opposite charge, the formulas of ionic compounds are always empirical formulas. No individual molecules can be recognized in a solid ionic compound (see Figure 4.12), and therefore no molecular formula can be assigned to these compounds. The empirical formulas of simple ionic compounds may be derived from the charges of the common ions of the main group elements, which are listed in Table 4.5.

The metals of groups 1, 2, and 3 form positive ions by loss of 1, 2, and 3 electrons, respectively. The nonmetals of groups 5, 6, and 7 form negative ions by the gain of 3, 2, and 1 electron, respectively. *The valence of an element that forms an ion is equal to the charge on the ion* because its charge determines the number of oppositely charged ions that it will combine with. The names of the negative ions formed by the common nonmetals are given in Table 4.6.

Note that the ions N^{3-}, O^{2-}, F^-, Na^+, Mg^{2+}, and Al^{3+} all have the same electron arrangement, namely that of neon, 2, 8. They differ only in their nuclear charge which ranges from $+7$ for N to $+13$ for Al. *Atoms and ions that have the same electron arrangement are said to be* **isoelectronic**.

We can readily understand the inertness of neon and the tendency of atoms such as Na and F to attain the same electron arrangement as in neon in compound formation by considering their shell structure. Starting at lithium electrons are added to the $n = 2$ shell, and at neon the $n = 2$ shell is filled with eight electrons. Although neon has a core charge of $+8$ and might therefore be expected to attract an electron even more strongly than fluorine (core charge = $+7$), there is no space for an additional electron in the valence shell of neon. An additional electron would have to enter the $n = 3$ shell where it would be only very weakly held. Thus a fluorine atom has a high electron affinity and a strong tendency to form the F^- ion, whereas neon has an extremely small electron affinity and no tendency to form an Ne^- ion. Moreover, because of its high core charge neon strongly attracts the electrons in its valence shell so that it is even more difficult to remove an electron from neon than from fluorine. In other words, neon has a very high ionization energy. We therefore do not expect neon to form either an Ne^- or an Ne^+ ion. We say that neon has an especially stable or inert electron arrangement; it has no tendency to attract an additional electron, and it is very difficult to remove an electron. Sodium (2, 8, 1), the element following neon, has a low ionization energy and its outer electron is easily removed. The sodium ion (2, 8) that is formed has eight electrons in its valence shell just like neon, so it is very difficult to remove a second electron, and it only weakly attracts an electron to reform a neutral sodium atom. Thus in almost all of its compounds sodium is in the form of the Na^+ ion. Similarly a fluorine atom (2, 7) has a strong tendency to gain an additional electron to form an F^- ion (2, 8), and this electron is difficult to remove to reform the F atom.

Helium has a filled $n = 1$ shell. The maximum number of electrons in this shell is only 2. But because the shell is filled, helium has little tendency to add an electron to give He^- (2, 1) and because helium has a high ionization energy it is difficult to remove an electron to give He^+. Thus, like neon, helium has a

TABLE 4.6 Names of the Negative Ions of the Common Nonmetals	
F^-	Fluoride
Cl^-	Chloride
Br^-	Bromide
I^-	Iodide
O^{2-}	Oxide
S^{2-}	Sulfide
N^{3-}	Nitride
P^{3-}	Phosphide

stable, inert electron arrangement. The ions H^-, Li^+, and Be^{2+} are isoelectronic with He; each has just two electrons in the $n = 1$ shell. These elements therefore do not obey the octet rule in their ionic compounds. They may be said to obey a duet rule.

As we shall see in more detail later the other noble gases which, like neon, have eight electrons in their valence shells have a high ionization energy and a very low electron affinity so they also have stable, inert electron arrangements. Other atoms frequently gain or lose electrons in compound formation until they have the same stable electron arrangements as these noble gases; in other words, they obey the octet rule, although, as we shall see, there are also many exceptions.

═ *Example 4.5* ══════════════════════════════════

EMPIRICAL FORMULAS AND LEWIS STRUCTURES FOR IONIC COMPOUNDS

Give the empirical formulas and draw Lewis structures for the ionic compounds formed by each of the following pairs of elements: Na, O; Mg, Br; K, S; Al, F; Na, P.

Solution

We can find the charge on each of the ions formed by these elements from Table 4.5 or directly from the position of the element in the periodic table. From the charges on the ions we can deduce the ratio of the numbers of positive and negative ions in the compound and hence write the empirical formula and the Lewis structure. The following table lists the results.

Elements	Ions	Empirical Formula	Lewis Structure
Na, O	Na^+, $:\overset{..}{\underset{..}{O}}:^{2-}$	Na_2O	$(Na^+)_2(:\overset{..}{\underset{..}{O}}:^{2-})$
Mg, Br	Mg^{2+}, $:\overset{..}{\underset{..}{Br}}:^-$	$MgBr_2$	$(Mg^{2+})(:\overset{..}{\underset{..}{Br}}:^-)_2$
K, S	K^+, $:\overset{..}{\underset{..}{S}}:^{2-}$	K_2S	$(K^+)_2(:\overset{..}{\underset{..}{S}}:^{2-})$
Al, F	Al^{3+}, $:\overset{..}{\underset{..}{F}}:^-$	AlF_3	$(Al^{3+})(:\overset{..}{\underset{..}{F}}:^-)_3$
Na, P	Na^+, $:\overset{..}{\underset{..}{P}}:^{3-}$	Na_3P	$(Na^+)_3(:\overset{..}{\underset{..}{P}}:^{3-})$

EXERCISE 4.7

Give the empirical formula and the Lewis structure for the compound formed by each of the following pairs of elements: Ca, F; Mg, N; Na, S.

COVALENT BONDS

We have seen that when a metal such as magnesium, from which the valence-shell electrons are easily removed, combines with a nonmetal such as chlorine, which has a strong tendency to gain an additional electron, an ionic compound is formed in which oppositely charged ions are held together by electrostatic attraction. But how are the two identical atoms held together in the chlorine molecule, Cl_2? If one chlorine atom were to gain an electron to become Cl^-, the other would have to lose an electron to become Cl^+. Because of its high core charge, the chlorine atom holds on to its valence electrons rather strongly, and it has a high ionization energy. In any case Cl^+ would have only six, rather than eight, electrons in its valence shell. Lewis suggested that since each chlorine atom in Cl_2 requires an additional electron to complete its octet they may be thought of as *sharing* a pair of electrons. Thus he viewed the formation of the Cl_2 molecule in the following way:

$$:\ddot{\text{C}}\text{l}\cdot + \cdot\ddot{\text{C}}\text{l}: \longrightarrow :\ddot{\text{C}}\text{l}\!:\!\ddot{\text{C}}\text{l}:$$

Shared pair of electrons

By sharing a pair of electrons, both atoms effectively acquire eight electrons in their valence shells (Figure 4.13).

The atoms in the molecules of many compounds formed between the non-metals can be thought of as being held together in the same way. For example, phosphorus in group 5 has a valence of 3 and forms the chloride PCl_3. Phosphorus needs three more electrons to complete its octet, and it can obtain these electrons by sharing a pair of electrons with each of three chlorine atoms:

$$\cdot\dot{\text{P}}\cdot + 3\cdot\ddot{\text{C}}\text{l}: \longrightarrow :\ddot{\text{C}}\text{l}\!:\!\dot{\text{P}}\!:\!\ddot{\text{C}}\text{l}:$$
$$:\ddot{\text{C}}\text{l}:$$

The diagrams

$$:\ddot{\text{C}}\text{l}\!:\!\ddot{\text{C}}\text{l}: \quad \text{and} \quad :\ddot{\text{C}}\text{l}\!:\!\dot{\text{P}}\!:\!\ddot{\text{C}}\text{l}:$$
$$:\ddot{\text{C}}\text{l}:$$

which show the arrangement of the electrons in the Cl_2 and PCl_3 molecules, are **Lewis structures**.

Whenever two atoms are held by a pair of shared electrons, we say that there is a **covalent bond** between them. In short, a covalent bond may be described as a pair of shared electrons. The Cl_2 molecule has one covalent bond. The PCl_3 molecule has three covalent bonds.

We can describe the bond in the hydrogen molecule, H_2, in the same way:

$$\text{H}\cdot + \cdot\text{H} \longrightarrow \text{H}\!:\!\text{H}$$

We note, however, that hydrogen has only one electron in the first ($n = 1$) shell; it only needs one more electron to fill this shell and thus to have the same electron arrangement as the noble gas helium.

The bonding in many molecules formed by hydrogen with other nonmetallic elements can be described in the same way. We can derive the Lewis structure

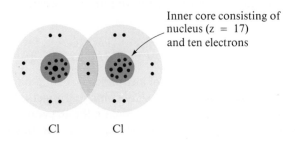

Inner core consisting of nucleus (z = 17) and ten electrons

Cl   Cl

FIGURE 4.13
Formation of a Covalent Bond by Sharing of an Electron Pair.

By sharing a pair of electrons, each Cl atom effectively acquires a valence shell of eight electrons in the Cl_2 molecule.

for the water molecule as follows:

$$2H\cdot + :\overset{\cdot\cdot}{\underset{\cdot}{O}}\cdot \longrightarrow :\overset{\cdot\cdot}{\underset{H}{O}}:H$$

Each hydrogen atom has effectively filled its valence shell with two electrons, and the oxygen atom has filled its valence shell with eight electrons.

It is sometimes convenient when you are learning to write Lewis structures to distinguish between the electrons that originally belonged to different atoms, as we have done above. However, because all electrons are identical, we do not usually make this distinction. We normally draw the Lewis structures of Cl_2, PCl_3, H_2, and H_2O as follows, where all the electrons are represented by identical dots:

$$:\overset{\cdot\cdot}{\underset{\cdot\cdot}{Cl}}:\overset{\cdot\cdot}{\underset{\cdot\cdot}{Cl}}: \qquad :\overset{\cdot\cdot}{\underset{\cdot\cdot}{Cl}}:\overset{}{\underset{}{P}}:\overset{\cdot\cdot}{\underset{\cdot\cdot}{Cl}}: \qquad H:H \qquad :\overset{\cdot\cdot}{\underset{\cdot\cdot}{O}}:H$$
$$:\overset{}{\underset{\cdot\cdot}{Cl}}: \qquad\qquad\qquad H$$

For the compounds of hydrogen with carbon, nitrogen, oxygen, and fluorine, we can write the following Lewis structures:

$$\begin{array}{cccc} H & & & \\ H:\overset{\cdot\cdot}{C}:H & H:\overset{\cdot\cdot}{N}:H & :\overset{\cdot\cdot}{O}:H & :\overset{\cdot\cdot}{\underset{\cdot\cdot}{F}}:H \\ H & H & H & \\ \text{Methane} & \text{Ammonia} & \text{Water} & \text{Hydrogen fluoride} \end{array}$$

The valence of hydrogen is 1, because it needs only one electron to complete its valence shell of two electrons, and it therefore forms one covalent bond. The valence of nitrogen is three, because it needs three electrons to complete its valence shell of eight electrons (an octet), and it therefore forms three covalent bonds.

Molecules are frequently represented by simplified forms of the Lewis structures. In particular, a bonding electron pair shared between two atoms is usually represented by a line drawn between the two atoms. In this convention the Lewis structures of water, ammonia, methane, and carbon tetrachloride molecules are written in the following manner:

$$\begin{array}{cccc} & & H & :\overset{\cdot\cdot}{Cl}: \\ & & | & | \\ :\overset{\cdot\cdot}{O}-H & H-\overset{\cdot\cdot}{N}-H & H-C-H & :\overset{\cdot\cdot}{\underset{\cdot\cdot}{Cl}}-C-\overset{\cdot\cdot}{\underset{\cdot\cdot}{Cl}}: \\ | & | & | & | \\ H & H & H & :\overset{}{\underset{\cdot\cdot}{Cl}}: \end{array}$$

Depending on the number of bonds that it forms, an atom with an octet of electrons may have one or more pairs of electrons that are not forming bonds; these pairs are called **unshared pairs**, **nonbonding pairs**, or **lone pairs**. In the ammonia molecule the nitrogen atom has one unshared pair of electrons. In the water molecule the oxygen atom has two unshared pairs of electrons. The chlorine atoms in carbon tetrachloride each have three unshared pairs of electrons.

LEWIS STRUCTURES OF COVALENT MOLECULES

In order to draw the Lewis structure of any molecule containing three or more atoms we need to know how the atoms are connected together. This can only be found with complete certainty by determining the structure of the molecule by a suitable experiment. However, for the present we will only consider some simple molecules in which H, halogen, O, or N are attached to a central atom. In these cases it is normally the atom with the highest valence, that is, the atom that needs the most electrons to complete its valence shell, that is the central atom in the molecule. Nitrogen is the central atom in ammonia, carbon in methane, oxygen in water, and so on. To draw the Lewis structure we write down the Lewis symbol of the central atom and then form bonds from the central atom to the other atoms until each atom has an octet.

─ *Example 4.6* ───────────────────────────

LEWIS STRUCTURES

Draw the Lewis structure for silicon tetrachloride, $SiCl_4$, a colorless, volatile liquid.

Solution

Silicon is in group 4 of the periodic table and has the Lewis symbol $\cdot\overset{\cdot}{Si}\cdot$. Silicon therefore can accomodate four more electrons in its valence shell to complete its octet. Chlorine is in group 7 and has the Lewis symbol $:\overset{\cdot\cdot}{Cl}\cdot$; it requires one more electron to complete its octet. Silicon can therefore use each of its four electrons to form four shared pairs with four chlorine atoms, as follows:

$$\cdot\overset{\cdot}{Si}\cdot + 4:\overset{\cdot\cdot}{Cl}\cdot \longrightarrow \quad :\overset{\overset{\displaystyle :\overset{\cdot\cdot}{Cl}:}{}}{\underset{\underset{\displaystyle :\overset{\cdot\cdot}{Cl}:}{}}{Cl}}:Si:\overset{\cdot\cdot}{Cl}: \quad \text{or} \quad :\overset{\overset{\displaystyle :\overset{\cdot\cdot}{Cl}:}{\mid}}{\underset{\underset{\displaystyle :\underset{\cdot\cdot}{Cl}:}{\mid}}{Cl}}-Si-\overset{\cdot\cdot}{Cl}:$$

───────────────────────────────────────

EXERCISE 4.8

Draw the Lewis structures of NF_3 and SCl_2.

MULTIPLE BONDS

When two atoms share one electron pair, the bond is called a **single bond**. Some elements such as carbon and oxygen can form **double bonds**, which consist of two pairs of shared electrons. For example, carbon normally forms four covalent bonds and oxygen two covalent bonds; therefore, the CO_2 molecule must have two covalent bonds between each oxygen atom and the carbon atom. Each oxygen atom shares *two* pairs of electrons with the carbon atom, so there are two double bonds in the CO_2 molecule:

$$:\overset{\cdot\cdot}{O}::C::\overset{\cdot\cdot}{O}: \quad \text{or} \quad :\overset{\cdot\cdot}{O}=C=\overset{\cdot\cdot}{O}:$$

A carbon atom can also form double bonds with other carbon atoms. Carbon–carbon double bonds are found in many molecules. For example, the hydrocarbon ethene (ethylene) has the following Lewis structure:

$$\overset{H}{\underset{H}{}}\!\!>\!\!C::C\!\!<\!\!\overset{H}{\underset{H}{}} \quad \text{or} \quad \overset{H}{\underset{H}{}}\!\!>\!\!C=C\!\!<\!\!\overset{H}{\underset{H}{}}$$

Here each carbon atom shares an electron pair with each of two hydrogen atoms, and it shares two electron pairs with the other carbon atom.

Sometimes, three pairs of electrons may be shared between two atoms, thereby forming a **triple bond**. Examples are found in the hydrocarbon ethyne (acetylene), C_2H_2, and in the nitrogen molecule, N_2:

$$H—C≡C—H \qquad :N≡N:$$

Ethyne   Nitrogen

$=$ *Example 4.7* $=$

LEWIS STRUCTURES

Draw the Lewis structure for the highly poisonous gas hydrogen cyanide, HCN.

Solution

The Lewis symbols for the atoms are H·, ·C·, and ·N·. Hydrogen can form one bond; carbon, four; and nitrogen, three. Carbon has the highest valence so we make it the central atom and then attach the other atoms. If we first form the CH bond we have

$$H· + ·C· \longrightarrow H:C·$$

We see that we can then form three bonds between carbon and nitrogen to give the final structure:

$$H:C· + ·N: \longrightarrow H:C:::N: \quad \text{or} \quad H—C≡N:$$

In drawing Lewis structures, particularly for molecules in which there are double and triple bonds, it is very useful to remember that the only possible combinations of bonds and unshared pairs of electrons for neutral H, F, O, N, and C atoms in covalent molecules are the following:

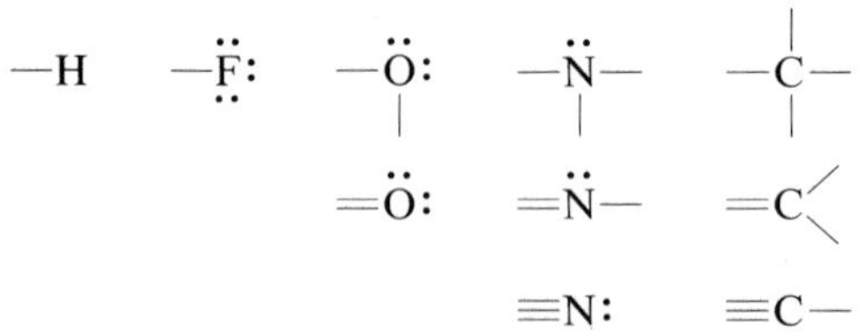

$$—H \quad —\ddot{\underset{\cdot\cdot}{F}}: \quad —\ddot{\underset{|}{O}}: \quad —\ddot{\underset{|}{N}}— \quad —\underset{|}{\overset{|}{C}}—$$

$$=\ddot{O}: \quad =\ddot{N}— \quad =C\!\!<$$

$$≡N: \quad ≡C—$$

Exactly similar arrangements hold for the heavier elements of groups 4–7 in those molecules in which these elements obey the octet rule.

Although we can very often write a Lewis structure rather easily by the method that we have outlined above, for some more complex molecules we will need to follow a more precisely defined set of rules which we give in Chapter 9.

DOUBLE AND TRIPLE BOND COVALENT RADII

When two electron pairs are shared between two atoms the atoms are held together more strongly than they are by a single pair and thus a double bond is shorter than a single bond between the same two atoms. Three shared pairs of

TABLE 4.7 Single, Double, and Triple Bond Covalent Radii (pm)						
	C	N	O	P	S	Cl
Single bond radius	77	70	66	110	104	99
Double bond radius	67	60	56	100	94	89
Triple bond radius	60	53				

electrons hold two atoms together still more strongly so a triple bond is shorter still. Some covalent radii for atoms forming double and triple bonds are given in Table 4.7.

POLYATOMIC IONS

We have seen in Figure 3.5 that $NH_3(g)$ reacts with $HCl(g)$ to form a white cloud of solid ammonium chloride, NH_4Cl. Ammonium chloride is an ionic substance containing a chloride ion, Cl^-, and an ammonium ion, NH_4^+. We can imagine that it is formed by the transfer of an H^+ from HCl to the ammonia molecule:

$$\text{H}-\overset{\overset{\displaystyle H}{|}}{\underset{\underset{\displaystyle H}{|}}{\text{N}}}: + \text{H}-\ddot{\ddot{\text{Cl}}}: \longrightarrow \left[\text{H}-\overset{\overset{\displaystyle H}{|}}{\underset{\underset{\displaystyle H}{|}}{\text{N}}}-\text{H} \right]^+ + :\ddot{\ddot{\text{Cl}}}:^-$$

The ammonium ion is an example of a **polyatomic ion**. A polyatomic ion is simply a molecule in which the atoms are held together by covalent bonds, but which has an overall positive or negative charge—in this case, $+1$. But since the term molecule is usually reserved for a neutral species, charged molecules are normally called polyatomic ions.

Another example of a polyatomic ion is the hydroxide ion. We saw in Experiment 4.1 that sodium reacts with water to form a solution of sodium hydroxide, NaOH. Sodium hydroxide is an ionic solid consisting of sodium ions, Na^+, and hydroxide ions, OH^-. First let us draw the Lewis structure.

Oxygen has six electrons, $:\overset{\cdot\cdot}{\underset{\cdot}{O}}\cdot$, and hydrogen, $\cdot$H, has one. We can form one covalent bond:

$$:\overset{\cdot\cdot}{\underset{\cdot}{O}}\cdot + \cdot\text{H} \longrightarrow :\overset{\cdot\cdot}{\underset{\cdot}{O}}:\text{H}$$

Oxygen has only seven electrons but we must add another electron to give the negative charge. We can imagine that this electron comes from the sodium atom, $Na\cdot$, when it forms the sodium ion, Na^+.

$$:\overset{\cdot\cdot}{\underset{\cdot}{O}}:\text{H} + \qquad \longrightarrow [:\overset{\cdot\cdot}{\underset{\cdot}{O}}:\text{H}]^-$$

FORMAL CHARGE

Where is the charge located on a polyatomic ion? We can obtain an approximate answer to this question by assuming that the electrons of each bond are shared *equally* between the two atoms that are bonded. Thus for the ammonium ion,

$$\left[\begin{array}{c} \text{H} \\ \text{H:N:H} \\ \text{H} \end{array} \right]^{+}$$

we assign one electron of each bond (red) to each hydrogen and one electron of each bond (black) to the nitrogen.

Of the total of eight electrons in the ion we have assigned one to each of the hydrogen atoms and four to the nitrogen atom. Since each hydrogen atom has a core charge of $+1$ and one electron, it has a zero charge. The nitrogen atom has a core charge of $+5$; but it has only four electrons assigned to it, so it has a charge of $+5 - 4 = +1$. Thus, on the basis of equal sharing of the covalent bond electrons, the positive charge of the ammonium ion is located on the nitrogen atom; this charge can be indicated in the Lewis structure as follows:

$$\begin{array}{c} \text{H} \\ \text{H:N:H} \\ \text{H} \end{array}$$

The $+1$ charge on the nitrogen atom is called its **formal charge**. We will see later that the formal charge is not necessarily the real charge on an atom in a molecule or polyatomic ion because the bond electrons are not always shared equally between the atoms forming the bond.

Notice that the ammonium ion and the methane molecule have the same Lewis structures. They have the same number of nuclei and the same number of electrons arranged in the same way. However, carbon has a core charge of $+4$, and since it is assigned four electrons, it has a zero charge:

$$\begin{array}{cc} \text{H} & \text{H} \\ \text{H:N:H} & \text{H:C:H} \\ \text{H} & \text{H} \end{array}$$

To find the formal charges in the hydroxide ion $[:\overset{..}{\text{O}}:\text{H}]^{-}$ we assign one of the electrons of the covalent bond to oxygen. The oxygen atom also has six more electrons—three unshared pairs—which are not shared with any atom and which therefore are all assigned to the oxygen atom, giving a total of seven electrons to oxygen (black). Since oxygen has a core charge of $+6$, it has a formal charge of $+6 - 7 = -1$. We assign one of the electrons of the covalent bond to hydrogen (red), which therefore has a zero formal charge. Thus we write the Lewis structure as follows:

$$^{\ominus}:\overset{..}{\underset{..}{\text{O}}}\cdot\text{H}$$

It is important to be able to correctly assign formal charges in Lewis structures, and you should always assign the formal charges when you write a Lewis structure.

The procedure that we have used for assigning formal charges can be summarized by the following rule:

$$\left(\begin{array}{c} \text{Formal} \\ \text{charge} \end{array} \right) = \left(\begin{array}{c} \text{Core} \\ \text{charge} \end{array} \right) - \left(\begin{array}{c} \text{Number of} \\ \text{unshared electrons} \end{array} \right) - \frac{1}{2}\left(\begin{array}{c} \text{Number of} \\ \text{shared electrons} \end{array} \right)$$

Alternatively, since the core charge is equal to the group number, and since $\frac{1}{2}$(number of shared electrons) = number of bonds, we have

$$\begin{pmatrix}\text{Formal}\\\text{charge}\end{pmatrix} = \begin{pmatrix}\text{Group}\\\text{number}\end{pmatrix} - \begin{pmatrix}\text{Number of}\\\text{unshared electrons}\end{pmatrix} - \begin{pmatrix}\text{Number}\\\text{of bonds}\end{pmatrix}$$

═ Example 4.8

POLYATOMIC IONS AND FORMAL CHARGES

When HCl(g) is dissolved in water it reacts to form the H_3O^+ ion and a Cl^- ion:

$$HCl(g) + H_2O(l) \longrightarrow H_3O^+ + Cl^-$$

Draw a Lewis structure for the H_3O^+ ion and assign formal charges to each of the atoms.

Solution

We can imagine that the H_3O^+ ion is formed in the same manner as the NH_4^+ ion, namely by transferring an H^+ from HCl to the water molecule:

$$H-\ddot{O}\!:\; +\; H-\ddot{\underset{\cdot\cdot}{Cl}}\!: \;\longrightarrow\; \left[H-\overset{\cdot\cdot}{O}-H \right]^+ +\; :\!\ddot{\underset{\cdot\cdot}{Cl}}\!:^-$$
$$\;\;\;\underset{H}{|} \qquad\qquad\qquad\;\; \underset{H}{|}$$

Or we can remove an electron from O to give O^+,

$$:\!\ddot{O}\!\cdot\; \longrightarrow\; \cdot\ddot{O}\!\cdot^+ + e^-$$

and then we can form bonds to three H atoms:

$$\cdot\ddot{O}\!\cdot^+ + 3H\cdot \;\longrightarrow\; H-\overset{\cdot\cdot}{\underset{|}{O}}{}^{\oplus}-H$$
$$\qquad\qquad\qquad\qquad\;\; \underset{H}{}$$

Constructing the Lewis structure in this way shows us that the formal charge is on oxygen. But let us check just to make sure. Oxygen has a core charge of $+6$. It has two unshared, nonbonding, electrons and a half share in three bonding pairs, that is, three electrons, for a total of five. Thus it has resultant charge of $+6 - 5 = +1$. Or using the formula

$$\begin{pmatrix}\text{Formal}\\\text{charge}\end{pmatrix} = \begin{pmatrix}\text{Core}\\\text{charge}\end{pmatrix} - \begin{pmatrix}\text{Number of}\\\text{unshared electrons}\end{pmatrix} - \frac{1}{2}\begin{pmatrix}\text{Number of}\\\text{shared electrons}\end{pmatrix}$$
$$\text{Formal charge} = 6 - 2 - 3 = +1$$

Each hydrogen has a half share in a pair, that is, one electron, so each hydrogen has a zero charge.

We note that the number of covalent bonds formed by an atom depends on its formal charge. A neutral nitrogen atom forms three bonds, as in NH_3, but a positively charged nitrogen atom forms four bonds, as in NH_4^+:

$$H-\ddot{N}-H \qquad\qquad H-\overset{\overset{\displaystyle H}{|}}{\underset{\underset{\displaystyle H}{|}}{N}}-H$$
$$\;\;\underset{H}{|}$$

A positively charged nitrogen has four electrons and is isoelectronic with a carbon atom:

$$:\ddot{N}\cdot \longrightarrow \cdot\ddot{N}^{\oplus} + e^{-} \qquad \cdot\dot{C}\cdot$$

$$
\begin{array}{cc}
\downarrow & \downarrow \\
\mathrm{H} & \mathrm{H} \\
| & | \\
\mathrm{H-\overset{\oplus}{N}-H} & \mathrm{H-C-H} \\
| & | \\
\mathrm{H} & \mathrm{H}
\end{array}
$$

Thus both a neutral carbon atom and a positively charged nitrogen atom have four electrons and can form four bonds. They both have a valence of 4. A neutral nitrogen atom has a valence of 3. Similarly a neutral oxygen atom forms two covalent bonds as in the water molecule, but a negatively charged oxygen atom forms only one covalent bond as in the hydroxide ion, OH^-, and a positively charged oxygen atom forms three covalent bonds as in H_3O^+. It is very useful in drawing Lewis structures to be able to recognize the possible combinations of bonds, nonbonding pairs, and formal charges for common atoms. Because carbon, nitrogen, oxygen, and fluorine always obey the octet rule, they always have four electron pairs in their valence shell. So the only possibilities are

$$
\begin{array}{ccccc}
\overset{|}{\underset{|}{-C-}} & -\overset{..}{C}{}^{\ominus} & & & \\[2mm]
\overset{|}{\underset{|}{-\overset{\oplus}{N}-}} & \overset{|}{\underset{|}{-\ddot{N}-}} & -\ddot{N}{:}^{\ominus} & -\ddot{N}{:}^{2\ominus} & :\ddot{N}{:}^{3\ominus} \\[2mm]
-\overset{\oplus}{\underset{|}{\ddot{O}}}{} & \overset{|}{-\ddot{O}{:}} & -\ddot{O}{:}^{\ominus} & :\ddot{O}{:}^{2\ominus} & \\[2mm]
& & -\ddot{F}{:} & :\ddot{F}{:}^{\ominus} &
\end{array}
$$

EXCEPTIONS TO THE OCTET RULE

Although, as we have seen, the octet rule is very useful in enabling us to understand how both covalent and ionic compounds are formed, it has many exceptions. One is obvious and has already been mentioned: the rule cannot apply to hydrogen, which has a valence shell that is filled by only two electrons. The elements on the left-hand side of the periodic table also depart from the octet rule in their covalent compounds. Strict application of the octet rule would predict that the elements in groups 1, 2, and 3 should not form covalent compounds because they have fewer than four electrons in their valence shells and cannot therefore complete their octets by electron sharing. However, these elements do form some covalent molecules. Boron trifluoride, BF_3, and boron trichloride, BCl_3, are two examples. In forming these molecules, boron uses all three electrons to form covalent bonds,

$$\cdot\dot{B}\cdot + 3:\!\ddot{F}\!\cdot \longrightarrow :\!\ddot{F}\!-\!B\!-\!\ddot{F}\!: \\ \qquad\qquad\qquad\quad |\\ \qquad\qquad\qquad\quad :\!\ddot{F}\!:$$

but it still has only six electrons in its valence shell. It is therefore an exception to the octet rule.

A similar exception is the $BeCl_2$ molecule

$$:\overset{..}{\underset{..}{Cl}}:Be:\overset{..}{\underset{..}{Cl}}:$$

in which there are only four electrons in the valence shell of beryllium.

We will encounter some further exceptions to the octet rule in Chapter 8.

BEYOND THE OCTET RULE

We have seen that Lewis's idea that a covalent bond is formed by a pair of shared electrons enables us to understand the number of bonds formed by an atom in a covalent compound and therefore to rationalize the formulas of covalent compounds. But we have not explained *why* a pair of shared electrons is able to hold two atoms together. We will take up this important topic in Chapter 7 where we will see that our current understanding of chemical bonding goes far beyond the simple ideas on which Lewis structures are based. Nevertheless, Lewis structures are a very useful way of approximately describing the arrangement of the electrons in molecules and polyatomic ions. We will make extensive use of them in this book.

One very important application of Lewis structures is for the prediction of molecular shape and we take up this topic in the next section.

4.4
Molecular Shape and the VSEPR Model

Molecules have a fascinating variety of different shapes. Some are long and thin, some are round, some are flat, some are rings, and others are spirals. We have already mentioned the shapes of several simple molecules. Water is an angular molecule, ammonia is a pyramidal molecule, and methane has a tetrahedral shape. Larger molecules may have much more complex shapes. In Chapter 1 (Figure 1.9) we illustrated the shape of the sucrose (sugar) molecule which, although it is considerably more complex than the water molecule, is simple compared with the DNA molecule (Figure 4.14). Although the detailed structure of DNA is very complicated its overall shape is a beautiful simple double spiral. The function of DNA, which is to transmit information from one generation to the next, is intimately related to its structure and shape. However, we shall see that no matter how complex a molecule is, the geometric arrangement of the covalent bonds around any particular atom can easily be deduced on the basis of some very simple principles. We use several different types of models to help us visualize the shapes of molecules. These are described in Box 4.3.

Ionic bonds *are nondirectional.* An ion attracts to itself as many ions of opposite charge as can be packed around it. Thus as we shall see in more detail in Chapters 5 and 11 the structure of an ionic crystal such as sodium chloride (Figure 4.12) depends on the charges and sizes of the ions of which it is composed. In contrast **covalent bonds** *are directional*; an atom forms covalent bonds in certain specific directions. Molecular geometry—that is, the geometric arrangement of the atoms in molecules and in covalently bonded crystals—depends on the directions of the covalent bonds formed by each atom. A simple

FIGURE 4.14
The DNA Molecule.

The DNA molecule consists of
two long chains that spiral around
each other.

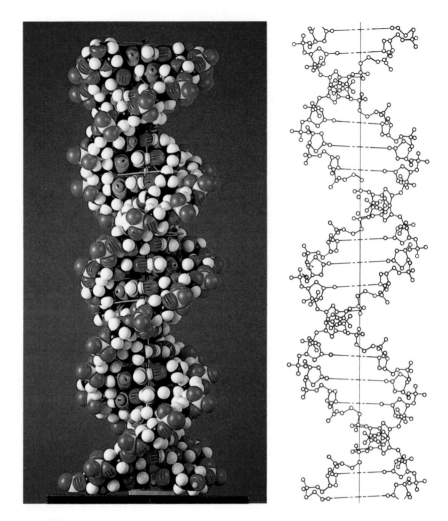

model that is based on Lewis structures enables us to predict the directions of
the bonds around any atom in a molecule. We introduce this model here. We
shall make considerable use of it in the following chapters and we take it up
again in more detail in Chapter 9.

THE VALENCE SHELL ELECTRON PAIR REPULSION (VSEPR) MODEL

According to the **VSEPR model** *the electron pairs in a valence shell are arranged
so as to keep as far apart as possible,* that is, *the electron pairs behave as if they
repel each other.* As we have just seen the atoms in many molecules have eight
electrons or four pairs in the valence shell. In other words, they obey the octet
rule. The arrangement of four electron pairs that keeps them as far apart as
possible is at the corners of a tetrahedron. Atoms that do not obey the octet
rule may have only two or three electron pairs in their valence shells and they
have a linear or equilateral triangular arrangement respectively (Figure 4.15).
A model that is very convenient for demonstrating the arrangements of electron
pairs in a valence shell is the electron pair sphere model described in Experiment
4.2.

BOX 4.3

MOLECULAR MODELS

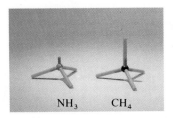

NH₃ CH₄

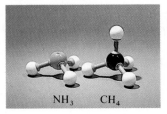

NH₃ CH₄

NH₃ CH₄

To visualize the shapes of molecules, you will find it very helpful to use a set of molecular models. These models do not have to be elaborate; they can be constructed even from gum drops and toothpicks. Relatively inexpensive sets are available commercially.

Stick models use plastic or metal forms consisting of a number of spokes radiating from a central point. Plastic tubes can be pushed onto the spokes to represent the bonds to other atoms and nonbonding electron pairs. The atoms are not directly represented in these models. Alternatively, plastic or wooden balls, drilled with holes at appropriate angles, can be used to represent atoms; wooden or plastic rods or stiff springs can be used to represent bonds. This kind of model is often called a *ball-and-stick model*. It represents only the directions of

the bonds in space, not the relative sizes of the atoms. Nevertheless, a ball-and-stick model gives a reasonably good representation of the general shape of a molecule and the relative orientations of the bonds.

Space-filling models are more elaborate. Since bonded atoms share electrons, they come together more closely than do nonbonded atoms; thus the atoms in a molecule are not spherical. In space-filling models they are represented by spheres with one or more slices cut off at suitable angles; each of the flat faces thus formed has a fastener, which enables it to be connected to similar spheres. These models give no direct representation of bonds, but compared with stick models they give a much more accurate representation of the shape of a molecule and of the relative sizes of its atoms.

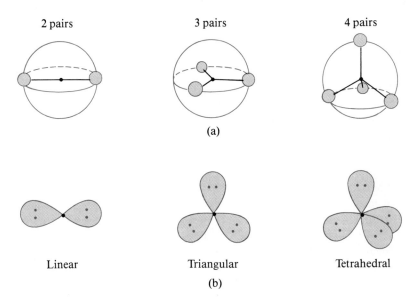

2 pairs 3 pairs 4 pairs

(a)

Linear Triangular Tetrahedral

(b)

FIGURE 4.15
Arrangements of 2, 3, and 4 Pairs of Electrons in a Valence Shell.

(a) Electron pairs represented by red spheres are arranged on the surface of a sphere representing the valence shell so as to be as far apart as possible. (b) Here each electron pair is represented by an egg-shaped charge cloud. They are arranged around the core of the atom so as to be as far apart as possible.

Experiment 4.3

Electron Pair Arrangements: Styrofoam Sphere Models

  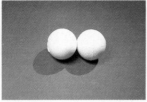 

Two spheres have a linear arrangement   Three spheres adopt a triangular arrangement

Four spheres adopt a tetrahedral arrangement

Electron pair arrangements can be demonstrated by joining pairs or threes of styrofoam spheres by elastic bands held in place by small nails or toothpicks. Each sphere represents a pair of electrons, or more exactly the space occupied by a pair of electrons. The elastic band provides a force of attraction between the spheres that corresponds to the electrostatic attraction between the electrons and a positive core situated at the mid-point of the elastic band. The spheres naturally adopt the arrangements shown. If they are forced into some other arrangement—such as the square planar arrangement for four spheres—they immediately adopt the tetrahedral arrangement when the restraining force is removed.

On the basis of these arrangements of electron pairs we can predict the shapes of a large number of molecules (Table 4.8). Let us first consider the hydrides of the elements of period 2. Their Lewis structures are as follows:

$$H\!:\!Be\!:\!H \qquad H\!:\!\overset{\displaystyle \cdot\cdot}{\underset{\displaystyle H}{B}}\!:\!H \qquad H\!:\!\overset{\displaystyle H}{\underset{\displaystyle H}{C}}\!:\!H \qquad H\!:\!\overset{\displaystyle \cdot\cdot}{\underset{\displaystyle H}{N}}\!:\!H \qquad H\!:\!\overset{\displaystyle \cdot\cdot}{\underset{\displaystyle H}{O}}\!: \qquad H\!:\!\overset{\displaystyle \cdot\cdot}{\underset{\displaystyle \cdot\cdot}{F}}\!:$$

AX$_2$ MOLECULES: LINEAR In BeH$_2$ there are two electron pairs in the valence shell of beryllium. Two electron pairs have a linear arrangement and since both of them are forming bonds to hydrogen atoms the hydrogen atoms have the same arrangement and the molecule is linear. The angle between the two bonds is 180°. All similar molecules such as BeCl$_2$ would be expected to have the same linear shape and we describe them as AX$_2$ molecules where A is the central atom and X are the atoms bonded to A.

Lewis structure	X:A:X	H:Be:H	:C̈l:Be:C̈l:
Shape	X—A—X	H—Be—H	Cl—Be—Cl

AX$_3$ MOLECULES: EQUILATERAL TRIANGULAR In BH$_3$ there are three electron pairs in the valence shell of boron. Three electron pairs have an equilateral triangular arrangement and since all of them are forming bonds to hydrogen atoms the three hydrogen atoms have the same arrangement and the molecule has an equilateral triangular shape. The angle between any pair of bonds is 120°. All similar molecules formed by the elements of group 3, such as BF_3, BCl_3, and $AlCl_3$, have the same shape. We describe them as AX_3 molecules.

TABLE 4.8
Electron Pair Arrangements and Shapes of Molecules

Number of Electron Pairs	Arrangement of Electron Pairs	Number of Nonbonding Electron Pairs	Class of Molecule	Arrangement of Bonds or Shape of Molecule	Examples	Predicted Bond Angle
2	Linear	0	AX_2	Linear X—A—X	BeH_2, $BeCl_2$	180°
3	Equilateral triangular	0	AX_3	Equilateral triangular	BF_3, $AlCl_3$	120°
		1	AX_2E	Angular	$SnCl_2$	120°
4	Tetrahedral	0	AX_4	Tetrahedral	CH_4, CCl_4, SiH_4, NH_4^+	109.5°
		1	AX_3E	Triangular pyramidal	NH_3, NF_3, PCl_3, H_3O^+	109.5°
		2	AX_2E_2	Angular	H_2O, F_2O, SCl_2	109.5°
		3	AXE_3	Linear	HF, HCl	—

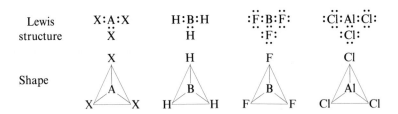

AX₄ MOLECULES: TETRAHEDRAL In CH_4 there are four electron pairs in the valence shell of carbon. Four electron pairs have a tetrahedral arrangement and so the CH_4 molecule has a tetrahedral shape. The angle between any pair of bonds is 109°. All similar molecules formed by the elements of group 4, such as CCl_4 and SiF_4, have the same shape. We describe them as AX_4 molecules.

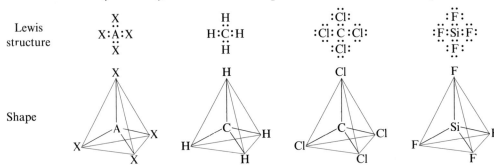

The ammonium ion also has a tetrahedral AX_4 geometry:

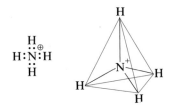

AX₃E MOLECULES: TRIANGULAR PYRAMIDAL In the valence shell of nitrogen in NH_3 there are four electron pairs with a tetrahedral arrangement but one of them is a nonbonding or lone pair and only three of them are bonding pairs. We describe such a molecule as an AX_3E molecule, where E represents the lone pair. It has a triangular pyramidal shape. It does *not* have a tetrahedral shape because it is the positions of the atomic nuclei A and X that determine the shape of a molecule and the nitrogen atom is at the apex of a triangular pyramid formed by the three hydrogen atoms. The arrangement of the four pairs of electrons is tetrahedral and the angle between any pair of bonds is predicted to be 109° as in CH_4 but the shape of the molecule is triangular pyramidal. All similar molecules formed by the elements of group 5 have the same shape.

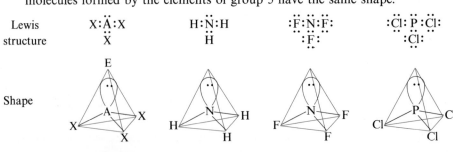

AX_2E_2 MOLECULES: ANGULAR There are four electron pairs in the valence shell of the oxygen atom in H_2O but two of them are nonbonding pairs and only two are bonding pairs. We describe such a molecule as an AX_2E_2 molecule. The four electron pairs have a tetrahedral arrangement but the molecule is angular because the shape describes the positions of the three atoms which form an angle—they do not have a linear arrangement as in an AX_2 molecule. The angle between the bonds is predicted to be 109° as in CH_4 and NH_3. All similar molecules formed by the elements of group 6 such as OF_2 and SCl_2 have the same angular shape.

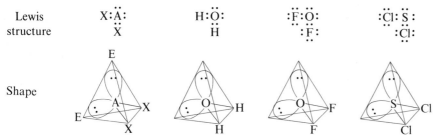

AXE_3 MOLECULES: LINEAR (DIATOMIC) Molecules such as HF have three nonbonding and one bonding pair in their valence shells. They may be described as AXE_3 molecules and because they are diatomic they are necessarily linear.

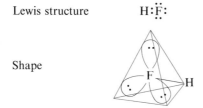

AX_2E MOLECULES: ANGULAR Group 4 elements form a few molecules in which they are bonded to only two other atoms rather than four. Examples include $SnCl_2$ and the very reactive molecule CH_2. The central atom has only three pairs of electrons in its valence shell and is therefore an exception to the octet rule. One of these pairs is a nonbonding pair and two are bonding pairs which therefore give an angular molecule with a predicted bond angle of 120°.

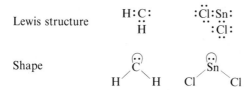

The molecular shapes of singly bonded molecules that we have described up to this point are summarized in Table 4.8. We next consider the shapes of some molecules containing double and triple bonds.

THE SHAPES OF MOLECULES CONTAINING DOUBLE AND TRIPLE BONDS

Carbon, nitrogen, and oxygen form many molecules in which there are double and triple bonds. We can readily extend the ideas in the previous section to enable us to predict the shapes of these molecules. The shapes of singly bonded molecules can be predicted on the basis that the bonds and the nonbonding pairs adopt the arrangement that maximizes their distance apart. Similarly the shapes of molecules in which there are double and triple bonds can be predicted on the basis that the bonds and the nonbonding pairs adopt that arrangement that keeps them as far apart as possible. For the molecules containing multiple bonds formed by carbon, nitrogen, and oxygen, there are three possible geometries.

LINEAR AX$_2$ MOLECULES   Examples include CO_2, HCN, and C_2H_2.

| Lewis structure | :Ö::C::Ö: | H:C:::N: | H:C:::C:H |
| Shape | O=C=O | H—C≡N | H—C≡C—H |

In each case the carbon atom is bonded to only two other atoms and it has no lone pairs in its valence shell, so the geometry around each carbon atom is AX$_2$ linear. We note that even if there is more than one "central" atom in a molecule we can predict the geometry at each atom, as in ethyne which has a linear geometry at each carbon atom and therefore has an overall linear shape.

TRIANGULAR AX$_3$ MOLECULES   An example is ethene, C_2H_4.

Lewis structure

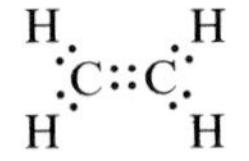

$$\begin{array}{cc} H & H \\ :C::C: \\ H & H \end{array}$$

Shape

$$\begin{array}{cc} H & H \\ \diagdown & \diagup \\ C=C \\ \diagup & \diagdown \\ H & H \end{array}$$

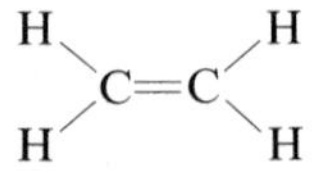

The geometry at each carbon atom is triangular with predicted bond angles of 120°. The molecule has an overall planar geometry (the shape of C_2H_4 and related molecules is discussed in more detail in Chapter 9).

We will encounter many more molecules in which there are multiple bonds in the following chapters and we will discuss some of the finer details of the geometry of these and other molecules in Chapter 9.

═ *Example 4.9* ═══

PREDICTING MOLECULAR SHAPES

Classify the following molecules and ions as AX$_4$, AX$_3$E, AX$_2$E$_2$, or AXE$_3$, and describe their shapes.

NCl_3 PF_3 OH^- $AsCl_3$ PH_4^+ SiO_4^{4-} OCl_2

Solution

	Lewis Structure	*Class of Molecule*	*Shape*
NCl_3	:Cl:N:Cl: :Cl:	AX$_3$E	Triangular pyramidal
PF_3	:F:P:F: :F:	AX$_3$E	Triangular pyramidal
OH^-	$^{\ominus}$:Ö—H	AXE$_3$	Linear

	Lewis Structure	Class of Molecule	Shape
$AsCl_3$	:Cl:As:Cl: :Cl:	AX_3E	Triangular pyramidal
$PH_4{}^+$	H H:P:H H	AX_4	Tetrahedral
$SiO_4{}^{4-}$	:O: :O:Si:O: :O:	AX_4	Tetrahedral
OCl_2	:O:Cl: :Cl:	AX_2E_2	Angular

IMPORTANT TERMS

The **alkali metals** are the elements of group 1.

The **alkaline earth metals** are the elements of group 2.

The **bond length** is the distance between the nuclei of two atoms that have a chemical bond between them.

The **chalcogens** are the elements of group 6.

A **chemical bond** is said to exist between any two atoms that are held strongly together.

The **core** of an atom is the nucleus plus the completed inner electron shells.

The **core charge** of an atom is equal to the atomic number Z minus the total number of electrons in the completed inner shells.

A **covalent bond** consists of a pair of electrons shared between two atoms.

The **covalent** (or atomic) **radius** is one-half the distance between two atoms of the same kind held together by a covalent bond.

A **double bond** consists of two pairs of shared electrons.

The **effective nuclear charge** is the apparent nuclear charge exerted on an electron in an atom. It is equal to the actual nuclear charge reduced by an amount that makes allowance for the repulsion of the other electrons.

The **formal charge** of an atom in a molecule or ion is given by either of the following equivalent rules:

Formal charge

$$= \begin{pmatrix} \text{Core} \\ \text{charge} \end{pmatrix} - \begin{pmatrix} \text{Number of} \\ \text{unshared electrons} \end{pmatrix} - \frac{1}{2}\begin{pmatrix} \text{Number of} \\ \text{shared electrons} \end{pmatrix}$$

$$= \begin{pmatrix} \text{Group} \\ \text{number} \end{pmatrix} - \begin{pmatrix} \text{Number of} \\ \text{unshared electrons} \end{pmatrix} - \begin{pmatrix} \text{Number} \\ \text{of bonds} \end{pmatrix}$$

It is the charge that the atom would have if the electrons in all of the covalent bonds that it forms were shared equally with its neighbors.

A **group** is a vertical column of the periodic table.

The **halogens** are the elements of group 7.

There is said to be an **ionic bond** between two oppositely charged ions that are held together by their electrostatic attraction.

Ionic compounds are composed of oppositely charged ions held together by electrostatic attraction, that is, by ionic bonds.

The **ionization energy** is the energy needed to remove an electron from an atom in the gas phase.

A **Lewis (electron dot) symbol** is the symbol of an element surrounded by a number of dots equal to the number of electrons in its valence shell.

A **negative ion** is a negatively charged atom or molecule. Negative ions may be formed by the addition of one or more electrons to a neutral atom or molecule.

A **lone pair** is a pair of electrons in the valence shell of an atom in a molecule that is not involved in the bonds in the molecule. It is also called a **nonbonding** or **unshared pair**.

The **noble gases** are the elements of group 8.

An **octet** is a valence shell containing eight electrons.

The **octet rule** states that when an atom forms a compound, it gains, loses, or shares electrons to obtain eight electrons in its valence shell.

A **period** is a horizontal row of the periodic table.

A **positive ion** is a positively charged atom or molecule. Positive ions may be formed by the removal of one or more electrons from a neutral atom or molecule.

A **semimetal (metalloid)** is an element whose properties are intermediate between those of a metal and those of a nonmetal.

A **shell** is one of the successive layers of electrons around an atom.

A **single bond** is formed by one pair of shared electrons.

A **transition element** is one of the ten elements in each period between groups 2 and 3 in the periodic table, excluding the 14 lanthanide and 14 actinide elements.

A **triple bond** is formed by three shared pairs of electrons.

Valence is the combining power of an element. In ionic compounds it is equal to the charge on an ion of the element. In covalent compounds it is the number of bonds formed by a neutral atom of the element.

The **valence shell** is the outer electron shell of an atom.

P R O B L E M S *

Periodic Table

1. Locate each of the following elements in the periodic table by group and by period. Name the element and classify it as a metal, a semimetal (metalloid), or a nonmetal:

He   P   K   Ca
Te   Br   Al   Sn

2. **(a)** To what period and group of the periodic table does the element with each of the following atomic numbers belong?

2   7   9   11   16   19

(b) Give the symbol of each of the above elements. From their positions in the periodic table, classify each as a metal or a nonmetal.

3. Identify three pairs of elements whose average atomic masses do not follow the order of their atomic numbers. Explain.

4. The element with atomic number 22 forms crystals that melt at 1668 °C to give a liquid that boils at 3313 °C. The crystals are hard, conduct heat and electricity, can be drawn into thin wires, and emit electrons when exposed to ultraviolet light. On the basis of these properties, classify the element as a metal or nonmetal. Which element is it?

5. By referring to the periodic table, classify each of the following elements as a main group element or a transition metal. For the main group elements, indicate the group to which the element belongs and whether it is a metal or a nonmetal.

Se   P   Mn   Kr   W   Al   Pb

6. By referring to the periodic table, classify each of the following elements as a main group element or a transition metal. For the main group elements, indicate the group to which the element belongs and whether it is a metal or a nonmetal.

Ar   Rb   V   Br   Ba   Fe   Au

7. Classify each of the following elements as a metal or a nonmetal. To which group of the periodic table does each belong? How many valence electrons does each have? What is the valence of each element, assuming that it obeys the octet rule?

Li   Mg   S   P   Br
Ne   As   Se   Cl   Ba

8. Classify each of the following elements as a metal or a nonmetal. To which group of the periodic table does each belong? How many valence electrons does each have? What is the valence of each element, assuming that it obeys the octet rule?

Ca   Al   F   Cs   N
I   K   Ar   O   Si

9. Write balanced equations for the reactions between the following, if any:

(a) Magnesium with steam

(b) Sulfur with hydrogen

(c) Sodium with iodine

(d) Potassium with water

(e) Hydrogen with chlorine

(f) Neon with water

10. Predict the empirical formulas of the chlorides formed by the elements in group 3 of the periodic table.

11. Francium, Fr, is an alkali metal. Tin, Sn, is in the same group of the periodic table as carbon. Astatine, At, is a halogen, and radon, Rn, is a noble gas. Write formulas for their hydrides (if any). What are the expected products (if any) of the reactions of these elements with chlorine?

12. Astatine is a radioactive halogen. Predict the following properties of astatine:

(a) physical state at 25 °C (solid, liquid, or gas)

* Answers to problems numbered in blue appear at the end of the book.

(b) formula of the potassium salt

(c) formula of a gaseous astatine molecule

(d) color of solid astatine

13. Assign each of the following elements to its appropriate group in the periodic table; classify each as a metal or a non-metal, and write the empirical formula of its hydride (if any):

C   Ca   He   B   Cl
Li   O   F   P   Mg

14. Complete and balance the following equations:

(a) $Mg(s) + Br_2(l) \longrightarrow$

(b) $Ca(s) + O_2(g) \longrightarrow$

(c) $Na(s) + I_2(s) \longrightarrow$

(d) $Mg(s) + N_2(g) \longrightarrow$

(e) $K(s) + H_2O(l) \longrightarrow$

15. Complete and balance the following equations:

(a) $Li(s) + S(s) \longrightarrow$

(b) $Ca(s) + H_2O(l) \longrightarrow$

(c) $Ne(g) + H_2O(l) \longrightarrow$

(d) $Sr(s) + Br_2(l) \longrightarrow$

(e) $Mg(s) + H_2(g) \longrightarrow$

16. **(a)** Arrange the following elements in pairs in terms of their greatest similarity of chemical and physical properties:

Na   Mg   C   Cl   Ca   Si   K   F

(b) Write balanced equations for their expected reactions with hydrogen and oxygen, respectively.

17. **(a)** To which group of the periodic table does each of the following belong?

Mg   K   Br   P   Si   Al   S   Pb

(b) What are the empirical formulas of the fluorides, and the sulfides, of the above elements?

18. In some versions of the periodic table, hydrogen is placed at the top of group 7, rather than at the top of group 1. Justify this location.

19. Give the empirical formula of the simplest compound formed between each of the following pairs of elements:

Fluorine and oxygen   Aluminum and sulfur
Boron and chlorine   Carbon and sulfur
Magnesium and nitrogen

Shell Model

20. The atomic numbers of phosphorus, carbon, and potassium, are, respectively, 15, 6, and 19. Without reference to any other information, deduce, for each of these elements, **(a)** the number of electrons in each shell and **(b)** the core charge.

21. What is meant by the *valence shell* of an atom? How many electrons are there in the valence shell of each of the following atoms?

(a) Boron   **(b)** A halogen   **(c)** Neon

(d) Magnesium   **(e)** Helium

22. Determine the number of electrons in each of the following, and in each case give the arrangement of the electrons in shells:

(a) A neutral chlorine atom

(b) A negatively charged chlorine atom (Cl^-)

(c) A silicon atom

(d) A positively charged neon atom (Ne^+)

23. **(a)** Which of the electron shells of an atom is called the valence shell?

(b) For the elements of a given group of the periodic table, what is the common feature of their valence shells?

(c) How many electrons are there in the valence shells of each of the following?

N N^{3-} Be Be^{2+} O
O^{2-} Na Na^+ Cl Cl^-

(d) Which of the species in part (c) are described as *isoelectronic*?

24. What is the core charge of each of the following atoms and ions?

C   Mg Mg^{2+} Sl
O O^{2-} S^{2-} Br

25. What are the core charges of atoms of each of the following?

(a) Fluorine   **(b)** Nitrogen   **(c)** Sulfur

(d) Lithium   **(e)** Oxygen   **(f)** Magnesium

(g) Cesium   **(h)** Silicon   **(i)** Iodine

Ionization Energies

26. Explain what is meant by the *ionization energy* of an element. Why is there a tendency for the ionization energy to increase in going from left to right across any *period*, and to decrease in going from the top to the bottom of any *group* in the periodic table?

27. Without consulting a table of ionization energies, deduce the order of ionization energies for the atoms F, Ne, and Na. Explain.

28. Without consulting a table of ionization energies, arrange the following atoms in order of the increasing value of their ionization energies.

   Ba   Cs   F   S   As

29. How much energy is needed to convert 1.00 g of sodium atoms to gaseous Na^+ ions?

30. Explain why the ionization energy of helium is greater than that of any other atom.

Lewis Structures

31. What relationship exists between the Lewis electron dot symbol of an element and its position in the periodic table?

32. Write Lewis (electron dot) symbols for each of the following elements:

   K   Ca   B   Sn   Sb
   Te   Br   Xe   As   Ge

33. Four elements have the following Lewis symbols:

   A·   ·Ḋ·   ·Ë·   :G̈·

   (a) Place each element in its appropriate group in the periodic table.

   (b) Which of these elements are expected to form an ion? What is the charge on each of these ions?

34. In order to attain a structure in which all of the electron shells are filled, how many electrons must be

   (a) *Lost* by each of the following atoms?

   Li   H   Al   Na
   Ca   Mg   Rb   Sc

   (b) *Gained* by each of the following atoms?

   O   H   Cl   N   P   S   F   I

35. On the basis of the octet rule and the group in the periodic table to which each belongs, predict the charges on the ions formed by each of the following elements, and write their symbols:

   (a) Magnesium   (b) Rubidium   (c) Bromine

   (d) Sulfur   (e) Aluminum   (f) Lithium

36. Draw the Lewis structure of each of the following ionic compounds:

   (a) LiCl   (b) Na_2O (c) AlF_3

   (d) CaS   (e) $MgBr_2$

37. What is the empirical formula of the compound composed of each of the following pairs of ions?

   (a) $NH_4{}^+$, S^{2-} (b) Fe^{3+}, O^{2-}

   (c) Cu^+, O^{2-} (d) Al^{3+}, Cl^-

38. Predict the empirical formula of the ionic compound formed by each of the following pairs of elements, and write their Lewis structures:

   (a) Li, S   (b) Be, O   (c) Mg, Br

   (d) Na, H   (e) Al, I

39. Predict the empirical formula of the ionic compound formed by each of the following pairs of elements, and write their Lewis structures:

   (a) Ca, I   (b) Ca, O   (c) Al, S

   (d) Ca, Br   (e) Rb, Se   (f) Ba, O

40. Write the empirical formula of each of the following compounds, and give the balanced equation for the formation of each from its elements:

   (a) Barium iodide   (b) Aluminum chloride

   (c) Calcium oxide   (d) Sodium sulfide

   (e) Aluminum oxide

41. Select from the following as many pairs of elements as possible that would be expected to form *ionic* binary compounds, and write the Lewis structures:

   H   O   F   Mg   Al   Ca

42. What is the Lewis (electron dot) symbol for each of the following atoms and ions? Select from this list two pairs of isoelectronic species:

   (a) Ca   (b) Ca^{2+} (c) Ne   (d) O^{2-}

   (e) S^{2-} (f) Cl   (g) Ar   (h) N^{3-}

43. Draw the Lewis structures for each of the following molecules:

   (a) H_2 (b) HCl   (c) HI   (d) PH_3

   (e) SiF_4 (f) F_2O (g) Cl_2

44. Draw Lewis structures for each of the following molecules and ions:

   H_2CO P_2 CN^-
   H_2NNH_2 HOOH

45. In one of its several different forms, the element sulfur consists of S_6 molecules, in which the sulfur atoms are joined in a ring. Draw a Lewis structure for S_6.

46. Use the periodic table to predict the empirical formulas of the hydrides of antimony, bromine, tellurium, and tin, and write their Lewis structures.

47. Find the formal charges on each of the atoms in each of the following:

NH_4^+ BH_4^- CH_3^+
CH_3^- SCl_3^+

48. Draw the Lewis structure of each of the following ions, and assign formal charges where appropriate:

NH_2^- H_3O^+ H_2F^+
PH_4^+ BF_4^- PCl_4^+

49. To which elements does the octet rule always apply in writing the Lewis structures of their compounds?

50. Draw Lewis structures for the hydrides and oxides of the elements in the third period.

51. Draw Lewis structures for each of the following, and assign formal charges to each of their atoms. The first atom in the formula is the central atom of the molecule in each case:

OH_3^+ OH^- NF_4^+ SF_2
CS_2 OF_2 BF_4^-

52. Which of the following compounds contains a central atom which is an exception to the octet rule?

Cl_2O $BeCl_2$ $AlCl_3$ PCl_3

Covalent Radii

53. Why do the covalent radii of the atoms in any *period* generally decrease in going from left to right in the periodic table, and those of the atoms in any *group* increase in going from the top to the bottom? Why are no covalent radii given for He, Ne, and Ar?

54. The bond length in Br_2 is 227 pm, and the covalent radius of carbon is 77 pm. Predict the carbon–bromine bond lengths in carbon tetrabromide, CBr_4.

55. From the data in the accompanying table, calculate the covalent radii of each atom, and predict the bonds lengths in Br_2, $BrCl$, and I_2.

Molecule	Bond Length (pm)
Cl_2	198
CCl_4	176
CBr_4	194
CI_4	215

56. Without consulting a table of covalent radii or the periodic table, predict which atom of the following pairs is expected to have the larger covalent (atomic) radius:

(a) F and Cl   **(b)** B and C   **(c)** C and Si
(d) P and Al   **(e)** Si and O

57. (a) On the basis of the C–C distance of 154 pm in diamond, the P–P distance of 220 pm in P_4, the S–S distance of 208 pm in S_8, and the Cl–Cl distance of 198 pm in Cl_2, what are the covalent (atomic) radii of C, P, S, and Cl, respectively?

(b) What are the expected lengths of the bonds formed by the central atoms in PCl_3, CCl_4, SCl_2, and $P(CH_3)_3$?

VSEPR Model and Molecular Geometry

58. Draw diagrams of each of the following types of molecule, and give the value of the XAX bond angle in each case:

(a) A linear AX_2 molecule
(b) An equilateral triangular AX_3 molecule
(c) A tetrahedral AX_4 molecule

59. Draw diagrams to illustrate the geometry of each of the following types of molecule. In each case name the geometric shape:

(a) AX_3E **(b)** AX_2E_2 **(c)** AXE_3
(d) AX_2E **(e)** AXE_2 **(f)** AX_2

60. BCl_3 is a planar molecule with 120° bond angles, while PCl_3 is pyramidal with 100° bond angles. Explain why these two tetraatomic molecules have such different shapes.

61. Using the AX_nE_m nomenclature, describe each of the following molecules and deduce their shapes:

H_2O H_3O^+ PCl_3 BCl_3 SiH_4

62. Using the AX_nE_m nomenclature, describe each of the following molecules and deduce their shapes:

BH_4^- H_2S NH_4^+ BeH_2 BeH_4^{2-}

63. Use the VSEPR model to predict the geometric shape of each of the following molecules and ions:

BO_3^{3-} O_3 $SnBr_2$ $PbCl_4$

64. Use the VSEPR model to predict the geometric shape of each of the following molecules and ions:

BF_4^- CF_4 NF_4^+

65. Draw Lewis structures for each of the chlorides of the elements of the second period from Be to F. Describe each using the AX_nE_m nomenclature, name the geometric shape, and draw a diagram to illustrate the shape.

Miscellaneous

*66. Draw a cube and connect the four appropriate corners to form a tetrahedron. Join the midpoint of the cube (the midpoint of the tetrahedron) to each of the four corners of the tetrahedron. These lines represent the bonds in a tetrahedral AX_4 molecule, such as CH_4. Using trigonometry, calculate the angle between the bonds.

*67. When Mendeleev proposed his periodic table in 1869, he predicted that an element unknown at the time would be discovered between calcium and titanium in period 4, and he predicted some of the expected properties. When this element was discovered in 1879, it was called scandium, Sc. By comparison with the following properties of calcium and titanium, make your own predictions about some of the properties of scandium and find some data in Chapter 21 to verify some of your predictions:

Element		Melting Point (K)	Boiling Point (K)	Density (g cm^{-3})
Ca	metal	1110	1757	1.55
Ti	metal	1941	3560	4.50
Sc	?	?	?	?

68. (a) Explain what is meant by the core charge of an atom.

(b) What is the relationship between the core charge of an atom and its position in the periodic table?

(c) What is the core charge of each of the following?

    H   C   N   F   Ne
    Al S   Cl Ar   Ca

(d) Which of the atoms in each of the following pairs has the higher ionization energy?

    Ne, Ar; Na, Cl; Be, Mg; F, Cl; N, P

* The asterisk denotes the more difficult problems.

*69. Indium oxide contains 82.7 mass % indium. It occurs naturally in ores containing zinc oxide, $ZnO(s)$, and it was therefore originally assumed to have the empirical formula InO. On this basis, calculate the atomic mass of indium, and predict its location in the periodic table. Explain why this location is unsuitable for indium. Mendeleev suggested that the empirical formula must be In_2O_3. Calculate the atomic mass of indium on this basis, and find its location in the periodic table. Is this location reasonable?

*70 An element A is a silvery white solid, melting point 845 °C, which reacts with water with the evolution of hydrogen. It reacts with $HCl(g)$ to give $H_2(g)$ and a solid chloride with empirical formula ACl_2. When 0.230 g of A was reacted completely with water, 144.1 mL of $H_2(g)$, measured at 25 °C and 740 mm Hg pressure, were evolved. The chloride was analyzed by dissolving 0.1456 g of ACl_2 in water and adding excess silver nitrate solution, to precipitate all the chlorine as insoluble silver chloride, $AgCl(s)$. The mass of AgCl obtained was 0.3760 g. What is the element A, and what is its atomic mass?

71. (a) Why are the ions O^{2-}, F^-, Na^+, and Mg^{2+} referred to as an isoelectronic series of ions?

(b) Explain the decrease in ionic radius in the following series of ions:

O^{2-} (140 pm) $>$ F^- (135 pm) $>$ Na^+ (102 pm) $>$ Mg^{2+} (72 pm)

72. With the aid of two appropriate examples in each case, explain each of the following terms:

(a) Bond angle   (b) Bond length
(c) Molecular geometry   (d) Bonding electron pair
(e) Nonbonding (lone) electron pair
(f) Double bond   (g) Triple bond

73. Write Lewis structures for each of the following:

    C_2H_4 C_2H_2 CN^-
    N_2 NO^+ P_2

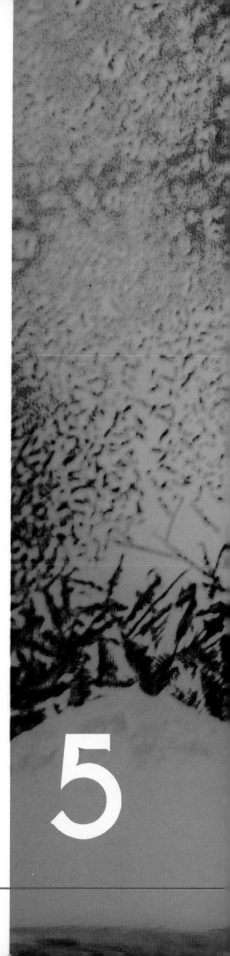

The Halogens

5

The periodic table showing H, He at top; metals, nonmetals, semimetals; transition elements:

Period	1	2										3	4	5	6	7	8
1	H																He
2	Li	Be		Metals		Nonmetals		Semimetals				B	C	N	O	F	Ne
3	Na	Mg				Transition Elements						Al	Si	P	S	Cl	Ar
4	K	Ca	Sc	Ti	V	Cr	Mn	Fe	Co	Ni	Cu	Zn	Ga	Ge	As	Se	Br	Kr
5	Rb	Sr	Y	Zr	Nb	Mo	Tc	Ru	Rh	Pd	Ag	Cd	In	Sn	Sb	Te	I	Xe
6	Cs	Ba	La	Hf	Ta	W	Re	Os	Ir	Pt	Au	Hg	Tl	Pb	Bi	Po	At	Rn
7	Fr	Ra	Ac	104	105	106	107											

The elements of group 7, fluorine, chlorine, bromine, iodine, and astatine, are collectively called the **halogens**. Astatine is an extremely rare, radioactive element, and we do not discuss it here. The halogens are a family of elements that have many of the characteristic properties of nonmetals and they provide an excellent example of the gradual change in properties that is often observed from one element to the next in a group of the periodic table. They form many important compounds and have many important uses. For example, chlorine is used to purify water supplies; the hypochlorite ion, OCl^-, is the active component of household bleach; fluoride ion is added to the water in many cities and to toothpaste to help prevent tooth decay; iodine is used as an antiseptic.

The halogens and their compounds provide many examples of both ionic and covalent substances. Discussion of the halogens will enable us to expand our knowledge and understanding of chemical bonding. The reactions of the halogens illustrate three very important types of reactions: oxidation–reduction, acid–base and precipitation reactions. We begin our study of these three reaction types in this chapter.

5.1
The Halogens

These elements are all composed of diatomic molecules in which the two atoms are held together by a single covalent bond. They all have the same Lewis structure:

$$:\ddot{F}-\ddot{F}: \quad :\ddot{Cl}-\ddot{Cl}: \quad :\ddot{Br}-\ddot{Br}: \quad :\ddot{I}-\ddot{I}:$$

From top to bottom in the group, the atoms follow the usual trend and become progressively larger. Therefore the molecules also increase in size, in the order $F_2 < Cl_2 < Br_2 < I_2$. The halogens are among the most reactive of the elements; this reactivity decreases from fluorine to iodine. The compounds of the halogens with another element are called fluorides, chlorides, bromides, and iodides, or, in general, **halides**; examples are sodium chloride, NaCl, and phosphorus trifluoride, PF_3.

Be careful to distinguish between the names halo*gen* for the elements and hal*ide* for the compounds. F_2 is the element fluo*rine*. NaF is the compound sodium fluo*ride*. You do not find the element fluo*rine*, F_2, in toothpaste, but you find the fluo*ride* ion F^-

TABLE 5.1
Some Properties of the Halogens

	Melting Point (°C)	Boiling Point (°C)	Color	Bond Length (pm)
F_2	−219	−188	Pale yellow	143
Cl_2	−101	−34	Yellow-green	199
Br_2	−7	59	Red-brown	228
I_2	113	185	Black	266

Table 5.1 summarizes some physical properties of the halogens. As the size of their molecules increases with increasing atomic number, so do their melting points and boiling points. At room temperature fluorine and chlorine are gases, whereas bromine is a red-brown liquid, and iodine is a black solid (see Experiment 5.1). These facts indicate that the strength of the intermolecular forces between the halogen molecules increases with increasing atomic number and molecular size, in the order $F_2 < Cl_2 < Br_2 < I_2$. This relationship between molecular size and the strength of intermolecular forces is typical of many substances, for reasons that we discuss in Chapter 12.

 ___ *Experiment 5.1*

Physical Properties of the Halogens

Chlorine is a pale yellow-green gas at room temperature. The tube inside the flask contains dry ice (solid carbon dioxide, −78°C). Yellow, liquid chlorine (b.p. −35°C) is condensing on the cold tube.

Bromine is red-brown liquid that is sufficiently volatile to fill the flask with vapor.

When solid iodine is gently heated it forms a violet vapor, which can be seen recrystallizing on the cold flask on top of the beaker.

FIGURE 5.1
A stockpile of pure salt.

The salt in 1 km³ of seawater would form a mound 100 m high and 250 m in diameter at its base.

SOURCES AND USES

Chlorine is the most abundant of the halogens, making up about 0.20% of the earth's crust (Table 3.2). Fluorine composes only 0.03% of the earth's crust; bromine and iodine are far less abundant. Because the halogens are very reactive, combining directly with many other elements and reacting with many compounds, they are not found in the free state on the earth. They occur in the form of compounds, particularly compounds of metals.

CHLORINE   Naturally occurring compounds of chlorine include alkali metal chlorides, such as sodium chloride and potassium chloride, and alkaline earth metal chlorides, such as magnesium chloride and calcium chloride. These chlorides occur in seawater and salt deposits that have been formed by the evaporation of ancient seas (Figure 5.1). The total amount of sodium chloride in the earth's crust is truly enormous. The oceans alone contain 3% sodium chloride, which amounts to a total of approximately 4.6×10^{19} kg of sodium chloride. This amount would form a solid cube of sides 277 km, or 172 miles.

Elemental chlorine, Cl_2, is used for bleaching wood pulp in the manufacture of paper, for bleaching textiles, for making plastics, such as poly(vinyl chloride) (PVC), insecticides, and dry-cleaning agents, and for the manufacture of bromine. It is widely used to kill bacteria in water for domestic use. The hypochlorite ion, OCl^-, is used for the same purpose in swimming pools.

BROMINE   Metal bromides occur in small amounts in seawater and in salt deposits as well as in the water from mineral springs. In the past seawater was used as a source of bromide ion, but the concentration of bromide in the oceans is only 65–70 ppm. Israel today produces a significant amount of bromine from the waters of the Dead Sea, which contains 4500–5000 ppm of bromine. In the United States most commercially produced bromine is obtained from subterranean *brines*, which are concentrated aqueous solutions of chloride and bromide that can be pumped to the surface. At the present time the most economical brine deposits are in Arkansas; they contain 3800–5000 ppm of bromide ion.

One of the most important uses for bromine is for the manufacture of bromine compounds used as gasoline additives. Bromine compounds are also widely used as pesticides and for treating plastic materials and textiles to make them fireproof. A considerable amount of silver bromide is used in the manufacture of photographic film.

IODINE   At one time seaweed (kelp), which concentrates the iodine in seawater, was an important source. When the seaweed is burned, its ashes contain as much as 1% of iodide ion. Until recently the most important commercial source of iodine was sodium iodate, $NaIO_3$, which occurs in small amounts in deposits of sodium nitrate found in Chile. Underground brines containing up to 100 ppm iodide have recently been discovered in Japan, in Michigan, and in Oklahoma. These are now the world's major source of iodine.

Iodine does not have as many important uses as the other halogens. Silver iodide is used in the manufacture of photographic film. Certain antiseptics are organic compounds containing iodine; a solution of the element in alcohol (tincture of iodine) is also a good, although rather old-fashioned, antiseptic. Iodine is an essential element in our diet because it is part of the structure of the growth-regulating hormone thyroxine, which is produced by the thyroid gland. Lack of iodine in our diet leads to enlargement of the thyroid gland, a condition called goiter. The necessary iodine can be obtained from fish, from sea salt, and from iodized table salt, which contains 0.01% of sodium iodide or potassium iodide.

FLUORINE Although it is less abundant than chlorine, fluorine is widely distributed in minerals such as fluorspar, CaF_2; cryolite, Na_3AlF_6; and fluorapatite, $Ca_5(PO_4)_3F$. Plants absorb small amounts of calcium from the soil and this calcium eventually passes into our bones and teeth, which are mainly hydroxyapatite, $Ca_5(PO_4)_3OH$. When OH^- is replaced by F^-, a harder, more acid-resistant layer of $Ca_5(PO_4)_3F$ is formed on the surface of teeth. Because this hard layer protects teeth from decay, most cities now add sodium fluoride to their water supply. A somewhat less effective protection is given by toothpaste containing fluoride ion. Other widely used compounds of fluorine are Teflon, which is used as a non-stick surface on cooking utensils, and Freons, such as CCl_2F_2, which are used as the working fluid in refrigerators and air conditioners.

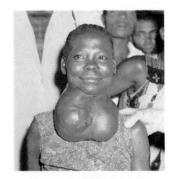
A woman suffering from goitre.

Teflon is a long-chain polymer molecule with the empirical formula CF_2:

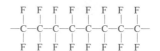

REACTIONS WITH NONMETALS

The halogens combine readily with almost all the other nonmetals to give compounds that consist of covalent molecules.

HYDROGEN HALIDES Hydrogen reacts with each of the halogens producing the **hydrogen halides** HF, HCl, HBr, and HI. For example,

$$H_2(g) + Cl_2(g) \longrightarrow 2HCl(g)$$

The reaction between hydrogen and fluorine is violent under all conditions. The reaction of hydrogen with chlorine is slow in the dark but becomes explosive in sunlight or other bright light or on heating to 250 °C. Bromine reacts more slowly with hydrogen, and iodine reacts still more slowly at r
ature. A temperature of at least 200 °C is needed to obtain a reasonably rapid reaction and in the industrial process for the production of HBr and HI by direct combination of the elements a catalyst is also used. If a stream of hydrogen is ignited in air the flame will continue to burn in an atmosphere of $Cl_2(g)$ or $Br_2(g)$, producing HCl or HBr (Figure 5.2). Thus although all the halogens react with hydrogen to form a hydrogen halide the reactivity of the halogens towards hydrogen decreases quite markedly from fluorine to iodine.

The hydrogen halides are all colorless gases that have pungent odors, and are all very soluble in water (see Experiment 5.2). Some of their properties are summarized in Table 5.2. The bond lengths of the molecules increase steadily from HF to HI as the size of the halogen atom increases. The melting points and boiling points increase from HCl to HI as the intermolecular forces become stronger with increasing atomic number and increasing size of the halogen atom. Hydrogen fluoride, however, has unexpectedly high melting and boiling points compared with the other halogen halides; these unusual properties will be discussed in Chapter 12.

NONMETAL HALIDES When chlorine is heated with carbon, carbon tetrachloride (tetrachloromethane) is produced:

$$C(s) + 2Cl_2(g) \longrightarrow CCl_4(l)$$

FIGURE 5.2
Hydrogen Burning in Chlorine.

Hydrogen burns with a blue-white flame in an atmosphere of chlorine to form HCl(g). The remaining yellow chlorine can be seen in the lower part of the beaker.

Experiment 5.2

Solubility of Hydrogen Chloride in Water

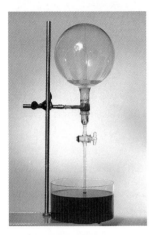

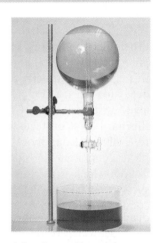

The flask contains colorless hydrogen chloride gas. When the tap is opened hydrogen chloride dissolves in the water, reducing the pressure in the flask so that water is drawn into the flask from the dish below, and creating a fountain.

Water is drawn into the flask so rapidly that it forms a jet that hits the top of the flask.

A few drops of bromothymol blue have been added to the water in the dish. The color changes to yellow as the water enters the flask because hydrogen chloride forms an acid solution in water. Hydrogen chloride is so soluble in water that the flask fills almost completely.

When red phosphorus is gently heated in a slow stream of chlorine, liquid phosphorus trichloride, PCl_3, is formed (see Figure 5.3):

$$2P(s) + 3Cl_2(g) \longrightarrow 2PCl_3(l)$$

If an excess of chlorine is used, pale yellow solid phosphorus pentachloride, PCl_5, is also produced (see Chapter 8).

When a stream of dry chlorine is passed over molten sulfur, an orange-red liquid is formed (Figure 5.3). It is a mixture of two chlorides of sulfur: S_2Cl_2,

	Melting Point (°C)	Boiling Point (°C)	Bond Length (pm)	Solubility in Water at 10 °C (g L^{-1})
HF	−83	20	92	Miscible
HCl	−115	−85	127	780
HBr	−89	−67	141	2100
HI	−51	−35	161	2340

TABLE 5.2
Some Properties of the Hydrogen Halides

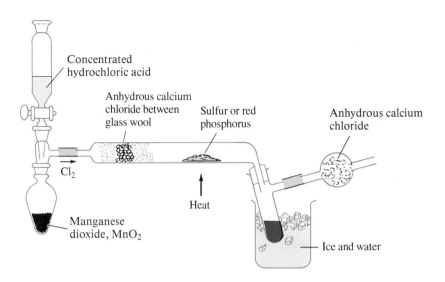

FIGURE 5.3
Preparation of Some
Nonmetal Chlorides.

The nonmetal chlorides PCl_3, S_2Cl_2, and SCl_2 may be prepared in this apparatus by passing chlorine over the heated element. The chlorine is first dried by passing it over anhydrous calcium chloride.

disulfur dichloride, which is a foul-smelling, orange liquid, and SCl_2, sulfur dichloride, which is a deep red liquid:

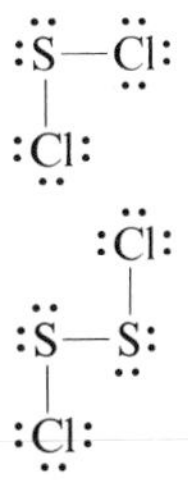

$$2S(s) + Cl_2(g) \longrightarrow S_2Cl_2(l)$$

$$S(s) + Cl_2(g) \longrightarrow SCl_2(l)$$

Many other nonmetal halides such as PF_3, CF_4, CBr_4, OF_2, and ICl can be made similarly. They are all covalent molecular compounds. The shapes of some nonmetal halides and their Lewis structures are given in Figure 5.4. Table 5.3 summarizes the properties of some nonmetal chlorides. Most of them are gases or liquids with low boiling points. The boiling points and melting points increase, with increasing atomic number, as we go down each group, so that a few of the halides of the heavier elements are solids.

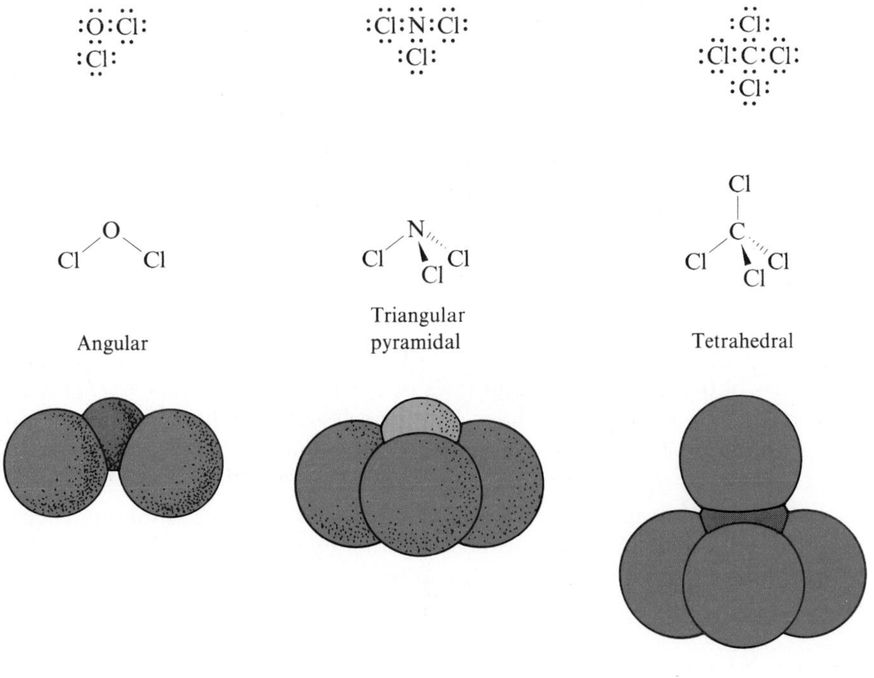

FIGURE 5.4
Lewis Structures and
Molecular Shapes of
Nonmetal Halides.

Halides of group 6 elements, such as Cl_2O, are all angular molecules like H_2O. The halides of group 5, such as PCl_3 and NCl_3, are all triangular pyramidal AX_3E molecules; they have the same shape as NH_3. The halides of group 4 are tetrahedral AX_4 molecules, with the group 4 atom at the center and a halogen atom at each of the four corners of the tetrahedron; CCl_4 is an example.

Angular

Triangular pyramidal

Tetrahedral

TABLE 5.3				
Properties of Some Chlorides of Nonmetals				
Period	Chloride	Melting Point (°C)	Boiling Point (°C)	Bond Length (pm)
2	CCl_4	−23	77	176
	NCl_3	−40	71	173
	OCl_2	−120	2	169
	FCl	−156	−100	163
3	$SiCl_4$	−68	57	201
	PCl_3	−94	76	204
	SCl_2	−122	59	200
	Cl_2	−101	−35	198
4	$GeCl_4$	−50	83	209
	$AsCl_3$	−16	130	216
	$SeCl_2{}^a$	—	—	—
	$BrCl$	−66	5	214
5	$SnCl_4$	−33	114	232
	$SbCl_3$	73	223	238
	$TeCl_2$	209	327	234
	ICl	27	97	232

^{a} $SeCl_2$ is an unstable compound that has never been isolated in the pure state.

Carbon tetrachloride is the best known and most widely used of these non-metal chlorides. It dissolves many of the other nonmetal halides and many other substances such as Br_2 and I_2 (see Experiment 1.6); it is therefore used as a solvent. Because it dissolves grease, it has been used as a cleaning fluid in dry cleaning. But recently CCl_4 has been replaced by other solvents since it has been found to be carcinogenic.

REACTIONS WITH METALS: METAL HALIDES

The halogens react with many metals to form halides (see Experiment 5.3). Some typical reactions are the following:

$$2Na(s) + Cl_2(g) \longrightarrow 2NaCl(s)$$
$$Ca(s) + F_2(g) \longrightarrow CaF_2(s)$$
$$Mg(s) + Cl_2(g) \longrightarrow MgCl_2(s)$$

The compounds formed in such reactions have quite different physical and chemical properties from the halides of the nonmetals. Some of the physical properties of the fluorides and chlorides of the alkali and alkaline earth metals are summarized in Table 5.4. They are all colorless crystalline solids (see Figure 5.5) with high melting points. In contrast, the nonmetal halides are generally gases, liquids, or solids of low melting point, as indicated in Table 5.5.

As we have seen in Chapter 4, the metal halides are ionic compounds consisting of oppositely charged ions held together by electrostatic attraction. The formation of these ionic compounds can be represented as follows:

$$\cdot Na + \cdot \ddot{\underset{..}{Cl}} : \qquad (Na^+)(:\ddot{\underset{..}{Cl}}:^-)$$
$$\cdot Ca \cdot + 2 \cdot \ddot{\underset{..}{F}} : \longrightarrow (Ca^{2+})(:\ddot{\underset{..}{F}}:^-)_2$$

FIGURE 5.5
Sodium Chloride Crystals.

Sodium chloride forms colorless cubic crystals. These crystals can be seen by looking at table salt under a microscope. Large crystals are sometimes found in salt deposits; they are called halite or rock salt crystals.

Experiment 5.3

Reactions of Metals with Chlorine

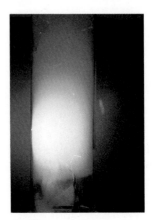

Sodium metal heated until it melts burns with a intense yellow flame when inserted into a jar of chlorine. A white smoke of sodium chloride particles is formed.

When steel wool is ignited in air and then inserted into a jar of chlorine it continues to burn, forming dense red-brown fumes of $FeCl_3$ which deposit on the sides of the jar. The sparks are white hot iron.

Small pieces of antimony react immediately, becoming white hot when dropped into a jar of chlorine. They can be seen here bouncing from the bottom of the jar. A white smoke of antimony trichloride $SbCl_3$ is formed.

Burning magnesium continues to burn with a bright white flame when inserted into a jar of chloride gas. White solid magnesium chloride is formed.

TABLE 5.4
Some Properties of the Alkali and Alkaline Earth Metal Fluorides and Chlorides

Compound	Melting Point (°C)	Boiling Point (°C)	Solubility in Water at 25 °C (g L^{-1})
LiF	845	1681	1.3
NaF	995	1704	40.1
KF	856	1501	1020[a]
RbF	775	1408	1310
CsF	682	1250	3700[a]
LiCl	610	1382	850[a]
NaCl	808	1465	360
KCl	772	1407	350
RbCl	777	1381	940
CsCl	645	1300	1900
BeF$_2$	—	800[b]	5500
MgF$_2$	1263	2227	0.13
CaF$_2$	1418	2500	0.016
SrF$_2$	1400	2460	0.12
BaF$_2$	1320	2260	1.6
BeCl$_2$	405	488	720[a]
MgCl$_2$	714	1418	550
CaCl$_2$	772	>1600	830[a]
SrCl$_2$	875	1250	560[a]
BaCl$_2$	1350	—	370[a]

[a] Solubilities of hydrates KF·2H$_2$O, CsF·H$_2$O, LiCl·H$_2$O, BeCl$_2$·4H$_2$O, CaCl$_2$·6H$_2$O, SrCl$_2$·6H$_2$O, BaCl$_2$·2H$_2$O.
[b] Sublimes (changes directly from solid to gas without forming a liquid).

TABLE 5.5
Melting Points (°C) of Some Chlorides

Metal Chlorides Ionic Bonds Solids			Nonmetal Chlorides Polar Covalent and Covalent Bonds Liquids and Gases				
LiCl 610	BeCl$_2$ 405		BCl$_3$ −107	CCl$_4$ −23	NCl$_3$ −40	OCl$_2$ −120	FCl −156
NaCl 808	MgCl$_2$ 714	AlCl$_3$ 192^a	SiCl$_4$ −68	PCl$_3$ −94	SCl$_2$ −122	Cl$_2$ −101	

^{a} Under pressure; sublimes at 180 °C.

The bonds in the metal halides are ionic, while those in the nonmetal halides are covalent. These are two extreme types of bonds; the bonds in many substances have an intermediate character. The nature of a bond between two atoms depends primarily on the *electronegativity* of the atoms. We take up this important concept in the next section.

EXERCISE 5.1

Write balanced equations for the reactions in Experiment 5.3.

═ *Example 5.1* ═══

REACTIONS OF THE HALOGENS WITH METALS AND NONMETALS

Write a balanced equation for each of the following reactions:

   (a) Barium with fluorine

   (b) Silicon with fluorine

   (c) Potassium with iodine

   (d) Phosphorus with bromine

In each case state whether you expect the product to be (i) a solid or (ii) a liquid or gas at room temperature.

Solution

   (a) Barium is in group 2 so the formula of barium fluoride is BaF$_2$. Therefore the equation for the reaction is

$$Ba(s) + F_2(g) \longrightarrow BaF_2(s)$$

   Barium fluoride is expected to be a solid because it is the halide of a metal and consists of Ba^{2+} and F$^-$ ions.

   (b) Silicon is in group 4 so the formula of silicon fluoride is SiF$_4$ and the equation for the reaction is

$$Si(s) + 2F_2(g) \longrightarrow SiF_4(g)$$

   Silicon tetrafluoride is expected to be either a gas or a liquid because it is the fluoride of a nonmetal and consists of covalent SiF$_4$ molecules. It is in fact a gas at room temperature.

(c) Potassium is an alkali metal in group 1 so the formula of potassium iodide is KI and the equation for the reaction is

$$2K(s) + I_2(s) \longrightarrow 2KI(s)$$

Potassium iodide is expected to be a solid because it is the halide of a metal and thus consists of K^+ and I^- ions.

(d) Phosphorus is in group 5 so the formula of its bromide is expected to be PBr_3 and the equation for the reaction is

$$2P(s) + Br_2(l) \longrightarrow 2PBr_3(l)$$

Phosphorus tribromide is expected to be a liquid or gas because it is the halide of a nonmetal and therefore consists of covalent molecules. It is in fact a liquid.

EXERCISE 5.2

Write a balanced equation for each of the following reactions:

(a) Strontium with bromine

(b) Arsenic with chlorine

(c) Rubidium with fluorine

(d) Bromine with chlorine

In each case state whether you expect the product to be (i) a solid or (ii) a gas or liquid at room temperature.

5.2
Electronegativity

In a molecule such as H_2, F_2, or Cl_2 the electron pair of the covalent bond is shared equally between the two atoms because they are identical and therefore attract the electrons of the bond equally strongly. However, in molecules such as HF, HCl, or ClF, in which the two atoms are not the same, one atom attracts electrons more strongly than the other. So the electron pair forming the bond is shared unequally between the two atoms. *The ability of an atom in a molecule to attract the electrons of a covalent bond to itself is called its* **electronegativity**. The greater the electronegativity of an atom, the more strongly it attracts the electrons of a bond. Electronegativity is a qualitative concept and is not a quantity that can be directly measured experimentally.

According to Coulomb's law, the force of attraction exerted by the positively charged core of an atom on an electron in its valence shell is given by

$$F = k\frac{Q_1 Q_2}{r^2} = k\frac{(Z_{core}e)(-e)}{r^2}$$

where $Q_1 = +Z_{core}e$, the core charge, $Q_2 = -e$, the charge on an electron, and r is the distance of the electron from the core. We see that this force is proportional to the core charge, $Z_{core}e$, and that it decreases with increasing distance of the electron from the core. We might expect that we could calculate a value for the electronegativity of an atom from this expression. However, the resultant force on any one electron in the valence shell depends not only on the attraction exerted on it by the positive core but also on the repulsions due to all the other electrons, which cannot be calculated in an exact manner.

Although the electronegativity of an atom cannot be measured experimentally, nor calculated accurately, electronegativity is a very useful concept, and approximate values have been estimated by several methods. Figure 5.6 gives values for the electronegativities of the elements of the main groups.

ELECTRONEGATIVITY AND THE PERIODIC TABLE

As we would expect from Coulomb's law, the electronegativity of an atom increases with increasing core charge and decreases with increasing distance of the valence shell from the core, that is, with increasing atomic size. Thus there are two important trends in electronegativity within the periodic table:

1. Electronegativity increases across a period as the core charge increases.
2. Electronegativity generally decreases from top to bottom in a group as atomic size increases and the bond electrons are further from the nucleus.

Figure 5.6 summarizes these trends.

The nonmetals on the right-hand side of the periodic table have high electronegatives, and the metals on the left-hand side have low electronegativities. To a good approximation, elements with an electronegativity, χ (Greek letter chi), of 2.0 or greater are nonmetals. Elements with an electronegativity χ of less than 2.0 are metals. Values are not given for helium, neon, and argon because there are no known compounds of these elements. The halogens are

FIGURE 5.6
Electronegativities of the Main Group Elements.

Across any period the electronegativity increases from left to right. Generally, the electronegativity decreases down any group from top to bottom, although there are some exceptions to this trend among the heavier elements.

Period	\multicolumn{8}{c}{Group}							
	1	2	3	4	5	6	7	8
1	H 2.2							He —
2	Li 1.0	Be 1.5	B 2.0	C 2.5	N 3.1	O 3.5	F 4.1	Ne —
3	Na 1.0	Mg 1.2	Al 1.3	Si 1.7	P 2.1	S 2.4	Cl 2.8	Ar —
4	K 0.9	Ca 1.0	Ga 1.8	Ge 2.0	As 2.2	Se 2.5	Br 2.7	Kr 3.1
5	Rb 0.9	Sr 1.0	In 1.5	Sn 1.7	Sb 1.8	Te 2.0	I 2.2	Xe 2.4

Electronegativity decreases

Electronegativity increases

among the most electronegative elements—indeed, fluorine is the most electronegative of all the elements ($\chi = 4.1$). The second most electronegative element is oxygen ($\chi = 3.5$), followed by krypton and nitrogen ($\chi = 3.1$), and chlorine ($\chi = 2.8$). The decrease in electronegativity on descending a group is largest in group 7 and smallest in group 1. There are some exceptions among the elements of period 4 to the general decrease in electronegativity on descending a group.

Because electronegativities cannot be determined in a quantitative manner, and because different methods of obtaining electronegativities give somewhat different values, the electronegativity values in Figure 5.6 have only a limited quantitative significance. Moreover, the electronegativity of a given element does not have a truly constant value. It varies somewhat from one molecule to another, depending on the number and the nature of the other atoms that are bonded to it. Nevertheless, electronegativity is an important and useful concept, as we will see in the next section and on many other occasions.

POLAR BONDS

If two atoms forming a diatomic molecule have exactly the same electronegativity, then the bonding electron pair will be shared exactly equally between the two atoms. In other words, the electron density of the bonding electron pair is distributed equally and symmetrically between the two atoms, as shown in Figure 5.7(a). When the two atoms forming the bond are identical, in molecules such as H_2, F_2 and Cl_2, for example, the sharing of the bonding electron density is always equal and symmetrical.

In contrast, if the atoms forming a bond have different electronegativities, the bonding pair is shared unequally. An example is the C—F bond: fluorine has a greater electronegativity than carbon and therefore the fluorine atom attracts the electrons of the bonding pair more strongly than carbon does. Thus the electron density of the bonding pair is not distributed equally between the valence shells of the two atoms but more of the density is in the valence shell of the fluorine atom than the carbon atom. As a result the fluorine atom has slightly more electron density in its valence shell than in the free atom and so it has a small negative charge. And the carbon atom has slightly less electron density in its valence shell than in the free atom so it has a small positive charge. These charges are of equal magnitude but opposite sign, and the overall charge is zero. Such small charges are denoted by the symbols $\delta+$ (delta plus) and $\delta-$ (delta minus), which indicate some fraction of the charge of one electron. The bond in a diatomic molecule such as HCl, in which the atoms carry small charges of equal magnitude but opposite sign,

$$\overset{\delta+}{H}-\overset{\delta-}{Cl}$$

is called a **polar covalent bond**. When the difference in the electronegativities of the two atoms forming a bond is zero, the charges on the two atoms are zero. The bonding electron pair is shared equally between the two atoms, and the bond is said to be a **nonpolar bond**, as in H_2, F_2, and Cl_2. The greater the difference in the electronegativities of two atoms forming a bond, the more polar is the bond; that is, the greater are the charges on its atoms.

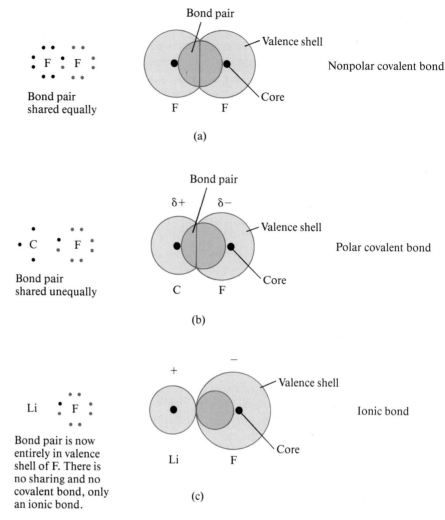

FIGURE 5.7
Nonpolar, Polar and Ionic Bonds.

(a) In a nonpolar bond such as that in the F_2 molecule the bond pair is shared equally between the two F atoms. Thus each atom effectively has seven electrons and since the core charge of F is $+7$ both fluorine atoms have a resultant charge of zero. (b) In a polar bond such as the C—F bond the bond electron pair is shared unequally between the carbon and fluorine atoms; the more electronegative fluorine atom acquires a greater share of the bond pair. It effectively has $7 + \delta$ electrons and therefore an overall negative charge of $-\delta$. (c) In an ionic bond such as that in LiF the "bond pair" has been acquired completely by the F atom so that it has eight electrons in its valence shell and charge of -1. It has become a fluoride ion and there is no longer a covalent bond but rather an ionic bond between Li^+ and F^-.

═ *Example 5.2* ═══════════════════════════════

POLAR AND NONPOLAR BONDS

Which of the following molecules have polar bonds and which have nonpolar bonds? For the polar bonds indicate the charges on the atoms.

H_2 FCl N_2 HCl O_2 ICl

Solution

H_2	Nonpolar	
FCl	Polar	$\overset{\delta-}{F}\!-\!\overset{\delta+}{Cl}$ since $\chi(F) > \chi(Cl)$
N_2	Nonpolar	
HCl	Polar	$\overset{\delta+}{H}\!-\!\overset{\delta-}{Cl}$ since $\chi(Cl) > \chi(H)$
O_2	Nonpolar	
ICl	Polar	$\overset{\delta+}{I}\!-\!\overset{\delta-}{Cl}$ since $\chi(Cl) > \chi(I)$

IONIC BONDS

If we consider the series of bonds F—F, O—F, N—F, C—F, B—F, Be—F, and Li—F, we expect the bond polarity to increase from zero in F_2 to a rather large value in LiF, as the electronegativity difference increases (see Figure 5.8). Indeed, the sharing of the electron pair between Li and F is so unequal that the electron pair is acquired almost completely by the fluorine atom. As a result, the fluorine atom has a negative charge that is nearly equal to -1. The fluorine atom has gained an electron to become the fluoride ion, F^-, and the lithium atom has lost an electron to become the lithium ion, Li^+. An electron has effectively been transferred from Li to F.

Thus lithium fluoride consists of positive lithium ions and negative fluoride ions held together by electrostatic attraction between the oppositely charged ions. Thus we no longer think of a lithium atom and a fluorine atom as being bonded by a shared electron pair. Rather, we think of the bond as arising from the electrostatic attraction between oppositely charged ions, and the bond is described as an **ionic bond**. An ionic bond can therefore be thought of as the extreme case of a polar bond, as indicated in Figure 5.8.

PREDICTING THE NATURE OF BONDS

It would be convenient if electronegativity values could be used to predict whether a particular bond is nonpolar covalent, polar covalent, or ionic. Unfortunately, the values are too approximate to enable us to make such predictions with complete confidence. If the electronegativity difference is zero, then

Period 2 fluorides	LiF	BeF_2	BF_3	CF_4	NF_3	OF_2	F_2
Electronegativity difference	3.1	2.6	2.1	1.6	1.0	0.6	0
Bond polarity	Highly polar		Polar				Non-polar
	Li^+ $:\ddot{F}:^-$		$\overset{\delta+}{C} : \overset{\delta-}{\ddot{F}}:$				$:\ddot{F}:\ddot{F}:$
Bond pair	Very unequally shared		Unequally shared				Equally shared
Type of bond	Ionic		Polar covalent				Covalent

FIGURE 5.8
Variation of Bond Polarity Across the Periodic Table.

The bonds in the fluorides of the second-period elements become increasingly polar from right to left. Fluorine, F_2, is nonpolar; OF_2, NF_3, CF_4, and BF_3 are covalent molecules with increasingly polar bonds; LiF and BeF_2 have highly polar bonds and are usually described as ionic. In the solid state LiF and BeF_2 consist of an infinite three-dimensional array of positive and negative ions.

the bond is certainly nonpolar covalent. If the electronegativity difference is 1.0 or smaller, the bond is almost certainly polar covalent. Examples include the bonds in PCl_3 (0.7), SCl_2 (0.4), and NF_3 (1.0). If the electronegativity difference is greater than 2.0, the bond is almost certainly ionic. Examples include the bonds in NaF (3.1) and K_2O (2.6). For electronegativity differences between 1.0 or 2.0, however, predictions are difficult to make. For example, AsF_3 (1.9) is a molecular compound with polar covalent bonds, but NaCl (1.8) is an ionic compound composed of Na^+ and Cl^- ions.

A simpler and more reliable guide to the nature of the bonds between two atoms of the main group elements is given by the following generalizations:

- Bonds between a group 1 or a group 2 metal atom and a nonmetal atom are ionic.
- Bonds between two different nonmetal atoms are polar covalent.
- Bonds between two identical nonmetal atoms are nonpolar covalent.

For polar covalent bonds we can, however, use the difference in the electronegativity of the two atoms to decide which bonds are the most polar. For example, reference to Figure 5.8 shows that the NF bond ($\chi(F) - \chi(N) = 1.0$) is expected to be less polar than the CF bond ($\chi(F) - \chi(C) = 1.6$).

═ *Example 5.3* ═══

PREDICTING THE NATURE OF BONDS

State whether the bond between each of the following pairs of atoms is ionic, polar covalent, or nonpolar covalent.

LiH   LiF   CH   NH   OH   NN   RbBr   SiH   CaO

Which of the polar covalent bonds would be expected to have the greatest polarity?

Solution

LiH	Metal–nonmetal	Ionic
LiF	Metal–nonmetal	Ionic
CH	Nonmetal–nonmetal	Polar covalent
NH	Nonmetal–nonmetal	Polar covalent
OH	Nonmetal–nonmetal	Polar covalent
NN	Identical nonmetals	Nonpolar covalent
RbBr	Metal–nonmetal	Ionic
SiH	Nonmetal–nonmetal	Polar covalent
CaO	Metal–nonmetal	Ionic

We can decide which of the polar covalent bonds will have the greatest polarity by considering the difference in the electronegativities of the two atoms. Reference to Figure 5.6 shows that the greatest difference (1.3) is between O and H, so that the OH bond is the most polar of the polar covalent bonds.

EXERCISE 5.3

State whether the bonds in each of the following substances are ionic, polar covalent, or nonpolar covalent:

CaH_2 $MgCl_2$ SiH_4 NH_3 P_4 SO_2 Na_2O Cl_2O

EXERCISE 5.4

Arrange the following bonds in order of increasing polarity:

C—O   C—S   P—P   B—F

5.3
Ionic Substances

We have seen that the nonmetal halides consist of molecules in which the atoms are held together by strong covalent bonds. In contrast, metal halides consist of ions held together by strong ionic bonds, that is, by the electrostatic attraction between oppositely charged ions. How do we know, in fact, that a substance such as lithium fluoride or sodium chloride has ionic bonds? One important piece of evidence is that such substances conduct an electric current when melted.

ELECTRICAL CONDUCTIVITY

If copper wires connected to a current source and a light bulb are dipped into molten lithium chloride, the bulb lights up (see Experiment 5.4). Molten (liquid) lithium chloride is therefore capable of conducting an electric current. Any substance through which an electric current will pass when an electric potential (voltage) is applied to it is said to be an *electrical conductor* and to have an **electrical conductivity**.

An electric current is composed of moving electric charges. In the copper wires the current is conducted by electrons (as we will see in Chapter 10). In molten lithium chloride the current is conducted by positive lithium ions and

Experiment 5.4

Electrical Conductivity of Molten Lithium Chloride

 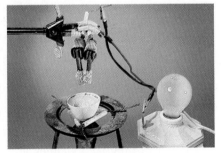

The crucible contains lithium chloride, which is heated to its melting point (610°C) to give a colorless liquid. The light bulb connected to a circuit with graphite electrodes shines brightly, showing that molten lithium chloride is an electrical conductor.

If the heating is stopped and the electrodes are removed from the molten lithium chloride, a plug of solid lithium chloride forms between the electrodes. The bulb does not light, showing that solid lithium chloride is not an electrical conductor.

negative chloride ions, which move in opposite directions through the liquid under the influence of the applied potential. In contrast, many liquid nonmetal halides, such as CCl_4, PCl_3, and SCl_2, do not conduct an electric current; they are nonconductors, or *insulators*.

A solid metal halide, unlike a solid metal, does not, however, conduct a current because the ions are fixed in position and are unable to move relative to each other. In a solid metal the conductivity is due to electrons which are free to move through the solid (see Chapter 10).

When a metal halide is dissolved in water, it gives a solution that also conducts an electric current. In the solution the ions separate from each other and move independently through the solution. In contrast, substances that are soluble in water but consist of covalent molecules, such as alcohol (ethanol) and sugar (sucrose), do not give conducting solutions. Substances that dissolve in water to give conducting solutions are called **electrolytes**. Substances that dissolve in water to give nonconducting solutions are called **nonelectrolytes**.

Other evidence that metal halides are ionic comes from their structures, which we consider in the next section.

THE STRUCTURES OF IONIC CRYSTALS

When one sodium atom reacts with one chlorine atom to form a sodium ion and a chloride ion,

$$\cdot Na + \cdot \ddot{\underset{..}{Cl}} : \longrightarrow Na^+ : \ddot{\underset{..}{Cl}} :^-$$

these two ions are held together by electrostatic attraction, giving a diatomic molecule in which the bond is ionic. Such a molecule is sometimes called an **ion pair**. This ionic molecule, or ion pair, occurs only in gaseous sodium chloride at very high temperature.

When a large number of positive ions and negative ions are formed, each positive ion tends to surround itself with as many negative ions as possible, and each negative ion tends to surround itself with as many positive ions as it can. The result is a structure consisting of a regular pattern of alternate positive and negative ions in which no individual molecules can be distinguished. In the structure of sodium chloride (see Figure 5.9), each sodium ion is surrounded by six chloride ions, and each chloride ion is surrounded by six sodium ions. The six chloride ions are arranged in the form of an **octahedron** (see Figure 5.10) around each Na^+ ion. Similarly, the six Na^+ ions surrounding each Cl^- ion have an octahedral arrangement. This regular arrangement continues indefinitely in three dimensions throughout the whole crystal. The structure is therefore described as an *infinite three-dimensional structure*. Thus an **ionic crystal** is composed of an infinite array of positive and negative ions in which no individual molecules can be distinguished.

Melting or vaporizing an ionic substance requires that the strong ionic bonds, which extend throughout the crystal, be broken, which takes a considerable amount of energy. Ionic substances, such as the alkali and alkaline earth metal halides, therefore generally have high melting points and boiling points, as we have seen in Table 5.4. In contrast, the nonmetal halides, such as CCl_4 and PCl_3, consist of molecules in which the atoms are held together by strong covalent bonds, but the molecules themselves are only held together by weak intermolecular forces. Only a small amount of energy is needed to overcome the weak

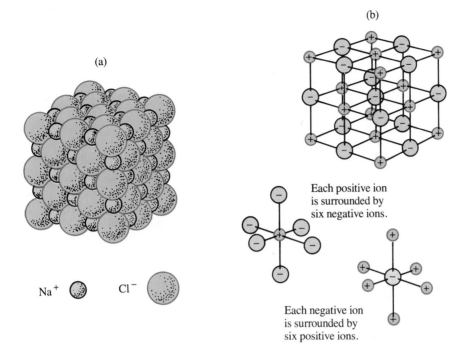

(a)

Na$^+$ ⚪   Cl$^-$ ⚪

(b)

Each positive ion
is surrounded by
six negative ions.

Each negative ion
is surrounded by
six positive ions.

FIGURE 5.9
The Structure of Sodium
Chloride.

The regular array of sodium ions
and chloride ions shown in (a) extends indefinitely throughout the
crystal. The crystal can therefore
be regarded as one huge molecule.
No individual NaCl molecules
can be recognized in the structure.
Each sodium ion is surrounded by
six chloride ions, which have an
octahedral arrangement. And each
chloride ion is surrounded by six
sodium ions, which also have an
octahedral arrangement. This arrangement is most easily seen in
the expanded view of the structure
shown in (b). Here the ions have
been separated from each other
but their arrangement has been
maintained. The lines joining the
ions do not represent bonds; they
are included only to assist in visualizing the structure. The attraction
of one ion for others of opposite
sign is exerted in all directions. In
contrast, covalent bonds form
only in specific directions.

forces *between* the molecules, so the nonmetal halides are generally gases and
liquids with low melting points (Table 5.3).

In the solid state the molecules of a covalent molecular compound such as
CCl_4 or H_2O are fixed in position. When these substances are melted and then
vaporized the molecules are free to move relative to each other, but they remain
intact because their atoms are held together by strong covalent bonds. Ice, liquid
water, and steam all consist of H_2O molecules; solid, liquid, and gaseous carbon tetrachloride all consist of CCl_4 molecules. The regular, three-dimensional
structure of sodium chloride or any other ionic substance cannot, however, be
retained in the liquid or gaseous state. Gaseous sodium chloride, which is only
formed at a very high temperature, consists of small molecules, such as Na^+Cl^-
and $(Na^+)_2(Cl^-)_2$, in which the bonding is predominantly ionic.

THE SODIUM CHLORIDE AND CESIUM
CHLORIDE STRUCTURES

Most of the alkali metal halides and many other ionic compounds have the
same arrangement of ions as sodium chloride. They are said to have the *sodium
chloride structure* (see Figure 5.9). However, cesium chloride and a few other
alkali metal halides have the *cesium chloride structure* (see Figure 5.11) in which
each Cs^+ is surrounded by eight Cl^- ions in a *cubic* arrangement, and each
Cl^- is surrounded by eight Cs^+ ions in a cubic arrangement. As in the NaCl
structure, no individual CsCl molecules can be distinguished. The fact that each
positive ion is surrounded by eight Cl^- ions, rather than six Cl^- ions, as in
the NaCl structure, is a consequence of the larger size of the Cs^+ ion. A positive
ion attracts as many negative ions as can pack around it. There is room for
eight Cl^- ions around a Cs^+ ion, but for only six Cl^- ions around the smaller
Na^+ ion.

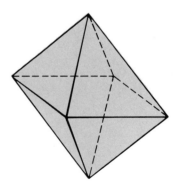

FIGURE 5.10
An Octahedron.

The octahedron is a very important
shape found commonly in molecules and crystal structures. It is
a regular polyhedron with eight
equilateral triangular faces, twelve
equivalent edges, and six equivalent corners. A template for
making an octahedron can be
found in the Study Guide.

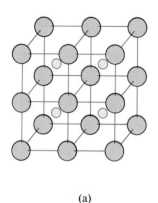

(a)

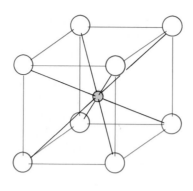

(b)

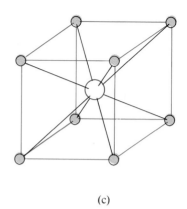

(c)

FIGURE 5.11
The Structure of Cesium Chloride.

(a) An expanded view of the cesium chloride structure. This arrangement of ions extends throughout the crystal. No individual cesium chloride molecules can be distinguished. (b) Each cesium ion is surrounded by a cubic arrangement of eight chloride ions. (c) Each chloride ion is surrounded by a cubic arrangement of eight cesium ions.

Li   Li⁺   F   F⁻

152 pm   74 pm   64 pm   133 pm

Na   Na⁺   Cl   Cl⁻

186 pm   102 pm   99 pm   181 pm

Positive ions are smaller and negative ions are larger than the corresponding covalently bound atoms

SIZES OF IONS: IONIC RADII

The sizes of ions can be obtained from the structures of crystals as determined by X ray crystallography (described in Chapter 11). Some values for the radii of the **cations** (positively charged ions) formed by the alkali and alkaline earth metals and of the **anions** (negatively charged ions) formed by the halogens and the group 6 elements are given in Table 5.6. By comparison with the covalent (atomic) radii in Figure 4.9, we see that *positive ions are smaller and negative ions are larger than the corresponding covalently bound atoms*. Removal of the valence electrons from an alkali or alkaline earth metal atom corresponds to the removal of its outer shell, so the positive ion that is formed is considerably smaller than the neutral atom. Addition of one or two electrons to a neutral

	TABLE 5.6 Ionic Radii (pm)			
	Group			
Period	*1*	*2*	*6*	*7*
2	Li⁺ 76	Be²⁺ 27	O²⁻ 138	F⁻ 133
3	Na⁺ 102	Mg²⁺ 72	S²⁻ 184	Cl⁻ 181
4	K⁺ 138	Ca²⁺ 100	Se²⁻ 198	Br⁻ 196
5	Rb⁺ 149	Sr²⁺ 116	Te²⁻ 221	I⁻ 216
6	Cs⁺ 170	Ba²⁺ 136		

atom to form a negative ion increases the number of repulsions between the valence-shell electrons and causes the valence shell to expand.

Note that within a period the alkaline earth M^{2+} ion is **isoelectronic** with the alkali metal M^+ ion; that is, it has the same electron arrangement. But because M^{2+} has a greater nuclear charge than M^+, its electrons are pulled closer to the nucleus, and the M^{2+} ion is therefore smaller than the M^+ ion. For example, Na^+ and Mg^{2+} both have the electron arrangement 2, 8; but Mg^{2+} ($Z = 12$) has a smaller radius (72 pm) than Na^+ ($Z = 11$), which has a radius of 102 pm. Similarly, within a period the group 6 X^{2-} ions and the halogen X^- ions are isoelectronic, but the X^- ion has a greater nuclear charge and therefore a smaller radius than the X^{2-} ion. For example, O^{2-} and F^- both have the electron arrangement 2, 8, but O^{2-} ($Z = 8$) has a radius of 140 pm, whereas F^- ($Z = 9$) has a radius of 133 pm.

$=$ Example 5.4

ION SIZES

In each of the following pairs, choose the larger ion without reference to Table 5.6:

(a) Na^+, F^- (b) Na^+, K^+ (c) F^-, Cl^-

(d) Na^+, Mg^{2+} (e) S^{2-}, Cl^-

Solution

(a) The ions Na^+ and F^- are isoelectronic; they have the same electron arrangement 2, 8. But these electrons are attracted by different nuclear charges. The atomic number Z of Na is 11; therefore the charge on the nucleus of both an Na atom and an Na^+ ion is $+11$. The atomic number of F is 9; therefore the charge on the nucleus of both F and F^- is $+9$. The electrons of Na^+ are attracted by a greater nuclear charge than are those of F^- and are therefore pulled in closer to the nucleus. Thus F^- is larger than Na^+.

(b) Although K^+ (2, 8, 8) has a nuclear charge of $+19$, and Na^+ (2, 8) has a nuclear charge of $+11$, the K nucleus is shielded by an additional shell of eight electrons. The core charge acting on the outer shell of eight electrons is $+19 - 8 - 2 = +9$ for K^+ and is $+11 - 2 = +9$ for Na^+. Thus the electrons are attracted by the same core charge, but since K^+ has three shells while Na^+ has only two, K^+ is the larger ion.

(c) The same type of argument we used in part (b) shows that Cl^- is the larger ion.

(d) The ions Na^+ and Mg^{2+} are isoelectronic; they have the electron arrangement 2, 8. But Mg^{2+} has a nuclear charge of $+12$ whereas Na^+ has a nuclear charge of $+11$. So the electrons of Na^+ are less strongly attracted than are those of Mg^{2+}, and therefore Na^+ is the larger ion.

(e) The same type of argument we used in part (d) shows that S^{2-} is the larger ion.

EXERCISE 5.5

In each of the following pairs choose the larger ion without reference to Table 5.6:

(a) K^+, Cl^- (b) K^+, Ca^{2+} (c) Mg^{2+}, Ca^{2+}

(d) O^{2-}, S^{2-} (e) O^{2-}, F^-

FORMULAS AND NAMES OF
IONIC COMPOUNDS

In addition to monatomic anions such as chloride and oxide there are several polyatomic anions containing oxygen (oxoanions) that we will encounter frequently throughout this book. They are included in Table 5.7 and are discussed in more detail in Chapters 8 and 9. All ionic compounds are neutral, that is, the total charge on the cations equals the total charge on the anions. Thus when we write the empirical formula of an ionic substance the charges of the ions must add up to zero. Some examples of ionic compounds between the magnesium ion Mg^{2+} and various anions are given in Table 5.8. The name of an ionic compound is simply a combination of the names of the ions in which the name of the cation is given first.

EXERCISE 5.6

Write the formula and give the name of the salt formed by each of the following pairs of ions:

(a) Na^+, SO_4^{2-} (b) Al^{3+}, O^{2-} (c) K^+, PO_4^{3-}

(d) Mg^{2+}, OH^- (e) Ba^{2+}, CO_3^{2-}

TABLE 5.7
Some Common Anions

F^-	Fluoride	OH^-	Hydroxide
Cl^-	Chloride	NO_3^-	Nitrate
Br^-	Bromide	ClO_4^-	Perchlorate
I^-	Iodide	CO_3^{2-}	Carbonate
O^{2-}	Oxide	SO_4^{2-}	Sulfate
S^{2-}	Sulfide	PO_4^{3-}	Phosphate
N^{3-}	Nitride	$CH_3CO_2^-$	Acetate

TABLE 5.8
Formulas and Names of Some Compounds of the Magnesium Ion, Mg^{2+}

Mg^{2+}	+	O^{2-}	$\longrightarrow$ MgO	Magnesium oxide
Mg^{2+}	+	$2Cl^-$	$\longrightarrow$ $MgCl_2$	Magnesium chloride
Mg^{2+}	+	$2NO_3^-$	$\longrightarrow$ $Mg(NO_3)_2$	Magnesium nitrate
Mg^{2+}	+	SO_4^{2-}	$\longrightarrow$ $MgSO_4$	Magnesium sulfate
$3Mg^{2+}$	+	$2PO_4^{3-}$	$\longrightarrow$ $Mg_3(PO_4)_2$	Magnesium phosphate

5.4
Oxidation–Reduction Reactions

In the reaction between sodium and chlorine to give sodium chloride,

$$2Na + Cl_2 \longrightarrow 2(Na^+Cl^-)$$

sodium atoms lose electrons to become sodium ions, and chlorine molecules gain electrons to become chloride ions. We can therefore conveniently imagine that the reaction takes place in two steps, called *half-reactions*:

$$Na \longrightarrow Na^+ + e^-$$
$$Cl_2 + 2e^- \longrightarrow 2Cl^-$$

The electrons produced in the first reaction must be used up in the second reaction, because no electrons appear in the overall reaction. To obtain the balanced equation for the overall reaction, we must multiply the first reaction by 2 and add it to the second reaction so that the electrons produced in the first reaction are used up in the second reaction:

$$2Na \longrightarrow 2Na^+ + 2e^-$$
$$\underline{Cl_2 + 2e^- \longrightarrow 2Cl^-}$$
$$2Na + Cl_2 \longrightarrow 2(Na^+Cl^-)$$

The overall reaction is an **electron transfer reaction**: electrons are transferred from sodium to chlorine. Such a reaction is commonly called an **oxidation–reduction reaction**, or a **redox** reaction.

We previously defined oxidation as the addition of oxygen to an element or compound and reduction as the removal of oxygen from a compound. The formation of sodium oxide, Na_2O, from sodium and oxygen can be represented by the equation

$$4Na + O_2 \longrightarrow 2(Na^+)_2O^{2-}$$

This equation can similarly be split into two halves to show that it is an electron transfer reaction. Each sodium atom loses an electron to form a sodium ion:

$$Na \longrightarrow Na^+ + e^-$$

And each oxygen molecule gains four electrons to form two oxide ions:

$$O_2 + 4e^- \longrightarrow 2O^{2-}$$

Combining the two equations

$$4(Na \longrightarrow Na^+ + e^-) \text{ oxidation}$$
$$O_2 + 4e^- \longrightarrow 2O^{2-} \text{ reduction}$$

gives the overall equation

$$4Na + O_2 \longrightarrow 2(Na^+)_2O^{2-}$$

We would describe this reaction by saying that sodium is oxidized by oxygen to sodium oxide. But we note that the reactions of sodium with oxygen and with chlorine involve the same half-reaction for sodium, namely

$$Na \longrightarrow Na^+ + e^-$$

So that we say that sodium is *oxidized* in both reactions and that in general

An element is oxidized when it loses electrons

and that

Oxidation is the loss of electrons

The half-reactions for chlorine and oxygen show that they both gain electrons:

$$Cl_2 + 2e^- \longrightarrow 2Cl^-$$
$$O_2 + 4e^- \longrightarrow 2O^{2-}$$

We say that both chlorine and oxygen are *reduced* and that in general

An element is reduced when it gains electrons

and that

Reduction is the gain of electrons

LEO the lion says GER. This phrase may help you to remember the terminology for redox reactions.
Loss of Electrons is Oxidation
Gain of Electrons is Reduction

Because electrons are neither produced nor used up in chemical reactions, an oxidation half-reaction must always be accompanied by an appropriate reduction half-reaction. So the overall reaction is an oxidation–reduction or redox reaction. Electrons are transferred from an element that is oxidized to an element that is reduced.

According to this more general definition of oxidation, many substances other than oxygen can oxidize other substances. They are said to be *oxidizing agents*. Any substance that tends to gain electrons—for example, nonmetals such as oxygen, sulfur, and the halogens—may behave as an oxidizing agent. Any substance from which electrons can be readily removed, such as metals, may behave as a *reducing agent*.

An **oxidizing agent** *is a substance that can gain electrons, and a* **reducing agent** *is a substance that can give up electrons.* Thus an oxidizing agent is reduced and a reducing agent is oxidized.

Oxidation–reduction reactions are a very important general class of reactions, and we will encounter many examples in later chapters. In general, the reaction between a metal and a nonmetal is an oxidation–reduction reaction in which the metal behaves as a reducing agent and is oxidized and the non-metal behaves as an oxidizing agent and is reduced.

THE HALOGENS AS OXIDIZING AGENTS

The halogens are all strong oxidizing agents, but their strength decreases in the series $F_2 > Cl_2 > Br_2 > I_2$. When an aqueous solution of chlorine, Cl_2, is added to a colorless aqueous solution of sodium bromide, NaBr, the solution rapidly becomes orange-red because of the formation of bromine, Br_2, in the reaction (see Experiment 5.5):

$$Cl_2 + 2Br^- \longrightarrow 2Cl^- + Br_2$$

This reaction is an oxidation–reduction reaction in which bromide ion is oxi-

Experiment 5.5

Oxidation–Reduction Reactions of the Halogens

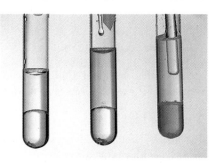 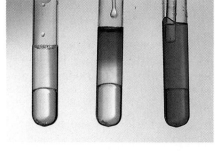

Left: The top layer is an aqueous solution of NaBr. The bottom layer is carbon tetrachloride, CCl_4. Center: When a few drops of a dilute aqueous solution of Cl_2 are added, the NaBr solution becomes yellow because Br_2 is formed. Right: Bromine is much more soluble in carbon tetrachloride than in water. On stirring, it is extracted into the carbon tetrachloride, forming a brown solution.

Left: The top layer is an aqueous solution of NaI. The bottom layer is CCl_4. Center: When a few drops of an aqueous solution of Cl_2 are added, iodine, I_2, is formed. Iodine reacts with iodide ion, I^-, to form a brown solution of the triiodide ion, I_3^-. Right: On stirring, I_2 is extracted into the CCl_4, forming a violet solution.

dized to bromine,

$$2Br^- \longrightarrow Br_2 + 2e^-$$

while chlorine is reduced to chloride ion,

$$Cl_2 + 2e^- \longrightarrow 2Cl^-$$

Similarly, if an aqueous solution of chlorine is added to a colorless aqueous solution of an iodide, or if gaseous chlorine is passed into an iodide solution, the solution becomes brown because of the formation of iodine, I_2:

$$Cl_2 + 2I^- \longrightarrow 2Cl^- + I_2$$

Chlorine oxidizes iodide ion to iodine. Iodine is insoluble in water, and the color is due to the brown triiodide ion, I_3^-, formed by combination of an iodine molecule and an iodide ion:

$$I_2(s) + I^-(aq) \longrightarrow I_3^-(aq)$$

When this solution is shaken with carbon tetrachloride, which is insoluble in water, molecular iodine is removed from I_3^- and forms a violet solution in the carbon tetrachloride.

If bromine is added to an aqueous solution of an iodide, iodine is formed according to the equation

$$Br_2 + 2I^- \longrightarrow 2Br^- + I_2$$

However, if bromine is added to an aqueous solution of a chloride, no reaction is observed. In other words, the reaction

$$Cl_2 + 2Br^- \longrightarrow 2Cl^- + Br_2$$

proceeds from left to right but not from right to left. Whereas chlorine can oxidize bromide and iodide, bromine can oxidize only iodide. Iodine, the weakest oxidizing agent among the halogens, oxidizes neither Br^- nor Cl^-.

Similar reactions occur between the halogens and the hydrogen halides in the gas phase (see Experiment 5.6). For example,

$$Cl_2(g) + 2HI(g) \longrightarrow I_2(s) + 2HCl(g)$$

These simple experiments show that the order of oxidizing strengths is $Cl_2 > Br_2 > I_2$. Fluorine is a much stronger oxidizing agent than chlorine. Fluorine, for example, oxidizes water to oxygen:

$$2F_2(g) + 2H_2O(l) \longrightarrow 4HF(aq) + O_2(g)$$

In contrast, chlorine and bromine dissolve largely unchanged in water, and iodine is insoluble. Fluorine reacts so vigorously with many substances that it can only be handled in the laboratory by using special apparatus and taking suitable precautions.

The order of oxidizing strengths $F_2 > Cl_2 > Br_2 > I_2$ is not unexpected when we recall that the electronegativities of F, Cl, Br, and I also decrease in the same sequence. The electronegativity of a halogen is a measure of the tendency of a halogen atom to attract the electrons of a covalent bond and therefore to acquire a partial negative charge. It is therefore not the same as the oxidizing strength of a halogen, which is a measure of the tendency of the diatomic molecule X_2 to acquire two electrons to become two X^- ions

$$X_2 + 2e^- \longrightarrow 2X^-$$

Nevertheless, both the oxidizing strength and the electronegativity of the halogens are related to the tendency of halogen atoms to acquire electrons. Therefore, not surprisingly, as the electronegativity of X decreases, the oxidizing strength

Experiment 5.6

Oxidation of Hydrogen Iodide by Chlorine

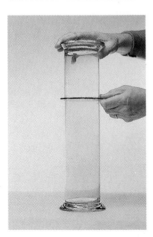

Chlorine is separated from twice its volume of hydrogen iodide by a glass plate.

When the plate is removed, chlorine oxidizes hydrogen iodide to iodine. Both the violet vapor and the black solid can be seen. Some brown iodine monochloride, ICl, is also observed on the sides of the jar.

of X_2 also decreases in the order $F_2 > Cl_2 > Br_2 > I_2$.

The ease with which halide ions can be oxidized increases in the order $F^- < Cl^- < Br^- < I^-$, which is the order of their strengths as reducing agents.

PREPARATION OF THE HALOGENS

Since the halogens occur naturally as halide ions, they are prepared by oxidation of halide ion to the free halogen. Although in principle Cl^- could be oxidized by fluorine, F_2, this method is impractical because fluorine is expensive to produce and reacts vigorously with water and many other substances. In the industrial manufacture of chlorine, chloride ion is oxidized by directly removing electrons in a process called **electrolysis**, in which an electric current is passed through molten sodium chloride or a concentrated aqueous solution of sodium chloride:

$$2Cl^-(aq) \xrightarrow{\text{Electrolysis}} Cl_2 + 2e^-$$

Enormous quantities of chlorine are made by this process which is discussed in detail in Chapter 17.

In the laboratory chloride ion can be oxidized to gaseous chlorine, Cl_2, by using a strong oxidizing agent. A suitable oxidizing agent is manganese dioxide, MnO_2. The reaction is carried out in an acidic solution:

$$MnO_2(s) + 2Cl^-(aq) + 4H^+(aq) \longrightarrow Mn^{2+}(aq) + Cl_2(g) + 2H_2O(l)$$

As we have seen, bromide ion is readily oxidized to bromine by chlorine. Because chlorine is manufactured in very large amounts and is relatively inexpensive, it is used to prepare bromine from the bromide ion present in subterranean brines:

$$Cl_2(g) + 2Br^-(aq) \longrightarrow Br_2(l) + 2Cl^-(aq)$$

Chlorine and steam are passed through the brine. The gaseous bromine that is formed is carried off from the solution with steam and excess chlorine. It is condensed from this mixture and purified by distillation. Iodine can be obtained from an aqueous iodide solution by the same method.

Because the fluoride ion, F^-, is so difficult to oxidize—it is a very weak reducing agent—the only practical method of making fluorine, F_2, is by electrolysis of molten sodium fluoride or a solution of sodium fluoride in liquid hydrogen fluoride:

$$2F^-(l) \longrightarrow F_2(g) + 2e^-$$

Despite the difficulties of handling fluorine because of its great reactivity, a large plant can produce about 9 tons of fluorine, F_2, a day. The annual U.S. and Canadian production is about 5000 tons. Much of this fluorine is used for the processing of uranium for nuclear power plants (see Chapter 25).

= *Example 5.5* =

OXIDIZING AND REDUCING AGENTS

Which of the following species is the strongest reducing agent?

I_2 I^- Cl^- F_2 F^-

Solution

I^-, Cl^-, and F^-, are reducing agents. I_2 and F_2 are oxidizing agents. The halide ions increase in reducing strength in the order $F^- < Cl^- < Br^- < I^-$. So in the list above I^- is the strongest reducing agent.

EXERCISE 5.7

Which of the following species is the strongest oxidizing agent?

Br_2 Br^- Cl^- I_2 F^-

≡ *Example 5.6* ≡

REDOX REACTIONS

In each of the following reactions, identify the oxidizing agent, the reducing agent, the species (molecule or ion) that is oxidized, and the species that is reduced:

   (a) $Ca(s) + Br_2(l) \longrightarrow Ca^{2+}(Br^-)_2(s)$

   (b) $4Li(s) + O_2(g) \longrightarrow 2(Li^+)_2O^{2-}(s)$

   (c) $Fe(s) + S(s) \longrightarrow Fe^{2+}S^{2-}(s)$

   (d) $2Fe^{2+}(aq) + Cl_2(g) \longrightarrow 2Fe^{3+}(aq) + 2Cl^-(aq)$

Solution

   (a) The Ca loses electrons to become Ca^{2+}. It is therefore a reducing agent and is itself oxidized:

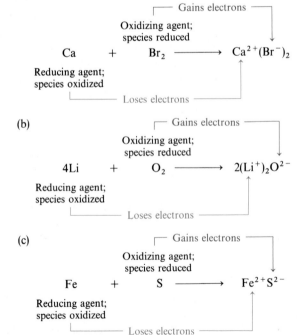

(d)

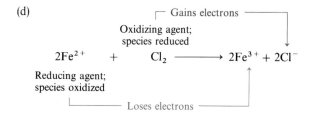

EXERCISE 5.8

In each of the following reactions, identify the oxidizing agent, the reducing agent, the species (molecule or ion) that is oxidized, and the species that is reduced:

(a) $2Cu(s) + O_2(g) \rightarrow 2CuO(s)$

(b) $2Cs(s) + I_2(s) \rightarrow 2CsI(s)$

(c) $Zn(s) + S(s) \rightarrow ZnS(s)$

(d) $Zn(s) + Cu^{2+}(aq) \rightarrow Zn^{2+}(aq) + Cu(s)$

5.5
Precipitation Reactions

When ionic compounds dissolve in water, the ions separate from each other and mix with the water molecules. A solution of sodium chloride in water consists simply of sodium ions, Na^+, and chloride ions, Cl^-, mixed with H_2O molecules. If potassium fluoride is dissolved in water, the solution contains potassium ions, K^+, and fluoride ions, F^-. If we mix equal amounts of solutions of NaCl and KF of equal concentrations, we obtain a solution containing equal concentrations of Na^+, K^+, Cl^-, and F^-. If we were to prepare a solution by mixing equal amounts of solutions of NaF and KCl, rather than NaCl and KF, we would again obtain a solution containing equal concentrations of Na^+, K^+, Cl^-, and F^-. These two solutions are identical in every way and cannot be distinguished. Thus in solutions of mixtures of ionic compounds the individual compounds lose their identity.

If we were to evaporate a solution containing equal amounts of Na^+, K^+, Cl^-, and F^-, would we obtain a mixture of all four possible compounds, NaF, NaCl, KF, and KCl, or would we obtain only some of them? The answer depends on the solubilities of the four possible compounds. If we allow the solution to evaporate, sodium fluoride, in fact, will crystallize first, since it has the smallest solubility (see Table 5.4).

If we start with solutions of KF and NaCl, we can represent this reaction by the equation

$$KF(aq) + NaCl(aq) \longrightarrow NaF(s) + KCl(aq)$$

where we have represented each substance by its empirical formula. The remaining solution still contains some Na^+ and F^- ions together with all the K^+ and Cl^- ions. No reaction occurs on simply mixing the solutions. When the solution is concentrated by evaporation until the solubility of NaF is exceeded, Na^+ ions combine with F^- ions to form a **precipitate** of insoluble NaF:

$$Na^+(aq) + F^-(aq) \longrightarrow NaF(s)$$

This reaction is an example of a **precipitation reaction**.

Precipitation of sodium fluoride only occurs from a rather concentrated solution, because sodium fluoride is relatively soluble. More common examples of precipitation reactions involve substances that have very small solubilities; they precipitate, therefore, even from dilute solutions.

Although the chlorides of the alkali and alkaline earth metals are all soluble in water, the chlorides of some metals such as silver and lead have very small solubilities. Thus if we mix dilute solutions of a soluble silver salt such as silver nitrate, $AgNO_3$, and sodium chloride, silver chloride precipitates immediately:

$$AgNO_3(aq) + NaCl(aq) \longrightarrow AgCl(s) + NaNO_3(aq)$$

Since these are all ionic compounds, we can also write the equation in terms of the ions:

$$Ag^+(aq) + NO_3^-(aq) + Na^+(aq) + Cl^-(aq) \longrightarrow (Ag^+Cl^-)(s) + Na^+(aq) + NO_3^-(aq)$$

We see that the ions Na^+ and NO_3^- appear on both sides of the equation; they take no part in the reaction and hence are said to be **spectator ions**. We can cancel them from the equation, leaving

$$Ag^+(aq) + Cl^-(aq) \longrightarrow (Ag^+Cl^-)(s)$$

This equation is called a **net ionic equation**. It is usually written as

$$Ag^+(aq) + Cl^-(aq) \longrightarrow AgCl(s)$$

where the ions in the solid are not shown. This equation represents the reaction that occurs when a solution of *any* soluble silver salt is mixed with a solution of *any* soluble chloride, for example,

$$AgClO_4(aq) + KCl(aq) \longrightarrow AgCl(s) + KClO_4(aq)$$

as we can see if we write this equation in the ionic form and cancel the spectator ions.

Silver bromide and silver iodide are also insoluble, and precipitates of silver bromide and silver iodide are obtained when silver ion is added to aqueous solutions containing bromide or iodide ions. Since most other silver salts are soluble, the formation of these precipitates is a good test for chloride, bromide, or iodide ions (see Experiment 5.7).

SOLUBILITY RULES

There is no general theory that enables us to predict the solubility of a substance. Fortunately there are some simple rules which enable us to make fairly reliable predictions about the solubilities of ionic substances in aqueous solution. These rules are summarized in Table 5.9. They apply to all the common anions and cations discussed in this book and you will find it useful to learn them. Although some substances are described in Table 5.9 as being insoluble you should realize that no substance is completely insoluble; at least a very small amount will always dissolve but the amount that dissolves is negligible for most purposes. So the definition of insoluble is somewhat arbitrary. We will describe any substance that has a solubility less than 0.01 mol L^{-1} as insoluble.

Experiment 5.7

Precipitation Reactions

When an aqueous NaCl solution is added to an aqueous AgNO$_3$ solution a white precipitate of insoluble AgCl is formed

When an aqueous solution of NaI is added to an aqueous solution of AgNO$_3$ a pale yellow precipitate of insoluble AgI is formed.

When an aqueous solution of NaI is added to an aqueous solution of Pb(NO$_3$)$_2$ a bright yellow precipitate of insoluble PbI$_2$ is formed.

TABLE 5.9
Solubilities of Some Common Salts and Hydroxides in Water at 25 °C

Soluble[a]	Exceptions	
	Insoluble	Sparingly Soluble
Na$^+$, K$^+$, and NH$_4$$^+$ salts		
Nitrates (NO$_3$$^-$)		
Perchlorates (ClO$_4$$^-$)		
Fluorides (F$^-$)	Mg^{2+}, Ca^{2+}, Sr^{2+}, Ba^{2+}, Pb^{2+}	
Chlorides, bromides, and iodides	Ag$^+$, Hg$_2$$^{2+}$, PbI$_2$	PbCl$_2$, PbBr$_2$
Sulfates (SO$_4$$^{2-}$)	Sr^{2+}, Ba^{2+}, Pb^{2+}	Ca^{2+}, Ag$^+$, Hg$_2$$^{2+}$
Acetates (CH$_3$CO$_2$$^-$)		Ag$^+$, Hg$_2$$^{2+}$

Insoluble[a]	Exceptions	
	Soluble	Sparingly Soluble
Carbonates (CO$_3$$^{2-}$)	Na$^+$, K$^+$, NH$_4$$^+$	
Phosphates (PO$_4$$^{3-}$)	Na$^+$, K$^+$, NH$_4$$^+$	
Sulfides (S^{2-})	Na$^+$, K$^+$, NH$_4$$^+$, Mg^{2+}, Ca^{2+}, Sr^{2+}, Ba^{2+}	
Hydroxides (OH$^-$)	Na$^+$, K$^+$, NH$_4$$^+$, Ba^{2+}	Ca^{2+}, Sr^{2+}

[a] The classifications *soluble*, *sparingly soluble*, and *insoluble* are somewhat arbitrary. Their approximate significance is as follows: soluble, ≥ 0.1 mol L^{-1}; sparingly soluble, ≤ 0.1 mol L^{-1} and ≥ 0.01 mol L^{-1}; insoluble, ≤ 0.01 mol L^{-1}.

═ *Example 5.7* ═══════════════════════════════════

SOLUBILITY RULES

Use the solubility rules in Table 5.9 to predict which of the following substances are insoluble and which are soluble:

KBr $NaNO_3$ MgF_2 $BaSO_4$ $FeCl_2$

Solution

KBr is soluble because all K^+ salts are soluble and all Br^- salts are soluble with a few exceptions that do not include K^+.

$NaNO_3$ is soluble because all Na^+ salts and all NO_3^- salts are soluble.

MgF_2 is insoluble because although most fluorides are soluble MgF_2 is one of the exceptions.

$BaSO_4$ is insoluble because although most sulfates are soluble $BaSO_4$ is one of the exceptions.

$FeCl_2$ is soluble because all chlorides are soluble with a few exceptions that do not include Fe^{2+}.

EXERCISE 5.9

Use the solubility rules to predict which of the following substances are soluble and which are insoluble:

$CaCO_3$ PbI_2 $LiBr$ CuO $MgSO_4$

═ *Example 5.8* ═══════════════════════════════════

PRECIPITATION REACTIONS

Use the solubility rules to predict whether a precipitate will form when the following pairs of aqueous solutions are mixed. Explain and write a net ionic equation for those cases in which a precipitate forms.

(a) $MgCl_2$ and Na_3PO_4

(b) $Al_2(SO_4)_3$ and $Ba(NO_3)_2$

(c) $Ba(NO_3)_2$ and $MgCl_2$

Solution

(a) Possible products are $Mg_3(PO_4)_2$ and $NaCl$. Since $Mg_3(PO_4)_2$ is insoluble, a precipitation reaction will occur:

$$3Mg^{2+} + 2PO_4^{3-} \longrightarrow Mg_3(PO_4)_2(s)$$

(b) Possible products are $Al(NO_3)_3$ and $BaSO_4$. Since $BaSO_4$ is insoluble, a precipitation reaction will occur:

$$Ba^{2+} + SO_4^{2-} \longrightarrow BaSO_4(s)$$

(c) Possible products are $BaCl_2$ and $Mg(NO_3)_2$. Both of these compounds are soluble, so there will be no precipitation reaction. The solution will consist of a mixture of Ba^{2+}, Cl^-, Mg^{2+}, and NO_3^-.

Use the solubility rules to predict whether a precipitate will form when the following pairs of aqueous solutions are mixed:

(a) $MgCl_2$ and NaF

(b) $MgSO_4$ and $NaCl$

(c) $FeCl_2$ and Na_2S

5.6
Acid–Base Reactions

AQUEOUS SOLUTIONS OF HYDROGEN HALIDES: THE HYDRONIUM ION

Hydrogen chloride and the other hydrogen halides are typical molecular covalent substances that have very low melting and boiling points and are nonconductors of electricity in both the liquid and the solid states. But the hydrogen halides are very soluble in water and form solutions that are very good conductors of electricity. Ions must be formed when the hydrogen halides are dissolved in water. What are these ions?

If a silver nitrate solution is added to an aqueous solution of hydrogen chloride, a white precipitate of silver chloride is obtained:

$$Ag^+(aq) + Cl^-(aq) \longrightarrow AgCl(s)$$

This reaction shows that one of the ions in the hydrogen chloride solution is the chloride ion. If a chloride ion is formed from hydrogen chloride, a hydrogen ion (a proton) must also be formed:

$$H\!:\!\overset{..}{\underset{..}{Cl}}\!:(aq) \longrightarrow H^+(aq) + :\overset{..}{\underset{..}{Cl}}\!:^{\ominus}(aq)$$

A proton has an empty valence shell and attracts electrons very strongly. In aqueous solution it combines with a water molecule to form the **hydronium ion**, H_3O^+:

Protons, H^+, are never found as free species in aqueous solution.

$$H^+ + :\overset{..}{\underset{H}{O}}\!:H \longrightarrow H\!:\!\overset{..}{\underset{H}{O}}\!:^{\oplus}\!H$$

Thus the net reaction that occurs when hydrogen chloride gas dissolves in water is the transfer of a hydrogen ion, H^+, from a hydrogen chloride molecule to a water molecule, with the formation of the hydronium ion, H_3O^+, and a chloride ion, Cl^-:

$$:\!\overset{..}{\underset{..}{Cl}}\!-H + :\!\overset{..}{\underset{H}{O}}\!-H \longrightarrow :\!\overset{..}{\underset{..}{Cl}}\!:^- + H\!-\!\overset{..}{\underset{H}{O}}\!\overset{\oplus}{}\!-H$$

Because of the large electronegativity difference between hydrogen and chlorine and between hydrogen and oxygen, both hydrogen chloride and water are polar molecules:

$$\overset{\delta-}{:\!\overset{..}{\underset{..}{Cl}}}\!-\!\overset{\delta+}{H}\overset{\curvearrowleft}{}\overset{2\delta-}{:\!\overset{..}{O}}\!-\!\overset{\delta+}{H} \longrightarrow :\!\overset{..}{\underset{..}{Cl}}\!:^- + H\!-\!\overset{..}{\underset{H}{O}}\!\overset{\oplus}{}\!-H$$
$$\overset{\delta+}{}H$$

The positive hydrogen atom of the hydrogen chloride molecule is attracted to the negative oxygen atom of the water molecule. A hydrogen ion (proton) then moves from the chlorine atom to the oxygen atom of the water molecule, leaving behind the electrons that it originally shared with the chlorine atom to give a $:\overset{..}{\underset{..}{Cl}}:^-$ ion. The hydrogen ion attaches itself to one of the unshared pairs of electrons of the oxygen atom, forming an additional O—H bond and converting the water molecule to a hydronium ion, H_3O^+.

If a concentrated solution of HCl in water is cooled to $-15\,°C$, colorless crystals of the compound $H_3O^+Cl^-$ separate. This is hydronium chloride, an ionic compound that is similar to other ionic chlorides such as sodium chloride and calcium chloride. It differs from these simple ionic compounds only in that it has a polyatomic cation instead of a simple monatomic ion such as Na^+ or Ca^{2+}. The H_3O^+ ion and the NH_3 molecule have the same Lewis structure,

$$H—\overset{..}{\underset{|}{N}}—H \qquad H—\overset{..}{\underset{|}{O}}—H$$
$$H \qquad\qquad H$$

in which both the nitrogen atom and the oxygen atom form three covalent bonds. In their valence shells they both have three shared (bonding) pairs of electrons and one unshared pair of electrons. The H_3O^+ also has the same triangular pyramidal AX_3E shape as the NH_3 molecule.

When a hydrogen ion becomes attached to one of the unshared pairs of electrons on the oxygen atom of a water molecule, it acquires a share in this electron pair and thus effectively acquires one electron. It therefore loses its charge. The oxygen atom at the same time effectively loses an electron when one of its unshared pairs is converted to a shared pair. Thus the oxygen atom becomes positively charged; it has a formal charge of $+1$. This formal charge can also be found by following the procedure for finding formal charges given in Chapter 4:

Formal charge = Group number − Number of bonds − Number of unshared electrons

$$= +6 - 3 - 2 = +1$$

Triangular Pyramidal AX_3E Molecules

ACIDS

Solutions of hydrogen chloride and the other hydrogen halides in water share certain properties with the aqueous solutions of a number of other substances. These properties have long been used to identify these substances as **acids**. Some of these characteristic properties are as follows:

· Aqueous solutions of acids have a sour taste. We are familiar with the taste of vinegar or lemon juice. Vinegar is an aqueous solution of acetic acid, $C_2H_4O_2$; lemon juice is an aqueous solution containing citric acid, $C_6H_8O_7$.

· Acids have the ability to change the colors of substances called *indicators*. For example, litmus, a substance extracted from certain lichens, has a characteristic red color in aqueous acid solutions. Grape juice, tea, and red cabbage also act as indicators (Experiment 5.8).

· When acids are added to a solution of a soluble carbonate such as sodium carbonate, Na_2CO_3, or to an insoluble solid carbonate such as calcium carbonate, $CaCO_3$, carbon dioxide is evolved. The solution bubbles vigorously—it is said to effervesce.

Experiment 5.8

Effect of Acids and Bases on some Natural Indicators

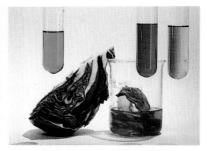

The middle tube contains purple grape juice. On the left acid has been added, changing the color to red. On the right base has been added, changing the color to an olive green.

When lemon juice is added to tea the color changes from brown to a yellow-orange.

In the beaker the natural color of red cabbage is being extracted into methanol. The middle tube contains the red methanol extract of red cabbage. On the left acid has been added, changing the color to pink. On the right base has been added, changing the color to green.

- Aqueous solutions of acids react with many metals, such as zinc and magnesium, producing hydrogen, which bubbles off from the solution. The solution effervesces (Experiment 5.9).

These common characteristic properties of aqueous solutions of acids result from the fact that, like a solution of hydrogen chloride in water, they all contain the hydronium ion, H_3O^+. These properties of aqueous acid solutions are the properties of the hydronium ion.

An acid has at least one hydrogen atom attached to an electronegative atom or group of atoms. We may write the formula of an acid as HA, where A is an electronegative atom or group of atoms (such as Cl or NO_3). The HA bond is therefore polar

$$\overset{\delta+}{H}\!-\!\overset{\delta-}{A}$$

and the hydrogen atom has a small positive charge. This positively charged H atom is attracted to the negatively charged oxygen atom of a water molecule and then H^+ is transferred to one of the unshared electron pairs of the water molecule to form H_3O^+.

We can represent the reaction of an acid with water by the general equation

$$HA + H_2O \longrightarrow H_3O^+ + A^-$$

For example,

$$HCl + H_2O \longrightarrow H_3O^+ + Cl^-$$
$$HNO_3 + H_2O \longrightarrow H_3O^+ + NO_3^-$$

In each of these reactions the acid gives up an H^+, that is, a proton. Water in turn accepts a proton. The reaction is thus a **proton transfer reaction**; one substance donates a proton and the other accepts a proton. In 1923 the Danish

Experiment 5.9

Reactions of Acids

Magnesium reacts with dilute hydrochloric acid producing bubbles of $H_2(g)$.

Zinc reacts with dilute hydrochloric acid producing bubbles of $H_2(g)$.

Marble chips ($CaCO_3$) react with dilute hydrochloric acid producing bubbles of $CO_2(g)$.

chemist Johannes Brønsted (1879–1947) and the English chemist Thomas Lowry (1874–1936) proposed that an **acid** may be defined as a **proton donor**. A **proton acceptor** is called a **base**. A proton transfer reaction is therefore called an **acid–base reaction**.

When an acid is dissolved in water, there is an acid–base reaction in which water behaves as a base, accepting a proton from the acid:

$$HA(aq) + H_2O(l) \longrightarrow H_3O^+(aq) + A^-(aq)$$

This equation is sometimes written in the form

$$HA(aq) \longrightarrow H^+(aq) + A^-(aq)$$

and the acid is commonly said to *ionize* in water. Remember, however, that in this equation $H^+(aq)$ is merely an abbreviation for $H_3O^+(aq)$. There is no evidence for the existence of free hydrogen ions, that is, free protons, in any solutions. Because the proton is a bare nucleus and is therefore extremely small, it can approach very closely to an unshared electron pair and is then held very strongly by the electron pair. Consequently, a proton formed in water becomes attached to an unshared pair of electrons on the oxygen atom of a water molecule, forming the hydronium ion, H_3O^+.

All acids react with water to form the hydronium ion, H_3O^+.

The reaction of acids with carbonate ions, CO_3^{2-}, is another example of an acid–base reaction. In this case the base is the carbonate ion. It can accept protons from two H_3O^+ ions to form carbonic acid, H_2CO_3:

$$CO_3^{2-}(aq) + 2H_3O^+(aq) \longrightarrow H_2CO_3(aq) + 2H_2O(l)$$

Carbonic acid, however, is unstable and decomposes to water and carbon dioxide,

$$H_2CO_3(aq) \longrightarrow H_2O(l) + CO_2(g)$$

which bubbles off from the solution.

However, the reactions of metals with the hydronium ion are *not* acid–base reactions. In the reaction of magnesium with an aqueous acid,

$$Mg(s) + 2H_3O^+(aq) \longrightarrow Mg^{2+}(aq) + H_2(g) + 2H_2O(l)$$

each magnesium atom loses two electrons to become Mg^{2+}:

$$Mg \longrightarrow Mg^{2+} + 2e^-$$

It is therefore oxidized. At the same time hydrogen in the hydronium ion is reduced to molecular hydrogen. That this is a reduction can be seen more clearly if we write the equation in the form

$$Mg(s) + 2H^+(aq) \longrightarrow Mg^{2+}(aq) + H_2(g)$$

Each hydrogen ion acquires an electron to become a hydrogen atom, and two hydrogen atoms form a hydrogen molecule:

$$2H^+(aq) + 2e^- \longrightarrow H_2(g)$$

The reactions of metals with the hydronium ion in aqueous acid solutions are therefore oxidation–reduction reactions, and $H^+(aq)$ is an oxidizing agent.

STRONG AND WEAK ACIDS

Hydrogen chloride reacts completely with water; it is completely converted to the corresponding ions:

$$HCl(g) + H_2O(l) \longrightarrow H_3O^+(aq) + Cl^-(aq)$$

Acids that react completely with water in this way are called **strong acids**. They are completely ionized and their solutions have a high electrical conductivity. Both HBr and HI are also strong acids.

In contrast, the reaction of HF with water,

$$HF(aq) + H_2O(l) \longrightarrow H_3O^+(aq) + F^-(aq)$$

is incomplete. This is shown by the fact that the electrical conductivity of a solution of HF is much smaller than that of a solution of HCl of the same concentration. There are therefore fewer ions in the HF solution than in the HCl solution. Much of the HF in the solution remains as un-ionized HF molecules. An acid that is incompletely ionized in an aqueous solution is called a **weak acid**.

The incomplete ionization of HF in water does not mean, however, that the reaction of HF with water simply stops after a small number of HF molecules have reacted to give H_3O^+ and F^- ions and that no more HF molecules react. In fact, this reaction continues indefinitely. However, at the same time a reaction occurs between the products H_3O^+ and F^- to give back the original reactants HF and H_2O. The reaction between HF and H_2O is called the **forward reaction**. The reaction between H_3O^+ and F^-, to give HF and H_2O, is called the **reverse reaction**. The rate of the reverse reaction increases as the concentration of the products increases and *when the rate of the reverse reaction is equal to the rate of the forward reaction, a state of* **dynamic equilibrium** *is reached.* There is then no further change in the concentrations of HF, H_3O^+, H_2O, or F^-; all four of these species are present in equilibrium. The two reactions proceed simultaneously but at the same rate, so the reaction appears to have stopped. The dynamic nature of the equilibrium is emphasized by writing the equation for the reaction with two arrows, one pointing in each direction:

$$HF(aq) + H_2O(l) \rightleftharpoons H_3O^+(aq) + F^-(aq)$$

The concept of dynamic equilibrium is extremely important in chemistry. All reactions eventually reach a state of equilibrium. In the ionization of strong acids such as HCl, HBr, and HI, there is, in principle, a reverse reaction. But the concentrations of un-ionized HCl, HBr, or HI that are present at equilibrium are infinitesimally small, so we may say that the reaction has proceeded to completion.

Another example of dynamic equilibrium is provided by the reactions between hydrogen and the halogens to form the hydrogen halides that we discussed above. The reaction between H_2 and Cl_2 proceeds with explosive violence and goes to completion. In other words, the amounts of unreacted H_2 and Cl_2 that remain when equilibrium is reached are infinitesimally small. In contrast, when I_2 and H_2 are heated together in the gas phase the reaction is slower and it reaches equilibrium before all the H_2 and I_2 have been used up. At equilibrium HI is decomposing to form H_2 and I_2 at the same rate as it is being formed:

$$H_2(g) + I_2(g) \rightleftharpoons 2HI(g)$$

We will meet many more examples of equilibria in the following chapters and in Chapters 13 and 14 we discuss equilibria in more detail and in a quantitative manner.

Table 5.10 lists some common strong and weak acids with which you should be familiar. Many of these, such as sulfuric acid, H_2SO_4, and nitric acid, HNO_3, are oxoacids which ionize in water to give the corresponding oxoanion, SO_4^{2-}, and NO_3^-. These oxoacids will be discussed in more detail in Chapter 8. The only common strong acids are the acids given in the table; almost all other acids are weak, and only the most common weak acids are listed in Table 5.10.

Hypochlorous acid, HOCl, is a weak acid. It is formed in a solution of chlorine in water. In such a solution most of the chlorine remains as Cl_2 molecules,

H:Ö:Cl:

TABLE 5.10 Some Common Acids			
Strong Acids		*Weak Acids*	
Hydrochloric acid	HCl	Hydrofluoric acid	HF
Hydrobromic acid	HBr	Carbonic acid	H_2CO_3
Hydroiodic acid	HI	Phosphoric acid	H_3PO_4
Sulfuric acid	H_2SO_4	Acetic acid	CH_3CO_2H
Nitric acid	HNO_3	Hypochlorous acid	HOCl
Perchloric acid	$HClO_4$	Boric acid	H_3BO_3

but small equilibrium amounts of HCl and the weak acid HOCl are formed:

$$Cl_2(aq) + H_2O(l) \rightleftharpoons HCl(aq) + HOCl(aq)$$

Acetic acid, $C_2H_4O_2$, is another example of a weak acid (see Experiment 5.10). Vinegar is a dilute solution of acetic acid in water. The ionization of acetic acid may be written as

⊕ H_3O^+ ⊖ Cl^-

A strong acid is completely ionized in water. For example HCl is completely converted to H_3O^+ and Cl^-.

or more simply as

$$CH_3CO_2H(aq) + H_2O(l) \rightleftharpoons CH_3CO_2^-(aq) + H_3O^+(aq)$$

The negative ion, $CH_3CO_2^-$, is called the *acetate ion*. Acetic acid is a member of a series of acids, called *carboxylic acids*. These acids are further discussed in Chapter 23. Only the hydrogen attached to the electronegative oxygen atom in the acetic acid molecule is acidic in water. Carbon has only a very slightly greater electronegativity than hydrogen. Thus C—H bonds normally have only a very small polarity, and the hydrogen atoms have only a very small positive charge. Hence they have no tendency to be transferred to a water molecule. These three hydrogens of the CH_3 group are therefore nonacidic. The hydrogen atoms in methane and almost all other hydrocarbons are also not acidic in water.

○ CH_3CO_2H

⊕ H_3O^+ ⊖ $CH_3CO_2^-$

A weak acid is only slightly ionized in water. For example, out of 100 molecules of acetic acid in a $0.1M$ solution of acetic acid in water only one is converted to H_3O^+ and $CH_3CO_2^-$.

We will see in Chapter 14 when we deal with the ionization of acids and bases in a quantitative manner that, for almost all the *weak* acids with which we will be concerned, the extent of ionization is very small and almost all the acid remains in solution in the un-ionized form. For example, in a $0.10M$ solution of acetic acid only 1% of the acid is ionized, that is, converted to H_3O^+ and $CH_3CO_2^-$ ions, and 99% remains as un-ionized CH_3CO_2H molecules:

$$CH_3CO_2H(aq) + H_2O(l) \rightleftharpoons CH_3CO_2^-(aq) + H_3O^+(aq)$$
$$\approx 99\% \qquad\qquad\qquad \approx 1\%$$

This behavior of weak acids is in contrast with that of strong acids which are essentially completely converted to H_3O^+ and the corresponding anion; the amount of un-ionized acid remaining in solution is negligibly small:

$$HCl(aq) + H_2O(l) \longrightarrow Cl^-(aq) + H_3O^+(aq)$$
$$\approx 0\% \qquad\qquad\qquad \approx 100\%$$

NAMING BINARY ACIDS

The solutions formed when the hydrogen halides are dissolved in water are given special names:

Hydrogen halide	Formula	Name of aqueous solution
Hydrogen fluoride	HF	Hydrofluoric acid
Hydrogen chloride	HCl	Hydrochloric acid
Hydrogen bromide	HBr	Hydrobromic acid
Hydrogen iodide	HI	Hydroiodic acid

Experiment 5.10

Electrical Conductivity of Solutions of Hydrochloric Acid, Hydrofluoric Acid and Acetic Acid

Two metal electrodes connected to a power supply are dipped into a beaker of water. When HCl(g) is passed into the water the light bulb in the circuit glows brightly. Water behaves as a base, accepting protons from HCl(g) to give H_3O^+(aq) and Cl^-(aq).

The left-hand beaker contains a 0.10M solution of HF and the right-hand beaker a 0.10M solution of acetic acid, CH_3CO_2H. The light bulb glows dimly showing that the solution has only a small electrical conductivity. Acetic acid and hydrofluoric acid are weak acids and are only slightly ionized in water.

BASES

A **base** is a proton acceptor. Water behaves as a base, accepting a proton from an acid to form H_3O^+. Ammonia, NH_3, is another base. It is a polar molecule like water, and the positively charged hydrogen of the hydrogen chloride molecule is attracted to the negatively charged nitrogen of the ammonia molecule and transferred to its unshared pair of electrons, forming a fourth N—H bond.

$$\overset{\delta-}{:\!\ddot{C}l}\!-\!\overset{\delta+}{H} + \overset{\overset{\delta+}{H}}{\underset{\underset{\delta+}{H}}{\overset{3\delta-}{:\!N}\!-\!H^{\delta+}}} \longrightarrow :\!\ddot{C}l\!:^{\ominus} + \underset{H}{\overset{H}{H\!-\!N^{\oplus}\!-\!H}}$$

When a stream of HCl(g) meets a stream of NH_3(g) a white cloud of solid ammonium chloride, NH_4Cl, is formed.

The products of this reaction are the ammonium ion and the chloride ion. Both hydrogen chloride and ammonia are gases. When they are allowed to mix, a dense white cloud of solid ammonium chloride is formed (see Experiment 3.7). Although acid–base reactions are commonly carried out in aqueous solution they are by no means confined to aqueous solution. They may occur in many other solvents and also in the gas phase as illustrated by the reaction between NH_3(g) and HCl(g).

Ammonia also reacts with water in a proton transfer reaction in which water acts as an acid and ammonia as a base. The reaction is incomplete, however, and the following equilibrium is set up:

$$\underset{\text{acid}}{H_2O(l)} + \underset{\text{base}}{NH_3(aq)} \rightleftharpoons NH_4^+(aq) + OH^-(aq)$$

The products of this reaction are the ammonium ion, NH_4^+, and the **hydroxide**

ion, OH⁻. This reaction can be represented in terms of Lewis structures in the following way:

$$
\overset{H}{\underset{\cdot\cdot}{:O}}{-}H + \overset{\overset{H}{|}}{:N}{-}H \;\longrightarrow\; \overset{\cdot\cdot}{:O:}_{\ominus} + H{-}\overset{\overset{H}{|}}{N}{\overset{\oplus}{-}}H
$$

A proton is transferred from the water molecule to the unshared pair of electrons on the nitrogen atom of the ammonia molecule. Since the reaction of ammonia with water is incomplete, ammonia is described as a **weak base**.

A base, B, is a proton acceptor. A base in water, therefore, gives rise to the hydroxide ion:

$$B + H_2O \longrightarrow BH^+ + OH^-$$

If this reaction is complete, the base is strong. A **strong base** is fully ionized in water to give OH⁻. A **weak base** is incompletely ionized in water to give OH⁻.

Most weak bases, like most weak acids, are only ionized to a very small extent in water. In a 0.10M solution of ammonia in water only about 1% of the ammonia molecules are converted to ammonium ion and 99% remain as NH_3 molecules:

$$NH_3(aq) + H_2O(l) \rightleftharpoons NH_4^+(aq) + OH^-(aq)$$
$$\approx 99\% \qquad\qquad\qquad \approx 1\%$$

Aqueous solutions of ammonia are often called ammonium hydroxide even though only a very small percentage of the ammonia is actually converted to ammonium and hydroxide ions.

An example of a strong base in water is the oxide ion, O^{2-}. The oxide ion accepts a proton from a water molecule and is completely converted to the hydroxide ion:

$$O^{2-}(aq) + H_2O(l) \longrightarrow OH^-(aq) + OH^-(aq)$$

The alkali metal oxides, for example, $(Na^+)_2O^{2-}$, which are ionic compounds, are strong bases in water:

$$(Na^+)_2O^{2-}(s) + H_2O(l) \longrightarrow 2Na^+(aq) + 2OH^-(aq)$$

The product of this reaction is a solution of sodium hydroxide, NaOH. All the alkali metals form similar hydroxides. They can also be prepared by the reaction of the alkali metal with water, as we saw in Experiment 4.1. For example,

$$2K(s) + H_2O(l) \longrightarrow 2K^+(aq) + 2OH^-(aq) + H_2(g)$$

The alkali metal hydroxides are all ionic compounds composed of alkali metal cations and hydroxide ions, for example, potassium hydroxide, K^+OH^-. They are soluble in water, and they behave as strong bases because they give a quantitative yield of hydroxide ion:

$$K^+OH^-(s) \longrightarrow K^+(aq) + OH^-(aq)$$

Another strong base in water is the hydride ion, H^-:

$$H^-(aq) + H_2O(l) \longrightarrow H_2(g) + OH^-(aq)$$

Pellets of solid potassium hydroxide, KOH.

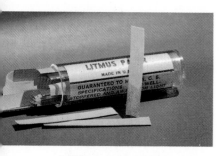

Red litmus paper turns mauve
when dipped into basic (alkaline)
solution.

Thus sodium hydride reacts with water to give hydrogen and a solution of
sodium hydroxide:

$$NaH(s) + H_2O(l) \longrightarrow NaOH(aq) + H_2(g)$$

All bases form the hydroxide ion, OH⁻, in water. Aqueous solutions of bases
therefore have several common properties by which they have long been recognized; these are the properties of the hydroxide ion, OH^-. These *basic solutions*
are also often called *alkaline solutions*; their properties include the following:

- Basic (alkaline) solutions have an unpleasant, bitter taste.
- Basic (alkaline) solutions have the ability to change the colors of indicators
  and, in particular, to reverse the effect of acids on these indicators.
- Basic (alkaline) solutions have the ability to destroy or neutralize the properties of acids.

ACID–BASE REACTIONS IN AQUEOUS SOLUTION

A reaction between an acid and a base is often called a **neutralization reaction**
because, when the reaction is complete, it tends to give a "neutral" solution,
that is, a solution that is neither acidic nor basic. The reaction of an aqueous
solution of HCl with an aqueous solution of NaOH is an example:

$$NaOH(aq) + HCl(aq) \longrightarrow NaCl(aq) + H_2O(l)$$

Another example is

$$KOH(aq) + HBr(aq) \longrightarrow KBr(aq) + H_2O(l)$$

The products of these reactions between acids and bases in aqueous solution
are a **salt** and water. A salt consists of a positive ion (cation) derived from a
base and a negative ion (anion) derived from an acid. The salts formed in the
above reactions are the alkali metal halides, sodium chloride, Na^+Cl^-, and
potassium bromide, K^+Br^-. By evaporating the solutions these salts could be
obtained as solids.

Colors of Some Common Indicators		
	Acidic Solution	*Basic Solution*
Methyl red	Red	Yellow
Bromo-thymol blue	Yellow	Blue
Phenol-phthalein	Colorless	Red

E X E R C I S E 5 . 1 1

Write an equation for an acid–base reaction by which each of the following salts
could be prepared:

LiI K_2SO_4 $Mg(NO_3)_2$

Since NaOH consists of Na^+ and OH^- ions and KOH consists of K^+ and
OH^- ions, and since HCl and HBr ionize in water to give the H_3O^+, Cl^-, and
Br^- ions, we may rewrite the above equations as

$$Na^+(aq) + OH^-(aq) + H_3O^+(aq) + Cl^-(aq) \longrightarrow Na^+(aq) + Cl^-(aq) + 2H_2O(l)$$

and

$$K^+(aq) + OH^-(aq) + H_3O^+(aq) + Br^-(aq) \longrightarrow K^+(aq) + Br^-(aq) + 2H_2O(l)$$

Canceling the spectator ions in both cases, we obtain the equation

$$H_3O^+(aq) + OH^-(aq) \longrightarrow 2H_2O(l)$$

The equation

$$H_3O^+(aq) + OH^-(aq) \longrightarrow 2H_2O(l)$$

represents the reaction that occurs when any acid reacts with any base in aqueous solution.

A common laboratory procedure is the determination of the concentration of a solution of a base by allowing it to react with a solution of an acid of known concentration, or vice versa. The procedure is called a **titration**. A solution of the acid is placed in a buret, and it is run into a known volume of the solution of the base until just sufficient acid solution has been added to react with all the base. This point is determined by observation of the change in color of a suitable indicator (see Figure 5.12).

An acid-base titration using methyl red as an indicator.

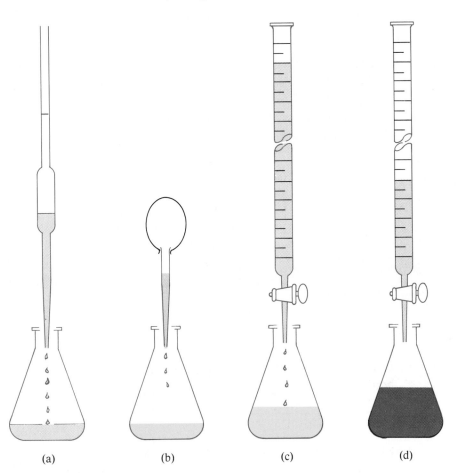

(a)   (b)   (c)   (d)

FIGURE 5.12
An Acid–Base Titration.

(a) A known volume of a solution of a base is placed in a flask using a pipet. (b) A few drops of a suitable indicator are added. (c) An acid solution of known concentration is placed in a buret and the level of the solution in the buret is noted. (d) The acid solution is run into the flask until the indicator just changes color. The level of the solution in the buret is then read again.

$=$ *Example 5.9* $=$

ACID–BASE TITRATION

In a titration 25.00 mL of sodium hydroxide, NaOH, solution were neutralized by 32.72 mL of hydrochloric acid, HCl. The HCl solution had a concentration of 0.129M. Find the concentration of the NaOH solution.

Solution

First, we must write the balanced equation for the reaction:

$$NaOH(aq) + HCl(aq) \longrightarrow NaCl(aq) + H_2O(l)$$

Now, since we know both the volume and the concentration of the HCl solution, we can determine how many moles of HCl took part in the reaction:

$$(32.72 \text{ mL})\left(\frac{1 \text{ L}}{1000 \text{ mL}}\right)\left(\frac{0.129 \text{ mol}}{1 \text{ L}}\right) = 4.22 \times 10^{-3} \text{ mol HCl}$$

The balanced equation shows that 1 mol of HCl reacts with 1 mol of NaOH. Therefore 4.22×10^{-3} mol HCl reacts with 4.22×10^{-3} mol NaOH. This amount of NaOH was in 25.00 mL of the NaOH solution. Therefore the concentration of the NaOH solution is

$$\left(\frac{4.22 \times 10^{-3} \text{ mol NaOH}}{25.00 \text{ mL}}\right)\left(\frac{1000 \text{ mL}}{1 \text{ L}}\right) = 0.169 \text{ mol L}^{-1} = 0.169M$$

Once we understand the principles, we can solve this problem more quickly in one step, as follows (concn is the abbreviation for concentration, and soln is the abbreviation for solution):

$$\text{Concn NaOH soln} = (32.72 \text{ mL HCl})\left(\frac{1 \text{ L}}{1000 \text{ mL}}\right)\left(\frac{0.129 \text{ mol HCl}}{1 \text{ L HCl}}\right)$$

$$\times \left(\frac{1 \text{ mol NaOH}}{1 \text{ mol HCl}}\right)\left(\frac{1}{25.00 \text{ mL}}\right)\left(\frac{1000 \text{ mL}}{1 \text{ L}}\right)$$

$$= 0.169 \text{ mol L}^{-1} = 0.169M$$

Still more time can be saved by working in millimoles (10^{-3} mol) rather than in moles. A 1M solution contains 1 mmol in 1 mL. Hence a 0.129M solution contains 0.129 mmol in 1 mL. Therefore

$$\text{Concn NaOH soln} = (32.72 \text{ mL HCl})\left(\frac{0.129 \text{ mmol HCl}}{1 \text{ mL}}\right)$$

$$\times \left(\frac{1 \text{ mmol NaOH}}{1 \text{ mmol HCl}}\right)\left(\frac{1}{25.00 \text{ mL}}\right)$$

$$= 0.169 \text{ mmol mL}^{-1} = 0.169 \text{ mol L}^{-1} = 0.169M$$

Let us summarize the steps that we followed in Example 5.9:

$$\text{Volume HCl} \xrightarrow[\text{HCl}]{\text{concn}} \text{moles HCl} \xrightarrow{\text{equation}} \text{moles NaOH} \xrightarrow[\text{NaOH}]{\text{volume}} \text{concn NaOH}$$

We follow a similar series of steps in any other calculation of this type.

EXERCISE 5.12

A 25.00-mL sample of an aqueous solution of KOH was neutralized by 38.60 mL of a 0.0500M solution of hydrobromic acid, HBr. What is the concentration of the KOH solution?

CONJUGATE ACID–BASE PAIRS

When an acid, HA, gives up a proton, the anion, A^-, that is formed is a base, because it can add a proton to re-form the acid HA:

$$HA \rightleftharpoons A^- + H^+$$

The anion A^- is called the **conjugate base** of the acid HA, and HA and A^- are described as a **conjugate acid–base pair**. Examples include the following:

$$HCl \rightleftharpoons Cl^- + H^+$$
$$H_2O \rightleftharpoons OH^- + H^+$$
$$NH_4^+ \rightleftharpoons NH_3 + H^+$$
$$H_3O^+ \rightleftharpoons H_2O + H^+$$

Acid   Conjugate base

When an acid is a cation, such as NH_4^+, its conjugate base is a neutral molecule, such as NH_3.

Similarly, when a base B accepts a proton, the cation, BH^+, that is formed is an acid, because it can lose a proton to give back the base B:

$$B + H^+ \rightleftharpoons BH^+$$

The cation BH^+ is called the **conjugate acid** of the base B, and BH^+ and B are described as a conjugate acid–base pair. For example,

$$H_2O + H^+ \rightleftharpoons H_3O^+$$
$$NH_3 + H^+ \rightleftharpoons NH_4^+$$
$$OH^- + H^+ \rightleftharpoons H_2O$$
$$F^- + H^+ \rightleftharpoons HF$$

Base   Conjugate acid

If the base is an anion, such as F^-, the conjugate acid is a neutral molecule, such as HF.

═ Example 5.10 ═══════════════════════════

CONJUGATE ACIDS AND BASES

(a) What is the conjugate acid of each of the following: CN^-, Br^-, and PH_3?

(b) What is the conjugate base of each of the following: HF, $HClO_4$, NH_3, and NH_4^+?

Solution

(a) The conjugate acid is the species obtained by adding a proton to a base. Thus we have

$$CN^- + H^+ \longrightarrow HCN$$
$$Br^- + H^+ \longrightarrow HBr$$
$$PH_3 + H^+ \longrightarrow PH_4^+$$

Base   Conjugate acid

(b) The conjugate base is the species formed by removing a proton from an acid. Thus we have

$$HF \longrightarrow F^- + H^+$$
$$HClO_4 \longrightarrow ClO_4^- + H^+$$
$$NH_3 \longrightarrow NH_2^- + H^+$$
$$NH_4^+ \longrightarrow NH_3 + H^+$$

Acid   Conjugate base

Since a base can obtain a proton only from an acid, and an acid can give up a proton only if there is a base to accept it, every acid–base reaction involves two conjugate acid–base pairs:

$$AH + B \rightleftharpoons BH^+ + A^-$$

Acid$_1$ + Base$_2$ $\rightleftharpoons$ Acid$_2$ + Base$_1$

Acid$_1$ and base$_1$ (AH and A$^-$) form one conjugate acid–base pair; acid$_2$ and base$_2$ (BH$^+$ and B) constitute another. For example,

Acid$_1$ + Base$_2$ $\rightleftharpoons$ Acid$_2$ + Base$_1$

$$HF + H_2O \rightleftharpoons H_3O^+ + F^-$$
$$H_2O + CN^- \rightleftharpoons HCN + OH^-$$
$$H_2O + NH_3 \rightleftharpoons NH_4^+ + OH^-$$

═ *Example 5.11*

CONJUGATE ACID–BASE PAIRS

In the following reaction identify the acid on the left and its conjugate base on the right and the base on the left and its conjugate acid on the right:

$$HF(aq) + NH_3(aq) \rightleftharpoons NH_4^+(aq) + F^-(aq)$$

Solution

HF is the acid on the left; it loses a proton forming its conjugate base F$^-$. NH$_3$ is the base on the left; it gains a proton forming its conjugate acid NH$_4^+$.

EXERCISE 5.13

In the following reaction identify two conjugate acid–base pairs:

$$CH_3CO_2H(aq) + H_2O(l) \rightleftharpoons CH_3CO_2^-(aq) + H_3O^+(aq)$$

We see from these examples that water can act both as an acid and as a base. It accepts a proton from an acid, such as HF, forming the conjugate acid H_3O^+:

$$HF + H_2O \rightleftharpoons H_3O^+ + F^-$$

But it also donates a proton to an ammonia molecule, forming its conjugate base OH^-:

$$H_2O + NH_3 \rightleftharpoons NH_4^+ + OH^-$$

Many molecules and ions can behave either as an acid or as a base depending on the circumstances, and are said to be *amphiprotic*, or *amphoteric*. **Amphiprotic** means having the ability to either accept or donate a proton. **Amphoteric** is a more general term that means having the ability to act either as an acid or as a base.

The existence of amphiprotic substances such as water serves to emphasize the relative nature of the terms *acid* and *base*. It is not strictly correct to state that water *is* an acid or that it *is* a base. Water may be said to *behave as an acid* or to *behave as a base*, depending on the circumstances. In the presence of HCl water behaves as a base, but in the presence of NH_3 it behaves as an acid.

Since water is amphiprotic, a proton exchange can occur between two water molecules. In this reaction one water molecule acts as an acid and the other as a base:

$$H_2O + H_2O \qquad H_3O^+ + OH^-$$
$$\text{Acid}_1 \quad \text{Base}_2 \qquad \text{Acid}_2 \quad \text{Base}_1$$

This reaction is known as the **self-ionization**, or **autoprotolysis, of water**. The concentrations of the ions H_3O^+ and OH^- in equilibrium with pure water are very small, but nevertheless, the self-ionization of water is important in a quantitative treatment of acid–base reactions in water, as we will see in Chapter 14.

STRENGTHS OF ACIDS AND BASES

STRONG ACIDS The strong acids listed in Table 5.10 are all completely ionized when dissolved in a large amount of water. Their reaction with water goes to completion. For example,

$$HCl + H_2O \longrightarrow H_3O^+ + Cl^-$$
$$HClO_4 + H_2O \longrightarrow H_3O^+ + ClO_4^-$$

Since all these acids are completely converted to H_3O^+, any intrinsic differences in their strengths cannot be detected. Their strengths are said to be leveled by the base, water, and this effect is referred to as the **leveling effect** of water. **The strongest acid that can exist in a dilute aqueous solution is H_3O^+.** Any acid that is intrinsically stronger than H_3O^+ is quantitatively converted to H_3O^+ in water.

The conjugate bases of strong acids are such weak bases that they have no tendency to remove a proton from an H_3O^+ ion. They therefore have no tendency to remove a proton from a water molecule, which is a much weaker acid than H_3O^+. In other words, they have no base properties in water. The equilibrium

$$Cl^- + H_2O \rightleftharpoons OH^- + HCl$$

lies very far over on the left-hand side. Although formally they may be described as bases because they have the potential to add a proton, anions such as Cl^-, Br^-, NO_3^-, and ClO_4^- are much too weak to act as bases in water (see Figure 5.13). Thus we do not normally consider the above reaction to be an equilibrium because the equilibrium concentration of HCl is much too small to be detected. In other words, the reverse reaction

$$HCl + OH^- \longrightarrow H_2O + Cl^-$$

goes to completion.

STRONG BASES Similar considerations apply to strong bases. They are completely converted to the hydroxide ion in water, and any intrinsic differences in their strengths cannot be detected. For example,

$$O^{2-} + H_2O \longrightarrow OH^- + OH^-$$
$$H^- + H_2O \longrightarrow H_2 + OH^-$$

The strongest base that can exist in a dilute aqueous solution is the hydroxide ion, OH^-. Water therefore also levels the strengths of strong bases, just as it levels the strengths of strong acids. The range of acid–base strengths that is accessible in aqueous solution therefore lies between the strong acid H_3O^+ and the strong base OH^-.

Since the reactions of strong bases with water proceed completely to the right, the reverse reactions show no tendency to proceed. Thus the conjugate acids of O^{2-} and H^-, in other words, OH^- and H_2, have no tendency to donate a proton to an OH^- ion. They also therefore have no tendency to donate a proton to the much weaker base, H_2O. In other words, neither OH^- nor H_2 behave as acids in water (see Figure 5.13).

WEAK ACIDS AND BASES A weak acid such as HF is incompletely ionized in water. In other words, it is only partially converted to H_3O^+, and the following equilibrium is set up:

$$HF + H_2O \rightleftharpoons H_3O^+ + F^-$$

In this case the reverse reaction does occur; F^- accepts a proton from H_3O^+ to a limited extent, and so it behaves as a base. The F^- ion also accepts a proton from an H_2O molecule but to an even more limited extent, to form very small equilibrium amounts of HF and OH^-:

$$F^- + H_2O \rightleftharpoons HF + OH^-$$

Thus the conjugate base of a weak acid is a weak base in water.

Similarly, NH_4^+, the conjugate acid of the weak base NH_3,

$$NH_3 + H_2O \rightleftharpoons NH_4^+ + OH^-$$

is a weak acid in water:

$$NH_4^+ + H_2O \rightleftharpoons H_3O^+ + NH_3$$

In a series of acids of increasing strength in water, the strength of the conjugate base decreases until finally, for a strong acid in water, the conjugate base shows no base properties at all in water. Similarly, in a series of bases of increasing strength in water, the conjugate acids decrease in strength until finally,

FIGURE 5.13
Strengths of Acids and Bases in Water.

	ACIDS			BASES	
	Acid	Conjugate Base		Base	Conjugate Acid
Strong acids 100% dissociated in water / Completely converted to H_3O^+	$HClO_4 + H_2O \longrightarrow H_3O^+ + ClO_4^-$		Too weak to behave as bases in water	ClO_4^-	
	$HCl + H_2O \longrightarrow H_3O^+ + Cl^-$			Cl^-	
	$HNO_3 + H_2O \longrightarrow H_3O^+ + NO_3^-$			NO_3^-	
Strongest acid	$H_3O^+ + H_2O \rightleftharpoons H_3O^+ + H_2O$		Weakest base	$H_2O + H_2O \rightleftharpoons OH^- + H_3O^+$	
Weak acids	$HF + H_2O \rightleftharpoons H_3O^+ + F^-$		Weak bases	$F^- + H_2O \rightleftharpoons OH^- + HF$	
	$CH_3CO_2H + H_2O \rightleftharpoons H_3O^+ + CH_3CO_2^-$			$CH_3CO_2^- + H_2O \rightleftharpoons OH^- + CH_3CO_2H$	
	$NH_4^+ + H_2O \rightleftharpoons H_3O^+ + NH_3$			$NH_3 + H_2O \rightleftharpoons OH^- + NH_4^+$	
Weakest acid	$H_2O + H_2O \rightleftharpoons H_3O^+ + OH^-$		Strongest base	$OH^- + H_2O \rightleftharpoons OH^- + H_2O$	
Too weak to behave as acids in water	NH_3		Strong bases 100% dissociated in water / Completely converted to OH^-	$NH_2^- + H_2O \longrightarrow OH^- + NH_3$	
	H_2			$H^- + H_2O \longrightarrow OH^- + H_2$	
	OH^-			$O^{2-} + H_2O \longrightarrow OH^- + OH^-$	

(left margin: Range of acid and base strengths in water)

for a strong base, the conjugate acid has no acid properties at all in water (see Figure 5.13). In general, the stronger the acid, the weaker its conjugate base and vice versa.

5.7
Reaction Types

In this chapter we have met three very important types of reactions:

1. In *oxidation–reduction reactions* electrons are transferred from one species (molecule or ion) to another species. For example, in the reaction

$$2Br^-(aq) + Cl_2(aq) \longrightarrow Br_2(aq) + 2Cl^-(aq)$$

electrons are transferred from Br^- ions (the reducing agent) to Cl_2 molecules (the oxidizing agent).

2. In *acid–base reactions* hydrogen ions (protons), H^+, are transferred from one species (molecule or ion) to another species. For example, in the reaction

$$HBr(aq) + H_2O(l) \longrightarrow H_3O^+(aq) + Br^-(aq)$$

protons are transferred from HBr molecules (the acid) to H_2O molecules (the base).

3. In *precipitation reactions* an insoluble solid is formed from a solution of soluble substances. For example, in the reaction

$$KBr(aq) + AgNO_3(aq) \longrightarrow AgBr(s) + KNO_3(aq)$$

insoluble silver bromide, AgBr, is formed on mixing aqueous solutions of KBr and $AgNO_3$.

It is important that you are able to recognize these three types of reactions. They are the most important types of reaction that occur in aqueous solution and you will meet them many times in the following chapters.

═ *Example 5.12* ═══════════════════════════════════

IDENTIFYING REACTION TYPES

Identify each of the following reactions as an oxidation–reduction, acid–base, or precipitation reaction

(a) $Mg(s) + F_2(g) \longrightarrow Mg^{2+}(F^-)_2(s)$

(b) $2Ca(s) + O_2(g) \longrightarrow 2CaO(s)$

(c) $Pb(NO_3)_2(aq) + 2NaCl(aq) \longrightarrow PbCl_2(s) + 2NaNO_3(aq)$

(d) $HNO_3(aq) + NH_3(aq) \longrightarrow NH_4^+(aq) + NO_3^-(aq)$

(e) $2Na(s) + S(s) \longrightarrow Na_2S(s)$

(f) $H_3O^+(aq) + CO_3^{2-}(aq) \longrightarrow HCO_3^-(aq) + H_2O(l)$

(g) $Zn(s) + 2HCl(aq) \longrightarrow Zn^{2+}(aq) + 2Cl^-(aq) + H_2(g)$

Solution

(a) Oxidation–reduction reaction. The Mg loses electrons to become Mg^{2+}; it is oxidized:

$$Mg \longrightarrow Mg^{2+} + 2e^-$$

The F_2 gains electrons to become $2F^-$; it is reduced:

$$F_2 + 2e^- \longrightarrow 2F^-$$

(b) Oxidation–reduction reaction. Calcium oxide, CaO, is a compound of a group 2 metal with a group 6 nonmetal. It is therefore ionic, $Ca^{2+}O^{2-}$. Thus Ca loses electrons to become Ca^{2+} and is oxidized, while O_2 gains electrons to become $2O^{2-}$ and is reduced.

(c) Precipitation reaction. An insoluble solid, $PbCl_2$, is formed from an aqueous solution of two soluble substances, $Pb(NO_3)_2$ and NaCl.

(d) Acid–base reaction. Nitric acid, HNO_3, loses a proton to form the nitrate ion, NO_3^-, and NH_3 gains a proton to give the ammonium ion, NH_4^+. Thus it is a proton transfer, or acid–base, reaction.

(e) Oxidation–reduction reaction. Sodium sulfide, Na_2S, is a compound of a group 1 metal and a group 6 nonmetal and is therefore ionic. Thus Na loses an electron to become Na^+, and S gains two electrons to become S^{2-}. It is an electron transfer, or oxidation–reduction, reaction.

(f) Acid–base reaction. The H_3O^+ ion loses a proton to become H_2O; it is an acid. The carbonate ion, CO_3^{2-}, gains a proton to become the hydrogen carbonate ion, HCO_3^-; CO_3^{2-} is a base.

(g) Oxidation–reduction reaction. Although this reaction involves the acid HCl(aq), it is not an acid–base reaction. The proton in HCl is not transferred to a base. In fact, Zn loses two electrons to become Zn^{2+}; it is oxidized. And two H^+ (H_3O^+) ions gain two electrons to become $H_2(g)$:

$$2H^+ + 2e^- \longrightarrow H_2$$

EXERCISE 5.14

Identify each of the following reactions as an oxidation–reduction, acid–base, or precipitation reaction. In each case give reasons for your choice:

(a) $Mg(s)$ $+ H_2(g)$ $\rightarrow MgH_2(s)$

(b) $Cu(s)$ $+ S(s)$ $\rightarrow CuS(s)$

(c) $NH_3(g)$ $+ HBr(g)$ $\rightarrow NH_4Br(s)$

(d) $CuSO_4(aq)$ $+ K_2S(aq)$ $\rightarrow CuS(s)$ $+ K_2SO_4(aq)$

(e) $Mg(s)$ $+ 2HBr(aq)$ $\rightarrow MgBr_2(aq) + H_2(g)$

(f) $NaH(s)$ $+ H_2O(l)$ $\rightarrow NaOH(aq) + H_2(g)$

(g) $Ba(OH)_2(aq) + H_2SO_4(aq) \rightarrow BaSO_4(s)$ $+ 2H_2O(l)$

IMPORTANT TERMS

An **acid** is a proton donor. In aqueous solution all acids transfer a proton to a water molecule, to give the hydronium ion, H_3O^+.

An **acid–base reaction** is a reaction in which protons are transferred; it is a proton transfer reaction.

An **amphiprotic substance** is one that can both add or donate a proton.

An **amphoteric substance** is one that can behave as both an acid and a base.

Autoprotolysis (self-ionization) is an acid–base reaction in which one solvent molecule acts as an acid, donating a proton to another solvent molecule, which acts as a base.

A **base** is a proton acceptor. In aqueous solution a base accepts a proton from a water molecule, forming the hydroxide ion, OH^-.

A **conjugate acid** is the acid formed by the addition of a proton to a base.

A **conjugate base** is the base formed by the loss of a proton from an acid.

Electronegativity is the ability of an atom in a molecule to attract the electrons of a covalent bond to itself.

Equilibrium is reached in a reaction when the rate of the forward reaction equals the rate of the back (reverse) reaction so that there is no change in the overall concentrations of the reactants and products.

The **halogens** are the elements fluorine, chlorine, bromine, iodine, and astatine; they form group 7 of the periodic table.

The **hydronium ion**, H_3O^+, is the ion that is formed by all *acids* in aqueous solution. It is the strongest acid that can exist in water.

The **hydroxide ion**, OH^-, is the ion that is formed by all *bases* in aqueous solution. It is the strongest base that can exist in water.

An **ionic bond** is a bond that results from the electrostatic attraction between oppositely charged ions.

An **ionic crystal** is a crystal that is composed of an infinite array of positive and negative ions and in which no individual molecules can be distinguished.

A **neutralization reaction** is a reaction between an acid and a base.

A **nonpolar bond** is a bond formed between atoms of equal electronegativity.

An **octahedron** is a regular solid with 8 equivalent equilateral triangular faces, 12 equivalent edges, and 6 equivalent vertices.

Oxidation is a process in which electrons are lost.

An **oxidation–reduction reaction** is a reaction in which electrons are transferred; it is an electron transfer reaction.

An **oxidizing agent** is a substance that has a tendency to gain electrons.

A **polar bond** is a covalent bond between two atoms of different electronegativity in which the atom of higher electronegativity has a partial negative charge and the other atom has a partial positive charge; a bond in which the bond electrons are unequally shared.

A **precipitation reaction** is one in which an insoluble substance separates from solution.

A **reducing agent** is a substance from which electrons are easily removed.

Reduction is a process in which electrons are gained.

A **salt** is an ionic compound consisting of a positive ion (cation) derived from a base and a negative ion (anion) derived from an acid. It is the product of the neutralization of a base such as NaOH with an acid such as an aqueous solution of HCl.

Spectator ions are ions that are present in a solution but which do not take part in any reactions that are occurring.

A **strong acid** in water is an acid that is quantitatively converted to H_3O^+.

A **strong base** in water is a base that is quantitatively converted to OH^-.

An acid–base **titration** is a procedure in which a measured volume of a solution of an acid (or a base) is added to a known volume of a solution of a base (or an acid) until the reaction is complete as indicated by the color change of a suitable indicator.

A **weak acid** in water is an acid that is only partially converted to H_3O^+.

A **weak base** in water is a base that is only partially converted to OH^-.

P R O B L E M S *

Halogens: Preparation and Properties

1. What is the most important source of bromine? Briefly describe how bromine is obtained from this natural source.

2. (a) Write the balanced equation for the preparation of chlorine from the reaction of manganese dioxide, $MnO_2(s)$, with hydrochloric acid, to give manganese dichloride and $Cl_2(g)$.

(b) How can hydrogen chloride be prepared from chlorine?

(c) Explain why hydrogen chloride gas gives a strongly acidic solution when it dissolves in water.

3. What is the advantage of using sea salt rather than pure sodium chloride to add flavor to our food? Why is table salt not pure sodium chloride?

4. Write a balanced equation for each of the following reactions:

(a) The reaction of white phosphorus with chlorine to give phosphorus trichloride

(b) The reaction of chlorine with sulfur to give **(i)** SCl_2, **(ii)** S_2Cl_2

(c) The reaction of fluorine with carbon to give carbon tetrafluoride

(d) The reaction of bromine with arsenic to give arsenic tribromide

5. Write Lewis structures for each of the products of the reactions in Problem 4.

6. What is the important mineral in bones and teeth? Why is a small amount of fluoride ion added to most water supplies?

7. Write balanced equations for each of the following reactions:

(a) Barium with chlorine

(b) Aluminum with bromine

(c) Potassium with iodine

(d) Phosphorus with chlorine

(e) Phosphorus with iodine

8. Name each of the products of the reactions in Problem 7, and write their Lewis structures.

9. Javex is a solution containing sodium hypochlorite, NaOCl(aq), and sodium chloride, NaCl(aq), obtained by passing chlorine into sodium hydroxide solution.

(a) Write the balanced equation for this reaction.

(b) How many grams of sodium hypochlorite would be obtained by passing 1.00 L of $Cl_2(g)$, at 25 °C and 1.00 atm pressure, into excess NaOH(aq)?

10. Complete and balance each of the following equations. If no reaction occurs, write NR. Identify the oxidizing agent and the reducing reagent for each reaction that occurs at room temperature.

(a) $Br_2(l) + NaCl(aq) \rightarrow$

(b) $NaCl(s) + F_2(g) \rightarrow$

(c) $I_2(s) + NaF(aq) \rightarrow$

(d) $CaBr_2(s) + F_2(g) \rightarrow$

(e) $KF(aq) + Br_2(l) \rightarrow$

(f) $MgI_2(aq) + Cl_2(g) \rightarrow$

(g) $LiI(aq) + Br_2(l) \rightarrow$

(h) $I_2(s) + MgCl_2(aq) \rightarrow$

* Answers to problems numbered in blue appear at the end of the text.

Electronegativity

11. What are the two principal properties of an atom that determine its electronegativity?

12. Describe how the electronegativities of the elements change in going across any period of the periodic table from left to right, and in descending any group. Explain these variations.

13. Without reference to Figure 5.6, select the element of higher electronegativity in each of the following pairs:

    (a) F, Cl   **(b)** F, O   **(c)** P, S   **(d)** C, Si

    (e) O, P   **(f)** Br, Se   **(g)** P, Al

14. Arrange the following elements in order of decreasing electronegativity:

    Al   S   F   Sr

Ionic, Polar, and Covalent Bonds

15. Classify the bonds in each of the following substances as ionic, nonpolar covalent, or polar covalent:

    Cl_2 PCl_3 LiCl   ClF $MgCl_2$

16. Classify the bonds in each of the following substances as ionic, nonpolar covalent, or polar covalent:

    Li_2O MgO O_2 SO_2 Cl_2O NO

17. Classify each of the following molecules as polar or nonpolar:

    HBr I_2 ClF H_2 LiH

18. Explain why sodium chloride forms a solid ionic compound at room temperature, rather than a gas consisting of Na^+Cl^- molecules (ion pairs).

19. Suggest a reason why ions with charges greater than 3 are rarely found in ionic compounds.

20. Explain on the basis of structure why fluorine is a gas, whereas lithium fluoride is a crystalline solid at room temperature.

21. Without reference to Table 5.6, select the larger ion of each of the following pairs of ions:

    (a) K^+, Ca^{2+} **(b)** S^{2-}, Cl^- **(c)** Cl^-, K^+

    (d) Na^+, Li^+ **(e)** I^-, Br^-

22. In each of the following pairs, select the species (atom or ion) with the larger radius (covalent or ionic).

    (a) Cl, Cl^- **(b)** Na, Na^+

    (c) Mg^{2+}, Na   **(d)** Cl^-, K^+

* The asterisk denotes the more difficult problems.

23. On the basis of the electronegativities of the atoms forming the bonds, deduce the expected polarities of the bonds in each of the nitrogen trihalides.

24. Deduce the ionic, polar covalent, or nonpolar covalent nature of the bonds formed between each of the elements of the second period and **(a)** hydrogen, **(b)** fluorine. In the case of the polar covalent bonds indicate the direction of polarity of the bonds by writing $\delta+$ and $\delta-$ on the appropriate atoms.

Names and Formulas of Ionic Compounds

25. Write the empirical formulas of the compounds composed of the following ions:

    (a) NH_4^+, PO_4^{3-} **(b)** Fe^{3+}, O^{2-}

    (c) Cu^+, O^{2-} **(d)** Al^{3+}, SO_4^{2-}

26. Write the empirical formulas of the following compounds:

    (a) Barium iodide   **(b)** Aluminum chloride

    (c) Ammonium perchlorate

    (d) Calcium nitrate   **(e)** Aluminum sulfate

    (f) Calcium carbonate   **(g)** Magnesium hydroxide

27. Give the empirical formula and name the ionic compound formed by each of the following pairs of elements:

    (a) Ca, I   **(b)** Be, O   **(c)** Al, S

    (d) Mg, Br   **(e)** Rb, Se   **(f)** Ba, O

Oxidation–Reduction Reactions

28. Would you expect chlorine to react with iodide ion, or iodine to react with chloride ion? In the reaction that does proceed, identify the oxidizing agent, the reducing agent, the substance oxidized, and the substance reduced.

29. Complete and balance each of the following equations. If no reaction occurs, write NR:

    (a) $Cl_2(g) + KI(aq) \rightarrow$

    (b) $I_2(s) + NaCl(aq) \rightarrow$

    (c) $Br_2(l) + NaI(aq) \rightarrow$

    (d) $F_2(g) + H_2O(l) \rightarrow$

30. Suggest a reason why samples of crystalline sodium iodide are sometimes a pale yellow color, although both the Na^+ and I^- ions are colorless.

*31. Chlorine, bromine, and iodine all oxidize hydrogen sulfide, $H_2S(g)$, to sulfur, $S(s)$. Chlorine and bromine oxidize nitrous acid, $HNO_2(aq)$, to nitric acid, $HNO_3(aq)$, but iodine does not, and chlorine will oxidize bromide ion to bromine.

Write balanced equations for each of these reactions, and use the results to arrange these halogens in order of *increasing* strength as oxidizing agents.

32. For each of the following reactions, identify the oxidizing agent, the reducing agent, the species oxidized, and the species reduced:

(a) $2Rb(s) + I_2(s) \rightarrow 2RbI(s)$

(b) $4Al(s) + 3O_2(g) \rightarrow 2(Al^{3+})_2(O^{2-})_3(s)$

(c) $2Cu^{2+}(aq) + 4I^-(aq) \rightarrow 2Cu^+I^-(s) + I_2(s)$

(d) $Zn(s) + S(s) \rightarrow Zn^{2+}S^{2-}(s)$

(e) $Mg(s) + 2HCl(aq) \rightarrow MgCl_2(aq) + H_2(g)$

33. Suppose that 2.45 g of magnesium are burned completely in chlorine. What mass of magnesium chloride is formed? Which of the reactants is oxidized and which is reduced?

34. Explain how you could identify each solution if you were given colorless aqueous solutions of an ionic chloride, an ionic bromide, and an ionic iodide. Write balanced equations for each reaction used in the identification.

Precipitation Reactions

35. Use the solubility rules (Table 5.9) for ionic compounds to predict which of the following are soluble, and which are insoluble, in water:

(a) $PbSO_4$ (b) AgI (c) Na_2CO_3

(d) FeS (e) $AgNO_3$ (f) $Cu(OH)_2$

36. Use the solubility rules (Table 5.9) for ionic compounds to predict which of the following are soluble, and which are insoluble, in water:

(a) $AlCl_3$ (b) $CaSO_4$ (c) $CuSO_4$

(d) $LiOH$ (e) $BaCO_3$ (f) Na_2S

37. Predict whether or not a precipitate will form when aqueous solutions of each of the following pairs of substances are mixed. Where a reaction occurs, write the balanced equation for the reaction, and the corresponding net ionic equation:

(a) $FeCl_3 + NaOH \rightarrow$

(b) $NaClO_4 + AgNO_3 \rightarrow$

(c) $BaCl_2 + KOH \rightarrow$

(d) $Pb(NO_3)_2 + H_2SO_4 \rightarrow$

(e) $AgNO_3 + Na_2S \rightarrow$

38. Predict whether or not a precipitate will form when aqueous solutions of each of the following pairs of substances are mixed. Where a reaction occurs, write the bal-

anced equation for the reaction, and the corresponding net ionic equation:

(a) $Na_2CO_3 + CaCl_2 \rightarrow$

(b) $NaNO_3 + CaBr_2 \rightarrow$

(c) $AgNO_3 + NaI \rightarrow$

(d) $BaCl_2 + MgSO_4 \rightarrow$

(e) $HCl + Pb(NO_3)_2 \rightarrow$

39. An aqueous solution of silver nitrate contains 2.50 g of silver nitrate. To this solution is added sufficient barium chloride solution to precipitate all the silver as silver chloride. What mass of dry silver chloride is obtained?

40. In attempting to prepare a $0.10M$ solution of silver nitrate in tap water, a student claimed that not all the silver nitrate could be dissolved, because the final solution remained cloudy. Explain this student's problem. The solubility of silver nitrate is 245 g in 100 g of water at 25 °C.

Acid–Base Reactions

41. A 35.00-mL sample of barium hydroxide solution required 32.12 mL of $0.112M$ aqueous sulfuric acid for neutralization. What was

(a) The molarity of the barium hydroxide solution?

(b) The mass of barium sulfate precipitated?

42. (a) What volume of $0.012M$ $HNO_3(aq)$ is required to neutralize 25.1 mL of $0.021M$ $Ca(OH)_2(aq)$?

(b) What volume of $0.112M$ $HCl(aq)$ is required to react completely with 1.234 g of calcium carbonate?

(c) If 24.1 mL of $0.056M$ $HCl(aq)$ are required to neutralize 25.00 mL of $Ca(OH)_2(aq)$, what volume of $0.032M$ $H_3PO_4(aq)$ will be required to neutralize 28.2 mL of the same solution of $Ca(OH)_2(aq)$?

43. Describe three tests that you could make to decide whether a given aqueous solution was acidic.

44. For each of the following acids, write a balanced equation for its reaction with water, and classify the acid as a strong acid or a weak acid:

(a) Nitric acid   (b) Phosphoric acid

(c) Hypochlorous acid   (d) Sulfuric acid

(e) Hydrofluoric acid   (f) Perchloric acid

45. Write balanced equations to show how each of the following bases is ionized in aqueous solution. Classify each base as a strong base or a weak base.

(a) Sodium oxide   (b) Potassium hydroxide

(c) Ammonia   (d) Lithium hydride

46. With respect to their behavior in aqueous solution, classify each of the following as *strong acid*, *weak acid*, *strong base*, *weak base*, or *salt*:

(a) Potassium hydroxide   (b) Water

(c) Acetic acid   (d) Sulfuric acid

(e) Ammonia   (f) Potassium iodide

(g) Calcium chloride   (h) Carbonic acid

(i) Sodium oxide   (j) Potassium acetate

47. Write the balanced equation for a suitable acid–base reaction for the preparation of each of the following salts:

(a) Calcium sulfate   (b) Lithium fluoride

(c) Sodium sulfide   (d) Ammonium nitrate

(e) Magnesium perchlorate

(f) Aluminum chloride

48. Would an aqueous solution of each of the following be expected to be acidic, neutral, or basic? Explain.

(a) RbF   (b) $CaCl_2$

(c) KNO_3 (d) NH_4Cl

49. Give the formulas and names of the conjugate bases of the following acids:

(a) HF   (b) HNO_3 (c) $HClO_4$

(d) H_2O (e) H_3O^+

50. Give the formulas and names of the conjugate acids of the following bases:

(a) NH_3 (b) F^- (c) OH^-

(d) H_2O (e) H^-

51. What are the conjugate bases of each of the following.

H_2O NH_3 HCl HF NH_4^+

52. What are the conjugate acids of each of the following:

H_2O NH_3 OH^- NH_2^- S^{2-}

53. (a) Classify each of the following as a strong acid, a weak acid, or as having no acidic properties in aqueous solution:

HCl HF HNO_3 CH_4 NH_4^+

(b) Classify each of the following as a strong base or a weak base, or as having negligible basicity, in aqueouse solution:

Cl^- F^- NO_3^- NH_3 O^{2-}

54. (a) What is the simplest equation that represents the reaction between a solution of a strong acid, such

as $HCl(aq)$ or $HNO_3(aq)$, and a solution of a strong base, such as $NaOH(aq)$ or $Ca(OH)_2(aq)$? Explain.

(b) When an aqueous solution of a strong base, such as $NaOH(aq)$, is added to a solution of a weak acid, such as $HF(aq)$, why does the reaction go quantitatively to completion to form the solution of a salt, such as, for example, $NaF(aq)$?

55. Write the formulas, draw the Lewis structures, and describe the shape of each of the following:

(a) Hydronium ion   (b) Ammonium ion

(c) Hydroxide ion

56. Write balanced equations for the ionization of each of the following in aqueous solution. Indicate which of the reactants and which of the products of these reactions are behaving as acids, and which are behaving as bases:

(a) HCl (b) CH_3CO_2H (c) $HClO_4$

(d) H_2SO_4 (e) $HOCl$ (f) NH_4^+

57. Write balanced equations for the reactions that occur when (a) sodium hypochlorite, $NaOCl(s)$, and (b) ammonium chloride, $NH_4Cl(s)$, form aqueous solutions. Are the resulting solutions acidic, basic, or neutral? Explain.

58. What is the strongest *acid* species, and the strongest *basic* species, that is found in aqueous solutions? The hydride ion, H^-, and the oxide ion, O^{2-}, are both strong bases in water. Write balanced equations for the reaction of each of $LiH(s)$, $CaH_2(s)$, $Li_2O(s)$, and $CaO(s)$ with water.

59. Write the simplest balanced equation that describes the neutralization of any acid by any base in aqueous solution. Suppose 250 mL of $1.00M$ $HCl(aq)$ are neutralized with 500 mL of $0.500M$ $NaOH(aq)$. What is the concentration of sodium chloride in the resulting solution? If it is evaporated to dryness, what mass of sodium chloride is obtained?

60. What volume of $0.100M$ $HCl(aq)$ is required to react completely with 5.00 g of calcium hydroxide, $Ca(OH)_2(s)$?

61. The primary active ingredient in an antacid is the salt $NaAl(OH)_2CO_3(s)$, which reacts with stomach acid according to the equation

$$NaAl(OH)_2CO_3(s) + 4HCl(aq) \longrightarrow$$
$$NaCl(aq) + AlCl_3(aq) + 3H_2O(l) + CO_2(g)$$

What mass of this salt is required to react with 2.00 L of $0.120M$ $HCl(aq)$?

62. What volume of $0.124M$ $HBr(aq)$ is required to neutralize 25.00 mL of $0.107M$ $NaOH(aq)$ solution?

63. What volume of $0.115M$ KOH(aq) solution is required to react completely with 100.0 mL of $0.211M$ HF(aq)?

Reaction Types

64. Using a compound of bromine as one of the reactants, give at least one example of each of the following types of reaction:

    (a) Acid–base   **(b)** Oxidation–reduction

    (c) Precipitation

65. Balance each of the following equations. Classify each as an acid–base, an oxidation–reduction, or a precipitation reaction.

    (a) $AgNO_3(aq) + BaCl_2(aq) \rightarrow Ba(NO_3)_2(aq) + AgCl(s)$

    (b) $NH_3(aq) + H_2SO_4(aq) \rightarrow (NH_4)_2SO_4(aq)$

    (c) $Na_2O(s) + H_2O(l) \rightarrow NaOH(aq)$

    (d) $Al(s) + Br_2(l) \rightarrow AlBr_3(s)$

    (e) $Ca(OH)_2(aq) + CO_2(g) \rightarrow CaCO_3(s) + H_2O(l)$

66. Suggest methods of preparing each of the following salts, using a reaction involving an acid. Write the balanced equation for each reaction and classify each as an acid–base, oxidation–reduction, or precipitation reaction.

    $CaCl_2$ $MgSO_4$ Na_2SO_4 $AgCl$

67. Which of the following reactions are acid–base reactions and which are oxidation–reduction reactions? For the acid–base reactions, identify the acids and the bases. For the oxidation–reduction reactions, identify the oxidizing agent and the reducing agent.

    (a) $Cl_2(aq) + 2I^-(aq) \rightarrow 2Cl^-(aq) + I_2(s)$

    (b) $HCl(aq) + H_2O(l) \rightarrow Cl^-(aq) + H_3O^+(aq)$

    (c) $Zn(s) + 2HCl(aq) \rightarrow ZnCl_2(aq) + H_2(g)$

    (d) $HCO_3^-(aq) + H_3O^+(aq) \rightarrow CO_2(aq) + 2H_2O(l)$

Miscellaneous

*__68.__ Element A is a metal and element B is a liquid. They react to give a compound of empirical formula AB, which is a colorless crystalline solid that melts at 760 °C and boils at 1380 °C. The solid is a nonconductor, but when molten, and when it is dissolved in water, it conducts electricity. The addition of an excess of an aqueous solution of silver nitrate to a solution containing 0.543 g of AB(s) dissolved in water gave 0.857 g of a pale yellow insoluble compound. When chlorine was bubbled through an aqueous solution of AB, the solution turned brown. Identify the elements A and B, and the compound AB, and explain the reasons for your conclusions.

69. The salt $MX_2(s)$ is a metal halide which gives a neutral solution when dissolved in water. The aqueous solution does not react with bromine, but turns brown when $Cl_2(g)$ is bubbled through it. A solution containing 5.35 g of MX_2 required 710 mL of chlorine, measured at 25 °C and 1.00 atm pressure, for complete reaction. Identify the salt MX_2, and explain how each of the above observations supports your identification.

70. (a) Explain in terms of its size and core charge, why fluorine is the most electronegative element.

    (b) Write balanced equations for the reactions of fluorine with

        (i) Water   **(ii)** Potassium

        (iii) Aluminum   **(iv)** Phosphorus

71. Astatine, At, is the last member of the halogen family of elements. Predict the following for astatine:

    (a) Physical state

    (b) Ionization energy and size (compared with iodine)

    (c) The formula and Lewis structure of a compound with bromine

    (d) The acid strength of HAt in water

    (e) The shape of the PAt_3 molecule

    (f) The type of bond in KAt

    (g) Balanced equations for the reaction of At with **(i)** Na, **(ii)** Ca, **(iii)** P, and **(iv)** H_2

*__72.__ Bleaching powder, $Ca(OCl)_2(s)$, is the calcium salt of hypochlorous acid, HOCl. When carbon dioxide is passed into an aqueous solution of this salt, HOCl and insoluble calcium carbonate are formed.

    (a) Write the balanced equation for this reaction.

    (b) In an experiment that started with 8.00 g of $Ca(OCl)_2(s)$, the final solution of HOCl in water had a volume of 200 mL. Assuming that the reaction with CO_2 goes to completion, what is the molar concentration of HOCl in the final solution?

*__73.__ Radium, Ra, a radioactive element, is the last element in group 2 of the periodic table, and was discovered by Marie Curie in France in 1898. When it reacts with dilute hydrochloric acid, and the solution is heated to dryness, a pure anhydrous yellow-white salt, melting at 1000 °C, is obtained. Analysis shows that this salt contains 70.1% Ra by mass. What is the atomic mass of radium? Write the balanced equation for the reaction of Ra with HCl(aq).

74. (a) Give the Brønsted–Lowry definitions of an acid and a base.

(b) What is a conjugate acid–base pair? Illustrate your answer by means of two examples.

75. (a) What are the empirical formulas of the simplest hydrides of

  (i) Carbon   **(ii)** Nitrogen

  (iii) Oxygen   **(iv)** Fluorine

(b) Write balanced equations for the reactions, if any, of each of the hydrides in part (a) with water.

(c) Describe the acid–base properties in water of each of these four hydrides, using one or more of the terms *strong acid, strong base, weak acid, weak base,* or *neither acid nor base.*

*76. Two gaseous elements, X and Y, have densities of 0.0658 g L^{-1} and 2.315 g L^{-1}, respectively, at 100 °C and 1.00 atm pressure. When 50 mL of X reacted with 100 mL of Y, the resulting volume was 150 mL (all volumes measured at the same temperature and pressure). After shaking with water to give a solution which turned blue litmus red, the volume of the remaining gas was 50 mL at the same temperature and pressure. The remaining gas did not support combustion, but it reacted vigorously with hot sodium to give a white solid. Identify X and Y, and write balanced equations for each of the reactions described above.

77. Name the compounds with each of the following formulas. State which are ionic substances and which are covalent substances. Draw the Lewis structure for each. For the covalent species, indicate the direction of the polarity of the bonds by writing $\delta+$ and $\delta-$, as appropriate.

(a) $MgCl_2$ **(b)** SCl_2 **(c)** PCl_3 **(d)** HF

(e) OCl_2 **(f)** CS_2 **(g)** NF_3 **(h)** LiH

78. A sample of a metal X, of mass 4.315 g, combined with 0.481 L of $Cl_2(g)$, measured at 1.00 atm pressure and 20 °C, to form a metal chloride of empirical formula XCl. Identify X. Is the salt XCl soluble in water?

79. The density of sodium chloride is 2.17 g cm^{-3}. How many sodium ions and chloride ions are there in an NaCl crystal that is a cube of side 1 mm?

80. Write balanced equations for the reactions of each of the following with water:

(a) Hydrogen bromide

(b) Carbon dioxide

(c) Ammonia

(d) Chlorine

(e) Fluorine

81. Approximately 70% of the total annual U.S. and Canadian production of fluorine is used for the manufacture of uranium hexafluoride, UF_6. Assuming that the theoretical yield of UF_6 is obtained, how much UF_6 is produced each year?

82. Describe what is observed when a few drops of an aqueous solution of chlorine are added to an aqueous solution of potassium iodide and the solution is then shaken with an equal volume of carbon tetrachloride. Explain the color changes and write balanced equations to describe the reactions that occur.

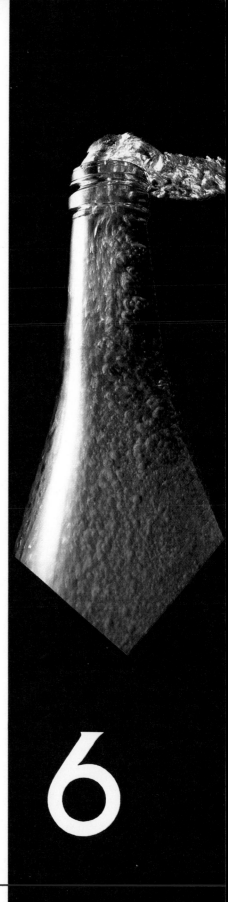

Carbon,
Energy, and
Thermochemistry

6

Periodic table showing Groups 1–8 and Periods 1–7, with a legend: Metals, Nonmetals, Semimetals, and a region labeled "Transition Elements."

Chemical reactions supply over 90% of the energy used in industrialized countries for industry and transportation. The chemical reactions that supply this energy are the combustion of fossil fuels such as coal, natural gas, and petroleum, all of which are composed of carbon compounds. Only relatively small amounts of energy are produced by hydroelectric power (4%) and nuclear reactors (5%). Our bodies too are made up of many different carbon compounds. It is the oxidation of some of these compounds, in particular carbohydrates and fats, that supplies the energy for life.

Chemical reactions may be exothermic or endothermic, that is, they occur with either the release or the absorption of energy. The combustion of fossil fuels is an exothermic process that releases energy in the form of heat. Whether energy is absorbed or released depends on whether the energy stored in the products is greater or less than the energy stored in the reactants. The energy stored in elements and compounds is called **chemical energy**.

When two atoms combine the resulting molecule is more stable than the free atoms. In other words, the formation of a bond occurs with the release of energy, so it is an exothermic process. The breaking of a bond requires energy, so it is an endothermic process. If, in a chemical reaction, more energy is released by the formation of new bonds in the products than is needed to break bonds in the reactants the reaction is exothermic. But if more energy is needed to break bonds than is released by the formation of new bonds the reaction is endothermic. Chemical energy is therefore the energy associated with chemical bonds.

The energy released or absorbed in chemical reactions is most frequently observed in the form of heat, but it may also take other forms such as light and electrical energy. For example, we have seen that when magnesium burns in air, a considerable amount of energy is evolved as both heat and light (see Experiment 1.3).

Clearly, it is important to have quantitative information about the energy changes associated with chemical reactions. For instance, an engineer designing a space rocket must know how much energy is released in the reaction between hydrogen and oxygen in order to calculate how much liquid oxygen and liquid hydrogen must be carried in order to lift the rocket clear of the earth. An engineer must know how much fuel oil is needed to operate an electricity generating station for a given period. A chemical engineer designing a chemical plant to operate as economically as possible needs to know how much heat each reaction in a process absorbs or evolves, so as to ensure that as far as possible the heat evolved in one reaction provides the heat necessary for another reaction. It is important for a biochemist to know the energy changes associated with the reactions in living cells to be able to understand the processes that occur in the cells and how a living organism transforms energy for its needs.

The quantitative study of the heat changes accompanying chemical reactions is called **thermochemistry**. This subject forms a major part of this chapter. But because combustion reactions of carbon compounds are the main source of the world's energy we begin by describing the element carbon and some of its simpler compounds and their reactions. We conclude with a discussion of alternative energy sources. The world's supply of carbon-based fossil fuels is limited, so if our present civilization is to survive we have to be concerned with possible energy sources other than the oxidation of carbon and its compounds.

6.1
Carbon and its Compounds

Although living matter is composed of an enormous variety of carbon compounds, carbon is only the fourteenth most abundant element. Only 0.08% of the earth's crust is carbon, and about half of this is in the form of the carbonate ion, CO_3^{2-}. The most common metal carbonates are calcium carbonate, $CaCO_3$, in its various forms such as chalk, limestone, and marble; magnesium carbonate, $MgCO_3$; and dolomite, $CaMg(CO_3)_2$. The remaining carbon is present in vegetable and animal matter, in coal and petroleum, and as carbon dioxide in the atmosphere and the oceans.

There is an enormous number of carbon compounds, and the vast majority of them are classified as **organic compounds**. At one time it was believed that many compounds of carbon could only be obtained from organic, or living, matter; hence these compounds were called organic compounds. Compounds occurring in the nonliving or mineral world were called **inorganic compounds**. Today we know that carbon compounds found in living matter can be synthesized from substances that are not found in living matter. Nevertheless, we retain the name *organic* to denote the compounds of carbon with the exception of CO_2, carbonates, and a few other compounds traditionally regarded as inorganic compounds. With a very few exceptions we may define **organic chemistry** as the chemistry of carbon compounds and **inorganic chemistry** as the chemistry of all the other elements and their compounds. The division between the inorganic and organic compounds of carbon is, however, an artificial one. In the process of photosynthesis carbon dioxide and water are converted by plants to organic compounds. We convert organic compounds in our bodies back to carbon dioxide and water. Indeed, life itself appears to have originated from simple inorganic compounds such as H_2, H_2O, NH_3, and HCN, as well as CH_4, which were present in the original atmosphere of our planet. Probably these compounds were combined under the action of lightning discharges or ultraviolet radiation to form the amino acids and other organic compounds that form the basic of life.

Very nearly all organic compounds contain hydrogen. The large number that contain carbon and hydrogen only are called **hydrocarbons**. All the others can be thought of as being derived from the hydrocarbons by replacing hydrogen atoms by other atoms or groups of atoms. Modern civilization is almost totally dependent on hydrocarbons, which occur in the earth's crust as *natural gas* and *petroleum*. Natural gas, and gasoline, diesel fuel, domestic heating oil, and industrial fuel oil, which are mixtures of hydrocarbons obtained by the distillation of petroleum, provide our major source of energy. Hydrocarbons are also the starting materials for the synthesis of a wide variety of organic compounds, ranging from drugs to plastics.

In this section we consider the element carbon, some of its more important inorganic compounds, and a few of the simpler hydrocarbons. The hydrocarbons and the organic compounds derived from them are discussed in more detail in Chapter 23.

CARBON

Carbon is in group 4 of the periodic table. The other elements in this group are silicon, germanium, tin, and lead. In this group in the middle of the periodic table there is a considerably greater change in properties from top to bottom than there is, for example, in the alkali metal group on the left and the halogen group on the right. Carbon is a typical nonmetal, silicon and germanium are semimetals, and tin and lead are metals. It is because of this considerable change in properties down the group that it is convenient to consider these elements separately. Silicon is discussed in Chapter 22 together with another semimetal, boron, and lead is discussed in Chapter 10 along with three other metals, aluminum, copper, and iron. However, the elements of group 4 do have one important feature in common, namely they each have four valence electrons. In its compounds carbon almost invariably completes its valence shell by forming four covalent bonds which, as we saw in Chapter 4, have a tetrahedral arrangement around the carbon atom. Silicon and germanium also usually form four covalent bonds with a tetrahedral arrangement, whereas tin and lead, which are larger and have smaller ionization energies than carbon and silicon, often lose just two of their valence electrons to form the Sn^{2+} and Pb^{2+} ions.

ALLOTROPES

Carbon occurs in several different solid forms, the most important of which are diamond and graphite. *The different forms of an element are called* **allotropes**. Thus diamond and graphite are allotropes of carbon.

Diamond and graphite are so different that if we did not know that they both consist of carbon, and carbon only, we would find it difficult to believe that they are forms of the same element. Pure diamond forms beautiful, transparent, colorless, very hard, and highly refractive crystals (see Figure 6.1). In contrast, graphite is a soft, black substance that is used as a lubricant, and mixed with varying amounts of clay it is used as the "lead" in lead pencils. Diamond has a much higher density (3.53 g cm^{-3}) than graphite (2.25 g cm^{-3}) and is an electrical insulator, whereas graphite is a fairly good conductor. Graphite is exceptional among the nonmetals in being a fairly good conductor of electricity. This unusual behavior of graphite is discussed in Chapter 10

That diamond consists only of carbon was conclusively proved by Faraday and Davy in 1813 when they focused sunlight onto a diamond by means of a large lens. The diamond ignited and burned completely to give carbon dioxide as the only product (see Box 17.3).

where we consider the conductivity of metals, which graphite resembles in this respect. At ordinary temperatures diamond is inert. Graphite, on the other hand, reacts at room temperature with several acids and other oxidizing agents.

In diamond each carbon atom forms four covalent bonds to four neighboring carbon atoms. These four bonds have a tetrahedral arrangement, that is, each carbon atom has a tetrahedral AX_4 geometry. Thus diamond has a three-dimensional network structure (Figure 6.2), and a crystal of diamond may therefore be regarded as one giant covalent molecule.

In contrast, graphite consists of planar sheets of carbon atoms in which each carbon atom is surrounded by only three other carbon atoms. Each carbon atom has a planar triangular AX_3 arrangement of bonds (see Figure 6.3a). The distance between adjacent carbon atoms in each sheet is 142 pm. The layers are stacked one on another with a separation of 334 pm (see Figure 6.3b). This distance between the sheets is much greater than the length of a single carbon–carbon bond. There are therefore no chemical bonds between the sheets of carbon atoms but only relatively weak intermolecular forces.

Although a carbon atom has four valence electrons and normally forms four bonds, in graphite each carbon atom is bonded to only three other atoms. We therefore expect that one of the three bonds will be a double bond. This double bond can be formed to any of the three neighboring carbon atoms. Thus for each carbon atom there are three possible bond arrangements (Figure 6.4). Because the sheets of carbon atoms are not held to each other by strong covalent bonds, but by relatively weak intermolecular forces, they can slide over each other rather easily. Thus graphite is a rather soft substance that feels quite slippery and is a useful lubricant. In contrast, diamond is very hard because, to move any of the carbon atoms relative to each other, strong covalent bonds must be broken which requires a considerable amount of energy. The remarkable differences in the properties of diamond and graphite, despite the fact that

FIGURE 6.1
Diamond and Graphite.

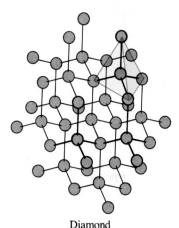

Diamond

FIGURE 6.2
Structure of Diamond.

Each carbon atom forms four bonds to neighboring carbon atoms. The four bonds formed by each carbon atom have a tetrahedral AX_4 arrangement. A three-dimensional network of covalent bonds extends throughout the crystal.

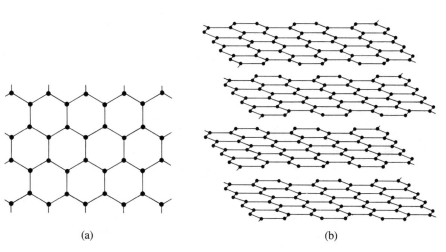

(a)

(b)

FIGURE 6.3
Structure of Graphite.

(a) Graphite is composed of flat sheets of carbon atoms in which each carbon atom is bonded to three neighboring carbon atoms in a planar AX_3 arrangement.
(b) These sheets of carbon atoms are stacked one on another and held together by relatively weak intermolecular forces.

FIGURE 6.4
Bonding in Graphite.

There are three possible arrangements for one double bond and two single bonds to each carbon atom:

Therefore there are three possible arrangements of the double bonds and single bonds in a sheet of carbon atoms, that is, three possible Lewis structures.

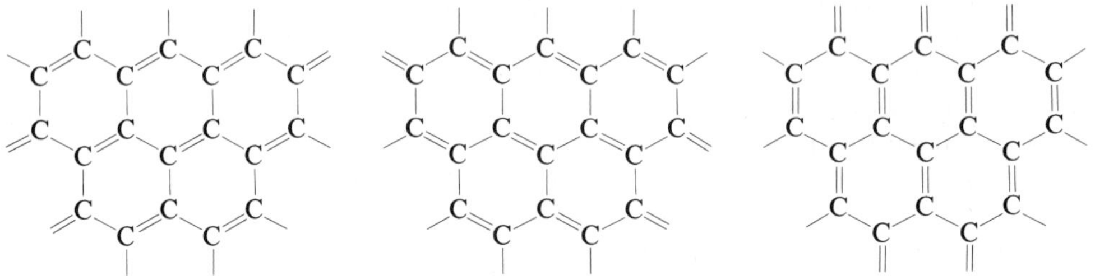

they have exactly the same composition, demonstrate that the properties of a solid depend very much on its structure. We shall take up this important subject again in Chapter 11.

Under ordinary conditions of temperature and pressure, graphite is the stable form of carbon. Diamond is unstable but, fortunately for the owners of diamond jewelry, the change to graphite is much too slow to detect under these conditions. The conversion of diamond to graphite necessitates a drastic rearrangement of the atoms which involves breaking many carbon–carbon bonds. This is why this conversion is normally exceedingly slow. At high pressures, however, diamond is more stable than graphite. Diamond has a greater density than graphite and therefore a smaller volume. Thus the equilibrium between diamond and graphite is shifted in favor of diamond at high pressure. Moreover the change becomes relatively rapid at a temperature of 1500 °C or higher. It is under these conditions of high temperature and pressure that diamonds were formed in the earth's crust. After many years of research scientists have been able to reproduce these conditions in the laboratory. Most of the small diamonds used in industry are now made by a high-pressure process developed by the General Electric Company.

There are several other forms of elemental carbon such as carbon black, charcoal, and coke that consist mainly of extremely small crystals of graphite.

Carbon black is a pure form of soot that is deposited when hydrocarbons are burned in a very limited supply of air. For example,

$$2C_2H_2(g) + O_2(g) \longrightarrow 4C(s) + 2H_2O(g)$$

Ethyne

Carbon black has a very intense color and is used in large quantities as a pigment for paint, paper, and printer's ink, and to reinforce and color the rubber used in automobile tires.

Charcoal is made by heating wood and other organic materials to a high temperature in the absence of air. We are familiar with the fact that charcoal is much lighter than the wood from which it is made; in other words, it appears to have a very low density. The density is low because charcoal is extremely porous; it has a structure resembling a sponge but with holes that are too small to be visible to the eye. This porous structure means that it has a very

Graphite

$d = 2.25$ g cm^{-3}
Atomic volume =
    5.34 cm^3

Increased ↑↓ Decreased
pressure   pressure

Diamond

$d = 3.53$ g cm^{-3}
Atomic volume =
    3.40 cm^3

The interconversion of diamond and graphite is very slow at ordinary temperatures but fairly rapid at 1500 °C.

large surface area relative to its volume. This large surface area can adsorb considerable quantities of other substances. Charcoal that has been thoroughly cleaned by heating with steam is known as *activated charcoal*. It has many applications, such as removing unburnt hydrocarbons from automobile exhaust, unpleasant and dangerous gases from the air, and impurities from water (see Experiment 6.1). Many municipal water treatment plants pass water through beds of activated charcoal.

Coke is made by heating coal in the absence of air. Coal is a very complex material consisting of many organic compounds. It contains 60%–90% C together with H, O, N, S, Al, Si, and some other elements. When coal is heated to a high temperature, it decomposes, producing a variety of gaseous and liquid products. The mixture of gases, which is largely methane and hydrogen, is known as *coal gas*. The mixture of liquid products, which includes many hydrocarbons and other organic compounds, is called *coal tar*. The solid residue contains 90%–98% C and is known as coke. Coke is used in enormous quantities in industry as a reducing agent for the production of metals, phosphorus, and other substances.

CARBON MONOXIDE

When carbon is burned in a limited supply of air, carbon monoxide, CO, is obtained:

$$2C(s) + O_2(g) \longrightarrow 2CO(g)$$

Industrially, carbon monoxide mixed with hydrogen is made on a large scale by passing steam over red-hot coke:

$$C(g) + H_2O(g) \longrightarrow CO(g) + H_2(g)$$

This mixture of hydrogen and carbon monoxide is called *water gas*.

Experiment 6.1

Adsorption by Activated Charcoal

The flask contains bromine vapor. The dish contains charcoal.

When the charcoal is added to the flask the color of the bromine vapor rapidly diminishes in intensity. After only 20 seconds it is substantially reduced.

After 1m 44s no more bromine vapor can be seen because it has been completely adsorbed by the charcoal.

Another increasingly important method of making carbon monoxide is the high-temperature reaction of methane (natural gas) with steam in the presence of a catalyst:

$$CH_4(g) + H_2O(g) \xrightarrow{\text{Catalyst}} CO(g) + 3H_2(g)$$

When prepared in this way, the mixture of carbon monoxide and hydrogen is known as *synthesis gas*, because of its importance as the basic material from which many organic compounds are synthesized. It is also used as a fuel.

Carbon monoxide is a colorless, odorless, and tasteless gas that has only a slight solubility in water. It is very toxic because it combines with hemoglobin in the blood, thus preventing hemoglobin from carrying out its function as an oxygen carrier. It is particularly dangerous because it is odorless and therefore not easily detected. Carbon monoxide is produced when tobacco burns in cigarettes and it is present in the exhaust gases from automobiles.

The Lewis structure of carbon monoxide is

$$:\overset{\ominus}{C}\equiv\overset{\oplus}{O}:$$

We can arrive at this structure as follows:

$$\cdot\dot{C}\cdot + \cdot\ddot{O}: \longrightarrow \cdot\dot{C}=\ddot{O}:$$

This completes an octet of electrons on oxygen but carbon still has only six electrons in its valence shell. Carbon can acquire an octet if one of the unshared pairs on oxygen is used to form a third bond:

$$:C\overset{\frown}{=}\ddot{O}: \longrightarrow :\overset{\ominus}{C}\equiv\overset{\oplus}{O}:$$

This results in oxygen acquiring a formal positive charge and carbon a formal negative charge. Each atom has an unshared pair and a half share in six bonding electrons; therefore each atom has effectively five electrons. Since carbon has a core charge of $+4$ it has a formal charge of $+4 - 5 = -1$, and since oxygen has a core charge of $+6$ it has a formal charge of $+6 - 5 = +1$.

Carbon monoxide is a reducing agent which can be oxidized to carbon dioxide. It burns in air to form carbon dioxide:

$$2CO(g) + O_2(g) \longrightarrow 2CO_2(g)$$

It reduces steam at high temperature, giving an equilibrium mixture of CO_2 and H_2:

$$CO(g) + H_2O(g) \rightleftharpoons CO_2(g) + H_2(g)$$

Carbon monoxide reduces metal oxides to metals. For example,

$$Fe_2O_3(s) + 3CO(g) \longrightarrow 2Fe(s) + 3CO_2(g)$$

CARBON DIOXIDE AND CARBONIC ACID

The combustion of carbon and carbon-containing compounds in excess oxygen leads to the formation of carbon dioxide, CO_2. It is also produced by heating alkaline earth metal carbonates, such as calcium carbonate:

$$CaCO_3(s) \longrightarrow CaO(s) + CO_2(g)$$

Beer and wine making also produce carbon dioxide as a by-product because

Because oxygen has a much higher electronegativity than carbon, each of the three C—O bonds is polar in the sense $\overset{\delta^+}{C}—\overset{\delta^-}{O}$. This means that the *real* charges on the carbon and oxygen atoms are much smaller than the formal charges of -1 and $+1$.

The shell of an egg consists mainly of calcium carbonate. When the egg is placed in aqueous hydrochloric acid, bubbles of CO_2 are evolved.

the fermentation of a sugar such as glucose gives ethanol and carbon dioxide:

$$C_6H_{12}O_6(aq) \longrightarrow 2C_2H_5OH(aq) + 2CO_2(g)$$

In the laboratory we can prepare small amounts of carbon dioxide by adding a dilute aqueous acid to a metal carbonate

$$CaCO_3(s) + 2H^+(aq) \longrightarrow Ca^{2+}(aq) + CO_2(g) + H_2O(l)$$

It is important to note that large scale industrial preparations of substances normally start from readily available, usually naturally occuring, and relatively cheap materials, and they are often carried out in the gas phase at high temperature and pressure. For example, carbon dioxide is made by heating chalk, and carbon monoxide is made by heating steam with coke. For the preparation of small amounts of substances in the laboratory, however, we are more concerned with convenience than with the availability of large quantities of cheap starting materials. Thus we use substances that are normally available in the laboratory, although they may be relatively expensive, and that will react to give the desired product in a convenient reaction, that is, one that can be carried out at or near room temperature, at atmospheric pressure, in solution (preferably in water), and in glass apparatus. Thus a convenient laboratory preparation of carbon dioxide is the reaction of a solid metal carbonate with a dilute aqueous acid such as HCl(aq). But this would be an expensive and not very convenient reaction to use on an industrial scale. For example, it would generate a large quantity of a waste solution of a calcium salt, such as $CaCl_2$.

Carbon dioxide is a linear AX_2 molecule with two double bonds. It is a colorless gas with a very slight odor. When cooled, it forms a white solid (dry ice) that sublimes at $-78\,°C$ at 1 atm pressure. Liquid carbon dioxide can be obtained only under pressure.

Carbon dioxide dissolves readily in water to give a solution that contains a small equilibrium amount of carbonic acid, H_2CO_3 (Experiment 6.2):

$$CO_2(aq) + H_2O(l) \rightleftharpoons H_2CO_3(aq)$$

This solution is commonly called soda water; it has a slightly acidic taste.

Carbonic acid is a weak acid that is partially ionized in two stages:

$$H_2CO_3 + H_2O \rightleftharpoons H_3O^+ + HCO_3^-$$
$$HCO_3^- + H_2O \rightleftharpoons H_3O^+ + CO_3^{2-}$$

Carbonic acid and its two ions, HCO_3^- and CO_3^{2-}, are represented by the following Lewis structures:

Carbonate ion, CO_3^{2-}, and hydrogen carbonate ion, HCO_3^-, being the anions of a weak acid, H_2CO_3, are weak bases. In the presence of an acid CO_3^{2-} adds a proton to give its conjugate acid HCO_3^-,

$$CO_3^{2-}(aq) + H^+(aq) \longrightarrow HCO_3^-(aq)$$

and in turn HCO_3^- adds a proton to give its conjugate acid H_2CO_3:

$$HCO_3^-(aq) + H^+(aq) \longrightarrow H_2CO_3(aq)$$

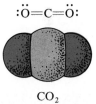

$:\ddot{O}{=}C{=}\ddot{O}:$

CO₂

Solid CO₂

Experiment 6.2

Carbon Dioxide Behaves as an Acid in Water

A piece of solid carbon dioxide (dry ice) is dropped into water to which a small amount of bromothymol blue indicator has been added.

As the CO_2 dissolves in the water it forms carbonic acid, which changes the color of the indicator from blue to yellow. The white smoke is a mist of water droplets condensed from the air by the cold CO_2 gas.

The original blue color of the indicator can be restored by adding sodium hydroxide solution, which converts the carbonic acid to sodium carbonate.

But H_2CO_3 is not very stable and it largely decomposes to CO_2 and water,

$$H_2CO_3(aq) \rightleftharpoons H_2O(l) + CO_2(aq)$$

and as CO_2 has only a limited solubility in water some of the CO_2 bubbles off as $CO_2(g)$:

$$CO_2(aq) \rightleftharpoons CO_2(g)$$

Thus, as we have seen above, addition of an acid to a carbonate gives bubbles of $CO_2(g)$.

Because carbon dioxide does not burn or easily support combustion of other substances and because it has a higher density than air, it is used as a fire extinguisher. It sinks down on the fire, forming a blanket of carbon dioxide that excludes air, extinguishing the fire. Only a few very reactive metals such as sodium, potassium, and magnesium will burn in carbon dioxide. For example, a previously ignited piece of magnesium will burn in carbon dioxide with a spluttering flame to produce white magnesium oxide and black carbon particles (Experiment 1.3):

$$CO_2(g) + 2Mg(s) \longrightarrow 2MgO(s) + C(s)$$

In photosynthesis carbon dioxide is combined with water to form many organic compounds (see Box 6.1).

CARBON DISULFIDE

Carbon combines with sulfur on heating to give carbon disulfide, CS_2:

$$C(s) + 2S(s) \longrightarrow CS_2(l)$$

Carbon disulfide is a linear AX_2 molecule with double bonds, just like carbon dioxide:

$$:\!\ddot{S}\!=\!C\!=\!\ddot{S}\!:$$

Carbon disulfide is a toxic, flammable liquid, but it is a useful solvent for sulfur and rubber.

TETRACHLOROMETHANE (CARBON TETRACHLORIDE)

When carbon disulfide is heated with chlorine, tetrachloromethane, CCl_4, and disulfur dichloride, S_2Cl_2, are formed:

$$CS_2(g) + 3Cl_2(g) \longrightarrow CCl_4(g) + S_2Cl_2(g)$$

The two products can be separated by distillation. Other ways of preparing tetrachloromethane are described in Chapter 23. Tetrachloromethane is a useful solvent. Because it is a good solvent for oils and grease, it may be used as a cleaning agent. But it should be used with suitable precautions because it is toxic and the liquid can pass through the skin.

HYDROGEN CYANIDE

When heated to approximately 1000–1200 °C in the presence of a catalyst, a mixture of methane and ammonia burns in air to give hydrogen cyanide in an exothermic reaction:

$$2CH_4(g) + 2NH_3(g) + 3O_2(g) \longrightarrow 2HCN(g) + 6H_2O(g)$$

Ammonia and methane also react in the absence of air. The reaction is endothermic and requires a temperature of 1200–1300 °C and a platinum catalyst:

$$CH_4(g) + NH_3(g) \xrightarrow[\text{Catalyst}]{1200\,°C} HCN(g) + 3H_2(g)$$

Hydrogen cyanide is a colorless liquid that boils just above room temperature (25.6 °C). It has the odor and taste of bitter almonds and is highly toxic. It is the gas used in execution chambers; a concentration of only 0.2% by volume causes death within minutes.

HCN can be conveniently prepared in the laboratory by adding acid to a metal cyanide such as NaCN or KCN:

$$NaCN(s) + HCl(aq) \longrightarrow NaCl(aq) + HCN(g)$$

HCN is a weak acid and therefore the position of the equilibrium

$$CN^- + H_3O^+ \rightleftharpoons HCN + H_2O$$

lies far to the right. Because of its low boiling point HCN is evolved as a gas. For this reason acids should not be added to a metal cyanide except in a fume hood. The HCN molecule has a triple CN bond. It has a linear geometry as expected for an AX_2 molecule.

$$H\!-\!C\!\equiv\!N\!:$$

BOX 6.1

THE CARBON CYCLE AND THE GREENHOUSE EFFECT

The carbon atoms on earth are continually being recycled through various compounds, producing what is called the *carbon cycle*. Although carbon dioxide constitutes only 0.05% of the earth's atmosphere by volume, it plays a vital role in this cycle and generally on earth.

There is a fine balance in nature between the processes that produce carbon dioxide and those that use it up. Plants take up carbon dioxide and use it in *photosynthesis*. In this process they use the energy provided by sunlight to convert carbon dioxide and water into carbohydrates and other organic compounds:

$$x CO_2 + y H_2O \longrightarrow C_x H_{2y} O_y$$

Animals consume these carbon-containing compounds and convert them to CO_2 in exothermic reactions with O_2; these reactions provide the energy needed to keep the body at its normal temperature and to enable the bodily functions to be carried out. The CO_2 is exhaled into the atmosphere by animals and used by plants. In addition, CO_2 in the atmosphere is in equilibrium with an enormous amount of CO_2 dissolved in the oceans and lakes. Some of this CO_2 is converted by marine animals into calcium carbonate, $CaCO_3$, the main component of their shells. The shells are eventually converted into limestone, which represents an enormous store of carbon on the earth. When the limestone weathers under the action of rain and surface water, some of the CO_2 is released into the atmosphere. Carbon dioxide is also produced by the decay of plants as well as by the respiration of animals.

Human activities are altering the balance established by nature's carbon cycle. The combustion of fossil fuels and industrial activities such as smelting add CO_2 to the atmosphere. Moreover, the amount of CO_2 removed from the atmosphere by photosynthesis is being reduced by the continued destruction of vast forested areas, particularly the tropical jungles. As a result of these activities, the CO_2 concentration in the atmosphere has been increasing slowly in the past hundred years, and the rate of increase is accelerating.

There is strong evidence that this accumulation of CO_2 in the atmosphere is slowly increasing the earth's temperature, because CO_2 absorbs some of the energy radiated from the earth's surface and converts it to heat. This phenomenon is known as the *greenhouse effect*. Although an increase in the earth's average temperature might appear to be a benefit to the many people who live in temperate climates with cold winters, it could have very serious consequences. For example, it would lead to extensive melting of the polar ice caps, which would cause extensive flooding of coastal areas, including major cities such as New York and Washington. A drastic reduction in the use of fossil fuels is essential if we are to slow up and eventually stop the warming of the earth caused by the accumulation of CO_2 in the atmosphere. Alternative energy sources is an increasingly important area of research for chemists.

Isoelectric
species

$:\overset{\cdot\cdot}{\underset{\cdot\cdot}{C}}{}^{\ominus}$

$:\overset{\cdot\cdot}{\underset{\cdot}{N}}\cdot$

$:\overset{\cdot\cdot}{\underset{\cdot\cdot}{O}}{}^{\oplus}$

N_2	$:N\equiv N:$
CN	$:\overset{\ominus}{C}\equiv N:$
C_2^{2-}	$:\overset{\ominus}{C}\equiv \overset{\ominus}{C}:$
CO	$:\overset{\ominus}{C}\equiv \overset{\oplus}{O}:$

Hydrogen cyanide behaves as a weak acid in water. An aqueous solution is called *hydrocyanic acid*. Its salts are the cyanides, such as sodium cyanide, NaCN. They contain the cyanide ion, $:\overset{\ominus}{C}\equiv N:$, which is isoelectronic with $:\overset{\ominus}{C}\equiv\overset{\oplus}{O}:$ and $:N\equiv N:$.

Hydrogen cyanide has many important applications in the plastics industry. Sodium cyanide is used for the extraction of gold and silver from their ores.

When an alkali metal cyanide is fused with sulfur, a *thiocyanate* is formed; for example,

$$S(s) + NaCN(s) \longrightarrow NaSCN(s)$$

The thiocyanate ion, SCN^-, has the Lewis structure $\overset{\ominus}{:}\overset{\cdot\cdot}{\underset{\cdot\cdot}{S}}-C\equiv N:$.

CARBIDES

In addition to its compounds with nonmetals such as oxygen, sulfur, and nitrogen, carbon reacts with many metals. The compounds of carbon with the

The Carbon Cycle in Nature. *Carbon dioxide in the atmosphere and dissolved in the ocean is converted by photosynthesis, P, into carbohydrates in plants. Oxygen is released in the process. Animals on land and in the sea use the carbohydrates for food, F, respiration, R, and decomposition, D, of animals and plants use up oxygen and produce carbon dioxide. In the oceans carbon dioxide reacts with water to give carbonate ion, $CO_3{}^{2-}$, which combines with calcium ion present in the sea to form deposits of calcium carbonate as sediments, S. The shells of certain aquatic animals also consist mainly of calcium carbonate and are formed* *from the calcium and carbonate ions in the ocean. These shells accumulate on the ocean floor, contributing to the sediment of calcium carbonate. Other sediments are formed when dead organic matter accumulates on the bed of lakes and seas in the absence of oxygen. These accumulations ultimately lead to coal and oil deposits. Combustion, C, of coal and oil regenerates carbon dioxide. The combustion of huge amounts of coal and oil in recent times has upset the balance in the carbon cycle, and the amount of carbon dioxide in the atmosphere is slowly increasing.*

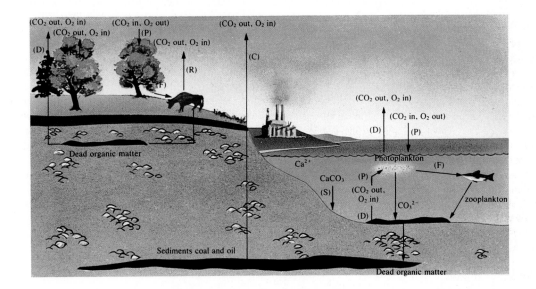

alkali and alkaline earth metals contain the carbide ion, $C_2{}^{2-}$; examples are Na_2C_2 and CaC_2. The carbide ion is isoelectronic with the nitrogen molecule and the cyanide ion and has a triple bond, $:\overset{\ominus}{C}{\equiv}\overset{\ominus}{C}:$.

Commercially, calcium carbide is made by heating lime with coke in a furnace:

$$CaO(s) + 3C(s) \xrightarrow{\text{Heat}} CaC_2(s) + CO(g)$$

Calcium carbide has a structure similar to that of sodium chloride (see Figure 6.5). It reacts with water to give ethyne (acetylene), C_2H_2:

$$CaC_2(s) + 2H_2O(l) \longrightarrow Ca(OH)_2(aq) + C_2H_2(g)$$

These carbides are salts of ethyne, which is too weak an acid to ionize in water. Therefore, the carbide ion, $C_2{}^{2-}$, is a strong base in water. It removes protons from water to give C_2H_2, which bubbles off as a gas:

$$C_2{}^{2-}(s) + 2H_2O(l) \longrightarrow C_2H_2(g) + 2OH^-(aq) \qquad\qquad H{-}C{\equiv}C{-}H$$

There are many other metal carbides that do not contain the $C_2{}^{2-}$ ion. Often

FIGURE 6.5
Structure of Calcium Carbide.

This structure is very similar to
that of sodium chloride.

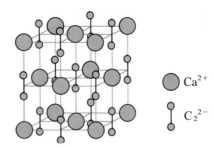

Ca^{2+}

C_2^{2-}

they have unexpected formulas, such as iron carbide, Fe_3C, which is called
cementite and is an important component of some steels. We cannot easily
explain the bonding and structures of these compounds by the simple bonding
models used in this book; they are treated in advanced texts of inorganic
chemistry.

With more electronegative elements such as silicon, carbon forms a covalent
rather than an ionic carbide. *Silicon carbide*, SiC, is a covalent compound with
the diamond structure, except that each alternate atom is a silicon atom rather
than a carbon atom (see Figure 6.6). Silicon carbide, commonly known as
carborundum, is almost as hard as diamond and is used as an abrasive; for
example, carbide sandpaper and grinding wheels are coated with silicon carbide.

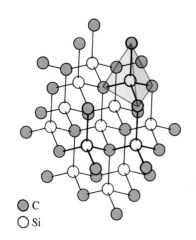

○ C
○ Si

FIGURE 6.6
Structure of Silicon Carbide.

The structure is very similar to
that of diamond; each alternate
atom in the diamond structure is
replaced by a silicon atom. Each
carbon atom and each silicon
atom has a tetrahedral geometry
in this three-dimensional,
covalently bonded network
structure.

═ *Example 6.1* ═══

REACTIONS OF CARBON COMPOUNDS

Predict the products of the following reactions and write a balanced equation in each
case:

(a) $CuO(s) + CO(g) \longrightarrow$

(b) $BaCO_3(s) \xrightarrow{\text{Heat}}$

(c) $Na_2CO_3(s) + HCl(aq) \longrightarrow$

(d) $HCN(g) + O_2(g) \longrightarrow N_2(g) +$
$\qquad\qquad$ Excess

Which of reactions (a), (c), and (d) are acid–base reactions, and which are redox
reactions?

Solution

(a) Carbon monoxide is a reducing agent that reduces metal oxides to the metal
(p. 266). Therefore

$$CuO(s) + CO(g) \longrightarrow Cu(s) + CO_2(g)$$

Oxygen is removed from CuO and added to CO, so this is a redox reaction.

(b) Alkaline earth metal carbonates decompose on heating to the oxide and
carbon dioxide (p. 266). Therefore

$$BaCO_3(s) \longrightarrow BaO(s) + CO_2(g)$$

(c) Carbonates react with acids to give carbon dioxide (p. 267).

$$Na_2CO_3(aq) + 2HCl(aq) \longrightarrow 2NaCl(aq) + CO_2(g) + H_2O(l)$$

This is an acid–base reaction as can be seen if we write it in the ionic form

$$CO_3^{2-}(aq) + 2H_3O^+(aq) \longrightarrow H_2CO_3(aq) + 2H_2O(l)$$

$$H_2CO_3(aq) \longrightarrow H_2O(l) + CO_2(g)$$

Protons are transferred from the acid H_3O^+ to the base CO_3^{2-}.

(d) Here we are told that nitrogen is one of the products. If we make the reasonable assumption that this is the only nitrogen-containing product we only have to worry about the C and H. We know that hydrocarbons are oxidized to CO_2 and H_2O in excess oxygen so it is reasonable to assume that this is also the case here. Thus the balanced equation is

$$4HCN(g) + 5O_2(g) \longrightarrow 2N_2(g) + 4CO_2(g) + 2H_2O(g)$$

This is not a completely certain prediction about the products of the reaction. Another product might have been an oxide of nitrogen and we cannot be absolutely sure that the C and H will be oxidized to CO_2 and H_2O as they are in hydrocarbons. Nevertheless it is a reasonable prediction that turns out to be correct. This is a redox reaction.

PREDICTING REACTION PRODUCTS

In one case in Example 6.1 we noted that it is was not possible to predict the products of the reaction with complete certainty. Nobody can be expected to remember all of the details of even the relatively small number of reactions that are dealt with in this book. Moreover, it is not practical to consult reference books in the library every time we wish to know the products of a reaction. Thus it is most important to be able to predict what are most likely to be the products of at least a number of common reactions. We can do this by being able to recognize the most important reaction types, and by knowing a few typical examples of each, with which we can compare the reaction that we are interested in. We can also get considerable help from the periodic table because we have seen that

1. Nonmetals have typical reactions that are different from the typical reactions of metals.
2. Elements in the same group have similar reactions.
3. There are often regular trends in chemical properties down any group.

By using this kind of knowledge you should be able to give a reasonably sensible answer to questions such as: What reaction, if any, is likely to occur when

Compound A is added to water?

Substance B is heated?

A mixture of compounds C and D is heated?

Aqueous solutions of compounds C and D are mixed?

when A, B, C, and D are simple substances that you have already studied, or are closely related substances.

Being able to answer questions like those above depends not so much on having a good memory, but rather on having a good understanding of chemistry, in particular, understanding similarities and trends in the periodic table and being able to recognize (a) the nature of the reacting substances, ionic, covalent or metallic, acid or base, oxidizing agent or reducing agent, and (b) the important reaction types such as acid–base, oxidation–reduction, and precipitation.

Each time you observe a reaction in the laboratory or as a lecture demonstration ask yourself: What kinds of substances are taking part? What type of reaction is it? Do I know any similar reactions? In this way you can build up and consolidate your knowledge of reaction chemistry. When you have to answer questions such as those in Example 6.1 or in Exercise 6.1 which follows, ask yourself the same questions. If you cannot think of an answer, look through the chapter again to try to find some similar reactions. All the problems of this type are based on information already given, so you should be able to come up with the answer if you have read or reread the text carefully enough. When asked to complete an equation do not be tempted to invent products that you have never heard of simply to balance the equation. Think carefully about the conditions under which a reaction is carried out. For example, if a reaction is occurring in an acidic solution, OH^- cannot be a reactant or a product. If a reaction is taking place in aqueous solution, O^{2-} cannot be a product because, being a strong base, it is completely converted to OH^-.

EXERCISE 6.1

Predict the products and write the balanced equation for each of the following reactions:

(a) $MgCO_3(s) \xrightarrow{\text{Heat}}$

(b) $PbO(s) + CO(g) \xrightarrow{\text{Heat}}$

(c) $H_2O(g) + CO(g) \xrightarrow{\text{Heat}}$

(d) $MgC_2(s) + H_2O(l) \longrightarrow$

(e) $C_2H_6(g) + H_2O(g) \xrightarrow{\text{Heat}}$

Which of the above are acid–base reactions and which are redox reactions?

EXERCISE 6.2

Which of the following anions are strong bases, which are weak bases, and which have no basic properties in water?

CN^- CO_3^{2-} Cl^- HCO_3^- C_2^{2-} OH^- F^-

EXERCISE 6.3

Draw the Lewis structures and give the names of three species (molecules or ions) that are isoelectronic with the cyanide ion.

ALKANES

One large and important group of hydrocarbons is the **alkanes**. They have single bonds between the carbon atoms and the general formula C_nH_{2n+2}, where n has all integral values from 1 up to a very large number. The simplest alkanes are *methane*, CH_4, which is the major constituent of natural gas; *ethane*, C_2H_6, a minor constituent of natural gas; and *propane*, C_3H_8 and *butane*, C_4H_{10}, both of which are used as fuels. Methane has $n = 1$, ethane has $n = 2$, and butane has $n = 4$. Naturally occurring petroleum is a liquid mixture of many alkanes with values of n from 5 to 40.

The alkanes up to butane are gases at ordinary temperatures and pressures; the higher members are liquids and solids. The melting points and the boiling points of the alkanes increase with increasing molecular size (see Table 6.1). Alkanes with more than 16 carbon atoms are waxy solids. Paraffin wax is a mixture of solid alkanes.

The Lewis structures for methane, ethane, and propane are as follows:

Methane

Ethane

Methane CH_4 Ethane C_2H_6 Propane C_3H_8

These structures are also called **structural formulas**. Frequently, structural formulas are further simplified by omitting the bond lines to the hydrogen atoms and collecting together the hydrogen atoms bonded to a given carbon atom. For example,

$$CH_4 \quad CH_3\text{---}CH_3 \quad CH_3\text{---}CH_2\text{---}CH_3$$

In these molecules each carbon atom forms four tetrahedrally arranged bonds, some to hydrogen atoms and some to other carbon atoms (see Figure 6.7).

Alkanes are rather unreactive compounds. At ordinary temperatures they do not react with oxidizing agents such as oxygen or chlorine. They do not react with reducing agents such as hydrogen, and they have no acid or base properties. An important reason for this lack of reactivity is that both the C—H bond and the C—C bond are strong, and a rather large amount of energy is needed to break them. Only at relatively high temperatures do the molecules have enough energy for the C—H or C—C bonds to be broken when molecules

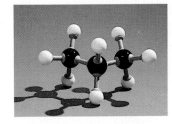

Propane

TABLE 6.1
Boiling Points and Melting Points of Alkanes, C_nH_{2n+2}

n	Name	Boiling Point (°C at 1 atm)	Melting Point (°C)	Formula
1	Methane	−162	−183	CH_4
2	Ethane	−89	−172	CH_3CH_3
3	Propane	−42	−188	$CH_3CH_2CH_3$
4	Butane	0	−138	$CH_3(CH_2)_2CH_3$
5	Pentane	36	−130	$CH_3(CH_2)_3CH_3$
6	Hexane	69	−95	$CH_3(CH_2)_4CH_3$
7	Heptane	98	−91	$CH_3(CH_2)_5CH_3$
8	Octane	126	−57	$CH_3(CH_2)_6CH_3$
9	Nonane	151	−54	$CH_3(CH_2)_7CH_3$
10	Decane	174	−30	$CH_3(CH_2)_8CH_3$
20	Icosane	343	−37	$CH_3(CH_2)_{18}CH_3$
30	Triacontane	446	66	$CH_3(CH_2)_{28}CH_3$

FIGURE 6.7
Structures of Methane,
Ethane, and Propane.

(a) Ball-and-stick models.
(b) Space-filling models.

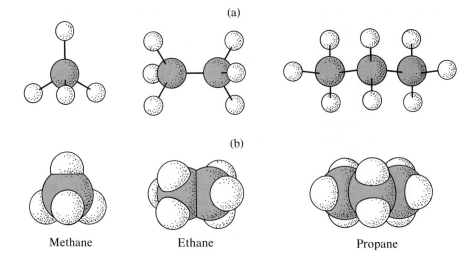

(a)

(b)

Methane   Ethane   Propane

collide. Also, unlike nitrogen in ammonia, for example, the carbon atom has
no unshared pairs of electrons to which a hydrogen ion can be added—alkanes
are not bases. Moreover, the C—H bond has very little polarity—alkanes are
not acids.

By far the most important reaction of alkanes is their reaction with oxygen
at elevated temperatures. If oxygen is in excess, complete combustion occurs
to give carbon dioxide and water; for example,

$$C_3H_8(g) + 5O_2(g) \longrightarrow 3CO_2(g) + 4H_2O(g)$$

ETHENE (ETHYLENE)

When alkanes are strongly heated, they decompose to give hydrogen and simpler hydrocarbons. One of the most important products of these reactions is
ethene (ethylene), C_2H_4:

$$C_2H_6(g) \longrightarrow C_2H_4(g) + H_2(g)$$
$$C_4H_{10}(g) \longrightarrow 2C_2H_4(g) + H_2(g)$$

In the *ethene* molecule the two carbon atoms share two pairs of electrons, that
is, they are joined by a double bond:

$$\begin{array}{ccc} H & & H \\ \diagdown & & \diagup \\ & C{=}C & \\ \diagup & & \diagdown \\ H & & H \end{array}$$

Figure 6.8 shows two models of the ethene molecule. Since the four bonds
around carbon have a tetrahedral orientation, the two carbon–carbon bonds
must be bent to join the carbon atoms. The molecule has a rigid structure, and
all six atoms lie in the same plane. Each carbon atom may be described as
having an AX_3 geometry; therefore there is a planar triangular arrangement of
the three attached atoms around each C atom.

Ethene is the simplest member of the group of hydrocarbons called **alkenes**
that contain one or more double bonds. Like all hydrocarbons, ethene burns
in air or oxygen:

$$C_2H_4(g) + 3O_2(g) \longrightarrow 2CO_2(g) + 2H_2O(g)$$

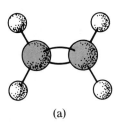

(a)

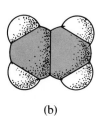

(b)

FIGURE 6.8
Structure of Ethene.

(a) A ball-and-stick model. (b) A
space-filling model. Both models
show that the molecule is planar.

In 1987 ethene was the fifth most important industrial chemical produced in the United States in terms of the amount manufactured per year. It is the starting material for the synthesis of many other important substances. Almost half the ethene produced is polymerized to give *polyethylene* (polythene):

$$\frac{n}{2}\,H_2C{=}CH_2 \xrightarrow{\text{Catalyst}} (CH_2)_n$$

Ethene   Polyethylene

In a **polymerization reaction** a large number of small molecules (*monomers*) are combined to form a single very large molecule (*polymer*). In the formation of polyethylene one of the bonds of the carbon–carbon double bond breaks, and new bonds form between the individual molecules:

$$\cdots + H_2C{=}CH_2 + H_2C{=}CH_2 + H_2C{=}CH_2 + H_2C{=}CH_2 + \cdots$$

$$\downarrow$$

$$\cdots CH_2{-}CH_2{-}CH_2{-}CH_2{-}CH_2{-}CH_2{-}CH_2{-}CH_2\cdots$$

Thousands of ethene molecules are joined together to form one molecule of polyethylene. Polymers are discussed in more detail in Chapter 24.

ETHYNE (ACETYLENE)

Hydrocarbons that contain a carbon–carbon triple bond are called **alkynes**. The simplest alkyne is *ethyne*, C_2H_2, which is commonly called *acetylene*.

As Figure 6.9 shows, ethyne is a linear molecule in which both carbon atoms have an AX_2 geometry. It is an important industrial chemical that is being made in increasing amounts by the high-temperature thermal decomposition of ethane:

$$C_2H_6(g) \xrightarrow{\text{Heat}} C_2H_2(g) + 2H_2(g)$$

It was formerly made by the reaction of calcium carbide with water (see Experiment 6.3):

$$CaC_2(s) + 2H_2O(l) \longrightarrow Ca(OH)_2(aq) + C_2H_2(g)$$

Ethyne burns in excess oxygen with a very hot bright flame:

$$2C_2H_2(g) + 5O_2(g) \longrightarrow 4CO_2(g) + 2H_2O(g)$$

In a limited supply of air or oxygen it gives a smoky flame because of the formation of carbon (Experiment 6.3).

One early form of portable lighting was the acetylene lamp in which a constant supply of ethyne was produced by water slowly dripping onto calcium carbide. A major use for ethyne is in the oxyacetylene torch for welding and cutting metals.

BENZENE

Another important class of hydrocarbons are the **arenes**, of which the simplest member is *benzene*, C_6H_6. Benzene and other arenes were first discovered among the products obtained when coal is heated to a high temperature in the absence of air. The products are *coke* and a complex mixture of liquids, called

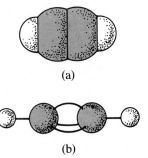

(a)

(b)

FIGURE 6.9
Structure of Ethyne.

Ethyne is a linear AX_2 molecule. (a) Space-filling model. (b) Ball-and-stick model.

Experiment 6.3

Preparation and Combustion of Ethyne

Ethyne is being prepared by the reaction of calcium carbide, CaC_2, with water. It burns with a smoky yellow flame, depositing soot (carbon) on the surface of the evaporating dish.

coal tar, that are mainly hydrocarbons, including benzene and other arenes, and a mixture of gases—mainly CH_4, H_2, and CO—called *coal gas*.

Coal is no longer the major source of benzene. Substantial amounts of benzene are present in petroleum from some sources, but large amounts are also made by heating alkanes to a high temperature in the presence of suitable catalysts.

Many of the arenes and their derivatives have pleasant odors; therefore they are often called *aromatic hydrocarbons*. Many are toxic and some are carcinogenic. The volatile arenes are very flammable and burn with a yellow sooty flame. In contrast, alkanes, alkenes, and alkynes in an adequate amount of air burn with a bluish flame, leaving little carbon residue. Benzene is a colorless liquid that boils at 80 °C and freezes at 5.5 °C. It is less dense than water and is insoluble in water.

Although benzene was first prepared by Michael Faraday in 1814, its structure remained a mystery until 1866. It was not easy to see how the formula C_6H_6 could be reconciled with a valence of 4 for carbon and a valence of 1 for hydrogen, and several structures were proposed. The structure that we now know to be nearly correct was suggested by Friedrich August Kekulé (1829–1896) in 1866. He proposed that the benzene molecule is a six-membered ring containing alternate double and single bonds:

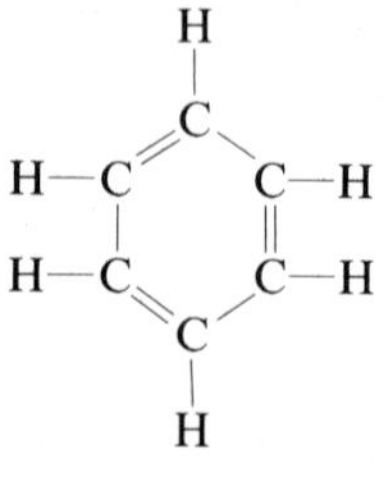

$$
\begin{array}{c}
\text{H} \\
\text{C} \\
\text{H—C} \quad \text{C—H} \\
\text{H—C} \quad \text{C—H} \\
\text{C} \\
\text{H}
\end{array}
$$

We will see in Chapter 9 that this structure has to be modified slightly.

Many other arenes can be derived from benzene by replacing one or more hydrogen atoms with alkyl groups. The simplest derivative is *methylbenzene*, or *toluene*:

$$
\begin{array}{c}
CH_3 \\
| \\
C \\
H-C \diagup \ \diagdown C-H \\
\| \qquad \| \\
H-C \diagdown \ \diagup C-H \\
C \\
| \\
H
\end{array}
$$

ORGANIC CHEMISTRY

An enormous number of compounds can be obtained from hydrocarbons by replacing one or more hydrogen atoms by other atoms or groups, either directly or indirectly. The study of these compounds constitutes the subject of organic chemistry, which we will study in Chapter 23. Because there are some very common and important organic compounds, in addition to the hydrocarbons, which we will encounter in the following chapters, it is useful to be familiar with their names and structures and to understand how they are related to the hydrocarbons. We will discuss the preparation and properties of these substances in Chapter 23.

ALCOHOLS If one or more hydrogen atoms in a hydrocarbon are replaced by hydroxyl, OH, groups, the compound that results is called an alcohol. Some examples are

$$
\begin{array}{cc}
H & H \\
| & | \\
H-C-H & H-C-OH \\
| & | \\
H & H \\
\text{Methane} & \text{Methanol}
\end{array}
$$

$$
\begin{array}{ccc}
H\ \ H & H\ \ H & H\ \ H \\
|\ \ \ | & |\ \ \ | & |\ \ \ | \\
H-C-C-H & H-C-C-OH & HO-C-C-OH \\
|\ \ \ | & |\ \ \ | & |\ \ \ | \\
H\ \ H & H\ \ H & H\ \ H \\
\text{Ethane} & \text{Ethanol} & \text{Ethanediol} \\
& & \text{(glycol)}
\end{array}
$$

CARBOXYLIC ACIDS These are an important group of weak acids among which we have already encountered acetic acid, the most common of the carboxylic acids.

$$
\begin{array}{cc}
H & O \\
| & \| \\
H-C-H & H-C-OH \\
| & \\
H & \\
\text{Methane} & \text{Methanoic acid} \\
& \text{(formic acid)}
\end{array}
$$

$$
\begin{array}{cc}
H\ \ H & H\ \ O \\
|\ \ \ | & |\ \ \ \| \\
H-C-C-H & H-C-C-OH \\
|\ \ \ | & | \\
H\ \ H & H \\
\text{Ethane} & \text{Ethanoic acid} \\
& \text{(acetic acid)}
\end{array}
$$

Many organic substances are known by two different names: a common name such as acetic acid and a systematic name such as ethanoic acid which is based on the name of the parent hydrocarbon.

AMINES These are a group of compounds related to ammonia, which like ammonia are weak bases.

$$
\begin{array}{ccc}
\underset{\substack{|\\ \text{H}}}{\overset{\substack{\text{H}\\ |}}{\text{H}-\text{C}-\text{H}}} & \underset{\substack{|\\ \text{H}}}{\overset{\substack{\text{H}\\ |}}{\text{H}-\text{C}-\text{NH}_2}} & \underset{\substack{|\quad|\\ \text{H}\ \ \text{H}}}{\overset{\substack{\text{H}\ \text{H}\ \text{H}\\ |\quad|\quad|}}{\text{H}-\text{C}-\text{N}-\text{C}-\text{H}}}\\
\text{Methane} & \text{Methylamine} & \text{Dimethylamine}
\end{array}
$$

Other common organic compounds that we will meet in later chapters include

$$
\begin{array}{ccc}
\underset{\substack{|\quad\quad|\\ \text{H}\ \ \ \ \text{H}}}{\overset{\substack{\text{H}\ \ \text{O}\ \ \text{H}\\ |\ \ \ ||\ \ \ |}}{\text{H}-\text{C}-\text{C}-\text{C}-\text{H}}} & \overset{\substack{\text{O}\\ ||}}{\text{H}-\text{C}-\text{H}} & \underset{\substack{|\quad\quad|\\ \text{H}\ \ \ \ \text{H}}}{\overset{\substack{\text{H}\ \ \ \ \text{H}\\ |\quad\quad|}}{\text{H}-\text{C}-\text{O}-\text{C}-\text{H}}}\\
\text{Propanone} & \text{Methanal} & \text{Diethyl ether}\\
\text{(acetone)} & \text{(formaldehyde)} &
\end{array}
$$

EXERCISE 6.4

Draw a structural formula and give the name of each of the following compounds, all of which have been described in this chapter:

C_2H_4 C_6H_6 C_3H_8 CH_2O C_2H_6O $C_2H_4O_2$.

6.2
Energy and the First Law of Thermodynamics

As we discussed at the beginning of the previous section, most of the world's energy is currently obtained from the combustion of fossil fuels, which are mainly hydrocarbons. We are all familiar with the idea of energy and we have a qualitative idea of what we mean by energy from everyday life. We will be particularly concerned in this chapter with the energy, usually in the form of heat, that can be obtained from chemical reactions. *The quantitative study of the heat changes associated with chemical reactions is called* **thermochemistry**. Thermochemistry is part of a subject of much wider scope called thermodynamics. **Thermodynamics** *is the science of the transformations of energy.* Before we begin our study of the heat changes associated with chemical reactions, however, we need to consider a few of the fundamental ideas of thermodynamics. We begin with energy. Energy, like matter, is a concept that is difficult to define precisely. We will use the following definition:

Energy is the capacity to do work or to transfer heat.

When a force displaces an object, we say that **work** *is done on the object. The* work done, w, is defined as the magnitude of the force, F, multiplied by the magnitude of the displacement, d:

$$w = F \times d$$

In SI force is measured in newtons and distance in meters so the SI unit of work is the newton meter (N m) which is called a joule (J):

$$1\,J = 1\,N\,m = 1\,kg\,m\,s^{-2} \times m = 1\,kg\,m^2\,s^{-2}$$

Heat *is energy that is transferred as the result of a temperature difference.* Heat always flows from a warmer object to a colder object. At the molecular level, when the molecules of the warmer object collide with those of the colder object, the molecules of the warmer object, which have a higher average kinetic energy, lose kinetic energy to those of the colder object. The average kinetic energy of the molecules of the warmer object decreases, and its temperature falls. The average kinetic energy of the molecules of the cooler object increases, and its temperature increases. Heat flows between the two objects until they have the same temperature.

FIRST LAW OF THERMODYNAMICS

One of the most important characteristics of energy is that it is conserved. The **law of conservation of energy** can be stated in several ways. In Chapter 2 we gave the following definition:

> **Energy can neither be created nor destroyed but only changed from one form to another.**

An alternative statement of the law is the following:

> **The energy of a system that is isolated from its surroundings is constant.**

By a **system** we mean any part of the universe in which we are particularly interested, for example the substances taking part in a reaction and the reaction vessel. By the **surroundings** we mean all the rest of the universe, but it can usually be taken to mean just the immediate surroundings of the system. By an **isolated system** we mean a system that cannot exchange matter or energy with its surroundings. We will be particularly concerned in what follows with a **closed system**, which is one in which exchange of energy with the surroundings may occur but in which there is no transfer of matter to or from the surroundings.

Energy can be transferred to a system from its surroundings or from a system to its surroundings in two ways: either by allowing heat to flow into or out of the system or by the surroundings doing work on the system or the system doing work on the surroundings. If an amount of heat q flows into a system from the surroundings and the surroundings do an amount of work w on the system, then the change in the internal energy of the system

$$E_{final} - E_{initial} = \Delta E$$

is given by the expression

$$\Delta E = q + w$$

This is a mathematical expression of the law of conservation of energy; in this form it is usually known as **the first law of thermodynamics**.

The change in energy resulting from doing work on a system or by a flow of heat into the system is measured in joules.

If 10 kJ of energy is transferred to a system as heat, we write $q = 10$ kJ. If the same amount of energy is transferred as heat from the system to the surroundings, we write $q = -10$ kJ. The negative sign indicates loss of energy from the system.

If 15 kJ of energy is transferred to the system by doing work on the system, we write $w = +15$ kJ. If the system does 15 kJ of work on the surroundings, we write $w = -15$ kJ, because the system has lost 15 kJ of energy.

If we do 15 kJ of work on a system and it simultaneously loses 10 kJ of energy as heat to the surroundings, the internal energy of the system increases by

$$\Delta E = q + w$$
$$= -10 \text{ kJ} + 15 \text{ kJ}$$
$$= 5 \text{ kJ}$$

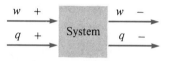

Sign convention for q and w

In general when heat flows into the system from the surroundings or when work is done on the system by the surroundings q and w have positive values indicating that the internal energy of the system increases. When heat flows from the system to the surroundings or when the system does work on the surroundings, q and w have negative values indicating that the internal energy of the system decreases.

KINETIC AND POTENTIAL ENERGY

There are two fundamentally different forms of energy: kinetic energy and potential energy.

- *Kinetic energy* is the energy possessed by an object by virtue of its motion. For a body of mass m with a speed v, the kinetic energy $(KE) = \frac{1}{2}mv^2$.
- *Potential energy* is the energy possessed by an object by virtue of its position in relation to another object that exerts a force upon it.

At this point it will be useful to examine the ideas of heat and work in a little more detail in terms of kinetic and potential energy. Suppose a book with a mass of 1.0 kg is raised to a height of 10 cm. The work required is equal to the force acting on the book times the distance that it is raised, h. The force is $F = mg$, where m is the mass of the book and g is the acceleration due to gravity ($g = 9.8$ m s^{-2}). The work done on the book is

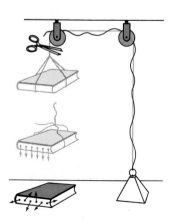

$$w = Fh = mgh$$
$$= 1.0 \text{ kg} \times 9.8 \text{ m s}^{-2} \times 0.10 \text{ m}$$
$$= 0.98 \text{ kg m}^2 \text{ s}^{-2}$$
$$= 0.98 \text{ J}$$

Thus 0.98 J of work is done on the book.

The potential energy of the book is therefore increased by the same amount, 0.98 J. If the book is allowed to fall back to the table, it could do this much work. If we were to attach it to a string running over a pulley and then to a weight, the book could raise the weight and thus do work on the weight. However, if the book is allowed to fall freely, it does no work. Its potential energy is converted to kinetic energy as it falls. And when it hits the table, this kinetic energy is converted to heat, which will slightly raise the temperature of the book and the table.

This example helps us to understand the essential difference between heat and work. When work is done on the book to lift it from the table, all its molecules are moved away from the table in an organized, coordinated way; we may say that work produces organized motion. When the book falls to the table, the force exerted by gravity causes all the molecules in the book to move in a coordinated, organized fashion toward the table. However, when the book hits the table, this organized molecular motion is transformed to the random or disorganized motion that is heat. The random, disorganized motion of the atoms and molecules of the book is increased, and its temperature is raised accordingly. We may say that *heat is associated with random, disorganized motion, whereas work is associated with organized motion.*

INTERNAL ENERGY AND STATE FUNCTIONS

The internal energy of a system is the sum of the kinetic and potential energies of all the particles in the system. We cannot measure the absolute internal energy of a system, but we can measure the changes in the internal energy in terms of the work done on a system and the energy added to the system as heat.

The internal energy of a system is a characteristic property of the system. Other characteristic properties of a system include pressure, volume, temperature, and composition. Once some of these properties have been specified for the substances in a system the state of the system has been completely specified. This means that the system can be exactly duplicated by anyone, anywhere, at any time. For example, if you report the results of an experiment on 100 mL of oxygen, O_2, at 25 °C and 1 atm pressure, the experiment can be set up identically at any time or place knowing only four pieces of information, namely: oxygen, O_2, 100 mL, 25 °C, and 1 atm. In other words, the state of the system has been exactly specified by these particular properties. Properties of this kind are called *state functions.* These properties of the state of a system depend only on the state itself and not on how that state was reached, that is, they do not depend on the history of the system. Thus *a **state function** is any property of a system that depends only on the state of the system and not on how that state was reached.*

For a system going from some initial state (i) to some final state (f), the change in internal energy, ΔE, is given by

$$\Delta E = E_f - E_i$$

And since both E_f and E_i are state functions characteristic of the final state and the initial state, ΔE is always exactly the same no matter how the change from the initial state to the final state is carried out. In other words, it is completely independent of the route by which the change from initial state to final state took place.

A given system may change from an initial state to some final state by many routes. Different amounts of heat q may be added to the system, and different amounts of work w may be done on the system, provided only that the sum of q and w is constant, that is, that they satisfy the relation

$$\Delta E = q + w$$

Do not be tempted to confuse the meaning of the word *state* as we use it here with the same word when it is used in speaking of the state of the economy or the state of our civilization today. You cannot understand the state of economics or of any human aspects of the state of the world without understanding the history of how the present state was reached. But you *can* understand completely the state of 100 mL of $O_2(g)$ at 25 °C and 1 atm without having to worry about where the oxygen came from or what its temperature was yesterday.

The amount of heat added to the system, q, and the work done on the system, w, are not state functions. The initial and final states of a system may be likened to the balance in a bank account in which deposits and withdrawals are made in two currencies, namely, *heat* and *work*. The balance at the end of the month depends only on the sum of all the deposits and withdrawals. A particular balance could have been achieved in innumerable different ways; it is independent of the number and order of the deposits and withdrawals and the currency in which they were made, provided only that they have the same sum and therefore give the same final balance.

One mole of water at 25 °C has a certain internal energy. One mole of ice at 0 °C has a lower internal energy. The difference in their internal energies is always the same irrespective of the path by which the water at 25 °C is converted to ice at 0 °C. For example, the water could be cooled directly; that is, energy could be transferred as heat to the surroundings. Or it could first be boiled, in which case it does work in pushing away the atmosphere as it expands. Then the steam could be converted directly to ice by letting it come into contact with a surface held at 0 °C, when energy would be transferred as heat from the steam to the surface (the surroundings).

The preceding discussion leads to another alternative statement of the *first law of thermodynamics*:

> The change in energy, ΔE, accompanying the change of a system from an initial state to a final state is determined only by the initial and final states and is independent of the path by which change is effected.

6.3
Enthalpy

Chemical reactions, particularly those involving gases, often occur with a change in volume. For example, in the combustion of propane

$$C_3H_8(g) + 5O_2(g) \longrightarrow 3CO_2(g) + 4H_2O(g)$$

6 mol of reactants give 7 mol of products, so the volume of the products is greater than the volume of the reactants if the pressure is kept constant. In other words, the system expands as the reaction proceeds. When a system expands it does work by pushing the atmosphere away. If the change in the volume of the system is ΔV and the pressure of the surroundings is P_{surr}, then the system does work $P_{surr}\Delta V$ on the surroundings, as shown in Figure 6.10. So the work done on the system is $w = -P_{surr}\Delta V$. From the first law we then have

$$\Delta E = q + w = q - P_{surr}\Delta V$$

or

$$q = \Delta E + P_{surr}\Delta V$$

If the reaction is carried out in a closed vessel so that there is no volume change and the system does no work, then, since $\Delta V = 0$,

$$q_V = \Delta E$$

where we have added the subscript V to indicate that the heat flow is at constant volume. *The heat, q_V, that flows into a system at constant volume is equal to the change in the internal energy of the system.*

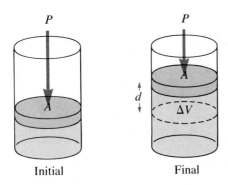

FIGURE 6.10
Work Done in a Reaction at Constant Pressure.

Consider a reaction carried out in a reaction vessel with a tightly fitted piston, acted upon by an external pressure P_{surr}, and in which there is an increase in volume, ΔV. Since pressure is force per unit area, the force acting on the piston is PA, where A is the area of the piston. If the piston moves out a distance, d, the work done by the system is force $\times$ distance $= PAd$. Now since Ad is the increase in the volume of the system, ΔV, the work done by the system is $PAd = P\Delta V$ at constant pressure P.

But many reactions are carried out in open vessels and therefore at a constant pressure $P = P_{surr}$. For these reactions the heat that flows into the system at constant pressure, q_P, is not equal to the internal energy change but to $\Delta E + P\Delta V$. In such cases it is convenient to make use of another property of the system, namely the enthalpy. Enthalpy is defined by the expression

$$H = E + PV$$

Since the internal energy, the pressure, and the volume are all state functions, enthalpy is also a state function and therefore

$$\Delta H = \Delta E + \Delta(PV)$$

Since P is equal to the constant pressure of the surroundings, the change in PV is due only to the change in volume, so

$$\Delta(PV) = P\Delta V$$

Thus

$$\Delta H = \Delta E + P\Delta V \qquad \text{at constant pressure}$$

But we have seen that the heat, q_P, that flows into the system at constant pressure is given by

$$q_P = \Delta E + P\Delta V$$

So we see that *the heat, q_P, that flows into the system at constant pressure is equal to the enthalpy change, ΔH:*

$$\Delta H = q_P$$

For a chemical reaction the enthalpy change is given by the equation

$$\Delta H = \sum H_{products} - \sum H_{reactants}$$

where $\sum H_{products}$ is the sum of the enthalpies of the products and $\sum H_{reactants}$ is the sum of the enthalpies of the reactants.

The symbol $\sum$ (capital Greek sigma) is used to denote "the sum of".

When the total enthalpy of the products, $\sum H_{products}$, is greater than the total enthalpy of the reactants, $\sum H_{reactants}$, the enthalpy change, ΔH, is positive. There is a flow of heat $q_P = \Delta H$ from the surroundings to the reaction system. In other words, the reaction is *endothermic*.

When the total enthalpy of the products is less than that of the reactants, the enthalpy change, ΔH, is negative. Thus, $q_p = \Delta H$ is also negative, and heat flows from the reaction system to the surroundings. In other words, the reaction is *exothermic*. Figure 6.11 illustrates the enthalpy changes for an endothermic and an exothermic reaction.

The combustion of methane is an example of an exothermic reaction:

$$CH_4(g) + 2O_2(g) \longrightarrow CO_2(g) + 2H_2O(l) \qquad \Delta H = -890.4 \text{ kJ}$$

ΔH is negative; 890.4 kJ of heat are evolved (pass from the system to the surroundings) for every mole of $CH_4(g)$ consumed in the reaction, at constant pressure. The reaction between hydrogen and iodine to form hydrogen iodide is endothermic:

$$H_2(g) + I_2(s) \longrightarrow 2HI(g) \qquad \Delta H = 52.2 \text{ kJ}$$

ΔH is positive; 52.2 kJ of heat are absorbed (flow from the surroundings to the system) for every mole of H_2 and I_2 that react.

Experiment 6.4 shows the exothermic reaction between phosphorus and bromine

$$2P(s) + 3Br_2(l) \longrightarrow 2PBr_3(l) \qquad \Delta H = -278 \text{ kJ}$$

Experiment 6.4

The Exothermic Reaction between Phosphorus and Bromine

The dish contains red phosphorus and the dropper contains brown liquid bromine.

When the bromine is added to the phosphorus it reacts immediately to form phosphorus tribromide, PBr, which is a colorless volatile liquid. The heat produced in this exothermic reaction vaporizes the PBr which forms a white mist on reacting with the moisture in the air.

Eventually the heat produced by the reaction causes the mixture to ignite.

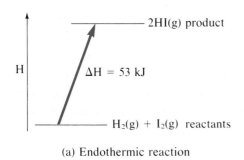

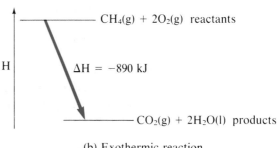

(a) Endothermic reaction

(b) Exothermic reaction

FIGURE 6.11

Enthalpy Changes in Exothermic and Endothermic Reactions.
(a) Heat is absorbed in an endothermic reaction. (b) Heat is evolved in an exothermic reaction.

6.4
Thermochemistry

CALORIMETRY

The measurement of the heat absorbed or evolved in chemical reactions is called **calorimetry**, after *caloric*, the old name for heat. The apparatus used for the measurement is known as a **calorimeter**. The amount of heat absorbed or evolved is calculated from the temperature change that it produces in the reacting system, or in some other system to which the heat is transferred. To do this calculation we need to know the temperature change that is produced in a given substance by the addition or the removal of a given amount of heat. Different substances need different amounts of heat to raise their temperature by a given amount.

The heat required to raise the temperature of an object or of a given amount of a substance by 1 kelvin (= 1 °C) is called the **heat capacity** of the object or of the particular amount of substance. The heat required to raise the temperature of 1 mole of a substance by 1 kelvin is the **molar heat capacity** of the substance. The molar heat capacity of liquid water is 75.4 J K^{-1} mol^{-1}. The heat required to raise the temperature of 1 gram of a substance is called its **specific heat capacity** or simply its **specific heat**. The specific heat of water is 4.18 J K^{-1} g^{-1}.

A very simple calorimeter can be made from two Styrofoam coffee cups (Figure 6.12a). Because Styrofoam is a very good heat insulator, a coffee cup calorimeter absorbs only a very small amount of the heat produced in a reaction carried out inside it. Its temperature is therefore not increased significantly, and so very little heat is transmitted to the surroundings.

This simple calorimeter is most convenient for measuring the heat evolved in reactions carried out in dilute aqueous solution. Almost all the heat evolved in the reaction goes to raise the temperature of the solution. If the solution is dilute, its heat capacity does not differ significantly from that of water, and the amount of heat evolved or absorbed in the reaction can be calculated from the change in the temperature of the solution. A reaction that is easily carried out in a coffee cup calorimeter is the neutralization of an acid with a base. Some typical results are given in Example 6.2.

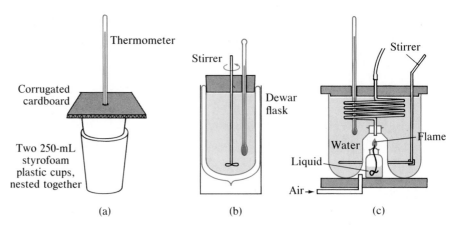

FIGURE 6.12
Calorimeters.

(a) A coffee cup calorimeter is suitable for a student laboratory experiment but not for accurate measurements. (b) A calorimeter of this type is suitable for making accurate measurements on reactions occurring in solution. (c) A flame calorimeter is used for measuring the enthalpies of combustion of gases and volatile liquids. The heat evolved in the combustion raises the temperature of the water in the calorimeter. From the known heat capacity of water the quantity of heat released in the reaction can be calculated.

Example 6.2

DETERMINATION OF THE ENTHALPY CHANGE FOR A NEUTRALIZATION REACTION

A sample of 50 mL of a $0.20M$ solution of HCl was mixed with 50 mL of a $0.20M$ solution of NaOH in a coffee cup calorimeter. The initial temperature of both solutions was 22.2 °C. After the two solutions were mixed, the temperature rose to 23.5 °C. What is the enthalpy change for the neutralization reaction that occurs:

$$H_3O^+(aq) + OH^-(aq) \rightleftharpoons 2H_2O(l)$$

Solution

The total volume of the solution is 100 mL, and since it is a dilute aqueous solution its density is 1.00 g mL^{-1}. Therefore the mass of the solution is 100 g. The temperature rise was 23.5 °C − 22.2 °C = 1.3 °C = 1.3 K. The specific heat capacity of water is $4.18 \text{ J K}^{-1}\text{g}^{-1}$.

Energy released as heat = specific heat capacity × mass of solution × temperature rise. Thus the energy released as heat by the reaction is

$$(4.18 \text{ J K}^{-1}\text{ g}^{-1})(100 \text{ g})(1.3 \text{ K}) = 540 \text{ J}$$

The HCl solution contained

$$(0.20 \text{ mol L}^{-1})(50 \text{ mL})\left(\frac{1 \text{ L}}{1000 \text{ mL}}\right) = 0.010 \text{ mol HCl}$$

The NaOH solution similarly contained 0.010 mol NaOH. Thus the heat evolved in the reaction of 0.01 mol of NaOH with 0.01 mol of HCl is 540 J. So for the reaction of 1.0 mol of NaOH with 1.0 mol of HCl we have

$$\text{NaOH(aq)} + \text{HCl(aq)} \longrightarrow \text{NaCl(aq)} + \text{H}_2\text{O(l)} \qquad \Delta H = -54 \text{ kJ}$$

Then writing the equation in terms of the ions present in the solution

$$H_3O^+(aq) + Cl^-(aq) + Na^+(aq) + OH^-(aq) \longrightarrow Na^+(aq) + Cl^-(aq) + 2H_2O(l)$$

and cancelling the spectator ions we have

$$H_3O^+(aq) + OH^-(aq) \longrightarrow 2H_2O(l) \qquad \Delta H = -54 \text{ kJ}$$

EXERCISE 6.5

A 25-mL sample of a $0.10M$ aqueous H_2SO_4 solution and 50 mL of an aqueous $0.10M$ KOH solution were mixed in a coffee cup calorimeter. The temperature of the solution rose from 21.20 to 22.10 °C as a result of the reaction that occurred on mixing the solutions. Calculate the enthalpy change for the reaction

$$H_2SO_4(aq) + 2KOH(aq) \longrightarrow K_2SO_4(aq) + 2H_2O(l)$$

Compare the value you obtain with that obtained from the neutralization of HCl(aq) in Example 6.2 and explain any difference.

A coffee cup calorimeter does not give very accurate results because the Styrofoam absorbs a small amount of heat and some heat is lost to the surroundings. A calorimeter constructed from a vacuum flask, as shown in Figure 6.12(b), gives more accurate results. Careful experiments using this type of calorimeter have given the accurate value of $\Delta H = -56.02$ kJ for the enthalpy change for the neutralization reaction in aqueous solution. The equation

$$H_3O^+(aq) + OH^-(aq) \longrightarrow 2H_2O(l) \qquad \Delta H = -56.02 \text{ kJ}$$

applies to the reaction of any strong acid with any strong base. For example,

$$HClO_4(aq) + KOH(aq) \longrightarrow KClO_4(aq) + H_2O(l) \qquad \Delta H = -56.02 \text{ kJ}$$
$$H_2SO_4(aq) + 2KOH(aq) \longrightarrow K_2SO_4(aq) + 2H_2O(l) \qquad \Delta H = 2(-56.02 \text{ kJ})$$
$$= -112.04 \text{ kJ}$$

For studying the enthalpies of combustion of gases and liquids, a flame calorimeter such as the one shown in Figure 6.12(c) can be used.

Example 6.3

DETERMINATION OF THE ENTHALPY CHANGE FOR A COMBUSTION REACTION

When 0.510 g of ethanol was burned in oxygen in a flame calorimeter containing 1200 g of water, the temperature of the water rose from 22.46 °C to 25.52 °C. What is the enthalpy change, ΔH, for the combustion of 1 mol of ethanol?

$$C_2H_5OH(l) + 3O_2(g) \longrightarrow 2CO_2(g) + 3H_2O(l)$$

Solution

The increase in the temperature of the water is

$$25.52 \text{ °C} - 22.46 \text{ °C} = 3.06 \text{ °C} = 3.06 \text{ K}$$

The specific heat capacity of water is 4.18 J K^{-1} g^{-1}. Therefore the amount of heat added to the water by the combustion of 0.510 g ethanol is

$$(4.18 \text{ J K}^{-1} \text{ g}^{-1})(1200 \text{ g})(3.06 \text{ K})\left(\frac{1 \text{ kJ}}{1000 \text{ J}}\right) = 15.3 \text{ kJ}$$

The heat *evolved* in the combustion of 1 mol of ethanol (molar mass 46.05 g) is

$$\left(\frac{15.3 \text{ kJ}}{0.510 \text{ g}}\right)\left(\frac{46.05 \text{ g}}{1 \text{ mol}}\right) = 1380 \text{ kJ mol}^{-1}$$

Therefore for the combustion of 1 mol of ethanol, $\Delta H = -1380$ kJ.

EXERCISE 6.6

When 1.3 g of butane, C_4H_{10}, were burned in oxygen in a flame calorimeter containing 1800 g of water, the temperature of the water rose from 20.2 to 28.2 °C. What is the enthalpy change for the combustion of 1 mol of butane?

The heat changes accompanying some reactions, for example the combustion of solids, are conveniently carried out in a bomb calorimeter (Figure 6.13). The reaction is then being carried out at constant volume rather than at constant pressure, so the heat absorbed, q, is equal to the energy change ΔE and not the enthalpy change ΔH. The value of ΔH can be calculated from ΔE but we will not consider that here. For many reactions, particularly those involving solids and liquids, which have small volume changes, the difference between ΔH and ΔE is very small.

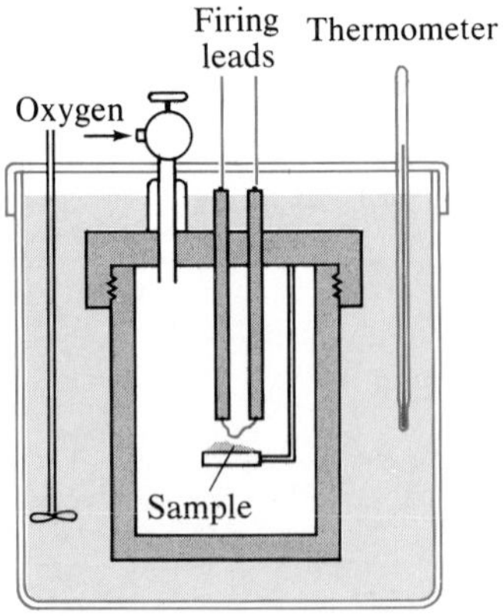

FIGURE 6.13
A Bomb Calorimeter.

A bomb calorimeter is particularly useful for measuring the enthalpy of combustion of solids. The reaction is initiated when a wire in contact with the solid is heated by passing an electric current through it. The reaction occurs at constant volume so that q, the heat absorbed, is equal to ΔE, the change in internal energy.

STANDARD ENTHALPY CHANGES

Since reaction enthalpies depend somewhat on the conditions under which the reaction is carried out, we must specify some standard conditions. The **standard enthalpy** of reaction is the enthalpy change for a reaction when all the reactants and all the products are in their standard states. A substance is in its *standard state* when it is at a pressure of 1 atm and at a specified temperature, which is usually chosen to be 25 °C. All standard enthalpy changes in this book are at 25 °C. Standard conditions are denoted by adding the superscript ° to the symbol ΔH, to give $\Delta H°$. For the combustion of 1 mol of methane under standard conditions, we write the equation

$$CH_4(g) + 2O_2(g) \longrightarrow CO_2(g) + 2H_2O(l) \qquad \Delta H° = -890.4 \text{ kJ}$$

This equation indicates that in the combustion of 1 mol of CH_4 in sufficient oxygen or air, 1 mol of CO_2 and 2 mol of H_2O are formed and 890.4 kJ of heat are released at a pressure of 1 atm and at 25 °C.

It might seem surprising that the combustion of methane can apparently be carried out at 25 °C. In fact, when methane is ignited, its temperature rises rapidly as it burns and produces heat, so the combustion is not in fact occurring at 25 °C. But if the reactants are at 25 °C at the start of the reaction, and the products cool to 25 °C at the conclusion of the reaction, all the heat produced in the reaction is eventually transferred to the calorimeter. It is immaterial that the reaction occurs at some temperature other than 25 °C, provided that the reactants are initially at 25 °C and the products are also at 25 °C. Because ΔH is a state function its value is independent of the path that the reaction takes in proceeding from reactants to products.

Although we usually avoid fractional coefficients in writing equations for reactions, for convenience when we discuss reaction enthalpies, we often write the equation for the reaction of 1 mol of the reactant under consideration. Therefore the equation for the combustion of ethane may be written as

$$C_2H_6(g) + \tfrac{7}{2}O_2(g) \longrightarrow 2CO_2(g) + 3H_2O(l) \qquad \Delta H° = -1560.4 \text{ kJ}$$

This equation tells us that the enthalpy change for the combustion of 1 mol of ethane under standard conditions is -1560.4 kJ.

Note that *standard enthalpy changes, $\Delta H°$, are for the numbers of moles of reactants and products as given in the equation for the reaction.* If we were to write the above equation in the more usual form, which involves 2 mol of ethane, we would have

$$2C_2H_6(g) + 7O_2(g) \longrightarrow 4CO_2(g) + 6H_2O(l)$$

and $\Delta H° = 2(-1560.4) \text{ kJ} = -3120.8 \text{ kJ}.$

HESS'S LAW

The combustion of ethane can be carried out directly, or alternatively in the following steps (see Figure 6.14).

1. If ethane is strongly heated, it is decomposed to ethene, C_2H_4, and hydrogen; 136.2 kJ is absorbed for each mole of ethane decomposed:

$$C_2H_6(g) \longrightarrow C_2H_4(g) + H_2(g) \qquad \Delta H° = +136.2 \text{ kJ}$$

2. If ethene is burned in air, the reaction is very exothermic:

$$C_2H_4(g) + 3O_2(g) \longrightarrow 2CO_2(g) + 2H_2O(l) \qquad \Delta H° = -1410.8 \text{ kJ}$$

3. The combustion of hydrogen also releases heat:

$$H_2(g) + \tfrac{1}{2}O_2(g) \longrightarrow H_2O(l) \qquad \Delta H° = -285.8 \text{ kJ}$$

Together, these three reactions give the same result as the direct combustion of ethane. This result can be seen by adding the left and right sides of the

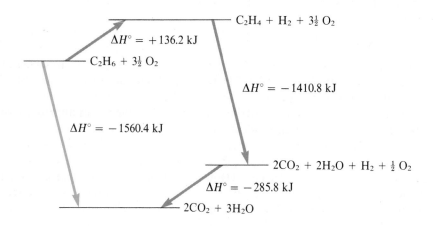

FIGURE 6.14
Enthalpy Changes in the Combustion of Ethane.

The combustion of ethane to carbon dioxide and water in the gas phase can be carried out either directly or by the intermediate formation of ethene and hydrogen. The overall enthalpy change for the process is the same for the two routes. Directly, we have $\Delta H° = -1560.4$ kJ; indirectly, we have $\Delta H° = +136.2 - 1410.8 - 285.8 = -1560.4$ kJ.

equations and then canceling terms that appear on each side:

$$C_2H_6(g) \longrightarrow C_2H_4(g) + H_2(g) \qquad \Delta H^\circ = +136.2 \text{ kJ}$$
$$C_2H_4(g) + 3O_2(g) \longrightarrow 2CO_2(g) + 2H_2O(l) \qquad \Delta H^\circ = -1410.8 \text{ kJ}$$
$$\underline{H_2(g) + \tfrac{1}{2}O_2(g) \longrightarrow H_2O(l) \qquad \Delta H^\circ = -285.8 \text{ kJ}}$$
$$C_2H_6(g) + \tfrac{7}{2}O_2(g) \longrightarrow 2CO_2(g) + 3H_2O(l) \qquad \Delta H^\circ = -1560.4 \text{ kJ}$$

We find also that if we add the enthalpy changes for each reaction, we obtain a value equal to the experimentally determined value of the enthalpy change for the overall reaction. Thus the net ΔH° for the combustion of C_2H_6 in O_2 carried out indirectly by the formation of C_2H_4 and H_2 and the combustion of each of these separately is *exactly* the same as ΔH° for the direct, one-stage combustion.

The Russian chemist Germain Hess (1802–1850) found experimentally that a similar statement could be made for a large number of different reactions. In fact, this conclusion is valid for *any* chemical reaction and is expressed as **Hess's law**:

> **The enthalpy change in a reaction is the same regardless of the path by which the reaction occurs.**

In other words, the enthalpy change in a reaction is simply the difference in the enthalpies of the reactants and the products; it depends *only* on the enthalpies of the reactants and the products and is independent of the way in which the reactants are transformed into the products (see Figure 6.14).

Hess's law is a direct consequence of the fact that enthalpy is a state function. The enthalpy of the products, under the specified conditions, does not depend on the route by which they were obtained.

We can use Hess's law to calculate enthalpy changes that are not easy to measure directly. *Enthalpies of combustion* are one of the simplest to measure experimentally, and they are often used to obtain values for other reaction enthalpies. For example, the enthalpy change for the formation of methane from carbon and hydrogen,

$$C(s) + 2H_2(g) \longrightarrow CH_4(g)$$

can be obtained from the ΔH° values for the combustion of carbon, hydrogen, and methane. These values are as follows:

$$(1) \qquad C(s) + O_2(g) \longrightarrow CO_2(g) \qquad \Delta H^\circ = -393.5 \text{ kJ}$$
$$(2) \qquad H_2(g) + \tfrac{1}{2}O_2(g) \longrightarrow H_2O(l) \qquad \Delta H^\circ = -285.8 \text{ kJ}$$
$$(3) \qquad CH_4(g) + 2O_2(g) \longrightarrow CO_2(g) + 2H_2O(l) \qquad \Delta H^\circ = -890.4 \text{ kJ}$$

To obtain the equation for the formation of methane from carbon and hydrogen, we first multiply equation (2) by 2 and add it to equation (1), because this gives us $C(s) + 2H_2(g)$ as reactants on the left-hand side as required. Then we reverse equation (3) to obtain $CH_4(g)$ as a product on the right-hand side, and add it to $(1) + (2) \times (2)$. *The values of ΔH° must be combined in exactly the same way.* Reaction (3) is exothermic; therefore the reverse reaction is endothermic and we must change the sign of ΔH accordingly.

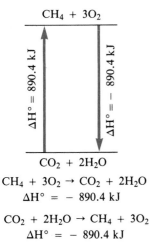

CH$_4$ + 3O$_2$

$\Delta H^\circ = 890.4$ kJ

$\Delta H^\circ = -890.4$ kJ

CO$_2$ + 2H$_2$O

$$CH_4 + 3O_2 \rightarrow CO_2 + 2H_2O$$
$$\Delta H^\circ = -890.4 \text{ kJ}$$

$$CO_2 + 2H_2O \rightarrow CH_4 + 3O_2$$
$$\Delta H^\circ = -890.4 \text{ kJ}$$

$$\Delta H^\circ$$
$$C(s) + O_2(g) \longrightarrow CO_2(g) \qquad -393.5 \text{ kJ}$$
$$2 \times [H_2(g) + \tfrac{1}{2}O_2(g) \longrightarrow H_2O(l)] \qquad 2(-285.8 \text{ kJ})$$
$$\underline{CO_2(g) + 2H_2O(l) \longrightarrow CH_4(g) + 2O_2(g) \qquad +890.4 \text{ kJ}}$$
$$C(s) + 2H_2(g) + 2O_2(g) + CO_2(g) + 2H_2O(l) \longrightarrow$$
$$CO_2(g) + 2H_2O(l) + CH_4(g) + 2O_2(g) \qquad -74.7 \text{ kJ}$$

Canceling terms that appear on both sides gives the desired reaction:

$$C(s) + 2H_2(g) \longrightarrow CH_4(g) \qquad -74.7 \text{ kJ}$$

EXERCISE 6.7

Given the standard enthalpy changes for the two reactions

$$2P(s) + 3Cl_2(g) \longrightarrow 2PCl_3(g) \qquad \Delta H° = -574 \text{ kJ}$$

$$PCl_3(g) + Cl_2(g) \longrightarrow PCl_5(s) \qquad \Delta H° = -87.9 \text{ kJ}$$

find the standard enthalpy change for the reaction

$$2P(s) + 5Cl_2(g) \longrightarrow 2PCl_5(s)$$

STANDARD ENTHALPIES OF FORMATION

By using Hess's law we can calculate the enthalpy change for any reaction from the enthalpy changes of an appropriate set of related reactions. So we do not need to list in a reference book the enthalpy change for every reaction. This would be an enormous task and even to look up a value that we might need would be difficult. We need only list a selected number of reactions from which the standard enthalpy change for any other reaction can be calculated. The most useful reactions for this purpose are the reactions in which a compound is formed from its elements in their standard states. *The enthalpy change for the reaction in which a compound is formed from its elements in their standard states is called the* **standard enthalpy of formation, $\Delta H_f°$.**

Because an element can exist in the gaseous, liquid, and solid forms—and often in several different solid forms called allotropes—we must specify the form of the element on which standard enthalpies of formation are based. This is the most stable form of the element under standard conditions, that is, at 1 atm and at a specified temperature, usually 25 °C. Thus for bromine, it is $Br_2(l)$, not $Br_2(g)$ or $Br(g)$; and for carbon, it is graphite, not diamond or $C(g)$.

Since to form the most stable form of an element in its standard state from the most stable form of the element in its standard state is to make no change at all, the corresponding standard enthalpy of formation is clearly zero. Thus by definition, *the enthalpy of formation of the most stable form of an element in its standard state is zero.* The standard states of the most stable forms of the elements are the arbitrary zero from which all enthalpy changes are measured. This choice is analogous to the choice of sea level as the arbitrary zero from which all land altitudes and ocean depths are measured. Thus the standard enthalpies of formation of carbon dioxide, water, and methane are the enthalpy changes for the following reactions at 25 °C (298 K) and 1 atm:

There are exceptions to this convention for a very few elements. For example, white phosphorus, rather than the more stable red phosphorus, is the form chosen for specifying the standard state of phosphorus, because white phosphorus is more easily obtained in a pure form.

$$C(\text{graphite}) + O_2(g) \longrightarrow CO_2(g) \qquad \Delta H_f°(CO_2, g) = -393.5 \text{ kJ mol}^{-1}$$

$$H_2(g) + \tfrac{1}{2}O_2(g) \longrightarrow H_2O(l) \qquad \Delta H_f°(H_2O, l) = -285.8 \text{ kJ mol}^{-1}$$

$$C(\text{graphite}) + 2H_2(g) \longrightarrow CH_4(g) \qquad \Delta H_f°(CH_4, g) = -74.8 \text{ kJ mol}^{-1}$$

Standard enthalpies of formation for some common substances at 25 °C are given in Table 6.2. A more extensive list is given in Appendix B, Table B.2.

REACTION ENTHALPIES FROM ENTHALPIES OF FORMATION From the $\Delta H_f°$ values for the reactants and products of a reaction, we can calculate $\Delta H°$ for

TABLE 6.2
Standard Enthalpies of Formation at 25 °C

Substance	ΔH_f° (kJ mol^{-1})	Substance	ΔH_f° (kJ mol^{-1})
$CH_4(g)$	-74.5	$SO_2(g)$	-296.8
$C_2H_2(g)$	$+226.8$	$SO_3(g)$	-395.7
$C_2H_4(g)$	$+52.3$	$O_3(g)$	$+142.7$
$C_2H_6(g)$	-84.7	$H_2SO_4(l)$	-814.0
$C_3H_8(g)$ (*n*-propane)	-103.8	$HNO_3(l)$	-174.1
$C_4H_{10}(g)$ (*n*-butane)	-126.1	$B_5H_9(s)$	$+73.2$
$C_5H_{12}(g)$ (*n*-pentane)	-146.4	$B_2O_3(s)$	-1273.5
$CO(g)$	-110.5	$NaOH(s)$	-425.6
$CO_2(g)$	-393.5	All elements	
$H_2O(l)$	-285.8	(most stable form)	0
$H_2O(g)$	-241.8	C(diamond)	-1.90
$NO(g)$	$+90.3$	H(g, atomic)	$+218.0$
$NO_2(g)$	$+33.2$	C(g, atomic)	$+716.7$
$NH_3(g)$	-46.2		

the reaction by using Hess's law. Consider the combustion of methane. This reaction may be carried out directly according to the equation

$$CH_4(g) + 2O_2(g) \longrightarrow CO_2(g) + 2H_2O(l) \qquad \Delta H_1^\circ \qquad (1)$$

Alternatively, at least in principle, the reaction may be carried out by first decomposing the reactants into the elements in their standard states (equation 2) and then recombining these elements to form the products (equation 3):

$$CH_4(g) + 2O_2(g) \longrightarrow C(s) + 2H_2(g) + 2O_2(g) \qquad \Delta H_2^\circ \qquad (2)$$

$$C(s) + 2H_2(g) + 2O_2(g) \longrightarrow CO_2(g) + 2H_2O(l) \qquad \Delta H_3^\circ \qquad (3)$$

From Hess's law we have

$$\Delta H_1^\circ = \Delta H_3^\circ + \Delta H_2^\circ$$

Reaction (3) corresponds to the sum of the formation reactions for all the products from their elements. Thus

$$\Delta H_3^\circ = \text{Sum of the enthalpies of formation, } \Delta H_f^\circ, \text{ of all the products}$$
$$= (1 \text{ mol } CO_2)[\Delta H_f^\circ(CO_2, g)] + (2 \text{ mol } H_2O)[\Delta H_f^\circ(H_2O, l)]$$

Reaction (2) is the reverse of the formation reactions for all the reactants. Thus

$$\Delta H_2^\circ = -(\text{Sum of the enthalpies of formation, } \Delta H_f^\circ, \text{ of all the reactants})$$
$$= -(1 \text{ mol } CH_4)[\Delta H_f^\circ(CH_4, g)] - (2 \text{ mol } O_2)[\Delta H_f^\circ(O_2, g)]$$

Hence

$$\Delta H_1^\circ = \Delta H_3^\circ + \Delta H_2^\circ$$
$$= (\text{Sum of the enthalpies of formation, } \Delta H_f^\circ, \text{ of the products})$$
$$\quad - (\text{Sum of the enthalpies of formation, } \Delta H_f^\circ, \text{ of the reactants})$$
$$= (1 \text{ mol } CO_2)[\Delta H_f^\circ(CO_2, g)] + (2 \text{ mol } H_2O)[\Delta H_f^\circ(H_2O, l)]$$
$$\quad - (1 \text{ mol } CH_4)[\Delta H_f^\circ(CH_4, g)] - (2 \text{ mol } O_2)[\Delta H_f^\circ(O_2, g)]$$

Using the values given in Table 6.2, we have

$$\Delta H^\circ = (1 \text{ mol})(-393.5 \text{ kJ mol}^{-1}) + (2 \text{ mol})(-285.8 \text{ kJ mol}^{-1})$$
$$- (1 \text{ mol})(-74.8 \text{ kJ mol}^{-1}) - (2 \text{ mol})(0)$$
$$= -890.3 \text{ kJ}$$

This approach is valid for any reaction, since any reaction can, in principle, be carried out by first decomposing all the reactants into their elements and then combining the elements to give the final products. Thus for any reaction

$$n_A A + n_B B + \cdots \longrightarrow n_X X + n_Y Y + \cdots$$

we have

$$\Delta H^\circ = [n_X(\Delta H_f^\circ)_X + n_Y(\Delta H_f^\circ)_Y + \cdots] - [n_A(\Delta H_f^\circ)_A + n_B(\Delta H_f^\circ)_B + \cdots]$$
$$= \sum n_p(\Delta H_f^\circ)_p - \sum n_r(\Delta H_f^\circ)_r$$

where p stands for products and r for reactants. In other words,

$$\Delta H^\circ = (\text{Sum of standard enthalpies of formation of products})$$
$$- (\text{Sum of standard enthalpies of formation of reactants})$$

Notice that the ΔH_f° values listed in Table 6.2 are for the formation of 1 mol of the substance. If n mol of a substance are involved in a reaction, the corresponding ΔH_f° value must be multiplied by the factor n.

Notice also that the standard enthalpies of formation of the elements are zero only for their most stable form. For example, the standard enthalpy of formation of gaseous molecular hydrogen, H_2, is zero, but the standard enthalpy of formation of gaseous *atomic* hydrogen is the enthalpy change for the reaction

$$\tfrac{1}{2}H_2(g) \longrightarrow H(g, \text{atomic}) \qquad \Delta H^\circ = +218.0 \text{ kJ}$$

and so

$$\Delta H_f^\circ(H, g, \text{atomic}) = +218.0 \text{ kJ mol}^{-1}$$

Similarly, the standard enthalpy of formation of solid carbon (graphite) is zero, but the standard enthalpy of formation of atomic carbon is the enthalpy change for the reaction

$$C(s, \text{graphite}) \longrightarrow C(g, \text{atomic}) \qquad \Delta H^\circ = +716.7 \text{ kJ}$$

and so

$$\Delta H_f^\circ(C, g, \text{atomic}) = +716.7 \text{ kJ mol}^{-1}$$

The standard enthalpy of formation of diamond is also not zero; it is the enthalpy change for the reaction

$$C(\text{graphite}) \longrightarrow C(\text{diamond}) \qquad \Delta H^\circ = +1.90 \text{ kJ}$$

so that $\Delta H_f^\circ(\text{diamond}) = +1.90 \text{ kJ mol}^{-1}$.

═ *Example 6.4* ═══════════════════════════════

USING STANDARD ENTHALPIES OF FORMATION

Find the standard enthalpy change for the oxidation of NH_3 according to the equation

$$4NH_3(g) + 5O_2(g) \longrightarrow 4NO(g) + 6H_2O(g)$$

from the standard enthalpies of formation given in Table 6.2.

Solution

$$\Delta H^\circ = \sum n_p (\Delta H_f^\circ)_p - \sum n_r (\Delta H_f^\circ)_r$$
$$= (4 \text{ mol NO})(+90.3 \text{ kJ mol}^{-1} \text{ NO}) + (6 \text{ mol H}_2\text{O})(-241.8 \text{ kJ mol}^{-1} \text{ H}_2\text{O})$$
$$- [(4 \text{ mol NH}_3)(-46.2 \text{ kJ mol}^{-1} \text{ NH}_3) + (5 \text{ mol O}_2)(0)]$$
$$= +361.2 \text{ kJ} - 1450.8 \text{ kJ} + 184.8 \text{ kJ} - 0$$
$$= -904.8 \text{ kJ}$$

═ *Example 6.5* ═

USING STANDARD ENTHALPIES OF FORMATION

The hydrides of boron, which have unexpected formulas such as B_5H_9 and B_4H_{10}, have very high enthalpies of combustion. Some years ago, an extensive study was made of the hydrides of boron with a view to their utilization as rocket fuels. The hydride B_5H_9 ignites spontaneously in air with a green flash to produce solid B_2O_3 and water. What is the enthalpy of combustion of B_5H_9 under standard conditions?

Solution

The unbalanced equation for the reaction is

$$B_5H_9(g) + O_2(g) \longrightarrow B_2O_3(s) + H_2O(l)$$

Two moles of B_5H_9 are needed for every 5 mol of B_2O_3 produced in order to balance the boron atoms. Hence the 18 hydrogen atoms must appear as 9 water molecules. There are then a total of 24 oxygen atoms on the right side of the equation, and so $12 O_2$ molecules are needed on the left side. The balanced equation is

$$2B_5H_9(g) + 12O_2(g) \longrightarrow 5B_2O_3(s) + 9H_2O(l)$$

or for 1 mol of B_5H_9

$$B_5H_9(g) + 6O_2(g) \longrightarrow \tfrac{5}{2}B_2O_3(s) + \tfrac{9}{2}H_2O(l)$$

Using the standard enthalpies of formation given in Table 6.2, we have

$$\Delta H^\circ = \sum n_p (\Delta H_f^\circ)_p - \sum n_r (\Delta H_f^\circ)_r$$
$$= (\tfrac{5}{2} \text{ mol})(-1273.5 \text{ kJ mol}^{-1})$$
$$+ (\tfrac{9}{2} \text{ mol})(-285.8 \text{ kJ mol}^{-1}) - (1 \text{ mol})(73.2 \text{ kJ mol}^{-1})$$
$$= -4543 \text{ kJ}$$

The standard enthalpy of combustion of $B_5H_9(g)$ is $-4543 \text{ kJ mol}^{-1}$.

EXERCISE 6.8

Calculate the standard enthalpy change, ΔH°, for the reaction

$$2SO_2(g) + O_2(g) + 2H_2O(g) \longrightarrow 2H_2SO_4(l)$$

from the standard enthalpies of formation in Table 6.2. How much heat would be liberated if 5.20 g of SO_2 were converted to H_2SO_4?

ENTHALPIES OF FORMATION FROM ENTHALPIES OF COMBUSTION Direct determination of the enthalpies of formation of many compounds is difficult. For example, pure hydrocarbons are not easily obtained by directly combining carbon and hydrogen. In general, the enthalpy of formation of a hydrocarbon may be obtained from its experimentally measured enthalpy of combustion and the tabulated enthalpies of formation of the combustion product, CO_2 and H_2O, as we have already seen in the case of CH_4.

=== Example 6.6 ===

ENTHALPY OF FORMATION FROM ENTHALPIES OF COMBUSTION

The combustion of 1 mol of liquid benzene, $C_6H_6(l)$, at 25 °C and 1.00 atm to produce $CO_2(g)$ and $H_2O(l)$ liberates 3267 kJ of heat when the products are also at 25 °C and 1.00 atm. What is the standard enthalpy of formation of $C_6H_6(l)$?

Solution

The equation for the combustion of 1 mol of C_6H_6 is

$$C_6H_6(l) + 7\tfrac{1}{2}O_2(g) \longrightarrow 6CO_2(g) + 3H_2O(l)$$

We are given that the combustion of 1 mol of C_6H_6 at standard conditions liberates 3267 kJ of heat. Therefore $\Delta H° = -3267 \text{ kJ mol}^{-1}$. Then

$$\Delta H° = \sum n_p(\Delta H_f°)_p - \sum n_r(\Delta H_f°)_r$$
$$-3267 \text{ kJ} = 6\,\Delta H_f°(CO_2, g) + 3\,\Delta H_f°(H_2O, l) - \Delta H_f°(C_6H_6, l) - 7\tfrac{1}{2}\,\Delta H_f°(O_2, g)$$

Using the values given in Table 6.2 we have

$$-3267 \text{ kJ} = (6 \text{ mol})(-393.5 \text{ kJ mol}^{-1}) + (3 \text{ mol})(-285.8 \text{ kJ mol}^{-1})$$
$$- (1 \text{ mol})[\Delta H_f°(C_6H_6, l)] - 0$$

Therefore

$$(1 \text{ mol})[\Delta H_f°(C_6H_6, l)] = [6(-393.5) + 3(-285.8) + 3267] \text{ kJ}$$
$$= (-2361.0 - 857.4 + 3267) \text{ kJ}$$
$$\Delta H_f°(C_6H_6, l) = 49 \text{ kJ mol}^{-1}$$

EXERCISE 6.9

The combustion of 1 mol of butane, $C_4H_{10}(g)$, in excess oxygen at 25 °C and 1.00 atm liberates 2877 kJ of heat. What is the standard enthalpy of formation of butane at 25 °C?

The standard enthalpies of formation provide a measure of the relative amounts of chemical energy stored in a compound. If the standard enthalpy of formation of a compound is negative, the compound contains less stored energy than the standard states of the elements from which it is formed. The $\Delta H_f°$ for CO_2 is -393 kJ mol^{-1}, so CO_2 contains less stored energy than C(graphite) and $O_2(g)$. If $\Delta H_f°$ is positive, the compound contains more stored energy than the elements from which it is formed. The $\Delta H_f°$ for ethyne, C_2H_2, is $+226.0 \text{ kJ mol}^{-1}$; C_2H_2 therefore contains a large amount of stored energy. If it is decomposed to the elements, this energy is released as heat. When ethyne is burned in oxygen, the reaction is highly exothermic, because the energy stored in C_2H_2 is released together with the energy that is released in the formation of CO_2 and H_2O, which both contain less stored energy than the elements from which they are made. We express the same idea in a different way when we say that C_2H_2 is less stable than the elements from which it is formed, whereas CO_2 and H_2O are more stable than the elements from which they are formed.

BOND ENERGIES

The atoms in a molecule are held together by strong forces of attraction that we call chemical bonds. We began our study of the nature of the forces operating in chemical bonds in Chapter 4 and we will continue this discussion in Chapter 7. Here we will be concerned with the energy changes associated with chemical bonds. In order to decompose a molecule into its constituent atoms work must be done against the forces of attraction, that is, energy must be supplied to break chemical bonds. The decomposition of a molecule into its constituent atoms is an endothermic process. The stronger the bonds in a molecule the more energy is needed to decompose the molecule into its atoms.

Thus the enthalpy change for the reaction

$$H_2(g) \longrightarrow 2H(g, \text{atomic}) \qquad \Delta H^\circ = 436.0 \text{ kJ}$$

in which a hydrogen molecule is dissociated into two hydrogen atoms is a measure of the strength of the bond in the hydrogen molecule. This value is the *bond dissociation enthalpy*, or *bond enthalpy*, of the hydrogen molecule; it is commonly, although strictly speaking incorrectly, called the *bond dissociation energy* or *bond energy* and the small difference between the enthalpy of dissociation and the energy of dissociation is ignored. This is generally justifiable because, as we shall see, only approximate average values can be given for most bond energies. Thus the bond energy of the hydrogen molecule is $436.0 \text{ kJ mol}^{-1}$:

$$BE(\text{H}-\text{H}) = 436 \text{ kJ mol}^{-1}$$

The **bond energy** *of a diatomic molecule is the energy needed to dissociate* 1 *mol of molecules into gaseous atoms*. Notice that the enthalpy of formation of 1 mol of hydrogen atoms, which is given in Table 6.2, is the ΔH° for the reaction:

$$\tfrac{1}{2}H_2(g) \longrightarrow H(g, \text{atomic}) \qquad \Delta H^\circ = \frac{436.0}{2} \text{ kJ} = 218.0 \text{ kJ}$$

that is,

$$\Delta H_f^\circ(\text{H, g, atomic}) = 218.0 \text{ kJ mol}^{-1}$$

The bond energies of diatomic molecules range from values of 149 kJ mol^{-1} and 155 kJ mol^{-1} for the rather weak bonds in I_2 and F_2 to higher values of 941 kJ mol^{-1} and 1070 kJ mol^{-1} for the very strong bonds in N_2 and CO (Table 6.3).

The bond energies in polyatomic molecules are not quite so simple to determine. We consider two possible cases: first, when all the bonds are the same and, second, when there are two or more different kinds of bonds in the molecule.

The energy needed to dissociate a CH_4 molecule completely into a C atom

TABLE 6.3
Bond Energies

Bond	Bond Energy[a] (kJ mol^{-1})	Bond	Average Bond Energy[b] (kJ mol^{-1})	Bond	Average Bond Energy[b] (kJ mol^{-1})
H—H	436	O—O	138	C—F	485
H—F	565	N—N	159	O—Cl	205
H—Cl	431	N=N	418	N—Cl	201
H—Br	364	C—C	348	C—Cl	326
H—I	297	C=C	619	P—Cl	326
F—F	155	C≡C	812	S—Cl	276
Cl—Cl	239	O—H	463	C—O	335
Br—Br	190	N—H	389	C=O	707
I—I	149	C—H	413	C≡O	1070
O=O	494	P—H	318	C—N	293
N≡N	941	S—H	364	C=N	616
		O—F	184	C≡N	879

[a] These bond energies are the dissociation energies of diatomic molecules that have only one bond; they are therefore exact values.

[b] These bond energies are obtained, as described in the text, from molecules that contain more than one bond; they are average, not exact, values therefore.

and four H atoms,

$$CH_4(g) \longrightarrow C(g, \text{atomic}) + 4H(g, \text{atomic})$$

represents the energy needed to break all four C—H bonds. One-quarter of this value is the *average bond energy* for a C—H bond in methane. It is not easy to determine the enthalpy change for this reaction experimentally, but we can calculate it from the enthalpies of formation given in Table 6.2.

$$\Delta H° = \Delta H_f°(C, \text{atomic}) + 4 \, \Delta H_f°(H, \text{atomic}) - \Delta H_f°(CH_4, g)$$
$$= (1 \text{ mol})(716.7 \text{ kJ mol}^{-1}) + (4 \text{ mol})(218.0 \text{ kJ mol}^{-1}) - (1 \text{ mol})(-74.5 \text{ kJ mol}^{-1})$$
$$= 1663 \text{ kJ}$$

since this enthalpy change is the energy required to break four C—H bonds, the average C—H bond energy is one-quarter of this value, that is, 415.8 kJ mol^{-1}.

When there are two or more kinds of bonds in a molecule, the determination of the bond energies is slightly more complicated and involves certain approximations. For example, from the appropriate enthalpies of formation we may obtain a value for the enthalpy change for the reaction

$$C_2H_6(g) \longrightarrow 2C(g, \text{atomic}) + 6H(g, \text{atomic})$$

in which six C—H bonds and one C—C bond are broken. Since we have two unknowns, namely, the C—H and C—C bond energies, we cannot determine values for both of them, and we have to make an assumption about one of them. If we assume, for example, that the C—H bond energy in C_2H_6 is the same as that in CH_4, we could then calculate a value for the C—C bond energy.

═ Example 6.7 ═══════════════════════════════

BOND ENERGIES

From the data in Table 6.2 calculate a value for the enthalpy change for the reaction

$$C_2H_6(g) \longrightarrow 2C(g, atomic) + 6H(g, atomic)$$

Then assuming that the C—H bond energy is the same as that in CH_4, calculate a value for the C—C bond energy in C_2H_6.

Solution

$$\Delta H° = (2 \text{ mol})(716.7 \text{ kJ mol}^{-1}) + (6 \text{ mol})(218.0 \text{ kJ mol}^{-1}) - (1 \text{ mol})(-84.7 \text{ kJ mol}^{-1})$$

$$= 2826.1 \text{ kJ}$$

Since C_2H_6 has six C—H bonds and one C—C bond, if BE(C—H) is the bond energy of a C—H bond and BE(C—C) is the bond energy of a C—C bond, then the energy needed to break all the bonds is

$$BE(\text{C—C}) + 6BE(\text{C—H}) = 2826.1 \text{ kJ mol}^{-1}$$

If we now assume that BE(C—H) is the same as it is in CH_4, that is, BE(C—H) = 415.8 kJ mol^{-1}, then

$$BE(\text{C—C}) = (2826.1 \text{ kJ mol}^{-1}) - 6(415.8 \text{ kJ mol}^{-1})$$

$$BE(\text{C—C}) = 331.3 \text{ kJ mol}^{-1}$$

EXERCISE 6.10

Calculate the standard enthalpy change for the reaction

$$C_3H_8(g) \longrightarrow 3C(g, atomic) + 8H(g, atomic)$$

from the standard enthalpies of formation in Table 6.2. Assuming that the C—H bond energy, BE(C—H), is the same as in methane, calculate a value for the average C—C bond energy in propane, C_3H_8, which has eight C—H and two C—C bonds. Compare this value with the value obtained in Example 6.7.

We obtained two slightly different values for the C—C bond energy in Example 6.7 and Exercise 6.10. A third value for BE(C—C) can be obtained from data for diamond. The enthalpy change for the vaporization of 1 mol of carbon atom is

$$C(\text{diamond}) \longrightarrow C(g, atomic) \qquad \Delta H° = 714.8 \text{ kJ}$$

In this process all the C—C bonds in the solid are broken. Since each carbon atom forms four bonds, but each bond is shared between two atoms, the $\Delta H°$ of this process is a measure of the energy required to break $\frac{4}{2} = 2$ bonds per carbon. Thus we obtain BE(C—C) = 714.8/2 kJ = 357.4 kJ mol^{-1}, which we can compare with the values of 335.8 kJ mol^{-1} from C_3H_8 (Exercise 6.10) and 331.3 kJ mol^{-1} from C_2H_6. It is not surprising that these values for the C—C bond energy do not agree exactly because we would not expect the C—H bond energy to be exactly the same in CH_4, C_2H_6, and C_3H_8, as we have assumed, nor would we expect a C—C bond to have exactly the same bond energy in all substances.

   Although the C—C bond energies obtained from different compounds differ somewhat, the average of the values obtained from a large number of compounds gives a reasonably good approximation for the energy of the C—C bond in any substance. This value is called the *average bond energy* of the C—C bond. Values for the average bond energies of some common bonds are listed in Table 6.3. These values give a good idea of the relative strengths of different bonds. We see that the C—C and C—H bonds are among the strongest of the single bonds, whereas the N—N, O—O, and F—F bonds are relatively weak. That the properties of a given bond, such as the bond energy, do not vary greatly from one substance to another is a very important concept in chemistry. It enables us to make many predictions about the properties of substances from the knowledge gained from the study of other substances that contain the same bonds.

DOUBLE AND TRIPLE BONDS

From the heats of formation of ethene and ethyne the energy needed to dissociate the molecule completely into atoms can be calculated. If we assume that the C—H bonds have the same strength as those in alkanes, values for the bond energies of the C=C double bond and the C≡C triple bond can be obtained in the same manner as for C—C single bonds.

⎓ Example 6.8 ⎓

BOND ENERGY OF THE C=C DOUBLE BOND

From the value for the enthalpy of formation of ethene, C_2H_4, in Table 6.2 and the bond energy of the C—H bond in CH_4 (415.8 kJ mol^{-1}), calculate a value for the C=C bond energy.

Solution

To find the C=C bond energy, we must first find the energy needed to decompose the C_2H_4 molecule completely into atoms. This energy is the enthalpy change for the reaction

$$C_2H_4(g) \longrightarrow 2C(g) + 4H(g)$$

Thus

$$\Delta H° = (2 \text{ mol})[\Delta H_f°(C, g)] + (4 \text{ mol})[\Delta H_f°(H, g)] - (1 \text{ mol})[\Delta H_f°(C_2H_4, g)]$$
$$= (2 \text{ mol})(716.7 \text{ kJ mol}^{-1}) + (4 \text{ mol})(218.0 \text{ kJ mol}^{-1}) - (1 \text{ mol})(52.3 \text{ kJ mol}^{-1})$$
$$= 2253.1 \text{ kJ}$$

This enthalpy change is the energy required to break four C—H bonds and one C=C bond. So

$$BE(C\!=\!C) + 4BE(C\!-\!H) = 2253.1 \text{ kJ mol}^{-1}$$

If $BE(C—H)$ is the same as it is in CH_4, that is, 415.8 kJ mol^{-1}, then

$$BE(C\!=\!C) = 2253.1 \text{ kJ mol}^{-1} - 4(415.8 \text{ kJ mol}^{-1}) = 589.9 \text{ kJ mol}^{-1}$$

E X E R C I S E   6 . 1 1

From the standard enthalpy of formation, ΔH_f°, of ethyne (acetylene), C_2H_2, given in Table 6.2 and the C—H bond energy in CH_4 of 415.8 kJ mol^{-1}, calculate a value for the C≡C triple bond energy.

The average bond energies of the C=C and C≡C bonds given in Table 6.3 are mean values calculated from a number of different compounds. The values for some other double and triple bonds are also given in the table.

We see from Table 6.3 that a double bond is very approximately twice as strong as the corresponding single bond, and a triple bond is three times as strong. But notice that whereas a C≡C bond is considerably less than three times as strong as a C—C bond, the N≡N triple bond is considerably more than three times as strong as the corresponding single bond. However, a discussion of these interesting differences would take us beyond the scope of this book. Because of the great strength of the N≡N triple bond, many reactions in which N_2 is formed are highly exothermic. Thus many nitrogen compounds decompose or burn in explosive reactions. Explosions are simply reactions in which a large amount of energy is released and a large volume of gaseous products is formed very rapidly (see Box 6.2 and Experiment 6.5).

Experiment 6.5

The Decomposition of Nitrogen Triiodide

The dark brown substance on the filter paper is a compound of nitrogen triiodide and ammonia, $NI_3 \cdot NH_3$.

When dry it is extremely shock sensitive and, when touched lightly with a feather, explodes violently, producing a cloud of purple-brown iodine vapor.

The explosion is violent enough to punch a hole through the filter papers and the asbestos mat on which they were resting.

Because of the great strength of the triple bond in the N_2 molecule, the decomposition of NI_3 to N_2 and I_2 is a strongly exothermic reaction. It is also very rapid so that a large amount of heat and a large volume of gaseous products are formed rapidly which produces a violent explosion.

BOX 6.2

EXPLOSIVES

An important application of chemical energy is the use of explosives. Any substance that undergoes a very rapid chemical reaction that is strongly exothermic—or that produces a large volume of gaseous products from a solid or a liquid—is potentially an explosive. The destructive power of an explosion is due to the shock wave caused by the rapid increase in volume from the gases formed or to the rapid expansion of the atmosphere as a consequence of the large amount of heat released in a short time, or to both of these circumstances.

The oldest known explosive is *gun powder*, which was used in ancient times in China, Arabia, and India. Gun powder is a mixture of approximately 75% KNO_3, 12% S, and 13% C. The products include a large volume of gases including CO_2, CO, and N_2, as well as a dense smoke that consists of fine particles of K_2CO_3, K_2SO_4, and K_2S.

For many purposes gun powder has been replaced by stronger explosives such as ammonium nitrate. When ammonium nitrate is detonated, it decomposes in a very exothermic reaction to give a large volume of gaseous products:

$$2NH_4NO_3(s) \longrightarrow 2N_2(g) + O_2(g) + 4H_2O(g)$$

The oxygen produced can also be used to oxidize other substances, thus increasing the energy released. A commonly used explosive for blasting in mines is composed of 95% NH_4NO_3 and 5% fuel oil. Ammonium nitrate is also used as a fertilizer. Normally, its use as a fertilizer is safe, because ammonium nitrate must be denoted before it will explode. However, the careless handling of large quantities of ammonium nitrate fertilizer can cause a massive explosion. For example, in 1947 a ship carrying ammonium nitrate fertilizer exploded and leveled a huge area of Texas City, Texas, claiming 576 lives.

If an explosive reaction is very rapid, the shock wave may travel at very high speeds, up to 6 km s^{-1}, and the explosive is classified as a high explosive. Trinitrotoluene (TNT), $C_7H_5O_6N_3$, and nitroglycerin, $C_3H_5O_9N_3$, are examples. The slower combustion that occurs in low explosives such as gun powder produces shock waves that travel at about 100 m s^{-1}.

The oxygen required for the very rapid combustion of high explosives cannot come from the air, because the oxidation is too rapid. For these high explosives the

Alfred Nobel
(1833–1896)

oxygen comes from the explosive itself. Often such explosives are mixed with other substances such as NH_4NO_3 in order to increase the amount of oxygen available. Many common explosives contain nitrogen compounds. Their combustion produces oxides of nitrogen and molecular nitrogen, which is formed in a very exothermic process because of the high bond energy of the nitrogen molecule.

Nitroglycerin is a liquid that explodes 25 times as fast as gun powder and with three times the energy per gram. Although it was used as an explosive soon after it was first prepared in 1866, nitroglycerin is much too unstable and sensitive to shock to be handled safely. For transportation the containers were packed into a type of clay called kieselguhr in order to cushion them from shocks as much as possible. Alfred Nobel, a Swedish inventor, noticed that when nitroglycerin leaked from the containers, it was soaked up by the clay. The nitroglycerin-soaked clay is much more stable and less sensitive to shock than pure nitroglycerin. Nobel called this safer explosive *dynamite*, and he made enough money from manufacturing it to found the Nobel prizes in peace, literature, physics, chemistry, and physiology and medicine.

APPROXIMATE REACTION ENTHALPIES FROM BOND ENERGIES

If the enthalpy change for a reaction is not known and the standard enthalpies of formation from which it could be calculated are either not known or are not readily available, bond energies can be used to obtain an approximate value for the enthalpy change. We may need just to know if a reaction is exothermic or endothermic and so a quick estimate of the enthalpy change is all we require. The basic idea is simple. The heat absorbed in a reaction, $q = \Delta H$, is equal to the energy needed to break all the bonds in the reactants minus the energy evolved in the formation of the bonds in the products. If more energy is needed to break the bonds in the reactants than is obtained by forming the bonds in the products, then the reaction is endothermic. If less energy is needed to break all the bonds in the reactants than is evolved in the formation of the bonds in the products, then the reaction is exothermic (see Figure 6.15). In summary

When we use bond energies, we find $\Delta H°$ by subtracting the values for the products from those for the reactants. When we use enthalpies of formation, however, we do the reverse—we subtract the values for the reactants from those for the products.

$$\Delta H° = \sum BE(\text{bonds broken}) - \sum BE(\text{bonds formed})$$

or

$$\Delta H° = \sum [\text{bond energies(reactants)}] - \sum [\text{bond energies(products)}]$$

Example 6.9

USING BOND ENERGIES TO ESTIMATE REACTION ENTHALPIES

Use the average bond energies in Table 6.3 to calculate an approximate $\Delta H°$ for the oxidation of HI(g) by $Cl_2(g)$ (Experiment 5.6):

$$Cl_2(g) + 2HI(g) \longrightarrow I_2(g) + 2HCl(g)$$

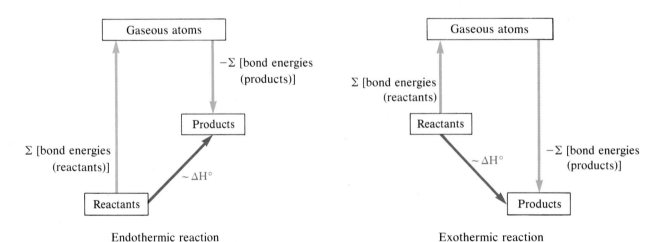

FIGURE 6.15
Approximate Reaction Enthalpies from Bond Energies.

Any reaction can, in principle, be carried out by first decomposing all the gaseous reactants into free atoms. This process is endothermic, and the heat needed is the sum of the bond energies of the reactants. Then the free atoms may be combined to give the products. This process is exothermic, and the heat evolved is the sum of the bond energies of the products. Hence $\Delta H = \sum [\text{bond energies(reactants)}] - \sum [\text{bond energies(products)}]$. Since the bond energies are approximate, this method gives only an approximate value for the enthalpy change of a reaction in the gas phase.

Solution

First, we dissociate the Cl_2 and HI molecules into free atoms, and then we combine the atoms to form the I_2 and HCl molecules:

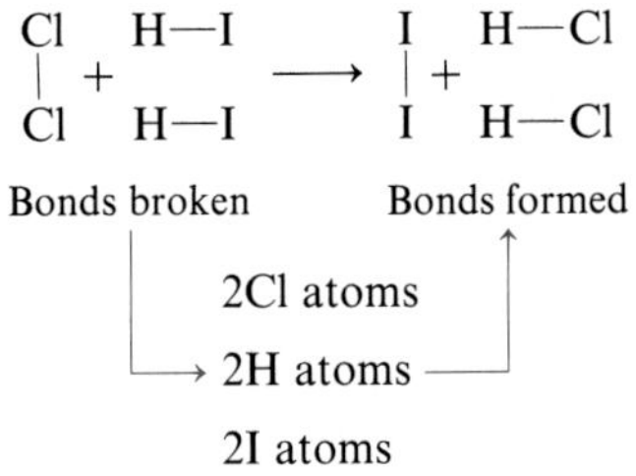

$$
\begin{array}{ccc}
\begin{matrix} Cl \\ | \\ Cl \end{matrix} + \begin{matrix} H—I \\ \\ H—I \end{matrix} & \longrightarrow & \begin{matrix} I \\ | \\ I \end{matrix} + \begin{matrix} H—Cl \\ \\ H—Cl \end{matrix}
\end{array}
$$

Bonds broken   Bonds formed

2Cl atoms

2H atoms

2I atoms

Bonds broken: 1 mol Cl—Cl, 2 mol H—I

To break 1 mol of Cl_2 bonds requires 239 kJ. To break 2 mol of HI bonds requires $2 \text{ mol} \times 297 \text{ kJ mol}^{-1} = 594 \text{ kJ}$. The total energy needed to break all the bonds in the reactants is therefore

$$\sum BE(\text{bonds broken}) = 239 \text{ kJ} + 594 \text{ kJ} = 833 \text{ kJ}$$

Bonds formed: 1 mol I—I, 2 mol H—Cl

The energy evolved in the formation of 1 mol of I—I bonds is 149 kJ. The energy evolved in the formation of 2 mol of H—Cl bonds is $2 \text{ mol} \times 431 \text{ kJ mol}^{-1} = 862 \text{ kJ}$. The total energy evolved in the formation of all the bonds in the products is

$$\sum BE(\text{bonds formed}) = 149 \text{ kJ} + 862 \text{ kJ} = 1011 \text{ kJ}$$

Thus for this reaction, 833 kJ is needed to break all the bonds in the reactants, but 1011 kJ is evolved in the formation of the new bonds in the products. Hence the overall reaction is exothermic:

$$\Delta H° = \sum BE(\text{bonds broken}) - \sum BE(\text{bonds formed})$$
$$= 833 \text{ kJ} - 1011 \text{ kJ}$$
$$= -178 \text{ kJ}$$

EXERCISE 6.12

Use bond energies to estimate a value for the standard enthalpy change, $\Delta H°$, for the addition of hydrogen to ethene to give ethane:

$$C_2H_4(g) + H_2(g) \longrightarrow C_2H_6(g)$$

6.5
Energy Sources

FOSSIL FUELS

Approximately 90% of the energy used in the industrialized world today comes from the combustion of the fossil fuels, coal, petroleum, and natural gas. These fossil fuels were formed when animal and vegetable matter decayed and became covered with other deposits and subjected to enormous pressure and high temperatures for millions of years. Thus the energy that we can obtain today by burning wood and fossil fuels originally came from the sun. By the process of photosynthesis, plants make carbohydrates and other organic molecules in which energy is stored until it is released on combustion.

Natural gas consists chiefly of methane and small amounts of ethane, propane, and butane. The propane and butane can be separated by compressing and cooling the gas until the propane and the butanes are liquefied. The liquefied propane and butane is sold as bottled gas.

Petroleum is a liquid mixture of many different alkanes and other hydrocarbons. It is normally separated by *fractional distillation* (see Figure 6.16), not into pure hydrocarbons but into mixtures of hydrocarbons, called fractions, each of which boils over a certain limited temperature range. The particular fractions that are collected depend partly on the source of the petroleum and partly on the proposed uses of the fractions. Typical fractions are shown in Table 6.4. So that the demand for gasoline can be met, much of the kerosene and higher-boiling-point fractions are decomposed by heating to a high temperature to form the shorter-chain alkanes of gasoline.

Petroleum Distallation Towers

FIGURE 6.16
Oil Refinery Distillation
Column.

Petroleum is heated with super-heated steam at the bottom of a tall distillation column. Most of the petroleum is vaporized; the higher-boiling-point components condense at a low point in the column, and the lower-boiling-point components move toward the top of the column. Fractions of different compositions are taken from the column at different heights.

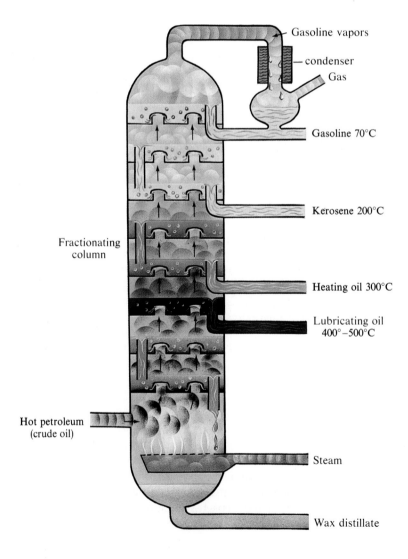

TABLE 6.4
Typical Fractions Obtained in the Distillation of Petroleum

Fraction	Boiling Point (°C)	Composition	Uses
Gas	Up to 20	Alkanes from CH_4 to C_4H_{10}	Synthesis of other carbon compounds; fuel
Petroleum ether	20–70	C_5H_{12}, C_6H_{14}	Solvent; gasoline additive for cold weather
Gasoline	70–180	Alkanes from C_6H_{14} to $C_{10}H_{22}$	Fuel for gasoline engines
Kerosene	180–230	$C_{11}H_{24}$, $C_{12}H_{26}$	Jet engine fuel
Light gas oil	230–305	$C_{13}H_{28}$ to $C_{17}H_{36}$	Fuel for furnaces and for diesel engines
Heavy gas oil and light lubricating distillate	305–405	$C_{18}H_{38}$ to $C_{25}H_{52}$	Fuel for generating stations; lubricating oil
Lubricants	405–515	Higher alkanes	Thick oils, greases, and waxy solids; lubricating grease; petroleum jelly
Solid residue			Pitch or asphalt for roofing and road material

Coal is a complex mixture of many compounds that contain a high percentage of carbon and hydrogen but also many other elements. The composition of coal varies considerably depending on its age and location. A typical bituminous coal has the approximate composition 80% C, 6% H, 8% O, 5% S, and 1% N. When this type of coal is burned a considerable amount of sulfur dioxide, SO_2, is formed, contributing to atmospheric pollution and acid rain (see Chapter 8).

Before the Industrial Revolution in Europe wood and peat were the only significant fuels, and they were used primarily for heating and cooking. During the latter half of the eighteenth century coal grew rapidly in importance because of the increasing use of energy for power. Power was generated by burning coal in a steam engine, which could be used for driving machinery, for generating electricity, and for transportation. Today petroleum and natural gas provide 70% of our energy requirements and coal only about 20%. The invention of the internal combustion engine led to the replacement of coal-burning steam-engines in trains and ships with the diesel engine, and to the development of the private automobile. These developments have in turn led to an increasing demand for petroleum. Hydrocarbons are also valuable as the raw materials for the manufacture of plastics and many other materials that have come to be a part of modern life. Despite the discovery of new deposits of petroleum, the world's total petroleum resources are limited and will last for only a relatively short time. Estimates of this time vary from 30 years to several hundred years. One of the challenges facing humanity is to learn how to utilize, efficiently and safely, alternative energy sources, such as coal, nuclear power, solar energy, and synthetic fuels, and to find alternative raw materials for all the substances now manufactured from petroleum.

ALTERNATIVE ENERGY SOURCES

COAL CONVERSION One alternative to the use of petroleum and natural gas would be to return to the use of coal, of which the United States and other countries have enormous reserves. Solid coal is difficult and expensive to transport and is of no use as a fuel for the internal combustion engine. So, much of it would have to be converted to a liquid fuel. If coal is heated at a high temperature with oxygen and steam it can be converted to a mixture of gases that contains principally methane, carbon monoxide, and hydrogen, all of which are useful as fuels. Carbon monoxide and hydrogen can also be converted to methanol,

$$CO(g) + 2H_2(g) \xrightarrow[\text{Catalyst}]{\text{Heat}} CH_3OH(g)$$

which is another useful fuel, as well as being the starting material for the production of synthetic fibres and plastics (see Chapter 24). However, coal suffers from the same disadvantage as petroleum in that its use as a fuel contributes to the increase in the concentration of CO_2 in the atmosphere and hence to the greenhouse effect (see Box 6.1).

HYDROGEN We have already mentioned the possibility of using hydrogen as a fuel in Box 3.2. Because of its very low mass and a high enthalpy of combustion it is ideal as a rocket fuel. But because hydrogen is not present in the atmosphere it must be made from one of its compounds. The obvious choice is water of which the oceans provide an almost inexhaustible supply. However, the decomposition of water to hydrogen and oxygen

$$H_2O(l) \longrightarrow H_2(g) + \tfrac{1}{2}O_2(g) \qquad \Delta H° = 286 \text{ kJ}$$

requires 286 kJ per mole of water. No way has yet been found to carry out this reaction economically. Electrolysis is a possibility, if electricity can be produced cheaply enough, or sunlight might be used as the energy source to decompose water in a photochemical reaction (see Chapter 7), if a suitable catalyst to speed up the reaction can be found. Research is also in progress to attempt to modify the photosynthetic process so that plants will release hydrogen from water instead of using it to produce carbohydrates and other organic compounds. Challenging problems await the research chemist in this area.

BIOMASS The term biomass refers to all animal and plant materials—both dead and alive. Wood, leaves, animal and human excreta, and waste food are all forms of biomass. Biomass is an important form of stored energy. Wood of course has long been used as an energy source for cooking and heating. Indeed, the destruction of forests over the past 1000 years or so has greatly changed the face of the earth and has led to the disappearance of many animal and plant species. Other forms of biomass are now being increasingly used as a source of two important fuels—biogas and ethanol.

If animal wastes and plants are allowed to rot in the absence of air, certain species of bacteria break down the waste to form **biogas** which, like natural gas, is predominately methane. The methane can be used directly for cooking and heating or used to generate electricity. The residue which has a high nitrogen content can be used as a fertilizer. Biogas production is particularly important in rural areas and in Third World countries. China has about 4 million biogas production plants.

Indian villager stirring dung in the intake tank of a biogas digester

The traditional method for the production of **ethanol**, C_2H_5OH, is fermentation. This is a reaction in which sugars and carbohydrates are converted to ethanol by the action of yeast:

$$C_6H_{12}O_6(aq) \longrightarrow 2C_2H_5OH(aq) + 2CO_2(g)$$

The sugar and carbohydrates can come from many different plants and fruits. A car engine needs little modification to enable it to burn ethanol, or no modification at all for an ethanol–gasoline mixture known as gasohol. Indeed, gasohol is already being sold in some parts of the United States, and ethanol is extensively used in Brazil.

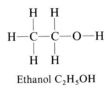

Ethanol C_2H_5OH

FOOD

Oxidation reactions are the major source not only of the energy for heating and power production but also of the energy that sustains all living creatures. We meet our energy needs by using the chemical energy that is released by the oxidation of carbohydrates, proteins, and fats in food. These oxidation reactions are typified by the oxidation of glucose:

$$C_6H_{12}O_6(s) + 6O_2(g) \longrightarrow 6CO_2(g) + 6H_2O(l) \qquad \Delta H^\circ = -2803.0 \text{ kJ}$$

In the body this reaction occurs in a relatively slow and controlled manner, by a series of rather complicated steps. In contrast, the combustion of glucose outside the body occurs only at a high temperature, with the rapid evolution of a considerable amount of heat. The total energy made available in the complex series of reactions that occur in the body is exactly the same as would be obtained if the glucose were simply burned in air.

Therefore the energy that can be obtained from any food can be measured by burning it in oxygen in a calorimeter. It must first be dried, because most foods contain considerable quantities of water. Since foods are usually not pure substances, we cannot express the results as enthalpies of combustion per mole. They are usually expressed as the enthalpy change per gram. Some typical average values for foods are given in Table 6.5. You are probably aware that the energy content of food is commonly measured in calories. The calorie is a

Sign advertising gasohol in New Mexico

TABLE 6.5
Enthalpies of Combustion of Various Foods

Food	$-\Delta H^\circ$ (kJ g^{-1})	$-\Delta H^\circ$ (kcal oz^{-1})
Onions	1.0	6.7
Beer	1.5	10
Apples	1.9	13
Milk	2.8	18
Potatoes	2.9	19
Beef	6.3	42
Bread	11	73
Sugar	17	110
Cheese	18	120
Cake	20	130
Butter	33	221

1 ounce (oz) = 28 g; 1 kcal = 4.18 kJ.

One calorie is the amount of heat required to raise the temperture of one gram of water by 1 °C.

unit of energy that was used in science until recently when it was replaced by the joule. However, many people including nutritionists and physicians continue to use the calorie (cal) and kilocalorie (kcal), so it is important to be able to convert from one unit to the other. The relationship between the two units is

$$4.184 \text{ J} = 1 \text{ cal}$$

The energy content of food is usually expressed in kilocalories but unfortunately this is sometimes called a Calorie which, although it is written with a capital C, can be a cause of confusion.

$$4.184 \text{ kJ} = 1 \text{ kcal} = 1 \text{ Calorie}$$

IMPORTANT TERMS

An **alkane** is a hydrocarbon in which all the C—C bonds are single bonds. Alkanes have the general formula C_nH_{2n+2}.

An **alkene** is a hydrocarbon that has one or more carbon–carbon double bonds.

An **alkyne** is a hydrocarbon that has one or more carbon–carbon triple bonds.

The **average bond energy** of a bond in a polyatomic molecule is the average energy needed to break 1 mol of bonds of a particular type (for example a C—H bond).

The **bond energy** of a diatomic molecule is the energy needed to dissociate 1 mol of molecules in the gas state into atoms in the gas state.

A **calorimeter** is a device used to measure the quantity of heat absorbed or evolved during a chemical reaction or a physical change.

Calorimetry is the measurement of the heat absorbed or evolved in chemical reactions or other changes.

Endothermic reactions absorb heat from the surroundings.

Enthalpy is a state function defined by the expression $H = E + PV$.

The **enthalpy change** ΔH is the difference in the enthalpy of the final and initial states of a system. It is equal to the heat absorbed by the system at constant pressure, q_P.

The **standard enthalpy change** $\Delta H°$ for a reaction is the enthalpy change for the reaction when all the reactants and the products are in their standard states.

The **standard enthalpy of formation** of a compound is the enthalpy change for the reaction in which 1 mol of the compound is formed from its elements in their standard states.

Exothermic reactions give off heat to the surroundings.

The **first law of thermodynamics** states that energy cannot be created or destroyed, or the energy of an isolated system

is constant. In mathematical form, $\Delta E = q + w$, where ΔE is the change in internal energy, q is the amount of heat added to the system, and w is the amount of work done on the system. It is an alternative name for the **law of conservation of energy**.

Heat is energy that is transferred as a result of a temperature difference.

The **heat capacity** of an object or of a given amount of a substance is the amount of heat required to raise the temperature of the object or of the substance by 1 K. **Molar heat capacity** is the amount of heat needed to raise the temperature of 1 mol of a substance by 1 K. The **specific heat capacity (specific heat)** is the amount of heat needed to raise the temperature of 1 g of a substance by 1 K.

Hess's law states that the enthalpy change for a reaction is the sum of the enthalpy changes for the individual steps of the reaction, or the enthalpy change for a reaction is independent of the path by which the reaction occurs.

The **internal energy** E is the sum of the kinetic and potential energies of all the particles in a system.

Hydrocarbons are compounds containing only carbon and hydrogen.

Inorganic chemistry is the chemistry of all the elements except carbon.

Organic chemistry is the chemistry of carbon compounds.

A **state function** is any property of a system that depends only on the state of the system and not on how that state was reached.

The **surroundings** represent that portion of the universe with which a system interacts.

A **system** is a portion of the universe that is chosen for study.

Thermochemistry is the quantitative study of heat changes in chemical reactions.

PROBLEMS *

Carbon and its Compounds

1. Write balanced equations for the reaction of carbon monoxide with each of the following, and state the conditions under which each reaction occurs:

(a) H_2 (b) O_2

(c) H_2O (d) Fe_2O_3

2. Write balanced equations for each of the following:

(a) Two ways of preparing carbon monoxide

(b) Two ways of preparing carbon dioxide

(c) One way of preparing carbon disulfide

3. Write Lewis structures for each of the following molecules and ions, and predict their shapes:

(a) CO_2 (b) CO (c) CN^-

(d) C_2^{2-} (e) HCN

Name each of the above species.

4. What is the chemical composition of each of the following: lime, soda water, natural gas, coke, carbon black, and chalk.

5. Write a balanced equation to describe the reaction that occurs on heating each of the following to a sufficiently high temperature: (a) $BaCO_3(s)$, (b) $H_2CO_3(aq)$, (c) $HI(g)$, (d) $C_2H_6(g)$.

6. Write a balanced equation to describe the reaction that occurs when carbon is heated with each of the following to a sufficiently high temperature: (a) $CuO(s)$, (b) $CaO(s)$, (c) $S(s)$, (d) $O_2(g)$

7. Write a balanced equation to describe the reaction that occurs when methane is strongly heated with each of the following: (a) $H_2O(g)$, (b) $NH_3(g)$, (c) $O_2(g)$

8. What is the chemical formula for each of the following: (a) limestone, (b) hydrocyanic acid, (c) acetylene, (d) calcium carbide, (e) carborundum?

9. What is the angle between the carbon–carbon bonds in diamond and in graphite? Explain why graphite is quite soft whereas diamond is very hard.

*10. Two important Lewis structures can be written for the thiocyanate ion, SCN^-. Draw both Lewis structures. On the basis that nitrogen forms multiple bonds more readily than does sulfur, which of the structures more closely represents the structure of the thiocyanate ion?

* Answers to problems numbered in blue appear at the end of the text.

*11. When concentrated sulfuric acid is dripped onto carbon tetrabromide at 160 °C, a compound containing 6.4% C, 85.0% Br, and 8.6% O, by mass, is obtained. At 25 °C and a pressure of 1.00 atm, 0.940 g of this compound has a volume of 112 mL. What is the molecular formula of the compound? Draw its Lewis structure and deduce its molecular shape.

*12. When aluminum oxide is heated with coke in an electric furnace, a yellow carbide of aluminum is obtained which is stable up to 1400 °C and reacts with water to give methane. A sample of the carbide of mass 0.500 g was reacted with excess water and the methane collected in a 250-mL bulb at 25 °C. The pressure in the bulb was measured as 1.02 atm. What is the empirical formula of the carbide? What mass of aluminum carbide is needed to produce 20.0 L of methane at 25 °C and 1.00 atm pressure?

*13. When malonic acid, $CH_2(CO_2H)_2$, is heated with a large excess of the dehydrating agent $P_4O_{10}(s)$, an oxide of carbon containing 53.0 mass % carbon is obtained, which boils at 6 °C and freezes to a white solid at −111 °C. The volume of 0.200 g of the oxide was found to be 74.3 mL at 27 °C and a pressure of 740 mm Hg.

(a) What is the molecular formula of the oxide?

(b) Write the balanced equation for its formation from malonic acid.

(c) Suggest a Lewis structure for this oxide.

*14. When mercury(II) cyanide is heated a gaseous compound X, containing only carbon and nitrogen, is obtained. Analysis shows that it contains 46.2% C by mass. At 100 °C and 0.950 atm pressure, 0.208 g of X has a volume of 126 mL.

(a) What are the empirical formula and the molecular formula of X?

(b) Draw a possible Lewis structure for X and deduce its molecular shape.

*15. A colorless, odorless gas Y is obtained when carbon monoxide and sulfur vapor are passed through a heated tube. Y burns in excess oxygen to give $CO_2(g)$ and $SO_2(g)$. At 25 °C and 1.00 atm pressure, 0.246 g of Y has a volume of 100 mL. When this sample of Y was mixed with 200 mL of oxygen and burned completely, the final volume of gases was 250 mL at the same temperature and pressure. When these gases were bubbled through a large volume of water, only 50 mL of the mixture remained undissolved at 25 °C and 1 atm pressure. A heated platinum wire decomposed another sample of the gas into sulfur and carbon monoxide without changing the volume at 25 °C and 1 atm. Another sample of Y was dissolved in water to give a 0.100M solu-

* The asterisk denotes the more difficult problems.

tion. 25.00 mL of this acidic solution reacted completely with 50.00 mL of 0.200M KOH(aq) to give a solution containing the ions K^+(aq), $CO_3{}^{2-}$(aq), and S^{2-}(aq). What is the molecular formula of Y? Write balanced equations for each of the above reactions.

16. Write balanced equations for the preparation of ethyne, starting with calcium oxide, coke, and water.

17. Give the names and Lewis structures of the alkanes with 1, 2, and 3 carbon atoms, respectively.

18. Why is the tetrafluoroethene molecule, C_2F_4, planar, and the difluoroethyne molecule, C_2F_2, linear?

First Law of Thermodynamics

19. Calculate the work done when 1 mol of liquid water is vaporized at 373 K and a pressure of 1 atm. (The volume of liquid water may be neglected in comparison with that of H_2O(g).)

20. Calculate the work done, in joules, when 1 g of zinc metal dissolves in HCl(aq) at 25 °C, when the reaction is carried out in a vessel with a tight-fitting piston acted upon by an external pressure of 1 atm.

Calorimetry

21. When 0.150 g of liquid octane, C_8H_{18}(l), was burned in a flame calorimeter containing 1.500 kg of water, the temperature of the water rose from 25.246 to 26.386 °C. What is the standard enthalpy of combustion of octane at 25 °C?

22. A volume of 50.0 mL of 0.400M NaOH(aq) was added to 20.0 mL of 0.500M H_2SO_4(aq) in a calorimeter of heat capacity 39.0 J K^{-1}. The temperature of the resulting solution rose by 3.60 °C. What is the standard enthalpy of neutralization of H_2SO_4(aq) with NaOH(aq)?

23. A sample of 25.0 mL of HCl(aq) was mixed with 25.00 mL of KOH(aq) of the same concentration in a calorimeter. As a result, the temperature rose from 25.00 to 26.60 °C. Given that $\Delta H° = -56.02$ kJ for the reaction

$$H_3O^+(aq) + OH^-(aq) \longrightarrow 2H_2O(l)$$

and that the heat capacity of dilute aqueous solutions is 75.4 J K^{-1} mol^{-1}, determine the concentration of the HCl(aq). (Assume the heat capacity of the calorimeter to be negligible.)

24. Calcium oxide (lime) reacts with water in an exothermic reaction to give calcium hydroxide:

$$CaO(s) + H_2O(l) \longrightarrow Ca(OH)_2(s)$$

A sample of 5.40 g of CaO(s) was added to 500 mL of water in a calorimeter of heat capacity 350 J K^{-1}. The observed

temperature increase was 2.60 K. What is the standard enthalpy change for the reaction of 1 mol of CaO(s) to give 1 mol of $Ca(OH)_2$(s)?

25. When 2.500 g of sulfur were burned completely in oxygen to SO_2(g), the temperature of 1.00 kg of water in a bomb calorimeter of negligible heat capacity was raised from 22.00 to 27.40 °C.

   (a) Calculate the heat evolved, q, per mole of SO_2(g) formed.

   (b) Is this heat evolved, q, equal to the ΔH for the reaction? Explain.

26. When 1.000 g of naphthalene, $C_{10}H_8$(s), was burned in excess oxygen in a bomb calorimeter of negligible heat capacity at 298 K, 40.10 kJ of heat were evolved. The combustion products were CO_2(g) and H_2O(l). Calculate $\Delta E°$ for the combustion of 1 mol of naphthalene.

27. A small well-insulated catalytic hydrogenation apparatus has a heat capacity of 1.500 kJ K^{-1}. When 1.500 g of ethylene are hydrogenated completely to ethane in the apparatus, what temperature rise should be observed? (Assume that the heat capacities of the gases are negligible compared with that of the apparatus.)

Hess's Law

28. Calculate the value of $\Delta H°$ for the reaction

$$CuCl_2(s) + Cu(s) \longrightarrow 2CuCl(s)$$

given the information

$Cu(s) + Cl_2(g) \longrightarrow CuCl_2(s)$	$\Delta H° = -206$ kJ
$2Cu(s) + Cl_2(g) \longrightarrow 2CuCl(s)$	$\Delta H° = -36$ kJ

29. Calculate $\Delta H°$ for the reaction

$$2F_2(g) + 2H_2O(l) \longrightarrow 4HF(g) + O_2(g)$$

given that

$H_2(g) + F_2(g) \longrightarrow 2HF(g)$	$\Delta H° = -542$ kJ
$2H_2(g) + O_2(g) \longrightarrow 2H_2O(l)$	$\Delta H° = -572$ kJ

30. Calculate $\Delta H°$ for the reaction

$$2CO(g) + O_2(g) \longrightarrow 2CO_2(g)$$

given that

$C(graphite) + O_2(g) \longrightarrow CO_2(g)$	$\Delta H° = -393.5$ kJ
$2C(graphite) + O_2(g) \longrightarrow 2CO(g)$	$\Delta H° = -221.0$ kJ

31. Calculate $\Delta H°$ for the reduction of FeO(s) to Fe(s) by CO(g), given that

$$Fe_2O_3(s) + CO(g) \longrightarrow 2FeO(s) + CO_2(g) \quad \Delta H° = 38 \text{ kJ}$$

$$Fe_2O_3(s) + 3CO(g) \longrightarrow 2Fe(s) + 3CO_2(g) \quad \Delta H° = -28 \text{ kJ}$$

32. Calculate the standard enthalpy change for the reaction of nitrogen dioxide with water,

$$3NO_2(g) + H_2O(l) \longrightarrow 2HNO_3(aq) + NO(g)$$

given that

$$2NO(g) + O_2(g) \rightarrow 2NO_2(g) \quad \Delta H° = -173 \text{ kJ}$$

$$2N_2(g) + 5O_2(g) + 2H_2O(l) \rightarrow 4HNO_3(aq) \quad \Delta H° = -255 \text{ kJ}$$

$$N_2(g) + O_2(g) \rightarrow 2NO(g) \quad \Delta H° = 181 \text{ kJ}$$

33. Calculate the standard enthalpy change for the reaction

$$2C(s) + H_2(g) \longrightarrow C_2H_2(g)$$

given that

$$2C_2H_2(g) + 5O_2(g) \rightarrow 4CO_2(g) + 2H_2O(l) \quad \Delta H° = -2600 \text{ kJ}$$

$$C(s) + O_2(g) \rightarrow CO_2(g) \quad \Delta H° = -390 \text{ kJ}$$

$$2H_2(g) + O_2(g) \rightarrow 2H_2O(l) \quad \Delta H° = -572 \text{ kJ}$$

34. Calculate the standard enthalpy change for the reaction

$$2H_2O_2(l) \longrightarrow 2H_2O(l) + O_2(g)$$

given that

$$2H_2(g) + O_2(g) \longrightarrow 2H_2O(g) \quad \Delta H° = -483.6 \text{ kJ}$$

$$H_2O(l) \longrightarrow H_2O(g) \quad \Delta H° = 44.0 \text{ kJ}$$

$$H_2(g) + O_2(g) \longrightarrow H_2O_2(l) \quad \Delta H° = -187.6 \text{ kJ}$$

35. It has been proposed that the following reaction might occur in the stratosphere:

$$HO(g) + Cl_2(g) \longrightarrow HOCl(g) + Cl(g)$$

Calculate the standard enthalpy change for this reaction from the following data:

$$Cl_2(g) \longrightarrow 2Cl(g) \quad \Delta H° = 242 \text{ kJ}$$

$$H_2O_2(g) \longrightarrow 2HO(g) \quad \Delta H° = 134 \text{ kJ}$$

$$H_2O_2(g) + 2Cl(g) \longrightarrow 2HOCl(g) \quad \Delta H° = -209 \text{ kJ}$$

36. When $PCl_3(g)$ is formed from *white* phosphorus and chlorine gas, $\Delta H°$ for the reaction is -306 kJ mol^{-1}, while for the formation of $PCl_3(g)$ from *red* phosphorus and chlorine gas, $\Delta H°$ is -288 kJ mol^{-1}. From this information calculate $\Delta H°$ for the conversion of red phosphorus to white phosphorus.

Enthalpies of Formation and Combustion

37. Calculate the standard enthalpy of formation of gaseous dinitrogen tetraoxide, $N_2O_4(g)$, from the following data:

$$N_2(g) + O_2(g) \longrightarrow 2NO(g) \quad \Delta H° = 180.6 \text{ kJ}$$

$$2NO(g) + O_2(g) \longrightarrow 2NO_2(g) \quad \Delta H° = -114.2 \text{ kJ}$$

$$2NO_2(g) \longrightarrow N_2O_4(g) \quad \Delta H° = -58.0 \text{ kJ}$$

38. What are the enthalpies of formation of each of $H_2O(l)$, $H_2O(g)$, and $NH_3(g)$, given that

$$H_2(g) + \tfrac{1}{2}O_2(g) \longrightarrow H_2O(l) \quad \Delta H° = -285.8 \text{ kJ}$$

$$H_2O(g) \longrightarrow H_2O(l) \quad \Delta H° = -44.0 \text{ kJ}$$

$$2NH_3(g) \longrightarrow N_2(g) + 3H_2(g) \quad \Delta H° = 92.4 \text{ kJ}$$

39. For $H_2(g)$, $N_2(g)$, $O_2(g)$, $F_2(g)$, and $Cl_2(g)$, but not for $Br_2(g)$ or $I_2(g)$, the standard enthalpies of formation are 0 kJ mol^{-1}. Explain why. In what physical states are the standard enthalpies of formation of bromine and iodine also 0 kJ mol^{-1}?

40. The standard enthalpy of combustion of liquid *n*-heptane, $C_7H_{16}(l)$, is -4816.9 kJ. The products of this combustion are liquid water and carbon dioxide gas. Calculate the standard enthalpy of formation of liquid *n*-heptane.

41. What is the enthalpy change when 1 mol of $SO_3(g)$ reacts with 1 mol of $H_2O(g)$ to give 1 mol of $H_2SO_4(l)$, with all the compounds in their standard states?

42. Calculate $\Delta H_f°$ for ethyne, $C_2H_2(g)$, from the standard enthalpy change of -312 kJ for the reaction

$$C_2H_2(g) + 2H_2(g) \longrightarrow C_2H_6(g)$$

and the standard enthalpy of formation of ethane, $C_2H_6(g)$, given in Table 6.2.

43. Calculate the standard enthalpy of formation of solid magnesium hydroxide from the following data:

$$2Mg(s) + O_2(g) \longrightarrow 2MgO(s) \quad \Delta H° = -1203.7 \text{ kJ}$$

$$MgO(s) + H_2O(l) \longrightarrow Mg(OH)_2(s) \quad \Delta H° = -36.7 \text{ kJ}$$

$$2H_2O(l) \longrightarrow 2H_2(g) + O_2(g) \quad \Delta H° = 571.6 \text{ kJ}$$

44. Many cigarette lighters contain liquid butane, $C_4H_{10}(l)$, for which $\Delta H_f° = -127 \text{ kJ mol}^{-1}$. Calculate the heat evolved by 1.00 g of liquid butane in the lighter when it is burned, assuming that the products of combustion are $CO_2(g)$ and $H_2O(g)$ under standard conditions.

45. Given that the standard enthalpy of combustion of ethanol, $C_2H_5OH(l)$, is $-1370 \text{ kJ mol}^{-1}$, and using data given in Table 6.2, calculate the standard enthalpy of formation of ethanol.

46. Calculate $\Delta H_f°$ for propane, $C_3H_8(g)$, from its standard enthalpy of combustion ($-2044 \text{ kJ mol}^{-1}$) and data given in Table 6.2.

47. Calculate ΔH° for the conversion of 3 mol of gaseous acetylene, $C_2H_2(g)$, to 1 mol of gaseous benzene, $C_6H_6(g)$, given that it takes 434.5 J to convert 1.00 g of liquid benzene to gaseous benzene. $\Delta H_f^\circ(C_6H_6) = 49$ kJ mol^{-1}

48. When 2 mol of gaseous hydrogen iodide are formed from 1 mol of gaseous hydrogen and 1 mol of solid iodine, 51.9 kJ of heat are absorbed. What is the standard enthalpy of formation of HI(g)?

49. Calculate the standard enthalpy change when 1 mol of n-octane, $C_8H_{18}(l)$, reacts with oxygen in a constant-pressure container to give $CO_2(g)$ and $H_2O(l)$. $\Delta H_f^\circ(C_8H_{18}) = -224$ kJ mol^{-1}

50. When 4.766 mL of liquid acetic acid, $CH_3CO_2H(l)$ (density 1.049 g mL^{-1}), are burned in excess oxygen to $CO_2(g)$ and $H_2O(l)$, the standard enthalpy change is -72.62 kJ. Calculate the standard enthalpy change for the oxidation of 1 mol of ethane to acetic acid:

$$C_2H_6(g) + \tfrac{3}{2}O_2(g) \longrightarrow CH_3CO_2H(l) + H_2O(l)$$

51. The standard enthalpies of combustion of graphite and diamond to $CO_2(g)$ are -393.7 kJ mol^{-1} and -395.6 kJ mol^{-1}, respectively. What is the enthalpy of formation of C(diamond) from C(graphite)? Which is the more stable, diamond or graphite?

Bond Energies

52. The standard enthalpies of formation of atomic oxygen gas and atomic nitrogen gas are 249 and 473 kJ mol^{-1}, respectively. What is the dissociation energy of **(a)** the double bond in $O_2(g)$ and **(b)** the triple bond in $N_2(g)$?

53. The standard enthalpy of formation of ClF(g) is -55.7 kJ mol^{-1}, and the dissociation energies of $F_2(g)$ and $Cl_2(g)$ are 155 kJ mol^{-1} and 239 kJ mol^{-1}, respectively. What is the bond dissociation energy of ClF(g)?

54. Calculate the average P—Cl bond energy in PCl_3, given the following thermochemical data. What is the average P—Cl bond energy in PCl_5? What is the average bond energy of the two additional P—Cl bonds in PCl_5? The ΔH_f° values (kJ mol^{-1}) are as follows:

   P(g), 316.2   Cl(g), 119.5

   $PCl_3(g)$, -287.0 $PCl_5(g)$, -374.7

55. (a) Calculate the average O—H bond energy in water.

   (b) Using the above result, calculate the O—O bond energy in hydrogen peroxide.

   (c) Is the O—O bond in H_2O_2 a strong or a weak bond compared, for example, to the C—C bond?

The ΔH_f° values (kJ mol^{-1}) are as follows:

   H(g), 218.0   O(g), 249.1

   $H_2O_2(g)$, -136.4 $H_2O(g)$, -241.8

56. Calculate the CO bond energy in CO and in CO_2 using the following ΔH_f° values (kJ mol^{-1}):

   CO(g), -110.5 $CO_2(g)$, -393.5

   O(g), 249.1   C(g), 716.7

57. Calculate the CO bond energy in methanal (formaldehyde), H_2CO, from the average C—H bond energy given in Table 6.3 and the ΔH_f° values given below. Compare this bond energy with that calculated for the bonds in CO_2 in Problem 56. The ΔH_f° values (kJ mol^{-1}) are as follows:

   $H_2CO(g)$, -115.9 C(g), 716.7

   H(g), 218.0   O(g), 249.1

58. Using the average C—C and C—H bond energies given in Table 6.3, estimate the standard enthalpy of formation of ethane, $C_2H_6(g)$.

59. Use the average bond energies of Table 6.3 to estimate ΔH° for each of the following gas-phase reactions:

   (a) $H_2S + Cl_2 \rightarrow SCl_2 + H_2$

   (b) $CH_4 + 2F_2 \rightarrow CH_2F_2 + 2HF$

   (c) $CH_2Cl_2 + CH_4 \rightarrow 2CH_3Cl$

60. Use the average bond energies of Table 6.3 to estimate ΔH° for each of the following gas-phase reactions:

   (a) $C_2H_2 + C_2H_6 \rightarrow 2C_2H_4$

   (b) $2H_2O_2 \rightarrow 2H_2O + O_2$

   (c) $CO + H_2 \rightarrow H_2CO$

61. Estimate the standard enthalpy of formation, ΔH_f°, of gaseous methylpropene, $(CH_3)_2C{=}CH_2(g)$, taking any required bond energy data from Table 6.3.

62. From the average bond energies of Table 6.3, and the S—S and P—P single bond energies of 255 kJ mol^{-1} and 172 kJ mol^{-1} respectively, estimate the standard enthalpies of formation of each of the following:

   (a) $H_2O(g)$ **(b)** $H_2S(g)$ **(c)** $NH_3(g)$

   (d) $PH_3(g)$ **(e)** $CH_4(g)$

63. Given that the standard enthalpy of dissociation of $H_2(g)$ is 436 kJ mol^{-1}, and that of $Cl_2(g)$ is 239 kJ mol^{-1}, what is the standard enthalpy of formation of HCl(g), if the standard enthalpy of dissociation of HCl(g) is 431 kJ mol^{-1}?

Miscellaneous

64. Define and explain each of the following terms:

(a) Standard enthalpy change, $\Delta H°$

(b) Standard enthalpy of formation, $\Delta H_f°$

(c) Standard enthalpy of neutralization

(d) The standard state of an element

(e) Average bond energy

*65. (a) Define standard enthalpy of formation, illustrating your answer by reference to $MgCO_3(s)$.

(b) When 0.203 g of magnesium was dissolved in an excess of dilute HCl(aq) in a vacuum flask, the temperature rose by 8.61 K. In another experiment, it was found that the vacuum flask and its contents required 506 J to raise the temperature by 1.02 K. Calculate the heat released in the first experiment, and hence find the enthalpy change per mole of magnesium for the reaction.

(c) In a similar experiment, using the same apparatus, $MgCO_3(s)$ was allowed to react with excess of dilute HCl(aq), and the enthalpy change was found to be -90.4 kJ per mole of $MgCO_3(s)$. Use the result from part (b) and the standard enthalpies of formation of $H_2O(l)$ and $CO_2(g)$ to find the standard enthalpy of formation of $MgCO_3(s)$.

66. Natural gas often contains unwanted $H_2S(g)$, which can be removed by reaction with $SO_2(g)$ according to the equation

$$2H_2S(g) + SO_2(g) \longrightarrow 3S(s) + 2H_2O(g)$$

(a) Identify the oxidizing agent and the reducing agent in this reaction.

(b) Calculate $\Delta H°$ at 25 °C for the reaction, assuming that the sulfur formed is orthorhombic sulfur consisting of S_8 molecules, which is the most stable form of sulfur at room temperature.

*67. The standard enthalpies of combustion of C(s, graphite), $H_2(g)$, $C_2H_6(g)$, and $C_3H_8(g)$ are -393.5, -285.8, -1559.8, and -2219.9 kJ mol^{-1}, respectively. Calculate the enthalpies of formation of $C_2H_6(g)$ and $C_3H_8(g)$. Use bond energies to predict approximate $\Delta H_f°$ and $\Delta H°$(combustion) values for butane, $C_4H_{10}(g)$.

68. Propane, $C_3H_8(g)$, is used widely as a domestic fuel and increasingly as a fuel for motor vehicles. What is its standard enthalpy of formation, given that the combustion of 1.00 g of propane releases 46.3 kJ of heat when it is burned to $CO_2(g)$ and $H_2O(g)$ at 298 K and 1 atm?

69. Given that the enthalpy of formation of glucose,

$C_6H_{12}O_6(s)$, is -1273 kJ mol^{-1}, and the overall photosynthesis reaction is

$$6CO_2(g) + 6H_2O(l) \longrightarrow C_6H_{12}O_6(s) + 6O_2(g)$$

calculate the $\Delta H°$ for this reaction per mole, and per gram, of glucose. (In nature, the large energy requirement for this reaction is provided for by light from the sun, rather than as heat.)

70. (a) Assuming that coke (graphite) and natural gas (methane) have identical costs per gram, which of these fuels is the more economical for heating a home?

(b) If the price per gram of the more economical fuel of part (a) is doubled, would that fuel still be the more economical?

71. The daily intake of calories for the average adult is about 2500 kcal, (1 cal = 4.184 J). Using the data in Table 6.5, calculate each of the following:

(a) How many pounds of cake that would have to be consumed to satisfy this daily requirement?

(b) Assuming a diet of only bread and butter, in equal parts by mass, how many pounds of bread and how many pounds of butter would have to be consumed to meet the daily requirement?

72. By comparing the cost per joule of gasoline (assumed to consist only of hexane, $C_6H_{14}(l)$) and electricity in your region, decide which form of energy is currently the best buy, using the fact that 1 watt (W) equals 1 J s^{-1} and calculating the number of joules in one kilowatt hour (kW h), the usual unit given on utility bills. Hexane has $\Delta H_f° = -199$ kJ mol^{-1} and a density of 1.38 g mL^{-1}. Assume that gasoline burns completely to $CO_2(g)$ and $H_2O(l)$. Take the price of gasoline as that of regular gasoline at a local service station.

73. Methanol, $CH_3OH(l)$, is a potential fuel supplement. Although it provides only one-half as much energy per liter as gasoline, it is clean burning and has a high octane number. It is made industrially from synthesis gas at high pressure using a catalyst at about 300 °C:

$$2H_2(g) + CO(g) \longrightarrow CH_3OH(l)$$

Use the following standard enthalpies of combustion to determine $\Delta H°$ for the synthesis of methanol by the above reaction.

$CH_3OH(l) + \frac{3}{2}O_2(g) \longrightarrow CO_2(g) + 2H_2O(l) \ \Delta H° = -726.6$ kJ

$C(graphite,s) + \frac{1}{2}O_2(g) \longrightarrow CO(g) \ \Delta H° = -110.5$ kJ

$C(graphite,s) + O_2(g) \longrightarrow CO_2(g) \ \Delta H° = -393.5$ kJ

$H_2(g) + \frac{1}{2}O_2(g) \longrightarrow H_2O(l) \ \Delta H° = -285.8$ kJ

Quantum Theory and the Electronic Structure of Atoms and Molecules

7

We have seen in Chapters 4, 5, and 6 that Lewis structures and the concepts of ionic and covalent bonds are very useful in enabling us to understand the compositions and properties of substances. We have described two important types of substances—ionic compounds and covalent molecular substances. Ionic compounds consist of oppositely charged ions held together by electrostatic attraction. Examples are sodium chloride, Na^+Cl^-, magnesium chloride, $Mg^{2+}(Cl^-)_2$, and magnesium oxide, $Mg^{2+}O^{2-}$. Covalent molecular substances consist of molecules in which the atoms are held together by covalent bonds. Examples are water, H_2O, chlorine, Cl_2, and methane, CH_4. Although we have described how oppositely charged ions are held together by electrostatic attraction, we have not yet answered the question "How does a pair of shared electrons hold two atoms together in a covalent bond?" To answer this question, and to generally extend our understanding of chemical bonds, we will first need to consider the properties of electrons and their arrangement in atoms in some detail.

Experiments carried out in the early years of this century showed that the laws of motion that apply to ordinary-sized objects do not apply to electrons. A new theory was developed to account for the behavior of electrons and other very small particles; it is known as **quantum mechanics**. Quantum mechanics had its origins in attempts to explain the nature of light and its interactions with matter, so we begin with a discussion of light.

7.1
Light and Electromagnetic Waves

For hundreds of years philosophers and scientists argued about the nature of light. Isaac Newton (1642–1727) believed that light consisted of a stream of particles. The Dutch physicist Christian Huygens (1629–1695) believed that light was a type of wave motion. The dispute over the nature of light was apparently resolved by the work of Scottish physicist James Clerk Maxwell (1831–1879). Maxwell showed that all the then-known properties of light could be accounted for by means of equations based on the hypothesis that visible light and other forms of radiation such as ultraviolet light and radio waves are all propagated through space as electromagnetic waves.

An **electromagnetic wave** consists of electric and magnetic fields that oscillate in directions perpendicular to each other and perpendicular to the direction in which the wave is traveling, as Figure 7.1 illustrates. When we say that an electric or magnetic field oscillates we mean that the strength of the field increases, reaches a maximum, decreases to zero, increases in the opposite direction, reaches a maximum, and again decreases to zero, and so on (Figure 7.2).

All waves can be described in terms of their velocity, frequency, wavelength, and amplitude (see Figure 7.3). The **wavelength**, λ (Greek letter lambda), is the distance between successive wave *crests*, or points of equal displacement on successive waves. The **frequency**, ν (Greek letter nu), is the number of wave crests that pass a given point in 1 second. It has the units second^{-1}, which is given the special name **hertz (Hz)**:

$$1\,\text{Hz} = 1\,\text{s}^{-1}$$

A frequency of 10 Hz means that ten wave crests pass a given point in 1 s. If

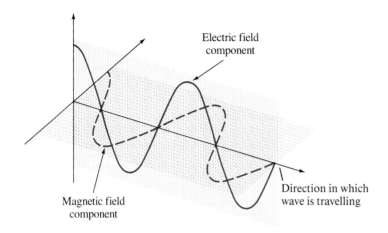

FIGURE 7.1
Electromagnetic Waves.

Electromagnetic waves consist of electric and magnetic fields that oscillate in directions perpendicular to each other and to the direction in which the wave travels.

there are v waves per second moving past a given point, and if the length of each wave is λ, the distance traveled by the wave in 1 s is λv, which is its **speed**, v:

$$v = \lambda v$$

Light and all other types of electromagnetic radiation have a constant speed of $2.997\,924\,6 \times 10^8$ m s^{-1} in a vacuum; this speed is given the symbol c:

$$c = \lambda v = 2.997\,924\,6 \times 10^8 \text{ m s}^{-1}$$

For most purposes in this book it will be sufficiently accurate to use this value rounded off to three significant figures, that is

$$c = 3.00 \times 10^8 \text{ m s}^{-1}$$

Radiation provides one way in which energy is transferred. For example, energy from the sun reaches the earth as ultraviolet, visible, and infrared radiation.

The **amplitude**, A, of a wave is the height of a crest or the depth of a trough. The energy per unit volume stored in a wave is proportional to A^2. In the case of light the **intensity**, or brightness, of light is proportional to A^2.

In ordinary conversation the words speed and velocity are used interchangeably. In physics, however, speed is a quantity that has magnitude only—it is a scalar—and it is given the symbol v, whereas velocity has both speed and direction—it is a vector—and it is given the symbol **v**. Here we use the terms speed and velocity in their strict sense although the difference between them will not be of great concern to us.

THE ELECTROMAGNETIC SPECTRUM

The complete range of electromagnetic waves is called the **electromagnetic spectrum** (see Figure 7.4). When we speak of visible light, or the *visible spectrum*, we are referring to radiation with wavelengths in the range 4×10^{-7} to 7.5×10^{-7} m. Our eyes are sensitive only to this very small part of the complete electromagnetic spectrum. X rays have wavelengths as short as 10^{-13} m, and ultraviolet, visible, and infrared radiation have increasingly longer wavelengths, in the range of 10^{-8} to 10^{-4} m. In contrast, radio waves have wavelengths as long as 1 km or more.

White light from the sun or an incandescent light bulb consists of all the wavelengths in the visible spectrum. When white light is passed through a glass prism, it is spread out into the band of colors shown in Figure 7.4, ranging from long-wavelength red light to shorter-wavelength violet light.

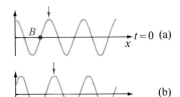

FIGURE 7.2
Series of "snapshots"
showing a sinusoidal wave
propagating to the right.

Note that the crest, marked by an
arrow, shifts its position as time
passes; it moves exactly one
wavelength while point B moves
through a complete cycle—first
down to a minimum, then back up
through zero to a maximum, and
back to zero displacement—as
shown in (i).

Example 7.1

FREQUENCY AND WAVELENGTH

A radio station broadcasts on a frequency of 900 kHz. What is the wavelength of the
electromagnetic radiation emitted by the transmitter?

Solution

We can rearrange the equation $c = v\lambda$ to give $\lambda = c/v$. Then inserting the values of c
and v and converting kilohertz (10^3 s^{-1}) to hertz (s^{-1}) we have

$$\lambda = \frac{c}{v} = \left(\frac{3.00 \times 10^8 \text{ m s}^{-1}}{900 \text{ kHz}}\right)\left(\frac{1 \text{ kHz}}{10^3 \text{ s}^{-1}}\right)$$

$$= 3.33 \times 10^2 \text{ m} = 0.333 \text{ km}$$

EXERCISE 7.1

The colors that make up visible light range in wavelength from 400 to 750 nm
(violet to red). What is the corresponding range of frequencies, in hertz?

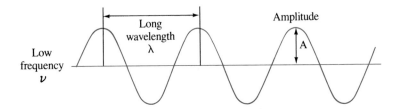
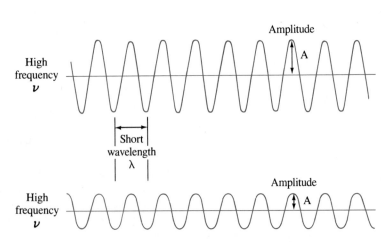

FIGURE 7.3
Properties of Waves.

Any wave can be described by its frequency, wavelength, amplitude, and speed.
The speed of a wave is the product of its wavelength and frequency. The energy of
a wave is related to its amplitude. For example, the greater the height of an ocean
wave, the greater is its destructive power.

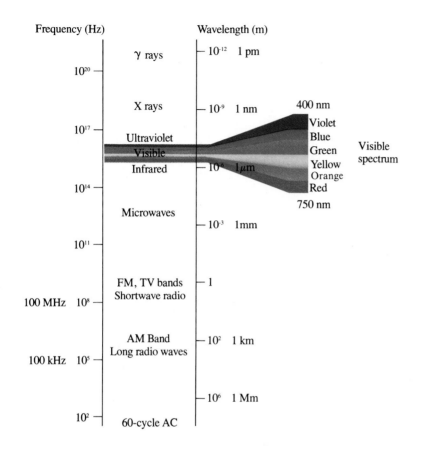

FIGURE 7.4
The Electromagnetic Spectrum.

All electromagnetic waves have the same speed in a vacuum, but their wavelengths and frequencies vary. Thus electromagnetic waves with a low frequency have a long wavelength; those with a high frequency have a short wavelength. Electromagnetic radiation with a wavelength of about 400 to 750 nm is detectable by the human eye and thus constitutes visible light.

INTERFERENCE AND DIFFRACTION PATTERNS

The view that light is an electromagnetic wave is supported by experimental observations that show that light behaves like other waves, such as waves on a water surface. In particular, light exhibits the property known as **interference**. When monochromatic light (light of a single frequency or color) is allowed to pass through a very narrow slit (for example, 0.1 mm in width), the slit behaves as a very small light source scattering light in all directions. When two such slits are placed very close together and the light coming from them falls onto a screen, we do not see an image of each slit but rather a pattern of alternate light and dark lines called a **diffraction pattern** (Figure 7.5). Because light waves from the two sources travel different distances to reach a given point on the screen, the crests and troughs of the two waves do not necessarily arrive together. Bright lines are formed on the screen at those positions where the waves that arrive from the two sources are *in phase*. In other words, the crests and troughs of the two waves coincide and combine to produce a resultant wave of greater amplitude. The dark lines are produced at those positions where waves arriving from the two sources are *out of phase*. The crest of one wave arrives at the same time as the trough of the other wave so that the resultant wave has a zero amplitude, in other words, there is zero light intensity at this point. Diffraction patterns are only produced by very narrow slits very close together, that is, when the dimensions of the slits are comparable with the wavelength

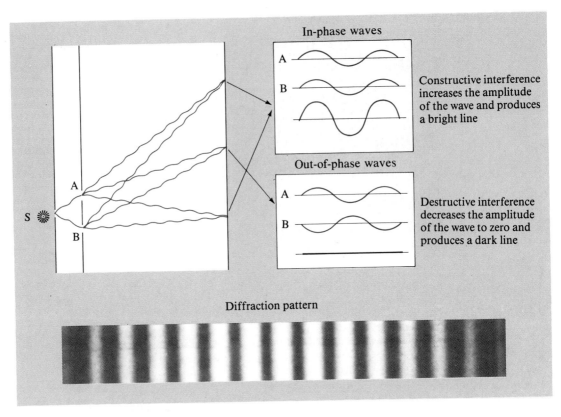

FIGURE 7.5
Diffraction Pattern Produced by Light Waves
Passing Through Two Slits.

Monochromatic light passing through two, narrow, closely spaced slits produces
a diffraction pattern. When the waves arrive in phase, reinforcing each other,
they produce a bright band. When the waves arrive out of phase, they cancel
each other, creating a dark line.

of the light. Two sharp images would be obtained of slits that are large compared with the wavelength of the light. In general light produces sharp images
of ordinary-sized objects, giving the impression that light travels in straight
lines like a beam of particles.

Any regular arrangement of tiny slits or holes will produce a diffraction
pattern with visible light. A particularly important type of diffraction pattern
can be obtained from the regularly arranged atoms in a crystal. Radiation is
scattered from atoms, that is, it bounces off in all directions. The interference
between the radiation scattered from each of the atoms in a crystal produces a
diffraction pattern when the wavelength of the radiation is comparable with the
distances between the atoms. It so happens that the wavelength of X rays is of
the right order of magnitude and so X rays of appropriate wavelength give a
diffraction pattern when scattered from crystals (Figure 7.6). From the distances
between the light and dark spots in the pattern and knowing the wavelength
of the X rays one can calculate the distances between the atoms in the crystal.
This is the basis of a very important method of determining the structures of
crystalline solids that is called X ray crystallography (see Chapter 11).

FIGURE 7.6
Diffraction Pattern Produced
when X Rays are Scattered by
a Sodium Chloride Crystal.

THE QUANTIZATION OF ENERGY

For several decades after Maxwell described radiation in terms of electromagnetic waves, the wave theory was able to provide an explanation for all the observations related to the transmission of light. But difficulties arose when attempts were made to use the wave theory to explain several observations concerning the interaction of light and matter. These difficulties were not resolved until a revolutionary idea, first proposed by Planck in 1900, and developed by Einstein and others in the next few years, was accepted (see Box 7.1). These physicists proposed (1) that when an atom emits or absorbs energy in the form of light or other electromagnetic radiation, the energy of the atom cannot change continuously but can only change by small finite amounts called **quanta** (singular quantum), and (2) that the energy of light or other electromagnetic radiation is quantized, that is, it is transmitted as quanta of radiation, called **photons**.

The energy of a photon is given by the expression

$$\text{Energy} = h\nu$$

where ν is the frequency of the radiation and h is a constant called the Planck constant. The value of the Planck constant is $6.626\,18 \times 10^{-34}$ J s. Thus the energy of a single photon of yellow light of frequency 5.50×10^{14} Hz is

$$E = (6.626 \times 10^{-34}\text{ J s})(5.50 \times 10^{14}\text{ s}^{-1})$$
$$= 3.64 \times 10^{-19}\text{ J}$$

This is an extremely small amount of energy, so that even a very feeble source of light emits an enormous number of photons every second, each of which has only a very small amount of energy. We are therefore never conscious of the fact that the energy of light is transmitted as particlelike photons.

$=$ *Example 7.2* $=$

ENERGY OF PHOTONS

What is the energy of one photon of red light of wavelength 650 nm?

BOX 7.1

ALBERT EINSTEIN (1879–1955)

Einstein was born in Germany in the old city of Ulm on the Danube. As a child, he was so slow at learning that his parents feared he might be retarded. At high school he disliked the harsh discipline, and when his family emigrated to Milan, Einstein left his school and joined them. He applied for admission to the Swiss Federal Polytechnical School in Zurich but was refused because he did not have a high school diploma and he failed the entrance examination, although he did very well in mathematics and physics. He then spent two years at a small college and finally was able to enter the Polytechnical School in Zurich. He did not particularly impress his teachers, and after graduation he had difficulty finding employment.

After taking several part-time positions, Einstein went to work as a junior official in the Swiss Patent Office in Berne. The work appears to have left Einstein lots of time to think about theoretical physics. In 1905, at the age of twenty-six, he published three articles, any one of which would have established him as one of the world's leading physicists. The first proposed that light has a particlelike, as well as a wavelike, nature and explained the photoelectric effect. The second paper explained Brownian motion, the random erratic motion of very small particles suspended in a liquid, as being due to collisions with the rapidly moving molecules of the liquid. The third showed that ideas of absolute space and time had to be replaced by the concept that space and time are relative to each other (the theory of relativity). It was in this paper that Einstein derived the famous equation $E = mc^2$.

Einstein reached his revolutionary conclusions by means of rather simple but uncompromising logic based on experimental observations. Remarkably, he did all this work without any contact with other important physicists of the time.

After the publication of these papers, the University of Zurich offered him a position, and Einstein quickly became an important figure in the world of theoretical physics. In 1914 he was persuaded to move to Berlin as the head of the physics department of the world-famous Kaiser-Wilhelm Institute. Despite his prestigious position he was not entirely happy under the militaristic Prussian rulers of Germany, but he continued to work in Germany throughout World War I and the difficult years that followed. Ultimately, Hitler's repression of Jews forced Einstein to leave. He arrived in New York in October 1933, and he stayed in the United States until his death in 1955.

Albert Einstein is universally recognized as the greatest physicist of our age. Some say that if someone else had discovered the theory of relativity, Einstein's other work would have made him the second greatest physicist of his time. His ideas radically changed our concepts of space and time. From the 1905 publication of the theory of relativity until the end of his long life, he concentrated on one main task: the attempt to find a single unifying theory that would explain all physical events.

Solution

The frequency of the red light is

$$\nu = \frac{c}{\lambda} = \frac{3.00 \times 10^8 \text{ m s}^{-1}}{650 \times 10^{-9} \text{ m}} = 4.61 \times 10^{14} \text{ s}^{-1} = 4.61 \times 10^{14} \text{ Hz}$$

The energy of one photon is

$$E = h\nu = 6.63 \times 10^{-34} \text{ J s} \times 4.61 \times 10^{14} \text{ Hz}$$
$$= 3.06 \times 10^{-19} \text{ J}$$

We will now look in a little more detail at just one of the experiments that led to these remarkable ideas concerning quanta and photons. One of the most striking observations that was not in accord with the wave theory was the photoelectric effect.

THE PHOTOELECTRIC EFFECT

When light, particularly ultraviolet light, shines on the surface of a metal, electrons are emitted from the surface (Figure 7.7). This phenomenon is known as the photoelectric effect. Experiments established the following:

1. No electrons are emitted unless the light has a frequency greater than a certain mininum value which is characteristic for each metal, no matter how intense the light.
2. When electrons are emitted the number of electrons is proportional to the light intensity.

Einstein showed that these observations could only be explained by using the revolutionary postulate that light consists of photons. A certain amount of energy is needed to pull an electron in an atom away from the nucleus. If a photon has at least this amount of energy, then on colliding with an atom it can knock out an electron, but if it has a lower energy it will not be able to do so. Since the energy of a photon is $h\nu$ it follows that the light must have a certain minimum frequency before it can eject an electron from an atom, which explains observation (1). Moreover, the greater the light intensity the greater the number of photons colliding with the metal surface and therefore the greater the number of electrons emitted, in accordance with observation (2).

According to the wave theory, however, the energy of light is related to its amplitude, that is, to its intensity. Thus the wave theory predicts that light of any frequency should be able to eject an electron from an atom provided it is intense enough, which is not what is observed.

The photoelectric effect was discovered by the German physicist Heinrich Hertz (1857–1894) in 1888 during the course of his experiments on generating radio waves. His name is commemorated in the unit for the frequency of a wave.

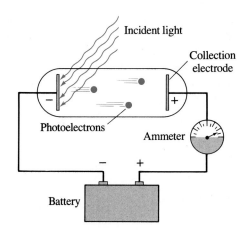

FIGURE 7.7
The Photoelectric Effect.

Detecting the photoelectric effect. The photoelectrons ejected from the irradiated metal plate are attracted to the positive collection electrode at the other end of the tube. The current that results is measured with an ammeter.

PHOTOCHEMICAL REACTIONS The description of light in terms of photons is also needed to understand chemical reactions that are caused by the absorption of light. These reactions are known as **photochemical reactions**. We saw in Chapter 5 that a bright light will initiate the reaction between hydrogen and chlorine, which then occurs with explosive violence. But experiments show that only blue-green light or light of higher frequency will cause the reaction. Red light, no matter how intense, has no effect. We can understand this in terms of the photon theory of light since it takes a certain amount of energy, and therefore a photon which has at least this energy, to dissociate a chlorine molecule into two chlorine atoms:

$$Cl_2 + h\nu \longrightarrow 2Cl$$

This dissociation of the chlorine molecule by a photon of sufficiently high frequency is followed by the reaction of a chlorine atom with a hydrogen molecule,

$$Cl + H_2 \longrightarrow HCl + H$$

to form a hydrogen chloride molecule and a hydrogen atom. The hydrogen atom then reacts with a chlorine molecule to form another hydrogen chloride molecule and another chlorine atom:

$$H + Cl_2 \longrightarrow HCl + Cl$$

This chlorine atom can react with another hydrogen molecule, and so these two successive reactions can continue indefinitely.

The overall reaction is the sum of these two reactions:

$$\begin{array}{c} Cl + H_2 \longrightarrow HCl + H \\ \underline{H + Cl_2 \longrightarrow HCl + Cl} \\ H_2 + Cl_2 \longrightarrow 2HCl \end{array}$$

The very reactive chlorine and hydrogen atoms are used up as fast as they are formed and do not appear as products of the overall reaction. This reaction is an example of a chain reaction. Chain reactions are discussed in more detail in Chapter 19.

= Example 7.3 =

PHOTOCHEMICAL REACTIONS

The energy needed to dissociate a chlorine molecule into chlorine atoms (the bond energy) is 243 kJ mol^{-1}. What is the maximum wavelength of light that will initiate the hydrogen–chlorine reaction?

Solution

The energy needed to dissociate one chlorine molecule is

$$E = \left(\frac{243 \text{ kJ}}{1 \text{ mol}}\right)\left(\frac{1 \text{ mol}}{6.022 \times 10^{23} \text{ molecules}}\right)\left(\frac{1000 \text{ J}}{1 \text{ kJ}}\right)$$

$$= 4.035 \times 10^{-19} \text{ J (molecule)}^{-1}$$

Now since one photon is needed to dissociate one chlorine molecule, the photon must have a minimum energy of 4.035×10^{-19} J. It must therefore have a minimum frequency of

$$v = \frac{E}{h} = \frac{4.035 \times 10^{-19} \text{ J}}{6.63 \times 10^{-34} \text{ J s}} = 6.09 \times 10^{14} \text{ s}^{-1}$$

We can now convert frequency to wavelength:

$$\lambda = \frac{c}{v} = \left(\frac{3.00 \times 10^8 \text{ m}}{1 \text{ s}}\right)\left(\frac{1 \text{ s}}{6.09 \times 10^{14}}\right)$$

$$= 4.93 \times 10^{-7} \text{ m} = (4.93 \times 10^{-7} \text{ m})\left(\frac{10^9 \text{ nm}}{1 \text{ m}}\right)$$

$$= 493 \text{ nm}$$

This value is the maximum wavelength of light that will dissociate a chlorine molecule; it is in the green region of the spectrum.

Many other reactions are caused by the absorption of light by molecules. In the upper atmosphere many photochemical reactions involving nitrogen and oxygen molecules occur as a result of the absorption of ultraviolet radiation from the sun. One of these reactions is the formation of ozone (see Chapters 3 and 18).

EXERCISE 7.2

Ozone is formed in the stratosphere by the following reactions:

$$O_2 \xrightarrow{h\nu} 2O$$
$$O + O_2 \longrightarrow O_3$$

If it takes ultraviolet light with a wavelength no longer than 240 nm to cause this reaction, what is the energy needed to dissociate 1 mol of oxygen molecules into oxygen atoms?

THE DUAL NATURE OF LIGHT

Since light behaves in some experiments as if it consisted of photons, but nevertheless also exhibits wavelike behavior, as in the phenomena of diffraction and interference, we have to accept the idea that light can behave like particles and like waves. Which behavior we observe depends on the kind of experiment being carried out. When light is passed through two adjacent slits, we observe an interference pattern that is characteristic of waves; when light interacts with matter and there is a transfer of energy from the light to electrons, we observe the particlelike behavior of light. This dual nature of light is something that is not familiar from everyday experience, but experiment shows us that this is how light behaves.

7.2
Atomic Spectra

Once the ideas of Planck and Einstein concerning quanta and photons had been accepted, studies of the absorption and emission of electromagnetic radiation by atoms and molecules began to provide important information about their structures. We consider now the spectra of atoms, that is, **atomic spectra**. The range of frequencies (in wavelengths) of light emitted or absorbed by an atom is called its **spectrum** (plural, spectra).

When the alkali metals or compounds of the alkali metals are heated to a high temperature in a very hot flame, they impart distinctive colors to the flame that are characteristic of a particular metal—for example, red for lithium, yellow for sodium, and lilac for potassium (Experiment 7.1). If the light emitted by the flame is passed first through a slit (to obtain a narrow beam) and then through a glass prism, a spectrum is obtained. Unlike the continuous spectrum obtained from sunlight, this spectrum consists of only a few sharp lines, as Figure 7.8 shows. Thus light emitted by an alkali metal flame does not contain all the frequencies in the visible region but only a few frequencies that are characteristic of the particular metal. Such a spectrum is called an emission **line spectrum**.

If an electric discharge is passed through helium, the gas glows with a characteristic blue-violet color. If the light is passed through a prism, the spectrum obtained again consists of only a limited number of sharp lines. It is a line spectrum like those given by the alkali metals (see Figure 7.8).

Experiment 7.1

Colored Flames and Atomic Spectra

Compounds of the alkali metals give characteristic colors to a flame. The color arises from alkali metal atoms that are raised to an excited state by the high temperature of the flame and then return to the ground state, emitting light of a characteristic frequency. The flames shown here were produced by holding a platinum wire on which a small amount of the alkali metal chloride had been placed in the flame of a bunsen burner.

Lithium—red   Sodium—yellow   Potassium—lilac

FIGURE 7.8
Atomic (Line) Spectra of (top to bottom) Hydrogen,
Sodium, Helium, and Neon.

What is the origin of these spectra? When an alkali metal is heated to a high temperature in a flame, some of its atoms are raised to high energy states by collisions with fast-moving atoms and molecules in the flame. When an alkali metal compound is heated in a flame, it is decomposed to give alkali metal atoms, which are similarly in high energy states. When an electric discharge is passed through helium gas, the very fast-moving electrons of the discharge collide with helium atoms, transferring some of their energy to the helium atoms and thus raising them to a higher energy state. Atoms that have been raised to a high energy state are said to be in an **excited state**. The normal, unexcited state of an atom is called its **ground state**. An excited atom is unstable, and may lose some or all of its excess energy by emitting light. This light constitutes the spectrum of the atom (Figure 7.8).

Hydrogen, helium, and
mercury discharge tubes.

The fact that excited atoms emit only certain frequencies of light is completely in accord with Planck and Einstein's postulate that when an atom emits or absorbs light its energy can only change by certain definite amounts or quanta. In other words an atom cannot have a continuous range of energies but only certain definite energies. We say that the energy of an atom is quantized. If an atom in an excited state has energy E_2, whereas in the ground state it has energy E_1, then the energy emitted when the atom returns from the excited state to the ground state is $E_2 - E_1$. This energy is emitted as a single photon with a frequency v:

$$E_2 - E_1 = hv$$

Each frequency in the emitted light, and thus each line in the spectrum, corresponds to the difference in energy between two different states (see Figure 7.9).

From the observed spectra of atoms we conclude that each atom has a set of definite **energy levels**, E_1, E_2, E_3, and so on (Figure (7.9)). An atom may have a rather large number of energy levels, but it cannot have an energy that does not correspond to one of these levels. A set of energy levels is rather like a set of shelves, each one at a certain height. An object can be placed on any one of the shelves but not at any position between the shelves. Thus the object can have the potential energy corresponding to the height of any particular shelf, but it cannot have any energy between these values.

When the energy of an atom has its lowest value, that is, when the atom is in its ground state, we say that the atom is in its lowest energy level. When it is in an excited state it is in one of a set of higher energy levels. When an atom in an excited state loses energy so that it is then in its ground state or another excited state of lower energy, we say that it moves from a higher energy level to a lower energy level. In so doing it emits a photon of light with an energy equal to the difference in energy between the two levels. Such a movement of an atom from one energy level to another is called a **transition** between the two levels. Thus from the spectrum of an atom we could in principle determine the set of energy levels for that particular atom. So let us now see what the spectrum of the simplest atom, hydrogen, tells us about the energy levels of the hydrogen atom.

THE SPECTRUM OF THE HYDROGEN ATOM

In the visible region of the spectrum of the hydrogen atom we see the lines shown in Figure 7.10. There are also lines in the infrared and ultraviolet regions of the spectrum. Although these lines cannot be seen by eye they can be recorded on suitable photographic film. The energy levels of the hydrogen atom that can be deduced from the spectrum are given by the very simple formula

$$E = -\frac{2.179 \times 10^{-18}}{n^2} \text{ J atom}^{-1} \qquad n = 1, 2, 3, \ldots$$

where n is a number that can have any positive integral value. The numbers n are called **quantum numbers**. As we will see shortly, this expression gives a set of energy levels that accounts for all the lines in the visible, infrared, and ultraviolet regions of the spectrum. The energy levels given by this expression are the energies for a single hydrogen atom. But it is often convenient to work on a molar basis, in which case we convert this expression to one that gives the possible energies for a mole of hydrogen atoms, as follows:

$$E = \left(-\frac{1}{n^2}\right)\left(\frac{2.179 \times 10^{-18} \text{ J}}{1 \text{ atom}}\right)\left(\frac{6.022 \times 10^{23} \text{ atoms}}{1 \text{ mol}}\right)\left(\frac{1 \text{ kJ}}{1000 \text{ J}}\right)$$

$$= \frac{-1312}{n^2} \text{ kJ mol}^{-1}$$

FIGURE 7.9
Energy Levels.

Each atom has a characteristic set of energy levels. An atom emits light when an electron returns from an excited state to a lower energy state. Each frequency in the line spectrum therefore corresponds to the difference in energy between two energy levels of the atom.

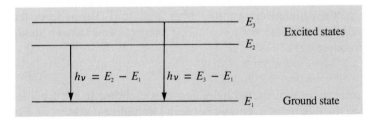

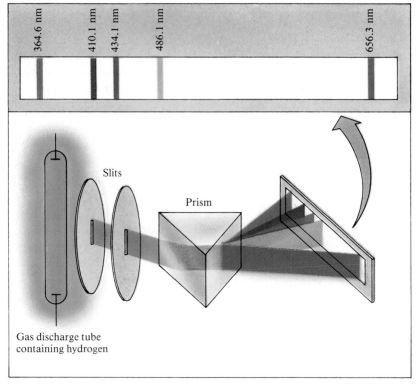

Hydrogen discharge tube

FIGURE 7.10
The Spectrum of the Hydrogen Atom.

When an electric discharge is passed through a tube containing H_2 gas the molecules are split up into atoms which are produced in excited states:

$$H_2 \xrightarrow{\text{Electric discharge}} 2H^*$$

$$H^* \longrightarrow H + h\nu$$

The light emitted by these atoms in returning to lower energy states produces the line spectrum of the hydrogen atom when it is passed through a prism or another device for separating the light into its component wavelengths, such as a diffraction grating.

In 1913 the young Danish physicist Niels Bohr (1885–1962) proposed a very simple model of the hydrogen atom which accounted quantitatively for the set of hydrogen-atom energy levels. According to his model, the electron in a hydrogen atom rotates around the nucleus in any one of a limited number of circular orbits, just as the earth rotates in its orbit around the sun. When the electron is in the orbit closest to the nucleus, it has the lowest possible energy and the hydrogen atom is in its ground state. When the electron is in one of the orbits further away from the nucleus, it has a higher energy because work must be done against the electrostatic attraction exerted by the nucleus in order to move the electron away from the nucleus. The electron is then in a higher energy level and we say that the atom is in an excited state. When the electron moves to a lower energy level it emits a photon of light of energy equal to the difference in energy between the two levels.

The energy levels of the hydrogen atom are shown in Figure 7.11. In the lowest energy level or the ground state the electron is in the orbit closest to

FIGURE 7.11
Energy Levels and Transitions
for the Hydrogen Atom.

Each line in the spectrum of the
hydrogen atom is produced by a
photon emitted when the electron
returns from an excited state to a
lower energy level. The energy of
the photon, and hence the frequency of the light emitted, depends on the difference in energy
between the two energy levels.
For example, the $n = 1$ state has
an energy of -1312 kJ mol^{-1};
the $n = 2$ state has an energy of
-328 kJ mol^{-1}. Thus the energy
change associated with a transition from the $n = 2$ to the $n = 1$
level is $-328 - (-1312) =$
984 kJ mol^{-1}, which is the energy
of the photons producing one of
the lines in the ultraviolet region.
All the transitions to the $n = 1$
level produce lines in the ultraviolet region. Transitions to the $n = 2$
level, which have smaller energies
than transitions to the $n = 1$ level,
produce the series of lines observed
in the visible spectrum, and transitions to the $n = 3$ level, which
have still smaller energies, produce
the series of lines in the infrared
region of the spectrum.

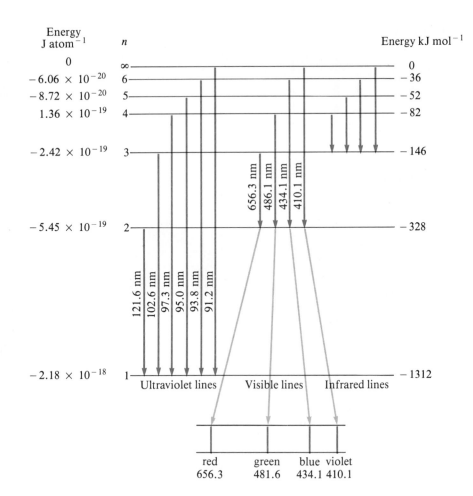

the nucleus. This state corresponds to $n = 1$ and has energy

$$E = \frac{-1312}{1^2} \text{ kJ mol}^{-1} = -1312 \text{ kJ mol}^{-1}$$

The next lowest energy level, in which the electron is in the next nearest orbit
to the nucleus, and which we call the first excited state, corresponds to $n = 2$
and has energy

$$E = \frac{-1312}{2^2} \text{ kJ mol}^{-1} = -328.0 \text{ kJ mol}^{-1}$$

The energy of the $n = 3$ state, or second excited state, is $-1312/3^2 =$
-145.8 kJ mol^{-1}, and so on.

To find the frequency or wavelength of the light emitted when an electron
moves from a higher to a lower energy level, we first find the difference in energy
between the two levels. This difference in energy is equal to the energy of 1 mol
of the emitted photons:

$$\text{E(1 mol photons)} = \Delta E = E_{\text{initial}} - E_{\text{final}} = (-1312/n_{\text{initial}}^2) - (-1312/n_{\text{final}}^2) \text{ kJ mol}^{-1}$$

$$= -1312 \left(\frac{1}{n_{\text{initial}}^2} - \frac{1}{n_{\text{final}}^2} \right) \text{ kJ mol}^{-1}$$

It is convenient to remove the negative sign by reversing the terms in the brackets to give

$$E(1 \text{ mol photons}) = 1312\left(\frac{1}{n_{\text{final}}^2} - \frac{1}{n_{\text{initial}}^2}\right) \text{kJ mol}^{-1}$$

Then we convert this energy to a frequency or wavelength, as shown in the next example.

═ *Example 7.4* ══

ENERGY OF A HYDROGEN SPECTRUM LINE

Find the energy change when an electron moves from the $n = 4$ level to the $n = 2$ level in the hydrogen atom. If this energy is emitted as a photon, what is the wavelength of the photon?

Solution

We have

$$\Delta E = 1312\left(\frac{1}{n_{\text{f}}^2} - \frac{1}{n_{\text{i}}^2}\right) \text{kJ mol}^{-1}$$

where ΔE is the energy difference between the initial and final states. Substituting $n_{\text{i}} = 4$ and $n_{\text{f}} = 2$ gives

$$\Delta E = 1312\left(\frac{1}{2^2} - \frac{1}{4^2}\right) = 246.0 \text{ kJ mol}^{-1}$$

This is the energy change for a *mole of hydrogen atoms*. But because we have to find the energy of the emitted photon and one hydrogen atom emits one photon we need the energy change for *one hydrogen atom*, which is

$$\Delta E = \left(\frac{246.0 \text{ kJ}}{1 \text{ mol}}\right)\left(\frac{1 \text{ mol}}{6.022 \times 10^{23} \text{ atoms}}\right)\left(\frac{1000 \text{ J}}{1 \text{ kJ}}\right)$$

$$= 4.085 \times 10^{-19} \text{ J atom}^{-1}$$

and since $E_{\text{photon}} = \Delta E$

$$E_{\text{photon}} = 4.085 \times 10^{-19} \text{ J atom}^{-1}$$

To find the wavelength of this photon, we use the relationships

$$E_{\text{photon}} = h\nu \quad \text{and} \quad \lambda\nu = c$$

Thus

$$E_{\text{photon}} = \frac{hc}{\lambda} \quad \text{or} \quad \lambda = \frac{hc}{E_{\text{photon}}}$$

$$\lambda = \frac{(6.63 \times 10^{-34} \text{ J s})(3.00 \times 10^8 \text{ m s}^{-1})}{4.085 \times 10^{-19} \text{ J}}$$

$$= 4.87 \times 10^{-7} \text{ m} = 487 \text{ nm}$$

───

EXERCISE 7.3

(a) What is the energy change associated with the transition of a hydrogen atom in the $n = 5$ excited state to the ground state?

(b) What are the frequency and the wavelength of the corresponding line in the spectrum?

(c) In what region of the spectrum is this line found?

We saw in Example 7.4 that the wavelength of the line corresponding to the transition of the electron in the hydrogen atom from the $n = 4$ level to the $n = 2$ level is 487 nm, which is in the blue-green region of the visible spectrum. In fact, as we can see in Figure 7.11, all the lines corresponding to transitions from higher levels to the $n = 2$ level are in the visible region of the spectrum. The lines arising from transitions to the $n = 1$ level, that is, to the ground state, are in the ultraviolet region, and those arising from transitions to the $n = 3$ level are in the infrared region. In principle there are other lines that are also in the infrared region that arise from transitions to the $n = 4$ level and so on, but they become increasingly difficult to observe.

You may have wondered how it is that the energy of an electron in a hydrogen atom can have *negative* values. There is nothing mysterious about this. We have seen that as the electron gets closer to the nucleus its energy *decreases*. In other words, the energy levels have lower energies the closer they are to the nucleus. The negative values arise simply because the zero of the energy scale is taken to be when the electron is at a very large distance from the nucleus, in other words, when it is in the $n = \infty$ level, which has

$$E = \frac{-1312}{\infty^2} \, \text{kJ mol}^{-1} = 0$$

So all the energy levels necessarily have negative values for their energies.

Because they have more than one electron, other elements have more complicated atomic spectra than hydrogen; however, these spectra can all be accounted for in terms of a set of energy levels characteristic of each element. The atomic spectrum of an element is unique; it therefore provides a very convenient method for identifying that element. For example, the sodium spectrum has two characteristic intense lines in the yellow region of the spectrum, which enable sodium and its compounds to be recognized very easily.

The emission of light by excited atoms also has some practical uses in everyday life. The yellow light of certain street and highway lamps is emitted by excited sodium atoms produced by bombardment of the sodium atoms in sodium vapor by the fast-moving electrons of an electric discharge. This type of lamp produces less heat than an incandescent light bulb and therefore wastes less energy. Another type of street lamp, the mercury lamp, gives a blue-green light, which arises from excited mercury atoms produced in mercury vapor by an electric discharge. The red light emitted by excited neon atoms is also familiar from neon signs, which are made from tubes that contain neon gas through which an electric discharge is passed.

We have not discussed how Bohr was able to calculate the exact values of the hydrogen-atom energy levels from his model. We will not do so, because it was found that Bohr's model did not predict the spectrum of any atom with more than one electron. As we will see, we no longer think of the electrons in an atom as rotating around the nucleus in orbits, although we still use the very important idea of electron energy levels. Nevertheless, at the time, Bohr's theory represented a great advance in our understanding of the properties of atoms.

We have seen in this section that atomic spectra provide clear evidence that the energies of the electrons in an atom are quantized, and we have looked at the energy levels of the single electron in the hydrogen atom in some detail. The spectra of the atoms of other elements are more complicated than the spectrum of the hydrogen atom because these atoms have more electrons occupying more energy levels. Although it is possible to deduce the energy level scheme of an atom from its atomic spectrum, we will find it simpler to use the ionization energies of an atom to provide us with information about the energy levels of atoms other than hydrogen.

7.3
Electron Configurations

The form of the periodic table and the values of ionization energies led us to conclude in Chapter 4 that the electrons in an atom are arranged in shells. Because of the increasing distance of successive shells from the nucleus, less energy is needed to remove an electron from the $n = 2$ shell of a given atom than from the $n = 1$ shell, and still less energy is needed to remove an electron from the $n = 3$ shell, and so on. But do all the electrons in a given shell have precisely the same energy? That is, are they all in the same energy level? In order to answer this question, we must consider ionization energies in a little more detail.

ENERGY LEVELS FROM IONIZATION ENERGIES

We described in Chapter 4 how ionization energies may be measured by the *electron impact method*, in which atoms in the gas phase are bombarded with fast-moving electrons. In general, electron impact experiments give a value for the ionization energy of the electron that is most easily removed from the atom—in other words, the ionization energy for an electron in the highest occupied energy level. An alternative, and generally more accurate, method that provides information on *all* the energy levels of an atom is known as **photoelectron spectroscopy**; this method uses photons to knock electrons out of atoms. Electrons obtained in this way are called *photoelectrons*.

Photoelectron spectroscopy is very similar to the photoelectric effect experiment, except that photons are used to knock electrons out of atoms in the gas phase instead of from the surface of a metal. And because electrons are usually less easily removed from the atoms of other elements than they are from the atoms of metals, very-high-energy photons such as the photons of very-short-wavelength ultraviolet radiation or even X rays must be used. If the energy of the photon is greater than the ionization energy of the electron, the excess energy appears as kinetic energy, $\frac{1}{2}mv^2$, of the electron, which is ejected from the atom with a speed v. In other words, the speed of the ejected electron depends on how much excess energy it has received. So, if IE is the ionization energy of the electron and KE is the kinetic energy with which it leaves the atom we have

$$E_{photon} = h\nu = IE + KE$$

Rearranging this equation gives

$$IE = h\nu - KE$$

Hence we can find the ionization energy IE if we know the frequency or wavelength of the photon and we can measure the kinetic energy of the photoelectron. The kinetic energy of the photons is measured in the photoelectron spectrometer as shown in Figure 7.12.

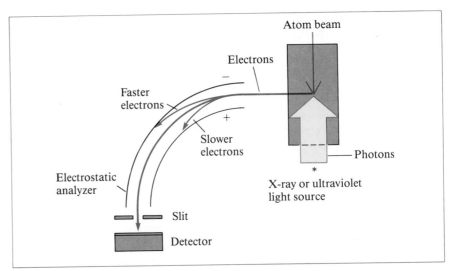

FIGURE 7.12
Schematic Diagram of a Photoelectron Spectrometer.

A beam of atoms is irradiated with ultraviolet light or X rays of a known frequency. The kinetic energies of the ejected electrons are measured by passing them into an electrostatic analyzer which consists of two curved plates, one charged positively and the other negatively. The electric field between the plates deflects each electron into a curved path, the curvature depending on the speed and therefore on the kinetic energy of the electron. Thus only electrons of one particular speed and therefore one particular kinetic energy will have a path of just the right curvature to pass right through the analyzer and through the slit to reach the detector. The detector counts the electrons as they arrive. By varying the charge on the plates, that is, the voltage between the plates, electrons of different energies can be detected.

If photons of sufficient energy are used, an electron may be ejected from *any* of the energy levels of an atom. Thus if we bombard a large number of atoms with photons of sufficient energy, electrons will be ejected from all the energy levels of the atom. If the photons all have the same energy, electrons ejected from a given energy level will all have the same energy so that electrons with only a few different energies will be obtained corresponding to the number of energy levels in the atom.

HELIUM The photoelectrons obtained from helium, using radiation of a single frequency, all have the same kinetic energy. Since helium has two electrons in the $n = 1$ shell, both these electrons must have the same energy. In other words, there is only one energy level for the $n = 1$ shell, and both electrons are in this level. From the known energy of the photons and the measured kinetic energy of the photoelectrons the ionization energy of an electron in this level is found to be 2.37 MJ mol^{-1}. The energy of this level is therefore 2.37 MJ mol^{-1} less than the energy of an ionized helium atom; it has an energy of -2.37 MJ mol^{-1}, with the energy of the ionized helium atom being taken to be zero. This value is the same as that obtained from electron impact experiments (Table 4.2).

NEON When neon atoms are irradiated with light of a sufficiently high frequency, electrons with three different kinetic energies are obtained. This result shows that the electrons in a neon atom occupy three energy levels. A given atom may lose an electron from any one of these levels if the energy of the photon with which it interacts is large enough. Let us suppose that we are using photons with an energy of 100.00 MJ mol^{-1}. The kinetic energies of the photoelectrons are then found to be 16.00, 95.32, and 97.93 MJ mol^{-1}. So the three ionization energies are

$$IE = 100.0 - 16.0 \quad = 84.0 \text{ kJ mol}^{-1}$$
$$IE = 100.00 - 95.32 = 4.68 \text{ kJ mol}^{-1}$$
$$IE = 100.00 - 97.93 = 2.07 \text{ kJ mol}^{-1}$$

The electrons with the very high ionization energy of 84.0 kJ mol^{-1} (the most difficult to remove) must come from the $n = 1$ shell. So, as for the helium atom, there is only one energy level for the $n = 1$ shell. In the neon atom this level has an energy of -84.0 kJ mol^{-1}, which is much lower than in the helium atom because of the higher nuclear charge of neon. The electrons with ionization energies of only 4.68 and 2.07 MJ mol^{-1} must come from the $n = 2$ shell. Thus there are two energy levels in the $n = 2$ shell which have energies of -4.68 kJ mol^{-1} and -2.07 kJ mol^{-1}. We shall see shortly how the eight electrons in the $n = 2$ shell of neon are distributed between these two levels. The levels are labeled 2s and 2p, respectively. The number 2 indicates that the electrons are in the $n = 2$ shell. The label s always indicates the lowest energy level in a given shell, and p denotes the next highest energy level. Thus the $n = 2$ shell consists of the 2s and 2p energy levels. The $n = 1$ shell consists of a single 1s energy level. Energy level diagrams for the helium and neon atoms are given in Figure 7.13.

The labels s and p—and also d and f which we will introduce shortly—were originally abbreviations for words such as sharp and principal which were used in the description of atomic spectra. However they have lost their original significance and are now merely labels for the energy levels within a shell, and, as we will see later, for the atomic orbitals corresponding to these energy levels.

PERIODS 1, 2, AND 3

The ionization energies of the electrons in the atoms from hydrogen to argon are given in Table 7.1 and in Figure 7.14. We see that lithium has two ionization energies: one for the two electrons in the 1s level and the second one for one electron in the 2s level. Similarly, beryllium has two ionization energies corresponding to two electrons in the 1s level and two in the 2s level. However, boron and all the following elements up to neon have three ionization energies, showing that there must be a second energy level in the $n = 2$ shell, namely,

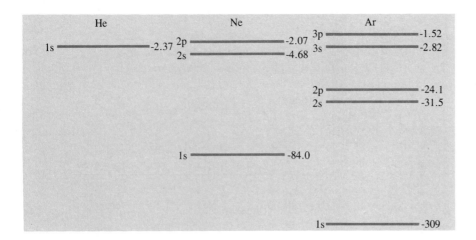

FIGURE 7.13
Energy Levels for the Helium, Neon, and Argon Atoms Obtained by Photoelectron Spectroscopy.

The energies are in megajoules per mole.

TABLE 7.1
Ionization Energies (MJ mol^{-1}) for the First 21 Elements

Element	1s	2s	2p	3s	3p	3d	4s
H	1.31						
He	2.37						
Li	6.26	0.52					
Be	11.5	0.90					
B	19.3	1.36	0.80				
C	28.6	1.72	1.09				
N	39.6	2.45	1.40				
O	52.6	3.04	1.31				
F	67.2	3.88	1.68				
Ne	84.0	4.68	2.08				
Na	104	6.84	3.67	0.50			
Mg	126	9.07	5.31	0.74			
Al	151	12.1	7.19	1.09	0.58		
Si	178	15.1	10.3	1.46	0.79		
P	208	18.7	13.5	1.95	1.06		
S	239	22.7	16.5	2.05	1.00		
Cl	273	26.8	20.2	2.44	1.25		
Ar	309	31.5	24.1	2.82	1.52		
K	347	37.1	29.1	3.93	2.38		0.42
Ca	390	42.7	34.0	4.65	2.90		0.59
Sc	433	48.5	39.2	5.44	3.24	0.77	0.63

FIGURE 7.14
Ionization Energies for the First 18 Elements.

The ionization energy of the 1s level increases very rapidly with increasing Z. The ionization energy of an electron in an s level is always slightly greater than that of an electron in the corresponding p level.

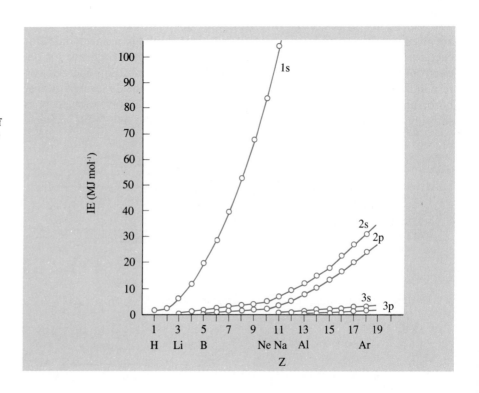

the 2p level. In boron one electron is in this level; in carbon two electrons are in this level. The number of electrons in the 2p level increases from element to element up to six in neon, when the 2p level is filled and the $n = 2$ shell is complete.

The arrangement of the electrons for the elements of the first two periods is therefore as follows:

	$n = 1$	$n = 2$
H	$1s^1$	
He	$1s^2$	
Li	$1s^2$	$2s^1$
Be	$1s^2$	$2s^2$
B	$1s^2$	$2s^2 2p^1$
C	$1s^2$	$2s^2 2p^2$
N	$1s^2$	$2s^2 2p^3$
O	$1s^2$	$2s^2 2p^4$
F	$1s^2$	$2s^2 2p^5$
Ne	$1s^2$	$2s^2 2p^6$

These representations of the electron arrangement are called **electron configurations**.

The element following neon is sodium. It is in the third period and has an electron in the $n = 3$ shell. Sodium is an alkali metal like lithium, and therefore it has a similar electron configuration, with just one electron in its valence shell. It has four ionization energies and the configuration $1s^2 2s^2 2p^6 3s^1$. Similarly, magnesium has four ionization energies corresponding to the 1s, 2s, 2p, and 3s levels. But the next element, aluminum, has five ionization energies; its 3s level is filled with two electrons and the remaining electron in the $n = 3$ shell must be in the 3p level. Like the 2p level, the 3p level can contain a maximum of six electrons; electrons are added to this level until it is filled at argon. Thus the electron configurations of the elements from sodium to argon are as follows:

Na	$1s^2 2s^2 2p^6 3s^1$
Mg	$1s^2 2s^2 2p^6 3s^2$
Al	$1s^2 2s^2 2p^6 3s^2 3p^1$
Si	$1s^2 2s^2 2p^6 3s^2 3p^2$
P	$1s^2 2s^2 2p^6 3s^2 3p^3$
S	$1s^2 2s^2 2p^6 3s^2 3p^4$
Cl	$1s^2 2s^2 2p^6 3s^2 3p^5$
Ar	$1s^2 2s^2 2p^6 3s^2 3p^6$

═ *Example 7.5* ══════════════════════════════════

ENERGY LEVELS FROM PHOTOELECTRON SPECTROSCOPY

When a beam of neon atoms was irradiated with a beam of X rays of wavelength 0.2291 nm, electrons with kinetic energies of 428.2 MJ mol^{-1}, 517.5 MJ mol^{-1}, and 520.1 MJ mol^{-1} were obtained. Calculate the corresponding ionization energies of neon and hence find the energy levels of the neon atom.

Solution

First, we need the energy of the X ray photons. For one photon $E = h\nu$ and $\nu = c/\lambda$, so

$$E_{photon} = \frac{hc}{\lambda} = \left(\frac{6.626 \times 10^{-34} \text{ J s}}{0.2291 \times 10^{-9} \text{ m}} \right) \left(\frac{2.998 \times 10^8 \text{ m}}{1 \text{ s}} \right)$$

$$= 8.671 \times 10^{-16} \text{ J per photon}$$

For 1 mol of photons

$$E_{photon}\ mol^{-1} = \left(\frac{8.671 \times 10^{-16}\ J}{1\ photon}\right)\left(\frac{6.022 \times 10^{23}\ photons}{1\ mol}\right)$$

$$= 5.222 \times 10^8\ J\ mol^{-1} = 522.2\ MJ\ mol^{-1}$$

Now, we can use the relationship between the energy of the photons, E_{photon}, the kinetic energy of the electrons, KE, and their ionization energy, IE,

$$IE = E_{photon} - KE$$

to find the ionization energies:

$$IE = 522.2\ MJ\ mol^{-1} - 438.2\ MJ\ mol^{-1} = 84.0\ MJ\ mol^{-1}$$

$$IE = 522.2\ MJ\ mol^{-1} - 517.5\ MJ\ mol^{-1} = 4.7\ MJ\ mol^{-1}$$

$$IE = 522.2\ MJ\ mol^{-1} - 520.1\ MJ\ mol^{-1} = 2.1\ MJ\ mol^{-1}$$

Hence the energy levels of neon are $-84.0\ MJ\ mol^{-1}$, $-4.7\ MJ\ mol^{-1}$, and $-2.1\ MJ\ mol^{-1}$.

EXERCISE 7.4

When lithium atoms were irradiated with X rays of wavelength 8.481 nm, electrons with kinetic energies of 7.85 MJ mol^{-1} and 13.59 MJ mol^{-1} were obtained. Calculate the two ionization energies of lithium and hence find the energy levels of the lithium atom.

The existence of the s and p energy levels in the $n = 2$ and $n = 3$ shells accounts for some of the small irregularities in the ionization energy of the most easily removed electron that we saw in Table 4.3 and Figure 4.5 but did not explain. We might have expected a steady increase in the ionization energy of an $n = 2$ shell electron from lithium to neon as the core charge increases, but this increase is interrupted at boron and at oxygen. The ionization energy of the electron that is most easily removed from boron (0.80 MJ mol^{-1}) is slightly lower, not higher, than that for beryllium (0.90 MJ mol^{-1}), even though boron has a higher core charge than beryllium. We now see that this energy is lower because the most easily removed electron comes from the 2s level in beryllium, but the most easily removed electron in boron comes from the 2p level. Electrons in the 2p level of boron have a slightly lower ionization energy than the 2s electrons of beryllium, despite the increase in the core charge.

A similar situation occurs in the third period. The ionization energy of the most easily removed electron in aluminum (0.58 MJ mol^{-1}) is less than the ionization energy of the preceding element magnesium (0.74 MJ mol^{-1}). The electron removed from aluminum is a 3p electron, whereas the electron removed from magnesium is a more tightly held 3s electron. The similar anomaly observed between group 5 and group 6 atoms of the same period is explained in Section 7.5.

PERIODS 4 TO 7

The element following argon is another alkali metal, potassium, and the next element is the alkaline earth metal calcium. Each of these elements has six different ionization energies, one of which has a very small value and therefore corresponds to an electron from the $n = 4$ shell. We may assign these elements the following electron configurations:

$$
\begin{array}{ll}
\text{K} & 1s^2 2s^2 2p^6 3s^2 3p^6 4s^1 \\
\text{Ca} & 1s^2 2s^2 2p^6 3s^2 3p^6 4s^2
\end{array}
$$

The next element, scandium, has seven ionization energies. However, the ionization energy for the most easily removed electron is higher than that for a calcium 4s electron. If this electron were being removed from the 4p level, then by analogy with boron and aluminum it should have a lower ionization energy than that for the 4s electron in calcium. Instead, the electron is in a third level of the $n = 3$ shell, the 3d level, which is higher in energy than the 3p level and may accommodate up to ten electrons. The elements from scandium to zinc all have electrons in a 3d level; these elements are the transition elements of the fourth period. Thus the electron configuration of scandium is $1s^2 2s^2 2p^6 3s^2 3p^6 3d^1 4s^2$; that of zinc is $1s^2 2s^2 2p^6 3s^2 3p^6 3d^{10} 4s^2$.

== *Example 7.6* ===

ELECTRON CONFIGURATIONS

Write the ground state electron configuration for an arsenic atom ($Z = 33$).

Solution

We first find arsenic in the periodic table. It is in period 4 of group 5. Using the periodic table in Figure 7.15 and working from top to bottom, we move across each period writing down the occupancies of the energy levels until we come to the element arsenic:

$$
\begin{array}{ll}
\text{Period 1} & 1s^2 \\
\text{Period 2} & 2s^2 2p^6 \\
\text{Period 3} & 3s^2 3p^6 \\
\text{Period 4} & 4s^2 3d^{10} 4p^3
\end{array}
$$

By writing these down in order we obtain the complete electron configuration, $1s^2 2s^2 2p^6 3s^2 3p^6 4s^2 3d^{10} 4p^3$.

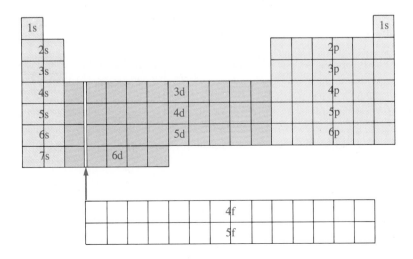

FIGURE 7.15
The Periodic Table and Electron Configurations.

This diagram of the periodic table shows the energy levels that are being filled with electrons in the different regions of the table. The s and p regions correspond to the main groups, the d region to the transition metals and the f region to the lanthanides and actinides.

Another method is to remember (1) the order in which the energy levels are filled (namely, 1s, 2s, 2p, 3s, 3p, 4s, 3d, 4p) and (2) the maximum number of electrons that can be accommodated in each level (namely, 2 in s, 6 in p, 10 in d). Then fill the energy levels until the number of electrons is equal to the atomic number of arsenic, $Z = 33$, which gives the same result: $1s^2 2s^2 2p^6 3s^2 3p^6 4s^2 3d^{10} 4p^3$.

Electron configurations are also often written with the energy levels listed strictly in order of the quantum number n; in this case we have $1s^2 2s^2 2p^6 3s^2 3p^6 3d^{10} 4s^2 4p^3$. Both ways of writing electron configurations are acceptable.

EXERCISE 7.5

Write the electron configuration of the tellurium atom ($Z = 52$).

The fact that electrons in the ground states of potassium and calcium occupy the 4s level, rather than the 3d level, shows that for these two elements the 3d level is of higher energy than the 4s level. Although the *average* energy of the electrons in all the energy levels of the $n = 3$ shell is lower than that of electrons in the $n = 4$ shell, the energies of the different levels in the two shells overlap for some elements. In general, for the $n = 3, 4, 5, \ldots$ shells the energy levels of neighboring shells overlap (see Figure 7.16), which complicates the electron configurations of the elements in periods 4 to 7.

Following zinc, electrons enter the 4p level, starting with gallium and finishing with krypton. Thus krypton has the valence-shell electron configuration $4s^2 4p^6$, which is like those of argon, $3s^2 3p^6$, and neon, $2s^2 2p^6$. The electron configurations are as follows:

Ga	$1s^2 2s^2 2p^6 3s^2 3p^6 3d^{10} 4s^2 4p^1$
Ge	$1s^2 2s^2 2p^6 3s^2 3p^6 3d^{10} 4s^2 4p^2$
As	$1s^2 2s^2 2p^6 3s^2 3p^6 3d^{10} 4s^2 4p^3$
Se	$1s^2 2s^2 2p^6 3s^2 3p^6 3d^{10} 4s^2 4p^4$
Br	$1s^2 2s^2 2p^6 3s^2 3p^6 3d^{10} 4s^2 4p^5$
Kr	$1s^2 2s^2 2p^6 3s^2 3p^6 3d^{10} 4s^2 4p^6$

FIGURE 7.16
Energy Levels.

The energy of the shells increases with increasing n, and the energy of the subshells increases in the order s, p, d, f. When $n \geq 3$, there is some overlap between the energy levels of one shell and the next highest shell.

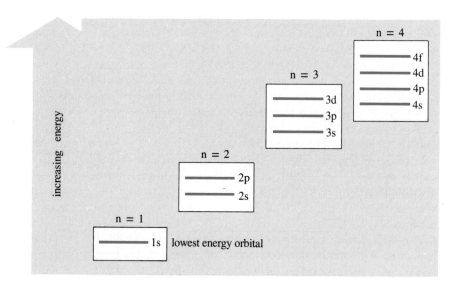

The six elements from Ga to Kr resemble the six elements from Al to Ar in which the 3p level is progressively filled, and the six elements from B to Ne in which the 2p level is filled.

Electron configurations of all the elements are given in Table 7.2. For convenience only the energy levels higher than those in the preceding noble gas are shown in full in the table. Thus the electron configuration of sodium, $1s^2 2s^2 2p^6 3s^1$, is written as [Ne] $3s^1$, and the electron configuration of bromine, $1s^2 2s^2 2p^6 3s^2 3p^6 4s^2 3d^{10} 4p^5$, is written as [Ar] $3d^{10} 4s^2 4p^5$. We will be concerned primarily with the electron configurations of the first 36 elements.

GENERAL RULES The order in which the various energy levels are occupied can be remembered by making use of the diagram in Figure 7.15 or that in Figure 7.16. For the first 36 elements the order is 1s, 2s, 2p, 3s, 3p, 4s, 3d, 4p. Each of the energy levels in a given shell is often called a **subshell**. Thus the $n = 2$ shell has a 2s and a 2p subshell, and the $n = 3$ shell has a 3s, a 3p, and a 3d subshell.

The arrangements of electrons in the shells and the subshells (energy levels) of an atom are summarized in Table 7.3. These arrangements are governed by several rules, as follows:

1. In the $n = 1$ shell there is only one subshell (1s); in the $n = 2$ shell there are two subshells (2s and 2p); in the $n = 3$ shell there are three subshells (3s, 3p, and 3d); and in the $n = 4$ shell there are four subshells (4s, 4p, 4d, and 4f). In general, *the number of subshells in any shell is equal to the quantum number n.*

2. The maximum number of electrons that can occupy the $n = 1$ shell is 2 ($1s^2$); in the $n = 2$ shell it is 8 ($2s^2 2p^6$); and in the $n = 3$ shell it is 18 ($3s^2 3p^6 3d^{10}$). In general, *the maximum number of electrons in a shell is $2n^2$.* For example, for $n = 3$, the maximum number of electrons is $2 \times 3^2 = 18$.

3. *The number of electrons in an s subshell is 2, in a p subshell is 6, in a d subshell is 10, in an f subshell is 14, and so on.* We will be concerned only with the s, p, and d subshells.

ENERGY LEVELS AND THE PERIODIC TABLE Figure 7.15 shows the relationship between the periodic table and the electron configurations. The alkali and alkaline earth metals are called the *s block elements*, because the highest-energy, most easily removed electrons for these elements are s electrons. The elements on the right, which are mainly nonmetals, are called the *p block elements* because the highest-energy electrons for these elements are p electrons. The elements in the middle, the transition elements, are called the *d block elements*.

We have seen that we can conclude from the form of the periodic table that electrons are arranged in shells around the nucleus. The measurement of ionization energies by photoelectron spectroscopy enables us to determine the energy levels of the electrons in each shell, and to find how the electrons are distributed among these levels, in other words, to deduce the electron configurations. But this does not give us a very clear picture of where the electrons are located and how they are moving around the nucleus. We have mentioned that Bohr's model of electrons rotating around the nucleus in circular orbits had to be modified for atoms other than hydrogen. Let us now see why this was necessary and how we can best describe the electrons in an atom.

TABLE 7.2
Electron Configurations of the Elements

Atomic Number	Element	Electron Configuration	Atomic Number	Element	Electron Configuration
1	H	$1s^1$	43	Tc	$[Kr]\,4d^5 5s^2$
2	He	$1s^2$	44	Ru	$[Kr]\,4d^7 5s^1$
3	Li	$[He]\,2s^1$	45	Rh	$[Kr]\,4d^8 5s^1$
4	Be	$[He]\,2s^2$	46	Pd	$[Kr]\,4d^{10}$
5	B	$[He]\,2s^2 2p^1$	47	Ag	$[Kr]\,4d^{10} 5s^1$
6	C	$[He]\,2s^2 2p^2$	48	Cd	$[Kr]\,4d^{10} 5s^2$
7	N	$[He]\,2s^2 2p^3$	49	In	$[Kr]\,4d^{10} 5s^2 5p^1$
8	O	$[He]\,2s^2 2p^4$	50	Sn	$[Kr]\,4d^{10} 5s^2 5p^2$
9	F	$[He]\,2s^2 2p^5$	51	Sb	$[Kr]\,4d^{10} 5s^2 5p^3$
10	Ne	$[He]\,2s^2 2p^6$	52	Te	$[Kr]\,4d^{10} 5d^2 5p^4$
11	Na	$[Ne]\,3s^1$	53	I	$[Kr]\,4d^{10} 5s^2 5p^5$
12	Mg	$[Ne]\,3s^2$	54	Xe	$[Kr]\,4d^{10} 5s^2 5p^6$
13	Al	$[Ne]\,3s^2 3p^1$	55	Cs	$[Xe]\,6s^1$
14	Si	$[Ne]\,3s^2 3p^2$	56	Ba	$[Xe]\,6s^2$
15	P	$[Ne]\,3s^2 3p^3$	57	La	$[Xe]\,5d^1 6s^2$
16	S	$[Ne]\,3s^2 3p^4$	58	Ce	$[Xe]\,4f^1 5d^1 6s^2$
17	Cl	$[Ne]\,3s^2 3p^5$	59	Pr	$[Xe]\,4f^3 \quad 6s^2$
18	Ar	$[Ne]\,3s^2 3p^6$	60	Nd	$[Xe]\,4f^4 \quad 6s^2$
19	K	$[Ar]\,4s^1$	61	Pm	$[Xe]\,4f^5 \quad 6s^2$
20	Ca	$[Ar]\,4s^2$	62	Sm	$[Xe]\,4f^6 \quad 6s^2$
21	Sc	$[Ar]\,3d^1 4s^2$	63	Eu	$[Xe]\,4f^7 \quad 6s^2$
22	Ti	$[Ar]\,3d^2 4s^2$	64	Gd	$[Xe]\,4f^7 5d^1 6s^2$
23	V	$[Ar]\,3d^3 4s^2$	65	Tb	$[Xe]\,4f^9 \quad 6s^2$
24	Cr	$[Ar]\,3d^5 4s^1$	66	Dy	$[Xe]\,4f^{10} \quad 6s^2$
25	Mn	$[Ar]\,3d^5 4s^2$	67	Ho	$[Xe]\,4f^{11} \quad 6s^2$
26	Fe	$[Ar]\,3d^6 4s^2$	68	Er	$[Xe]\,4f^{12} \quad 6s^2$
27	Co	$[Ar]\,3d^7 4s^2$	69	Tm	$[Xe]\,4f^{13} \quad 6s^2$
28	Ni	$[Ar]\,3d^8 4s^2$	70	Yb	$[Xe]\,4f^{14} \quad 6s^2$
29	Cu	$[Ar]\,3d^{10} 4s^1$	71	Lu	$[Xe]\,4f^{14} 5d^1 6s^2$
30	Zn	$[Ar]\,3d^{10} 4s^2$	72	Hf	$[Xe]\,4f^{14} 5d^2 6s^2$
31	Ga	$[Ar]\,3d^{10} 4s^2 4p^1$	73	Ta	$[Xe]\,4f^{14} 5d^3 6s^2$
32	Ge	$[Ar]\,3d^{10} 4s^2 4p^2$	74	W	$[Xe]\,4f^{14} 5d^4 6s^2$
33	As	$[Ar]\,3d^{10} 4s^2 4p^3$	75	Re	$[Xe]\,4f^{14} 5d^5 6s^2$
34	Se	$[Ar]\,3d^{10} 4s^2 4p^4$	76	Os	$[Xe]\,4f^{14} 5d^6 6s^2$
35	Br	$[Ar]\,3d^{10} 4s^2 4p^5$	77	Ir	$[Xe]\,4f^{14} 5d^7 6s^2$
36	Kr	$[Ar]\,3d^{10} 4s^2 4p^6$	78	Pt	$[Xe]\,4f^{14} 5d^9 6s^1$
37	Rb	$[Kr]\,5s^1$	79	Au	$[Xe]\,4f^{14} 5d^{10} 6s^1$
38	Sr	$[Kr]\,5s^2$	80	Hg	$[Xe]\,4f^{14} 5d^{10} 6s^2$
39	Y	$[Kr]\,4d^1 5s^2$	81	Tl	$[Xe]\,4f^{14} 5d^{10} 6s^2 6p^1$
40	Zr	$[Kr]\,4d^2 5s^2$	82	Pb	$[Xe]\,4f^{14} 5d^{10} 6s^2 6p^2$
41	Nb	$[Kr]\,4d^4 5s^1$	83	Bi	$[Xe]\,4f^{14} 5d^{10} 6s^2 6p^3$
42	Mo	$[Kr]\,4d^5 5s^1$	84	Po	$[Xe]\,4f^{14} 5d^{10} 6s^2 6p^4$

TABLE 7.2
Electron Configurations of the Elements (*continued*)

Atomic Number	Element	Electron Configuration	Atomic Number	Element	Electron Configuration
85	At	$[Xe] 4f^{14}5d^{10}6s^26p^5$	95	Am	$[Rn] 5f^7 \quad 7s^2$
86	Rn	$[Xe] 4f^{14}5d^{10}6s^26p^6$	96	Cm	$[Rn] 5f^76d^17s^2$
87	Fr	$[Rn] \quad 7s^1$	97	Bk	$[Rn] 5f^9 \quad 7s^2$
88	Ra	$[Rn] \quad 7s^2$	98	Cf	$[Rn] 5f^{10} \quad 7s^2$
89	Ac	$[Rn] 6d^17s^2$	99	Es	$[Rn] 5f^{11} \quad 7s^2$
90	Th	$[Rn] 6d^27s^2$	100	Fm	$[Rn] 5f^{12} \quad 7s^2$
91	Pa	$[Rn] 5f^26d^17s^2$	101	Md	$[Rn] 5f^{13} \quad 7s^2$
92	U	$[Rn] 5f^36d^17s^2$	102	No	$[Rn] 5f^{14} \quad 7s^2$
93	Np	$[Rn] 5f^46d^17s^2$	103	Lr	$[Rn] 5f^{14}6d^17s^2$
94	Pu	$[Rn] 5f^6 \quad 7s^2$			

TABLE 7.3
Arrangement of Electrons in Shells and Subshells

Shell n	Energy Levels	Number of Electrons	Total Number of Electrons in Shell ($2n^2$)
1	1s	2	2
2	2s	2	8
	2p	6	
3	3s	2	18
	3p	6	
	3d	10	
4	4s	2	32
	4p	6	
	4d	10	
	4f	14	

7.4
The Wave Properties of the Electron

In 1924 French physicist Louis de Broglie (Figure 7.17) made the bold suggestion that since light exhibits both wave and particle characteristics, particles like electrons, protons, and atoms should show wavelike properties. De Broglie went further and proposed that the wavelength associated with a particle of mass m and velocity $\mathbf{v}$ is

$$\lambda = \frac{h}{m\mathbf{v}}$$

The product $m\mathbf{v}$ is called **momentum.**

FIGURE 7.17
Louis Victor, Prince de
Broglie (1892–1977).

When de Broglie put forward his hypothesis, there were no experimental observations to support it, and it was not taken seriously. Then in 1927 Clinton Davisson in the United States and George Thomson in Britain independently demonstrated that beams of electrons are diffracted by crystals. Figure 7.18 shows a diffraction pattern produced by passing a beam of electrons through very thin aluminum foil. Subsequently, diffraction patterns were observed for other particles such as neutrons, protons, and helium nuclei.

═ *Example 7.7* ═══════════════════════════════

THE WAVELENGTH OF AN ELECTRON

The mass of an electron is 9.11×10^{-31} kg. What is the wavelength of an electron with a velocity of 6.12×10^6 m s^{-1}?

Solution

We use the equation

$$\lambda = \frac{h}{m\mathbf{v}}$$

The value of the Planck constant h is 6.63×10^{-34} J s. We must convert this constant to basic SI units by using the unit conversion factor 1 kg m^2 s^{-2}/1 J in order to obtain the wavelength in meters. We then have

$$\lambda = \frac{h}{m\mathbf{v}} = \frac{6.63 \times 10^{-34} \text{ J s}}{(9.11 \times 10^{-31} \text{ kg})(6.12 \times 10^6 \text{ m s}^{-1})} \left(\frac{1 \text{ kg m}^2 \text{ s}^{-2}}{1 \text{ J}} \right)$$

$$= 1.19 \times 10^{-10} \text{ m} = 119 \text{ pm}$$

Thus electrons moving with a velocity of 6.12×10^6 m s^{-1} have a wavelength of 119 pm, which, as we can see from Figure 7.4, is comparable with the wavelength of X rays.

───

Example 7.7 shows that the wavelength of a rapidly moving electron is in the X ray region and is considerably shorter than that of visible light. This fact is put to good use in the electron microscope, as Figure 7.19 shows.

FIGURE 7.18
Diffraction Patterns Produced by Electrons and by X Rays.

Diffraction patterns produced by (a) a beam of fast-moving electrons and (b) a beam of X rays passing through a thin aluminum foil. The foil consists of many tiny aluminum crystals each of which produces a diffraction pattern and these combine to form the pattern shown. The similarity of the two patterns shows that electrons behave like X rays and can have wavelike properties.

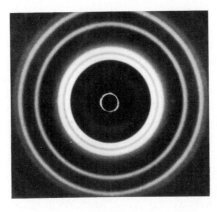

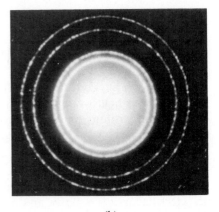

(a)   (b)

The wave behavior of matter can only be observed for very small particles such as electrons and atoms. For an ordinary-sized object the associated wavelength is much too small to be observed. For example, the wavelength of a 1500-kg car moving at 30 m s^{-1} ($\approx$ 60 miles h^{-1}) can be calculated from de Broglie's equation: it is only 1.5 × 10^{-38} m. This wavelength is much too small to be measured and therefore has no physical significance.

FIGURE 7.19
The Electron Microscope.

EXERCISE 7.6

What is the wavelength of (a) a proton (mass 1.6726 × 10^{-27} kg) traveling at 3.46 × 10^{10} m s^{-1} and (b) a baseball (mass 0.10 kg) traveling at 30 m s^{-1} (approximately 60 miles h^{-1})?

The size of an object that can be distinguished in an ordinary microscope is limited by the wavelength of visible light. Because diffraction blurs the image, a clear and accurate image of an object that is smaller than the wavelength of the light used cannot be obtained. Thus the lower limit to the size of objects that can be distinguished in an ordinary optical microscope is about 500 nm. However, objects as small as 0.1 nm can be observed in a microscope that uses beams of electrons instead of visible light, because the wavelength of electrons traveling at high speed is much shorter than that of visible light. Here we see an electron microscope photograph of a piece of graphite. The bright bands are layers of carbon atoms that are only 341 pm apart. This corresponds to a magnification of about 15 million times.

Thus we have seen that in certain aspects of their behavior electrons resemble particles, while in others they resemble waves. Similarly, in some aspects of its behavior light resembles particles, and in other aspects it resembles waves. Neither the wave model nor the particle model can be used exclusively to describe either matter or light. Some phenomena are best described by one model, while other phenomena are best described by the other model.

UNCERTAINTY PRINCIPLE

Whereas a particle is localized in space and its position can be accurately defined, a wave is spread out in space. If an electron has both wave and particle properties, can its position in space be defined? If it is a particle, we should be able to find precisely where it is located. But if it is a wave, we cannot do so. By considering the various ways in which one might try, in principle, to find the position of an electron, the German physicist Werner Heisenberg (Figure 7.20) came to the conclusion in 1927 that there are definite limitations on the accuracy with which one can define the position of an electron. He proposed that the behavior of electrons was governed by the uncertainty principle, which states that

The uncertainty Δx in the position of an electron times the uncertainty $\Delta(mv)$ in its momentum must be greater than the Planck constant, or expressed mathematically

$$\Delta x \, \Delta(mv) > h$$

This means that the more accurately we know the momentum of an electron (the product of its mass and velocity) the less accurately can we know its position and vice versa. This limitation applies in principle to all particles but because the Planck constant is so small the uncertainty in the position of a particle with a mass of even 1 mg is much too small to be of any significance. However, for an electron it implies that we cannot simultaneously know both the position and the velocity of an electron as it moves around the nucleus. In other words, Bohr's model of an electron moving with a known velocity in a well-defined orbit cannot be correct. How then are we going to describe how the electrons move in atoms and molecules? To do so we must use quantum mechanics.

FIGURE 7.20
Werner Heisenberg (1901–1976).

7.5
Quantum Mechanics

The ideas of Einstein, de Broglie, Bohr, Heisenberg, and others led to the development, during the period 1924–1928, of a new theory that could describe the behavior of electrons in atoms and molecules. This theory is called **quantum mechanics**. No attempt is made in this theory to describe the position or the path of an electron precisely. Instead, the theory gives the *probability* of finding an electron at some specified point in an atom or a molecule. This probability can be obtained from an equation proposed in 1925 by the Austrian physicist Erwin Schrödinger (1887–1961) (Figure 7.21). This equation is often called the **Schrödinger wave equation** because it is based on the description of an electron as a wave. It is not a wave that travels through space like the electromagnetic waves used to describe light, but it is a *standing wave* that is confined to an atom or molecule. A familiar example of standing waves is the waves formed in the strings of a violin or other string instrument when they are plucked or bowed.

ONE- AND TWO-DIMENSIONAL STANDING WAVES

Because both ends of a violin string are fixed, only certain standing-wave patterns are possible. Only integral or half-integral wavelengths can fit into the length of the string as shown in Figure 7.22. This is a macroscopic analogy to quantization in a microscopic system. The vibrations of a violin string are quantized in the sense that only certain vibrations are allowed. The points at which the amplitude of the wave is zero are called **nodes**. Excluding the fixed ends of the string, the lowest-energy vibration has zero nodes; the next lowest has one node; and so on. The possible vibrations are designated by a number n. For the lowest-energy vibration (called the fundamental), $n = 1$, and there are zero nodes. For the vibration of next-highest energy (called the first overtone), $n = 2$, and there is one node. The second overtone, $n = 3$, has two nodes, and so on. In general, a vibration has $n - 1$ nodes.

The vibrations of a violin string are in one dimension. Now consider the vibrations of a circular drumhead, which are in two dimensions. They are illus-

FIGURE 7.21
Erwin Schrödinger (1887–1961).

Schrödinger was born and educated in Austria, obtaining his Ph.D. at the University of Vienna in 1910. After serving in the Austrian army in World War I, he held positions at several universities. While at the University of Zurich, in 1925, he published his famous wave equation that made such an important contribution to the new field of quantum mechanics, which was also being developed at the same time by several other physicists, including Heisenberg and Pauli. In 1928 he succeeded Planck as professor of theoretical physics at the University of Berlin. Subsequently, his life reflected the troubled times in Europe. In 1933, disgusted with the advent of Nazism, he left Germany and went to Oxford. In 1936 he returned to Austria, but only two years later Hitler annexed Austria to Germany and he was forced to move again. He arrived in Rome with only the possessions that he could carry in a backpack and found asylum in the Vatican. From there he went to the Institute of Advanced Studies in Dublin, where he remained until 1955, when he retired to his native Austria.

trated in Figure 7.23. The nodes are not points but lines on the surface. They are of two kinds: straight lines and circles. The total number of nodes is again given by $n - 1$, while the number of straight-line nodes is given by a second number, l, which cannot be greater than the total number of nodes, $n - 1$. Therefore the value of l is limited to $0, 1, 2, \ldots, n - 1$. For example, for $n = 2$ the possible values of l are 0 or 1; thus there are two different types of vibration possible, one having a circular node (for which $l = 0$) and the other a linear node ($l = 1$). The $l = 1$ vibration that has a linear node can occur in two mutually perpendicular directions. There are only two such vibrations because a vibration in any other direction can always be represented as a sum of the two independent vibrations.

These examples of standing waves in musical instruments show us that only certain standing-wave patterns and energy states are possible. They also introduce us to the very important idea of nodes which are an essential feature of standing waves. The form of the waves can be described by integral numbers which give the number and nature of the nodes; one number for a one-dimensional system and two for a two-dimensional system. Let us now consider the standing waves that describe electrons in atoms and molecules, which are three dimensional.

HYDROGEN ATOMIC ORBITALS

We will confine our discussion to the very simple case of the hydrogen atom where there is only one electron. The standing waves associated with each of the energy levels for the electron in the hydrogen atom can be obtained by solving the appropriate form of the Schrödinger equation for the hydrogen atom. Although we will not do this we can obtain some appreciation of these standing-wave patterns by extrapolating from the one- and two-dimensional cases that we have just considered. For atoms these standing waves are called **orbitals**. The orbitals for the lower energy levels of the hydrogen atom are shown in Figure 7.24.

Because we are now dealing with a three-dimensional system we need three integral numbers n, l and m, which are called **quantum numbers**, to describe the orbitals. The circular and linear nodes of the two-dimensional system are now replaced by **nodal surfaces** which may be **planar** or **spherical**. The total number of nodes is $n - 1$, as for one- and two-dimensional waves. The number of planar nodes is given by the quantum number l, which, as before, can have all integral values from 0 to $n - 1$. The number of spherical nodes is therefore $n - 1 - l$.

Orbitals for which $l = 0$ have only spherical nodes and are called **s orbitals**. Thus for $n = 1$ there is a 1s orbital with no nodes, for $n = 2$ a 2s orbital with one node, for $n = 3$ a 3s orbital with two nodes, and so on.

Orbitals for which $l = 1$ have one planar node and are called **p orbitals**. Thus there are 2p orbitals with one planar node and no spherical nodes and 3p orbitals with one planar node and one spherical node, and so on.

Orbitals for which $l = 2$ have two planar nodes and are called **d orbitals**. The 3d orbitals have two planar nodes and no spherical nodes.

The energy of the hydrogen-atom orbitals depends only on the value of n, and they therefore increase in energy in the order $1s < 2s = 2p < 3s = 3p = 3d$.

The orbitals with planar nodes can have different relative orientations in space. For example, the planar node in a p orbital can be perpendicular to

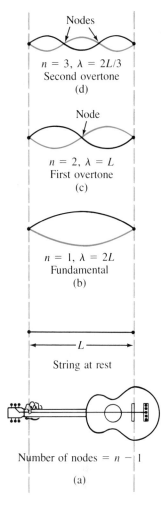

FIGURE 7.22
Vibrations of a String, such as a Guitar or Violin String, Fixed at Both Ends.

(a) String of length L at rest.
(b) The lowest-energy vibration ($n = 1$) is called the fundamental.
(c) The next-highest-energy vibration ($n = 2$) is called the first overtone. (d) The next-highest-energy vibration ($n = 3$) is called the second overtone. In general, there are $n - 1$ points of zero amplitude (nodes) for each value of n, and the wavelength λ is given by $2L/n$.

Within the figure:
Nodes
$n = 3$, $\lambda = 2L/3$
Second overtone
(d)

Node
$n = 2$, $\lambda = L$
First overtone
(c)

$n = 1$, $\lambda = 2L$
Fundamental
(b)

L
String at rest

Number of nodes $= n - 1$
(a)

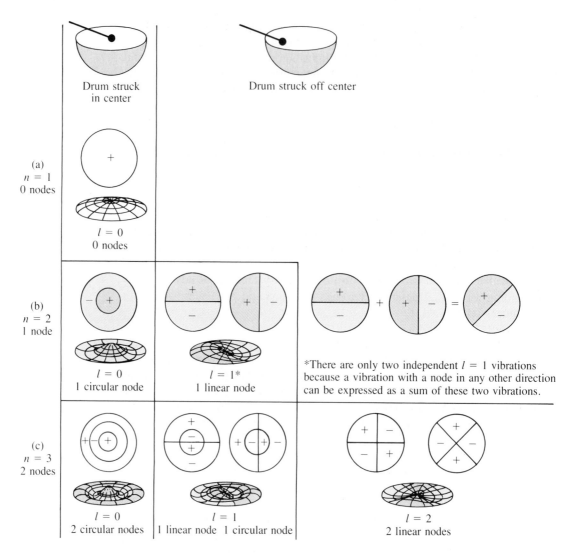

FIGURE 7.23
Vibrations of a Circular Surface, such as a Drumhead.

(a) There is only one vibration ($n = 1$) with no nodes; this is the lowest-energy vibration. (b) There are three vibrations with one node ($n = 2$); one has a circular node and the other two have linear nodes. There are only two vibrations with a linear node, because a vibration with a node in any other direction can be expressed by the appropriate sum of the two independent vibrations with nodes at right angles. (c) There are five vibrations with two nodes ($n = 3$); one has two circular nodes, two have one circular and one linear node, and two have two linear nodes.

any one of the x, y, or z axes. Thus there are three independent p orbitals, designated p_x, p_y, and p_z, with their planar node perpendicular to the x, y, or z axis, respectively. For any orbital the number of possible independent orientations is given by the number of values of the third quantum number m, which has all integral values from $-l$ through 0 to $+l$. Hence for $l = 0$, there is only one possible value of m, namely zero, so there is only one possible orientation of an s orbital. This is clear if we consider that if we rotate a sphere we cannot distinguish any new orientation that is produced from the original orientation. For $l = 1$ (one planar node), m has the three values, -1, 0, and $+1$, so there are three possible orientations for a p orbital as we have just seen. For a d orbital which has $l = 2$ there are five possible orientations corresponding to the five possible m values of -2, -1, 0, $+1$, and $+2$. The possible values of the quantum numbers n, l, and m are summarized in Table 7.4.

The diagrams in the top half of Figure 7.24 show where the amplitude of the wave has a positive value and where it has a negative value, but they do not completely describe the electron wave because they do not show how the amplitude of the wave varies in the regions between the nodes. This is difficult to do on the same diagrams so we also show how the amplitude of the wave varies along a line radiating from the nucleus. In the Schrödinger equation the amplitude of the wave is given the symbol ψ (psi) and ψ is called the **wave function**. Since s orbitals are spherical, the variation in ψ is exactly the same in any direction from the nucleus. In the 1s orbital ψ has its maximum value at the nucleus and decreases with increasing distance from the nucleus, eventually reaching an infinitesimally small value. In the 2s orbital ψ again has its maximum value at the nucleus; but since there is a spherical node, ψ passes through a value of zero, then decreases to a minimum value, and finally increases again to an infinitesimally small negative value.

The overall shape of a 2p orbital may be approximately described as a sphere sliced in half by the planar node through the nucleus. In one hemisphere ψ has a positive value, and in the other hemisphere it has a negative value; at the nucleus the value of ψ is zero. In the positive half of the orbital ψ increases to a maximum value and then decreases to an infinitesimally small value. In the negative half of the orbital ψ decreases to a minimum value and

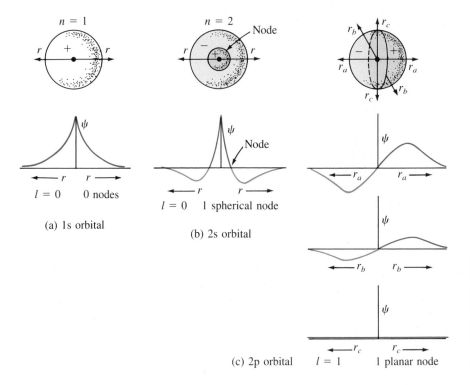

(a) 1s orbital

(b) 2s orbital

(c) 2p orbital

FIGURE 7.24
Hydrogen-Atom Orbitals.

(a) The spherical 1s orbital and the variation of ψ in any direction from the nucleus in a 1s orbital.
(b) The spherical 2s orbital and the variation of ψ in any direction from the nucleus in a 2s orbital.
(c) The 2p orbital and the variation of ψ in the directions r_a, r_b, and r_c in the 2p orbital.

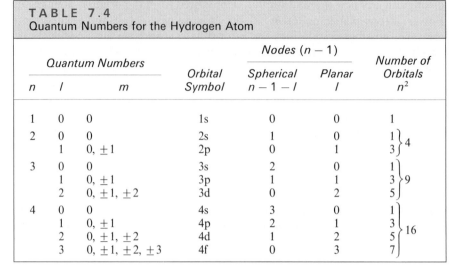

TABLE 7.4
Quantum Numbers for the Hydrogen Atom

Quantum Numbers			Orbital Symbol	Nodes (n − 1)		Number of Orbitals n^2
n	l	m		Spherical $n − 1 − l$	Planar l	
1	0	0	1s	0	0	1
2	0	0	2s	1	0	1 } 4
	1	0, ±1	2p	0	1	3
3	0	0	3s	2	0	1 } 9
	1	0, ±1	3p	1	1	3
	2	0, ±1, ±2	3d	0	2	5
4	0	0	4s	3	0	1 } 16
	1	0, ±1	4p	2	1	3
	2	0, ±1, ±2	4d	1	2	5
	3	0, ±1, ±2, ±3	4f	0	3	7

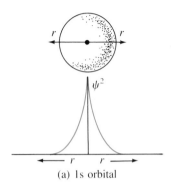

(a) 1s orbital

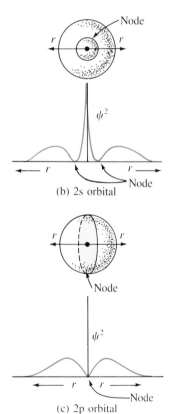

(b) 2s orbital

(c) 2p orbital

FIGURE 7.25
Variation of ψ^2 (Probability of Finding an Electron) in 1s, 2s, and 2p Orbitals of the Hydrogen Atom.

(a) The 1s orbital; (b) the 2s orbital; (c) the 2p orbital.

then increases to an infinitesimally small negative value. The orbital is not truly spherical, and ψ does not attain the same maximum value in every direction. The largest maximum and minimum values of ψ are along the directions (r_a) perpendicular to the nodal plane. In any other direction the maxima and minima in ψ are smaller, and in any direction along the nodal plane ψ is always zero. There are three independent 2p orbitals with their nodal planes perpendicular to the x, y, and z axes, respectively; they are designated $2p_x$, $2p_y$, and $2p_z$. A 3d orbital has two planar nodes. But the shapes of 3d orbitals will not be of any concern to us until much later.

In the one- and two-dimensional standing waves that we described above the amplitude is a measure of the displacement of the violin string or the surface of the drum from its equilibrium position. What then is the meaning of the amplitude of the electron wave? Surprisingly, perhaps, ψ has no physical significance but *the value of ψ^2 at any particular point gives the probability of finding the electron at that point.* The variation of ψ^2 is quite similar to the variation of ψ except that ψ^2 is everywhere either positive or zero. We can use diagrams similar to those in Figure 7.24 to show the variation of ψ^2, but we can leave out the positive and negative signs because ψ^2 is everywhere positive except at the nodes, where it is zero, as shown in Figure 7.25.

We can conveniently think of the electron as spread out in the form of a negative charge cloud surrounding the nucleus. The density of the charge cloud is not uniform; the value of ψ^2 at any point represents the density of the charge cloud at that point. We can represent the charge cloud by a dot density diagram in which the density of dots in any particular small volume represents the density of the charge cloud, that is, the value of ψ^2 in the small volume (see Figure 7.26). In the 1s orbital the density of the charge cloud is greatest at the nucleus and decreases to an infinitesimally small value with increasing distance from the nucleus. In the 2s orbital the maximum density is again at the nucleus, but it decreases to zero at the node and then increases again and passes through a maximum before decreasing to an infinitesimally small value.

The charge density for a $2p_z$ orbital consists of two clouds that have the shape of flattened spheres, one above and one below the nodal plane (Figure

7.26). The density of these charge clouds is again not uniform, but they are more dense toward the center of each cloud, the point of maximum density being on the z axis above and below the nodal plane. The charge density distributions for the other 2p orbitals have the same shape; but in the $2p_y$ orbital the two maximum density regions are along the y axis, and in the $2p_x$ orbital they are along the x axis.

Although the electron density of an isolated atom in principle extends to infinity, it rapidly becomes insignificant at quite a small distance from the nucleus. It is therefore convenient to choose a surface of constant, but small, electron density to represent the size and shape of the orbital. This boundary surface is usually chosen so that it encloses 90% of the electron density. It is therefore a good representation of the overall size and shape of the electron density distribution. As Figure 7.27 shows, an s orbital has a spherical boundary surface and a p orbital has a boundary surface that can be approximately described as two flattened spheres, one on each side of the nodal plane.

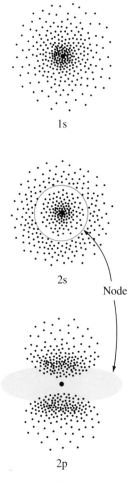

1s

2s

Node

2p

FIGURE 7.26
Dot Density Diagrams Showing Charge Clouds for 1s, 2s, and 2p Orbitals of the H Atom.

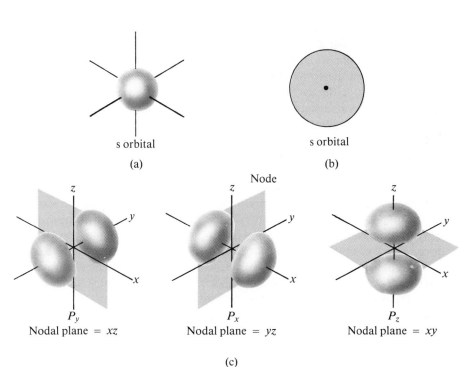

s orbital

(a)

s orbital

(b)

P_y
Nodal plane $= xz$

P_x
Nodal plane $= yz$

Node

P_z
Nodal plane $= xy$

(c)

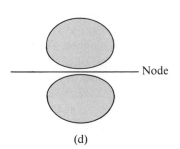

Node

(d)

FIGURE 7.27
Boundary Surfaces for s and p Orbitals.

Boundary surfaces are drawn so as to enclose approximately 90% of the electron density. (a) Boundary surface for an s orbital. (b) Cross-section through the boundary surface of an s orbital. (c) Boundary surfaces of the three p orbitals. (d) Cross-section through the boundary surface of a p orbital. In these boundary-surface representations of orbitals any spherical nodes inside the boundary surface are not shown.

The orbitals that we have described represent the probability of finding the electron at any point in the space around the nucleus, or the distribution of electron density around the nucleus, for the single electron of the hydrogen atom in the ground (1s) state and for the various excited states of the hydrogen atom. The electron density is closest to the nucleus in the ground (1s) state. In the excited states there is more electron density further away from the nucleus, and there are one or more nodes in the distribution.

POLYELECTRONIC ATOMS

The Schrödinger equation becomes too complicated to solve exactly for any atom or molecule containing more than one electron, although for most atoms and many simple molecules, given the power of modern computers, a very accurate result can be obtained by successive approximations. However, for most purposes it is a sufficiently good approximation to assume that the standing-wave patterns, that is, the orbitals, for an atom with two or more electrons have the same general form as the orbitals for the hydrogen atom, and in particular that the orbitals are described by the same three quantum numbers n, l, and m. Now in the hydrogen atom ψ^2 gives the probability distribution of the single electron in its different orbitals: 1s, 2s, 2p, etc. In a polyelectronic atom we are concerned with many electrons. So the question arises: Can all the electrons in a given atom be described by the same orbital? Clearly this is not the case, because if it were, all the electrons in an atom would have the same energy and would be in the lowest energy level, that is, the 1s level. The electron configurations that we obtained from the form of the periodic table, and from ionization energies, show that there are at most only two electrons in the 1s level, two in the 2s level, and six in the 2p level. Since there is only one 1s orbital, one 2s orbital, and three 2p orbitals, it is clear that *the maximum number of electrons associated with any orbital is two*. It is customary to state that the maximum number of electrons that can *occupy* any given orbital is two. This description is used because it is often convenient to think of an orbital as a "box" into which we can put a maximum of two electrons. The box is then full and any more electrons must be placed in other boxes, that is, in other orbitals. Nevertheless, we should remember that an orbital does not exist in the absence of the electron—an orbital is the standing-wave pattern of an electron in a given energy state.

In order to understand why a maximum of two electrons may be described by the same orbital, we need to consider another property of an electron, namely its spin.

ELECTRON SPIN

Several experiments carried out in the 1920s (Box 7.2) showed that an electron has magnetic properties; in other words, when it is placed in a magnetic field, it behaves rather like a small bar magnet. The magnetic properties of the electron can be understood if we think of it as a charged sphere rotating about an axis through its center (Figure 7.28). A rotating charge generates a magnetic field, and therefore the electron behaves like a magnet.

An ordinary bar magnet may be placed in any orientation in a magnetic field, but its energy is lowest when it is pointing in the direction of the field, as when a compass needle, which is free to rotate, points toward the magnetic

BOX 7.2

EXPERIMENTAL EVIDENCE FOR ELECTRON SPIN

Experiments carried out by Otto Stern and Walter Gerlach in Germany beginning in 1921 provided the first direct evidence for electron spin. They passed a beam of alkali metal or silver atoms between the poles of a magnet designed to give a very nonuniform field. In each case the beam of atoms was split into two. Since the atoms were not charged, they could be deflected only if they behaved like magnets.

A magnet situated in a nonuniform magnetic field experiences a resultant force that displaces it, because one pole is situated in a stronger magnetic field than the other pole. In a uniform magnetic field both poles experience the same force, and there is no resultant force to displace the magnet. The amount by which a moving magnet is deflected from its original path in a nonuniform field depends on the orientation of the magnet. When it is lined up along the direction in which the field is changing the most (the greatest field gradient), it experiences the greatest force. When it is at right angles to this direction (zero field gradient), it experiences no deflecting force.

A beam of tiny magnets having random orientations, when entering the magnetic field, would be spread out into a continuous band. The fact that a beam of alkali metal or silver atoms is split up only into two beams shows not only that the atoms are magnetic but also that the atomic magnets can only have two orientations with respect to the magnetic field.

The explanation of the magnetic behavior of these atoms was given by two Dutch physicists, George Uhlenbeck and Sam Goudsmit. To explain certain fine details of atomic spectra, they proposed in 1925 that an electron has spin. The alkali metal atoms and the silver atom each have one electron in their valence shells. It is the spin of this single unpaired electron that is responsible for the magnetic properties of these atoms.

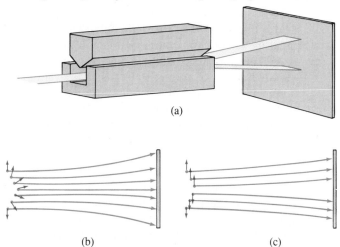

The Experimental Arrangement for the Stern–Gerlach Experiment. (a) A beam of atoms is passed through an inhomogeneous magnetic field produced by the specially shaped pole pieces. (b) The expected result for magnets that can take up any orientation with respect to the field: the beam of atoms is spread out uniformly. (c) The experimental result for silver atoms which have only one unpaired electron in the valence shell. The beam of atoms is split up into two distinct beams. This result shows that the magnetic moment due to the single valence-shell electron can only take up two orientations with respect to the field.

north pole. The electron, however, differs from a compass needle in that it can have only two orientations in a magnetic field, either along the direction of the field or in the opposite direction. When the axis of rotation is pointing in the direction of the field, the electron has a lower energy than when the axis is pointing in the opposite direction. The energy of the electron in a magnetic

FIGURE 7.28
Electron Spin.

A spinning electron generates a magnetic field and behaves like a bar magnet. However, unlike a compass needle, it can only have two orientations when placed in a magnetic field. Either its spin axis points in the direction of the magnetic field (clockwise spin), or its spin axis points in the opposite direction (counter-clockwise spin). (a) A compass needle can have any orientation in a magnetic field. If left free to rotate it takes up the lowest-energy orientation on the left. But by applying a force to the needle it can be rotated into another orientation. Work must be done to rotate the needle so these other orientations all have a higher energy. (b) In contrast an electron has only two possible orientations in a magnetic field. All other orientations are forbidden. Thus we say that the orientation of the spinning electron is quantized; it can only line up either with or against the field.

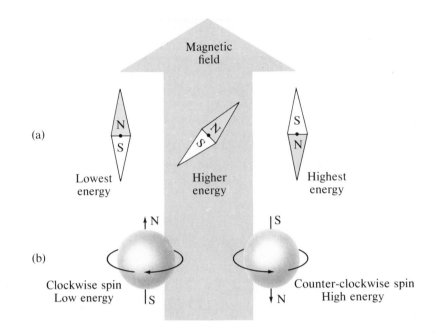

field is quantized, and there are only two corresponding energy levels. In contrast, the energy of a compass needle in the earth's magnetic field is not quantized. The compass needle can have any orientation, and there is essentially an infinite number of corresponding energy levels.

When the electron moves from the lower to the upper of its two energy levels, the orientation of the axis of the spinning electron changes from being along the direction of the magnetic field to being in the opposite direction. In other words, the direction of rotation of the electron is reversed. The difference in energy between the two levels for an electron in a magnetic field is very small. In the field of a strong electromagnet the difference in energy between the two orientations of the electron is about 10^6 times smaller than the ionization energy of an electron in an atom. Thus it takes very little energy to reverse the spin of an electron. In a system containing more than one electron, any two electrons may be spinning in the same direction—we say that they have the same spin—or in opposite directions—we say that they have opposite spins.

The experiment described in Box 7.2 shows that a hydrogen atom behaves like a very small magnet because of the spin of its single electron. The same experiment shows that a helium atom, which has two electrons in its 1s orbital, does *not* behave like a magnet. The only way to explain that helium does not behave like a magnet is to assume that the two electrons have opposite spins so that the magnetic fields that they produce are equal and opposite and thus cancel each other. This behavior of electrons was summarized by the Swiss physicist Wolfgang Pauli (1900–1958) in 1925 in the statement

> **An orbital can accommodate no more than two electrons and these electrons must have opposite spins.**

This statement is called the **Pauli exclusion principle**. When an orbital contains two electrons of opposite spin it excludes all other electrons. Because the orbital describes the spatial distribution of the two electrons this is somewhat equivalent to saying that two electrons in an orbital occupy a certain region of space and exclude all other electrons from this space. However, this is not as rigid an exclusion as it is for macroscopic objects because orbitals can overlap in space.

ORBITAL BOX DIAGRAMS AND ELECTRON CONFIGURATIONS

As we mentioned above, it is often convenient to represent an orbital as a "box" in which we place arrows pointing either up or down to represent the spin of the electron. If there is only one electron in the orbital it may have its spin either "up" or "down":

If there are two electrons in the orbital their spins must be opposite:

We can now use box diagrams to write out the ground state electron configurations of the elements in a little more detail than we have done so far. In the ground state of any atom the electrons will be in the lowest-energy orbitals that are available. Thus for hydrogen and helium the electrons are in the 1s orbital:

H $1s^1$ ↑

He $1s^2$ ↑↓

The orbital box diagrams for the next elements Li ($1s^2 2s^1$) and Be ($1s^2 2s^2$) are shown in Figure 7.29. In these orbital box diagrams we put the orbital boxes in order of increasing energy but we make no attempt to indicate their relative energies.

The 2p level has three 2p orbitals, $2p_x$, $2p_y$, and $2p_z$, which all have the same energy, so we can put the fifth electron of the boron atom into any of the three boxes representing these orbitals and it is immaterial whether the spin of the electron is "up" or "down". Thus all the following diagrams are equivalent,

and it does not matter which we use. However, it is conventional to put the first electron to be placed in these orbitals in the first box with its spin "up",

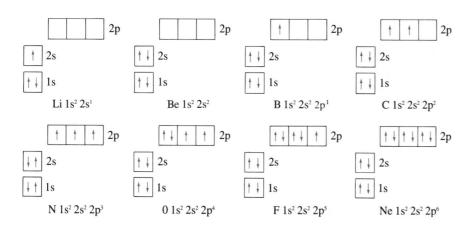

FIGURE 7.29
Orbital Diagrams for the Elements Lithium to Neon.

These diagrams are the ground state configurations. In accordance with Hund's rule, electrons in the same energy level as far as possible occupy separate orbitals and have the same spin. When electrons must occupy the same orbital, they have opposite spins, as the arrows indicate.

as shown in Figure 7.29. When we draw the box diagram for carbon we have a choice among the three 2p boxes in which to place the sixth electron. It could go into the box already containing an electron, in which case the two electrons would have to have opposite spins, or it could go into one of the unoccupied boxes, in which case it could have either spin:

$$\boxed{\uparrow\downarrow}\;\boxed{}\;\boxed{} \qquad \boxed{\uparrow}\;\boxed{\uparrow}\;\boxed{} \qquad \boxed{\uparrow}\;\boxed{}\;\boxed{\uparrow} \qquad \boxed{\uparrow}\;\boxed{\downarrow}\;\boxed{} \qquad \boxed{\uparrow}\;\boxed{}\;\boxed{\downarrow}$$

In fact the electron occupies one of the empty orbitals and has the same spin as the electron in the occupied orbital. This is a general rule known as **Hund's rule** which can be stated as follows:

> In the lowest-energy (ground state) electron configuration electrons in the same energy level as far as possible occupy separate orbitals and have the same spin.

Following this rule we can represent the electron configurations of carbon and the other elements of the second period by the orbital box diagrams shown in Figure 7.29. When the number of electrons in an energy level does not exceed the number of available orbitals, then, following Hund's rule, all the electrons occupy separate orbitals and have the same spin, as in the cases of boron, carbon, and nitrogen. When there are more electrons than available orbitals, then at least some of the orbitals must contain two electrons of opposite spin, as in the case of oxygen, fluorine, and neon.

Hund's rule is a consequence of the fact that two electrons in the same orbital have the same distribution in space—they are as close together as they can be—whereas electrons in different orbitals have different distributions in space and they are therefore, on average, further apart. Because they are further apart electrons in different orbitals have a lower electrostatic repulsion energy than they would have if they were in the same orbital. But why should two electrons in separate orbitals also have the same spin in the lowest-energy (ground state) configuration? The reason is that in the three possible arrangements of two electrons with the same spin among three orbitals of the same energy,

$$\boxed{\uparrow}\;\boxed{\uparrow}\;\boxed{} \qquad \boxed{\uparrow}\;\boxed{}\;\boxed{\uparrow} \qquad \boxed{}\;\boxed{\uparrow}\;\boxed{\uparrow}$$

the two electrons are necessarily always in separate orbitals and therefore have a minimum repulsion energy, whereas, if they have opposite spin, the possible arrangements of the two electrons include those in which the electrons are in the same orbital,

$$\boxed{\uparrow\downarrow}\;\boxed{}\;\boxed{} \quad \boxed{}\;\boxed{\uparrow\downarrow}\;\boxed{} \quad \boxed{}\;\boxed{}\;\boxed{\uparrow\downarrow} \quad \boxed{\uparrow}\;\boxed{\downarrow}\;\boxed{} \quad \boxed{\uparrow}\;\boxed{}\;\boxed{\downarrow} \quad \boxed{}\;\boxed{\uparrow}\;\boxed{\downarrow}$$

so the average energy of repulsion is greater than for electrons with the same spin.

═ *Example 7.8* ══════════════════════════════════

ORBITAL BOX DIAGRAMS

Write the electron configuration and draw an orbital box diagram for the ground state of each of the following atoms and ions:

$$O \qquad F^- \qquad Na \qquad Na^+ \qquad Si \qquad S^{2-}$$

Solution

First, we write the electron configuration:

$$O \quad 1s^2 2s^2 2p^4$$

Then we draw one box for each s level and three boxes for each p level:

1s 2s 2p

Now we place two electrons of opposite spin in each of the 1s and 2s levels:

1s 2s 2p

Then we place one electron in each of the 2p orbitals, following Hund's rule:

1s 2s 2p

Finally, we add the fourth electron to one of the 2p orbitals.

O

1s 2s 2p

The same procedure gives the other orbital diagrams:

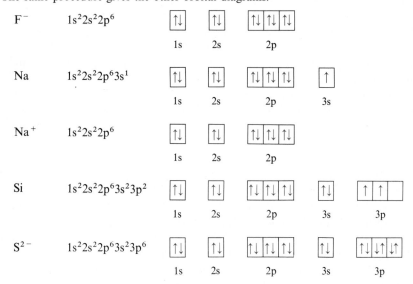

		1s	2s	2p	3s	3p
F^-	$1s^2 2s^2 2p^6$					
Na	$1s^2 2s^2 2p^6 3s^1$					
Na^+	$1s^2 2s^2 2p^6$					
Si	$1s^2 2s^2 2p^6 3s^2 3p^2$					
S^{2-}	$1s^2 2s^2 2p^6 3s^2 3p^6$					

EXERCISE 7.7

Write the electron configurations and draw an orbital box diagram for each of the following atoms and ions:

$$Mg \quad Al \quad Cl \quad Cl^- \quad K^+$$

The orbital diagrams in Figure 7.29 represent ground state electron configurations. There are many other possible electron configurations for these atoms, but they all represent higher-energy excited states. The following diagrams show a few of the many possible excited state electron configurations of boron, carbon, and nitrogen:

B (excited state) $\boxed{\uparrow\downarrow}$ $\boxed{\uparrow}$ $\boxed{\uparrow\,|\,\uparrow\,|\,}$

                          1s   2s   2p

C (excited state) $\boxed{\uparrow\downarrow}$ $\boxed{\uparrow}$ $\boxed{\uparrow\,|\,\uparrow\,|\,\uparrow}$

                          1s   2s   2p

C (excited state) $\boxed{\uparrow\downarrow}$ $\boxed{\uparrow}$ $\boxed{\uparrow\,|\,\uparrow\,|\,}$ $\boxed{\uparrow}$

                          1s   2s   2p   3s

N (excited state) $\boxed{\uparrow\downarrow}$ $\boxed{\uparrow}$ $\boxed{\uparrow\downarrow\,|\,\uparrow\,|\,\uparrow}$

                          1s   2s   2p

FIRST IONIZATION ENERGIES

Earlier, in Chapter 4, we predicted that the energy needed to remove the most easily removed electron (the first ionization energy) should increase across any period of the periodic table, because the core charge increases from left to right across any period. As Table 7.1 shows, this is the observed trend for periods 2 and 3 with minor deviations at boron and oxygen in period 2 and at aluminum and sulfur in period 3.

The energy of the hydrogen-atom orbitals depends only on the quantum number n, but the small decrease in the ionization energy from beryllium to boron and from magnesium to aluminum indicates that, in a many-electron atom, an electron in a p subshell is more easily removed than one in an s subshell. In other words, the energy of the orbitals in a many-electron atom depends on both the n and l quantum numbers. Let us consider why this should be.

We assumed previously that the electrons in the valence shell of an atom have their probability density located very largely outside the inner-core electrons, so that for an atom with a nuclear charge Z, the valence electrons are attracted by a charge of $+Z$, but are repelled by the core electrons. To a first approximation, the valence-shell electrons experience the resultant attraction of a core charge of $Z - 2$ for the second-period elements and $Z - 2 - 8$ for the third-period elements. The inner electrons shield, or screen, the outer-shell valence electrons from the full effect of the nuclear charge $+Z$. However, this is a rather crude approximation for two reasons: (1) the valence electrons also repel each other, and (2) the valence-shell electrons are not totally outside the core electrons. Thus, in fact, the valence electrons feel the attraction of an *effective nuclear charge* of $Z - S$, where S is called the *screening constant*, rather than the core charge. The core charge is only a rough approximation to the effective nuclear charge but it is adequate for many purposes and is very useful because its value for any atom is easily found simply by counting electrons. Values of the screening constant, S, and therefore of the effective nuclear charge, $Z - S$, are not easily determined. However, we can see that the screening constant is not the same for valence electrons in different subshells; in particular, it is greater for

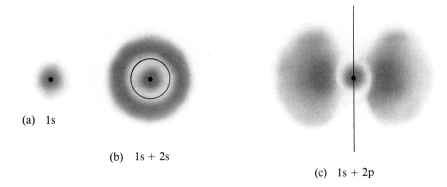

(a)   1s

(b)   1s + 2s

(c)   1s + 2p

FIGURE 7.30
Effective Nuclear Charge.

A 2s electron penetrates inside the 1s shell and has some probability of being found right at the nucleus. This part of the electron density experiences a greater attractive force than the rest of the electron density, which is "outside" the 1s shell. Thus a 2s electron experiences the attraction of a greater charge than the core charge of $Z - 2$. This charge is called the effective nuclear charge. In contrast, the 2p electron has no density at the nucleus and only a very small amount of density "inside" the 1s shell. Thus it experiences the attraction of a charge only very slightly greater than $Z - 2$. A 2s electron experiences a greater effective nuclear charge than a 2p electron does, so an electron in a 2s orbital has a slightly lower energy than one in a 2p orbital.

p electrons than for s electrons. A 2s electron has an appreciable probability of being found very close to the nucleus, and indeed right at the nucleus, whereas a 2p electron has a node at the nucleus, and therefore a zero probability of being found there and only a low probability of being found close to the nucleus (see Figure 7.30). Thus a 2s electron has more density "inside" the 1s electrons than a 2p electron and is therefore less shielded by the 1s electrons than is a 2p electron. In other words, the effective nuclear charge for a 2s electron is slightly greater than that for a 2p electron. As a consequence the 2p energy level is slightly higher than the 2s energy level, and so a 2p electron has a correspondingly lower ionization energy than a 2s electron. Thus, the first ionization energy of boron, $1s^2 2s^2 2p^1$, in which the 2p electron is removed, is lower than that of beryllium, $1s^2 2s^2$, in which a 2s electron is removed, despite the increase in nuclear charge from beryllium to boron. The first ionization energy of aluminium, which has a $3s^2 3p^1$ valence shell, is, for a similar reason, lower than that of magnesium, which has a $3s^2$ valence shell.

Oxygen has four 2p electrons, two of which must therefore occupy the same orbital, while the three 2p electrons of nitrogen all occupy separate orbitals. The small decrease in the ionization energy from nitrogen to oxygen is due to the fact that it is easier to remove an electron from a filled orbital than from a singly occupied orbital, because of the repulsion between the two electrons in the filled orbital. Thus the ionization energy of oxygen is slightly lower than that of nitrogen despite the fact that the core charge of oxygen is greater than that of nitrogen. For the same reason, in the third period, the ionization energy of sulfur is slightly less than that of phosphorus. The anomalies in the electron affinites of the elements that we noted in Chapter 4 have a similar origin. Thus beryllium has a smaller electron affinity than lithium, despite the increased nuclear charge, because the added electron must occupy a higher-energy 2p orbital rather than the 2s orbital. The electron affinity of nitrogen is smaller than that of carbon, despite the increased nuclear charge, because the electron must be added to a 2p orbital that already contains an electron rather than to an empty 2p orbital as for carbon.

7.6
The Covalent Bond

Having discussed electron configurations in some detail, we are now in a position to look a little more closely at the covalent bond. In particular, we can begin to answer the question: How does a shared pair of electrons hold two nuclei together?

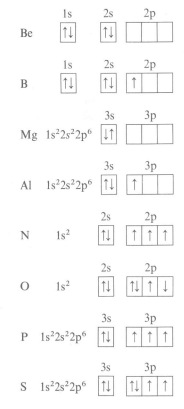

THE HYDROGEN MOLECULE

As two hydrogen atoms approach each other the nucleus of one atom begins to exert an attractive force on the electron of the other atom. Each hydrogen atom has space in its 1s orbital for a second electron so that, provided the two electrons have opposite spin, they are both drawn towards the region between the nuclei where the two orbitals overlap (Figure 7.31). As a result the electron

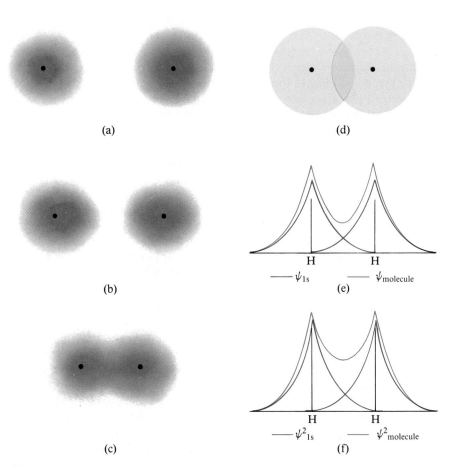

FIGURE 7.31
Electron Density Changes in the Formation of an H_2 Molecule.

(a) When the two hydrogen atoms are far apart they only attract each other very weakly. (b) As the two atoms get closer the electron in one atom begins to feel the attraction of the nucleus of the other atom. (c) As the two hydrogen atoms get closer still their electron densities are able to overlap because each hydrogen has space for another electron in its 1s orbital. The electrostatic attraction between this increased electron density and the nuclei holds them together and thus a stable molecule is formed. (d) An approximate representation of the overlap of two hydrogen 1s orbitals. (e) This shows how $\psi_{molecule}$ varies along the line joining the nuclei. Constructive interference between the two H 1s orbitals (wave functions) leads to an increased value of $\psi_{molecule}$ in the region between the nuclei in the molecule. (f) Corresponding to the increased value of $\psi_{molecule}$ between the nuclei there is an increase in the value of $\psi^2_{molecule}$. In other words, there is an increased electron density in the internuclear region. It is this increase in the electron density between the two nuclei that provides the attractive force holding the two nuclei together.

density increases in the region between the two nuclei and it is basically this increased electron density between the positive nuclei that pulls them together by electrostatic attraction. But as the nuclei and electrons come closer together, there is also an increase in the nucleus–nucleus and electron–electron repulsions. At a certain distance between the nuclei these repulsive forces equal the attractive forces between the nuclei and electrons, and the total energy of the system is then a minimum (see Figure 7.32). If the nuclei were to come closer, the repulsive forces would dominate, and the total energy of the system would therefore increase. The internuclear separation at which the energy is a minimum is the bond length of the hydrogen molecule in its ground state.

We can represent the formation of the hydrogen molecule in several different ways. For example, we can use an orbital box diagram,

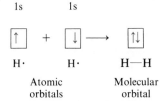

or we can represent the approximate shapes of the orbitals,

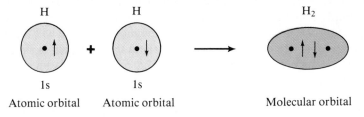

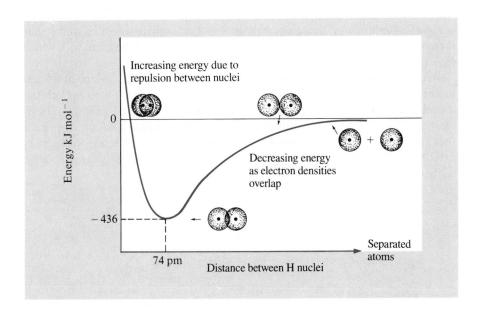

FIGURE 7.32
Energy Changes in the Formation of an H_2 Molecule.

The curve shows the change in energy as the distance between two hydrogen nuclei changes. The distance at which the energy is a minimum is the distance between the nuclei in the H_2 molecule, 74 pm.

or we can use an energy level diagram,

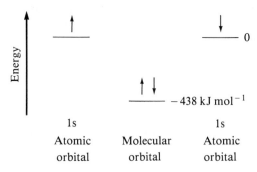

THE MOLECULAR ORBITAL AND LOCALIZED ORBITAL MODELS

Our description of the hydrogen molecule is the simplest example of a general bonding model called the **molecular orbital model**. In this model, two electrons of opposite spin are assigned, in accordance with the Pauli exclusion principle, to each of a set of orbitals that extend over all the nuclei in the molecule. These orbitals are therefore called **molecular orbitals**. Although this is an important and useful model, for all except the simplest molecules it requires some rather sophisticated mathematics. Moreover, the picture that it gives of a molecule is not simply related to Lewis structures which are a convenient and simple description of the electronic structure of many molecules. We will therefore use an alternative model, called the **localized orbital model**, which is simple to use and is closely related to Lewis structures. This model retains one of the basic ideas of the molecular orbital model, namely that each bond can be represented by an orbital such as that which we have described for the hydrogen molecule. But this orbital is restricted to two nuclei only, even in a molecule that contains three or more nuclei. Such an orbital is sometimes called a *localized molecular orbital*. We will refer to it simply as a **localized orbital**, and specifically as a **bonding orbital**. Any electrons in the molecule which are not involved in bonding are assumed to remain associated with only one atom as localized nonbonding electron pairs (lone pairs), and to occupy localized orbitals that we will call **nonbonding orbitals** or **lone-pair orbitals**. For the present we will assume that these nonbonding orbitals are simply the atomic orbitals of the atom with which they are associated.

According to the localized orbital model an atomic orbital on one atom can be combined with a suitable orbital on another atom to give a bonding orbital, if each atomic orbital is occupied by only one electron. For example, the F atom has the electron configuration $1s^2 2s^2 2p^5$; it therefore has one 2p orbital that contains only one electron. This 2p orbital may be combined with the 1s orbital of an H atom to form a bonding orbital. Using orbital box diagrams, we can represent the formation of the HF molecule as follows:

We obtain a rather complicated diagram if we attempt to show the approximate shapes of all the orbitals in a case like this so we will further simplify the model by assuming that both the bonding and the nonbonding orbitals are roughly spherical in shape. Such a diagram closely resembles a Lewis structure and we will see in Chapter 9 that this is a better description than might at first appear to be the case.

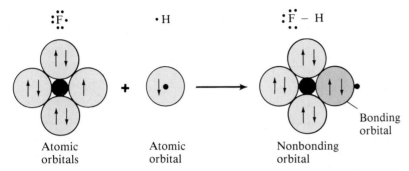

The orbital box diagram becomes complicated and less useful for a triatomic molecule such as water so we illustrate the localized orbital model for water using the spherical orbital picture only:

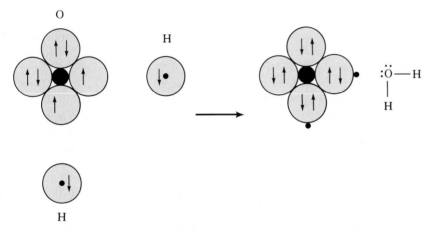

We see that the Pauli exclusion principle provides the explanation for why we can consider that the electrons in a valence shell are arranged in localized pairs. The approximate model that we have arrived at here, according to which each electron pair can be regarded as occupying a roughly spherical localized orbital, is essentially identical with the model that we used in Experiment 4.1 in Chapter 4 to demonstrate the tetrahedral arrangement of four electron pairs in a valence shell, where we represented the space occupied by an electron pair as a sphere.

VALENCE AND ELECTRON CONFIGURATIONS

We can describe any single bond in terms of a bonding orbital formed by the combination of two atomic orbitals. Thus the number of bonds formed by an atom depends on the number of electrons in singly occupied orbitals, that is, on the number of *unpaired electrons*. Table 7.5 shows the valences we would predict on this basis for the elements in the second period.

TABLE 7.5
Ground State Electron Configurations and Predicted Valences for the Second-Period Elements

Element	1s	2s	2p	Number of Unpaired Electrons	Predicted Valence	Observed Valence
Li	↑↓	↑		1	1	1
Be	↑↓	↑↓		0	0	2
B	↑↓	↑↓	↑	1	1	3
C	↑↓	↑↓	↑ ↑	2	2	4
N	↑↓	↑↓	↑ ↑ ↑	3	3	3
O	↑↓	↑↓	↑↓ ↑ ↑	2	2	2
F	↑↓	↑↓	↑↓ ↑↓ ↑	1	1	1
Ne	↑↓	↑↓	↑↓ ↑↓ ↑↓	0	0	0

These valences are in accordance with experiment and with the positions of the elements in the periodic table for Li, N, O, F, and Ne, but not for Be, B, and C, which have valences of 2, 3, and 4, not 0, 1, and 2 as predicted from the number of unpaired electrons. However, if we were to excite an electron from the 2s orbital to a vacant 2p orbital, the number of unpaired electrons in Be, B, and C would correspond to the observed valence in each case, as shown in Table 7.6. In this case an electron is said to have been *promoted* from the 2s orbital to a 2p orbital. This excited state of the atom is often called the **valence state**.

We can then imagine that in the formation of the methane molecule each of the singly occupied orbitals of the carbon atom is used to form a bonding orbital with a hydrogen 1s orbital. But it is not correct to think that, in the formation of methane from a carbon atom and four hydrogen atoms, this excited state of the carbon atom is actually formed. Rather, the four bonding orbitals are formed directly from the carbon atom in its ground state and the four hydrogen atoms (Figure 7.33).

TABLE 7.6
Valence State Electron Configurations and Predicted Valences for Be, B, and C

Element	1s	2s	2p	Number of Unpaired Electrons	Predicted Valence	Observed Valence
Be	↑↓	↑	↑	2	2	2
B	↑↓	↑	↑ ↑	3	3	3
C	↑↓	↑	↑ ↑ ↑	4	4	4

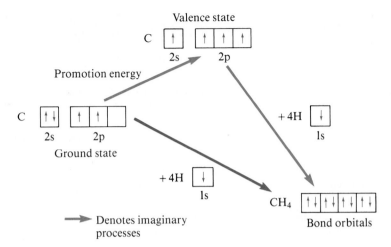

FIGURE 7.33
The Formation of the Methane Molecule.

In order to understand how carbon can form four bonds as in methane rather than just two as expected from its ground state electron configuration, it is convenient to imagine that an electron is first promoted from the 2s orbital to the 2p orbital to give an excited or valence state of the carbon atom in which there are four singly occupied orbitals each of which can be used to form a C—H bond. The energy needed for promotion of the electron is provided by the energy released in the formation of four bonds. In fact the four bonding orbitals in methane are formed directly from carbon in its ground state. We can imagine that the processes of electron promotion and bond formation occur simultaneously.

Nevertheless, imagining the formation of methane in two stages is often convenient. First, the promotion of a 2s electron into a vacant 2p orbital gives the electron configuration $2s^1 2p^3$ (the valence state), which is followed by the formation of four bonding orbitals from these carbon orbitals and the four hydrogen 1s orbitals. A certain amount of energy is needed to promote the 2s electron to a 2p orbital. But, because the formation of a C—H bond liberates a considerable amount of energy, this is more than compensated for by the energy evolved in the formation of two additional C—H bonds.

The advantage of thinking about the process in this way is that we can easily count the number of unpaired electrons in the valence state, and we can therefore easily predict how many bonds the atom will form. In particular, we note that Be, B, and C all form as many bonds as possible; that is, they use *all* their valence-shell electrons in bond formation.

For Li, N, O, and F promotion of a 2s electron to a 2p orbital does not change the number of unpaired electrons, so promotion does not allow the formation of additional bonds. In the case of Ne, since all the orbitals are filled, an electron cannot be promoted from a 2s orbital to a 2p orbital. Thus neon has no unpaired electrons and forms no compounds. Promotion of an electron to an $n = 3$ orbital would need more energy than could be obtained by bond formation.

Similar considerations apply to the elements of period 3. For example, silicon in group 4 has a valence of 4 in compounds such as $SiCl_4$ and SiH_4. We can imagine that these compounds are formed via the valence state of silicon, which has the electron configuration $3s^1 3p^3$ with four unpaired electrons, rather than via the ground state $3s^2 3p^2$ configuration, which has only two unpaired electrons.

IMPORTANT TERMS

A **bonding orbital** is the orbital occupied by the two electrons of a single bond.

A **diffraction pattern** is obtained when waves from two or more regularly arranged sources interfere with each other.

The **electron configuration** of an atom is a description of how the electrons are distributed among the energy levels of the atom.

Electron spin is a concept used to explain the magnetic properties of electrons. It is imagined that a spherical electron can spin around its own axis, thus generating a magnetic field.

In an **excited state** an atom has one or more of its electrons in energy levels other than the lowest available level.

The **frequency** of a wave is the number of wave crests that pass a given point in one second.

In the **ground state** of an atom all the electrons are in their lowest-possible energy levels.

Hund's rule states that in the lowest-energy (ground state) electron configuration electrons in the same energy level as far as possible occupy separate orbitals and have the same spin.

Interference occurs between waves when waves from two different sources arrive at the same point.

A **localized orbital** is an orbital that is confined to only one or two nuclei in a molecule. It may be a bonding orbital or a nonbonding (or lone-pair) orbital.

A **molecular orbital** is an orbital that extends over all the nuclei in a molecule.

Momentum is the product of the mass and the velocity of an object:

$$Momentum = mv$$

A **node** is a point, line, or surface at which the amplitude of a wave or the wave function, ψ, is zero.

A **nonbonding orbital** is an orbital that is associated with only one nucleus in a molecule. It is also called a **lone-pair orbital**.

An **orbital** is the standing electron wave associated with an energy level of an atom or molecule. The form of the orbital can be obtained by solving the Schrödinger equation for the atom or molecule.

The **Pauli exclusion principle** states that no orbital can contain more than two electrons and these electrons must have opposite spins.

A **photochemical reaction** is a reaction that is caused by the absorption of light by a molecule.

The **photoelectric effect** is the emission of electrons by metals when light with a frequency greater than a certain minimum frequency shines upon them.

In **photoelectron spectroscopy** photons are used to knock electrons out of atoms. From the energy of the photons and the kinetic energies of the electrons thus obtained, the ionization energies of the electrons in an atom can be determined.

A **photon** is a quantum of light; its energy is given by $E = hv$, where v is the frequency of the light and h is the Planck constant.

A **quantized** system is one that can have only certain energies. The energies of the electrons in atoms are quantized.

A **quantum** is the smallest amount of energy that can be emitted or absorbed by an atom as radiation. If v is the frequency of the radiation emitted or absorbed, the energy of the quantum is $E = hv$.

Quantum numbers are the integers used to describe the orbitals of the hydrogen atom.

A **spectrum** is the range of frequencies of radiation emitted by an atom, a heated filament, the sun, or other energy source. A *line spectrum* consists of only a few sharp lines; it consists of radiation of only a few definite wavelengths.

Subshell is an alternative name for each of the energy levels associated with a particular shell.

The **uncertainty principle** states that the uncertainty Δx in the position of an electron times the uncertainty $\Delta(mv)$ in its momentum must be greater than the Planck constant, h.

$$\Delta x \, \Delta(mv) > h$$

The **wavelength** is the distance between successive wave crests or points of equal displacement on successive waves.

PROBLEMS *

Electromagnetic Radiation

1. Radio station CBC in Toronto, Ontario, broadcasts its FM signal at 94.1 MHz and its AM signal at 740 kHz. What are the wavelengths in meters of these signals?

2. Mercury lamps used for street and highway lighting emit the atomic spectrum of mercury. One of the lines in this spectrum is in the blue region and has a wavelength of 435.8 nm. Express this wavelength in **(a)** meters, **(b)** micrometers, and **(c)** angstroms. What is the frequency of this radiation?

3. Citizens' band (CB) radio operates at a frequency of 27.3 MHz. What is the wavelength of this radiowave?

* Answers to problems numbered in blue appear at the end of the text.

4. Calculate the range of frequencies associated with each of the regions of the electromagnetic spectrum from the following wavelengths:

(a) Radio (1 km to 30 cm)

(b) Microwave (30 cm to 2 mm)

(c) Far infrared (2 mm to 30 μm)

(d) Near infrared (30 μm to 710 nm)

(e) Visible (710 nm to 400 nm)

(f) Ultraviolet (400 nm to 4 nm)

(g) X rays (4 nm to 30 pm)

(h) γ rays (30 pm to 0.1 pm)

5. A helium–neon laser produces light of wavelength 633 nm. What is the color and frequency of this light?

6. A sodium vapor street lamp emits radiation of wavelength 589.2 nm. What is the color and the frequency of this radiation?

7. The atomic spectrum of lithium has a strong red line at 670.8 nm. What is the energy of each photon of this wavelength? What is the energy of 1 mol of these photons?

8. The first ionization energy of potassium is 0.42 MJ mol^{-1}. What is the maximum wavelength of light that will ionize a potassium atom in the gas phase?

9. The yellow color of a sodium vapor street lamp is due to lines at 589.6 nm and 589.0 nm. What are the frequencies of these transitions, and what is their energy difference?

10. By use of a suitable filter, the green mercury emission line of wavelength 546.1 nm can be isolated. Calculate the energy of

(a) One photon of light of this wavelength

(b) One mole of photons of light of this wavelength

Photoelectric Effect and Photochemistry

11. The longest-wavelength light that causes an electron to be emitted from a gaseous lithium atom is 520 nm. Gaseous lithium atoms are irradiated with light of wavelength 360 nm. What is the kinetic energy of the emitted electrons in kilojoules per mole?

12. Photons of minimum energy 486 kJ mol^{-1} are needed to ionize sodium atoms. Calculate the lowest frequency of light that will ionize a sodium atom. What is the the color of this light? If light of energy 600 kJ mol^{-1} is used, what is the velocity of the emitted electrons?

13. When light of wavelength 470.0 nm falls on the surface of potassium metal, electrons are emitted with a velocity of 6.4×10^4 m s^{-1}.

(a) What is the kinetic energy of the emitted electrons?

(b) What is the energy of a 470.0-nm photon?

(c) What is the minimum energy required to remove an electron from potassium?

14. When light of frequency 1.30×10^{15} Hz shines on the surface of cesium metal, photoelectrons are ejected with a kinetic energy of 5.2×10^{-19} J. What is the longest wavelength of light that will cause the removal of electrons from cesium?

15. One type of burglar alarm uses the photoelectric effect. Provided that visible light falling on a metal plate causes the emission of photoelectrons, the alarm is inactive. When the light beam is blocked by an intruder, the alarm is set off. Would magnesium metal be a suitable material for the metal plate, given that the lowest frequency that can cause the emission of an electron is 8.95×10^{14} Hz?

16. Light of minimum wavelength 439 nm dissociates chlorine molecules into chlorine atoms. Will light of the same wavelength dissociate bromine molecules into bromine atoms, given that the dissociation energy of $Br_2(g)$ is 190 kJ mol^{-1}? What is the maximum wavelength of light that will dissociate $Br_2(g)$ molecules into bromine atoms?

17. Using the wavelengths listed in Problem 4, calculate the energy range, in kilojoules per mole, of the photons of each type of radiation in the electromagnetic spectrum. What type of radiation will supply just sufficient energy to cause each of the following dissociations?

(a) $H_2(g) \rightarrow 2H(g)$ $\Delta H_f^\circ(H, g) = 218$ kJ mol^{-1}

(b) $O_2(g) \rightarrow 2O(g)$ $\Delta H_f^\circ(O, g) = 247$ kJ mol^{-1}

(c) $Cl_2(g) \rightarrow 2Cl(g)$ $\Delta H_f^\circ(Cl, g) = 120$ kJ mol^{-1}

(d) $F_2(g) \rightarrow 2F(g)$ $\Delta H_f^\circ(F, g) = 78$ kJ mol^{-1}

(e) $N_2(g) \rightarrow 2N(g)$ $\Delta H_f^\circ(N, g) = 471$ kJ mol^{-1}

18. Nitrogen dioxide, $NO_2(g)$, is one of the components of photochemical smog. The energy required to dissociate $NO_2(g)$ molecules into NO(g) and oxygen atoms is 305 kJ mol^{-1}.

(a) What maximum wavelength of light will cause this dissociation?

(b) What type of radiation is it?

(c) If the minimum wavelength of light that strikes the earth's surface at sea level is 320 nm, does the dissociation of $NO_2(g)$ to NO(g) and O(g) atoms occur near to the surface of the earth?

The Atomic Hydrogen Spectrum

19. What wavelength of light is emitted when an electron moves from the $n = 6$ to the $n = 2$ energy level of a hydrogen

atom? In what region of the electromagnetic spectrum is the corresponding spectral line found?

20. Calculate the energy required to excite an electron from the $n = 2$ to the $n = 4$ energy level of the hydrogen atom. What maximum wavelength of light will cause this excitation?

21. Lines in the Lyman series in the spectrum of atomic hydrogen arise from transitions to the $n = 1$ level. One of these lines has a wavelength of 103 nm. What is the n quantum number of the electrons in the excited atoms that give rise to this line?

22. How much energy, in megajoules per mole, is needed to ionize hydrogen atoms, starting from **(a)** the ground state and **(b)** the first excited state? In each case, find the maximum wavelength of light that will ionize a hydrogen atom.

23. The Balmer series of emission lines from atomic hydrogen end in the $n = 2$ energy level. Calculate, the nanometers, the longest-wavelength line of the Balmer series.

24. Considering only the $n = 1, 2, 3, 4$, and 5 energy levels of the hydrogen atom, how many spectral lines are possible using only this set of levels? How many are in the ultraviolet region, and how many are in the visible spectrum?

25. The Paschen series of emission lines from atomic hydrogen occur in the infrared spectrum. They arise from transitions to the $n = 3$ level. One of these lines has a wavelength of 1094 nm. Determine the value of the n quantum number for the upper level involved in this transition.

Matter Waves

26. What is the wavelength associated with **(a)** an electron traveling with 20.0% of the speed of light and **(b)** a lithium atom of mass 7.02 u moving at the same speed? (Remember $1 \text{ J} = 1 \text{ kg m}^2 \text{ s}^{-2}$.)

27. What wavelength is associated with an electron moving with a speed of $5.00 \times 10^7 \text{ km s}^{-1}$?

28. What wavelength is associated with a neutron moving with a speed of $1.00 \times 10^2 \text{ m s}^{-1}$?

29. A beam of neutrons has a wavelength of 300 pm. What is the speed of the neutrons?

30. Major league pitchers can throw a baseball with a maximum speed of 95 miles h^{-1}. What de Broglie wavelength is associated with a 5.0 oz baseball thrown at this speed?

31. The mass of an alpha particle (helium nucleus) is

6.65×10^{-24} g. What is the wavelength of an alpha particle that has a kinetic energy of 4.0×10^{-12} J? To what fraction of the speed of light does this kinetic energy correspond?

32. (a) What is the speed of a rapidly moving electron with a wavelength of 0.100 nm?

   (b) To what region of the electromagnetic spectrum does this wavelength correspond?

Ionization Energies

33. Which of the following groups of elements is arranged correctly in order of increasing first ionization energy?

   (a) C < Si < Li < Ne   **(b)** Ne < C < Si < Li

   (c) Li < Si < C < Ne   **(d)** Ne < Si < C < Li

34. In each of the following sets, which atom or ion has the smallest first ionization energy?

(a) H, He, Li   **(b)** Mg, Al, Si   **(c)** Cs, Ba, La

(d) P, S, Cl   **(e)** O, O^-, O^{2-}

35. Place the following species in order of increasing first ionization energy: Ca, K, Ca^{2+}, and Cl. Briefly explain your answer.

36. Arrange the following species in order according to the energy needed to remove the most easily removed electron:

   He Li^+ Ar   Ne Be^{2+}

37. Which species of each of the following pairs has the higher ionization energy?

   (a) C, Si   **(b)** Ca, Mg   **(c)** Be, B

   (d) Br, S   **(e)** Ar, K^+

38. Does it take more energy, or less energy, to remove an electron from a K^+ ion than from a Ca^{2+} ion in the gas phase? Explain.

39. (a) Explain briefly how ionization energies are obtained by using a photoelectron spectrometer.

   (b) Argon has ionization energies of -1.52, -2.82, -24.1, -31.5, and -309 MJ mol. Interpret these ionization energies in terms of the electronic configuration of argon.

40. When a beam of argon atoms was irradiated with photons, no 2s photoelectrons were observed unless the incident photons had a wavelength of at least 3.80 nm. In another experiment, higher-energy photons were used and 2s electrons with kinetic energies of 1.12×10^{-16} J were ejected. What was the wavelength of these higher-energy photons?

41. The ionization energy of helium is 2.37 MJ mol^{-1}. What is the kinetic energy of the photoelectrons produced when helium gas is irradiated with radiation of wavelength 40.0 nm?

Electron Configurations and Quantum Numbers

42. What is the total number of electrons associated with each of the $n = 1$, $n = 2$, and $n = 3$ energy levels?

43. Which of the following atomic orbital designations are not possible?

  (a) 6s   **(b)** 1p   **(c)** 4d   **(d)** 2d

  (e) 3p   **(f)** 4d   **(g)** 5p   **(h)** 2s

44. Which of the following is not an allowable set of quantum numbers? Explain why.

	n	l	m
(a)	2	0	0
(b)	1	1	0
(c)	2	1	-1
(d)	6	5	5

45. In terms of the Pauli exclusion principle explain why beryllium cannot have the electron configuration $1s^4$.

46. Without reference to Table 7.2, decide which of the following electron configurations are not allowed by the Pauli exclusion principle, and explain why:

  (a) $1s^2 2s^2 2p^4$ **(b)** $1s^2 2s^2 2p^6 3s^3$

  (c) $1s^2 3p^1$ **(d)** $1s^2 2s^2 2p^6 3s^2 3p^{10}$

47. What is the value of the l quantum number associated with each of the following energy levels?

  (a) 1s   **(b)** 3p   **(c)** 4d

  (d) 6s   **(e)** 5d   **(f)** 5p

48. Without reference to Table 7.2, select the electron configurations among the following that are not ground state configurations:

  (a) $1s^2 2p^1$ **(b)** $1s^2 2s^2 3p^2$

  (c) $1s^2 2s^2 2p^5$ **(d)** $1s^2 2s^2 3p^6 3d^3$

49. Without reference to Table 7.2, give the ground state electron configurations of each of the following atoms using the orbital box notation:

  (a) K   **(b)** Al

  (c) Cl   **(d)** Ti $(Z = 22)$

  (e) Zn $(Z = 30)$ **(f)** As $(Z = 33)$

50. Give the ground state electron configurations of each of the following atoms and ions using the orbital box notation:

  (a) Be   **(b)** N   **(c)** F   **(d)** Mg

  (e) Cl$^+$ **(f)** Ne$^+$ **(g)** Al^{3+}

51. Identify each of the elements that have the following ground state electron configurations:

  (a) $1s^2 2s^1$ **(b)** $1s^2 2s^2 2p^3$

  (c) $[Ar]4s^2$ **(d)** $[Ar]4s^2 3d^2$

  (e) $[Ar]4s^2 3d^{10} 4p^3$ **(f)** $[Kr]5s^2 3d^{10} 5p^5$

  (g) $[Xe]6s^2$

52. How many unpaired electrons are in the ground states of each of the following: O, O$^-$, O^{2-}, S, F, Ar, Fe

53. On which quantum numbers does the energy of an electron depend in each of the following?

  (a) a one-electron atom or ion

  (b) a many-electron atom or ion.

54. Give the corresponding atomic orbital designation (1s, 3p, and so on) for electrons with the following quantum numbers:

	(a)	**(b)**	**(c)**	**(d)**
n	2	4	3	2
l	1	0	2	0

55. What arrangements of valence-shell electrons are associated with

  (a) The main group element with the lowest first ionization energy in any period?

  (b) The main group element with the highest first ionization energy in any period?

56. Without reference to Table 7.2, draw orbital box diagrams for the ground state valence shells of each of the following atoms:

  (a) P   **(b)** Ca   **(c)** V $(Z = 23)$

  (d) O   **(e)** Br $(Z = 35)$

57. Use Hund's rule to decide which of the following electron configurations are not ground states:

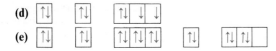

Identify each element.

58. By drawing orbital box diagrams, determine how many unpaired electrons there are for the ground state electron configurations of each of the following atoms:

(a) P   (b) Si   (c) I   (d) Se   (e) Sc

59. Categorize each of the following electron configurations as that of a ground state, an excited state, or not possible. For the allowed configurations identify the element; for those that are not possible explain why.

(a) $1s^2 2s^2$ (b) $1s^2 3s^1$

(c) $1s^2 2d^3$ (d) $[Ne]3s^2 3d^1$

(e) $[Ar]4s^2 3d^2$ (f) $1s^2 2s^2 2p^6 3s^1$

(g) $[Ne]3s^2 3d^{12}$ (h) $[Ar]3s^2 3p^1$

60. To what group of the periodic table, and to which period, does arsenic belong? How many orbitals are occupied in the ground state of arsenic? How many of these orbitals is only singly occupied? Name the elements in the same group as arsenic in periods 2 and 3 and write their ground state electron configurations.

Atomic and Molecular Orbitals

61. For the hydrogen atom:

(a) For $n = 3$, how many spherical nodes and how many planar nodes are there for $l = 0$, $l = 1$, and $l = 2$?

(b) Is $l = 3$ possible for $n = 3$?

(c) How are the orbitals with $n = 3$ and $l = 0$, and $n = 3$ and $l = 1$, usually designated?

62. What is the relationship between the number of planar nodes in an atomic orbital and its quantum number l?

63. What is the relationship between the orbital shape and each of the permissible values of the l quantum number for $n = 3$?

64. (a) Draw rough sketches of each of the 1s, 2s, and 3s orbitals showing the nodes.

(b) Draw rough sketches of each of the 2p and 3p orbitals showing the nodes.

65. What is the qualitative relationship between the energy of an atomic orbital and its number of nodes? What is the

relationship between the number of nodes of an atomic orbital and its quantum number n?

66. Explain what is meant by each of the following terms:

(a) atomic orbital   (b) molecular orbital

(c) localized orbital   (d) bonding orbital

(e) nonbonding orbital

67. Explain why two hydrogen atoms attract each other when brought together and combine to form the H_2 molecule, while two helium atoms repel each other when brought together and do not form a molecule.

Miscellaneous

68. Excited barium atoms in a bunsen burner flame can return to their ground states by emitting photons of energy 3.62×10^{-19} J. What color is imparted to the bunsen burner flame due to light of this wavelength?

69. When an excited aluminum atom with an electron in a 4s orbital returns to the ground state, it emits light of wavelength 395 nm. When an excited aluminum atom with an electron in a 3d orbital returns to the ground state, it emits light of wavelength 319 nm.

(a) Calculate the energy separation between the 3d and 4s energy levels of aluminum.

(b) Which is higher in energy, the 3d energy level or the 4s energy level?

70. Write the ground state electron configuration of silicon. How many unpaired electrons are there in the ground state? What valence would you predict for silicon? How do you explain the fact that silicon forms the compounds $SiCl_4$ and SiH_4?

71. Write the ground state electron configuration of boron. How many unpaired electrons are there in this ground state? How many covalent bonds would you expect boron to form? How do you explain that boron forms molecules such as BF_3 and the ion BF_4^-?

72. Some sunglasses have special lenses that darken on exposure to strong light and become paler in the shade. These lenses contain a small amount of silver chloride. Light causes the following reaction to occur

$$AgCl(s) \longrightarrow Ag(s) + Cl$$

The silver that is formed darkens the lens. In the absence of light the reverse reaction occurs. The standard enthalpy change for this reaction is 310 kJ mol^{-1}. What is the maximum wavelength of light that can cause this reaction?

73. Elements that have large ionization energies also usually have large electron affinities. Explain this statement. Which group of elements would you expect to be an exception?

Phosphorus and Sulfur

8

Among the 20 most abundant elements in the earth's crust (Table 3.2), 10 are nonmetals: O, Si, H, Cl, P, C, S, Ar, N, and F. We have already discussed oxygen and hydrogen (Chapter 3), fluorine and chlorine (Chapter 5), and carbon (Chapter 6). In this chapter we consider two more common nonmetals, phosphorus and sulfur. They are neighboring elements in the third period. Phosphorus is in group 5 and sulfur is in group 6. They are relatively abundant elements, twelfth and fifteenth, respectively. The elements and their compounds have many important practical uses. Sulfuric acid, H_2SO_4, is produced by the chemical industry in a larger quantity than any other substance. Enormous quantities of phosphorus compounds are used as fertilizers, and phosphorus plays a vital role in life processes. Indeed, after carbon, hydrogen, oxygen, and nitrogen, phosphorus and sulfur are the two most abundant nonmetals in living matter. For example phosphorus is an essential part of adenosine triphosphate (ATP) and sulfur is a component of two of the amino acids from which proteins are built (Chapter 24).

Phosphorus in group 5 has the valence-shell electron configuration $3s^2 3p^3$. It has five electrons in its valence shell and therefore is expected to complete its octet by forming three covalent bonds. Similarly sulfur, which has the valence-shell electron configuration $3s^2 3p^4$, with six electrons in its valence shell, is expected to complete its octet by forming two covalent bonds. However, in addition to the expected valences of 2 and 3 respectively for sulfur and phosphorus, these elements have other important valences. For example, sulfur forms the oxides SO_2 and SO_3, in which it has valences of 4 and 6 respectively. We will see that these additional valences arise because in the $n = 3$ quantum shell, but not in the $n = 2$ shell, there are d orbitals which can be used in the formation of covalent bonds. Thus, although sulfur resembles oxygen, the first element in group 6, in having a valence of 2, sulfur differs from oxygen in that it has many important compounds in which it has a higher valence. Similarly phosphorus resembles nitrogen in that they both have a valence of 3, but phosphorus differs from nitrogen in that it also forms many important compounds in which it has a higher valence.

These two elements form a relatively large number of important compounds and in order to help us discuss and understand their chemistry we will find it convenient to introduce the concepts of oxidation state and oxidation number. We will thus extend and make more quantitative our discussion of oxidation–reduction reactions.

Among the compounds of these elements are two important acids—phosphoric acid, H_3PO_4, and sulfuric acid, H_2SO_4—that are examples of an important class of acids, the oxoacids. We discuss the structures and strengths of these acids in this chapter, thereby extending our knowledge of acids and bases and their reactions.

8.1
The Elements and their Properties

OCCURRENCE AND PRODUCTION OF THE ELEMENTS

SULFUR Sulfur has been known since the beginning of recorded history. In the Bible it is called *brimstone*, which means "the stone that burns". It has been known for so long because in volcanic regions it is found as the free element on the earth's surface, although in small amounts. Important deposits of sulfur are found at depths of 300 m or more in association with sodium chloride and the mineral anhydrite—calcium sulfate, $CaSO_4$. Reduction of calcium sulfate by bacteria probably formed the elemental sulfur in these deposits.

Today the most productive deposits are in Louisiana and Texas. Until recently, they produced about 80% of the world's sulfur, but this percentage is declining. When these sulfur deposits were first discovered, they were considered inaccessible. They are covered by layers of sand, gravel, and mud, which would have made normal mining difficult and dangerous because of the possibility of excavations collapsing. Moreover, the deposits contain considerable quantities of the poisonous gases SO_2 and H_2S.

These problems were solved by the development of the **Frasch process**. A bore is made down to the sulfur, and three concentric pipes are pushed down the bore (see Figure 8.1). Superheated water (150 °C) under pressure is pumped

Jupiter and two of its satellites, Io (left) and Europa. The surface composition of Io is uncertain, but scientists believe it may be a mixture of salts and sulfur.

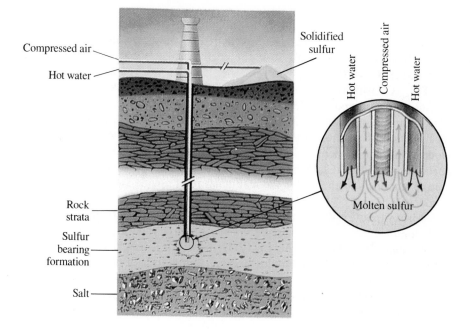

FIGURE 8.1
Extraction of Sulfur by the Frasch Process.

A bore is made by a drilling rig similar to the rigs used for drilling oil wells; then three concentric pipes are pushed down the bore. Superheated water at about 150 °C is forced down the outer pipe. When it has melted the sulfur around the end of the pipe, compressed air is used to blow the mixture of hot water and molten sulfur to the surface.

FIGURE 8.2
A pile of pure sulfur awaiting shipment.

down one pipe, melting the sulfur around the bottom of the pipe. Compressed air is then sent down the inner pipe, and a bubbly froth of air, water, and molten sulfur is forced to the surface through the third pipe. At the surface the molten sulfur solidifies. The sulfur obtained by this process is 99.5% pure and is suitable for most commerical purposes (Figure 8.2).

Compounds of sulfur are also widespread. There are many sulfide minerals such as pyrite, FeS_2, galena, PbS, and chalcocite, Cu_2S, and several common sulfates such as gypsum, $CaSO_4 \cdot 2H_2O$. These minerals are not important sources of sulfur but the sulfides are important sources of metals such as lead and copper. During the smelting operations that extract the metal from the sulfide ore, sulfur dioxide is produced:

$$Cu_2S(s) + O_2(s) \longrightarrow 2Cu(s) + SO_2(g)$$

When this sulfur dioxide is allowed to escape into the atmosphere, severe pollution and acid rain can result (see Box 14.1). For example, in the past sulfur dioxide produced by the nickel smelters at Sudbury, Ontario, destroyed the vegetation for many miles around. The landscape was so denuded of vegetation that it was used as a training ground for the astronauts who made the first landing on the moon.

Controls on SO_2 emission are now being introduced, and at least some of the SO_2 is removed from the gases evolved in smelting operations. One method for removing SO_2 uses the reaction of SO_2 with H_2S to produce elemental sulfur:

$$2H_2S(g) + SO_2(g) \longrightarrow 3S(s) + 2H_2O(g)$$

Another source of sulfur of increasing importance is the H_2S present in some natural gas wells. It is removed from the natural gas and converted to sulfur by using its reaction with SO_2. Considerable quantities of sulfur are now produced as a by-product of the purification of natural gas and of smelting operations, so that the Frasch process is becoming less important.

Apparently complicated formulas such as $Ca_5(PO_4)_3(OH)$ are easier to remember if you recall that in any ionic compound the sum of the positive charges on the cations must be equal and opposite to the sum of the negative charges on the anions. The charge on five Ca^{2+} is $+10$. The total charge on three PO_4^{3-} is -9 and the -1 charge on the OH^- ion makes a total of -10. The singly charged OH^- can be replaced by other singly charged ions such as Cl^- and F^-.

PHOSPHORUS Phosphorus is too reactive to occur in nature in the free state. Most of the phosphorus in the earth's crust occurs as the minerals fluorapatite, $Ca_5(PO_4)_3F$, hydroxyapatite, $Ca_5(PO_4)_3(OH)$, and chlorapatite, $Ca_5(PO_4)_3Cl$, which are collectively known as phosphate rock. These minerals are all ionic compounds containing calcium ions, Ca^{2+}, phosphate ions, PO_4^{3-} and fluoride, F^-, chloride, Cl^-, or hydroxide, OH^-, ions. Phosphate rock and sulfur are two of the basic substances on which a very large part of modern inorganic chemical industry is based. Other substances of basic importance are salt, NaCl, limestone, $CaCO_3$, and sand, SiO_2. These compounds are the starting substances for the manufacture of a wide variety of important products.

Hydroxyapatite is the main constituent of the bones and teeth of animals. Fluoridation of water leads to the replacement of hydroxyapatite, $Ca_5(PO_4)_3(OH)$, in teeth enamel by fluorapatite, $Ca_5(PO_4)_3F$, which is less basic and more resistant to attack by acids. Complex organic compounds of phosphorus are essential constituents of many proteins and of nerve and brain tissue. Phosphate groups are an essential part of the structure of DNA, the storehouse of genetic information, and they play a vital role in the transfer of energy in our bodies.

FIGURE 8.3
Discovery of Phosphorus.

German alchemist Hennig Brand discovered phosphorus in 1669 when he strongly heated the residue left by the evaporation of urine and observed the striking blue-green light that is emitted by phosphorus vapor when it comes in contact with air. The urine residue contained organic compounds and phosphate and when it was heated the organic compounds were decomposed to carbon which reduced the phosphate to phosphorus.

Thus phosphorus is an essential component of our diet. An adult excretes daily phosphorus compounds containing about 2 g of phosphorus. Indeed, phosphorus was discovered by German alchemist Hennig Brand, who obtained it by strongly heating the residue obtained on the evaporation of urine (see Figure 8.3).

The industrial manufacture of phosphorus is based on the reaction that led to the discovery of phosphorus, namely, the reduction of phosphate by carbon. A mixture of phosphate rock, coke, and silica sand is fed into an electric furnace and heated to 1400–1500 °C by passing a large electric current through it (see Figure 8.4). The reactions that occur in the molten mixture are complex but may be summarized by two equations:

$$2Ca_3(PO_4)_2(l) + 6SiO_2(l) \longrightarrow P_4O_{10}(g) + 6CaSiO_3(l)$$
$$P_4O_{10}(g) + 10C(s) \longrightarrow P_4(g) + 10CO(g)$$

FIGURE 8.4
Preparation of Phosphorus.

A mixture of phosphate rock, coke, and silica sand is fed into the top of the furnace and heated by a large electric current that is passed between two large graphite rods in the lower part of the furnace. The phosphorus vapor passes out of the furnace through an electrostatic precipitator to remove dust and is then condensed by a spray of water at about 70 °C. The liquid phosphorus is run off at the bottom of the spray tower into large storage tanks.

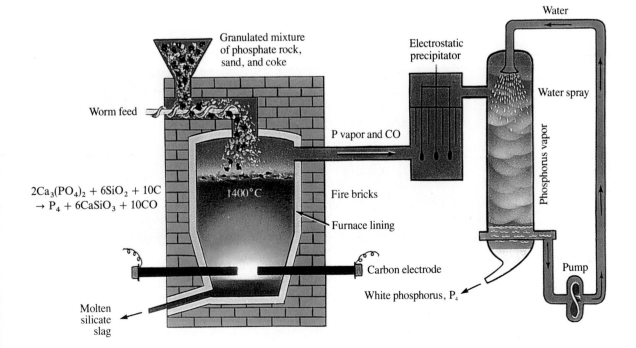

Granulated mixture of phosphate rock, sand, and coke

Electrostatic precipitator

Water

Worm feed

P vapor and CO

Water spray

$2Ca_3(PO_4)_2 + 6SiO_2 + 10C$
$\rightarrow P_4 + 6CaSiO_3 + 10CO$

1400°C

Fire bricks

Phosphorus vapor

Furnace lining

Carbon electrode

White phosphorus, P_4

Pump

Molten silicate slag

The oxide, P_4O_{10}, formed in the first reaction is reduced to elemental phosphorus by carbon in the second. The gaseous phosphorus emerging from the furnace is condensed by a spray of warm water, and the liquid is pumped to storage tanks through steam-heated pipes, which keep it above its melting point of 44 °C.

PROPERTIES OF THE ELEMENTS

SULFUR Sulfur can be obtained in several different solid forms. When sulfur is allowed to crystallize at room temperature from a suitable solvent such as carbon disulfide, it is obtained in the form of brilliant yellow crystals of *orthorhombic sulfur* (Figure 8.5). If orthorhombic sulfur is heated to 95.5 °C, it changes to another crystalline form, *monoclinic sulfur*. However, the rate of this transformation is quite slow, and it is simpler to obtain monoclinic sulfur by cooling molten sulfur. Monoclinic sulfur crystallizes from molten sulfur at 119.3 °C in the form of long needles (Figure 8.5). But monoclinic sulfur is not stable at room temperature; it slowly changes back to orthorhombic sulfur. As we saw in Chapter 6, the different forms of an element, such as diamond and graphite, are called *allotropes*. Thus orthorhombic sulfur and monoclinic sulfur are two allotropes of sulfur. Their names refer to the different structures of their crystals. Both crystals contain S_8 molecules, which consist of a zigzag ring of eight sulfur atoms. Each sulfur atom has an angular AX_2E_2 geometry with a bond angle of 108° (Figure 8.6). The arrangement of the S_8 molecules in orthorhombic sulfur is shown in Figure 8.7. The arrangement of the molecules in monoclinic sulfur is slightly different.

Another form of sulfur is obtained by quickly cooling molten sulfur by pouring it into cold water. A brown rubbery material known as *plastic sulfur* is obtained, as described in Experiment 8.1 (on p. 382). It is not stable and within a few hours it is transformed back into crystalline orthorhombic sulfur.

FIGURE 8.5
Crystals of Orthorhombic and Monoclinic Sulfur.

(a) Orthorhombic sulfur is the stable allotrope of sulfur at room temperature.
(b) The needle-like crystals of monoclinic sulfur are stable only above 119 °C. They are sometimes found in the vicinity of volcanoes where they are formed from the hot gases emitted by the volcano.

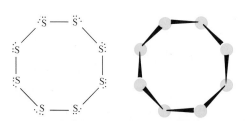

FIGURE 8.6
The Crown-Shaped S_8
Molecule.

Each sulfur atom has an angular
AX_2E_2 geometry.

Plastic sulfur consists of very long chains of sulfur atoms rather than S_8 rings. When liquid sulfur is heated to about 160 °C, one of the S—S bonds in some of the rings breaks, and the rings open to give chains of eight sulfur atoms:

The sulfur atoms at each end of the chain have only seven valence electrons. Therefore they have a strong tendency to attract an additional electron and are very reactive. An S_8 chain reacts with an S_8 ring causing it to open up and thereby forming a 16-atom chain:

But the sulfur atoms at the ends of the chain still have only seven electrons, so the process continues, leading to the formation of very long chains of thousands of atoms. As the temperature is raised from 160 °C to 190 °C, the chains

FIGURE 8.7
The Structure of
Orthorhombic Sulfur.

Both orthorhombic and monoclinic sulfur consist of crown-shaped rings of eight sulfur atoms. In orthorhombic sulfur these molecules are packed together as shown here. In monoclinic sulfur the arrangement of the S_8 molecules is slightly different.

Experiment 8.1

Plastic Sulfur

Sulfur is heated until it melts to form an orange liquid.

On further heating the color becomes dark red and the liquid becomes very viscous.

When the liquid is rapidly cooled by pouring it into cold water, a rubbery brown solid—plastic sulfur—is formed.

Plastic sulfur is elastic, like rubber.

become longer and more tangled, and the liquid becomes increasingly thick and sticky, like maple syrup or molasses. It is described as being very viscous. A liquid like water that flows easily has a low **viscosity**, while a liquid like molasses or heavy engine oil that flows slowly is said to have a high viscosity.

The behavior of liquid sulfur on heating is unusual. Most liquids become less viscous on heating, because the increased thermal motion of the molecules enables them to move past each other more easily. Liquid sulfur, however, becomes more viscous. When this viscous liquid sulfur is cooled rapidly—for example, by pouring it into water, with which it does not react—the long tangled chains do not have time to rearrange to the more stable cyclic S_8 molecules. So the rubbery solid that is obtained is not crystalline. Its molecules do not have the ordered arrangement that is characteristic of a crystalline solid. They still retain the random arrangement that is characteristic of liquids. Solids that are not crystalline are called **amorphous solids**. Plastic sulfur exhibits properties similar to those of rubber, which also consists of long-chain molecules. Long-chain molecules tend to coil up into compact shapes. If plastic sulfur or rubber is stretched, the coiled-up molecules straighten out a little, and the material stretches. If the stretching force is removed, the molecules tend to resume their coiled-up structure and the material contracts again (see Figure 8.8).

There are still other, less important allotropes of sulfur that contain rings of six, seven, twelve, and more sulfur atoms.

PHOSPHORUS Like sulfur, phosphorus also has several allotropes. *White phosphorus* is obtained when phosphorus vapor is condensed. It is a colorless crystalline solid, with a melting point of 44 °C, but it rapidly becomes white

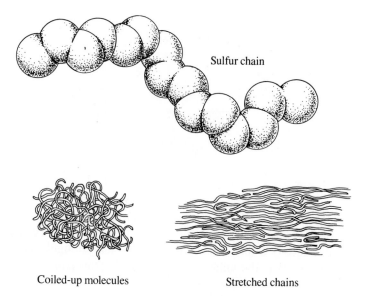

Sulfur chain

Coiled-up molecules

Stretched chains

FIGURE 8.8
Plastic Sulfur.

Plastic sulfur is an amorphous (noncrystalline) allotrope of sulfur consisting of long chains of sulfur atoms. These chains have an irregular, disordered arrangement rather than the regular arrangement characteristic of crystalline substances. The chains tend to coil up and form a tangled mass. When a force is applied, the molecules straighten out a little but coil up again when the force is removed. So the solid can stretch and contract like rubber.

and opaque unless stored under nitrogen in the dark. White phosphorus is very toxic. The vapor causes decay of cartilage and bones, particularly of the nose and jaw.

White phosphorus reacts with oxygen at room temperature, emitting a bluish green light (see Experiment 8.2). Some of the energy produced in the reaction is emitted as electromagnetic radiation. This phenomenon is known as **chemiluminescence**, which is the emission of energy released by a reaction as light rather than as heat. At temperatures above 40 °C the oxidation becomes quite rapid, and phosphorus ignites. In order to avoid the danger of fire, white phosphorus is stored away from contact with the air; usually, it is stored under water, in which it is insoluble (Figure 8.9). It is soluble, however, in certain

The caribbean flashlight fish is chemiluminescent. Chemical reactions are the source of the light it emits.

Experiment 8.2

Oxidation of White Phosphorus: Chemiluminescence

When water containing a piece of white phosphorus is boiled, the jet of steam coming from the tube at the top of the flask glows in a subdued light with a blue-green color, as the phosphorus vapor in the steam is oxidized by the oxygen in the air. Some of the energy produced by the oxidation of phosphorus is in the form of light rather than heat.

FIGURE 8.9
Red and White Phosphorus.

White phosphorus rapidly oxidizes
in the air so it is usually stored
under water in which it is insoluble.

other solvents such as carbon disulfide, CS_2. White phosphorus consists of
tetraatomic P_4 molecules that have a tetrahedral structure (Figure 8.10). Each
of the phosphorus atoms has a pyramidal AX_3E geometry. As the angles at
the corners of a face of a tetrahedron are only $60°$ it is thought that the P—P
bonds are bent, so that the angles between the electron pairs in the valence
shell of phosphorus are considerably larger than $60°$.

There are several other allotropes of phosphorus; the best known are *red
phosphorus* and *black phosphorus*. Red phosphorus (Figure 8.9) can be obtained
by heating white phosphorus in the absence of air at atmospheric pressure.
Black phosphorus is obtained by heating white or red phosphorus under a
very high pressure. Both allotropes have much higher melting points and
boiling points than white phosphorus. For example, red phosphorus sublimes
(changes directly to vapor) at $417 °C$ and only melts under pressure at $540 °C$.
Neither red nor black phosphorus ignites spontaneously in air, and they are
much less reactive and less poisonous than white phosphorus. These differences
result from differences in their structures. Red and black phosphorus do not
consist of P_4 molecules but have polymeric network structures. The structure
of black phosphorus is shown in Figure 8.11. The structure of red phosphorus
is more complicated. To melt and vaporize a network structure requires that
covalent bonds be broken, so the melting point and boiling point are high, as
we saw in Chapter 6 for diamond and graphite. The vapor obtained from both
red and black phosphorus consists of P_4 molecules; when the vapor is condensed, white phosphorus is obtained. Red phosphorus is an essential component of the striking surface on a box of safety matches (Box 8.1).

EXERCISE 8.1

On what properties of sulfur does its extraction by the Frasch process depend?

FIGURE 8.10
The Tetrahedral P_4 Molecule.

(a) Ball-and-stick model. (b) Space-
filling model. (c) In this ball-and-
stick model the angles between the
bonds at each phosphorus atom are
considerably larger than $60°$ so
that the bonds are bent.

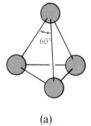

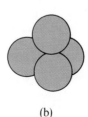

(a)   (b)   (c)

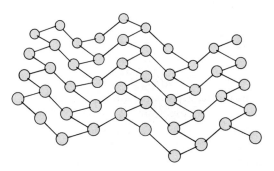

FIGURE 8.11
The Structure of Black Phosphorus.

Each P atom has a pyramidal AX_3E geometry, but the atoms are not connected so as to form P_4 molecules but instead form a corrugated sheet of atoms, that is, a two-dimensional network or giant two-dimensional molecule. These sheets of atoms are then stacked one upon another in the crystal. When black phosphorus is vaporized, this structure is broken up and P_4 molecules are formed.

BOX 8.1

THE CHEMISTRY OF MATCHES

The heads of safety matches are made of an oxidizing agent such as potassium chlorate, $KClO_3$, mixed with sulfur, fillers, and glass powder. The side of the box contains red phosphorus, binder, and powdered glass or sand. The heat generated by friction when the match is struck causes a minute amount of red phosphorus to be converted to white phosphorus, which ignites spontaneously in the air. This sets off the decomposition of the potassium chlorate to give oxygen. The sulfur then catches fire and ignites the wood.

$$2KClO_3(s) \longrightarrow 2KCl(s) + 3O_2(g)$$

$$S(s) + O_2(g) \longrightarrow SO_2(g)$$

The heads of "strike anywhere" matches contain an oxidizing agent, such as $KClO_3$, or manganese dioxide, MnO_2, together with tetraphosphorus trisulfide, P_4S_3, glass and binder. The phosphorus sulfide is easily ignited, the $KClO_3$ decomposes to give oxygen, which in turn causes the P_4S_3 to burn more vigorously.

P_4S_3

The red tips on the strike anywhere matches on the left are P_4S_3. The remainder of the tip is composed of $KClO_3$ or MnO_2, sulfur, glue, and a coloring agent. The head of the safety matches on the right contain $KClO_3$, sulfur, powdered glass, glue, and a coloring agent.

The striking surface on the side of the box of safety matches consists of red phosphorus, ground glass, and glue.

The striking surface on the side of a box of strike anywhere matches is simply a rough surface of powdered glass and glue.

8.2
Reactions and Compounds of Sulfur

OXIDES

From its position in the periodic table we expect sulfur to have a valence of 2 as it does in many of its compounds. We would therefore expect it to form the oxide SO. Indeed, this oxide is formed in the gas phase when an electric discharge is passed through sulfur dioxide but, rather surprisingly, it is a very unstable and reactive substance. The much more stable and better-known oxides of sulfur are sulfur dioxide, SO_2, and sulfur trioxide, SO_3.

Thus when sulfur is heated in air, it ignites at 350 °C and burns with a blue flame to produce *sulfur dioxide*, SO_2 (Experiment 3.1). Sulfur dioxide is a colorless gas with a pungent, choking odor. It destroys bacteria and is used as a preservative in the storage of fruits, such as apples, and in the preparation of dried fruits, such as prunes and apricots. It condenses to a liquid at −10 °C; at 20 °C it can be liquefied by a pressure of about 3 atm. It is usually sold as a liquid under pressure in metal cylinders.

When sulfur dioxide is heated with oxygen in the presence of finely divided platinum metal or vanadium pentoxide, V_2O_5, another oxide, sulfur trioxide, SO_3, is formed:

$$2SO_2(g) + O_2(g) \longrightarrow 2SO_3(g)$$

The platinum or vanadium pentoxide acts as a *catalyst* for the reaction. Sulfur trioxide condenses to a colorless liquid at 44.5 °C and freezes to transparent crystals at 16.8 °C.

HIGHER VALENCES OF SULFUR AND PHOSPHORUS

The formulas SO_2 and SO_3 show that sulfur can exhibit valences of 4 and 6 in addition to the expected valence of 2, which it has in compounds such as H_2S and SCl_2. Indeed, sulfur forms many compounds in which it exhibits these *higher valences*. These compounds include the sulfite ion, SO_3^{2-}, the sulfate ion, SO_4^{2-}, sulfuric acid, H_2SO_4, and the fluorides SF_4 and SF_6.

From its position in the periodic table we expect phosphorus to have a valence of 3, which it does in the oxide P_4O_6 (empirical formula P_2O_3) and the chloride PCl_3. But phosphorus also forms an oxide P_4O_{10} (empirical formula P_2O_5), the chloride PCl_5, and phosphoric acid, H_3PO_4. In each of these compounds phosphorus has a valence of 5.

To understand the higher valences of phosphorus and sulfur, we must look at their electron configurations (Table 8.1). They are just like those of the corresponding elements of the second period, nitrogen and oxygen, except that the electrons occupy 3s and 3p rather 2s and 2p orbitals.

Thus sulfur has the following ground state electron configuration:

3s   3p

Like oxygen, sulfur is therefore expected to form two covalent bonds as it does

TABLE 8.1 Ground State Electron Configurations of Third-Period Elements	
Na	[Ne] $3s^1$
Mg	[Ne] $3s^2$
Al	[Ne] $3s^2 3p^1$
Si	[Ne] $3s^2 3p^2$
P	[Ne] $3s^2 3p^3$
S	[Ne] $3s^2 3p^4$
Cl	[Ne] $3s^2 3p^5$
Ar	[Ne] $3s^2 3p^6$

in H_2S. However, the $n = 3$ shell also has five 3d orbitals which are not occupied in the ground state of sulfur:

If an electron is promoted from a 3p orbital to one of the 3d orbitals we obtain the excited or valence state configuration:

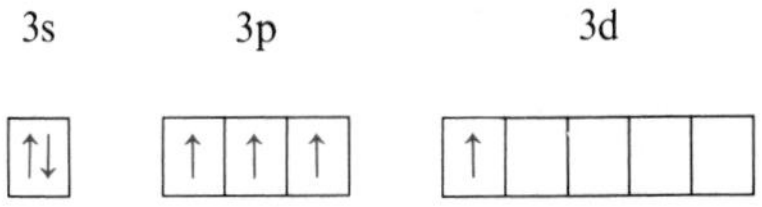

3s   3p   3d

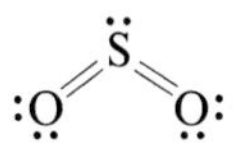

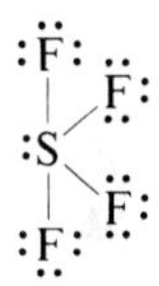

In this valence state there are four electrons in singly occupied orbitals, that is, four unpaired electrons. These electrons can be used to form four covalent bonds, as in SO_2 and SF_4. There is also an unshared pair of electrons in a nonbonding orbital. Thus SO_2 has two double bonds and an unshared pair and is therefore an AX_2E type molecule, and it has the expected angular shape with a bond angle of 120°. We will discuss the structure of SF_4 in Chapter 9.

Promotion of both a 3s and a 3p electron to a 3d orbital gives another valence state in which there are six electrons in singly occupied orbitals:

3s   3p   3d

These six unpaired electrons can be used to form six covalent bonds, as in the compounds SO_3 and SF_6. Sulfur trioxide is a planar AX_3 molecule and sulfur hexafluoride, SF_6, is an AX_6 octahedral molecule. (See Chapter 9.)

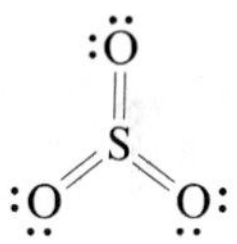

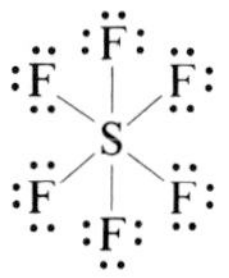

Thus, because the $n = 3$ shell can contain a maximum of 18 rather than 8 electrons as in the $n = 2$ shell, sulfur can form compounds such as SO_2 and SF_4 in which it has 10 electrons in its valence shell and SO_3 and SF_6 in which it has 12 electrons in its valence shell. As we have already discussed in Chapter 7 in the case of carbon, these excited or valence states of the sulfur atom are not formed before compound formation begins but rather four or six bonding orbitals are formed as the oxygen or fluorine atoms combine with the sulfur atom. Nevertheless, it is convenient to think of these higher-valence compounds as being formed from the corresponding valence state because we can then readily see how many bonds will be formed.

Phosphorus has the following ground state configuration:

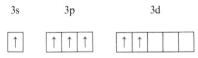

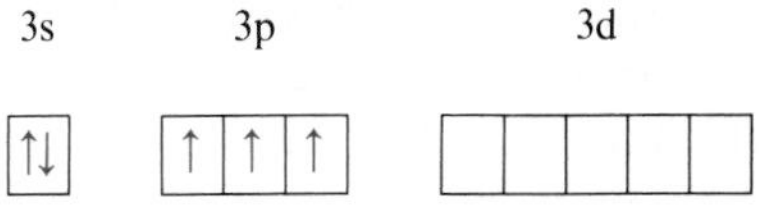

If an electron is promoted from the 3s orbital to a 3d orbital, the electron configuration becomes

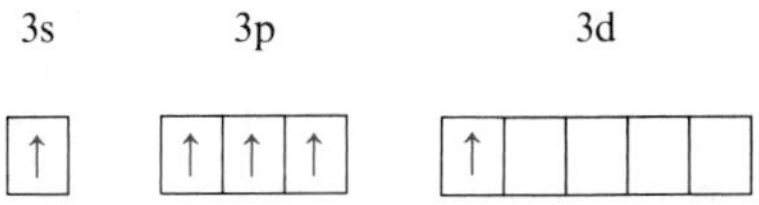

3s   3p   3d

This valence state has five unpaired electrons. Phosphorus in this state can therefore form five covalent bonds, as in PCl_5 and P_4O_{10}.

In writing Lewis structures for the higher-valence compounds of sulfur and phosphorus we must remember that they do not obey the octet rule; they may have 10 or 12 electrons in their valence shells. For example, in the case of SO_2 we can use 2 of the 6 sulfur electrons to form two bonds to an oxygen atom and 2 more electrons to form two bonds to the second oxygen atom, which leaves the 2 remaining electrons as an unshared pair. Thus the sulfur atom in SO_2 has a pair of nonbonding electrons and four shared pairs of electrons, making a total of five pairs, or 10 electrons, in its valence shell:

$$:\ddot{S}\cdot + 2:\ddot{O}\cdot \longrightarrow :\ddot{O}::\ddot{S}::\ddot{O}: \quad \text{or} \quad :\ddot{O}=\ddot{S}=\ddot{O}:$$

In the molecule SO_3 sulfur uses all 6 of its valence electrons in the formation of three double bonds to three oxygen atoms:

It therefore has six pairs, or 12 electrons, in its valence shell.

Gaseous sulfur trioxide consists partly of this molecule and partly of a trimeric molecule, S_3O_9, which has the ring structure

in which each sulfur atom forms six bonds. The liquid consists very largely of S_3O_9 molecules. Sulfur trioxide reacts with water to give sulfuric acid, H_2SO_4,

$$H_2O(l) + SO_3(l) \longrightarrow H_2SO_4(aq)$$

in which sulfur again forms six bonds:

H_2SO_4

In both $(SO_3)_3$ and H_2SO_4 the sulfur atom forms two double and two single bonds and has no nonbonding pairs of electrons. Thus the sulfur atom has an AX_4 type geometry and therefore the oxygen atoms have a tetrahedral arrangement around the sulfur atom (Figure 8.12).

In both SO_2 and SO_3, and indeed in all its other covalent compounds, oxygen forms only two bonds and always has a valence shell of eight electrons. The valence shell of oxygen, the $n = 2$ shell, is completely filled by eight electrons, and oxygen always obeys the octet rule. In fact, the only elements that invariably obey the octet rule in compound formation are carbon, nitrogen, oxygen, and fluorine. All heavier elements obey the octet rule only in some of their compounds. In other compounds they may have 10, 12, or even 14 electrons in their valence shells. In the latter compounds the atoms attached to sulfur, phosphorus, etc. are usually very electronegative atoms such as fluorine,

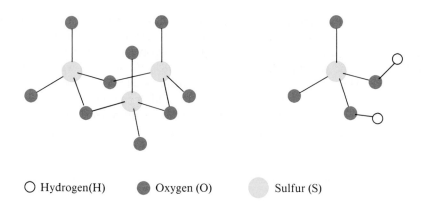

FIGURE 8.12
The Structures of H_2SO_4 and $(SO_3)_3$.

In both molecules there is an AX_4 tetrahedral arrangement of four oxygen atoms around each sulfur.

○ Hydrogen(H)   ● Oxygen (O)   ○ Sulfur (S)

oxygen, and chlorine. To understand why, recall that a d orbital has a higher energy than an s or a p orbital in the same shell, and an electron in a d orbital is at a correspondingly greater distance from the nucleus. Energy must be supplied to move the electrons to a higher energy level, that is, to pull the electron away from the nucleus. Only the atoms of elements with high electronegativities, such as fluorine, oxygen, and chlorine, appear to be able to attract the valence-shell electrons of phosphorus and sulfur sufficiently strongly to promote one or more of these electrons from an s or p orbital into a d orbital.

INDUSTRIAL PREPARATION OF SULFURIC ACID: CONTACT PROCESS

Sulfuric acid may be described as the world's most important industrial chemical; it is produced in larger quantities than any other substance and has many uses. The uses of sulfuric acid are so important and so varied that it has been said that a country's sulfuric acid production is a good measure of its industrial development. The annual U.S. production, which is the largest in the world, is approximately 40 million tons. About half of this amount is used for the production of phosphate fertilizers. Other uses include the manufacture of paints, dyes, explosives, detergents, and synthetic fibers.

The industrial preparation of sulfuric acid involves three reactions that we have already encountered:

1. The burning of sulfur or a metal sulfide in air to give SO_2:

$$S + O_2 \longrightarrow SO_2$$
$$CuS + O_2 \longrightarrow Cu + SO_2$$

2. The oxidation of SO_2 to SO_3:

$$2SO_2 + O_2 \xrightarrow{\text{Catalyst}} 2SO_3$$

3. The combination of SO_3 with water to give H_2SO_4:

$$SO_3 + H_2O \longrightarrow H_2SO_4$$

The oxidation of SO_2 to SO_3 is very slow at ordinary temperatures. To increase the reaction rate this oxidation is carried out at approximately 400 °C in the presence of a *catalyst*, usually vanadium pentaoxide, V_2O_5. The process is known as the **contact process** because the reaction takes place when SO_2

Catalysts are substances that speed up reactions but remain unchanged at the end of the reaction. Vanadium penta-oxide is an example of a common type of catalyst called a neterogeneous catalyst. This type of catalyst is a solid which adsorbs the reactant molecules on its surface so that they are held more closely and for a longer time than during a fleeting contact in the gas phase. Thus they have a greater probability of reacting than they do in a collision of the reactant molecules in the gas phase. Catalysts are discussed in more detail in Chapter 19.

and O_2 molecules come into contact on the surface of the solid V_2O_5 catalyst. The SO_3 is not allowed to react directly with water, because this is a violent reaction that produces a dense mist of H_2SO_4 droplets that is not easily handled. Instead, the gaseous SO_3 is absorbed in 98% H_2SO_4, and water is added at a controlled rate to keep the concentration of H_2SO_4 at approximately 98%. This solution is the acid that is normally sold as *concentrated sulfuric acid*.

If no water is added, the concentration of the sulfuric acid rises to 100%, and then a series of *polysulfuric acids* are formed:

$$H_2SO_4 + SO_3 \longrightarrow H_2S_2O_7$$

Disulfuric acid

$$H_2SO_4 + 2SO_3 \longrightarrow H_2S_3O_{10}$$

Trisulfuric acid

A mixture of sulfuric acid and polysulfuric acids is often called *oleum* or *fuming sulfuric acid*.

PROPERTIES AND REACTIONS OF SULFURIC ACID: LE CHÂTELIER'S PRINCIPLE

SOLUTIONS IN WATER Sulfuric acid is a strong acid and its reaction with water is highly exothermic:

$$H_2O(l) + H_2SO_4(l) \longrightarrow H_3O^+(aq) + HSO_4^-(aq)$$

Consequently, the dilution of concentrated sulfuric acid must be carried out with care. If water is added to the concentrated acid, the heat of the reaction may be sufficient to raise the temperature of the water to its boiling point and cause drops of the acid to be thrown violently out of the container. The only safe procedure is to add the concentrated acid slowly, with constant stirring, to a large amount of cold water.

Since sulfuric acid has two OH groups, it has two ionizable hydrogen atoms and it can donate a second proton to another water molecule:

$$HSO_4^- + H_2O \rightleftharpoons H_3O^+ + SO_4^{2-}$$

An acid that has two ionizable hydrogen atoms is known as a **diprotic acid**. Another example of a diprotic acid is carbonic acid, H_2CO_3, that we discussed in Chapter 6.

The anions formed in these two reactions are the hydrogen sulfate ion, HSO_4^-, and the sulfate ion, SO_4^{2-}. The HSO_4^- ion is a weak acid and its reaction with water is not complete. The Lewis structures for the hydrogen sulfate ion and the sulfate ion are

A solution of sulfuric acid in water contains H_3O^+, HSO_4^-, and SO_4^{2-}. The acidic properties of such a solution are those of H_3O^+, and they are the same therefore as those of an aqueous solution of any strong acid such as hydrochloric acid.

When we speak of hydrochloric acid we always mean a solution of hydrogen chloride in water. Hydrogen chloride is a gas, which condenses to a liquid only at $-85\,°C$. The pure liquid is therefore not commonly encountered and is not of any great importance.

The situation is very different, however, for sulfuric acid. The name *sulfuric acid* is itself ambiguous since it is used both for the pure substance H_2SO_4 and for solutions of H_2SO_4 in water. Unlike pure liquid HCl, pure liquid H_2SO_4 is easily obtained. It is a colorless liquid with a melting point of $10.4\,°C$. The 98% H_2SO_4 that is produced industrially and is commonly used in the laboratory is a mixture of water and sulfuric acid, but since the latter is in a large excess, it is best regarded as a solution of water in sulfuric acid rather than as a solution of sulfuric acid in water. It contains H_2SO_4 molecules and some H_3O^+ and HSO_4^-, but no un-ionized H_2O molecules. Its properties are quite similar to those of the 100% pure H_2SO_4 that consists only of H_2SO_4 molecules.

Concentrated (98%) sulfuric acid has four important properties, which are utilized in a number of important ways:

1. It has a high boiling point of $338\,°C$.
2. It is a strong acid.
3. It is a dehydrating agent.
4. It is a strong oxidizing agent.

First we consider some uses of sulfuric acid that depend on its boiling point and the fact that it is a strong acid.

PREPARATION OF HYDROGEN CHLORIDE Concentrated sulfuric acid reacts with sodium chloride to give hydrogen chloride, which is evolved as a gas from the reaction mixture:

$$NaCl(s) + H_2SO_4(\text{conc aq}) \rightleftharpoons HCl(g) + NaHSO_4(s)$$

In this reaction sulfuric acid *protonates*—that is, adds a proton to—the chloride ion:

$$Cl^- + H_2SO_4 \rightleftharpoons HCl + HSO_4^-$$

We saw in Chapter 5 that chloride ion, Cl^-, is not a base in water and is not protonated by the hydronium ion, H_3O^+, as it is the anion (conjugate base) of HCl, a strong acid in water. But in concentrated sulfuric acid there is not enough water to ionize all the sulfuric acid and thus in addition to H_3O^+ and HSO_4^- ions in the solution there are also H_2SO_4 molecules. Un-ionized sulfuric acid is a stronger acid than H_3O^+ and it can protonate Cl^- to give its conjugate acid HCl.

In principle, the reaction between NaCl and H_2SO_4 should come to equilibrium, but since hydrogen chloride is a gas, if the reaction is not carried out in a closed container, hydrogen chloride is evolved as it is formed and is lost from the reaction mixture. Thus the concentration of HCl in the system never reaches its equilibrium value and so more and more hydrogen chloride is formed until the reaction has gone essentially to completion (see Experiment 8.3).

Experiment 8.3

Preparation of Volatile Acids

When sodium nitrate is heated with concentrated sulfuric acid, nitric acid is formed. The nitric acid distills into the test tube and collects as a pale yellow liquid. Some brown NO_2 vapor is also formed by decomposition of the nitric acid in the hot flask.

This is a close-up view of the pale yellow liquid nitric acid. Pure nitric acid is colorless but here it is colored yellow by the small amount of nitrogen dioxide, NO_2, which was formed.

When concentrated sulfuric acid is added to sodium chloride, hydrogen chloride gas is evolved, causing effervescence. When the HCl(g) comes in contact with a swab soaked in aqueous ammonia solution that is held at the mouth of the tube, white fumes of solid ammonia chloride are formed.

This is not the reaction that is used for the industrial preparation of nitric acid. As we shall see in Chapter 18 nitric acid is prepared on a large scale by the catalytic oxidation of ammonia. It is important to remember that a reaction that is convenient for the preparation of a substance on a small scale is not always the most suitable and the most economical for the industrial preparation of this substance on a large scale. The ready availability of the starting products in large amounts and their cost are some of the most important factors to be taken into account in the design of an industrial process. But for small-scale laboratory preparations these factors are of lesser importance.

PREPARATION OF NITRIC ACID Nitric acid, HNO_3, can be prepared in a similar manner by heating sodium nitrate with concentrated sulfuric acid:

$$NaNO_3(s) + H_2SO_4(conc\ aq) \rightleftharpoons HNO_3(g) + NaHSO_4(s)$$

In this reaction sulfuric acid protonates the nitrate ion:

$$NO_3^- + H_2SO_4 \rightleftharpoons HNO_3 + HSO_4^-$$

The nitrate ion, like the chloride ion, is not a base in aqueous solution, but it is protonated by concentrated sulfuric acid.

Unlike hydrogen chloride, which is a gas, nitric acid is a liquid with a boiling point of 86 °C. It can be distilled from the reaction mixture by heating and is collected as a pale yellow liquid. Because sulfuric acid has a very high boiling point, it does not distill at the same time (see Experiment 8.3).

LE CHÂTELIER'S PRINCIPLE The preparations of hydrogen chloride and nitric acid are examples of the effect of removing one of the substances taking part in an equilibrium. In any chemical equilibrium the reaction between the reactants to give the products, the forward reaction, is proceeding at the same rate as the reaction between the products to give back the reactants. If one of the products is removed, or it its concentration is decreased in some way, then the rate of the back reaction is decreased. The rate of the forward reaction is unchanged and is therefore greater than the rate of the back reaction. Thus, the products continue to form and, if they are continually removed, the system never comes to equilibrium and the reaction continues until all the reactants are used up. In other words, the reaction is driven essentially to completion.

In general we can state that:

For any system in chemical equilibrium, the effect of a change in the concentration of any of the substances taking part in the equilibrium is to cause the position of the equilibrium to shift so as to minimize the change in concentration.

Thus, if the concentration of one of the products is decreased, then the position of the equilibrium shifts so that more product forms until equilibrium is once again established. If one or more products are continually removed then equilibrium is never reached and the reaction goes to completion.

This statement is a special case of a general principle first stated by French chemist Henri Le Châtelier (1850–1936), which describes how systems at equilibrium react to changes in any of the factors that affect the equilibrium. **Le Châtelier's principle** is as follows:

When any of the conditions that affect the position of a dynamic equilibrium are changed, the position of the equilibrium shifts so as to minimize the change.

These conditions include the concentrations of the reactants and the products, the temperature, and the pressure. We will study the effect of changing the concentrations of the substances taking part in an equilibrium in a quantitative way in Chapters 13, 14, and 15. In this chapter we will see several more examples of how we can qualitatively predict the effect of changing the concentrations of the substances involved in an equilibrium by using Le Châtelier's principle.

SULFATES Sulfuric acid is a diprotic acid which reacts with bases to form two series of salts, the hydrogen sulfates and the sulfates:

$$H_2SO_4(aq) + KOH(aq) \longrightarrow KHSO_4(aq) + H_2O(l)$$
$$\text{Potassium hydrogen sulfate}$$

$$KHSO_4(aq) + KOH(aq) \longrightarrow K_2SO_4(aq) + H_2O(l)$$
$$\text{Potassium sulfate}$$

The overall reaction is

$$H_2SO_4(aq) + 2KOH(aq) \longrightarrow K_2SO_4(aq) + 2H_2O(l)$$

Metal hydrogen sulfates and sulfates are ionic compounds containing metal ions and HSO_4^- or SO_4^{2-}. They are generally soluble in water but the sulfates of Ca^{2+}, Sr^{2+}, Ba^{2+}, Pb^{2+}, and Ag^+ are insoluble (Table 5.9). Some insoluble sulfates—such as $CaSO_4 \cdot 2H_2O$ (gypsum), $BaSO_4$ (barite), and $PbSO_4$—occur as minerals. Barium sulfate is very insoluble, and its formation as a white precipitate on addition of an acidic aqueous solution of $BaCl_2$ is often used as a test for the presence of sulfate in a solution:

$$Ba^{2+}(aq) + SO_4^{2-}(aq) \longrightarrow BaSO_4(s)$$

Common soluble sulfates include $Na_2SO_4 \cdot 10H_2O$, $(NH_4)_2SO_4$, $MgSO_4 \cdot 7H_2O$ (Epsom salts), $CuSO_4 \cdot 5H_2O$, $ZnSO_4 \cdot 7H_2O$, and some sulfates containing two different positive ions, such as $(NH_4)_2Fe(SO_4)_2 \cdot 6H_2O$ and $KAl(SO_4)_2 \cdot 12H_2O$ (alum), which are known as double salts.

The crystals that form when many salts crystallize from water often contain water molecules and are therefore called **hydrates**. They have formulas such as $CuSO_4 \cdot 5H_2O$. These salts are said to be hydrated. The water that is contained in hydrated salts is called **water of crystallization**. In most cases the water molecules, or at least most of them, are bonded to the metal ion to form **hydrated ions** such as $Cu(H_2O)_4^{2+}$. Because the oxygen atom in the water molecule has a higher electronegativity than the hydrogen atoms the bonds are polar; the oxygen atom carries a small negative charge and the hydrogen atoms a small positive charge. The positive charge of the metal atom attracts an unshared electron pair on the negatively charged oxygen atom of a water molecule. Metal atoms, particularly if they have a $+2$ or $+3$ charge, often combine with 4 or even 6 water molecules in this way. In $CuSO_4 \cdot 5H_2O$ only four of the water molecules are bonded to the Cu^{2+} ion. Thus the crystalline solid is composed of $Cu(H_2O)_4^{2+}$ ions, SO_4^{2-} ions, and H_2O molecules. Its formula could be written as $Cu(H_2O)_4^{2+} \cdot SO_4^{2-} \cdot H_2O$.

EXERCISE 8.2

Write balanced equations for the neutralization of the following bases with sulfuric acid in aqueous solution:

$CsOH$ NH_3 $Ca(OH)_2$ MgO

EXERCISE 8.3

What volume of $0.112M$ $H_2SO_4(aq)$ must be added to 100.0 mL of $0.056M$ $Ba(OH)_2(aq)$ to precipitate all the barium as $BaSO_4(s)$?

SULFURIC ACID AS A DEHYDRATING AGENT Water is completely ionized in solution in sulfuric acid. It behaves as a strong base:

$$H_2O(l) + \underset{\text{Concentrated 98\%}}{H_2SO_4(l)} \longrightarrow H_3O^+ + HSO_4^-$$

This property makes sulfuric acid a very good **dehydrating agent**. Gases that do not react with sulfuric acid, such as O_2, N_2, CO_2, and SO_2, may be dried by bubbling them through concentrated sulfuric acid.

Hydrated salts, such as $CuSO_4 \cdot 5H_2O$, lose their water of crystallization when stored in a closed container (*desiccator*) with concentrated sulfuric acid. A small amount of water vapor is in equilibrium with the hydrated salt, and as this vapor is absorbed by the sulfuric acid, more of the salt gives up its water of crystallization. Thus slowly all the water is removed:

$$\underset{\text{Blue}}{CuSO_4 \cdot 5H_2O} \rightleftharpoons \underset{\text{White}}{CuSO_4} + \underset{\substack{\text{Absorbed by sulfuric acid}}}{5H_2O}$$

$$(H_2O + H_2SO_4 \longrightarrow H_3O^+ + HSO_4^-)$$

This reaction is another example of the operation of Le Châtelier's principle. The copper sulfate that results is a white powder and is called *anhydrous* copper sulfate. It can be reconverted to the blue hydrated salt simply by addition of water. This reaction can be used as a test for water.

The tendency of sulfuric acid to combine with water is so strong that it will remove hydrogen and oxygen in a 2:1 atomic ratio from many compounds that do not contain water in a molecular form. For example, it removes hydrogen and oxygen as water from carbohydrates and many other organic compounds, leaving behind a charred residue of carbon. Wood, paper, starch, cotton, and sugar (sucrose) are all dehydrated in this way (Experiment 8.4):

$$C_{12}H_{22}O_{11}(s) + 11H_2SO_4(l) \longrightarrow 12C(s) + 11H_3O^+ + 11HSO_4^-$$

SULFURIC ACID AS AN OXIDIZING AGENT Concentrated aqueous sulfuric acid is a strong oxidizing agent. For example it will oxidize the bromide ion in solid NaBr to bromine, Br_2, and the iodide ion in solid sodium iodide to iodine, I_2 (Experiment 8.5, p. 396). It also oxidizes carbon to CO_2 and copper to the Cu^{2+} ion.

$$Br^- \longrightarrow Br_2$$
$$I^- \longrightarrow I_2$$
$$C \longrightarrow CO_2$$
$$Cu \longrightarrow Cu^{2+}$$

Experiment 8.4

Sulfuric Acid as a Dehydrating Agent

Concentrated sulfuric acid is added to sugar, $C_{12}H_{22}O_{11}$.

The concentrated sulfuric acid removes water from the sugar to form black carbon and steam.

The steam mixes with the carbon to form a black pillar which rises high above the beaker as the mixture heats and expands.

Experiment 8.5

Sulfuric Acid as an Oxidizing Agent

Concentrated sulfuric acid is added to white solid sodium bromide and white solid sodium iodide.

The sulfuric acid oxidizes the sodium bromide to reddish-brown bromine and the sodium iodide to violet iodine.

In these reactions the sulfuric acid must have been reduced. It is found by experiment that in each case another product of the reaction is sulfur dioxide, SO_2. So sulfuric acid must have been reduced to sulfur dioxide and we can write the following unbalanced equations

$$Br^- + H_2SO_4(conc) \longrightarrow Br_2 + SO_2$$
$$C + H_2SO_4(conc) \longrightarrow CO_2 + SO_2$$
$$Cu + H_2SO_4(conc) \longrightarrow Cu^{2+} + SO_2$$

We recognize the conversion of copper metal to copper ion and the conversion of bromide ion to bromine as oxidations because they involve the loss of electrons:

$$2Br^- \longrightarrow Br_2 + 2e^-$$
$$Cu \longrightarrow Cu^{2+} + 2e^-$$

But it is not so clear that the conversion of sulfuric acid to sulfur dioxide is a reduction, because it is not obvious that this reaction involves a gain of electrons. Similarly, although the conversion of carbon to carbon dioxide is an oxidation in the original sense of addition of oxygen, it is not obvious that this reaction involves a loss of electrons. To help us to understand how these reactions may be regarded as redox reactions and to help us balance the equations for these reactions we use the concept of oxidation numbers.

OXIDATION NUMBERS

The **oxidation number** is a measure of the extent of oxidation of an element in its compounds and is assigned by using a set of simple rules. The free element is the standard state from which the extent of oxidation and reduction is measured. *An element in any of its allotropic forms is assigned an oxidation number of 0.*

A singly charged positive ion formed by the loss of one electron from a neutral atom is assigned an oxidation number of $+1$. Thus sodium in Na^+ has an oxidation number of $+1$. A doubly charged positive ion formed by the loss of two electrons from an atom is assigned an oxidation number of $+2$, and so on. Singly charged, monatomic, negative ions, such as Cl^-, are assigned an oxidation number of -1, and doubly charged negative ions, such as O^{2-}, have an oxidation number of -2. *The oxidation number of a monatomic ion is equal to its charge.*

But how do we assign oxidation numbers in covalent compounds such as SO_2 and polyatomic ions such as SO_4^{2-}? Consider the reaction

$$H_2 + Cl_2 \longrightarrow 2HCl$$

Because chlorine is more electronegative than hydrogen, hydrogen chloride is a polar molecule. The chlorine atom has a small negative charge, and the hydrogen atom has a small positive charge:

$$\overset{\delta+}{H} - \overset{\overset{\cdot\cdot}{}\,\delta-}{\underset{\cdot\cdot}{Cl}}:$$

There has been some transfer of electron density from hydrogen to chlorine; the hydrogen has been oxidized and the chlorine reduced, but less than one electron has been transferred. Assigning fractional oxidation numbers to the hydrogen and chlorine atoms would be complicated and not very useful. So *the oxidation number is defined as the charge that would be associated with a particular atom if both electrons of each bond were transferred completely to the atom of higher electronegativity.* Thus in hydrogen chloride the bonding electron pair is assigned to chlorine because it is more electronegative than hydrogen. Consequently, for the purpose of calculating oxidation numbers, the Cl atom is considered to be Cl^- and is assigned an oxidation number of -1. The H atom then is thought of as having lost an electron, becoming H^+, and is assigned an oxidation number of $+1$:

$$H:\overset{\cdot\cdot}{\underset{\cdot\cdot}{Cl}}: \xrightarrow[\text{to chlorine}]{\text{Transfer bond electrons}} \overset{+1}{H} \;\; \overset{\cdot\cdot}{\underset{\cdot\cdot}{:Cl}}:^{-1}$$

Similarly, in PCl_3 the shared electrons are assigned to Cl, and the oxidation number of each Cl atom is -1. The oxidation number of P, which is thought of as having lost three electrons, is $+3$:

$$:\overset{\cdot\cdot}{\underset{\cdot\cdot}{Cl}}:\overset{\cdot\cdot}{\underset{\cdot\cdot}{P}}:\overset{\cdot\cdot}{\underset{\cdot\cdot}{Cl}}: \xrightarrow[\text{to chlorine}]{\text{Transfer bond electrons}} \overset{-1}{:\overset{\cdot\cdot}{\underset{\cdot\cdot}{Cl}}:} \overset{+3}{\overset{\cdot\cdot}{P}} \overset{-1}{:\overset{\cdot\cdot}{\underset{\cdot\cdot}{Cl}}:}$$
$$\underset{\;\;\;\;\;\;:\overset{\cdot\cdot}{\underset{\cdot\cdot}{Cl}}:}{\;}\qquad\qquad\qquad \underset{:\overset{\cdot\cdot}{\underset{\cdot\cdot}{Cl}}:_{-1}}{\;}$$

In assigning oxidation numbers in covalent compounds, we imagine electrons to be transferred from one atom to another, but no electrons are added to or removed from the molecule. Thus the oxidation numbers of all the atoms must add up to zero, as is the case in HCl and PCl_3. Similarly, in any polyatomic ion the sum of the oxidation numbers must equal the charge on the ion. We therefore have the following rule:

In any neutral molecule the sum of the oxidation numbers is zero; in an ion the sum of the oxidation numbers is equal to the charge on the ion.

Because fluorine is the most electronegative element in any covalent or ionic compound, it always has an oxidation number of -1. Because the other halogens also have high electronegativities, they too have an oxidation number of -1 in most, but not all, of their compounds. Similarly, in its compounds oxygen is normally assigned an oxidation number of -2. In its covalent compounds hydrogen is usually less electronegative than the atom to which it is attached, so hydrogen usually has an oxidation number $+1$.

We can assign oxidation numbers for atoms in a great many compounds if we combine these generalizations about the halogens, oxygen, and hydrogen with the rule that the sum of the oxidation numbers of the atoms must equal zero for a neutral compound or the charge of an ion. For example, in sulfur dioxide each oxygen has an oxidation number of -2. The sum of their oxidation numbers is -4, and since the sum of all the oxidation numbers must be zero, the oxidation number of the sulfur atom is $+4$:

$$\overset{-2}{O}\overset{+4}{S}\overset{-2}{O}$$

In sulfuric acid, H_2SO_4, the sum of the oxidation numbers of hydrogen and oxygen is

$$2H = 2 \times (+1) = +2$$
$$4O = 4 \times (-2) = \underline{-8}$$
$$-6$$

Therefore since the sum of the oxidation numbers must be zero, sulfur in this case has an oxidation number of $+6$:

$$\overset{+1}{H_2}\ \overset{+6}{S}\ \overset{-2}{O_4}$$
$$2(+1) + (+6) + 4(-2) = 0$$

In the hydrogen sulfate ion, HSO_4^-, the sum of the oxidation numbers must be -1:

$$\overset{+1}{H}\ \overset{+6}{S}\ \overset{-2}{O_4}$$
$$(+1) + (+6) + 4(-2) = -1$$

One situation that has not been covered by the previous rules is that in which an atom is bonded to another of identical electronegativity and, in particular, *when it is bonded to one of the same kind. In this case the bond electrons are divided equally between the two atoms.* Thus in the case of the H_2 molecule each atom is assigned one electron, and therefore each H atom has an oxidation number of zero. Similar arguments apply to the eight sulfur atoms in S_8 and the four phosphorus atoms in P_4. As previously stated, *the oxidation number of an atom in any of the allotropic forms of an element is 0.*

Another possibility arises when there is a bond between two like atoms in a molecule that also contains other atoms. For example, in hydrogen peroxide, H_2O_2, each hydrogen is assigned an oxidation number of $+1$. And from the rule that the sum of the oxidation numbers must be zero, each oxygen atom must have an oxidation number of -1. Alternatively, we see that if the electrons of the O—O bond are divided equally between the two oxygen atoms, each oxygen is assigned seven electrons. It would then have a charge of -1 and therefore an oxidation number of -1. This compound is one of the rare cases in which oxygen does not have an oxidation number of -2:

$$\overset{+1}{H}\overset{-1}{:\ddot{O}}\overset{-1}{:\ddot{O}}\overset{+1}{:H}$$

In summary, the rules for assigning oxidation numbers are:

1. In any of the allotropic forms of an element each atom is assigned an oxidation number of 0.

2. Fluorine, the most electronegative element, has an oxidation number of -1 in all its compounds.

3. Hydrogen in its compounds usually has an oxidation number of $+1$. The only exceptions occur in metal hydrides, in which hydrogen has an oxidation number of -1, as, for example, in Na^+H^-.

4. Oxygen usually has oxidation number -2 in its compounds. There are two important exceptions:

   · When oxygen is combined with fluorine, as in F_2O, its oxidation number is $+2$.

   · In compounds containing oxygen–oxygen bonds, such as hydrogen peroxide, the oxidation number of oxygen is -1.

5. The halogens other than fluorine have an oxidation number of -1 except when they are combined with a more electronegative element, that is, with oxygen or a more electronegative halogen.

6. The sum of the oxidation numbers of all the atoms in a neutral compound is 0 and in an ion is equal to the charge on the ion. As a result, the oxidation number of a monatomic ion is equal to its charge, for example, $+2$ for Cu^{2+} and -1 for Cl^-.

═ Example 8.1 ═

ASSIGNING OXIDATION NUMBERS

Assign oxidation numbers (abbreviated as O.N.) to each of the elements in the following compounds:

(a) NaCl   (b) BaF_2 (c) CO_2 (d) SCl_2 (e) H_2S

Solution

(a) NaCl is ionic: Na^+Cl^-. The O.N.s are equal to the charges. Therefore O.N.(Na) $= +1$; O.N.(Cl) $= -1$. Or, using rule 5, O.N.(Cl) $= -1$, so O.N.(Na) $= +1$.

(b) BaF_2 is ionic: $Ba^{2+}(F^-)_2$. Therefore O.N.(Ba) $= +2$; O.N.(F) $= -1$. Or, using rule 2, O.N.(F) $= -1$, so O.N.(Ba) $= +2$.

(c) Each O has O.N. $= -2$, and therefore C has O.N. $= +4$.

(d) Each Cl has O.N. $= -1$, and therefore S has O.N. $= +2$.

(e) Each H has O.N. $= +1$, and therefore S has O.N. $= -2$.

EXERCISE 8.4

What is the oxidation number of each element in each of the following compounds?

H_3PO_4 PH_3 H_2CO_3 ClF

= Example 8.2

ASSIGNING OXIDATION NUMBERS

Assign oxidation numbers to each of the atoms in the following molecules and ions:

(a) HS^- (b) S_2Cl_2 (c) SO_4^{2-} (d) HSO_3^-

Solution

(a) The sum of the oxidation numbers must be -1 (rule 6). Hydrogen has O.N. = $+1$. So $+1 + O.N.(S) = -1$, and hence O.N.(S) = -2.

(b) Chlorine has O.N. = -1. So

$$2(-1) + 2[O.N.(S)] = 0$$

Hence O.N.(S) = $+1$. This oxidation state is unusual for sulfur. It arises because the molecule has an S—S bond:

$$:\overset{..}{\underset{..}{Cl}}-\overset{..}{\underset{..}{S}}-\overset{..}{\underset{..}{S}}-\overset{..}{\underset{..}{Cl}}:$$

The S—S bond electrons are shared equally between the S atoms, and the S—Cl bond electrons are assigned to Cl. Thus each S atom is assigned five electrons so each has a charge of $+1$. Hence O.N.(S) = $+1$.

(c) The sum of the O.N.s must be -2 (rule 6). Oxygen has O.N. = -2, so we have $4(-2) + O.N.(S) = -2$. Hence, O.N.(S) = $+6$.

(d) Taking the oxidation numbers of oxygen and hydrogen to be -2 and $+1$, respectively, we have

$$3(-2) + (+1) + O.N.(S) = -1$$
$$O.N.(S) = +4$$

EXERCISE 8.5

Assign oxidation numbers to each of the elements in each of the following:

SO_3^{2-} ClO_4^- S_2^{2-} $Al_2(SO_4)_3$ NaH_2PO_4

OXIDATION NUMBERS AND FORMAL CHARGES

Oxidation numbers do not correspond to the actual distribution of charge in a molecule or polyatomic ions. They are useful, however, because they enable us to follow the state of oxidation of an element in its reactions and its compounds and to keep count of electrons in oxidation–reduction reactions. Thus oxidation numbers resemble formal charges in being arbitrary, but note that the two numbers are not the same. Oxidation numbers are assigned by assuming that the electrons of bonds are shared *unequally* between atoms, while formal charges are found by assuming that bonding electrons are shared *equally* between atoms. In other words, oxidation numbers are based on the assumption that all bonds can be regarded as ionic, while formal charges are based on the assumption that all bonds are purely covalent and have no polarity.

For the chloride ion, ammonium ion, and sulfate ion, the formal charges and oxidation numbers are as follows:

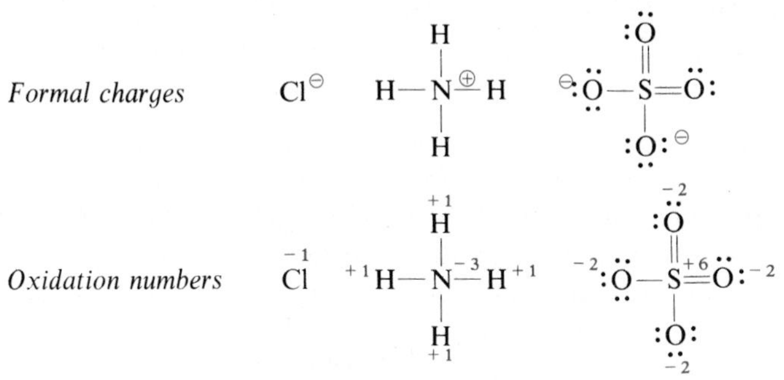

Neither formal charges nor oxidation numbers are equal to the real charges on atoms in ions and molecules except for monatomic ions such as Cl^-. Depending on whether the bonds are more covalent (equal sharing of bonding electrons) or more ionic (unequal sharing of bonding electrons) the real charges will be closer to the formal charges or to the oxidation numbers respectively. However, the main purpose of formal charges is to help us write correct Lewis structures and to keep proper account of electrons in molecules and ions. The main purpose of oxidation numbers is to help us follow the changes in the state of oxidation of an element in its different compounds and hence to better understand oxidation–reduction reactions.

Let us now see how oxidation numbers are useful to us in describing the compounds of sulfur and their reactions.

OXIDATION STATES OF SULFUR

The common oxidation numbers found for sulfur in its compounds are $+6$, $+4$, $+2$, 0, and -2, as shown in Figure 8.13. These numbers describe the states of oxidation of sulfur. When it has an oxidation number of $+6$, as in H_2SO_4, sulfur is said to be in the $+6$ oxidation state, which is the most highly oxidized state of sulfur. When it has an oxidation number of -2, as in H_2S, sulfur is said to be in the -2 oxidation state, which is the least highly oxidized, or the most highly reduced, state of sulfur. In its elemental forms, S_n (usually S_8), sulfur is in the 0 oxidation state. When sulfur is converted to H_2S, it is reduced to its lowest (-2) oxidation state. When it is oxidized to H_2SO_4, it is converted to its highest $(+6)$ oxidation state. Note that the ground state of sulfur in which there are two unpaired electrons gives rise to both the -2 and the $+2$ oxidation states in compound formation depending on whether sulfur is combined with

Oxidation state	Compounds
+6	SO_3, H_2SO_4, SO_4^{2-}, $H_2S_2O_7$
+4	SO_2, HSO_3^-, SO_3^{2-}
+2	SCl_2
0	S_8 and all other forms of elemental sulfur
−2	H_2S, S^{2-}

Oxidation Oxidation number increases →

Reduction Oxidation number decreases →

a less or a more electronegative element. But the valence states in which there are four or six unpaired electrons only give rise to the +4 and +6 oxidation states, because only electronegative elements such as fluorine and oxygen can cause the promotion of electrons to the 3d level.

The concept of oxidation numbers leads to another useful definition of oxidation and reduction: **Oxidation** *may be defined as any change in which the oxidation number of an atom increases.* **Reduction** *may be defined as any change in which the oxidation number of an atom decreases.*

Let us now see how these definitions are equivalent to the definitions that we gave in Chapter 5 in terms of electron loss and gain. Every oxidation reaction is accompanied by a reduction reaction and vice versa. Thus any redox reaction can be split into two half-reactions. For example, for the reaction

$$2Na(s) + Br_2(l) \longrightarrow 2Na^+(s) + 2Br^-(s)$$

the oxidation half-reaction is

$$Na \longrightarrow Na^+ + e^- \quad \text{1 electron lost}$$
$$\text{O.N.s} \quad 0 \qquad +1 \qquad \text{Increase in O.N.} = 1$$

and the reduction half-reaction is

$$Br_2 + 2e^- \longrightarrow 2Br^- \quad \text{2 electrons gained}$$
$$\text{O.N.s} \quad 0 \qquad \quad 2(-1) \quad \text{Decrease in O.N.} = 2$$

Thus we see that *electron loss is accompanied by an increase in the oxidation number* and *electron gain is accompanied by a decrease in the oxidation number* equal to the number of electrons lost or gained.

Now we can apply these ideas to the reduction of sulfuric acid to sulfur dioxide:.

$$H_2 \quad S \quad O_4 \longrightarrow S \quad O_2$$
$$\text{O.N.s} \quad 2(+1) \quad +6 \quad 4(-2) \qquad +4 \quad 2(-2)$$

The oxidation number of sulfur decreases from +6 to +4; therefore two electrons are involved in this reduction. So for the half-reaction for the reduction of H_2SO_4 to SO_2 we have

$$H_2SO_4 + 2e^- \longrightarrow SO_2$$

But this equation does not balance for atoms or for charge so there must be some other reactants and/or products. In this case the complete half-reaction is

$$H_2SO_4 + 2H^+ + 2e^- \longrightarrow SO_2 + 2H_2O$$

In any acid solution H^+ is always available (from H_3O^+ or in concentrated sulfuric acid from H_2SO_4) and another product in any aqueous solution may be H_2O. The complete equation for the oxidation of Br^- by concentrated aqueous sulfuric acid can be obtained by combining the equation for the sulfuric acid reduction half-reaction with the equation for the oxidation half-reaction of bromide ion to bromine:

$$2Br^- \longrightarrow Br_2 + 2e^-$$
$$H_2SO_4 + 2H^+ + 2e^- \longrightarrow SO_2 + 2H_2O$$

Adding these equations gives

$$2Br^- + H_2SO_4 + 2H^+ \longrightarrow Br_2 + SO_2 + 2H_2O$$

Similarly we can combine the equation for the oxidation of copper to Cu^{2+},

$$Cu \longrightarrow Cu^{2+} + 2e^-$$

with the equation for the reduction of sulfuric acid to give SO_2 to obtain the complete equation

$$Cu + H_2SO_4 + 2H^+ \longrightarrow Cu^{2+} + SO_2 + 2H_2O$$

Unlike some metals such as zinc and magnesium, copper does not dissolve in dilute aqueous HCl or H_2SO_4. The hydronium ion is not a strong enough oxidizing agent to oxidize copper. However, copper *is* oxidized by the stronger oxidizing agent H_2SO_4, which is present in concentrated aqueous sulfuric acid.

THE +4 OXIDATION STATE: SULFUR DIOXIDE AND SULFITES

In addition to sulfur dioxide, the most important compounds in which sulfur is in the +4 oxidation state are the metal **sulfites**. Sulfur dioxide is very soluble in water (9.4 g L^{-1} at STP), and it has generally been assumed that *sulfurous acid*, H_2SO_3, is formed in such solutions. The reaction would be

$$H_2O + SO_2 \rightleftharpoons H_2SO_3$$

In fact, there is no firm evidence that the molecule H_2SO_3 exists. Nevertheless, the anions HSO_3^- and SO_3^{2-} are formed in small amounts:

$$SO_2(aq) + 2H_2O \rightleftharpoons H_3O^+ + HSO_3^-$$
$$HSO_3^- + H_2O \rightleftharpoons H_3O^+ + SO_3^{2-}$$

Thus sulfur dioxide behaves as a weak diprotic acid in water, and a solution of SO_2 in water is usually called sulfurous acid.

The sulfite and hydrogen sulfite ions have the following Lewis structures:

They both have the expected pyramidal AX_3E geometry.

We can think of the reaction between SO_2 and water as taking place as follows:

AX₃E pyramidal geometry of SO_3^{2-} and HSO_3^-

Metal sulfites are all soluble in water. Addition of an aqueous acid to a solution of a sulfite causes the above reactions to proceed from right to left and SO_2 is formed:

$$SO_3^{2-}(aq) + 2H^+(aq) \qquad SO_2(aq) + H_2O(l)$$

When the solution becomes saturated with SO_2 it then escapes into the gas phase. The high concentration of H^+ (H_3O^+) and the escape of the SO_2 into the gas phase both shift the position of the above equilibrium to the right (Le Châtelier's principle). Small amounts of SO_2 can be prepared in the laboratory in this way.

Both SO_2 and SO_3^{2-} are good reducing agents. They are oxidized to SO_3 and sulfate ion, SO_4^{2-}, by many oxidizing agents, although, as we have seen, the reaction with O_2 is very slow in the absence of a catalyst such as V_2O_5. For example, an orange-red solution of Br_2 in water is decolorized by SO_2 or by a sulfite solution, because Br_2 is reduced to Br^- (see Experiment 8.6):

$$Br_2(aq) + SO_2(aq) + 2H_2O(l) \longrightarrow 2Br^-(aq) + SO_4^{2-}(aq) + 4H^+(aq)$$

This reaction is simply the reverse of the reaction that occurs when concentrated sulfuric acid oxidizes bromide ion to bromine. The above reaction is driven to the right because the reaction is in dilute aqueous solution and all the sulfuric acid is ionized—to SO_4^{2-} and $H^+(aq)$—so that there is no un-ionized H_2SO_4 which is needed to oxidize Br^- to Br_2. The high concentration of SO_2 also drives the reaction to the right. Both Cl_2 and I_2 are reduced in a similar way.

The reducing properties of an aqueous solution of sulfur dioxide make it a useful bleaching agent. For example, colored compounds in wool, silk, paper, and straw become colorless when reduced by sulfur dioxide.

Because the sulfur in SO_2 is in an intermediate oxidation state, it may either be oxidized to the $+6$ state or be reduced to a lower oxidation state. Thus SO_2 can act not only as a reducing agent but also as an oxidizing agent. For example, it is an oxidizing agent in the important reaction with H_2S used in the purification of natural gas:

$$2H_2\overset{-2}{S} + \overset{+4}{S}O_2 \longrightarrow 3\overset{0}{S} + 2H_2O$$

In this reaction sulfur in the -2 oxidation state in H_2S is oxidized to sulfur in the 0 oxidation state by SO_2 in which sulfur is in the $+4$ oxidation state. The sulfur in SO_2 is, at the same time, reduced to the 0 oxidation state.

When an aqueous solution of a sulfite is boiled with sulfur, the thiosulfate ion, $S_2O_3^{2-}$, is formed:

$$S(s) + SO_3^{2-}(aq) \longrightarrow S_2O_3^{2-}(aq)$$

The Lewis structure for this ion is the same as that of the sulfate ion, except that a sulfur atom replaces one of the oxygen atoms. The replacement of an oxygen atom by a sulfur atom in a compound is often denoted by the prefix *thio-*, as in thiosulfate.

When dilute acid is added to an aqueous solution of a thiosulfate, it is decomposed to sulfur and sulfur dioxide:

$$S_2O_3^{2-}(aq) + 2H_3O^+(aq) \longrightarrow S(s) + SO_2(g) + 3H_2O(l)$$

This reaction is the reverse of the reaction by which the thiosulfate ion is formed. Addition of acid converts the small equilibrium amount of SO_3^{2-} to

Sulfur dioxide reduces the red pigment in a dampened rose to a colorless substance.

Thiosulfate ion

H_2O and SO_2 and thus drives the equilibrium back to the left:

$$S(s) + SO_3^{2-}(aq) \rightleftharpoons S_2O_3^{2-}(aq)$$
$$\downarrow + 2H^+(aq)$$
$$H_2O(l) + SO_2(g)$$

This reaction is another example of the operation of Le Châtelier's principle.

THE −2 OXIDATION STATE:
HYDROGEN SULFIDE AND METAL SULFIDES

HYDROGEN SULFIDE Unlike water, which is a liquid at room temperature, H_2S is a gas (melting point $-85.6\ °C$; boiling point $-60.7\ °C$). It has a powerful, unpleasant odor, and it is very poisonous.

Experiment 8.6

Sulfur Dioxide as a Reducing Agent

When sulfur dioxide is passed into a purple solution of potassium permanganate it rapidly reduces it to a colorless solution of manganous ion, Mn^{2+}.

When sulfur dioxide is passed into an orange-red solution of bromine it reduces it more slowly.

After 2 min 10 s the bromine color has noticeably decreased as bromine, Br_2, is reduced to colorless bromide ion, Br^-.

After 2 m 41 s the reduction of bromine to bromide ion is almost complete.

AX$_2$E$_2$ angular geometry of H$_2$S

The Lewis structure of H$_2$S is just like that of water. So it is an AX$_2$E$_2$ molecule and it has the expected angular shape although the bond angle is smaller than in H$_2$O (see Chapter 9). Hydrogen sulfide is soluble in water; it is a weak diprotic acid:

$$H_2S + H_2O \rightleftharpoons H_3O^+ + HS^- \qquad \text{Hydrogen sulfide ion}$$
$$HS^- + H_2O \rightleftharpoons H_3O^+ + S^{2-} \qquad \text{Sulfide ion}$$

In hydrogen sulfide sulfur is in its lowest oxidation state (-2). It is therefore a good reducing agent; it is usually oxidized to sulfur. The half-reaction for this oxidation in aqueous solution is

$$H_2S(aq) \longrightarrow S(s) + 2H^+(aq) + 2e^-$$

If an aqueous solution of H$_2$S is exposed to the air, the H$_2$S is oxidized by the oxygen of the air and a precipitate of sulfur is slowly formed:

$$2H_2S(aq) + O_2(g) \longrightarrow 2S(s) + 2H_2O(l)$$

The same reaction occurs when H$_2$S burns in air; sulfur is deposited on a cold surface held near the flame (see Experiment 8.7). Some SO$_2$ is formed at the same time:

$$2H_2S(g) + 3O_2(g) \longrightarrow 2H_2O(l) + 2SO_2(g)$$

If H$_2$S is passed into an aqueous solution of bromine, Br$_2$, the red-brown color of the Br$_2$ disappears as it is reduced to Br$^-$. At the same time a milky precipitate of finely divided sulfur is formed. If we add the equation for the reduction of bromine to bromide ion

$$Br_2(aq) + 2e^- \longrightarrow 2Br^-(aq)$$

to the equation for the oxidation of H$_2$S to sulfur

$$H_2S(aq) \longrightarrow S(s) + 2H^+(aq) + 2e^-$$

we obtain the overall equation for the reaction:

$$H_2S(g) + Br_2(aq) \longrightarrow 2H^+(aq) + 2Br^-(aq) + S(s)$$

Chlorine and iodine are reduced in the same way.

SULFIDES AND POLYSULFIDES   The **sulfides** of the alkali and alkaline earth metals are colorless solids that are soluble in water. The sulfides of most other metals are often colored and insoluble. They are precipitated when H$_2$S is passed through a solution of a salt of the metal (see Experiment 8.7):

$$Zn^{2+}(aq) + H_2S(g) \longrightarrow ZnS(s) + 2H^+(aq)$$
$$Pb^{2+}(aq) + H_2S(g) \longrightarrow PbS(s) + 2H^+(aq)$$

We can think of this reaction as occurring in two stages. First the ionization of H$_2$S as a weak acid,

$$H_2S(aq) \rightleftharpoons 2H^+(aq) + S^{2-}(aq)$$

followed by the combination of the metal cation with S^{2-} to form the insoluble sulfide:

$$Pb^{2+}(aq) + S^{2-}(aq) \longrightarrow PbS(s)$$

Experiment 8.7

Properties of Hydrogen Sulfide

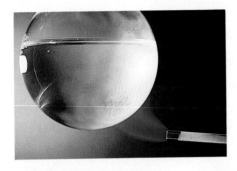

Hydrogen sulfide burns in air with a blue flame, depositing yellow sulfur on a cool surface.

When H_2S is passed into a colorless aqueous solution of zinc sulfate, $ZnSO_4$, a white precipitate of insoluble zinc sulfide, ZnS, is formed.

When H_2S is passed into a colorless aqueous solution of lead nitrate, $Pb(NO_3)_2$, a dark brown precipitate of lead sulfide, PbS, is formed.

Removal of sulfide ion by the precipitation of the insoluble metal sulfide forces the ionization of H_2S to the right:

$$H_2S(aq) \longrightarrow 2H^+(aq) + \underset{\substack{\text{Removed as insoluble}\\\text{metal sulfide}}}{S^{2-}(aq)}$$

The equilibrium between H_2S and S^{2-} can also be forced to the left by using an excess of $H^+(aq)$ and allowing the H_2S to escape:

$$\underset{\substack{\text{Allowed to escape}\\\text{as gas}}}{H_2S(aq)} \longleftarrow \underset{\text{Excess acid}}{2H^+(aq)} + S^{2-}(aq)$$

Thus small amounts of H_2S can be prepared in the laboratory by adding a dilute aqueous acid such as hydrochloric acid to a metal sulfide. For example

$$FeS(s) + 2H^+(aq) \longrightarrow Fe^{2+}(aq) + H_2S(g)$$
$$FeS(s) + 2HCl(aq) \longrightarrow FeCl_2(aq) + H_2S(g)$$

Many insoluble metal sulfides occur in nature and are important ores. Examples include NiS, CuS, Ag_2S, HgS, and PbS.

A colorless solution of an alkali or alkaline earth metal sulfide such as Na_2S can dissolve considerable amounts of sulfur to give a bright yellow solution containing a mixture of polysulfides:

$$S^{2-} + S \longrightarrow S_2{}^{2-} \qquad \text{Disulfide} \qquad {}^{\ominus}:\!\ddot{S}\!-\!\ddot{S}\!:^{\ominus}$$

$$S^{2-} + 2S \longrightarrow S_3{}^{2-} \qquad \text{Trisulfide} \qquad {}^{\ominus}:\!\ddot{S}\!-\!\ddot{S}\!-\!\ddot{S}\!:^{\ominus}$$

$$S^{2-} + 3S \longrightarrow S_4{}^{2-} \qquad \text{Tetrasulfide} \qquad {}^{\ominus}:\!\ddot{S}\!-\!\ddot{S}\!-\!\ddot{S}\!-\!\ddot{S}\!:^{\ominus}$$

These polysulfide ions consist of chains of sulfur atoms. The common mineral pyrite, FeS_2, is a salt of the disulfide ion, $S_2{}^{2-}$.

═ Example 8.3 ═

REACTIONS OF SULFUR COMPOUNDS

Complete and balance each of the following equations:

(a) $H_2SO_4(\text{conc aq}) + Ag(s) \rightarrow Ag^+(aq) +$

(b) $H_2SO_4(aq) + CuS(s) \rightarrow$

(c) $Na_2S_2O_3(aq) + H_2SO_4(aq) \rightarrow$

Solution

(a) $Ag(s)$ is oxidized to $Ag^+(aq)$. The equation for the half-reaction is

$$Ag(s) \longrightarrow Ag^+(aq) + e^-$$

The equation for the reduction of concentrated H_2SO_4 has been given above:

$$H_2SO_4(\text{concn aq}) + 2H^+(aq) + 2e^- \longrightarrow SO_2(g) + 2H_2O(l)$$

Multiplying the first equation by 2 and adding to the second equation gives the balanced equation for the overall reaction:

$$H_2SO_4(aq) + 2Ag(s) + 2H^+(aq) \longrightarrow 2Ag^+(aq) + SO_2(g) + 2H_2O(l)$$

(b) Sulfuric acid like other acids reacts with metal sulfides to give H_2S. The equation can be written by analogy with the equation for FeS,

$$CuS(s) + H_2SO_4(aq) \longrightarrow CuSO_4(aq) + H_2S(g)$$

or by using the equation

$$2H^+(aq) + S^{2-}(aq) \longrightarrow H_2S(aq)$$

and adding the spectator ions Cu^{2+} and $SO_4{}^{2-}$.

(c) This is simply the empirical formula version of the ionic equation given in the text above, namely

$$S_2O_3{}^{2-}(aq) + 2H^+(aq) \longrightarrow S(s) + SO_2(g) + H_2O(l)$$

so the required equation is obtained by adding $2Na^+$ and $SO_4{}^{2-}$ to both sides of the equation

$$Na_2S_2O_3(aq) + H_2SO_4(aq) \longrightarrow Na_2SO_4(aq) + S(s) + SO_2(g) + H_2O(l)$$

Dilute sulfuric acid behaves only as an acid whereas concentrated sulfuric acid also behaves as an oxidizing agent and a dehydrating agent.

EXERCISE 8.6

Complete and balance the following equations:

(a) $H_2S(aq) + Cl_2(aq) \rightarrow$

(b) $H_2S(aq) + Ag^+(aq) \rightarrow$

(c) $H_2SO_4(conc\ aq) + I^-(aq) \rightarrow$

EXERCISE 8.7

Write balanced equations for the following reactions:

(a) The production of ammonium sulfate fertilizer from ammonia

(b) The reaction of dilute sulfuric acid with magnesium

(c) The formation of a precipitate when dilute sulfuric acid is added to an aqueous solution of lead nitrate, $Pb(NO_3)_2$

8.3
Reactions and Compounds of Phosphorus

The most important oxidation states of phosphorus are $+5$, $+3$, 0, and -3. Figure 8.14 gives examples of compounds in which phosphorus is in these oxidation states.

OXIDES

When phosphorus is burnt in air or oxygen, two different oxides may be formed; they have the empirical formulas P_2O_3 and P_2O_5 (see Experiment 8.8 on p. 410). Their molecular masses indicate that their molecular formulas are P_4O_6 and P_4O_{10}, respectively. If the supply of oxygen is limited, P_4O_6, tetraphosphorus hexaoxide, is the major product. In excess oxygen, P_4O_{10}, tetraphosphorus decaoxide, is formed:

$$P_4(s) + 3O_2(g) \longrightarrow P_4O_6(s)$$
$$P_4(s) + 5O_2(g) \longrightarrow P_4O_{10}(s)$$

Figure 8.15 shows how the structures of these oxides are related to that of the P_4 tetrahedron. In P_4O_6 phosphorus exhibits its normal valence of 3 and

Oxidation state	Compounds
$+5$	P_4O_{10}, H_3PO_4, PO_4^{3-}
$+3$	P_4O_6, H_3PO_3, HPO_3^{2-}
0	P_4
-3	PH_3, PH_4^+

Oxidation — Oxidation number increases

Reduction — Oxidation number decreases

FIGURE 8.14
Classification of Compounds of Phosphorus by Oxidation State.

Experiment 8.8

Reaction of White Phosphorus with Oxygen

When a piece of white phosphorus is inserted into a flask of oxygen it ignites, burns, and produces a white smoke of P_4O_{10}.

As the phosphorus burns it produces a brilliant white light.

When the reaction is complete, the inside of the flask is coated with a layer of white P_4O_{10}.

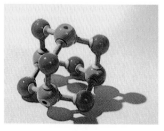

P_4O_6

P_4O_{10}

is in the $+3$ oxidation state. Each phosphorus atom forms three bonds and has one unshared pair to complete an octet of electrons. Each phosphorus atom has a pyramidal AX_3E geometry. The structure of P_4O_{10} is similar, except that an additional oxygen atom is bonded to each phosphorus atom by a double bond. Phosphorus is in the $+5$ oxidation state. Each phosphorus atom uses all its five valence electrons to form five covalent bonds, three single bonds and one double bond, and has an AX_4 tetrahedral geometry.

The oxide P_4O_{10} reacts vigorously with excess water, liberating a considerable amount of heat and forming H_3PO_4, *orthophosphoric acid*, which is usually called simply *phosphoric acid* (see Experiment 8.9.)

$$P_4O_{10}(s) + 6H_2O(l) \longrightarrow 4H_3PO_4(aq)$$

If a limited quantity of water is used, *metaphosphoric acid*, HPO_3, is formed:

$$P_4O_{10}(s) + 2H_2O(l) \longrightarrow 4HPO_3(s)$$

The oxide P_4O_{10} has such a strong tendency to react with water that it is a very powerful dehydrating agent. It can even remove water from H_2SO_4. When a mixture of concentrated H_2SO_4 and P_4O_{10} is heated, sulfur trioxide, SO_3, is formed:

$$2H_2SO_4(l) + P_4O_{10}(s) \longrightarrow 2SO_3(g) + 4HPO_3(s)$$

PHOSPHORIC AND PHOSPHOROUS ACIDS

PHOSPHORIC ACID Phosphoric acid is a colorless solid with a melting point of 42 °C. A concentrated (82% by mass) solution in water is usually used in the laboratory.

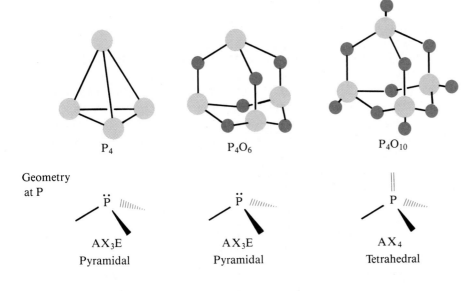

P_4 P_4O_6 P_4O_{10}

FIGURE 8.15
Structures of P_4, P_4O_6, and P_4O_{10}.

In P_4O_6 each edge of the P_4 tetrahedron is bridged by an oxygen atom. The structure of P_4O_{10} is the same, except that an additional oxygen atom is bonded to each phosphorus atom by a double bond.

Geometry at P

AX_3E AX_3E AX_4
Pyramidal   Pyramidal   Tetrahedral

Phosphoric acid is **triprotic** because it has three OH groups and thus three ionizable hydrogen atoms. It ionizes in water in three stages:

$$H_3PO_4 + H_2O \rightleftharpoons H_3O^+ + H_2PO_4^-$$
$$H_2PO_4^- + H_2O \rightleftharpoons H_3O^+ + HPO_4^{2-}$$
$$HPO_4^{2-} + H_2O \rightleftharpoons H_3O^+ + PO_4^{3-}$$

Experiment 8.9

The Reaction of P_4O_{10} with Water

The glass spoon contains red phosphorus. The beaker of water is colored yellow because methyl red indicator has been added.

After igniting the phosphorus in a bunsen flame it is placed in the beaker just above the surface of the water. The white fumes are P_4O_{10}.

When the P_4O_{10} dissolves in the water it forms phosphoric acid, H_3PO_4, and the methyl red indicator turns red.

$$:\ddot{O}H$$
$$:\ddot{O}=\overset{\displaystyle |}{\underset{\displaystyle |}{P}}-\ddot{O}H$$
$$:\ddot{O}H$$

Phosphoric acid

$$:\ddot{O}:^{\ominus}$$
$$:\ddot{O}=\overset{\displaystyle |}{\underset{\displaystyle |}{P}}-\ddot{O}H$$
$$:\ddot{O}H$$

Dihydrogen phosphate ion

$$\ddot{O}:^{\ominus}$$
$$:\ddot{O}=\overset{\displaystyle |}{\underset{\displaystyle |}{P}}-\ddot{O}:^{\ominus}$$
$$:\ddot{O}H$$

Hydrogen phosphate ion

$$:\ddot{O}:^{\ominus}$$
$$:\ddot{O}=\overset{\displaystyle |}{\underset{\displaystyle |}{P}}-\ddot{O}:^{\ominus}$$
$$:\ddot{O}:^{\ominus}$$

Phosphate ion

$$\overset{\displaystyle O}{\underset{\displaystyle O}{\overset{\displaystyle \|}{\underset{\displaystyle |}{P}}}}$$
$$^{-}O \qquad O^{-}$$

Tetrahedral AX_4 geometry

Even the first of these ionizations is incomplete. Thus unlike sulfuric acid, which is a strong acid, phosphoric acid is a weak acid.

Notice that there is no relationship between the *number* of ionizable hydrogen atoms and the *strength* of an acid. Hydrochloric acid, HCl, which has one ionizable hydrogen, and H_2SO_4, which has two ionizable hydrogens, are both strong acids. But carbonic acid, H_2CO_3, which has two ionizable hydrogens, and H_3PO_4, which has three ionizable hydrogens, are both weak acids.

In its reactions with bases H_3PO_4 forms three series of salts. These salts contain the dihydrogen phosphate ion, $H_2PO_4^-$, the hydrogen phosphate ion, HPO_4^{2-}, and the phosphate ion, PO_4^{3-}. For example, 1 mol of H_3PO_4 reacts with 1, 2, or 3 mol of potassium hydroxide, KOH, to give KH_2PO_4, K_2HPO_4, or K_3PO_4, respectively:

$$KOH + H_3PO_4 \longrightarrow KH_2PO_4 + H_2O$$
$$2KOH + H_3PO_4 \longrightarrow K_2HPO_4 + 2H_2O$$
$$3KOH + H_3PO_4 \longrightarrow K_3PO_4 + 3H_2O$$

Phosphoric acid and its anions all have a tetrahedral AX_4 geometry.

EXERCISE 8.8

What volume of $0.113M$ KOH solution must be added to 50.0 mL of $0.105M$ H_3PO_4 solution to prepare a solution of K_3PO_4?

PHOSPHATES AS FERTILIZERS Phosphorus is essential for plant life. Although most soils contain an appreciable amount of phosphate, it is often present in the form of insoluble salts and hence is unavailable to plants. Phosphate rock can be ground up and added to phosphorus-deficient soil, but it is not soluble enough to be a very good fertilizer. A better fertilizer can be obtained by treating phosphate rock with aqueous sulfuric acid to convert it to the more soluble calcium dihydrogen phosphate. The amount and concentration of the acid are adjusted so that all the water is used up in the formation of hydrated salts:

$$Ca_3(PO_4)_2(s) + 2H_2SO_4(aq) + 5H_2O(l) \longrightarrow Ca(H_2PO_4)_2 \cdot H_2O(s) + 2CaSO_4 \cdot 2H_2O(s)$$

If this equation is written in the ionic form and all the spectator ions are eliminated, we see that the reaction is simply

$$PO_4^{3-}(aq) + 2H^+(aq) \longrightarrow H_2PO_4^-(aq)$$

in which phosphate ion, which is a base, combines with hydrogen ions from the sulfuric acid. The solid mixture of hydrated calcium sulfate (gypsum), $CaSO_4 \cdot 2H_2O$, and hydrated calcium dihydrogen phosphate, $Ca(H_2PO_4)_2 \cdot H_2O$, that results is called *superphosphate of lime*.

A more concentrated phosphorus fertilizer called *triple phosphate* is made by treating phosphate rock with phosphoric acid to give a product that is essentially pure calcium dihydrogen phosphate and contains no calcium sulfate. The phosphoric acid is made by treating phosphate rock with a larger quantity of sulfuric acid than is needed for the production of superphosphate of lime:

$$Ca_3(PO_4)_2(s) + 3H_2SO_4(aq) + 6H_2O(l) \longrightarrow 2H_3PO_4(aq) + 3CaSO_4 \cdot 2H_2O(s)$$

This equation can be written more simply as:

$$PO_4{}^{3-}(aq) + 3H^+(aq) \longrightarrow H_3PO_4(aq)$$

The insoluble calcium sulfate is removed by filtration, leaving phosphoric acid, which is then used to treat more phosphate rock to form calcium dihydrogen phosphate (triple phosphate), $Ca(H_2PO_4)_2$:

$$Ca_3(PO_4)_2 + 4H_3PO_4 \longrightarrow 3Ca(H_2PO_4)_2$$

Another important fertilizer is made by treating phosphoric acid with ammonia:

$$H_3PO_4 + NH_3 \longrightarrow NH_4H_2PO_4$$

The product, ammonium dihydrogen phosphate, $NH_4{}^+H_2PO_4{}^-$, supplies the soil with nitrogen as well as phosphorus.

POLYMERIC PHOSPHORIC ACIDS AND CONDENSATION Phosphoric acid readily undergoes **condensation**, *the reaction of two or more molecules to form a larger molecule by the elimination of small molecules, such as water.* Two hydroxyl groups on different molecules frequently eliminate a molecule of water in this way; the product is a new molecule containing a bridging oxygen atom:

$$X{-}OH + HO{-}X \longrightarrow X{-}O{-}X + H_2O$$

This process can produce polymeric chains or rings. For example, when phosphoric acid is heated, it loses water and condenses to give diphosphoric acid (pyrophosphoric acid), $H_4P_2O_7$:

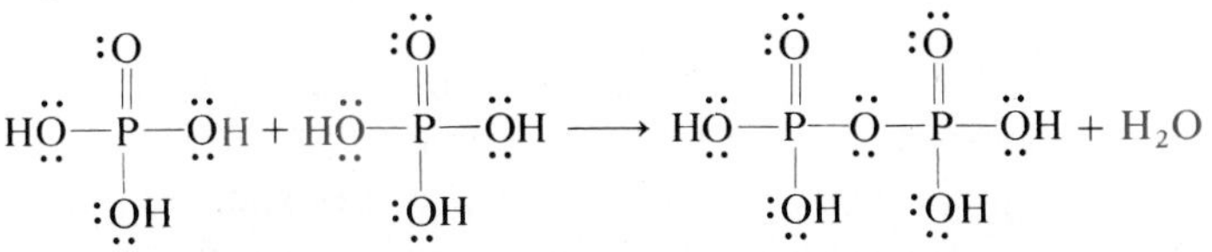

Continued heating leads to the formation of triphosphoric acid (often called tripolyphosphoric acid), $H_5P_3O_{10}$, and acids with still longer chains.

The interconversion of triphosphates, diphosphates, and phosphates, such as adenosine triphosphate (ATP) and adenosine diphosphate (ADP), is very important in many biochemical processes. These reactions occur at body temperature under the influence of catalysts known as enzymes.

Further dehydration of phosphoric acid leads eventually to metaphosphoric acid:

$$H_3PO_4 \longrightarrow HPO_3 + H_2O$$

As we have seen, HPO_3 can also be made by addition of a limited amount of water to P_4O_{10}.

The formula HPO_3 is the empirical formula of metaphosphoric acid, which is in fact polymeric. It has many different forms depending on the degree of polymerization. Thus the general formula is $(HPO_3)_n$, where $n = 3, 4, 5, 6, \ldots$. These molecules are rings, such as $(HPO_3)_3$ and $(HPO_3)_4$, or very long chains, $(HPO_3)_n$. Salts of the metaphosphoric acids, *metaphosphates*, are used extensively as water softeners.

PHOSPHOROUS ACID AND PHOSPHITES When P_4O_6 reacts with water, it forms phosphorous acid, H_3PO_3:

$$P_4O_6 + 6H_2O \longrightarrow 4H_3PO_3$$

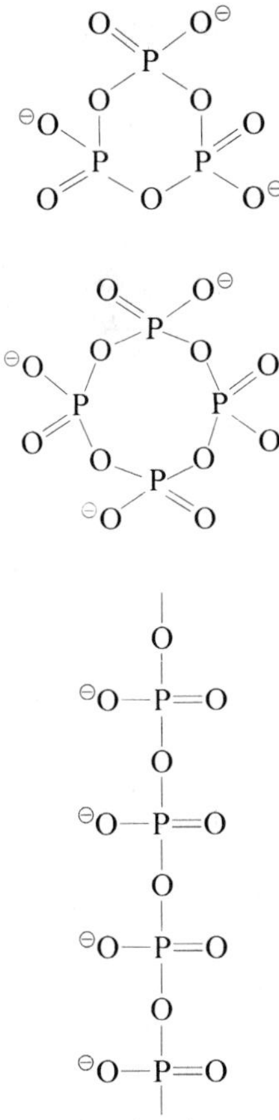

Metaphosphates

Phosphorous acid is a white solid with a melting point of 74 °C. In aqueous solution it behaves as a weak acid. It has only two replaceable hydrogen atoms; it is a *diprotic acid*. At first sight we might expect phosphorous acid to be triprotic, like phosphoric acid. But H_3PO_3 has only two OH groups; the third hydrogen is bonded directly to phosphorus. The P—H bond is nonpolar, because hydrogen and phosphorus have essentially the same electronegativity ($\chi_P = 2.1$, $\chi_H = 2.2$), so this third hydrogen does not ionize in water. Phosphorous acid is therefore diprotic and forms two series of salts, for example, sodium dihydrogen phosphite, NaH_2PO_3, and sodium hydrogen phosphite, Na_2HPO_3.

Just as SO_2 and sulfites are rather easily oxidized to the $+6$ oxidation state, so phosphorous acid and the phosphites, in which phosphorus is in the $+3$ oxidation state, are readily oxidized to phosphates, in which phosphorus is in the $+5$ oxidation state. They are therefore good reducing agents. For example, phosphorous acid precipitates copper from a solution of a copper salt:

$$Cu^{2+}(aq) + H_3PO_3(aq) + 3H_2O(l) \longrightarrow Cu(s) + H_3PO_4(aq) + 2H_3O^+(aq)$$

The names, formulas, and structures of the more important oxoacids of phosphorus are summarized in Table 8.2.

PHOSPHIDES AND PHOSPHINE

The lowest oxidation state of phosphorus is the -3 state. When phosphorus is heated with some metals, **phosphides**, such as calcium phosphide, Ca_3P_2, and aluminum phosphide, AlP, are formed. If these phosphides are treated with dilute aqueous acid, *phosphine*, PH_3, is formed:

$$Ca_3P_2(s) + 6HCl(aq) \longrightarrow 2PH_3(g) + 3CaCl_2(aq)$$

This reaction is quite analogous to that of a metal sulfide with an aqueous acid to give hydrogen sulfide, for example,

$$CaS(s) + 2HCl(aq) \longrightarrow CaCl_2(aq) + H_2S(g)$$

Phosphine is an unpleasant-smelling toxic gas. The phosphine molecule has a pyramidal AX_3E structure like ammonia. Phosphine is much less soluble in water than ammonia, and it is an exceedingly weak base. In other words, the equilibrium

$$PH_3(g) + H_2O(l) \rightleftharpoons PH_4^+(aq) + OH^-(aq)$$

lies far to the left. Because PH_3 is an exceedingly weak base, the phosphonium ion, PH_4^+, is almost a strong acid and it therefore reacts with water almost quantitatively liberating gaseous phosphine:

$$PH_4^+(aq) + H_2O(l) \longrightarrow H_3O^+(aq) + PH_3(g)$$

Thus very few phosphonium salts are known as they can only be prepared in the absence of water, for example by the reaction

$$PH_3(g) + HCl(g) \longrightarrow PH_4Cl(s)$$

Phosphorous acid

Phosphine
Pyramidal AX_3E geometry

Phosphonium ion
Tetrahedral AX_4 geometry

═ *Example 8.4* ═══════════════════

REACTIONS OF PHOSPHORUS COMPOUNDS

Starting with white phosphorus and a solution of sodium hydroxide explain, giving suitable equations, how a sample of NaH_2PO_3 could be made.

TABLE 8.2
Oxoacids of Phosphorus

Oxidation State	Name	Formula	Structure
3	Phosphor*ous* acid	H_3PO_3	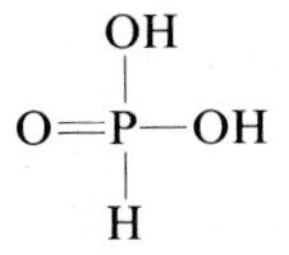
5	Phosphor*ic* acid	H_3PO_4	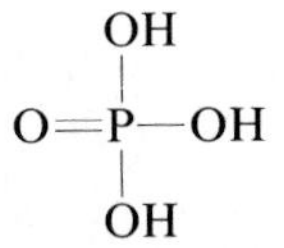
5	*Di*phosphor*ic* acid	$H_4P_2O_7$	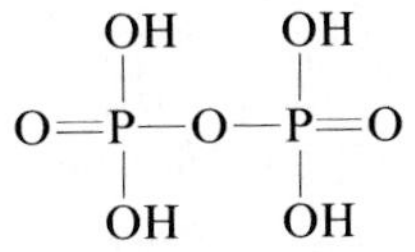
5	*Tri*phosphor*ic* acid	$H_5P_3O_{10}$	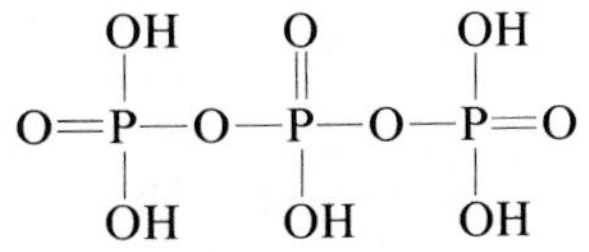
5	*Meta*phosphor*ic* acid	$(HPO_3)_n$	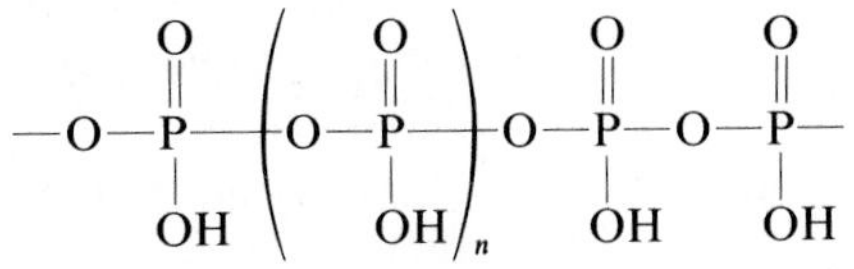

Solution

NaH_2PO_3 is a salt of phosphorous acid, H_3PO_3. If we first prepare H_3PO_3 we can neutralize it with the NaOH solution to form the salt. H_3PO_3 can be made by adding P_4O_6 to water. The P_4O_6 is made by burning P_4 in a limited supply of air. The equations are

$$P_4(s) + 3O_2(g) \longrightarrow P_4O_6(s)$$

$$P_4O_6(s) + 6H_2O(l) \longrightarrow 4H_3PO_3(aq)$$

$$H_3PO_3(aq) + NaOH(aq) \longrightarrow NaH_2PO_3(aq) + H_2O(l)$$

EXERCISE 8.9

Starting with Ca(s), P(s), HCl(g), and H_2O explain, giving suitable equations, how solid PH_4Cl could be made.

8.4
Structure and Acid–Base Strengths

In Chapters 5 and 6 and in this chapter we have described several different acids and bases. We are now in a position to give some answers to these questions: What makes a substance an acid or a base? What determines the strength of an acid or a base? We first consider the nonmetal hydrides.

NONMETAL HYDRIDES

As Figure 8.16 shows, acid–base strengths vary in a regular way in the periodic table. In groups 4–7 the acid strength of the hydrides increases from left to right and decreases from top to bottom.

Two factors appear to be important in determining the acid strength of the nonmetal hydrides: (1) the electronegativity of the nonmetal, in other words *the polarity of the X—H bond*, and (2) *the strength of the X—H bond* as measured for example by the bond energy.

In the series CH_4, NH_3, H_2O, HF the X—H bond becomes more polar with increasing electronegativity of X. Thus the hydrogen atom acquires an increasingly large positive charge and therefore becomes easier to remove as a proton (H^+). The increase in acid strength from SiH_4 to HCl can similarly be attributed to the increase in the electronegativity from Si to Cl. Neither CH_4 nor SiH_4 behaves as an acid in water because the C—H and Si—H bonds have very little polarity.

In going down a group it appears that the bond strength is the dominant factor. The bond energies of the X—H bonds (Table 6.3) decrease from period 2 to period 3 with increasing size of X. The bond electron density is at a greater distance from the nuclei that it is holding together and so the bond is weaker. Thus it becomes easier to break the X—H bond and the acid strength increases correspondingly. Thus HCl is a strong acid whereas HF is a weak acid.

However, these explanations are undoubtedly an oversimplification of a rather complex situation. Another factor that probably contributes to the weak acid behavior of HF will be mentioned in Chapter 12.

The base strengths of the hydrides vary in just the opposite manner from their acid strengths. Their basicity decreases from left to right across the periodic table and down a group. Basicity depends on the availability of an unshared electron pair to which a proton can be added. Neither CH_4 nor SiH_4 have an unshared pair so they do not behave as bases. From NH_3 to HF the increasing electronegativity of X causes the unshared electron pairs to be held more strongly and thus makes them less available for sharing with a proton. Thus the base strength decreases. NH_3 is a weak base. Water is a much weaker base

Period	Group 4	Group 5	Group 6	Group 7
Period 2	CH₄ — No acid or base properties	NH₃ — Weak base	H₂O — Amphoteric	HF — Weak acid
3	SiH₄ — No acid or base properties	PH₃ — Weak base	H₂S — Weak acid	HCl — Strong acid

Acid strength increases →
Base strength increases →

Base strength increases (vertical) / Acid strength increases (vertical)

FIGURE 8.16
Acid and Base Strengths of Nonmetal Hydrides.

than NH_3 and HF does not behave as a base. Note that the base strength of a molecule does not depend on how many unshared pairs it has—NH_3 has one and HF has three. Base strength depends only on the availability of the unshared pairs for sharing with a hydrogen ion. Note also that, although the water molecule has two unshared pairs of electrons, it adds only one proton to give the hydronium ion, H_3O^+. The positive charge thus produced on the oxygen atom strongly attracts the remaining unshared pair and makes it much less available for sharing with a proton. Thus the H_4O^{2+} ion has never been observed.

With increasing size of the atom X an unshared pair becomes larger and less localized. It therefore attracts a proton less strongly than a smaller and more concentrated unshared pair. Thus PH_3 is a much weaker base than NH_3 and, whereas water is a weak base, H_2S has no basic properties in water.

OXOACIDS

Phosphoric acid, phosphorous acid, and sulfuric acid are members of an important class of compounds known as **oxoacids**. Oxoacids have one or more OH groups attached to an electronegative atom. They have the general formula $XO_m(OH)_n$, where $m = 0, 1, 2, \ldots$ and $n = 1, 2, 3, \ldots$. For example, H_3PO_4 can be written as $PO(OH)_3$, where $m = 1$ and $n = 3$; H_2SO_4 can be written as $SO_2(OH)_2$, where $m = 2$ and $n = 2$.

Other important oxoacids include perchloric acid, $HClO_4$ or $ClO_3(OH)$, and nitric acid, HNO_3 or $NO_2(OH)$.

The oxoacids $XO_m(OH)_n$ have one or more hydroxyl, OH, groups. But we know that some compounds containing OH groups, such as NaOH and $Ca(OH)_2$, are ionic hydroxides, Na^+OH^- and $Ca^{2+}(OH^-)_2$, and are therefore strong bases. What, then, determines if a compound containing OH groups is an acid or a base?

The structures and acid–base properties of the hydroxides of the elements of the third period are shown in Figure 8.17. We see that **the hydroxides of metals are bases, while the hydroxides of nonmetals are acids**. More specifically,

FIGURE 8.17
Acid and Base Strengths of the Hydroxyl Compounds of the Elements of Period 3.

Group						
1	2	3	4	5	6	7
NaOH	$Mg(OH)_2$	$Al(OH)_3$	$Si(OH)_4$	$[P(OH)_5]$	$[S(OH)_6]$	$[Cl(OH)_7]$
				$-H_2O$	$-2H_2O$	$-3H_2O$
Na^+OH^-	$Mg^{2+}(OH^-)_2$	$Al(OH)_3$	$Si(OH)_4$	$O{=}P(OH)_3$	$O{=}S(OH)_2$	$O{=}Cl{-}OH$
Strong base	Weak base	Amphoteric	Weak acid	Weak acid	Strong acid	Strong acid

Acid strength increases →

← Base strength increases

we see that the acid strength of the hydroxides increases from left to right across each period, and the base strength decreases. Sodium hydroxide, NaOH, and magnesium hydroxide, $Mg(OH)_2$, are strong bases, $Al(OH)_3$ is amphoteric (it is both a weak base and a weak acid), while $Si(OH)_4$ and the other nonmetal hydroxides are all acids, their strength increasing from left to right. Silicic acid, $Si(OH)_4$, and phosphoric acid, $PO(OH)_3$, are weak, while sulfuric acid, $SO_2(OH)_2$, and perchloric acid, $ClO_3(OH)$, are strong.

If the atom X in XOH has a low electronegativity, the X—O bond is polar, with the oxygen having a negative charge and X a positive charge:

$$\overset{\delta+}{X}-\overset{\delta-}{O}-H$$

When this is the case, XOH ionizes to give X^+ and OH^- as in NaOH and $Mg(OH)_2$. When X has a high electronegativity, the X—O bond is less polar and so XOH has a decreased tendency to ionize as X^+ and OH^-. But the O—H bond becomes more polar,

$$X-\overset{\delta-}{O}-\overset{\delta+}{H}$$

So there is an increased tendency for the hydrogen to be donated as a proton to a base to give XO^- and BH^+. Thus acid strength increases with increasing electronegativity of X, and base strength decreases.

The oxoacids (the hydroxides of the nonmetals) often do not have the maximum number of hydroxyl groups that would be expected from their valences. Instead, pairs of —OH groups are often replaced by a doubly bonded oxygen, =O. They can be imagined as being derived from the hypothetical hydroxides with the maximum number of OH groups by loss of one or more water molecules. For example,

$$HO-\underset{\underset{OH}{|}}{\overset{\overset{OH}{|}}{C}}-OH \longrightarrow O{=}C\overset{OH}{\underset{OH}{\diagdown}} + H_2O$$

and

$$\underset{HO}{\overset{HO}{\diagdown}}\underset{\underset{OH}{|}}{\overset{\overset{OH}{|}}{S}}\overset{OH}{\diagup} \longrightarrow O{=}\underset{\underset{OH}{|}}{\overset{\overset{O}{\|}}{S}}-OH + 2H_2O$$

The strength of an oxoacid depends on the value of m in the formula $XO_m(OH)_n$. The larger the value of m, the number of doubly bonded oxygen atoms, the stronger is the acid. Each oxygen atom attracts electrons away from the central atom X which in turn attracts electrons more strongly away from the O—H groups, thus increasing the polarity of the O—H bonds and therefore the strength of the acid. In general, for an acid $XO_m(OH)_n$, if $m = 0$ or 1 the acid is weak, but if $m = 2$ or 3 the acid is strong. We see from Figure 8.17 that silicic acid, with $m = 0$, and phosphoric acid, with $m = 1$, are weak. But sulfuric acid, with $m = 2$, and perchloric acid, with $m = 3$, are strong.

This correlation of acid strength with the number of doubly bonded oxygen atoms also implies that acid strength increases with the increasing oxidation state of the central atom. Sulfurous acid, H_2SO_3 (oxidation number S = +4), is a weak acid, while sulfuric acid, H_2SO_4 (oxidation number S = +6), is a strong acid. Hypochlorous acid, ClOH (oxidation number Cl $= +1$), is a weak acid, while perchloric acid, $HClO_4$ (oxidation number Cl $= +7$), is a strong acid.

═ Example 8.5 ═══════════════════════════════

STRENGTHS OF OXOACIDS

Classify the following acids as either strong or weak:

HNO_2 $HBrO_4$ H_4SiO_4 HNO_3

Solution

First we write each formula in the form $XO_m(OH)_n$. Then we have

HNO_2	$NO(OH)$	$m = 1$	Weak
$HBrO_4$	$BrO_3(OH)$	$m = 3$	Strong
H_4SiO_4	$Si(OH)_4$	$m = 0$	Weak
HNO_3	$NO_2(OH)$	$m = 2$	Strong

EXERCISE 8.10

Classify the following acids as either strong or weak:

HOCl $HClO_3$ H_2SO_3 H_5IO_6 H_2SeO_4

8.5
Names of Compounds

Some of the compounds we have encountered so far have names that give no clue to their composition; water, ammonia, methane, and acetic acid are examples. If all compounds were named in such an unsystematic way, chemists would have to memorize hundreds of thousands of names. Fortunately, the names of most compounds are related to their composition in a systematic way. We have already described in Chapter 4 how to name simple ionic compounds. Now that we have discussed oxidation states and electronegativity we can give the rules used to name both covalent and ionic inorganic compounds. The names of organic compounds are discussed in Chapter 23.

BINARY COMPOUNDS

The simplest compounds are **binary compounds**; they are composed of only two elements. We name them by giving first the name of the less electronegative (more metallic) element and then the stem of the name of the other element (almost always a nonmetal), to which the suffix *ide* has been added. The following examples give the names of some binary compounds:

NaCl	Sodium chloride	$CaCl_2$	Calcium chloride
KBr	Potassium bromide	Na_2S	Sodium sulfide
MgO	Magnesium oxide	Mg_3N_2	Magnesium nitride
HBr	Hydrogen bromide		

When two elements form more than one binary compound, we need some way of distinguishing the two compounds. The simplest method uses a Greek prefix to indicate the numbers of each atom in the formula. The prefixes are *mono* for one, *di* for two, *tri* for three, *tetra* for four, *penta* for five, and *hexa* for six. By this method we have the following names:

Carbon monoxide	CO	Carbon dioxide	CO_2
Sulfur dioxide	SO_2	Sulfur trioxide	SO_3
Phosphorus trichloride	PCl_3	Phosphorus pentachloride	PCl_5
Sulfur tetrafluoride	SF_4	Tetraphosphorus hexaoxide	P_4O_6

The prefix *mono* is generally omitted before the first element in the name.

Alternatively, we can distinguish two compounds of the same elements by specifying the oxidation state of the less electronegative element with a Roman numeral. For example, phosphorus in PCl_3 has an oxidation number of $+3$, while in PCl_5 it has an oxidation number of $+5$. Thus according to this system, the names of PCl_3 and PCl_5 are phosphorus(III) chloride and phosphorus(V) chloride. Other examples are as follows:

SO_2	Sulfur(IV) oxide	SO_3	Sulfur(VI) oxide
$FeCl_2$	Iron(II) chloride	$FeCl_3$	Iron(III) chloride

This method is more commonly used for the compounds of metals with nonmetals than for compounds between nonmetals. Thus the name phosphorus trichloride is more commonly used than phosphorus(III) chloride.

Note that when the stoichiometry of a compound is not indicated in its name, as, for example, in aluminum chloride, we must know that aluminum is in group 3 and has a valence of 3 in order to write the correct formula, $AlCl_3$. The names aluminum trichloride or aluminum(III) chloride are more specific, but they are not always used.

BINARY ACIDS

The aqueous solutions of nonmetal hydrides that behave as acids are named by adding the prefix *hydro* and the suffix *ic* to the stem of the name of the nonmetal:

Binary hydride	*Formula*	*Name of aqueous solution*
Hydrogen fluoride	HF	Hydrofluoric acid
Hydrogen chloride	HCl	Hydrochloric acid
Hydrogen bromide	HBr	Hydrobromic acid
Hydrogen iodide	HI	Hydroiodic acid
Hydrogen cyanide	HCN	Hydrocyanic acid

OXOACIDS

If an element forms only one oxoacid, it is named by adding *ic acid* to the stem of the name of the element. For example:

H_2CO_3	Carbonic acid
H_3BO_3	Boric acid

When an element has more than one oxidation state, it may form more than one oxoacid. If the element forms only two oxoacids, the acid containing the element in the higher oxidation state is given the suffix *ic*; the acid having the element in the lower oxidation state is given the ending *ous*. Thus we have

$$H_3PO_4 \quad \text{Phosphoric acid}$$
$$H_3PO_3 \quad \text{Phosphorous acid}$$

The name of the anion derived from the *ic* acid ends in *ate*, while the name of the anion derived from the *ous* acid ends in *ite*:

H_2SO_4	Sulfuric acid	SO_4^{2-}	Sulfate
H_2SO_3	Sulfurous acid	SO_3^{2-}	Sulfite
HNO_3	Nitric acid	NO_3^-	Nitrate
HNO_2	Nitrous acid	NO_2^-	Nitrite
H_3PO_4	Phosphoric acid	PO_4^{3-}	Phosphate
H_2CO_3	Carbonic acid	CO_3^{2-}	Carbonate

A few elements form oxoacids in more than two oxidation states. In this case the prefixes *hypo* and *per* are used to designate a lower and a higher oxidation state, respectively. The oxoacids of chlorine, of which we have already mentioned HClO and $HClO_4$, provide a good example:

HClO	Hypochlorous acid	ClO^-	Hypochlorite
$HClO_2$	Chlorous acid	ClO_2^-	Chlorite
$HClO_3$	Chloric acid	ClO_3^-	Chlorate
$HClO_4$	Perchloric acid	ClO_4^-	Perchlorate

Many acids are *polyprotic*; they have more than one ionizable hydrogen and therefore give two or more anions. The three anions derived from phosphoric acid are dihydrogen phosphate, $H_2PO_4^-$, hydrogen phosphate, HPO_4^-, and phosphate, PO_4^{3-}. When it reacts with sodium hydroxide, phosphoric acid gives the following salts:

NaH_2PO_4	Sodium dihydrogen phosphate
Na_2HPO_4	Disodium hydrogen phosphate (or sodium hydrogen phosphate)
Na_3PO_4	Trisodium phosphate (or sodium phosphate)

The terminations "ous" and "ic" were formerly also used to distinguish between the salts of a metal in two oxidation states. Thus copper(I) chloride was called cuprous chloride and copper(II) chloride was called cupric chloride. Similarly iron(II) chloride was called ferrous chloride and iron(III) chloride was called ferric chloride. It would seem to be logical also to replace the "ous" and "ic" terminology for acids by a system based on the use of a Roman numeral to designate the oxidation state of the central atom. Indeed such a system has been suggested and is being used in some European countries but it has not yet been adopted in North America.

═ *Example 8.6* ════════════════════════════

NAMING COMPOUNDS

Name each of the following compounds:

(a) $ZnCO_3$ (b) SiF_4 (c) Na_2SO_3 (d) $HBrO_4$ (e) Li_3N

Solution

(a) Zinc carbonate   (b) Silicon tetrafluoride   (c) Sodium sulfite

(d) Perbromic acid   (e) Lithium nitride

EXERCISE 8.11

Name each of the following compounds:

(a) $MgHPO_4$ (b) $HClO_3$ (c) P_4S_3 (d) Mg_3N_2 (e) N_2O_5

═ *Example 8.7* ═══

WRITING FORMULAS FROM NAMES

Write the formulas of each of the following compounds:

(a) Phosphorus triiodide   (b) Barium iodide

(c) Magnesium phosphate (d) Manganese(IV) oxide

(e) Calcium hydrogen carbonate (f) Hypobromous acid

Solution

(a) PI_3.

(b) BaI_2. It is not BaI because barium is in group 2 and has a valence of 2 and forms the Ba^{2+} ion. Since BaI_2 is ionic we can also write the formula as $Ba^{2+}(I^-)_2$.

(c) $Mg_3(PO_4)_2$. Magnesium is in group 2 and therefore forms the Mg^{2+} ion. The phosphate ion is PO_4^{3-}. Thus we can write the formula as $(Mg^{2+})_3(PO_4^{3-})_2$. The sum of the positive charges must equal the sum of the negative charges, so the formula is *not* $MgPO_4$.

(d) MnO_2. The Roman numeral IV indicates that manganese is in the oxidation state $+4$ and has an oxidation number of $+4$. Therefore the formula must be MnO_2, assuming, as is usual, that oxygen has oxidation number -2.

(e) $Ca(HCO_3)_2$. The hydrogen carbonate ion is HCO_3^-, and calcium forms the Ca^{2+} ion.

(f) $HOBr$. This compound is an oxoacid of bromine in which bromine is in the $+1$ oxidation state. It is named in the same way as $HOCl$, hypochlorous acid.

───

EXERCISE 8.12

Write the formula of each of the following compounds:

(a) Aluminum sulfate   (b) Strontium bromide

(c) Hydrocyanic acid   (d) Potassium dihydrogen phosphate

(e) Chromium(VI) oxide (f) Phosphorus acid

═══

IMPORTANT TERMS

A **binary compound** is composed of only two elements.

Condensation is the reaction of two or more molecules to form a larger molecule with the elimination of small molecules, such as water.

A **diprotic acid** has two ionizable hydrogen atoms. A **triprotic acid** has three ionizable hydrogen atoms.

A **half-reaction** describes either the oxidation or the reduction part of an oxidation–reduction reaction. Electrons appear on the right side of the equation describing the oxidation half-reaction and on the left side of the equation describing the reduction half-reaction.

Le Châtelier's principle states that when one of the conditions that affect the position of a dynamic equilibrium is changed, the position of the equilibrium shifts in such a way as to minimize the effect of the change.

Oxidation is any change in which the oxidation number of an atom increases.

The **oxidation number** is a measure of the extent of oxidation of an element in its compounds. It is assigned by using a set of rules. For a monatomic ion it is equal to the charge on the ion. For an atom in a covalent molecule it is equal to the charge that the atom would have if both electrons of each covalent bond were assigned to the more electronegative element.

Oxoacids have one or more OH groups attached to an electronegative atom. They have the general formula $XO_m(OH)_n$.

Reduction is any change in which the oxidation number of an atom decreases.

PROBLEMS *

Sulfur

1. On what properties of sulfur does its extraction by the Frasch process depend? What other industrial method is used to prepare sulfur?

2. Name and briefly describe the forms and molecular structures of three *allotropes* of sulfur. Describe and explain what happens when sulfur is gradually heated to its boiling point.

3. Draw orbital box diagrams showing the valence state of sulfur when it has valences of 2, 4, and 6. Explain why sulfur cannot have a valence of 8, and why oxygen cannot have valences of 4 and 6.

4. (a) Suggest a reason why the industrial process for manufacturing sulfuric acid is called the *contact* process.

(b) Explain why sulfuric acid should always be diluted by adding the acid to water, rather than the other way round.

5. Give two examples of the use of concentrated sulfuric acid to prepare acids that are more volatile than sulfuric acid. Why is this method unsuitable for the preparation of hydrogen iodide?

6. Write balanced equations for the reaction of sulfuric acid with each of the following. In each case, state whether the reaction occurs at room temperature or only at high temperature, and whether dilute or concentrated acid should be used:

(a) An aqueous solution of $NaOH(aq)$

(b) An aqueous solution of $BaCl_2(aq)$

(c) $NaF(s)$ (d) $NaNO_3(s)$ (e) Carbon

7. Write a balanced equation for a reaction in which sulfuric acid acts in each of the following ways:

(a) As an oxidizing agent   (b) As an acid

(c) As a dehydrating agent

* Answers to problems numbered in blue appear at the end of the text.

8. Write Lewis structures for the following molecules and ions and then use the VSEPR model to predict their shapes:

SO_2, SO_3, H_2SO_4, SO_3^{2-}, $S_2O_3^{2-}$

9. Write a balanced equation for each of the following reactions:

(a) The reaction of zinc with dilute aqueous sulfuric acid.

(b) The reaction of concentrated sulfuric acid with sodium iodide

(c) The reaction of concentrated sulfuric acid with silver

(d) The reaction of dilute sulfuric acid with solid magnesium hydroxide

10. Describe how you would prepare a solution of sodium thiosulfate starting with $NaOH(aq)$, $S(s)$, and $SO_2(g)$. Write a balanced equation for each of the reactions that you see.

11. A compound containing only iron and sulfur was analyzed by heating in air to convert all the sulfur to sulfur dioxide. At a pressure of 750 mm Hg and 25 °C the volume of SO_2 obtained from 0.4203 g of the compound was 173.6 mL. What is the empirical formula of the compound? Is it likely to be an ionic compound or a covalent compound?

12. An aqueous solution is known to contain either $SO_4^{2-}(aq)$ or $SO_3^{2-}(aq)$, but not both. Suggest two tests that could be used to identify the anion in the solution.

13. One method used for the removal of SO_2 from the flue gases of power plants is to react it with an aqueous solution of H_2S. One product of this reaction is sulfur. Write a balanced equation for the reaction. What volume of $H_2S(g)$ at 25 °C and 750 mm Hg pressure is needed to remove all the SO_2 formed by burning 1 metric ton (10^3 kg) of coal containing 4.00% sulfur by mass? What mass of elemental sulfur is formed?

14. Water from springs and wells is often contaminated with small concentrations of hydrogen sulfide, which gives it a bad smell. The H_2S can be removed by treatment of the water with chlorine, which oxidizes the H_2S to sulfur. Write the balanced equation for this reaction. If the H_2S content of the water from a particular source is 22 ppm by mass, how much chlorine will be needed to remove the H_2S from 5000 L of water?

*15. A compound contains 23.7% S, 52.6% Cl, and 23.7% O, by mass. The volume occupied by 0.337 g of the gaseous compound at 100 °C and 770 mm Hg pressure is 75.6 mL. Calculate the empirical and molecular formulas, and suggest a possible Lewis structure for the compound, which has a boiling point of 69 °C.

16. (a) How is hydrogen sulfide gas prepared in the laboratory?

   (b) Hydrogen sulfide burns in air with a blue flame to give sulfur dioxide, but in a limited supply of air, sulfur is formed. Write balanced equations for each of these reactions.

   (c) Write equations to show H_2S acting as a diprotic acid in water.

   (d) Give *three* examples of reactions in which H_2S behaves as a reducing agent.

17. Write the formula for each of the following compounds:

   (a) potassium sulfate   (b) calcium hydrogen sulfate

   (c) calcium sulfite   (d) potassium tetrasulfide

   (e) sodium disulfate   (f) aluminum sulfate

18. Give an example of reactions in which sulfuric acid (a) behaves as an acid, (b) behaves as an oxidizing agent, (c) behaves as a dehydrating agent, and (d) takes part in a precipitation reaction.

19. Classify each of the following compounds as ionic or covalent and write their Lewis structures. Which of the species contain double bonds, and which of the bonds are polar?

   (a) Na_2S (b) Na_2S_2 (c) MgS (d) S_8

   (e) CS_2 (f) SO_2 (g) SO_3 (h) Cl_2S

Name each of the above compounds.

Phosphorus

20. Explain the following properties of the *allotropes* of phosphorus in terms of their structures:

   (a) White phosphorus has a lower melting point than black phosphorus and is more volatile

   (b) White phosphorus is soluble in carbon disulfide, $CS_2(l)$, while red phosphorus is insoluble

21. Write a balanced equation for the combustion of phosphine, $PH_3(g)$, in air to give $P_4O_{10}(s)$.

* The asterisk denotes the more difficult problems.

22. Show by means of balanced equations how phosphorus rock can be used as a source of

   (a) Phosphoric acid   (b) Sodium phosphate

   (c) White phosphorus

   (d) Superphosphate-of-lime fertilizer

23. One of the components of a match head is phosphorus sulfide, $P_4S_3(s)$. Propose a structure for P_4S_3 based on the structures of P_4 and P_4O_6.

24. Suggest possible explanations for the fact that PF_5 and PCl_5 are known compounds but it has not proved possible to make either PH_5 or NCl_5.

25. Striking a match involves the combustion of $P_4S_3(s)$ to produce a white smoke of $P_4O_{10}(s)$ and sulfur dioxide gas. Write a balanced equation for this reaction. Calculate the maximum volume of $SO_2(g)$ at 20 °C and 772 mm Hg pressure that results from the combustion of 0.157 g of P_4S_3.

*26. Dry hydrogen fluoride, $HF(g)$, reacts with $P_4O_{10}(s)$ to give a gas containing 29.8% P, 54.8% F, and 15.4% O, by mass. This gas has a density of 4.64 g L^{-1} at STP. What is the molecular formula of the gas? Draw a possible Lewis structure for this compound and write down the oxidation numbers of each of its elements.

*27. When aluminum phosphide, $AlP(s)$, is dissolved in dilute sulfuric acid, a gas X, with a density of 1.531 g L^{-1} at STP, is evolved. Evaporation of the remaining aqueous solution to dryness, followed by recrystallization of the solid product from water, and drying, gives a salt containing 15.8% Al, 28.1% S, and 56.1% O, by mass. When 0.158 g of the gas X is reacted completely with hydrogen iodide, 0.752 g of a white crystalline compound containing 19.1% P, 2.5% H, and 78.4% I, by mass, is obtained. What is the gas X? Write balanced equations for all the above reactions.

*28. Write a balanced equation for each of the following reactions:

   (a) The reaction of red phosphorus with concentrated nitric acid to give orthophosphoric acid, nitrogen dioxide, and water

   (b) The preparation of white phosphorus, $P_4(s)$, from calcium phosphate by reaction with carbon (coke) and sand, $SiO_2(s)$, to give calcium silicate, carbon monoxide and phosphorus

   (c) The formation of disodium hydrogen phosphate from phosphoric acid and aqueous sodium hydroxide

   (d) The reduction of $Cu^{2+}(aq)$ with phosphorus acid to give $Cu(s)$ and phosphoric acid

(e) The reaction of white phosphorus with aqueous sodium hydroxide to give sodium dihydrogen hypophosphite, $NaH_2PO_2(aq)$, and phosphine

29. A 3.064-g sample of phosphorus was burned completely in air to give 7.020 g of an oxide, which reacted with 2.671 g of water to give 9.691 g of an oxoacid of phosphorus. Determine the empirical formulas of the oxide and the oxoacid.

30. Starting with white phosphorus, P_4, and NaOH(aq) how would you prepare an aqueous solution of Na_3PO_4? Write balanced equations for any reactions that you use.

31. A sodium salt of a phosphorus oxoacid contains 25.3% P, 43.5% O, and 31.2% Na, by mass. What is its empirical formula, and what is the probable Lewis structure of its anion?

32. (a) Phosphorus pentachloride exists as PCl_5 molecules in the gas phase, but as $PCl_4{}^+PCl_6{}^-$ ions in the solid. Draw Lewis structures for each of these three species.

(b) Red phosphorus reacts with $Cl_2(g)$ to give a mixture of PCl_3 and PCl_5. If 5.43 g of red phosphorus reacts with chlorine to give 28.05 g of a mixture of the phosphorus chlorides, what is its composition?

33. Write a formula for each of the following compounds:

(a) tetraphosphorus hexaoxide

(b) calcium dihydrogen phosphate

(c) calcium phosphide

(d) phosphorous acid

(e) phosphonium iodide

(f) sodium metaphosphate

34. Calcium hydrogen phosphate, $CaHPO_4(s)$, is produced commercially from calcium carbonate and phosphoric acid. Write the balanced equation for the reaction. How many grams of calcium carbonate are required to react with 48.0 mL of an aqueous solution that is 85.5 mass % phosphoric acid, and what is the maximum mass of $CaHPO_4(s)$ that could be obtained from this reaction? The density of 85.5 mass % phosphoric acid is 1.70 g mL^{-1}.

35. Calculate the volume in milliliters needed for the complete neutralization of 25.00 mL of $0.200M$ $H_3PO_3(aq)$ by $0.150M$ NaOH(aq).

Oxidation–Reduction

36. Define *oxidation* and *reduction* in terms of **(a)** electron transfer and **(b)** change in oxidation number.

37. What is the range of possible oxidation states for phosphorus? Give two examples of compounds containing phosphorus in each of these oxidation states.

38. What are the oxidation states of each of the elements in each of the following?

(a) PH_3 **(b)** AsH_3 **(c)** PF_3

(d) $K^+MnO_4{}^-$ **(e)** SiO_2 **(f)** $PH_4{}^+$

39. What is the oxidation number of oxygen in each of the following?

(a) O_2 **(b)** O_3 **(c)** OF_2 **(d)** H_2O_2

(e) $O_2{}^{2-}$ **(f)** O^{2-}

40. What are the oxidation numbers of each element in each of the following:

(a) CH_4 **(b)** $CO_3{}^{2-}$ **(c)** $C_2O_4{}^{2-}$

(d) CH_3CO_2H **(e)** $SO_4{}^{2-}$ **(f)** $Cr_2O_7{}^{2-}$

41. What are the oxidation numbers for each of the following?

(a) C in CO   **(b)** S in SO_2

(c) P in $Ca_5(PO_4)_3(OH)$ **(d)** S in SF_6

(e) S in $S_2{}^{2-}$ **(f)** P in $H_2PO_3{}^-$

42. What are the oxidation numbers of each of the elements in each of the following:

(a) N_2H_4 **(b)** Li_2O **(c)** PH_4I

(d) $NaClO_4$ **(e)** NaOCl   **(f)** CaS_2

(g) AlP   **(h)** S_8 **(i)** $BaSO_3$

43. Assign oxidation numbers to each element in each of the following:

(a) N_2 **(b)** H_2CO_3 **(c)** NH_3

(d) PH_3 **(e)** VO^+ **(f)** $VO_2{}^+$

(g) MnO_2 **(h)** HNO_3 **(i)** HNO_2

44. Write equations for each of the following half-reactions in acidic aqueous solution:

(a) The oxidation of H_2S to S

(b) The oxidation of SO_2 to $SO_4{}^{2-}$

(c) The oxidation of H_3PO_3 to H_3PO_4

45. (a) If an atom in a molecular species is *oxidized*, does its oxidation number increase or decrease?

(b) If an atom in a molecular species is *reduced*, does its oxidation number increase or decrease?

(c) Write half-equations for each of the following in aqueous acid:

   (i) The oxidation of Cu^+ to Cu^{2+}

   (ii) The reduction of S_8 to S^{2-}

   (iii) The reduction of NO_3^- to NO

   (iv) The oxidation of $S_2O_3^{2-}$ (thiosulfate ion) to SO_4^{2-}

46. When an ionic bromide is heated with concentrated sulfuric acid, bromide ion is oxidized to bromine and the sulfuric acid is reduced to sulfur dioxide. Write the balanced equation for this reaction. When sulfur dioxide is bubbled through an aqueous solution of bromine, the bromine is reduced to bromide ion, and the sulfur dioxide is oxidized to sulfate ion. Write the balanced equation for this reaction. Explain why this reaction proceeds in one direction under one set of conditions and in the reverse direction under another set of conditions.

47. Sulfur dioxide can behave both as an oxidizing agent and as a reducing agent. Write a balanced equation to show SO_2 acting as **(a)** an oxidizing agent and **(b)** a reducing agent.

Lewis Structures

48. Draw Lewis structures for each of the following species, and use the VSEPR theory to predict the expected molecular geometries:

   (a) H_2SO_4 **(b)** SO_2 **(c)** SO_4^{2-}

   (d) HSO_3^- **(e)** H_2S

49. Draw Lewis structures for each of the following species, and use the VSEPR model to predict the expected molecular geometries:

   (a) Phosphoric acid, $(HO)_3PO$

   (b) Phosphorous acid, $HP(O)(OH)_2$

   (c) Phosphate ion, PO_4^{3-}

   (d) Phosphite ion, HPO_3^{2-}

*50. Draw Lewis structures for each of the following:

   (a) Disulfuric acid

   (b) Diphosphoric acid

   (c) Fluorosulfate ion, FSO_3^-

   (d) Fluorophosphate ion, FPO_3^{2-}

51. Draw Lewis structures for each of the following:

   (a) SO_3 **(b)** SO_3^{2-}

   (c) $S_2O_3^{2-}$ **(d)** $S_2O_7^{2-}$

52. Draw Lewis structures for **(a)** the triphosphate ion, $P_3O_{10}^{5-}$, and **(b)** the trisulfate ion, $S_3O_{10}^{2-}$. What is the oxidation number of phosphorus in (a), and the oxidation number of sulfur in (b)?

Acids and Bases

53. Classify each of the following acids as either strong or weak:

   (a) $B(OH)_3$ (boric acid)   **(b)** Phosphorous acid

   (c) H_2SeO_3 (selenous acid)   **(d)** Nitric acid

   (e) Phosphoric acid

54. Explain the acid and base properties of the following period 2 hydrides listed in aqueous solution:

CH_4	NH_3	H_2O	HF
No acid or base properties	Weak base	Very weak base	No base properties
	No acid properties	Very weak acid	Weak acid

55. Explain the acid and base properties of the following period 3 hydrides in aqueous solution:

SiH_4	PH_3	H_2S	HCl
No acid or base properties	Very weak base	No base properties	No base properties
	No acid properties	Weak acid	Strong acid

56. Explain why **(a)** HCl is a stronger acid in water than HF and **(b)** H_2S is a stronger acid in water than H_2O.

57. Explain why an anion $H_{m-1}XO_n^-$ is always a weaker acid than the parent acid H_mXO_n in aqueous solution (for example HSO_4^- versus H_2SO_4).

58. What are the relationships between the following mostly unknown hydroxides and the known oxoacids of these elements?

   (a) $B(OH)_3$ **(b)** $C(OH)_4$ **(c)** $N(OH)_5$

   (d) $Si(OH)_4$ **(e)** $P(OH)_5$ **(f)** $S(OH)_6$

59. Explain why KOH and $Ca(OH)_2$ are strong bases in aqueous solution, while $Si(OH)_4$ is a very weak acid.

Nomenclature

60. Write the formulas of each of the following compounds:

   (a) Hydrofluoric acid   **(b)** Phosphoric acid

   (c) Phosphorous acid   **(d)** Hypochlorous acid

   (e) Sulfuric acid   **(f)** Sulfurous acid

61. Write the formulas of each of the following compounds:

   (a) Calcium sulfate

(b) Phosphorus pentabromide

(c) Phosphorus(III) iodide

(d) Ammonium hydrogen phosphate

(e) Calcium nitride

(f) Sulfur tetrafluoride

(g) Chromium(III) chloride

62. Name each of the compounds with the following formulas:

(a) K_3PO_4 **(b)** CF_4 **(c)** $ZnCO_3$

(d) $Ca(OCl)_2$ **(e)** $CaSO_3$ **(f)** PH_4I

(g) $Na_2S_2O_7$ **(h)** $Na_5P_3O_{10}$

63. Name each of the compounds with the following formulas:

(a) K_2SeO_4 **(b)** H_2Te **(c)** Na_2S_4

(d) FeS_2 **(e)** $RbHSO_4$ **(f)** P_4O_6

(g) Na_2HPO_4 **(h)** Na_2HPO_3

64. What is the name of each of the oxoacids with the following molecular formulas:

(a) H_2SO_4 **(b)** H_3PO_4 **(c)** H_3PO_3

(d) H_2CO_3 **(e)** HNO_3.   **(f)** H_3SO_3

(g) H_4SiO_4 **(h)** $HOCl$ **(i)** $HClO_4$

65. Write balanced equations for all the possible reactions of each the following oxoacids with sodium hydroxide in aqueous solution, and name each of the possible salts that could be obtained from the solutions:

(a) Nitric acid, HNO_3

(b) Sulfuric acid, H_2SO_4

(c) Phosphoric acid, H_3PO_4

(d) Phosphorus acid, H_3PO_3

66. What species are present in aqueous solutions of each of the following? Give both names and formulas.

(a) Sulfuric acid   **(b)** Sulfur dioxide

(c) Phosphoric acid   **(d)** Carbonic acid

(e) Sodium phosphate

(f) Calcium hydrogen phosphate

(g) Ammonium nitrate   **(h)** Ammonium sulfate

Miscellaneous

67. Classify each of the following reactions in aqueous solution as an acid–base reaction or an oxidation–reduction reaction, and balance each equation. For the acid–base reactions, identify all of the conjugate acid–base pairs. For the oxidation–reduction reactions, identify the elements that are oxidized and those that are reduced:

(a) $CO_3^{2-} + H_2SO_4 \rightarrow HSO_4^- + HCO_3^-$

(b) $HSO_4^- + H_2O \rightarrow H_3O^+ + SO_4^{2-}$

(c) $Cu + H_2SO_4 \rightarrow Cu^{2+} + SO_4^{2-} + SO_2 + H_2O$

(d) $Br^- + H_2SO_4 \rightarrow Br_2 + SO_4^{2-} + SO_2 + H_2O$

(e) $CaCO_3 + H_3O^+ \rightarrow Ca^{2+} + HCO_3^- + H_2O$

(f) $Mg + H_3O^+ \rightarrow Mg^{2+} + H_2 + H_2O$

(g) $Ca_3P_2 + H_2O \rightarrow Ca(OH)_2 + PH_3$

68. Explain in terms of Le Châtelier's principle why sulfuric acid and phosphoric acid can be used to prepare $HCl(g)$ from sodium chloride, but nitric acid is not suitable for this purpose.

69. Each of the following groups of compounds has a common structural feature. Explain what it is in each case.

(a) P_4, P_4O_6, and P_4O_{10}

(b) $(HPO_3)_3$ and $(SO_3)_3$

(c) $H_4P_2O_7$, $H_2S_2O_7$, and $(SO_3)_n$

(d) PO_4^{3-}, HPO_3^{2-}, PH_4^+, SO_4^{2-}, and $S_2O_3^{2-}$

70. Draw orbital box diagrams for the ground state electron configurations of arsenic and selenium. Predict the formulas of the hydrides of these elements. Arsenic forms an unstable chloride of molecular formula $AsCl_5$, and selenium forms the fluorides SeF_4 and SeF_6. Give the orbital box diagrams for the valence states of these elements on which these higher-valence compounds are based.

***71.** Phosphine may be prepared by heating white phosphorus with aqueous sodium hydroxide:

$$P_4(s) + 3NaOH(aq) + 3H_2O \longrightarrow PH_3(g) + 3NaH_2PO_2(aq)$$

(a) Give the formula of the acid from which the salt NaH_2PO_2 could be obtained.

(b) Name this acid.

(c) The acid is monoprotic. Draw its Lewis structure.

(d) Assign oxidation numbers to phosphorus in each of its compounds in the above equation.

(e) In the reaction, which species is oxidized and which is reduced?

(f) Suggest why ammonia cannot be prepared from nitrogen by a similar reaction.

(g) What is the minimum mass of phosphorus in grams that would be needed to give 100 mL of phosphine gas at 25 °C and 0.95 atm pressure?

72. A sample of coal contains 5.00% sulfur by mass. When burned in a power plant it will give sulfur dioxide in the stack gas, which will be converted in the atmosphere to sulfuric acid and deposited as "acid rain". Write the simplest equations to show how sulfur dioxide is converted to sulfuric acid in the atmosphere. Calculate the mass of sulfuric acid that could result from burning 1000 kg of this coal. How much $Ca(OH)_2$ would be required to neutralize this sulfuric acid?

73. A yellow solid A was heated in air and gave a colorless gas B, which was dissolved in water to give a colorless solution C. Another sample of the yellow solid was mixed with iron powder and heated strongly to produce a black solid D. When dilute sulfuric acid was added to D, a colorless gas E was formed. When E was bubbled through solution C, a finely divided yellow precipitate was formed, which was found to be identical to A in all its chemical and physical properties. Identify A, B, C, D, and E, and write balanced equations for all the above reactions.

*74. An element X is a reactive white solid, insoluble in water but soluble in organic solvents, such as carbon disulfide, from which it can be recovered by evaporating the solvent. It melts at 44 °C and boils at 280 °C. X as vapor has a density of 2.242 g L^{-1} at 400 °C and 1 atm pressure. 25.0 mL of X(g) react completely with 150 mL of chlorine at a given temperature and pressure to give 100 mL of a gaseous chloride at the same temperature and pressure. This chloride contains 77.4% Cl by mass, boils below 100 °C, and is a liquid at room temperature. Its molar mass was found to be 137.3 g mol^{-1}. Identify X. Draw its Lewis structure, and that of its chloride. Write the balanced equation for the reaction of X(g) with $Cl_2(g)$.

75. Excess sulfur was heated with 4.00 g of red phosphorus to give a compound X, which after removal of unreacted sulfur had a mass of 14.35 g. Analysis showed that X con-

tained only phosphorus and sulfur. Deduce the empirical formula of X and suggest a molecular structure.

*76. A compound X contains phosphorus, sulfur, and chlorine. A 0.9209-g sample of X reacted completely with water to give phosphoric acid, hydrogen sulfide, and chloride ion. The aqueous solution was made up to exactly 250 mL. 25.00 mL of this solution reacted with 32.47 mL of $0.0500M$ $AgNO_3(aq)$ to precipitate all the chloride ion as AgCl(s). In another experiment, 0.4437 g of X was dissolved in water, and all the phosphate ion precipitated as magnesium ammonium phosphate, $MgNH_4PO_4(s)$, which when heated strongly to constant mass gave 0.2932 g of magnesium diphosphate, $Mg_2P_2O_7(s)$. Finally, in a molar mass determination, 0.6775 g of X gave a pressure of 446.5 mm Hg when vaporized in a 300-mL flask at 250 °C. Deduce the empirical and molecular formulas of X, and draw its Lewis structure.

77. (a) Why is sulfuric acid referred to as a diprotic acid, and phosphoric acid as a triprotic acid?

(b) Why is sulfuric acid a strong acid and phosphoric acid a weak acid in aqueous solution?

(c) What are the products of dehydration of sulfuric acid and phosphoric acid, respectively?

78. You are provided with test tubes containing each of the following substances, all unlabeled. Explain how each could be identified, using only their physical properties, or such tests that could be performed with a bunsen burner, wooden splints, water, and litmus paper:

(a) Hydrogen   **(b)** Oxygen

(c) Helium   **(d)** Chlorine

(e) Sulfur dioxide   **(f)** Carbon dioxide

(g) Ammonia   **(h)** Hydrogen chloride

Molecular Geometry

9

The shape or geometry of the molecules of a substance is often of fundamental importance in determining some of its properties. The ability of enzymes, the catalysts in reactions in living systems, to increase the rates of reactions thousands or even millions of times is intimately related to their shape (see Chapter 24). It is believed too that differences in the smell of various substances is also related to the shapes of their molecules. An interesting example of the importance of molecular shape is provided by the molecule retinal which, like many other molecules, can adopt two or more different shapes. Retinal, which is found in the retina of the eye, has two shapes known as *cis* and *trans* (see Chapter 23). When a molecule of retinal absorbs a quantum of light it changes from the *cis* form to the *trans* form. This change of shape causes a signal to be sent from the retina along the optic nerve to the brain.

cis-Retinal   trans-Retinal

The shapes of molecules are determined by the number and geometric arrangement of the bonds formed by each atom in a molecule. The properties of solids with covalent network structures similarly depend on the number and geometric arrangement of the bonds around each atom. We only have to recall the differences in the properties of allotropes, such as white and red phosphorus, orthorhombic and plastic sulfur, and diamond and graphite to appreciate how important the geometric arrangement of the atoms and bonds in a substance is in determining its properties.

Although chemists realized as long ago as the middle of the past century that atoms in molecules must have definite arrangements in space, they had no direct methods for determining the geometric arrangement of the atoms. Nevertheless by studying the properties of substances and their reactions and by using some ingenious arguments, they were able to deduce much information about the geometric arrangement of atoms in molecules. They concluded that the four single bonds formed by a carbon atom must have a tetrahedral arrangement rather than, for example, a square arrangement. Today chemists have many different methods for studying the structures of molecules and we have detailed information about the lengths of all the bonds and the angles between the bonds for many thousands of molecules (Box 9.1).

We have seen in Chapter 4 that we can predict the general shapes of a large number of molecules simply on the basis of the Lewis structure for the molecule by using the VSEPR model—that electron pairs in the valence shell of an atom adopt the arrangement that keeps them as far apart as possible. In that chapter we restricted our attention to molecules in which the atoms had no more than four electron pairs in their valence shell. But we have seen in Chapter 8 that in their higher oxidation states phosphorus and sulfur form many molecules in which they have five or six electron pairs in their valence shells. So we must extend the VSEPR model to cover these and similar cases. We have also seen that the bond angle in the water molecule is 104.5° whereas in Chapter 4, using the VSEPR model, we predicted an angle of 109°. Although this is a small difference, we will show that the VSEPR model can account in a qualitative way for this and similar small differences from the predicted ideal angles.

In this type of formula, which is often used for organic molecules, the carbon and hydrogen atoms are omitted and only the C—C bonds are shown. Thus

$$\begin{array}{c} CH_3 \\ CH_3 \end{array}\!\!\!\!\!>\!\!C=\!CH_2-\!CH=\!CH_2$$

would be represented as

BOX 9.1

THE EXPERIMENTAL DETERMINATION OF MOLECULAR GEOMETRY

Until the present century chemists had no direct methods for determining the shapes of molecules. Nevertheless they were able to reach some important conclusions concerning the arrangements of atoms in space by means of arguments such as the following.

If we were to successively replace the hydrogen atoms in methane by halogen atoms, we could make the molecules CH_3F, CH_2FCl, and $CHFClBr$. Suppose that these molecules have a *square planar shape*; we predict that the following different molecules would exist:

CH₃F   CH₂FCl

CHFClBr

On the basis of a square planar shape, we expect only one type of molecule with the formula CH_3F but two molecules with the formula CH_2FCl and three molecules with the formula $CHFClBr$. In fact, there is only one substance with the formula CH_3F and only one with the formula CH_2FCl, but there are two substances with the formula $CHFClBr$. Evidence like this led chemists to conclude that the four bonds formed by a carbon atom do not have a planar arrangement but, rather, a tetrahedral arrangement, because this arrangement leads to the prediction of the correct number of compounds with the same formula (*isomers*):

CH₃F   CH₂FCl   CHFClBr

The two isomers of CHFClBr differ in a rather subtle way; they are like right- and left-handed gloves. Right- and left-handed molecules, called *chiral molecules*, can be recognized by the fact that they rotate the plane of polarization of light in opposite directions.

Today we have available several methods by which the distances between the atoms in a molecule or in a network solid can be measured and thus the geometry of the molecule or the structure of the solid completely defined. These methods are based on the interaction of different kinds of electromagnetic radiation with atoms and molecules. The most important and generally useful of these methods is the diffraction of X rays by crystalline solids (See Chapter 11). Other methods, which are particularly applicable to gases, are the diffraction of a beam of electrons (electron diffraction) and the absorption of infrared and microwave radiation by molecules (infrared and microwave spectroscopy).

By means of these different experimental methods it is often possible to determine distances between the atoms in a molecule or in a crystalline solid to an accuracy of 0.1 pm, or better, and the angles between bonds (bond angles) to 0.1°. In fact the atoms in a molecule are not rigidly fixed in position but each atom is vibrating around its average position. Thus it is the distances between these mean positions that are measured. The frequencies of the vibrations can also be measured, for example by infrared spectroscopy, and these frequencies give us information about bond strengths—strong bonds have high vibrational frequencies and weak bonds have low vibrational frequencies.

That we have been able to obtain so much detailed information about the structure of something so incredibly small as a molecule is one of the most striking features of modern chemistry. This detailed knowledge has enabled us greatly to improve our understanding of the properties of substances, and throughout this book we give many examples of the dependence of the properties of substances on their molecular and solid state structures.

In Chapter 7 we went beyond Lewis structures and showed how we can understand the covalent bond in terms of the quantum mechanical description of atoms and molecules. In that chapter we considered only diatomic molecules. Here we will discuss how we can describe the bonds in polyatomic molecules.

Because Lewis structures are the starting point for the prediction of molecular geometry and because they are the simplest and commonest method of illustrating the manner in which electrons are arranged in molecules and ions, it is essential to be able to draw them correctly. So we will first look at some rules which will enable us to draw Lewis structures correctly even in apparently difficult cases.

9.1
Lewis Structures

In Chapter 4 we saw how to draw Lewis structures for simple molecules. Knowing the valences of the elements and taking account of the octet rule, when it is valid, we can often arrive at a correct Lewis structure quite easily. However, some cases are more difficult, and it is for dealing with such cases that we now describe a general set of rules for drawing Lewis structures. With practice you may not always need to work systematically through these rules, but they are useful when you are uncertain how to proceed. The rules are summarized below:

1. Before any Lewis structure can be drawn, we must know how the atoms are connected in the molecule. This information can only be obtained with complete certainty by experiment. If we do not know the arrangement of the atoms, often we can deduce a probable arrangement of the atoms as follows: (a) The structure assumed must be consistent with the known valences of the elements; (b) in a molecule with the formula AX_n the unique atom A is usually the central atom to which the ligands X are attached; (c) atom A is also normally the least electronegative atom or the atom with the highest valence. So the first step is to draw a diagram of the molecule or polyatomic ion, showing the atoms connected by single bonds.

2. Add the number of valence electrons on each atom in the molecule to find the total number of valence electrons. If the molecule is charged—that is, if it is a polyatomic ion—add one electron for each negative charge or subtract one electron for each positive charge.

3. Subtract the number of electrons needed to form the single bonds between the atoms in the diagram that you drew in rule 1 from the total number of electrons and use the remainder to complete octets around each atom except hydrogen. If there are insufficient electrons to complete all the octets, complete those of the more electronegative atoms first. If there are more than enough electrons to complete all the octets, add the remaining electrons in pairs to the central atom which must be from period 3 or a later period. Then assign formal charges.

4. If any atom still has an incomplete octet, convert nonbonding electron pairs to bonding pairs. In other words, use nonbonding pairs to form double or triple bonds until each atom has an octet. Then reassign formal charges.

5. If rule 4 creates additional formal charges, use the structure given by rule 3.

6. If the central atom is from period 3 or a later period, the octet rule may not apply. Form additional multiple bonds in order to remove as many formal charges as possible.

To illustrate the application of the rules, we now use them to deduce the Lewis structures of PCl_3 and the carbonate ion CO_3^{2-}. The arrangement of the atoms in PCl_3 is known to be

$$Cl\text{—}P\text{—}Cl$$
$$\overset{|}{Cl}$$

1. We could in any case have deduced this arrangement because it is the only one consistent with a valence of 3 for P and 1 for Cl and because P is the least electronegative of the atoms in the molecule and it is the unique atom.

2. The total number of electrons in the valence shell of an atom is given by its electron configuration or simply by its position in the periodic table:

Phosphorus (group 5)	$5e^-$	1P	$5e^-$
Chlorine (group 7)	$7e^-$	3Cl	$21e^-$
			$\overline{26e^-}$

3. The three single bonds in the structure use up 6 of these electrons. Thus $26 - 6 = 20$ electrons remain to be allocated. To complete their octets, 6 electrons are added to each Cl and 2 to the P atom.

$$:\overset{\cdot\cdot}{\underset{\cdot\cdot}{Cl}}\text{—}\overset{\cdot\cdot}{P}\text{—}\overset{\cdot\cdot}{\underset{\cdot\cdot}{Cl}}:$$
$$\overset{|}{\underset{\cdot\cdot}{:Cl:}}$$

This structure uses up all 20 electrons. Each atom has a zero formal charge.

4. All atoms have an octet, so no further adjustment is necessary. Rules 5 and 6 are not applicable in this case.

Now we deduce the Lewis structure for CO_3^{2-}.

1. The arrangement of the atoms in CO_3^{2-} is

$$O$$
$$\overset{|}{O\text{—}C\text{—}O}$$

2. The total number of electrons is found as follows:

Carbon (group 4)	$4e^-$	1C	$4e^-$
Oxygen (group 6)	$6e^-$	3O	$18e^-$
Negative charges			$2e^-$
			$\overline{24e^-}$

3. Six of these electrons are used for the three single bonds, so $24 - 6 = 18$ electrons remain to be distributed. Three pairs of electrons can be added to each O (the most electronegative atom) to complete an octet on each one. Thus we obtain

$$:\overset{\cdot\cdot}{O}:^{\ominus}$$
$$^{\ominus}:\overset{\cdot\cdot}{\underset{\cdot\cdot}{O}}\text{—}\underset{\oplus}{C}\text{—}\overset{\cdot\cdot}{\underset{\cdot\cdot}{O}}:^{\ominus}$$

This uses up all 18 electrons. Formal charges are then assigned. Each O atom (core charge $+ 6$) is assigned $6 + 1 = 7$ electrons and therefore has a -1 formal charge. The C atom (core charge $+4$) is assigned 3 electrons and therefore has a $+1$ formal charge.

4. In this structure C does not have an octet. One of the nonbonding electron pairs is therefore moved from an O to form a double bond with C, thus removing the formal charges on these atoms.

$$:\ddot{O}:^{\ominus} \qquad\qquad :\ddot{O}:^{\ominus}$$
$$^{\ominus}:\ddot{O}-\underset{\oplus}{C}-\ddot{O}:^{\ominus} \longrightarrow :O=C-\ddot{O}:^{\ominus}$$

Rules 5 and 6 are not applicable in this case.

═ Example 9.1 ═

DRAWING LEWIS STRUCTURES

Draw the Lewis structure for the nitrate ion, NO_3^-.

Solution

1. The arrangement of the atoms is known to be

$$\begin{array}{c} O \\ | \\ O-N-O \end{array}$$

   It would in any case have been a plausible assumption to put the less electronegative and unique atom N in the center of the three O atoms.

2. The total number of electrons is found as follows:

Nitrogen (group 5)	$5e^-$	1N	$5e^-$
Oxygen (group 6)	$6e^-$	3O	$18e^-$
One negative charge			$1e^-$
			$24e^-$

3. Six of these electrons are used for the three single bonds, so $24 - 6 = 18$ remain to be distributed. Three pairs can be added to each O to give

$$:\ddot{O}:^{\ominus}$$
$$^{\ominus}:\ddot{O}-\underset{\textcircled{+2}}{N}-\ddot{O}:^{\ominus}$$

   Each O has a -1 formal charge, and the N (core charge $+5$) has a $+2$ formal charge.

4. In this structure the N atom does not have an octet; therefore one of the nonbonding electron pairs on O is moved to form a double bond to N:

$$:\ddot{O}:^{\ominus} \qquad\qquad :\ddot{O}:^{\ominus}$$
$$^{\ominus}:\ddot{O}-\underset{\textcircled{+2}}{N}-\ddot{O}:^{\ominus} \longrightarrow :\ddot{O}=\underset{\oplus}{N}-\ddot{O}:^{\ominus}$$

   Rules 5 and 6 are not applicable in this case.

If we had not known the structure of NO_3^-, we might have arranged the atoms in the following way: O—N—O—O. If we follow the rules, we arrive at the Lewis structure:

$$:\ddot{O}=\ddot{N}-\ddot{O}-\ddot{O}:^{\ominus}$$

This is a correct Lewis structure and is, in fact, a known ion called peroxonitrite. It is not, however, the nitrate ion. There are two ions with the formula NO_3^-: the nitrate ion and the peroxonitrite ion, each with a corresponding Lewis structure. This example emphasizes the need to know how the atoms are arranged before we can draw a correct Lewis structure for a molecule or ion.

For atoms that can have more than eight electrons in their valence shell, we need to use rule 6. As an example, let us consider the sulfate ion, SO_4^{2-}.

1. The S atom is in the middle of four O atoms:

$$
\begin{array}{c}
\text{O} \\
| \\
\text{O—S—O} \\
| \\
\text{O}
\end{array}
$$

Note that S is the less electronegative atom and the unique atom.

2. The total number of electrons is found as follows:

Sulfur (group 6)	$6e^-$	1S	$6e^-$
Oxygen (group 6)	$6e^-$	4O	$24e^-$
Negative charges			$2e^-$
			$32e^-$

3. The four bonds use up 8 of these electrons, leaving 24 to distribute. These 24 can be allocated 6 to each O to complete the octet on each atom:

$$
\begin{array}{c}
:\ddot{\text{O}}:^{\ominus} \\
| \\
{}^{\ominus}:\ddot{\text{O}}\text{—S}^{(2+)}\ddot{\text{O}}:^{\ominus} \\
| \\
:\ddot{\text{O}}:_{\ominus}
\end{array}
$$

The S atom has a formal charge of $+2$, and each O atom has a formal charge of -1.

4. Each atom has an octet, so no more bonds need to be formed.

5. Rule 5 is not applicable.

6. The structure in step 3 is a correct Lewis structure, but we can draw a better structure by applying rule 6, which recognizes the fact that S is not restricted to eight electrons in its valence shell. The formal charges on the S atom and two of the O atoms can be removed by forming two double bonds. This step does not change the number of electrons in the valence shell of each of the O atoms, but it increases the number in the valence shell of S to 12. We then obtain

$$
\begin{array}{c}
:\ddot{\text{O}} \\
\| \\
:\ddot{\text{O}}\text{=S—}\ddot{\text{O}}:^{\ominus} \\
| \\
:\ddot{\text{O}}:^{\ominus}
\end{array}
$$

This is the structure we gave in Chapter 8.

In general, when two or more Lewis structures can be written, the structure with the fewest formal charges is the preferred structure. Removing formal charges moves electrons from negatively charged atoms toward positively charged atoms, thereby decreasing the energy of the structure. (See Box 9.2)

═ *Example 9.2* ═══════════════════════════════════

DRAWING LEWIS STRUCTURES

Draw the Lewis structure of SO_2.

B O X   9 . 2

MORE ON LEWIS STRUCTURES

Some of the Lewis structures that we described in Chapter 8 and in this chapter may differ from those that you learned in another chemistry course in high school or elsewhere. For example, we have given the structures

whereas you may have learned the structures

for these molecules. So you may be wondering if the structures that you learned previously are wrong.

They are not wrong, but they are not the best structures that can be given for these molecules. It is important to remember that Lewis structures are only very approximate representations of the electron distributions in molecules. This electron distribution is a continuous charge cloud in which individual electrons cannot be recognized. Determining this electron distribution experimentally is very difficult, although it has been done in a few cases by X ray crystallography. In relatively simple molecules the electron distribution can be calculated with some accuracy on the basis of better models for the electron distribution than are provided by Lewis structures. However, we cannot discuss these calculations at the level of this book. We have to make do with Lewis structures, which are, in any case, widely used by chemists because of their simplicity. We must be aware of their limitations, however. A continuous electron distribution is only very crudely represented by dots and lines representing bonding and nonbonding electron pairs.

The structures originally proposed by Lewis have subsequently been modified in two important ways to make them more consistent with experimental observations, particularly bond lengths.

First, to account for the fact, for example, that the three CO bonds in the carbonate ion, CO_3^{2-}, all have the same length, the concept of *resonance* has been introduced. Resonance is a way of allowing for the fact that in many molecules not all electrons can be regarded as strictly localized nonbonding pairs associated with only one atom or as bonding pairs associated with just

two atoms. Some electrons may behave as both bonding and nonbonding pairs, and some bonding pairs may be delocalized over more than two atoms.

Second, to account for the existence of molecules such as PCl_5 and SF_6 and for the short SO bonds in SO_2, SO_3, and SO_4^{2-}, we recognized that the octet rule is not necessarily valid for the elements of period 3 and beyond. These elements may have more than 8 electrons in their valence shell in some of their compounds, because the $n = 3$ shell can accommodate a maximum of 18 electrons. The structures that you may have learned previously, such as IV, V, and VI, are the structures that Lewis wrote for these molecules and ions and they are therefore correct Lewis structures. Indeed, we deduce their structures following our rules 1 to 4, but these structures have subsequently been modified so that they are in better agreement with experimental data, such as bond lengths, and because it has been recognized that elements such as sulfur do not need to obey the octet rule in all their compounds. We allow for this modification in rule 6.

Thus the structure III and the equivalent resonance structures are preferred to the structure VI because the second structure has many formal charges and is not consistent with the length of the SO bonds. Both structures are only a very crude representation of the electron distribution in these molecules, but most chemists now consider that the double bond structures for SO_2, SO_3, and SO_4^{2-}, for example, are better representations of the electron distributions than Lewis's original structures.

If we wish, we can go one step further and say that SO_4^{2-}, for example, is best represented by a combination of all the possible Lewis structures, giving more weight to those structures that we consider to be of greatest importance—that is, the structures of lower energy. For SO_4^{2-} we would then have the following resonance structures, in order of decreasing importance.

| 6 equivalent structures | 4 equivalent structures | 1 structure |

However, this begins to get too complicated to be worthwhile; we are trying to push our simple model beyond its limits. We will normally just use the single structure or the equivalent resonance structures that most nearly represent the electron distribution as far as we can tell from the available experimental information.

Solution

1. The arrangement of atoms is O—S—O.

2. There are 18 electrons and two bonds. Therefore 14 electrons are available to complete the octets.

3. Because oxygen is more electronegative than sulfur we complete the octets on oxygen first. This uses up six pairs of electrons. The remaining pair is placed on sulfur which still has only three pairs

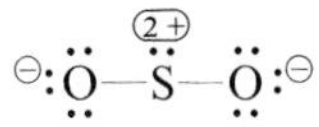

$$\overset{\ominus}{:}\!\overset{..}{\underset{..}{O}}\!\!-\!\!\overset{\overset{\textcircled{\scriptsize 2+}}{..}}{S}\!\!-\!\!\overset{..}{\underset{..}{O}}\!\overset{\ominus}{:}$$

4. Forming a double bond completes the octet on sulfur

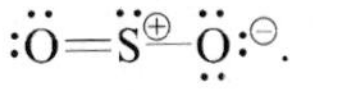

$$:\!\overset{..}{O}\!\!=\!\!\overset{..}{S}{}^{\oplus}\!\overset{..}{\underset{..}{O}}\!\overset{\ominus}{:}.$$

5. Rule 5 is not applicable.

6. The structure in step 4 is a correct Lewis structure. But since S is a third-period element, we can eliminate the formal charges by forming another double bond, thereby giving S a valence shell of ten electrons. The preferred Lewis structure is therefore

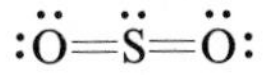

$$:\!\overset{..}{O}\!\!=\!\!\overset{..}{S}\!\!=\!\!\overset{..}{O}:$$

Note that the purpose of a Lewis structure is to show how the electrons are arranged in bonding pairs and nonbonding pairs in an ion or a molecule. Although the shape of a molecule can be predicted from the Lewis structures, the Lewis structure need not show the shape. Thus the Lewis structure of SO_2, which is an angular AX_2E molecule, may be correctly drawn as

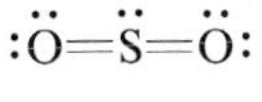

$$:\!\overset{..}{O}\!\!=\!\!\overset{..}{S}\!\!=\!\!\overset{..}{O}:\quad \text{or}\quad$$

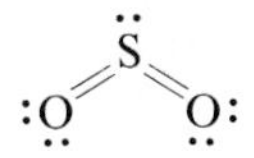

The Lewis structure of a more complicated molecule is often most easily drawn without attempting to show the shape. Thus we draw the Lewis structure of $SO_4{}^{2-}$ as follows:

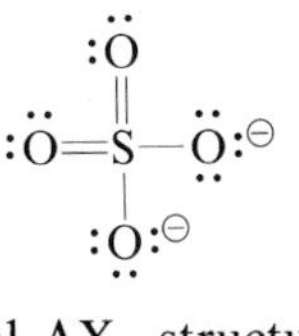

The sulfate ion has a tetrahedral AX_4 structure and we can show this more conveniently by means of a separate diagram which emphasizes the shape but does not attempt to show all the electrons:

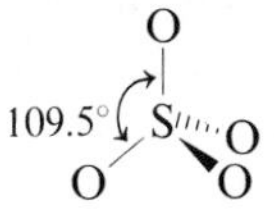

We saw in Chapter 4 that boron which is in group 3 and has only three electrons in its valence shell ($2s^2 2p^1$) does not have an octet of electrons in compounds such as BF_3. So let us see how our rules enable us to obtain the Lewis structure of such a molecule.

1. In BCl_3 the boron atom is in the center of the three chlorine atoms:

$$
\begin{array}{c}
Cl \\
| \\
Cl-B-Cl
\end{array}
$$

2. There are a total of $3 + 21 = 24$ electrons.
3. Six electrons are used for the three bonds, therefore 18 remain. Six can be added to each chlorine atom to complete their octets, thus using up all 18 electrons.

$$
\begin{array}{c}
:\ddot{Cl}: \\
| \\
:\ddot{Cl}-B-\ddot{Cl}:
\end{array}
$$

This structure leaves the boron atom without an octet.
4. Following rule 4 we would write the structure

$$
\begin{array}{c}
:\ddot{Cl}: \\
| \\
{}^{\oplus}:\ddot{Cl}=B{}^{\ominus}-\ddot{Cl}:
\end{array}
$$

However, we find that we have created formal charges. This structure is expected to be less stable than that in which all the atoms have a zero formal charge, particularly since there is a positive charge on the very electronegative chlorine atom and a negative charge on the much less electronegative boron. We therefore apply rule 5 and return to the structure obtained in step 3.

When we draw Lewis structures, it is useful to be able to recognize the possible combinations of bonds, nonbonding pairs, and formal charges for common atoms. Because carbon, nitrogen, oxygen, and fluorine always obey the octet rule, they always have four electron pairs in their valence shell. So the only possibilities are

$$
\begin{array}{cccccc}
-\overset{|}{\underset{|}{C}}- & -\overset{..}{\underset{|}{C}}{}^{\ominus} & & & & \\
& & & & & \\
-\overset{\oplus}{\underset{|}{N}}- & -\overset{..}{\underset{|}{N}}- & -\overset{..}{\underset{|}{N}}:{}^{\ominus} & -\overset{..}{\underset{..}{N}}:{}^{\text{\textcircled{2}}\ominus} & :\overset{..}{\underset{..}{N}}:{}^{\text{\textcircled{3}}\ominus} & \\
-\overset{..}{\underset{|}{O}}{}^{\oplus} & -\overset{..}{\underset{|}{O}}: & -\overset{..}{\underset{..}{O}}:{}^{\ominus} & :\overset{..}{\underset{..}{O}}:{}^{\text{\textcircled{2}}\ominus} & & \\
& & -\overset{..}{\underset{..}{F}}: & :\overset{..}{\underset{..}{F}}:{}^{\ominus} & &
\end{array}
$$

When you have drawn a Lewis structure, particularly if you have not followed the above rules in detail, remember to check the following:

1. Hydrogen forms only one bond, —H.
2. Period 2 elements never exceed the octet but occasionally, as in some compounds of boron and aluminum, they may have fewer than eight electrons in the valence shell.

3. As many formal charges as possible have been removed by the formation of double and triple bonds, provided that the period 2 elements do not exceed the octet.

9.2
The VSEPR Model

In Chapter 7 we saw that the electrons in most molecules can be described as occupying either bonding orbitals or nonbonding orbitals. In accordance with the Pauli principle each orbital can accomodate two electrons, provided that they are of opposite spin. In other words only two electrons can have the same standing wave pattern and these electrons must have opposite spins. The square of the function, ψ, that describes the wave pattern or orbital gives the probability that the electron will be found at any particular point in the atom. It is convenient to think of an electron as a cloud of negative charge which is most dense where the electron is most probably to be found and less dense where it is less likely to be found. In other words the charge cloud is a time-average picture of the electron. Thus according to the localized orbital model there is a charge cloud corresponding to each of the bonds in the molecule and one corresponding to each of the nonbonding or lone pairs. Each charge cloud represents the distribution of the total electron density of two electrons of opposite spin. In fact localized orbitals, and therefore the corresponding charge clouds, are not fully localized; they overlap somewhat. Nevertheless, it is a convenient approximation to assume that the charge clouds occupy completely separate regions of space. We have in fact already made use of this approximation in the electron-pair sphere model which we introduced in discussing the VSEPR model in Chapter 4. We saw that by representing each pair of electrons by a sphere we could account for the valence shell electron pair arrangements that are the basis of the VSEPR theory.

The electron pair charge clouds share equally all the space available to them in the valence shell, so that the most probable location of the electron pairs is where they are as far apart as possible (Figure 9.1 on p. 438). Electron pairs behave as if they repel each other; this is a consequence not only of electrostatic repulsion but also of the operation of the Pauli exclusion principle. The arrangements of two to six electron pair charge clouds in the valence shell of an atom are summarized in Table 9.1. We have already discussed the shapes of molecules based on valence shells containing two to four electron pairs and we discuss five and six electron pairs later in this section. First we will see how we can use the VSEPR model to help us understand some of the finer details of molecular geometry.

NONEQUIVALENT ELECTRON PAIRS

It is clearly only a very rough approximation to assume, as we have done so far, that all the electron pair charge clouds in a valence shell are spherical and have the same size. In fact they differ in size and shape for at least three important reasons:

1. The charge clouds of nonbonding pairs are larger and take up more space in the valence shell of an atom than the charge clouds of bonding pairs.

FIGURE 9.1
Arrangements of Charge
Clouds for Two, Three, and
Four Electron Pairs.

(a) The electron pairs share the
valence shell, each occupying its
own segment and thus keeping
the regions of maximum charge
density as far apart as possible.
(b) These diagrams show the most
probable locations of the electron
pairs on the surface of a sphere
which represents the valence shell.

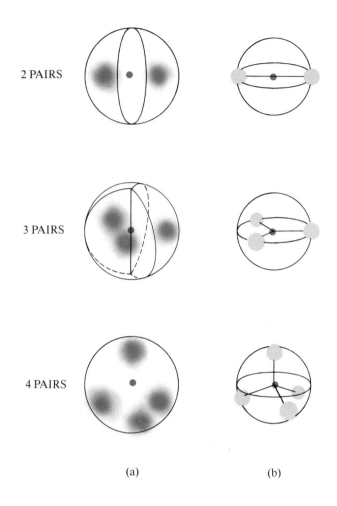

2 PAIRS

3 PAIRS

4 PAIRS

(a)   (b)

2. The space occupied by a bonding pair charge cloud in the valence shell of
    the central atom A increases with increasing electronegativity of the cen-
    tral atom A and decreases with increasing electronegativity of the ligand
    X.

3. The two electron pairs of a double bond have a combined charge cloud
    that is larger than the charge cloud of a single bond and the combined
    charge cloud of the three electron pairs of a triple bond is still larger.

BONDING AND NONBONDING PAIRS A lone pair charge cloud is pulled in
towards the core of the central atom and tends to spread out around the core.
In contrast, the charge cloud of a bonding pair is pulled away from the core
of the central atom by the attraction of the core of the ligand, and is shared between the valence shells of the central atom and the ligand. In short, a nonbonding electron pair charge cloud takes up more space in the valence shell of
an atom than does a bonding electon pair charge cloud (Figure 9.2).

The difference in the space occupied by nonbonding and bonding electron
pairs affects bond angles at any atom with one or more nonbonding electron
pairs in its valence shell. Four equivalent bonding pairs adopt a perfect tetrahedral arrangement, and all molecules of the type AX_4 with four identical
ligands X have bond angles of exactly 109.5°. However, because nonbonding

TABLE 9.1
Arrangements of Electron Pair Charge Clouds in a Valence Shell

Number of Electron Pairs	Arrangement	
2	Linear	
3	Equilateral triangular	
4	Tetrahedral	
5	Trigonal bipyramidal	
6	Octahedral	

electron pairs take up more space in the valence shell than bonding pairs, the bonding pairs in AX_3E and AX_2E_2 molecules are pushed closer together than those in an AX_4 molecule and the angles between the bonds in AX_3E and AX_2E_2 molecules are predicted to be smaller than the ideal angle of 109.5° for a tetrahedron (Figure 9.3). For example the bond angle in the NH_3 molecule is 107°. Further examples are given in Table 9.2.

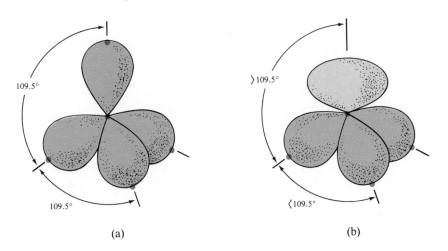

109.5°

109.5°

(a)

〉109.5°

〈109.5°

(b)

FIGURE 9.2
Bonding and Nonbonding Electron Pair Charge Clouds.

(a) Four equivalent bonding pair charge clouds. (b) One nonbonding (lone) pair charge cloud and three bonding pair charge clouds. The nonbonding pair charge cloud is a larger cloud and takes up more space in the valence shell than the bonding pair charge clouds.

FIGURE 9.3
Bond Angles in AX$_3$E and
AX$_2$E$_2$ Molecules.

Because a lone pair charge cloud
occupies more space in the valence
shell of A than a bond pair charge
cloud, bond angles in these molecules are smaller than in AX$_4$
molecules. (a) The bond angle in
NH$_3$ is 107°. (b) The bond angle
in H$_2$O is 104°.

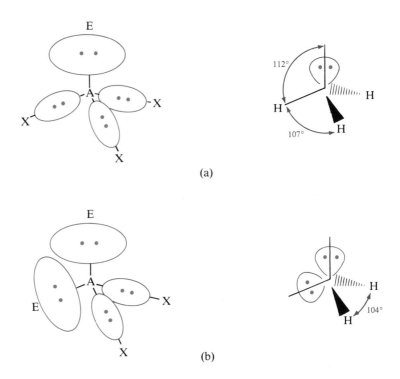

(a)

(b)

ELECTRONEGATIVITY Even in AX$_4$ molecules, if all the ligands X are not
identical, the bond angles deviate from the tetrahedral angle because the space
occupied by a bonding electron pair depends to some extent on the ligand. In
particular the space occupied by the charge cloud of a bonding electron pair
in the valence shell of the central atom A decreases with increasing electronegativity of a ligand X, because more of the charge cloud is pulled away from
the core of A towards the ligand X. Conversely the bonding electron pair charge
cloud occupies an increasing amount of space in the valence shell of A with

TABLE 9.2
Observed Bond Angles in Some
AX$_3$E and AX$_2$E$_2$ Molecules

AX$_3$E Molecules		AX$_2$E$_2$ Molecules	
NH$_3$	107°	H$_2$O	104°
NF$_3$	102°	F$_2$O	103°
PH$_3$	93°	H$_2$S	92°
PI$_3$	102°	SCl$_2$	100°
PBr$_3$	101°	SF$_2$	98°
PCl$_3$	100°		
PF$_3$	97°		

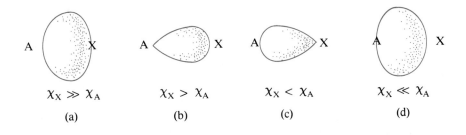

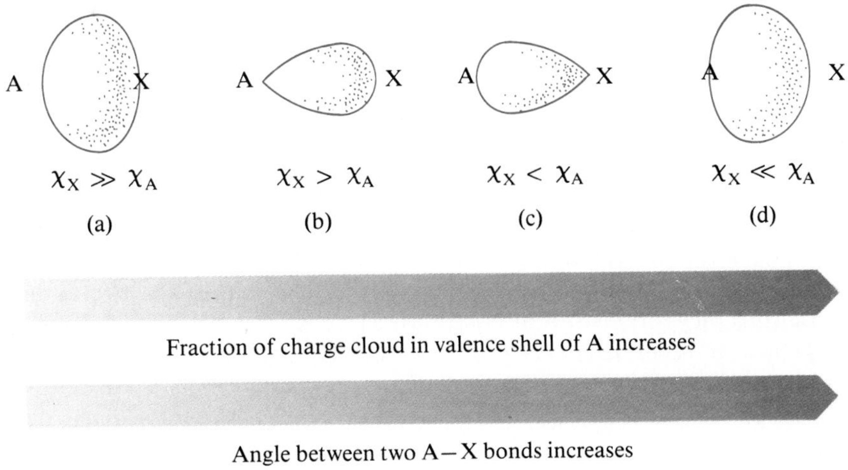

$\chi_X \gg \chi_A$ $\chi_X > \chi_A$ $\chi_X < \chi_A$ $\chi_X \ll \chi_A$

(a)   (b)   (c)   (d)

Fraction of charge cloud in valence shell of A increases

Angle between two A—X bonds increases

FIGURE 9.4
Effect of Electronegativity on Bond Angles.

The space occupied by a bonding electron pair charge cloud in the valence shell of A increases with increasing electronegativity of A and decreases with increasing electronegativity of X. (a) When the electronegativity of X is much greater than that of A the electron pair is a lone pair on X and the AX bond is ionic, A^+X^-. (b) With increasing electronegativity of A the electron pair becomes a bonding pair. Most of the charge cloud is still in the valence shell of X and rather little is in the valence shell of A, and the AX bond is polar $\overset{\delta+ \quad \delta-}{A-X}$. (c) With a further increase in the electronegativity of A the charge cloud occupies an increasing amount of space in the valence shell of A and the AX bond is polar $\overset{\delta- \quad \delta+}{A-X}$. (d) When the electronegativity of X is negligibly small the charge cloud is entirely in the valence shell of A and so becomes a lone pair on A, and the AX bond is ionic, A^-X^+. As the fraction of the charge cloud in the valence shell of A increases, so the angle between two AX bonds increases.

increasing electronegativity of A (Figure 9.4). Thus we expect bond angles between electronegative ligands to be smaller than those between less electronegative ligands. For example, as we see in Table 9.2 the bond angle decreases in the series PI_3, PBr_3, PCl_3, PF_3, with increasing electronegativity of the ligand. Moreover because nitrogen is more electronegative than phosphorus the bond angle in NF_3 (102°) is larger than that in PF_3 (97°). Further examples are given in Table 9.2 and Figure 9.5.

DOUBLE AND TRIPLE BONDS We have already briefly discussed the prediction of the shapes of molecules containing double and triple bonds in Chapter 4 but now we can look at these molecules in a little more detail in terms of the

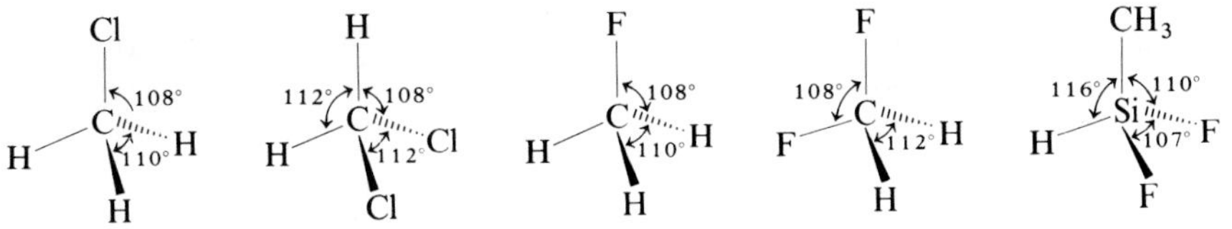

FIGURE 9.5
Effect of Electronegativity on Bond Angles.

Angles between the bonds to more electronegative ligands are smaller than those between the bonds to less electronegative ligands.

$$:\overset{..}{\underset{..}{O}}:^{\ominus}$$
$$\underset{\underset{..}{C}}{|}$$
$$:\overset{..}{O}\qquad\overset{..}{\underset{..}{O}}:^{\ominus}$$

$$CO_3{}^{2-}$$

$$:\overset{..}{O}$$
$$\|$$
$$:\overset{..}{O}=S-\overset{..}{\underset{..}{O}}:^{\ominus}$$
$$|$$
$$:\overset{..}{\underset{..}{O}}:_{\ominus}$$

$$SO_4{}^{2-}$$

$$:\overset{..}{\underset{}{O}}:^{\ominus}$$
$$|$$
$$:\overset{..}{O}=P-\overset{..}{\underset{..}{O}}:^{\ominus}$$
$$|$$
$$:\overset{..}{\underset{..}{O}}:_{\ominus}$$

$$PO_4{}^{3-}$$

$$^{\ominus}:\overset{..}{\underset{..}{O}}-\overset{..}{\underset{}{S}}=\overset{..}{\underset{..}{O}}:$$
$$|$$
$$:\overset{..}{\underset{..}{O}}:_{\ominus}$$

$$SO_3{}^{2-}$$

sizes and shapes of electron pair charge clouds. Consider the ethene molecule in which there is a double bond between the two carbon atoms:

Three of the four tetrahedrally arranged electron pairs around each carbon atom are shared between the two carbon atoms, thus forming three bent bonds which lie around the line joining the two carbon atoms (Figure 9.7).

If the charge clouds of the four electron pairs in the valence shell of each carbon atom have a tetrahedral arrangement, then the two pairs shared between the carbon atoms must lie one on each side of the line joining the two carbon atoms. Thus they form two bent bonds just as we see in a ball-and-stick model of the ethene molecule (Figure 9.6). The four C—H bonds then all lie in the same plane so that the ethene molecule is planar.

In the ethyne molecule, C_2H_2, there is a triple bond between the two carbon atoms:

Three of the four tetrahedrally arranged electron pairs around each carbon atom are shared between the two carbon atoms, thus forming three bent bonds which lie around the line joining the two carbon atoms (Figure 9.7).

We must be careful not to interpret our bent-bond model of the double bond as meaning that there are two quite distinct regions of electron density with a "hole" in the middle. Each stick in fact represents a large electron density cloud, and these overlap sufficiently that the *total* electron density has a maximum along the C—C direction. There is one large cloud of electron density in which the individual electron pairs cannot be distinguished. Similar considerations apply to the three electron pairs of a triple bond (see Figure 9.8). For many purposes therefore it is convenient to consider the two combined electron pair charge clouds of a double bond as one (two-electron-pair) charge cloud and similarly the three combined electron pair charge clouds of a triple bond as one (three-electron-pair) charge cloud. We can then predict the shapes of molecules containing double and triple bonds on the basis that the double bond and triple bond charge clouds are arranged at a maximum distance from the other single bond and lone pair charge clouds. This is in fact what we already have done in Chapter 4 when we briefly considered the application of the VSEPR model to molecules with multiple bonds. Thus to derive the arrangement of the ligands around any atom we simply count the number of charge clouds, whether they correspond to single bonds, double bonds, triple bonds, or lone pairs. This is equivalent to counting the total number of ligands X and lone pairs E. For example in ethene there are three charge clouds in the valence

FIGURE 9.6
The Ethene Molecule.

The arrangement of the electron pairs around each carbon atom is tetrahedral. Two pairs are shared between the two carbon atoms giving the molecule an overall planar shape. (a) Ball-and-stick model. (b) Charge cloud model. Each charge cloud is approximately represented by a sphere. (c) Balloon model.

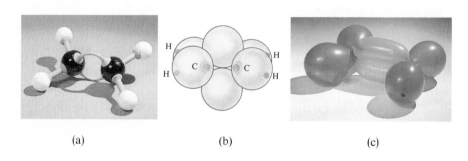

(a)   (b)   (c)

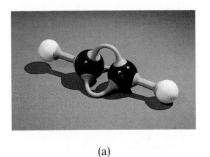

(a)

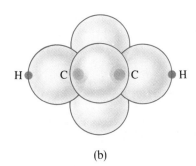

(b)

(c)

FIGURE 9.7
The Ethyne Molecule.

The arrangement of the electron pairs around each carbon atom is tetrahedral. Three pairs are shared between the two carbon atoms giving the molecule a linear shape. (a) Ball-and-stick model. (b) Charge cloud model. Each charge cloud is approximately represented by a sphere. (c) Balloon model.

shell of each carbon atom, two single bond charge clouds and one double bond charge cloud, and there are no lone pair charge clouds. Thus, each carbon atom is described as having an AX_3 geometry, which is planar triangular. Alternatively we see that there are three ligands X (two H and one C atom) and no lone pairs on each carbon atom, so again we conclude that the geometry around each carbon atom is triangular AX_3.

The Lewis structures for CO_2 and SO_2 are

$$\ddot{O}=C=\ddot{O}: \qquad :O \overset{\overset{\textstyle \ddot{S}}{\diagup \diagdown}}{} O:$$

Thus CO_2 is an AX_2 molecule and is therefore linear. SO_2 is an AX_2E molecule and is therefore angular with a bond angle close to $120°$ (see Figure 9.9 on p. 444).

An important class of ions containing multiple bonds are the anions such as carbonate, CO_3^{2-}, and sulfate, SO_4^{2-}, derived from oxoacids. We can derive their shapes in the manner just outlined. For example, from the Lewis structure for the carbonate ion we see that it is an AX_3 molecule. Accordingly, it has a planar triangular structure. The sulfate and phosphate ions are both AX_4 molecules with four ligands (two double and two single bonds in SO_4^{2-} and one double and three single bonds in PO_4^{3-}) and no unshared pairs of electrons.

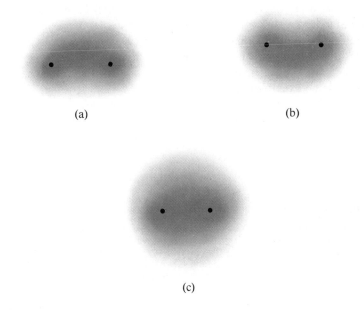

(a)

(b)

(c)

FIGURE 9.8
Charge Cloud Representation of a Double Bond.

The overlap of the electron densities of the charge clouds corresponding to the two individual bent bonds forms a large charge cloud. (a) This charge cloud corresponds to one of the bonds making up the double bond. (b) This charge cloud corresponds to the other half of the double bond. (c) The total electron density of the combined charge clouds. The maximum is along the C—C direction, and the individual electron pair densities cannot be distinguished.

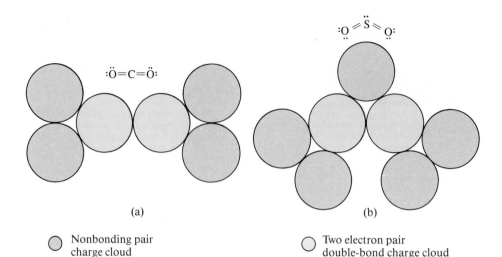

Nonbonding pair
charge cloud

Two electron pair
double-bond charge cloud

FIGURE 9.9
Multiple Bonds in the CO_2, SO_2, and HCN Molecules.

(a) CO_2—an AX_2 molecule. Since the carbon atom has two double bonds and no unshared electron pairs, CO_2 can be classified as an AX_2 molecule. (b) SO_2—an AX_2E molecule. Since the sulfur atom has two double bonds and an unshared electron pair, SO_2 can be classified as an AX_2E molecule. (c) HCN—an AX_2 molecule. Since HCN has a single bond and a triple bond on the central carbon atom, HCN can be classified as an AX_2 molecule.

They are therefore tetrahedral in shape. The sulfite ion, SO_3^{2-}, has three ligands and one unshared pair of electrons. Therefore it is an AX_3E molecule and has a triangular pyramidal shape like ammonia.

Table 9.3 summarizes the shapes of molecules and ions containing double and triple bonds.

Since a double bond charge cloud is larger than a single bond charge cloud, we expect deviations from ideal bond angles in all molecules having both single and double bonds. For molecules having an AX_3 geometry around a central atom, the ideal bond angles are 120°. We therefore expect angles involving a double bond to be somewhat greater than 120° and angles involving only single bonds to be somewhat less than 120°, as in ethene. For AX_4 molecules we expect angles involving double bonds to be somewhat larger than the tetrahedral angle of 109.5° and angles involving only single bonds to be somewhat smaller than 109.5°. Some examples are given in Figure 9.10.

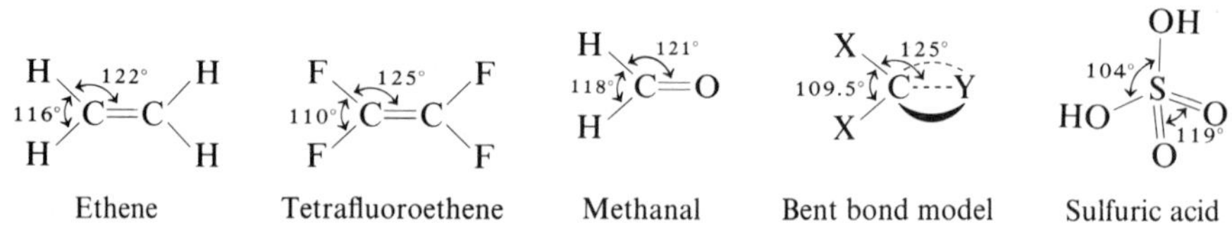

Ethene   Tetrafluoroethene   Methanal   Bent bond model   Sulfuric acid

FIGURE 9.10
Bond Angles in Some Molecules Containing Double Bonds.

The ideal bond angles in AX_3 and AX_4 molecules are 120° and 109.5°, respectively. Because the electron cloud of a double bond is larger than that of a single bond, angles involving double bonds are larger than those between single bonds. Therefore in AX_3 and AX_4 molecules involving double bonds, the angles involving double bonds are larger than the ideal angles, and those involving single bonds only are smaller than the ideal angles.

TABLE 9.3
Shapes of Molecules Containing Double and Triple Bonds

Total Number of Ligands and Nonbonding Pairs	Arrangement of Ligands and Nonbonding Pairs	Number of Ligands	Number of Nonbonding Pairs	Class of Molecule	Shape of Molecule	Examples
2	Linear	2	0	AX_2	Linear	$O=C=O$ $H-C\equiv N$
3	Triangular	3	0	AX_3	Triangular	
		2	1	AX_2E	Angular	
4	Tetrahedral	4	0	AX_4	Tetrahedral	
		3	1	AX_3E	Triangular pyramidal	
		2	2	AX_2E_2	Angular	

To predict the departure of the bond angle in a molecule such as SO_2 from the ideal angle of $120°$, we need to know whether a double bond charge cloud is larger or smaller than a lone pair charge cloud. Unfortunately there is no simple way to predict their relative sizes. The bond angle in SO_2 which has two double bonds and a lone pair is $119°$, which is very close to the ideal angle of $120°$ for an AX_2E molecule, so it appears that, in this case at least, the double bond charge cloud and the lone pair charge cloud are very nearly the same size.

= Example 9.3 =

SHAPES OF MOLECULES AND IONS WITH MULTIPLE BONDS

Classify the following molecules and ions as AX_2, AX_2E, AX_3, and so on, and describe their shapes:

$$SeO_3^{2-} \quad ClO_4^- \quad CS_2 \quad H_3PO_3$$

Selenium is in group 6.

Solution

	Lewis structure	Class of molecule	Shape
SeO_3^{2-}	$^{\ominus}\ddot{\text{O}}$—Se—$\ddot{\text{O}}:^{\ominus}$ with $\|$:Ö: below	AX_3E	Triangular pyramidal
ClO_4^-	:O: $\|$:Ö=Cl—Ö:$^{\ominus}$ $\|$:Ö	AX_4	Tetrahedral
CS_2	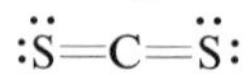 :S̈=C=S̈:	AX_2	Linear
H_3PO_3	H $\|$ HÖ—P=Ö: $\|$:OH	AX_4	Tetrahedral

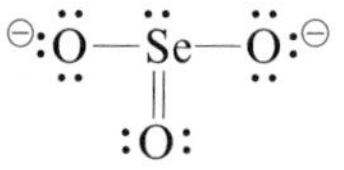

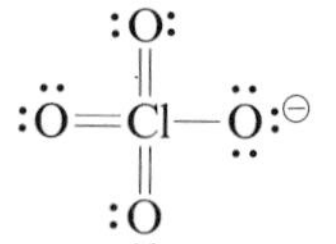

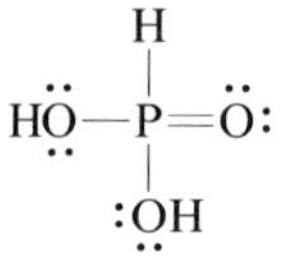

EXERCISE 9.2

Draw a Lewis structure for each of the following species. Classify each one in terms of the AX_nE_m nomenclature and describe its shape.

$H_2CO \quad HCO_3^- \quad HPO_4^{2-} \quad ClO_3^- \quad CS_3^{2-} \quad HCO_2H$

VALENCE SHELLS CONTAINING FIVE AND SIX ELECTRON PAIRS

In Chapter 8 we saw that in their high-oxidation-state compounds the period 3 elements do not obey the octet rule; they have more than four electron pairs in their valence shells. For example, in PCl_5 phosphorus has a valence shell containing five electron pairs and in SF_6 sulfur has a valence shell containing six electron pairs. The shapes of these and similar molecules are based on the arrangements of five and six electron pair charge clouds in a valence shell.

The arrangement that keeps five electron pairs as far apart as possible is the trigonal bipyramid, which consists of two triangular pyramids sharing a face (Figure 9.11). The trigonal bipyramid has six equivalent triangular faces, nine edges and *five vertices*. The PCl_5 molecule and almost all other AX_5 molecules have this shape, with a bonding electron pair occupying each vertex.

The arrangement that keeps six electron pairs as far apart as possible is the octahedron, which is a regular solid with *six equivalent vertices*, 12 equivalent edges and 8 equilateral triangular faces (Figure 9.12). The SF_6 molecule and all other AX_6 molecules have this shape, with a bonding electron pair occupying each vertex.

In PCl_5 and SF_6 all the electron pairs are bonding pairs. Several other molecular shapes arise when some of the electron pairs are nonbonding. Examples of such molecules are found among the fluorides of the halogens and xenon, such as ClF_3 and XeF_4. They are all covalent molecules in which the central atom is in a higher oxidation state which can be considered to arise from a valence state in which one or more of the electrons have been promoted to d orbitals (Figure 9.13 on p. 448). These compounds can all be prepared by direct reaction of the element with fluorine and we will discuss them in more detail in Chapter 20. Here we can illustrate the usefulness of the VSEPR model by using it to predict the shapes of these molecules (see Experiment 9.1, p. 449).

AX_6, AX_5E, AND AX_4E_2 MOLECULES AX_6 molecules such as SF_6 are octahedral (Figure 9.12). The six bonds all have the same length and they make an angle of 90° with their neighbors. Many hydrated ions such as $Al(H_2O)_6^{3+}$ and $Fe(H_2O)_6^{2+}$ also have this shape (see Chapter 10).

If one of the electron pairs of an octahedral arrangement is a nonbonding pair E, the molecule is of the AX_5E type and has a *square pyramidal shape*, as Figure 9.14 illustrates. The halogen pentafluorides, ClF_5, BrF_5, and IF_5, are examples of this type of molecule. Because the nonbonding electron pair charge cloud is larger than the bonding pair charge clouds the molecules are slightly

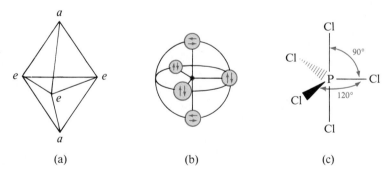

(a)   (b)   (c)

FIGURE 9.11
The Trigonal Bipyramidal Arrangement of Five Valence-Shell Electron Pairs.

(a) A trigonal bipyramid has six equivalent triangular faces, nine edges and five vertices. The two axial vertices *a* are not equivalent to the three equatorial vertices *e*. (b) The charge clouds of five electron pairs in a valence shell have a trigonal bipyramidal arrangement. This arrangement gives the largest possible distances between the maxima of each of the charge clouds. (c) PCl_5 is a trigonal bipyramidal AX_5 molecule. The angle between an axial and an equatorial bond is 90°. The angle between two equatorial bonds is 120°.

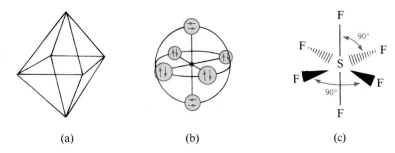

| | (a) | (b) | (c) |

FIGURE 9.12
The Octahedral Arrangement of Six Valence-Shell Electron Pairs.

(a) An octahedron has eight equilateral triangular faces, twelve edges and six equivalent vertices. (b) The charge clouds of six valence-shell electron pairs have an octahedral arrangement. This arrangement gives the largest possible distances between the maxima of each of the charge clouds. (c) SF_6 is an octahedral AX_6 molecule. The angle between any adjacent pair of bonds is 90°.

	3s	3p	3d		Oxidation states	Example compounds
P				Ground state	−3, +3	PCl_3, PH_3
				Valence state	+5	PCl_5, H_3PO_4
S				Ground state	−2, +2	SCl_2, H_2S
				Valence state	+4	SF_4, SO_2
				Valence state	+6	SF_6, SO_3, H_2SO_4
Cl				Ground state	−1, +1	ClF, HOCl
				Valence state	+3	ClF_3, $HClO_2$
				Valence state	+5	ClF_5, $HClO_3$
				Valence state	+7	$HClO_4$

	5s	5p	5d			
Xe				Ground state	0	—
				Valence state	+2	XeF_2
				Valence state	+4	XeF_4
				Valence state	+6	XeF_6, XeO_3
				Valence state	+8	XeO_4

FIGURE 9.13
Oxidation States and Ground State and Valence State Electron Configurations for Phosphorus, Sulfur, Chlorine, and Xenon.

Experiment 9.1

AX₆, AX₅, AX₄, and AX₃ Molecules: Balloon Models

If six balloons are tied together they automatically adopt an AX₆ octahedral arrangement.

If one of the balloons is popped, the remaining five adopt an AX₅ trigonal bipyramidal arrangement.

If another balloon is popped, the remaining four adopt an AX₄ tetrahedral arrangement.

If a third balloon is popped, the remaining three adopt an AX₃ triangular arrangement.

distorted from the ideal square pyramidal shape. All the bond angles are slightly smaller than the ideal angle of 90°, and the bonds in the square base are slightly longer than the axial bond. The large nonbonding pair pushes adjacent bonding pairs away from the core, thus increasing these bond lengths, but the nonbonding pair is too far away from the axial bond pair to affect it appreciably (Figure 9.14).

In an AX_4E_2 molecule such as XeF_4 the two nonbonding pairs are at opposite corners of the octahedron so that the molecule has a *square planar shape*. Because of their large size, the lone pair charge clouds keep as far apart as possible. They therefore occupy opposite (*trans*) positions of the octahedron rather than adjacent (*cis*) positions (Figure 9.15 on p. 450).

AX_5, AX_4E, AX_3E_2, AND AX_2E_3 MOLECULES The most probable arrangement of five electron pairs is at the corners of a trigonal bipyramid. An AX_5 molecule therefore has a trigonal bipyramidal shape. Phosphorus pentachloride and phosphorus pentafluoride are examples of trigonal bipyramidal AX_5 molecules (see Figure 9.11 on p. 447).

Of the five bonds, three are equivalent *equatorial* bonds in the same plane, and the other two are equivalent *axial* bonds. The equatorial bonds make angles

FIGURE 9.14
Square Pyramidal AX₅E Molecules.

Because the lone pair charge cloud occupies more space in the valence shell of A than the bond pair charge clouds the bond angles are smaller than the ideal angle of 90° and the bonds in the base of the square pyramid are longer than the bond to the apex, as illustrated for BrF₅.

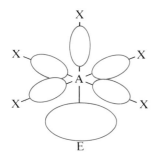 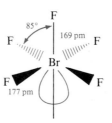

FIGURE 9.15
Square AX_4E_2 Molecules.

The two lone pair charge clouds can each take up more space if they occupy opposite vertices (*trans* positions) of the octahedron, rather than adjacent vertices (*cis* positions). Thus AX_4E_2 molecules have a regular square shape as shown for XeF_4.

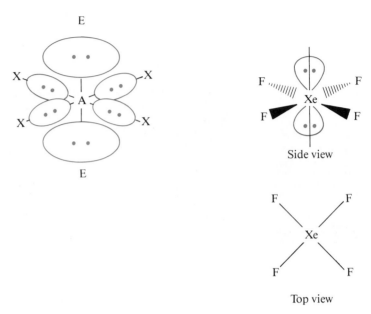

Side view

Top view

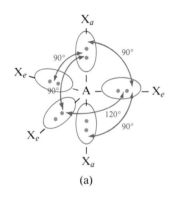

(a)

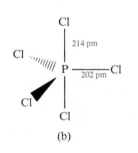

(b)

FIGURE 9.16
Bond Lengths in AX_5 Molecules.

The axial bond pairs have three neighboring pairs at 90° whereas the equatorial bond pairs have only two neighbors at 90°. Thus the axial pairs are more crowded than the equatorial pairs and are pushed further from the nucleus than the equatorial pairs, so that the axial bonds are longer than the equatorial bonds, as shown for PCl_5.

of 120° with each other, and the axial bonds are colinear and make angles of 90° with the equatorial bonds. The axial and equatorial bonds are not geometrically equivalent. The two axial bonds are longer than the three equatorial bonds, because the axial positions, which have three nearest neighbors at 90°, are more crowded than the equatorial positions, which have only two nearest neighbors at 90°. The axial pairs therefore tend to be forced further away from the central core than the equatorial pairs (Figure 9.16).

Because the equatorial positions are less crowded than the axial positions, there is a strong preference for the large lone pair charge clouds to occupy the equatorial rather than the axial positions of a trigonal bipyramidal arrangement. Thus in AX_4E, AX_3E_2, and AX_2E_3 molecules the nonbonding pairs of electrons always occupy the equatorial positions. As a result, an AX_4E molecule such as SF_4 has one lone pair in an equatorial position, two S—F equatorial bonds and two S—F axial bonds. The shape of the molecule which is determined by the positions of the sulfur and fluorine atoms is that of a disphenoid but it is often called a seesaw shape. The alternative shape of an AX_4E molecule with the lone pair in the less favorable axial position has never been observed.

An AX_3E_2 molecule such as ClF_3 is planar and is usually described as T-shaped. There are two lone pairs in equatorial positions, one Cl—F equatorial bond and two Cl—F axial bonds. The two possible alternative shapes for an AX_3E_2 molecule with either one or two lone pairs in the axial positions have not been found for any molecule of this type.

An AX_2E_3 molecule such as XeF_2 has all three equatorial positions occupied by lone pairs and it is therefore a linear molecule with two axial Xe—F bonds (Figure 9.17). We saw in Chapter 5 that, although iodine has a very small solubility in water, it dissolves in an aqueous solution of I^- to form the brown I_3^- ion. This ion has a linear AX_2E_3 geometry as shown in Example 9.4.

Because lone pair electron clouds are larger than bonding pair electron clouds, the observed bond angles in AX_4E and AX_3E_2 molecules are smaller than the 90° and 120° angles in the ideal shapes. The difference in the lengths of the axial and equatorial bonds that is observed whenever there are five electron pairs in the valence shell of the central atom is accentuated in the

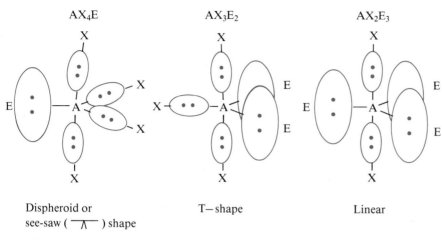

FIGURE 9.17
AX_4E, AX_3E_2, and AX_2E_3
Molecules.

Because lone pair charge clouds
are larger than bond pair charge
clouds they always occupy the
less crowded equatorial positions
rather than the axial positions.
The larger space occupied by the
lone pair charge clouds also causes
the bond angles in these molecules
to be smaller than the ideal angles
of 90° and 120°, as shown for SF_4
and ClF_3.

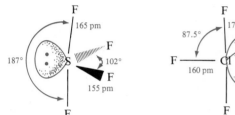

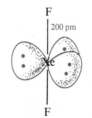

AX_3E_2 and AX_4E molecules because the nonbonding pair is closer to the axial
bonding pairs than to the equatorial pairs (Figure 9.17).

The three nonbonding pairs in AX_2E_3 molecules have a symmetrical arrangement, so they do not distort the predicted linear geometry, although they
cause the A—X bonds to be longer than they would be in the absence of the
unshared pairs.

═ *Example 9.4* ═══

SHAPES OF MOLECULES WITH FIVE AND SIX ELECTRON PAIRS

What are the expected shapes of each of the following species?

   SeF_6 AsF_5 BrF_5 BrF_3 I_3^-

Se is in group 6, and As is in group 5.

Solution

First we draw each of the Lewis structures, then express each in terms of the AX_nE_m
nomenclature, and hence deduce the shape:

SeF_6	AsF_5	BrF_5	BrF_3	I_3^-
AX_6	AX_5	AX_5E	AX_3E_2	AX_2E_3
Octahedron	Trigonal bipyramid	Square pyramid	T-shaped	Linear

EXERCISE 9.3

What are the expected shapes of each of the following species?

PCl_6^- SiF_5^- IF_3 BrF_4^- ICl_2^-

9.3
Resonance Structures

The Lewis structure of the planar triangular carbonate ion,

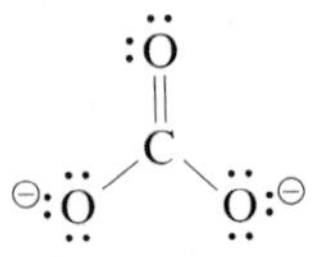

indicates that the central carbon atom forms one double bond and two single bonds to oxygen atoms. We saw in Chapter 4 that double bonds are shorter than single bonds, and in this chapter that the angles formed by double bonds are larger than those formed by single bonds. We therefore expect one of the bonds in this ion to be shorter than the other two and two of the angles to be larger than the third. However, determination of the structures of this ion by X ray crystallography has shown that all three bonds have the same length (131 pm) and that the bond angles are all exactly 120° (see structure at left).

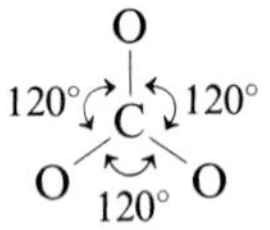

Similarly, the Lewis structures of the phosphate and sulfate ions have both single and double bonds. Yet again, all the bonds are found to have the same length and all the bond angles are equal; these ions have a regular tetrahedral shape:

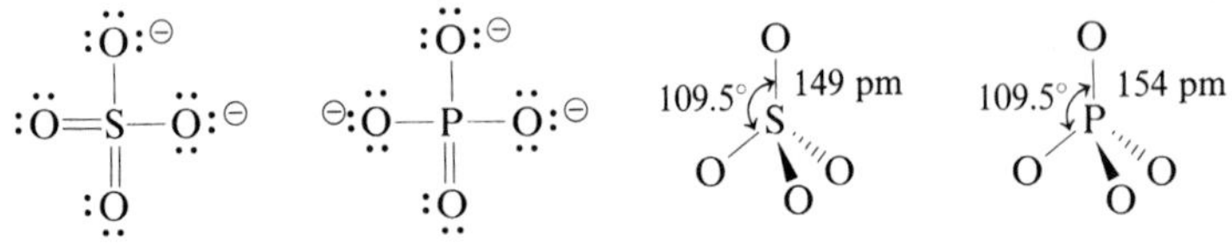

Carbonate ion

DELOCALIZED ELECTRONS

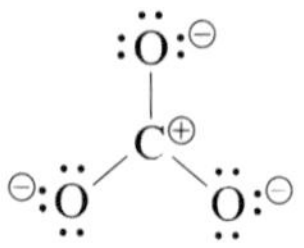

Clearly, the Lewis structures do not give a completely accurate description of the arrangement of the electrons in these ions. Let us see why they do not by returning to the derivation of the Lewis structure of the carbonate ion. Rule 3 leads to the structure shown to the left. There are then three ways to complete the octet on carbon, following rule 4:

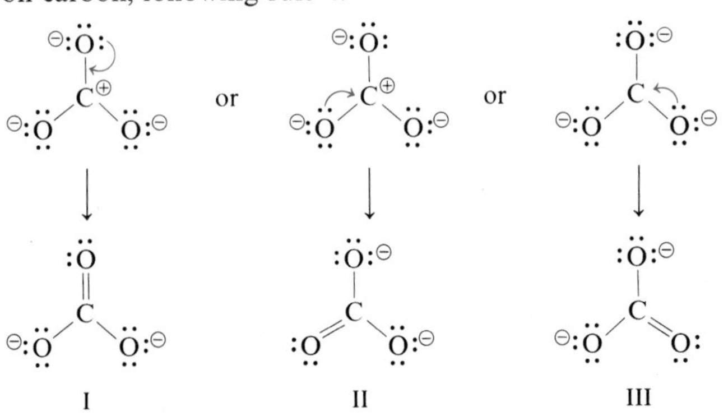

I   II   III

So there are three possible Lewis structures, I, II, and III. The carbon atom cannot discriminate among the three oxygen atoms. It does not accept a complete electron pair from any one of them but, rather, one-third of a pair of electrons from each of the three oxygen atoms. Each of these three electron pairs is partially a nonbonding pair and partially a bonding pair. Two-thirds of the charge cloud is on an oxygen atom and one-third is in the bonding region. Clearly, we cannot draw a single conventional Lewis structure to describe the arrangement of these electron pairs because Lewis structures depict electrons as being completely localized either as bond pairs or as nonbonding pairs.

There are many molecules in which at least some of the electrons cannot be described as localized bond or lone pair electrons—they are said to be **delocalized**. One way to describe the electron arrangement in such a molecule is to use molecular orbitals which extend over the whole molecule. But a convenient alternative in many cases is to use a combination of several Lewis structures that are called **resonance structures**. Thus the carbonate ion is not correctly described by any one of the structures I, II, or III, but it may be described by a combination of these structures which we now call resonance structures. We indicate that they are resonance structures by means of double-headed arrows between them:

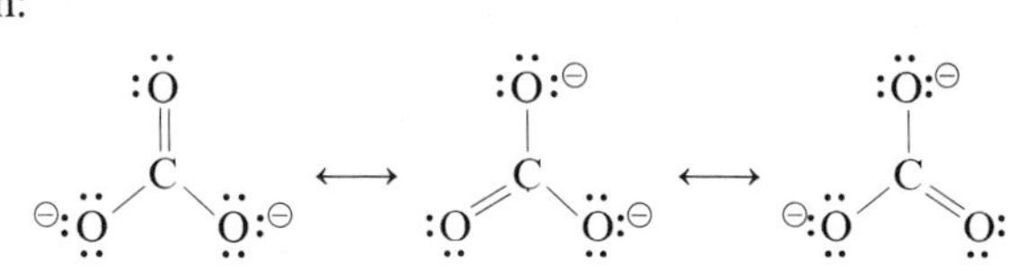

Of the four electron pairs in the valence shell of each oxygen atom, one is a bonding pair and two are nonbonding pairs in each of the above structures. But the fourth pair is nonbonding in two of the structures and bonding in the third structure. In other words one electron pair on each oxygen atom is partly bonding and partly nonbonding. All three bonds are equivalent and each may be described as a $1\frac{1}{3}$ bond; each is said to have a **bond order** of $1\frac{1}{3}$.

BOND ORDER

The **order of a bond** is *defined as the total number of electron pairs that constitute the bond.* A triple bond has a bond order of 3, a double bond has a bond order of 2, and a single bond has a bond order of 1. In the carbonate ion the bonds are intermediate between single and double bonds and have a bond order of $1\frac{1}{3}$. The observed bond length of 131 pm for each bond is consistent with this bond order; it is intermediate between that for the CO double bond in methanal (121 pm) and that for the CO single bond in methanol (143 pm):

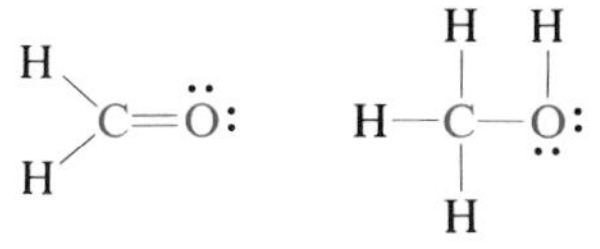

| CO bond length | 121 pm | 143 pm |
| | Methanal (formaldehyde) | Methanol |

To find the bond order from a set of resonance structures, we take the same bond in each of the resonance structures, sum the bond order for this bond in all the structures, and then divide by the number of structures. To determine the bond order of each CO bond in the carbonate ion, we may use the top CO bond in structures I, II, and III. The sum of the bond orders is $2 + 1 + 1 = 4$. The number of structures is 3, so the average bond order is $\frac{4}{3}$, or $1\frac{1}{3}$. Note that it does not matter which bond we select; for any bond the result is the same.

The formal charge on each oxygen atom in the carbonate ion may be obtained by a similar procedure. The sum of the charges on any given oxygen atom for all three structures is $0 - 1 - 1 = -2$. The average charge on oxygen is then obtained by dividing -2 by the number of structures; it is $-\frac{2}{3}$. We see that the -2 charge of the ion is spread over the three oxygen atoms.

RESONANCE

It is often said that there is **resonance** between the three structures I, II, and III for the carbonate ion. However, we must be careful not to interpret such a statement as meaning that resonance is a phenomenon. There is no rearrangement of electrons corresponding to the change of one structure to another. The molecule never has any of the three individual Lewis structures, all of which localize the electrons. It has only one structure in which some of the electrons are less localized—in other words, more spread out—than is represented by any single resonance structure.

We use resonance structures because no individual Lewis structure is capable of accurately representing the arrangement of electrons in some molecules. Lewis structures depict each electron pair as either a bonding pair or a nonbonding pair, and they do not allow for the possibility that a pair of electrons may be partially bonding and partially nonbonding. The concepts of resonance and of resonance structures are necessary because of the inadequacies of Lewis structures for representing the structures of some molecules (see Box 9.2). We emphasize again that *resonance is a concept and not a phenomenon*. A phenomenon is some change in nature that we can observe. A concept is a model or theory proposed by us to explain phenomena that we observe.

Whenever we can write more than one equivalent Lewis structure for a polyatomic ion or molecule, as in the case of the carbonate ion, some of the electrons are more delocalized than any one of the individual structures indicates. When we can only write one Lewis structure, as in the case of methane, we may reasonably suppose that the electrons are localized approximately in the manner indicated by the Lewis diagram.

The Lewis structure for the phosphate ion has one double bond and three single PO bonds. The following four resonance structures may be written:

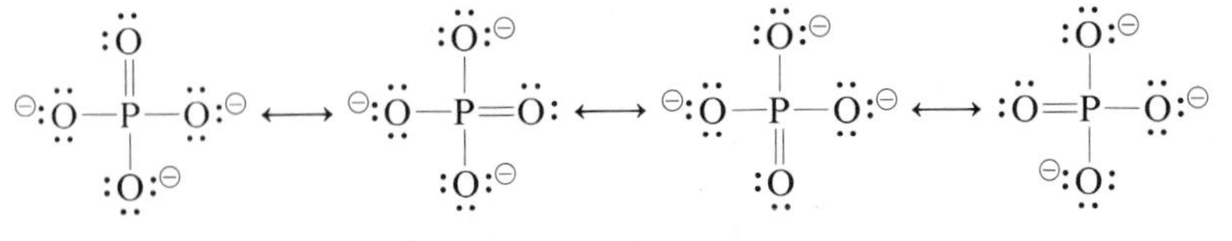

They show that each bond has an order of $1\frac{1}{4}$. Each oxygen atom has an electron pair that is 25% a bonding pair and 75% a nonbonding pair. Each oxygen has a charge of $-\frac{3}{4}$, because a total of three negative charges are shared between four oxygen atoms. Or we may sum the charge on any one oxygen for all the structures $(-1 - 1 - 1 + 0 = -3)$ and divide by the total number of structures, to give $-\frac{3}{4}$.

═ *Example 9.5* ═══════════════════════════════

RESONANCE STRUCTURES

The sulfate ion has a tetrahedral structure in which all four SO bonds have the same length (149 pm). Draw resonance structures for the sulfate ion, and determine the bond order and the charge on each oxygen atom.

Solution

The Lewis structure of the sulfate ion is

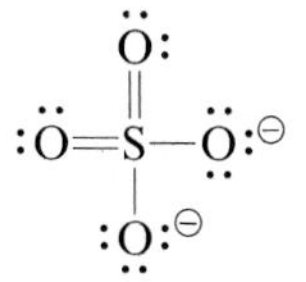

There are six possible combinations of double and single bonds, as follows:

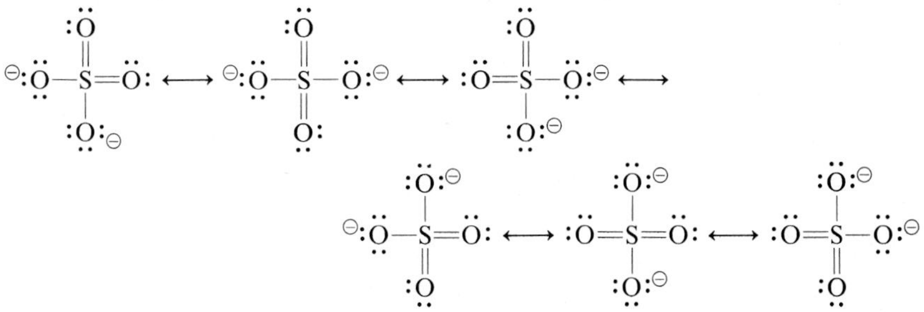

The sum of the bond orders for any given SO bond is $2 + 2 + 2 + 1 + 1 + 1 = 9$. Hence each bond has an order of $\frac{9}{6} = 1.5$. The sum of the charges on any given oxygen atom is $0 + 0 + 0 - 1 - 1 - 1 = -3$. Hence each oxygen atom has a charge of $-\frac{3}{6} = -\frac{1}{2}$.

──

EXERCISE 9.4

Draw all the resonance structures for the sulfite ion, $SO_3{}^{2-}$. Find the SO bond order and the charge on each oxygen atom.

The sulfur–oxygen bonds in the following molecules and ions illustrate the relationship between bond order and bond length:

Structure	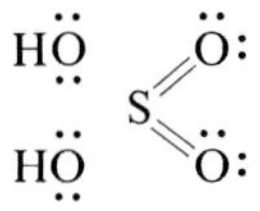	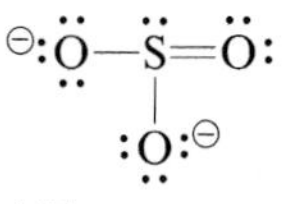	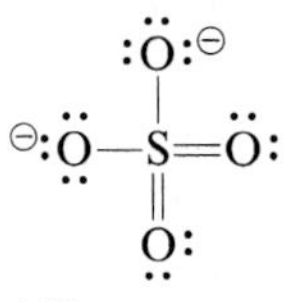
Bond length (pm)	157 (S—O)	151	149
Bond order	1.00	1.33	1.50

Structure			
Bond length (pm)	143	143	142 (S=O)
Bond order	2	2	2

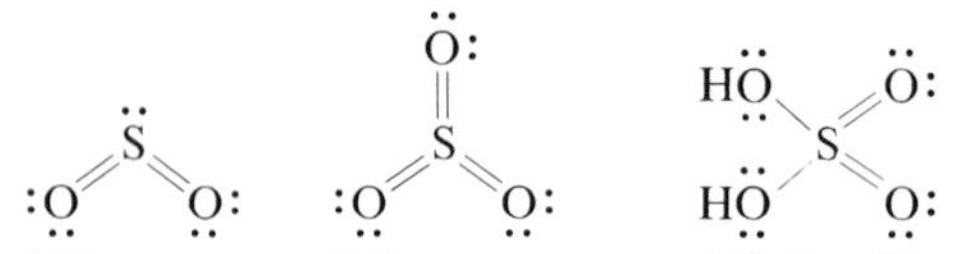

In general, bond length decreases as bond order increases. The more electron pairs that form the bond, the more strongly they hold the two nuclei together.

≡ *Example 9.6* ═══════════════════════════

RESONANCE STRUCTURES

Draw resonance structures for the methanoate (formate) ion, HCO_2^-. What is the formal charge on each O atom? What is the bond order of the CO bonds? Predict an approximate value for the CO bond length.

Solution

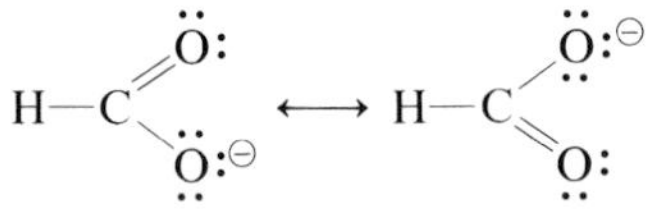

The formal charge on each O atom is $-\frac{1}{2}$. The CO bond order is $\frac{3}{2} = 1.5$. The bond length is predicted to be intermediate between that of CO_3^{2-} (bond order 1.33, bond length 131 pm) and methanal H_2CO (bond order 2.0, bond length 121 pm). Taking the mean of these two bond lengths as a rough approximation, we predict a value of 126 pm. The observed value in sodium methanoate is 127 pm.

───────────────────────────────────────

EXERCISE 9.5

Draw all the resonance structures for each of the following species, assigning formal charges where appropriate. Find the bond order in each case.

(a) HCO_3^- (b) HSO_4^- (c) HPO_4^{2-}

9.4
Hybrid Orbitals

We saw in Chapter 7 that the approximate form of the molecular orbital occupied by the two electrons in the H_2 molecule can be obtained by taking the sum of the two hydrogen 1s atomic orbitals, that is, by adding the standing-wave patterns of the 1s electron in each atom to form a new standing-wave pattern for each of the electrons in the molecule (Figure 9.18). In the region between the two nuclei where there is appreciable overlap of the two 1s waves the resultant wave, ψ, has an increased amplitude. Therefore there is a corresponding increase in ψ^2, that is, in the electron density in this region, and it is this increased electron density that is responsible for holding the two hydrogen nuclei together.

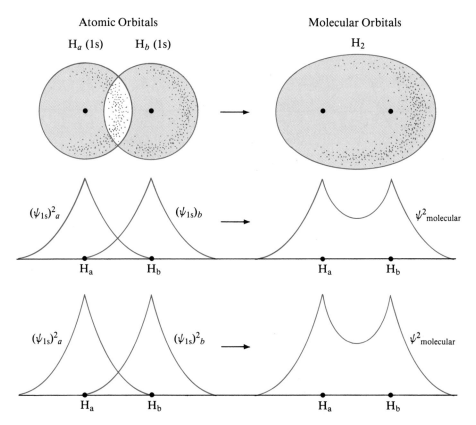

Atomic Orbitals

H_a (1s) H_b (1s)

Molecular Orbitals

H_2

$(\psi_{1s})^2_a$ $(\psi_{1s})_b$ $\psi^2_{molecular}$

H_a H_b H_a H_b

$(\psi_{1s})^2_a$ $(\psi_{1s})^2_b$ $\psi^2_{molecular}$

H_a H_b H_a H_b

FIGURE 9.18
Formation of the H_2 Molecule.

Constructive interference between the standing-wave patterns, ψ, for each of the hydrogen atoms H_a and H_b produces a standing-wave pattern for the molecule which has an increased amplitude in the region between the nuclei. There is therefore a corresponding increase in the probability, ψ^2, of finding an electron in this region, that is, there is an increase in the density of the charge cloud in the internuclear region. It is this increased electron density that is responsible for holding the two nuclei together.

Any two singly occupied orbitals that have a significant overlap can be combined in the same way to give an approximate representation of a localized bonding orbital. For example, to obtain the approximate form of the bonding orbital in the HF molecule we combine the singly occupied 2p orbital of the F atom with the H 1s orbital (Figure 9.19 on p. 458). Whereas a 2s orbital is spherical, a 2p orbital has directional character. Therefore, maximum overlap between the hydrogen 1s orbital and the fluorine 2p orbital, and hence the greatest increase in electron density in the bonding region and the strongest bond, is obtained if the overlap occurs along the axis of the p orbital, that is, in the direction in which ψ has its greatest value.

To describe the bonds in the water molecule we can use the two singly occupied 2p orbitals of an oxygen atom $(2s^2 2p_x^2 2p_y^1 2p_z^1)$ to form localized bonding orbitals by combining each of them with a hydrogen 1s atomic orbital. These bond orbitals and therefore the corresponding charge clouds are at 90° to each other (Figure 9.20). Clearly this is not the best picture that we can obtain of the

FIGURE 9.19
Formation of the Localized
Molecular Orbital or Bonding
Orbital in the HF Molecule.

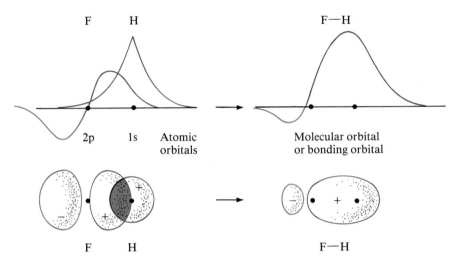

bonds in the water molecule because the observed bond angle is 104.5°, rather than 90°.

If we use the same method to obtain the localized orbitals corresponding to the bonds formed by the carbon atom $(2s^1 2p_x^1 2p_y^1 2p_z^1)$ in the methane molecule we obtain three bond orbitals that make angles of 90° with each other and a fourth bond orbital in an unspecified direction formed by the 2s orbital, which is spherical and therefore can form a bond of the same strength in any direction (Figure 9.21). It is clear, therefore, that we cannot construct localized orbitals corresponding to the tetrahedrally arranged electron pair charge clouds of the VSEPR model by using s and p atomic orbitals.

In order to obtain bond orbitals that have the tetrahedral arrangement of the bond electron pair charge clouds in methane and similar AX_4 molecules we have to use a set of four equivalent orbitals on the carbon atom that have a tetrahedral arrangement. Such a set of four tetrahedrally directed orbitals, that are called sp^3 **hybrid orbitals**, can be obtained from one s and three p orbitals by a mathematical operation called **hybridization**. By combining each of these hybrid orbitals with an appropriate orbital on a ligand atom we obtain a localized bond orbital corresponding to each of the four tetrahedral bonds. The square of the wave function, ψ^2, for each of these bond orbitals gives an electron density distribution or charge cloud concentrated primarily along one

FIGURE 9.20
Formation of Two Localized
Molecular Orbitals or Bonding
Orbitals in the H_2O Molecule.

According to this model the angle
between the two orbitals and therefore between the two corresponding
charge clouds is 90°.

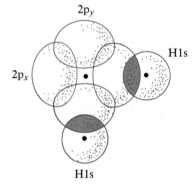

Atomic orbitals

Two localized molecular
orbitals or bonding orbitals

(The small negative "tails" of
the orbitals have been
omitted for simplicity.)

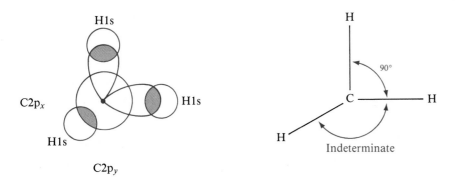

FIGURE 9.21
A Localized Orbital Model for the Methane Molecule.

In this model the carbon 2s and 2p orbitals are used to form localized bonding orbitals by combining them with hydrogen 1s orbitals. The $2p_z$ orbital which is perpendicular to the plane of the paper is not shown. It would give a bonding orbital at 90° to the bonding orbitals formed from the $2p_x$ and $2p_y$ orbitals. Thus this model predicts three C—H bonds at 90° to each other and a fourth bond that is not equivalent to the other three and whose direction cannot be predicted.

of the four tetrahedral directions. A set of hybrid orbitals can also be obtained that corresponds to the geometric arrangement of the bonds in each of the AX_2, AX_3, AX_5, and AX_6 molecular geometries, as shown in Table 9.4.

AX_2 MOLECULES: sp HYBRID ORBITALS

To understand the concept of hybridization let us consider first the linear BeH_2 molecule. Beryllium has the valence state configuration $2s^1 2p^1$. If we were to use these orbitals to form bonding orbitals by overlapping them with H 1s orbitals

---

TABLE 9.4
Hybrid Orbitals

Valence State	Hybrid Orbitals	Arrangement of Orbitals	
s p	sp	Linear	
s p	sp^2	Triangular	
s p	sp^3	Tetrahedral	
s p d	sp^3d	Trigonal bipyramidal	
s p d	sp^3d^2	Octahedral	

we would obtain two nonequivalent orbitals whose relative orientation we could not specify because an s orbital can form a bond of equal strength in any direction (Figure 9.22).

However, if we take the sum and difference of these two orbitals we generate two new equivalent orbitals called sp hybrid orbitals which are at 180° with respect to each other (Figure 9.23). By overlapping each of these orbitals with an H 1s orbital we generate two equivalent bond orbitals oriented at 180° to each other (Figure 9.24). In this and similar figures the shapes of the hybrid orbitals have been distorted by emphasizing their directional character and reducing their overlap in order to make the figures as clear as possible.

The concept of hybridization arises from the fact that the s, p, d set of orbitals that we have used to describe polyelectronic atoms is not unique. Certain combinations of these orbitals are an equally valid description of the electrons in an atom. Because orbitals are standing electron waves, combinations of them are like the in-phase and out-of-phase combinations of two standing waves in a vibrating string (Figure 9.25 on p. 462). It is important to understand that hybridization is not a physical phenomenon but simply a mathematical operation by means of which a set of atomic orbitals is converted into an equivalent set of hybrid orbitals. Such a set of hybrid orbitals gives exactly the same total electron density distribution for the atom as the set of atomic orbitals from which it is formed. It is only the electron density distribution that can be observed—the orbitals, that is, the electron waves cannot be observed. Thus any set of orbitals that leads to the same total electron density distribution is a valid description of the atom. We can use whichever set of orbitals is the most convenient for a particular purpose. For the beryllium atom the 2s and 2p orbitals are appropriate for describing the energies of the two electrons, but the sp hybrid orbitals, which are strongly directed and more localized than the 2s and 2p atomic orbitals are more suitable for constructing the localized bonding orbitals corresponding to the two Be—H bonds.

AX$_3$ MOLECULES: sp^2 HYBRID ORBITALS

To describe the bonds in a triangular molecule such as BF$_3$ we can combine one 2s orbital and two 2p orbitals to form three equivalent sp^2 orbitals that are directed toward the corners of an equilateral triangle (Figure 9.26 on p. 462). These sp^2 orbitals have a similar shape to an sp orbital but here, in order to simplify the figure, we show only their very approximate shape as we did in Figure 9.24. Each of these hybrid orbitals can be overlapped with a 2p orbital

FIGURE 9.22
The BeH$_2$ Molecule.

The formation of localized bonding orbitals by overlapping H 1s orbitals with the 2s and 2p orbitals of beryllium gives nonequivalent bonding orbitals whose relative directions cannot be predicted.

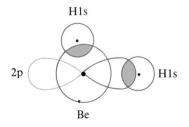

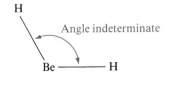

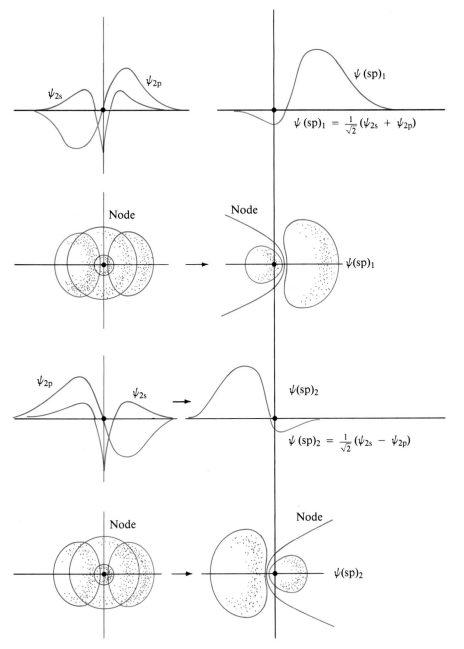

FIGURE 9.23
sp Hybrid Orbitals.

The two sp hybrid orbitals are constructed by taking the sum and difference of a 2s and a 2p orbital. They have their maxima in opposite directions at 180° from each other. Where the wave function, ψ, is positive it is shown in red and where it is negative it is shown in blue.

$$\psi(sp)_1 = \frac{1}{\sqrt{2}}(\psi_{2s} + \psi_{2p})$$

$$\psi(sp)_2 = \frac{1}{\sqrt{2}}(\psi_{2s} - \psi_{2p})$$

on a fluorine atom to form a localized bonding orbital. The set of three localized bonding orbitals is oriented towards the corners of an equilateral triangle.

AX_4 MOLECULES: sp^3 HYBRID ORBITALS

To describe the bonds in methane or any other molecule in which a carbon atom forms four equivalent bonds we can construct four equivalent combinations of the s and three p orbitals that are called tetrahedral sp^3 hybrid orbitals. They are strongly directed toward the corners of a tetrahedron (Figure 9.26 on p. 462),

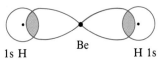

sp orbitals

FIGURE 9.24
The Formation of Bonding Orbitals for the BeH_2 Molecule Using sp Hybrid Orbitals.

The shapes of the sp hybrid orbitals have been greatly simplified by reducing the extent to which they spread out "sideways", thus emphasizing their directional character.

FIGURE 9.25
Combinations of Two
Standing Waves in a String.

The sum and difference of two
standing waves in a string gives
two more "localized" waves.

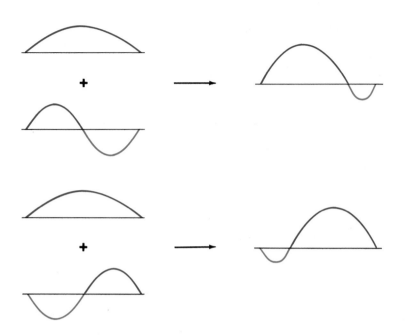

and have a shape that is similar to the shapes of sp and sp^2 hybrid orbitals. We can then use these sp^3 hybrid orbitals to form the localized orbitals that describe the bonds. We can think of each of the sp^3 hybrid orbitals as being composed of a "mixture" of one-quarter of the s orbital and one-quarter of each of the p orbitals.

We must be very careful not to take these approximate orbital shapes too literally as they could give the false impression that the total electron density distribution of four electrons in these orbitals is different from that of four electrons in a 2s and three 2p orbitals. In fact the electron density distribution of the carbon atom is spherical, and any orbital set that is chosen to describe the electrons in the carbon atom must give a spherical distribution. In the methane molecule the electron density distribution is no longer spherical. This is because the formation of four C—H bonds causes a concentration of the electron density along the four tetrahedral bond directions. It is not the formation of hybrid orbitals that causes this change in the electron density—this mathematical operation can never change the total electron density—but rather the formation of four localized bonding orbitals on the approach of the four hydrogen atoms. Hybridization is a mathematical operation and not a physical phenomenon.

FIGURE 9.26
Triangular sp^2 Hybrid Orbitals
and Tetrahedral sp^3 Hybrid
Orbitals.

sp^2 hybrid orbitals

We have seen that the concept of resonance was invented to enable us to continue to use Lewis structures for molecules for which they would otherwise be inadequate because some of the electrons are delocalized. Similarly hybrid orbitals were invented so that the bonds in a molecule can be described in terms of localized bond orbitals formed by combining a ligand orbital with an orbital on the central atom, because the s, p, and d orbitals that are used to describe the electrons in free atoms are inadequate for this purpose.

AX$_5$ AND AX$_6$ MOLECULES: sp^{3}d AND sp^{3}d^2 HYBRIDIZATION

In molecules in which the central atom has more than four pairs of electrons in its valence shell we imagine that the molecule is formed from the central atom in a valence state in which one or more electrons are promoted to the d orbitals. The hybrid orbitals that are used to describe the bonding in such molecules are therefore formed from s, p, and d orbitals. The molecule PF$_5$, for example, can be thought of as being formed from the phosphorus atom in the valence state s^{1}p^{3}d^1 and therefore the corresponding hybrid orbitals on the phosphorus atom are called sp^{3}d hybrid orbitals. They have a trigonal bipyramidal geometry. Similarly, the hybrid orbitals on the sulfur atom that can be used to construct the bonding orbitals for a molecule such as SF$_6$ are called sp^{3}d^2 hybrid orbitals. They have an octahedral geometry (see Table 9.4).

HYBRID ORBITALS AND THE VSEPR MODEL

Our discussion of hybrid orbitals has necessarily been incomplete and oversimplified because a more thorough discussion would have taken us too deeply into quantum mechanics. If you take a course in physical chemistry or quantum mechanics you will have the opportunity to improve your understanding of hybrid orbitals and of the chemical bond in general. In the meantime you can, for many purposes, simply think of the terms sp, sp^2, sp^3, sp^{3}d, and sp^{3}d^2 as convenient labels for the orbitals corresponding to the localized electron pair charge clouds of the VSEPR model in AX$_2$, AX$_3$, AX$_4$, AX$_5$, and AX$_6$ molecules, respectively, as shown in Table 9.4.

AX$_n$E$_m$ MOLECULES: LONE PAIR ORITALS

We can use the same set of sp^3 orbitals to form the bonding orbitals in AX$_3$E and AX$_2$E$_2$ molecules as we used for AX$_4$ molecules. In AX$_4$ molecules each of the sp^3 orbitals is used to form a bonding orbital. But in AX$_3$E molecules such as NH$_3$ only three of the sp^3 orbitals are used to form bonding orbitals and the other contains the lone pair of electrons—it is a nonbonding orbital. Similarly in AX$_2$E$_2$ molecules such as H$_2$O two of the sp^3 orbitals are used to form bonding orbitals while the other two contain the lone pairs—they are nonbonding orbitals (Figure 9.27). In discussing the VSEPR model we saw that the lone pair charge clouds are larger and occupy more space than the bonding pair charge clouds and that therefore the angle between the bonds in less than 109.5°. Therefore it is only an approximation to describe each of the bonding orbitals in an AX$_3$E or AX$_2$E$_2$ molecule as being derived from an sp^3 orbital. The set of sp^3 hybrid orbitals may be modified to take this into account but we cannot go into these finer details here.

FIGURE 9.27
Approximate Bonding and
Lone Pair Orbitals Based on
sp^3 Hybrid Orbitals for the
NH_3 and H_2O Molecules.

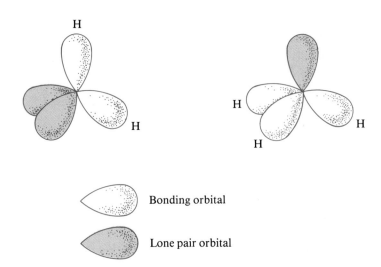

Bonding orbital

Lone pair orbital

We can similarly use sp^3d and sp^3d^2 orbitals to give approximate descriptions of the bonds in AX_4E, AX_5E, and similar molecules.

═ *Example 9.7* ═

HYBRID ORBITALS

What is the shape of each of the following molecules? In each case what set of hybrid orbitals on the central atom may be used to construct an appropriate set of localized bonding orbitals?

   SiF_4 $AlCl_3$ $BeCl_2$ PF_3 PF_5 ClF_3 IF_5

Solution

In each case write the Lewis structure and hence predict the shape. The number of atomic orbitals needed to construct the set of hybrid orbitals is equal to the number of electron pairs in the valence shell. Appropriate hybrid orbitals are formed by taking the requisite number of atomic orbitals in the order s, p, d.

	SiF_4	$AlCl_3$	$BeCl_2$	PF_3	PF_5	ClF_3	BrF_5
Number of valence-shell electron pairs	4	3	2	4	5	5	6
Hybrid orbitals	sp^3	sp^2	sp	sp^3	sp^3d	sp^3d	sp^3d^2
	AX_4	AX_3	AX_2	AX_3E	AX_5	AX_3E_2	AX_5E
Shape	Tetrahedral	Trigonal planar	Linear	Trigonal pyramidal	Trigonal bipyramidal	T-shaped	Square pyramidal

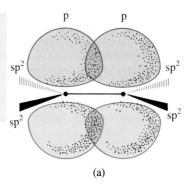

(a)

EXERCISE 9.6

For each of the following species predict the molecular shape and name the set of hybrid orbitals that may be used to construct an appropriate set of bonding orbitals.

NH_4^+ $\quad$ H_3O^+ $\quad$ $SnCl_2$ $\quad$ PCl_6^- $\quad$ XeF_2 $\quad$ XeF_4

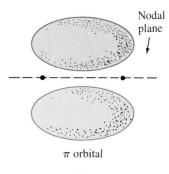

Nodal plane

π orbital

(b)

THE HYBRID ORBITAL DESCRIPTION OF MULTIPLE BONDS

A widely used description of multiple bonds is based on hybrid orbitals. Consider, for example, the ethene molecule which has the following planar geometry:

$$\underset{H}{\overset{H}{}}\,116°\,\overset{122°}{\underset{H}{C}}=\underset{H}{\overset{H}{C}}$$

Each carbon atom is bonded to three other atoms which lie in a plane and make angles of approximately 120° with each other. The bonds may be approximately described by using a set of three equivalent sp^2 hybrid orbitals on each of the carbon atoms to form localized bonding orbitals with two hydrogen 1s orbitals and with an sp^2 orbital on the other carbon atom (see Figure 9.28). A single electron remains in the $2p_z$ orbital on each carbon atom. These two orbitals are combined in a "sideways" overlap. The result is a bonding orbital that has a nodal plane in the plane of the molecule. It is called a **π (pi) orbital**. Bonding orbitals that are formed by "end-on" overlap of atomic orbitals are symmetrical around the bond axis and are called **σ (sigma) orbitals**.

DOUBLE BONDS The C=C double bond is thus approximately described as consisting of a σ orbital formed by "end-on" overlap of sp^2 orbitals and a π orbital formed by "sideways" overlap of two 2p orbitals. According to this description, the double bond consists of a **σ bond** and a **π bond**. In contrast, our previous description suggested that the double bond consists of two equivalent bent bonds. Neither description is exact and, as we will see, they are in fact equivalent.

We may put our previous description of the double bond, which was based on a tetrahedral arrangement of four electron pairs around each carbon atom, into orbital language by using four equivalent tetrahedral sp^3 hybrid orbitals on each carbon atom. Two of these hybrid orbitals are combined with hydrogen 1s orbitals to form the two C—H bond orbitals and two are combined with similar orbitals on the other carbon atom to form the double bond orbitals (see Figure 9.29). According to this description, the double bond consists of two charge clouds, one on either side of the C—C axis, in other words, two "bent" bonds. But the sum of these two charge clouds is a four-electron cloud that has its maximum along the C—C axis, as shown in Figure 9.30. In contrast, according to the alternative description the double bond consists of an axially symmetric σ charge cloud and a π charge cloud that has electron density on both sides of a nodal plane through the C atoms (Figure 9.29). But again

FIGURE 9.28
Description of the Ethene Molecule in Terms of σ and π Orbitals.

The C—H and C—C σ bond orbitals formed by a set of sp^2 hybrid orbitals on each carbon atom are shown here only as lines. The "sideways" overlap of two p orbitals gives a π orbital. A σ orbital is symmetrical around the bond direction, but a π orbital has a planar node in the bond direction.

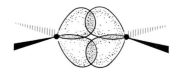

FIGURE 9.29
Bent-Bond Description of the Ethene Molecules.

Here the double bond is represented by two bent bonds, each of which is formed by the overlap of a pair of sp^3 orbitals.

FIGURE 9.30
Equivalence of the σ–π
and Bent-Bond Models for
Double Bonds.

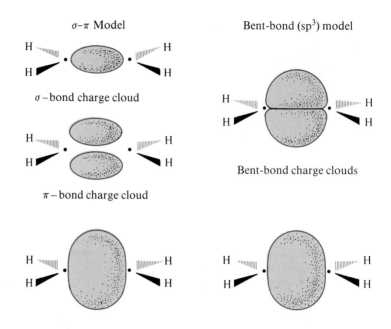

σ–π Model

σ–bond charge cloud

π–bond charge cloud

Bent-bond (sp³) model

Bent-bond charge clouds

Double-bond (4 electron) charge cloud

the sum of these two charge clouds gives a four-electron cloud that has a maximum along the C—C axis (Figure 9.30). Both methods for describing the double bond lead to essentially the same description of the bonding around each carbon atom as that given by the VSEPR model, that is, two two-electron C—H bond charge clouds and a four-electron C=C double bond charge cloud. These charge clouds keep as far apart as possible. If they were all equivalent they would be in a plane at 120°, but since the double bond charge cloud is larger than the single bond charge clouds the HCC angle is expected to be larger than 120° and the HCH bond angle smaller than 120°, in agreement with the experimental data. Since there is no way that we can observe the two individual charge clouds of the double bond we cannot distinguish between these two descriptions. The bent-bond model predicts an HCH bond angle of 109° while the σ–π model predicts an HCH bond angle of 120°, compared with the observed angle of 116°. Thus neither is strictly correct; they are both only approximate descriptions of the double bond. We can use whichever is the most convenient for a particular purpose.

TRIPLE BONDS A triple bond, such as that in ethyne, is frequently described by using a set of two sp hybrid orbitals and a $2p_y$ and a $2p_z$ orbital (see Figure 9.31). The sp hybrid orbitals are used to form the localized orbitals describing the C—H and C—C bonds, and the $2p_y$ and $2p_z$ orbitals are combined with similar orbitals on the other carbon atom to form two π-type orbitals. Again, however, this description is not unique.

An alternative description can be based on a set of four tetrahedral sp³ orbitals on each carbon atom. One of these orbitals is used to form the C—H bonding orbital, and the other three are used to form three bent bonds between the two carbon atoms (Figure 9.32). Both models predict a linear molecule. But the combination of the three bent-bond charge clouds and the combination

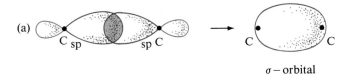

FIGURE 9.31
σ–π Model for the Triple Bond.

(a) A σ orbital is formed by overlap of an sp hybrid orbital on each carbon atom.
(b) Two π orbitals are formed by "sideways" overlap of the $2p_y$ and $2p_z$ orbitals on each carbon atom.

of the two π charge clouds and the σ charge cloud both give essentially the same six-electron triple bond charge cloud as is used in the VSEPR model. The C—H bond charge cloud and the six-electron triple bond charge cloud keep as far apart as possible, again leading to the prediction of a linear molecule.

BENZENE In Chapter 6 we described the hydrocarbon benzene, C_6H_6, and we gave the structure

$$
\begin{array}{c}
\text{H} \\
\text{H} \quad \text{C} \quad \text{H} \\
\text{C} \quad \text{C} \\
\text{C} \quad \text{C} \\
\text{H} \quad \text{C} \quad \text{H} \\
\text{H}
\end{array}
$$

first proposed by Kekulé. According to this structure there are alternate single and double bonds in the ring of six carbon atoms. By comparison with the

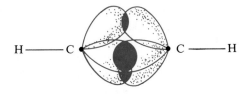

sp³ orbitals

FIGURE 9.32
The Bent-Bond Model of the Triple Bond.

bond lengths in ethane and ethene we would expect the single bonds in benzene to have a length of approximately 154 pm and the double bonds to have
a length of approximately 134 pm. In fact all six carbon–carbon bonds have
the same length of 140 pm. There are two alternative ways in which we can
place the double bonds in the benzene ring:

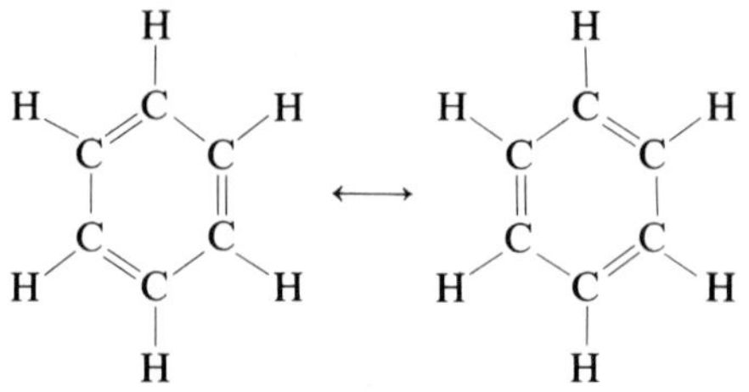

These represent two resonance structures. These structures are called the Kekulé
structures for benzene. For simplicity just the C—C bonds are often shown,
with the C and H atoms and the C—H bonds omitted.

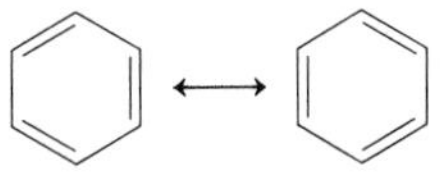

Taken together these two structures describe the fact that three of the electron
pairs are not as localized as implied by a single Kekulé structure. If we calculate
the bond order for any of the carbon–carbon bonds by taking the sum of the
bond orders for any particular bond in the two structures $(1 + 2 = 3)$ and
dividing by the number of structures (2), we obtain a bond order for each bond
of 1.5. The observed bond length of 140 pm is consistent with this bond order.

The three delocalized electron pairs are often depicted by a circle drawn
inside the hexagon representing the six C—C bonds, which allows us to use
one structural formula to represent benzene:

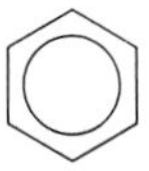

Each carbon atom forms three localized bonds with an AX_3 geometry, two to
adjacent carbon atoms and one to a hydrogen atom. In addition, there are six
delocalized electrons that are spread around the ring of six carbon atoms.
Benzene is often represented simply by one Kekulé structure; we must then
remember that the double bond electrons are in fact delocalized around the
ring.

Alternatively we may use a hybrid orbital description of the bonding in
benzene. Benzene has a regular hexagonal structure and so the bond angles at
each carbon atom are exactly $120°$. Thus we may use a set of sp^3 hybrid orbitals
on each carbon atom to form three σ-bonds (the CH and two CC bonds). This
leaves a singly occupied 2p orbital on each carbon atom (Figure 9.33). These
2p orbitals may be combined by sideways overlap to form π-type molecular
orbitals which extend over all six carbon atoms. Three such orbitals are each
occupied by two electrons. The sum of the electron densities in these three
orbitals gives a ring of electron density above and below the plane of the carbon
atoms. (Figure 9.33). The six π-electrons are delocalized over the six carbon
atoms. It is this delocalization of six of the electrons in benzene that we alternatively describe by two resonance structures.

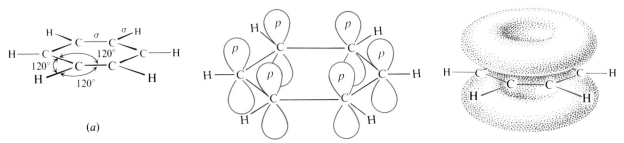

(a)

FIGURE 9.33
The σ-π model of the
bonding in benzene.

(a) Only the σ bonds are shown.
(b) π-molecular orbitals can be
formed by the sideways overlap of
a 2p orbital on each carbon atom.
(c) The π-charge cloud forms a
ring above and below the nodal
plane through the carbon atoms.

9.5
Molecular Polarity and Dipole Moments

The geometric arrangement of the bonds in a molecule and the electronegativities of the atoms determine the distribution of charge in the molecule. The distribution of charge in a molecule has an important effect on its properties and on the forces between molecules (intermolecular forces), as we will discuss in Chapter 12.

We have seen that when two atoms in a diatomic molecule have different electronegativities, the bond is polar covalent; HCl is an example. Because it has a greater electronegativity than H and because the bond electrons are shared unequally between Cl and H, the Cl atom in HCl has a small negative charge ($\delta-$) and the H atom has a small positive charge ($\delta+$):

$$\overset{\delta+}{H}-\overset{\delta-}{Cl}$$

Two separated, equal, and opposite charges constitute a **dipole**.

A dipole is described quantitatively by its **dipole moment**, μ, which is defined as the product of the magnitude of the charge, Q, at each end of the dipole and the distance, r, between the charges:

$$+Q \quad -Q$$
$$\bullet\!\!-\!\!r\!\!\to\!\!\bullet$$
$$\mu = Qr$$

In SI units the charge Q is measured in coulombs; the distance r is measured in meters. So the dipole moment is measured in coulomb meters, C m.

As Figure 9.34 describes, the dipole moment of a molecule is found experimentally by measuring how the substance composed of these molecules affects the ability of a capacitor to store electric charge.

Table 9.5 lists the dipole moments of several molecules. Note that the dipole moment of the hydrogen halides increases with an increase in the difference in

FIGURE 9.34
Measurement of Dipole
Moments.

A substance is placed between the plates of a capacitor, and the plates are charged as shown. (a) Nonpolar molecules orient themselves randomly between the plates. (b) In contrast, polar molecules, with a dipole moment, tend to line up with their + ends pointing toward the negative plate and their − ends pointing toward the positive plate. This lining up of the polar molecules enables the plates to accept a greater electric charge. The greater the dipole moment, the greater is the charge that the plates can hold, that is, the greater the capacitance of the capacitor in (b).

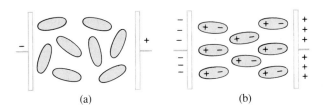

(a)　　(b)

TABLE 9.5
Dipole Moments of Some Molecules

Molecule	Dipole Moment (Units of 10^{-30} C m)
HF	6.36
HCl	3.43
HBr	2.63
HI	1.27
H_2O	6.17
H_2S	3.12
NH_3	4.88
PCl_3	2.30
BCl_3	0
SO_2	5.42
CO_2	0

electronegativity between the halogen and hydrogen because the charges on the atoms increase with increasing electronegativity difference.

A molecule that has a dipole moment is described as **polar**. All *diatomic* molecules with polar covalent bonds have dipole moments and thus are polar. But not all polyatomic molecules with polar covalent bonds are polar. The geometry as well as the polarity of its bonds determines whether a polyatomic molecule is polar.

Consider SO_2 and CO_2. Because oxygen has a greater electronegativity than either carbon or sulfur, we expect both a CO bond and an SO bond to be polar. But we see from Table 9.5 that SO_2 has a dipole moment of 5.42×10^{-30} C m, whereas the dipole moment of CO_2 is 0. The important difference between the two molecules lies in their geometry. The CO_2 molecule is a linear AX_2 molecule, whereas SO_2 is an angular AX_2E molecule:

$$\overset{\delta-\quad 2\delta+\quad \delta-}{O=C=O} \qquad \overset{..}{\underset{\delta-O\diagup\quad\diagdown O^{\delta-}}{S^{2\delta+}}}$$

Linear AX_2 Angular AX_2E

The dipole associated with a polar bond is usually represented by an arrow pointing from the positive charge to the negative charge

$$\overset{\delta+\qquad \delta-}{\underset{\longrightarrow}{+}}$$

In both CO_2 and SO_2 there are two bond dipoles which we can represent as follows:

$$\overset{\delta-\quad 2\delta+\quad \delta-}{\underset{\underset{\mu=0}{\longleftarrow + \quad + \longrightarrow}}{O=C=O}} \qquad \underset{\underset{\mu\neq0}{}}{\overset{2\delta+}{\underset{\delta-O\quad\quad O^{\delta-}}{S}}} \quad \text{Sum of SO dipole moments}$$

We can see that the two dipoles in CO_2 point in opposite directions away from the carbon atom so that they cancel each other, in other words, their vector sum is zero. But in SO_2 the two dipoles both point toward the same side of the sulfur atom so that their vector sum is not zero. There is a resultant dipole that points from the sulfur atom along the bisector of the OSO angle, so this molecule has a nonzero dipole moment.

We can reach the same conclusions by noting that in CO_2 the center of the negative charges coincides with the center of the positive charges, so that there is no resultant displacement of charge in the molecule and therefore the dipole moment is zero. But in SO_2 the center of negative charge does not coincide with the center of positive charge, so that there is a resultant displacement of charge in the molecule and therefore the molecule has a dipole moment.

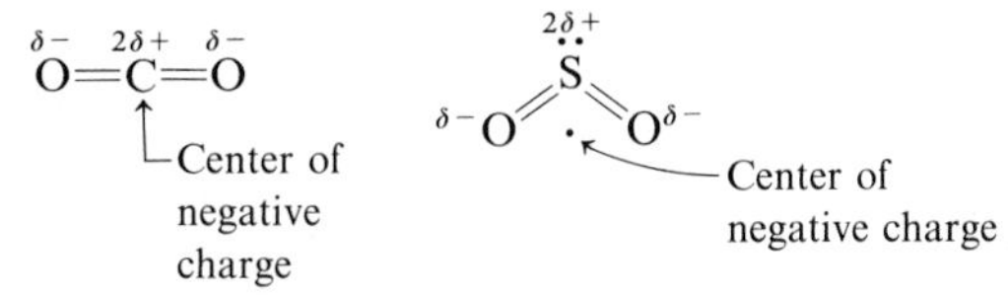

The water molecule, which is an angular AX_2E_2 molecule, is also a polar molecule with a dipole moment:

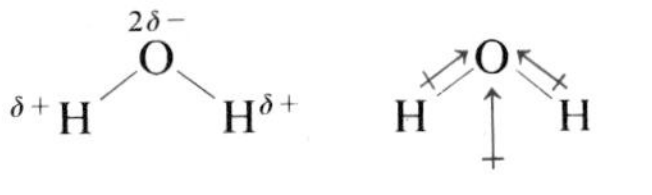

For a triatomic molecule consisting of a central atom A with two identical ligands X, the measurement of the dipole moment can tell us immediately whether the molecule is linear or angular. If it has a dipole moment it is angular not linear. AX_2 molecules do not have a dipole moment, AX_2E and AX_2E_2 do.

== *Example 9.8* ==

DIPOLE MOMENTS AND MOLECULAR GEOMETRY

Which of the following molecules will have a dipole moment and which will not?

CS_2 $BeCl_2$ $SnCl_2$ H_2S SCl_2

Solution

First, we draw the Lewis structures for the molecules, and then we classify them as AX_2 linear, AX_2E angular, or AX_2E_2 angular.

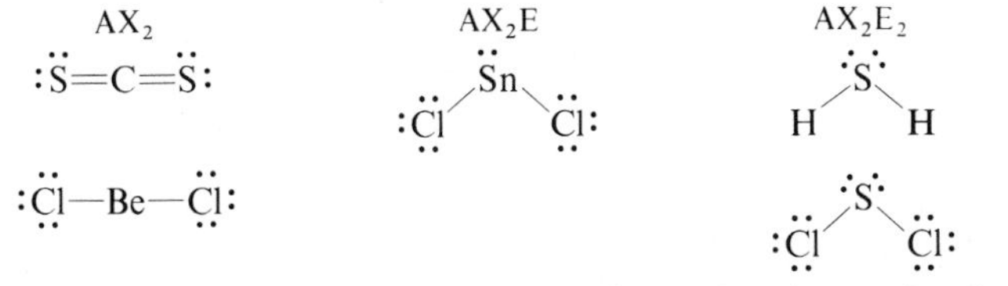

| AX₂ | AX₂E | AX₂E₂ |
| Linear, $\mu = 0$, nonpolar | Angular, $\mu \neq 0$, polar | Angular, $\mu \neq 0$, polar |

The linear CS_2 and $BeCl_2$ molecules have no dipole moment and are nonpolar, while $SnCl_2$, H_2S, and SCl_2, which are all angular molecules, have dipole moments and are polar.

We can similarly compare molecules in which three ligands are attached to a central atom. For example, Figure 9.35 compares PCl_3 and BCl_3. As the figure shows, PCl_3, which is an AX_3E molecule with a triangular pyramidal shape, has a dipole moment ($\mu = 2.30 \times 10^{-30}$ C m); BCl_3, which is a planar

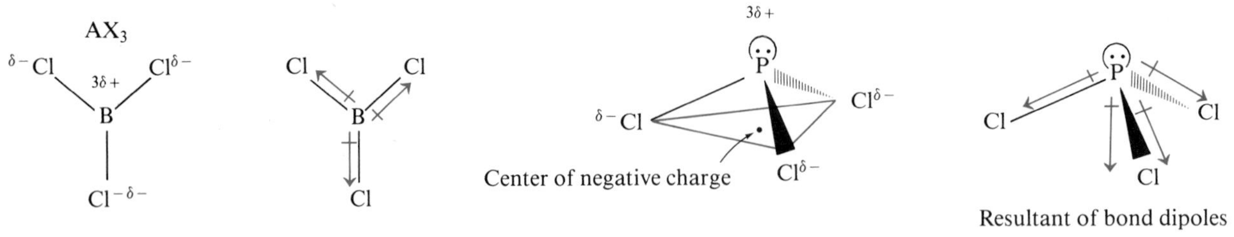

AX₃

FIGURE 9.35
Dipole Moments of AX_3 and AX_3E Molecules.

(a) In BCl_3 the center of the negative charges on the Cl atoms coincides with the positive charge on the B atom. Therefore the molecule has no dipole moment. In other words, the three BCl bond dipoles, which have a symmetrical planar arrangement, have a vector sum of zero. (b) In PCl_3 the center of the negative charges does not coincide with the positive charge on the phosphorus atom. So the molecule has a dipole moment and is polar. The vector sum of the three bond dipoles in PCl_3 is a dipole that has its positive end on the P atom and points toward the midpoint of the triangular base of the pyramid.

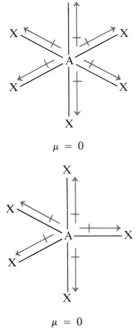

$\mu = 0$

$\mu = 0$

FIGURE 9.36
Dipole Moment of CCl_4—an AX_4 Molecule.

The sum of the upper C—Cl dipoles is a vector that points from the C atom to the midpoint of the upper face of the cube; the sum of the lower C—Cl dipoles points toward the midpoint of the lower face of the cube. These two vectors have equal magnitudes and point in opposite directions; therefore their sum is zero. The center of the four negative charges coincides with the positive charge on the C atom. The dipole moment of CCl_4 is zero.

AX_3 molecule, does not ($\mu = 0$). In general, AX_3 molecules, like AX_2 molecules, are nonpolar; AX_3E molecules, like AX_2E and AX_2E_2 molecules, have dipole moments and are polar.

Figure 9.36 shows CCl_4, a tetrahedral AX_4 molecule. The C—Cl bonds are polar, but the center of the four negative charges on the Cl atoms lies in the center of the tetrahedron and thus coincides with the positive charge on the C atom. Therefore the molecule has no dipole moment and is nonpolar. We reach the same conclusion if we consider the bond dipoles; they have a vector sum of zero. This can be seen if we represent the tetrahedron by four corners of a cube and consider the sum of the vectors two at a time, as shown in Figure 9.36. Thus AX_4 molecules, like AX_3 and AX_2 molecules, have a zero dipole moment. Similarly, as Figure 9.37 shows, trigonal bipyramidal AX_5 molecules and octahedral AX_6 molecules also have zero dipole moments. But if the valence shell of the central atom contains a lone pair of electrons the molecule may or may not have a dipole moment, depending on its geometry. Thus AX_2E and AX_2E_2 molecules are both angular and have dipole moments, as we have seen. But an AX_2E_3 molecule, such as XeF_2, is linear and therefore has a zero dipole moment.

═ *Example 9.9* ═══════════════

DIPOLE MOMENTS AND MOLECULAR GEOMETRY

Show by means of appropriate diagrams that SF_4 is a polar molecule ($\mu \neq 0$) whereas XeF_4 is nonpolar ($\mu = 0$).

Solution

SF_4 is an AX_4E molecule which has a seesaw (disphenoid) shape. The axial dipoles cancel each other but the equatorial dipoles have a resultant that points along the bisector of the FSF angle so that the molecule is polar.

XeF_4 is a square planar AX_4E_2 molecule. The four Xe—F dipoles therefore cancel each other and the molecule is nonpolar. Alternatively we can see that the center of

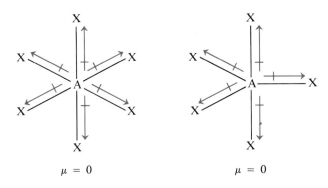

FIGURE 9.37
Trigonal Bipyramidal AX_5 Molecules and Octahedral AX_6 Molecules Have Zero Dipole Moments.

$\mu = 0$ $\mu = 0$

the four negative charges on the fluorine atoms coincides with the positive charge on the xenon atom so that there is no resultant displacement of charge and therefore XeF_4 has a zero dipole moment.

AX_4E 　　AX_4E_2
$\mu \neq 0$ 　　$\mu = 0$

EXERCISE 9.7

Which of the following molecules will have a dipole moment, and which will not?

SF_2 　 PCl_3 　 BrF_3 　 BrF_5 　 SF_6 　 OCS

IMPORTANT TERMS

In a **bent bond** the maximum of the electron density does not lie on the straight line between two bonded atoms.

A **bond angle** is the angle formed by two bonds to the same atom.

A **bond dipole** is formed by the partial charges on two atoms that are bonded together.

The **bond length** is the distance between the nuclei of two atoms that are bonded together.

The **bond order** is the number of electron pairs involved in a bond between two atoms. For molecules and ions that are described by resonance structures, the bond order may be fractional.

A localized **bonding orbital** is formed by the overlap of an atomic or hybrid orbital on one atom with an atomic or hybrid orbital on an adjacent atom. It is occupied by a bonding pair of electons.

A **bonding pair** is a pair of electrons that is shared between two atoms and that bonds them together.

A **dipole** is formed by two separated charges of equal magnitude and opposite sign.

The **dipole moment**, μ, is the product of the magnitude of the charges forming a dipole and the distance between the charges: $\mu = Qr$.

A **double bond** is a covalent bond of order 2; a bond formed by the sharing of two electron pairs.

A set of **hybrid orbitals** is obtained by taking an appropriate combination of a set of atomic orbitals. A hybrid orbital is used to form a localized bonding orbital by combining it with an atomic or hybrid orbital on an adjacent atom.

Hybridization is the mathematical process of forming a set of hybrid orbitals; it is not a physical phenomenon.

A **ligand** is a single atom, or a group of atoms, such as OH, that is attached to a central atom.

A **nonbonding orbital** is an atomic or hybrid orbital that is occupied by a nonbonding pair of electrons.

A **nonbonding pair** is a pair of electrons in the valence shell of an atom that is not forming a bond; also called an *unshared pair* or a *lone pair*.

A **nonpolar bond** is formed between two atoms that have the same electronegativity. A **nonpolar molecule** is a molecule that has no dipole moment.

A **pi (π) bond** is a bond that is described by a π bonding orbital.

A **pi (π) orbital** has a nodal plane through the atoms of a planar molecule. Its approximate shape may be obtained by combining two or more atomic p orbitals in a "sideways" manner.

A **polar bond** is a bond formed between two atoms with different electronegativities. A **polar molecule** is a molecule that has a dipole moment.

Resonance is a method for describing the structures of molecules that cannot be adequately represented by a single Lewis structure because the electrons are not as localized as shown in a Lewis structure. Such a molecule may be represented by a combination of two or more Lewis structures called **resonance structures**.

A **sigma (σ) bond** is a bond that is described by a σ bonding orbital.

A **sigma (σ) orbital** is a localized molecular (or bonding) orbital that is symmetrical around the straight line joining two atoms. It may be obtained by combining two atomic or hybrid orbitals in an "end-on" manner.

A **triple bond** is a covalent bond of order 3; a bond formed by the sharing of three electron pairs.

PROBLEMS *

Lewis Structures

1. Draw Lewis structures for each of the following ions:

$$S_2^{2-} \qquad SO_3^{2-} \qquad ClO^- \qquad ClO_4^-$$

2. In each of the following species, the central atom is a sulfur atom. Draw the Lewis structure of each:

$$S_2O_3^{2-} \qquad SO_3F^- \qquad SOF_2 \qquad SO_2F_2 \qquad SNF_3$$

3. Draw Lewis structures for each of the following oxo-acids:

$$HNO_2 \qquad HNO_3 \qquad H_2SO_4 \qquad HClO_3 \qquad H_2CO_3$$

4. Draw Lewis structures for each of the following species:

$$CN^- \qquad NO_2^- \qquad O_3$$

5. Dinitrogen monoxide, N_2O, is one of several oxides of nitrogen. Draw Lewis structures for both of the atom arrangements NON and NNO. By comparing the formal charges on atoms in these Lewis structures, decide which structure is the most probable.

6. In addition to ammonia, there are two other hydrides of nitrogen. They are hydrazine, N_2H_4, and diazine, N_2H_2. Draw Lewis structures for each of these molecules.

7. Complete the Lewis structures of the following molecules, adding unshared electron pairs and extra bonds, as required. Give approximate values for each bond angle.

(a)
$$\begin{array}{c} Cl \\ \diagdown \\ \diagup \quad C{-}O \\ Cl \end{array}$$
(b) $F{-}N{-}N{-}F$

(c) $O{-}C{-}S$
(d)
$$\begin{array}{c} H \qquad\qquad H \\ \diagdown \qquad \diagup \\ C{-}C \\ \diagup \qquad \diagdown \\ H \qquad\qquad C{-}N \end{array}$$

(e)
$$\begin{array}{c} F \\ \diagdown \\ \diagup \quad S{-}O \\ F \end{array}$$

8. The Lewis structure of the hydrocarbon allene is

$$\begin{array}{c} H \qquad\qquad\qquad H \\ \diagdown \qquad\qquad\qquad \diagup \\ C{=}C{=}C \\ \diagup \qquad\qquad\qquad \diagdown \\ H \qquad\qquad\qquad H \end{array}$$

(a) Are all three carbon atoms colinear? Explain.

(b) Are all of the hydrogen atoms in the same plane? If not, how would you expect them to be arranged? Explain why.

9. Which of the following molecules and ions are exceptions to the octet rule?

$$BCl_3 \qquad NO_2^+ \qquad NO_3^- \qquad BrO_2^- \qquad PF_3 \qquad XeF_2$$

The VSEPR Model

10. Describe the shape that you would expect for each of the following molecules and ions:

BH_3 BO_3^{3-} $SnCl_3^-$ H_2Se SiF_5^- BrF_5

11. Four electron pairs in the valence shell of an atom have a tetrahedral arrangement and an AX_4 molecule is described as having a tetrahedral shape. But AX_3E and AX_2E_2 molecules, which also have four electron pairs in the valence shell of A, are described as having triangular pyramidal and angular shapes respectively. Explain why.

12. Classify each of the following molecules in terms of the AX_nE_m nomenclature, and describe their approximate shapes:

SO_2 SO_3 F_2SO_2 F_2SO F_3SN

13. An AX_4 type molecule has regular tetrahedral geometry with XAX angles of 109.5°, while the bond angles in ammonia, NH_3, and water, H_2O, are 107° and 104.5° respectively. Explain the reasons for these differences.

14. Account for the deviation of the following bond angles from the ideal tetrahedral angle for each pair of molecules:

(a) H_2O, 104.5°; F_2O, 101°

(b) NH_3, 107°; PH_3, 94°

(c) NH_3, 107°; NF_3, 102°

15. In an AX_5 trigonal bipyramidal molecule such as PF_5, there are three equatorial bonds and two axial bonds. Explain why the axial bonds are always found to be longer than the equatorial bonds in such molecules.

16. Explain why, in molecules based upon a central atom with five pairs of electrons in its valence shell, some of which are lone pairs, i.e. molecules of the type AEX_4, AE_2X_3, and AE_3X_2, the lone pairs of electrons are always found in equatorial positions, rather than in axial positions. Give an example of each of these types of molecules.

Resonance

17. Why is it necessary to introduce the concept of *resonance* to account for the observed structures of species such as CO_3^{2-} and SO_3^{2-}, in which all the bonds have the same length, intermediate between the length of a single bond and a double bond?

18. Draw Lewis structures, including resonance structures where appropriate, for each of the following molecules and ions:

$CH_3CO_2^-$ CH_3OH ClO_3^- $H_2PO_4^-$

19. Draw the resonance structures for each of the following and calculate the formal charge on each atom and the bond orders of the bonds in the actual molecular species:

(a) NO_3^- (b) O_3 (c) PO_4^{3-}

(d) SO_4^{2-} (e) ClO_4^-

20. Would you expect the ClO bond length in the ClO^- ion to be longer or shorter than that in the ClO_4^- ion? Explain.

21. The nitric acid molecule, HNO_3, has two different NO bond lengths of 121 pm and 136 pm. All the bonds in the nitrate ion, NO_3^-, have the same length of 126 pm, as do the bonds in the nitrite ion, NO_2^-, 121 pm. Account for these differences in terms of the orders of the NO bonds in these molecules.

22. The observed bond lengths in phosphorus(V) oxide, P_4O_{10}, are 160 and 140 pm. On the basis of these two values, account for the following observed PO bond lengths:

PO_4^{3-}, 154 pm HPO_3^{2-}, 151 pm
$P_2O_7^{4-}$, 152 and 161 pm

23. Draw the two Lewis (resonance) structures of nitryl chloride, O_2NCl, in which nitrogen is the central atom. Show any formal charges, and predict the bond orders of the NO bonds and the approximate bond angles in this molecule.

Hybrid Orbitals

24. What set of hybrid orbitals on the central atom would it be appropriate to use to describe the bonds in (a) CF_4 and (b) SF_6?

25. What set of hybrid orbitals on the central atom would it be appropriate to use to describe the σ-bonds in

(a) H_2O (b) H_3O^+ (c) CO_2

(d) CO_3^{2-} (e) PF_5 (f) ClF_3

26. Describe the bonds in ethene in terms of the bent bond model and in terms of the σ–π model. Explain why these are both only approximate descriptions of the bonding.

27. Describe the bonds in propene, $CH_3CH=CH$, in terms of the σ–π model.

28. What set of hybrid orbitals would it be appropriate to use to describe the σ-bonds formed by each of the carbon atoms in propyne, $CH_3C\equiv CH$?

29. Describe (a) a double bond, and (b) a triple bond, in terms of σ and π orbitals.

30. Arsenic, As, is a member of the nitrogen family of elements:

(a) Use the VSEPR model to predict the shape of AsF_5

(b) Describe the bonding in AsF_5 in terms of hybrid orbitals

31. Describe the bonding in each of the following nitrogen-containing molecules in terms of hybrid orbitals on the nitrogen atom:

(a) NF_3 (b) H_3CCN (c) NH_4^+

(d) NO^+ (e) NO_2^+ (f) $(CH_3)_4N^+$

32. Use the VSEPR model to predict the shape of each of the following species. In each case give the set of hybrid orbitals on the central atom that would be appropriate for describing the bonds and lone pairs.

BrF_3 SbF_6^- BrF_4^- SiF_5^- SiF_6^{2-}

Dipole Moments

33. A molecule ACl_3 is found to have a dipole moment. Predict the group of the periodic table to which the central atom A belongs. What would be your answer if ACl_3 had no dipole moment ($\mu = 0$)?

34. Indicate the expected polarity of each of the bonds in each of the following molecules. Which molecules would have a dipole moment?

(a) CO_2 (b) SO_2 (c) H_2O

(d) NH_3 (e) SO_3 (f) BeH_2

35. Draw the Lewis structures of each of the following molecules and indicate the expected polarity of each bond. Hence, deduce which of the molecules will have a dipole moment:

(a) NF_3 (b) OF_2 (c) H_2S (d) PH_3

(e) CCl_4 (f) SCO   (g) SO_3 (h) BF_3

*36. In each of the following groups of molecules, arrange the molecules in order of the expected increase in their dipole moments:

(a) HF, HCl, HBr, HI   (b) AsH_3, NH_3, PH_3

(c) Cl_2O, F_2O, H_2O (d) H_2O, H_2S

37. Explain each of the following statements:

(a) The dipole moment of F_2O is much smaller than that of H_2O, even though both molecules have rather similar bond angles

* The asterisk denotes the more difficult problems.

(b) OCS has a dipole moment, while CS_2 has no dipole moment

38. Explain why sulfur hexafluoride, SF_6, is a nonpolar molecule, even though fluorine is much more electronegative than sulfur.

39. (a) Explain why a tetrahedral molecule such as CCl_4 has no dipole moment, while a molecule such as CF_2Cl_2 has a dipole moment.

(b) Explain why a trigonal bipyramidal molecule such as PF_5 has no dipole moment.

*40. The dipole moments of HF and H_2O, in units of 10^{-30} C m, are 6.36 and 6.17, respectively. The observed bond lengths are HF, 91.7 pm; and H_2O, 95.7 pm. The HOH bond angle in the latter molecule is 104.5°. Calculate the partial charge on each atom in each of these polar molecules.

*41. The dipole moments of HCl and H_2S, in units of 10^{-30} C m, are 3.43 and 3.12. The observed bond lengths are HCl, 127.4 pm; and H_2S, 135 pm. The bond angle in the latter molecule is 92°. Calculate the partial charge on each atom in each of these polar molecules.

Miscellaneous

42. In the usual Lewis structure for BF_3, the boron atom is an exception to the octet rule. Structures may be written in which boron completes its octet. Write three such structures for BF_3 with their formal charges. Why are such structures expected to be less important than the structure in which boron has an incomplete octet?

43. Draw Lewis structures for each of the following, giving all the possible Lewis structures, where applicable. Which will have the longest NO bond? Explain.

NO_2^+ NO_3^- NO_2^- FNO_2

44. (a) Define bond order.

(b) How does the length of a particular bond type (for example carbon–carbon, or carbon–oxygen) change with increasing bond order?

45. The noble gas xenon, Xe, was discovered by Sir William Ramsay in 1898, and identified as an element of group 8. Since no compounds of group 8 elements had been discovered at the time, it was assumed that xenon was also inert. In 1962, xenon compounds were made for the first time. In particular Xe was found to react with $F_2(g)$ under pressure to give the covalent compounds XeF_2, XeF_4, and XeF_6.

(a) What are the valence state electron configurations of Xe in each of these compounds?

(b) What are the expected shapes of XeF_2 and XeF_4?

(c) Is XeF_6 expected to be an octahedral molecule?

46. With the aid of two suitable examples in each case, explain each of the following terms:

(a) Polar bond   **(b)** Bond dipole

(c) Dipole moment   **(d)** Bent bond

(e) Ligand   **(f)** Resonance structure

(g) Bond order

47. (a) What is the Lewis octet rule?

(b) What is a Lewis structure?

(c) What is a resonance structure?

(d) Give two examples of molecules where the central atom obeys the octet rule.

(e) Give two examples of molecules where the central atom has less than an octet of electrons.

(f) Give two examples of molecules where the central atom has more than an octet of electrons.

48. For each of the following give examples of two molecules or ions having the specified structural features:

(a) A double bond

(b) A triple bond

(c) Linear geometry

(d) Trigonal bipyramidal geometry

(e) Octahedral geometry

(f) A formal charge on the central atom

49. (a) Write a Lewis structure for the nitrate ion, NO_3^-, in which the central atom is N. What geometry and approximate bond angles are expected?

(b) What concept is used to explain why the nitrate ion is found to be planar with the shape of an equilateral triangle, with an O atom at each corner and the N atom at the center, with three equal N—O bond lengths of 126 pm, intermediate in length between an NO single bond (136 pm) and an NO double bond (116 pm)?

50. (a) Write two resonance structures for the ozone molecule, O_3.

(b) What concepts are used to explain **(i)** why O_3 is an angular molecule, and **(ii)** why it has two OO bonds of length 128 pm, intermediate in length between the OO bond in hydrogen peroxide (147 pm) and the OO bond in oxygen (121 pm)?

51. Complete the following table:

Species	Molecular Shape	Hybrid Orbitals on the Central Atom	Approximate Bond Angle
(a) CH_4			
(b) PH_4^+			
(c) NF_3			
(d) F_2O			
(e) H_3O^+			

52. Give examples of molecules with structures that fit each of the following classifications of the VSEPR model:

(a) AX_2 **(b)** AX_3 **(c)** AX_2E **(d)** AX_4

(e) AX_3E **(f)** AX_2E_2 **(g)** AXE_3

Which of these types of molecule have dipole moments?

53. Give examples of molecules with structures that fit each of the following classifications of the VSEPR model:

(a) AX_5 **(b)** AX_4E **(c)** AX_3E_2 **(d)** AX_2E_3

(e) AX_6 **(f)** AX_5E **(g)** AX_4E_2

Which of these types of molecule have dipole moments?

54. Peroxoacetyl nitrate, a toxic compound that is present in photochemical smog (see Box 18.1), has the structure

(a) Use the VSEPR model to determine the geometry around each of the two carbon atoms, the two non-terminal oxygen atoms, and the nitrogen atom. Give the appropriate AX_nE_m designation for each of these atoms and give approximate values for the bond angles at each of the atoms 1 to 5.

(b) Use the σ–π model of find the number of σ bonds and the number of π bonds formed by each of the carbon atoms and the nitrogen atom.

55. Using Lewis structures explain why F_3^- has never been made, although I_3^- forms readily when iodine is dissolved in an aqueous KI solution (see Experiment 5.5).

56. Use the bond energies in Table 6.3 to calculate a value for the standard enthalpy of formation of benzene assuming that it has a single Kékulé structure with alternate single and double bonds. Compare the value that you obtain with the experimentally determined value for benzene which is $\Delta H_f^\circ = +83 \text{ kJ mol}^{-1}$. Explain the large difference and suggest why this difference is called the resonance or delocalization energy of benzene.

57. Lewis structures are based on the assumption that all the electrons in a molecule are localized in bonds or lone pairs. Explain, with two suitable examples, what is meant by delocalized electrons and describe the concept that was introduced to enable such delocalized electrons to be described in terms of Lewis structures.

58. Explain why the bonds in methane cannot be conveniently described in terms of localized bond orbitals formed from the 2s and 2p orbitals of the carbon atom. What concept is used to enable us to describe each of the bonds in methane in terms of a localized bond orbital on carbon and a 1s orbital of a hydrogen atom?

59. Discuss two alternative models for describing the bonding in benzene and explain how they are both consistent with its regular hexagonal structure.

Some Common Metals: Aluminum, Iron, Copper, and Lead

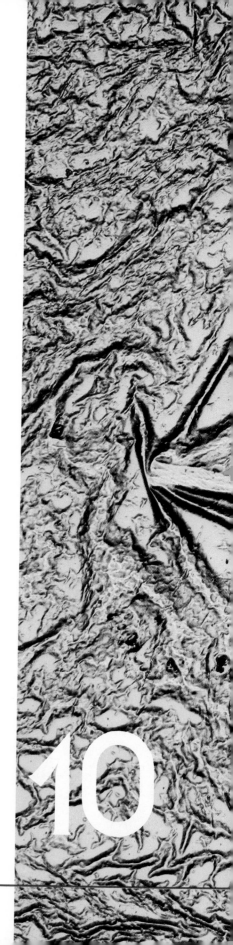

10

■ Metals   ■ Nonmetals   ■ Semimetals

Transition Elements

Period	1	2											3	4	5	6	7	8
1	H																	He
2	Li	Be											B	C	N	O	F	Ne
3	Na	Mg											Al	Si	P	S	Cl	Ar
4	K	Ca	Sc	Ti	V	Cr	Mn	Fe	Co	Ni	Cu	Zn	Ga	Ge	As	Se	Br	Kr
5	Rb	Sr	Y	Zr	Nb	Mo	Tc	Ru	Rh	Pd	Ag	Cd	In	Sn	Sb	Te	I	Xe
6	Cs	Ba	La	Hf	Ta	W	Re	Os	Ir	Pt	Au	Hg	Tl	Pb	Bi	Po	At	Rn
7	Fr	Ra	Ac	104	105	106	107											

In Chapters 5, 6, and 8 we discussed the chemistry of the halogens, carbon, phosphorus, and sulfur, which are all nonmetals. Of the approximately 100 elements about 80 are metals. This chapter is devoted to the chemistry of four typical metals, all of which are familiar to you: aluminum, iron, copper, and lead. These metals are all produced in very large quantities and have many important uses. We first describe the reactions by which the metals are obtained from their naturally occurring compounds. Then we discuss the structures of metals, that is, the arrangements of the atoms in solid metals and the nature of the forces holding the atoms together, which are called metallic bonds. We will see that the metals have particularly simple structures, which we will use later as a basis for the discussion of the structures of many other solids. Finally, we describe some of the reactions of the metals and their compounds. Consideration of these reactions will enable us to further extend our knowledge and understanding of acid–base, oxidation–reduction, and precipitation reactions.

10.1
Properties and Uses

Aluminum is in group 3 below boron in the periodic table. Lead is the last element in group 4. In this group the properties of the elements change gradually from those of carbon, a typical nonmetal, to those of lead, a typical metal. Copper and iron are members of the first transition series in the fourth period.

Some of the properties of these four elements are summarized in Table 10.1. All four show the typical properties of metals; they are good *conductors of heat and electricity*, they are *malleable* and *ductile*, and they have a shiny surface that reflects light, a so-called *metallic luster*. The shiny surface is not always apparent, because many metals become coated with a layer of oxide or carbonate when they are exposed to the air. If the metal is cut or the surface layers are scraped away, the bright shiny surface is exposed.

For many practical purposes **alloys** are often more useful than pure metals. An alloy is a solid that has metallic properties and that is usually composed

Samples of (clockwise) copper, lead, aluminum, and iron

				Melting	Boiling		Electrical
Element	Z	Electron Configuration	Abundance (%)	Point (°C)	Point (°C)	Density (g cm^{-3})	Conductivity (A m^{-1} V^{-1})
Aluminum	13	[Ne] $3s^2 3p^1$	7.5	660	2467	2.71	37.7×10^6
Iron	26	[Ar] $3d^6 4s^2$	4.7	1535	2750	7.86	9.9×10^6
Copper	29	[Ar] $3d^{10} 4s^1$	0.0058	1083	2567	8.97	59.6×10^6
Lead	82	[Xe] $4f^{14} 5d^{10} 6s^2 6p^2$	0.002	327	1740	11.40	4.8×10^6

TABLE 10.1
Some Properties of Aluminum, Iron, Copper, and Lead

of two or more metals. The alloys of copper include *brass*, an alloy of copper and zinc, and *bronze*, an alloy of copper and tin. An alloy may also be composed of a metal and a nonmetal, such as carbon, silicon, or nitrogen. In a few cases alloys are compounds, but more frequently they are solid solutions or simply mixtures. Alloys are often far stronger and harder than their constituent metallic elements.

ALUMINUM

Because aluminum is a rather reactive metal, it is not found in nature in the free state; however, its compounds are very common. After oxygen and silicon, aluminum ranks as the third most abundant element on the earth's surface (Table 3.2) and it is the most abundant of all the metals. It occurs as aluminum oxide, Al_2O_3, in the minerals corundum and bauxite, as spinel, $MgAl_2O_4$, and as many complex aluminosilicates such as beryl, $Be_3Al_2Si_6O_{18}$.

Although pure aluminum is a rather soft and weak metal, it has several hard, strong alloys, such as duralumin, which contains copper, manganese, and magnesium. Because these aluminum alloys have a low density and are not subject to corrosion, and because the metal surface is protected by a thin hard unreactive film of aluminum oxide, Al_2O_3, they have many important uses. For example, they are used in aircraft and space vehicles and for garden furniture, door and window frames, and kitchen utensils. Billions of beer and soft drink cans are manufactured each year from aluminum, and in the form of foil it is used as a wrapping material. Other applications of aluminum are based on the fact that it is an excellent conductor of electricity. Although a wire of aluminum has only one-third the conductivity of a copper wire of the same diameter, the aluminum wire is much lighter because of the low density of aluminum. Since it is also considerably cheaper than copper, aluminum is being used increasingly for electric wiring and is widely used in high-voltage transmission lines.

IRON

Iron ranks as the fourth most abundant element in the earth's crust and the second most abundant metal (see Table 3.2). It is found only in very small amounts on the earth's surface in the free state, usually in association with nickel. But it is believed to form a large fraction of the earth's core, and it is

Articles made from aluminum

Jules Verne in his prophetic story *From Earth to the Moon* (1865) chose aluminum as the perfect material for his projectile. As one of his characters remarked, "It is easily wrought, it is very widely distributed forming the basis of most rocks, is three times lighter than iron, and seems to have been created for the express purpose of furnishing us with the material for our projectile." It is interesting that at that time aluminum was a rare, little-known metal as no practical method for extracting it from its ores had been developed.

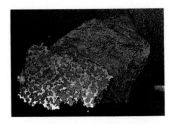

An iron-nickel meteorite

Pyrite

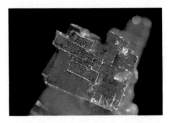

Cuprite

Galena

commonly found in meteorites. Most of the iron in the earth's crust is combined with oxygen, silicon, and sulfur. Important iron minerals include the oxides *hematite*, Fe_2O_3, and *magnetite*, Fe_3O_4, and the carbonate $FeCO_3$, *siderite*. The sulfide *pyrite*, FeS_2, also called "fool's gold" because of a superficial resemblance to gold, is another abundant iron mineral. But it is not used as a source of iron because of the difficulty of removing the large amount of sulfur.

Pure iron is a rather soft silvery metal that very few people have ever seen. Obtaining it free of carbon, oxygen, nitrogen, phosphorus, sulfur, and other metals is difficult; and pure iron so obtained has no use because of its poor mechanical properties. However, the alloys of iron have a greater industrial importance than any other metal. Almost all iron is produced in the form of steel, which is an alloy of iron containing carbon and certain transition metals, such as nickel, chromium, manganese, and vanadium. Steel is much harder and stronger than pure iron. It is used in buildings, bridges, ships, cars, machinery of all kinds, and many other familiar items.

COPPER

Copper is a shiny red metal that is resistant to corrosion. Very few metals are found on the surface of the earth in the uncombined state, and then only in small amounts. Among these copper, silver, and gold are by far the most important. They were the earliest known metals because they are found in the free state, whereas other metals must be obtained from their compounds, and it took time to develop the processes for doing this (see Box 10.1).

Because of their availability as the free elements and their resistance to corrosion, these metals have been used since antiquity for coins, and they are called the *coinage metals*. However, as the price of these metals has increased dramatically in recent times, their use in coins has decreased. Today gold and silver are not so used at all, and even copper is being replaced by other metals.

Because copper has a very high electrical conductivity, second only to silver, and because it is rather soft and very ductile and thus easily drawn out into thin wires, one of its most important uses is for electric wiring. Elemental copper occurs in amounts too small to be a useful source of the metal, which is therefore obtained from its minerals, which include the sulfides *chalcocite*, Cu_2S, and *chalcopyrite*, $CuFeS_2$, the oxide *cuprite*, Cu_2O, and the hydroxide carbonate *malachite*, $Cu_2(CO_3)(OH)_2$.

LEAD

Pure lead is a soft metal with a high density and a relatively low melting point. When freshly cut, it has a silvery luster. But when it is exposed to the air, a surface layer of oxide and carbonate forms, and the lead turns a dull blue-gray.

Although it is not abundant, lead is easily extracted from its most important ore, *galena*, PbS. As a result, it has been known for a long time. In ancient Rome lead pipes were used for the water supply of villas; buildings such as the Pantheon had roofs sheathed in lead; and even wine casks were lined with lead. In the Middle Ages lead was used extensively in the building of the great cathedrals, for roofs, gutters, and stained glass windows. And until quite recently lead was extensively used for domestic water pipes.

Lead forms slightly soluble lead hydroxide, $Pb(OH)_2$, in contact with water and oxygen. It is also soluble in dilute acids such as acetic acid in the presence

B O X 1 0 . 1

DISCOVERY AND USE OF METALS IN HISTORY

Articles of copper, silver, and gold have been found in the remains of the earliest civilizations. These objects were made from the naturally occurring metals simply by shaping them with stone hammers and chisels. Other metals were not available until people had learned how to extract a metal from one of its ores; we do not know how this momentous discovery was first made. The ores of copper, lead, tin, and mercury are easily reduced to the metal by heat or by heating with charcoal. Thus it seems likely that the formation of a metal was first observed when a fire was built on an ore-bearing rock or when an ore was dropped accidentally into a fire. One plausible suggestion is that copper was obtained accidentally from its ore malachite, $Cu_2(CO_3)(OH)_2$. In early Egypt malachite, which is green, was used for facial adornment. When it is heated, it decomposes to give copper oxide, which, when heated with carbon, is reduced to the metal. If malachite had been accidentally dropped into a charcoal fire, metallic copper would have been formed.

Before 3000 B.C., perhaps because copper and tin ores frequently occur together, people discovered that copper could be hardened by alloying it with tin. The resulting bronze could be forged into utensils, tools and weapons so superior to others then available that this discovery produced a major revolution in technology, initiating the period known as the Bronze Age.

The next great step forward was made about 1000 B.C. with the discovery of iron smelting, the production of iron from iron ore. This discovery marked the beginning of the Iron Age. The rise and fall of early civilizations was closely tied to progress and discoveries in metallurgy.

By the time of Christ, people had learned how to make and to use the metals iron, copper, silver, tin, gold, mercury, and lead. But the nature of the processes involved remained an engima until the end of the eighteenth century. Until then, the extraction of metals from their ores was an art and certainly not a science. Lack of understanding of the processes involved in making metals encouraged the belief that one metal might be changed into another, leading to the alchemists' futile attempts to transmute lead into gold.

All the metals known to early humans, except iron, are rare elements, but they either occur in the uncombined state or can be obtained easily from their ores. The most abundant metal, aluminum, is difficult to obtain from its ores and was not prepared as the pure element until 1825; a practical method for the large-scale industrial production of aluminum was not devised until 1886. (See Chapter 17)

of oxygen. Lead compounds are toxic and measures are now being taken to reduce our exposure to them. The Romans were unaware of the toxicity of lead compounds. Not only did they store wine in lead containers but they even added lead acetate to wine to sweeten it. Thus it seems that many Romans, particularly the rich governing class, suffered from lead poisoning, which led to both madness and sterility, and some historians claim that lead poisoning was one cause of the decline of the Roman Empire.

Today lead is used for making storage batteries, for covering electric cables, and as an important component of several alloys such as type metal and solder. Large quantities have been used for the production of the gasoline additive tetraethyl lead, $Pb(C_2H_5)_4$. But this use is now rapidly declining, since the health hazards associated with "leaded" gasoline have been recognized.

10.2
Metallurgy: Extraction of Metals

Metallurgy is the science of metals. It had its origins in the development of methods of extracting metals from their ores. An **ore** is a mineral deposit that may be profitably treated for the extraction of one or more metals.

Most metals occur in the earth's crust in an oxidized form, that is, as positive ions. The most important ores of many metals are either oxides or sulfides.

Metal	Ore (oxidized form of metal)
Al	Bauxite $Al_2O_3 \cdot xH_2O$
Fe	Hematite $Fe_2O_3 \cdot xH_2O$
	Pyrite FeS_2
Cu	Chalcopyrite $CuFeS_2$
	Cuprite Cu_2O
Pb	Galena PbS

Thus if we wish to obtain a metal from one of its ores, a reducing agent must be used. In order of increasing cost, the more commonly used reducing agents are carbon (usually in the form of coke), carbon monoxide, hydrogen, free electrons (in electrochemical processes; see Chapter 17), and other metals, for example, sodium and aluminum.

EXTRACTION OF METALS FROM OXIDE ORES

Often metals can be extracted from their oxide ores simply by heating the oxide with carbon, usually in the form of coke, or with carbon monoxide. This process is used, for example, with cuprite, Cu_2O:

$$Cu_2O(s) + C(s) \longrightarrow 2Cu(l) + CO(g)$$

Aluminum is sufficiently reactive to be used for reducing oxides of some less reactive metals to the metal. Some manganese and chromium are produced industrially by using aluminum powder as the reducing agent:

$$3MnO_2(s) + 2Al(s) \longrightarrow 3Mn(s) + 2Al_2O_3(s)$$

In the **thermite process** aluminum powder is used for producing small quantities of molten iron for special purposes such as the welding and repair of railway lines:

$$Fe_2O_3(s) + 2Al(s) \longrightarrow Al_2O_3(s) + 2Fe(l)$$

This reaction is strongly exothermic, producing sufficient heat to melt the iron (see Experiment 10.1).

MANUFACTURE OF IRON AND STEEL

In prehistoric times iron was prepared simply by mixing iron ore with charcoal and heating the mixture in a fired clay pot. The modern process uses the same reaction. Iron ore, coke (carbon), and limestone, $CaCO_3$, are added at the top of a blast furnace while preheated air (or pure oxygen) is blown in at the bottom (see Figure 10.1 on p. 486). The reducing agent is the carbon monoxide that is produced from the coke at the high temperature in the furnace:

$$2C(s) + O_2(g) \longrightarrow 2CO(g)$$

The iron ore, which is an impure mixture of Fe_2O_3, Fe_3O_4, and silicates, is reduced as it moves down the furnace—first to FeO and then, in the lower and hottest part of the furnace, to liquid iron, which runs down to the bottom of the furnace. The overall process is

$$Fe_2O_3(s) + 3CO(g) \longrightarrow 2Fe(l) + 3CO_2(g)$$

Since the majority of the rocks on the earth's surface are composed of silicon dioxide, SiO_2 (silica), and silicates (substances containing oxoanions of silicon such as SiO_3^{2-}), these substances are almost always present in an ore, and they must be removed before or during the reduction of the ore. Various mechanical methods are used, but an important chemical method is to add a material called a *flux* that combines with the silica and the silicates to produce a material called a *slag*. The slag is liquid at the temperature of the molten metal and does not dissolve in the molten metal; it forms a separate liquid layer, which can easily be removed. In the production of iron in the blast furnace, limestone, $CaCO_3$, is added along with the coke and ore. The limestone decomposes to CaO, which

Experiment 10.1

The Reduction of Iron(III) Oxide to Iron Using Aluminum as the Reducing Agent: The Thermite Process

The crucible contains a mixture of Fe_2O_3 and aluminum powder. A small amount of a mixture of potassium chlorate and sugar is placed on the top of the Fe_2O_3–Al mixture and a few drops of concentrated sulfuric acid are added to start the reaction.

The heat of the strongly exothermic reaction of H_2SO_4 with the $KClO_3$–sugar mixture ignites the Fe_2O_3–Al mixture, which reacts violently in a strongly exothermic reaction. A shower of white hot sparks is emitted and the crucible becomes red hot.

A ball of white hot iron can be seen glowing in the bottom of the crucible.

When the crucible has cooled, a magnet can be used to pick up the ball of iron.

combines with silica impurities in the ore to form a layer of molten slag, $CaSiO_3$, which collects on top of the iron. This slag protects the molten iron from reoxidation. Solid slag is used in building materials such as cement, concrete, and cinder blocks.

$$CaCO_3(s) \longrightarrow CaO(s) + CO_2(g)$$
$$CaO(s) + SiO_2(s) \longrightarrow CaSiO_3(l)$$

The product of the blast furnace is called *pig iron*. It contains about 4% carbon, 2% silicon, and small amounts of other elements such as phosphorus and sulfur. Most of the carbon and nearly all the other impurities are removed by melting the pig iron and blowing oxygen through it for a short time (Figure 10.2 on p. 487) in a process called the basic oxygen process. The impurities are converted to their oxides by this treatment and either are evolved as gases, such as SO_2 and CO_2, or form a slag with added calcium oxide:

$$Si \xrightarrow{O_2} SiO_2 \xrightarrow{CaO} CaSiO_3$$
$$P_4 \xrightarrow{O_2} P_4O_{10} \xrightarrow{CaO} Ca_3(PO_4)_2$$

FIGURE 10.1
Blast Furnace.

Iron ore, coke, and limestone are added at the top of the furnace, and air is blown in at the bottom. The oxygen of the air reacts with the hot coke to form carbon monoxide, which passes up the furnace and reduces the iron oxide as it moves down, first to iron oxide, FeO, and finally to iron. Liquid iron collects at the bottom of the furnace covered by a layer of molten slag, $CaSiO_3$. The iron is tapped off at the bottom of the furnace and allowed to solidify, forming pig iron.

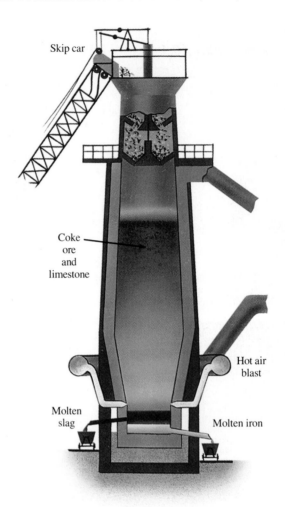

Skip car

Coke
ore
and
limestone

Hot air
blast

Molten
slag

Molten iron

The product of this process is ordinary *carbon steel*, of which there are many different kinds. They differ primarily in the amount of carbon that they contain, from less than 0.2% up to 1.5%. Addition of other metals such as nickel, chromium, manganese, tungsten, and vanadium gives a very large number of alloy steels, of which stainless steel is the most familiar; it typically contains 18% chromium and 8% nickel.

EXTRACTION OF METALS FROM SULFIDE ORES

Sulfide ores are first heated strongly in air to remove the sulfur as sulfur dioxide, SO_2, and to form the oxide. The process is known as **roasting**. It produces enormous quantities of sulfur dioxide, much of which is allowed to escape into the atmosphere and is a major source of acid rain. For example, galena, PbS, is roasted in the air to give the oxide, which is then reduced to lead with coke in a small blast furnace:

$$2PbS(s) + 3O_2(g) \longrightarrow 2PbO(s) + 2SO_2(g)$$
$$2PbO(s) + C(s) \longrightarrow 2Pb(l) + CO_2(g)$$

Metal sulfide + O_2(g)
↓
Metal oxide + SO_2(g)

Metal oxide + C(coke)
↓
Metal + CO_2(g)

FIGURE 10.2
Basic Oxygen Furnace for the Production of Steel.

A typical basic oxygen furnace is charged with about 200 tons of molten pig iron, 100 tons of scrap iron, and 20 tons of limestone. A stream of hot oxygen gas is blown through the mixture to oxidize the impurities, which either escape as their gaseous oxides or form a slag with the calcium oxide formed from the limestone.

The copper ore $CuFeS_2$ is first roasted in a limited supply of air. This roasting converts the iron to FeO and removes some of the sulfur as SO_2 but leaves the copper as Cu_2S:

$$2CuFeS_2(s) + 4O_2(g) \longrightarrow Cu_2S(s) + 2FeO(s) + 3SO_2(g)$$

The FeO forms a slag of $FeSiO_3$ with any silica that is present or is added, and this slag is removed, leaving a relatively pure molten copper sulfide, Cu_2S. Air is then blown through the molten Cu_2S, converting it to the metal:

$$Cu_2S(l) + O_2(g) \longrightarrow 2Cu(l) + SO_2(g)$$

This copper is then further purified by electrolysis (Chapter 17).

EXERCISE 10.1

Assign oxidation numbers to each of the elements in the following reactions and hence decide which elements are oxidized and which are reduced. Which of the reactions is not an oxidation–reduction reaction?

(a) $Cu_2O(s) + C(s) \rightarrow 2Cu(s) + CO(g)$

(b) $Fe_2O_3(s) + 2Al(s) \rightarrow Al_2O_3(s) + 2Fe(l)$

(c) $CaO(s) + SiO_2(s) \rightarrow CaSiO_3(l)$

(d) $2PbS(s) + 3O_2(g) \rightarrow 2PbO(s) + 2SO_2(g)$

(e) $Cu_2S(l) + O_2(g) \rightarrow 2Cu(l) + SO_2(g)$

10.3
Structure of Metals

Copper, aluminum, iron, and lead—and indeed all metals except mercury—are crystalline solids at room temperature. Usually, a metal consists of many tiny crystals that are not visible to the eye, although they can often be seen under the microscope. Occasionally, small gold and silver crystals are found in nature (see Figure 10.3).

CLOSE-PACKED ARRANGEMENTS OF SPHERES: HEXAGONAL AND CUBIC CLOSE-PACKED STRUCTURES

The crystalline form of metals implies that the atoms in the structure are packed together in a regular manner. Because metals consist of identical spherical atoms, their structures are relatively simple: they are almost all based on two ways that spheres may be packed together as closely as possible. If we place identical spherical balls in a slightly inclined tray and gently push them together as closely as possible, we obtain a **close-packed arrangement** (see Figure 10.4 and Experiment 10.2). The spheres are arranged in straight rows that make angles of 60° with each other. Each sphere is surrounded by a regular hexagonal arrangement of six other spheres. Figure 10.4 also shows how the rows of atoms define the edges of "two-dimensional" crystals of different shapes. Although these two-dimensional crystals may have various shapes, the angles between the edges are always 60° or 120°.

When layers of close-packed atoms are stacked so that the atoms of one layer nestle in the hollows between the atoms in the adjacent layer, a three-dimensional, close-packed structure is obtained. There are two ways in which close-packed layers can be stacked, and thus there are two closely related but distinct *close-packed structures*. In **hexagonal close packing** the spheres in the third layer lie directly over the spheres of the first layer. The sequence of layers in the hexagonal close-packed structure is therefore described as ABABAB . . . (Figure 10.4). In **cubic close packing** the spheres in the third layer do *not* lie above the spheres of the first layer. The sequence of layers in the cubic close-packed structure is therefore described as ABCABCABC . . . . It is not until the fourth layer is reached that the atoms lie directly above those in the first layer (Figure 10.5). In both forms of close packing each sphere is in contact with twelve neighboring spheres, six in its own layer, three in the layer above, and three in the layer below (Figure 10.6 on p. 490).

FIGURE 10.3
Crystals of Metals.

(a) Gold and (b) silver are sometimes found naturally in crystalline forms. (c) This section through a piece of zinc that has been allowed to solidify very slowly shows a mass of interlocking crystals.

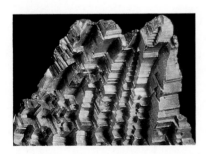

(a)

(b)

(c)

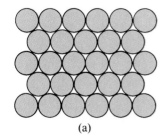

(a)

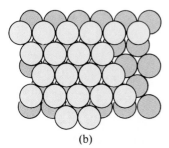

(b)

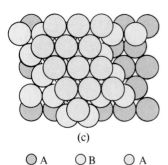

(c)

 A   ◯ B   ◯ A

FIGURE 10.4
Hexagonal Close Packing.

(a) A close-packed layer of spheres.
(b) A second close-packed layer can be placed so that it nestles in the depressions in the first layer.
(c) A third layer of spheres may be added in two ways. We may place the third layer in the depressions of the second layer so that it lies directly above the spheres in the first layer. If we call the bottom layer an A layer and the second layer a B layer, then since the third layer is an exact replica of the first layer, we may also label it an A layer. Thus we have an arrangement in which the layers alternate in position, and so it is an ABABAB... arrangement, which is called hexagonal close packing.

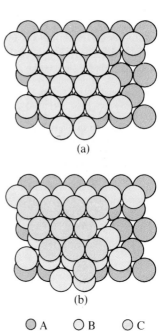

(a)

(b)

◯ A   ◯ B   ◯ C

FIGURE 10.5
Cubic Close Packing.

(a) Two close-packed layers. (b) A third layer may be placed so that it does *not* lie above the spheres in the bottom layer. Only when a fourth layer is added do the spheres lie directly above the spheres in the first layer. This arrangement is therefore different from the ABABA... arrangement. It is an ABCABCA... arrangement, which is called cubic close packing.

___ *Experiment 10.2*

Close Packing of Spheres

When a fine jet of air is bubbled through a soap solution a layer of soap bubbles forms on the surface of the solution.

Each bubble is surrounded by a hexagonal arrangement of six more bubbles giving a two-dimensional close-packed arrangement.

FIGURE 10.6
Cubic and Hexagonal Close Packing.

In both forms of close packing each sphere is in contact with twelve neighboring spheres, six in its own layer, three in the layer above and three in the layer below. No more than twelve spheres all of the same size can be placed so that they all touch a single sphere, and there are two ways of arranging these twelve spheres. In (a) we see how the twelve spheres are arranged in hexagonal close packing and in (b) we see how they are arranged in cubic close packing.

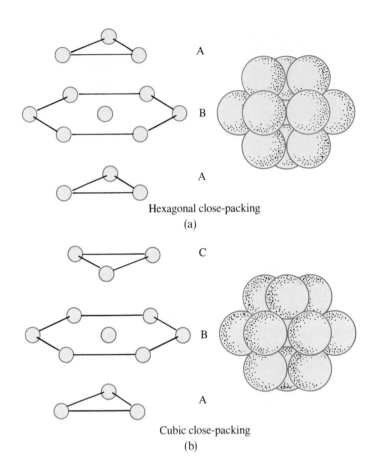

Hexagonal close-packing
(a)

Cubic close-packing
(b)

BODY-CENTERED CUBIC STRUCTURE

A majority of the metals have one of these two close-packed structures. Most of the remaining metals have a third type of structure, called the **monatomic body-centered cubic structure**. To understand this structure, consider first a single layer of spheres packed in a square arrangement (Figure 10.7). Although the spheres are packed in a regular way, they are not close packed; each sphere has only four rather than six close neighbors, and a given number of spheres occupy a greater total area than do the same number in the close-packed arrangement. If layers of this type are stacked directly one on another, simple cubic-packed arrangement is obtained. If such an arrangement is expanded slightly so that the spheres no longer quite touch each other, an additional sphere can be inserted in the middle of each cube. This arrangement is the monatomic body-centered cubic structure, in which each atom is in contact with only 8 neighbors rather than 12, as in a close-packed structure (Figure 10.7).

In the two close-packed arrangements, the spheres occupy 74% of the total volume; the rest of the space is the interstices ("holes") between the spheres. In the body-centered cubic structure the spheres occupy only 68% of the total volume.

Copper, aluminum, and lead have the cubic close-packed structure, whereas iron has the body-centered cubic structure. Common metals having the hexagonal close-packed structure include zinc and magnesium. A few metals have different, more complicated structures.

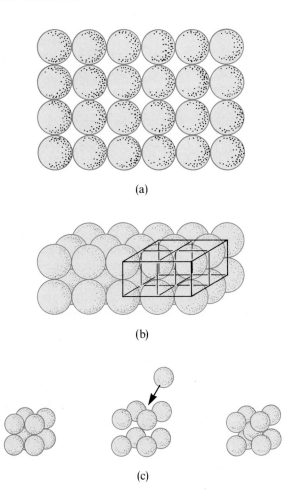

FIGURE 10.7
Simple Cubic and Body-
Centered Cubic Packing.

(a) A square-packed layer. (b) A
second layer is stacked directly
above the first layer to give a
simple cubic packing. (c) The large
holes remaining in the middle of
each cube are not large enough
to accommodate another sphere.
But if each cube is expanded by
15% another sphere can be inserted in the center of each cube,
thereby giving a body-centered
cubic packing.

10.4
Metallic Bonding

Metal atoms have low ionization energies; one, two, and sometimes three electrons can be rather easily removed to give positive ions. Because of their small
core charges, metals have little tendency to accept electrons to form negative
ions. Hence we do not expect a metal to have an ionic structure consisting of
positive and negative ions. Covalent bonding also is not possible in solid metals,
because a metal atom does not have enough valence electrons to form covalent
bonds to all its 12 or 8 neighboring atoms. A sodium atom, for example, has
only one valence electron. A sodium atom could therefore form a localized
covalent bond to only one of its neighbors (see Figure 10.8 on p. 492). Even in
aluminum each atom could form bonds to only 3 of the neighboring atoms.

Metals, in fact, have a type of bonding that differs from both ionic and
covalent bonding; it is called **metallic bonding**. We will describe metallic bonding
initially by using sodium as an example. Sodium is simple to discuss because
it has only one valence electron, but the model that we will develop applies to
all metals.

The 3s orbital of a sodium atom can overlap with the 3s orbital of another
sodium atom to form a covalent bond and give a diatomic Na_2 molecule, as

FIGURE 10.8
Covalent Bond
Representation of Sodium
Crystal.

A sodium atom can form a bond
to only one of its four nearest
neighbors in the two-dimensional
representation of sodium shown
here.

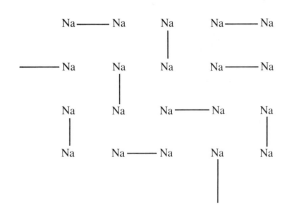

shown in Figure 10.9, just as the 1s orbital of a hydrogen atom overlaps with that of another hydrogen atom in the formation of the H_2 molecule. But the Na_2 molecule is found only in the gas phase. In the solid state the 3s orbital of each sodium atom overlaps with those of all its close neighbors to give an electron charge cloud that is spread continuously over all the sodium atoms in the crystal (Figure 10.9).

Each sodium atom in metallic sodium may be considered to have given up its single 3s electron to give a sodium ion, Na^+, and a "free" electron. These electrons form a cloud of negative charge, a kind of electron gas, that occupies the space between the positive ions and can move freely between them. The electrostatic attraction between the cloud of negative electrons and the positive ions holds them together. The **metallic bond** may be defined as the force of attraction between a mobile charge cloud (electron gas) and positively charged atomic cores. In any metal each atom may be considered to have contributed some or all of its valence electrons to this mobile electron cloud. For example, aluminum consists of Al^{3+} ions held together by an electron cloud made up of three electrons from each atom (see Figure 10.10).

As we have seen before, resonance structures are used to describe the distribution of electrons in ions and molecules when they are not localized, as implied by a normal Lewis structure. A metal represents an extreme case of electron delocalization; the electrons may be regarded as being delocalized over the whole metal crystal. We could, in principle, represent a metal by a very large number of resonance structures, such as the structures given in Figure 10.8, each one having a different arrangement of bonds, but this model is clearly

FIGURE 10.9
Metallic Bond in Sodium.

(a) The 3s orbitals of two isolated sodium atoms overlap in the formation of a diatomic Na_2 molecule. (b) In solid sodium the 3s orbital of any one sodium atom overlaps with those of all its neighbors. (c) Thus the electron density is spread out among all the sodium atoms to form a cloud of freely moving electrons—a negative charge cloud or electron gas. The electrostatic attraction between the electron cloud and the positive ions constitutes the metallic bond.

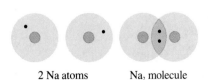

2 Na atoms Na_2 molecule

(a)

(b)

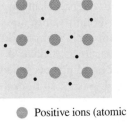

● Positive ions (atomic cores)

· Electrons

(c)

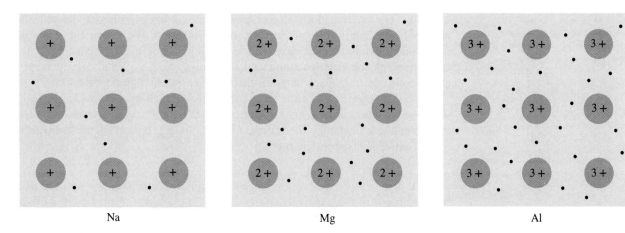

Na Mg Al

not very practical. The delocalized electron cloud model is much more convenient, and it provides a satisfactory explanation for most of the properties of metals.

Just as many compounds have bonds that are intermediate in character between covalent and ionic, so other compounds have bonds that can be described as intermediate between metallic and ionic or between metallic and covalent. Each of the three types of chemical bond—covalent, ionic, and metallic—represents an extreme limiting case. In most compounds the bonds have an intermediate character, although it is often convenient to classify compounds according to their predominant bonding type.

PROPERTIES OF METALS

The charge cloud (electron gas) model of metallic bonding is a very simplified description of a metal. Nevertheless, we can use this model to provide an elementary explanation of most of the typical properties of metals.

MELTING POINT AND STRENGTH OF THE METALLIC BOND Consider the three metals sodium, magnesium, and aluminum in period 3. Since sodium has one electron in its valence shell, magnesium two, and aluminum three, there are twice as many electrons in the charge cloud of magnesium and three times as many in the charge cloud of aluminum as in sodium (see Figure 10.10). Since the core charge also increases from sodium, $+1$, to aluminum, $+3$, the strength of the attraction between the atomic cores and the charge cloud increases from sodium to magnesium to aluminum. Therefore we expect the strength of the metallic bonding to increase in the same order.

To melt a metal, the atoms must be separated slightly so that they have sufficient space to move around each other. The stronger the bonds between the atoms, the more energy will be needed to separate them and therefore the higher will be the melting point. Other factors such as the structure of the metal also affect the melting point, but melting point does give us a rough indication of the strength of the metallic bonding. The melting points of the alkali and alkaline earth metals and aluminum are given in Table 10.2. In accordance with our model, the melting points increase in the series sodium, magnesium, aluminum, and in general group 2 metals have higher melting points than the group

FIGURE 10.10
Metallic Bonds in Sodium, Magnesium, and Aluminum.

The number of electrons contributed to the metallic bond increases from one per sodium atom, to two for each magnesium atom, to three for each aluminum atom. Thus the strength of the metallic bond increases from sodium to magnesium to aluminum.

TABLE 10.2
Melting Points (°C) of Some Metals

Li	Be	
180	1280	
Na	Mg	Al
98	650	660
K	Ca	
64	838	
Rb	Sr	
39	770	
Cs	Ba	
29	725	

1 metals. The melting point decreases with increasing atomic size down both groups 1 and 2, although magnesium is an exception. We expect the strength of the metallic bond to decrease with increasing atomic size as the average distance between the electron cloud and the center of each atomic core increases and the electrostatic force between them decreases correspondingly.

Note that copper and iron have much higher melting points (Table 10.1) than the groups 1 and 2 metals, indicating that the metallic bonds are very strong in these elements. Strong metallic bonds are typical of the transition metals as we will see in Chapter 21.

ELECTRICAL CONDUCTIVITY Since the valence electrons in a metal are free to move between the positively charged ions, an electric current, which consists of a stream of negative electrons, will flow through a metal under the influence of an applied voltage, provided, for example, by a battery. This type of conduction, which is due to the movement of electrons, is called **metallic conduction**. Silver is the most highly conducting metal, followed by the other coinage metals, copper and gold, and then aluminum.

A metal conducts electricity in both the solid and liquid states. In contrast, most ionic substances conduct electricity only in the liquid state, and covalent substances do not conduct either in the solid or in the liquid state. They are said to be *insulators*. In a solid ionic crystal the ions are not free to move; in the liquid state the ions can move, and the current consists of moving positive and negative ions. Covalent solids or liquids have no ions, and the electrons are not free to move because they are held firmly to the atoms as bonding or nonbonding pairs, so they are nonconductors. Graphite is an exception because there are delocalized electrons in each layer (Figure 6.3) that can move just like the electrons in a metal. So graphite is a good conductor in the directions along its planes, but it is a poor conductor in the direction perpendicular to its planes.

Whereas the conductivity of an ionic conductor decreases with decreasing temperature, the conductivity of a metallic conductor increases as the temperature is decreased. As a molten ionic substance or a solution of an ionic substance is cooled, the speeds of the ions decrease. Thus under the influence of an applied voltage, the distance moved by the ions in a given time decreases with decreasing temperature; hence the conductivity decreases. In a metal, on the other hand, the small and extremely light electrons move at very high speeds, even at low temperatures. They are hindered in their motion by collisions with the vibrating positive ions (see Figure 10.11). Because the amplitude of the vibrations of the ions decreases with decreasing temperature, they interfere to a lesser extent with the motion of the electrons. Thus the conductivity of the metal increases as the temperature is decreased.

HEAT CONDUCTION The transfer of heat energy results from collisions of the fast-moving particles in the hotter part of a material with the slower-moving particles in the cooler part. The slower particles are speeded up by these collisions, and the temperature of this part of the material therefore rises. At a given temperature all the particles—electrons and nuclei—in a solid have the same average kinetic energy. But because electrons have a much smaller mass than nuclei, they move at much higher speeds than the nuclei. Thus the free

Copper tubing

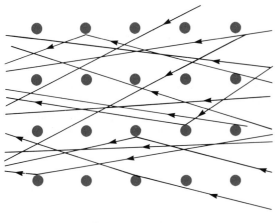

Low temperature

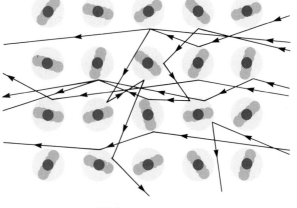

High temperature

FIGURE 10.11
Effect of Temperature on
Electrical Conductivity.

The movement of electrons through a metal is impeded by their collisions with the atomic cores. The atomic cores are vibrating, and the amplitude of the vibrations increases with increasing temperature. The greater the amplitude of these vibrations, the greater is the effective volume occupied by the core, and the greater is the chance that an electron will collide with an atomic core and that its motion through the solid will be impeded. With decreasing temperature the amplitude of the vibrations of the atomic cores decreases, and the chance of an electron colliding with a core decreases correspondingly. Hence the electrons are less impeded by the atomic cores, and the conductivity increases correspondingly.

electrons move rapidly from one part of a metal to another, making metals very good conductors of heat. Covalent and ionic substances are generally poor conductors of heat because they have no free electrons and because their atoms and ions move only relatively slowly and to a limited extent about fixed positions.

MECHANICAL PROPERTIES Because the electrons that form the bonds in a metal are constantly moving and are not fixed in position between any particular atoms, the atoms in a solid metal can be moved with respect to each other without breaking the bonds between them. For this reason a metal can be distorted in shape relatively easily; that is, it is malleable and ductile. For example, steel can be rolled out into thin sheets for making automobiles and household appliances, and copper can be drawn out into thin wires for use in electric circuits. Most covalent and ionic solids cannot be distorted in this way; they break into pieces if they are placed under stress in an attempt to alter their shape, because bonds must be broken if the atoms or ions are moved relative to each other.

LUSTER Light falling on a metal surface causes the loosely held electrons of the metal to vibrate at the frequency of the light. A vibrating charge, such as an electron, emits electromagnetic waves, and therefore it emits light at the same frequency as the incident light. Thus light is reflected from a metal surface, which therefore has a shiny appearance.

THERMIONIC EFFECT The free electrons in a metal have a distribution of speeds just like the molecules of a gas (Chapter 3). When a metal is heated, the speeds of the electrons increase; at a certain temperature a few acquire enough energy to escape from the metal. If they are attracted away from the metal by a positive potential, a flow of electrons—in other words, an electric current—is obtained. This phenomenon is known as the **thermionic effect**. In a television picture tube the electrons emitted by a heated metal wire are accelerated and focused into a beam, which moves across the coated end of the tube to produce the visible pattern that is the television picture.

10.5

Reactions and Compounds of Aluminum, Iron, Copper, and Lead

In this section we describe some of the reactions of the metals aluminum, iron, copper, and lead and some of their more important compounds. Very many of these reactions are acid–base, oxidation–reduction, or precipitation reactions. In considering these reactions, we outline a systematic method for balancing oxidation–reduction equations, and we introduce the concept of Lewis acids and bases.

OXIDATION STATES OF ALUMINUM, IRON, COPPER, AND LEAD

As we saw in Chapter 8, classifying the compounds of an element according to the oxidation state of the element in the compound is very convenient. Table 10.3 shows the principal oxidation states of aluminum, iron, copper, and lead. Because they are metals, they have only positive oxidation states. Aluminum has only one oxidation state, $+3$, while the other metals each have two common oxidation states, $+2$ and $+3$ for iron, $+1$ and $+2$ for copper, and $+2$ and $+4$ for lead. The $+3$ oxidation state of aluminum is expected from its position in group 3 of the periodic table and its valence-shell electron configuration $3s^2 3p^1$. Similarly the $+4$ oxidation state of lead is expected from its position in group 4 and its valence-shell electron configuration $6s^2 6p^2$, but it also has a common oxidation state of $+2$ in which only the 6p electrons are used in compound formation. The oxidation states of iron and copper cannot be so easily predicted from their positions in the periodic table. All the elements in the first series of transition metals have an oxidation state of $+2$. In addition iron has a common oxidation state of $+3$ and copper has a $+1$ oxidation state. We look in more detail at the relationship between the oxidation states of iron and copper and their electron configurations in Chapter 21. For the present it is important to remember these common oxidation states.

TABLE 10.3
Oxides and Chlorides of Aluminum, Iron, Copper, and Lead

Oxidation State	Al	Fe	Cu	Pb
$+1$			Cu_2O; $CuCl$	
$+2$		FeO; $FeCl_2$; Fe_3O_4[a]	CuO; $CuCl_2$	PbO; $PbCl_2$; Pb_3O_4[a]
$+3$	Al_2O_3; $AlCl_3$	Fe_2O_3; $FeCl_3$		
$+4$				PbO_2; $PbCl_4$

Note: Except for $PbCl_4$, which is a covalent liquid, all these compounds are solids and have three-dimensional network structures with predominantly ionic bonding. In the gas phase, however, $FeCl_3$ and $AlCl_3$ are covalent.

[a] Fe_3O_4 and Pb_3O_4 are mixed-oxidation-state oxides; they have metal atoms in two different oxidation states.

REACTIONS OF METALS

Metals are readily oxidized to positive ions such as M^+, M^{2+}, or M^{3+}. In their reactions with nonmetals the metal is oxidized and the nonmetal is reduced to give a product that is usually an ionic compound. For example,

$$2Na(s) + Cl_2(g) \longrightarrow 2NaCl(s)$$
$$2Mg(s) + O_2(g) \longrightarrow 2MgO(s)$$

Aluminum, iron, copper, and lead undergo similar reactions. We will consider their reactions with the important nonmetals oxygen, sulfur, and the halogens, to give oxides, sulfides, and halides, respectively. For example,

$$Cu(s) + S(s) \longrightarrow CuS(s)$$
$$2Fe(s) + 3Cl_2(g) \longrightarrow 2FeCl_3(s)$$

From the solubility rules in Table 5.9 we see that the oxides and sulfides of aluminum, iron, copper, and lead, like the oxides and sulfides of most other metals except the alkali and alkaline earth metals, are insoluble in water, but most of the halides of these metals are soluble in water. Remembering this information about solubilities will help us understand the reactions by which these compounds may be prepared in solution. In addition to the compounds that can be prepared by direct reaction of a metal with a nonmetal the other important compounds of metals that we will consider are the hydroxides and the salts of oxoacids, in particular the sulfates and nitrates. The hydroxides of most metals, including copper, iron, aluminum, and lead, are insoluble in water (Table 5.9) and they can therefore be prepared by precipitation reactions such as

$$Cu^{2+}(aq) + 2OH^-(aq) \longrightarrow Cu(OH)_2(s)$$

Sulfates and nitrates are mostly soluble in water (Table 5.9). They may be prepared by the reaction of the metal with the appropriate oxoacid. For example, as we saw in Chapter 8,

$$Cu(s) + 2H_2SO_4(aq, concn) \xrightarrow{\text{Heat}} CuSO_4(aq) + 2H_2O(l) + SO_2(g)$$

Some halides can also be prepared by the reaction of the appropriate hydrohalic acid with the metal. For example,

$$2Al(s) + 6HCl(aq) \longrightarrow 2AlCl_3(aq) + 3H_2(g)$$

We have previously seen (Chapter 5) that many other metals, but by no means all metals, react with the hydrohalic acids in this way. For example, magnesium reacts with hydrochloric acid as follows:

$$Mg(s) + 2HCl(aq) \longrightarrow MgCl_2(aq) + H_2(g)$$

We will next consider in more detail the reactions of copper, iron, aluminum, and lead with acids, and then we will discuss some of the important compounds of each of these metals.

REACTIONS WITH ACIDS

Many metals react with *dilute* aqueous solutions of acids; hydrogen is evolved and a solution of an ionic salt of the metal is formed. For example, aluminum and iron react with hydrochloric acid in this way (see Experiment 10.3):

$$Fe(s) + 2HCl(aq) \longrightarrow FeCl_2(aq) + H_2(g)$$

$$2Al(s) + 6HCl(aq) \longrightarrow 2AlCl_3(aq) + 3H_2(g)$$

Aluminum and iron react in a similar way with other acids, such as HBr and H_2SO_4, to give hydrogen and the corresponding salt of the acid:

$$2Al(s) + 6HBr(aq) \longrightarrow 2AlBr_3(aq) + 3H_2(g)$$

$$Fe(s) + H_2SO_4(aq) \longrightarrow FeSO_4(aq) + H_2(g)$$

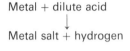

Metal + dilute acid

↓

Metal salt + hydrogen

The reactions of iron and aluminum with dilute acids are examples of a general redox reaction in which a metal is oxidized to a positive ion by the hydrogen in hydronium ion, H_3O^+, which is reduced to elemental hydrogen, H_2:

$$2H_3O^+ + 2e^- \longrightarrow 2H_2O + H_2(g)$$

We have previously seen that magnesium and zinc also react in this way. Since acids such as HCl and H_2SO_4 are ionized to give H_3O^+ in dilute aqueous solution, the reaction between a divalent metal and an aqueous acid may be

Experiment 10.3

Reactions of Metals with Acids

When dilute hydrochloric acid is added to zinc (left) and iron (right) the metals dissolve and bubbles of hydrogen gas are evolved.

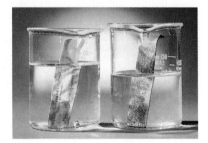

When dilute hydrochloric acid is added to copper (left) and lead (right) the metals do not react.

Reactions with nitric acid. Left: Copper dissolves in concentrated nitric acid, forming a dark green solution of $Cu(NO_3)_2$ and brown NO_2 gas. Right: Copper dissolves in dilute nitric acid, forming a blue solution of $Cu(NO_3)_2$ and colorless NO gas.

described by the general equation

$$M(s) + 2H_3O^+(aq) \longrightarrow M^{2+}(aq) + H_2(g) + 2H_2O(l)$$

For example,

$$Fe(s) + 2H_3O^+(aq) \longrightarrow Fe^{2+}(aq) + H_2(g) + 2H_2O(l)$$

This equation is frequently written in the simpler form

$$M(s) + 2H^+(aq) \longrightarrow M^{2+}(aq) + H_2(g)$$

However, we must remember that H^+ is always associated with a water molecule to form H_3O^+. The essential reaction in each case is the oxidation of the metal to M^{2+} and the reduction of H^+ (H_3O^+) to hydrogen, H_2. There is a transfer of electrons from the metal to H^+ (H_3O^+). Similar equations can be written for the formation of M^+ and M^{3+}.

Example 10.1

THE REACTIONS OF DILUTE AQUEOUS ACIDS WITH METALS

Write balanced equations for the following reactions:

(a) Iron with dilute sulfuric acid

(b) Magnesium with dilute sulfuric acid

(c) Aluminum with dilute sulfuric acid

Solution

(a) From the general reaction above we can write

$$Fe(s) + 2H^+(aq) \longrightarrow Fe^{2+}(aq) + H_2(g)$$

If we add sulfate ion, $SO_4{}^{2-}$, to each side of this equation we obtain the equation that we require:

$$Fe(s) + H_2SO_4(aq) \longrightarrow FeSO_4(aq) + H_2(g)$$

(b) Mg is in group 2 and therefore forms the Mg^{2+} ion. Thus the equation is exactly similar to that for iron:

$$Mg(s) + H_2SO_4(aq) \longrightarrow MgSO_4(aq) + H_2(g)$$

(c) Aluminum forms the Al^{3+} ion; therefore we have

$$2Al(s) + 6H^+(aq) \longrightarrow 2Al^{3+}(aq) + 3H_2(g)$$

Adding $3SO_4{}^{2-}$ to each side of the equation gives

$$2Al(s) + 3H_2SO_4(aq) \longrightarrow Al_2(SO_4)_3(aq) + 3H_2(g)$$

EXERCISE 10.2

Write balanced equations for the following reactions:

(a) The reaction of dilute hydrochloric acid with chromium to give $CrCl_2(aq)$

(b) The reaction of zinc with dilute sulfuric acid

(c) The reaction of aluminum with dilute perchloric acid, $HClO_4$

Copper, lead, and some other metals do not react with *dilute* aqueous acid solutions because the hydronium ion, H_3O^+, is not a sufficiently strong oxidizing agent to oxidize these metals. But, as we saw in Chapter 8, copper is oxidized to Cu^{2+} by hot *concentrated* sulfuric acid, which is reduced to sulfur dioxide. In addition to H_3O^+ and HSO_4^- ions concentrated sulfuric acid contains molecular sulfuric acid, H_2SO_4, which at high temperature is a stronger oxidizing agent than H_3O^+. Lead is similarly oxidized by hot concentrated sulfuric acid to $PbSO_4$:

$$Pb(s) + 2H_2SO_4(aq, conc) \longrightarrow PbSO_4(s) + SO_2(g) + 2H_2O(l)$$

Nitric acid and the nitrate ion are strong oxidizing agents even at room temperature. Thus copper is oxidized by both concentrated nitric acid, in which the oxidizing agent is molecular HNO_3, and dilute nitric acid, in which the nitrate ion, NO_3^-, is the oxidizing agent. The reduction products of nitric acid depend on the concentration of the acid and the temperature. We may represent the reactions of copper with nitric acid by the following simplified equations (the abbreviation "dil" represents "dilute"):

$$Cu(s) + 4HNO_3(aq, conc) \longrightarrow Cu(NO_3)_2(aq) + 2NO_2(g) + 2H_2O(l)$$
$$3Cu(s) + 8HNO_3(aq, dil) \longrightarrow 3Cu(NO_3)_2(aq) + 2NO(g) + 4H_2O(l)$$

In these reactions molecular nitric acid, HNO_3, and the nitrate ion, NO_3^-, not H_3O^+, are the oxidizing agents. Copper is oxidized to copper(II) nitrate. Nitric acid, in which nitrogen is in the $+5$ oxidation state, is reduced to NO_2, in which it is in the $+4$ oxidation state, or to NO, in which it is in the $+2$ oxidation state.

BALANCING OXIDATION–REDUCTION EQUATIONS

The somewhat complicated equations describing the reactions between copper and nitric acid are not simple to balance unless we use a systematic approach. If we know all the reactants and the products for an oxidation–reduction reaction, then we can balance the equation by following five steps, as we show now for the oxidation of copper by concentrated nitric acid.

1. *Write down the reactants and products of the reaction and inspect them to determine which elements have undergone changes in their oxidation numbers.* We have

$$Cu + HNO_3 \longrightarrow Cu(NO_3)_2 + NO_2$$
$$\text{Reactants} \qquad\qquad \text{Products}$$

For reactions in aqueous solution, it is convenient to rewrite the equation in the ionic form:

$$Cu + H_3O^+ + NO_3^- \longrightarrow Cu^{2+} + 2NO_3^- + NO_2$$

Examining the oxidation numbers, we find

$$\overset{+5\ -2}{NO_3^-} \xrightarrow{\text{Reduction}} \overset{+4\ -2}{NO_2}$$
$$\overset{0}{Cu} \xrightarrow{\text{Oxidation}} \overset{+2}{Cu^{2+}}$$

The reaction of concentrated
nitric acid with a copper penny

Only N and Cu have undergone changes in oxidation numbers.

2. *Write a half equation for the reduction process and another half equation for the oxidation process. Balance all atoms except O and H, and then balance the oxidation numbers by adding the requisite number of electrons to the appropriate side of the equation.* In our example we have

$$\text{Reduction}\quad \overset{+5}{N}O_3^- + e^- \longrightarrow \overset{+4}{N}O_2$$

$$\text{Oxidation}\quad \overset{0}{Cu} \longrightarrow \overset{+2}{Cu}^{2+} + 2e^-$$

Nitrogen is balanced, and in the change from NO_3^- to NO_2 the oxidation number of N changes from $+5$ to $+4$, so one electron must be involved in the reaction. Similarly, Cu is balanced, and the oxidation number of Cu changes from 0 for the metal to $+2$ for Cu^{2+}. Two electrons must be added to the right side of the equation for this half-reaction. At this point the half equations are not necessarily balanced with respect to charges or H and O atoms.

3. *Balance each half-reaction for charge.* The next step is to *balance charges*. How we do this depends on whether the reaction is taking place in an acidic solution or a basic solution. In an acidic aqueous solution H^+ (H_3O^+) ions are always present, and in a basic aqueous solution OH^- ions are always present. Thus for acidic solutions we add H^+ (H_3O^+) to balance charges, and for basic solutions we add OH^- to balance charges. In the half equation for the reduction of NO_3^- to NO_2, there are two negative charges on the left side of the equation and no charges on the right side, since NO_2 is a neutral molecule. The reaction is an *acidic* solution, so H_3O^+ ions can participate in the reaction. For simplicity we will write H_3O^+ as H^+. We balance charges by adding the requisite number of H^+ to the appropriate side of the equation:

$$NO_3^- + e^- + 2H^+ \longrightarrow NO_2$$

The equation is now balanced with respect to the changes in both the oxidation numbers and the charges. The equation $Cu \rightarrow Cu^{2+} + 2e^-$ is already balanced.

4. *Balance each half equation with respect to H and O atoms.* If the H and O atoms do not balance, because we are concerned with aqueous solutions, we can add an appropriate number of H_2O molecules to either side of the equation. The left side of our equation has two more H atoms and one more O atom than the right side; it can be balanced by adding one H_2O to the right side:

$$NO_3^- + e^- + 2H^+ \longrightarrow NO_2 + H_2O$$

The half-reaction for reduction of NO_3^- to NO_2 is now balanced in all respects. In general, for reactions carried out in acidic aqueous solution, H_2O molecules and H_3O^+ (H^+) ions may participate in the reaction as either reactants or products.

5. *Multiply the two half equations by appropriate coefficients so that, when the two half equations are added, the electrons on one side of the equation cancel those on the other side.* In our example this step can be accomplished by adding the equation for the oxidation of Cu to twice the equation for the reduction of NO_3^-:

$$\begin{array}{l} Cu \longrightarrow Cu^{2+} + 2e^- \\ 2NO_3^- + 2e^- + 4H^+ \longrightarrow 2NO_2 + 2H_2O \\ \hline Cu + 2NO_3^- + 4H^+ \longrightarrow Cu^{2+} + 2NO_2 + 2H_2O \end{array}$$

This equation is the required balanced equation for the oxidation–reduction reaction.

The final equation should be checked to see whether it is balanced for atoms and for charges:

	Left side	Right side
Cu atoms	1	1
N atoms	2	2
O atoms	6	6
H atoms	4	4
Charge	2+	2+

We see that atoms and charges balance.

We can also check changes in the oxidation numbers:

Oxidation number change in element(s) reduced

Oxidation number change in element(s) oxidized

$$2N(+5) \longrightarrow 2N(+4) = -2 \qquad Cu(0) \longrightarrow Cu^{2+}(+2) = +2$$

We see that the total *decrease* in the oxidation number of N is equal to the total *increase* in the oxidation number of Cu, so the overall sum of the oxidation number changes is zero.

The equation can also be written in terms of the empirical and molecular formulas of the reactants and products. In this case addition of two NO_3^- ions to each side of the equation gives

$$Cu + 4NO_3^- + 4H^+ \longrightarrow Cu^{2+} + 2NO_3^- + 2NO_2 + 2H_2O$$

or

$$Cu + 4HNO_3 \longrightarrow Cu(NO_3)_2 + 2NO_2 + 2H_2O$$

═ *Example 10.2* ═══

BALANCING EQUATIONS FOR REDOX REACTIONS (ACIDIC SOLUTION)

Copper reacts with dilute nitric acid to give nitrogen monoxide, NO. Write a balanced equation for the reaction.

Solution

$$Cu + H_3O^+ + NO_3^- \longrightarrow Cu^{2+} + NO$$

Reactants   Products

We have

$$\text{Oxidation} \qquad \overset{0}{Cu} \longrightarrow \overset{+2}{Cu^{2+}}$$

$$\text{Reduction} \quad \overset{+5}{NO_3^-} \longrightarrow \overset{+2}{NO}$$

Adding electrons, we have

$$\text{Oxidation} \qquad \overset{0}{Cu} \longrightarrow \overset{+2}{Cu^{2+}} + 2e^-$$

$$\text{Reduction} \quad \overset{+5}{NO_3^-} + 3e^- \longrightarrow \overset{+2}{NO}$$

and the oxidation equation is balanced. But the reduction equation is not balanced.

We first balance charges by adding H^+:

$$NO_3^- + 3e^- + 4H^+ \longrightarrow NO$$

We then balance atoms by adding $2H_2O$ to the right-hand side of the equation:

$$NO_3^- + 3e^- + 4H^+ \longrightarrow NO + 2H_2O$$

The equations for both half-reactions are now balanced.

We add three times the first equation to twice the second equation to eliminate the electrons:

$$
\begin{aligned}
3Cu &\longrightarrow 3Cu^{2+} + 6e^- \\
2NO_3^- + 6e^- + 8H^+ &\longrightarrow 2NO + 4H_2O \\
\hline
3Cu + 2NO_3^- + 8H^+ &\longrightarrow 3Cu^{2+} + 2NO + 4H_2O
\end{aligned}
$$

This equation is the balanced equation for the oxidation–reduction reaction between Cu and dilute HNO_3 to give Cu^{2+} and NO.

Finally, we check for atoms and charges:

Left side	Right side
3Cu	3Cu
2N	2N
6O	6O
8H	8H
6+	6+

And we check for oxidation number changes:

$$2N(+5) \longrightarrow 2N(+2) = -6 \quad \text{and} \quad 3Cu(0) \longrightarrow 3Cu^{2+}(+2) = +6$$

The balanced equation for any oxidation–reduction reaction can be obtained in the same way, provided we know all the reactants and all the products. In *acidic solutions* we balance charges by adding H^+ as appropriate. In *basic solutions* only insignificant concentrations of H^+ ions are present; so we balance charges by adding an appropriate number of OH^- to the equation. In general, in a basic solution H_2O molecules and OH^- ions may take part in a reaction as either reactants or products.

═ Example 10.3 ═

BALANCING EQUATIONS FOR REDOX REACTIONS (BASIC SOLUTION)

Sodium nitrate reacts with zinc in sodium hydroxide solution to give ammonia and the tetrahydroxozinc(II) ion, $Zn(OH)_4^{2-}$. Write a balanced equation for this reaction.

Solution

$$\underset{\text{Reactants}}{Zn + NO_3^-} \longrightarrow \underset{\text{Products}}{Zn(OH)_4^{2-} + NH_3}$$

$$\text{Oxidation} \quad \overset{0}{Zn} \longrightarrow \overset{+2}{Zn(OH)_4^{2-}} + 2e^-$$

We balance charges by adding $4OH^-$ to the left side:

$$Zn + 4OH^- \longrightarrow Zn(OH)_4^{2-} + 2e^-$$

Atoms are then also balanced. For the reduction half-reaction we have

$$\text{Reduction} \quad \overset{+5}{N}O_3{}^- + 8e^- \longrightarrow \overset{-3}{N}H_3$$

We balance charges by adding $9OH^-$ to the right side of the equation:

$$NO_3{}^- + 8e^- \longrightarrow NH_3 + 9OH^-$$

We balance atoms by adding $6H_2O$ to the left side of the equation:

$$NO_3{}^- + 8e^- + 6H_2O \longrightarrow NH_3 + 9OH^-$$

The electrons are eliminated by adding four times the first equation to the second equation:

$$
\begin{array}{l}
4Zn + 16OH^- \longrightarrow 4Zn(OH)_4{}^{2-} + 8e^- \\
\underline{NO_3{}^- + 8e^- + 6H_2O \longrightarrow NH_3 + 9OH^-} \\
4Zn + NO_3{}^- + 6H_2O + 7OH^- \longrightarrow 4Zn(OH)_4{}^{2-} + NH_3
\end{array}
$$

Checking the overall equation for atoms and charges, we have

Left side	Right side
4Zn	4Zn
1N	1N
16O	16O
19H	19H
8−	8−

Checking for oxidation number changes we have

$$4Zn(0) \longrightarrow 4Zn(+2) = +8 \quad \text{and} \quad N(+5) \longrightarrow N(-3) = -8$$

The steps to follow for balancing all oxidation–reduction equations are:

1. Write down the reactants and products of the reaction, and inspect them to determine which elements have undergone changes in their oxidation numbers.
2. Write half equations representing the changes in the oxidized and reduced species, and balance the oxidation number changes by adding the requisite number of electrons to the appropriate side of each equation.
3. Balance each half equation for charge by adding the requisite number of hydrogen ions (in acidic solution) or hydroxide ions (in basic solution) to the appropriate side of the equation.
4. Balance each half equation with respect to atoms by adding the requisite number of water molecules to the appropriate side of the equation.
5. Multiply the two half equations by appropriate coefficients so that, when the two half equations are added, the electrons on one side of the equation cancel those on the other side.
6. Check the final equation to see that it is balanced for atoms and for charges.

EXERCISE 10.3

Balance the following equations for redox reactions in acidic solution:

(a) $Br^-(aq) + MnO_4{}^-(aq) \rightarrow Br_2(l) + Mn^{2+}(aq)$

(b) $PbS(s) + NO_3{}^-(aq) \rightarrow PbSO_4(s) + NO(g)$

EXERCISE 10.4

Balance the following equations for redox reactions in basic solution:

(a) $Al(s) + MnO_4^-(aq) \rightarrow MnO_2(s) + Al(OH)_4^-(aq)$

(b) $Al(s) + OH^-(aq) \rightarrow Al(OH)_4^-(aq) + H_2(g)$

There are several ways of balancing oxidation–reduction equations. Sometimes, one is more convenient than the others, but beginning students are advised to adopt one method and stick to it. We will illustrate a variation in the method we have described by again using the reaction between copper and concentrated nitric acid as an example.

1. Write the reactants and products of the reaction and inspect them to determine which elements have undergone changes in their oxidation numbers:

$$\text{Oxidation} \quad \overset{0}{Cu} \longrightarrow \overset{+2}{Cu^{2+}}$$

$$\text{Reduction} \quad \overset{+5}{NO_3^-} \longrightarrow \overset{+4}{NO_2}$$

2. Write two half equations and balance them for atoms. In an acid solution we may use hydrogen ions, H^+, and H_2O molecules as necessary:

$$\text{Oxidation} \qquad Cu \longrightarrow Cu^{2+}$$

$$\text{Reduction} \quad NO_3^- + 2H^+ \longrightarrow NO_2 + H_2O$$

3. Balance for charges by adding electrons as necessary:

$$Cu \longrightarrow Cu^{2+} + 2e^-$$

$$NO_3^- + 2H^+ + e^- \longrightarrow NO_2 + H_2O$$

At this point we can check that the number of electrons in each equation is equal to the change in oxidation number.

4. Multiply the equations by appropriate coefficients so that, on adding, we have the same number of electrons on each side, and they cancel:

$$Cu \longrightarrow Cu^{2+} + 2e^-$$
$$\underline{2NO_3^- + 2e^- \quad + 4H^+ \longrightarrow 2NO_2 + 2H_2O}$$
$$Cu \quad + 2NO_3^- + 4H^+ \longrightarrow Cu^{2+} + 2NO_2 + 2H_2O$$

The final equation should be checked as before for atom balance, for charge balance, and for oxidation number changes.

≡ *Example 10.4* ═══════════════════════════════

**BALANCING EQUATIONS FOR REDOX REACTIONS
(ALTERNATIVE METHOD)**

Hot, concentrated H_2SO_4 reacts with Cu to give $CuSO_4$ and SO_2. Write the balanced equation for this reaction using the preceding method.

Solution

$$Cu + H_2SO_4 \longrightarrow CuSO_4 + SO_2$$

or writing it in ionic form, we have

$$Cu + 2H^+ + SO_4^{2-} \longrightarrow Cu^{2+} + SO_4^{2-} + SO_2$$

1. We see that Cu is oxidized to Cu^{2+}:

$$\overset{0}{Cu} \longrightarrow \overset{+2}{Cu^{2+}}$$

Sulfuric acid is reduced to SO_2:

$$\overset{+6}{SO_4^{2-}} \longrightarrow \overset{+4}{SO_2}$$

2. Balance for atoms:

$$Cu \longrightarrow Cu^{2+}$$
$$4H^+ + SO_4^{2-} \longrightarrow SO_2 + 2H_2O$$

3. Balance for charges by adding electrons:

$$Cu \longrightarrow Cu^{2+} + 2e^-$$
$$4H^+ + SO_4^{2-} + 2e^- \longrightarrow SO_2 + 2H_2O$$

4. Adding the two equations gives the overall balanced equation:

$$Cu + 4H^+ + SO_4^{2-} \longrightarrow Cu^{2+} + SO_2 + 2H_2O$$

Adding one SO_4^{2-} to each side, we can write the equation in the following alternative form:

$$Cu + 2H_2SO_4 \longrightarrow CuSO_4 + SO_2 + 2H_2O$$

The final equation should be checked for atom balance, charge balance, and oxidation number changes.

REACTIVITY OF METALS

On the basis of the above discussion of the reactions of metals with dilute aqueous acids and our discussion in Chapter 4 of the reactions of the alkali and alkaline earth metals with water we can classify metals according to their reactivity toward the hydronium ion, H_3O^+, that is, the ease with which they are oxidized by H_3O^+. The alkali metals and some of the alkaline earth metals are extremely reactive and are so readily oxidized that they react even with water in which the concentration of H_3O^+ is very small. Other metals such as Zn, Fe, and Al are less reactive, but they are oxidized by H_3O^+ in aqueous acid solutions. Metals such as Cu, Ag, and Au are still less reactive and are not oxidized by H_3O^+:

Na, K, Ca	Mg, Zn, Al, Fe, Pb	Cu, Ag, Au
Very reactive	Less reactive	Unreactive
React with water	React with H_3O^+ in aqueous acids	Do not react with H_3O^+ in aqueous acids

This is called the *activity series* of the metals. The metals on the left are strong reducing agents. Those on the right are weak reducing agents. In Chapter 17 we will see how to put this series on a quantitative basis.

═ *Example 10.5* ═══════════════════════════════

REACTIONS OF METALS

Write balanced equations for the following reactions:

(a) The reaction of aluminum with bromine

(b) The reaction of iron with excess oxygen

Solution

(a) The only oxidation state of aluminum is $+3$ so the formula of the product aluminum bromide is $AlBr_3$, and the equation for the reaction is therefore

$$2Al(s) + 3Br_2(l) \longrightarrow 2AlBr_3(s)$$

(b) Iron has two oxidation states $+2$ and $+3$, and it forms the two corresponding oxides FeO and Fe_2O_3. In excess oxygen we would expect the higher oxide Fe_2O_3 to be formed and we saw in Experiment 3.1 that this oxide is indeed formed. The equation is

$$4Fe(s) + 3O_2(g) \longrightarrow 2Fe_2O_3(s)$$

EXERCISE 10.5

Write balanced equations for the following reactions:

(a) The reaction of copper with excess oxygen

(b) The reaction of aluminum with sulfur

(c) The reaction of excess lead with oxygen

(d) The reaction of an aqueous solution of copper sulfate with a solution of sodium hydroxide

COMPOUNDS OF ALUMINUM; LEWIS ACIDS AND BASES

Aluminum is in group 3 and has the electron configuration $[Ne]\, 3s^2 3p^1$. The majority of aluminum compounds are ionic and contain Al^{3+}, formed by the loss of the three valence-shell electrons. However, there are also some important covalent compounds of aluminum.

The ground state electron configuration of aluminum is

It has only one unpaired electron, and we would therefore expect aluminum to be univalent in any covalent compounds that it might form. But like beryllium, boron, and carbon, aluminum normally forms compounds in which it uses all its valence-shell electrons. So we can conveniently imagine that such compounds are formed not from the ground state of aluminum but from a valence state in which a 3s electron has been promoted to a 3p orbital and in which there are therefore three unpaired electrons:

$$3s \qquad 3p$$

[Ne]   ↑   ↑ ↑

By sharing these electrons, it forms three covalent bonds. The fact that aluminum forms covalent as well as ionic bonds reflects the fact that it lies near the borderline between the metals and the nonmetals of the periodic table.

ALUMINUM OXIDE Although aluminum is rather reactive, it does not corrode in the atmosphere because the surface of the metal quickly becomes covered by a thin, hard layer of aluminum oxide, Al_2O_3. In nature aluminum oxide, also called *alumina*, occurs as *bauxite*, $Al_2O_3 \cdot xH_2O$, and in the anhydrous form, Al_2O_3, as *corundum*. To a first approximation, Al_2O_3 can be regarded as consisting of the ions Al^{3+} and O^{2-}. The force between two ions at a given distance is, according to Coulomb's law, proportional to the product of their charges. Thus if we neglect differences in the sizes of the ions, the force between an Al^{3+} and an O^{2-} would be six times that between singly charged ions such as Na^+ and Cl^-. For this reason Al_2O_3 is very hard and has a very high melting point (approximately 2000 °C). However, the small highly charged Al^{3+} attracts electrons very strongly. It therefore attracts the electron pairs of adjacent O^{2-} ions so that they are partially shared with the Al^{3+} ion. In other words, the bonds have some covalent character; they may be described as very polar covalent bonds.

Transparent crystals of corundum containing traces of other metal ions, which impart colors to the crystals, occur naturally; they are used as gemstones. For example, sapphires are corundum crystals that have a blue color because a small number of Al^{3+} ions are replaced by Fe^{2+}, Fe^{3+}, and Ti^{3+}. Rubies are red because a small number of Al^{3+} ions are replaced by Cr^{3+} ions. Artificial rubies and sapphires are produced by melting Al_2O_3 with small amounts of other suitable oxides in the flame of an oxyhydrogen torch (Figure 10.12).

Like many other metal oxides, aluminum oxide is insoluble in water. Its insolubility can be attributed to the very strong ionic bonding in the solid that results from the high charges on the ions.

ALUMINUM CHLORIDE, $AlCl_3$, AND ALUMINUM BROMIDE, $AlBr_3$ When aluminum is heated in chlorine or hydrogen chloride, aluminum chloride is formed:

$$2Al(s) + 3Cl_2(g) \longrightarrow 2AlCl_3(s)$$
$$2Al(s) + 6HCl(g) \longrightarrow 2AlCl_3(s) + 3H_2(g)$$

Aluminum reacts vigorously with liquid bromine at room temperature to form aluminum bromide, $AlBr_3$ (see Experiment 10.4):

$$2Al(s) + 3Br_2(l) \longrightarrow 2AlBr_3(s)$$

Aluminum chloride and aluminum bromide are white crystalline substances. They are very **hygroscopic**, strongly absorbing water from the atmosphere and eventually dissolving in the absorbed water. When heated at atmospheric pressure, $AlCl_3$ sublimes at 183 °C without melting. Aluminum chloride may be purified by heating it until it sublimes and then condensing the vapor, thus leaving behind all the less volatile impurities.

In a formula such as $Al_2O_3 \cdot xH_2O$, x denotes that the ratio of H_2O to Al_2O_3 is uncertain and may vary from sample to sample.

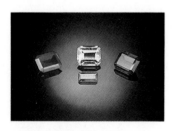

FIGURE 10.12
Synthetic Gemstones.

Ruby, diamond, sapphire, and emerald. Ruby and sapphire are impure forms of corundum. Emerald is an impure form of beryl, $Be_3Al_2Si_6O_{18}$.

Experiment 10.4

The Preparation of Aluminum Bromide and Aluminum Chloride

Aluminum pellets can be seen on the watch glass on top of the beaker, which contains bromine. The aluminum pellets are tipped into the bromine

and after a few minutes the aluminum begins to react vigorously with the bromine. The aluminum pellets ignite, burning with a bright flame, and skate around the surface of the bromine.

The product of the reaction is a white solid, aluminum bromide, which coats the beaker and the watch glass.

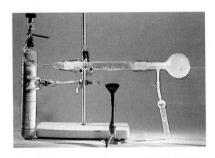

When aluminum is heated in a stream of chlorine, it forms aluminum chloride, which condenses as a white powder in the flask on the right. The reaction is exothermic, and round drops of red-hot molten aluminum (mp 660°C) can be seen in the tube. The yellow-green color is due to the chlorine.

Solid aluminum chloride is an ionic solid consisting of Al^{3+} and Cl^-. But the gas, which is obtained when aluminum chloride sublimes, consists of covalent Al_2Cl_6 molecules:

$$\ddot{:}\overset{..}{\underset{..}{Cl}} \quad \overset{\oplus}{\overset{..}{Cl}} \quad \overset{..}{\underset{..}{Cl}}\ddot{:} \\ \qquad Al^{\ominus} \qquad Al^{\ominus} \\ \ddot{:}\overset{..}{\underset{..}{Cl}} \quad \underset{\oplus}{\overset{..}{Cl}} \quad \overset{..}{\underset{..}{Cl}}\ddot{:}$$

Both aluminum atoms have a tetrahedral AX_4 geometry, so the overall geometry of the molecule is

$$\begin{array}{ccc} Cl & Cl & Cl \\ \diagdown & \diagdown & \diagup \\ & Al \qquad Al \\ \diagup & \diagup & \diagdown \\ Cl & Cl & Cl \end{array}$$

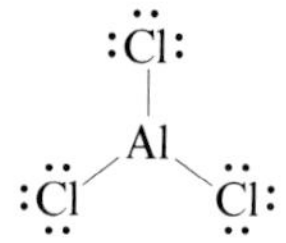

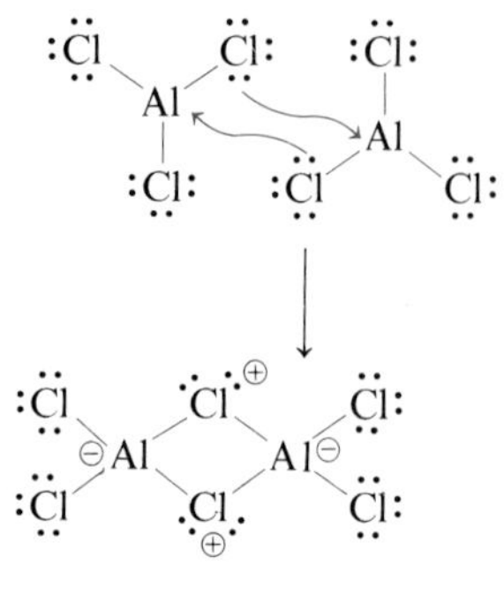

The four *terminal* chlorine atoms are in a plane at 90° to the plane of the two *bridging* chlorine atoms. Above 750 °C, Al_2Cl_6 molecules dissociate into $AlCl_3$ molecules, which have a planar triangular AX_3 geometry.

Unlike $AlCl_3$, aluminum bromide is not ionic in the solid state but consists of Al_2Br_6 molecules, just like aluminum chloride in the gaseous state. At high temperatures in the gas phase Al_2Br_6 molecules dissociate to $AlBr_3$ molecules.

In the $AlCl_3$ molecule the Al atom has only six electrons in its valence shell and thus has space for more electrons. In the Al_2Cl_6 molecule each Al atom completes its octet by accepting an electron pair from a Cl atom. When an Al—Cl bond is formed in this way, formal charges are generated, + on Cl and − on Al, because a pair of electrons initially belonging only to a Cl atom is shared with the Al atom. In general, whenever an Al atom forms four bonds, it must carry a formal negative charge, whenever a Cl atom forms two covalent bonds, it must carry a formal positive charge.

In forming a covalent chloride in the vapor state, aluminum is behaving like a nonmetal. This is consistent with its position in the periodic table on the borderline between metals and nonmetals.

LEWIS ACIDS AND BASES The formation of Al_2Cl_6 by the donation of an unshared electron pair on a Cl atom into the valence shell of the Al atom is an example of a type of behavior that is common for molecules in which the central atom has only six or fewer electrons in its valence shell. Such molecules often combine with other molecules or ions by accepting an electron pair, thus converting a lone pair into a shared pair in order to complete an octet of electrons in their valence shells. Thus $AlCl_3$ also combines with a Cl^- ion to give the tetrahedral $AlCl_4^-$ ion:

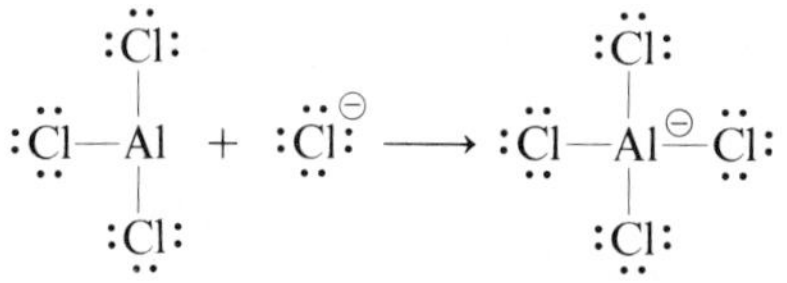

Boron, the first element in group 3, forms covalent trihalides, such as boron trichloride and boron trifluoride, in which the boron atom has only six electrons in its valence shell. Unlike aluminum chloride, boron trichloride and boron trifluoride do not form dimeric molecules, but they do combine with a chloride ion or a fluoride ion to give the tetrahedral BCl_4^- and BF_4^- ions and with other neutral molecules, such as ammonia, that have unshared electron pairs:

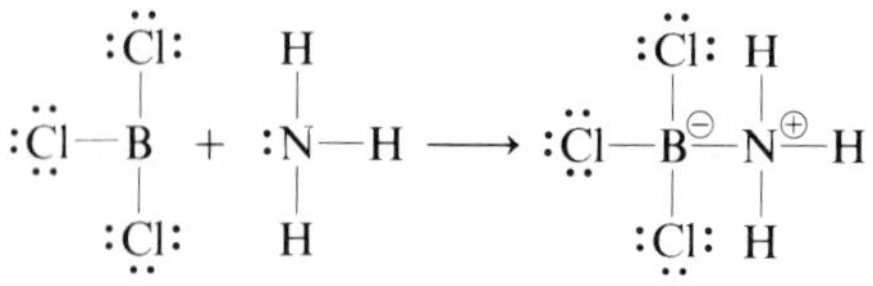

Such a compound between two neutral molecules is called an **adduct**, in this case the boron trichloride–ammonia adduct.

A proton adds to a molecule or ion with an unshared pair of electrons (a base), forming a covalent bond in an acid–base reaction:

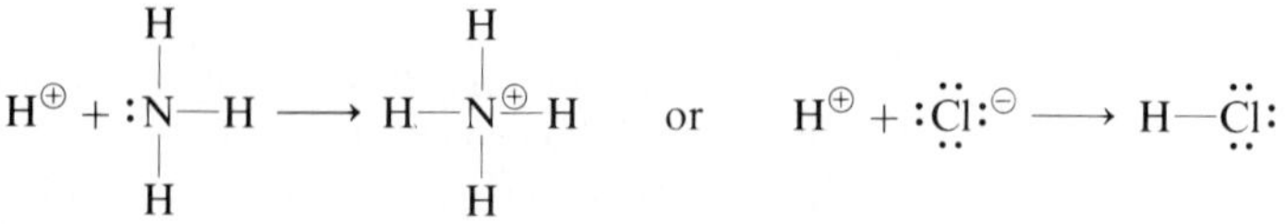

Thus by forming a covalent bond by sharing a lone pair of another molecule or ion, BCl_3 and $AlCl_3$ behave like a proton in an acid–base reaction.

Lewis (Figure 4.11) therefore suggested that all molecules and ions that combine with other molecules and ions by accepting an electron pair should be regarded as acids. Acids thus defined are called **Lewis acids**, to differentiate them from acids as defined by Brønsted and Lowry. Molecules that provide an electron pair are called **Lewis bases**.

A Lewis acid is an electron pair acceptor.

A Lewis base is an electron pair donor.

Bases are the same according to both the Brønsted–Lowry and the Lewis definitions; in both cases the prerequisite for base behavior is the availability of an unshared electron pair, which can be shared either with a proton or with a Lewis acid. A prerequisite for Lewis acid behavior is that a molecule or ion has space available in its valence shell, in other words a vacant orbital, that can accommodate an electron pair. For example, when boron is forming only three bonds, as in BCl_3, it has a vacant 2p orbital. It is this orbital that accepts the electron pair from the donor atom and the valence shell of boron is then complete. Boron is then forming four bonds and it has a formal negative charge.

In forming Al_2Cl_6, each $AlCl_3$ molecule behaves as both a Lewis acid and a Lewis base. The Al atom accepts an electron pair, and a Cl atom donates an electron pair.

ACIDIC AND BASIC OXIDES Another example of a Lewis acid–base reaction is provided by the reaction in which slag is formed when ores are reduced to obtain metals. The silica impurity in the ore reacts with lime, CaO, to form calcium silicate, $CaSiO_3$:

$$CaO + SiO_2 \longrightarrow CaSiO_3$$

Or since CaO and $CaSiO_3$ are ionic,

$$O^{2-} + SiO_2 \longrightarrow SiO_3{}^{2-}$$

Similar examples include

$$Na_2O + CO_2 \longrightarrow Na_2CO_3 \quad \text{or} \quad O^{2-} + CO_2 \longrightarrow CO_3{}^{2-}$$

and

$$MgO + SO_3 \longrightarrow MgSO_4 \quad \text{or} \quad O^{2-} + SO_3 \longrightarrow SO_4{}^{2-}$$

In each case an unshared pair of electrons of the oxide ion of an ionic metal oxide forms a bond with the central atom of a nonmetal oxide:

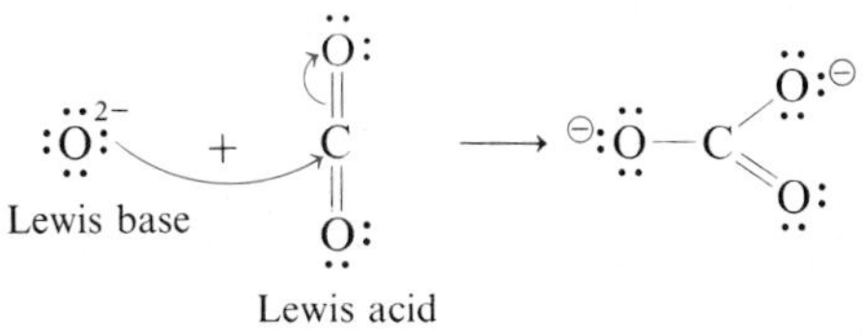

Because the C atom in CO_2 has four pairs of electrons in its valence shell, we might think that it would not be able to accept an electron pair. But one of the electron pairs of a CO double bond shifts to the O atom to make room for the incoming electron pair from the oxide ion. The reaction between an

The curved arrows in this and similar formulas and equations show the movement of electron pairs. Here one of the unshared pairs of the oxide ion is shared with the carbon atom to form a bond and at the same time one of the electron pairs of one of the double bonds moves onto the oxygen atom.

oxide ion and SO_3 is similar:

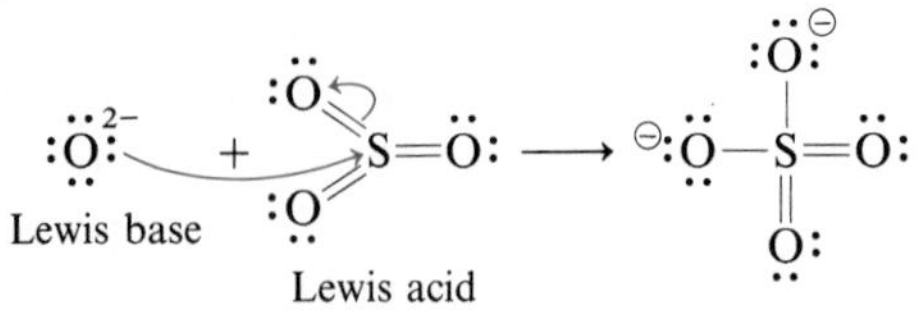

From the above examples we see that it is not necessary that one of the atoms in a molecule should have an incomplete valence shell, that is an empty orbital, in order that it can behave as a Lewis acid. It is only necessary that an empty orbital can become available, as for example, by the conversion of one of the electron pairs of a double bond to a nonbonding pair as in CO_2 and SO_3.

In summary, the oxide ion is a Lewis base, and ionic metal oxides are therefore called **basic oxides**. Covalent nonmetal oxides are Lewis acids and are called **acidic oxides**. An acidic oxide reacts with a basic oxide to form a salt:

$$CaO \quad + \quad SiO_2 \quad \longrightarrow \quad CaSiO_3$$
$$CaO \quad + \quad CO_2 \quad \longrightarrow \quad CaCO_3$$

| Basic oxide | Acidic oxide | Salt |
| Lewis base | Lewis acid | |

The acid and base character of oxides can also be seen in their reactions in aqueous solution. Basic oxides react with water to give basic solutions; for example,

$$Na_2O(s) + H_2O(l) \longrightarrow 2Na^+(aq) + 2OH^-(aq)$$

Acidic oxides react with water to give an acidic solution; for example,

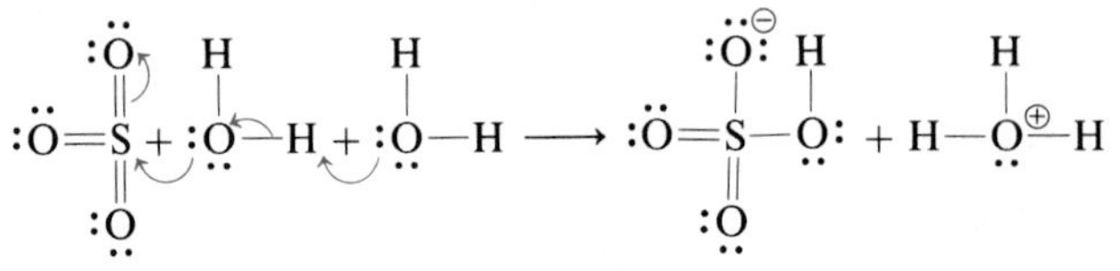

$$SO_3 + 2H_2O \longrightarrow HSO_4^- + H_3O^+$$

This reaction is another example of a Lewis acid–base reaction. Here the Lewis acid SO_3 accepts an electron pair from the Lewis base H_2O, forming an S—O bond. At the same time the H_2O molecule loses a proton to another H_2O molecule to give HSO_4^- and H_3O^+.

Because they react with water to give acids, acidic oxides are often called **acid anhydrides**. Thus an acidic oxide can be thought of as being derived from an acid by the removal of water—although, in practice, in some cases it is very difficult to dehydrate the acid to give the oxide. For example,

$$H_2SO_4 - H_2O \longrightarrow SO_3 \quad \text{(Occurs only with difficulty)}$$

$$H_2CO_3 - H_2O \longrightarrow CO_2 \quad \text{(Occurs readily)}$$

We saw in Chapter 8 that P_4O_{10} reacts with excess water to give orthophosphoric acid, H_3PO_4:

$$P_4O_{10} + 6H_2O \longrightarrow 4H_3PO_4$$

Thus P_4O_{10} is the anhydride of phosphoric acid.

Similarly, the anhydride of nitric acid is the oxide N_2O_5:

$$N_2O_5 + H_2O \longrightarrow 2HNO_3 \qquad 2HNO_3 - H_2O \longrightarrow N_2O_5$$

Visualizing the dehydration of nitric acid as follows is helpful:

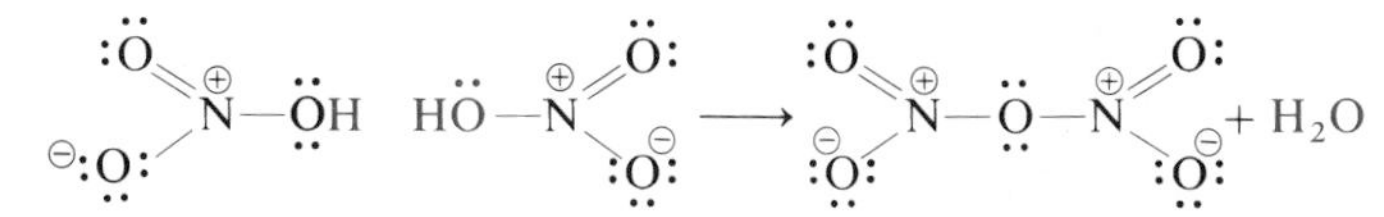

Thus acidic oxides react with water to give an oxoacid, and they react with an oxide ion to give the anion of the corresponding acid. For example,

$$SO_3 + H_2O \longrightarrow H_2SO_4$$

and

$$SO_3 + O^{2-} \longrightarrow SO_4^{2-}$$

Some oxides such as MgO and SiO_2 react only very slowly and incompletely with water. Their acid–base properties are, however, shown by their reactions with acids and bases. For example, MgO reacts with hydrochloric acid to give $MgCl_2$; it is therefore a basic oxide:

$$MgO(s) + 2HCl(aq) \longrightarrow MgCl_2(aq) + H_2O(l)$$

And silicon dioxide reacts with an aqueous solution of NaOH to give Na_2SiO_3; it is therefore an acidic oxide:

$$SiO_2(s) + 2NaOH(aq) \longrightarrow Na_2SiO_3(aq) + H_2O(l)$$

Some oxides, such as aluminum oxide, Al_2O_3, have both acid and base properties; they are called **amphoteric oxides**. Although corundum is very inert and insoluble in both acids and bases, other forms of Al_2O_3 that have slightly different structures are soluble in both acidic and basic aqueous solutions:

$$Al_2O_3(s) + 6HCl(aq) \longrightarrow 2Al^{3+}(aq) + 6Cl^-(aq) + 3H_2O(l)$$
$$Al_2O_3(s) + 2NaOH(aq) + 3H_2O(l) \longrightarrow 2Na^+(aq) + 2Al(OH)_4^-(aq)$$

In general, the oxides of the main group metals are basic, because they are ionic and the oxide ion is a Lewis base (and a Brønsted–Lowry base). However, oxides of elements such as aluminum that are near the borderline between metals and nonmetals in the periodic table may be amphoteric. The relationship between the acid and base character of an oxide and the position of the element in the periodic table is shown in Figure 10.13.

Period	Basic oxides	Amphoteric oxides	Acidic oxides			Oxides with no acidic or basic properties	
2	Li_2O	BeO	B_2O_3	CO_2	N_2O_3	O_2	F_2O
					N_2O_5		
3	Na_2O	MgO	Al_2O_3	SiO_2	P_2O_3	SO_2	Cl_2O
					P_2O_5	SO_3	Cl_2O_7

FIGURE 10.13
Acid and Base Properties of the Oxides of Periods 2 and 3.

On the left of the periodic table the oxides are basic. On the right they are acidic (except O_2 and F_2O).

═ *Example 10.6* ══════════════════════════════

ACIDIC AND BASIC OXIDES

Write a balanced equation for the reaction of the oxide ion with

(a) SiO_2 (b) N_2O_5

Solution

(a) The equation is $SiO_2 + O^{2-} \longrightarrow SiO_3^{2-}$.

(b) N_2O_5 is the anhydride of nitric acid,

$$N_2O_5 + H_2O \longrightarrow 2HNO_3$$

so an oxide ion will react with an N_2O_5 molecule to give two nitrate ions:

$$N_2O_5 + O^{2-} \longrightarrow 2NO_3^-$$

E X E R C I S E   1 0 . 6

Write a balanced equation for the reaction of the oxide ion with

(a) CO_2 (b) P_4O_{10}

═ *Example 10.7* ══════════════════════════════

ACIDIC AND BASIC OXIDES

Write balanced equations for the reactions of the following oxides with water. In each case state whether the oxide is acidic or basic.

(a) CO_2 (b) BaO

Solution

(a) Since CO_2 is a nonmetal oxide, it is acidic. It reacts with water to give the weak acid, carbonic acid:

$$H_2O(l) + CO_2(aq) \rightleftharpoons H_2CO_3(aq)$$
$$H_2CO_3(aq) + H_2O(l) \rightleftharpoons H_3O^+(aq) + HCO_3^-(aq)$$

(b) Group 1 and 2 metal oxides are basic:

$$BaO(s) + H_2O(l) \longrightarrow Ba(OH)_2(aq)$$

E X E R C I S E   1 0 . 7

Write equations for the reactions of the following oxides with water. In each case state whether the oxide is acidic or basic:

(a) SO_2 (b) Rb_2O

ALUMINUM SALTS In addition to the direct reaction between aluminum and chlorine, $AlCl_3$ can also be prepared by the reaction of aluminum or the oxide Al_2O_3 with hydrochloric acid:

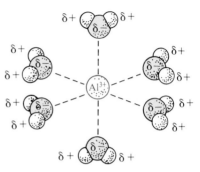

$$2Al(s) + 6HCl(aq) \longrightarrow 2AlCl_3(aq) + 3H_2(g)$$

$$Al_2O_3(s) + 6HCl(aq) \longrightarrow 2AlCl_3(aq) + 3H_2O(l)$$

Other salts can be prepared similarly by using the appropriate acid; for example,

$$2Al(s) + 3H_2SO_4(aq) \longrightarrow Al_2(SO_4)_3(aq) + 3H_2(g)$$

Aluminum chloride crystallizes from solution as $AlCl_3 \cdot 6H_2O$. It is said to be **hydrated** and the water that it contains is called **water of crystallization**. Six water molecules surround each aluminum ion in an octahedral arrangement (Figure 10.14). It is preferable therefore to write the formula as $[Al(H_2O)_6]Cl_3$ to show that the six water molecules are bonded to the aluminum ion. The bonds between the water molecules and the aluminum ion can be described in two ways: (1) the bond can be considered to arise from the electrostatic attraction between the negatively charged oxygen atom of the polar water molecule and the positive aluminum ion; or (2) the bonds can be considered to be covalent bonds formed by a water molecule donating an electron pair to the empty valence shell of the aluminum ion.

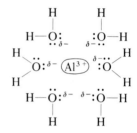

$Al(H_2O)_6^{3+}$ —ionic bonding

The actual bonding between the aluminum ion and a water molecule is best regarded as being intermediate between these two extremes. In other words, an electron pair on an oxygen atom is donated to the aluminum ion, but this donation is only partial, so the resulting bond is intermediate between a covalent bond and an ionic bond. The formal charge on aluminum is $+3$ in the ionic description and -3 in the covalent description. The real situation is intermediate between these two extremes and the real charge on aluminum is probably close to zero.

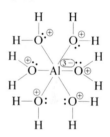

$Al(H_2O)_6^{3+}$ —covalent bonding

The formation of the hydrated aluminum ion, $Al(H_2O)_6{}^{3+}$, provides another example of a Lewis acid–base reaction. The aluminum ion behaves as a Lewis acid, accepting not one but six electron pairs from water molecules. In general, because they have empty valence shells, metal ions, particularly those with high charges, behave as electron pair acceptors, or Lewis acids.

FIGURE 10.14
The Hydrated Aluminum Ion $Al(H_2O)_6{}^{3+}$.

Six water molecules are arranged octahedrally around the aluminum ion. The bonding can be best regarded as intermediate between the two extremes of ionic bonding and covalent bonding.

Aluminum sulfate crystallizes from aqueous solution as the hydrate $Al_2(SO_4)_3 \cdot 18H_2O$. Six water molecules are associated with each aluminum ion, and six more form part of the crystal structure. If an equimolar mixture of potassium sulfate and aluminum sulfate is allowed to crystallize, a salt with the composition $KAl(SO_4)_2 \cdot 12H_2O$ is obtained as large octahedral-shaped crystals (see Experiment 10.5). This salt is called **alum**, or, more specifically, potassium alum. Other related salts such as ammonium alum, $(NH_4)Al(SO_4)_2 \cdot 12H_2O$, can be prepared. They are all ionic solids consisting of ammonium, potassium, or another singly charged cation, aluminum ions, sulfate ions, and water molecules.

ALUMINUM HYDROXIDE If hydroxide ion is added to a solution of an aluminum salt, a white gelatinous precipitate of aluminum hydroxide, $Al(OH)_3$, is obtained:

$$Al^{3+}(aq) + 3OH^-(aq) \longrightarrow Al(OH)_3(s)$$

Except for the hydroxides of the alkali metals and barium, most metal hydroxides are insoluble in water so they can be prepared by a similar precipitation reaction.

Experiment 10.5

Growing Alum Crystals

A crystal of potassium alum growing in a saturated solution.

Octahedral crystals of potassium alum.

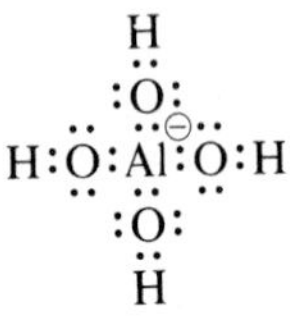

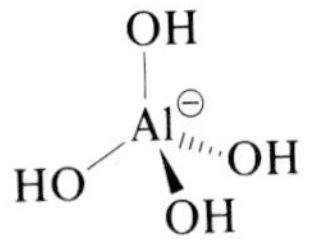

If excess OH^- ion is added to the precipitate of $Al(OH)_3$, it dissolves because of the formation of the soluble tetrahydroxoaluminum ion, $Al(OH)_4^-$:

$$Al(OH)_3(s) + OH^-(aq) \longrightarrow Al(OH)_4^-(aq)$$

This reaction is analogous to the reaction between $AlCl_3$ and Cl^- to give $AlCl_4^-$. It is another example of Lewis acid–base behavior, in which $Al(OH)_3$ is the Lewis acid and OH^- is the Lewis base.

Aluminum hydroxide also dissolves in acids to give an aluminum salt:

$$Al(OH)_3(s) + 3HCl(aq) \longrightarrow AlCl_3(aq) + 3H_2O(l)$$

Thus aluminum hydroxide behaves as both a base and an acid. It is amphoteric, like the oxide.

Like most metal hydroxides, except those of the alkali metals, aluminum hydroxide decomposes when strongly heated to give the oxide and water:

$$2Al(OH)_3(s) \longrightarrow Al_2O_3(s) + 3H_2O(g)$$

The reactions of aluminum and its compounds are summarized in Figure 10.15.

Tetrahydroxoaluminate ion

EXERCISE 10.8

Complete and balance each of the following equations:

(a) $Al(s) \rightarrow AlBr_3 \cdot 6H_2O(s)$

(b) $Al_2O_3(s) \rightarrow Al(NO_3)_3(aq)$

(c) $Al(OH)_3(s) \rightarrow Al_2O_3(s)$

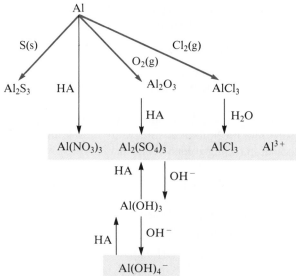

FIGURE 10.15
Preparation and Reactions of
Aluminum Compounds.

All reactants are in aqueous solution unless otherwise specified. All
products are solids except those in
blue boxes which are in aqueous
solution. All reactions which require heat are denoted by red
arrows. All other reactions are at
room temperature.

COMPOUNDS OF LEAD

Lead is in group 4 and has the electron configuration $[Xe]\ 4f^{14}5d^{10}6s^26p^2$, with a valence shell consisting of the four $6s^26p^2$ electrons. Its valence-shell electron configuration is similar to that of carbon, $2s^22p^2$, and it forms a few compounds in which all four electrons are used to form four covalent bonds. In these compounds, which are predominately covalent and resemble the corresponding compounds of carbon and silicon, lead is in the +4 oxidation state.

However, we have pointed out previously that elements in the central groups of the periodic table resemble each other less than those in the groups at either end. The elements in group 4 all have a common valence of 4, but there is a distinct trend in the group from the typically nonmetallic element carbon at the top to the typically metallic element lead at the bottom. Because of the large size of the lead atom, its valence-shell electrons are much less strongly held than are those of carbon. Thus lead loses two electrons from its ground state configuration quite readily to form the Pb^{2+} ion in which lead is in the +2 oxidation state:

	6s	6p
Pb ground state	↓↑	↑ ↑ ☐
Pb^{2+}	↑↓	☐ ☐ ☐

These Pb(II) compounds are typical ionic compounds like those formed by other metals and they are the most common compounds of lead. The less numerous Pb(IV) compounds are much less ionic and are best described as covalent.

LEAD(II) OXIDE When lead is heated in air, it forms the yellow oxide Pb(II)O.

LEAD(II) CHLORIDE, PbCl₂ Boiling, concentrated hydrochloric acid slowly dissolves lead to form $PbCl_2$:

$$Pb(s) + 2HCl(aq) \longrightarrow PbCl_2(aq) + H_2(g)$$

Lead(II) chloride is more easily prepared by addition of a soluble chloride to a solution of a lead(II) salt (see Experiment 10.6):

$$Pb^{2+}(aq) + 2Cl^-(aq) \longrightarrow PbCl_2(s)$$

It is only sparingly soluble in cold water but is more soluble in hot water. When the hot solution is cooled, white needlelike crystals of $PbCl_2$ separate out.

LEAD(II) IODIDE, PbI_2 Lead(II) iodide can be precipitated from a solution of a Pb(II) salt. The precipitate is somewhat soluble in hot water. When the solution is cooled, sparkling golden yellow crystals are obtained (see Experiment 10.6):

$$Pb^{2+}(aq) + 2I^-(aq) \longrightarrow PbI_2(s)$$

LEAD(II) HYDROXIDE, $Pb(OH)_2$ When NaOH(aq) is added to a solution of a soluble Pb(II) salt a white precipitate of $Pb(OH)_2$ is obtained:

$$Pb^{2+}(aq) + 2OH^-(aq) \longrightarrow Pb(OH)_2(s)$$

But this precipitate redissolves as more OH^- is added because of the formation

Experiment 10.6

Reactions of Pb^{2+}(aq)

Left to right: Pb^{2+} forms an insoluble precipitate of white $PbSO_4$ with SO_4^{2-}, an insoluble precipitate of yellow PbI_2 with I^-, and a white insoluble precipitate of $PbCl_2$ with Cl^-.

On cooling, the solution PbI_2 slowly precipitates as shiny golden-yellow spangles.

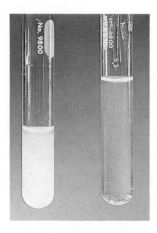

Left: When an aqueous NaOH solution is added to a solution of a soluble Pb^{2+} salt, a white insoluble precipitate of $Pb(OH)_2$ is formed. Right: When an excess of sodium hydroxide is added, the white precipitate dissolves to give a solution of $Pb(OH)_4^{2-}$.

of soluble $Na_2Pb(OH)_4$, which contains the $Pb(OH)_4^{2-}$ ion:

$$Pb(OH)_2(s) + 2OH^-(aq) \longrightarrow Pb(OH)_4^{2-}(aq)$$

Lead hydroxide is also soluble in aqueous acids; it is amphoteric:

$$Pb(OH)_2(s) + 2HCl(aq) \longrightarrow PbCl_2(aq) + 2H_2O(l)$$

LEAD(II) SULFIDE When lead is heated with sulfur $PbS(s)$ is obtained. When $H_2S(g)$ is passed into an aqueous solution of a lead(II) salt a black precipitate of PbS is formed (Experiment 10.6).

LEAD(II) SULFATE AND NITRATE These salts can be prepared from PbO or $Pb(OH)_2$ and the appropriate acid. Lead(II) sulfate is not very soluble so it is conveniently prepared by addition of a soluble sulfate to a solution of a soluble Pb(II) salt such as $Pb(NO_3)_2$. Lead(II) nitrate decomposes on heating to give nitrogen dioxide, NO_2, and oxygen:

$$2Pb(NO_3)_2(s) \xrightarrow{\text{Heat}} 2PbO(s) + 4NO_2(g) + O_2(g)$$

This reaction is often used to prepare small amounts of NO_2 (see Chapter 18).

The preparation and reactions of lead(II) compounds are summarized in Figure 10.16.

LEAD(IV) COMPOUNDS If lead is heated in oxygen, or if PbO is strongly heated in air, it is further oxidized to Pb_3O_4, *red lead*:

$$6PbO(s) + O_2(g) \xrightarrow{\text{Heat}} 2Pb_3O_4(s)$$

The oxide Pb_3O_4 is the basis of a red paint that is widely used for protecting iron and steel structures against rust; apparently, it forms an oxidized, unreactive layer on the surface of the iron. Red lead, Pb_3O_4, contains both Pb(II) and Pb(IV) and may be written as $Pb(II)_2Pb(IV)O_4$.

Red lead reacts with concentrated nitric acid to give soluble lead(II) nitrate and lead dioxide, $Pb(IV)O_2$, a brown insoluble solid in which lead is in the $+4$ oxidation state:

$$Pb_3O_4(s) + 4HNO_3(aq, concn) \longrightarrow 2Pb(NO_3)_2(aq) + PbO_2(s) + 2H_2O(l)$$

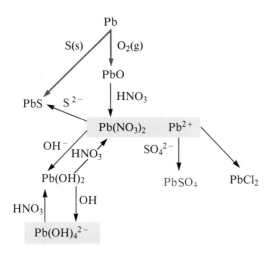

FIGURE 10.16
Preparation and Reactions of Lead(II) Compounds.

All reactants are in aqueous solution except where otherwise specified. All products are solids except where they are in a blue box when they are in aqueous solution. Reactions that require heat are denoted by red arrows. All other reactions occur at room temperature.

The reaction of Pb_3O_4 with HNO_3 is consistent with the formulation of this oxide as $Pb(II)_2Pb(IV)O_4$, because the Pb(II) atoms are converted to soluble $Pb(II)(NO_3)_2$, lead nitrate, leaving the Pb(IV) as the insoluble oxide PbO_2. Lead dioxide is an important component of lead storage batteries (Chapter 17).

Lead(IV) chloride (lead tetrachloride), $PbCl_4$, is a yellow liquid that decomposes when warmed to give chlorine and $PbCl_2$:

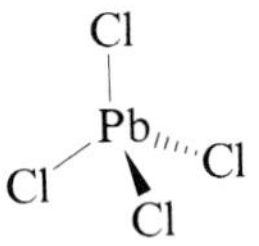

$$PbCl_4(l) \longrightarrow PbCl_2(s) + Cl_2(g)$$

It is a covalent compound consisting of tetrahedral $PbCl_4$ molecules, analogous to carbon tetrachloride, CCl_4, and silicon tetrachloride, $SiCl_4$.

EXERCISE 10.9

Write balanced equations for the following reactions:

(a) The preparation of $Pb(NO_3)_2(aq)$ from PbO(s)

(b) The preparation of $Pb(OH)_2(s)$ from $Pb(NO_3)_2(aq)$

(c) The preparation of $PbSO_4(s)$ from $Pb(NO_3)_2(aq)$

EXERCISE 10.10

Explain, giving appropriate balanced equations, how you would prepare pure samples of PbO(s) and $PbO_2(s)$ from red lead, $Pb_3O_4(s)$.

COMPOUNDS OF IRON

Iron is a member of the first transition metal series of elements. A characteristic feature of these elements is that they have several different oxidation states.

As Table 10.4 shows, most of the elements in the first transition series have two electrons in the 4s orbital, which they can lose to form M^{2+} ions, such as Fe^{2+}. Because the 3d and 4s electrons have very similar energies, these elements also have oxidation states in which they utilize some or all of their 3d electrons in addition to the 4s electrons. In particular, iron has a common +3 oxidation state.

OXIDES We expect two oxides of iron, FeO and Fe_2O_3, corresponding to the +2 and +3 oxidation states of iron, respectively. In fact there is a third oxide of iron, namely Fe_3O_4, which contains iron in both the +2 and +3 oxidation states and its formula can therefore be written as $Fe(II)Fe(III)_2O_4$. As we have seen both Fe_2O_3, hematite, and Fe_3O_4, magnetite, are important ores of iron. When iron burns in air the main product is Fe_3O_4 with some Fe_2O_3. Fe_3O_4 is also formed when iron is heated in steam in the large-scale preparation of hydrogen (Chapter 3):

$$3Fe(s) + 4H_2O(g) \longrightarrow Fe_3O_4(s) + 4H_2(g)$$

Fe_2O_3 can be made in a pure state by heating the hydroxide $Fe(OH)_3$:

$$2Fe(OH)_3(s) \longrightarrow Fe_2O_3(s) + H_2O(g)$$

TABLE 10.4
Electron Configurations of the Elements of the First Transition Series

Sc	[Ar] $3d^14s^2$
Ti	[Ar] $3d^24s^2$
V	[Ar] $3d^34s^2$
Cr	[Ar] $3d^54s^1$
Mn	[Ar] $3d^54s^2$
Fe	[Ar] $3d^64s^2$
Co	[Ar] $3d^74s^2$
Ni	[Ar] $3d^84s^2$
Cu	[Ar] $3d^{10}4s^1$
Zn	[Ar] $3d^{10}4s^2$

The oxide FeO can be made by reducing Fe_2O_3 with hydrogen at 300 °C:

$$Fe_2O_3(s) + H_2(g) \longrightarrow 2FeO(s) + H_2O(g)$$

IRON(II) CHLORIDE, $FeCl_2$ As previously mentioned, iron reacts with aqueous solutions of acids to form iron(II) salts. With hydrochloric acid iron(II) chloride, $FeCl_2$, is formed. It crystallizes as the tetrahydrate, $FeCl_2 \cdot 4H_2O$:

$$Fe(s) + 2HCl(aq) \longrightarrow FeCl_2(aq) + H_2(g)$$

IRON(II) SULFATE Iron dissolves in sulfuric acid to form iron(II) sulfate,

$$Fe(s) + H_2SO_4(aq) \longrightarrow FeSO_4(aq) + H_2(g)$$

which crystallizes from solution as $FeSO_4 \cdot 7H_2O$. Six water molecules are arranged octahedrally around Fe^{2+}, and there is an additional water molecule forming part of the crystal structure. Hydrated Fe(II) salts are generally pale green.

IRON(II) HYDROXIDE, $Fe(OH)_2$ When an alkali metal hydroxide is added to a solution of an Fe(II) salt, a white precipitate, $Fe(OH)_2$, is obtained:

$$Fe^{2+}(aq) + 2OH^-(aq) \longrightarrow Fe(OH)_2(s)$$

The precipitate rapidly becomes a dirty green and finally brown, because the oxygen in the air oxidizes iron(II) hydroxide to iron(III) hydroxide, $Fe(OH)_3$ (see Experiment 10.7).

Experiment 10.7

Reactions of Fe^{2+}(aq)

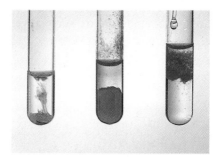

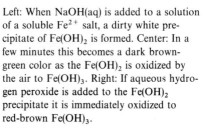

Left: When NaOH(aq) is added to a solution of a soluble Fe^{2+} salt, a dirty white precipitate of $Fe(OH)_2$ is formed. Center: In a few minutes this becomes a dark brown-green color as the $Fe(OH)_2$ is oxidized by the air to $Fe(OH)_3$. Right: If aqueous hydrogen peroxide is added to the $Fe(OH)_2$ precipitate it is immediately oxidized to red-brown $Fe(OH)_3$.

When an Fe^{2+} solution is added to a purple solution of potassium permanganate, it is immediately decolorized as purple MnO_4^- is reduced to colorless Mn^{2+}. An aqueous solution of Mn^{2+} is actually pale pink, but here the solution is too dilute for the color to be seen.

IRON(II) SULFIDE, FeS When iron and sulfur are heated together, black FeS is obtained (see Experiment 3.2). It is often used for the preparation of hydrogen sulfide by treating it with an aqueous acid; for example,

$$FeS(s) + H_2SO_4(aq) \longrightarrow FeSO_4(aq) + H_2S(g)$$

When hydrogen sulfide is passed through a solution of an iron(II) salt, the reverse reaction occurs and iron(II) sulfide is precipitated:

$$Fe^{2+}(aq) + H_2S(g) \longrightarrow FeS(s) + 2H^+(aq)$$

The reversibility of this reaction provides another example of Le Châtelier's principle. Excess hydrogen sulfide and a low $H^+(aq)$ concentration cause iron(II) sulfide to precipitate. But a high $H^+(aq)$ concentration and a low hydrogen sulfide concentration cause iron(II) sulfide to dissolve with the formation of hydrogen sulfide.

IRON(II) DISULFIDE Another important sulfide of iron, which occurs widely as a mineral, is pyrite, FeS_2. It is an ionic compound consisting of Fe^{2+} and disulfide ions, $[:\ddot{S}—\ddot{S}:]^{2-}$.

IRON(III) CHLORIDE, FeCl₃ When iron is heated in chlorine, $FeCl_3$ is obtained. Like $AlCl_3$, the deep red-black $FeCl_3$ sublimes on heating and forms covalent dimeric molecules, Fe_2Cl_6, in the vapor state. Also like $AlCl_3$, it behaves as a Lewis acid and combines with Cl^- to give the tetrahedral $FeCl_4^-$ ion:

$$FeCl_3 + Cl^- \longrightarrow FeCl_4^-$$

Iron(III) chloride is soluble in water, and it crystallizes as yellow $FeCl_3 \cdot 6H_2O$.

IRON(III) HYDROXIDE, Fe(OH)₃ When an alkali metal hydroxide is added to a solution of $FeCl_3$, a brown gelatinous precipitate of $Fe(OH)_3$ is obtained. When this is heated, it loses water to give iron(III) oxide, Fe_2O_3:

$$2Fe(OH)_3 \longrightarrow Fe_2O_3 + 3H_2O$$

The more important reactions of iron and its compounds are summarized in Figure 10.17.

FIGURE 10.17
Preparation and Reactions of Iron Compounds.

All reactants are in aqueous solution unless otherwise specified. All products are solids unless they are enclosed in a blue box when they are in aqueous solution. Reactions denoted by red arrows require heat. All other reactions are at room temperature.

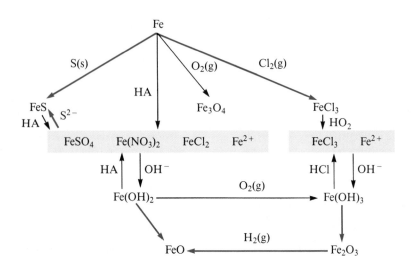

RUSTING OF IRON Although iron, particularly in the form of steel, is a very useful metal, it has the disadvantage that it corrodes or rusts rather easily when exposed to the atmosphere. Rust is brown hydrated iron(III) oxide, $Fe_2O_3 \cdot H_2O$. Both water and oxygen are required to form rust; iron does not rust in dry air or in water that is free of O_2.

The initial step in aqueous solution is

$$2Fe(s) + 4H^+(aq) + O_2(g) \longrightarrow 2Fe^{2+}(aq) + 2H_2O(l)$$

In the second step Fe^{2+} is oxidized by oxygen to insoluble Fe_2O_3 in which iron is in the $+3$ oxidation state:

$$4Fe^{2+}(aq) + O_2(g) + 4H_2O(l) \longrightarrow 2Fe_2O_3(s) + 8H^+(aq)$$

The overall reaction, the sum of these two reactions, is

$$4Fe(s) + 3O_2(g) \longrightarrow 2Fe_2O_3(s)$$

Aluminum also forms an oxide in the air, but it is in the form of a tough continuous skin over the surface of the metal and thus protects it from further reaction with oxygen. Thus although aluminum is a rather reactive metal, it does not corrode. In contrast, iron(III) oxide does not form a continuous skin over the surface of the iron. Rather, it flakes off, exposing more metal to attack by oxygen, and therefore iron continues to rust.

Iron may be protected against corrosion in various ways. It may be coated with a thin layer of a metal, such as tin, that is not attacked by the atmosphere or with a thin layer of chromium, which is protected itself by a thin layer of the oxide, or with a layer of zinc on which an adhering, insoluble layer of zinc carbonate hydroxide, $Zn_2CO_3(OH)_2$, is formed. Iron that is protected by a layer of zinc is called *galvanized iron*. Painting iron also protects it; red lead paint is often used. Many alloys of iron such as stainless steels, which contain nickel and chromium, are more resistant to corrosion than ordinary steel.

OXIDATION AND REDUCTION REACTIONS OF IRON Because they can be oxidized to iron(III) compounds, iron(II) compounds can behave as reducing agents. An important reaction in volumetric analysis is the reduction of permanganate ion, MnO_4^-, to manganous ion by Fe^{2+} in an acidic solution:

$$5Fe^{2+}(aq) + MnO_4^-(aq) + 8H^+(aq) \longrightarrow 5Fe^{3+}(aq) + Mn^{2+}(aq) + 4H_2O(l)$$

Iron(III) compounds can behave as oxidizing agents because they can be reduced to iron(II) compounds. For example, I^- is oxidized to I_3^- by Fe^{3+}:

$$2Fe^{3+}(aq) + 3I^-(aq) \longrightarrow 2Fe^{2+}(aq) + I_3^-(aq)$$

We saw in Chapter 5 that the strength of the halide ions as reducing agents increases in the order $F^- < Cl^- < Br^- < I^-$. Of the halide ions only I^- is strong enough to reduce Fe^{3+} to Fe^{2+}. Thus FeI_3 cannot be prepared, although FeF_3, $FeCl_3$, and $FeBr_3$ can.

═ *Example 10.8* ═══════════════════════════

BALANCING EQUATIONS FOR REDOX REACTIONS

Derive the balanced equation for the reduction of MnO_4^- to Mn^{2+} by Fe^{2+}.

Solution

$$\text{Oxidation} \qquad Fe^{2+} \longrightarrow Fe^{3+} + e^-$$

$$\text{Reduction} \quad MnO_4^- \longrightarrow Mn^{2+}$$

Assiging oxidation numbers, we have

$$\overset{+7}{Mn}O_4^- \longrightarrow \overset{+2}{Mn}^{2+}$$

Therefore we add five electrons to the left side:

$$MnO_4^- + 5e^- \longrightarrow Mn^{2+}$$

Next, we add $8H^+$ to the left side to balance charges:

$$MnO_4^- + 8H^+ + 5e^- \longrightarrow Mn^{2+}$$

Finally, we add $4H_2O$ to the right side to balance atoms:

$$MnO_4^- + 8H^+ + 5e^- \longrightarrow Mn^{2+} + 4H_2O$$

Adding this equation to $5(Fe^{2+} \rightarrow Fe^{3+} + e^-)$ gives the required equation:

$$5Fe^{2+} + MnO_4^- + 8H^+ \longrightarrow 5Fe^{3+} + Mn^{2+} + 4H_2O$$

COMPOUNDS OF COPPER

Copper, like iron, is a transition element. It has two common oxidation states, +1 and +2. The compounds of copper are predominately ionic and contain Cu^+ or Cu^{2+}. The electron configuration of copper is $[Ar]\, 3d^{10}4s^1$. It loses the 4s electron to give Cu^+, and it may also lose one of the 3d electrons to give Cu^{2+}.

COPPER(II) COMPOUNDS When copper is heated strongly in air or oxygen, black *copper(II) oxide*, CuO, is formed. It can also be prepared by heating copper(II) nitrate:

$$2Cu(NO_3)_2(s) \longrightarrow 2CuO(s) + 4NO_2(g) + O_2(g)$$

As we have seen, lead nitrate, $Pb(NO_3)_2$, decomposes on heating in the same way.

⎯ *Example 10.9* ⎯⎯⎯⎯⎯⎯⎯⎯⎯⎯⎯⎯⎯⎯⎯⎯⎯⎯⎯⎯⎯⎯⎯⎯⎯

OXIDATION NUMBERS

Assign oxidation numbers to all the elements in the reaction

$$2Cu(NO_3)_2(s) \longrightarrow 2CuO(s) + 4NO_2(g) + O_2(g)$$

and find which elements are oxidized and which are reduced.

Solution

$$2\overset{+2}{Cu}(\overset{+5}{N}\overset{-2}{O_3})_2 \longrightarrow 2\overset{+2}{Cu}\overset{-2}{O} + 4\overset{+4}{N}\overset{-2}{O_2} + \overset{0}{O_2}$$

The oxidation state of copper remains +2 throughout the reaction. Nitrogen oxidizes some of the oxygen from the −2 state to the element (oxidation state 0) and is itself reduced to the +4 state.

Copper(II) nitrate can be obtained as blue hydrated crystals, $Cu(NO_3)_2 \cdot 6H_2O$, by evaporating a solution made by dissolving copper in nitric acid.

The most common salt of copper(II) is *copper(II) sulfate*, $CuSO_4$. It can be prepared by the reaction of CuO(s) with dilute sulfuric acid:

$$CuO(s) + H_2SO_4(aq) \longrightarrow CuSO_4(aq) + H_2O(l)$$

It crystallizes from solution in large blue crystals of the hydrate $CuSO_4 \cdot 5H_2O$. White $CuSO_4$, formed when the hydrate is heated strongly, is called anhydrous copper(II) sulfate. It can be used to detect small amounts of water in solvents such as alcohol and ether; when it is added to these liquids, it becomes blue if water is present because the hydrate $CuSO_4 \cdot 5H_2O$ is formed.

Copper(II) sulfide is a black solid formed by heating copper with excess sulfur or by passing H_2S into a solution of a Cu^{2+} salt (see Experiment 10.8):

$$Cu(s) + S(s) \longrightarrow CuS(s)$$
$$Cu^{2+}(aq) + H_2S(g) \longrightarrow CuS(s) + 2H^+(aq)$$

Copper(II) hydroxide is formed as a pale blue precipitate when OH^- is added to a solution of Cu^{2+} salt:

$$Cu^{2+}(aq) + 2OH^-(aq) \longrightarrow Cu(OH)_2(s)$$

This precipitate dissolves in an aqueous NH_3 solution to give a deep blue

Blue crystals of $CuSO_4 \cdot 5H_2O$ and white anhydrous $CuSO_4$

Experiment 10.8

Reactions of Cu^{2+}(aq)

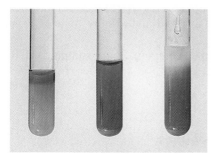

Left: When an aqueous solution of KI is added to a blue Cu^{2+} solution, I^- is oxidized to brown I_3^-, Cu^{2+} is reduced to Cu^+, and a white precipitate of CuI is formed. Center: When sufficient KI has been added to react with all the Cu^{2+}, the solution is brown due to I_3^- and contains a white precipitate of CuI. Right: When a solution of sodium thiosulfate, $Na_2S_2O_3$, is added, it reduces the I_3^- to colorless I^-, leaving the white precipitate of CuI.

When H_2S(g) is bubbled into a blue solution of a soluble Cu^{2+} salt, a black precipitate of copper (II) sulfide, CuS, is formed.

solution of the tetraamminecopper(II) ion, $Cu(NH_3)_4^{2+}$. If $Cu(OH)_2$ is heated, black CuO is formed:

$$Cu(OH)_2(s) \longrightarrow CuO(s) + H_2O(l)$$

COPPER(I) COMPOUNDS If CuO is heated strongly with copper filings or copper powder, red *copper(I) oxide*, Cu_2O, is formed:

$$CuO(s) + Cu(s) \longrightarrow Cu_2O(s)$$

Black *copper(I) sulfide*, Cu_2S, can be obtained simply by heating copper(II) sulfide, CuS:

$$2CuS(s) \longrightarrow Cu_2S(s) + S(s)$$

It can also be made by heating the requisite amounts of copper and sulfur:

$$2Cu(s) + S(s) \longrightarrow Cu_2S(s)$$

White *copper(I) chloride*, CuCl, is formed when HCl gas is passed over heated copper:

$$2Cu(s) + 2HCl(g) \longrightarrow 2CuCl(s) + H_2(g)$$

Copper(I) iodide can be obtained by the reduction of Cu^{2+} with I^- in a reaction similar to that between Fe^{3+} and I^- (see Experiment 10.8):

$$2Cu^{2+}(aq) + 5I^-(aq) \longrightarrow 2CuI(s) + I_3^-(aq)$$

Copper(I) iodide is insoluble in water. The only copper(I) compounds that can be made in aqueous solution are insoluble compounds like CuI, because Cu^+ is not stable in aqueous solution; it reacts to give Cu and Cu^{2+}:

$$2Cu^+(aq) \longrightarrow Cu^{2+}(aq) + Cu(s)$$

In this reaction half of the copper(I) is oxidized to copper(II) and the other half is reduced to copper(0), that is, elemental copper. This type of reaction in which an element oxidizes itself to give a more oxidized and a more reduced form is called a **disproportionation reaction**.

The preparation and reactions of copper compounds are summarized in Figure 10.18.

$=$ *Example 10.10* $=$

REACTIONS OF COPPER AND ITS COMPOUNDS

In an experiment 1.50 g of copper were dissolved in concentrated nitric acid, and the solution was evaporated to dryness to give blue crystals. When these crystals were heated strongly, a black powder was obtained. When the black powder was dissolved in dilute aqueous sulfuric acid, a blue solution was obtained, which on evaporation yielded blue crystals. Write balanced equations for all the reactions, and calculate the maximum amount of the final product that could be obtained.

Solution

$$Cu(s) + 4HNO_3(aq) \longrightarrow Cu(NO_3)_2(aq) + 4NO_2(g) + 2H_2O(l)$$

$$Cu(NO_3)_2(aq) + 6H_2O(l) \xrightarrow{\text{Evaporate}} Cu(NO_3)_2 \cdot 6H_2O(s)$$
$$\text{Blue crystals}$$

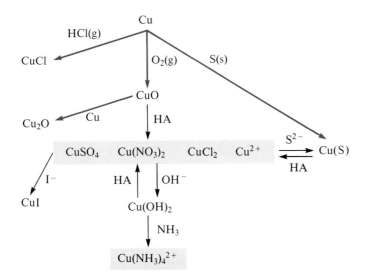

FIGURE 10.18
Preparation and Reactions of Copper Compounds.

All reactants are in aqueous solution unless otherwise specified. All products are solids except those in blue boxes which are in aqueous solution. Reactions which require heat are denoted by red arrows. All other reactions are at room temperature.

$$Cu(NO_3)_2 \cdot 6H_2O \xrightarrow{\text{Heat}} CuO(s) + 2NO_2(g) + O_2(g) + 6H_2O(g)$$
Black

$$CuO(s) + H_2SO_4(aq) \longrightarrow CuSO_4(aq) + H_2O(l)$$

$$CuSO_4(aq) + 5H_2O(l) \xrightarrow{\text{Evaporate}} CuSO_4 \cdot 5H_2O(s)$$

We see from these equations that 1 mol of Cu(s) gives 1 mol of $CuSO_4 \cdot 5H_2O(s)$. Hence 63.5 g Cu give 249.6 g $CuSO_4 \cdot 5H_2O$ and therefore

$$1.50 \text{ g Cu give } \frac{1.50 \text{ g} \times 249.6 \text{ g}}{63.5 \text{ g}} = 5.90 \text{ g } CuSO_4 \cdot 5H_2O$$

Or using unit factors

$$(1.50 \text{ g Cu})\left(\frac{1 \text{ mol Cu}}{63.5 \text{ g Cu}}\right)\left(\frac{1 \text{ mol } CuSO_4 \cdot 5H_2O}{1 \text{ mol Cu}}\right)\left(\frac{249.6 \text{ g}}{1 \text{ mol } CuSO_4 \cdot 5H_2O}\right)$$
$$= 5.90 \text{ g } CuSO_4 \cdot 5H_2O$$

SUMMARY OF THE REACTIONS OF Al, Cu, Fe, AND Pb

The large number of reactions and compounds mentioned in this section may appear confusing at first sight. But there are many similarities between the reactions and compounds of the four metals, and it will make it easier to remember and understand these reactions if we look at these similarities. The important compounds of the metals that we have considered are the oxides, hydroxides, halides, sulfides, sulfates, and nitrates. These compounds are summarized in Table 10.5 on p. 528. The preparation and reactions of the compounds of a typical metal M that has a +2 oxidation state (except the group 2 metals) are summarized in Figure 10.19 on p. 528. These reactions also apply to other oxidation states such as +1 (except group 1 metals) and +3, although not to the +4 state of lead which forms essentially covalent compounds.

TABLE 10.5
Compounds of Aluminum, Iron, Lead, and Copper

	Oxides (insoluble)	Hydroxides (insoluble)	Sulfides (insoluble)	Chlorides (soluble)	Sulfates (soluble)	Nitrates (soluble)
Al(III)	Al_2O_3	$Al(OH)_3$	—	$AlCl_3$	$Al_2(SO_4)_3$	$Al(NO_3)_3$
Fe(II)	FeO	$Fe(OH)_2$	FeS	$FeCl_2$	$FeSO_4$	—
	Fe_3O_4					
Fe(III)	Fe_2O_3	$Fe(OH)_3$	—	$FeCl_3$	$Fe_2(SO_4)_3$	$Fe(NO_3)_3$
Pb(II)	PbO	$Pb(OH)_2$	PbS	$PbCl_2$[a]	$PbSO_4$[a]	$Pb(NO_3)_2$
	Pb_3O_4					
Pb(IV)	PbO_2	—	—	$PbCl_4$[b]	—	—
Cu(I)	Cu_2O	—	Cu_2S	$CuCl$	—	—
Cu(II)	CuO	$Cu(OH)_2$	CuS	$CuCl_2$	$CuSO_4$	$Cu(NO_3)_2$

[a] Insoluble.   [b] Reacts with water.

With the exception of the metals of groups 1 and 2, the oxides, hydroxides, and sulfides of metals are generally insoluble in water, whereas the halides, sulfates, and nitrates are generally soluble—but note that $PbCl_2$, $PbBr_2$, PbI_2, and $PbSO_4$ are insoluble (see Table 5.9).

Insoluble compounds can usually be prepared by precipitation. For example,

$$Al^{3+}(aq) + 3OH^-(aq) \longrightarrow Al(OH)_3(s)$$
$$Fe^{2+}(aq) + S^{2-}(aq) \longrightarrow FeS(s)$$

The oxide ion, O^{2-}, is too strong a base to exist in water, so oxides cannot be prepared in this way, but they can be prepared by direct combination of the elements. They can also be prepared by dehydrating a hydroxide; for example,

$$Cu(OH)_2(s) \xrightarrow{\text{Heat}} CuO(s) + H_2O(g)$$

Sulfides can similarly be prepared by direct combination.

Soluble compounds such as halides and sulfates cannot be prepared by precipitation, but they can be prepared by the reaction of the oxide, and in some

FIGURE 10.19
Preparative Methods for the +2 Oxidation State Compounds of a Metal.

All reactants are in aqueous solution unless otherwise specified. All products are solids unless they are enclosed in a blue box when they are in aqueous solution. Reactions denoted by red arrows require heat. All other reactions are at room temperature.

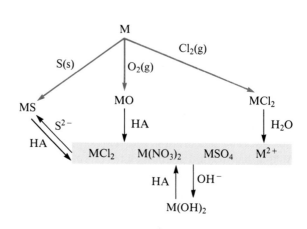

TABLE 10.6
Preparative Methods for the Compounds of Aluminum, Lead, Iron, and Copper

Direct Combination of Elements Redox	Precipitation	Metal + Acid Redox	Oxide + Acid Acid–Base
Oxides	Hydroxides	Sulfates	Sulfates
Sulfides	Sulfides	Nitrates	Nitrates
Halides	Pb Halides	Halides	Halides

cases the metal, with an aqueous solution of the appropriate acid:

$$Fe(s) + H_2SO_4(aq) \longrightarrow Fe^{2+}(aq) + SO_4^{2-}(aq) + H_2(g)$$

$$CuO(s) + H_2SO_4(aq) \longrightarrow Cu^{2+}(aq) + SO_4^{2-}(aq) + H_2O(l)$$

Halides can also be prepared by direct combination of the elements; for example,

$$Fe(s) + 3Cl_2(g) \longrightarrow 2FeCl_3(s)$$

Because the halogens are strong oxidizing agents, direct combination usually gives a halide of the metal in one of its higher oxidation states, for example Fe(III) and Cu(II). The methods of preparation are summarized in Table 10.6.

IMPORTANT TERMS

An **acidic oxide** is an oxide of a nonmetal that dissolves in water to give an acidic solution or that reacts with a base or a basic oxide to give a salt.

An **alloy** is a solid that has metallic properties and is usually composed of two or more metals; it may also be composed of a metal and a nonmetal such as carbon, silicon, or nitrogen.

An **alum** is a double salt, $MAl(SO_4)_2 \cdot 12H_2O$, where M is an alkali metal ion or the ammonium ion.

An **amphoteric oxide** is an oxide that has both acid and base properties.

A **basic oxide** is an oxide of a metal that dissolves in water to give a basic solution or that reacts with an acid or an acidic oxide to give a salt.

In **body-centered cubic packing** each spherical atom is surrounded by eight neighbors that are arranged at the corners of a cube.

Cubic close packing is an arrangement of spherical atoms to give close-packed layers in which each atom is surrounded by a hexagon of nearest neighbors. The layers are stacked on each other in an ABCABC... arrangement so that the atoms in any layer lie directly above those three layers below.

A **disproportionation reaction** is one in which an element in a given oxidation state reacts to give compounds containing the element in two different oxidation states.

Hexagonal close packing is an arrangement of spherical atoms in close-packed layers in which each atom is surrounded by a hexagon of nearest neighbors. The layers are stacked one on another in an ABAB... arrangement so that the atoms in alternate layers are directly above each other.

A **hydrated salt** is one that crystallizes from water with molecules of *water of crystallization*.

A **hygroscopic compound** is one that will absorb water directly from the atmosphere.

A **Lewis acid** is an electron pair acceptor. A **Lewis base** is an electron pair donor.

A **metallic bond** is the type of bond that holds the atoms together in a metal. It is the force of attraction between a mobile charge cloud (electron gas) and positively charged atomic cores.

Water of crystallization. See hydrated salt.

PROBLEMS *

Metallurgy

1. **(a)** Give four different reducing agents that are used to prepare metals from their compounds.

   (b) Write balanced equations for the preparation of **(i)** copper, **(ii)** iron, **(iii)** lead, and **(iv)** aluminum, using a different reducing agent in each case.

2. A copper ore contains 1.60% as Cu_2S. What volume of $SO_2(g)$, measured at 20 °C and 1.00 atm pressure, is obtained when 1.00 metric ton (10^3 kg) of the ore is roasted?

3. What minimum mass of limestone, containing 96 mass % $CaCO_3$, is required in a blast furnace charged with 2.00 metric tons (2×10^3 kg) of iron ore containing 12.2 mass % silica, SiO_2?

4. What is the difference between a mineral and an ore? How do you account for the fact that the commonest ores are oxides, carbonates, and sulfides?

5. **(a)** Write balanced equations to show how a metal may be extracted from each type of ore mentioned in Problem 4.

   (b) Write balanced equations for the extraction of a metal from an oxide ore, using as the reducing agent **(i)** another metal and **(ii)** a nonmetal.

6. **(a)** What mass of an ore containing 30 mass % Fe_2O_3 is required to produce 10^6 metric tons of steel?

   (b) How many kilograms of iron are present in each metric ton of the ore?

   (c) Calculate the mass of coke (assuming that it is pure carbon and that only carbon monoxide is formed) that is needed to reduce 1 metric ton of the ore to iron.

7. Describe the function of each of the following in the manufacture of pig iron:

   (a) $Fe_3O_4(s)$ **(b)** Coke

   (c) Carbon monoxide   **(d)** Calcium oxide

8. **(a)** How can aluminum be obtained from one of its compounds using another metal as the reducing agent?

   (b) How is lead obtained from its sulfide ore galena, $PbS(s)$?

   (c) What function does the limestone added during the smelting of iron play in the process?

* Answers to problems numbered in blue appear at the end of the book.

9. Describe how iron is produced in a blast furnace, and how it is converted to steel by the basic oxygen process. Write balanced equations for each of the important chemical reactions involved.

10. What is slag, and how and why is it formed in a blast furnace?

Properties and Structures of Metals

11. Which of the elements of the third period (sodium to argon) and of group 4 (carbon to lead) are metals? Describe in general how the metallic character of the elements changes throughout the periodic table.

12. Draw diagrams to illustrate **(a)** closest packing and **(b)** square packing of spheres in a single two-dimensional layer of spheres. What is the number of nearest neighbors in (a) and (b) in **(i)** two dimensions and **(ii)** three dimensions?

13. Describe and contrast the type of bonding in **(a)** an Na_2 molecule in the gas phase and **(b)** solid sodium metal. What is the essential difference between the bonding in the two cases? Which would you expect to be greater, the Na–Na distance in Na_2 or the Na–Na distance in sodium metal?

14. Explain why **(a)** a metal is a good conductor of heat and electricity and **(b)** the electrical conductivity of a metal increases with decrease in temperature.

15. Draw a diagram to illustrate the body-centered cubic structure, showing the eight nearest neighbors, and the six next-nearest neighbors, of each atom in the structure.

16. The metals gold, silver, copper, tin, lead, and iron are mentioned in the Bible, but lithium, sodium, potassium, and aluminum, for example, were not isolated as the elements until the nineteenth century. What differences in properties between the "biblical" elements and the alkali metals and aluminum account for the late discovery of the latter elements?

17. Considering the properties discussed in this chapter, which metal would you consider the best choice for each of the following?

   (a) The main structure of a space vehicle

   (b) A statue for a city park

Justify your choice by citing some advantageous and some disadvantageous properties of each metal. Are there advantages in using alloys of these metals, rather than the pure metals?

18. Explain why:

    (a) Copper was one of the earliest known metals

    (b) Copper-covered roofs acquire a green deposit

    (c) Although aluminum is a rather reactive metal, it has many uses, including cooking utensils

    (d) Aluminum was a precious rare metal until the beginning of the twentieth century

19. Explain the following increase in melting point:

$$\text{Na } 98\,°C < \text{Mg } 650\,°C < \text{Al } 660\,°C$$

Iron melts at $1535\,°C$. Suggest a possible reason why the melting point of iron is so much higher than the three metals above.

20. Explain in terms of their structures why carbon in the form of diamond is very hard and fractures under a large stress, whereas copper is relatively soft and deforms when subjected to a stress.

Reactions of Metals with Acids

21. (a) Name two metals that react with dilute hydrochloric acid, and write the balanced equations for the reactions that occur.

    (b) Name two metals that *do not* react with dilute hydrochloric acid.

    (c) Why is it that many metals that do not react with dilute hydrochloric acid nevertheless react with dilute nitric acid?

22. Write balanced equations for each of the following reactions:

    (a) Aluminum with dilute sulfuric acid

    (b) Silver with hot concentrated sulfuric acid

23. Place the metals Al, Cu, Mg, and Na in order of decreasing reactivity towards $H_2O(aq)$. What are the experimental observations that provide the evidence for this order?

Oxidation–Reduction

24. Assign oxidation numbers to each of the atoms in each of the following:

    (a) $Al_2O_3 \cdot xH_2O$ **(b)** $AlCl_4^-$ **(c)** SiO_2

    (d) SiO_3^{2-} **(e)** $Pb(OH)_4^{2-}$ **(f)** $PbCl_4$

25. Assign oxidation numbers to each of the atoms in each of the following:

    (a) $Al(OH)_4^-$ **(b)** Al_2Cl_6

    (c) $Fe_2(SO_4)_3$ **(d)** $CuCl$

    (e) $CaCO_3$ **(f)** $KAl(SO_4)_2 \cdot 12H_2O$

    (g) FeS_2 **(h)** $Zn_2CO_3(OH)_2$

26. Define *oxidation* and *reduction* in terms of the transfer of electrons. Explain each of the following reactions in aqueous solution in terms of electron transfer, indicating whether each reactant is oxidized, reduced, or takes part in neither oxidation nor reduction. Balance each of the equations.

    (a) $Zn(s) + H_2SO_4(aq) \rightarrow ZnSO_4(aq) + H_2(g)$

    (b) $I_3^-(aq) + S_2O_3^{2-}(aq) \rightarrow I^-(aq) + S_4O_6^{2-}(aq)$

27. Write balanced equations for each of the following oxidation–reduction reactions in acidic aqueous solution:

    (a) $SO_3^{2-} + MnO_4^- \rightarrow SO_4^{2-} + Mn^{2+}$

    (b) $H_2O_2 + MnO_4^- \rightarrow Mn^{2+} + O_2(g)$

    (c) $Zn(s) + NO_3^- \rightarrow Zn^{2+} + N_2O(g)$

    (d) $P_4(s) + NO_3^- \rightarrow H_2PO_4^- + NO(g)$

    (e) $NO_3^- + I_2(s) \rightarrow IO_3^- + NO_2(g)$

28. Write balanced equations for each of the following oxidation–reduction reactions in acidic aqueous solution:

    (a) $MnO_4^- + Fe^{2+} \rightarrow Mn^{2+} + Fe^{3+}$

    (b) $ClO_3^- + Cl^- \rightarrow Cl_2 + ClO_2$

    (c) $Cu(s) + HNO_3 \rightarrow Cu^{2+} + NO(g)$

    (d) $MnO_4^- + SO_2(g) \rightarrow Mn^{2+} + SO_4^{2-}$

    (e) $HI + HNO_3 \rightarrow I_2 + NO$

29. Write balanced equations for each of the following oxidation–reduction reactions in basic aqueous solution:

    (a) $MnO_4^- + I^- \rightarrow MnO_2(s) + IO^-$

    (b) $Br_2 \rightarrow Br^- + BrO_3^-$

    (c) $I^- + ClO^- \rightarrow IO_3^- + Cl^-$

    (d) $SO_3^{2-} + MnO_4^- \rightarrow SO_4^{2-} + MnO_2(s)$

    (e) $Mn^{2+} + H_2O_2 \rightarrow MnO_2(s)$

30. Write balanced equations for each of the following oxidation–reduction reactions in basic aqueous solution:

    (a) $CN^- + MnO_4^- \rightarrow CNO^- + MnO_2$

    (b) $Cr^{3+} + OCl^- \rightarrow CrO_4^{2-} + Cl^-$

    (c) $I^- + ClO_3^- \rightarrow I_3^- + Cl^-$

    (d) $NH_3 + ClO^- \rightarrow N_2H_4 + Cl^-$

    (e) $MnO_2(s) + O_2 \rightarrow MnO_4^-$

Acidic and Basic Oxides

31. What are the *anhydrides* of the following oxoacids?

    H_2SO_4 HNO_3 H_2CO_3 $HClO_4$

32. Write an equation for the reaction of each of the following oxides with water. Classify each oxide as an acidic or basic oxide.

K_2O SrO SO_2 SO_3

CO_2 P_4O_6 Cl_2O_7

33. Which of Al_2O_3 and SiO_2 would you describe as the more basic? Which of SiO_2 and SO_2 would you describe as the more acidic? Explain why.

34. Write balanced equations for each of the following reactions:

(a) Lithium oxide with silica

(b) Sodium oxide with dinitrogen pentaoxide

(c) Calcium oxide with phosphorus(V) oxide

Lewis Acids and Bases

35. (a) What are the similarities and the differences between the Brønsted–Lowry and the Lewis acid–base concepts?

(b) Why is the reaction between calcium oxide and silica, to give calcium silicate, $CaSiO_3$, classified as an acid–base reaction?

36. (a) In terms of their positions in the periodic table, which elements from periods 2 and 3 are expected to form acidic oxides, and which are expected to form amphoteric oxides?

(b) Give two examples each of basic oxides, acidic oxides, and amphoteric oxides.

(c) Write balanced equations for the reactions with water of each of the oxides selected in part (b).

37. Write the balanced equations for reactions of each of the following which support their classification as Lewis acids:

$AlCl_3$ $AlBr_3$ BF_3 BCl_3

38. Which of the following may behave as Lewis acids and which as Lewis bases?

NH_3 Cu^{2+} Al^{3+} SiO_2 Cl^-

Compounds and Reactions of Al, Fe, Cu, and Pb

39. Describe and write balanced equations for the reactions that occur when an aqueous solution of sodium hydroxide is added to aqueous solutions of each of the following:

(a) $AlCl_3$ **(b)** $CuSO_4$

(c) $FeSO_4$ **(d)** $Pb(NO_3)_2$

40. Give three examples taken from this chapter of metal hydroxides that decompose on heating to give the corresponding oxides. Write a balanced equation for each of the reactions.

41. Solutions of $CuSO_4(aq)$, $Al_2(SO_4)_3(aq)$, $Pb(NO_3)_2(aq)$, and $FeSO_4(aq)$ were prepared but not labeled. Describe some simple tests that could be used to identify each of the solutions.

42. Write balanced equations for each of the following reactions:

(a) The reaction of powdered aluminum with $NaOH(aq)$ to give sodium aluminate, $NaAl(OH)_4$, and hydrogen.

(b) The reaction of aluminum oxide with carbon and chlorine to give aluminum trichloride and carbon monoxide.

(c) The decomposition of ammonium alum on heating, to give ammonia, sulfuric acid, aluminum oxide, and water.

43. Give the names and formulas of all the iron and copper salts that may be produced from each of the following acids:

(a) Nitric acid   **(b)** Sulfuric acid

(c) Phosphoric acid

44. Give the systematic name and the formula for each of the following:

(a) Limestone   **(b)** Alumina

(c) Magnetite   **(d)** Pyrite

(e) Coke   **(f)** Red lead

(g) Rust   **(h)** Alum

45. Describe the reactions that occur in the rusting of iron.

46. What is the mass percentage of aluminum, and the mass percentage of silicon, in common clay, assuming the composition $H_2Al_2(SiO_4)_2·H_2O$?

47. When 0.250 g of aluminum metal was heated in a stream of dry chlorine gas, 1.236 g of a white hygroscopic solid, containing only aluminum and chlorine, were formed. Above 183 °C, the solid changed from solid to vapor, and this sample gave a pressure of 720 torr when contained in a 210-mL flask at 250 °C. What is the empirical formula and the molecular formula of the aluminum chloride? Write the Lewis structure.

48. Soft solder is an alloy of lead and tin. When 1.00 g of soft solder was dissolved in concentrated nitric acid by warming, and the solution was diluted with water to precipitate insoluble $SnO_2 \cdot 2H_2O$, the mass of dry hydrated tin oxide was found to be 0.778 g. What is the composition of soft solder?

49. A red powder resulted from strongly heating 2.00 g of lead in excess oxygen. When this powder was treated with concentrated nitric acid, a brown powder was formed which was filtered off and dried. When a solution of potassium iodide was added to the filtrate, a bright yellow precipitate was formed. Write balanced equations for each of the above reactions and identify all the products. What is the maximum amount of the brown powder, and of the dry yellow precipitate, that could be obtained?

50. A deep red-black solid resulted from heating 1.50 g of iron in excess chlorine. When the solid was dissolved in water and excess of aqueous sodium hydroxide was added, a gelatinous brown precipitate resulted. When this precipitate was strongly heated, it formed a red-brown powder. Write balanced equations for each of the above reactions and identify all the products. What is the maximum mass of red-brown powder that could be obtained?

51. Aluminum has many important commercial uses, including structural materials and cooking utensils. Yet, it dissolves readily in aqueous acids and bases, can be used to reduce the oxides of other metals, as in the thermite process, and burns readily in chlorine. Explain why aluminum has so many important practical uses despite its evident reactivity. Under what conditions does it become dangerous to use?

*52. Aluminum brass contains copper, zinc, and aluminum. When 1.000 g was reacted with $0.100M$ $H_2SO_4(aq)$, 149.3 mL of $H_2(g)$ was evolved, measured at 25 °C and 1.00 atm pressure. When an identical sample was dissolved in hot concentrated sulfuric acid, 411.1 mL of $SO_2(g)$ were obtained at 25 °C and 1.00 atm pressure. What is the composition of aluminum brass?

53. White phosphorus causes severe burns. A first-aid procedure to render harmless any phosphorus that has accidently contacted the skin is to treat it with an aqueous solution of copper sulfate, $CuSO_4(aq)$, which is reduced to a mixture of copper metal and copper(I) phosphide. Write balanced equations for

(a) The reaction of $P_4(s)$ with $CuSO_4(aq)$ to give copper(I) phosphide, phosphorous acid, and sulfuric acid

(b) The reaction of $P_4(s)$ with $CuSO_4(aq)$ to give copper, phosphoric acid, and sulfuric acid

54. What is observed when solid lead nitrate is heated? Write a balanced equation for the reaction that occurs.

* The asterisk denotes the more difficult problems.

55. What qualitative experiments could you perform to show that ammonium alum, $(NH_4)Al(SO_4)_2 \cdot 12H_2O$, is composed of ammonium, aluminum, and sulfate ions, and water molecules?

*56. Calculate the empirical formula of hydrated iron(II) ammonium sulfate from the following experimental data: **(i)** When heated strongly to constant mass, 0.7840 g of the salt gave 0.1600 g of iron(III) oxide. **(ii)** Addition of excess of aqueous barium chloride to a solution of 0.7840 g of the salt dissolved in water gave 0.9336 g of barium sulfate. **(iii)** When a solution containing 0.3920 g of the salt was boiled with excess $NaOH(aq)$, ammonia gas was liberated. When absorbed in 50.0 mL of $0.10M$ $HCl(aq)$, the excess acid that remained after reaction with the ammonia required 30.0 mL of $0.10M$ $NaOH(aq)$ for neutralization.

57. Copper(II) ammonium sulfate was found to contain 27.3 mass % of water of crystallization, and on strongly heating gave 19.89% of its mass of copper(II) oxide. What is the empirical formula of hydrated copper(II) ammonium sulfate?

58. Describe and explain the reactions that could be used to prepare each of the following substances from copper(II) sulfate pentahydrate (more than one step may be required):

(a) Copper

(b) Copper(II) chloride

(c) Copper(I) chloride

(d) Copper(II) tetraammine sulfate monohydrate, $Cu(NH_3)_4SO_4 \cdot H_2O$

59. Starting with the metal, how would you prepare $Al_2(SO_4)_3(aq)$, $CuSO_4(aq)$, $PbSO_4(s)$?

60. Starting with the oxides which are insoluble in water, how would you prepare $Fe(OH)_3(s)$ and $Cu(OH)_2(s)$?

61. Many insoluble compounds can be made by precipitation from aqueous solution. Explain why Fe_2O_3 and CuO cannot be made by this method. Describe how you would prepare these oxides starting with aqueous solutions of $FeSO_4$ and $CuSO_4$ respectively.

62. You are provided with a yellow solid which may be either AgI or PbI_2. Describe a simple physical test that you could carry out to determine which it is.

63. Describe what is observed when an aqueous solution of KI is added to a blue solution of $CuSO_4$. Write balanced equations for the reactions that occur.

64. You are provided with a black solid which might be either CuO or CuS. Describe a simple chemical test that you could cary out to decide which it is.

Miscellaneous

65. The mineral *atacamite* has the formula $CuCl_2 \cdot xCu(OH)_2$, where x is an integer. In a titration experiment it was found that 21.45 mL of $0.4071M$ HCl(aq) were required to react completely with 0.6217 g of atacamite. What is x in the empirical formula of this mineral?

66. Write Lewis structures for each of the following molecules and ions:

(a) $AlCl_3$

(b) Al_2Cl_6

(c) $Al(H_2O)_6{}^{3+}$

(d) $Al(OH)_4{}^-$

67. When a mixture of aluminum powder and Fe_2O_3 is ignited, a highly exothermic reaction occurs as was demonstrated in Experiment 10.1. Write the equation for the reaction and calculate the standard enthalpy change from the standard enthalpies of formation in Table B.2 Appendix B.

68. The booster rockets of the space shuttle use a mixture of aluminum and ammonium perchlorate as fuel. The equation for the reaction is

$$3Al(s) + 3NH_4ClO_4(s) \longrightarrow$$
$$Al_2O_3(s) + AlCl_3(s) + 3NO(g) + 6H_2O(g)$$

Using data from Table B.2 Appendix B calculate the standard enthalpy change, $\Delta H°$, for the reaction.

The Solid State

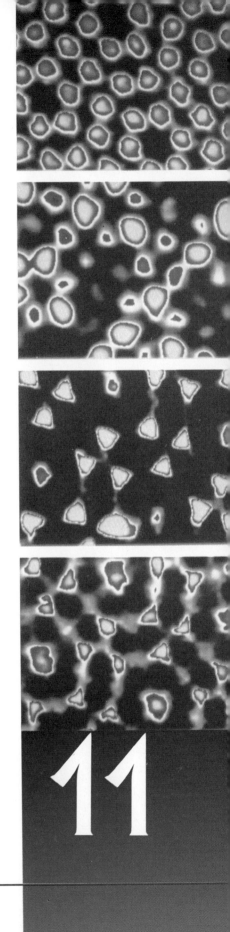

11

The vast majority of substances are solids under ordinary conditions. The importance in our daily lives of metals, glass, concrete, plastics, silicon chips, and gemstones hardly needs emphasizing.

In earlier chapters we have given considerable emphasis to gases and to solutions, for two important reasons. First, chemical reactions occur much more readily between gases and between substances in solution than between solids, because in the gas state and in solution molecules are free to move and to collide with each other. Second, unlike solids, all gases have similar properties; they obey the ideal gas law, at least approximately. This is because they all consist of widely separated, rapidly moving molecules between which there are only relatively weak forces.

In contrast, the forces holding the atoms and molecules together in a solid are relatively strong, and the structures of solids are varied and often complex. As a result, different types of solids often have very different properties. We have already described some of the differences between molecular, ionic, and metallic solids. In this chapter we give a more comprehensive account of the different types of solids and their properties.

The properties of a solid depend on its composition and its structure. The allotropes of an element, which, of course, all have the same composition, provide striking illustrations of the effect of structure on properties. We have only to recall the differences between white and red phosphorus or between diamond and graphite to appreciate the profound effect of structure on properties.

Most solids are **crystalline**; that is, they have a regular periodic arrangement of their atoms or ions, which is reflected in the fact that they form crystals with flat faces and definite angles between the faces. We will be mainly concerned with crystalline solids in this chapter. But as we will see, some solids do not have a regular periodic arrangement of their atoms; they are called **amorphous solids**.

11.1
Crystals

The most obvious characteristic of a crystalline substance is that it forms crystals. The large crystals of some substances that occur naturally have a fascination and beauty that has been recognized throughout history (Figure 11.1), and it

FIGURE 11.1
A Natural Specimen of Quartz.

Quartz is one of the many crystalline forms of silicon dioxide.

is not difficult to grow large well-formed crystals of some substances in the laboratory (Experiment 10.5). A crystal has flat faces that are at characteristic angles to each other. The flat faces of a crystal are a consequence of the regular ordered arrangement of the atoms, ions, or molecules in the crystal (Figure 11.2). If a crystal is broken into two or more pieces it breaks along the planes formed by the regularly arranged atoms, ions, or molecules to give smaller crystals that have a similar shape to the original crystal, and in particular have the same angles between their faces (Figure 11.2). Even if a crystal is damaged the characteristic angles between its faces are retained and it can be recognized by measuring these angles (Figure 11.3). Crystals of a given substance grown under different conditions may have different shapes because the various crystal faces may grow at different rates under different conditions. However, these shapes are all related to each other as we see in Figure 11.4. In this particular case the crystals are cubic or octahedral or have an intermediate shape, but in all cases the characteristic angles between the faces are either 90° or 135°. In an amorphous substance in which the atoms or molecules have a disordered, rather than an ordered, arrangement there are no planes of atoms to give rise to flat crystal faces and so amorphous substances do not form crystals with characteristic shapes. But this has the advantage that an amorphous substance like glass can be worked into an infinite variety of different shapes which is not possible with a crystalline substance.

The structures of crystalline solids can be determined with considerable accuracy from the patterns produced when a beam of X rays is diffracted by a crystal (see Box 11.1). This technique, known as **X ray crystallography**, was invented and developed by William and Lawrence Bragg (Figure 11.5).

Because the properties of solids depend so strongly on their structure, it is very useful to classify solids according to their structures. This we do in the next section.

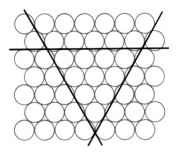

FIGURE 11.2
Cleavage Planes in a Crystal.

A crystal will break relatively easily—it is said to cleave—along any of the planes shown by thick lines to give a smaller crystal with the same angles between the faces as the original crystal. It will only break with great difficulty—it does not cleave—along a plane such as that shown by the dotted line.

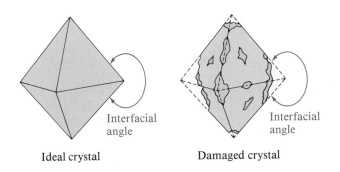

Ideal crystal   Damaged crystal

FIGURE 11.3
Constant Interfacial Angles.

Even if a crystal is damaged it may be recognized by measuring the angles between its faces which are always the same for the same crystalline form of a substance.

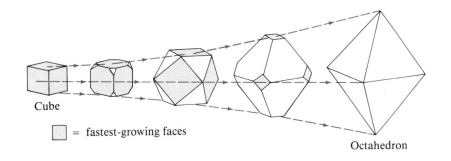

Cube

☐ = fastest-growing faces

Octahedron

FIGURE 11.4

Crystals of a given substance may have different shapes when grown under different conditions. If some faces grow faster than others the shape of the crystal may change as shown here. However, the characteristic angles between the faces are always retained.

BOX 11.1

X RAY DIFFRACTION AND THE STRUCTURES OF SOLIDS

Before 1912 chemists had no direct way of determining how the atoms are arranged in molecules or crystalline solids. The discovery by German physicist von Laue, in 1912, that a beam of X rays is diffracted by a crystalline solid led to the rapid development of a new branch of science called *X ray crystallography*, which has had an enormous effect on the development of chemistry. By studying the diffraction of X rays by crystalline solids, we can obtain precise information about the arrangement of the atoms in solids.

We cannot see atoms and molecules even with the most powerful optical microscope because the wavelength of visible light is much longer than their dimensions, so light waves pass over them undisturbed, much as ocean waves are undisturbed by a small floating cork. Only if we could use waves with a wavelength that is much smaller than the dimensions of atoms could we see atoms directly just as we can see ordinary-sized objects with visible light. But the energy of such very short wavelength radiation would be so great that the substance being studied would be decomposed. However, if the wavelength of 'he waves were at least comparable with the distances between the atoms in a crystal, we could obtain a diffraction pattern similar to that obtained when visible light is passed through two or more very narrow slits, as we described in Chapter 7. The distances between the lines in the diffraction pattern may be calculated from the distance between the slits if the wavelength of the

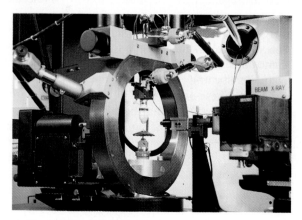

An X-ray diffractometer.

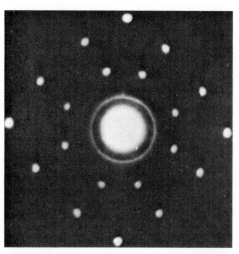

Diffraction pattern produced by a sodium chloride crystal

light is known. Thus we would expect to be able to calculate the distances between the atoms in a crystal if we could use radiation of wavelength suitable to give such a diffraction pattern. X rays in fact have wavelengths of the same order of magnitude as the distances between the atoms in a crystal. So a crystalline solid gives a diffraction pattern when X rays are passed through it.

X rays interact with the electrons in an atom and are scattered in all directions. This is quite similar to what happens when waves on a lake pass a post—ripples spread out in all directions from the post. The ripples from two or more such posts at equal distances apart would interfere with each other, giving a diffraction pattern. In the same way when X rays of suitable wavelength pass a line of atoms, the scattered X rays are in phase and reinforce each other in certain directions, whereas they are out of phase and cancel each other in other directions. In other words, a diffraction pattern is produced which may be recorded, either on photographic film or electronically. Of course there are many rows of atoms in three dimensions in a crystal so that the diffraction pattern may be complicated. Nevertheless, it is possible to calculate the distances between the atoms in the crystal from the diffraction pattern if the wavelength of the X rays is known. Not only can the distances between the atoms be obtained but each of the atoms can be identified because the intensity of the X rays scattered from an atom depends directly on the number of electrons in the atom, that is, on its atomic number. In order

BOX 11.1 (cont.)

to obtain a complete diffraction pattern the crystal must be rotated around its axes so that different rows and planes of atoms are presented to the beam. In a modern X ray diffractometer this is done automatically under the control of a computer, which also records the positions and intensities of all the diffracted beams, of which there may be several thousand or more. From this information the structure of the crystal can usually be obtained

Before the advent of computers the collection of all the data and the calculation of the structure was a tedious and time-consuming process and many months of work were often required. But with a modern diffractometer and a powerful computer a structure can often be completely determined in a few days. The structures of increasingly large and complex molecules such as proteins are now being routinely determined by X ray diffraction.

FIGURE 11.5
W. Lawrence Bragg (1890–1971) and William H. Bragg (1862–1942).

William Bragg left England in 1885, where he had graduated from the University of Cambridge, to become professor of physics at the University of Adelaide, Australia, at the age of twenty-three. While at Adelaide he made some important discoveries in the field of radioactivity. His son Lawrence was born in 1890. The family returned to England when William Bragg was offered a position as professor of physics at the University of Leeds, England, in 1909. In 1912 he became intrigued with von Laue's discovery that a beam of X rays could be diffracted by a crystal, because, like many other physicists, he believed that X rays consisted of fast-moving particles. He discussed this discovery with his son Lawrence, who had graduated from the University of Adelaide at the age of eighteen and was studying at the University of Cambridge. During the summer of 1912 Lawrence worked out the theory of the diffraction of X rays by crystals, deducing what is now known as the Bragg equation. Father and son became convinced that X rays were indeed a form of electromagnetic radiation, and working together, they went on to deduce the structures of several ionic crystals. Thus they founded the science of X ray crystallography, to which they were both to devote a major part of their scientific lives. In 1915 they jointly received the Nobel Prize for physics for their work on X rays and the determination of the structures of crystals. They are the only father-and-son team to have received this honor, and Lawrence is also the youngest person to have received it.

11.2
Classification of Solids

MOLECULAR AND NETWORK SOLIDS

A simple classification of the most important types of solids is given in Table 11.1. We first divide solids into two main types, molecular solids and network solids.

Molecular solids consist of individual molecules in which the bonds are covalent held together by relatively weak intermolecular forces. Some molecular substances, such as I_2, S_8, P_4, and P_4O_6, are solids at room temperature. But because of the weakness of intermolecular forces, many are liquids, including H_2O, Br_2, CCl_4, and H_2SO_4, and a large number are gases, for example, CO_2, CH_4, HCl, and NH_3. Molecular solids are often soft and have no great mechanical strength.

Network solids do not contain finite individual molecules. They have a continuous network of atoms or ions in which each atom or ion is bound strongly to its neighbors, and the regular network extends indefinitely throughout the crystal.

We can distinguish three main types of network solids depending on the nature of the bonds between the atoms or ions: covalent solids, ionic solids, and metallic solids. In **covalent network solids** the atoms are held together by covalent bonds, and the covalently bound network of atoms extends throughout the crystal. Diamond and silicon dioxide, SiO_2, are examples (Figure 11.6). In **ionic solids** oppositely charged ions are held together by their mutual electrostatic attraction, that is, by ionic bonds. The regular arrangement of ions extends continuously throughout the crystal, as in sodium chloride (Figure 11.6). In **metallic solids** positive ions are held together in a regular arrangement by free electrons, that is, by a mobile electron cloud. The regular arrangement of positive ions extends throughout the crystal (Figure 11.6).

TABLE 11.1
Classification of Solids

	Molecular	Network					
		1D chains	2D layers (sheets)		3D networks		
Structural units	Molecules	Atoms	Atoms	Ions of opposite charge	Atoms	Ions of opposite charge	Positive ions and electrons
Type of interaction between structural units	Intermolecular forces	Covalent bonds	Covalent bonds	Ionic bonds	Covalent bonds	Ionic bonds	Metallic bonds
Type of solid	Covalent molecular	Covalent chain	Covalent layer	Ionic	Covalent network	Ionic	Metallic
Examples	CH_4, CO_2, HCl, NH_3, I_2, S_8, P_4	S_n $(SO_3)_n$	C(graphite) P(black)	$MgCl_2$	C(diamond) SiO_2 SiC	NaCl MgO CaF_2	Ca Fe Al

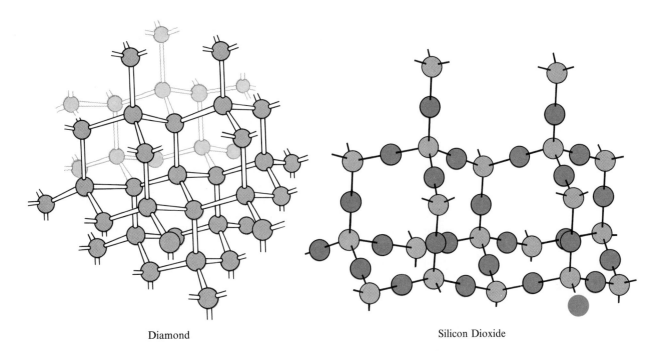

Diamond

Silicon Dioxide

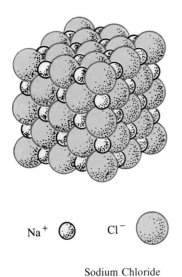

Na^+ ◯ Cl^- ◯

Sodium Chloride

Copper

FIGURE 11.6
Three-Dimensional Network Solids.

In these solids the regular arrangement of atoms or ions extends indefinitely in three dimensions and no individual molecules can be recognized. The bonding may be covalent as in diamond, polar covalent as in silicon dioxide, ionic as in sodium chloride, or metallic as in copper. The structures of diamond and silicon dioxide are shown in expanded form so that the directions of the covalent bonds can be seen. In diamond each carbon atom is bonded to four other carbon atoms in a tetrahedral arrangement; each carbon atom has an AX_4 geometry. Silicon dioxide occurs in several different forms. The form shown here is called β-cristobalite. In all these forms each silicon atom is surrounded by a tetrahedral arrangement of four oxygen atoms; each silicon atom has a tetrahedral AX_4 geometry.

No sharp dividing line can be drawn between ionic and covalent network solids. When the electronegativity difference between the atoms is small, the bonds can be regarded as covalent. If the electronegativity difference is large, as in silicon dioxide, the bonds are polar. If the electronegativity difference is very large, the bonds can be regarded as ionic, as in sodium chloride. Again, in many solids with metallic properties the bonding may not be purely metallic but may be intermediate between covalent and metallic or ionic and metallic.

In network solids individual molecules cannot be distinguished. A network structure clearly cannot be maintained in the gas phase. Because the particles of a gas are far apart, all gases consist of molecules or individual atoms. If a network solid such as silicon dioxide, SiO_2, is converted to a gas, it undergoes a profound change in structure. In the solid state each silicon atom forms four single bonds to four different oxygen atoms, which surround it in a tetrahedral arrangement (see Figure 11.6). In the gas, silicon dioxide consists of linear SiO_2 molecules, similar to CO_2. In solid sodium chloride, each Cl^- is surrounded by six Na^+ ions, and each Na^+ is surrounded by six Cl^- ions. In contrast, in the gas there are small clusters of two or more ions—in other words, small molecules such as NaCl and Na_2Cl_2.

Note that solids in which the bonding is metallic or ionic are necessarily network solids, but solids in which the bonds are covalent may be either molecular or network solids. In a molecular solid there are also weak intermolecular forces holding the molecules together. In a two-dimensional network solid there are covalent bonds in each layer and intermolecular forces between the layers, but in a covalent three-dimensional network solid there are only covalent bonds.

MELTING POINTS OF MOLECULAR AND NETWORK SOLIDS Molecular substances usually consist of the same molecules in the solid, liquid, and gaseous states. In most cases no chemical bonds are broken when a molecular solid is melted or vaporized. The individual molecules remain intact. Water consists of the same H_2O molecules in ice, liquid water, and steam. They have a regular arrangement in ice, but are moving in a disordered way in liquid water, and are much further apart and moving more rapidly in the gas state. Only the relatively weak intermolecular forces *between* the molecules have to be overcome to separate the molecules. Molecular solids are converted to the corresponding liquid and gas at relatively low temperatures because only a small amount of energy is needed to overcome the weak intermolecular forces. All molecular substances either have a melting point below 400 °C or decompose below this temperature, in which case they cannot be obtained in the liquid or gaseous state.

For the vast majority of molecular solids no chemical bonds are broken when they are melted or vaporized. Nevertheless, there are a few exceptions, one of which we have already met. When sulfur is melted one of the bonds in the eight-atom rings breaks and the rings join to form long chains. In this process, however, there is no change in the total number of bonds as each sulfur atom is still forming two bonds. However, on vaporizing the liquid the chains break up to give S_2 molecules in the gas phase.

In contrast, many covalent, metallic, and ionic network solids have very high melting points and still higher boiling points. For example, the melting point of the covalent solid diamond is 3600 °C, the melting point of iron is

1540 °C, and the melting point of the ionic solid magnesium oxide is 2800 °C. These substances have high melting points because a large amount of energy is needed to break the strong covalent, metallic, or ionic bonds in order to destroy the regular structure of the network solid, allowing the atoms to move relative to each other to form a liquid. Not all ionic solids have very high melting points, however. For example, potassium nitrate melts at 334 °C, and potassium hydrogen sulfate, $KHSO_4$, melts at 210 °C. The melting point depends on the strength of the ionic bonds. In accordance with Coulomb's law, the strength of ionic bonds increases with the charge on the ions and with decreasing distance between the ions. Thus the strongest bonds are between small highly charged ions such as Mg^{2+} and O^{2-}, whereas the bonds between large, singly charged ions such as K^+ and NO_3^- or HSO_4^- are relatively weak.

The melting points of metals cover a wide range, from 3400 °C for tungsten to as low as -39 °C for mercury. Mercury is the only metal that melts below room temperature, but cesium (mp 29 °C) and gallium (mp 30 °C) melt just above room temperature. There are two important reasons that metals may have relatively low melting points. First, metallic bonds are not broken when a metal melts. The metal ions are held together by a mobile electron cloud, which can adjust to the movement of the positive ions and still keep them strongly bound together. Second, some metallic bonds are relatively weak because only a small number of electrons are available for metallic bonding. For example, in the alkali metal cesium the metallic bonds are formed by only one electron from each metal atom. Nevertheless the boiling points of these metals are relatively high because in forming the gas phase the metallic bonds must be broken to give the free atoms (Hg 357 °C, Cs 705 °C, Ga 2403 °C).

In summary, the melting points of molecular solids are usually low, often well below room temperature. No molecular solid is known to have a melting point above 400 °C. In contrast, network solids usually have high melting points, often as high as several thousand degrees; with very few exceptions network solids melt above room temperature.

Gallium, with a melting point of 30 °C, melts when held in the hand.

= Example 11.1 =

MELTING POINTS OF MOLECULAR AND NETWORK SOLIDS

What types of forces (bonds and intermolecular forces) must be overcome in order to melt each of the following solids:

CS_2 SiC Br_2 Ca Na_2SO_4

Solution

CS_2 consists of covalent CS_2 molecules held together by weak *intermolecular forces*, which are the only forces that have to be overcome when $CS_2(s)$ melts. *SiC* resembles diamond in that each Si atom and each C atom is part of a covalent network solid in which every atom is part of a tetrahedral AX_4 arrangement. On melting, many *covalent bonds* have to be broken. *Br_2* consists of covalent Br_2 molecules held together by only weak *intermolecular forces*. *Ca* is a metal in which the forces are those of *metallic bonding*. *Na_2SO_4* consists of Na^+ and SO_4^{2-} ions held together by *ionic bonds*.

EXERCISE 11.1

What kinds of chemical bonds (if any) are broken

(a) When solid iodine is melted?

(b) When solid copper is vaporized?

(c) When graphite is vaporized?

(d) When quartz, SiO_2, is melted?

(e) When solid carbon dioxide, CO_2, is vaporized?

(f) When solid phosphorus(III) oxide, P_4O_6, is melted?

ONE-, TWO-, AND THREE-DIMENSIONAL NETWORKS

We can also classify network solids according to the type of network rather than the type of bonding. We can distinguish three types of network, one dimensional (1D), two dimensional (2D), and three dimensional (3D).

A **three-dimensional network solid** is one in which strong bonds extend indefinitely in three dimensions so that the whole crystal may be regarded as one giant, almost infinite molecule. Examples include diamond (covalent bonding), silicon dioxide, SiO_2 (polar covalent bonding), sodium chloride (ionic bonding), and copper (metallic bonding) (see Figure 11.6).

A **two-dimensional network solid** has a network that extends indefinitely in only two dimensions; therefore the atoms or ions form an infinite layer, or sheet, of atoms. Examples include graphite (covalent and metallic bonding), black phosphorus (covalent bonding), and magnesium chloride (ionic bonding) (see Figure 11.7). The sheets of atoms are not necessarily just one atom thick, as in graphite; they may have two or more layers of atoms or ions, but they do not extend indefinitely in the third dimension. The sheets are stacked one on another and are held together by relatively weak intermolecular forces.

Recall that three-dimensional covalent and ionic network solids, such as diamond and aluminum oxide, are very hard, because strong bonds must be broken if atoms are displaced. In contrast, two-dimensional solids are often quite soft, because the sheets of atoms can slide over each other relatively easily. They also tend to form flaky, platelike crystals, as found, for example, for graphite and the silicate minerals mica and talc (Chapter 22). Two-dimensional network solids are intermediate between three-dimensional solids, in which all the atoms are held together by strong bonds, and molecular solids, in which the atoms are held together in small groups (molecules) by strong bonds and the molecules are held together by weak intermolecular forces. A two-dimensional network solid consists of extremely large (virtually infinite), sheetlike molecules held together in the crystal by intermolecular forces.

One-dimensional network solids consist of infinitely long chain molecules held together by weak intermolecular forces. These chains may be arranged in a regular fashion to give a crystalline solid, as in the polymeric form of sulfur trioxide $(SO_3)_n$ (Figure 11.8). The crystals of such solids are often thin and fibrous. Frequently, however, long-chain molecules get tangled up with each other and

Mica

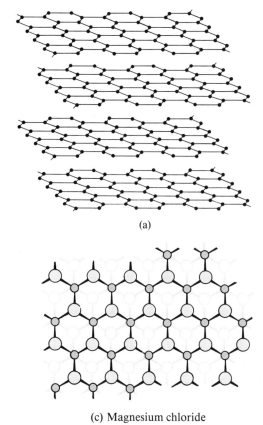

(a)

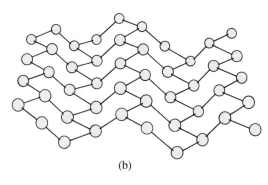

(b)

(c) Magnesium chloride

FIGURE 11.7
Two-Dimensional Network Solids, or Layer Molecules.

These solids have a network of bonded atoms that extends in two dimensions only, forming sheets of atoms or giant two-dimensional molecules. The sheets are stacked on each other in the solids. (a) The bonding in the sheets may be covalent, as in graphite, which consists of planar, two-dimensional carbon atoms. (b) The bonding in black phosphorus is also covalent; it consists of buckled two-dimensional sheets of phosphorus atoms. (c) The bonding may also be ionic, as in magnesium chloride, which consists of Mg^{2+} ions in the octrahedral holes between two planar layers of Cl^- ions.

- Lower level of Cl^- ions
- ◯ Upper layer of Cl^- ions
- ◉ Mg^{2+} ions

form an amorphous solid, as in the case of plastic sulfur. Polyethylene and many other familiar plastics consist of long-chain molecules forming an amorphous solid. A long-chain molecule that extends indefinitely in one dimension is not normally considered to constitute a network. So we will refer to such molecules simply as long-chain molecules and not as one-dimensional networks.

═ *Example 11.2* ═══════════════════════

MOLECULAR AND NETWORK SOLIDS

State whether each of the following solids is a network solid or a molecular solid. In each case name the type of chemical bonds that are present.

    (a) Brass   (b) White phosphorus

    (c) Black phosphorus   (d) Iron(III) oxide, Fe_2O_3

Solution

    (a) Brass is an alloy of copper and zinc (Chapter 10)—it is a network solid with metallic bonds.

    (b) White phosphorus is a molecular solid containing P_4 molecules in which there are P—P covalent bonds.

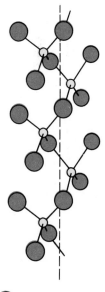

Oxygen
◯ Sulfur

FIGURE 11.8
The Long-Chain Molecule of $(SO_3)_n$.

It has an overall spiral shape.

(c) Black phosphorus is a two-dimensional network solid in which the bonds are covalent.

(d) Fe_2O_3 is a network solid with ionic bonds between Fe^{3+} and O^{2-} ions.

EXERCISE 11.2

State whether each of the following solids is a network solid or a molecular solid. In each case name the types of chemical bonds that are present.

(a) Diamond

(b) Monoclinic sulfur

(c) Aluminum chloride hexahydrate, $AlCl_3 \cdot 6H_2O$

(d) Solid carbon dioxide

AMORPHOUS SOLIDS AND GLASSES

Silicon dioxide, silica, SiO_2, crystallizes as the mineral quartz and in other forms such as β-cristobalite, but if molten silica is cooled rapidly it forms a transparent, noncrystalline solid that is called an amorphous solid or glass. Amorphous solids are hard and rigid like crystalline solids but they do not have sharp melting points. When heated they gradually soften and become increasingly more fluid without ever exhibiting any sharp melting point. Even when apparently solid they may flow, although very slowly, so that it may be hundreds of years before any perceptible motion is observed. For example, some very old glass windows have been found to be slightly thicker at the bottom than at the top. The basic difference between amorphous and crystalline solids is in their structures. Amorphous solids lack the regular, repeating arrangement of atoms or ions that is characteristic of crystalline solids, and so they do not have the plane faces at definite angles to each other that we find in crystals. For example, in the amorphous form of silica each silicon atom is surrounded by a tetrahedral arrangement of four oxygen atoms, as in the crystalline forms, but in the amorphous form these tetrahedra are linked together in a random rather than a regular manner. Figure 11.9 illustrates, by a two-dimensional diagram, the difference between an ordered crystalline structure and a disordered amorphous solid.

When amorphous solids are heated they soften gradually and become more fluid without exhibiting any sharp melting point, because the solid already has the random arrangement characteristic of the liquid state. With increasing temperature the motion of the atoms increases, and there is a smooth transition from the solid to the liquid state through an intermediate stage that may be described as a very soft solid or a very viscous liquid. In contrast, in a crystalline solid the regular arrangement of the atoms must change to the random arrangement of the liquid. This change can only occur when the atoms of the solid acquire enough energy to be able to move past each other. The regular arrangement then changes abruptly to the random arrangement characteristic of the liquid, and a sharp melting point is observed.

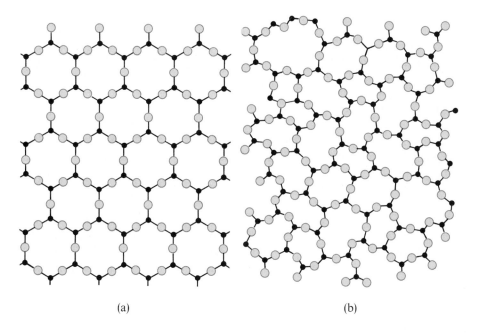

(a) (b)

FIGURE 11.9
Two-Dimensional Models
for a Crystalline Solid and an
Amorphous Solid.

(a) In the crystalline solid the arrangement of the atoms is regular throughout the crystal. (b) In the amorphous solid the arrangement of the atoms is irregular. But in both cases there is an equilateral triangular arrangement of the ○ atoms around the ● atoms.

The amorphous solid that is most familiar to us is ordinary glass such as we find in windows, laboratory apparatus, electric light bulbs, mirrors, and so on. It has a multitude of uses in science, industry, and the arts. Glass has the approximate composition of a sodium calcium silicate with the formula $Na_2CaSi_6O_{14}$, but its composition can vary widely depending on the use for which it is made (Chapter 22). It is the fact that it softens over a range of temperature and does not melt sharply that enables it to be molded and blown into an enormous variety of shapes.

Generally, the amorphous state of a solid is less stable than the crystalline state, and the amorphous state may change spontaneously to the crystalline state. We saw in Chapter 8 that plastic sulfur changes within a few hours to crystalline orthorhombic sulfur. But the transformation of an amorphous solid to a crystalline solid may be extremely slow, because at ordinary temperatures the atoms do not have sufficient kinetic energy to undergo the complicated rearrangement needed to give the ordered structure.

Glass can be decorative as well as functional

11.3
Lattices

The atoms, ions, or molecules of a crystalline substance are arranged in a regular manner and form a repeating, three-dimensional pattern. Because there are an enormous number of different structures, we need a systematic method of describing and classifying them. We base our classification of structure on the concept of the **space lattice**, which is a regular, repeating arrangement of points in space. But before considering space lattices, it will be helpful to examine two-dimensional arrangements of points called *two-dimensional lattices* or *nets*.

TWO-DIMENSIONAL LATTICES, OR NETS

Two-dimensional patterns are familiar from the designs on many wallpapers, carpets, and tiled floors. Any two-dimensional pattern can be described by a **net**, or **two-dimensional lattice**, which is a regular repeating arrangement of points in a plane. Figure 11.10(a) provides an example.

To construct the net corresponding to this two-dimensional pattern, we lay tracing paper on it and put a dot at some chosen point of the pattern, for example, at the eye of each white horse (Figure 11.10). The regular arrangement of points that we obtain in this way is the net on which the pattern is based. The whole pattern can be constructed by placing the white horse, which is called the **motif** of the pattern, in the same position with respect to each point of the net.

There are five types of two-dimensional lattices, as Figure 11.11 shows. They are the hexagonal, square, rectangular, rhombic, and parallelogram lattices. They differ in the symmetry of the arrangement of the points; the hexagonal lattice has the most symmetrical arrangement of points and the parallelogram has the least symmetrical.

UNIT CELLS

Because of the regular repeating arrangement of the points in a net or a two-dimensional lattice, we need only describe a small part of the lattice in order to specify it completely. We choose four points and connect them to give a parallelogram (see Figure 11.11). This figure is known as a **unit cell**. We can generate the complete lattice by repeatedly moving the unit cell in the directions of its edges by a distance equal to the cell edge (Figure 11.12).

As Figure 11.12 shows, unit cells for any given lattice can be chosen in many different ways, and may include some cells that have interior points. The most convenient cell is the smallest cell that shows the full symmetry of the lattice. For the square, rectangular, and parallelogram lattices the unit cells normally chosen are the square, the rectangle, and the parallelogram. Any other cells are either larger or are not as symmetrical as the lattice. For the hexagonal lattice the unit cell is a rhombus with an angle of 60°. For the rhombic lattice a

The fact that there are only five possible two-dimensional lattices can be understood in terms of the possible shapes of tiles that can be used to cover a floor completely without changing the orientation of the tiles or leaving gaps. There are just five possible shapes that can be used: a hexagon, a square, a rectangle, a rhombus, and a parallelogram. No other shapes, such as a pentagon or an octagon, can be used. Triangles can only be used if they are placed in two different orientations or used in pairs and this is equivalent to using a parallelogram.

FIGURE 11.10
A Two-Dimensional Pattern.

(a) This two-dimensional pattern is by the Dutch artist M. C. Escher. (b) This two-dimensional lattice or net is the basis of the pattern. The lattice may be obtained by placing a dot at the same position with respect to each white horse, for example, at the eye of the horse. The white horse is the motif of the pattern which can be reconstructed by placing the horse in the same position at each lattice point.

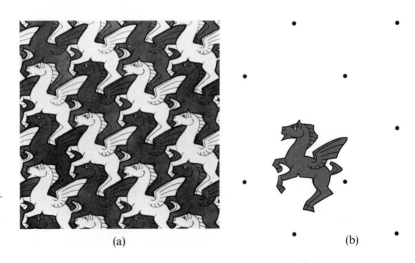

(a)   (b)

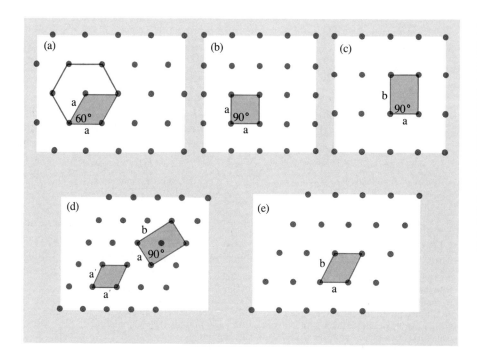

FIGURE 11.11
The Five Two-Dimensional
Lattices or Nets.

(a) The hexagonal lattice: the unit
cell is a rhombus with equal sides
and an angle of 60°. (b) The square
lattice: the unit cell is a square.
(c) The rectangular lattice: the unit
cell is a rectangle. (d) The rhombic
lattice, or centered-rectangular
lattice: the unit cell that is normally
chosen is the centered rectangle.
The alternative primitive cell is a
rhombus with equal sides and an
angle that is not equal to 60° or
90°. (e) The parallelogram lattice:
the unit cell is a parallelogram
with unequal sides and an angle
that is not 90°.

rectangular cell with an interior point is normally chosen. A cell with an interior
point is called a **centered cell**—in this case a *centered rectangular cell*. The other
four cells that do not contain any interior points are known as **primitive cells**.

To describe any two-dimensional pattern, we must specify the unit cell by
the lengths of its edges and the angle between them, and we must specify the
motif and its position with respect to the lattice point. Figure 11.13 on p. 550
shows some imaginary two-dimensional structures of molecular, ionic, and
metallic crystals. The description of a metallic crystal (Figure 11.13a) is particularly simple since the motif is a single atom that may be placed at a lattice point.
Thus a close-packed layer of metal atoms can be described by a hexagonal
lattice with a motif consisting of a single atom at a lattice point. The simple
two-dimensional ionic crystal shown in Figure 11.13(b) can be described by a
square lattice. In this case the motif consists of two ions, a negative ion and a
positive ion. If the negative ion is placed at a lattice point, the positive ion is
located in the middle of the square unit cell. In a molecular crystal the motif
is often a single molecule (Figure 11.13c).

FIGURE 11.12
Unit Cells.

(a) A unit cell is constructed by
choosing four points of the lattice
and connecting them to give a
four-sided figure with pairs of
parallel sides. The complete lattice
can be generated by repeatedly
moving the unit cell in the directions of its edges by a distance
equal to the cell edge. (b) A unit cell
can be chosen in many different
ways. The most convenient cell is
the smallest cell that has the full
symmetry of the lattice. The most
symmetrical cell may be a centered
cell which has an additional lattice
point at its center.

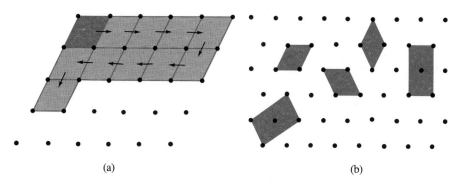

(a)   (b)

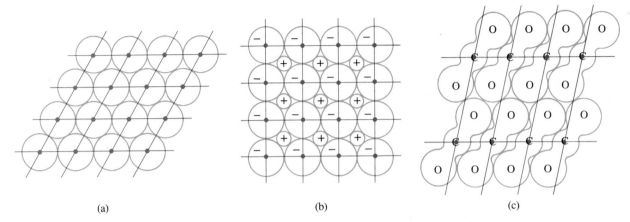

(a)   (b)   (c)

FIGURE 11.13
Structures of Some Imaginary Two-Dimensional Crystals.

(a) This metallic crystal can be described by a hexagonal lattice. The motif is a single metal atom that is placed at a lattice point. (b) This ionic crystal can be described by a square lattice. The motif is a single anion and cation. If the anion is placed at a lattice point, the cation is at the center of the unit cell. (c) This molecular crystal can be described by a parallelogram lattice. The motif is the CO_2 molecule, which has been placed with the C atom at a lattice point.

SPACE LATTICES

A regular three-dimensional arrangement of points in space is called a *space lattice*. Just as there are only 5 possible two-dimensional lattices, there are only 14 possible three-dimensional space lattices. They are known as *Bravais lattices* after the French mathematician who first described them. The most symmetrical of the 14 three-dimensional lattices are the 3 cubic lattices shown in Figure 11.14. Many structures can be described in terms of the primitive cubic lattice

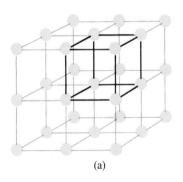

(a)

FIGURE 11.14
The Three Cubic Lattices

(a) The primitive cubic lattice. One unit cell is outlined with heavy lines. (b) The unit cells of the primitive, body-centered, and face-centered cubic lattices. The body-centered cell has an additional point at the center of the cube. The face-centered cell has additional points at the center of each face.

Primitive   Body-centered   Face-centered
cubic   cubic   cubic

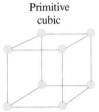 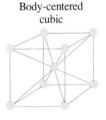

(b)

TABLE 11.2
The Cubic Unit Cells

Type of Cell	Number of Points at Corners	Number of Points in Faces	Number of Points in Center of Cube	Total
Primitive	$8 \times \frac{1}{8}$	0	0	1
Body-centered	$8 \times \frac{1}{8}$	0	1	2
Face-centered	$8 \times \frac{1}{8}$	$6 \times \frac{1}{2}$	0	4

and two centered versions of this lattice, the body-centered cubic lattice and the face-centered cubic lattice.

The unit cell of a space lattice is obtained by joining eight points to give a solid with pairs of parallel faces—a parallelepiped. The unit cell of the **primitive cubic lattice** is a cube. The unit cell of the **body-centered cubic lattice** is a cube with an additional point in the middle of the cube. The unit cell of a **face-centered cubic lattice** is a cube with an additional point in the center of each face.

As we will see in the next section, it is important to recognize how many lattice points are associated with each unit cell. The primitive cubic cell has a point at each corner, but each point is shared with seven other unit cells that meet at that point. Thus only one-eighth of the point belongs to any particular cell. Since the unit cell has eight corners, then $8 \times \frac{1}{8} = 1$ point is associated with each primitive unit cell. The body-centered cubic cell has an additional point at its center, which is not shared with other cells. Thus there are two lattice points associated with this unit cell. The face-centered cubic unit cell has an additional point at the center of each face, each of which is shared between two adjacent unit cells. Since the cube has six faces, $6 \times \frac{1}{2} = 3$ additional points are added to the unit cell, to make a total of four points. These conclusions are summarized in Table 11.2.

We can now describe the structures of some different types of network solids in terms of their space lattices.

11.4
Metallic Crystals

Because they consist of a network of single identical atoms, metals generally have very simple structures. We discussed the structures of metals in Chapter 10 in terms of close packing of spherical atoms. As Figure 11.15 shows, almost all the metals have a cubic close-packed, hexagonal closed-packed, or monatomic body-centered cubic structure. We can now describe these structures in terms of their space lattices.

THE IRON (MONATOMIC BODY-CENTERED CUBIC) STRUCTURE

We can describe any three-dimensional repeating structure in terms of the appropriate space lattice and the motif that is associated with each lattice point.

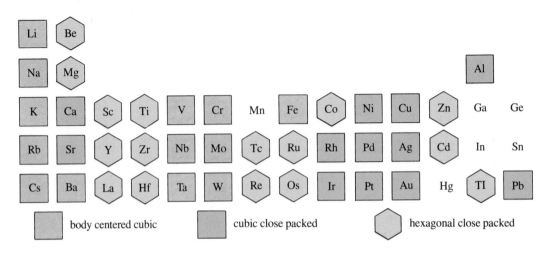

| body centered cubic | cubic close packed | hexagonal close packed |

FIGURE 11.15
Structures of the Metals.

All the metals except Mn, Hg, Ga, In, Ge, and Sn have one of the three types of structures shown in the key.

For iron the lattice is body-centered cubic and the motif is a single iron atom which can be conveniently placed at each lattice point (Figure 11.16). Hence we can call the structure of iron a *monatomic body-centered cubic structure*, that is, a structure in which a single atom is associated with each lattice point.

Since the unit cell of the body-centered cubic lattice contains two lattice points, in the iron structure it contains two iron atoms. We can reach the same conclusion by counting the contents of a cell. As shown in Figure 11.16(a), there are eight corner atoms that are shared between eight unit cells, plus the atom in the center of the cell. Therefore there are

$$8 \text{ corner atoms} \times \tfrac{1}{8} \text{ atom per cell} = \tfrac{8}{8} = 1 \text{ atom}$$

$$1 \text{ center atom per cell} = 1 \text{ atom}$$

$$\text{Total} = 2 \text{ atoms}$$

Although it might appear that the lattice point in the middle of the cell is different from the lattice points at the corners, such is not the case. In any space lattice every point is equivalent to every other point. In this case every point is surrounded by eight other points in the form of a cube. Thus every iron atom is surrounded by eight other iron atoms in a cubic arrangement. The number of neighboring atoms that are in contact with a given atom is called the **coordination number** of that atom. Each iron atom has a coordination number of eight.

Other metals that have the iron structure include the alkali metals, and barium, chromium, and vanadium.

FIGURE 11.16
Iron (Monatomic Body-Centered Cubic) Structure.

(a) The lattice is body-centered cubic. (b) One atom (the motif) is situated at each lattice point. (c) There are two atoms per unit cell, $1 + 8(\tfrac{1}{8})$. (d) Each atom has a cubic eight coordination.

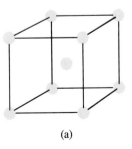

(a)

(b)

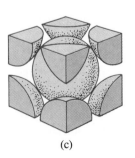

(c)

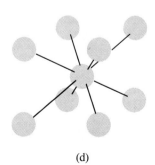

(d)

Example 11.3

UNIT CELL DIMENSIONS AND ATOMIC RADIUS

The length of the unit cell edge in the iron structure is 286 pm. What is the radius of the iron atom?

Solution

In calculations on structures involving cubic unit cells, it is useful to know the relationships between a cube edge, a face diagonal, and a body diagonal. These relationships can be obtained from trigonometry or from the Pythagorean theorem.

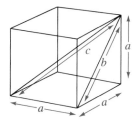

If the length of the unit cell edge is a, if b is a face diagonal, and if c is a body diagonal, then

$$b^2 = a^2 + a^2 = 2a^2 \quad \text{or} \quad b = \sqrt{2}\,a$$
$$c^2 = a^2 + b^2 = 3a^2 \quad \text{or} \quad c = \sqrt{3}\,a$$

From Figure 11.16 we see that for the iron structure the length of the body diagonal c is equal to $4r$, where r is the radius of the iron atom. Hence

$$r = \frac{c}{4} = \frac{\sqrt{3}}{4}\,a = \frac{\sqrt{3} \times 286}{4}\ \text{pm} = 124\ \text{pm}$$

The radius of the iron atom is 124 pm.

Example 11.4

UNIT CELL DIMENSIONS AND DENSITY

Vanadium has the iron (monatomic body-centered cubic) structure. The length of the edge of the unit cell is found by X ray diffraction to be 305 pm. What is the density of vanadium?

Solution

We can find the density by finding the mass and the volume of a unit cell:

$$\text{Volume} = (\text{Cell edge})^3 = (3.05 \times 10^{-8}\ \text{cm})^3$$

The iron structure has two atoms per unit cell. Therefore the mass of the unit cell is the mass of two vanadium atoms:

$$\text{Mass of 2 mol of vanadium atoms} = 2 \times 50.94\ \text{g}$$

$$\text{Hence, mass of two vanadium atoms} = \frac{2 \times 50.94\ \text{g}}{6.022 \times 10^{23}}$$

$$\text{Density} = \frac{\text{Mass}}{\text{Volume}} = \frac{2 \times 50.94\ \text{g}}{6.022 \times 10^{23} \times 3.05^3 \times 10^{-24}\ \text{cm}^3} = 5.96\ \text{g cm}^{-3}$$

EXERCISE 11.3

Titanium has the iron (monatomic body-centered cubic) structure, and its density is 4.50 g cm^{-3}. Calculate the length of the edge of the unit cell and the atomic radius of titanium.

THE COPPER (CUBIC CLOSE-PACKED) STRUCTURE

In Chapter 10 we described copper as having a cubic close-packed structure. We can now describe it in more detail in terms of its space lattice, which is face-centered cubic (see Figure 11.17a). The motif is a single copper atom that is located at each lattice point (Figure 11.17b). Because there are four lattice points associated with the face-centered cubic unit cell, there are four copper atoms per unit cell (Figure 11.17c). We can also reach this conclusion by counting the copper atoms inside a unit cell. Those at the corners are shared between eight unit cells, while those at the faces are shared between two cells. Therefore

$$8 \text{ corner atoms} \times \tfrac{1}{8} \text{ atom per cell} = \tfrac{8}{8} = 1 \text{ atom}$$
$$6 \text{ face atoms} \times \tfrac{1}{2} \text{ atom per cell} = \tfrac{6}{2} = 3 \text{ atoms}$$

giving a total of four atoms. This structure is the cubic close-packed structure, which, as we saw in Chapter 10, consists of close-packed layers stacked one on another in the sequence ABCABCABC . . . .

The close-packed layers of atoms in this structure are not immediately apparent because they are not parallel to the faces of the unit cell. Rather, they are perpendicular to the body diagonals of the cell, as shown in Figure 11.18. The name *cubic* close packing refers to the fact that the structure is based on a face-centered cubic lattice. The motif is a single atom situated at each lattice point, so it may also be called the *monatomic face-centered cubic structure*.

Many other metals, including Al, Ca, Sr, Co, Ni, Pb, Ag, and Au, have the cubic close-packed structure. The noble gases Ne, Kr, Ar, and Xe, which consist of single atoms, also have this structure in the solid state.

═ *Example 11.5* ═══

LATTICE TYPE FROM DENSITY AND UNIT CELL DIMENSIONS

Silver has a structure based on a cubic lattice. The edge of the unit cell is found to have a length of 408 pm by X ray diffraction. The density of silver is 10.6 g cm^{-3}. How many atoms of silver are there in the unit cell? What type of cubic lattice is the structure of silver based on?

FIGURE 11.17
Copper (Cubic Close-Packed) Structure.

(a) The lattice is face-centered cubic. (b) The motif is a single copper atom situated at a lattice point. (c) There are four atoms per unit cell, $6(\tfrac{1}{2}) + 8(\tfrac{1}{8})$.

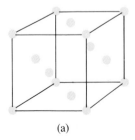

(a)

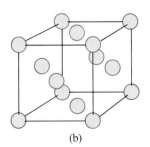

(b)

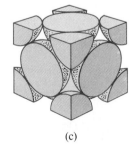

(c)

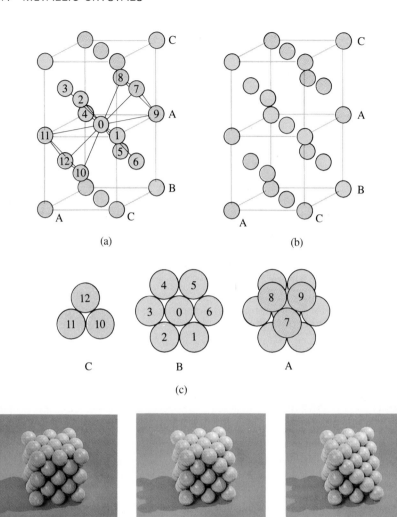

FIGURE 11.18
The Cubic Close-Packed Structure.

(a) Two unit cells are shown, so that we can see that the 12 nearest neighbors of the atom 0 are atoms 1 to 12. (b) The close-packed layers ABC are perpendicular to the body diagonal of the face-centered cubic cell. (c) In the close-packed layers ABC, layer B contains the atom 0 surrounded by atoms 1 to 6; layer A contains atoms 7, 8, and 9; and layer C contains atoms 10, 11, and 12. (d) The cubic close-packed structure showing the removal of successive layers of atoms, exposing the close-packed layers perpendicular to the body diagonal.

Solution

To find how many atoms of silver there are in the unit cell, we will need to find the mass of the unit cell. We can find the mass from the density and the volume:

$$\text{Volume of unit cell} = (\text{Cell edge})^3 = (408 \text{ pm})^3 = (4.08 \times 10^{-8} \text{ cm})^3$$

$$\text{Mass of unit cell} = \text{Density} \times \text{Volume}$$

$$= 10.6 \text{ g cm}^{-3} \times 4.08^3 \times 10^{-24} \text{ cm}^3$$

If the cell contains x atoms, its mass is

$$x(\text{Atomic mass}) = \frac{x(107.9 \text{ g})}{6.022 \times 10^{23}}$$

Hence we have

$$\frac{x(107.9 \text{ g})}{6.022 \times 10^{23}} = 10.6 \times 4.08^3 \times 10^{-24} \text{ g}$$

$$x = 4.02$$

There are four atoms per unit cell. Therefore it is the unit cell of a face-centered cubic lattice.

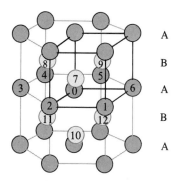

FIGURE 11.19
The Hexagonal Close-Packed
Structure.

The 12 nearest neighbors of atom
0 are atoms 1 to 12. A unit cell is
shown in heavy lines (it is not a
cubic cell).

EXERCISE 11.4

Aluminum has the cubic close-packed structure, and its atomic radius is 143 pm.
Calculate (a) the length of the unit cell edge, (b) the volume of the unit cell, and
(c) the density of aluminum.

THE HEXAGONAL CLOSE-PACKED STRUCTURE

Some other metals, such as magnesium, titanium, and zinc, have the second
type of close-packed structure—hexagonal close-packed. As we saw in Chapter
10 this structure consists of close-packed layers stacked one on another in the
sequence ABABAB . . . . It is based not on a cubic lattice but on a hexagonal
lattice (see Figure 11.19). The unit cell is not cubic so we will not consider it
further. Models of the cubic close-packed and hexagonal close-packed structures are shown in Figure 11.20.

Accurate determination of the size of the unit cell for a metal structure provides a method for the experimental determination of the Avogadro constant,
as shown in the next example.

= *Example 11.6* ==

**THE AVOGADRO CONSTANT FROM DENSITY AND
UNIT CELL DIMENSIONS**

Chromium has the iron (monatomic body-centered cubic) structure. Its density
is 7.19 g cm^{-3}, and the length of the edge of the unit cell is 288.4 pm. Use these data
to calculate a value for the Avogadro constant.

Solution

To solve this problem, we first find the mass of the unit cell from its volume and density:

$$\text{Mass of unit cell} = \text{Volume} \times \text{Density}$$
$$= (2.884 \times 10^{-8} \text{ cm})^3 \times 7.19 \text{ g cm}^{-3}$$
$$= 2.884^3 \times 7.19 \times 10^{-24} \text{ g}$$

FIGURE 11.20
Models of the Hexagonal
Close-Packed and Cubic
Close-Packed Structures.

In the hexagonal close-packed
structure ABABA . . . the close-
packed layers are perpendicular
to the vertical axis. In the
cubic close-packed structure
ABCABC . . . , of which two unit
cells are shown here, the close-
packed layers are perpendicular
to a body diagonal of the unit
cell.

The unit cell of the iron structure contains two atoms. The mass of two Cr atoms is therefore $2.884^3 \times 7.19 \times 10^{-24}$ g. Hence the mass of 2 mol of chromium atoms is

$$2.884^3 \times 7.19 \times 10^{-24} N_A \text{ g} = 2(\text{molar mass Cr}) = 2 \times 52.00 \text{ g}$$

Hence

$$\text{Avogadro constant } N_A = 6.03 \times 10^{23}$$

This value is not exactly equal to the more accurate value (6.022×10^{23}) that we have been using, because the density was given only to three significant figures.

EXERCISE 11.5

Iridium has the cubic close-packed structure. The edge of the unit cell has a length of 383.3 pm. The density of iridium is 22.61 g cm^{-3}. Calculate a value for the Avogadro constant.

11.5
Covalent Molecular Crystals

The structure of a molecular crystal is determined primarily by the size and shape of the molecules, which determines how they may be most efficiently packed together. Thus methane, the molecules of which have a nearly spherical shape, has a cubic close-packed structure. The structure of any molecular crystal can be described in terms of the appropriate space lattice and a motif which is usually one molecule. X ray crystallography enables the positions of *all* the atoms in the unit cell to be determined. Thus we find not only the positions of the molecules in the unit cell but also the distances between all the atoms in the molecule, and hence the complete structure of the molecule. By operating at a sufficiently low temperature even the structures of substances that are gases or liquids at room temperature can be determined by X ray crystallography. Figure 11.21 shows the structures of CO_2 and of Cl_2, Br_2, and I_2. In these

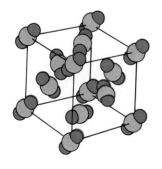

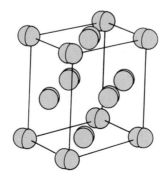

(a)   (b)

FIGURE 11.21
The Structures of CO_2 and of Cl_2, Br_2, and I_2.

(a) The unit cell of CO_2 is face-centered cubic. The molecule is linear and the length of the CO bond is found to be 116 pm.
(b) Cl_2, Br_2 and I_2 all have the same structure. The unit cell is face-centered but it is not cubic. The bond lengths are Cl_2, 198 pm; Br_2, 227 pm; and I_2, 276 pm.

diagrams the molecules have been reduced in size for clarity. They would in fact be closely packed together and fill almost all the space, as shown in the hypothetical two-dimensional structure of CO_2 in Figure 11.13(c).

11.6
Covalent Network Crystals

The structures of covalent network crystals are determined primarily by the geometry of the bonds formed by each atom, as in the diamond structure.

THE DIAMOND STRUCTURE

The diamond structure was briefly described in Chapter 6. We can now describe it in terms of its space lattice, which is face-centered cubic. The unit cell is shown in Figure 11.22. A carbon atom may be located at each lattice point, and there are also four more carbon atoms located at a position one-fourth of the way along each body diagonal of the unit cell. Since there are four lattice points associated with the unit cell of a face-centered cubic lattice, there are a total of $4 + 4 = 8$ carbon atoms in each unit cell of the diamond structure. Each atom in this structure has an AX_4 geometry and forms four covalent bonds to four nearest neighbors in a tetrahedral arrangement around it (Figure 11.22). This network of tetrahedrally coordinated carbon atoms extends throughout the crystal.

Among the other elements of group 4, silicon, germanium, and one form of tin have the diamond structure.

THE β-CRISTOBALITE STRUCTURE

One of the many forms of silica, SiO_2, namely β-cristobalite, has a structure very closely related to that of silicon (Figure 11.23). The silicon atoms have just the same cubic close-packed arrangement as in the element. An oxygen atom is located between each pair of silicon atoms so that each silicon is surrounded by a tetrahedral AX_4 arrangement of oxygen atoms. There are eight silicon atoms and sixteen oxygen atoms in the unit cell.

TWO-DIMENSIONAL NETWORK (LAYER) STRUCTURES

Arsenic and phosphorus each form three bonds with a triangular pyramidal AX_3E geometry. One allotrope of each element contains the tetraatomic molecules P_4 or As_4; another form of each element has a layer structure. The layer structure of As can be derived by breaking all the vertical bonds between the buckled sheets in the diamond structure (Figure 11.6). It consists of buckled layers of atoms in which the atoms are all joined in hexagonal rings, as Figure 11.24(a) shows. The structure of black phosphorus is very similar, but the hexagonal rings are joined to each other in a slightly different manner (Figure 11.24b).

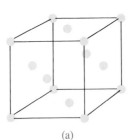
(a)

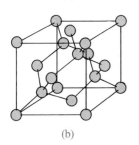
(b)

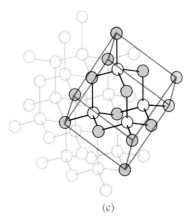
(c)

FIGURE 11.22
The Diamond Structure.

(a) The lattice is face-centered cubic. (b) There is an atom at each lattice point and at one-fourth the distance along each body diagonal. (c) This shows another view of the structure in which a body diagonal of the outlined unit cell is in a vertical position. The atoms at the face-centered lattice points are shaded; the other four atoms are shown as open circles.

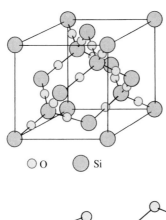

○ O   ● Si

FIGURE 11.23
The Structure of Silicon Dioxide, SiO_2.

Silicon dioxide (silica) crystallizes in a number of different forms. Each Si atom is surrounded by a tetrahedron of O atoms, each of which in turn is bonded to another Si atom. In the different forms of silica the SiO_4 tetrahedra are arranged in different ways. The structure shown here is the form of silica called β-cristobalite. Note that the O atom of each SiO_4 tetrahedron is shared with another Si atom, so each Si atom has a half share in four O atoms and the overall composition is SiO_2.

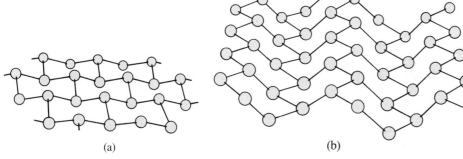

(a)   (b)

FIGURE 11.24
Layer Structures.

(a) The structure of arsenic is based on a pyramidal AX_3E geometry at each arsenic atom. The three bonds at each arsenic atom link the atoms into six-membered rings. (b) The structure of black phosphorus is similar, except that the six-membered rings are joined to each other in a slightly different way.

11.7
Ionic Crystals

The structures of ionic crystals are determined by the tendency of each ion to attract to itself as many ions of opposite charge as possible. As a consequence all ionic substances have network structures, whereas covalent substances may have either molecular or network structures.

Ionic crystals that are composed of equal numbers of ions of opposite charge have the simplest structures. They have the general formula MX, where M is a positive ion and X is a negative ion. These structures are called **1:1 structures**. We have briefly described some simple structures of this type in Chapter 5, and we can now consider them in more detail in terms of their space lattices.

THE SODIUM CHLORIDE STRUCTURE

Figure 11.25 shows the unit cell of the sodium chloride structure, which is based on a face-centered cubic lattice. The motif is one sodium ion and one chloride ion separated by half the length of the unit cell edge. If a sodium ion is placed at each lattice point, there is a chloride ion associated with each sodium ion

Space-filling model of sodium chloride (NaCl)

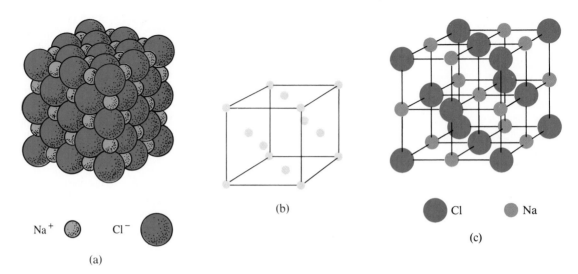

Na⁺ ●   Cl⁻ ●

(a)

(b)

● Cl   ● Na

(c)

FIGURE 11.25
The Sodium Chloride
Structure.

(a) A general view of the structure showing how the ions are closely packed
together. (b) The lattice is face-centered cubic. (c) On this lattice we place a positive
ion at each lattice point and a negative ion to the right of each positive ion at a
distance equal to one-half of the cell edge. The unit cell contains four cations
and four anions.

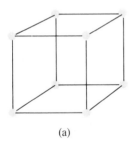

(a)

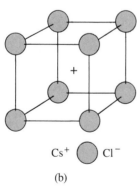

Cs⁺ ●   Cl⁻

(b)

FIGURE 11.26
The Cesium Chloride
Structure.

(a) The lattice is a primitive cubic
lattice. (b) The anions are at the
lattice points, and the cation is at
the center of the cell. The unit cell
contains one cation and one anion.

at a distance of one-half the unit cell edge away. Since the unit cell of the
face-centered cubic lattice has four lattice points associated with it, there are
four Na^+ ions and four Cl^- ions in the unit cell, in other words, four NaCl
formula units per unit cell. Each positive ion is surrounded by an octahedral
arrangement of six negative ions, and each negative ion is surrounded by an
octahedral arrangement of six positive ions. Each ion has a coordination number
of 6.

The sodium chloride structure is very common. It is found for all the alkali
metal halides (except CsCl, CsBr, and CsI), the oxides and sulfides of Mg, Ca,
Sr, and Ba, and the oxides of a number of transition metals such as iron(II)
oxide, FeO.

THE CESIUM CHLORIDE STRUCTURE

Cesium chloride, cesium bromide, and cesium iodide have a structure that is
different from sodium chloride; it is called the cesium chloride structure. In this
structure, each Cs^+ ion is surrounded by a cubic arrangement of eight Cl^- ions,
and each Cl^- ion is surrounded by a cubic arrangement of eight Cs^+ ions (see
Figure 11.26). The lattice is the primitive cubic lattice. If the Cl^- ions are placed
at each lattice point, then there is a Cs^+ at the center of each cell. Since this is a
primitive cell, it has only one lattice point associated with it. There is therefore
only one Cs^+ and one Cl^- associated with each unit cell, or in other words,
there is one CsCl formula unit per unit cell.

THE SPHALERITE STRUCTURE

A third important 1:1 structure is the structure of sphalerite, one form of zinc
sulfide, ZnS. In this structure the ions are only four-coordinated. The structure

is based on the face-centered cubic lattice, and the unit cell is shown in Figure 11.27. If the sulfide ions are placed at the lattice points, there are zinc ions one-fourth of the distance along each body diagonal. This structure is the same as the diamond structure, except that alternate atoms are zinc and sulfur. Since there are four lattice points associated with the face-centered cubic lattice, each unit cell contains four zinc ions and four sulfide ions, or four ZnS formula units. By connecting each ion to its nearest neighbors, we see that each ion has a tetrahedral coordination. Because of the large size of the sulfide ion compared with the size of the zinc ion, only four, rather than six or eight, sulfide ions can be packed around a zinc ion.

It is only a rather crude approximation to describe ZnS as an ionic crystal. Many crystals that are commonly described as consisting of doubly charged ions have some covalent character, particularly when the difference in the electronegativities of the two elements is not very large. In the present case $\chi_{Zn} = 1.6$ and $\chi_S = 2.5$, so the difference is only 0.9. In other words, there are covalent bonds between Zn and S, although the bonds have a considerable polarity. Whenever a small highly charged positive ion is situated next to a large negative ion, we expect some electron density to be transferred from the electron pairs of the negative ion to the empty valence shell of the positive ion, giving some covalent character to the bonds. Assuming that there are polar covalent bonds between Zn and S, they would each have four electron pairs in their valence shells and would therefore be expected to have an AX_4 geometry. Thus the structure of ZnS is consistent with a polar covalent description of the bonding.

Many substances, including CuCl, CuBr, CuI, BeS, HgS, and SiC, possess the sphalerite structure. The amount of covalent character in the bonds varies from one substance to another; in SiC the bonds are almost nonpolar covalent bonds.

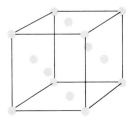

(a)

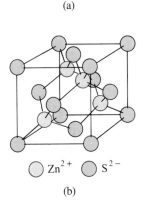

$\bigcirc$ Zn^{2+} $\bigcirc$ S^{2-}

(b)

FIGURE 11.27
The Sphalerite (ZnS) Structure.

(a) The lattice is face-centered cubic. (b) There are sulfide ions, S^{2-}, at the lattice points and zinc ions, Zn^{2+}, a fourth of the distance along each body diagonal. The unit cell contains four zinc ions and four sulfide ions.

=== *Example 11.7* ===

STRUCTURE OF AN IONIC SOLID FROM DENSITY AND CELL DIMENSIONS

Potassium iodide, KI, has a cubic unit cell with a cell edge of 705 pm. The density of KI is 3.12 g cm^{-3}. How many K^+ ions and I^- ions are contained in the unit cell? What is the structure of KI?

Solution

We can find the number of formula units per unit cell by finding the mass of the unit cell from its volume and density:

$$\text{Volume of unit cell} = (7.05 \times 10^{-8} \text{ cm})^3$$

$$\text{Mass of unit cell} = \text{Volume} \times \text{Density}$$

$$= (7.05 \times 10^{-8})^3 \text{ cm}^3 \times 3.12 \text{ g cm}^{-3}$$

$$= 1.09 \times 10^{-21} \text{ g}$$

This mass is the mass of a certain number of K^+ and I^-. We can find the total mass of one K^+ and one I^- from the molar mass of KI:

$$\text{Molar mass KI} = 166.0 \text{ g}$$

$$\text{Mass of one empirical formula} = \frac{166.0 \text{ g}}{6.022 \times 10^{23}} = 2.76 \times 10^{-22} \text{ g}$$

$$\text{Number of empirical formulas per unit cell} = \frac{\text{Mass of unit cell}}{\text{1 empirical formula mass}}$$

$$= \frac{1.09 \times 10^{-21} \text{ g}}{2.76 \times 10^{-22} \text{ g}} = 3.95$$

There are four empirical formulas per unit cell, that is, four K^+ and four I^- ions per unit cell. Thus there are probably four lattice points in the unit cell, which is therefore a face-centered cubic lattice. We conclude that KI has the NaCl structure.

= *Example 11.8* =

IONIC RADIUS FROM THE DENSITY OF AN IONIC CRYSTAL

The density of CsCl is 3.99 g cm^{-3}.

(a) From this value and the structure of CsCl (Figure 11.26), calculate the length of the edge of the unit cell.

(b) What is the distance between the centers of a Cs^+ ion and a neighboring Cl^- ion?

(c) The radius of Cl^- is 180 pm. What is the radius of Cs^+?

Solution

(a) We first calculate the mass of the unit cell. The unit cell contains one Cs^+ and one Cl^-. Hence

$$\text{Mass of unit cell} = \text{Mass}(Cs^+ + Cl^-)$$

$$= \frac{(132.9 + 35.5) \text{ g}}{6.022 \times 10^{23}} = 2.796 \times 10^{-22} \text{ g}$$

Next, we calculate the volume of the cell from the relationship

$$\text{Volume} = \frac{\text{Mass}}{\text{Density}} = \frac{2.796 \times 10^{-22} \text{ g}}{3.99 \text{ g cm}^{-3}} = 7.01 \times 10^{-23} \text{ cm}^3$$

If we let the edge of the unit cell be a, then the volume of the unit cell is a^3. Hence

$$a^3 = 7.01 \times 10^{-23} \text{ cm}^3$$

and therefore

$$a = 4.12 \times 10^{-8} \text{ cm} = 412 \text{ pm}$$

(b) We see from Figure 11.26 that the Cs^+ ion at the center of the unit cell is in contact with eight Cl^- ions at the corners. Therefore the distance between the center of the Cs^+ ion and the center of a Cl^- ion is the distance from the center of the unit cell to one corner, that is, half the length of a body diagonal. We saw in Example 11.3 that the length of the body diagonal is $\sqrt{3}\,a$. Therefore the distance between a Cs^+ and a Cl^- is

$$\frac{\sqrt{3}\,a}{2} = \frac{\sqrt{3} \times 412}{2} = 357 \text{ pm}$$

(c) Let r_{Cl^-} be the radius of Cl^- and r_{Cs^+} be the radius of Cs^+. Then

$$r_{Cl^-} + r_{Cs^+} = 357 \text{ pm}$$

If $r_{Cl^-} = 180$ pm, then $r_{Cs^+} = 177$ pm.

IONIC STRUCTURES BASED ON CLOSE PACKING OF ANIONS

Many ionic crystal structures can be described as close-packed arrangements of the larger anions with the smaller cations occupying the holes in the close-packed arrangement. Consider sodium chloride and sphalerite, which both have structures based on the face-centered cubic lattice. If the anions are large enough to touch each other, they have a cubic close-packed arrangement. The holes in such a cubic close-packed structure are of two types, **tetrahedral holes**, which have four neighboring anions in a tetrahedral arrangement, and **octahedral holes**, which are surrounded by six anions in an octahedral arrangement (Figure 11.28).

As shown in Figure 11.28 there are four octahedral holes per unit cell—one in the center of the unit cell and $12 \times \frac{1}{4} = 3$ holes situated at the middle of each edge of the unit cell. Thus we can describe the sodium chloride structure as a cubic close-packed arrangement of chloride ions, with all the octahedral holes occupied by sodium ions. There are eight tetrahedral holes per unit cell situated

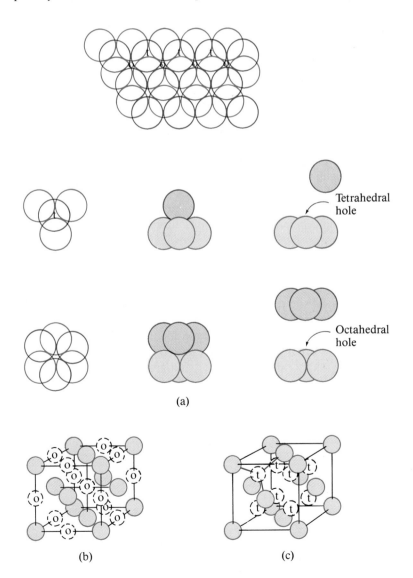

(a)

(b)　　　　(c)

FIGURE 11.28
Tetrahedral and Octahedral Holes in a Cubic Close-Packed Structure.

(a) Tetrahedral holes and octahedral holes between two close-packed layers. (b) The positions of the octahedral holes (o) in the face-centered cubic lattice of the cubic close-packed structure. (c) The positions of the tetrahedral holes (t).

in a cubic arrangement in the interior of the unit cell. We can describe the sphalerite structure as a cubic close-packed arrangement of sulfide ions with half the tetrahedral holes occupied by cations.

The next simplest types of ionic crystal structures are the **1:2 and 2:1 structures.** Two of these structures that can be described in terms of the face-centered cubic lattice or cubic close packing of one of the ions are the fluorite and the antifluorite structures.

THE FLUORITE STRUCTURE

Fluorite is the mineral name for calcium fluoride, CaF_2. The fluorite structure is shown in Figure 11.29. The calcium ions are located at the face-centered cubic lattice points and therefore have a cubic close-packed arrangement. The fluoride ions then occupy *all* the eight tetrahedral holes. There are therefore eight fluoride ions and four calcium ions in the unit cell, which is consistent with the empirical formula CaF_2. Each fluoride ion is surrounded by four calcium ions in a tetrahedral arrangement, while each calcium ion is surrounded by eight fluoride ions, which have a cubic arrangement. There are four calcium ions and eight fluoride ions per unit cell.

Other ionic compounds that have the fluorite structure include SrF_2, BaF_2, PbF_2, and $BaCl_2$.

THE ANTIFLUORITE STRUCTURE

Several M_2X compounds have a structure closely related to the fluorite structure; it is called the antifluorite structure. In this structure the smaller cations occupy the position of the fluoride ions, and the larger anions occupy the positions of the calcium ions in the fluorite structure. The following oxides and sulfides of the alkali metals, Li_2O, Na_2O, K_2O, Rb_2O, Li_2S, K_2S, and Rb_2S, have this structure. In these structures the oxide or sulfide ions have a cubic close-packed arrangement, with the cations occupying all the tetrahedral holes.

IONIC RADII

The number of negative ions (anions) that will fit around a given positive ion (cation) depends on the relative sizes of the two ions, that is, on the ratio of the radius of the positive ion to that of the negative ion, or r_+/r_-. Table 11.3 gives the possible range of the **radius ratio** for tetrahedral four coordination, octahedral six coordination, and cubic eight coordination. It shows that if we imagine starting with a very small cation and gradually increase its size, when the radius of the cation r_+ is 0.225 times the radius of the anion, r_-, that is, when

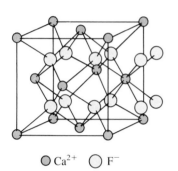

⦿ Ca^{2+} ◯ F^-

FIGURE 11.29
The Fluorite (CaF_2) Structure.

The lattice is face-centered cubic. The calcium ions are at the lattice points and have a cubic close-packed arrangement. The fluoride ions occupy all the tetrahedral holes.

TABLE 11.3 Radius Ratios		
Coordination Number	*Radius Ratio r_+/r_-*	*Coordination Polyhedron*
4	0.225–0.414	Tetrahedron
6	0.414–0.732	Octahedron
8	0.732–1.00	Cube

$r_+ = 0.225r_-$ or $r_+/r_- = 0.225$, we can place just four anions in a tetrahedral arrangement around the cation, with each anion touching its three neighbors. If the size of the cation is increased further, the anions no longer touch each other but there is insufficient room for additional anions until $r_+ = 0.414r_-$. Six anions can then be grouped around the cation in an octahedral arrangement, with each anion just touching its four neighbors. If the size of the cation is further increased, the anions no longer touch each other; but when $r_+ = 0.732r_-$, eight anions can be grouped around the cation in a cubic arrangement. Note that $r_+ = 0.225r_-$ is the radius of a tetrahedral hole, and $r_+ = 0.414r_-$ is the radius of an octahedral hole in a close-packed arrangement of anions.

These limiting values of the radius ratio are useful in enabling us to understand the structures of ionic crystals, but to use them, we need to know the radii of ions. The distances between the centers of adjacent ions in an ionic crystal can be measured by X ray diffraction with considerable accuracy. Values of these interionic distances for the alkali halides are given in Table 11.4. These distances vary in the expected manner; that is, we expect the halide ions to increase in size from F^- to I^- and the alkali metal ions to increase in size from Li^+ to Cs^+. But unfortunately, there is no completely reliable way of dividing up these distances to give values characteristic of the individual ions.

Chemists, physicists, and geologists have all suggested different ways of dividing up interionic distances to obtain the radii of ions. A recent set of values based on the analysis of a very large number of crystal structures is given in Table 11.5. The values in this table—or, indeed, any other table of ionic radii—must be regarded as approximate since they vary somewhat with the method by which they are determined. Such a set of ionic radii assumes that an ion always has exactly the same size, in other words, that each ion behaves as a hard sphere. But ions, like neutral atoms, are somewhat compressible. They may be compressed to different extents in different crystals, which means that the radius of an ion may vary somewhat from one crystal to another. Nevertheless, the values in Table 11.5 are reasonably accurate for most crystals containing these ions.

As we saw in Chapter 5, both cations and anions increase in size down any group of the periodic table. Comparison of the ionic radii in Table 11.5 with the atomic (covalent) radii in Figure 4.9 shows that anions are, in general, larger than the corresponding neutral atoms in covalent molecules, while cations are, in general, smaller than the corresponding neutral atoms in covalent molecules.

From the ionic radii in Table 11.5 and the radius ratios for different coordination numbers given in Table 11.3, we might expect to be able to predict

TABLE 11.4
Interionic Distances (pm) in Alkali Metal Halide Crystals

	Li^+	Na^+	K^+	Rb^+	Cs^+
F^-	201	231	266	282	300
Cl^-	257	281	314	327	356
Br^-	275	298	329	343	371
I^-	300	323	353	366	395

TABLE 11.5
Ionic Radii (pm)

Li^+	Be^{2+}												O^{2-}	F^-
74	27												138	133
Na^+	Mg^{2+}	Al^{3+}											S^{2-}	Cl^-
102	72	53											184	181
K^+	Ca^{2+}	Sc^{3+}	Ti^{2+}	V^{2+}	Cr^{2+}	Mn^{2+}	Fe^{2+}	Co^{2+}	Ni^{2+}	Cu^{2+}	Zn^{2+}		Se^{2-}	Br^-
138	100	73	86	79	82	82	77	74	70	73	75		198	196
Rb^+	Sr^{2+}												Te^{2-}	I^-
149	116												221	216
Cs^+	Ba^{2+}													
170	136													

the coordination number for any given pair of ions and therefore to make a prediction about the crystal structure. Some examples of such predictions for 1:1 structures are given in Table 11.6.

Although the ionic radius ratios are consistent with the coordination numbers of the ions in many crystal structures, there are also a large number of exceptions, only a few of which are shown in Table 11.6. Cesium chloride, bromide, and iodide have radius ratios consistent with the eight coordination observed in their structures, and many of the other alkali halides that have the NaCl structure have radius ratios consistent with six coordination. However, others have the NaCl structure, even though their radius ratios are too large, as in the case of RbCl, or too small, as in the case of LiI. Although ZnS has radius ratio consistent with the four coordination found in its structure, BeS also has this structure, despite the very small value of its radius ratio

These exceptions to predictions made from the radius ratios are not very surprising in view of the somewhat approximate nature of the values of the ionic radii. These exceptions also lead us to suspect that there are other factors that

TABLE 11.6
Radius Ratios and 1:1 Structures

r_+/r_-	Coordination Number	Structure	Examples			
			Compound	r_+/r_-	Compound	r_+/r_-
0.225–0.414	4	Sphalerite	ZnS	0.41	BeS	0.19
0.414–0.732	6	Sodium chloride	LiCl	0.41	LiI	0.34
			MgO	0.51	LiBr	0.38
			NaBr	0.52	KCl	0.77
			CaS	0.54	RbCl	0.83
			FeO	0.55	BaO	0.97
			NaCl	0.57		
			MnO	0.59		
			KBr	0.71		
			CaO	0.71		
0.732–1.00	8	Cesium chloride	CsI	0.79		
			CsBr	0.87		
			CsCl	0.94		

influence the structures of ionic crystals. For example, we know that many "ionic crystals" have considerable covalent character in their bonding. The structures of such substances are influenced by the geometry of the covalent bonds, as is certainly the case for ZnS and BeS, where each atom forms a set of four tetrahedral AX_4 bonds.

IMPORTANT TERMS

An **amorphous solid** is non-crystalline; it does not have plane faces with characteristic angles between them, and it does not have a sharp melting point. The atoms, molecules, or ions of an amorphous solid have a random, disordered arrangement rather than the regular periodic arrangement characteristic of crystalline solids.

A **body-centered cubic cell** has a lattice point at each corner of a cube and one in the center of the cube, giving a total of two points.

The **coordination number** of an ion is the number of ions of opposite sign that are in contact with it in the crystal structure.

Crystalline solids form crystals with flat faces and definite angles between the faces and they have sharp melting points. The atoms, molecules, or ions in a crystalline solid have a regular periodic arrangement.

A **face-centered cubic cell** has one lattice point at each corner of a cube and one at the center of each face, giving a total of four points.

Molecular solids consist of individual molecules held together by relatively weak intermolecular forces.

A **motif** is the object, which for a crystal is the atom, ion or molecule, or arrangement of atoms, ions or molecules that, when associated with each lattice point, produces the complete pattern or crystal structure.

A **net**, or **two-dimensional lattice**, is a regular arrangement of points in a plane.

Network solids have a continuous network of atoms or ions in which each atom or ion is bound strongly to each of its near neighbors, and the regular network extends indefinitely throughout the crystal.

A **primitive cubic unit cell** has lattice points at each corner of a cube. It has one point associated with it.

The **radius ratio** is the ratio of the radius of the cation to the radius of the anion in an ionic crystal.

A **space lattice** is a regular, three-dimensional arrangement of points in space.

The **unit cell** of a two-dimensional lattice is obtained by choosing four points of the lattice and joining them to give a parallelogram. The unit cell of a three-dimensional lattice is obtained by joining eight points of the lattice to give a parallelepiped. Many different unit cells are possible. The cell usually chosen is the smallest that shows the full symmetry of the lattice. It is possible, by repeatedly moving the cell in the directions of its edges by a distance equal to the cell edge, to generate the complete lattice.

PROBLEMS *

Types of Solids

1. Which of the following solids are molecular solids, and which are network solids?

   C S_8 CO_2 P_4O_6 NaCl   MgO   Al

2. Explain why molecular solids usually melt below 400 °C, while network solids often have melting points higher than 400 °C.

3. (a) What type of force holds the molecules together in orthorhombic sulfur?

   (b) What type of bond holds the atoms together in black phosphorus?

(c) What type of bond holds the ions together in magnesium oxide?

(d) What type of bond holds the atoms together in silicon dioxide?

(e) What type of bond holds the atoms together in copper?

4. Explain why (a) polymeric sulfur trioxide, $(SO_3)_n$, forms needlelike crystals while (b) plastic sulfur is a soft rubberlike solid.

5. What is an amorphous solid? Give an example and explain why an amorphous solid does not have a sharp melting point.

* Answers to problems numbered in blue appear at the end of the text.

6. Explain in terms of their structures why each of the following pairs of substances have very different chemical and physical properties:

(a) Silica and carbon dioxide

(b) Oxygen and sulfur

(c) Diamond and graphite

7. Explain what is meant by a two-dimensional network solid. Give an example.

8. Which of the following substances form a network solid and which form a molecular solid?

CO   SiC Cl_2 Mg $MgCl_2$

9. What types of interactions (bonds and intermolecular forces) must be broken to melt each of the following solids:

BaO   Diamond I_2
P_4O_{10} Cu   Graphite

10. Describe the properties of each of the following types of solids in terms of their melting points and solubilities in water:

(a) Molecular   (b) Ionic

(c) Metallic   (d) Covalent network

11. Both sodium chloride and magnesium oxide are ionic solids, yet MgO melts at a temperature about 2000 K higher than NaCl. Explain why.

12. Explain why NaCl(s) has a three-dimensional network structure and does not consist of NaCl molecules as in the gas phase?

Lattices

13. What is a space lattice?

14. What is a unit cell?

15. Describe each of the three cubic unit cells, and draw a diagram of each.

16. How many lattice points are associated with each of the following?

(a) A primitive cubic unit cell

(b) A body-centered cubic unit cell

(c) A face-centered cubic unit cell

Structures of Solids

17. What are the three common structures for metals? In each case give the corresponding space lattice and the motif.

18. Describe the structure of each of the following:

(a) Copper   (b) Sodium   (c) Diamond

19. Describe the structure of each of the following:

(a) Potassium chloride   (b) Barium oxide

(c) Copper(I) chloride

20. Explain the difference between a cubic close-packed structure and a hexagonal close-packed structure.

21. Draw diagrams to show the relationship of the close-packed layers in a cubic close-packed structure to the unit cell.

22. Aluminum has a structure based on a cubic lattice. The length of the edge of the unit cell is 405 pm at 25 °C, and the density of aluminum is 2.70 g cm^{-3} at the same temperature. How many aluminum atoms are there in the unit cell? On what type of cubic lattice must the structure of aluminum be based?

23. The density of platinum at 25 °C is 21.5 g cm^{-3} and it has the cubic close-packed structure with a unit cell edge of 392 pm at 25 °C. Calculate the atomic mass of platinum from these data.

24. Aluminum has the cubic close-packed structure. The length of the unit cell edge is 405 pm. Calculate the length of the diagonal of a face of the unit cell, and hence find the radius of an aluminum atom.

25. The compound copper(I) chloride, CuCl, has the sphalerite (zinc sulfide) structure. Its density is 3.41 g cm^{-3}.

(a) What is the length of the edge of the unit cell?

(b) What is the shortest distance between the centers of a Cu$^+$ ion and Cl$^-$ ion?

(c) The radius of the Cl$^-$ ion is 180 pm. What is the radius of the Cu$^+$ ion?

26. At 24 K, neon has a cubic lattice with a unit cell edge of 450 pm. The density of solid neon is 1.45 g cm^{-3} at this temperature. How many neon atoms are in the unit cell? What type of cubic lattice does crystalline neon have? What is the radius of the neon atom?

27. Solid krypton has a cubic close-packed lattice with a cell edge of 559 pm.

(a) Sketch the unit cell, showing the positions of the atoms.

(b) How many krypton atoms are there in the unit cell?

(c) Calculate the density of solid krypton.

(d) Find the radius of the krypton atom.

(e) Compare the radius of krypton found here with the covalent radius given in Figure 4.9, and explain the difference between the two values.

28. Gold has a cubic close-packed lattice and a density of 9.329 g cm^{-3}. The length of the unit cell edge is found by X ray crystallography to be 407 pm. Calculate a value for the Avogadro constant.

29. In crystalline sodium chloride the distance between the centers of neighboring Na$^+$ and Cl$^-$ ions is 281 pm, and the density is 2.165 g cm^{-3}. Calculate a value for the Avogadro constant.

30. Copper has a density of 8.930 g cm^{-3}. The edge of the unit cell is 361.5 pm. Calculate a value for the Avogadro constant.

31. Calcium fluoride, CaF$_2$, has a cubic lattice. The length of the unit cell edge is 546.3 pm. The density of CaF$_2$ is 3.180 g cm^{-3}. How many formula units of CaF$_2$ are there per unit cell?

32. Potassium fluoride has the sodium chloride structure and a density of 2.481 g cm^{-3}.

(a) What is the length of the edge of the unit cell?

(b) What is the distance between the center of a potassium ion and the center of a neighboring fluoride ion?

33. On the basis of the diagram of the unit cell of β-cristobalite, one of the forms of silica, given in Figure 11.23, identify the space lattice and describe the positions of the Si and O atoms in the unit cell.

34. The bromide ion, Br$^-$, has an ionic radius of 195 pm. If bromide ions are packed closely together in a three-dimensional network, so that octahedral holes are formed, what is the maximum radius for the cations that can fit into these octahedral holes?

35. The element with atomic number 73 is tantalum, a metal with a crystalline structure based upon one of the cubic lattices. Its density is 16.6 g cm^{-3}, and the unit cell edge is 328 pm. How many tantalum atoms are there in the unit cell? What is the structure of tantalum?

36. Which unit cell has the higher mass, that of cesium chloride or that of sodium chloride? Explain.

37. Molybdenum has the iron (monatomic body-centered cubic) structure. Its density is 10.22 g cm^{-3}. What is the radius of the molybdenum atom?

*__38.__ Mercury(II) sulfide has the sphalerite (ZnS) structure.

* The asterisk denotes the more difficult problems.

The shortest distance between the center of an Hg^{2+} ion and the center of an S^{2-} ion is 253 pm. Calculate the density of mercury(II) sulfide.

39. Cesium bromide, CsBr, crystallizes with the cesium chloride structure. If the closest distance between the centers of oppositely charged ions is 371 pm, calculate the density of cesium bromide.

40. (a) How many ions are associated with a unit cell for each of the following?

(i) Barium oxide   (ii) Cesium iodide

(iii) Lithium sulfide

(b) How many lattice points belong to each of the above unit cells?

41. What is the coordination number of Zn^{2+} in zinc sulfide? What is the coordination number of S^{2-} in zinc sulfide? How is the zinc sulfide structure related to the diamond structure?

42. (a) What is the coordination number of Ca^{2+} in fluorite, CaF$_2$(s)?

(b) What is the coordination number of F$^-$ in fluorite?

(c) How is the fluorite structure related to that of zinc sulfide?

43. At a temperature of about 1000 °C, iron undergoes a transition from the monatomic body-centered cubic structure to the face-centered cubic structure. The length of the unit cell edge increases from 286 pm to 363 pm. Calculate the density of this high temperature form of iron.

44. Consider the unit cell of sodium chloride. How many cations, and how many anions, are wholly within the unit cell? How many are partially in the unit cell (shared with other unit cells)?

45. Barium has the monatomic body-centered cubic structure, with a unit cell edge of 502 pm. What is the density of barium?

*__46.__ Barium fluoride has the fluorite structure with a unit cell edge of 618 pm.

(a) How many BaF$_2$ formula units are in the unit cell?

(b) What is the ionic radius of Ba^{2+}, given that the ionic radius of F$^-$ is 135 pm?

(c) How does this calculated value for the radius of Ba^{2+} compare with that given in Table 11.5?

47. Silver has the copper (monatomic face-centered cubic) structure, and a density of 10.50 g cm^{-3}. How many silver atoms are contained in a cube of silver with an edge length of 1.00 mm?

Miscellaneous

48. Which ion in each of the following pairs is the larger ion?

    (a) F^-, I^- **(b)** Na^+, K^+ **(c)** O^{2-}, F^-

    (d) Be^{2+}, Cl^- **(e)** K^+, Cl^-

49. Describe the structure of diamond and the nature of the bonding in diamond.

50. Describe the structure of graphite and the nature of the bonding in graphite.

51. Account for the physical properties of diamond and graphite in terms of their structures.

52. From data collected for many molecules, the average carbon–carbon single bond length is 154 pm and the carbon–carbon double bond length is 134 pm. In a layer of graphite, the distance between adjacent carbon atoms is 142 pm. Explain this observed carbon–carbon bond length in graphite in terms of its resonance structures and the carbon° carbon bond order.

53. In terms of the radii of the ions involved, explain why sodium chloride has a different structure from that of cesium chloride.

54. Why is it not possible to obtain precise values for ionic radii?

55. Why are anions generally larger than the corresponding neutral atoms in covalent molecules, while cations are in general smaller than the corresponding neutral atoms?

Water, Liquids, Solutions, and Intermolecular Forces

12

We discussed gases in Chapter 3 and solids in Chapter 11. We have left liquids to last because they are the most difficult to treat theoretically and are the least well understood. The simplifying assumptions that apply to the gaseous state—namely, that the volume of the atoms or molecules is negligible compared with the volume occupied by the gas, and that the forces between the molecules are also negligible—are not applicable to liquids. The molecules in a liquid are packed closely together, and the forces between them are relatively strong. In contrast to the properties of a gas, the properties of a liquid depend very much on the size and shape of the molecules and on the forces between them. More-over, the molecules in a liquid are not arranged in the regular patterns that we find in most solids, so it is much more difficult to describe the structure of a liquid than the structure of a solid.

Yet liquids are extremely important. Water is the most common liquid and the most common substance on earth. It covers 72% of the earth's surface. It is the only substance that is found naturally as a liquid, as a solid (snow and ice), and as a gas (water vapor) in the atmosphere (Figure 12.1). Water is es-sential to life; the human body is 65% water by mass, and a jellyfish is 97% water. Blood, which is 83% water, carries dissolved nutrients and oxygen to the body cells and transports waste products from the cells. Urine is an aqueous solution of the waste products from body processes. Water is the solvent in which chemical reactions occur in the body. For example, in the digestive tract, proteins and carbohydrates are broken down into smaller molecules which can be absorbed into the blood. Water is also an important reactant. For example, in photosynthesis it combines with carbon dioxide to form carbohydrates.

In this chapter we discuss the properties of liquids, with particular emphasis on water. Although water is colorless, odorless, and tasteless, as well as being quite commonplace, compared with most other liquids it has, as we shall see, some remarkable and unusual properties. It is a good solvent for many other substances and we will consider some of the properties of aqueous solutions as well as solutions in other solvents. In particular we will discuss those proper-ties of solutions called colligative properties—vapor pressure, freezing point, boiling point, and osmotic pressure.

The solubility of substances in water and other liquids is strongly dependent on the strength of the forces acting between the solvent molecules, between the

FIGURE 12.1
Solid, Liquid, and Gaseous Water.

Water occurs naturally in all its three forms: solid ice, liquid water, and gaseous water (water vapor). The water vapor in the atmosphere condenses to form clouds which consist of very small droplets of liquid water.

solute molecules, and between the solvent and the solute molecules. We have frequently mentioned these intermolecular forces, and we now discuss their nature and strength in some detail. One type of intermolecular force, the hydrogen bond, is of particular importance in water and plays a vital role in living matter.

Not only is water a good solvent for many reactions, but it also undergoes many important reactions. Some of these we have encountered in previous chapters. In this chapter we review these reactions.

12.1
Natural Waters

Of the vast amount of water on the earth's surface, 97% is in the oceans. The remainder is fresh water, but 2.1% is in the form of ice caps and glaciers, and only 0.7% is readily available in rivers and lakes and as underground water. Water is continually being redistributed. It evaporates from the lakes and oceans into the atmosphere and returns in the form of rain and snow. This water runs off the surface or percolates through the ground into rivers and lakes and finally returns to the sea. In doing so, it dissolves many substances from the earth's crust, which accumulate in the oceans. This redistribution of water makes life on the land possible, and the dissolved substances that accumulate in the ocean are important for marine life.

SEAWATER

The 97% of all the earth's water that is found in the seas and oceans contains too much sodium chloride and other dissolved substances to be useful as drinking water or for many other purposes. Fresh water can be obtained from seawater by distillation but this is an expensive process because a large amount of energy is needed. It is only practical in regions with a hot sunny climate when solar energy can be used. Another process that is used is reverse osmosis which we discuss later in this chapter.

Although the oceans are not generally useful as a source of water, they are a vast storehouse of many other substances. Table 12.1 gives the concentrations of the most common of these. Each cubic meter of seawater contains 1.5 kg of dissolved substances, but the low concentrations of most of them make any process for isolating them expensive and uneconomical at present. However,

TABLE 12.1 Major Constituents of Seawater (ppm)	
Sodium, Na^+	10 561
Magnesium, Mg^{2+}	1272
Calcium, Ca^{2+}	400
Potassium, K^+	380
Chloride, Cl^-	18 980
Sulfate, SO_4^{2-}	2649
Hydrogen carbonate, HCO_3^-	142
Bromide, Br^-	65
Other substances	34

two important substances are recovered commercially from seawater—sodium chloride and magnesium. In the past, bromine was also obtained from seawater, but it is now mostly obtained from underground brines (Chapter 5).

Sodium chloride is the most abundant dissolved substance in seawater, and much of it is obtained from this source. In hot climates seawater trapped in shallow ponds is allowed to evaporate in the sun until the solubility of sodium chloride is exceeded.

Magnesium is removed from seawater by adding lime, CaO, which precipitates magnesium hydroxide:

$$CaO(s) + H_2O(l) \longrightarrow Ca^{2+}(aq) + 2OH^-(aq)$$
$$Mg^{2+}(aq) + 2OH^-(aq) \longrightarrow Mg(OH)_2(s)$$

The magnesium hydroxide is filtered off and converted to magnesium chloride with hydrochloric acid:

$$Mg(OH)_2(s) + 2HCl(aq) \longrightarrow MgCl_2(aq) + 2H_2O(l)$$

The solution is evaporated to give solid magnesium chloride which, after being dried, is melted and then electrolyzed to give magnesium and chlorine (see Chapter 17).

EXERCISE 12.1

Classify the three reactions used in the extraction of magnesium from seawater as acid–base, redox, or precipitation.

PURIFICATION OF WATER

The water needed for homes, agriculture, and industry is taken from lakes, rivers, and underground sources. Much of this water must be treated to remove bacteria and other dangerous impurities. Figure 12.2 illustrates a typical purification process. After a preliminary filtration, the water is allowed to stand in large tanks so that fine sand and other very small particles can settle. This process is aided by first making the water slightly alkaline by the addition of lime,

$$CaO(s) + H_2O(l) \longrightarrow Ca^{2+}(aq) + 2OH^-(aq)$$

and then adding aluminum sulfate or alum to give a gelatinous precipitate of $Al(OH)_3$:

$$Al^{3+}(aq) + 3OH^-(aq) \longrightarrow Al(OH)_3(s)$$

This precipitate settles out slowly and carries with it much of the suspended matter, including most of the bacteria. The water is then again passed through a filter and often through activated charcoal, which adsorbs most of the impurities that still remain. Then the water is sprayed into the air to speed up the oxidation of dissolved organic substances.

In the final stage of purification the water is treated with an oxidizing agent to complete the destruction of bacteria. Ozone, O_3, is very effective, but it must be generated on the site. In North America chlorine is most frequently used because it can be shipped in tanks in liquid form and dispensed from the tanks

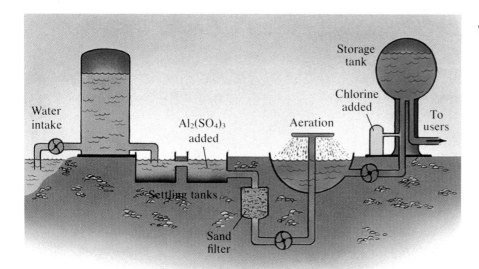

FIGURE 12.2
Water Purification System.

directly into the water supply. The hypochlorous acid, HOCl, formed by the chlorine in water is the effective sterilizing agent:

$$Cl_2(g) + 2H_2O(l) \longrightarrow HOCl(aq) + H_3O^+(aq) + Cl^-(aq)$$

A very small concentration of approximately 1 ppm of hypochlorous acid is sufficient to kill bacteria.

Even after this treatment water is not completely pure. It contains small amounts of dissolved salts, particularly sodium, potassium, magnesium, and calcium, in the form of chlorides, sulfates, fluorides, and hydrogen carbonates. These salts have no harmful effects in the low concentrations usually present; indeed, they provide essential "minerals" for the body. Water also contains dissolved gases, in particular, oxygen, nitrogen, and carbon dioxide. The dissolved oxygen is essential to aquatic life.

12.2
Phase Changes

With the exception of helium, all liquids, when cooled sufficiently at atmospheric pressure, freeze to form a solid. Also, all liquids are volatile to some extent; that is, they tend to evaporate to form gases. At atmospheric pressure, a liquid is completely transformed to gas when heated to a particular temperature called its *boiling point*.

The term *vapor* is frequently used to refer to a gas that is formed by the evaporation of a liquid or a gas that is easily condensed to give a liquid. Thus one usually refers to water vapor or to gasoline vapor. However, there is no difference between a vapor and a gas; every vapor is a gas, and every gas can be condensed to a liquid. It is merely a matter of habit that we speak of water vapor rather than gaseous water.

The transformation of a substance from one state to another is called a **phase change**. Figure 12.3 summarizes the possible phase changes and terms used to describe them.

FIGURE 12.3
Phase Changes.

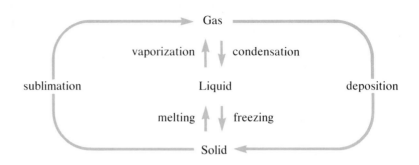

MELTING AND FREEZING

When a solid is heated, the kinetic energy of the atoms or molecules increases so that the molecules vibrate about their fixed positions more vigorously. With increasing temperature this motion eventually becomes sufficiently violent that the atoms or molecules break away from their fixed positions and the ordered arrangement of the solid is replaced by the more random arrangement of the liquid. In other words, the solid melts. At the melting point of the solid, which is also the freezing point of the liquid, melting and freezing are occurring at the same rate, and the system is in dynamic equilibrium.

Because the molecules, atoms, or ions in a liquid are moving faster and have a less ordered arrangement than those in a solid, they usually take up a slightly greater volume. Thus the volume of a solid substance is normally smaller than that of the same substance in the liquid state at the melting point and therefore the density of the solid is normally greater than that of the liquid. But, as we shall see, water is an exception to this general rule.

Because the energy of the molecules must be increased in order to break up their arrangement in a solid, melting (or fusion) is an endothermic process. The **molar enthalpy of fusion**, ΔH_{fus}, is the heat required to melt 1 mol of a substance at 1 atm pressure. For example, the molar enthalpy of fusion of ice at 0 °C is the enthalpy change for the process

$$H_2O(s) \longrightarrow H_2O(l) \qquad \Delta H_{fus} = 6.01 \text{ kJ mol}^{-1}$$

Thus 6.01 kJ are required to melt 1 mol of ice at 0 °C. Values for the molar enthalpy of fusion of some solids are given in Table 12.2. In a molecular solid the molecules are held together by relatively weak intermolecular forces whereas in a network solid they are held together by strong chemical bonds—covalent,

TABLE 12.2 Molar Enthalpies of Fusion			
Substance	*Formula*	ΔH_{fus} *(kJ mol^{-1})*	*Melting Point (°C)*
Hydrogen	H_2	0.12	−259
Methane	CH_4	0.94	−164
Mercury	Hg	2.3	−38.9
Tetrachloromethane	CCl_4	2.51	−22.9
Ethanol	C_2H_5OH	5.01	−114.6
Water	H_2O	6.01	0.0
Benzene	C_6H_6	10.6	5.5
Silver	Ag	11.3	961
Sodium chloride	NaCl	27.2	801

metallic, or ionic. Thus it takes considerably more energy to break enough of the chemical bonds in a network solid to enable the atoms or ions to move freely with respect to each other than it does to overcome the weak intermolecular forces holding the molecules together in a molecular solid. Thus the melting points and molar enthalpies of fusion of molecular solids are generally considerably lower than those of network solids.

A few substances exhibit an unusual state of matter that is intermediate between a liquid and a crystalline solid; it is called a *liquid crystal* (see Box 12.1).

BOX 12.1

LIQUID CRYSTALS

The displays in many calculators, wristwatches, readout meters, and temperature-measuring strips use substances called *liquid crystals*. These substances are unusual in that they have a structure intermediate between that of a liquid and that of a crystalline solid. In a liquid the molecules have a random arrangement, and they are able to move past each other. In a crystalline solid the molecules have an ordered arrangement and are in fixed positions. In a small temperature range just above the melting point, the molecules of a liquid crystal have an ordered arrangement, but they are able to move past each other, so the substance is a liquid. Molecules that form liquid crystals have rather special shapes; they are either long and cylindrical—rodlike—or large and flat—platelike. Rodlike molecules can form arrangements like those shown in the accompanying figure. They can rotate around their own axis, and they can slide past each other, but they remain parallel to each other, like soda straws or matches when they are packed parallel to each other.

The practical applications of liquid crystals depend on their rather remarkable optical properties. Because of the ordered arrangement of the molecules, they can diffract light, just as the planes of atoms in a crystal diffract X rays (Box 11.1). Only one of the wavelengths of white light is reflected by a liquid crystal, which therefore appears colored. As the temperature is changed, the distance between the layers of molecules changes, and therefore the color of the reflected light changes correspondingly. Liquid crystals can therefore also be used as sensitive temperature-measuring devices. A liquid crystal film can be used to map the temperature on the surface of an object. One medical application is for locating veins by the slightly higher temperature of the skin above a vein.

The optical properties of liquid crystals are also affected by an electric field. When a thin film of a liquid crystal is placed between two electrodes and an electric potential is applied, a rearrangement of the structure occurs, and the transparent liquid crystal becomes opaque. This property is used in the number displays of digital watches and other instruments. Only very small potentials are needed, so these devices consume very little power.

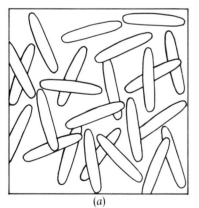
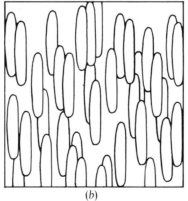
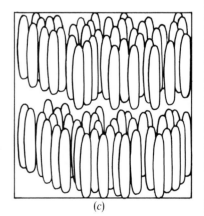

(a) (b) (c)

The arrangement of molecules in an ordinary liquid and in two types of liquid crystals. (a) An ordinary liquid. (b) A liquid crystal: the molecules can freely rotate and slip past each other, but retain their parallel orientation. (c) Another type of liquid crystal: the molecules are arranged in layers. They can rotate and move past each other in the layers, and the layers can also slide over each other, but the layers are retained.

EVAPORATION AND CONDENSATION

The molecules of a liquid have a distribution of energies similar to those in a gas (Figure 12.4). Molecules with enough kinetic energy to enable them to overcome the intermolecular forces holding them together may leave the liquid to form the vapor. This process is called **evaporation**. As the more energetic molecules leave the liquid, the average kinetic energy of the remaining molecules must decrease, so the temperature of the liquid must decrease, unless heat flows in from the surrounding to keep the temperature constant.

Vaporization is an endothermic process. The **molar enthalpy of vaporization**, ΔH_v, is the heat required to transform 1 mol of liquid into vapor at 1 atm pressure and at a specified temperature. Values of the molar enthalpy of vaporization of some liquids at their boiling points are given in Table 12.3. Values range from as low as 0.9 kJ mol^{-1} for hydrogen, H_2, to 612 kJ mol^{-1} for carbon, C.

In order to convert a covalent liquid to a gas, all the molecules must be widely separated, so the enthalpy of vaporization is a good measure of the strength of the intermolecular forces that hold the molecules together in the liquid. As we see in Table 12.3 the enthalpies of vaporization of network substances are much higher than those of molecular substances because a large number of strong bonds, whether ionic, covalent, or metallic, must be broken to convert the liquid to vapor. Both diamond and graphite vaporize to give a gas consisting mainly of carbon atoms; red and black phosphorus give a vapor consisting of P_4 molecules. Ionic substances such as NaCl and CaO give a vapor consisting of small ionic molecules such as the diatomic molecules Na^+Cl^- and $Ca^{2+}O^{2-}$. Metals vaporize to give single atoms or small clusters of atoms.

Although the exact nature of network substances in the liquid state is not entirely clear, their high enthalpies of vaporization show that they still contain many unbroken bonds—presumably bonds are continually being broken and new ones formed as the atoms or ions move around in the liquid.

One of the important roles played by water in the body is that of maintaining the body temperature close to 37 °C. The metabolic processes in the body gen-

FIGURE 12.4
Distribution of Kinetic Energies of Molecules in a Liquid.

The average kinetic energy of the molecules increases with increasing temperature. With increasing temperature a rapidly increasing number of molecules have more than the minimum energy, E_{min}, needed for a molecule to escape from the liquid into the vapor phase. The shaded area shows the fraction of the total number of molecules that have an energy greater than E_{min}.

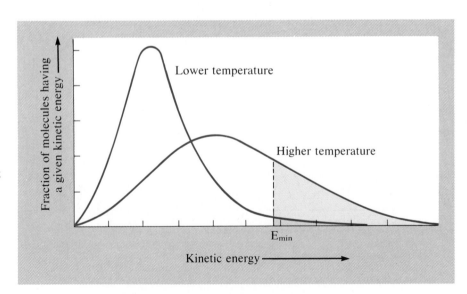

TABLE 12.3
Molar Enthalpies of Vaporization

Substance	Formula	ΔH_v (kJ mol^{-1})	Boiling Point (°C)
Hydrogen	H_2	0.9	−253
Methane	CH_4	10.4	−164
Pentane	C_5H_{12}	27.0	36.1
Tetrachloromethane	CCl_4	30.0	76.7
Benzene	C_6H_6	30.8	80.2
Ethanol	C_2H_5OH	38.6	78.5
Water	H_2O	40.7	100
Mercury	Hg	59.3	357
Sodium chloride	NaCl	207	1465
Carbon (graphite)	C	612	4830

erate heat, and any heat in excess of that required to maintain the body temperature must be dissipated. If the air temperature is lower than the body temperature, we lose heat by conduction. But if we are exercising vigorously or if the air temperature is very high, we cannot get rid of the heat rapidly enough by conduction. We then rely on the evaporation of water, that is, on sweating, to remove heat and maintain the body temperature. Because of its relatively high enthalpy of vaporization, water is very effective for this purpose.

VAPOR PRESSURE

We are familiar with the fact that if water or gasoline is left in an open container or spilled on the ground, it gradually evaporates until all the liquid has been converted to vapor. But if a sufficient amount of liquid is present in a *closed* container, it does not all evaporate. Instead, an equilibrium is established between the liquid and the vapor, and a constant pressure is established in the container, as Figure 12.5 illustrates. At first the amount of vapor increases, and the pressure that it exerts increases correspondingly. As the amount of vapor increases, the chance that a molecule in the vapor phase will collide with the surface of the liquid and return to the liquid phase also increases. Eventually, the number of molecules hitting the surface and returning to the liquid is equal

FIGURE 12.5
Vapor Pressure.

(a) A bottle completely filled with water is placed in a container that can be evacuated by pumping out all the air. When all the air has been removed and the pressure gauge registers zero, the tap is turned to close the container, and the stopper of the flask is removed by remote control. (b) The water then begins to evaporate into the container, and the pressure gauge begins to register the pressure of the water vapor. As the amount of water vapor increases, the chance that a gaseous molecule in the vapor phase will collide with the surface of the water and return to the liquid phase also increases. (c) Eventually, the number of molecules returning to the liquid phase is equal to the number leaving. No further changes in the amounts of liquid and vapor are then observed. The vapor exerts a constant pressure (the vapor pressure of water at the temperature of the experiment), which is registered on the pressure gauge as a constant value.

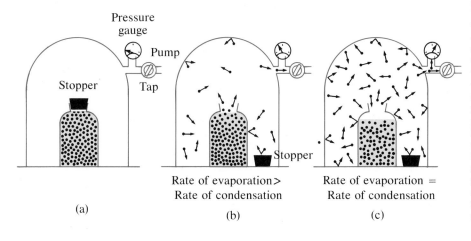

Pressure gauge · Pump · Stopper · Tap

Rate of evaporation >
Rate of condensation

Rate of evaporation =
Rate of condensation

(a)   (b)   (c)

to the number leaving. A state of dynamic equilibrium is reached, which can be summarized by the equation

$$H_2O(l) \rightleftharpoons H_2O(g)$$

No further changes in the amounts of water in the liquid phase and in the vapor phase are then observed, and the constant amount of vapor in the container exerts a constant pressure. The **vapor pressure** is the constant pressure exerted by the vapor above a liquid when equilibrium is established. The vapor pressure of a liquid is independent of the volume of the container in which it is confined, and it has a constant, characteristic value at a given temperature (see Figure 12.6).

The only liquids that have appreciable vapor pressures at ordinary temperatures and pressures are those consisting of covalent molecules and so we will be primarily concerned with such liquids in what follows.

The vapor pressure of a liquid depends on the tendency of the molecules to escape into the vapor phase, which in turn depends on the strength of the intermolecular forces in the liquid. When the intermolecular forces are weak, the molecules escape easily, and the vapor pressure is high. Thus the vapor pressure of a liquid at a particular temperature provides a measure of the magnitude of the intermolecular forces.

If a liquid is not confined in a container, it still evaporates, but the vapor diffuses away. The rate of condensation is thus always less than the rate of evaporation, and the liquid continues to evaporate until it has all been converted to vapor.

The vapor pressure of a liquid depends on the temperature, as Table 12.4 demonstrates. As the temperature increases, a larger fraction of the molecules

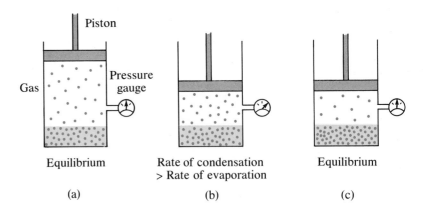

FIGURE 12.6
The Vapor Pressure of a Liquid is Independent of the Volume of the Container.

(a) A liquid and its vapor are confined in a container with a tight-fitting piston.
(b) If we push in the piston, thus decreasing the volume of the vapor, the concentration of the molecules in the gas phase will momentarily increase. In turn, the rate of collisions of gas molecules with the surface will increase, and thus the rate at which they return to the liquid will increase. Since this rate will be greater than the rate at which the molecules evaporate, the system will no longer be in equilibrium. Vapor will condense until the concentration of gas molecules has decreased to its original value. (c) The system is then again in equilibrium, and the vapor pressure again has its original value.

TABLE 12.4
Vapor Pressure of Water

T (°C)	P (mm Hg)	T (°C)	P (mm Hg)
0	4.6	23	21.0
5	6.5	24	22.3
10	9.2	25	23.8
15	12.8	30	31.8
16	13.6	40	55.3
17	14.5	50	92.3
18	15.5	60	147.4
19	16.5	70	233.5
20	17.5	80	355.1
21	18.6	90	525.8
22	19.8	100	760.0

have sufficient energy to overcome the intermolecular forces and escape from the liquid phase. Thus the vapor pressure increases with increasing temperature.

Figure 12.7 shows how the vapor pressures of pentane, tetrachloromethane, and water vary with temperature. At any given temperature pentane has a higher vapor pressure than tetrachloromethane, which in turn has a higher vapor pressure than water. Thus the intermolecular forces are stronger in water than in tetrachloromethane, which in turn has stronger intermolecular forces than pentane.

BOILING POINT When the temperature of a liquid is raised to the point at which the vapor pressure is equal to the atmospheric pressure, bubbles of vapor form in the liquid, and it is said to boil. A bubble of vapor can only form and expand in a liquid when the pressure in the bubble is sufficient to push the liquid away against the atmospheric pressure acting on the liquid—that is, when the vapor pressure of the liquid is equal to the atmospheric pressure. The **normal boiling point** of a liquid is defined as the temperature at which the vapor pressure equals 1 atm (760 mm Hg or 101.3 kPa). The temperature of the liquid

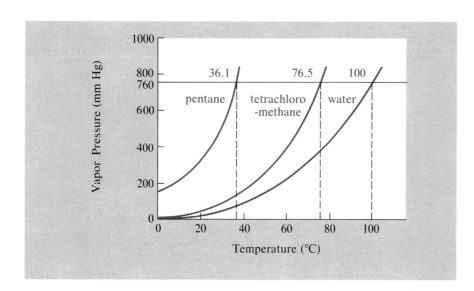

FIGURE 12.7
Variation of Vapor Pressure with Temperature.

The vapor pressure of a liquid increases with increasing temperature. The temperature at which the vapor pressure equals the pressure exerted by the atmosphere is the boiling point of the liquid.

A pressure cooker is used in the kitchen to decrease cooking time. A pressure cooker is a sealed container which allows steam to escape only when it reaches some predetermined pressure. The pressure inside the pressure cooker is the sum of the atmospheric pressure plus the pressure due to the steam. Consequently the water boils at a temperature above 100 °C at sea level. Because the process of cooking involves chemical reactions and reactions are faster at higher temperatures, the cooking time is decreased.

cannot be raised above the boiling point, because if more heat is supplied, it simply causes bubbles to be formed more rapidly.

Water boils at 100 °C at 1 atm pressure; therefore 100 °C is its normal boiling point. Tetrachloromethane, which has a higher vapor pressure, boils at 76.5 °C at 1 atm pressure, and pentane boils at 36.1 °C. If the atmospheric pressure is lower, as it is at higher altitude, the boiling point of a liquid is correspondingly reduced. In Boulder, Colorado, at an altitude of 5430 ft, the atmospheric pressure is 610 mm Hg and water boils at the temperature at which its vapor pressure reaches 610 mm Hg, that is, at 94 °C.

$=$ Example 12.1 $=$

BOILING POINTS AT REDUCED PRESSURE

Using Figure 12.7, estimate the boiling points of C_5H_{12}, CCl_4, and H_2O at a pressure of 400 mm Hg.

Solution

If we draw a horizontal line on Figure 12.7 at a pressure of 400 mm Hg, we can read off the temperature at which this line cuts each vapor pressure curve. Thus we find these boiling points:

C_5H_{12}, 25 °C CCl_4, 60 °C H_2O, 80 °C

EXERCISE 12.2

At what pressure will (a) tetrachloromethane and (b) water boil at a temperature of 60 °C?

VAPOR PRESSURE AND PARTIAL PRESSURE When a liquid evaporates in the presence of other gases, such as air, the total pressure of the gas phase is the sum of several partial pressures: the pressure caused by the vapor molecules and the pressure caused, for example, by the N_2 and O_2 of the air. The pressure exerted by the water vapor is of importance when a gas is collected over water, as shown in Figure 12.8. The total pressure is the pressure of the gas plus the

FIGURE 12.8
Collecting an Insoluble Gas over Water.

(a) The inverted cylinder is filled completely with water. (b) The gas displaces the water in the cylinder. (c) The height of the cylinder is adjusted so that the water levels inside and outside the cylinder are equal. The total pressure in the cylinder, which is the sum of the partial pressure of the gas that has been collected, p_{gas}, and the vapor pressure of water, p_{H_2O}, at the temperature of the water, is then equal to the atmospheric pressure. Thus $p_{gas} = p_{atm} - p_{H_2O}$.

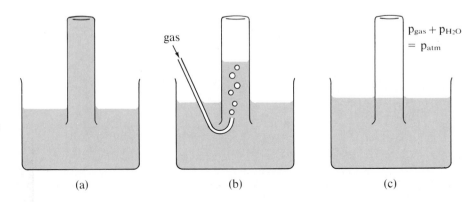

(a)   (b)   (c)

vapor pressure of water at the particular temperature. Thus to find the pressure of the gas, we must subtract the vapor pressure of water from the total pressure, as shown in Example 12.2.

═ Example 12.2 ═

CORRECTING FOR THE VAPOR PRESSURE OF WATER IN A GAS COLLECTED OVER WATER

When 2.050 g of a mixture of magnesium metal and magnesium oxide were treated with an excess of dilute hydrochloric acid, 510 mL of hydrogen were collected over water at 20.0 °C and a pressure of 742 mm Hg. How many moles of H_2 were produced? What was the percentage by mass of the Mg metal in the mixture?

Solution

The total pressure of 742 mm Hg is equal to the sum of the vapor pressure of water and the pressure of hydrogen:

$$P_{total} = p_{H_2} + p_{H_2O}$$

From Table 12.4 the vapor pressure of water at 20.0 °C is 17.5 mm Hg. Therefore

$$p_{H_2} = P_{total} - p_{H_2O} = 742 - 17.5 = 724 \text{ mm Hg}$$

$$n_{H_2} = \frac{p_{H_2} V}{RT} = \frac{(724/760) \text{ atm} \times 0.510 \text{ L}}{0.0821 \text{ L atm mol}^{-1} \text{ K}^{-1} \times 293 \text{ K}} = 0.0202 \text{ mol } H_2$$

Magnesium reacts with dilute HCl according to the equation

$$\text{Mg(s)} + 2\text{HCl(aq)} \longrightarrow \text{MgCl}_2\text{(aq)} + H_2\text{(g)}$$

Magnesium oxide reacts with dilute HCl according to the equation

$$\text{MgO(s)} + 2\text{HCl(aq)} \longrightarrow \text{MgCl}_2\text{(aq)} + H_2O\text{(l)}$$

so only the reaction with the Mg forms H_2. Hence 1 mol of Mg gives 1 mol of H_2, and 0.0202 mol of H_2 are produced from 0.0202 mol of Mg. Therefore the mass of Mg in the mixture is

$$(0.0202 \text{ mol Mg}) \left(\frac{24.31 \text{ g Mg}}{1 \text{ mol Mg}} \right) = 0.491 \text{ g Mg}$$

$$\text{Mass percent Mg in mixture} = \frac{0.491}{2.050} \times 100 = 24.0\%$$

EXERCISE 12.3

A 10.0-L sample of air saturated with water vapor at 25 °C and 1.00 atm pressure was bubbled through concentrated sulfuric acid to remove the water. The increase in the mass of the sulfuric acid was found to be 0.2307 g. What is the vapor pressure of water at 25 °C? What volume of dry air would have been collected at 25 °C and 1.00 atm pressure?

VAPOR PRESSURE OF SOLIDS A solid also exerts a vapor pressure, although this pressure is often very small because very few molecules in a solid normally have enough energy to break loose from the solid to form a vapor. The process by which a solid is converted directly to vapor is called **sublimation**. If a solid has a relatively high vapor pressure, sublimation is a convenient procedure for its purification. For example, we saw in Experiment 5.1 that iodine is easily sublimed; it can therefore be conveniently purified in this way. The vapor pressure of solid carbon dioxide (dry ice) is also rather high; it is equal to 1 atm at $-78\ °C$. In other words, at this temperature the solid is transformed directly to the vapor without melting. Liquid carbon dioxide can be obtained only under an external pressure of at least 5.2 atm; at this pressure carbon dioxide melts at $-57\ °C$.

12.3
Intermolecular Forces

The fact that all gases condense to liquids at a sufficiently low temperature shows that there must be attractive forces between molecules that are strong enough to hold the molecules together in the liquid and solid states. These forces are called **intermolecular forces**. We have mentioned these forces many times in previous chapters, and in this chapter we have seen that the relative magnitudes of these forces can be estimated from melting and boiling points and enthalpies of fusion and vaporization. We now consider the origin of these forces in more detail.

Although it is convenient to distinguish several different types of intermolecular forces, they are all electrostatic. In fact, all the different types of interactions among atoms, molecules, and ions, whether we describe them as covalent bonds, ionic bonds, metallic bonds, or intermolecular forces, result from electrostatic attractions and repulsions between positive nuclei and negative electrons.

DIPOLE–DIPOLE FORCES

We saw in Chapter 9 that many molecules containing polar bonds behave as electric dipoles, that is, as if they have a center of positive charge that is separated from the center of negative charge. Such molecules are called polar molecules and they attract each other because of the interaction of their dipoles. When two dipoles are arranged head to tail, they attract each other, because the sum of the attractive forces between opposite charges is greater than the sum of the repulsive forces between like charges. If two dipoles are arranged head to head or tail to tail, they repel each other. Since the head-to-tail arrangements have a lower energy than the head-to-head arrangements, the head-to-tail arrangements predominate in any collection of polar molecules. Therefore there is an overall attraction between such molecules (see Figure 12.9).

All polar molecules therefore attract each other by dipole–dipole interactions. Dipole–dipole attractions are relatively weak compared with ion–ion attractions, because the charges on polar molecules are generally quite small. Moreover, this attractive force decreases extremely rapidly with increasing distance; it is proportional to $1/r^7$, whereas the force between two charges is proportional to $1/r^2$ (Coulomb's law). We can understand this because when two dipoles are more than a very short distance apart, the distance between opposite charges is not significantly shorter than the distance between like charges, so that the resultant attractive force between them is almost zero; it is only significant when the two dipoles are close together (see Figure 12.9).

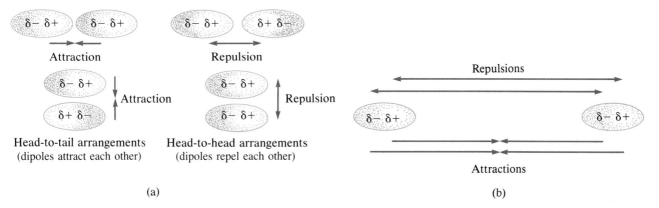

FIGURE 12.9
Intermolecular Forces: Dipole–Dipole Forces.

(a) Two polar molecules can have many different relative arrangements. When they have a head-to-tail arrangement, their dipoles attract each other. When they have a head-to-head arrangement, the two dipoles repel each other. The head-to-tail arrangements have a lower energy, and are therefore more common, than the head-to-head arrangements. So there is an overall attraction between the polar molecules. (b) When two dipoles are not very close together the forces of attraction are almost exactly balanced by the forces of repulsion so that the resultant attraction is very small. Thus the force of attraction between two dipoles decreases much more rapidly with increasing distance than the force of attraction between two opposite charges.

EXERCISE 12.4

In which of the following liquids would you expect that there will be dipole–dipole forces between the molecules?

(a) H_2O (b) CCl_4 (c) CH_2Cl_2 (d) BCl_3 (e) SO_2

LONDON (DISPERSION) FORCES

There must be forces of attraction even between the nonpolar molecules of gases such as O_2, N_2, and H_2 since these gases condense to liquids at sufficiently low temperatures. In 1926 German physicist Fritz London (1900–1954) proposed an explanation for these forces. They are therefore often called **London forces**, but they are also known as **dispersion forces**. London postulated that they result from the fact that even a nonpolar molecule or an atom may have an *instantaneous* dipole moment. The electrons in a molecule are in constant motion, and their *average* distribution is such that in a nonpolar molecule the center of negative charge of the electrons coincides with the center of positive charge of the nuclei. But at any one instant the centers of positive and negative charges do not, in general, coincide, and the molecule will have an **instantaneous dipole moment**, which is constantly changing in magnitude and direction with the movement of electrons in the molecule (see Figure 12.10). This fluctuating dipole attracts and repels the electrons in an adjacent molecule; in other words, it induces a fluctuating dipole in this second molecule which tends to fluctuate in phase with the dipole in the first molecule. Thus two adjacent molecules have

FIGURE 12.10
Intermolecular Forces: London (Dispersion) Forces.

(a) The movement of the electrons in a nonpolar molecule gives rise to a small instantaneous dipole, which continually changes in magnitude and direction. (b) This fluctuating dipole can induce a dipole in a neighboring molecule. This induced dipole changes in magnitude and direction following the changes in the fluctuating dipole in the first molecule. Thus there is always an attraction between the two molecules.

a head-to-tail arrangement of their fluctuating dipoles more often than a head-to-head arrangement. Because of this interaction between fluctuating dipoles on adjacent molecules, even nonpolar molecules attract each other.

London showed that the force of attraction between two nonpolar molecules is inversely proportional to the seventh power of the distance and proportional to a property of each molecule called the *polarizability*, α. Thus for two molecules A and B

$$F \propto \frac{\alpha_A \alpha_B}{r^7}$$

We see that London forces fall off very rapidly with distance, like dipole–dipole forces.

POLARIZABILITY   *The* **polarizability** *of an atom or a molecule is a measure of the ease with which the electrons and nuclei can be displaced from their average positions.* The more easily the electrons and nuclei can be displaced, the greater the polarizability. The greater the polarizability, the greater the magnitude of the instantaneous dipole and therefore the stronger the London forces. Since the mass of an electron is much less than the mass of any nucleus, electrons are displaced much more easily; they therefore make the largest contribution to the polarizability.

The electrons that are most easily displaced are the valence electrons since they are furthest from the nucleus and are held less strongly than the other electrons. Therefore they make the greatest contribution to the polarizability, and to a first approximation we may consider that the polarizability is due to just these electrons. The force acting on the valence electrons depends on their distance from the nucleus and on the core charge. In any group of the periodic table the core charge is constant, so we expect polarizability to increase as atomic size increases from the top to the bottom of the group. For example, we see from the polarizabilities given in Table 12.5 that the polarizability of the hydrogen halides increases from HF to HI. The units of polarizability are the units of volume, cubic meters, m^3.

In a molecule a dipole is induced in each atom, and the total induced-dipole moment is the resultant of all these small dipoles. Hence large molecules with many atoms have larger polarizabilities than small molecules. The polarizability of a molecule increases with both increasing size and increasing number of the atoms in the molecule. So we expect the magnitude of the fluctuating dipoles, and therefore the strength of the London forces, to be greater the greater the number of atoms in a molecule and the larger the atoms. We see from the values in Table 12.5 that, as we expect, the polarizability of nitrogen, N_2, is greater than that of hydrogen, H_2, the polarizability of carbon tetrachloride is greater than that of methane, and that of carbon dioxide is greater than that of carbon monoxide. These differences in the polarizabilities and therefore in the strength of the intermolecular (London) forces in these substances is reflected in their boiling points, as we can see from the data in Table 12.5. The boiling point of N_2 (-196 °C) is higher than that of H_2 (-253 °C), the boiling point of CCl_4 is higher than that of CH_4, and the boiling point of CO_2 is higher than that of CO.

Because of its smaller size, fluorine has a much smaller polarizability than chlorine. Molecular fluorides, therefore, usually have much lower boiling points than the corresponding chlorides. For example, the boiling point of CCl_4 is $+77$ °C, whereas that of CF_4 is -128 °C. In fact, the boiling points of fluorides are generally not much higher than those of the corresponding hydrides; the boiling point of methane is -161 °C. Although fluorine has seven valence electrons whereas hydrogen has only one, the fluorine electrons are held very tightly by the high core charge, and so the polarizability of fluorine is not much greater than that of hydrogen.

The boiling point increases in the series HCl (-85 °C), HBr (-67 °C), HI (-35 °C), despite the fact that the dipole moment decreases from HCl to HI and therefore the strength of the dipole–dipole forces decreases correspondingly. The increase in the boiling point is due to the increasing polarizability of the molecules in the series from HCl to HI (Table 12.5). We see that, particularly for larger molecules, the London forces make an important contribution to the overall intermolecular forces, even for molecules that have fairly large dipole moments.

We have seen that there are two important contributions to the intermolecular forces between covalent molecules: (1) dipole–dipole forces which are only

	TABLE 12.5 Dipole Moments and Polarizabilities		
Substance	Dipole Moment μ (10^{-30} C m)	Polarizability α (10^{-31} m^3)	Boiling Point (°C)
He	0	2.0	-268.9
Ar	0	16.6	-185.9
H_2	0	8.2	-252.9
N_2	0	17.7	-195.8
CO	0.33	19.8	-190
CO_2	0	26.3	-78
HF	6.37	5.1	19
HCl	3.60	26.3	-85
HBr	2.67	30.1	-67
HI	1.40	54.5	-35
H_2O	6.17	14.8	100
NH_3	4.90	22.2	-34.4
CCl_4	0	105	76.5
CH_4	0	26.0	-161.5
SO_2	5.42	43.4	-10

present if the molecules are polar, that is, if they have a dipole moment; (2) London (dispersion) forces which act between all molecules and are particularly important for large molecules. London forces are very often stronger than the dipole–dipole forces between polar molecules. In fact it is only rarely that dipole–dipole forces are the dominating influence on properties such as the boiling point, except for the special type of very strong dipole–dipole force known as a hydrogen bond that we will consider in the next section. As we can see in Table 12.5 the boiling points of H_2O and HF are quite out of line with those of other substances with similar dipole moments and polarizabilities. Obviously there must be very strong forces acting between H_2O molecules and between HF molecules. As we will see in Section 12.5 these strong forces are the special type of dipole–dipole forces called hydrogen bonds.

= *Example 12.3* ══

INTERMOLECULAR FORCES

Which of the following substances is most likely to exist as a gas at room temperature and normal atmospheric pressure: PCl_3, Cl_2, $MgCl_2$, or Br_2? Why?

Solution

The weakest intermolecular forces will be found in the substance with the smallest nonpolar molecules. Of the substances listed, $MgCl_2$ is ionic and PCl_3 consists of polar molecules. Both Cl_2 and Br_2 consist of nonpolar molecules, but since Cl_2 molecules are smaller than Br_2 molecules, they will be less polarizable and therefore have the smaller intermolecular forces. Thus Cl_2 is the substance most likely to be a gas at room temperature and normal atmospheric pressure. Even if the dipole–dipole forces in PCl_3 were small, we note that it is a larger molecule than Cl_2 (three Cl atoms and one P atom rather than two Cl atoms), so we expect the intermolecular forces in PCl_3 to be larger.

EXERCISE 12.5

By considering the relative magnitudes of the intermolecular forces expected for the following substances, place them in order of increasing boiling point: C_2H_5Cl, $AsCl_3$, SCl_2, CH_4, C_2H_6.

= *Example 12.4* ══

INTERMOLECULAR FORCES

Explain why CH_3F (bp $-78\,°C$) has a much lower boiling point than CCl_4 (bp $76.5\,°C$) even though CH_3F is a polar molecule and CCl_4 is nonpolar.

Solution

CCl_4 has four large polarizable chlorine atoms and therefore the London forces in CCl_4 are stronger than the combined effect of the much weaker London forces and the dipole–dipole forces in CH_3F.

NONIDEAL BEHAVIOR OF GASES AND VAN DER WAALS FORCES

The existence of attractive forces between molecules is one important reason why real gases do not obey the ideal gas equation exactly (Chapter 3). The greater the intermolecular forces between the molecules of a gas, the more the gas deviates from the ideal gas law. Gases with very small intermolecular forces obey the ideal gas law most closely; small molecules with no dipole moment, such as He, Ne, H_2, and N_2, are examples. Larger molecules, and particularly those with dipole moments, such as SO_2 and HI, show greater deviations from the ideal gas law.

Dutch physicist Johannes van der Waals (1837–1923) was the first to study deviations from the ideal gas law in some detail. Hence intermolecular forces are frequently called **van der Waals forces**.

VAN DER WAALS RADII

Although all molecules attract each other by intermolecular (van der Waals) forces, when two molecules come very close together their electron clouds begin to overlap and repel each other strongly, so that the attraction between the molecules is opposed by a strong repulsion which increases rapidly as the molecules come closer together. When the molecules are at some particular distance apart the repulsive force equals the attractive force and the total energy of the two molecules is a minimum (Figure 12.11). The two molecules may then be regarded as "touching" each other (see Figure 12.12). By measuring the distances

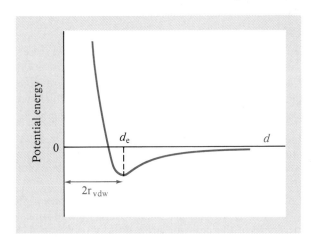

FIGURE 12.11
Potential Energy of Interaction of Two Nonbonded Atoms.

At large separations d, two atoms of the same element that are not able to form a bond because they are already part of a molecule or are noble gas atoms, attract each other as a consequence of the intermolecular forces between them. So the potential energy of the pair of atoms decreases slowly as the distance d decreases. But as their electron clouds begin to overlap and to repel each other strongly the potential energy increases sharply if they are brought still closer together. So the potential energy has a minimum value at some particular distance d_e between their nuclei. The van der Waals radius is taken as one-half the distance d_e when the two atoms are at their equilibrium separation, $r_{vdW} = \frac{1}{2}d_e$.

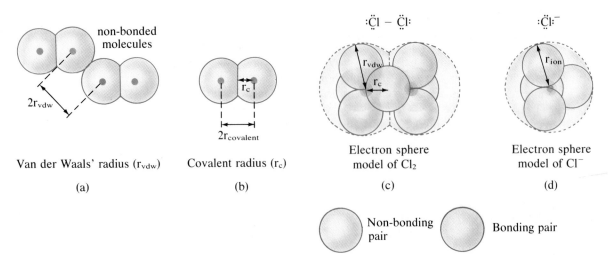

FIGURE 12.12
Van der Waals Radii, Covalent Radii, and Ionic Radii.

(a) The van der Waals radius, r_{vdW}, is half the distance between two similar atoms in separate molecules in a solid. The molecules are close-packed in the solid but are not bonded together. (b) The covalent radius, r_c, is half the distance between two similar atoms joined by a covalent bond in the same molecule. (c) According to the electron sphere model of Cl_2, r_c is equal to the *radius* of the sphere occupied by the bonding electron pair, but r_{vdW} is equal to the *diameter* of the sphere occupied by a nonbonding pair, which shows why $r_{vdW} \approx 2r_c$ for the same atom. (d) In the electron sphere model of Cl^- the radius of the ion, r_{ion}, is also equal to the *diameter* of an electron pair. So for a monatomic anion $r_{ion} \approx r_{vdW}$ for the same atom.

between molecules that are touching each other in the solid state, we can obtain a set of radii, called **van der Waals radii**, that *are a measure of the size of an atom in a molecule, in those directions in which it is not forming a bond.*

For example, from the structure of solid chlorine which we described in Chapter 11 (see Figure 11.21), we can measure two distances:

1. The distance between the chlorine nuclei in the *same* molecule. One-half this distance gives the *covalent radius* of the chlorine atom, as discussed in Chapter 4 (Figure 4.9). The covalent radius is a measure of the size of the chlorine atom when it is forming a bond.

2. The distance between the chlorine nuclei in two *adjacent* molecules. One-half this distance is the *van der Waals radius* of chlorine. This radius is the effective size of the chlorine atom in any direction in which it is *not* forming a bond (see Figure 12.12).

Van der Waals radii are considerably larger than covalent radii. For most atoms van der Waals radii are approximately twice the covalent radii (see Figure 12.12) and are approximately equal to the radii of the corresponding anions (Table 11.5).

Some van der Waals radii for atoms are listed in Table 12.6. Like covalent radii, van der Waals radii are approximate and are not strictly constant from one molecule to another. They depend somewhat on how the molecules are packed together in the solid state.

TABLE 12.6
Van der Waals Radii of Atoms (pm)

H	110	N	150	O	140	F	135
		P	190	S	185	Cl	180
		As	200	Se	200	Br	195
		Sb	220	Te	220	I	215

12.4
Some Unusual Properties of Water

Although water is the most common liquid on the earth, it is also one of the most unusual. It has several properties that at first sight are unexpected and unlike those of most other liquids.

One unusual property is that the density of solid water (ice) is less than that of liquid water at the melting point. For almost all other substances the solid is more dense than the liquid. As a liquid is cooled, the kinetic energy of the molecules decreases, their movements become more restricted, and they pack more closely together. This contraction and increase in density continues until the liquid freezes to a solid, at which point the density increases abruptly, because the molecules are packed still more closely together. However, water behaves in a different manner when it is cooled. Its density increases in the normal way until a temperature of 3.98 °C is reached, when its density is 1.000 00 g cm^{-3}. Then the density *decreases* very slightly with decreasing temperature to 0.00 °C, when it has a density of 0.999 87 g cm^{-3}. Then when water freezes, its density again *decreases* abruptly, to 0.917 g cm^{-3}, rather than increasing (Figure 12.13).

This unusual variation in the density of water with temperature makes aquatic life possible in cold climates. When the temperature of the atmosphere falls to near or below 0 °C, the water on the surface of lakes, ponds, and rivers cools first. Since it is then more dense, it sinks to the bottom, replacing warmer

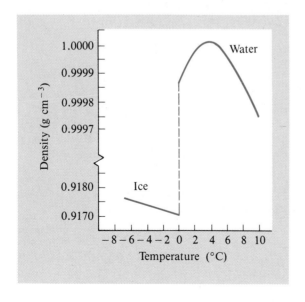

FIGURE 12.13
Temperature Dependence of the Densities of Ice and Liquid Water.

water, which in its turn is cooled and sinks to the bottom. But when all the water has been cooled to the temperature of maximum density (3.98 °C), any further cooling produces a surface layer, which since it is less dense, remains on top, eventually freezing. Because ice is considerably less dense than water, the ice remains on the surface. If ice were more dense than water, it would sink to the bottom of a lake, and eventually the whole lake would freeze. Clearly, fish could not survive under such conditions. In many parts of the world the ice at the bottom of a lake would not melt in the summer, and life on the lake bed would not be possible.

We are used to the fact that water freezes at 0 °C and boils at 100 °C, and so at first these properties do not seem unusual. But if we compare the boiling point and melting point of water with those of similar substances, we find that both are far higher than expected. For example, the boiling points of the hydrides of the elements of group 6 decrease from H_2Te to H_2S as the size of the molecules decreases, and therefore their polarizability decreases, as Figure 12.14 illustrates. Extrapolation of these boiling points would lead us to expect that water would have a boiling point of approximately -100 °C, that is, about 200 °C lower than the observed boiling point.

In contrast, the boiling points of the hydrides of the elements of group 4, namely SnH_4, GeH_4, SiH_4, and CH_4, which are nonpolar, decrease in the expected manner with decreasing size and the consequent decrease in the strength of the London forces. The boiling points of the hydrides of group 6 are somewhat higher than those of the hydrides of group 4 because the group 6 hydrides are polar molecules. Therefore in addition to the London forces there are also dipole–dipole forces.

But the dipole moment of water (6.17×10^{-30} C m) is quite inadequate to account for its exceptionally high boiling point. The dipole moment of water is only slightly larger than that of SO_2 ($\mu = 5.42 \times 10^{-30}$ C m), which also has a

FIGURE 12.14
Boiling Points of the Hydrides of Groups 4, 5, 6 and 7.

The boiling points of the nonpolar group 4 hydrides decrease from SnH_4 with decreasing molecular size and, therefore, decreasing intermolecular (London) forces. Group 6 hydrides have higher boiling points than hydrides of group 4 because they are polar molecules, and therefore there are additional dipole–dipole attractions. Nevertheless, London forces make the largest contribution to the overall intermolecular forces. Therefore the boiling points decrease from H_2Te to H_2S with decreasing molecular size. But the boiling point of H_2O is very much higher than expected. This high boiling point is attributed to a strong additional intermolecular force called the *hydrogen bond*, which is not present in the group 4 hydrides and is much weaker in H_2S, H_2Se, and H_2Te. Ammonia in group 5 and HF in group 7 also have unexpectedly high boiling points, for the same reason.

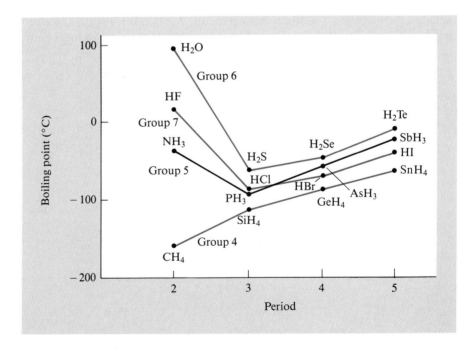

considerably greater polarizability than that of water (see Table 12.5), yet SO_2 has a boiling point of only $-10\,°C$. The exceptionally high boiling point of water must be due to an additional stronger force between H_2O molecules that is not present between SO_2 molecules nor between H_2S, H_2Se, or H_2Te molecules or the hydrides of group 4. This strong force is called the *hydrogen bond*.

12.5
The Hydrogen Bond

The hydrogen bond is found not only in water but rather generally in substances that have N—H, O—H, and F—H bonds. For example, although the boiling points of hydrogen fluoride and ammonia are not as high as that of water they are much higher than expected from the extrapolation of the boiling points of the other hydrides in the same groups (see Figure 12.14).

What, then, is a hydrogen bond? Like all other attractive forces between atoms and molecules, it is electrostatic in origin. Because the water molecule is polar, water molecules attract each other by dipole–dipole forces. But the comparison of the dipole moments of SO_2 and H_2O seems to show that dipole–dipole forces cannot account for the very strong attractions between water molecules. However, we must take into account the unique character of the hydrogen atom, which consists only of a proton and an electron and which has no inner shell of nonbonding electrons. When an H atom is bonded to a very electronegative atom, a significant amount of electron density is removed from the H atom. Moreover there are no inner-shell electrons to shield the nucleus (proton) of the hydrogen atom, so it can get very close to the negative end of a dipole and the electrostatic attraction between them is therefore unusually strong (see Figure 12.15). Indeed, the H atom of a water molecule gets so close to the O atom of another water molecule that it interacts almost exclusively with one particular unshared pair of the O atom. The interaction of a positively charged H atom with an unshared electron pair is strongest when the unshared electron pair is rather small and localized, as it is in the N, O, and F atoms. The larger and more diffuse unshared pairs on longer atoms such as Cl, S, and P do not interact as strongly with positively charged H atoms.

We see therefore that a hydrogen bond is a dipole–dipole force that is exceptionally strong because the proton of a positively charged H atom that is attached to an electronegative atom can get very close to an unshared pair of electrons in another molecule. The proton is particularly strongly attracted by this unshared pair of electrons if it is a small localized pair such as is found on the small electronegative N, O, and F atoms. A **hydrogen bond** *may be defined as an intermolecular attraction in which a hydrogen atom that is bonded to an electronegative atom is attracted to an unshared electron pair on another small electronegative atom.*

By far the most common and strongest hydrogen bonds are therefore the following combinations, in which the hydrogen bond is depicted by a dashed line:

$$
\begin{array}{lll}
\text{F—H---:F} & \text{F—H---:O} & \text{F—H---:N} \\
\text{O—H---:F} & \text{O—H---:O} & \text{O—H---:N} \\
\text{N—H---:F} & \text{N—H---:O} & \text{N—H---:N}
\end{array}
$$

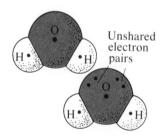

Unshared electron pairs

FIGURE 12.15
The Hydrogen Bond between Two Water Molecules.

In the water molecule the very small positive core of the hydrogen atom (the proton) is surrounded by very little electron density—a total of less than one electron. Therefore the proton can get very close to an unshared electron pair of the oxygen atom of another H_2O molecule and so it is strongly attracted by this unshared pair, thus forming a hydrogen bond.

The hydrogen bond is intermediate in strength between other intermolecular forces and the much stronger covalent and ionic bonds. The energies of most hydrogen bonds range from about 5 to 25 kJ mol^{-1}, whereas the energies of most covalent bonds range from 100 to over 500 kJ mol^{-1} (Chapter 6).

The relative weakness of the hydrogen bond compared with other bonds is also seen in its length. For example, Figure 12.16 shows that in ice the H---O hydrogen bond has a length of 177 pm, whereas the O—H covalent bond has a length of 99 pm. Although the O---H hydrogen bond is longer than the O—H covalent bond, it is nevertheless much shorter than the distance of 250 pm that we would expect between an H atom and an O atom if there were not a very strong attraction between them. This distance between two atoms (H and O) that are *not* bonded together is obtained by adding the van der Waals radii of H and O given in Table 12.6: $r_{vdw}(O) + r_{vdw}(H) = 140$ pm $+ 110$ pm $= 250$ pm.

The hydrogen bond can be regarded as either a very strong intermolecular force or a very weak chemical bond. Its name implies that it is more often regarded as a weak bond than an intermolecular force. It resembles a covalent bond in that it is localized between two particular atoms. In contrast, intermolecular forces normally act between molecules as a whole rather than between specific atoms.

If we think of the hydrogen bond as a chemical bond, we might ask whether it is covalent or ionic. As is often the case, the answer to this question is not clear-cut. Since the valence shell of the hydrogen atom can contain only two electrons, we normally consider that the hydrogen atom can form only one covalent bond. However, when hydrogen is bonded to a very electronegative element, some of its electron density is pulled away by the electronegative atom, leaving the hydrogen deficient in electron density. It can therefore accept some electron density from the unshared electron pair to which it is attracted. Thus in some cases the hydrogen atom and the atom with the unshared pair share some electron density. In these cases the hydrogen bond can be considered to have some covalent character.

THE STRUCTURE OF ICE

The strong hydrogen bonding between water molecules accounts not only for the exceptionally high boiling and melting points of water but also for the structure and low density of ice and the expansion of water below 4 °C. Ice has a network structure similar to that of diamond; each oxygen atom is surrounded by four other oxygen atoms in a tetrahedral arrangement (see Figure 12.16). Each oxygen has two hydrogen atoms covalently bound to it, and it has two unshared pairs of electrons, each of which forms a hydrogen bond with the hydrogen atoms of neighboring water molecules.

Each water molecule could have up to 12 nearest neighbors if they were packed closely together. But the number of neighbors of any water molecule is limited to 4 because each oxygen atom can form only two covalent O—H bonds and two O---H hydrogen bonds. As a result, the structure is an open one in which there is room for more molecules.

When ice melts, the regular structure collapses to some extent because some of the hydrogen bonds are broken, and consequently, some of the water molecules pack more closely together. Hence the density increases upon melting. However, only about 15% of the hydrogen bonds are broken when ice melts.

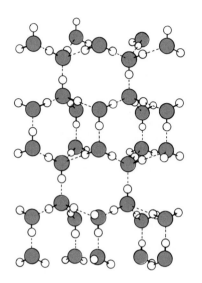

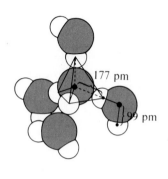

177 pm

99 pm

FIGURE 12.16
The Structure of Ice.

Each water molecule is surrounded by a tetrahedral arrangement of four other water molecules to which it is bound by hydrogen bonds. Each oxygen atom forms covalent bonds to two hydrogen atoms and two hydrogen bonds with hydrogen atoms of neighboring water molecules. The structure is a three-dimensional network rather similar to the structures of diamond and silica. It is a very open structure. When ice melts and some of the hydrogen bonds are broken, the molecules pack more closely together and so the density of water is greater than that of ice.

Thus in liquid water at 0 °C many water molecules are still hydrogen-bonded in groups that retain more or less the same structure they have in ice. With increasing temperature more hydrogen bonds are broken and more molecules pack more closely together, so the density continues to increase. Normally as a liquid is heated, the increased motion of the molecules causes them to take up more space. At 3.98 °C this increased motion becomes more important than the continued collapse of the structure. So above 3.98 °C the density of water begins to decrease in the normal way.

Hydrogen bonds are of widespread importance. For example, they hold long-chain protein molecules in certain definite shapes; the shape of a protein molecule plays an important role in its function. As we will see in the next section, hydrogen bonds also play an important role in determining the solvent properties of water.

WHY HF IS A WEAK ACID

In Chapter 8 we explained the fact that HF is a weak acid in water, whereas HCl is strong, in terms of the fact that the HF bond is stronger and more difficult to break than the HCl bond. However, hydrogen bonding is another important, and perhaps the most important, factor. The ions H_3O^+ and F^- that are formed by the ionization of HF in water are held together by a strong hydrogen bond $F^- \text{---} H_3O^+$, and this hydrogen-bonded complex is incompletely dissociated to free H_3O^+ and F^- ions (Figure 12.17). Another complication is that HF also forms a hydrogen-bonded complex with F^- so that not all the

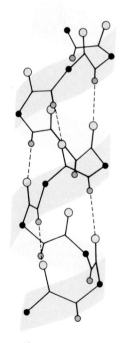

---Hydrogen bond

The spiral shape of a protein molecule

HF ionizes in water to form H_3O^+ and F^- but these two ions form a hydrogen-bonded complex so that the concentration of free H_3O^+ is less than the amount formed by complete ionization and therefore HF is a weak acid. HF also forms a hydrogen-bonded complex with F^- so that not all the HF reacts with water to form H_3O^+, which further reduces the equilibrium concentration of H_3O^+ and thus the acid strength of HF.

$$H-\overset{\cdot\cdot}{\underset{H}{O}}: + H-F \longrightarrow H-\overset{\oplus}{\underset{H}{O}}-H\text{-}\text{-}F^- \rightleftharpoons H-\overset{\oplus}{\underset{H}{O}}-H + F^-$$

$$F^- + H-F \rightleftharpoons F^-\text{-}\text{-}\text{-}H-F$$

HF reacts with water to form H_3O^+ (Figure 12.17). For both these reasons HF is a much weaker acid than HCl. The Cl^- ion forms much weaker hydrogen bonds than F^- and so there is no appreciable concentration of the hydrogen-bonded complexes $Cl\text{-}\text{-}\text{-}H_3O^+$ and $Cl^-\text{-}\text{-}\text{-}H-Cl$ to reduce the concentration of free H_3O^+.

12.6
Solutions in Water and Other Solvents

In the process of forming a solution, one substance mixes completely with another to form a homogeneous mixture. Two gases always form a gaseous mixture or solution. When two gases are placed in contact, the rapid, random motions of the molecules of each gas cause them to form a random, disordered mixture. However, when two liquids are brought into contact, they may or may not mix. Hydrocarbons are insoluble in water, whereas ethanol, C_2H_5OH, is soluble in water in all proportions.

What then determines whether one substance will dissolve in another or not? There are two important factors to be considered:

1. The tendency of all systems to move towards a state of disorder;
2. The relative strengths of the intermolecular forces in the solvent and the solute and between the solvent and the solute.

The first of these factors is illustrated by the fact that when two gases are placed in contact, because of the rapid random motion of their molecules, they always mix with each other, that is, they form a solution. The molecules in the mixture have a more disordered arrangement than when the two gases were separated. This tendency of systems to become more disordered is a one-way process. Left to itself the mixture of two gases will never separate into two pure gases. We would certainly be very surprised if the air in a room were to separate into oxygen at one end and nitrogen at the other! In principle this could occur but in practice the chance that it would occur is infinitesimally small. Although this example may seem rather obvious and trivial it illustrates a very important law of nature, namely that systems tend spontaneously toward a state of greater disorder. Figure 12.18 shows that a disordered state can be achieved by millions of different arrangements of the molecules in a system whereas the ordered state corresponds to only one unique arrangement of the molecules. Thus the probability of this ordered arrangement being reached spontaneously is extremely small. But, because a disordered arrangement can be obtained in millions of different ways, there is a very high probability of obtaining such an arrangement. We measure the disorder in a system by a property called **entropy**. We say that a mixture of gases has a higher entropy than the separate pure gases. We will discuss the concept of entropy in more detail in Chapter 16 but we will have occasion to mention it again in this chapter.

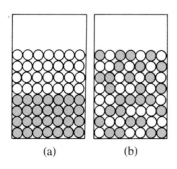

(a)   (b)

(a) Red and black spheres in an ordered arrangement in a closed container. (b) After shaking for some time the spheres are thoroughly mixed up and have a disordered arrangement. There are millions of such disordered arrangements but only one ordered arrangement. So a disordered arrangement is much more probable than an ordered arrangement and the chance of the ordered arrangement being formed on continuing to shake the container is negligible. The disordered arrangement has a higher entropy than the ordered arrangement.

Ethanol and water mix completely with each other just as two gases do. Ethanol and water form a solution: they are said to be **miscible** with each other. In contrast, hexane and water do not mix. Hexane is not soluble in water and water is not soluble in hexane. Hexane and water are **immiscible**. Since the random motions of the molecules, or in other words the tendency for the system to attain greater disorder or a higher entropy, would be expected to cause the liquids to mix with each other, something must be preventing the mixing of the hexane molecules with the water molecules. It is the hydrogen bonds between water molecules that prevent the mixing. If hexane molecules were to mix with water molecules, many of the hydrogen bonds between water molecules would have to be broken. They are not broken because there are only weak, London forces between the hexane molecules and the water molecules, and these forces are not strong enough to pull the water molecules apart. The strong attractions between water molecules due to hydrogen bonding thus prevent water molecules from mixing with hexane molecules.

Ethanol, C_2H_5OH, is soluble in water because the OH group can form hydrogen bonds. Thus hydrogen bonds between ethanol molecules and water molecules replace some of the hydrogen bonds between water molecules. The energy needed to break the hydrogen bonds between water molecules is compensated by the energy evolved in the formation of hydrogen bonds between ethanol and water molecules, and so ethanol dissolves in water. Hydrocarbons are also soluble in each other. They mix easily because there are only weak London forces of attraction between hydrocarbon molecules.

These examples illustrate an often quoted generalization: "**Like dissolves like.**" In other words, chemically similar substances are soluble in each other, whereas dissimilar substances are insoluble. Ethanol is like water in that both contain OH groups and are hydrogen-bonded. Different hydrocarbons are very similar to each other, and they are also soluble in each other. But water and a hydrocarbon are not at all similar; one is polar and the other is nonpolar, and they are not soluble in each other.

We can apply the same considerations to understand the solubility of **solids in liquids**. Nonpolar substances, such as I_2 and S_8, are soluble in a nonpolar liquid such as tetrachloromethane, CCl_4, or carbon disulfide, CS_2, but they are insoluble in water, which is polar (Experiment 1.6). Sugar contains many OH groups that can form hydrogen bonds with water. It is therefore very soluble in water, but it is insoluble in CCl_4.

Substances having **covalent network structures**—such as diamond, graphite, red phosphorus, silicon dioxide, and silicon carbide—are insoluble in water. They are insoluble in water because the strong covalent bonds holding the atoms together prevent them from separating to mix with the H_2O molecules.

Similarly **metals** are insoluble in water because the strong metallic bond prevents the metal atoms from separating to mix with the water molecules.

Many **ionic solids** such as MgO, AgCl, CuO, CaF_2, and $BaSO_4$ are also insoluble in water because the ionic bonds are too strong to allow the ions to separate and mix with the H_2O molecules.

However, many other ionic compounds, such as NaCl, KNO_3, $MgSO_4$, and $CuCl_2$, are soluble in water despite the strong forces holding the ions together. Why? There must be a relatively strong attractive force between the ions and the H_2O molecules so that the H_2O molecules can pull the ions away from the crystal. This force is the electrostatic attraction between an ion and a dipole. Positive ions attract the negative end of the H_2O dipole, and consequently,

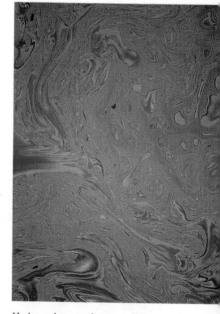

Hydrocarbons such as gasoline and oil are insoluble in water and have a lower density than water. Thus a small amount of oil forms a film on the surface of water. Interference between the light reflected from the top surface of the oil and that reflected from the interface between the oil and the water gives this colored diffraction pattern.

each positive ion becomes surrounded by H_2O molecules, as we described in Chapter 10. Similarly, negative ions attract the positive end of the H_2O dipole, and consequently, each negative ion also becomes surrounded by H_2O molecules (see Figure 12.19). The ions are said to be *hydrated*. In the case of the F^- ion and oxoanions such as SO_4^{2-}, NO_3^-, CO_3^{2-}, and PO_4^{3-}, the ion–dipole interaction is enhanced by hydrogen bonding between H_2O molecules and F^- ions or the O atoms of the oxoanions.

There are, therefore, two important factors determining solubility. One is the tendency of the random, thermal motions of atoms and molecules to cause them to become mixed up with each other, as always happens in the case of gases. The second is the relative strengths of the attractive forces between the solute particles, the attractive forces between the solvent particles, and the attractive forces between the solute and the solvent particles. The first factor is an example of the tendency of all systems to become more mixed up or disordered as a result of the thermal motions of atoms and molecules—in other words, for the *entropy* to increase. This tendency toward disorder is opposed by attractive forces between the atoms or molecules of the solvent or solute that may prevent them from mixing. The forces between the molecules in gases are always weak, so they mix easily. But the forces between the atoms and molecules in liquids and solids are often much stronger, so they may or may not mix easily, depending on the relative magnitudes of these forces.

As we shall see in more detail in Chapter 16, the direction in which a change occurs in any isolated system is determined by the tendency for the disorder or entropy to increase and for the energy of the system to decrease: the system tends toward maximum entropy and minimum energy. Thus there is a tendency for one substance to dissolve in another, for the molecules to mix up with each other, because of the resulting increase in the disorder or entropy of the system. But they will not do so if this causes the energy of the system to increase by too much, that is, if too much energy is needed to overcome the forces of attraction between the solvent or solute molecules. So let us look now at the energy changes involved in the process of forming a solution.

FIGURE 12.19
An Ionic Crystal
Dissolving in Water.

(a) A small crystal before it dissolves in the surrounding water. (b) Ions are pulled away from the crystal by the attraction of the polar water molecules. (c) Hydrated ions are formed. (d) The crystal has dissolved, giving a solution of hydrated positive and negative ions.

(a)　　　　　(b)　　　　　(c)　　　　　(d)

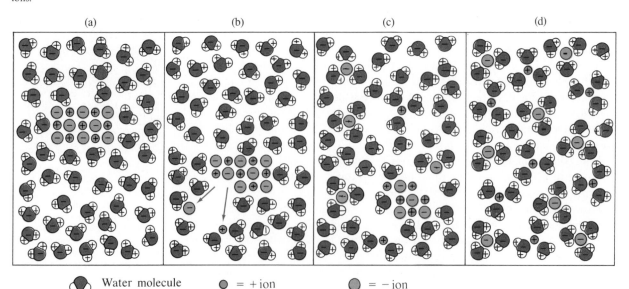

Water molecule　　● = + ion　　● = − ion

ENTHALPY OF SOLUTION

We can conveniently imagine the process of forming a solution of a liquid or a solid as taking place in three steps: (1) the solvent molecules are separated from each other, (2) the solute molecules are separated from each other, and (3) the solvent and solute molecules are allowed to mix with each other. Because there are forces of attraction between the solvent molecules, energy is required to separate them and so step 1 is endothermic. Similarly, step 2 is endothermic; energy is needed to separate the solute molecules. Because there are forces of attraction between the solvent and solute molecules, energy is liberated when they are attracted to each other, so step 3 is exothermic. The enthalpy change associated with the formation of the solution, the **molar enthalpy of solution** ΔH_{soln}, is the sum of the enthalpy changes for the three steps:

$$\Delta H_{soln} = \Delta H_1 + \Delta H_2 + \Delta H_3$$

Thus the enthalpy of solution may have a positive sign (energy absorbed) or negative sign (energy liberated), depending on the relative magnitudes and signs of ΔH_1, ΔH_2, and ΔH_3. In other words, the formation of a solution may be an exothermic or an endothermic process (Figure 12.20). In the case of hexane and water ΔH_1 for the separation of the water molecules is large and positive because hydrogen bonds must be broken. ΔH_2 is positive but small because only relatively weak London forces must be overcome to separate the hexane molecules. ΔH_3 is small and negative because there are only weak London forces of attraction between hexane and water molecules. Thus ΔH_{soln} is large and positive. To dissolve hexane in water would require a large input of energy and such processes tend not to occur. The tendency for mixing to occur, that is, for entropy to increase, is not sufficient in this case to cause hexane to dissolve in water.

As a second example, let us consider the solubility of an ionic solid in water (Figure 12.21). The term ΔH_2 is large and positive because a large amount of energy is needed to pull the ions apart. This is called the **lattice energy**. ΔH_1, the energy needed to pull the water molecules apart, is also large and positive

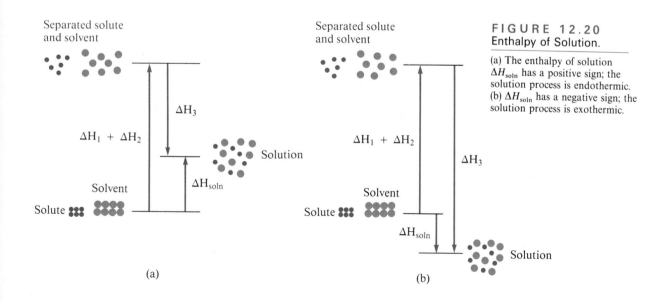

(a)

(b)

FIGURE 12.20
Enthalpy of Solution.

(a) The enthalpy of solution ΔH_{soln} has a positive sign; the solution process is endothermic. (b) ΔH_{soln} has a negative sign; the solution process is exothermic.

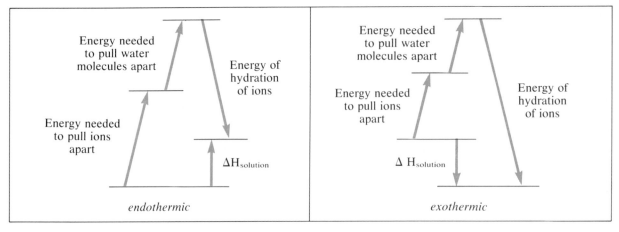

FIGURE 12.21
Enthalpy of Solution of an Ionic Solid in Water.

The enthalpy of solution may be positive or negative, depending on the relative magnitudes of three energy terms: the energy needed to pull the ions apart, the energy needed to pull the solvent molecules apart, and the energy of hydration of the ions.

because hydrogen bonds must be broken. Finally, ΔH_3 is large and negative because of the strong attractions between the ions and the polar water molecules. Thus the enthalpy of solution of an ionic salt may be positive or negative (see Figure 12.21) and Experiment 12.1). When the enthalpy of solution is positive (endothermic) a salt will be soluble if the enthalpy of solution is not too large and the entropy increase on forming the solutions is sufficiently large. In the case of sodium chloride the exothermic and endothermic terms very nearly cancel each other so that the resulting enthalpy of solution has only a very small value of $+3$ kJ mol^{-1}. Let us look at this case in a little more detail. The lattice energy of sodium chloride, ΔH_2, is known to be 786 kJ mol^{-1}:

$$NaCl(s) \longrightarrow Na^+(g) + Cl^-(g) \qquad \Delta H_2 = 786 \text{ kJ mol}^{-1}$$

The enthalpy change for dissolving the gaseous ions in water is called the **enthalpy of hydration** of the ions: it is the sum of ΔH_1 and ΔH_3 and it has been found to have the value -783 kJ mol^{-1}:

$$Na^+(g) + Cl^-(g) + H_2O(l) \longrightarrow Na^+(aq) + Cl^-(aq)$$

$$\Delta H_{\text{hyd}} = \Delta H_1 + \Delta H_3 = -783 \text{ kJ mol}^{-1}$$

Hence

$$\Delta H_{\text{soln}} = \Delta H_1 + \Delta H_2 + \Delta H_3 = 786 - 783 = 3 \text{ kJ mol}^{-1}$$

Although the enthalpy of solution of sodium chloride in water is positive (endothermic), sodium chloride is soluble in water because of the large entropy change associated with mixing sodium ions and chloride ions with water.

Table 12.7 summarizes the various possibilities for the formation of a solution for different types of solutes and solvents. For the cases of polar solvent with polar solute and nonpolar solvent with nonpolar solute, the enthalpy of solution is small and the solution forms because of the increase in disorder (entropy). For the cases of a nonpolar solvent with polar solute and polar solvent with nonpolar solute, no solution forms because the enthalpy of solution is large and

Experiment 12.1

Enthalpy of Solution

Ammonium nitrate dissolves in water endothermically. The beaker contains water at room temperature, and the dish contains ammonium nitrate; there is a small pool of water on the wooden block. The ammonium nitrate is added to the water and the beaker is placed on the wooden block.

So much heat is absorbed that the temperature of the solution decreases to below 0 °C as the ammonium nitrate dissolves and the water on the block freezes solid. The block can then be lifted with the beaker.

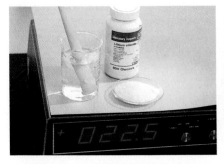

The beaker contains water at 22.5 °C and the dish contains solid lithium chloride.

In the experiment shown here, the temperature rose from 21 to 46 °C when lithium chloride was dissolved in the water in the beaker.

TABLE 12.7
Enthalpy of Solution

Solvent	Solute	ΔH_1	ΔH_2	ΔH_3	ΔH_{soln}	Solution Forms
Polar	Polar	+	+	−	Small	Yes
Polar	Nonpolar	+	Small	Small	+	No
Nonpolar	Nonpolar	Small	Small	Small	Small	Yes
Nonpolar	Polar	Small	+	Small	+	No

ΔH_1, separation of solvent molecules—endothermic.
ΔH_2, separation of solute molecules—endothermic.
ΔH_3, mixing of solvent and solute molecules—exothermic.
$\Delta H_{soln} = \Delta H_1 + \Delta H_2 + \Delta H_3$.

positive—the increase in the energy of the system that the solution process would involve prevents the mixing that would otherwise occur. These considerations help us to understand the rule that "like dissolves like".

─── *Example 12.5* ──────────────────────────────

ENTHALPY OF SOLUTION

Ammonium nitrate has a reasonably large positive enthalpy of solution ($\Delta H = 26.4$ kJ mol^{-1}). This property is exploited in cold packs, which are used to treat athletic injuries. Cold packs contain bags of ammonium nitrate and water; these bags are broken when the pack is kneaded, allowing the ammonium nitrate to dissolve in the water. What is the final temperature reached when 10.0 g of ammonium nitrate dissolve in 50.0 g of water, initially at room temperature (20 °C)?

Solution

$$\text{Heat absorbed} = (10.00 \text{ g})\left(\frac{1 \text{ mol}}{80.1 \text{ g}}\right)\left(\frac{26.4 \text{ kJ}}{1 \text{ mol}}\right) = 3.30 \text{ kJ}$$

We can calculate the temperature decrease caused by removing 3300 J by using the heat capacity of the solution which we will assume is the same as that of water, 4.2 J K^{-1} g^{-1} (Chapter 6).

The removal of 4.2 J from 1 g of water will decrease its temperature by 1 °C. So the removal of 3.30 kJ from 50 g of water will decrease its temperature by

$$\left(\frac{3.30 \text{ kJ}}{50 \text{ g}}\right)\left(\frac{1000 \text{ J}}{1 \text{ kJ}}\right)\left(\frac{1 \text{ g}}{4.2 \text{ J K}^{-1}}\right) = 16 \text{ K}$$

The temperature is lowered from 20 to 4 °C.

This calculation is only approximate since we have used the enthalpy of solution for a 1M solution. When a concentrated solution is formed, the enthalpy of solution is somewhat smaller. The approximate calculated value nevertheless demonstrates that a considerable cooling effect may be achieved by dissolving some salts in water.

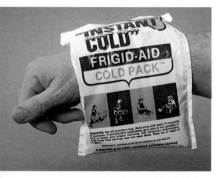

When the pouch of water inside this portable cold pack is broken, the solid ammonium nitrate dissolves in the water endothermically and the pack is cooled to below 0 °C.

12.7
Colligative Properties

Antifreeze is added to an automobile radiator to prevent water from freezing in the winter. Salt is placed on roads in the winter to prevent the formation of ice. A cucumber shrivels when placed in a concentrated salt solution. Red blood cells placed in water burst.

At first, there does not appear to be much connection between these varied facts, but they all depend on some related properties of solutions called colligative properties. **Colligative properties** *are properties of solutions that depend on the number of solute and solvent molecules (or ions), but not on the nature of the solute or of the solvent or on the nature of the forces between them.* These properties are the vapor pressure, the boiling point elevation, the freezing point depression, and the osmotic pressure.

VAPOR PRESSURE OF SOLUTIONS

Measurement of the vapor pressure of solutions shows that the vapor pressure of the solvent in a solution is less than that of the pure solvent. The French chemist François Raoult (1830–1901) showed that the vapor pressure of the solvent in a solution is independent of the nature of the solute and depends only on its concentration. The vapor pressure of the solvent is given by the expression

$$p_{solvent} = x_{solvent} p_{solvent}^{\circ}$$

where $p_{solvent}$ is the vapor pressure of the solvent in the solution, $p_{solvent}^{\circ}$ is the vapor pressure of the pure solvent, and $x_{solvent}$, which is called the **mole fraction** of the solvent, is given by the expression

$$x_{solvent} = \frac{\text{moles solvent}}{\text{moles solvent} + \text{moles solute}} \quad \text{Similarly,} \quad x_{solute} = \frac{\text{moles solute}}{\text{moles solvent} + \text{moles solute}}$$

The mole fraction has no units; it represents the fraction of the total number of molecules in a solution that are solvent molecules, or are solute molecules.

If the solute is nonvolatile, then the vapor pressure of the solution is due only to the solvent in the solution. Thus for a nonvolatile solute

$$p_{solution} = p_{solvent} = x_{solvent} p_{solvent}^{\circ}$$

An approximate explanation of why the vapor pressure of a solution is lower than that of the pure solvent is given in Figure 12.22. When the pure solvent is in equilibrium with its vapor, solvent molecules are leaving the surface of the solvent (evaporating) at exactly the same rate as they are returning to the liquid from the vapor by colliding with the surface (condensing). In the solution some of the surface of the liquid is occupied by solute molecules so the chance that a solvent molecule leaves the surface is reduced and the rate at which it leaves is correspondingly reduced. At equilibrium the rate at which the molecules are returning to the liquid from the vapor must also be decreased correspondingly and this means that the concentration of molecules in the vapor phase must be decreased; in other words the vapor pressure is decreased.

However, a better and more exact explanation can be given in terms of the disorder or entropy of the system. We have seen that a solute dissolves in a solvent because the random thermal motions of the solvent and the solute molecules cause them to mix, that is to become more disordered. Thus solutions have a higher entropy than the pure solvent. The molecules of a vapor have more random motion than those of a liquid, so a vapor has a higher entropy than the corresponding liquid. Thus the tendency of a liquid to evaporate is

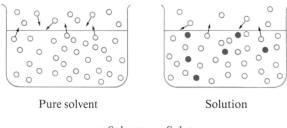

Pure solvent   Solution

○ Solvent ● Solute

FIGURE 12.22
Vapor Pressure of a Solution.

The vapor pressure of a solution is lower than that of the pure solvent. Fewer molecules escape from the surface of the solution than from the pure solvent in the same time interval.

another example of a system tending to increase its entropy. Because a solution has a higher entropy than a pure solvent, the difference in entropy between a solution and the vapor is less than that between the pure solvent and the vapor. Thus solvent molecules have a smaller tendency to leave a solution than to leave the pure solvent to become vapor. So the vapor pressure of a solution is less than that of the solvent.

⹀ Example 12.6

VAPOR PRESSURE OF A SOLUTION

What is the vapor pressure at 20 °C of a solution of 24.00 g of sucrose ($C_{12}H_{22}O_{11}$) in 150.0 g of water if the vapor pressure of pure water at 20 °C is 17.54 torr?

Solution

Raoult's law states that

$$p_{solution} = x_{H_2O}p^{\circ}_{H_2O}$$

In order to find the mole fraction of water, x_{H_2O}, we must calculate the number of moles of water and the number of moles of sucrose in the solution:

$$n_{sucrose} = \frac{24.00 \text{ g}}{342 \text{ g mol}^{-1}} = 0.0702 \text{ mol}$$

$$n_{H_2O} = \frac{150.0 \text{ g}}{18.02 \text{ g mol}^{-1}} = 8.324 \text{ mol}$$

$$x_{H_2O} = \frac{n_{H_2O}}{n_{H_2O} + n_{sucrose}} = \frac{8.324 \text{ mol}}{(8.324 + 0.0702) \text{ mol}} = 0.9916$$

$$p_{solution} = 0.9916 \times 17.54 \text{ torr} = 17.39 \text{ torr}$$

EXERCISE 12.6

An aqueous solution contains 10.0 mass % of urea, $(NH_2)_2CO$. If the vapor pressure of pure water is 55.3 mm Hg at 40 °C, what is the vapor pressure of the urea solution at the same temperature?

A direct consequence of the lowering of the vapor pressure of the solvent in a solution is that the boiling point of the solution is higher than that of the pure solvent.

ELEVATION OF THE BOILING POINT

Figure 12.23 shows that if the vapor pressure of a solution is lower than the vapor pressure of the pure solvent, then in order to increase the vapor pressure of the solution to 1 atm, we must increase its temperature to above the boiling point of the pure solvent. In other words, the boiling point of the solution is higher than that of the pure solvent.

The elevation of the boiling point is proportional to the number of moles

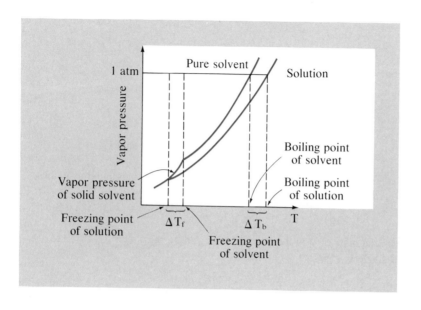

FIGURE 12.23
Elevation of Boiling Point and Depression of Freezing Point.

A solution has a lower vapor pressure than the pure solvent; it must be heated to a *higher* temperature before its vapor pressure reaches the atmospheric pressure. The boiling point of a solution is greater than that of the solvent by ΔT_b. At the freezing point the vapor pressure of the solvent in a solution is lower than that of the pure solvent. Therefore the vapor pressure of the solid in equilibrium with the solution must also be lower. The vapor pressure of the solid decreases with decreasing temperature. So the pure solid and a solution are in equilibrium at a *lower* temperature than the solid and pure solvent. The freezing point of a solution is lower than that of the pure solvent by ΔT_f.

of solute in a given amount of solvent. We can write

$$\Delta T = K_b m$$

where m is the number of moles of solute in 1 kg of solvent, called the **molality** of the solute, and K_b is the **boiling point elevation constant**. Values of K_b for some solvents are given in Table 12.8. Note that *molality* is not the same as *molarity. The **molality**, m, is the number of moles of solute in 1 kg of solvent*; the molarity, M, is the number of moles of solvent in *1 L of solution*.

The elevation of the boiling point of a solvent by a solute can be used to determine the molar mass of the solute.

≡ *Example 12.7*

MOLAR MASS FROM BOILING POINT ELEVATION

A sample of 1.20 g of a nonvolatile organic compound was dissolved in 60.0 g of benzene. The boiling point of the solution was found to be 80.96 °C. Pure benzene boils at 80.08 °C. What is the molar mass of the solute?

TABLE 12.8
Boiling Point Elevation Constants

Solvent	Boiling Point (°C)	K_b (°C kg mol^{-1})
Water	100	0.52
Ethanol	79	1.19
Acetic acid	118	2.93
Benzene	80	2.53
Cyclohexane	81	2.79
Carbon disulfide	46	2.37
Carbon tetrachloride	76.5	5.03

Solution

$$\Delta T = 80.96\ {}^{\circ}\text{C} - 80.08\ {}^{\circ}\text{C} = 0.88\ {}^{\circ}\text{C}$$

From Table 12.8 we have

$$K_b(\text{benzene}) = 2.53\ {}^{\circ}\text{C kg mol}^{-1}$$

Therefore

$$m = \frac{\Delta T}{K_b} = \frac{0.88\ {}^{\circ}\text{C}}{2.53\ {}^{\circ}\text{C kg mol}^{-1}} = 0.35\ \text{mol kg}^{-1}$$

The number of moles of solute in 60.0 g solvent is then found as follows:

$$\left[\frac{0.35\ \text{mol (solute)}}{1\ \text{kg (solvent)}}\right]\left(\frac{1\ \text{kg}}{1000\ \text{g}}\right)[60.0\ \text{g (solvent)}] = 0.021\ \text{mol (solute)}$$

Then the mass of 1 mol, the molar mass, can be calculated:

$$\text{Molar mass} = \frac{1.20\ \text{g}}{0.021\ \text{mol}} = 57\ \text{g mol}^{-1}$$

EXERCISE 12.7

What would be the boiling point of the solution in Exercise 12.6?

DEPRESSION OF THE FREEZING POINT

At the freezing point of a solvent the solid and the liquid are in equilibrium. They must therefore have the same vapor pressure. If the solid had a lower vapor pressure, all the liquid would be transformed into solid—in other words, they could not be in equilibrium. If the vapor pressure of the solid were higher than that of the liquid, all the solid would be transformed into liquid. So *at equilibrium the vapor pressure of the solid and liquid must be equal.*

The solid that separates from a solution is in most cases the pure solid solvent. Because the vapor pressure of a solution is less than that of the pure solvent, a solution can only be in equilibrium with the pure solid solvent if the vapor pressure of the solid solvent is decreased by decreasing its temperature. Thus a solution and the pure solid solvent are in equilibrium at a lower temperature than the pure liquid and the solid solvent. In other words, the freezing point of a solution is lower than that of the pure solvent as Figure 12.23 shows.

The relationship between the depression of the freezing point and the number of molecules of solute in 1 kg of solvent is similar to that for the boiling point elevation:

$$\Delta T = K_f m$$

Values for the **freezing point depression constant**, K_f, for some solvents are given in Table 12.9.

The measurement of the freezing point of a solution can be used, like the measurement of the boiling point, to determine the molar mass of a solute. If we know the empirical formula of a solute, we can then find its molecular formula, as the following example shows.

TABLE 12.9 Freezing Point Depression Constants		
Solvent	Freezing Point (°C)	K_f (°C kg mol^{-1})
Water	0.0	1.86
Acetic acid	16.7	3.90
Sulfuric acid	10.4	5.98
Benzene	5.5	5.10
Cyclohexane	6.5	20.2

= Example 12.8 =

MOLAR MASS FROM FREEZING POINT DEPRESSION

A solution of 2.95 g of sulfur in 100 g of cyclohexane had a freezing point of 4.18 °C. The freezing point of pure cyclohexane is 6.50 °C. What is the molecular formula of sulfur in this solvent?

Solution

$$\Delta T = 6.50 - 4.18 = 2.32 \ ^\circ C$$

From Table 12.9 we have

$$K_f = 20.2 \ ^\circ C \ kg \ mol^{-1}$$

Therefore

$$m = \frac{\Delta T}{K_f} = \frac{2.32 \ ^\circ C}{20.2 \ ^\circ C \ kg \ mol^{-1}} = 0.115 \ mol \ kg^{-1}$$

The number of moles of solute in 100 g of solvent is then found as follows:

$$\left[\frac{0.115 \ mol \ (solute)}{1 \ kg \ (solvent)} \right] \left(\frac{1 \ kg}{1000 \ g} \right) [100 \ g \ (solvent)] = 0.0115 \ mol \ (solute)$$

Since we know the mass of the solute, we can find the mass of 1 mol, that is, the molar mass:

$$Molar \ mass = \frac{2.95 \ g}{0.0115 \ mol} = 257 \ g \ mol^{-1}$$

Hence the mass of one molecule of sulfur is 257 u.
    Since the atomic mass of sulfur is 32.06 u, there must be

$$\frac{257 \ u \ molecule^{-1}}{32.06 \ u \ atom^{-1}} = 8.02 \ atoms \ molecule^{-1}$$

In other words, there are eight sulfur atoms in a sulfur molecule. Thus the molecular formula of sulfur in solution in cyclohexane is S_8.

= Example 12.9 =

FREEZING POINT DEPRESSION

How much of the antifreeze ethylene glycol (1,2-ethanediol), $C_2H_6O_2$, must be added to 1.00 L of water so that the solution does not freeze above $-20.0 \ ^\circ C$?

Solution

From Table 12.9 the freezing point depression constant for water is 1.86 °C kg mol^{-1}. We can find the molality of a solution with a freezing point of -20.0 °C as follows:

$$m = \frac{\Delta T}{K_b} = \frac{20.0\ °C}{1.86\ °C\ kg\ mol^{-1}} = 10.8\ mol\ kg^{-1}$$

Since 1.00 L of water has a mass of 1.00 kg, we need 10.8 mol of ethylene glycol. Since the molar mass of ethylene glycol is 62.1 g mol^{-1}, we need

$$10.8\ mol \times \left(\frac{62.1\ g}{1\ mol}\right) = 671\ g\ ethylene\ glycol$$

EXERCISE 12.8

The freezing point of a solution of 1.50 g of glutamic acid in 100 g of water is -0.189 °C. What is the molar mass of glutamic acid? Glutamic acid is an α-amino acid (see Chapter 24).

OSMOSIS AND OSMOTIC PRESSURE

Osmosis is another colligative property that is not only very useful for the measurement of molar mass but also of great biological importance. A *membrane* is a thin liquid or solid film; paper, cellophane, and parchment are examples. The walls of the living cell are membranes. Most membranes are *semipermeable*; that is, they allow the passage of some molecules or ions but not others. Some important membranes allow the passage of water but not of ions or other solutes.

If a solution is separated from pure solvent by a semipermeable membrane that allows the passage of solvent molecules but not solute molecules, more solvent molecules pass through the semipermeable membrane from the solvent to the solution than pass in the other direction. Thus there is a net flow of solvent from the pure solvent to the solution, which is called **osmosis** (Figure 12.24 and Experiment 12.2). The volume of the solution increases and that of the pure solvent decreases. This flow of solvent continues until the level of the solution is sufficiently higher than that of the pure solvent that the hydrostatic pressure due to this height of solution is sufficient to force the solvent molecules back through the membrane, at the same rate as they come through the membrane from the solvent. Equilibrium is then reached and osmosis stops. The pressure that is just sufficient to stop osmosis is called the **osmotic pressure**. Alternatively as shown in Figure 12.24 an external pressure can be applied to the solution until there is no resultant flow through the membrane. This pressure is the osmotic pressure.

Solvent molecules leave a pure solvent and pass through a semipermeable membrane into a solution because they have a greater entropy in the solution than in the pure solvent. They do not move in the reverse direction because this would lead to a decrease in entropy. It is for just the same reason that, if we were to have a pure solvent and a solution in separate containers in an enclosed space, there would be a transfer of solvent from the pure solvent to the solution (Figure 12.24).

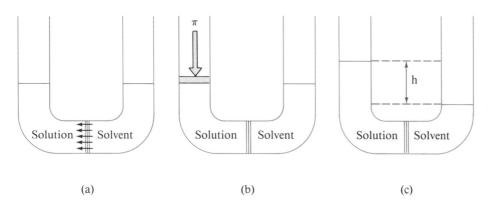

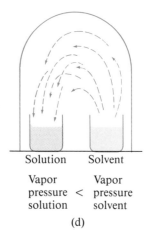

(a)   (b)   (c)

Solution   Solvent

Vapor   Vapor
pressure < pressure
solution   solvent

(d)

Osmosis is a colligative property because the osmotic pressure of a solution depends on the number of solute molecules or ions but not on their nature. The osmotic pressure is related to the number of moles of solute by an equation that closely resembles the ideal gas equation:

$$\Pi V = nRT$$

where Π is the osmotic pressure, V is the volume of the solution, n is the number of moles of solute, T is the Kelvin temperature, and R is the gas constant.

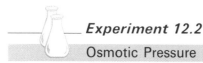

Experiment 12.2

Osmotic Pressure

A cucumber is placed in a graduated cylinder containing a concentrated aqueous solution of sodium chloride. The cell walls of the cucumber act as a semipermeable membrane and water flows from the cucumber into the sodium chloride solution.

After several days the originally firm cucumber has become soft and spongy because of the loss of water.

FIGURE 12.24
Osmotic Pressure.

(a) When a solution and the solvent are separated by a semi-permeable membrane there is a transfer of solvent from the pure solvent to the solution. (b) This transfer of solvent can be stopped by exerting a sufficient pressure on the solution, by means of a suitable piston for example. The pressure needed to just stop the flow of solvent is called the osmotic pressure. (c) Or the solvent can be allowed to flow until the height of the solution above that of the pure solvent gives a sufficient hydrostatic pressure to stop the flow of solvent. (d) The flow of solvent through a semipermeable membrane is analogous to the flow of solvent in the vapor phase that occurs when separate samples of solvent and solution are left in a closed container. Because the vapor pressure of the solvent is higher than that of the solution, there is an overall transfer of solvent to the solution.

Osmosis is a very sensi.ive method for measuring molar masses, and it is widely used for proteins, polymers, and other large molecules.

═ Example 12.10 ═

MOLAR MASS FROM OSMOTIC PRESSURE

An aqueous solution containing 1.10 g of a protein in 100 mL of the solution had an osmotic pressure at 20 °C of 3.93×10^{-3} atm. What is the molar mass of the protein?

Solution

From the equation $\Pi V = nRT$ we calculate the number of moles of protein in the solution:

$$n = \frac{\Pi V}{RT} = \frac{(3.93 \times 10^{-3} \text{ atm})(0.100 \text{ L})}{(0.0821 \text{ L atm mol}^{-1} \text{ K}^{-1})(293 \text{ K})}$$

$$= 1.63 \times 10^{-5} \text{ mol}$$

Hence

$$\text{Molar mass} = \frac{1.10 \text{ g}}{1.63 \times 10^{-5} \text{ mol}} = 67\,500 \text{ g mol}^{-1}$$

ˈBecause of the very high molar mass, the solution is very dilute, 1.63×10^{-4} mol L^{-1}, and therefore the osmotic pressure is very small, only 3.93×10^{-3} atm. However, this pressure corresponds to a difference in height between the solvent and the solution of 2.99 cm, which is easily measured.

In contrast, the freezing point depression of this solution is only

$$\Delta T = 1.86 \text{ °C kg mol}^{-1} \times 1.63 \times 10^{-4} \text{ mol kg}^{-1} = 0.0003 \text{ °C}$$

This freezing point depression is much too small to be measured with any accuracy. Thus freezing point depression, and similarly boiling point elevation, cannot be used for measuring the molar masses of very large molecules. But osmotic pressure is much more sensitive and is widely used for this purpose.

═ Example 12.11 ═

MOLAR MASS FROM OSMOTIC PRESSURE

(a) Analysis of hemoglobin from red blood cells showed that it contains 0.328% by mass of iron. From this fact and the atomic mass of iron deduce the minimum molar mass of the hemoglobin.

(b) An aqueous solution containing 80.0 g of hemoglobin in 1.00 L had an osmotic pressure of 0.0260 atm at 4 °C. What is the molar mass of hemoglobin?

(c) How many iron atoms are in a hemoglobin molecule?

Solution

(a) The analytical data show that 100 g of hemoglobin contains

$$0.328 \text{ g Fe} = \frac{0.328 \text{ g}}{55.85 \text{ g mol}^{-1}} = 5.87 \times 10^{-3} \text{ mol Fe}$$

One molecule of hemoglobin must contain at least one iron atom. Therefore in 100 g of hemoglobin there can be at most 5.87×10^{-3} mol of hemoglobin. Thus

the minimum molar mass of hemoglobin is

$$\frac{100 \text{ g}}{5.87 \times 10^{-3} \text{ mol}} = 17\ 100 \text{ g mol}^{-1}$$

(b) We can find the number of moles of hemoglobin in solution as follows:

$$n = \frac{\Pi V}{RT} = \frac{(0.0260 \text{ atm})(1.00 \text{ L})}{(0.0821 \text{ atm L mol}^{-1} \text{ K}^{-1})(277 \text{ K})}$$
$$= 1.14 \times 10^{-3} \text{ mol}$$

Hence the molar mass is

$$\frac{80.0 \text{ g}}{1.14 \times 10^{-3} \text{ mol}} = 70\ 200 \text{ g mol}^{-1}$$

(c) The actual molar mass is approximately four times the minimum molar mass that we found in (a), assuming that the molecule contains only one iron atom. Therefore we conclude that the hemoglobin molecule contains four iron atoms.

The Ghar-Lapsi, Malta, plant provides fresh water from seawater using reverse osmosis

EXERCISE 12.9

What osmotic pressure will be exerted by a 0.0020M solution of sucrose ($C_{12}H_{22}O_{11}$) at 25 °C?

EXERCISE 12.10

A solution containing 0.8330 g of a protein of unknown structure in 170.0 mL of water was found to have an osmotic pressure of 5.20 mm Hg at 25 °C. What is the molar mass of the protein?

If a pressure greater than the osmotic pressure is applied to a solution in contact with the solvent across a semipermeable membrane, the flow of solvent into the solution is not only stopped but actually reversed, and solvent is forced through the membrane from the solution leaving the solute behind. This process is known as **reverse osmosis** (Figure 12.25). It has some potential for the purification of water, particularly for obtaining drinking water from seawater. The method has already been developed to a limited extent for ocean-going sail boats and for life rafts but it is at present too expensive for large-scale use.

ELECTROLYTE SOLUTIONS

Colligative properties depend on the numbers of solvent and solute molecules, not on their nature. One mole of NaCl and 1 mol of HCl both give 2 mol of ions, ($Na^+ + Cl^-$) and ($H_3O^+ + Cl^-$), respectively, in solution. These solutes therefore give twice the freezing point depression, boiling point elevation, or osmotic pressure of a nonelectrolyte solution of the same concentration. If we call i the number of moles of solute particles produced in a solution by 1 mol

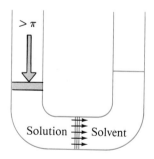

FIGURE 12.25
Reverse Osmosis.

If a pressure greater than the osmotic pressure is applied to the solution the solvent can be forced out of the solution leaving the solute behind.

FIGURE 12.26
Svante Arrhenius.

Arrhenius was born near Uppsala, Sweden, in 1859, and became one of the most brilliant physical chemists of his time. He was the first to propose, in his Ph.D. thesis of 1884, that substances such as NaCl exist as ions in aqueous solution. This was a revolutionary suggestion, since the electron had not yet been discovered and chemists had great difficulty in understanding how sodium and chlorine atoms could become charged. Arrhenius barely passed his Ph.D. oral, and only after evidence for the electron had accumulated in the 1890s did Arrhenius's ideas concerning the nature of solutions of salts in water come to be generally accepted. Finally, in 1903, for the same thesis that had barely been accepted for the Ph.D., Arrhenius was awarded the Nobel Prize in chemistry. His other great contribution to chemistry was the idea of activation energy in reactions, which we consider in Chapter 19.

of solute, then $i = 2$ for HCl and NaCl and $i = 3$ for Na_2SO_4. The equations for the colligative properties are then as follows:

Elevation of boiling point	$\Delta T = iK_b m$
Depression of freezing point	$\Delta T = iK_f m$
Osmotic pressure	$\Pi V = inRT$

The fact that salts and acids give larger freezing point depressions than nonelectrolytes in aqueous solution provided Swedish chemist Svante Arrhenius (1859–1927) with strong evidence for his theory that these substances are electrolytes that form ions in aqueous solution (see Figure 12.26).

=== *Example 12.12* ===

FREEZING POINT OF AN ELECTROLYTE SOLUTION

How many grams of NaCl must be added to 1.00 L (1.00 kg) of water to decrease the freezing point to $-6.00\ °C$?

Solution

$$NaCl(s) \xrightarrow{H_2O} Na^+(aq) + Cl^-(aq)$$

Therefore $i = 2$. Using the equation

$$\Delta T = iK_f m \qquad \text{or} \qquad m = \frac{\Delta T}{iK_f}$$

and substituting $i = 2$, $\Delta T = 6.00\ °C$, and $K_f = 1.86\ °C\ kg\ mol^{-1}$ (from Table 12.9), we have

$$m = \frac{6.00\ °C}{2 \times 1.86\ °C\ kg\ mol^{-1}\ (\text{solvent})} = 1.61\ mol\ kg^{-1}\ (\text{solvent})$$

Therefore $1.61\ mol \times 58.44\ g\ mol^{-1} = 94.1\ g$ NaCl must be added to $1.00\ kg = 1.00\ L$ of water.

EXERCISE 12.11

What concentration of calcium chloride is required to prevent ice forming in an aqueous solution down to $-10.0\ °C$?

12.8
Reactions of Water

Water is by far the most widely used solvent, and so reactions in aqueous solution are very common. Water is often a reactant as well as a solvent and it will be useful to review the reactions of water and to classify them according to reaction type. We have already introduced the three most important reaction types, namely acid–base (both Brønsted–Lowry and Lewis), precipitation, and oxidation–reduction. If you learn to recognize these reaction types you will have a basis for a good knowledge and understanding of chemical reactions. As we have already seen water acts as both an acid and a base in the Brønsted sense and as a base in the Lewis sense; thus it can behave as a proton donor, a proton acceptor, and an electron pair donor. It can also act as an oxidizing agent or as a reducing agent toward appropriate reactants. We will now consider each of these different types of reaction in turn.

BRØNSTED ACID–BASE REACTIONS

Water can donate a proton to a base such as ammonia and thus behave as an acid:

$$H_2O + NH_3 \rightleftharpoons NH_4^+ + OH^-$$

Also, one of its unshared electron pairs can accept a proton from an acid such as hydrochloric acid, and thus it behaves as a base:

$$H_2O + HCl \longrightarrow H_3O^+ + Cl^-$$

Recall from Chapter 5 that water exerts a *leveling effect* on the strengths of acids and bases. All acids that are quantitatively converted to H_3O^+ in aqueous solution are said to be strong. They may have different intrinsic strengths, that is, different proton-donating abilities, but in solution in water their strengths cannot be differentiated. The only common strong acids are HCl, HBr, HI, HNO_3, H_2SO_4 and $HClO_4$. All other common acids are weak. Similarly all bases that are quantitatively converted to OH^- are said to be strong. The only common strong bases are O^{2-}, H^-, and NH_2^-, and metal hydroxides $M(OH)_n$; all other common bases are weak. We also saw in Chapter 5 that the conjugate acids of weak bases are weak acids and the conjugate bases of weak acids are weak bases. Thus the ammonium ion, NH_4^+, the conjugate acid of the weak base ammonia, NH_3, is a weak acid, and the hydrogen carbonate ion, HCO_3^-, the conjugate base of the weak acid carbonic acid, H_2CO_3, is a weak base. The range of acid–base strengths that is available in water lies between the acid H_3O^+, the strongest acid that can exist in water, and OH^-, the strongest base that can exist in water.

The reaction between *any* acid and *any* base in aqueous solution is always exactly the same reaction, namely the reaction between hydronium ions and hydroxide ions to give water:

$$H_3O^+(aq) + OH^-(aq) \rightleftharpoons 2H_2O(l)$$

This reaction occurs essentially quantitatively because water is only very slightly ionized; the equilibrium lies very far to the right so that only a *very* small concentration of H_3O^+ and OH^- ions can exist together in water. If we write the equations for acid-base reactions using the empirical formulas of the reactants, the reactions all *look* different; for example

$$NaOH(aq) + HCl(aq) \longrightarrow NaCl(aq) + H_2O(l)$$

$$2KOH(aq) + H_2SO_4(aq) \longrightarrow K_2SO_4(aq) + 2H_2O(l)$$

But the only reaction in every case is the reaction of H_3O^+ with OH^-. The other ions in the solution are spectator ions. Even if the acid or the base is weak, the neutralization reaction is just the same, and the reactions go to completion because the ionization of the weak acid or base is forced to the right

Common Strong Acids
HCl
HBr
HI
HNO_3
H_2SO_4
$HClO_4$

Common Strong Bases
O^{2-}
OH^-
H^-
NH_2^-

by the removal of H_3O^+ or OH^- which are in equilibrium with the weak acid or base. So we can write equations such as

$$2NH_3(aq) + H_2SO_4(aq) \longrightarrow (NH_4)_2SO_4(aq) + H_2O(l)$$

and

$$H_2CO_3(aq) + 2NaOH(aq) \longrightarrow Na_2CO_3(aq) + H_2O(l)$$

where the arrow indicates that these reactions go essentially to completion. We are now in a position to predict the products and write the equation for the reaction of any acid or of any base with water and of any acid with any base in aqueous solution.

═ *Example 12.13* ═══════════════════════════

ACID–BASE REACTIONS IN AQUEOUS SOLUTION

Write balanced equations for each of the following reactions:

(a) The reaction of nitrite ion, NO_2^-, with water

(b) The reaction of ammonium ion with water

(c) The reaction of sulfurous acid, H_2SO_3, with excess cesium hydroxide in aqueous solution

(d) The reaction of hypochlorous acid, HOCl, with calcium hydroxide in aqueous solution

Solution

(a) Nitrite ion is the anion (conjugate base) of nitrous acid. Because nitrite ion is not on our list of strong bases, and clearly cannot be an acid, it must be a weak base. It is the conjugate base of the weak acid, HNO_2. The equation for its reaction with water, which acts as an acid here, is

$$NO_2^-(aq) + H_2O(l) \rightleftharpoons HNO_2(aq) + OH^-(aq)$$

(b) Ammonium ion is a weak acid. It is not on our list of strong acids; it is the conjugate acid of the weak base ammonia, NH_3. The equation for its reaction with water is

$$NH_4^+(aq) + H_2O(l) \rightleftharpoons NH_3(aq) + H_3O^+(aq)$$

(c) Sulfurous acid is not on our list of strong acids so it must be weak. However, it is not essential to know this because the reaction of any acid (strong or weak) with a strong base such as cesium hydroxide, CsOH, goes to completion:

$$H_2SO_3(aq) + 2CsOH(aq) \longrightarrow Cs_2SO_3(aq) + 2H_2O(l)$$

Excess CsOH was specified so both hydrogens of sulfurous acid will be removed by reaction with OH^-. Recall that cesium is in group 1 so its hydroxide has the formula CsOH. The product is cesium sulfite.

(d) Calcium is in group 2; therefore its hydroxide has the formula $Ca(OH)_2$. Its reaction with hypochlorous acid can be represented by the equation

$$2HOCl(aq) + Ca(OH)_2(aq) \longrightarrow Ca(OCl)_2(aq) + 2H_2O(l)$$

EXERCISE 12.12

Write balanced equations for each of the following reactions:

(a) The reaction of cyanide ion with water

(b) The reaction of carbide ion, C_2^{2-}, with water

(c) The reaction of sodium acetate with perchloric acid in aqueous solution

(d) The reaction of calcium hydroxide with excess phosphoric acid in aqueous solution

LEWIS ACID–BASE REACTIONS

A water molecule has two unshared pairs of electrons that are available to be shared with an atom that can accommodate more electrons in its valence shell. Thus a water molecule can behave as a Lewis base.

REACTIONS OF NONMETAL OXIDES WITH WATER Water reacts with nonmetal oxides to give oxoacids. For example,

$$SO_3 + H_2O \longrightarrow H_2SO_4$$
$$CO_2 + H_2O \rightleftharpoons H_2CO_3$$
$$P_4O_{10} + 6H_2O \longrightarrow 4H_3PO_4$$

Experiment 12.3

Reactions of Metal and Nonmetal Oxides with Water

The glass spoon contains sodium metal. Phenolphthalein indicator has been added to the water in the beaker.

When sodium metal is ignited it burns with a yellow flame, producing a white smoke of sodium oxide that dissolves in water to form sodium hydroxide. Phenolphthalein indicator turns pink, showing that the solution is now alkaline.

Sulfur burns with a blue flame, producing sulfur dioxide. The water in the beaker is colored yellow because methyl red indicator has been added.

When the sulfur dioxide produced by the burning sulfur dissolves in water the methyl red indicator turns red, showing that the solution formed is acidic.

These oxides have polar bonds; therefore the nonmetal atoms have a positive charge, and the oxygens have a negative charge. The reaction is initiated by attack of the negative oxygen atom of a water molecule on the positive non-metal atom of the oxide. A proton is then transferred from the water molecule to a negative oxygen atom. For example

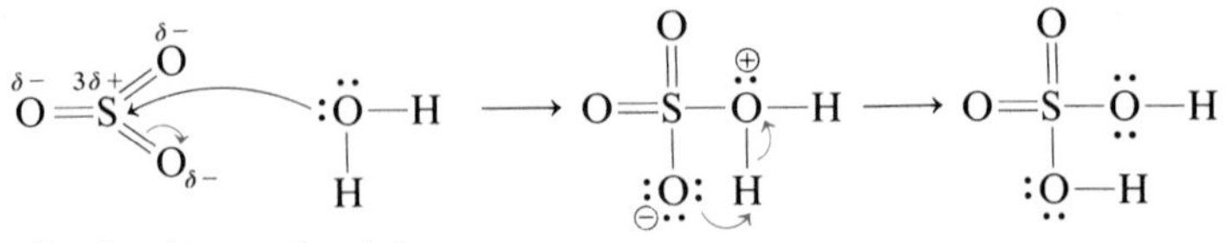

The sand that we find on the beach is mainly silicon dioxide so the fact that silicon dioxide does not dissolve in water under or-dinary conditions is a matter of everyday experience. Its insolubility, like that of diamond, is a consequence of its three-dimensional network structure and the strong covalent bonds that hold the structure together.

In these reactions water is a Lewis base, and the oxide is a Lewis acid. Other nonmetal oxides that react in the same way with water are listed in Table 12.10. Silicon dioxide, SiO_2, is included in this list although it does not react with water except at high temperature and pressure. Carbon monoxide is not in the list as it is an exception in that it does not react with water under ordinary conditions.

REACTIONS OF NONMETAL HALIDES WITH WATER Most of the halides of the nonmetals of period 3 and beyond react with water to give the correspond-ing oxoacid and hydrogen halides (see Experiment 12.4). For example,

$$PCl_3 + 3H_2O \longrightarrow HPO(OH)_2 + 3HCl$$

$$SiCl_4 + 4H_2O \longrightarrow Si(OH)_4 + 4HCl$$

Other halides that react in this way are PF_3, PF_5, PCl_5, and SF_4. Sulfur hexafluoride, SF_6, does not react with water and is an exception. These reac-tions resemble the reactions of water with the oxides of the nonmetals. The negatively charged O atom of the water molecule attacks the positively charged nonmetal atom. Then an H atom is transferred to the halogen atom, with the elimination of the hydrogen halide:

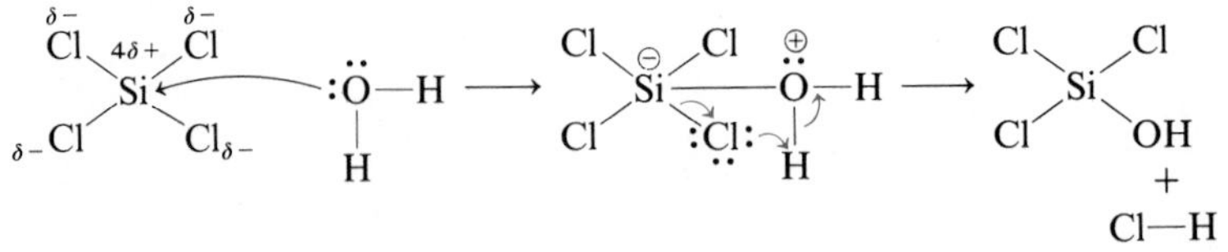

Repetition of this reaction leads eventually to the formation of silicic acid, $Si(OH)_4$. Again, the water molecule is behaving as a Lewis base and the metal halide as a Lewis acid. Silicon tetrachloride can behave as a Lewis acid because the valence shell of silicon can accommodate more than eight electrons. In contrast carbon tetrachloride neither reacts with nor dissolves in water, as we saw previously in Experiment 1.6. The valence shell of carbon, like that of the other period 2 elements, is restricted to eight electrons—carbon obeys the octet rule—and so carbon tetrachloride cannot accept an unshared pair of electrons from a water molecule. In other words, carbon tetrachloride cannot behave as a Lewis acid. Moreover, the C atom is completely surrounded by four large Cl atoms, which block the access of a water molecule to the carbon atom.

Sulfur hexafluoride is like carbon tetrachloride in that it does not react with water. It appears that the valence shell of sulfur can contain a maximum of six electron pairs. As there is no space available for another electron pair, it does not behave as a Lewis acid.

TABLE 12.10
Nonmetal Oxides that Behave as Lewis Acids toward Water

CO_2	P_4O_{10}
SiO_2	SO_2
N_2O_3	SO_3
N_2O_5	Cl_2O
P_4O_6	Cl_2O_7

Experiment 12.4

Reactions of Metal and Nonmetal Chlorides with Water

The four test tubes contain water to which methyl red indicator has been added. Below the tubes are samples of four nonmetal halides. From left to right these are, carbon tetrachloride, CCl_4, silicon tetrachloride $SiCl_4$, and phosphorus trichloride PCl_3, which are all colorless liquids, and phosphorus pentachloride PCl_5, which is pale yellow.

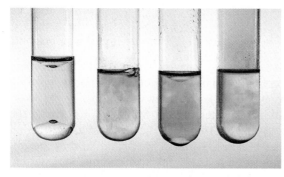

When the halide samples are added to the water in the test tubes, all but the carbon tetrachloride react, forming acidic solutions, thus causing the methyl red indicator to change color.

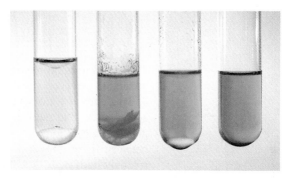

From left to right we note (a) that carbon tetrachloride forms an immiscible layer on the bottom of the tube, (b) that some insoluble silicic acid, $Si(OH)_4$ is precipitated, (c) that some of PCl_3 has not yet reacted and forms an immiscible layer on the bottom of the tube, (d) that PCl_5 has reacted completely to give an acidic solution of H_3PO_4 and HCl.

EXERCISE 12.13

Write balanced equations for the reactions of the following halides with water:

(a) PCl_5 (b) PF_3 (c) SF_4

HYDRATION OF METAL IONS Small metal ions with two or more positive charges, particularly Al^{3+}, Mg^{2+}, and transition metal ions such as Fe^{3+}, Fe^{2+}, and Cu^{2+}, are strongly hydrated in aqueous solution. Although we can think of this hydration as resulting from the electrostatic attraction of an ion for a dipole, in so far as there is some donation of the electron density of an oxygen lone pair into the empty valence shell of the metal ion, we can also regard this attraction as a Lewis acid–base reaction in which water is behaving as a Lewis base and the metal ion as a Lewis acid:

$$Al^{3+} + :\overset{..}{O}-H \longrightarrow \overset{(2+)}{Al}-\overset{\oplus}{O}-H$$

Six water molecules are bound to the aluminum ion in this way, so that the overall equation for the formation of the hydrated ion is

$$Al^{3+} + 6H_2O \longrightarrow Al(H_2O)_6{}^{3+}$$

Because the aluminum ion attracts electrons away from the oxygen atom of a water molecule, the oxygen atom acquires a positive charge that is greater than in an isolated water molecule. This positively charged oxygen attracts electrons away from the hydrogen atoms which acquire a greater positive charge and become correspondingly more acidic than in an isolated water molecule. Such strongly hydrated ions therefore behave as weak acids in aqueous solution:

$$Al(OH_2)_6{}^{3+} + H_2O \longrightarrow H_3O^+ + Al(OH)(OH_2)_5{}^{2+}$$

The metal ions of group 1 have only a +1 charge. Therefore the electrostatic attraction between such an ion and a polar water molecule is not very strong and so they are not strongly hydrated. Hence they do not behave as weak acids—they give neutral solutions in water. Except for Be^{2+} and Mg^{2+}, the group 2 metal ions are relatively large. Therefore the electrostatic attraction between these ions and a water molecule is rather weak and so they are not strongly hydrated. Thus they do not behave as weak acids in aqueous solution. Of the group 2 ions only Be^{2+} and Mg^{2+} are small enough to be strongly hydrated and they give acidic solutions in water.

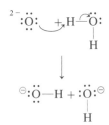

REACTIONS OF METAL OXIDES WITH WATER Water reacts with some metal oxides to give metal hydroxides. For example,

$$Na_2O + H_2O \longrightarrow 2NaOH$$

This is a Brønsted acid–base reaction between the oxide ion, which is a strong base, and water, which is acting as an acid:

$$O^{2-} + H_2O \longrightarrow 2OH^-$$

Many metal oxides such as MgO and CuO are, however, insoluble and therefore they do not react with water.

$=$ *Example 12.14* $=$

REACTIONS OF NONMETAL OXIDES WITH WATER

Write balanced equations for the reaction of each of the following oxides with water and name the product.

(a) N_2O_5 (b) Cl_2O

Solution

(a) Nitrogen is in the $+5$ oxidation state in N_2O_5 and it must be in the same oxidation state in the product since this is not a redox reaction. Nitric acid, HNO_3, is the oxoacid of nitrogen in the $+5$ oxidation state. If we simply add H_2O to N_2O_5 we get $H_2N_2O_6$, which we can see is $2HNO_3$. The required equation is

$$N_2O_5 + H_2O \longrightarrow 2HNO_3$$

(b) Chlorine is in the $+1$ oxidation state in Cl_2O, so it must be in the $+1$ oxidation state in the product, since this is not a redox reaction. Hypochlorous acid, $HOCl$, is the oxoacid of chlorine in the $+1$ oxidation state. If we add H_2O to Cl_2O we obtain $H_2Cl_2O_2$, which we recognize as $2HOCl$. The required equation is

$$Cl_2O + H_2O \longrightarrow 2HOCl$$

EXERCISE 12.14

Write balanced equations for the reaction of each of the following oxides with water:
(a) P_4O_6 (b) Cl_2O_7 (c) CaO (d) K_2O

OXIDATION–REDUCTION REACTIONS

Water is also an oxidizing agent and a reducing agent, although its oxidizing, and particularly its reducing, properties are rather weak. When it behaves as an *oxidizing* agent, that is, when it gains electrons, it is reduced to hydrogen:

$$2H_2O(l) + 2e^- \longrightarrow 2OH^-(aq) + H_2(g)$$

We are familiar with its reaction with sodium and other reactive metals, which are strong reducing agents:

$$2Na(s) + 2H_2O(l) \longrightarrow 2NaOH(aq) + H_2(g)$$

Similarly, water reacts with the hydride ion, H^-, which is oxidized to elemental hydrogen:

$$H^-(aq) + H_2O(l) \longrightarrow H_2(g) + OH^-(aq)$$

This reaction is also an example of an acid–base reaction, since the hydride ion is a strong base that adds a proton to give hydrogen.

Water is a very weak *reducing* agent:

$$2H_2O(l) \longrightarrow 4H^+(aq) + O_2(g) + 4e^-$$

It will reduce only very strong oxidizing agents such as fluorine:

$$2F_2(g) + 2H_2O(l) \longrightarrow 4HF(aq) + O_2(g)$$

Chlorine reacts in the same way with water, but the reaction is very slow, although it is speeded up somewhat by strong light:

$$2Cl_2(g) + 2H_2O(l) \longrightarrow 4HCl(aq) + O_2(g)$$

The main reaction between chlorine and water is the formation of a small equilibrium amount of hypochlorous acid, HOCl:

$$Cl_2(g) + 2H_2O(l) \rightleftharpoons HOCl(aq) + H_3O^+(aq) + Cl^-(aq)$$

This reaction is just like the reactions of nonmetal halides with water in which the nonmetal halide behaves as a Lewis acid. In this case Cl_2 behaves as a Lewis acid, one of the chlorine atoms accepting an electron pair from water:

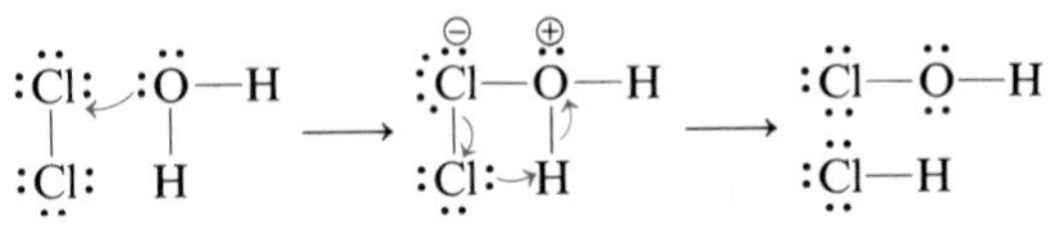

The reactions of water that we have discussed in this section are summarized in Table 12.11. These reactions are for liquid water at ordinary temperatures, and the reactions of steam at high temperature and pressure have been excluded. We can use this table to predict the products of reactions in which water is one of the reactants. We first identify the nature of the other reactant. Is it an element, an oxide, a halide, or a salt? We then need to decide whether the element is a metal or a nonmetal or whether the oxide or the halide is a compound of a metal or a nonmetal. In considering a salt, we need to remember that either the anion or the cation, or both, may react with water. This scheme does not cover all possible inorganic compounds, but it does cover a very large number of the most important types of compounds, in particular those that we have been emphasizing in this book. Its function is to help you to organize your knowledge of reactions and thereby understand them better, but not to provide a comprehensive scheme for predicting the products of *all* the reactions of water.

Important Nonmetals

3	4	5	6	7	8
B	C	N	O	F	Ne
	Si	P	S	Cl	Ar
				Br	Kr
				I	Xe

Important Metals

1	2	3	4
Li	Be		
Na	Mg	Al	
K	Ca		
Rb	Sr		
Cs	Ba		Pb

Transition metals including Cu and Fe

═ Example 12.15 ═

IDENTIFYING REACTION TYPES

Write a balanced equation for the reaction of each of the following with water at ordinary temperature and pressure. In each case state whether the reaction is a Brønsted acid–base reaction (BAB), a Lewis acid–base reaction (LAB), an oxidation–reduction reaction (OR), or if there is no reaction.

    (a) HF   (b) KF   (c) P_4O_6 (d) Na_2S (e) CO   (f) K

Solution

    (a) HF is a weak acid and H_2O behaves as a base:

$$HF + H_2O \rightleftharpoons H_3O^+ + F^- \qquad (BAB)$$

    (b) K^+ is not strongly hydrated and therefore does not behave as an acid. F^- is a weak base (HF is a weak acid) and H_2O therefore behaves as an acid:

$$F^- + H_2O \rightleftharpoons HF + OH^-$$

$$KF + H_2O \rightleftharpoons HF + KOH \qquad (BAB)$$

    (c) P_4O_6 is a nonmetal oxide. It therefore behaves as a Lewis acid, reacting with water (a Lewis base) to give an oxoacid (phosphorous acid):

$$P_4O_6 + 6H_2O \longrightarrow 4H_3PO_3 \qquad (LAB)$$

TABLE 12.11
Reactions of Water

Reactant		Reaction
Element	Metal	Metals of groups 1 and 2 (except Be) reduce water to H_2. Other metals do not react. $$2Na + 2H_2O \longrightarrow 2NaOH + H_2$$
	Nonmetal	Only F_2 is a strong enough oxidizing agent to oxidize water to O_2: $$2F_2 + 2H_2O \longrightarrow O_2 + 4HF$$ Cl_2 also oxidizes water but very slowly. Cl_2 and Br_2 also behave as Lewis acids and form ClOH and BrOH. The other nonmetals are insoluble and do not react.
Oxide	Metal	Ionic. The oxide ion is a strong base: $$O^{2-} + H_2O \longrightarrow 2OH^-$$ Only the oxides of groups 1 and 2 (except BeO and MgO) are soluble in water and give basic solutions. Other metal oxides are insoluble.
	Nonmetal	Covalent. Lewis acids. Give oxoacids: $$CO_2 + H_2O \rightleftharpoons H_2CO_3$$
Halide	Metal	Ionic. F^- is a weak base: $$F^- + H_2O \rightleftharpoons HF + OH^-$$ Cl^-, Br^-, I^- are not bases, no reaction.
	Nonmetal	Covalent. Lewis acids (except period 2). Give oxoacid and HX: $$PCl_3 + 3H_2O \rightleftharpoons H_3PO_3 + 3HCl$$
Salt	Cation	NH_4^+ is a weak acid: $$NH_4^+ + H_2O \rightleftharpoons NH_3 + H_3O^+$$ Cations of groups 1 and 2 (except Be^{2+} and Mg^{2+}), no reaction. Be^{2+}, Mg^{2+}, Al^{3+}, and transition metal ions are strongly hydrated and behave as weak acids: $$Al(OH_2)_6^{3+} + H_2O \rightleftharpoons H_3O^+ + Al(OH_2)_5(OH)^{2+}$$
	Anion	Anions of weak acids are weak bases: $$CO_3^{2-} + H_2O \rightleftharpoons HCO_3^- + OH^-$$ Anions of strong acids, no reaction.

(d) Sulfide ion, S^{2-}, is a weak base (H_2S is a weak acid) and H_2O behaves as an acid:

$$S^{2-} + H_2O \rightleftharpoons HS^- + OH^-$$

or

$$S^{2-} + 2H_2O \rightleftharpoons H_2S + 2OH^-$$
$$K_2S + 2H_2O \rightleftharpoons H_2S + 2KOH \quad \text{(BAB)}$$

(e) No reaction. Nonmetal oxides usually react with water to give oxoacids as in (c) but CO is an exception.

(f) K is a metal and therefore a reducing agent. It is a sufficiently strong reducing agent to reduce water to hydrogen. In this reaction water is an oxidizing agent oxidizing potassium to K^+.

$$2K + 2H_2O \longrightarrow 2KOH + H_2 \qquad (OR)$$

EXERCISE 12.15

Write a balanced equation for the reaction of each of the following with water at ordinary temperature and pressure:

(a) F^- (b) PF_3 (c) KNO_3 (d) KNO_2 (e) Cl_2O (f) SF_6.

In each case state whether the reaction is a Brønsted acid–base reaction, a Lewis acid–base reaction, a redox reaction, or if there is no reaction.

IMPORTANT TERMS

A **colligative property** is a property of a solution that depends only on the number of solute particles (molecules or ions) and not on their nature.

A **hydrogen bond** is an intermolecullar attraction in which a hydrogen atom that is bonded to an electronegative atom is attracted to an unshared electron pair on another electronegative atom.

An **instantaneous dipole moment** is the fluctuating, temporary dipole moment that results from the motions of the electrons in an atom or molecule.

Intermolecular forces are the forces of attraction between molecules.

London forces are intermolecular forces that result from fluctuating instantaneous dipoles. They are the only intermolecular forces between nonpolar molecules. They are also called **dispersion forces**.

The **molality, m** of a solute in a solution is the number of moles of the solute in 1 kg of solvent.

The **molar enthalpy of fusion**, ΔH_{fus}, is the heat required to melt 1 mol of a solid substance at 1 atm pressure.

The **molar enthalpy of solution** is the heat absorbed at constant pressure when 1 mol of a solute dissolves in a solvent at a given temperature. The enthalpy of solution depends on the concentration of the final solution, but it has a constant value for sufficiently dilute solutions, that is, for a sufficiently large amount of solvent.

The **molar enthalpy of vaporization**, ΔH_v, is the heat required to vaporize 1 mol of a liquid substance at 1 atm pressure.

The **mole fraction** of the solvent in a solution is the fraction of the total number of molecules in the solution that are solvent molecules. The **mol fraction** of a solute is the fraction of the total number of molecules that are solute molecules.

Osmosis is the flow of solvent through a semipermeable membrane from a more dilute to a more concentrated solution.

Osmotic pressure is the pressure that must be exerted on a solution to prevent osmosis.

The **polarizability** of an atom or a molecule is a measure of the ease with which the electrons and nuclei can be displaced from their equilibrium positions by an external force so that a dipole is created.

Sublimation is the process by which a solid is converted directly to vapor.

Van der Waals force is an alternative name for intermolecular force.

The **van der Waals radius** is a measure of the size of an atom in the directions in which it is not forming a bond. It is equal to one-half the distance between adjacent atoms of the same kind in molecules that are not bonded together.

Vapor pressure is the pressure exerted by a vapor in equilibrium with a liquid or a solid.

Molal and Molar Concentrations and Mole Fraction

1. The mol fraction of water in an aqueous ethanol solution is 0.925. What is the molality of ethanol in the solution?

2. An aqueous solution of sodium chloride contains 9.65 mass % NaCl. What is the mole fraction of water in the solution?

3. Concentrated aqueous nitric acid is 69.0 mass % HNO_3 and has a density of 1.41 g mL^{-1}. What is the molarity, the molality, and the mole fraction of nitric acid in concentrated nitric acid?

4. Concentrated aqueous hydrochloric acid contains 37.0 mass % HCl and has a density of 1.18 g mL^{-1}. What is the molarity, the molality, and the mole fraction of HCl in concentrated hydrochloric acid?

5. A solution is prepared by dissolving 22.40 g of magnesium chloride, $MgCl_2$, in 200 mL of water. Taking the density of pure water to be 1.000 g mL^{-1}, and that of the solution to be 1.089 g mL^{-1}, calculate the mole fraction, the molarity, and the molality of $MgCl_2$ in this solution.

6. A solution prepared from 95.94 g of water and 10.66 g of pure H_2SO_4 had a volume of 100.0 mL. Calculate the density of the solution and the molality and the molarity of sulfuric acid. What is the mole fraction of water in the solution?

Vapor Pressure and Enthalpies of Change of State

7. Why is the molar enthalpy of sublimation of a solid always greater than the molar enthalpy of vaporization of the corresponding liquid?

8. (a) Why does the vapor pressure of a liquid increase with increasing temperature?

(b) At 20 °C, the vapor pressure of benzene is 75.0 mm Hg, and that of toluene is 50.0 mm Hg. Which of these hydrocarbons is expected to have the higher boiling point?

(c) The normal boiling points of diethyl ether, $(C_2H_5)_2O$, methanol, CH_3OH, and propanone (acetone), $(CH_3)_2CO$, are, respectively, 34.5, 64.5, and 56.1 °C. Which of these liquids will have the highest vapor pressure, and which will have the lowest vapor pressure, at 25 °C?

9. Before the invention of the refrigerator, butter and milk were stored in the summer in porous clay pots standing in water. Why would a clay pot standing in water keep the contents cool, even on a hot day?

10. Calculate the standard enthalpy change for the combustion of methane in air to give $CO_2(g)$ and $H_2O(l)$. Use this value to calculate the amount of methane (natural gas) that would have to be burned to change 1000 kg of liquid water at 100 °C to steam at 100 °C.

11. Define each of the following:

    (a) condensation   **(b)** evaporation   **(c)** sublimation

    (d) boiling point

12. Explain what effect (if any) each of the following has on the vapor pressure of a liquid: **(a)** surface area of the liquid **(b)** volume of the container   **(c)** temperature **(d)** intermolecular forces   **(e)** volume of the liquid.

Partial Pressure of Water Vapor

13. A student heated potassium chlorate, $KClO_3(s)$, mixed with a little manganese dioxide, $MnO_2(s)$, catalyst, to prepare a sample of oxygen. The gas was collected in a gas jar over water. After adjusting the water levels inside and outside the jar until they were the same, the volume of gas inside the jar was 365 mL. Atmospheric pressure was 745 torr, and the temperature was 21 °C.

    (a) What was the partial pressure of $O_2(g)$ in the sample collected?

    (b) What would the volume of dry $O_2(g)$ have been if it had been collected over mercury under the same conditions of temperature and pressure?

    (c) What would the volume of dry $O_2(g)$ have been at STP?

    (d) What minimum amount of potassium chlorate was required to give this amount of oxygen?

14. 5.00 L of air were saturated with water vapor at 25 °C and then completely dried by bubbling through concentrated sulfuric acid, to give an increase in mass of the sulfuric acid of 0.115 g. What is the vapor pressure of water at 25 °C?

15. When 0.540 g of lithium hydride reacts completely with water, what volume of gas can be collected over water at 25 °C and a pressure of 754 torr? If the volume of the resulting solution is 50.5 mL, what concentration of LiOH(aq) does it contain? What volume of dry $H_2(g)$ would have been collected at STP?

16. In an experiment 0.1022 g of aluminum was completely dissolved in excess of dilute sulfuric acid, and the resulting hydrogen was collected over water at 27 °C and a pressure of 740 mm Hg. What was the volume of the gas collected?

Intermolecular Forces

17. Elemental boron melts at 2300 °C and is almost as hard as diamond. It is a poor electrical conductor in both solid and liquid states. On the basis of these properties, is boron expected to be a molecular solid, an ionic solid, or a covalent network solid? Would you expect boron to be soluble in water?

18. Tin tetrachloride, $SnCl_4$, is a liquid boiling at 114 °C and melting at -33 °C, while tin dichloride, $SnCl_2$, is a solid with a melting point of 246 °C. Suggest a reason for these differences in terms of intermolecular forces and structure.

19. Is each of the following expected to form a molecular solid or a network solid? What is expected to be the dominant intermolecular force in each case. Explain your choices.

(a) Nitrogen   (b) Hydrogen sulfide

(c) Chromium   (d) Calcium oxide

(e) Silane, SiH_4 (f) Silica, SiO_2

(g) Potassium hydroxide   (h) Sulfuric acid

20. Which substance in each of the following pairs would be expected to have the higher boiling point? Give reasons for your choice.

(a) ClF and BrF   (b) BrCl and Cl_2

(c) KBr and BrCl   (d) K and Br_2

21. Explain why chlorobenzene, C_6H_5Cl, has a higher boiling point (132 °C) than benzene, C_6H_6 (80 °C), and why the boiling point of hexachlorobenzene, C_6Cl_6 (310 °C), is so much higher than that of benzene.

22. What types of intermolecular attractions are important in each of the following?

(a) Solid iodine   (b) Calcium oxide

(c) Carbon dioxide gas

(d) Liquid methyl chloride, CH_3Cl

(e) Liquid hydrogen fluoride

23. The boiling points of the fluorides of the second period elements are as follows: LiF, 1717 °C; BeF_2, 1175 °C; BF_3, -101 °C; CF_4, -128 °C; NF_3, -120 °C; OF_2, -145 °C; F_2, -188 °C. Account for the very large change in boiling point between LiF and BeF_2 and the remainder, in terms of the nature and the strengths of the intermolecular forces in these substances.

24. Why is the vapor pressure of water at 25 °C less than the vapor pressure of gasoline at the same temperature?

25. Which substance of each of the following pairs will have the greater enthalpy of vaporization? Justify your choices.

(a) Cl_2O or H_2O (b) CCl_4 or CBr_4

(c) He or Ar

26. Which of the following properties would you expect to depend on the strength of intermolecular forces? Explain why.

(a) Boiling point   (b) Enthalpy of vaporization

(c) Molar mass   (d) Solubility

(e) Viscosity   (f) Covalent radius

(g) Electronegativity   (h) Bond energy

27. Make qualitative predictions about each of the following solubilities and explain your predictions:

(a) HCl(g) in water and in pentane, C_5H_{12}(l)

(b) Water in liquid HF and in gasoline

(c) Chloroform (trichloromethane), $CHCl_3$, in water and in tetrachloromethane (carbon tetrachloride), CCl_4

(d) Naphthalene, $C_{10}H_8$, in water and in benzene, C_6H_6

(e) N_2(g) and hydrogen cyanide, HCN(g), in water

(f) Benzene, C_6H_6(l), in toluene (methylbenzene), C_7H_8, and in water

28. What is the most important kind of intermolecular force in each of the following substances?

(a) Solid argon   (b) Cl_2(g)

(c) Molten lithium fluoride

(d) Solid aluminum chloride hexahydrate, $AlCl_3 \cdot 6H_2O$

(e) Liquid methanol   (f) Liquid sulfuric acid

(g) Hydrogen chloride gas

(h) Hexafluoroethane, $C_2^{:}F_6$(g)

29. Identify the most important intermolecular force between the molecules of each of the following substances at room temperature:

(a) Carbon dioxide   (b) Water

(c) Hydrogen fluoride   (d) Iodine

(e) Iodine chloride, ICl   (f) Helium

30. What is the strongest intermolecular attraction, or bond, that must be broken when each of the following substances is melted?

(a) Benzene   (b) Ethanol

(c) Ethane   (d) Barium oxide

(e) Chlorine   (f) Hydrogen chloride

31. For each of the following pairs of substances, predict which substance will have the higher boiling point, and justify your choice:

(a) Calcium or sodium

(b) Silane, SiH_4, or methane

(c) Ethane or propane

(d) Ammonia or phosphine

(e) Fluorine or chlorine

(f) Sulfur dioxide or silica, SiO_2

32. In the substances with the following molecular formulas, which will have dipole–dipole interactions between the molecules?

CO H_2S CH_2Cl_2 CCl_4

CH_4 Cl_2 HBr SO_2

33. Which substance of each of the following pairs is likely to be the more soluble in water? Explain each choice. For the less soluble of each pair, suggest a better solvent than water.

(a) Hydrogen peroxide or benzene

(b) Ethane or ethanediol, $HOCH_2CH_2OH$

(c) Sugar or hexane, C_6H_{12}

(d) Chloroform (trichloromethane) or magnesium chloride

(e) Iodine or hydrogen iodide

(f) Lithium chloride or tetrachloromethane

(g) Methanol, CH_3OH, or ethane

34. Which of each of the following pairs of gases would be expected to show the greater deviation from ideal gas behavior? Explain.

(a) Cl_2 and F_2 (b) Cl_2 and BrCl

(c) CH_4 and CF_4 (d) CO_2 and SO_2

35. (a) Explain why carbon dioxide is a molecular solid while silica is a covalent network solid.

(b) Why do the halogens form molecular solids?

(c) Why are many ionic solids soluble in water but no simple ionic solid is soluble in tetrachloromethane?

(d) Why is the difference in the melting points of water and hydrogen sulfide much greater than that between ammonia and phosphine?

36. Nonpolar tetrachloromethane boils at 76.5 °C, while polar trichloromethane (chloroform) has the lower boiling point of 61.7 °C. Explain why the boiling point of polar chloroform is lower than that of nonpolar tetrachloromethane.

37. (a) What are the requirements for a substance to be capable of forming hydrogen bonds?

(b) Which of the following substances are hydrogen bonded in the liquid state? Explain your answers.

(i) Ammonia   (ii) Sodium chloride

(iii) Hydrogen fluoride   (iv) Hydrogen

(v) Methane   (vi) Lithium hydride

(vii) Methanol   (viii) Acetic acid

38. Arrange the following substances in order of the expected increase in their boiling points, and account for the order you select in terms of intermolecular forces:

(a) Chlorine   (b) Sodium chloride

(c) Hydrogen chloride   (d) Neon

(e) Oxygen   (f) Water

(g) Argon   (h) Hydrogen fluoride

39. Explain the differences in the boiling points of each of the following pairs of substances:

	Substance	Bp (°C)	Substance	Bp (°C)
(a)	$(CH_3)_2O$	35	C_2H_5OH	79
(b)	HF	20	HCl	−85
(c)	CCl_4	76	LiCl	1360
(d)	HCl	−85	LiCl	1360

40. Define each of the following and clearly explain the difference between each pair:

(a) Polarity and polarizability

(b) London forces and dipole-dipole forces

(c) Van der Waals radius and covalent radius

(d) Polar covalent bond and hydrogen bond

41. Water has a greater density at 0 °C than does ice, while liquid bromine has a smaller density than solid bromine at its melting point. Account for these differences.

42. (a) Explain why real gases deviate from ideal behavior, especially at high pressures and low temperatures. To what factors can this deviation from ideality be ascribed?

(b) Which gas of each of the following pairs would be expected to show the greater deviation from ideal gas behavior, and why?

(i) Bromine and fluorine

(ii) Carbon monoxide and nitrogen

(iii) Acetic acid and acetyl chloride, CH_3COCl

Colligative Properties

43. What mass of sugar, $C_{12}H_{22}O_{11}(s)$, must be dissolved in a 250-mL cup of water in order to raise the boiling point to 101.0 °C at a pressure of 1 atm?

44. Both camphor, $C_{10}H_{16}O(s)$, and naphthalene, $C_{10}H_8(s)$, are used to make mothballs. A solution in 100.0 g of ethanol contained 5.00 g of mothballs and had a boiling point of 78.89 °C, compared with the boiling point of the pure solvent of 78.41 °C. Were the mothballs made from camphor or from naphthalene?

*45. A 1.25-g sample of urea dissolved in 17.0 g of water gave a boiling point elevation of 0.65 °C. What is the molar mass of urea? Given that the composition of urea is 20.0% C, 46.7% N, 6.7% H, and 26.6% O, by mass, what is its molecular formula? Urea also dissolves in organic solvents. Suggest a possible Lewis structure for the urea molecule. How could boiling point elevation measurements be used to distinguish a sample of urea from a sample of ammonium cyanate, which has the same empirical formula? Write the Lewis structure of ammonium cyanate.

46. When 2.60 g of sulfur were dissolved in 200 g of diethyl ether, a rise in the boiling point of 0.105 °C was observed. What is the molar mass of sulfur dissolved in ether? What is the molecular structure of sulfur in ether? ($K_b^{ether} = 2.10$ °C kg mol^{-1}.)

*47. (a) What is the relationship between the vapor pressure of a pure solvent, $p°$, the mole fraction of an involatile solute, x_{solute}, and the vapor pressure of the solution, p?

(b) The hydrocarbon limonene is the major constituent of lemon oil. A solution of limonene in 78.0 g of benzene had a vapor pressure of 90.6 mm Hg at 25 °C, and the vapor pressure of pure benzene at 25 °C is 95.2 mm Hg. What is the approximate molar mass of limonene? What is its molecular formula and its accurate molar mass?

48. A solution containing 1.00 g of aluminum bromide in 100 g of benzene had a freezing point 0.099 °C lower than that of pure benzene. What is the molar mass of aluminum bromide dissolved in benzene? What is its molecular formula? Draw a Lewis structure for this molecule.

* The asterisk denotes the more difficult problems.

49. Down to approximately what temperature is salt effective in melting the ice on streets and sidewalks? (The solubility of sodium chloride in water is 5.8 molal at 0 °C.)

50. An aqueous solution containing 89.0 g of sucrose in 1.00 kg of solution had a freezing point of −0.48 °C. What is the molar mass of sucrose?

51. The main ingredient of commercial antifreeze is ethylene glycol (ethane-1,2-diol), $HOCH_2CH_2OH$. What concentration of aqueous glycol is needed to prevent an automobile radiator from freezing down to a temperature of −30 °C?

52. Calculate the freezing point and the boiling point of each of the following solutions:

(a) 50.0 g of urea, $(NH_2)_2CO$, in 750 g of water

(b) 17.1 g of sucrose, $C_{12}H_{22}O_{11}$, in 500 g of water

(c) 25.0 g of glycerol (1,2,3-trihydroxypropane), in 100 g of water

53. What is the freezing point of each of the following?

(a) A 0.10 molal solution of sucrose in water

(b) A 0.10 molal solution of $NaNO_3$ in water

(c) A 0.10 molal solution of $Ca(NO_3)_2$ in water

54. Arrange the following aqueous solutions in order of decreasing freezing point relative to that of pure water:

(a) 0.10 molal NaCl　　　(b) 0.40 molal sucrose

(c) 0.10 molal $BaCl_2$　　(d) 0.10 molal sucrose

55. (a) What equation relates the osmotic pressure of a dilute solution to its concentration at a given temperature?

(b) What osmotic pressure is expected to be exerted by an 0.001 molar solution of a nonelectrolyte at 25 °C?

(c) A solution contained 0.012 79 g of an aromatic hydrocarbon dissolved in 25.00 mL of cyclohexane and exerted an osmotic pressure of 74.7 mm Hg at 27 °C. What is the molar mass of the aromatic hydrocarbon?

56. Insulin is a protein that regulates carbohydrate metabolism by decreasing blood sugar levels. A deficiency of insulin leads to diabetes. An aqueous solution containing 5.00 g of insulin in 50 mL of solution was found to have an osmotic pressure of 0.427 atm at 25 °C. What is the molecular mass of insulin?

57. A 1.0 mass % solution of the polymer Gum Arabic (empirical formula $C_{12}H_{22}O_{11}$) was found to have an osmotic pressure of 7.22 mm Hg at 25 °C. What is the average

molar mass of Gum Arabic, and the number of monomer units per polymer molecule?

Reactions of Water

58. Write balanced equations for the reactions, if any, of each of the following substances with water. Classify each reaction as an acid–base reaction or as an oxidation–reduction reaction, as appropriate. In each case state whether the reaction occurs at 25 °C or only at high temperature.

(a) Calcium carbide   (b) Sulfur dioxide

(c) Sulfur trioxide   (d) Magnesium nitride

(e) Carbon monoxide   (f) Sodium amide

(g) White phosphorus   (h) Phosphorus trichloride

59. Write balanced equations for the reactions, if any, of each of the following substances with water. Classify each reaction as an acid–base or as an oxidation–reduction reaction, as appropriate. In each case state whether the reaction occurs at 25 °C or only at high temperature.

(a) Phosphorus pentachloride

(b) Calcium   (c) Sodium oxide

(d) Fluorine   (e) Chlorine

(f) Methane   (g) Magnesium hydride

60. Write balanced equations for two reactions in which water behaves as a Lewis base.

61. Write balanced equations for the half-reactions in which water behaves as (a) an oxidizing agent and (b) a reducing agent.

62. Explain why water is described as an amphoteric substance.

63. Balance each of the following equations and classify the reactant water as an acid, a base, an oxidizing agent, or a reducing agent:

(a) $Al(s) + H_2O(g) \rightarrow Al_2O_3(s) + H_2(g)$

(b) $KNH_2(s) + H_2O(l) \rightarrow KOH(aq) + NH_3(aq)$

(c) $F_2(g) + H_2O(l) \rightarrow HF(aq) + O_2(g)$

(d) $P_4O_{10}(s) + H_2O(l) \rightarrow H_3PO_4(aq)$

(e) $BaO_2(s) + H_2O(l) \rightarrow Ba(OH)_2(aq) + H_2O_2(aq)$

(f) $Cl_2O_7(l) + H_2O(l) \rightarrow HClO_4(aq)$

(g) $CaH_2(s) + H_2O(l) \rightarrow Ca(OH)_2(aq) + H_2(g)$

64. Give two examples of each of the following:

(a) A salt of a metal that gives a basic solution in water

(b) A salt of a metal that gives a neutral solution in water

(c) A salt of a metal that gives an acidic solution in water

65. Why is the reaction of an acid anhydride, such as SO_2, SO_3, or N_2O_5, described as a Lewis acid–base reaction?

66. Write balanced equations for each of the following:

(a) The behavior of $FeCl_3 \cdot 6H_2O$ as a weak acid in water

(b) The reactions in which $Al_2O_3(s)$ dissolves in (i) $HCl(aq)$ and (ii) $NaOH(aq)$

(c) The reaction that occurs when NH_4NO_3 is dissolved in water

Miscellaneous

67. Write balanced equations to describe each of the following processes:

(a) The industrial preparation of hydrogen

(b) A reaction for the preparation of pure oxygen

(c) A reaction in which water behaves as an oxidizing agent

(d) A reaction in which water behaves as a reducing agent

(e) The reaction of an acidic oxide with water

(f) The reaction of a basic oxide with water

68. In terms of the kinetic molecular theory, explain each of the following:

(a) Why a liquid evaporates below its boiling point

(b) Why ice disappears on a very cold sunny day in winter without melting

69. A given volume of acetic acid vapor at 150 °C and 1 atm pressure diffuses through a porous plug in 9.58 min. Repetition of the same experiment with the same volume of oxygen under the same conditions gave a diffusion time of 5.00 min. What is the molar mass of acetic acid under these conditions? In terms of intermolecular forces, account for the fact that this molar mass is not that calculated from the formula CH_3CO_2H.

70. Explain each of the following:

(a) Why it is possible for some substances to go from solid to vapor without going through the intermediate liquid stage

(b) Why the vapor pressure of a solution is less than that of the pure solvent

(c) Why the density of ice at 0 °C is less than that of water at the same temperature

(d) Why the freezing point of a solution is less than that of the pure solvent

71. Explain the principle on which each of the following is based, and in each case give at least one example of a practical application:

  (a) Fractional distillation

  (b) Depression of freezing point

  (c) Reverse osmosis

*72. Measurement of the freezing points of 0.010 molal solutions of water, potassium sulfate, nitric acid, and dinitrogen pentaoxide in *pure sulfuric acid* gave freezing point depressions of 0.122 °C, 0.245 °C, 0.245 °C, and 0.367 °C, respectively. Spectroscopic measurements showed that the last two solutions contain the nitronium ion, NO_2^+. Write balanced equations for the reaction of each of the solutes with sulfuric acid. ($K_f^{H_2SO_4} = 6.12$ °C kg mol^{-1}.)

73. Explain each of the following facts.

  (a) Water has a boiling point some 200 °C higher than would be expected in comparison with the boiling points of H_2S, H_2Se, and H_2Te.

  (b) The boiling point of water is even more anomalous than that of ammonia.

  (c) The density of ice is less than that of liquid water at the temperature of melting.

  (d) Water has its maximum density at 3.98 °C.

  (e) Methanol, CH_3OH, is miscible with water, while hexane, C_6H_{14}, is immiscible.

  (f) Sodium oxide is readily soluble in water, while magnesium oxide is insoluble.

74. Predict the products of the following reactions, balance each of the equations, and state whether the water behaves as a Brønsted acid, Brønsted base, Lewis acid, Lewis base, oxidizing agent, or reducing agent:

  (a) $P_4O_{10}(s) + H_2O(l) \rightarrow$

  (b) $PCl_3(l) + H_2O(l) \rightarrow$

  (c) $Ca(s) + H_2O(l) \rightarrow$

  (d) $F_2(g) + H_2O(l) \rightarrow$

  (e) $CO_3^{2-}(aq) + H_2O(l) \rightarrow$

  (f) $CaO(s) + H_2O(l) \rightarrow$

  (g) $HNO_3(aq) + H_2O(l) \rightarrow$

  (h) $NH_4Br(aq) + H_2O(l) \rightarrow$

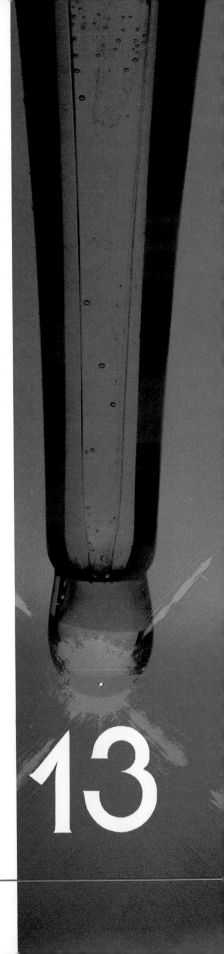

Chemical Equilibrium

The Nature of Equilibrium
The Equilibrium Constant
Heterogeneous Equilibria
The Effect of Changing Conditions
on Equilibria
Le Châtelier's Principle

13

A very important question concerning any reaction is the extent to which it proceeds under particular conditions. Does it proceed to completion, that is, until one of the reactants is completely used up? Does it proceed to only a limited extent, or perhaps not at all? In other words, do we obtain the maximum possible yield of the products, only a small yield, or perhaps no measurable yield? For example, we have seen in Chapter 5 that if hydrogen and chlorine are exposed to a bright light, an explosive reaction occurs in which the hydrogen and chlorine combine completely; in other words, a quantitative yield of HCl is obtained. The reaction is said to proceed to completion. On the other hand, if hydrogen and iodine are heated together, the reaction is much less vigorous and does not go to completion; some hydrogen and iodine remain unreacted.

We have also seen in Chapters 5 and 12 that a strong acid such as HCl ionizes completely in water; that is, all the HCl molecules donate their protons to H_2O molecules to give H_3O^+ and Cl^-. In contrast, a weak acid such as HF does not ionize completely in water. The HF molecules, H_2O molecules, H_3O^+ ions, and F^- ions are all present in a state of *equilibrium* in which their concentrations do not change. But the fact that the concentrations do not change does not mean that the reaction of HF with H_2O stops. In fact, HF molecules continue to donate protons to H_2O molecules, but at the same time H_3O^+ ions donate protons to F^- ions. At equilibrium the rates of these two reactions are equal, so the concentrations of all the species taking part in the reaction remain constant. We write the equation for such a reaction as follows:

$$HF(aq) + H_2O(l) \rightleftharpoons H_3O^+(aq) + F^-(aq)$$

where the double arrow indicates that the reaction is taking place in both directions simultaneously. Similarly, we write the equation for the reaction of H_2 with I_2, in which hydrogen, iodine, and hydrogen iodide are all present together when equilibrium is reached, as follows:

$$H_2(g) + I_2(g) \rightleftharpoons 2HI(g)$$

In Chapter 12 we met several examples of *physical equilibria*. For example, when a liquid such as water is kept in a closed container, water molecules evaporate at a certain rate which depends on the temperature, but they also return to the liquid (condense) at a rate which depends on their concentration in the gas phase. When the rate at which they condense equals the rate at which they evaporate, equilibrium is established and the constant pressure exerted by the water vapor in the gas phase is the vapor pressure of the water. A physical equilibrium differs from a chemical equilibrium in that in a physical process no bonds are broken and there are no changes in chemical composition. But physical and chemical equilibria have a very important feature in common, namely that at equilibrium a forward process and a reverse process are proceeding at the same rate so that there is no overall change in the system.

In this chapter we show how **chemical equilibria** can be treated in a quantitative manner. We will see that we can assign to a reaction a quantity called the *equilibrium constant* that enables us to predict how far a reaction will proceed toward completion before it comes to equilibrium. We will also see how the concentrations of the reactants and products that are present at equilibrium in a reaction depend on several factors, such as initial concentration of the reactants, pressure, and temperature. We will see how in an industrial process, such as the synthesis of ammonia, it is possible to adjust the conditions so as to maximize the yield of ammonia.

A physical equilibrium

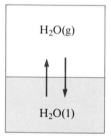

Rate ($H_2O(g) \rightarrow H_2O(l)$) = Rate ($H_2O(l) \rightarrow H_2O(g)$)

A chemical equilibrium

H₂	HI	I₂
I₂	H₂ I₂	H₂
I₂	HI	I₂
HI	H₂	HI
H₂		

Rate ($H_2 + I_2 \rightarrow HI + HI$) = Rate ($HI + HI \rightarrow H_2 + I_2$)

In this chapter, in which we deal with the basic principles of chemical equilibrium, we concentrate mainly on reactions involving gases. We will apply these principles to acid–base reactions in solution in the next chapter, to precipitation reactions in Chapter 15 and to oxidation–reduction reactions in Chapter 17.

13.1
The Equilibrium Constant

All chemical reactions carried out in a *closed system* eventually reach a state of **equilibrium** in which the concentrations of all the reactants and products do not change with time. By "in a closed system" we mean that the total mass of the reactants and products remains constant; in other words, as the reaction proceeds, none of the products is removed or allowed to escape from the reaction mixture, and no additions of any of the reactants are made. Some reactions proceed very rapidly; others are much slower. But in all cases a reacting system eventually reaches equilibrium. In some cases, such as the reaction between H_2 and Cl_2 to give HCl, the concentrations of H_2 and Cl_2 that remain when equilibrium is reached are too small to be measured. In such a case we say that the reaction has gone to completion, and we usually write a single arrow between the reactants and the products in the equation for the reaction:

$$H_2(g) + Cl_2(g) \longrightarrow 2HCl(g)$$

In contrast, in the reaction between H_2 and I_2 in the gas phase there are measurable concentrations of all the reactants and products when equilibrium is reached. When I_2 vapor, which has a violet color, is mixed with colorless H_2, colorless hydrogen iodide, HI, is formed. As the reaction proceeds and I_2 is used up, the color of the I_2 vapor fades, eventually reaching a constant intensity that does not change further; the reaction has then reached equilibrium. The rate of the forward reaction between H_2 and I_2 is then equal to the rate of the reverse reaction between HI molecules to give back H_2 and I_2, so the concentrations of all three species, H_2, I_2, and HI, remain constant:

$$H_2(g) + I_2(g) \rightleftharpoons 2HI(g)$$

If we start with H_2 and I_2, then their concentrations decrease and the concentration of HI increases with time, until all three concentrations become constant when equilibrium is reached, as illustrated in Figure 13.1 on p. 632.

Equilibrium is reached whether we start with the reactants or the products. For example, a colorless sample of HI gradually becomes violet as I_2 is formed. The concentration of HI decreases, and the concentrations of H_2 and I_2 increase. Eventually, the intensity of the violet color due to the I_2 becomes constant as the concentrations of H_2, I_2, and HI reach their equilibrium values. Figure 13.1 shows that whether we start with 1 mol of I_2 and 1 mol of H_2 or with 2 mol of HI in a vessel of the same volume, the same equilibrium mixture of H_2, I_2, and HI is obtained. Table 13.1 gives the results of three experiments. They show that if we start with different initial concentrations of H_2 or I_2 or HI, we obtain different equilibrium concentrations. In fact every different set of initial concentrations gives a different equilibrium mixture. Table 13.2 gives the concentrations of a number of such equilibrium mixtures. Here and throughout this chapter we use [ ]$_{eq}$ to denote the concentrations of reactants or products *at equilibrium*.

FIGURE 13.1
Concentration Changes
During Approach to
Equilibrium.

$$H_2(g) + I_2(g) \rightleftharpoons 2HI(g)$$

On the left, we start with equal
initial concentrations of H_2 and
I_2. As the reaction proceeds, the
concentrations of H_2 and I_2 decrease and the concentration of
HI increases, until each attains a
constant value at equilibrium. *On
the right*, we start with HI alone.
As the reaction to give H_2 and I_2
proceeds, the concentration of HI
decreases and the concentrations
of H_2 and I_2 increase, until they
all become constant when equilibrium is reached. The initial concentration of HI on the right is
twice the initial concentration of
H_2 and I_2 on the left. Under these
circumstances the same equilibrium
mixture is attained whether we start
on the left or on the right.

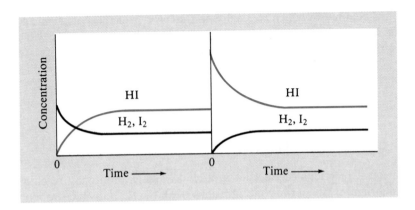

These equilibrium concentrations are all different but there is an important relationship between them. We see in the fourth column of the table that the
quantity K given by the expression

$$K = \frac{[HI]_{eq}^2}{[H_2]_{eq}[I_2]_{eq}}$$

has a constant value that is independent of the actual concentrations of the
reactants and the products at equilibrium. The constant K is called the **equilibrium constant** for the reaction. It has a value of 54.4 at 698 K for the H_2–I_2
reaction. It has a different value at another temperature, so the temperature
must always be stated.

Experiments have shown that for any reaction *at equilibrium* a similar expression involving the concentrations of the products and the reactants at
equilibrium has a characteristic value. We can write a general equation for a
reaction as follows:

$$aA + bB + cC + \cdots \rightleftharpoons pP + qQ + rR + \cdots$$

TABLE 13.1
Initial and Equilibrium Concentrations for the Reaction
$H_2(g) + I_2(g) \rightleftharpoons 2HI(g)$ at 698 K

	Concentration (mol L^{-1})		
	H_2	I_2	HI
Experiment 1			
Initial	4.46×10^{-2}	4.46×10^{-2}	0
Equilibrium	9.46×10^{-3}	9.46×10^{-3}	7.04×10^{-2}
Experiment 2			
Initial	1.68×10^{-2}	2.11×10^{-2}	9.40×10^{-2}
Equilibrium	1.20×10^{-2}	1.64×10^{-2}	0.104
Experiment 3			
Initial	0.811	0.394	1.362
Equilibrium	0.540	0.123	1.904

TABLE 13.2
Equilibrium Concentrations in the H_2–I_2 System at 698 K

$[I_2]_{eq}$ (mol L^{-1})	$[H_2]_{eq}$ (mol L^{-1})	$[HI]_{eq}$ (mol L^{-1})	$K = \left(\dfrac{[HI]^2}{[H_2][I_2]}\right)_{eq}$
0.47×10^{-3}	0.48×10^{-3}	3.5×10^{-3}	54.3
1.14×10^{-3}	1.14×10^{-3}	8.4×10^{-3}	54.3
1.71×10^{-3}	2.91×10^{-3}	16.5×10^{-3}	54.7
1.25×10^{-3}	3.56×10^{-3}	15.6×10^{-3}	54.7
0.74×10^{-3}	4.56×10^{-3}	13.5×10^{-3}	54.0
2.34×10^{-3}	2.25×10^{-3}	16.9×10^{-3}	54.2
3.13×10^{-3}	1.83×10^{-3}	17.7×10^{-3}	54.7
		Average value	54.4

In this equation A, B, C, and so on, are the reactants; and P, Q, R, and so on, are the products. The letters $a, b, c, \ldots, p, q, r, \ldots$ represent the *number of moles* of each substance involved in the balanced equation for the reaction. For this general reaction the equilibrium constant is

$$K = \left(\frac{[P]^p[Q]^q[R]^r \cdots}{[A]^a[B]^b[C]^c \cdots}\right)_{eq}$$

where $[A], [B], [C], \ldots, [P], [Q], [R], \ldots$ are the concentrations of the reactants and the products *at equilibrium*. To simplify the writing of the expression, we have dropped the subscript eq on each concentration term and have given the expression in brackets a subscript eq to remind us that the concentrations are equilibrium concentrations. In other words:

To obtain the expression for the equilibrium constant for any reaction, we raise the equilibrium concentration of each product to the power given by the number of moles of that product in the balanced equation for the reaction, and we multiply these. We then multiply the concentrations of each reactant, similarly raised to the power given by the number of moles of the reactant in the balanced equation. Finally, we divide the resulting expression for the products by that for the reactants.

For example, for the H_2–I_2 reaction $p = 2$, $a = 1$, and $b = 1$; so

$$K = \left(\frac{[HI]^2}{[H_2][I_2]}\right)_{eq}$$

For the reaction between N_2 and H_2 to give ammonia, NH_3,

$$N_2 + 3H_2 \rightleftharpoons 2NH_3$$

$p = 2$, $a = 1$, and $b = 3$; so

$$K = \left(\frac{[NH_3]^2}{[N_2][H_2]^3}\right)_{eq}$$

Some data for this reaction are given in Table 13.3. We see that K has a constant value independent of the concentrations of the reactants and products at equilibrium.

The expression for the equilibrium constant can be derived by the methods of thermodynamics and also by considering how the rates of the forward and reverse reactions depend on the concentrations of the reactants and products, as we will see in Chapter 19.

TABLE 13.3
Equilibrium Concentrations in the $N_2 + 3H_2 \rightleftharpoons 2NH_3$
System at 500 °C

$[H_2]_{eq}$ (mol L^{-1})	$[N_2]_{eq}$ (mol L^{-1})	$[NH_3]_{eq}$ (mol L^{-1})	$K = \left(\dfrac{[NH_3]^2}{[N_2][H_2]^3}\right)_{eq}$ (mol^{-2} L^2)
1.15	0.75	0.261	5.98×10^{-2}
0.50	1.00	0.087	6.05×10^{-2}
1.35	1.15	0.412	6.00×10^{-2}
2.43	1.85	1.27	6.08×10^{-2}
1.47	0.750	0.376	5.93×10^{-2}
		Average value	6.0×10^{-2}

Example 13.1

EQUILIBRIUM CONSTANT EXPRESSIONS

What is the expression for the equilibrium constant of each of the following reactions?

$$2SO_2(g) + O_2(g) \rightleftharpoons 2SO_3(g)$$
$$2H_2(g) + O_2(g) \rightleftharpoons 2H_2O(g)$$
$$CH_4(g) + H_2O(g) \rightleftharpoons CO(g) + 3H_2(g)$$

Solution

$$K_1 = \left(\frac{[SO_3]^2}{[SO_2]^2[O_2]}\right)_{eq} \qquad K_2 = \left(\frac{[H_2O]^2}{[H_2]^2[O_2]}\right)_{eq} \qquad K_3 = \left(\frac{[CO][H_2]^3}{[CH_4][H_2O]}\right)_{eq}$$

Measurements of the equilibrium concentrations of the reactants and the products have given the following values for the equilibrium constants in Example 13.1:

$$K_1 = 287 \text{ mol}^{-1} \text{ L at } 1000 \text{ °C}$$
$$K_2 = 5.9 \times 10^{40} \text{ mol}^{-1} \text{ L at } 25 \text{ °C}$$
$$K_3 = 3.73 \times 10^{-3} \text{ mol}^2 \text{ L}^{-2} \text{ at } 1000 \text{ K}$$

Note that the units of the equilibrium constant depend on the form of the equilibrium constant expression.

EXERCISE 13.1

Write equilibrium constant expressions for the following:

(a) The reaction of ethane with steam at high temperature,

$$C_2H_6(g) + 2H_2O(g) \rightleftharpoons 2CO(g) + 5H_2(g)$$

(b) The decomposition of NO_2 at high temperature,

$$2NO_2(g) \rightleftharpoons 2NO(g) + O_2(g)$$

Example 13.2

CALCULATION OF K FROM EXPERIMENTAL DATA

At equilibrium for the reaction

$$2SO_2(g) + O_2(g) \rightleftharpoons 2SO_3(g)$$

the concentrations of the reactants and products at 1000 K were found to be $[SO_2] = 0.27$ mol L^{-1}, $[O_2] = 0.40$ mol L^{-1}, and $[SO_3] = 0.33$ mol L^{-1}. What is the value of the equilibrium constant K at this temperature?

Solution

$$K = \frac{[SO_3]^2}{[SO_2]^2[O_2]} = \frac{(0.33 \text{ mol L}^{-1})^2}{(0.27 \text{ mol L}^{-1})^2(0.40 \text{ mol L}^{-1})} = 3.7 \text{ mol L}^{-1}$$

EXERCISE 13.2

At equilibrium for the reaction

$$H_2(g) + Br_2(g) \rightleftharpoons 2HBr(g)$$

a 10.0-L reaction vessel was found to contain 2.5×10^{-3} mol of H_2, 0.150 mol of HBr and 2.8×10^{-3} mol of Br_2. What is the value of K at this temperature?

EQUILIBRIUM CONSTANT EXPRESSIONS INVOLVING PRESSURES

For reactions involving gases, we will often find it convenient to express the concentrations in terms of the partial pressures of the gases. Consider a mixture of n_A moles of A, n_B moles of B, ..., in a volume V. If the partial pressure of A is p_A, then from the ideal gas law we may write

$$p_A V = n_A RT$$

or

$$p_A = \frac{n_A}{V} RT$$

Now n_A/V is the molar concentration of A, that is, $[A]$. Hence $p_A = [A]RT$. Similarly, we may write $p_B = [B]RT$, and so on, for every gas in the mixture.

At a constant temperature RT is a constant, so $p_A \propto [A]$. Similarly, for every gas in the mixture the partial pressure is proportional to the concentration. Therefore the equilibrium constant expression written in terms of partial pressures is also a constant for the reaction at a given temperature. For example, for the synthesis of NH_3,

$$N_2(g) + 3H_2(g) \rightleftharpoons 2NH_3(g)$$

we have

$$K_p = \left\{ \frac{(p_{NH_3})^2}{(p_{N_2})(p_{H_2})^3} \right\}_{eq}$$

where p_{NH_3} is the partial pressure of NH_3, and so on. The experimental value of K_p at 400 K is 55 atm^{-2}. Note that the units of K_p are pressure units. The value of K_p depends on the units used to express pressure, which are usually atmospheres. Note also that we write K_p with a subscript p to denote that the equilibrium constant has pressure units. Similarly, we may write K_c with a subscript c to denote that the equilibrium constant has concentration (mol L^{-1}) units.

─ *Example 13.3* ═══

CONVERSION OF K_p TO K_c

The value of K_p for the reaction

$$N_2(g) + 3H_2(g) \rightleftharpoons 2NH_3(g)$$

is 55 atm^{-2} at 400 K. What is the value of K_c at the same temperature?

Solution

As we have seen above, $p_{NH_3} = [NH_3]RT$, etc., so we may write

$$K_p = \frac{(p_{NH_3})^2}{(p_{N_2})(p_{H_2})^3} = \frac{([NH_3]RT)^2}{([N_2]RT)([H_2]RT)^3} = \frac{[NH_3]^2}{[N_2][H_2]^3}(RT)^{-2}$$

$$= K_c(RT)^{-2}$$

Hence

$$K_c = K_p(RT)^2 = (55 \text{ atm}^{-2})[(0.0821 \text{ L atm K}^{-1} \text{ mol}^{-1})(400 \text{ K})]^2$$

$$= 5.9 \times 10^4 \text{ mol}^{-2} \text{ L}^2$$

───

In general, K_p and K_c are related by the formula

$$K_p = K_c(RT)^{(p+q+r\cdots)-(a+b+c\cdots)} = K_c(RT)^{\Delta n}$$

where Δn is the difference between the total number of moles of products and the total number of moles of reactants in the balanced equation for the reaction. For example, in Example 13.3, $a = 1$, $b = 3$, and $p = 2$, so that $\Delta n = -2$ and $K_p = K_c(RT)^{-2}$ as deduced in the example.

EXERCISE 13.3

Calculate K_c at 900 K for the reaction

$$N_2(g) + 3F_2(g) \rightleftharpoons 2NF_3(g)$$

for which $K_p = 4.72$ atm^{-2}.

THE EFFECT OF MULTIPLYING AN EQUATION BY A CONSTANT

The value and the units of the equilibrium constant for a reaction depend on the form of the equation that we use to represent the reaction. For example, we sometimes write the equation for the reaction of hydrogen and oxygen to give water in the form

$$H_2(g) + \tfrac{1}{2}O_2(g) \rightleftharpoons H_2O(g)$$

for which the equilibrium constant expression is

$$K = \frac{[H_2O]}{[H_2][O_2]^{1/2}} = 2.4 \times 10^{20} \text{ mol}^{-1/2} \text{ L}^{1/2}$$

We see that K for the reaction written this way is related to K_2 for the reaction written as in Example 13.1 by the expression $K = K_2^{1/2}$. If, as in this case, we multiply the coefficients in an equation by $\tfrac{1}{2}$, then we must raise the equilibrium constant K to the power $\tfrac{1}{2}$. Or in general, if we multiply the coefficients in the equation by a factor n, then we must raise the equilibrium constant for the original equilibrium to the power n to obtain the new equilibrium constant.

EXERCISE 13.4

For the reaction

$$H_2(g) + Br_2(g) \rightleftharpoons 2HBr(g)$$

$K_p = 3.5 \times 10^4$ at 1495 K. What is the value of K_p for the following reactions at 1495 K?

(a) $2H_2(g) + 2Br_2(g) \rightleftharpoons 4HBr(g)$

(b) $\tfrac{1}{2}H_2(g) + \tfrac{1}{2}Br_2(g) \rightleftharpoons HBr(g)$

SIMULTANEOUS REACTIONS

Sometimes we need the equilibrium constant for a reaction that is the sum of two or more reactions. For example, by adding the two equations

$$2HCl(g) \rightleftharpoons H_2(g) + Cl_2(g) \qquad K_1 = 4.2 \times 10^{-34} \text{ at } 25\,^{\circ}C$$

and

$$I_2(g) + Cl_2(g) \rightleftharpoons 2ICl(g) \qquad K_2 = 2.1 \times 10^5 \text{ at } 25\,^{\circ}C$$

we obtain the equation for the reaction of HCl with I_2, namely

$$2HCl(g) + I_2(g) \rightleftharpoons 2ICl(g) + H_2(g)$$

We can obtain the equilibrium constant expression K_3 for this reaction by simply multiplying together K_1 and K_2:

$$K_1 K_2 = \frac{[H_2][Cl_2]}{[HCl]^2} \times \frac{[ICl]^2}{[I_2][Cl_2]} = \frac{[H_2][ICl]^2}{[HCl]^2[I_2]} = K_3$$

$$= (4.2 \times 10^{-34})(2.1 \times 10^5) = 8.8 \times 10^{-29} \text{ at } 25\,^{\circ}C$$

In general, if a reaction is the sum of two other reactions, 1 and 2, then the equilibrium constant for the overall reaction is the product of the equilibrium constants of reactions 1 and 2:

$$K_{overall} = K_1 K_2$$

THE POSITION OF EQUILIBRIUM

An important use of the value of the equilibrium constant for a reaction is to estimate approximately where the position of equilibrium lies, that is, whether the concentrations of the products are greater than the concentrations of the reactants or vice versa. For example, for the reaction between H_2 and Cl_2 in the gas phase,

$$H_2(g) + Cl_2(g) \rightleftharpoons 2HCl(g) \qquad K_c = \left(\frac{[HCl]^2}{[H_2][Cl_2]}\right)_{eq}$$

the equilibrium constant K at 25 °C has the very large value of 2.5×10^{33}. The large value of K indicates that at equilibrium the concentration of HCl is very large compared with the concentrations of the reactants H_2 and Cl_2. In other words, the reaction goes very nearly to completion. The position of the equilibrium is said to lie "far to the right", and the equation for the reaction is normally written with a single arrow:

$$H_2(g) + Cl_2(g) \longrightarrow 2HCl(g)$$

In contrast, the equilibrium constant for the decomposition of water to H_2 and O_2,

$$2H_2O(g) \rightleftharpoons 2H_2(g) + O_2(g) \qquad K_c = \left(\frac{[H_2]^2[O_2]}{[H_2O]^2}\right)_{eq}$$

has the very small value of 1.7×10^{-41} mol L^{-1} at 25 °C. This small value indicates that the position of the equilibrium lies far to the left. At equilibrium there is a very large concentration of H_2O and only small concentrations of O_2 and H_2. Thus if we start with H_2 and O_2 instead of H_2O, then the reaction goes almost to completion. That is, at equilibrium there is a very large concentration of H_2O and only very small concentrations of H_2 and O_2.

The *formation* of H_2O from H_2 and O_2,

$$2H_2(g) + O_2(g) \rightleftharpoons 2H_2O(g)$$

has the equilibrium constant

$$K_c = \left(\frac{[H_2O]^2}{[H_2]^2[O_2]}\right)_{eq}$$

If we call the equilibrium constant for the decomposition of water K_{decomp} and the equilibrium constant for the formation of water K_{form}, we see that

$$K_{form} = \left(\frac{[H_2O]^2}{[H_2]^2[O_2]}\right)_{eq} = \frac{1}{\left(\frac{[H_2]^2[O_2]}{[H_2O]^2}\right)_{eq}} = \frac{1}{K_{decomp}}$$

Therefore from the value for K_{decomp} we can find the value for K_{form}:

$$K_{form} = \frac{1}{K_{decomp}} = \frac{1}{1.7 \times 10^{-41} \text{ mol L}^{-1}} = 5.9 \times 10^{40} \text{ mol}^{-1} \text{ L}$$

As we expect, the equilibrium constant for the formation of water has a very large value; the reaction goes very nearly to completion.

In general, the equilibrium constant for a reaction as written from right to left is the reciprocal of the equilibrium constant for the reaction written from left to right:

$$K_{\text{right to left}} = \frac{1}{K_{\text{left to right}}}$$

EXERCISE 13.5

Using the data in Exercise 13.4 find the value of K_p for each of the following reactions:

(a) $2HBr(g) \rightleftharpoons H_2(g) + Br_2(g)$

(b) $HBr(g) \rightleftharpoons \frac{1}{2}H_2(g) + \frac{1}{2}Br_2(g)$

EXERCISE 13.6

Equilibrium constants for the following two reactions are given at 25 °C:

$$N_2(g) + \tfrac{1}{2}O_2(g) \rightleftharpoons N_2O(g) \qquad K = 7.1 \times 10^{-19} \text{ atm}^{-1/2}$$

$$N_2(g) + O_2(g) \rightleftharpoons 2NO(g) \qquad K = 4.2 \times 10^{-31}$$

What is the value of the equilibrium constant at 25 °C for the following reaction?

$$N_2O(g) + \tfrac{1}{2}O_2(g) \rightleftharpoons 2NO(g)$$

THE REACTION QUOTIENT

We can also use the equilibrium constant to predict in which direction a reaction will proceed for any given initial—that is, nonequilibrium—concentrations of reactants and products. Suppose we mix certain amounts of H_2, I_2, and HI. Will there be more or less HI when equilibrium is reached? For example, if 1.0×10^{-2} mol each of H_2 and I_2 and 2.0×10^{-3} mol of HI are placed in a 1-L flask at 698 K, will more HI or more H_2 and I_2 be formed as the system comes to equilibrium?

To answer this question, we calculate the **reaction quotient**, Q. It has the same form as the expression for the equilibrium constant, but the concentrations of reactants and products are *not* the equilibrium concentrations. We first find the value of the reaction quotient,

$$Q = \frac{[HI]^2}{[H_2][I_2]}$$

for the *initial concentrations*. Substituting, we have

$$Q = \frac{(2.0 \times 10^{-3})^2 \text{ mol}^2 \text{ L}^{-2}}{(1.0 \times 10^{-2}) \text{ mol L}^{-1} \times (1.0 \times 10^{-2}) \text{ mol L}^{-1}} = 0.040$$

We then compare this value with the value of K_c:

$$K_c = \left(\frac{[HI]^2}{[H_2][I_2]} \right)_{eq} = 54.4$$

We see that Q is much smaller than K_c. If the value of the reaction quotient

$$Q = \frac{[HI]^2}{[H_2][I_2]}$$

is to increase until it equals K_c, the concentration of HI must increase and the concentrations of H_2 and I_2 must decrease. Thus we conclude that, if we start with these initial concentrations of reactants and products, more HI will be formed until equilibrium is established.

If instead the initial concentration of HI is 2.0×10^{-1} mol L^{-1}, then we have

$$Q = \frac{(2.0 \times 10^{-1})^2 \text{ mol}^2 \text{ L}^{-2}}{(1.0 \times 10^{-2}) \text{ mol L}^{-1} \times (1.0 \times 10^{-2}) \text{ mol L}^{-1}} = 400$$

In this case the reaction quotient is greater than K_c. In order to establish equilibrium the concentration of HI must decrease and the concentrations of H_2 and I_2 must increase; more H_2 and I_2 will be formed until equilibrium is reached.

Finally, suppose the initial concentration of HI is 7.38×10^{-2} mol L^{-1} and the H_2 and I_2 concentrations are each 1.00×10^{-2} mol L^{-1}. In this case the reaction quotient is

$$Q = \frac{(7.38 \times 10^{-2})^2}{(1.00 \times 10^{-2})^2} = 54.4$$

which is equal to K_c. Thus the system is at equilibrium, and there will be no change in the concentrations of H_2, I_2, or HI.

In short, if we know the initial concentrations of the reactants and the products, we can predict the direction in which a reaction will proceed, as follows:

- If $Q < K$, the concentrations of the products will increase and the concentrations of the reactants will decrease until equilibrium is reached.
- If $Q > K$, the concentrations of the reactants will increase and the concentrations of the products will decrease until equilibrium is reached.
- If $Q = K$, the reactants and the products are at equilibrium, and there will be no change in the concentrations of reactants or products.

═ Example 13.4 ═

THE REACTION QUOTIENT AND POSITION OF EQUILIBRIUM

The equilibrium constant for the reaction

$$2SO_2(g) + O_2(g) \rightleftharpoons 2SO_3(g)$$

is $K_p = 0.14$ atm^{-1} at 900 K. If a reaction vessel is filled with SO_3 at a partial pressure of 0.10 atm and with O_2 and SO_2 each at a partial pressure of 0.20 atm, is the reaction at equilibrium? If not, in what direction does it proceed?

Solution

The value of the reaction quotient Q for the initial conditions is

$$Q = \frac{(p_{SO_3})^2}{(p_{SO_2})^2(p_{O_2})} = \frac{(0.10 \text{ atm})^2}{(0.20 \text{ atm})^2(0.20 \text{ atm})} = 1.25 \text{ atm}^{-1}$$

Since $Q > K_p$, the reaction is not at equilibrium. The reaction will proceed so as to decrease the value of Q. Thus more SO_2 and O_2 will be formed, and the amount of SO_3 will decrease. In other words, the reaction will proceed to the left.

EXERCISE 13.7

The equilibrium constant for the reaction

$$2NO(g) + O_2(g) \rightleftharpoons 2NO_2(g)$$

is 1.20 L mol^{-1} at 1000 K. The following concentrations were found in a reaction vessel: $[O_2] = 1.25$ mol L^{-1}, $[NO] = 2.25$ mol L^{-1}, and $[NO_2] = 3.25$ mol L^{-1}. Is the system at equilibrium? If not, will the concentration of NO_2 be greater or smaller than 3.25 mol L^{-1} when equilibrium is attained?

CALCULATION OF EQUILIBRIUM CONCENTRATIONS

The most important use of the equilibrium constant is to calculate the concentrations of reactants and products that will be present at equilibrium starting with some initial concentrations. We illustrate this use of the equilibrium constant in Examples 13.5 and 13.6.

=== Example 13.5 ===

CALCULATION OF EQUILIBRIUM CONCENTRATIONS

Suppose we introduce 0.100 mol of H_2 and 0.100 mol of I_2 into a 10.0-L flask at 698 K. What will be the concentrations of H_2, I_2, and HI at equilibrium?

Solution

The first step in solving any equilibrium problem is to write the equation for the equilibrium reaction. In this case the reaction is

$$H_2 + I_2 \rightleftharpoons 2HI$$

The second step is to write the expression for the equilibrium constant and look up its value at the temperature of the reaction if the value is not given with the data for the problem. In this case the value $K_c = 54.4$ at 698 K has been given in Table 13.2. Thus we can write

$$K_c = \left(\frac{[HI]^2}{[H_2][I_2]}\right)_{eq} = 54.4$$

The third step is to write expressions for the concentrations of each of the substances present at equilibrium. We do not know these concentrations, but they are all related to each other through the balanced equation for the reaction. Suppose that when equilibrium is reached, x moles of H_2 have reacted. Then the equation shows that they must have combined with x moles of I_2 to form $2x$ moles of HI. We can conveniently write this information under the equation for the reaction:

	H_2	$+ I_2$	$\rightleftharpoons 2HI$
Initial amounts	0.100 mol	0.100 mol	0 mol
Equilibrium amounts	$(0.100 - x)$ mol	$(0.100 - x)$ mol	$2x$ mol
Equilibrium concentrations	$\dfrac{(0.100 - x)\ \text{mol}}{10.0\ \text{L}}$	$\dfrac{(0.100 - x)\ \text{mol}}{10.0\ \text{L}}$	$\dfrac{2x\ \text{mol}}{10.0\ \text{L}}$

The fourth step is to substitute these concentrations in the expression for K_c and to solve for x. We have

$$\frac{(2x/10.0)^2 \text{ mol}^2 \text{ L}^{-2}}{[(0.100-x)/10.0 \text{ L}][(0.100-x)/10.0 \text{ L}] \text{ mol}^2 \text{ L}^{-2}} = 54.4$$

Taking the square root of both sides of the equation we have

$$\frac{2x}{0.100-x} = 7.38$$

Solving for x gives

$$x = 0.0787$$

Substituting this value of x in the equilibrium concentrations gives

$$[H_2] = \frac{(0.100-0.0787) \text{ mol}}{10.0 \text{ L}} = 0.002\,13 \text{ mol L}^{-1}$$

Similarly,

$$[I_2] = 0.002\,13 \text{ mol L}^{-1} \qquad [HI] = 0.0157 \text{ mol L}^{-1}$$

We can check these results by calculating K_c:

$$K_c = \frac{[HI]^2}{[H_2][I_2]} = \frac{(0.0157)^2}{(0.002\,13)^2} = 54.3$$

≡ Example 13.6 ≡

CALCULATION OF EQUILIBRIUM CONCENTRATIONS

A 10.0-L flask is filled with 0.200 mol of HI at 698 K. What will be the concentrations of H_2, I_2, and HI at equilibrium?

Solution

Following the same steps as in Example 13.5 we have:

First step

$$2HI \rightleftharpoons H_2 + I_2$$

Second step

$$K_c = \left(\frac{[H_2][I_2]}{[HI]^2}\right)_{eq}$$

What is the value of the equilibrium constant for this reaction? In Example 13.5 we used the value for the equilibrium constant K_{form} for the formation of HI,

$$K_{form} = \left(\frac{[HI]^2}{[H_2][I_2]}\right)_{eq} = 54.4$$

If we write K_{decomp} for the equilibrium constant for the decomposition of HI, we have

$$K_{decomp} = \frac{1}{K_{form}} = \frac{1}{54.4} = 0.0184 = K_c$$

For the *third step*, we assume that at equilibrium x moles of H_2 are formed. The equation for the reaction shows that x moles of I_2 are also formed and that $2x$ moles of HI have reacted, so the amount present at equilibrium is $0.20 - 2x$.

	2HI	$\rightleftharpoons$	H_2	$+$	I_2
Initial amounts	0.200 mol		0 mol		0 mol
Equilibrium amounts	(0.200 − 2x) mol		x mol		x mol
Equilibrium concentrations	$\dfrac{(0.200 - 2x)\ \text{mol}}{10.0\ \text{L}}$		$\dfrac{x\ \text{mol}}{10.0\ \text{L}}$		$\dfrac{x\ \text{mol}}{10.0\ \text{L}}$

The fourth step is to substitute these values in the expression for the equilibrium constant:

$$K_c = \frac{(x/10.0)^2\ \text{mol}^2\ \text{L}^{-2}}{[(0.200 - 2x)/10.0]^2\ \text{mol}^2\ \text{L}^{-2}} = 0.0184$$

Taking the square root of both we have

$$\frac{x}{0.200 - 2x} = 0.136$$

Solving for x gives

$$x = 0.0214$$

Since the volume is 10.0 L, the concentrations at equilibrium are

$$[H_2] = 0.002\ 14\ \text{mol L}^{-1} \qquad [I_2] = 0.002\ 14\ \text{mol L}^{-1}$$

$$[HI] = \frac{0.200 - 2(0.0214)}{10.0} = 0.0157\ \text{mol L}^{-1}$$

Note that the equilibrium concentrations are exactly the same as the concentrations we obtained in Example 13.5 which demonstrates again that the same equilibrium position can be approached from both sides. In other words, the equilibrium mixture is the same whether we start with 0.01 mol L^{-1} of H_2 and 0.01 mol of I_2 or with 0.02 mol L^{-1} of HI in 10.0 L.

EXERCISE 13.8

The equilibrium constant for the reaction

$$CO(g) + H_2O(g) \rightleftharpoons CO_2(g) + H_2(g)$$

is 5.10 at 700 K. If 1 mol each of CO, H_2O, CO_2, and H_2 are sent into a 10.0-L reaction vessel at 700 K, what is the concentration of each substance when equilibrium is reached?

= *Example 13.7* =

CALCULATION OF EQUILIBRIUM PARTIAL PRESSURES

At 1000 K the equilibrium constant for the reaction

$$2SO_2(g) + O_2(g) \rightleftharpoons 2SO_3(g)$$

is $K_p = 3.50\ \text{atm}^{-1}$. If the total pressure in the reaction vessel is 1.00 atm, and the partial pressure of O_2 at equilibrium is 0.10 atm, what are the partial pressures of SO_2 and SO_3?

Solution

If we let the partial pressure of $SO_2 = x$ atm, then the partial pressure of $SO_3 = (1.00 - p_{O_2} - x)$ atm $= (0.90 - x)$ atm.

$$K = \frac{(p_{SO_3})^2}{(p_{SO_2})^2(p_{O_2})} = \frac{(0.90 - x)^2}{x^2 \times 0.1} \text{ atm}^{-1} = 3.50 \text{ atm}^{-1}$$

$$\frac{0.90 - x}{x} = (0.35)^{1/2}$$

$$x = 0.57$$

$$p_{SO_2} = 0.57 \text{ atm} \qquad p_{SO_3} = 0.33 \text{ atm}$$

EXERCISE 13.9

Phosphorus pentachloride dissociates to phosphorus trichloride and chlorine on heating:

$$PCl_5(g) \rightleftharpoons PCl_3(g) + Cl_2(g)$$

When 2.42 g of PCl_5 were placed in a 2.00-L reaction vessel and heated to 250 °C they were completely vaporized, and the pressure in the vessel was found to be 359 mm Hg when equilibrium was reached. What was the partial pressure of each gas at equilibrium? What is the value of the equilibrium constant K_p at 250 °C?

13.2
Heterogeneous Equilibria

In the equilibria that we have discussed so far in this chapter, all the reactants and products have been gases. Equilibria in which all the substances involved are in the same phase are called **homogeneous equilibria**. However, many equilibria involve solids and gases, solids and liquids, or liquids and gases. These equilibria are called **heterogeneous equilibria**.

An example is the decomposition of calcium carbonate, which when heated gives calcium oxide and carbon dioxide:

$$CaCO_3(s) \rightleftharpoons CaO(s) + CO_2(g)$$

The equilibrium constant for this reaction is

$$K'_c = \left(\frac{[CaO][CO_2]}{[CaCO_3]}\right)_{eq}$$

But $CaCO_3$ and CaO are pure solids. What do we mean by the concentration of a pure solid? The number of molecules or ions in a given volume of a pure solid is fixed and cannot vary. Thus the "concentration" of a pure solid, the number of moles per liter, is constant and is independent of the amount of pure solid that is present. So both $[CaO]$ and $[CaCO_3]$ are constant. We can rearrange the expression for K'_c to give

$$\frac{K'_c[CaCO_3]}{[CaO]} = [CO_2]_{eq}$$

and since $[CaCO_3]$ and $[CaO]$ are constant, $K'_c[CaCO_3]/[CaO]$ is a constant, which we will write as K_c. Thus

$$K_c = [CO_2]_{eq}$$

or in terms of partial pressures

$$K_p = (p_{CO_2})_{eq}$$

At 900 °C, $K_p = 1.04$ atm; thus if $CaCO_3$ is heated in a closed vessel, it will decompose to CaO and CO_2 until the pressure of CO_2 reaches 1.04 atm. The system will then have reached equilibrium, and the amounts of $CaCO_3$, CaO, and CO_2 will remain constant. However, if $CaCO_3$ is heated in a vessel open to the atmosphere, the CO_2 diffuses away, and it never attains a partial pressure of 1.04 atm. So the reaction is driven to completion, and all the $CaCO_3$ decomposes to CaO (see Figure 13.2).

The same considerations apply to pure liquids so we can state:

The pure solids and liquids taking part in heterogeneous equilibria are not included in the equilibrium constant expression.

$=$ *Example 13.8* $=$

EQUILIBRIUM CONSTANT EXPRESSIONS FOR HETEROGENEOUS EQUILIBRIA

Write the expressions for the equilibrium constants for the following reactions in terms of concentrations and in terms of partial pressures:

(a) $$C(s) + CO_2(g) \rightleftharpoons 2CO(g)$$

(b) $$FeO(s) + CO(g) \rightleftharpoons Fe(s) + CO_2(g)$$

(c) $$PCl_5(s) \rightleftharpoons PCl_3(l) + Cl_2(g)$$

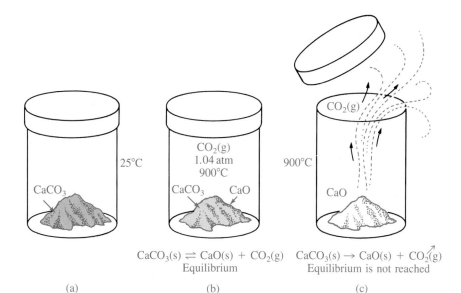

$CaCO_3(s) \rightleftharpoons CaO(s) + CO_2(g)$ $CaCO_3(s) \rightarrow CaO(s) + CO_2(g)$
Equilibrium   Equilibrium is not reached

(a)   (b)   (c)

FIGURE 13.2
Heterogeneous Equilibrium.

(a) At 25 °C there is no observable tendency for solid $CaCO_3$ to decompose. (b) When solid $CaCO_3$ is heated in a closed vessel at 900 °C, it decomposes to give solid CaO and gaseous CO_2. When the pressure of CO_2 reaches 1.04 atm, equilibrium is established. (c) When $CaCO_3$ is heated in an open vessel, the CO_2 diffuses away. The equilibrium pressure of CO_2 is never reached, and $CaCO_3$ is completely converted to CaO.

Solution

(a) C(s) is not included in the equilibrium constant expression.

$$K_c = \left(\frac{[CO]^2}{[CO_2]}\right)_{eq} \qquad K_p = \left(\frac{(p_{CO})^2}{p_{CO_2}}\right)_{eq}$$

(b) FeO(s) and Fe(s) are not included in the equilibrium constant expression.

$$K_c = \left(\frac{[CO_2]}{[CO]}\right)_{eq} \qquad K_p = \left(\frac{p_{CO_2}}{p_{CO}}\right)_{eq}$$

(c) $PCl_5(s)$ and $PCl_3(l)$ are not included in the equilibrium constant expression.

$$K_c = [Cl_2]_{eq} \qquad K_p = (p_{Cl_2})_{eq}$$

EXERCISE 13.10

Write expressions for the equilibrium constants K_c and K_p for the following equilibria:

(a) $NH_4Cl(s) \rightleftharpoons NH_3(g) + HCl(g)$

(b) $3Fe(s) + 4H_2O(g) \rightleftharpoons Fe_3O_4(s) + 4H_2(g)$

$=$ *Example 13.9* $=$

HETEROGENEOUS EQUILIBRIA

Carbon dioxide is reduced to carbon monoxide by heating with carbon at a high temperature:

$$C(s) + CO_2(g) \rightleftharpoons 2CO(g) \qquad K_p = 1.90 \text{ atm}$$

In a particular experiment the total pressure at equilibrium was found to be 2.00 atm. What were the partial pressures of CO and CO_2?

Solution

If we let the partial pressure of CO be x atm then since the total pressure is 2.00 atm the partial pressure of CO_2 is $(2.00 - x)$ atm.

$$K = \frac{(p_{CO})^2}{p_{CO_2}} = \frac{x^2}{2.00 - x} \text{ atm} = 1.90 \text{ atm}$$

Rearranging gives

$$x^2 + 1.90x - 3.80 = 0$$

Solving this quadratic equation (see Appendix A) gives

$$x = 1.22 \text{ or } x = -3.12$$

We can ignore the negative value of x as it has no physical significance, so

$$p_{CO} = x \text{ atm} = 1.22 \text{ atm}$$

and

$$p_{CO_2} = (2.00 - x) \text{ atm} = 0.78 \text{ atm}$$

EXERCISE 13.11

The equilibrium constant for the reaction

$$NH_4Cl(s) \rightleftharpoons NH_3(g) + HCl(g)$$

is 1.04×10^{-2} atm^2 at 275 °C. What will be the partial pressures of NH_3 and HCl in equilibrium with $NH_4Cl(s)$ in a closed vessel at 275 °C?

13.3
The Effect of Changing Conditions on an Equilibrium: Le Châtelier's Principle

We are often concerned with how to maximize the amount of product that can be obtained from an equilibrium reaction. For example, in the industrial manufacture of ammonia from nitrogen and hydrogen (Chapter 3),

$$N_2(g) + 3H_2(g) \rightleftharpoons 2NH_3(g)$$

it is important to be able to adjust the conditions so as to obtain the largest possible yield of ammonia. The position of an equilibrium, that is, the concentrations of reactants and products at equilibrium, is determined (1) by the initial concentrations or partial pressures of the reactants and products, and (2) by the temperature. We have seen how we can calculate the equilibrium amounts of the reactants and products if we know the equilibrium constant at any given temperature. We have already seen in Chapter 8 and some of the following chapters that Le Châtelier's principle is very useful for predicting how the position of an equilibrium can be shifted by changing the concentration of one of the reactants or products. Now we will look more closely at this very useful principle to see how we can use it to predict qualitatively the effect on the position of an equilibrium of changing concentrations, pressures, or temperature, even if we do not know the equilibrium constant for a reaction.

Recall from Chapter 8 that Le Châtelier's principle states that:

If any of the conditions affecting a system at equilibrium is changed, the position of the equilibrium shifts so as to minimize the change.

CONCENTRATION CHANGES

We can restate Le Châtelier's principle for the special case of concentration changes:

When the concentrations of any of the reactants or products in a reaction at equilibrium are changed, the position of the equilibrium shifts so as to reduce the change in concentration that was made.

For example, if to the reaction

$$H_2(g) + I_2(g) \rightleftharpoons 2HI(g)$$

at equilibrium we add H_2, the concentration of H_2 is momentarily increased so that it is no longer the equilibrium concentration. According to Le Châtelier's principle, a new equilibrium will be reached in which the concentration of H_2 is less than it was after the addition of H_2 although more than in the original mixture (see Figure 13.3). The adjustment of the equilibrium never fully compensates for the change in concentration that was made. We can summarize the changes that occur in the system by stating that the addition of H_2 causes the position of the equilibrium to shift to the right.

We could arrive at the same conclusion by considering the reaction quotient,

$$Q = \frac{[HI]^2}{[H_2][I_2]}$$

After the addition of H_2 the concentration of H_2 is greater than the equilibrium value, so Q is less than K. The concentration of HI will therefore increase and the concentrations of H_2 and I_2 will decrease until $Q = K$ and equilibrium is reestablished. Again, we conclude that the addition of H_2 causes the position of equilibrium to shift to the right.

Similarly, we can predict that if we were to remove H_2 from the reaction mixture at equilibrium, more H_2 would be formed to replace the H_2 removed. A new equilibrium would be set up with a decreased concentration of HI and an increased concentration of I_2.

Although removing H_2 from this particular reaction mixture would not be very practical, the shift in an equilibrium caused by the removal of a product often has great practical importance. When a product is removed, a reaction that would otherwise come to equilibrium without going to completion can be driven to completion. This circumstance is favored, for example, when one of the products of a reaction is a gas or a substance that can easily be vaporized. We saw in Chapter 8 that, when sodium chloride reacts with concentrated sulfuric acid,

$$NaCl(s) + H_2SO_4(conc, aq) \rightleftharpoons HCl(g) + NaHSO_4(aq)$$

the equilibrium is shifted to the right, because HCl is a gas that escapes from the reaction mixture. Therefore the equilibrium concentration of HCl in the reaction mixture is never reached, and the reaction goes very nearly to completion. The effects of concentration changes on an equilibrium are demonstrated in Experiment 13.1.

FIGURE 13.3
Le Châtelier's Principle: Effect of Concentration Changes.

When H_2 is added to the system

$$H_2 + I_2 \rightleftharpoons 2HI$$

at equilibrium at time t_1 the equilibrium shifts to the right. So the concentrations of H_2 and I_2 both decrease and the concentration of HI increases, until a new position of equilibrium is reestablished at time t_2. The final equilibrium concentration of H_2 is less than it was after the addition of H_2 but greater than it was in the original equilibrium mixture.

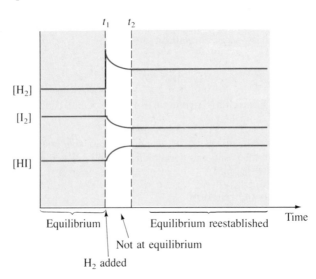

Experiment 13.1

Le Châtelier's Principle

The dish contains a concentrated aqueous solution of blue $CoCl_4^{2-}$ ions. They are in equilibrium with pink $Co(H_2O)_6^{2+}$ ion.

$$CoCl_4^{2-} + 6\,H_2O \rightleftharpoons Co(H_2O)_6^{2+} + 4Cl^-$$

$$\text{blue} \qquad\qquad\qquad \text{pink}$$

As water is added the equilibrium shifts to the right and the blue solution becomes pink.

Sufficient water has been added to turn the solution completely pink.

When a concentrated solution of chloride ion is added, the equilibrium shifts back to the left and the solution again becomes blue.

The formation of pink $Co(H_2O)_6^{2+}$ is an exothermic reaction. Thus when the solution is cooled the equilibrium shifts to the right and the solution turns pink. When the solution is heated the equilibrium shifts to the left and the solution turns blue.

PRESSURE CHANGES

The total pressure of a system at equilibrium can be changed in two ways:

1. By adding an inert gas (a gas that does not react with any of the reactants and products) while keeping the volume of the reaction vessel constant. The total pressure is increased, but although the total pressure is increased, the partial pressures of the reactants and products are not changed so *increasing the pressure by the addition of an inert gas has no effect on the position of an equilibrium.*

2. By decreasing the volume of the reaction vessel so that the total pressure is increased. In this case the partial pressures, and therefore the concentrations, of all the reactants and products are increased. The system is no longer at equilibrium and reaction occurs until a new equilibrium is reached.

Let us see why, in the Haber process for synthesis of ammonia (Chapter 3), a high pressure is used. Suppose that the equilibrium

$$N_2(g) + 3H_2(g) \rightleftharpoons 2NH_3(g)$$

has been established in a vessel with a variable volume. What is the effect of decreasing the volume of the vessel, thereby increasing the concentrations, or partial pressures, of all the gases and, therefore, the total pressure? (see Figure 13.4). Le Châtelier's principle predicts that the position of equilibrium will shift so as to reduce the pressure increase. Thus we need to know whether a shift in the position of the equilibrium to the left or to the right will cause a pressure decrease. Because the pressure is proportional to the total number of molecules in a given volume the equilibrium will shift in the direction that reduces the total number of molecules. Thus in the present case the position of the equilibrium will shift to the right because the formation of two molecules of product ($2NH_3$) causes the disappearance of four molecules of reactants ($N_2 + 3H_3$) and thus a decrease in the total number of molecules. The amount of NH_3 at equilibrium is therefore increased by increasing the pressure. Hence, the Haber process for the synthesis of ammonia employs a pressure in the range of 100–300 atm.

Consider now the decomposition of ethane to ethene and hydrogen:

$$C_2H_6(g) \rightleftharpoons C_2H_4(g) + H_2(g)$$

There are more molecules on the right than on the left. Therefore in this case a pressure increase will shift the equilibrium to the left. In other words, the amounts of C_2H_4 and H_2 at equilibrium will be reduced by increasing the pressure. The reaction is therefore carried out at atmospheric pressure.

We can summarize the effect of a pressure increase on any gas phase equilibrium as follows:

$$\text{Reactants} \underset{\text{Fewer molecules on left}}{\overset{\text{Fewer molecules on right}}{\rightleftharpoons}} \text{Products}$$

If, as in the case of the equilibrium

$$H_2(g) + I_2(g) \rightleftharpoons 2HI(g)$$

the reaction is not accompanied by any change in the total number of molecules, a change of pressure has no effect on the position of the equilibrium.

FIGURE 13.4
Le Châtelier's Principle:
Effect of Pressure.

(a) In the reaction

$$N_2 + 3H_2 \rightleftharpoons 2NH_3$$

if the volume of the reaction vessel is decreased to half its original value, (b), the concentrations (partial pressures) of N_2, H_2, and NH_3 are all doubled, and the total pressure is therefore doubled. The equilibrium thus shifts to the right, (c), which decreases the total number of molecules and thereby decreases the total pressure. When equilibrium is reestablished, the final pressure P_f is less than $2P_i$ but greater than P_i.

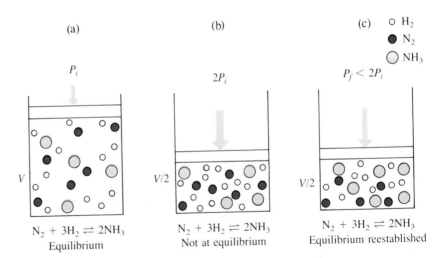

EXERCISE 13.12

Predict how the position of the equilibrium shifts when the volume of the system is reduced, thus increasing the total pressure, for each of the following reaction equilibria:

(a) $$PCl_3(g) + Cl_2(g) \rightleftharpoons PCl_5(g)$$

(b) $$N_2(g) + O_2(g) \rightleftharpoons 2NO(g)$$

(c) $$CaCO_3(s) \rightleftharpoons CaO(s) + CO_2(g)$$

TEMPERATURE CHANGES

It is important to understand that although changes in concentrations and pressure may affect the position of an equilibrium they do *not* affect the value of the equilibrium constant. After a change in conditions has been made the new equilibrium concentrations or pressure still satisfy the *same* equilibrium constant. But the effect of temperature is different because *the value of the equilibrium constant depends on the temperature.* Thus for the same initial concentrations of reactants and products a change in temperature will shift the position of equilibrium because the value of the equilibrium constant changes. If we know how the equilibrium constant changes with temperature we can, of course, calculate the equilibrium concentrations at a new temperature. But Le Châtelier's principle enables us to make a qualitative prediction of the effect of temperature provided we know whether the reaction is exothermic or endothermic.

The formation of hydrogen iodide from hydrogen and iodine is an exothermic reaction. For the purpose of applying Le Châtelier's principle, we can conveniently write this equilibrium as

$$H_2(g) + I_2(g) \rightleftharpoons 2HI(g) + Heat$$

When this reaction proceeds to the right, energy is liberated in the form of heat, which raises the temperature of the reaction mixture. According to Le Châtelier's principle, changing the temperature causes the position of the equilibrium to shift in the direction that minimizes the temperature change. Thus if the temperature is decreased, the hydrogen–iodine equilibrium will shift to the right, increasing the concentration of hydrogen iodide, because the heat thus produced will tend to raise the temperature, thus counteracting the initial temperature decrease. Conversely, if the temperature is increased, the equilibrium will shift to the left, because in this direction the reaction is endothermic, and the heat absorbed will tend to lower the temperature of the system, thus counteracting the initial temperature increase.

We can come to the same conclusion if we know the value of the equilibrium constant at two temperatures. In this case the equilibrium constant,

$$K_c = \left(\frac{[HI]^2}{[H_2][I_2]} \right)_{eq}$$

decreases from 67.5 at 325 °C to 54.4 at 425 °C. Thus the yield of HI decreases with increasing temperature, as we predicted from Le Châtelier's principle.

The synthesis of ammonia is an exothermic reaction:

$$N_2(g) + 3H_2(g) \rightleftharpoons 2NH_3(g) \qquad \Delta H° = -92.38 \text{ kJ}$$

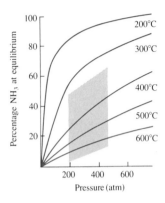

FIGURE 13.5
Effect of Temperature and Pressure on the Equilibrium $N_2 + 3H_2 \rightleftharpoons 2NH_3$.

Increasing pressure increases the percentage of NH_3 at equilibrium; increasing temperature decreases the percentage of NH_3 at equilibrium. The range of conditions normally used in the Haber process is shown by the shaded area.

We may therefore write

$$N_2(g) + 3H_2(g) \rightleftharpoons 2NH_3(g) + \text{Heat}$$

According to Le Châtelier's principle, raising the temperature shifts the equilibrium to the left and decreases the equilibrium concentration of ammonia. Thus for a maximum yield of ammonia the reaction should be carried out at as low a temperature as possible. A very low temperature cannot be used, however, because the reaction is then too slow. In practice, a catalyst is used to speed up the reaction, and even then a temperature in the range 400–500 °C is necessary in order to obtain a reasonable reaction rate (Figure 13.5). *Catalysts do not change the position of an equilibrium, but they do increase the rate at which equilibrium is attained.* We will discuss catalysts in more detail in Chapter 19. The effect of a temperature change on the position of an equilibrium is demonstrated in Experiment 13.1.

We can summarize the effect of a *temperature increase* on an equilibrium as follows:

Exothermic reactions	*Endothermic reactions*
Equilibrium concentration of products decreases	Equilibrium concentration of products increases
Position of equilibrium shifts to left	Position of equilibrium shifts to right
K decreases	K increases

The effects of a temperature decrease are just the reverse of the effects of a temperature increase.

══ *Example 13.10* ══════════════════

USING LE CHÂTELIER'S PRINCIPLE

What conditions of pressure and temperature give a high equilibrium yield of products in the following reactions?

(a) $2SO_2(g) + O_2(g) \rightleftharpoons 2SO_3(g)$ 　　 $\Delta H° = -198$ kJ

(b) $H_2(g) + CO_2(g) \rightleftharpoons H_2O(g) + CO(g)$ 　　 $\Delta H° = +41$ kJ

Solution

(a) This reaction is accompanied by a decrease in the number of gaseous molecules; hence a high equilibrium concentration of the product, SO_3, will be favored by a high pressure. The reaction is exothermic, so the equilibrium yield of SO_3 will be increased by carrying out the reaction at a low temperature.

(b) This reaction is not accompanied by any changes in the number of molecules, so it is not affected by a change in pressure. The reaction is endothermic, so the yield of H_2O and CO will be increased at a high temperature.

E X E R C I S E   1 3 . 1 3

For each of the following reactions predict which conditions of pressure and temperature will favor a high yield of product:

(a) $N_2(g) + O_2(g) \rightleftharpoons 2NO(g)$ 　　 $\Delta H° = 173$ kJ

(b) $CO(g) + 3H_2(g) \rightleftharpoons H_2O(g) + CH_4(g)$ 　　 $\Delta H° = -206$ kJ

EXERCISE 13.14

The reaction

$$NH_4HS(s) \rightleftharpoons NH_3(g) + H_2S(g)$$

is at equilibrium in a vessel of volume V at 200 °C. The reaction is endothermic. For each of the following changes, state whether the partial pressure of NH_3 will increase, decrease, or remain the same when equilibrium has been reestablished.

(a) $NH_3(g)$ is added

(b) $H_2S(g)$ is added

(c) $NH_4HS(s)$ is added

(d) The temperature is increased

(e) The total pressure is increased by adding $Ar(g)$

(f) The volume of the reaction vessel is increased to $2V$

IMPORTANT TERMS

Equilibrium is said to have been reached in a chemical reaction when the rate of the forward reaction is equal to the rate of the back reaction and no further changes in the concentrations of the reactants or the products take place.

The **equilibrium constant** is an expression involving the concentrations of the reactants and the concentrations of the products; it has a constant value for a system at equilibrium. For the reaction

$$aA + bB + \cdots \rightleftharpoons pP + qQ + \cdots$$

the equilibrium constant expression is

$$K = \left(\frac{[P]^p[Q]^q \cdots}{[A]^a[B]^b \cdots} \right)_{eq}$$

In a **heterogeneous equilibrium** some of the participating substances are in different phases. For example some may be gases and the others solids.

In a **homogeneous equilibrium** all the participating substances are in the same phase. Commonly they are all in the gas phase or all in solution.

Le Châtelier's principle states that if any of the conditions that affect a system at equilibrium are changed, the position of the equilibrium shifts so as to minimize the change.

The **reaction quotient**, Q, is a ratio of concentration terms that has the same form as the equilibrium constant expression, but the concentrations are *not* the equilibrium concentrations. Its value, compared to the equilibrium constant, K, indicates the direction in which a reaction will proceed in order to establish equilibrium.

PROBLEMS *

Equilibrium Constant Expressions

1. (a) Explain the term equilibrium.

   (b) What is the difference between static equilibrium and dynamic equilibrium?

   (c) What is the difference between a physical equilibrium and a chemical equilibrium? Give two examples of each.

2. (a) What is the difference between the equilibrium constant expression and the reaction quotient?

   (b) If the value of the reaction quotient for a reaction is greater than that of the equilibrium constant expres-

sion do you expect that the concentrations of the reactants will be greater or smaller when equilibrium is reached? Under what conditions is the value of the reaction quotient equal to that of the equilibrium constant expression?

3. Write the equilibrium constant expressions K_c and K_p for each of the following reactions:

   (a) $2N_2O_5(g) \rightleftharpoons 4NO_2(g) + O_2(g)$

   (b) $2SO_2(g) + O_2(g) \rightleftharpoons 2SO_3(g)$

   (c) $SO_2(g) + \frac{1}{2}O_2(g) \rightleftharpoons SO_3(g)$

   (d) $P_4(g) + 5O_2(g) \rightleftharpoons P_4O_{10}(g)$

   (e) $PCl_5(g) \rightleftharpoons PCl_3(g) + Cl_2(g)$

4. For concentrations expressed in **(a)** moles per liter and **(b)** atmospheres, what are the units of the equilibrium constants K_c and K_p, respectively, for each of the equilibria in Problem 3?

5. Write the equilibrium constant expressions K_c and K_p for each of the following reactions:

 (a) $2NOBr(g) \rightleftharpoons 2NO(g) + Br_2(g)$

 (b) $Fe_2O_3(s) + 3CO(g) \rightleftharpoons 2Fe(s) + 3CO_2(g)$

 (c) $N_2O(g) + 4H_2(g) \rightleftharpoons 2NH_3(g) + H_2O(g)$

 (d) $2KNO_3(s) \rightleftharpoons 2KNO_2(s) + O_2(g)$

 (e) $2Pb(NO_3)_2(s) \rightleftharpoons 2PbO(s) + 4NO_2(g) + O_2(g)$

6. For each of the equilibria in Problem 5, what are the units of K_c, with concentrations expressed in moles per liter, and K_p, with concentration units expressed in atmospheres?

7. The equilibrium constant, K_c, for the equilibrium

$$N_2O_4(g) \rightleftharpoons 2NO_2(g)$$

has a value 0.212 mol L^{-1} at 100 °C. What is the value of K_p?

8. What is the value of K_p for each of the following equilibria?

 (a) $N_2(g) + O_2(g) \rightleftharpoons 2NO(g)$
  $K_c = 2.5 \times 10^{-3}$ at 2100 °C

 (b) $2H_2(g) + CO(g) \rightleftharpoons CH_3OH(g)$
  $K_c = 300$ mol^{-2} L^2 at 425 °C

 (c) $PCl_5(g) \rightleftharpoons PCl_3(g) + Cl_2(g)$
  $K_c = 6.35 \times 10^{-2}$ mol L^{-1} at 250 °C

9. The reaction of nitrogen with oxygen to give nitrogen monoxide,

$$N_2(g) + O_2(g) \rightleftharpoons 2NO(g)$$

has an equilibrium constant of 2.5×10^{-3} at 2100 °C.

 (a) What is the equilibrium constant expression?

 (b) What are the units of the equilibrium constant when **(i)** the concentrations are in moles per liter and **(ii)** the concentrations are expressed as partial pressures?

10. The equilibrium constant for the reaction

$$N_2O_4(g) \rightleftharpoons 2NO_2(g)$$

is 0.212 mol L^{-1} at 100 °C. What is the value of the equilibrium constant for the same reaction written as

 (a) $2NO_2(g) \rightleftharpoons N_2O_4(g)$

 (b) $NO_2(g) \rightleftharpoons \frac{1}{2}N_2O_4(g)$

Homogeneous Gas Phase Equilibria

11. At 425 °C, $K_c = 300$ mol^{-2} L^2 for the reaction in which methanol, $CH_3OH(g)$, is synthesized from hydrogen and carbon monoxide:

$$2H_2(g) + CO(g) \rightleftharpoons CH_3OH(g)$$

If the concentrations of H_2, CO, and CH_3OH are each 0.10 mol L^{-1}, is the system at equilibrium at 425 °C? If the reaction is not at equilibrium, will the concentration of CH_3OH be greater than, or less than, 0.10 mol L^{-1} when equilibrium is achieved?

12. When the reaction

$$2SO_2(g) + O_2(g) \rightleftharpoons 2SO_3(g)$$

had attained equilibrium at a particular temperature, the concentrations of the reactants and products were found to be $[SO_2] = 0.010$ mol L^{-1}, $[O_2] = 0.20$ mol L^{-1}, and $[SO_3] = 0.100$ mol L^{-1}. What is the value of K_c?

13. For the reaction

$$H_2(g) + CO_2(g) \rightleftharpoons H_2O(g) + CO(g)$$

at 600 K, the following concentrations, in moles per liter, were found at equilibrium: $[H_2] = 0.600$; $[CO_2] = 0.459$; $[H_2O] = 0.500$; $[CO] = 0.425$. Calculate the value of K_c for this equilibrium. What is the value of K_p?

14. For the reaction in Problem 13, at the same temperature, what would be the equilibrium concentrations if initially 1.00 mol of $H_2(g)$ and 1.00 mol of $CO_2(g)$ were placed in a sealed 5.00-L vessel?

15. At 1000 K, iodine molecules dissociate into iodine atoms with $K_c = 3.76 \times 10^{-5}$ mol L^{-1} for the equilibrium

$$I_2(g) \rightleftharpoons 2I(g)$$

What will be the equilibrium concentration of $I_2(g)$ and $I(g)$ at this temperature after initially introducing 1.00 mol of I_2 into a 2.00-L flask? What is the percent dissociation of I_2 at this temperature?

16. Phosphorus pentachloride, $PCl_5(g)$, dissociates at high temperature into phosphorus trichloride, $PCl_3(g)$, and chlorine, $Cl_2(g)$. Initially, 0.200 mol of PCl_5 were placed in a 5.00-L flask at 200 °C, and at equilibrium the concentration of PCl_5 was found to be 0.015 mol L^{-1}. Calculate the value of the equilibrium constant, K_c, for this reaction at 200 °C.

17. At high temperature, $H_2(g)$ and $CO_2(g)$ react to form $H_2O(g)$ and $CO(g)$. When 0.100 mol of H_2 and 0.200 mol of CO_2 were mixed in a 2.00-L flask at 2000 °C, the amount of water in the equilibrium mixture was found to be 1.541 g. What is the equilibrium constant, K_c, for this reaction at 2000 °C?

18. The equilibrium constant at 490 °C for the reaction

$$2HI(g) \rightleftharpoons H_2(g) + I_2(g)$$

is $K_c = 0.022$. What are the equilibrium concentrations of HI, H_2, and I_2, when an initial amount of 2.00 mol of HI(g) is placed in a 4.3-L flask at 490 °C?

19. The equilibrium constant K_c at 698 K is 54.4 for the reaction

$$H_2(g) + I_2(g) \rightleftharpoons 2HI(g)$$

A reaction mixture contains 0.10 mol L^{-1} each of H_2 and I_2, and 1.0 mol L^{-1} of HI.

(a) Has the reaction yet reached equilibrium?

(b) If not, calculate the concentrations of H_2, I_2, and HI at equilibrium.

(c) What would be the concentrations at equilibrium if one started with no H_2 or I_2 but 1.2 mol L^{-1} of HI?

20. For the reaction

$$N_2(g) + 3H_2(g) \rightleftharpoons 2NH_3(g)$$

the equilibrium constant, K_c, is 6.0×10^{-2} mol^{-2} L^2 at 500 °C. Initially, the concentrations of N_2 and H_2 were each 1.0 mol L^{-1}, and that of NH_3 was 0.10 mol L^{-1}. Answer each of the following questions, without explicitly solving for the equilibrium concentrations.

(a) Is the initial system at equilibrium?

(b) A student calculated that, in order to achieve equilibrium, 0.010 mol of the N_2 would be converted to ammonia. Was this student correct? If not, was the student's estimate too large, or too small?

21. The reaction

$$NO(g) + NO_2(g) \rightleftharpoons N_2O_3(g)$$

has the equilibrium constant $K_p = 2.00$ atm^{-1} at 20 °C.

(a) If, in an equilibrium mixture at 20 °C, the partial pressures of NO and NO_2 are 1.00 atm and 0.500 atm, respectively, what is the partial pressure of N_2O_3?

(b) Another mixture at 20 °C, but not at equilibrium, contains no N_2O_3, but has partial pressures of NO and NO_2 of 1.00 atm and 0.500 atm, respectively. What will be the partial pressure of each gas when this mixture has achieved equilibrium at 20 °C?

22. A 5.00-L flask containing 1 mol of HI(g) is heated to 800 °C. What is the percentage dissociation of the HI, given that the value of the equilibrium constant K_c, is, 6.34×10^{-4} at 800 °C, for the reaction

$$2HI(g) \rightleftharpoons H_2(g) + I_2(g)$$

23. Dinitrogen tetraoxide dissociates into nitrogen dioxide according to the equation

$$N_2O_4(g) \rightleftharpoons 2NO_2(g)$$

In a mixture of the two gases at 100 °C, the concentrations were found to be $[N_2O_4] = 0.10$ mol L^{-1} and $[NO_2] = 0.12$ mol L^{-1}.

(a) What is the value of the reaction quotient Q for this mixture?

(b) Given that $K_c = 0.212$ mol L^{-1} at 100 °C for the reaction, is the mixture at equilibrium?

(c) If not, will $[NO_2]$ increase, or decrease, as equilibrium is reached?

(d) What are the equilibrium concentrations of NO_2 and N_2O_4?

(e) What is the value of K_p at 100 °C for this reaction?

(f) What are the partial pressures of NO_2 and N_2O_4 in the equilibrium mixture?

Le Châtelier's Principle

24. The value of the equilibrium constant for a reaction increases as the temperature is increased. Is the forward reaction exothermic or endothermic? What can one say about the enthalpy change for the reverse reaction?

25. The dissociation of $PCl_5(g)$ to $PCl_3(g)$ and $Cl_2(g)$ is an endothermic reaction. What is the effect on the percentage dissociation of PCl_5 of each of the following changes?

(a) Compressing the gaseous mixture

(b) Increasing the volume of the gaseous mixture

(c) Decreasing the temperature

(d) Adding $Cl_2(g)$ to the equilibrium mixture at constant volume

26. The endothermic reaction

$$2NO(g) + Br_2(g) \rightleftharpoons 2NOBr(g)$$

has an equilibrium constant $K_p = 116.6$ atm^{-1} at 25 °C.

(a) If 0.108 atm of NOBr, 0.100 atm of NO, and 0.010 atm of Br_2 are mixed together at 0 °C, what reaction will occur, if any? Explain your answer.

(b) When 5.00 atm of NOBr is injected into a container at 50 °C, the equilibrium mixture of gases is found to contain 4.30 atm of NOBr. Calculate the value of K_p for the reaction at 50 °C. Compare the value of K_p at 50 °C with that at 25 °C and explain why it increases (or decreases).

27. For each of the following reactions, calculate the standard enthalpy change for the forward reaction, from the data in Appendix B. In each case, what is the effect on the position of equilibrium of changing temperature, and changing pressure? What combination of temperature and pressure will maximize the yield of product(s)?

(a) $2NO(g) + Cl_2(g) \rightleftharpoons 2NOCl(g)$

(b) $2SO_2(g) + O_2(g) \rightleftharpoons 2SO_3(g)$

(c) $N_2(g) + 3H_2(g) \rightleftharpoons 2NH_3(g)$

(d) $CO(g) + H_2O(g) \rightleftharpoons CO_2(g) + H_2(g)$

28. For the equilibrium

$$2SO_2(g) + O_2(g) \rightleftharpoons 2SO_3(g) \qquad \Delta H° = -198 \text{ kJ}$$

what will be the effect on the equilibrium concentration of SO_3 of each of the following changes?

(a) Doubling the volume of the reaction vessel

(b) Increasing the temperature at constant volume

(c) Adding more O_2 to the reaction vessel

(d) Adding helium to the reaction vessel at constant volume

29. In which direction will the following equilibrium shift in response to each of the following changes in conditions:

$$C(s) + 2H_2(g) \rightleftharpoons CH_4(g) \qquad \Delta H° = -75 \text{ kJ}$$

(a) Increasing the temperature

(b) Increasing the volume of the reaction vessel

(c) Increasing the partial pressure of hydrogen

(d) Adding more carbon

30. The equilibrium constant is $K_p = 3.2 \times 10^2 \text{ atm}^{-1/2}$ at 425 °C for the reaction

$$SO_2(g) + \tfrac{1}{2}O_2(g) \rightleftharpoons SO_3(g)$$

At 525 °C, the value of K_p decreases to 33 atm$^{-1/2}$. Is the reaction as written exothermic or endothermic?

31. Determine the effect of (i) an increase in temperature, and (ii) an increase in pressure, on the position of equilibrium of each of the following:

(a) $2NO_2(g) \rightleftharpoons N_2O_4(g)$

(b) $2NO_2(g) \rightleftharpoons 2NO(g) + O_2(g)$

Use such data as you need from Appendix B.

Heterogeneous Equilibria

32. Write the expressions for the equilibrium constants K_c and K_p for each of the following reactions:

(a) $H_2O(g) \rightleftharpoons H_2(g) + \tfrac{1}{2}O_2(g)$

(b) $H_2O(l) \rightleftharpoons H_2(g) + \tfrac{1}{2}O_2(g)$

(c) $2H_2O(l) \rightleftharpoons 2H_2(g) + O_2(g)$

(d) $H_2(g) + O_2(g) \rightleftharpoons H_2O_2(l)$

(e) $2HgO(s) \rightleftharpoons 2Hg(l) + O_2(g)$

33. What are the equilibrium constant expressions K_p and

K_c for each of the following reactions?

(a) $C(s) + O_2(g) \rightleftharpoons CO_2(g)$

(b) $MgCO_3(g) \rightleftharpoons MgO(s) + CO_2(g)$

(c) $2NaHCO_3(s) \rightleftharpoons Na_2CO_3(s) + CO_2(g) + H_2O(g)$

(d) $ZnO(s) + CO(g) \rightleftharpoons Zn(s) + CO_2(g)$

(e) $3Fe(s) + 4H_2O(g) \rightleftharpoons Fe_3O_4(s) + 4H_2(g)$

34. The following reaction is at equilibrium at 298 K in a closed 1.00-L vessel:

$$C(\text{graphite, s}) + O_2(g) \rightleftharpoons CO_2(g) \qquad \Delta H° = -393.5 \text{ kJ}$$

What will be the effect of each of the following changes on the equilibrium concentration of $O_2(g)$?

(a) Addition of graphite

(b) Addition of $CO_2(g)$

(c) Addition of $O_2(g)$

(d) Lowering the temperature

(e) Addition of a catalyst

35. At 25 °C, $K_p = 9.1 \text{ atm}^{-2}$ for the reaction

$$NH_3(g) + H_2S(g) \rightleftharpoons NH_4HS(s)$$

(a) Calculate the value of K_c for the reaction at 25 °C.

(b) How does the value of K_c change if the above equation is multiplied throughout by 2?

(c) If 1.00 mol of solid ammonium hydrogen sulfide, NH_4HS, was placed in an evacuated 1.00-L flask at 25 °C, what would be the total pressure of gases at equilibrium?

(d) Does your answer to part (c) depend on which balanced equation you use? Explain.

36. Suppose 1.00 mol each of $NH_3(g)$, $H_2S(g)$, and $NH_4HS(s)$ are placed in an evacuated 1.00-L flask at 25 °C. Use data from Problem 35 to decide whether there will be more than 1 mol of $NH_4HS(s)$ at equilibrium.

37. Calculate the equilibrium constant K_c for the reaction

$$MnO_2(s) + 2CO(g) \rightleftharpoons Mn(s) + 2CO_2(g)$$

from the equilibrium constants of the following reactions:

$$MnO_2(s) + 2H_2(g) \rightleftharpoons Mn(s) + 2H_2O(g) \qquad K_c = 182$$

$$CO(g) + H_2O(g) \rightleftharpoons CO_2(g) + H_2(g) \qquad K_c = 0.052$$

What is the value of K_p for the reaction?

Miscellaneous

38. For the gas phase reaction

$$CO(g) + H_2(g) \rightleftharpoons H_2CO(g)$$

(a) Use bond energy data from Table 6.3 to estimate $\Delta H°$ for this reaction

(b) Decide whether high or low temperature, and high or low pressure, will favor a high yield of methanal (formaldehyde), $H_2CO(g)$

* **39.** A volume of 1 L of dinitrogen tetraoxide, $N_2O_4(g)$, has a mass of 2.50 g at 60 °C and 1.00 atm pressure.

(a) What is the percentage dissociation of N_2O_4 to NO_2 at 60 °C?

(b) What is the value of K_p for the reaction $N_2O_4(g) \rightleftharpoons 2NO_2(g)$?

(c) What is the value of K_c?

40. Sulfur dioxide and oxygen in the mole ratio 2:1 were allowed to reach equilibrium in the presence of a catalyst, at a pressure of 5.00 atm. At equilibrium, 33% of the SO_2 was converted to $SO_3(g)$. What is the value of the equilibrium constant K_p for the reaction?

$$2SO_2(g) + O_2(g) \rightleftharpoons 2SO_3(g)$$

41. Dinitrogen tetraoxide, N_2O_4, dissociates in the gas phase according to

$$N_2O_4(g) \rightleftharpoons 2NO_2(g)$$

The reaction is endothermic, $\Delta H° = 57.1$ kJ.

(a) When 0.800 mol of N_2O_4 was introduced into a

* The asterisk denotes the more difficult problems.

5.00-L flask at 100 °C, the equilibrium concentration of NO_2 was found to be 0.140 mol L^{-1}. What is the value of K_c at 100 °C?

(b) If 0.400 mol of N_2O_4 is introduced into a 5.00-L flask at 100 °C, what are the equilibrium concentrations of NO_2 and N_2O_4?

(c) What effect would each of the following changes have on the equilibrium concentration of NO_2? Explain.

(i) Increasing the temperature

(ii) Decreasing the volume of the vessel

*42. Anhydrous aluminum chloride was found to have the following densities in the gas phase at a pressure of 1 atm:

$T(°C)$	200	600	800
Density (g mL^{-1})	6.87×10^{-3}	2.65×10^{-3}	1.51×10^{-3}

(a) What is the molecular formula of aluminum chloride at 200 °C, and at 800 °C?

(b) What species are in equilibrium at 600 °C?

(c) Remembering that the total pressure is 1.00 atm, what are the partial pressures of the species in equilibrium at 600 °C?

(d) What are the values of K_p and K_c for the equilibrium at 600 °C?

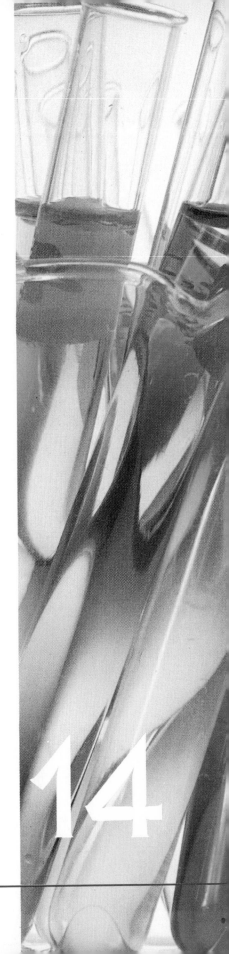

Acid–Base Equilibria

14

In earlier chapters, particularly Chapters 5, 8, and 12, we have discussed many examples of acids and bases and we have seen that acid–base reactions are one of the most common and important reaction types. We encounter many examples of such reactions in everyday life. We know that the acidity of the water in a swimming pool or in an aquarium must be carefully controlled. The acidity of our own blood has to stay constant if we are to remain healthy. Acid rain has become a topic of widespread concern in both North America and Europe. Sulfuric acid, which is produced by chemical industry in a larger amount than any other substance, is used for the manufacture of fertilizers, polymers, iron and steel, and a wide variety of other important substances. Ammonia and other household cleaners are bases. Other bases such as lime (calcium oxide) and caustic soda (sodium hydroxide) are extensively used in the chemical industry.

In this chapter we will be studying acid–base reactions in a quantitative way using the basic principles of equilibrium that were introduced in the previous chapter. We have seen that in water some acids and bases are strong while many others are weak. We are now in a position to describe the strengths of acids and bases in a quantitative manner. We will introduce the pH scale for measuring the acidity of a solution, and we will see how indicators work and how we can use buffer solutions to maintain a constant acidity or pH.

First we recall from Chapter 5 that, according to the definition of acids and bases given in 1923 by the Danish chemist J. N. Brønsted and the English chemist T. M. Lowry, an acid is a proton donor and a base is a proton acceptor. In water an acid donates a proton to the water molecule to give the hydronium ion H_3O^+:

$$HCl + H_2O \longrightarrow H_3O^+ + Cl^-$$

Thus any substance that forms the hydronium ion in water behaves as an acid. A base accepts a proton from a water molecule to give the hydroxide ion OH^-:

$$NH_3 + H_2O \rightleftharpoons NH_4^+ + OH^-$$

Thus any substance that forms the hydroxide ion in water behaves as a base, including the metal hydroxides such as NaOH.

14.1
Acid and Base Strengths

STRONG ACIDS AND BASES

A strong acid such as hydrochloric acid, HCl, or nitric acid, HNO_3, reacts very nearly completely with water:

$$HCl + H_2O \longrightarrow H_3O^+ + Cl^-$$

The equilibrium lies far to the right; we say that the acid is fully ionized or that it is fully dissociated into hydronium ions and the conjugate base of the acid. The equilibrium constant,

$$K = \left(\frac{[H_3O^+][Cl^-]}{[H_2O][HCl]} \right)_{eq}$$

has a very large value, that is, $K \gg 1$.

$=$ *Example 14.1* $=\!=\!=\!=\!=\!=\!=\!=\!=\!=\!=\!=$

ION CONCENTRATIONS IN A SOLUTION OF A STRONG ACID

What are the concentrations of H_3O^+ and Cl^- in 250 mL of an aqueous HCl solution containing 0.01 mol of HCl?

Solution

The equation shows that 0.01 mol of HCl is converted to 0.01 mol of H_3O^+ and 0.01 mol of Cl^-. We then convert these amounts to concentrations by using the volume of the solution. Note that there must be some HCl in equilibrium with H_3O^+ and Cl^-, but the amount is so small that it is effectively zero.

	HCl	$+ H_2O \longrightarrow$	H_3O^+	$+$	Cl^-
Initial amounts	0.01 mol		0 mol		0 mol
Equilibrium amounts	0 mol		0.01 mol		0.01 mol
Equilibrium concentrations	0 mol L^{-1}		$\left(\dfrac{0.01\ \text{mol}}{250\ \text{mL}}\right)\left(\dfrac{10^3\ \text{mL}}{1\ \text{L}}\right)$		$\left(\dfrac{0.01\ \text{mol}}{250\ \text{mL}}\right)\left(\dfrac{10^3\ \text{mL}}{1\ \text{L}}\right)$
	0 mol L^{-1}		0.04 mol L^{-1}		0.04 mol L^{-1}

Thus

$$[H_3O^+] = [Cl^-] = 0.04\ \text{mol L}^{-1}$$

Example 14.1 is simple but it illustrates two important points. First, a strong acid is quantitatively ionized. Second, $[H_3O^+] = [Cl^-]$. The second conclusion comes from the balanced equation, and it exemplifies an important principle: *All solutions are electrically neutral.* The sum of the charges on all the positive ions must always be equal to the sum of the charges on all the negative ions in a solution.

As we saw in Chapter 5, a strong base is either an ionic substance containing OH^- or an ion or molecule that is completely protonated in water, thus producing an equal amount of OH^-. The most common strong bases are soluble hydroxides such as NaOH and KOH, which are ionic compounds consisting of metal ions and hydroxide ions. They therefore give a quantitative amount of OH^- when dissolved in water. Two examples of species that are completely protonated in water are the oxide ion, O^{2-}, and the hydride ion, H^-:

$$O^{2-} + H_2O \longrightarrow OH^- + OH^-$$
$$H^- + H_2O \longrightarrow H_2 + OH^-$$

In both cases the position of equilibrium lies far to the right.

WEAK ACIDS

The only common strong acids are HCl, HBr, HI, HNO_3, H_2SO_4, and $HClO_4$. Almost all other acids are weak (see Experiment 5.9). Weak acids do not react completely with water. For example, in a solution of acetic acid in water the following equilibrium is established:

$$CH_3CO_2H + H_2O \rightleftharpoons CH_3CO_2^- + H_3O^+$$

The equilibrium constant expression for this reaction is

$$K_c = \left(\frac{[CH_3CO_2^-][H_3O^+]}{[CH_3CO_2H][H_2O]}\right)_{eq}$$

We will be concerned only with dilute solutions of acids and bases in which there is an enormous excess of water compared with the solute species. The concentration of water is very large and is hardly changed by changing the amounts of the solute species. Thus in the above expression for the equilibrium constant we can take $[H_2O]$ to be constant. We can therefore modify this expression by multiplying both sides by $[H_2O]$:

$$K_c[H_2O] = \left(\frac{[H_3O^+][CH_3CO_2^-]}{[CH_3CO_2H]}\right)_{eq}$$

Since $[H_2O]$ is constant in dilute solutions, the product $K_c[H_2O]$ is also constant. It is called the **acid dissociation constant** and is given the symbol K_a:

$$K_a(CH_3CO_2H) = K_c[H_2O] = \left(\frac{[H_3O^+][CH_3CO_2^-]}{[CH_3CO_2H]}\right)_{eq}$$

The expressions for the acid dissociation constants of other weak acids are written in the same way. For example, HF is a weak acid in water:

$$HF + H_2O \rightleftharpoons H_3O^+ + F^-$$

Thus the equilibrium constant $K_a(HF)$ is given by the expression

$$K_a(HF) = \left(\frac{[H_3O^+][F^-]}{[HF]}\right)_{eq}$$

In general, for any weak acid, HA, in water

$$HA + H_2O \rightleftharpoons H_3O^+ + A^-$$

and

$$K_a(HA) = \left(\frac{[H_3O^+][A^-]}{[HA]}\right)_{eq}$$

Values of the acid dissociation constants of some weak acids are given in Table 14.1. Like other equilibrium constants, K_a values depend on the temperature. They are normally quoted at 25 °C. These values show the extent to which the acids are dissociated in water. The strong acids are essentially completely converted to H_3O^+ and an anion. The position of equilibrium lies very far to the right. These acids have dissociation constants that are much larger than 1.0 and that cannot be measured with any accuracy because of the difficulty of measuring the very small equilibrium concentration of un-ionized acid HA. The strongest acid that can exist in appreciable concentrations in water is the H_3O^+ ion which has a dissociation constant of 1.0. Weak acids have dissociation constants that are less than 1.0, and for most weak acids very much smaller than 1.0. In Table 14.1 the weak acids are listed in order of decreasing strength. Thus phosphoric acid, $K_a = 7.5 \times 10^{-3}$ mol L^{-1}, is much more extensively ionized than hydrocyanic acid, $K_a = 4.9 \times 10^{-10}$ mol L^{-1}. Phosphoric acid is a stronger acid than hydrocyanic acid, although both are weak. The position of the equilibrium lies to the left in both cases, but it is much further to the left for hydrocyanic acid than for phosphoric acid. From the value of

TABLE 14.1
Dissociation Constants for Some Acids in Water at 25 °C

Acid	Proton Transfer Reaction		K_a (mol L^{-1})	pK_a ($= -\log_{10} K_a$)
	ACID	CONJUGATE BASE		
Perchloric acid	$HClO_4$	$+ H_2O \longrightarrow H_3O^+ + ClO_4^-$	Very large	—
Hydrochloric acid	HCl	$+ H_2O \longrightarrow H_3O^+ + Cl^-$	Very large	—
Sulfuric acid	H_2SO_4	$+ H_2O \longrightarrow H_3O^+ + HSO_4^-$	Very large	—
Nitric acid	HNO_3	$+ H_2O \longrightarrow H_3O^+ + NO_3^-$	Very large	—
Hydronium ion	H_3O^+	$+ H_2O \rightleftharpoons H_3O^+ + H_2O$	1.0	0.00
Sulfurous acid[a]	SO_2	$+ 2H_2O \rightleftharpoons H_3O^+ + HSO_3^-$	1.2×10^{-2}	1.92
Hydrogen sulfate ion	HSO_4^-	$+ H_2O \rightleftharpoons H_3O^+ + SO_4^{2-}$	1.2×10^{-2}	1.92
Phosphoric acid	H_3PO_4	$+ H_2O \rightleftharpoons H_3O^+ + H_2PO_4^-$	7.5×10^{-3}	2.12
Nitrous acid	HNO_2	$+ H_2O \rightleftharpoons H_3O^+ + NO_2^-$	4.5×10^{-4}	3.35
Hydrofluoric acid	HF	$+ H_2O \rightleftharpoons H_3O^+ + F^-$	3.5×10^{-4}	3.45
Acetic acid	CH_3CO_2H	$+ H_2O \rightleftharpoons H_3O^+ + CH_3CO_2^-$	1.8×10^{-5}	4.74
Hydrated aluminum ion	$Al(H_2O)_6^{3+}$	$+ H_2O \rightleftharpoons H_3O^+ + Al(H_2O)_5OH^{2+}$	7.2×10^{-6}	5.14
Carbonic acid[b]	CO_2	$+ 2H_2O \rightleftharpoons H_3O^+ + HCO_3^-$	4.3×10^{-7}	6.37
Hydrogen sulfide	H_2S	$+ H_2O \rightleftharpoons H_3O^+ + HS^-$	9.1×10^{-8}	7.04
Dihydrogen phosphate ion	$H_2PO_4^-$	$+ H_2O \rightleftharpoons H_3O^+ + HPO_4^{2-}$	6.2×10^{-8}	7.21
Hypochlorous acid	$HOCl$	$+ H_2O \rightleftharpoons H_3O^+ + OCl^-$	3.1×10^{-8}	7.51
Ammonium ion	NH_4^+	$+ H_2O \rightleftharpoons H_3O^+ + NH_3$	5.6×10^{-10}	9.25
Hydrocyanic acid	HCN	$+ H_2O \rightleftharpoons H_3O^+ + CN^-$	4.9×10^{-10}	9.31
Hydrogen phosphate ion	HPO_4^{2-}	$+ H_2O \rightleftharpoons H_3O^+ + PO_4^{3-}$	2.1×10^{-13}	12.68
Hydrogen sulfide ion	HS^-	$+ H_2O \rightleftharpoons H_3O^+ + S^{2-}$	1.3×10^{-13}	12.88
Water	H_2O	$+ H_2O \rightleftharpoons H_3O^+ + OH^-$	1.0×10^{-14}	14.00
	CONJUGATE ACID	BASE		

[a] There is no evidence for the formation of the undissociated acid H_2SO_3.
[b] The equilibrium given here is the sum of the two equilibria $H_2O + CO_2 \rightleftharpoons H_2CO_3$ and $H_2CO_3 + H_2O \rightleftharpoons H_3O^+ + HCO_3^-$ because the amount of H_2CO_3 formed is small and not accurately known.

the dissociation constant K_a of a weak acid, we can calculate the extent of its dissociation in water.

CALCULATING THE EXTENT OF DISSOCIATION OF A WEAK ACID

As an example, we will calculate the concentration of H_3O^+ in a $0.20M$ solution of acetic acid.

The first step in solving any equilibrium problem is to write the equation for the equilibrium reaction. In this case the reaction is

$$CH_3CO_2H(aq) + H_2O(l) \rightleftharpoons H_3O^+(aq) + CH_3CO_2^-(aq)$$

The second step is to write the expression for the equilibrium constant and look up its value. From Table 14.1, K_a for acetic acid is 1.8×10^{-5} mol L^{-1}. Thus we can write

$$K_a = \left(\frac{[H_3O^+][CH_3CO_2^-]}{[CH_3CO_2H]}\right)_{eq} = 1.8 \times 10^{-5} \text{ mol L}^{-1}$$

The third step is to write expressions for the concentrations of each of the species present at equilibrium. We first imagine that when acetic acid is added to water, no proton transfer occurs, so the *initial* concentration of acetic acid is 0.20M. Then we imagine that the proton transfer occurs to give a certain concentration of H_3O^+, which we wish to find. Let us call the concentration of H_3O^+ x mol L^{-1}. The concentration of acetate ion, $CH_3CO_2^-$, is then also x mol L^{-1}, since 1 mol of $CH_3CO_2^-$ is formed for each mole of H_3O^+ formed. The concentration of undissociated acetic acid, CH_3CO_2H, that remains at equilibrium is therefore $(0.20 - x)$ mol L^{-1}. Thus

$$CH_3CO_2H + H_2O \rightleftharpoons H_3O^+ + CH_3CO_2^-$$

Initial concentrations	0.20		0	0	mol L^{-1}
Equilibrium concentrations	$0.20 - x$		x	x	mol L^{-1}

The fourth step is to substitute these concentrations in the expression for K_a and to solve for x, the H_3O^+ concentration:

$$K_a = \left(\frac{[H_3O^+][CH_3CO_2^-]}{[CH_3CO_2H]}\right)_{eq} = \frac{(x \text{ mol } L^{-1})(x \text{ mol } L^{-1})}{(0.20 - x) \text{ mol } L^{-1}}$$

$$= \frac{x^2}{0.20 - x} \text{ mol } L^{-1} = 1.8 \times 10^{-5} \text{ mol } L^{-1}$$

Since this expression can be rearranged to give a quadratic equation, it can be solved by the formula given in Appendix A. However, a simpler and much quicker approximate method can be used. Because the value of K_a is very small, we know that the position of the equilibrium is far to the left; only a very small amount of acetic acid is dissociated. Therefore x, the H_3O^+ concentration, will be very small compared with the concentration of CH_3CO_2H. We will assume then that x is much smaller than 0.20 and can be neglected with respect to 0.20. So we can write

$$0.20 - x \approx 0.20$$

Using this approximation, we have

$$\frac{x^2}{0.20} \text{ mol } L^{-1} = 1.8 \times 10^{-5} \text{ mol } L^{-1}$$

$$x^2 = 0.20 \times 1.8 \times 10^{-5} = 0.36 \times 10^{-5} = 3.6 \times 10^{-6}$$

Taking the square root of both sides, we have $x = 1.9 \times 10^{-3}$, and

$$[H_3O^+] = 1.9 \times 10^{-3} \text{ mol } L^{-1}$$

We see that x is indeed much smaller than 0.20, so our approximation is justified. The assumption that x is very small compared with 0.20 is equivalent to assuming that the concentration of un-ionized acetic acid at equilibrium is the same as the initial concentration of acetic acid. When is it valid to make this assumption? The answer to this question depends on the accuracy to which we wish to calculate the H_3O^+ concentration. But since most dissociation constants are known to an accuracy of about 5% we cannot expect to be able to calculate the H_3O^+ concentration to any greater accuracy. Thus working to this accuracy the approximation is valid if x, that is, $[H_3O^+]$, is 5% or less of the initial acid concentration. In the example we have just discussed $[H_3O^+]$ is

$$\frac{0.0019 \text{ mol } L^{-1}}{0.20 \text{ mol } L^{-1}} \times 100\% = 1\%$$

of the initial acid concentration. In all the examples given in this book, except for a few concerned with polyprotic acids, the approximation gives a sufficiently accurate result.

We have just seen that the concentration of H_3O^+ in a 0.20M solution of acetic acid is only 1% of the initial concentration of the acid. This means that 1% of the acid is dissociated. The **percent dissociation** of an acid is defined as follows:

The percentage dissociation of a weak acid increases with increasing dilution of the solution.

$$\text{Percent dissociation} = \frac{[H_3O^+]}{[HA]_{\text{initial}}} \times 100\%$$

Example 14.2

CALCULATING THE EXTENT OF DISSOCIATION OF A WEAK ACID

What is the H_3O^+ concentration in a 0.20M solution of hydrocyanic acid, HCN? What is the percent dissociation of the acid?

Solution

First, we write the equation for the equilibrium:

$$HCN(aq) + H_2O(l) \rightleftharpoons H_3O^+(aq) + CN^-(aq)$$

Second, we write the expression for the equilibrium constant and find the value of K_a from Table 14.1:

$$K_a = \left(\frac{[H_3O^+][CN^-]}{[HCN]}\right)_{eq} = 4.9 \times 10^{-10} \text{ mol L}^{-1}$$

Third, we let x mol $L^{-1} = [H_3O^+]$ at equilibrium. Then we have

$$HCN(aq) + H_2O(l) \rightleftharpoons H_3O^+(aq) + CN^-(aq)$$

Initial concentrations	0.20		0	0	mol L^{-1}
Equilibrium concentrations	0.20 − x		x	x	mol L^{-1}

Substituting into the expression for K_a, we have

$$K_a = \frac{x^2}{0.20 - x} \text{ mol L}^{-1} = 4.9 \times 10^{-10} \text{ mol L}^{-1}$$

Since K_a is very small, HCN is only very slightly dissociated and the concentration of H_3O^+ will be very small. Therefore we assume that x is much smaller than 0.20 and can be neglected with respect to 0.20. In other words, $0.20 - x \approx 0.20$. We then have

$$\frac{x^2}{0.20} = 4.9 \times 10^{-10}$$

$$x^2 = 0.98 \times 10^{-10} = 98 \times 10^{-12}$$

Taking the square root of each side, we obtain

$$x = 9.9 \times 10^{-6}$$

Therefore

$$[H_3O^+] = 9.9 \times 10^{-6} \text{ mol L}^{-1}$$

We see that our assumption that $x \ll 0.2$ is certainly justified.

Since the concentration of HCN that is dissociated is equal to the concentration of $[H_3O^+]$ that is formed, 9.9×10^{-6} mol L^{-1}, the percent dissociation of the acid is

$$\frac{9.9 \times 10^{-6} \text{ mol L}^{-1}}{0.20 \text{ mol L}^{-1}} \times 100\% = 0.005\%$$

At a concentration of $0.20M$, HCN is ionized to only a very small extent, namely, 0.005%. Thus it is an extremely weak acid. Only 5 molecules in 100 000 are ionized; the rest remain as un-ionized HCN molecules.

EXERCISE 14.1

What is the H_3O^+ concentration and the percent dissociation of HF in a $0.50M$ solution?

WEAK BASES

Methylamine

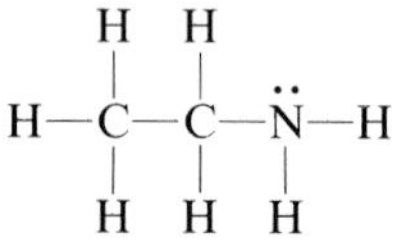

Ethylamine

Aniline

In addition to ammonia there are many organic molecules derived from ammonia such as methylamine, CH_3NH_2, ethylamine, $C_2H_5NH_2$, and aniline, $C_6H_5NH_2$, that are also weak bases. Another important type of weak base is the anion of a weak acid. Thus fluoride ion, F^-, and acetate ion, $CH_3CO_2^-$, and the anions of all other weak acids are weak bases.

A weak base is incompletely protonated by water:

$$B + H_2O \rightleftharpoons BH^+ + OH^-$$

The solution contains the base, B, the protonated base, BH^+, hydroxide ion, OH^-, and water in equilibrium. The equilibrium constant for this reaction is

$$K_c = \left(\frac{[BH^+][OH^-]}{[B][H_2O]}\right)_{eq}$$

As we did when considering weak acids, we assume that $[H_2O]$ is constant. Hence we may incorporate $[H_2O]$ into the equilibrium constant and write

$$K_c[H_2O] = K_b = \left(\frac{[BH^+][OH^-]}{[B]}\right)_{eq}$$

The constant K_b is called the **base dissociation constant**. A list of K_b values for some weak bases is given in Table 14.2.

The strongest base that can exist in an appreciable concentration in aqueous solution is the hydroxide ion, OH^-, which has a dissociation constant of 1.0. Stronger bases are essentially completely converted to OH^- in water and they have very large dissociation constants. Weak bases have dissociation constants that are less than 1.0. They are listed in order of decreasing strength in Table 14.2. Phosphate ion is a stronger base than ammonia but both are weak and the position of equilibrium lies well over on the left-hand side in both cases, but further to the left for ammonia than for phosphate ion.

TABLE 14.2
Dissociation Constants for Some Weak Bases at 25 °C

Base	Proton Transfer Reaction		K_b (mol L^{-1})	pK_b ($= -\log_{10} K_b$)
	BASE	CONJUGATE ACID		
Hydride ion	H^-	$+ H_2O \rightleftharpoons OH^- + H_2$	Very large	—
Amide ion	NH_2^-	$+ H_2O \rightleftharpoons OH^- + NH_3$	Very large	—
Oxide ion	O^{2-}	$+ H_2O \rightleftharpoons OH^- + OH^-$	Very large	—
Hydroxide ion	OH^-	$+ H_2O \rightleftharpoons OH^- + H_2O$	1.00	0.0
Sulfide ion	S^{2-}	$+ H_2O \rightleftharpoons OH^- + HS^-$	7.7×10^{-2}	1.11
Phosphate ion	PO_4^{3-}	$+ H_2O \rightleftharpoons OH^- + HPO_4^{2-}$	4.8×10^{-2}	1.32
Ethylamine	$C_2H_5NH_2$	$+ H_2O \rightleftharpoons OH^- + C_2H_5NH_3^+$	4.7×10^{-4}	3.33
Methylamine	CH_3NH_2	$+ H_2O \rightleftharpoons OH^- + CH_3NH_3^+$	3.9×10^{-4}	3.41
Carbonate ion	CO_3^{2-}	$+ H_2O \rightleftharpoons OH^- + HCO_3^-$	2.1×10^{-4}	3.68
Cyanide ion	CN^-	$+ H_2O \rightleftharpoons OH^- + HCN$	2.0×10^{-5}	4.70
Ammonia	NH_3	$+ H_2O \rightleftharpoons OH^- + NH_4^+$	1.8×10^{-5}	4.74
Hydrogen phosphate ion	HPO_4^{2-}	$+ H_2O \rightleftharpoons OH^- + H_2PO_4^-$	1.6×10^{-7}	6.79
Hydrogen carbonate ion	HCO_3^-	$+ H_2O \rightleftharpoons OH^- + H_2CO_3$	2.5×10^{-8}	7.60
Aniline	$C_6H_5NH_2$	$+ H_2O \rightleftharpoons OH^- + C_6H_5NH_3^+$	4.3×10^{-10}	9.36
Dihydrogen phosphate ion	$H_2PO_4^-$	$+ H_2O \rightleftharpoons OH^- + H_3PO_4$	1.3×10^{-12}	11.88
	CONJUGATE BASE	ACID		

We can find the OH^- concentration in an aqueous solution of a weak base by a method similar to the method we used to find the H_3O^+ concentration in a solution of a weak acid, as the following example shows.

═ *Example 14.3* ═══════════════════════════════

CALCULATING THE EXTENT OF DISSOCIATION OF A WEAK BASE

What is the concentration of OH^- in a 0.10M aqueous solution of ammonia, and what is its percent dissociation?

Solution

The first step is to write the equation for the equilibrium reaction:

$$NH_3(aq) + H_2O(l) \rightleftharpoons NH_4^+(aq) + OH^-(aq)$$

The second step is to write the expression for the equilibrium constant and look up its value. From Table 14.2 we see that K_b for NH_3 is 1.8×10^{-5} mol L^{-1}. Therefore

$$K_b = \left(\frac{[NH_4^+][OH^-]}{[NH_3]}\right)_{eq} = 1.8 \times 10^{-5} \text{ mol } L^{-1}$$

The third step is to write expressions for the concentration of each of the species in solution. We let the equilibrium concentration of OH^- be x mol L^{-1}. Thus

	NH_3	$+ H_2O \rightleftharpoons$	NH_4^+	$+ OH^-$	
Initial concentrations	0.10		0	0	mol L^{-1}
Equilibrium concentrations	$0.10 - x$		x	x	mol L^{-1}

The fourth step is to substitute these values in the expression for the equilibrium constant and then solve for x. We have

$$K_b = \frac{[NH_4^+][OH^-]}{[NH_3]} = \frac{x^2}{0.10 - x} \text{ mol L}^{-1} = 1.8 \times 10^{-5} \text{ mol L}^{-1}$$

We could solve this equation for x by using the formula for a quadratic equation (Appendix A). However, because K_b is small, we know that the extent of ionization will be small and therefore that x is much less than 0.10. Hence we assume that $0.10 - x \approx 0.10$, and the equation becomes

$$\frac{x^2}{0.10} = 1.8 \times 10^{-5}$$

$$x^2 = 1.8 \times 10^{-6}$$

$$x = 1.3 \times 10^{-3}$$

Hence

$$[OH^-] = 1.3 \times 10^{-3} \text{ mol L}^{-1}$$

Thus $[OH^-]$ is 1.3% of the initial concentration of NH_3, 0.10 mol L^{-1}; hence the assumption that we made in solving the equation was justified. The *percent dissociation* of the base is

$$\frac{[OH^-]}{[base]_{initial}} \times 100\% = \frac{1.3 \times 10^{-3}}{0.1} \times 100\% = 1.3\%$$

EXERCISE 14.2

What is the OH^- concentration in a $0.10M$ solution of methylamine, CH_3NH_2? What is the percent dissociation of the methylamine?

14.2
The Self-Ionization of Water

We saw in Chapters 5 and 12 that water can act as a very weak acid and also as a very weak base. Very small concentrations of H_3O^+ and OH^- are therefore formed in water by the reaction

$$H_2O(l) + H_2O(l) \rightleftharpoons H_3O^+(aq) + OH^-(aq)$$

in which one water molecule behaves as an acid and the other as a base. This reaction is called the **self-ionization**, or **autoprotolysis**, of water, and the equilibrium constant is

$$K_c = \left(\frac{[H_3O^+][OH^-]}{[H_2O]^2}\right)_{eq}$$

Because the concentration of water is very nearly constant in any dilute solution, we can multiply K_c by $[H_2O]^2$ to give a new constant, K_w, called the **ionic product constant of water**:

$$K_w = K_c[H_2O]^2 = ([H_3O^+][OH^-])_{eq}$$

Measurement of the electrical conductivity of carefully purified water has shown that, at 25 °C, $[H_3O^+] = [OH^-] = 1.00 \times 10^{-7}$ mol L^{-1}. Thus

$$K_w = (1.00 \times 10^{-7} \text{ mol } L^{-1})(1.00 \times 10^{-7} \text{ mol } L^{-1})$$
$$= 1.00 \times 10^{-14} \text{ mol}^2 \text{ } L^{-2} \text{ at } 25 \text{ °C}$$

This equilibrium constant applies not only to pure water but also to the self-ionization of water in any aqueous solution. Hydronium ions and hydroxide ions are present in any aqueous solution, and they are always in equilibrium with water molecules.

Let us calculate the concentration of OH^- in a 0.010M solution of HCl. The concentration of OH^- will be less than it is in pure water because the H_3O^+ from the ionization of the HCl will shift the position of the equilibrium of the self-ionization reaction to the left. If the concentration of OH^- is x mol L^{-1}, then the concentration of H_3O^+ *from the self-ionization of water* must also be x mol L^{-1}. The concentration of H_3O^+ from the ionization of HCl is 0.010 mol L^{-1}, so the *total* concentration of H_3O^+ is $(0.010 + x)$ mol L^{-1}. Thus

$$2H_2O \rightleftharpoons H_3O^+ + OH^-$$

Equilibrium concentrations $\qquad\qquad 0.010 + x \qquad x \qquad$ mol L^{-1}

Substituting into the equilibrium constant for the self-ionization of water, we have

$$K_w = ([H_3O^+][OH^-])_{eq} = (0.010 + x)x \text{ mol}^2 \text{ } L^{-2} = 1.0 \times 10^{-14} \text{ mol}^2 \text{ } L^{-2}$$

Since x must be very small, we can assume that $x \ll 0.010$, and therefore that $0.010 + x \approx 0.010$. So

$$0.010x = 1.0 \times 10^{-14}$$

Hence

$$x = 1.0 \times 10^{-12}$$
$$[OH^-] = 1.0 \times 10^{-12} \text{ mol } L^{-1}$$

The H_3O^+ from the ionization of an acid shifts the position of the water self-ionization equilibrium to the left so that both $[OH^-]$ and $[H_3O^+]$ from the self-ionization are much smaller than 10^{-7} mol L^{-1} (in the above case they are both 10^{-12} mol L^{-1}). The acid is said to *repress* the self-ionization of water. We see that

1. Even in a solution of an acid there is still a very small concentration of OH^- in equilibrium with the H_3O^+. The concentration of OH^- is never zero in any aqueous solution but it is very small in an acid solution.
2. The concentration of H_3O^+ *arising from the self-ionization of water* is negligible compared with the concentration of H_3O^+ from the dissociation of an acid.

In calculating the degree of dissociation of a weak acid (Example 14.2) we ignored any H_3O^+ arising from the self-ionization of water. We see now that we were justified in doing so. We will continue to make use of this approximation which is valid in all cases except those in which $[H_3O^+]$ arising from the acid is extremely small ($< 10^{-6}$ mol L^{-1}) and we will not be concerned with such cases.

Similar considerations apply to solutions of bases in water.

1. There is always some H_3O^+ in equilibrium with OH^- in any solution of a base, although its concentration is very small.

2. The concentration of OH^- arising from the self-ionization of water is negligible compared with the concentration of OH^- due to a base.

═ *Example 14.4* ═══

CALCULATING THE HYDROXIDE ION CONCENTRATION IN A SOLUTION OF A STRONG ACID

What is the concentration of OH^- in a $1.0 \times 10^{-5} M$ solution of nitric acid?

Solution

Nitric acid is a strong acid and is therefore completely dissociated. The concentration of H_3O^+ from the acid is therefore $1.0 \times 10^{-5} M$. We can neglect the H_3O^+ from the self-ionization of water, so the total H_3O^+ concentration in the solution is also 10^{-5} mol L^{-1}. Thus

$$K_w = ([H_3O^+][OH^-])_{eq} = 10^{-14} \text{ mol}^2 \text{ L}^{-2}$$

Substituting $[H_3O^+] = 10^{-5}$ mol L^{-1} gives

$$(10^{-5} \text{ mol L}^{-1})[OH^-] = 10^{-14} \text{ mol}^2 \text{ L}^{-2}$$

Thus

$$[OH^-] = 10^{-9} \text{ mol L}^{-1}$$

14.3
The pH Scale

Solutions of acids and bases in water may have H_3O^+ and OH^- concentrations that vary over a very wide range, that is, from greater than 1 mol L^{-1} down to 10^{-14} mol L^{-1} or lower. The relationship between the H_3O^+ and OH^- concentrations in aqueous solutions is shown in Table 14.3. *Acidic solutions* have $[H_3O^+]$ greater than 10^{-7} mol L^{-1} and $[OH^-]$ less than 10^{-7} mol L^{-1} at 25 °C. *Basic (alkaline) solutions* have $[OH^-]$ greater than 10^{-7} mol L^{-1} and $[H_3O^+]$ less than 10^{-7} mol L^{-1} at 25 °C. A *neutral solution*, such as pure water, has $[H_3O^+] = [OH^-] = 10^{-7}$ mol L^{-1}.

The concentration of H_3O^+ or OH^- in aqueous solutions has many important consequences. Plants tolerate only limited ranges of H_3O^+ concentration in the soil. Thus an important factor determining the distribution of a plant is the soil acidity or basicity. Many reactions in living systems occur only in very narrow ranges of H_3O^+ concentration. Blood has a constant H_3O^+ concentration very close to 2.5×10^{-7} mol L^{-1}. A solution with an H_3O^+ concentration of 10^{-3} mol L^{-1} has a sour or tart flavor, which is quite pleasant, especially if it is sweetened with sugar, as in many soft drinks. But a solution having an H_3O^+ concentration of 10^{-1} mol L^{-1} or higher is not only unpleasant to taste but also dangerous because it burns the skin.

In expressing the concentration of H_3O^+ in a solution, we can avoid the use of negative powers of 10 by using the pH scale. The pH of a solution is the

TABLE 14.3 H_3O^+ and OH^- Concentrations of Common Substances and the pH Scale				
pH	$[H_3O^+]$	$[OH^-]$	pH *of Some Common Substances*	
-1	10	10^{-15}	Concentrated HCl (37%)	
0	1	10^{-14}	$1M$ HCl solution	
1	10^{-1}	10^{-13}		
2	10^{-2}	10^{-12}	Stomach acid Lemon juice	Acid
3	10^{-3}	10^{-11}	Orange juice	pH < 7
4	10^{-4}	10^{-10}	Wine Soda water	
5	10^{-5}	10^{-9}	Tomato juice Rainwater	
6	10^{-6}	10^{-8}		
7	10^{-7}	10^{-7}	Milk	Neutral
8	10^{-8}	10^{-6}	Blood Seawater	
9	10^{-9}	10^{-5}	Baking soda solution	
10	10^{-10}	10^{-4}	Borax solution Toilet soap	
11	10^{-11}	10^{-3}	Milk of magnesia	Basic
12	10^{-12}	10^{-2}	Household ammonia	pH > 7
13	10^{-13}	10^{-1}		
14	10^{-14}	1	$1M$ NaOH solution	
15	10^{-15}	10	Drain cleaner	

negative logarithm to the base 10 of the hydrogen ion concentration, $[H_3O^+]$. Thus

$$pH = -\log_{10}[H_3O^+]$$

The units of $[H_3O^+]$ are moles per liter, but when the logarithm is taken, the units are dropped because it is impossible to take the logarithm of a quantity with units. Therefore pH has no units.

If, as in pure water, the H_3O^+ concentration is 10^{-7} mol L^{-1}, then

$$pH = -\log_{10}[H_3O^+] = -\log_{10}(10^{-7}) = -(-7.0) = +7.0$$

ACIDIC SOLUTIONS In a $0.10M$ solution of a strong acid such as HCl, $[H_3O^+]$ is 1.0×10^{-1} mol L^{-1}. Therefore

$$pH = -\log(1.0 \times 10^{-1}) = -(-1.0) = +1.0$$

═ Example 14.5 ═══════════════════════════════════

pH CALCULATIONS: ACID SOLUTIONS

What is the pH of each of the following aqueous solutions?

(a) $1 \times 10^{-3}M$ HCl   (b) $5 \times 10^{-3}M$ HCl   (c) $1M$ HCl

Solution

Because HCl is a strong acid, we have the following concentrations:

(a) $[H_3O^+] = 1 \times 10^{-3} \text{ mol L}^{-1}$.

(b) $[H_3O^+] = 5 \times 10^{-3} \text{ mol L}^{-1}$.

(c) $[H_3O^+] = 1 \text{ mol L}^{-1}$.

Dropping the units and taking the negative logarithm of $[H_3O^+]$, we have the following pH values:

(a) $pH = -\log[H_3O^+] = -\log(1 \times 10^{-3}) = 3$.

(b) $pH = -\log[H_3O^+] = -\log(5 \times 10^{-3}) = 2.3$.

(c) $pH = -\log[H_3O^+] = -\log(1) = 0$.

$=$ *Example 14.6* $=$

pH CALCULATIONS: ACID SOLUTIONS

What is the pH of a 0.20M solution of acetic acid?

Solution

Acetic acid is a weak acid. We saw previously (page 664) that for a 0.20M solution of acetic acid $[H_3O^+]$ is $1.9 \times 10^{-3} \text{ mol L}^{-1}$. So

$$pH = -\log(1.9 \times 10^{-3}) = -(-2.72) = 2.72$$

EXERCISE 14.3

What is the pH of a 0.10M solution of hypochlorous acid, HOCl?

BASIC SOLUTIONS For any solution of a base we can calculate the OH^- concentration, as we saw in Example 14.3. The H_3O^+ concentration in basic solutions is very small, but it can be easily calculated from the OH^- concentration and the ionic product constant for water,

$$K_w = [H_3O^+][OH^-] = 1.00 \times 10^{-14} \text{ mol}^2 \text{ L}^{-2}$$

Thus we may also conveniently describe basic solutions in terms of their pH. In a basic solution $[OH^-] > 10^{-7}$, so $[H_3O^+] < 10^{-7}$ and therefore pH > 7 (see Table 14.3). There are no upper or lower limits to the pH scale, but for the vast majority of practical applications pH values are in the range 0–14.

pOH

It is convenient for some purposes to define pOH in an analogous way to the definition of pH:

$$pOH = -\log_{10}[OH^-]$$

Then taking negative logarithms of the above expression for the ionic product constant of water we have

$$-\log[H_3O^+] - \log[OH^-] = -\log(1.00 \times 10^{-14}) = 14.00$$

Hence

$$pH + pOH = 14.00$$

This expression is very convenient for calculating pH if we know pOH or vice versa.

═ Example 14.7 ══════════════════════════════════════

pH CALCULATIONS: BASIC SOLUTIONS

Calculate the pH of the following aqueous solutions:

(a) 0.0100M NaOH   (b) 0.134M NaOH

Solution

Since NaOH is a strong base, it is fully ionized to give Na^+ and OH^-.

(a) $[OH^-] = 1.00 \times 10^{-2}$ mol L^{-1}

$$pOH = -\log(1.00 \times 10^{-2}) = -(-2.00) = 2.00$$

and therefore pH = 14.00 − 2.00 = 12.00

(b) $[OH^-] = 0.134$ mol L^{-1}

$$pOH = -\log(0.134) = -(-0.87) = 0.87$$

$$pH = 14.00 - 0.87 = 13.13$$

═ Example 14.8 ══════════════════════════════════════

pH CALCULATIONS: BASIC SOLUTIONS

Aniline, $C_6H_5NH_2$, is a much weaker base than ammonia and has a dissociation constant, $K_b = 4.3 \times 10^{-10}$ mol L^{-1}. What is the pH of a 0.010M solution of aniline in water?

Solution

We first write the equation for the equilibrium:

$$C_6H_5NH_2 + H_2O \rightleftharpoons C_6H_5NH_3^+ + OH^-$$

Next, we write the expression for the equilibrium constant:

$$K_b = \frac{[C_6H_5NH_3^+][OH^-]}{[C_6H_5NH_2]} = 4.3 \times 10^{-10} \text{ mol L}^{-1}$$

We then let the equilibrium concentration of OH^- be x mol L^{-1} and write

$$C_6H_5NH_2 + H_2O \rightleftharpoons C_6H_5NH_3^+ + OH^-$$

Initial concentrations	0.010		0	0	mol L^{-1}
Equilibrium	0.010 − x		x	x	mol L^{-1}

We can now substitute these values in the equilibrium constant expression, so we have

$$\frac{x^2}{0.010 - x} \text{ mol L}^{-1} = 4.3 \times 10^{-10} \text{ mol L}^{-1}$$

Since the value of the equilibrium constant is very small, we can reasonably assume that x is very small compared with 0.010. Thus we have

$$\frac{x^2}{0.010} = 4.3 \times 10^{-10}$$

$$x^2 = 4.3 \times 10^{-12}$$

$$x = 2.1 \times 10^{-6}$$

$$[\text{OH}^-] = 2.1 \times 10^{-6} \text{ mol L}^{-1}$$

We see that x is indeed very small compared with 0.010, and therefore our assumption that $0.010 - x \approx 0.010$ was justified.

$$\text{pOH} = -\log(2.1 \times 10^{-6}) = -(-5.68) = 5.68$$

$$\text{pH} = 14.00 - 5.68 = 8.32$$

EXERCISE 14.4

What is the pH of a $0.05M$ solution of methylamine, CH_3NH_2?

pK VALUES Because the dissociation constants of weak acids and bases are normally expressed in terms of negative powers of 10 (see Tables 14.1 and 14.2), it is convenient to define pK_a and pK_b in the same way that we defined pH:

> The stronger the acid, the larger is K_a and the smaller is pK_a.

$$pK_a = -\log K_a \quad \text{and} \quad pK_b = -\log K_b$$

pK_a and pK_b values are listed in the last column of Tables 14.1 and 14.2 respectively.

EXERCISE 14.5

For phosphoric acid, H_3PO_4, $K_a = 7.5 \times 10^{-3}$ mol L^{-1}. What is the pK_a value for phosphoric acid?

RELATIONSHIP BETWEEN K_a AND K_b FOR AN ACID AND ITS CONJUGATE BASE

Recall from Chapter 5 that an acid and a base related by the equation

$$\text{base} + \text{H}^+ \rightleftharpoons \text{acid}$$

are called a conjugate acid–base pair. Thus HF and F^- constitute a conjugate acid–base pair because

$$\text{F}^- + \text{H}^+ \rightleftharpoons \text{HF}$$

We may say either that "F^- is a base and HF is its conjugate acid" or that "HF is an acid and F^- is its conjugate base". Ammonia, NH_3, and the ammonium ion, NH_4^+, also constitute a conjugate acid–base pair, since

$$NH_3 + H^+ \rightleftharpoons NH_4^+$$

In Chapter 5 we saw also that the greater the strength of an acid the weaker is its conjugate base and the greater the strength of a base the weaker is its conjugate acid. We can now put this relationship on a quantitative basis in terms of the acid and base dissociation constants K_a and K_b. For the ionization of a weak acid, HA, we have

$$HA + H_2O \rightleftharpoons H_3O^+ + A^- \qquad K_a(HA) = \left(\frac{[H_3O^+][A^-]}{[HA]}\right)_{eq}$$

Its conjugate base, A^-, behaves as a weak base in water:

$$A^- + H_2O \rightleftharpoons HA + OH^- \qquad K_b(A^-) = \left(\frac{[HA][OH^-]}{[A^-]}\right)_{eq}$$

For convenience we omit the parentheses $(\ )_{eq}$, but we must remember that these expressions for K_a and K_b are only valid when the [ ] refers to an *equilibrium* concentration.

If we now multiply $K_a(HA)$ and $K_b(A^-)$, we have

$$K_a(HA)K_b(A^-) = \frac{[H_3O^+][A^-]}{[HA]} \times \frac{[HA][OH^-]}{[A^-]} = [H_3O^+][OH^-]$$

Hence,

$$K_a(HA)K_b(A^-) = [H_3O^+][OH^-] = K_w = 10^{-14}\ mol^2\ L^{-2} \text{ at } 25\ °C$$

In general, for any acid

$$K_a(\text{acid})K_b(\text{conjugate base}) = K_w = 10^{-14}\ mol^2\ L^{-2}$$

Taking negative logarithms of both sides gives

$$pK_a(\text{acid}) + pK_b(\text{conjugate base}) = pK_w = 14$$

where we have written pK_w for $-\log K_w$. Thus the larger is K_a (that is, the stronger the acid), the smaller is K_b (that is, the weaker the conjugate base), their product always being equal to $10^{-14}\ mol^2\ L^{-2}$. In the limit of a strong acid, such as HCl, for which $K_a \gg 1$, $K_b(Cl^-) \ll 10^{-14}$. In other words, Cl^- is a weaker base than water and does not therefore behave as a base in water (Table 14.4).

It is not necessary to list values of both K_a and K_b for conjugate acid–base pairs, because one can always be obtained from the other. Many reference books give only K_a values but for convenience we have listed both K_a and K_b values for some common acids and bases in Tables 14.1 and 14.2.

$=$ *Example 14.9* $=$

CALCULATING K_b FROM K_a FOR A CONJUGATE ACID–BASE PAIR

What is the base dissociation constant for the fluoride ion, F^-? From Table 14.1 $K_a(HF) = 3.5 \times 10^{-4}\ mol\ L^{-1}$.

TABLE 14.4
Strengths of Conjugate Acid–Base Pairs

	K_a	Acid	Base	K_b	
Strong leveled to strength of H_3O^+	$\gg 1.0$	$HCl + H_2O \longrightarrow H_3O^+ + Cl^-$	$Cl^- + H_2O \longleftarrow HCl + OH^-$	$\ll 10^{-14}$	Not bases
	$\gg 1.0$	$HNO_3 + H_2O \longrightarrow H_3O^+ + NO_3^-$	$NO_3^- + H_2O \longleftarrow HNO_3 + OH^-$	$\ll 10^{-14}$	
	1.0	$H_3O^+ + H_2O \rightleftharpoons H_3O^+ + H_2O$	$H_2O + H_2O \rightleftharpoons H_3O^+ + OH^-$	10^{-14}	
Weak	1.2×10^{-2}	$HSO_4^- + H_2O \rightleftharpoons H_3O^+ + SO_4^{2-}$	$SO_4^{2-} + H_2O \rightleftharpoons HSO_4^- + OH^-$	8.3×10^{-13}	Weak
	3.5×10^{-4}	$HF + H_2O \rightleftharpoons H_3O^+ + F^-$	$F^- + H_2O \rightleftharpoons HF + OH^-$	2.9×10^{-10}	
	9.1×10^{-8}	$H_2S + H_2O \rightleftharpoons H_3O^+ + HS^-$	$HS^- + H_2O \rightleftharpoons H_2S + OH^-$	1.1×10^{-7}	
	5.6×10^{-10}	$NH_4^+ + H_2O \rightleftharpoons H_3O^+ + NH_3$	$NH_3 + H_2O \rightleftharpoons NH_4^+ + OH^-$	1.8×10^{-5}	
	10^{-14}	$H_2O + H_2O \rightleftharpoons H_3O^+ + OH^-$	$OH^- + H_2O \rightleftharpoons H_2O + OH^-$	1.0	Strong leveled to strength of OH^-
Not acids	$\ll 10^{-14}$	$NH_3 + H_2O \longrightarrow H_3O^+ + NH_2^-$	$NH_2^- + H_2O \longrightarrow NH_3 + OH^-$	$\gg 1.0$	
	$\ll 10^{-14}$	$OH^- + H_2O \longrightarrow H_3O^+ + O^{2-}$	$O^{2-} + H_2O \longrightarrow OH^- + OH^-$	$\gg 1.0$	

Solution

$$K_a(HF)K_b(F^-) = K_w$$

$$K_b(F^-) = \frac{1.0 \times 10^{-14} \text{ mol}^2 \text{ L}^{-2}}{K_a(HF)}$$

$$= \frac{1.0 \times 10^{-14} \text{ mol}^2 \text{ L}^{-2}}{3.5 \times 10^{-4} \text{ mol L}^{-1}} = 2.9 \times 10^{-11} \text{ mol L}^{-1}$$

EXERCISE 14.6

Given that K_b for the carbonate ion, CO_3^{2-}, has a value of 2.1×10^{-4} mol L^{-1}, calculate the value of K_a for its conjugate acid, HCO_3^-.

14.4
Acid–Base Properties of Salts

When an acid reacts with a base to give a salt, the acid is often said to neutralize the base. Thus we might think that the solution of the salt that is formed is neutral, that is, that it has a pH of 7. Although many salts do give neutral solutions in water, a large number do not, because many cations and anions behave as acids or bases. The acid–base properties of some common anions and cations are summarized in Table 14.5.

ANIONS

Anions (conjugate bases) of strong acids such as Cl^- have no basic properties in water. They give neutral solutions.

Anions (conjugate bases) of weak acids such as CN^- and CO_3^{2-} are weak bases. They give basic solutions in water. For example,

$$CN^- + H_2O \rightleftharpoons HCN + OH^-$$

TABLE 14.5
Acid–Base Properties of Some Common Ions

	Cations	*Anions*
Acidic	H_3O^+, NH_4^+, $Al(H_2O)_6^{3+}$, $Fe(H_2O)_6^{3+}$	HSO_4^-, $H_2PO_4^-$
Neutral	Mg^{2+}, Ca^{2+}, Sr^{2+}, Ba^{2+}, Li^+, Na^+, K^+, Rb^+, Cs^+, Ag^+	NO_3^-, ClO_4^-, Cl^-, Br^-, I^-
Basic	None	PO_4^{3-}, CO_3^{2-}, SO_3^{2-}, F^-, CN^-, OH^-, S^{2-}, $CH_3CO_2^-$, HCO_3^-, NO_2^-, HS^-, HPO_4^{2-}, SO_4^{2-} (very weak, almost neutral)

Anions containing hydrogen that are derived from polyprotic acids may be either acids or bases because they may either lose or gain a proton. For example, HSO_4^- and $H_2PO_4^-$ are acids,

$$HSO_4^- + H_2O \rightleftharpoons SO_4^{2-} + H_3O^+$$
$$H_2PO_4^- + H_2O \rightleftharpoons HPO_4^{2-} + H_3O^+$$

but HPO_4^{2-} and HCO_3^- are bases,

$$HPO_4^{2-} + H_2O \rightleftharpoons H_2PO_4^- + OH^-$$
$$HCO_3^- + H_2O \rightleftharpoons H_2CO_3 + OH^-$$

How do we know whether a given anion of this type will behave as an acid or as a base in solution in water? We simply compare its K_a and K_b values (or its pK_a and pK_b values). If $K_a > K_b$ ($pK_a < pK_b$), it behaves as an acid. If $K_b > K_a$ ($pK_a > pK_b$) it behaves as a base. For example, we see from Table 14.1 that $pK_a(H_2PO_4^-) = 7.21$, and from Table 14.2 that $pK_b(H_2PO_4^-) = 11.88$, so $pK_a(H_2PO_4^-) < pK_b(H_2PO_4^-)$, and therefore $H_2PO_4^-$ behaves as an acid and not as a base in aqueous solution.

EXERCISE 14.7

By comparing the K_a and K_b values for HPO_4^{2-} from Tables 14.1 and 14.2 decide whether HPO_4^{2-} behaves as a base or as an acid in aqueous solution.

CATIONS

Metal ions are generally hydrated, as we have described in Chapters 10 and 12. Many hydrated metal ions behave as acids, particularly when the metal has a positive charge of 2 or greater. For example,

$$Al(H_2O)_6^{3+} + H_2O \rightleftharpoons H_3O^+ + Al(OH)(H_2O)_5^{2+}$$

The charge on the metal ion attracts some of the electron density of an unshared pair of electrons on the oxygen atom into the valence shell of the metal atom, thereby producing a small positive charge on the oxygen atom. This in turn attracts electrons more strongly from the hydrogen atoms so that they acquire a greater positive charge than in the free water molecule, and hence become more acidic. The only common hydrated metal ions that do *not* behave as acids are Li^+, Na^+, K^+, Rb^+, Cs^+, Mg^{2+}, Ca^{2+}, Sr^{2+}, Ba^{2+}, and Ag^+. They give neutral solutions in water. All other hydrated metal ions give acidic solutions in water. *The conjugate acids of weak bases* are weak acids. The most common example is the ammonium ion,

$$NH_4^+ + H_2O \rightleftharpoons H_3O^+ + NH_3$$

Most other acidic cations of this type are the conjugate acids of organic bases related to ammonia, such as the conjugate acid of methylamine, $CH_3NH_3^+$, and the conjugate acid of aniline, $C_6H_5NH_3^+$.

TABLE 14.6
Acid–Base Properties of Aqueous Solutions of Some
Common Salts

Basic Solutions, pH > 7	Neutral Solutions, pH = 7	Acidic Solutions, pH < 7
Neutral cation	Neutral cation	Acidic cation
Basic anion	Neutral anion	Neutral anion
NaCN	KCl	NH_4Cl
KF	$BaCl_2$	$Al(H_2O)_6Cl_3$
$Na(CH_3CO_2)$	$Ca(NO_3)_2$	$Fe(H_2O)_6(NO_3)_3$
Na_2CO_3	$Mg(ClO_4)_2$	$C_6H_5NH_3Cl$
Na_3PO_4		
		Neutral cation
		Acidic anion
		$KHSO_4$

SALTS

We can now predict whether a salt will give an acidic, basic, or neutral solution in water by considering the acid–base properties of both the anion and the cation. There are a large number of salts that do not give neutral solutions. As we have seen, there are many metal cations that give an acidic solution and many anions that behave as weak bases because they are the conjugate bases of weak acids (Table 14.6).

Neutral salts contain a neutral cation and a neutral anion. They include salts of Li^+, Na^+, K^+, Rb^+, Cs^+, Mg^{2+}, Ca^{2+}, Sr^{2+}, Ba^{2+}, and Ag^+, with anions of strong acids, such as Cl^- and NO_3^-—for example, KCl, $BaCl_2$, and $AgNO_3$.

Acidic salts contain an acidic cation and a neutral anion or a neutral cation and an acidic anion. They include the following:

- Salts of metal cations, except Li^+, Na^+, K^+, Rb^+, Cs^+, Mg^{2+}, Sr^{2+}, Ba^{2+}, and Ag^+, with anions of strong acids, for example, $AlCl_3$ and $Fe_2(SO_4)_3$.
- Ammonium salts of strong acids, for example, NH_4Cl.
- Some salts of polyprotic acids, for example, $NaHSO_4$.

Basic salts contain a neutral cation and a basic anion. They include salts of Li^+, Na^+, K^+, Rb^+, Cs^+, Mg^{2+}, Ca^{2+}, Sr^{2+}, Ba^{2+}, and Ag^+, with anions of weak acids, such as CN^-, F^-, and CO_3^{2-}—for example, NaCN, KF, and Na_2CO_3.

For a salt of an acidic cation, such as NH_4^+, and a basic anion, such as CN^-, we cannot predict whether the solution will be acidic, basic, or neutral unless we know K_a for the cation and K_b for the anion. If $K_a > K_b$ the solution will be acidic, and if $K_b > K_a$ the solution will be basic.

═ *Example 14.10* ═══════════════════════════════════

ACID–BASE PROPERTIES OF SALTS

Predict whether the following salts give acidic, basic, or neutral solutions when dissolved in water: NaBr, K_2CO_3, $AlCl_3$, NH_4ClO_4, and $(NH_4)_2S$.

Solution

NaBr	Neutral cation, neutral anion	Therefore the solution is neutral.
K_2CO_3	Neutral cation, basic anion	Therefore the solution is basic.
$AlCl_3$	Acidic cation, neutral anion	Therefore the solution is acidic.
NH_4ClO_4	Acidic cation, neutral anion	Therefore the solution is acidic.
$(NH_4)_2S$	Acidic cation, basic anion	We cannot make a prediction without information on $K_a(NH_4^+)$ and $K_b(S^{2-})$.

EXERCISE 14.8

Predict whether the pH of an aqueous solution of each of the following salts is less than, greater than, or equal to 7:

$FeCl_3$ NH_4NO_3 $NaHCO_3$ K_2HPO_4

If we know the appropriate K_a or K_b value, we can calculate the pH of a solution of a salt, as the following examples show.

Example 14.11

pH OF A SALT SOLUTION

What is the pH of a 0.10*M* solution of sodium cyanide?

Solution

The first step is to recognize whether the cation or the anion reacts with water and then to write the equation for that reaction. Sodium is one of the cations that is not sufficiently strongly hydrated to give an acidic solution—it gives a neutral solution. But cyanide is the conjugate base of the weak acid HCN so it is a weak base and therefore reacts with water according to the following equation:

$$CN^- + H_2O \rightleftharpoons HCN + OH^-$$

The second step is to write the expression for the equilibrium constant and to look up the value of $K_b(CN^-)$ in Table 14.2 or to calculate it from $K_a(HCN)$:

$$K_b = \frac{[HCN][OH^-]}{[CN^-]} = 2.0 \times 10^{-5} \text{ mol L}^{-1}$$

The third step is to write an expression for the concentration of each of the species in solution. Because NaCN is a salt and is fully dissociated in aqueous solution, the initial concentrations of Na^+ and CN^- are both 0.10*M*. If we let $[OH^-] = x$ mol L^{-1}, we have

	CN^-	$+ H_2O \rightleftharpoons$	HCN	$+ OH^-$	
Initial concentrations	0.10		0	0	mol L^{-1}
Equilibrium concentrations	$0.10 - x$		x	x	mol L^{-1}

Hence

$$\frac{x^2}{0.10 - x} \text{ mol L}^{-1} = 2.0 \times 10^{-5} \text{ mol L}^{-1}$$

Since K_b is small, we assume that $x \ll 0.10$ and hence $0.10 - x \approx 0.10$. Thus we have

$$\frac{x^2}{0.10} = 2.0 \times 10^{-5}$$

$$x^2 = 2.0 \times 10^{-6}$$

$$x = 1.4 \times 10^{-3}$$

$$[OH^-] = 1.4 \times 10^{-3} \text{ mol L}^{-1}$$

Since x is less than 5% of the initial concentration of CN^-, the assumption that $0.10 - x \approx 0.10$ was justified. Thus

$$pH = -\log(1.4 \times 10^{-3}) = -(-2.85) = 2.85$$

$$pH = 14.00 - 2.85 = 11.15$$

= *Example 14.12* =

pH OF A SALT SOLUTION

What is the pH of a $0.20M$ solution of NH_4Cl?

Solution

Ammonium chloride, NH_4Cl, is an ionic solid consisting of ammonium ions, $NH_4{}^+$, and chloride ions, Cl^-. The chloride ion is not a base (it is the anion of a strong acid), but the ammonium ion is a weak acid. We follow exactly the same steps as in Example 14.11. First, we write the equation for the reaction of ammonium ion with water:

$$NH_4{}^+ + H_2O \rightleftharpoons H_3O^+ + NH_3$$

The equilibrium constant is the acid dissociation constant, $K_a(NH_4{}^+)$. From Table 14.1 we see that $K_a(NH_4{}^+) = 5.6 \times 10^{-10}$ mol L^{-1}. Thus

$$K_a(NH_4{}^+) = \frac{[H_3O^+][NH_3]}{[NH_4{}^+]} = 5.6 \times 10^{-10} \text{ mol L}^{-1}$$

Now we obtain an expression for each of the concentrations by letting $[H_3O^+] = x$ mol L^{-1}:

	$NH_4{}^+$	$+ H_2O$	$\rightleftharpoons$	H_3O^+	$+ NH_3$	
Initial concentrations	0.20			0	0	mol L^{-1}
Equilibrium concentrations	$0.20 - x$			x	x	mol L^{-1}

Therefore

$$\frac{x^2}{0.20 - x} \text{ mol L}^{-1} = 5.6 \times 10^{-10} \text{ mol L}^{-1}$$

Making the usual approximation that $x \ll 0.20$, we have

$$\frac{x^2}{0.20} = 5.6 \times 10^{-10}$$

Solving for x gives

$$x = 1.1 \times 10^{-5} \text{ and } [H_3O^+] = 1.1 \times 10^{-5} \text{ mol L}^{-1}$$

Therefore

$$pH = -\log(1.1 \times 10^{-5}) = 4.96$$

Whenever you calculate the pH of an aqueous solution think about the reaction(s) involved and make a qualitative estimate of the pH; for example, is it expected to be less than, equal to, or greater than 7? Use this qualitative answer to check that your numerical answer is reasonable.

EXERCISE 14.9

What is the pH of a $0.30M$ NaF solution?

14.5
pH: Applications and Measurement

The measurement of the pH of aqueous solutions has many important applications. The rates of chemical reactions involved in biochemical processes are often very sensitive to the hydronium ion concentration of the medium. In fermentation, for example, control of the pH is very important. In fact, the concept of pH was invented in 1909 by the Danish chemist Søren Sørensen (1868–1939) while he was working on problems connected with the brewing of beer at the Carlsberg Brewery in Copenhagen. Many body fluids have well-defined pH values that must be maintained if the body is to function in a normal way. For example, the fluid in the stomach has a pH of approximately 1.4. This rather high acidity is important for the proper digestion of food. But if the stomach fluid becomes much more acidic, we are soon made aware of it by the pain and discomfort. Blood has a constant pH of 7.4.

Pure water has a pH of 7.0. Ordinary rainwater and drinking water are normally very slightly acidic (pH < 7.0), mainly because they contain a small amount of dissolved CO_2. This dissolved CO_2, which reacts with water to some extent, produces H_3O^+ and hydrogen carbonate ion, HCO_3^-:

$$CO_2(aq) + 2H_2O(l) \rightleftharpoons H_3O^+(aq) + HCO_3^-(aq)$$

Rainwater also contains very small amounts of sulfuric and nitric acids. These acids are formed from SO_2 emitted by volcanoes and NO formed in lightning discharges. However, when we speak of *acid rain*, we mean rain that is much more acidic than it has normally been in the past or than it is in areas where the atmosphere is not polluted with SO_2 and other acid-producing substances resulting from industrial processes and automobile emissions (see Box 14.1).

Our sense of taste is remarkably sensitive to pH. Indeed, taste was one of the earliest ways in which acids and bases were distinguished. We are able to detect a sour, tart, or acidic taste in a solution with a pH between 4 and 5. Soda water, which is a solution of carbon dioxide in water, has a pH of about 4 and tastes quite definitely acidic. Most fruit juices and soft drinks have a pH in the range of 2–3. Their familiar acid taste arises from the weak acids that they contain. For example, lemon juice contains citric acid (see Chapter 23). A solution with a pH of 1 or less is not only unpleasant to taste but is dangerous because it burns the skin.

Basic solutions with a pH greater than 7 have a bitter taste and a characteristic slippery or soapy feel. They cause a characteristic wrinkling of the skin. A 2% solution of sodium hydrogen carbonate, $NaHCO_3$, is an effective mouthwash, but it has a very unpleasant taste. Very basic solutions of pH 12 or 13 are dangerous because they attack the skin quite rapidly. Some drain cleaners are a concentrated NaOH solution with added scent and coloring. They have a pH as high as 14 or 15 and should be handled with great care.

INDICATORS

A convenient way of determining the approximate pH of a solution is by the use of indicators. *An* **indicator** *is a weak acid that has a conjugate base with a different color from that of the acid.* Colored substances have been used as indicators for a long time. Litmus, which is extracted from certain lichens, is an example. In basic solutions, in which it has a blue color, it is in its conjugate base form. In acid solutions, in which it has a red color, it is in its acid form. We saw in Experiment 5.8 that tea, red cabbage, and grape juice are indicators. Experiment 14.1 on p. 686 shows that the colored substance that gives a rose its red color is an indicator. In the laboratory we use various dyes such as phenolphthalein, which is colorless in acid solution and pink in basic solution, and methyl red, which is red in acid solution and yellow in basic solution (see Figure 14.1).

The pH at which the color change occurs depends on the indicator. It is determined by the position of the equilibrium between the acid form of the indicator, denoted by HIn, and its conjugate base, denoted by In^-:

$$HIn + H_2O \rightleftharpoons In^- + H_3O^+$$

According to Le Châtelier's principle, this equilibrium is shifted to the left if the H_3O^+ concentration is increased, for example, by adding acid to the solution. If the indicator is phenolphthalein, it is then very largely in its colorless form,

Vegetables such as red cabbage can serve as indicators.

PHENOLPHTHALEIN

Acid form: Colorless   Conjugate base form: Pink

FIGURE 14.1
The Indicators Phenolphthalein and Methyl Red.

METHYL RED

Acid form: Red   Conjugate base form: Yellow

HIn. But if the solution is made basic, the H_3O^+ concentration is much reduced, the equilibrium shifts to the right, and the indicator is converted almost entirely to the In^- form, which in the case of phenolphthalein is pink. Thus we can distinguish between an acidic solution and a basic solution by adding a small amount of phenolphthalein. The solution will be pink if it is basic but colorless if it is acidic.

The exact pH at which the color change of an indicator occurs can be found from the acid dissociation constant of the indicator:

$$K_a(HIn) = \frac{[In^-][H_3O^+]}{[HIn]}$$

BOX 14.1

ACID RAIN

Rain is normally slightly acidic, with a pH of 5.6, because it contains dissolved carbon dioxide from the atmosphere. But the rain now falling in many parts of the industrialized world, even in regions remote from industry, is often much more acidic. The average pH of rain in the northeastern United States has decreased over the past years and is now 4.3. Rain with a pH as low as 1.5—which is only three times less concentrated than the $0.1M$ acid often used for titrations in the laboratory—has been recorded at Wheeling, West Virginia. The Los Angeles basin routinely has fogs made up of suspended water droplets with a pH of 2.2–4.0.

Acid rain is caused by oxides of sulfur and nitrogen which are present in the atmosphere as a result of combustion processes. Coal and oil typically contain 1%–3% sulfur and this is converted into SO_2 when the coal and oil are burned. Sulfur dioxide is also produced when sulfide ores are roasted to convert them to oxides that can be reduced to the metal (see Chapter 10). For example

$$Cu_2S + 2O_2 \longrightarrow 2CuO + SO_2$$

Sulfur dioxide is slowly oxidized to sulfur trioxide in the atmosphere and this dissolves in water droplets to give a dilute solution of sulfuric acid. Of course, some sulfur dioxide arises from natural sources. Scientists have estimated that the eruption of Mount St. Helens in the state of Washington in 1980 blew out approximately 400 000 tons of sulfur dioxide. But that is only about 1.5% of the estimated total sulfur dioxide from man-made sources in the United States for the same year. Nitrogen monoxide, NO, is formed during electrical storms, but it is also formed in combustion processes, particularly in internal combustion engines, as a result of the combination of

atmospheric nitrogen and oxygen. Nitrogen monoxide reacts with oxygen in the atmosphere to give nitrogen dioxide, NO_2, which in turn reacts with water in clouds and raindrops to give a solution of nitric acid:

$$3NO_2(g) + H_2O(l) \longrightarrow 2HNO_3(aq) + NO(g)$$

Some of the harmful effects of acid rain are now well established. Because marble—a form of calcium carbonate—is soluble in acid (see Chapter 5), Greek and Italian monuments that have withstood many centuries of natural weathering are now rapidly deteriorating. There are even more serious harmful ecological effects. In limestone areas acid rain is largely neutralized, but in other regions this does not happen and lakes and rivers have consequently become acidified. It is estimated that 20 000 lakes in Sweden have now become too acidic for fish and other life. The same effects are being observed in lakes in the United States and Canada. Acid rain dissolves aluminum compounds from the soil and washes them into lakes where they poison the fish. Plant life is also affected because acid rain kills microorganisms in the soil that are responsible for nitrogen fixation, and it dissolves and washes away essential magnesium, calcium, and potassium compounds. Moreover, acid rain can dissolve the waxy coating that protects leaves from fungi and bacteria. The recent deterioration of some coniferous forests in Germany and Canada has been attributed to the effects of acid rain.

What can be done to combat the effects of acid rain? A local and limited solution that has been used in Sweden is to add lime, CaO, to lakes to neutralize the acid, but this is expensive and must be repeated annually. In the past, very high smoke stacks were built at smelting plants

Rearranging this equation, we have

$$[H_3O^+] = K_a \frac{[HIn]}{[In^-]}$$

Taking negative logarithms of both sides gives

$$-\log[H_3O^+] = -\log K_a - \log \frac{[HIn]}{[In^-]}$$

$$pH = pK_a - \log \frac{[HIn]}{[In^-]} = pK_a + \log \frac{[In^-]}{[HIn]}$$

Left: This photo taken in 1910 shows the effect of 400 years of weathering on a grotesque decorating Lincoln Cathedral in England. Right: In 1984, only 74 years later, acid rain and other atmospheric pollution have worn the figure to a barely recognizable remnant.

to disperse the sulfur dioxide as high in the atmosphere as possible. This certainly benefitted the vegetation around the smelting plants which in many cases had been completely killed for many miles around. However, sulfur dioxide may be carried in air currents many hundreds of miles to be deposited, often in another country, as acid rain. The Swedes blame the British for their acid rain, and controversy rages between Canada and the United States over the acid rain believed by both countries to originate on the other side of the border. The obvious solution to the problem is to drastically reduce emissions of the oxides of sulfur and nitrogen. Sulfur dioxide could be removed from the gases produced by combustion and by smelting, for example, by passing them over lime which reacts with sulfur dioxide to produce solid calcium sulfite:

$$CaO(s) + SO_2(g) \longrightarrow CaSO_3(s)$$

Other solutions include removing sulfur from oil and coal or using only low sulfur content fuels. Or we could convert to alternative energy sources such as solar power or nuclear power. Most solutions are expensive or bring problems of their own—such as disposal of nuclear wastes. Because these solutions are expensive, and often unpopular with certain groups, such as coalminers who might lose their jobs, governments are reluctant to act. They use the fact that, neither all the effects of acid rain, nor all the details of its formation are yet well understood, as an excuse to delay. Although scientists will continue to search for a better understanding of the phenomenon of acid rain and for better and less expensive solutions to the problems that it causes, we already know enough about its harmful effects to warrant drastic steps to reduce emissions of the oxides of sulfur and nitrogen.

Experiment 14.1

A Natural Indicator

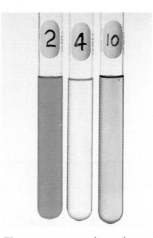

The compound that gives the color to a red rose can be dissolved in methanol to give a red solution, leaving the rose in the beaker on the left a pale pink color.

The rose extract can be used as an indicator. A small amount of the extract has been added to each of these tubes. The tube on the left is an aqueous solution at pH = 2, the middle tube is at pH = 4, and the right-hand tube is at pH = 10.

The indicator will be half in its acid form and half in its base form—in other words, in the middle of its color change—when

$$[HIn] = [In^-] \quad \text{or} \quad \frac{[In^-]}{[HIn]} = 1$$

and the pH will then be

$$pH = pK_a + \log 1 = pK_a + 0 = pK_a$$

TABLE 14.7
Properties of Some Indicators

	pK_a	*Effective* pH *Range*	Color Acid Form	Base Form
Methyl violet	1.6	0.0–3.0	Yellow	Violet
Methyl orange	4.2	3.1–4.4	Red	Yellow
Methyl red	5.2	4.2–6.2	Red	Yellow
Bromothymol blue	7.1	6.0–7.8	Yellow	Blue
Thymol blue	8.2	7.9–9.4	Yellow	Blue
Phenolphthalein	9.5	8.3–10.0	Colorless	Red
Alizarine yellow	11.0	10.1–12.1	Yellow	Red

FIGURE 14.2
Colors of Some Indicators.

From left to right: Phenolphthalein, bromothymol blue, and methyl red. The numbers on the tubes denote the pH.

Thus the pH at which an indicator changes color depends on its pK_a. For phenolphthalein $K_a = 3 \times 10^{-10}$, so the pH at which it is half in its base form and half in its acid form is

$$pH = pK_a = -\log(3 \times 10^{-10}) = 9.5$$

Thus we expect phenolphthalein to change color in the vicinity of pH 9.5. In fact, the color change occurs over a range of pH, and this range is partly determined by the sensitivity of the eye to the color change. In general, when $[In^-]/[HIn]$ or $[HIn]/[In^-]$ reaches a value of approximately 10, the eye cannot detect any further color change if the concentration ratio is further increased. Thus the visible color change occurs approximately over the range

$$pH = pK_a(HIn) \pm \log 10 \quad \text{or} \quad pH = pK_a(HIn) \pm 1$$

and the color change is half complete when $pH = pK_a(HIn)$. For example, methyl red has a pK_a of 5.2. It is clearly red in a solution of pH 4.2 and clearly yellow in a solution of pH 6.2. Thus as the acidity increases over this range, the indicator changes color from yellow through orange to red.

Table 14.7 lists several indicators together with their useful pH ranges. The colors of some of these indicators are shown in Figure 14.2. Notice that indicators do not in general change color at pH 7. If we compare the color of a solution of unknown pH containing a suitable indicator with the colors of that indicator in a number of solutions of known pH, we can make a fairly accurate determination of the pH. The smallest amount of indicator that will give a clearly observable color should be used. Adding too much indicator disturbs the acid–base equilibrium in the solution and changes the pH.

Universal indicator is a mixture of several indicators that has several color changes over a wide pH range. With such a solution one can find the approximate pH of any solution within this range. So-called *pH paper* is impregnated with a universal indicator. When a strip of this paper is immersed in a solution, the pH can be judged from the resulting color.

For accurate pH measurements a pH meter such as the one shown in Figure 14.3 is used. The principles on which the pH meter operates are described in

Most indicators have a useful pH range of approximately 2 pH units.

FIGURE 14.3
pH Meter.

A pH meter is an instrument that employs two electrodes to measure the potential difference between the solution to be tested and a standard solution of known pH. This potential difference is related to the pH of the solution being tested. The pH is read directly from the dial of the instrument.

Experiment 14.2

The pH of Some Common Solutions

The color of universal indicator shows that solutions of lime juice in water, soda water, and vinegar are all acidic.

The color of the universal indicator shows that Drāno, household ammonia, and Milk of Magnesia are all basic (alkaline).

pH can be measured more accurately with a pH meter. Vinegar has a pH of 2.3.

A suspension of Milk of Magnesia in water has a pH of 9.4.

The color produced when a strip of pH paper is dipped into a solution gives an approximate value for the pH of the solution.

Chapter 17. Experiment 14.2 demonstrates the measurement of the pH of some common solutions.

DETERMINATION OF K_a AND K_b

One important application of pH measurements is the determination of the dissociation constants of acids and bases. By measuring the pH of a solution of an acid or base of known concentration the value of K_a or K_b can be calculated as shown in the following examples.

═ Example 14.13 ═══

DETERMINATION OF K_a FROM pH

A 0.125M solution of benzoic acid, $C_6H_5CO_2H$, a monoprotic acid, has a pH of 2.56. What is the value of K_a for benzoic acid?

Solution

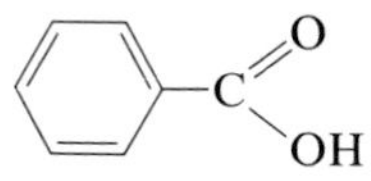

We will represent the formula of benzoic acid as HBz. The concentration of H_3O^+ in the solution can be obtained from the pH:

$$[H_3O^+] = 10^{-pH} = 10^{-2.56} = 2.8 \times 10^{-3} \text{ mol L}^{-1}$$

Now we can find the concentrations of each of the species present in the solution:

	HBz	$+ H_2O \rightleftharpoons$	H_3O^+	$+$	Bz^-	
Initial concentrations	0.125		0		0	mol L^{-1}
Equilibrium concentrations	$0.125 - 2.8 \times 10^{-3}$		2.8×10^{-3}		2.8×10^{-3}	mol L^{-1}

From these equilibrium concentrations we can calculate K_a:

$$K_a = \frac{[H_3O^+][Bz^-]}{[HBz]} = \frac{(2.8 \times 10^{-3})(2.8 \times 10^{-3})}{0.122} = 6.4 \times 10^{-5} \text{ mol L}^{-1}$$

EXERCISE 14.10

Lactic acid is a monoprotic acid that accumulates in the muscles during strenuous exercise and can lead to cramps. The pH of a $0.10M$ solution is found to be 2.43. What is the K_a of lactic acid?

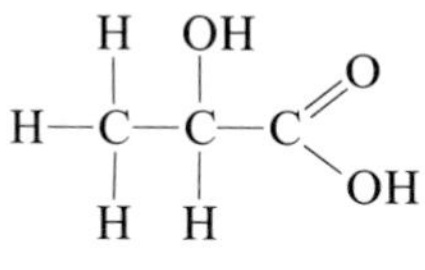

Lactic acid

Example 14.14

DETERMINATION OF K_b FROM pH

A $0.20M$ solution of trimethylamine, $(CH_3)_3N$, has a pH of 11.51. What is the value of K_b for trimethylamine?

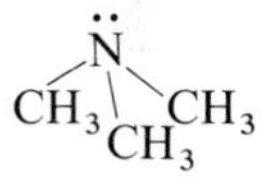

Trimethylamine

Solution

We can find the OH^- concentration from the pH as follows:

$$pOH = 14.00 - pH = 2.49 \quad \text{hence, } [OH^-] = 3.2 \times 10^{-1}$$

Now we can find the concentrations of each of the species present in the solution:

	$(CH_3)_3N$	$+ H_2O \rightleftharpoons$	$(CH_3)_3NH^+$	$+$	OH^-	
Initial concentrations	0.20		0		0	mol L^{-1}
Equilibrium concentrations	$0.20 - 3.2 \times 10^{-3}$		3.2×10^{-3}		3.2×10^{-3}	mol L^{-1}

Hence

$$K_b = \frac{[(CH_3)_3NH^+][OH^-]}{[(CH_3)_3N]} = \frac{(3.2 \times 10^{-3})(3.2 \times 10^{-3})}{0.20} \text{ mol L}^{-1}$$
$$= 5.1 \times 10^{-5} \text{ mol L}^{-1}$$

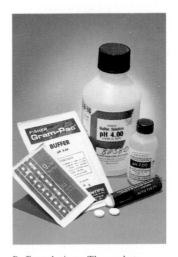

Dimethylamine

Dimethylamine, $(CH_3)_2NH$, is a weak base used in the manufacture of detergents. The pH of a $1.00M$ aqueous solution of dimethylamine is 12.36. What is K_b for dimethylamine?

14.6
Buffer Solutions

In many chemical processes and in many reactions that occur in living systems, the pH must be held at a constant value. For example, the control of the pH is very important in the treatment of sewage, in electroplating, and in the manufacture of photographic materials. Maintaining a constant pH is important in many metabolic processes, because the function of proteins depends on their structures and their structures depend on the pH. Proteins are both acids and bases, and as their charge changes on gaining or losing protons, their shape changes. Therefore the pH must remain constant so that they maintain a particular shape. For example, the pH of the blood has to be close to 7.4; the pH of saliva is close to 6.8; and the enzymes in the stomach function only at a pH of approximately 1.5.

A solution that has the ability to maintain a very nearly constant pH even when moderate amounts of acid or base are added to it is called a **buffer solution**. In order to react with either added acid or added base, a buffer solution must contain appreciable concentrations of both a base and an acid. The usual and most convenient way to prepare such a solution is from roughly equal quantities of a weak acid and its conjugate base, or a weak base and its conjugate acid.

> **Buffer solutions contain either a weak acid and a salt of the weak acid, or a weak base and a salt of the weak base, in approximately equal proportions.**

For example, a buffer solution prepared from approximately equal amounts of acetic acid and sodium acetate contains roughly equal concentrations of CH_3CO_2H and $CH_3CO_2^-$. If H_3O^+ is added to the solution, it combines almost completely with $CH_3CO_2^-$ to form CH_3CO_2H:

$$CH_3CO_2^- + H_3O^+ \longrightarrow CH_3CO_2H + H_2O$$

Buffer solutions. The packets contain a solid mixture of an acid and its conjugate base which, when added to a suitable amount of water, produces a buffer solution with the pH stated on the packet.

Therefore there is no appreciable change in the H_3O^+ concentration. That this reaction does go almost to completion can be seen from the value of its equilibrium constant, which is the inverse of the acid dissociation constant of acetic acid:

$$K = \frac{[CH_3CO_2H]}{[CH_3CO_2^-][H_3O^+]} = \frac{1}{\dfrac{[H_3O^+][CH_3CO_2^-]}{[CH_3CO_2H]}} = \frac{1}{K_a(CH_3CO_2H)}$$

$$= \frac{1}{1.8 \times 10^{-5}} = 5.6 \times 10^4 \text{ mol}^{-1}\text{ L}$$

If OH^- is added to the acetic acid–acetate buffer solution, it combines almost completely with acetic acid, converting it to acetate ion:

$$CH_3CO_2H + OH^- \longrightarrow CH_3CO_2^- + H_2O$$

So again there is almost no change in the OH^- (or H_3O^+) concentration and therefore in the pH. That this reaction also goes very nearly to completion is shown by its very large equilibrium constant, which is the inverse of the base dissociation constant of $CH_3CO_2^-$:

$$K = \frac{[CH_3CO_2^-]}{[CH_3CO_2H][OH^-]} = \frac{1}{\dfrac{[CH_3CO_2H][OH^-]}{[CH_3CO_2^-]}} = \frac{1}{K_b(CH_3CO_2^-)}$$

$$= \frac{1}{5.6 \times 10^{-10} \text{ mol}^{-1} \text{ L}} = 1.8 \times 10^9 \text{ mol}^{-1} \text{ L}$$

Another way of looking at the consequences of adding acid or base to an acetic acid–acetate buffer solution is to apply le Châtelier's principle to the equilibrium:

$$CH_3CO_2H + H_2O \rightleftharpoons CH_3CO_2^- + H_3O^+$$

Addition of acid, that is, of H_3O^+, drives the equilibrium to the left. Since acetic acid is a weak acid, only a very small concentration of H_3O^+ can remain when the new equilibrium is reached. Therefore virtually all the added H_3O^+ is removed, and there is almost no decrease in the pH. If OH^- is added to the buffer solution, it reacts with H_3O^+ to form water. In accordance with Le Châtelier's principle more acetic acid must then dissociate to give more H_3O^+ to replace that which is removed. The overall result is the removal of the added OH^- and the simultaneous conversion of CH_3CO_2H to acetate ion, $CH_3CO_2^-$. Essentially all the added OH^- is removed in this way, and there is almost no increase in the pH of the solution.

An acetic acid–acetate buffer solution will therefore maintain an almost constant pH provided that the amounts of added acid or base are not large enough to remove most of the $CH_3CO_2^-$ or CH_3CO_2H originally present in the buffer solution. As long as both $CH_3CO_2^-$ and CH_3CO_2H are present in appreciable amounts, a small and nearly constant equilibrium concentration of H_3O^+ is maintained, and there is little change in the pH.

A buffer solution composed of a weak base and a salt of the weak base behaves in a similar manner. For example, if we have a solution of ammonia and ammonium chloride, added acid reacts with the weak base, NH_3, and added base reacts with the ammonium ion, which is a weak acid:

$$NH_3 + H_3O^+ \longrightarrow NH_4^+ + H_2O$$
$$NH_4^+ + OH^- \longrightarrow NH_3 + H_2O$$

Again, there is no significant change in either the H_3O^+ or OH^- concentrations and thus no appreciable change in the pH of the solution.

CALCULATION OF THE pH OF A BUFFER SOLUTION

To make up a buffer solution for some particular pH, we must know how to calculate the pH of any given buffer solution. As an example, we will calculate the pH of a solution of 0.050 mol of acetic acid and 0.050 mol of sodium acetate in 1 L.

The first step is to write the equation for the equilibrium in the solution. The solution contains both acetate ions and acetic acid molecules, and the equilibrium is

$$CH_3CO_2H + H_2O \rightleftharpoons CH_3CO_2^- + H_3O^+$$

This is the equilibrium for the ionization of acetic acid in water. However, the solution that we are considering differs from a solution of acetic acid in water in that it has a large concentration of acetate ion as well as acetic acid.

The second step is to write the expression for the equilibrium constant and look up its value. In this case we have

$$K_a = \frac{[H_3O^+][CH_3CO_2^-]}{[CH_3CO_2H]} = 1.8 \times 10^{-5} \text{ mol L}^{-1}$$

The third step is to write an expression for the concentration of each of the species, remembering that 0.050 mol $CH_3CO_2^-$ comes from the sodium acetate:

$$CH_3CO_2H + H_2O \rightleftharpoons CH_3CO_2^- + H_3O^+$$

Initial concentrations	0.050	0.050	0	mol L^{-1}
Equilibrium concentrations	$0.050 - x$	$0.050 + x$	x	mol L^{-1}

The initial concentrations of the acetic acid and the acetate ion are both 0.050 mol L^{-1}, but a small amount of the acetic acid must ionize in order to establish the equilibrium with the acetate ion. We let the concentration of hydronium ion that is formed be x mol L^{-1}. Then at equilibrium the concentration of acetic acid will be $0.050 - x$ mol L^{-1}, and the concentration of acetate ion will be $0.050 + x$ mol L^{-1}. Hence

$$K_a = \frac{[H_3O^+][CH_3CO_2^-]}{[CH_3CO_2H]} = \frac{x(0.050 + x)}{0.050 - x} \text{ mol L}^{-1} = 1.8 \times 10^{-5} \text{ mol L}^{-1}$$

This is a quadratic equation and may be solved by using the formula in Appendix A. However, because acetic acid is a weak acid, we know that x will be small compared with 0.050, and we can see, by using Le Châtelier's principle, that the large concentration of acetate ion will further decrease the ionization of acetic acid; thus x must be very small indeed. Therefore we make the approximation that $0.050 - x \approx 0.050 + x \approx 0.050$. Thus

$$\frac{0.050x}{0.050} \text{ mol L}^{-1} = 1.8 \times 10^{-5} \text{ mol L}^{-1}$$

$$x = 1.8 \times 10^{-5}$$

$$[H_3O^+] = 1.8 \times 10^{-5} \text{ mol L}^{-1}$$

$$pH = -\log(1.8 \times 10^{-5}) = 4.74$$

We see that x is indeed very small compared with 0.050, and therefore our approximation was justified.

In order to demonstrate the buffering action of this solution we will see how the pH is affected by the addition of 0.001 mol of HCl. HCl is a strong acid so we are adding 0.001 mol H_3O^+ and 0.001 mol Cl$^-$ to the solution. The added H_3O^+ reacts almost completely with the acetate ion, converting it to acetic acid:

$$CH_3CO_2^- + H_3O^+ \longrightarrow CH_3CO_2H + H_2O$$

Before reaction with HCl	0.050	0.001	0.050	mol L^{-1}
After reaction with HCl	0.049	0	0.051	mol L^{-1}

Now we recalculate the pH using the new concentrations of CH_3CO_2H and $CH_3CO_2^-$. As before we let the concentration of H_3O^+ in equilibrium with CH_3CO_2H and $CH_3CO_2^-$ be x mol L^{-1}.

$$CH_3CO_2H + H_2O \rightleftharpoons CH_3CO_2^- + H_3O^+$$

Initial concentrations	0.051	0.049	0	mol L^{-1}
Equilibrium concentrations	0.051 − x	0.049 + x	x	mol L^{-1}

$$K_a = \frac{[CH_3CO_2^-][H_3O^+]}{[CH_3CO_2H]} = 1.8 \times 10^{-5} \text{ mol L}^{-1}$$

Hence

$$x = [H_3O^+] = K_a \frac{[CH_3CO_2H]}{[CH_3CO_2^-]} = (1.8 \times 10^{-5}) \frac{0.051 - x}{0.049 + x}$$

Making the usual assumption that $x \ll 0.049$

$$x = (1.8 \times 10^{-5}) \frac{0.051}{0.049} = 1.9 \times 10^{-5}$$

$$[H_3O^+] = 1.9 \times 10^{-5} \text{ mol L}^{-1}$$

Therefore

$$pH = -\log(1.9 \times 10^{-5}) = 4.72$$

This pH differs by only 0.02 from the pH of the original buffer solution.

We see that the buffer works because the concentration of H_3O^+ depends on the ratio $[CH_3CO_2H]/[CH_3CO_2^-]$ and this hardly changes on the addition of HCl provided that the amounts of CH_3CO_2H and $CH_3CO_2^-$ are relatively large compared with the amount of HCl added. In this case $[CH_3CO_2H]/[CH_3CO_2^-]$ changes from 0.050/0.050 = 1.00 to 0.051/0.049 = 1.04

It may seem that we added only a small amount of HCl. But if the same amount of HCl had been added to a solution of pH 4.7 that was *not* a buffer solution, the pH would have changed by a much larger amount. For example, an HCl solution of concentration 0.000 020 mol L^{-1} has a pH of 4.7. If we add 0.001 mol of HCl to 1 L of this solution, the total $[H_3O^+]$ is 0.001 020, and the pH of the solution is 3.0. Thus in this case the pH changes from 4.7 to 3.0—a much greater change than we found for the buffer solution. Or if we add 0.001 mol of HCl to 1 L of pure water, the pH changes from 7 to 3.

─ Example 14.15 ─────────────────────

BUFFERING ACTION

What is the pH change when 0.01 mol of HCl is added to a solution containing 0.05 mol of acetate ion and 0.05 mol of acetic acid in 1.0 L of solution?

Solution

Following the previous argument, we find that the initial concentrations after the addition of HCl are

$$[CH_3CO_2H] = 0.06 \text{ mol L}^{-1} \quad \text{and} \quad [CH_3CO_2^-] = 0.04 \text{ mol L}^{-1}$$

If we let the equilibrium concentration of H_3O^+ be x mol L^{-1}, and assuming that $x \ll 0.04$, we have

$$x = [H_3O^+] = K_a \frac{[CH_3CO_2H]}{[CH_3CO_2^-]} = (1.8 \times 10^{-5})\frac{0.06}{0.04}$$

Therefore

$$x = 2.7 \times 10^{-5} \quad [H_3O^+] = 2.7 \times 10^{-5} \text{ mol L}^{-1}$$

and

$$pH = 4.57$$

Thus the pH changes from 4.74 to 4.57.

We see that even on adding ten times as much HCl as in the preceding discussion, the change in pH is still very small. If this much HCl had been added to 1.0 L of solution of pH 4.7 that was *not* a buffer, the pH would have changed from 4.7 to 2.0.

$=$ *Example 14.16* $=$

pH OF A BUFFER SOLUTION

What is the pH of a buffer solution prepared from 1.00 mol of NH_3 and 0.40 mol of NH_4Cl in 1.0 L of solution?

Solution

The equilibrium is between NH_3 and the NH_4^+:

$$NH_3 + H_2O \rightleftharpoons NH_4^+ + OH^-$$

The appropriate equilibrium constant is therefore the base dissociation constant of NH_3 (Table 14.2):

$$K_b = \frac{[NH_4^+][OH^-]}{[NH_3]} = 1.8 \times 10^{-5} \text{ mol L}^{-1}$$

We then obtain the concentrations of the various species in solution:

	NH_3	$+ H_2O \rightleftharpoons$	NH_4^+	$+ OH^-$	
Initial concentrations	1.0		0.4	0	mol L^{-1}
Equilibrium concentrations	$1.0 - x$		$0.4 + x$	x	mol L^{-1}

Substituting in the equilibrium constant expression gives

$$K_b = \frac{(0.4 + x)x}{1.0 - x} \text{ mol L}^{-1} = 1.8 \times 10^{-5} \text{ mol L}^{-1}$$

Making the usual assumption that x is very small compared with the initial concentrations, we have

$$\frac{0.4x}{1.0} = 1.8 \times 10^{-5}$$

$$x = 4.5 \times 10^{-5}$$

Thus

$$[\text{OH}^-] = 4.5 \times 10^{-5} \text{ mol L}^{-1}$$

$$\text{pOH} = 4.35$$

$$\text{pH} = 14.00 - 4.35 = 9.65$$

The problem in Example 14.16 can also be solved using the equilibrium

$$\text{NH}_4{}^+ + \text{H}_2\text{O} \rightleftharpoons \text{NH}_3 + \text{H}_3\text{O}^+$$

and the equilibrium constant

$$K_a(\text{NH}_4{}^+) = \frac{[\text{NH}_3][\text{H}_3\text{O}^+]}{[\text{NH}_4{}^+]}$$

Since the concentrations of NH_3 and $\text{NH}_4{}^+$ must satisfy both equilibria the same answer is obtained whichever method is used.

EXERCISE 14.12

Calculate the pH of the buffer solution in Example 14.16 using $K_a(\text{NH}_4{}^+) = 5.6 \times 10^{-10} \text{ mol L}^{-1}$.

THE HENDERSON–HASSELBALCH EQUATION

We can derive a general equation for calculating the pH of any buffer solution that is a mixture of a weak acid and its conjugate base. We have

$$\text{HA} + \text{H}_2\text{O} \rightleftharpoons \text{A}^- + \text{H}_3\text{O}^+$$

$$K_a = \left(\frac{[\text{H}_3\text{O}^+][\text{A}^-]}{[\text{HA}]} \right)_{eq}$$

This equation can be rearranged to give

$$[\text{H}_3\text{O}^+] = \frac{K_a[\text{HA}]}{[\text{A}^-]}$$

Taking negative logarithms of both sides, we obtain

$$-\log[\text{H}_3\text{O}^+] = -\log K_a - \log \frac{[\text{HA}]}{[\text{A}^-]}$$

$$\text{pH} = \text{p}K_a + \log \left(\frac{[\text{A}^-]}{[\text{HA}]} \right)_{eq}$$

We have seen that the equilibrium concentrations of HA and A^- do not differ significantly from the concentrations of the weak acid and its conjugate base used to make up the buffer solution. Hence to a very good approximation we may write

$$pH = pK_a + \log\left(\frac{[A^-]}{[HA]}\right)_{initial}$$

or

$$pH = pK_a + \log\frac{[base]}{[acid]}$$

where [acid] and [base] are the concentrations of acid and conjugate base used to make up the buffer solution. This equation is called the **Henderson–Hassel-balch equation**.

A buffer is most efficient in resisting changes in pH when both the acid and its conjugate base are present in approximately equal amounts—in other words, when $[A^-] = [HA]$. In this case

$$pH = pK_a + \log\frac{[A^-]}{[HA]} = pK_a + \log 1 = pK_a$$

Thus we can choose a buffer for a given pH range on the basis of the pK_a value of the acid. For a buffer solution to operate around pH 9, we might choose an NH_3–NH_4^+ buffer, since $pK_a(NH_4^+) = 9.3$. For a buffer to operate around a pH of 5, we might choose a CH_3CO_2H–$CH_3CO_2^-$ buffer, since $pK_a(CH_3CO_2H) = 4.7$.

The constant pH of blood is maintained by several conjugate acid–base pairs including $H_2PO_4^-$, HPO_4^{2-}, and H_2CO_3, HCO_3^-. Several proteins also contribute to the buffering action.

≡ Example 14.17

USING THE HENDERSON–HASSELBALCH EQUATION

Blood has a pH of 7.40. Use the Henderson–Hasselbalch equation to calculate the ratio of hydrogen carbonate ion, HCO_3^-, to carbonic acid, H_2CO_3, in blood; $pK_a(H_2CO_3) = 6.37$.

Solution

The Henderson–Hasselbalch equation is

$$pH = pK_a + \log\frac{[base]}{[acid]}$$

Rearranging to solve for the ratio of concentrations, we obtain

$$\log\frac{[base]}{[acid]} = pH - pK_a$$

$$= 7.40 - 6.37 = 1.03$$

Therefore

$$\frac{[base]}{[acid]} = 10.7$$

Thus the mole ratio of HCO_3^- to H_2CO_3 must be 10.7 to 1.

━ *Example 14.18* ━━━━━━━━━━━━━━━━━━━━━━━━━━━━━

USING THE HENDERSON–HASSELBALCH EQUATION

What is the pH of a buffer solution obtained by adding 25.00 mL of a 0.10M sodium hydroxide solution to 75.00 mL of a 0.10M solution of acetic acid?

Solution

The NaOH and CH_3CO_2H react to form sodium acetate, CH_3CO_2Na. We need first to calculate the amounts of CH_3CO_2H and CH_3CO_2Na in the buffer solution. Initially, we have

$$\text{moles of NaOH} = (25 \text{ mL})(0.10 \text{ mol L}^{-1})\left(\frac{1 \text{ L}}{10^3 \text{ mL}}\right) = 2.5 \times 10^{-3} \text{ mol}$$

$$\text{moles of } CH_3CO_2H = (75 \text{ mL})(0.10 \text{ mol L}^{-1})\left(\frac{1 \text{ L}}{10^3 \text{ mL}}\right) = 7.5 \times 10^{-3} \text{ mol}$$

	$CH_3CO_2H +$	OH^-	$\longrightarrow$	$CH_3CO_2^- + H_2O$	
Initial amounts	7.5×10^{-3}	2.5×10^{-3}		0	moi
After mixing	5.0×10^{-3}	0		2.5×10^{-3}	mol

Note that to apply the Henderson–Hasselbalch equation, we do not need to change these amounts into concentrations, because the ratio of concentrations, [base]/[acid], is equal to the ratio of the number of moles. That is,

$$pH = pK_a + \log \frac{[\text{base}]}{[\text{acid}]} = 4.74 + \log \frac{2.5 \times 10^{-3} \text{ mol}}{5.0 \times 10^{-3} \text{ mol}}$$

$$= 4.74 + \log 0.5 = 4.44$$

EXERCISE 14.13

What is the pH of each of the following?

(a) A solution prepared by mixing 10.0 mL of 0.10M NH_4Cl with 20.0 mL of 0.10M NH_3

(b) A solution prepared by dissolving 15.0 g of $NaHCO_3$ and 18.0 g of Na_2CO_3 in sufficient water to make 1.00 L of solution

EXERCISE 14.14

What mole ratio of acetic acid and sodium acetate would you need to prepare a buffer solution with a pH of 5.00?

14.7
Acid–Base Titrations

Acids and bases are such common and important substances that it is frequently necessary to determine the concentration of a solution of an acid or a base.

One convenient and widely used procedure for doing this is called a **titration**. A titration can be based on any reaction that is rapid and goes to completion, for example, the reaction between an acid and a base. A solution of known concentration of a base is added to a known volume of a solution of an acid of unknown concentration, and the volume of the solution of the base needed to just react completely with the acid is determined.

The calculation of the concentration of the acid solution from the measured volume of the base solution was explained in Chapter 5. We will be concerned here with how we find the **equivalence point**, that is, the point during the titration at which just enough base has been added to react completely with all the acid. The usual way of determining when we have reached the equivalence point during a titration is by measuring the pH of the solution with a pH meter, or by using an indicator that changes color at the pH of the equivalence point. We will see now how we can calculate the pH at any point during an acid–base titration, and in particular at the equivalence point.

TITRATION OF A STRONG ACID WITH A STRONG BASE

We consider first the change in the pH during the titration of a strong acid with a strong base. As an example, we will calculate the change in pH during the titration of 25 mL of $0.100M$ HCl solution with $0.100M$ NaOH solution. In these calculations we will find it convenient to work in millimoles (abbreviated mmol):

$$1 \text{ mmol} = 10^{-3} \text{ mol}$$

Because the volumes of the solutions used are relatively small and are usually measured in milliliters the mole is an inconveniently large unit. We will also find it convenient to define molarity in terms of millimoles per milliliter:

$$\text{Molarity} = \frac{\text{mol of solute}}{\text{L of solution}} = \frac{\dfrac{\text{mol of solute}}{1000}}{\dfrac{\text{L of solution}}{1000}} = \frac{\text{mmol of solute}}{\text{mL of solution}}$$

A $1.0M$ solution contains 1.0 mol of solute per liter of solution, or 1.0 mmol of solute per milliliter of solution.

In this titration the reaction is

$$\text{HCl} + \text{NaOH} \longrightarrow \text{NaCl} + \text{H}_2\text{O}$$

At the equivalence point the solution contains only NaCl and it has a pH of 7.00. At any other point in the titration we can calculate the pH as shown in Table 14.8 where n_a is the number of millimoles of HCl initially present in the solution and n_b is the number of millimoles of NaOH added. Before the titration there are $25.0 \text{ mL} \times 0.1 \text{ mmol mL}^{-1} = 2.50 \text{ mmol HCl} = 2.50 \text{ mmol}$ H_3O^+. When, for example, 20.0 mL of $0.100M$ NaOH = 2.00 mmol NaOH = 2.00 mmol OH^- have been added, the solution contains 2.50 mmol − 2.00 mmol H_3O^+ = 0.50 mmol H_3O^+. Since the total volume of the solution at this point is 20.0 mL + 25.0 mL = 45.0 mL the concentration of H_3O^+ is 0.50 mmol/ 45.0 mL = 0.011 mmol mL^{-1} = 0.011 mol L^{-1} and the pH = $-\log 0.011$ = 1.95. The pH at other points in Table 14.8 is calculated in the same way. If the

TABLE 14.8
Change in pH During Titration of 25.0 mL of 0.100M HCl with 0.100M NaOH at 25 °C

V_a (HCl) (mL)	V_b (NaOH) (mL)	Total Volume: $V_a + V_b$ (mL)	Initial Acid (n_a) (mmol)	Added Base (n_b) (mmol)	$n_a + n_b$ $n_{H_3O^+}$ (mmol)	$[H_3O^+]$ (mmol mL^{-1}) (mol L^{-1})	pH
25.0	0.0	25.0	2.50	0.00	2.50	0.100	1.00
	10.0	35.0		1.00	1.50	4.29×10^{-2}	1.37
	20.0	45.0		2.00	0.50	1.11×10^{-2}	1.95
	24.5	49.5		2.45	0.05	1.0×10^{-3}	3.00
	24.9	49.9		2.49	0.01	2.0×10^{-4}	3.70
	25.0	50.0		2.50	—	10^{-7}	7.00[a]

					$n_{OH^-}, n_b - n_a$ (mmol)	$[OH^-]$ (mmol mL^{-1}) (mol L^{-1})	
	25.1	50.1		2.51	0.01	2.0×10^{-4}	10.30
	25.5	50.5		2.55	0.05	1.0×10^{-3}	11.00
	30.0	55.0		3.00	0.50	9.1×10^{-3}	11.96
	40.0	65.0		4.00	1.50	2.31×10^{-2}	12.36
	50.0	75.0		5.00	2.50	3.33×10^{-2}	12.52

[a] Equivalence point: solution contains only NaCl.

pH is plotted against the volume of NaOH solution added, the curve in Figure 14.4 is obtained. This is called a **titration curve** or **pH curve**. We see that there is a very rapid increase in the pH very close to the equivalence point. It increases from 3.7 when 24.9 mL of NaOH have been added, to 7.0 when exactly 25.0 mL of NaOH have been added, to 10.3 when 25.1 mL have been added. Thus it is simple to detect the equivalence point in this titration by measuring the pH or by using any indicator that changes color between pH 4 and pH 10.

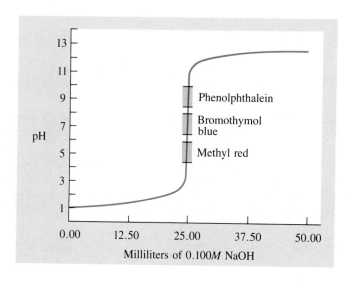

FIGURE 14.4
Titration of 25.00 mL of 0.100M HCl with 0.100M NaOH.

Any one of these three indicators could be used to determine the equivalence point of this titration.

TITRATION OF A WEAK ACID WITH A STRONG BASE

The choice of indicator for the titration of a weak acid such as acetic acid with a strong base such as sodium hydroxide is more limited. The variation in the pH is not the same as in a strong-acid–strong-base titration, because at the equivalence point of a titration of acetic acid with sodium hydroxide, the solution contains sodium acetate. This solution is basic and has pH > 7.0.

We consider now the change in pH during the titration of 25 mL of 0.100M acetic acid solution with a 0.100M sodium hydroxide solution. We can conveniently consider four stages in the titration.

1. *Initially*, we have a 0.100M solution of CH_3CO_2H. We can calculate the pH of this solution as described in Example 14.6, and we find pH = 2.9.

2. *Between the initial acid solution and the equivalence point*, insufficient NaOH has been added to neutralize all the CH_3CO_2H, so the solution contains both CH_3CO_2H and CH_3CO_2Na. These solutions are therefore *buffer solutions*, and their pH may be calculated as shown in Examples 14.17 and 14.18.

TABLE 14.9
Change in pH During Titration of 25.0 mL of 0.100M CH_3CO_2H with 0.100M NaOH at 25 °C

V_a CH_3CO_2H (mL)	V_b NaOH (mL)	Total Volume, $V_a + V_b$ (mL)	n_b CH_3CO_2Na (mmol)	n_a CH_3CO_2H (mmol)	$\dfrac{n_b}{n_a}$	$[H_3O^+]$ (mol L^{-1})	pH
25.0	0.00	25.0	0.00	2.50	—	1.34×10^{-3}	2.87[a]
	5.00	30.0	0.50	2.00	0.25	7.20×10^{-5}	4.14[b]
	10.0	35.0	1.00	1.50	0.67	2.69×10^{-5}	4.57[b]
	12.5	37.5	1.25	1.25	1.00	1.80×10^{-5}	4.74[b]
	15.0	40.0	1.50	1.00	1.50	1.20×10^{-5}	4.92[b]
	20.0	45.0	2.00	0.50	4.00	4.5×10^{-6}	5.35[b]
	22.0	47.0	2.20	0.30	7.40	2.4×10^{-6}	5.61[b]
	23.0	48.0	2.30	0.20	11.50	1.6×10^{-6}	5.81[b]
	24.0	49.0	2.40	0.10	34.00	8.0×10^{-7}	6.12[b]
	25.0	50.0	25.0	0.00	—	1.9×10^{-9}	8.72[c]
				Excess OH$^-$ (mmol)		[OH$^-$] (mol L^{-1})	
	26.0	51.0	2.50	0.10		1.96×10^{-3}	11.29[d]
	27.0	52.0	2.50	0.20		3.85×10^{-3}	11.59[d]
	30.0	55.0	2.50	0.50		9.09×10^{-3}	11.96[d]
	35.0	60.0	2.50	1.00		1.67×10^{-2}	12.22[d]
	40.0	65.0	2.50	1.50		2.31×10^{-2}	12.36[d]
	50.0	75.0	2.50	2.50		3.33×10^{-2}	12.52[d]

[a] 0.100M solution of CH_3CO_2H.
[b] Buffer solutions.
[c] 0.050M solution of CH_3CO_2Na.
[d] Solutions containing excess OH$^-$ and CH_3CO_2Na.

3. *At the equivalence point* the addition of 25.00 mL of $0.100M$ NaOH solution to 25.00 mL of $0.100M$ CH_3CO_2H solution gives a $0.050M$ solution of CH_3CO_2Na. The pH of this solution is calculated to be 8.87 using the method given in Example 14.11.

4. *After the equivalence point is reached*, we are adding strong base, NaOH, to a solution of CH_3CO_2Na. The excess OH^- represses the ionization of $CH_3CO_2^-$ as a weak base:

$$CH_3CO_2^- + H_2O \rightleftharpoons CH_3CO_2H + OH^-$$

So the additional OH^- from this equilibrium is negligible compared with the added OH^-. Therefore the pH can be calculated simply from the concentration of OH^- added after the equivalence point.

The calculations of the pH for selected points in the titration of 25 mL of $0.100M$ CH_3CO_2H with $0.100M$ NaOH are summarized in Table 14.9. The complete curve is shown in Figure 14.5.

From the figure we see that methyl red changes color long before the equivalence point, so methyl red is of no use for detecting the equivalence point. In contrast, phenolphthalein changes color at the equivalence point. Furthermore, at this point the pH changes very rapidly with the volume of added NaOH. Therefore the volume of NaOH solution needed to reach the equivalence point can be determined accurately, that is, to within one drop of the added solution. Thus phenolphthalein, but not methyl red, is a suitable indicator for this titration. *The only suitable indicators are those that have a color change over a pH range that falls within the steep part of the titration curve close to the equivalence point.*

Figure 14.6 shows how the shape of the titration curve changes with the dissociation constant of the acid being titrated. When the acid is very weak, $K_a < 10^{-8}$, the slope of the curve in the vicinity of the equivalence point is not steep enough to cause any indicator to change color rapidly. So titration using an indicator is not a good method for determining the concentration of a solution of a very weak acid. But for strong acids and for many common weak acids and bases, titrations are a widely used method for determining the concentrations of solutions of acids and bases.

Although we often speak of an acid–base reaction as a neutralization reaction the solution at the equivalence point is not necessarily neutral. At the equivalence point a stoichiometric amount of acid has reacted with a base. The solution is neutral, pH = 7, if both acid and base are strong. But if the base is strong and the acid is weak the solution at the equivalence point is basic, pH > 7, and if the base is weak and the acid is strong the solution at the equivalence point is acidic, pH < 7.

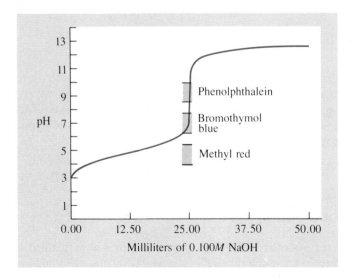

FIGURE 14.5
Titration of 25.00 mL of $0.100M$ CH_3CO_2H with $0.100M$ NaOH.

Of the three indicators, only phenolphthalein could be used to determine the equivalence point of this titration.

FIGURE 14.6
Titration of 25.00 mL of
0.100M Solutions of Weak
Acids with 0.100M NaOH.

In the titration of very weak acids,
$K_a \leqslant 10^{-8}$, the pH changes only
slowly in the vicinity of the equivalence point, so no indicator will
give a sharp color change.

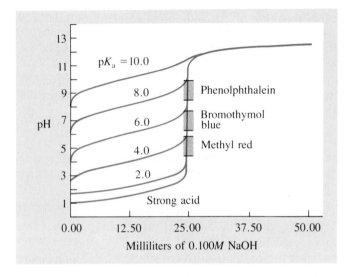

FIGURE 14.7
Titration of 25.00 mL of
0.100M NH$_3$ with 0.100M
HCl.

Of the three indicators only
methyl red could be used to
determine the equivalence point
of this titration.

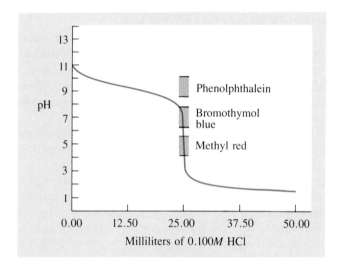

TITRATION OF A WEAK BASE WITH A STRONG ACID

The change in the pH during a titration of a weak base, such as NH_3, with a
strong acid, such as HCl, can be calculated in just the same way as illustrated
for the titration of a weak acid with a strong base. At the equivalence point
we have a solution of NH_4Cl, and the pH is less than 7. A typical titration
curve is shown in Figure 14.7.

14.8
Polyprotic Acids

An acid that can donate more than one proton to a base is called a **polyprotic
acid**. In Chapter 8 we saw that sulfuric acid, H_2SO_4, is a *diprotic acid* and that

phosphoric acid, H_3PO_4, is a *triprotic acid*. In Chapter 6 we saw that carbonic acid, H_2CO_3, is a diprotic acid.

Phosphoric acid has three distinct acid dissociation constants.

1. Dissociation of phosphoric acid:

$$H_3PO_4(aq) + H_2O(l) \rightleftharpoons H_3O^+(aq) + H_2PO_4^-(aq)$$

$$K_{a_1}(H_3PO_4) = \frac{[H_3O^+][H_2PO_4^-]}{[H_3PO_4]} = 7.1 \times 10^{-3} \text{ mol L}^{-1}$$

2. Dissociation of the dihydrogen phosphate ion:

$$H_2PO_4^-(aq) + H_2O(l) \rightleftharpoons H_3O^+(aq) + HPO_4^{2-}(aq)$$

$$K_{a_2}(H_3PO_4) = K_a(H_2PO_4^-) = \frac{[H_3O^+][HPO_4^{2-}]}{[H_2PO_4^-]} = 6.2 \times 10^{-8} \text{ mol L}^{-1}$$

3. Dissociation of the monohydrogen phosphate ion:

$$HPO_4^{2-}(aq) + H_2O(l) \rightleftharpoons H_3O^+(aq) + PO_4^{3-}(aq)$$

$$K_{a_3}(H_3PO_4) = K_a(HPO_4^{2-}) = \frac{[H_3O^+][PO_4^{3-}]}{[HPO_4^{2-}]} = 4.4 \times 10^{-13} \text{ mol L}^{-1}$$

Each successive acid dissociation constant of a polyprotic acid is approximately 10^{-5} times the value of the preceding one. Each successive proton is more difficult to remove because of the increased charge of the anion from which it is being removed.

Let us see now how we calculate the concentrations of all the species in a $0.100M$ aqueous solution of H_3PO_4. We consider first the first dissociation:

$$H_3PO_4 + H_2O \rightleftharpoons H_3O^+ + H_2PO_4^-$$

Initial concentrations	0.100		0	0	mol L^{-1}
Equilibrium concentrations	0.100 − x		x	x	mol L^{-1}

$$K_{a_1}(H_3PO_4) = \frac{[H_3O^+][H_2PO_4^-]}{[H_3PO_4]} = \frac{x^2}{0.100 - x} \text{ mol L}^{-1} = 7.1 \times 10^{-3} \text{ mol L}^{-1}$$

Solving the quadratic equation for x (see Appendix A) gives $x = 0.023$, and therefore $[H_3O^+] = 0.023$ mol L^{-1}. Hence

$$[H_2PO_4^-] = 0.023 \text{ mol L}^{-1} \quad \text{and} \quad [H_3PO_4] = 0.077 \text{ mol L}^{-1}$$

Note that the approximation $x \ll 0.1$ is not valid in this case because $x > (5\%)[H_3PO_4]$.

You might object that these concentrations cannot be correct because we have not taken into account that some of the $H_2PO_4^-$ is dissociated to HPO_4^{2-}, and in principle your objection is justified. However, the differences in the dissociation constants K_{a_1}, K_{a_2}, and K_{a_3} are so great that the concentrations of HPO_4^{2-} and PO_4^{3-} formed are negligibly small, as we will now show. We have

$$K_{a_2}(H_3PO_4) = K_a(H_2PO_4^-) = \frac{[H_3O^+][HPO_4^{2-}]}{[H_2PO_4^-]} = 6.2 \times 10^{-8} \text{ mol L}^{-1}$$

$$H_2PO_4^- + H_2O \rightleftharpoons H_3O^+ + HPO_4^{2-}$$

Initial concentrations	0.023		0.023	0	mol L^{-1}
Equilibrium concentrations	0.023 − x		0.023 + x	x	mol L^{-1}

FIGURE 14.8
Distribution Diagram for
H_3PO_4.

This diagram shows how the
amount of each species expressed
as a fraction, α, of the total varies
with the pH of the aqueous solu-
tion, for a total concentration of
$0.10M$.

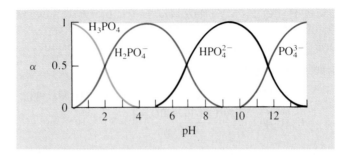

Since x is very small, we can write

$$0.023 - x \approx 0.023 + x \approx 0.023$$

Hence

$$[HPO_4^{2-}] = K_{a_2} \frac{[H_2PO_4^-]}{[H_3O^+]} = \frac{(6.2 \times 10^{-8})(0.023)}{0.023} \text{ mol L}^{-1}$$

$$= 6.2 \times 10^{-8} \text{ mol L}^{-1}$$

Thus we see that $[HPO_4^{2-}]$ is negligible compared with $[H_2PO_4^-] = 0.023$ mol L^{-1}.

Similarly, we have

$$[PO_4^{3-}] = \frac{K_a[HPO_4^{2-}]}{[H_3O^+]} = \frac{(4.4 \times 10^{-13})(6.2 \times 10^{-8})}{0.023} \text{ mol L}^{-1}$$

$$= 1.2 \times 10^{-18} \text{ mol L}^{-1}$$

which is negligible compared with $[H_2PO_4^-] = 0.023$ mol L^{-1} and $[HPO_4^{2-}] = 6.2 \times 10^{-8}$ mol L^{-1}.

Thus the concentrations of the various species in a $0.100M$ H_3PO_4 solution are

$$[H_3PO_4] = 0.077M \qquad [H_3O^+] = 0.023M \qquad [H_2PO_4^-] = 0.023M$$

$$[HPO_4^{2-}] = 6.2 \times 10^{-8}M \qquad [PO_4^{3-}] = 1.2 \times 10^{-18}M$$

A simple way to show how the concentrations of the species in a H_3PO_4 solution depend on the pH is by means of a *distribution diagram*, such as the one shown in Figure 14.8 in which the concentrations of the species are plotted against the pH. This diagram shows that, although three separate dissociation equilibria are involved, only one of the dissociation equilibria is important at any given pH, because the three acid dissociation constants have very different values.

=== *Example 14.19* ===

pH OF A POLYPROTIC ACID SOLUTION

Hydrogen sulfide is a diprotic acid:

$$H_2S + H_2O \rightleftharpoons H_3O^+ + HS^- \qquad K_{a_1} = 9.1 \times 10^{-8} \text{ mol L}^{-1}$$

$$HS^- + H_2O \rightleftharpoons H_3O^+ + S^{2-} \qquad K_{a_2} = 1.3 \times 10^{-13} \text{ mol L}^{-1}$$

A saturated solution of hydrogen sulfide in water has a concentration of approximately $0.10M$. What is the pH of this solution and what are the equilibrium concentrations of H_2S, HS^-, and S^{2-}?

Solution

Because the second dissociation constant is much smaller than the first the equilibrium concentration of H_3O^+ is determined almost entirely by the first dissociation:

$$H_2S + H_2O \rightleftharpoons H_3O^+ + HS^-$$

Initial concentrations	0.10		0 0	mol L^{-1}
Equilibrium concentrations	$0.10 - x$		x x	mol L^{-1}

$$K = \frac{[H_3O^+][HS^-]}{[H_2S]} = 9.1 \times 10^{-8} = \frac{x^2}{0.10 - x}$$

Assuming that $x \ll 0.10$

$$x^2 = 9.1 \times 10^{-9}$$

so

$$x = 9.5 \times 10^{-5}$$

Therefore

$$[H_3O^+] = [HS^-] = 9.5 \times 10^{-5} \text{ mol L}^{-1}$$

and

$$pH = 4.02$$
$$[H_2S] = 0.10 - 9.5 \times 10^{-5} = 0.10 \text{ mol L}^{-1}$$

To find the S^{2-} concentration we need to consider the second dissociation:

$$HS^- + H_2O \rightleftharpoons H_3O^+ + S^{2-}$$

Initial concentrations	9.5×10^{-5}		9.5×10^{-5} 0	mol L^{-1}
Equilibrium concentrations	$9.5 \times 10^{-5} - x$		$9.5 \times 10^{-5} + x$ x	mol L^{-1}

Assuming that x is very small

$$9.5 \times 10^{-5} - x \approx 9.5 \times 10^{-5} + x \approx 9.5 \times 10^{-5}$$

$$K = \frac{[H_3O^+][S^{2-}]}{[HS^-]} = \frac{(9.5 \times 10^{-5})x}{9.5 \times 10^{-5}} \text{ mol L}^{-1} = 1.3 \times 10^{-13} \text{ mol L}^{-1}$$

Hence

$$x = 1.3 \times 10^{-13}$$
$$[S^{2-}] = 1.3 \times 10^{-13} \text{ mol L}^{-1}$$

Thus we have

$$[H_3O^+] = 9.5 \times 10^{-5} \text{ mol L}^{-1} \qquad pH = 4.02$$
$$[HS^-] = 9.5 \times 10^{-5} \text{ mol L}^{-1}$$
$$[S^{2-}] = 1.3 \times 10^{-13} \text{ mol L}^{-1}$$
$$[H_2S] = 0.10 \text{ mol L}^{-1}$$

IMPORTANT TERMS

The **acid dissociation constant** for a weak acid in water is given by the expression

$$K_a = \left(\frac{[H_3O^+][A^-]}{[HA]}\right)_{eq}$$

The **base dissociation constant** for a weak base in water is given by the expression

$$K_b = \left(\frac{[BH^+][OH^-]}{[B]}\right)_{eq}$$

A **buffer solution** has the ability to resist changes in pH. A buffer solution can be prepared from a weak acid and a salt of the weak acid or from a weak base and a salt of the weak base.

An **indicator** is a weak acid or base whose color depends on

the pH of the solution, because its acid and base forms have different colors.

The **ionic product constant of water**, K_w, is the product of the equilibrium concentrations of H_3O^+ and OH^-. $K_w = ([H_3O^+][OH^-])_{eq}$. This is also called the **autoprotolysis constant** of water.

A **titration** is a procedure in which the volume of a solution needed to react completely with a known volume of another solution may be determined.

pH is a convenient method for expressing the concentration of H_3O^+ in a solution. It is defined as

$$pH = -\log_{10}[H_3O^+]$$

A **polyprotic acid** has more than one ionizable hydrogen.

PROBLEMS *

Acids and Bases and pH

1. What are the molar concentrations of the ions in each of the following aqueous solutions?

   (a) $10^{-5}M$ HNO$_3$ **(b)** $0.0023M$ HCl

   (c) $0.113M$ HClO$_4$ **(d)** $0.034M$ HBr

   (e) $10^{-3}M$ NaOH   **(f)** $0.145M$ Ba(OH)$_2$

2. What are the molar concentrations of the ions in each of the following aqueous solutions?

   (a) $0.0234M$ HI   **(b)** $10^{-5}M$ Ca(OH)$_2$

   (c) $0.204M$ LiOH   **(d)** $0.342M$ HBr

   (e) $0.0023M$ Sr(OH)$_2$ **(f)** $10^{-4}M$ Na$_2$O

3. Name five strong acids and four weak acids.

4. Name three strong bases and three weak bases.

5. What is the H_3O^+ ion concentration and the percent dissociation of a 0.010M solution of hydrocyanic acid, HCN(aq)?

6. What is the H_3O^+ ion concentration and the percent dissociation of a 0.060M solution of hydrofluoric acid, HF(aq)?

7. Which of a 0.0010M solution of HNO$_3$(aq) and a 0.200M solution of acetic acid contains the higher concentration of H_3O^+ ions?

* Answers to problems numbered in blue appear at the end of the text.

8. What is the OH$^-$ ion concentration and the percent dissociation of a 0.040M solution of aqueous ammonia?

9. What is the OH$^-$ ion concentration and the percent dissociation of a 0.080M solution of aniline?

10. What is the pH of each of the solutions in Problem 1?

11. What is the pH of each of the solutions in Problem 2?

12. What is the pH of a 0.0050M solution of HF?

13. The pH of a 0.100M solution of hypochlorous acid, HOCl(aq), is 4.2. What are the K_a and the pK_a of hypochlorous acid?

14. Muscles may ache after strenuous exercise, because lactic acid, p$K_a = 3.08$, is formed faster than it is metabolized to CO$_2$ and water. What is the pH of the fluid in a muscle when the lactic acid concentration reaches 1.0×10^{-3} mol L^{-1}?

15. What is the pH of each of the following solutions?

   (a) A solution prepared by adding 25.0 mL of 0.10M KOH(aq) to 50.0 mL of 0.080M HNO$_3$(aq)

   (b) A solution prepared by adding 25.0 mL of 0.13M HCl(aq) to 35.0 mL of 0.12M NaOH(aq)

   (c) A solution prepared by adding 35.0 mL of 0.050M HNO$_3$(aq) to 70.0 mL of 0.025M NaOH(aq)

16. The pK_a values of acetic acid, monochloroacetic acid, $ClCH_2CO_2H$, dichloroacetic acid, Cl_2CHCO_2H, and trichloroacetic acid, Cl_3CCO_2H, are 4.8, 2.9, 1.3, and 0.7, respectively. Arrange these acids in order of their decreasing acid strength in water. What is the pH of a 0.10M solution of each?

17. What is the pH of a 0.0050M aqueous solution of KF?

18. Arrange the following 0.10M aqueous solutions in order of increasing pH: No calculations are needed.

(a) NaCN   (b) KCl   (c) LiOH   (d) HBr

(e) NH_3 (f) NH_4Cl (g) K_2O

19. What are the equilibrium concentrations of NH_4^+, NH_3, OH^-, and H_3O^+, in a 0.020M solution of $NH_4Cl(aq)$?

20. What is the pH of each of the following aqueous solutions?

(a) 0.010M acetic acid   (b) 0.10M hydrofluoric acid

(c) 0.0030M ammonia   (d) 0.10M sodium acetate

(e) 0.20M ammonium chloride

21. Suggest a suitable acid–base reaction by which each of the following salts could be prepared. Would you expect 0.10M solutions of each salt to be acidic, basic, or neutral? Explain your answer in each case.

(a) Ammonium nitrate   (b) Ammonium chloride

(c) Calcium sulfate   (d) Potassium acetate

(e) Aluminum chloride   (f) Lithium iodide

22. Write balanced equations for the reactions that occur when each of the following is dissolved in water:

(a) Sodium oxide   (b) Potassium sulfide

(c) Sodium amide

23. Write a balanced equation for the reaction of each of the following with water, if any. State whether the resulting solution is acidic, neutral, or basic.

(a) Carbon dioxide   (b) Sulfur trioxide

(c) Calcium oxide   (d) Phosphorus trichloride

(e) Sulfur   (f) Lithium hydride

24. Equal volumes of a hydrochloric acid solution of pH 2.00 and hydrochloric acid of pH 3.00 are mixed. What is the pH of the resulting solution?

25. Nitrous acid, HNO_2, is a weak acid in water (p$K_a = 3.35$). An aqueous solution of nitrous acid with pH 2.00 is required. What stoichiometric concentration of nitrous acid is needed?

26. The value of K_w, the ionic product of water, depends, as do all equilibrium constants, on temperature. From the following data, calculate the pH of pure water at the temperatures indicated:

$t(°C)$	$K_w \times 10^{14}$ (mol^2 L^{-2})
0	0.115
10	0.293
30	1.471
50	5.476
100	51.3

27. Citric acid has a pK_a of 3.10. What is the pH of a 0.10M solution of citric acid?

28. What is the pH of each of the following aqueous solutions?

(a) 0.20M KCN   (b) 0.50M NH_3

(c) 0.10M HBr   (d) 0.10M NH_4ClO_4

(e) 0.30M $NaNH_2$ (f) 0.33M LiI

29. 30.0 g of phosphoric acid, H_3PO_4, are dissolved in water to give 500 mL of solution. What volume of 0.200M NaOH(aq) is required to form $Na_3PO_4(aq)$ quantitatively? Will the resulting solution be acidic, basic, or neutral?

30. Sodium carbonate is often added to swimming pools to change the pH. Does this raise, or lower, the pH of the pool? Explain.

31. For each of the following acids, name the conjugate base, give its formula and calculate its pK_b:

Acid	K_a (mol L^{-1})
(a) H_2CO_3	3.3×10^{-7}
(b) H_2S	9.1×10^{-8}
(c) HNO_2	4.5×10^{-4}
(d) H_2SO_3	1.7×10^{-2}

32. Formic acid (methanoic acid), HCO_2H, has a pK_a of 3.68. Calculate the pH of (a) a 0.10M solution of formic acid and (b) a 0.10M solution of sodium formate.

33. What is the pH of a 0.100M solution of $AlCl_3(aq)$?

34. What is the pH of a 0.100M solution of sodium hydrogen sulfate?

35. Phosphorous acid, H_3PO_3, is a diprotic acid with K_a values of 5×10^{-2} mol L^{-1} and 2×10^{-5} mol L^{-1}, respectively, for its two stages of ionization.

   (a) Calculate the pH of a $0.10M$ solution of H_3PO_3(aq).

   (b) Calculate the pH of a $0.10M$ solution of NaH_2PO_3(aq).

Indicators

36. (a) In using an indicator, why is it important to add the smallest amount possible to the solution being investigated?

   (b) At what pH value does the ratio of the basic form In^- to the acidic form HIn of methyl orange indicator ($pK_a = 4.2$) have the following values?

   (i) 5:1   **(ii)** 1:1   **(iii)** 1:5

37. Estimate the pH of a colorless aqueous solution that turns yellow when methyl red is added to it and yellow when bromothymol blue is added.

38. What color would you expect a solution containing methyl red to be when the pH is 5.0?

39. From Table 14.7 select a suitable indicator for detecting the equivalence point in each of the following titrations of $0.1M$ acid with $0.1M$ base:

   (a) HCl(aq) with NaOH(aq)

   (b) HF(aq) with KOH(aq)

   (c) NH_3(aq) with HCl(aq)

   (d) CH_3NH_2(aq) with HCl(aq)

40. When each of the indicators listed below is added to a $0.10M$ solution of a weak acid, the colors are as shown. What is the approximate K_a of the acid?

Indicator	Color
Methyl violet	Violet
Methyl orange	Yellow
Methyl red	Orange
Bromothymol blue	Yellow

41. What would be the color of each of the following indicators in a $0.10M$ solution of potassium fluoride?

   (a) Thymol blue

   (b) Phenolphthalein

   (c) Bromothymol blue

42. What would be the color of each of the following indicators in a $0.10M$ solution of ammonium chloride?

   (a) Methyl red   **(b)** Methyl orange

   (c) Bromothymol blue

43. The indicator thymol blue turns from yellow to blue over the pH range 7.9–9.4. Estimate the pK_a of thymol blue.

Buffer Solutions

44. A buffer solution contains $0.30M$ ammonia and $0.25M$ ammonium chloride. What is its pH?

45. A buffer solution is made from equal volumes of $0.10M$ hydrocyanic acid and $0.10M$ sodium cyanide. What is its pH?

46. What ratio of masses of ammonia and ammonium chloride is required to prepare a buffer solution with a pH of 9.0?

47. What is the pH of a solution prepared by mixing 25.0 mL of $0.100M$ NaOH(aq) with 50.0 mL of $0.100M$ acetic acid?

48. What is the pH of a solution prepared by mixing 15.0 mL of $0.0100M$ NH_3(aq) and 25.0 mL of $0.0100M$ NH_4Cl(aq)?

49. What is the change in pH when 1.00 mL of $1.00M$ NaOH(aq) is added to 100 mL of a solution containing $0.18M$ NH_3(aq) and $0.10M$ NH_4Cl(aq)?

50. Use the Henderson–Hasselbalch equation to calculate the mole ratio of acetic acid to acetate ion that is required to prepare a buffer solution of pH 4.50.

51. (a) How many grams of sodium acetate must be added to 1.00 L of $0.200M$ CH_3CO_2H(aq) to give a solution of pH 5.00?

   (b) What is the pH of the above solution after 1.00 mL of $12.0M$ HCl(aq) has been added to it?

52. Formic acid (methanoic acid), HCO_2H (pK_a 3.68), has one ionizable proton. What is the pH of a solution prepared by dissolving 1.00 g each of formic acid and sodium formate in sufficient water to give 100.0 mL of solution? What would be the pH of the solution after addition of another 100.0 mL of distilled water?

53. You have available cyanic acid, HOCN (pK_a 3.66), sodium cyanate, sodium hydroxide, and distilled water. Describe two different ways of making a buffer solution with a pH of approximately 3.7.

54. A phosphate buffer is frequently used to maintain a constant pH in biological experiments. What is the pH of the phosphate buffer solution prepared by dissolving 3.40 g of potassium dihydrogen phosphate and 3.55 g of disodium hydrogen phosphate in sufficient water to give 500 mL of solution? What is the pH if this solution is diluted with water to a volume of 600 mL?

55. What is the pH of a buffer solution prepared by mixing 25.0 mL of a 0.020M solution of aniline, $C_6H_5NH_2$, with 10.0 mL of a 0.030M solution of the salt anilinium chloride, $C_6H_5NH_3{}^+Cl^-$? What is the pH of the solution after addition of 1.00 mL of 0.040M $HNO_3(aq)$? What would be the pH of the solution if 2.00 mL of 0.030M KOH(aq) is added rather than the nitric acid?

Titrations

56. Explain each of the following terms and the differences between them:

(a) The strength of an acid and its concentration

(b) The equivalence point and the end point of a titration

57. From Table 14.7, select suitable indicators for detecting the equivalence points in each of the following titrations:

(a) 0.10M NaOH(aq) with 0.10M HCl(aq)

(b) 0.10M acetic acid with 0.10M NaOH(aq)

(c) 0.10M NH_3(aq) with 0.10M HCl(aq)

58. From Table 14.7, select suitable indicators for detecting the equivalence points in each of the following titrations:

(a) 0.10M KOH(aq) with 0.05M HNO_3(aq)

(b) 0.20M HF(aq) with 0.10M NaOH(aq)

(c) 0.10M CH_3NH_2(aq) with 0.10M HCl(aq)

59. 25.00 mL of 0.10M acetic acid were titrated with 0.08M KOH(aq). What is the pH at the following points in the titration?

(a) Initially

(b) After 15.63 mL of base have been added

(c) At the equivalence point

(d) After addition of 50.00 mL of base

60. What would be the approximate pK_a of a suitable indicator that could be used to detect the equivalence point of the titration in Problem 59?

61. 25.00 mL of 0.10M NH_3(aq) were titrated with 0.10M HCl(aq). What is the pH at the following points in the titration?

(a) Initially

(b) After 10.00 mL of acid have been added

(c) After 12.50 mL of acid have been added

(d) At the equivalence point

(e) After addition of 40.00 mL of acid

62. Which of methyl red and thymol blue would be a suitable indicator to detect the equivalence point in the titration in Problem 61? Explain.

63. Consider the titration of 10.0 mL of 0.010M hypochlorous acid, HOCl(aq), with 0.010M KOH(aq). Calculate the pH of each solution after addition of the following volumes of base:

(a) 0.00 mL   (b) 5.00 mL   (c) 9.50 mL

(d) 10.00 mL   (e) 10.50 mL   (f) 15.00 mL

64. 50.00 mL of an aqueous solution of acetic acid were titrated with NaOH(aq). The initial pH before addition of base was 2.72. At the equivalence point in the titration the pH was 8.92. Calculate each of the following:

(a) The initial concentration of acid

(b) The volume of base added to reach the equivalence point

(c) The concentration of the NaOH(aq)

65. Prove the following statement: "The pK_a of a weak acid is the same as the pH at the half-equivalence point in its titration with a strong base such as NaOH(aq)." (The half-equivalence point is the point where half as much base has been added as is needed to attain the equivalence point.)

*66. 10.00 mL of 0.100M H_3PO_4(aq) are titrated with 0.125M NaOH(aq). Calculate the pH of the solution after the following volumes of base have been added:

(a) 0.00 mL   (b) 4.00 mL   (c) 8.00 mL

(d) 12.00 mL   (e) 16.00 mL   (f) 20.00 mL

Miscellaneous

67. Calculate the equilibrium concentrations of each of the ions $CO_3{}^{2-}$, $HCO_3{}^-$, OH^-, and H_3O^+ in a 0.200M solution of sodium carbonate in water.

68. What is the pH of 0.010M aqueous solutions of each of the following:

(a) KOH   (b) KCl   (c) NH_3

(d) HF   (e) KF   (f) HNO_3

* The asterisk denotes the more difficult problems.

69. Which of the following act as buffer solutions? Explain.

   (a) 25 mL of $0.10M$ HNO_3(aq) and 25 mL of $0.10M$ $NaNO_3$(aq)

   (b) 25 mL of $0.10M$ HNO_2(aq) and 25 mL of $0.10M$ $NaNO_2$(aq)

   (c) 25 mL of $0.10M$ aqueous acetic acid and 25 mL of $0.15M$ KOH(aq)

   (d) 25 mL of $0.10M$ aqueous acetic acid and 25 mL of $0.05M$ KOH(aq)

70. Phenol, C_6H_5OH, is a weak acid ($pK_a = 9.80$):

$$C_6H_5OH + H_2O \rightleftharpoons H_3O^+ + C_6H_5O^-$$

What is the solubility of phenol in water if the pH of a saturated solution is 4.90?

*71. A few milliliters of an aqueous solution of a monobasic acid were titrated with a strong base. If the pH after addition of 5.00 mL of base was 5.0, and the equivalence point was reached after addition of a further 5.00 mL of strong base, what was the pK_a of the acid?

*72. A student titrated a few milliliters of an unknown acid, HA, with an NaOH(aq) solution of unknown concentration. After addition of 5.00 mL of NaOH(aq) the pH was 6.0, and the equivalence point was reached after adding an additional 7.00 mL of base. What is the K_a of the acid?

The Alkali and Alkaline Earth Metals

15

Group 1

Period	1	2											3	4	5	6	7	8
1	H																	He
2	Li	Be											B	C	N	O	F	Ne
3	Na	Mg											Al	Si	P	S	Cl	Ar
4	K	Ca	Sc	Ti	V	CR	Mn	Fe	Co	Ni	Cu	Zn	Ga	Ge	As	Se	Br	Kr
5	Rb	Sr	Y	Zr	Nb	Mo	Tc	Ru	Rh	Pd	Ag	Cd	In	Sn	Sb	Te	I	Xe
6	Cs	Ba	La	Hf	Ta	W	Re	Os	Ir	Pt	Au	Hg	Tl	Pb	Bi	Po	At	Rn
7	Fr	Ra	Ac	104	105	106	107											

■ Metals ■ Nonmetals ■ Semimetals

Transition Elements

The ten most abundant elements in the earth's crust

O	Na
Si	K
Al	Mg
Fe	H
Ca	Ti

From left to right, the metals magnesium, sodium, and calcium

The elements of group 1, the *alkali metals*, are lithium, sodium, potassium, rubidium, cesium, and francium. The elements of group 2, the *alkaline earth metals*, are beryllium, magnesium, calcium, strontium, barium, and radium. Four of these elements, namely calcium, sodium, potassium, and magnesium, are among the ten most abundant in the earth's crust (Table 3.2). In this chapter we will concentrate mainly on these four elements.

We have previously mentioned that sodium chloride and limestone, $CaCO_3$, are two of the most important raw materials on which a large part of the chemical industry is based. They are used, for example, for the production of calcium oxide (lime), sodium hydroxide, and sodium carbonate, which are among the top twenty chemicals produced by the chemical industry (see Table 15.1) and which have a multitude of uses. Because they are soluble in water and are readily available, the sodium and potassium salts of the common acids are widely used compounds.

Potassium is essential to nearly all forms of life, both plant and animal. The major use of potassium compounds, particularly potassium chloride, is as a fertilizer. Sodium is an essential component of the diet of both humans and animals because many biological processes are controlled by the concentration of sodium ions. Farmers and ranchers often place large blocks of salt in the fields

TABLE 15.1
Top 20 Industrial Chemicals Produced in the United States (1987)

1. Sulfuric acid	11. Sodium carbonate
2. Nitrogen	12. Urea
3. Ethene	13. Nitric acid
4. Ammonia	14. Ethylene dichloride
5. Oxygen	15. Ammonium nitrate
6. Calcium oxide (lime)	16. Benzene
7. Sodium hydroxide	17. Carbon dioxide
8. Chlorine	18. Ethylbenzene
9. Phosphoric acid	19. Vinyl chloride
10. Propene	20. Terephthalic acid

for cattle to lick. Magnesium is required for nerve impulse transmissions, muscle contraction, and carbohydrate metabolism. Magnesium is part of the chlorophyll molecule which is vital to the process of photosynthesis in green plants. Calcium is an essential element in the formation of bones and teeth (Chapter 8) and is required for maintaining the rhythm of the heart, and in blood clotting. A healthy human excretes each day an average of 20–30 g of the chloride, sulfate, and phosphate salts of sodium, potassium, magnesium, and calcium.

In this chapter we consider the properties and reactions of the alkali and alkaline earth metals and their compounds. Almost all the salts of the alkali metals are soluble in water, but several common salts of the alkaline earth metals have rather low solubilities. To discuss these solubilities in a quantitative way, we use another type of equilibrium constant called the solubility product constant.

The word "salary" comes from the Latin word *salarium*, which originally meant "salt money" or money paid to Roman soldiers for guarding salt deposits.

15.1
The Elements

PHYSICAL PROPERTIES

The alkali and alkaline earth metals have typical metallic properties. They are good conductors of heat, have high electrical conductivities, which increase with decreasing temperature, are mostly malleable and ductile, and have a shiny metallic luster when freshly cut (see Experiment 15.1). All the elements in these two groups, with the exception of magnesium and beryllium, are rather soft and mechanically weak, and they react with the atmosphere; consequently, they are of no importance as structural materials. However, magnesium is an important constituent of many lightweight structural alloys.

The alkali metals have several unusual properties. Unlike the alkaline earth metals and most other typical metals, the alkali metals have low densities and low melting points (see Tables 15.2 and 15.3). In fact, lithium is the lightest of all metals. Lithium, sodium, and potassium are all less dense than water, which is very unusual for a metal. The melting points of all the alkali metals are considerably lower than those of almost all other metals. The melting point of cesium is so low (29 °C) that it becomes a liquid on a hot day. The only metal that has a lower melting point is mercury (−30 °C). Alloys of two or more alkali metals have melting points that are even lower than those of the pure metals. Sodium–potassium alloys containing 40%–90% potassium are liquid at room temperature. They are used in certain types of nuclear reactors as coolants. Liquid metals are particularly useful as coolants because they have a high heat conductivity, a high heat capacity, and very high boiling points.

Many metals have close-packed structures (Chapters 10 and 11). Among the alkaline earth metals, beryllium and magnesium have the hexagonal close-packed structure and calcium and strontium have the cubic close-packed structure, with each atom having twelve close neighbors. In contrast, the alkali metals and barium have the same structure as iron, namely the monatomic body-centered cubic structure, in which each atom has only eight close neighbors.

The metallic bonding in the alkali metals is relatively weak, because they have only one valence electron per atom available for metallic bonding. In contrast, the alkaline earth metals have two valence electrons, and other metals

Densities of Some Common Metals (g cm^{-3}) at 25 °C

Li	0.53	Al	2.71
K	0.86	Fe	7.86
Na	0.97	Cu	8.94
Ca	1.55	Pb	11.40
Mg	1.74	Hg	13.55

Monatomic body—centered cubic

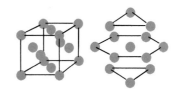

Cubic close packing

Hexagonal close packing

Experiment 15.1

Physical Properties of the Alkali Metals

Sodium is soft enough to be cut with a knife. The bright metallic luster soon dulls as the surface is oxidized.

Sodium is good conductor of electricity.

Sodium melts at 97.5°C and potassium melts at 62.3°C. The tubes contain sodium (left) and potassium (right) covered with paraffin to protect them from oxidation by the oxygen in the atmosphere. When the tubes are heated in boiling water, both sodium and potassium melt to form a shiny metallic ball that looks like mercury.

have two or more valence electrons available for metallic bonding. When a metal is melted, the metallic bonds holding the atoms together are not broken, because the mobile electron cloud that holds the atomic cores together can move with the moving atoms. The melting of a metal therefore requires only a slight weakening of the bonds as the atoms move a little further apart than they were in the solid. For the alkali metals, which have relatively weak metallic bonding, melting requires only a small amount of energy, and their melting points and enthalpies of fusion are quite low (see Table 15.2). However, when a liquid metal is vaporized, individual atoms or small molecules are formed, and the majority of the metallic bonds are broken. Thus a rather large amount of energy is needed to vaporize a metal, and the boiling points and enthalpies of vaporization are always high (see Table 15.2), although lower for the alkali metals, because of their relatively weak bonding, than for other metals.

The alkali metals have only one electron in their valence shells, whereas the alkaline earth metals have two (Tables 15.4 and 15.5). The first column in Table 15.4 gives the energy needed to remove the most easily removed electron to give

TABLE 15.2
Properties of the Alkali Metals

Element	Density $(g \, cm^{-3})$	Melting Point $(°C)$	Boiling point $(°C)$	ΔH_{fus} $(kJ \, mol^{-1})$	ΔH_{vap} $(kJ \, mol^{-1})$
Lithium	0.53	186	1336	2.9	148
Sodium	0.97	97.5	880	2.6	99
Potassium	0.86	62.3	760	2.4	79
Rubidium	1.53	38.5	700	2.2	76
Cesium	1.87	28.5	670	2.1	66

TABLE 15.3
Properties of the Alkaline Earth Metals

Element	Density $(g \, cm^{-3})$	Melting Point $(°C)$	Boiling Point $(°C)$	ΔH_{fus} $(kJ \, mol^{-1})$	ΔH_{vap} $(kJ \, mol^{-1})$
Beryllium	1.86	1280	2970	12	309
Magnesium	1.74	650	1100	9	128
Calcium	1.55	850	1490	8	151
Strontium	2.6	770	1380	9	139
Barium	3.6	710	1140	7.7	151

an M^+ ion. This energy is called the *first ionization energy*. For the alkali metals, the most easily removed electron is an s electron. The second column gives the energy needed to remove the next most easily removed electron to give an M^{2+} ion. This energy is called the *second ionization energy*. For an alkali metal atom this second electron must come from a completed inner shell, and therefore it has a much higher ionization energy. Thus one, but not two, electrons may be rather easily removed from an alkali metal atom. For the alkaline earth metals the two s electrons can be rather easily removed—the first and second ionization energies are relatively low. But a third electron can only be removed with much greater difficulty, because it must come from a completed inner shell. Thus

TABLE 15.4
Properties of the Alkali Metal Atoms

Element	Ionization Energies $(MJ \, mol^{-1})$ 1st $M \rightarrow M^+$	2nd $M^+ \rightarrow M^{2+}$	Atomic Radius (pm)	Ionic Radius (M^+) (pm)	Electron Configuration
Lithium	0.52 (2s)	7.28 (1s)	152	74	[He] $2s^1$
Sodium	0.51 (3s)	4.56 (2p)	186	102	[Ne] $3s^1$
Potassium	0.42 (4s)	3.06 (3p)	231	138	[Ar] $4s^1$
Rubidium	0.40 (5s)	2.65 (4p)	244	149	[Kr] $5s^1$
Cesium	0.38 (6s)	2.42 (5p)	267	170	[Xe] $6s^1$

TABLE 15.5
Properties of the Alkaline Earth Metal Atoms

| Element | Ionization Energies (MJ mol^{-1}) | | | Atomic Radius (pm) | Ionic Radius (M^{2+}) (pm) | Electron Configuration |
	1st $M \rightarrow M^+$	2nd $M^+ \rightarrow M^{2+}$	3rd $M^{2+} \rightarrow M^{3+}$			
Beryllium	0.90 (2s)	1.75 (2s)	14.81 (1s)	111	27	[He] $2s^2$
Magnesium	0.74 (3s)	1.45 (3s)	7.72 (2p)	160	72	[Ne] $3s^2$
Calcium	0.59 (4s)	1.15 (4s)	4.93 (3p)	197	100	[Ar] $4s^2$
Strontium	0.55 (5s)	1.06 (5s)	4.14 (4p)	215	126	[Kr] $5s^2$
Barium	0.50 (6s)	0.96 (6s)	3.47 (5p)	222	136	[Xe] $6s^2$

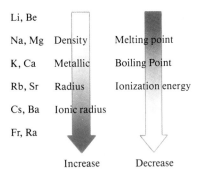

Li, Be
Na, Mg
K, Ca
Rb, Sr
Cs, Ba
Fr, Ra

Density · Metallic · Radius · Ionic radius — Increase

Melting point · Boiling Point · Ionization energy — Decrease

the third ionization energy is much larger than the first and second (Table 15.5). Hence the alkali and alkaline earth metals are readily oxidized to M^+ and M^{2+}, respectively. The ionization energies decrease from the top to the bottom of each group as the size of the atoms increases and the valence electrons are therefore at an increasingly greater distance from the core. The increasing size of the atoms and the decreasing ionization energy with increasing atomic number means that the electrons in the mobile charge cloud are at a greater distance from the nuclei and are therefore less strongly attracted. Hence the strength of the metallic bonding and the melting and boiling points decrease from the top to the bottom of both groups.

Recall that when we previously discussed ionization energies in Chapter 7, we considered the removal of *one* electron only, but not necessarily the most easily removed electron. The ionization energies in Table 7.1 are for the removal of one electron from the various orbitals of a neutral atom. These neutral-atom ionization energies also show that removing the 3s electron of sodium (0.50 MJ mol^{-1}) is much easier than removing a 2p electron (3.67 MJ mol^{-1}). The ionization energies in Tables 15.4 and 15.5 are for the successive removal of electrons from the same atom. The most weakly held electron is removed first from the neutral atom, then the next most weakly held electron from an M^+ ion, and so on.

EXERCISE 15.1

Explain why the melting points and the boiling points of the groups 1 and 2 metals decrease with increasing atomic mass and why the melting points and boiling points of the group 2 metals are higher than those of the group 1 metals.

FLAME TESTS

When the alkali and alkaline earth metals and their compounds are strongly heated in a flame, they impart characteristic bright colors to the flame (see Table 15.6). Only beryllium and magnesium do not give colored flames. As we saw in Chapter 7, the colors arise from atoms that have been raised to excited

TABLE 15.6
Flame Colors Produced by the Alkali and
Alkaline Earth Metals

Metal	Flame Color	Metal	Flame Color
Li	Red	Cs	Blue
Na	Yellow	Ca	Orange red
K	Lilac	Sr	Deep red
Rb	Purple	Ba	Pale green

states at the high temperature of the flame; atoms in these excited states may lose their excess energy by emitting light of characteristic wavelengths. Compounds of these elements contain the metals as positive ions in the solid state; but when they are heated to a high temperature in a flame, they dissociate into gaseous atoms, not ions. For example, at high temperatures sodium chloride dissociates into Na and Cl atoms, not into Na^+ and Cl^- ions. Hence the compounds exhibit the same characteristic flame colors as the elements. These colored flames provide a very convenient qualitative test for these elements in mixtures and compounds (see Experiments 7.1 and 15.2). Many of the brilliant colors of fireworks and aerial rockets are provided by salts of the alkali and alkaline earth metals (Box 15.1).

OCCURRENCE

The alkali metals commonly occur as chlorides in enormous salt deposits that were formed by the evaporation of ancient seas. They are also present in vast quantities, but in less accessible form, in the world's oceans (see Table 15.7).

Experiment 15.2

Flame Tests for the Alkaline Earth Metals

Calcium compounds give an orange red color.

Strontium compounds give a deep red color.

Barium compounds give a pale green color.

BOX 15.1

FIREWORKS

Alkali and alkaline earth metal salts are used to produce many of the color effects in fireworks. The invention of fireworks over one thousand years ago is usually attributed to the Chinese. The basis of many fireworks is a mixture of potassium nitrate, charcoal, and sulfur that is called *black powder*. The Chinese found that if the mixture was ignited in a sealed container it produced an explosion and a loud bang. They also used black powder to make the first rockets. The hot gases—mainly CO_2, SO_2, and oxides of nitrogen—produced by the reaction act as the rocket propellant. The potassium nitrate is called the oxidizer because it provides the oxygen needed to convert the carbon and sulfur, which are called the fuels, to CO_2 and SO_2. Potassium chlorate, $KClO_3$, and potassium perchlorate, $KClO_4$, are also used as oxidizers. Potassium salts are used rather than sodium salts for two reasons. (1) Many sodium salts are hygroscopic, that is they absorb moisture from the atmosphere. (2) As we saw in Experiment 7.1, sodium salts impart a brilliant yellow color to a flame that is so intense that it masks the colors produced by any other salts that are present. The lilac color produced by potassium is much less intense. Some of the substances used to produce special colors are

Red: $Sr(NO_3)_2$ and $SrCO_3$ and also Ca and Li salts

Green: $Ba(NO_3)_2$ and $Ba(ClO_4)_2$ and also some Cu(II) salts

Blue: $CuCO_3$, $CuSO_4$, CuO, and Cu(I)Cl

Yellow: Nonhygroscopic Na salts such as Na_3AlF_6 and sodium oxalate

Violet/purple: Cesium, potassium, and rubidium salts

White: Magnesium and aluminum metals

Gold sparks: Iron filings

White smoke: KNO_3/S mixture

Colored smoke: KNO_3/S/volatile organic dye mixture

Many of the techniques invented by the Chinese are still used today in the manufacture of multistage fireworks involving numerous special effects. Despite many improvements in methods of handling the materials used in fireworks, their manufacture is still a hazardous operation that should be left to the experts who take numerous safety precautions. In spite of these precautions we occasionally hear of massive explosions at fireworks factories.

Some inland seas from which there is no outlet contain high concentrations of sodium chloride. For example, the Dead Sea between Israel and Jordan contains about 20% by mass of sodium chloride plus large amounts of other salts.

When an inland sea evaporates, the solid that is deposited is not a uniform mixture. Instead, because the least soluble salts are deposited first, followed by the more soluble ones, a typical salt deposit consists of several layers. Each of these layers consists of a relatively pure salt such as *rock salt*, NaCl, *sylvite*, KCl, and *carnallite*, $KMgCl_3 \cdot 6H_2O$. These salt deposits are mined either by conventional methods or by forcing water into them to form a concentrated salt solution, *brine*, which is pumped to the surface and evaporated.

Magnesium occurs not only as the chloride in seawater and in salt deposits

TABLE 15.7	
Concentrations of Ions in Seawater	
Element	Concentration (mol L^{-1})
Na^+	0.48
Mg^{2+}	0.05
Ca^{2+}	0.01
K^+	0.01
Cl^-	0.51
SO_4^{2-}	0.07
HCO_3^-	0.002

but also as the carbonate in the minerals *magnesite*, $MgCO_3$, and *dolomite*, $MgCa(CO_3)_2$, and as the sulfate *epsomite*, $MgSO_4 \cdot 7H_2O$.

Calcium occurs widely and abundantly as the carbonate, phosphate, sulfate, and fluoride. Naturally occurring calcium carbonate, $CaCO_3$, exists in a wide variety of forms including *chalk*, *limestone*, and *marble*. Chalk has been subjected to the least pressure during its formation and is the softest $CaCO_3$ rock. Limestone is a more highly compressed form, while marble, which is the hardest, has been subjected to such high temperatures and pressures that it has melted and subsequently recrystallized. Marble is an impure form of the crystalline form of calcium carbonate, called *calcite*. Marble may contain many other minerals, and it consequently varies greatly in color and texture. Pure colorless crystals of calcite also occur naturally.

A salt mine in Great Saline, Texas

The enormous deposits of these forms of calcium carbonate that we find on the earth all had their origin in the sea. The weathering of calcium silicate rocks over millions of years by water and wind gradually changed insoluble calcium silicate into soluble calcium salts, which were carried by rivers into the sea. The calcium dissolved in the sea was used by small sea creatures to form their shells, which are chiefly calcium carbonate. As these animals died, deposits of calcium carbonate were laid down on the ocean floor. Eventually, the deposits were compressed by layers of other deposits above them to form sedimentary rock. Calcium carbonate deposits continue to be laid down on the ocean floor today. Phosphate rock, a mixture of *fluorapatite*, $Ca_5(PO_4)_3F$, *chlorapatite*, $Ca_5(PO_4)_3Cl$, and *hydroxyapatite*, $Ca_5(PO_4)_3(OH)$, occurs widely; as we saw in Chapter 8, it is the most important source of phosphorus. Hydroxyapatite is the principal constituent of bones in humans and animals.

The alkali and alkaline earth metals also occur widely as complex silicates and aluminosilicates (see Chapter 22). But because it is difficult to obtain the elements or other useful compounds of the alkali or alkaline earth metals from these minerals, they are not important as sources of the elements.

PREPARATION

Since the elements occur exclusively in the form of the M^+ and M^{2+} ions, these ions must be reduced in order to prepare the elements. The very low ionization energies of the alkali and alkaline earth metals indicate that they are very easily oxidized to their positive ions; thus the ions are difficult to reduce to the elemental form.

The world-famous Taj Mahal in India is built entirely of marble.

For many years sodium was obtained by reducing sodium carbonate with carbon at a high temperature:

$$Na_2CO_3(s) + 2C(s) \longrightarrow 2Na(l) + 3CO(g)$$

But today sodium is prepared entirely by electrolysis of molten sodium chloride, as described in Chapter 17. In this process sodium ions are reduced by free electrons (the strongest possible reducing agent):

$$Na^+ + e^- \xrightarrow{\text{electrolysis}} Na$$

Potassium can be prepared by electrolysis of molten potassium chloride but this is not a suitable large-scale process for several reasons. For example, potassium is rather soluble in molten potassium chloride and is therefore not easily separated. Potassium is usually prepared by using sodium as a reducing agent:

$$KCl(l) + Na(l) \rightleftharpoons NaCl(l) + K(g)$$

The position of equilibrium in this reaction lies well to the left because potassium is a stronger reducing agent than sodium. Molten sodium is introduced at the bottom of a column of molten potassium chloride at about 850 °C. At this temperature potassium is a gas while sodium is still liquid (Table 15.2). Potassium vapor distills off from the top of the column and is constantly removed so equilibrium is never reached. Thus the reaction is forced to the right in accordance with Le Châtelier's principle. The alkaline earth metals are usually prepared by electrolysis (see Chapter 17).

REACTIONS

The chemistry of the alkali and alkaline earth metals is relatively simple. Because one or two electrons are easily removed from their atoms to form the M^+ and M^{2+} ions, respectively, there is only one stable oxidation state for each group, +1 for the alkali metals and +2 for the alkaline earth metals. With the exception of a very few compounds, primarily those of lithium and beryllium, all their compounds are ionic.

As we might expect from their low ionization energies, all these elements are very reactive. The metals react directly with many other elements and compounds. In Chapter 5 we saw that they react with the halogens to give ionic halides, such as NaCl and $CaBr_2$. Except for beryllium, they react with hydrogen on heating to give ionic hydrides, such as NaH and CaH_2. Lithium is the only element that reacts with nitrogen at room temperature, and magnesium, calcium, strontium, and barium react on heating, to give ionic nitrides containing the N^{3-} ion. Because of its strong triple bond N_2 is rather unreactive, so the formation of these nitrides shows the great reactivity and strong reducing properties of these alkali and alkaline earth metal elements.

The alkali and alkaline earth metals all react readily with oxygen, and except for beryllium and magnesium they must be stored out of contact with the atmosphere, usually under kerosene or some other hydrocarbon. The reactions with oxygen are more complicated than we might have expected. Only lithium, beryllium, magnesium, and calcium give the expected oxides, Li_2O, BeO, MgO, and CaO, in a pure state when they react with excess oxygen under ordinary conditions. The other metals give peroxides such as Na_2O_2, which contain the

$Ca^{2+}(:\ddot{B}r:^-)_2$

$Ca^{2+}(:H^-)_2$

$(Li^+)_3:\ddot{N}:^{3-}$

$(Ca^{2+})_3(:\ddot{N}:^{3-})_2$

$(Li^+)_2:\ddot{O}:^{2-}$

Lithium oxide

$(Na^+)_2{}^-:\ddot{O}- \ddot{O}:^-$

Sodium peroxide

TABLE 15.8
Some Reactions of the Alkali Metals[a]

With oxygen
$4Li(s) + O_2(g) \longrightarrow 2Li_2O(s)$
$2Na(s) + O_2(g) \longrightarrow Na_2O_2(s)$
$K(s) + O_2(g) \longrightarrow KO_2(s)$
$Rb(s) + O_2(g) \longrightarrow RbO_2(s)$
$Cs(s) + O_2(g) \longrightarrow CsO_2(s)$

With nitrogen
$6Li(s) + N_2(g) \longrightarrow 2Li_3N(s)$

With halogens
$2M(s) + X_2(g) \longrightarrow 2MX(s)$

With water
$2M(s) + 2H_2O(l) \longrightarrow 2MOH(aq) + H_2(g)$

With hydrogen
$2M(s) + H_2(g) \longrightarrow 2MH(s)$

Note: Except for the reactions with water, most of these reactions only proceed at a reasonable rate at temperatures above room temperature.

[a] M denotes an alkali metal and X a halogen.

TABLE 15.9
Some Reactions of the Alkaline Earth Metals[a]

With oxygen
$2M(s) + O_2(g) \longrightarrow 2MO(s)$
$Ba(s) + O_2(g) \longrightarrow BaO_2(s)$

With halogens
$M(s) + X_2 \longrightarrow MX_2(s)$

With water
$Mg(s) + H_2O(g) \longrightarrow MgO(s) + H_2(g)$
$M(s) + 2H_2O(l) \longrightarrow M(OH)_2(s) + H_2(g)$

With hydrogen
$M(s) + H_2(g) \longrightarrow MH_2(s)$

With nitrogen
$3M(s) + N_2(g) \longrightarrow M_3N_2(s)$

Note: Except for the reactions of Ca, Sr, and Ba with water, most of these reactions only proceed at a reasonable rate at temperatures above room temperature.

[a] M denotes Mg, Ca, Sr, or Ba and X a halogen.

For many years flashbulbs for photography contained magnesium foil in an atmosphere of oxygen. When the magnesium was heated by an electric current it burned rapidly, producing a brilliant white light.

O_2^{2-} ion, and superoxides such as KO_2, which contain the O_2^- ion. The peroxide and superoxide ions are discussed in Chapter 18.

Some of the most important reactions of the alkali and alkaline earth metals are summarized in Tables 15.8 and 15.9. In each case the element is converted to an ionic compound containing an M^+ or M^{2+} ion, so that these are all oxidation–reduction reactions.

In the next two sections we discuss some of the more important compounds of the alkali and alkaline earth metals.

15.2
Compounds of the Alkali Metals

HYDROXIDES

The alkali metals reduce water to hydrogen and are oxidized to the corresponding M^+ cation. For example,

$$2Na(s) + 2H_2O(l) \longrightarrow 2Na^+(aq) + 2OH^-(aq) + H_2(g)$$

The reactions of the alkali metals with water are quite vigorous and become increasingly so from lithium to cesium (see Experiments 4.1 and 15.3). A small piece of sodium floats when dropped onto water and moves around on the surface in a spectacular way, propelled by the hydrogen evolved and the currents set up in the water by the large amount of heat produced. With potassium, which also floats on the surface, the reaction is even more spectacular, since the heat evolved is sufficient to ignite the hydrogen, which burns with a flame colored violet by the potassium. The reactions of cesium and rubidium are still more violent and are dangerous to demonstrate.

The alkali metal hydroxides can also be produced by reaction of the metal oxides with water; for example,

$$Li_2O(s) + H_2O(l) \longrightarrow 2Li^+(aq) + 2OH^-(aq)$$

This is an acid–base reaction, not an oxidation–reduction reaction, since there is no change in the oxidation number of any of the elements involved. If we write the equation in terms of ions, we see that the reaction is a simple transfer of a proton from a water molecule to an oxide ion:

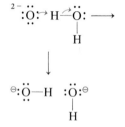

$$O^{2-} + H_2O \longrightarrow 2OH^-$$

The oxide ion behaves as a strong base in water and is quantitatively converted to the hydroxide ion (Chapters 5 and 14).

However, because preparing the alkali metals and their oxides is expensive and difficult, the hydroxides are made commerically by the electrolysis of an aqueous solution of the alkali metal chloride. This process is described in Chapter 17.

Experiment 15.3

Reactions of Some Alkali and Alkaline Earth Metals with Water

Lithium—the lightest of all metals—floats on water, reacting less vigorously than sodium and producing hydrogen.

Calcium is heavier than water and sinks to the bottom of the beaker. A rapid stream of hydrogen bubbles rises to the surface.

Calcium hydroxide, $Ca(OH)_2$, is the other product of the reaction. Because it is not very soluble, a white precipitate is soon formed.

Magnesium does not react with cold water but it does react with hot water. The hydrogen produced collects at the top of the tube.

The alkali metal hydroxides are all white crystalline ionic solids containing the metal ion, M^+, and the hydroxide ion, OH^-. They are all *deliquescent*; that is, they absorb water very readily from the atmosphere to form a concentrated aqueous solution of the hydroxide. They behave as strong bases in water, dissolving to give a solution of M^+ and OH^- ions.

Sodium hydroxide is an important strong base. It is used extensively in the production of many chemicals, textiles, cleaners, soap, and paper, and in petroleum refining. In industry and commerce sodium and potassium hydroxides are frequently called *caustic soda* and *caustic potash*.

Many salts of the alkali metals are easily and conveniently prepared by the reaction of an alkali metal hydroxide with the appropriate acid. For example, lithium nitrate may be prepared by the reaction between lithium hydroxide, $LiOH$, and nitric acid, HNO_3:

$$HNO_3(aq) + LiOH(aq) \longrightarrow LiNO_3(aq) + H_2O(l)$$
$$\text{Acid} \qquad\qquad \text{Base} \qquad\qquad\qquad \text{Salt}$$

An important property of the alkali metal salts—in particular, the halides, sulfates, carbonates, nitrates, and the phosphates—is that they are all soluble in water. Since sodium and potassium salts are relatively inexpensive, they are the salts normally used to carry out reactions involving these anions in aqueous solution.

EXERCISE 15.2

Write balanced equations for the preparation of aqueous solutions of each of the following salts from the appropriate alkali metal hydroxide and acid:

K_2SO_4 CsI $NaClO_4$ Na_3PO_4 $RbNO_3$

HALIDES

The alkali metal halides are all colorless, crystalline ionic compounds that are soluble in water. Some of the properties of the fluorides and chlorides were given in Table 5.4. All the alkali metal halides, except $CsCl$, $CsBr$, and CsI, have the sodium chloride structure. Cesium chloride, bromide, and iodide all have the cesium chloride structure. These structures were discussed in Chapter 11 (Figures 11.25 and 11.26).

The halides of the alkali metals are readily available compounds, most of which are found in natural salt deposits. They can be obtained by the direct reaction between the alkali metal and a halogen (see Experiment 5.3); for example,

$$2K(s) + Br_2(l) \longrightarrow 2KBr(s)$$

But they would not normally be made in this way, because the starting materials are expensive and the reactions are inconveniently vigorous. They may be more easily prepared by the reaction between the hydroxide and the appropriate acid; for example,

$$CsOH(aq) + HI(aq) \longrightarrow CsI(aq) + H_2O(l)$$

Sodium chloride is the usual source of almost all other compounds containing sodium or chlorine. Its most important uses are for the manufacture of hydrochloric acid (see Chapter 8) and of sodium hydroxide and chlorine (see Chapter 17).

HYDRIDES

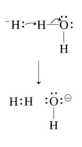

The hydrides of alkali metals are white crystalline ionic compounds containing M^+ and H^- ions; they have the sodium chloride structure. They can be made by heating the element in hydrogen.

These hydrides are all powerful reducing agents. They react vigorously with water, liberating H_2 and forming OH^-:

$$H^- + H_2O \longrightarrow OH^- + H_2$$

The hydride ion, H^-, accepts a proton from water to become H_2. It is therefore a strong base in water. But this reaction is also an oxidation–reduction reaction. The hydride ion (oxidation number -1) is oxidized by the hydrogen in water (oxidation number $+1$) to give H_2 (oxidation number 0).

CARBONATES

Potassium and sodium carbonates have been known since very early times. Potassium carbonate was obtained from the ashes left after the burning of wood and other land plants—hence the common name *potash*. Land plants make little use of sodium, but it is present in considerable quantities in marine plants. Thus the ashes from the burning of seaweed (kelp) were an early source of sodium carbonate (soda ash). They were also a source of iodine (see Chapter 5).

Sodium carbonate, Na_2CO_3, is a very important industrial chemical, which in 1987 ranked eleventh among the industrial chemicals produced in the United States (see Table 15.1). It is used in the manufacture of glass, paper, detergents, and soap. In the home it is used for washing, for cleaning, and for softening "hard" water.

Most of the sodium carbonate used in North America comes from large deposits of the mineral *trona*, $Na_5(CO_3)_2(HCO_3)\cdot 2H_2O$, found in Wyoming. The trona is simply crushed and heated. Heating causes the reaction

$$2HCO_3^- \longrightarrow CO_3^{2-} + H_2O + CO_2$$

to proceed to the right as gaseous H_2O and CO_2 are evolved. Thus the equation for the overall reaction is

$$2Na_5(CO_3)_2(HCO_3)\cdot 2H_2O(s) \longrightarrow 5Na_2CO_3(s) + CO_2(g) + 5H_2O(g)$$

The crude sodium carbonate obtained in this way is dissolved in water, the insoluble impurities are filtered off, and the sodium carbonate is crystallized, filtered, and heated to give pure anhydrous sodium carbonate, Na_2CO_3.

At temperatures below 35.2 °C sodium carbonate crystallizes from water as the hydrated form, $Na_2CO_3\cdot 10H_2O$, and is familiar in this form under the name *soda crystals* or *washing soda*. Above this temperature the monohydrate

$Na_2CO_3 \cdot H_2O$ crystallizes. On standing in the air, $Na_2CO_3 \cdot 10H_2O$ slowly loses water and crumbles to a white powder of $Na_2CO_3 \cdot H_2O$.

Because the carbonate ion is the anion of a weak acid, HCO_3^-, it is a weak base (Chapters 5 and 14):

$$CO_3^{2-} + H_2O \rightleftharpoons HCO_3^- + OH^-$$

Solutions of sodium carbonate are therefore basic (see Excercise 15.3).

EXERCISE 15.3

An aqueous solution is prepared by adding 100 g of washing soda, $Na_2CO_3 \cdot 10H_2O$ to sufficient water to obtain 1.00 L of solution. What is the pH of the solution? (See Table 14.2 on p. 667 for any data needed.)

Sodium hydrogen carbonate, $NaHCO_3$ (sodium bicarbonate, bicarbonate of soda, or baking soda), can be made by passing CO_2 into a solution of NaOH:

$$CO_2(g) + NaOH(aq) \longrightarrow NaHCO_3(aq)$$

When $NaHCO_3(s)$ is heated, H_2O and CO_2 are driven off, and $Na_2CO_3(s)$ is formed as we described above for the preparation of sodium carbonate from the ore trona.

$$2NaHCO_3(s) \longrightarrow Na_2CO_3(s) + H_2O(g) + CO_2(g)$$

Sodium hydrogen carbonate is used in cooking and in medicine and for the manufacture of baking powder, which is a mixture of sodium hydrogen carbonate and a weak acid, for example, sodium dihydrogen phosphate. Baking powder is a leavening agent used in the making of biscuits, cakes, and other baked goods. In the solid mixture there is no reaction but, on adding water and warming during cooking, carbon dioxide is evolved:

$$H_3O^+ + HCO_3^- \rightleftharpoons CO_2 + 2H_2O$$

The reaction is driven to the right by the evolution of CO_2. The carbon dioxide causes the baked goods to "rise", giving them a light texture because of the many "holes" formed by the escaping carbon dioxide.

Solutions of $NaHCO_3$ in water are basic, but they have a lower pH than solutions of Na_2CO_3 because HCO_3^- is a weaker base than CO_3^{2-}.

SULFATES

Sodium sulfate, Na_2SO_4, and *sodium hydrogen sulfate*, $NaHSO_4$, are by-products of the manufacture of hydrogen chloride by the reaction of sulfuric acid with sodium chloride (Chapter 8):

$$NaCl + H_2SO_4 \longrightarrow NaHSO_4 + HCl$$

$$2NaCl + H_2SO_4 \longrightarrow Na_2SO_4 + 2HCl$$

Large amounts of sodium sulfate are used in the paper industry. Sodium hydrogen sulfate gives acidic solutions in water:

$$HSO_4^- + H_2O \rightleftharpoons H_3O^+ + SO_4^{2-}$$

EXERCISE 15.4

Starting with the element, explain by means of suitable balanced equations how the following compounds of lithium could be prepared: the oxide, hydride, and bromide in the solid state, and the carbonate and sulfate in aqueous solution.

EXERCISE 15.5

Give the formulas and an alternative name for each of the following substances: caustic soda, potash, rock salt, chalk, limestone, dolomite.

EXERCISE 15.6

Aqueous solutions of each of the following substances are prepared by adding 1 mol to 1.00 L of water: KCl, NaH, K_2CO_3, $NaHSO_4$. List the substances in order of decreasing pH of their solutions. (No calculations are required.)

15.3
Compounds of the Alkaline Earth Metals

The alkaline earth metals received this name because their oxides are very stable to heat and they react with water to give alkaline (basic) solutions. The alchemists referred to any nonmetallic solid that was unaffected by strong heat as an *earth*. The oxides were therefore called *alkaline earths*. The name *alkaline earth metals* for the family of elements was then derived from the name for the oxides which, because of the difficulty of obtaining the metals, were known long before the elements themselves.

We will not consider the compounds of beryllium, because it is a rather rare element whose compounds are not of great practical importance. Moreover, like the first element in other groups, its properties and behavior are not always typical of the group as a whole.

OXIDES AND HYDROXIDES

The oxides of the alkaline earth metals are formed when the elements burn in air (Experiment 15.4), although for magnesium this reaction is accompanied by the formation of the nitride, Mg_3N_2:

$$2Mg(s) + O_2(g) \longrightarrow 2MgO(s)$$
$$3Mg(s) + N_2(g) \longrightarrow Mg_3N_2(s)$$

Both these reactions are highly exothermic, and energy is released as both heat and light (see Experiment 1.3). When barium is heated in air or oxygen at 500–600 °C, the peroxide, BaO_2, is formed (see Chapter 18).

Experiment 15.4

Oxidation of Calcium to Calcium Oxide

When calcium is strongly
heated in the air it burns with
a red flame.

The product is white solid
calcium oxide.

The oxides are normally made by decomposing the carbonates by heat; for example,

$$MgCO_3(s) \longrightarrow MgO(s) + CO_2(g)$$

$$CaCO_3(s) \longrightarrow CaO(s) + CO_2(g)$$

Calcium oxide is made by heating limestone in a *limekiln* (Figure 15.1 on p. 728); the product, CaO, is known as *lime* (or *quicklime*).

The oxides of magnesium, calcium, strontium, and barium all have the sodium chloride crystal structure. Magnesium oxide is insoluble in water, and its melting point is very high because of the strength of the interionic forces. The electrostatic force of attraction between an M^{2+} ion and an X^{2-} ion is four times the force between an M^+ ion and an X^- ion at the same distance apart.

The alkaline earth metal oxides are basic. They react with acids to give salts; for example

$$MgO(s) + 2HCl(aq) \longrightarrow MgCl_2(aq) + H_2O(l)$$

Magnesium oxide reacts very slowly with water to give insoluble magnesium hydroxide:

$$MgO(s) + H_2O(l) \longrightarrow Mg(OH)_2(s)$$

Calcium oxide reacts exothermically with water to give *calcium hydroxide* $Ca(OH)_2$, or *slaked lime*, which has a low solubility in water.

$$CaO(s) + H_2O(l) \longrightarrow Ca(OH)_2(s) \qquad \Delta H° = -65.2 \text{ kJ}$$

Calcium hydroxide can also be made by precipitation from a solution of a soluble calcium salt:

$$CaCl_2(aq) + 2NaOH(aq) \longrightarrow Ca(OH)_2(s) + 2NaCl(aq)$$

FIGURE 15.1
Limekiln.

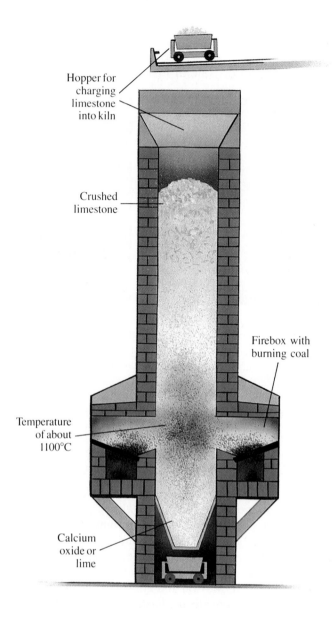

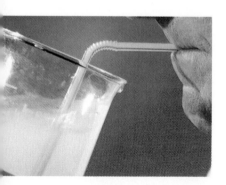

The presence of CO_2 in exhaled air can be demonstrated by blowing into a solution of calcium hydroxide (limewater). A white precipitate of calcium carbonate is formed.

A third method of preparation is by the direct reaction of the element with water (see Experiment 15.3):

$$Ca(s) + 2H_2O(l) \longrightarrow Ca(OH)_2(s) + H_2(g)$$

A saturated solution of $Ca(OH)_2$ is called *limewater* and is commonly used as a test for CO_2. When CO_2 is passed through limewater, the solution turns cloudy, because insoluble $CaCO_3$ precipitates:

$$Ca(OH)_2(aq) + CO_2(aq) \longrightarrow CaCO_3(s) + H_2O(l)$$

But continued passage of CO_2 causes the precipitate to dissolve to give a clear solution again. The calcium carbonate precipitate dissolves because carbonate ion is removed from the solution due to the formation of hydrogen carbonate ion by the reaction

$$CO_3^{2-} + CO_2 + H_2O \rightleftharpoons 2HCO_3^-$$

Since CO_2 and H_2O are in equilibrium with carbonic acid H_2CO_3, we can also write this reaction as

$$CO_3^{2-} + H_2CO_3 \rightleftharpoons 2HCO_3^-$$

The removal of carbonate ion by this reaction displaces the equilibrium

$$CaCO_3(s) \rightleftharpoons Ca^{2+}(aq) + CO_3^{2-}(aq)$$

to the right and so the $CaCO_3$ dissolves. Adding the equations for these two reactions gives the equation for the overall reaction:

$$CaCO_3(s) + CO_2(g) + H_2O(l) \longrightarrow Ca(HCO_3)_2(aq)$$

Calcium hydroxide is a strong base. It is completely dissociated into Ca^{2+} and OH^- in solution:

$$Ca(OH)_2(s) \rightleftharpoons Ca^{2+}(aq) + 2OH^-(aq)$$

Lime and slaked lime are very important commercial bases because of their ready availability and cheapness. Lime is used in making glass and cement. It is used to remove SiO_2 and other acidic oxides in metallurgical processes, such as the smelting of iron (Chapter 10), by combining with them to form a slag, such as calcium silicate, $CaSiO_3$ (Box 15.2).

Magnesium hydroxide, $Mg(OH)_2$, is only slightly soluble in water. It is conveniently made by adding a soluble hydroxide, such as NaOH, to a solution of a soluble magnesium salt, such as $MgCl_2$ or $MgSO_4$, to give a white precipitate of $Mg(OH)_2$:

$$MgSO_4(aq) + 2NaOH(aq) \longrightarrow Mg(OH)_2(s) + Na_2SO_4(aq)$$

Although magnesium reacts only very slowly with water, it reacts rapidly with steam to give the hydroxide:

$$Mg(s) + 2H_2O(g) \longrightarrow Mg(OH)_2(s) + H_2(g)$$

A burning piece of magnesium will continue to burn brightly in an atmosphere of steam (Experiment 1.3).

A suspension of magnesium hydroxide in water is called *milk of magnesia*; it is used as a medicine to correct excess acidity in the stomach (Box 15.3).

CARBONATES

Calcium and magnesium carbonates are minerals and are the most important sources of these elements. Unlike the alkali metal carbonates (other than Li_2CO_3), they decompose when heated, giving the oxides and CO_2:

$$CaCO_3 \xrightarrow{\text{Heat}} CaO + CO_2$$

As we saw in Chapter 13, if $CaCO_3$ is heated in a closed system, it does not completely decompose but comes into equilibrium with CaO and CO_2. When it is heated in the atmosphere, however, the CO_2 escapes and never reaches the equilibrium pressure, so the reaction is driven to completion.

BOX 15.2

LIME AND LIMESTONE

The transformation of limestone to lime by heating and the setting of lime mortar are among the first chemical reactions, other than fire itself, to be utilized by humans. Lime mortar was used by the Greeks in the building of their temples, by the Romans in the building of their roads, and by the Chinese in the building of the Great Wall. Mortar consists of one part lime to three parts sand (silicon dioxide) mixed with water to make a thick paste. This paste was placed between blocks of stone or bricks to hold them together in constructing roads and buildings. The first reaction that occurs is the conversion of lime to slaked lime:

$$CaO(s) + H_2O(l) \longrightarrow Ca(OH)_2(s)$$

$$\text{Lime} \qquad\qquad\qquad \text{Slaked lime}$$

The slaked lime then absorbs CO_2 from the air and is converted to calcium carbonate:

$$Ca(OH)_2(s) + CO_2(g) \longrightarrow CaCO_3(s) + H_2O(l)$$

The sand does not react, but the sand particles are bound together by the calcium carbonate as it forms to give a hard solid that holds together the stone blocks or bricks between which it is placed.

At the present time cement from which concrete is made is more widely used in building than lime mortar. Cement is made by heating a mixture of lime, sand, and clay to about 1500 °C. In this case the sand and the clay react with the lime to give calcium aluminosilicate (see Chapter 22).

Limestone is also one of the oldest agricultural chemicals. It is spread on the ground to neutralize undesirable acids. It also supplies essential Ca^{2+} ions and, since limestone almost always contains some $MgCO_3$, Mg^{2+} ions as well.

As we saw in Chapter 10, an important use of limestone is in the manufacture of iron in the blast furnace. Limestone is added to the blast furnace with the iron ore and coke to remove silica and other impurities as a slag:

$$CaCO_3(s) \xrightarrow{\text{Heat}} CaO(s) \xrightarrow{SiO_2} CaSiO_3(l)$$

Another very important use of lime is for the manu-

The Great Wall of China

facture of calcium carbide, CaC_2 (Chapter 6), by heating it with coke at 2000 °C:

$$CaO(s) + 3C(s) \longrightarrow CaC_2(s) + CO(g)$$

Calcium carbide was formerly the principal source of ethyne (acetylene) which results from its reaction with water (Chapter 6):

$$CaC_2(s) + 2H_2O(l) \longrightarrow Ca(OH)_2(s) + C_2H_2(g)$$

Other important substances used in the manufacture of certain plastics are made from calcium carbide.

Whenever an inexpensive, readily available, base is needed in a chemical process, lime is most often used. For example, it is used to maintain an optimum pH for the biological oxidation of sewage and for the removal of SO_2 and H_2S from the stack gas of electricity generating stations that use fossil fuels and metallurgical smelters:

$$CaO(s) + SO_2(g) \longrightarrow CaSO_3(s)$$

$$CaO(s) + H_2S(g) \longrightarrow CaS(s) + H_2O(g)$$

Enormous quantities of purified calcium carbonate are used in the paper industry for enhancing the brightness, opacity, and smoothness of paper. Lime is used in the dairy industry for reducing acidity prior to pasteurization and in the sugar industry in the refining process.

LIMESTONE CAVES, STALACTITES, AND STALAGMITES Large and beautiful caves are a feature of limestone country in many parts of the world. These caves have been formed by water running through the limestone in small streams and occasionally large underground rivers. Although $CaCO_3$ is insoluble in pure water, it is slightly soluble in natural water because of the presence of a small amount of carbon dioxide which, as we have seen above, converts insoluble calcium carbonate to soluble calcium hydrogen carbonate:

$$CaCO_3(s) + CO_2(aq) + H_2O(l) \rightleftharpoons Ca^{2+}(aq) + 2HCO_3^-(aq)$$

In an underground stream the dissolved calcium hydrogen carbonate is continually washed away. Thus equilibrium is never reached, so the calcium carbonate continues to dissolve, and over a very long time underground caves may be formed.

A common feature of limestone caves are the stalactites and stalagmites, which are often found in impressively large and beautiful shapes (Figure 15.2). They are formed by the reversal of the reaction that creates the caves. Water that has passed through limestone rock becomes saturated with $Ca(HCO_3)_2$, seeps through to the roof of a cave, and gathers as a drop hanging from the roof. The water slowly evaporates, and some of the CO_2 is lost from the solution. In accordance with Le Châtelier's principle, the equilibrium shifts to the left to produce more CO_2 and H_2O, and some of the Ca^{2+} in solution is redeposited

---

BOX 15.3

ANTACIDS

Calcium carbonate, magnesium carbonate, and magnesium hydroxide are bases that are frequently used as antacids, that is, for neutralizing excess stomach acid. The mucous membrane lining the stomach secretes gastric juices, which contain hydrochloric acid and pepsinogen, which in the stomach is converted to the enzyme pepsin. The hydrochloric acid and the enzyme pepsin play essential roles in the digestive process. The normal pH of the stomach ranges from 1.2 to 3.0. This concentration of acid is sufficient to dissolve zinc! It also dissolves the stomach lining, which is replaced at the rate of about half a million cells a minute! If we overeat, the stomach responds by producing an excessive amount of

hydrochloric acid, and we suffer from indigestion. When this happens, we may take an antacid to neutralize the excess stomach acid. Some common antacids are as follows:

Commercial name	Active ingredients
Tums	$CaCO_3$
Milk of Magnesia	$Mg(OH)_2$
Alka Seltzer	$NaHCO_3$, citric acid
Rolaids	$NaAl(OH)_2CO_3$
Bisodol	$NaHCO_3$, $MgCO_3$

Some antacids, such as Alka Seltzer, fizz when added to water because the citric acid that they contain reacts with HCO_3^- and CO_3^{2-} to produce CO_2:

$$NaHCO_3 + HC_6H_7O_7 \rightarrow H_2O + CO_2 + NaC_6H_7O_7$$

Citric acid   Sodium citrate

$$\begin{array}{c} H \\ | \\ H-C-CO_2H \\ | \\ HO-C-CO_2H \\ | \\ H-C-CO_2H \\ | \\ H \end{array}$$

Citric acid

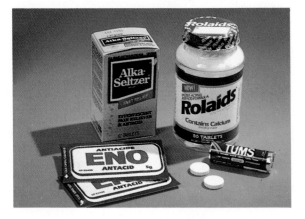

FIGURE 15.2
Stalactites and Stalagmites
in a Limestone Cave.

as $CaCO_3$. The evaporation of many drops from the same place over many thousands of years gradually leads to the formation of a conical-shaped stalactite. Moreover, a drop hanging from the roof may fall to the floor, where it evaporates slowly, again depositing a very small amount of $CaCO_3$. If drops continue to hit the same spot, a stalagmite is built up. It may grow sufficiently to join the stalactite, forming a column.

HARDNESS OF WATER   Water containing appreciable amounts of Ca^{2+} and Mg^{2+} is called *hard water*. The anions that are usually present with these cations are Cl^-, SO_4^{2-}, and HCO_3^-. Water containing only very small concentrations of Ca^{2+} and Mg^{2+} is called *soft water*.

Most natural water, particularly in limestone regions, is hard. The use of hard water for domestic purposes and in industry presents several problems. For example, soap consists of the sodium salts of long-chain carboxylic acids such as sodium stearate, $C_{17}H_{35}CO_2^-Na^+$. In hard water soap forms an insoluble scum of the insoluble calcium salts of these acids, which has no cleansing power. When hard water is heated, the reaction by which $CaCO_3$ is dissolved in cave formation is reversed as CO_2 is driven off and $CaCO_3$ is precipitated:

$$Ca^{2+}(aq) + 2HCO_3^-(aq) \longrightarrow CaCO_3(s) + CO_2(aq) + H_2O(l)$$

Calcium sulfate, which is less soluble in hot water than in cold, is also precipitated. A mixture of $CaCO_3$ and $CaSO_4$ is the familiar "scale" that forms inside teakettles and in pipes in hot-water boilers and heating systems. After a sufficiently long time enough deposit may form on the inside of a pipe to block it completely.

Calcium and magnesium can be removed from water—in other words, the water can be "softened"—by adding Na_2CO_3 to precipitate the insoluble carbonates (Experiment 15.5):

$$Ca^{2+}(aq) + CO_3^{2-}(aq) \longrightarrow CaCO_3(s)$$
$$Mg^{2+}(aq) + CO_3^{2-}(aq) \longrightarrow MgCO_3(s)$$

If the hardness is due only to $Ca(HCO_3)_2$, it can be removed by simply boiling

Experiment 15.5

Hardness of Water

Left: Two drops of soap solution form a thick lather in distilled water. Center: Two drops of soap solution in hard water give no lather, but the solution becomes cloudy because insoluble calcium salts of the soap are precipitated. Right: If sodium carbonate from the dish is added to the water, calcium ion is precipitated as calcium carbonate. Two drops of soap solution then give a small amount of lather.

water to precipitate $CaCO_3$ or by adding enough lime to convert all the $Ca(HCO_3)_2$ to $CaCO_3$:

$$Ca^{2+}(aq) + 2HCO_3^-(aq) + Ca(OH)_2(s) \longrightarrow 2CaCO_3(s) + 2H_2O(l)$$

HALIDES

The halides of the alkaline earth metals are all rather soluble in water, with the exception of CaF_2, SrF_2, and BaF_2, which have very low solubilities. Calcium fluoride occurs widely as the mineral *fluorite*, which is an important source of fluorine. The solubility of CaF_2 is only 2.1×10^{-4} mol L^{-1}, but when it is added to drinking water, it gives the low concentration of F^- that helps prevent tooth decay (Chapter 5).

Magnesium and calcium chlorides crystallize from water as $MgCl_2 \cdot 6H_2O$ and $CaCl_2 \cdot 6H_2O$. Anhydrous calcium chloride can be made by heating the hydrate. It combines strongly with water to re-form the hydrate, and it is often used for drying gases. However, when $MgCl_2 \cdot 6H_2O$ is heated, the $MgCl_2$ and water react to form MgO and HCl:

$$MgCl_2 \cdot 6H_2O(s) \longrightarrow MgO(s) + 2HCl(g) + 5H_2O(g)$$

Because of the small size of Mg^{2+}, the $Mg(H_2O)_6^{2+}$ ion is relatively acidic and donates a proton to the Cl^- ion to give HCl, which is evolved as a gas, driving the reaction to the right. Hydrated aluminum chloride, $AlCl_3 \cdot 6H_2O$, and some other hydrated metal halides behave in the same way on heating.

The fluorides of calcium, strontium, and barium have a relatively simple crystal structure known as the *calcium fluoride*, or *fluorite*, *structure* (Figures 11.25 and 15.3a). Like the sodium chloride structure, it is based on a face-centered cubic lattice. Magnesium chloride has a layer structure (Figures 11.7 and 15.3b).

SULFATES

Magnesium sulfate is soluble in water and crystallizes as the hydrate $MgSO_4 \cdot 7H_2O$. It occurs naturally in this form as *epsomite* and is commonly called *Epsom salts*. In medicine it is useful as a purgative.

Calcium sulfate occurs in nature as the mineral *gypsum*, $CaSO_4 \cdot 2H_2O$. When gypsum is heated to a little above 100 °C, it loses three-quarters of its water of

Scale buildup in a boiler pipe

The reaction of the halides of metals having small highly charged ions with water on heating is similar to the reaction of nonmetal halides with water. The metal halides give the metal oxide and HCl while the nonmetal halides give an oxoacid and HCl (Chapter 12).

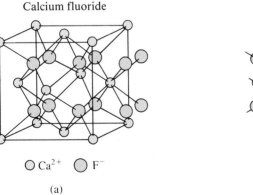

Calcium fluoride

Ca²⁺   F⁻

(a)

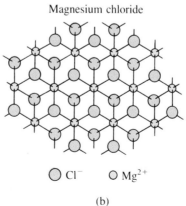

Magnesium chloride

Cl⁻   Mg²⁺

(b)

FIGURE 15.3
Structures of Calcium Fluoride and Magnesium Chloride.

(a) The calcium fluoride structure is based on a face-centered cubic lattice. The calcium ions are at the lattice points and have a cubic close-packed arrangement. The fluoride ions occupy all the tetrahedral holes. (b) Magnesium chloride has a layer structure. Each magnesium ion is surrounded by an octahedral arrangement of six chloride ions. The chloride ions form layers, with the magnesium ions sandwiched between them.

Experiment 15.6

Some Insoluble Alkaline Earth Metal Compounds

A solution of ammonium carbonate added to a solution of calcium chloride produces a white precipitate of calcium carbonate.

A solution of sodium sulfate added to a solution of barium chloride produces a heavy white precipitate of barium sulfate.

A solution of sodium hydroxide added to a solution of magnesium chloride produces a gelatinous precipitate of magnesium hydroxide.

crystallization, giving a white powder of $CaSO_4 \cdot \frac{1}{2}H_2O$, which is called _plaster of paris_:

$$CaSO_4 \cdot 2H_2O \xrightarrow{\text{Heat}} CaSO_4 \cdot \tfrac{1}{2}H_2O + \tfrac{3}{2}H_2O$$

A paste of plaster of paris and a small amount of water rapidly sets to a solid, which consists of a mass of interlocking crystals of the dihydrate. The formation of the dihydrate is accompanied by an increase in volume, which is very valuable in making plaster casts, since the increase in volume forces the plaster into all the crevices of any mold into which it is poured; thus all the details are faithfully reproduced. Plaster of paris is used in the manufacture of wallboard and as a component of the plaster used for the surface of interior walls and ceilings.

The solubilities of the alkaline earth metal sulfates decrease steadily from $BeSO_4$ and $MgSO_4$, which are very soluble in water, to $CaSO_4$, which is only slightly soluble, to $SrSO_4$ and $BaSO_4$, which are insoluble. The formation of a precipitate of $BaSO_4$, when a solution of a soluble barium salt, such as $BaCl_2$, is added to a solution suspected to contain sulfate, confirms the presence of sulfate in the solution (Experiment 15.6):

$$Ba^{2+}(aq) + SO_4{}^{2-}(aq) \longrightarrow BaSO_4(s)$$

Like all elements of high atomic number, barium absorbs X rays strongly. If a suspension of barium sulfate is swallowed, it coats the alimentary canal. Then X ray photographs of the alimentary canal, which is normally transparent to X rays, can be obtained (Figure 15.4). Unlike calcium and magnesium, barium is poisonous, but the solubility of barium sulfate is so small that it passes through the body without being absorbed or having any harmful effects.

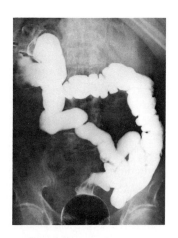

FIGURE 15.4
X Ray Photograph of the Large Intestine.

The intestine has been coated with barium sulfate to render it opaque to X rays.

EXERCISE 15.7

Explain the following observations, giving appropriate balanced equations.

(a) When an aqueous solution of sodium hydroxide is added to an aqueous solution of magnesium chloride a white precipitate is formed.

(b) When carbon dioxide is bubbled into a suspension of calcium carbonate in water the calcium carbonate dissolves to give a clear solution.

(c) When a solution of sodium flouride is added to a solution of calcium chloride a white precipitate forms.

(d) When a solution of sodium carbonate is added to hard water a white precipitate is formed.

15.4
Solubility and Precipitation Reactions

We have seen that the hydroxides and all the salts of the alkali metals are soluble in water, whereas many of the hydroxides and salts of the alkaline earth metals have very small solubilities; these insoluble compounds include $Mg(OH)_2$, $CaCO_3$, CaF_2, $CaSO_4$, $BaCO_3$, and $BaSO_4$ (see Experiment 15.6). In previous chapters we encountered other insoluble salts such as $AgCl$, $PbSO_4$, and FeS and insoluble hydroxides such as $Cu(OH)_2$ and $Pb(OH)_2$. In Chapter 5 (Table 5.9) we gave some useful rules for the solubilities of ionic compounds in water. Unfortunately we cannot give any simple general explanation for these solubility rules. However, the following observations may be helpful.

In general, as a consequence of Coulomb's law, ions having multiple charges are more strongly held together than ions having single charges, and smaller ions are more strongly held together than larger ions. Thus ionic substances with small, highly charged ions tend to be less soluble than ionic substances composed of large, singly charged ions. For example, the compounds of singly charged cations—the alkali metals and NH_4^+—are all soluble. And the salts of Cl^-, Br^-, I^-, NO_3^-, and ClO_4^- are all soluble, except for their salts with some M^{2+} ions and Ag^+. However, most hydroxides are insoluble. Also, we note that with few exceptions salts of the doubly charged ions CO_3^{2-}, S^{2-}, and PO_4^{3-} are insoluble. However, most sulfates are soluble. Note that although the salts of the multiply charged CO_3^{2-} and PO_4^{3-} ions are insoluble, many salts of the singly charged HCO_3^- and $H_2PO_4^-$ ions are soluble.

SOLUBILITY PRODUCT CONSTANT

When we use the solubility rules to predict whether or not a precipitate will form when two solutions are mixed we should strictly speaking say that precipitate *may* form, rather than it *will* form, because if the solutions were very dilute a precipitate might not form. What concentrations of ions will, in fact, give a precipitate? We can answer this question by considering the equilibrium between a solid salt and its ions in a saturated solution of the salt in a more quantitative manner.

Even if a salt is described as insoluble, there must be small, perhaps very

The formation of a precipitate of red silver chromate, Ag_2CrO_4

small, concentrations of ions in equilibrium with the undissolved salt. For example, if we have a saturated solution of $CaSO_4$ in equilibrium with some undissolved $CaSO_4$, the Ca^{2+} and $SO_4{}^{2-}$ ions are continually leaving the crystals and passing into solution, while ions from the solution are continually attaching themselves to the crystals. At equilibrium the rates of the two opposing processes are equal:

$$CaSO_4(s) \rightleftharpoons Ca^{2+}(aq) + SO_4{}^{2-}(aq)$$

The expression for the equilibrium constant is

$$K_c = \left(\frac{[Ca^{2+}][SO_4{}^{2-}]}{[CaSO_4(s)]} \right)_{eq}$$

The equilibrium between $CaSO_4(s)$ and $Ca^{2+}(aq)$ and $SO_4{}^{2-}(aq)$ is an example of a heterogeneous equilibrium because it involves both a solid and a solution. As we saw in Chapter 13, the concentration of a pure solid is a constant, which does not vary, and so it can be incorporated in the equilibrium constant to give another constant, K_{sp}, which is called the **solubility product constant**:

$$K_c[CaSO_4(s)] = K_{sp} = ([Ca^{2+}][SO_4{}^{2-}])_{eq}$$

In general, for an ionic compound with the formula A_xB_y the equilibrium in a saturated solution can be written as

$$A_xB_y(s) \rightleftharpoons xA^{m+}(aq) + yB^{n-}(aq)$$

The solubility product is then

$$K_{sp} = ([A^{m+}]^x[B^{n-}]^y)_{eq}$$

In other words, the solubility product constant is equal to the product of the concentrations of the ions involved in the equilibrium, each raised to the power of its coefficient in the equation for the equilibrium.

EXERCISE 15.8

Write equations for the equilibrium between the following solid ionic compounds and their ions in solution: MgF_2, $Fe(OH)_3$, $Mg_3(PO_4)_2$, Ag_2S. Hence write an expression for the solubility product constant for each compound.

CALCULATING THE SOLUBILITY PRODUCT CONSTANT OF AN IONIC COM-POUND FROM ITS SOLUBILITY We can obtain the value of the solubility product constant for an ionic compound from the measured value of its solubility. For example, the solubility of $CaSO_4$ in water at 25 °C is 4.9×10^{-3} mol L^{-1}. This is the concentration of the dissolved $CaSO_4$. But since the $CaSO_4$ is fully ionized to give equal amounts of Ca^{2+} and $SO_4{}^{2-}$, the concentration of each of these ions is also 4.9×10^{-3} mol L^{-1}. In other words,

$$[Ca^{2+}] = [SO_4{}^{2-}] = 4.9 \times 10^{-3} \text{ mol } L^{-1}$$

Therefore

$$K_{sp} = ([Ca^{2+}][SO_4{}^{2-}])_{eq} = (4.9 \times 10^{-3} \text{ mol } L^{-1})^2$$
$$= 2.4 \times 10^{-5} \text{ mol}^2 \text{ L}^{-2}$$

> **TABLE 15.10**
> Solubility Product Constants, K_{sp} of Some Slightly Soluble Salts at 25 °C

Bromides		Ca(OH)$_2$	6.5×10^{-6}
PbBr$_2$	2.1×10^{-6}	Cu(OH)$_2$	4.8×10^{-20}
Hg$_2$Br$_2$	5.6×10^{-23}	Fe(OH)$_2$	7.9×10^{-16}
AgBr	5.0×10^{-13}	Fe(OH)$_3$	1.6×10^{-39}
Carbonates		Pb(OH)$_2$	6×10^{-16}
BaCO$_3$	5.0×10^{-9}	Mg(OH)$_2$	7.1×10^{-12}
CaCO$_3$	4.7×10^{-9}	Sr(OH)$_2$	3.2×10^{-4}
CuCO$_3$	2.3×10^{-10}	Zn(OH)$_2$	3.0×10^{-16}
FeCO$_3$	2.1×10^{-11}	Iodides	
PbCO$_3$	7.4×10^{-14}	PbI$_2$	7.9×10^{-9}
MgCO$_3$	1×10^{-5}	Hg$_2$I$_2$	1.1×10^{-28}
Ag$_2$CO$_3$	8.1×10^{-12}	AgI	8.3×10^{-17}
SrCO$_3$	9.3×10^{-10}	Sulfates	
ZnCO$_3$	1×10^{-10}	BaSO$_4$	3.2×10^{-7}
Chlorides		CaSO$_4$	2×10^{-5}
PbCl$_2$	1.7×10^{-5}	PbSO$_4$	6.3×10^{-7}
Hg$_2$Cl$_2$	1.2×10^{-18}	Ag$_2$SO$_4$	1.5×10^{-5}
AgCl	1.8×10^{-10}	SrSO$_4$	3.2×10^{-7}
Fluorides		Sulfides	
BaF$_2$	1.7×10^{-6}	CuS	6×10^{-36}
CaF$_2$	3.9×10^{-11}	FeS	8×10^{-19}
PbF$_2$	3.6×10^{-8}	PbS	3×10^{-28}
MgF$_2$	6.6×10^{-9}	HgS	5×10^{-54}
SrF$_2$	2.9×10^{-9}	Ag$_2$S	8×10^{-51}
Hydroxides		ZnS	3×10^{-24}
Al(OH)$_3$	3×10^{-34}		
Ba(OH)$_2$	3×10^{-4}		

Solubility product constants for some common salts and hydroxides are given in Table 15.10. Very insoluble salts have very small values; more soluble salts have larger values. For example, the solubility product constant of the divalent metal hydroxides ranges from 3×10^{-4} mol^3 L^{-3} for the relatively soluble Ba(OH)$_2$ to 4.8×10^{-20} mol^3 L^{-3} for the much less soluble Cu(OH)$_2$. Many metal sulfides are very insoluble; for example, HgS has the solubility product constant 5×10^{-54} mol^2 L^{-2}.

═ *Example 15.1* ═════════════════════════════════════

CALCULATING THE SOLUBILITY PRODUCT CONSTANT FROM SOLUBILITY

The solubility of CaF$_2$ in water at 25 °C is found to be 2.14×10^{-4} mol L^{-1}. What is the value of K_{sp} at this temperature?

Solution

The equation for the equilibrium is

$$CaF_2(s) \rightleftharpoons Ca^{2+}(aq) + 2F^-(aq)$$

Hence each mole of CaF_2 that dissolves produces 1 mol of Ca^{2+} and 2 mol of F^-. Thus

$$[Ca^{2+}] = 2.14 \times 10^{-4} \text{ mol L}^{-1}$$
$$[F^-] = 4.28 \times 10^{-4} \text{ mol L}^{-1}$$

From the equation for the equilibrium we have

$$K_{sp}(CaF_2) = ([Ca^{2+}][F^-]^2)_{eq}$$

Hence

$$K_{sp} = (2.14 \times 10^{-4})(4.28 \times 10^{-4})^2 \text{ mol}^3 \text{ L}^{-3}$$
$$= 3.9 \times 10^{-11} \text{ mol}^3 \text{ L}^{-3}$$

EXERCISE 15.9

The solubility of Ag_2SO_4 in water at 25 °C is 1.55×10^{-2} mol L^{-1}. What is the value of K_{sp} at this temperature?

CALCULATING SOLUBILITY FROM THE SOLUBILITY PRODUCT CONSTANT
We can calculate the solubility of an ionic compound if we know its solubility product constant, as the next example shows.

= Example 15.2 =

CALCULATING SOLUBILITY FROM THE SOLUBILITY PRODUCT CONSTANT

The solubility product constant of lead(II) chloride, $PbCl_2$, is 1.7×10^{-5} mol^3 L^{-3} at 25 °C. What is the solubility of $PbCl_2$ in water at this temperature?

Solution

The equilibrium is

$$PbCl_2(s) \rightleftharpoons Pb^{2+}(aq) + 2Cl^-(aq)$$

We have to find the solubility of $PbCl_2$, so we will let this be x mol L^{-1}. The equation shows that every mole of $PbCl_2$ that dissolves gives 1 mol of Pb^{2+} and 2 mol of Cl^-. Therefore

$$[Pb^{2+}] = x \text{ mol L}^{-1} \quad \text{and} \quad [Cl^-] = 2x \text{ mol L}^{-1}$$

From the equation we can write

$$K_{sp} = ([Pb^{2+}][Cl^-]^2)_{eq}$$

And since we know that $K_{sp} = 1.7 \times 10^{-5}$ mol^3 L^{-3}, we have

$$[Pb^{2+}][Cl^-]^2 = 1.7 \times 10^{-5} \text{ mol}^3 \text{ L}^{-3}$$

Substituting gives

$$x(2x)^2 = 4x^3 = 1.7 \times 10^{-5}$$
$$x^3 = 4.3 \times 10^{-6}$$

Taking the cube root of each side yields

$$x = 1.6 \times 10^{-2}$$

$$[Pb^{2+}] = 1.6 \times 10^{-2} \text{ mol L}^{-1}$$

Thus the solubility of $PbCl_2$ is 1.6×10^{-2} mol L^{-1}.

Example 15.2 is somewhat artificial because the solubility product constant is normally obtained from the measured solubility, as we saw in Example 15.1. More important uses of the solubility product constant include predicting whether a precipitate will form when two solutions are mixed; the solubility of a substance in the presence of another having a common ion; and the dependence of solubility on pH. We take up these applications of the solubility product constant in the following sections.

EXERCISE 15.10

The solubility product constant of LiF is 1.7×10^{-3} mol L^{-1} at 25 °C. What is the solubility of LiF at this temperature?

PRECIPITATION REACTIONS

We can use the solubility product constant to predict whether a precipitate will form when two solutions are mixed. To do this, we calculate the reaction quotient, Q, from the ion concentrations. Recall that Q is calculated from an expression that has the form of the equilibrium constant but in which the concentrations are the initial concentrations, *not* the equilibrium concentrations. In the present case Q has the form of the solubility product constant and is called the **ion product**. It is the product of the initial concentrations of the ions, each raised to the same power as in the K_{sp} expression. We then compare the value of Q with the value of the solubility product, K_{sp}. There are three possibilities:

- If $Q < K_{sp}$, the solution is not saturated, and no precipitate forms.
- If $Q = K_{sp}$, the solution is just saturated.
- If $Q > K_{sp}$, the solution is supersaturated, and a precipitate will form until the ion product becomes equal to K_{sp}.

For example, if we mix a $2 \times 10^{-4} M$ solution of $AgNO_3$ with an equal volume of a $2 \times 10^{-4} M$ solution of NaCl, will a precipitate form? Because the volume of the solution is doubled on mixing equal volumes, the concentrations of each of the ions will be halved. Therefore, initially

$$[Ag^+] = 1 \times 10^{-4} \text{ mol L}^{-1} \quad \text{and} \quad [Cl^-] = 1 \times 10^{-4} \text{ mol L}^{-1}$$

Hence

$$Q = [Ag^+][Cl^-] = 1 \times 10^{-8} \text{ mol}^2 \text{ L}^{-2}$$

From Table 15.10 we find

$$K_{sp}(\text{AgCl}) = 1.8 \times 10^{-10} \text{ mol}^2 \text{ L}^{-2}$$

Thus $Q > K_{sp}$, so we conclude that a precipitate will form.

$=$ *Example 15.3* $=$

PREDICTING THE FORMATION OF A PRECIPITATE

Equal volumes of a $2 \times 10^{-3}M$ $\text{Pb(NO}_3)_2$ solution and a $2 \times 10^{-3}M$ NaI solution are mixed. Will a precipitate of PbI_2 form?

Solution

Since equal volumes of the two solutions are mixed, the resulting concentration of Pb^{2+} will be $1 \times 10^{-3}M$ and the resulting concentration of I^- will be $1 \times 10^{-3}M$. Hence

$$Q = [\text{Pb}^{2+}][\text{I}^-]^2 = (1 \times 10^{-3})(1 \times 10^{-3})^2 = 1 \times 10^{-9} \text{ mol}^3 \text{ L}^{-3}$$

From Table 15.10

$$K_{sp} = 7.9 \times 10^{-9} \text{ mol}^3 \text{ L}^{-3}$$

Therefore

$$Q < K_{sp}$$

and we conclude that no precipitate will form.

EXERCISE 15.11

A 25.0-mL portion of a $0.02M$ Na_2SO_4 aqueous solution is added to 50.0 mL of $0.01M$ $\text{Ba(NO}_3)_2$ solution. Will a precipitate of BaSO_4 form?

COMMON-ION EFFECT

Suppose we have a saturated solution of lead chloride in equilibrium with the solid salt:

$$\text{PbCl}_2(s) \rightleftharpoons \text{Pb}^{2+}(aq) + 2\text{Cl}^-(aq)$$

If we now add sodium chloride or some other soluble chloride to the solution and thus increase the concentration of chloride ion, then according to Le Châtelier's principle, the equilibrium will shift to the left. In other words, more lead chloride will precipitate, and the concentration of lead ions in solution will decrease. The solubility of lead chloride is less in a solution containing chloride ion than it is in pure water. The solubility of any salt is decreased in the presence of another salt that has a common ion. This decrease in solubility is an example of the **common-ion effect**. Any equilibrium in which ions are formed is shifted to the left in a solution that contains an ion common with one of those formed in the equilibrium.

= *Example 15.4* ==

COMMON-ION EFFECT

What is the solubility of $PbCl_2$ in a $1.00M$ HCl solution?

Solution

Let the solubility of $PbCl_2$ in $1.00M$ HCl be x mol L^{-1}. The concentration of Pb^{2+} in the solution is then x mol L^{-1}, and the concentration of Cl^- is 1.00 mol L^{-1} (from the HCl) plus $2x$ (from the $PbCl_2$). Thus we have

$$PbCl_2(s) \rightleftharpoons Pb^{2+}(aq) + 2Cl^-(aq)$$

Equilibrium concentrations $\qquad\qquad\qquad x \qquad\quad 1.00 + 2x \qquad$ mol L^{-1}

The solubility product expression is

$$K_{sp} = ([Pb^{2+}][Cl^-]^2)_{eq}$$

and the value for the solubility product, from Table 15.10, is 1.7×10^{-5} mol^3 L^{-3}. Substituting this value and the equilibrium concentrations, we have

$$x(1.00 + 2x)^2 \text{ mol}^3 \text{ L}^{-3} = 1.7 \times 10^{-5} \text{ mol}^3 \text{ L}^{-3}$$

Since x will be very small, we can assume that $2x \ll 1.00$ and therefore, to a good approximation, $1.00 + 2x \approx 1.00$. Hence

$$x(1.00)^2 = 1.7 \times 10^{-5}$$
$$x = 1.7 \times 10^{-5}$$
$$[Pb^{2+}] = 1.7 \times 10^{-5} \text{ mol L}^{-1}$$

The Pb^{2+} concentration is decreased from $1.6 \times 10^{-2}M$ in water (see Example 15.2) to $1.7 \times 10^{-5}M$ in $1M$ HCl. Because of the common-ion effect, the solubility of $PbCl_2$ in $1.00M$ HCl is reduced to about a thousandth of its solubility in pure water. Note that our assumption that $2x(= 3.4 \times 10^{-5}) \ll 1.0$ is justified.

EXERCISE 15.12

What is the solubility of PbI_2 in a $0.100M$ HI solution?

The common-ion effect can be useful if we wish to ensure that we have almost completely removed an ion from solution. Radium, the heaviest of the alkaline earth metals, is very rare and highly radioactive. It could therefore be important to remove as much Ra^{2+} as possible from a solution. Since $RaSO_4$ has a very low solubility (6×10^{-6} mol L^{-1}) the concentration of Ra^{2+} could be decreased to 6×10^{-6} mol L^{-1} by precipitating $RaSO_4$. Although this solubility is small, it is not negligible, particularly in view of the fact that radium is highly radioactive. But the concentration of Ra^{2+} in solution is easily reduced to a much lower value by using an excess of SO_4^{2-} to precipitate $RaSO_4$.

Let us assume that we use sulfuric acid to precipitate $RaSO_4$ and that we add sufficient sulfuric acid so that the final concentration of SO_4^{2-} in solution is 1 mol L^{-1}. We can calculate the concentration of Ra^{2+} in solution if we know the solubility product constant for $RaSO_4$. We can find it from the solubility of $RaSO_4$ in water, which is 6.0×10^{-6} mol L^{-1}. Therefore

$$[Ra^{2+}] = [SO_4{}^{2-}] = 6.0 \times 10^{-6} \text{ mol } L^{-1}$$

Hence

$$K_{sp} = [Ra^{2+}][SO_4{}^{2-}] = (6.0 \times 10^{-6})^2 \text{ mol}^2 \text{ } L^{-2} = 3.6 \times 10^{-11} \text{ mol}^2 \text{ } L^{-2}$$

We have supposed that the final concentration of $SO_4{}^{2-}$ is 1.00 mol L^{-1} after the $RaSO_4$ is precipitated by adding excess H_2SO_4. Therefore

$$[Ra^{+2}] = \frac{K_{sp}}{[SO_4{}^{2-}]} = \frac{3.6 \times 10^{-11} \text{ mol}^2 \text{ } L^{-2}}{1.00 \text{ mol } L^{-1}} = 3.6 \times 10^{-11} \text{ mol } L^{-1}$$

Because of the common-ion effect of the excess $SO_4{}^{2-}$, the solubility of $RaSO_4$ is reduced to 4×10^{-11} mol L^{-1}, which is approximately a hundred thousandth of its solubility in pure water.

DEPENDENCE OF SOLUBILITY ON pH

An insoluble precipitate may be dissolved by decreasing the concentration of one of the ions in equilibrium so that the ion product becomes less than the solubility product constant. One way of removing one of the ions is by the formation of a weak acid or base. For example, insoluble carbonates, such as $CaCO_3$, are soluble in acids because $CO_3{}^{2-}$ combines with the added H_3O^+ to form $HCO_3{}^-$ and H_2CO_3:

$$CO_3{}^{2-} + H_3O^+ \rightleftharpoons HCO_3{}^- + H_2O$$
$$HCO_3{}^- + H_3O^+ \rightleftharpoons H_2CO_3 + H_2O$$

Then H_2CO_3 decomposes to give H_2O and CO_2:

$$H_2CO_3 \rightleftharpoons H_2O + CO_2$$

The combined effect of these reactions is to decrease the concentration of $CO_3{}^{2-}$ so that the ion product, $Q = [Ca^{2+}][CO_3{}^{2-}]$, becomes smaller than the solubility product constant, $K_{sp}(CaCO_3)$. Hence the equilibrium

$$CaCO_3(s) \rightleftharpoons Ca^{2+}(aq) + CO_3{}^{2-}(aq)$$

shifts to the right. In other words, $CaCO_3$ dissolves until equilibrium is reestablished with a higher concentration of Ca^{2+}. We can see from the distribution diagram for carbonic acid (Figure 15.5) that $CO_3{}^{2-}$ is completely converted to $HCO_3{}^-$ and H_2CO_3 in an acid solution (pH < 7) and the $CO_3{}^{2-}$ concentration

FIGURE 15.5
Distribution Diagram for
Carbonic Acid.

Carbonic acid, H_2CO_3, dissociates in water to produce the $HCO_3{}^-$ and $CO_3{}^{2-}$ ions. The distribution diagram shows how the concentration of each of these species, expressed as a fraction of the total concentration of all three species, varies with the pH of the solution.

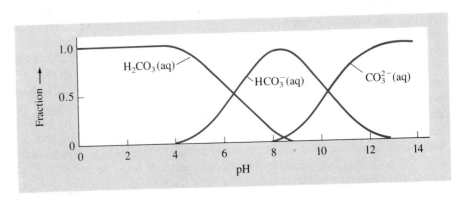

is negligibly small. Thus $CaCO_3$ is completely soluble in an acid solution and CO_2 bubbles off from the solution as a consequence of the decomposition of H_2CO_3. We recall from Chapter 5 that the liberation of carbon dioxide from a carbonate is one of the characteristic properties of an acid.

The solubility of a salt of any weak acid such as a fluoride or a sulfide depends on the pH of the solution in a similar way. The solubility of a salt of a strong acid such as a chloride, a bromide, or a sulfate does not depend on the pH.

═ Example 15.5 ═

SOLUBILITY OF THE SALT OF A WEAK ACID IN AN ACIDIC SOLUTION

Calcium fluoride is insoluble in water but dissolves in dilute hydrochloric acid. Explain why and write an equation for the overall reaction.

Solution

In order for CaF_2 to dissolve, either the F^- or Ca^{2+} must be removed by reaction with H_3O^+ or Cl^-, thus shifting the equilibrium.

$$CaF_2(s) \rightleftharpoons Ca^{2+}(aq) + 2F^-(aq)$$

to the right. Because Ca^{2+} is not a base—it has no unshared electron pairs to accept a proton—it does not react with H_3O^+. But because HF is a weak acid, F^- is a weak base; therefore it reacts with H_3O^+ to form HF:

$$F^-(aq) + H_3O^+(aq) \rightleftharpoons HF(aq) + H_2O(l)$$

The position of this equilibrium lies to the right as we can see from its equilibrium constant

$$K = \frac{1}{K_a(HF)} = \frac{1}{3.5 \times 10^{-4}} = 2.9 \times 10^3 \text{ mol L}^{-1}$$

So the F^- concentration is reduced to a much smaller value and therefore the equilibrium

$$CaF_2(s) \rightleftharpoons Ca^{2+}(aq) + 2F^-(aq)$$

shifts to the right and more CaF_2 dissolves. Adding the above two equations (after multiplying the first by 2) gives the equation for the overall reaction:

$$CaF_2(s) + 2H_3O^+(aq) \longrightarrow 2HF(aq) + 2H_2O(l) + Ca^{2+}(aq)$$

or, alternatively:

$$CaF_2(s) + 2HCl(aq) \longrightarrow 2HF(aq) + CaCl_2(aq)$$

SELECTIVE PRECIPITATION

Because they are the salts of the weak acid H_2S, metal sulfides can be dissolved in acid:

$$MS(s) + 2H_3O^+(aq) \longrightarrow M^{2+}(aq) + H_2S(g) + 2H_2O(l)$$

The more insoluble the sulfide—in other words, the smaller its solubility product constant—the higher the hydronium ion concentration that will be needed to dissolve it. Figure 15.6 shows the dependence of the solubilities of FeS

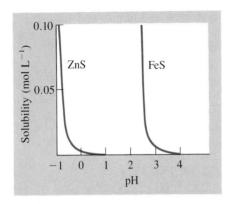

FIGURE 15.6
Solubilities of ZnS and FeS
as a Function of pH.

FeS dissolves at a higher pH
than that which is needed to
dissolve ZnS.

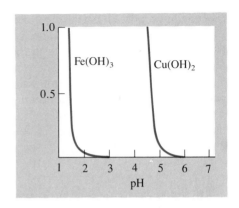

FIGURE 15.7
Solubilities of Cu(OH)$_2$ and
Fe(OH)$_3$ as a Function of pH.

Cu(OH)$_2$ dissolves at a higher
pH than that which is needed to
dissolve Fe(OH)$_3$.

($K_{sp} = 8 \times 10^{-19}$) and ZnS ($K_{sp} = 3 \times 10^{-24}$) on the pH of the solution. If acid is added to a mixture of ZnS(s) and FeS(s), the FeS will dissolve first at a pH between 3 and 4. Or if we were to pass H$_2$S into a solution of Fe^{2+} and Zn^{2+} at a pH of 2.5, for example, only ZnS would be precipitated, leaving Fe^{2+} in solution. The separation of metal ions by selective precipitation of metal sulfides and other insoluble salts in this way is an important technique in *qualitative analysis*, which is part of the laboratory work in many introductory chemistry courses.

The solubility of metal hydroxides similarly depends on the pH. The equilibrium between a solid hydroxide such as Cu(OH)$_2$ and its ions,

$$Cu(OH)_2(s) \rightleftharpoons Cu^{2+}(aq) + 2OH^-(aq)$$

is shifted to the right by the removal of OH$^-$ on the addition of an acid. We can find how the solubility of Cu(OH)$_2$ depends on the pH of the solution as follows. The solubility equilibrium for Cu(OH)$_2$ is

$$Cu(OH)_2(s) \rightleftharpoons Cu^{2+}(aq) + 2OH^-(aq)$$

The solubility product constant is

$$K_{sp} = ([Cu^{2+}][OH^-]^2)_{eq} = 4.8 \times 10^{-20} \text{ mol}^3 \text{ L}^{-3}$$

which we can rearrange as

$$[Cu^{2+}] = \frac{4.8 \times 10^{-20} \text{ mol}^3 \text{ L}^{-3}}{[OH^-]^2}$$

Now since $K_w = [H_3O^+][OH^-] = 1.0 \times 10^{-14} \text{ mol}^2 \text{ L}^{-2}$, then

$$[OH^-] = \frac{1.0 \times 10^{-14}}{[H_3O^+]}$$

Therefore

$$[Cu^{2+}] = \frac{(4.8 \times 10^{-20} \text{ mol}^3 \text{ L}^{-3})[H_3O^+]^2}{(1.0 \times 10^{-14})^2 \text{ mol}^4 \text{ L}^{-4}} = (4.8 \times 10^8 \text{ mol}^{-1} \text{ L})[H_3O^+]^2$$

At pH = 6, $[H_3O^+] = 10^{-6}$ mol L^{-1}, and therefore $[Cu^{2+}] = 4.8 \times 10^{-4}$ mol L^{-1}. In other words, the solubility of $Cu(OH)_2$ is only 4.8×10^{-4} mol L^{-1}. But at pH = 5.0, $[H_3O^+] = 10^{-5}$ mol L^{-1}, and therefore $[Cu^{2+}] = 4.8 \times 10^{-2}$ mol L^{-1}. The solubility of $Cu(OH)_2$ at pH 5 is therefore 0.048 mol L^{-1}, which is much greater than the solubility at pH = 6.0. Figure 15.7 shows a plot of the solubility of $Cu(OH)_2$ and of $Fe(OH)_3$ against the pH. We can see that if we were to adjust the pH to 4.0, $Cu(OH)_2$ would dissolve, but $Fe(OH)_3$, which has a smaller solubility product, would remain undissolved. In this way we could separate a mixture containing Cu^{2+} and Fe^{3+}.

═ *Example 15.6* ═══════════════════════════

DEPENDENCE OF SOLUBILITY ON pH

What is the solubility of $Fe(OH)_3$ in an aqueous solution at pH = 3.0 and at pH = 1.2?

Solution

For $Fe(OH)_3$ the solubility equilibrium is

$$Fe(OH)_3(s) \rightleftharpoons Fe^{3+}(aq) + 3OH^-(aq)$$

From Table 15.10 the solubility product constant is

$$K_{sp} = ([Fe^{3+}][OH^-]^3)_{eq} = 1.6 \times 10^{-39} \text{ mol}^3 \text{ L}^{-3}$$

Following the same argument we used for copper hydroxide, we deduce the equation

$$[Fe^{3+}] = (1.6 \times 10^3 \text{ mol}^{-2} \text{ L}^2)[H_3O^+]^3$$

At pH 3.0, $[H_3O^+] = 1.0 \times 10^{-3}$ mol L^{-1}, and therefore

$$[Fe^{3+}] = 1.6 \times 10^{-6} \text{ mol L}^{-1}$$

At pH 1.2, $[H_3O^+] = 6.3 \times 10^{-2}$, and therefore

$$[Fe^{3+}] = 0.40 \text{ mol L}^{-1}$$

Thus at pH 3.0 the solubility of $Fe(OH)_3$ is only 1.6×10^{-6} mol L^{-1}, but at pH 1.2 it is 0.40 mol L^{-1}.

───

We saw in Chapter 10 that a precipitate of $Cu(OH)_2$ will dissolve in concentrated aqueous ammonia to give a deep blue solution that contains the $Cu(NH_3)_4^{2+}$ ion. This ion is analogous to a hydrated ion, except that NH_3 molecules replace the water molecules. It is called a *complex ion* and is discussed in more detail in Chapter 21. In this case the equilibrium

$$Cu(OH)_2(s) \rightleftharpoons Cu^{2+}(aq) + 2OH^-(aq)$$

is shifted to the right because of the decrease in the concentration of $Cu^{2+}(aq)$ caused by the formation of $Cu(NH_3)_4^{2+}$ on the addition of NH_3 to the solution.

EXERCISE 15.13

What is the solubility of $Pb(OH)_2$ in aqueous solutions of pH 7.5 and pH 6.5?

IMPORTANT TERMS

The **common-ion effect** is the effect on a system at equilibrium caused by the addition of a substance with an ion common with one involved in the equilibrium.

The **solubility product constant**, K_{sp}, for a slightly soluble ionic substance is the equilibrium constant for the equilib-

rium between the solid and the dissolved ions in a saturated solution. It is the product of the concentrations of the ions in the saturated solution raised to the powers indicated by the coefficients in the balanced equation for the equilibrium.

PROBLEMS *

Group 1 and Group 2 Metals: Properties and Reactions

1. Why does the reactivity of the alkali metals increase with increasing atomic number? Comment on the relative reactivities of the alkali metals in their reactions with water.

2. Write balanced equations for the reactions of lithium and calcium, respectively, with **(a)** bromine, **(b)** sulfur, and **(c)** nitrogen. Name each of the products and write their Lewis structures.

3. Complete and balance each of the following equations, and name each of the products:

(a) $K(s) + Br_2(l) \rightarrow$ (b) $Li(s) + O_2(g) \rightarrow$

(c) $Na(s) + H_2(g) \rightarrow$ (d) $Li(s) + N_2(g) \rightarrow$

(e) $LiH(s) + H_2O(l) \rightarrow$ (f) $Li_3N(s) + H_2O(l) \rightarrow$

(g) $K(s) + H_2O(l) \rightarrow$

4. Explain why an aqueous solution of sodium hydrogen carbonate can act as a buffer solution. What is the common name for sodium hydrogen carbonate?

5. What is lime? What happens when water is added to lime? Why is lime such an important industrial chemical?

6. Write balanced equations for the reactions of magnesium with **(a)** oxygen, **(b)** sulfur, and **(c)** nitrogen.

7. What volume of hydrogen gas at STP would result from the reaction of 2.00 g of calcium hydride with excess water?

8. Explain why the first ionization energy of sodium (0.50 MJ mol^{-1}) is lower that the first ionization energy of magnesium (0.74 MJ mol^{-1}), but the second ionization

energy of sodium (4.56 MJ mol^{-1}) is much greater than the second ionization energy of magnesium (1.45 MJ mol^{-1}).

9. Which of the following ionic compounds of sodium react with water? Write a balanced equation for each reaction.

(a) Sodium chloride   (b) Sodium hydride

(c) Sodium hydroxide   (d) Sodium oxide

(e) Sodium sulfate   (f) Sodium carbonate

10. Complete and balance each of the following equations:

(a) $Mg(s) + Cl_2(g) \rightarrow$ (b) $Ca(s) + O_2(g) \rightarrow$

(c) $Sr(s) + H_2(g) \rightarrow$ (d) $Mg(s) + H_2(g) \rightarrow$

(e) $Ca(s) + H_2O(l) \rightarrow$

11. How do the solubilities of the hydroxides of the alkaline earth metals vary from the top to the bottom of the group, and why?

12. Explain with the aid of appropriate balanced equations how limestone caverns, and the stalactites and stalagmites found in these caverns, are formed.

13. Write balanced equations for the reactions that occur when each of the following compounds is heated, and name the products:

(a) Calcium carbonate   (b) Calcium hydroxide

(c) Sodium hydrogen carbonate

(d) Magnesium chloride hexahydrate

14. A mixture contains calcium carbonate, calcium hydrogen carbonate, and calcium oxide. When 10.00 g of this mixture are heated to constant mass, 0.200 g of water and 1.500 g of CO_2 are obtained. What is the composition of the mixture?

15. Seawater contains 0.05 mol L^{-1} of $Mg^{2+}(aq)$. How many liters of seawater would have to be processed to yield 1 metric ton (10^3 kg) of magnesium, if the extraction process is 70% efficient?

16. What is the maximum amount of washing soda that can be obtained from 1 metric ton (10^3 kg) of trona ore?

17. Write equations for the acid–base reactions that occur when soluble compounds containing the following anions are dissolved in water:

(a) O^{2-} (b) H^- (c) N^{3-} (d) CO_3^{2-}

18. Give the empirical formulas and write the Lewis structures of

(a) The hydride of francium

(b) An oxide of cesium

(c) An oxide of barium

(d) The sulfate of radium

(e) The sulfides of sodium and lithium

19. When the salt $MgCl_2 \cdot 6H_2O$ is heated, the products are $MgO(s)$, $HCl(g)$, and $H_2O(g)$, while the action of heat on the salt $BaCl_2 \cdot 2H_2O$ gives anhydrous $BaCl_2(s)$ and water. Account for this difference in behavior.

20. When 1.00 g of a white solid, A, was strongly heated, another white solid, B, and 125 mL of a gas, C, were obtained at 750 mm Hg and 25 °C. When the gas, C, was bubbled into a solution of $Ca(OH)$, another white solid, D, was obtained. A solution of the white solid B in water turned red litmus paper blue. When dilute aqueous HCl was added to the solution until it turned blue litmus paper red and the solution then evaporated to dryness, a white solid, E, was obtained. When the white solid E was placed in a bunsen flame it colored the flame green. Finally, if dilute sulfuric acid was added to an aqueous solution of E a white precipitate, F, was formed. Identify the compounds, A,B,C,D,E and F.

21. An approximately circular lake with a diameter of 4.2 km, and an average depth of 8.2 m, has a pH of 3.9. How much limestone should be added to the lake to raise its pH to 6.3, assuming that no carbon dioxide is lost to the atmosphere?

22. (a) Give the names and empirical formulas of three compounds of calcium that occur naturally.

(b) Under what conditions does calcium metal react with (i) hydrogen, (ii) water, and (iii) nitrogen?

(c) Write a balanced equation for each of the reactions in part (b), name the products, and write their Lewis structures.

23. Explain each of the following characteristic properties of the group 1 alkali metals:

(a) They are reducing agents.

(b) Their compounds are ionic.

(c) Their oxides behave as strong bases in water

(d) Their salts of strong acids give neutral aqueous solutions (pH 7).

24. Explain each of the following characteristic properties of the group 2 alkaline earth metals:

(a) They are reducing agents, and they increase in reactivity down the group.

(b) With the exception of some beryllium compounds, nearly all their compounds are ionic.

(c) Their hydroxides increase in solubility in water in descending the group, from $Be(OH)_2$ and $Mg(OH)_2$, which are very insoluble to $Ba(OH)_2$, which is soluble.

(d) With the exception of some beryllium and magnesium salts, their salts with the anions of strong acids give neutral solutions (pH 7).

25. Among the alkaline earth metals, beryllium is exceptional in that many of its simple compounds are covalent, its salts, such $BeCl_2 \cdot 4H_2O$, give acidic aqueous solutions, and its hydroxide, $Be(OH)_2$, is amphoteric. Account for these properties of beryllium.

26. Why is it that magnesium, in contrast with calcium, reacts very slowly with cold water, but both metals react rapidly with dilute hydrochloric acid?

27. Explain why magnesium carbonate, which is insoluble in water, dissolves readily when carbon dioxide is bubbled through a suspension of $MgCO_3(s)$ in water.

28. Write balanced equations for the reaction of calcium with each of the following reactants. Describe the conditions for reaction, and name each of the products:

(a) Water   (b) Chlorine   (c) Hydrogen

(d) Oxygen   (e) Hydrogen bromide

Solubility and Solubility Product Constants

29. Write the solubility product constant expressions for each of the following in aqueous solution, and state the units in each case: (a) $Fe(OH)_3(s)$ (b) $Ca_3(PO_4)_2(s)$

30. What are the solubility product constant expressions, and the units of K_{sp}, for each of the following?

(a) $AgCl(s)$ **(b)** $BaF_2(s)$ **(c)** $Cr(OH)_3(s)$

(d) $Bi_2S_3(s)$ **(e)** $Cu(OH)_2(s)$

31. Write the solubility product constant expressions for the following:

(a) The fluorides of silver, calcium, and lead

(b) The hydroxides of silver, magnesium, and aluminium

(c) The sulfates of silver and strontium

32. The amount of lead sulfate that dissolves in 2.00 L of water at 25 °C is 0.060 g. What is the solubility product constant of lead sulfate?

33. The solubility of magnesium fluoride in water is 0.0012 mol L^{-1}. What is the solubility product constant for magnesium fluoride?

34. The solubilities of lead sulfide, calcium fluoride, and chromium(III) hydroxide in water at 25 °C are 4.41 pg L^{-1}, 16.8 mg L^{-1}, and 5.62 µg L^{-1}, respectively. What are the values of their solubility product constants at 25 °C?

35. Use the data of Table 15.10 to calculate the solubility in water at 25 °C of each of the following:

(a) Magnesium carbonate   **(b)** Silver chloride

(c) Aluminum hydroxide   **(d)** Lead(II) iodide

36. The solubility product constant for silver chloride is 1.8×10^{-10} mol^2 L^{-2} at 25 °C. What mass of silver chloride will give a saturated solution of silver chloride in 250 mL of water at 25 °C?

37. The solubility product constant of lead(II) hydroxide is 6×10^{-16} mol^3 L^{-3} at 25 °C. What is the molar concentration of $Pb^{2+}(aq)$ in a saturated solution of lead(II) hydroxide at 25 °C?

38. Is a precipitate of lead(II) chloride expected to form at 25 °C when 50 mL of $0.10M$ $Pb(NO_3)_2(aq)$ is mixed with 100 mL of $0.05M$ $NaCl(aq)$?

39. Using data from Table 15.10, answer each of the following:

(a) Will $Mg(OH)_2(s)$ be precipitated when 10.0 mL of $0.10M$ $MgCl_2(aq)$ is mixed with 10.0 mL of $2.0M$ $NH_3(aq)$?

(b) Will $Sr(OH)_2(s)$ be precipitated when 10.0 mL of $0.10M$ $SrCl_2(aq)$ is mixed with 10.0 mL of $2.0M$ $NH_3(aq)$?

(c) What are the solubilities of $Mg(OH)_2(s)$ and $Sr(OH)_2(s)$ in a $0.10M$ $NH_4Cl(aq)$–$0.10M$ $NH_3(aq)$ buffer solution?

40. The solubility product constants of Fe(II) hydroxide and Fe(III) hydroxide are 7.9×10^{-15} mol^3 L^{-3} and 1.6×10^{-39} mol^4 L^{-4}, respectively. Which of these hydroxides is the least soluble in water?

41. Will a precipitate form when 50 mL of $1.00M$ $KOH(aq)$ is mixed with each of the following:

(a) 50 mL of $0.0010M$ $BaCl_2(aq)$

(b) 50 mL of $1.00M$ $BaCl_2(aq)$

42. 50 mL of $2.00M$ $NaCl(aq)$ is mixed with 50 mL of $0.020M$ $AgNO_3(aq)$.

(a) What mass of silver chloride is precipitated from the solution?

(b) What concentration of silver ions remains in solution after the precipitate has formed?

43. (a) Is a precipitate formed when 50 mL of $2.00M$ $Pb(NO_3)_2(aq)$ is mixed with 50 mL of $4.00 \times 10^{-3}M$ $NaI(aq)$?

(b) If a precipitate is formed, what is it, and what mass of solid is obtained?

(c) What are the concentrations of $Pb^{2+}(aq)$, $I^-(aq)$, $NO_3^-(aq)$, and $Na^+(aq)$ that remain in the solution?

Common-Ion Effect

44. Calculate the solubility of calcium fluoride at 25 °C in each of the following:

(a) Pure water   **(b)** $0.010M$ $CaCl_2(aq)$

(c) $0.100M$ $NaF(aq)$

45. What mass of barium fluoride will dissolve at 25 °C in 1.00 L of a solution that is already $0.20M$ in $Ba^{2+}(aq)$ ions?

46. An aqueous solution is prepared by dissolving 1.0×10^{-8} mol of $MgCO_3(s)$ in 2.00 L of water at 25 °C. How many moles of the soluble salt $MgCl_2(s)$ can be dissolved in this solution before $MgCO_3(s)$ begins to precipitate?

47. Predict whether the amount of $Fe(OH)_3(s)$ in equilibrium with its saturated solution will increase, or decrease, under each of the following conditions:

(a) Addition of $KOH(aq)$

(b) Addition of $HCl(aq)$

48. What is the solubility of calcium phosphate in each of the following solutions?

(a) $0.10M$ $CaCl_2(aq)$ (b) $0.10M$ $Na_3PO_4(aq)$

(c) Pure water

49. An aqueous solution is $0.0010M$ in fluoride ion and $0.010M$ in carbonate ion. A concentrated solution of magnesium chloride is added. Which precipitates first, magnesium fluoride or magnesium carbonate?

50. Initially a solution is $0.10M$ in both fluoride ion and sulfate ion. If Pb^{2+}(aq) is added in the form of a soluble salt, which precipitates first, lead sulfate or lead fluoride? What percentage of one anion remains in solution when the other begins to precipitate as its lead salt?

51. The solubility of lead(II) chloride in water is 0.016 mol L^{-1} at 25 °C. Why, in separating lead from other cations in qualitative analysis, is $PbCl_2(s)$ precipitated in $3M$ $HCl(aq)$?

Solubility and pH

52. What is the minimum pH at which a precipitate of $Fe(OH)_2(s)$ will form at 25 °C in a $0.005M$ $FeCl_2(aq)$ solution?

53. Milk of magnesia is an aqueous suspension of magnesium hydroxide. What is the pH of milk of magnesia?

54. For which of the following compounds does the solubility increase when the pH of the solution is decreased?

(a) Copper(II) sulfide   (b) Silver iodide

(c) Magnesium carbonate   (d) Copper(II) hydroxide

(e) Calcium fluoride   (f) Lead(II) sulfate

55. What is the solubility of lead(II) hydroxide in an aqueous solution buffered at a pH of 8.0 at 25 °C?

56. At pH = 8.0 at 25 °C, the solubility of nickel(II) hydroxide is $0.0020M$. What is the solubility product constant of nickel(II) hydroxide? What is the solubility of the hydroxide at a pH of 0?

57. What are the solubilities at 25 °C of copper(II) hydroxide and lead(II) hydroxide in an aqueous solution buffered at a pH of 5.5? Could $Cu(OH)_2(s)$ be separated from $Pb(OH)_2(s)$ at this pH?

58. Calculate the molar solubility of zinc hydroxide in water and in $0.100M$ $NaOH(aq)$ at 25 °C.

59. A saturated solution of hydrogen sulfide in water has a concentration of 0.10 mol L^{-1} at 25 °C, and the acid dissociation constant for the equilibrium

$$H_2S(aq) + 2H_2O \rightleftharpoons 2H_3O^+(aq) + S^{2-}(aq)$$

* The asterisk denotes the more difficult problems.

has the value 1.0×10^{-19} mol^2 L^{-2} at 25 °C. What are the maximum concentrations of (a) Cu^{2+}(aq), (b) Fe^{2+}(aq), (c) Pb^{2+}(aq), and (d) Ag^+(aq) that can exist in solutions saturated with H_2S and maintained at a pH of 4.00?

60. Calculate the pH of saturated solutions of each of the following at 25 °C:

(a) $Al(OH)_3(s)$ (b) $Ca(OH)_2(s)$ (c) $Sr(OH)_2(s)$

(d) $Mg(OH)_2(s)$ (e) $Fe(OH)_2(s)$

Miscellaneous

***61.** An aqueous solution containing 0.203 g of hydrated magnesium chloride, $MgCl_2 \cdot xH_2O$, was titrated with $AgNO_3(aq)$. Complete precipitation of all the chloride ion in solution required 20.0 mL of $0.10M$ $AgNO_3(aq)$. On heating another sample of the hydrated magnesium chloride to constant mass in a stream of $HCl(g)$, the reduction in mass was 53.2% of the original mass. On heating to constant mass *in air*, the reduction in mass of another sample was 80.2% of the original mass.

(a) What is the empirical formula of hydrated magnesium chloride?

(b) What is the product when the hydrated salt is heated in $HCl(g)$?

(c) What is the product when the hydrated salt is heated in air?

(d) Explain the reason for the difference between the reactions in part (b) and part (c).

62. A sample of limestone contained both calcium carbonate and magnesium carbonate. On heating 2.634 g of the limestone to constant mass, the residue had a mass of 1.288 g. What was the mass percent of $CaCO_3(s)$ in the limestone sample?

63. How would you prepare a sample of each of the following in the laboratory?

(a) $MgSO_4 \cdot 7H_2O(s)$ from $MgCO_3(s)$

(b) $BaSO_4(s)$ from $BaCO_3(s)$

***64.** The following are among the methods commonly used for the extraction of metals from their ores:

(a) Reduction of an oxide with carbon (coke)

(b) Electrolysis of an aqueous solution of a chloride

(c) Electrolysis of a molten chloride

(d) Strongly heating an oxide in air

(e) Reduction of a chloride with a more reactive metal

Discuss the preparation of each of sodium and calcium with reference to each of the above methods.

65. Among the alkali metals, soluble lithium and sodium salts are often hydrated, while those of the heavier alkali metals are usually anhydrous. In contrast, nearly all the salts of alkaline earth metals are hydrated at room temperature. Discuss the above observations in terms of ionic radii and charges on the alkali and alkaline earth metal cations.

66. Unlike the other sulfates of the group 2 metals, beryllium sulfate gives a solution in water that is acidic and from which the salt $BeSO_4 \cdot 4H_2O$ can be crystallized. Explain these observations.

*67. A sample of each of the following is provided in separate unlabeled bottles:

   (a) Ammonium carbonate

   (b) Barium carbonate

   (c) Sodium hydrogen sulfate

   (d) Sodium carbonate

   (e) Ammonium chloride

   (f) Calcium nitrate

   (g) Copper(II) nitrate

   (h) Sodium chloride

   (i) Silver nitrate

You are provided with test tubes, indicator paper, water, and a bunsen burner, but no reagents except the samples themselves. Describe a series of procedures that would enable you to label all the bottles correctly.

68. (a) Write a balanced equation for the reaction that occurs when limestone dissolves in rainwater.

   (b) What reaction occurs, and what is observed, when the above solution is boiled?

   (c) To what is the "hardness" of water due?

   (d) How can "hard" water be "softened"?

69. **(a)** $Na_2SO_4(aq)$ is added to a solution containing $1.00M$ $BaCl_2(aq)$ and $0.001M$ $SrCl_2(aq)$. Which will precipitate first, $BaSO_4(s)$ or $SrSO_4(s)$?

   (b) $Na_2SO_4(aq)$ is added to a solution containing $0.001M$ $BaCl_2(aq)$ and $1.00M$ $SrCl_2(aq)$. Which will precipitate first, $BaSO_4(s)$ or $SrSO_4(s)$?

70. Which alkali and alkaline earth compounds can be identified with respect to the cations they contain from the colors that they impart to flames?

*71. When a pellet of solid sodium hydroxide is placed on a watch glass and left exposed to the atmosphere, it gradually

changes to a colorless liquid. After a longer exposure, it first changes to colorless transparent crystals and then into a white opaque solid with a mass 55% greater than the original pellet. Explain all the above changes.

72. **(a)** In which of the following solutions will lead(II) chloride be the *least* soluble at 25 °C?

   (i) Pure water   (ii) $0.10M$ $NaCl(aq)$

   (iii) $0.20M$ $Pb(NO_3)_2(aq)$

   (b) What is meant by the common-ion effect?

   (c) Explain the common-ion effect in terms of Le Châtelier's principle.

73. The solubility product constants of silver chloride are 2.15×10^{-10} $mol^2\,L^{-2}$ at 100 °C and 2.10×10^{-11} $mol^2\,L^{-2}$ at 5 °C. If exactly 1 L of AgCl(aq) solution saturated with silver chloride at 100 °C is cooled to 5 °C, what mass of AgCl(s) will crystallize out?

74. Explain each of the following:

   (a) The reactivities of the group 1 and group 2 metals increase in descending each group.

   (b) The melting points of the alkali metals are rather low, but the boiling points are much higher than the melting points.

   (c) The alkali metals have much lower densities than the alkaline earth metals and metals such as aluminum and copper.

   (d) Almost all the alkali metal salts are soluble in water, but the least soluble of these salts is lithium fluoride.

75. Write balanced equations for the reaction of calcium oxide with each of the following oxides, and name the products:

   (a) Water   **(b)** Carbon dioxide

   (c) Sulfur dioxide   **(d)** Silica

   (e) Phosphorus(V) oxide

76. Explain why magnesium oxide dissolves readily in dilute HCl(aq) or dilute $HNO_3(aq)$ whereas it is insoluble in water.

77. In the qualitative analysis of cations, $Pb^{2+}(aq)$, $Ag^+(aq)$, and $Hg_2^{2+}(aq)$ are precipitated from solution as their chlorides by adding $2M$ HCl(aq).

   (a) Compare the solubilities of these chlorides in (i) water and (ii) $2M$ HCl(aq), at 25 °C.

   (b) Why is the precipitation in the analysis scheme carried out using $2M$ HCl(aq)?

Thermodynamics, Entropy, and Free Energy

16

In the three preceding chapters we have studied chemical equilibria in some detail. We have seen that at any given temperature each reaction is characterized by an equilibrium constant that tells us to what extent the reactants are converted to products when equilibrium is reached. Some reactions go almost to completion; for others substantial amounts of reactants are present together with the products when equilibrium is reached; and other reactions hardly proceed at all. We have seen that the equilibrium constant is a very important and useful property of a reaction, but we have not yet attempted to answer the question: Why do some reactions have a very large equilibrium constant and therefore go to completion, whereas others have a very small equilibrium constant and hardly proceed at all? Thermodynamics provides us with an answer to this question and we take up this subject in this chapter.

One of humankind's greatest achievements was the discovery of how sources of energy other than our own muscles or those of animals can be used. The development of this knowledge created modern civilization. Most of our energy comes from the burning of fossil fuels and from nuclear reactions; these processes supply energy as heat. But most energy is needed in the form of mechanical energy to do work in order to operate a machine or propel a vehicle. The transformation of heat to work is achieved by the use of devices called engines, which use the energy stored in a fuel to perform mechanical work to drive, for example, an electric generator, an automobile, or a ship.

The science of energy and its transformations is known as *thermodynamics*. It grew out of the Industrial Revolution and the need to understand how to convert as much as possible of the heat obtained by burning a fuel into mechanical energy—in other words, how to improve the efficiency of an engine. In engineering the main use of thermodynamics is therefore in connection with engines. In chemistry, however, thermodynamics has another very important application; it helps us to answer the question: Why do some reactions proceed to completion, others proceed only partially, and still others hardly at all?

We will see that the direction in which a reaction proceeds is determined by both the change in *energy* and the change in *entropy* that is associated with the reaction. We studied energy changes in chemical reactions in some detail in Chapter 6 where we made use of the *first law of thermodynamics*, that is, the *law of conservation of energy*, according to which the energy of an isolated system is constant—energy can neither be created nor destroyed. In Chapter 12 we briefly mentioned that entropy is a measure of the disorder of a system and that the tendency for disorder to increase is responsible for the process of forming a solution, that is, for one substance to become mixed up with another. In this chapter we consider entropy in more detail, and then show how a quantity called the *Gibbs free energy*, which combines the energy change and the entropy change for a reaction, enables us to predict the direction in which a reaction will proceed under some given conditions.

16.1
Spontaneous Processes

Some changes in nature occur spontaneously, and some do not. A gas expands to fill the space available to it, but it does not spontaneously contract to a smaller volume. A hot object cools down to the temperature of the environment, but an object does not spontaneously get hotter than its environment. Many chemical reactions proceed spontaneously in one direction only. Hydrogen combines explosively with oxygen to give water when the reaction is initiated by a spark, but water does not decompose spontaneously to hydrogen and oxygen. Burning diamonds in oxygen gives carbon dioxide, but carbon dioxide, even when strongly heated, does not decompose to give diamonds and oxygen. Clearly, something causes some chemical reactions and certain physical processes to proceed in one direction only. Of course, these changes can be reversed: a gas *can* be compressed to a smaller volume; an object *can* be heated to a temperature above that of its environment; and water *can* be decomposed to hydrogen and oxygen by passing an electric current through it. However, in each case work must be done to accomplish these changes. They do not occur spontaneously; rather, it is the reverse process that occurs spontaneously. What is it that causes all these changes to occur spontaneously in one particular direction?

Common experience shows that spontaneous processes involving macroscopic objects proceed with a decrease in potential energy. A rock spontaneously rolls down a mountain if it is dislodged, but it never rolls up to the top again. A skier slides spontaneously down a ski run, but work must be done by the skier, or by the ski lift, if the skier is to reach the top of the run again.

Although we take it as almost self-evident that an object tends to achieve a position of minimum energy, we must remember that if something loses energy spontaneously, something must at the same time gain energy, because according to the first law of thermodynamics, the total energy of a system and its surroundings remains constant (Chapter 6). The potential energy of the skier descending the ski run is converted to kinetic energy as his speed increases and to heat because of friction between the skis and the snow. Finally, when the skier stops, all the kinetic energy is converted to heat. Although the skier has lost energy, the surroundings have gained energy, so the *total* energy has not decreased. But, in this spontaneous process, there has been a transfer of energy from the system to the surroundings.

We might therefore expect that a chemical reaction would proceed spontaneously if the reacting system decreases in energy by transferring heat to its surroundings. In other words, we might expect spontaneous reactions to be exothermic, and this is frequently the case. But there are many endothermic chemical reactions and physical processes that proceed spontaneously. For example, ammonium nitrate dissolves spontaneously in water in an endothermic process (Experiment 12.2):

$$NH_4NO_3(s) \longrightarrow NH_4^+(aq) + NO_3^-(aq) \qquad \Delta H° = 27.4 \text{ kJ}$$

Dinitrogen pentaoxide, N_2O_5, decomposes spontaneously at room temperature into NO_2 and O_2 in an endothermic reaction (Chapter 18):

$$2N_2O_5(s) \longrightarrow 4NO_2(s) + O_2(g) \qquad \Delta H° = 219.0 \text{ kJ}$$

Water left in an open container spontaneously evaporates, although the vaporization of water is an endothermic process:

$$H_2O(l) \longrightarrow H_2O(g) \qquad \Delta H° = 44.0 \text{ kJ}$$

In all these examples the enthalpy of the system increases rather than

Recall from Chapter 6 that the change in internal energy of a system, ΔE, is not in general exactly equal to the change in the enthalpy, ΔH. But in almost all cases the difference between ΔE and ΔH is quite small, and it can be neglected for our present purposes.

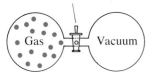

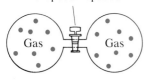

FIGURE 16.1
The Expansion of an Ideal
Gas is a Spontaneous Process.

decreases. Clearly, the direction of spontaneous change is not always determined by the tendency of a system to achieve minimum energy.

It will help us to zero in on the real reason for spontaneous processes if we consider a process in which there is no energy change at all. An ideal gas expands spontaneously at constant temperature into any space made available to it. Since the energy of an ideal gas depends only on the temperature, its energy does not change during the process, and therefore no heat is evolved or absorbed. Moreover, if the gas expands into a vacuum, no work is done by, or on, the system (see Figure 16.1). That this spontaneous expansion of a gas should occur does not surprise us because it is a familiar phenomenon; indeed, we would be very surprised if it did not occur. But *why* does it occur? It occurs because, according to the kinetic molecular theory, the molecules of a gas are moving randomly and at high speeds. They move through the valve connecting the two flasks in Figure 16.1 and thus move into the empty flask. Some molecules eventually return to the original flask, but more move from left to right than from right to left, until there are equal numbers of molecules in both flasks and equilibrium is established. Thus the spontaneity of this process is associated with the molecular nature of matter and specifically with the random motions of large numbers of molecules. This random molecular motion causes the molecules of a gas to spread out in space as much as possible. Not only does a single gas diffuse into an empty space, but two gases diffuse into each other to form a uniform mixture. They become completely mixed up with each other as a result of their random thermal motions.

In general, because of the random motions of molecules, there is a tendency for all molecules to become more dispersed in space and more mixed up with each other. In other words, there is a tendency for molecules to become more disordered. If we start with two separate gases and allow them to mix, the final state of the system is more mixed-up—that is, is more disordered—than the original state. This is quite similar to the shuffling of a pack of cards that was initially ordered according to the four suits. The process of shuffling mixes them up —that is, it increases their disorder—and continued shuffling is very unlikely to return them to the original order. The mixing up of the cards is a spontaneous process that occurs in only one direction; we never observe the "unmixing" of the cards. This "unmixing" does not occur because there are millions of ways in which a mixed-up disordered arrangement can be obtained but only one way in which the completely ordered arrangement can be obtained. Thus the probability of obtaining a disordered arrangement is very great while the probability of obtaining the disordered arrangement is very small. Thus a spontaneous process in an *isolated* system is one in which there is an increase in disorder in the system (Experiment 16.1).

For a system that is not isolated from its surroundings, we must also take into account what happens in the surroundings. For example, in an exothermic process the total energy of the system and its surroundings is constant. But energy is transferred as heat from the system to the surroundings; that is, it is dispersed in the surroundings. For example, a hot block of metal spontaneously cools down as energy is transferred as heat from the metal to its surroundings. The random thermal motion of the molecules of the surroundings is increased by an amount corresponding to the heat transferred from the system to the surroundings. The energy that was originally concentrated in the metal block becomes much more dispersed because of the very large—essentially infinite—size of the surroundings. In this spontaneous process there is no change in the

Experiment 16.1

Entropy

A red liquid and a green liquid and separate layers of red and green balls represent ordered states.

When the red and green liquids are mixed and the red and green balls are mixed, we obtain a more disordered state in each case, that is, a state of higher entropy.

No matter how long the brown mixture of the red and green liquids is stirred it will never separate into the two original red and green liquids. Similarly, if the red and green balls are stirred or shaken they will never separate into the two original layers. The two liquids or the red and green balls mix spontaneously, but they will never unmix spontaneously. Spontaneous change in an isolated system is always accompanied by an increase in disorder, randomness, or entropy.

total energy of the system and its surroundings, but there is an increase in the disorder of the surroundings.

The example of the book that is dropped on the table, which we discussed in Chapter 6, also illustrates that a spontaneous process is accompanied by an increase in disorder. When the book hits the table, the ordered motion of its molecules, which are all moving toward the table as it falls, is replaced by an increase in the chaotic, random motion of the molecules of the book and the table, which both become a little warmer. It would be quite consistent with the law of conservation of energy (the first law of thermodynamics) for the book and table to become a little cooler and for the book to jump back off the table! We know that this never happens. The change that occurs when the book drops to the table and its energy is converted to heat is a spontaneous process, whereas the reverse process is not spontaneous. There has been no change in the total energy of the book and the table in this process. What has changed is the way the energy is distributed. The energy that was originally concentrated in the book has become dispersed over the table and the book as random thermal motions. The reverse process—which would involve the spontaneous ordering of the random motions of the molecules to cause them all to move in the same direction so that the book jumps off the table—is so improbable that it is never observed, although it would not be contrary to the law of conservation of energy.

If we consider a system and its surroundings, then whether or not a change in the system is spontaneous does not depend on energy changes, because the total energy of the system and its surroundings is constant, but it depends on the change in the disorder of the system and its surroundings. *In any spontaneous process the total disorder in the system and its surroundings increases.*

16.2
Entropy

To put the above ideas on a quantitative basis, we need a quantity that is a measure of the disorder of the particles (atoms and molecules) that make up the system and the dispersal of energy associated with these particles. This quantity is called **entropy** and it is given the symbol S. The disorder in a system depends only on the conditions that determine the state of the system, such as composition, temperature, and pressure. The change in entropy therefore depends only on the initial and final states of the system. *Entropy*, like enthalpy, *is a state function.*

THE SECOND LAW OF THERMODYNAMICS

In any spontaneous process the total entropy of a system and its surroundings increases.

This law of nature is called the **second law of thermodynamics**. It follows that *the entropy of the universe is increasing.* This is an alternative statement of the second law. An analogous statement of the *first law* is that *the energy of the universe is constant.* For any spontaneous process we may write

$$\Delta S_{universe} = \Delta S_{system} + \Delta S_{surroundings} > 0$$

In other words, for any spontaneous process the total entropy change must be positive.

STANDARD MOLAR ENTROPIES

So far we have given only a qualitative description of the quantity we have called entropy. Although we cannot discuss here how quantitative values for the entropies of substances can be obtained, we can increase our understanding of the concept of entropy by considering the results of the measurement of the entropy of a substance as a function of temperature.

The entropy of oxygen as a function of temperature is shown in Figure 16.2.

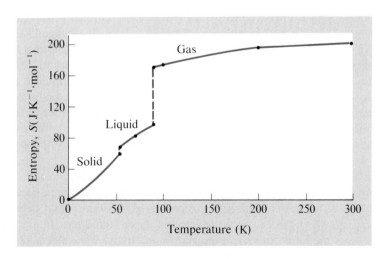

FIGURE 16.2
Entropy of Oxygen as a Function of Temperature.

At 0 K the entropy of any pure crystalline substance is zero—the molecules have a perfectly regular arrangement and no thermal motion, so there is no disorder. As the temperature increases, the molecules begin to vibrate about their mean positions—their random thermal motion begins to increase—and therefore the entropy of oxygen increases. At the melting point the regular arrangement of the molecules in the solid changes to the more random arrangement of the molecules in the liquid, so there is an abrupt increase in the entropy. Then the entropy again steadily increases with increasing temperature until the liquid boils. There is then a large increase in the volume of the oxygen and therefore in the random distribution of the molecules in space. Therefore there is a correspondingly large, abrupt increase in the entropy. The entropy of gaseous oxygen then continues to increase slowly with increasing temperature.

Values of the standard molar entropies of some substances at 25 °C are given in Table 16.1. (Additional values are given in Appendix B.) The **standard molar entropy**, $S°$, is the entropy of 1 mol of the substance in its standard state—that is, at 1 atm pressure—and at 25 °C. The units of entropy are joules per kelvin per mole ($J K^{-1} mol^{-1}$). Note that entropy values are normally given in joules, not kilojoules.

The data in Table 16.1 shows that gases in general have larger entropies than solids. In a gas, the molecules have a more random motion than in a solid, and their energy is spread over a much larger volume; that is, the energy of the molecules of a gas is more dispersed than the energy associated with the motions

Because absolute values of the entropies of substances can be measured experimentally we can use absolute rather than relative entropies and we do not need to assume that the entropies of the elements are zero. In contrast, because absolute values of the enthalpies of substances cannot be obtained we use enthalpies of formation, $\Delta H_f°$, based on the assumption that the enthalpies of formation of the elements are zero.

TABLE 16.1
Standard Molar Entropies, $S°$, at 25 °C ($J K^{-1} mol^{-1}$)*

Solids		Liquids		Gases	
C (diamond)	2.4	H_2O	70.0	H_2	130.6
C (graphite)	5.8	Hg	75.9	N_2	191.5
Fe	27.3	Br_2	152.2	F_2	202.7
S (rhombic)	32.0	CH_3OH	126.8	O_2	205.0
Cu	33.2	HNO_3	155.6	Cl_2	223.0
P (white)	41.1	CH_3CO_2H	159.8	I_2	260.6
Ag	42.6	C_2H_5OH	160.7	CH_4	86.1
I_2	116.1	CH_2Cl_2	177.8	HCl	186.8
MgO	27.0	$CHCl_3$	201.7	H_2O	188.7
CaO	38.1	CCl_4	216.4	NH_3	192.7
SiO_2 (quartz)	41.5	$SiCl_4$	239.7	H_2S	205.6
NaF	51.3	C_5H_{12}ᵃ	263.3	HI	206.5
NaCl	72.5	C_6H_6ᵇ	172.2	CO	197.6
NaBr	87.2	C_6H_{14}ᵃ	295.9	CO_2	213.7
Fe_2O_3	87.4	C_7H_{16}ᵃ	328.5	C_2H_6	229.5
$CaCO_3$	92.9	C_8H_{18}ᵃ	361.2	C_3H_8	269.9
AgCl	96.2			C_4H_{10}ᵃ	310.1
NaI	98.5			C_5H_{12}ᵃ	348.9
Glucose ($C_6H_{12}O_6$)	182.4			C_6H_6ᵇ	269.2
P_4O_{10}	231				
Sucrose ($C_{12}H_{22}O_{11}$)	360				

* Additional values are given in Tables B.2 and B.3.
ᵃ Straight-chain alkanes. ᵇ Benzene.

of the molecules in a solid. The entropies of liquids, in general, fall between those of gases and solids. We note also that substances consisting of large molecules usually have higher entropies than substances with smaller molecules, because energy is shared between more atoms and is therefore more dispersed.

ENTROPY CHANGES IN REACTIONS

$\Delta S°$ refers to the standard entropy change for the reacting system and not to any entropy change that there might be in the surroundings.

The **standard entropy change** for a reaction is easily calculated from the standard molar entropies, using the expression

$$\Delta S° = \sum S°(\text{products}) - \sum S°(\text{reactants})$$

As an example, let us calculate the standard entropy change for the rusting of iron:

$$4Fe(s) + 3O_2(g) \longrightarrow 2Fe_2O_3(s)$$

For this reaction we may write

$$\Delta S° = 2S°(Fe_2O_3) - [4S°(Fe) + 3S°(O_2)]$$

Using the values given in Table 16.1, we obtain

$$\begin{aligned}
\Delta S° &= (2 \text{ mol})(87.4 \text{ J K}^{-1} \text{ mol}^{-1}) - [(4 \text{ mol})(27.3 \text{ J K}^{-1} \text{ mol}^{-1}) \\
&\quad + (3 \text{ mol})(205.0 \text{ J K}^{-1} \text{ mol}^{-1})] \\
&= -549.4 \text{ J K}^{-1}
\end{aligned}$$

= Example 16.1 =

STANDARD ENTROPY CHANGE FOR A REACTION

Use the data in Table 16.1 to calculate $\Delta S°$ for the reaction

$$CaCO_3(s) \longrightarrow CaO(s) + CO_2(g)$$

Solution

$$\begin{aligned}
\Delta S° &= S°(CaO) + S°(CO_2) - S°(CaCO_3) \\
&= (1 \text{ mol})(38.1 \text{ J K}^{-1} \text{ mol}^{-1}) + (1 \text{ mol})(213.7 \text{ J K}^{-1} \text{ mol}^{-1}) \\
&\quad - (1 \text{ mol})(92.9 \text{ J K}^{-1} \text{ mol}^{-1}) \\
&= 158.9 \text{ J K}^{-1}
\end{aligned}$$

EXERCISE 16.1

Use the data in Table 16.1 to find the entropy change for the reduction of iron(III) oxide to iron using carbon monoxide as the reducing agent:

$$Fe_2O_3(s) + 3CO(g) \longrightarrow 2Fe(s) + 3CO_2(g)$$

We see that there is a large decrease in entropy in the rusting of iron because the highly dispersed oxygen gas reacts to form a compact, ordered solid.

Conversely, in the decomposition of calcium carbonate there is a large increase in entropy because a gas is produced from a solid. Whenever a reaction involves one or more gases, we can easily make a qualitative prediction of the entropy change for the reaction.

If the number of moles of gas *increases* during a reaction, the entropy change for the reaction is *positive*.

If the number of moles of gas *decreases* during a reaction, the entropy change for the reaction is *negative*.

--- *Example 16.2* ==

ENTROPY CHANGES IN REACTIONS INVOLVING GASES

Predict whether the entropy increases or decreases in the following reaction:

$$N_2(g) + 3H_2(g) \longrightarrow 2NH_3(g)$$

Solution

Two moles of gas are produced from 4 mol, so we expect a decrease in the entropy in this reaction; in other words, we predict that ΔS will be negative.

EXERCISE 16.2

Predict the sign of the entropy change for each of the following reactions:

(a) $NH_3(g) + HCl(g) \longrightarrow NH_4Cl(s)$

(b) $BaO(s) + CO_2(g) \longrightarrow BaCO_3(s)$

(c) $CH_4(g) + 2O_2(g) \longrightarrow CO_2(g) + 2H_2O(l)$

16.3
Gibbs Free Energy

The rusting of iron is accompanied by a large decrease in the entropy of the system. Nevertheless, the rusting of iron is a spontaneous process. To understand why, we must also take into account the entropy change in the surroundings, because according to the second law of thermodynamics it is the total entropy change $\Delta S_{system} + \Delta S_{surroundings}$ that must increase during a spontaneous process. The entropy change in the surroundings results from the exchange of energy as heat between the system and its surroundings. An exothermic reaction transfers energy to its surroundings in the form of heat. The energy transferred from the system to the surroundings at constant pressure is $-\Delta H_{system}$ where ΔH_{system} is the enthalpy change in the reacting system. This energy becomes dispersed in the surroundings; therefore the entropy of the surroundings is increased. Conversely, an endothermic reaction withdraws energy in the form of heat from its surroundings. Therefore, in this case, the entropy of the surroundings decreases. The rusting of iron is an exothermic process so the entropy of the surroundings increases and, as we will now see, the total entropy change is also positive.

From thermodynamics it can be shown that the quantitative relationship between the energy transferred as heat to the surroundings at constant pressure and the entropy increase in the surroundings is

$$\Delta S_{surroundings} = -\frac{\Delta H_{system}}{T}$$

In calculating $\Delta S_{surroundings}$ it is important to remember that the units of S are joules per kelvin, whereas ΔH is expressed in kilojoules which must therefore be converted to joules.

The enthalpy change at 25 °C for the rusting of iron

$$4Fe(s) + 3O_2(g) \longrightarrow 2Fe_2O_3(s)$$

is simply twice the molar enthalpy of formation of $Fe_2O_3(s)$ (see Appendix Table B.2):

$$\Delta H° = 2\ \Delta H_f°(Fe_2O_3) = (2\ mol)(-824.2\ kJ\cdot mol^{-1}) = -1648.4\ kJ$$

The release of this energy to the surroundings increases the entropy of the surroundings by the amount $-\Delta H_{system}/T$:

$$\Delta S_{surroundings} = -\frac{\Delta H_{system}}{T} = \frac{-1648.4\ kJ}{298.15\ K}\left(\frac{1000J}{1\ kJ}\right)$$

$$= +5529\ J\ K^{-1}$$

This large increase in the entropy of the surroundings completely outweighs the entropy decrease in the reacting system, which we calculated in Section 16.2 to be $-549\ J\ K^{-1}$, so there is a large net increase in the total entropy:

$$\Delta S_{total}° = \Delta S_{system}° + \Delta S_{surroundings}°$$

$$= -549\ J\ K^{-1} + 5529\ J\ K^{-1} = +4980\ J\ K^{-1}$$

Thus, although in the rusting of iron there is a decrease in the entropy of the reacting system, there is a larger increase in the entropy of the surroundings. The rusting of iron is therefore a spontaneous process. We can never hope to stop this process, only to slow it up as much as possible.

The importance of both $\Delta S_{surroundings}$ and ΔS_{system} in determining the spontaneity of a process can also be seen in solution formation. As we discussed in Chapter 12 the entropy of mixing of the solute and solvent is always positive. But a solution will not form if the enthalpy of mixing is too large and positive. In this case a large amount of heat flows from the surroundings to the system and so the change in the entropy of the surroundings is large and negative and thus ΔS_{total} is negative and no solution is formed, or in other words a solution does not form spontaneously.

Instead of calculating ΔS_{system} and $\Delta S_{surroundings}$ in order to predict whether or not a reaction is spontaneous, it would be much more convenient if we could make the prediction on the basis of a property of the system only. If fact, we can do so by making use of the relationship $\Delta S_{surroundings} = -\Delta H_{system}/T$. For a spontaneous reaction

$$\Delta S_{total} = \Delta S_{system} + \Delta S_{surroundings} > 0$$

Substituting $\Delta S_{surroundings} = -\Delta H_{system}/T$ gives

$$\Delta S_{total} = \Delta S_{system} - \frac{\Delta H_{system}}{T} > 0$$

FIGURE 16.3
Josiah Willard Gibbs (1839–1903).

Gibbs, the son of a Yale professor, was the first person to be awarded a Ph.D. in science from an American university. After a period of study in France and Germany, he returned to New Haven in 1869. In 1871 he became professor of theoretical physics at Yale, retaining that position until his death. He is regarded by many as the most brilliant of native-born American scientists. He was a modest, reserved person who was said to be a poor teacher. He traveled very little, and he had almost no contact with the great scientists of the time in Europe. Working by himself, he applied the principles of thermodynamics to chemical reactions in a thorough and rigorous mathematical fashion. He published this work from 1876 to 1878 in the *Transactions of the Connecticut Academy of Sciences*. Unfortunately, this journal was not well known in Europe, and he wrote in such a concise and abstract style that most other scientists found his writing difficult to understand. Even Einstein said of a book that Gibbs wrote later, "It is a masterpiece but it is hard to read." It was not until the 1890s that his work was discovered by physical chemists in Europe. He was then universally recognized for his outstanding contribution to thermodynamics, and today his work remains the foundation of modern chemical thermodynamics.

Multiplying through by T, we have

$$T \Delta S_{\text{total}} = T \Delta S_{\text{system}} - \Delta H_{\text{system}} > 0$$

or

$$- T \Delta S_{\text{total}} = - T \Delta S_{\text{system}} + \Delta H_{\text{system}} = \Delta H_{\text{system}} - T \Delta S_{\text{system}} < 0$$

Thus an alternative criterion for a spontaneous reaction is that

$$\Delta H_{\text{system}} - T \Delta S_{\text{system}} < 0$$

or in other words

$$\Delta H_{\text{system}} - T \Delta S_{\text{system}} \text{ must be negative}$$

Now we have a criterion for a spontaneous reaction that is expressed *only* in terms of the properties of the system and we do not need to consider also what is happening in the surroundings. Gibbs (Figure 16.3) showed that it was very convenient to define a new function

$$G = H - TS$$

where we have now dropped the subscript system because all the quantities in this equation refer to the system only. G is called the **Gibbs free energy**. Because H and S are state functions G is also a state function. For any change at constant temperature and pressure we have

$$\Delta G = \Delta H - T \Delta S$$

Since we have seen above that for a spontaneous process $\Delta H - T \Delta S$ must be negative, it follows that the change in the Gibbs free energy, ΔG, must be negative. In other words, the **Gibbs free energy decreases during a spontaneous process**.

For many reactions ΔH is much larger than $T \Delta S$, so as a rough approximation, $\Delta G \approx \Delta H$. Since ΔG is negative in a spontaneous process, we see why so many spontaneous reactions are exothermic (ΔH is negative). Nevertheless, there are some endothermic reactions that are spontaneous because they are accompanied by a large increase in the entropy of the system. Many thermal decompositions that form gases, such as the decomposition of calcium carbonate, are of this type.

There are four possible combinations of ΔH and ΔS:

	ΔH	ΔS	ΔG	*Spontaneous Reaction*
(1)	−	+	−at all T	Yes
(2)	−	−	−at low T	Yes
			+at high T	No
(3)	+	−	+at all T	No
(4)	+	+	+at low T	No
			−at high T	Yes

1. This first case is an exothermic reaction (ΔH is negative) in which there is an increase in entropy in the system (ΔS is positive). The heat emitted to the surroundings causes an increase in the entropy of the surroundings, and since the entropy of the system also increases, the total entropy increases and the process is spontaneous.

2. If the entropy of the system decreases (ΔS is negative), then *at a sufficiently high temperature $T \Delta S$ will be large and negative and therefore $\Delta H - T \Delta S$ will be positive, and the process will not be spontaneous. At low temperature*, however, $T \Delta S$ is small. In this case $\Delta G = \Delta H - T \Delta S$ is negative, so the process is spontaneous. At low temperature the surroundings are relatively ordered, so the transfer of a given amount of heat causes a relatively large increase in the disorder. But at high temperature the surroundings are more disordered, and the same amount of heat produces a much smaller increase in the entropy. Thus at low temperature a given amount of heat transferred to the surroundings produces a large entropy increase in the surroundings, $\Delta S_{surroundings} = -\Delta H/T$, which counterbalances the entropy decrease in the system, so the total entropy increase in the universe is positive. However, the transfer of the same amount of heat at high temperature causes only a small increase in the entropy of the surroundings, which is insufficient to overcome the negative entropy change in the system.

3. A reaction that is endothermic and has a negative entropy change can never be spontaneous because ΔG is always positive.

4. An endothermic reaction may be spontaneous if it is accompanied by a large enough increase in the entropy of the system. Since the term $T \Delta S$ increases with increasing T, the reaction is more likely to be spontaneous at high temperature than at low temperature. In an endothermic reaction the surroundings lose heat, so the entropy of the surroundings must decrease, and this entropy change for a given amount of heat is larger at

A spontaneous reaction is one that proceeds from the reactants on the left of the equation for the reaction to the products on the right until equilibrium is reached.

lower temperatures than at higher temperatures, when the system is more highly disordered. Thus only at some sufficiently high temperature is the negative entropy change in the surroundings sufficiently small that it is outweighed by the positive entropy change in the system.

Notice that a reaction for which ΔG is positive and that is therefore not spontaneous in the forward direction—that is, from left to right—has a negative ΔG in the reverse direction—that is, from right to left. Thus a reaction that does not proceed spontaneously in the forward direction is spontaneous in the reverse direction. Finally, if $\Delta G = 0$ the reaction has no tendency to proceed in either direction; in other words, the system must be at equilibrium. In summary:

ΔG	*Reaction*
Negative	Spontaneous
Positive	Not spontaneous (spontaneous in reverse direction)
Zero	At equilibrium

CALCULATING STANDARD FREE ENERGY CHANGES

The **standard free energy change**, $\Delta G°$, for any reaction may be found from the standard free energies of formation, $\Delta G_f°$, of the reactants and products in just the same way as a standard enthalpy change is calculated

$$\Delta G° = \sum \Delta G_f°(\text{products}) - \sum \Delta G_f°(\text{reactants})$$

The **standard free energy of formation**, $\Delta G_f°$, is the free energy change for the formation of 1 mol of a compound from its elements in their standard states. The standard free energies of formation, $\Delta G_f°$, of the elements in their standard states are taken to be zero, as for $\Delta H_f°$ values. But notice that the standard entropies of the elements are *not* zero. As we have seen, absolute values of S can be obtained, but only relative, not absolute, values of G and H can be obtained. Value of $\Delta G_f°$ can be calculated from the corresponding $\Delta H_f°$ and $S°$ values. However, it is a great convenience not to have to do these calculations each time a $\Delta G°$ value is required but simply to refer to a table of $\Delta G_f°$ values, such as Table 16.2.

We will now calculate the entropy, enthalpy, and free energy changes associated with the following reactions:

(a) $CaCO_3(s) \longrightarrow CaO(s) + CO_2(g)$

(b) $N_2(g) + 3H_2(g) \longrightarrow 2NH_3(g)$

(c) $H_2(g) + Cl_2(g) \longrightarrow 2HCl(g)$

First, we calculate the entropy changes for these reactions, using the data in Table 16.1.

(a) $\Delta S° = S°(CaO) + S°(CO_2) - S°(CaCO_3)$

$= (1 \text{ mol})(38.1 \text{ J K}^{-1} \text{ mol}^{-1}) + (1 \text{ mol})(213.7 \text{ J K}^{-1} \text{ mol}^{-1})$

$-(1 \text{ mol})(92.9 \text{ J K}^{-1} \text{ mol}^{-1})$

$= 158.9 \text{ J K}^{-1}$

TABLE 16.2
Standard Free Energies of Formation, ΔG_f°, 25 °C

	ΔG_f° (kJ mol^{-1})		ΔG_f° (kJ mol^{-1})
$AgCl(s)$	−109.8	$CH_3CO_2H(l)$	−390
$CO(g)$	−137.2	$C_6H_6(l)$	124.7
$CO_2(g)$	−394.4	$CaO(s)$	−603.5
$CH_4(g)$	−50.8	$CaCO_3(s)$	−1128.8
$C_2H_6(g)$	−32.9	$Fe_2O_3(s)$	−742.2
$C_3H_8(g)$	−23.4	$H_2S(g)$	−33.4
$C_4H_{10}(g)^a$	−17.2	$HCl(g)$	−95.3
$C_5H_{12}(l)^a$	−9.6	$HI(g)$	1.6
$C_6H_{14}(l)^a$	−4.4	$HNO_3(l)$	−80.8
$C_7H_{16}(l)^a$	1.0	$H_2SO_4(l)$	−690.1
$C_8H_{18}(l)^a$	6.4	$H_2O(l)$	−237.2
$CH_2O(g)$	−113	$H_2O(g)$	−228.6
$CH_3OH(l)$	−116.4	$NH_3(g)$	−16.4
$C_2H_5OH(l)$	−174.9	$NO(g)$	86.6
$CH_2Cl_2(l)$	−67.3	$NO_2(g)$	51.3
$CH_3Cl(g)$	−57.4	$N_2O_4(g)$	97.8
$CCl_4(l)$	−65.3	$NaCl(s)$	−384.3
$C_6H_{12}O_6(s)$ (glucose)	−919.2	$NaBr(s)$	−349.1
		$NaI(s)$	−282.4

^{a} Straight-chain alkanes.

(b) $\Delta S^\circ = 2S^\circ(NH_3) - [S^\circ(N_2) + 3S^\circ(H_2)]$

$= (2 \text{ mol})(192.7 \text{ J K}^{-1} \text{ mol}^{-1}) - [(1 \text{ mol})(191.5 \text{ J K}^{-1} \text{ mol}^{-1})$

$+ (3 \text{ mol})(130.6 \text{ J K}^{-1} \text{ mol}^{-1})]$

$= -197.9 \text{ J K}^{-1}$

(c) $\Delta S^\circ = 2S^\circ(HCl) - [S^\circ(H_2) + S^\circ(Cl_2)]$

$= (2 \text{ mol})(186.8 \text{ J K}^{-1} \text{ mol}^{-1}) - [(1 \text{ mol})(130.6 \text{ J K}^{-1} \text{ mol}^{-1})$

$+ (1 \text{ mol})(223.0 \text{ J K}^{-1} \text{ mol}^{-1})]$

$= 20.0 \text{ J K}^{-1}$

Note that in the third reaction the entropy change is small because 2 mol of gaseous reactants form 2 mol of gaseous products.

Let us now find the enthalpy changes in these reactions so that we can finally calculate the free energy changes and thus predict whether these reactions will occur spontaneously. Using the data in Tables 6.1, B.2, and B.3, we have

(a) $\Delta H^\circ = \Delta H_f^\circ(CaO) + \Delta H_f^\circ(CO_2) - \Delta H_f^\circ(CaCO_3)$

$= (1 \text{ mol})(-635.1 \text{ kJ mol}^{-1}) + (1 \text{ mol})(-393.5 \text{ kJ mol}^{-1})$

$- (1 \text{ mol})(-1206.9 \text{ kJ mol}^{-1})$

$= 178.3 \text{ kJ}$

(b) $\Delta H^\circ = 2 \Delta H_f^\circ(NH_3) - [\Delta H_f^\circ(N_2) - 3 \Delta H_f^\circ(H_2)]$

$= (2 \text{ mol})(-45.9 \text{ kJ mol}^{-1}) - (0 + 0)$

$= -91.8 \text{ kJ}$

(c)
$$\Delta H° = 2 \Delta H_f°(HCl) - [\Delta H_f°(H_2) + \Delta H_f°(Cl_2)]$$
$$= (2 \text{ mol})(-92.3 \text{ kJ mol}^{-1}) - (0 + 0)$$
$$= -184.6 \text{ kJ}$$

We see that reaction (a), the thermal decomposition of calcium carbonate, is endothermic, ΔH is positive, and that ΔS is also positive. Thus this reaction is an example of case 4 above. It will only be spontaneous at high temperature when $T\Delta S$ is large enough that $\Delta G = \Delta H - T\Delta S$ is negative.

Reaction (b), the synthesis of ammonia, is exothermic, ΔH is negative, but it is accompanied by a large negative entropy change so it is an example of case 2 above. This reaction will only be spontaneous at low temperature when $T \Delta S$ is small enough that $\Delta G = \Delta H - T \Delta S$ is negative.

Reaction (c), the synthesis of hydrogen chloride, is strongly exothermic, ΔH is negative, and it is accompanied by a positive entropy change so it is an example of case 1 above and is therefore spontaneous.

Let us now calculate the free energy changes for these reactions so that we can examine our qualitative predictions more closely. We can calculate the free energy change either from the expression

In calculating ΔG from $\Delta H - T\Delta S$ it is important to remember than values of ΔS are given in joules and they must be converted to kilojoules because ΔG values are always expressed in kilojoules.

$$\Delta G° = \Delta H° - T \Delta S°$$

or directly from the values of the free energies of formation listed in Table 16.2.

For reaction (a), $\Delta S° = 159 \text{ J K}^{-1} = 0.159 \text{ kJ K}^{-1}$ and $\Delta H° = 178.3 \text{ kJ}$. Therefore

$$\Delta G° = 178.3 \text{ kJ} - (298 \text{ K})(0.159 \text{ kJ K}^{-1})$$
$$= 130.9 \text{ kJ}$$

Alternatively, using the data in Table 16.2

$$\Delta G° = \Delta G_f°(CaO) + \Delta G_f°(CO_2) - \Delta G_f°(CaCO_3)$$
$$= (1 \text{ mol})(-603.5 \text{ kJ mol}^{-1}) + (1 \text{ mol})(-394.4 \text{ kJ mol}^{-1})$$
$$- (1 \text{ mol})(-1128.8 \text{ kJ mol}^{-1})$$
$$= 130.9 \text{ kJ}$$

This large positive $\Delta G°$ value indicates that this reaction has no tendency to proceed at 25 °C under standard conditions.

For reaction (b), $\Delta S° = -198 \text{ J K}^{-1} = -0.198 \text{ kJ K}^{-1}$. Therefore

$$\Delta G° = -91.8 \text{ kJ} - (298 \text{ K})(-0.198 \text{ kJ K}^{-1})$$
$$= -32.8 \text{ kJ}$$

or alternatively,

$$\Delta G° = 2 \Delta G_f°(NH_3) - [\Delta G_f°(N_2) + 3 \Delta G_f°(H_2)]$$
$$= -(2 \text{ mol})(-16.4 \text{ kJ mol}^{-1}) - (0 + 0)$$
$$= -32.8 \text{ kJ}$$

In this case the negative $\Delta G°$ value indicates that the reaction will, in principle, proceed spontaneously at room temperature. However, we should be careful to note that the statement that a reaction proceeds spontaneously does not mean that it necessarily proceeds rapidly. In fact, a spontaneous reaction may be very slow. For example, the synthesis of ammonia is a very slow reaction at room temperature. In order to speed up the reaction, one must increase the temperature and employ a catalyst (see Chapters 3 and 19).

For reaction (c), $\Delta S° = 20.0$ J K^{-1} = 0.0200 kJ K^{-1}. Therefore

$$\Delta G° = -184.6 \text{ kJ} - (298 \text{ K})(0.020 \text{ kJ K}^{-1})$$
$$= -190.6 \text{ kJ}$$

or alternatively,

$$\Delta G° = 2 \Delta G_f°(HCl) - [\Delta G_f°(H_2) + \Delta G_f°(Cl_2)]$$
$$= (2 \text{ mol})(-95.3 \text{ kJ mol}^{-1}) - (0 + 0)$$
$$= -190.6 \text{ kJ}$$

This large negative $\Delta G°$ value indicates that this reaction has a strong tendency to proceed at room temperature. As we saw in Chapter 5, if this reaction is initiated by a spark or by light of suitable frequency, it proceeds explosively to give a quantitative yield of the products.

═ *Example 16.3* ══════════════════════════════

STANDARD FREE ENERGY CHANGE FOR A REACTION

Using the data in Table 16.2, calculate $\Delta G°$ for the reaction

$$CH_3OH(l) + \tfrac{3}{2}O_2(g) \longrightarrow CO_2(g) + 2H_2O(l)$$

Solution

We have

$$\Delta G° = \Delta G_f°(CO_2) + 2 \Delta G_f°(H_2O) - [\Delta G_f°(CH_3OH) + \tfrac{3}{2}\Delta G_f°(O_2)]$$
$$= (1 \text{ mol})(-394.4 \text{ kJ mol}^{-1}) + (2 \text{ mol})(-237.2 \text{ kJ mol}^{-1})$$
$$\quad - [(1 \text{ mol})(-116.4 \text{ kJ mol}^{-1}) + (\tfrac{3}{2} \text{ mol})(0 \text{ kJ mol}^{-1})]$$
$$= -752.4 \text{ kJ}$$

EXERCISE 16.3

Using the data in Table 16.2 calculate the value of the standard free energy change for the combustion of methane in excess oxygen.

SPONTANEITY AND REACTION RATE

We have mentioned several times that the fact that a reaction has a negative ΔG and therefore occurs spontaneously does not necessarily mean that it occurs rapidly. We have seen that the rusting of iron is a spontaneous reaction but it is quite slow. The standard free energy of formation of water is -228.6 kJ mol^{-1}. So the reaction of hydrogen with oxygen to give water is a spontaneous process for which the position of equilibrium lies very far to the right. But oxygen and hydrogen can be mixed at room temperature without any observable reaction taking place. However, if the mixture is ignited, an extremely rapid explosive reaction occurs (Experiment 16.2). We have also seen that the synthesis of ammonia from nitrogen and hydrogen is a spontaneous reaction with a negative ΔG, but at room temperature it is extremely slow. In the

Experiment 16.2

A Spontaneous Reaction

The balloon contains hydrogen.

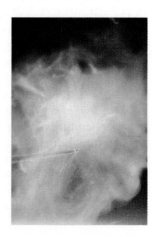

When a flame is touched to the balloon the hydrogen combines with the oxygen of the air in a rapid, explosive reaction. The position of equilibrium for the reaction

$$2H_2(g) + O_2(g) \rightarrow 2H_2O(g)$$

lies far to the right—it is a spontaneous reaction.

industrial process the reaction is speeded up by using high temperature and a catalyst. Although thermodynamics can tell us the direction in which a reaction will occur and how far it will proceed before equilibrium is reached, it can tell us nothing about the rate of the reaction. Thermodynamics is concerned only with the initial and final states of a system. It is not concerned with how the system changes from the initial to the final state. In contrast, the rate of a reaction depends very much on how it proceeds from the reactants to the products. The study of the rates of reactions, why some are fast while others are slow, is called *kinetics*. We will take up this subject in Chapter 19.

TEMPERATURE DEPENDENCE OF $\Delta G°$

We have seen that the reaction

$$CaCO_3(s) \longrightarrow CaO(s) + CO_2(g)$$

has a large positive $\Delta G°$ value at room temperature, indicating that it has no tendency to proceed under these conditions; indeed, it tends to proceed in the reverse direction. However, we can see from the equation

$$\Delta G° = \Delta H° - T\Delta S°$$

that because $\Delta S°$ is positive, $\Delta G°$ will have a negative value if the temperature is increased sufficiently. Let us calculate the value of $\Delta G°$ for $T = 1000$ K, assuming that ΔS and ΔH do not change very much with temperature:

$$\Delta G^\circ = 178.3\ \text{kJ} - (1000\ \text{K})(0.159\ \text{kJ K}^{-1}) = 19\ \text{kJ}$$

The value is still positive. But at 1200 K

$$\Delta G^\circ = 178.3\ \text{kJ} - (1200\ \text{K})(0.159\ \text{kJ K}^{-1}) = -13\ \text{kJ}$$

The value of ΔG° is negative, so we expect the reaction to proceed spontaneously at this temperature.

This prediction agrees with predictions made previously on the basis of Le Châtelier's principle. Since the decomposition of calcium carbonate is an endothermic reaction, Le Châtelier's principle predicts that the equilibrium is shifted to the right if the temperature is raised; that is, the products are favored. At higher temperatures the transfer of a given amount of energy as heat from the surroundings to the system produces a smaller decrease in the entropy of the surroundings than at lower temperatures. At a sufficiently high temperature the entropy decrease in the surroundings becomes smaller than the entropy increase in the reacting system, and the reaction becomes spontaneous.

Since ΔG° is negative at 1200 K but positive at 1000 K, we could guess that, at a temperature of approximately 1100 K, ΔG° would be zero. Thus at this temperature the reaction would have no tendency to proceed in either direction; it would be at equilibrium. We can calculate this temperature from the equation

$$\Delta G^\circ = \Delta H^\circ - T\,\Delta S^\circ$$

by setting $\Delta G^\circ = 0$. In this case we have

$$0 = 178.3 - T(0.159\ \text{kJ K}^{-1})$$

$$T = 1121\ \text{K} = 848\ ^\circ\text{C}$$

Thus at 848 °C and at standard conditions—that is, a pressure of 1 atm—the system is at equilibrium. At a higher temperature it proceeds to the right, ΔG negative; at a lower temperature it proceeds to the left, ΔG positive. However, such a calculation is only approximate, because it is based on the assumption that ΔH° and ΔS° do not vary with temperature, which is only an approximation.

Many thermal decompositions are of this type. Solids that are stable at room temperature often decompose in an endothermic reaction at high temperature, if the decomposition produces one or more gases and is therefore accompanied by a large increase in entropy of the system. For example, nitrates of many metals decompose on heating to give gaseous NO_2 and O_2 (Chapters 10 and 18):

$$2\text{Pb}(\text{NO}_3)_2(s) \longrightarrow 2\text{PbO}(s) + 4\text{NO}_2(g) + \text{O}_2(g)$$

For most reactions ΔH and ΔS vary only very slowly with temperature. Experiment shows that for a temperature change of approximately 100 °C ΔH usually changes by less than 1% and ΔS by 2%–3%. But ΔG depends markedly on temperature because of the term T in $\Delta G = \Delta H - T\,\Delta S$.

GIBBS FREE ENERGY AND THE EQUILIBRIUM CONSTANT

In Chapter 13 we introduced the reaction quotient Q, which for the gas-phase reaction $a\text{A} + b\text{B} + \cdots \rightarrow p\text{P} + q\text{Q} + \cdots$ can be written in the form

$$Q = \frac{(p_\text{P})^p(p_\text{Q})^q \cdots}{(p_\text{A})^a(p_\text{B})^b \cdots}$$

	Reaction Quotient Equilibrium Constant	Gibbs Free Energy Change	Total Entropy Change,
Process is	Q/K	ΔG	ΔS_{total}
Spontaneous	<1	$-$	$+$
Not spontaneous	>1	$+$	$-$
At equilibrium	$=1$	0	0

TABLE 16.3
Criteria for Spontaneity of Reactions

where p_A, p_B, ... are the partial pressures of A, B, ... and so on, and in general are *not* the equilibrium partial pressures. We also saw in Chapter 13 that:

- If $Q < K_p$ or $Q/K_p < 1$, the reaction proceeds from left to right; in other words, the reaction is *spontaneous*.
- If $Q > K_p$ or $Q/K_p > 1$, the reaction proceeds from right to left; in other words, it is *not spontaneous* in the direction written (left to right), but it is spontaneous in the reverse direction.
- If $Q = K_p$ or $Q/K_p = 1$, the reaction is at *equilibrium*.

The three criteria that we have now discussed for deciding whether or not a reaction will proceed spontaneously are summarized in Table 16.3. The relationship between ΔG and ΔS_{total} is $\Delta G = -T\Delta S_{total}$ and there must also be a relationship between Q/K_p and ΔG. Thermodynamics shows that this relationship is

$$\Delta G = RT \ln(Q/K_p)$$

or

$$\Delta G = RT \ln Q - RT \ln K_p$$

If all the reactants and products are in their standard states—that is, 1 atm pressure—$\Delta G = \Delta G°$ and $Q = 1$, or $\ln Q = 0$, so that

$$\Delta G° = -RT \ln K_p$$

This equation shows that the equilibrium constant is determined by the standard Gibbs free energy change for the reaction. If $\Delta G°$ is negative, $K_p > 1$ and the reaction is spontaneous under standard conditions. If $\Delta G°$ is positive, $K_p < 1$ and the reaction is not spontaneous under standard conditions, but is spontaneous in the reverse direction. If $\Delta G° = 0$, $K_p = 1$ and the reaction is at equilibrium under standard conditions.

As an example, let us calculate the equilibrium constant for the formation of NO(g) from the elements at 25 °C. The equation for the reaction is

$$N_2(g) + O_2(g) \longrightarrow 2NO(g)$$
$$\Delta G° = 2 \Delta G_f°(NO, g) - [\Delta G_f°(N_2, g) + \Delta G_f°(O_2, g)]$$
$$= (2 \text{ mol})(86.6 \text{ kJ mol}^{-1}) - (0 + 0) = 173.2 \text{ kJ mol}^{-1}$$

Then we can find K_p from the expression

$$\ln K_p = \frac{-\Delta G^\circ}{RT} = -\frac{173.2 \text{ kJ mol}^{-1}}{(0.00831 \text{ kJ mol}^{-1} \text{ K}^{-1})(298 \text{ K})} = -69.9$$

$$K_p = 4.25 \times 10^{-31}$$

This very small equilibrium constant shows that the reaction has very little tendency to proceed at room temperature. At standard conditions, because all the gas pressures are 1 atm, $Q = 1$, so $Q \gg K$, or $Q/K \gg 1$. Thus the reaction is not spontaneous in the direction written but proceeds spontaneously in the reverse direction, although at 25 °C it is exceedingly slow.

═ *Example 16.4* ═══════════════════════════════════

CALCULATING EQUILIBRIUM CONSTANTS FROM ΔG° VALUES

The standard free energy change at 25 °C for the reaction

$$N_2(g) + 3H_2(g) \rightleftharpoons 2NH_3(g)$$

is $\Delta G^\circ = -32.8$ kJ. What is the equilibrium constant, K_p, for this reaction?

Solution

$$\Delta G^\circ = -RT \ln K_p$$

Therefore,

$$\ln K_p = \frac{-\Delta G^\circ}{RT} = \frac{32.8 \text{ kJ mol}^{-1}}{(0.00831 \text{ kJ mol}^{-1} \text{ K}^{-1})(298 \text{ K})} = 13.25$$

$$K_p = 5.68 \times 10^5$$

This large value of the equilibrium constant tells us that the position of the equilibrium is far to the right; that is, at equilibrium there will be a high pressure (or concentration) of NH_3 and low pressures (or concentrations) of N_2 and H_2. We can reach essentially the same conclusion from the negative ΔG° value, which tells us that at standard conditions—all the gases at 1 atm pressure—the reaction is spontaneous, that is, it proceeds to the right until equilibrium is reached.

_____

Now we can see one reason for the importance of the Gibbs free energy. From a table of standard free energies of formation we can calculate the equilibrium constant for many thousands of reactions. In considering possible processes by which a substance might be synthesized we can immediately discard any processes which have a large positive ΔG° and therefore a very small equilibrium constant. We know that at equilibrium we will have only a very small concentration of the desired product.

The standard free energy change for a reaction, ΔG°, is the free energy change for the particular case in which all the reactants and products are in their standard state—1 atm pressure for gases. The value of ΔG° will give us a good idea whether the reaction is likely to proceed or not under other conditions but we can easily calculate the free energy change for nonstandard conditions by using the equation

$$\Delta G = -RT \ln \frac{Q}{K_p}$$

to evaluate ΔG for any initial concentrations of reactants and products, as the next example shows.

═ *Example 16.5* ═══════════════════════════════

CALCULATING ΔG FOR A REACTION FROM Q AND K_p

Calculate ΔG at 298 K for the reaction

$$N_2(g) + 3H_2(g) \rightleftharpoons 2NH_3(g)$$

if the reaction mixture consists of 10 atm N_2, 10 atm H_2, and 1 atm NH_3.

Solution

$$Q = \frac{(p_{NH_3})^2}{(p_{N_2})(p_{H_2})^3} = \frac{1^2}{10 \times 10^3} = \frac{1}{10^4} \text{ atm}^{-2}$$

and

$$K_p = 5.68 \times 10^5 \text{ atm}^{-2} \text{ (see Example 16.4)}$$

$$\Delta G = -RT \ln \frac{Q}{K_p} = (8.31 \text{ J K}^{-1})(298 \text{ K})\left(\ln \frac{1}{5.68 \times 10^9}\right)$$

$$= (8.31 \text{ J K}^{-1})(298 \text{ K})(-22.5) = -55.7 \text{ kJ}$$

Increasing the pressure of both N_2 and H_2 from 1 to 10 atm changes ΔG from -32.8 to -55.7 kJ. Thus the reaction has a still greater tendency to proceed to the right, in other words the position of equilibrium shifts to the right, as we could have predicted using Le Châtelier's principle.

─────────────────────────────────────

GIBBS FREE ENERGY AND WORK

Another important property of the free energy change for a reaction is that it can tell us how much useful work can be derived from the reaction. In fact *the maximum possible useful work obtainable from a process at constant pressure and temperature is equal to the change in free energy for the process*:

$$w_{max} = -\Delta G$$

G is called the Gibbs *free* energy because $-\Delta G$ for a spontaneous process represents the energy that is *free* to do useful work. For a nonspontaneous process ΔG tells us the minimum amount of work that must be done in order to make the process occur. Any process that occurs spontaneously can in principle be utilized for the performance of work. The greater the free energy change, the greater is the tendency for the reaction to occur spontaneously and the greater is the amount of work that can be obtained from the process.

If we have some gas compressed in a cylinder fitted with a piston, then the spontaneous expansion of the gas will drive the piston, which in turn could be used to turn a wheel, for example. In an automobile engine the spontaneous combustion of gasoline produces a large amount of gas, which drives the pistons of the engine. The amount of work that can be derived from a particular process depends on how it is carried out. If gasoline is burned in an open container, some of the energy of the reaction is used to push away the atmosphere, and the rest is converted to heat so that no work can be obtained. But if gasoline is burned in an automobile engine, some work can be obtained. In practice, the theoretically possible maximum amount of work can never be obtained—all engines are somewhat inefficient. In an automobile engine only about 20% of the maximum possible work is obtained. Nevertheless, it is very useful to know the maximum amount of work that can theoretically be obtained from any chemical reaction or other process.

In our bodies we oxidize foodstuffs to obtain energy to do work. For the oxidation of 1 mol of glucose,

$$C_6H_{12}O_6(s) + 6O_2(g) \longrightarrow 6CO_2(g) + 6H_2O(l)$$

$$\Delta H^\circ = -2808 \text{ kJ} \quad \text{and} \quad \Delta G^\circ = -2870 \text{ kJ}$$

Hence 2870 kJ is the maximum amount of work that can be done by a person as a result of metabolizing 1 mol (180.2 g) of glucose. The work that a person must do in climbing a height h is given by $w = mgh$, where m is the mass of the person and g is the acceleration due to gravity. Thus in order to climb a height of 100 m, a 60-kg woman would need to do $(60 \text{ kg})(9.8 \text{ m s}^{-2})(100 \text{ m}) = 60\,000 \text{ J} = 60 \text{ kJ}$ of work. In order to do this much work, she would need to metabolize a minimum of $(60 \text{ kJ}/2870 \text{ kJ})(180.2 \text{ g}) = 3.8 \text{ g}$ of glucose. In practice, because the conversion of energy to work in the body is not 100% efficient, she would need more than this amount of glucose.

The reaction between hydrogen and oxygen to produce water is a spontaneous process ($\Delta G^\circ = -474 \text{ kJ}$) that can be utilized to do work. The reverse reaction, the decomposition of water to hydrogen and oxygen, is a nonspontaneous reaction with $\Delta G^\circ = 474 \text{ kJ}$. Work must be done to decompose water to hydrogen and oxygen. For the decomposition of 1 mol of water at 25 °C and 1 atm pressure, 474 kJ of work must be done, by passing an electric current through the water, for example. Another way in which work can be done on a reacting system in order to drive a reaction in the nonspontaneous direction is to couple the reaction with another reaction that can do work, that is, one that has a negative ΔG.

COUPLED REACTIONS

Consider, for example, the extraction of copper from the ore Cu_2S. For the decomposition

$$Cu_2S(s) \longrightarrow 2Cu(s) + S(s) \tag{1}$$

$\Delta H^\circ = +79.5 \text{ kJ}$ and $\Delta S^\circ = -22.4 \text{ J K}^{-1}$. The reaction is strongly endothermic and has a negative entropy change. Therefore ΔG° is positive, and the reaction is nonspontaneous at all temperatures. At 25 °C we calculate the value of $\Delta G^\circ = \Delta H^\circ - T\Delta S^\circ$ to be $+86.2 \text{ kJ mol}^{-1}$. For this reaction to proceed, work must be done on the system in some way. We can couple it with another reaction so that the overall process is spontaneous. Consider, for example, the reaction

$$S(s) + O_2(g) \longrightarrow SO_2(g) \tag{2}$$

for which

$$\Delta H^\circ = -296.8 \text{ kJ} \quad \text{and} \quad \Delta S^\circ = 11.1 \text{ kJ K}^{-1}$$

This reaction is exothermic and is accompanied by an increase in entropy. It is therefore spontaneous under all conditions. At 25 °C, $\Delta G^\circ = -300.1 \text{ kJ}$.

By adding the equations (1) and (2), we obtain the equation for the overall reaction. By adding the values of $\Delta H°$ and $\Delta G°$ for the two reactions, we obtain values for the overall reaction:

$$Cu_2S(s) \longrightarrow 2Cu(s) + S(s) \qquad \Delta G° = +86.2 \text{ kJ} \qquad \Delta H° = +79.5 \text{ kJ}$$
$$S(s) + O_2(g) \longrightarrow SO_2(g) \qquad \Delta G° = -300.1 \text{ kJ} \qquad \Delta H° = -296.8 \text{ kJ}$$
$$\overline{Cu_2S(s) + O_2(g) \longrightarrow 2Cu(s) + SO_2(s) \qquad \Delta G° = -213.9 \text{ kJ} \qquad \Delta H° = -217.3 \text{ kJ}}$$

We see that the overall reaction is exothermic. More importantly, it has a negative ΔG value and it is spontaneous, because the negative free energy change for the second reaction is larger than the positive free energy change for the first reaction. The second reaction may be said to drive the first reaction.

The coupling of reactions to cause a nonspontaneous reaction to occur is very important in biochemical systems. Many of the reactions that are essential to life do not occur spontaneously in the human body. These reactions are made to occur by coupling them with reactions that are spontaneous. An example of such a reaction is the hydrolysis of ATP to ADP.

$$ATP^{3-} + H_2O \longrightarrow ADP^{2-} + H_2PO_4^-$$

At body temperature (37 °C) the standard entropy, enthalpy, and free energy changes for this reaction are

$$\Delta G° = -30 \text{ kJ} \qquad \Delta H° = -20 \text{ kJ} \qquad \Delta S° = +34 \text{ J K}^{-1}$$

The hydrolysis of ATP can therefore be used to drive any reaction for which $\Delta G < 30$ kJ. For example, the synthesis of a protein from its constituent amino acids is accompanied by a large increase in free energy, not only because it is endothermic, but also because the building of a single protein molecule from a large number of separate amino acid molecules (Chapter 24) is accompanied by a large decrease in entropy. Nevertheless this process occurs in the body because it is coupled to the hydrolysis of ATP. For example, the synthesis of a myoglobin molecule from approximately 150 amino acids is accompanied by the hydrolysis of about 450 ATP molecules so that the overall process has a negative ΔG. The ADP thus formed is then converted back to ATP by coupling this reaction with the oxidation of glucose in a multistep process in which 38 ATP molecules are generated for each molecule of glucose consumed. In this way the energy that we obtain from food such as glucose is used, for example, for the synthesis of proteins.

IMPORTANT TERMS

Entropy, S, is a state function that measures the extent of disorder or randomness in a system and the dispersal of energy in the system.

The **Gibbs free energy**, G, is a state function related to the enthalpy, H, the temperature, T, and the entropy, S, of a system by the expression: $G = H - TS$. ΔG is a measure of the spontaneity of a reaction.

The **second law of thermodynamics** states that the entropy of the universe (a system plus surroundings) increases in any spontaneous process.

The **standard entropy change**, $\Delta S°$, of a reaction is the change in entropy under standard conditions (1 atm and a specified temperature, usually 25 °C); it is given by

$$\Delta S° = \sum S°(\text{products}) - \sum S°(\text{reactants})$$

The **standard free energy change**, $\Delta G°$, for a reaction is the change in free energy under standard conditions (1 atm and a specified temperature, usually 25 °C); it is given by

$$\Delta G° = \sum \Delta G_f°(\text{products}) - \sum \Delta G_f°(\text{reactants})$$

The **standard free energy of formation**, $\Delta G_f°$, is the free energy change when 1 mol of a substance is formed from its elements in their standard states.

The **standard molar entropy**, $S°$, of a substance is the entropy of 1 mol of the substance in its standard state (1 atm) and at a specified temperature, usually 25 °C.

PROBLEMS *

Entropy

1. For each of the following, use qualitative reasoning to decide which system will have the larger entropy:

   (a) A mole of ice at 0 °C, or a mole of water at the same temperature

   (b) A pack of cards arranged in suits, or a pack of cards randomly shuffled

   (c) A collection of jigsaw pieces, or the completed puzzle

   (d) Solid ammonium chloride, or an aqueous solution of NH_4Cl

2. Does the degree of disorder in each of the following processes increase or decrease as the process proceeds? Is the entropy change for each process positive or negative?

   (a) The evaporation of 1 mol of ethyl alcohol (ethanol)

   (b) $2Mg(s) + O_2(g) \rightarrow 2MgO(s)$

   (c) $XeO_3(s) \rightarrow Xe(g) + \frac{3}{2}O_2(g)$

   (d) $N_2(g) + 3H_2(g) \rightarrow 2NH_3(g)$

   (e) $BaCl_2 \cdot H_2O(s) \rightarrow BaCl_2(s) + H_2O(g)$

* Answers to problems numbered in blue appear at the end of the text.

3. Predict the sign of the entropy change for each of the following reactions:

   (a) $CaCO_3(s) \rightarrow CaO(s) + CO_2(g)$

   (b) $NH_3(g) + HCl(g) \rightarrow NH_4Cl(s)$

   (c) $BaO(s) + CO_2(g) \rightarrow BaCO_3(s)$

4. Does an aqueous solution of Al^{3+} ions have a larger entropy before, or after, hydration of the ions? Why then are the ions hydrated?

5. Under what conditions is each of the following statements true?

   (a) In a spontaneous process, the system moves toward a state of lower energy.

   (b) In a spontaneous process, the system moves toward a state of greater entropy.

6. Predict the sign of the entropy change, ΔS, for each of the following reactions:

   (a) $2CO(g) + O_2(g) \rightarrow 2CO_2(g)$

   (b) $Mg(s) + Cl_2(g) \rightarrow MgCl_2(s)$

(c) $2C_2H_6(g) + 7O_2(g) \rightarrow 4CO_2(g) + 6H_2O(g)$

(d) $CH_4(g) + 2O_2(g) \rightarrow CO_2(g) + 2H_2O(l)$

(e) $Al_2Cl_6(g) \rightarrow 2AlCl_3(g)$

7. Predict the sign of the entropy change, ΔS, for each of the following reactions:

(a) $H_2(g) + Br_2(l) \rightarrow 2HBr(g)$

(b) $ZnO(s) + H_2S(g) \rightarrow ZnS(s) + H_2O(l)$

(c) $2H_2(g) + O_2(g) \rightarrow 2H_2O(l)$

(d) $2C_2H_6(g) + 7O_2(g) \rightarrow 4CO_2(g) + 6H_2O(l)$

(e) $2N_2O_5(g) \rightarrow 4NO_2(g) + O_2(g)$

8. Predict the sign of the entropy change, ΔS, for each of the following processe .

(a) $He(g, 298\ K, 1\ atm) \rightarrow He(g, 298\ K, 0.5\ atm)$

(b) $I_2(g, 298\ K, 1\ atm) \rightarrow I_2(s, 298\ K, 1\ atm)$

(c) $H_2O(s, 273\ K, 1\ atm) + 10H_2O(l, 373\ K, 1\ atm) \rightarrow 11H_2O(l, <373\ K, 1\ atm)$

9. Calculate the standard entropy change, $\Delta S°$, associated with each of the following reactions at 298 K:

(a) $C(s, graphite) + O_2(g) \rightarrow CO_2(g)$

(b) $C_2H_5OH(l) + 3O_2(g) \rightarrow 2CO_2(g) + 3H_2O(l)$

(c) $C_6H_{12}O_6(s) + 6O_2(g) \rightarrow 6CO_2(g) + 6H_2O(l)$

(d) $H_2(g) + I_2(s) \rightarrow 2HI(g)$

10. What is the standard entropy change, $\Delta S°$, associated with each of the following reactions at 298 K? Explain qualitatively the sign of each of the entropy changes by comparing the extent of molecular disorder in the reactants and the products:

(a) $CaCO_3(s) \rightarrow CaO(s) + CO_2(g)$

(b) $Br_2(l) + 3F_2(g) \rightarrow 2BrF_3(g)$

(c) $2CO(g) + O_2(g) \rightarrow 2CO_2(g)$

(d) $C(s, graphite) + H_2O(l) \rightarrow CO(g) + H_2(g)$

(e) $2Na(s) + Cl_2(g) \rightarrow 2NaCl(s)$

11. Find the standard entropy change, $\Delta S°$, associated with each of the following reactions:

(a) $S(s, rhombic) + O_2(g) \rightarrow SO_2(g)$

(b) $N_2(g) + O_2(g) \rightarrow 2NO(g)$

(c) $P_4(s) + 5O_2(g) \rightarrow P_4O_{10}(s)$

(d) $4Fe(s) + 3O_2(g) \rightarrow 2Fe_2O_3(s)$

(e) $2H_2(g) + O_2(g) \rightarrow 2H_2O(l)$

12. Calculate the standard entropy change, $\Delta S°$, associated with each of the following reactions at 298 K:

(a) $PCl_3(g) + Cl_2(g) \rightarrow PCl_5(g)$

(b) $H_2O(l) + SO_3(g) \rightarrow H_2SO_4(l)$

(c) $C_2H_2(g) + 2H_2(g) \rightarrow C_2H_6(g)$

(d) $2C(s, graphite) + O_2(g) \rightarrow 2CO(g)$

(e) $H_2(g) + Br_2(l) \rightarrow 2HBr(g)$

Gibbs Free Energy

13. Using the data in Tables 16.1, 16.2, and Appendix B2, calculate the standard enthalpy change, $\Delta H°$, the standard free energy change, $\Delta G°$, and the standard entropy change, $\Delta S°$, for the reaction

$$Fe_2O_3(s) + 3C(s, graphite) \longrightarrow 2Fe(s) + 3CO(g)$$

Is this reaction spontaneous at 25 °C? Verify that $\Delta G° = \Delta H° - T\Delta S°$, and discuss whether the enthalpy change and the entropy change, respectively, work for or against the spontaneity of the reaction. Which factor dominates?

14. Calculate the standard free energy change at 25 °C for the reaction

$$H_2(g) + Cl_2(g) \longrightarrow 2HCl(g)$$

Is the reaction as written spontaneous? What are the relative contributions of the enthalpy change and the entropy change to the spontaneity of the reaction? Which factor dominates?

15. What is the value of $\Delta G°$ for each of the following reactions? Indicate which reactions are spontaneous under standard conditions.

(a) $C_3H_8(g) + 5O_2(g) \rightarrow 3CO_2(g) + 4H_2O(g)$

(b) $N_2O_4(g) \rightarrow 2NO_2(g)$

(c) $CH_4(g) + CCl_4(l) \rightarrow 2CH_2Cl_2(l)$

16. The $\Delta G_f°$ values for $SO_2(g)$, $H_2S(g)$, and $NO_2(g)$ are, respectively, -300.1, -33.4, and $+51.3$ kJ mol^{-1}. Which of these gases has the greatest tendency to decompose into its elements at 298 K?

17. We saw in Chapter 6 that HCN is produced industrially by the reaction

$$2NH_3(g) + 3O_2(g) + 2CH_4(g) \xrightarrow[\text{catalyst}]{1200°} 2HCN(g) + 6H_2O(g)$$

Is the high temperature needed in order to obtain a high equilibrium yield of the products or to increase the speed of the reaction?

18. Predict the sign of the Gibbs free energy change for reactions at *low* temperature, for which

(a) ΔH is positive and ΔS is positive

(b) ΔH is negative and ΔS is positive

(c) ΔH is negative and ΔS is negative

(d) ΔH is positive and ΔS is negative

19. Repeat Problem 18 for *high*-temperature conditions.

20. Calculate the standard free energy change for each of the following reactions. Comment on the relative oxidizing powers of F_2, Cl_2, and Br_2, and on the relative strengths of zinc and iron as reducing agents:

(a) $2NaF(s) + Cl_2(g) \rightarrow 2NaCl(s) + F_2(g)$

(b) $2NaBr(s) + Cl_2(g) \rightarrow 2NaCl(s) + Br_2(l)$

(c) $PbO_2(s) + 2Zn(s) \rightarrow Pb(s) + 2ZnO(s)$

(d) $Al_2O_3(s) + 2Fe(s) \rightarrow 2Al(s) + Fe_2O_3(s)$

21. (a) Calculate $\Delta G°$ for the formation of 1 mol of $CO_2(g)$ from C(s, diamond) and $O_2(g)$ at 298 K and 1 atm pressure.

(b) What does the sign of this $\Delta G°$ signify?

(c) Should the owners of diamonds be concerned about the conversion of diamond to carbon dioxide? Explain.

Gibbs Free Energy and the Equilibrium Constant

22. For the reaction

$$2SO_2(g) + O_2(g) \longrightarrow 2SO_3(g)$$

what is the value of the equilibrium constant K_p at 298 K? Is this reaction spontaneous at 298 K?

23. For the reaction

$$C(s, graphite) + CO_2(g) \longrightarrow 2CO(g)$$

assuming that $\Delta H°$ and $\Delta S°$ are independent of temperature, what is the value of the equilibrium constant K_p at 700 °C?

24. For the reaction

$$PCl_5(g) \longrightarrow PCl_3(g) + Cl_2(g)$$

what is the value of the equilibrium constant K_p at 298 K?

25. For the reaction

$$N_2O_4(g) \longrightarrow 2NO_2(g)$$

what is the value of the equilibrium constant K_p at 25 °C?

26. Consider the reaction

$$CH_4(g) + 2O_2(g) \longrightarrow CO_2(g) + 2H_2O(g)$$

(a) Is this reaction spontaneous at 298 K?

(b) What is the value of the equilibrium constant K_p at 298 K?

(c) How do you account for the fact that a mixture of methane and oxygen shows a negligible extent of reaction at room temperature, even after a very long time?

27. Calculate the standard free energy change associated with the combustion of liquid methanol:

$$2CH_3OH(l) + 3O_2(g) \longrightarrow 2CO_2(g) + 4H_2O(l)$$

(a) Is this reaction spontaneous under standard conditions?

(b) What is the value of the equilibrium constant K_p at 298 K?

(c) Does the value of the equilibrium constant favor the formation of reactants or of products?

(d) What effect would an increase in pressure have on the spontaneity of the reaction?

(e) What effect would an increase in temperature have on the spontaneity of the reaction?

28. For the reaction:

$$2C(s, graphite) + H_2(g) \longrightarrow C_2H_2(g)$$

the standard free energy change is 209 kJ.

(a) Is this reaction a practical route for the synthesis of ethyne (acetylene), C_2H_2, at room temperature?

(b) Would the reaction be expected to be spontaneous at high temperature?

(c) Calculate the equilibrium constant K_p at 1200 K, assuming that $\Delta H°$ and $\Delta S°$ are independent of temperature.

29. Using your answer to Problem 25, deduce whether a mixture in which the partial pressures of $NO_2(g)$ and $N_2O_4(g)$ are 0.020 atm and 0.040 atm, respectively, is at equilibrium at 25 °C. If not, in which direction will the reaction proceed to establish equilibrium?

30. Calculate the standard free energy change for both of the following reactions. State which is spontaneous under standard conditions, and calculate the equilibrium constant K_p for each at 25 °C:

(a) $Fe_2O_3(s) + 13CO(g) \rightarrow 2Fe(CO)_5(g) + 3CO_2(g)$

(b) $CH_4(g) + N_2(g) \rightarrow HCN(g) + NH_3(g)$

Use $\Delta G_f°$ values from Appendix B, and $\Delta G_f°(HCN, g) = 124.7$ kJ mol^{-1}, and $\Delta G_f°(Fe(CO)_5, g) = -697.3$ kJ mol^{-1}.

Miscellaneous

31. For the reaction

$$2CO(g) + O_2(g) \longrightarrow 2CO_2(g)$$

what is ΔG for the reaction of a mixture where in which the partial pressure of each gas is 0.020 atm? Is ΔG larger or smaller than $\Delta G°$? In which direction will the reaction proceed in this mixture?

32. What is $\Delta G°$ for each of the following reactions? If these reactions were coupled, what would be the overall reaction, and what would be its $\Delta G°$ value? Would the coupled reaction be spontaneous?

    (a) $2CO_2(g) + 4H_2O(l) \rightarrow 2CH_3OH(l) + 3O_2(g)$

    (b) $2C(s) + 2O_2(g) \rightarrow 2CO_2(g)$

33. Explain, in terms of the changes in the entropy of the system and of the surroundings, why endothermic reactions are favored by an increase in temperature.

34. Explain, in terms of the natural tendency for energy to disperse, why chemical reactions take place spontaneously in the direction corresponding to a decrease in the Gibbs free energy.

35. For the formation of $CO(g)$ and $CO_2(g)$ from their elements at 1500 °C, the Gibbs free energies of formation are -250 kJ mol^{-1} and -380 kJ mol^{-1}, respectively. On the basis of the following information, discuss the feasibility of reducing each of the specified metal oxides to the metal with carbon at 1500 °C:

    (a) $4Al(s) + 3O_2(g) \rightarrow Al_2O_3(s)$
                                $\Delta G°(1500\ °C) = -2250$ kJ mol^{-1}

    (b) $2Fe(s) + O_2(g) \rightarrow 2FeO(s)$
                                  $\Delta G°(1500\ °C) = -250$ kJ mol^{-1}

    (c) $2Pb(s) + O_2(g) \rightarrow 2PbO(s)$
                                  $\Delta G°(1500\ °C) = -120$ kJ mol^{-1}

    (d) $2Cu(s) + O_2(g) \rightarrow 2CuO(s)$
                                  $\Delta G°(1500\ °C) = 0$ kJ mol^{-1}

36. Consider the formation of acetylene by two processes: (1) the formation of 1 mol of $C_2H_2(g)$ from graphite and $H_2(g)$, and (2) the formation of $C_2H_2(g)$ from methane gas and oxygen, according to

$$4CH_4(g) + 3O_2(g) \longrightarrow 2C_2H_2(g) + 6H_2O(g)$$

    (a) From a thermodynamic point of view, are either of these reactions feasible for the production of acetylene at room temperature? Explain.

    (b) Would the addition of a suitable catalyst to each of the reactions change your answer to part (a)?

37. In Chapter 10, the production of lead from the conversion of $PbS(s)$ to $PbO(s)$ followed by reduction with coke (carbon) was mentioned.

    (a) Calculate $\Delta G°$ at 298 K for the reduction of $PbO(s)$ to $Pb(s)$ by $Zn(s)$ (a different reducing agent) to give

* The asterisk denotes the more difficult problems.

$ZnO(s)$. Use the values $S°(PbO, s) = 69.5$ J K^{-1} mol^{-1}; $\Delta H_f°(PbO, s) = -217.9$ kJ mol^{-1}; $S°(Pb, s) = 64.9$ J K^{-1} mol^{-1}, and data from Appendix B.

    (b) Is the reduction of PbO to Pb by Zn a spontaneous reaction?

38. Although both $N_2(g)$ and acetylene, $C_2H_2(g)$, have strong triple bonds, nitrogen is very unreactive, while acetylene is very reactive.

    (a) Using bond energy data, calculate $\Delta H°$ for the hydrogenation of 1 mol of $N_2(g)$ to form hydrazine gas, $N_2H_4(g)$, and $\Delta H°$ for the hydrogenation of 1 mol of $C_2H_2(g)$ to form ethane gas, $C_2H_6(g)$.

    (b) Determine the signs of the $\Delta G°$ values for the two hydrogenation reactions in part (a), by qualitative consideration of the sign of $\Delta S°$ for each reaction.

    (c) Are the signs of the $\Delta G°$ values for these hydrogenation reactions consistent with the difference in the reactivities of $N_2(g)$ and $C_2H_2(g)$?

39. For the reaction

$$2SO_2(g) + O_2(g) \longrightarrow 2SO_3(g)$$

    (a) Calculate the value of the equilibrium constant K_p at 298 K

    (b) Calculate the effect of raising the temperature from 298 to 500 K on the value of the equilibrium constant, assuming that both ΔS and ΔH for the reaction are independent of temperature

    (c) Calculate the value of the equilibrium constant K_c at 298 K and 500 K

40. (a) Using only $\Delta G_f°$ values from Appendix B, calculate the value of the equilibrium constant K_p at 25 °C for the reaction

$$CO(g) + 2H_2(g) \longrightarrow CH_3OH(l)$$

    (b) Repeat the above calculation using only $\Delta H_f°$ and $S°$ data from Appendix B.

    (c) Does each method give the same value for K_p?

41. (a) At 25 °C and 1 atm pressure, calculate $\Delta H°$ for the reaction

$$2O_3(g) \longrightarrow 3O_2(g)$$

    (b) Calculate $\Delta G°$ for this reaction under the same conditions.

*42. When liquid water and water vapor are in equilibrium, the following equilibrium holds:

$$H_2O(l) \rightleftharpoons H_2O(g)$$

(a) Write the expression for the equilibrium constant K_p.

(b) Using the $\Delta G°$ value, calculate the pressure of $H_2O(g)$ in equilibrium with $H_2O(l)$ at 25 °C. This is the vapor pressure of water. Compare your calculated value with the value in Table 12.4.

(c) Calculate the vapor pressure of water at 30 °C, assuming that both ΔS and ΔH are independent of temperature.

43. Consider the reaction

$$H_2(g) + Cl_2(g) \longrightarrow 2HCl(g)$$

(a) Calculate the value of the equilibrium constant K_p at 25 °C.

(b) If 1.00 atm of each of these three gases is placed in a closed vessel at 25 °C, what will be the direction of spontaneous change?

(c) Is your prediction consistent with the experimental observation? Explain your answer.

*44. Consider the two reactions

$$S(s) + O_2(g) \longrightarrow SO_2(g)$$
$$SO_2(g) + \tfrac{1}{2}O_2(g) \longrightarrow SO_3(g)$$

(a) Calculate $\Delta G°$ for each of the reactions, and for the reaction given by the sum of these two reactions.

(b) Are the individual reactions of part (a) spontaneous?

(c) Is the reaction given by the sum of these reactions spontaneous?

(d) If sulfur-burning coal is burned, which of SO_2 and SO_3 should predominate in the products?

(e) What is the relationship between the above reactions and the production of acid rain?

Electrochemistry

17

Oxidation–reduction reactions involve the transfer of electrons. Since a flow of electrons constitutes an electric current, we can use chemical reactions to produce electrical energy. Conversely we can use electrical energy to carry out chemical reactions that do not proceed spontaneously.

The batteries for flashlights, radios, pocket calculators, watches, and automobiles, to name just a few of their many applications, use chemical reactions to produce an electric current. The industrial production of many metals, such as sodium, aluminum, and copper, depends on the use of electrical energy to reduce the positive metal ions to the metal.

In this chapter we consider first some examples of the use of electrical energy to carry out chemical reactions. This process is called *electrolysis*. Then we discuss the use of chemical reactions for producing electricity in *electrochemical cells*. We will see that an understanding of electrochemical cells will improve our understanding of the chemistry of metals and allow us to answer a question such as: Why do iron and magnesium dissolve in dilute hydrochloric acid, whereas copper and silver do not, although they will dissolve in concentrated nitric acid? In fact, we will be able to put oxidation–reduction reactions on a quantitative basis, just as we did for acid–base reactions in Chapter 14.

Corrosion of metals occurs as a result of oxidation–reduction reactions, and we will see that the chemistry of corrosion is closely related to the chemistry of electrochemical cells.

17.1
Electrolysis

We mentioned in Chapter 15 that certain metal ions, such as Na^+, K^+, and Ca^{2+}, are difficult to reduce to the metal, and that in many cases the most practical method is to use electrons directly, that is, to use an electric current. **Electrolysis** may be defined as *a process in which electrical energy is used to produce chemical change*. We consider first the preparation of sodium and chlorine by the electrolysis of molten sodium chloride.

ELECTROLYSIS OF MOLTEN SODIUM CHLORIDE

The electrolysis of molten sodium chloride is carried out in an **electrolytic cell**, which consists of a container to hold the molten sodium chloride and two *electrodes* that are made of a solid conducting material, usually a metal or graphite. When a battery or other direct current (dc) source is connected to the two electrodes by conducting wires, it pushes electrons into one electrode, which becomes negatively charged, and it withdraws electrons from the other electrode, which becomes positively charged (see Figure 17.1). The negatively charged electrode attracts positive sodium ions to its surface. Here each sodium ion acquires an electron from the electrode and is reduced to a sodium atom. These sodium atoms combine to form sodium metal, which rises to the surface of the molten sodium chloride. Negative chloride ions are attracted to the positive electrode, and each chloride ion gives up an electron to the electrode and is oxidized to a chlorine atom. These chlorine atoms combine to form chlorine molecules, which form bubbles of gas that rise to the surface.

Thus the reactions taking place at the electrode surfaces are as follows:

| Positive electrode: | anode | $2Cl^- \longrightarrow Cl_2 + 2e^-$ | Oxidation |
| Negative electrode: | cathode | $Na^+ + e^- \longrightarrow Na$ | Reduction |

The electrode at which oxidation occurs is called the **anode**. The electrode at which reduction occurs is called the **cathode**. To remember at which electrode oxidation occurs and at which reduction occurs note that both *a*node and *o*xidation begin with a vowel while both *c*athode and *r*eduction begin with a consonant.

The overall reaction that takes place in the electrolytic cell is called the *cell reaction*. The equation for the complete cell reaction is obtained by combining the anode half-reaction with the cathode half-reaction. It may be necessary to multiply either or both of the equations by appropriate coefficients so that the same number of electrons as are produced in the anode reaction are used up in the cathode reaction. Recall that we used this process before for balancing oxidation–reduction equations. In this case we must multiply the equation for the reduction half-reaction by 2 to give

Cathode reaction	$2Na^+(l) + 2e^- \longrightarrow 2Na(l)$
Anode reaction	$2Cl^-(l) \longrightarrow Cl_2(g) + 2e^-$
Overall reaction	$2Na^+(l) + 2Cl^-(l) \longrightarrow Cl_2(g) + 2Na(l)$

We saw in Chapter 5 that sodium and chlorine combine spontaneously in a highly exothermic reaction to give sodium chloride. The equilibrium

$$2Na(s) + Cl_2(g) \longrightarrow 2NaCl(s) \qquad \Delta H° = -822 \text{ kJ} \qquad \Delta G° = -769 \text{ kJ}$$

lies far to the right. A considerable amount of energy is therefore needed to carry out the reverse reaction—the decomposition of sodium chloride into its elements. In the electrolysis of sodium chloride this energy is supplied by a battery, or other source of direct current, which "pumps" electrons through the electrolytic cell.

The industrial preparation of sodium is carried out in a cell such as the one shown in Figure 17.2. The cell is designed to prevent the sodium and chlorine produced from coming into contact and re-forming sodium chloride. This process is the most important method for making sodium and is also a source of chlorine.

In the course of some of the earliest experiments on the effects of electric currents on substances, the elements sodium and potassium were first prepared by Humphrey Davy (Box 17.1) in 1806 by the electrolysis of their molten carbonates.

In any electrolysis electric current is supplied by an external source. The electrons enter and leave the material being electrolyzed—the *electrolyte*—by means of suitable conducting electrodes. At the surface of each electrode, electrons are transferred from the electrode to the electrolyte, or from the electrolyte to the electrode, by means of chemical reactions. *Electrons are transferred to the electrolyte from the cathode in a reduction reaction*, for example, the reduction of Na^+ to Na. *Electrons are transferred from the electrolyte to the anode in an oxidation reaction*, for example, the oxidation of Cl^- to Cl_2. Charge is carried across the electrolyte not by the movement of free electrons but by the movement of positive ions toward the negative electrode, or cathode, and negative ions toward the positive electrode, or anode.

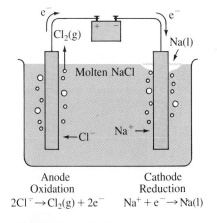

Anode
Oxidation
$2Cl^- \rightarrow Cl_2(g) + 2e^-$

Cathode
Reduction
$Na^+ + e^- \rightarrow Na(l)$

FIGURE 17.1
Electrolysis of Molten Sodium Chloride.

The electric current outside the electrolytic cell is carried by the electrons, which are pushed around the circuit by the battery. Inside the molten electrolyte the current is carried by the movement of the positive and negative ions toward the electrodes. At the anode Cl^- ions give up electrons to the electrode, to give Cl atoms, which combine to give Cl_2 molecules. At the cathode Na^+ ions accept electrons from the electrode, to give Na atoms, which form liquid sodium metal.

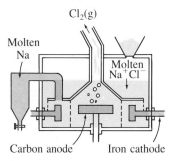

FIGURE 17.2
Cell for Commercial Production of Sodium.

A section through the cell is shown. The iron cathode in the form of a ring surrounds the carbon anode. The cell is designed to prevent the sodium produced at the cathode from coming into contact with the chlorine formed at the anode to re-form sodium chloride.

BOX 17.1

HUMPHREY DAVY (1778–1829)

Davy was brought up on a farm in Cornwall, England. His father had lost much of his money in a mining speculation and died when Davy was only sixteen. Forced to choose a profession that would enable him to help support the family, Davy was apprenticed to a surgeon and apothecary. During his apprenticeship he read widely and studied hard. He set himself a course of study that included theology, geography, medicine, logic, languages (English, French, Latin, Greek, Italian, Spanish, and Hebrew), physics, mechanics, rhetoric and oratory, history, and mathematics.

Just before his ninéteenth birthday he began his study of chemistry by reading Lavoisier's famous *Traité Elémentaire de Chimie* (Box 3.1). Almost immediately, without having had any formal training in chemistry, he began to test Lavoisier's theories with homemade apparatus. Among his many experiments during this period was one showing that plants absorb carbon dioxide in the presence of light and produce oxygen. Despite his youth he had the confidence to challenge established ideas, even Lavoisier's.

In 1797 Davy became an assistant at an institute for the study of the curative powers of the various gases that were being discovered at that time. Among these gases were oxygen, hydrogen, carbon dioxide, and carbon monoxide, not all of which had beneficial effects on the patients at the institute. Davy breathed all these gases and appears to have nearly killed himself with carbon monoxide. In early experiments he noticed that dinitrogen monoxide, N_2O, had an intoxicating effect. It was commonly called "laughing gas", and laughing gas parties became fashionable among the rich. Davy suggested that it be used during operations, which at the time were carried out without anesthetics. Indeed, N_2O later became widely used as the first chemical anesthetic.

In 1801, at the age of twenty-three, Davy was made director of the recently formed Royal Institution in London. His public lectures on chemistry were phenomenally successful, and Davy became a popular celebrity; he has been called the most handsome of all the great scientists.

In one of his greatest experiments, Davy prepared sodium and potassium by the electrolysis of their molten carbonates, using an enormous battery containing 250 metal plates.

In 1812, suffering from overwork and almost certainly from poisoning caused by tasting and smelling all the substances with which he worked and even breathing large quantities of gases, Davy needed a vacation. He resigned from the Royal Institution, married a rich Scottish widow, and set out on a grand tour of Europe, with Michael Faraday as his assistant (Box 17.3). As he traveled, he had discussions with many famous scientists, made geological and biological observations, and continued to do many experiments in chemistry and physics. He applied his ability to many practical problems. One of his most important inventions was the safety lamp for miners. It has been said, perhaps unjustly, that despite his many achievements his greatest discovery was Michael Faraday.

This type of conduction in which the charge carriers are ions is called **ionic conduction**. In the electrodes and in the external circuit (the wires connecting the electrodes to the source of power), the charge carriers are free electrons. This is *electronic, or metallic, conduction* (Chapter 10). The charge carriers change from electrons to ions, or vice versa, at the surfaces between the electrodes and the electrolyte. It is this change that produces the chemical reactions at the electrodes—oxidation at the anode and reduction at the cathode.

PREPARATION OF METALS BY ELECTROLYSIS

Magnesium, calcium, strontium, and barium are prepared by electrolysis of their molten chlorides. Magnesium is the most important of these metals and is produced in the largest quantities. Magnesium chloride is obtained from seawater as described in Chapter 12. It is then melted and electrolyzed to give magnesium and chlorine. The overall reaction is

$$MgCl_2(l) \xrightarrow{\text{Electrolysis}} Mg(l) + Cl_2(g)$$

Probably the most important electrolytic process for the preparation of a metal is that for aluminum (Chapter 10). This process is called the *Hall process* in North America after its inventor Charles M. Hall (Box 17.2). Aluminum is produced by the electrochemical reduction of a molten mixture of Al_2O_3 and *cryolite*, Na_3AlF_6, an ionic compound composed of Na^+ and octahedral AlF_6^{3-} ions. Cryolite is a rare mineral found in appreciable quantities only in Greenland, but the large quantities required for the Hall process are now made synthetically. The aluminum oxide is obtained from the mineral bauxite, $Al_2O_3 \cdot xH_2O$, which generally contains only 35%–60% Al_2O_3 together with oxides of iron, silicon, and titanium as well as small amounts of clay and other silicates. The melting point of Al_2O_3 (2050 °C) is much too high to allow it to be conveniently used for electrolysis. But it dissolves in molten cryolite, Na_3AlF_6, at about 1000 °C.

Ionic solids do not conduct electricity because the ions are not free to move but in the liquid state in which the ions are free to move they are very good conductors.

$$(Na^+)_3 \begin{bmatrix} F & F \\ F & \diagdown \; | \; \diagup & F \\ & Al & \\ F & \diagup \; | \; \diagdown & F \\ & F & \end{bmatrix}^{3-}$$

BOX 17.2

CHARLES MARTIN HALL (1863–1914)

The process we use today for manufacturing aluminium was invented by a young American while he was still an undergraduate at Oberlin College. Inspired by a professor's remark that anyone who could invent a cheap process for producing aluminum would make a fortune, Charles Hall set out to try in 1885. At the time, aluminum cost $90 a pound and was more expensive than either silver or gold. It is said that the very rich flaunted their wealth by dining with aluminum knives and forks.

Hall worked in a woodshed using homemade and borrowed equipment. After about a year he found that Al_2O_3 dissolves in molten cryolite to give a conducting solution, from which aluminum can be deposited by passing an electric current. He used an iron frying pan as a container for the molten cryolite–alumina mixture, which he melted over a blacksmith's forge. The electric current came from electrochemical cells that he made from jars that his mother used to can fruit.

By an odd coincidence Paul Héroult, who was the same age as Hall, made the same discovery independently in France about the same time. As a result of the discovery of Hall and Héroult, the large-scale production of aluminum became economically feasible for the first time, and it became a common and familiar metal.

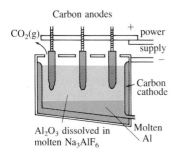

Carbon anodes

FIGURE 17.3
Electrolytic Cell for the
Production of Aluminum by
the Hall Process.

A section through the cell is
shown. The carbon anodes are oxidized to $CO_2(g)$ and as they are
consumed they are lowered further
into the molten mixture of Al_2O_3
and Na_3AlF_6. Aluminum ions are
reduced to aluminum at the carbon cathode that lines the tank.
Because molten aluminum is more
dense than the molten Al_2O_3–
Na_3AlF_6 mixture, aluminum collects at the bottom of the cell and
may be run off.

A typical cell is shown in Figure 17.3. The anodes are graphite rods dipping
into the molten electrolyte. The cathode is a steel vessel, lined with graphite,
that contains the molten electrolyte. Neither the nature of the species present
in solution nor the electrode reactions are completely understood. For simplicity we will assume that the Al_2O_3 is ionized to Al^{3+} and O^{2-}. We can represent the reaction at the cathode by the equation

$$\text{Cathode} \qquad Al^{3+}(l) + 3e^- \longrightarrow Al(l)$$

At the anode several reactions occur in which carbon dioxide, oxygen, and
fluorine are formed. The principal reaction at this electrode may be approximately represented by the equation

$$\text{Anode} \qquad C(s) + 2O^{2-}(l) \longrightarrow CO_2(g) + 4e^-$$

The approximate overall reaction is therefore

$$4Al^{3+}(l) + 6O^{2-}(l) + 3C(s) \longrightarrow 4Al(l) + 3CO_2(g)$$

The graphite electrodes are gradually converted to carbon dioxide and must
be replaced from time to time. The energy consumption is high, and the process
is economically feasible only when carried out near a cheap source of electric
power, for example, at a site where hydroelectric power is produced.

QUANTITATIVE ASPECTS OF ELECTROLYSIS

The principal cost in the electrolytic preparation of aluminum and other metals
is the cost of electricity. Therefore it is important to know how much electricity
is needed to produce a certain amount of metal. The amount of electricity needed
can be calculated from the equation for the appropriate electrode reaction. In
the case of the electrolysis of molten sodium chloride, the reactions are

$$\text{Cathode} \qquad Na^+(l) + e^- \longrightarrow Na(l)$$
$$\text{Anode} \qquad 2Cl^-(l) \longrightarrow Cl_2(g) + 2e^-$$

The passage of one electron produces one sodium atom; the passage of 1 mol
of electrons produces 1 mol of sodium. The passage of two electrons produces
one molecule of Cl_2; the passage of 2 mol of electrons produces 1 mol of Cl_2.
In summary:

Electrode reaction		Product
$Na + e^- \longrightarrow Na$	1 mol electrons	1 mol Na = 23.0 g Na
$2Cl^- \longrightarrow Cl_2 + 2e^-$	2 mol electrons	1 mol Cl_2 = 70.9 g Cl_2

The charge on one electron is $1.602\ 19 \times 10^{-19}$ coulomb (C). Therefore the
charge on 1 mol electrons = $6.022\ 05 \times 10^{23}$ electrons, is $(6.022\ 05 \times 10^{23}$ electrons $mol^{-1})(1.602\ 19 \times 10^{-19}$ C $electron^{-1}) = 96\ 485$ C mol^{-1}. The charge of
1 mol of electrons ($96\ 485$ C mol^{-1}) is called the Faraday constant, F, in honor
of Michael Faraday, who discovered the quantitative laws governing electrolysis
(Box 17.3):

$$F = 96\ 485 \text{ C mol}^{-1}$$

or $96\ 500$ C mol^{-1} when rounded off to three significant figures, which is accurate enough for most purposes. Thus we can find the charge Q of n moles of
electrons by using the expression

$$Q = nF$$

BOX 17.3

MICHAEL FARADAY (1791–1867)

Michael Faraday was one of ten children of a blacksmith in London. He had only a basic elementary education before being apprenticed to a bookbinder at the age of fourteen. His employer, who was uncharacteristically lenient for the times, allowed Faraday to read the books in his shop, and Faraday educated himself. In 1812 he was given tickets to attend lectures given by Humphrey Davy (Box 17.1) at the Royal Institution. He wrote up careful, complete notes of the lectures and bound them in a book. Friends persuaded Faraday to send the notes to Davy in support of his application for a position as an assistant to Davy, who was director of the Royal Institution. Faraday obtained the position, beginning not only his prolific scientific career but also the very fruitful collaboration with Davy. Soon after he was appointed, Faraday left with Davy on a grand tour of Europe, which did much to broaden Faraday's scientific education and gave him the opportunity to meet many famous scientists.

When they were in Florence, Davy and Faraday were able to use a very large lens belonging to the Duke of Tuscany to prove conclusively that diamond consists only of carbon—an idea that was difficult for many scientists at that time to accept. Davy and Faraday used the lens to focus the sun's rays on a diamond enclosed in a bulb of pure oxygen. After about an hour the diamond began to burn. "The diamond", Faraday wrote in his journal, "glowed brilliantly with a scarlet light and when placed in the dark continued to burn for about four minutes." They burned the diamond completely and showed that the bulb then contained nothing but carbon dioxide and excess oxygen.

In 1825 Faraday replaced Davy as director of the Royal Institution, and his reputation soon began to rival that of Davy. Faraday was the first to liquefy several gases, including carbon dioxide, hydrogen sulfide, hydrogen bromide, and chlorine; he discovered benzene and determined its composition; and he discovered the quantitative laws of electrolysis. The Faraday constant, F, was named in his honor. Faraday made even greater contributions to physics. He found that an electric current could be induced in a wire by a moving magnet. He provided both the experimental basis and basic ideas for the theory of electromagnetism, which was later developed by Maxwell. Albert Einstein ranked Faraday with Newton, Galileo, and Maxwell as the greatest physicists of all time.

Faraday was a member of a religious sect so strict that for a time he was denied membership of the church because he had accepted an invitation to have lunch with Queen Victoria on a Sunday! In accordance with his religious beliefs, he tried to live a simple life, accepting rather reluctantly the many honors that came to him. His beliefs enabled him to solve without any uncertainty a moral problem that still faces scientists. During the Crimean War, in the 1850s, the British government asked him to head an investigation of the possibility of preparing large quantities of poison gas for use on the battlefield. Faraday refused to have anything to do with the project, and nothing came of the idea at that time.

Since the production of 1 mol, or 23.0 g, of sodium by reduction of sodium ions requires 1 mol of electrons, it requires an amount of charge

$$Q = nF = (1 \text{ mol})(96\,500 \text{ C mol}^{-1})$$
$$= 96\,500 \text{ C}$$

The production of 1 mol Cl_2 requires 2 mol of electrons and therefore an amount of charge equal to $2 \times 96\,500$ C. In summary:

Electrode reaction	Charge	Product
$Na^+ + e^- \longrightarrow Na$	96 500 C	1 mol Na = 23.0 g Na
$2Cl^- \longrightarrow Cl_2 + 2e^-$	$2 \times 96\,500$ C	1 mol Cl_2 = 70.9 g Cl_2

In practice, charge is usually determined by measuring a current flow for a given time. A charge of one coulomb passes a given point when a current of one ampere (A) flows for one second:

$$1 \text{ coulomb} = 1 \text{ ampere} \times 1 \text{ second}$$

$$1 \text{ C} = 1 \text{ A s}$$

Thus if molten NaCl is electrolyzed for 1.00 h with a current of 50.0 A, the number of coulombs passed through it is

$$50.0 \text{ A} \times 3600 \text{ s} = 180\,000 \text{ A s} = 180\,000 \text{ C}$$

Thus the number of moles of electrons passing through the sodium chloride is

$$n = \frac{Q}{F} = \frac{180\,000 \text{ C}}{96\,500 \text{ C mol}^{-1}} = 1.87 \text{ mol}$$

1.87 mol electrons produces

$$1.87 \text{ mol electrons} \left(\frac{1 \text{ mol Na}}{1 \text{ mol electrons}} \right) = 1.87 \text{ mol Na}$$

1.87 mol electrons produces

$$1.87 \text{ mol electrons} \left(\frac{1 \text{ mol Cl}_2}{2 \text{ mol electrons}} \right) = 0.935 \text{ mol Cl}_2$$

☰ Example 17.1

CALCULATING THE AMOUNTS OF ELECTROLYSIS PRODUCTS

What mass of aluminum will be produced in 1.00 h by the electrolysis of molten $AlCl_3$, using a current of 10.0 A?

Solution

First, from the current and the time, we calculate the number of coulombs:

$$Q = 10.0 \text{ A} \left(\frac{1 \text{ C}}{1 \text{ A s}} \right) \times 1.00 \text{ h} \left(\frac{3600 \text{ s}}{1 \text{ h}} \right) = 3.60 \times 10^4 \text{ C}$$

We then find the number of moles of electrons:

$$n = \frac{Q}{F} = \frac{3.60 \times 10^4 \text{ C}}{96\,500 \text{ C mol}^{-1}} = 0.373 \text{ mol electrons}$$

The half-reaction for the reduction of aluminum is

$$Al^{3+} + 3e^- \longrightarrow Al$$

Thus 3 mol of electrons are needed to produce 1 mol (27.0 g) of Al. Hence 0.373 mol of electrons produces

$$0.373 \text{ mol of electrons} \left(\frac{1 \text{ mol Al}}{3 \text{ mol electrons}} \right) \left(\frac{27.0 \text{ g Al}}{1 \text{ mol Al}} \right) = 3.36 \text{ g Al}$$

$\equiv$ *Example 17.2* $\equiv$

CALCULATING THE AMOUNTS OF ELECTROLYSIS PRODUCTS

What volume of chlorine at STP is produced when a current of 20.0 A is passed through molten sodium chloride for 2.00 h?

Solution

$$\text{Number of coulombs} = Q = 20.0 \text{ A} \left(\frac{1 \text{ C}}{1 \text{ A s}} \right) \times 2.00 \text{ h} \times \frac{3600 \text{ s}}{1 \text{ h}}$$

$$= 1.44 \times 10^5 \text{ C}$$

$$n = \frac{Q}{F} = \frac{1.44 \times 10^5 \text{ C}}{96\,500 \text{ C mol}^{-1}} = 1.49 \text{ mol electrons}$$

The anode reaction is

$$2Cl^- \longrightarrow Cl_2 + 2e^-$$

Therefore 2 mol of electrons produce 1 mol of Cl_2. Hence 1.49 mol of electrons produce

$$\frac{1.49 \text{ mol } Cl_2}{2} = 0.745 \text{ mol } Cl_2$$

From the ideal gas law, we know that 1 mol of an ideal gas occupies 22.4 L at STP. Therefore

$$\text{Volume of } Cl_2 \text{ at STP} = 0.745 \text{ mol } Cl_2 \left(\frac{22.4 \text{ L}}{1 \text{ mol } Cl_2} \right)$$

$$= 16.7 \text{ L}$$

EXERCISE 17.1

What mass of magnesium and what volume of chlorine at STP will be produced by the electrolysis of molten magnesium chloride using a current of 3.00 A for 24.0 h?

ELECTROLYSIS OF AQUEOUS SOLUTIONS

In any electrolysis there must be an oxidation reaction at the anode and a reduction reaction at the cathode. But these reactions are not necessarily the oxidation of the negative ions and the reduction of the positive ions in the solution. In aqueous solution, water may be oxidized or reduced or the electrodes may be oxidized or reduced. For the moment we will consider only inert electrodes. We first discuss the electrolysis of an aqueous sodium chloride solution, which is an important industrial process for the manufacture of sodium hydroxide and chlorine.

ELECTROLYSIS OF AQUEOUS SODIUM CHLORIDE If we electrolyze a concentrated aqueous sodium chloride solution, we find that chlorine is produced at the anode, but at the cathode we obtain hydrogen and not sodium as in the electrolysis of molten sodium chloride (see Figure 17.4). Hydrogen is produced at the cathode because the cathode reaction is the reduction of water,

$$2H_2O + 2e^- \longrightarrow H_2 + 2OH^-$$

rather than the reduction of sodium ion to sodium metal:

$$Na^+ + e^- \longrightarrow Na$$

We conclude that water is more easily reduced than the sodium ion.

Chlorine is produced at the anode so the anode reaction is

$$\text{Anode} \qquad 2Cl^- \longrightarrow Cl_2(g) + 2e^- \qquad \text{(Oxidation)}$$

as in the electrolysis of molten sodium chloride. The alternative possibility is the oxidation of water:

$$2H_2O \longrightarrow O_2 + 4H^+ + 4e^- \qquad \text{(Oxidation)}$$

But the production of chlorine rather than oxygen at the anode shows that Cl^- is more easily oxidized than H_2O.

The overall equation for the cell reaction is

$$2Cl^-(aq) \longrightarrow Cl_2(g) + 2e^-$$
$$\underline{2H_2O(l) + 2e^- \longrightarrow H_2(g) + 2OH^-(aq)}$$
$$2H_2O(l) + 2Cl^-(aq) \longrightarrow Cl_2(g) + H_2(g) + 2OH^-(aq)$$

Since the sodium ion is not reduced at the cathode, it takes no part in the reaction—it is a spectator ion. By adding sodium ion to both sides we can rewrite this equation in the form

$$2H_2O(l) + 2NaCl(aq) \longrightarrow Cl_2(g) + H_2(g) + 2NaOH(aq)$$

The products of the electrolysis of an aqueous solution of sodium chloride are chlorine at the anode and sodium hydroxide and hydrogen at the cathode.

ELECTROLYSIS OF AQUEOUS SODIUM SULFATE If an aqueous solution of sodium sulfate is electrolyzed, the products are oxygen at the anode and hydrogen at the cathode (see Figure 17.5). At the anode the possibilities are either that sulfate ion is oxidized or that water is oxidized. Since sulfur in the sulfate ion is in its highest oxidation state, that is, $+6$, it is not surprising that it is the

FIGURE 17.4
Electrolysis of Aqueous
Sodium Chloride Solution.

The Cl^- ions are attracted to the anode, to which they give up electrons to form Cl atoms, which combine to form $Cl_2(g)$. At the cathode electrons are transferred to H_2O molecules; water is reduced to OH^- ions and H atoms, and the latter combine to give $H_2(g)$. The Na^+ ions attracted to the cathode are not reduced but keep the solution neutral; in other words, a solution of NaOH is formed at the cathode.

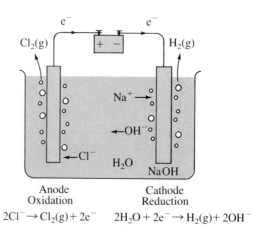

Anode   Cathode
Oxidation   Reduction
$2Cl^- \rightarrow Cl_2(g) + 2e^-$ $2H_2O + 2e^- \rightarrow H_2(g) + 2OH^-$

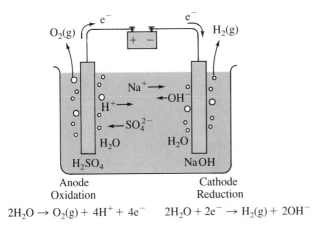

FIGURE 17.5
Electrolysis of Aqueous
Sodium Sulfate Solution.

Water molecules give up two
electrons at the anode to give O
atoms and H^+ ions; the former
combine to give O_2 molecules.
Water molecules accept electrons
at the cathode to give H atoms
and OH^- ions; the former combine
to give H_2 molecules. The Na^+
and SO_4^{2-} ions transport the
current through the cell, and by
their motion in opposite directions
they keep the solution neutral at
every point. Thus a solution of
$H_2SO_4(H_3O^+, SO_4^{2-}$ is formed
at the anode, and a solution of
$NaOH (Na^+OH^-)$ is formed at the
cathode.

water that is oxidized rather than the sulfate ion. The anode reaction is

$$\text{Anode} \quad 2H_2O(l) \longrightarrow O_2(g) + 4H^+(aq) + 4e^- \quad \text{(Oxidation)}$$

We have seen in discussing the electrolysis of aqueous sodium chloride that water, not sodium ion, is reduced at the cathode:

$$\text{Cathode} \quad 2H_2O(l) + 2e^- \longrightarrow H_2(g) + 2OH^-(aq) \quad \text{(Reduction)}$$

If we multiply this equation by 2 and add it to the equation for the anode reaction, we obtain the equation for the overall cell reaction:

$$6H_2O(l) \longrightarrow O_2(g) + 2H_2(g) + 4OH^-(aq) + 4H^+(aq)$$

This equation can be simplified, because H^+ and OH^- combine to give water. We can then subtract $4H_2O$ from each side to give

$$2H_2O(l) \longrightarrow O_2(g) + 2H_2(g)$$

This is the overall equation for the electrolysis of water.

Neither the sodium ion nor the sulfate ion participate in the electrode reactions. What then is their function? By their movement within the solution, they serve to conduct the current between the electrodes and to maintain the electrical neutrality of the solution. Positive hydronium (hydrogen) ions are produced at the anode. The negative sulfate ions arriving at the anode compensate for the charge of the hydronium ions so that the solution has no overall charge. Thus sulfuric acid ($H_2SO_4(aq) = H_3O^+(aq) + SO_4^{2-}(aq)$) is produced in the vicinity of the anode. Similarly positive sodium ions arriving at the cathode compensate for the negative charge of the hydroxide ions that are produced at the cathode, so that the solution has no overall charge. Thus a solution of sodium hydroxide is produced in the vicinity of the cathode ($NaOH(aq) = Na^+(aq) + OH^-(aq)$). Unless these solutions are kept apart by some barrier between them they will diffuse into each other and react to form sodium sulfate, and so overall the sodium sulfate remains unchanged and the only reaction is the electrolysis of water as we saw above.

Nevertheless the sodium sulfate plays an important role in the electrolysis in that its ions carry the current through the solution. Electrolysis can only occur as fast as the sulfate ions arrive at the anode and the sodium ions arrive at the cathode. Electrolysis of a very dilute sodium sulfate solution would be very slow. In principle we could even electrolyze pure water but the concentrations of H_3O^+ and OH^- ions resulting from the self-dissociation of water are each only 10^{-7} mol L^{-1} so electrolysis would be extremely slow.

EXERCISE 17.2

What volume of hydrogen and what volume of oxygen at STP will be produced by the electrolysis of an aqueous solution of sodium sulfate by a current of 0.10 A for 1.0 h?

ELECTROLYSIS OF AQUEOUS COPPER(II) SULFATE   In the electrolysis of a copper sulfate solution, oxygen is evolved at the anode and copper is deposited on the cathode. Thus as we saw for sodium sulfate, sulfate ion is not oxidized at the anode; rather, water is oxidized to oxygen. The anode reaction is

$$\text{Anode} \qquad 2H_2O(l) \longrightarrow O_2(g) + 4H^+(aq) + 4e^- \qquad \text{(Oxidation)}$$

In solutions of sodium chloride and sodium sulfate, water is reduced at the cathode, rather than Na^+, but in the present case Cu^{2+} is reduced to copper. We conclude that Cu^{2+} is more easily reduced than water, although Na^+ is less easily reduced than water. The cathode reaction is

$$\text{Cathode} \qquad Cu^{2+}(aq) + 2e^- \longrightarrow Cu(s) \qquad \text{(Reduction)}$$

The equation for the overall cell reaction can be obtained by multiplying the equation for the cathode reaction by 2 and adding it to the equation for the anode reaction, to give

$$2H_2O(l) + 2Cu^{2+}(aq) \longrightarrow O_2(g) + 4H^+(aq) + 2Cu(s)$$

The electrolysis of some aqueous solutions is demonstrated in Experiment 17.1.

ELECTROLYTIC PREPARATION OF CHLORINE AND SODIUM HYDROXIDE

The electrolysis of aqueous sodium chloride is an extremely important reaction because it is the principal method by which chlorine and sodium hydroxide are made industrially. Indeed, more than 90% of the world's chlorine is made by this process. Sodium hydroxide and chlorine are two of the basic chemicals produced by the chemical industry. In 1987 they ranked seventh and eighth, respectively, among all the chemicals produced in the United States (Table 15.1). From the equation for the overall cell reaction,

$$2NaCl(aq) + 2H_2O(l) \longrightarrow Cl_2(g) + H_2(g) + 2NaOH(aq)$$

we see that hydrogen is another important product of the process, although as we saw in Chapter 3, there are other important methods for making hydrogen.

The chlorine gas must be kept separated from the hydrogen and from the sodium hydroxide so that it cannot react with them. The cell shown in Figure 17.6 accomplishes this separation by using a porous diaphragm, usually made of asbestos, through which the solution can be made to flow but which prevents

Experiment 17.1

Electrolysis of Some Aqueous Solutions

Electrolysis of a colorless aqueous solution of potassium iodide with inert (copper) electrodes produces brown I_3^- at the anode and H_2 and OH^- at the cathode. The bubbles of hydrogen can be seen rising from the cathode, and the OH^- ion causes phenophthalein in the solution to turn pink.

The beaker contains a solution of copper sulfate and a graphite electrode which is connected to the positive terminal of a battery. The other electrode is a strip of silver which is connected to the negative terminal of the battery.

When the silver electrode is dipped into the solution, electrolysis commences and bubbles of oxygen can be seen rising from the graphite electrode.

After a few minutes the silver electrode is removed from the solution. It can be seen to be coated with a layer of red-brown copper.

the passage of the gases produced at the electrodes. This diaphragm also prevents OH^- produced at the cathode from diffusing to the anode compartment, where it would react with the chlorine. The solution that flows out of the cathode compartment contains 11% NaOH and 16% NaCl. This solution is concentrated by evaporation. The less soluble NaCl crystallizes and can be filtered off. When the concentration of NaOH reaches 50% by mass, only about 1% NaCl remains in solution. For many applications this small impurity of NaCl is not important.

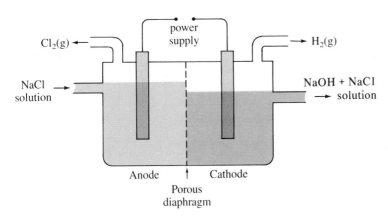

FIGURE 17.6
Diaphragm Cell for Production of Cl_2 by Electrolysis of Aqueous Sodium Chloride Solution.

The anode and cathode compartments are separated by a porous asbestos diaphragm. A difference in level between the solutions in the anode and cathode compartments keeps the solution flowing from the anode to the cathode compartment, which prevents movement of OH^- ions formed at the cathode into the anode compartment, where they would react with the chlorine.

In the type of cell shown in Figure 17.7 a stream of mercury is used as the negative electrode. Hydrogen is not formed at this electrode, but sodium ions are reduced to sodium, which dissolves in the mercury. The mercury electrode is not inert; it takes part in the electrode reaction by dissolving the sodium that is formed by reduction of Na^+:

$$Na^+ + Hg + e^- \longrightarrow \quad Na(Hg)$$

Sodium amalgam

The solution of sodium in mercury, which is called *sodium amalgam*, flows out of the cell and is subsequently decomposed by being mixed with water:

$$2Na(amalgam) + 2H_2O \longrightarrow 2Na^+ + 2OH^- + H_2 + (mercury)$$

Thus the hydrogen and the sodium hydroxide are produced separately from the chlorine, and very pure sodium hydroxide is obtained. The mercury is returned to the main cell and used again, but some mercury is always lost. A serious disadvantage of this type of cell is that a very small amount of mercury tends to find its way into the waste water from the plant, creating an environmental hazard (see Box 21.3).

ELECTROLYTIC REFINING OF COPPER

In most of the examples of electrolysis that we have considered so far, the electrodes were inert; in other words, they did not react under the conditions of the electrolysis. However, not all electrodes are inert. We have already seen that in the manufacture of aluminum by the electrolysis of a solution of Al_2O_3 in molten Na_3AlF_6, the carbon anodes are oxidized to CO_2. Oxidation of the anode also occurs when copper is used as an anode in the electrolysis of an aqueous solution. Copper dissolves to give Cu^{2+}, rather than water being oxidized. The anode reaction is

$$\text{Anode} \qquad Cu(s) \longrightarrow Cu^{2+}(aq) + 2e^-$$

In a cell in which we have a copper anode in a solution of copper(II) sulfate,

FIGURE 17.7
Mercury Cell for Production of Chlorine by Electrolysis of Aqueous Sodium Chloride Solution.

At the anode Cl^- ions are oxidized to $Cl_2(g)$. At the mercury cathode Na^+ ions are reduced to sodium, which dissolves in the mercury. This Na–Hg amalgam is pumped into the tank at the top of the figure, where it comes into contact with water and reacts to form sodium hydroxide and hydrogen. The mercury is then pumped back into the electrolytic cell.

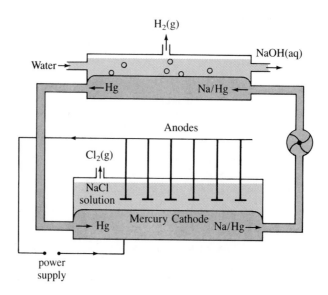

copper goes into solution at the anode, and at the cathode copper is deposited. The cathode reaction is

$$\text{Cathode} \qquad Cu^{2+}(aq) + 2e^- \longrightarrow Cu(s)$$

This type of cell is the basis for an important method for the purification (refining) of crude copper (see Figure 17.8). Impure copper is the anode of the cell. Unreactive metals that are not as easily oxidized as copper, such as silver and gold, fall to the bottom of the cell as a sludge. Other metals that are easily oxidized, such as zinc and iron, go into solution as their positive ions, together with the copper. But if the potential at which the cell operates is suitably adjusted, ions such as Fe^{2+} and Zn^{2+}, which are less readily reduced than Cu^{2+}, are not reduced to the metal at the cathode. Thus only copper is deposited at the cathode, which may be a thin sheet of pure copper or another metal on which a layer of pure copper is built up. The sludge is treated to isolate the less easily oxidized metals such as Ag, Au, and Pt which are valuable by-products of the process.

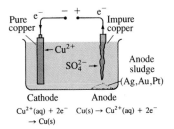

FIGURE 17.8
Electrolytic Refining of Copper.

The impure copper dissolves at the anode. Pure copper is deposited at the cathode.

ELECTROPLATING

We have just seen that copper can be "plated out" on the cathode during electrolysis. Other metals can be deposited on a cathode in the same way. This process is called *electroplating*. The object to be plated is the cathode and the plating metal is the anode of an electrolytic cell. Many automobile parts are plated with chromium to improve their appearance and to protect them against corrosion (Experiment 17.2). Knives, forks, spoons, trays, jugs, and many other objects in the home are often silverplated.

Example 17.3

CALCULATING THE AMOUNTS OF ELECTROLYSIS PRODUCTS

How much copper would be deposited during the electrolysis of an aqueous copper sulfate solution if a current of 1.30 A was passed for 1.50 h?

Solution

$$\text{Number of coulombs} = (1.30 \text{ A})\left(\frac{1 \text{ C}}{1 \text{ A s}}\right)(1.50 \text{ h})\left(\frac{3600 \text{ s}}{1 \text{ h}}\right)$$

$$= 7.02 \times 10^3 \text{ C}$$

$$n = \frac{Q}{F} = \frac{7.02 \times 10^3 \text{ C}}{96\ 500 \text{ C mol}^{-1}}$$

$$= 0.0727 \text{ mol electrons}$$

Copper is produced at the cathode according to the equation

$$Cu^{2+} + 2e^- \longrightarrow Cu$$

$$\text{Moles of Cu} = 0.0727 \text{ mol electrons}\left(\frac{1 \text{ mol Cu}}{2 \text{ mol of electrons}}\right)$$

$$= 0.0364 \text{ mol Cu}$$

$$= 0.0364 \text{ mol Cu}\left(\frac{63.55 \text{ g Cu}}{1 \text{ mol Cu}}\right) = 2.31 \text{ g Cu}$$

Experiment 17.2

Electroplating

A current is passed between strips of copper in an aqueous solution of dichromic acid $H_2Cr_2O_7$, obtained by adding chromium trioxide, CrO_3, to water.

After a few minutes the copper strip has become plated with a shiny layer of chromium.

EXERCISE 17.3

By measuring the amount of a metal deposited at the cathode by weighing the cathode before and after an electrolysis, we can find the amount of charge passed through a cell. An apparatus for measuring charge in this way is called a coulometer. If 0.5230 g of silver was deposited from a silver nitrate solution in a coulometer, how many coulombs of charge were passed through the coulometer?

SUMMARY OF ELECTRODE REACTIONS

There are three different types of reactions that may occur at the electrodes in electrolysis:

Reaction involves	Anode	Cathode
1. Electrolyte	Anions oxidized	Cations reduced
2. Solvent	Solvent oxidized	Solvent reduced
3. Electrode	Anode oxidized	Cathode reduced

1. The simplest situation is the electrolysis of a *molten* electrolyte with inert electrodes. In such cases the cation is always reduced and the anion is always oxidized. For example:

Electrolyte	Anode reaction	Cathode reaction
Molten NaCl	$2Cl^- \longrightarrow Cl_2 + 2e^-$	$Na^+ + e^- \longrightarrow Na$
Molten $MgCl_2$	$2Cl^- \longrightarrow Cl_2 + 2e^-$	$Mg^{2+} + 2e^- \longrightarrow Mg$

2. In the electrolysis of *aqueous* electrolyte solutions water is oxidized to oxygen at the anode if the anion is less easily oxidized than water; water is reduced to hydrogen at the cathode if the cation is less easily reduced

than water. From the results of experiments we can classify the common anions and cations according to whether they are more or less easily oxidized than water:

Cations	*Anions*
Less easily reduced than water:	Less easily oxidized than water:
Na^+, K^+, Mg^{2+}, Ca^{2+}, Al^{3+},	F^-, SO_4^{2-}, NO_3^-, ClO_4^-,
Fe^{2+}, Pb^{2+}	CO_3^{2-}, PO_4^{3-}
More easily reduced than water:	More easily oxidized than water:
Cu^{2+}, Ag^+, $H_3O^+(H^+)$	Cl^-, Br^-, I^-

Some examples are as follows:

Electrolyte	Anode reaction	Cathode reaction
Aqueous Na_2SO_4	$2H_2O \longrightarrow O_2 + 4H^+ + 4e^-$	$2H_2O + 2e^- \longrightarrow H_2 + 2OH^-$
Aqueous KI	$2I^- \longrightarrow I_2 + 2e^-$	$2H_2O + 2e^- \longrightarrow H_2 + 2OH^-$
Aqueous $CuSO_4$	$2H_2O \longrightarrow O_2 + 4H^+ + 4e^-$	$Cu^{2+} + 2e^- \longrightarrow Cu$

Chloride ion is a borderline case in that it is not much more easily oxidized than water. In a concentrated solution, chloride ion is oxidized to chlorine; but in a dilute solution of Cl^- there is such a large excess of water that water is preferentially oxidized to oxygen.

3. Mostly we are concerned with electrolysis using inert electrodes, but there are a few important cases where the electrode undergoes reaction. For example, copper is more easily oxidized than water. So if a copper anode is used in the electrolysis of a solution having an anion that is not easily oxidized, the copper anode goes into solution as Cu^{2+}. In the electrolysis of molten Al_2O_3, oxide ions react with the graphite anode to give carbon dioxide. In the electrolysis of aqueous sodium chloride with a mercury cathode, sodium ions are reduced to sodium, which dissolves in the mercury electrode. Some examples are as follows:

Electrolyte	Anode material	Anode reaction
Aqueous Na_2SO_4	Cu	$Cu \longrightarrow Cu^{2+} + 2e^-$
Aqueous $CuSO_4$	Cu	$Cu \longrightarrow Cu^{2+} + 2e^-$
Molten Al_2O_3	C	$C + 2O^{2-} \longrightarrow CO_2 + 4e^-$
Aqueous NaCl	Inert	$2Cl^- \longrightarrow Cl_2 + 2e^-$

Electrolyte	Cathode material	Cathode reaction
Aqueous Na_2SO_4	Inert	$2H_2O + 2e^- \longrightarrow H_2 + 2OH^-$
Aqueous $CuSO_4$	Inert	$Cu^{2+} + 2e^- \longrightarrow Cu$
Molten Al_2O_3	Inert	$Al^{3+} + 3e^- \longrightarrow Al$
Aqueous NaCl	Hg	$Na^+ + Hg + e^- \longrightarrow Na–Hg(amalgam)$

═ *Example 17.4* ═══════════════════════════════

PREDICTING THE PRODUCTS OF ELECTROLYSIS

Predict the products of electrolysis of each of the following:

(a) A dilute aqueous solution of sulfuric acid with inert electrodes

(b) Molten lead bromide with inert electrodes

(c) An aqueous solution of silver nitrate with inert electrodes

(d) An aqueous solution of silver nitrate with silver electrodes

Solution

(a) At the cathode $H^+(aq)$ is more easily reduced than H_2O. Thus $H^+(aq)$ is reduced to hydrogen:

$$2H^+(aq) + 2e^- \longrightarrow H_2$$

At the anode SO_4^{2-} is less easily oxidized than H_2O. Thus water is oxidized to oxygen:

$$2H_2O \longrightarrow O_2 + 4H^+ + 4e^-$$

(b) At the cathode Pb^{2+} is reduced to lead:

$$Pb^{2+} + 2e^- \longrightarrow Pb$$

At the anode Br^- is oxidized to bromine:

$$2Br^- \longrightarrow Br_2 + 2e^-$$

(c) At the cathode Ag^+ is more easily reduced than water, so silver is deposited:

$$Ag^+ \longrightarrow Ag(s) + e^-$$

At the anode water is more easily oxidized than NO_3^-. Thus water is oxidized to oxygen.

(d) At the cathode Ag^+ is more easily reduced than water, so silver is deposited. At the anode Ag is more easily oxidized than water, so the silver electrode dissolves to give $Ag^+(aq)$:

$$Ag(s) \longrightarrow Ag^+ + e^-$$

EXERCISE 17.4

Predict the products of electrolysis of each of the following:

(a) Molten calcium iodide with inert electrodes

(b) A dilute aqueous solution of phosphoric acid with inert electrodes

(c) A dilute aqueous solution of copper(II) nitrate with inert electrodes

(d) A dilute aqueous solution of copper(II) nitrate with copper electrodes

17.2
Electrochemical Cells

In principle any spontaneous oxidation–reduction reaction, that is, any reaction for which ΔG is negative, can be used to produce an electric current if an appropriate cell can be set up. To understand this, we consider the reaction that occurs when we place a piece of zinc in a solution of copper(II) sulfate. The zinc soon becomes coated with a reddish deposit of metallic copper, and the blue color of the solution due to the hydrated copper ion begins to fade. If enough zinc is present and enough time is allowed, the solution eventually becomes colorless. These changes result from a reaction in which electrons are transferred from zinc to copper; each Cu^{2+} ion deposited as metallic copper takes two electrons from a zinc atom, which dissolves as Zn^{2+} ion (Experiment 17.3):

$$Zn(s) + Cu^{2+}(aq) \longrightarrow Zn^{2+}(aq) + Cu(s)$$

In other words, zinc is oxidized to $Zn^{2+}(aq)$, while $Cu^{2+}(aq)$ is reduced to copper metal. This reaction can be used as the basis of an **electrochemical cell** to produce an electric current.

An electrochemical cell, which is also called a *voltaic cell*, or a *galvanic cell*, employs a spontaneous chemical reaction to produce an electric current. In contrast, in an electrolytic cell an electric current is used to provide the energy needed to force a nonspontaneous reaction to occur. We have seen that if an electric current is passed through molten sodium chloride it can be decomposed into sodium and chlorine. In contrast, if sodium and chlorine were allowed to combine in a suitable electrochemical cell, an electric current could be produced.

When the reaction between zinc and copper ions occurs on the surface of the zinc the flow of electrons from the zinc to the copper ions cannot be used. In order to produce an electric current the oxidation and reduction reactions must be separated so that the flow of electrons can be arranged to occur through an external circuit. This can be done by setting up a cell such as that shown in Figure 17.9.

A piece of zinc, called the zinc electrode, is immersed in a solution of zinc sulfate, and a piece of copper, the copper electrode, is immersed in a solution of copper(II) sulfate. The two solutions are kept apart by a porous barrier that permits ions to move from one solution to the other but prevents the mixing of the two solutions by diffusion. A conducting metal wire connects the two electrodes. At the surface of the zinc electrode, Zn^{2+} ions are formed and pass

Experiment 17.3

The Reaction of Zinc with an Aqueous Solution of Copper Sulfate

A strip of zinc and an aqueous solution of copper sulfate.

When the zinc strip is placed in the copper sulfate solution it rapidly becomes coated with copper.

When the zinc strip is removed from the solution, the characteristic red-brown color of the copper deposit is clearly seen.

FIGURE 17.9
An Electrochemical Cell.

This cell is based on the overall reaction

$$Zn(s) + Cu^{2+}(aq) \longrightarrow$$
$$Zn^{2+}(aq) + Cu(s)$$

(a) shows the experimental arrangement. The voltage of the cell is 1100 mV = 1.100 V. (b) is a schematic diagram showing the movement of the ions and the reactions that take place at the electrodes.

(a)

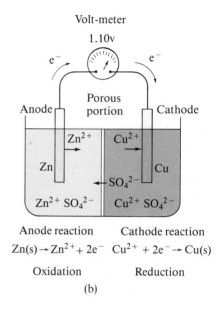

Anode reaction   Cathode reaction
$Zn(s) \rightarrow Zn^{2+} + 2e^-$ $Cu^{2+} + 2e^- \rightarrow Cu(s)$

Oxidation   Reduction

(b)

into the solution, leaving excess electrons in the electrode. These electrons flow through the wire to the copper electrode, where they combine with Cu^{2+} ions arriving at the electrode surface. These Cu^{2+} ions are thereby converted to copper atoms, which are deposited on the electrode surface. (Any metal could be used for this electrode and it would become coated with a layer of copper.) The electrode at which oxidation occurs, in this case the zinc electrode, is called the *anode*. The electrode at which reduction occurs, in this case the copper electrode, is the *cathode*. These definitions are the same as those we gave previously for an electrolytic cell.

In the copper–zinc cell the electrode reactions are as follows:

Anode	$Zn(s) \longrightarrow Zn^{2+}(aq) + 2e^-$	(Oxidation)
Cathode	$Cu^{2+}(aq) + 2e^- \longrightarrow Cu(s)$	(Reduction)
Overall reaction	$Zn(s) + Cu^{2+}(aq) \longrightarrow Zn^{2+}(aq) + Cu(s)$	

The two halves of the cell are called half cells. The zinc electrode dipping into a zinc sulfate solution is one of the half cells, and the copper electrode dipping into a copper sulfate solution is the other half cell.

Neither half-cell reaction can take place by itself. Each half-cell reaction must be accompanied by another half-cell reaction that can use up or supply the necessary electrons. These electrons constitute the electric current that flows around the external circuit. Not only must the circuit be complete outside the cell, but it must be complete inside as well; that is, ions must be able to move from one half cell to the other. In the cell in Figure 17.9, the ions move through the porous partition.

Another method is to connect the two half cells by a *salt bridge*, as shown in Figure 17.10. If the two solutions are not in contact in this way, the electrode reactions would quickly cease. Zinc ions going into solution at the zinc electrode would give the solution in the zinc half cell a positive charge, and the attraction of this charge for the electrons in the zinc electrode would prevent them from leaving. Similarly, at the copper electrode the removal of copper ions from the solution would leave the solution with an overall negative charge.

This negative charge would prevent more electrons from entering the copper electrode, so the electrode reaction would stop. When the porous partition is in place, sulfate ions can move from the copper half cell to the zinc half cell. The loss of the sulfate ions from the copper half cell keeps this solution neutral, and the sulfate ions that pass through the barrier neutralize the charge of the zinc ions that are formed in the zinc half cell, so this solution also remains neutral. A salt bridge serves the same purpose (see Figure 17.10). The movement of sulfate ions in one direction and of zinc and copper ions in the opposite direction constitutes the current that flows through the cell.

Experiment 17.4 shows that an electrochemical cell is formed whenever two different metals are inserted into a conducting medium.

CELL POTENTIALS

The electric current produced by an electrochemical cell is the result of electrons being pushed out of the anode and around the external circuit to the cathode, where they are used up. When a current flows between two points, we say that there is a *potential difference* between the two points: The current flows from the higher potential to the lower potential. This potential difference is measured in volts and is often called the *voltage* of the cell. When we speak of a 6-volt (V) battery, we mean a battery that has a potential difference of 6 V between its terminals. The potential difference is a measure of the work that the battery can do by pushing charge around an external circuit. If 1 J of work is done by moving a charge of 1 C through a potential difference, the potential difference is 1 V:

$$1\ V = 1\ J\ C^{-1}$$

The potential or voltage between the electrodes of an electrochemical cell is called the *cell potential*, E_{cell}. For a given cell reaction the cell potential depends on the concentrations of the ions in the cell, the temperature, and the partial

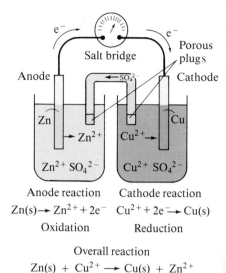

Anode reaction   Cathode reaction

$$Zn(s) \rightarrow Zn^{2+} + 2e^- \quad Cu^{2+} + 2e^- \rightarrow Cu(s)$$

Oxidation   Reduction

Overall reaction

$$Zn(s) + Cu^{2+} \rightarrow Cu(s) + Zn^{2+}$$

FIGURE 17.10
Electrochemical Cell with Salt Bridge.

The two half cells are connected by a salt bridge, which consists of an inverted U tube containing a conducting solution, such as $NaNO_3(aq)$ or $Na_2SO_4(aq)$. The salt bridge allows movement of ions between the two half cells but prevents the solution in the anode and cathode compartments from mixing. (a) shows the experimental arrangement. (b) is a schematic diagram showing the movement of the ions and the reactions occurring at the electrodes.

Experiment 17.4

A Simple Electrochemical Cell

Strips of copper and zinc inserted into a lemon form an electrochemical cell. The measured voltage is 0.9 V.

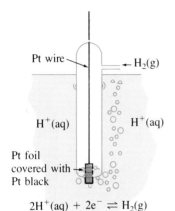

Pt wire

H₂(g)

H⁺(aq)   H⁺(aq)

Pt foil
covered with
Pt black

$$2H^+(aq) + 2e^- \rightleftharpoons H_2(g)$$

FIGURE 17.11
A Hydrogen Electrode.

This consists of a piece of platinum foil covered with a layer of finely divided platinum (platinum black), attached to a Pt wire, and surrounded by an atmosphere of hydrogen. The whole is immersed in an aqueous acidic solution.

pressures of any gases that might be involved in the cell reactions. When all the concentrations are 1 mol L^{-1}, all partial pressures of gases are 1 atm, and the temperature is 25 °C, the cell potential is called the **standard cell potential**, E°_{cell}. The cell potential can be measured by connecting a high-resistance voltmeter across the cell. The standard cell potential for the zinc–copper cell (Figure 17.9) is found to be 1.10 V.

In principle, any spontaneous oxidation–reduction reaction can be used as the basis of an electrochemical cell. For example, zinc dissolves in dilute aqueous acids:

$$Zn(s) + 2H^+(aq) \longrightarrow Zn^{2+}(aq) + H_2(g)$$

How do we set up an electrode involving hydrogen? We use a piece of platinum foil coated with a black layer of finely divided platinum that serves as a catalyst for the reaction

$$2H^+(aq) + 2e^- \longrightarrow H_2(g)$$

but which is otherwise inert. This electrode is immersed in a solution of $H^+(aq)$ ions, for example, a sulfuric acid solution. Hydrogen ions combine with electron from the metal to give hydrogen gas, which forms bubbles on the surface of the platinum. Thus a hydrogen electrode consists of bubbles of hydrogen gas on the surface of platinum (see Figure 17.11).

A cell based on the reaction between zinc and dilute acid is shown in Figure 17.12. It is composed of a half cell consisting of a hydrogen electrode immersed in a solution of sulfuric acid and a half cell consisting of a zinc electrode in a solution of zinc sulfate. Zinc dissolves to give Zn^{2+} ions, and the electrons released flow around the wire of the external circuit and enter the platinum electrode, where they combine with hydrogen ions at its surface to give H_2. The standard cell potential is 0.76 V.

Anode	$Zn(s) \longrightarrow Zn^{2+}(aq) + 2e^-$	(Oxidation)
Cathode	$2H^+(aq) + 2e^- \longrightarrow H_2(g)$	(Reduction)
Overall reaction	$Zn(s) + 2H^+(aq) \longrightarrow Zn^{2+}(aq) + H_2(g)$	$E^\circ = 1.10$ V

Rather than represent a cell by a detailed drawing, as in Figures 17.9, 17.10, and 17.12, we can use a shorthand representation called a **cell diagram**. For example, the cell in Figure 17.9 would be represented by

$$Zn(s)|ZnSO_4(aq)||CuSO_4(aq)|Cu(s)$$

and the cell shown in Figure 17.12 would be represented as

$$Zn(s)|ZnSO_4(aq)||H_2SO_4(aq)|H_2(g), Pt(s)$$

In these diagrams the single vertical line separates the electrode from the solution with which it is in contact, and the double vertical lines indicate a salt bridge or porous barrier between the two solutions. In such a cell diagram the anode—the electrode at which oxidation occurs—is on the left, and the cathode—the electrode at which reduction occurs—is on the right.

Just as we can think of the overall cell reaction as the sum of two half-reactions, we can think of the cell potential as the sum of the two half-cell potentials: E_{ox} due to the oxidation half-reaction and E_{red} due to the reduction half-reaction:

$$E_{cell} = E_{ox} + E_{red}$$

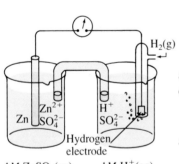

Zn²⁺
Zn SO₄²⁻

H⁺
SO₄²⁻

H₂(g)

Hydrogen
electrode

1M ZnSO₄(aq)   1M H⁺(aq)

Anode   Cathode

$Zn(s) \rightarrow$ $2H^+ + 2e^- \rightarrow$
$Zn^{2+} + 2e^-$ $H_2(g)$

FIGURE 17.12
Standard Electrochemical Cell Using a Hydrogen Electrode.

This cell is based on the reaction

$Zn(s) + 2H^+(aq) \longrightarrow$
$Zn^{2+}(aq) + H_2(g)$

STANDARD REDUCTION POTENTIALS

It would be useful to have values for half-cell potentials, but no potential can be measured unless two half cells are connected to give a cell in which an overall reaction can occur. Therefore we cannot measure experimentally the potential associated with any individual half-reaction; we can measure only the *sum* of two half-cell potentials.

However, we can obtain relative values of standard half-cell potentials by making an arbitrary assumption about the potential of one particular half cell. The half cell chosen as the standard with which all other half-cell potentials are compared is the hydrogen half cell. This half cell is arbitrarily assigned a potential of 0 V.

$$E^{\circ}_{2H^+ + 2e^- \to H_2} = 0$$

The standard cell potential for the zinc–hydrogen cell is 0.76 V. Now since

$$E^{\circ}_{cell} = E^{\circ}_{ox} + E^{\circ}_{red}$$
$$0.76 \text{ V} = E^{\circ}_{Zn \to Zn^{2+} + 2e^-} + E^{\circ}_{2H^+ + 2e^- \to H_2}$$

and since we have assumed that

$$E^{\circ}_{2H^+ + 2e^- \to H_2} = 0$$
$$0.76 \text{ V} = E^{\circ}_{Zn \to Zn^{2+} + 2e^-} + 0$$

or

$$E^{\circ}_{Zn \to Zn^{2+} + 2e^-} = 0.76 \text{ V}$$

Thus we see that the standard potential for the half-cell reaction

$$Zn(s) \longrightarrow Zn^{2+} + 2e^-$$

is $E^{\circ}_{ox} = 0.76$ V.

Now we can use this value to find the half-cell potential for the copper electrode by combining this electrode with the zinc electrode. We have already seen that the standard potential of the zinc–copper cell is 1.10 V. Therefore

$$1.10 \text{ V} = E^{\circ}_{ox} + E^{\circ}_{red}$$
$$= E^{\circ}_{Zn(s) \to Zn^{2+} + 2e^-} + E^{\circ}_{Cu^{2+} + 2e^- \to Cu(s)}$$
$$= 0.76 \text{ V} + E^{\circ}_{Cu^{2+} + 2e^- \to Cu(s)}$$
$$E^{\circ}_{Cu^{2+} + 2e^- \to Cu(s)} = 0.34 \text{ V}$$

Thus for the reaction

$$Cu^{2+} + 2e^- \longrightarrow Cu(s)$$

we have

$$E^{\circ}_{red} = 0.34 \text{ V}$$

We could alternatively have obtained the standard potential for the copper half cell by combining it with a hydrogen electrode. However, we know that copper does not dissolve in dilute aqueous acids. The equilibrium

$$Cu(s) + 2H^+ \rightleftharpoons Cu^{2+} + H_2(g)$$

lies far to the left. In other words, it is the reverse reaction that occurs when we combine a hydrogen half cell with a copper half cell; that is,

$$Cu^{2+} + H_2(g) \longrightarrow Cu(s) + 2H^+$$

The experimentally measured voltage of this cell is 0.34 V. Thus

$$0.34 \text{ V} = E^\circ_{ox} + E^\circ_{red}$$

$$= E^\circ_{H_2(g) \rightarrow 2H^+ + 2e^-} + E^\circ_{Cu^{2+} + 2e^- \rightarrow Cu(s)}$$

$$= 0 + E^\circ_{Cu^{2+} + 2e^- \rightarrow Cu(s)}$$

So for the reaction

$$Cu^{2+} + 2e^- \longrightarrow Cu(s)$$

we have

$$E^\circ_{red} = 0.34 \text{ V}$$

This is the same value we obtained above.

By combining any half cell with a hydrogen electrode, we can determine the standard half-cell potential. Some of these potentials are for oxidation reactions, and others are for reduction reactions. It is convenient, in tabulating the values, to write them all the same way. By convention, *all standard half-cell potentials are listed as reductions* and are called **standard reduction potentials**.

The sign of the potential of an oxidation reaction is reversed when it is written as a reduction reaction.

═ *Example 17.5* ═══════════════════════════════════

STANDARD REDUCTION POTENTIALS

Iron dissolves in dilute acid to give Fe^{2+}:

$$Fe(s) + 2H^+(aq) \longrightarrow Fe^{2+}(aq) + H_2(g)$$

The potential of a cell constructed from a standard hydrogen electrode and a half cell consisting of an iron electrode in a $1M$ solution of $FeSO_4$ is 0.44 V. What is the standard reduction potential of the Fe half cell?

Solution

$$E^\circ_{cell} = E^\circ_{ox} + E^\circ_{red}$$

$$0.44 \text{ V} = E^\circ_{Fe \rightarrow Fe^{2+} + 2e^-} + 0$$

Therefore

$$Fe \longrightarrow Fe^{2+} + 2e^- \qquad E^\circ_{ox} = 0.44 \text{ V} \quad \text{and} \quad E^\circ_{red} = -0.44 \text{ V}$$

EXERCISE 17.5

The standard cell potential for the cell

$$Zn(s)|ZnSO_4(aq)||AgNO_3(aq)|Ag(s)$$

is 1.56 V. What is the standard reduction potential for the reaction $Ag^+(aq) + e^- \rightarrow Ag(s)$?

Table 17.1 gives the value of the standard reduction potentials for a number of half-reactions. Recall that these are the half-cell potentials when the partial

TABLE 17.1
Standard Reduction Potentials

Oxidizing Agents	Reaction[a]	Reducing Agents	E°_{red} (V)
	Acidic Solution		
Very weak	$Li^+ + e^- \longrightarrow Li(s)$	Very strong	-3.05
	$K^+ + e^- \longrightarrow K(s)$		-2.93
	$Ca^{2+} + 2e^- \longrightarrow Ca(s)$		-2.87
	$Na^+ + e^- \longrightarrow Na(s)$		-2.71
	$Mg^{2+} + 2e^- \longrightarrow Mg(s)$		-2.36
	$H_2(g) + 2e^- \longrightarrow 2H^-$		-2.25
	$Al^{3+} + 3e^- \longrightarrow Al(s)$		-1.66
	$2H_2O + 2e^- \longrightarrow H_2(g) + 2OH^-$		-0.83
	$Zn^{2+} + 2e^- \longrightarrow Zn(s)$		-0.76
	$Cr^{3+} + 3e^- \longrightarrow Cr(s)$		-0.74
	$Fe^{2+} + 2e^- \longrightarrow Fe(s)$		-0.44
	$Cr^{3+} + e^- \longrightarrow Cr^{2+}$		-0.41
	$V^{3+} + e^- \longrightarrow V^{2+}$		-0.26
	$Ni^{2+} + 2e^- \longrightarrow Ni(s)$		-0.25
	$Sn^{2+} + 2e^- \longrightarrow Sn(s)$		-0.16
	$Pb^{2+} + 2e^- \longrightarrow Pb(s)$		-0.13
Increasing Oxidizing strength	$2H^+ + 2e^- \longrightarrow H_2(g)$	Increasing reducing strength	0
	$AgBr(s) + e^- \longrightarrow Ag(s) + Br^-$		$+0.10$
	$S(s) + 2H^+ + 2e^- \longrightarrow H_2S(aq)$		$+0.14$
	$Cu^{2+} + e^- \longrightarrow Cu^+$		$+0.15$
	$AgCl(s) + e^- \longrightarrow Ag(s) + Cl^-$		$+0.22$
	$Cu^{2+} + 2e^- \longrightarrow Cu(s)$		$+0.34$
	$Cu^+ + e^- \longrightarrow Cu(s)$		$+0.52$
	$I_2(s) + 2e^- \longrightarrow 2I^-$		$+0.54$
	$O_2(g) + 2H^+ + 2e^- \longrightarrow H_2O_2(aq)$		$+0.68$
	$Fe^{3+} + e^- \longrightarrow Fe^{2+}$		$+0.77$
	$Ag^+ + e^- \longrightarrow Ag(s)$		$+0.80$
	$NO_3^- + 2H^+ + e^- \longrightarrow NO_2(g) + H_2O$		$+0.80$
	$2Hg^{2+} + 2e^- \longrightarrow Hg_2^{2+}$		$+0.92$
	$NO_3^- + 4H^+ + 3e^- \longrightarrow NO(g) + 2H_2O$		$+0.97$
	$Br_2 + 2e^- \longrightarrow 2Br^-$		$+1.09$
	$O_2(g) + 4H^+ + 4e^- \longrightarrow 2H_2O$		$+1.23$
	$Cr_2O_7^{2-} + 14H^+ + 6e^- \longrightarrow 2Cr^{3+} + 7H_2O$		$+1.33$
	$Cl_2(g) + 2e^- \longrightarrow 2Cl^-$		$+1.36$
	$MnO_4^- + 8H^+ + 5e^- \longrightarrow Mn^{2+} + 4H_2O$		$+1.49$
	$Au^{3+} + 3e^- \longrightarrow Au(s)$		$+1.50$
	$MnO_2 + 4H^+ + 2e^- \longrightarrow Mn^{2+} + 4H_2O$		$+1.61$
	$H_2O_2(aq) + 2H^+ + 2e^- \longrightarrow 2H_2O$		$+1.78$
	$Co^{3+} + e^- \longrightarrow Co^{2+}$		$+1.81$
Very strong	$F_2 + 2e^- \longrightarrow 2F^-$	Very weak	$+2.87$
	Basic Solution		
	$2H_2O + 2e^- \longrightarrow H_2(g) + 2OH^-$		-0.83
	$Fe(OH)_3(s) + e^- \longrightarrow Fe(OH)_2(s) + OH^-$		-0.56
	$O_2(g) + e^- \longrightarrow O_2^-$		-0.56
	$O_2(g) + 2H_2O + 4e^- \longrightarrow 4OH^-$		$+0.40$

[a] All ions are in aqueous solution, and H_2O is in the liquid state.

pressures of any gases are 1 atm and the concentrations of ions are $1M$. By convention, the most negative reduction potentials are listed at the top of the table, the most positive at the bottom.

The significance of a negative sign is that, relative to H_2, the substances on the right tend to give up electrons and the half-reaction tends to proceed from right to left. The larger the negative value, the greater is the tendency of the reaction to proceed from right to left. The positive sign indicates that, relative to H^+, the substances on the left tend to acquire electrons, and the half-reaction tends to proceed from left to right. The larger the positive value, the greater is the tendency of the reaction to proceed to the right.

The substances on the right in the table are listed in order of decreasing strength as reducing agents. At the top of the table substances such as Li, K, Mg, and H^- are strong reducing agents. At the bottom of the table the substances on the right are very poor reducing agents, but those on the left are strong oxidizing agents, for example, NO_3^-, O_2, MnO_4^-, and Cl_2. The substances on the left are listed in order of increasing strength as oxidizing agents. Lithium is the strongest reducing agent listed in the table, and F_2 is the strongest oxidizing agent.

Notice that the table includes a number of half-reactions that do not involve the reduction of a metal ion to the corresponding metal, for example, the reduction of MnO_4^- to Mn^{2+}, the reduction of the halogens to the corresponding halide ions, and the reduction of Fe^{3+} to Fe^{2+}. In principle, at least, the electrode potentials corresponding to these reactions can be determined by setting up a half cell consisting of a solution of the ions (and gases) involved in the equilibrium and a platinum electrode to supply or remove the electrons. This half cell can then be combined with a hydrogen electrode or any other electrode of known potential.

RELATIVE STRENGTHS OF OXIDIZING AND REDUCING AGENTS

We can use the table of standard reduction potentials to predict the relative oxidizing or reducing strengths of any of the substances listed in the table. For example, since zinc comes above iron and has a more negative reduction potential, it is the stronger reducing agent. Since Br_2 comes below Fe^{3+} in the table and has a larger positive potential, it is the stronger oxidizing agent.

=== *Example 17.6* ===

RELATIVE STRENGTHS OF OXIDIZING AGENTS

Place the following in order of their strengths as oxidizing agents:

Cu^{2+} MnO_4^- Br_2 Zn^{2+}

Solution

If we list the reduction potentials, we have

$$Zn^{2+} + 2e^- \longrightarrow Zn \qquad\qquad -0.76 \text{ V}$$
$$Cu^{2+} + 2e^- \longrightarrow Cu \qquad\qquad +0.34 \text{ V}$$
$$Br_2 + 2e^- \longrightarrow 2Br^- \qquad\qquad +1.09 \text{ V}$$
$$MnO_4^- + 8H^+ + 5e^- \longrightarrow Mn^{2+} + 4H_2O \qquad +1.49 \text{ V}$$

The substance with the highest positive potential is the strongest oxidizing agent. Therefore the order of oxidizing strength is

$$MnO_4^- > Br_2 > Cu^{2+} > Zn^{2+}$$

EXERCISE 17.6

Place the following in order of their strengths as oxidizing agents in aqueous solution:

Sn^{2+} Co^{3+} H_2O NO_3^-

EXERCISE 17.7

Place the following in order of their strengths as reducing agents in aqueous solution:

H_2S $Sn(s)$ $Al(s)$ H_2O_2 Fe^{2+}

DIRECTION OF OXIDATION–REDUCTION REACTIONS

We can predict which way the overall reaction will proceed when we combine any two half-reactions. The reactions at the top of the table have a greater tendency than those at the bottom to proceed in the reverse direction, that is, to the left. Therefore, *when we combine any two half-reactions, the one that is higher in the table will go to the left, driving the other reaction to the right.* Consider the combination of two half-reactions:

$$Zn^{2+} + 2e^- \longrightarrow Zn(s) \quad \text{and} \quad Fe^{2+} + 2e^- \longrightarrow Fe(s)$$

The zinc reaction is the higher in the table, so it will go to the left, forcing the other reaction to the right. Thus we must reverse the first reaction,

$$Zn(s) \longrightarrow Zn^{2+} + 2e^-$$

and then add it to the second,

$$Fe^{2+} + 2e^- \longrightarrow Fe(s)$$

to get the overall reaction:

$$Zn(s) + Fe^{2+} \longrightarrow Fe(s) + Zn^{2+}$$

Zinc is the stronger reducing agent and it reduces Fe^{2+} to Fe.

A simple way to remember how to combine two half-reactions is to write them down in the order in which they are listed in the table:

$$Zn^{2+} + 2e^- \longrightarrow Zn(s)$$
$$Fe^{2+} + 2e^- \longrightarrow Fe(s)$$

Then connect the two reactions by a counterclockwise arrow—it has the form of a letter C (combine counterclockwise)—that shows the direction in which the reactions proceed when they are combined.

$$Zn^{2+} + 2e^- \longrightarrow Zn(s)$$

$$Fe^{2+} + 2e^- \longrightarrow Fe(s)$$

Thus if we wish to combine the two reactions

$$Cu^{2+} + 2e^- \longrightarrow Cu(s) \quad \text{and} \quad Ag^+ + e^- \longrightarrow Ag(s)$$

we write them in the order in which they are found in the table and we connect them with a ↄ to show us which way the reactions will proceed when combined. We then multiply the second reaction by 2 and add the two reactions to give

$$Cu(s) + 2Ag^+ \longrightarrow Cu^{2+} + Ag(s)$$

Copper reduces silver ion to silver.

We can similarly predict that iron will reduce Cu^{2+} to Cu and that zinc will reduce Pb^{2+} to Pb and Sn^{2+} to Sn (see Experiment 17.5).

CALCULATION OF CELL POTENTIALS

We can calculate the voltage of a cell made up of any two half cells by adding the half-cell voltages. Consider a cell made by combining the two half cells

$$Zn^{2+} + 2e^- \longrightarrow Zn(s) \quad \text{and} \quad Fe^{2+} + 2e^- \longrightarrow Fe(s)$$

Experiment 17.5

Metal Displacement Reactions

An iron nail dipped in copper sulfate solution becomes coated with a reddish brown layer of copper.

A coil of copper wire suspended in silver nitrate solution becomes covered with "whiskers" of silver, and the solution slowly turns blue as copper ions are formed.

A zinc rod in lead nitrate solution becomes coated with a spongy layer of lead.

Pieces of granulated zinc in an acidic solution of tin(II) chloride, $SnCl_2$ become coated with fine crystals of tin.

We have seen that the first reaction will proceed in the reverse direction and that the overall reaction in the cell will be

$$Zn(s) + Fe^{2+} \longrightarrow Zn^{2+} + Fe(s)$$

The cell diagram is

$$Zn(s)|Zn^{2+}(aq)||Fe^{2+}(aq)|Fe(s)$$

The half-reactions and the corresponding potentials are

| Anode | $Zn(s) \longrightarrow Zn^{2+} + 2e^-$ | $E^\circ_{ox} = +0.76$ V |
| Cathode | $Fe^{2+} + 2e^- \longrightarrow Fe(s)$ | $E^\circ_{red} = -0.44$ V |

The potential for the oxidation of zinc is found by changing the sign of the potential for the reduction reaction given in Table 17.1. Adding the half-cell reactions and potentials gives, for the complete cell,

$$Zn(s) + Fe^{2+} \longrightarrow Zn^{2+} + Fe(s) \qquad E^\circ_{cell} = +0.32 \text{ V}$$

Now consider the cell

$$Cu(s)|Cu^{2+}(aq)||Ag^+(aq)|Ag(s)$$

obtained by combining the two half cells

$$Cu^{2+} + 2e^- \longrightarrow Cu(s) \quad \text{and} \quad Ag^+ + e^- \longrightarrow Ag(s)$$

The copper half cell is higher in the table, and therefore the half-reaction must be reversed and written as an oxidation before combining it with the silver half-cell reaction:

Anode	$Cu(s) \longrightarrow Cu^{2+} + 2e^-$	$E^\circ_{ox} = -E^\circ_{red} = -0.34$ V
Cathode	$2[Ag^+ + e^- \longrightarrow Ag(s)]$	$E^\circ_{red} = +0.80$ V
Overall	$Cu(s) + 2Ag^+ \longrightarrow Cu^{2+} + 2Ag(s)$	$E^\circ_{cell} = +0.46$ V

Notice that when we multiply the equation for a half-reaction by an appropriate coefficient to obtain an equation for the overall reaction, we do *not* multiply the half-cell potential by this coefficient. The half-cell potential is always the value given in the table and is independent of any coefficient by which it may be necessary to multiply the corresponding equation in order to obtain a balanced equation for the overall reaction. The standard potential does not depend on the amounts of the reactants and the products but only on their concentrations, which are 1 mol L^{-1}. It therefore does not depend on how many electrons are transferred. A potential difference may be regarded as a difference in level between which the electrons flow; it does not depend on how many electrons flow between the levels.

Let us calculate the cell potential for the cell

$$Ag(s)|Ag^+(aq)||Cu^{2+}(aq)|Cu(s)$$

that is, for the reaction

$$2Ag(s) + Cu^{2+} \longrightarrow 2Ag^+ + Cu(s)$$

which is the reverse of the reaction that we have just considered. The half-cell reactions and their appropriate potentials are as follows:

Anode	$2[Ag(s) \longrightarrow Ag^+ + e^-]$	$E^\circ_{ox} = -E^\circ_{red} = -0.80$ V
Cathode	$Cu^{2+} + 2e^- \longrightarrow Cu(s)$	$E^\circ_{red} = +0.34$ V
Overall	$2Ag(s) + Cu^{2+} \longrightarrow 2Ag^+ + Cu(s)$	$E^\circ_{cell} = -0.46$ V

The overall cell potential has a negative value, whereas in the previous case it had a positive value. This negative value indicates that the reaction does *not* proceed in the direction written but in the reverse direction. The reaction actually proceeds in the direction

$$2Ag^+ + Cu(s) \longrightarrow Cu^{2+} + 2Ag(s)$$

as we have seen previously. *The calculated cell potential for a reaction that proceeds spontaneously as written from left to right is always positive. A calculated cell potential that is negative indicates that the reaction proceeds spontaneously in the reverse direction.*

Example 17.7

CALCULATING STANDARD CELL POTENTIALS

What will be the spontaneous reaction when the following half-reactions are combined? What is the value of E°_{cell}?

(a) $Fe^{3+} + e^- \rightarrow Fe^{2+}$

(b) $MnO_4^- + 8H^+ + 5e^- \rightarrow Mn^{2+} + 4H_2O$

Solution

From Table 17.1 we see that reaction (a) is higher in the table, so it proceeds to the left, driving reaction (b) to the right. We therefore reverse reaction (a) and change the sign of its potential. After multiplying by the appropriate coefficient, we add it to reaction (b). We then add the half-cell potentials to obtain E°_{cell}:

$$\begin{array}{ll} 5[Fe^{2+} \longrightarrow Fe^{3+} + e^-] & E^\circ_{ox} = -0.77 \text{ V} \\ MnO_4^- + 8H^+ + 5e^- \longrightarrow Mn^{2+} + 4H_2O & E^\circ_{red} = +1.49 \text{ V} \\ \hline 5Fe^{2+} + MnO_4^- + 8H^+ \longrightarrow 5Fe^{3+} + Mn^{2+} + 4H_2O & E^\circ_{cell} = +0.72 \text{ V} \end{array}$$

The overall reaction is the reduction of MnO_4^- to Mn^{2+} and the oxidation of Fe^{2+} to Fe^{3+}. The cell voltage has a positive value, confirming that we combined the two half-cell reactions correctly.

EXERCISE 17.8

What will be the spontaneous reaction, and what is the value of E°_{cell}, when the following half-reactions are combined?

(a) $Cr_2O_7^{2-} + 14H^+ + 6e^- \rightarrow 2Cr^{3+} + 7H_2O$

(b) $Cu^{2+} + 2e^- \rightarrow Cu(s)$

Example 17.8

CELL DIAGRAMS AND STANDARD CELL POTENTIALS

Using the half-reactions

(a) $Fe^{2+} + 2e^- \rightarrow Fe$

(b) $Cu^{2+} + 2e^- \rightarrow Cu$

construct an electrochemical cell and predict the standard cell potential. Give the cell diagram. Draw a sketch of the cell, labeling the anode and the cathode, and showing the direction of current flow.

Solution

The data in Table 17.1 show that Fe is a better reducing agent than Cu and Cu^{2+} is a better oxidizing agent than Fe^{2+}. Therefore Fe will reduce Cu^{2+}. Thus the anode and cathode reactions and the overall reaction are

Anode	$Fe \longrightarrow Fe^{2+} + 2e^-$	$E_{ox} = -E^\circ_{red} = +0.44$ V
Cathode	$Cu^{2+} + 2e^- \longrightarrow Cu$	$E^\circ_{red} = +0.34$ V
Overall	$Fe + Cu^{2+} \longrightarrow Fe^{2+} + Cu$	$E^\circ_{cell} = +0.78$ V

The cell diagram is $Fe(s)|Fe^{2+}(aq)\,\|\,Cu^{2+}(aq)|Cu(s)$

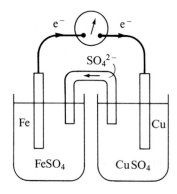

EXERCISE 17.9

Construct an electrochemical cell based on the half-reactions

(a) $Al^{3+} + 3e^- \rightarrow Al$

(b) $Mg^{2+} + 2e^- \rightarrow Mg$

Calculate the standard cell potential Give the cell diagram.

GIBBS FREE ENERGY AND CELL POTENTIAL

The calculated potential for a cell gives us another criterion for deciding whether or not a given reaction, in this case the oxidation–reduction reaction occurring in a cell, is spontaneous or not. We saw in Chapter 16 that the free energy change and the equilibrium constant for a reaction also tell us whether or not the reaction occurs spontaneously. So we expect that the cell potential will be related to the free energy change ΔG and the equilibrium constant for the reaction occurring in the cell. The relationship is

$$\Delta G = -nFE_{cell} = -RT \ln(K/Q)$$

or, if all substances are in their standard states,

$$\Delta G^\circ = -nFE^\circ_{cell} = -RT \ln K$$

= *Example 17.9* =

CALCULATION OF ΔG AND K FROM E_{cell}

Is Fe^{2+}(aq) spontaneously oxidized by the oxygen of the air in acidic solution? Calculate $\Delta G°$ and K for the reaction.

Solution

First write the two half-reactions, then combine them to give the overall reaction:

Reduction	$O_2(g) + 4H^+(aq) + 4e^- \rightarrow 2H_2O$	$E°_{red} = \quad 1.23$ V
Oxidation	$4[Fe^{2+} \rightarrow Fe^{3+} + e^-]$	$-E_{red} = -0.77$ V
Overall	$4Fe^{2+} + O_2(g) + 4H^+ \rightarrow 4Fe^{3+} + 2H_2O$	$E°_{cell} = \quad 0.46$ V

$E_{ox} = $

Since $E°_{cell}$ is positive, the reaction will proceed spontaneously from left to right if all substances are in their standard states. Since 1 V $= 1$ J C^{-1} we have

$$\Delta G° = -nFE°_{cell} = -4(96\ 500\ \text{C})(0.46\ \text{V}) = -1.8 \times 10^5\ \text{J} = -180\ \text{kJ}$$

The large negative value of $\Delta G°$ indicates that the reaction proceeds spontaneously to the right.

$$\ln K = \frac{\Delta G°}{RT} = \frac{(180\ \text{kJ mol}^{-1})\left(\dfrac{1000\ \text{J}}{1\ \text{kJ}}\right)}{(8.31\ \text{J mol}^{-1}\ \text{K}^{-1})(298\ \text{K})} = 73$$

$$K = 1.0 \times 10^{31}$$

This is a very large equilibrium constant so again we see that the reaction is spontaneous and the position of equilibrium lies far to the right.

EXERCISE 17.10

What is the standard free energy change associated with the reaction in the zinc–copper cell? What is the equilibrium constant for this reaction?

We can summarize the various criteria that we have used to decide whether or not a given reaction occurs spontaneously as follows:

A reaction will proceed spontaneously from left to right as written if	$\Delta G < 0$	$Q < K$	$E_{cell} > 0$
A reaction will not proceed spontaneously but will proceed in the reverse direction right to left if	$\Delta G > 0$	$Q > K$	$E_{cell} < 0$
The system is at equilibrium and no overall reaction will occur if	$\Delta G = 0$	$Q = K$	$E_{cell} = 0$

PRODUCTS OF ELECTROLYSIS

We saw earlier that in the electrolysis of aqueous solutions the reactions at the electrodes might be oxidation or reduction of ions, of water, or of the electrode. The table of standard reduction potentials enables us to predict which reactions will occur at the electrode, although, as we will see, there are some limitations to these predictions.

The substance that is reduced at the cathode will be the one that is most easily reduced. All those metal ions that are above the reaction for the reduction of water in the table are more difficult to reduce than water, so it is water that is preferentially reduced in a solution containing any of these ions. Thus Na^+, Ca^{2+}, and Al^{3+}, for example, cannot be reduced in aqueous solution because water is preferentially reduced. But ions such as Cr^{3+}, Cu^{2+}, and Ag^+, which come below water in the table, are more easily reduced; they are therefore reduced to the metal in the electrolysis of aqueous solutions of their cations. Similarly, any reaction that comes above the reaction for the oxidation of water,

$$2H_2O(l) \longrightarrow 4e^- + 4H^+(aq) + O_2(g)$$

will occur more easily than the oxidation of water. Thus we see that in aqueous solution Br^- will be preferentially oxidized to Br_2, and I^- will be oxidized to I_2. But F^- will not be oxidized to F_2, and Mn^{2+} will not be oxidized to MnO_4^-; instead, water will be oxidized.

However, there is a complication. We have already seen that chloride ion is oxidized to chlorine when a sodium chloride solution is electrolyzed, although the table predicts that water should be oxidized preferentially. This unexpected result is due to the fact that a higher voltage is sometimes needed for electrolysis than is indicated by the reduction potential. This additional voltage required to cause electrolysis is called *overvoltage*. Its origin is connected with the rates of the electrode reactions. If these reactions are slow, as is often the case when a gas is involved, then an additional voltage is required for the electrode reaction to take place. The overvoltage is considerably higher for the oxidation of water than for chloride ion, so chloride ion is more easily oxidized. As we will see, there is also a concentration effect. In very dilute solutions of chloride ion, water is more easily oxidized, and oxygen rather than chlorine is obtained.

Unfortunately, we cannot therefore always make reliable predictions about the products of electrolysis from standard reduction potentials. Nevertheless, we understand the general principles that govern which electrode reactions will occur and therefore why different electrode reactions are observed. Overvoltages are of considerable practical importance, but further discussion of them is beyond the scope of this book.

NONSTANDARD CONDITIONS: THE NERNST EQUATION

The values of the standard reduction potentials listed in Table 17.1 are for solutions in which the dissolved species all have concentrations of $1M$ and the partial pressures of any gases involved are 1 atm. In practice the concentrations of the species taking part in the cell reaction are not always $1M$, nor are their partial pressures 1 atm. Even if we start with $1M$ concentrations, as the cell reaction proceeds the concentrations of the reactants necessarily decrease. When equilibrium is reached there is no further tendency for the cell reaction to occur. The cell then produces no current and the cell potential is zero. Thus, as the cell reaction proceeds and the concentrations of the reactants decrease, the cell potential decreases and eventually becomes zero. A flashlight battery, for example, has a limited life. After it has been used for some time its voltage becomes too small to be useful and we say that the battery is "dead".

The quantitative relationship between the cell potential and the concentrations of the reactants is called the **Nernst equation**. It was first derived by

German chemist Walter Nernst (1864–1941) in 1899. The potential E of a cell is given by the expression

$$E = E° - \frac{RT}{nF} \ln Q$$

In this equation, $E°$ is the standard potential and Q is the reaction quotient, that is, an expression that has the same form as the expression for the equilibrium constant but in which the concentrations are not the equilibrium concentrations (see Chapter 13). In this equation, F is the Faraday constant and n is the number of moles of electrons transferred in the reaction. Note that ln means logarithm to the base e, that is, the natural logarithm, not logarithm to the base 10, which we write as log. The relationship between $\log x$ and $\ln x$ is $\ln x = 2.303 \log x$. At 25 °C

$$\frac{RT}{F} = \frac{(8.31 \text{ J K}^{-1})(298.2 \text{ K})}{96\,500 \text{ C mol}^{-1}} = 0.0257 \text{ J C}^{-1} = 0.0257 \text{ V}$$

Thus at 25 °C the Nernst equation becomes

$$E = E° - \frac{0.0257}{n} \ln Q \quad \text{or} \quad E = E° - \frac{0.0592}{n} \log Q$$

Using this equation, we can calculate the potential of a cell if we know the concentrations of the reactants and the standard cell potential, $E°$. Consider, as an example, a cell based on the reaction

$$Zn(s) + Cu^{2+}(aq) \longrightarrow Zn^{2+}(aq) + Cu(s)$$

for which $E° = 1.10$ V. In this reaction 2 mol of electrons are transferred for each mole of Cu^{2+} that is reduced, so $n = 2$. Hence the Nernst equation becomes

$$E = 1.10 - \frac{0.0257}{2} \ln \frac{[Zn^{2+}]}{[Cu^{2+}]}$$

Recall from Chapter 13 that Q includes the concentrations of species in solution but not the solids taking part in a reaction.

Consider a case in which the cell started to operate with $[Zn^{2+}] = [Cu^{2+}] = 1.00M$, and after a certain time the concentrations had changed so that $[Cu^{2+}] = 0.10M$ and $[Zn^{2+}] = 1.90M$. Then

$$E = 1.10 - \frac{0.0257}{2} \ln \frac{1.9}{0.1} = 1.10 - \frac{0.0257}{2} \ln 19$$

$$= 1.10 - 0.038 = 1.06 \text{ V}$$

We see that the cell voltage has decreased but only by a rather small amount.

Let us calculate the voltage after a further period of time when the concentrations have reached the values $[Cu^{2+}] = 0.001$ and $[Zn^{2+}] = 1.999$:

$$E = 1.10 - \frac{0.0257}{2} \ln \frac{1.999}{0.001}$$

$$= 1.10 - 0.10 = 1.00 \text{ V}$$

The cell voltage continually decreases as the cell reaction proceeds.

EQUILIBRIUM CONSTANTS FROM THE NERNST EQUATION The potential of a cell becomes zero when the cell reaction has attained equilibrium. The reaction quotient, Q, is then equal to the equilibrium constant, K, and the Nernst equation may be written in the form

$$E = E° - \frac{RT}{nF} \ln K = 0$$

This can be rearranged to

$$\ln K = \frac{nFE°}{RT} = \frac{nE°}{0.0257} \quad \text{or} \quad \log K = \frac{nE°}{0.0592} \quad \text{at 25°C}$$

This equation enables us to calculate the equilibrium constant for any oxidation-reduction reaction in aqueous solution from the corresponding standard cell potential, which in turn can often be obtained from the tabulated standard reduction potentials of the two half-cell reactions (Table 17.1).

$=$ Example 17.10 ==

EQUILIBRIUM CONSTANT AND THE NERNST EQUATION

What is the equilibrium constant for the following reaction at 25 °C?

$$Cu(s) + Br_2(aq) \longrightarrow Cu^{2+} + 2Br^-$$

Solution

First we write the two half-cell reactions and calculate the standard cell potential:

$$
\begin{array}{lll}
Br_2(aq) + 2e^- \longrightarrow 2Br^-(aq) & E°_{red} = & 1.09 \text{ V} \\
\underline{Cu(s) \longrightarrow Cu^{2+}(aq) + 2e^-} & \underline{E°_{ox} = -E°_{red} = -0.34 \text{ V}} \\
Br_2(aq) + Cu(s) \longrightarrow Cu^{2+}(aq) + 2Br^-(aq) & E°_{cell} = & 0.75 \text{ V}
\end{array}
$$

For this reaction $n = 2$ so we have

$$\log K = \frac{nE°}{0.0592} = \frac{2(0.75)}{0.0592} = 25$$

$$K = 1.0 \times 10^{25}$$

EXERCISE 17.11

What is the equilibrium constant at 25 °C for the reaction

$$5Fe^{2+} + MnO_4^- + 8H^+ \longrightarrow 5Fe^{3+} + Mn^{2+} + 4H_2O$$

Example 17.10 and Exercise 17.11 illustrate an important use of the Nernst equation, namely, to obtain values of equilibrium constants for reactions that go essentially to completion—that is, for reactions that have very large (or very small) equilibrium constants—and that are difficult to determine in any other way. Solubility product constants, which often have very small values, can be similarly obtained by using the Nernst equation.

Example 17.11

SOLUBILITY PRODUCT CONSTANT AND THE NERNST EQUATION

Calculate a value for the solubility product constant of AgCl at 25 °C from standard reduction potentials.

Solution

The overall reaction for which we require the cell potential is

$$AgCl(s) \longrightarrow Ag^+ + Cl^-$$

We can obtain this reaction by adding the half-reactions:

$$Ag(s) \longrightarrow Ag^+ + e^- \qquad E_{ox}^\circ = -0.80 \text{ V}$$
$$AgCl(s) + e^- \longrightarrow Ag(s) + Cl^- \qquad E_{red}^\circ = +0.22 \text{ V}$$

Adding the half-cell potentials gives the standard cell potential:

$$AgCl(s) \longrightarrow Ag^+ + Cl^- \qquad E^\circ = -0.58 \text{ V}$$

The negative potential shows that the reaction is spontaneous in the reverse direction from that written. We know that addition of chloride ion to a solution of silver ion gives a precipitate of silver chloride. The equilibrium position of the reaction lies far to the left. We can calculate the equilibrium constant from the expression

$$\ln K = \frac{nE^\circ}{0.0257} = \frac{1(-0.58)}{0.0257} = -23$$

Hence

$$K = [Ag^+][Cl^-] = 1.5 \times 10^{-10} = K_{sp}(AgCl)$$

EXERCISE 17.12

Calculate a value for the solubility product constant of AgBr at 25 °C from standard reduction potentials.

THE pH METER Another important application of the Nernst equation is for determining the concentration of an ion in solution by measuring the potential of a suitably designed cell. Consider a cell composed of a copper electrode and a hydrogen electrode:

$$Pt, H_2(g)|H^+(aq)||Cu^{2+}(aq)|Cu(s)$$

The cell reaction is

$$Cu^{2+}(aq) + H_2(g) \longrightarrow Cu(s) + 2H^+(aq)$$

and the Nernst equation for this reaction is

$$E = E^\circ - \frac{0.0592}{2} \log \frac{[H^+]^2}{[Cu^{2+}]p_{H_2}}$$

If the concentration of Cu^{2+} is 1 mol L^{-1} and the pressure of H_2 is 1 atm, then

$$E = E° - \frac{0.0592}{2} \log [H^+]^2$$

or

$$E = E° - 0.0592 \log [H^+] = E° + 0.0592(pH)$$

Thus the potential of the cell is proportional to the pH. By measuring the cell potential, E, we can obtain the pH of the solution in which the hydrogen electrode is immersed. A pH meter is an instrument based on such a cell, although it does not use a hydrogen electrode but a more practical electrode called a *glass electrode*. This electrode makes use of the fact that a potential difference is developed across a thin glass membrane that depends on the concentrations of hydrogen ions on the two sides of the membrane. If the hydrogen ion concentration inside a thin-walled glass bulb is kept constant, the hydrogen ion concentration of the solution into which the electrode is dipped can be measured from the potential developed across the thin glass wall (Figure 17.13).

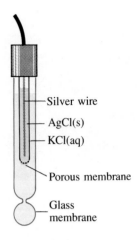

FIGURE 17.13
Glass Electrode.

The potential of this electrode depends on the concentration of hydrogen ion in any solution into which it is immersed.

BATTERIES

A **battery** is an electrochemical cell that is used as an energy source. It stores energy in the form of oxidizing and reducing agents and then releases the energy as electricity when it is needed. In principle, any oxidation–reduction reaction can be used as the basis of an electrochemical cell, but there are many limitations to the use of most reactions as the basis of a practical battery. A useful battery should be reasonably light and compact and easily transported; it should have a reasonably long life, both when it is being used and when it is not. Also, it is essential for many applications that the voltage of the cell stay constant during use.

There are two main types of batteries: *primary cells*, in which the reaction occurs only once and the battery is then dead and cannot be used again, and *secondary cells*, which can be recharged by passing a current through them so that they can be used again and again.

PRIMARY CELLS The most familiar battery is the type that is widely used, for example, in portable radios and flashlights (Figure 17.14). It is called a *Leclanché cell*, after its inventor, or a *dry cell*. The anode consists of a zinc can; the cathode is a graphite rod surrounded by powdered MnO_2, which is a conductor. The space between the electrodes is filled with a moist paste of NH_4Cl and $ZnCl_2$. Strictly speaking, the cell is not dry; it contains a moist paste in place of an aqueous solution. The electrode reactions are complex but they may be written approximately as

Anode	$Zn(s) \longrightarrow Zn^{2+} + 2e^-$
Cathode	$MnO_2(s) + NH_4^+ + e^- \longrightarrow MnO(OH)(s) + NH_3(aq)$

In the cathode reaction manganese is reduced from the +4 oxidation state to the +3 oxidation state. Ammonia is not liberated as a gas, but it combines with Zn^{2+} to form the ion, $Zn(NH_3)_4^{2+}$. The dry cell does not have an indefinite life, even when not used, because the acidic NH_4Cl corrodes the zinc can.

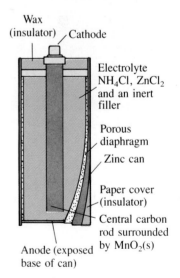

FIGURE 17.14
The Dry Cell.

The cell consists of a zinc anode (the battery container) and a cathode that is a graphite rod surrounded by powdered solid MnO_2. The electrolyte is a moist paste of NH_4Cl and $ZnCl_2$. The cell has a voltage of 1.5 V.

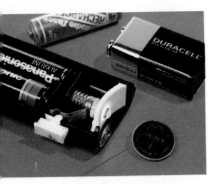

Alkaline cells, a lithium cell,
and a rechargeable Nicad cell

An improved version of this battery, known as an *alkaline battery*, has a zinc rod anode and a cathode of MnO_2. The electrolyte is KOH in the form of a thick gel. The battery is enclosed in a steel container. The alkaline battery has a longer life than the dry cell but is more expensive. Both cells have a potential of 1.5 V.

Another important battery is the *mercury cell*. It can be made in very small sizes and has many uses, for example, in hearing aids, watches, and cameras. A zinc–mercury amalgam is the anode; a paste of HgO and carbon is the cathode. The electrolyte is a paste of KOH and ZnO (Figure 17.15). The electrode reactions are as follows:

Anode	$Zn(amalgam) + 2OH^- \longrightarrow ZnO(s) + H_2O + 2e^-$
Cathode	$HgO(s) + H_2O + 2e^- \longrightarrow Hg(l) + 2OH^-$
Overall	$Zn(amalgam) + HgO(s) \longrightarrow ZnO(s) + Hg(l)$

The overall cell reaction does not involve any ions in solution whose concentrations can change. As a result, this battery has the advantage of maintaining a constant voltage (1.34 V) throughout its life.

A mercury cell in a wrist
watch.

SECONDARY CELLS   The best known and most important secondary cell is the *lead storage battery* (Figure 17.16). The anode is a grid of lead alloy packed with finely divided spongy lead. The cathode is a grid of lead alloy packed with lead(IV) oxide, PbO_2. The electrolyte is a solution of sulfuric acid (38% by mass). At the anode lead is oxidized to Pb^{2+} and insoluble $PbSO_4$ is formed. At the cathode PbO_2 is reduced to Pb^{2+} and $PbSO_4$ is formed. The reactions are as follows:

Anode	$Pb(s) + SO_4^{2-} \longrightarrow PbSO_4(s) + 2e^-$
Cathode	$PbO_2(s) + SO_4^{2-} + 4H^+ + 2e^- \longrightarrow PbSO_4(s) + 2H_2O$
Overall	$Pb(s) + PbO_2(s) + 2H_2SO_4(aq) \longrightarrow 2PbSO_4(s) + 2H_2O$

Thus lead sulfate is formed at each electrode, and sulfuric acid is used up. The sulfuric acid solution thereby becomes more dilute as the reaction proceeds. The extent to which the battery has been discharged can be checked by measuring the density of the acid, which decreases with decreasing concentration (see Figure 17.17).

An important advantage of a secondary cell is that it can be recharged. A potential that is larger than the cell potential is applied to the cell so as to force a current through the cell and reverse the electrode reactions. Therefore during recharging the $PbSO_4$ is reduced to Pb at one electrode, and at the other electrode Pb is oxidized to PbO_2. The overall reaction during charge is

$$2PbSO_4(s) + 2H_2O \longrightarrow Pb(s) + PbO_2(s) + 2H_2SO_4(aq)$$

A single lead cell has a potential of 2.0 V. Normally, three or six such cells are connected in series to give a 6-V or a 12-V battery. The lead storage battery has proved to be a very practical source of electric power, particularly for applications in which it is used many times but can be easily recharged between. It is used, for example, to start the engine of an automobile and then is recharged by the generator as soon as the engine is running.

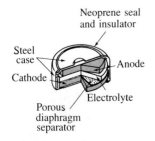

Neoprene seal
and insulator

Steel
case

Anode

Cathode

Electrolyte

Porous
diaphragm
separator

FIGURE 17.15
Mercury Cell.

If a lead storage battery is recharged too rapidly, however, hydrogen and oxygen may be formed at the electrodes:

Anode $\qquad$ $2H_2O \longrightarrow O_2(g) + 4H^+ + 4e^-$

Cathode $\qquad$ $2H^+ + 2e^- \longrightarrow H_2(g)$

There is then some danger of an explosion because of the possible reaction

$$2H_2(g) + O_2(g) \longrightarrow 2H_2O(g)$$

The oxygen for this reaction may be the oxygen produced in the cell or may come from the air. Thus sparks and flames should not be brought near a lead storage battery when it is being charged. The anode reaction that produces oxygen uses up water, which may also be lost by evaporation. Thus water must be added to the battery from time to time. Maintenance-free batteries use lead–calcium alloy electrodes at which the evolution of hydrogen and oxygen is very slow. Because very little water is electrolyzed, there is no need to add water to the cell.

The *nickel–cadmium* (Nicad) *battery* is smaller and lighter than the lead battery and it can be made as a sealed unit. It is used in portable, cordless appliances such as electric razors and calculators. It consists of a cadmium anode and a cathode that is a metal grid containing NiO(OH). The electrolyte is KOH. The reactions during discharge and charge are

Anode $\qquad$ $Cd(s) + 2OH^- \underset{\text{Charge}}{\overset{\text{Discharge}}{\rightleftharpoons}} Cd(OH)_2(s) + 2e^-$

Cathode $\qquad$ $NiO(OH)(s) + H_2O + e^- \underset{\text{Charge}}{\overset{\text{Discharge}}{\rightleftharpoons}} Ni(OH)_2(s) + OH^-$

The overall reaction is

$$Cd(s) + 2NiO(OH)(s) + 2H_2O \underset{\text{Charge}}{\overset{\text{Discharge}}{\rightleftharpoons}} Cd(OH)_2(s) + 2Ni(OH)_2(s)$$

Since all the reactants and products in the overall reaction are in the solid state, there is no change in the concentration of any ions in solution during the discharge reaction. Therefore the voltage of the cell (1.35 V) remains very constant.

In recent years much research has been done to develop secondary cells that will deliver large currents for a lengthy period. Such a cell would be particularly useful for an electrically powered automobile.

FUEL CELLS The oxidizing and reducing agents in a secondary cell can be regenerated by recharging, but batteries can also be made in which the reactants are fed continuously to the electrodes. Such batteries are called **fuel cells**. One of the most successful fuel cells uses the reaction of hydrogen with oxygen to form water. The cell is illustrated in Figure 17.18. The electrodes are hollow tubes made of porous compressed carbon impregnated with a catalyst. The electrolyte is a concentrated aqueous KOH solution. At the anode hydrogen is oxidized to water. At the cathode oxygen is reduced to the -2 oxidation state in the hydroxide ion:

Anode $\qquad$ $2[H_2(g) + 2OH^- \longrightarrow 2H_2O + 2e^-]$

Cathode $\qquad$ $O_2(g) + 2H_2O + 4e^- \longrightarrow 4OH^-$

Overall $\qquad$ $2H_2(g) + O_2(g) \longrightarrow 2H_2O$

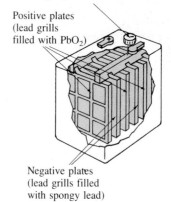

Capped hole for testing and replenishing electrolyte of H_2SO_4 and distilled water.

Positive plates (lead grills filled with PbO_2)

Negative plates (lead grills filled with spongy lead)

FIGURE 17.16
Lead Storage Battery.

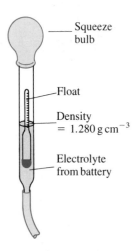

Squeeze bulb

Float

Density $= 1.280 \, g \, cm^{-3}$

Electrolyte from battery

FIGURE 17.17
Measuring the Density of Sulfuric Acid Electrolyte in a Lead Storage Battery by Using a Hydrometer.

The depth to which the float is submerged depends on the density of the liquid.

This cell employs the reaction

$$2H_2(g) + O_2(g) \longrightarrow 2H_2O(g)$$

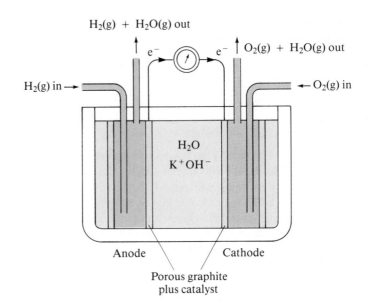

Such a cell runs continuously as long as the reactants are supplied. Because fuel cells convert the energy of a fuel directly to electricity, they are potentially more efficient than conventional methods of generating electricity on a large scale by burning hydrocarbon fuels or by using nuclear reactors. At present fuel cells are too expensive for largescale use, but their importance is growing. They have been used very successfully in some special applications, for example, in spacecraft.

17.3
Corrosion

Corrosion is the deterioration of metals as a result of their reactions with the environment. The rusting of iron, the tarnishing of silver, the development of a green coating on copper, brass, and bronze are all familiar examples of corrosion. It causes enormous damage to bridges, ships, and cars, for example. The damage and the efforts taken to prevent it cost billions of dollars a year.

Many metals corrode because they are relatively easily oxidized by the atmosphere to form the oxide, hydroxide or carbonate. The formation of an oxidized layer is not always harmful, however. Aluminum forms a tough, dense transparent layer of Al_2O_3, which protects the metal underneath from further oxidation. Iron, on the other hand, forms an oxide coating that is relatively porous and easily cracks as it thickens. As a result, oxygen and moisture can continue to reach the metal, and it continues to oxidize until it is completely destroyed.

Rust is hydrated iron(III) oxide, $Fe_2O_3 \cdot xH_2O$. It forms only in the presence of oxygen and water. Rusting is an electrochemical process. The reactions involved are complex and not completely understood, but the main steps are thought to be as follows. The iron in one part of an iron object behaves as an anode, and the iron is oxidized to Fe^{2+}:

$$Fe(s) \longrightarrow Fe^{2+} + 2e^- \qquad E^{\circ}_{ox} = 0.44 \text{ V}$$

The electrons produced in this half-reaction flow through the metal to another part of the object, which acts as a cathode. Here atmospheric oxygen is reduced in the presence of $H^+(aq)$ supplied by H_2CO_3 formed from dissolved CO_2:

$$O_2(g) + 4H^+ + 4e^- \longrightarrow 2H_2O \qquad E^\circ_{red} = 1.23 \text{ V}$$

Figure 17.19 shows the formation of rust in the vicinity of a drop of water on an iron surface.

The circuit is completed by the movement of ions through water on the surface of the iron. This explains why rusting is particularly rapid in salt water, which has a high concentration of ions. Salt spread on roads in the winter similarly speeds up the rusting of cars. The overall reaction is the sum of the cathode and anode reactions:

$$2Fe(s) + O_2(g) + 4H^+ \longrightarrow 2Fe^{2+} + 2H_2O \qquad E^\circ_{cell} = 1.67 \text{ V}$$

The Fe^{2+} is further oxidized by atmospheric oxygen to rust:

$$4Fe^{2+} + O_2(g) + 4H_2O \longrightarrow 2Fe_2O_3(s) + 8H^+$$

Note that this reaction also provides hydrogen ions that are needed in the reduction of oxygen.

Silver does not react with the oxygen of the air; it tarnishes due to the layer of black Ag_2S formed by reaction with traces of H_2S in the atmosphere and on contact with sulfur-containing foods such as eggs. Experiment 17.6 shows how a simple electrochemical cell can be constructed in the home to remove the tarnish on silver.

INHIBITING CORROSION

The most obvious way of minimizing corrosion is to cover the surface of the metal with a coating that keeps out air and moisture. This coating can be provided in several ways. Steel used in bridges and buildings is often painted with red lead paint, which contains Pb_3O_4. The Pb_3O_4 apparently oxidizes the surface of the iron to form a tough continuous layer of the oxide, which resists further oxidation.

Another method of protecting iron is to coat it with a layer of another metal. Zinc, tin, and chromium are often used for this purpose. The use of a zinc coating to protect iron is called *galvanizing*. Since zinc is above iron in the table of reduction potentials, it is more easily oxidized. It would not appear

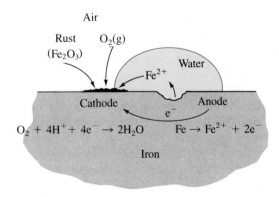

FIGURE 17.19
The Rusting of Iron.

Iron in contact with water forms the anode, where iron is oxidized to Fe^{2+}. Iron in contact with air forms the cathode, where oxygen is reduced to water. Thus a cell is set up and iron continues to go into solution as Fe^{2+}. Where the Fe^{2+} solution is in contact with the air, the Fe^{2+} is oxidized to insoluble Fe_2O_3 (rust).

Experiment 17.6

Removing the Tarnish on Silver Objects

(a) Silver objects slowly tarnish on the surface due to the formation of a dark layer of silver sulfide, Ag_2S. This tarnish can be removed by placing the silver objects in an aluminum dish containing a warm solution of sodium hydrogen carbonate (otherwise known as sodium bicarbonate, or baking soda).

(b) An electrochemical cell is thus set up in which the following reaction occurs:

$$3(Ag^+)_2(S^{2-}) + 2Al \rightarrow 6Ag + 2Al^{3+} + 3S^{2-}$$

The tarnish is thus removed with less loss of silver than occurs with the use of silver polish.

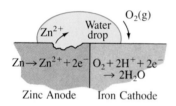

FIGURE 17.20
Galvanized Iron.

Cathodic protection of iron in contact with zinc is provided by galvanizing.

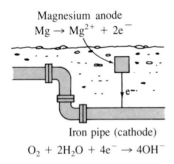

$O_2 + 2H_2O + 4e^- \rightarrow 4OH^-$

FIGURE 17.21
Cathodic Protection.

When a block of magnesium or zinc is attached to an iron pipe or tank buried underground, the magnesium forms the anode in an electrolytic cell and the iron becomes the cathode. Thus the magnesium or zinc is slowly oxidized and goes into solution as Mg^{2+} or Zn^{2+}. Oxygen is reduced at the iron cathode leaving the iron unattacked.

therefore to be very useful for protecting iron. However, zinc is oxidized to $Zn(OH)_2$, which reacts with the CO_2 in the atmosphere to form a tough layer of $Zn(OH)_2 \cdot xZnCO_3$, which strongly adheres to the surface, protecting the metal beneath. Even if the zinc layer is cracked or broken, it continues to protect the iron, as it becomes the anode in a cell in which the iron is the cathode. Hence zinc is oxidized rather than iron (see Figure 17.20).

A layer of tin, which becomes coated with a layer of oxide, can also be used to protect iron, as, for example, in tin cans. Tin comes below iron in the table of standard reduction potentials and is therefore less easily oxidized. But if the tin coating is cracked or scratched, the tin becomes the cathode in a cell and the iron becomes the anode. The oxidation of Fe to Fe^{3+}—rusting—will then be speeded up.

CATHODIC PROTECTION   We have seen that iron corrodes by becoming the anode of an electrochemical cell. However, if we connect the iron to a more easily oxidized metal such as zinc or magnesium, then this metal becomes the anode of a cell and *it* corrodes instead of the iron. The iron becomes the cathode at which oxygen is reduced to water. This method of preventing the corrosion of iron is called **cathodic protection** (Experiment 17.7). For example, if a magnesium or zinc rod is connected to an underground tank or steel pipe, then the magnesium or zinc is oxidized instead of the pipe (see Figure 17.21). The more easily oxidized metal is called a *sacrificial anode*. Although it is gradually dissolved, it is easier to replace than the pipe or the tank. This method also provides excellent protection for bridge foundations and steel ships.

Experiment 17.7

Corrosion and Cathodic Protection

An iron nail is placed in aqueous agar gel to which $K_3Fe(CN)_6$, potassium hexacyanoferrate(III), and phenolphthalein indicator have been added. When the nail corrodes in contact with water and the oxygen in the air, Fe^{2+} ions are formed at the ends of the nail which act as the anode. The formation of the Fe^{2+} ion is shown by the blue color, which is due to a compound called prussian blue formed between Fe^{2+} and $Fe(CN)_6^{3-}$. Along the length of the nail, which behaves as the cathode, OH^- ions are formed, which change the color of the phenolphthalein indicator to pink.

A small piece of zinc is attached to the nail. Zinc is more easily oxidized than iron so it is the anode in an electrochemical cell and the iron is the cathode. Thus it is the zinc that is oxidized and therefore no Fe^{2+} ions are formed and no blue color is observed. Attaching a piece of zinc to an iron object thus prevents the corrosion of the iron.

A piece of copper attached to the nail does not prevent corrosion because copper is less easily oxidized than iron. Thus Fe^{2+} is formed at the ends of the nail as before as shown by the formation of the blue color.

IMPORTANT TERMS

An **anode** is the electrode at which oxidation occurs.

A **battery** is an electrochemical cell that is used as an energy source.

A **cathode** is the electrode at which reduction occurs.

Cathodic protection is a method of protecting a metal from corrosion by making it the cathode of a cell in which some other, more easily oxidized metal acts as the anode and is oxidized preferentially.

The **coulomb** (C) is the SI unit of electric charge. It is the quantity of electric charge carried by 1 ampere (A) in 1 second (s): $1 \text{ C} = 1 \text{ A s}$.

An **electrochemical cell** is a device for producing an electric current. It consists of two half cells, in one of which a substance is oxidized and in the other a substance is reduced.

An **electrode** is a strip of metal, or other conducting material, by means of which electrons are transferred to or from a solution in an electrochemical or electrolytic cell.

Electrolysis is a process in which electrical energy is used to produce chemical change.

An **electrolytic cell** is an apparatus in which electrolysis occurs, that is, in which a chemical reaction is carried out by using an external source of electrical energy.

The **Faraday constant** is the charge associated with 1 mol of electrons, 96 485 C mol^{-1}.

A **fuel cell** is an electrochemical cell in which the substances that are oxidized and reduced are fed continuously to the anode and the cathode, respectively.

In **ionic conduction**, an electric current is carried by the movement of ions.

The **Nernst equation** is an expression that gives the relation between the voltage of a cell and the concentrations of the reactants.

The **standard cell potential**, E°_{cell}, is the potential of a cell when all the concentrations of solutes are 1 mol L^{-1}, all the partial pressures of gases are 1 atm, and the temperature is 25 °C.

The **standard reduction potential**, E°, is the reduction potential of a half-reaction with all the ions at $1M$ concentration and all the gases at 1 atm, measured relative to the hydrogen electrode, for which E° is exactly 0 V at 25 °C.

PROBLEMS *

Electrolysis

1. **(a)** Explain each of the following terms:

  (i) Electrolysis   **(ii)** Electrolyte

  (iii) Nonelectrolyte

  (b) In what way does ionic conduction differ from electronic conduction?

2. Define each of the following:

  (i) Anode   **(ii)** Cathode
  (iii) Electrolyte   **(iv)** The Faraday constant, F

3. For inert electrodes, write equations for the electrode reactions that occur during the electrolysis of each of the following:

  (a) Molten aluminum chloride

  (b) A dilute aqueous solution of aluminum chloride

Explain why the products of electrolysis in (a) are different from those in (b).

4. What are the products of the electrolysis of each of the following using inert electrodes?

  (a) Molten magnesium bromide

  (b) Aqueous copper(II) nitrate

  (c) Aqueous hydrogen iodide

  (d) Molten iron(III) chloride

5. Draw a sketch of a cell for the electrolysis of HBr(aq) using platinum electrodes. Label the anode and the cathode. Write equations for the electrode reactions and the overall cell reaction. Indicate the direction of electron flow in the

external circuit and the directions in which ions move in the solution.

6. Draw a sketch of a cell for the electrolysis of $CuSO_4$(aq) using copper electrodes. Label the anode and cathode. Show the direction in which electrons flow in the external circuit and the directions in which the ions move in the solution. Write equations for the electrode reactions, and for the overall cell reaction.

7. How many coulombs of electricity are required for each of the following reductions?

  (a) 1 mol of Cu^{2+}(aq) to Cu(s)

  (b) 1 mol of Fe^{3+}(aq) to Fe^{2+}(aq)

  (c) 1 mol of MnO_4^-(aq) to Mn^{2+}(aq)

  (d) 1 mol of ClO_3^-(aq) to Cl^-(aq)

8. How many coulombs of electricity are required for each of the following oxidations?

  (a) 1 mol of H_2O(l) to O_2(g)

  (b) 1 mol Cl_2(g) to ClO_3^-(aq)

  (c) 1 mol Pb(s) to PbO_2(s)

  (d) 1 mol FeO(s) to Fe_2O_3(s)

9. How many coulombs are required to produce each of the following:

  (a) 50.0 mL of O_2(g), at STP, from Na_2SO_4(aq)

  (b) 50.0 kg of Al(s) from molten Al_2O_3

  (c) 20.0 g of calcium from molten calcium chloride

  (d) 5.00 g of silver from aqueous silver nitrate

* Answers to problems numbered in the blue appear at the end of the text.

10. Using a current of 2.00 A, how long will it take to produce 1.00 kg of nickel by the electrolysis of a solution of nickel(II) chloride?

11. What mass of sodium hydroxide, and what mass of chlorine gas, will be produced by electrolysis of aqueous sodium chloride for 3.00 h using a current of 0.200 A?

12. What volume of $H_2(g)$, and what volume of $O_2(g)$, at 27 °C and a pressure of 740 torr, are produced by electrolysis of $Na_2SO_4(aq)$ for 30 min, using a current of 4.00 A? (Vapor pressure of H_2O at 27 °C = 26.7 torr.)

13. How long will it take to deposit 16.0 g of silver from an aqueous solution of silver nitrate, using a current of 6.00 A?

14. A metal tray has dimensions of 24 cm × 12 cm. How long will it take to plate the tray with a layer of silver of thickness 0.020 mm, using a current of 7.65 A? (Neglect the amount of silver needed to plate the edges of the tray; density of silver = 10.54 g cm^{-3}.)

15. For how long would an electric current of 1.50 A have to be passed through a solution containing chromium(III) sulfate for a steel object of surface area 0.10 m^2 to be plated with a layer of chromium 0.10 mm thick? (Density of chromium = 7.1 g cm^{-3}.)

16. The same quantity of electricity that plates out 10.0 g of copper from a $CuSO_4(aq)$ solution is passed through a solution of $NiCl_2(aq)$. What mass of nickel will be plated out?

17. When a current of 1.487 A was passed through aqueous hydroiodic acid for 1 h 0 min 1.5 s, 7.2428 g of iodine were formed at the anode. From these data, determine a value for the Faraday constant and compare this value with the accepted value.

18. When a current of 0.500 A was passed through a solution containing an unknown $M^{2+}(aq)$ ion for exactly 2 h, 1.98 g of the metal M were deposited. What is the atomic mass of M, and what is the metal?

19. A Hall cell for the production of aluminum from alumina, Al_2O_3, operates at a current of 1.300×10^5 A. What mass of aluminum will be produced in exactly 1 min? A hundred such cells are connected in series. How much aluminum can be produced per day? What mass of carbon will be consumed per day at the anodes?

20. In an electrolysis cell, AgCl(s) is reduced to Ag(s) and $Cl^-(aq)$ at the cathode and the copper anode is oxidized to $Cu^{2+}(aq)$. A current of 0.700 A was passed through the cell for 1.600 h. How many grams of silver were deposited at the

cathode, and how many grams of copper were dissolved at the anode?

21. Two electrolysis cells are connected in series. In the first $Fe^{3+}(aq)$ is reduced to Fe(s), and in the second $Cu^{2+}(aq)$ is reduced to Cu(s). After passage of an electric current for 30.00 min, 1.030 g of iron were deposited in the first cell. How many grams of copper were deposited in the second cell?

22. Assuming that the reduction of $Zn^{2+}(aq)$ to zinc metal is the only cathode reaction, how long will it take to plate 1.000 g of zinc onto a suitable cathode in an electrolysis cell containing $ZnSO_4(aq)$, using a current of 8.000 A?

23. When an aqueous solution of sodium chloride is electrolyzed for 1.500 h at a current of 2.500 A, $Cl_2(g)$ is produced at the anode and $H_2(g)$ and $OH^-(aq)$ are produced at the cathode. At the end of the electrolysis the cathode compartment contains 1.000 L of solution. What is its pH?

24. In the electrolysis of aqueous potassium iodide with inert electrodes, a brown color is observed in the solution at one of the electrodes. To what species is the brown color attributed? Does it occur at the cathode or at the anode? Write balanced equations for the reaction at each electrode, and for the overall cell reaction.

25. The same electric current was passed for 1.00 h through each of three electrolysis cells fitted with platinum electrodes. The cells contained aqueous solutions of $CuSO_4$, $AgNO_3$, and H_2SO_4, respectively. During this time, 0.106 g of copper was deposited at the cathode in the first cell. Calculate

(a) The average current in milliamps

(b) The mass of silver deposited at the cathode in the second cell

(c) The total volume of gas, measured at 20 °C and a pressure of 750 mm Hg, liberated in the second cell, given that the vapor pressure of water is 17.5 mm Hg

*26. A mass of 0.5305 g of an unknown alkali metal chloride was dissolved in water and an electric current was passed through the solution until no more chlorine gas was evolved at the anode. The volume of chlorine, collected over water at 25 °C and 762.1 mm Hg pressure, was 114.2 mL.

(a) Assuming that all the chloride ion was oxidized to chlorine, identify the alkali metal.

(b) If the current used was 0.200 A, for how many minutes was it passed to liberate all the chloride ion in solution as $Cl_2(g)$?

(c) What reaction occurred at the cathode, and what was the anode reaction in this reaction?

* The asterisk denotes the more difficult problems.

(d) In practice, would the production of $Cl_2(g)$ be expected to be quantitative? Explain.

(Vapor pressure of water at 25 °C = 23.8 mm Hg.)

Standard Reduction Potentials

27. Under standard conditions, which of the following will be oxidized by dichromate ion, $Cr_2O_7^{2-}$, in acidic aqueous solution?

(a) $F^-(aq)$ **(b)** $Cl^-(aq)$ **(c)** $Br^-(aq)$

(d) $I^-(aq)$ **(e)** $Hg_2^{2+}(aq)$ **(f)** $Mn^{2+}(aq)$

(g) $Fe^{2+}(aq)$ **(h)** $H_2S(aq)$ to sulfur

28. Use standard half-cell reduction potentials to predict which of the following reactions will occur in acid solution (with all soluble substances at a concentration of $1M$). For each reaction that is predicted to occur, write the balanced equation.

(a) $H_2O_2(aq) + Cu^{2+}(aq) \rightarrow Cu(s) + O_2(g)$

(b) $Ag^+(aq) + Fe^{2+}(aq) \rightarrow Ag(s) + Fe^{3+}(aq)$

(c) $I^-(aq) + NO_3^-(aq) \rightarrow I_2(s) + NO(g)$

29. From a consideration of the appropriate standard reduction potentials, predict whether or not $Cr_2O_7^{2-}(aq)$ in acid solution will oxidize water to $O_2(g)$. Why does a solution of $Cr_2O_7^{2-}(aq)$ in acid solution remain unchanged for a long time?

30. Under standard conditions, which of the following equations will occur as written? Balance each equation.

(a) $Mg(s) + Cr^{3+}(aq) \rightarrow Mg^{2+}(aq) + Cr(s)$

(b) $I_2(aq) + H^+(aq) \rightarrow H_2(g) + I^-(aq)$

(c) $Cu(s) + NO_3^-(aq) \rightarrow Cu^{2+}(aq) + NO_2(g) + H_2O$
(in acid solution)

31. Arrange the following metals in the order in which they displace each other from aqueous solution:

Aluminum   Copper   Iron   Magnesium   Zinc

32. Which of the following species are reduced by $Fe^{2+}(aq)$ under standard conditions in acid solution?

(a) $Ag^+(aq)$ **(b)** $Cr^{3+}(aq)$ **(c)** $Zn^{2+}(aq)$

(d) $I_2(s)$ **(e)** $Br_2(aq)$ **(f)** $MnO_4^-(aq)$

33. Which of the following substances are oxidized by manganese dioxide, $MnO_2(s)$, under standard conditions in acid solution?

(a) $Br^-(aq)$ **(b)** $Ag^+(aq)$

(c) $I^-(aq)$ **(d)** $Cl^-(aq)$

34. Use standard reduction potentials to predict the reaction that occurs, if any, between the following substances under standard conditions in acid solution:

(a) $Fe^{3+}(aq)$ and $I^-(aq)$ **(b)** $Ag^+(aq)$ and $Cu(s)$

(c) $Fe^{3+}(aq)$ and $Br^-(aq)$ **(d)** $Ag(s)$ and $Fe^{3+}(aq)$

(e) $Br_2(aq)$ and $Fe^{2+}(aq)$

35. Use standard reduction potentials to predict whether each of the following reactions will occur in acid solution with all soluble substances at a concentration of $1M$. Complete and balance the equation for each reaction that occurs.

(a) $Mn^{2+}(aq) + Cr_2O_7^{2-}(aq)$
    $\rightarrow MnO_4^-(aq) + Cr^{3+}(aq)$

(b) $O_2(g) + Br^-(aq) \rightarrow Br_2(aq)$

(c) $Au(s) + Cl_2(aq) \rightarrow Au^{3+}(aq) + Cl^-(aq)$

36. Predict which of the following metals should be oxidized by 1 atm of $O_2(g)$ in $1M$ acid solution at room temperature. The reduction product in each case is water.

(a) Silver   **(b)** Copper   **(c)** Gold

(d) Zinc   **(e)** Tin

37. Predict which of the following metal ions would be reduced by 1 atm of $H_2(g)$ in acid aqueous solution at room temperature:

(a) Au^{3+} **(b)** Ni^{2+} **(c)** Pb^{2+}

(d) Ag^+ **(e)** Fe^{2+}

38. Considering all the ions to be at a concentration of $1M$, which of the following reactions are spontaneous at 25 °C and 1 atm pressure? Balance each equation.

(a) $Br^-(aq) + Cl_2(g) \rightarrow Br_2(aq) + Cl^-(aq)$

(b) $MnO_4^-(aq) + H_3O^+(aq) + Au(s)$
    $\rightarrow Mn^{2+}(aq) + Au^{3+}(aq) + H_2O(l)$

(c) $Cr^{3+}(aq) + H_2(g) \rightarrow Cr^{2+}(aq) + H_3O^+(aq)$

(d) $AgBr(s) + Sn(s) \rightarrow Ag(s) + Sn^{2+}(aq) + Br^-(aq)$

(e) $Cu^+(aq) \rightarrow Cu^{2+}(aq) + Cu(s)$

39. Use standard reduction potentials to predict whether a spontaneous reaction occurs on mixing each of the following pairs of reactants to give solutions where all the concentrations are $1M$. Write balanced equations for the reactions that are spontaneous.

(a) $Ag(s)$ and $Pb^{2+}(aq)$ **(b)** $Mg(s)$ and $Fe^{2+}(aq)$

(c) $Cu(s)$ and $H_3O^+(aq)$ **(d)** $Fe(s)$ and $Fe^{3+}(s)$

(e) $Pb(s)$ and $Ni^{2+}(aq)$

*40. Suppose the standard silver chloride electrode

$$AgCl(s) + e^- \longrightarrow Ag(s) + Cl^-(aq, 1M)$$

rather than the standard hydrogen electrode, had been assigned a potential of 0.00 V at 25 °C.

(a) What would then be the numerical value of the standard reduction potential at 25 °C for each of the following reactions:

(i) $Cl_2(g) + 2e^- \rightarrow 2Cl^-(aq)$

(ii) $2H^+(aq) + 2e^- \rightarrow H_2(g)$

(b) With this new standard, calculate the standard cell potential for

$$Cl_2(g) + H_2(g) \longrightarrow 2H^+(aq) + 2Cl^-(aq)$$

(c) Does a different choice of which standard half-cell potential is assigned a value of zero have any effect on a standard cell potential, $E°_{cell}$?

41. Select a suitable reagent that under standard conditions will

(a) Oxidize tin to $Sn^{2+}(aq)$ but not oxidize $H_2(g)$ to $H^+(aq)$

(b) Reduce $Cr^{3+}(aq)$ to $Cr^{2+}(aq)$ but not all the way to $Cr(s)$

42. Using only the data from Table 17.1, for basic aqueous solutions, how many spontaneous redox reactions can you devise? Write each as a balanced equation.

43. Describe simple experiments that you could perform to place the metals zinc, magnesium, calcium, and copper in their correct order in the electrochemical series.

44. A book describing the construction of stained glass windows comments that the metallic framework holding the glass pieces can be given a bronzelike finish, instead of the usual gray solder appearance, by wiping the solder (usually about 60 mass % lead) with a dilute copper sulfate solution.

(a) Explain how this process works.

(b) What would you expect to happen when the bronzelike finish is wiped with a dilute solution of aqueous silver nitrate?

45. Explain each of the following.

(a) A collar of zinc is placed on the steel propeller of a sea-going yacht.

(b) A galvanized iron garbage can may be used for years, but an opened "tin-can" rusts in a few days.

(c) The black tarnish (silver sulfide) on a silver spoon may be removed by placing the spoon in contact with a piece of aluminum foil in a sodium carbonate solution.

Electrochemical Cells

46. Draw a sketch of an electrochemical cell in which the reaction is

$$Zn(s) + 2Ag^+(aq) \longrightarrow Zn^{2+}(aq) + 2Ag(s)$$

(a) Show the cathode and the anode, the directions in which the ions move, the direction in which the electrons move in the external circuit, and write the electrode reactions.

(b) Calculate the standard cell voltage.

47. **(a)** Draw a sketch to illustrate the construction of the electrochemical cell

$$Zn(s)|Zn^{2+}(aq)\,\|\,Ni^{2+}(aq)|Ni(s)$$

showing the direction of electron flow in the external circuit.

(b) What is the standard cell voltage, $E°_{cell}$?

48. For each of the following standard electrochemical cells, write the two half-reactions and the overall cell reaction, and calculate the standard cell voltage, $E°_{cell}$:

(a) $Al(s)|Al^{3+}(aq)\,\|\,Cu^{2+}(aq)|Cu(s)$

(b) $Pb(s)|Pb^{2+}(aq)\,\|\,Ag^+(aq)|Ag(s)$

(c) $Ag(s)|Ag^+(aq)\,\|\,Cl^-(aq)|Cl_2(g),\,Pt(s)$

49. Consider the reaction

$$Fe(s) + 2H^+(g) \longrightarrow Fe^{2+}(aq) + H_2(g)$$

(a) Draw a sketch of an electrochemical cell in which this reaction occurs.

(b) What are the charge carriers in the wire that connects the two electrodes?

(c) At which electrode does reduction occur? Is this the anode or the cathode?

(d) Write the equation for the reaction that occurs at the cathode.

(e) Since electrons are involved in this cathodic half-reaction, where do they come from (or to where are they moving)?

50. Draw a sketch of the experimental arrangements for each of the standard electrochemical cells for which the overall cell reactions are as follows. In each case, place the

anode compartment on the left, and indicate the direction of electron flow in the external circuit.

(a) $Zn(s) + Br_2(aq) \rightarrow Zn^{2+}(aq) + 2Br^-(aq)$

(b) $Pb(s) + 2Ag^+(aq) \rightarrow Pb^{2+}(aq) + 2Ag(s)$

(c) $Cu^+(aq) + Fe^{3+}(aq) \rightarrow Cu^{2+}(aq) + Fe^{2+}(aq)$

51. Write balanced equations for the reactions that occur at the anode and at the cathode of each of the following cells. What is the standard potential of each cell? For each example, draw a sketch of the cell, label the anode and the cathode, and show the direction in which electrons move in the external circuit.

(a) A lead wire dipping into $1M$ $PbCl_2(aq)$, and a copper wire dipping into $1M$ $CuSO_4(aq)$.

(b) $Cl_2(g)$ bubbling over a platinum wire in $1M$ $NaCl(aq)$, and a silver wire coated with $AgCl(s)$ dipping into a similar solution

(c) Two platinum wires dipping into $1M$ solutions of $HI(aq)$, with one of the wires having $H_2(g)$ at a pressure of 1 atm bubbling over it

52. A cell based on the reaction

$$5Fe^{2+} + MnO_4^- + 8H^+ \longrightarrow 5Fe^{3+} + Mn^{2+} + 4H_2O$$

may be set up as follows. A platinum electrode is dipped into a beaker containing $KMnO_4(aq)$ acidified with $H_2SO_4(aq)$; another platinum electrode is dipped into a second beaker containing $FeSO_4(aq)$, and the two solutions are connected by a salt bridge. The electrodes are then connected to a voltmeter.

(a) What reactions occur at the cathode and anode?

(b) In which direction do electrons move in the external circuit?

(c) In which directions do the ions move through the solutions?

(d) What is the initial reading on the voltmeter?

Nernst Equation

53. Predict qualitatively the effect of each of the following changes on the cell voltage of the following cell:

$$Zn(s)|Zn^{2+}(aq, 1M)\,\|\,Cu^{2+}(aq, 1M)|Cu(s)$$

(a) Adding $2M$ $Zn^{2+}(aq)$ to the $Zn(s)|Zn^{2+}(aq)$ half cell

(b) Adding $Zn(s)$ to the $Zn(s)|Zn^{2+}(aq)$ half cell

(c) Adding a few drops of dilute $NaOH(aq)$ to the $Zn(s)|Zn^{2+}(aq)$ half cell

(d) Adding concentrated $NH_3(aq)$ to the $Cu(s)|Cu^{2+}(aq)$ half cell

54. Consider a room temperature electrochemical cell composed of a $Cu(s)|Cu^{2+}(aq)$ half cell and an $Ag(s)|Ag^+(aq)$ half cell.

(a) Calculate the cell voltage when all ions are at $1M$ concentrations.

(b) Calculate the cell voltage when $[Cu^{2+}]$ is $2.0M$ and $[Ag^+]$ is $0.05M$.

55. Predict the effect on the voltage of an electrochemical cell utilizing the reaction

$$Cl_2(aq) + 2I^-(aq) \longrightarrow 2Cl^-(aq) + I_2(s)$$

of increasing the iodide ion concentration.

*56. The standard reduction potential for the reaction

$$2HOCl(aq) + 2H^+ + 2e^- \longrightarrow Cl_2(aq) + 2H_2O$$

is $E^\circ = 1.63$ V. What is the equilibrium constant for the following reaction

$$Cl_2(aq) + H_2O \rightleftharpoons H^+ + Cl^- + HOCl(aq)$$

57. An electrochemical cell is constructed on the basis of the reaction

$$Sn^{2+} + Pb(s) \longrightarrow Pb^{2+} + Sn(s)$$

When the concentration of $Sn^{2+}(aq)$ in the cathode compartment is $1.00M$ and the voltage of the cell is 0.22 V at 25 °C, what is the concentration of $Pb^{2+}(aq)$ ions in the anode compartment? If under these conditions the concentration of SO_4^{2-} ions in the anode compartment is $1.00M$, what is the value of the solubility product constant of $PbSO_4$?

58. Separate beakers contain a piece of iron immersed in $1.00M$ $FeSO_4(aq)$ and a piece of copper immersed in $1.00M$ $CuSO_4(aq)$. The solutions are then connected by a salt bridge, and the metals are connected by a conducting wire.

(a) Which metal will dissolve and which will increase in mass?

(b) What will be the initial voltage between the metals?

(c) Which of the two solutions will increase in concentration as the reaction proceeds?

(d) Will the initial voltage increase, or decrease, as the reaction proceeds?

59. The voltage of the electrochemical cell

$$Sn(s)|Sn^{2+}(aq, xM)\,\|\,Pb^{2+}(aq, 0.001M)|Pb(s)$$

is 0.00 V at 25 °C. What is the $Sn^{2+}(aq)$ concentration, xM, in the cell under these conditions?

60. For the electrochemical cell

$$Pb(s)|Pb^{2+}(aq)\,\|\,Cu^{2+}(aq)|Cu(s)$$

calculate the cell voltages for each of the following ion concentrations:

	$[Pb^{2+}]$ (mol L^{-1})	$[Cu^{2+}]$ (mol L^{-1})
(a)	1.0	1.0
(b)	1.0	1.0×10^{-5}
(c)	1.0×10^{-3}	1.0×10^{-2}
(d)	6.0×10^{-5}	2.0×10^{-2}

61. For the following electrochemical cell at 25 °C

$$Ag(s)|Ag^+(aq, 0.01M)||Br^-(aq, 0.50M)|Br_2(l), Pt$$

(a) Calculate the cell voltage

(b) Write the equilibrium constant expression for the cell reaction

(c) Calculate the value of the equilibrium constant

62. For the following cell at 25 °C

$$Pt|Fe^{2+}(aq, 0.10M), Fe^{3+}(aq, 0.01M)||Cu^{2+}(aq, 0.50M)|Cu(s)$$

(a) Calculate the cell voltage

(b) Write the equilibrium constant expression for the cell reaction

(c) Calculate the value of the equilibrium constant

63. An electrochemical cell is composed of an $Ag(s)|Ag^+(aq, 0.10M)$ half cell and a hydrogen electrode dipping into a solution of unknown pH, with an $H_2(g)$ pressure of 1 atm. When the two half cells are connected by a salt bridge, the measured cell voltage is 0.859 V. What is the pH of the solution?

*64. The *calomel* half cell, in which the reaction is

$$Hg_2Cl_2(s) + 2e^- \longrightarrow 2Hg(l) + 2Cl^-(aq)$$

has a standard reduction potential of 0.285 V. In many analytical chemistry applications, the pH of a solution is found by measuring the voltage of an electrochemical cell composed of a standard calomel half cell and a standard hydrogen electrode dipping into the solution of unknown pH. At the equivalence point in the titration of aqueous acetic acid with NaOH(aq) a cell voltage of 0.823 V was measured. What was the pH of the solution? How could this pH be used to calculate a value of K_a for acetic acid?

*65. Use the standard reduction potential data from Table 17.1 to calculate

(a) The solubility product constant for AgBr(s)

(b) The standard cell voltage for the cell reaction in which 1 mol of AgBr(s) is formed from its elements in their standard states

66. Calculate the equilibrium constant for the reaction in which $I^-(aq)$ is oxidized by $Fe^{3+}(aq)$, to give $I_2(s)$ and $Fe^{2+}(aq)$.

67. Calculate the voltage of the cell composed of an $Mg(s)|Mg^{2+}(aq)$ half cell in which $[Mg^{2+}]$ is 0.0500M, and an $Ni(s)|Ni^{2+}(aq)$ half cell in which $[Ni^{2+}]$ is 1.50M. Write the equation for the overall cell reaction, and draw a sketch of the cell, indicating the anode and cathode and the direction of current flow.

68. Given that the reaction in an electrochemical cell is

$$Fe^{2+}(aq) + Ag^+(aq) \longrightarrow Ag(s) + Fe^{3+}(aq)$$

predict the effect on the cell voltage of each of the following changes:

(a) An increase in $[Ag^+]$

(b) An increase in $[Fe^{3+}]$

(c) A twofold increase in both $[Fe^{3+}]$ and $[Fe^{2+}]$

(d) A decrease in the amount of Ag(s)

(e) A decrease in $[Fe^{2+}]$

(f) Addition of NaCl(aq) to the $Ag^+(aq)$ solution

69. The measured voltage at 25 °C of the cell in which the following reaction takes place, at the concentrations shown, is initially 0.293 V:

$$Cd(s) + Pb^{2+}(aq, 0.150M) \longrightarrow Pb(s) + Cd^{2+}(aq, 0.0250M)$$

(a) Calculate E°_{cell}, the standard cell voltage for a similar standard cell $(Cd^{2+}(aq) + 2e^- \rightarrow Cd(s)$, standard reduction potential $= -0.400$ V).

(b) Calculate the equilibrium constant for the overall cell reaction.

*70. Calculate the Gibbs free energy change associated with the reactions occurring in each of the following electrochemical cells:

(a) $Fe(s)|Fe^{2+}(aq, 1.0M)||Cu^{2+}(aq, 0.1M)|Cu(s)$

(b) $Zn(s)|Zn^{2+}(aq, 0.05M)||H^+(aq, 0.10M)|H_2(g, 1 \text{ atm}), Pt(s)$

(c) $Cr(s)|Cr^{3+}(aq, 0.003M)||Ni^{2+}(aq, 0.02M)|Ni(s)$

(d) $Pt(s)|V^{3+}(aq, 0.10M), V^{2+}(aq, 0.01M)||Ag^+(aq, 0.001M)|Ag(s)$

*71. Using the following standard reduction potentials, determine whether any precipitate will form when 10^{-6} mol each of silver nitrate, potassium iodate, KIO_3, and potassium iodide, are dissolved in 1 L of water at 25 °C:

$Ag^+(aq) + e^- \longrightarrow Ag(s)$	$E^\circ = +0.80$ V
$AgIO_3(s) + e^- \longrightarrow Ag(s) + IO_3^-(aq)$	$E^\circ = +0.35$ V
$AgI(s) + e^- \longrightarrow Ag(s) + I^-(aq)$	$E^\circ = -0.15$ V

Miscellaneous

A possible source of power for a heart pacemaker is to implant a zinc electrode and a platinum electrode into the body tissues. These electrodes inserted into the oxygen-containing body fluid form a "biogalvanic" cell, in which zinc is oxidized and oxygen is reduced.

(a) Write the equations for the anode and cathode reactions.

(b) Estimate the standard voltage of such a cell.

(c) If a current of 40 μA is drawn from the cell, how often will a zinc electrode of mass 5.0 g have to be replaced?

73. (a) Write the two half-reactions for a fuel cell in which ethane, $C_2H_6(g)$, is oxidized in acid solution by $O_2(g)$ to give $CO_2(g)$ and water.

(b) How many liters of ethane at STP would be needed to generate a current of 0.50 A for 6.00 h? How many liters of oxygen at STP would be consumed at the cathode?

Further Chemistry of Nitrogen and Oxygen

18

In earlier chapters we described some of the simpler compounds of the very common and important elements nitrogen and oxygen. But some important aspects of their chemistry could not be discussed adequately in terms of the ideas and theories described in those chapters so in this chapter we extend our discussion of these two elements.

The oxides of nitrogen are of importance from several points of view. Nitrogen monoxide, NO, is formed from lightning discharges in the atmosphere and then reacts with oxygen to form nitrogen dioxide, NO_2. The same reaction occurs in the internal combustion engine and NO and NO_2 have become well known in recent years as serious air pollutants. These oxides are also of interest because, unlike the vast majority of stable molecules, both NO and NO_2 have an odd number of electrons and therefore we cannot write conventional Lewis structures for them in which all the atoms obey the octet rule. Most molecules with an odd number of electrons are very reactive and they play an important role in many chemical reactions. Many studies of the rates of reactions have been based on the reactions of the oxides of nitrogen and it will be useful, therefore, to describe their structures and properties before we take up our study of the rates of reactions in the next chapter.

Oxygen, too, has some unusual and unexpected compounds. Some of the alkali and alkaline earth metals react with oxygen to form peroxides, such as $(Na^+)_2O_2^{2-}$, and superoxides, such as $K^+O_2^-$. The superoxide ion, O_2^-, is similar to NO in that it has an odd number of electrons and we cannot write a normal Lewis structure for it. When treated with acids, peroxides give hydrogen peroxide, H_2O_2. This is a strong and reactive oxidizing agent that in dilute solution in water is used as an antiseptic and as a bleach. The pure substance is important as a rocket propellant. We also discuss ozone, O_3, an allotrope of oxygen, which is a more reactive oxidizing agent than oxygen and is an important constituent of the upper atmosphere.

18.1
Oxides of Nitrogen

2s   2p

N can form three covalent bonds (as in NH_3) or accept three electrons to form N^{3-}.

Nitrogen is the first member of group 5 of the periodic table. It has the electron configuration $1s^2 2s^2 2p^3$ with a single electron in each 2p orbital. Therefore it has a valence of 3 and is expected to form the oxide N_2O_3. This oxide can be made at low temperatures, but it readily decomposes to NO and NO_2, which are much more stable. The oxides of nitrogen and some of their physical properties are listed in Table 18.1 Nitrogen exhibits positive oxidation states only in its compounds with oxygen and fluorine, because only these elements are more electronegative than nitrogen. In its oxides nitrogen has oxidation numbers from +1 to +5 (Table 18.2).

NITROGEN MONOXIDE, NO

This can also be called *nitrogen(II) oxide*, and it is often known by its older name of *nitric oxide*. It is colorless, reactive gas that is only slightly soluble in water.

TABLE 18.1
Properties of the Oxides of Nitrogen

Oxidation State	Formula	Name	Physical State at 25 °C	Mp (°C)	Bp (°C)
+5	N_2O_5	Dinitrogen pentaoxide	White solid	30	47*
+4	N_2O_4	Dinitrogen tetraoxide	White solid		
+4	NO_2	Nitrogen dioxide	Brown gas	−11.2	21.2
+3	N_2O_3	Dinitrogen trioxide	Deep blue liquid (−80 °C)	−102	—*
+2	NO	Nitrogen monoxide (nitric oxide)	Colorless gas	−90.8	−88.5
+1	N_2O	Dinitrogen monoxide (nitrous oxide)	Colorless gas	−163.7	−151.8

* Decomposes

PREPARATION Nitrogen combines with oxygen slowly and at very high temperatures to give a small equilibrium yield of NO:

$$N_2(g) + O_2(g) \rightleftharpoons 2NO(g) \qquad \Delta H° = 180 \text{ kJ} \qquad \Delta G° = 173 \text{ kJ}$$

The reaction is endothermic, and the equilibrium constant is only 10^{-30} at 25 °C. In accordance with Le Châtelier's principle, the equilibrium shifts to the right with increasing temperature, but even at 2400 K the equilibrium constant is only 2.5×10^{-2}; hence the equilibrium concentration of NO is small.

In nature nitrogen combines with oxygen to give NO in lightning discharges in electric storms. As the NO leaves the discharge, it is rapidly cooled to ordinary temperatures. The decomposition of NO is then very slow, so the equilibrium

TABLE 18.2
Oxidation States of Nitrogen

Oxidation State	Example
+5	HNO_3, N_2O_5, NO_3^-
+4	NO_2, N_2O_4
+3	HNO_2, N_2O_3, NO_2^-
+2	NO
+1	N_2O
0	N_2
−1	NH_2OH
−2	N_2H_4
−3	NH_3

does not shift back to the left to any appreciable extent. In other words, the concentration of NO is "frozen" at the high-temperature value.

A small but significant amount of NO is formed by the reaction between the nitrogen and the oxygen in the air at the high temperature (≈ 2300 °C) of the spark in the cylinders of internal combustion engines. Automobiles are therefore a major contributor to the pollution of the atmosphere caused by NO and NO_2 (Box 18.1).

Nitrogen monoxide is prepared on a large scale in the first step of the *Ostwald process* for the manufacture of nitric acid. Ammonia and oxygen are passed over a platinum–rhodium wire gauze that has been heated to approximately 900 °C by an electric current. The hot wire gauze catalyzes the oxidation of ammonia to nitrogen monoxide:

$$4NH_3(g) + 5O_2(g) \longrightarrow 4NO(g) + 6H_2O(g) \qquad \Delta H° = -906 \text{ kJ}$$

Once the reaction has started, because it is strongly exothermic, it supplies the heat needed to keep the catalyst at the operating temperature (see Experiment 18.1).

Nitrogen monoxide can be conveniently prepared in the laboratory by reducing *dilute* nitric acid with copper (see Experiment 18.2):

$$3Cu(s) + 8HNO_3(aq) \longrightarrow 3Cu(NO_3)_2(aq) + 2NO(g) + 4H_2O(l)$$

A rhodium–platinum catalyst gauze being installed in a plant for producing nitric acid by the oxidation of ammonia.

PROPERTIES Because the nitrogen in NO is in a low ($+2$) oxidation state, NO behaves as a good reducing agent. With oxygen it forms nitrogen dioxide, NO_2:

$$2NO(g) + O_2(g) \longrightarrow 2NO_2(g)$$

Nitrogen monoxide reduces ozone to oxygen and is oxidized to NO_2,

$$NO(g) + O_3(g) \longrightarrow O_2(g) + NO_2(g)$$

and at high temperatures it reduces CO_2 to CO:

$$NO(g) + CO_2(g) \xrightarrow{\text{Heat}} CO(g) + NO_2(g)$$

With chlorine it forms nitrosyl chloride, NOCl,

$$2NO(g) + Cl_2(g) \longrightarrow 2NOCl(g)$$

Nitrosyl chloride

an orange-yellow gas that contains nitrogen in the $+3$ state. When bubbled into aqueous solutions of oxidizing agents, NO is usually oxidized to nitrate ion. For example, it reduces iodine to iodide ion,

$$2NO(g) + 3I_2(s) + 4H_2O(l) \longrightarrow 2NO_3^-(aq) + 8H^+(aq) + 6I^-(aq)$$

and MnO_4^- to Mn^{2+}.

Because there are stable lower oxidation states of nitrogen, NO will also react as an oxidizing agent. For example, it is reduced to nitrogen by hydrogen, which is oxidized to water,

$$2NO(g) + 2H_2(g) \qquad N_2(g) + 2H_2O(l)$$

and by phosphorus, which is oxidized to P_4O_6,

$$P_4(s) + 6NO(g) \longrightarrow P_4O_6(s) + 3N_2(g)$$

BOX 18.1

SMOG

An unpleasant and dangerous combination of smoke and fog that we now call smog was a feature of many industrial cities in Europe for hundreds of years. It was a product of a damp climate, and smoke and sulfur dioxide produced by the burning of coal for heat and power. This type of smog has become much less common in recent years, since the widespread use of coal has decreased. It has been replaced by a different type of smog called photochemical smog, which is primarily the product of a combination of automobile exhaust and a sunny climate. The process of photochemical smog formation is complex and all its details are not yet well understood. It begins with the combination of atmospheric nitrogen and oxygen to form NO at the high temperature of an automobile engine or the furnaces of coal- or oil-burning electric power plants. When NO is emitted into the atmosphere, it is slowly oxidized by oxygen to NO_2. The concentration of NO_2 is sometimes high enough for it to be clearly visible as a brown layer in the atmosphere, particularly from an airplane.

Although NO_2 is a health hazard, primarily because it attacks the lungs, its concentration usually does not reach a dangerous level. However, as we have explained in Box 14.1, it leads to the formation of acid rain. More-over, it initiates a complex series of reactions that lead to other, more dangerous atmospheric pollutants. The NO_2 is dissociated by ultraviolet radiation ($\lambda \leqslant 392$ nm) from the sun to give NO molecules and O atoms:

$$NO_2 \longrightarrow NO + O$$

Free oxygen atoms are extremely reactive and combine with molecular oxygen to give ozone:

$$O + O_2 \longrightarrow O_3$$

Because ozone is a powerful oxidizing agent, it reacts destructively with many materials, including rubber, paint, and vegetation. Moreover, it is very irritating to the eyes and it may damage the lungs. It also reacts with hydrocarbons that are released into the atmosphere from automobile exhausts as a result of the incomplete combustion of gasoline. These reactions produce a variety of organic molecules such as organic peroxides and peroxoacetylnitrates (abbreviated PAN), which are strong eye irritants and are damaging to plant life.

Los Angeles was the first city to achieve notoriety because of its photochemical smog, which results from its huge number of automobiles, its sunny climate, and its geographic situation. Normally, the temperature of the air decreases with increasing altitude (Chapter 3), and as dirty, hot air rises from the surface, cleaner, cool air descends to take its place. In certain locations such as Los Angeles a condition known as a temperature inversion can arise. Cool air flows down from the nearby mountains to form a pool on the surface underneath a mass of hot air. Here it is trapped, and atmospheric pollutants accumulate in this pool of cool air rather than being removed by rising hot air currents.

In recent years serious efforts have been made to reduce the smog problem in many cities. For example, standards have been set for automobile exhaust emissions. Automobile manufacturers have attempted to reduce the concentrations of undesirable components of automobile exhaust in two ways: first, by altering the conditions of combustion to reduce NO formation and the concentration of unburnt fuel and, second, by adding catalytic converters to the exhaust system to remove unburnt hydrocarbons and to decompose NO to N_2 and O_2. But a really efficient catalyst for this decomposition has still to be found.

Experiment 18.1

Preparation of NO by the Catalytic Oxidation of NH₃

When a coil of copper wire is heated in a flame and then suspended over a warm, concentrated, aqueous solution of ammonia, the copper coil continues to glow brightly for a long time as the ammonia is oxidized on the metal surface in a strongly exothermic reaction. The heat produced in the reaction is usually sufficient to melt the copper wire so that molten copper drops into the ammonia solution producing a blue color.

Experiment 18.2

Nitrogen Monoxide, NO

Nitrogen monoxide may be prepared by reacting copper with 6 M nitric acid. In this experiment a jar containing 6 M nitric acid has been inverted over some pieces of copper in a trough of water. Colorless NO is evolved and collects in the jar as it is only slightly soluble in water. The Cu²⁺ ion formed causes the nitric acid to turn blue.

A jar of colorless NO.

When the stopper is removed, the NO reacts with oxygen of the atmosphere to form brown NO₂.

If the jar of NO₂ is then inverted over water containing a little methyl red indicator, the water rises rapidly in the jar as the NO₂ dissolves to form a solution of HNO₂ and HNO₃, which changes the color of the indicator from yellow to red.

NITROGEN DIOXIDE, NO$_2$

The product of the reaction between NO and oxygen is nitrogen dioxide, NO$_2$. It is a red-brown gas. It may also be called nitrogen(IV) oxide. Two molecules of NO$_2$ combine with each other to form colorless *dinitrogen tetraoxide*, N$_2$O$_4$. At ordinary temperatures the gas is an equilibrium mixture of NO$_2$ and N$_2$O$_4$:

$$2NO_2 \rightleftharpoons N_2O_4 \qquad \Delta H° = -57 \text{ kJ}$$

Because the formation of N$_2$O$_4$ is an exothermic reaction, the equilibrium shifts to the right with decreasing temperature, and more N$_2$O$_4$ is formed. At 21 °C the gas condenses to a deep red-brown liquid, which is a mixture of NO$_2$ and N$_2$O$_4$ in equilibrium. As the temperature is decreased, the color of the liquid mixture becomes less intense. Finally, at -11 °C it freezes to a white solid that consists entirely of N$_2$O$_4$ molecules (see Experiment 18.3).

EXERCISE 18.1

Use Le Châtelier's principle to predict whether the equilibrium constant for the reaction

$$N_2O_4(g) \rightleftharpoons 2NO_2(g)$$

increases or decreases with increasing temperature.

Nitrogen dioxide can be conveniently made in the laboratory by reducing *concentrated* nitric acid with copper:

$$Cu(s) + 4HNO_3(aq) \longrightarrow Cu(NO_3)_2(aq) + 2NO_2(g) + 2H_2O(l)$$

It can also be produced by heating certain metal nitrates (see Experiment 18.4). For example,

$$2Pb(NO_3)_2(s) \longrightarrow 2PbO(s) + 4NO_2(g) + O_2(g)$$

Experiment 18.3

Nitrogen Dioxide and Dinitrogen Tetraoxide

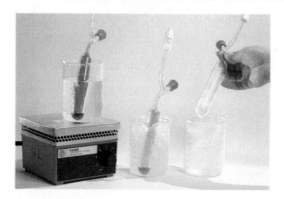

The equilibrium N$_2$O$_4$ $\leftrightarrows$ 2NO$_2$ shifts to the left with decreasing temperature. A large amount of dark brown NO$_2$ vapor can be seen in the tube on the left which is in hot water. The tube in the middle, which is in ice, is much paler because the equilibrium has been shifted to the left and there is therefore less NO$_2$ vapor in the tube. The yellow-brown liquid at the bottom of the tube consists of N$_2$O$_4$ containing a little NO$_2$. The tube on the right has been kept in a freezing mixture at a temperature of approximately -15°C. It contains white solid N$_2$O$_4$.

Experiment 18.4

Preparation of Nitrogen Dioxide

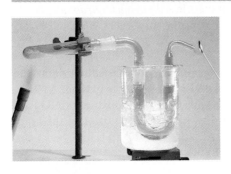

Nitrogen dioxide may be prepared by heating lead nitrate, $Pb(NO_3)_2$. Here the NO_2 may be seen condensing as a pale brown liquid in the U-tube cooled in ice. Oxygen, which is also produced in the reaction, ignites a glowing splint held at the exit tube.

Since nitrogen is in a high (+4) oxidation state, NO_2 is a good oxidizing agent. For example, it oxidizes SO_2 to SO_3:

$$SO_2(g) + NO_2(g) \longrightarrow SO_3(g) + NO(g)$$

It oxidizes CO to CO_2,

$$CO(g) + NO_2(g) \longrightarrow CO_2(g) + NO(g)$$

and I^- to I_2. But since it can be oxidized to the +5 oxidation state it can also act as a reducing agent and, for example, it will reduce MnO_4^- to Mn^{2+} in aqueous solution (Experiment 18.5).

EXERCISE 18.2

Write an appropriate balanced equation to show each of the following acting (i) as a reducing agent and (ii) as an oxidizing agent:
  (a) NO(g)   (b) $NO_2(g)$

STRUCTURES OF NO AND NO_2: FREE RADICALS

Nitrogen monoxide and nitrogen dioxide are unusual molecules in that they have an odd number of valence electrons: 11 for NO and 17 for NO_2. The vast majority of stable molecules have an even number of valence electrons, which may be described either as shared (bonding) pairs or as unshared (nonbonding) pairs.

NITROGEN MONOXIDE Because NO and NO_2 have an odd number of valence electrons, we cannot write normal Lewis structures for them in which all the electrons are paired and in which all the atoms have an octet of electrons in their valence shells. For example, we can write the following structure for NO:

$$:\overset{\textstyle .}{N}=\overset{\textstyle ..}{O}:$$

Experiment 18.5

Properties of Nitrogen Dioxide

A colorless aqueous solution of KI turns brown when NO_2 gas is passed into it, The NO_2 oxidizes colorless I^- to I_2, forming brown I_3^-.

A pink aqueous solution of $KMnO_4$ becomes colorless as NO_2 gas is passed into it. The NO_2 reduces MnO_4^- to Mn^{2+} and is itself oxidized to NO_3^-.

Here oxygen has eight electrons in its valence shell, but nitrogen has only seven, and the odd or unpaired electron is located on nitrogen. We can write another possible Lewis structure for NO by placing the odd electron on oxygen rather than on nitrogen, so oxygen has only seven electrons in its valence shell and nitrogen has eight:

$$\overset{\ominus}{:}\!\!N\!=\!\overset{\oplus}{\dot{O}}\!:$$

This second structure has formal charges; it is therefore somewhat less important than the previous structure. The two structures may be regarded as resonance structures:

$$\dot{N}\!=\!\ddot{O}\!: \longleftrightarrow \overset{\ominus}{:}\!\dot{N}\!=\!\overset{\oplus}{\dot{O}}\!:$$

Thus the odd electron is found on both the nitrogen atom and the oxygen atom. Hence we can alternatively describe NO by assuming that the odd electron is shared between the nitrogen atom and the oxygen atom:

$$:\dot{N}\!\!\doteq\!\!\dot{O}:$$

In this structure both the nitrogen atom and the oxygen atom have a valence shell of eight electrons. The bond may be described as consisting of two shared pairs and a single shared electron. It has a bond order of 2.5. The NO bond length is consistent with this description, since it is intermediate between that of the triple bond in N_2 and that of the double bond in O_2:

	$:N\!\!\equiv\!\!N:$	$:\dot{N}\!\!\doteq\!\!\dot{O}:$	$:\ddot{O}\!\!=\!\!\dot{O}:$
Bond length	109 pm	115 pm	121 pm
Bond order	3	2.5	2

FREE RADICALS Molecules containing an odd number of electrons are quite rare. They are usually exceedingly reactive and have only very short lifetimes because they react with almost any other substance with which they come in contact. Odd-electron molecules are called **free radicals**. Examples include the methyl radical and the hydroxyl radical, H—O·. Although free radicals are often formed in small amounts during the course of many reactions, they get used up again—they are said to be intermediates—and they cannot generally be isolated in appreciable amounts. The methyl radical, for example, reacts rapidly with many other molecules. In the absence of other molecules it combines with another methyl radical to give ethane, C_2H_6.

Although NO reacts with oxygen and with some other substances, it is relatively unreactive for an odd-electron molecule or free radical. At ordinary temperatures two NO molecules do not combine with each other. That the odd electron in NO is shared between the nitrogen and oxygen atoms rather than localized on one atom, as in the methyl radical, may be the reason that NO is considerably less reactive than most other free radicals.

NITROGEN DIOXIDE The NO_2 molecule is also an odd-electron molecule or free radical; it has 17 valence electrons. We can write several Lewis structures for NO_2. There are two structures with the odd electron on nitrogen and a total of only 7 electrons in the valence shell of nitrogen:

There are also two structures with the odd electron on oxygen and only 7 electrons in the valence shell of one of the oxygen atoms:

Again, the odd electron can be considered to be shared between oxygen and nitrogen, and we can then write the following two structures in which all the atoms have octets:

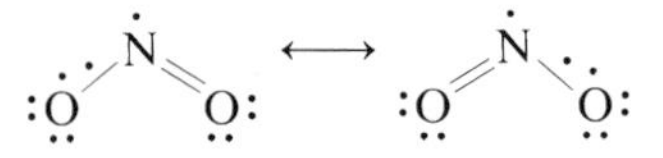

The molecule is angular with a bond length of 119 pm and a bond angle of 134°. Because a single unshared electron occupies a smaller amount of space in the valence shell than an unshared pair of electrons, the bond angle is considerably larger than the ideal angle of 120° for an AX_2E molecule.

MOLECULAR ORBITALS The difficulty of formulating satisfactory Lewis structures for NO and NO_2 again illustrates the limitations of Lewis structures. They are very convenient for the large number of molecules in which all the electrons can be described with reasonable accuracy as localized pairs, either bonding or nonbonding. However, for molecules in which some of the electrons are delocalized or for molecules with odd numbers of electrons, such as NO and NO_2, Lewis structures are less satisfactory. In such cases molecular orbital theory provides an alternative to the use of the concept of resonance to improve the description of molecules in terms of Lewis structures. In **molecular orbital theory** electrons are not described as localized bonding and nonbonding pairs; instead, they are allocated to orbitals, called **molecular orbitals**, that are associated with the entire molecule (see Chapter 7).

DINITROGEN TETRAOXIDE, N_2O_4, AND
DINITROGEN TRIOXIDE, N_2O_3

Two NO_2 molecules combine to give dinitrogen tetraoxide, N_2O_4, a white solid that melts at -11 °C to give an equilibrium mixture of NO_2 and N_2O_4. The strong oxidizing power of N_2O_4 has made it useful as a rocket fuel. A Lewis structure for N_2O_4 may be written by combining two NO_2 structures in which the odd electron is on nitrogen. A total of four resonance structures can be obtained in this way (see Figure 18.1).

Nitrogen monoxide and nitrogen dioxide also combine at low temperature by pairing their odd electrons to give *dinitrogen trioxide*, N_2O_3 (see Figure 18.2). This deep blue liquid decomposes completely at room temperature to NO and NO_2 (see Experiment 18.6).

At very low temperatures two NO molecules also combine in the same way to give the N_2O_2 molecule in the solid state (see Figure 18.3). This molecule is very unstable and is not known in the liquid or gaseous states. In other words, at room temperature the equilibrium lies far to the right:

$$N_2O_2(g) \rightleftharpoons 2NO(g)$$

In each of the molecules N_2O_4, N_2O_3, and N_2O_2 the N—N bond is much longer than the value of 140 pm predicted from the covalent radius of nitrogen (Figure 4.9), and the bond length increases from N_2O_4 to N_2O_3 to N_2O_2. With

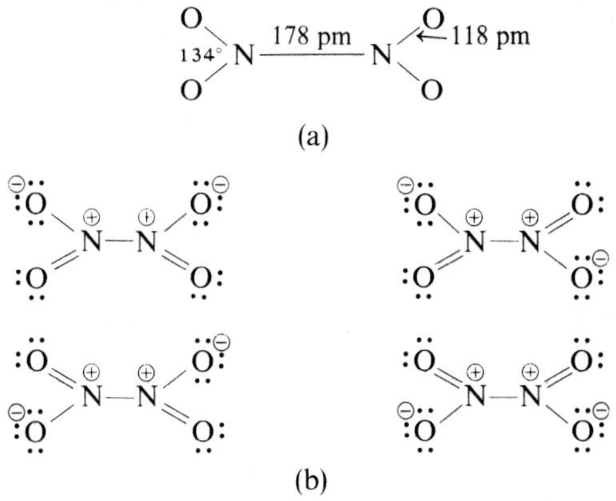

(a)

(b)

FIGURE 18.1
Structure of Dinitrogen Tetraoxide.

Dinitrogen tetraoxide is a planar molecule with the dimensions shown in (a). It can be represented by the four resonance structures in (b).

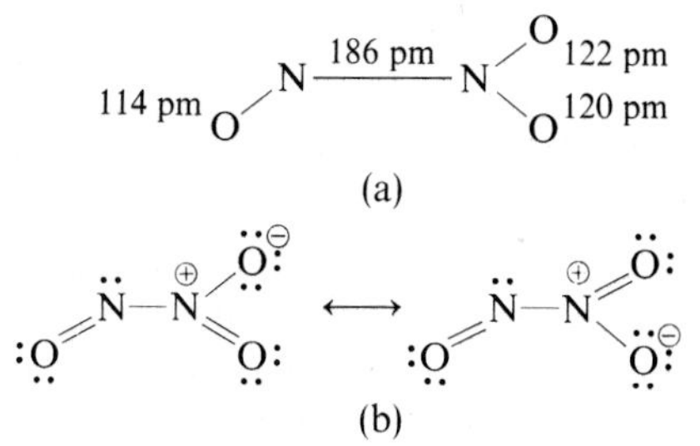

(a)

(b)

FIGURE 18.2
Structure of Dinitrogen Trioxide.

(a) Dimensions; (b) resonance structures.

Experiment 18.6

Dinitrogen Trioxide, N₂O₃

The tube contains pale brown liquid N₂O₄ at 0°C.

When colorless NO is bubbled into the N₂O₄ it is converted to deep blue liquid N₂O₃.

increasing length the N—N bond becomes weaker and the molecules therefore dissociate with increasing ease. No simple theory can describe satisfactorily the bonding in these molecules.

DINITROGEN MONOXIDE, N₂O

When solid ammonium nitrate is gently heated, it melts and decomposes to give N_2O, dinitrogen monoxide,

$$\overset{-3}{N}\overset{+5}{H_4}NO_3(l) \longrightarrow \overset{+1}{N_2}O(g) + 2H_2O(g) \quad \Delta H° = -37 \text{ kJ}$$

which is often called by its older name, *nitrous oxide.* In this oxidation–reduction reaction, nitrogen in the -3 oxidation state in NH_4^+ is oxidized by nitrogen in the $+5$ state in the nitrate ion, to give N_2O in which nitrogen is in the $+1$ oxidation state. Since the reaction forms 3 mol of gaseous products and is exothermic, it can occur with explosive violence if the solid is heated too strongly (see Box 6.2).

Dinitrogen monoxide may be described by two resonance structures:

$$:\overset{..}{O}=\overset{\oplus}{N}=\overset{..}{N}:^{\ominus} \longleftrightarrow {}^{\ominus}:\overset{..}{O}-\overset{\oplus}{N}\equiv N:$$

Dinitrogen monoxide is isoelectronic with carbon dioxide,

$$:\overset{..}{O}=C=\overset{..}{O}:$$

and has the same linear AX_2 structure.

In contrast to NO and NO_2, dinitrogen monoxide has an even number of electrons and is a rather unreactive substance. It is a colorless gas with a pleasant smell and a sweet taste. When breathed in small amounts, it acts as a mild

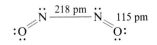

FIGURE 18.3
Structure of N₂O₂ in the Solid State.

intoxicant. Because of this property, it has been called laughing gas. In large amounts N_2O acts as a general anesthetic and is sometimes used for this purpose in dental and other minor surgery. Dinitrogen monoxide also has the useful property of being rather soluble in fats, a property exploited in making self-whipping cream. Cream is packaged with N_2O under pressure to increase its solubility. When the pressure is released, some the N_2O escapes to form tiny bubbles, which produce whipped cream.

DINITROGEN PENTAOXIDE, N_2O_5

The oxide N_2O_5 is made by dehydrating 100% nitric acid, HNO_3, with phosphorus(V) oxide, P_4O_{10}:

$$4HNO_3(l) + P_4O_{10}(s) \longrightarrow 2N_2O_5(g) + 4HPO_3(s)$$

It has the structure shown in Figure 18.4. Because it is formed by the removal of water from nitric acid, dinitrogen pentaoxide may be described as the *anhydride of nitric acid*. In general, an *acid anhydride* is the oxide obtained by removing water from an oxoacid.

$$2HNO_3 \longrightarrow N_2O_5 + \qquad H_2O$$
$$\text{Combines with } P_4O_{10}$$

Dinitrogen pentaoxide is a white volatile solid that decomposes slowly at room temperature to NO_2 and O_2:

$$2N_2O_5(s) \longrightarrow 4NO_2(l) + O_2(g)$$

It reacts readily with water to give nitric acid.

Dinitrogen pentaoxide is an interesting example of a substance that has a different structure in the solid and gaseous states. In the gas phase dinitrogen pentaoxide consists of covalent N_2O_5 molecules. In contrast, the white solid is an ionic compound composed of nitronium ions, NO_2^+, and nitrate ions, NO_3^-. It might therefore be called nitronium nitrate. The nitronium ion is isoelectronic with CO_2 and N_2O and has a linear AX_2 structure (see Figure 18.4). We have previously seen in Chapter 10 that aluminum trichloride is ionic in the solid state but consists of covalent Al_2Cl_6 molecules in the gaseous state.

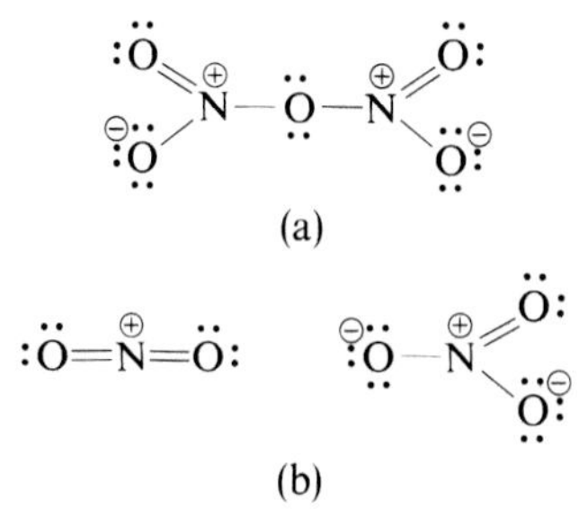

(a)

(b)

FIGURE 18.4
Structure of Dinitrogen Pentaoxide.

(a) In the gas state, dinitrogen pentaoxide consists of N_2O_5 molecules. One of four equivalent resonance structures is shown here. (b) In the solid state, dinitrogen pentaoxide is an ionic compound composed of the ions NO_2^+ and NO_3^-.

EXERCISE 18.3

Give the names and structures of two acid anhydrides, other than dinitrogen pentaoxide, and the names and structures of the acids from which they are derived.

18.2
Oxoacids of Nitrogen

There are two important oxoacids of nitrogen: nitrous acid, HNO_2, and nitric acid, HNO_3. They contain nitrogen in the $+3$ and $+5$ oxidation states, respectively. Nitric acid is a very important industrial chemical that ranked thirteenth in annual U.S. production in 1987 (see Table 15.1). It is used in the manufacture of explosives, dyes, and fertilizers. It is both a strong acid and a strong oxidizing agent.

NITRIC ACID, HNO₃

Nitric acid is prepared from ammonia by the **Ostwald process**, which was devised by German chemist Wilhelm Ostwald (1853–1932). The first step in this process is the catalyzed oxidation of NH_3 to NO (see page 832). The NO then reacts with excess oxygen to form NO_2,

$$2NO(g) + O_2(g) \longrightarrow 2NO_2(g)$$

which is passed into water, with which it reacts to give nitric acid and NO, which is recycled:

$$3NO_2(g) + H_2O(l) \longrightarrow 2HNO_3(aq) + NO(g)$$

In this reaction nitrogen in the $+4$ oxidation state is converted to nitrogen in the $+5$ state in nitric acid and nitrogen in the $+2$ state in NO. In other words, nitrogen is both oxidized and reduced. This reaction is another example of self-oxidation–reduction, or *disproportionation* (see Chapter 10).

The aqueous solution of nitric acid prepared in this way contains dissolved NO and NO_2, which are blown out by a stream of air and recycled. The acid is normally sold as "concentrated nitric acid", which has a concentration of $16M$, or 68% HNO_3 by mass. Concentrated nitric acid is colorless, but with time it usually develops a slight yellow color, because it decomposes when exposed to light, forming yellow-brown NO_2:

$$4HNO_3(aq) \xrightarrow{h\nu} 4NO_2(g) + O_2(g) + 2H_2O(l)$$

Anhydrous (100%) HNO_3 is obtained by mixing aqueous nitric acid with concentrated sulfuric acid and distilling off the volatile HNO_3. It can also be made by heating a metal nitrate with concentrated sulfuric acid (see Experiment 18.7):

$$NaNO_3(s) + H_2SO_4(l) \rightleftharpoons NaHSO_4(s) + HNO_3(g)$$

The equilibrium is continually shifted to the right by the removal of the volatile HNO_3 from the reaction mixture.

PROPERTIES Pure nitric acid is a colorless liquid that boils at 84.1 °C and freezes at −41.6 °C. The nitric acid molecule can be represented by two resonance structures:

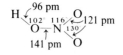

Nitric acid is a strong acid that is 100% dissociated in dilute aqueous solution:

$$HNO_3(l) + H_2O(l) \longrightarrow H_3O^+(aq) + NO_3^-(aq)$$

NITRATES There are many salts of nitric acid and almost all of them are soluble in water. The nitrate ion, NO_3^-, is isoelectronic with the carbonate ion and has the same AX_3 planar triangular shape. The bonding in the nitrate ion can be described by three resonance structures:

All three NO bonds have a length of 126 pm and a bond order of $1\frac{1}{3}$.

Colorless nitric acid (left) becomes yellow (right) when exposed to light.

Dimensions of HNO_3 molecule

Experiment 18.7

Nitric Acid

Nitric acid may be prepared by heating sodium nitrate with concentrated sulfuric acid. Nitric acid is volatile and distills over to the tube on the right, where it can be seen condensing as a pale yellow liquid. Pure liquid nitric acid is colorless. Here it is colored yellow by NO_2 formed by some decomposition of the hot HNO_3 vapor.

OXIDIZING PROPERTIES Nitric acid and the nitrate ion both contain nitrogen in the $+5$ oxidation state and are good oxidizing agents. Depending on the concentration of nitric acid and the nature of the reducing agent, they may be reduced to any of the other oxidation states of nitrogen. The corresponding half-reactions are given in Table 18.3.

The two most common of these reactions and their corresponding standard reduction potentials are as follows:

$$NO_3^-(aq) + 2H^+(aq) + e^- \longrightarrow NO_2(g) + H_2O(l) \qquad E° = 0.78 \text{ V}$$
$$NO_3^-(aq) + 4H^+(aq) + 3e^- \longrightarrow NO(g) + 2H_2O(l) \qquad E° = 0.96 \text{ V}$$

Concentrated aqueous nitric acid oxidizes sulfur to sulfuric acid and carbon to carbon dioxide and is reduced to nitrogen dioxide:

$$S(s) + 6HNO_3(\text{conc, aq}) \longrightarrow H_2SO_4(aq) + 6NO_2(g) + 2H_2O(l)$$
$$C(s) + 4HNO_3(\text{conc, aq}) \longrightarrow CO_2(g) + 4NO_2(g) + 2H_2O(l)$$

TABLE 18.3
Half-Reactions for the Reduction of Nitrate Ion

Reaction	Oxidation State of Nitrogen in Reduction Product
$NO_3^-(aq) + 2H^+(aq) + e^- \longrightarrow NO_2(g) + H_2O(l)$	$+4$
$NO_3^-(aq) + 3H^+(aq) + 2e^- \longrightarrow HNO_2(aq) + H_2O(l)$	$+3$
$NO_3^-(aq) + 4H^+(aq) + 3e^- \longrightarrow NO(g) + 2H_2O(l)$	$+2$
$2NO_3^-(aq) + 10H^+(aq) + 8e^- \longrightarrow N_2O(g) + 5H_2O(l)$	$+1$
$2NO_3^-(aq) + 12H^+(aq) + 10e^- \longrightarrow N_2(g) + 6H_2O(l)$	0
$NO_3^-(aq) + 10H^+(aq) + 8e^- \longrightarrow NH_4^+(aq) + 3H_2O(l)$	-3

Dilute aqueous nitric acid oxidizes H_2S to sulfur and SO_2 to sulfuric acid. In these reactions, nitric acid is reduced to NO:

$$3H_2S(aq) + 2HNO_3(aq) \longrightarrow 3S(s) + 2NO(g) + 4H_2O(l)$$

$$3SO_2(aq) + 2HNO_3(aq) + 2H_2O(l) \longrightarrow 3H_2SO_4(aq) + 2NO(g)$$

In the table of standard reduction potentials (Table 17.1) the half-reactions for most of the metals are above the half-reactions for the reduction of nitric acid to NO_2 and NO. Thus we expect nitric acid to dissolve all these metals, including metals such as copper and silver which are not oxidized by the H_3O^+ ion. Only a few metals such as gold and platinum, which have more positive reduction potentials than NO_3^- are not dissolved by nitric acid.

We can see from these half-reactions that the oxidizing strength of NO_3^- depends on the hydrogen ion concentration. According to Le Châtelier's principle, increasing the hydrogen ion concentration should cause the reactions to proceed more to the right. In other words, the oxidizing strength of NO_3^- should increase with increasing hydrogen ion concentration. Thus nitric acid is a much stronger oxidizing agent than a neutral solution of a metal nitrate. We can also arrive at this conclusion by using the Nernst equation (see Exercise 18.4).

Because of the variety of reduction products of nitric acid, its reactions with reducing agents are often quite complicated and give several reduction products. The relative amounts of the possible products depend on the temperature, the concentration of the nitric acid, and the nature of the reducing agent. As we saw in Chapter 10, when copper reacts with concentrated nitric acid, the main reduction product is NO_2, in which nitrogen is in the $+4$ oxidation state; with dilute nitric acid the main reduction product is NO, in which nitrogen is in the $+2$ oxidation state:

$$Cu(s) + 4HNO_3(aq, \text{concn}) \longrightarrow Cu(NO_3)_2(aq) + 2NO_2(g) + 2H_2O(l)$$

$$3Cu(s) + 8HNO_3(aq, \text{dil}) \longrightarrow 3Cu(NO_3)_2(aq) + 2NO(g) + 4H_2O(l)$$

The standard reduction potentials show that the reduction of NO_3^- with copper should not stop at NO_2 but should continue to NO. However, the reaction of concentrated nitric acid with copper is very rapid, and NO_2 escapes from the solution before it can be reduced to NO.

EXERCISE 18.4

Use the Nernst equation to calculate the half-cell potential for

$$NO_3^- + 4H^+ + 3e^- \longrightarrow NO + 2H_2O$$

for (a) $[H^+] = 10$ mol L^{-1} and (b) a neutral solution with $[H^+] = 10^{-7}$ mol L^{-1}, assuming $[NO_3^-] = 1$ mol L^{-1} and $p_{NO} = 1$ atm. Explain why an aqueous solution of nitric acid is a much stronger oxidizing agent than an aqueous solution of sodium nitrate.

USES OF NITRIC ACID   Much of the nitric acid produced is combined with ammonia to give ammonium nitrate:

$$NH_3(g) + HNO_3(l) \longrightarrow NH_4NO_3(s)$$

Ammonium nitrate is widely used as a fertilizer. Sodium nitrate and potassium nitrate are also important fertilizers. Extensive deposits of sodium nitrate are found in the desert regions of Chile.

Nitric acid is also used in the manufacture of explosives. Two important explosives are trinitrotoluene (TNT), and nitroglycerin. Nitroglycerin, the explosive component of dynamite (Box 6.2), is made by the reaction of glycerol (1,2,3-propanetriol) with nitric acid:

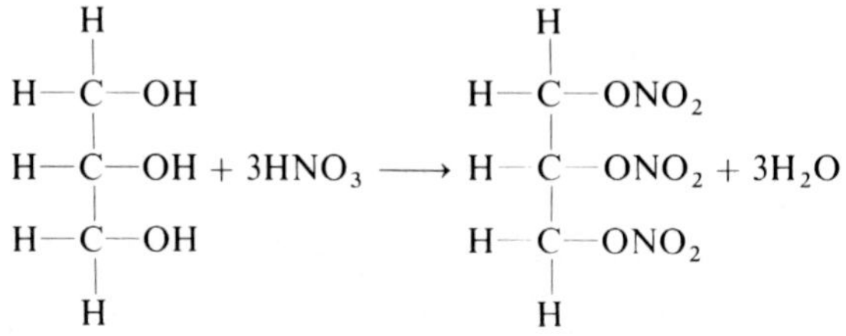

$$
\begin{array}{l}
\text{H} \\
| \\
\text{H—C—OH} \\
| \\
\text{H—C—OH} + 3HNO_3 \longrightarrow \\
| \\
\text{H—C—OH} \\
| \\
\text{H}
\end{array}
\qquad
\begin{array}{l}
\text{H} \\
| \\
\text{H—C—ONO}_2 \\
| \\
\text{H—C—ONO}_2 + 3H_2O \\
| \\
\text{H—C—ONO}_2 \\
| \\
\text{H}
\end{array}
$$

Trinitrotoluene is made by heating toluene with a mixture of concentrated nitric and sulfuric acids. Nitric acid is protonated by concentrated sulfuric acid:

$$HNO_3 + H_2SO_4 \longrightarrow H_2NO_3^+ + HSO_4^-$$

Although both acids are strong in water, sulfuric acid is a stronger acid than nitric acid, and in the absence of water, nitric acid acts as a base toward sulfuric acid. The ion $H_2NO_3^+$ is then dehydrated by sulfuric acid, producing the nitronium ion, NO_2^+:

$$H_2NO_3^+ + H_2SO_4 \longrightarrow NO_2^+ + H_3O^+ + HSO_4^-$$

Thus the overall reaction of nitric acid with concentrated sulfuric acid is

$$HNO_3 + 2H_2SO_4 \longrightarrow NO_2^+ + H_3O^+ + 2HSO_4^-$$

The nitronium ion reacts with toluene to give trinitrotoluene:

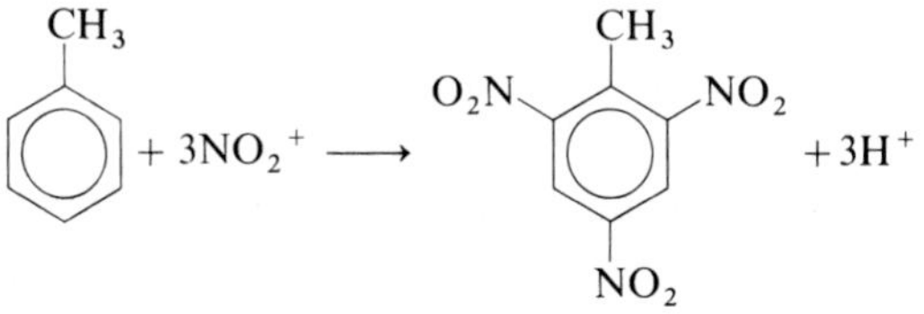

When TNT and nitroglycerin are heated or subjected to shock, they decompose rapidly, giving a large quantity of gaseous products in exothermic reactions. For example, nitroglycerin decomposes according to the following equation:

$$4C_3H_5N_3O_9(l) \longrightarrow 6N_2(g) + 12CO_2(g) + 10H_2O(g) + O_2(g)$$

NITROUS ACID, HNO$_2$

Unlike nitric acid, nitrous acid is a weak acid in water. It can be obtained by dissolving either N_2O_3 or a mixture of NO and NO_2 in water:

$$N_2O_3 + H_2O \longrightarrow 2HNO_2$$

Dinitrogen trioxide, N_2O_3, is the anhydride of nitrous acid. Only dilute solutions of nitrous acid can be obtained. If we attempt to concentrate a solution by heating, the nitrous acid decomposes to give NO and NO_2:

$$2HNO_2(aq) \longrightarrow NO(g) + NO_2(g) + H_2O(l)$$

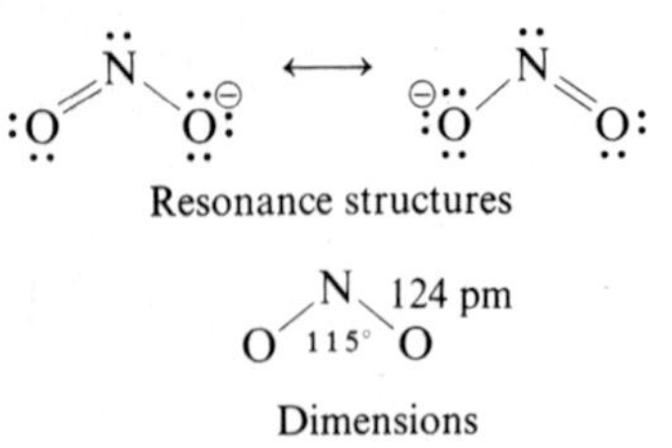

Resonance structures

Dimensions

FIGURE 18.5
Structure of the Nitrite Ion.

The Lewis structure of nitrous acid is

$$:\ddot{O}=\ddot{N}-\ddot{O}-H$$

The salts of nitrous acid are the *nitrites* (Figure 18.5). Sodium nitrite can be made by passing a mixture of NO and NO_2 into aqueous sodium hydroxide solution:

$$NO(aq) + NO_2(g) + 2NaOH(aq) \longrightarrow 2NaNO_2(aq) + H_2O(l)$$

Nitrites can also be made by heating alkali metal nitrates above their melting point:

$$2NaNO_3(l) \longrightarrow 2NaNO_2(l) + O_2(g)$$

Because they are readily oxidized to nitric acid and nitrate ion, nitrous acid and the nitrites are good reducing agents:

$$NO_2^- + H_2O \longrightarrow NO_3^- + 2e^- + 2H^+$$

Sodium nitrite and sodium nitrate have long been used in the preservation of meat. Meat darkens when it is stored because the blood is oxidized. Sodium nitrate and nitrite are reduced to NO by compounds in the meat. The NO combines with the hemoglobin and retards its oxidation, thus maintaining a fresh red appearance to the meat. Nitrites and nitrates also prevent the growth of a bacterium that causes botulism, a dangerous and sometimes fatal form of food poisoning. In the past few years, however, concern has risen about the possibility of a reaction between nitrite ion and certain organic compounds in the meat or in the digestive system to form nitrosamines, which are carcinogenic.

18.3
Hydrazine and Hydrazoic Acid

HYDRAZINE, N_2H_4

It is useful to think of hydrazine as being derived from ammonia by replacement of a hydrogen atom by an NH_2 group:

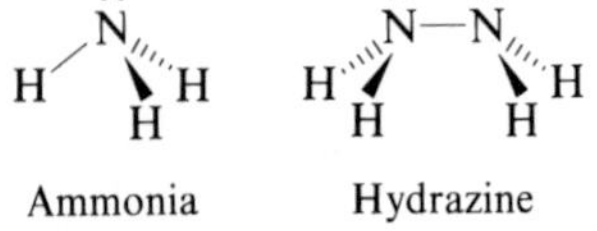

Ammonia   Hydrazine

Both nitrogen atoms in N_2H_4 have the expected pyramidal AX_3E geometry.

An aqueous solution of hydrazine can be prepared by oxidizing ammonia with a solution of sodium hypochlorite:

$$2NH_3(aq) + OCl^-(aq) \longrightarrow N_2H_4(aq) + H_2O(l) + Cl^-(aq)$$

The chloramines, H_2NCl and $HNCl_2$, are also produced in this reaction. They are both toxic and explosive. Thus it is dangerous to mix household bleach and an ammonia cleaning solution.

Hydrazine is a colorless liquid that melts at 2 °C and boils at 114 °C. It is a weak base, which is protonated to give the $N_2H_5^+$ and $N_2H_6^{2+}$ ions. Hydrazine has an endothermic enthalpy of formation ($\Delta H_f^\circ = 51$ kJ mol^{-1}); it is therefore a somewhat unstable compound.

Hydrazine burns in air with considerable evolution of heat:

$$N_2H_4(l) + O_2(g) \longrightarrow N_2(g) + 2H_2O(g) \qquad \Delta H° = -622 \text{ kJ}$$

Hydrazine is a good reducing agent and reacts vigorously with strong oxidizing agents such as the halogens, nitric acid, dinitrogen tetraoxide, and hydrogen peroxide with the evolution of a large amount of heat and the generation of a large volume of gaseous products:

$$2N_2H_4(l) + N_2O_4(l) \longrightarrow 3N_2(g) + 4H_2O(g)$$
$$N_2H_4(l) + 2H_2O_2(l) \longrightarrow N_2(g) + 4H_2O(g)$$

The oxidation of N_2H_4 or substituted hydrazines, where one or more of the hydrogens are replaced by methyl groups, is used to power rockets. For example on the Apollo missions to the moon, the rocket engines of the command module and the lunar landing vehicles used N_2O_4 as an oxidizer and a mixture of hydrazine and dimethylhydrazine as fuel. The U.S. space shuttle orbiter also used N_2O_4 and monomethylhydrazine. They react to give water, nitrogen, and carbon dioxide:

$$5N_2O_4(l) + 4N_2H_3(CH_3)(l) \longrightarrow 12H_2O(g) + 9N_2(g) + 4CO_2(g)$$

This reaction begins immediately the reactants are mixed which enables the rocket engines to be started and stopped as required. Since the reaction is highly exothermic and gives a large number of gaseous molecules, a high thrust is obtained for minimum mass of fuel.

Appollo lunar landing module.

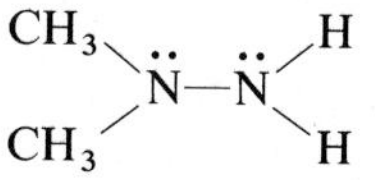

Dimethylhydrazine

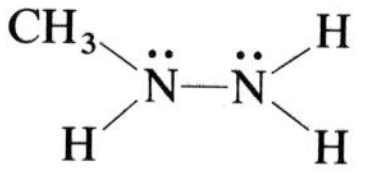

Monomethylhydrazine

EXERCISE 18.5

N_2O_4 (molar mass 92) boils at 21 °C. N_2H_4 (molar mass 32) boils at 114 °C. Explain why the boiling point of hydrazine is so much higher than that of dinitrogen tetraoxide.

EXERCISE 18.6

Using the enthalpies of formation in Table 6.2 and $\Delta H_f°(H_2O_2) = -188 \text{ kJ mol}^{-1}$ and $\Delta H_f°(N_2H_4) = 51 \text{ kJ mol}^{-1}$, find the enthalpy changes for the following reactions:

(a) $N_2H_4(l) + 2H_2O_2(l) \longrightarrow N_2(g) + 4H_2O(g)$

(b) $2N_2H_4(l) + N_2O_4(l) \longrightarrow 3N_2(g) + 4H_2O(g)$

EXERCISE 18.7

Calculate an approximate value for $\Delta H°$ for the reaction of dinitrogen tetraoxide and monomethylhydrazine from the bond energies given in Table 6.3.

═ *Example 18.1* ═════════════════════════════

HYDRAZINE AS A REDUCING AGENT

Write a balanced equation for the reaction between hydrazine and Fe^{3+} in which hydrazine is oxidized to N_2.

Solution

The oxidation number of each nitrogen in hydrazine is -2. Therefore the oxidation half-reaction involves 4 electrons and is

$$N_2H_4(aq) \longrightarrow N_2(g) + 4H^+(aq) + 4e^-$$

The reduction half-reaction is

$$Fe^{3+}(aq) + e^- \longrightarrow Fe^{2+}(aq)$$

Combining the equations for the two half-reactions gives the equation for the overall reaction

$$4Fe^{3+}(aq) + N_2H_4(aq) \longrightarrow 4Fe^{2+}(aq) + N_2(g) + 4H^+(aq)$$

EXERCISE 18.8

Write a balanced equation for the reduction of $MnO_4^-(aq)$ to Mn^{2+} by hydrazine in which it is oxidized to N_2.

HYDRAZOIC ACID, HN_3

:Ö=C=Ö:

:Ö=N⁺=Ö:

⁻:N̈=N⁺=Ö:

⁻:N̈=N⁺=N̈:⁻

Isoelectronic species

Hydrazoic acid is a liquid that boils at 37 °C. It is a weak acid in water. It is dangerously explosive. Salts of hydrazoic acid are called *azides*. The azides of heavy metals such as lead, mercury, and barium explode on being struck sharply and are used as detonators.

The azide ion, N_3^-, is isoelectronic with CO_2, N_2O, and NO_2^+, and it has the expected linear AX_2 structure with a bond length of 116 pm. Hydrazoic acid has the structure shown in Figure 18.6. It can be represented by the following two resonance structures:

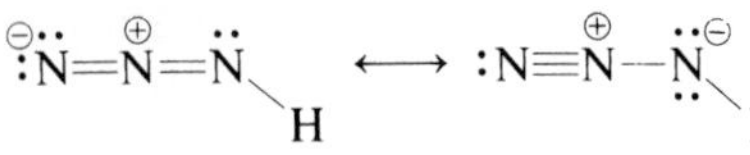

$$\overset{\ominus}{:}\!\!\ddot{N}\!\!=\!\!\overset{\oplus}{N}\!\!=\!\!\ddot{N}\diagdown_H \quad \longleftrightarrow \quad :N\!\!\equiv\!\!\overset{\oplus}{N}\!\!-\!\!\overset{\ddots\ominus}{N}\diagdown_H$$

Sodium azide, NaN_3, is used as the gas source in automobile air safety bags (see Box 18.2).

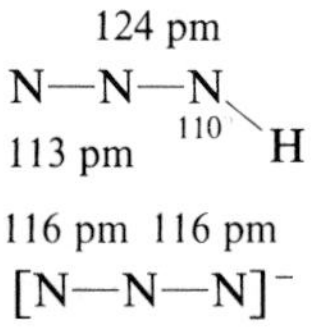

FIGURE 18.6
Structures of Hydrazoic Acid and the Azide Ion.

124 pm
N—N—N
113 pm   ¹¹⁰   H

116 pm 116 pm
[N—N—N]⁻

BOX 18.2

AUTOMOBILE AIR BAGS

Within the next five years air bags will be widely used in automobiles in the United States to protect the occupants in the event of a front-end collision. It is essential that the bag inflate very quickly, that is, within a few one-hundredths of a second. Thus a gas must be produced by some very rapid reaction. It is highly desirable that the gas should be nontoxic and noninflammable.

Most current designs of air bag use sodium azide, NaN_3, as the source of the inflating gas. Sodium azide is used because it contains 65% nitrogen by mass and it decomposes cleanly and rapidly, but without any flame or explosion, at temperatures of 350 °C or higher:

$$2NaN_3(s) \longrightarrow 2Na(l) + 3N_2(g)$$

The decomposition is initiated by a device which produces red-hot sparks. The temperature is then maintained by the exothermic reaction of iron(III) oxide, Fe_2O_3, with the sodium that is produced to give Na_2O and Fe:

$$6Na(l) + Fe_2O_3(s) \longrightarrow 3Na_2O(s) + 2Fe(s)$$

Because sodium azide is poisonous all the reactants are kept in a sealed container and only the nitrogen gas is allowed to enter the "air" bag. Typically the sodium azide mixture is triggered to react at about 10 milliseconds into the crash. The bag is completely filled within 30 milliseconds and porous sections in the bag allow it to deflate in 100–200 milliseconds.

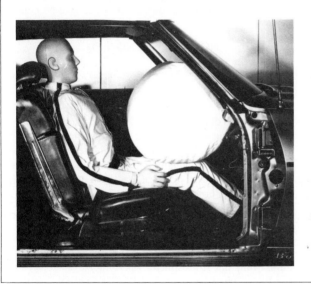

18.4
OZONE

Ozone or trioxygen is an allotrope of oxygen consisting of O_3 molecules. It is a pale blue gas with a characteristic odor. When an electric discharge is passed through gaseous oxygen about 10% of the oxygen is converted to ozone:

$$3O_2(g) \longrightarrow 2O_3(g)$$

The smell of ozone is often noticed during electric storms and in the vicinity of electric motors. It is an important constituent of the stratosphere (Box 18.3).

B O X   1 8 . 3

THE OZONE LAYER

Ozone is an important constituent of the stratosphere, principally at altitudes between 15 and 25 km, where it may reach a concentration of 10 ppm, compared with 0.04 ppm at lower altitudes. Ozone is formed by the decomposition of oxygen by ultraviolet radiation from the sun having wavelengths shorter than 240 nm:

$$O_2 \xrightarrow{h\nu} 2O$$

At a height of 200 km or more above the earth's surface, the pressure is extremely low, and the concentration of oxygen atoms is so low that they rarely recombine to form O_2 molecules but remain as oxygen atoms, making up the atomic oxygen layer of the homosphere (Chapter 3). But at lower altitudes, oxygen atoms immediately combine with oxygen molecules to form ozone:

$$O + O_2 \longrightarrow O_3$$

But this ozone absorbs ultraviolet light ($\lambda = 240 - 310$ nm), and is decomposed back to O and O_2. Thus an equilibrium is set up.

$$O_3 \underset{}{\overset{h\nu}{\rightleftharpoons}} O_2 + O$$

The concentration of ozone remains constant at its equilibrium value, but ultraviolet light is continuously absorbed and is converted to kinetic energy of O atoms and O_2 molecules, that is, into heat. As a consequence, most of the ultraviolet radiation from the sun is absorbed before it reaches the earth's surface. Since living cells can be destroyed by ultraviolet radiation, the ozone layer protects us from its damaging effects. If there were no ozone in the stratosphere, the intensity of the ultraviolet radiation reaching the surface would be such that life as we know it would not be possible. Even a small decrease in the concentration of ozone could lead to an increased incidence of skin cancer.

Although too much ozone in the atmosphere at low altitudes as in photochemical smog, is harmful to us, too little in the stratosphere could be even more serious. There has been concern in recent years that certain atmospheric pollutants may be decreasing the concentration of ozone in the ozone layer. Supersonic aircraft (SSTs) fly at altitudes up to 18 km, that is, at the height of the ozone layer. The NO they emit in their exhaust reduces ozone to oxygen and is oxidized to NO_2:

$$NO + O_3 \longrightarrow O_2 + NO_2$$

The NO_2 combines with oxygen atoms to give NO and O_2:

$$NO_2 + O \longrightarrow NO + O_2$$

The net reaction is

$$O_3 + O \longrightarrow 2O_2$$

Thus ozone is decomposed and NO is regenerated. Here NO is behaving as a catalyst: it increases the rate of decomposition of O_3 but is regenerated in the reaction and undergoes no overall change.

Another cause of concern has been the release of chlorofluorocarbons, such as CF_2Cl_2 and $CFCl_3$, into the atmosphere. These substances have been widely used as propellants in spray cans and as refrigerant gases. They are unreactive compounds and do not appear to undergo any reactions in the lower atmosphere. They diffuse into the stratosphere, where they are subject to

It condenses to a deep blue liquid at -112 °C. The liquid and even the gas, if it is not at low pressure or diluted with an inert gas such as nitrogen, may decompose explosively in a strongly exothermic reaction:

$$2O_3 \longrightarrow 3O_2 \qquad \Delta H° = -285 \text{ kJ}$$

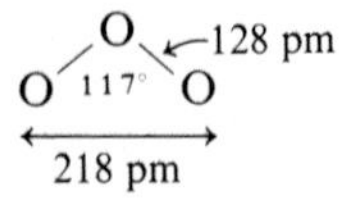

FIGURE 18.7
Structure of Ozone.

Ozone is isoelectronic with the nitrite ion, NO_2^-, and is an angular AX_2E molecule (see Figure 18.7).

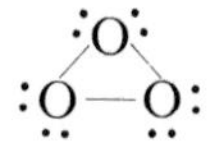

We might have expected ozone to be described by the Lewis structure ultraviolet radiation, which breaks a C—Cl bond, giving a free Cl atom:

$$CF_2Cl_2 \xrightarrow{h\nu} CF_2Cl + Cl$$

These Cl atoms can catalyze the decomposition of ozone in a manner analogous to NO:

$$O_3 + Cl \longrightarrow O_2 + ClO$$
$$ClO + O \longrightarrow O_2 + Cl$$

Chlorine atoms are regenerated, and the overall reaction is

$$O + O_3 \longrightarrow 2O_2$$

In the past few years direct measurements of the ozone concentration in the ozone layer have confirmed that it has decreased by about 3%, which, although small, could be very significant, particularly if the decrease continues. However, a much more dramatic, although temporary, reduction in the concentration of ozone in the ozone layer has been observed in the early spring (September) each year in the Antarctic, giving rise to what has been called the "ozone hole" as shown in the accompanying diagram. The interpretation of these results shows that the mechanism of the destruction of ozone in this part of the atmosphere is more complicated than outlined above. It seems that much of the chlorine in Freons is slowly converted in the atmosphere to HCl(g) and chlorine nitrate, $ClNO_3(g)$, which normally decompose quite slowly to yield Cl and ClO radicals which are catalysts for the decomposition of ozone. However, because of the very low temperatures, the atmosphere over the Antarctic in winter contains a considerable concentration of ice crystals and when HCl and $ClNO_3$ are adsorbed onto the surface of these crystals they are catalytically

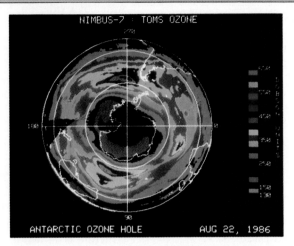

Map of total ozone in the Southern Hemisphere illustrating the Antarctic ozone "hole," which is the oval feature generally covering the Antarctic, portrayed in gray and violet colors. On August 22, 1986, the ozone "hole" is forming.

decomposed to HOCl and Cl_2:

$$H_2O + ClNO_3 \longrightarrow HNO_3 + HOCl$$
$$HCl + ClNO_3 \longrightarrow HNO + Cl_2$$

When sunlight returns to Antarctica in the spring both Cl_2 and HOCl are rapidly decomposed photochemically to produce relatively large concentrations of Cl and ClO radicals which catalyze the decomposition of ozone. The study of these reactions is still far from complete and we see that there is a need for more research on the reactions of even quite simple compounds of chlorine. We note also the importance of surfaces as catalysts which we study in Chapter 19.

However, this structure implies bond angles of 60° and equal distances between all the oxygen atoms, which is not in agreement with the experimentally determined structure (Figure 18.7). Since the covalent radius of oxygen is 66 pm (Figure 4.9), the length of an oxygen–oxygen single bond is predicted to be 132 pm. The observed distance of 218 pm between two of the oxygen atoms in ozone is so much larger than this that there can be no bond between these two oxygen atoms. Ozone is therefore represented by the following two resonance structures:

Ozone is an extremely powerful oxidizing agent. The only common oxidizing agent that is stronger is fluorine. Ozone can be used for destroying bacteria in water by oxidation. Unlike chlorine, it leaves no taste in the water.

EXERCISE 18.9

What is the bond order of the bonds in the ozone molecule? Show that the observed bond length of 128 pm is consistent with this bond order.

18.5
Hydrogen Peroxide, Peroxides, and Superoxides

Hydrogen peroxide, H_2O_2, can be thought of as being derived from water by replacement of a hydrogen atom by an OH group (see Experiment 18.8). Its structure is shown in Figure 18.8.

Although the angular geometry at each oxygen atom can be predicted by the VSEPR model, there is no simple theory that enables one to predict the

Experiment 18.8

Hydrogen Peroxide

An interesting although not very practical method of preparing a dilute aqueous solution of hydrogen peroxide is to allow a hydrogen flame to play on an ice cube. As the flame is rapidly cooled by the ice, the hydrogen reacts with oxygen to produce H_2O_2 as well as H_2O. The H_2O_2 dissolves in the water formed by the melting ice so that a dilute solution of H_2O_2 collects in the dish.

That the solution thus formed contains hydrogen peroxide can be demonstrated by adding it to a pink aqueous solution of $KMnO_4$. The $KMnO_4$ is decolorized as the hydrogen peroxide reduces the MnO_4^- ion to Mn^{2+} and is itself oxidized to oxygen.

When an aqueous solution of hydrogen peroxide is added to solid black lead sulfide, PbS, the PbS is oxidized to white $PbSO_4$ and the H_2O_2 is reduced to H_2O.

overall shape of the molecule—that is, whether it will be planar or nonplanar. In the solid state it is nonplanar (see Figure 18.8). The O—O bond (147 pm) is somewhat longer than the single-bond distance of 132 pm predicted from the covalent radius of 66 pm (Figure 4.9), which suggests that the O $\cdots$ O bond is rather weak and therefore rather easily broken to give two $\cdot \ddot{O}H$ radicals. Thus H_2O_2 is a reactive substance.

In the laboratory hydrogen peroxide can be prepared by stirring solid barium peroxide, BaO_2, with a cold aqueous solution of sulfuric acid. Insoluble barium sulfate is precipitated and can be filtered off. If stoichiometric amounts of BaO_2 and sulfuric acid are used, a pure aqueous solution of hydrogen peroxide can be obtained:

$$BaO_2(s) + H_2SO_4(aq) \longrightarrow BaSO_4(s) + H_2O_2(aq)$$

Hydrogen peroxide is a colorless, viscous liquid that boils at 150 °C. Like water, it is strongly associated by hydrogen bonding (Chapter 12). When the pure liquid is heated, it decomposes rapidly and even explosively in a disproportionation reaction:

$$2H_2O_2(l) \longrightarrow 2H_2O(l) + O_2(g) \qquad \Delta H° = -196 \text{ kJ}$$

Normally, it is used as an aqueous solution (for example, 30% by mass for use in the laboratory and 3% by mass for pharmaceutical use). Such aqueous solutions decompose very slowly at room temperature, but the decomposition is catalyzed by many different substances, including Fe(II) salts, manganese dioxide, powdered platinum, and blood (Experiment 1.5). Hydrogen peroxide is a very weak acid in aqueous solution.

Hydrogen peroxide is an important industrial chemical that has many applications; for example, it is used as a bleaching agent for textiles and for wood pulp and waste paper in paper making. Its use for bleaching hair is well known. Its bleaching action is due to its strong oxidizing properties. The oxidation number of oxygen in hydrogen peroxide is -1. It is reduced to H_2O, in which oxygen has an oxidation number of -2. The corresponding half-reaction is

$$H_2O_2(aq) + 2H^+(aq) + 2e^- \longrightarrow 2H_2O(l) \qquad E° = +1.77 \text{ V}$$

The large positive value of the standard reduction potential shows that it is a stronger oxidizing agent than NO_3^- or MnO_4^-. Hydrogen peroxide oxidizes Fe^{2+} to Fe^{3+}, I^- to I_2, SO_2 (or SO_3^{2-}) to SO_4^{2-}, and PbS to $PbSO_4$ (see Experiment 18.8). This last reaction is used to restore the original white color to old paintings in which white lead pigment, $Pb_3(OH)_2(CO_3)_2$, has become converted to dark brown PbS in an urban atmosphere containing H_2S.

Because it can be oxidized to oxygen, in which the oxidation number of oxygen is zero, H_2O_2 can also behave as a reducing agent:

$$H_2O_2(aq) \longrightarrow O_2(g) + 2H^+(aq) + 2e^-$$

For example, H_2O_2 will reduce MnO_4^- to Mn^{2+}.

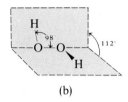

FIGURE 18.8
Structure of Hydrogen Peroxide.

(a) Lewis structure. (b) Dimensions. The two OH bonds in the H_2O_2 molecule are not in the same plane. In the solid state, there is an angle of 112° between them.

EXERCISE 18.10

Write a balanced equation for the oxidation of PbS(s) to $PbSO_4$(s) with H_2O_2(aq).

When sodium is heated in a limited supply of air, it forms the expected oxide Na_2O. But in an excess of air it forms the pale yellow peroxide Na_2O_2, which contains the **peroxide ion**, O_2^{2-}:

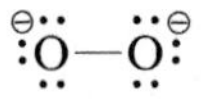

$$\overset{\ominus}{\underset{..}{\overset{..}{O}}}-\overset{..}{\underset{..}{\overset{..}{O}}}\overset{\ominus}{}$$

Barium behaves in the same way to give barium peroxide, BaO_2.

Potassium, rubidium, and cesium give the expected oxides K_2O, Rb_2O, and Cs_2O in a limited amount of air or oxygen. But when they are heated in excess oxygen, they form the oxides KO_2, RbO_2, and CsO_2. These orange-red compounds contain the **superoxide ion**, O_2^-, which has 13 electrons and is a stable, free-radical ion. It can be represented by two resonance structures,

$$:\overset{..}{\underset{..}{O}}-\overset{..}{\underset{..}{\overset{.}{O}}}\overset{\ominus}{}\quad\longleftrightarrow\quad\overset{\ominus}{:}\overset{..}{\underset{..}{O}}-\overset{.}{\underset{..}{O}}:$$

or by one structure in which the odd electron is shared between the two oxygen atoms, as in NO:

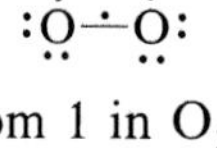

$$:\overset{.}{\underset{..}{O}}\cdot\cdot\overset{.}{\underset{..}{O}}:$$

As the bond order increases from 1 in O_2^{2-} to 1.5 in O_2^- and to 2 in the oxygen molecule, O_2, the bond length decreases correspondingly:

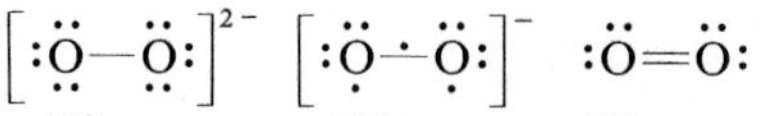

$$\left[:\overset{..}{\underset{..}{O}}-\overset{..}{\underset{..}{O}}:\right]^{2-}\quad\left[:\overset{.}{\underset{..}{O}}\cdot\cdot\overset{.}{\underset{..}{O}}:\right]^-\quad:\overset{..}{O}=\overset{..}{O}:$$
$$\quad149\text{ pm}\qquad\qquad133\text{ pm}\qquad121\text{ pm}$$

Potassium superoxide KO_2 is used in self-contained breathing apparatus to remove moisture and CO_2 from the exhaled air and replace it by oxygen. The reactions are

$$2KO_2(s) + 2H_2O(l) \longrightarrow 2KOH + O_2(g) + H_2O_2(l)$$
$$4KO_2(s) + 2CO_2(g) \longrightarrow 2K_2CO_3(s) + 2O_2(g)$$

EXERCISE 18.11

The reactions of KO_2 with water and carbon dioxide are redox reactions. Assign oxidation numbers to show which elements are reduced and which are oxidized. Then write an equation for the reduction half-reaction and an equation for the oxidation half-reaction in both cases.

IMPORTANT TERMS

An **acid anhydride** is the oxide obtained from an oxoacid by removing water.

A **free radical** is a molecule, an atom, or an ion containing an odd number of electrons.

P R O B L E M S *

Names and Formulas

1. What is the formula of each of the following compounds?

   (a) Nitric acid   (b) Nitrous acid

   (c) Potassium nitrite   (d) Nitrogen monoxide

   (e) Dinitrogen pentaoxide   (f) Hydrazine

   (g) Sodium azide

2. What is the formula of each of the following compounds?

   (a) Ozone   (b) Sodium peroxide

   (c) Barium peroxide   (d) Potassium superoxide

   (e) Hydrogen peroxide

3. What are the names of the compounds with each of the following formulas?

   (a) $NaNO_3$ (b) KNO_2 (c) N_2O_4

   (d) HN_3 (e) BaO_2 (f) KO_2

   (g) N_2H_4 (h) LiN_3 (i) Li_3N

4. Write the formula, and name each of the anhydrides of the following oxoacids:

   (a) HNO_3 (b) HNO_2 (c) H_3PO_3

   (d) H_2SO_4 (e) H_2CO_3

5. Give the molecular formula of each of the following compounds and state the oxidation number of nitrogen in each:

   (a) Ammonia   (b) Hydrazine

   (c) Nitrogen   (d) Nitric acid

   (e) Dinitrogen monoxide   (f) Nitrogen monoxide

   (g) Nitrous acid   (h) Dinitrogen tetraoxide

Reactions

6. Each of the following reactions is a method by which nitrogen may be prepared in the laboratory. In each case, write the balanced equation and indicate which substances are reduced and which oxidized:

   (a) The reaction of ammonia with copper(II) oxide, to give nitrogen, water, and copper

   (b) The reaction of an aqueous solution of sodium nitrite with a solution of ammonium chloride, to give, on gentle heating, nitrogen and sodium chloride

   (c) The reaction of nitrogen monoxide with ammonia over a red-hot copper catalyst, to give nitrogen and water

   (d) The decomposition of solid ammonium dichromate, $(NH_4)_2Cr_2O_7$, on heating, to give chromium(III) oxide, nitrogen, and water

7. Write balanced equations and identify the oxidation number of nitrogen in the products formed when each of the following nitrates is heated:

   (a) Potassium nitrate   (b) Ammonium nitrate

   (c) Lead(II) nitrate

8. Write balanced equations for each of the following reactions:

   (a) The decomposition of nitrous acid to nitric acid and nitrogen monoxide in aqueous solution

   (b) The decomposition of nitric acid to nitrogen dioxide and oxygen

   (c) The reaction of benzene with a mixture of sulfuric acid and nitric acid, to give nitrobenzene

9. Write balanced equations for the reactions of (a) dilute $HNO_3(aq)$ with $Cu(s)$, with $NH_3(g)$, and with $Mg(OH)_2(s)$; (b) concentrated $HNO_3(aq)$ with $Cu(s)$ and with $Mg(OH)_2(s)$; (c) anhydrous (100%) HNO_3 with $P_4O_{10}(s)$ and with $H_2SO_4(l)$.

10. Describe, using suitable balanced equations, how nitric acid is prepared from ammonia.

11. When sulfur dioxide gas is bubbled through an acidic solution of barium nitrate, a white precipitate is formed. What is the white precipitate? Explain by means of suitable balanced equations how it is formed.

12. Write a balanced equation for each of the following reactions. State whether each reaction is an acid–base or an oxidation–reduction reaction. For the acid–base reactions, indicate which reactant is the acid and which is the base. For the oxidation–reduction reactions, state which reactant is oxidized and which is reduced.

   (a) The reaction of sodium nitrate with concentrated sulfuric acid on heating

   (b) The photochemical decomposition of nitric acid

   (c) The decomposition of lead(II) nitrate on heating

   (d) The reaction of nitric acid with anhydrous (100%) sulfuric acid

   (e) The reaction of nitrous acid with water

   (f) The reaction of ammonia with hypochlorite ion in aqueous solution

(g) The catalyzed decomposition of hydrogen peroxide

(h) The reaction of barium peroxide with dilute sulfuric acid

13. How can **(a)** nitrogen monoxide, **(b)** ammonia, and **(c)** nitrogen dioxide be prepared from nitric acid?

14. Describe how **(a)** ozone and **(b)** hydrogen peroxide may be prepared from oxygen.

15. How may a sample of each of the following be prepared in the laboratory?

(a) Dinitrogen monoxide, $N_2O(g)$

(b) Nitrogen monoxide, $NO(g)$

(c) Dinitrogen tetraoxide, $N_2O_4(l)$

16. Draw the Lewis structure of each of the oxides in Problem 15.

17. What reaction occurs between N_2O_4 and water?

18. Write balanced reactions for each of the following reactions of ozone:

(a) Oxidation of black lead(II) sulfide to white lead(II) sulfate

(b) Oxidation of iodide ion to iodine in basic aqueous solution

(c) Oxidation of $Fe^{2+}(aq)$ to $Fe^{3+}(aq)$ in acidic solution

19. Write balanced equations for the action of heat on each of the following solids:

(a) HgO   **(b)** PbO_2 **(c)** BaO_2 **(d)** KNO_3

20. Balance each of the following equations, and classify each as an acid–base or oxidation–reduction reaction:

(a) $HNO_3 \xrightarrow{\text{light}} NO_2 + H_2O + O_2$

(b) $HNO_3 + H_2O \rightarrow H_3O^+ + NO_3^-$

(c) $HNO_3 + H_2SO_4 \rightarrow NO_2^+ + H_3O^+ + HSO_4^-$

(d) $S + HNO_3 \rightarrow H_2SO_4 + NO$

(e) $I_2 + HNO_3 \rightarrow HIO_3 + NO + H_2O$

(f) $P_4 + HNO_3 + H_2O \rightarrow H_3PO_4 + NO + NO_2$

(g) $Zn + HNO_3 \rightarrow Zn(NO_3)_2 + N_2O + H_2O$

(h) $Fe^{2+} + HNO_3 \rightarrow Fe^{3+} + NO + H_2O$

21. Balance each of the following equations, and in each case classify NO_2 as an oxidizing agent or a reducing agent:

(a) $H_2S + NO_2 \rightarrow NO + H_2O + S_8$

(b) $NO_2 + I^- + H_2O \rightarrow NO + I_3^- + OH^-$

(c) $MnO_4^- + NO_2 + H_2O \rightarrow Mn^{2+} + H_3O^+ + NO_3^-$

(d) $NO_2 + H_2O \rightarrow HNO_2 + HNO_3$

22. When an alkali metal nitrate such as potassium nitrate or sodium nitrate is strongly heated, it melts, evolving oxygen, and a nitrite is formed. Alternatively, the nitrite is formed by heating the nitrate with lead or copper, to give also PbO(s) or CuO(s). Write a balanced equation for each of these reactions.

23. A useful method for the determination of nitrate in aqueous solution is to determine the amount of nitrogen monoxide produced by the reaction

$$2NO_3^- + 3H_2SO_4 + 3Hg(l) \longrightarrow$$
$$3HgSO_4 + 2H_2O + 2OH^- + 2NO(g)$$

in aqueous solution. If 3.26 L of $NO(g)$ at STP were obtained from 2.00 L of a nitrate solution, what was the concentration of nitrate ion in the solution?

Lewis Structures and Molecular Geometry

24. Draw Lewis structures for each of the following species:

(a) NH_4^+ **(b)** $N_2H_5^+$ **(c)** HNO_2

(d) H_2O_2 **(e)** O_2^{2-} **(f)** ClNO

25. Draw Lewis structures for each of the following molecules and ions. Indicate in each case whether a single Lewis structure is sufficient for a description of the bonding, or whether several resonance structures are required.

(a) Nitric acid   **(b)** Nitrate ion

(c) Nitrous acid   **(d)** Nitrite ion

(e) Nitrogen monoxide   **(f)** Dinitrogen tetraoxide

(g) Dinitrogen pentaoxide (g)

26. Using the VSEPR model, categorize each of the following (where nitrogen is the central atom) in terms of the AX_nE_m nomenclature, and predict the approximate geometry of each:

(a) NO_2 **(b)** NO_2^+ **(c)** NO_3^-

(d) $HONO_2$ **(e)** N_2O **(f)** ClNO

(g) NF_3 **(h)** HONO   **(i)** CH_3NO_2

27. Draw a Lewis structure for the oxide that is the anhydride of each of the following oxoacids:

(a) Nitric acid   **(b)** Nitrous acid

(c) Phosphoric acid   **(d)** Sulfuric acid

28. Which of the molecules with the following molecular formulas has a dipole moment?

(a) NO   **(b)** NO_2 **(c)** N_2

(d) O_3 (e) N_2O (f) N_2O_4

(g) H_2O_2 (h) N_2H_4 (i) HNO_3

(j) $NOCl$

29. Which three species mentioned in this chapter are isoelectronic with CO_2?

30. Write Lewis structures for the ozone, O_3, molecule. Explain why ozone is not given the Lewis structure

$$:\overset{\displaystyle :O:}{\underset{\displaystyle :O\mathbf{-}O:}{}}$$

Equilibrium

31. In the gas phase, NO_2 is in equilibrium with N_2O_4. What will be the effect on the equilibrium concentration of N_2O_4 of each of the following changes?

(a) Increasing the total pressure of the system by adding an inert gas

(b) Increasing the volume of the system at constant temperature

(c) Decreasing the temperature at constant volume (use data from Appendix B to calculate ΔH° for the reaction)

32. Calculate the pH of each of the following solutions:

(a) $0.020M$ $HNO_2(aq)$ ($pK_a = 3.14$)

(b) $0.020M$ $NaNO_2(aq)$

(c) A solution that is $0.010M$ in $NaNO_2(aq)$ and $0.010M$ in $HNO_2(aq)$

Thermochemistry and Thermodynamics

33. Calculate the standard enthalpy change for the reaction

$$4NH_3(g) + 5O_2(g) \longrightarrow 4NO(g) + 6H_2O(g)$$

Under what conditions of temperature and pressure would the equilibrium concentration of $NO(g)$ be maximized?

34. The standard enthalpy of formation of ozone, $O_3(g)$, is $142\ kJ\ mol^{-1}$, and the dissociation energy of $O_2(g)$ is $498\ kJ\ mol^{-1}$. What is the average bond energy of the two bonds in the O_3 molecule?

35. Calculate the standard enthalpy change for the reaction

$$CO(g) + NO_2(g) \longrightarrow CO_2(g) + NO(g)$$

36. Use standard enthalpies of formation and the dissociation energy of $O_2(g)$ ($498\ kJ\ mol^{-1}$) to find the enthalpy change for both steps in the conversion of $O_3(g)$ to $O_2(g)$ catalyzed by $NO(g)$:

$$NO(g) + O_3(g) \longrightarrow NO_2(g) + O_2(g)$$
$$NO_2(g) + O_3(g) \longrightarrow 2O_2(g) + NO(g)$$

37. The standard enthalpies of formation of $NO(g)$, $NO_2(g)$, and $N_2O_3(g)$ are 90.3, 33.2, and $83.7\ kJ\ mol^{-1}$, respectively, and their standard entropies are 210.7, 239.9, and $312.2\ J\ K^{-1}\ mol^{-1}$, respectively.

(a) Calculate the standard free energy change for the reaction

$$N_2O_3(g) \longrightarrow NO(g) + NO_2(g)$$

(b) Is this reaction spontaneous under standard conditions?

(c) What is the equilibrium constant K_p at $25\ ^\circ C$?

(d) Below what temperature is N_2O_3 stable?

(e) Write Lewis structures for each of the species N_2O_3, NO, and NO_2.

38. On heating, liquid hydrazine, $N_2H_4(l)$, can decompose *either* to nitrogen and hydrogen *or* to ammonia and nitrogen:

$$N_2H_4(l) \longrightarrow N_2(g) + 2H_2(g)$$
$$3N_2H_4(l) \longrightarrow 4NH_3(g) + N_2(g)$$

(a) Qualitatively, without doing any calculations, decide which of these two reactions has the largest entropy change per mole of hydrazine decomposed.

(b) Calculate ΔG° for the decomposition of 1 mol of hydrazine into $N_2(g)$ and $H_2(g)$ at $25\ ^\circ C$. (At this temperature, for liquid hydrazine, ΔH_f° is $50.5\ kJ\ mol^{-1}$ and S° is $121.2\ J\ K^{-1}\ mol^{-1}$.)

(c) Is the above reaction spontaneous at $25\ ^\circ C$?

Oxidation–Reduction

39. What is the oxidation state of nitrogen in each of the following?

(a) NH_4^+ (b) N_2H_4 (c) KNO_2

(d) NO_2 (e) NH_2OH (f) N_2O

(g) NH_4NO_3 (h) LiN_3 (i) Li_3N

40. Balance each of the following equations:

(a) $H_2S(aq) + NO_3^-(aq) \rightarrow S(s) + NO(aq)$ in acid solution

(b) $Zn(s) + NO_3^-(aq) \rightarrow Zn(OH)_4^{2-} + NH_3(aq)$ in basic solution

(c) $P_4(s) + NO_3^-(aq) \rightarrow H_3PO_4(aq) + NO(aq)$ in acid solution

41. Write equations for the half-reactions for the reduction of $NO_3^-(aq)$ to each of the following, in acid solution:

(a) NO_2 (b) NO (c) N_2 (d) NH_4^+

42. Explain why, in aqueous solution, hydrogen peroxide can act both as an oxidizing agent and as a reducing agent.

Write the balanced equation for each of the following reactions:

(a) Oxidation of sulfur dioxide to sulfate ion by hydrogen peroxide in acid solution

(b) Reduction of ozone to water by hydrogen peroxide in acid solution

(c) Oxidation of iodide ion to iodine by hydrogen peroxide in acid solution

(d) Oxidation of chromium(III) hydroxide, $Cr(OH)_3$, to chromate ion, CrO_4^{2-}, by hydrogen peroxide in basic solution

(e) Oxidation of nitrite ion to nitrate ion by hydrogen peroxide in acid solution

43. Explain why nitrous acid can behave both as an oxidizing agent and as a reducing agent. Write balanced equations for one reaction where it behaves as an oxidizing agent and one reaction where it behaves as a reducing agent.

44. The standard reduction potential for the half-reaction

$$O_3(g) + 2H^+ + 2e^- \longrightarrow O_2(g) + H_2O$$

is 2.07 V. On the basis of standard reduction potentials from Table 17.1, decide which of the following oxidations could be achieved using ozone as the oxidizing agent in $1.00M$ acid, and for which $O_2(g)$ would be ineffective:

(a) H_2O to H_2O_2 (b) Mn^{2+} to MnO_4^-

(c) Cr^{3+} to $Cr_2O_7^{2-}$ (d) I_2 to IO_3^-

(e) Co^{2+} to Co^{3+}

Miscellaneous

*45. A gas X, which is an oxide of nitrogen, supports the combustion of a variety of substances. X may be prepared by gently heating ammonium nitrate. When 0.1020 g of white phosphorus was burned completely in X, 0.2337 g of an oxide of phosphorus was obtained, and another gas, Y. Starting with a given volume of X, the volume of Y produced was identical at a given temperature and pressure, but Y was found not to support combustion. The relative densities of X and Y were 1.571:1.000. Identify X and Y, give their Lewis structures, and explain the above reaction.

46. (a) How may an aqueous solution of hydrogen peroxide be prepared in the laboratory?

(b) Balance each of the following equations, indicating for each whether hydrogen peroxide acts as an oxidizing agent or a reducing agent:

(i) $PbS(s) + H_2O_2(aq) \rightarrow PbSO_4(s) + H_2O$

(ii) $Fe^{2+}(aq) + H_2O_2(aq) \rightarrow Fe^{3+}(aq)$ (acid solution)

(iii) $I^-(aq) + H_2O_2(aq) \rightarrow I_3^-(aq)$ (acid solution)

* The asterisk denotes the more difficult problems.

(iv) $HOCl(aq) + H_2O_2(aq) \rightarrow HCl(aq) + O_2(g)$

(v) $MnO_4^-(aq) + H_2O_2(aq) \rightarrow Mn^{2+}(aq) + O_2(g)$

47. Hydrazine can decompose either to nitrogen and hydrogen, or to nitrogen and ammonia.

(a) Write the balanced equation for each of these reactions.

(b) Show by calculation which of these reactions is favored thermodynamically.

48. A hot coil of platinum wire is inserted into a flask containing a mixture of ammonia and oxygen. Brown fumes are observed forming at the surface of the wire. Write balanced equations for the reactions taking place.

49. Explain each of the following:

(a) Phosphorus forms a trichloride and a pentafluoride, but nitrogen forms only a trichloride

(b) Ammonia is a stronger base than phosphine

(c) Nitrogen is less reactive than white phosphorus

(d) White phosphorus is more reactive than red phosphorus

50. In terms of the bond orders in the following species, account for the trends in the observed bond lengths, and name each species:

Species	Bond Length (pm)
O_2^+	112
O_2	121
O_3	128
O_2^-	128
O_2^{2-}	149

51. Account for the trend in the observed bond angles in the following species:

Species	ONO Bond Angle
NO_2^+	180°
NO_2	135°
NO_2^-	115°

52. What volume of nitrogen, at 60 °C and 750 torr, is formed by the thermal decomposition of 10.0 g of ammonium nitrite?

53. A sample of impure sodium nitrate of mass 1.354 g was heated, and the evolved oxygen was collected over water at 25 °C and a pressure of 760 torr. (The vapor pressure of water at 25 °C is 20.8 torr.) Assume that all the oxygen comes from the sodium nitrate, and calculate the purity of the sample if 131.8 mL of oxygen were collected.

Rates of Chemical Reactions

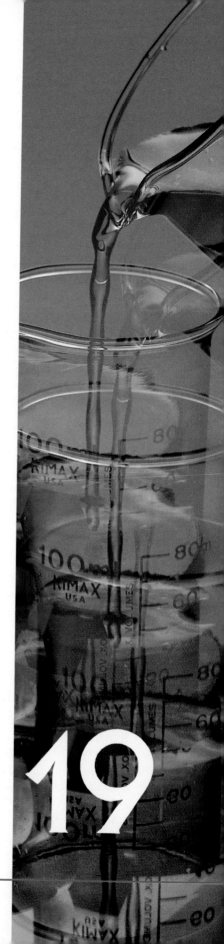

19

In Chapters 13–16 we studied in some detail the factors that influence the final position of equilibrium reached by a reaction. We saw that a reaction with a very large equilibrium constant has a strong tendency to go to completion, in other words, the position of equilibrium lies far over toward the products so that very little of the reactants remain when equilibrium is reached. Such a reaction is said to be spontaneous. However, as we have already pointed out, spontaneous does not mean fast. A spontaneous reaction is simply a reaction that proceeds as written from left to right but it may be slow or it may be fast. The reaction between hydrogen and oxygen to give water is a spontaneous reaction for which the position of equilibrium lies very far to the right:

$$2H_2(g) + O_2(g) \rightleftharpoons 2H_2O(l)$$

But at 25 °C hydrogen and oxygen can be mixed together and no reaction is apparent even over a long period of time because the reaction is very slow at this temperature.

The rates of reactions vary widely from those that reach equilibrium in a millisecond or less to those that take millions of years. The explosive reaction between oxygen and a hydrocarbon in the cylinder of an automobile, the explosive decomposition of TNT, and the reaction between a strong acid and a strong base, which occurs as fast as acid is added to base, are all examples of very fast reactions. Although slower than explosions, the reactions that lead to the contraction of muscles and that transmit nervous impulses are still very rapid. In contrast, the rusting of iron in air is rather slow, although usually it is not as slow as we would like. The weathering of rocks is still slower and may take millions of years to complete. Although we would like to be able to further slow up reactions such as the rusting of iron and the deterioration of foods, we are more often concerned with how we could speed up slow reactions so that we can obtain a useful amount of a desired product in a reasonable time. Experiment 19.1 shows that permanganate ion is reduced very rapidly by Fe^{2+} ion but more slowly by oxalate ion $C_2O_4{}^{2-}$.

In order to be able to control the rates of reactions, we need to understand the factors that influence the rate. In this chapter we consider the effect of the concentrations of the reactants, the temperature, and the presence of catalysts on the rate of a reaction. In order for a reaction to take place between two molecules, the molecules must collide. The probability of such a collision increases with increasing concentrations of the reactants. Thus we expect reaction rates to depend on the concentrations of the reactants. However, we will see that if every collision led to a reaction, then almost all reactions would be extremely fast. In fact, in most cases only a very small fraction of the collisions actually lead to a reaction, and we will discuss why this is so. We will see that increasing the temperature increases the rate of almost all reactions because it increases the number of effective collisions in a given time.

Many reactions of industrial importance employ catalysts, and most reactions in living organisms are catalyzed. We will discuss how a catalyst can increase the rate of a reaction without being used up.

First we need to consider just how we express and measure the rate of a reaction.

Experiment 19.1

Rates of Oxidation

A solution of oxalic acid has been added to a purple solution of potassium permanganate. Even after 6 min 20 s the solution is still purple because the reduction of MnO_4^- to Mn^{2+} is slow at 20 °C.

After 7 min 35 s the solution becomes colorless as the reaction finally goes to completion.

In contrast, the reduction of MnO_4^- by Fe^{2+} in acidic solution is very rapid. At 20 °C the reaction is already complete within 1 s.

19.1
Reaction Rate

The rate (or speed) of a chemical reaction is defined in the same way that we define the speed of an automobile or any other object. The speed of an automobile is equal to the change in its position, or the distance traveled, divided by the time taken to travel the distance between its initial and final positions:

$$\text{Speed of automobile} = \frac{\text{Change in position}}{\text{Time for the change}} = \frac{\text{Distance traveled}}{\text{Time taken}}$$

The **rate of a chemical reaction** is defined as the change in the concentration of a reactant (or product) in a given time interval. Consider the reaction between nitrogen monoxide and ozone that is probably one of the reactions responsible for decreasing the concentration of ozone in the ozone layer of the upper atmosphere (Box 18.3):

$$NO(g) + O_3(g) \longrightarrow NO_2(g) + O_2(g)$$

If the concentration of NO_2 at time t_1 is $[NO_2]_1$, and if at time t_2 it has increased to $[NO_2]_2$, then the concentration of NO_2 has changed by

$$\Delta[NO_2] = [NO_2]_2 - [NO_2]_1$$

in the time interval $\Delta t = t_2 - t_1$. The rate of the reaction is then

$$\frac{\text{Change in concentration of } NO_2}{\text{Time taken for change}} = \frac{\Delta[NO_2]}{\Delta t}$$

The rate is given by the change in the concentration divided by the time needed for the change. Since the concentrations of the reactants decrease, $\Delta[\text{reactant}]$ is a negative quantity. Therefore if we measure the rate in terms of the change in concentration of one of the reactants, we define the rate as $-\Delta[\text{reactant}]/\Delta t$ so that the rate is always a positive quantity. The rate of the reaction is given by the change in the concentration of any of the reactants or products in a certain time interval. For the reaction of NO with O_3 to give NO_2 and O_2, 1 mol of NO reacts with 1 mol of O_3 to give 1 mol of NO_2 and 1 mol of O_2. Thus

$$\text{Rate of reaction} = \frac{\Delta[\text{NO}_2]}{\Delta t} = \frac{\Delta[\text{O}_2]}{\Delta t} = -\frac{\Delta[\text{NO}]}{\Delta t} = -\frac{\Delta[\text{O}_3]}{\Delta t}$$

For a reaction involving different numbers of moles of reactants and products, we must be careful to specify which species is being used to measure the rate of the reaction. For example, in the reaction

$$N_2(g) + 3H_2(g) \longrightarrow 2NH_3(g)$$

two ammonia molecules are formed for every nitrogen molecule that is used up. In other words, the rate at which ammonia is formed is twice the rate at which N_2 disappears, and hydrogen disappears three times as fast. Therefore if we express the rate of reaction in terms of the rate at which nitrogen disappears, we have

$$\text{Rate of reaction} = -\frac{\Delta[\text{N}_2]}{\Delta t} = \frac{1}{2}\frac{\Delta[\text{NH}_3]}{\Delta t} = -\frac{1}{3}\frac{\Delta[\text{H}_2]}{\Delta t}$$

Speed has the units of distance divided by time and may be expressed, for example, as miles per hour (miles h^{-1}), kilometers per hour (km h^{-1}), or meters per second (m s^{-1}). Reaction rate has the units of concentration divided by time. We express concentrations in moles per liter (mol L^{-1}), but time may be given in any convenient unit—seconds, minutes, hours, days, or possibly years. Therefore the units of reaction rate may be mol L^{-1} s^{-1}, mol L^{-1} min^{-1}, mol L^{-1} h^{-1}, and so on.

=== Example 19.1 ===

REACTION RATE IN TERMS OF CONCENTRATIONS OF REACTANTS AND PRODUCTS

The rate of the reaction

$$2N_2O_5(g) \longrightarrow 4NO_2(g) + O_2(g)$$

can be expressed as $\Delta[O_2]/\Delta t$. Write an expression for the rate in terms of each of the other molecules involved in the reaction.

Solution

The equation shows that the N_2O_5 is used up twice as fast as O_2 is formed. Therefore

$$\text{Rate} = \frac{\Delta[\text{O}_2]}{\Delta t} = -\frac{1}{2}\frac{\Delta[\text{N}_2\text{O}_5]}{\Delta t}$$

The equation also shows that NO_2 is formed four times as fast as oxygen is formed. Therefore

$$\text{Rate} = \frac{\Delta[\text{O}_2]}{\Delta t} = -\frac{1}{2}\frac{\Delta[\text{N}_2\text{O}_5]}{\Delta t} = \frac{1}{4}\frac{\Delta[\text{NO}_2]}{\Delta t}$$

To measure the rate of a reaction experimentally, we can select any one of the reactants or products and measure the change in its concentration with time. Any convenient method of determining concentration can be used. A small sample could be withdrawn from the reaction mixture and the reaction in the sample stopped—for example, by lowering the temperature or by diluting it in a large amount of solvent. The sample could then be analyzed for one or more of the products or reactants.

A measurement that can be made directly on the reaction mixture is often more convenient. For example, in the reaction between colorless NO and O_3, the change in the concentration of NO_2 can be followed by a measurement of the increase in the intensity of the red-brown color of NO_2. If the hydrogen ion concentration changes during a reaction, this can be followed by measuring the pH, for example, with a pH meter. Reactions involving gases are often accompanied by a change in volume or pressure. For example, in the reaction of N_2 with H_2 to give NH_3, the total number of molecules decreases with time, so the pressure decreases if the reaction is carried out in a vessel of constant volume. Thus the rate of the reaction can be measured by the rate of change of the pressure.

Table 19.1 shows the concentrations of the reactants during the reaction

$$CO(g) + NO_2(g) \longrightarrow CO_2(g) + NO(g)$$

TABLE 19.1
Rate Data for the Reaction $CO(g) + NO_2(g) \rightarrow CO_2(g) + NO(g)$

[CO] (mol L^{-1})	[NO$_2$] (mol L^{-1})	$t(s)$	$Rate = -\dfrac{\Delta[CO]}{\Delta t} = -\dfrac{\Delta[NO_2]}{\Delta t}$ (mol L^{-1} s^{-1})	Slope = Rate at Time t (mol L^{-1} s^{-1})
0.100	0.100	0		4.9×10^{-3}
			3.3×10^{-3}	
0.067	0.067	10		2.2×10^{-3}
			1.7×10^{-3}	
0.050	0.050	20		1.2×10^{-3}
			1.0×10^{-3}	
0.040	0.040	30		0.8×10^{-3}
			0.7×10^{-3}	
0.033	0.033	40		0.5×10^{-3}
			0.3×10^{-3}	
0.017	0.017	100		0.1×10^{-3}

We can find the reaction rate for each 10-s interval by dividing the change in concentration over the 10-s interval by 10 s. For example, for the period $t = 0$ to $t = 10$ s the rate is given by

$$-\frac{\Delta[CO]}{\Delta t} = -\frac{0.067 - 0.100}{10 \text{ s}} \text{ mol L}^{-1} = 3.3 \times 10^{-3} \text{ mol L}^{-1} \text{ s}^{-1}$$

Values of the rate obtained in this way are given in column 4 of Table 19.1. Because the reaction rate is changing with time, these values are only average rates over each 10-s interval.

We can obtain a more exact value for the rate by reducing the size of the time interval. In the limit when t is very small, $-\Delta[CO]/\Delta t$ is the slope of the tangent to the curve at the time t as shown in Figure 19.1. Thus

$$\text{Rate} = \text{Limiting value} \left(-\frac{\Delta[CO]}{\Delta t} \right) = -\text{ Slope of curve}$$

This is the *instantaneous rate* at a particular time t. Values of the rate obtained from the slope of the curve at a given time are given in column 5 of Table 19.1. Notice from columns 4 and 5 that the rate of the reaction continually decreases as the reaction proceeds because the concentrations of the reactants are continually decreasing.

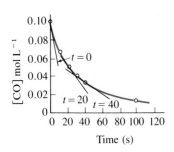

FIGURE 19.1
Concentration of CO as a Function of Time for the Reaction

$$CO(g) + NO_2(g) \longrightarrow$$
$$CO_2(g) + NO(g)$$

19.2
Rate Laws and Reaction Order

We have mentioned that we expect the rate of a reaction to depend on the concentrations of the reactants. For the reaction

$$NO(g) + O_3(g) \longrightarrow NO_2(g) + O_2(g)$$

experiment shows that

Rate of reaction is proportional to the product $[NO][O_3]$

or

Rate of reaction $= k[NO][O_3]$

where k is a constant called the **rate constant**. At a given temperature k is a constant characteristic of the reaction. Its value is independent of the concentrations of the reactants, although it does depend on the temperature. For the reaction between NO and O_3, $k = 1.6 \times 10^7 \text{ mol}^{-1} \text{ L s}^{-1}$ at 25 °C. The rate constant k is a measure of the intrinsic rate of the reaction: it is the rate when the concentrations of all the reactants are 1 mol L^{-1}. Fast reactions have large k values, while slow reactions have small k values.

An expression of this type, which relates the rate of a reaction to the concentrations of the reactants, is called a **rate law**. In general, for a reaction

$$aA + bB + cC + \cdots \longrightarrow \text{Products}$$

the rate law often has the form

$$\text{Rate} = k[A]^x[B]^y[C]^z \ldots$$

where x is called the **order** of the reaction with respect to A, y is the order with respect to B, z is the order with respect to C, and the sum, $x + y + z + \cdots$, is called the **overall order**.

For the reaction between NO and O_3 the rate law is

$$\text{Rate} = k[\text{NO}][\text{O}_3]$$

We say that the reaction is first-order with respect to NO, first-order with respect to O_3, and second-order overall. This means that the rate of the reaction is directly proportional to the concentration of NO and also directly proportional to the concentration of O_3. If we double the concentration of either NO or O_3, the rate will double. If we double the concentrations of both NO and O_3, the rate increases by four times.

The rate law for any reaction must be determined experimentally, because, as we will see, *there is no necessary connection between the order with respect to any given reactant and the coefficient for the reactant in the balanced equation for the reaction*. In other words, x, y, z, ... are not necessarily the same as a, b, c, ...

How do we find experimentally how the rate depends on the concentration of each of the reactants? For a reaction in which there is only one reactant, the problem is fairly simple. Consider, for example, the decomposition of dinitrogen pentaoxide:

$$2\text{N}_2\text{O}_5(g) \longrightarrow 4\text{NO}_2(g) + \text{O}_2(g)$$

Table 19.2 and Figure 19.2 show experimental data for this reaction. The reaction was followed by measuring the increase in pressure accompanying the reaction, from which the partial pressure of N_2O_5 (column 2) and hence the concentration of N_2O_5 (column 3) were calculated. The values of the rate given in column 4 were obtained by drawing tangents to the curve in Figure 19.2 on p. 866. By plotting the rate against the concentration of N_2O_5 (Figure 19.3 on p. 866), we obtain a straight line, showing that the rate is proportional to the concentration of N_2O_5. Therefore the rate law is

$$\text{Rate} = k[\text{N}_2\text{O}_5]$$

We can obtain the value of the rate constant from the slope of the straight line:

$$k = \frac{\text{Rate}}{[\text{N}_2\text{O}_5]} = \text{Slope} = 0.030 \text{ min}^{-1}$$

In contrast, a plot of the rate against $[\text{N}_2\text{O}_5]^2$ does *not* give a straight line

TABLE 19.2
Rate Data for Decomposition of $\text{N}_2\text{O}_5(g)$

Time (min)	$p_{\text{N}_2\text{O}_5}$ (mm Hg)	$[\text{N}_2\text{O}_5]$ (mol L^{-1})	Rate (mol L^{-1} min^{-1})
0	301.6	0.0152	
10	224.8	0.0113	3.4×10^{-4}
20	166.7	0.0084	2.5×10^{-4}
30	123.2	0.0062	1.8×10^{-4}
40	92.2	0.0046	1.3×10^{-4}
50	69.1	0.0035	1.0×10^{-4}
60	51.1	0.0026	0.8×10^{-4}
70	37.5	0.0019	0.6×10^{-4}
80	27.4	0.0014	0.4×10^{-4}

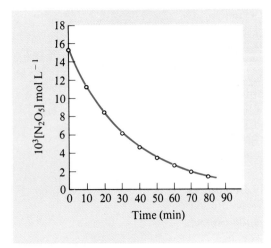

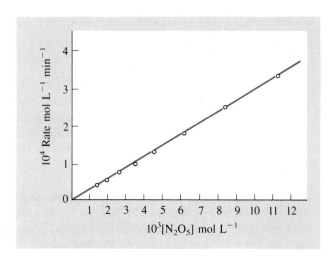

FIGURE 19.2
Concentration of N_2O_5 as a
Function of Time for the
Decomposition of N_2O_5.

FIGURE 19.3
Rate of Decomposition of
N_2O_5 as a Function of N_2O_5
Concentration.

(Figure 19.4), showing that the rate law is *not*

$$\text{Rate} = k[N_2O_5]^2$$

In other words, the reaction is *not* second-order, even though there are two
molecules of N_2O_5 on the left side of the equation.

INITIAL-RATE METHOD

For most reactions there is more than one reactant, all of whose concentrations
are changing, so the previous method cannot be used. In such cases a simple
method is the *initial-rate method*. In this method the rate of the reaction is measured at the beginning of the reaction, that is, before the concentrations have
had time to change significantly. The initial concentration of only one reactant
is then changed, while all the other initial concentrations are kept constant, and

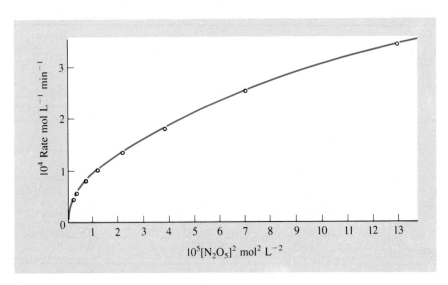

FIGURE 19.4
Rate of Decomposition of
N_2O_5 as a Function of
$[N_2O_5]^2$.

TABLE 19.3			
Rate Data for the Reaction $2NO(g) + Cl_2(g) \rightarrow 2NOCl(g)$ at 300 K			
	Initial Concentrations (mol L^{-1})		
Experiment	*[NO]*	*[Cl_2]*	*Initial Rate* (mol $L^{-1} s^{-1}$)
1	0.010	0.010	1.2×10^{-4}
2	0.010	0.020	2.3×10^{-4}
3	0.020	0.020	9.6×10^{-4}

the rate is measured again. This procedure is repeated for each reactant, and the initial rate is measured each time.

For example, the data in Table 19.3 were obtained for the reaction between NO and Cl_2:

$$2NO(g) + Cl_2(g) \longrightarrow 2NOCl(g)$$

The rate law can be written in the general form

$$\text{Rate} = k[NO]^x[Cl_2]^y$$

where x and y are the orders with respect to NO and Cl_2, respectively. We note that when [NO] is kept constant and [Cl_2] is doubled, as between experiments 1 and 2, the rate doubles; thus the rate is proportional to [Cl_2] and $y = 1$. When [Cl_2] is kept constant and [NO] is doubled, as between experiments 2 and 3, the rate quadruples; thus the rate is proportional to $[NO]^2$ and $x = 2$. Therefore the rate law is

$$\text{Rate} = k[NO]^2[Cl_2]$$

The reaction is second-order in NO, first-order in Cl_2, and third-order overall.

≡ *Example 19.2* ≡══

RATE CONSTANT FROM THE RATE EQUATION

What is the value of the rate constant at 300 K for the following reaction?

$$2NO(g) + Cl_2(g) \longrightarrow 2NOCl(g)$$

Solution

We have seen that the rate law for this reaction is

$$\text{Rate} = k[NO]^2[Cl_2]$$

Therefore

$$k = \frac{\text{Rate}}{[NO]^2[Cl_2]}$$

Substituting the concentrations from experiment 1 (Table 19.3), we have

$$\frac{1.2 \times 10^{-4} \text{ mol}^{-1} s^{-1}}{(0.010 \text{ mol } L^{-1})^2(0.010 \text{ mol } L^{-1})} = 1.2 \times 10^2 \text{ mol}^{-2} L^2 s^{-1}$$

Substituting the concentrations and rates for experiments 2 and 3 also gives $k = 1.2 \times 10^2$ mol^{-2} L^2 s^{-1} in both cases.

$=$ *Example 19.3* $=$

REACTION RATE FROM THE RATE EQUATION

What is the rate of the reaction in Example 19.2 when $[NO] = 0.030$ mol L^{-1} and $[Cl_2] = 0.040$ mol L^{-1}?

Solution

We have seen that the rate law is rate $= k[NO]^2[Cl_2]$ and that k is 1.2×10^2 mol^{-2} L^2 s^{-1}. Therefore, if

$$[NO] = 0.030 \text{ mol } L^{-1} \quad \text{and} \quad [Cl_2] = 0.040 \text{ mol } L^{-1}$$

then

$$\text{Rate} = (1.2 \times 10^2 \text{ mol}^{-2} \text{ L}^2 \text{ s}^{-1})(0.030 \text{ mol } L^{-1})^2(0.040 \text{ mol } L^{-1})$$
$$= 4.3 \times 10^{-3} \text{ mol } L^{-1} \text{ s}^{-1}$$

$=$ *Example 19.4* $=$

RATE EQUATION FROM INITIAL-RATE DATA

For the reaction between nitrogen monoxide and hydrogen,

$$2NO(g) + 2H_2(g) \longrightarrow N_2(g) + 2H_2O(g)$$

the following initial-rate data were obtained:

| Experiment | Initial Pressures (mm Hg) | | Initial Rate (mm Hg s^{-1}) |
	NO	H_2	
1	359	300	1.50
2	300	300	1.03
3	152	300	0.25
4	300	289	1.00
5	300	205	0.71
6	300	147	0.51

What is the order of the reaction with respect to NO and with respect to H_2, and what is the rate law?

Solution

In determining the order we are concerned only with *ratios* of concentrations; so we can use the pressure data directly, without converting them to concentrations, because at constant V and T pressure is directly proportional to concentration. If we consider experiments 2 and 3, we see that when the pressure of NO is reduced by a factor of approximately 2 while keeping p_{H_2} constant, the rate decreases by a factor of approximately 4. We conclude that the reaction is second-order with respect to NO. By considering experiments 4 and 6, we see that reducing the pressure of H_2 by a factor of 2 while keeping p_{NO} constant reduces the rate from 1.00 to 0.51 mm Hg s^{-1}, that is, by a factor of approximately 2. Thus the rate is directly proportional to the H_2 concentration; in other words, the reaction is first-order with respect to H_2. Therefore the overall rate law is

$$\text{Rate} = k[NO]^2[H_2]$$

Experiment	[NO] (mol L^{-1})	[O$_2$] (mol L^{-1})	Initial Rate (mol L^{-1} s^{-1})
1	0.0010	0.0010	7×10^{-6}
2	0.0010	0.0020	1.4×10^{-5}
3	0.0010	0.0030	2.1×10^{-5}
4	0.0020	0.0030	8.4×10^{-5}
5	0.0030	0.0030	1.9×10^{-4}

INTEGRATED RATE LAW METHOD

Another generally useful method is the integrated rate law method. We have seen that we can determine the rate at different times from the slope of a plot of the concentration of a reactant or a product against time; and then by plotting the rate against concentration, we can find the order. It is generally more convenient, however, to determine the order directly from a suitable plot of concentration against time. This can be done using an integrated rate law. A rate law describes how the rate varies with the concentrations of the reactants. By the methods of the calculus, the rate law can be transformed to an *integrated rate law*, which *shows how the concentrations of the reactants vary with time.*

For a *first-order reaction*, if [A] is the concentration of a reactant A and k_1 is the rate constant, then

$$\text{Rate} = -\frac{\Delta[A]}{\Delta t} = k_1[A]$$

From the methods of the calculus, it can be shown that

$$\ln[A]_t = -k_1 t + \ln[A]_0$$

where $[A]_t$ is the concentration of A at time t and $[A]_0$ is the initial concentration of A. The equation has the form of an equation for a straight line,

$$y = ax + b$$

where a is the slope of the line. Thus, for a first-order reaction, a plot of $\ln[A]_t$ against t is a straight line with a slope of $-k_1$.

Table 19.4 gives data for the decomposition of N_2O_5 at 67 °C, and a plot of $\ln[N_2O_5]$ against t is given in Figure 19.5 on p. 870. Since this plot is a straight line, we conclude that the reaction is first-order. From the slope of the line, we get the rate constant $k_1 = 0.0345$ min^{-1}.

For a *second-order reaction* that follows the rate law

$$\text{Rate} = k_2[A]^2$$

TABLE 19.4
Rate Data for Decomposition of N_2O_5

t (min)	$[N_2O_5]$ (mol L^{-1})	ln $[N_2O_5]$
0	1.000	0.000
10	0.705	−0.350
20	0.497	−0.699
30	0.349	−1.053
40	0.246	−1.402
50	0.173	−1.754

it can be shown by calculus that the concentration of A varies with time according to the equation

$$\frac{1}{[A]_t} = k_2 t + \frac{1}{[A]_0}$$

This again has the form of an equation for a straight line. In this case a plot of $1/[A]$ against t gives a straight line. The slope of this line is k_2, where k_2 is the rate constant.

EXERCISE 19.3

By means of a suitable plot, show that the data in Table 19.1 are consistent with a second-order rate law, and determine the value of the rate constant.

HALF-LIFE

A simple method for determining the order of a reaction is to measure the *half-life*. The half-life of a reaction, $t_{1/2}$, is the time required for the concentration of

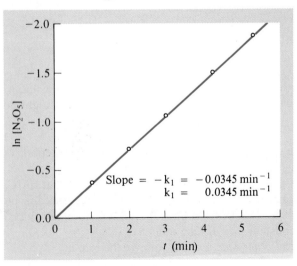

FIGURE 19.5
A First-Order Reaction.

Plot of ln $[N_2O_5]$ versus time for the decomposition of N_2O_5.

a reactant to decrease to half its initial value. If the initial concentration is $[A]_0$, then $[A] = \frac{1}{2}[A]_0$ at $t = t_{1/2}$.

The integrated rate law for a first-order reaction,

$$\ln [A]_t = -k_1 t + \ln [A]_0$$

can be written in the form

$$\ln \left(\frac{[A]_0}{[A]_t} \right) = k_1 t$$

If we substitute $[A] = \frac{1}{2}[A]_0$, we get

$$\ln \left(\frac{[A]_0}{\frac{1}{2}[A]_0} \right) = k_1 t_{1/2}$$

$$\ln 2 = k_1 t_{1/2}$$

or

$$t_{1/2} = \frac{\ln 2}{k_1} = \frac{0.693}{k_1}$$

For a first-order reaction, $t_{1/2}$ is independent of the concentration of A. Thus if the concentration of the reactant decreases to half its initial value in 10 min, then it will decrease again by a factor of 2 in the next 10 min, so after 20 min it will be a fourth of its initial value, and so on (see Figure 19.6). In fact, at any time t during a first-order reaction, the time needed for the concentration at that time to drop to half its value is equal to the half-life, $t_{1/2}$ (see Figure 19.6).

For a second-order reaction, by substituting $[A] = \frac{1}{2}[A]_0$ at $t = t_{1/2}$, we find that

$$t_{1/2} = \frac{1}{k_2 [A]_0}$$

so $t_{1/2}$ is *not* independent of the initial concentration. If we double the initial concentration, the half-life is halved.

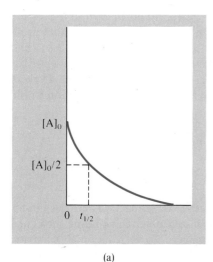

(a)

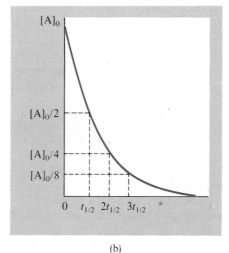

(b)

FIGURE 19.6
Half-Life of a First-Order Reaction.

The half-life is independent of the initial concentration. In (a) the initial concentration of the reactant $[A]_0$ is less than it is in (b), but the time $t_{1/2}$ needed for the initial concentration to decrease to half its initial value is the same. The time needed for the initial concentration to decrease to a quarter of its initial value is $2t_{1/2}$ and to an eighth of its initial value is $3t_{1/2}$.

═ *Example 19.5* ═════════════════════════════════════

FIRST-ORDER REACTION AND HALF-LIFE

The hydrolysis of sucrose to glucose and fructose is catalyzed by the enzyme sucrase. It is first order in the concentration of sucrose. If the half-life is 80 min at 20 °C, what proportion of the initial sucrose will remain after 160 and 320 min?

Solution

Since the reaction is first-order, the concentration is halved each 80 min. And since 160 min is two half-life periods, the initial concentration is reduced to $\frac{1}{2} \times \frac{1}{2} = \frac{1}{4}$ of the original value. After 320 min, or four half-life periods, the concentration is reduced to $(\frac{1}{2})^4 = \frac{1}{16}$ of its original value.

EXERCISE 19.4

From the data of Table 19.4, deduce an approximate value for the rate constant using the half-life method.

19.3
Factors Influencing Reaction Rates

As we have stated earlier, a reaction can only occur when the reacting molecules collide. Thus the rate of a reaction depends on the rate at which collisions occur and on what fraction of these collisions are effective in leading to reaction.

COLLISION RATES

The rate at which molecules of two substances, A and B, collide in the gas phase can be calculated from a quantitative treatment of the molecular kinetic theory of gases (Chapter 3). If we have two gases, A and B, each at a pressure of 0.5 atm, in a total volume of 1 L and at a temperature of 298 K, then from the ideal gas equation we can calculate that there is 0.02 mol each of gas A and of gas B. Kinetic theory shows that under these conditions there will be approximately 10^7 mol of collisions per second—a truly enormous number. If every collision of an A molecule with a B molecule led to reaction, the reaction would be complete in approximately

$$\frac{0.02 \text{ mol}}{10^7 \text{ mol s}^{-1}} \approx 10^{-9} \text{ s}$$

In other words, the reaction would be over almost instantaneously if products were formed on every collision.

In fact, very few reactions occur this rapidly. Most reactions are very much slower because *most collisions do not result in reaction*; the colliding molecules frequently just bounce off each other unchanged. For a reaction to occur on the collision of two molecules, the molecules must collide (1) with a certain orientation and (2) with sufficient energy.

Lycopodium powder, the spores of a common moss, consists of very small particles that burn very rapidly when sprayed into a flame. The rate at which a solid reacts increases with increasing surface area. The very fine particles of lycopodium powder have a very large surface area for a given mass.

ORIENTATION, OR STERIC, EFFECT

The necessity for two molecules to collide with the correct relative orientation in order that the reaction can occur is described as a **steric**, or **orientation**, **effect**. We can see the importance of the correct orientation of the reacting molecules by considering again the reaction of nitrogen monoxide with ozone:

$$NO(g) + O_3(g) \longrightarrow NO_2(g) + O_2(g)$$

In this reaction an oxygen atom must be transferred from an ozone molecule to a nitrogen monoxide molecule. Therefore it is reasonable to assume that, for reaction to occur during a collision, one of the terminal oxygen atoms of an O_3 molecule must collide with the nitrogen of the NO molecule, as shown in Figure 19.7. We would not expect collisions between the central oxygen atom of the O_3 molecule and the nitrogen atom of the NO molecule, or between the O_3 molecule and the oxygen atom of the NO molecule, to lead readily to the transfer of an oxygen atom to the nitrogen atom. In other words, in such collisions the reaction would be unlikely to occur.

In many reactions, the required orientation of the molecules occurs only in a very small fraction of the collisions, particularly if the molecules are large and complicated.

ENERGY OF ACTIVATION

The second reason why not every collision leads to a reaction is that a collision must occur with a certain minimum energy before a reaction can take place. When the atoms of the reactant molecules are rearranged to form the new product molecules, bonds must be broken and new bonds formed. For example, in the reaction of NO with O_3, one of the OO bonds in the O_3 molecule is broken, and a new NO bond is formed to give the NO_2 molecule (see Figure 19.7).

The OO bond must first be stretched—that is, the OO bond must be partially broken—before the oxygen atom can begin to form a bond with the nitrogen

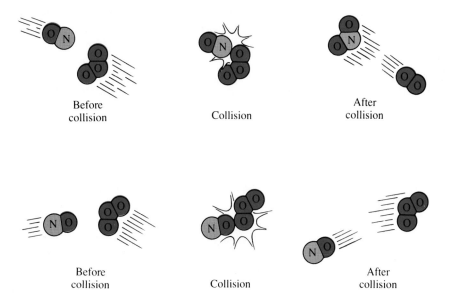

Before collision

Collision

After collision

Before collision

Collision

After collision

FIGURE 19.7
The Orientation Effect.

(a) A collision between an NO and an O_3 molecule in this orientation may lead to reaction. (b) A collision between an NO and an O_3 molecule in this orientation does *not* lead to reaction.

atom of NO. Energy must be supplied to accomplish this stretching. Partial formation of the new NO bond liberates some energy, and this partially compensates for the energy needed to break the OO bond. A point is reached eventually at which the energy liberated by the formation of the new NO bond is greater than the energy needed to stretch the OO bond further, and there is an overall liberation of energy.

Figure 19.8 shows how the total potential energy of the system changes as an oxygen atom moves from the O_3 molecule to the NO molecule. At first, the energy of the system increases as the OO bond stretches. The energy eventually reaches a maximum and finally decreases again as the new NO bond is formed.

The intermediate state of the reacting system at which its energy is a maximum is called the **transition state**, or **activated complex**, of the reaction. The energy needed to pass from the reactants to the transition state is the **activation energy**, E_a. It is the difference in energy between the reactants and the transition state.

The activation energy is provided by the kinetic energy of the reacting molecules. Unless the kinetic energy of their relative motion is at least equal to E_a, no reaction will occur, and the molecules simply bounce off each other unchanged. The activation energy represents a barrier that must be overcome in order for the reaction to occur.

As an analogy, imagine the movement of a marble on the surface shown in Figure 19.9. The movement of the marble from left to right represents the progress of the reaction, and the barrier corresponds to the activation energy. As the marble climbs up the barrier, its potential energy increases. If it is rolled at a low velocity and therefore has low kinetic energy, it will climb only partway up the barrier and then roll back again (Figure 19.9a). Its kinetic energy is too low, and its energy is all converted to potential energy before it reaches the top of the barrier. This situation corresponds to a collision in which the kinetic

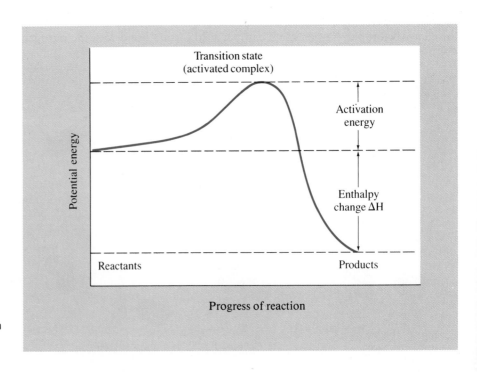

FIGURE 19.8
Change of Potential Energy in a Collision of O_3 with NO to Give O_2 and NO_2.

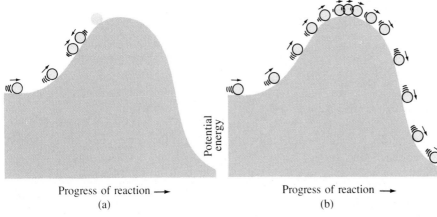

Progress of reaction ⟶
(a)

Progress of reaction ⟶
(b)

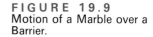

FIGURE 19.9
Motion of a Marble over a Barrier.

(a) Here the marble has insufficient energy to climb the potential energy barrier; it rolls back down the same side of the barrier—no reaction has occurred. (b) Here the marble has sufficient energy to get to the top of the potential energy barrier and roll down the other side—reaction has occurred.

energy of the reacting molecules is too small for reaction to occur; the molecules simply bounce apart unchanged. The marble must have sufficient initial kinetic energy to enable it to roll up the side of the barrier and reach the top, which in this analogy corresponds to the formation of the transition state. It may then roll down the other side and finish on the right-hand side, which corresponds to the completion of the reaction (Figure 19.9b). As the marble rolls down the other side, the potential energy that it gained on moving to the top of the energy barrier is reconverted to kinetic energy. The corresponding situation in a chemical reaction is when the product molecules move apart as the reaction is completed.

Although energy is always needed to reach the transition state, the overall reaction may be exothermic or endothermic, depending on whether more or less energy is needed to reach the transition state than is obtained in passing from the transition state to the products (see Figure 19.10). The energy of the reactant molecules depends on the temperature and, as we will see in the next section, raising the temperature increases the number of molecules with sufficient energy to react—and therefore increases the reaction rate.

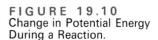

FIGURE 19.10
Change in Potential Energy During a Reaction.

(a) An endothermic reaction; (b) an exothermic reaction.

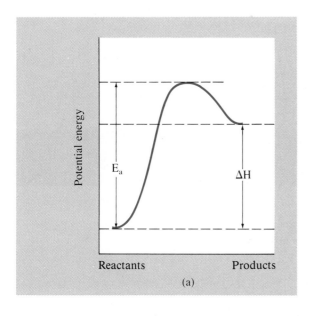

Reactants   Products
(a)

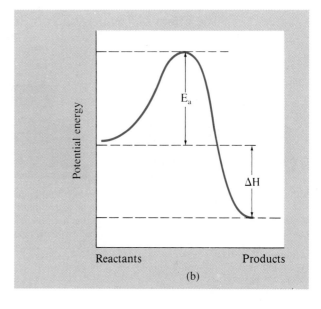

Reactants   Products
(b)

TEMPERATURE AND REACTION RATE: THE ARRHENIUS EQUATION

The rates of almost all chemical reactions increase with increasing temperature (see Experiment 19.2). The effect is quite marked, and the rate often increases by a factor of between 2 and 4 for every 10 K rise in temperature. Many industrial processes, such as the synthesis of ammonia, are carried out at a rather high temperature in order to speed up the reaction.

The average kinetic energy of the molecules in a gas increases with increasing temperature. Not all the molecules are moving at the same speed, but at a given temperature they have a certain average speed and a corresponding average kinetic energy. As we saw in Chapter 3, the distribution of speeds and kinetic energies of the molecules has the form shown in Figure 19.11(a). Only a very few molecules have very low energies or very high energies; the majority have energies close to the average. At a higher temperature the distribution of energies becomes wider, the number of molecules having rather high energies increases, and the average energy also increases.

The energies associated with the collisions between two molecules of a gas have a distribution similar to the kinetic energies of the individual molecules (see Figure 19.11b). In some cases the total energy of two colliding molecules will be small, and in a few cases it will be very high. In a large number of collisions the molecules will have an energy close to the average value. If the minimum energy for a reaction to occur—that is, the activation energy, E_a—is much higher than the average energy of the colliding molecules, only a very small fraction of the collisions can lead to reaction, and the reaction will be very slow. If the activation energy is small, however, a large fraction of the collisions that have a suitable orientation will lead to reaction, and the reaction will be fast. Thus the rate constant of a reaction depends on the activation energy for the reaction.

We see from Figure 19.11(b) that the fraction of the total collisions that can lead to reaction increases rather rapidly with increasing temperature. Thus the

Experiment 19.2

The Effect of Temperature on Reaction Rate

At 20 °C after 6 min 21 s the reaction between permanganate ion MnO_4^- and oxalic acid is still incomplete.

At 34.7 °C after only 40 s the pale pink color of the solution shows that the reaction has proceeded much further. Thus the reaction is much faster at 34.7 °C than at 20 °C.

Finally, in less than another second the reaction is complete.

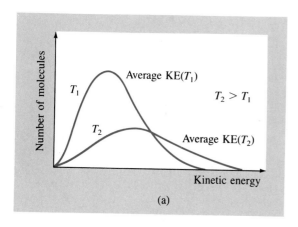

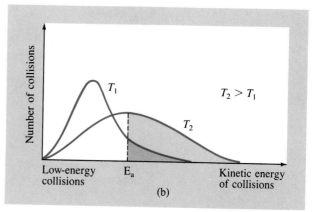

(a)

(b)

FIGURE 19.11
Kinetic Energy and Collision
Energy Distributions.

(a) Distribution of kinetic energies
of molecules at temperatures T_1
and T_2; (b) distribution of collision
energies at temperatures T_1 and
T_2. The shaded areas show that
the fraction of molecules with
kinetic energies of collision greater
than the activation energy, E_a,
increases with an increase in
temperature.

rate of most reactions increases greatly as the temperature is raised. The Swedish chemist Svante Arrhenius (1859–1927) (Figure 12.27) first proposed the equation that shows how the rate constant of a reaction depends on the temperature. The **Arrhenius equation** is

$$k = A \exp(-E_a/RT)$$

where k is the rate constant, A is a constant (the Arrhenius parameter) that is relatively independent of temperature, E_a is the activation energy, R is the gas constant, and T is the temperature in kelvins. The Arrhenius parameter A is the total number of collisions per second that have the correct orientation. The exponential factor $\exp(-E_a/RT)$ represents the fraction of the total number of collisions that have sufficient energy to react, that is, in which the molecules have at least the energy of activation E_a.

The Arrhenius parameter A increases with increasing temperature but its rate of increase is relatively small. In contrast, the $\exp(-E_a/RT)$ factor in the Arrhenius equation, increases rapidly with increase in temperature and is responsible for a large increase in reaction rate, as Example 19.6 demonstrates.

For many purposes, we will find it useful to write the Arrhenius equation in logarithmic form. Taking natural logarithms of both sides gives

$$\ln k = \ln A - \frac{E_a}{RT}$$

so that a plot of $\ln k$ versus $1/T$ should give a straight line with a slope equal to $-E_a/R$. Such a plot provides a method for determining the activation energy of a reaction from values of its rate constant at different temperatures, as is shown in Example 19.7.

═ Example 19.6 ═══════════════════════════════

INCREASE OF THE RATE OF A REACTION WITH INCREASE IN TEMPERATURE

By what factor does the rate of a reaction increase if its activation energy is 50 kJ mol^{-1} and the temperature is increased from (a) 300 K to 320 K, and (b) 300 K to 400 K?

Solution

For two temperatures T_1 and T_2 we can write

$$\ln k_{T_1} = \ln A - \frac{E_a}{RT_1} \tag{1}$$

and

$$\ln k_{T_2} = \ln A - \frac{E_a}{RT_2} \tag{2}$$

where k_{T_1} and k_{T_2} are the rate constants at T_1 and T_2, respectively. Subtracting equation (1) from equation (2) gives

$$\ln k_{T_2} - \ln k_{T_1} = \ln\left(\frac{k_{T_2}}{k_{T_1}}\right) = \frac{E_a}{R}\left(\frac{1}{T_1} - \frac{1}{T_2}\right)$$

(a) Substituting $E_a = 50\ \text{kJ mol}^{-1}$, $R = 8.31\ \text{J K}^{-1}\ \text{mol}^{-1}$, $T_1 = 300\ \text{K}$, and $T_2 = 320\ \text{K}$ in the above equation gives

$$\ln\left(\frac{k_{T_2}}{k_{T_1}}\right) = \frac{(50\ \text{kJ mol}^{-1})\left(\dfrac{1000\ \text{J}}{1\ \text{kJ}}\right)}{8.31\ \text{J K}^{-1}\ \text{mol}^{-1}}\left(\frac{1}{300} - \frac{1}{320}\right)\text{K}^{-1} = 1.25$$

$$\frac{k_{T_2}}{k_{T_1}} = e^{1.25} = 3.49$$

(b) Substituting $E_a = 50\ \text{kJ mol}^{-1}$, $R = 8.31\ \text{J K}^{-1}$, $T_1 = 300\ \text{K}$, and $T_2 = 400\ \text{K}$ in the above equation gives

$$\ln\left(\frac{k_{T_2}}{k_{T_1}}\right) = \frac{(50\ \text{kJ mol}^{-1})\left(\dfrac{1000\ \text{J}}{1\ \text{kJ}}\right)}{8.31\ \text{J K}^{-1}\ \text{mol}^{-1}}\left(\frac{1}{300} - \frac{1}{400}\right)\text{K}^{-1} = 5.01$$

$$\frac{k_{T_2}}{k_{T_1}} = e^{5.01} = 150$$

For the same initial concentrations of reactants, the ratio of the rates is the same as the ratio of the rate constants. Thus, for a reaction with an activation energy of 50 kJ mol^{-1}, an increase in the temperature from 300 K to 320 K more than trebles the rate, and an increase from 300 K to 400 K increases the rate 150-fold.

$=$ *Example 19.7* $=$

DETERMINATION OF THE ACTIVATION ENERGY E_a FROM THE ARRHENIUS EQUATION

When cyclopropane is heated it is converted to propene:

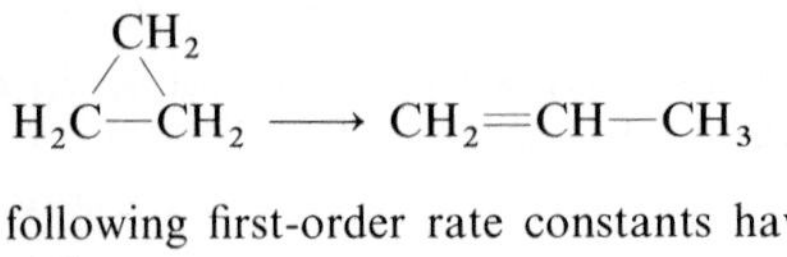

$$\underset{\text{H}_2\text{C}-\text{CH}_2}{\overset{\text{CH}_2}{\diagup\diagdown}} \longrightarrow \text{CH}_2{=}\text{CH}{-}\text{CH}_3$$

For this reaction, the following first-order rate constants have been measured, at the temperatures indicated:

$T\ (°C)$	$k_1\ (s^{-1})$
300	2.39×10^{-10}
320	1.64×10^{-9}
340	9.91×10^{-9}
360	5.34×10^{-8}
380	2.60×10^{-7}
400	1.15×10^{-6}

By plotting $\ln k$ against $1/T$, determine the value of the activation energy, E_a, for this reaction.

Solution

The Arrhenius equation has the form $\ln k = \ln A - E_a/R$ and the plot of $\ln k$ versus $1/T$ should be a straight line with a slope of $-E_a/R$. First we calculate $1/T$, remembering that T is in kelvins, and $\ln k$ for each experiment and then plot $1/T$ versus $\ln k$. The converted data give

$T\ (K)$	$\dfrac{1}{T} \times 10^3\ (K^{-1})$	$\ln k$
573	1.745	-22.15
593	1.686	-20.23
613	1.631	-18.43
633	1.580	-16.75
653	1.531	-15.16
673	1.486	-13.68

You should obtain a plot like that in Figure 19.12:

$$\text{Slope} = -3.31 \times 10^4\ \text{K} = -\frac{E_a}{R}$$

Hence

$$E_a = (3.31 \times 10^4\ \text{K})(8.31\ \text{J K}^{-1}\ \text{mol}^{-1})\left(\frac{1\ \text{kJ}}{1000\ \text{J}}\right)$$

$$= 275\ \text{kJ mol}^{-1}$$

As we saw in Example 19.6, for two temperatures T_1 and T_2, the Arrhenius equation may be written in the form

$$\ln\left(\frac{k_{T_2}}{k_{T_1}}\right) = \frac{E_a}{R}\left(\frac{1}{T_1} - \frac{1}{T_2}\right)$$

We may use this equation to calculate the activation energy for a reaction using rate constants measured experimentally at *two* different temperatures. Alternatively, if we know the activation energy for a reaction and the rate constant at one temperature, we can calculate the rate constant at any other temperature.

FIGURE 19.12
Arrhenius Plot (ln k versus
$1/T$) for the Data in Example
19.7.

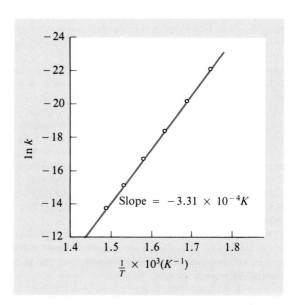

$$\frac{1}{T} \times 10^3(K^{-1})$$

EXERCISE 19.5

For the reaction in Example 19.7, at what temperature will the rate of the reaction be 100 times as fast as it is at 400 °C?

═ *Example 19.8* ═══════════════════════════════════════

RATE CONSTANT AT T_2 FROM ACTIVATION ENERGY AND THE RATE CONSTANT AT T_1

The reaction

$$2NOCl(g) \longrightarrow 2NO(g) + Cl_2(g)$$

has an activation energy of 100 kJ mol^{-1} and the rate constant at 350 K is 8.0×10^{-6} mol^{-1} L s^{-1}. What is the value of the rate constant at 400 K?

Solution

We substitute the given values in the equation

$$\ln\left(\frac{k_{T_2}}{k_{T_1}}\right) = \frac{E_a}{R}\left(\frac{1}{T_1} - \frac{1}{T_2}\right)$$

to give

$$\ln\left(\frac{k_{400}}{8.0 \times 10^{-6}}\right) = \frac{(100 \text{ kJ mol}^{-1})\left(\dfrac{1000 \text{ J}}{1 \text{ kJ}}\right)}{8.31 \text{ J K}^{-1} \text{ mol}^{-1}}\left(\frac{1}{350} - \frac{1}{400}\right)K^{-1}$$

$$= 4.30$$

Therefore

$$\frac{k_{400}}{8.0 \times 10^{-6} \text{ mol}^{-1} \text{ L s}^{-1}} = e^{4.30} = 73.7$$

$$k_{400} = 5.9 \times 10^{-4} \text{ mol}^{-1} \text{ L s}^{-1}$$

EXERCISE 19.6

For the reaction

$$2HI(g) \longrightarrow H_2(g) + I_2(g)$$

the rate constant is 3.91×10^{-4} mol L^{-1} s^{-1} at 370 °C, and 4.05×10^{-2} mol L^{-1} s^{-1} at 470 °C. What is the activation energy for this reaction?

EXERCISE 19.7

What is the rate constant at 450 °C for the reaction in Example 19.8?

EXERCISE 19.8

What is the activation energy of a reaction for which the rate constant is observed to double from 25 °C to 35 °C?

19.4
Reaction Mechanisms

The study of the factors that influence the rate of a reaction, such as temperature and concentration, is important if we wish to control the rates of reactions—to speed them up or slow them down as necessary. But the study of reaction rates is also important in helping us understand in detail how reactions occur. The balanced equation does not tell us *how* a reaction occurs; it merely summarizes the numbers of moles of reactants and products. But how exactly are the reactants transformed to the products? It is often difficult to answer this question. But we can get much useful information, at least for some reactions, by studying their rates and how the rates are influenced by concentration and temperature changes.

The details of the process by which the reactants are converted to the products is called the **mechanism** of the reaction. The study of the mechanism of reactions, including many of biological interest, is a very active area of research in chemistry today.

SINGLE-STEP REACTIONS

The simplest reactions are those that occur in a single step. They may be classified according to the number of molecules that are involved in the single step.

BIMOLECULAR REACTIONS The most common type of single-step reactions are those that occur on the collision of two reactant molecules. These are called **bimolecular reactions**. Examples include

$$NO(g) + O_3(g) \longrightarrow NO_2(g) + O_2(g)$$

and

$$NOCl(g) + NOCl(g) \longrightarrow 2NO(g) + Cl_2(g)$$

As we have seen, we expect the rate of collisions between two molecules to be proportional to the concentrations of each of the molecules. That is,

$$\text{Rate} = k[NO][O_3] \quad \text{and} \quad \text{Rate} = k[NOCl][NOCl] = k[NOCl]^2$$

In other words, we expect these bimolecular reactions to have second-order rate laws. Second-order rate laws are, in fact, observed for these reactions.

TERMOLECULAR REACTIONS If a reaction occurs by the simultaneous collision of three molecules, we call it a **termolecular reaction**. The probability of a simultaneous collision between three molecules is much smaller than the probability of a collision between two molecules; therefore such reactions are uncommon. The simultaneous collision of four molecules would be so rare that it is safe to say that no reaction actually occurs in this way. If the equation for a reaction shows four molecules of reactants, as, for example, in

$$2NO(g) + 2H_2(g) \longrightarrow N_2(g) + 2H_2O(g)$$

we presume that the reaction does not occur by the simultaneous collision of four molecules. Instead, we assume that it occurs in two or more simpler steps. However, before considering multistep reactions, we must discuss one further important type of single-step reaction—unimolecular reactions.

UNIMOLECULAR REACTIONS Not all reactions are the direct result of collisions between molecules. A reaction may simply be the consequence of the decomposition or internal rearrangement of a single molecule. Such a reaction is called **unimolecular**. An example of a unimolecular reaction is the conversion of cyclopropane to propene that we discussed in Example 19.7:

$$\overset{\displaystyle CH_2}{\underset{\displaystyle H_2C-CH_2}{\diagup\diagdown}} \longrightarrow CH_3-CH=CH_2$$

In this reaction a carbon–carbon bond in cyclopropane must first be broken. Where does this energy come from? We know that as a result of random collisions between the molecules a very small fraction of the molecules have a much higher energy than the average energy. Part of this energy is the kinetic energy associated with the motion of the molecules in space (translational and rotational motion) and part of it is the kinetic and potential energy associated with the vibrations of the atoms in the molecule (Figure 19.13). There are random fluctuations of this energy between the different vibrational motions of the molecule. Only when sufficient energy is concentrated in the vibration of one of the carbon–carbon bonds is it possible for this bond to break. At a given temperature a very small but constant fraction of the molecules have their energy concentrated in this way. Thus if the total concentration of molecules is doubled the concentration of molecules with sufficient energy in a carbon–carbon bond to cause it to break will also double. In other words the rate of reaction doubles. Thus the rate of the reaction is proportional to the concentration of cyclopropane: the reaction is first-order.

$$\text{Rate} = k[\text{cyclopropane}]$$

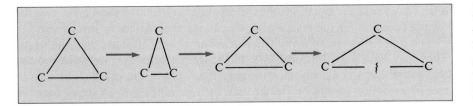

FIGURE 19.13
Vibration of a C—C Bond in the Cyclopropane Molecule.

The bond alternately contracts and expands. The greater the vibrational energy associated with the bond the larger is the amplitude of vibration, and if the energy is large enough the bond may break.

These arguments apply for any unimolecular reaction,

$$A \longrightarrow \text{Products}$$

so that the rate law is

$$\text{Rate} = k[A]$$

The rate is proportional simply to the concentration of one species; this is a *first-order rate law.*

Thus *for single-step reactions:*

- A unimolecular reaction has a first-order rate law.
- A bimolecular reaction has a second-order rate law.
- A termolecular reaction has a third-order rate law.

MULTISTEP REACTIONS

If a reaction occurs in a single step, the order for each reactant in the rate law corresponds to the coefficient of that reactant in the balanced equation for the reaction. However, the rate law for many reactions does not correspond to the balanced equation for the reaction.

For example, the rate law for the reaction

$$2N_2O_5(g) \longrightarrow 4NO_2(g) + O_2(g)$$

is found by experiment to be

$$\text{Rate} = k[N_2O_5]$$

Thus although the reaction could, in principle, take place by collisions between two N_2O_5 molecules, we can conclude that it does not, because such a single-step bimolecular reaction would lead to a second-order rate law:

$$\text{Rate} = k[N_2O_5]^2$$

It must therefore proceed by a more complicated process in which there are two or more successive steps.

The decomposition of N_2O_5 is, in fact, believed to proceed as follows:

$$2[N_2O_5 \longrightarrow NO_2 + NO_3] \quad \text{(Slow)}$$
$$NO_2 + NO_3 \longrightarrow NO + NO_2 + O_2 \quad \text{(Fast)}$$
$$NO + NO_3 \longrightarrow 2NO_2 \quad \text{(Fast)}$$

Each of the steps is called an *elementary process.* Elementary processes are almost always either unimolecular or bimolecular. Occasionally, they may be termolecular. All the steps taken together constitute the **mechanism of the reac-**

tion. The sum of the equations for the separate consecutive steps must give the overall equation for the reaction, as we can see that it does in this case.

The different steps of a multistep reaction do not all have the same rate. They may have very different rates; in the simple cases that we will consider, one of the steps is usually much slower than any of the others. The overall rate of a reaction cannot be greater than the rate of the slowest step. Therefore if one of the steps in a reaction mechanism is much slower than the others, it is called the **rate-determining step**, or the **rate-limiting step**. In such cases the rate law is determined by the rate law for the slowest step of the reaction. For example, in the decomposition of N_2O_5 the rate law for the slow unimolecular step is

$$\text{Rate} = k[N_2O_5]$$

which is the rate law for the overall reaction.

Even when the exponents of the rate law match the coefficient of the balanced equation, we cannot be certain that the reaction occurs in one step. In fact, if the balanced equation involves more than two molecules of reactants, it probably occurs in several steps. An example is the reaction between NO and Cl_2,

$$2NO(g) + Cl_2(g) \longrightarrow 2NOCl(g)$$

for which the rate law is

$$\text{Rate} = k[NO]^2[Cl_2]$$

Thus, the reaction could be termolecular. But the reaction might occur by the following two steps: first, a rapid equilibrium with an equilibrium constant K_1,

$$2NO \underset{}{\overset{K_1}{\rightleftharpoons}} N_2O_2 \quad \text{(Fast equilibrium)}$$

and second, a slow bimolecular reaction with a rate constant k_2,

$$N_2O_2 + Cl_2 \xrightarrow{k_2} 2NOCl \quad \text{(Slow)}$$

The rate law for the slow second step is

$$\text{Rate} = k_2[N_2O_2][Cl_2] \tag{1}$$

In this form the rate law is not very useful because it involves N_2O_2, which is present in only a very small concentration during the reaction and is not one of the reactants nor one of the final products. We require the rate law in a form that can be measured experimentally, in other words, in a form that involves only the concentrations of the reactants—in this case NO and Cl_2.

We can eliminate the concentration of N_2O_2 from the rate law by using the fact that the first step of the mechanism is an equilibrium for which

$$K_1 = \left(\frac{[N_2O_2]}{[NO]^2}\right)_{\text{eq}}$$

where K_1 is the equilibrium constant. Hence

$$[N_2O_2] = K_1[NO]^2 \tag{2}$$

Substituting (2) in (1) gives

$$\text{Rate} = k_2K_1[NO]^2[Cl_2]$$

If we write $k_2K_1 = k$, then

$$\text{Rate} = k[NO]^2[Cl_2]$$

This is the observed rate law. Therefore this two-step mechanism is consistent with the rate law. But we cannot conclude with complete certainty that this *is* the mechanism of the reaction. We need to test other plausible mechanisms to make sure that they are not also consistent with the observed rate law.

Another possible mechanism is

$$2NO \xrightarrow{k_1} N_2O_2 \qquad \text{(Slow)}$$

$$N_2O_2 + Cl_2 \xrightarrow{k_2} 2NOCl \qquad \text{(Fast)}$$

The rate law for the first step is

$$\text{Rate} = k_1[NO]^2$$

Since this step is the slow, rate-limiting step, this rate law is also the rate law for the overall reaction:

$$\text{Rate} = k[NO]^2$$

But this is not the observed rate law, so we conclude that this mechanism is not the actual mechanism of the reaction.

 = *Example 19.9* =

RATE LAW FROM REACTION MECHANISM

What is the rate law for the following mechanism?

$$NO + Cl_2 \xrightarrow{k_1} NOCl + Cl \qquad \text{(Slow)}$$

$$NO + Cl \xrightarrow{k_2} NOCl \qquad \text{(Fast)}$$

Solution

For the first step, the rate law is

$$\text{Rate} = k_1[NO][Cl_2]$$

Since this step is the slow, rate-limiting step, this rate law is also the rate law for the overall reaction:

$$\text{Rate} = k[NO][Cl_2]$$

Again, this differs from the experimentally observed rate law, so we can conclude that this also is not the mechanism of the reaction between NO and Cl_2.

E X E R C I S E   1 9 . 9

What is the rate law for the reaction

$$H_2(g) + Br_2(g) \longrightarrow 2HBr(g)$$

if the mechanism is

$$Br_2 \rightleftharpoons 2Br \qquad \text{(Fast equilibrium)}$$

$$H_2 + Br \longrightarrow HBr + H \qquad \text{(Slow)}$$

$$Br_2 + H \longrightarrow HBr + Br \qquad \text{(Fast)}$$

Another reaction that is more complicated than we might expect from the rate law is the gas-phase reaction between hydrogen and iodine to give hydrogen iodide:

$$H_2(g) + I_2(g) \longrightarrow 2HI(g)$$

The second-order rate law,

$$Rate = k[H_2][I_2]$$

was first established experimentally in 1894. For over seventy years chemists thought that the reaction occurred by the simple bimolecular collision of an H_2 molecule with an I_2 molecule. But in 1967 it was shown that the reaction is considerably speeded up when exposed to an intense visible light.

Because the iodine molecule has a relatively weak bond (bond energy 149 kJ mol^{-1}), visible light is capable of dissociating iodine molecules to iodine atoms:

$$I_2 \xrightarrow{hv} 2I$$

This suggests that iodine atoms are involved in the reaction, which is now believed to occur by the following sequence of steps:

$$I_2 \rightleftharpoons 2I \qquad \text{(Fast equilibrium)} \qquad \text{(1)}$$

$$I + H_2 \rightleftharpoons H_2I \qquad \text{(Fast equilibrium)} \qquad \text{(2)}$$

$$H_2I + I \longrightarrow 2HI \qquad \text{(Slow)} \qquad \text{(3)}$$

In the first reaction a small equilibrium concentration of iodine atoms is established by the dissociation of iodine molecules. Some of these iodine atoms react rapidly with hydrogen molecules in the second step to form a small equilibrium concentration of H_2I molecules. Finally, iodine atoms and H_2I molecules react relatively slowly to give HI. As the H_2I and I are used up in reaction (3) they are replenished by the shifting of the equilibria (1) and (2).

Step (3) is the slowest, so it is the rate-limiting reaction. It determines the rate of formation of HI and therefore the rate of the overall reaction. Since the third reaction is a bimolecular reaction, its rate law is

$$Rate = k_3[H_2I][I] \qquad \text{(4)}$$

and because reaction (3) is the rate-determining step, this rate law is also the rate law for the overall reaction. We may relate the concentrations of both H_2I and I to the concentrations of H_2 and I_2 by means of the equilibrium constants for reactions (1) and (2). These are

$$K_1 = \left(\frac{[I]^2}{[I_2]}\right)_{eq} \qquad K_2 = \left(\frac{[H_2I]}{[I][H_2]}\right)_{eq}$$

Hence

$$[I]^2 = K_1[I_2] \quad \text{and} \quad [H_2I] = K_2[I][H_2]$$

Substituting for $[H_2I]$ in the rate law (4), we obtain

$$Rate = k_3K_2[I][H_2][I]$$

Then substituting for $[I]^2$ gives

$$Rate = k_3K_2K_1[H_2][I_2]$$

If we then define a new constant $k = k_3 K_2 K_1$, we may write

$$\text{Rate} = k[\text{H}_2][\text{I}_2]$$

This is the observed rate law. It shows that the proposed mechanism is a possible mechanism for the reaction. Since it also accounts for the effect of visible light on the reaction, which increases the concentration of iodine atoms and therefore speeds up the reaction, it is now the accepted mechanism.

The determination of the rate law for a reaction is a very important part of the study of any reaction. Although several mechanisms may be consistent with the rate law, there will be many that are not consistent and can therefore be eliminated. Further experiments can then often be made to obtain the additional information necessary to decide between those mechanisms that are consistent with the rate law. However, we can only say that a given mechanism is *consistent* with the rate law and with all the other information about the reaction and is therefore an acceptable mechanism. It is conceivable that another mechanism will be discovered that also is consistent with the rate law and all other available experimental data. New experimental data will then be needed to decide between these two possible mechanisms.

REACTION INTERMEDIATES

In most multistep reactions, species formed in one or more of the steps are used up in the following steps and do not appear in the final products. Such species are called **reaction intermediates**. They are often unfamiliar molecules such as N_2O_2, NO_3, and H_2I or free atoms such as I. Frequently, as in the cases of NO_3, H_2I, and I, they are odd-electron species, that is, free radicals. They are usually very reactive and play an important role in reactions, but they are normally present only in very small concentrations during the reaction.

EQUILIBRIUM CONSTANT AND REACTION MECHANISM

When a reacting system reaches equilibrium, the rate of the forward reaction equals the rate of the reverse reaction. For the single-step bimolecular reaction

$$\text{NO(g)} + \text{O}_3\text{(g)} \rightleftharpoons \text{NO}_2\text{(g)} + \text{O}_2\text{(g)}$$

the rate laws for the forward and back reactions are

$$\text{Rate forward reaction} = k_f[\text{NO}][\text{O}_3]$$
$$\text{Rate reverse reaction} = k_r[\text{NO}_2][\text{O}_2]$$

At equilibrium the rate of the forward reaction equals the rate of the reverse reaction. Thus

$$k_f([\text{NO}][\text{O}_3])_{eq} = k_r([\text{NO}_2][\text{O}_2])_{eq}$$

so that

$$\left(\frac{[\text{NO}_2][\text{O}_2]}{[\text{NO}][\text{O}_3]}\right)_{eq} = \frac{k_f}{k_r}$$

Since k_f/k_r is a constant, which we may write as K, we have

$$\frac{k_f}{k_r} = K = \left(\frac{[NO_2][O_2]}{[NO][O_3]}\right)_{eq}$$

We recognize this expression as the equilibrium constant expression for the reaction. Thus the equilibrium constant expression for a single-step reaction is a consequence of the rate laws for the forward and back reactions and the fact that at equilibrium the rates of the forward and the back reactions are equal.

We have seen in Chapter 13 that we can always write the correct expression for the equilibrium constant from the balanced equation for a reaction, although we cannot always deduce the rate law from the balanced equation. For example, from the balanced equation for the reaction

$$H_2(g) + I_2(g) \rightleftharpoons 2HI(g)$$

we may write

$$K = \frac{[HI]^2}{[H_2][I_2]}$$

If we were to assume, incorrectly, that the reaction is a single-step bimolecular reaction, then we could write

$$\text{Rate forward reaction} = k_f[H_2][I_2]$$

$$\text{Rate reverse reaction} = k_r[HI]^2$$

At equilibrium

$$k_f([H_2][I_2])_{eq} = k_r([HI]^2)_{eq}$$

$$\frac{k_f}{k_r} = \left(\frac{[HI]^2}{[H_2][I_2]}\right)_{eq} = K$$

This is the correct expression for the equilibrium constant, even though we did not use the correct mechanism. Let us now derive the equilibrium constant expression from the accepted mechanism.

When equilibrium is reached between the reactants and the final products of a reaction, each step in the reaction mechanism must also have reached equilibrium. In the present case three equilibria will be established:

$$I_2 \rightleftharpoons 2I \tag{1}$$

$$I + H_2 \rightleftharpoons H_2I \tag{2}$$

$$H_2I + I \rightleftharpoons 2HI \tag{3}$$

Step (3) is written as an equilibrium here, whereas in deriving the rate law only the forward reaction was considered. The rate law thus applies only to the early stages of the reaction, when it is far from equilibrium. Derivation of the rate law for the system as it approaches equilibrium is complex and beyond the scope of this book.

The three equilibrium constants for the three successive steps are

$$K_1 = \left(\frac{[I]^2}{[I_2]}\right)_{eq} \qquad K_2 = \left(\frac{[H_2I]}{[I][H_2]}\right)_{eq} \qquad K_3 = \left(\frac{[HI]^2}{[H_2I][I]}\right)_{eq}$$

Since the sum of equations (1), (2), and (3) is the overall equation

$$H_2 + I_2 \rightleftharpoons 2HI$$

the product of the equilibrium constants $K_1K_2K_3$ is the equilibrium constant for the overall reaction (see Chapter 13):

$$K_1K_2K_3 = \frac{[I]^2[H_2I][HI]^2}{[I_2][I][H_2][H_2I][I]} = \frac{[HI]^2}{[I_2][H_2]} = K$$

Thus we see that taking the individual steps of the reaction into account gives the same expression for the equilibrium constant as is obtained from the balanced equation for the equilibrium between the reactants and the final products of the reaction. In fact, for any reaction *the equilibrium constant expression is independent of the actual mechanism by which the reaction occurs.* Thus we can always write the correct expression for the equilibrium constant for a reaction simply from the balanced equation for the reaction, whether or not we know the detailed mechanism by which the reaction occurs.

19.5
Catalysis

In previous chapters we have met many examples of the use of catalysts to increase the rate of a reaction. We are now in a position to look at how a catalyst works. A **catalyst** is a substance that increases the rate of a chemical reaction although it is not used up in the reaction. In other words, its final concentration is equal to its initial concentration. *A catalyst does not affect the equilibrium position of a reaction, but it does increase the rate at which equilibrium is reached.* A catalyst does not appear in the overall equation for the reaction because it is regenerated during the reaction. It may be regarded as both a reactant and a product, and it therefore cancels out of the balanced equation. Experiment 19.3 demonstrates a reaction that is catalyzed by water.

There are two different types of catalysts: homogeneous and heterogeneous.

Experiment 19.3
Catalysis

  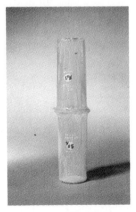

The upper jar contains SO_2 gas. The lower jar contains H_2S gas. The two gases are separated by a glass plate.

The plate is removed but no reaction is observed. A few drops of water are then added and the pair of jars is inverted a few times to mix the two gases quickly.

Water catalyzes the reaction between H_2S and SO_2 so that the sides of the jars are rapidly coated with yellow sulfur.

FIGURE 19.14
Activation Energy for the
Iodide-Ion-Catalyzed
Decomposition of Hydrogen
Peroxide.

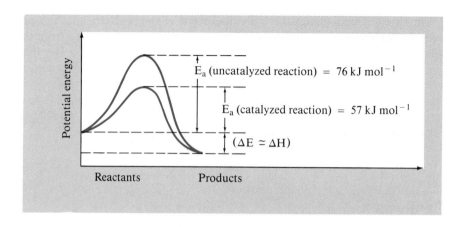

A **homogeneous catalyst** *is in the same phase as the reactants*; usually both are in solution or both are in the gas phase. *A* **heterogeneous catalyst** *is in a different phase from the reactants.* Frequently the catalyst is a solid and the reactants are in solution or in the gas phase. In Experiment 1.5 we saw that solid manganese dioxide catalyzes the decomposition of an aqueous solution of hydrogen peroxide. This is an example of heterogeneous catalysis. The decomposition of hydrogen peroxide in aqueous solution is also catalyzed by a small amount of a soluble iodide. This is an example of homogeneous catalysis. The reaction takes place in two steps:

$$H_2O_2(aq) + I^-(aq) \longrightarrow IO^-(aq) + H_2O(l) \qquad \text{(Slow)}$$
$$H_2O_2(aq) + IO^-(aq) \longrightarrow H_2O(l) + O_2(g) + I^-(aq) \qquad \text{(Fast)}$$

Overall reaction $\qquad\qquad 2H_2O_2(aq) \longrightarrow 2H_2O(l) + O_2(g)$

The iodide ion takes part in the first step but is regenerated in the second so it is not used up in the reaction and it does not therefore appear in the overall equation; in other words, it behaves as a catalyst. This catalyzed reaction has an activation energy of 57 kJ mol^{-1} and is therefore much faster than the uncatalyzed reaction which has an activation energy of 76 kJ mol^{-1}. *The function of a catalyst is to provide an alternative path for the reaction which has a lower activation energy than that of the uncatalyzed reaction* (Figure 19.14).

$=$ *Example 19.10* $================================$

RATE LAW FOR A CATALYZED REACTION

Deduce the rate law for the iodide-catalyzed decomposition of hydrogen peroxide in aqueous solution.

Solution

The slow step of the reaction is

$$H_2O_2(aq) + I^-(aq) \longrightarrow IO^-(aq) + H_2O(l)$$

So the rate law is

$$\text{Rate} = k[H_2O_2][I^-]$$

The reaction is second-order. Notice that the catalyst appears in the rate law although it does not appear in the overall equation for the reaction.

EXERCISE 19.10

Dinitrogen monoxide, N_2O, is relatively stable at room temperature, but at 600 °C it decomposes,

$$2N_2O(g) \longrightarrow 2N_2(g) + O_2(g)$$

according to the mechanism

$$N_2O(g) \longrightarrow N_2(g) + O(g) \qquad \text{(Slow)}$$
$$O(g) + N_2O(g) \longrightarrow N_2(g) + O_2(g) \qquad \text{(Fast)}$$

The reaction is catalyzed by a trace of $Cl_2(g)$, and the catalyzed reaction follows the mechanism

$$Cl_2 \rightleftharpoons 2Cl \qquad \text{(Fast equilibrium)}$$
$$N_2O + Cl \longrightarrow N_2 + ClO \qquad \text{(Slow)}$$
$$ClO + ClO \longrightarrow Cl_2 + O_2 \qquad \text{(Fast)}$$

Confirm that Cl_2 behaves as a catalyst for this reaction and derive the rate laws for the uncatalyzed and the catalyzed reactions.

In the contact process for the industrial manufacture of sulfuric acid solid vanadium pentaoxide, V_2O_5, is used as a heterogeneous catalyst for the reaction between SO_2 and O_2 (Chapter 8). Heterogeneous catalysts are used in many industrial processes. For example, a mixture of iron and iron oxide is used in the Haber process for the synthesis of ammonia (Chapter 3), and platinum is the catalyst for the oxidation of ammonia to nitrogen monoxide in the Ostwald process for the manufacture of nitric acid (Chapter 18). In these and many similar cases, the catalyst is a solid, the reactants are gases, and the rate-limiting step occurs on the surface of the solid catalyst. Hence heterogeneous catalysis is often also called **surface catalysis**.

Although heterogeneous catalysis is of enormous importance, the detailed mechanism by which many heterogeneous catalysts work is not well understood. But we do know that reactant molecules become attached to the surface of the catalyst; this weakens some of the bonds in the reactant molecules and enables them to take part much more easily in a reaction with another molecule.

An important example is the use of the surface of a metal, such as platinum or nickel, as a catalyst for the addition of hydrogen to a carbon–carbon double bond, as in the conversion of ethene to ethane. In the absence of a catalyst this reaction does not occur readily. But in the presence of finely divided nickel or platinum, it occurs rapidly at room temperature and at a high pressure of several hundred atmospheres:

(a), (b) Hydrogen and ethene
molecules are adsorbed onto a
platinum surface. The hydrogen
molecules dissociate into hydrogen
atoms. (c) The hydrogen atoms
move over the surface of the platinum and one of them combines
with the ethene molecule to form
C_2H_5 which remains attached to
the platinum surface. (d) Another
hydrogen atom moves over the
surface and combines with the
C_2H_5 forming ethane, C_2H_6,
which leaves the surface.

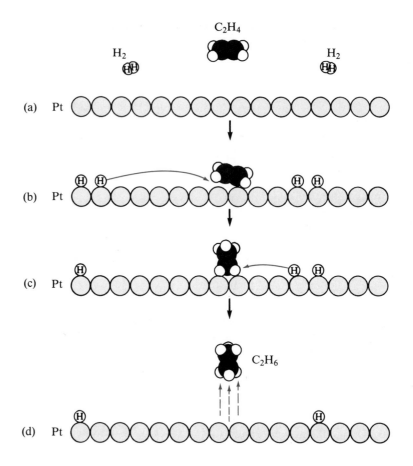

Both the C_2H_4 and the H_2 molecules become attached to the metal surface;
they are said to be *adsorbed* (see Figure 19.15). In this process the H_2 molecules
are dissociated into H atoms, which become attached to platinum atoms on
the surface. But these hydrogen atoms can move rather easily across the metal
surface from one platinum atom to another. When they encounter an ethene
molecule, they combine with it readily. The successive addition of two hydrogen
atoms gives the final product, ethane, which escapes from the surface.

An important application of heterogeneous catalysts is in the catalytic converters in the exhaust systems of automobiles. The exhaust gases contain unburnt
hydrocarbons and CO and the catalyst converts them to CO_2 and water. Newer
converters also contain a catalyst to speed up the decomposition of NO which
is formed from N_2 and O_2 at the high temperature reached in the cylinders of
an automobile engine. The catalysts in a catalytic converter are destroyed by lead
and this led to the introduction in 1974 of unleaded gasoline, which in turn
has considerably reduced the environmental pollution formerly caused by the
lead in leaded gasoline.

ENZYMES

Enzymes, which catalyze reactions in living organisms, are another very important class of catalysts. They are complex proteins that are very specific catalysts

for biochemical reactions. For example, a solution of sugar at 37 °C does not oxidize at a significant rate. Yet in the body at this temperature sugar is rapidly oxidized to CO_2 and H_2O:

$$C_{12}H_{22}O_{11}(aq) + 12O_2(aq) \longrightarrow 12CO_2(aq) + 11H_2O(l)$$

This reaction is accomplished by means of an enzyme catalyst. Enzymes are amazingly efficient catalysts that increase the rates of reactions by factors as great as 10^{20} so that they occur rapidly at body temperature. They are discussed in more detail in Chapter 24.

19.6
Chain Reactions

We have seen in Chapter 5 that the reaction between hydrogen and chlorine at room temperature is extremely slow; no measurable amount of hydrogen chloride is formed, even over a long period. However, if the mixture of gases is exposed to a bright light, a very rapid reaction occurs, which we observe as an explosion. This reaction is an example of an important class of reactions called **chain reactions**. A chain reaction is a multistep reaction in which a free radical is formed and the free radical initiates a series of two or more reactions in which the products are formed and the free radical is regenerated. Because one free radical may thus cause the formation of a large number of product molecules, chain reactions may be very fast. As an example, we will consider the mechanism of the reaction between hydrogen and chlorine.

Absorption of light of a certain minimum frequency (see Example 7.3) causes chlorine molecules to dissociate to atoms:

$$Cl_2 \xrightarrow{\ hv\ } 2Cl \qquad \qquad \textbf{(1)}$$

This first step is called *chain initiation*. Chlorine atoms are very reactive, and a Cl atom reacts rapidly with an H_2 molecule, producing an HCl molecule and an H atom:

$$Cl + H_2 \longrightarrow HCl + H \qquad \qquad \textbf{(2)}$$

The very reactive H atom then attacks a Cl_2 molecule, forming an HCl molecule and regenerating a Cl atom:

$$H + Cl_2 \longrightarrow HCl + Cl \qquad \qquad \textbf{(3)}$$

The Cl atom can attack another H_2 molecule, as in step 2, producing another H atom, which can react with a Cl_2 molecule, as in step 3. These two reactions can repeat themselves many times, giving a chain of successive reactions. They are called *chain-propagating steps*. The overall reaction

$$H_2 + Cl_2 \longrightarrow 2HCl$$

is the sum of reactions (2) and (3). These reactions continue indefinitely until the Cl atoms are removed in some way, for example, by combining with each other:

$$Cl + Cl \longrightarrow Cl_2$$

This reaction is called a *chain termination reaction*. Because the concentration of Cl atoms is very low, this reaction is relatively slow. Thus a Cl atom may

Both the hydrogen atom H·, and the chlorine atom, :Cl·, have a single unpaired electron so they are free-radicals.

cause the rapid formation of many thousands of HCl molecules before it is re-moved from the reacting system (see Figure 19.16).

The mechanism of the H_2–Cl_2 reaction can therefore be summarized by the following equations:

$$Cl_2 \longrightarrow 2Cl \qquad \text{(Chain initiation)}$$

$$\left.\begin{array}{l} Cl + H_2 \longrightarrow HCl + H \\ H + Cl_2 \longrightarrow HCl + Cl \end{array}\right\} \quad \text{(Chain propagation)}$$

$$2Cl \longrightarrow Cl_2 \qquad \text{(Chain termination)}$$

Such a sequence of reactions is typical of a chain reaction. A chain reaction is initiated by a relatively slow step that produces a highly reactive intermedi-ate. In subsequent steps this reactive intermediate attacks a reactant molecule to form a product molecule and either regenerates itself or produces another reactive intermediate, which can attack another reactant molecule to form the product, and so on. Eventually, the reactive intermediate is removed in some chain termination step.

Chain reactions can lead to *explosions*, as in the hydrogen–chlorine reaction, if the products of the reaction are gases and if the chain-propagating steps pro-duce heat faster than it can be conducted away. In such a case the increasing temperature continually increases the rate of the reaction, which produces heat still faster, and so on. The extremely rapid increase in the volume of the gaseous products then produces an explosion (Box 6.2).

A very important application of chain reactions is in *polymerization*, the type of reaction used in the manufacture of plastics, such as polyethylene (Chapter 24).

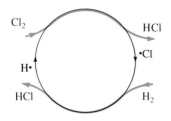

FIGURE 19.16
Hydrogen–Chlorine Chain Reaction.

A Cl atom reacts with an H_2 molecule to form an HCl molecule and an H atom. The H atom re-acts with a Cl_2 molecule to form another HCl molecule and another Cl atom, which reacts with an-other H_2 molecule, and so on. One Cl atom can cause the for-mation of a very large number of HCl molecules.

IMPORTANT TERMS

The **activated complex** is the particular arrangement of at-oms that has the highest energy during the rearrangement of the atoms of the reactants to those of the products. It is also called the **transition state**.

The **activation energy** is the minimum energy that the reac-tant molecules must possess in order to be able to react. It is the difference in energy between the reactants and the transition state.

The **Arrhenius equation** is

$$k = A \exp(-E_a/RT)$$

where k is the rate constant at temperature T and E_a is the activation energy. For two temperatures T_1 and T_2, and rate constants k_{T_1} and k_{T_2}, it takes the form

$$\ln\left(\frac{k_{T_2}}{k_{T_1}}\right) = \frac{E_a}{R}\left(\frac{1}{T_1} - \frac{1}{T_2}\right)$$

A **bimolecular reaction** is a single-step reaction between two molecules.

A **catalyst** is a substance that increases the rate of a reaction without being used up in the reaction. A catalyst changes the mechanism of a reaction and decreases the activation energy, E_a.

A **chain reaction** is a multistep reaction that begins with an initial step, which produces a reactive intermediate (that is usually a free-radical), continues with chain-propagating steps, which form the products and regenerate the reactive intermediate, and is completed by one or more chain termination steps, in which the reactive intermediate is removed.

The **half-life** of a reaction, $t_{1/2}$, is the time required for the concentration of a reactant to decrease to half its initial value.

A **heterogeneous catalyst** is in a different phase from the reactants.

A **homogeneous catalyst** is in the same phase as the reactants.

The **order of a reaction** with respect to a given reactant is the exponent of the concentration of that reactant in the rate law.

The **orientation (or steric) effect** describes the fact that colliding molecules must have a suitable relative orientation in order to react.

The **overall order of a reaction** is the sum of the exponents of the concentration terms in the rate law for the reaction.

The **rate constant** is the proportionality constant, k, that appears in a rate law: rate $= k[A]^x[B]^y[C]^z\ldots$

The **rate law** is the equation that relates the rate of a reaction to the concentrations of the reactants; it has the form

$$\text{rate} = k[A]^x[B]^y[C]^z\ldots$$

A **rate-limiting reaction** (or **rate-determining reaction**) is the slowest step in the series of reaction steps that constitute the mechanism of a reaction. It is the step that determines the rate of the overall reaction.

Reaction intermediates are species formed in one step of a multistep reaction and used up in a following step, so they do not appear in the final products.

A **reaction mechanism** is the series of successive reaction steps (elementary processes) by which reactants are converted to products in a reaction.

The **reaction rate** is the change in the concentration of one of the reactants, or one of the products, of a reaction in a given time interval divided by the time interval.

A **surface catalyst** provides a reactive surface on which the rate-limiting step of a reaction can take place. It is a heterogeneous catalyst.

A **termolecular reaction** is a reaction that occurs by the simultaneous collision of three molecules.

Transition state (see activated complex).

In a **unimolecular reaction** a single molecule undergoes decomposition or rearrangement.

PROBLEMS *

Reaction Rates

1. The reaction

$$CH_3OH(aq) + HCl(aq) \longrightarrow CH_3Cl(aq) + H_2O(l)$$

was followed by measuring the change in hydronium ion concentration with time, to give the results in the accompanying table. Calculate the average reaction rate for each time interval.

Time (min)	$[H_3O^+]$ (mol L^{-1})
0	1.85
80	1.66
159	1.53
314	1.31
628	1.02

2. On heating, gaseous sulfuryl chloride, SO_2Cl_2, decomposes to $SO_2(g)$ and $Cl_2(g)$. The accompanying table gives data at 320 °C for the concentration of the remaining SO_2Cl_2. How could the progress of this reaction be followed experimentally? Calculate the average rate of the reaction for each time interval.

Time (min)	$[SO_2Cl_2]$ (mol L^{-1})
0	0.010 00
20	0.009 70
50	0.009 28
100	0.008 61
200	0.007 41
400	0.005 49
700	0.003 50
1000	0.002 23

3. For each of the following reactions, express the rate in terms of the change in concentration with time of the reactant, and relate this to the rate of formation of each of the products.

(a) $2HI(g) \rightarrow H_2(g) + I_2(g)$

(b) $2NOCl(g) \rightarrow 2NO(g) + Cl_2(g)$

(c) $2N_2O_5(g) \rightarrow 4NO_2(g) + O_2(g)$

4. For each of the following reactions, express the rate in terms of the disappearance of each reactant. Give the relationship between these rates and the rate of appearance of each of the products.

* Answers to problems numbered in blue appear at the end of the text.

(a) $2C_4H_{10}(g) + 13O_2(g) \rightarrow 8CO_2(g) + 10H_2O(g)$

(b) $C_2H_6(g) + 2H_2O(g) \rightarrow 2CO(g) + 5H_2(g)$

(c) $5Br^-(aq) + BrO_3^-(aq) + 6H_3O^+(aq) \rightarrow 3Br_2(aq) + 9H_2O$

5. For the reaction

$$4NH_3(g) + 3O_2(g) \longrightarrow 2N_2(g) + 6H_2O(g)$$

it was found that at a given temperature and at a certain time the rate of formation of $N_2(g)$ was 0.27 mol L^{-1} s^{-1}.

(a) At what rate was water being formed?

(b) At what rate was NH_3 being consumed?

(c) At what rate was O_2 being consumed?

6. For each of the following reactions, express the rate in terms of the change in concentration of oxygen, and relate this to the rate of change of the concentrations of each of the other reactants and products.

(a) $H_2(g) + O_2(g) \rightarrow H_2O_2(g)$

(b) $2H_2(g) + 2O_2(g) \rightarrow 2H_2O_2(g)$

(c) $2NO(g) + O_2(g) \rightarrow 2NO_2(g)$

(d) $4PH_3(g) + 8O_2(g) \rightarrow P_4O_{10}(s) + 6H_2O(g)$

7. In a reaction of ozone with ammonia at 500 K, according to the equation

$$5O_3(g) + 6NH_3(g) \longrightarrow 6NO(g) + 9H_2O(g)$$

the rate of increase in the pressure of NO in a given time interval was 1095 mm Hg s^{-1}. What was the rate of disappearance of ozone, and the rate of disappearance of ammonia, in moles per liter per second, for the same time interval?

Rate Laws

8. (a) How is the rate of a reaction defined?

(b) What is the rate law for a reaction that is first-order in a reactant X and second-order in a reactant Y?

(c) Express the rate constant for the reaction in (b) in terms of the reaction rate.

(d) What are the units of the rate constant for the rate law described in part (b)?

9. What is the rate law for a reaction with the balanced equation

$$aA + bB + cC \longrightarrow Products$$

if the reaction is found experimentally to be first-order in A, first-order in B, and second-order in C. What are the units of the rate constant?

10. The rate constant for a first-order reaction at a certain temperature is 3.7×10^{-2} s^{-1}. For an initial concentration of reactant of 0.040 mol L^{-1}, what is the initial rate

(a) In moles per liter per second?

(b) In moles per liter per hour?

11. The rate law for a reaction is found to be

$$\text{Rate} = k[B]^2[C]$$

with $k = 4.0 \times 10^{-3}$ mol^{-2} L^2 s^{-1}. What is the reaction rate when

(a) $[B] = [C] = 0.010$ mol L^{-1}?

(b) $[B] = [C] = 0.050$ mol L^{-1}?

12. The experimental data for the rate of decomposition of nitrosyl bromide, NOBr(g), at a given temperature is given below. What is the order of the reaction with respect to NOBr and what is the rate law?

[NOBr] (mol L^{-1})	Rate (mol L^{-1} s^{-1})
0.14	0.50
0.31	1.90
0.45	4.00
0.54	5.70
0.71	9.80

13. Plot the rates that you obtained in Problem 2 against the concentration of SO_2Cl_2, and against this concentration squared. Is the reaction first- or second-order in SO_2Cl_2?

14. For the gas-phase reaction

$$2NO(g) + Cl_2(g) \longrightarrow 2NOCl(g)$$

(i) doubling the initial concentration of each reactant increases the initial rate by a factor of 8, and (ii) doubling only the initial concentration of Cl_2 increases the initial rate by only a factor of 2. What is the order of the reaction with respect to NO, with respect to Cl_2, and overall? What is the rate law?

15. When ethanal (acetaldehyde), CH_3CHO, is heated, it decomposes to methane and carbon monoxide. At a given temperature, doubling the initial concentration of ethanal increases the initial rate by a factor of 2.83. What is the order with respect to ethanal, and what is the rate law?

16. For the reaction

$$2NO(g) + H_2(g) \longrightarrow N_2O(g) + H_2O(g)$$

the following data were obtained at a given temperature. What is the rate law for the reaction, and what is the value of the rate constant?

[NO] (mol L^{-1})	[H$_2$] (mol L^{-1})	Initial Rate (mol L^{-1} min^{-1})
0.150	0.800	0.500
0.075	0.800	0.125
0.150	0.400	0.250

[N$_2$O$_5$] (mol L^{-1})	Initial Rate (mol L^{-1} min^{-1})
2.00	1.26×10^{-3}
1.80	1.13×10^{-3}
1.51	0.94×10^{-3}
0.92	0.57×10^{-3}

17. In the reaction of $H_2(g)$ with $Br_2(g)$ to give $HBr(g)$, doubling the initial concentration of H_2 doubles the initial rate, but tripling the initial concentration of Br_2 increases the rate by a factor of 1.75. What are the orders of the reaction with respect to H_2 and Br_2, respectively? What is the rate law? Is the reaction an elementary single-step bimolecular reaction?

18. The decomposition of gaseous hydrogen peroxide to $O_2(g)$ and $H_2O(g)$ is a first-order reaction. Experimentally, at a given temperature, the initial concentration of H_2O_2 was found to decrease to one-half in 17.0 min.

  (a) What is the half-life of the reaction?

  (b) What fraction of the initial H_2O_2 would remain after **(i)** 51.0 min, and **(ii)** ten half-life periods?

19. In another experiment at a higher temperature than in Problem 18, one-fourth of the initial H_2O_2 was found to remain after 8.0 min. What is the half-life of H_2O_2 under these conditions? How long will it take for the concentration of H_2O_2 to decrease to 3.125% of its initial value?

20. Using the data given in Problem 2 for the decomposition of SO_2Cl_2, determine the rate law for the reaction by plotting the data using the integrated first-order rate equation, and the integrated second-order rate equation. Determine the value of the rate constant from the appropriate linear plot.

21. The rate of decomposition of H_2O_2 in aqueous solution at a particular temperature was measured by titrating samples of the solution withdrawn at given times with $KMnO_4(aq)$ under acidic conditions. The following results were obtained.

t (min)	0	10	20
mL KMnO$_4$(aq)	22.8	13.8	8.3

Show that the reaction is first-order in H_2O_2 and find its half-life and the value of the rate constant.

22. At 45 °C, N_2O_5 decomposes in solution in tetrachloromethane according to the equation

$$2N_2O_5 \longrightarrow 4NO_2 + O_2$$

The following rate data were obtained.

(a) Determine the rate law for the reaction.

(b) Calculate the value of the rate constant.

(c) How long would it take for $[N_2O_5]$ to decrease to 0.50 mol L^{-1}?

(d) How could this reaction be followed experimentally to obtain the initial rates?

23. Annual production of the insecticide DDT amounted to about 7.5×10^7 kg in the 1960s. In 1972, DDT was banned for general use in the United States by the Environmental Protection Agency. At ordinary temperatures, the half-life of DDT in soil is about 10 years. In years, how long will it take for 1000 kg of DDT originally sprayed on the ground in 1965 to decrease to 1 g?

24. At 570 K, azomethane, H_3CNNCH_3, decomposes to ethane and nitrogen. The reaction is first-order, with a rate constant of 2.50×10^{-4} s^{-1}. If azomethane, initially at a pressure of 200 torr at 570 K, is allowed to decompose for 30 min, what will be the resulting partial pressure of azomethane, and the total pressure?

25. For the room temperature reaction between nitrogen monoxide and oxygen

$$2NO(g) + O_2(g) \longrightarrow 2NO_2(g)$$

the following initial rate data were obtained.

[NO] (mol L^{-1})	[O$_2$] (mol L^{-1})	Initial Rate (mol L^{-1} s^{-1})
0.010	0.020	0.014
0.010	0.010	0.007
0.020	0.040	0.114
0.040	0.020	0.227

Determine the order of the reaction with respect to NO and O_2, and the value of the rate constant.

26. In acid solution, bromate ions, $BrO_3{}^-$, slowly oxidize bromide ions to bromine.

  (a) Write the balanced equation for this reaction.

(b) From the following initial rate data, deduce the rate equation.

$[BrO_3^-]$ (mol L^{-1})	$[Br^-]$ (mol L^{-1})	$[H_3O^+]$ (mol L^{-1})	Relative Rate
0.05	0.25	0.30	1
0.05	0.25	0.60	4
0.10	0.25	0.60	8
0.05	0.25	0.60	2
0.05	0.50	0.30	2

27. The thermal decomposition of phosphine to phosphorus and hydrogen is a first-order reaction:

$$4PH_3(g) \longrightarrow P_4(g) + 6H_2(g)$$

The half-life of the reaction is 35.0 s at 680 °C. Calculate **(a)** the rate constant and **(b)** the time required for 90% of the phosphine to decompose.

Arrhenius Equation and Activation Energy

28. Rate constants at several different temperatures for the reaction

$$2HI(g) \longrightarrow H_2(g) + I_2(g)$$

are given below. Plot ln k versus $1/T$, and obtain the value for the activation energy of the reaction, E_a, from the slope of the plot. What is the value of the rate constant at 400 °C?

Temperature (°C)	k (mol^{-1} L s^{-1})
302	1.18×10^{-6}
356	3.33×10^{-5}
374	8.96×10^{-5}
410	5.53×10^{-4}
427	1.21×10^{-3}

29. From standard enthalpies of formation, calculate the standard enthalpy change for the reaction in Problem 28. Draw a diagram (like those in Figure 19.10) showing the $\Delta H°$ and E_a for the reaction. What is the activation energy for the reaction $H_2(g) + I_2(g) \rightarrow 2HI(g)$?

30. Explain how the rate of a reaction is affected by each of the following: **(a)** the frequency of collisions, **(b)** the kinetic energy of collisions, **(c)** the orientation of the molecules during a collision.

31. The activation energy of a reaction is 80 kJ mol^{-1}. Calculate the temperature at which the rate of the reaction will be ten times the rate at 0 °C.

32. Given that the activation energy of a reaction is

100 kJ mol^{-1}, to what temperature must the reaction mixture be raised for its rate constant to have exactly twice the value it has at 27 °C?

33. The rate of a reaction at 50 °C is three times the rate at 25 °C. What is its activation energy?

34. By what factor will a catalyst increase the rate of a reaction if it decreases the activation energy from 200 kJ mol^{-1} to 100 kJ mol^{-1} at 100 °C?

35. Nitrogen dioxide decomposes to nitrogen monoxide and oxygen at high temperature. Find the activation energy for this reaction from the following rate constant data.

T (K)	k (mol^{-1} L s^{-1})
650	3.16
730	28.2
800	1.58×10^2
900	1.12×10^3
1000	5.01×10^3

36. For the reaction

$$2N_2O(g) \longrightarrow 2N_2(g) + O_2(g)$$

the rate constant is 1.1×10^{-3} mol^{-1} L s^{-1} at 565 °C, and 3.8×10^{-3} mol^{-1} L s^{-1} at 728 °C. What is the value of the rate constant at 780 °C?

37 Why does it take longer to boil an egg high in the Rocky Mountains, in comparison with the cooking time on the plains of the Midwestern United States?

38. For the reaction

$$CO(g) + NO_2(g) \longrightarrow CO_2(g) + NO(g)$$

the rate constant at 425 °C has the value 1.3 mol^{-1} L s^{-1}, and at 525 °C the value is 23 mol^{-1} L s^{-1}. Calculate the activation energy of the reaction and the rate constant at 298 °C.

39. Explain the difference between a bimolecular reaction and a second-order reaction.

40. Define each of the following:

(a) elementary process (step),

(b) reaction mechanism,

(c) rate-limiting reaction

(d) reaction intermediate

Reaction Mechanisms

41. For the reaction

$$Cl_2(g) + CO(g) \longrightarrow COCl_2(g)$$

the following mechanism has been proposed.

$$Cl_2 \rightleftharpoons 2Cl \qquad \text{(Fast equilibrium)}$$
$$Cl + CO \rightleftharpoons COCl \qquad \text{(Fast equilibrium)}$$
$$COCl + Cl_2 \longrightarrow COCl_2 + Cl \qquad \text{(Slow)}$$

In terms of the reactants CO and Cl_2, what is the expected rate law? What name is given to species such as COCl and Cl?

42. The reaction

$$NO_2(g) + CO(g) \longrightarrow NO(g) + CO_2(g)$$

is believed to occur via the following two bimolecular reactions.

$$NO_2 + NO_2 \longrightarrow NO_3 + NO \qquad \text{(Slow)}$$
$$NO_3 + CO \longrightarrow NO_2 + CO_2 \qquad \text{(Fast)}$$

What rate law is expected for this mechanism? What rate law would be expected if the reaction occurred directly as an elementary process in a single step?

43. The conversion of ozone to molecular oxygen in the upper atmosphere

$$2O_3(g) \longrightarrow 3O_2(g)$$

is thought to occur via the mechanism

$$O_3 \rightleftharpoons O_2 + O \qquad \text{(Fast equilibrium)}$$
$$O + O_3 \longrightarrow 2O_2 \qquad \text{(Slow)}$$

What rate law is consistent with this mechanism? Explain why the reaction rate decreases as the concentration of O_2 increases, in other words, why O_2 appears in the rate law with a negative order.

44. For the reaction

$$2NO(g) + O_2(g) \longrightarrow 2NO_2(g)$$

the rate law is

$$\text{Rate} = k[NO]^2[O_2]$$

(a) Explain why this reaction is unlikely to occur by an elementary termolecular process.

(b) Devise two mechanisms for this reaction that are consistent with the observed rate law that do not involve the simultaneous collision of three molecules.

45. By considering bond energies, explain why the observation that visible light increases the rate of the reaction

$$H_2(g) + I_2(g) \longrightarrow 2HI(g)$$

cannot be explained by a mechanism that involves the formation of H atoms to initiate the reaction.

46. One mechanism for the first-order decomposition of $N_2O_5(g)$ has been proposed in the text. Another proposed mechanism consists of the following steps:

* The asterisk denotes the more difficult problems.

$$N_2O_5 \longrightarrow NO_2 + NO_3 \qquad \text{(Slow)}$$
$$NO_3 \longrightarrow NO + O_2 \qquad \text{(Fast)}$$
$$NO + N_2O_5 \longrightarrow NO_2 + N_2O_4 \qquad \text{(Fast)}$$
$$N_2O_4 \rightleftharpoons 2NO_2 \qquad \text{(Fast equilibrium)}$$

(a) What is the molecularity of each step in this reaction?

(b) Is the proposed mechanism consistent with the observed rate law?

47. In acid solution, iodide ion is oxidized to iodine by hydrogen peroxide according to the equation

$$2I^-(aq) + H_2O_2(aq) + 2H_3O^+(aq) \longrightarrow I_2(s) + 4H_2O(l)$$

The following mechanism has been proposed.

$$H_2O_2 + I^- \longrightarrow H_2O + IO^- \qquad \text{(Slow)}$$
$$H_3O^+ + IO^- \longrightarrow HOI + H_2O \qquad \text{(Fast)}$$
$$HOI + H_3O^+ + I^- \longrightarrow I_2 + 2H_2O \qquad \text{(Fast)}$$

(a) For this mechanism, what is the predicted order of the reaction in H_2O_2, I^-, and H_3O^+, respectively?

(b) If the mechanism is correct, what should be the effect on the rate of changing the pH of the solution from 2.00 to 4.00?

Catalysis

48. State the effect that a catalyst has on each of the following.

(a) The rate of a reaction

(b) The activation energy of a reaction

(c) The enthalpy change of a reaction

(d) The temperature required for a reaction to proceed at a given rate

(e) The position of equilibrium of a reaction

49. Define each of the following:

(a) homogeneous catalyst, (b) heterogeneous catalyst.

50. Does the concentration of a homogeneous catalyst appear in the rate law for a catalyzed reaction? Explain.

51. Explain how a catalyst increases the rate of a reaction.

*52. The iodination of acetone by iodine in aqueous solution proceeds according to the equation

$$CH_3COCH_3(aq) + I_2(aq) \longrightarrow CH_3COCH_2I(aq) + H^+(aq) + I^-(aq)$$

In the presence of a weak base, B, the rate law is

$$\text{Rate} = k[CH_3COCH_3][B]$$

Suggest a three-step mechanism that is consistent with this rate law. What is the function of the weak base B?

*53. Peroxodisulfate ions, $S_2O_8^{2-}$, oxidize iodide ions to iodine according to the balanced equation

$$S_2O_8^{2-}(aq) + 2I^-(aq) \longrightarrow 2SO_4^{2-}(aq) + I_2(s)$$

The reaction is catalyzed by $Fe^{3+}(aq)$ ions, but not by $Cr^{3+}(aq)$ ions. On the basis of the following standard reduction potentials, suggest a possible mechanism for the catalyzed reaction.

$$Fe^{3+}(aq) + e^- \longrightarrow Fe^{2+}(aq) \qquad E^\circ = +0.77 \text{ V}$$
$$Cr^{3+}(aq) + e^- \longrightarrow Cr^{2+}(aq) \qquad E^\circ = -0.41 \text{ V}$$
$$S_2O_8^{2-}(aq) + 2e^- \longrightarrow 2SO_4^{2-}(aq) \qquad E^\circ = +2.01 \text{ V}$$
$$I_2(s) + 2e^- \longrightarrow 2I^-(aq) \qquad E^\circ = +0.54 \text{ V}$$

Chain Reactions

54. Describe each step in the following mechanism for the bromination of methane as *chain initiation, chain propagation,* or *chain termination.*

$$Br_2 \longrightarrow 2Br$$
$$Br + CH_4 \longrightarrow CH_3 + Br$$
$$CH_3 + Br_2 \longrightarrow CH_3Br + Br$$
$$2Br \longrightarrow Br_2$$

What is the maximum wavelength of light that could catalyze this reaction?

55. The chlorination of methane proceeds by a chain mechanism, and the overall reaction is

$$CH_4(g) + Cl_2(g) \longrightarrow CH_3Cl(g) + HCl(g)$$

It is believed that methyl radicals, CH_3, and chlorine atoms are involved in the mechanism, which is initiated by light of maximum wavelength 480 nm.

(a) Write a series of steps that constitute a reasonable mechanism for this reaction.

(b) Identify each step in your mechanism as *chain initiation, chain propagation,* or *chain termination.*

(c) Some ethane, $C_2H_6(g)$, is also formed in this reaction. Explain why.

Miscellaneous

56. (a) Distinguish the terms *order* and *molecularity.*
   (b) How can one analyze experimental results of concentration versus time to show whether a reaction is first-order in a reactant A, or second-order in a reactant A?
   (c) For what kinds of reactions are the order and molecularity the same?

57. (a) Why is it, in a bimolecular reaction in the gas phase, that not all collisions between reactant molecules lead to the formation of products?

(b) For an elementary one-step reaction, how can one distinguish between a unimolecular mechanism and a bimolecular mechanism?

(c) What difference is there between the way that reaction rate is increased by an increase in temperature, and the way that it is increased by a catalyst? Explain.

*58. Bromide ions are oxidized by hydrogen peroxide to bromine in acidic aqueous solution.

(a) Write the balanced equation for this reaction.

(b) In separate experiments, where first the initial concentration of H_2O_2 and then the initial concentration of bromide ion were doubled, keeping the other concentrations constant, the rate was found to double at constant pH. When the pH alone was changed from 1.00 to 0.400, the rate increased fourfold. Deduce the rate law for the reaction.

(c) If under certain conditions the rate of disappearance of Br^- is 7.2×10^{-3} mol L^{-1} s^{-1}, what is the rate of disappearance of H_2O_2, and what is the rate of appearance of Br_2?

(d) What is the effect on the rate constant of increasing the pH?

(e) If the initial solution is diluted with water so that its volume is doubled, what would be the effect on the initial reaction rate?

(f) Suggest a possible mechanism for this reaction.

59. The following three-step mechanism has been proposed for the formation of tetrachloromethane from chlorine and trichloromethane (chloroform):

$$CHCl_3(g) + Cl_2(g) \longrightarrow CCl_4(g) + HCl(g)$$
$$Cl_2 \rightleftharpoons 2Cl \qquad \text{(Fast equilibrium) (1)}$$
$$Cl + CHCl_3 \longrightarrow CCl_3 + HCl \qquad \text{(2)}$$
$$CCl_3 + Cl \longrightarrow CCl_4 \qquad \text{(3)}$$

The observed rate law is

$$\text{Rate} = k[CHCl_3][Cl_2]^{1/2}$$

Is this mechanism consistent with the rate law if (i) step (2) is much slower than step (3), or (ii) step (3) is much slower than step (2)?

60. Define each of the following terms:

(a) reaction rate,

(b) reaction order,

(c) rate constant,

(d) half-life,

(e) molecularity.

The Noble Gases: More Chemistry of the Halogens

Halides of Nonmetals in their Higher
Oxidation States
Oxides and Oxoacids of the Halogens
The Noble Gases and their Compounds

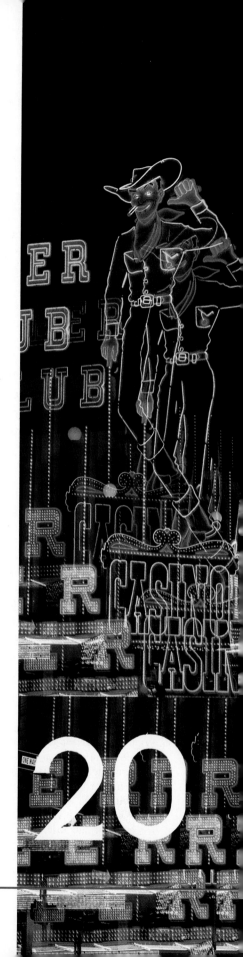

20

One of the most startling discoveries in chemistry in recent times occurred in 1962, when Neil Bartlett prepared the first compound of xenon. This discovery greatly surprised most chemists, since it had become generally accepted that the elements of group 8 could not form compounds. For this reason they were generally referred to as the *inert gases*. This supposed lack of compound formation by the noble gases, all of which, except helium, have eight electrons in their valence shells, had led to the idea that a valence shell of eight electrons is a particularly stable arrangement. This idea was the basis of Lewis's octet rule. The success of the octet rule appeared to confirm the idea that the noble gases could not form compounds and so very few attempts were made to prepare such compounds for nearly 50 years.

In Chapter 5 we discussed some of the chemistry of the halogens, in particular that of the elements, the hydrogen halides, and the halide ions. We restricted our attention almost exclusively to compounds with halogens in the -1 or $+1$ oxidation states. Like other elements of the *third period*, such as phosphorus and sulfur, chlorine forms many compounds in higher oxidation states, such as chlorates and perchlorates, in which the valence shell of chlorine is not limited to an octet of electrons, and the same is true for bromine and iodine. Formation of noble gas compounds also requires expansion of their valence shells beyond the octet, and the chemistry of the noble gases is closely related to the chemistry of the halogens in their higher oxidation states.

We now take up some further aspects of the chemistry of the halogens in their higher valence states and the chemistry of the noble gases. Before they were discovered, some chemists had in fact predicted the probable existence of compounds of the noble gases by extrapolating from the known higher-oxidation-state compounds of the halogens and the elements of groups 5 and 6.

20.1
Halides of Nonmetals in their Higher Oxidation States

In Chapter 8 we discussed the higher oxidation states of phosphorus and sulfur, namely, the $+5$ state of phosphorus and the $+4$ and $+6$ states of sulfur. These higher oxidation states are possible because these elements can have more than eight electrons in their valence shells. Recall that compounds containing the element in these higher oxidation states can be regarded as being formed from a valence state in which one or more electrons are promoted to the 3d orbitals (see Figure 20.1). Only the most electronegative ligands such as oxygen, fluorine, and chlorine can form these higher-oxidation-state compounds because only these ligands are sufficiently electronegative to pull one or more electrons out of the 3s and 3p orbitals into the 3d orbitals in which they are on average further from the nucleus than in the 3s and 3p orbitals. In addition to the -1 and $+1$ oxidation states that arise from the ground state electron configuration, chlorine, bromine, and iodine have $+3$, $+5$, and $+7$ oxidation states, which correspond to valence state electron configurations in which one, two, or three electrons have been promoted to d orbitals (see Figure 20.1 and Table 20.1).

Fluorine is such a strong oxidizing agent that it can oxidize most nonmetals to their highest oxidation states. Thus fluorine reacts with phosphorus to give PF_3 and PF_5 and with sulfur to give SF_4 and SF_6. Fluorine oxidizes even the other halogens, chlorine, bromine, and iodine, to higher oxidation states. It oxidizes chlorine and bromine to the $+1$, $+3$, and $+5$ oxidation states in ClF, ClF_3, and ClF_5, and BrF, BrF_3, and BrF_5; and iodine to IF_3 and IF_5, and to the $+7$ state in IF_7. Chlorine reacts with phosphorus to give PCl_3 and PCl_5 (see Experiment 20.1), but since it is a weaker oxidizing agent than fluorine, it

	3s	3p	3d			Number of unpaired electrons	Example compounds
P	↑↓	↑ ↑ ↑			Ground state	3	PCl_3, PH_3
	↑	↑ ↑ ↑	↑		Valence state	5	PCl_5, H_3PO_4
S	↑↓	↑↓ ↑ ↑			Ground state	2	SCl_2, H_2S
	↑↓	↑ ↑ ↑	↑		Valence state	4	SF_4, SO_2
	↑	↑ ↑ ↑	↑ ↑		Valence state	6	SF_6, SO_3, H_2SO_4
Cl	↑↓	↑↓ ↑↓ ↑			Ground state	1	ClF, $HOCl$
	↑↓	↑↓ ↑ ↑	↑		Valence state	3	ClF_3, $HClO_2$
	↑↓	↑ ↑ ↑	↑ ↑		Valence state	5	ClF_5, $HClO_3$
	↑	↑ ↑ ↑	↑ ↑ ↑		Valence state	7	$HClO_4$

FIGURE 20.1
Ground State and Valence State Electron Configurations for Phosphorus, Sulfur, and Chlorine.

TABLE 20.1 Classification of Compounds of the Halogens by Oxidation State	
Oxidation State	*Typical Compounds*
−1	NaCl, HCl, HI
0	F_2, Cl_2, Br_2, I_2
+1	HClO, NaClO, ClF
+3	$HClO_2$, $NaClO_2$, ClF_3, BrF_3
+5	$HClO_3$, $NaClO_3$, BrF_5, IF_5
+7	$HClO_4$, $NaClO_4$, H_5IO_6, IF_7

gives chlorides of sulfur only in lower oxidation states, namely, S_2Cl_2 and SCl_2 (Chapter 8). It oxidizes iodine to ICl and ICl_3 but not to the +5 and +7 states (see Experiment 20.2). These compounds of fluorine and chlorine with the other halogens are known as **interhalogen compounds**.

As we shall see in Section 20.3 fluorine is the only halogen that is a sufficiently strong oxidizing agent to oxidize xenon to its higher oxidation states of +2, +4, and +6.

The nonmetal halides are typical covalent substances, consisting of covalent molecules and having relatively low melting points and boiling points. Some of their physical properties are summarized in Table 20.2. Phosphorus pentachloride resembles $AlCl_3$ in that it is ionic in the solid state but covalent in the gas phase (Chapter 10). It consists of covalent PCl_5 molecules in the gas phase but is an ionic crystal composed of PCl_4^+ and PCl_6^- ions in the solid state (Figure 20.2 on p. 906).

Experiment 20.1

Preparation of Phosphorus Pentachloride, PCl_5

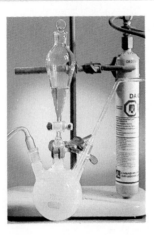

Phosphorus pentachloride may be prepared from the reaction of phosphorus trichloride with chlorine. Here, colorless liquid PCl_3 is being slowly dripped into a flask into which gaseous chlorine is being passed. White (very pale yellow) PCl_5 is formed immediately and may be seen coating the sides of the flask.

Experiment 20.2

Preparation of Iodine Monochloride
and Iodine Trichloride

Chlorine is passed into a flask
containing solid iodine.

Red liquid ICl and a small amount
of orange solid ICl_3 are rapidly
formed.

TABLE 20.2
Halides of Phosphorus, Sulfur, and the Halogens

Phosphorus

PF_3	Colorless gas	bp $= -95\,°C$
PF_5	Colorless gas	bp $= -84\,°C$
PCl_3	Colorless liquid	bp $= 76\,°C$
PCl_5	White solid	Sublimes at 163 °C

Sulfur

SF_4	Colorless gas	bp $= -38\,°C$
SF_6	Colorless gas	bp $= -64\,°C$
S_2Cl_2	Yellow liquid	bp $= 138\,°C$
SCl_2	Red liquid	bp $= 59\,°C$

Chlorine

ClF	Colorless gas	bp $= -100\,°C$
ClF_3	Colorless gas	bp $= 12\,°C$
ClF_5	Colorless gas	bp $= -14\,°C$

Bromine

BrF	Pale brown gas	bp $= 20\,°C$
BrF_3	Yellow liquid	bp $= 126\,°C$
BrF_5	Colorless liquid	bp $= 41\,°C$

Iodine

IF_3	Yellow solid	bp $= 28\,°C$ (decomposes)
IF_5	Colorless liquid	bp $= 100\,°C$
IF_7	Colorless gas	Solid sublimes at 5 °C
ICl	Red solid	mp $= 27\,°C$
ICl_3	Orange solid	Sublimes at 64 °C

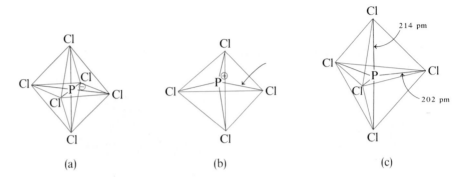

(a)   (b)   (c)

FIGURE 20.2
Structure of Phosphorus Pentachloride.

(a) In the gas phase phosphorus pentachloride consists of PCl_5 molecules. In the solid state it is an ionic crystal composed of (b) tetrahedral PCl_4^+ ions and (c) octahedral PCl_6^- ions.

PROPERTIES

Most nonmetal halides are rather reactive compounds. As we saw in Chapter 12, they almost all react very readily with water to give a hydrogen halide and either an oxoacid or an oxide; or example,

$$PCl_5(s) + 4H_2O(l) \longrightarrow H_3PO_4(aq) + 5HCl(aq)$$

$$SF_4(g) + 2H_2O(l) \longrightarrow SO_2(aq) + 4HF(aq)$$

$$ClF_5(g) + 3H_2O(l) \longrightarrow HClO_3(aq) + 5HF(aq)$$

These reactions between nonmetal halides and water are all Lewis acid–base reactions (Chapter 10). The nonmetal halide is the Lewis acid and water is the Lewis base. The reaction can be thought of as occurring in a number of steps, each of which is a Lewis acid–base reaction. The first step in the reaction of PCl_5 with water involves the donation of an unshared electron pair on the oxygen atom of a water molecule into the valence shell of the phosphorus atom followed by the loss of an HCl molecule:

Repetition of this reaction four more times gives $P(OH)_5$ which loses a water molecule to give H_3PO_4.

SF_6 is exceptional in that it is not attacked by water even at 500 °C. It appears that six fluorine atoms completely fill all the space around the sulfur atom. There is therefore no space for a water molecule to approach close enough to the sulfur atom to donate an electron pair into its valence shell. In its inertness to water SF_6 resembles CCl_4 and CF_4, two other nonmetal halides that do not react with water. Four chlorine or fluorine atoms completely surround the smaller carbon atom, leaving no room for attack by a water molecule.

EXERCISE 20.1

Write balanced equations for the reactions of the following nonmetal halides with water.

(a) PCl_3 (b) BrF (c) BrF_5

Another property of many nonmetal halides is their ability either to lose or to gain halide ions to form cations and anions; for example,

$$PCl_5 \longrightarrow PCl_4^+ + Cl^-$$
$$PCl_5 + Cl^- \longrightarrow PCl_6^-$$

The formation of the ionic solid $PCl_4^+PCl_6^-$ from gaseous PCl_5 molecules can be regarded as a transfer of a chloride ion from one PCl_5 molecule to another. Other examples include

$$BrF_3(l) + KF(s) \longrightarrow K^+BrF_4^-(s)$$
$$ICl_3(s) + AlCl_3(s) \longrightarrow ICl_2^+AlCl_4^-(s)$$
$$ICl_3(s) + KCl(s) \longrightarrow K^+ICl_4^-(s)$$
$$ICl(s) + KCl(s) \longrightarrow K^+ICl_2^-(s)$$

Anions such as BrF_4^-, ICl_4^-, and ICl_2^- are called **polyhalide ions**. Other anions of this type are I_3^-, which is formed by the addition of I^- to I_2, and Br_3^-, which is formed by the similar reaction of Br^- with Br_2. The formation of I_3^- was demonstrated in Experiment 5.5. The above reactions in which polyhalide ions are formed are further examples of Lewis acid–base reactions.

EXERCISE 20.2

In the above reactions of ICl_3 with $AlCl_3$ and KCl identify the Lewis acid and the Lewis base in each case.

MOLECULAR SHAPES

As we have seen in Chapter 9, the shapes of the molecules and ions of the interhalogen compounds are readily predicted using the VSEPR model. Recall that to deduce the shape of a molecule we write the Lewis structure, and use it to classify the species according to the AX_nE_m nomenclature, where m is the number of unshared (lone pair) electron charge clouds and n is the number of attached atoms (ligands), which is equal to the number of single, double, or triple bond charge clouds. The shapes of the interhalogen molecules and ions are based on the trigonal bipyramidal arrangement of five, or the octahedral arrangement of six, bonding and nonbonding charge clouds ($m + n = 5$ or 6). As we have already indicated in Chapter 9, the bonds in AX_5, AX_4E, AX_3E_2, and AX_2E_3 molecules can be approximately described as being formed from trigonal bipyramidal sp^3d hybrid orbitals on the central atom. The bonds in AX_6, AX_5E, and AX_4E_2 molecules can similarly be described as being formed from octahedral sp^3d^2 hybrid orbitals on the central atom (see Table 9.4). The

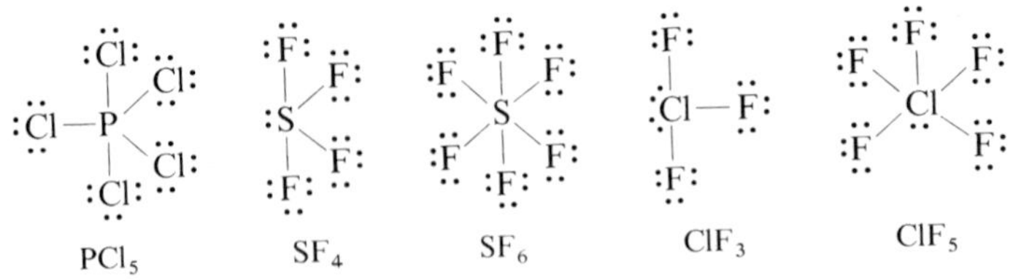

FIGURE 20.3
Lewis Structures of Some
Nonmetal Halides.

Lewis structures of some interhalogen molecules and some other halides of non-metals in their higher valence states are given in Figure 20.3 and their shapes are summarized in Table 20.3 and Figure 20.4. Because unshared pair charge clouds occupy more space in the valence shell than single bond charge clouds the unshared pairs are located in those positions where there is more space available to them. In a trigonal bipyramidal arrangement the equatorial positions are less crowded than the axial positions because an equatorial position has only two close neighbors at 90° whereas an axial position has three close neighbors at 90°. Thus in AX_4E, AX_3E_2, and AX_2E_3 molecules, in which the ligands X are attached by single bonds, the unshared pairs are always located in the equatorial positions of the trigonal bipyramidal arrangement giving disphenoid, T-shape, and linear molecules respectively. AX_5E molecules have a square pyramidal shape based on the octahedral arrangement of six charge clouds and AX_4E_2 molecules have a square planar shape with the two large lone pair charge clouds located as far apart as possible, that is, at 180° rather than at 90° to each other, so that each one has the maximum amount of space available to it.

Because nonbonding electron pair charge clouds are larger than single bond charge clouds, molecular species containing nonbonding electron pairs in the valence shell of the central atom are often slightly distorted from the ideal shapes. Some examples are shown in Figure 20.4. In AX_5E molecules all the bond angles are slightly smaller than the ideal of 90°, and the bonds in the square base are slightly longer than the axial bond. The difference in the lengths of the axial and equatorial bonds that is observed whenever there are five pairs of electrons in the valence shell of the central atom is accentuated in the AX_4E and AX_3E_2 molecules because a nonbonding pair is closer to the axial bonding pairs than to the equatorial pairs; all the bond angles are smaller than the 90°

TABLE 20.3
Shapes of Molecules Based on Valence Shells Containing Five and Six Electron Pairs

Number of Electron Pairs	Number of Unshared Pairs	Type of Molecule	Molecular Shape	Example
5	0	AX_5	Trigonal bipyramid	PF_5, PCl_5
	1	AX_4E	Disphenoid	SF_4
	2	AX_3E_2	T shape	ClF_3
	3	AX_2E_3	Linear	ICl_2^-, XeF_2
6	0	AX_6	Octahedron	SF_6
	1	AX_5E	Square pyramid	BrF_5, IF_5
	2	AX_4E_2	Square planar	ICl_4^-, XeF_4

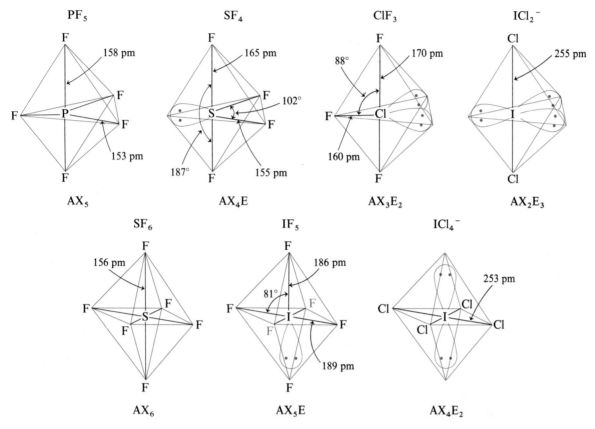

FIGURE 20.4
Examples of AX_5, AX_4E, AX_3E_2, AX_2E_3, AX_6, AX_5E, and AX_4E_2
Molecules.

and 120° angles of the ideal shapes. AX_4E_2 and AX_2E_3 molecules have the lone pairs arranged symmetrically, so they do not distort the predicted geometries.

EXERCISE 20.3

Classify each of the following molecules and ions in terms of the AX_nE_m nomenclature and deduce their shapes.

BrF_3 BrF_4^+ ICl_2^+ I_3^- IF_6^+

20.2
Oxides and Oxoacids of the Halogens

The most important compounds containing the halogens in positive oxidation states are the oxoacids. These are summarized in Table 20.4.

Recall that when there are only two oxoacids of a given element, the *-ous* ending is used to indicate the lower-oxidation-state acid and *-ic* the higher-oxidation-state acid. For elements such as the halogens that form oxoacids in

TABLE 20.4
Oxoacids and Oxoanions of the Halogens

Oxidation	Formula	Name	Formula	Name	Formula	Name
Oxoacids						
+1	$HClO$	Hypochlorous	$HBrO$	Hypobromous	HIO	Hypoiodous
+3	$HClO_2$	Chlorous				
+5	$HClO_3$	Chloric	$HBrO_3$	Bromic	HIO_3	Iodic
+7	$HClO_4$	Perchloric	$HBrO_4$	Perbromic	HIO_4	Periodic
					H_5IO_6	Paraperiodic
Anions						
+1	ClO^-	Hypochlorite	BrO^-	Hypobromite	IO^-	Hypoiodite
+3	ClO_2^-	Chlorite				
+5	ClO_3^-	Chlorate	BrO_3^-	Bromate	IO_3^-	Iodate
+7	ClO_4^-	Perchlorate	BrO_4^-	Perbromate	IO_4^-	Periodate
					IO_6^{5-}	Paraperiodate

as many as four oxidation states, this simple terminology must be modified by the addition of prefixes to the names. The prefix *hypo-* signifies a lower oxidation state and *per-* a higher oxidation state, as Table 20.5 shows. In the IUPAC system the oxidation state is designated by a Roman numeral, and the names of all oxoacids end in *-ic*, as shown in Table 20.5. It would seem logical to use the systematic IUPAC nomenclature for the halogen oxoacids, but the traditional names continue to be widely used, and it will be a long time before the change to the new names is complete. In this book we use the traditional names because they are used much more frequently at present than the IUPAC names.

OXOACIDS OF CHLORINE

The oxoacids of chlorine and their salts are particularly important and have several useful applications. There are oxoacids corresponding to each of the positive oxidation states of chlorine (Table 20.4).

HYPOCHLOROUS ACIDS AND HYPOCHLORITES Hypochlorous acid is a weak acid in water ($K_a = 3.1 \times 10^{-8}$ mol L^{-1}). It exists only in dilute aqueous solution. The pure acid, or even concentrated solutions, cannot be made because the acid decomposes when dilute solutions are concentrated:

$$2HOCl(aq) + 2H_2O(aq) \longrightarrow 2H_3O^+(aq) + 2Cl^-(aq) + O_2(g)$$

TABLE 20.5
Names of Chlorine Oxoacids

Common Name	Formula	Systematic Name
Hypochlorous	$HClO$	Chloric(I) acid
Chlorous	$HClO_2$	Chloric(III) acid
Chloric	$HClO_3$	Chloric(V) acid
Perchloric	$HClO_4$	Chloric(VII) acid

In this reaction hypochlorous acid, in which chlorine is in the $+1$ oxidation state, oxidizes water to oxygen and is itself reduced to Cl^- (oxidation number $= -1$). Even fairly dilute solutions of hypochlorous acid are not completely stable; they decompose slowly—more rapidly in sunlight.

Hypochlorous acid is formed in equilibrium with molecular chlorine in solutions of chlorine in water:

$$Cl_2 + 2H_2O \rightleftharpoons HOCl + H_3O^+ + Cl^-$$

About a third of the chlorine in a $0.1M$ solution is converted to HOCl. We can increase the concentration of HOCl in the solution by removing the Cl^- and thus shifting the equilibrium to the right. Chloride ion can be removed by adding silver oxide, $(Ag^+)_2O^{2-}$. The Ag^+ ion reacts with Cl^- to form insoluble AgCl, which can be removed by filtration, and the O^{2-} neutralizes the H_3O^+. The overall equation for this reaction is

$$2Cl_2(g) + Ag_2O(s) + H_2O(l) \longrightarrow 2AgCl(s) + 2HOCl(aq)$$

Hypochlorites can be made by passing chlorine into a cold solution of a soluble hydroxide:

$$Cl_2 + 2OH^- \longrightarrow Cl^- + ClO^- + H_2O$$

This is essentially the same as the reaction of Cl_2 with water. The equilibrium

$$Cl_2 + 2H_2O \rightleftharpoons Cl^- + HOCl + H_3O^+$$

is shifted to the right by removal of HOCl and H_3O^+ by reaction with the hydroxide ion, which converts them to OCl^- and H_2O, respectively. The reactions of chlorine with water and with OH^- are disproportionation reactions in which chlorine in the zero oxidation state is converted to chlorine in the $+1$ and -1 oxidation states.

Hypochlorite solutions are prepared on a large scale by electrolyzing a *cold* aqueous solution of sodium chloride. The solution is vigorously stirred to mix the chlorine produced at the anode (see Chapter 17),

$$2Cl^- \longrightarrow Cl_2 + 2e^-$$

with the hydroxide ion produced at the cathode,

$$2H_2O + 2e^- \longrightarrow H_2 + 2OH^-$$

so that the reaction

$$Cl_2 + 2OH^- \longrightarrow Cl^- + ClO^- + H_2O$$

takes place.

Hypochlorous acid and hypochlorites are strong oxidizing agents:

$$OCl^-(aq) + 2H^+(aq) + 2e^- \longrightarrow Cl^-(aq) + H_2O(l)$$

Sodium hypochlorite, NaOCl, and calcium hypochlorite, $Ca(OCl)_2$, have many important uses. Calcium hypochlorite is known as bleaching powder, and solutions of sodium hypochlorite are familiar as household bleach (Experiment 20.3). Hypochlorites are used in industry to bleach cotton and linen materials and paper pulp. They kill microorganisms and are therefore used as disinfectants and to purify water.

Experiment 20.3

Properties of Sodium Hypochlorite

Left to right: (a) Tetrachloroethene, C_2Cl_4 is added to a colorless aqueous solution of potassium iodide. The C_2Cl_4 is immiscible with water and sinks to the bottom of the tube, forming a separate layer. (b) A few drops of a dilute aqueous solution of sodium hypochlorite, NaOCl, are added. This oxidizes iodide ion to iodine, which forms a purple solution in the C_2Cl_4 and a brown solution of I_3^- in the aqueous layer. (c) When the tube is shaken, all the iodine is extracted into the C_2Cl_4. (d) When an excess of sodium hypochlorite solution is added, iodine is further oxidized to iodate ion, IO_3^-, and when the tube is shaken, both layers again become colorless.

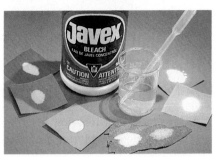

Household bleach is a solution of sodium hypochlorite. Here its ability to bleach different samples of colored paper is shown. The bleaching action results from the oxidation of the colored dye to a colorless compound.

CHLORIC ACID AND CHLORATES   In aqueous solution hypochlorite ion decomposes slowly in two ways. As we have stated, it decomposes to H_3O^+, Cl^-, and O_2, and it also disproportionates to chlorate ion, ClO_3^-, and Cl^-:

$$3ClO^- \longrightarrow 2Cl^- + ClO_3^-$$

The rate of this reaction increases if the solution is warmed. Chlorates can therefore be prepared by passing chlorine into a *hot* hydroxide solution. The overall equation for the reaction is

$$3Cl_2 + 6OH^- \longrightarrow 5Cl^- + ClO_3^- + 3H_2O$$

In the commercial preparation a hot solution of potassium chloride is electrolyzed with vigorous stirring so that the chlorine produced at the anode mixes with the hydroxide produced at the cathode. Potassium chlorate has a solubility of only 3 g in 100 g of water at 0 °C, but potassium chloride has a solubility of 28 g in 100 g. The solubility of $KClO_3$ is further decreased by the common-ion effect (Chapter 15) of the potassium chloride produced in the reaction. Hence when the solution resulting from the electrolysis is cooled, potassium chlorate crystallizes.

Chloric acid, $HClO_3$, is a strong acid in water, but neither the pure substance nor concentrated aqueous solutions can be obtained. If solutions of $HClO_3$ are evaporated, the acid decomposes, sometimes explosively.

The chlorate ion may be described by three equivalent resonance structures. It is an AX_3E molecule and has the expected triangular pyramidal structure (see Figure 20.5).

Chloric acid and chlorates are powerful oxidizing agents. Chlorates form sensitive explosive mixtures with reducing agents such as carbon, sulfur, and many organic compounds, which are converted in exothermic reactions to gaseous products such as CO_2 and SO_2. Potassium chlorate finds extensive use in matches (Box 8.1), fireworks (Box 15.1), and various types of explosives because of its strength as an oxidizing agent. Sodium chlorate destroys microorganisms and bacteria in the soil by oxidation. It is therefore used to kill weeds and other vegetation.

When potassium chlorate is heated it decomposes in two ways:

(1) The reaction

$$2KClO_3(s) \longrightarrow 2KCl(s) + 3O_2(g)$$

is catalyzed by various substances; manganese dioxide is frequently used (see Experiment 20.4).

(2) In the absence of a catalyst, gently heating potassium chlorate gives potassium perchlorate as the main product:

$$4KClO_3(s) \longrightarrow 3KClO_4(s) + KCl(s)$$

This reaction is an example of a disproportionation, in which potassium chlorate may be said to oxidize itself. Although almost all perchlorates are rather soluble in water, the solubility of potassium perchlorate is only 0.75 g in 100 g of water at 0 °C; therefore it is easily separated from the soluble KCl.

PERCHLORIC ACID AND PERCHLORATES Perchlorates are strong oxidizing agents but are somewhat less sensitive and less dangerous than chlorates. Potassium perchlorate and ammonium perchlorate mixed with carbon or various organic compounds are used as explosives and as rocket fuels. The solid booster rockets of the space shuttle use a solid propellant that consists of 70% ammonium perchlorate, 16% aluminum powder, and a polymer known as PBAN (a polybutadiene–acrylic acid–acrylonitrile copolymer), which binds the mixture into a fairly hard material that looks and feels like a hard rubber eraser. This propellant is quite safe to prepare and handle. Perchlorates are used extensively in fireworks and flares (Box 15.1).

The perchlorate ion, $ClO_4{}^-$, may be represented by four equivalent resonance structures. It is an AX_4 species and therefore has a tetrahedral structure (see

Resonance structures

Dimensions

FIGURE 20.5
Structure of the Chlorate
Ion: An AX_3E Molecule.

Experiment 20.4

Thermal Decomposition of Potassium Chlorate

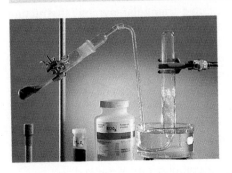

When white potassium chlorate, $KClO_3$, is heated in the presence of black manganese dioxide, MnO_2, which acts as a catalyst, it decomposes to potassium chloride and oxygen. Here we see the oxygen being collected over water.

Figure 20.6). It is isoelectronic with the tetrahedral PO_4^{3-} and SO_4^{2-} ions. The observed bond length (144 pm) is much smaller than the sum of the single bond, covalent radii of chlorine and oxygen, which is 165 pm. This bond length is consistent with the bond order of 1.75 implied by the resonance structures.

Anhydrous perchloric acid, $HClO_4$, can be prepared by distilling a mixture of potassium perchlorate and concentrated sulfuric acid under reduced pressure. It is a somewhat hazardous substance as it is a very powerful oxidizing agent that explodes violently on contact with organic matter such as wood and cloth. Perchloric acid is a strong acid in water and its dilute aqueous solutions are much less reactive than the pure acid and are safe to handle. Aqueous solutions of perchloric acid and perchlorates are potentially good oxidizing agents but the reactions are usually quite slow at ordinary temperatures.

Although all the halogen oxoacids and their anions are good oxidizing agents, their oxidizing ability and, particularly, the rates at which they react depend very much on the conditions. Their oxidizing strengths do not simply increase with increasing number of oxygen atoms.

CHLOROUS ACID AND CHLORITES There is a fourth oxoacid of chlorine, chlorous acid, $HClO_2$, which is known only in aqueous solution, in which it behaves as a weak acid ($K_a = 1.1 \times 10^{-2}$ mol L^{-1}). Its salts are the chlorites. They are good oxidizing agents but are not very important. The chlorite ion is an angular AX_2E_2 molecule. It can be represented by two resonance structures (see Figure 20.7). In the series of anions ClO_2^-, ClO_3^-, ClO_4^-, the bond orders obtained from the resonance structures are 1.50, 1.67, and 1.75, and the bond lengths decrease correspondingly from 157 to 148 to 144 pm.

STRENGTHS OF CHLORINE OXOACIDS The oxoacids of chlorine form an interesting series,

One of 4 equivalent resonance structures

Dimensions

FIGURE 20.6
Structure of the Perchlorate Ion: An AX_4 Molecule.

Acid strength increases

in which the nonbonding pairs are successively replaced by oxygen atoms. This causes a steady increase in the acid strength from hypochlorous acid, which is the weakest, to chlorous acid, which is still a weak acid, and to chloric acid and perchloric acid, which are both strong acids. We have previously seen that sulfuric acid, H_2SO_4, is stronger than sulfurous acid, H_2SO_3, and that nitric acid, HNO_3, is stronger than nitrous acid, HNO_2. The increasing number of oxygen atoms in the series HOCl, HOClO, HOClO₂, and HOClO₃ increases the effective electronegativity of the chlorine atom by withdrawing electrons from the chlorine, thereby increasing the ability of chlorine to withdraw electrons from the OH group (Chapter 8).

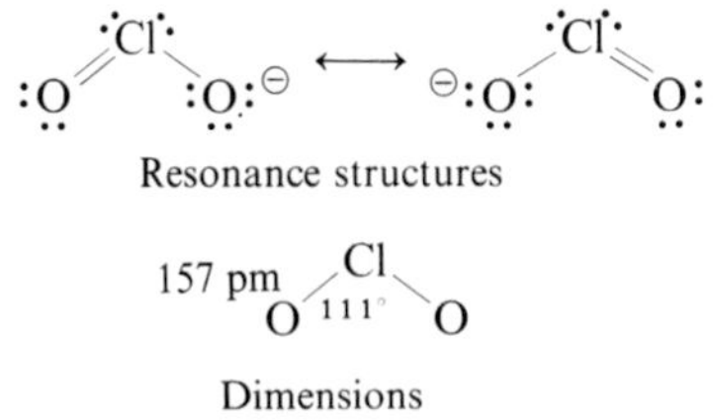

We have seen previously (Chapter 8) that if the formulas of oxoacids are written in the form $XO_m(OH)_n$ the strength of the acid increases with the value of m (see Table 20.6). According to this classification, perchloric acid is a stronger acid than sulfuric acid, nitric acid, or chloric acid. However, all four acids are completely ionized in aqueous solution. They are all strong acids, and we cannot distinguish between their strengths by studying their solutions in water. Measurements on the ionization of these acids in solvents that are less basic than water have shown that perchloric acid is indeed a stronger acid than nitric acid or sulfuric acid.

Resonance structures

157 pm

Dimensions

FIGURE 20.7
Structure of the Chlorite Ion: An AX₂E₂ Molecule.

CHLORINE OXIDES

The oxides of chlorine, unlike those of phosphorus and sulfur, are all somewhat unstable and explosive. They include Cl_2O, Cl_2O_3, ClO_2, Cl_2O_6, and Cl_2O_7. The structures of some of these oxides are shown in Figure 20.8. Of these only chlorine dioxide, ClO_2, is of any importance. It is prepared by reducing $NaClO_3$ with SO_2 in acid solution:

$$2NaClO_3(aq) + SO_2(g) + H_2SO_4(aq) \longrightarrow 2ClO_2(g) + 2NaHSO_4(aq)$$

Chlorine dioxide is a yellow-brown gas that condenses to a red liquid at $-59\,°C$. The ClO_2 molecule, which has 33 electrons, is an odd-electron molecule. Like NO_2, it is a stable free radical, and it shows even less tendency to dimerize than NO_2. It can be represented by the Lewis structures in Figure 20.8. They indicate that the odd electron is delocalized over the chlorine and oxygen atoms. It is an angular AX_2E_2 molecule.

Despite its instability and the fact that it can form explosive mixtures with air, ClO_2 is produced in large quantities. It is used as a bleaching agent for

TABLE 20.6 Strengths of Some Oxoacids $XO_m(OH)_n$				
	$X(OH)_n$	$XO(OH)_n$	$XO_2(OH)_n$	$XO_3(OH)_n$
	ClOH	ClO(OH)	ClO₂(OH)	ClO₃(OH)
	BrOH	PO(OH)₃	SO₂(OH)₂	
	IOH	IO(OH)₅	NO₂(OH)	
	Si(OH)₄	NO(OH)		
	B(OH)₃	HPO(OH)₂		
		CO(OH)₂		
Strength in water K_a (mol L⁻¹)	Very weak $<10^{-7}$	Weak 10^{-2}–10^{-4}	Strong >1	Very strong $\gg 1$

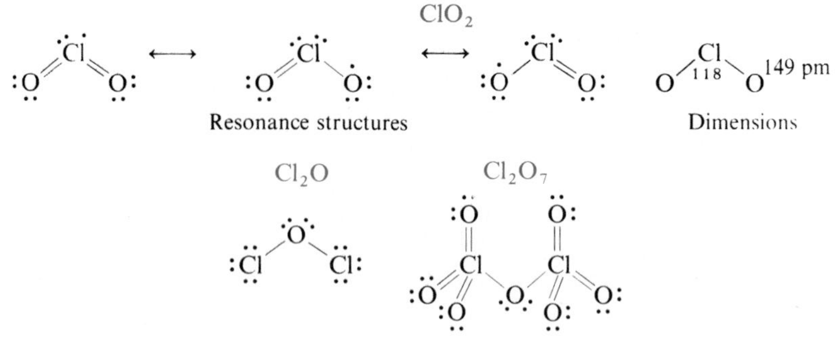

FIGURE 20.8
Structures of Some Oxides
of Chlorine.

ClO_2

Resonance structures   Dimensions

Cl_2O Cl_2O_7

cellulose in the pulp and paper industry and in the manufacture of white flour. It is also used for water purification.

The chlorine oxides are unusual compounds that are still not well understood. No one has yet explained satisfactorily the fact that both ClO_2 and NO_2 are very unreactive for free radicals.

OXOACIDS OF BROMINE AND IODINE

The oxoacids of bromine are $HOBr$, $HBrO_3$, and $HBrO_4$. They resemble the corresponding chlorine oxoacids. Bromous acid, $HBrO_2$, is not known.

The oxoacids of iodine are hypoiodous acid, HOI, iodic acid, HIO_3, and the periodic acids, HIO_4 and H_5IO_6. Sodium iodate occurs naturally in small amounts in sodium nitrate, deposits of which are found in certain desert regions of Chile. Iodine is obtained from sodium iodate by reduction with sulfur dioxide or with sodium hydrogen sulfite, $NaHSO_3$:

$$2IO_3^-(aq) + 5SO_2(g) + 4H_2O(l) \longrightarrow I_2(s) + 8H^+(aq) + 5SO_4^{2-}(aq)$$

Experiment 20.5 shows the reduction of potassium iodate to iodine by sulfur dioxide.

Iodine is insoluble in water but dissolves in a solution of hydroxide ion with the formation of hypoiodite ion, IO^-. This reaction is quite similar to the reaction of Cl_2 with OH^-:

$$I_2(s) + 2OH^-(aq) \longrightarrow IO^-(aq) + I^-(aq) + H_2O(l)$$

However, hypoiodite ion is considerably less stable than hypochlorite ion, and it disproportionates rather rapidly:

$$3IO^-(aq) \longrightarrow IO_3^-(aq) + 2I^-(aq)$$

This reaction is further accelerated by warming the solution. Thus when iodine is dissolved in *warm* hydroxide solution, iodate and iodide are formed:

$$3I_2(s) + 6OH^-(aq) \longrightarrow 5I^-(aq) + IO_3^-(aq) + 3H_2O(l)$$

We have seen that chlorine reacts similarly to give chlorate and chloride in hot NaOH solution.

Whereas chloric acid is unstable and cannot be isolated in a pure form, iodic acid is a stable, white crystalline solid. It can be made by oxidizing iodine with nitric acid (see Experiment 20.6):

$$I_2(s) + 10HNO_3(aq) \longrightarrow 2HIO_3(aq) + 10NO_2(g) + 4H_2O(l)$$

Experiment 20.5

Reduction of Potassium Iodate

Sulfur dioxide is passed into
a colorless solution of
potassium iodate

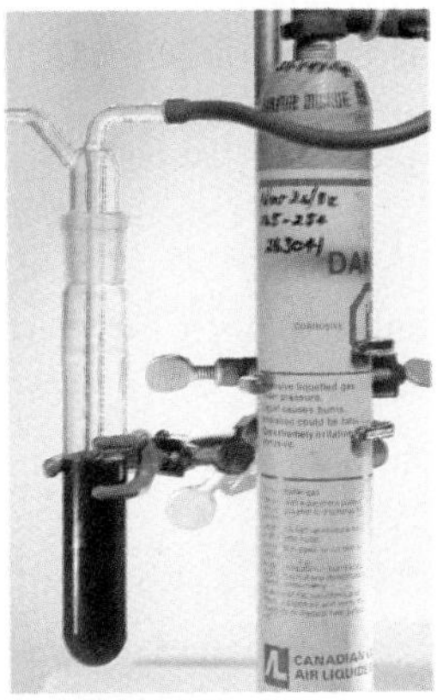

The potassium iodate is rapidly reduced to black
solid iodine.

Experiment 20.6

Preparation of Iodic Acid

Iodine can be oxidized to
iodic acid, HIO_3, by
concentrated nitric acid.
Here, pale yellow
concentrated nitric acid has
been added to solid iodine
in the bottom of the flask.

On warming, the flask is
filled with brown NO_2
vapor resulting from the
reduction of nitric acid.

After some time, the iodic
acid formed can be seen as a
white crystalline solid in the
bottom of the flask and
coating the sides.

Like the chlorate and bromate ions, iodic acid and the iodate ion have a triangular pyramidal AX_3E geometry (see Figure 20.9).

Periodic acid and the periodates differ from perchloric acid and perbromic acid in that they exist in two forms (see Figure 20.10). Periodic acid is a solid that, when crystallized from water, has the formula H_5IO_6. This form is known as *paraperiodic acid*. It is a weak acid (see Table 20.6). If it is heated to 80 °C under vacuum, it loses water to give HIO_4, which is known as *metaperiodic acid*:

$$H_5IO_6 \longrightarrow HIO_4 + 2H_2O$$

Salts of both forms of the acid are known, such as $NaIO_4$, $Na_2H_3IO_6$, and Ag_5IO_6.

EXERCISE 20.4

The standard electrode potentials for the reduction of ClO^-, ClO_2^-, ClO_3^-, and ClO_4^-, respectively, to Cl^- ion in basic aqueous solution are 0.89, 0.78, 0.63, and 0.56 V, respectively. Arrange these oxochloro anions in order of increasing strength as oxidizing agents in basic solution, and write the balanced equations for the half-reactions.

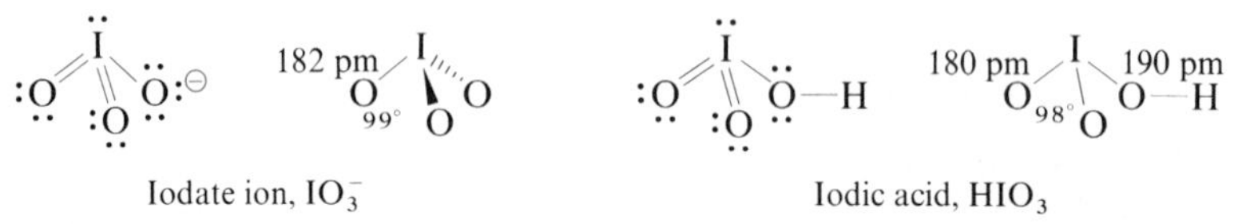

FIGURE 20.9
Structures of Iodic Acid and the Iodate Ion.

Iodate ion, IO_3^- Iodic acid, HIO_3

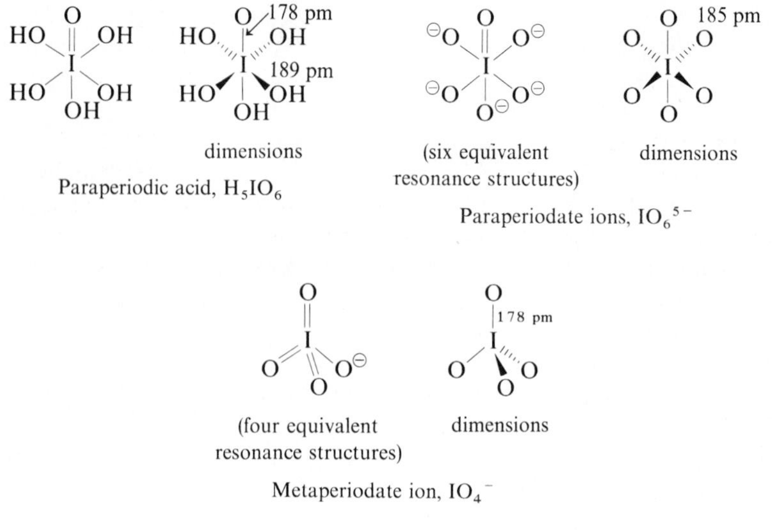

FIGURE 20.10
Structures of Periodic Acid and the Periodate Ion.

dimensions

Paraperiodic acid, H_5IO_6

(six equivalent resonance structures)   dimensions

Paraperiodate ions, IO_6^{5-}

(four equivalent resonance structures)   dimensions

Metaperiodate ion, IO_4^-

EXERCISE 20.5

Write balanced equations for each of the following reactions.

(a) The formation of perbromate ion and fluoride ion from the reaction of bromate ion and fluorine gas in basic aqueous solution

(b) The reaction of calcium hypochlorite with potassium iodide in HCl(aq) to give I_3^-(aq)

(c) The reaction of potassium chlorate with sulfuric acid to give potassium hydrogen sulfate, chlorine dioxide, and oxygen

20.3
The Noble Gases

OCCURRENCE AND USES

Helium was the first noble gas to be discovered. It is formed in the sun by the fusion of hydrogen nuclei (protons); this reaction provides the sun with its energy (Chapter 25). Indeed, helium was identified on the sun before it was discovered on the earth. Certain lines in the spectrum of the sun were attributed to an element not known on earth; hence it was named helium after the Greek word *helios*, for "sun". The discovery of the other noble gases is described in Box 20.1 on p. 920.

The noble gases occur in the atmosphere as monatomic gases. Argon constitutes approximately 1% of the atmosphere, neon only about 0.002%, and the other noble gases are present in still smaller amounts. Helium, formed from the α particles emitted in the disintegration of uranium and other radioactive elements (see Chapter 25), is found trapped in uranium minerals, from which it can be liberated by heating. Some of the helium formed in this way has gradually diffused into pockets of natural gas, which often contains as much as 1% helium. Although helium is rather rare on the earth, it is the second most abundant element in the sun and in the solar system as a whole. Like hydrogen, it is too light to have remained in appreciable amounts in the earth's atmosphere (Chapter 3).

The fact that the noble gases occur in the uncombined monatomic state in the atmosphere is a clear indication of their lack of reactivity. All their important uses depend on this lack of reactivity.

Helium is the second lightest of all the elements and is therefore used for filling balloons and airships in place of the much more dangerous hydrogen (Box 3.4). It is also used, mixed with oxygen, to provide a gas for divers to breathe. If they breathe ordinary air, nitrogen dissolves in the blood under the high pressures at which divers work. When a diver comes to the surface, the dissolved nitrogen is released into the blood in the form of bubbles, a painful and dangerous condition. A helium–oxygen mixture can be breathed safely because helium is much less soluble in the blood than is nitrogen.

The most important use for neon is in neon signs. When an electric current is passed through neon at low pressure, excited neon atoms emit the characteristic atomic spectrum of neon, which has strong lines in the red part of the spectrum (Chapter 7).

BOX 20.1

DISCOVERY OF THE NOBLE GASES

The story of the discovery of the noble gases illustrates the importance of paying attention to minor discrepancies in experimental results. For over a hundred years it had been thought that air consisted solely of nitrogen and oxygen, together with small, variable amounts of water vapor and carbon dioxide. Then in 1785 English scientist Henry Cavendish (1731–1810) made some new investigations of the atmosphere. He added oxygen to the air and passed an electric spark through the mixture. This caused the formation of NO_2:

$$N_2 + O_2 \longrightarrow 2NO \quad \text{and} \quad 2NO + O_2 \longrightarrow 2NO_2$$

Cavendish continued the sparking until no further change in the volume occurred, that is, until all the nitrogen had been removed. He then removed the NO_2 by reacting it with a solution of KOH, and he removed the excess of oxygen by allowing it to react with a solution of potassium pentasulfide:

$$2K_2S_5 + 3O_2 \longrightarrow 2K_2S_2O_3 + 6S$$

After this treatment a small bubble of gas remained; it had a volume of not more than $\frac{1}{120}$ of the original volume of the air. Cavendish concluded that if there were another component of the air, other than nitrogen and oxygen, it constituted less than 1% by volume. But most chemists apparently assumed that the remaining bubble was simply nitrogen that had not been removed, perhaps because the sparking had not been carried on for a

sufficient time, and Cavendish's experiment was forgotten.

More than a hundred years later Lord Rayleigh (1842–1919), an English physicist, was making careful measurements of the densities of several gases, including nitrogen. He prepared nitrogen from ammonia by passing a mixture of oxygen with excess ammonia over red-hot copper; the copper catalyzes the oxidation of ammonia to nitrogen:

$$4NH_3 + 3O_2 \longrightarrow 6H_2O + 2N_2$$

After it was dried, the nitrogen was found to have a density of 1.2505 g L^{-1} at 0 °C and 1 atm. Rayleigh also made nitrogen by passing dry air, from which the CO_2 had been removed, over red-hot copper, which removed the oxygen by combining with it to form copper oxide. The remaining gas, after it was dried, was presumed to be nitrogen. But it was found to have a density of 1.2672 g L^{-1} at 0 °C and 1 atm. In other words, nitrogen prepared from air had a density 0.5% greater than that of nitrogen prepared from ammonia. Rayleigh was convinced that this difference was greater than the experimental error in his measurements, and he suggested several tentative explanations. For example, he proposed that atmospheric nitrogen might contain N_3 molecules analogous to O_3 molecules.

But it was not until William Ramsay (1852–1916), professor of chemistry at University College, London, became interested in the problem, that the true reason for these different densities was found. He believed that

---

Argon, by far the most abundant of the noble gases, has some important applications. It is used to increase the useful life of incandescent light bulbs because it reduces the rate at which the filament sublimes, hence permitting the filament to be heated to a higher temperature and thus to produce a whiter light. Argon is extensively used in industry to provide a chemically inert atmosphere for welding and in the manufacture of metals and alloys.

No significant use has been found for krypton and xenon. Somewhat surprisingly, xenon is a good anesthetic, but it is too expensive for general use. The radioactive gas radon is used in the treatment of cancer. Radon is sealed into a small tube and placed close to the cancerous tissues to be destroyed.

COMPOUNDS OF THE NOBLE GASES

After the discovery of the noble gases, many attempts were made to prepare compounds of these elements. But none was successful, and the elements became known as the *inert gases*. In fact, no compound of any of these elements was made until 1962, when Neil Bartlett prepared the first compound of xenon (see

William Ramsay.

ably higher density than nitrogen. Ramsay tried to cause this new element to react with many other substances, but he had no success. He therefore called the element *argon*, after the Greek word for "lazy".

Ramsay also identified the gas given off by some radioactive minerals as helium, whose atomic spectrum had previously been observed in the light from the sun. Helium, like argon, proved to be inert. Because there was no place for these two elements in the periodic table which, as originally formulated by Mendeleev, had only seven groups, Ramsay made the bold suggestion that there must be a whole family of inert monatomic gases like argon and helium. He considered it likely that these gases were also present in the atmosphere, so he set out with his assistant, William Travers, to find them. They prepared large amounts of liquid air and then fractionally distilled it. In this way they separated argon and the new element krypton in May 1898, followed by neon in June, and by xenon in July of the same year. Because they appeared to be inert and therefore had a zero valence he placed these new elements in group 0 on the left side of the table preceding the alkali metals. Today we usually consider them to be group 8 following the halogens.

Both Cavendish's and Rayleigh's experiments illustrate the importance of knowing the accuracy of measurements, so that discrepancies greater than the possible experimental error can be recognized. Rayleigh's attention to a small discrepancy that was greater than his experimental error led to the discovery of the whole family of noble gases.

nitrogen prepared from the atmosphere might contain a small amount of another, unidentified gas, as suggested by Cavendish's experiment. He therefore repeated Cavendish's experiment, using several different methods to remove both the oxygen and the nitrogen from the air. Like Cavendish, he found that approximately 1% of the air remained after both the oxygen and the nitrogen had been removed. By examining the atomic spectrum of the residual gas, Ramsay then showed that it was not nitrogen but a new element and that it had a consider-

Box. 20.2 on p. 922). Since that time more compounds of xenon have been made, as well as a few compounds of krypton and radon, and we can no longer regard these elements as truly inert. They are now called the *noble gases*.

Although before 1962 it was generally accepted that the inert gases were incapable of forming compounds, a few chemists continued to believe that synthesizing compounds of the noble gases should be possible. In 1933 Linus Pauling (Box 20.3 on p. 924) went so far as to predict the formulas of several compounds of the noble gases that he felt should exist.

A consideration of the periodic table shows us the basis of Pauling's predictions. We saw in Chapter 4 that if we consider the normal valences of the nonmetals, we conclude that the noble gases would have a valence of zero (see Table 20.7 on p. 922). A valence of zero is consistent with their inertness. But the nonmetals in the third and subsequent periods also have higher valences, because they can expand their valence shells beyond 8 electrons; there are d orbitals in these shells that can be used in compound formation. For example, in some fluorides of the nonmetals the central atom has a valence shell of 12 electrons and in others a valence shell of 10 electrons (see Tables 20.8 and 20.9). Comparison of the formulas of these compounds leads to the prediction that

BOX 20.2

DISCOVERY OF NOBLE GAS COMPOUNDS

The first compound of the noble gases was prepared in 1962 at the University of British Columbia by Neil Bartlett, who was born in England in 1932. At that time Bartlett was studying the reactions of platinum hexafluoride, PtF_6, which is a volatile red solid. He found that the vapor of PtF_6 reacts rapidly with oxygen at room temperature to form the yellow ionic solid $O_2^+PtF_6^-$. This surprising reaction implies that PtF_6 is a very strong oxidizing agent that can remove an electron from an oxygen molecule to give the ion O_2^+. Bartlett argued that if PtF_6 could remove an electron from O_2, it should be able to remove an electron from Xe, since the ionization energy of xenon (1.167 MJ mol^{-1}) is slightly lower than the ionization energy of O_2 (1.176 MJ mol^{-1}). He therefore predicted that xenon and PtF_6 would react to form $Xe^+PtF_6^-$:

$$Xe + PtF_6 \longrightarrow Xe^+PtF_6^-$$

Bartlett carried out the reaction and found that xenon does indeed react with the red vapor of PtF_6 to give a yellow solid, which Bartlett believed to be $Xe^+PtF_6^-$.

Bartlett's discovery caused great interest and excitement. His reaction was immediately repeated by others, and other reactions of xenon were tried. In particular, xenon was found to react directly with fluorine under pressure to give XeF_2, XeF_4, and XeF_6.

It is interesting to note that William Ramsay (Box 20.1) sent a sample of argon to the French chemist Henri Moissan (1852–1907), who had recently discovered

Neil Bartlett

flourine, to see whether it would react with fluorine. Moissan did not succeed. Unfortunately, he was unable to try the reaction between fluorine and xenon because Ramsay had prepared such a small amount of xenon that he was unable to send a sample to Moissan. Had Moissan been able to attempt this reaction, the compounds of

TABLE 20.7
Predicted Valences of Nonmetals Based on a Valence Shell of Eight Electrons

Group	4	5	6	7	8
Valence	4	3	2	1	0
	CF_4	NF_3	OF_2	FF	Ne
	SiF_4	PF_3	SF_2	ClF	Ar

TABLE 20.8
Nonmetal Fluorides with a Valence Shell of Twelve Electrons

Known Compounds		Predicted by Extrapolation
Group 6	Group 7	Group 8
SF_6	ClF_5	(ArF_4), still unknown
SeF_6	BrF_5	(KrF_4), still unknown
TeF_6	IF_5	XeF_4, now known

xenon might have been discovered 60 years earlier. Such a discovery might have greatly influenced the development of the theory of chemical bonding. It was only because the noble gases were found by Ramsay and others to be inert that Lewis was led to formulate his octet rule in 1916. Had Moissan been able to make XeF_4, for example, the octet rule might never have been proposed. The success of this rule reinforced the idea that the noble gases were unreactive and, indeed, gave some apparent theoretical foundation to the concept of the inertness of the noble gases. It became generally accepted that the noble gases did not form compounds because they had a stable octet of electrons, and chemists began to forget that Lewis was only led to formulate his rule because the noble gases had been found by experiment to be unreactive. Only a very few chemists, such as Linus Pauling, continued to think about the possible existence of noble gas compounds.

Bartlett not only made a logical and apparently simple, but nevertheless brilliant, deduction from his experiment with oxygen and PtF_6, but he also had sufficient confidence in the correctness of his argument to question the orthodox view that the noble gases did not form compounds.

As a postscript to this story, it is interesting to point out that the true composition of the xenon compound prepared by Bartlett has never been established with certainty. Probably it was not the simple compound $Xe^+PtF_6^-$ but rather a complex mixture containing the compounds $XeF^+ PtF_6^-$ and $XeF^+ Pt_2F_{11}^-$. Despite the apparent logic of Bartlett's argument that PtF_6 should

SCIENTIFIC AMERICAN

NOBLE-GAS CHEMISTRY   SIXTY CENTS

May 1964

Bartlett's apparatus for the reaction between xenon and platinum hexafluoride. The flask on the left contains PtF_6 vapor, the flask on the right contains the yellow solid product "XePtF_6."

be able to oxidize Xe to Xe^+, there is no firm evidence that Xe^+ exists as a stable species. Because Xe^+ is a free radical, we expect it to be very reactive, and if it is formed in the reaction with PtF_6, it probably combines immediately with a fluorine atom to give XeF^+. Nevertheless, Bartlett's brilliantly conceived experiment opened a new field of chemistry—the chemistry of the noble gases.

TABLE 20.9
Nonmetal Fluorides with a Valence Shell of Ten Electrons

Known Compounds			Predicted by Extrapolation
Group 5	Group 6	Group 7	Group 8
PF_5	SF_4	ClF_3	(ArF_2), still unknown
AsF_5	SeF_4	BrF_3	KrF_2, now known
SbF_5	TeF_4	IF_3	XeF_2, now known

argon, krypton, and xenon should form tetrafluorides and difluorides. Of these possible compounds only XeF_4, XeF_2, and KrF_2 have so far been prepared.

Similarly, we might predict the formation of the trioxides and tetraoxides of the noble gases by extrapolation from the known formulas of the isoelectronic anions of the preceding nonmetals (see Table 20.10). Of these possible compounds only XeO_3 and XeO_4 have so far been prepared.

Since the valence shells of heliums and neon are completely filled with two and eight electrons respectively, we do not expect to find any compounds of

BOX 20.3

LINUS PAULING

Linus Pauling was born in Portland, Oregon, in 1901. He graduated from Oregon State College in 1922 and obtained his Ph.D. in chemistry in 1925 from the California Institute of Technology. After a period of study

in Europe, he became a professor at the California Institute of Technology in 1927, and he remained there for the rest of his academic career.

Pauling did much to develop our understanding of the chemical bond. Among the important concepts that he introduced were electronegativity and resonance structures. He was one of a very small number of chemists who seriously considered the possibility that the noble gases might form compounds. In 1939 he published his ideas on chemical bonding in a book entitled *The Nature of the Chemical Bond*. This book proved to be one of the most influential chemistry books of the present century. In 1954 Pauling was awarded the Nobel Prize in chemistry for his work on molecular structure.

In the 1950s, Pauling turned to studying the structures of biopolymers (Chapter 24). He was one of the first to suggest that protein molecules have a helical shape. Pauling's claim that large doses of vitamin C are effective in preventing the common cold has attracted attention since 1970.

After World War II, Pauling became a passionate supporter of nuclear disarmament. In 1962 he was awarded the Nobel Peace Prize, thus becoming the second person in history to win two Noble Prizes. He is still actively pursuing his research in chemistry today (1988).

TABLE 20.10
Isoelectronic Oxoanions and Oxides

Known Anions			Predicted by Extrapolation
Group 5	Group 6	Group 7	Group 8
—	SO_3^{2-}	ClO_3^-	(ArO_3), still unknown
AsO_3^{3-}	SeO_3^{2-}	BrO_3^-	(KrO_3), still unknown
SbO_3^{3-}	TeO_3^{2-}	IO_3^-	XeO_3, now known
PO_4^{3-}	SO_4^{2-}	ClO_4^-	(ArO_4), still unknown
AsO_4^{3-}	SeO_4^{2-}	BrO_4^-	(KrO_4), still unknown
—	—	IO_4^-	XeO_4, now known

these two elements. However, argon, krypton, xenon, and radon can all have more than eight electrons in their valence shells, and thus they could, in principle, form compounds. No compounds of argon have so far been made, and although there appears to be at least one compound of radon, the chemistry of this element is very difficult to study because of its radioactivity. Our knowledge of the chemistry of the noble gases is therefore confined to krypton and xenon.

The noble gases in their ground states do not have any unpaired electrons, that is, electrons in singly occupied orbitals. Consequently, they can form compounds only via a valence state in which one or more s or p electrons is promoted to an unoccupied d orbital, as illustrated in Figure 20.11 on p. 926.

Only the most electronegative elements—in particular, fluorine and oxygen—exert sufficient attraction on the valence-shell electrons to move some of them into the d orbitals, where they are on average further from the nucleus than in the s and p orbitals. Except for one compound that contains Xe—N bonds, all the known compounds of krypton and xenon have Kr—F, Xe—F, and Xe—O bonds.

FLUORIDES When xenon and fluorine are heated together, XeF_2 and XeF_4 are formed; the relative amounts of these compounds depend on whether xenon or fluorine is in excess. If xenon is heated with an excess of fluorine under pressure, XeF_6 is obtained:

$$Xe(g) + F_2(g) \longrightarrow XeF_2(s)$$
$$Xe(g) + 2F_2(g) \longrightarrow XeF_4(s)$$
$$Xe(g) + 3F_2(g) \longrightarrow XeF_6(s)$$

Xenon difluoride can also be prepared more simply by exposing a glass bulb containing equal amounts of fluorine and xenon gases to strong sunlight or ultraviolet light. Over a few hours, colorless crystals of xenon difluoride form on the sides of the bulb (Figure 20.12 on p. 926).

The fluorides XeF_2, XeF_4, and XeF_6 are all colorless, crystalline and stable but very reactive, solids. They are all strong oxidizing agents. For example, a solution of XeF_2 in water rapidly oxidizes HCl to Cl_2 and OH^- to O_2:

$$2HCl(aq) + XeF_2(aq) \longrightarrow Xe(g) + Cl_2(g) + 2HF(aq)$$
$$4OH^-(aq) + 2XeF_2(aq) \longrightarrow 2Xe(s) + O_2(g) + 4F^-(aq) + 2H_2O(l)$$

In each case the XeF_2 is reduced to xenon. Only one fluoride of krypton, KrF_2, is known.

OXIDES The oxides of xenon, XeO_3 and XeO_4, cannot be prepared by the direct reaction of xenon with oxygen. The trioxide, which is a white explosive solid, is obtained by the reaction of either XeF_4 or XeF_6 with water. Xenon hexafluoride reacts very rapidly with water to form xenon trioxide:

$$XeF_6(s) + 3H_2O(l) \longrightarrow XeO_3(s) + 6HF(aq)$$

This is a typical reaction of a nonmetal halide with water to give a hydrogen halide and a nonmetal oxide or oxoacid. For example, it is analogous to the reaction of PCl_5 with water:

$$PCl_5(s) + 4H_2O(l) \longrightarrow H_3PO_4(aq) + 5HCl(aq)$$

Hydrolysis of XeF_4 does not yield XeO_2 but rather XeO_3 and Xe by a disproportionation reaction:

$$6XeF_4(s) + 12H_2O(l) \longrightarrow 2XeO_3(s) + 4Xe(g) + 3O_2(g) + 24HF(aq)$$

Xenon tetraoxide is obtained from the reaction of barium perxenate,

FIGURE 20.11
Ground State and Valence State Electron Configurations for Xenon.

5s	5p	5d		Number of unpaired electrons	Typical compounds
[↑↓]	[↑↓][↑↓][↑↓]	[ ][ ][ ][ ][ ]	Ground state	0	—
[↑↓]	[↑↓][↑↓][↑]	[↑][ ][ ][ ][ ]		2	XeF_2
[↑↓]	[↑↓][↑][↑]	[↑][↑][ ][ ][ ]		4	XeF_4
[↑↓]	[↑][↑][↑]	[↑][↑][↑][ ][ ]	Valence states	6	XeF_6, XeO_3
[↑]	[↑][↑][↑]	[↑][↑][↑][↑][ ]		8	XeO_4

Ba_2XeO_6, with concentrated sulfuric acid (see Exercise 20.7). It is a highly unstable and explosive gas.

Xenon trioxide behaves as a weak acid in aqueous solution and its reaction with alkali metal hydroxides gives hydrogen xenates of formula $MHXeO_4$, where M is potassium, rubidium, or cesium. Solutions of hydrogen xenates are unstable and slowly disproportionate in basic solution to give xenon and perxenate ion:

$$2HXeO_4^-(aq) + 2OH^-(aq) \longrightarrow XeO_6^{4-}(aq) + Xe(g) + O_2(g) + 2H_2O(l)$$

Aqueous perxenate ion is also formed when a dilute solution of XeO_3 is oxidized with ozone under basic conditions:

$$3XeO_3(aq) + O_3(g) + 12OH^-(aq) \longrightarrow 3XeO_6^{4-}(aq) + 6H_2O(l)$$

Insoluble salts such as Ba_2XeO_6 may be precipitated from these solutions. Perxenate ion is a very strong oxidizing agent with a standard reduction potential of +2.1 V. Thus, for example, $Mn^{2+}(aq)$ is oxidized to $MnO_4^-(aq)$ by perxenate ion in aqueous solution.

FIGURE 20.12
Preparation of Xenon Difluoride, XeF_2.

Left: When a mixture of equal volumes of xenon and fluorine is sealed in a glass bulb and placed in strong sunlight or under an ultraviolet lamp, xenon difluoride, XeF_2, is formed as colorless crystals which coat the sides of the flask. Right: A close-up view of the XeF_2 crystals.

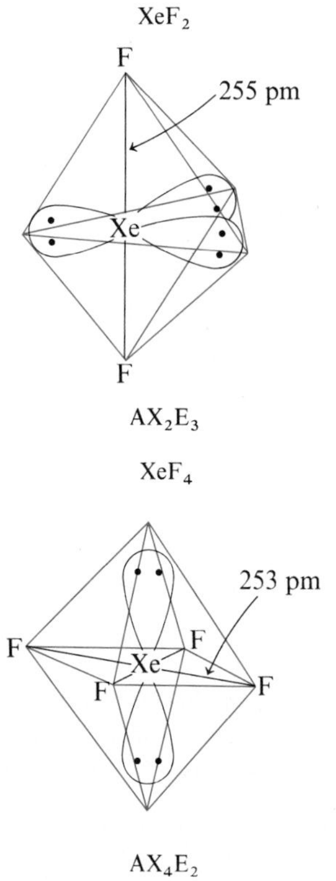

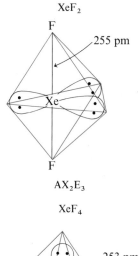

XeF₂

255 pm

Xe

F

AX₂E₃

XeF₄

253 pm

F   F
  Xe
F   F

AX₄E₂

FIGURE 20.13
Structure of XeF₂, an AX₂E₃
Molecule, and XeF₄, an
AX₄E₂ Molecule.

STRUCTURES OF NOBLE GAS COMPOUNDS The structures of the noble gas compounds provide some interesting examples of the application of the VSEPR model. Xenon difluoride is a linear AX_2E_3 molecule, while XeF_4 is a square planar AX_4E_2 molecule with the two nonbonding electron pairs in *trans* positions (see Figure 20.13). Xenon hexafluoride, XeF_6, has seven electron pairs in the valence shell of xenon and is an AX_6E molecule, a type we have not yet discussed. Because of the presence of the nonbonding pair, the arrangement of the six bonding pairs would not be expected to be octahedral. It may be described as a distorted octahedron. The presence of the unshared electron pair forces the fluorine atoms surrounding the unshared pair further apart than in a regular octahedron (see Figure 20.14).

The two oxides XeO_3 and XeO_4 are isoelectronic with iodate and metaperiodate and have the same pyramidal AX_3E and tetrahedral AX_4 structures, as Figure 20.15 shows.

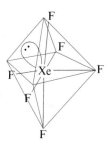

FIGURE 20.14
Structure of XeF₆: An AX₆E
Distorted Octahedral
Molecule.

FIGURE 20.14
Structure of XeF₆: An AX₆E
Distorted Octahedral
Molecule.

EXERCISE 20.6

What are the Lewis structures of the hydrogen xenate and the perxenate ions?

EXERCISE 20.7

Write the balanced equations for (a) the oxidation of $Mn^{2+}(aq)$ by $XeO_6^{4-}(aq)$ in acidic solution, to give $MnO_4^-(aq)$ and $HXeO_4^-(aq)$, and (b) the reaction of concentrated sulfuric acid with Ba_2XeO_6 to give XeO_4.

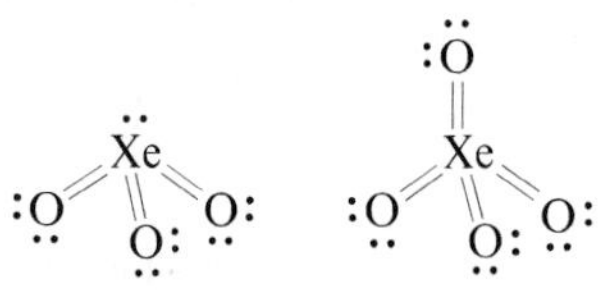

:Ö:
  Xe   Xe
:O   O: :O   O:
  Ö: Ö:

FIGURE 20.15
Structure of XeO₃, an AX₃E
Molecule, and XeO₄, an AX₄
Molecule.

IMPORTANT TERMS

An **interhalogen compound** is a compound formed between two different halogens. Examples are ClF and BrF_5.

PROBLEMS *

Molecular Geometry and Bonding

1. Draw orbital box diagrams for each of the following:

   (a) The $+1$, $+3$, $+5$, and $+7$ oxidation states of iodine

   (b) The $+2$, $+4$, and $+6$ oxidation states of xenon

2. (a) On the basis of the VSEPR model, what are expected shapes of AX_5 and AX_6 molecules?

   (b) Give two examples each of AX_5 molecules and AX_6 molecules

   (c) What are the expected shapes of each of the following?

   SeF_4 BrF_3 XeF_4 ICl_2^-

3. Compare the structure of phosphorus pentachloride in the gaseous and solid phases.

4. (a) Why is it possible for the elements of the third and subsequent periods to form compounds in which they have more than eight electrons in their valence shells?

   (b) Describe the shapes of the molecules formed between (i) phosphorus and fluorine, and (ii) sulfur and fluorine.

5. Molar mass determination shows that the molecular formula of iodine trichloride is I_2Cl_6, rather than ICl_3. By analogy with the structures of $AlCl_3$ and Al_2Cl_6, predict the structure of I_2Cl_6.

6. Chlorine, bromine, and iodine form covalent oxofluorides of general molecular formulas AO_2F and AO_3F. Draw the Lewis diagrams of these molecules and predict their shapes.

7. The covalent fluorides XeF_4 and XeF_6 react with SbF_5 to give the ionic compounds $XeF_3^+SbF_6^-$ and $XeF_5^+SbF_6^-$. Would these be classified as acid–base reactions? Explain. Draw a Lewis structure for each of these ions and use the VSEPR model to predict their shapes. What are the expected approximate bond angles?

8. Which of the following molecules is expected to have a dipole moment?

   (a) Phosphorus pentafluoride

   (b) Selenium tetrafluoride

* Answers to problems numbered in blue appear at the end of the text.

   (c) Chlorine trifluoride

   (d) Bromine pentafluoride

   (e) Xenon difluoride

   (f) Xenon tetrafluoride

9. Use the VSEPR model to predict the shape of each of the following ions.

   (a) SiF_6^{2-} (b) SiF_5^- (c) PF_6^-

   (d) SeF_5^- (e) BrF_4^- (f) IF_4^-

10. Use the VSEPR model to predict the shape of each of the following oxofluorides of xenon.

   (a) $XeOF_4$ (b) XeO_2F_2 (c) XeO_3F_2

What is the oxidation state of xenon in each of these compounds?

11. What compounds are known to result from the reaction of each of the halogens chlorine, bromine, and iodine with fluorine? Deduce the molecular geometry of each. Which of these molecules is expected to have a dipole moment?

12. (a) In AX_5 molecules such as PCl_5, PF_5, or AsF_5, two of the bonds are described as axial and the other three are described as equatorial. Explain why the axial bonds in such molecules are invariably found to be slightly longer than the equatorial bonds.

   (b) In AX_4E, AX_3E_2, and AX_2E_3 molecules, explain why the unshared electron pairs on the central atom A are invariably found in equatorial positions, rather than in axial positions.

   (c) In what way do the lone pairs in equatorial positions in AX_4E, AX_3E_2, and AX_2E_3 molecules affect the ideal geometry and/or bond lengths?

The Halogens

13. Arrange the following oxoacids in order of increasing acid strength, and explain your choice of order.

   (a) Hypochlorous acid   (b) Phosphoric acid

   (c) Perchloric acid   (d) Sulfuric acid

14. Write balanced equations for each of the following reactions.

(a) The oxidation of a solution of chromium(III) sulfate, $Cr_2(SO_4)_3(aq)$, with aqueous sodium hypochlorite to give sodium chromate, $Na_2CrO_4(aq)$, and $NaCl(aq)$ in basic solution

(b) The oxidation of potassium iodide by sodium chlorate to give iodine and sodium chloride in basic aqueous solution

(c) The oxidation of iodide by iodate ion in acidic aqueous solution

15. Write a balanced equation for the oxidation of bromate ion by xenon difluoride to give perbromate ion, xenon gas, and fluoride ion in basic aqueous solution. What is the expected geometry of the perbromate ion?

16. Write balanced equations for the reactions that occur when chlorine gas is bubbled into **(a)** cold aqueous $NaOH(aq)$, and **(b)** hot aqueous $NaOH(aq)$.

17. What reaction occurs when potassium chlorate is gently heated? What reaction occurs if manganese dioxide is present when potassium chlorate is heated? Write balanced equations for each reaction.

18. Write balanced equations for the reaction of each of the following with water.

(a) Phosphorus pentachloride

(b) Iodine pentafluoride

(c) Sulfur tetrafluoride

(d) Bromine

19. Describe by means of appropriate balanced equations the preparation of a dilute solution of hypochlorous acid. What happens when an attempt is made to concentrate this solution by evaporation?

20. Write balanced equations for each of the following reactions:

(a) Bromine is added to a hot concentrated solution of $NaOH(aq)$

(b) Bromine is added to an aqueous solution of hydrogen peroxide, and oxygen is liberated

(c) Potassium bromide is heated with concentrated aqueous phosphoric acid

(d) Sulfur dioxide is bubbled through an aqueous solution of potassium iodate

21. Name each of the compounds with the following formulas:

(a) $HOBr$ **(b)** $Ca(OCl)_2$ **(c)** $KBrO_3$

(d) $KClO_2$ **(e)** $Mg(ClO_4)_2$ **(f)** HIO_3

(g) $HBrO_4$ **(h)** H_5IO_6

22. What is the oxidation state of the halogen in each of the compounds in Problem 21?

23. Write equations for the half-reactions that describe the behavior of each of ClO_3^-, ClO_2^-, and ClO^- when they are reduced to Cl^- in aqueous acidic solution.

24. When heated, iodic acid is first dehydrated to give the oxide $I_2O_5(s)$, and then it further decomposes to oxygen and iodine. Write balanced equations for each of these reactions.

25. Explain why sulfur forms the fluorides SF_4 and SF_6, but oxygen does not form the corresponding fluorides OF_4 and OF_6.

26. Iodine reacts with liquid chlorine at $-40\,°C$ to give an orange compound containing 54.5% iodine and 45.5% chlorine by mass. What is the empirical formula of the orange compound? If the molar mass is $467\ g\ mol^{-1}$, what is the molecular formula? What is the Lewis structure of this compound?

27. What is the oxidation number of the halogen in each of the following molecules and ions?

(a) ClO_3^- **(b)** BrF_3 **(c)** $HClO_4$

(d) ClO_2^- **(e)** ClO_2 **(f)** H_5IO_6

28. Write the equation for the half-reaction in which $ClO_3^-(aq)$ is reduced to $Cl_2(g)$ in acidic solution. This reaction can be carried out using $FeSO_4(aq)$. Write the balanced equation for the complete oxidation–reduction reaction.

29. What is a disproportionation reaction? Give two examples from the chemistry of the halogens, and another example from the chemistry of the noble gases.

30. Iodine is prepared from sodium iodate by adding a controlled amount of sodium hydrogen sulfite, $NaHSO_3$, which is oxidized to sodium sulfate. Why is it important that $NaHSO_3$ is not added in excess? How much sodium hydrogen sulfite is required to prepare iodine from $50.0\ kg$ of sodium iodate, and what is the maximum possible yield of iodine?

31. By means of balanced equations, and a brief description of the reaction conditions, indicate methods by which each of the following may be prepared:

(a) Lead(II) chloride from lead

(b) Anhydrous iron(III) chloride from iron

(c) Phosphorus trichloride from phosphorus

(d) Tetrachloromethane from methane

(e) Aluminum chloride hexahydrate from aluminum

32. When 0.3574 g of potassium iodate was heated to constant mass, 0.2772 g of a white solid residue remained, and during the reaction 61.5 mL of dry oxygen, measured at 25 °C and a pressure of 755 mm Hg, were evolved. What is the balanced equation for the reaction?

33. (a) What is the maximum possible yield of sodium hypochlorite from electrolyzing a cold solution of NaCl(aq) for 60 min, using a current of 4.50 A?

(b) What is the maximum possible yield of potassium chlorate from electrolyzing a hot solution of KCl(aq) for 145 min, using a current of 3.05 A?

34. Explain and write balanced equations for each of the following.

(a) Addition of 2 mL of concentrated sulfuric acid to 0.5 g of sodium bromate gives a brown solution. On adding 1 mL of tetrachloroethane and shaking, the brown color is absorbed into the layer.

(b) When $H_2S(g)$ is bubbled through an aqueous solution of potassium bromate, finely divided sulfur is precipitated. After filtering off the sulfur, the clear filtrate, acidified with $HNO_3(aq)$, reacts with $AgNO_3(aq)$ to give a pale yellow precipitate that is readily soluble in $NH_3(aq)$.

(c) Heating solid potassium bromate gives a colorless gas which reignites a glowing splint.

(d) Addition of KBr(aq) to a solution of potassium bromate, followed by acidification with dilute acid, gives a brown solution. On shaking with 1 mL of tetrachloroethane the brown color is extracted into the layer.

35. From the standard reduction potentials for the reduction of the halogens to their halide ions (Table 17.1), calculate the equilibrium constants for each of the following reactions at 25 °C:

(a) $Cl_2(g) + 2Br^-(aq) \rightarrow Br_2(l) + 2Cl^-(aq)$

(b) $Cl_2(g) + 2I^-(aq) \rightarrow I_2(s) + 2Cl^-(aq)$

(c) $Br_2(l) + 2I^-(aq) \rightarrow I_2(s) + 2Br^-(aq)$

*36. (a) An aqueous solution of sodium iodate was treated with chlorine gas in the presence of excess sodium hydroxide, to give a white precipitate of the salt $Na_3H_2IO_6(s)$. Write a balanced equation for the reaction.

(b) On treating the above salt with excess $HNO_3(aq)$ and evaporating the solution until crystals formed, another white compound was formed. After drying these crystals at 110 °C, their composition was found to be 10.8% Na, 59.3% I, and 29.9% O, by

mass. Calculate the empirical formula for this compound and write the balanced equation for its formation.

(c) Draw a Lewis structure for the salt $Na_3H_2IO_6$, and for the compound prepared in part (b). Describe the molecular geometry of the anion in each in terms of the VSEPR model.

37. The pK_a values for hypochlorous acid, hypobromous acid, and hypoiodous acid are, respectively, 7.51, 8.62, and 10.64, at 25 °C.

(a) Give an explanation for the trend in acidities.

(b) What is the pH of a 0.010M solution of each of these acids?

*38. Iodine was heated to 70–80 °C with fuming nitric acid until it had all dissolved and the solution became a light yellow. During the reaction a colorless gas was evolved, which reacted with air to give brown fumes. The solution was evaporated to dryness on a steam bath, to give a white solid. This was recrystallized from concentrated nitric acid by dissolving it and rapidly cooling the solution. White crystals containing 72.1% I, 27.3% O, and 0.6% H, by mass, were obtained. These melted at 110 °C, and at 220 °C, on heating, lost water to give another white compound containing 76.0% I and 24.0% O, by mass.

(a) Identify the two compounds.

(b) Draw their Lewis structures and deduce their molecular geometries.

(c) Write balanced equations for each of the reactions involved in their preparation.

(d) Use the balanced equations for the reaction of chlorine with hot and cold NaOH(aq) to explain what is meant by a disproportionation reaction.

*39. Write balanced equations for each of the following methods for preparing the yellow explosive gas chlorine dioxide:

(a) The reaction of potassium chlorate with concentrated sulfuric acid to give potassium perchlorate, chlorine dioxide, and potassium hydrogen sulfate

(b) The reaction of oxalic acid, $H_2C_2O_4$, with potassium chlorate in aqueous solution, to give carbon dioxide, potassium oxalate, $K_2C_2O_4$, and chlorine dioxide

(c) The reaction of chlorine gas with silver chlorate to give silver chloride, oxygen, and chlorine dioxide

40. Hypofluorous acid, HOF, was first prepared in 1971 by the reaction between fluorine gas and ice at −40 °C. It is a white solid that melts at −11 °C to a pale yellow liquid.

* The asterisk denotes the more difficult problems.

It is a very reactive substance that decomposes below room temperature to give HF and O_2. It reacts with dilute aqueous acid to give hydrogen peroxide, with dilute base to give oxygen, and with HF to give fluorine. Write balanced equations for the preparation of HOF and for each of the reactions described above.

41. Calculate the average bond energies of the bonds to fluorine in each of the following interhalogen compounds from the standard enthalpy of formation data for the gaseous compound,

ΔH_f° (kJ mol^{-1})		ΔH_f° (kJ mol^{-1})	
ClF	-54	ClF$_3$	-163
BrF	-58	BrF$_3$	-256
BrF$_5$	-429	IF$_5$	-838
IF$_7$	-954		

and the following standard enthalpies of formation of the gaseous halogen atoms: F(g), 79.4 kJ mol^{-1}; Cl(g), 119.5 kJ mol^{-1}; Br(g), 111.9 kJ mol^{-1}; and I(g), 106.8 kJ mol^{-1}.

42. Calculate values for the bond energies of ClF(g), Cl_2(g), ClBr(g), and ICl(g) from the following standard enthalpy of formation data:

ΔH_f° (kJ mol^{-1})		ΔH_f° (kJ mol^{-1})	
ClF(g)	-54	F(g)	79.4
Cl$_2$(g)	0	Cl(g)	119.5
ClBr(g)	15	Br(g)	111.9
ICl(g)	18	I(g)	106.8

*43. The dipole moments of HF, HCl, HBr, and HI in units of 10^{-30} C m, are 6.36, 3.43, 2.63, and 1.27, respectively. The observed bond lengths are HF, 91.7; HCl, 127.4; HBr, 141.4; and HI, 160.9 pm. Calculate the partial charges on each atom in each of these polar covalent bonds.

*44. Bleaching powder is obtained by reacting slaked lime, $Ca(OH)_2$, with chlorine. It has the composition 36.6% Ca, 43.2% Cl, 19.5% O, and 0.6% H, by mass. When excess $AgNO_3$(aq) is added to 1.000 g of bleaching powder dissolved in water, a mass of 0.874 g of AgCl(s) is precipitated, and when an acidified aqueous solution containing 1.000 g of bleaching powder is titrated with 0.100M KI(aq), 121.9 mL of the KI(aq) are needed to reduce all the hypochlorite ion, OCl$^-$(aq), in solution to chloride ion, Cl$^-$(aq).

(a) What is the empirical formula of bleaching powder?

(b) Write the balanced equation for the preparation of bleaching powder.

Noble Gases

45. In terms of intermolecular forces, explain each of the following:

(a) The enthalpies of vaporization (in kJ mol^{-1}): He, 0.092; Ne, 1.84; Ar, 6.28; Kr, 9.67; Xe, 13.68; and Rn, 18.0

(b) Why the solubilities of the noble gases in water increase progressively from helium to radon

46. What is the oxidation number of xenon in each of the following molecules and ions?

(a) XeF$_2$ **(b)** XeF$_4$ **(c)** XeO$_2$F$_2$

(d) XeF$_3^+$ **(e)** XeO$_3$ **(f)** XeO$_4$

47. Why do we not expect to find compounds of neon analogous to those formed by krypton and xenon?

*48. Colorless crystals of a compound A containing only xenon and fluorine reacted with a limited amount of water to give a compound B, which was a colorless liquid containing xenon, fluorine, and oxygen. Compound B reacted with excess water to give a colorless solution from which a compound C, a colorless crystalline solid containing only xenon and oxygen, was isolated when the solution was evaporated to dryness. Analysis and molar mass determinations gave the following results: A, 46.5% F; B, 34.0% F, 7.2% O; C, 26.8% O. Molar masses: A, 245; B, 223; C, 179 g mol^{-1}.

(a) Determine the molecular formulas of A, B, and C.

(b) Explain each of the above reactions.

(c) Use the VSEPR model to predict the shapes of A, B, and C.

49. The standard enthalpy change for the reaction

$$XeO_3(s) \longrightarrow Xe(g) + \tfrac{3}{2}O_2(g)$$

is -402 kJ mol^{-1}, and the enthalpy of sublimation of XeO$_3$(s) is 80 kJ mol^{-1}. Given that ΔH_f°(O, g) is 249.1 kJ mol^{-1}, calculate the XeO bond energy. Compare the calculated value with some of the bond energies given in Table 6.3, and comment on the stability of XeO$_3$.

50. Xenon trioxide added to an aqueous solution containing manganese(II) ion, Mn^{2+}(aq), immediately oxidizes it to permanganate ion, MnO$_4^-$(aq), and is itself reduced to xenon. Write a balanced equation for this reaction. What can be said about the standard reduction potential of XeO$_3$?

*51. The reaction of 0.2100 g of XeF_4(s) with NaI(aq) to give Xe(g), F^-(aq), and I_3^-(aq) ions, gave a solution which required 40.00 mL of 0.0100M sodium thiosulfate solution, $Na_2S_2O_3$(aq), for complete reaction of all the I^-(aq). Calculate

(a) The purity of the XeF_4 sample

(b) The volume of Xe(g) liberated, at STP

($S_2O_3^{2-}$(aq) is oxidized by iodine to $S_4O_6^{2-}$(aq).)

*52. In the periodic table, the noble gases are sometimes designated as group 0, and sometimes as group 8. Give arguments for and against each designation, and give your opinion as to which you consider the more appropriate.

53. Explain why xenon and fluorine combine to give XeF_2 when placed in strong sunlight but do not combine in the dark. What is the maximum wavelength of light that could be used to cause this reaction to occur?

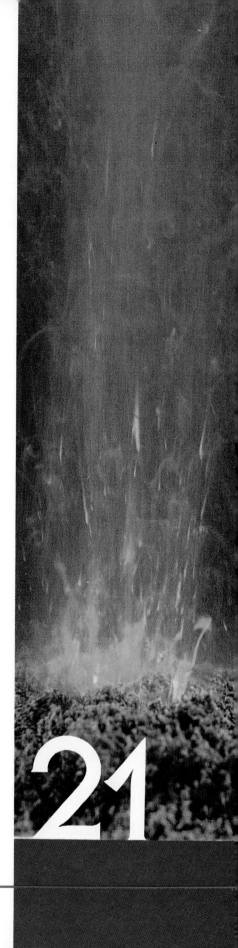

The Transition Metals

Properties of the Transition Metals
The First Series of Transition Elements
Coordination Compounds: Structure, Color, and Magnetism
Silver, Gold, and Mercury

21

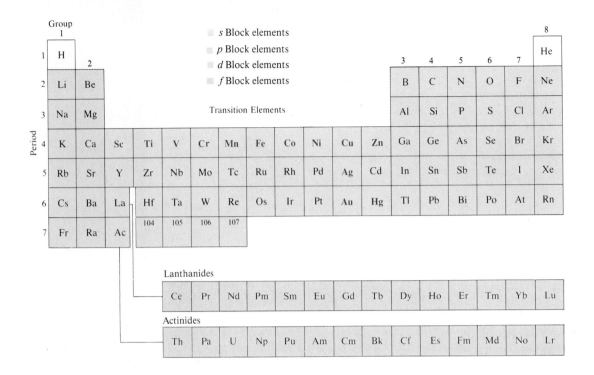

Some colored compounds of the transition metals (top to bottom) V, Cr, Mn, Fe, Co, Ni, and Cu.

More than three-quarters of the elements are metals. Among these are the elements known as the **transition elements** or **transition metals**. They occupy the middle portion of the periodic table between the main groups 2 and 3 in periods 4, 5, 6 and 7. They include some very important metals, such as iron and copper, that we discussed in Chapter 10, as well as vanadium, chromium, cobalt, nickel, silver, gold, and mercury.

They exhibit a wide variety of properties, but they have several common characteristics. These include the following:

1. *Multiple oxidation states*: Unlike the metals in groups 1 and 2 and aluminum in group 3, the majority of the transition metals form compounds in several different oxidation states.

2. *Complex ions*: The transition metals form a large number of complex ions. In this respect they again differ from the metals of groups 1 and 2, which form relatively few complex ions.

3. *Color*: The majority of the compounds of the transition metals are colored. We have already met purple $KMnO_4$, yellow-brown $FeCl_3$, and blue $CuSO_4 \cdot 5H_2O$. Again, they differ from the main group metals, the majority of which form colorless compounds.

4. *Magnetism*: Many of the compounds of the transition metals are paramagnetic—they are attracted by a magnetic field. In contrast, the compounds of the main group metals are almost always diamagnetic—they are weakly repelled by a magnetic field.

Many of the transition metals have important practical applications. Some of them are very hard and strong and are important structural materials. The prime examples are the many different steels, which are alloys of iron with other transition metals such as vanadium, chromium, manganese, cobalt, and nickel, as well as other elements such as carbon. Silver, gold, and platinum are important for their lack of reactivity and resistance to corrosion. Platinum and some other transition metals such as rhodium and iron are also important as catalysts in industrial processes such as the manufacture of nitric acid and of ammonia.

Several of the transition metals are important as trace elements in living systems. In addition to H, C, N, O, P, S, Na, K, and Ca, which are required by the human body in relatively large amounts, we also require smaller amounts of other elements including several of the transition metals. Iron is an essential part of hemoglobin, the oxygen carrier in the blood. Cobalt is the crucial element in vitamin B_{12}. Molybdenum and iron are essential components of nitrogenase, a biological catalyst used by plants to convert atmospheric nitrogen into ammonia. Copper and zinc are important in other biological catalysts that we call enzymes.

The transition elements are sometimes called the **d-block elements** because they have d electrons among their valence electrons. As we shall see it is the presence of d electrons in the valence shells of these elements that is responsible for most of their characteristic properties. Similarly the elements in groups 1 and 2 are called **s-block elements** and the elements in groups 3–8 are called **p-block elements**. In the same way the elements between lanthanum and hafnium and between actinium and element 104 are called **f-block elements** because they have f electrons in their valence shells. These elements are also called the **lanthanides** and **actinides** respectively.

The four rows of transition elements are often called the first, second, third, and fourth transition series. We will restrict our discussion mainly to the first series because this includes most of the familiar and more important of the transition metals, although we will also briefly discuss silver, gold, and mercury.

21.1
Properties of the Transition Metals

We begin by reviewing the electron configurations of the transition metals because they determine their characteristic properties.

ELECTRON CONFIGURATIONS AND IONIZATION ENERGIES

The electron configurations of the first 38 elements are given in Table 21.1. Recall from Chapter 7 that for some elements the sublevels of the $n = 3$ shell and the $n = 4$ shell overlap (see Figure 21.1 on p. 937). In particular, this is the case for potassium, calcium, scandium, and the succeeding elements. Thus for potassium and calcium the 4s level is lower in energy than the 3d level and is therefore filled before the 3d level. Then at scandium the 3d level begins to fill. Scandium is the first element in the first transition series, and zinc, which has both 3d and 4s levels filled, is the last element. In the second series the 4d level is filled, and in the third series the 5d level is filled.

As we saw in Chapter 7, we can obtain considerable information on electron configurations from ionization energies. Table 21.2 on p. 938 gives the first

TABLE 21.1
Electron Configurations of the First 38 Elements

Period		$n=1$ s	2 s	p	3 s	p	d	4 s	p	d	f	5 s
1	H	$1s^1$										
	He	$1s^2$										
2	Li	$1s^2$	$2s^1$									
	Be	$1s^2$	$2s^2$									
	B	$1s^2$	$2s^2$	$2p^1$								
	C	$1s^2$	$2s^2$	$2p^2$								
	N	$1s^2$	$2s^2$	$2p^3$								
	O	$1s^2$	$2s^2$	$2p^4$								
	F	$1s^2$	$2s^2$	$2p^5$								
	Ne	$1s^2$	$2s^2$	$2p^6$								
3	Na	$1s^2$	$2s^2$	$2p^6$	$3s^1$							
	Mg	$1s^2$	$2s^2$	$2p^6$	$3s^2$							
	Al	$1s^2$	$2s^2$	$2p^6$	$3s^2$	$3p^1$						
	Si	$1s^2$	$2s^2$	$2p^6$	$3s^2$	$3p^2$						
	P	$1s^2$	$2s^2$	$2p^6$	$3s^2$	$3p^3$						
	S	$1s^2$	$2s^2$	$2p^6$	$3s^2$	$3p^4$						
	Cl	$1s^2$	$2s^2$	$2p^6$	$3s^2$	$3p^5$						
	Ar	$1s^2$	$2s^2$	$2p^6$	$3s^2$	$3p^6$						
4	K	$1s^2$	$2s^2$	$2p^6$	$3s^2$	$3p^6$		$4s^1$				
	Ca	$1s^2$	$2s^2$	$2p^6$	$3s^2$	$3p^6$		$4s^2$				
	Sc	$1s^2$	$2s^2$	$2p^6$	$3s^2$	$3p^6$	$3d^1$	$4s^2$				
	Ti	$1s^2$	$2s^2$	$2p^6$	$3s^2$	$3p^6$	$3d^2$	$4s^2$				
	V	$1s^2$	$2s^2$	$2p^6$	$3s^2$	$3p^6$	$3d^3$	$4s^2$				
	Cr	$1s^2$	$2s^2$	$2p^6$	$3s^2$	$3p^6$	$3d^5$	$4s^1$				
	Mn	$1s^2$	$2s^2$	$2p^6$	$3s^2$	$3p^6$	$3d^5$	$4s^2$				
	Fe	$1s^2$	$2s^2$	$2p^6$	$3s^2$	$3p^6$	$3d^6$	$4s^2$				
	Co	$1s^2$	$2s^2$	$2p^6$	$3s^2$	$3p^6$	$3d^7$	$4s^2$				
	Ni	$1s^2$	$2s^2$	$2p^6$	$3s^2$	$3p^6$	$3d^8$	$4s^2$				
	Cu	$1s^2$	$2s^2$	$2p^6$	$3s^2$	$3p^6$	$3d^{10}$	$4s^1$				
	Zn	$1s^2$	$2s^2$	$2p^6$	$3s^2$	$3p^6$	$3d^{10}$	$4s^2$				
	Ga	$1s^2$	$2s^2$	$2p^6$	$3s^2$	$3p^6$	$3d^{10}$	$4s^2$	$4p^1$			
	Ge	$1s^2$	$2s^2$	$2p^6$	$3s^2$	$3p^6$	$3d^{10}$	$4s^2$	$4p^2$			
	As	$1s^2$	$2s^2$	$2p^6$	$3s^2$	$3p^6$	$3d^{10}$	$4s^2$	$4p^3$			
	Se	$1s^2$	$2s^2$	$2p^6$	$3s^2$	$3p^6$	$3d^{10}$	$4s^2$	$4p^4$			
	Br	$1s^2$	$2s^2$	$2p^6$	$3s^2$	$3p^6$	$3d^{10}$	$4s^2$	$4p^5$			
	Kr	$1s^2$	$2s^2$	$2p^6$	$3s^2$	$3p^6$	$3d^{10}$	$4s^2$	$4p^6$			
5	Rb	$1s^2$	$2s^2$	$2p^6$	$3s^2$	$3p^6$	$3d^{10}$	$4s^2$	$4p^6$			$5s^1$
	Sr	$1s^2$	$2s^2$	$2p^6$	$3s^2$	$3p^6$	$3d^{10}$	$4s^2$	$4p^6$			$5s^2$

ionization energy, that is, the ionization energy of the most easily removed electron, for the first 36 elements. Let us compare the ionization energies for the third- and fourth-period elements.

In the third period the ionization energy increases from Na ($3s^1$) to Mg ($3s^2$) and then decreases for Al ($3s^2 3p^1$), despite the increased core charge. The

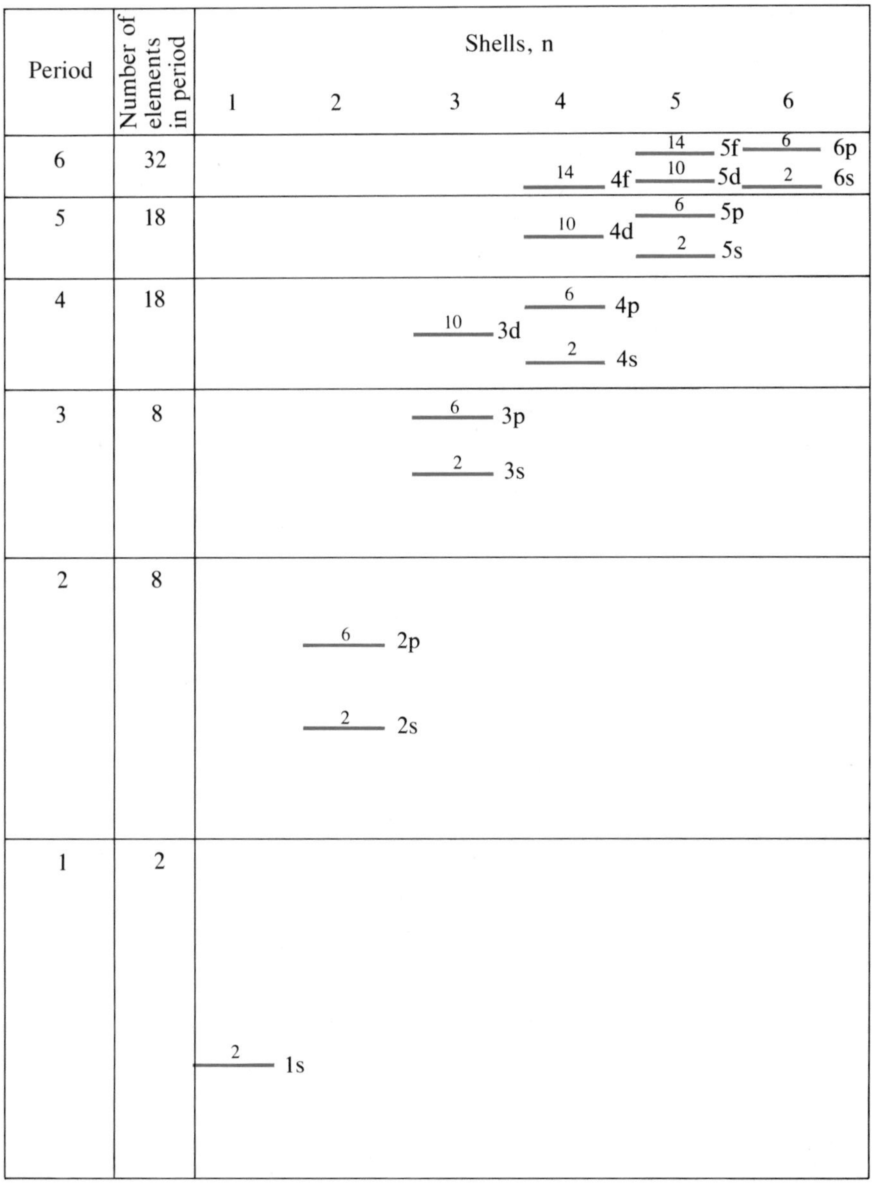

Period	Number of elements in period	Shells, n					
		1	2	3	4	5	6
6	32				14 —4f	14 —5f; 10 —5d	6 —6p; 2 —6s
5	18				10 —4d	6 —5p; 2 —5s	
4	18			10 —3d	6 —4p; 2 —4s		
3	8			6 —3p; 2 —3s			
2	8		6 —2p; 2 —2s				
1	2	2 —1s					

FIGURE 21.1
Energy Level Diagrams.

This diagram is not to scale and it is not meant to represent the energy levels for any particular element. By filling each of the levels in order of increasing energy, we can, however, correctly predict the electron configurations of most of the elements. In particular, it shows that for the neutral atoms of transition elements, the 4s level is below the 3d level and is therefore filled first. However, the 4s level is not below the 3d level for all the elements or, in fact, for the ions of the transition metals.

reason for the decrease is that in Al there is an electron in a 3p level, which has a higher energy than the 3s level, so this electron has a lower ionization energy. In the fourth period there is a similar increase in ionization energy from K ($4s^1$) to Ca ($4s^2$), but at Sc there is a small increase in the ionization energy rather than the decrease observed for Al in period 3. The increase in ionization energy clearly shows that the electron configuration of scandium is not analogous to that of aluminum. As we have seen the 3d level, which has an energy that is very close to the 4s level, begins to be occupied at this point. Thus scandium has the electron configuration $3d^1 4s^2$. The ionization energy then increases slowly with increasing nuclear charge up to Zn. Only then do we observe a marked decrease in the ionization energy at Ga ($3d^{10} 4s^2 4p^1$), when the 4p level starts to fill.

TABLE 21.2
First Ionization Energies for the First 36 Elements

Z	Element	Ionization Energy (MJ mol^{-1})	Z	Element	Ionization Energy (MJ mol^{-1})
1	H	1.312	19	K	0.419
2	He	2.373	20	Ca	0.590
3	Li	0.520	21	Sc	0.631
4	Be	0.899	22	Ti	0.658
5	B	0.809	23	V	0.650
6	C	1.086	24	Cr	0.653
7	N	1.400	25	Mn	0.717
8	O	1.314	26	Fe	0.759
9	F	1.680	27	Co	0.758
10	Ne	2.080	28	Ni	0.737
11	Na	0.496	29	Cu	0.745
12	Mg	0.732	30	Zn	0.906
13	Al	0.578	31	Ga	0.579
14	Si	0.786	32	Ge	0.762
15	P	1.012	33	As	0.946
16	S	1.000	34	Se	0.941
17	Cl	1.251	35	Br	1.139
18	Ar	1.521	36	Kr	1.303

DENSITIES, MELTING POINTS, RADII, AND METALLIC BONDING

Some properties of potassium, calcium, and the first series of transition metals are listed in Table 21.3. Since the atoms in a metallic crystal are packed closely together and are touching each other, the distance between the nuclei of two adjacent atoms is twice the radius of the atom. These radii, which are called **metallic radii**, are listed in Table 21.3 and are also shown in Figure 21.2.

We recall from Chapter 10 that we expect the melting point of a metal to increase with increasing strength of the metallic bond. The strength of the bond-

TABLE 21.3
Properties of Potassium, Calcium, and the First Series of Transition Metals

Metal	Atomic Number	Atomic Mass	Melting Point (K)	Density (g cm^{-3})	r_{metallic} (pm)
K	19	39.10	337	0.86	235
Ca	20	40.08	1110	1.55	197
Sc	21	44.96	1812	3.0	187
Ti	22	47.90	1941	4.51	147
V	23	50.94	2173	6.1	134
Cr	24	52.00	2148	7.19	128
Mn	25	54.94	1518	7.43	127
Fe	26	55.85	1809	7.86	126
Co	27	58.93	1768	8.9	125
Ni	28	58.70	1726	8.9	125
Cu	29	63.55	1356	8.96	128
Zn	30	65.38	693	7.14	134

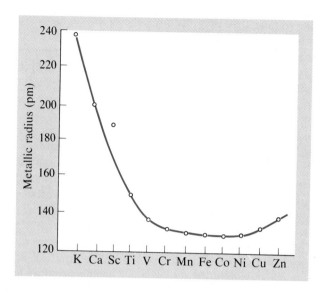

FIGURE 21.2
Metallic Radii of the
Elements K to Zn.

These radii are obtained by taking
one-half the distance between adjacent atoms in the crystal of the
metal.

ing in a metal is determined primarily by the number of electrons that contribute to the metallic bonds. Potassium $(4s^1)$ has only one electron available
for metallic bonding, while calcium $(4s^2)$ has two electrons available for metallic
bonding. Thus calcium has a higher melting point than potassium (Table 21.3
and Figure 21.3). Since scandium has a still higher melting point, and the melting point continues to increase up to vanadium, we conclude that the strength
of the metallic bonding also increases up to vanadium. An increasing number of
electrons, therefore, must contribute to the metallic bonding; these must be the
4s and 3d electrons. Presumably, scandium $(3d^14s^2)$ has three metallic bonding
electrons, titanium $(3d^24s^2)$ four, and vanadium $(3d^34s^2)$ five. However, after
vanadium the melting point decreases slightly to nickel and then more markedly
to copper and zinc. Manganese has an unexpectedly low melting point, which
reflects the fact that it has an unusual structure that is not close-packed.

There is also a marked increase in the density from 0.86 g cm^{-3} for potassium to 7.19 g cm^{-3} for chromium. The density depends on the mass of the

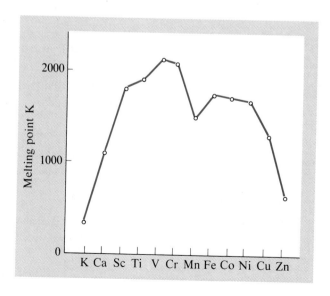

FIGURE 21.3
Melting Points of the
Elements K to Zn.

atoms and the volume that they occupy. The atomic mass increases by only 20% from potassium to chromium but the density increases by over 800%, so the volume that the atoms occupy must be decreasing. The increasingly strong metallic bonds pull the atoms more closely together, so their radii decrease from potassium to chromium (see Table 21.3 and Figure 21.2). The metallic radius is then almost constant from chromium to nickel and finally increases slightly to copper and then zinc.

We conclude from the melting points and the densities (or metallic radii) that after vanadium, which can have a maximum of five metallic bonding electrons, or chromium, which can have a maximum of six, the number of metallic bonding electrons remains constant at five or six up to nickel, despite the fact that the total number of 3d electrons is increasing. After nickel the number of metallic bonding electrons decreases. We expect all the electrons to be more tightly held with increasing atomic number and core charge. Apparently the 3d electrons become so tightly held after vanadium or chromium that not all of them are used to form metallic bonds. We will see that we reach a similar conclusion about the availability of these electrons for bonding from considering the oxidation states of these elements.

OXIDATION STATES

We saw in the previous section that the 3d electrons contribute to the metallic bonding in the metals of the first transition series. Not surprisingly, therefore, the 3d electrons are also used in compound formation by these elements; the 4s electrons are always used, and one or more of the 3d electrons may be used. This possibility of using different numbers of electrons in compound formation is responsible for the variety of oxidation states exhibited by these elements. The common oxidation states of the transition metals are summarized in Table 21.4.

The maximum observed oxidation state increases from +3 for scandium to +7 for manganese, but thereafter it decreases. Although the +6 oxidation state is known for iron, its most common oxidation states are +2 and +3. For nickel, copper, and zinc, the highest common oxidation state is only +2.

TABLE 21.4
Oxidation States of the First Series of Transition Metals

	Sc	Ti	V	Cr	Mn	Fe	Co	Ni	Cu	Zn	
Total number of 3d and 4s electrons	3	4	5	6	7	8	9	10	11	12	
									+1	—	} Ionic compounds
	—	+2	+2	+2	+2	+2	+2	+2	+2	+2	
	+3	+3	+3	+3	+3	+3	+3	+3	+3		
		+4	+4	+4	+4	+4	+4	+4			} Covalent compounds
			+5	+5	+5	—	+5				
				+6	+6	+6					
					+7						

The more common oxidation states are in red.

The variation in the maximum oxidation state shows that in compound formation all the 3d electrons are available for bonding up to manganese, but for the following elements, although the total number of 3d electrons increases, the number of 3d electrons available for bonding decreases, as the core charge increases, from 4 in iron, in the $+6$ oxidation state, to none in the $+2$ oxidation state of zinc. In zinc the 3d electrons may be considered to have become part of the core. The $+2$ oxidation state is found for all the elements except scandium. It corresponds to the loss of the two 4s electrons to form dipositive ions such as Mn^{2+}, Fe^{2+}, and Ni^{2+}.

21.2
The First Series of Transition Elements

In this section we consider some of the more important compounds of the first series of transition elements. As the valence electrons become increasingly strongly held with increasing core charge from left to right in the series, these elements become stronger oxidizing agents in their higher oxidation states and weaker reducing agents in their lower oxidation states. For example, Ti(IV) compounds are only weak oxidizing agents whereas Mn(VII) compounds, such as $KMnO_4$, are strong oxidizing agents. In contrast, Ti(II) and V(II) compounds are strong reducing agents whereas Co(II) compounds are only weak reducing agents. Just as for the main group elements, compounds formed by the transition elements in oxidation states of $+4$ or higher are predominately covalent in nature, and they resemble the compounds of the nonmetals in many of their properties.

SCANDIUM ($3d^1 4s^2$)

Scandium is widely distributed in nature, principally in the form of Sc_2O_3. The three valence electrons are easily removed to give Sc^{3+}. All the compounds of scandium contain this ion; they resemble the corresponding Al^{3+} compounds. They have no important uses.

TITANIUM ($3d^2 4s^2$)

Titanium is the tenth most abundant element in the earth's crust. After iron it is the second most abundant transition metal. The principal ores are *rutile*, TiO_2, and *ilmenite*, $FeTiO_3$.

PREPARATION AND USES The element is difficult to prepare in the pure state because at high temperatures it reacts readily with carbon, hydrogen, nitrogen, and oxygen. The most important method for preparing the metal involves heating rutile or ilmenite with coke and chlorine. The volatile covalent compound $TiCl_4$ distills off, leaving behind involatile, ionic $FeCl_2$:

$$FeTiO_3(s) + 3C(s) + 3Cl_2(g) \longrightarrow FeCl_2(s) + 3CO(g) + TiCl_4(g)$$

The $TiCl_4$ is condensed, purified, and then vaporized into a vessel filled with red-hot magnesium shavings in an atmosphere of argon in which it is reduced to titanium metal, $TiCl_4(g) + 2Mg(s) \rightarrow 2MgCl_2(l) + Ti(s)$. Most of the liquid

$MgCl_2$ is drained off, and the remainder is removed by washing it out with water or by vaporizing it in a vacuum at about 900 °C:

Titanium is becoming an increasingly important metal because it is very strong, has a low density, is highly resistant to corrosion, and has a high melting point. It has many important uses in rockets, jet engines, and aircraft, where its very high melting point makes it much more useful than aluminum. It is also widely used in chemical plants because of its great resistance to corrosion; for example, it is not attacked by any of the common acids or by chlorine. Steel containing titanium is particularly hard and very resistant to abrasion.

OXIDATION STATES Titanium can use two, three, or four valence electrons to give compounds containing titanium in the $+2$, $+3$, and $+4$ oxidation states (see Table 21.5). The $+4$ state is the most stable oxidation state. Compounds of titanium in the $+3$ and $+2$ states are very strong reducing agents that are readily oxidized to the $+4$ state.

COMPOUNDS The most important Ti(IV) compound is the dioxide TiO_2, which occurs naturally as the mineral rutile. Titanium dioxide is very important as a white pigment that is widely used in paints, paper, and many other products. Titanium dioxide is an acidic oxide that reacts with molten metal hydroxides to give titanates such as Na_2TiO_3.

Titanium(IV) chloride consists of covalent $TiCl_4$ molecules. It is a colorless liquid that boils at 136 °C and has a pungent odor. It fumes strongly in moist air, forming a dense white cloud of titanium dioxide (Experiment 21.1):

$$TiCl_4(l) + 2H_2O(s) \longrightarrow TiO_2(s) + 4HCl(g)$$

It is used in producing smoke screens and in skywriting. In its volatility and reaction with water, it resembles a nonmetal halide such as $SiCl_4$ or PCl_3. Titanium tetrachloride, $TiCl_4$, can be prepared by heating the element in chlorine or by heating the more readily available dioxide with carbon and chlorine:

$$TiO_2(s) + 2C(s) + 2Cl_2(g) \longrightarrow TiCl_4(g) + 2CO(g)$$

Titanium(III) chloride, $TiCl_3$, can be made by heating $TiCl_4$ with hydrogen:

$$2TiCl_4(l) + H_2(g) \longrightarrow 2TiCl_3(s) + 2HCl(g)$$

It is an ionic compound that is soluble in water to give violet solutions containing the hydrated titanium(III) ion, $Ti(H_2O)_6{}^{3+}$.

In the $+3$ oxidation state titanium has one unused valence electron; that is, it has the valence-shell electron arrangement $3d^1$, which is usually written simply as d^1. We emphasize that titanium uses all its valence electrons in the $+4$ state by describing it as a d^0 state (see Table 21.5).

TABLE 21.5
Oxidation States of Titanium

Oxidation State	Unused d-Electrons	Typical Examples
$+4$	d^0	TiO_2, $TiCl_4$
$+3$	d^1	Ti^{3+}
$+2$	d^2	$TiCl_2$

Experiment 21.1

Hydrolysis of Titanium Tetrachloride

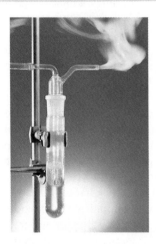

When dry air is bubbled through colorless liquid titanium tetrachloride, $TiCl_4$, a dense white smoke of TiO_2 is formed as soon as the $TiCl_4$ vapor comes into contact with atmospheric moisture.

When $TiCl_4$ is heated with titanium, titanium(II) chloride, $TiCl_2$, can be prepared:

$$TiCl_4(l) + Ti(s) \longrightarrow 2TiCl_2(g)$$

All Ti(II) compounds are very strong reducing agents. For example, they reduce water to hydrogen:

$$TiCl_2(s) + 2H_2O(l) \longrightarrow TiO_2(s) + 2HCl(aq) + H_2(g)$$

There is therefore no aqueous solution chemistry of Ti^{2+}.

VANADIUM ($3d^3 4s^2$)

OCCURRENCE AND USES Although vanadium is very widely distributed, it constitutes only 0.02% of the earth's crust. It occurs mainly in the form of the vanadate ion, VO_4^{3-}. Like titanium, pure vanadium is very difficult to prepare because it combines readily at high temperature with oxygen, nitrogen, and carbon. Because it is used mainly as a component of alloy steels, it is usually made in the form of ferrovanadium, an iron–vanadium alloy, rather than as the pure metal.

COMPOUNDS The oxidation states of vanadium are summarized in Table 21.6. In its highest ($+5$) oxidation state vanadium behaves more like a non-metal than a metal.

The most important compound of vanadium is vanadium(V) oxide, V_2O_5, an orange-red solid that can be prepared by heating vanadium in oxygen. It is important as a catalyst in the contact process for the manufacture of sulfuric acid (Chapter 8). V_2O_5 dissolves in a strongly basic solution to give the vanadate ion, VO_4^{3-}:

$$V_2O_5(s) + 6OH^-(aq) \longrightarrow 2VO_4^{3-}(aq) + 3H_2O(l)$$

TABLE 21.6
Oxidation States of Vanadium

Oxidation State	Unused d-Electrons	Typical Examples
+5	d^0	V_2O_5, VO_4^{3-}, VO_2^+
+4	d^1	VO^{2+}
+3	d^2	V^{3+}
+2	d^3	V^{2+}

Except in strongly basic solutions vanadium in the +5 and +4 oxidation states is found in aqueous solution in the form of the oxovanadium ions VO_2^+ and VO^{2+}. The ions V^{5+} and V^{4+} have charges that are too high for them to be able to exist in aqueous solution. Hydrated V^{5+} and V^{4+} would be highly acidic (recall that Al gives acidic solutions in water—Chapter 10) and would lose protons to give the hydrated oxo ions. For example

$$V(H_2O)_6^{4+} + 2H_2O \longrightarrow VO(H_2O)_5^{2+} + 2H_3O^+$$

A yellow solution of $VO_2^+(aq)$ can be reduced with a suitable reducing agent such as zinc, first to a blue solution of $VO(H_2O)_5^{2+}$, then to a green solution of $V(H_2O)_6^{3+}$, and finally to violet $V(H_2O)_6^{2+}$ (Experiment 21.2). Both V^{2+} and V^{3+} are strong reducing agents, and their aqueous solutions are oxidized in the air to vanadium(IV).

CHROMIUM ($3d^54s^1$)

OCCURRENCE AND USES Chromium is familiar as the protective coating applied to automobile parts and to many other steel objects to improve their appearance and increase their resistance to corrosion. The commonest ore of chromium is *chromite*, $FeCr_2O_4$. This ore is reduced by carbon in an electric furnace to give ferrochrome, an iron–chromium alloy, that is used to make a variety of stainless steels.

OXIDATION STATES The principal oxidation states of chromium are summarized in Table 21.7.

The unexpected electron configuration $3d^54s^1$, rather than $3d^44s^2$, again emphasizes that the 3d and 4s levels are very close in energy and that alternative electron arrangements may differ only very slightly in energy. When the 3d and 4s levels are sufficiently close, the repulsion between the two electrons in the 4s orbital can cause one of them to move into a 3d orbital.

The +6 oxidation state is the maximum possible oxidation state for chromium, because in this state it uses all six valence electrons in the formation of bonds. Chromium(VI) compounds are generally strong oxidizing agents and are readily reduced to the +3 state, which is the most stable oxidation state of chromium. In the +2 state chromium is a reducing agent and is readily oxidized to the +3 state.

THE +6 OXIDATION STATE In its highest oxidation state chromium has a high electronegativity and behaves like a nonmetal. Chromium(VI) compounds are predominantly covalent. The yellow chromate ion, CrO_4^{2-}, resembles the sul-

Experiment 21.2

Oxidation States of Vanadium

A yellow acidified solution of ammonium vanadate, $(NH_4)_3VO_4$, in which vanadium is in the $+5$ oxidation state, can be reduced with zinc amalgam first to a blue solution of VO^{2+}, in which vanadium is in the $+4$ oxidation state, then to a green solution of V^{3+}, and finally to a violet solution of V^{2+}.

If a layer of purple solution of potassium permanganate is gently added on top of a violet solution of V^{2+}, it slowly oxidizes the V^{2+} over a period of a few hours, forming successive layers of the different oxidation states of vanadium. From bottom to top: violet V^{2+}, green V^{3+}, blue VO^{2+}, yellow VO_4^{3-}, brown MnO_2 (produced by reduction of MnO_4^-), and purple $KMnO_4$.

fate ion and has the same tetrahedral structure. Metal chromates often resemble the corresponding sulfates. For example, like $BaSO_4$ and $PbSO_4$, the chromates $BaCrO_4$ and $PbCrO_4$ are insoluble in water.

The corresponding acid, chromic acid, H_2CrO_4, cannot be made, because when a yellow chromate solution is acidified, the solution becomes orange-red owing to the formation of the dichromate ion, $Cr_2O_7^{2-}$ (see Experiment 21.3):

$$2CrO_4{}^{2-}(aq) + 2H_3O^+(aq) \longrightarrow Cr_2O_7{}^{2-}(aq) + 3H_2O(l)$$

TABLE 21.7
Oxidation States of Chromium

Oxidation State	Unused d-Electrons	Typical Examples
$+6$	d^0	CrO_3, $CrO_4{}^{2-}$, $Cr_2O_7{}^2$
$+3$	d^3	Cr_2O_3, Cr^{3+}
$+2$	d^4	Cr^{2+}

Experiment 21.3

Chromium(VI) Compounds

The dish on the left contains a solution of yellow chromate ion, CrO_4^{2-}. When HCl(aq) is added to the chromate solution in the middle dish we see the formation of orange-red dichromate ion, $Cr_2O_7^{2-}$. On the right sufficient HCl(aq) has been added to complete the conversion of chromate to dichromate.

We can think of this reaction as occurring by the protonation of the chromate ion to give the hydrogen chromate ion, $HCrO_4^-$, followed by elimination of water from two $HCrO_4^-$ ions:

$$\underset{\ominus O}{\overset{O}{\underset{}{}}}\!\!Cr\!\!\overset{O}{\underset{OH}{}} + \underset{HO}{\overset{O}{}}\!\!Cr\!\!\overset{O}{\underset{O\ominus}{}} \longrightarrow \underset{\ominus O}{\overset{O}{}}\!\!Cr\!\!\overset{O}{\underset{O}{}}\!\!Cr\!\!\overset{O}{\underset{O\ominus}{}} + H_2O$$

This is an example of a condensation reaction (Chapter 8). The dichromate ion has a structure similar to that of the disulfate ion (Chapter 8). Potassium dichromate, $K_2Cr_2O_7$, crystallizes from solution as bright orange-red crystals. The formation of the dichromate ion is reversible, and if base is added to a dichromate solution, the color changes back to the yellow color of the chromate ion:

$$Cr_2O_7{}^{2-}(aq) + 2OH^-(aq) \longrightarrow 2CrO_4{}^{2-}(aq) + H_2O(l)$$

If potassium dichromate is dissolved in concentrated sulfuric acid, it forms dichromic acid, which is dehydrated to give red solid chromium(VI) oxide (chromium trioxide), CrO_3:

$$Cr_2O_7{}^{2-}(aq) + 2H^+(aq) \rightleftharpoons H_2Cr_2O_7(aq) \longrightarrow 2CrO_3(s) + H_2O(l)$$

An acidic dichromate solution is often used in the laboratory as a strong oxidizing agent:

$$Cr_2O_7{}^{2-} + 14H^+ + 6e^- \longrightarrow 2Cr^{3+} + 7H_2O \qquad E° = 1.33 \text{ V}$$

THE +3 OXIDATION STATE The Cr^{3+} ion, which has a violet color in aqueous solution, forms many salts. These include $CrCl_3 \cdot 6H_2O$, $Cr_2(SO_4)_3 \cdot 18H_2O$, and *chrome alum*, $KCr(SO_4)_2 \cdot 12H_2O$, which forms large, violet octahedral crystals. In these salts Cr^{3+} is hydrated by six water molecules to give the octahedral ion $Cr(H_2O)_6{}^{3+}$. An aqueous solution of $CrCl_3$ is green rather than violet (Experiment 21.4) because of the formation of the green complex ion $CrCl_2(H_2O)_4{}^+$.

When an aqueous solution of ammonia or sodium hydroxide is added to a solution of a chromium(III) salt, a gray-green precipitate of chromium(III) hydroxide, $Cr(OH)_3$, is obtained (Experiment 21.4):

$$Cr^{3+}(aq) + 3OH^-(aq) \longrightarrow Cr(OH)_3(s)$$

If the hydroxide is strongly heated, it is dehydrated to give chromium(III) oxide, Cr_2O_3:

$$2Cr(OH)_3(s) \longrightarrow Cr_2O_3(s) + 3H_2O(g)$$

Chromium(III) oxide can also be made conveniently in a rather spectacular reaction by heating ammonium dichromate:

$$(NH_4)_2Cr_2O_7(s) \longrightarrow N_2(g) + 4H_2O(g) + Cr_2O_3(s)$$

In this reaction the dichromate ion oxidizes the ammonium ion to nitrogen and is itself reduced to Cr_2O_3 (see Experiment 21.5 on p. 948).

Chromium(III) oxide is a rather inert substance of high melting point. When exposed to the air, chromium becomes coated with a very thin, but very hard, unreactive layer of Cr_2O_3, which prevents any further attack on the metal; thus chromium is used as a protective and decorative coating for other metals. The oxide is used as a green pigment in paint.

Acidified Cr^{3+} solutions can be reduced with zinc or other reducing agents to Cr^{2+}. In aqueous solution this ion is hydrated, $Cr(H_2O)_6^{2+}$, and is blue in color (see Experiment 21.4). The chromium(II) ion is a very strong reducing agent, and its solutions must be protected from oxygen, which oxidizes it rapidly to Cr^{3+}. Oxygen can be conveniently removed from a mixture of gases by bubbling it through a chromium(II) solution.

Experiment 21.4

Reactions of Chromium and its Compounds

 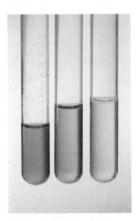 

Chromium dissolves in dilute hydrochloric acid to give hydrogen and a blue solution of $CrCl_2$, which is rapidly oxidized by oxygen in the air to a green solution of $CrCl_3$.

When sodium peroxide, Na_2O_2, solution is added to a green solution of $CrCl_3$, it is oxidized to a yellow solution of CrO_4^{2-}.

When aqueous sodium hydroxide is added to a green solution of $CrCl_3$, a grey-green precipitate of $Cr(OH)_3$ is formed.

Experiment 21.5

Thermal Decomposition of Ammonium Dichromate: The Volcano Experiment

A pile of orange ammonium dichromate can be ignited with a flame.

It then burns spontaneously, producing a shower of sparks, in a very exothermic reaction.

The pile becomes red hot, giving the appearance of a volcano in eruption. The product is dark green Cr_2O_3, which is a light, fluffy powder with a considerably greater volume than the original ammonium dichromate.

EXERCISE 21.1

Write balanced equations for

(a) The reduction of an acidic solution of $Cr_2O_7^{2-}$ to Cr^{3+} with $SO_2(g)$

(b) The oxidation of Cr^{2+} to Cr^{3+} by $O_2(g)$ in acid solution

MANGANESE ($3d^5 4s^2$)

OCCURRENCE AND USES The most important ore of manganese is *pyrolusite*, MnO_2, from which the metal is obtained by reduction with carbon or carbon monoxide. Since the pure metal is not widely used, it is mostly prepared in the form of ferromanganese, an iron–manganese alloy, by the reduction of a mixture of MnO_2 and iron oxides. Manganese is used to produce steel, containing 12% Mn and 1% C, that is very hard and tough and resists abrasion extremely well. It is used, for example, for naval armor plate, bulldozer blades, and dredger buckets.

The principal oxidation states of manganese are summarized in Table 21.8.

COMPOUNDS The only common compound of manganese in the **+4 oxidation state** is manganese(IV) oxide, which occurs naturally as pyrolusite and is commonly known as manganese dioxide. Its formula is usually written as MnO_2, but careful analysis shows that it is usually closer to $MnO_{1.85}$. Although it has considerable covalent character, this oxide may be approximately described as consisting of Mn^{4+} ions and oxide ions. The unusual variable composition arises because some of the oxide ions are missing from the crystal structure, leaving a number of holes. A corresponding number of Mn^{4+} ions are replaced by Mn^{3+} ions in order to compensate for the deficiency of negative charge caused by the missing oxide ions. In this structure an oxide ion can move from its position in the lattice to an adjacent hole, which thus creates a new hole into which another oxide ion can move, and so on. Thus oxide ions can move through solid MnO_2, and it is therefore an electrical conductor. The fact that MnO_2 is an electrical conductor as well as an oxidizing agent is the basis for its use in the dry cell (Chapter 17).

A compound that has a formula such as $MnO_{1.85}$, in which the subscripts are not whole numbers, is called a **nonstoichiometric compound**. Such compounds are only possible in the solid state and only when a metal has at least two stable oxidation states. Nonstoichiometric compounds are common among the oxides and other compounds of the transition metals in their lower oxidation states.

Manganese dioxide oxidizes hydrochloric acid to chlorine and is reduced to manganese(II) ion, Mn^{2+}, in which manganese is in the **+2 oxidation state**:

$$MnO_2(s) + 4HCl(aq) \longrightarrow Cl_2(g) + Mn^{2+}(aq) + 2Cl^-(aq) + 2H_2O(l)$$

This reaction provides a convenient method for preparing chlorine in the laboratory (Chapter 5). Hydrated $MnCl_2 \cdot 6H_2O$ can be crystallized from this solution. It contains the pale pink hydrated Mn^{2+} ion, $Mn(H_2O)_6^{2+}$. There are many other salts of Mn^{2+} such as $MnSO_4 \cdot 7H_2O$ and $Mn(NO_3)_2 \cdot 6H_2O$, which are all pale pink and also contain the hydrated ion $Mn(H_2O)_6^{2+}$.

When hydroxide ion is added to an aqueous solution containing Mn^{2+}, a white precipitate of $Mn(OH)_2$ is obtained. In the presence of air this precipiate rapidly turns brown as it is oxidized to $MnO(OH)$, one of the few compounds that contain manganese in the **+3 oxidation state** (see Experiment 21.6 on p. 950).

When H_2S is passed into a basic solution of an Mn^{2+} salt, a pale pink precipitate of MnS, manganese(II) sulfide, is obtained:

$$Mn^{2+}(aq) + H_2S(aq) \longrightarrow MnS(s) + 2H^+(aq)$$

Because manganese in MnO_2 is in the +4 oxidation state it can behave not only as an oxidizing agent, being reduced to Mn^{2+}, but also as a reducing agent. For example, if MnO_2 is heated with solid KOH in the presence of air, it reduces oxygen from the 0 to the -2 oxidation state and is itself oxidized to the manganate ion, MnO_4^{2-}, which contains manganese in the **+6 oxidation state** (see Experiment 21.7 on p. 951):

$$2MnO_2(s) + 4KOH(s) + O_2(g) \longrightarrow 2K_2MnO_4(s) + 2H_2O(l)$$

Potassium manganate is an ionic compound containing the green manganate ion, MnO_4^{2-}.

The manganate ion, MnO_4^{2-}, is stable only in basic solution. If a solution of manganate, MnO_4^{2-}, is acidified, it gives permanganate, MnO_4^-, and

TABLE 21.8
Oxidation States of Manganese

Oxidation State	Unused d-Electrons	Typical Examples
+7	d^0	MnO_4^-
+6	d^1	MnO_4^{2-}
+4	d^3	MnO_2
+3	d^4	$MnO(OH)$
+2	d^5	Mn^{2+}

Experiment 21.6

Reactions of Manganese(II)

Left to right: (a) An aqueous solution of the manganese(II) cation, Mn^{2+}, is very pale pink. (b) When aqueous sodium hydroxide solution is added, a gelatinous precipitate of $Mn(OH)_2$ is obtained. (c) On standing, this slowly becomes brown as it is oxidized to the manganese(III) compound, $MnO(OH)$. (d) If hydrogen peroxide is added to the $Mn(OH)_2$ precipitate it is rapidly oxidized to dark brown manganese(IV) oxide, MnO_2.

manganese dioxide, MnO_2. In this reaction the $+6$ oxidation state disproportionates to the **$+7$ oxidation state** and the $+4$ state:

$$3MnO_4^{2-} + 4H^+ \rightleftharpoons 2MnO_4^- + MnO_2 + 2H_2O$$

$\quad\quad$ Green $\quad\quad\quad\quad\quad\quad\quad\quad$ Purple $\quad\quad$ Black

The permanganate ion, MnO_4^-, has a deep purple color. This reaction is reversible. If OH^- is added to the purple solution of MnO_4^- containing black insoluble MnO_2, a clear green solution of MnO_4^{2-} is obtained.

A solution of potassium permanganate is a powerful oxidizing agent that is used as a disinfectant and bleaching agent and for purifying water. Potassium permanganate is used extensively as an oxidizing agent in analytical chemistry. Under acidic conditions MnO_4^- is reduced to Mn^{2+}:

$$MnO_4^- + 8H^+ + 5e^- \longrightarrow Mn^{2+} + 4H_2O \qquad E^\circ = 1.51 \text{ V}$$

EXERCISE 21.2

Write balanced equations for the following oxidations by MnO_4^- in acid solution:

(a) Oxidation of I^- to I_2

(b) Oxidation of Fe^{2+} to Fe^{3+}

(c) Oxidation of oxalic acid, $(COOH)_2$, to CO_2

EXERCISE 21.3

What is the concentration of an acidic potassium permanganate solution if exactly 25.0 mL of it are completely reduced to Mn^{2+} by 73.2 mL of a 0.106M solution of $FeSO_4$?

Experiment 21.7

Preparation of Potassium Manganate, K_2MnO_4, and Potassium Permanganate, $KMnO_4$

The dish on the left contains a mixture of white pellets of KOH and black solid MnO_2. When this mixture is heated, a very dark green molten mass of K_2MnO_4 is obtained (right).

A solution of dark green K_2MnO_4 is obtained when water is added to the K_2MnO_4 in the dish. This dark green solution is shown in the left test tube. When dilute HCl is added to the dark green K_2MnO_4, the K_2MnO_4 disproportionates to pink $KMnO_4$ and black MnO_2. In the middle tube the reaction is incomplete and some dark green K_2MnO_4 solution remains at the bottom of the tube. In the right tube the reaction has gone to completion, and the MnO_2 can be seen at the bottom of the tube below the pink solution of $KMnO_4$.

IRON ($3d^6 4s^2$), COBALT ($3d^7 4s^2$), NICKEL ($3d^8 4s^2$)

OCCURRENCE AND USES Iron was considered in Chapter 10. **Cobalt** is a rather scarce element that occurs principally as the sulfide CoS. It is usually found together with nickel sulfide, NiS, and copper sulfide, CuS. In the extraction of the metal, the ore is roasted in air to give the impure oxide Co_2O_3, which is then dissolved in sulfuric acid, and the hydroxide, $Co(OH)_3$, is precipitated by adding $Ca(OH)_2$ or Na_2CO_3. The hydroxide is heated to give the oxide, which is then reduced with hydrogen or carbon. Cobalt is used in the manufacture of a number of special alloys, for example, Alnico, an alloy of Co, Ni, Al, and Cu, which can be very strongly magnetized and is therefore used for making magnets.

Nickel occurs as NiS together with CuS and FeS. It occurs in sufficient concentrations to be worth mining in only a few places. The most important nickel deposits are in Sudbury, Ontario, and central Africa. The metal is prepared by roasting the ore to give the oxide and then reducing the oxide with carbon to give impure nickel containing copper and iron. Pure nickel is obtained by passing CO(g) over the impure nickel at 80 °C. The nickel reacts to give nickel tetracarbonyl, $Ni(CO)_4$, which is a volatile covalent compound that boils at 43 °C and is purified by distillation. It is then decomposed to pure nickel and CO(g) by heating at 200 °C:

$$Ni(s) + 4CO(g) \rightleftharpoons Ni(CO)_4(g)$$

Many transition metals form compounds with carbon monoxide with intriguing formulas such as $Fe(CO)_5$ and $Mn_2(CO)_{10}$. Their structures and properties are discussed in inorganic chemistry courses.

The properties of nickel are much like those of cobalt; it is almost entirely used in the manufacture of stainless steel and special alloys. An important alloy is Monel (72% Ni, 25% Cu, and 3% Fe), which is resistant to oxidation and corrosion and therefore has important applications in chemical plants. Monel is also the sheath metal on electric heating elements in kitchen ranges and ovens. Another important alloy is coinage metal, which is an alloy of 70% Cu and 30% Ni. Nichrome (60% Ni, 25% Fe, and 15% Cr) has a high melting point and a relatively high resistance and is widely used for electric resistance heaters.

COMPOUNDS OF IRON   The common oxidation states of iron are $+2$ (d^6) and $+3$ (d^5). Thus in aqueous solution iron exists in the form of the green Fe^{2+}(aq) and violet Fe^{3+}(aq) ions. The compounds formed by these ions, including the insoluble oxide, hydroxide, and sulfide and the soluble sulfate, nitrate, and halides, were summarized in Table 10.5. Fe^{3+}(aq) is a good oxidizing agent and is reduced to Fe^{2+}(aq).

As we described in Chapter 10, iron forms three oxides, FeO, Fe_2O_3, and Fe_3O_4. Like MnO_2, these oxides have a variable composition and are nonstoichiometric. The oxide FeO has the sodium chloride structure and may be regarded as a close-packed arrangement of oxide ions with Fe^{2+} ions occupying the octahedral holes (Chapter 11). If a small number of Fe^{2+} ions were replaced by two-thirds as many Fe^{3+} ions, we would have an iron-deficient but electrically neutral crystal. The actual composition of iron(II) oxide is usually $Fe_{0.95}O$. Conversion of three-quarters of the Fe^{2+} to Fe^{3+} would give the composition $FeO \cdot Fe_2O_3$, or Fe_3O_4. Finally, replacement of all the Fe^{2+} by two-thirds as many Fe^{3+} gives the composition $Fe_{0.67}O$, that is, Fe_2O_3.

COMPOUNDS OF COBALT   The common oxidation states of cobalt are $+2$ (d^7) and $+3$ (d^6). $Co(H_2O)_6^{3+}$ is a very strong oxidizing agent and most of the simple compounds of cobalt are compounds of $Co(H_2O)_6^{2+}$. But there are many complex ions in which cobalt is in the $+3$ oxidation state; they are described later in this chapter.

The cobalt(II) ion is hydrated in aqueous solution and has the formula $Co(H_2O)_6^{2+}$. It has a pale pink color. Most salts crystallize from solution in a hydrated form and are pink or red. Examples are $CoCl_2 \cdot 6H_2O$, $Co(NO_3)_2 \cdot 6H_2O$, and $CoSO_4 \cdot 6H_2O$. They all contain $Co(H_2O)_6^{2+}$.

Addition of sodium hydroxide solution to a solution of a cobalt(II) compound gives a pink precipitate of cobalt(II) hydroxide, $Co(OH)_2$ (see Experiment 21.8). If $Co(OH)_2$ is heated, it loses water to give the olive green oxide CoO, an ionic compound that has the sodium chloride structure. It is a basic oxide that dissolves in acids to give cobalt(II) salts.

If a solution of ammonium sulfide is added to a solution of a cobalt(II) salt, or if H_2S is passed into a basic solution of a cobalt(II) salt, a black precipitate of cobalt(II) sulfide, CoS, is obtained:

$$Co^{2+}(aq) + S^{2-}(aq) \longrightarrow CoS(s)$$

COMPOUNDS OF NICKEL   All the common compounds of nickel contain the element in the $+2$ oxidation state. Nickel(II) salts such as $NiCl_2 \cdot 6H_2O$ and $NiSO_4 \cdot 7H_2O$ contain the green hydrated nickel ion, $Ni(H_2O)_6^{2+}$. Addition of hydroxide ion to a solution of an Ni^{2+} salt gives a green precipitate of $Ni(OH)_2$ (see Experiment 21.9). Aqueous ammonia also gives the same precipitate, but with excess ammonia this precipitate dissolves to give a deep blue solution containing the complex ion $Ni(NH_3)_6^{2+}$.

Experiment 21.8

Reactions of Cobalt(II)

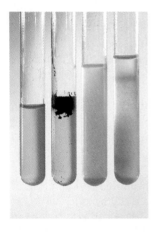

Left to right: (a) A pink solution of $CoCl_2$. (b) When a solution of ammonium sulfide is added, a black precipitate of cobalt(II) sulfide, CoS, is obtained. (c) When excess aqueous NaOH solution is added, a pink precipitate of $Co(OH)_2$ is obtained. (d) When a solution of hydrogen peroxide is added on top of the suspended $Co(OH)_2$ precipitate, the $Co(OH)_2$ is oxidized to $Co(OH)_3$, which forms a brown ring on the interface between the two solutions.

COPPER ($3d^{10}4s^1$)

The element following nickel in the first transition metal series is copper, which we discussed in Chapter 10. Its electron configuration is $3d^{10}4s^1$. As with chromium, this differs from the electron configuration of most of the other transition metals of the first series in that it only has one 4s electron. Copper may lose the single 4s electron to give Cu(I) compounds in which it has a $3d^{10}$ electron configuration and therefore has a completed $n = 3$ shell. In addition, it may also lose one of the 3d electrons to give the Cu(II) oxidation state, which has a d^9 electron configuration.

Experiment 21.9

Reactions of Nickel(II)

Left to right: (a) A solution of nickel(II) chloride, $NiCl_2$, is green. (b) When a solution of ammonium sulfide is added, a black precipite of nickel(II) sulfide, NiS, is obtained. (c) When NaOH(aq) is added, a pale green precipitate of $Ni(OH)_2$ is obtained.

The common compounds of Cu(II) including the soluble sulfate, nitrate, and halides and the insoluble oxide, hydroxide, and sulfide were summarized in Table 10.5.

ZINC ($3d^{10}4s^2$)

The next element, and the last in the first transition metal series, is zinc. The principal ore of zinc is zinc sulfide, ZnS, which is known as *sphalerite*. The structure of ZnS was described in Chapter 11. When ZnS is roasted in a furnace, it is converted to the oxide ZnO, which is then reduced to the metal by strongly heating it with carbon (coke).

Zinc has only one oxidation state, the $+2$ state. Its valence-shell electron configuration is $3d^{10}4s^2$. It can lose the two 4s electrons to give the Zn^{2+} ion, which has a d^{10} arrangement of unused valence electrons. Typical salts of Zn^{2+} are $ZnCl_2 \cdot H_2O$, $ZnSO_4 \cdot 7H_2O$, and $Zn(NO_3)_2 \cdot 6H_2O$. In aqueous solution the zinc(II) ion is in the hydrated form, $Zn(H_2O)_6^{2+}$.

Addition of an alkali metal hydroxide or aqueous ammonia to a solution of a zinc(II) salt gives a white gelatinous precipitate of zinc(II) hydroxide, $Zn(OH)_2$:

$$Zn^{2+}(aq) + 2OH^-(aq) \longrightarrow Zn(OH)_2(s)$$

The precipitate dissolves in excess hydroxide because of the formation of the $Zn(OH)_4^{2-}$ ion,

$$Zn(OH)_2(s) + 2OH^-(aq) \longrightarrow Zn(OH)_4^{2-}(aq)$$

and in excess ammonia because of the formation of the $Zn(NH_3)_4^{2+}$ ion.

If hydrogen sulfide is passed into a basic solution of a zinc(II) salt, a white precipitate of zinc sulfide, ZnS, is obtained.

SUMMARY OF THE PROPERTIES OF THE TRANSITION METALS Sc TO Zn

Now that we have considered the chemistry of the elements from scandium to zinc, we can review the trends in the properties of these elements. The highest oxidation state increases from $+3$ for scandium to $+7$ for manganese. The compounds of these elements in oxidation states of $+4$ or higher are predominantly covalent, and in these oxidation states the elements behave more like nonmetals than metals. Their oxides are acidic, and they form oxoacids and oxoanions as the elements in the main groups 4–7 do—for example, titanates, TiO_3^{2-} (like silicates, SiO_3^{2-}), vanadates, VO_4^{3-} (like phosphates, PO_4^{3-}), and chromates, CrO_4^{2-} (like sulfates, SO_4^{2-}). However, after manganese the theoretically possible maximum oxidation states, which would be $+8$ for iron, $+9$ for cobalt, and so on, are not known. Because the core charge increases from $+4$ for titanium, to $+7$ for manganese, and to $+9$ for cobalt, the 3d electrons are increasingly strongly held, so that after manganese not all of them are available for compound formation.

Thus from titanium to manganese, the highest oxidation state becomes increasingly less stable, and the oxoanions become increasingly strong oxidizing agents. As the higher oxidation states become less stable, the lower oxidation states become more stable. Thus whereas Ti^{2+} is a strong reducing agent that is easily oxidized to Ti(IV) compounds, Ni^{2+}, Cu^{2+}, and Zn^{2+} are the stable oxidation states of these elements, and they cannot easily be oxidized to higher oxidation states. The redox behavior of the first series of transition metals is summarized in Table 21.9. In their lower oxidation states these elements show

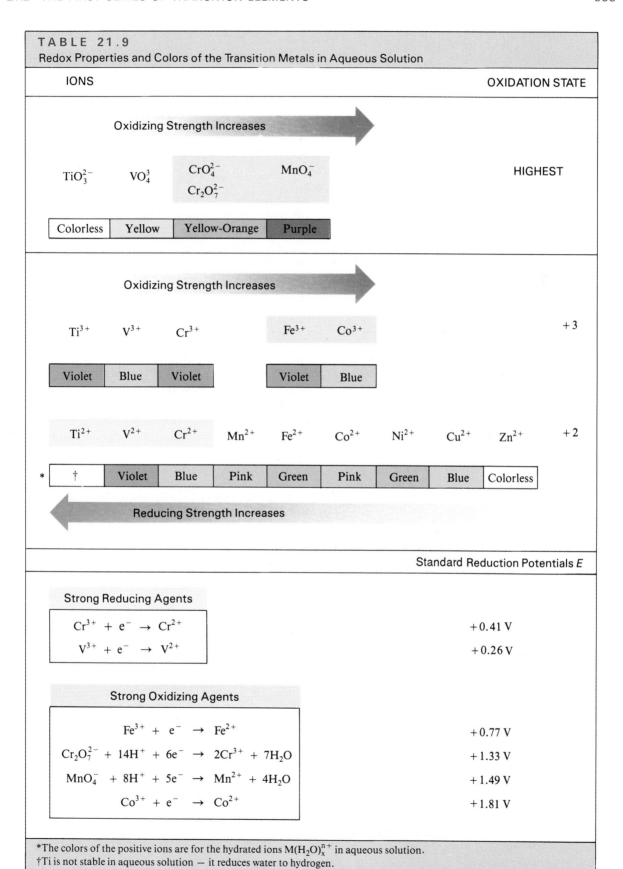

TABLE 21.9
Redox Properties and Colors of the Transition Metals in Aqueous Solution

IONS   OXIDATION STATE

Oxidizing Strength Increases →

TiO_3^{2-} VO_4^{3} CrO_4^{2-} MnO_4^- HIGHEST
                $Cr_2O_7^{2-}$

| Colorless | Yellow | Yellow-Orange | Purple |

Oxidizing Strength Increases →

Ti^{3+} V^{3+} Cr^{3+} Fe^{3+} Co^{3+} +3

| Violet | Blue | Violet | | Violet | Blue |

Ti^{2+} V^{2+} Cr^{2+} Mn^{2+} Fe^{2+} Co^{2+} Ni^{2+} Cu^{2+} Zn^{2+} +2

* | † | Violet | Blue | Pink | Green | Pink | Green | Blue | Colorless |

← Reducing Strength Increases

Standard Reduction Potentials E

Strong Reducing Agents

| $Cr^{3+} + e^- \rightarrow Cr^{2+}$ | +0.41 V |
| $V^{3+} + e^- \rightarrow V^{2+}$ | +0.26 V |

Strong Oxidizing Agents

$Fe^{3+} + e^- \rightarrow Fe^{2+}$	+0.77 V
$Cr_2O_7^{2-} + 14H^+ + 6e^- \rightarrow 2Cr^{3+} + 7H_2O$	+1.33 V
$MnO_4^- + 8H^+ + 5e^- \rightarrow Mn^{2+} + 4H_2O$	+1.49 V
$Co^{3+} + e^- \rightarrow Co^{2+}$	+1.81 V

*The colors of the positive ions are for the hydrated ions $M(H_2O)_x^{n+}$ in aqueous solution.
†Ti is not stable in aqueous solution — it reduces water to hydrogen.

TABLE 21.10
Solubility and Preparation of M^{2+} and M^{3+} Compounds of the Transition Metals

Soluble compounds
Nitrates, sulfates, and halides
Preparation   Oxide, hydroxide, or metal + acid
                  $Cr(OH)_3(s) + 3HCl(aq) \longrightarrow CrCl_3(aq) + 3H_2O(l)$
                  $Cr(s) + 2HCl(aq) \longrightarrow CrCl_2(aq) + H_2(g)$

Insoluble compounds
Hydroxides and sulfides
Preparation   Precipitation
                  $Co^{2+}(aq) + 2OH^-(aq) \longrightarrow Co(OH)_2(s)$
                  $Zn^{2+}(aq) + S^{2-}(aq) \longrightarrow ZnS(s)$

Oxides
Preparation   Metal + oxygen
                  $2Cu(s) + O_2(g) \xrightarrow{\text{Heat}} 2CuO(s)$
                  Heat hydroxide
                  $2Fe(OH)_3(s) \xrightarrow{\text{Heat}} Fe_2O_3(s) + H_2O(l)$

typical metal behavior; they form basic oxides and hydroxides, such as FeO, NiO, $Fe(OH)_2$, and $Ni(OH)_2$, that dissolve in acids to form ionic salts. The methods of preparing the compounds of these elements in their +2 and +3 oxidation states are summarized in Table 21.10.

An obvious question that comes to mind is, why are the 4s electrons always used before the 3d electrons in compound formation, although we stated earlier in discussing electron configurations that the 4s level has a lower energy than the 3d level? The reason is that although the 3d level is above the 4s level for the neutral atoms of the transition metals, the 3d level is below the 4s level for their ions. The greater resultant charge acting on the electrons in the ions than in the neutral atom decreases the energy of the 3d level more than that of the 4s level; as a result, in the ions the 3d level has a slightly lower energy than the 4s level. Thus when electrons are removed from a neutral transition metal atom, the first two always *appear to have come* from the 4s level (see Figure 21.4). Thus the electron configurations of the first transition series metal ions can be obtained from the electron configurations of the elements by first removing the 4s electrons and then as many d electrons as necessary.

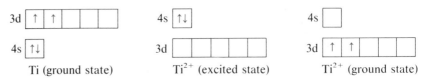

FIGURE 21.4
Orbital Energies for Ti and Ti^{2+}.

Although the 3d orbitals have a higher energy than the 4s orbital in the neutral Ti atom, they have a lower energy in the Ti^{2+} ion. Thus the ground state configuration of Ti^{2+} is $3d^2 4s^0$, not $4s^2 3d^0$. If we imagine that two electrons are removed from the 3d orbital of Ti ($3d^2 4s^2$), we obtain an excited state of Ti^{2+}, $3d^0 4s^0$, which reverts to the ground state, $3d^2 4s^2$.

21.3
Coordination Compounds

A characteristic property of transition metals is their ability to form a large number of stable complexes. An inorganic molecule or ion that contains several atoms, including one or more metal atoms, is called a **coordination compound** or **coordination complex**; it most commonly consists of a single metal atom surrounded by several atoms or groups of atoms. When the coordination complex carries a charge, it is called a **complex ion**. Hydrated ions such as $Al(H_2O)_6^{3+}$ and $Fe(H_2O)_4^{2+}$ and the ions $AlCl_4^-$, $Al(OH)_4^-$, and $FeCl_4^-$, which we discussed in Chapter 10, are all examples of complex ions. In this chapter we have encountered a variety of other complex ions such as $Cr(H_2O)_6^{3+}$, $Ni(H_2O)_6^{2+}$, $Ni(NH_3)_6^{2+}$, $Zn(NH_3)_4^{2+}$, and $Zn(OH)_4^{2-}$. The formation of $Co(NH_3)_6^{3+}$ and the similar complex ammines $Ni(NH_3)_6^{2+}$ and $Cu(NH_3)_4^{2+}$ is demonstrated in Experiment 21.10.

We have seen in Chapter 10 that a metal ion, particularly if it is small and multiply charged, strongly attracts an unshared electron pair in a molecule such as H_2O or NH_3 or an anion such as Cl^- or OH^-. A molecule or ion that becomes bonded to a metal atom in this way is called a **ligand**. The transition metals form complexes with a wide variety of molecules and anions. The ligands that are most commonly found in transition metal complexes are listed in Table 21.11 on p. 958).

As we discussed in Chapter 10 the bond between a metal atom and a ligand is intermediate in character between an ionic bond and a covalent bond. In some complexes it is largely ionic, while in others it is largely covalent. At the ionic extreme the bond can be considered to be due to the electrostatic attraction between the metal ion and the negative charge of the anion or the negative end of a polar molecule. At the covalent extreme the originally unshared pair of the ligand is shared with the metal atom. If a considerable amount of charge

Experiment 21.10

Copper(II), Cobalt(III), and Nickel(II) Ammines

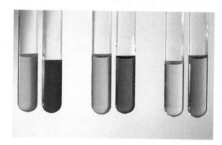

Left to right: (a) When an excess of aqueous ammonia is added to a blue solution of $Cu^{2+}(aq)$, the very deep blue $Cu(NH_3)_4^{2+}$ complex ion is formed. (b) When an excess of ammonia is added to a pink solution of $Co^{2+}(aq)$, it is oxidized to Co^{3+} by the oxygen in the air and becomes deep red in color as the $Co(NH_3)_6^{3+}$ complex ion is formed. (c) A green solution of $Ni^{2+}(aq)$ becomes violet in color when an excess of aqueous ammonia is added because the $Ni(NH_3)_6^{2+}$ complex ion is formed.

TABLE 21.11
Some Common Ligands

	Ligand	Name in Complex
Water	H_2O	Aquo
Ammonia	NH_3	Ammine[a]
Bromide	Br^-	Bromo
Chloride	Cl^-	Chloro
Cyanide	CN^-	Cyano
Hydroxide	OH^-	Hydroxo
Oxalate	$C_2O_4{}^{2-}$	Oxalato
Ethylenediamine	$H_2NCH_2CH_2NH_2$	Ethylenediamine (en)

[a] Note the double *mm*, rather than a single *m*, as in the amines RNH_2 (Chapter 23).

is donated from the ligand to the metal atom, then the bond is largely covalent. But if only a small amount of charge is donated from the ligand to the metal atom, then the bond is largely ionic.

The ligands are said to be *coordinated* to the metal atom. The number of bonds formed to ligands is called the **coordination number** of the metal atom. The bond between a ligand and a metal is sometimes called a **coordinate bond**. But a coordinate bond consists of a pair of shared electrons, like any other single bond, and like other single bonds it may be predominantly covalent or predominantly ionic.

The coordination number of transition metal atoms is very frequently 6; less commonly, it is 4. Other coordination numbers such as 2, 3, 5, 8, and 9 are sometimes found. If the ligand is a negative ion, then the charge on the complex ion differs from the charge on the uncoordinated metal ion; for example,

$$Fe^{3+} + 4Cl^- \longrightarrow FeCl_4^-$$

A complex ion may have both neutral molecules and negative ions as ligands. For example, Cr^{3+} forms the ion $Cr(H_2O)_4Cl_2^+$.

The tendency of the transition metals to form complex ions is so strong that in aqueous solution they are always either hydrated or in the form of some other complex ion. Because of the large number and variety of these complex ions, the transition metals often have a more complicated and more interesting aqueous solution chemistry than that of the main group metals. For example, when ammonium chloride and aqueous ammonia are added to a solution of cobalt(II), and the solution is treated with an oxidizing agent such as air, iodine, or hydrogen peroxide and then acidified with hydrochloric acid, four compounds can be obtained:

1. A yellow compound with the composition $CoCl_3 \cdot 6NH_3$
2. A purplish red compound $CoCl_3 \cdot 5NH_3$
3. A green compound $CoCl_3 \cdot 4NH_3$
4. A violet compound with the same composition, $CoCl_3 \cdot 4NH_3$

For a long time the bewildering variety of similar, closely related compounds of cobalt and many other transition metals posed a difficult problem. The structures of these compounds were not understood until Alfred Werner, a 26-year-old Swiss chemist (1866–1919), suggested in 1893 that ammonia molecules, chloride ions, and other molecules and ions could be strongly attached to a metal ion to give complex ions. Werner proposed that all these compounds of cobalt chloride and ammonia should be formulated as ionic compounds containing complex ions of cobalt.

Werner based his theory on a great variety of experimental evidence. One example of the evidence that he used to support his theory is provided by the reaction of these cobalt compounds with excess silver nitrate. He found the following results:

- One mole of $CoCl_3 \cdot 6NH_3$ reacts immediately with 3 mol of $AgNO_3$ to give a precipitate of 3 mol of AgCl.
- One mole of $CoCl_3 \cdot 5NH_3$ reacts immediately with only 2 mol of $AgNO_3$ to give a precipitate of 2 mol of AgCl.
- One mole of $CoCl_3 \cdot 4NH_3$ reacts immediately with only 1 mol of $AgNO_3$ to give a precipitate of 1 mol of AgCl.

He concluded that $CoCl_3 \cdot 6NH_3$ contains three Cl^- ions, whereas $CoCl_3 \cdot 5NH_3$ contains only two Cl^- ions and $CoCl_3 \cdot 4NH_3$ contains only one Cl^- ion. In order to account for these results, he proposed that all these compounds contain complex ions. The first compound contains the complex ion $[Co(NH_3)_6]^{3+}$ and should therefore be formulated as $[Co(NH_3)_6]^{3+}(Cl^-)_3$. The second contains the complex ion $[Co(NH_3)_5Cl]^{2+}$ and should therefore be formulated as $[Co(NH_3)_5Cl]^{2+}(Cl^-)_2$, in which there are only two chloride ions and one chlorine that is covalently bound to the metal atom and does not, therefore, give a precipitate with silver chloride. Finally, $CoCl_3 \cdot 4NH_3$ contains the complex ion $[Co(NH_3)_4Cl_2]^+$ and should be formulated as $[Co(NH_3)_4Cl_2]^+Cl^-$, in which there is only one chloride ion and two chlorine atoms covalently bound to the cobalt atom.

Notice that in order to indicate clearly which ions or molecules are attached to the metal atom, we enclose the formula of the complex ion in square brackets. The charge on the complex ion is written outside the square brackets. These brackets should not be confused with the brackets that we used in previous chapters to indicate the concentration of a species.

Werner also proposed that when there are six ligands they have an octahedral arrangement around the central metal atom (Figure 21.5). This enabled him to explain the existence of two forms of the complex ion $[Co(NH_3)_4Cl_2]^+$. The two chlorine ligands can either occupy adjacent corners of the octahedron or they can occupy opposite corners (Figure 21.5 on p. 960). The form in which the two chlorines are adjacent is called the *cis* form and the form in which they are at opposite corners is called the *trans* form. Molecules and ions which have the same composition and in which the atoms are all connected to each other in the same way but which differ only in the positions of identical atoms or groups in space are called **geometric isomers**. Further examples of geometric isomers are described in Chapter 23.

GEOMETRY OF COORDINATION COMPOUNDS

Werner proposed shapes for the various coordination complexes in order to explain the numbers of isomers that he found, but he had no direct experimental evidence for the shapes. Today the results of X ray crystallographic studies have provided us with information on the shapes of a very large number of coordination compounds. The most common geometries observed for coordination compounds are linear for two ligands, square planar or tetrahedral for

FIGURE 21.5
Octahedral Co(III)
Complexes.

Werner proposed that transition metal complexes with six ligands have an octahedral shape as we see here for $[Co(NH_3)_6]^{3+}$. This enabled him to explain the existence of two forms of the complex $[Co(NH_3)_4Cl_2]^+$. They are the *cis* and *trans* geometric isomers.

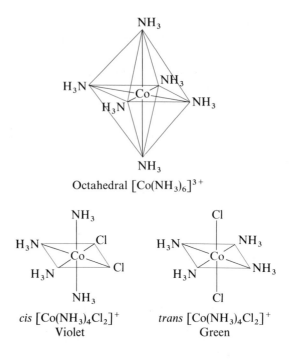

Octahedral $[Co(NH_3)_6]^{3+}$

cis $[Co(NH_3)_4Cl_2]^+$
Violet

trans $[Co(NH_3)_4Cl_2]^+$
Green

four ligands, and octahedral for six ligands (Figure 21.6). We recognize that the linear, tetrahedral, and octahedral geometries are just the geometries predicted by the VSEPR model for AX_2, AX_4, and AX_6 molecules, that is, for molecules with 2, 4, and 6 bonding electron pairs in their valence shells and no nonbonding pairs, but the VSEPR model does not predict a square planar shape for an AX_4 species.

The most important difference between the coordination compounds of the transition metals and the compounds of the main group metals is that in many

$H_3N—Ag—NH_3$

(a) Linear $[Ag(NH_3)_2]^+$

(b) Tetrahedral $[FeCl_4]^-$; Square planar $[Cu(NH_3)_4]^{2+}$

FIGURE 21.6
Some Common Shapes for
Inorganic Complexes.

(a) AX_2 complexes are linear.
(b) AX_4 complexes are tetrahedral or square planar. (c) AX_6 complexes are octahedral.

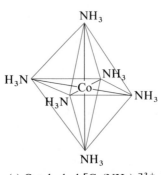

(c) Octahedral $[Co(NH_3)_6]^{3+}$

coordination compounds the central metal atom has some unused nonbonding d electrons. These nonbonding d electrons can be thought of as being largely "inside" the electron pairs associated with the metal–ligand bonds. For a main group element the bonding electron pairs and any nonbonding pairs surround a spherical core consisting of completed electron shells. This spherical core has no influence on the arrangement of the surrounding electron pairs. But in a transition metal complex the bonding electron pairs surround a d shell which may be incomplete and therefore nonspherical. The interaction between a nonspherical d shell and the surrounding bonding electron pairs of the ligands may distort the arrangement of the ligands predicted by the VSEPR model so that, for example, an AX_4 compound of a transition metal may have a square planar rather than a tetrahedral shape. We will look at the interaction between the d electrons and the ligands in a little more detail in a following section where we also show that these interactions are of great importance in determining the color and magnetic properties of transition metal complexes.

CHELATES

The atom in a ligand that is bound directly to the metal is called the *donor atom*. For example, oxygen is the donor atom in $Cr(H_2O)_6^{3+}$. Most of the ligands in Table 21.11 have only one donor atom. They are called **monodentate ligands**. But two ligands in the table have two donor atoms. Ethylenediamine (1,2-diaminoethane), $H_2NCH_2CH_2NH_2$, has two donor nitrogen atoms, and in the oxalate ion, $C_2O_4^{2-}$, two of the oxygen atoms behave as donor atoms. These ligands are called **bidentate ligands**.

In an octahedral complex ethylenediamine (en) occupies two adjacent octahedral positions. It forms complexes such as those shown in Figure 21.7. The formation of some ethylenediamine complexes of Ni(II) is demonstrated in Experiment 21.11 on p. 962. The oxalate ion, $C_2O_4^{2-}$, forms similar complexes (Figure 21.8 on p. 962).

Other ligands can occupy three or more coordination sites. They are called **polydentate ligands**. An interesting example is the ethylenediaminetetraacetate ion, $EDTA^{4-}$ (Figure 21.9 on p. 962), which has six donor atoms and can therefore occupy all six sites in an octahedrally coordinated ion, as in the complex ion $Co(EDTA)^-$. Complexes containing bidentate or polydentate ligands are often known as **chelates**, a name derived from the Greek word for "claw", because the polydentate ligands appear to grasp the metal atom between two or more donor atoms. Such ligands are called *chelating agents*.

In general, polydentate ligands form more stable complexes than do related monodentate ligands. This property makes chelating agents very useful for removing metal ions from solution. For example, Ca^{2+} and Mg^{2+} ions, which

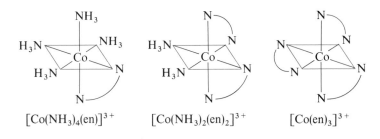

$[Co(NH_3)_4(en)]^{3+}$ $[Co(NH_3)_2(en)_2]^{3+}$ $[Co(en)_3]^{3+}$

FIGURE 21.7
Some Chelate Complexes of Cobalt Formed by the Bidentate Ligand $H_2NCH_2CH_2NH_2$.

The ligand $H_2NCH_2CH_2NH_2$ (en) is shown as N—N.

Experiment 21.11

Ethylenediamine Complexes of Ni(II)

When diaminoethane (ethylene diamine), $H_2NCH_2CH_2NH_2$, is added to a green aqueous solution of Ni^{2+}, the solution changes first to green-blue, then to purple-blue, and finally to violet as the chelate complexes $Ni(H_2O)_4(en)^{2+}$, $Ni(H_2O)_2(en)_2^{2+}$, and $Ni(en)_3^{2+}$ are successively formed.

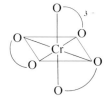

FIGURE 21.8
The Chelate Complex
$Cr(C_2O_4)^{3-}$.

The oxalate ion,

is abbreviated as O—O.

cause hardness in water (Chapter 15), can be removed by forming complex ions with the triphosphate anion:

Certain metal ions such as Hg^{2+} and Pb^{2+}, which tend to accumulate in the body and act as poisons, can be removed by forming a chelate with EDTA, thus converting them into harmless soluble forms that are eliminated from the body. Various complexes of transition metals with chelating agents play vital roles in reactions in living systems. One example is discussed in Box 21.1.

FORMATION CONSTANTS

The stabilities of complex ions in aqueous solution are measured by determining the equilibrium constant for the reaction of the hydrated ion with the

EDTA

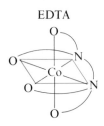

$[Co(EDTA)]^-$

FIGURE 21.9
The Chelate Complex of
Cobalt(III) Formed by the
Hexadentate Ligand
Ethylenediaminetetraacetate
(EDTA).

ligand. For example, for the reaction

$$Co(H_2O)_6{}^{2+}(aq) + 6NH_3(aq) \rightleftharpoons Co(NH_3)_6{}^{2+}(aq) + 6H_2O(l)$$

the expression for the equilibrium constant is

$$K_c = \frac{[Co(NH_3)_6{}^{2+}][H_2O]^6}{[Co(H_2O)_6{}^{2+}][NH_3]^6}$$

Since for dilute solutions $[H_2O]$ is very nearly constant, we can rearrange this equation as

$$K_f = \frac{K_c}{[H_2O]^6} = \frac{[Co(NH_3)_6{}^{2+}]}{[Co(H_2O)_6{}^{2+}][NH_3]^6}$$

where K_f is the **formation constant** of the complex ion.

For $Co(NH_3)_6{}^{2+}$, $K_f = 1 \times 10^5$ mol^{-6} L^6. Table 21.12 gives values of formation constants for some other complex ions. A large value for K_f indicates a very stable complex. For example, the formation constant of $Ni(NH_3)_6{}^{2+}$ is 6×10^8 mol^{-6} L^6, whereas that for the chelate $Ni(en)_3{}^{2+}$ is 4×10^{18} mol^{-3} L^3, indicating that the bidentate ligand $NH_2CH_2CH_2NH_2$ forms a considerably more stable complex than NH_3.

BOX 21.1

HEMOGLOBIN

Among the transition metals that are essential to life, iron is extremely important as an essential component of the oxygen carrier *hemoglobin*. If our diet is deficient in iron, we may suffer from anemia and feel tired and weak. Hemoglobin contains four complex protein molecules, each of which has a heme molecule as part of its structure. Each heme molecule contains an iron atom coordinated by four nitrogen atoms in a square. Octahedral coordination around the iron is completed by a nitrogen atom from another part of the protein molecule and an oxygen molecule or a water molecule:

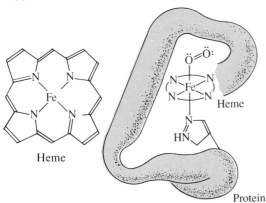

Heme

Protein

Hemoglobin picks up oxygen and forms *oxyhemoglobin* in the lungs. Oxygen is released in the tissues, where it is needed for cell metabolism, and replaced by a water molecule to form *deoxyhemoglobin*.

Carbon monoxide also behaves as a ligand. It forms a complex with hemoglobin that is about two hundred times as strong as the complex with oxygen and makes the hemoglobin useless as an oxygen carrier. If we breathe carbon monoxide, some of the oxyhemoglobin in our blood is converted to carboxyhemoglobin; breathing air containing only 0.1% CO converts about 60% of the hemoglobin to carboxyhemoglobin in a few hours. If a substantial part of the hemoglobin in the blood is converted to carboxyhemoglobin, the body suffers from acute oxygen deficiency and death soon results.

Long exposure to even small concentrations of carbon monoxide can have serious consequences. For example, air inhaled through a lighted cigarette contains about 400 ppm of carbon monoxide. Consequently, heavy smokers have as much as 6% of the hemoglobin in their blood constantly converted to carboxyhemoglobin. As a result, their blood is not as efficient at carrying oxygen as the blood of nonsmokers, and their hearts must work harder. This is probably a contributing factor in heart disease and heart attacks.

TABLE 21.12
Formation Constants for Some Complex Ions at 25 °C

Ion	K_f	Ion	K_f
$Cu(NH_3)_4^{2+}$	1×10^{12}	$Cu(en)_2^{2+}$	1.6×10^{20}
$Ni(NH_3)_6^{2+}$	6×10^8	$Ni(en)_3^{2+}$	4×10^{18}
$Zn(NH_3)_4^{2+}$	5×10^8	$Co(en)_3^{2+}$	8×10^{13}
$Ag(NH_3)_2^+$	1×10^8	$Zn(en)_3^{2+}$	1.2×10^{13}
$Co(NH_3)_6^{2+}$	1×10^5	$Fe(en)_3^{3+}$	4×10^9
$Fe(CN)_6^{3-}$	1×10^{31}	$Mn(en)_3^{2+}$	5×10^5
$Fe(CN)_6^{4-}$	1×10^{24}	$Al(EDTA)^-$	1.4×10^{16}
$Ni(CN)_4^{2-}$	1×10^{30}	$Co(EDTA)^-$	2×10^{16}
$Zn(CN)_4^{2-}$	5×10^{16}	$Cu(EDTA)^{2-}$	6.3×10^{18}
$Hg(CN)_4^{2-}$	4×10^{41}	$Fe(EDTA)^{2-}$	2.1×10^{14}
$Ag(CN)_2^-$	1×10^{21}	$Fe(EDTA)^-$	1.3×10^{25}
$Al(C_2O_4)_2^-$	1.0×10^{13}	$Mn(EDTA)^{2-}$	1.1×10^{14}
$Co(C_2O_4)_2^{2-}$	1.3×10^7	$Hg(EDTA)^{2-}$	6.3×10^{21}
$Cu(C_2O_4)^{2-}$	2.0×10^{10}	$Ni(EDTA)^{2-}$	4.2×10^{18}
$Fe(C_2O_4)_2^{2-}$	4.0×10^9	$Ag(EDTA)^{3-}$	2.1×10^7
$Mn(C_2O_4)_2^{2-}$	6.3×10^5	$Zn(EDTA)^{2-}$	3.2×10^{16}
$Ni(C_2O_4)_2^{2-}$	3.3×10^6		
$Zn(C_2O_4)_2^{2-}$	2.3×10^7		

NOMENCLATURE

The rules for the systematic naming of complex compounds are as follows:

1. The common ligands have the names given in Table 21.11.
2. The number of any particular ligand is specified by di = 2, tri = 3, tetra = 4, penta = 5, hexa = 6, and so on. When confusion might result from the use of these prefixes, the alternative prefixes bis = 2, tris = 3, tetrakis = 4, and so on, are used.
3. The name of a negative (anionic) complex always ends in the suffix *-ate*, which is appended to the name of the metal or the stem of the name of the metal. For some metals the Latin stem is used. For example, in a negative complex iron is named ferrate and lead is named plumbate.
4. The oxidation state of the metal is indicated by a Roman numeral in parenthesis following the name of the metal.

Some examples are given in Table 21.13.

COLORS OF TRANSITION METAL COMPLEXES

We have seen that most transition metal compounds are colored. For example, hydrated Cu^{2+} salts are blue, hydrated Ni^{2+} salts are green, MnO_4^- is purple, and Fe_2O_3 is a deep red brown. The colors of the aqueous solutions of some transition metal ions are shown in Figure 21.10. In contrast, most of the compounds of the main group metals are colorless. Substances are colored because they absorb certain wavelengths of visible light while transmitting (or reflecting)

TABLE 21.13
Names of Some Complexes

Complex	Name
$[Co(H_2O)_6]^{3+}$	Hexaaquocobalt(III) ion
$[CoCl_6]^{3-}$	Hexachlorocobaltate(III) ion
$[Co(NH_3)_4Cl_2]^+$	Dichlorotetraamminecobalt(III) ion
$[Ag(NH_3)_2]^+$	Diamminesilver(I) ion
$[Ag(CN)_2]^-$	Dicyanoargentate(I) ion
$[Cr(NH_3)_3Cl_3]$	Trichlorotriamminechromium(III)

others. Leaves of plants are green because they contain the molecule chlorophyll, a complex of magnesium, which absorbs red and blue light but transmits (or reflects) green and yellow light. Figure 21.11 shows the absorption spectrum of chlorophyll.

A substance absorbs light when one or more of its electrons are excited from their ground state energy levels to higher energy levels. If the difference in energy of the levels between which the electron is excited is between 170 and 290 kJ mol^{-1}, then as we saw in Chapter 7, the substance absorbs visible light, that is, light of wavelengths between 700 and 400 nm. For many substances this energy difference is greater than 290 kJ mol^{-1}, so they absorb ultraviolet light, rather than visible light, and are therefore colorless.

In contrast to the spectra of free atoms, which, as we saw in Chapter 7, consist of sharp lines, the spectra of molecules are broad bands. Molecules absorb not one wavelength but a range of wavelengths (see Figure 21.11) because the energy levels of an atom in a molecule are affected by the surrounding atoms. And because atoms in molecules are in constant motion, vibrating around their equilibrium position, their effect on the energy levels of an atom

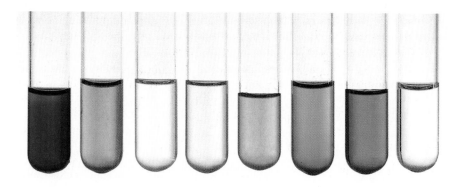

FIGURE 21.10
The Colors of Some Transition Metal Ions.

Hydrated ions are shown here in solution and in the solid state.

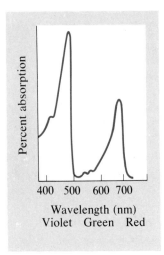

Percent absorption

400 500 600 700

Wavelength (nm)
Violet Green Red

FIGURE 21.11
Absorption Spectrum of Chlorophyll.

This shows the percentage of the incident light that is absorbed at each wavelength. Chlorophyll absorbs red and blue light; therefore the transmitted (or reflected) light is green.

to which they are bonded is constantly varying. Thus the transition of an electron when it absorbs light occurs not between energy levels of precisely fixed energy but between energy levels that have a range of energies.

In transition metal compounds it is the d electrons that are responsible for the absorption of light and therefore for the color. In a free metal atom the five d orbitals all have the same energy but in a coordination compound the interaction between the ligands and the d electrons causes the d orbitals to have different energies and transitions can occur between these d orbitals. The shapes of the d orbitals are shown in Figure 21.12, where we have placed them inside an octahedral AX_6 arrangement of six ligands, each with a bonding electron pair directed toward the metal atom. Because of the repulsion between the ligand electron pairs and the electrons in the d orbitals all the d orbitals increase in energy. But we can see that the regions of maximum electron density of the d_{xy}, d_{yz}, and d_{xz} orbitals point in directions *between* the ligands while the regions of maximum electron density of the d_{z^2} and $d_{x^2-y^2}$ orbitals point directly *at* the ligands. Thus there will be a greater repulsion between the ligand electron pairs and the electrons in the d_{z^2} and $d_{x^2-y^2}$ orbitals than between the ligand electron pairs and electrons in the d_{xy}, d_{yz}, and d_{xz} orbitals. This means that electrons in the d_{xy}, d_{yz}, and d_{xz} orbitals have a lower energy than electrons in the $d_{x^2-y^2}$ and d_{z^2} orbitals. In other words the five d orbitals are split into two sets as shown in Figure 21.13. The energy difference between the two sets of levels is given the symbol Δ_o, where the subscript o indicates that the splitting is caused by an octahedral arrangement of ligands.

The simplest case that we can consider is the $Ti(H_2O)_6^{3+}$ ion which has a violet color because it has an absorption band with a maximum at 500 nm. The Ti^{3+} ion has a d^1 configuration and in the ground state the d electron is in one of the lower energy d_{xy}, d_{yz}, or d_{xz} orbitals. Absorption of a quantum of light gives an excited state with the electron in one of the d_{z^2} or $d_{x^2-y^2}$

FIGURE 21.12
The Five d Orbitals and their Relation to Six Ligands in an Octahedral Arrangement around the Central Metal Ion.

The $d_{x^2-y^2}$ and d_{z^2} orbitals point toward the ligands whereas the d_{xy}, d_{yz}, and d_{xz} orbitals point between the ligands.

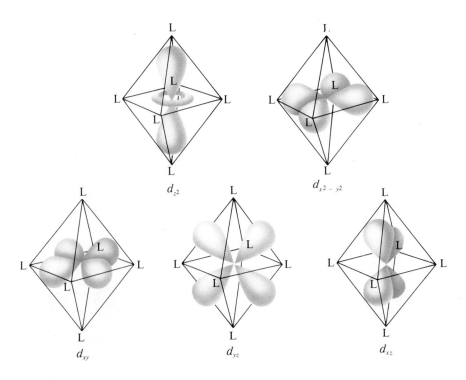

d_{z^2} $d_{x^2-y^2}$

d_{xy} d_{yz} d_{xz}

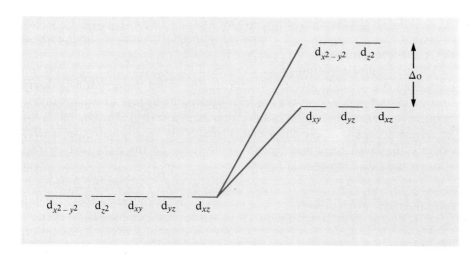

FIGURE 21.13
Energies of the 3d Orbitals
for a Metal Ion in an
Octahedral Complex.

In the free metal ion all five d
orbitals have the same energy but
in an octahedral complex their
energies are increased and split
into two sets. The difference in
energy between the two sets is designated Δ_o.

orbitals (Figure 21.14). From the wavelength of the maximum of the absorption band we can calculate the energy of the light quantum absorbed which
is equal to the energy difference between the two sets of d orbitals, which in
this case is 240 kJ mol^{-1}. This energy difference depends on the strength of
the interaction between the ligands and the d electrons. If the interaction is
weak the splitting, Δ_o, i small, but if the interaction is strong the splitting is
large. Thus we can see that the energy of the absorbed light and therefore the
color of the complex will depend very much on how strongly the ligands interact with the metal atom and therefore on the nature of the ligands. Thus
$[Cu(H_2O)_4]^{2+}$ is pale blue, but $[Cu(NH_3)_4{}^{2+}]$ is an intense deep blue;
$[Fe(H_2O)_6]^{3+}$ is a pale violet, but $[Fe(H_2O)_5SCN]^{2+}$ is deep red; and
$[Co(NH_3)_6]^{3+}$ is yellow, but $[Co(NH_3)_5Cl]^{2+}$ is purple. However, in these
cases, and in fact in almost all the other cases, the metal atom has more than
one d electron and a quantitative treatment is considerably more complicated
because we have to take into account not only the splitting of the energies of
the d orbitals but also the repulsion energies between two or more d electrons.
Nevertheless we can get an approximate idea of the strength of the interaction
between the ligands and transition metal ions from the colors of the complexes
of a given ion. In general it is found that the splitting between the d levels
increases in the order

$$Cl^- < F^- < H_2O < NH_3 < en < CN^-$$

The ordering of ligands in terms of their ability to cause a splitting of the d levels

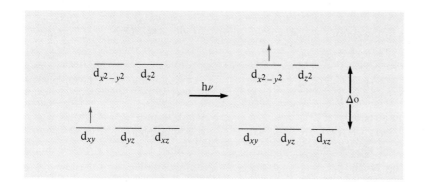

FIGURE 21.14
Absorption Spectrum of
$Ti(H_2O)_6{}^{3+}$.

The absorption spectrum of
$Ti(H_2O)_6{}^{3+}$ has a broad band with
a maximum at 500 nm. This corresponds to a d orbital splitting
of 240 kJ mol^{-1}.

is called the **spectrochemical series**. We can see from Table 21.14 that the colors of several Co^{3+} complexes are consistent with the ordering of the ligands in the spectrochemical series.

The model of the splitting of the energies of the d orbitals in a coordination compound that we have just described is usually called the **crystal field theory** because in its simplest form the model considers the ligands to be monatomic negative ions surrounding the metal cation as in a crystal. The ligands at the left of the spectrochemical series that interact only weakly with the metal d orbitals are called **weak-field ligands**, while those on the right that interact strongly are called **strong-field ligands**.

MAGNETIC PROPERTIES OF COORDINATION COMPOUNDS

Many transition metal complexes are **paramagnetic**, that is, they are attracted by a magnetic field. This is in contrast with almost all compounds of the main group elements which are slightly repelled by a magnetic field and are said to be **diamagnetic**. We saw in Chapter 7 that an electron has magnetic properties; it can be imagined to be spinning around its own axis thereby generating a magnetic field. In almost all compounds of the main group elements the electrons are present in pairs of opposite sign. The magnetic field generated by one electron in a pair is cancelled by the opposite magnetic field of the electron of opposite spin. Thus almost all compounds of the main group elements are nonmagnetic—they are said to be diamagnetic. The only exceptions are the small number of stable free radicals such as we discussed in Chapter 18 which, because they have an odd number of electrons, must have one unpaired electron. However, the situation is different for the transition metals, because in many of their complexes the metal atom has unpaired d electrons. Such compounds are therefore paramagnetic. The strength of their paramagnetism, which is a function of the number of unpaired electrons, can be found experimentally by measuring the force by which they are attracted into a magnetic field. Hence we can count the number of unpaired electrons in a complex ion. This gives us further useful information about the splitting of the d levels in such complexes.

As an example let us consider the complexes of $Fe^{2+}(d^6)$. Some of these, such as $Fe(H_2O)_6^{2+}$ have four unpaired electrons and are paramagnetic, where-

TABLE 21.14
Colors of Some Cobalt Complexes

Co^{3+} Complex	Wavelength Absorbed (nm)	Color of Light Absorbed	Color of Complex
CoF_6^{3-}	700	Red	Green
$[Co(CO_3)_3]^{3-}$	640	Red orange	Green blue
$[Co(H_2O)_6]^{3+}$	600	Orange	Blue
$[Co(NH_3)_5Cl]^{2+}$	535	Yellow	Purple
$[Co(NH_3)_5OH]^{2+}$	500	Blue green	Red
$[Co(NH_3)_6]^{3+}$	475	Blue	Yellow orange
$[Co(CN)_5Br]^{3-}$	415	Violet	Yellow

as others, such as $Fe(CN)_6^{2-}$, have no unpaired electrons, and are therefore diamagnetic. This difference in magnetic properties results from a difference in the magnitude of the d orbital splitting (Figure 21.15). Water is a weak-field ligand and causes only a small splitting of the d orbitals, while CN^- is a strong-field ligand which causes a large splitting of the d orbitals. We know from Hund's rule that electrons occupy a set of orbitals of the same energy one at a time keeping their spins parallel. In order to minimize their mutual repulsion energy they only pair up after each orbital contains one electron. The same rule applies to the d orbitals, even if they do not all have the same energy provided that the splitting is not too large. In such a case the total energy of the d electrons is lower when they occupy separate orbitals than it would be if they were paired up in the set of orbitals of lowest energy. This is the case in $Fe(H_2O)_6^{2+}$ which, as we see in Figure 21.15, has four unpaired electrons. However, in $Fe(CN)_6^{2-}$, because CN^- is a strong-field ligand, the separation between the two sets of d levels is large. In this case the total energy of the d electrons is lower if they are all in the lowest energy set, despite the increased energy of repulsion, than if two of the electrons were in the higher energy set. So in this case there are six electrons in three orbitals and there are therefore no unpaired electrons.

$Fe(H_2O)_6^{2+}$ which has a maximum number of unpaired electrons is called a **high-spin complex**, whereas $Fe(CN)_6^{4-}$, which has fewer unpaired electrons, is called a **low-spin complex**.

=== Example 21.1 ===================================

PREDICTING THE NUMBER OF UNPAIRED ELECTRONS IN A COMPLEX

Predict the number of unpaired electrons in the complex ion $[Cr(CN)_6]^{4-}$.

Solution

Because each CN^- in the complex has a charge of -1 and the overall charge on the complex ion is -4, the charge on the metal ion must be $+2$ ($-6 + 2 = -4$). It is therefore Cr^{2+}, which has a d^4 configuration. Since CN^- is a strong-field

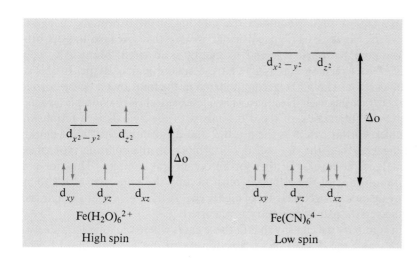

FIGURE 21.15
Effect of the Ligand on the Electron Configuration of Fe^{2+}.

H_2O is a weak-field ligand that causes only a small splitting of the d orbitals: the electrons occupy all the orbitals according to Hund's rule, giving a maximum number of unpaired electrons. CN^- is a strong-field ligand that causes a large splitting of the d orbitals: the electrons occupy only the low energy set of three orbitals so that in this case there are no unpaired electrons.

ligand we expect that the splitting of the energies of the d orbitals will be large. So the energy level diagram will be

$$
\begin{array}{ccc}
\underline{} & \underline{} & \\
d_{x^2-y^2} & d_{z^2} & \qquad \uparrow \\[4pt]
\underline{\uparrow\downarrow} \quad & \underline{\uparrow} \quad & \underline{\uparrow} \qquad \Delta_0 \text{ (large)} \\
d_{xy} & d_{yz} & d_{xz} \qquad \downarrow
\end{array}
$$

All the electrons will be in the lower set of orbitals, thus leaving 2 unpaired electrons.

EXERCISE 21.4

Predict the number of unpaired electrons in the complex ion CoF_6^{3-}.

DEVIATIONS FROM THE STRUCTURES PREDICTED BY THE VSEPR MODEL

As we have mentioned, many AX_4 complexes are square planar in shape rather than tetrahedral. For example, several complexes of Cu^{2+}, such as $CuCl_4^{2-}$, are square planar. The Cu^{2+} ion has a d^9 configuration. Let us first consider the Cu^{2+} ion in an octahedral complex. We can see from Figure 21.16 that either the d_{z^2} or the $d_{x^2-y^2}$ orbital has only one electron. If there is only one electron in the $d_{x^2-y^2}$ orbital there will be less electron density in the xy plane so that the four ligands in this plane will suffer less repulsion than the two ligands in the z direction. The octahedral shape of the complex will therefore be distorted because the ligands in the z direction will be pushed further away from the metal ion than the four ligands in the xy plane. In other words, the shape will be an elongated "octahedron". Alternatively, if an electron is missing from the d_{z^2} orbital we can see that the "octahedron" will be flattened (Figure 21.16). It is not possible to predict which of these distortions will occur, but if the interaction between the ligands and the d electrons is strong enough the complex will be distorted in one of these two ways. Almost all complexes of this type in fact have an elongated "octahedral" shape. Moreover, the interaction with the d electrons is often strong enough that two ligands along the z axis are lost, giving a square planar AX_4 complex.

In a square planar complex the splitting of the d orbitals changes to that shown in Figure 21.17. In particular we see that because the d_{z^2} orbital no longer points toward a ligand its energy is lowered. Many AX_4 complexes of Ni^{2+} (d^8) such as $Ni(CN)_4^{2-}$ have a square planar shape. In this case the eight electrons can all be accommodated in the four lowest energy orbitals. We can contrast this with the situation in a tetrahedral Ni^{2+} coomplex. Figure 21.17 shows the interaction of the d orbitals with the ligands in a tetrahedral AX_4 complex. In this case it is the d_{xy}, d_{yz}, and d_{xz} orbitals that point more toward the ligands than the d_{z^2} and $d_{x^2-y^2}$ orbitals so the splitting pattern of the d orbitals is the reverse of that for an octahedral complex. Thus for a d^8 configuration four of the electrons must occupy the upper set of d orbitals. For many ligands this geometry is of higher energy than the square planar geometry, so the square planar geometry is preferred.

In summary we can say that in those cases where the interaction between the ligands and the d shell of a transition metal ion is weak the d electrons do

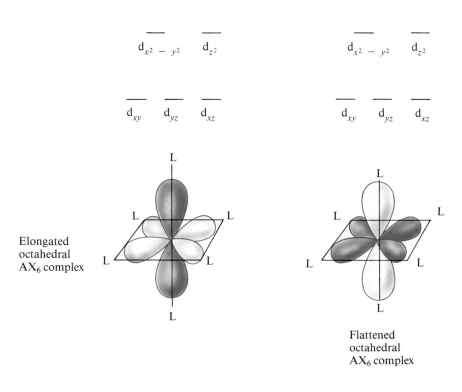

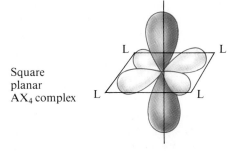

FIGURE 21.16
Effect of a Nonspherical d Shell on the Shape of a Complex.

In an octahedral complex of a d^9 ion such as Cu^{2+} one of the upper set of d orbitals has only one electron. Either the four ligands in the xy plane experience a smaller repulsion giving an elongated "octahedral" shape or the two ligands in the z direction experience a smaller repulsion giving a flattened "octahedral" shape. Most complexes of this type have the elongated shape and in many cases two ligands are lost giving a square planar AX_4 geometry.

not distort the shape and the VSEPR rules are obeyed. But when the interaction between the d shell and the ligands is strong the VSEPR predicted shapes may be distorted. In particular an AX_6 complex may be distorted so that it has an elongated "octahedral" shape and an AX_4 complex may have a square planar rather than a tetrahedral shape.

FIGURE 21.17
d Orbital Splittings in Square Planar and Tetrahedral AX₄ Complexes.

In a square planar complex the lowest energy orbitals are d_{xz} and d_{yz} and the highest energy orbital is $d_{x^2-y^2}$. In a tetrahedral complex the d orbitals are split into a low energy $d_{x^2-y^2}$, d_{z^2} set and a high energy d_{xy}, d_{yz}, and d_{xz} set.

AX₄ SQUARE PLANAR

AX₄ TETRAHEDRAL

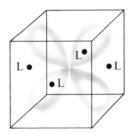

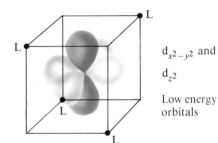

d_{yz}

Lowest energy orbital

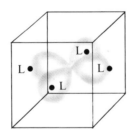

$d_{x^2-y^2}$ and d_{z^2}

Low energy orbitals

$d_{x^2-y^2}$

Highest energy orbital

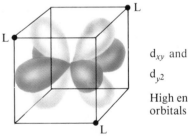

d_{xy} and d_{y^2}

High energy orbitals

21.4
Second and Third Series of Transition Metals

We discuss the second and third series of transition elements in periods 5 and 6 only very briefly, considering in detail only silver, gold, and mercury, which are metals of significant historical and practical importance.

In the fifth period, after two electrons enter the 5s subshell to give the alkali metal rubidium, Rb (5s¹), and the alkaline earth metal strontium, Sr (5s²), the elements from yttrium, Y, to cadmium, Cd, constitute a *second series of ten transition metals* in which the 4d subshell is filled with electrons (Table 21.15).

The second series of ten transition elements is followed by six main group elements, In, Sn, Sb, Te, I, and Xe, completing the eighteen elements in the fifth

TABLE 21.15
Electron Configurations for Elements from Rb to Rn

Period		n =	4				5				6	
		s	p	d	f	s	p	d	f	s	p
5	Rb	$4s^2$	$4p^6$			$5s^1$					
	Sr	$4s^2$	$4p^6$			$5s^2$					
	Y	$4s^2$	$4p^6$	$4d^1$		$5s^2$					
	Zr	$4s^2$	$4p^6$	$4d^2$		$5s^2$					
	Nb	$4s^2$	$4p^6$	$4d^4$		$5s^1$					
	Mo	$4s^2$	$4p^6$	$4d^5$		$5s^1$					
	Tc	$4s^2$	$4p^6$	$4d^5$		$5s^2$					
	Ru	$4s^2$	$4p^6$	$4d^7$		$5s^1$					
	Rh	$4s^2$	$4p^6$	$4d^8$		$5s^1$					
	Pd	$4s^2$	$4p^6$	$4d^{10}$							
	Ag	$4s^2$	$4p^6$	$4d^{10}$		$5s^1$					
	Cd	$4s^2$	$4p^6$	$4d^{10}$		$5s^2$					
	In	$4s^2$	$4p^6$	$4d^{10}$		$5s^2$	$5p^1$				
	Sn	$4s^2$	$4p^6$	$4d^{10}$		$5s^2$	$5p^2$				
	Sb	$4s^2$	$4p^6$	$4d^{10}$		$5s^2$	$5p^3$				
	Te	$4s^2$	$4p^6$	$4d^{10}$		$5s^2$	$5p^4$				
	I	$4s^2$	$4p^6$	$4d^{10}$		$5s^2$	$5p^5$				
	Xe	$4s^2$	$4p^6$	$4d^{10}$		$5s^2$	$5p^6$				
6	Cs	$4s^2$	$4p^6$	$4d^{10}$		$5s^2$	$5p^6$			$6s^1$	
	Ba	$4s^2$	$4p^6$	$4d^{10}$		$5s^2$	$5p^6$			$6s^2$	
	La	$4s^2$	$4p^6$	$4d^{10}$		$5s^2$	$5p^6$	$5d^1$		$6s^2$	
	Ce	$4s^2$	$4p^6$	$4d^{10}$	$4f^2$	$5s^2$	$5p^6$			$6s^2$	
	Pr	$4s^2$	$4p^6$	$4d^{10}$	$4f^3$	$5s^2$	$5p^6$			$6s^2$	
	Nd	$4s^2$	$4p^6$	$4d^{10}$	$4f^4$	$5s^2$	$5p^6$			$6s^2$	
	Pm	$4s^2$	$4p^6$	$4d^{10}$	$4f^5$	$5s^2$	$5p^6$			$6s^2$	
	Sm	$4s^2$	$4p^6$	$4d^{10}$	$4f^6$	$5s^2$	$5p^6$			$6s^2$	
	Eu	$4s^2$	$4p^6$	$4d^{10}$	$4f^7$	$5s^2$	$5p^6$			$6s^2$	
	Gd	$4s^2$	$4p^6$	$4d^{10}$	$4f^7$	$5s^2$	$5p^6$	$5d^1$		$6s^2$	
	Tb	$4s^2$	$4p^6$	$4d^{10}$	$4f^9$	$5s^2$	$5p^6$			$6s^2$	
	Dy	$4s^2$	$4p^6$	$4d^{10}$	$4f^{10}$	$5s^2$	$5p^6$			$6s^2$	
	Ho	$4s^2$	$4p^6$	$4d^{10}$	$4f^{11}$	$5s^2$	$5p^6$			$6s^2$	
	Er	$4s^2$	$4p^6$	$4d^{10}$	$4f^{12}$	$5s^2$	$5p^6$			$6s^2$	
	Tm	$4s^2$	$4p^6$	$4d^{10}$	$4f^{13}$	$5s^2$	$5p^6$			$6s^2$	
	Yb	$4s^2$	$4p^6$	$4d^{10}$	$4f^{14}$	$5s^2$	$5p^6$			$6s^2$	
	Lu	$4s^2$	$4p^6$	$4d^{10}$	$4f^{14}$	$5s^2$	$5p^6$	$5d^1$		$6s^2$	
	Hf	$4s^2$	$4p^6$	$4d^{10}$	$4f^{14}$	$5s^2$	$5p^6$	$5d^2$		$6s^2$	
	Ta	$4s^2$	$4p^6$	$4d^{10}$	$4f^{14}$	$5s^2$	$5p^6$	$5d^3$		$6s^2$	
	W	$4s^2$	$4p^6$	$4d^{10}$	$4f^{14}$	$5s^2$	$5p^6$	$5d^4$		$6s^2$	
	Re	$4s^2$	$4p^6$	$4d^{10}$	$4f^{14}$	$5s^2$	$5p^6$	$5d^5$		$6s^2$	
	Os	$4s^2$	$4p^6$	$4d^{10}$	$4f^{14}$	$5s^2$	$5p^6$	$5d^6$		$6s^2$	
	Ir	$4s^2$	$4p^6$	$4d^{10}$	$4f^{14}$	$5s^2$	$5p^6$	$5d^7$		$6s^2$	
	Pt	$4s^2$	$4p^6$	$4d^{10}$	$4f^{14}$	$5s^2$	$5p^6$	$5d^9$		$6s^1$	
	Au	$4s^2$	$4p^6$	$4d^{10}$	$4f^{14}$	$5s^2$	$5p^6$	$5d^{10}$		$6s^1$	
	Hg	$4s^2$	$4p^6$	$4d^{10}$	$4f^{14}$	$5s^2$	$5p^6$	$5d^{10}$		$6s^2$	
	Tl	$4s^2$	$4p^6$	$4d^{10}$	$4f^{14}$	$5s^2$	$5p^6$	$5d^{10}$		$6s^2$	$6p^1$
	Pb	$4s^2$	$4p^6$	$4d^{10}$	$4f^{14}$	$5s^2$	$5p^6$	$5d^{10}$		$6s^2$	$6p^2$
	Bi	$4s^2$	$4p^6$	$4d^{10}$	$4f^{14}$	$5s^2$	$5p^6$	$5d^{10}$		$6s^2$	$6p^3$
	Po	$4s^2$	$4p^6$	$4d^{10}$	$4f^{14}$	$5s^2$	$5p^6$	$5d^{10}$		$6s^2$	$6p^4$
	At	$4s^2$	$4p^6$	$4d^{10}$	$4f^{14}$	$5s^2$	$5p^6$	$5d^{10}$		$6s^2$	$6p^5$
	Rn	$4s^2$	$4p^6$	$4d^{10}$	$4f^{14}$	$5s^2$	$5p^6$	$5d^{10}$		$6s^2$	$6p^6$

period. Xenon, the last element in the period, has the electron arrangement $[Kr]4d^{10}5s^25p^6$. Neither the $n = 4$ shell nor the $n = 5$ shell is yet complete, since the 4f subshell and the 5d, 5f, and 5g subshells are still empty. In the two elements Cs and Ba, the 6s shell is completed. The 4f shell is then filled, which gives rise to the lanthanide series of 14 elements following lanthanum. The filling of the 4f shell completes the $n = 4$ shell, which then has a maximum of 32 electrons. The lanthanides are followed by the third series of transition metals in which the 5d subshell is filled. The most familiar of these metals are tungsten, platinum, gold, and mercury.

SILVER AND GOLD

Roman coins were made from a natural mixture of gold and silver.

Copper, silver, and gold are in the same group of transition metals and are known as the *coinage metals*. They all have a $d^{10}s^1$ outer-electron configuration. Because of the high core charges of these elements, we expect only a few of the d electrons to be available for compound formation. In fact, all three elements form compounds in the $+1$ oxidation state, in which only the single s electron has been used, leaving a complete d^{10} subshell. However, these elements also form compounds in higher oxidation states. Indeed, the most common oxidation state of copper is Cu(II). Silver forms both Ag(II) and Ag(III) compounds, but these compounds are rather uncommon. Gold has a common Au(III) oxidation state in addition to Au(I).

Both silver and gold are rare elements, but they have long been known in their elemental forms. Some of their properties are summarized in Table 21.16. Like copper, they are relatively soft metals that are very malleable and ductile, and they have relatively low melting points compared with many other metals. Silver, like most metals, has a silvery white color, but the red color of copper and the yellow of gold are unusual. Because they occur in the free state in nature, they are easily worked into different shapes, and because they are rather inert, they have had many uses since ancient times (Box 10.1). Gold, because it is completely resistant to the atmosphere and because it has a fine yellow color and luster and is rare, has always been valued for jewelry and other ornamental purposes. It is the most malleable and most ductile of all metals. It can, for example, be hammered into sheets only 10 nm (10^{-6} cm) thick. Very thin sheets of gold—gold leaf—have been used for many purposes, from decorating books to decorating large buildings.

The three metals, copper, silver, and gold, are known as the coinage metals because of their traditional use for coins. However, as gold and silver have become more valuable, their use for this purpose has greatly decreased. Except for special commemorative coins, gold and silver coins are no longer in current use. Up to the mid 1960s U.S. silver coins contained 90% silver and 10% copper.

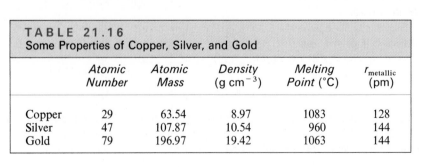

TABLE 21.16
Some Properties of Copper, Silver, and Gold

	Atomic Number	Atomic Mass	Density (g cm^{-3})	Melting Point (°C)	$r_{metallic}$ (pm)
Copper	29	63.54	8.97	1083	128
Silver	47	107.87	10.54	960	144
Gold	79	196.97	19.42	1063	144

This composition is known as coin silver. Sterling silver is an alloy of 92% silver and 8% copper used in tableware. "Silver" coins are now made from copper–nickel alloy. Pure gold is too soft for normal use in jewelry, so it is alloyed with copper, silver, and other metals.

COMPOUNDS OF SILVER Silver occurs not only as the free metal but also as *argentite*, Ag_2S, and *chlorargyrite*, $AgCl$. To extract silver from the sulfide it is treated with a solution of sodium cyanide, which dissolves the silver as the complex ion $Ag(CN)_2{}^-$. Addition of zinc to this solution then precipitates metallic silver:

$$Ag_2S(s) + 4CN^-(aq) \longrightarrow 2Ag(CN)_2{}^-(aq) + S^{2-}(aq)$$

$$2Ag(CN)_2{}^-(aq) + Zn(s) \longrightarrow 2Ag(s) + Zn(CN)_4{}^{2-}(aq)$$

Silver tarnishes in the atmosphere because of the presence of traces of hydrogen sulfide, which reacts with silver in the presence of oxygen to form a thin film of black silver sulfide:

$$4Ag(s) + 2H_2S(g) + O_2(g) \longrightarrow 2Ag_2S(s) + 2H_2O(l)$$

Sulfur-containing proteins in eggs and other foods stain silver spoons by a similar reaction (see Experiment 17.6).

Silver, like copper, is insoluble in dilute hydrochloric and sulfuric acids but dissolves in dilute nitric acid to give silver nitrate, $AgNO_3$. This is a colorless crystalline salt and is the most common compound of silver. Silver nitrate is easily reduced to metallic silver by organic matter. If a drop of silver nitrate solution is spilled on the skin, it leaves a brown-black stain of finely divided metallic silver.

Silver oxide, Ag_2O, is obtained as a dark brown precipitate when sodium hydroxide is added to a solution of silver nitrate:

$$2Ag^+(aq) + 2OH^-(aq) \rightleftharpoons Ag_2O(s) + H_2O(l)$$

It is slightly soluble in water, producing a weakly basic solution because it is in equilibrium with Ag^+ and OH^-.

When Cl^-, Br^-, or I^- is added to an aqueous solution of silver ion, a precipitate of the insoluble silver halide is obtained. Silver chloride is white, $AgBr$ is pale yellow, and AgI is bright yellow. In contrast, silver fluoride, AgF, is very soluble in water. The formation of these precipitates is often used as a test for Cl^-, Br^-, and I^- and also for Ag^+ (see Chapters 5 and 15). The test can be confirmed and I^- can be distinguished from Cl^- and Br^- by noting the solubility of the precipitates in aqueous ammonia solution. Silver chloride and $AgBr$ both dissolve because of the formation of the $Ag(NH_3)_2{}^+$ complex ion, but AgI remains insoluble.

Silver chloride, bromide, and iodide slowly turn black when exposed to ordinary daylight, because they undergo a photochemical decomposition to give silver and the corresponding halogen:

$$2AgBr \xrightarrow{hv} 2Ag + Br_2$$

This reaction is the basis of the photographic process (Experiment 21.12 and Box 21.2). Thirty percent of the silver used by industry in the United States goes into the manufacture of photographic film.

Experiment 21.12

The Photographic Process

A key is placed on a filter paper covered with powdered white silver chloride.

The silver chloride is then exposed to a strong light. The silver chloride slowly darkens as grey-black silver is formed.

When the key is removed, a white image of the key is left because the silver chloride underneath the key was protected from the light.

COMPOUNDS OF GOLD Gold is a very unreactive metal, and most of its compounds can be reduced or decomposed back to gold rather easily. It does not react with oxygen under any conditions or with any of the common acids. Gold does dissolve in a mixture of concentrated nitric and hydrochloric acids, which has long been known as *aqua regia*, forming a yellow solution of the strong acid $HAuCl_4$. Yellow crystals of the hydrate $(H_3O^+)(AuCl_4^-) \cdot 3H_2O$ can be obtained from this solution. When heated, this hydrate loses water and HCl to give yellow $Au(III)Cl_3$, which then loses Cl_2 to give yellow $Au(I)Cl$ and finally elemental gold.

When $OH^-(aq)$ is added to a solution of $HAuCl_4$, $Au(OH)_3$ is precipitated. When $Au(OH)_3$ is heated, it gives brown Au_2O_3, which decomposes at 150 °C to gold and oxygen.

EFFECT OF COMPLEX-ION FORMATION ON SOLUBILITY

Silver bromide and silver chloride are soluble in an aqueous ammonia solution, whereas silver iodide is insoluble because of the much lower solubility in water of silver iodide than silver chloride and bromide. Let us look at this difference in solubility in a quantitative way.

The equilibria that we are concerned with are

$$AgCl(s) \rightleftharpoons Ag^+(aq) + Cl^-(aq) \qquad K_{sp} = 1.8 \times 10^{-10} \text{ mol}^2 \text{ L}^{-2} \text{ (see Table 15.10)}$$

and

$$Ag^+(aq) + 2NH_3(aq) \rightleftharpoons Ag(NH_3)_2^+(aq) \qquad K_f = 1.0 \times 10^8 \text{ mol}^{-2} \text{ L}^2$$

(see Table 21.12)

These equations may be added to give the equation for the overall equilibrium:

$$AgCl(s) + 2NH_3(aq) \rightleftharpoons Ag(NH_3)_2^+(aq) + Cl^-(aq)$$

BOX 21.2

PHOTOGRAPHY

Black and white photographic film is composed of very small crystals of a silver halide (usually silver bromide) suspended in gelatin—called photographic emulsion—coated on a cellulose acetate plastic. When the film is briefly exposed to light, a very small fraction of the silver ions in some of the silver halide crystals undergo photochemical decomposition, producing silver atoms. This process is called *photosensitization*. The crystals that are sensitized in this way are much more easily reduced completely to silver than the crystals that have not been sensitized. The reason sensitized crystals are much more easily reduced is not well understood, but it is a very important feature of the photographic process.

The process of reducing the sensitized crystals is known as *developing*, and the reducing agent is the *developer*. Organic reducing agents are used that penetrate the gelatin and rapidly reduce the sensitized crystals but only reduce the nonsensitized crystals very slowly. A common developer is 1,4-dihydroxybenzene (hydroquinone), which is oxidized to quinone, a ketone containing two carbonyl groups:

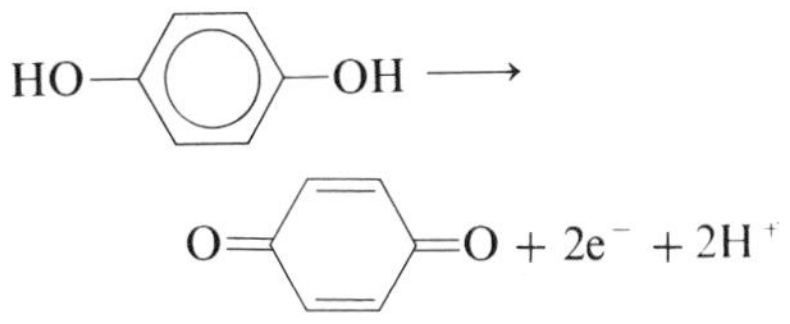

When the sensitized film is developed, the sensitized crystals are reduced to silver; the amount of elemental silver is thereby increased by a factor of about 10^{10}. The image on the film is therefore intensified by the same factor. In the next step, called *fixing*, undeveloped crystals of silver halide are dissolved by a solution of sodium thiosulfate, commonly called *hypo*. Without this step nonsensitized crystals would be slowly reduced by light, and the whole film would eventually turn black. The sodium thiosulfate solution dissolves the nonsensitized silver halide crystals by forming the complex ion $Ag(S_2O_3)_2{}^{3-}$:

$$AgBr(s) + 2S_2O_3{}^{2-} \longrightarrow$$
$$Ag(S_2O_3)_2{}^{3-} + Br^-$$

After the film has been thoroughly washed, all that remains on the film is metallic silver. It is denser and therefore darker in the parts of the film that were exposed to the most light; it is less dense, and therefore not so dark, in those parts of the film that were exposed to less light. The film is then called a *negative*. When light is shone through the negative onto photographic paper—that is, paper covered with photographic emulsion—a positive print is obtained by essentially the same process.

The corresponding equilibrium constant is obtained by multiplying the equilibrium constants K_{sp} and K_f. Therefore we have

$$K = K_{sp}K_f = \left(\frac{[Ag^+][Cl^-][Ag(NH_3)_2{}^+]}{[Ag^+][NH_3]^2}\right)_{eq} = \left(\frac{[Ag(NH_3)_2{}^+][Cl^-]}{[NH_3]^2}\right)_{eq}$$

$$= (1.8 \times 10^{-10}\ mol^2\ L^{-2})(1 \times 10^8\ mol^{-2}\ L^2) = 1.8 \times 10^{-2}$$

If silver chloride is added to a $1M$ NH_3 solution, and if we let x mol L^{-1} be the concentration of $Ag(NH_3)_2{}^+$ that is formed, we may write

$$AgCl(s) + 2NH_3(aq) \rightleftharpoons Ag(NH_3)_2{}^+(aq) + Cl^-(aq)$$

Initial concentrations	1	0	0	mol L^{-1}
Equilibrium concentrations	$1 - 2x$	x	x	mol L^{-1}

$$K = \left(\frac{[Ag(NH_3)_2{}^+][Cl^-]}{[NH_3]^2}\right)_{eq} = \frac{x^2}{(1 - 2x)^2} = 1.8 \times 10^{-2}$$

Taking the square root of both sides, we have

$$\frac{x}{1 - 2x} = 0.13$$

Solving for x gives

$$x = 0.10 \quad or \quad [Ag(NH_3)_2{}^+] = 0.10\ mol\ L^{-1}$$

The concentration of the complex ion $Ag(NH_3)_2{}^+$ in equilibrium with solid silver chloride is quite high, showing that silver chloride has a considerable solubility in ammonia. We can find the concentration of Ag^+ in solution before the addition of ammonia as follows:

$$[Ag^+][Cl^-] = K_{sp} = 1.8 \times 10^{-10}\ mol^2\ L^{-2}$$

But $[Ag^+] = [Cl^-]$. Therefore,

$$[Ag^+]^2 = 1.8 \times 10^{-10}\ mol^2\ L^{-2}$$

$$[Ag^+] = 1.3 \times 10^{-5}\ mol\ L^{-1}$$

The concentration of silver in solution as Ag^+ is only 1.3×10^{-5} mol L^{-1}, but after ammonia is added, the concentration of Ag in solution as $Ag(NH_3)_2{}^+$ is 0.10 mol L^{-1}.

EXERCISE 21.5

Using data from Tables 15.10 and 21.12 find the equilibrium constant for the reaction

$$AgI(s) + 2NH_3(aq) \rightleftharpoons Ag(NH_3)_2{}^+(aq) + I^-(aq)$$

Hence show that the concentration of $Ag(NH_3)_2{}^+$ in a $1M$ aqueous ammonia solution is 9×10^{-5} mol L^{-1}.

In Exercise 21.5 we saw that the concentration of silver in solution in the form of the complex ion $Ag(NH_3)_2{}^+$ is very small. In other words, no significant amount of silver iodide dissolves in $1M$ ammonia solution.

In general, an insoluble precipitate of a metal salt will dissolve in a solution of a complexing agent if the formation constant of the complex is large enough, that is, if the position of the overall equilibrium between the insoluble salt and the complexing agent to give the complex ion lies toward the right. The equilibrium constant for this reaction depends on the relative magnitudes of the solubility product constant of the insoluble salt and the formation constant of the complex. A very insoluble salt will be dissolved only by a complexing agent that forms a very strong complex. Thus ammonia forms a strong enough complex with the silver ion to dissolve silver chloride and bromide, but it will not dissolve the much less soluble silver iodide, which has a much smaller solubility product. Similarly, cyanide ion is used in extracting silver from the ore argentite, Ag_2S. Although Ag_2S is very insoluble in water, it can be dissolved in a solution of CN^- because of the formation of the complex ion $Ag(CN)_2^-$.

MERCURY

Mercury is in the same group of transition metals as zinc and cadmium. Some of the properties of zinc, cadmium, and mercury are summarized in Table 21.17. We discussed zinc earlier in this chapter. Cadmium is sometimes used as a protective coating for iron and steel. It is also used for the control rods in nuclear reactors (Chapter 25) and as a component of the nickel–cadmium cell (Chapter 17). The chemistry of cadmium resembles that of zinc rather closely. The three elements Zn, Cd, and Hg have the same $d^{10}s^2$ valence-shell electron configuration. Only the s electrons are used in compound formation. Thus zinc and cadmium form the Zn^{2+} and Cd^{2+} ions and many complexes of these ions. Mercury also forms a similar Hg^{2+} ion as well as the unusual $[Hg—Hg]^{2+}$ ion in which two mercury atoms are joined by a covalent bond.

The symbol for mercury comes from the Latin *hydrargyrum*, meaning "liquid silver". Mercury is the only metal that is liquid at room temperature (mp = -38.9 °C), although the melting points of two other metals are only just above room temperature (Cs, mp 28.5 °C; Ga, mp 29.8 °C). Some of mercury's many uses depend on this property. It is used, for example, in electric switches, thermometers, and barometers. It is also used in mercury vapor lamps and in fluorescent lighting tubes (Chapter 7). Mercury vapor and soluble mercury compounds are toxic (see Box 21.3).

Mercury forms an alloy with almost every metal except iron. These mercury alloys are known as **amalgams**. The metal in these amalgams is often less reactive than it is in the free state. For example, sodium amalgam reacts only slowly with water (Experiment 21.13). A mercury amalgam called dental amalgam is

TABLE 21.17 Some Properties of Zinc, Cadmium, and Mercury					
	Atomic Number	*Atomic Mass*	*Density* (g cm^{-3})	*Melting Point* (°C)	$r_{metallic}$ (pm)
Zinc	30	65.37	7.14	419	138
Cadmium	48	112.41	8.64	321	154
Mercury	80	200.59	13.55	-39	157

BOX 21.3

MERCURY: POISONING AND POLLUTION

Mercury, lead, and arsenic are commonly known to be toxic, but there are many popular misconceptions about the extent and dangers of environmental pollution by mercury and other toxic elements. It must be emphasized that the toxicity of the element and of various compounds of the element may be very different. For mercury to exert its toxic effect, it must enter the blood, either in the digestive system or in the lungs. Mercury amalgams, because of their insolubility, have long been used in dentistry without any harmful effects. Mercury has only a very small vapor pressure, but if mercury vapor is breathed over a long time, small amounts enter the body through the lungs, and a toxic concentration may be slowly built up. More volatile mercury compounds such as dimethyl mercury are much more dangerous. Soluble mercury compounds are equally dangerous because they can enter the body through the digestive system.

Mercury(II) nitrate, which is soluble in water, was at one time used to treat the fur used to make felt hats. Persons working in this trade who had long exposure to the mercury(II) nitrate often displayed symptoms of mercury poisoning, which include the loss of teeth and hair, and nervous disorders such as tremors, loss of memory, paralysis, and insanity. Hence the origin of the phrase

"mad as a hatter", familiar from the mad hatter in *Alice in Wonderland*.

In recent years the pollution of the environment by mercury has aroused concern. The burning of coal, which contains about 1 ppm Hg, releases about 5000 tons of mercury into the atmosphere every year. In the past paper mills have used a mercury-containing fungicide to

used to fill teeth. It consists of about 50% mercury and 50% "dental alloy", which is mainly silver and tin.

The most important ore of mercury is the sulfide *cinnabar*, HgS. Mercury is easily obtained from the ore by heating it in the air:

$$HgS(s) + O_2(g) \longrightarrow Hg(l) + SO_2(g)$$

Although mercury is a very rare element, it has been known since ancient times, because it is so easily obtained from it ores.

Mercury does not react with dilute acids, but it does react with concentrated nitric acid and, if heated, with concentrated sulfuric acid to give mercury(II) nitrate and mercury(II) sulfate, respectively:

$$Hg(l) + 4HNO_3(conc) \longrightarrow Hg(NO_3)_2(aq) + 2NO_2(g) + 2H_2O(l)$$
$$Hg(l) + 2H_2SO_4(conc) \longrightarrow HgSO_4(aq) + SO_2(g) + 2H_2O(l)$$

Mercury(II) chloride, $HgCl_2$, can be made by heating mercury(II) sulfate with sodium chloride:

$$HgSO_4(s) + 2NaCl(s) \longrightarrow Na_2SO_4(s) + HgCl_2(g)$$

The volatile $HgCl_2$ distills off. Mercury(II) chloride is a covalent compound that even in aqueous solution is present mainly as un-ionized covalent molecules. Since there are only two bonding electron pairs in the valence shell of the mercury atom, $HgCl_2$ is a linear AX_2 molecule.

prevent the growth of fungi during pulping. Mercury compounds therefore occurred in the waste products of the process and even in small amounts in the paper. Large amounts of mercury are used in the production of sodium hydroxide and chlorine by electrolysis of aqueous sodium chloride (see Chapter 17). Although in principle this mercury is recovered, small amounts find their way into the waste products of the process, which are frequently discharged into rivers.

It was once believed that the discharge of mercury compounds into rivers and lakes was safe because they are mostly insoluble and were thought to be converted slowly to very insoluble mercury(II) sulfide. However, it is now clear that this practice constitutes a very considerable hazard. Elemental mercury and mercury(I) compounds become slowly converted to mercury(II) compounds. Bacteria in the water then convert the Hg(II) compounds to CH_3Hg^+ and $(CH_3)_2Hg$. These compounds gradually accumulate in the plants and small organisms growing in the water. Fish feed on these plants and organisms and in turn tend to accumulate the mercury. This mercury then passes into larger fish, which eat the small fish, and in fish such as sharks and swordfish mercury can reach dangerously high levels. Before this danger was realized, there were several cases of mass mercury poisoning. For example, at Minimata, Japan, industrial mercury wastes accumulated in the sediment at the bottom of an ocean bay over many years. Over a period of ten years more than fifty people in Minimata died of mercury poisoning as a result of eating fish caught in the bay. Many others suffered from mercury poisoning, and many children were born with serious deformities and brain damage.

We should note, however, that high levels of mercury in fish are not necessarily always to be attributed to human carelessness. Natural sources account for the mercury found in fish in some isolated mountain lakes and in arctic regions.

Because mercury is eliminated from the body, we can tolerate small amounts. The half-life of inorganic mercury compounds in the body is about six days; that is, half of the mercury that is ingested is eliminated in about six days. If only small amounts enter the body at infrequent intervals, a dangerous concentration of mercury is not built up. In contrast, the half-life of organomercury compounds, such as $(CH_3)_2Hg$, in the body averages about 70 days and is even longer in some organs such as the brain. Thus even if a person regularly ingests even very small amounts of organomercury compounds, the concentration of mercury in the body continues to rise for some months until it reaches a constant level, which may be a toxic level.

Mercury combines with oxygen when heated in the air to give mercury(II) oxide:

$$2Hg(l) + O_2(g) \longrightarrow 2HgO(s)$$

However, when HgO is heated more strongly, it decomposes to mercury and oxygen (Experiment 21.14). Mercury(II) oxide can also be obtained by adding hydroxide ion to a solution of an Hg^{2+} salt:

$$Hg^{2+}(aq) + 2OH^-(aq) \longrightarrow HgO(s) + H_2O(l)$$

When H_2S is passed into a solution of a mercury(II) salt, insoluble black HgS is precipitated.

A very unusual property of mercury is that it forms the Hg_2^{2+} ion in which mercury atoms are joined by a covalent bond. Cadmium is the only other metal known to form an analogous ion. This ion can be obtained, for example, by the reaction between mercury and a solution of mercury(II) nitrate:

$$Hg(l) + Hg(NO_3)_2(aq) \longrightarrow Hg_2(NO_3)_2(aq)$$

The Hg_2^{2+} ion contains mercury in the +1 oxidation state and is called the mercury(I) ion.

If chloride ion is added to a solution of mercury(I) nitrate, a white insoluble precipitate of mercury(I) chloride, Hg_2Cl_2, is obtained. Like mercury(II) chloride, Hg_2Cl_2 is a covalent molecule with a linear structure in which both mercury atoms have the expected linear AX_2 geometry.

Experiment 21.13
Sodium Amalgam

A piece of sodium floats on the surface of mercury, which has a much higher density.

When the tube is gently warmed, a solid compound of sodium and mercury (sodium amalgam) is formed. Here this solid amalgam is being lifted out of the tube.

Sodium amalgam reacts considerably more slowly with water than does sodium itself to form hydrogen and an NaOH solution.

Experiment 21.14
The Thermal Decomposition of Mercury(II) Oxide

When red mercury(II) oxide is heated it darkens and decomposes to give oxygen and mercury, which condenses as silver droplets on the cooler part of the tube.

When the tube is allowed to cool, the mercury oxide returns to its original red color.

Dimethyl mercury, CH_3—Hg—CH_3, is a volatile covalent compound. It is an example of an **organometallic compound**, a compound in which alkyl groups, or substituted alkyl groups, are attached to a metal atom. Dimethyl mercury can be prepared by the reaction of mercury–sodium amalgam with chloromethane:

$$Hg(l) + 2Na(l) + 2CH_3Cl(g) \longrightarrow Hg(CH_3)_2(l) + 2NaCl(s)$$

It is a strong-smelling, very toxic, volatile substance (bp 96 °C). It has a linear covalent structure. Reaction of $(CH_3)_2Hg$ with $HgCl_2$ gives methylmercury chloride, CH_3HgCl, another linear covalent molecule:

$$(CH_3)_2Hg + HgCl_2 \longrightarrow 2CH_3HgCl$$

If the Cl is replaced by sulfate or nitrate, an ionic salt that contains the covalent methylmercury cation, CH_3—Hg^+, is obtained, for example, $CH_3Hg^+NO_3^-$ (see Box 21.3).

All the transition metals form similar organometallic compounds. The study of the preparation and reactions of these compounds is a very active field of research at the present time, partly because many of them have unusual and intriguing structures and partly because some of them are very effective and convenient catalysts for some important organic reactions.

IMPORTANT TERMS

The **actinides** are a group of 14 elements in period 7 between actinium and element 104. In this series of elements the 5f orbitals are being filled.

An **amalgam** is a liquid or solid alloy of mercury.

A **bidentate ligand** is a molecule or ion containing two atoms that can each donate an electron pair to a metal atom, such as ethylenediamine and the oxalate ion.

A **chelate** is a ligand with two or more donor atoms.

A **complex (coordination compound)** is a molecule containing one or more ligands bonded to a metal atom.

A **complex ion** is a complex that has an overall charge.

A **coordinate bond** is the bond between a ligand and a metal atom in a complex. It does not differ from an ordinary single bond.

The **coordination number** is the number of donor atoms bonded to a metal atom.

Crystal field theory is a model for explaining the splitting of the d electron energy levels in coordination compounds. It assumes that the ligands can be represented by monatomic negative ions surrounding the metal cation in a definite geometric arrangement as in an ionic crystal.

Diamagnetic substances are weakly repelled by a magnetic field. All their electrons are paired.

The **formation constant** is the equilibrium constant for the equilibrium between a hydrated ion and a complex ion in aqueous solution.

Geometric isomers are molecules which have the same composition and in which all the atoms are connected to each other in the same way but which differ in the positions of atoms or groups of atoms in space.

High-spin complexes have their d electrons distributed

through all the d orbitals in accordance with Hund's rule to give the maximum number of unpaired electrons.

The **lanthanides** are the 14 metals in period 6 between lanthanum and hafnium.

A **ligand** is a molecule or an ion that is attached to a central metal atom in a complex.

In a **low-spin complex** the d electrons are as far as possible accommodated in the set of d orbitals of lowest energy. A low-spin complex has fewer unpaired electrons than the corresponding high-spin complex.

A **monodentate** ligand is a ligand with only one donor atom.

A **nonstoichiometric compound** is a solid with a formula in which one of the subscripts is nonintegral, for example, $MnO_{1.85}$.

An **organometallic compound** is a compound containing metal–carbon bonds.

Paramagnetic substances are attracted by a magnetic field. Their molecules or ions have one or more unpaired electrons.

A **polydentate ligand** is a ligand in which the number of donor atoms is greater than two, for example, $(EDTA)^{4-}$, which is hexadentate.

The **spectrochemical series** is an ordering of ligands in terms of their ability to cause a splitting of the d electron energy levels.

Strong-field ligands cause a large splitting of the energies of the d orbitals.

The **transition metals** are a group of thirty-five elements that occupy the middle portion of the periodic table between groups 2 and 3 in periods 4, 5, 6, and 7.

Weak-field ligands cause a small splitting of the energies of the d orbitals.

PROBLEMS *

Electron Configurations

1. Write the ground state electron configurations of each of the following ions:

Cr^{3+} Ni^{2+} Cu^+ Co^{2+} Au^{3+} Fe^{3+}

2. What is the ground state electron configuration of each of the following?

Ti   Cr   Fe   Ni   Zn
Au Hg^{2+} Cu^{2+} Ag^+ Mn^{2+}

3. How many 3d electrons are there in each of the following ions?

Cu^{2+} Fe^{3+} Mn^{2+} Ag^+ V^{3+} Ti^{2+}

Oxidation States

4. What is the oxidation number of the transition metal in each of the following compounds and ions?

(a) $KMnO_4$ (b) $K_2Cr_2O_7$

(c) $Ag(CN)_2^-$ (d) $[Co(NH_3)_4Cl_2]^+$

(e) $K_3[Cr(CN)_6]$ (f) Na_2CoCl_4

(g) K_2MnO_4 (h) $MnO(OH)$

(i) VO_2Cl (j) $TiO \cdot SO_4$

5. What is the oxidation number of the metal in each of the following complex ions?

(a) $FeCl_4^-$ (b) $Co(H_2O)_6^{3+}$ (c) $Al(OH)_4^-$

(d) $Ag(NH_3)_2^+$ (e) $Fe(CN)_6^{4-}$ (f) $Fe(CN)_6^{3-}$

6. By writing out their electron configurations, decide which oxidation states are possible in principle for each of the elements with atomic numbers 22, 25, and 27. Which of these oxidation states are in fact found in known compounds of these elements?

7. What is the oxidation state of the transition metal in each of the following ions?

(a) CrO_4^{2-} (b) $Cr_2O_7^{2-}$ (c) MnO_4^-

(d) VO_4^{3-} (e) VO^{2+} (f) FeO_4^{2-}

8. What is the oxidation number of chromium in each of the following?

(a) CrO_3 (b) Cr_2O_3

(c) CrF_2 (d) $PbCrO_4$

(e) $Cr_2(SO_4)_3$ (f) $[Cr(H_2O)_5NH_3]Cl_2$

* Answers to problems numbered in blue appear at the end of the text.

Complexes

9. Consider the complex ion $[Co(NH_3)_3(H_2O)_2Cl]^+$.

(a) Identify each of the ligands and their charges.

(b) What is the oxidation state of cobalt in this ion?

(c) What would be the charge on the complex ion if Cl^- ligands were to replace the water molecules?

(d) What is the geometric arrangement of the ligands around the central Co atom?

10. What is the coordination number of the central metal atom in each of the following compounds?

(a) $[Zn(NH_3)_4]Cl_2$ (b) $[Co(NH_3)_3Cl_3]$

(c) $[Co(NH_3)_5Cl]Cl_2$ (d) $K_2[FeCl_4]$

11. Name each of the complex ions in Problem 10.

12. Draw a diagram of the structure of each of the following complex ions:

(a) $trans[Cr(NH_3)_4Cl_2]^+$ (b) $[Co(C_2O_4)_3]^{3-}$

(c) $[Cr(C_2O_4)Br_4]^{3-}$ (d) $cis[Pt(en)_2(CN)_2]^{2-}$

13. Name each of the complex ions in Problem 12.

14. Addition of $AgNO_3(aq)$ to aqueous solutions of each of the platinum compounds; $PtCl_4 \cdot 6NH_3$, $PtCl_4 \cdot 5NH_3$, $PtCl_4 \cdot 4NH_3$, $PtCl_4 \cdot 3NH_3$ and $PtCl_4 \cdot 2NH_3$ were found by Werner to give 4, 3, 2, 1, and 0 moles of AgCl per mole of complex respectively. How did Werner explain these observations?

15. A compound with the empirical formula $Co(NH_3)_5SO_4Br$ exists in a red form and a violet form. Solutions of the red form give a precipitate of AgBr(s) on addition of $AgNO_3(aq)$, but no precipitate on addition of $BaCl_2(aq)$. In contrast, solutions of the violet compound give a white precipitate of $BaSO_4(s)$ on addition of $BaCl_2(aq)$, but no precipitate on addition of $AgNO_3(aq)$. From these observations draw the structure of each compound.

16. Write balanced equations to represent each of the following observations.

(a) Silver chloride dissolves in an aqueous solution of of sodium thiosulfate, $Na_2S_2O_3$ (see Box 21.2).

(b) The green complex of formula $[Cr(en)_2Cl_2]Cl$ reacts slowly with water to give an orange-brown compound. When an aqueous solution of the orange-brown compound is treated with $AgNO_3(aq)$, 3 mol of $AgCl(s)$ is precipitated per mole of compound.

(c) Insoluble $Ni(OH)_2(s)$ dissolves in an excess of $NH_3(aq)$.

(d) A pink solution of $CoSO_4(aq)$ turns deep blue on addition of concentrated $HCl(aq)$.

17. Explain why calcium oxalate, CaC_2O_4, which is insoluble in water, dissolves in a solution of EDTA.

18. Draw a Lewis structure for the tetrahydroxozincate(II) ion, $Zn(OH)_4{}^{2-}$. What is the formal charge on the zinc atom, and what is the shape of this ion?

19. Four complexes of cobalt have the following compositions by mass: A, 25.22% Co, 45.99% Cl, 29.19% NH_3; B, 25.22% Co, 45.59% Cl, 29.19% NH_3; C, 23.53% Co, 42.46% Cl, 34.01% NH_3; D, 22.03% Co, 39.76% Cl, 38.21% NH_3.

(a) Calculate the empirical formulas of A, B, C and D.

(b) When excess $AgNO_3(aq)$ was added to an aqueous solution of each complex 0.232 g of A gave 0.143 g of $AgCl(s)$; 0.255 g of B gave 0.157 g of $AgCl(s)$; 0.226 g of C gave 0.258 g of $AgCl(s)$; and 0.348 g of D gave 0.559 g of $AgCl(s)$. From these results and the empirical formulas calculated above, suggest possible structures for A, B, C and D.

20. $Fe^{3+}(aq)$ ion forms a deep red complex ion of formula $Fe(SCN)^{2+}(aq)$ with thiocyanate ion, SCN^-, for which the formation constant is $1 \times 10^3 \text{ mol}^{-1} \text{ L}$. In a given solution, the equilibrium concentrations of $Fe^{3+}(aq)$ and $SCN^-(aq)$ were $0.010M$ and $0.0003M$, respectively. What was the concentration of the complex ion?

21. Give the formula of each of the following:

(a) Sodium tetrahydroxozincate(II)

(b) Sodium tetrahydroxoaluminate(III)

(c) Bromotriaquoplatinum(II) chloride

(d) Tetrachloroferrate(II) ion

(e) Trisethylenediamminenickel(II) bromide

(f) Potassium tetracyanocobaltate(II)

22. Name the complexes with each of the following formulas:

(a) $KMn(CN)_5$ **(b)** $[Fe(CN)_6]^{4-}$

(c) $[Co(NH_3)_5NO_2]^{2+}$ **(d)** $Cu(NH_3)_4SO_4$

(e) $Co(NH_3)_6Cl_2$ **(f)** $Ni(CO)_4$

23. Would you expect the complex $Zn(NH_3)_4Cl_2$ to be colored? Explain.

* The asterisk denotes the more difficult problems.

24. A solution of cobalt(III) in aqueous ammonia is yellow.

(a) Estimate the wavelength of maximum absorption for this solution.

(b) When acid is added to the solution, would the wavelength of maximum absorption increase, or decrease? Explain.

25. An aqueous solution of manganese(II) nitrate is pale pink, while an aqueous solution of potassium hexacyanomanganate(II) is deep blue. Account for this difference.

*26. The complex ion $[Co(NH_3)_4Cl_2]^+$ exists in *two* isomeric forms. Show that this is consistent with an octahedral structure for the ion, and that it is inconsistent with alternative shapes, such as a planar hexagon, or a triangular prism, in which the cobalt atom is at the center of the hexagon, or prism.

*27. There are three different compounds, A, B, and C, with the empirical formula $CrCl_3 \cdot 6H_2O$. When 10.00 g of the dark green compound A were kept in a closed container with a dehydrating agent, the mass eventually decreased to the constant value of 8.65 g. When 10.00 g of the light green compound B were kept in a closed container with a dehydrating agent, the mass decreased to 9.32 g. When the violet compound C was kept in a closed container with a dehydrating agent, its mass did not change.

(a) What are the structures of A, B, and C?

(b) If an excess of $AgNO_3(aq)$ is added to aqueous solutions containing 1.00 g of each of these compounds, respectively, what mass of $AgCl(s)$ will be precipitated in each case?

28. Calculate the formation constant, K_f, of the complex ion $Cu(NH_3)_4{}^{2+}(aq)$ at 25 °C from the following data:

$$Cu(NH_3)_4{}^{2+} + 2e^- \longrightarrow Cu(s) + 4NH_3 \quad E° = -0.03 \text{ V}$$
$$Cu(H_2O)_4{}^{2+} + 2e^- \longrightarrow Cu(s) + 4H_2O \quad E° = 0.35 \text{ V}$$

29. Explain why the d_{xy}, d_{yz}, and d_{xz} orbitals lie lower in energy than the $d_{x^2-y^2}$ and d_{z^2} orbitals in an octahedral complex of a transition metal ion.

30. The complex ion $[Co(NH_3)_6]^{3+}$ contains no unpaired electrons, whereas $[Mn(NH_3)_6]^{2+}$ contains five. Account for this difference in terms of the splitting of the energies of the 3d orbitals. Which of these ions is paramagnetic?

31. The complex ion $NiCl_4{}^{2-}$ has a tetrahedral shape, whereas the $Ni(CN)_4{}^{2-}$ ion has a square planar shape.

(a) Explain this difference in shape.

(b) Which of these complex ions is paramagnetic?

(c) How many unpaired electrons are there in the paramagnetic ion?

32. Draw diagrams to show the occupation of the d orbitals in each of the following complex ions: $[Fe(CN)_6]^{4-}$, $[Co(NH_3)_6]^{3+}$, and $[Fe(H_2O)_6]^{3+}$. Which of these ions is paramagnetic? How many unpaired electrons are there in each of the paramagnetic ions?

Properties of Transition Metal Compounds

33. How does the acidic character of a transition metal oxide change with change in its oxidation state? Discuss the oxides of a particular transition metal to illustrate your answer.

34. How does the ionic–covalent character of the compounds formed by a transition metal change with change in its oxidation state? Illustrate your answer by describing some of the compounds of one particular transition metal.

35. Explain why the melting points of the first series of transition metals increase along the series to a maximum at chromium and then decrease again. Is your answer consistent with the observed trend in the densities of these transition metals? Explain.

36. In their common compounds, the early members of the first transition series are commonly in their respective highest possible oxidation states, while this is not true for the later members of the series. Discuss this statement and include in your answer reference to each of the following:

(a) The common oxidation states of titanium and vanadium and examples of their compounds

(b) The common oxidation states of cobalt and nickel and examples of their compounds

37. Explain why the compounds of Sc^{3+} are colorless but those of Ti^{3+} and V^{3+} are colored.

38. Explain why the compounds of Cu(II) are generally blue or green, while those of Cu(I) are colorless.

Reactions of Transition Metal Compounds

39. In what form does copper exist in each of the following?

(a) Aqueous copper(II) sulfate solution

(b) After addition of excess of concentrated HCl(aq) to the solution in (a)

(c) After the addition of excess concentrated NH_3(aq) to the solution in (b)

(d) After addition of excess KCN(aq) to the solution in (a)

40. Consider Cr, Mn, Fe, Co, Ni, Cu, and Zn.

(a) Which of these elements form ions that give a colorless aqueous solution?

(b) Which oxidation state is common to all these metals?

(c) Which hydroxides of these metals of general formula $M(OH)_2$ are **(i)** blue, **(ii)** white, **(iii)** green?

(d) Which ion of which metal gives a pale green aqueous solution which forms a red-brown precipitate on the addition of NaOH(aq) and H_2O_2(aq)?

(e) Which ion of which metal gives a pale pink aqueous solution that turns blue on addition of excess of concentrated HCl(aq)?

(f) Which metals form compounds in which their oxidation state is $+6$? Give examples of compounds of this oxidation state for two different metals.

(g) An aqueous solution containing manganese has a pale pink color, which might be due to Mn^{2+}(aq) or to very dilute MnO_4^-(aq). How would you distinguish between these two possibilities?

41. Give reasons for classifying each of the following oxides as acidic, basic, or amphoteric, supporting your answers wherever possible by relevant balanced equations:

(a) MgO   **(b)** SO_3 **(c)** CrO_3 **(d)** ZnO

(e) Cr_2O_3 **(f)** CuO   **(g)** TiO_2 **(h)** V_2O_5

42. Explain each of the following observations.

(a) When H_2S(g) is bubbled through a solution containing both $ZnCl_2$(aq) and $CuCl_2$(aq), and moderately concentrated HCl(aq), only CuS(s) is precipitated.

(b) Silver chloride is insoluble in water but readily soluble in a solution of sodium thiosulfate, $Na_2S_2O_3$(aq).

(c) A solution of iron(III) chloride has a pH <7.

(d) When NH_3(aq) is added to a solution of $CuSO_4$(aq), a pale blue precipitate is formed which dissolves in excess NH_3(aq) to give a deep blue solution.

43. What is the pH of a 0.050M solution of $Cr(NO_3)_3$(aq), given that the pK_a of the $Cr(H_2O)_6^{3+}$(aq) ion is 3.70, and assuming that only the first stage of ionization of the hydrated Cr^{3+} ion is important?

44. A violet aqueous solution of a chromium(III) salt is treated with NaOH(aq) to give a gelatinous gray-green precipitate, which dissolves in excess of the reagent to give a clear green solution. When sodium peroxide, Na_2O_2(s), is added and the solution is boiled, the color changes from green to yellow. Addition of H_2SO_4(aq) to the solution until its pH is less than 7 changes the color from yellow to orange. Explain, using appropriate balanced equations, the above sequence of reactions.

45. Use standard reduction potentials from Table 17.1 to predict what might happen when

(a) A piece of zinc is placed in **(i)** 1.0M $NiCl_2$(aq), **(ii)** 1.0M HCl(aq)

(b) A piece of nickel is placed in **(i)** $1.0M$ $ZnCl_2(aq)$, **(ii)** $1.0M$ $HCl(aq)$

46. By means of appropriate balanced equations, explain each of the following observations.

(a) Addition of acid to a yellow solution of potassium chromate changes the color to orange.

(b) Addition of base to the above orange solution changes the color back to the original yellow.

(c) Heating orange ammonium dichromate crystals causes them to decompose, rather spectacularly, with the evolution of gas and the formation of a bulky green powder.

(d) When $SO_2(g)$ is bubbled into an acidified orange solution of potassium dichromate, a green solution results.

(e) When $H_2S(g)$ is bubbled into an acidified potassium dichromate solution, a green solution with a finely divided precipitate suspended in it is obtained.

47. Explain why oxalic acid can be used to remove rust stains. Write a balanced equation for the reaction involved.

48. Using standard reduction potentials from Table 17.1, place $1.00M$ solutions of each of the following ions in order of increasing oxidizing strength, assuming that in each case the reduction is from $M^{3+}(aq)$ to $M^{2+}(aq)$:

$$Cr^{3+} \quad V^{3+} \quad Fe^{3+} \quad Co^{3+} \quad Mn^{3+} \quad Ti^{3+}$$

Which of these ions is stable in $1.00M$ acid? ($E°$ for $Mn^{3+} + e^- \rightarrow Mn^{2+}$ is 1.50 V, and $E°$ for $Ti^{3+} + e^- \rightarrow Ti^{2+}$ is 2.00 V.)

Miscellaneous

49. Chromium(VI) dioxodichloride, CrO_2Cl_2, is a deep red liquid boiling at 117 °C. Draw its Lewis structure and predict its geometry.

50. A mixture of 0.540 g of pure iron powder and excess sulfur was heated to give a gray-black compound. After burning off excess sulfur, the mass of the gray-black compound was found to be 0.905 g. Calculate the empirical formula of this sulfide of iron. Do you think that this result is accurate, or might it reflect a large experimental error? Explain.

51. What volume of nickel tetracarbonyl gas measured at 80 °C and 2.00 atm pressure results from the complete reaction of 2.50 g of nickel with carbon monoxide? What volume of $CO(g)$, measured at 25 °C and 1 atm, would result from the decomposition of this amount of nickel tetracarbonyl at 200 °C?

52. Brass is a zinc–copper alloy. When 0.50 g of brass was reacted completely with dilute sulfuric acid, 102.8 mL of hydrogen were collected over water at 25 °C and 756 mm Hg

pressure. What is the mass percent composition of brass? (The vapor pressure of water at 25 °C is 23.8 mm Hg.)

53. One important type of stainless steel contains 18 mass % nickel. What minimum mass of nickel sulfide ore, NiS, needs to be processed to give 1 metric ton (10^3 kg) of stainless steel?

54. Vitamin B_{12} is a coordination complex of cobalt with the molecular formula $C_{63}H_{90}O_{14}N_{14}PCo$. What is the mass percent Co in vitamin B_{12}? If the minimum daily requirement of vitamin B_{12} is 1 μg, how much cobalt per day does the human body need to remain healthy?

55. The solubility of $Cr(OH)_3(s)$ in water at 25 °C is 5.6 g L^{-1}. Calculate the value of the solubility product constant of chromium(III) hydroxide at 25 °C.

56. For how long would a current of 1.5 A have to be passed through a solution of $Cr_2(SO_4)_3(aq)$ in order to coat a metal object of surface area 1.00 m^2 with a 0.10-mm layer of chromium. Why do you suppose that the thickness of the coating on chromium-plated steel is generally very thin?

57. The solubility product constant of silver phosphate is 1.8×10^{-18} mol^4 L^{-4} at 25 °C. What is the solubility of silver phosphate in **(a)** water and **(b)** $0.0010M$ silver nitrate solution?

*__58.__ A solution of 0.545 g of mercury(II) nitrate in 100 g of water depresses the freezing point by 0.093 °C, while a solution of 1.084 g of mercury(II) chloride in 100 g of water depresses the freezing point by 0.075 °C. What do these data tell you about the species present in each of these solutions? ($K_f(H_2O) = 1.86$ °C mol^{-1} kg.)

59. A sample of hydrated iron(II) sulfate of mass 0.673 g was dissolved in water to give 250 mL of solution. On titration with $0.020M$ $KMnO_4(aq)$, 25.00 mL of the iron sulfate solution required 24.00 mL $KMnO_4(aq)$ in acid solution to give the first pale pink coloration.

(a) Calculate the mass percent of $FeSO_4$ in hydrated iron(II) sulfate, and hence the value of x in its formula, $FeSO_4 \cdot xH_2O$.

*__60.__ The red color of ruby is due to the presence of a small number of Cr^{3+} ions in octahedral sites in the close-packed arrangement of oxide ions in the mineral corundum, Al_2O_3. Draw a diagram showing the splitting of the energies of the 3d orbitals of the Cr^{3+} ion. If the ruby crystal was subjected to a very high pressure how would you expect the wavelength of the absorption maximum and therefore the color of the crystal to change?

61. Silver chromate has a solubility product constant of 1.7×10^{-12} mol^3 L^{-3} at 25 °C. What is the solubility of silver chromate in $0.10M$ $K_2CrO_4(aq)$?

62. The standard reduction potentials for the reduction of $V^{3+}(aq)$ to $V^{2+}(aq)$, $VO_2^+(aq)$ to $V^{3+}(aq)$, and $Fe^{3+}(aq)$ to $Fe^{2+}(aq)$ in acid solution are, respectively, $+0.20$ V, $+1.00$ V, and $+0.77$ V. Using these data, predict what (if anything) is expected to occur when equal amounts of each of the following are mixed:

(a) $2.0M$ $Fe^{3+}(aq)$ and $2.0M$ $V^{3+}(aq)$

(b) $2.0M$ $Fe^{2+}(aq)$ and $2.0M$ $V^{3+}(aq)$

(a) $2.0M$ $Fe^{3+}(aq)$ and $2.0M$ $V^{2+}(aq)$

*63. A sample of liquid titanium(IV) chloride was dissolved in concentrated HCl(aq) and excess ammonium chloride was added. After boiling, evaporation, and cooling, yellow crystals were deposited. A solution was prepared containing 0.1664 g of the dry yellow crystals. On heating, this solution precipitated 0.0400 g of white titanium dioxide, which was filtered off. The filtrate and washings were diluted to 250 mL of solution in a volumetric flask. When excess of silver nitrate solution was added to 100 mL of this solution, 0.1708 g of silver chloride was precipitated. Analysis of the yellow crystals showed that they contained 8.4 mass % N. Assuming that the yellow crystals contained water of crystallization, determine their empirical formula, and suggest a Lewis structure for this compound.

64. The value of Δ_o for the $[CrF_6]^{3-}$ complex ion is 182 kJ mol^{-1}. What is the expected wavelength of maximum absorbtion for this ion?

65. Standard reduction potentials for the reaction $M^{2+}(aq) + 2e^- \rightarrow M(s)$ for a number of the metals of the first transition metal series are given in Table 17.1. The remainder are

$$Sc^{2+} + 2e^- \longrightarrow Sc(s) \qquad E^\circ = 2.12 \text{ V}$$
$$Ti^{2+} + 2e^- \longrightarrow Ti(s) \qquad E^\circ = 1.63 \text{ V}$$
$$V^{2+} + 2e^- \longrightarrow V(s) \qquad E^\circ = -1.18 \text{ V}$$
$$Mn^{2+} + 2e^- \longrightarrow Mn(s) \qquad E^\circ = -1.18 \text{ V}$$
$$Co^{2+} + 2e^- \longrightarrow Co(s) \qquad E^\circ = -0.28 \text{ V}$$

Predict which of the metals of the first transition metal series should be soluble in $1.00M$ HCl(aq).

*66. (a) Explain why an almost colorless solution of FeSO$_4$(aq) turns yellow-brown when exposed to air, and white Fe(OH)$_2$(s) precipitated from such a solution by adding excess NaOH(aq) rapidly changes to a dirty brown color in air.

(b) Can the above reactions be reversed by bubbling H$_2$(g) through the solutions?

(Assume $1.00M$ solutions and a pressure of H$_2$(g) of 1 atm; use standard reduction potentials from Chapter 17, and $E^\circ = -0.56$ V for the half-reaction $Fe_2O_3(s) + 3H_2O + 2e^- \rightarrow 2Fe(OH)_2(s) + 2OH^-.$)

67. Name each of the following compounds and describe their structures:

(a) $CoCl_2 \cdot 6H_2O$ **(b)** $Co(NH_3)_6Cl_3$

(c) $Co(NH_3)_5Cl_3$ **(d)** $Co(NH_3)_4Cl_3$

68. (a) Write the expression for the formation of the complex ion $Ag(CN)_2^-$ in aqueous solution.

(b) What is the solubility of Ag$_2$S(s) in **(i)** water and **(ii)** $3.0M$ NaCN(aq)? ($K_{sp}(Ag_2S) = 8 \times 10^{-51}$ mol^3 L^{-3}; $K_f(Ag(CN)_2^-) = 1.0 \times 10^{21}$ mol^{-2} L^2.)

*69. When aqueous solutions of ammonia and potassium iodide are added to an aqueous solution of Cr(NO$_3$)$_3$ a solid compound A is obtained that analysis shows to contain 73.53% iodine by mass. From the results of the following experiments deduce the formula of the compound and predict the structure of the complex ion present.

(a) On strongly heating 0.270 g of the compound A in excess oxygen 0.0365 g of CrO$_3$ were obtained.

(b) It took 21.95 mL of $0.100M$ HCl(aq) to react completely with all the ammonia in 0.229 g of the compound A.

(c) The freezing point of water was lowered by 1.08 °C when 0.401 g of the compound A were dissolved in 10.00 g of water. $K_f(H_2O) = 1.86$ °C kg mol^{-1}.

Boron and Silicon: Two Semimetals

22

Group 1	2					Transition Elements						3	4	5	6	7	8
1	H																He
2	Li	Be											B	C	N	O	F	Ne
3	Na	Mg											Al	Si	P	S	Cl	Ar
4	K	Ca	Sc	Ti	V	Cr	Mn	Fe	Co	Ni	Cu	Zn	Ga	Ge	As	Se	Br	Kr
5	Rb	Sr	Y	Zr	Nb	Mo	Tc	Ru	Rh	Pd	Ag	Cd	In	Sn	Sb	Te	I	Xe
6	Cs	Ba	La	Hf	Ta	W	Re	Os	Ir	Pt	Au	Hg	Tl	Pb	Bi	Po	At	Rn
7	Fr	Ra	Ac	104	105	106	107											

Metals   Nonmetals   Semimetals

Silicon is the second most abundant element in the earth's crust. *Silicates*, which are compounds of silicon, oxygen, and many different metals, comprise 95% of the rocks that make up the earth's crust. It is not surprising, therefore, that silicates have played a very significant role in history. One feature that distinguishes *Homo sapiens* from all other mammals is a highly developed ability to use tools. The first tools were pebbles of flint, a form of silicon dioxide, some of which were fashioned into sharp cutting edges for axes and hunting weapons. The Latin word for flint is *silex*, and from it comes our words *silicon*, *silica*, *silicate*, *silicone*, and so on. Another important step in human development occurred when pottery was first made by heating clay, which is a mixture of certain types of silicates. Pottery making ranks with brewing as one of the oldest chemical technologies.

As civilization evolved, silicate glazes and enamels for decorating pottery and metal objects, and glass for making bottles and ornamental objects, were developed. These are all made from molten silicates, which do not crystallize readily on cooling but, instead, form a hard, transparent, amorphous solid that we call a glass. The chemistry of silicates, however, is only one aspect of the chemistry of silicon. Today the element itself is assuming great importance because the silicon chip is the basis of the microcomputer revolution.

In contrast to silicon, boron is rare; it constitutes only 0.0003% of the earth's crust. However, because it occurs in rather concentrated deposits in a few areas, it is a relatively well-known element with some important uses.

Boron and silicon are neighboring elements in the periodic table, and although boron is in group 3 and silicon is in group 4, they have many similar properties. Both are **semimetals**, or *metalloids*, that is, elements with some of the properties of a metal and some of a nonmetal. The semimetals are found in a band running diagonally across the periodic table from top left to bottom right.

22.1
Silicon

Silicon follows carbon in group 4. Whereas carbon is a typical nonmetal, silicon and the next element, germanium, are semimetals, and they are followed by tin and lead, which are metals. Unlike the elements in the groups on the left and right sides of the periodic table, those in the main groups in the middle of the table show a considerable variation in properties on descending the group. Silicon is a shiny, silvery solid that looks like a metal but has only a low electrical conductivity; moreover, its conductivity increases with increasing temperature, whereas the conductivity of a metal decreases with increasing temperature (Chapter 10). Substances that have small electrical conductivity in the solid state that increases appreciably with increasing temperature are known as **semiconductors**. We will discuss them and their important role in electronics at the end of this chapter.

Some Group 4 elements. From left to right: carbon, silicon, tin, and lead.

Silicon occurs not only as silicates but also as *silicon dioxide*, SiO_2, which has been known for centuries as *silica*, which is familiar, in an impure form, as sand.

The element can be prepared from silica by heating it with coke to a temperature of about 3000 °C in an electric arc furnace:

$$SiO_2(s) + 2C(s) \longrightarrow Si(l) + 2CO(g)$$

The reactants are added continuously at the top of the furnace. Carbon monoxide escapes from the furnace and burns to give carbon dioxide, while the molten silicon (mp 1414 °C) runs out from the bottom of the furnace and solidifies. This silicon is pure enough for many purposes, such as the manufacture of alloys with metals, but ultrapure silicon, needed in many electronic devices is obtained by first converting impure silicon to silicon tetrachloride by heating it in chlorine:

$$Si(s) + 2Cl_2(g) \longrightarrow SiCl_4(l) \qquad (bp\ 57.6\ °C)$$

The resulting $SiCl_4$ is then purified by distillation and reduced to silicon by heating with hydrogen or magnesium:

$$SiCl_4(g) + 2H_2(g) \longrightarrow Si(s) + 4HCl(g)$$
$$SiCl_4(g) + 2Mg(s) \longrightarrow Si(s) + 2MgCl_2(s)$$

The magnesium chloride is removed from the silicon by washing it out with hot water. The silicon can be further purified by **zone refining** (Figure 22.1). In this process, a short segment of a rod of silicon is heated until it melts. The impurities are more soluble in the molten silicon than in the solid and therefore concentrate in the liquid. The rod is slowly moved through the heater so that the molten zone traverses the length of the rod, removing impurities as it moves. When the impure molten zone reaches the end of the rod, it is allowed to solidify and is cut off. This process is repeated as often as necessary. In the laboratory silica can be reduced to silicon by heating it with magnesium (Experiment 22.1).

Silicon has the diamond structure (Chapter 11), but the Si—Si bonds are longer and weaker than the C—C bonds, and so the melting point and boiling point of silicon are lower than those of carbon (see Table 22.1). Silicon has no allotropic form analogous to graphite. This is consistent with the fact that a

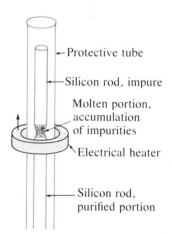

← Protective tube

← Silicon rod, impure

Molten portion, accumulation of impurities

Electrical heater

Silicon rod, purified portion

FIGURE 22.1
Purification of Silicon by Zone Melting.

A rod of impure silicon is drawn very slowly through the electric heater. The molten zone in which the impurities dissolve is thus moved to one end of the rod, where, after solidifying, it can be cut off.

Experiment 22.1

Reduction of Silica to Silicon

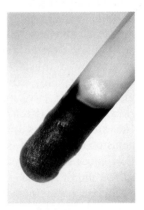

When silica, SiO_2 (sand), is heated with magnesium powder, a strongly exothermic reaction occurs and the mixture begins to glow brightly.

It continues to glow very brightly even when the tube is removed from the bunsen burner.

When the tube is allowed to cool, it is seen to have become distorted by the heat of the reaction and the inside is coated with a shiny metallic-looking deposit of silicon.
The other product of the reaction, magnesium oxide, can be seen as a white powder higher up in the tube.

third-period element such as silicon has a much smaller tendency to form multiple bonds than a second-row element such as carbon.

Silicon is a rather unreactive element and is not attacked by acids. It reacts with hot, concentrated aqueous hydroxide solutions and with molten hydroxides to give silicates; it reacts with oxygen at high temperatures to give silicon dioxide.

TABLE 22.1
Properties of Boron, Carbon, and Silicon

	Atomic Number	Atomic Mass	Density $(g\ cm^{-3})$	Melting Point $(^{\circ}C)$	Boiling Point $(^{\circ}C)$	Covalent Radius (pm)
B	5	10.81	2.54	2300	2550	90
C (diamond)	6	12.01	3.52	3550	4827	77
Si	14	28.08	2.36	1414	2355	117

22.2
Silicon Dioxide, SiO$_2$

Silicon dioxide, or *silica*, occurs in several crystalline forms, including *quartz*, *cristobalite*, and *tridymite*, and in several amorphous forms, such as agate and flint. Quartz is the best known of the silica minerals. It is one of the commonest minerals in the earth's crust and is often found in the form of colorless, transparent crystals, which are sometimes beautifully formed and of enormous size. Quartz is a constituent of many rocks, which are often complex mixtures of different minerals. Granite, for example, is a mixture of the three silicon minerals, quartz, mica, and feldspar. Some forms of quartz that contain traces of impurities are beautifully colored and are often used as semiprecious gemstones. Amethyst, for example, is quartz that is colored violet by traces of Fe(III). Onyx, jasper, carnelian, and flint are colored forms of noncrystalline silicon dioxide.

Silicon dioxide also forms the cell walls of tiny one-cell plants called *diatoms*, which are the most abundant components of marine plankton. When the diatom dies, the silica cell walls sink to the bottom of the ocean to form deposits of a material called *diatomaceous earth*. This is used in many commercial preparations, including polishes and paint removers, and also for decolorizing oils.

The structures of all the crystalline forms of silicon dioxide are based on the SiO$_4$ tetrahedron. Each silicon atom forms four Si—O bonds, which have a tetrahedral AX$_4$ arrangement around the silicon. Each oxygen is bonded to another silicon atom to give an infinite three-dimensional structure. The structures of the different forms differ in the ways that the tetrahedra are arranged. Cristobalite has the simplest structure, closely related to that of diamond (see Figure 22.2). Each silicon atom occupies the position of a carbon atom in the diamond structure. But the silicon atoms are not connected directly to each other as in elemental silicon; rather, they are connected via an oxygen atom, which is situated between each pair of silicon atoms (see Figure 22.2). In quartz the SiO$_4$ tetrahedra are joined in a more complicated manner.

The contrast between silicon dioxide and carbon dioxide is striking. Carbon dioxide is a molecular substance consisting of simple triatomic CO$_2$ molecules and is a gas at room temperature. Silicon dioxide has an infinite three-dimensional structure and is a hard crystalline solid. As previously noted, second-period elements, such as carbon, readily form multiple bonds, as in the CO$_2$ molecule, whereas silicon, a third-period element, has less tendency to form multiple bonds, and thus SiO$_2$ has an infinite three-dimensional network structure that involves only single bonds. Because of the considerable difference in electronegativity between silicon and oxygen, SiO bonds are polar, and the structure may be regarded as being intermediate between a purely covalent three-dimensional network and an ionic crystal consisting of Si^{4+} and O^{2-} ions.

If any form of silica is melted (mp $\approx$ 1600 °C) and the liquid is then cooled, it normally does not crystallize at the original melting point; instead, it becomes more and more viscous as the temperature is lowered. At about 1500 °C it is so viscous that it no longer flows, and has many of the characteristics of a solid. It is an amorphous rather than a crystalline solid, and it is called *silica glass*. It has a structure related to that of quartz and the other crystalline forms of silicon dioxide in that each silicon is surrounded by a tetrahedral arrangement

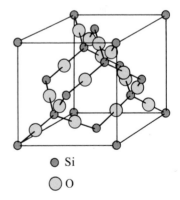

● Si

◯ O

FIGURE 22.2
Structure of Silicon Dioxide.

The structures of the three forms of silicon dioxide, quartz, cristobalite, and tridymite, all have a tetrahedral arrangement of four Si—O bonds around each silicon atom. The structure of cristobalite is based on a face-centered cubic lattice. Each silicon atom occupies the position of a carbon atom in the structure of diamond. An oxygen atom is located between each pair of silicon atoms.

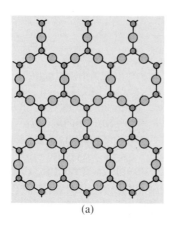
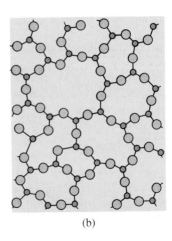

(a)

(b)

FIGURE 22.3
Two-Dimensional Representation of the Structures of Crystalline SiO_2 and Amorphous SiO_2.

(a) In crystalline SiO_2, the SiO_4 tetrahedra have a regular arrangement. (b) In amorphous SiO_2 (silica glass), the SiO_4 tetrahedra have an irregular, random arrangement.

of four oxygen atoms and each tetrahedron is joined to four others by the oxygen atoms at their corners. However, the tetrahedra have a random arrangement rather than one of the regular arrangements of quartz, cristobalite, or tridymite (see Figure 22.3).

Silica glass is transparent and has many of the properties of ordinary glass, but it is more difficult to work; it becomes soft at a much higher temperature

Experiment 22.2

Formation of Metal Silicates: A Chemical Garden

The bottle contains an aqueous solution of sodium silicate to which crystals of various colored transition metal salts have been added. The metal ions react with the sodium silicate to form a thin membrane of insoluble metal silicate. Water penetrates the membrane by osmosis, causing it to expand and then burst. A new membrane then forms and the process repeats, causing a column of colored metal silicate to form. Here the crystals are resting on a layer of sand, and the columns growing up from the crystals form what is sometimes called a "chemical garden."

than ordinary glass and has a smaller range of temperature over which it remains soft and able to be shaped. Silica glass expands very little when it is
heated, and it can therefore be heated and cooled very rapidly without breaking.
For this reason it is useful for certain laboratory apparatus. In addition, it is
transparent to ultraviolet light and is therefore used for making mercury vapor
ultraviolet lamps and for windows and lenses in instruments that employ ultraviolet light.

Silicon dioxide is an acidic oxide that reacts with NaOH or Na_2CO_3 on
heating to give silicates:

$$SiO_2(s) + 4NaOH(s) \longrightarrow Na_4SiO_4(s) + 2H_2O(g)$$
$$SiO_2(s) + Na_2CO_3(s) \longrightarrow Na_2SiO_3(s) + CO_2(g)$$

These are only simplified representations of the reactions; there are many silicates with complex structures, as we will see in the next section.

A concentrated solution of sodium silicate in water is called *water glass*. It is
used for fireproofing wood and cloth, for adhesives, and for preserving eggs. It
can be used to prepare other metal silicates, as demonstrated in Experiment 22.2.

22.3
Silicic Acid and Silicates

Silicon dioxide is the anhydride of silicic acid. When finely powdered silicon
dioxide is shaken with water for a long time, a very slightly acidic solution is
obtained, due to the formation of a very small amount of silicic acid, $Si(OH)_4$:

$$SiO_2(s) + 2H_2O(l) \rightleftharpoons Si(OH)_4(aq)$$

The very small solubility of silicon dioxide in water is increased at high
pressure and temperature, which accounts for the deposits of silica found
around many hot geysers. The silicon dioxide that is dissolved in water at high
pressure below the surface comes out of solution when the water rises to the
surface, where the pressure and temperature are much lower.

More concentrated solutions of silicic acid can be obtained by the reaction
of silicon tetrachloride with water (see Experiment 12.3),

$$SiCl_4(l) + 4H_2O(l) \longrightarrow Si(OH)_4(aq) + 4HCl(aq)$$

or by the reaction of an aqueous solution of sodium silicate with hydrochloric
acid,

$$Na_2SiO_3(aq) + H_2O(l) + 2HCl(aq) \longrightarrow Si(OH)_4(aq) + 2NaCl(aq)$$

However, the product of these reactions is not simply a solution of $Si(OH)_4$.
A gelatinous solid is formed that consists of polymeric silicic acids formed by
condensation reactions such as

Continuation of such reactions leads to a complex mixture of many polymeric
acids. In Chapter 8 we noted a similar tendency of phosphoric acid to form

A dessicator containing silica gel, which strongly absorbs water and thus keeps dry whatever is stored in the dessicator. The silica gel is mixed with a small amount of cobalt(II) chloride which is blue when anhydrous but pink when hydrated, showing that the silica gel is no longer absorbing any water.

polymeric acids by condensation, although the tendency is less marked for phosphorus than for silicon. Phosphoric acid gives stable solutions in water that contain only H_3PO_4 and its ions, whereas $Si(OH)_4$, although probably present in aqueous solutions, cannot be separated from the complicated mixture of the condensed forms of the acid that is always formed.

Silicic acid is very weak ($K_a = 1 \times 10^{-10}$ mol L^{-1}). It is a member of the class of very weak acids such as HOCl and $B(OH)_3$, which have the general formula $X(OH)_n$ (Chapter 20). It is the first member of the series of third-period oxoacids. These acids increase in strength from the very weak silicic acid to the very strong perchloric acid (see Figure 22.4).

If the mixture of polymeric silicic acids is heated, causing further condensation reactions to occur, a hard, granular, translucent substance called *silica gel* is obtained. It has a large surface area and readily adsorbs water and other substances. It is widely used as a drying agent and as a catalyst. Small bags of silica gel are frequently packed with delicate scientific apparatus, cameras, and electronic equipment to protect them from damage by moisture during transport.

SILICATES

Over one thousand silicates occur naturally. Their number and complexity result from the many different ways in which SiO_4 tetrahedra can be linked together. The simplest silicates contain the anion SiO_4^{4-}, derived from the acid $Si(OH)_4$. *Olivine* is an important mineral of this type; it is the principal component of the earth's mantle, which lies between the core and the crust (see Box 22.1). Olivine is an iron magnesium silicate that is often represented by the formula $FeMgSiO_4$, although its composition may vary from Mg_2SiO_4 to Fe_2SiO_4. The SiO_4^{4-} tetrahedra are packed together in a compact way in olivine, giving it a greater density than most other silicate minerals.

Many silicates are the salts of the many polymeric forms of silicic acid. The condensation of two silicic acid molecules gives the acid $H_6Si_2O_7$:

$$\underset{\substack{| \\ OH}}{\overset{\substack{OH \\ |}}{HO-Si-OH}} + \underset{\substack{| \\ OH}}{\overset{\substack{OH \\ |}}{HO-Si-OH}} \longrightarrow \underset{\substack{| \\ OH}}{\overset{\substack{OH \\ |}}{HO-Si-O}}-\underset{\substack{| \\ OH}}{\overset{\substack{OH \\ |}}{Si-OH}} + H_2O$$

$XO_m(OH)_n$ $m =$	$X(OH)_4$ 0	$XO(OH)_3$ 1	$XO_2(OH)_2$ 2	$XO_3(OH)$ 3
	$HO-Si-OH$ (with OH above and below)	$HO-P-OH$ (with O above, OH below)	$O=S-OH$ (with O above, OH below)	$O=Cl-OH$ (with O above, O below)
	Very weak			Very strong

Acid strength increases →

Tendency to polymerize decreases

FIGURE 22.4
Oxoacids of the Third Period.

The formula of an oxoacid may be written $XO_m(OH)_n$. Acid strength increases with increasing values of m (Chapters 8 and 20). Silicic acid, $m = 0$, is very weak. Perchloric acid, $m = 3$, is very strong.

The anion derived from this acid is $Si_2O_7^{6-}$. It consists of two SiO_4 tetrahedra sharing a corner:

BOX 22.1

MINERALS AND THE COMPOSITION AND STRUCTURE OF THE EARTH

The earth consists principally of three concentric layers. There is a central sphere consisting mainly of iron and nickel, which has a radius of approximately 3500 km. Surrounding the core is a mantle of silicate rock, mostly olivine, $FeMgSiO_4$, approximately 3000 km thick. On top of the mantle is a very thin crust ranging from about 5 km in thickness under the oceans to as much as 100 km under the continents. The temperature of the mantle is such that the rocks are soft and somewhat plastic and flow slowly. As a result, pieces of the crust, called *plates*, can be regarded as "floating" on the mantle, like icebergs in the oceans. Collisions between these plates produce mountain ranges, volcanic activity, and earthquakes.

Soon after the formation of our planet, most of the interior was molten; thus the heaviest substances—in particular, iron and nickel—sank to the center, forming the core. Around this core various metal silicates stratified according to their density. One of the most dense of the silicates, olivine, now constitutes the mantle. The less dense silicates floated to the top and formed the crust.

The substances that constitute the solid part of the earth's crust (the lithosphere) are called *minerals*. Minerals may be classified into three major groups: native elements, silicate minerals, and nonsilicate minerals. The most important of the native elements are copper, silver, gold, and sulfur. Some of the more important nonsilicate miner-

A sample of olivine, a silicate rock found in the earth's mantle.

als that have been mentioned in earlier chapters are the following:

- The oxides Al_2O_3 (bauxite), Cu_2O (cuprite), and Fe_2O_3 (hematite).
- The carbonates $CaCO_3$ (limestone, calcite) and $MgCO_3$ (magnesite).
- The sulfates $BaSO_4$ (barite) and $CaSO_4 \cdot 2H_2O$ (gypsum).
- The halides $NaCl$ (halite, rock salt) and KCl (sylvite).
- The phosphate $Ca_5(PO_4)_3F$ (fluorapatite).

Despite their abundance, the silicate minerals are not useful as sources of metals, because it is more difficult to obtain a metal from a silicate than from other compounds such as oxides and sulfides. Thus we rely on these much rarer minerals as sources of metals. The silicates are used mainly for making pottery, glass, and cement.

The condensation of three silicic acid molecules gives the acid $H_8Si_3O_{10}$:

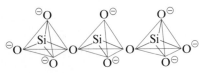

The anion derived from this acid is $Si_3O_{10}{}^{8-}$, and it consists of three SiO_4 tetrahedra sharing corners:

Continued condensation reactions of this type can give still longer chains, which may also join head to tail to give rings. A typical example is the cyclic anion $Si_6O_{18}{}^{12-}$ found in the mineral *beryl*, $Be_3Al_2Si_6O_{18}$ or $(Be^{2+})_3(Al^{3+})_2$ $(Si_6O_{18}{}^{12-})$. In this anion there are six SiO_4 tetrahedra, each of which shares two corners with neighboring SiO_4 tetrahedra:

Each of the oxygen atoms that is not shared with another silicon carries a negative charge, giving a total charge of $12-$ to the ion. Rings of other sizes are also known, as well as the infinitely long chain

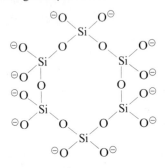

In this chain, each silicon is bonded to four oxygens, two of which are not shared and two of which are shared with other silicon atoms. Thus there is a total of 3 oxygen atoms $[(2 \times 1) + (2 \times \frac{1}{2})]$ for each silicon atom. The empirical formula for the chain is $SiO_3{}^{2-}$ since each of the unshared oxygen atoms carries a negative charge. Silicates that contain this infinite-chain anion are called *pyroxenes*. An example is *diopside*, $CaMg(SiO_3)_2$, which is composed of Ca^{2+} and Mg^{2+} ions and the infinite-chain anion $(SiO_3{}^{2-})_n$ (see Figure 22.5).

Two $(SiO_3{}^{2-})_n$ chains may be joined together by sharing oxygen atoms on alternate tetrahedra, giving a double chain with the empirical formula $Si_4O_{11}{}^{6-}$ (see Figure 22.5). Silicates that contain this double-chain anion are called *amphiboles*.

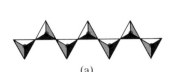

(a)

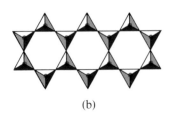

(b)

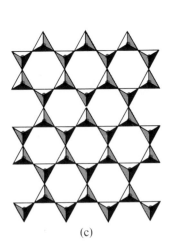

(c)

FIGURE 22.5
Polymeric Silicate Anions.

(a) Pyroxenes have single chains of linked SiO_4 tetrahedra. (b) Amphiboles have double chains of linked SiO_4 tetrahedra. (c) Clays and mica have sheets of linked SiO_4 tetrahedra. Each tetrahedron represents an SiO_4 group.

The mineral *tremolite*, $Ca_2Mg_5(Si_4O_{11})_2(OH)_2$, is an example of an amphibole. Tremolite is composed of Ca^{2+} ions, Mg^{2+} ions, OH^- ions, and the infinite, double-chain anion $(Si_4O_{11}{}^{6-})_n$. It is one of a group of minerals called *asbestos*. Asbestos minerals have a fibrous structure that reflects the chain structure of the anion. Fibers of asbestos are very strong and have many uses. Asbestos is a very effective insulating and fire-resistant material. The fibers can be woven into blankets that are used for insulation and extinguishing fires and into cloth for making protective clothing. Unfortunately, small particles of asbestos suspended in the air and in water supplies are a health hazard (see Box 22.2).

If a large number of chains are joined side by side, a layer of SiO_4 tetrahedra results (see Figure 22.5). Each tetrahedron then shares three corners with neighboring tetrahedra. Each silicon atom is bonded to three oxygen atoms that it shares with other silicon atoms and to one oxygen atom that carries a negative charge. Thus each silicon atom is associated with $2\frac{1}{2}$ oxygen atoms $[(1 \times 1) + (3 \times \frac{1}{2})]$, one of which carries a negative charge. Therefore the empirical formula of this sheet anion is $Si_2O_5{}^{2-}$. Several minerals contain this sheet anion,

BOX 22.2

ASBESTOS

The ready availability of the fibrous silicate minerals known as asbestos and their nonflammability, flexibility, mechanical strength, and inertness to chemical attack have led to their widespread use. Asbestos has many applications in the building industry as a fire retardant and as a heat and sound insulating material. It is an important component of solid materials used for pipes, brake pads, and many other applications.

When airborne asbestos particles reach the lungs, they do not dissolve

Crude asbestos fibers

and are not expelled: they tend to remain, irritating and scarring the lungs. Extensive exposure to airborne asbestos in appreciable concentrations can lead to the disease asbestosis and to a greatly increased risk of cancer, especially for those who smoke. Many workers who have been exposed to asbestos in its mining and processing have contracted asbestosis, and there is concern that the general public is being exposed to levels of asbestos that after 20–30 years may give some of them the same health problems as those experienced by asbestos workers.

Asbestos fibers magnified 100 times

Inuit soapstone carvings.

for example, *talc*, $Mg_3(Si_2O_5)_2(OH)_2$, which is used for making talcum powder, and *kaolinite*, $Al_2Si_2O_5(OH)_4$, which is a type of clay that is particularly valuable for pottery making. Talc and kaolinite consist of layers, each of which is composed of two $(Si_2O_5^{2-})_n$ sheet anions together with hydroxide ions and sufficient cations to neutralize the total charge of the $(Si_2O_5^{2-})_n$ anions and the hydroxide ions (Figure 22.6). These neutral layers slide easily over each other, giving these minerals a characteristic soft, soapy feel. *Soapstone*, an easily carved material, is talc in compact form. Soapstone objects made by the Inuit inhabitants of Canada and Alaska have become well known.

The layer structure of a clay such as kaolinite can be penetrated by water molecules, which provide a lubricant between the layers, enabling them to slip easily over each other. Thus wet clay is pliable and slippery, but when it is baked or fired in a kiln to 1100 °C or more, the water molecules are driven out, and the layers lock into a rigid structure.

Because molecules can be inserted between the layers of a clay mineral, these substances have enormous effective surface areas. They can absorb large amounts of water and many organic substances. A dried clay called Fuller's earth is often used for removing stains and grease from fabrics, for removing some of the natural oil from wool in the textile industry, and for clarifying and deodorizing oils.

We have seen then that, when SiO_4 tetrahedra share two corners, we get chain and ring silicates; and when they share three corners, we get sheet silicates. When SiO_4 tetrahedra share all four corners, the various three-dimensional structures of silicon dioxide result.

EXERCISE 22.1

Bearing in mind that the oxygen atom joining two Si atoms (Si—O—Si) has no formal charge, but an oxygen atom bonded to just one Si atom (Si—O^-) has a formal charge of -1, what are the empirical formulas of the species formed by

(a) Two SiO_4 tetrahedra sharing one corner

(b) Three SiO_4 tetrahedra sharing two corners each to form a six-membered ring

(c) An infinite number of SiO_4 tetrahedra sharing two corners each

(d) An infinite number of SiO_4 tetrahedra sharing three corners each

(e) An infinite number of SiO_4 tetrahedra sharing four corners each

(f) Infinite chains of the type formed in (c) sharing corners on alternate tetrahedra

ALUMINOSILICATES

In many silicate minerals, some of the silicon atoms in the SiO_4 tetrahedra are replaced by aluminum atoms to give *aluminosilicates*. But to form four bonds like the silicon atom it replaces, the aluminum atom must have four electrons; in other words, Si is replaced by an Al^- ion. There are numerous aluminosilicate anions. We consider just a few examples.

In the mineral *albite*, $NaAlSi_3O_8$, one-quarter of the silicon atoms in silicon dioxide are replaced by Al^-, giving the three-dimensional framework anion $(AlSi_3O_8^-)_n$. Albite consists of this anion and sodium ions, Na^+. Minerals in

A sample of albite from Brazil. The white plates are partially stained by iron oxides.

		K^+ K^+ K^+
$(Si_2O_2^{5-})_n$	$(Si_2O_2^{5-})_n$	$(AlSi_3O_{5-}^{10})_n$
$Mg^{2+}OH^-Mg^{2+}OH^-$	$Al^{3+}OH^-Al^{3+}OH^-$	$Al^{3+}OH^-Al^{3+}OH^-$
$(Si_2O_2^{5-})_n$	$(Si_2O_2^{5-})_n$	$(AlSi_3O_{5-}^{10})_n$
		K^+ K^+ K^+
$(Si_2O_2^{5-})_n$	$(Si_2O_2^{5-})_n$	$(AlSi_3O_{5-}^{10})_n$
$Mg^{2+}OH^-Mg^{2+}OH^-$	$Al^{3+}OH^-Al^{3+}OH^-$	$Al^{3+}OH^-Al^{3+}OH^-$
$(Si_2O_2^{5-})_n$	$(Si_2O_2^{5-})_n$	$(AlSi_3O_{5-}^{10})_n$
		K^+ K^+ K^+
Talc	Kaolinite	Muscovite (mica)

FIGURE 22.6
Structures of Some Layer Silicates and Aluminosilicates.

Edge view of the layer structures of talc, $Mg_3(Si_2O_5)_2(OH)_2$, kaolinite, $Al_2(Si_2O_5)(OH)_4$, and muscovite (mica), $KAl_2(AlSi_3O_{10})$. In talc and kaolinite each layer consists of a sandwich of two $(Si_2O_5^{2-})_n$ sheet anions held together by metal ions and hydroxide ions. Each layer has a zero overall charge and there are only weak van der Waals forces between the layers which therefore slide over each other easily so that these minerals are soft. In muscovite each layer consists of two $(AlSi_3O_{10}^{5-})_n$ anions held together by Al^{3+} and OH^- ions. In this case, however, the layers have an overall negative charge and they are held strongly together by potassium ions. Muscovite is therefore a relatively hard but flaky mineral that is commonly known as mica.

which up to one-half of the silicon atoms in silicon dioxide are replaced by aluminum are *feldspars*. Feldspar minerals are one of the essential constituents of granite.

= Example 22.1

EMPIRICAL FORMULA OF AN ALUMINOSILICATE MINERAL

The mineral *anorthite* is a feldspar mineral containing Ca^{2+} ions and an anion formed by replacing half of the silicon atoms in SiO_2 by aluminum. What is the empirical formula of this mineral?

Solution

If we double the empirical formula SiO_2 and then replace one of the Si atoms by Al^-, we obtain the empirical formula $AlSiO_4^-$ for the infinite, three-dimensional aluminosilicate anion. For electrical neutrality, two of these empirical formula units are required by each Ca^{2+}, so the empirical formula of the mineral anorthite is $Ca(AlSiO_4)_2$.

Aluminosilicates can also be derived from silicate anions by replacing some of the Si atoms by Al^- ions. Muscovite, $KAl_2(AlSi_3O_{10})(OH)_2$, has a structure based on the amphibole layer anion with one-quarter of the silicon atoms replaced by aluminum. If we double the empirical formula $Si_2O_5^{2-}$ of the amphibole layer anion, we obtain $Si_4O_{10}^{4-}$. Then replacing one of the silicon atoms by Al^- gives the empirical formula $AlSi_3O_{10}^{5-}$. Each layer of the structure consists of two of these infinite sheet $(AlSi_3O_{10}^{5-})_n$ anions held together by OH^- and Al^{3+} ions. Each layer has an overall negative charge and the layers are held together by layers of potassium ions (Figure 22.6). The layers do not, therefore, slide over each other as readily as the neutral layers in talc. So muscovite is not as soft as talc. Muscovite, which is commonly known as mica, cleaves readily into thin transparent sheets that are used for windows in stoves and furnaces. Small flecks are used to add the glitter to "metallic" paints.

A sample of mica. Note the cleavage into thin, flat sheets.

EXERCISE 22.2

What is the empirical formula of the mica mineral phlogopite in which one-quarter of the silicon atoms in talc, $Mg_3Si_4O_{10}(OH)_2$, are replaced by Al^- and the additional negative charge is balanced by K^+ ions?

GLASS

The manufacture and use of **glass** is another chemical technology that was developed early in history. We have already mentioned that when molten silicon dioxide is cooled it does not crystallize but gradually becomes more viscous and eventually gives a glass or amorphous solid, in which the SiO_4 tetrahedra are arranged in a random manner. Many silicates and aluminosilicates have the same property. Since they usually melt at lower temperatures and stay soft over a wider temperature range, they are easier to work with than silicon dioxide. For example, ordinary window and bottle glass is a mixture of sodium and calcium silicates.

The composition of glass is normally expressed in terms of the oxides of the elements that it contains, even though these elements are actually present in the form of complex sodium and calcium silicates. Ordinary glass has the approximate composition by mass of 12% Na_2O, 12% CaO, and 76% SiO_2. It is made by heating together sodium carbonate, calcium oxide (lime) or calcium carbonate (limestone), and silicon dioxide in the form of white sand. Numerous other oxides can be added to glass to give it special properties. For example, adding boron oxide, B_2O_3, gives borosilicate glass, which expands very little on being heated. This type of glass is used in cooking utensils and laboratory ware and is sold under trade names such as Pyrex. Addition of PbO gives a glass with a high refractive index and a high density; it is used for the manufacture of cut glass and for lenses.

Glass may also be colored by the addition of other oxides. Indeed, ordinary glass often contains a small amount of Fe(II), from an impurity in the sand, which gives it a green color. This color can be removed by the addition of MnO_2. Addition of CoO gives a blue glass; Cr_2O_3 gives a deep green glass; and SnO_2 gives a white opaque glass.

A recent development has been that of "photochromic" glass for use in sunglasses. This glass darkens on exposure to bright sunlight but becomes clear again in a subdued indoor light. The glass contains AgCl (or AgBr), which is decomposed by light (Chapter 21) to give finely divided black silver:

$$AgCl \xrightarrow{h\nu} Ag + Cl$$

Since the Ag and Cl atoms are trapped in adjacent positions in the solid glass, they recombine in the absence of light to re-form AgCl.

Glass provides the glaze on pottery that makes it impervious to water. A glaze is simply a coating of glass that is formed by covering the pottery with a thin paste of the appropriate oxides and heating to a high temperature. A similar coating of glass, called *vitreous enamel*, can also be applied to metal surfaces in much the same way. Ornamental objects made of enameled metal were found in the tomb of Tutankhamen, who died in 1350 B.C.

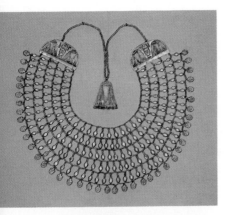

Enamelled gold jewelry from ancient Egypt.

CEMENT

Cement is an aluminosilicate made by heating a powdered mixture of limestone ($CaCO_3$), sand (SiO_2), and clay to about 1500 °C. The resulting solid mass is powdered and mixed with a little gypsum, $CaSO_4 \cdot 2H_2O$. Cement has the approximate composition 60% CaO, 20% SiO_2, 10% Al_2O_3, and 10% other oxides. When cement is mixed with water, a thick viscous material is obtained, which slowly hardens as interlocking crystals of hydrated aluminosilicates are formed. Cement is usually mixed with sand, gravel, or crushed rock to give a hard, strong material known as *concrete*.

22.4
Silicone Polymers

The wide range of applications of the mineral silicates depends on their great thermal stability and inertness toward other substances. These properties are related to the great strength of the silicon–oxygen bond, which has an average bond energy (464 kJ mol^{-1}) that is much greater than the silicon–silicon bond energy (196 kJ mol^{-1}) or even the carbon–carbon, single-bond energy (348 kJ mol^{-1}). Chemists have been able to combine the strength and inertness of the silicon–oxygen bond with some of the useful properties of organic polymers (Chapter 24) in the synthetic polymers known as *silicones*.

The simplest silicones are chain polymers that have the general formula $(R_2SiO)_n$, where R is an alkyl or aryl group such as CH_3, C_2H_5, or C_6H_5 (see Figure 22.7). There are also cyclic polymers (see Figure 22.7) and cross-linked polymers in which chains are held together by sharing oxygen atoms.

Silicones are prepared by first heating silicon with chloroalkanes such as chloromethane, CH_3Cl. The main product of the reaction with CH_3Cl is dimethyl silicon dichloride, $(CH_3)_2SiCl_2$:

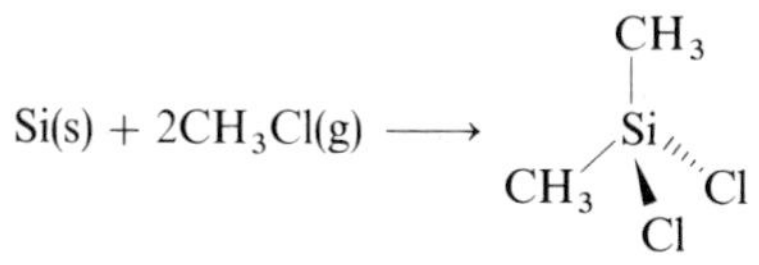

$$Si(s) + 2CH_3Cl(g) \longrightarrow \begin{array}{c} CH_3 \\ | \\ Si \\ CH_3 \diagdown \quad Cl \\ Cl \end{array}$$

But small amounts of $(CH_3)_3SiCl$ and CH_3SiCl_3 are also formed. The reaction

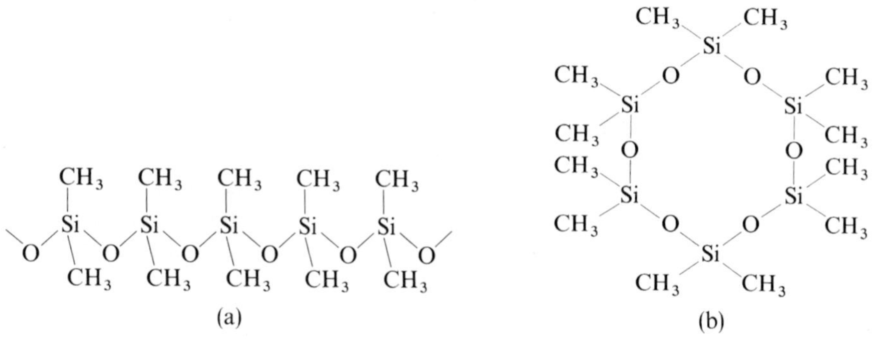

FIGURE 22.7
Structures of Some Silicone Polymers.

(a) A long-chain polymer. (b) A cyclic polymer.

of dimethyl silicon dichloride with water gives dimethyl silicon dihydroxide, $(CH_3)_2Si(OH)_2$:

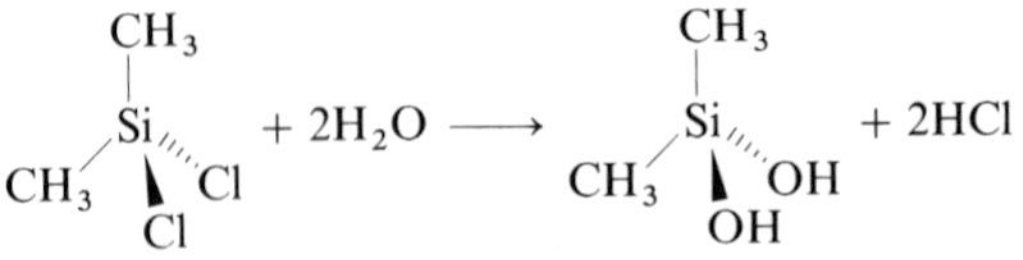

This polymerizes in a condensation reaction to give a chain polymer:

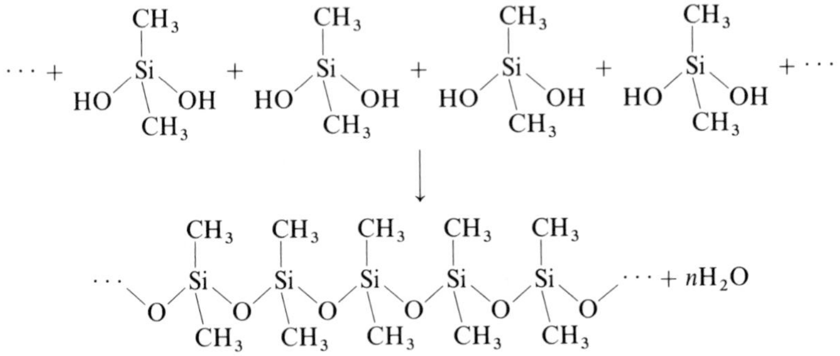

The length of the polymer chain can be controlled by introducing a certain amount of $(CH_3)_3SiCl$ into the reaction mixture to produce $(CH_3)_3SiOH$. When $(CH_3)_3SiOH$ condenses with another Si—OH group, it terminates the chain:

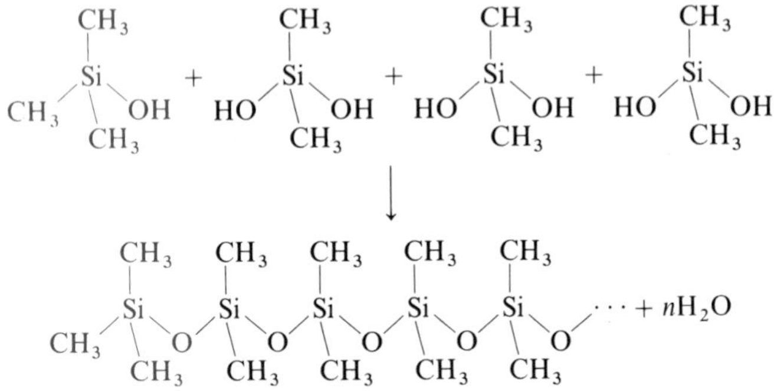

Silicones consisting of chains with up to about ten silicon atoms are liquids, while those with longer chains are greases and waxes.

If a certain amount of CH_3SiCl_3 is introduced into the reaction mixture, cross-links between the chains are produced. The reaction of CH_3SiCl_3 with water gives $CH_3Si(OH)_3$. When this product condenses with other Si—OH groups, it can form three Si—O—Si links and it can therefore join the SiOSiOSi chains together (see Figure 22.8). In this way a variety of solids can be produced, ranging from hard materials to soft, rubbery materials.

A drop of water placed on the untreated half of a cloth (left) wets it. On the half treated with a silicone polymer a drop of water does not spread and wet the cloth.

Silicones have several useful properties. They are excellent electrical insulators, good lubricants, water-repellent, and nontoxic. Because they are very resistant to heat, they can be used as replacements for rubber and lubricants in many high-temperature applications. The viscosity of the liquid silicones changes very little with temperature, so they retain their fluidity down to very low temperatures. They are often used in place of hydrocarbon oils for low-temperature applications. The solids also retain their rubbery properties at low temperatures and do not become hard and brittle, like many rubbers based on organic molecules.

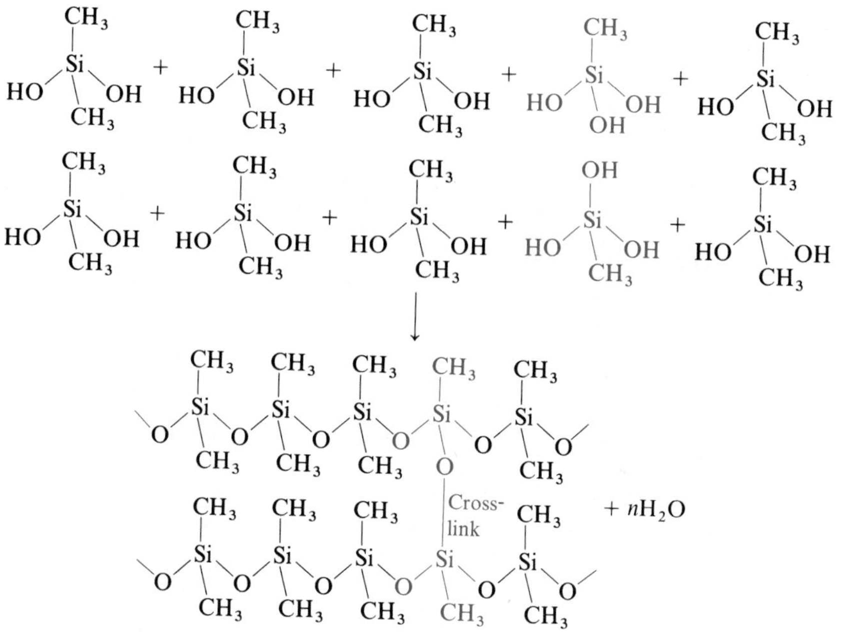

FIGURE 22.8
Formation of Cross-Linked Silicone Polymer.

Introduction of a small amount of $CH_3Si(OH)_3$ into $(CH_3)_2Si(OH)_2$ causes the formation of cross-links between the chains.

Silicones with the formula $(R_2SiO)_n$ are the silicon analogues of ketones. However, whereas ketones are simple molecules with $C=O$ double bonds, the corresponding silicones are all polymers in which there are only single bonds. As we have seen before, carbon, a second-period element, has a much stronger tendency to form multiple bonds than does silicon, a third-period element.

22.5
Other Silicon Compounds

SILANES

The great difference between the chemistry of silicon and the chemistry of carbon is also illustrated by the hydrides of silicon, which are called *silanes*. In their general formula Si_nH_{2n+2}, the silanes are analogous to the alkanes, C_nH_{2n+2}, but only a few silanes are known. They are extremely reactive and are spontaneously flammable in the air (see Experiment 22.3):

$$SiH_4(g) + 2O_2(g) \longrightarrow SiO_2(s) + 2H_2O(l)$$

In contrast, the alkanes burn only when ignited. The silanes react rapidly with basic aqueous solutions to give hydrogen:

$$SiH_4(g) + OH^-(aq) + 3H_2O(l) \longrightarrow SiO(OH)_3{}^-(aq) + 4H_2(g)$$

SILICON HALIDES

Silicon tetrachloride, which is a colorless liquid (bp 57 °C), can be made by passing chlorine through a red-hot mixture of sand and coke:

$$SiO_2(s) + 2C(s) + 2Cl_2(g) \longrightarrow SiCl_4(l) + 2CO(g)$$

Experiment 22.3

Preparation of Silane, SiH$_4$

A mixture of silica, SiO$_2$, is heated with magnesium in a 1:2 mole ratio.

The reaction is highly exothermic and the tube glows brightly even when it is removed from the flame. The product contains magnesium silicide, Mg$_2$Si.

When the magnesium silicide is tipped into dilute hydrochloric acid it reacts to form silane, SiH$_4$. When the bubbles of silane reach the surface of the solution, they ignite spontaneously with an explosive pop.

Silicon tetrafluoride, which is a colorless gas, can be made by the reaction of silicon with fluorine,

$$Si(s) + 2F_2(g) \longrightarrow SiF_4(g)$$

or by heating calcium fluoride with sand and concentrated sulfuric acid:

$$2CaF_2(s) + 2H_2SO_4(concn) + SiO_2(s) \longrightarrow SiF_4(g) + 2CaSO_4(s) + 2H_2O(l)$$

In this reaction, hydrogen fluoride is produced by the reaction of calcium fluoride with concentrated sulfuric acid, and then HF reacts with silicon dioxide:

$$SiO_2(s) + 4HF(aq) \rightleftharpoons SiF_4(g) + 2H_2O(l)$$

The position of the equilibrium is shifted to the right by the removal of water by sulfuric acid.

Both SiCl$_4$ and SiF$_4$ are covalent tetrahedral AX$_4$ molecules. In contrast to the carbon compounds CF$_4$ and CCl$_4$, which are stable in the presence of water, both SiCl$_4$ and SiF$_4$ react rapidly with water to give silicic acid, Si(OH)$_4$, and a hydrogen halide:

$$SiCl_4(l) + 4H_2O(l) \longrightarrow Si(OH)_4(aq) + 4HCl(aq)$$

$$SiF_4(g) + 4H_2O(l) \longrightarrow Si(OH)_4(aq) + 4HF(aq)$$

In reacting in this way with water the silicon halides resemble most other nonmetal halides (Chapters 12 and 19).

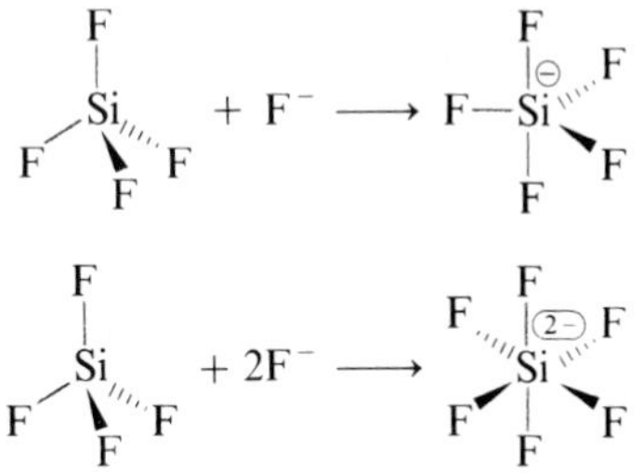

FIGURE 22.9
Formation of SiF_5^- and SiF_6^{2-} Ions.

Because it can accept more electrons into its valence shell, the silicon atom in SiF_4 can combine with one or two F^- ions to form the SiF_5^- and SiF_6^{2-} ions. In these reactions SiF_4 is behaving as a Lewis acid.

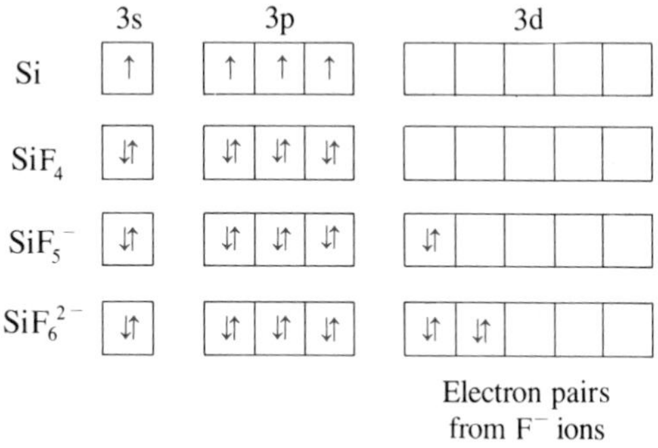

Electron pairs
from F^- ions

Silicon tetrafluoride reacts with fluoride ion to form the ions SiF_5^- and SiF_6^{2-}. These ions are isoelectronic with PF_5 and SF_6 and have the same trigonal bipyramidal AX_5 and octahedral AX_6 structures (see Figure 22.9).

22.6
Comparison of Carbon and Silicon

We have mentioned that there are considerable differences in the chemistry of the elements in the second period of the periodic table and the elements of the same groups in the third period. These differences are particularly evident for carbon and silicon. In this section we review some of these differences and give some explanations for them. There are thousands of alkanes and other hydrocarbons but only a very few silanes are known and, as we have seen, they are very reactive. However, long chains of silicon atoms are known in compounds which have methyl groups attached to silicon rather than hydrogen atoms. Nevertheless these compounds are much more reactive than their carbon analogues. One reason for this difference in reactivity is the difference in the strengths of the C—C and Si—Si bonds. As we saw in Chapter 6, one useful measure of bond strength is the bond energy. Table 22.2 lists bond energies for boron and most of the elements in groups 4–7. The C—C bond is indeed the strongest of all the single covalent bonds between like atoms, an important factor in the lack of reactivity of compounds containing chains and rings of carbon atoms.

Another important difference between silicon and carbon is that the valence shell of carbon is completely filled in all its compounds, whereas silicon has the possibility of adding extra electrons to its valence shell by utilizing its 3d orbitals. Carbon tetrafluoride does not combine with fluoride ion, but SiF_4 forms SiF_5^- and SiF_6^{2-}, because silicon can exceed the octet by making use of its 3d orbitals

TABLE 22.2 Bond Energy Values (kJ mol^{-1})				
B—B 301	C—C 348	N—N 159	O—O 138	F—F 155
	Si—Si 196	P—P 197	S—S 266	Cl—Cl 237
	Ge—Ge 163	As—As 177	Se—Se 193	Br—Br 190
	Sn—Sn 152	Sb—Sb 142	Te—Te 126	I—I 149
B—O 523	C—O 335	N—O 113	O—O 143	F—O 184
	Si—O 464	P—O 368		Cl—O 207

to accept F$^-$ ions (see Figure 22.9). Similarly, SiH_4, $SiCl_4$, and SiF_4 react rapidly with water, whereas the alkanes and CCl_4 and CF_4 do not. The empty 3d orbitals on the silicon atom provide an easy route for the water molecule to attack a silicon atom by donating one of its unshared electron pairs to a vacant 3d orbital.

Although the Si—Si bond is not so strong as the C—C bond (Table 22.2), the Si—O bond is stronger than the C—O bond and indeed second only to the B—O bond in strength among the energies of bonds formed by oxygen (Table 22.2). The strength of the Si—O bond is certainly consistent with the great stability of silicon dioxide and the mineral silicates, but it does not completely explain why the silicates are relatively unreactive since the bond energy of the Si—F bond in SiF_4 is even greater (598 kJ mol^{-1}) and yet SiF_4 reacts rapidly with water. The reasons for reactivity or lack of reactivity of compounds are many and complex, and they cannot be fully discussed in terms of a few simple concepts such as bond energies. Further discussion of these interesting questions, however, would take us well beyond the scope of this book.

Finally, it is interesting that, despite many attempts, a complete lack of success in making compounds containing Si=Si double bonds had led chemists to conclude that such a bond does not form. However, Robert West found in 1981 that compounds containing Si=Si bonds can be made if the groups attached to the silicon atoms are large enough that they prevent other molecules from approaching the very reactive silicon atoms (see Figure 22.10). Thus we see that the ability of silicon to accept extra electron pairs into its valence shell is responsible for the lack of compounds containing Si=Si bonds. The Si=Si bond can be formed, but molecules with an Si=Si bond are too reactive to be isolated unless the molecule has a structure that prevents other reactants, such as water molecules, from coming close to the silicon atoms.

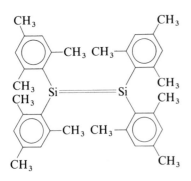

FIGURE 22.10
The First Compound Containing an Si=Si Double Bond.

It was prepared by Robert West at the University of Michigan in 1981. The bulky organic groups prevent other molecules from reaching and reacting with the Si=Si double bond.

22.7
Boron and its Compounds

Boron is the first element in group 3. It differs in many of its properties from the next element in the group, aluminum, which is a metal. Boron is a very hard, black, shiny solid. Although it is somewhat metallic in appearance, it is a very poor conductor of electricity. It is best regarded as a semimetal, like silicon.

BORIC ACID AND BORATES

Boric acid and the borates are among the simplest and most important of the compounds of boron. Boric acid, $B(OH)_3$, is a stable, colorless crystalline compound that forms thin, platelike crystals. It consists of planar molecules with an equilateral triangular AX_3 geometry around boron. The molecules are held together in flat sheets by hydrogen bonds, as shown in Figure 22.11.

Boric acid is a very weak monoprotic acid ($K_a = 6.0 \times 10^{-10}$ mol L^{-1}). It ionizes in water in an unusual way. Instead of donating one of its hydrogen atoms to a water molecule; it removes an OH^- from a water molecule, leaving an H^+ ion, which combines with another water molecule to give an H_3O^+ ion:

$$B(OH)_3(aq) + 2H_2O(l) \rightleftharpoons B(OH)_4{}^-(aq) + H_3O^+(aq)$$

Thus $B(OH)_3$ behaves as a Lewis acid (Chapter 10). The boron atom has a vacant 2p orbital, which can accept an electron pair from a water molecule. This water molecule, simultaneously or in a subsequent step, loses a proton to another water molecule:

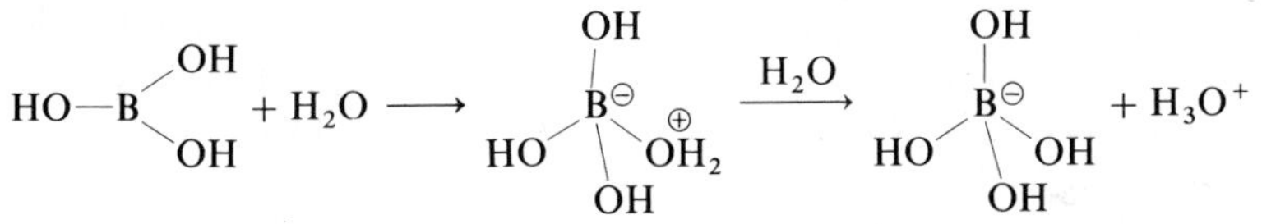

Lewis acid   Brönsted acid

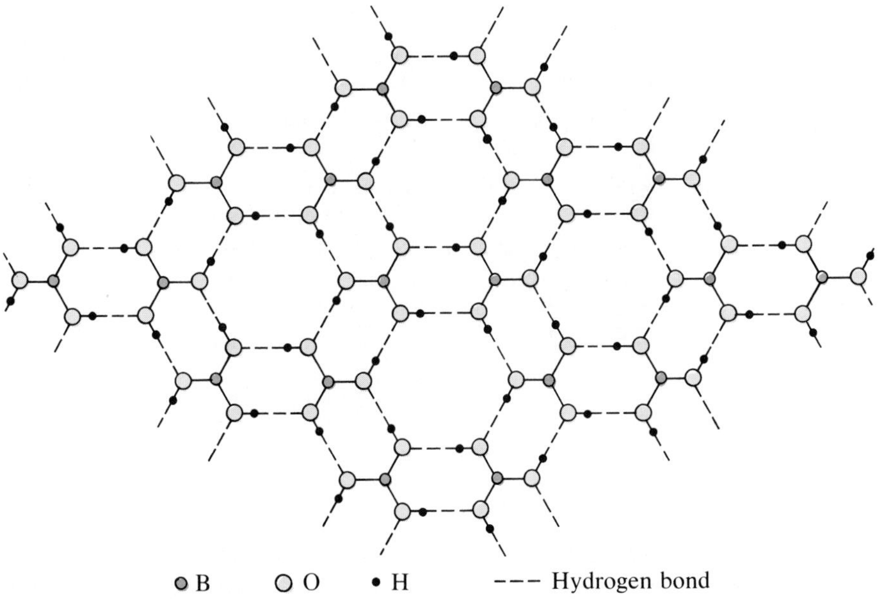

● B   ◯ O   • H   - - - Hydrogen bond

FIGURE 22.11
Structure of Boric Acid.

The planar triangular AX_3 molecules are held together in flat sheets by hydrogen bonds.

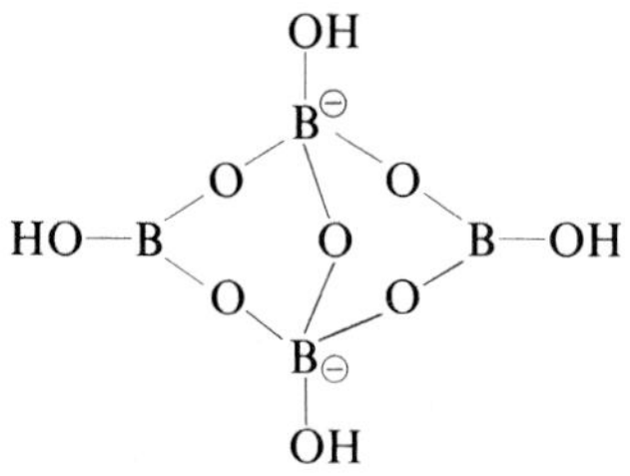

FIGURE 22.12
Structure of Tetraborate
Anion, $B_4O_5(OH)_4{}^{2-}$, in
Sodium Tetraborate (Borax).

Two of the boron atoms have a
tetrahedral AX_4 geometry, and
the other two have a planar AX_3
geometry. The two tetrahedrally
coordinated boron atoms each
have a formal negative charge.

In this way, boron completes its octet by forming the tetrahedral borate ion $B(OH)_4{}^-$.

When it is heated, boric acid loses water to form various condensed boric acids, such as cyclic metaboric acid:

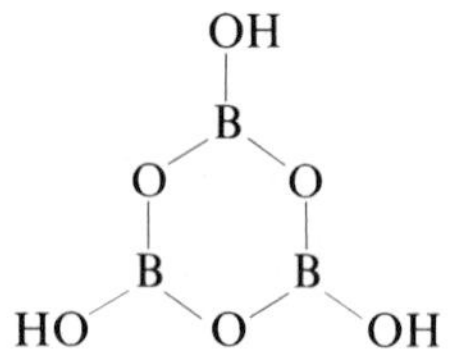

Further heating gives boric oxide, B_2O_3, which, like silicon dioxide, is often formed in an amorphous form, or glass. The crystalline form has a complex network structure based on planar BO_3 groups.

A few of the salts of boric acid, the borates, contain the simple anion $BO_3{}^{3-}$, but most are derived from condensed forms of the acid. The most familiar and most important is sodium tetraborate, $Na_2[B_4O_5(OH)_4]\cdot 8H_2O$, which is commonly called *borax* and for which the formula is often written as $Na_2B_4O_7\cdot 10H_2O$. The structure of the tetraborate anion, $B_4O_5(OH)_4{}^{2-}$, is shown in Figure 22.12.

Borax and other borates are the only important boron minerals. They are found in only a very few places, for example, California and Turkey, but the deposits in those places are extremely large. Figure 22.13 shows a view of the enormous open pit mine in the Mohave desert near the town of Boron.

Aqueous solutions of borax are basic, because tetraborate is the anion of a weak acid and is therefore a weak base:

$$B_4O_5(OH)_4{}^{2-}(aq) + 5H_2O(l) \rightleftharpoons 4H_3BO_3(aq) + 2OH^-(aq)$$

Because the borates of calcium and magnesium are insoluble, borax is used as a water softener and as a component of washing powders.

FIGURE 22.13
A Borax Mine.

The open pit mine of U.S. Borax
at Boron, California, is the world's
principal source of borates. To
expose the ore, which lies approximately 300–600 feet below the
surface, the overburden is removed by using explosives,
electric shovels, and gigantic
trucks.

BORON HALIDES

The boron halides are typical covalent nonmetal halides. Boron trifluoride, BF_3, and boron trichloride, BCl_3, are gases at room temperature; BBr_3 is a liquid, and BI_3 is a solid. They all consist of molecules with the expected AX_3 planar triangular structure. Although the electronegativity difference between boron and fluorine is 2.1, boron trifluoride is a covalent molecular compound with polar B—F bonds rather than an ionic crystal containing B^{3+} and F^- ions. We see again that the electronegativity difference between two elements is not always a reliable guide to whether they will form an ionic crystal or a molecular compound (Chapter 5).

Because of the presence of the vacant 2p orbital on the boron atom in the boron halides, they are rather reactive compounds, unlike the carbon tetrahalides. For example, BF_3 reacts with an F^- ion to form BF_4^- in which the valence shell of boron is completed. Boron trifluoride reacts in a similar way with many other molecules and ions that have unshared electron pairs that can be donated to its vacant 2p orbital. Thus it forms compounds with ammonia and other amines and with ethers and alcohols. In all these cases BF_3 is behaving as a Lewis acid, that is, as an electron pair acceptor (Chapter 10):

$$
\begin{array}{cc}
\text{F} & \text{H} \\
\text{F—B} + :\text{N—H} \\
\text{F} & \text{H}
\end{array}
\longrightarrow
\begin{array}{cc}
\text{F} & \text{H} \\
\text{F—}\overset{\ominus}{\text{B}}\text{—}\overset{\oplus}{\text{N}}\text{—H} \\
\text{F} & \text{H}
\end{array}
\qquad
\begin{array}{cc}
\text{F} & \text{CH}_3 \\
\text{F—B} + \overset{..}{\underset{..}{\text{O}}} \\
\text{F} & \text{CH}_3
\end{array}
\longrightarrow
\begin{array}{cc}
\text{F} & \text{CH}_3 \\
\text{F—}\overset{\ominus}{\text{B}}\text{—}\overset{\oplus}{\underset{..}{\text{O}}} \\
\text{F} & \text{CH}_3
\end{array}
$$

The boron halides react with water to give boric acid and the hydrogen halides. For example,

$$BCl_3(g) + 3H_2O(l) \longrightarrow B(OH)_3(aq) + 3HCl(aq)$$

This reaction is similar to the reaction of $SiCl_4$ with water. It occurs readily because the vacant 2p orbital on the boron can accept an electron pair from a water molecule. In contrast, the carbon tetrahalides have no vacant orbital, and they do not react with water.

BORANES

At least twenty-five compounds of boron and hydrogen have been prepared. They are known as the *boranes*. They all have unexpected formulas such as B_2H_6, B_4H_{10}, B_5H_9, and $B_{10}H_{14}$. We would expect the formula of the simplest borane to be BH_3, but this molecule is not known as a stable species. The simplest borane that can be isolated is diborane, B_2H_6; its structure is shown in Figure 22.14.

The unusual and unexpected feature of this structure is that there are two hydrogen atoms, called *bridging hydrogens*, shared between the two borons. However, there are not enough electrons for each of the lines shown in the structure to represent an electron pair. Each boron contributes 3 electrons and each hydrogen 1 electron, making a total of 12 electrons, or six pairs, for the molecule. Thus there can be a maximum of only six ordinary covalent bonds, whereas the structure appears to have eight bonds. Because B_2H_6 has too few electrons for all the atoms to be held together by normal electron pair bonds between two nuclei, it is often described as an **electron-deficient molecule**. The bonding in diborane is best described as involving two **three-center bonds**, in which one electron pair holds together three rather than two nuclei. The ar-

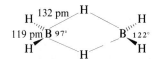

FIGURE 22.14
Structure of Diborane, B_2H_6.

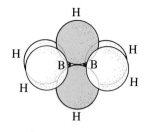

FIGURE 22.15
Electron Pair Sphere Model of
Diborane.

○ Two-center bonds

● Three-center bonds

There is a tetrahedral arrangement
of electron pairs around each of
the boron atoms. Each three-
center bond is formed by one
electron pair shared between two
boron atoms and a hydrogen
atom.

rangement of the electron pairs is shown in Figure 22.15. Each boron atom is surrounded by four electron pairs which have the expected tetrahedral arrangement. But two of these electron pairs form three-center bonds in which one electron pair holds together two boron nuclei and a hydrogen nucleus. In the following diagram of the structure the two electron pairs forming the three-center bonds are denoted by ⊙:

The following alternative but equivalent representation of the bonding in diborane emphasizes its close relationship to the bent-bond model of ethene:

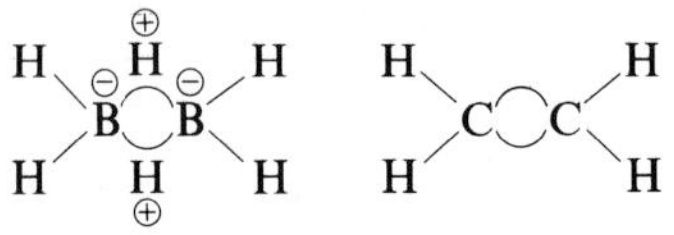

In this way each boron atom completes on octet in its valence shell, whereas in a BH_3 molecule the boron atom would have only six electrons in its valence shell and an empty 2p orbital (see Figure 22.16). Another way to represent the structure of this molecule is by means of the two resonance structures.

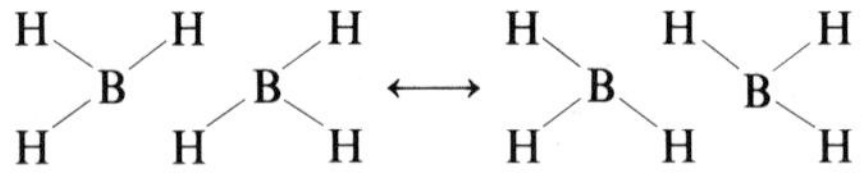

A molecule of diborane readily adds two hydride ions to give two borohydride ions, BH_4^-: $B_2H_6 + 2H^- \rightarrow 2BH_4^-$. This ion has the expected tetrahedral AX_4 structure and ordinary two-center, electron pair bonds. Salts such as lithium borohydride, $LiBH_4$, and sodium borohydride, $NaBH_4$, are good reducing agents with many applications in organic chemistry.

The higher boranes have unusual and fascinating structures (see Figure 22.17) that cannot be explained in terms of simple bonding theories, so they are usually discussed in terms of the molecular orbital theory.

There are also many anions derived from the boranes that are more complicated than BH_4^-. A particularly fascinating example is $B_{12}H_{12}^{2-}$ (Figure 22.18). It has the shape of an icosahedron, which has 12 equivalent vertices, 20 equilateral triangular faces, and 30 equivalent edges. The bonding in the electron-deficient $B_{12}H_{12}^{2-}$ can be satisfactorily described only by using molecular orbital theory.

The structures of these remarkable compounds serve to illustrate the fact that although the simple ideas of chemical bonding presented in this book enable us to understand the structures of many compounds, they are inadequate for many others. A more detailed treatment of the chemical bond is needed to be able to understand such compounds. Although our understanding of chem-

FIGURE 22.16
Formation of B_2H_6 from Two
BH_3 Molecules.

The molecule BH_3 is not known
under ordinary conditions. Two
BH_3 molecules combine to give a
B_2H_6 molecule.

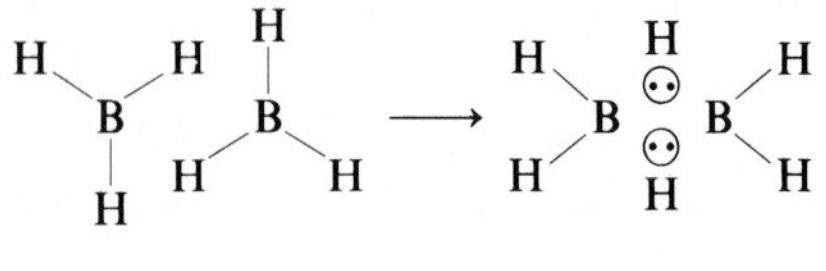

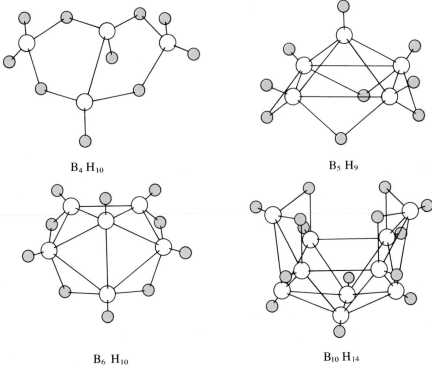

B_4H_{10}

B_5H_9

B_6H_{10}

$B_{10}H_{14}$

FIGURE 22.17
Structures of Some Higher Boranes.

They are all electron-deficient molecules in which there are not enough electrons to form ordinary single bonds between adjacent atoms. The bonding in these molecules can only be described in terms of three-center and other multicenter bonds.

ical bonding has progressed considerably since Lewis first proposed the idea of the shared electron pair, many aspects of this subject are still not fully understood. Chemists continue to prepare new compounds, the structures of which present a challenge for even the most sophisticated theories.

22.8
Semiconductors

We are in the midst of a technological revolution, the computer revolution, which is no less significant than the Industrial Revolution of the nineteenth century. The computer revolution is based on the semiconductor properties of silicon and other semimetals. A *semiconductor* is not as conducting as a metal, but it does have a significant electrical conductivity, which unlike the conductivity of a metal, increases with increasing temperature.

Semiconductors have electrical properties that are intermediate between those of nonconducting covalent solids and highly conducting metals. In order to understand semiconductors, let us first review our model of the bonding in metals. In the charge cloud or electron gas model we imagine that each metal atom provides one or more of its valence electrons to a mobile charge cloud that can move freely between the positively charged metal atom cores. These mobile electrons hold the positively charged cores together and constitute the metallic bond. In a covalent solid all the valence electrons are held tightly by the atoms in localized bonds or as unshared pairs. They are not free to move through the crystal, so a covalent crystal is a nonconductor. We can think of a semiconductor as a covalent substance in which the valence electrons are

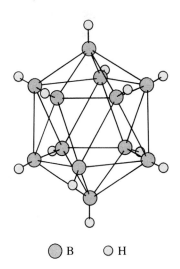

○ B   ○ H

FIGURE 22.18
Structure of the $B_{12}H_{12}^{2-}$ Anion.

Each boron atom is at one of the corners of an icosahedron. A hydrogen atom is attached to each boron atom. The icosahedron is a regular solid with 12 equivalent vertices, 20 equilateral triangular faces, and 30 equivalent edges.

not held very strongly; as a consequence, a few of them have enough energy to break away from the atom with which they were associated and move freely through the crystal. With increasing temperature more electrons acquire sufficient energy to break away from their atoms, and so the conductivity increases with increasing temperature.

This is a very oversimplified picture of a semiconductor, but a full discussion of semiconductors is beyond the scope of this book. Nevertheless, we can go a little more deeply into the properties of semiconductors by starting with an alternative view of the bonding in metals.

ENERGY LEVELS IN SOLIDS

In a crystal of sodium, for example, the 3s orbital of each sodium atom overlaps extensively with that of its neighbors. As a consequence the individual 3s orbitals may be considered to be replaced by a set of molecular orbitals or energy levels equal in number to the number of atoms in the crystal. The set of energy levels is called a *band*, in this case a 3s band (see Figure 22.19). Since each orbital, in accordance with the Pauli exclusion principle, may contain two electrons, and since each sodium atom contributes only one electron, only half the levels are occupied. Thus there are many empty levels into which an electron may be excited. The movement of electrons from one orbital to another constitutes an electric current; thus sodium is an electrical conductor.

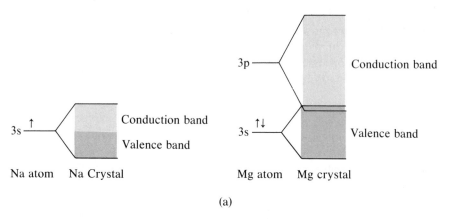

(a)

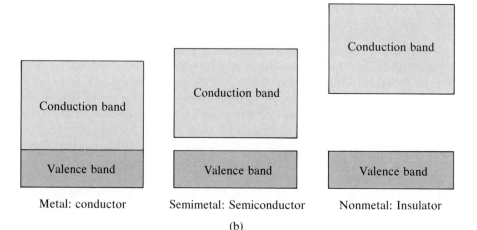

(b)

FIGURE 22.19
Valence and Conduction Bands in Metals, Semiconductors, and Insulators.

(a) In a sodium or a magnesium crystal, the single 3s orbital of an isolated metal atom is replaced by numerous closely spaced energy levels, which are said to form a band. In sodium, this band is half-filled. In magnesium, this band is full but overlaps the empty 3p band. (b) The separation of the valence band and the conduction band increases from a metal, to a semiconductor, to an insulator.

If we apply the same arguments to magnesium, which has two electrons in a 3s orbital, we conclude that the 3s band will be completely filled. There are no empty orbitals in the 3s band into which electrons can be excited by the application of an electric field. In other words, an electric field cannot cause any movement of the electrons, and therefore magnesium might be expected to be a nonconductor. However, the 3p orbitals also form a band of closely spaced energy levels, and this 3p band overlaps the 3s band (see Figure 22.19). Thus the application of an electric field can move electrons into this empty band, which is called a *conduction band*. The filled 3s band is called the *valence band*.

In an insulator such as diamond the valence band is completely filled, and the conduction band is at a much higher energy; so the electrons from the valence band cannot move into the conduction band, and diamond is a nonconductor. In a semiconductor such as silicon the conduction band does not overlap the valence band as in a metal but is sufficiently close to the valence band that a few electrons have sufficient thermal energy to jump the gap and occupy the conduction band. Thus silicon is a conductor, but because there are only very few electrons in the conduction band, it is a much poorer conductor than a metal.

The conductivity of a semiconductor increases with increasing temperature because an increasing number of electrons have sufficient thermal energy to occupy the conduction band. Thus the difference between a metal, a semiconductor, and a nonconductor depends on the energy gap between the valence band and the conduction band. Figure 22.20 shows the energy gaps for carbon, silicon, germanium, and tin. Carbon is an insulator; silicon and germanium are semiconductors; and tin is a metal. In a metal the conduction band overlaps the valence band or is very close to the valence band.

APPLICATIONS OF SEMICONDUCTORS

The important applications of semiconductors depend on the discovery that the conductivity of a semiconductor can be changed in a dramatic but controllable way by the addition of very small amounts of certain impurities. If a very

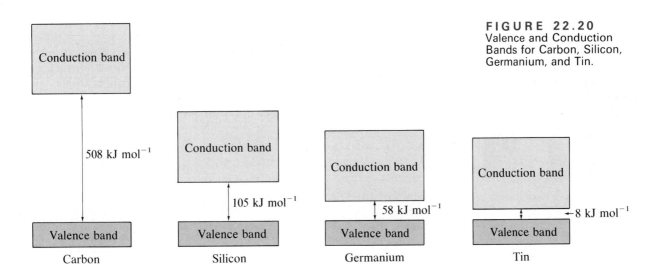

FIGURE 22.20
Valence and Conduction Bands for Carbon, Silicon, Germanium, and Tin.

FIGURE 22.21
An n–p Semiconductor Junction.

Current is carried in a p-type semiconductor by the motion of holes and in an n-type semiconductor by the motion of electrons. Left: With the potential in this direction, a large current can flow. Right: With the potential applied in this direction, little current flows.

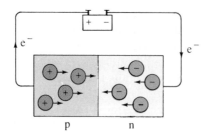
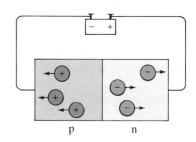

small amount of phosphorus or arsenic is added to crystalline silicon, the phosphorus or arsenic atoms are incorporated into the silicon structure. However, only four electrons are needed to form bonds to the surrounding four silicon atoms, so one electron is left over. This electron occupies an energy level in the conduction band, and under the influence of an applied electric potential it can move through the crystal. The number of electrons that move into the conduction band and, therefore, the conductivity are proportional to the number of phosphorus or arsenic atoms that are added.

If a boron atom is added to silicon, it contributes only three electrons to the structure. Because it needs four electrons to form four bonds, the boron steals an electron from an adjacent silicon atom, leaving this atom short of one electron and therefore positively charged. This vacancy in the valence shell of silicon is called a *positive hole*. An electron from an adjacent silicon atom can move into this hole, thus transferring the positive hole to this adjacent atom, and so on. Under the influence of an applied potential the positive hole moves through the crystal. Because the charge is carried by the positive holes, this type of conductor is called a p-type conductor. In contrast, if arsenic or phosphorus is added to the silicon, the conductor is called an n-type conductor, because conduction is due to negative electrons moving in the conduction band.

If a p-type semiconductor is placed in contact with an n-type semiconductor, a very useful device called a p–n junction is obtained. Figure 22.21 shows what happens if a potential is applied across a p–n junction. If the n-type conductor is made negative, electrons flow into the conductor and continue toward the junction. Electrons are simultaneously removed from the p-type conductor, producing more holes, which move toward the junction. At the junction the electrons fall into the holes—the holes and electrons may be said to neutralize each other—and the current continues to flow.

If the potential is applied in the opposite direction, electrons flow out of the n-type conductor and electrons enter the p-type conductor, thus removing the holes, which flow away from the junction. Therefore in the region near the junction there are no longer holes in the p conductor nor any free electrons in the n conductor. Since there is no mechanism for producing more holes or electrons, the current ceases to flow.

Thus a p–n junction conducts current in only one direction. It can therefore convert alternating current to direct current. Such a device is called a *rectifier*. The p–n junction has replaced the vacuum tube diode in electronics for this purpose and is often called a diode.

In addition to rectifying current, n- and p-type conductors can be combined into devices, called *transistors*, that amplify currents and voltages. Since their invention in 1948, transistors and other semiconductor devices have completely revolutionized the electronics industry.

FIGURE 22.22
Integrated Circuit on a Silicon Chip.

A chip this size can contain many hundreds of transistors.

The most important development in recent years has been the production of *integrated circuits* (Figure 22.22) consisting of thousands of resistors, transistors, rectifiers, and capacitors constructed from n- and p-type conductors on a single piece of silicon having dimensions of no more than a few millimeters. Integrated circuit chips are at the heart of digital wristwatches, handheld calculators, and personal computers.

IMPORTANT TERMS

An **electron-deficient molecule** is a molecule in which there are insufficient electrons for all the atoms to be held together by two-center, electron pair bonds.

A **glass** is a solid that has an amorphous rather than a crystalline structure. One of its characteristic properties is that it softens over a temperature range rather than melting sharply at a well-defined temperature.

A **semiconductor** is a solid that has an appreciable electrical conductivity, which increases with increasing temperature rather than decreasing, as in the case of a metal.

A **semimetal** (or *metalloid*) is an element that has properties intermediate between those of a metal and a nonmetal.

In a **three-center bond** one pair of electrons holds three nuclei together.

Zone refining is a technique for purifying solids in which a molten zone is moved along a bar of a solid material. Impurities concentrate in the molten zone and are thus removed to one end of the bar, which can then be cut off.

PROBLEMS *

Silicon

1. (a) What is the fundamental structural unit on which the structures of all silicates is based?

   (b) How is this unit modified in the aluminosilicates?

2. Give an example of a silicate mineral with a layer structure. Explain the empirical formula of the silicate anion in this structure.

3. On what structure is an amphibole mineral based, and what is the empirical formula of the silicate anion in the amphiboles?

4. Draw the Lewis structure of the cyclic $Si_4O_{12}^{8-}$ anion.

5. Draw the Lewis structure of the $Si_6O_{18}^{12-}$ anion. Which silicate mineral contains this anion?

6. Predict the structure of the silicate anion (chain, ring, double chain, sheet, etc.) in each of the following silicate minerals:

   (a) Gillespite, $BaFeSi_4O_{10}$

   (b) Dentitoite, $BaTiSi_3O_9$

   (c) Chrysolite, $Mg_3Si_2O_5(OH)_4$

   (d) Rhodonite, $CaMn_4Si_5O_{15}$

   (e) Vermiculite, $Mg_3Si_4O_{10}(OH)_4 \cdot xH_2O$

7. Predict the structure of the aluminosilicate anion (chain, ring, sheet, network, etc.) in each of the following aluminosilicate minerals:

   (a) Anorthite, $Ca(Al_2Si_2O_8)$

   (b) Muscovite, $KAl_2(Si_3AlO_{10})(OH)_2$

   (c) Amesite, $Mg_2Al(SiAlO_5)(OH)_4$

   (d) Thomsonite, $NaCa_2(Si_5Al_5O_{20}) \cdot 6H_2O$

8. Write a reasonable ionic formulation for each of the following minerals, and describe the structure of its silicate, or aluminosilicate, anion:

   (a) Diopside, $CaMgSi_2O_6$

   (b) Orthoclase, $KAlSi_3O_8$

   (c) Hardystonite, $Ca_2ZnSi_2O_7$

   (d) Denitoite, $BaTiSi_3O_9$

9. What are the three principal components of common glass?

10. What is photochromic glass, and how does it work?

11. Which properties of glass are similar to those of (a) a crystalline solid and (b) a liquid?

12. The crystal structure of silicon is similar to that of diamond.

(a) Draw a sketch of the unit cell.

(b) How many atoms are there in the unit cell?

(c) How many nearest neighbors has each Si atom in silicon?

(d) If the length of the edge of the unit cell is 545 pm, calculate **(i)** the density of silicon, **(ii)** the length of the Si—Si bond, and **(iii)** the covalent radius of silicon.

13. Explain the process of *zone refining*.

14. Compare the structures of graphite, diamond, and silicon. Explain why there is no allotrope of silicon with a structure analogous to that of graphite.

15. Draw Lewis structures for each of the Si_2Cl_6 and Si_2Cl_6O molecules, and predict which molecule would be expected to have a dipole moment.

16. What is a silicone? Draw the Lewis structure of part of the structure of a typical silicone. List some of the important properties and uses of silicones.

17. What carbon compounds are the analogues of the silicones? How do their structures differ from those of the silicones? Explain this difference in structure.

18. Describe the structures of the silicone polymer resulting from the hydrolysis of each of the following:

(a) $(CH_3)_3SiCl$ **(b)** $(CH_3)_2SiCl_2$ **(c)** CH_3SiCl_3

19. What is the general formula of a silane containing n silicon atoms? Why are silanes very reactive in comparison with their carbon analogues?

20. Draw the Lewis structures and deduce the molecular geometries of each of the following:

(a) SiO_4^{4-} **(b)** $(H_3Si)_2O$ **(c)** $(CH_3)_2SiCl_2$
(d) SiF_4^- **(e)** SiF_5^- **(f)** SiF_6^{2-}

21. An organic silicon compound A containing chlorine was reacted with water, and another organic silicon compound B was isolated from the aqueous solution by distillation. Compound B contained carbon, hydrogen, silicon, and oxygen. When 0.1803 g of B was burned completely in oxygen, 0.2931 g of carbon dioxide, 0.1800 g of water, and 0.1334 g of silica were obtained. At 200 °C and 755 mm Hg pressure, 0.3345 g of gaseous compound B occupied a volume of 80.5 mL. Identify the compounds A and B, name them, and write their Lewis structures.

22. Give the formula of the methyl silicon chloride that on hydrolysis gives the polymer $[Si(CH_3)_2O]_n$.

* The asterisk denotes the more difficult problems.

*23. When gaseous silicon tetrafluoride is bubbled into water, silicic acid precipitates and the strong acid $H_2SiF_6(aq)$ is formed. After filtering off the silica and adding excess $BaCl_2(aq)$ to the filtrate, barium hexafluorosilicate precipitates.

(a) What is the molarity of the $H_2SiF_6(aq)$ solution formed from the reaction of 50 mL of $SiF_4(g)$, at 10 °C and a pressure of 756 mm Hg, with 50 mL of water?

(b) If 20 mL of $0.010M$ $BaCl_2(aq)$ is added to the above solution, what mass of barium hexafluorosilicate is precipitated, and what is the final molarity of the solution with respect to the strong acid H_2SiF_6, and its pH?

Boron

24. One of the gaseous boranes has a density of 1.23 g L^{-1} at STP. Calculate its molar mass and propose a plausible molecular formula.

25. Calculate the bond energy of each of the B—H bridge bonds in diborane, $B_2H_6(g)$, given that the average B—H bond energy in the ion BH_4^- is 389 kJ mol^{-1} and the following standard enthalpies of formation. ΔH_f°(kJ mol^{-1}): $B_2H_6(g)$, 36; B(g), 560; H(g), 218. Assume that each of the B—H terminal bonds have the same energy as the B—H bonds in BH_4^-. Compare the value of the B—H bridge bond energy with that of the B—H bonds in BH_4^-, and account for the difference.

26. Explain why boric acid, $B(OH)_3$, is a weak acid, while aluminum hydroxide, $Al(OH)_3$, is amphoteric in aqueous solution.

27. Define the terms *Lewis acid* and *Lewis base*. Write three balanced equations for reactions in which boron compounds behave as Lewis acids.

28. Write Lewis structures and predict the molecular shapes of each of the following molecules and ions:

BF_3 BF_4^- $BF_3 \cdot NH_3$
$AlCl_3$ $AlCl_4^-$ Al_2Cl_6

29. Draw the Lewis structures and deduce the molecular geometries of each of the following:

BH_4^- $B_2O_5^{4-}$ $B_3O_6^{3-}$

30. Account for each of the following observations.

(a) Boric acid is a very weak monoprotic acid ($pK_a = 9$) in aqueous solution.

(b) An aqueous solution of $BCl_3(g)$ is highly acidic.

(c) Sodium borate gives a basic aqueous solution.

31. Boron nitride, empirical formula BN, has a structure analogous to that of graphite under ordinary conditions but is converted to a very hard form with a diamondlike structure at high temperature and very high pressure. Describe the bonding in each of these two forms of boron nitride.

32. By comparison with the isoelectronic carbon compound, suggest a structure for borazine, $B_3N_3H_6$, a colorless liquid boiling at 55 °C. How many isomers with the molecular formula $B_3N_3H_4Cl_2$ should exist?

*33. Reaction of boric acid with ethanol and concentrated sulfuric acid, followed by distillation, gives a liquid boiling at 118 °C with the composition 7.4% B, 32.9% O, 49.3% C, and 10.4% H, by mass. At 200 °C and a pressure of 740 mm Hg, 0.371 g of this compound was found to occupy a volume of 101.3 mL. Calculate the empirical and molecular formulas of the compound, and suggest a possible structure.

Miscellaneous

34. Describe the chemistry of carbon, silicon, and lead, with reference to each of the following:

(a) The possible valence states and the stability and covalent or ionic nature of compounds in these valence states

(b) The structures and reactions of the oxides

(c) The trends in the covalent and ionic character of the chlorides of these elements

35. What is the chemical composition of each of the following minerals?

(a) Limestone   (b) Gypsum   (c) Quartz

(d) Bauxite   (e) Talc   (f) Beryl

36. Name each of the oxoacids with the following molecular formulas. Place them in order of increasing strength and explain this order.

$Si(OH)_4$ $OP(OH)_3$ $O_2S(OH)_2$ $HOClO_3$

37. Explain with the aid of a suitable example what is meant by the term *three-center bond*.

38. From the following data, calculate and compare the average bond energies of the C—H and Si—H bonds in CH_4 and SiH_4, respectively. Using these values, calculate and compare the C—C and Si—Si bond energies in $C_2H_6(g)$ and $Si_2H_6(g)$, respectively:

	ΔH_f° (kJ mol^{-1})		ΔH_f° (kJ mol^{-1})
$CH_4(g)$	-75	$C(g)$	716
$SiH_4(g)$	34	$Si(g)$	450
$C_2H_6(g)$	-84	$H(g)$	218
$Si_2H_6(g)$	80		

39. Diborane, B_2H_6, reacts with water to give boric acid and hydrogen gas. Write the balanced equation for this reaction. What would be the pH of the solution that results from reaction of 1.50 g of diborane with water, given that the pK_a of boric acid is 9.22?

40. Germanium is the element below silicon in group 4. What properties would you predict for the oxide GeO_2 and the chloride $GeCl_4$?

Organic Chemistry

23

Organic chemistry is the chemistry of all the compounds of carbon except for the few that are considered to be inorganic compounds which we discussed in Chapter 6. It may seem surprising that a complete branch of chemistry is devoted to just one element but 90% of all known compounds contain carbon. Although organic compounds were originally obtained from plants and animals the great majority known today do not exist in nature. The major sources of organic compounds at the present time are petroleum and natural gas which provide the hydrocarbons from which all the other organic compounds are synthesized. Although we tend to think of hydrocarbons primarily as fuels, they are no less important as the source of the enormous variety of organic compounds used in the manufacture of many materials essential to modern life, such as synthetic fabrics, rubber, plastics, medicines, drugs, and solvents. Substances derived from petroleum are frequently referred to as *petrochemicals*, and the industry that produces them is called the *petrochemical industry*.

Organic compounds either are hydrocarbons or can be conveniently regarded as derived from hydrocarbons by replacing one or more of the hydrogen atoms with other atoms or groups of atoms called functional groups.

We will first review and extend the discussion of hydrocarbons that we gave in Chapter 6 and then we will discuss several classes of compounds containing some of the more important functional groups.

23.1
Alkanes

The alkanes are a class of hydrocarbons in which all the carbon–carbon bonds are single bonds. The simplest alkanes are methane, CH_4, which is the major constituent of natural gas, ethane, C_2H_6, propane, C_3H_8, and butane, C_4H_{10}. Alkanes have the general formula C_nH_{2n+2}.

The melting points and boiling points of the alkanes increase with increasing molecular size because of the increasing strength of the intermolecular (London)

TABLE 23.1
Boiling Points and Melting Points of n-Alkanes, C_nH_{2n+2}

n	Name	Boiling Point (°C at 1 atm)	Melting Point (°C)	Formula
1	Methane	−162	−183	CH_4
2	Ethane	−89	−183	CH_3CH_3
3	Propane	−42	−188	$CH_3CH_2CH_3$
4	Butane	0	−138	$CH_3(CH_2)_2CH_3$
5	Pentane	36	−130	$CH_3(CH_2)_3CH_3$
6	Hexane	69	−95	$CH_3(CH_2)_4CH_3$
7	Heptane	98	−91	$CH_3(CH_2)_5CH_3$
8	Octane	126	−57	$CH_3(CH_2)_6CH_3$
9	Nonane	151	−54	$CH_3(CH_2)_7CH_3$
10	Decane	174	−30	$CH_3(CH_2)_8CH_3$
20	Eicosane	343	37	$CH_3(CH_2)_{18}CH_3$
30	Triacontane	446	66	$CH_3(CH_2)_{28}CH_3$

forces (Table 23.1). The alkanes up to butane are gases at ordinary temperatures and pressures. The alkanes from $n = 5$ to $n = 16$ are liquids and those with more than 16 carbon atoms are waxy solids. Paraffin wax is a mixture of solid alkanes.

STRUCTURES OF ALKANES

In Chapters 6 and 9 the structures of some of the simpler alkanes were discussed. Both ball-and-stick and space-filling models for methane, ethane, and propane are illustrated in Figure 23.1. The four bonds formed by each carbon atom in an alkane have a geometry that is close to tetrahedral and they can be described in terms of a tetrahedral set of sp^3 hybrid orbitals on each carbon atom. By analogy with methane, ethane, and propane we expect the structural formulas for butane, C_4H_{10}, and pentane, C_5H_{12}, to be

$$H-\overset{\displaystyle H}{\underset{\displaystyle H}{C}}-\overset{\displaystyle H}{\underset{\displaystyle H}{C}}-\overset{\displaystyle H}{\underset{\displaystyle H}{C}}-\overset{\displaystyle H}{\underset{\displaystyle H}{C}}-H \qquad \text{or} \qquad CH_3-CH_2-CH_2-CH_3$$

$$H-\overset{\displaystyle H}{\underset{\displaystyle H}{C}}-\overset{\displaystyle H}{\underset{\displaystyle H}{C}}-\overset{\displaystyle H}{\underset{\displaystyle H}{C}}-\overset{\displaystyle H}{\underset{\displaystyle H}{C}}-\overset{\displaystyle H}{\underset{\displaystyle H}{C}}-H \qquad \text{or} \qquad CH_3-CH_2-CH_2-CH_2-CH_3$$

Such alkanes are often called *straight-chain alkanes*, because all the carbon atoms are connected in one continuous chain. In fact, the carbon chains are not straight but have a zigzag shape with an angle between the C—C bonds at each carbon atom of approximately 109°, as Figure 23.2 shows.

STRUCTURAL ISOMERS

There are two different substances with slightly different properties that have the molecular formula C_4H_{10}; one has a boiling point of 0 °C and the other has a boiling point of -10 °C. Similarly, there are three different substances with the formula C_5H_{12}; they have boiling points of 10 °C, 28 °C, and 36 °C.

There are two different butanes and three different pentanes, because carbon atoms in an alkane with four or more carbon atoms can be joined in different

(a)

(b)

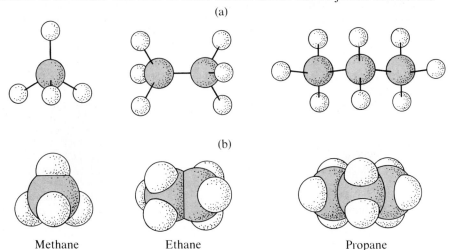

Methane   Ethane   Propane

FIGURE 23.1
Structures of Methane, Ethane, and Propane.

(a) Ball-and-stick models. (b) Space-filling models.

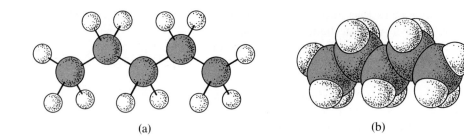

(a)

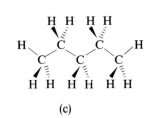

(b)   (c)

FIGURE 23.2
Structure of Pentane.

Pentane is called a straight-chain alkane because all the carbon atoms are connected in a continuous chain. But the chain actually has a zigzag shape; the angles between the bonds at each carbon atom are approximately 109°. (a) Ball-and-stick model. (b) Space-filling model. (c) Structural formula.

ways. The four carbon atoms of C_4H_{10} may be joined consecutively in one chain as we have seen. But we can draw another structure in which there is a continuous chain of only three carbon atoms; the fourth carbon is attached to the middle carbon of the chain, forming a **side chain** (Figure 23.3):

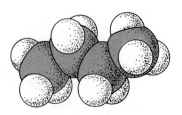

Isobutane

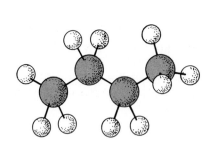

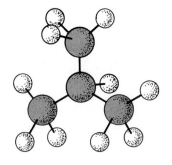

FIGURE 23.3
n-Butane and Isobutane.

(a) Ball-and-stick models. (b) Space-filling models.

n-Butane

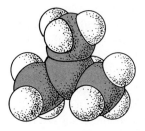

Isobutane

This is an example of a **branched-chain alkane**. It is called *isobutane*, whereas that with a single four-carbon chain is called normal butane, or *n-butane*.

Notice that the middle carbon in isobutane is attached to three other carbons, whereas all the carbons in *n*-butane are bonded at most to two others. Both butanes have the same molecular formula, C_4H_{10}, and the same molar mass; however, since their structures differ, they have slightly different properties. Molecules that have the same molecular formula but different structures are called **structural isomers**. In other words, structural isomers have the same number of each kind of atom, and thus the same molecular formula, but the atoms are arranged in different ways.

For pentane three unique arrangements of carbon atoms are possible. Therefore, there are three isomers with the following structures:

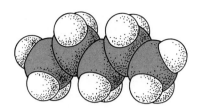

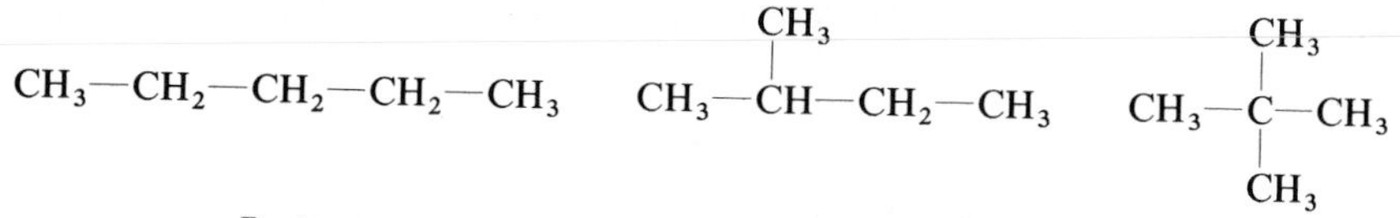

n-Pentane

$$CH_3—CH_2—CH_2—CH_2—CH_3 \qquad CH_3—\overset{\displaystyle CH_3}{\underset{|}{CH}}—CH_2—CH_3 \qquad CH_3—\overset{\displaystyle CH_3}{\underset{|}{\overset{|}{C}}}—CH_3$$
$$CH_3$$

n-Pentane   Isopentane   Neopentane
(bp = 36 °C)   (bp = 28 °C)   (bp = 10 °C)

We might think that there is a fourth possibility:

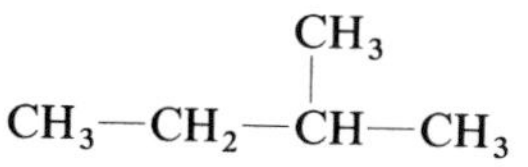

$$CH_3—CH_2—\overset{\displaystyle CH_3}{\underset{|}{CH}}—CH_3$$

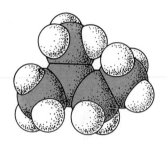

Isopentane

But this structure is identical with isopentane, as we can see by rotating the structure about an axis passing through the middle C—C bond,

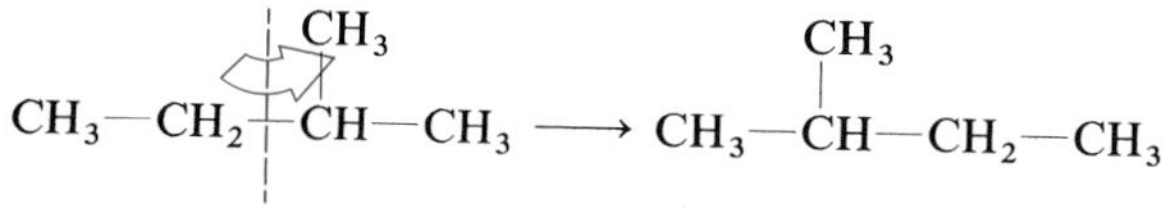

$$CH_3—CH_2—\overset{\displaystyle CH_3}{\underset{|}{CH}}—CH_3 \longrightarrow CH_3—\overset{\displaystyle CH_3}{\underset{|}{CH}}—CH_2—CH_3$$

or by comparing models of the two structures.

Structural isomers of alkanes differ in the way in which the atoms are connected together. For example, the carbon atoms in *n*-pentane are joined in a continuous chain, whereas in isopentane the chain is not continuous but is branched. Remember that structural formulas do not give any information about the three-dimensional geometry of molecules; they merely show how the atoms are connected. Thus we can write isopentane as

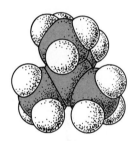

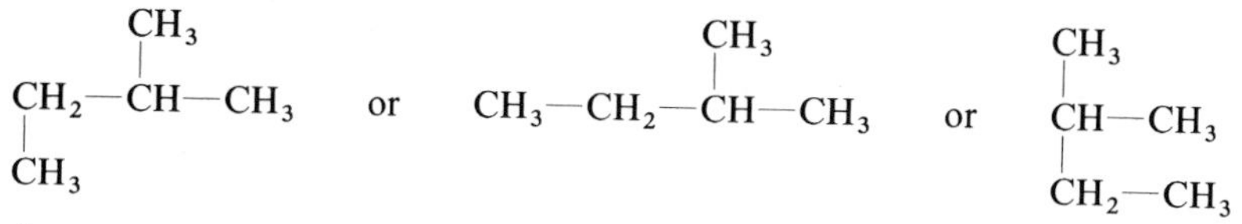

Neopentane

$$\overset{\displaystyle CH_3}{\underset{|}{\underset{CH_3}{CH_2—CH—CH_3}}} \quad \text{or} \quad CH_3—CH_2—\overset{\displaystyle CH_3}{\underset{|}{CH}}—CH_3 \quad \text{or} \quad \overset{\displaystyle CH_3}{\underset{|}{\underset{CH_2—CH_3}{CH—CH_3}}}$$

All three structures represent the same molecule. They do not represent different structural isomers because the atoms are all connected in the same way. It is irrelevant whether the angle between two C—C bonds is shown as 90° or as 180°, because structural formulas are not meant to give any information on bond angles.

Following the pentanes, C_5H_{12}, we have the hexanes, C_6H_{14}, the heptanes, C_7H_{16}, and so on. As the number of carbon atoms increases, the number of isomers increases rapidly. There are 3 isomeric pentanes, 5 hexanes, 9 heptanes, and 75 decanes, $C_{10}H_{22}$. The names and formulas of the first ten *n*-alkanes are given in Table 23.1, together with their boiling points and melting points.

It takes practice to be able to recognize whether two structural formulas that look different actually represent different isomers and to find the correct number of isomers of a particular alkane. To identify the isomers systematically, first find the longest continuous chain of carbon atoms, and then look for the location of the side chains along the principal chain. *n*-Pentane has a continuous chain of five carbon atoms. Isopentane has a continuous chain of four carbon atoms with one CH_3 side chain. Neopentane has a continuous chain of only three carbon atoms with two CH_3 side chains. No other arrangements of five carbon atoms are possible.

═ *Example 23.1* ═══

THE STRUCTURES OF ISOMERIC HYDROCARBONS

What are the structures of the five isomers of hexane, C_6H_{14}?

Solution

First, we write a structure with a continuous chain of six carbon atoms:

$$CH_3-CH_2-CH_2-CH_2-CH_2-CH_3$$

(1)

This structure is the only possibility for a six-carbon chain; simply bending the chain cannot give a different isomer.

Second, we consider structures containing a continuous chain of five carbon atoms. There are two such isomers:

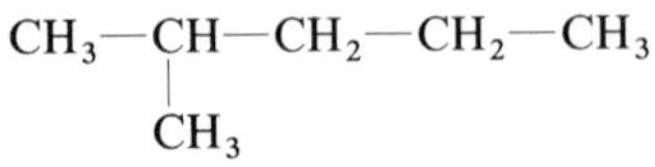

$$CH_3-\underset{\underset{CH_3}{|}}{CH}-CH_2-CH_2-CH_3 \quad \text{and} \quad CH_3-CH_2-\underset{\underset{CH_3}{|}}{CH}-CH_2-CH_3$$

(2)   (3)

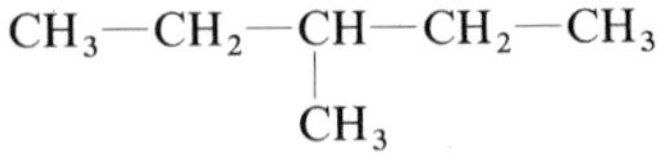

There are no others. For example,

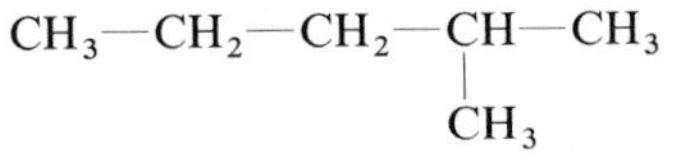

$$CH_3-CH_2-CH_2-\underset{\underset{CH_3}{|}}{CH}-CH_3$$

is the same as (2), because there is a $-CH_3$ group attached to a carbon atom one from the end of a continuous chain of five C atoms.

We must also consider structures in which there is a continuous chain of four C atoms. There are two isomers of this kind:

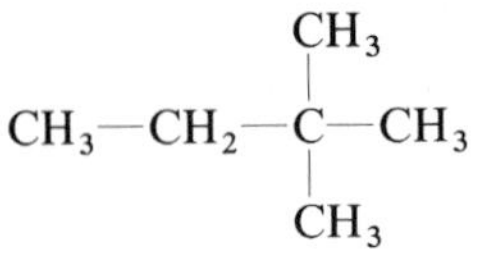

$$CH_3-CH_2-\underset{\underset{CH_3}{|}}{\overset{\overset{CH_3}{|}}{C}}-CH_3 \quad \text{and} \quad CH_3-\underset{\underset{CH_3}{|}}{CH}-\underset{\underset{CH_3}{|}}{CH}-CH_3$$

(4)   (5)

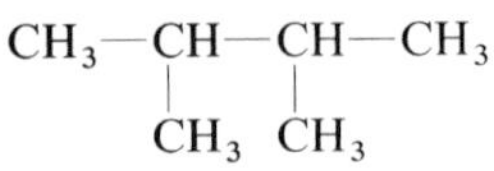

We could try writing structures with a continuous chain of only three carbon atoms, but this does not lead to any different structures. For example, in the structure

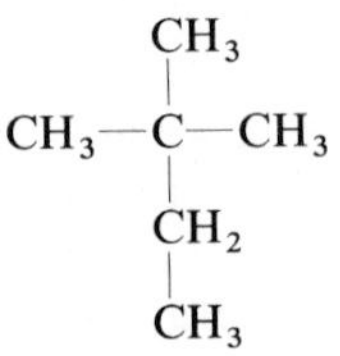

$$CH_3-\underset{\underset{\underset{\underset{CH_3}{|}}{CH_2}}{|}}{\overset{\overset{CH_3}{|}}{C}}-CH_3$$

the longest continuous chain contains four C atoms. The structure is the same as (4). We have simply drawn a structure that at first sight looks different but, in fact, is not.

CONFORMATIONS

Although we know that the four bonds around each carbon atom in ethane have a tetrahedral arrangement, this does not tell us how the C—H bonds at one end of the molecule are oriented with respect to those at the other end. Indeed, it might seem that there could be two ethane molecules with rather different overall shapes, as shown in Figure 23.4. In one form of the molecule, the C—H bonds at one end of the molecule are directly opposite the C—H bonds at the other end. The C—H bonds are said to have an **eclipsed arrangement**. In the other form the C—H bonds at one end of the molecule, rather than being exactly opposite the bonds at the other end, lie exactly between them. The C—H bonds are said to have a **staggered arrangement**.

Different arrangements of atoms that can be converted into one another by rotation about single bonds are called **conformations**. Ethane can have an eclipsed conformation, a staggered conformation, and an infinite number of conformations between these two that are called **skew conformations**.

As we might expect, there are small differences in energy between the different conformations. The eclipsed conformation has a slightly higher energy and is therefore slightly less stable than the staggered conformation. But because the energy difference between the different conformations is small, even at low temperatures the conformation of an ethane molecule is continually changing. In effect, one end of the molecule is continually rotating with respect to the other end. We say that there is **free rotation** around the C—C bond. Any orientation of the C—H bonds at one end with respect to those at the other end is possible. This free rotation is easily demonstrated by using a ball-and-stick model of ethane.

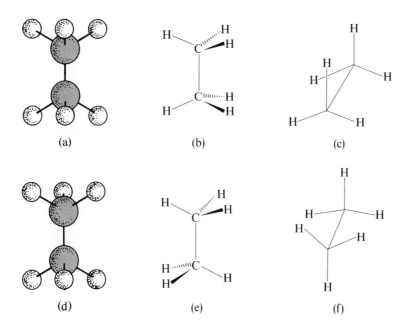

(a) (b) (c)

(d) (e) (f)

FIGURE 23.4
Eclipsed and Staggered Conformations of Ethane.

(a), (b), and (c) are representations of the eclipsed conformation. (d), (e), and (f) are representations of the staggered conformation. In (c) and (f) the carbon atoms are not shown specifically; they are represented by the junction of the four bonds.

CYCLOALKANES

Another important series of hydrocarbons is the **cycloalkanes**. They have $(CH_2)_n$ chains in which the two ends are joined, forming a closed loop or ring. The simplest members of this series are the following:

	Melting point (°C)	Boiling point (°C)
Cyclopropane, C_3H_6	-127	-33
Cyclobutane, C_4H_8	-80	13
Cyclopentane, C_5H_{10}	-94	49
Cyclohexane, C_6H_{12}	7	81

An unusual feature of the structures of cyclopropane and cyclobutane is that the bond angles are unexpectedly small. If the C—C bonds are drawn as straight lines, the angles between them are only 60° and 90°, respectively. Since the angles between four single bonds on a carbon atom are always close to 109.5°, the C—C bonds must be bent, as shown in Figure 23.5. The region where the bond electron density is concentrated, as shown for example by the region

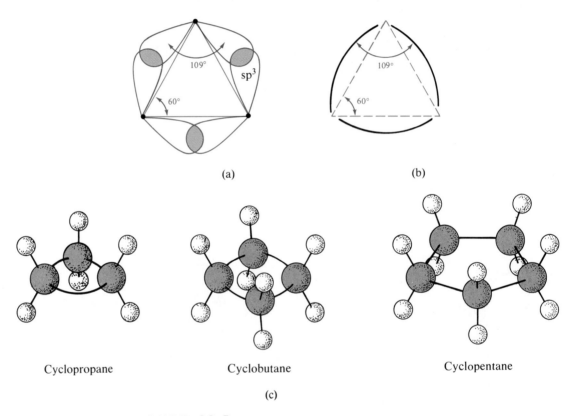

(a)   (b)

Cyclopropane   Cyclobutane   Cyclopentane

(c)

FIGURE 23.5
Structures of Cycloalkanes: Bent Bonds.

(a) The region of maximum overlap of the sp^3 orbitals on adjacent carbon atoms and therefore the region of greatest concentration of electron density is not along the internuclear axis but displaced to the outside of the ring. Thus the carbon–carbon bonds are described as "bent" as shown in (b). The angle between the bonds at each carbon atom is 109° as expected for an AX_4 geometry (sp^3 hybrid orbitals) although the angle of the equilateral triangular shape of the cyclopropane ring is only 60°. (c) The amount of bond bending is less in cyclobutane and negligible in cyclopentane. (The internal angle of a regular pentagon is 108° which is very close to the tetrahedral angle of 109°.)

of maximum overlap of the sp³ orbitals, is not along the internuclear axis as it normally is but is displaced to one side. Such bonds are said to be bent or strained and they are weaker than normal carbon–carbon single bonds. So cyclopropane and cyclobutane are more reactive than the corresponding straight-chain alkanes, propane and butane. Because the angle in a pentagon is 108° the bonds in cyclopentane are not strained and it behaves like a normal alkane. Cyclohexane is also not strained but it is not a planar ring like cyclo-pentane. If it were planar the bond angles in the ring would have to be 120° which is appreciably larger than the expected bond angle of 109°. However, normal tetrahedral bond angles can be maintained if cyclohexane adopts a non-planar conformation (Figure 23.6).

REACTIONS OF ALKANES

At ordinary temperatures alkanes are rather unreactive but they undergo several important reactions at higher temperatures. As we have seen in Chapter 6 their most important reaction is combustion in air or oxygen; for example

$$C_3H_8(g) + 5O_2(g) \longrightarrow 3CO_2(g) + 4H_2O(g)$$

Alkanes also react with the halogens at high temperatures or in a photo-chemical reaction to give products in which one or more of the hydrogen atoms

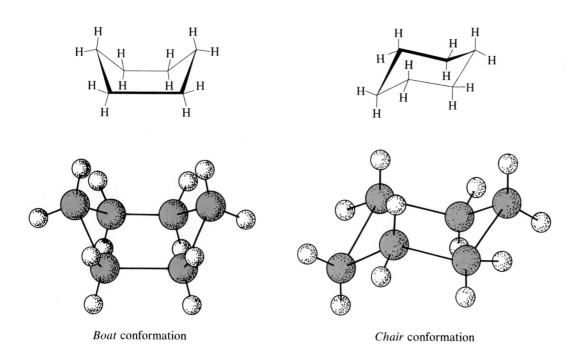

Boat conformation   *Chair* conformation

FIGURE 23.6
Structure of Cyclohexane.

Cyclohexane is not planar, because a planar arrangement would require 120° bond angles at each carbon atom. There are two nonplanar conformations of cyclohexane, called the *boat conformation* and the *chair conformation*. The bond angles at each carbon atom are approximately 109°.

are replaced by halogen atoms:

$$CH_4 + Cl_2 \xrightarrow{\text{Heat or } hv} CH_3Cl + HCl$$
Chloromethane

$$C_2H_6 + Cl_2 \longrightarrow C_2H_5Cl + HCl$$
Chloroethane

When alkanes are heated by themselves to a sufficiently high temperature, they decompose to a mixture of molecules including smaller alkanes, other hydrocarbons such as ethene, and hydrogen. For example,

$$C_4H_{10} \xrightarrow{\text{Heat}} CH_2{=}CH_2 + C_2H_6$$

$$C_2H_6 \xrightarrow{\text{Heat}} CH_2{=}CH_2 + H_2$$

In the petroleum industry such reactions are called **cracking reactions**, because they make small molecules out of larger ones.

PETROLEUM AND NATURAL GAS

Animal and vegetable matter that has been covered with other deposits and then subjected to enormous pressure and high temperature for millions of years has been transformed into a complex mixture of alkanes and other hydrocarbons. The more volatile alkanes in the mixture constitute *natural gas*, and the liquid mixture of all the others is called *petroleum*.

Natural gas consists chiefly of methane and small amounts of ethane, propane, and the two butanes. It also contains small amounts of helium. The propane and butane can be separated by compressing and cooling the gas until the propane and the butanes are liquefied. The liquefied propane and butane is sold as bottled gas.

Petroleum is normally separated by *fractional distillation* (see Figure 23.7) not into pure hydrocarbons but into mixtures of hydrocarbons, called fractions, each of which boils over a certain limited temperature range. The particular fractions that are collected depend on the source of the petroleum and on the proposed uses of the fractions. Typical fractions are shown in Table 23.2. So that the demand for gasoline can be met, much of the kerosene and higher-boiling-point fractions are decomposed in cracking reactions to form the shorter-chain alkanes of gasoline.

The industrialized world has come to rely heavily on petroleum as an energy source. The invention of the internal combustion engine (see Box 23.1) led to the replacement of coal-burning steam engines in trains and ships with the diesel engine and to the development of the private automobile. These developments have in turn led to an increasing demand for petroleum. Hydrocarbons are also valuable as the raw materials for the manufacture of plastics and many other materials that have come to be a part of modern life. As a consequence the search for new oil deposits continues all over the globe, and deposits that are more difficult of access, such as those under the oceans and in the Arctic, are now being exploited. Despite the discovery of new deposits, the world's total petroleum resources are limited and will last for only a relatively short time. Estimates of this time vary from 30 years to several hundred years. One of the challenges facing humanity is to learn how to utilize, efficiently and safely, alternative energy sources, such as coal, nuclear power, and solar energy, and to find alternative raw materials for all the substances now manufactured from petroleum.

Offshore drilling allows us to reach the large deposits of oil that lie under water.

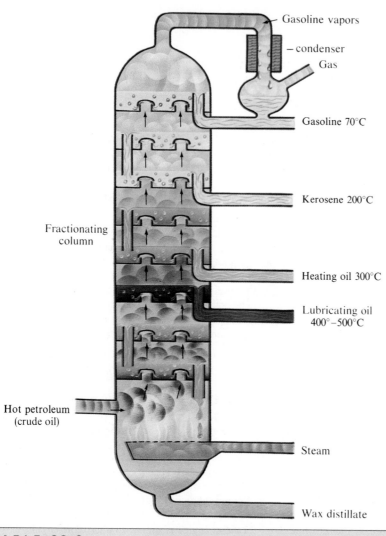

Gasoline vapors

— condenser

Gas

Gasoline 70°C

Kerosene 200°C

Fractionating
column

Heating oil 300°C

Lubricating oil
400°–500°C

Hot petroleum
(crude oil)

Steam

Wax distillate

FIGURE 23.7
Oil Refinery Distillation
Column.

Petroleum is heated with superheated steam at the bottom of a
tall distillation column. Most of the
petroleum is vaporized; the higher-
boiling-point components condense at a low point in the column,
and the lower-boiling-point components move toward the top of
the column. Fractions of different
compositions are taken from the
column at different heights.

TABLE 23.2
Typical Fractions Obtained in Distillation of Petroleum

Fraction	Boiling Point (°C)	Composition	Uses
Gas	Up to 20°	Alkanes from CH_4 to C_4H_{10}	Synthesis of other carbon compounds; fuel
Petroleum ether	20°–70°	C_5H_{12}, C_6H_{14}	Solvent; gasoline additive for cold weather
Gasoline	70°–180°	Alkanes from C_6H_{14} to $C_{10}H_{22}$	Fuel for gasoline engines
Kerosene	180°–230°	$C_{11}H_{24}$, $C_{12}H_{26}$	Jet engine fuel
Light gas oil	230°–305°	$C_{13}H_{28}$ to $C_{17}H_{36}$	Fuel for furnaces and for diesel engines
Heavy gas oil and light lubricating distillate	305°–405°	$C_{18}H_{38}$ to $C_{25}H_{52}$	Fuel for generating stations; lubricating oil
Lubricants	405°–515°	Higher alkanes	Thick oils, greases, and waxy solids; lubricating grease; petroleum jelly
Solid residue			Pitch or asphalt for roofing and road material

THE INTERNAL COMBUSTION ENGINE

The old-fashioned steam engine is an *external combustion engine*, because the fuel is burned outside the engine and the heat produced is used to convert water into steam. A wide variety of fuels could be used, of which coal was the most common. But steam engines are inconveniently heavy and require a driving fluid (water) that must be carried around. For all forms of transportation, from private automobiles to trains and airplanes, the steam engine has been replaced by the *internal combustion engine* in which the fuel is burned inside the engine. There are two important types of internal combustion engine: the diesel engine and the gasoline engine.

In the internal combustion engine, the hydrocarbon fuel is burned in air, yielding mainly carbon dioxide and water as products. For example, for the representative alkane nonane the principal reaction is

$$C_9H_{20}(g) + 14O_2(g) \longrightarrow 9CO_2(g) + 10H_2O(g)$$

Some oxides of nitrogen are also formed by the reaction of oxygen with the nitrogen in the air.

In the operation of the gasoline engine a mixture of gasoline and air is sucked into a cylinder as a piston descends, thereby creating a partial vacuum by increasing the cylinder volume. The valve on the cylinder then closes and the piston returns, compressing the air–fuel mixture. An electric spark is then passed through the compressed air–fuel mixture to ignite it. Since the number of moles of gaseous products exceeds the number of moles of reactants, and since the reaction products are much hotter than the reactants because the reaction is exothermic, the gas pressure on the piston increases, forcing its descent and thus delivering power to the engine.

The burning of the fuel–air mixture in the gasoline engine must occur at a rate that delivers a smooth thrust to the descending piston. Too rapid a reaction causes a distinct explosive noise known as engine knocking or pinging, and some of the power is wasted. Most of the hydrocarbons obtained by distillation of petroleum are unbranched alkanes, which tend to explode too rapidly and cause knock. The highly branched alkane isooctane causes little knocking and is arbitrarily given an octane rating of 100:

$$CH_3 - \underset{\underset{\displaystyle CH_3}{|}}{\overset{\overset{\displaystyle CH_3}{|}}{C}} - CH_2 - \underset{}{\overset{\overset{\displaystyle CH_3}{|}}{CH}} - CH_3$$

Isooctane, C_8H_{18}

At the other end of the scale is *n*-heptane which causes considerable knocking and is given an octane rating of 0. The octane rating of any fuel is then established by determining the ratio of isooctane and *n*-heptane needed to produce the same amount of knocking. For example, a 50–50 mixture has an octane rating of 50.

The octane number of gasoline is increased by heating it and passing it over a catalyst (catalytic re-forming) to convert some of the less-branched alkanes to more highly branched isomers. For example,

$$CH_3 - CH_2 - CH_2 - CH_2 - CH_3 \xrightarrow[\text{Heat}]{\text{Catalyst}}$$

$$CH_3 - \overset{\overset{\displaystyle CH_3}{|}}{CH} - CH_2 - CH_3$$

The octane rating of gasolines can also be increased by using various additives. In recent years the most common additive was tetraethyllead, $Pb(C_2H_5)_4$. But because lead compounds are toxic, leaded gasoline is gradually being phased out.

A diesel engine is similar to a gasoline engine, except that the fuel–air mixture is more highly compressed so that its temperature rises sufficiently to ignite the mixture without the use of a spark. Thus unlike the more familiar gasoline engine, the diesel engine does not have any spark plugs.

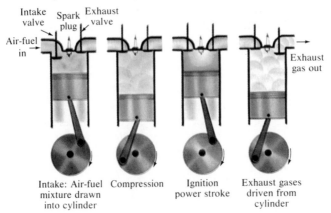

Intake: Air-fuel mixture drawn into cylinder   Compression   Ignition power stroke   Exhaust gases driven from cylinder

Four-Cycle Internal Combustion Engine.

23.2
Alkenes and Alkynes

Alkenes are hydrocarbons which contain at least one carbon–carbon double bond. Alkynes are hydrocarbons which contain at least one carbon–carbon triple bond. We have already discussed ethene, the simplest alkene (Figure 23.8), and ethyne, the simplest alkyne (Figure 23.9), in Chapter 6.

In Chapter 9 we saw that the double bond in ethene, or in any alkene, can be described in two different but equivalent ways, as we see in Figure 23.10. (1) The double bond can be considered to consist of two bent single bonds which can be represented by the overlap of two tetrahedral sp^3 hybrid orbitals on each carbon atom. (2) The double bond can be considered to consist of a sigma (σ) bond and a pi (π) bond. The formation of the σ bond can be represented by the end-on overlap of two sp^2 hybrid orbitals while the π bond can be represented by the sideways overlap of two p orbitals. Either model is consistent with the observed planar geometry.

A carbon–carbon triple bond can similarly be described either as consisting of three bent single bonds or as consisting of a σ bond and two π bonds (Figure 23.11). Both models are consistent with the observed linear geometry.

The second member of the alkene series is *propene*, C_3H_6. Like propane, it has three carbon atoms, but in propene one of the carbon–carbon bonds is a double bond. It is also known as *propylene*:

$$ \underset{H}{\overset{H}{}}\!\!\!\!\!\diagdown \!\!\! \underset{}{C} \!\! = \!\! \underset{}{C} \!\!\!\!\!\diagup\!\!\!\!\!\overset{H}{\underset{CH_3}{}} \qquad \text{or} \qquad CH_2\!\!=\!\!CH\!\!-\!\!CH_3 $$

Figure 23.12 shows two models of the molecule. Propene can be regarded as a derivative of ethene, in which one of the H atoms has been replaced by a —CH_3 group. It does not matter which of the four H atoms in the planar CH_2=CH_2 molecule is replaced by a —CH_3 group since all four are equivalent; regardless of which one is chosen, exactly the same molecule results. The chief industrial use of propene is in the production of *polypropylene*, a polymer used in the manufacture of packing materials and synthetic fibers for ropes and carpets.

Butene, C_4H_8, contains four carbon atoms and one C=C bond. There are four isomers of butene:

$$ \underset{H}{\overset{H}{}}\!\!\!\diagdown \!\! C\!\!=\!\!C\!\!\diagup\!\!\!\!\overset{CH_2\!-\!CH_3}{\underset{H}{}} \qquad\qquad \underset{H}{\overset{H}{}}\!\!\!\diagdown \!\! C\!\!=\!\!C\!\!\diagup\!\!\!\!\overset{CH_3}{\underset{CH_3}{}} $$

(1)   (2)

$$ \underset{H_3C}{\overset{H}{}}\!\!\!\diagdown \!\! C\!\!=\!\!C\!\!\diagup\!\!\!\!\overset{CH_3}{\underset{H}{}} \qquad\qquad \underset{H}{\overset{H_3C}{}}\!\!\!\diagdown \!\! C\!\!=\!\!C\!\!\diagup\!\!\!\!\overset{CH_3}{\underset{H}{}} $$

(3)   (4)

Structures (1), (2), and (3) are structural isomers because their atoms are connected in different ways. But (3) and (4) differ in a more subtle way: they both have one CH_3 group on each carbon atom and they differ only in the position of these groups relative to each other. These isomers exist because *free rotation does not occur around a* C=C *bond*. In order for rotation to occur

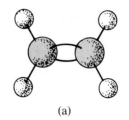
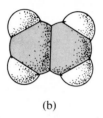

FIGURE 23.8
Structure of Ethene.

(a) A ball-and-stick model. (b) A space-filling model. Both models show that the molecule is planar.

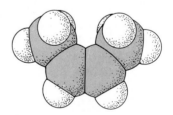

cis-2-butene

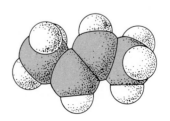

trans-2-butene

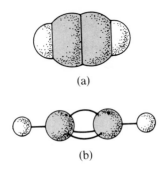

FIGURE 23.9
Structure of Ethyne.

Both carbon atoms in ethyne have a linear AX_2 geometry so ethyne is a linear molecule. (a) Space-filling model. (b) Ball-and-stick model.

around a double bond, one of the two bonds in the double bond would have to be broken, which requires a considerable amount of energy.

Isomers in which all the atoms are connected in the same way and that differ only with respect to the positions of identical groups in space are known as **geometric isomers**. Structure (3) is named the **trans isomer**, and (4) is called the *cis* isomer. *Trans* means across (as in transatlantic). In *trans*-butene the CH_3 groups are across from each other, while in *cis*-butene the CH_3 groups are on the *same* side of the double bond.

There are also many alkenes containing two or more double bonds. They are called alka*dienes*, alka*trienes*, and so on. One very important compound of this type is *butadiene*, CH_2=CH—CH=CH_2 (see Figure 23.13). When butadiene is polymerized, a synthetic rubber is produced (see Chapter 24). The product is similar in many respects to natural rubber, which is obtained from the sap of the rubber tree and is a polymer of *isoprene*. Isoprene is very similar to butadiene but has the hydrogen atom on one of the inner carbon atoms replaced by a —CH_3 group (see Figure 23.13).

REACTIONS OF ALKENES AND ALKYNES

Like all hydrocarbons, alkenes and alkynes burn in excess air or oxygen to give carbon dioxide and water. As a consequence of the fact that they have double or triple bonds, alkenes and alkynes are much more reactive than alkanes because they can undergo addition reactions. In an addition reaction molecules such as Cl_2, HCl, or H_2 are added to a double bond which is thereby converted to a single bond, or to a triple bond, which is converted to a double bond. The halogens or hydrogen halides react readily under ordinary conditions but the addition of hydrogen requires high pressure and a catalyst. Some typical addition reactions of alkenes and alkynes are

$$H_2C=CH_2 + H—Cl \longrightarrow CH_3—CH_2Cl$$
$$H_2C=CH_2 + Br—Br \longrightarrow CH_2Br—CH_2Br$$
$$H_2C=CH_2 + H—H \longrightarrow CH_3—CH_3$$
$$HC\equiv CH + H—Br \longrightarrow H_2C=CHBr$$
$$HC\equiv CH + Cl—Cl \longrightarrow HClC=CHCl$$
$$HC\equiv CH + H—H \longrightarrow H_2C=CH_2$$

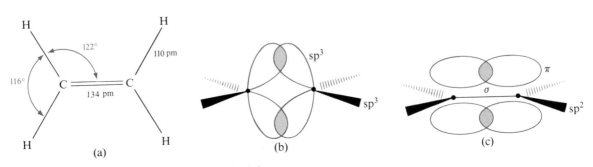

FIGURE 23.10
The Ethene Molecule.

(a) Dimensions of the planar ethene molecule. (b) Bent-bond model of the double bond based on tetrahedral sp^3 hybrid orbitals. (c) σ-π model of the double bond based on sp^2 hybrid orbitals and a p orbital on each carbon atom.

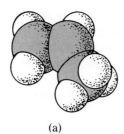

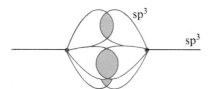

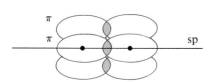

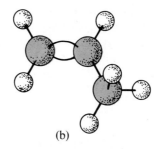

(a)

(b)

FIGURE 23.11
The Ethyne Molecule.

(a) Dimensions of the linear ethyne molecule. (b) Bent-bond model of the triple bond based on tetrahedral sp³ hybrid orbitals. (c) σ–π model of the triple bond based on sp hybrid orbitals and two p orbitals on each carbon atom.

FIGURE 23.12
Structure of Propene.

(a) A space-filling model. (b) A ball-and-stick model.

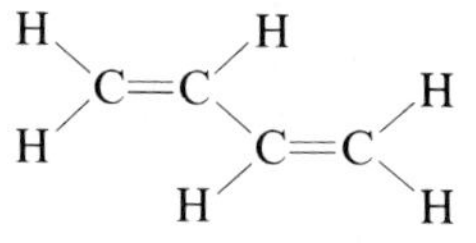

Butadiene

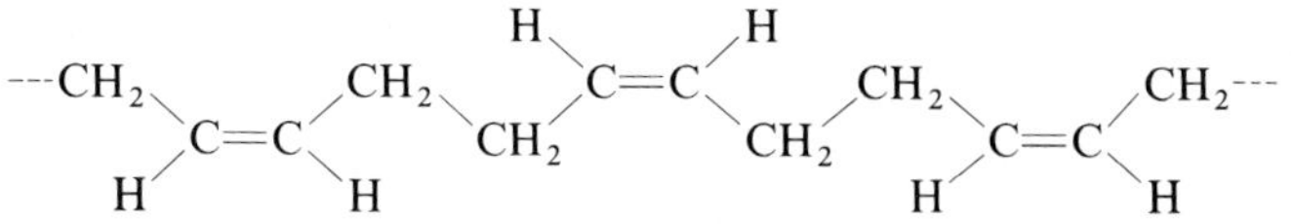

Polybutadiene (a synthetic rubber)

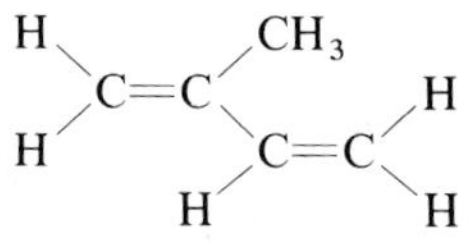

Isoprene

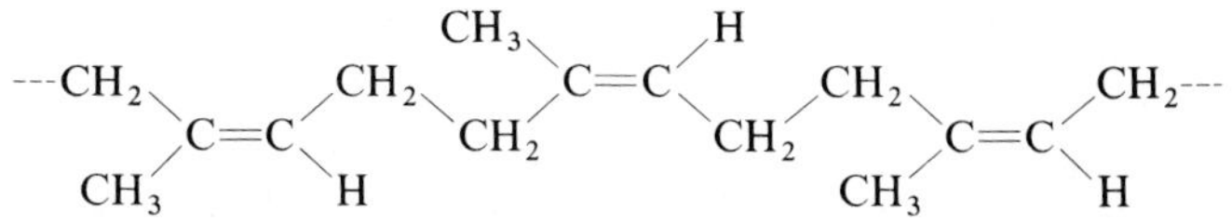

Polyisoprene (natural rubber)

FIGURE 23.13
Structures of Butadiene, Isoprene, and Their Polymers.

Butadiene and isoprene (2-methyl-butadiene) are dienes; they have two double bonds. When they are polymerized, one double bond per molecule is lost. Polybutadiene is a synthetic rubber, and poly-isoprene is natural rubber.

In general an **addition reaction** is any reaction in which two molecules combine to give a third molecule.

An **elimination reaction** is the opposite of an addition reaction. It involves the decomposition of a molecule into two molecules, one of which, normally the smaller one, is said to be eliminated. For example, the cracking reaction of ethane involves the elimination of a molecule of hydrogen and the formation of ethene:

$$C_2H_6 \longrightarrow C_2H_4 + H_2$$

Hydrocarbons that can undergo addition reaction, namely alkenes and alkynes, are called **unsaturated hydrocarbons**. Alkanes, which cannot undergo addition reactions, are called **saturated hydrocarbons**.

23.3
Naming of Hydrocarbons

The naming of individual compounds and of classes of related compounds has been an important but somewhat controversial subject during the development of chemistry. The names of compounds, particularly those discovered or first prepared many years ago, were chosen for a variety of reasons. Many names, particularly of substances that have been known for a long time, contain no structural information. For example, acetic acid, CH_3CO_2H, was named from the Latin word *acetum*, for "sour wine or vinegar". With the enormous and ever-increasing number of known compounds, particularly organic compounds, devising a systematic way of naming them has become essential so that their formulas and their structures can be deduced from their names. The most complete and useful rules were evolved during several international conferences; they are known as the *IUPAC* (International Union of Pure and Applied Chemistry) *rules*. We will show how to use these rules for naming hydrocarbons.

The naming of hydrocarbons is of particular importance because their names form the basis for naming all other organic compounds. The names and formulas of some of the alkanes are given in Table 23.1. For the first four hydrocarbons the common (nonsystematic) names are used:

$$CH_4 \qquad CH_3-CH_3 \qquad CH_3-CH_2-CH_3 \qquad CH_3-CH_2-CH_2-CH_3$$
Methane   Ethane   Propane   Butane

The higher members, beginning with pentane, are named systematically by denoting the number of carbon atoms with a numerical prefix (*pent-*, *hex-*, *hept-*, and so on) and adding the ending *-ane* to classify the compound as an alkane. Thus we have

$$CH_3-CH_2-CH_2-CH_2-CH_3$$
Pentane

and

$$CH_3-CH_2-CH_2-CH_2-CH_2-CH_3$$
Hexane

Alkanes are classified as *continuous-chain* (that is, unbranched) if all the carbon atoms in the chains are linked to no more than two other carbon atoms. They are called *branched-chain* if one or more carbon atoms are linked to more than two other carbon atoms. Thus

$$CH_3-CH_2-CH_2-CH_2-CH_2-CH_3$$

Continuous-chain hydrocarbon

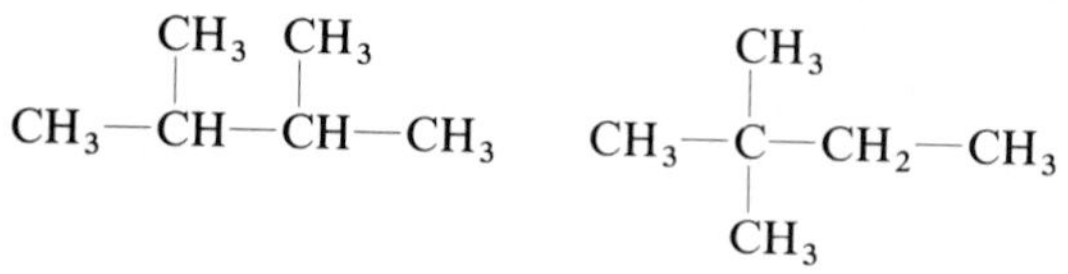

$$CH_3-\underset{\underset{CH_3}{|}}{CH}-\underset{\underset{CH_3}{|}}{CH}-CH_3 \qquad CH_3-\underset{\underset{CH_3}{\overset{CH_3}{|}}}{\overset{|}{C}}-CH_2-CH_3$$

Branched-chain hydrocarbons

The rules for the systematic naming of hydrocarbons are as follows:

1. The *longest* continuous chain of carbon atoms is taken to be the parent hydrocarbon. Single carbon atoms or shorter chains of carbon atoms may then be attached to this longest chain at various places. They substitute for hydrogen atoms of the parent hydrocarbon and are therefore called **substituent groups**. Thus the following hydrocarbon is regarded as a substituted pentane, rather than as a substituted butane, because the longest continuous chain has five carbon atoms:

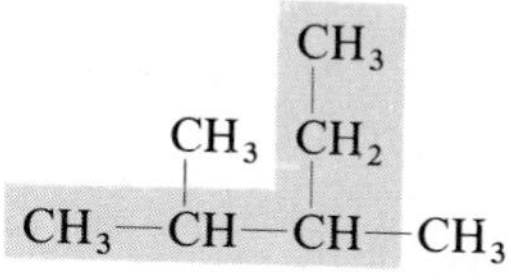

$$CH_3-\underset{\underset{\underset{CH_3}{|}}{CH}}{}-\underset{\underset{\underset{CH_3}{|}}{\overset{CH_2}{\overset{|}{}}}}{CH}-CH_3$$

2. The substituent groups attached to the main chain are named by replacing the ending *-ane* of the alkane by *-yl*. Hence the substituent groups are known as **alkyl groups**; the simplest examples are methyl, $-CH_3$, and ethyl, $-C_2H_5$.

3. The parent hydrocarbon is numbered starting from the end of the chain, and the substituent groups are assigned numbers corresponding to their position on the chain. The direction of numbering is chosen so as to give the lowest numbers to the side-chain substituents. When there are two or more substituents of the same kind, the number is denoted by the prefixes *di* (2), *tri* (3), *tetra* (4), *penta* (5), and so on. Thus the hydrocarbon illustrated above is 2,3-dimethylpentane:

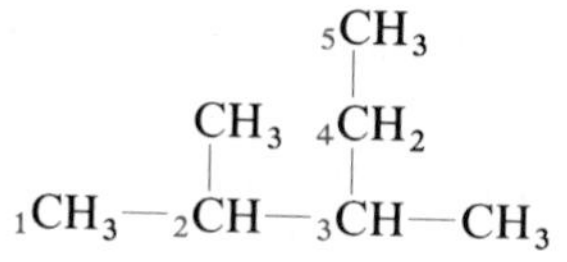

$$_1CH_3-_2CH-_3CH-CH_3 \qquad not \qquad _5CH_3-_4CH-_3CH-CH_3$$

2,3-Dimethylpentane   3,4-Dimethylpentane

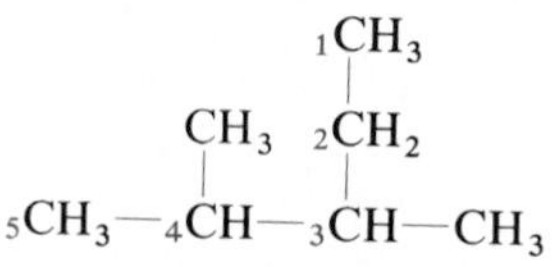

Normally, the structure of this hydrocarbon would be written as follows, that is, with the longest continuous chain displayed horizontally:

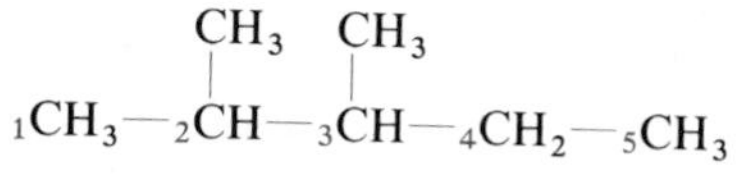

$$_1CH_3-_2CH-_3CH-_4CH_2-_5CH_3$$

═ *Example 23.2* ═══════════════════════════════════

THE NAMING OF ALKANES

What are the systematic names of the following alkanes?

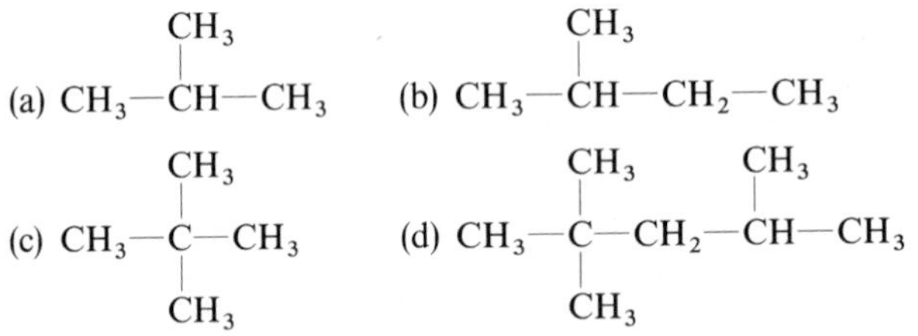

$$
\begin{array}{ll}
& \quad CH_3 \qquad\qquad\qquad CH_3 \\
& \quad | \qquad\qquad\qquad\quad | \\
(a)\ CH_3\!-\!CH\!-\!CH_3 & (b)\ CH_3\!-\!CH\!-\!CH_2\!-\!CH_3 \\[4pt]
& \quad CH_3 \qquad\qquad\quad CH_3 \qquad CH_3 \\
& \quad | \qquad\qquad\qquad\ | \qquad\quad | \\
(c)\ CH_3\!-\!C\!-\!CH_3 & (d)\ CH_3\!-\!C\!-\!CH_2\!-\!CH\!-\!CH_3 \\
& \quad | \qquad\qquad\qquad\ | \\
& \quad CH_3 \qquad\qquad\quad CH_3
\end{array}
$$

Solution

(a)
$$
\begin{array}{c}
CH_3 \\
| \\
{}_1CH_3-{}_2CH-{}_3CH_3
\end{array}
$$

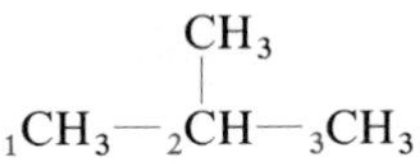

We have previously called this compound isobutane. But according to the systematic rules, it is named as a derivative of propane, since the longest continuous chain has three carbon atoms. Since the methyl group is on the second carbon atom in the chain, the systematic name of the hydrocarbon is 2-methylpropane.

(b)
$$
\begin{array}{c}
CH_3 \\
| \\
{}_1CH_3-{}_2CH-{}_3CH_2-{}_4CH_3
\end{array}
$$

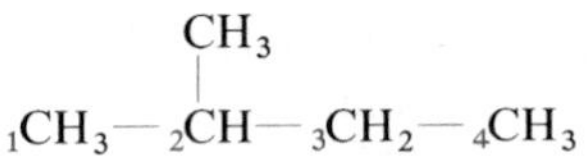

This alkane is named as a derivative of butane and numbered from the left so that the substituent —CH_3 group comes at carbon atom 2. It is 2-methylbutane.

(c)
$$
\begin{array}{c}
CH_3 \\
| \\
{}_1CH_3-{}_2C-{}_3CH_3 \\
| \\
CH_3
\end{array}
$$

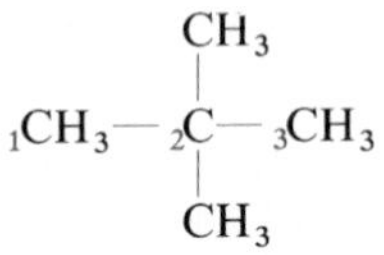

This alkane is a derivative of propane and has two methyl substituents at carbon atom 2. It is named 2,2-dimethylpropane; its common name is neopentane. When there is more than one of the same kind of substituent attached to the same carbon atom, we repeat the number of this carbon atom as many times as there are attached groups.

(d)
$$
\begin{array}{c}
CH_3 \qquad\quad CH_3 \\
| \qquad\qquad\ | \\
{}_1CH_3-{}_2C-{}_3CH_2-{}_4CH-{}_5CH_3 \\
| \\
CH_3
\end{array}
$$

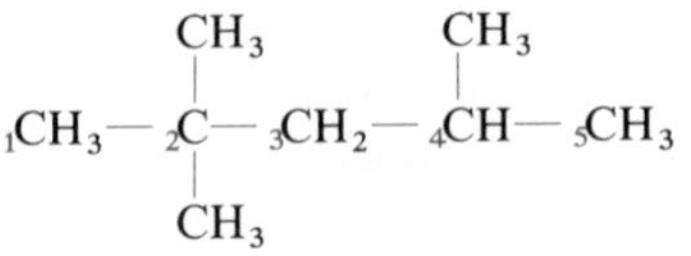

Whichever end of the chain we number from, there are substituents at carbon atoms 2 and 4, but we choose the numbering that gives the most *highly substituted* carbon—that is, the carbon with the greatest number of substituent groups—the lowest number; this is carbon atom 2. The compound is 2,2,4-trimethylpentane.

──

E X E R C I S E   2 3 . 1

What are the systematic names of the five hexane isomers in Example 23.1?

Now that we have seen how to write the names of alkanes given their structural formulas, we can readily draw the structural formulas from the systematic name.

═ Example 23.3

DRAWING STRUCTURES OF ALKANES

Draw the structures of the following compounds:

(a) Hexane　　　　　　　(b) 2-Methylpentane　　　(c) 3-Methylpentane

(d) 2,2-Dimethylbutane　　(e) 2,3-Dimethylbutane

Solution

(a) According to systematic nomenclature, hexane has a six-carbon continuous chain:

$$CH_3-CH_2-CH_2-CH_2-CH_2-CH_3$$

(b) The longest continuous chain of 2-methylpentane contains 5 C atoms, and there is a methyl ($-CH_3$) substituent at C_2:

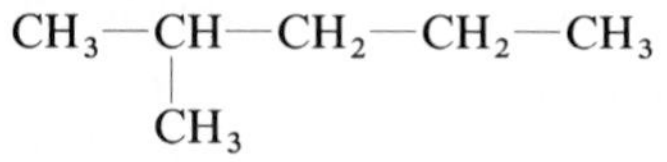

$$CH_3-CH-CH_2-CH_2-CH_3$$
$$\ \ \ \ \ \ \ \ |$$
$$\ \ \ \ \ \ \ CH_3$$

(c) The longest continuous chain of 3-methylpentane contains 5 C atoms, and there is a $-CH_3$ substituent at C_3.

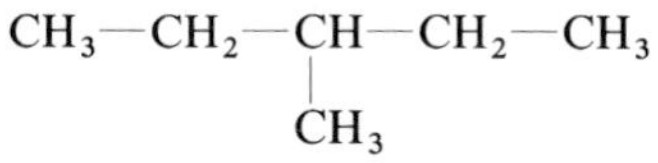

$$CH_3-CH_2-CH-CH_2-CH_3$$
$$\ \ \ \ \ \ \ \ \ \ \ \ \ \ |$$
$$\ \ \ \ \ \ \ \ \ \ \ \ \ CH_3$$

(d) The longest continuous chain of 2,2-dimethylbutane contains 4 C atoms, and there are two methyl substituents at C_2.

$$CH_3$$
$$|$$
$$CH_3-C-CH_2-CH_3$$
$$|$$
$$CH_3$$

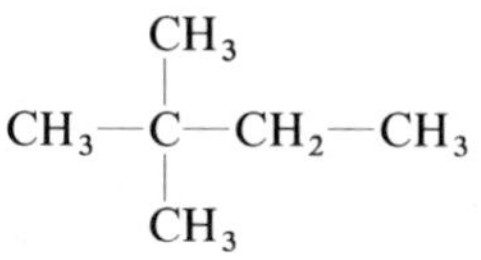

(e) The longest continuous chain of 2,3-dimethylbutane contains 4 C atoms, and there are methyl substituents at C_2 and C_3.

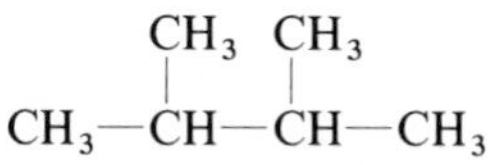

$$CH_3\ \ CH_3$$
$$|\ \ \ \ \ |$$
$$CH_3-CH-CH-CH_3$$

We have encountered all these five compounds previously. They are the structural isomers of hexane, C_6H_{14}.

We can also use the systematic method for naming compounds to decide whether two structures are in fact different structural isomers. For example, we saw previously that the two structures

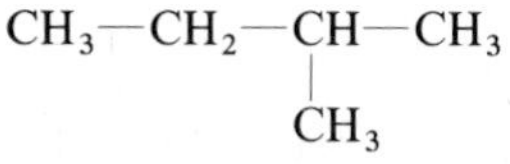

$$CH_3-CH_2-CH-CH_3 \quad \text{and} \quad CH_3-CH-CH_2-CH_3$$
$$\ \ \ \ \ \ \ \ \ \ \ \ \ |\ |$$
$$\ \ \ \ \ \ \ \ \ \ \ CH_3 \qquad\qquad\qquad\qquad CH_3$$

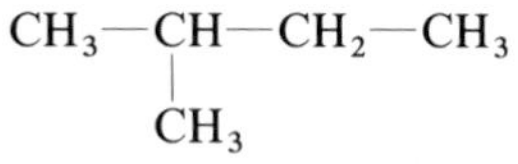

were identical by rotating one structure about an axis passing through the center of the middle C—C bond. However, if we name each of them system-atically, we see that they are the same compound because they have the same name, 2-methylbutane.

Two more rules are needed for naming hydrocarbons that contain multiple bonds.

4. Hydrocarbons containing one or more double bonds are *alkenes*. The longest continuous chain *containing a double bond* is given the name of the corresponding alkane but with the termination *-ane* changed to *-ene*. The chain is numbered so that the *first carbon atom* of the double bond is indicated by the *lowest possible number*. This number is usually omitted if the chain is so short that no ambiguity thereby arises; for example, CH_2=CH—CH_3, 1-propene, may be called simply propene.

	Systematic name	Common name
CH_2=CH_2	Ethene	Ethylene
CH_2=CH—CH_3	Propene	Propylene
CH_2=CH—CH_2—CH_3	1-Butene	
CH_3—CH=CH—CH_3	2-Butene	
CH_3—C=CH_2 with CH_3 branch	2-Methylpropene	Isobutylene

When compounds contain more than one C=C bond, a numerical prefix (*di-*, *tri-*, and so on) is used before the *-ene* suffix. For example,

$$CH_2=CH—CH=CH_2$$

1,3-Butadiene

= Example 23.4 =

THE NAMING OF ALKENES

Name the following compounds:

(a) CH_3—CH_2—C—CH_2—CH_2—CH_3 with CH_2 double-bonded below

(b) CH_3—C=CH—CH=C—CH_3 with CH_3 branches below

Solution

(a) The name is 2-ethyl-1-pentene:

$$CH_3—CH_2—\overset{2}{C}—\overset{3}{C}H_2—\overset{4}{C}H_2—\overset{5}{C}H_3$$
with $\overset{1}{C}H_2$ double-bonded below

(We do not use the longer chain of six C atoms to name the compound, since that chain does not incorporate the double bond.)

(b) The name is 2,5-dimethyl-2,4-hexadiene:

$$\overset{1}{C}H_3—\overset{2}{C}=\overset{3}{C}H—\overset{4}{C}H=\overset{5}{C}—\overset{6}{C}H_3$$
with CH_3 branches below

EXERCISE 23.2

Name the compound

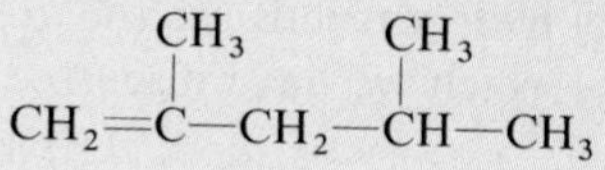

$$CH_2\text{=}\underset{\underset{CH_3}{|}}{C}\text{--}CH_2\text{--}\underset{\underset{CH_3}{|}}{CH}\text{--}CH_3$$

5. Hydrocarbons that contain triple bonds are *alkynes*. Their names are obtained by replacing the *-ane* ending of the corresponding alkane by *-yne*. The numbering system used for locating the triple bonds and any substituents is the same as that for the alkenes:

$$H\text{--}C\equiv C\text{--}H \qquad CH_3\text{--}C\equiv C\text{--}H \qquad CH_3\text{--}C\equiv C\text{--}CH_3$$

Ethyne (acetylene)   Propyne (methylacetylene)   2-Butyne (dimethylacetylene)

A final rule is needed for cyclic alkanes.

6. Cycloalkanes are named by adding the prefix *cyclo-* to the name of the corresponding noncyclic alkane having the same number of carbon atoms as the ring:

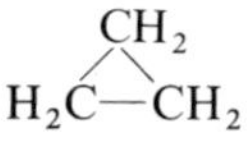

Cyclopropane 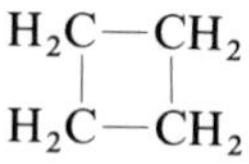   Cyclobutane

The carbon atoms are numbered around the ring so as to give the lowest possible numbers for the substituent positions. For example,

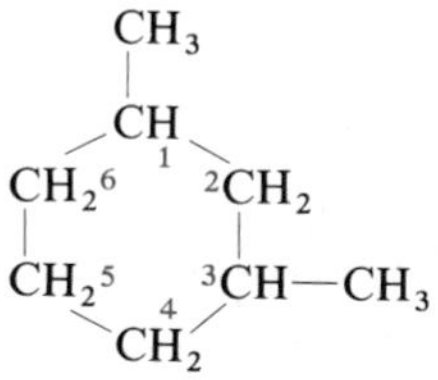

1,3-Dimethylcyclohexane

The systematic names for $CH_2\text{=}CH_2$ and $H\text{--}C\equiv C\text{--}H$ are ethene and ethyne, respectively. But the common and long-established names, ethylene and acetylene, continue to be widely used. Although chemists appreciate the utility of the IUPAC rules for naming more complex hydrocarbons, like most people, they are reluctant to change familiar things. They continue to use the well-established names for many simple and well-known compounds. Although it would be logical and in some ways simpler to use only IUPAC nomenclature through this book, you would be at a serious disadvantage in reading newspapers, magazines, and other books if you did not know the nonsystematic names that are widely used. Throughout the history of chemistry old names have slowly given way to new, more systematic names. For example, hydrochloric acid was at one time known as muriatic acid, but this name is now rarely used. Presumably ethene will gradually replace ethylene as the name for C_2H_4.

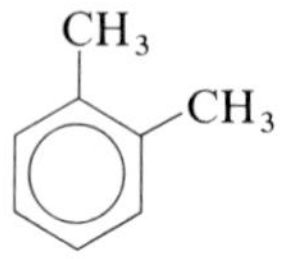

1,2-Dimethylbenzene
o-Xylene

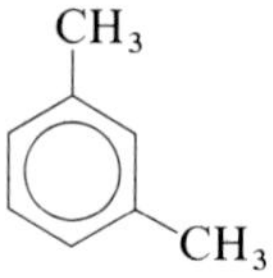

1,3-Dimethylbenzene
m-Xylene

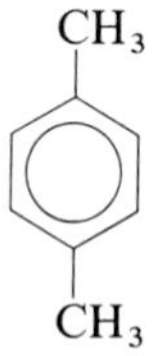

1,4-Dimethylbenzene
p-Xylene

FIGURE 23.14
The Three Isomeric Xylenes

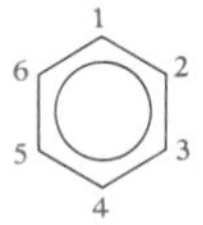

Ethylbenzene

23.4
Arenes

Another important class of hydrocarbons are the **arenes**, of which the simplest member is *benzene*, C_6H_6, which we have discussed previously in Chapters 6 and 9.

Many other hydrocarbons can be derived from benzene by replacing one or more hydrogen atoms with alkyl groups. These compounds are all *arenes*. The simplest derivative is *methylbenzene*, or *toluene*:

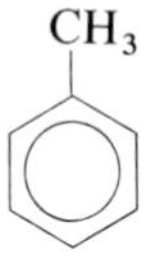

Methylbenzene (toluene)
(bp = 111 °C)

Substituting two methyl groups gives three *dimethylbenzenes* (*xylenes*) that are structural isomers. These three compounds differ only in the relative positions of the methyl groups (see Figure 23.14). According to the IUPAC rules of nomenclature, these isomeric molecules are named by numbering the carbon atoms of the ring so that the substituent groups have the lowest possible numbers:

Another compound that has the same molecular formula as the dimethylbenzenes, C_8H_{10}, is ethylbenzene.

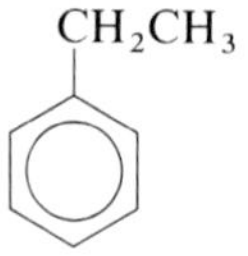

═ *Example 23.5* ═══════════════════════════════════════

THE STRUCTURES OF ISOMERIC ARENES

How many trimethylbenzenes, $C_6H_3(CH_3)_3$ are there?

Solution

We have three identical substituent groups. Only one compound results from placing them adjacent to each other, 1,2,3-trimethylbenzene. Another compound results from placing two groups adjacent to each other with the third group separated from this pair by one carbon atom, 1,2,4-trimethylbenzene. A final possibility is to have each of the —CH_3 groups separated from each other by one carbon atom, 1,3,5-trimethylbenzene.

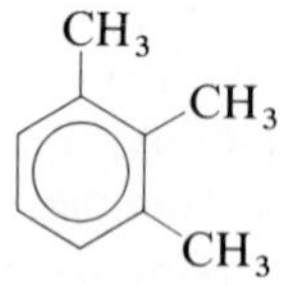

1,2,3-Trimethylbenzene 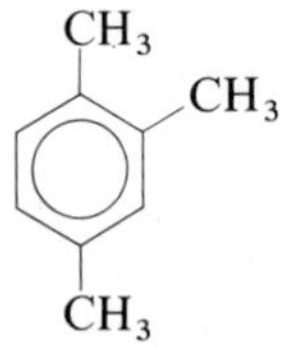   1,2,4-Trimethylbenzene 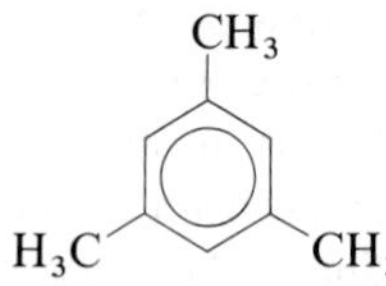   1,3,5-Trimethylbenzene

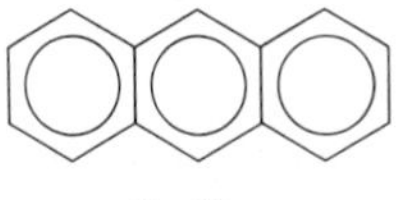

$C_{14}H_{10}$,
Anthracene (colorless)

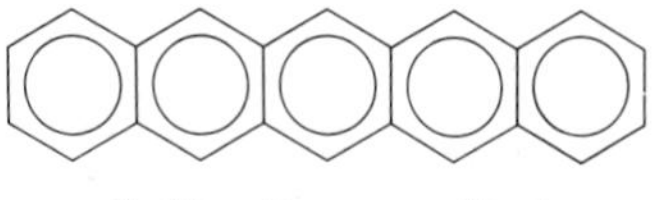

$C_{22}H_{14}$, Pentacene (blue)

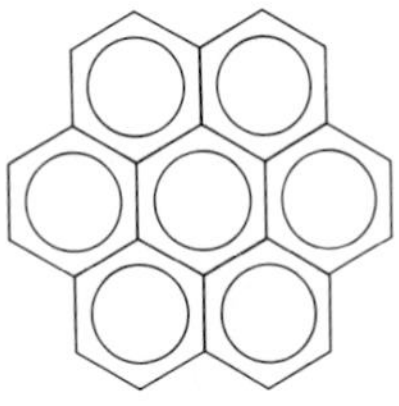

$C_{24}H_{12}$,
Coronene (pale yellow)

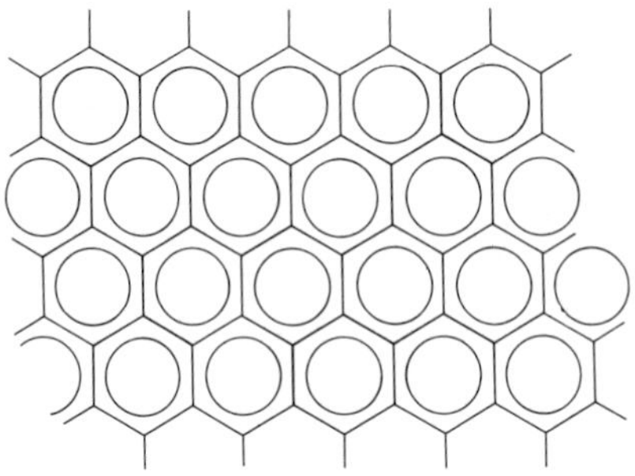

Graphite (black)

FIGURE 23.15
Structures of Some Arenes with Fused Rings.

As the number of rings increases, the arenes become more intensely colored. Graphite may be regarded as an arene of essentially infinite size; it is black.

Naphthalene consists of two benzene rings fused together, that is, two rings sharing two carbon atoms. It has the formula $C_{10}H_8$ rather than $C_{10}H_{10}$ because the two carbon atoms at the point of fusion do not have hydrogen atoms bonded to them:

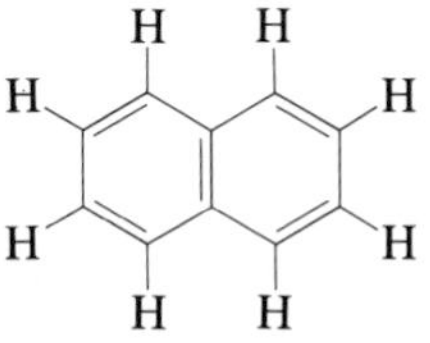

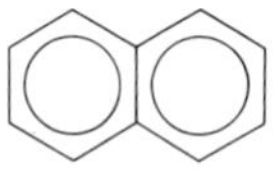

Naphthalene
(mp = 90 °C)

A large number of arenes are known that contain different numbers of benzene rings fused in different ways. Typical examples include anthracene, pentacene, and coronene, which are shown in Figure 23.15. Although the vast majority of hydrocarbons are colorless, arenes that contain several fused rings are often colored. When the number of rings becomes essentially infinite, a planar infinite sheet of carbon atoms is obtained (Figure 23.15). Graphite, which is black, consists of these sheets stacked one on another (Figure 6.2).

23.5 Identification and Analysis of Hydrocarbons

The hydrocarbons illustrate a common problem in organic chemistry, namely, that of identifying substances that have very similar compositions and very similar properties. For example, all the alkanes are rather unreactive, and except for the simplest members of the series, they have rather similar compositions. Therefore the identification of an alkane depends mainly on its physical properties, such as its boiling point, and on the determination of its molar mass. The molar mass of a volatile compound can be determined by measuring the volume occupied by a known mass of the gaseous compound at a known temperature and pressure, as we described in Chapter 3.

A modern method for determining molar mass and for identifying a substance is *mass spectrometry*. A substance in the gas state is bombarded with high-speed electrons. These electrons may knock an electron from a molecule of the substance to give a positive ion called the *parent ion*, which may also break up to give smaller positive ions called *fragment ions*. For example,

$$
\begin{array}{c}
CH_3 \\
| \\
CH_3-C-CH_3 + e^- \longrightarrow 2e^- + C_5H_{12}{}^+ \\
| \\
CH_3
\end{array}
\begin{array}{l}
\nearrow C_4H_9{}^+ \quad \text{Mass 57} \\
\longrightarrow C_3H_5{}^+ \quad \text{Mass 41} \\
\searrow C_2H_5{}^+ \quad \text{Mass 29} \\
\searrow C_2H_3{}^+ \quad \text{Mass 27}
\end{array}
$$

2,2dimethylpropane   Parent ion
                            Mass 72

Fragment ions

The positive ions are accelerated by an electric field and pass into a mass spectrometer, where their masses can be determined by the amount by which they are deflected by a magnetic field, as described in Box 2.3. A plot of the relative intensities of the peaks against the mass is called the mass spectrum of the substance (Figure 23.16). The line of highest mass in the spectrum, M^+,

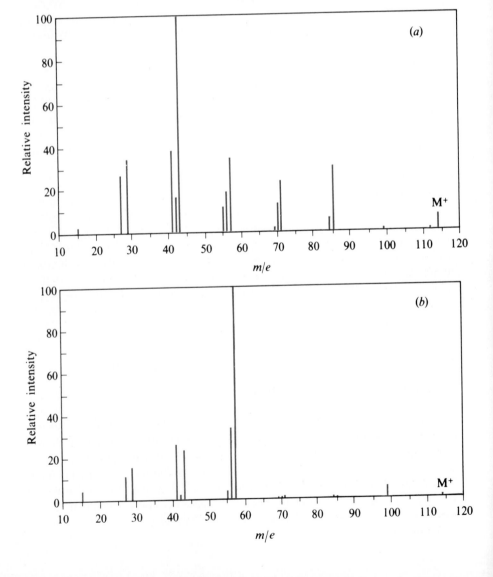

FIGURE 23.16
Mass Spectra of Two
Isomeric Alkanes.

(a) *n*-Octane. (b) 2,2,4-Trimethyl-pentane.

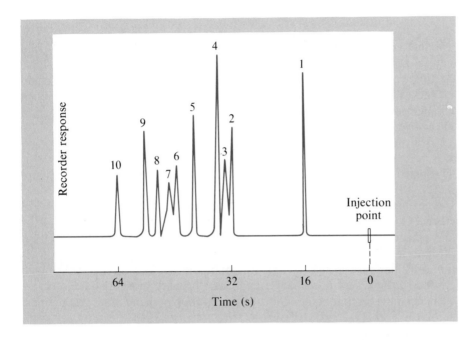

FIGURE 23.17
Gas Chromatogram
Obtained from a Mixture
of Nine Heptane Isomers.

Each of the isomers takes a different time to pass through the chromatographic column and is detected separately on emerging from the column. Peak 1 is a reference peak.

gives the mass of the parent ion and therefore of the molecule from which it is derived. The masses of the fragment ions often enable the chemist to determine the structure of the substance and therefore to identify it. Each substance has its own characteristic mass spectrum as can be seen, for example, by comparing the spectra of octane, $CH_3(CH_2)_6CH_3$ and the isomeric hydrocarbon 2,2,4-trimethylpentane (Figure 23.16).

Complex mixtures of very similar substances, such as hydrocarbons, present formidable problems for analysis. One useful method for volatile substances is *gas–liquid chromatography*, which we described in Chapter 1 (Figure 1.21). Figure 23.17 shows the results achieved in the separation of a mixture of nine heptanes. As indicated in the figure, each isomer emerges from the long separating tube at a different time. The area under each peak corresponds to the relative amount of each substance in the mixture.

The chart in Figure 23.17 is called a **chromatogram**. The peaks in the chromatogram can be identified by comparison with a chromatogram obtained from a known mixture of substances. Alternatively, if the apparatus is large enough, samples of each substance can be collected as they emerge. They can then be separately identified by means of, for example, a mass spectrometer.

23.6
The Petrochemical Industry

The raw materials of the petrochemical industry are natural gas and petroleum. Natural gas consists primarily of methane and small amounts of ethane, propane, and the butanes, the exact amounts of which vary with the source of the natural gas. Petroleum consists of a complex mixture of alkanes containing five or more carbon atoms. In the petrochemical industry these hydrocarbons are converted into a variety of important organic compounds. Two primary processes of the petrochemical industry are thermal cracking and the manufacture of synthesis gas.

THERMAL CRACKING

Thermal cracking is a process in which a mixture of alkanes obtained from natural gas or petroleum is heated to approximately 1400 °C for a short time. At this temperature longer-chain alkanes are broken down to shorter-chain alkanes, alkenes, and hydrogen. The reactions that occur are complex. When ethane is a major component of the hydrocarbon mixture, an important reaction in the overall process is its decomposition to ethene and hydrogen:

$$CH_3-CH_3 \longrightarrow CH_2=CH_2 + H_2$$

This reaction is believed to take place by a free-radical mechanism (see Figure 23.18).

The important products of the thermal cracking of alkane mixtures are ethene, C_2H_4, and propene, $CH_3-CH=CH_2$, together with hydrogen, methane, ethane, propane, and other hydrocarbons, depending on the composition of the original mixture. The products are separated by low-temperature fractional distillation. The higher-boiling-point fractions are further treated to make gasoline. Ethene and propene are important primary materials of the petrochemical industry that are used for the preparation of many other compounds.

SYNTHESIS GAS

Methane and other alkanes are converted to a mixture of hydrogen and carbon monoxide by heating them with steam at 700–800 °C in the presence of a nickel catalyst; for example,

$$CH_4(g) + H_2O(g) \longrightarrow CO(g) + 3H_2(g)$$
$$C_9H_{20}(g) + 9H_2O(g) \longrightarrow 9CO(g) + 19H_2(g)$$

The mixture of carbon monoxide and hydrogen obtained by this process is called **synthesis gas**. It is used as a fuel for domestic and industrial heating and as a starting material for the manufacture of many organic compounds.

$$
\begin{aligned}
CH_3-CH_3 &\longrightarrow \cdot CH_3 + \cdot CH_3 \\
\cdot CH_3 + H-CH_2-CH_3 &\longrightarrow CH_4 + \cdot CH_2CH_3
\end{aligned}
\quad\Big\} \text{ Chain initiation}
$$

$$
\begin{aligned}
\cdot CH_2CH_3 &\longrightarrow \cdot H + H_2C=CH_2 \\
\cdot H + CH_3-CH_3 &\longrightarrow H_2 + \cdot CH_2-CH_3
\end{aligned}
\quad\Big\} \text{ Chain propagation}
$$

$$
\begin{aligned}
\cdot H + \cdot H &\longrightarrow H_2 \\
\cdot CH_3 + \cdot CH_3 &\longrightarrow C_2H_6
\end{aligned}
\quad\Big\} \text{ Chain termination}
$$

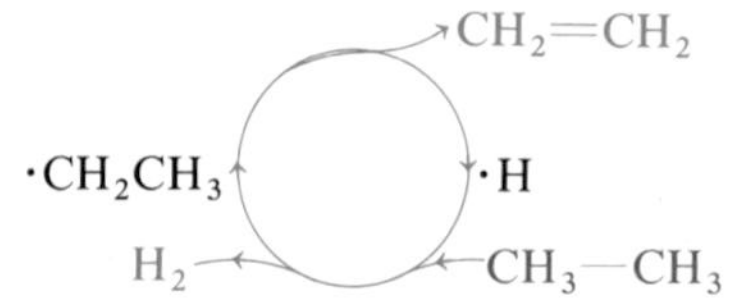

FIGURE 23.18
Thermal Cracking of Ethane.

The cracking of ethane is a chain reaction involving free-radical intermediates. The chain propagation steps are represented also in the "circle diagram", which shows that ethane is consumed and both ethene and molecular hydrogen are produced by these steps; the H· and ·CH₂CH₃ free radicals are seen to be produced and consumed within each cycle.

23.7
Functional Groups

Organic compounds either are hydrocarbons or can be imagined to be derived from hydrocarbons by replacing one or more of the hydrogen atoms with other atoms or groups of atoms called **functional groups**. For example, replacing a hydrogen atom in ethane with an OH group gives ethanol:

$$
\begin{array}{ccc}
\text{H}\ \text{H} & & \text{H}\ \text{H} \\
|\ \ | & & |\ \ | \\
\text{H}-\text{C}-\text{C}-\text{H} & \longrightarrow & \text{H}-\text{C}-\text{C}-\text{OH} \\
|\ \ | & & |\ \ | \\
\text{H}\ \text{H} & & \text{H}\ \text{H}
\end{array}
$$

Organic compounds with OH as the functional group are called **alcohols** (or phenols if they are derived from an arene). Alcohols have many common properties associated with the OH group.

The vast majority of organic compounds contain oxygen or nitrogen, or both, in addition to carbon and hydrogen; the most important functional groups are those containing oxygen or nitrogen or both (see Table 23.3). We may conveniently consider organic compounds as composed of one or more alkyl groups,

TABLE 23.3
Important Functional Groups

Functional Group	General Formula for Compound	Name	Suffix used in Systematic Name
—OH	R—OH[a]	Alcohol (phenol)	-ol
—OR	R—O—R	Ether	alkoxy (prefix)
$-\text{C}\overset{\displaystyle O}{\underset{\displaystyle H}{<}}$	$\text{R}-\text{C}\overset{\displaystyle O}{\underset{\displaystyle H}{<}}$	Aldehyde	-al
$-\text{C}\overset{\displaystyle O}{\underset{\displaystyle R}{<}}$	$\text{R}-\text{C}\overset{\displaystyle O}{\underset{\displaystyle R}{<}}$	Ketone	-one
$-\text{C}\overset{\displaystyle O}{\underset{\displaystyle OH}{<}}$	$\text{R}-\text{C}\overset{\displaystyle O}{\underset{\displaystyle OH}{<}}$	Carboxylic acid	-oic acid
$-\text{C}\overset{\displaystyle O}{\underset{\displaystyle OR}{<}}$	$\text{R}-\text{C}\overset{\displaystyle O}{\underset{\displaystyle OR}{<}}$	Ester	-oate
—X[b]	R—X	Haloalkane	halo (prefix)
—NH_2	R—NH_2	Amine	amino (prefix)
$-\text{C}\overset{\displaystyle O}{\underset{\displaystyle NH_2}{<}}$	$\text{R}-\text{C}\overset{\displaystyle O}{\underset{\displaystyle NH_2}{<}}$	Amide	-amide

[a] R is an alkyl or an aryl group. ROH is an alcohol when R is an alkyl group and a phenol when R is an aryl group.
[b] X is F, Cl, Br, or I.

denoted R, and one or more functional groups, such as those in Table 23.3. Thus the general formula for an alcohol is written as ROH. The alkane part of an organic molecule is relatively unreactive, and most of the reactions of organic molecules involve the formation or removal of functional groups and the conversion of one functional group into another.

In the following sections we consider the characteristic properties and reactions of several important functional groups.

23.8
Alcohols

INDUSTRIAL PREPARATION

When synthesis gas having the composition $2H_2 : 1CO$ is heated at 240–260 °C under a pressure of 50–100 atm in the presence of a $ZnO/CuO/Al_2O_3$ catalyst, a good yield of methanol is obtained:

$$CO(g) + 2H_2(g) \rightleftharpoons CH_3OH(g) \qquad \Delta H° = -91 \text{ kJ}$$

The position of the equilibrium is very favorable at room temperature, but even in the presence of a catalyst, the reaction is very slow at ordinary temperatures. Because the reaction is highly exothermic, the equilibrium constant decreases with increasing temperature (Table 23.4). The operating temperature of approximately 250 °C is as low as can be used to obtain a reasonable reaction rate, even with a catalyst. High pressure increases the yield because the reaction is accompanied by a decrease in the number of molecules.

Methanol is a primary product of the petrochemical industry and is used in the manufacture of many other substances including plastics, fertilizers, and pharmaceuticals. A minor but familiar use is as a component of windshield washer liquid, because it is miscible with water and has a low freezing point of −98 °C. Thus solutions of methanol and water will not freeze on windshields in cold weather, as would water.

Methanol can be imagined as being derived from methane by replacement of a hydrogen atom by a hydroxyl group, OH:

Methane　　　Methanol

TABLE 23.4
Temperature Variation of the Equilibrium Constant for the Reaction $CO(g) + 2H_2(g) \rightleftharpoons CH_3OH(g)$

Temperature (°C)	K (mol^{-2} L^2)
0	5.27×10^5
100	1.08×10
200	1.70×10^{-2}
300	2.31×10^{-4}
400	1.09×10^{-5}

It is a member of the class of compounds called alcohols, which contain the OH functional group and have the general formula ROH.

As we mentioned in Chapter 6, alkenes readily undergo *addition reactions*. An important reaction of this type is the addition of water to a double bond. Addition of water to ethene gives ethanol, according to the reaction

This reaction is extremely slow at room temperature, so a temperature of approximately 300 °C and a catalyst, such as phosphoric acid, are used to obtain a sufficiently rapid reaction. And a high pressure of 70 atm is used to increase the equilibrium yield.

Ethanol is commonly called *ethyl alcohol*, and is also known simply as *alcohol*. It can be imagined as being derived from ethane by replacing an H atom by a hydroxyl group, —OH. Only one alcohol is obtained from ethane because replacement of any of the H atoms gives the same molecule, C_2H_5OH.

However, two *different* propanols can be derived from propane since the hydrogen atoms of the central carbon are not equivalent to those of the terminal carbons; in other words, there are two isomers of propanol:

Propane   1-Propanol   2-Propanol

NOMENCLATURE

Alcohols are named by replacing the -*e* in the name of the alkane from which they are derived by the ending -*ol*. For example,

$$Methane \longrightarrow Methanol$$

$$Ethane \longrightarrow Ethanol$$

More complicated structures are dealt with by the same rules that we described for the hydrocarbons. *The longest carbon chain in a molecule is numbered in the direction that gives the substituent groups the smallest possible numbers.* Thus $CH_3CH_2CH_2OH$ is called 1-propanol, not 3-propanol.

There are four alcohols, C_4H_9OH. They can be derived from the two isomeric alkanes, C_4H_{10},

An alcohol containing a —CH_2OH group is known as a *primary alcohol*. An alcohol containing a $>$CHOH group is a *secondary alcohol*, and an alcohol containing a $\geq$COH group is a *tertiary alcohol*.

— *Example 23.6* ═══

CLASSIFICATION AND NAMING OF ALCOHOLS

Classify the four alcohols, C_4H_9OH, as primary, secondary, and tertiary, and name them.

Solution

Primary $CH_3CH_2CH_2CH_2$—OH

$$CH_3-\overset{\overset{\displaystyle CH_3}{|}}{\underset{\underset{\displaystyle H}{|}}{C}}-CH_2OH$$

1-Butanol

2-Methyl-1-propanol

Secondary $CH_3CH_2\overset{}{\underset{\underset{\displaystyle OH}{|}}{C}}HCH_3$

2-Butanol

Tertiary $CH_3-\overset{\overset{\displaystyle CH_3}{|}}{\underset{\underset{\displaystyle OH}{|}}{C}}-CH_3$

2-Methyl-2-propanol

EXERCISE 23.3

Classify and name the two alcohols of formula C_3H_7OH.

FERMENTATION

Throughout history, ethanol has been made by fermentation of carbohydrates and sugars (Figure 23.19). Beer and wine are still made in this way. **Fermentation** is a reaction, catalyzed by certain enzymes found in yeast, in which sugars and carbohydrates are broken down into ethanol and carbon dioxide (Experiment 23.1). For example,

$$C_6H_{12}O_6 \xrightarrow{\text{Yeast}} 2C_2H_5OH + 2CO_2$$

Glucose   Ethanol

FIGURE 23.19
Wine Making in Ancient Egypt.

Experiment 23.1

Preparation of Ethanol by Fermentation

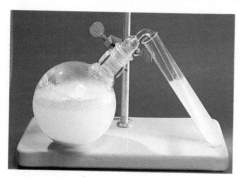

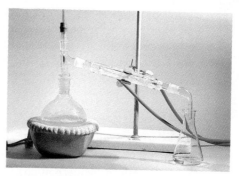

The flask contains a aqueous solution of sugar to which a little yeast has been added. The enzymes in the yeast catalyze the decomposition of sugar to ethanol and carbon dioxide. The CO_2 can be observed bubbling through lime water, a solution of $Ca(OH)_2$, producing a precipitate of insoluble calcium carbonate. After several days the evolution of CO_2 ceases and the reaction is complete

When the aqueous solution produced in this way is distilled, a more concentrated ethanol solution is obtained by collecting the fraction boiling at approximately 78°C.

When this concentrated ethanol solution is poured into a dish it burns when ignited.

If we begin with an aqueous solution of sugar, fermentation proceeds until the ethanol concentration reaches 13%, at which point the yeast no longer functions and fermentation ceases. Wine contains up to about 13% ethanol and beer usually 4%–6%. If wine is left exposed to the air, bacteria enter and catalyze the oxidation of ethanol to acetic acid (vinegar), CH_3CO_2H:

$$C_2H_5OH + O_2 \longrightarrow CH_3CO_2H + H_2O$$

This deterioration of wine by oxidation is prevented if the ethanol content is increased to approximately 20%. The ethanol content of fortified wines such as port, sherry, and vermouth is increased to this concentration by adding more alcohol after fermentation has ceased. Beverages such as whisky that contain more alcohol than fortified wines are made by distillation, which raises the ethanol content to about 50% (Experiment 23.1).

The intoxicating effect of ethanol is well known. Prolonged and excessive consumption of ethanol can lead to permanent liver damage. Methanol is considerably more toxic; small amounts can lead to blindness and even death.

DIOLS AND TRIOLS

Alcohols containing two or more OH groups are called *diols*, *triols*, and so on. Two important compounds of this type are 1,2-ethanediol, also known as *ethylene glycol*, or simply *glycol*,

$$\begin{array}{cc} CH_2 - CH_2 \\ | \quad\quad | \\ OH \quad OH \end{array}$$

and 1,2,3-propanetriol, which is also known as *glycerol*, or *glycerin*,

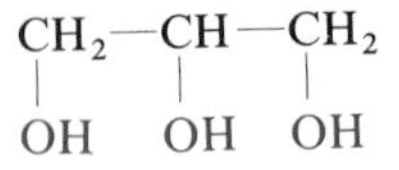

$$CH_2-CH-CH_2$$
$$\quad|\qquad|\qquad|$$
$$OH\quad OH\quad OH$$

Ethylene glycol is the main component of permanent antifreeze used in automobiles. Another important use is for the manufacture of polyester fibers. Glycerol is used in the manufacture of the explosives nitroglycerin and dynamite, in cosmetics, as a sweetening agent, and in the manufacture of plastics and synthetic fibers.

Antifreeze is ethylene glycol (1,2-ethanediol).

PROPERTIES

At 25 °C and 1 atm methanol, ethanol, and other alcohols containing up to about twelve carbon atoms are liquids. The boiling point increases regularly as the length of the carbon chain increases; the strength of the intermolecular forces increases as the number of carbon atoms increases (see Table 23.5). The boiling points are all much higher than for the corresponding alkanes. For example, methane boils at −164 °C, whereas methanol boils at 65 °C; ethane boils at −89 °C, but ethanol boils at 78 °C. The high boiling points of the alcohols result from hydrogen bonding. The hydrogen atom bonded to the highly electronegative oxygen atom of one molecule is hydrogen bonded to the oxygen of a neighboring molecule:

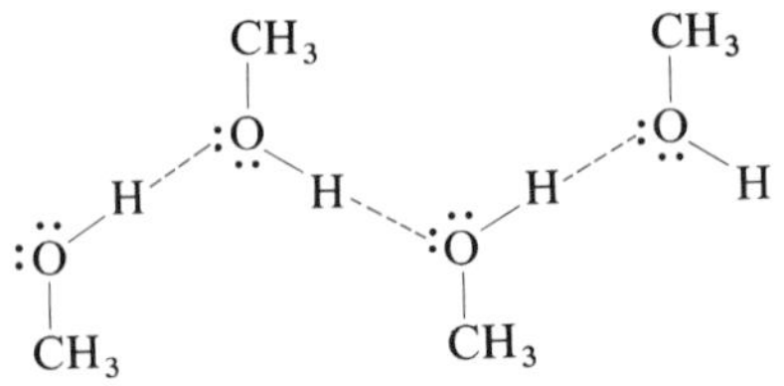

Methanol, ethanol, and the propanols are soluble in water at room temperature in all proportions. In contrast, the higher alcohols have only a limited solubility. Although the —OH group forms a hydrogen bond with water, as the hydrocarbon chains become longer, the solubility progressively decreases.

REACTIONS

Like water, alcohols behave as very weak acids and bases. They are too weakly acidic to donate a proton from their OH group to water, but they donate an

TABLE 23.5
Properties of Some Alcohols

Alcohol	Name	Bp (°C)	Solubility (g/100 g H_2O), 25 °C
CH_3OH	Methanol	65	Soluble in all proportions; miscible
CH_3CH_2OH	Ethanol	78	
$CH_3CH_2CH_2OH$	1-Propanol	97	
$CH_3CH_2CH_2CH_2OH$	1-Butanol	117	9.0
$CH_3CH_2CH_2CH_2CH_2OH$	1-Pentanol	138	2.7
$CH_3CH_2CH_2CH_2CH_2CH_2OH$	1-Hexanol	158	0.6

H^+ to the OH^- ion in concentrated aqueous hydroxide solutions, forming H_2O and the ethoxide ion:

$$C_2H_5OH + OH^- \rightleftharpoons C_2H_5O^- + H_2O$$

Ethoxide ion

The reaction of sodium with ethanol is similar to its reaction with water (Experiment 23.2). The products are hydrogen and sodium ethoxide, $C_2H_5O^-Na^+$:

$$2C_2H_5OH + 2Na \longrightarrow 2C_2H_5O^-Na^+ + H_2$$

As bases, alcohols are similar in strength to water and are protonated by strong acids. For example,

$$CH_3-\overset{..}{\underset{..}{O}}-H + H-Br \longrightarrow CH_3-\overset{\overset{H}{|}}{\underset{\oplus}{O}}-H + Br^-$$

Most alcohols can be dehydrated by heating them with concentrated H_2SO_4 or H_3PO_4 to give alkenes:

$$C_2H_5OH(l) \xrightarrow[\text{Heat}]{H_2SO_4(\text{conc})} CH_2{=}CH_2 + H_2O\ (+H_2SO_4 \rightarrow H_3O^+HSO_4^-)$$

This reaction is the reverse of the formation of an alcohol from an alkene, that is, the addition of H_2O to a double bond. The reaction is reversible, and the direction in which it proceeds is determined by the conditions. Concentrated sulfuric acid removes water and drives the reaction to the right.

Experiment 23.2

Reaction of Sodium with Ethanol

Sodium reacts with ethanol producing sodium ethoxide and hydrogen. When sodium is added to ethanol it sinks to the bottom and reacts less vigorously than it reacts with water.

The hydrogen produced may be easily collected as shown. Because ethoxide ion is basic, a few drops of phenolphthalein added to the ethanol color it pink.

Another important reaction of alcohols is oxidation to give aldehydes and ketones. This reaction is discussed in Section 23.10.

ALCOHOLS AS FUELS

Complete oxidation of a hydrocarbon gives carbon dioxide and water. The alcohols represent an intermediate or incomplete stage of oxidation of a hydrocarbon, so their enthalpies of combustion, although high, are not as high as those of the corresponding alkanes:

$$CH_4(g) + 2O_2(g) \longrightarrow CO_2(g) + 2H_2O(l) \qquad \Delta H° = -891 \text{ kJ mol}^{-1}$$

$$CH_3OH(l) + 1\tfrac{1}{2}O_2(g) \longrightarrow CO_2(g) + 2H_2O(l) \qquad \Delta H° = -726 \text{ kJ mol}^{-1}$$

$$C_2H_6(g) + 3\tfrac{1}{2}O_2(g) \longrightarrow 2CO_2(g) + 3H_2O(l) \qquad \Delta H° = -1560 \text{ kJ mol}^{-1}$$

$$C_2H_5OH(l) + 3O_2(g) \longrightarrow 2CO_2(g) + 3H_2O(l) \qquad \Delta H° = -1367 \text{ kJ mol}^{-1}$$

Because of an increase in the cost of petroleum products in recent years, there has been interest in replacing gasoline with alcohol produced by fermentation of agricultural waste. Mixtures of ethanol and gasoline—gasohol—are now sold as automobile fuel in some parts of the United States and in other countries, and methanol, which has a high octane rating of 110, is used in racing cars. Ethanol is used as a fuel in some camp stoves and in the home for heating and cooking food at the table. Unlike hydrocarbons, it normally does not form explosive mixtures with air, and any fire caused by burning alcohol can be easily extinguished with water.

PHENOLS

When a hydrogen atom in an arene (aromatic hydrocarbon) is replaced by an OH group, the resulting compound is called a **phenol**, not an alcohol. Replacing one of the hydrogen atoms of benzene with an —OH group gives the compound C_6H_5OH, which is called *benzenol* but is commonly known as *phenol*. It has the structure

Phenol is a weak acid, ($K_a = 1.3 \times 10^{-10}$ mol L^{-1} at 25 °C) though it is a stronger acid than the alcohols. It is a much weaker base than the alcohols.

Originally called *carbolic acid*, phenol is used as an antiseptic, a disinfectant, and a starting material in the manufacture of dyes and plastics. It is very effective in killing bacteria and was first used in medicine by Sir Joseph Lister in 1867.

23.9
Ethers

Ethers have the general formula

where R and R′ are alkyl groups. They may be prepared by dehydrating alcohols; for example.

$$C_2H_5OH + HOC_2H_5 \longrightarrow C_2H_5—O—C_2H_5 + H_2O$$
$$\text{Diethyl ether}$$

This reaction can be carried out by using a dehydrating agent such as concentrated sulfuric acid. We have seen above that with excess sulfuric acid ethene is obtained, but with excess ethanol the main product is diethyl ether.

Another method of preparing ethers is by the reaction of a sodium salt of an alcohol with an alkyl halide. For example,

$$C_2H_5—O^-Na^+ + CH_3—Br \longrightarrow C_2H_5OCH_3 + Na^+Br^-$$

The common names for ethers are based on the two alkyl groups attached to the oxygen atom. For example, $CH_3OC_2H_5$ is methyl ethyl ether. The systematic name is based on the alkyl group that has the longest carbon chain. The alkyl group with the shorter carbon chain and the oxygen atom are then called an *alkoxy group*; for example, $CH_3OC_2H_5$ is methoxyethane. These systematic names are not widely used.

Both alcohols and ethers are related to water and have the same angular structure at the oxygen atom. They are all examples of angular AX_2E_2 molecules. For example,

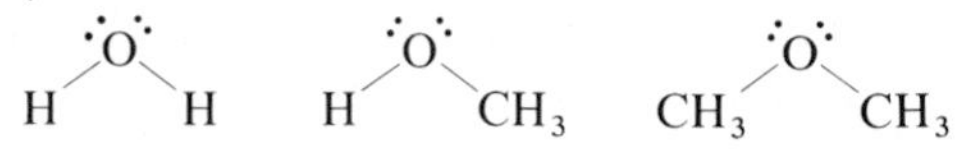

Water	Methanol	Dimethyl ether
(bp = 100 °C)	(bp = 65 °C)	(bp = −25 °C)

Both methanol and water are liquids, but dimethyl ether is a gas, because unlike water and methanol, dimethyl ether cannot form hydrogen bonds. For that reason, the boiling point of dimethyl ether is not very different from that of a simple alkane with a similar molecular mass ($CH_3—O—CH_3$, bp −25 °C; $CH_3—CH_2—CH_3$, bp −42 °C).

Diethyl ether (ethoxyethane), $C_2H_5OC_2H_5$ (bp = 35 °C), has only a limited solubility in water and is less dense than water. It is widely used as a solvent for organic compounds and for extracting organic compounds from water. In recent years MTB, methyl-*tert*butyl ether $CH_3—O—\overset{\displaystyle CH_3}{\underset{\displaystyle CH_3}{\overset{|}{\underset{|}{C}}}}—CH_3$ has become important as a replacement for lead tetraethyl $Pb(CH_3)_4$ as an antiknock agent in gasoline.

═ *Example 23.7* ═══════════════════════════════════

THE STRUCTURE AND PREPARATION OF ETHERS

Write the structure, and give a preparation, for ethoxybutane (ethylbutyl ether).

Solution

From the alkyl group and alkane names given in the name of the ether, the groups R and R′ must be ethyl and *n*-butyl, respectively, so the structure is $CH_3CH_2OCH_2CH_2CH_2CH_3$. This ether could be prepared by reacting the sodium salt of ethanol with an *n*-butyl halide such as 1-bromobutane:

$$CH_3CH_2O^-Na^+ + CH_3CH_2CH_2CH_2Br \longrightarrow CH_3CH_2OCH_2CH_2CH_2CH_3$$

(Attempts to dehydrate a mixture of ethanol and 1-butanol would give a mixture of diethyl ether, di-*n*-butyl ether, and the desired compound, and thus this route should be avoided.)

EXERCISE 23.4

Write the structure, and give a preparation, for methoxypropane (methylpropyl ether).

23.10
Aldehydes and Ketones

Primary and secondary alcohols decompose at 550–600 °C in the presence of a suitable catalyst (such as copper or silver) to form hydrogen and an aldehyde or ketone. This is called a **dehydrogenation reaction** and is an industrial method of preparation for aldehydes and ketones. Some examples are

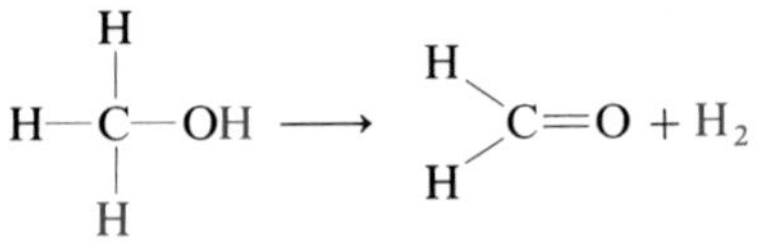

Methanol   Methanal
(formaldehyde)
(bp = −21 °C)

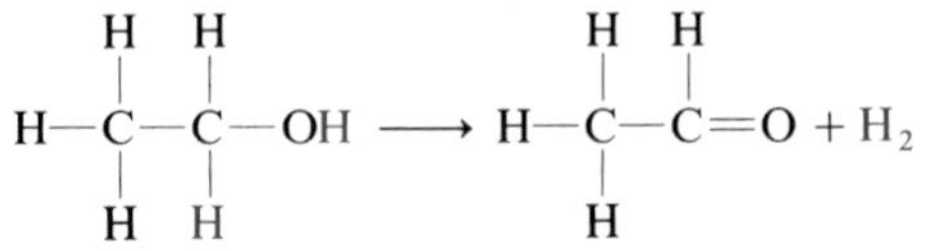

Ethanol   Ethanal
(acetaldehyde)
(bp = 21 °C)

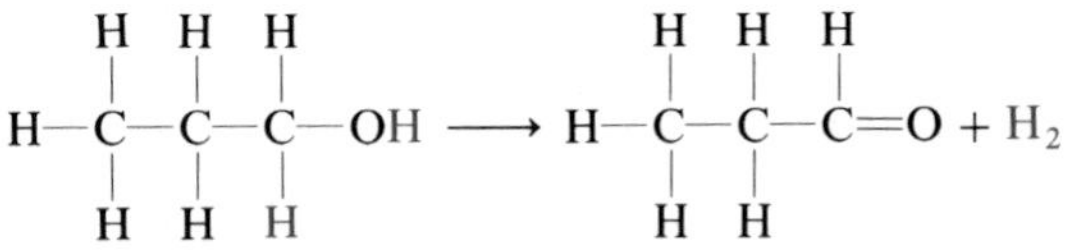

1-Propanol   Propanal
(bp = 50 °C)

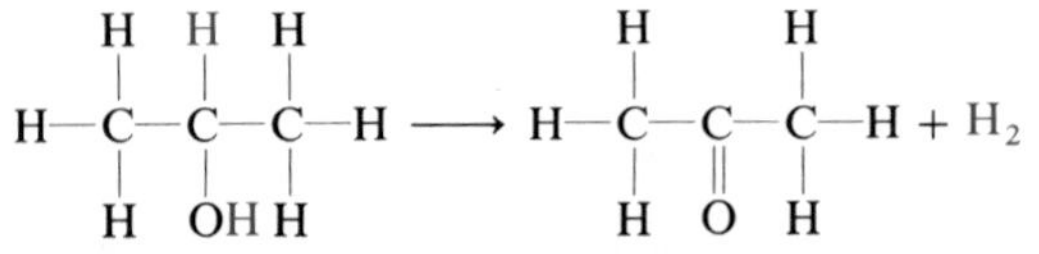

2-Propanol   2-Propanone
(acetone)
(bp = 56 °C)

Methanal, ethanal, and propanal all contain the

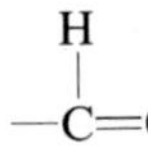

group and are the simplest members of the series of compounds known as **aldehydes**, which have the general formula

$$\underset{\underset{H}{|}}{R-C}=O$$

2-Propanone (acetone) is the simplest of a series of compounds known as **ketones**, all of which contain the

$$>C=O$$

group and have the general formula

$$\underset{R'}{\overset{R}{>}}C=O$$

Some examples are given in Figure 23.20.

Primary alcohols give aldehydes, while secondary alcohols give ketones on dehydrogenation. A tertiary alcohol cannot be dehydrogenated in this way, because it has no H atom on the C atom to which the OH group is attached.

PREPARATION

As we have seen, aldehydes and ketones are prepared on an industrial scale by the dehydrogenation of alcohols. In the laboratory they may be prepared by the oxidation of alcohols using oxidizing agents such as potassium permanganate or potassium dichromate in acid solution. Two hydrogen atoms are removed from the alcohol as water, and $KMnO_4$ is reduced to Mn^{2+} and $K_2Cr_2O_7$ is reduced to Cr^{3+}:

$$\underset{\underset{OH}{|}}{CH_3CHCH_3} \xrightarrow{K_2Cr_2O_7,\ H^+(aq)} \underset{\underset{O}{\|}}{CH_3CCH_3}$$

This method, however, is less convenient for the preparation of aldehydes as they are rather easily further oxidized to carboxylic acids as we shall see in the following section. Experiment 23.3 demonstrates the oxidation of ethanal by copper oxide which is reduced to copper.

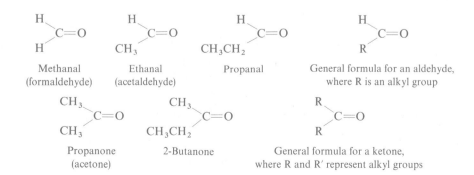

Methanal (formaldehyde)	Ethanal (acetaldehyde)	Propanal	General formula for an aldehyde, where R is an alkyl group
Propanone (acetone)	2-Butanone	General formula for a ketone, where R and R' represent alkyl groups	

FIGURE 23.20
Aldehydes and Ketones.

Experiment 23.3

Oxidation of Ethanol to Ethanal with Copper Oxide

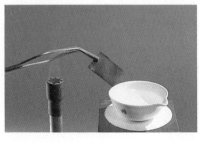

A piece of copper sheet is heated in a bunsen flame.

It rapidly becomes coated with a black layer of copper(II) oxide, CuO.

When the hot, oxide-coated copper is dipped into ethanol, it rapidly oxidizes the ethanol to ethanal and is itself reduced back to shiny metallic copper.

NOMENCLATURE

Aldehydes are named by replacing the *-e* of the corresponding alkane by *-al.* Since aldehydes are derived from primary alcohols, the

$$\begin{array}{c} H \\ | \\ -C=O \end{array}$$

group must always be at the end of the chain, so its C atom must always be carbon atom 1. Therefore we do not need to indicate its position in the name.

Ketones are named by replacing the *-e* in the name of the corresponding alkane by *-one.* The longest chain of C atoms is numbered from the end that gives the C atom of the C=O group the lowest possible number. A $>$C=O group is called a *carbonyl group.*

EXERCISE 23.5

Write a balanced equation for the preparation of propanone from 2-propanol using an acidic solution of potassium permanganate as the oxidizing agent.

═ *Example 23.8* ═══════════════════════════════════

THE STRUCTURE AND PREPARATION OF KETONES

Write the structure, and give a preparation, for 3-pentanone.

Solution

From the name, we deduce that the ketone group occurs as the third carbon in a five-carbon chain:

$$\begin{array}{c} O \\ || \\ CH_3CH_2CCH_2CH_3 \end{array}$$

This ketone can be prepared by oxidation of the corresponding alcohol, 3-pentanol:

$$CH_3CH_2\overset{\displaystyle OH}{\underset{\displaystyle |}{CH}}CH_2CH_3 \xrightarrow[H_3O^+]{KMnO_4} CH_3CH_2\overset{\displaystyle O}{\overset{\displaystyle \|}{C}}CH_2CH_3$$

EXERCISE 23.6

Draw the structure and give a preparation for 2-pentanone.

The common names *formaldehyde*, *acetaldehyde*, and *acetone* are very commonly used and should be remembered.

STRUCTURES

Ketones and aldehydes are examples of AX_3 molecules and therefore have a planar geometry around the carbon atom of the C=O group, with approximately 120° bond angles (Figure 23.21). If we describe the C=O double bond in terms of a σ and a π bond then we consider that the three σ bonds around the carbon atom are formed from sp^2 hybrid orbitals and that the π bond is formed from p orbitals on the carbon and oxygen atoms. Because oxygen is more electronegative than carbon, the oxygen atom of the carbonyl group carries a small negative charge, and the carbon of the C=O group carries a small positive charge. In other words, the carbonyl group is polar:

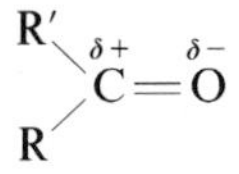

$$\overset{\displaystyle R'}{\underset{\displaystyle R}{>}}\overset{\delta+\quad\delta-}{C=O}$$

The polarity of the carbonyl group can also be represented by the two resonance structures

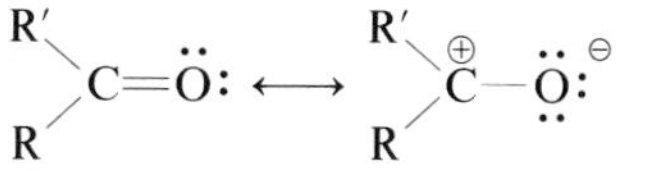

$$\overset{\displaystyle R'}{\underset{\displaystyle R}{>}}C=\ddot{O}: \longleftrightarrow \overset{\displaystyle R'}{\underset{\displaystyle R}{>}}\overset{\oplus}{C}-\ddot{\underset{\displaystyle \cdot\cdot}{O}}{:}^{\ominus}$$

PROPERTIES

Methanal (formaldehyde) is a gas with a penetrating odor and is very soluble in water. It is normally stored and sold as a 40% aqueous solution known as *formalin*. This is used as a disinfectant and for preserving biological specimens. A major use of formaldehyde is for the manufacture of resins, plastics, and adhesives.

Ethanal (acetaldehyde), CH_3CHO, is a liquid with a low boiling point of 21 °C. It is used largely for the synthesis of other organic compounds.

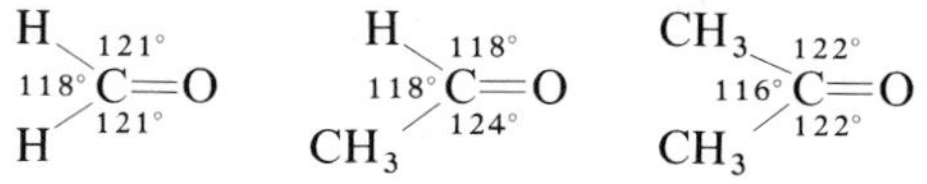

$$\overset{\displaystyle H}{\underset{\displaystyle H}{>}}\overset{118°\ 121°}{\underset{121°}{C=O}} \qquad \overset{\displaystyle H}{\underset{\displaystyle CH_3}{>}}\overset{118°\ 118°}{\underset{124°}{C=O}} \qquad \overset{\displaystyle CH_3}{\underset{\displaystyle CH_3}{>}}\overset{116°\ 122°}{\underset{122°}{C=O}}$$

Methanal   Ethanal   Propanone

FIGURE 23.21
Structures of Some Aldehydes and Ketones.

These are all AX_3 molecules with a planar triangular structure and approximately 120° bond angles.

FIGURE 23.22
Some Aromatic Aldehydes.

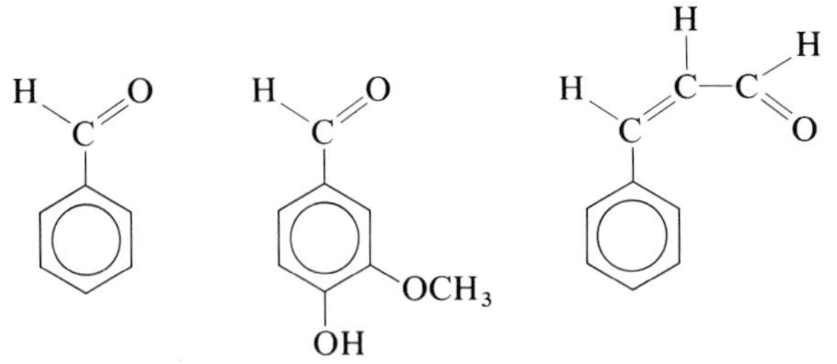

Benzaldehyde
(bitter almonds)

Vanillin
(vanilla)

Cinnamaldehyde
(cinnamon)

Propanone (acetone), CH_3COCH_3, is a volatile liquid boiling at 56 °C. It is a good solvent for many compounds, and it is completely miscible with water. Many aromatic aldehydes have pleasant odors (Figure 23.22).

23.11
Carboxylic Acids

In Chapters 5 and 14 we discussed acetic acid as an example of a weak acid in aqueous solution. It is a member of the class of compounds known as **carboxylic acids**, all of which have the general formula

$$R-\overset{\overset{\displaystyle O}{\|}}{C}-OH$$

For example, acetic acid is

$$CH_3-\overset{\overset{\displaystyle O}{\|}}{C}-OH$$

The $-\overset{\overset{\displaystyle O}{\|}}{C}-OH$ group is called the *carboxyl group*.

Carboxylic acids are named systematically by replacing the *-e* in the name of the parent alkane by *-oic acid*. Thus methane becomes methanoic acid, ethane becomes ethanoic acid, propane becomes propanoic acid, and so on. Examples are given in Table 23.6. These systematic names, however, have not been widely accepted, and the common names continue to be used extensively.

PREPARATION

A general method for preparing carboxylic acids in the laboratory is the oxidation of a primary alcohol or an aldehyde with an oxidizing agent such as potassium dichromate, $K_2Cr_2O_7$, in acid solution (see Experiment 23.5):

$$CH_3CH_2CH_2OH \xrightarrow{K_2Cr_2O_7,\ H_3O^+} CH_3CH_2C\overset{\nearrow O}{\underset{\searrow H}{}} \xrightarrow{K_2Cr_2O_7,\ H_3O^+} CH_3CH_2C\overset{\nearrow O}{\underset{\searrow OH}{}}$$

1-Propanol   Propanal   Propanoic acid

The relationships between alcohols, aldehydes, and carboxylic acids can be sum-

TABLE 23.6
Carboxylic Acids

Hydrocarbon	Carboxylic Acid	Common Name of Carboxylic Acid
H | H—C—H | H Methane	O || H—C—OH Methanoic acid	Formic acid
CH_3—CH_3 Ethane	O || CH_3—C—OH Ethanoic acid	Acetic acid
CH_3—CH_2—CH_3 Propane	O || CH_3—CH_2—C—OH Propanoic acid	Propionic acid
CH_3—CH_2—CH_2—CH_3 Butane	O || CH_3—CH_2—CH_2—C—OH Butanoic acid	Butyric acid
CH_3 | CH_3—C—CH_3 | H 2-Methylpropane	CH_3 O | || CH_3—C—C—OH | H 2-Methylpropanoic acid	Isobutyric acid

marized as follows:

$$\text{Primary alcohol} \underset{\text{Reduction}}{\overset{\text{Oxidation}}{\rightleftarrows}} \text{Aldehyde} \underset{\text{Reduction}}{\overset{\text{Oxidation}}{\rightleftarrows}} \text{Carboxylic acid}$$
$$\qquad RCH_2OH \qquad\qquad RCHO \qquad\qquad RCO_2H$$

$$\text{Secondary alcohol} \underset{\text{Reduction}}{\overset{\text{Oxidation}}{\rightleftarrows}} \text{Ketone}$$
$$\qquad RCH(OH)R \qquad\qquad R_2CO$$

Even a rather weak oxidizing agent, such as Ag^+, can oxidize an aldehyde to a carboxylic acid. Silver ion is reduced to metallic silver, which under the right conditions forms a shiny mirror on the sides of the reaction tube. This reaction is a useful test for aldehydes and serves to distinguish them from ketones and other related compounds (Experiment 23.4).

ACETIC ACID One industrial method for preparing acetic acid is the oxidation of ethanal (acetaldehyde) with air at 60–80 °C and 5 atm pressure in the presence of a catalyst:

$$2CH_3CHO + O_2 \longrightarrow 2CH_3CO_2H$$

Acetic acid can also be made by the enzyme-catalyzed oxidation of ethanol by air. This process has been used, since ancient times, to produce vinegar from

Experiment 23.4

Silver Mirror Test for Aldehydes

A clean beaker (or test tube) contains a solution of silver nitrate in aqueous ammonia. An aqueous solution of ethanal is added from a dropper and the solution is stirred.

The solution rapidly darkens in color as ethanal is oxidized to ethanoic acid and Ag⁺ is reduced to silver.

The inside of the beaker finally becomes coated with a shiny metallic silver mirror. Other aldehydes will give the same result.

wine and cider, but today much of the vinegar sold is made by diluting manufactured acetic acid. Vinegar is an approximately 5% solution of acetic acid in water. The oxidation of ethanol occurs in two stages; the second stage is more rapid than the first:

$$CH_3-CH_2-OH \xrightarrow{O_2} CH_3-\overset{\overset{\displaystyle H}{|}}{C}=O \xrightarrow{O_2} CH_3-\overset{\overset{\displaystyle OH}{|}}{C}=O$$

Ethanol   Ethanal (acetaldehyde)   Ethanoic acid (acetic acid)

FORMIC ACID The first member of the carboxylic acid series is *formic acid* (methanoic acid), HCO_2H:

$$H-C\overset{\displaystyle \ddot{O}:}{\underset{\displaystyle \ddot{O}-H}{}}$$

It was first prepared by the distillation of ants (Latin, *formica*). The swelling and irritation associated with ant bites and bee stings is partly due to formic acid.

OTHER CARBOXYLIC ACIDS In addition to formic acid, many other carboxylic acids can be isolated from natural sources. *Butyric acid* (*1-butanoic acid*), $CH_3CH_2CH_2CO_2H$, is obtained from butterfat and is responsible for the odor of rancid butter. The long-chain acids *palmitic acid*, $CH_3(CH_2)_{14}CO_2H$, and *stearic acid*, $CH_3(CH_2)_{16}CO_2H$, are obtained from many animal and vegetable fats (Chapter 24). Their sodium salts are important ingredients of soap.

There are also important carboxylic acids that contain more than one carboxyl group. The simplest is *oxalic acid* (*ethanedioic acid*), $(CO_2H)_2$, which consists of two carboxyl groups joined together. Oxalic acid occurs in the leaves of rhubarb and related plants and is poisonous. It removes calcium from the body as insoluble calcium oxalate, $Ca^{2+}(C_2O_4)^{2-}$.

Other important carboxylic acids contain hydroxyl groups as well as the carboxylic acid group. *Lactic acid* (2-hydroxypropanoic acid), $CH_3CH(OH)CO_2H$, is found in sour milk. It accumulates in the muscles of the body after strenuous exercise and is responsible for the soreness in the muscles. *Citric acid*, $C_3H_5O(CO_2H)_3$, is found in the juice of citrus fruits. The simplest aromatic carboxylic acid is *benzoic acid*, $C_6H_5CO_2H$. It is made by the oxidation of methylbenzene (toluene) with oxygen in the presence of a catalyst:

Rhubarb leaves contain oxalic acid. Pure oxalic acid is a white crystalline solid.

Benzoic acid

Citric acid and benzoic acid are commonly used as food preservatives. The dicarboxylic acid $C_6H_4(CO_2H)_2$ is called *phthalic acid*. Phthalic acid is used in the synthesis of dyes and in the manufacture of paints and varnishes. The aromatic hydroxycarboxylic acid $C_6H_4(OH)(CO_2H)$ is known as *salicylic acid*. *Aspirin* is acetylsalicylic acid (see Box 23.2).

STRUCTURE

The carboxylic acid group has a planar AX_3 geometry around the carbon atom (see Figure 23.23).

PROPERTIES

Carboxylic acids are weak acids in water; for example,

Acetic acid   Acetate ion

Table 23.7 gives pK_a values for some carboxylic acids. They are much stronger acids than the corresponding alcohols.

The acid strength of the carboxylic acids can be attributed to the presence of the polar carbonyl group, $C=O$. The electronegative oxygen atom attracts electrons away from the carbon atom to which it is attached, making this carbon atom more electronegative. The carbon atom in turn attracts electrons from the OH group, making the O—H bond more polar than it is in an alcohol and facilitating the donation of the proton to a water molecule (see Figure 23.24).

In the anion, $CH_3CO_2^-$, the negative charge is delocalized over both oxygen atoms. This delocalization of charge contributes to the stability of the anion and, therefore, to the tendency of the acid to ionize:

A similar delocalization of negative charge is not possible in the anion of an

Oxalic acid

Lactic acid

Citric acid

Phthalic acid

Salicylic acid

FIGURE 23.23
Structure of Methanoic (Formic) Acid.

The CO_2H group has a planar AX_3 geometry with bond angles that are close to 120°.

BOX 23.2

ASPIRIN

From early times people have sought substances that would relieve pain. Ethanol, opium, cocaine, and marijuana were all used by early societies for this purpose. Such pain-relieving substances are generally known as *analgesics*. In the eighteenth century it was found that an extract of willow bark was a good analgesic. In 1860 salicylic acid was isolated from this willow bark extract and found to be an efficient analgesic and an anti-inflammatory drug. About the same time chemists discovered that salicylic acid could be synthesized conveniently and cheaply by heating the sodium salt of phenol with carbon dioxide under pressure:

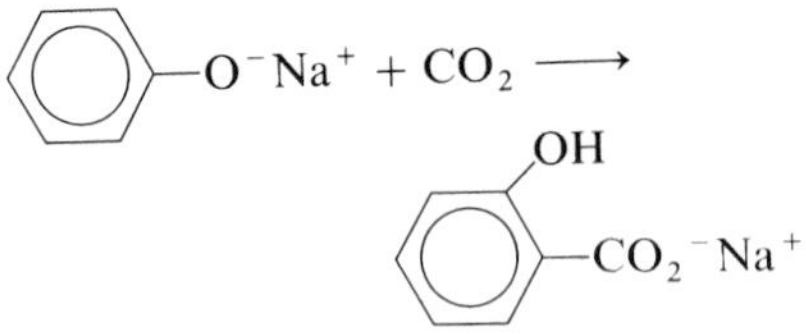

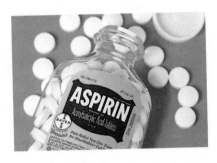

Salicylic acid then began to be widely used. But salicylic acid has a very sour taste and damages the tissue of the mouth and throat. Acetylsalicylic acid—aspirin—was found to be equally effective and not to have the disadvantages of salicylic acid. It was put into large-scale production in 1899 and soon became the largest-selling drug in the world, a position it retains today.

Aspirin not only relieves pain but also reduces fever and local inflammation. The only undesirable side effect is gastrointestinal bleeding. While the bleeding is usually slight and unimportant, in some persons it may be serious. Other persons have an allergic hypersensitivity to aspirin. Aspirin also inhibits the clotting of blood and should not be used by persons facing surgery or women awaiting childbirth.

More than $100 million is spent every year in the United States on thousands of tons of aspirin; three times as much is spent on products that are combinations of other drugs with aspirin. These products generally contain small amounts of the stimulant caffeine and other pain relievers that are probably no more effective than aspirin but are considerably more expensive. Despite its long use and intensive investigation, only in the past few years have we begun to understand how aspirin works.

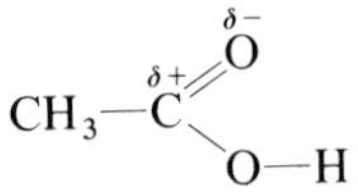

FIGURE 23.24
Acidity of Carboxylic Acids.

The acidity of the OH group is enhanced by the presence of the polar $\overset{\delta+}{C}=\overset{\delta-}{O}$ group.

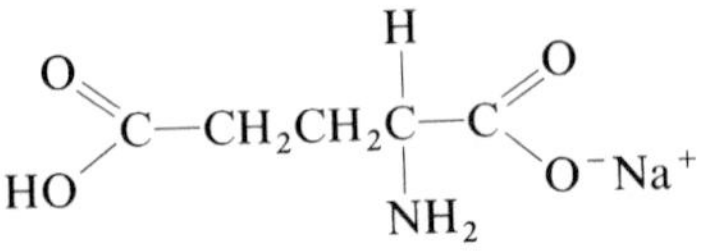

FIGURE 23.25
Monosodium Glutamate.

This compound is used as a meat tenderizer and flavor enhancer.

alcohol, such as CH_3O^-, and therefore alcohols are weaker acids than carboxylic acid. Both CO bonds in the acetate ion are expected to have a bond order of $1\frac{1}{2}$. Their length is consistent with this bond order, as can be seen from the data in Table 23.8.

Important salts of carboxylic acids are sodium benzoate, which is used as a food preservative, and monosodium glutamate (MSG), which is used as a flavor enhancer (Figure 23.25).

TABLE 23.7
Melting Points, Boiling Points, and pK_a Values for Some Carboxylic Acids

Acid	Formula	Mp (°C)	Bp (°C)	pK_a
Methanoic (formic)	HCO_2H	8.4	110.5	3.77
Ethanoic (acetic)	CH_3CO_2H	16.6	118	4.76
Propanoic (propionic)	$CH_3CH_2CO_2H$	−22	141	4.88
1-Butanoic (butyric)	$CH_3CH_2CH_2CO_2H$	−5	163	4.82
Ethanedioic (oxalic)	$(CO_2H)_2$	187	—	1.46
2-Hydroxypropanoic (lactic)	$CH_3CH(OH)CO_2H$	18	—	3.87
Benzoic	$C_6H_5CO_2H$	122	249	4.17
Phthalic	$C_6H_4(CO_2H)_2$	200[a]	—	3.00
Salicylic	$C_6H_4(OH)(CO_2H)$	159	—	3.00

[a] Decomposes.

TABLE 23.8
Bond Lengths and Bond Orders in
Some Molecules Containing CO Bonds

Molecule	Bond Order	Bond Length (pm)
CH_3-OH	1.00	143
CO_3^{2-}	1.33	129
$CH_3CO_2^-$	1.50	127
H_2CO	2.00	122
CO	3.00	113

23.12
Esters

Carboxylic acids react with alcohols in the presence of an acid as a catalyst to form compounds called **esters**. For example, acetic acid reacts with ethanol to give ethyl acetate (ethyl ethanoate) and water:

$$CH_3\overset{\overset{\textstyle O}{\|}}{C}-OH + H-OC_2H_5 \longrightarrow CH_3\overset{\overset{\textstyle O}{\|}}{C}-O-C_2H_5 + H_2O$$

Butanoic acid reacts with ethanol to give *ethylbutanoate*:

$$CH_3-CH_2-CH_2-\overset{\overset{\textstyle O}{\|}}{C}-OH + H-O-C_2H_5 \longrightarrow$$

$$CH_3-CH_2-CH_2-\overset{\overset{\textstyle O}{\|}}{C}-O-C_2H_5 + H_2O$$

In general, a carboxylic acid RCO_2H reacts with an alcohol $R'OH$ to give the ester

$$R-\overset{\overset{\textstyle O}{\|}}{C}-OR'$$

The systematic names for esters are obtained from the names of the acid and the alcohol from which they may be prepared. The ending *-oic* of the acid is replaced by *-oate* and the name is preceded by that of the alkyl group in the alcohol. For example, *methylpropanoate* is prepared from methanol and propanoic acid:

$$CH_3-CH_2-\overset{\overset{\textstyle O}{\|}}{C}-OH + HO-CH_3 \longrightarrow CH_3-CH_2-\overset{\overset{\textstyle O}{\|}}{C}-OCH_3 + H_2O$$

Example 23.9

THE PREPARATION OF AN ESTER

Write an equation for the preparation of pentylbutanoate from the appropriate alcohol and carboxylic acid.

Solution

$$CH_3CH_2CH_2CO_2H + HOCH_2CH_2CH_2CH_2CH_3 \longrightarrow$$

Butanoic acid   1-Pentanol

$$CH_3CH_2CH_2\overset{\overset{\textstyle O}{\|}}{C}OCH_2CH_2CH_2CH_2CH_3 + H_2O$$

Pentylbutanoate

EXERCISE 23.7

Write an equation for the preparation of ethylpropanoate from the appropriate alcohol and carboxylic acid.

The pleasant taste and odor of many fruits is due to the esters they contain.

Many esters have pleasant fruity odors. Indeed, they play an important role in determining the taste and fragrance of many fruits and flowers (see Table 23.9). In general, the flavor and odor of a fruit is due to a combination of many substances, not one compound. For example, it is estimated that more than 250 substances determine the flavor of strawberries. These include alcohols, carboxylic acids, esters, aldehydes, and ketones. Synthetic esters are widely used in perfumes and for artificial flavoring in foods and candies.

Since salicylic acid has both a carboxylic acid group and an —OH group, it can form esters in two ways. It can react, as an alcohol, with another carboxylic acid such as acetic acid to give the ester acetylsalicylic acid, which is aspirin (see Box 23.2):

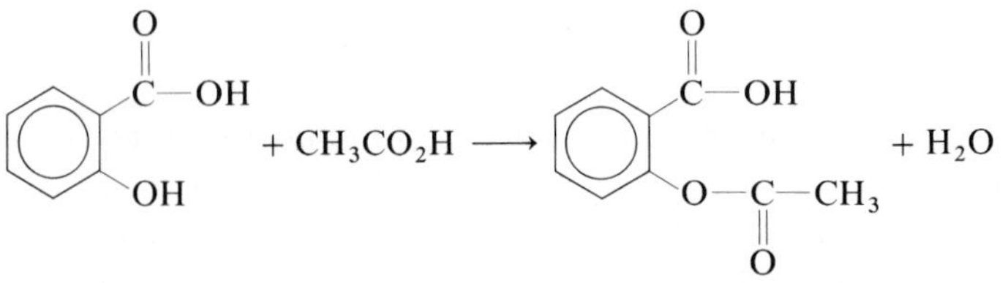

$$\underset{\text{Salicylic acid}}{\begin{array}{c}\overset{\overset{\textstyle O}{\|}}{C}-OH\\\\OH\end{array}} + CH_3CO_2H \longrightarrow \underset{\text{Aspirin}}{\begin{array}{c}\overset{\overset{\textstyle O}{\|}}{C}-OH\\\\O-\overset{\overset{\textstyle }{}}{C}-CH_3\\\overset{}{\|}\\O\end{array}} + H_2O$$

Salicylic acid   Aspirin

TABLE 23.9
Odors of Some Esters

Formula	Name	Odor
$\overset{O}{\overset{\|}{CH_3C}}OCH_2CH_2CH_2CH_2CH_3$	Pentylethanoate (amyl acetate)	Banana
$CH_3CH_2CH_2\overset{O}{\overset{\|}{C}}OCH_2CH_3$	Ethylbutanoate (ethyl butyrate)	Pineapple
$CH_3\overset{O}{\overset{\|}{C}}OCH_2(CH_2)_6CH_3$	Octylethanoate	Orange
$CH_3CH_2CH_2\overset{O}{\overset{\|}{C}}OCH_2CH_2CH_2CH_2CH_3$	Pentylbutanoate (amyl butyrate)	Apricot
Methyl anthranilate structure (COCH₃, NH₂)	Methyl anthranilate	Grape
Methyl salicylate structure (COCH₃, OH)	Methyl salicylate	Oil of wintergreen

Salicylic acid can also react as an acid when it is in the presence of an alcohol. For example, with methanol it is the methyl ester, methyl salicylate, that is formed. This ester is also known as *oil of wintergreen* and is used in the manufacture of perfumes and artificial flavors:

$$\text{(salicylic acid structure, C—OH, OH)} + CH_3OH \longrightarrow \text{(methyl salicylate structure, C—O—CH}_3\text{, OH)} + H_2O$$

Methyl salicylate
(oil of wintergreen)

The synthesis of aspirin and oil of wintergreen starting with the hydrocarbon benzene is a simple illustration of the important procedure called *organic synthesis* in which functional groups are added to a hydrocarbon and modified in a step-by-step procedure until the final desired product is obtained (Box 23.3).

The esters of diols and triols with carboxylic acids are an important group of natural compounds. Animal fats (such as lard and butter) and vegetable oils

BOX 23.3

ORGANIC SYNTHESIS

The production of organic compounds from readily available materials usually requires a number of steps. Each of these steps normally involves either the insertion of a functional group into a hydrocarbon or the conversion of one functional group to another. Each step produces an intermediate product that is used in the next step. The final step produces the desired product. The synthesis of large, complex organic molecules occurring in plants and animals may require a very large number of steps.

There may be many different possible routes to a given compound, and the best choice has to be made considering the following factors:

1. The cost and availability of the starting materials.

2. The number of separate steps needed.

3. The yield and the purity of the product obtained in each step.

4. The ease of separating and purifying the product of each step.

A simple example of a multistep organic synthesis is provided by the synthesis of aspirin and oil of wintergreen starting from methylbenzene (toluene). In this scheme we have not written a balanced equation for each reaction but we have merely shown each organic substance and the other reactant and conditions needed to bring about the reaction.

An organic synthesis in progress.

It has sometimes been said that organic synthesis is more of an art than a science. Designing a good organic synthesis, like all creative activity, requires imagination, extensive background knowledge, and experience. It is a challenging field for the chemist and one that has contributed enormously to our material well-being. Chemists have synthesized a truly enormous number of organic compounds—well over 2 million are known today and new ones are being made every day. These include plastics, synthetic fibers, paints, dyes, pesticides, medicines, drugs, perfumes, flavors, preservatives, lubricants, and solvents, all of which are prepared by synthesis from coal, petroleum, and natural gas.

$$\text{Methylbenzene (toluene)} \xrightarrow[\text{Catalyst}]{O_2} \text{Phenol} \xrightarrow{\text{NaOH}} \text{Sodium phenoxide} \xrightarrow[\substack{\text{Heat}\\\text{Pressure}}]{CO_2} \text{Sodium salicylate}$$

Methylbenzene (toluene) — Phenol — Sodium phenoxide — Sodium salicylate

$\xrightarrow{\text{HCl}}$ Salicylic acid

Methyl salicylate $\xleftarrow[\substack{H_2SO_4\\\text{Heat}}]{CH_3OH}$ Salicylic acid

$CH_3-\overset{O}{\overset{\|}{C}}-O-\overset{O}{\overset{\|}{C}}-CH_3$ Acetic anhydride

Aspirin $\xleftarrow{H_2SO_4}$

(such as corn oil, soybean oil, linseed oil, olive oil, and peanut oil) are all esters of 1,2,3-propanetriol (glycerol) and are called *glycerides*. They have the general formula

$$RCO_2CH_2$$
$$R'CO_2CH$$
$$R''CO_2CH_2$$

The R, R', and R'' groups are hydrocarbon chains with from 3 to 21 carbon atoms. These R groups may be saturated alkyl groups or they may contain one or more C=C double bonds. In general, animal fats contain saturated hydrocarbon chains, whereas vegetable oils have unsaturated hydrocarbon chains with one or more double bonds. In the manufacture of margarine from vegetable oils, some of the double bonds are converted to single bonds by adding hydrogen to them. This raises the melting point and converts an oil to a fat.

Inorganic acids also form esters with alcohols. For example, glycerol reacts with nitric acid to form the nitrate ester glyceroltrinitrate, commonly called nitroglycerin (Box 6.2). The esters of adenosine with diphosphoric and triphosphoric acid, ADP and ATP (Figure 23.26), are important energy storage compounds in biochemical processes (Chapters 16 and 24).

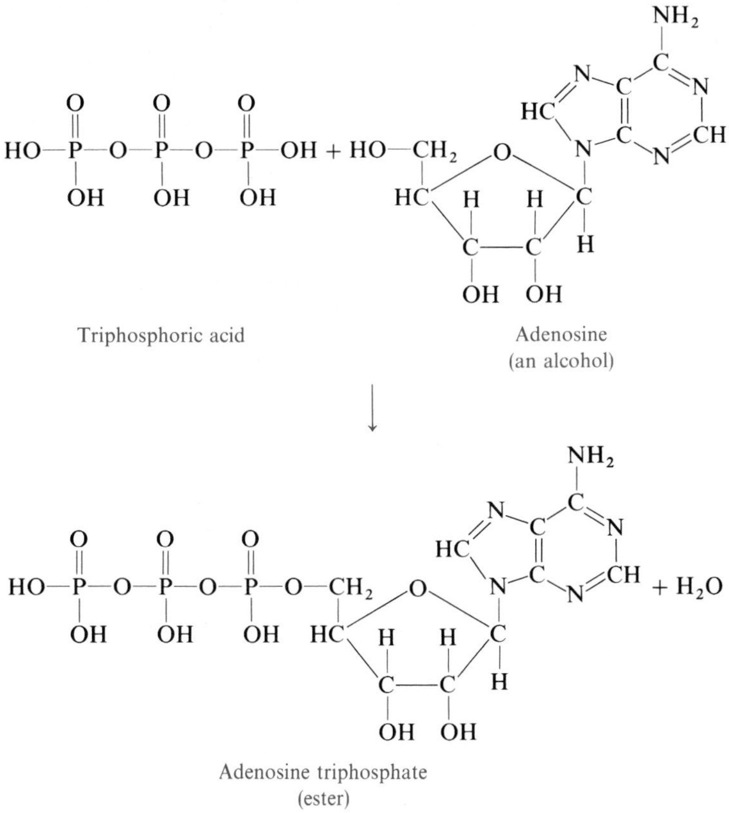

FIGURE 23.26
Adenosine Triphosphate.

ATP is an ester of triphosphoric acid and adenosine (an alcohol).

23.13
Amines

In the same way that alcohols and ethers may be regarded as derivatives of water, there are analogous compounds derived from ammonia. They are called **amines**. We can distinguish primary, secondary, and tertiary amines, depending on the number of alkyl groups (one, two, or three) attached to the nitrogen atom; for example,

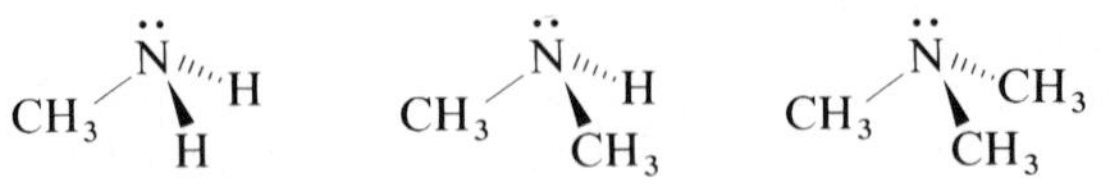

Ammonia   Primary amine   Secondary amine   Tertiary amine

Examples are given in Figure 23.27. All the amines have the same pyramidal AX_3E geometry around the nitrogen as the ammonia molecule.

The methylamines are obtained industrially by the reaction of methanol with ammonia at 400 °C in the presence of aluminum oxide as a catalyst:

$$CH_3OH + NH_3 \longrightarrow CH_3NH_2 + H_2O$$

$$CH_3OH + CH_3NH_2 \longrightarrow (CH_3)_2NH + H_2O$$

$$CH_3OH + (CH_3)_2NH \longrightarrow (CH_3)_3N + H_2O$$

The simplest aromatic amine is *aminobenzene*, $C_6H_5NH_2$, commonly known as *aniline*. It is the starting material for the preparation of a number of important dyes called azo dyes.

Most amines have unpleasant odors. The stench of decaying flesh is due to amines such as putrescine produced by the decomposition of proteins. Some examples of amines are given in Figure 23.28.

The simple aliphatic amines are generally soluble in water. Like ammonia, amines are weak bases (see Table 23.10):

$$CH_3-NH_2 + H_2O \rightleftharpoons CH_3-NH_3^+ + OH^-$$

Methylammonium ion

Methylamine   Dimethylamine   Trimethylamine
(primary amine)   (secondary amine)   (tertiary amine)

$$NH_2CH_2CH_2CH_2CH_2NH_2$$ Putrescine

$$NH_2CH_2CH_2CH_2CH_2CH_2NH_2$$ Cadaverine

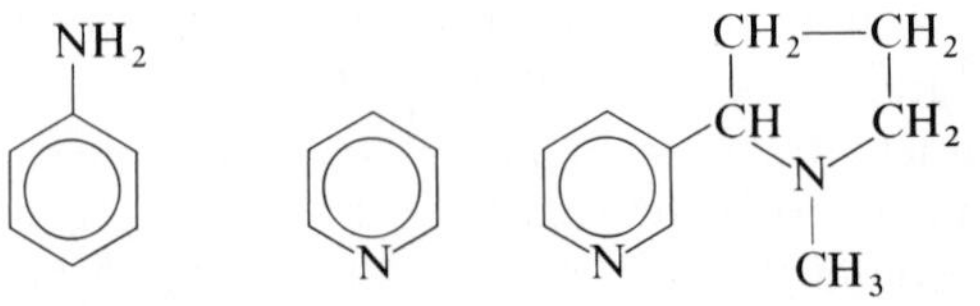

NH_2

Aminobenzene   Pyridine   Nicotine
(Aniline)

TABLE 23.10				
Properties of Ammonia and Some Amines				
Compound	Formula	Mp (°C)	Bp (°C)	pK_b
Ammonia	NH_3	−77.7	−33.4	4.75
Methylamine	CH_3NH_2	−92.5	−6.5	3.43
Dimethylamine	$(CH_3)_2NH$	−96	7.4	3.27
Trimethylamine	$(CH_3)_3N$	−124	3.5	4.19
Aniline	$C_6H_5NH_2$	−6	184	9.37

They react with acids, such as HCl, to give salts:

$$C_6H_5NH_2 + HCl \longrightarrow C_6H_5NH_3{}^+Cl^-$$

Anilinium chloride

23.14
Amides

Replacement of the OH group of a carboxylic acid by an NH_2 group gives a compound called an **amide**:

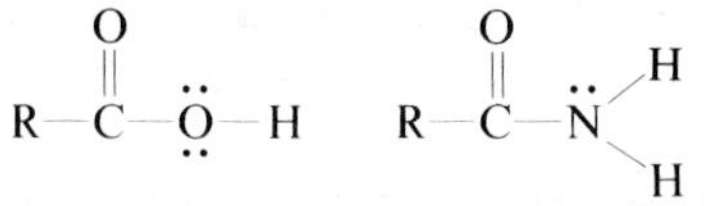

$$R-\overset{\overset{\displaystyle O}{\|}}{C}-\overset{\cdot\cdot}{\underset{\cdot\cdot}{O}}-H \qquad R-\overset{\overset{\displaystyle O}{\|}}{C}-\overset{\cdot\cdot}{N}\overset{\diagup H}{\diagdown H}$$

Carboxylic acid   Amide

The simplest members of this family of compounds are

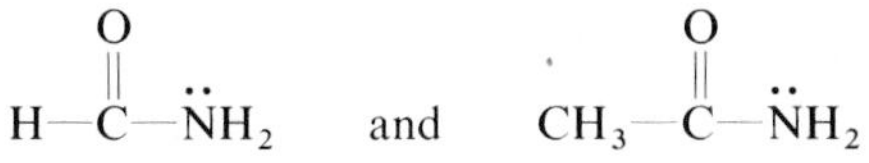

$$H-\overset{\overset{\displaystyle O}{\|}}{C}-\overset{\cdot\cdot}{N}H_2 \qquad \text{and} \qquad CH_3-\overset{\overset{\displaystyle O}{\|}}{C}-\overset{\cdot\cdot}{N}H_2$$

Methanamide   Ethanamide
(formamide)   (acetamide)

Amides can be prepared by the reaction of a carboxylic acid with ammonia or an amine to form a salt, which, when heated, eliminates water to give the amide. For example,

$$CH_3COOH + NH_3 \longrightarrow CH_3CO_2{}^-NH_4{}^+ \longrightarrow CH_3CONH_2 + H_2O$$

Acetic acid   Ammonia   Ammonium acetate   Acetamide

The amide group

$$-\overset{\overset{\displaystyle O}{\|}}{C}-\overset{\overset{\displaystyle H}{|}}{\underset{\cdot\cdot}{N}}-$$

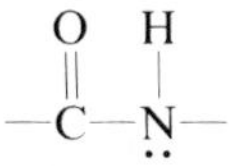

is the key functional group in the structure of proteins and some important polymers. They are discussed in Chapter 24.

An important feature of the amide group is that it has a planar geometry at both the carbon and the nitrogen atoms (see Figure 23.29). We expect a planar AX_3 geometry around the carbon atom, but we might have expected a pyramidal AX_3E geometry at the nitrogen atom. The nitrogen atom has a planar geometry because the lone pair on the nitrogen atom is delocalized into the

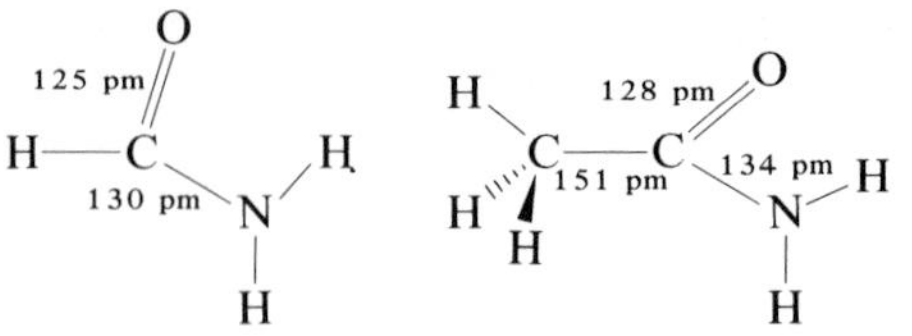

FIGURE 23.29
Structures of Methanamide and Ethanamide.

All the atoms of the amide group $CONH_2$ are in the same plane. Both the C and N atoms of the amide group have a planar AX_3 geometry.

C–N bonding region forming a partial CN double bond. Thus the amide group can be represented by the following two resonance structures:

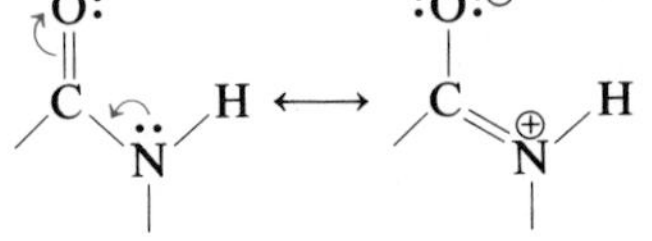

Hence both the carbon and the nitrogen have a planar AX_3 geometry. The C—N bond length of 130 pm in methanamide is intermediate between the C—N single bond length of 147 pm and the C=N double bond length of 124 pm. This short CN bond length is further evidence that the amide group is best described by the two resonance structures above.

Urea,

is a major animal waste product that is found in urine. In 1828 the German chemist Friedrich Wöhler (1800–1882) made urea by heating ammonium cyanate:

$$NH_4{}^+OCN^- \xrightarrow{\text{Heat}} NH_2CONH_2$$

This experiment led to the overthrow of the idea, widely accepted at the time, that organic compounds could only be obtained from living matter.

In 1987 urea ranked twelfth among the industrial chemicals produced in the United States. It is made by heating ammonia together with carbon dioxide at 175–200 °C and a pressure 200–400 atm:

Because of its high nitrogen content (46% by mass), urea is an important fertilizer. In the soil it decomposes slowly to ammonia. It is also used in manufacturing plastics, foams, and adhesives.

23.15
Halogen Derivatives of Alkanes and Alkenes

We have seen that alkanes are relatively unreactive compounds. Their carbon atoms have complete valence shells and cannot accept additional electron pairs. They also have no unshared electron pairs to donate to other atoms. For an alkane to undergo reaction, C—H bonds must be broken; but because a C—H bond is strong, a large amount of energy is needed to break it. However, at high temperatures the alkanes do undergo some important reactions. Combustion is one. Another very important reaction is that between alkanes and the halogens. Particularly important is the reaction of methane with chlorine.

CHLOROMETHANES

In the industrial process the reaction of methane and chlorine is carried out at a temperature between 350 and 750 °C:

$$CH_4 + Cl_2 \longrightarrow CH_3Cl + HCl$$

The product is *chloromethane* (methyl chloride), CH_3Cl. However, chloromethane is more reactive than methane, so the reaction continues to give *dichloromethane* (methylene dichloride), CH_2Cl_2, *trichloromethane* (chloroform), $CHCl_3$, and *tetrachloromethane* (carbon tetrachloride), CCl_4:

$$CH_3Cl + Cl_2 \longrightarrow CH_2Cl_2 + HCl$$
$$CH_2Cl_2 + Cl_2 \longrightarrow CHCl_3 + HCl$$
$$CHCl_3 + Cl_2 \longrightarrow CCl_4 + HCl$$

These reactions are chain reactions (see Figure 23.30). The relative amounts of the various chloromethanes in the final product can be adjusted by changing the proportions of the reactants. The HCl is removed by dissolving it in water, and the chloromethanes are then condensed and separated by fractional distillation. The boiling points of the chloromethanes increase from 24 °C for CH_3Cl to 79 °C for CCl_4 with increasing molecular size and the consequent increase in the strength of the intermolecular (London) forces.

The main use of chloromethane is in the preparation of chloromethyl silanes, such as $(CH_3)_2SiCl_2$, which are important intermediates in the manufacture of silicones (see Chapter 22).

Dichloromethane is a very good solvent, and nearly all its commercial uses depend on this property. For example, it is an important component of paint strippers. In the past *trichloromethane* (*chloroform*) was widely used as an anesthetic, but it is no longer an important anesthetic because it is believed to be carcinogenic. Its main use is as an intermediate in the manufacture of chlorofluoromethanes.

Tetrachloromethane (*carbon tetrachloride*) was formerly used as a dry-cleaning agent and in fire extinguishers. It has been replaced by other substances for these purposes because it is toxic and at high temperatures it reacts with the oxygen of the air to form the highly toxic gas phosgene, $Cl_2C{=}O$.

$$Cl_2 \xrightarrow{\text{Heat or light}} 2Cl\cdot \qquad \text{Chain initiation}$$

$$\left.\begin{array}{l}\cdot Cl + CH_4 \longrightarrow H{-}Cl + \cdot CH_3 \\ \cdot CH_3 + Cl_2 \longrightarrow CH_3Cl + \cdot Cl\end{array}\right\} \text{Chain propagation}$$

$$\left.\begin{array}{l}\cdot Cl + \cdot Cl \longrightarrow Cl_2 \\ \cdot CH_3 + \cdot CH_3 \longrightarrow C_2H_6\end{array}\right\} \text{Chain termination}$$

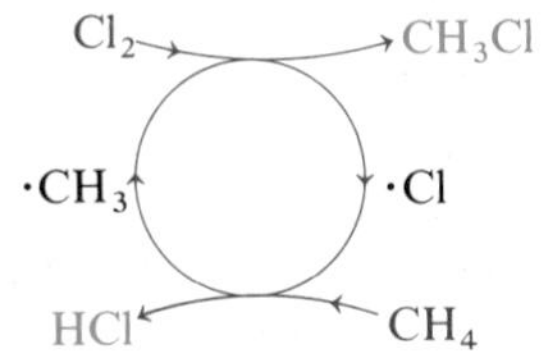

FIGURE 23.30 The Reaction Between Methane and Chlorine.

This is a chain reaction involving free-radical intermediates. It can be initiated by heat or light.

OTHER CHLOROALKANES AND CHLOROALKENES

Chloroethane, C_2H_5Cl, can be made by the chlorination of ethane, but it is more commonly made by adding HCl to ethene:

$$H_2C{=}CH_2 + HCl \longrightarrow C_2H_5Cl$$

Addition of chlorine to ethene gives *1,2-dichloroethane*:

$$H_2C{=}CH_2 + Cl_2 \longrightarrow ClCH_2{-}CH_2Cl$$

In 1987 1,2-dichloroethane ranked fourteenth among the industrial chemicals produced in the United States. Almost all of it is used to produce *chloroethene* (*vinyl chloride*), another very important industrial chemical. On heating dichloroethane, HCl is eliminated and a double bond is formed:

$$ClCH_2{-}CH_2Cl \xrightarrow[\text{Heat}]{} H_2C{=}CHCl + HCl$$

Chloroethene
(vinyl chloride)

Vinyl chloride is used mainly for the manufacture of poly (vinyl chloride), PVC, (see Chapter 24).

Several *chlorofluoromethanes* and *chlorofluoroethanes* have important uses as refrigerator fluids and as aerosol propellants in cans of spray polish, shaving cream, and so on. These compounds are commonly called Freons and include CF_2Cl_2 (bp = -30 °C) and $CClF_2CClF_2$ (bp = 24 °C). As we discussed in Chapter 18, there is increasing evidence that these substances are decreasing the concentration of ozone in the upper atmosphere. Their use is now being controlled and gradually reduced in the United States and some other countries.

Tetrachloroethene, $Cl_2C{=}CCl_2$, has replaced CCl_4 in "dry" cleaning because of its lower volatility and lower toxicity. *Tetrafluoroethene*, C_2F_4, on polymerization yields the plastic Teflon, which is widely used because of its inertness and resistance to heat. Teflon-lined nonstick pans are familiar in the kitchen.

23.16
Review of Functional Groups and Reactions

In this final section we review the functional groups and some of the reactions by which they are inserted into alkanes and by which they may be interconverted.

The simple functional groups OH and OR may be thought of as being derived from water and NH_2 as being derived from ammonia. Table 23.11 shows that five more important functional groups contain the carbonyl group, $C{=}O$. The carbonyl group may be combined with H or R to give aldehydes and ketones and with the simple functional groups OH, OR, and NH_2 to give carboxylic acids, esters, and amides, respectively.

The various reactions by which simple one-, two-, and three-carbon compounds containing these functional groups may be obtained from methane, ethene, and propene are summarized in Figures 23.31 and 23.32.

TABLE 23.11
Relationships Between Functional Groups

—H		$-\overset{\overset{\displaystyle O}{\parallel}}{C}-H$		Aldehyde
—R	Alkyl	$-\overset{\overset{\displaystyle O}{\parallel}}{C}-R$		Ketone
—OH	Alcohol	$-\overset{\overset{\displaystyle O}{\parallel}}{C}-OH$		Carboxylic acid
—OR	Ether	$-\overset{\overset{\displaystyle O}{\parallel}}{C}-OR$		Ester
$-NH_2$	Amine	$-\overset{\overset{\displaystyle O}{\parallel}}{C}-NH_2$		Amide

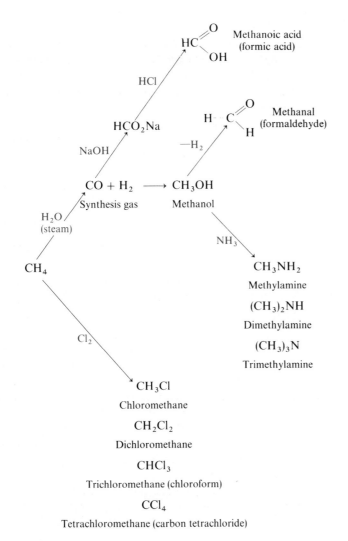

FIGURE 23.31
Some Products Made from Methane and Methanol.

FIGURE 23.32
Some Products Made from
Ethene and Propene.

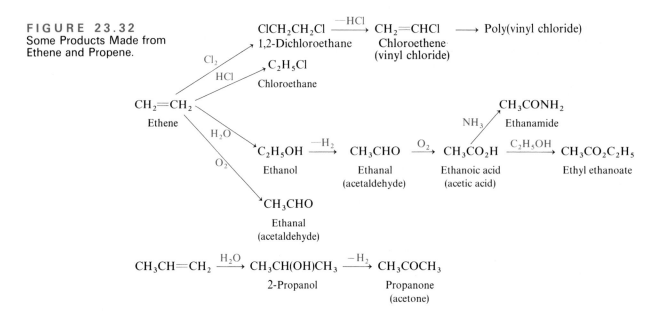

An **addition reaction** is a reaction in which two molecules combine to give a third molecule.

An **alkane** is a hydrocarbon in which all the carbon–carbon bonds are single bonds.

An **alkene** is a hydrocarbon that has one or more carbon–carbon double bonds.

An **alkyne** is a hydrocarbon that has one or more carbon–carbon triple bonds.

An **arene** is a hydrocarbon derived from benzene either by substituting its hydrogen atoms with alkyl groups or by fusing two or more benzene rings.

A *cis* **isomer** is a structural isomer of an alkene in which substituents at each end of the double bond are on the same side of the double bond.

Conformations are the different arrangements of the atoms and the groups in a molecule that are possible as a consequence of rotation about single bonds.

Cracking is a process in which alkanes are converted to shorter-chain alkanes and alkenes by heating.

Cycloalkanes are alkanes in which the carbon chain forms a ring.

A **dehydrogenation reaction** is one in which a hydrogen molecule is eliminated from a molecule.

The **eclipsed conformation** of a molecule is one in which the atoms or groups attached to the two ends of a single bond are in positions directly opposite each other.

An **elimination reaction** is the opposite of an addition reaction. It involves the decomposition of a molecule into two

molecules, one of which, normally the smaller one, is said to be eliminated.

Fermentation is a reaction catalyzed by certain enzymes in which sugars and carbohydrates are broken down into ethanol and carbon dioxide.

Free rotation is a property of a single bond that allows the atoms and the groups attached at one end of the bond to rotate freely with respect to those at the other end.

A **functional group** is an atom or group of atoms that replaces a hydrogen atom of a hydrocarbon; for example, the hydroxyl group, OH, and the carboxyl group, CO_2H, are functional groups.

Geometric isomers are isomers in which all the atoms are connected in the same way and that differ only in how identical groups are arranged in space.

Hydrocarbons are compounds containing only carbon and hydrogen.

Organic chemistry is the chemistry of carbon and its compounds.

Organic compounds are the compounds of carbon. But there are a few compounds such as carbon dioxide that are normally regarded as inorganic compounds.

A **saturated hydrocarbon** is one in which every carbon atom is bonded to four other atoms by single bonds; an alkane or cycloalkane.

A **staggered conformation** is one in which the atoms or the groups attached to one end of a single bond are in positions which lie between those at the other end.

Structural isomers have the same molecular formula but they have different structures. In other words, they have the same number of different kinds of atoms but the atoms are arranged in different ways.

Synthesis gas is a mixture of hydrogen and carbon monoxide that is obtained by heating methane with steam at 700–800 °C in the presence of a nickel catalyst.

Thermal cracking is a process in which alkanes are broken down on heating to give shorter-chain alkanes, alkenes, and hydrogen.

A *trans* **isomer** is an isomer of an alkene in which two substituent groups are on opposite sides of the double bond.

An **unsaturated hydrocarbon** is a hydrocarbon containing one or more double or triple bonds; an alkene or alkyne.

PROBLEMS *

Names, Formulas and Isomers

1. In naming hydrocarbons, and other organic compounds, the last part of the name gives information about the longest continuous chain of carbon atoms in the molecular structure. How many carbon atoms are in the longest continuous chain of carbon atoms in each of the following?

   (a) Butane

   (b) 2-Methylpropane

   (c) 2,2-Dimethyloctane

   (d) 2,3-Dimethylpentane

   (e) 2,2,5,5-Tetramethylhexane

2. Give the molecular formula of each of the following:

   (a) An alkane with 8 carbon atoms

   (b) An alkene with 6 carbon atoms and one double bond

   (c) An alkyne with 5 carbon atoms and one triple bond

   (d) A cycloalkane containing a six-membered ring of carbon atoms

3. Classify each of the following as an alkane, an alkene, or an alkyne, draw the Lewis structure, and give the molecular formula:

   (a) Methane   (b) Ethene

   (c) Propyne   (d) Cyclobutane

   (e) Cyclopropene

4. Which of the following names do not conform to the IUPAC rules for naming compounds? Rename those that do not conform.

   (a) 2-Ethylbutane   (b) 3,3-Dimethylbutane

   (c) 1-Ethylpropane   (d) 2,2-Dimethylpropane

   (e) 1,2-Dimethylpropane

5. Draw the structures and give the IUPAC names of each of the isomers with the molecular formula $C_2H_2Cl_2$.

6. Explain why 2-butene exists as *cis* and *trans* isomers, while 2-butyne does not.

7. Draw the structural formulas and give the IUPAC names of each of the nine isomers of heptane.

8. Draw the structures and give the IUPAC names of each of the six isomers with the molecular formula C_4H_8.

9. Draw a structural formula for each of the following:

   (a) 1,3-Cyclopentadiene

   (b) 1,3-Cyclohexadiene

   (c) 3-Chloro-2, 4-hexadiene

10. What is the difference between each of the following?

   (a) A saturated and an unsaturated hydrocarbon

   (b) A straight-chain and a branched-chain alkane

   (c) An aliphatic hydrocarbon and an aromatic hydrocarbon

   (d) A molecular conformation and an isomer

11. Give the IUPAC name of each of the following:

   (a)
   $$CH_3-\overset{\overset{\displaystyle H}{|}}{\underset{\underset{\displaystyle CH_3}{|}}{C}}-CH_3$$

   (b)
   $$CH_3\overset{\overset{\displaystyle CH_3}{|}}{C}HCH\underset{\underset{\displaystyle CH_2CH_2CH_2CH_3}{|}}{}CH_2CH_2CH_3$$

   (c) $CH_3-CH=CH_2$

   (d) $CH_3CH=C(CH_3)_2$

   (e) $CH_3CH_2CH_2CH_2CH=CHCH_3$

   (f)
   $$H_3C-\overset{\overset{\displaystyle H}{|}}{C}-\overset{\overset{\displaystyle H}{|}}{\underset{\underset{\displaystyle H_2C-C(CH_3)_2}{|}}{C}}-CH_3$$

   (g) $CH_2=CHCH_2CH=CHCH_3$

12. Give the systematic name of each of the following:

    (a) $CH_3CH_2CH{=}CH_2$

    (b) $CH_3CH_2C{\equiv}CH$

    (c) $(CH_3)_2C{=}CHCH_3$

    (d) $CH_2{=}CHCH_2CH_2CH{=}CH_2$

    (e)
$$\begin{array}{ccc} CH_3 & & CH_3 \\ & C{=}C & \\ H & & H \end{array}$$

    (f) $CH_3CH{=}CHCH{=}CH_2$

13. Which of the compounds with the following molecular formulas is *not* an isomer of heptane?

    (a) 2-Methylhexane   **(b)** 2,2-Dimethylpentane

    (c) 2,3-Dimethylbutane   **(d)** 2,3-Dimethylpentane

14. Draw the structures of as many isomers of formula C_5H_{10} as you can, and give the IUPAC name of each.

15. Write structural formulas for each of the following:

    (a) 2,2,4-Trimethylpentane

    (b) 1,3-Cyclobutadiene

    (c) 4,4-Dimethyl-1-pentyne

    (d) *Trans*-2-pentene

    (e) 1,3-Butadiene

    (f) Methyl-2,3-diethylbenzene

    (g) 1,3,5-Trichlorobenzene

    (h) *Cis*-3,4-Dichloro-2-heptene

16. Write structural formulas for each of the following:

    (a) 1,2,2-Trichlorobutane

    (b) 1,3,5-Tribromobenzene

    (c) 1,2-Dichlorocyclohexane

17. Complete combustion of a sample of a hydrocarbon gave 0.318 g of CO_2 and 0.163 g of H_2O. The mass of the hydrocarbon that occupied a 250-mL flask at 100 °C and 1 atm pressure was 0.4743 g. Determine the empirical and molecular formulas of the hydrocarbon. Write structural formulas for, and name, all the isomeric hydrocarbons with this molecular formula.

18. Complete combustion of 0.1540 g of a hydrocarbon gave 0.4832 g of CO_2. The mass of hydrocarbon that filled a 250-mL flask at 100 °C and 1 atm pressure was 0.4580 g. Determine the empirical and molecular formulas of the hydrocarbon. Write structural formulas for, and name, at least four isomers that have this molecular formula. How would you decide which of these isomers is the hydrocarbon under investigation?

19. Identify and name the *functional groups* in each of the following molecules:

    (a) CH_3CHO

    (b) $CH_3CH_2{-}\underset{\underset{O}{\|}}{C}{-}CH_3$

    (c) $CH_3CH_2{-}\underset{\underset{O}{\|}}{C}{-}OH$

    (d) $(CH_3)_2\underset{\underset{H}{|}}{C}{-}\underset{\underset{O}{\|}}{C}{-}NH_2$

    (e) $(CH_3)_3CNH_2$

    (f) $CH_3CH_2\underset{\underset{OH}{|}}{CH}CH_2CH_3$

    (g) $CH_3\underset{\underset{NH_2}{|}}{CH}{-}\underset{\underset{O}{\|}}{C}{-}OH$

    (h)

20. Name each of the compounds in Problem 19.

21. Name each of the compounds with the following molecular formulas:

    (a) CH_3OH **(b)** CH_3CH_2OH

    (c) HCO_2H **(d)** $CH_3CH_2\underset{\underset{OH}{|}}{CH}CH_3$

    (e) CH_3CH_2CHO **(f)** $HOCH_2CH_2OH$

    (g) CH_3COCH_3 **(h)** $(CH_3)_2CH{-}CHO$

    (i) $ClCH_2\underset{\underset{Br}{|}}{CH}CH_3$

22. Write structural formulas for each of the following compounds:

    (a) 2-Methyl-2-butanol   **(b)** 5-Methyl-2-hexanol

    (c) 2-Chloro-1-propanol   **(d)** 3-Chlorocyclohexanol

    (e) 1,3-Propanediol

23. Write structural formulas for each of the following aldehydes and ketones:

    (a) Butanal   **(b)** 2-Pentanone

    (c) 3-Methyl-2-butanone   **(d)** 3,3-Dimethylhexanal

    (e) Acetone   **(f)** Formaldehyde

24. Write the structural formulas and name of each of the following noncyclic compounds containing three carbon atoms:

    (a) A primary alcohol   **(b)** A secondary alcohol

    (c) An aldehyde   **(d)** A ketone

    (e) A carboxylic acid   **(f)** An ether

    (g) An amide   **(h)** A primary amine

    (i) A secondary amine   **(j)** An alkyne

25. Classify each of the following alcohols as a primary, a secondary, or a tertiary alcohol, and give the systematic name:

(a) $(CH_3)_2C(OH)CH_2CH_3$

(b) $CH_3CH(OH)CH_2CH_3$ (c) $(CH_3)_3CCH_2OH$

(d) $(CH_3)_2CH(OH)$ (e) $(CH_3CH_2)_2CH(OH)$

26. Name the ester that could be obtained from each of the following reactions, and write its structure:

(a) Methanoic acid and methanol

(b) Butanoic acid and ethanol

(c) Ethanoic acid and butanol

(d) Propionic acid and 2-propanol

(e) One mole of phosphoric acid and 1 mol of ethanol

Reactions

27. Give an example, by writing an appropriate balanced equation, of each of the following reaction types:

(a) Combustion of an alkane

(b) An elimination reaction of an alkane

(c) An elimination reaction of a primary alcohol

(d) An addition reaction of an alkene

28. With the aid of an appropriate example in each case, give a short definition or description of each of the following terms:

(a) A cracking reaction

(b) A polymerization reaction

(c) An addition reaction

(d) An elimination reaction

(e) A substitution reaction

(f) A free-radical mechanism

29. Write balanced equations for the reactions of ethanol with each of the following, and name the products:

(a) $OH^-(aq)$ (b) $H_3O^+(aq)$

(c) Sodium   (d) $H_2SO_4(conc)$

30. Explain why alkenes and alkynes are very reactive, while the alkanes are rather unreactive compounds.

31. Complete and balance each of the following equations, and name the products:

(a) C_6H_{10} $+ Br_2 \longrightarrow$
        Cyclohexene

(b) $C_3H_8 + O_2 \xrightarrow{\text{Heat}}$

(c) $C_2H_5OH_2 \xrightarrow[\text{Catalyst}]{\text{Heat}}$

(d) $CH_3OH + NH_3 \xrightarrow{\text{Heat}}$

32. Describe a simple chemical test to distinguish between each of the following pairs of gases:

(a) Ethane and ethyne

(b) Carbon dioxide and propane

(c) Ethene and propane

In each case describe what is observed.

33. The following aldehydes and ketones may each be prepared by the oxidation of a suitable alcohol. In each case, name the alcohol and draw its structural formula.

(a) Ethanal   (b) Propanone

(c) 2-Methylpropanal   (d) 2-Pentanone

34. Draw the structures for, and name, each of the esters formed from the reactions of the following pairs of substances:

(a) Ethanoic acid and 2-propanol

(b) Ethanoic acid and 1-butanol

(c) Benzoic acid and methanol

(d) Methanoic acid and ethanol

35. Describe, in terms of suitable balanced equations, each of the following preparations:

(a) Synthesis gas

(b) Methanol from synthesis gas

(c) Methanal from methanol

(d) Methylamine from methanol

36. Complete and balance each of the following equations:

(a) $C_2H_5OH(l) + O_2(g) \xrightarrow{\text{Catalyst}}$

(b) $CH_3(CH_2)_2OH(l) + NaOH(s) \longrightarrow$

(c) $CH_3OH(l) + Na(s) \longrightarrow$

(d) $C_2H_5OH(l) + H_2SO_4(concn) \longrightarrow$

(e) $CH_3CH{=}CH_2(g) + H_2O(g) \xrightarrow[\text{Catalyst}]{\text{Heat, Pressure}}$

37. Complete and balance each of the following equations:

(a) $CH_3CH_2CHO + K_2Cr_2O_7(aq) + H_3O^+(aq) \xrightarrow{\text{Heat}}$

(b) $CH_3CH_2CO_2H + (CH_3)_2CHOH \xrightarrow{\text{Heat}}$

(c) $CH_3OH + NH_3 \xrightarrow{\text{Heat}}$

(d) $CH_3NH_2 + H_2O(l)$ $\xrightarrow{\text{Room temperature}}$

(e) $CH_3NH_2 + HCl(aq) \longrightarrow$

(f) $CH_3CO_2H + C_2H_5NH_2$ $\xrightarrow{\text{Heat}}$

38. Write the overall balanced equation for the production of methanol from methane and water, via synthesis gas.

39. Write the balanced equation for the reaction by which the ester methyl methanoate (methyl formate) can be obtained from the appropriate carboxylic acid and alcohol.

40. Give the structural formula and name the organic product of each of the following reactions:

(a) $CO(g) + 2H_2(g)$ $\xrightarrow{\text{ZnO catalyst}}$

(b) 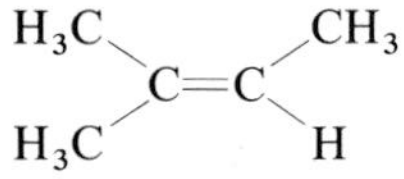 $\begin{array}{c} H_3C \\ H_3C \end{array} C=C \begin{array}{c} CH_3 \\ H \end{array} + Br_2 \longrightarrow$

(c) 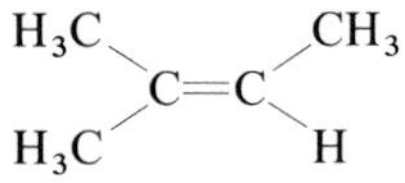 $\begin{array}{c} H_3C \\ H_3C \end{array} C=C \begin{array}{c} CH_3 \\ H \end{array} + H_2 \longrightarrow$

(d) $CH_3OH + C_2H_5OH$ $\xrightarrow{H_2SO_4}$

(e) $CH_3CH_2CH_2OH + Na \longrightarrow$

41. How is vinyl chloride (chloroethene) prepared from ethyne and chlorine?

42. Describe by suitable equations how ethene may be converted to ethyl ethanoate.

43. Each of the compounds with the following structures may be synthesized from an alkene, or an alkyne, and another reactant. In each case give the name and structure of the alkene and alkyne, and the other reactant:

(a) $CH_3CHOHCH_3$ **(b)** $CH_3CBr_2CBr_2CH_3$
(c) $CH_3CH=CHCH_3$ **(d)** $CH_3C(Br)=CH_2$

44. How would you prepare each of the following?

(a) 1,2-Dichloroethane from ethene

(b) Propanoic acid from 1-propanol

(c) Ethyl ethanoate from ethanal

Lewis Structures, Molecular Geometry, and Bonding

45. In terms of the AX_nE_m nomenclature of the VSEPR theory, classify each of the following molecules and ions in terms of the geometry around the starred atom, and draw a diagram to show the geometry:

(a) $\overset{*}{C}H_3OH$ **(b)** $CH_3\overset{*}{O}CH_3$ **(c)** $H\overset{*}{C}O_2H$

* The asterisk denotes the more difficult problems.

(d) $H\overset{*}{C}O_2^-$ **(e)** $CH_3\overset{*}{C}O_2H$ **(f)** $(CH_3)_2\overset{*}{C}O$
(g) $CH_3\overset{*}{N}H_2$ **(h)** $(CH_3)_4\overset{*}{N}^+$ **(i)** $(CH_3)_3\overset{*}{N}$

*__46.__ Suggest an explanation for the observed bond angles in tetrafluoroethene and why they differ from the observed bond angles in ethene.

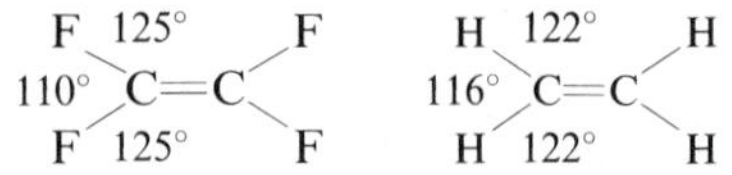

$$\begin{array}{c} F \quad 125° \quad F \\ 110° \; C=C \\ F \quad 125° \quad F \end{array} \qquad \begin{array}{c} H \quad 122° \quad H \\ 116° \; C=C \\ H \quad 122° \quad H \end{array}$$

47. Explain what is meant by a *molecular conformation.* Draw diagrams to illustrate two different conformations of **(a)** ethane and **(b)** cyclohexane.

48. Explain each of the following:

(a) Why all five atoms of the amide group, $-CONH_2$, are in the same plane

(b) Why all four atoms of the carboxylate group, $-COOH$ are in the same plane

49. Write Lewis structures for each of the following, including all the possible resonance structures, where appropriate:

(a) CO **(b)** CO_3^{2-}
(c) HCO_2^- **(d)** HCO_2H
(e) $HCONH_2$ **(f)** C_6H_6 (benzene)

50. Name the set of hybrid orbitals that can be used to approximately describe the bonds formed by the starred atoms in each of the following:

(a) $\overset{*}{C}H_4$ **(b)** $H_2\overset{*}{C}=\overset{*}{C}H_2$
(c) $H\overset{*}{C}\equiv\overset{*}{C}H$ **(d)** $H_3C-\overset{*}{O}-CH_3$

Properties of Organic Compounds

51. Explain why the structural isomers dimethyl ether (bp = -23 °C) and ethanol (bp = 78.3 °C) differ significantly in their boiling points? What simple chemical test could be used to distinguish ethanol from dimethyl ether?

52. In terms of intermolecular forces, account for the differences in the boiling points of the following substances:

Water, 100 °C   Methanol, 65 °C   Ethanol, 78 °C

Methane, -162 °C   Ethane, -89 °C

Dimethyl ether, -25 °C

53. The representative alkane in gasoline is nonane, C_9H_{20}, which has a density of 0.72 g cm^{-3} and a standard enthalpy of formation of -275 kJ mol^{-1}. Ethanol has a density of 0.79 g cm^{-3} and $\Delta H_f^\circ = -277$ kJ mol^{-1}. Compare the energy obtained under standard conditions from burning gasoline with that from burning ethanol, on the basis of (a) equal volumes and (b) equal masses. If gasoline costs 30¢ per liter and ethanol costs 50¢ per liter, which of these fuels is the more economical?

54. (a) What is the pH of a 0.10M aqueous solution of methylamine (pK_a = 3.34)?

(b) What is the pK_a of the methylammonium ion?

(c) What is the pH of a solution obtained by mixing 24.5 mL of 0.090M aqueous methylamine with 10.0 mL of 0.100M HCl(aq)?

Miscellaneous

55. A hydrocarbon contains 82.6 mass % carbon, and 0.470 g of it filled a 100 mL flask at 25 °C and a pressure of 750 mm Hg. What are the empirical and molecular formulas of the hydrocarbon? Can you write a unique structural formula for the hydrocarbon?

56. A small quantity of a liquid hydrocarbon containing 85.6 mass % carbon was placed in a 250 mL flask and heated to 100 °C on a boiling water bath. Under these conditions, all the hydrocarbon vaporized, and excess vapor escaped from the flask into the atmosphere. On cooling to room temperature, the amount of liquid in the flask was found to be 0.687 g. Atmospheric pressure was 760 mm Hg. What is the molar mass of the hydrocarbon? What is its molecular formula? Suggest a possible structure.

57. A sample of 0.200 g of a hydrocarbon containing 85.71 mass % carbon occupies a volume of 95.3 mL at 0.921 atm pressure and 27 °C. Draw possible structures for the hydrocarbon that are consistent with the molecular formula, and name them.

*58. An unsaturated hydrocarbon containing 88.8 mass % carbon was allowed to react with excess hydrogen over a palladium catalyst. The amount of H_2 consumed in reaction with 1.00 g of the hydrocarbon was 906 mL, measured at 25 °C and 1 atm pressure. In a molar mass determination, 0.1200 g of the hydrocarbon occupied a volume of 67.9 mL at 100 °C and 1.00 atm.

(a) What are the empirical and molecular formulas of the hydrocarbon?

(b) Write structures for the isomers that have this molecular formula.

(c) Discuss how the actual structure of the unsaturated hydrocarbon could be found by studying the products of its reaction with Br_2.

*59. A gaseous mixture of methane and an aliphatic alkene of volume 1.00 L has a mass of 0.882 g at 25 °C and 744 mm Hg pressure. When burned in excess oxygen it gave 2.641 g of carbon dioxide and 1.442 g of water. Identify the aliphatic alkene and calculate the mass percent composition of the mixture.

60. A compound was found on analysis and molar mass determination to have the molecular formula $C_5H_{12}O$. Oxidation converts it to another compound with the molecular formula $C_5H_{10}O$, which gives the characteristic reactions of a ketone. Suggest two or more reasonable structural formulas for the original compound.

*61. A mass of 156 mg of a compound containing carbon, hydrogen, and oxygen was burned in excess oxygen to give 512 mg of CO_2 and 209 mg of H_2O. At 100 °C, 156 mg of the compound occupied a volume of 93.3 mL at a pressure of 882 torr. Suggest a suitable structure for the compound, identify its functional group, and suggest tests that would confirm its presence in the compound.

*62. An organic compound contains 52.1% C, 13.1% H, and 34.8% O, by mass. At 100 °C and 1.00 atm pressure, 0.230 g of the compound occupies a volume of 153 mL. Calculate the empirical and molecular formulas of the compound and draw the possible structures. The compound was found to react with sodium to give hydrogen and a sodium salt. Write the balanced equation for this reaction, and name the compound and its salt. What volume of hydrogen, at 25 °C and a pressure of 730 torr, would result from the reaction of 0.250 g of the compound with excess sodium?

*63. Analysis of a gaseous organic compound gave 31.86% C, 5.31% H, and 62.8% Cl, by mass. In a molar mass determination, 0.113 g of the compound was found to occupy a volume of 23.75 mL at 15 °C and a pressure of 768.7 torr. Deduce the molar mass and the molecular formula of the compound, and suggest possible structural formulas.

*64. (a) A mass of 1.00 g of a primary alcohol was completely oxidized to a carboxylic acid, which required 83.3 mL of 0.20M NaOH(aq) to reach the equivalence point in a titration. Calculate the molar mass of the alcohol and deduce its molecular formula.

(b) Write the structural formula of the primary alcohol in part (a), and give its systematic name. What are the possible products of its oxidation?

(c) Name and give the structures of two other alcohols that are isomeric with the above alcohol, and discuss the possible products of their oxidation.

65. Write structures for the products of the oxidation of each of the following with acidic $KMnO_4$(aq):

(a) Butanol   **(b)** Propanal

(c) 2-Methyl-1-butanol   **(d)** 3-Methyl-2-butanol

(e) 3,3-Dimethylbutanal

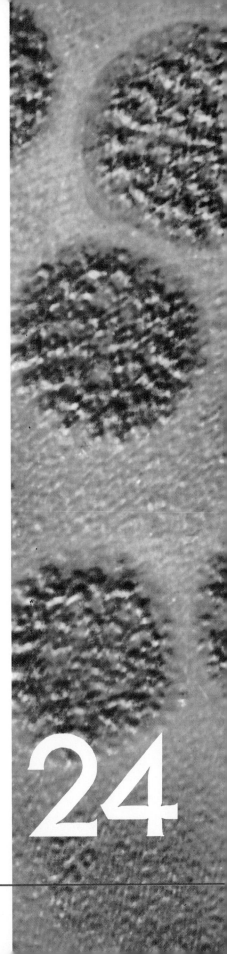

Polymers: Synthetic and Natural

Synthetic Polymers
Addition Polymers
Condensation Polymers
Biopolymers
Carbohydrates
Proteins
Nucleic Acids

24

We cannot go through a single day without using a dozen or more materials based on *synthetic polymers*. The materials that we commonly call **plastics**, which are used for cups and dishes, combs, telephones, pens, bags, pipes, paints, synthetic fibers, and kitchen counter tops, are all composed of synthetic polymers. The names of many of these materials are well known to most of us: polyethylene, polystyrene, polyurethane, Teflon, Formica, Saran, and so on. **Polymers** are very large molecules that are formed by the combination of many relatively small molecules called **monomers**. In synthetic polymers there is often only one type of monomer unit—or at the most a small number of different monomer units. In Chapter 6 we mentioned polyethylene, which is a polymer of composition $(CH_2)_n$ formed from ethene. Nylon is a polymer formed from two different monomers, hexanedioic acid and diaminohexane. There are also very many natural polymers, often called *biopolymers*, such as carbohydrates and proteins, that often contain many different monomer units. Proteins are made from the 20 natural amino acids. Silk, cotton, wool, and rubber are all polymers. The cellulose that gives strength to a tree trunk is a polymer.

Polymers may be made from both inorganic and organic molecules. We have previously mentioned a number of inorganic polymers such as $(SO_3)_n$ and metaphosphoric acid, $(HPO_3)_n$ (Chapter 8), and the silicates and silicones (Chapter 22). Polymer molecules may have many different forms; in particular, they may be chains, like polyethylene, or sheets, like talc and mica, or three-dimensional giant molecules, like quartz. But most synthetic polymers are long-chain organic molecules that typically contain thousands of monomer units. Such molecules have very high masses, of the order of 100 000 u or more. They are also often called **macromolecules**.

Although today polymers are familiar to almost everybody, only in the past 40 years have chemists learned how to synthesize them. Their enormous importance at the present time can be judged from the fact that half of the professional organic chemists employed by industry in the United States are engaged in research and development related to polymers. But nature has been using polymers, often far more complex than those synthesized by chemists, since the beginning of life. In this chapter we consider first a few examples of synthetic polymers and then some naturally occurring polymers of organic compounds.

24.1
Synthetic Polymers

The two main types of synthetic polymers are addition and condensation polymers.

ADDITION POLYMERS

Addition polymerization requires monomers which have a double bond; during polymerization the double bond becomes a single bond, releasing electrons to form new bonds (Figure 24.1). Some examples of addition polymers prepared by the successive addition of monomer units are given in Table 24.1.

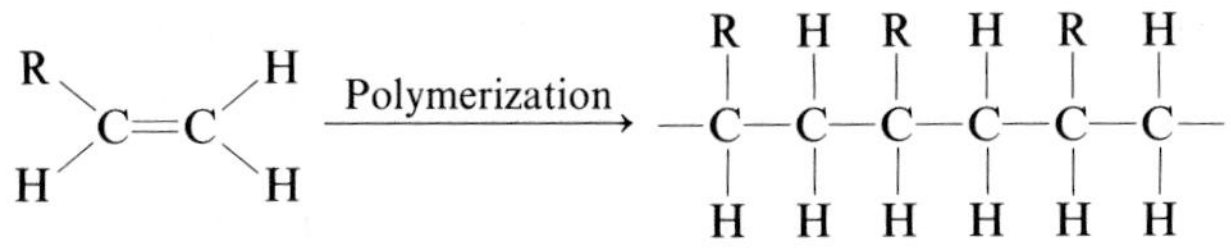

FIGURE 24.1
The Addition Polymerization
of an Alkene.

TABLE 24.1
Some Addition Polymers Produced from Substituted Ethenes

Monomer	Polymer	Typical Uses
$CH_2{=}CH_2$ Ethene	$-(CH_2{-}CH_2)_n$ Polyethylene	Containers, pipes, bags, toys, wire insulation
$CH_2{=}CHCH_3$ Propene	$-(CH_2{-}CH)_n$ $\quad\quad CH_3$ Polypropylene	Fibers for carpets, artificial turf, rope, fishing nets
$CH_2{=}CHCl$ Chloroethene (vinyl chloride)	$-(CH_2{-}CH)_n$ $\quad\quad Cl$ Poly(vinyl chloride) (PVC)	Garden hoses, floor tiles, plumbing, records, laboratory tubing
$CH_2{=}CHCN$ Propenenitrile (acrylonitrile)	$-(CH_2{-}CH)_n$ $\quad\quad CN$ Polyacrylonitrile (Orlon, Acrilan)	Fibers for cloth, carpets, upholstery
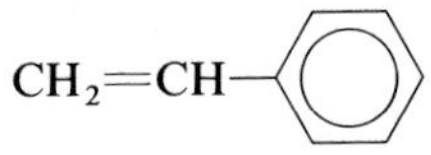 $CH_2{=}CH{-}\bigcirc$ Styrene	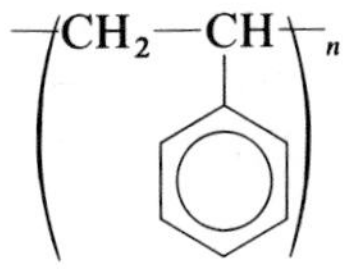 $-(CH_2{-}CH)_n$ $\quad\quad\bigcirc$ Polystyrene	Styrofoam, hot-drink cups, insulation
$CF_2{=}CF_2$ Tetrafluoroethene	$-(CF_2{-}CF_2)_n$ Teflon	Nonstick coating for kitchen utensils
$\quad\quad CH_3$ $CH_2{=}C{-}CO_2CH_3$ Methylmethacrylate	$\quad\quad CH_3$ $-(CH_2{-}C)_n$ $\quad\quad CO_2CH_3$ Polymethylmethacrylate	Plexiglas, Lucite, headlamp lenses, sunglasses, aircraft windows

POLYETHYLENE Polyethylene is formed by joining a large number of ethene molecules to form a long hydrocarbon chain:

$$\cdots CH_2{=}CH_2 + CH_2{=}CH_2 + CH_2{=}CH_2 \cdots \longrightarrow -CH_2{-}CH_2{-}CH_2{-}CH_2{-}CH_2{-}CH_2{-}$$

Monomer $\qquad\qquad\qquad\qquad\qquad\qquad\qquad$ Addition polymer

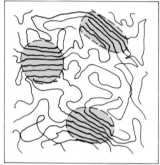

Crystalline polyethylene

Crystalline and amorphous
regions in polyethylene

FIGURE 24.2
Structure of Polyethylene.

In a polymer such as polyethylene the polymer chains are like so many pieces of spaghetti, some long and some short, coiling and twisting in all directions. This material cannot be expected to form regular crystals in the solid state. However, experiment has shown the existence of crystalline regions in the solid, called **crystallites**.

Ethene molecules are not easily polymerized, even at high temperature and pressure; the activation energy for the polymerization is high. The reaction must be initiated by introducing a reactive free radical (Chapters 18 and 19), usually provided by thermal decomposition of an organic peroxide, R—O—O—R, to give two free radicals, R—O·. The first step in the polymerization is then

$$R—O· + CH_2{=}CH_2 \longrightarrow R—O—CH_2—CH_2·$$

The product is a free radical that can react with another ethene molecule to give another free radical,

$$R—O—CH_2—CH_2· + CH_2{=}CH_2 \longrightarrow R—O—CH_2—CH_2—CH_2—CH_2·$$

and so on. The chain can continue to grow in this way until a termination reaction occurs, in which, for example, two chains add to each other or another free radical adds to the chain. The length of the chains and the degree to which the chains are branched can be controlled by varying the conditions of the polymerization and using different catalysts. Thus polyethylenes with a variety of different properties can be obtained. They may be viscous liquids or solids of varying degrees of hardness. The solid polymers are generally amorphous, but they may be at least partly crystalline (see Figure 24.2).

The polymerization of ethene can be written as

$$nCH_2{=}CH_2 \longrightarrow {+}CH_2CH_2{+}_n$$

The formula ${+}CH_2CH_2{+}_n$ means that the group in brackets is repeated n times; the value of n is usually several thousand or larger and varies from one chain to another. If we measure the chain length or the molecular mass of a polymer, we obtain an average value. The chains must be terminated by some other group or groups, but since there is only one such group for several thousand monomer units, these end groups have no appreciable effect on the composition or the properties.

Polyethylene with a molar mass of less than about 300 000 is called low density polyethylene and is used to make squeeze bottles, bags for food, and other flexible items. High density polyethylene which can have a relative molar mass of up to 3 000 000 is used to make more rigid items (see Figure 24.3).

═ *Example 24.1* ═══════════════════════════════════════

MOLAR MASS OF A POLYMER

What is the molar mass of polystyrene if a single molecule contains 2000 monomer units? Styrene has the molecular formula $C_6H_5—CH{=}CH_2$.

FIGURE 24.3
A Selection of Articles
Made from Polyethylene.

Solution

The molar mass of styrene is 104.2 g mol^{-1}. The polymer molecule has the molecular formula $(C_8H_8)_{2000}$. The molar mass is therefore

$$(2000)(104.2) \text{ g mol}^{-1} = 2.084 \times 10^5 \text{ g mol}^{-1}$$

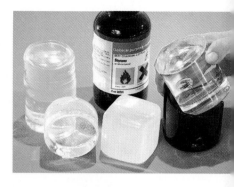

Blocks of solid transparent polystyrene that were formed by spontaneous polymerization in bottles of styrene that had been kept for a long time.

NATURAL AND SYNTHETIC RUBBER   Natural rubber is an addition polymer of 2-methyl-1,3-butadiene (isoprene) with an average chain length of about 5000 monomer units:

$$n \text{ CH}_2{=}\text{CH}{-}\overset{\overset{\displaystyle \text{CH}_3}{|}}{\text{C}}{=}\text{CH}_2 \longrightarrow {+}\text{CH}_2{-}\text{CH}{=}\overset{\overset{\displaystyle \text{CH}_3}{|}}{\text{C}}{-}\text{CH}_2{+}_n$$

Rubber was discovered by the native peoples of Central and South America who used rubber balls in their games. They obtained rubber by collecting the milky white sap called latex from rubber trees and allowing it to harden in the air. It was first taken to Europe from Haiti by Columbus in 1496. Much later it was named rubber by Joseph Priestley (Chapter 3) who found that it could be used to rub out pencil marks. However, natural rubber becomes soft and sticky in warm weather and brittle when it is cold so it remained a curiosity without any important practical applications until in 1839 the American inventor Charles Goodyear discovered the process which he called *vulcanization*. He found that if latex is heated with sulfur a rubber-like material is obtained which retains its elasticity and flexibility over a wide temperature range. In the vulcanization process the hydrocarbon chains become linked together by C—S—S—C bonds. The polymer is then said to be **cross-linked** (Figure 24.4). In a polymer the chains are normally tangled up with each other, rather like a bowl of spaghetti. When solid rubber is stretched, the chains straighten out to some extent. When the tension on the rubber is released, the chains tend to coil up again (see Experiment 24.1). The straighter, more ordered chains have a lower entropy than the coiled up, more disordered chains. When rubber is cross-linked by sulfur chains, the extent to which the chains can be straightened is limited, and they have a greater tendency to resume their original shape when the tension is released. Thus vulcanized rubber is stronger, harder, less sticky, and more "rubbery" than natural rubber.

As the demand for rubber grew, huge plantations of rubber trees were established in Sri Lanka, Malaysia, and Indonesia where they grew even better than in their native habitat. These sources of natural rubber, however, were cut off from Europe and North America during World War II and intensive efforts were then made to develop synthetic rubbers. An example of a synthetic rubber is polybutadiene which is made from butadiene:

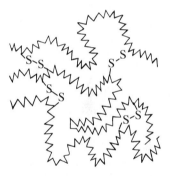

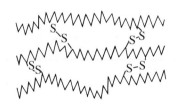

FIGURE 24.4
Vulcanized Rubber.

The hydrocarbon chains are held together by cross-linking chains of sulfur atoms. When tension is applied, the chains can straighten out, but they cannot slip past each other because of the polysulfide bridges. Thus rubber can be stretched only to a limited extent. When the tension is removed, the chains tend to coil up again, and the rubber resumes its original shape.

$$n(\text{CH}_2{=}\text{CH}{-}\text{CH}{=}\text{CH}_2) \longrightarrow {+}\text{CH}_2{-}\text{CH}{=}\text{CH}{-}\text{CH}_2{+}_n$$
$$\text{Butadiene} \qquad\qquad\qquad \text{Polybutadiene}$$

Materials like rubber that return to their original shape after stretching are called **elastomers**. Many different elastomers have been developed for special applications. For example, the elastomer obtained from chloroprene, in which the methyl group of isoprene is replaced by a chlorine atom, is much more resistant to gasoline and oil than natural rubber.

Experiment 24.1

Rubber

A long sharp-pointed needle can be pushed slowly through an inflated rubber balloon without causing it to burst because the long flexible rubber molecules move around the hole and seal it.

Example 24.2

STRUCTURE OF A SYNTHETIC RUBBER

Draw the structure of the addition polymer (neoprene rubber) formed from 2-chloro-1,3-butadiene (chloroprene).

Solution

The structure of 2-chloro-1,3-butadiene is

$$CH_2=\overset{\overset{\displaystyle Cl}{|}}{C}-CH=CH_2$$

so the polymer has the structure

$$\cdots-CH_2\overset{Cl}{\underset{}{\diagdown}}C=C\overset{H}{\underset{CH_2-CH_2}{\diagup}}\overset{Cl}{\diagdown}C=C\overset{H}{\underset{CH_2-CH_2}{\diagup}}\overset{Cl}{\diagdown}C=C\overset{H}{\underset{CH_2}{\diagup}}-\cdots$$

Neoprene rubber

Most synthetic elastomers, however, are made from two alkene monomers. Polymers of this type are called **copolymers**. For example, Saran, the film used to cover food in the kitchen, is a copolymer of 1,1-dichloroethene and chloroethene (vinyl chloride):

$$Cl_2C=CH_2 + CH_2=CHCl \longrightarrow \left(CH_2-\overset{\overset{\displaystyle Cl}{|}}{\underset{\underset{\displaystyle Cl}{|}}{C}}-CH_2-\overset{\overset{\displaystyle Cl}{|}}{CH}\right)_n$$

CONDENSATION POLYMERS

Addition polymers contain all the atoms of the original monomers. Another important class of polymers are the **condensation polymers** that are formed through condensation reactions in which monomers are joined into polymer chains by the elimination of small molecules such as water.

In Chapter 23 we saw that when a carboxylic acid and an amine are heated together, a condensation reaction occurs to give an amide and water:

$$\underset{\text{Carboxylic acid}}{R—\overset{\displaystyle O}{\overset{\|}{C}}—OH} + \underset{\text{Amine}}{H—\overset{\displaystyle H}{\overset{|}{N}}—R'} \longrightarrow \underset{\text{Amide}}{R—\overset{\displaystyle O}{\overset{\|}{C}}—\overset{\displaystyle H}{\overset{|}{N}}—R'} + \underset{\text{Water}}{H_2O}$$

We also saw that an ester is formed from a carboxylic acid and an alcohol in a condensation reaction:

$$R—CO_2H + R'—OH \longrightarrow R—\overset{\displaystyle O}{\overset{\|}{C}}—OR' + H_2O$$

Polyamides and *polyesters* are among the most important condensation polymers.

NYLON   Nylon-66 is a **polyamide** that is prepared by heating 1,6-hexanedioic acid (adipic acid) and 1,6-hexanediamine at 270 °C under pressure:

$$+ \underset{\text{Hexanedioic acid}}{HO—\overset{\displaystyle O}{\overset{\|}{C}}—(CH_2)_4—\overset{\displaystyle O}{\overset{\|}{C}}—OH} + \underset{\text{1,6-Hexanediamine}}{H—\overset{\displaystyle H}{\overset{|}{N}}—(CH_2)_6—\overset{\displaystyle H}{\overset{|}{N}}—H} + HO—\overset{\displaystyle O}{\overset{\|}{C}}—(CH_2)_4—\overset{\displaystyle O}{\overset{\|}{C}}—OH +$$

$$\downarrow$$

$$—\overset{\displaystyle H}{\overset{|}{N}}—\overset{\displaystyle O}{\overset{\|}{C}}—(CH_2)_4—\overset{\displaystyle O}{\overset{\|}{C}} \quad \overset{\displaystyle H}{\overset{|}{N}}—(CH_2)_6—\overset{\displaystyle H}{\overset{|}{N}}—\overset{\displaystyle O}{\overset{\|}{C}}—(CH_2)_4—\overset{\displaystyle O}{\overset{\|}{C}} \quad \overset{\displaystyle H}{\overset{|}{N}}— + nH_2O$$

The structure of nylon-66 may be written more briefly as

$$\left(\overset{\displaystyle H}{\overset{|}{N}}—\overset{\displaystyle O}{\overset{\|}{C}}—(CH_2)_4—\overset{\displaystyle O}{\overset{\|}{C}}—\overset{\displaystyle H}{\overset{|}{N}}—(CH_2)_6 \right)_n$$

Nylon produced in this way has an average molar mass of about 10 000 u and a melting point of 250 °C. While molten, it can be extruded into fibers. Hydrogen bonds between the NH groups in one chain and the CO groups in an adjacent chain hold the molecules together strongly enough to give nylon considerable tensile strength but not so strongly that it cannot be pulled out into thin fibers. Nylon is used in hosiery and other clothing. It resembles silk in its structure and properties but, being cheaper to produce, it has almost entirely replaced

Experiment 24.2

Synthesis of Nylon-610

When a solution of 1,6-diaminohexane, $H_2N(CH_2)_6NH_2$, in aqueous sodium hydroxide is poured gently onto a solution of decanedioyl chloride, $COCl(CH_2)_8COCl$, a white film of nylon-610 forms between the two layers. The film can be grasped with tweezers and pulled up as a nylon string, which can be wound on a glass rod as shown.

silk. The name nylon-66 refers to the fact that it is made from a six-carbon-atom carboxylic acid and a six-carbon-atom amine. Many other nylons have been made from other dicarboxylic acids and diamines (see Experiment 24.2 and Box 24.1).

B O X   2 4 . 1

WALLACE H. CAROTHERS AND NYLON

Wallace Carothers, born in Burlington, Iowa, in 1896, was responsible for the development of neoprene, the first synthetic rubber, and nylon. He joined the DuPont Company in 1928 as head of the organic chemistry division and was the first industrial chemist to be elected to the National Academy of Science. Unfortunately he became depressed by a family death and committed suicide by drinking cyanide at the age of 41. Although his career was brief he made enormous contributions to our understanding of the structure and properties of macromolecules.

Nylon, the first truly man-made fiber, was first marketed in 1938. It was used for toothbrush bristles, fishing line, and surgical sutures. Nylon stockings first went on sale in 1939 in Wilmington, Delaware, the home of the Du-

Pont Company, and women queued for hours to buy them. When World War II cut off all supplies of silk from the Far East, parachutes were made from nylon and it is still used for this purpose.

In recent times a new nylon polymer has been developed in The Netherlands. This new polymer has a better temperature stability and is stronger than traditional nylon polymers. Trade-named Stanyl it is nylon-46 and is made by reacting 1,4-diaminobutane and 1,6-hexanedioic acid (adipic acid). Although not yet manufactured on a large scale, it appears highly suitable for manufacturing both strong fibers and impact-resistant molded articles. It is interesting to note that nylon-46 was originally prepared by Wallace Carothers but was not fully investigated at the time.

EXERCISE 24.2

If 1.30×10^6 kg of nylon are made annually in the United States, how many grams of hexamethylenediamine must be consumed annually in the manufacture of nylon?

If each monomer molecule has only two active functional groups, then a linear polymer will be formed. If one of the monomers has three or more active functional groups, then a three-dimensional polymer may be formed. The first completely synthetic polymer, Bakelite, a phenol–formaldehyde polymer, is an example of a three-dimensional polymer. Its formation is illustrated in Experiment 24.3 and Figure 24.5. Highly branched or cross-linked polymers like this often do not melt when heated and are called thermosetting polymers. They are hard plastics which can be molded and machined.

POLYESTERS When a diol reacts with a dicarboxylic acid, a **polyester** is formed. For example, Dacron is a condensation polymer formed from 1,2-ethanediol and terephthalic acid (1,4-benzene dicarboxylic acid):

$$\cdots + HOCH_2CH_2OH + HO-\overset{\overset{\displaystyle O}{\|}}{C}-\bigcirc-\overset{\overset{\displaystyle O}{\|}}{C}-OH + HOCH_2CH_2OH + \cdots$$

$$\downarrow$$

$$-CH_2CH_2-O-\overset{\overset{\displaystyle O}{\|}}{C}-\bigcirc-\overset{\overset{\displaystyle O}{\|}}{C}-O-CH_2CH_2-O-\overset{\overset{\displaystyle O}{\|}}{C}-\bigcirc-\overset{\overset{\displaystyle O}{\|}}{C}-O-$$

$$+ n\,H_2O$$

Experiment 24.3

The Formation of Phenol–Formaldehyde Polymer

Concentrated HCl is poured into a solution containing aqueous HCHO (formaldehyde), glacial acetic acid, and phenol.

As the mixture is stirred it warms up and turns pink when the polymer begins to form.

After about one minute of stirring a solid phenol–formaldehyde condensation polymer forms, which is removed on the stirring stick.

The structure of Dacron may be written

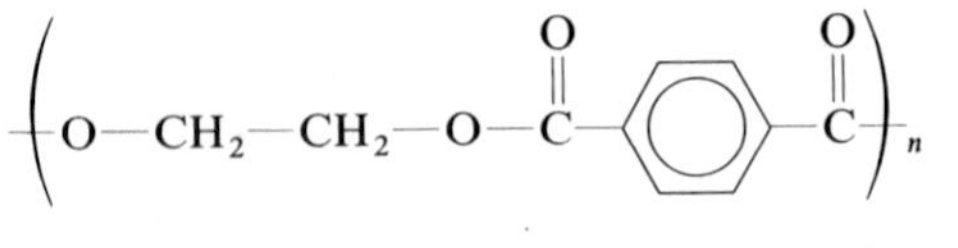

$$\left(O-CH_2-CH_2-O-\overset{O}{\underset{\|}{C}}-\bigcirc-\overset{O}{\underset{\|}{C}} \right)_n$$

Dacron forms strong fibers. It is used as a blend with cotton in clothing, and it has specialized uses, such as for seat belts and sails. In the form of thin sheets, this polymer is called Mylar.

═ Example 24.3 ═

STRUCTURE OF A POLYESTER

Kodel is a polyester made from terephthalic acid (1,4-benzene dicarboxylic acid) and the diol

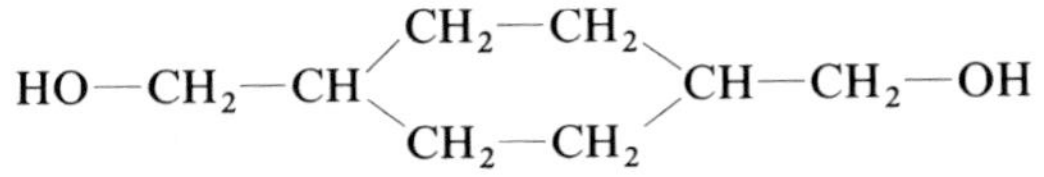

$$HO-CH_2-CH\overset{CH_2-CH_2}{\underset{CH_2-CH_2}{<>}}CH-CH_2-OH$$

What is the structure of Kodel?

Solution

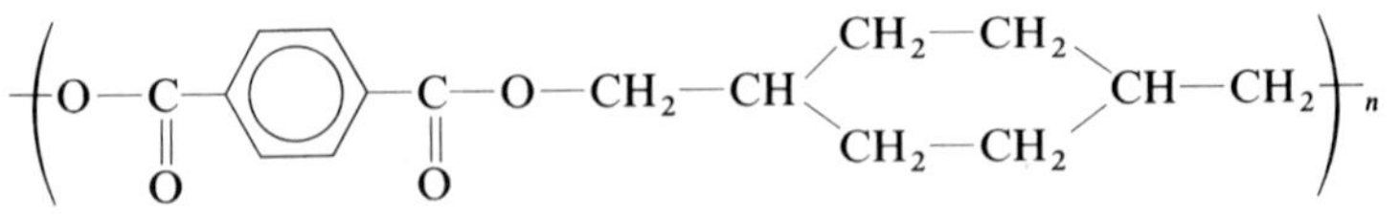

$$\left(O-\overset{}{\underset{O}{C}}-\bigcirc-\overset{}{\underset{O}{C}}-O-CH_2-CH\overset{CH_2-CH_2}{\underset{CH_2-CH_2}{<>}}CH-CH_2 \right)_n$$

═ Example 24.4 ═

CLASSIFICATION OF POLYMERS

Poly-4-methyl-1-pentene is a solid transparent polymer used in the manufacture of laboratory ware such as flasks and beakers. Is this an addition or a condensation polymer? Draw the structure of the polymer.

Solution

It is an addition polymer:

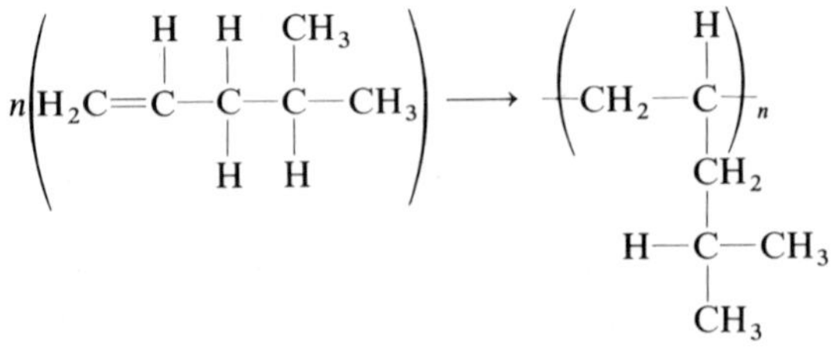

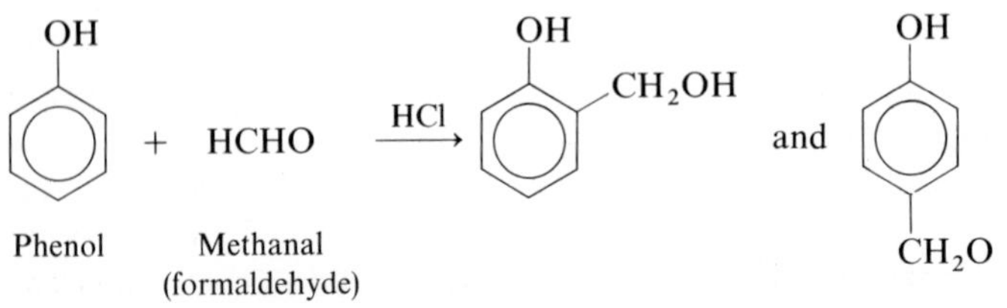

$$n\left(\overset{H\ \ H\ \ CH_3}{H_2C=\underset{H\ \ H}{C-C-C}-CH_3} \right) \longrightarrow \left(CH_2-\overset{H}{\underset{\underset{\underset{CH_3}{H-C-CH_3}}{CH_2}}{C}} \right)_n$$

FIGURE 24.5
The Formation of a Phenol–
Formaldehyde Plastic
(Bakelite) by Condensation.

$$\underset{\text{Phenol}}{\overset{OH}{\bigcirc}} + \underset{\substack{\text{Methanal} \\ \text{(formaldehyde)}}}{HCHO} \xrightarrow{HCl} \overset{OH}{\underset{}{\bigcirc}}^{CH_2OH} \quad \text{and} \quad \underset{CH_2O}{\overset{OH}{\bigcirc}}$$

Kevlar is an exceptionally strong polymer used in making canoes and bullet-proof vests (Figure 24.6). It is a condensation polymer with the following structure:

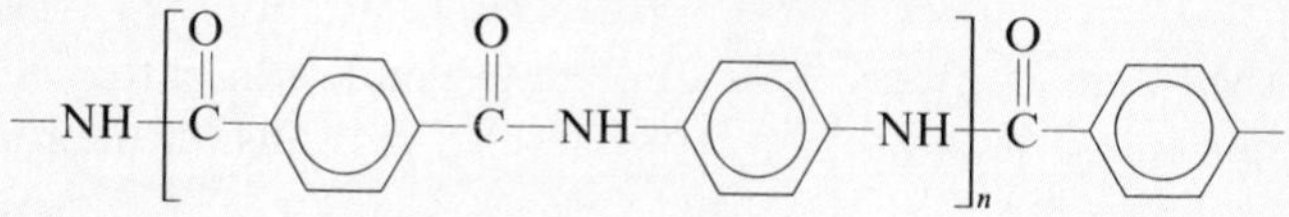

$$-NH-\overset{\overset{O}{\|}}{C}-\left[\bigcirc\right]-\overset{\overset{O}{\|}}{C}-NH-\bigcirc-NH-\overset{\overset{O}{\|}}{C}-\bigcirc-$$

What monomers are used to produce Kevlar?

FIGURE 24.6
A Bullet-Proof Vest
Made from Kevlar.

24.2
Biopolymers

Many biologically important substances are polymers. In this section we consider carbohydrates, proteins, and nucleic acids. Carbohydrates serve as energy sources and as the structural material of plants. Proteins are found in all parts of the body, and they have an enormous variety of functions. Some proteins are the structural components of skin, muscle, and hair; others control the transmission of nerve impulses; still others are enzymes that catalyze reactions. The nucleic acid, DNA, is the molecule in which an organism stores genetic information and through which it passes this information from generation to generation.

CARBOHYDRATES

Carbohydrates are synthesized by green plants from CO_2 and H_2O in the presence of sunlight in a process called *photosynthesis*:

$$xCO_2 + yH_2O \longrightarrow C_xH_{2y}O_y + xO_2$$

Because their empirical formulas can be written as $C_x(H_2O)_y$, carbohydrates were originally thought to be hydrates of carbon (hence the name). Carbohydrates can be classified into three main groups: monosaccharides, disaccharides, and polysaccharides.

About twenty monosaccharides occur naturally. They are commonly known as *sugars*. Two important examples are *glucose* and *fructose*. They both have the molecular formula $C_6H_{12}O_6$, but they have different structures. Glucose has a six-membered ring of five carbon atoms and an oxygen atom and has five OH groups, which accounts for its solubility in water (see Figure 24.7). It exists in two forms called α-*glucose* and β-glucose, which differ only in the orientation of one of the OH groups with respect to the ring. Fructose has a five-membered ring of four carbon atoms and one oxygen atom (see Figure 24.8).

A common disaccharide is sucrose (common table sugar). It is formed from an α-glucose molecule and a fructose molecule condensed together with the elimination of a water molecule (see Figure 24.9). The two monosaccharide units are joined by an ether linkage. The condensation reaction by which sucrose is formed from glucose and fructose is reversed in the stomach in a reaction that is catalyzed by the enzyme *sucrase*. Thus when we digest sucrose, glucose and fructose are formed, which are absorbed into the blood. The oxidation of glucose in living cells (aerobic metabolism) is an important source of energy

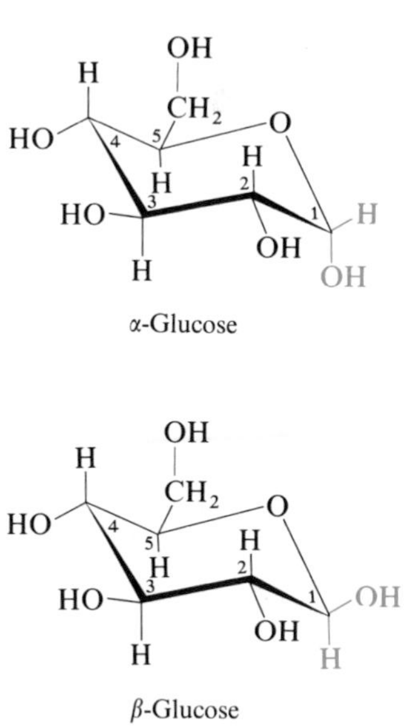

α-Glucose

β-Glucose

FIGURE 24.7
Structure of Glucose.

The α and β forms of glucose differ in the orientation of the OH group on C-1. In α-glucose this OH group is perpendicular to a plane through the ring. In β-glucose this OH group lies approximately in a plane through the ring.

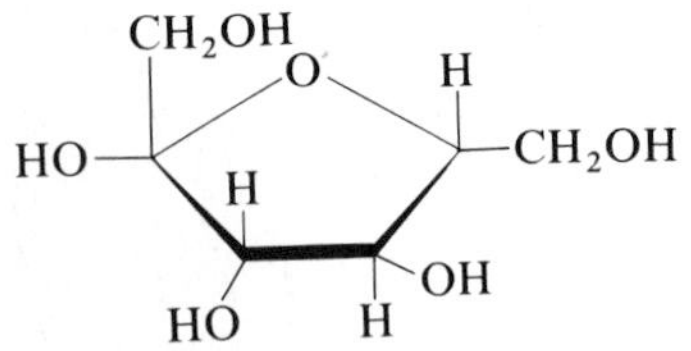

FIGURE 24.8
Structure of Fructose.

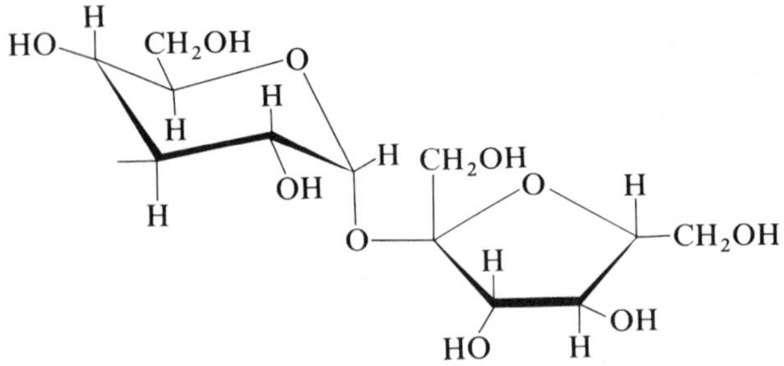

FIGURE 24.9
Structure of Sucrose, a
Disaccharide.

for all animals. It occurs in many steps catalyzed by enzymes and ultimately results in the formation of CO_2 and water:

$$C_6H_{12}O_6 + 6O_2 \longrightarrow 6CO_2 + 6H_2O \qquad \Delta H° = -2880 \, kJ$$

Two important polysaccharides are starch and cellulose. *Starch* is a mixture of polymers of α-glucose, and cellulose is a polymer of β-glucose. Starch consists mainly of *amylose*, which is a straight-chain (unbranched) polymer of α-glucose containing approximately 1000 to 4000 glucose units (Figure 24.10). Starch is broken down in the digestive tract in a series of steps that are catalyzed by enzymes.

Polysaccharides have two main functions: storage of excess glucose and the provision of mechanical support for plants. The structures and functions of polysaccharides are summarized in Table 24.2.

Cellulose is a straight-chain polymer of β-glucose containing on average about 3000 glucose units (Figure 24.11). Cellulose is the major structural component of wood and other plants. It accounts for more than one-half of all living matter. Humans do not possess the enzymes necessary to break down cellulose into glucose. Thus we are unable to digest cellulose. Animals such as cows and deer have intestinal bacteria that have the necessary enzymes for breaking cellulose down to glucose. If chemists could find a simple way to break cellulose down to glucose, we would have another important source of food.

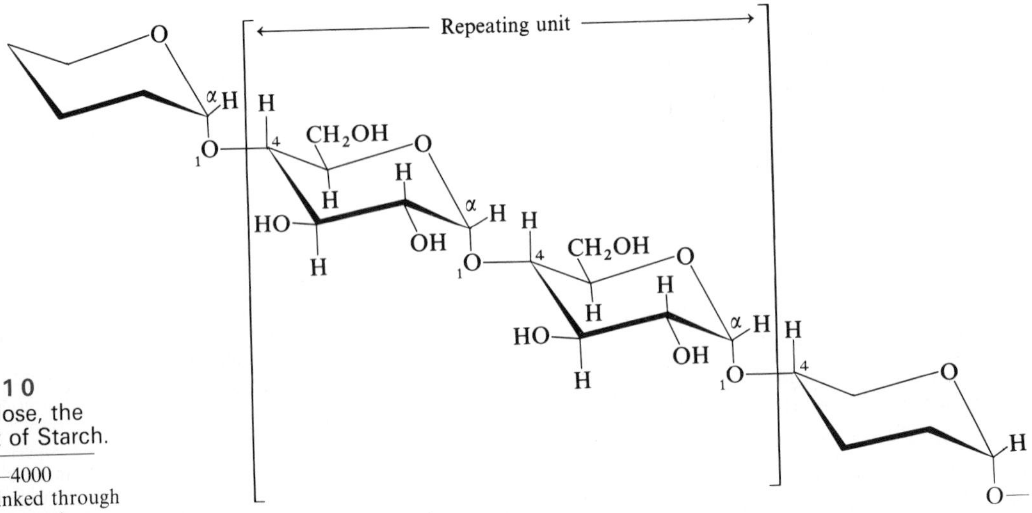

FIGURE 24.10
Structure of Amylose, the
Main Component of Starch.

Approximately 1000–4000
α-glucose units are linked through
oxygen atoms to form amylose.

TABLE 24.2
The Structures and Functions of Polysaccharides .

Polysaccharide	Monomer	Structure of Polymer Chain	Function
Cellulose	β-glucose	Unbranched	Structural material of plant cell wall—provides support to plants (especially trees)
Starch			
amylose (soluble starch)	α-glucose	Unbranched	Excess glucose in plants is stored as starch
amylopectin (insoluble starch)	α-glucose	Branched chains	
Glycogen	α-glucose	Branched chains (more highly branched than amylopectin)	Excess glucose in animals is stored as glycogen in the liver and muscles
Chitin	β-N-acetylglucosamine	Unbranched	Exoskeleton of insects, lobsters, etc.

═ *Example 24.5* ═

THE HYDROLYSIS OF CARBOHYDRATES

Write equations for the hydrolysis of sucrose and starch.

Solution

$$C_{12}H_{22}O_{11} + H_2O \longrightarrow C_6H_{12}O_6 + C_6H_{12}O_6$$

Sucrose $\qquad\qquad$ Glucose $\quad$ Fructose

$$(C_6H_{10}O_5)_n + nH_2O \longrightarrow nC_6H_{12}O_6$$

Starch $\qquad\qquad$ Glucose

═ *Example 24.6* ═

SIZE OF POLYSACCHARIDES

Amylose has a molar mass of about 3.0×10^5 g mol^{-1}. Approximately how many glucose units does an amylose molecule contain?

Solution

Each glucose unit in amylose has the formula $C_6H_{12}O_6 - H_2O = C_6H_{10}O_5$ and therefore has a formula mass of 162 g mol^{-1}. Hence

$$\text{Number of glucose units} = \frac{3.0 \times 10^5 \text{ g mol}^{-1}}{162 \text{ g mol}^{-1}} = 1900$$

FIGURE 24.11
Structure of Cellulose.

On average about 3000 β-glucose units are linked through oxygen atoms to form a cellulose molecule.

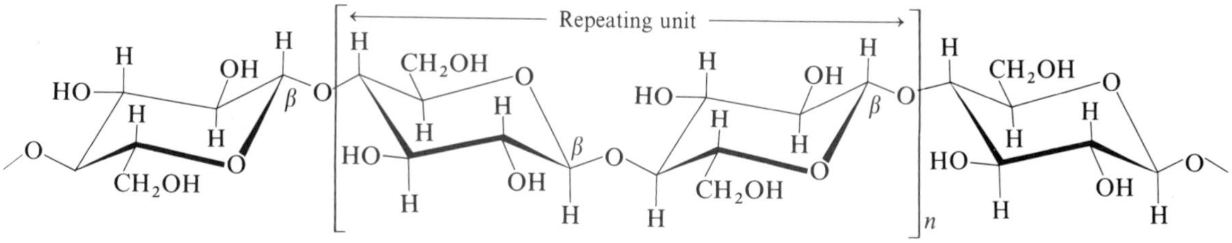

SEMISYNTHETIC POLYMERS FROM CELLULOSE

Cellulose contains a very large number of —OH groups that can react with carboxylic acids to give esters. Especially important are the *acetates*, such as cellulose triacetate, in which all three —OH groups of each glucose unit are replaced by acetate groups.

$$-O-\overset{\overset{\displaystyle O}{\|}}{C}-CH_3$$

This class of polymers is referred to collectively as *acetate polymers*. An example is Arnel, a strong fiber used in clothing, fabrics, and electric insulators.

Cellulose in the form of cotton or wood pulp also reacts with nitric acid to give *nitrocellulose*. Depending on the conditions of the reaction, the —OH groups of each glucose unit may be partially or fully replaced by nitrate groups, —ONO_2. The partially nitrated form, with camphor added to make it softer and more malleable, is called *celluloid*. It was patented as early as 1869 and was the first synthetic plastic. It was used for many years to make articles as diverse as baby rattles, shirt collars, and photographic film, but its use has diminished in recent years because of its flammability. Fully nitrated cellulose is known as *guncotton* and is used as an explosive and a propellant.

PROTEINS

Thousands of different proteins go into the makeup of a living cell. They take part in the thousands of reactions that take place in a living cell. **Proteins** are very complex, giant molecules, and one of the greatest achievements of modern science has been determining the structure of numerous proteins. But the exact details of all the complex processes in which they are involved will defy our understanding for some time to come. Here we can only give a glimpse of this fascinating and challenging area of chemistry.

Proteins are polyamides formed from amino acids. **Amino acids** have both an NH_2 group and a carboxyl group, CO_2H. The amino acids in proteins are called **α-amino acids** because they have the NH_2 group on the same carbon atom as the carboxylic acid group.

An amide is formed by condensation of an NH_2 group with a carboxylic acid group. Because an amino acid has both these functional groups, a polymer— a *polyamide*—can be formed by condensation:

$$H_2NCHC\overset{\displaystyle O}{\underset{\displaystyle R_1}{<}}\!\!OH \;+\; H_2NCHC\overset{\displaystyle O}{\underset{\displaystyle R_2}{<}}\!\!OH \;+\; H_2NCHC\overset{\displaystyle O}{\underset{\displaystyle R_3}{<}}\!\!OH$$

$$\downarrow$$

$$-NH-CH-\overset{\overset{\displaystyle O}{\|}}{C}-NH-CH-\overset{\overset{\displaystyle O}{\|}}{C}-NH-CH-\overset{\overset{\displaystyle O}{\|}}{C}-NH-\;+\;n H_2O$$
$$\quad\;\;\; R_1 \qquad\qquad\;\;\; R_2 \qquad\qquad\;\;\; R_3$$

The
$$\overset{\displaystyle O}{\overset{\displaystyle \|}{-C}}-NH-$$
groups linking the R groups in the polymer are called peptide links, and a polyamide is also called a *polypeptide*. Proteins are naturally occurring polypeptides. The general formula of a protein as we have written it looks simple, but there are 20 amino acids that are commonly found in nature and a given protein may contain many or all of them. These different amino acids are listed in Table 24.3. They differ in the nature of the group R. They are usually denoted by the three-letter abbreviations given in Table 24.3 such as *ala* for alanine and *gly* for glycine. Since a protein commonly contains as many as fifty or more monomer units, the number of possibilities for different proteins is truly enormous. Fortunately, in nature we find only a very small fraction of all these possibilities.

The order in which the amino acids occur in a protein is called the *primary structure*. The first primary structure of a protein was determined by British chemist Frederick Sanger in 1953. He determined the complete structure of the protein insulin (Box 24.2). For this he was awarded the Nobel Prize for chemistry in 1958. He also jointly won the 1980 Nobel Prize for chemistry for work

BOX 24.2

INSULIN AND DIABETES

The complete sequence of amino acid units in several hundred protein molecules is now known. An example is *insulin*, the hormone produced in the pancreas that is essential to the metabolism of carbohydrates in the body and the lack of which leads to diabetes. Diabetes is a serious and widespread disease, and many diabetics have to be injected with insulin daily to regulate their condition. Insulin contains 51 amino acids arranged in two chains and cross-linked in two places by the disulfide bond of cysteine. One chain contains 21 amino acid units and the other has 30 amino acid units (Figure 24.12). The polymer has a molecular mass of 5733. The amino acid sequence was determined by British biochemist Frederick Sanger, who received a Nobel Prize for the work in 1958, and insulin was synthesized in the laboratory for the first time in 1963.

The principal source of insulin for medical use since the 1920s, when it was discovered by Canadian scientist Frederick Banting (1891–1941) at the University of Toronto, had been the

Genetic engineering laboratory

pancreases of cattle, from which it had to be extracted. Today, however, synthetic insulin is available. Very recently, it has become possible to make insulin by the methods of *genetic engineering*. In this technique, genes are made—either synthetically or by modifying existing genes—and then added to simple organisms, such as bacteria, that can use the information from the genetic code to "manufacture" proteins. In 1978 chemically synthesized genes were added to the bacteria *E. coli*, and human insulin resulted. The use of biotechnology to produce medically important materials like insulin has enormous commercial possibilities.

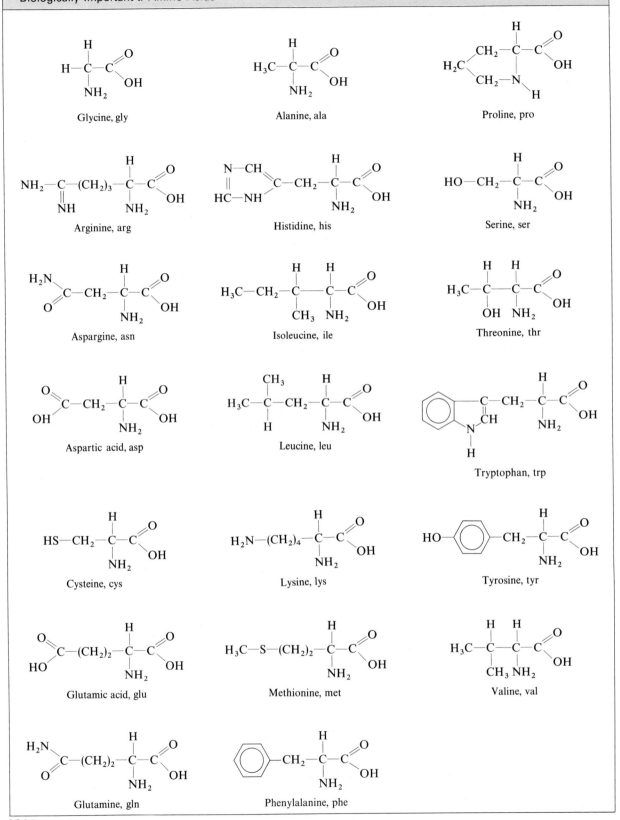

Glycine, gly

Alanine, ala

Proline, pro

Arginine, arg

Histidine, his

Serine, ser

Aspargine, asn

Isoleucine, ile

Threonine, thr

Aspartic acid, asp

Leucine, leu

Tryptophan, trp

Cysteine, cys

Lysine, lys

Tyrosine, tyr

Glutamic acid, glu

Methionine, met

Valine, val

Glutamine, gln

Phenylalanine, phe

on various aspects of DNA. Frederick Sanger was the third person to win two Nobel Prizes for scientific work. Now the primary structures of several hundred proteins have been determined.

The primary structure of beef insulin is shown in Figure 24.12. It has 51 amino acids that are linked into two polypeptide chains held together by S—S links. Some proteins consist of only a single polypeptide chain.

A very long protein chain can have an enormous number of different conformations. But a protein does not have a floppy structure that is continually changing; rather, it adopts a very definite conformation, called the *secondary structure*. This structure is very largely a consequence of hydrogen bonding. If two protein chains are laid parallel to each other but running in opposite directions, a very large number of N—H---O=C hydrogen bonds can be formed between them (see Figure 24.13). In fact, numerous parallel chains can be bonded together in this way to form a sheet (Figure 24.14). These sheets are then stacked one upon another to form a three-dimensional structure. Silk has a structure of this type, with the protein chains running in the direction of the silk fibers.

Wool and hair have a different type of structure, in which hydrogen bonds are formed between CO and NH groups in a single chain. These hydrogen bonds cause the chain to coil up into a spiral, called an α-helix, in which there are 3.6 amino acids for each turn of the helix (Figure 24.15). Three α-helices are then twisted together as in a rope to give a structure called a *protofibril*. The protofibrils are then packed together in parallel bundles in a wool or hair fiber. Wool fibers are somewhat elastic because only the weak hydrogen bonds must be broken in order to allow the helix to increase in length, much as a spring stretches when it is pulled. Then when the tension is released, the hydrogen bonds re-form, pulling the fiber back into its original helical shape.

Other proteins, such as myoglobin and hemoglobin in the blood and those proteins that behave as enzymes, have a still more complex structure. They consist of helical chains, but the chain is folded up in a complex way to give a much more compact structure. These proteins are called *globular proteins*. Parts of the chain have a helical structure, but at the bends in a chain the regular helical structure is disrupted (Figure 24.16). The protein is held in this folded shape mainly by interactions between the side groups R, some of which are ionic or polar, and also by disulfide bridges —S—S— between cysteine residues. The form of the folded chain is called the *tertiary* structure of the protein. Different globular proteins have very different and very characteristic folded shapes.

Enzymes are globular proteins that catalyze chemical reactions in living systems. More than a thousand enzymes have been identified, and the amino acid sequence in over a hundred of them has been determined. Two remarkable properties of enzymes are their extraordinary specificity—each enzyme catalyzes only one reaction or one group of closely related reactions—and their amazing efficiency—they may speed up reactions by factors of up to 10^{20}. Enzymes provide a very effective method for the control of reactions in living systems. Biochemical reactions do not take place in the body at an appreciable rate in the absence of the appropriate enzyme catalyst. Thus the presence or absence of the enzyme at a particular site enables reactions to be switched on or off. Enzymes work best within certain temperature and pH ranges; outside these ranges, the enzyme activity slows down or stops. At high temperatures or extreme pH, proteins may be denatured—the characteristic shape of the

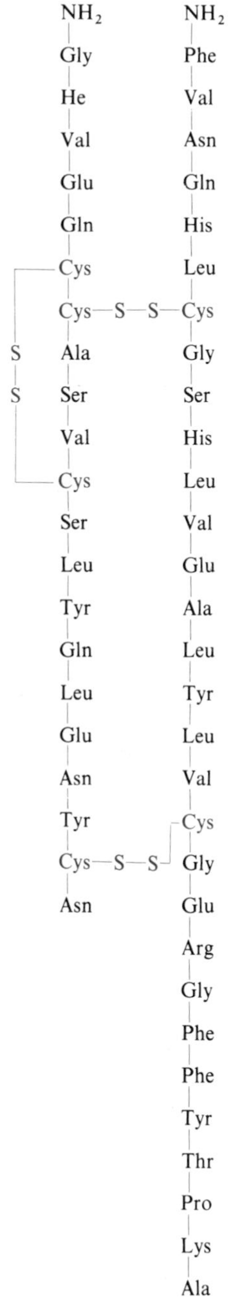

FIGURE 24.12
Primary Structure of Beef Insulin.

There are two polypeptide chains held together by disulfide linkages between cysteine residues.

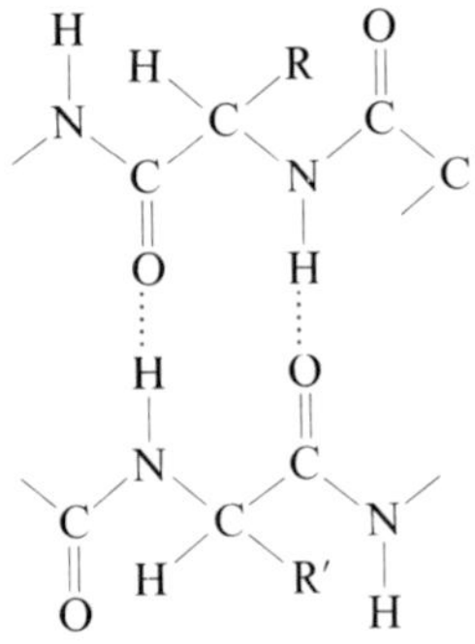

FIGURE 24.13
Hydrogen Bonding Between
Amino Groups and C=O
Groups.

protein is changed and its biological activity is destroyed. A common example is the heating of egg white in which the protein, albumin, is denatured.

The mechanism by which an enzyme acts has been the subject of intense research. A simple and popular theory is the lock-and-key theory, according to which the reactant molecule or molecules, called the *substrate*, fit into a pocket or cavity in the complex folded structure of the enzyme (Figure 24.17). The pocket in any particular enzyme has a very specific shape that can only accommodate one particular molecule or group of similar molecules. When the reactant molecules are held in the correct orientation for the reaction to occur, the reaction is much more rapid than it would be if the correct orientation were achieved only in a small percentage of random collisions. In other words, the enzyme behaves as a catalyst.

The lack of even one of the many enzymes in the body can cause serious disease. For example, some mentally retarded children suffer from the disease known as PKU (phenylketonuria). They lack the enzyme that converts phenylalanine to tyrosine:

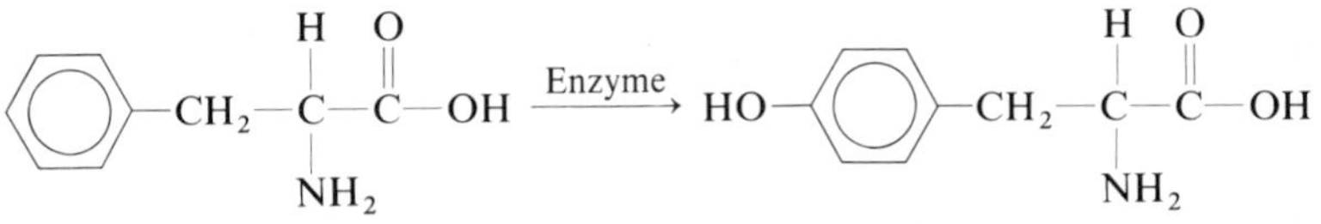

$$\text{Phenylalanine} \xrightarrow{\text{Enzyme}} \text{Tyrosine}$$

Instead, phenylalanine is converted to phenylpyruvic acid, and high levels of phenylpyruvic acid can lead to mental retardation in some way that is not well understood:

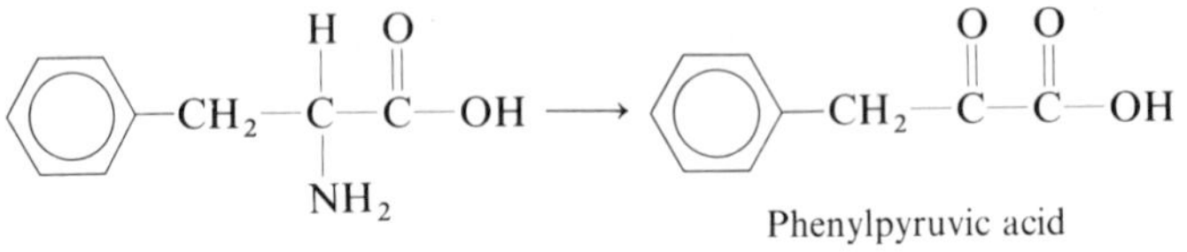

Phenylpyruvic acid

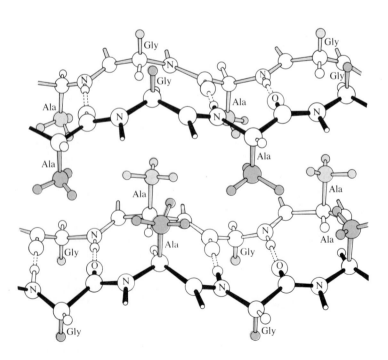

FIGURE 24.14
Sheets of Hydrogen-Bonded
Protein Chains as Found in
Silk.

Phenylpyruvic acid is easily detected in the urine, and infants are now routinely tested for its presence. If it is found in large amounts, the child can be given a special diet low in phenylalanine, and so the mental retardation can be prevented. The wide range of functions carried out by proteins in living organisms is summarized in Table 24.4.

NUCLEIC ACIDS

One of the most amazing aspects of life is the ability of living organisms to transmit their characteristics to their progeny. The observation that organisms reproduce their own species is widely known and self-evident. Yet the mystery of how this happens is one of the most challenging problems facing science today. Although we have a fair understanding of this process, our knowledge of many of its details is far from complete. Differences in species appear to result from differences in proteins. For example, the hemoglobin of cats differs slightly in amino acid sequence from the hemoglobin of mice and of humans.

How does an organism synthesize correctly its own characteristic proteins? We know that the information that is necessary to guide the correct synthesis of proteins is stored in the molecule **deoxyribonucleic acid**—usually abbreviated as **DNA**—which is found in the nuclei of all cells. Deoxyribonucleic acid is an example of a **nucleic acid**. Nucleic acids are polymers of nucleotides. A **nucleotide** is made by the condensation of a molecule of phosphoric acid, a molecule of a sugar—deoxyribose—and a molecule of a nitrogen compound called a *nitrogen*

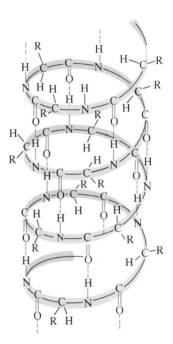

FIGURE 24.15
α-Helix Structure of Protein.

The helix conformation of the molecule is called the secondary structure.

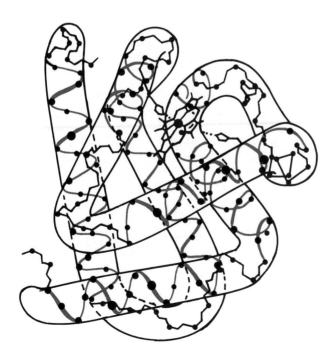

FIGURE 24.16
Tertiary Structure of the Globular Protein Myoglobin.

The heme group (colored) can bond readily to oxygen; myoglobin transports oxygen in muscle tissue.

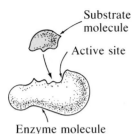

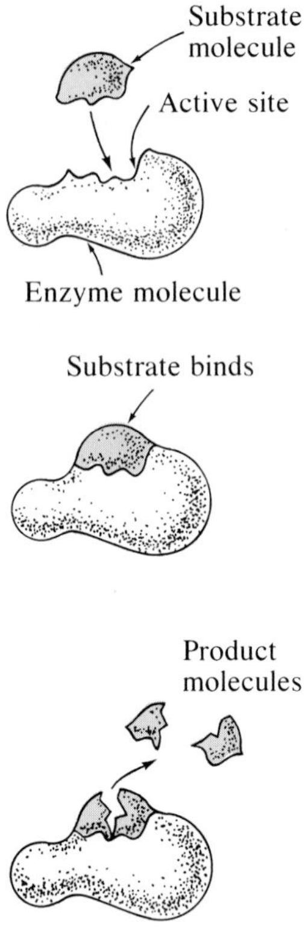

Substrate
molecule

Active site

Enzyme molecule

Substrate binds

Product
molecules

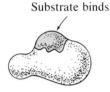

FIGURE 24.17
Lock-and-Key Theory of
Enzyme Action.

The substrate fits the active site of
an enzyme as a key fits a lock. The
bond-breaking and bond-making
processes that transform a substrate (reactant) to products occur
while the substrate is bound to the
active site of the enzyme.

TABLE 24.4
Protein Functions

Type	Examples	Specific Function of Example
Enzymes	Amylase	Catalyzes hydrolysis of starch in the mouth
	DNA polymerase	Catalyzes synthesis of DNA from nucleotide triphosphates in nucleus
Hormones	Insulin	Regulates glucose metabolism
Contractile proteins	Actin, myosin	Work together to contract muscles
Transport proteins	Hemoglobin	Transports oxygen in the blood
Structural proteins	α-keratin	Forms skin, nails, feathers
	Collagen	Forms cartilage, tendons
	Elastin	Forms ligaments
	Fibroin	Forms silk (from silkworm cocoon), spider webs
Protective proteins	Antibodies	Destroy foreign protein
	Fibrinogen	Forms an insoluble material in a blood clot, preventing bleeding

base. Four different nitrogen bases are found in DNA; they are adenine, guanine, cytosine, and thymine (see Figure 24.18). Thus there are four different nucleotides in DNA. These nucleotides are condensed into a polynucleotide, which consists of a sugar–phosphate backbone with a nitrogen base attached to each sugar (see Figure 24.19). An enormous number of different sequences of nitrogen bases are possible. In a typical small DNA polymer strand containing 1500 nucleotide units, there are 4^{1500}, or 10^{900}, different possible sequences.

The key to understanding how DNA works lies in its three-dimensional structure. In 1953 James Watson, an American biologist, and Francis Crick, an English biophysicist, working together in Cambridge, England, proposed

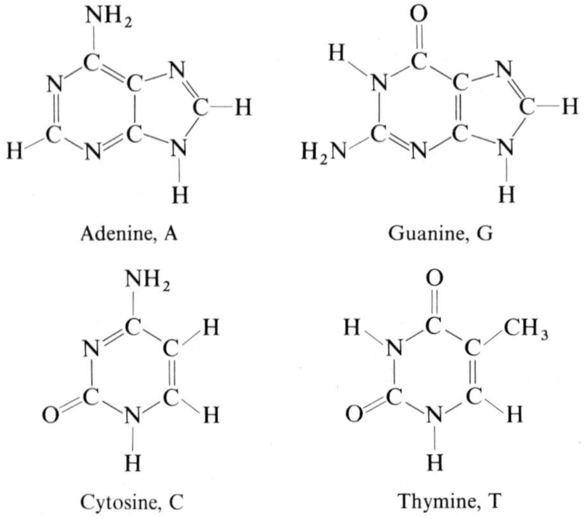

FIGURE 24.18
The Four Nitrogen Bases
Found in the Polynucleotide
DNA.

Adenine, A

Guanine, G

Cytosine, C

Thymine, T

FIGURE 24.19
Structure of a Nucleic Acid.

that DNA consists of two polynucleotides in the form of a **double helix** (see Figure 24.20). They based this suggestion on two observations: (1) X ray diffraction patterns of DNA indicated that it has a helical structure; (2) chemical analysis of DNA had shown that although the amounts of the different nitrogen bases in DNA vary from species to species, the amount of adenine is always equal to the amount of thymine and the amount of guanine is always equal to the amount of cytosine. Thus it seemed that these bases must somehow always be paired together.

Working with ball-and-stick molecular models (see Figure 24.21), they found that adenine, A, and thymine, T, were just the right size and shape to be linked together by two hydrogen bonds and that guanine, G, and cytosine, C, can join together at exactly the same distance (1.1 nm) by forming three hydrogen bonds (see Figure 24.20). Held together by hydrogen bonds in this way, the two strands have a constant separation of 1.1 nm. No other nitrogen base pairs have the right size and shape to form hydrogen bonds with the same separation between the chains. The two chains are said to be *complementary* to each other because the sequence of nitrogen bases in one chain completely determines the sequence in the other chain. Thus if the hydrogen bonds break, the helix can uncoil, and each of the separate chains can then act as a template for the formation of a new complementary chain. Thus we have an explanation of how DNA can replicate itself.

The sequence of bases in the DNA molecule is a code for the synthesis of all the proteins characteristic of a given organism. Each sequence of three bases along a DNA chain—for example, CGT—is the code for the synthesis of a

A space-filling model of DNA

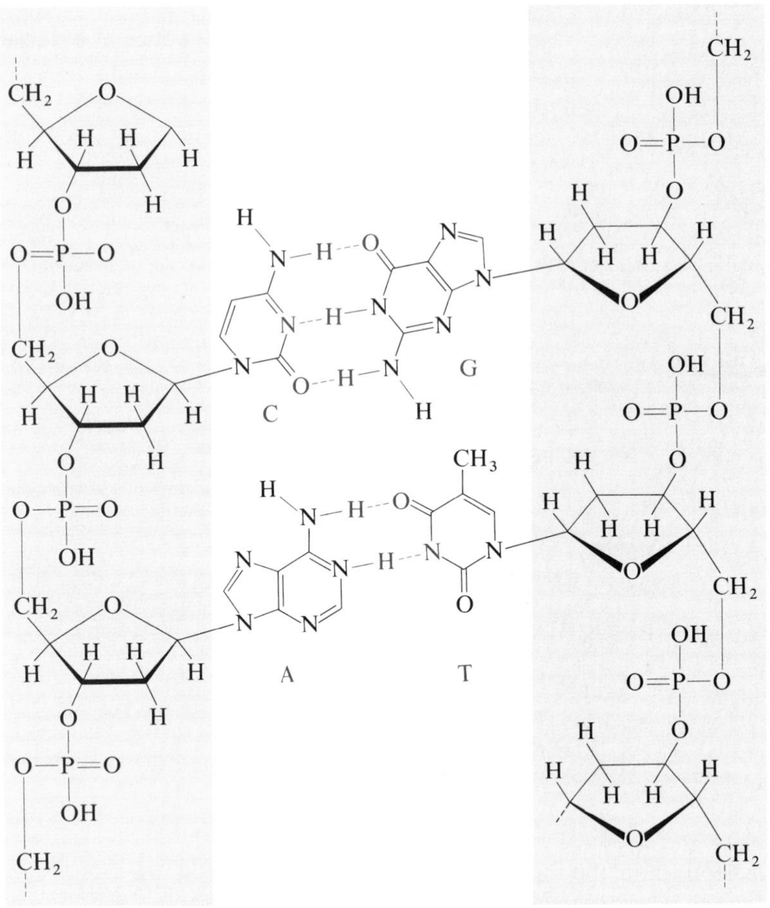

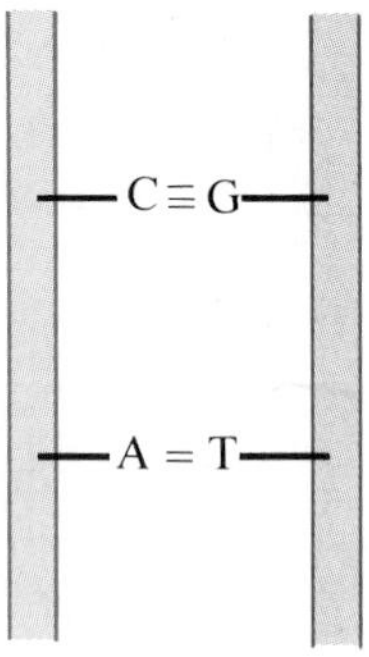

FIGURE 24.20
The Double-Helix Structure of DNA.

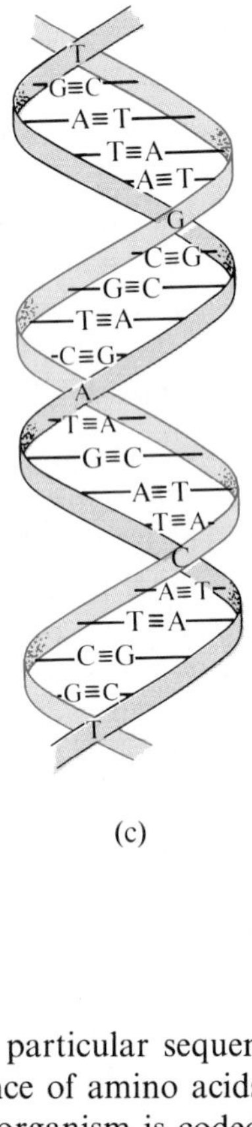

(a) A small portion of the two strands of the double helix showing how they are held together by hydrogen bonds between the nitrogen bases. There are three hydrogen bonds between cytosine, C, and guanine, G, and two hydrogen bonds between adenine, A, and thymine, T. (b) A simplified version of the same portion of the double helix. (c) This shows how the two strands are held together in a double helix by the hydrogen bonds between the nitrogen bases.

particular amino acid—in this case alanine. Thus the particular sequence of bases in a segment of DNA corresponds to the sequence of amino acids in a particular protein. Each of the proteins needed by an organism is coded into the DNA in this way. The triplet AAA is the code for phenylalanine, so the sequence CGTAAA corresponds to an ala–phe segment of a polypeptide, and so on. Each segment of the DNA molecule that codes the synthesis of one particular protein is called a *gene*. The exact mechanism by which the nitrogen base sequence along a DNA strand is used to build up a protein is complicated but reasonably well understood. However, consideration of this mechanism would take us too deeply into molecular biology and must be left for other courses.

FIGURE 24.21
Ball-and-Stick Model of DNA Constructed by Watson and Crick.

Watson (*left*) and Crick are examining the model of DNA that they built at Cambridge in 1953. James Watson was born in 1928 in Chicago, and he graduated from the University of Chicago at the age of nineteen. He obtained his Ph.D. in zoology from the University of Indiana in 1950, when he was only twenty-two. In 1951 he went to the University of Cambridge, where he worked with Francis Crick on the structure of DNA. Francis Crick was born in Northampton, England, in 1916. He graduated in physics, later joining a group of physicists and other scientists at Cambridge who had turned their attention to solving the challenging problems of the new science of molecular biology. Watson and Crick's proposal of the double-helix structure of DNA is regarded as one of the most significant break-throughs in science in recent times. For their work they were awarded the Nobel Prize in medicine and physiology in 1962. Watson later wrote a popular and highly successful account of the work leading up to their discovery in his book *The Double Helix*.

Mutations in DNA can cause errors in the biosynthesis of proteins with serious consequences for the organism. A well-known example is sickle-cell anemia. The red blood cells of persons afflicted with this disease have an unusual shape. This unusual shape results simply from the replacement of one glutamic acid monomer in the hemoglobin protein by a valine monomer. The unusual shape of these cells causes them to clump together and to block capillaries, thus preventing oxygen-carrying cells from reaching the tissues. Sickle-cell anemia is one example of many molecular diseases that are now being identified.

IMPORTANT TERMS

An **addition polymer** is a polymer than contains all the atoms of the monomer units from which it is composed.

An **α-amino acid** is an amino acid in which the —CO₂H and —NH₂ groups are attached to the same carbon atom.

An **amino acid** is an organic molecule containing both a carboxylate group and an amino group.

The **amino acid sequence** is the sequence of amino acids found in proteins (polypeptides).

A **carbohydrate** is a compound of carbon, hydrogen, and oxygen in which the ratio of H to O is the same as in water.

Cellulose is a polysaccharide polymer formed by the condensation of β-glucose units.

A **condensation polymer** is a polymer formed from monomers by a condensation reaction in which molecules of small molecular mass, such as H₂O, are eliminated.

A **copolymer** is a polymer produced by the polymerization of two or more types of monomers.

A **cross-linked polymer** is a polymer in which polymer chains are joined by cross-links.

DNA, deoxyribonucleic acid, is a polynucleotide consisting of two polynucleotide strands forming a double helix.

The **double helix** is the structure of DNA in which two helical monomer strands (polynucleotides) are held together by hydrogen bonding.

An **elastomer** is a polymer with elastic properties.

A **macromolecule** is a polymer.

A **monomer** is the basic repeating unit of a polymer.

A **nucleic acid** is a polymer of nucleotides—a polynucleotide.

A **nucleotide** is the basic building block of nucleic acids. It is formed by the condensation of an organic nitrogen base, a sugar, and a phosphoric acid molecule.

Plastic is the common name for a synthetic polymer.

A **polyamide** is a condensation polymer formed from an amino acid or from a dicarboxylic acid and a diamine.

A **polyester** is a condensation polymer formed from a dicarboxylic acid and a diol.

A **polymer** is a molecule consisting of a number of repeating units called monomers.

A **protein** is naturally occurring polyamide (polypeptide).

P R O B L E M S *

Synthetic Polymers

1. Define each of the following terms, and give one example of each type of substance.

    (a) Addition polymer   **(b)** Condensation polymer

    (c) α-amino acid   **(d)** Sugar

    (e) Polypeptide

2. What addition polymers result from polymerization of each of the following monomers? Draw a diagram of a segment of the structure of each polymer.

    (a) Chloroethene (vinyl chloride)

    (b) Phenylethene (styrene)   **(c)** Tetrafluoroethene

3. From what monomers are each of the following formed?

    (a) Teflon   **(b)** Saran   **(c)** PVC

    (d) Nylon   **(e)** Dacron

4. What are the structures of the polymers formed from each of the following monomers?

    (a) Propene   **(b)** 1,3-Butadiene

    (c) 1,6-Hexanediamine and 1,4-hexanedioic acid

    (d) 1,2-Ethanediol and terephthalic acid

5. The *trans* isomer of 2-methyl-1,3-butadiene forms a hard natural polymer known as gutta-percha. What is the structure of this polymer?

6. *Orlon* has the polymeric chain structure

$$-[-CH_2-CH-CH_2-CH-CH_2-CH-]-_n$$
$$\quad\quad\quad\; |\quad\quad\quad\quad |\quad\quad\quad\quad |$$
$$\quad\quad\quad CN\quad\quad\; CN\quad\quad\; CN$$

From what monomer is this synthesized?

7. Draw the structure of the repeating unit in condensation polymers made by condensing together each of the following pairs of monomers, with the elimination of methanol, CH_3OH:

    (a) 1,2-Ethanediol and dimethylpropanedioate

    (b) 1,3-Propanediol and diethyl-1,4-butanedioate

8. Nylon stockings dissolve readily in concentrated hydrochloric acid or sulfuric acid. Suggest a possible explanation.

9. What structural feature must an organic molecule have if it is to undergo addition polymerization?

* Answers to problems numbered in blue appear at the end of the text.

10. What intermolecular forces are present between the polymer chains in

    (a) An addition polymer such as polyethylene?

    (b) A condensation polymer, such as Terylene?

    (c) A polyamide?

Biopolymers

11. What molecular units combine to form a nucleotide?

12. What is the structural difference between α-glucose and β-glucose?

13. (a) What is an α-aminocarboxylic acid?

    (b) Draw the structures of aminoethanoic acid (glycine), and 2-aminopropanoic acid (alanine).

14. (a) The sugars are examples of carbohydrates. Why was the name carbohydrate originally given to such compounds, and how appropriate is it?

    (b) Give an example of each of the following.

       (i) A monosaccharide   **(ii)** A disaccharide

       (iii) A polysaccharide

    (c) Glucose and fructose have the same molecular formula, $C_6H_{12}O_6$. How do they differ structurally?

15. In what structural way does starch differ from cellulose?

16. How many glucose residues are there in a starch molecule of molecular mass 2.57×10^5 u?

17. Draw structures for each of the four nitrogen bases found in DNA, and name each of them.

18. Explain why two strands of DNA are always found to be linked together by interactions between two specific pairs of bases, rather than via all six possible combinations of these four bases.

19. What is meant by the primary structure, the secondary structure, and the tertiary structure of a protein?

20. Explain how an enzyme is believed to increase the rate of a specific biochemical reaction.

Miscellaneous

21. How much sucrose would have to be dissolved in 100 g of water so that the vapor pressure of the resulting solution was 90% of that of pure water at the same temperature?

22. A solution containing 1.772 g of nylon-66 in 100.0 mL of solution has an osmotic pressure of 10.1 mm Hg at 25 °C. What is the average molar mass of the nylon-66?

23. Name the functional groups in each of the following.

(a) Methylmethacrylate polymer (Lucite)

(b) Nylon-66   (c) 1,2-Ethanediol

(d) Terephthalic acid   (e) Fructose

(f) Glycine

24. Ethylamine, ethanoic acid, and aminoethanoic acid (glycine) have melting points of -81 °C, 17 °C, and 233 °C, respectively. In terms of intermolecular forces, how can these differences be explained?

25. What type of bonding is responsible for holding together the double helix of DNA?

26. (a) What nitrogen bases are found in a DNA molecule?

(b) Which combinations of these base pairs are found in DNA?

(c) Are the base pairs in DNA found inside the helix, outside the helix, or do they constitute part of the backbone structure?

27. (a) What is an enzyme?

(b) Why are enzymes important?

(c) What effect does an enzyme have on the activation energy of a reaction?

(d) Explain how the lock-and-key model describes the mode of action of an enzyme.

28. Draw the structure of the triglyceride formed between glycerol and stearic acid, $CH_3(CH_2)_{18}CO_2H$.

29. An amino acid isolated from a piece of animal tissue was believed to be glycine (aminoethanoic acid), $NH_2CH_2CO_2H$. When 0.0500 g of the amino acid was completely converted to ammonia, which was absorbed into 50.0 mL of 0.0500M HCl(aq), the excess HCl(aq) required 30.57 mL of 0.0600M NaOH(aq) for neutralization.

(a) How many moles of HCl were neutralized by the NH_3?

(b) How many grams of nitrogen were in the 0.500-g sample of the amino acid?

(c) Does the mass percentage of nitrogen in the sample conform to that expected for glycine?

Nuclear and Radiochemistry

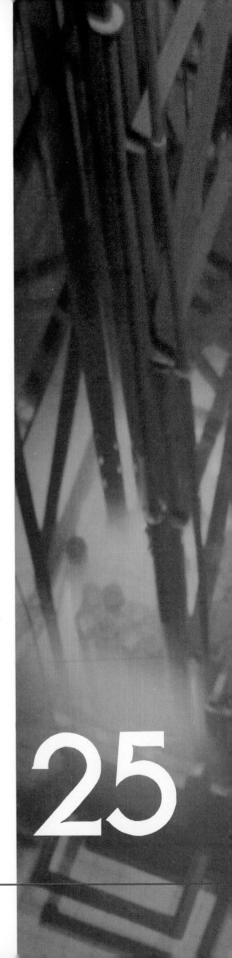

Radioactivity
Nuclear Stability
Radioactive Decay Rates
Artificial Radioisotopes
Nuclear Energy

25

Marie Curie (1867–1934) and her daughter, Irene (1897–1956). Marie Curie shared the 1903 Nobel Prize in physics with her husband, Pierre, and Antoine Becquerel for their research on radioactivity. In 1911 she won the Nobel Prize in chemistry for the discovery of the elements polonium and radium. Irene Curie shared the 1935 Nobel Prize in chemistry with her husband, Frederic Joliot.

We discussed the structure of the nuclei of atoms very early in this book because the charge of the nucleus—the atomic number Z—determines the number of surrounding electrons and therefore the chemical properties of the atom. Subsequently, however, we have paid little attention to the nucleus because nuclei remain unchanged in chemical reactions. Any process in which a nucleus undergoes a change is called a **nuclear reaction**, and such changes are usually considered to form part of physics rather than chemistry. Nevertheless, nuclear reactions have some very important applications in chemistry as well as being sources of enormous amounts of energy. Life as we know it exists because of the nuclear reactions that provide the energy for the sun. In this chapter we discuss nuclear reactions and some of their important applications. Nuclear reactions can be used in destructive ways, but electric power generation, medical diagnosis, and a variety of industrial applications have enriched our lives.

Some nuclei are unstable and spontaneously change into other nuclei by emitting electrons, positrons, or other particles, such as helium nuclei (α particles). Such nuclei are said to be radioactive. Many stable nuclei can also be transformed into unstable radioactive nuclei by bombarding them with other particles such as α particles and neutrons.

25.1
Radioactivity

All elements with atomic numbers of 83 or less, with the exception of technetium ($Z = 43$) and promethium ($Z = 61$), have one or more stable isotopes, but the nuclei of all the isotopes of the elements with atomic numbers greater than 83 (bismuth) are unstable. The spontaneous disintegration of a nucleus is called **radioactivity**, and an unstable nucleus that decomposes spontaneously is said to be **radioactive**. An isotope that is radioactive is usually called a **radio-isotope**. For example, uranium-238 nuclei emit helium nuclei, which for historical reasons are called α particles. A uranium nucleus is thus transformed into a thorium nucleus:

$$^{238}_{92}\text{U} \longrightarrow {}^{234}_{90}\text{Th} + {}^{4}_{2}\text{He}$$

In such nuclear reactions the total number of nucleons (protons and neutrons), and therefore their total charge, remains constant. The sum of the mass numbers and the sum of the atomic numbers of the products must equal the mass number and the atomic number of the disintegrating nucleus respectively. Thus whenever a helium nucleus is emitted, the mass number of the disintegrating nucleus decreases by 4 and its charge (atomic number) decreases by 2.

Not all radioactive nuclei emit α particles; some emit electrons. In the early studies of radioactivity, before these particles were identified as electrons, they were called β particles and the emission of electrons is still often called β emission. Thus thorium-234, which is produced by the radioactive disintegration of uranium-238, emits electrons and is transformed into protactinium-234:

$$^{234}_{90}\text{Th} \longrightarrow {}^{234}_{91}\text{Pa} + {}^{0}_{-1}\text{e}$$

In equations for nuclear reactions the electron is written as $_{-1}^{0}\text{e}$. The superscript refers to the very small mass of the electron relative to that of a proton or a neutron, and the subscript refers to the charge on the electron. Hydrogen has a radioactive isotope called tritium, $^3_1$H, which emits an electron to give an isotope of helium:

$$^3_1\text{H} \longrightarrow {}^3_2\text{He} + {}^{0}_{-1}\text{e}$$

When an electron is emitted, the mass number does not change, but the atomic number increases by 1. There are no electrons in nuclei. The emission of an electron results from the transformation of a neutron into a proton, which can be represented by the equation

$$^1_0\text{n} \longrightarrow {}^1_1\text{H} + {}^{0}_{-1}\text{e}$$

Other nuclei emit positrons. A **positron** is a particle with the same mass as an electron but with a positive charge. The symbol for a positron in an equation for a nuclear reaction is $^0_1$e. Two examples of positron emission are

$$^{39}_{19}\text{K} \longrightarrow {}^{39}_{18}\text{Ar} + {}^0_1\text{e}$$
$$^{11}_{6}\text{C} \longrightarrow {}^{11}_{5}\text{B} + {}^0_1\text{e}$$

There are no positrons in nuclei. The emission of a positron can be considered to result from the conversion of a proton to a neutron:

$$^1_1\text{H} \longrightarrow {}^1_0\text{n} + {}^0_1\text{e}$$

Positrons exist only for a very short time. Within about 10^{-9} s each positron combines with an electron and is converted to high-energy radiation called γ radiation that has a shorter wavelength than X rays.

Some nuclei undergo radioactive transformation without emitting any particles. Instead, one of the inner electrons of an atom, for example a 1s electron, enters the nucleus. This is called an **electron capture process**. Rubidium-81 is transformed in this way to krypton-81:

$$^{81}_{37}\text{Rb} + {}^{0}_{-1}\text{e} \longrightarrow {}^{81}_{36}\text{Kr}$$

Positron emission and electron capture lead to the same result; they both decrease the nuclear charge by 1 and leave the mass number unchanged.

The new nucleus formed in a radioactive-decay process may be in an excited state—in other words, its constituent neutrons and protons do not have their most stable arrangement—and it decays to the ground state by emitting high-energy γ radiation. For example, in the decay of $^{234}_{92}$U to $^{230}_{90}$Th by α particle emission, 77% of the $^{230}_{90}$Th nuclei are produced in the ground state by the emission of an α particle with an energy of 6.69×10^{-13} J, but 23% of the nuclei are produced in an excited state by the emission of an α particle with an energy of only 6.61×10^{-13} J. This excited state thorium atom, ^{230}Th, then emits γ radiation of energy 8×10^{-15} J and returns to the ground state (Figure 25.1).

The radioactive isotope cobalt-60 is used in the treatment of cancer. It emits an electron to give $^{60}_{28}$Ni in an excited state which decays to the ground state by two successive γ ray emissions

$$^{60}_{27}\text{Co} \longrightarrow {}^{60}_{28}\text{Ni}\left(\begin{array}{c}\text{excited}\\\text{state 1}\end{array}\right) + {}^{0}_{-1}\text{e}$$

$$^{60}_{28}\text{Ni}\left(\begin{array}{c}\text{excited}\\\text{state 1}\end{array}\right) \longrightarrow {}^{60}_{28}\text{Ni}\left(\begin{array}{c}\text{excited}\\\text{state 2}\end{array}\right) + \gamma(8.7 \times 10^{-13}\text{ J})$$

$$^{60}_{28}\text{Ni}\left(\begin{array}{c}\text{excited}\\\text{state 2}\end{array}\right) \longrightarrow {}^{60}_{28}\text{Ni}\left(\begin{array}{c}\text{ground}\\\text{state}\end{array}\right) + \gamma(2.1 \times 10^{-13}\text{ J})$$

Table 25.1 summarizes the various types of radioactive decay.

Do not confuse the excited states of nuclei with those of the electrons surrounding the nucleus. Particles are emitted from nuclei in excited nuclear states. Visible and ultraviolet photons are emitted by electrons in excited states.

^{60}Co is also used to irradiate food as a means of preserving it. The radiation kills the microorganisms which lead to spoilage but the food itself is largely unaffected.

FIGURE 25.1
Production of Gamma Rays.

In the decay of $^{234}_{92}U$ by α particle emission, α particles of two different energies are produced. Emission of α particles of energy 6.61×10^{-13} J leads to the formation of $^{230}_{90}Th$ in an excited state. This excited state decays to the ground state by emission of a γ ray photon of energy 8×10^{-15} J.

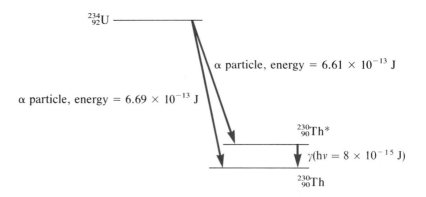

$^{234}_{92}U$

α particle, energy $= 6.61 \times 10^{-13}$ J

α particle, energy $= 6.69 \times 10^{-13}$ J

$^{230}_{90}Th^*$

$\gamma(h\nu = 8 \times 10^{-15}$ J)

$^{230}_{90}Th$

═ *Example 25.1* ═

WRITING EQUATIONS FOR NUCLEAR REACTIONS

Complete the following nuclear equations:

(a) $^{32}_{15}P \rightarrow ? + ^{0}_{-1}e$

(b) $^{43}_{19}K \rightarrow ^{43}_{20}Ca + ?$

(c) $^{210}_{84}Po \rightarrow ^{206}_{82}Pb + ?$

(d) $^{17}_{9}F \rightarrow ? + ^{0}_{1}e$

Solution

(a) The emission of an electron does not change the mass number but it increases Z by 1, so the nucleus that is formed is $^{32}_{16}S$.

(b) The mass number remains unchanged but the atomic number increases by 1, so the particle emitted must be an electron, $^{0}_{-1}e$.

(c) The mass number decreases by 4 and the atomic number decreases by 2, so the particle emitted must be a helium nucleus, $^{4}_{2}He$ (α particle).

(d) The emission of a positron means that the mass number does not change but the charge decreases by 1. Therefore the nucleus that is formed is $^{17}_{8}O$.

TABLE 25.1
Radioactive-Decay Processes

Particle Emitted		Change in		Example
		Mass Number	*Atomic Number*	
Helium nucleus (α particle)	$^{4}_{2}He$	Decreases by 4	Decreases by 2	$^{238}_{92}U \longrightarrow ^{234}_{90}Th + ^{4}_{2}He$
Electron (β particle)	$^{0}_{-1}e$	No change	Increases by 1	$^{14}_{6}C \longrightarrow ^{14}_{7}N + ^{0}_{-1}e$
Positron	$^{0}_{+1}e$	No change	Decreases by 1	$^{64}_{29}Cu \longrightarrow ^{64}_{28}Ni + ^{0}_{+1}e$
Electron capture		No change	Decreases by 1	$^{195}_{79}Au + ^{0}_{-1}e \longrightarrow ^{195}_{78}Pt$
γ ray photon	γ	No change	No change	$^{87}_{38}Sr^* \longrightarrow ^{87}_{38}Sr + \gamma$

* The asterisk denotes a nucleus in an excited state.

EXERCISE 25.1

Complete the following nuclear equations:

(a) $^{22}_{11}\text{Na} \rightarrow ^{0}_{1}\text{e} + ?$

(b) $^{27}_{12}\text{Mg} \rightarrow ^{0}_{-1}\text{e} + ?$

(c) $^{36}_{17}\text{Cl} \rightarrow ^{36}_{18}\text{Ar} + ?$

(d) $^{239}_{94}\text{Pu} \rightarrow ^{235}_{92}\text{U} + ?$

A variety of methods can be used to detect the emissions from radioactive materials. Photographic film and plates are sensitive not only to light but also to the energetic α particles, electrons, and γ rays emitted by radioactive substances. The greater the exposure to radioactive emissions, the greater is the blackening of the negative. Persons who work with radioactive substances carry a film badge to record the extent of their exposure to radiation.

An important instrument for detecting and measuring radioactivity is the **Geiger counter**. A Geiger counter (Figure 25.2) consists of a metal tube filled with a gas such as argon. One end of the tube has a thin window that allows fast-moving electrons, α particles and γ rays to pass through. In the center of the tube is a wire electrode. A potential difference of about 1000 V is maintained between the metal tube and the central wire. If a high-energy electron, α particle, or γ ray photon enters the tube through the window, it knocks electrons out of the atoms in its path. The electrons and ions that are thus formed are accelerated to high speeds by the high voltage between the central wire and the tube, and they in turn ionize other atoms, producing more electrons and ions, which in turn ionize more atoms, and so on. Thus a single, high-energy α particle or γ ray photon entering the tube produces an avalanche of ions and electrons. These give a brief pulse of electric current in the external circuit. This current pulse is amplified and recorded or is made audible as a click. The number of clicks in a given time is a direct measure of the number of particles entering the Geiger counter. Because a Geiger counter can record individual particles, it is an extremely sensitive device.

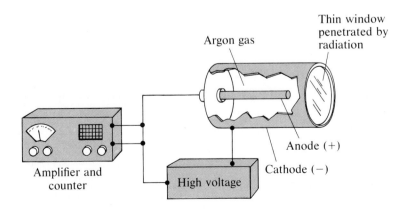

Argon gas

Thin window penetrated by radiation

Anode (+)

Cathode (−)

Amplifier and counter

High voltage

FIGURE 25.2
Schematic Representation of a Geiger Counter.

25.2
Nuclear Stability

What makes some nuclei stable and others unstable? If we plot the number of neutrons against the number of protons, that is, the atomic number, for all the known stable nuclei, we find that they all fall in a narrow band called the *band of stability* (see Figure 25.3). For stable nuclei with low atomic numbers, the number of protons is equal to the number of neutrons, but as Z increases, the number of neutrons exceeds the number of protons and the ratio of the number of neutrons to the number of protons reaches a value of approximately 1.5 for the heaviest stable nuclei. The distances between the particles in nuclei are very small—less than 10^{-15} m. At these distances the electrostatic repulsion between the protons is extremely large. Nuclei are only stable because there are very strong attractive forces between the nucleons. We call these attractive forces **nuclear forces**. Their nature has been under investigation for many years but is

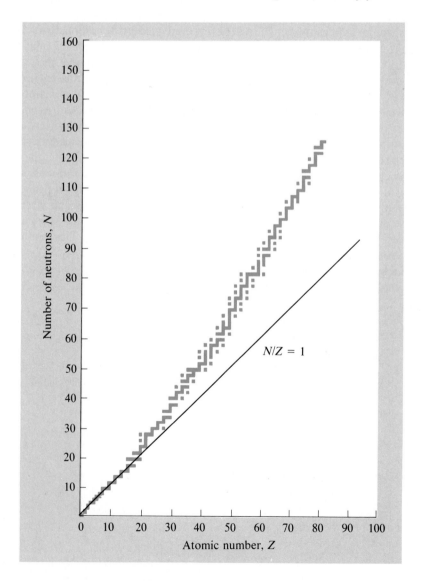

FIGURE 25.3
Plot of Number of Neutrons Against Number of Protons (Atomic Number, *Z*) for Stable Nuclei.

All the stable nuclei lie in a narrow band called the band of stability. For the light nuclei $N = Z$, but heavier stable nuclei contain more neutrons than protons.

still not fully understood. As the number of protons increases, the electrostatic repulsion between them increases, and a higher proportion of neutrons is needed to provide additional attractive forces to hold the nucleus together. But after bismuth, $Z = 83$, no increase in the number of neutrons is sufficient to hold the nucleus together, and all the nuclei with more protons than the bismuth nucleus are unstable.

Nuclei that lie outside the band of stability are unstable and decompose to give a nucleus with a more stable neutron-to-proton ratio. For example, a nucleus that lies above the band of stability must either gain protons or lose neutrons to become more stable. Thus we can understand why a nucleus such as ^{14}C, which lies above the band of stability, decays by the emission of an electron, because this process converts a neutron into a proton:

$$^{1}_{0}n \longrightarrow {}^{1}_{1}H + {}^{0}_{-1}e$$

For ^{14}C we have

$$^{14}_{6}C \longrightarrow {}^{14}_{7}N + {}^{0}_{-1}e$$

Nuclei located below the band of stability must increase their neutron-to-proton ratio to achieve stability. They can do so either by positron emission or by electron capture. Positron emission converts a proton into a neutron. An example is the decay of ^{11}C:

$$^{11}_{6}C \longrightarrow {}^{11}_{5}B + {}^{0}_{1}e$$

Electron capture similarly converts a proton to a neutron. For example,

$$^{7}_{4}Be + {}^{0}_{-1}e \longrightarrow {}^{7}_{3}Li$$

The electron is captured from an inner shell of the atom, which leaves the resulting atom in an electronically excited state. An electron then quickly drops from the valence shell to fill the vacant orbital, and a corresponding amount of energy in the form of X rays is emitted. An example is

$$^{40}_{19}K + {}^{0}_{-1}e \longrightarrow {}^{40}_{18}Ar$$

Nuclei with atomic numbers greater than 83 cannot achieve stability by electron or positron emission or electron capture. As a result, they often decay by emission of a helium nucleus (α particle), which removes two protons and two neutrons simultaneously. We have seen that ^{238}U decays to ^{234}Th, which is also radioactive and decays to ^{234}Pa. This isotope is unstable and decays to ^{234}U, which is also radioactive, and so on. Such a series of radioactive disintegrations continues until a stable (nonradioactive) isotope of an element is formed. Such a series of nuclear reactions is called a **radioactive-decay series**. For example, ^{238}U decays in a series of 14 nuclear reactions that eventually lead to the stable isotope ^{206}Pb. Another possibility for decomposition of heavy elements is fission, which we discuss later.

25.3
Radioactive-Decay Rates

HALF-LIFE

We cannot predict when an individual radioactive nucleus will decay, but each nucleus of the same kind in a sample has the same probability of decaying in a certain interval of time as any other. As a result, the rate of decay of a sample of radioactive material, that is, the number of disintegrations per unit time, is directly proportional to the number of radioactive nuclei, N, that the sample contains. The rate of decay of a given sample of radioactive material is therefore not constant, but it decreases with time since the number of nondisintegrated nuclei keeps decreasing. A characteristic property of each radioactive isotope is the time needed for half a given sample to disintegrate. This is called the **half-life**, $t_{1/2}$, of the particular radio isotope (Table 25.2). Some radioactive isotopes have very long half-lives. Others have very short half-lives. For example, $t_{1/2}$ for uranium-238 is 4.5×10^9 years, that of radon-222 is 3.8 days, while that of sodium-25 is only 1.0 min. Thus if we have a sample of 1 mg of radon-222 in a container, $\frac{1}{2}$ mg will remain after 3.8 days. After another 3.8 days one-half of the $\frac{1}{2}$ mg—or $\frac{1}{4}$ mg—will be left. Thus after 7.6 days only $\frac{1}{4}$ mg radon will remain, and after $3(3.8)$ days $= 11.4$ days, $\frac{1}{8}$ mg will remain, and so on. In other words, the amount of material which remains after the passage of a half-life period is one-half of that present at the beginning of the period. Figure 25.4 shows a plot of the decay of $^{222}_{86}$Rn. Since the number of disintegrations in a given time is proportional to the number of radioactive nuclei present, radioactive decay is an example of a first-order rate process (Chapter 19).

If there are N_0 nuclei at $t = 0$, and N at time t, then $\Delta N = N_0 - N$ nuclei have disintegrated in a time interval $\Delta t = t - t_0$. So

$$\text{Rate of decay} = \frac{\Delta N}{\Delta t} = kN$$

where k is a first-order rate constant. By comparison with the analogous equation for a first-order rate process that we discussed in Chapter 19, the number of nuclei N remaining after a time t can be calculated from the equation

$$\ln \frac{N_0}{N} = kt$$

where N_0 is the number of nuclei at $t = 0$. The time $t_{1/2}$ needed for half of the radioactive nuclei in any sample to decay—that is, the half-life—can be found

TABLE 25.2 Half-Lives of Some Radioisotopes		
Isotope	*Half-Life*	*Mode of Decay*
$^{214}_{84}$Po	164 s	α
$^{25}_{11}$Na	1.0 min	β
$^{131}_{53}$I	8.0 days	β
$^{222}_{86}$Rn	3.8 days	α
$^{32}_{15}$P	14.3 days	β
$^{60}_{27}$Co	5.3 years	β
$^{90}_{38}$Sr	28.8 years	β
$^{14}_{6}$C	5730 years	β
$^{230}_{94}$Pu	2.4×10^4 years	α
$^{40}_{19}$K	1.3×10^9 years	α
$^{238}_{92}$U	4.5×10^9 years	α
$^{232}_{90}$Th	1.4×10^{10} years	α

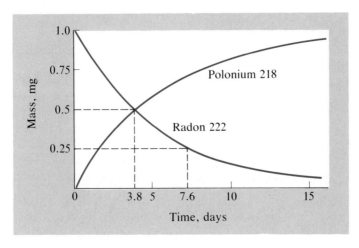

FIGURE 25.4
Plot of the Amount of
Radon-222 Against Time for
a Sample with an Initial
Mass of 1.0 mg.

Radon-222 decays to polonium-
218 by α particle emission with a
half-life of 3.8 days.

by writing $N = \frac{1}{2}N_0$ at $t = t_{1/2}$, which gives

$$\ln \frac{N_0}{\frac{1}{2}N_0} = kt_{1/2} \quad \text{or} \quad kt_{1/2} = \ln 2$$

Thus

$$t_{1/2} = \frac{0.693}{k}$$

═ Example 25.2 ═

RADIOACTIVE DECAY

The half-life of $^{222}_{86}$Rn is 3.8 days. How much $^{222}_{86}$Rn will remain after 8.5 days in a sample initially containing 45 μg of $^{222}_{86}$Rn?

Solution

We determine the rate constant k from the expression derived above:

$$k = \frac{0.693}{t_{1/2}} = \frac{0.693}{3.8 \text{ days}} = 0.18 \text{ day}^{-1}$$

Since for any given radioactive nucleus the number of atoms is proportional to the mass, we have

$$\ln \frac{N_0}{N} = \ln \frac{\text{Initial mass Rn}}{\text{Mass Rn after 8 days}} = kt$$

$$= 0.18 \text{ day}^{-1} \times 8.5 \text{ days} = 1.5$$

So

$$\ln \left(\frac{45 \text{ μg Rn}}{\text{Mass Rn after 8 days}} \right) = 1.5$$

and hence

$$\frac{45 \text{ μg Rn}}{\text{Mass Rn after 8 days}} = 4.5$$

Therefore,

$$\text{Mass Rn after 8 days} = \frac{45 \text{ μg Rn}}{4.5} = 10 \text{ μg Rn}$$

EXERCISE 25.2

How much $^{222}_{86}$Rn will remain after 12.0 days in a sample initially containing 30 µg of $^{222}_{86}$Rn?

Example 25.3

CALCULATING THE HALF-LIFE OF A RADIOISOTOPE

The radioisotope $^{131}_{53}$I is used in studies and tests on the thyroid gland. A sample that originally contained 1.00 mg of $^{131}_{53}$I contained 0.32 mg of $^{131}_{53}$I after 13.3 days. What is the half-life of $^{131}_{53}$I?

Solution

From the equation above for $t_{1/2}$, we obtain

$$k = 0.693/t_{1/2}$$

Substituting in the equation for $\ln(N_0/N)$ then gives

$$\ln \frac{N_0}{N} = \frac{0.693t}{t_{1/2}}$$

So

$$\ln \frac{1.00 \text{ mg } ^{131}_{53}\text{I}}{0.32 \text{ mg } ^{131}_{53}\text{I}} = \frac{0.693 \times 13.3 \text{ days}}{t_{1/2}}$$

$$t_{1/2} = \frac{0.693 \times 13.3}{1.14} \text{ days} = 8.08 \text{ days}$$

EXERCISE 25.3

A sample which originally contained 0.30 mg of ^{60}Co was found to contain only 0.25 mg of ^{60}Co 1.40 years later. What is the half-life of ^{60}Co?

You may wonder how we can possibly measure the half-life of a nucleus that disintegrates as slowly as uranium-238 ($t_{1/2} = 4.51 \times 10^9$ years) because the change in the mass of a sample of uranium over any measurable period of time, even as long as 10 years, is completely negligible. But because a Geiger counter is so sensitive that it can record individual particles, we can measure the half-life of uranium by measuring the rate at which it emits α particles. Let us assume that we have a 1.0-mg sample of uranium. This sample contains

$$\left(\frac{1.0 \times 10^{-3} \text{ g}}{238 \text{ g mol}^{-1}}\right)(6.022 \times 10^{23} \text{ nuclei mol}^{-1}) = 2.5 \times 10^{18} \text{ nuclei}$$

The first-order rate constant for the disintegration of uranium is

$$k = \frac{0.693}{t_{1/2}} = \frac{0.693}{4.51 \times 10^9 \text{ years}}$$

$$= \frac{0.693}{4.51 \times 10^9 \times 365 \times 24 \times 3600 \text{ s}}$$

$$= 4.9 \times 10^{-18} \text{ s}^{-1}$$

Rate of disintegration $= \dfrac{\Delta N}{\Delta t} = kN$

$$= 4.9 \times 10^{-18} \text{ s}^{-1} \times 2.5 \times 10^{18} \text{ nuclei}$$

$$= 12 \text{ nuclei s}^{-1}$$

Thus we see that the disintegration of uranium proceeds at a rate which can be relatively easily measured with a Geiger counter.

RADIOCHEMICAL DATING

An important application of radioactive decay is the dating of rocks, fossils, and ancient objects. Naturally occurring radioactive uranium-238 decays in a series of steps, the first of which gives ^{234}Th by α particle emission. This process has a half-life of 4.51×10^9 years and is by far the slowest of the steps that finally lead to the stable isotope ^{206}Pb. If we measure the amount of ^{206}Pb, for example with a mass spectrometer, we can find how much uranium-238 was present initially and, therefore, knowing the rate constant, how long it has taken this much ^{206}Pb to form. If we assume that the lead begins to accumulate once the rock has formed and solidified, the time it has taken the lead to form is equal to the age of the solid rock.

For rocks that contain potassium the reaction

$$^{40}_{19}\text{K} \xrightarrow{\text{Electron capture}} {}^{40}_{18}\text{Ar} \qquad t_{1/2} = 1.3 \times 10^9 \text{ years}$$

can also be used for dating.

$=$ *Example 25.4* $=$

CALCULATING THE AGE OF ROCKS

A sample of rock is found to contain 13.2 µg of uranium-238 and 3.42 µg of lead-206. If the half-life of $^{238}_{92}$U is 4.51×10^9 years, what is the age of the rock?

Solution

We first need to find how many grams of uranium-238 have decayed. The number of micrograms of $^{238}_{92}$U transformed into $^{206}_{82}$Pb is

$$(3.42 \text{ µg } ^{206}\text{Pb}) \left(\frac{238 \text{ g U mol}^{-1}}{206 \text{ g Pb mol}^{-1}} \right) = 3.95 \text{ µg } ^{238}\text{U}$$

The initial amount of uranium-238 in the ore was 13.2 µg plus the 3.95 µg that has been transformed into lead, that is

$$13.2 \text{ µg} + 3.95 \text{ µg} = 17.2 \text{ µg}$$

Then we have, from Example 25.3,

$$\ln \frac{N_0}{N} = \frac{0.693}{t_{1/2}} t$$

$$\ln \frac{17.2 \text{ µg}}{13.2 \text{ µg}} = \frac{0.693}{4.51 \times 10^9 \text{ years}} t$$

$$t = \frac{0.26 \times 4.51 \times 10^9 \text{ years}}{0.693} = 1.7 \times 10^9 \text{ years}$$

Thus we conclude that the uranium mineral crystallized from the molten magma 1.7 billion years ago. The oldest rocks on earth analyzed by this method solidified about 3.6 billion years ago. Clearly the earth is older than this, and other information indicates that the age of the earth is about 4.5 billion years.

EXERCISE 25.4

A sample of rock contains 17.4 µg of ^{238}U and 1.45 µg of ^{206}Pb. Given that the half-life of ^{238}U is 4.51×10^9 years, how old is the rock?

Another important radiodating method is that based on carbon-14, which decays by the reaction

$$^{14}_{6}C \longrightarrow ^{14}_{7}N + ^{0}_{-1}e$$

The half-life of carbon-14 is 5730 years. Carbon-14 is produced in the atmosphere by bombardment of nitrogen with cosmic rays that are high-energy particles such as protons and neutrons originating from the sun and other parts of the universe. Carbon-14 is formed by the reaction

$$^{14}_{7}N + ^{1}_{0}n \longrightarrow ^{14}_{6}C + ^{1}_{1}H$$

Because ^{14}C is produced in the upper atmosphere at a constant rate and because it decays at a constant rate, there is a small but constant concentration of $^{14}CO_2$ in the atmosphere. Plants use atmospheric CO_2 to make carbohydrates in photosynthesis, so there is the same small concentration of carbon-14 in all living plants and animals. But when a plant or animal dies, it no longer incorporates carbon-14, so the amount of carbon-14 in it decreases gradually with the passage of time. Thus by measuring the amount of carbon-14 left in any formerly living material, we can determine the time that has passed since it died.

Because the amount of carbon-14 in living matter is extremely small, it would be very difficult to measure the amount present with any accuracy. It is much simpler and much more accurate to measure the rate at which the carbon-14 disintegrates by counting the number of disintegrations per second per gram of material. In the atmosphere and in all living organisms, there are 15.3 disintegrations of carbon-14 per minute per gram of carbon. When the organism dies, this rate of disintegration decreases with a half-life of 5730 years. The rate of disintegration, R, at time t is proportional to N, the number of radioactive nuclei at time t. Thus we can transform the equation, obtained in Example 25.3,

$$t = \frac{t_{1/2}}{0.693} \ln \frac{N_0}{N}$$

to the form

$$t = \frac{t_{1/2}}{0.693} \ln \frac{R_0}{R}$$

For carbon, $R_0 = 15.3$ disintegrations per minute per gram and $t_{1/2} = 5730$ years. So we have

$$t = \frac{5730}{0.693} \ln \frac{15.3}{R} \text{ years} = 8.27 \times 10^3 \ln \frac{15.3}{R} \text{ years}$$

The time that has elapsed since any living material died can be determined by using this equation, as illustrated in the following example.

= *Example 25.5* =

CALCULATING AGES FROM CARBON-14 DATA

A sample of charcoal from one of the earliest Polynesian settlements in Hawaii had a disintegration rate of 13.6 disintegrations per minute per gram. What is the age of the charcoal?

Solution

The time since the tree providing the charcoal was cut can be obtained from the equation

$$t = 8.27 \times 10^3 \ln \frac{15.3}{R} \text{ years}$$

In this case $R = 13.6$ disintegrations per minute per gram, so that

$$t = 8.27 \times 10^3 \ln \frac{15.3}{13.6} \text{ years} = 974 \text{ years}$$

This result suggests that the Polynesians first arrived in Hawaii around the year 1010 A.D.

EXERCISE 25.5

What is the age of a sample of charcoal in which the rate of disintegration is 11.2 per minute per gram?

25.4
Artificial Radioisotopes

About one-half of the uranium-238 present when the earth's crust solidified still remains; the rest has been transformed into lead-206. But most radioactive nuclei decay much more rapidly, so even if they were present when the earth was formed, they would have completely disappeared many years ago and we would know nothing about them. However, a large number of radioactive nuclei have been made by nuclear reactions in recent times.

SYNTHESIS OF RADIOISOTOPES

Rutherford was the first to carry out a nuclear reaction in the laboratory. He observed that when he bombarded nitrogen with a beam of α particles he produced $^{17}_8O$ and protons by the reaction

$$^{14}_7N + {}^4_2He \longrightarrow {}^{17}_8O + {}^1_1H$$

This nuclear reaction gives the stable isotope $^{17}_8O$, but many radioactive nuclei can be produced by similar reactions. For example, $^{27}_{13}Al$ can be transformed to $^{30}_{15}P$ by bombardment with α particles:

$$^{27}_{13}Al + {}^4_2He \longrightarrow {}^{30}_{15}P + {}^1_0n$$

In these experiments α particles were obtained by the disintegration of radioactive elements such as uranium. But α particles obtained in this way do not have enough energy to react with many heavy nuclei. The high charge of a heavy nucleus repels a positively charged α particle so that it cannot enter the nucleus unless the particle has a very high energy. If α particles are to react with heavy nuclei, their energy must be increased by accelerating them to a high velocity in a machine called a particle accelerator, such as a cyclotron. For example, uranium-238 is converted to plutonium-239 when it is bombarded with high-energy α particles:

$$^{238}_{92}U + {}^4_2He \longrightarrow {}^{239}_{94}Pu + 3{}^1_0n$$

An alternative method for producing new nuclei is to use neutrons as the bombarding particles. Because they are neutral, neutrons are not repelled electrostatically by nuclei and they do not therefore need to be moving at a high speed to enter a nucleus. Neutrons can be obtained from neutron-producing reactions such as the conversion of $^{27}_{13}Al$ to $^{30}_{15}P$ or the conversion of $^{238}_{92}U$ to $^{239}_{94}Pu$. Nuclear reactors, in which many neutron-producing reactions occur, are often used for bombarding nuclei with neutrons to produce new nuclei. For example, cobalt-60, used in radiation therapy for cancer, is produced by the reaction:

$$^{59}_{27}Co + {}^1_0n \longrightarrow {}^{60}_{27}Co$$

In discussing the halogens in Chapters 5 and 20, we only very briefly mentioned the last element in the halogen family, astatine, because it has been much less studied than the other halogens. Astatine is radioactive and can only be produced in very small amounts. It was first made by the reaction

$$^{209}_{83}Bi + {}^4_2He \longrightarrow {}^{211}_{85}At + 2{}^1_0n$$

The isotope $^{211}_{85}At$ has a half-life of only 7.5 h, and even the most stable isotope, $^{210}_{85}At$, has a half-life of only 8.5 h, so large amounts cannot be accumulated.

Another radioactive element that is not found in nature is technetium, Tc, which is in the second series of transition metals below manganese. Technetium can be made by neutron bombardment of molybdenum:

$$^{98}_{42}Mo + {}^1_0n \longrightarrow {}^{99}_{42}Mo$$
$$^{99}_{42}Mo \longrightarrow {}^{99}_{43}Tc + {}^0_{-1}e$$

USES OF RADIOISOTOPES

Many artificially produced radioisotopes have important applications in medicine, agriculture, oil exploration, and many other fields. Because the radiation emitted from a radioisotope is easily detected, the movement of an element in the human body is easily followed. The radiation emitted by a radioisotope can give an image of any organ in which the radioisotope concentrates. Thus sodium-24 is used to follow blood circulation, technetium-99 is used for brain and liver scans, and iodine-123 is used for thyroid imaging.

In chemistry one of the important applications of radioisotopes is to study reaction mechanisms. We saw in Chapter 19 that most reactions take place in a series of steps. Radioisotopes can often be used to help work out the exact sequence of steps in a complex reaction mechanism. If a very small percentage of the atoms of an element in a sample are exchanged for a radioactive isotope, the element is said to be *labeled*. Since a radioactive isotope has the same chemical properties as the stable isotopes of an element, the radioactivity of the labeled element can be used to detect the movement of the element through a complex series of reactions.

Melvin Calvin (born 1911, St. Paul, Minnesota) and his fellow researchers exposed growing plants to an atmosphere of carbon dioxide labeled with radioactive carbon-14 for just a few seconds. They then extracted and separated as many of the compounds in the plant as possible. Those compounds that were found to contain carbon-14 could then be supposed to be involved in the early stages of photosynthesis. By much painstaking work from 1949 to 1957 they were able to work out a mechanism for the very complex reaction by which plants convert carbon dioxide and water to sugars and carbohydrates in the presence of light.

Until the development in the 1940s of accelerators for producing high-speed particles, the last element in the periodic table was element 92, uranium. Since that time the periodic table has been extended up to at least element 106. These transuranium elements have. been produced by nuclear reactions. We have already mentioned the formation of plutonium by α particle bombardment of uranium. Some other examples are given in Table 25.3. Some of these transuranium elements can be produced in small but nevertheless commercially useful quantities. Americium-241 is used in one type of home smoke detector (Figure 25.5). The α particles emitted by americium-241, like all α particles produced by radioactive nuclei, have a very low penetrating power, so they do not escape from the detector and thus do not constitute a health hazard.

EFFECTS OF RADIATION

The increasing use of radioisotopes in medicine and industry and the increasing number of nuclear reactors have led to increased concern over the biological effects of radiation. Electrons, α particles, and γ rays emitted by radioactive

---

TABLE 25.3
Synthesis of Some Transuranium Elements

Atomic Number	Name	Symbol	Reaction
93	Neptunium	Np	$^{238}_{92}U + ^{1}_{0}n \longrightarrow ^{239}_{93}Np + ^{0}_{-1}e$
94	Plutonium	Pu	$^{238}_{92}U + ^{2}_{1}H \longrightarrow ^{238}_{93}Np + 2^{1}_{0}n$
			$^{238}_{93}Np \longrightarrow ^{238}_{94}Pu + ^{0}_{-1}e$
95	Americium	Am	$^{239}_{94}Pu + ^{1}_{0}n \longrightarrow ^{240}_{95}Am + ^{0}_{-1}e$
96	Curium	Cm	$^{239}_{94}Pu + ^{4}_{2}He \longrightarrow ^{242}_{96}Cm + ^{1}_{0}n$

FIGURE 25.5
Home Smoke Detector.

The ionization chamber contains a small quantity of americium-241, which decays by α particle emission with a half-life of 432 years. The α particles ionize the air in the ionization chamber. The ions are accelerated by a potential provided by a battery so that an electric current flows across the ionization chamber. This current can be detected in an external circuit. When smoke enters the chamber, the smoke particles impede the movement of ions, and the current is reduced. This decrease in current is detected electronically.

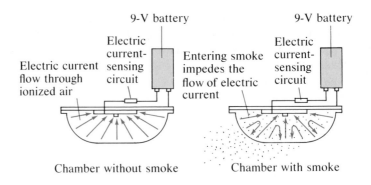

nuclei have energies far in excess of that needed to break chemical bonds. When these high-energy particles and gamma rays pass through matter, they break up some molecules, forming free radicals and ions. For example, water may be split into hydrogen atoms and hydroxyl radicals:

$$H_2O \longrightarrow H\cdot + \cdot OH$$

Many of these free radicals and ions are very reactive. In a biological system free radicals may disrupt the normal operation of the cell and may even kill the cell. Indeed, γ rays are used routinely to destroy cancerous cells.

The damage caused by a radiation source outside the body depends on the penetrating ability of the radiation. Gamma rays are particularly dangerous because, like X rays, they penetrate human tissue very effectively. In contrast, α particles are stopped by the skin, and electrons do not penetrate far below the skin. However, if a radiation source enters the body, it can be particularly dangerous. For example, α emitters are generally the nuclei of heavy elements that tend to concentrate in the bones where they may cause considerable damage to the bone and surrounding tissues. However, we should be aware that we are subject to radiation at all times. Radioactive minerals, such as those of uranium, have been present on the earth since its formation. Cosmic radiation causes nuclear reactions in the atmosphere which produce radioactive nuclei such as those of carbon and potassium. Since these are essential elements in living organisms, humans and all other organisms are continually subjected to radiation from the disintegration of these nuclei. The normal low level of radiation to which we are all exposed is called background radiation. This background radiation causes mutations in living cells and thus has actually contributed to the mechanism by which the great variety of living organisms has been produced. It is usually considered to be exposure to radiation that is many times in excess of normal background radiation that is dangerous.

25.5
Nuclear Energy

We saw in Chapter 2 that the mass of a helium-4 atom is not equal to the mass of its constituents. The mass of two protons and two neutrons is 4.031 88 u, but the mass of one $^4_2$He nucleus is only 4.001 50 u. The difference of 0.030 38 u arises because a very large amount of energy is released when protons and neutrons combine to form a nucleus. The amount of energy is so large that it has a significant mass equivalent, given by the Einstein equation

$$E = mc^2$$

where E is the energy, m is the mass, and c is the velocity of light. For a change in mass, Δm, the energy change is $\Delta E = c^2 \, \Delta m$. Thus the energy change for the formation of a helium nucleus from two protons and two neutrons is

$$E = (2.998 \times 10^8 \text{ m s}^{-1})^2(0.030\,38 \text{ u})\left(\frac{1.000 \text{ g}}{6.022 \times 10^{23} \text{ u}}\right)\left(\frac{1 \text{ kg}}{1000 \text{ g}}\right)$$

$$= 4.534 \times 10^{-12} \text{ kg m}^2 \text{ s}^{-2} = 4.534 \times 10^{-12} \text{ J}$$

The formation of 1 mol of helium-4 nuclei would produce an enormous amount of energy:

$$\left(\frac{6.022 \times 10^{23} \text{ nuclei}}{1 \text{ mol nuclei}}\right)(4.534 \times 10^{-12} \text{ J nuclei}^{-1}) = 2.730 \times 10^{12} \text{ J mol}^{-1}$$

Conversely, this amount of energy would be needed to break 1 mol of helium nuclei into their constituent protons and neutrons. The energy required to decompose a nucleus into protons and neutrons is called the **binding energy** of the nucleus; it is the energy needed to overcome the very strong nuclear forces that hold the nucleus together.

For comparison of the binding energies of different nuclei, it is convenient to quote the value of the *binding energy per nucleon*:

$$\text{Binding energy per nucleon} = \frac{\text{Binding energy}}{\text{Number of nucleons}} = \frac{\text{Binding energy}}{\text{Mass number}}$$

═ *Example 25.6* ═════════════════════════════════

CALCULATING THE BINDING ENERGY OF NUCLEI AND NUCLEONS

The mass of $^{56}_{26}$Fe is 55.906 38 u. What are the binding energy and the binding energy per nucleon?

To obtain the mass of the nucleus, subtract the total mass of the electrons from the mass of an atom of the isotope.

Solution

The masses of the proton and the neutron are given in Table 2.1. They are 1.007 28 and 1.008 66 u, respectively. The mass difference between the mass of the $^{56}_{26}$Fe nucleus and its constituent particles (26 protons and 30 neutrons) is

$$\Delta m = (26 \times 1.007\,28 \text{ u}) + (30 \times 1.008\,66 \text{ u}) - 55.906\,38 \text{ u}$$

$$= 56.4491 \text{ u} - 55.906\,38 \text{ u} = 0.5427 \text{ u}$$

Hence

$$\Delta E = c^2 \, \Delta m = (2.998 \times 10^8 \text{ m s}^{-1})^2(0.5427 \text{ u})\left(\frac{1.000 \text{ g}}{6.022 \times 10^{23} \text{ u}}\right)\left(\frac{1 \text{ kg}}{1000 \text{ g}}\right)$$

$$= 8.100 \times 10^{-11} \text{ kg m}^2 \text{ s}^{-2} = 8.100 \times 10^{-11} \text{ J}$$

This is the binding energy of $^{56}_{26}$Fe. There are 56 nucleons (26 protons and 30 neutrons) in $^{56}_{26}$Fe, so the binding energy per nucleon is

$$\frac{8.100 \times 10^{-11} \text{ J}}{56 \text{ nucleons}} = 1.446 \times 10^{-12} \text{ J nucleon}^{-1}$$

EXERCISE 25.6

The mass of ^{16}O atom is 15.994 91 u. Hence the mass of the ^{16}O nucleus is 15.990 53 u. What is its binding energy, and what is its binding energy per nucleon?

FIGURE 25.6
Plot of Binding Energy per
Nucleon Against Mass
Number.

The most stable nucleus is $^{56}_{29}$Fe.
If lighter nuclei are combined in a
fusion reaction, energy is released.
If nuclei heavier than $^{56}_{29}$Fe are
split in a fission reaction, energy
is released.

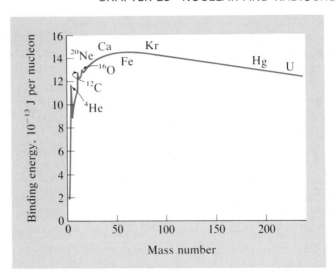

Figure 25.6 shows a plot of the binding energy per nucleon against the mass number. The binding energy per nucleon increases up to ^{56}Fe and then decreases slowly. Thus ^{56}Fe is the most stable nucleus. This curve shows that energy is released when very heavy nuclei are broken up into lighter nuclei. This process is called **fission**. A fission reaction was first discovered during the search for transuranium elements in the 1930s. If a uranium-235 nucleus absorbs a neutron, it breaks into two nuclei and at the same time several neutrons are ejected (Figure 25.7). Uranium-235 nuclei can split into several different pairs of nuclei. Two of these reactions are

$$^{235}_{92}U + ^{1}_{0}n \quad\left\langle\begin{array}{l} ^{137}_{52}Te + ^{97}_{40}Zr + 2\,^{1}_{0}n \\ ^{142}_{56}Ba + ^{91}_{36}Kr + 3\,^{1}_{0}n \end{array}\right.$$

The average number of neutrons produced in the fission of a uranium-235 nucleus is 2.4. Suppose that two of these neutrons each cause the fission of another uranium nucleus. These two fissions produce four neutrons, which can cause the fission of four more nuclei, which produces eight more neutrons, which can cause the fission of eight nuclei, and so on (Figure 25.8). The number of

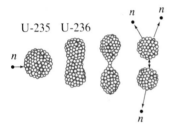

FIGURE 25.7
The Neutron-Induced Fission
of Uranium-235.

Absorption of a neutron produces
the $^{236}_{92}$U nucleus in an excited
state. This deforms and splits in
two in the same way as a vibrating
drop of liquid might split in two.
Simultaneously, several neutrons
are emitted.

FIGURE 25.8
Branching Chain Reaction
Produced When
Uranium-235 Undergoes
Fission.

If each fission produces two neutrons, these neutrons cause the fission of two uranium-235 nuclei,
each of which produces two neutrons. The four neutrons thus
produced cause the fission of four
more nuclei, thus producing eight
neutrons, which can cause the
fission of eight nuclei, and so on.
The number of fission reactions
increases very rapidly, resulting in
an explosive release of energy.

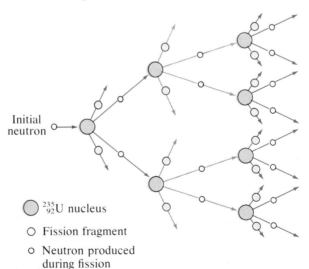

fissions rapidly escalates, and this process is called a **branching chain reaction**. The associated liberation of a large amount of energy causes an enormous explosion. But for a chain reaction to occur, the sample of uranium must have a minimum mass. Otherwise, neutrons escape from the surface of the sample before they have a chance to cause fission. The minimum mass of fissionable material needed to ensure that each fission process causes one further fission is called the **critical mass**. The critical mass depends on the shape and purity of a particular sample. If the critical mass is exceeded, a branching chain reaction occurs. One of the ways that a critical mass can be obtained is to very rapidly combine two pieces of fissionable material, each of which has less than the critical mass, for example, by firing one piece into the other by using a conventional explosive. This method is used in one type of nuclear (atom) bomb (Figure 25.9).

NUCLEAR REACTORS

In a nuclear reactor the fission reaction is carried out in a controlled manner so that just *one* of the 2.4 neutrons emitted in an average fission is captured by another fissionable nucleus. Achieving this precise control is one of the main problems of nuclear reactors. Some neutrons, of course, are lost through the reactor surface and by absorption into the material of which the reactor is constructed. But most important is the fact that natural uranium contains only 0.7% uranium-235. The much more abundant uranium-238 readily captures the fast neutrons liberated in fission but does not usually undergo fission itself. The neutrons absorbed by ^{238}U are therefore wasted. However, ^{238}U has little ability to capture *slow* neutrons, which readily cause fission in ^{235}U. Thus if the neutrons are slowed down they will not be absorbed by ^{238}U and therefore wasted, and more will be available to cause fission in ^{235}U.

A substance called a *moderator* is used to slow down the neutrons. When a fast neutron collides with a nucleus, some of its kinetic energy is imparted to the nucleus. When two particles collide, the greatest transfer of energy occurs if the particles have equal masses. Thus hydrogen nuclei would be the most efficient moderator for neutrons because they have very nearly the same mass. Indeed, ordinary ("light") water is used as a moderator in one type of nuclear reactor. The water is also the coolant in such reactors. However, protons also combine with neutrons to form deuterons

$$^1_1H + ^1_0n \longrightarrow ^2_1H$$

so that many neutrons are lost by reaction of neutrons with the H atoms of H_2O. Thus in such a reactor it is necessary to use uranium enriched to about 3% in ^{235}U. Uranium is enriched by converting it to uranium hexafluoride, UF_6, and using gaseous diffusion to separate $^{235}UF_6$ from $^{238}UF_6$ (see Chapter 3). This process requires a large plant and much energy and is therefore expensive.

Another important type of nuclear reactor uses "heavy" water, D_2O, as a moderator because deuterium nuclei have a much smaller tendency than protons to combine with neutrons. In this type of reactor natural uranium can be used as a fuel. The production of heavy water is an energy-consuming and expensive process, but once the reactor has been constructed only small amounts are needed to make up for small losses.

The fuel of a nuclear reactor consists of uranium oxide, UO_2, pellets packed into a zirconium alloy tube to form what is called a fuel rod. A number of these

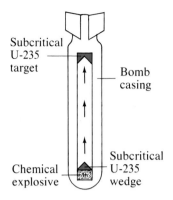

FIGURE 25.9
Diagram of an Atomic Bomb.

One piece of uranium-235 is fired into the other piece by means of a conventional explosive. When the two masses collide, the critical mass is exceeded, and a neutron chain reaction ensues with a corresponding release of energy as an explosion.

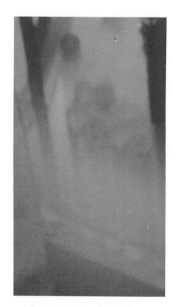

Looking down into a "light" water moderated nuclear reactor. The intense blue glow is due to radiation emitted during the fission process.

fuel rods are then packed together to form the reactor core. Control rods made of substances such as cadmium or boron, which strongly absorb neutrons, are inserted into spaces between the fuel rods. The control rods are moved in and out of the reactor core to adjust the rate of the fission reaction. The energy of the fission process appears as heat. The heat is transferred outside the reactor by means of a suitable fluid—often water. In some types of reactors it may be liquid sodium, or a sodium–potassium alloy, or a gas such as carbon dioxide or helium. This heat is used to boil water to form steam which is used to drive a turbine as in a conventional power plant (Figure 25.10).

NUCLEAR FUSION

We can see from Figure 25.6 that the conversion of very light nuclei to heavier nuclei also results in a release of large amounts of energy, even larger, in fact, than the energy obtained from fission reactions. Such reactions are called **fusion reactions** because small nuclei fuse to form a larger nucleus.

The basic energy-producing process in stars—and hence the source of nearly all the energy in the universe—is the fusion of hydrogen nuclei into helium nuclei. In the sun the following reactions predominate:

$$^1_1H + ^1_1H \longrightarrow ^2_1H + ^0_1e$$

$$^1_1H + ^2_1H \longrightarrow ^3_2He$$

$$^3_2He + ^3_2He \longrightarrow ^4_2He + 2\,^1_1H$$

In order to initiate and continue such reactions, a very high temperature is needed so that the nuclei collide with sufficient kinetic energy to overcome their mutual electrostatic repulsion. Indeed, a temperature of approximately 10^7 K is needed for fusion reactions to occur in such quantity that a substantial amount of energy is produced.

In order to produce a fusion reaction, we must somehow reproduce on earth the extremely high temperature of the sun. In the hydrogen or thermonuclear bomb this high temperature is produced by a fission reaction. In one such device a fission bomb is surrounded by lithium deuteride, $Li^2H(LiD)$. Neutrons from the fission reaction are captured by lithium nuclei:

$$^6_3Li + ^1_0n \longrightarrow ^4_2He + ^3_1H$$

And at the very high temperature produced by the fission reaction, tritium, ^3_1H, undergoes fusion with deuterium:

$$^3_1H + ^2_1H \longrightarrow ^4_2He + ^1_0n$$

But controlling nuclear fusion so that it can be used for the production of usable power is a problem that has challenged scientists and engineers for over twenty years. A fundamental difficulty is that the temperature is too high for the fusion reaction to be confined to any solid container. One approach involves using powerful laser beams to heat tiny pellets containing deuterium to a very high temperature to produce what are in effect miniature hydrogen bomb explosions. A succession of such explosions could furnish a steady supply of energy. Another method that has been extensively studied is to contain an extremely hot mass of gaseous atoms stripped of their electrons—called a plasma—by means of a very strong magnetic field. Intensive research continues, but practical fusion reactors are probably still far in the future.

FIGURE 25.10
A Nuclear Reactor.

The heat produced in the core is transferred to water, which flows in a closed loop through a vessel containing water that is heated to its boiling point, forming steam. The steam is used to drive a turbine, which produces electricity. The steam from the turbine is condensed to water in a cooling tower, and the resulting water is pumped back into the steam generator.

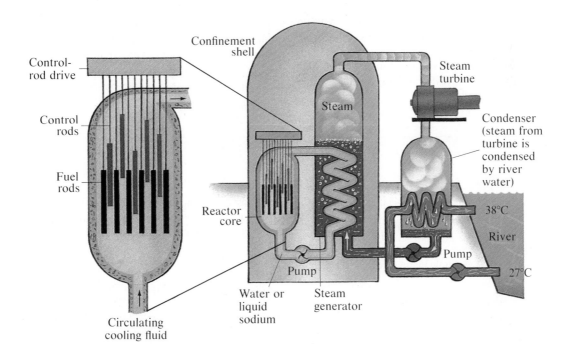

A nuclear power plant

IMPORTANT TERMS

The **binding energy** of a nucleus is the energy needed to overcome the very strong nuclear forces that hold the protons and neutrons together in the nucleus.

A **branching chain reaction** is a fission reaction caused by a neutron, which produces more than one neutron, which can in turn cause the fission of more nuclei, and so on.

The **critical mass** is the minimum mass of fissionable material for which a branching chain reaction can be sustained.

Electron capture is a process in which a nucleus absorbs an inner electron and is thus transformed to a new nucleus.

A **fission reaction** is a nuclear reaction in which a heavy nucleus splits into two lighter nuclei with the production of a large amount of energy.

A **fusion reaction** is a nuclear reaction in which two light nuclei combine to form a heavier nucleus with the production of a large amount of energy.

A **Geiger counter** is an instrument for detecting the α particles and γ rays emitted by radioactive nuclei.

The **half-life** of a radioactive nucleus is the time it takes for one-half of the nuclei in a given sample to disintegrate.

Nuclear forces are the strong attractive forces that hold the protons and the neutrons together in a nucleus.

A **nuclear reaction** is a process in which nuclei are changed into other nuclei and particles.

A **positron** is a particle with the same mass as an electron but with a positive charge.

A **radioactive-decay series** is the series of nuclear reactions by which an unstable radioactive nucleus decays to a stable nucleus.

Radioactivity is the spontaneous decomposition of unstable nuclei.

A **radioisotope** is a radioactive isotope.

PROBLEMS *

Composition of Nuclei

1. How many protons and how many neutrons are there in each of the following nuclei?

   (a) $^{6}_{3}\text{Li}$ **(b)** $^{13}_{6}\text{C}$ **(c)** $^{94}_{40}\text{Zr}$ **(d)** $^{137}_{56}\text{Ba}$

2. How many protons and how many neutrons are there in each of the following nuclei?

   (a) ^{22}Ne **(b)** ^{88}Sr **(c)** ^{92}Sr

   (d) ^{180}W **(e)** ^{242}Cm

Radioactivity

3. What is the mass (in atomic mass units) and the charge (in units of the charge on an electron) of each of the following **(a)** an α particle, **(b)** a β particle, and **(c)** a positron?

4. An early law that was formulated to express the results of the radioactive decay of nuclei is the Russell–Soddy group displacement law. This states that, relative to the position of a radioactive element in the periodic table,

   (a) Emission of an α particle results in a product isotope located two groups to the left in the periodic table, and

   (b) Emission of a β particle results in a product isotope located one group to the right in the periodic table.

Explain this law in terms of the nature of the α particle and the β particle, and give specific examples to illustrate the law.

* Answers to problems numbered in blue appear at the end of the text.

5. How could the Russell–Soddy group displacement law (see Problem 4) be extended to include **(a)** positron emission and **(b)** electron capture?

6. An $^{80}_{35}\text{Br}$ nucleus can decay by **(a)** electron emission, **(b)** positron emission, or **(c)** electron capture. What is the product nucleus in each case?

7. The $^{233}_{90}\text{Th}$ nucleus undergoes two successive electron emissions to give an isotope of uranium. Which isotope of uranium is formed?

8. Complete each of the following equations for radioactive-decay processes:

   (a) $^{32}_{15}\text{P} \rightarrow ^{32}_{16}\text{S} + ?$ **(b)** $^{15}_{8}\text{O} \rightarrow ^{15}_{7}\text{N} + ?$

   (c) $^{52}_{26}\text{Fe} \rightarrow ^{52}_{25}\text{Mn} + ?$ **(d)** $^{218}_{87}\text{Fr} \rightarrow ? + ^{4}_{2}\text{He}$

   (e) $^{50}_{26}\text{Fe} \rightarrow ? + ^{0}_{-1}\text{e}$ **(f)** $^{122}_{53}\text{I} \rightarrow ^{122}_{54}\text{Xe} + ?$

9. The $^{207}_{84}\text{Po}$ isotope of polonium can decay in three ways: **(a)** by electron capture, **(b)** by positron emission, and **(c)** by α particle emission. Write balanced nuclear equations for each of these nuclear reations.

10. Thorium-231 decays to lead-207 in a stepwise fashion by emitting the following particles in successive steps: β; α; α; β; α; α; α; β; β; α. Write the symbol for the isotope formed in each step.

11. Uranium-235 undergoes radioactive decay with the emission of an α particle. The product decays by the emission of an electron. This product emits an α particle, and the next product emits an electron. Write balanced equations for these first four steps in the decay scheme for uranium-235.

Nuclear Reactions

12. Fill in the missing symbols in each of the following equations for nuclear reactions:

(a) $^{35}_{17}Cl + ? \rightarrow ^{32}_{16}S + ^{4}_{2}He$

(b) $^{15}_{7}N + ? \rightarrow ^{12}_{6}C + ^{4}_{2}He$

(c) $^{12}_{6}C + ^{12}_{6}C \rightarrow ? + ^{1}_{1}H$

(d) $? + ^{4}_{2}He \rightarrow ^{7}_{4}Be + \gamma$

13. Complete and balance each of the following equations for nuclear reactions:

(a) $^{32}_{16}S + ^{1}_{0}n \rightarrow ^{1}_{1}H + ?$

(b) $^{7}_{4}Be + ^{0}_{-1}e$ (1s electron) $\rightarrow ?$

(c) $^{81}_{37}Rb \rightarrow ? + ^{0}_{+1}e$

(d) $^{1}_{1}H + ^{11}_{5}B \rightarrow 3?$

(e) $^{235}_{92}U + ^{1}_{0}n \rightarrow ^{135}_{54}Xe + ? + 2^{1}_{0}n$

(f) $^{249}_{98}Cf + ^{18}_{8}O \rightarrow ? + 4^{1}_{0}n$

14. Complete and balance each of the following equations for nuclear reactions:

(a) $^{179}_{75}Re + ^{0}_{-1}e \rightarrow ?$ 　　(b) $^{149}_{62}Sm \rightarrow ? + ^{4}_{2}He$

(c) $^{228}_{88}Ra \rightarrow ? + ^{0}_{-1}e$ 　　(d) $? + ^{25}_{12}Mg \rightarrow ^{26}_{13}Al + \gamma$

(e) $? + ^{26}_{12}Mg \rightarrow ^{26}_{13}Al + ^{1}_{0}n$

Nuclear-Decay Rates and Half-Lives

15. Germanium-66 decays by positron emission with a half-life of 2.50 h.

(a) Write the balanced equation for the nuclear reaction.

(b) How much ^{66}Ge remains from a 50.0 mg sample after 12.5 h?

16. Cesium-137 is produced from fission reactions in nuclear reactors. As a result of the Chernobyl accident in Russia, ^{137}Cs was spread over large areas of Europe. The half-life is 30.2 years. How long will it take for the activity to decrease to 1.0% of its value immediately after the accident?

17. Radioactive fallout from nuclear explosions or accidents contains ^{131}I, with a half-life of 8.0 days, and ^{90}Sr, with a half-life of 19.9 years. How long will it take each isotope to decay to (a) 10% and (b) 1.0% of its initial concentration? Which isotope will have the most serious long-term effects?

18. Determine the half-life of a radioactive isotope which gave the following disintegrations per minute at the times indicated:

Time (min)	Counts min^{-1}
0	668
2	611
4	534
8	429
16	277
24	179
40	75
60	25

19. A sample was known to contain originally 3.40 mg of radioactive ^{32}P. After 10.0 days it was found to contain only 2.09 mg of ^{32}P. What is the half-life of ^{32}P?

20. One of the principal sources of radioactivity inside your body is the disintegration of radioactive ^{40}K in your bones, since this is one of the naturally occuring isotopes of potassium. What is the half-life of ^{40}K, given that only 7.0% of the ^{40}K that was present in the earth when it was formed 4.5 billion years ago still remains.

21. A sample of charcoal from the Lascaux cave in France had a count rate of 2.4 disintegrations per minute per gram. Assuming that the fire that produced the charcoal was lit by the artists of the renowned cave paintings, when did these artists live?

22. A sample containing ^{42}K, a radioactive isotope of potassium with a half-life of 12.4 h, has an initial activity of 1.10×10^9 disintegrations per minute.

(a) Calculate the activity after 30.0 h.

(b) How long will it take for the initial activity to diminish to 1.0×10^5 disintegrations per minute?

23. A wooden artifact from an ancient Chinese temple contained 25.0 g of carbon and had a disintegration rate of 110 counts per minute. Determine the age of the artifact.

24. A sample of wood containing 50.0 g of carbon, from an ancient casket, had a disintegration rate of 95 disintegrations per minute. Approximately when did the person buried in that casket die?

Binding Energy

25. The mass of the ^{35}Cl isotope is 34.9689 u. Calculate the binding energy, and the binding energy per nucleon.

26. Calculate the total binding energy, and the binding energy per nucleon, for each of the following:

	Atom	Atomic mass (u)
(a)	$^{20}_{10}Ne$	19.992 44
(b)	$^{64}_{30}Zn$	63.929 14
(c)	$^{61}_{28}Ni$	60.930 06
(d)	$^{226}_{88}Ra$	226.025 4

27. Calculate the binding energy per nucleon for each of the following nuclei:

	Atom	Atomic mass (u)
(a)	$^{2}_{1}H$	2.014 10
(b)	$^{16}_{8}O$	15.994 91
(c)	$^{56}_{26}Fe$	55.934 9
(d)	$^{125}_{52}Te$	124.904 4
(e)	$^{235}_{92}U$	238.050 8

28. Plot the results of Problems 26 and 27 versus mass number. Compare your graph with Figure 25.6.

29. Calculate the binding energy per nucleon for each of ^{14}N and ^{15}N, for which the atomic masses are 14.003 07 u and 15.000 11 u, respectively. What do you conclude about the relative stabilities of these two isotopes? Which isotope has the greater natural abundance?

30. Is the reaction

$$^{7}_{3}Li + {}^{1}_{1}H \longrightarrow {}^{4}_{2}He + {}^{4}_{2}He$$

exothermic, or endothermic? What is the amount of energy absorbed or released in this nuclear reaction?

Nuclear Fusion and Fission

31. For a controlled fusion process, a possible nuclear reaction is the conversion of deuterium, $^{2}_{1}H$, to the helium isotope $^{3}_{2}He$:

$$^{2}_{1}H + {}^{2}_{1}H \longrightarrow {}^{3}_{2}He + {}^{1}_{0}n$$

Compare the energy released per mole of deuterium consumed in this nuclear reaction with that obtained by burning deuterium in oxygen to give heavy water, $^{2}_{1}H_2O$. Use data from Table 2.1 and assume that the enthalpy of combustion of deuterium is the same as that of natural H_2.

* The asterisk denotes the more difficult problems.

32. Explain the importance of each of the following in nuclear reactors:

    (a) Critical mass   **(b)** The moderator

    (c) The control rods

Miscellaneous

33. By consulting a reference book, such as the *Handbook of Physics and Chemistry*, give examples of stable nuclei with each of the following characteristics:

    (a) An *even* number of neutrons, and an *even* number of protons

    (b) An *even* number of neutrons, and an *odd* number of protons

    (c) An *odd* number of neutrons, and an *even* number of protons

    (d) An *odd* number of neutrons, and an *odd* number of protons

34. The sun radiates 3.9×10^{23} J of energy into space every second. By how much does the mass of the sun decrease each year?

***35.** For a system at equilibrium, no change in the concentrations of any of the reactants, or any of the products, is observed with time. Explain how you could use the radioactive isotope ^{131}I ($t_{1/2} = 8.0$ days) to show that the equilibrium

$$H_2(g) + I_2(g) \rightleftharpoons 2HI(g)$$

is a dynamic process; in other words, the forward and reverse reactions are occurring, but at the same rate.

36. The artificial isotope ^{18}F decays by two processes: (1) electron capture, and (2) positron emission. Find the difference in the energy released in the two processes. Any data required may be found in Table 2.1.

37. The heavy radioactive isotope curium, $^{240}_{96}Cm$, was synthesized by bombarding $^{232}_{90}Th$ nuclei with the nuclei of carbon atoms. Write a balanced equation for this reaction.

38. Explain each of the following terms:

    (a) Binding energy

    (b) A branching chain reaction

    (c) Critical mass

    (d) A fission reaction

    (e) A fusion reaction

APPENDIX A
MATHEMATICAL REVIEW

Scientific (Exponential) Notation

As we have seen in Chapter 1, it is usual in chemical calculations to express all numbers in a standard form. In scientific notation, all numbers, however large or small, are expressed as a number between $1.000\ldots$ and $9.999\ldots$, multiplied or divided by 10 an approximate number of times. For example

$$138.42 = 1.3842 \times 10 \times 10$$

is written in the form 1.3842×10^2, where 10^2 means 10×10. Here 2 is the power, or exponent, to which 10 is raised.

In general in scientific notation a number is expressed in the form

$$a.bcd \ldots \times 10^n$$

where $a.bcd \ldots$ is a number between $1.000 \ldots$ and $9.999 \ldots$ and n is a number (not necessarily a single digit) called an *exponent*.

To express a number smaller than $1.000 \ldots$ in scientific notation, the appropriate number between $1.000 \ldots$ and $9.999 \ldots$ is divided by 10 an appropriate number of times. For example

$$0.000\,138\,42 = \frac{1.3842}{10 \times 10 \times 10 \times 10} = 1.3842 \times 10^{-4}$$

Here the exponent -4 means that the number 1.3842 has been divided by 10 four times or, in other words, multiplied by $\frac{1}{10}$ four times. In general

$$10^{-n} = \frac{1}{10^n}$$

To transform a number larger than $9.999 \ldots$ to scientific notation, the decimal point is moved to the left until there is only *one* nonzero digit before the decimal point. If the decimal point is moved x places, the exponent $n = x$.

Thus in transforming 138.42 to scientific notation the decimal point is moved to the left two places

$$138.42$$

so the exponent $n = 2$ and we can write

$$138.42 = 1.3842 \times 10^2$$

To transform a number smaller than $1.000 \ldots$ to scientific notation, the decimal point is moved to the right until there is one nonzero digit before the decimal point. If the decimal point is moved y places the exponent $n = -y$.

Thus in transforming 0.000 138 42 to scientific notation the decimal point is moved to the right four places

$$0.000\,138\,142$$

so that the exponent $n = -4$ and we can write

$$0.000\,138\,42 = 1.3842 \times 10^{-4}$$

To transform a number from scientific notation to the ordinary form, we must move the decimal point in the opposite direction. If n is positive, the number is larger than $9.9999\ldots$ and the decimal point is moved to the right. Thus to transform 4.21×10^4 we move the decimal point four places to the right

$$4.2100_{\underset{\longrightarrow}{\rule{1.5em}{0pt}}}$$

to give 42 100. If n is negative, the number is smaller than $1.000\ldots$ so the decimal point is moved to the left. Thus to transform 6.2×10^{-3} we move the decimal point three places to the left

$$0\,006.2$$

to give 0.0062. Note that if the number we are dealing with is negative, the negative sign is retained when the number is transformed to or from scientific notation. Thus

$$-42100 = -4.21 \times 10^4$$

$$-0.003\,94 = -3.94 \times 10^{-3}$$

To add or subtract numbers in scientific notation the power of 10—that is, the exponent, n—must be the same in both numbers. For example, to add 6.234×10^4 and 1.203×10^3 we must first express both numbers so that they have the same exponent, n. If we transform 1.203×10^3 to 0.1203×10^4 we can then add the two numbers

$$6.234 \times 10^4 + 0.1203 \times 10^4 = 6.354 \times 10^4$$

To multiply two numbers in scientific notation we make use of the relation

$$(10)^x(10)^y = 10^{x+y}$$

In other words, we add the exponents. For example

$$(3.025 \times 10^3)(6.217 \times 10^{-6}) = 18.81 \times 10^{3-6}$$
$$= 18.81 \times 10^{-3}$$
$$= 1.881 \times 10^{-2}$$

To divide two numbers in scientific notation we make use of the relation

$$\frac{10^x}{10^y} = 10^{x-y}$$

In other words, we subtract the exponent of the number in the denominator from the exponent of the number in the numerator. For example

$$\frac{3.81 \times 10^{12}}{6.22 \times 10^{23}} = \frac{3.81}{6.22} \times 10^{12-23}$$
$$= 0.613 \times 10^{-11}$$
$$= 6.13 \times 10^{-12}$$

To raise a number to a power we make use of the relation

$$(10^x)^y = 10^{xy}$$

In other words, we multiply the exponents. For example

$$(3.142 \times 10^3)^4 = (3.142)^4 \times 10^{12}$$
$$= 97.46 \times 10^{12}$$
$$= 9.746 \times 10^{13}$$

To take a root of a number we make use of the relation

$$\sqrt[y]{10^x} = (10^x)^{1/y} = 10^{x/y}$$

In other words, we divide the exponents. For example

$$\sqrt[3]{6.22 \times 10^{23}} = (6.22 \times 10^{23})^{1/3}$$
$$= (0.622 \times 10^{24})^{1/3}$$
$$= (0.622)^{1/3} \times 10^8$$
$$= 0.854 \times 10^8$$

Note that in order to take the y'th root of a number it must be written in such a form that the exponent is divisible by y. In this case, therefore, we transformed 6.22×10^{23} to 0.622×10^{24} so that it could be divided by 3.

Using a Calculator

It is essential to be able to enter numbers expressed in scientific (exponential) notation into your calculator. Most calculators have an EXP or EE button to assist with such operations. To enter a number such as 2.5×10^7, first make the three touches required for 2.5, then press the EXP key, and finally press the key for 7. In other words, the keys you press after you have touched EXP correspond to the exponent of 10.

Notice that the "EXP" or "10" does not appear in the readout on your calculator. The blank spaces left between the 2.5 and the 7 (or 07) are meant to imply that the latter number is an exponent of 10.

Practice by entering 6.20×10^9 and 3.83×10^6 into your calculator.

Once a number has been entered in this manner, it can be used as a complete unit in any arithmetic operation. For example, multiplying the numbers 6.20×10^9 and 3.826×10^6 together should yield 2.37×10^{16}; most calculators use the following sequence:

6.20   EXP (or EE)   9 $\times$ 3.83   EXP (or EE)   6   =

To enter a number that has a *negative* exponent of 10, use the $+/-$ key on your calculator. For example, to enter 9.7×10^{-11}, first press the keys for 9.7, then EXP (or EE), then the $+/-$ key which changes the sign of the exponent from $+$ to $-$, and finally press the keys for 11. (Some calculators allow you to press the $+/-$ key after the numerical part of the exponent has been punched in.) To divide 3.92×10^{-6} by 4.44×10^{-8} use the following sequence:

3.92   EXP $+/-$ 6 $\div$ 4.44   EXP $+/-$ 8   =

The answer is 88.3.

Significant Figures

INTRODUCTION

Two types of numbers are used in chemistry

1. Exact numbers
2. Inexact numbers

Examples of exact numbers are numbers of things, such as 10 beakers of 12 pencils, and numbers whose values are precisely fixed by definition (60 minutes = 1 hour; the mass of exactly 12 atomic mass units for 1 atom of carbon-12).

Every measurement (other than that of counting) gives an inexact number because every such measurement is uncertain to some extent. The precision of a measurement is indicated by the number of figures used to record it. The digits in a properly recorded measurement are known as *significant figures*.

Use the following rules to determine the proper number of significant figures to be recorded for the result of a calculation.

LOCATION OF THE DECIMAL POINT

Zeros which are used to locate the decimal point, that is, zeros *before* the first nonzero digit, are *not* significant. For example, suppose that the distance between two points is measured as 3 cm. This measurement could also be expressed as 0.03 m.

$$3 \text{ cm} = 0.03 \text{ m}$$

Both values contain only one significant figure. The zeros in the second value are not significant since they only serve to locate the decimal point.

Zeros that arise as part of a measurement are significant. For example,

1. The number 106.540 has 6 significant figures
2. The number 0.000 5030 has 4 significant figures
3. The number 6.02×10^{23} has 3 significant figures

Numbers are often expressed in scientific (exponential) notation, using the appropriate number of significant figures, in order to avoid ambiguity concerning the number of significant figures. Thus, a number ending in zero such as 1580 is best expressed as 1.580×10^3 or 1.58×10^3 depending upon whether or not the measurement merits four or three significant figures.

ROUNDING OFF

If the calculated answer to a problem contains more figures than are significant, it should be rounded off.

1. If the figures following the last number to be retained are 4999 . . . or less, they are discarded and the last number is left unchanged. For example,

$$3.624 \text{ is } 3.62 \text{ to 3 significant figures}$$

2. If the figures following the last number to be retained are 5000 ... or greater, they are discarded and the last number is increased by one. For example,

> 7.635 becomes 7.64 if there are 3 significant figures
>
> 28.7257 becomes 28.726 if there are 5 significant figures

CALCULATIONS

ADDITION AND SUBTRACTION The result of an addition or subtraction should be reported to the same number of decimal places as that of the term with the least number of decimal places. For example, the answer to the addition

$$
\begin{array}{r}
28.16 \\
5.423 \\
0.0004 \\
\hline
33.5834
\end{array}
$$

should be reported as 33.58 (four significant figures).

MULTIPLICATION AND DIVISION The result of a multiplication or division is rounded off to the number of significant figures in the least precise term used in the calculation. The result of the multiplication

$$52.064 \times 1.24 = 64.5594$$

would be reported as 64.6 since the least precise term (1.24) has three significant figures. The result of the division

$$\frac{0.24}{1.346} = 0.1783$$

would be reported as 0.18 since the least precise term (0.24) has only two significant figures.

The presence of exact numbers in a mathematical expression does not affect the number of significant figures in the answer. Thus the result of the following calculation, in which the exact number 2 is used,

$$\frac{2.58 \times 0.1056 \times 2}{0.0267} = 20.4081$$

would be expressed as 20.4, that is, to three significant figures in keeping with the data. In other words, an exact number is considered to have an infinite number of significant figures.

MULTISTEP CALCULATIONS When a calculation involves several steps, small errors often are introduced by rounding off at the intermediate stages. Sometimes these errors become important if several rounding off corrections all happen to be in the same direction. Such errors may be avoided by doing all the numerical calculations in a problem in one operation, or by carrying an extra digit on the intermediate figures in a calculation and rounding off only the final answer.

Logarithms, Exponents, and Exponentials

Certain relationships in the study of chemistry involve exponents, exponentials, and/or logarithms. Here we review some properties of these operations.

If a number y is expressed in the form a^b, then we say that the exponent b is the *logarithm of y to the base a*. Thus,

if
$$y = a^b$$

then
$$\log_a y = b$$

The most common base used in science is the natural number $e\ (= 2.718\ldots)$; logarithms to base e often are abbreviated ln. Thus,

if
$$y = e^b$$

then
$$\log_e y = \ln y = b$$

Values for the ln (natural logarithm) of numbers can be obtained from your calculator by first entering the number and then pressing the LN button.

> **Exercise** Obtain ln(38.43), and ln(8.4×10^{-2}).
> **Answers** 3.649, -2.48

Given the value x for the ln of a number y, it is possible, using your calculator, to establish the value of the number y; here y is called the antilogarithm of x. If your calculator has an e^x button, enter the value (x) of the logarithm and then press this button. For example, entering 3.649 and pressing e^x yields 38.436. Thus we have reversed the process used in the exercise above. If your calculator has no e^x button, the value can be obtained by entering x and pressing INV, then LN; that is, "inverting" the ln operation.

> **Exercise** Given that ln $y = -3.58$, find the value of y.
> **Answer** 2.79×10^{-2}

In some applications it is required to raise some number y (which we will call the base) to the exponent of ("power of") another number x, for example, $0.5^{3.2}$. If your calculator has a y^x button, first enter the base y (0.5), press the y^x button, then enter 3.2 and press =. For example, by this procedure $0.5^{3.2}$ is found to equal 0.109.

In some applications, the value b of the exponent of a term in an equation is unknown; it is then convenient to rewrite the equation in logarithmic form in order to solve it. This can be done readily by realizing that if

$$y = a^b$$

then it follows that

$$\ln y = \ln(a^b)$$

In other words, we can equate the logarithms of the two sides of any equation. Further, recall that

$$\ln(a^b) = b \ln a$$

Thus our equation relating ln y to ln a can be transformed to

$$\ln y = b \ln a$$

Solving for b, we obtain

$$b = \frac{\ln y}{\ln a}$$

The value of b is then obtained by dividing the numerical value for $\ln y$ by the value for $\ln a$ (but *not* the $\ln$ of y/a).

For example, let us find the value of the exponent b for which the following equation is true.

$$1.342 = 1.800^b$$

First rewrite the equation in logarithmic form:

$$\ln 1.342 = b \ln 1.800$$

Therefore

$$b = \frac{\ln 1.342}{\ln 1.800} = \frac{0.2942}{0.5878} = 0.50$$

In other applications we may need to relate the logarithm of a product of two numbers to the logs of the individual numbers:

If $y = ab$

then $\ln y = \ln a + \ln b$

   Similarly

if $y = ae^x$

then $\ln y = \ln a + \ln e^x$

$$= \ln a + x$$

   Similarly

if $y = c/d$

then $\ln y = \ln c - \ln d$

Thus if we encounter equations which involve sums of differences of logarithmic terms, it often is convenient to re-express them as logarithms of products or quotients. We could rewrite

$$\ln a + \ln b \quad \text{as} \quad \ln(ab)$$

and $\ln c - \ln d \quad \text{as} \quad \ln(c/d)$

In chemistry, some quantities are defined as the logarithms to base 10 of other quantities. For example, $pH = -\log_{10} [H^+]$. If a number, y, is expressed as a power of 10, that is, as

$$y = 10^b$$

then the logarithm to base 10 of y is b; thus

$$\log_{10} y = b$$

The value of the $\log_{10}$ of any number can be obtained by entering the number in your calculator and pressing the LOG button; on some calculators this is a second function and another button must first be pressed—see your instructions.

Exercise Evaluate $\log_{10}$ of 38.43 and of 8.4×10^{-2}.
Answers 1.585, -1.076

In the expression

$$\log_{10} z = x$$

x is called the logarithm of z and z is called the antilogarithm of x. Thus for the expression

$$\log_{10} z = -1.076$$

which is equivalent to

$$z = 10^{-1.076}$$

if we wish to evaluate z, we say that we need the antilog of -1.076. To find this, we enter -1.076 in the calculator and press the 10^x or **INV LOG** button. This gives the result

$$8.4 \times 10^{-2}$$

Exercise Given that $\log_{10} z = -0.843$, evaluate z.
Answer 0.14

In summary, it is useful to memorize these relationships:

If $y = e^x$, $\log_e y = \ln y = x$

If $y = 10^x$, $\log_{10} y = x$

If $y = a^b$, $\ln y = b \ln a$, $\log_{10} y = b \log_{10} a$

If $y = ab$, $\ln y = \ln a + \ln b$, $\log_{10} y = \log_{10} a + \log_{10} b$

If $y = c/d$, $\ln y = \ln c - \ln d$, $\log_{10} y = \log_{10} c - \log_{10} d$

Solving Mathematical Equations

Many quantitative problems in introductory chemistry involve solving an algebraic equation. An equation may be simple in appearance,

$$4y = 15 \tag{1}$$

or rather complicated

$$ay^2 = be^{-cx} \tag{2}$$

In most cases, however, solving an equation requires only two steps:

1. Rearrange the equation. On the left side, isolate the variable (y or a or b or ...) whose numerical value is *not* known and for which you are trying to determine a value. After this step, the equation will have the form

   Left side   *Right side*
   unknown variable   =   algebraic expression involving numbers and
                                        variables whose values *are* known to you

2. Substitute numerical values for variables in the right-hand side of the new equation. Using your calculator determine the value for the unknown variable.

For example consider equation (1) above. We can isolate the unknown, y, on the left side by dividing both sides of the equation by the constant on the left, that is, by 4:

$$\frac{4y}{4} = \frac{15}{4}$$

Thus,

$$y = 15/4 = 3.75$$

Next consider equation (2). Let us assume first that the quantity whose value we wish to determine is b. To obtain it on the left side, switch sides of the equation:

$$be^{-cx} = ay^2$$

Then divide both sides by the terms *other* than b on the left, that is, divide by e^{-cx}:

$$b = \frac{ay^2}{e^{-cx}}$$

If we are supplied with numerical values for a, y, c, and x, the value of b can be computed.

As another example, suppose that the unknown variable in equation (2) is y rather than b. Dividing both sides of the original equation by a isolates y^2 on the left:

$$y^2 = \frac{be^{-cx}}{a}$$

To obtain y, take the square root of both sides:

$$y = \sqrt{\frac{be^{-cx}}{a}}$$

Finally, consider a case in which the unknown variable in equation (2) is x. We can isolate (on the left) the exponential term involving x by switching sides of the equation,

$$be^{-cx} = ay^2$$

and then dividing both sides by b to obtain

$$e^{-cx} = ay^2/b$$

Since x occurs in an exponent, this equation must be manipulated further so that it takes the form

$$x = \ldots\ldots$$

To accomplish this, take the logarithm to base e, that is, ln, of both sides of the equation:

Since

$$\ln(e^{-cx}) = -cx$$

therefore

$$-cx = \ln(ay^2/b)$$

Dividing by $-c$, we obtain our final result

$$x = \frac{-\ln(ay^2/b)}{c}$$

QUADRATIC EQUATIONS

In some equilibrium calculations, you may need to solve equations of the form

$$\frac{y^2}{M-y} = K \quad \text{and also} \quad \frac{y(N+y)}{M-y} = K$$

By multiplying both sides by $M-y$ and rearranging, either type can be recast into the standard form for a quadratic equation, that is,

$$ay^2 + by + c = 0$$

The general solutions to this equation are

$$y = \frac{-b \pm \sqrt{b^2 - 4ac}}{2a}$$

To solve for y, given values of a, b, and c, follow the following steps:

1. Evaluate $b^2 - 4ac$
2. Obtain the square root of $b^2 - 4ac$ (press $\sqrt{}$ on your calculator)
3. From this result, subtract b
4. Divide your answer by 2, then by a

For example, to solve

$$\frac{y^2}{10^{-2} - y} = 5 \times 10^{-4}$$

we multiply both sides by $10^{-2} - y$ and obtain

$$y^2 = 5 \times 10^{-4}(10^{-2} - y)$$

that is,

$$y^2 = 5 \times 10^{-6} - 5 \times 10^{-4}y$$

All terms are now brought to the left side:

$$y^2 + 5 \times 10^{-4}y - 5 \times 10^{-6} = 0$$

Comparing with the standard form, we make the following identifications:

$$a = 1$$
$$b = 5 \times 10^{-4}$$
$$c = -5 \times 10^{-6}$$

Thus

$$b^2 - 4ac = 25 \times 10^{-8} + 20 \times 10^{-6} = 2.025 \times 10^{-5}$$
$$\sqrt{b^2 - 4ac} = 4.50 \times 10^{-3}$$
$$-b \pm \sqrt{b^2 - 4ac} = 4.00 \times 10^{-3} \quad \text{or} \quad -5.00 \times 10^{-3}$$

Therefore

$$y = (-b \pm \sqrt{b^2 - 4ac})/2a = 2.00 \times 10^{-3} \quad \text{or} \quad -2.00 \times 10^{-3}$$

In the cases with which we will be concerned only the positive solution has physical significance so the negative solution is ignored.

> **Exercise** Solve the equation
>
> $$\frac{y^2}{2 \times 10^{-4} - y} = 2 \times 10^{-6}$$
>
> **Answer** $y = 1.90 \times 10^{-5}$

Appendix B: Tables

TABLE B.1
Atomic and Molar Masses of the Elements[a]

ELEMENT	SYMBOL	ATOMIC NUMBER	ATOMIC MASS (u) MOLAR MASS (g mol^{-1})	ELEMENT	SYMBOL	ATOMIC NUMBER	ATOMIC MASS (u) MOLAR MASS (g mol^{-1})
Actinium	Ac	89	227.0278[b]	Iodine	I	53	126.9045
Aluminum	Al	13	26.981 54	Iridium	Ir	77	192.22
Americium	Am	95	(243)[c]	Iron	Fe	26	55.847
Antimony	Sb	51	121.75	Krypton	Kr	36	83.80
Argon	Ar	18	39.948	Lanthanum	La	57	138.9055
Arsenic	As	33	74.9216	Lawrencium	Lr	103	(260)[c]
Astatine	At	85	(210)[c]	Lead	Pb	82	207.2
Barium	Ba	56	137.33	Lithium	Li	3	6.941
Berkelium	Bk	97	(247)[c]	Lutetium	Lu	71	174.967
Beryllium	Be	4	9.012 18	Magnesium	Mg	12	24.305
Bismuth	Bi	83	208.9804	Manganese	Mn	25	54.9380
Boron	B	5	10.81	Mendelevium	Md	101	(258)[c]
Bromine	Br	35	79.904	Mercury	Hg	80	200.59
Cadmium	Cd	48	112.41	Molybdenum	Mo	42	95.94
Calcium	Ca	20	40.08	Neodymium	Nd	60	144.24
Californium	Cf	98	(249)[c]	Neon	Ne	10	20.179
Carbon	C	6	12.011	Neptunium	Np	93	237.0482[b]
Cerium	Ce	58	140.12	Nickel	Ni	28	58.69
Cesium	Cs	55	132.9054	Niobium	Nb	41	92.9064
Chlorine	Cl	17	35.453	Nitrogen	N	7	14.0067
Chromium	Cr	24	51.996	Nobelium	No	102	(259)[c]
Cobalt	Co	27	58.9332	Osmium	Os	76	190.2
Copper	Cu	29	63.546	Oxygen	O	8	15.9994
Curium	Cm	96	(247)[c]	Palladium	Pd	46	106.42
Dysprosium	Dy	66	162.50	Phosphorus	P	15	30.973 76
Einsteinium	Es	99	(252)[c]	Platinum	Pt	78	195.08
Erbium	Er	68	167.26	Plutonium	Pu	94	(244)[c]
Europium	Eu	63	151.96	Polonium	Po	84	(209)[c]
Fermium	Fm	100	(257)[c]	Potassium	K	19	39.0983
Fluorine	F	9	18.998 403	Praseodymium	Pr	59	140.9077
Francium	Fr	87	(223)[c]	Promethium	Pm	61	(145)[c]
Gadolinium	Gd	64	157.25	Protactinium	Pa	91	231.0359[b]
Gallium	Ga	31	69.72	Radium	Ra	88	226.0254[b]
Germanium	Ge	32	72.59	Radon	Rn	86	(222)[c]
Gold	Au	79	196.9665	Rhenium	Re	75	186.207
Hafnium	Hf	72	178.49	Rhodium	Rh	45	102.9055
Helium	He	2	4.002 60	Rubidium	Rb	37	85.4678
Holmium	Ho	67	164.9304	Ruthenium	Ru	44	101.07
Hydrogen	H	1	1.007 94	Samarium	Sm	62	150.36
Indium	In	49	114.82	Scandium	Sc	21	44.9559

[a] The atomic masses of many elements are not invariant but depend on the origin and treatment of the material; the values given here apply to elements as they exist naturally on the earth and to certain artificial elements.

[b] For these radioactive elements the mass given is that for the longest-lived isotope.

[c] Atomic masses for these radioactive elements cannot be quoted precisely without knowledge of the origin of the elements; the value given is the atomic mass number of the isotope of that element of longest known half-life.

TABLE B.1 (continued)

ELEMENT	SYMBOL	ATOMIC NUMBER	ATOMIC MASS (u) MOLAR MASS (g mol^{-1})	ELEMENT	SYMBOL	ATOMIC NUMBER	ATOMIC MASS (u) MOLAR MASS (g mol^{-1})
Selenium	Se	34	78.96	Thulium	Tm	69	168.9342
Silicon	Si	14	28.0855	Tin	Sn	50	118.69
Silver	Ag	47	107.8682	Titanium	Ti	22	47.88
Sodium	Na	11	22.989 77	Tungsten	W	74	183.85
Strontium	Sr	38	87.62	Uranium	U	92	238.0289
Sulfur	S	16	32.06	Vanadium	V	23	50.9415
Tantalum	Ta	73	180.9479	Xenon	Xe	54	131.29
Technetium	Tc	43	(98)	Ytterbium	Yb	70	173.04
Tellurium	Te	52	127.60	Yttrium	Y	39	88.9059
Terbium	Tb	65	158.9254	Zinc	Zn	30	65.38
Thallium	Tl	81	204.37	Zirconium	Zr	40	91.22
Thorium	Th	90	232.0381[a]				

TABLE B.2
Thermodynamic Data: Elements and Inorganic Compounds at 25 °C

COMPOUND	ΔH_f° kJ mol^{-1}	S° J K^{-1} mol^{-1}	ΔG_f° kJ mol^{-1}	COMPOUND	ΔH_f° kJ mol^{-1}	S° J K^{-1} mol^{-1}	ΔG_f° kJ mol^{-1}
Ag(s)	0.0	42.6	0.0	N$_2$(g)	0.0	191.5	0.0
AgCl(s)	−127.1	96.2	−109.8	N$_2$O$_4$(g)	9.3	304.2	97.8
AlCl$_3$(s)	−704.2	110.7	−628.8	Na(s)	0.0	51.3	0.0
Al$_2$O$_3$(s)	−1676	50.9	−1582	NaF(s)	−573.7	51.3	−546.3
B$_5$H$_9$(s)	73.2	276	175	NaCl(s)	−411.2	72.5	−384.3
B$_2$O$_3$(s)	−1273.5	54.0	−1194.4	NaBr(s)	−361.1	87.2	−349.1
Br$_2$(l)	0.0	152.2	0.0	NaI(s)	−287.8	98.5	−282.4
BrF$_3$(g)	−255.6	292.4	−229.5	NaOH(s)	−425.6	64.5	−379.7
CaO(s)	−635.1	38.1	−603.5	Na$_2$O$_2$(s)	−511.7	104	−451.0
CaCO$_3$(s) (calcite)	−1206.9	92.9	−1128.8	NH$_3$(g)	−46.2	192.7	−16.4
Cl$_2$(g)	0.0	223.0	0.0	N$_2$H$_4$(l)	50.6	121.2	149.2
Cu(s)	0.0	33.2	0.0	NH$_4$ClO$_4$(s)	−295	186	−89
F$_2$(g)	0.0	202.7	0.0	NO(g)	90.3	210.6	86.6
Fe(s)	0.0	27.3	0.0	NO$_2$(g)	33.2	240.0	51.3
Fe$_2$O$_3$(s) (hematite)	−824	87.4	−742.2	HNO$_3$(l)	−174.1	155.6	−80.8
H(g)	218.0	114.6	203.3	NOCl(g)	51.7	261.6	66.1
H$_2$(g)	0.0	130.6	0.0	O$_2$(g)	0.0	205.0	0.0
HCl(g)	−92.3	186.8	−95.3	O$_3$(g)	142.7	238.8	163.2
HF(g)	−271.1	173.8	−273.2	P(s) (white)	0.0	41.1	0.0
HI(g)	26.4	206.5	1.6	P$_4$O$_{10}$(s)	−3010	231	−2724
HBr(g)	−36.4	198.6	−53.4	PCl$_3$(g)	−287.0	311.7	−267.8
HCN(g)	135.1	201.7	1247	PCl$_5$(g)	−374.9	364.5	−305.0
H$_2$O(g)	−241.8	188.7	−228.6	PbO$_2$(s)	−277.4	68.6	−217.4
H$_2$O(l)	−258.8	70.0	−237.2	S(s) (rhombic)	0.0	32.0	0.0
H$_2$O$_2$(l)	−187.8	109.6	−120.4	H$_2$S(g)	−20.6	205.6	−33.4
Hg(l)	0.0	75.9	0.0	SiO$_2$ (quartz)	−910.7	41.5	−856.3
I$_2$(s)	0.0	116.1	0.0	SiCl$_4$(l)	−687.0	239.7	−619.9
I$_2$(g)	62.4	260.6	19.4	SO$_2$(g)	−296.8	248.1	−300.1
MgO(s)	−601.5	27.0	−569.2	SO$_3$(g)	−395.7	256.6	−371.1
MnO$_2$(s)	−520.0	53.1	−465.2	H$_2$SO$_4$(l)	−814.0	145.9	−690.1
				ZnO(s)	−350.5	43.6	−320.5

TABLE B.3

Thermodynamic Data: Carbon and Carbon Compounds at 25 °C

COMPOUND	ΔH_f° kJ mol^{-1}	S° J K^{-1} mol^{-1}	ΔG_f° kJ mol^{-1}
C(g)	716.7	158.0	671.3
C (graphite)	0.0	5.8	0.0
C (diamond)	1.9	2.4	2.9
CO(g)	−110.5	197.6	−137.2
CO$_2$(g)	−393.5	213.7	−394.4
CH$_4$(g)	−74.5	186.1	−50.8
C$_2$H$_2$(g)	228.0	200.8	209.2
C$_2$H$_4$(g)	52.3	219.4	68.1
C$_2$H$_6$(g)	−84.7	229.5	−32.9
C$_3$H$_6$(g) (cyclopropane)	53.3	237	104
C$_3$H$_8$(g)	−103.8	269.9	−23.4
C$_4$H$_8$(g) (cyclobutane)	28.4	265	100
C$_4$H$_{10}$(g) (n-butane)	−126.1	310.1	−17.2
C$_5$H$_{10}$(g) (cyclopentane)	−78.4	293	39
C$_5$H$_{12}$(g) (n-pentane)	−146.4	348.9	−8.4
C$_5$H$_{12}$(l) (n-pentane)	−173.2	263.3	−9.6
C$_6$H$_6$(l) (benzene)	49.0	172.2	124.7
C$_6$H$_{12}$(g) (cyclohexane)	−123.3	298	32
n-C$_6$H$_{14}$(l)	−198.6	295.9	−4.4
n-C$_7$H$_{16}$(l)	−224.0	328.5	1.0
n-C$_8$H$_{18}$(l)	−250.0	361.2	6.4
CH$_2$O(g)	−108.7	218.7	−113
CH$_3$OH(l)	−239.1	126.8	−166.4
C$_2$H$_5$OH(l)	−277.1	160.7	−174.9
CH$_3$CO$_2$H(l)	−484.3	159.8	−390
C$_6$H$_{12}$O$_6$(s) (glucose)	−1273.3	182.4	−919.2
C$_{12}$H$_{12}$O$_{11}$(s) (sucrose)	−2226.1	360	
CH$_2$Cl$_2$(l)	−124.1	177.8	−67.3
CHCl$_3$(l)	−135.1	201.7	−73.7
CCl$_4$(l)	−129.6	216.4	−65.3

The International System of Units (SI)

TABLE B.4

SI Base Units

PHYSICAL QUANTITY	NAME	SYMBOL
Length	meter	m
Mass	kilogram	kg
Time	second	s
Electric current	ampere	A
Thermodynamic temperature	kelvin	K
Amount of substance	mole	mol
Luminous intensity	candela	cd

The above base units are defined as follows.

1. The **meter** is the length equal to 1 650 763.73 wavelength in vacuum of the radiation corresponding to the transition between the levels 2p$_{10}$ and 5d, of the krypton-86 atom.

2. The **kilogram** is the unit of mass; it is equal to the mass of the international proto-type of the kilogram.

3. The **second** is the duration of 9 192 631 770 periods of the radiation corresponding to the transition between the two hyperfine levels of the ground state of the cesium-133 atom.

4. The **ampere** is that constant current which, if maintained in two straight parallel conductors of infinite length, of negligible circular cross section, and placed 1 meter apart in a vacuum, would produce between these conductors a force equal to 2×10^{-7} newton per meter of length.

5. The **kelvin** unit of thermodynamic temperature is the fraction 1/273.16 of the thermodynamic temperature of the triple point of water.

6. The **mole** is the amount of substance of a system which contains as many elementary entities as there are atoms in 0.012 kilogram of carbon-12. When the mole is used, the elementary entities must be specified and may be atoms, molecules, ions, electrons, other particles, or specified groups of such particles.

7. The **candela** is the luminous intensity, in the perpendicular direction, of a surface of 1/600 000 square meter of black body at the temperature of freezing platinum under a pressure of 101 325 newtons per square meter.

T A B L E   B . 5
SI Derived Units

PHYSICAL QUANTITY	NAME	SYMBOL	DEFINITION
Frequency	hertz	Hz	s^{-1}
Energy	joule	J	$kg\ m^2\ s^{-2}$
Force	newton	N	$kg\ m\ s^{-2} = J\ m^{-1}$
Power	watt	W	$kg\ m^2\ s^{-3} = J\ s^{-1}$
Pressure	pascal	Pa	$kg\ m^{-1}\ s^{-2} = N\ m^{-2} = J\ m^{-3}$
Electric charge	coulomb	C	$A\ s$
Electric potential difference	volt	V	$kg\ m^2\ s^{-3}\ A^{-1} = J\ A^{-1}\ s^{-1}$
Electric resistance	ohm	Ω	$kg\ m^2\ s^{-3}\ A^{-2} = V\ A^{-1}$

ANSWERS TO EXERCISES

1.1 2.08×10^{-10} m, 2.40×10^{-7} m, 9.5×10^{-5} m, 9.7×10^{-11} m   **1.2** 2.5×10^4 g, 2.50×10^{-1} g, 2.5×10^{-6} g, 1.00×10^8 g.   **1.3** 39 mile gal^{-1}.
1.4 11 g cm^{-3}.   **1.5** 29.9 mass %.   **2.1** $^1_1$H 1, 1, 0, 1; $^4_2$He 4, 2, 2, 2; $^{19}_9$F 19, 9, 10, 9; $^{118}_{50}$Sn 118, 50, 68, 50; $^{238}_{92}$U 238, 92, 146, 92.   **2.2** ^{69}Ga, 60.3%; ^{71}Ga, 39.7%.
2.3 (a) 1.64×10^{-4} mol Si,   (b) 270 g Al, (c) 0.2621 mol NaCl,   (d) 132 g CO_2.   **2.4** (a) 100.1 u, 100.1 g;   (b) 183.2 u, 183.2 g.   **2.5** (a) 74.88% C, 25.14% H;   (b) 1.60% H, 22.23% N, 76.17% O; (c) 29.45% Ca, 23.56% S, 47.02% O.   **2.6** H_3PO_4.
2.7 Al_2O_3.   **2.8** C_6H_6O.   **2.9** CH_2O, $C_2H_4O_2$.
2.10 (a) $C_2H_6O + 3O_2 \longrightarrow 2CO_2 + 3H_2O$
(b) $N_2 + 3H_2 \longrightarrow 2NH_3$ (c) $4Fe + 3O_2 \longrightarrow 2Fe_2O_3$
2.11 $C_3H_8 + 5O_2 \longrightarrow 3CO_2 + 4H_2O$
2.12 (a) $C_2H_6O(l) + 3O_2(g) \longrightarrow 2CO_2(g) + 3H_2O(l)$
(b) $N_2(g) + 3H_2(g) \longrightarrow 2NH_3(g)$
(c) $4Fe(s) + 3O_2(g) \longrightarrow 2Fe_2O_3(s)$
(d) $C_3H_8(g) + 5O_2(g) \longrightarrow 3CO_2(g) + 4H_2O(g)$
2.13 4.62 g CO_2, 2.38 g H_2O.   **2.14** (a) Al(s), (b) 70.0 g,   (c) 11 g Fe_2O_3(s).   **2.15** 1.68 metric ton, 88.7%.   **2.16** (a) $0.967M$,   (b) $0.19M$,   (c) $0.639M$.
2.17 (a) 8.50 g,   (b) 51.5 g,   (c) 20.8 g.   **2.18** 1.62 mL.
2.19 2.34 g.
3.1 (a) $2Mg(s) + CO_2(g) \longrightarrow 2MgO(s) + C(s)$
(b) $CuO(s) + H_2(g) \longrightarrow Cu(s) + H_2O(g)$
(c) $N_2(g) + 3H_2(g) \longrightarrow 2NH_3(g)$
(d) $P_4(s) + 5O_2(g) \longrightarrow P_4O_{10}(s)$
3.2 (a) Sulfur,   (b) copper,   (c) helium, neon, or argon, (d) oxygen,   (e) aluminum.   **3.3** 6.51 atm.
3.4 116 K (-157 °C).   **3.5** 224 L.   **3.6** 3.00×10^4 L, 3.86 m.   **3.7** (a) 118 g mol^{-1},   (b) CF_2Cl_2.
3.8 120 g mol^{-1}, $CHCl_3$.   **3.9** 28.1 g mol^{-1}, C_2H_4.
3.10 300 L, 100 L.   **3.11** O_2, 8.3 atm; He, 10.6 atm.
3.12 26.7 cm from the HI end of the tube.
3.13 88.6 g mol^{-1}.   **4.1** Metals: Cu, K, Cs, Ga, Mg. Nonmetals: O, S, F, Ar, P. Semimetals: Ge, B.   **4.2** LiBr, $BaBr_2$, PBr_3; Li_3N, Ba_3N_2, PN:
$2Li(s) + Br_2(l) \longrightarrow 2LiBr$, $Ba(s) + Br_2(l) \longrightarrow BaBr_2(s)$
$P_4(s) + 6Br_2(l) \longrightarrow 4PBr_3(l)$,
$6Li(s) + N_2(g) \longrightarrow 2Li_3N(s)$
$3Ba(s) + N_2(g) \longrightarrow Ba_3N_2(s)$,
$P_4(s) + 2N_2(g) \longrightarrow 4PN(s)$
4.3 $+5$, $+7$, $+2$, $+2$.   **4.4** Be < O < F < He.
4.5 141 pm, 107 pm, 209 pm, 181 pm.

4.6 Rb· ·Ba· ·Pb· :Ï· :Xe:

4.7 CaF_2, Ca^{2+}[:F̈:$^-$]$_2$; Mg_3N_2, [Mg^{2+}]$_3$[:N̈:$^{3-}$]$_2$; [Na$^+$]$_2$:S̈:$^{2-}$ **4.8** :F̈—N̈—F̈: :C̈l—S̈:
     :F̈:   :C̈l:

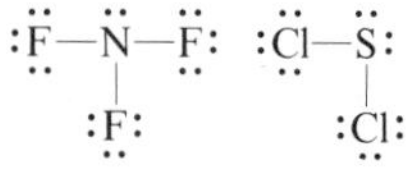

5.1 $2Na(l) + Cl_2(g) \longrightarrow 2NaCl(s)$;
$2Fe(s) + 3Cl_2(g) \longrightarrow 2FeCl_3(s)$;
$2Sb(s) + 3Cl_2(g) \longrightarrow 2SbCl_3(s)$;
$Mg(s) + Cl_2(g) \longrightarrow MgCl_2(s)$.
5.2 (a) $Sr(s) + Br_2(l) \longrightarrow SrBr_2(s)$
(b) $2As(s) + 3Cl_2(g) \longrightarrow AsCl_3(l)$
(c) $2Rb(s) + F_2(g) \longrightarrow 2RbF(s)$
(d) $Br_2(l) + Cl_2(g) \longrightarrow 2BrCl(g)$
5.3 CaH_2, ionic; $MgCl_2$, ionic; SiH_4, polar covalent; NH_3, polar covalent; P_4, nonpolar covalent; SO_2, polar covalent; Na_2O, ionic; Cl_2O, polar covalent.
5.4 P—P < C—S < C—O < B—F.   **5.5** (a) Cl$^-$, (b) K$^+$,   (c) Ca^{2+},   (d) S^{2-},   (e) O^{2-}.
5.6 (a) Na_2SO_4, sodium sulfate;   (b) Al_2O_3, aluminum oxide;   (c) K_3PO_4, potassium phosphate;   (d) $Mg(OH)_2$, magnesium hydroxide;   (e) $BaCO_3$, barium carbonate.
5.7 Br_2.

	Oxidizing Agent	Reducing Agent	Species Oxidized	Species Reduced
5.8				
(a)	O_2	Cu	Cu	O_2
(b)	I_2	Cs	Cs	I_2
(c)	S	Zn	Zn	S
(d)	Cu^{2+}	Zn	Zn	Cu^{2+}

5.9 Insoluble: $CaCO_3$, PbI_2, CuO. Soluble: LiBr, $MgSO_4$.
5.10 (a) Precipitate,   (b) no precipitate,   (c) precipitate.
5.11 $LiOH(aq) + HI(aq) \longrightarrow LiI(aq) + H_2O(l)$
$2KOH(aq) + H_2SO_4(aq) \longrightarrow K_2SO_4(aq) + 2H_2O(l)$
$Mg(OH)_2(s) + 2HNO_3(aq) \longrightarrow Mg(NO_3)_2(aq) + 2H_2O(l)$
5.12 $0.0772M$.   **5.13** $CH_3CO_2H(aq)$ and $CH_3CO_2^-(aq)$; $H_3O^+(aq)$ and $H_2O(l)$.   **5.14** (a) Oxidation–reduction, (b) oxidation–reduction,   (c) acid–base, (d) precipitation,   (e) oxidation–reduction,   (f) acid–base and oxidation–reduction,   (g) precipitation and acid–base.

6.1 (a) $MgCO_3(s) \xrightarrow{\text{Heat}} MgO(s) + CO_2(g)$
(b) $PbO(s) + CO(g) \xrightarrow{\text{Heat}} Pb(s) + CO_2(g)$ (redox)
(c) $H_2O(g) + CO(g) \xrightarrow{\text{Heat}} H_2(g) + CO_2(g)$ (redox)
(d) $MgC_2(s) + 2H_2O(l) \longrightarrow Mg(OH)_2(s) + C_2H_2(g)$ (acid–base)
(e) $C_2H_6(g) + 2H_2O(g) \xrightarrow{\text{Heat}} 2CO(g) + 5H_2(g)$ (redox)

6.2 Strong bases, OH$^-$, C_2^{2-}; weak bases, CN$^-$, CO_3^{2-}, HCO_3^-, F$^-$; no basic properties, Cl$^-$.   **6.3** Carbon monoxide, $^-$:C≡O:$^+$; nitrogen, :N≡N:; carbide ion, $^-$:C≡C:$^-$.

6.4

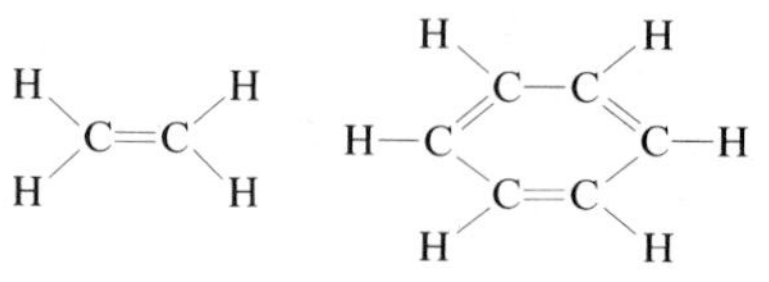

Ethene   Benzene

A-16

H H H
H—C—C—C—H
H H H
Propane

H
C=O
H
Methanal

H H
—H—C—C—OH *or*
H H
Ethanol

H H
H—C—O—C—H
H H
Diethyl ether

H O
H—C—C—OH
H
Ethanoic acid

6.5 $\Delta H = -110$ kJ, approximately twice that for the neutralization of the monoprotic acid HCl(aq), because H_2SO_4 is a diprotic acid. (The difference from $2(-54)$ kJ $= -108$ kJ is attributable to experimental error, because a coffee cup calorimeter is very crude.)　**6.6** $\Delta H = -2.7 \times 10^3$ kJ.
6.7 $\Delta H = -750$ kJ.　**6.8** $\Delta H° = -550.8$ kJ, -22.4 kJ.
6.9 $\Delta H_f° = -126$ kJ mol^{-1}.　**6.10** 3997.9 kJ,
BE(C—C) $= 335.8$ kJ mol^{-1}.　**6.11** BE(C≡C) $=$
811 kJ mol^{-1}.　**6.12** $\Delta H° = -119$ kJ.
7.1 4.0×10^{14} Hz to 7.5×10^{14} Hz.　**7.2** 499 kJ.
7.3 (a) 1259 kJ mol^{-1}; (b) 3.15×10^{15} Hz, 95.2 nm;
(c) ultraviolet.　**7.4** 6.25 MJ mol^{-1}, 0.51 MJ mol^{-1};
-6.25 MJ mol^{-1}, -0.51 MJ mol^{-1}.
7.5 $1s^2 2s^2 2p^6 3s^2 3p^6 3d^{10} 4s^2 4p^6 5s^2 4d^{10} 5p^4$.
7.6 1.15×10^3 m, 2.2×10^{-34} m.

		3s	3p
7.7 Mg,	$1s^2 2s^2 2p^6 3s^2$; [Ne]	↑↓	☐ ☐ ☐
	Al, $1s^2 2s^2 2p^6 3s^2 3p^1$; [Ne]	↑↓	↑ ☐ ☐
	Cl, $1s^2 2s^2 2p^6 3s^2 3p^5$; [Ne]	↑↓	↑↓ ↑↓ ↑
	Cl$^-$, $1s^2 2s^2 2p^6 3s^2 3p^6$; [Ne]	↑↓	↑↓ ↑↓ ↑↓
	K$^+$, $1s^2 2s^2 2p^6 3s^2 3p^6$; [Ne]	↑↓	↑↓ ↑↓ ↑↓

8.1 Orthorhombic sulfur (the stable form) is liquid at the temperature of superheated steam and occurs naturally in relatively pure deposits.
8.2 $2CsOH(aq) + H_2SO_4(aq) \longrightarrow Cs_2SO_4(aq) + 2H_2O(l)$
$2NH_3(aq) + H_2SO_4(aq) \longrightarrow (NH_4)_2SO_4(aq)$
$Ca(OH)_2(aq) + H_2SO_4(aq) \longrightarrow CaSO_4(s) + 2H_2O(l)$
$MgO(s) + H_2SO_4(aq) \longrightarrow MgSO_4(aq) + H_2O(l)$
8.3 25.0 mL.　**8.4** H_3PO_4: H, $+1$; O, -2; P, $+5$, PH_3:
H, $+1$; P, -3. H_2CO_3: H, $+1$; O, -2; C, $+4$. ClF: F, -1;
Cl, $+1$.　**8.5** SO_3^{2-}: O, -2; S, $+4$. ClO_4^-: O, -2; Cl,
$+7$. S^{2-}: S, -1. $Al_2(SO_4)_3$: Al, $+3$; S, $+6$; O, -2.
NaH_2PO_4: Na, $+1$; H, $+1$; P, $+5$; O, -2.
8.6 (a) $H_2S(aq) + Cl_2(aq) \longrightarrow 2HCl(aq) + S(s)$
(b) $H_2S(aq) + 2Ag^+(aq) + 2H_2O(l) \longrightarrow$
$Ag_2S(s) + 2H_3O^+(aq)$
(c) $5H_2SO_4(aq) + 2I^-(aq) \longrightarrow$
$I_2(s) + SO_2(g) + 2H_3O^+(l) + 4HSO_4^-(aq)$
8.7 (a) $2NH_3(aq) + H_2SO_4(aq) \longrightarrow (NH_4)_2SO_4(aq)$
(b) $Mg(s) + H_2SO_4(aq) \longrightarrow MgSO_4(aq) + H_2(g)$
(c) $Pb(NO_3)_2(aq) + H_2SO_4(aq) \longrightarrow$
$PbSO_4(s) + 2HNO_3(aq)$
8.8 139 mL.

8.9 $2Ca(s) + 2P(s) \longrightarrow Ca_3P_2(s)$
$Ca_3P_2(s) + 6HCl(aq) \longrightarrow 2PH_3(g) + 3CaCl_2(aq)$
8.10 HOCl, weak; $HClO_3$, strong; H_2SO_3, weak; H_5IO_6, weak; H_2SeO_4, strong.　**8.11** (a) Magnesium hydrogen phosphate, (b) chloric acid, (c) tetraphosphorus trisulfide, (d) magnesium nitride, (e) dinitrogen pentaoxide.　**8.12** (a) $Al_2(SO_4)_3$, (b) $SrBr_2$, (c) HCN(aq), (d) KH_2PO_4, (e) CrO_3, (f) H_3PO_3.

9.1

O—S—O (sulfate structure)　 O—P=O (phosphate structure)

9.2 H_2CO　　　HCO_3^-　　　HPO_4^{2-}

H₂CO: AX$_3$, Trigonal planar
HCO₃⁻: AX$_3$, Trigonal planar
HPO₄²⁻: AX$_4$, Tetrahedral

ClO_3^-　　　CS_3^{2-}　　　HCO_2H

ClO₃⁻: AX$_3$E, Trigonal pyramidal
CS₃²⁻: AX$_3$, Trigonal planar
HCO₂H: AX$_3$, Trigonal planar

9.3 PCl_6^-　　　SiF_5^-　　　IF_3

PCl₆⁻: AX$_6$, Octahedral
SiF₅⁻: AX$_5$, Trigonal bipyramidal
IF₃: AX$_3$E$_2$, T-shaped

BrF_4^-　　　ICl_2^-

BrF₄⁻: AX$_4$E$_2$, Square planar
ICl₂⁻: AX$_2$E$_3$, Linear

9.4 $:O=S—O:^- \longleftrightarrow ^-:O—S=O: \longleftrightarrow ^-:O—S—O:^-$

SO bond order $1\frac{1}{3}$; charge on each O atom $-\frac{2}{3}$.

9.5 (a)
$$H-\overset{\cdot\cdot}{\underset{\cdot\cdot}{O}}-C\overset{\overset{\displaystyle\ddot{O}:^-}{|}}{\underset{\displaystyle\ddot{O}:}{\|}} \longleftrightarrow H-\overset{\cdot\cdot}{\underset{\cdot\cdot}{O}}-C\overset{\overset{\displaystyle\ddot{O}:}{\|}}{\underset{\displaystyle\ddot{O}:^-}{|}}$$

(b)
$$H-\overset{\cdot\cdot}{\underset{\cdot\cdot}{O}}-\overset{\overset{\displaystyle\ddot{O}:}{\|}}{\underset{\displaystyle:\ddot{O}:^-}{S}}=\ddot{O}: \longleftrightarrow H-\overset{\cdot\cdot}{\underset{\cdot\cdot}{O}}-\overset{\overset{\displaystyle\ddot{O}:}{\|}}{\underset{\displaystyle\ddot{O}:}{S}}-\ddot{O}:^- \longleftrightarrow$$

$$H-\overset{\cdot\cdot}{\underset{\cdot\cdot}{O}}-\overset{\overset{\displaystyle:\ddot{O}:^-}{|}}{\underset{\displaystyle\ddot{O}:}{S}}=\ddot{O}:$$

S—OH bond order 1.00

S—OH bond order 1.67

(c)
$$H-\overset{\cdot\cdot}{\underset{\cdot\cdot}{O}}-\overset{\overset{\displaystyle:\ddot{O}:^-}{|}}{\underset{\displaystyle\ddot{O}:}{P}}-\ddot{O}:^- \longleftrightarrow H-\overset{\cdot\cdot}{\underset{\cdot\cdot}{O}}-\overset{\overset{\displaystyle:\ddot{O}:^-}{|}}{\underset{\displaystyle:\ddot{O}:^-}{P}}=\ddot{O}: \longleftrightarrow$$

$$H-\overset{\cdot\cdot}{\underset{\cdot\cdot}{O}}-\overset{\overset{\displaystyle\ddot{O}:}{\|}}{\underset{\displaystyle:\ddot{O}:^-}{P}}-\ddot{O}:^-$$

P—OH bond order 1.00

P—OH bond order 1.33

9.6 NH_4^+, AX_4, tetrahedron, sp^3; H_3O^+, AX_3E, trigonal pyramid, sp^3; $SnCl_2$, AX_2E, angular, sp^2; PCl_6^-, AX_6, octahedron, sp^3d^2; XeF_2, AX_2E_3, linear, sp^3d; XeF_4, AX_4E_2, square planar, sp^3d^2.   **9.7** Yes: SF_2, PCl_3, BrF_3, BrF_5, OCS. No: SF_6.

10.1

Reactants	Products	Oxidized	Reduced
(a) Cu +1, O −2, C 0	Cu 0, O −2, C +2	C	Cu
(b) Fe +3, O −2, Al 0	Fe 0, O −2, Al +3	Al	Fe
(c) Ca +2, O −2, Si +4	Ca +2, O −2, Si +4	Not a redox reaction	
(d) Pb +2, S −2, O 0	Pb +2, O −2, S +4	S	O
(e) Cu +1, S −2, O 0	Cu 0, O −2, S +4	S	O Cu

10.2 (a) $Cr(s) + 2HCl(aq) \longrightarrow CrCl_2(aq) + H_2(g)$
(b) $Zn(s) + H_2SO_4(aq) \longrightarrow ZnSO_4(aq) + H_2(g)$
(c) $2Al(s) + 6HClO_4(aq) \longrightarrow 2Al(ClO_4)_3(aq) + 3H_2(g)$
10.3 (a) $5[2Br^- \longrightarrow Br_2 + 2e^-]$ oxidation
$2[MnO_4^- + 5e^- + 8H^+ \longrightarrow$
$Mn^{2+} + 4H_2O]$ reduction
$\overline{10Br^- + 2MnO_4^- + 16H^+ \longrightarrow}$
$5Br_2 + 2Mn^{2+} + 8H_2O$

(b) $3[PbS + 4H_2O \longrightarrow$
$PbSO_4 + 8e^- + 8H^+]$ oxidation
$8[NO_3^- + 3e^- + 4H^+ \longrightarrow$
$NO + 2H_2O]$ reduction
$\overline{3PbS + 8H^+ + 8NO_3^- \longrightarrow}$
$3PbSO_4 + 8NO + 4H_2O$

10.4 (a) $Al + MnO_4^- + 2H_2O \longrightarrow MnO_2 + Al(OH)_4^-$
(b) $2Al + 2OH^- + 6H_2O \longrightarrow 2Al(OH)_4^- + 3H_2$
10.5 (a) $2Cu(s) + O_2(g) \longrightarrow 2CuO(s)$

A-18

(b) $2Al(s) + 3S(s) \longrightarrow Al_2S_3(s)$
(c) $2Pb(s) + O_2(g) \longrightarrow 2PbO(s)$
(d) $CuSO_4(aq) + 2NaOH(aq) \longrightarrow$
$Cu(OH)_2(s) + Na_2SO_4(aq)$
10.6 (a) $CO_2 + O^{2-} \longrightarrow CO_3^{2-}$,
(b) $P_4O_{10} + 6O^{2-} \longrightarrow 4PO_4^{3-}$
10.7 (a) $SO_2(g) + H_2O(l) \longrightarrow H_2SO_3(aq)$; *acidic*
(b) $Rb_2O(s) + H_2O(l) \longrightarrow 2RbOH(aq)$; *basic*
10.8 (a) $2Al(s) + 6HBr(aq) + 12H_2O(l) \longrightarrow$
$2AlBr_3 \cdot 6H_2O(aq) + 3H_2(g)$
(b) $2Al_2O_3(s) + 6HNO_3(aq) \longrightarrow$
$2Al(NO_3)_3(aq) + 3H_2O(l)$

(c) $2Al(OH)_3(s) \xrightarrow{\text{Heat}} Al_2O_3(s) + 3H_2O(g)$
10.9 (a) $PbO(s) + 2HNO_3(aq) \longrightarrow$
$Pb(NO_3)_2(aq) + H_2O(l)$
(b) $Pb(NO_3)_2(aq) + 2NaOH(aq) \longrightarrow$
$Pb(OH)_2(s) + 2NaNO_3(aq)$
(c) $Pb(NO_3)_2(aq) + Na_2SO_4(aq) \longrightarrow$
$PbSO_4(s) + 2NaNO_3(aq)$
10.10 $Pb_3O_4(s)$ is $Pb(II)_2Pb(IV)O_4$.
$Pb_3O_4(s) + 4HNO_3(aq, conc) \longrightarrow$
$2Pb(NO_3)_2(aq) + PbO_2(s) + 2H_2O(l)$

$2Pb(NO_3)_2(s) \xrightarrow{\text{Heat}} 2PbO(s) + 4NO_2(g) + O_2(g)$
11.1 (a) No bonds broken. Only weak induced London forces between I_2 molecules have to be overcome; (b) metallic bonds between Cu atoms; (c) strong covalent bonds between carbon atoms; (d) covalent Si—O bonds; (e) No bonds broken. Only weak London forces between CO_2 molecules; (f) No bonds broken. Only weak intermolecular forces between P_4O_{10} molecules.   **11.2** (a) *Network solid*, covalent bonds; (b) *molecular solid* in which covalently bonded S_8 molecules are held together by weak London forces; (c) ionic *network solid* with ionic bonds between Cl^- and $Al(H_2O)_6^{3+}$ ions, in which the H_2O molecules are bonded by polar covalent Al—O bonds; (d) *molecular solid* in which CO_2 molecules with covalent bonds are held together by weak London forces.
11.3 382 pm, 165 pm,   **11.4** (a) 404 pm,
(b) 6.594×10^{-23} cm^3,   (c) 2.72 g cm^{-3}.
11.5 6.052×10^{23}.
12.1 $CaO(s) + H_2O(l) \longrightarrow Ca^{2+}(aq) + 2OH^-(aq)$;
acid–base

$Mg^{2+}(aq) + 2OH^-(aq) \longrightarrow Mg(OH)_2(s)$; *precipitation*
$Mg(OH)_2(s) + 2HCl(aq) \longrightarrow MgCl_2(aq) + 2H_2O(l)$;
acid–base
12.2 (a) 210–220 mm Hg, (b) 400 mm Hg.
12.3 0.0313 atm (23.8 mm Hg), 9.67 L.   **12.4** (a), (c), and (e).   **12.5** $CH_4 < C_2H_6 < C_2H_5Cl < SCl_2 < AsCl_3$.
12.6 53.5 mm Hg.   **12.7** 100.95 °C.
12.8 147 g mol^{-1}.   **12.9** 0.05 atm.
12.10 1.75×10^4 g mol^{-1}.   **12.11** 1.79 mol kg^{-1}.
12.12 (a) $CN^-(aq) + H_2O(l) \rightleftharpoons HCN(aq) + OH^-(aq)$
(b) $C_2^{2-}(aq) + 2H_2O(l) \longrightarrow C_2H_2(g) + 2OH^-(aq)$
(c) $CH_3CO_2Na(aq) + HClO_4(aq) \longrightarrow$
$CH_3CO_2H(aq) + NaClO_4(aq)$
(d) $2H_3PO_4(aq) + Ca(OH)_2(aq) \longrightarrow$
$Ca(H_2PO_4)_2(aq) + 2H_2O(l)$

12.13 (a) $PCl_5(s) + 4H_2O(l) \longrightarrow H_3PO_4(aq) + 5HCl(aq)$
(b) $PF_3(g) + 3H_2O(l) \longrightarrow H_3PO_3(aq) + 3HF(aq)$
(c) $SF_4(g) + 3H_2O(l) \longrightarrow H_2SO_3(aq) + 4HF(aq)$
12.14 (a) $P_4O_6(s) + 6H_2O(l) \longrightarrow 4H_3PO_3(aq)$
(b) $Cl_2O_7(l) + H_2O(l) \longrightarrow 2HClO_4(aq)$
(c) $CaO(s) + H_2O(l) \longrightarrow Ca(OH)_2(aq)$
(d) $K_2O(s) + H_2O(l) \longrightarrow 2KOH(aq)$
12.15 (a) $F^-(aq) + H_2O(l) \longrightarrow HF(aq) + OH^-(aq)$;
Brønsted acid–base
(b) $PF_3(g) + 3H_2O(l) \longrightarrow H_3PO_3(aq) + 3HF(aq)$;
Lewis acid–base
(c) $KNO_3(s) \longrightarrow K^+(aq) + NO_3^-(aq)$; *no reaction*
(d) $NO_2^-(aq) + H_2O(l) \rightleftharpoons HNO_2(aq) + OH^-(aq)$
Brønsted acid–base
(e) $Cl_2O(aq) + H_2O(l) \longrightarrow 2HOCl(aq)$;
Lewis acid–base
(f) *No reaction.*

13.1 (a) $K = \left(\dfrac{[CO]^2[H_2]^5}{[C_2H_6][H_2O]^2}\right)$ (b) $K = \left(\dfrac{[NO]^2[O_2]}{[NO_2]^2}\right)_{eq}$

13.2 3.2×10^3. **13.3** $4.38 \times 10^4 \text{ mol}^{-2} \text{ L}^{-2}$.
13.4 (a) 1.23×10^9, (b) 1.9×10^2.
13.5 (a) 1.23×10^9 (b) 1.9×10^2
13.6 $5.9 \times 10^{-13} \text{ atm}^{1/2}$
13.7 $Q = 1.67 \text{ L mol}^{-1} > K_{eq}$. No; smaller.
13.8 $[CO] = [H_2O] = 0.061 \text{ mol L}^{-1}$,
$[CO_2] = [H_2] = 0.139 \text{ mol L}^{-1}$.
13.9 $p_{PCl_5} = 21 \text{ mm Hg}, p_{PCl_3} = p_{Cl_2} = 169 \text{ mm Hg}$.
$K_p = 1.36 \times 10^3 \text{ mm Hg}$. **13.10** (a) $K_c = [NH_3][HCl]$,
$K_p = p_{NH_3}p_{HCl}$. (b) $K_c = \dfrac{[H_2]^4}{[H_2O]^4}$; $K_p = \dfrac{p_{H_2}^4}{p_{H_2O}^4}$

13.11 $p_{NH_3} = p_{HCl} = 0.102 \text{ atm}$. **13.12** (a) In favor of formation of $PCl_5(g)$; (b) no change; (c) in favor of formation of $CaCO_3(s)$. **13.13** (a) High T, pressure not critical except in as far as it affects rate; (b) low T and high P.
13.14 (a) Increase, (b) decrease, (c) no change,
(d) increase, (e) no change, (f) increase.
14.1 $[H_3O^+] = 4.0 \times 10^{-3} \text{ mol L}^{-1}, 8.0\%$.
14.2 $[OH^-] = 6.2 \times 10^{-3} \text{ mol L}^{-1}, 6.2\%$. **14.3** pH = 4.25. **14.4** pH = 11.63. **14.5** $pK_{a_1} = 2.12$.
14.6 $K_a(HCO_3^-) = 4.8 \times 10^{-11} \text{ mol L}^{-1}$. **14.7** A base.
14.8 pH < 7, pH < 7, pH > 7, pH > 7. **14.9** pH = 8.47.
14.10 $K_a = 1.44 \times 10^{-6} \text{ mol L}^{-1}$.
14.11 $K_b = 5.4 \times 10^{-4} \text{ mol L}^{-1}$. **14.12** pH = 9.65.
14.13 (a) 9.65; (b) 10.34; **14.14** Ratio = 1.82.
15.1 Metal bonding strength decreases with increasing atomic mass down each group because as the size of the metal atom increases the distance between the nucleus and the bonding electrons increases and the strength of the electrostatic force of attraction decreases correspondingly. Hence the melting and boiling points decrease down each group. Group 2 metals have higher melting points and boiling points than group 1 metals because they have two rather than one metallic bonding electron per atom, and therefore the metallic bonding is considerably stronger for the group 2 metals than for the group 1 metals.
15.2 $2KOH(aq) + H_2SO_4(aq) \longrightarrow K_2SO_4(aq) + 2H_2O(l)$

$CsOH(aq) + HI(aq) \longrightarrow CsI(aq) + H_2O(l)$
$NaOH(aq) + HClO_4(aq) \longrightarrow NaClO_4(aq) + H_2O(l)$
$3NaOH(aq) + H_3PO_4(aq) \longrightarrow Na_3PO_4(aq) + 3H_2O(l)$
$RbOH(aq) + HNO_3(aq) \longrightarrow RbNO_3(aq) + H_2O(l)$
15.3 11.93.
15.4 $4Li(s) + O_2(g) \xrightarrow{\text{Heat}} 2Li_2O(s)$
$2Li(s) + H_2(g) \xrightarrow{\text{Heat}} 2LiH(s)$
$2Li(s) + Br_2(l) \xrightarrow{\text{Heat}} 2LiBr(s)$
$2Li(s) + 2H_2O(l) \longrightarrow 2LiOH(aq)$;
$2LiOH(aq) + CO_2(g) \longrightarrow Li_2CO_3(aq)$
$2Li(s) + H_2SO_4(aq) \longrightarrow Li_2SO_4(aq) + H_2(g)$
15.5 *Caustic soda*, sodium hydroxide, $NaOH(s)$; *potash*, potassium carbonate, $K_2CO_3(s)$; *rock salt*, sodium chloride, $NaCl(s)$; *chalk*, calcium carbonate, $CaCO_3(s)$; *limestone*, calcium carbonate, $CaCO_3(s)$; *dolomite*, calcium magnesium carbonate, $CaMg(CO_3)_2(s)$. **15.6** $NaH > K_2CO_3 >$ $KCl > NaHSO_4$. **15.7** (a) The white precipitate is insoluble magnesium hydroxide: $MgCl_2(aq) +$ $2NaOH(aq) \longrightarrow Mg(OH)_2(s) + 2NaCl(aq)$. (b) Soluble $Ca(HCO_3)_2$ is formed: $CaCO_3(s) + CO_2(aq) +$ $H_2O(l) \longrightarrow Ca(HCO_3)_2(aq)$. (c) The white precipitate is insoluble calcium fluoride: $CaCl_2(aq) +$ $2NaF(aq) \longrightarrow CaF_2(s) + 2NaCl(aq)$. (d) The white precipitate is insoluble calcium carbonate, because hard water contains calcium ions, $Ca^{2+}(aq)$: $Ca^{2+}(aq) +$ $CO_3^{2-}(aq) \longrightarrow CaCO_3(s)$.
15.8 $MgF_2(s) \rightleftharpoons Mg^{2+}(aq) + 2F^-(aq)$;
$K_{sp} = [Mg^{2+}][F^-]^2$
$Fe(OH)_3(s) \rightleftharpoons Fe^{3+}(aq) + 3OH^-(aq)$;
$K_{sp} = [Fe^{3+}][OH^-]^3$
$Mg_3(PO_4)_2(s) \rightleftharpoons 3Mg^{2+}(aq) + 2PO_4^{3-}(aq)$;
$K_{sp} = [Mg^{2+}]^3[PO_4^{3-}]^2$
$Ag_2S(g) \rightleftharpoons 2Ag^+(aq) + S^{2-}(aq); K_{sp} = [Ag^+]^2[S^{2-}]$
15.9 $K_{sp}(Ag_2SO_4, s) = 1.49 \times 10^{-5} \text{ mol}^3 \text{ L}^{-3}$.
15.10 $4.1 \times 10^{-2} \text{ mol L}^{-1}$. **15.11** $Q > K_{sp}$; therefore a precipitate will form. **15.12** $8.3 \times 10^{-7} \text{ mol L}^{-1}$.
15.13 $0.21 \text{ mol L}^{-1}, 21 \text{ mol L}^{-1}$.
16.1 15.5 J K^{-1}. **16.2** (a) negative, (b) negative,
(c) negative. **16.3** $\Delta G° = 919.6 \text{ kJ}$.
17.1 32.6 g Mg, 30.0 L $Cl_2(g)$. **17.2** 42 mL $H_2(g)$,
21 mL $O_2(g)$. **17.3** 468 C. **17.4** (a) Ca, I_2; (b) H_2, O_2; (c) Cu, O_2; (d) *anode*, Cu deposited; *cathode*, Cu dissolves. **17.5** $E°$ for $Ag^+(aq) + e^- \longrightarrow Ag(s)$ is 0.80 V. **17.6** *Oxidizing strength*: $H_2O(l) < Sn^{2+}(aq) <$ $NO_3^-(aq) < Co^{3+}(aq)$. **17.7** *Reducing strength*:
$Fe^{2+}(aq) < H_2O_2(aq) < H_2S(g) < Sn(s) < Al(s)$.
17.8 $3Cu(s) + Cr_2O_7^{2-}(aq) + 14H^+(aq) \longrightarrow$
$2Cu^{2+}(aq) + 2Cr^{3+}(aq) + 7H_2O(l); E°_{cell} = 0.99 \text{ V}$.
17.9 $Mg(s)|Mg^{2+}(aq)||Al^{3+}(aq)|Al(s), E°_{cell} = 0.70 \text{ V}$.
17.10 $E°_{cell} = 1.10 \text{ V}, \Delta G° = -212 \text{ kJ}, K = e^{85.6} =$ 1.5×10^{37}. **17.11** $E°_{cell} = 0.72 \text{ V}, K = e^{140} = 6.3 \times 10^{60}$.
17.12 $K_{sp} = 1.9 \times 10^{-12} \text{ mol}^2 \text{ L}^{-2}$
18.1 $\Delta H° = 57.1 \text{ kJ}$ (endothermic); K_{eq} increases with increase in temperature.
18.2 (a) (i) $NO(g) + CO_2(g) \xrightarrow{\text{Heat}} CO(g) + NO_2(g)$
$2NO(g) + O_2(g) \longrightarrow 2NO_2(g)$
$2NO(g) + Cl_2(g) \longrightarrow 2NOCl(g)$

A-19

(ii) $2NO(g) + 2H_2(g) \longrightarrow N_2(g) + 2H_2O(l)$
$P_4(s) + 6NO(g) \longrightarrow P_4O_6(s) + 3N_2(g)$

(b) (i) $MnO_4^-(aq) + 5NO_2(g) + H_2O(l) \longrightarrow$
$Mn^{2+}(aq) + 5NO_3^-(aq) + 2H^+(aq)$

(ii) $SO_2(g) + NO_2(g) \longrightarrow SO_3(g) + NO(g)$
$CO(g) + NO_2(g) \longrightarrow CO_2(g) + NO(g)$

18.3 For example, SO_3, Cl_2O_7, B_2O_3, P_4O_{10}, P_4O_6 derived from the respective acids:

HO—S(=O)(=O)—OH HO—Cl(=O)(=O)... HO—B(OH)—OH

Sulfuric acid   Perchloric acid   Boric acid

HO—P(OH)(=O)—... (phosphoric) HO—P(H)(=O)—... (phosphorous)

Phosphoric acid   Phosphorous acid

18.4 $E_{red}^\circ = 0.97$ V; (a) $E_{red} = 1.05$ V, (b) $E_{red} = 0.40$ V, so that solution (a) is a much stronger oxidizing agent than solution (b). Since solution (a) corresponds to a concentrated solution of $HNO_3(aq)$ and solution (b) corresponds to a neutral solution of a nitrate, such as $NaNO_3(aq)$, $HNO_3(aq)$ is a much stronger oxidizing medium than $NaNO_3(aq)$. **18.5** Like ammonia, N_2H_4 contains highly polar N—H bonds and each N atom has a highly localized lone pair, so that it forms strong intermolecular hydrogen bonds, which accounts for its high boiling point. In contrast, the intermolecular forces between N_2O_4 molecules are confined to weak London forces.
18.6 (a) -642.2 kJ, (b) -1049.7 kJ. **18.7** $\Delta H^\circ = -4930$ kJ. **18.8** $4MnO_4^-(aq) + 5N_2H_4(aq) + 12H^+(aq) \longrightarrow 4Mn^{2+}(aq) + 5N_2(g) + 16H_2O(l)$.
18.9 Bond order $1\frac{1}{2}$, intermediate between the single O—O bond length of 147 pm in H_2O_2 and the O=O bond length of 121 pm in O_2. **18.10.** $PbS(s) + 4H_2O_2(aq) \longrightarrow PbSO_4(s) + 4H_2O(l)$.

19.1 $-\dfrac{\Delta[C_3H_8]}{\Delta t}$; $-\dfrac{\Delta[H_2O]}{\Delta t}$; $\dfrac{\Delta[CO]}{\Delta t}$; $\dfrac{\Delta[H_2]}{\Delta t}$.

$-\dfrac{\Delta[C_3H_8]}{\Delta t} = -\dfrac{1}{3}\dfrac{\Delta[H_2O]}{\Delta t} = \dfrac{1}{3}\dfrac{\Delta[CO]}{\Delta t} = \dfrac{1}{7}\dfrac{\Delta[H_2]}{\Delta t}$

19.2 Second order in NO and first order in O_2; rate $= k[NO]^2[O_2]$; $k = 7 \times 10^3$ mol^2 L^{-2} s^{-1}. **19.3** The plot of $1/[NO]$ versus time t is a straight line showing that the reaction is second order, rate $= k_2[CO][NO_2]$, and the slope of the straight line gives $k = 0.488$ mol^{-1} L s^{-1}.
19.4 $k = 0.035$ min^{-1}. **19.5** 470°C
19.6 $E_a = 184$ kJ. **19.7** $k_{450} = 4.0 \times 10^2$ mol^{-1} Ls^{-1}
19.8 $E_a = 53$ kJ mol^{-1}. **19.9** Rate $= k[H_2][Br_2]^{1/2}$.
19.10 Addition of the equations for the three steps of the catalyzed reaction gives $2N_2O \longrightarrow 2N_2 + O_2$, in which Cl_2 appears as neither a reactant or a product, showing that it behaves as a catalyst: *uncatalyzed reaction*, rate $= k[N_2O]$; *catalyzed reaction*, rate $= k[N_2O][Cl_2]^{1/2}$.
20.1 (a) $PCl_3(l) + 3H_2O(l) \longrightarrow H_3PO_3(aq) + 3HCl(aq)$

(b) $BrF(g) + H_2O(l) \longrightarrow HOBr(aq) + HF(aq)$
(c) $BrF_5(g) + 3H_2O(l) \longrightarrow HBrO_3(aq) + 5HF(aq)$
20.2 $AlCl_3$, *Lewis acid*; ICl_3, *Lewis base*. ICl_3, *Lewis acid*; Cl^-, *Lewis base*. **20.3** BrF_3, AX_3E_2, T-shaped; BrF_4^+, AX_4E, disphenoid shape; ICl_2^+, AX_2E_2, angular; I_3^-, AX_2E_3, linear; IF_6^+, AX_6, octahedron.
20.4 $ClO^- > ClO_2^- > ClO_3^- > ClO_4^-$.
$ClO^-(aq) + 2e^- + H_2O(l) \longrightarrow Cl^-(aq) + 2OH^-(aq)$
$ClO_2^-(aq) + 4e^- + 2H_2O(l) \longrightarrow Cl^-(aq) + 4OH^-(aq)$
$ClO_3^-(aq) + 6e^- + 3H_2O(l) \longrightarrow Cl^-(aq) + 6OH^-(aq)$
$ClO_4^-(aq) + 8e^- + 4H_2O(l) \longrightarrow Cl^-(aq) + 8OH^-(aq)$
20.5 (a) $BrO_3^-(aq) + F_2(g) + 2OH^-(aq) \longrightarrow BrO_4^-(aq) + 2F^-(aq) + H_2O(l)$
(b) $OCl^-(aq) + 3I^-(aq) + 2H_3O^+(aq) \longrightarrow I_3^-(aq) + Cl^-(aq) + 3H_2O(l)$
(c) $4KClO_3(s) + 6H_2SO_4(l) \longrightarrow 4ClO_2(g) + O_2(g) + 4KHSO_4 + 2H_3O^+ + 2HSO_4^-$

20.6

$HXeO_4^-$ XeO_6^{4-}

20.7 (a) $Mn^{2+}(aq) + 5XeO_6^{4-}(aq) + 9H_3O^+(aq) \longrightarrow 2MnO_4^-(aq) + 5HXeO_4^{2-}(aq) + 11H_2O(l)$
(b) $Ba_2XeO_6(s) + 4H_2SO_4(l) \longrightarrow 2BaSO_4 + XeO_4(s) + 2H_3O^+ + 2HSO_4^-$
21.1 (a) $Cr_2O_7^{2-}(aq) + 3SO_2(g) + 2H_3O^+(aq) \longrightarrow 2Cr^{3+}(aq) + 3SO_4^{2-}(aq) + 3H_2O(l)$
(b) $4Cr^{2+}(aq) + O_2(g) + 4H_3O^+(aq) \longrightarrow 4Cr^{3+}(aq) + 6H_2O(l)$
21.2 (a) $2MnO_4^- + 10I^- + 16H^+ \longrightarrow 2Mn^{2+} + 5I_2 + 8H_2O$
(b) $5Fe^{2+} + MnO_4^- + 8H^+ \longrightarrow Mn^{2+} + 5Fe^{3+} + 4H_2O$
(c) $2MnO_4^- + 5(COOH)_2 + 6H^+ \longrightarrow 2Mn^{2+} + 10CO_2 + 8H_2O$
21.3 $0.0621M$. **21.4** CoF_6^{3-} is a complex ion of $Co^{3+}(d^4)$. Since F^- is a low field ligand, CoF_6^{3-} will be a high spin complex with four unpaired electrons.
21.5 For $AgI(s) + 2NH_3(aq) \rightleftharpoons Ag(NH_3)^+(aq) + I^-(aq)$, the equilibrium constant is $K_{sp}(AgI, s) \times K_f(Ag(NH_3)_2^+) = 8 \times 10^{-9}$; then for $[NH_3] = 1M$, $[Ag(NH_3)_2^+] = 9 \times 10^{-5}M$.
22.1 (a) $Si_2O_7^{6-}$, (b) SiO_3^{2-}, (c) SiO_3^{2-}, (d) $Si_2O_5^{2-}$, (e) SiO_2, (f) $Si_4O_{11}^{4-}$.
22.2 $KMg_3AlSi_3O_{10}(OH)_2$.
23.1 Hexane, 2-methylpentane, 3-methylpentane, 2,2-dimethylbutane, and 2,3-dimethylbutane.
23.2 2,4-Dimethyl-1-pentene. **23.3** 1-Propanol (primary), 2-propanol (secondary).
23.4 $CH_3OCH_2CH_2CH_3$, prepared from the reaction of the sodium salt of 1-propanol with methylbromide.
23.5 $5CH_3CH(OH)CH_3 + 2MnO_4^- + 6H^+ \longrightarrow 5(CH_3)_2CO + 2Mn^{2+} + 8H_2O$.
23.6 $CH_3C(O)CH_2CH_2CH_3$, prepared by oxidation of 2-pentanol.

23.7 $CH_3CH_2CO_2H + HOCH_2CH_3 \longrightarrow$

     Propanoic acid   Ethanol

$CH_3CH_2COCH_2CH_3 + H_2O$
      $\|$
      O

24.1 $[CH_2-CH=CH-CH_2-CH-CH_2]_n$

24.2 6.67×10^8 g.   **24.3** 1,4-benzene dicarboxylic acid and 1,4-benzene diamine.   **24.4** (a) (i) Amylose,

cellulose; (ii) glycogen, chitin, (b) (i) Starch or glycogen; (ii) cellulose.   **24.5** (i) Fibers, e.g. in silk; (ii) protofibrils, e.g. in wool and hair; (iii) enzymes, e.g. sucrase; (iv) hormones, e.g. insulin. Heating an enzyme to 100 °C results in its destruction (denaturation).
25.1 (a) $^{22}_{11}Na \longrightarrow {}^{0}_{1}e + {}^{22}_{10}Ne$.
(b) $^{27}_{12}Mg \longrightarrow {}^{0}_{-1}e + {}^{27}_{13}Al$. (c) $^{36}_{17}Cl \longrightarrow {}^{36}_{18}Ar + {}^{0}_{-1}e$.
(d) $^{239}_{94}Pu \longrightarrow {}^{235}_{92}U + {}^{4}_{2}He$.
25.2 3.5 μg.   **25.3** 5.3 years.   **25.4** 6.1×10^8 years.
25.5 2.58×10^3 years.   **25.6** 2.0446×10^{-11} J; 1.278×10^{-12} J nucleon^{-1}.

ANSWERS TO SELECTED PROBLEMS

CHAPTER 1

2. (a) Fe, (b) Ni, (c) Hg, (d) Si, (e) Cr, (f) He, (g) Ba, (h) Pb, (i) U, (j) Ca.   **4.** (a) Calcium, (b) bromine, (c) iron, (d) manganese, (e) copper, (f) silver, (g) gold, (h) zinc, (i) arsenic, (j) platinum.
6. (b) and (g) are *elements*; (a), (d), and (j) are *compounds*; the remainder are *mixtures*.   **8.** Soda water, filtered coffee, vinegar, and air are homogeneous mixtures; snow (a form of solid water) and "dry ice" (solid carbon dioxide) are homogeneous "pure" substances; wood and soil are heterogeneous mixtures.   **11.** The following are typical chemical properties (many other chemical reactions are given in the textbook): *water* reacts with many metals to give hydrogen gas; *copper* reacts with nitric acid to give a blue solution of copper nitrate and oxides of nitrogen (NO or NO_2 depending on the conditions); *iron* reacts with dilute sulfuric acid to give a pale green solution of iron(II) sulfate and hydrogen gas; *magnesium* burns brilliantly in air or oxygen to give white solid magnesium oxide, MgO; *hydrogen* burns in air or oxygen with a pale blue flame, to give water; *hydrogen peroxide* decomposes on heating to give water and oxygen gas.   **13.** (a) As_4 and As, (b) C_3H_6 and CH_2, (c) P_4O_{10} and P_2O_5, (d) XeF_4 and XeF_4.   **15.** (a) HO, (b) H_2O, (c) Li_2CO_3, (d) CH_2O, (e) S, (f) C_3H_7, (g) BH_3, (h) O.   **17.** (a) Magnesium, (b) magnesium oxide, (c) sodium chloride, (d) phosphorus (white), (e) sulfur (orthorhombic or monoclinic), (f) nitric acid, (g) sulfuric acid, (h) potassium nitrate, (i) sodium sulfate, (j) barium sulfate.   **19.** (a) MgO, (b) CO, (c) CO_2, (d) Br_2, (e) P_4, (f) S_8, (g) NH_3, (h) CH_4, (i) C_2H_6, (j) NaCl.   **21.** 190 pm.   **23.** H_2O, angular: XeF_2, linear; PH_3, trigonal pyramidal; BCl_3, trigonal planar.   **25.** 96.9 pm.   **27.** (a), (b), (e), and (f) are exact numbers.   **29.** (a) 300 (2 significant figures); (b) 116 200 (4 significant figures), (c) 0.0048 (2 significant figures), (d) −0.064 40 (4 significant figures).
31. (a) 2×10^{-4}, (b) 2×10^{-3}, (c) 2.4×10^5,

(d) 4×10^2, (e) 2.4×10.   **33.** (a) meter, m; (b) kilogram, kg; (c) meter cubed, m^3; (d) kilogram per cubic meter, kg m^{-3}.   **35.** (a) kilo, (b) deci, (c) centi, (d) micro, (e) pico.   **37.** (a) 1 ft/12 inch, (b) 12 inch/1 ft, (c) 1 km/0.6214 mile, (d) 0.6214 mile/1 km, (e) 10^6 mL/1 m^3, (f) 2.590 km^2/1 mile^2, (g) 10^{-4} m^2/1 cm^2.
39. (a) 3.9×10^7 m, (b) 68 kg, (c) 9.8×10^3 kg m^{-2}, (d) 2.7×10^{-2} km s^{-1}.   **41.** (a) 1.998×10^8 m s^{-1}
(b) 1.007×10^{-26} kg, (c) 5.358×10^{-8} m s^{-1}, (d) 2.241×10^{-5} m^{-3}.   **43.** (a) 2.998×10^{10} m, (b) 1.43×10^{-10} m, (c) 1×10^{-12} m, (d) 1.54×10^{-10} m.   **45.** 62.1 mile h^{-1}
47. 4 min 7 s.   **49.** 9.2 L.   **51.** Jupiter revolves around the sun with a period of 3.74×10^8 s at a distance of approximately 7.79×10^8 km. The average density of Jupiter is 1.330×10^3 kg m^{-3}.   **54.** 10^{-3}.   **55.** (a) 50.3 cm^3, (b) 186 g.   **57.** (a) 6.20×10^{-4} cm^3, (b) 1.57 cm^3, (c) 1.14×10^3 cm^3, (d) 943 cm^3.   **59.** 11 g cm^{-3}.
61. 2.96×10^4 kg.   **63.** 3.8×10^2 kg.
65. 1.9×10^{-29} m^3 atom^{-1}, 170 pm.   **68.** 1.28 g mL^{-1}.
69. 50 g acetic acid kg^{-1}.   **70.** (a) 4.21 mass %, (b) 7.248 mass %, (c) 10.6 mass % of HCl, 1.21 mass % of NaCl, (d) 52.7 mass %, (e) 82.41 mass %.
71. 874 g H_2SO_4.   **72.** 26 cm^3 or 26 mL.
74. 6.4×10^9 kg Au.   **75.** 2×10^7 kg seawater.
80. 3.6 g K_2SO_4, 11 mass %.   **81.** 6.0 g, 26.5 mass%.

CHAPTER 2

2. (a) H, 1; (b) O, 8; (c) F, 9; (d) Ne, 10; (e) Mg, 12; (f) P, 15; (g) Cl, 17; (h) Ca, 20; (i) Zn, 30.   **4.** (a) 5, (b) 7, (c) 1, (d) 10, (e) 17, (f) 8, (g) 16, (h) 19, (i) 26.   **6.** (a) $^{40}_{19}K$, (b) $^{30}_{14}Si$, (c) $^{40}_{18}Ar$, (d) $^{15}_{7}N$, (e) $^{32}_{16}S$, (f) $^{23}_{11}Na$, (g) $^{27}_{13}Al$.   **8.** $^{2}_{1}H$ 1, 1, 1; $^{19}_{9}F$ 9, 10, 9; $^{40}_{20}Ca$ 20, 20, 20; $^{112}_{48}Cd$ 48, 64, 48; $^{117}_{50}Sn$ 50, 67, 50; $^{131}_{54}Xe$ 54, 77, 54.

10.

Atomic symbol	$^{24}_{12}Mg$	$^{106}_{47}Ag$	$^{137}_{56}Ba$
Mass number	24	106	137
Atomic number	12	47	56
Number of protons	12	47	56
Number of electrons	12	47	56
Number of neutrons	12	59	81

12. 10.81 u.　**15.** $^{69}_{31}Ga$, 31 protons, 38 neutrons, 60.30% abundance; $^{71}_{31}Ga$, 31 protons, 40 neutrons, 39.70% abundance.　**16.** 0.70%.　**18.** (a) ^{79}Br and ^{81}Br; (b) $^{79}Br^{35}Cl$, $^{79}Br^{37}Cl$, $^{81}Br^{35}Cl$, and $^{81}Br^{37}Cl$.
20. 19.00 u, ^{19}F.　**22.** (a) (i) $1\ g/6.022 \times 10^{23}$ u, (ii) $1\ kg/6.022 \times 10^{26}$ u,　(iii) $1\ lb/2.732 \times 10^{26}$ u. (b) 7.122×10^{26} mol　**24.** 3.002×10^{21} atoms.
26. (a) 0.55 mol,　(b) 0.16 mol,　(c) 0.17 mol, (d) 0.10 mol.　**28.** 159.8 g mol^{-1}.
30. (a) 4.037×10^{-23} g atom^{-1},　(b) 13.99 cm^3 mol^{-1}, (c) 2.323×10^{-23} cm^3 atom^{-1},　(d) 177.0 pm.
32. 7.68×10^{23} molecules.　**33.** 18.02, 34.02, 58.44, 184.1, 28.01, 44.01, 16.04, 32.08, 17.03, 36.46 g mol^{-1}.
35. (a) 30.03 u, 60.05 u;　(b) 46.03 u, 46.03 u; (c) 342.3 u, 342.3 u;　(d) 29.06 u, 58.12 u;　(e) 13.84 u, 27.67 u;　(f) 17.01 u, 34.02 u.　**37.** (a) 11.2% H, 88.8% O; (b) 39.43% Na, 60.66% Cl;　(c) 79.88% C, 20.12% H; (d) 13.20% Mg, 86.80% Br;　(e) 27.29% C, 72.71% O.
39. (a) 85.63% C, 14.37% H;　(b) 92.24% C, 7.76% H; (c) 11.96% Mg, 34.87% Cl, 5.952% H, 47.22% O; (d) 20.09% Fe, 11.53% S, 5.08% H, 63.31% O; (e) 24.09% Cu, 21.24% N, 12.15% S, 36.39% O, 6.11% H.
41. 21.21% N, 471 g $(NH_4)_2SO_4$.　**43.** CBr_2, 171.8 u.
45. SF_4.　**47.** $C_4H_5N_2O$, 97.1 u.　**49.** $C_3H_8O_3$.
51. CH_2.　**53.** A, VO;　B, V_2O_3; C, VO_2; D, V_2O_5.
55. C_5H_7N, $C_{10}H_{14}N_2$.　**57.** $x = 1$, $Li_2SO_4 \cdot H_2O$.
59. $m = 4$, $C_{12}H_4Cl_6$.
60. (a) $2SO_2 + 2H_2O + O_2 \longrightarrow 2H_2SO_4$
(b) $2CH_3OH + 3O_2 \longrightarrow 2CO_2 + 4H_2O$
(c) $2H_2O_2 \longrightarrow 2H_2O + O_2$
(e) $Zn + 2HCl \longrightarrow ZnCl_2 + H_2$
62. (a) $2S + 3O_2 \longrightarrow 2SO_3$
(b) $2C_2H_2 + 3O_2 \longrightarrow 4CO + 2H_2O$
(c) $Na_2CO_3 + Ca(OH)_2 \longrightarrow 2NaOH + CaCO_3$
(d) $Na_2SO_4 + 4H_2 \longrightarrow Na_2S + 4H_2O$
(e) $2Cu_2S + 3O_2 \longrightarrow 2Cu_2O + 2SO_2$
(f) $2Cu_2O + Cu_2S \longrightarrow 6Cu + SO_2$
64. (a) $2S + 3O_2 \longrightarrow 2SO_3$
(b) $2CH_4 + 3O_2 \longrightarrow 2CO + 4H_2O$
(c) $Mg + H_2O \longrightarrow MgO + H_2$
(d) $C_4H_{10} + 4H_2O \longrightarrow 4CO + 9H_2$
66. 0.603 g Mg.　**68.** 1.45 g F_2, 3.95 g XeF_4.
70. 22.6 g phosphorus.
72. $3Ca(OH)_2 + 2H_3PO_4 \longrightarrow Ca_3(PO_4)_2 + 6H_2O$; 34.0 g $Ca(OH)_2$;　47.5 g $Ca_3(PO_4)_2$.　**74.** BaO_2.
76. 326 g CO_2.
78. $6Mg + B_2O_3 \longrightarrow 3MgO + Mg_3B_2$; 3.812 g B_4H_{10}.　**80.** 50 mass % Mg, 50 mass % Zn; 1.03M and 0.38M.　**81.** 5.6 mol H_2SO_4.　**82.** 7.81 g PCl_5.
84. 4.89 g $AlCl_3$, 1.71 g Al.　**86.** 77.93%.　**88.** 67.2%.
90. (a) 0.0100 mol NaOH,　(b) 2.50×10^{-3} mol NaOH,

(c) 2.52×10^{-4} mol NaOH,　(d) 3.37×10^{-4} mol NaOH.
92. 0.0380M $KMnO_4$.　**94.** 4.75 mL H_2SO_4.
96. (a) Add distilled water to 189 mL 0.100M $Ba(OH)_2$(aq) to give a total volume of 6.30 L.　(b) Add distilled water to 14.9 cm^3 of 35 mass % $Cr_2(SO_4)_3$ to give a total volume of 750 mL.　**98.** 0.24 g AgCl(s).
100. 850 mL 0.0120M HCl(aq).
102. (a) $2UF_5 + 2H_2O \longrightarrow UO_2F_2 + UF_4 + 4HF$; (b) 4.71 g UF_4.　**103.** (a) 3.51×10^{-3} u, (b) 7.83×10^{10} J,　(c) 5.97×10^4 g water.

CHAPTER 3

1. (a) N_2, O_2, Ar;　(b) O_2, CO_2, CH_4;　(c) in photosynthesis.　**3.** (a) Noble gases such as Ar and He,　(b) N_2 and O_2,　(c) CO_2,　(d) Ar,　(e) Au,　(f) H_2.
5. (a) A process in which an element or compound combines with oxygen (see page 106);　(b) a process in which oxygen is removed partially or completely from a compound.
7. (a) $Fe_2O_3(s) + 3CO(g) \longrightarrow 2Fe(s) + 3CO_2(g)$
(b) $Fe_2O_3(s) + 3H_2(g) \longrightarrow 2Fe(s) + 3H_2O(g)$
(c) $CuO(s) + CO(g) \longrightarrow Cu(s) + CO_2(g)$
(d) $Mg(s) + H_2O(g) \longrightarrow MgO(s) + H_2(g)$
9. (a) $3H_2(g) + N_2(g) \longrightarrow 2NH_3(g)$
(b) $N_2(g) + O_2(g) \longrightarrow 2NO(g)$
(c) $3Mg(s) + N_2(g) \longrightarrow Mg_3N_2(s)$
11. Neither CO(g) nor H_2(g) is appreciably soluble in water; conversion of the CO(g) to CO_2(g), by reaction with more steam, enables soluble CO_2(g) to be separated from insoluble H_2(g).
13. (a) $2N_2(g) + 5O_2(g) + 2H_2O(l) \longrightarrow 4HNO_3(aq)$; (b) 0.38 g HNO_3.　**15.** (a) Hydrogen chloride, HCl; (b) iron, Fe;　(c) ammonia, NH_3;　(d) methane, CH_4; (e) hydrogen, H_2.
17. (a) $2C(s) + O_2(g) \longrightarrow 2CO(g)$
(b) $2Ca(s) + O_2(g) \longrightarrow 2CaO(s)$
(c) $2KClO_3(s) \longrightarrow 2KCl(s) + 3O_2(g)$
(d) $C_3H_8(g) + 5O_2(g) \longrightarrow 3CO_2(g) + 4H_2O(g)$
(e) $C_2H_6O(l) + 3O_2(g) \longrightarrow 2CO_2(g) + 3H_2O(g)$
19. (a) Metal, solid;　(b) nonmetal, gas;　(c) nonmetal, gas;　(d) nonmetal, gas;　(e) nonmetal, solid; (f) nonmetal, solid;　(g) metal, solid;　(h) nonmetal, gas; (i) nonmetal, liquid.　**21.** 2.5 atm.　**23.** 0.28 atm.
25. $-258\ °C$ (15 K).　**27.** 1.50 atm.　**29.** 174 L.
31. 52.0 atm.　**33.** (a) 2.59 atm,　(b) 2.24 atm.
35. 0.163 mol.　**37.** 6.11 L.　**39.** 1.18 g.
41. (a) 28 kg BaO_2,　(b) 6.83 days.　**43.** 1.36 g.
45. 0.731 g L^{-1}.　**47.** 5.03 g L^{-1}.　**49.** O_3.
51. 140 g mol^{-1}, CCl_3F.　**53.** 34.6 g mol^{-1}.
55. (a) 116 g mol^{-1},　(b) $C_6H_{12}O_6$.　**57.** B_2H_6.
59. 2.08 atm, 0.881 atm, 2.96 atm.
61. (a) 7.1×10^{-4} mm Hg,　(b) 2.3×10^{19} molecules m^{-3}.
63. $p_{N_2} = 9.72 \times 10^{-3}$ atm, $p_{H_2O} = 2.92 \times 10^{-2}$ atm, $p_{NH_3} = 0.981$ atm.　**65.** (a) 1793 m s^{-1},　(b) 302 m s^{-1}, (c) 481 m s^{-1},　(d) 600 m s^{-1},　(e) 384 m s^{-1}, (f) 436 m s^{-1}.　**67.** 1.414.　**69.** 0.267 mile s^{-1}.
71. F_2.　**73.** B_4H_7, B_8H_{14}.　**75.** 262 L.　**77.** 224 L.
79. (a) $CaH_2(s) + 2H_2O(l) \longrightarrow Ca(OH)_2(s) + 2H_2(g)$; (b) 9.39 g CaH_2.　**81.** (a) 276 L,　(b) 1.24×10^3 L,

(c) 1.41×10^3 L.   **83.** 87.4%.
85. C_3H_6; $C_3H_6(g) + 9O_2(g) \longrightarrow 6CO_2(g) + 6H_2O(g)$.
87. 1630 times.   **89.** 30.29 g $CO_2(g)$.

CHAPTER 4

1.

Element	Group	Period	Type
He	8	1	Nonmetal
P	5	3	Nonmetal
K	1	4	Metal
Ca	2	4	Metal
Te	6	5	Nonmetal
Br	7	4	Nonmetal
Al	3	3	Metal
Sn	4	5	Metal

3. Ar and K, Co and Ni, Te and I, Th and Pa; position in the periodic table depends on the atomic number Z; average atomic mass depends on isotopic composition.   **5.** Transition metals: Mn and W. Main group elements: Se, group 6, nonmetal; P, group 5, nonmetal; Kr, group 8, nonmetal; Al, group 3, metal; Pb, group 4, metal.

7.

Symbol	Element	Type	Group	Valence Electrons	Common Valence
Li	Lithium	Metal	1	1	1
Mg	Magnesium	Metal	2	2	2
S	Sulfur	Nonmetal	6	6	2
P	Phosphorus	Nonmetal	5	5	3
Br	Bromine	Nonmetal	7	7	1
Ne	Neon	Nonmetal	8	8	0
As	Arsenic	Nonmetal	5	5	3
Se	Selenium	Nonmetal	6	6	2
Cl	Chlorine	Nonmetal	7	7	1
Ba	Barium	Metal	2	2	2

9. (a) $Mg(s) + H_2O(g) \longrightarrow MgO(s) + H_2(g)$
(b) $S(s) + H_2(g) \longrightarrow H_2S(g)$
(c) $2Na(s) + I_2(s) \longrightarrow 2NaI(s)$
(d) $2K(s) + 2H_2O(l) \longrightarrow 2KOH(aq) + H_2(g)$
(e) $H_2(g) + Cl_2(g) \longrightarrow 2HCl(g)$ (f) No reaction
11. FrH, SnH_4, AtH, no hydride; $FrCl$, $SnCl_4$, $AtCl$, no chloride.

13.

Element	Group	Valence	Type	Hydride
C	4	4	nm[a]	CH_4
Ca	2	2	m[b]	CaH_2
He	8	0	nm	—
B	3	3	nm	BH_3
Cl	7	1	nm	HCl
Li	1	1	m	LiH
O	6	2	nm	H_2O
F	7	1	nm	HF
P	5	3	nm	PH_3
Mg	2	2	m	MgH_2

[a] nm, nonmetal; [b] m, metal.

15. (a) $2Li(s) + S(s) \longrightarrow Li_2S(s)$
(b) $Ca(s) + 2H_2O(l) \longrightarrow Ca(OH)_2(s) + H_2(g)$
(c) No reaction   (d) $Sr(s) + Br_2(l) \longrightarrow SrBr_2(s)$
(e) $Mg(s) + H_2(g) \longrightarrow MgH_2(s)$

16. (a) Na, K; Mg, Ca; C, Si; Cl, F.
(b) $2Na(s) + H_2(g) \longrightarrow 2NaH(s)$;
$2K(s) + H_2(g) \longrightarrow 2KH(s)$
$4Na(s) + O_2(g) \longrightarrow 2Na_2O(s)$;
$4K(s) + O_2(g) \longrightarrow 2K_2O(s)$
$Mg(s) + H_2(g) \longrightarrow MgH_2(s)$;
$Ca(s) + H_2(g) \longrightarrow CaH_2(s)$
$2Mg(s) + O_2(g) \longrightarrow 2MgO(s)$;
$2Ca(s) + O_2(g) \longrightarrow 2CaO(s)$
$C(s) + 2H_2(g) \longrightarrow CH_4(g)$
$Si(s) + 2H_2(g) \longrightarrow SiH_4(g)$
$C(s) + O_2(g) \longrightarrow CO_2(g)$
$Si(s) + O_2(g) \longrightarrow SiO_2(s)$
$Cl_2(g) + H_2(g) \longrightarrow 2HCl(g)$;
$F_2(g) + H_2(g) \longrightarrow 2HF(g)$
$2Cl_2(g) + O_2(g) \longrightarrow 2Cl_2O(g)$;
$2F_2(g) + O_2(g) \longrightarrow 2F_2O(g)$
19. F_2O, Al_2S_3, BCl_3, CS_2, Mg_3N_2.   **21.** The valence shell is the outermost shell of electrons;   (a) 3,   (b) 7,   (c) 8,   (d) 2,   (e) 2.   **23.** (a) The outermost shell of electrons;   (b) the number of valence electrons is the same as the group number;   (c) 5, 8, 2, 2, 6, 8, 1, 8, 7, 8;   (d) N^{3-}, O^{2-}, and Na^+.   **25.** (a) +7,   (b) +5,   (c) +6,   (d) +1,   (e) +6,   (f) +2,   (g) +1,   (h) +4,   (i) +7.
27. Na < F < Ne.   **29.** 0.022 MJ.   **32.** $K\cdot$ $\cdot Ca\cdot$
$\cdot \overset{..}{B}\cdot$ $\cdot \overset{.}{S}n\cdot$ $:\overset{..}{S}b\cdot$ $:\overset{..}{T}e\cdot$ $:\overset{..}{B}r\cdot$ $:\overset{..}{X}e:$ $\cdot \overset{..}{A}s\cdot$ $\cdot \overset{.}{G}e\cdot$

33. (a) A, group 1; D, group 4; E, group 2; G, group 7;
(b) A^+, E^{2+}; G^-.   **35.** (a) Mg^{2+},   (b) Rb^+,   (c) Br^-,
(d) S^{2-},   (e) Al^{3+},   (f) Li^+.   **37.** (a) $(NH_4)_2S$,
(b) Fe_2O_3,   (c) Cu_2O,   (d) $AlCl_3$.

39. (a) CaI_2, $Ca^{2+}[:\overset{..}{\underset{..}{I}}:^-]_2$;   (b) CaO, $Ca^{2+}[:\overset{..}{\underset{..}{O}}:^{2-}]$;
(c) Al_2S_3, $[Al^{3+}]_2[:\overset{..}{\underset{..}{S}}:^{2-}]_3$;   (d) $CaBr_2$, $Ca^{2+}[:\overset{..}{\underset{..}{B}r}:^-]_2$;
(e) Rb_2Se, $[Rb^+]_2:\overset{..}{\underset{..}{S}e}:^{2-}$;   (f) BaO, $Ba^{2+}[:\overset{..}{\underset{..}{O}}:^{2-}]$.

41. The possible ions are Mg^{2+}, Al^{3+}, Ca^{2+}, and H^-, O^{2-}, and F^-, to give the following possible combinations: MgH_2, $Mg^{2+}[:H^-]_2$; MgO, $Mg^{2+}:\overset{..}{\underset{..}{O}}:^{2-}$; MgF_2, $Mg^{2+}[:\overset{..}{\underset{..}{F}}:^-]_2$; AlH_3, $Al^{3+}[:H^-]_3$; Al_2O_3, $[Al^{3+}]_2[:\overset{..}{\underset{..}{O}}:^{2-}]_3$; AlF_3, $Al^{3+}[:\overset{..}{\underset{..}{F}}:^-]_3$; CaH_2, $Ca^{2+}[:H^-]_2$; CaO, $Ca^{2+}:\overset{..}{\underset{..}{O}}:^{2-}$; CaF_2, $Ca^{2+}[:\overset{..}{\underset{..}{F}}:^-]_2$.   **43.** (a) H—H,   (b) $H-\overset{..}{\underset{..}{C}l}:$,
(c) $H-\overset{..}{\underset{..}{I}}:$,   (d) $H-\overset{H}{\underset{}{\overset{|}{P}}}-H$,   (e) $:\overset{..}{\underset{..}{F}}-\overset{:\overset{..}{F}:}{\underset{:\overset{..}{F}:}{\overset{|}{Si}}}-\overset{..}{\underset{..}{F}}:$,
(f) $:\overset{..}{\underset{..}{F}}-\overset{..}{\underset{..}{O}}-\overset{..}{\underset{..}{F}}:$,   (g) $:\overset{..}{\underset{..}{C}l}-\overset{..}{\underset{..}{C}l}:$ **45.**

47. $H-\overset{H}{\underset{H}{\overset{|}{N^+}}}-H$ $H-\overset{H}{\underset{H}{\overset{|}{B^-}}}-H$ $H-\overset{H}{\underset{H}{\overset{|}{C^+}}}-H$

$H-\overset{H}{\underset{H}{\overset{|}{C^-}}}-H$ $:\overset{..}{\underset{..}{C}l}-\overset{:\overset{..}{C}l:}{\overset{|}{S^+}}-\overset{..}{\underset{..}{C}l}:$

49. The elements of the second period (Li to Ne).

51. $H-\overset{\cdot\cdot}{\underset{H}{O}}{}^{+}-H$ $H-\overset{\cdot\cdot}{\underset{\cdot\cdot}{O}}:^{-}$ $:\overset{\cdot\cdot}{F}-\overset{\overset{\displaystyle:\overset{\cdot\cdot}{F}:}{|}}{\underset{\underset{\displaystyle:\overset{\cdot\cdot}{F}:}{|}}{N}}{}^{+}-\overset{\cdot\cdot}{F}:$ $:\overset{\cdot\cdot}{F}-\overset{\cdot\cdot}{\underset{\cdot\cdot}{S}}-\overset{\cdot\cdot}{F}:$

$:\overset{\cdot\cdot}{S}=C=\overset{\cdot\cdot}{S}:$ $:\overset{\cdot\cdot}{F}-\overset{\cdot\cdot}{\underset{\cdot\cdot}{O}}-\overset{\cdot\cdot}{F}:$ $:\overset{\cdot\cdot}{F}-\overset{\overset{\displaystyle:\overset{\cdot\cdot}{F}:}{|}}{\underset{\underset{\displaystyle:\overset{\cdot\cdot}{F}:}{|}}{B}}{}^{-}-\overset{\cdot\cdot}{F}:$

54. 192 pm.   **55.** Cl, 99 pm; C, 77 pm; Br, 117 pm; I, 138 pm; Br_2, 234 pm; BrCl, 216 pm; I_2, 276 pm.
57. (a) C, 77 pm; P, 110 pm; S, 104 pm; Cl, 99 pm;   (b) PCl, 209 pm; CCl, 176 pm; SCl, 203 pm; PC, 187 pm.

59.

	Molecular Type	Geometry	Shape
(a)	AX_3E		Trigonal pyramidal
(b)	AX_2E_2		Angular
(c)	AXE_3	:A—X	Linear
(d)	AX_2E		Angular
(e)	AXE_2	:A—X	Linear
(f)	AX_2	X—A—X	Linear

61. H_2O, AX_2E_2, angular; H_3O^+, AX_3E, trigonal pyramidal; PCl_3, AX_3E, trigonal pyramidal; BCl_3, AX_3, trigonal planar; SiH_4, AX_4, tetrahedral.   **63.** AX_3, trigonal planar; AX_2E, angular; AX_2E, angular; AX_4, tetrahedral.

65.

Lewis Structure	Type	Shape
$:\overset{\cdot\cdot}{Cl}-Be-\overset{\cdot\cdot}{Cl}:$	AX_2	Linear
$:\overset{\cdot\cdot}{Cl}-\overset{\underset{\displaystyle:\overset{\cdot\cdot}{Cl}:}{\mid}}{B}-\overset{\cdot\cdot}{Cl}:$	AX_3	Trigonal planar
$:\overset{\cdot\cdot}{Cl}-\overset{\overset{\displaystyle:\overset{\cdot\cdot}{Cl}:}{\mid}}{\underset{\underset{\displaystyle:\overset{\cdot\cdot}{Cl}:}{\mid}}{C}}-\overset{\cdot\cdot}{Cl}:$	AX_4	Tetrahedral
$:\overset{\cdot\cdot}{Cl}-\overset{\underset{\displaystyle:\overset{\cdot\cdot}{Cl}:}{\mid}}{N}-\overset{\cdot\cdot}{Cl}:$	AX_3E	Trigonal pyramidal
$:\overset{\cdot\cdot}{Cl}-\overset{\cdot\cdot}{\underset{\cdot\cdot}{O}}-\overset{\cdot\cdot}{Cl}:$	AX_2E_2	Angular
$:\overset{\cdot\cdot}{Cl}-\overset{\cdot\cdot}{F}:$	AXE_3	Linear

67. Sc should also be a metal and its valence should be 3, since it is found between Ca and Ti. Thus, it would be expected to resemble Al in group 3 in forming *ionic* compounds

such as $Sc_2O_3(s)$ and $ScCl_3(s)$. Similarly, its physical properties should be intermediate between those of Ca and Ti. Data in Chapter 21 confirm this: m.p. 1812 °C, density 3.0 g cm^{-3} (b.p. 2730 K).   **69.** On the basis of InO the molar mass of In is 76.6 g mol^{-1}, which places In in the periodic table in the vicinity of As or Se, where there is no vacancy. Moreover such a position would suggest that In should be a nonmetal. If it was in group 6 it could form an oxide of empirical formula InO, but this would have to be a volatile solid and would not resemble the ionic oxide ZnO(s). On the basis of In_2O_3 the molar mass is 115 g mol^{-1}, which places In in group 3 and period 5, between Cd and Sn, two metals, which is consistent with its resemblance to Zn (another metal) and the valence of 3 in the oxide $In_2O_3(s)$.
71. (a) Each contains a total of 10 electrons; species with the same number of electrons constitute an *isoelectronic series*.   (b) Each ion has the same outer shell of 8 electrons but the core charges increase in the order O^{2-} (+6), F^- (+7), Na^+ (+9), Mg^{2+} (+10), so that the ionic radii (sizes) are expected to decrease in the same order.

73. $\begin{matrix} H \\ \\ H \end{matrix}\!\!>\!C=C\!<\!\!\begin{matrix} H \\ \\ H \end{matrix}$ $H-C\equiv C-H$

$^-:C\equiv N:$ $:N\equiv N:$ $:N\equiv O:^+$ $:P\equiv P:$

CHAPTER 5

2. (a) $MnO_2(s) + 4HCl(aq) \longrightarrow MnCl_2(aq) + 2H_2O(l) + Cl_2(g)$ (b) $H_2(g) + Cl_2(g) \xrightarrow{\text{Sunlight}} 2HCl(g)$
(c) Polar $^{\delta+}H-Cl^{\delta-}$ is attracted to polar water molecules and transfers a proton to a water molecule to give $H_3O^+(aq)$ and $Cl^-(aq)$.   **4.** (a) $P_4(s) + 6Cl_2(g) \longrightarrow 4PCl_3(l)$
(b) (i) $S_8(s) + 8Cl_2(g) \longrightarrow 8SCl_2(l)$
(ii) $S_8(s) + 4Cl_2(g) \longrightarrow 4S_2Cl_2(l)$
(c) $C(s) + 2F_2(g) \longrightarrow CF_4(g)$
(d) $As_4(s) + 6Br_2(l) \longrightarrow 4AsBr_3(l)$
7. (a) $Ba(s) + Cl_2(g) \longrightarrow BaCl_2(s)$
(b) $2Al(s) + 3Br_2(l) \longrightarrow 2AlBr_3(s)$
(c) $2K(s) + I_2(s) \longrightarrow 2KI(s)$
(d) $P_4(s) + 6Cl_2(g) \longrightarrow 4PCl_3(l)$, *and*
$P_4(s) + 10Cl_2(g) \longrightarrow 4PCl_5(s)$
(e) $P_4(s) + 6I_2(s) \longrightarrow 4PI_3(s)$
9. (a) $Cl_2(g) + 2NaOH(aq) \longrightarrow NaOCl(aq) + NaCl(aq) + H_2O(l)$ (b) 3.04 g.   **12.** As the core charge increases in going from left to right across any *period*, electronegativity increases; in going down any *group*, the core charge remains constant but the distance of the valence-shell electrons from the nucleus increases as the number of filled shells of electrons increases, and the electronegativity progressively decreases.   **13.** (a) F, (b) F, (c) S, (d) C, (e) O, (f) Br, (g) P.   **15.** Cl_2, covalent; PCl_3, ClF, polar covalent; LiCl, $MgCl_2$, ionic.   **17.** I_2, H_2, covalent; HBr, ClF, polar covalent; LiH, ionic.   **19.** It becomes progressively more difficult to remove successive electrons from an atom to form a multicharged cation, such as M^{4+}. Moreover, if such an ion were formed the high formal charge would strongly attract electron pairs from neighboring anions to form polar covalent bonds.   **21.** (a) K^+,

(b) S^{2-}, (c) Cl^-, (d) Na^+, (e) I^-.   **23.** $^{\delta+}N-F^{\delta-}$, $^{\delta-}N-Cl^{\delta+}$, $^{\delta-}N-Br^{\delta+}$, $^{\delta-}N-I^{\delta+}$.

25. (a) $(NH_4)_3PO_4$, ammonium phosphate; (b) Fe_2O_3, iron(III) oxide; (c) Cu_2O, copper(I) oxide; (d) $Al_2(SO_4)_3$, aluminum sulfate.   **27.** (a) CaI_2, (b) BeO, (c) Al_2S_3, (d) $MgBr_2$, (e) Rb_2Se, (f) BaO.

29. (a) $Cl_2(g) + 3I^-(aq) \longrightarrow I_3^-(aq) + 2Cl^-(aq)$
(b) No reaction
(c) $Br_2(l) + 3I^-(aq) \longrightarrow I_3^-(aq) + 2Br^-(aq)$
(d) $2F_2(g) + 2H_2O(l) \longrightarrow O_2(g) + 4HF(aq)$

31. $X_2 + H_2S \longrightarrow 2HX + S$ (X = Cl, Br, or I);
$X_2 + HNO_2 + H_2O \longrightarrow 2HX + HNO_3$ (X = Cl or Br);
$Cl_2 + 2Br^- \longrightarrow 2Cl^- + Br_2$; $I_2 < Br_2 < Cl_2$.

33. 9.52 g $MgCl_2$; Mg is *oxidized* and Cl_2 is *reduced*.

35. (a) Insoluble, (b) insoluble, (c) soluble, (d) insoluble, (e) soluble, (f) insoluble.

37. (a) $FeCl_3(aq) + 3NaOH(aq) \longrightarrow$
$Fe(OH)_3(s) + 3NaCl(aq)$
$Fe^{3+}(aq) + 3OH^-(aq) \longrightarrow Fe(OH)_3(s)$
(d) $Pb(NO_3)_2(aq) + H_2SO_4(aq) \longrightarrow$
$PbSO_4(s) + 2HNO_3(aq)$
$Pb^{2+}(aq) + SO_4^{2-}(aq) \longrightarrow PbSO_4(s)$
(e) $2AgNO_3(aq) + Na_2S(aq) \longrightarrow Ag_2S(s) + 2NaNO_3(aq)$
$2Ag^+(aq) + S^{2-}(aq) \longrightarrow Ag_2S(s)$

39. 2.11 g $AgCl(s)$.   **41.** (a) $0.103M$ $Ba(OH)_2(aq)$,
(b) 0.840 g $BaSO_4(s)$.   **43.** (i) Action on an indicator, (ii) reaction with a solid carbonate to give $CO_2(g)$ (lime-water test), (iii) reaction with an active metal such as Mg or Zn to give $H_2(g)$.

45. (a) $Na_2O(s) + H_2O(l) \longrightarrow 2Na^+(aq) + 2OH^-(aq)$; strong base
(b) $KOH(s) \longrightarrow K^+(aq) + OH^-(aq)$; strong base
(c) $NH_3(aq) + H_2O(l) \rightleftharpoons NH_4^+(aq) + OH^-(aq)$; weak base
(d) $LiH(s) + H_2O(aq) \longrightarrow Li^+(aq) + OH^-(aq) + H_2(g)$; strong base

47. (a) $Ca(OH)_2(s) + H_2SO_4(aq) \longrightarrow$
$CaSO_4(aq) + 2H_2O(l)$
(b) $LiOH(s) + HF(aq) \longrightarrow LiF(aq) + H_2O(l)$
(c) $2NaOH(aq) + H_2S(g) \longrightarrow Na_2S(aq) + 2H_2O(l)$
(d) $NH_3(aq) + HNO_3(aq) \longrightarrow NH_4NO_3(aq)$
(e) $MgO(s) + 2HClO_4(aq) \longrightarrow Mg(ClO_4)_2(aq) + H_2O(l)$
(f) $Al_2O_3(s) + 6HCl(aq) \longrightarrow 2AlCl_3(aq) + 3H_2O(l)$

48. (a) Basic, (b) neutral, (c) neutral, (d) acidic.

49. (a) F^-, fluoride ion; (b) NO_3^-, nitrate ion; (c) ClO_4^-, perchlorate ion; (d) OH^-, hydroxide ion; (e) H_2O, water.   **51.** OH^-, NH_2^-, Cl^-, F^-, NH_3.

53. (a) HCl, HNO_3, H_3O^+ strong acids; HF, NH_4^+, weak acids; CH_4, no acidic properties. (b) Cl^-, NO_3^-, negligible basicity; F^-, NH_3, weak bases; O^{2-}, strong base.

55. (a) H_3O^+ $\overset{+}{H-\overset{\displaystyle ..}{\underset{\displaystyle |}{O}}-H}$ AX_3E, trigonal pyramidal;
$\qquad\qquad\quad H$

(b) NH_4^+ $\quad H$ $\qquad AX_4$, tetrahedral;
$\qquad\qquad H-\overset{|}{\underset{|}{N^+}}-H$
$\qquad\qquad\qquad H$

(c) OH^- $:\overset{..}{\underset{..}{O}}-H$ AXE_3, linear.

57. (a) $NaOCl(aq) \longrightarrow Na^+(aq) + OCl^-(aq)$
$OCl^-(aq) + H_2O(l) \rightleftharpoons HOCl(aq) + OH^-(aq)$; *basic*
(b) $NH_4Cl(aq) \longrightarrow NH_4^+(aq) + Cl^-(aq)$
$NH_4^+(aq) + H_2O(l) \rightleftharpoons NH_3(aq) + H_3O^+(aq)$; *acidic*

59. $H_3O^+(aq) + OH^-(aq) \longrightarrow 2H_2O(l)$; $0.333M$; 14.6 g NaCl.   **61.** 8.64 g.   **63.** 183 mL.

65. (a) $2AgNO_3(aq) + BaCl_2(aq) \longrightarrow Ba(NO_3)_2(aq) + 2AgCl(s)$; *precipitation*
(b) $2NH_3(aq) + H_2SO_4(aq) \longrightarrow (NH_4)_2SO_4(aq)$; *acid–base*
(c) $Na_2O(s) + H_2O(l) \longrightarrow 2NaOH(aq)$; *acid–base*
(d) $2Al(s) + 3Br_2(l) \longrightarrow 2AlBr_3(s)$; *oxidation–reduction*
(e) $Ca(OH)_2(aq) + CO_2(g) \longrightarrow CaCO_3(s) + H_2O(l)$; *acid–base* and *precipitation*

67. (a) Oxidation–reduction, oxidizing agent Cl_2, reducing agent I^-. (b) Acid–base, acid HCl, base H_2O. (c) Oxidation–reduction, oxidizing agent H_3O^+, reducing agent $Zn(s)$. (d) Acid–base, acid H_3O^+, base HCO_3^-.

69. $MgBr_2$.   **71.** (a) Solid; (b) ionization energy less than that of iodine, covalent radius greater than that of iodine; (c) AtBr, $:\overset{..}{\underset{..}{At}}-\overset{..}{\underset{..}{Br}}:$; (d) strong acid; (e) AX_3E, trigonal pyramidal; (f) ionic;
(g) (i) $2Na + At_2 \longrightarrow 2NaAt$,
(ii) $Ca + At_2 \longrightarrow CaAt_2$, (iii) $P_4 + 6At_2 \longrightarrow 4PAt_3$,
(iv) $H_2 + At_2 \longrightarrow 2HAt$.

73. $Ra(s) + 2HCl(aq) \longrightarrow RaCl_2(aq) + H_2(g)$; atomic mass 226 u.   **75.** (a) (i) CH_4, (ii) NH_3, (iii) H_2O, (iv) HF. (b) (i) No reaction except at high temperature, $CH_4 + H_2O \longrightarrow CO + 3H_2$;
(ii) $NH_3(aq) + H_2O(l) \rightleftharpoons NH_4^+(aq) + OH^-(aq)$;
(iii) $2H_2O(l) \rightleftharpoons H_3O^+(aq) + OH^-(aq)$;
(iv) $HF(aq) + H_2O(l) \rightleftharpoons H_3O^+(aq) + F^-(aq)$.
(c) (i) Neither acid nor base, (ii) weak base, (iii) weak acid and weak base, (iv) weak acid.

77.

	Formula	Name	Type	Lewis Structure
(a)	$MgCl_2$	Magnesium chloride	Ionic	$Mg^{2+}[:\overset{..}{\underset{..}{Cl}}:^-]_2$
(b)	SCl_2	Sulfur dichloride	Covalent	$^{\delta-}:\overset{..}{\underset{..}{Cl}}-S^{2\delta+}-\overset{..}{\underset{..}{Cl}}:^{\delta-}$
(c)	PCl_3	Phosphorus trichloride	Covalent	$^{\delta-}:\overset{..}{\underset{..}{Cl}}-P^{3\delta+}-\overset{..}{\underset{..}{Cl}}:^{\delta-}$ $:\overset{..}{\underset{..}{Cl}}:^{\delta-}$
(d)	HF	Hydrogen fluoride	Covalent	$^{\delta+}H-\overset{..}{\underset{..}{F}}:^{\delta-}$
(e)	OCl_2	Oxygen dichloride	Covalent	$^{\delta+}:\overset{..}{\underset{..}{Cl}}-\overset{..}{\underset{..}{O}}^{2\delta-}-\overset{..}{\underset{..}{Cl}}:^{\delta+}$
(f)	CS_2	Carbon disulfide	Covalent	$^{\delta+}:\overset{..}{S}=C^{2\delta-}=\overset{..}{S}:^{\delta+}$
(g)	NF_3	Nitrogen trifluoride	Covalent	$^{\delta-}:\overset{..}{\underset{..}{F}}-N^{3\delta+}-\overset{..}{\underset{..}{F}}:^{\delta-}$ $:\overset{..}{\underset{..}{F}}:^{\delta-}$
(h)	LiH	Lithium hydride	Ionic	$Li^+:H^-$

79. 2.24×10^{19} Na^+ ions and 2.24×10^{19} Cl^- ions.
80. (a) $HBr(g) + H_2O(l) \longrightarrow H_3O^+(aq) + Br^-(aq)$
(b) $CO_2(g) + H_2O(l) \rightleftharpoons H_2CO_3(aq)$;
$H_2CO_3(aq) + H_2O(l) \rightleftharpoons H_3O^+(aq) + HCO_3^-(aq)$
(c) $NH_3(aq) + H_2O(l) \rightleftharpoons NH_4^+(aq) + OH^-(aq)$
(d) $Cl_2(g) + H_2O(l) \rightleftharpoons HCl(aq) + HOCl(aq)$
(e) $2F_2(g) + 2H_2O(l) \longrightarrow 4HF(aq) + O_2(g)$

CHAPTER 6

1. (a) $CO(g) + H_2(g) \xrightarrow[\text{High } T]{\text{Catalyst}} H_2CO(g) \xrightarrow{H_2} H_3COH$
$\qquad\qquad\qquad\qquad\qquad$ Methanal $\qquad$ Methanol

(b) $2CO(g) + O_2(g) \xrightarrow{\text{Combustion}} 2CO_2(g)$

(c) $CO(g) + H_2O(g) \xrightarrow[\text{High } T]{\text{Catalyst}} CO_2(g) + H_2(g)$

(d) $Fe_2O_3(s) + 3CO(g) \xrightarrow{\text{High } T} 2Fe(s) + 3CO_2(g)$

3. (a) $:\ddot{O}=C=\ddot{O}:$, AX_2, linear, carbon dioxide.
(b) $^-:C\equiv O:^+$, linear, carbon monoxide.
(c) $^-:C\equiv N:$, linear, cyanide ion.
(d) $^-:C\equiv C:^-$, linear, carbide ion.
(e) $H-C\equiv N:$, AX_2, linear, hydrogen cyanide.
4. $CaO(s)$, $CO_2(aq)$, $CH_4(g)$, $C(s)$, $C(s)$, $CaCO_3(s)$.
6. (a) $CuO(s) + C(s) \longrightarrow Cu(s) + CO(g)$
(b) $CaO(s) + 3C(s) \longrightarrow CaC_2(s) + CO(g)$
(c) $C(s) + 2S(s) \longrightarrow CS_2(l)$
(d) $C(s) + O_2(g) \longrightarrow CO_2(g)$
8. (a) $CaCO_3(s)$, (b) $HCN(aq)$, (c) $C_2H_2(g)$,
(d) $CaC_2(s)$, (e) $SiC(s)$ (silicon carbide).

10. $^-:\ddot{S}-C\equiv N:$ and $:\ddot{S}=C=\ddot{N}:^-$, of which the first is preferred, since N forms multiple bonds more readily than the third-period element S. **12.** Al_4C_3, 39.2 g.
14. (a) CN, C_2N_2; (b) $:N\equiv C-C\equiv N:$

16. $CaO(s) + 3C(s) \xrightarrow{\text{Heat}} CaC_2(s) + CO(g)$;
$\qquad\qquad\qquad\qquad$ Calcium carbide

$CaC_2(s) + 2H_2O(l) \longrightarrow Ca(OH)_2(s) + C_2H_2(g)$
$\qquad\qquad\qquad\qquad\qquad\qquad\qquad$ Acetylene

17.

Methane $\qquad$ Ethane $\qquad$ Propane

19. 3.10 kJ. **21.** -5450 kJ mol^{-1}.
22. -119.5 kJ mol^{-1}. **24.** -66.5 kJ mol^{-1}.
26. -5146 kJ. **28.** 170 kJ. **30.** -566.0 kJ.
32. -49.0 kJ. **34.** -196.4 kJ. **36.** -18 kJ.
37. 8.4 kJ. **38.** -285.8 kJ, -241.8 kJ, -46.2 kJ.
40. -221.3 kJ. **42.** 227 kJ. **44.** -45.7 kJ.
46. -104 kJ. **48.** 26.4 kJ mol^{-1}. **50.** -686.9 kJ.
52. (a) 498 kJ mol^{-1}, (b) 946 kJ mol^{-1}.
54. 321 kJ mol^{-1}, 258 kJ mol^{-1}, 163 kJ mol^{-1}.
56. 1076.3 kJ mol^{-1}, 804.2 kJ mol^{-1}.
58. -84.6 kJ mol^{-1}. **60.** (a) -78 kJ, (b) -218 kJ,
(c) -27 kJ. **61.** -8 kJ. **63.** -94 kJ.
65. (a) $\Delta H°$ for the reaction $Mg(s) + C(graphite, s) + \frac{3}{2}O_2(g) \longrightarrow MgCO_3(s)$; (b) $q = -4270$ J, -511 kJ mol^{-1}; (c) -1100 kJ mol^{-1}.

67. -84.6 kJ mol^{-1}, -103.8 kJ mol^{-1}, -123 kJ mol^{-1}, -2880 kJ mol^{-1}. **69.** 2539 kJ mol^{-1}, 14.09 kJ g^{-1}.
71. (a) 1.20 lb, (b) 0.53 lb bread and 0.53 lb butter.
73. -128.0 kJ mol^{-1}

CHAPTER 7

1. 3.19 m, 4.05×10^2 m. **3.** 11.0 m.
5. 4.74×10^{14} Hz, red light in the visible spectrum.
7. 2.97×10^{-19} J, 1.79×10^2 kJ mol^{-1}.
9. 5.085×10^{14} Hz, 5.090×10^{14} Hz, 3.32×10^{-22} J.
11. 102 kJ mol^{-1}. **13.** (a) 1.9×10^{-21} J,
(b) 4.23×10^{-19} J, (c) 4.21×10^{-19} J. **15.** Maximum wavelength 335 nm; No.

17.

Type of Radiation	Energy Range (kJ mol^{-1})
Radio waves	1.20×10^{-7} to 3.99×10^{-4}
Microwave	3.99×10^{-4} to 5.98×10^{-2}
Far infrared	5.98×10^{-2} to 3.99
Near infrared	3.99 to 1.69×10^2
Visible	1.69×10^2 to 3.00×10^2
Ultraviolet	3.00×10^2 to 3.00×10^4
X rays	3.00×10^4 to 3.99×10^6
γ rays	3.99×10^6 to 1.20×10^9

(a) Ultraviolet, (b) ultraviolet, (c) visible, (d) near infrared, (e) ultraviolet. **19.** 411 nm (violet), visible spectrum. **21.** $n = 3$. **23.** 657 nm. **25.** $n = 6$.
27. 14.6 pm. **29.** 1.32×10^3 m s^{-1}.
31. 2.88×10^{-3} pm; 11.5%. **33.** (c)
35. $K < Ca < Cl < Ca^{2+}$. **37.** (a) C, (b) Mg,
(c) Be, (d) S, (e) K^+. **39.** (a) See text. (b) The observed peaks can be identified with the six electronic energy levels of Ar, as follows:

Peak (MJ mol^{-1})	Energy Level
-1.52	3p
-2.82	3s
-24.1	2p
-31.5	2s
-309	1s

There is an outermost ($n = 3$) shell containing two relatively closely spaced energy levels (3s and 3p), an inner ($n = 2$) shell containing two energy levels (2s and 2p), and an innermost ($n = 1$) shell with one energy level (1s).
41. 620 kJ mol^{-1}. **43.** (b) and (d). **45.** $1s^4$ would suggest that electrons of the same spin can occupy the same orbital, which is contrary to the Pauli exclusion principle.
47. (a) $l = 0$, (b) $l = 1$, (c) $l = 2$, (d) $l = 0$, (e) $l = 2$,
(f) $l = 1$. **49.** (a) $1s^2 2s^2 2p^6 3s^2 3p^6 4s^1$,
(b) $1s^2 2s^2 2p^6 3s^2 3p^1$, (c) $1s^2 2s^2 2p^6 3s^2 3p^5$,
(d) $1s^2 2s^2 2p^6 3s^2 3p^6 4s^2 3d^2$, (e) $1s^2 2s^2 2p^6 3s^2 3p^6 4s^2 3d^{10}$,
(f) $1s^2 2s^2 2p^6 3s^2 3p^6 3d^{10} 4s^2 4p^3$. **51.** (a) Li, (b) N,
(c) Ca, (d) Ti, (e) As, (f) I, (g) Ba. **53.** (a) The n quantum number, (b) the n and l quantum numbers.
55. (a) ns^1 (the alkali metal); (b) $ns^2 np^6$ (the noble gas).
57. (b) and (e) are not ground states. **59.** (a) Ground state of Be, (b) excited state of Li, (c) not possible, (d)

excited state of Al, (e) ground state of Ti, (f) ground state of Na, (g) not possible, (h) not possible.

61. (a) $l = 0$, two spherical nodes; $l = 1$, one spherical and one planar node; $l = 2$, two planar nodes. (b) No. (c) 3s, 3p. **63.** 3s is spherical with two spherical nodes. 3p has a nodal plane dividing it into two equal *lobes* which are intersected by a spherical node and $3p_x$, $3p_y$, and $3p_z$ differ only in their orientations in space. Each of the five 3d orbitals have two nodal planes. **65.** The energy of an orbital increases as the number of nodes increases; an atomic orbital with a quantum number n has a total of $n - 1$ nodes.

68. Green. **69.** (a) 72.3 kJ mol^{-1}, (b) 3d > 4s.

71. Ground state: $[He]2s^22p^1$, with one unpaired electron (valence 1), so that B should form only one covalent bond. To form three covalent bonds, as in BF_3, boron must have the excited state, $[He]\ 2s^12p_x^12p_y^1$, with three unpaired electrons. Boron in BF_3 still has an empty $2p_z$ orbital and can behave as a Lewis acid, accepting an unshared electron pair from a $:\overset{..}{\underset{..}{F}}:^-$ ion to form BF_4^-. **73.** Large ionization energies are associated with high core charges, i.e. with atoms of the elements in groups on the right-hand side of the periodic table, such as groups 6 and 7. For these elements the electron affinities are also large (and exothermic) because their high core charges strongly attract additional electrons to form negative ions, provided there are unfilled orbitals in the valence shell to accommodate them. An exception is the elements of group 8, the noble gases, with filled valence shells, that can accept additional electrons only into high energy orbitals outside the valence shell.

CHAPTER 8

3.

		3s	3p			3d				
S(II)	[Ne]	↑↓	↑↓ ↑ ↑							
S(IV)	[Ne]	↑↓	↑ ↑ ↑	↑						
S(VI)	[Ne]	↑	↑ ↑ ↑	↑ ↑						

Sulfur cannot have a valence of 8 because this would require promotion of an electron from a 2p orbital to a 3d orbital, which requires too much energy. For oxygen with the ground state $[He]2s^22p^4$ and a valence of 2, to have valences of 4 and 6, electrons would have to be promoted from the $n = 2$ to the $n = 3$ shell, which is energetically impossible.

5. The very high boiling point of H_2SO_4 makes it suitable for the preparation of more volatile acids such as HCl and HNO_3,

$NaCl(s) + H_2SO_4(l) \longrightarrow NaHSO_4(s) + HCl(g)$,

$KNO_3(s) + H_2SO_4(l) \longrightarrow KHSO_4(s) + HNO_3(g)$,

and the reactions are driven to completion by removal of the HCl(g) or HNO_3(g). HI(g) cannot be prepared by the same method because it is oxidized by H_2SO_4 to iodine:

$2NaI(s) + 5H_2SO_4(l) \longrightarrow$
$I_2(s) + SO_2(g) + 2H_3O^+ + 2Na^+ + 4HSO_4^-$.

7. (a) Examples include

$C(s) + 4H_2SO_4 \longrightarrow$
$CO_2(g) + 2SO_2(g) + 2H_3O^+ + 2HSO_4^-$

$2X^- + 5H_2SO_4 \longrightarrow$
$X_2 + SO_2 + 2H_3O^+ + 2HSO_4^-$ (X = Br or I)

$Cu(s) + 5H_2SO_4 \longrightarrow Cu^{2+} + SO_2 + 2H_3O^+ + 4HSO_4^-$

(b) H_2SO_4 is a diprotic acid which, depending on the conditions, donates either one or two protons to a base. Examples include

$H_2O + H_2SO_4 \longrightarrow H_3O^+ + HSO_4^-$ (H_2SO_4 in excess)

$2H_2O + H_2SO_4 \longrightarrow 2H_3O^+ + SO_4^{2-}$ (H_2O in excess)

$NaOH + 2H_2SO_4 \longrightarrow$
$Na^+ + H_3O^+ + 2HSO_4^-$ (concentrated acid)

$2NaOH + H_2SO_4 \longrightarrow$
$2Na^+ + SO_4^{2-} + 2H_2O$ (dilute acid)

(c) Examples include

$CuSO_4 \cdot 5H_2O \longrightarrow CuSO_4 + 5H_2O$

$C_{12}H_{22}O_{11} \longrightarrow 12C(s) + 11H_2O$

$HCO_2H \longrightarrow CO(g) + H_2O$

The water reacts with the H_2SO_4 to form $H_3O^+HSO_4^-$

9. (a) $Zn(s) + H_2SO_4(aq) \longrightarrow$
$Zn^{2+}(aq) + SO_4^{2-}(aq) + H_2(g)$

(b) $2NaI(s) + 5H_2SO_4(l) \longrightarrow$
$I_2(s) + SO_2(g) + 2H_3O^+(aq) + 4HSO_4^-(aq) + 2Na^+(aq)$

(c) $2Ag(s) + 5H_2SO_4(l) \longrightarrow$
$2Ag^+(aq) + SO_2(g) + 2H_3O^+(aq) + 4HSO_4^-(aq)$

(d) $Mg(OH)_2(s) + H_2SO_4(aq) \longrightarrow$
$Mg^{2+}(aq) + SO_4^{2-}(aq) + 2H_2O(l)$

11. FeS_2; ionic, $Fe^{2+}\ ^-:\overset{..}{\underset{..}{S}}-\overset{..}{\underset{..}{S}}:^-$

13. $SO_2(g) + 2H_2S(g) \longrightarrow 3S(s) + 2H_2O(g)$;
6.20×10^4 L; 1.20×10^2 kg.

15. SCl_2O_2, SCl_2O_2,

$$:\overset{\overset{\textstyle \overset{..}{O}:}{\|}}{\underset{\underset{\textstyle \overset{..}{O}:}{\|}}{Cl-S-Cl}}:$$

17. (a) K_2SO_4, (b) $Ca(HSO_4)_2$, (c) $CaSO_3$, (d) K_2S_4, (e) $Na_2S_2O_7$, (f) $Al_2(SO_4)_3$.

19. (a) Ionic, $[Na^+]_2\ :\overset{..}{\underset{..}{S}}:^{2-}$, sodium sulfide;

(b) ionic, $[Na^+]_2\ ^-:\overset{..}{\underset{..}{S}}-\overset{..}{\underset{..}{S}}:^-$, sodium disulfide;

(c) ionic, $Mg^{2+}\ :\overset{..}{\underset{..}{S}}:^{2-}$, magnesium sulfide;

(d) covalent, [ring of eight S atoms], sulfur;

(e) covalent, $:\overset{..}{\underset{..}{S}}=C=\overset{..}{\underset{..}{S}}:$, carbon disulfide;

(f) covalent, $:\overset{..}{O}=\overset{..}{S}=\overset{..}{O}:$, sulfur dioxide;

(g) covalent, $:\overset{..}{O}=\overset{\overset{\textstyle S}{\|}}{\underset{\underset{\textstyle \overset{..}{O}:}{}}{}}=\overset{..}{O}:$, sulfur trioxide;

(h) covalent, $:\overset{..}{\underset{..}{Cl}}-\overset{..}{\underset{..}{S}}-\overset{..}{\underset{..}{Cl}}:$, sulfur dichloride.

21. $4PH_3(g) + 8O_2(g) \longrightarrow P_4O_{10}(s) + 6H_2O(g)$

23. [cage structure of P and S atoms]

25. $P_4S_3(s) + 8O_2(g) \longrightarrow P_4O_{10}(s) + 3SO_2(g)$; 50.7 mL.

27. X is PH_3(g), $2AlP(s) + 3H_2SO_4(aq) \longrightarrow$
$Al_2(SO_4)_3(aq) + 2PH_3(g)$; $PH_3(g) + HI(g) \longrightarrow PH_4I(s)$.

29. P_2O_5, H_3PO_4.

31. $Na_5P_3O_{10}$,

$$:\overset{..}{\underset{..}{O}}: ^- \quad :\overset{..}{\underset{..}{O}}: ^- \quad :\overset{..}{\underset{..}{O}}: ^-$$
$$^-:\overset{..}{\underset{..}{O}}-\overset{|}{\underset{|}{P}}-\overset{..}{\underset{..}{O}}-\overset{|}{\underset{|}{P}}-\overset{..}{\underset{..}{O}}-\overset{|}{\underset{|}{P}}-\overset{..}{\underset{..}{O}}: ^-$$
$$\overset{..}{\underset{..}{O}}: \quad \overset{..}{\underset{..}{O}}: \quad \overset{..}{\underset{..}{O}}:$$

33. (a) P_4O_6, (b) $Ca(H_2PO_4)_2$, (c) Ca_3P_2, (d) H_3PO_3, (e) PH_4I, (f) $NaPO_3$. **35.** 66.7 mL.
37. *0*, elemental phosphorus (white, red, or black); -3, PH_3, P^{3-}, PH_4^+; $+3$, P_4O_6, PCl_3, PF_3, H_3PO_3; $+5$, P_4O_{10}, PF_5, PCl_5, $POCl_3$, H_3PO_4. **39.** (a) 0, (b) 0, (c) $+2$, (d) -1, (e) -1, (f) -2. **41.** (a) $+2$, (b) $+4$, (c) $+5$, (d) $+6$, (e) -1, (f) $+3$. **43.** (a) N(0); (b) H($+1$), C($+4$), O(-2); (c) N(-3), H($+1$); (d) P(-3), H($+1$); (e) V($+3$), O(-2); (f) V($+5$), O(-2); (g) Mn($+4$), O(-2); (h) H($+1$), N($+5$), O(-2); (i) H($+1$), N($+3$), O(-2). **45.** (a) Increase. (b) Decrease. (c) (i) $Cu^+ \longrightarrow Cu^{2+} + e^-$; (ii) $S_8 + 16e^- \longrightarrow 8S^{2-}$; (iii) $NO_3^- + 3e^- + 4H^+ \longrightarrow NO + 2H_2O$; (iv) $S_2O_3^{2-} + 5H_2O \longrightarrow 2SO_4^{2-} + 8e^- + 10H^+$. **47.** (a) $2H_2S + SO_2 \longrightarrow 3S + 2H_2O$, for example; (b) $2SO_2 + O_2 \longrightarrow 2SO_3$, for example.

49. (a)

$$H-\overset{..}{\underset{..}{O}}\diagdown \overset{\overset{..}{O}-H}{\underset{..}{P}} \diagup$$
$$:\overset{..}{\underset{..}{O}} \quad \overset{..}{\underset{..}{O}}-H$$

AX_4
Tetrahedral

(b)

$$H-\overset{..}{\underset{..}{O}}\diagdown \overset{\overset{..}{O}-H}{\underset{..}{P}} \diagup$$
$$H \quad \overset{..}{\underset{..}{O}}:$$

AX_4
Tetrahedral

(c)

$$:\overset{..}{\underset{..}{O}}\diagdown \overset{\overset{..}{O}:^-}{\underset{..}{P}} \diagup$$
$$^-:\overset{..}{\underset{..}{O}} \quad :\overset{..}{\underset{..}{O}}:^-$$

AX_4
Tetrahedral

(d)

$$H\diagdown \overset{\overset{..}{O}:^-}{\underset{..}{P}} \diagup$$
$$:\overset{..}{\underset{..}{O}} \quad :\overset{..}{\underset{..}{O}}:^-$$

AX_4
Tetrahedral

51. (a) $:\overset{..}{O}=S=\overset{..}{O}:$ (b) $:\overset{..}{O}=S-\overset{..}{\underset{..}{O}}:^-$ (c)
$$\underset{\overset{..}{\underset{..}{O}}:}{} \quad \underset{:\overset{..}{\underset{..}{O}}:^-}{}$$

(c)
$$:\overset{..}{\underset{..}{O}}:^-$$
$$:\overset{..}{O}=S=\overset{..}{O}:$$
$$:\overset{..}{\underset{..}{S}}:^-$$

(d)
$$\overset{..}{\underset{..}{O}}: \quad \overset{..}{\underset{..}{O}}:$$
$$^-:\overset{..}{\underset{..}{O}}-S-\overset{..}{\underset{..}{O}}-S-\overset{..}{\underset{..}{O}}:^-$$
$$\overset{..}{\underset{..}{O}}: \quad \overset{..}{\underset{..}{O}}:$$

53. (a), (b), (c), (e), weak; (d) strong. **55.** *Base properties*: SiH_4 has no lone pairs of electrons and therefore no basic properties. The lone pair of PH_3 is larger (more diffuse) than that of the weak base NH_3, and it is a weaker base than NH_3. Similar considerations apply to H_2S versus H_2O and HCl versus HF. *Acid properties*: acidity is expected to increase from left to right in the period, as the X—H bonds become more polar, but H_2S is a stronger acid than H_2O, and HCl a stronger acid than HF, because the strength of the X—H bond decreases down any group. **57.** The charge on the anion makes it more difficult to remove a proton from the anion than from the uncharged parent acid. **59.** KOH and $Ca(OH)_2$ are ionic hydroxides that dissociate completely in aqueous solution to their ions. $Si(OH)_4$, in contrast, is a covalent hydroxide that belongs to the weakest class of acids. **61.** (a) $CaSO_4$, (b) PBr_5, (c) PI_3, (d) $(NH_4)_2HPO_4$, (e) Ca_3N_2, (f) SF_4, (g) $CrCl_3$.

63. (a) Potassium selenate, (b) hydrogen telluride, (c) sodium tetrasulfide, (d) iron(II) disulfide, (e) rubidium hydrogen sulfate, (f) phosphorus(III) oxide (tetraphosphorus hexaoxide), (g) disodium hydrogen phosphate, (h) disodium hydrogen phosphite.
65. (a) $HNO_3(aq) + NaOH(aq) \longrightarrow$
$NaNO_3(aq) + H_2O(l)$; sodium nitrate.
(b) $H_2SO_4(aq) + NaOH(aq) \longrightarrow$
$NaHSO_4(aq) + H_2O(l)$; sodium hydrogen sulfate
$H_2SO_4(aq) + 2NaOH(aq) \longrightarrow Na_2SO_4(aq) + 2H_2O(l)$;
sodium sulfate (c) $H_3PO_4(aq) + NaOH(aq) \longrightarrow$
$NaH_2PO_4(aq) + H_2O(l)$; sodium dihydrogen phosphate
$H_3PO_4(aq) + 2NaOH(aq) \longrightarrow$
$Na_2HPO_4(aq) + 2H_2O(l)$; sodium hydrogen phosphate
$H_3PO_4(aq) + 3NaOH(aq) \longrightarrow Na_3PO_4(aq) + 3H_2O(l)$;
sodium phosphate (d) $H_3PO_3(aq) + NaOH(aq) \longrightarrow$
$NaH_2PO_3(aq) + H_2O(l)$; sodium dihydrogen phosphite
$H_3PO_3(aq) + 2NaOH(aq) \longrightarrow$
$Na_2HPO_3(aq) + 2H_2O(l)$; sodium hydrogen phosphite
67. (a) $\underset{base_1}{CO_3^{2-}} + \underset{acid_2}{H_2SO_4} \longrightarrow$

$\underset{base_2}{HSO_4^-} + \underset{acid_1}{HCO_3^-}$; *acid–base*

(b) $\underset{acid_1}{HSO_4^-} + \underset{base_2}{H_2O} \longrightarrow \underset{acid_2}{H_3O^+} + \underset{base_1}{SO_4^{2-}}$; *acid–base*

(c) $Cu + 2H_2SO_4 \longrightarrow Cu^{2+} + SO_4^{2-} + SO_2 + H_2O$; *oxidation–reduction*; Cu is oxidized and S in H_2SO_4 is reduced to SO_2. (d) $2Br^- + 2H_2SO_4 \longrightarrow$
$Br_2 + SO_2 + SO_4^{2-} + 2H_2O$; *oxidation–reduction*; Br^- is oxidized to Br_2 and S in H_2SO_4 is reduced to SO_2.
(e) $\underset{base_1}{Ca^{2+}CO_3^{2-}} + \underset{acid_2}{H_3O^+} \longrightarrow$

$Ca^{2+} + \underset{acid_1}{HCO_3^-} + \underset{base_2}{H_2O}$; *acid–base*

(f) $Mg + 2H_3O^+ \longrightarrow Mg^{2+} + H_2 + 2H_2O$; *oxidation–reduction*; Mg is oxidized to Mg^{2+} and H in H_3O^+ is reduced to H_2.
(g) $\underset{base_1}{[Ca^{2+}]_3[P^{3-}]_2} + \underset{acid_2}{6H_2O} \longrightarrow$

$3Ca^{2+} + \underset{base_2}{6OH^-} + \underset{acid_1}{2PH_3}$; *acid–base*

69. (a) Each contains a tetrahedral arrangement of four P atoms. (b) Each contains a cyclic six-membered ring, X—O—X—O—X—O. (c) Each contains X—O—X bridge bonds. (d) Each has AX_4 tetrahedral geometry for the bonds to the S or P atom. (e) Each has AX_4 tetrahedral geometry. **71.** (a) $H_2P(O)OH$.

(b) Hypophosphorous acid (I) (c)
$$\overset{H}{\underset{\overset{\|}{\underset{..}{O}}:}{\overset{|}{H-P-\overset{..}{\underset{..}{O}}-H}}}$$

(d) P_4, (0); PH_3, (-3); $H_2PO_2^-$, ($+1$). (e) P_4 is reduced to PH_3 and oxidized to $H_2PO_2^-$. (f) N_2 is very unreactive because of its strong triple bond. (g) 0.481 g.

73. A, sulfur; B, SO_2; C, $H_2SO_3(aq)$; D, $FeS(s)$; E, $H_2S(g)$.

$S(s) + O_2(g) \longrightarrow SO_2(g)$;

$SO_2(g) + H_2O(l) \longrightarrow H_2SO_3(aq)$

$Fe(s) + S(s) \longrightarrow FeS(s)$;

$Fe(s) + H_2SO_4(aq) \longrightarrow FeSO_4(aq) + H_2S(g)$

$2H_2S(g) + H_2SO_3(aq) \longrightarrow 3S(s) + 3H_2O(l)$

75. P_2S_5, and, by analogy with P_4O_{10},

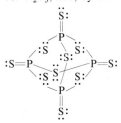

77. (a) Sulfuric acid, $(HO)_2SO_2$, has *two* ionizable protons and phosphoric acid, $(HO)_3PO$, has *three* ionizable protons. (b) In terms of the $XO_n(OH)_m$ nomenclature, H_2SO_4 has $n = m = 2$, and H_3PO_4 has $n = 1$ and $m = 3$. Thus, H_2SO_4 is a strong acid and H_3PO_4 is a weak acid in terms of their first ionizations in water. (c) SO_3, P_4O_{10}.

CHAPTER 9

1. $^-{:}\ddot{S}{-}\ddot{S}{:}^-$ $:\ddot{O}{=}\ddot{S}{-}\ddot{O}{:}^-$ $:\ddot{C}l{-}\ddot{O}{:}^-$ $:\ddot{O}{=}\ddot{C}l{-}\ddot{O}{:}^-$

3. $H{-}\ddot{O}{-}\ddot{N}{=}\ddot{O}{:}$ $H{-}O{-}N^+{=}\ddot{O}{:}$ $H{-}\ddot{O}{-}\overset{\ddot{O}:}{\underset{:O:}{S}}{-}\ddot{O}{-}H$

$H{-}\ddot{O}{-}\ddot{C}l{=}\ddot{O}{:}$ $H{-}\ddot{O}{-}\overset{O:}{C}{-}\ddot{O}{-}H$

5. *NON*, $^-{:}\ddot{N}{=}O^{2+}{=}\ddot{N}{:}^-$ *or* $:N{\equiv}O^{2+}{-}\ddot{N}{:}^{2-}$ *NNO*,

$^-{:}\ddot{N}{=}N^+{=}\ddot{O}{:}$ *or* $^{2-}{:}\ddot{N}{-}N^+{\equiv}O{:}^+$ *or*

$:N{\equiv}N^+{-}\ddot{O}{:}^-$ Of these structures only $^-{:}\ddot{N}{=}N^+{=}\ddot{O}{:}$ and $:N{\equiv}N^+{-}\ddot{O}{:}^-$, with the smallest formal charges, make the most important contributions.

7. $:\ddot{C}l{-}\overset{\ddot{O}:}{C}{-}\ddot{C}l:$ $:\ddot{F}{-}\ddot{N}{=}\ddot{N}{-}\ddot{F}:$ $:\ddot{O}{=}C{=}\ddot{S}:$

$\overset{H}{\underset{H}{}}C{=}C\overset{H}{\underset{C{\equiv}N:}{}}$ $:\ddot{F}:\overset{\ddot{S}{=}\ddot{O}:}{\underset{:\ddot{F}:}{}}$

9. BCl_3, $BrO_2{}^-$, XeF_2. **11.** Although the arrangement of the electron pairs in each case is tetrahedral, the shape of the molecule is determined by the positions of the ligands X. Thus we have

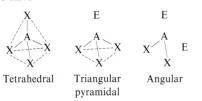

Tetrahedral   Triangular pyramidal   Angular

13. A bonding electron pair is attracted by two nuclei and takes up less space in the valence shell of a central atom A than does an unshared (lone) pair of electrons, which is attracted only by A. Thus, one lone pair in an AX_3E type molecule such as NH_3 causes the HNH angles to be slightly less than the tetrahedral angle, while the effect of two lone pairs in an AX_2E_2 type molecule such as H_2O causes the HOH angles to be even smaller than those in ammonia. **15.** In an AX_5 molecule, each of the axial AX bond pairs is repelled by three equatorial AX bond pairs making angles of 90° to it, while each equatorial bond pair is repelled by two axial bond pairs at 90° and two equatorial bond pairs at 120°. Because electron pair repulsions fall off rapidly with increasing distance apart, the electron pair repulsions at 90° predominate over all others, so that the axial bonds are longer than the equatorial bonds to balance the repulsions between all five bonds. **17.** When we draw the Lewis structures, we first write, respectively,

$\overset{:\ddot{O}:^-}{\underset{^-{:}\ddot{O}:\quad:\ddot{O}:^-}{C^+}}$ and $\overset{^-{:}\ddot{O}{-}\ddot{S}^+{-}\ddot{O}:^-}{\underset{:\ddot{O}:^-}{}}$ and then complete the

valence shell of the central atom by delocalizing an electron pair from *one* of the O atoms. In molecules with a number of oxygen atoms, for example, there is no unique way of achieving this and all the oxygen atoms delocalize an electron pair to the same extent. In the above examples each O atom delocalizes one-third of an electron pair. Since there is no simple way to depict this, we achieve the same result by writing all three possible Lewis structures, none of which represents the actual structure, and we call these *resonance structures*. The structure obtained by averaging all the resonance structures then represents the actual structure.

19. (a) $\overset{\ddot{O}:}{\underset{^-{:}\ddot{O}:\quad:\ddot{O}:^-}{N^+}}$ ⟷ $\overset{:\ddot{O}:^-}{\underset{^-{:}\ddot{O}:\quad:\ddot{O}:}{N^+}}$ ⟷ $\overset{:\ddot{O}:^-}{\underset{:\ddot{O}:\quad:\ddot{O}:^-}{N^+}}$

Formal charge $\tfrac{2}{3}-$ on O, $^+1$ on N; Bond order $1\tfrac{1}{3}$

(b) $:\ddot{O}{=}\ddot{O}^+{-}\ddot{O}:^-$ ↔ $^-{:}\ddot{O}{-}\ddot{O}^+{=}\ddot{O}:$

Formal charge $\tfrac{1}{2}-$ on each terminal O, $^+1$ on the central 0; Bond order $1\tfrac{1}{2}$

(c) $^-{:}\ddot{O}{-}\overset{:\ddot{O}:^-}{\underset{:\ddot{O}:^-}{P}}{=}\ddot{O}:$ ⟷ $^-{:}\ddot{O}{-}\overset{:\ddot{O}:^-}{\underset{O:}{P}}{-}\ddot{O}:^-$ ⟷

$:\ddot{O}{=}\overset{\ddot{O}:}{\underset{:\ddot{O}:^-}{P}}{-}\ddot{O}:^-$ ⟷ $^-{:}\ddot{O}{-}\overset{:\ddot{O}:}{\underset{:\ddot{O}:^-}{P}}{=}\ddot{O}:$

Formal charge on each O atom $= \tfrac{3}{4}-$; PO bond order $= 1\tfrac{1}{4}$.

(d) Six structures of the type $:\ddot{O}{=}\overset{:\ddot{O}:^-}{\underset{:\ddot{O}:^-}{S}}{=}\ddot{O}:$

Formal charge on each O atom $= \tfrac{1}{2}-$; SO bond order $= 1\tfrac{1}{2}$.

(e) Four structures of the type

$$:\overset{\displaystyle \overset{..}{\overset{..}{O}}:}{\underset{\displaystyle \underset{..}{\overset{..}{O}}:}{:\overset{..}{O}=Cl-\overset{..}{\overset{..}{O}}:^-}}$$

Formal charge on each O atom $= \frac{1}{4}-$; ClO bond order $1\frac{3}{4}$.

21. Nitric acid has two important resonance structures

$$H-\overset{..}{\underset{..}{O}}-\overset{+}{N}\overset{\displaystyle \overset{..}{O}:^-}{\underset{\displaystyle \underset{..}{\overset{..}{O}}:}{}} \longleftrightarrow H-\overset{..}{\underset{..}{O}}-\overset{+}{N}\overset{\displaystyle \overset{..}{O}:}{\underset{\displaystyle \underset{..}{\overset{..}{O}}:^-}{}}$$

with an N—OH bond of order 1.00 (136 pm) and two NO bonds of order $1\frac{1}{2}$ (121 pm). The nitrate ion has three resonance structures (see Problem 19) with each NO bond of order $1\frac{1}{3}$. The nitrite ion has two resonance structures,

$:\overset{..}{O}=\overset{..}{N}-\overset{..}{\underset{..}{O}}:^-$ and $^-:\overset{..}{\underset{..}{O}}-\overset{..}{N}=\overset{..}{O}:$, with a bond order of $1\frac{1}{2}$.

As expected, the NO bond length decreases as the NO bond order increases.

23.

$$:\overset{..}{Cl}-\overset{+}{N}\overset{\displaystyle \overset{..}{O}:}{\underset{\displaystyle \underset{..}{\overset{..}{O}}:^-}{}} \longleftrightarrow :\overset{..}{Cl}-\overset{+}{N}\overset{\displaystyle \overset{..}{O}:^-}{\underset{\displaystyle \underset{..}{\overset{..}{O}}:}{}}$$ with NO bond order

$1\frac{1}{2}$. NO_2Cl is an AX_3 type trigonal planar molecule. For three equivalent ligands, all the bond angles would be 120°. Here the NO bonds each have bond order $1\frac{1}{2}$ and the NCl bond has a bond order of 1. Thus, the greatest repulsions are between the two NO bonds, to give an ONO bond angle greater than 120° (135° observed). **25.** (a) sp^3, (b) sp^3, (c) sp, (d) sp^2, (e) sp^3d, (f) sp^3d.

27.

The C—H σ bonds of the CH_3 group are formed from overlap of carbon sp^3 hybrid orbitals with H 1s orbitals; the H_3C—C σ bond is formed from overlap of carbon sp^3 and C* sp^2 hybrid orbitals; the C*—C* σ bond is formed from overlap of sp^2 hybrid orbitals, and the C*—H σ bonds are formed from overlap of carbon sp^2 and H 1s orbitals. The π bond results from sideways overlap of a p orbital on each C* atom. **29.** (a) A double bond consists of a σ bond formed from overlap between sp^2 hybrid orbitals on each atom and a π bond formed from sideways overlap between the remaining p orbital on each atom. (b) A triple bond consists of a σ bond formed from overlap between sp hybrid orbitals on each atom and two mutually perpendicular π bonds formed from sideways overlap of the two remaining p orbitals on each atom. **31.** (a) NF_3 is an AX_3E type molecule with four tetrahedral sp^3 orbitals on N. (b) The N atom in H_3C—C≡N: has AXE linear geometry; two sp hybrid orbitals on N accommodate the nitrogen lone pair and form an N—C σ bond, leaving unpaired electrons in each of two mutually perpendicular p orbitals on N and C which form two π bonds. (c) NH_4^+ has AX_4 tetrahedral geometry, which requires four sp^3 hybrid orbitals on N. (d) NO^+ has the Lewis structure $:N≡O:^+$ in which the N atom has AXE linear geometry, as in (a) above. Two sp

A-30

orbitals are required to accommodate the lone pair and to form an N—O σ bond, leaving two mutually perpendicular p orbitals on N which can be used to form two π bonds.

(e) NO_2^+ has the Lewis structure $:\overset{..}{O}=\overset{+}{N}=\overset{..}{O}:$, with AX_2 linear geometry at N. Two sp hybrid orbitals are used to form σ bonds to the two O atoms, leaving two singly occupied p orbitals on N from which a π bond to each O atom can be formed. (f) $(CH_3)_4N^+$ is similar to NH_4^+ with AX_4 tetrahedral geometry at N. A set of four sp^3 hybrid orbitals on N is utilized to form four N—C σ bonds. **33.** A must be from group 3 or group 5, since its valence is 3. For A from group 3, ACl_3 would be a trigonal planar molecule with AX_3 geometry with $\mu = 0$. Thus, A must be from *group 5*, so that ACl_3 is an AX_3E type molecule with $\mu \neq 0$.

35. (a) $\delta^-:\overset{\displaystyle \overset{..}{N}^{3\delta+}}{F:}\overset{}{|}\,:\overset{..}{F}:^{\delta-}$ AX₃E, trigonal pyramidal, $\mu \neq 0$

(b) $\delta^-:\overset{..}{F}:\,\overset{..}{\overset{..}{O}}{}^{2\delta+}\,:\overset{..}{F}:^{\delta-}$ AX_2E_2, angular, $\mu \neq 0$

(c) $\delta^+ H\,\overset{\overset{..}{S}{}^{2\delta-}}{}\,H^{\delta+}$ AX_2E_2, angular, $\mu \neq 0$

(d) $\delta^- H\,\overset{\overset{..}{P}{}^{3\delta+}}{}\,H^{\delta-}$, $H^{\delta-}$ AX_3E, trigonal pyramidal, $\mu \neq 0$

(e) $\delta^-:\overset{..}{Cl}-\overset{\displaystyle :\overset{..}{Cl}:^{\delta-}}{\underset{\displaystyle :\overset{..}{Cl}:^{\delta-}}{C^{4\delta+}}}-\overset{..}{Cl}:^{\delta-}$ AX_4, tetrahedral, $\mu = 0$

(f) $\delta^-:\overset{..}{S}=\overset{*}{C}=\overset{..}{O}:^{\delta-}$ AXY, linear, $\mu \neq 0$
*Charge on C, $(\delta+)(+\bar{\delta}+)$

(g) $\delta^-:\overset{..}{O}\,\overset{\displaystyle \overset{..}{O}:^{\delta-}}{\underset{\displaystyle \underset{..}{O}:^{\delta-}}{S^{3\delta+}}}$ AX_3, trigonal planar, $\mu = 0$

(h) $\delta^-:\overset{..}{F}\,\overset{\displaystyle :\overset{..}{F}:^{\delta-}}{\underset{\displaystyle }{B^{3\delta+}}}\,:\overset{..}{F}:^{\delta-}$ AX_3, trigonal planar, $\mu = 0$

37. (a) The H—O bonds in water are more polar than are the F—O bonds in F_2O (electronegativity differences: F—O, 0.6; H—O, 1.3). For similar bond angles (and bond lengths) $\mu(H_2O) > \mu(F_2O)$. (b) Both molecules are AX_2 linear. For O=C=O the center of negative charge coincides with the center of positive charge (at carbon). For S=C=S, the bonds have different polarities and the center of positive charge does not coincide with the center of negative charge. **39.** (a) CCl_4 is regular tetrahedral AX_4 and the centers of positive and negative charge coincide at the C atom (see text). CCl_2F_2 is distorted tetrahedral AX_2Y_2; the center of negative charge does not coincide with the center of positive charge because the C—F bonds are more polar than the C—Cl bonds. (b) The bonds are polar, δ^+P—Fδ^-. The axial bonds are colinear, so for these bonds the center of negative charge coincides with the center of positive charge at the P atom. Similarly for the equatorial bonds the centers of positive and negative charge also coincide at the P atom, because the equatorial F atoms are at the corners of an equilateral triangle with P at the center. Thus, $\mu = 0$.

41. $^{+0.17e}H-Cl^{-0.17e}$

$$S^{-0.20e}$$
$$_{+0.10e}H \diagdown H^{+0.10e}$$

44. (a) Bond order is the number of electron pairs forming a bond. (b) Bond length decreases with increasing bond order as a bond of a given type becomes stronger.

45. (a) Xe(II) in XeF_2, Xe(IV) in XeF_4, and Xe(VI) in XeF_6, with the following valence-shell configurations:

	5s	5p	5d
Xe(II)	$\uparrow\downarrow$	$\uparrow\downarrow$ $\uparrow\downarrow$ $\uparrow$	$\uparrow$
Xe(IV)	$\uparrow\downarrow$	$\uparrow\downarrow$ $\uparrow$ $\uparrow$	$\uparrow$ $\uparrow$
Xe(VI)	$\uparrow\downarrow$	$\uparrow$ $\uparrow$ $\uparrow$	$\uparrow$ $\uparrow$ $\uparrow$

(b) XeF_2, AX_2E_3, linear; XeF_4, AX_4E_2, square planar. (c) No; it is an AX_6E molecule with seven pairs of electrons in the valence shell of the central Xe atom.

49. (a)

$$:\ddot{O}:^-$$
$$|$$
$$N^+$$
$$:\overset{..}{O} \quad \overset{..}{O}:^-$$

$\langle ONO \approx 120\,°C$; AX_3 trigonal planar geometry (b) Resonance; NO_3^- has three equivalent resonance structures; the NO bonds each have an order of $1\frac{1}{3}$, intermediate in length between an NO single bond and an NO double bond; the ion is exactly equilateral triangular with bond angles of $120°$.

51.

Species	Molecular Shape	Hybrid Orbitals on Central Atom	Approximate Bond Angle
CH_4	AX_4, tetrahedral	sp^3	$109.5°$ [a]
PH_4^+	AX_4, tetrahedral	sp^3	$109.5°$ [a]
NF_3	AX_3E, trigonal pyramidal	sp^3	$< 109.5°$
F_2O	AX_2E_2, angular	sp^3	$< 109.5°$
H_3O^+	AX_3E, trigonal pyramidal	sp^3	$< 109.5°$

[a] Exactly the tetrahedral angle.

53. (a) PCl_5, PF_5, AsF_5, SiF_5^-; (b) SF_4, SeF_4, $SeCl_4$; (c) ClF_3, ICl_3; (d) XeF_2, ICl_2^-; (e) SF_6, PF_6^-, SiF_6^{2-}; (f) BrF_5, IF_5; (g) XeF_4, ICl_4^-. (b), (c) and (f) have dipole moments.

CHAPTER 10

1. (a) C(coke), CO(g), H_2(g), reactive metals such as the alkali metals and aluminum, electrons (in electrolysis).
(b) (i) $Cu_2O(s) + CO(g) \longrightarrow 2Cu(s) + CO_2(g)$
(ii) $Fe_2O_3(s) + 2Al(s) \xrightarrow{Heat} 2Fe(l) + Al_2O_3(s)$, thermite process
(iii) $PbO(s) + C(s) \xrightarrow{Heat} Pb(l) + CO(g)$

(iv) $AlCl_3(s) + 3K(s) \xrightarrow{Heat} Al(s) + 3KCl(s)$
3. 0.423 metric ton.
5. (a) $Cu_2O + C \longrightarrow 2Cu + CO$;
$Fe_2O_3 + 3CO \longrightarrow 2Fe + 3CO_2$
$2PbS + 3O_2 \longrightarrow 2PbO + 2SO_2$;
$PbO + C \longrightarrow Pb + CO$
$Cu_2(CO_3)(OH)_2 + C \longrightarrow 2Cu + 2CO_2 + H_2O$
(b) (i) $Fe_2O_3 + 2Al \longrightarrow 2Fe + Al_2O_3$, thermite process
$AlCl_3 + 3K \longrightarrow Al + 3KCl$
(ii) $Fe_2O_3 + 3C \longrightarrow 2Fe + 3CO$;
$PbO + C \longrightarrow Pb + CO$
6. (a) 4.8×10^6 metric ton; (b) 210 kg Fe; (c) 68 kg coke.
8. (a) $AlCl_3 + 3K \longrightarrow Al + 3KCl$
(b) $2PbS + 3O_2 \longrightarrow 2PbO + 2SO_2$;
$PbO + C \longrightarrow Pb + CO$
(c) $CaCO_3 \longrightarrow CaO + CO_2$;
$CaO + SiO_2 \longrightarrow CaSiO_3$ (slag)
11. The majority of the elements are metals located on the left-hand side of the periodic table in groups 1, 2, and 3, the three series of transition metals in periods 4, 5, and 6, and the lanthanide and actinide series. The metals and nonmetals are divided by a diagonal band of elements, starting at beryllium (group 2, period 2) and extending to polonium (group 6, period 6), which exhibit both metallic and non-metallic properties and are called semimetals (metalloids). The nonmetals are located on the right-hand side of the periodic table in groups 4–8. In the third period, Na, Mg, and Al are metals, while Si is a semimetal and P, S, Cl (and Ar) are nonmetals. In group 4, C is a metal, Si and Ge are semimetals and Sn and Pb are metals. Metallic character increases down the group from Si to Pb. **13.** (a) A single bond formed by the overlap of two 3s orbitals. (b) De-localized metallic bonding to which each Na atom contributes one electron; Na_2 has the stronger bond between two Na atoms and has a shorter Na–Na distance than the Na–Na distance in the metal. **15.** See Figure 10.7. **17.** (a) Low density, to minimize the mass, high tensile strength, high melting point, and relatively low chemical reactivity are desirable properties. A good choice would be aluminum, or better one of its alloys such as duralumin. Although Al is relatively reactive it is protected by a hard thin layer of the oxide Al_2O_3 at all but very high temperatures. (b) Ease of casting and resistance to weathering would be the most desirable properties. Gold would be ideal but too expensive. Copper or one of its alloys, such as bronze, is commonly used. It rapidly acquires a brown protective tarnish of oxide, sulfide, and the basic sulfate $Cu_2(OH)_2SO_4$ which we generally regard as aesthetically pleasing. **21.** (a) Mg, Al, Fe, or Zn, for example:
$Mg + 2H_3O^+ \longrightarrow Mg^{2+} + 2H_2O + H_2$;
$2Al + 6H_3O^+ \longrightarrow 2Al^{3+} + 6H_2O + 3H_2$;
$Fe + 2H_3O^+ \longrightarrow Fe^{2+} + 2H_2O + H_2$;
$Zn + 2H_3O^+ \longrightarrow Zn^{2+} + 2H_2O + H_2$.
(b) Cu and Ag, for example, which are not oxidized by H_3O^+(aq). (c) HNO_3(aq) contains NO_3^-(aq) which is a stronger oxidizing agent than H_3O^+ and readily reduced to gases such as NO_2 and NO.
22. (a) $2Al(s) + 3H_2SO_4(aq) \longrightarrow Al_2(SO_4)_3(aq) + 3H_2(g)$;

(b) $2Ag(s) + 2H_2SO_4(l) \longrightarrow$
$Ag_2SO_4(s) + SO_2(g) + 2H_2O(l)$.
24. (a) H, $+1$; O, -2; Al, $+3$. (b) Cl, -1; Al, $+3$.
(c) O, -2; Si, $+4$. (d) O, -2; Si, $+4$. (e) O, -2; H, $+1$;
Pb, $+2$. (f) Cl, -1; Pb, $+4$.   **26.** *Oxidation* involves
the *loss* of electrons by the species that is oxidized; *reduction*
involves the *gain* of electrons by the species that is reduced.
(a) Zn(0) in Zn(s) is oxidized to Zn($+2$) in $Zn^{2+}SO_4^{2-}$;
H($+1$) in H_2SO_4 is reduced to H(0) in H_2 (S($+6$) in H_2SO_4
and $ZnSO_4$ remains unchanged). The equation is balanced
as written. (b) I($-\frac{1}{3}$) in I_3^- is reduced to I(-1) in I^-;
S($+2$) in $S_2O_3^{2-}$ is oxidized to S($+\frac{5}{2}$) in $S_4O_6^{2-}$:
$I_3^- + 2S_2O_3^{2-} \longrightarrow 3I^- + S_4O_6^{2-}$.
28. (a) $MnO_4^- + 5Fe^{2+} + 8H^+ \longrightarrow$
$Mn^{2+} + 5Fe^{3+} + 4H_2O$
(b) $2ClO_3^- + 2Cl^- + 4H^+ \longrightarrow Cl_2 + 2ClO_2 + 2H_2O$
(c) $3Cu + 2HNO_3 + 6H^+ \longrightarrow 3Cu^{2+} + 2NO + 4H_2O$
(d) $2MnO_4^- + 5SO_2 + 2H_2O \longrightarrow$
$2Mn^{2+} + 5SO_4^{2-} + 4H^+$
(e) $6HI + 2HNO_3 \longrightarrow 3I_2 + 2NO + 4H_2O$
30. (a) $3CN^- + 2MnO_4^- + H_2O \longrightarrow$
$3CNO^- + 2MnO_2 + 2OH^-$
(b) $2Cr^{3+} + 3OCl^- + 10OH^- \longrightarrow$
$2CrO_4^{2-} + 3Cl^- + 5H_2O$
(c) $9I^- + ClO_3^- + 3H_2O \longrightarrow 3I_3^- + Cl^- + 6OH^-$
(d) $2NH_3 + OCl^- \longrightarrow N_2H_4 + Cl^- + H_2O$
(e) $2MnO_2 + 4OH^- + O_2 \longrightarrow 2MnO_4^{2-} + 2H_2O$
32. $K_2O + H_2O \longrightarrow 2K^+ + OH^-$; basic oxide
$SrO + H_2O \longrightarrow Sr^{2+} + 2OH^-$; basic oxide
$SO_2 + H_2O \longrightarrow H_2SO_3$; acidic oxide
$SO_3 + H_2O \longrightarrow H_2SO_4$; acidic oxide
$CO_2 + H_2O \longrightarrow H_2CO_3$; acidic oxide
$P_4O_6 + H_2O \longrightarrow 2H_3PO_3$; acidic oxide
$Cl_2O_7 + H_2O \longrightarrow 2HClO_4$; acidic oxide
34. (a) $Li_2O + SiO_2 \longrightarrow Li_2SiO_3$ (lithium silicate)
(b) $Na_2O + N_2O_5 \longrightarrow 2NaNO_3$ (sodium nitrate)
(c) $6CaO + P_4O_{10} \longrightarrow 2Ca_3(PO_4)_2$ (calcium phosphate)
36. Period 2: BeO, *amphoteric*; B_2O_3, CO_2, N_2O_3, N_2O_5,
acidic. Period 3: Al_2O_3, *amphoteric*; SiO_2, P_4O_6, P_4O_{10},
SO_2, SO_3, Cl_2O_7, *acidic*.   **38.** NH_3, Cl^-, Lewis bases;
Cu^{2+}, Al^{3+}, SiO_2, Lewis acids.
40. $2Al(OH)_3(s) \longrightarrow Al_2O_3(s) + 3H_2O(g)$
$2Fe(OH)_3(s) \longrightarrow Fe_2O_3(s) + 3H_2O(g)$
$Cu(OH)_2(s) \longrightarrow CuO(s) + H_2O(g)$
42. (a) $2Al(s) + 2NaOH(aq) + 6H_2O(l) \longrightarrow$
$2NaAl(OH)_4^-(aq) + 3H_2(g)$
(b) $Al_2O_3(s) + 3C(s) + 3Cl_2(g) \longrightarrow 2AlCl_3(s) + 3CO(g)$
(c) $2(NH_4)Al(SO_4)_2 \cdot 12H_2O(s) \longrightarrow$
$2NH_3(g) + 4H_2SO_4(l) + Al_2O_3(s) + 21H_2O(g)$
44. (a) Calcium carbonate, $CaCO_3$; (b) aluminum oxide,
Al_2O_3; (c) iron(II) iron(III) oxide, Fe_3O_4 *or*
$Fe(II)[Fe(III)]_2O_4$; (d) iron(II) disulfide, FeS_2;
(e) carbon, C; (f) lead(II) lead(IV) oxide, Pb_3O_4 *or*
$[Pb(II)]_2Pb(IV)O_4$; (g) hydrated iron(III) oxide,
$Fe_2O_3 \cdot xH_2O$; (h) hydrated ammonium aluminum sulfate,
$(NH_4)Al(SO_4)_2 \cdot 12H_2O$.   **46.** 20.90% Al, 21.76% Si.
48. 49.5% tin, 50.5% lead.   **50.** $2Fe(s) + 3Cl_2(g) \longrightarrow$
$2FeCl_3(s)$ (red-black solid iron(III) chloride);

A-32

$FeCl_3(aq) + 3NaOH(aq) \longrightarrow Fe(OH)_3(s) + 3NaCl(aq)$
(brown Fe(III) hydroxide); $2Fe(OH)_3(s) \xrightarrow{\text{Heat}} Fe_2O_3(s) +$
$3H_2O(g)$ (red-brown Fe(III) oxide). The mass of
$Fe_2O_3(s)$ is 2.14 g.   **52.** 68.0% Cu, 19.2% Zn, 12.8%
Al.   **53.** (a) $3P_4 + 24H_2O + 12CuSO_4 \longrightarrow$
$4Cu_3P + 8H_3PO_3 + 12H_2SO_4$;
(b) $P_4 + 16H_2O + 10CuSO_4 \longrightarrow 4H_3PO_4 + 10Cu +$
$10H_2SO_4$.   **56.** $Fe(NH_4)_2(SO_4)_2 \cdot 6H_2O$.
58. (a) Electrolysis of $CuSO_4(aq)$, or by the series of
reactions
$CuSO_4(s) + 2OH^-(aq) \longrightarrow Cu(OH)_2(s)$
$Cu(OH)_2(s) \xrightarrow{\text{Heat}} CuO(s) + H_2O(g)$
$CuO(s) + H_2(g) \xrightarrow{\text{Heat}} Cu(s) + H_2O(g)$
(b) Starting with $Cu(OH)_2(s)$ or CuO(s), prepared as in part
(a), $Cu(OH)_2(s) + 2HCl(aq) \longrightarrow CuCl_2(aq) + 2H_2O(l)$
and the $CuCl_2(aq)$ solution can then be concentrated and
crystals of $CuCl_2 \cdot 2H_2O(s)$ obtained. (c) Starting with
Cu(s) obtained in part (a), $2Cu(s) + 2HCl(g) \xrightarrow{\text{Heat}}$
$2CuCl(s) + H_2(g)$. (d) $Cu(NH_3)_4SO_4 \cdot H_2O(s)$ can be crystallized from a solution obtained by adding the appropriate
amount of concentrated $NH_3(aq)$ to an aqueous solution
of $CuSO_4 \cdot 5H_2O$: $Cu^{2+}(aq) + 4NH_3(aq) \longrightarrow$
$Cu(NH_3)_4^{2+}(aq)$.   **60.** The oxides are soluble in dilute
acid, such as HCl(aq), from which solutions the hydroxides
can be precipitated by adding NaOH(aq):
$Fe^{3+}(aq) + 3OH^-(aq) \longrightarrow Fe(OH)_3(s)$;
$Cu^{2+}(aq) + 2OH^-(aq) \longrightarrow Cu(OH)_2(s)$.   **62.** PbI_2 is
relatively soluble in hot water, from which it separates as
golden-yellow spangles on cooling; AgI is very insoluble in
both hot and cold water.
64. $CuO(s) + 2HCl(aq) \longrightarrow CuCl_2(aq) + H_2O(l)$
$CuS(s) + 2HCl(aq) \longrightarrow CuCl_2(aq) + H_2S(g)$

66. (a) (b) (c) (d)

CHAPTER 11

1. Molecular solids: S_8, CO_2, and P_4O_6. Network solids:
C, NaCl, MgO, and Al.   **3.** (a) Weak London forces,
(b) covalent bonds, (c) ionic bonds, (d) covalent bonds,
(e) metallic bonds.   **5.** Amorphous solids lack the regular
repeating pattern of atoms that is found in crystalline solids.
An amorphous solid, such as glass, has a structure that is
intermediate between those of a crystalline solid and a liquid.
As the temperature is raised, the randomness of the molecules
in an amorphous solid gradually increases; there is no sharp
melting point, but rather the material gradually softens.
7. One in which the network extends indefinitely only in

two dimensions, to give a layer structure in which the layers are attracted only by weak intermolecular forces, e.g. graphite, black phosphrous, and magnesium chloride.
9. BaO, ionic; diamond, covalent bonds; I_2, weak London forces; P_4O_{10}, weak London forces; copper, metallic bonds; graphite, covalent bonds within each layer and London forces between molecular layers.   **11.** Melting points of ionic solids are related to the sizes of their ions, their charges, and the interionic distances. NaCl consists of Na^+ and Cl^- ions, while MgO consists of Mg^{2+} and O^{2-} ions. In terms of their ionic radii, Mg^{2+} is smaller than Na^+, and O^{2-} is smaller than Cl^-. In addition, the force of attraction between doubly charged ions is four times that between singly charged ions. Thus, the ionic bonds in MgO are much stronger than those in NaCl, and the melting point of MgO is much higher than that of NaCl.   **13.** A regular three-dimensional array of points in space.   **15.** See Figure 11.14.
17. Hexagonal close-packed, face-centered cubic, body-centered cubic. In each case the motif is a single metal atom at a lattice point.   **19.** (a) KCl has the NaCl lattice with the cations and anions arranged in a face-centered lattice, with the Cl^- ions situated halfway between the K^+ ions, and vice versa. (b) BaO consists of Ba^{2+} and O^{2-} ions and has the NaCl lattice. (c) CuCl, composed of Cu^+ and Cl^- ions, has the sphalerite (ZnS) structure based on the face-centered cubic lattice, with Cl^- ions at the lattice points and the Cu^+ ions located one-quarter of the distance along each body diagonal.   **21.** See Figure 11.18.
23. 195 g mol^{-1}.   **25.** (a) 578 pm,   (b) 250 pm,   (c) 70 pm.   **27.** (a) See Figure 11.17(a);   (b) 4;   (c) 3.18 g cm^{-3};   (d) 198 pm;   (e) 198 pm is the van der Waals radius of Kr; 111 pm is the covalent radius.
29. 6.08×10^{23} units mol^{-1}.   **31.** 4.   **33.** The unit cell is similar to that of diamond (face-centered cubic) with Si atoms at the lattice points and O atoms between each pair of Si atoms, at one-quarter of the distance along each body diagonal, so that each Si atom is surrounded by four O atoms at the corners of a tetrahedron.   **35.** 2, body-centered cubic.   **37.** 136 pm   **39.** 4.51 g cm^{-3}.
41. Zn^{2+}, coordination number 4; S^{2-}, coordination number 4. ZnS and diamond have the same structure; in ZnS, four Zn^{2+} ions replace the four C atoms in diamond that are situated one-quarter of the distance along each body diagonal, with the S^{2-} ions at the lattice points of a face-centered cubic unit cell.   **43.** 7.76 g cm^{-3}.
45. 3.60 g cm^{-3}.   **47.** 5.86×10^{19} Ag atoms.
48. (a) $I^- > F^-$,   (b) $K^+ > Na^+$,   (c) $O^{2-} > F^-$,   (d) $Cl^- \gg Be^{2+}$,   (e) $Cl^- > K^+$.   **53.** The cation: anion radius ratio is 0.567 in NaCl, and 0.944 in CsCl, while the limiting radius ratio for 6 coordination is 0.414–0.732, and that for 8 coordination is 0.732–1.00. Thus Na^+ ions in NaCl are surrounded by six Cl^- ions, and Cs^+ ions in CsCl are surrounded by eight Cl^- ions.   **55.** Anions have one or more electrons in excess of those in the neutral atom, while cations have one or more electrons fewer than the neutral atom, but the core charge remains the same as that of the free atom. The size of the species is related to the magnitude of the electron–electron repulsions in the valence shell; as the number of electrons increases the valence-shell electrons become more crowded and repel each other more strongly. Thus, the radius of the species is expected to increase in the order cation < neutral atom < anion.

CHAPTER 12

1. 4.5 m.   **3.** 15.4M, 35.3 m, 0.389.   **5.** 0.0208, 1.18 m, 1.15M.   **7.** In sublimation a solid passes directly from solid to gas (vapor). Thus the enthalpy of sublimation is the sum of the enthalpies of fusion and vaporization.
10. -890.6 kJ mol^{-1}, 2.54×10^3 mol CH_4 (40.7 kg).
12. (a) None; (b) none; (c) increase; (d) the greater intermolecular forces the smaller is the vapor pressure; (e) none.   **13.** (a) 726 torr,   (b) 356 mL,   (c) 324 mL, (d) 1.19 g.   **15.** 1.73 L, 1.34M, 1.52 L.   **17.** Covalent network solid; insoluble.   **19.** (a) Molecular solid, London forces;   (b) molecular solid, dipole–dipole and London forces;   (c) network solid, metallic bonds;   (d) network solid, ionic bonds;   (e) molecular solid, London forces; (f) network solid, covalent bonds;   (g) network solid, ionic bonds;   (h) molecular solid, hydrogen bonds.   **21.** Cl is more polarizable than H; thus the strength of the London forces increases from benzene to chlorobenzene to hexachlorobenzene.   **23.** LiF and BeF_2 are ionic substances and the remainder are covalent substances. Ionic substances have strong ionic bonds and covalent substances have relatively weak intermolecular forces, which accounts for the large differences between LiF and BeF_2 and the remainder. Among the covalent substances the strength of the London forces will depend primarily on the polarizabilities of the molecules and on the polarizabilities of their atoms and on the shape of the molecules (which decrease from B to F). The order of boiling points is $F_2 < OF_2 < CF_4 < NF_3 < BF_3$. For F_2, OF_2, and CF_4 the boiling points are related to the number of atoms per molecules. $CF_4 < NF_3 < BF_3$ because the average distance between molecules decreases despite a small decrease in polarizability.
25. (a) $H_2O > Cl_2O$ because of strong hydrogen bonding. (b) $CBr_4 > CCl_4$ because Br is more polarizable than Cl. (c) Ar > He because Ar is more polarizable than He.
27. (a) HCl(g) is more soluble in polar water than in nonpolar pentane. (b) Water is more soluble in liquid HF than in nonpolar gasoline because of hydrogen bonding. (c) Nonpolar CCl_4 is more soluble in $CHCl_3$ than in polar water. (d) Nonpolar naphthalene is more soluble in nonpolar benzene than in polar water. (e) HCN is more soluble in water than in nonpolar N_2 because of hydrogen bonding. (f) Nonpolar benzene is more soluble in nonpolar toluene than in polar water.   **29.** (a) London forces, (b) hydrogen bonding, (c) hydrogen bonding, (d) London forces, (e) London forces and dipole–dipole forces, (f) London forces.   **31.** (a) Ca > Na, because each Ca atom contributes two electrons and each Na atom contributes one electron to metallic bonding, and Ca^{2+} is a smaller ion than Na^+. (b) $SiH_4 > CH_4$, because of its higher polarizability. (c) $C_2H_6 > CH_4$, because of its greater polarizability. (d) $NH_3 > PH_3$, because of stronger hydrogen bonding. (e) $Cl_2 > F_2$, because of its greater polarizability.

(f) $SiO_2 > SO_2$, because the former is a covalent network solid. **32.** $CO, H_2S, CH_2Cl_2, HBr, SO_2$. **34.** (a) Cl_2, on account of both its larger size and greater polarizability. (b) $BrCl$, on account of its larger size, greater polarizability, and dipole moment. (c) CF_4, on account of its larger size and greater polarizability. (d) SO_2, on account of its larger size, greater polarizability, and dipole moment. **36.** The greater polarizability of CCl_4 compared with $CHCl_3$ leads to greater London forces, which outweigh the additional effects due to the small dipole moment of $CHCl_3$.
37. (a) Highly polar X—H bonds and strongly localized unshared pairs of electrons on X; thus, X = N, O, or F. (b) NH_3, HF, $H_3CC(O)OH$. **39.** (a) $C_2H_5OH >$ $(CH_3)_2O$, because only the former can form hydrogen bonds. (b) $HF > HCl$, because only the former can form strong hydrogen bonds. (c) $LiCl \gg CCl_4$, because $LiCl(l)$ is an ionic melt while $CCl_4(l)$ is a covalent molecular liquid where only weak London forces have to be overcome. (d) $LiCl \gg$ HCl, for the same reason as in (c). **42.** The definition of an ideal gas (kinetic molecular theory) assumes that the gaseous molecules have a negligible volume and that there are no intermolecular forces between gas molecules. Molecules of real gases have finite volumes and attract each other through intermolecular forces. At high pressure the volume of the molecules dominates, causing the volume to be greater than ideal. At low temperature intermolecular forces dominate, causing the observed volume to be smaller than the ideal volume. (b) (i) $Br_2 > F_2$, on account of both size and the relative magnitudes of the London forces; (ii) $CO >$ N_2, because, although the molar volumes and polarizabilities are similar, CO has a dipole moment; (iii) $CH_3CO_2H > CH_3COCl$, because, although the molar volumes and polarizabilities are expected to be rather similar, only acetic acid is capable of forming hydrogen bonds. **43.** 164 g sugar.
45. 58.7 g mol^{-1}, CN_2H_4O, $H_2N\!-\!\overset{\displaystyle \|}{\underset{\displaystyle O}{C}}\!-\!NH_2,$

ammonium cyanate, $H\!-\!\overset{\displaystyle H}{\underset{\displaystyle H}{\overset{\displaystyle |}{\underset{\displaystyle |}{N^+}}}}\!-\!H:\quad N\!\equiv\!C\!-\!\overset{..}{\underset{..}{O}}\!:^-$

Ammonium cyanate would give twice the boiling point elevation of a urea solution of the same concentration.
47. (a) $p = p°(1 - x_{solute})$; (b) 136 g mol^{-1}, $C_{10}H_{16}$, 136.2 g mol^{-1}. **49.** -22 °C. **51.** 16 m.
53. (a) -0.19 °C. (b) -0.37 °C, (c) -0.56 °C.
55. (a) $\pi = MRT$, where π is the osmotic pressure, M is the molarity of the solute species, R is the gas constant, and T is the absolute temperature. (b) 0.0245 atm. (c) 128 g mol^{-1}. **57.** 2.6×10^4 g mol^{-1}· 76 monomer units. **59.** (a) $PCl_5(s) + 9H_2O(l) \xrightarrow{25\,°C}$ $H_3PO_4(aq) + 5H_3O^+(aq) + 5Cl^-(aq)$; *acid–base*
(b) $2Li(s) + S(s) \xrightarrow{Heat} Li_2S(s)$; lithium sulfide, *oxidation–reduction*
(c) $Na_2O(s) + H_2O(l) \xrightarrow{25\,°C} 2Na^+(aq) + 2OH^-(aq)$; *acid–base*

A-34

(d) $2F_2(g) + 2H_2O(l) \xrightarrow{25\,°C} 4HF(aq) + O_2(g)$; *oxidation–reduction*
(e) $Cl_2(g) + H_2O(l) \xrightarrow{25\,°C} HOCl(aq) + HCl(aq)$; *oxidation–reduction*
(f) $CH_4(g) + H_2O(g) \xrightarrow{High\,T} CO(g) + 3H_2(g)$; *oxidation–reduction*
(g) $MgH_2(s) + 2H_2O(l) \xrightarrow{25\,°C} Mg(OH)_2(s) + H_2(g)$; *acid–base* and *oxidation–reduction*
61. (a) $2H_2O + 2e^- \longrightarrow H_2(g) + 2OH^-$;
(b) $2H_2O \longrightarrow 4H^+ + O_2(g) + 4e^-$
63. (a) $2Al(s) + 3H_2O(g) \longrightarrow Al_2O_3(s) + 3H_2(g)$; *oxidizing agent*
(b) $KNH_2(s) + H_2O(l) \longrightarrow KOH(aq) + NH_3(aq)$; *acid*
(c) $2F_2(g) + 2H_2O(l) \longrightarrow 4HF(aq) + O_2(g)$; *reducing agent*
(d) $P_4O_{10}(s) + 6H_2O(i) \longrightarrow 4H_3PO_4(aq)$; Lewis *base*
(e) $BaO_2(s) + 2H_2O(l) \longrightarrow Ba(OH)_2(aq) + H_2O_2(aq)$; *acid*
(f) $Cl_2O_7(l) + H_2O(l) \longrightarrow 2HClO_4(aq)$; *Lewis base*
(g) $CaH_2(s) + 2H_2O(l) \longrightarrow Ca(OH)_2(s) + 2H_2(g)$; *acid*
65. In each case the reaction is initiated by donation of a lone pair of electrons from a water molecule to the central atom of the anhydride, which is followed by transfer of a proton from the attached water to an oxygen of the anhydride (to give two —OH groups). Thus, in the initial step, water is an electron pair donor (Lewis base), and the anhydride is an electron pair acceptor (Lewis acid).
69. 120 g mol^{-1} corresponding to the dimer $(CH_3CO_2H)_2$ due to the association of acetic acid molecules through

hydrogen bonding: $H_3C\!-\!C\overset{\displaystyle \overset{..}{O}:\cdots\cdot H\!-\!\overset{..}{O}}{\underset{\displaystyle \underset{..}{O}\!-\!H\cdots\cdot:\overset{..}{O}}{}}C\!-\!CH_3$

72. $H_2O + H_2SO_4 \longrightarrow H_3O^+ + HSO_4^-$;
$K_2SO_4 + H_2SO_4 \longrightarrow 2K^+ + 2HSO_4^-$
$HNO_3 + 2H_2SO_4 \longrightarrow NO_2^+ + H_3O^+ + 2HSO_4^-$;
$N_2O_5 + 3H_2SO_4 \longrightarrow H_3O^+ + 2NO_2^+ + 3HSO_4^-$
74. (a) $P_4O_{10}(s) + 6H_2O(l) \longrightarrow 4H_3PO_4(aq)$; Lewis *base*
(b) $PCl_3(l) + 3H_2O(l) \longrightarrow H_3PO_3(aq) + 3HCl(aq)$; Lewis *base*
(c) $Ca(s) + 2H_2O(l) \longrightarrow Ca(OH)_2(s) + H_2(g)$; *oxidizing agent*
(d) $2F_2(g) + 2H_2O(l) \longrightarrow 4HF(aq) + O_2(g)$; *reducing agent*
(e) $CO_3^{2-}(aq) + H_2O(l) \longrightarrow HCO_3^-(aq) + OH^-(aq)$; *Brønsted acid*
(f) $CaO(s) + H_2O(l) \longrightarrow Ca(OH)_2(s)$; *Brønsted acid*
(g) $HNO_3(aq) + H_2O(l) \longrightarrow H_3O^+(aq) + NO_3^-(aq)$; *Brønsted base*
(h) $NH_4Br(aq) + H_2O(l) \rightleftharpoons$ $NH_3(aq) + H_3O^+(aq) + Br^-(aq)$; *Brønsted base*

CHAPTER 13

2. (a) The equilibrium constant K_{eq} is the value of the equilibrium constant expression when all of the concentration terms have their equilibrium values. The reaction quotient, Q, is the value of the equilibrium constant expression when the concentration terms are those for any nonequilibrium situation. (b) For $Q > K_{eq}$, the concentrations of products have to decrease in order for equilibrium to be achieved. For $Q = K_{eq}$, the system is at equilibrium.

3. (a) $K_c = \dfrac{[NO_2]^4[O_2]}{[N_2O_5]^2}$; $K_p = \dfrac{p_{NO_2}^4 p_{O_2}}{p_{N_2O_5}^2}$

(b) $K_c = \dfrac{[SO_3]^2}{[SO_2]^2[O_2]}$; $K_p = \dfrac{p_{SO_3}^2}{p_{SO_2}^2 p_{O_2}}$

(c) $K_c = \dfrac{[SO_3]}{[SO_2][O_2]^{1/2}}$; $K_p = \dfrac{p_{SO_3}}{p_{SO_2} p_{O_2}^{1/2}}$

(d) $K_c = \dfrac{[P_4O_{10}]}{[P_4][O_2]^5}$; $K_p = \dfrac{p_{P_4O_{10}}}{p_{P_4} p_{O_2}^5}$

(e) $K_c = \dfrac{[PCl_3][Cl_2]}{[PCl_5]}$; $K_p = \dfrac{p_{PCl_3} p_{Cl_2}}{p_{PCl_5}}$

5. (a) $K_c = \dfrac{[NO]^2[Br_2]}{[NOBr]^2}$; $K_p = \dfrac{p_{NO}^2 p_{Br_2}}{p_{NOBr}^2}$

(b) $K_c = \dfrac{[CO_2]^3}{[CO]^3}$; $K_p = \dfrac{p_{CO_2}^3}{p_{CO}^3}$

(c) $K_c = \dfrac{[NH_3]^2[H_2O]}{[N_2O][H_2]^4}$; $K_p = \dfrac{p_{NH_3}^2 p_{H_2O}}{p_{N_2O} p_{H_2}^4}$

(d) $K_c = [O_2]$; $K_p = p_{O_2}$. (e) $K_c = [NO_2]^4[O_2]$; $K_p = p_{NO_2}^4 p_{O_2}$. **7.** $K_p = K_c(RT) = 6.49$ atm.

9. (a) $K_c = \dfrac{[NO]^2}{[N_2][O_2]}$ or $K_c = \dfrac{p_{NO}^2}{p_{N_2} p_{O_2}}$. (b) $K_c = K_p = 2.5 \times 10^{-3}$ (unitless). **11.** System is initially *not* at equilibrium; to attain equilibrium, $[CH_3OH]$ has to increase; $[CH_3OH] > 0.10$ mol L^{-1}. **13.** $K_c = 0.772 = K_p$. **14.** $[H_2] = [CO_2] = 0.106$ mol L^{-1}; $[H_2O] = [CO] = 0.094$ mol L^{-1}. **16.** $K_c = 0.042$ mol L^{-1}. **18.** $[H_2] = [I_2] = 0.053$ mol L^{-1}; $[HI] = 0.36$ mol L^{-1}. **20.** (a) $Q = 1.0 \times 10^{-2}$ mol^{-2} L^{-2} < K_c; system *not* at equilibrium. (b) Too small. **22.** 4.80%.
24. Forward reaction *endothermic*; reverse reaction *exothermic* (ΔH negative). **26.** (a) $Q = 116.6$ atm^{-1} (0 °C) and $K_p(0$ °C$) < K_p(25$ °C$)$, so NOBr(g) would have to react to give more NO(g) and Br$_2$(g) for equilibrium to be achieved at 0 °C. (b) $K_p(50$ °C$) = 179$ atm$^{-1} > K_p(25$ °C$)$, as expected for endothermic forward reaction.
28. (a) $[SO_3]$ decreases; (b) $[SO_3]$ decreases; (c) $[SO_3]$ increases; (d) no change. **30.** Exothermic.

32. (a) $K_p = \dfrac{p_{H_2} p_{O_2}^{1/2}}{p_{H_2O}}$; $K_c = \dfrac{[H_2][O_2]^{1/2}}{[H_2O]}$.

(b) $K_p = p_{H_2} p_{O_2}^{1/2}$; $K_c = [H_2][O_2]^{1/2}$. (c) $K_p = p_{H_2}^2 p_{O_2}$; $K_c = [H_2]^2[O_2]$. (d) $K_p = p_{H_2}^{-1} p_{O_2}^{-1}$; $K_c = [H_2]^{-1}[O_2]^{-1}$. (e) $K_p = p_{O_2}$; $K_c = [O_2]$.
34. (a) No effect; (b) $[O_2]$ increases; (c) $[O_2]$ decreases; (d) $[O_2]$ decreases; (e) increases rate of reaction but has no effect on $[O_2]$. **36.** 1.99 mol NH$_4$HS(s).
38. (a) -27 kJ; (b) low temperature and high pressure. **40.** $K_p = 0.200$ atm^{-1}. **42.** (a) Al$_2$Cl$_6$(200 °C), AlCl$_3$(800 °C); (b) Al$_2$Cl$_6$(g) and 2AlCl$_3$(g); (c) $p_{AlCl_3} = 0.58$ atm, $p_{Al_2Cl_6} = 0.42$ atm; (d) $K_p = 0.80$ atm, $K_c = K_p RT = 1.1 \times 10^{-2}$ mol L^{-1}.

CHAPTER 14

1. (a) $[H_3O^+] = [NO_3^-] = 10^{-5}M$; (b) $[H_3O^+] = [Cl^-] = 0.0023M$; (c) $[H_3O^+] = [ClO_4^-] = 0.113M$; (d) $[H_3O^+] = [Br^-] = 0.034M$; (e) $[Na^+] = [OH^-] = 10^{-3}M$; (f) $[Ba^{2+}] = 0.145M$, $[OH^-] = 0.290M$.
3. Strong acids: HCl, HBr, HI, HNO$_3$, HClO$_4$, H$_2$SO$_4$, for example. Weak acids: HF, CH$_3$CO$_2$H, HCN, HOCl, H$_3$PO$_4$, for example. **5.** $[H_3O^+] = 2.2 \times 10^{-6}$ mol L^{-1}; 0.022%. **7.** 0.200M CH$_3$CO$_2$H.
9. $[OH^-] = 5.9 \times 10^{-6}$ mol L^{-1}, 0.007%. **11.** (a) 1.63, (b) 9.30, (c) 13.31, (d) 0.47, (e) 11.66, (f) 10.30.
13. $K_a = 4.0 \times 10^{-8}$ mol L^{-1}, p$K_a = 7.40$. **15.** (a) 1.70, (b) 12.20, (c) 7.00. **17.** 7.58.
19. $[NH_4^+] = 0.020M$, $[H_3O^+] = [NH_3] = 3.4 \times 10^{-6}M$, $[OH^-] = 2.9 \times 10^{-9}M$.
21. (a) NH$_3$(aq) + HNO$_3$(aq) $\longrightarrow$ NH$_4$NO$_3$(aq); *acidic* (b) NH$_3$(aq) + HCl(aq) $\longrightarrow$ NH$_4$Cl(aq); *acidic* (c) Ca(OH)$_2$(s) + H$_2$SO$_4$(aq) $\longrightarrow$ CaSO$_4$(aq) + 2H$_2$O(l); *basic* (d) KOH(aq) + CH$_3$CO$_2$H(aq) $\longrightarrow$ CH$_3$CO$_2$K(aq) + H$_2$O(l); *basic* (e) Al(OH)$_3$(s) + 3HCl(aq) $\longrightarrow$ AlCl$_3$(aq) + 3H$_2$O(l); *acidic* (f) LiOH(aq) + HI(aq) $\longrightarrow$ LiI(aq) + H$_2$O(l); *neutral*
23. (a) CO$_2$(g) + H$_2$O(l) $\rightleftharpoons$ H$_2$CO$_3$(aq); H$_2$CO$_3$(aq) + H$_2$O(l) $\rightleftharpoons$ H$_3$O$^+$(aq) + HCO$_3^-$(aq); HCO$_3^-$(aq) + H$_2$O(l) $\rightleftharpoons$ H$_3$O$^+$(aq) + CO$_3^{2-}$(aq); *acidic* (b) SO$_3$(g) + H$_2$O(l) $\longrightarrow$ H$_2$SO$_4$(aq); H$_2$SO$_4$(aq) + H$_2$O(l) $\longrightarrow$ H$_3$O$^+$(aq) + HSO$_4^-$(aq); HSO$_4^-$(aq) + H$_2$O(l) $\rightleftharpoons$ H$_3$O$^+$(aq) + SO$_4^{2-}$(aq); *acidic* (c) CaO(s) + H$_2$O(l) $\longrightarrow$ Ca^{2+}(aq) + 2OH$^-$(aq); *basic* (d) PCl$_3$(l) + 6H$_2$O(l) $\longrightarrow$ H$_3$PO$_3$(aq) + 3H$_3$O$^+$(aq) + 3Cl$^-$(aq), *acidic* (e) No reaction, *neutral* (f) LiH(s) + H$_2$O(l) $\longrightarrow$ Li$^+$(aq) + OH$^-$(aq) + H$_2$(g), *basic*
25. 0.23M. **27.** pH = 2.07. **29.** 4.59 L, *basic*.
31. (a) HCO$_3^-$, p$K_b = 7.52$; (b) HS$^-$, p$K_b = 6.96$; (c) NO$_2^-$, p$K_b = 10.65$; (d) HSO$_3^-$, p$K_b = 12.23$.
33. pH = 3.07. **35.** (a) 1.30, (b) 2.85.
37. 6.0 > pH > 4.4. **39.** (a) Bromothymol blue, (b) bromothymol blue, (c) methyl red, (d) methyl red.
41. (a) Green, (b) colorless, (c) blue. **43.** 8.2.
44. 9.35. **46.** 0.178. **48.** 9.03. **50.** 1.74.
52. 3.51, 3.51. **54.** 7.21, 7.21. **56.** (a) Strength refers to the extent of ionization, which increases with the K_a of the acid; concentration is the amount of acid per unit volume, which is independent of K_a. (b) Equivalence point is the point in the titration where moles of acid equals moles of base, corresponding to a salt solution. The end-point is the point in a titration where a particular indicator changes color, which occurs at a pH close to the value of pK_a for the indicator. **58.** (a) Bromothymol blue, (b) thymol blue, (c) methyl red. **59.** (a) 2.87, (b) 4.74, (c) 8.70, (d) 12.30. **60.** Thymol blue or phenolphthalein, p$K_a = 8.7 \pm 1$. **61.** (a) 11.13, (b) 9.43, (c) 9.25, (d) 5.28, (e) 1.64. **63.** (a) 4.75, (b) 7.51, (c) 8.79, (d) 9.60,

(e) 10.40, (f) 12.30. **64.** (a) 0.202M, (b) 28.6 mL, (c) 0.341 M. **66.** (a) 1.08, (b) 2.12, (c) 4.23 (d) 7.21. (e) 9.89, (f) 12.68. **68.** (a) 12.00 (b) 7.00 (c) 10.63 (d) 2.73 (e) 7.73 (f) 2.00 **69.** (a) Not a buffer (strong acid plus a salt of a strong acid); (b) buffer solution (weak acid plus a salt of the weak acid); (c) not a buffer solution (salt of a weak acid plus a strong base); (d) buffer solution (salt of a weak acid plus the weak acid). **71.** pK_a = 5.0 **72.** K_a = 7.1 × 10^{-7} mol L^{-1}.

CHAPTER 15

2. (a) 2Li(s) + Br$_2$(l) ⟶ 2LiBr(s); lithium bromide, Li$^+$:Br:$^-$

Ca(s) + Br$_2$(l) ⟶ CaBr$_2$(s); calcium bromide, Ca^{2+} [:Br:$^-$]$_2$

(b) 2Li(s) + S(s) $\xrightarrow{\text{Heat}}$ Li$_2$S(s); lithium sulfide, [Li$^+$]$_2$:S:$^{2-}$

Ca(s) + S(s) $\xrightarrow{\text{Heat}}$ CaS(s); calcium sulfide, Ca^{2+} :S:$^{2-}$

(c) 6Li(s) + N$_2$(g) $\xrightarrow{\text{Heat}}$ 2Li$_3$N(s); lithium nitride, [Li$^+$]$_3$:N:$^{3-}$

3Ca(s) + N$_2$(g) $\xrightarrow{\text{Heat}}$ Ca$_3$N$_2$(s); calcium nitride, [Ca^{2+}]$_3$ [:N:$^{3-}$]$_2$

3. (a) 2K(s) + Br$_2$(l) ⟶ 2KBr(s); potassium bromide (b) 4Li(s) + O$_2$(g) ⟶ 2Li$_2$O(s); lithium oxide (c) 2Na(s) + H$_2$(g) ⟶ 2NaH(s); sodium hydride (d) 6Li(s) + N$_2$(g) ⟶ 2Li$_3$N(s); lithium nitride (e) 2K(s) + 2H$_2$O(l) ⟶ 2KOH(aq) + H$_2$(g); potassium hydroxide, hydrogen (f) LiH(s) + H$_2$O(l) ⟶ LiOH(aq) + H$_2$(g); lithium hydroxide, hydrogen (g) Li$_3$N(s) + 3H$_2$O(l) ⟶ 3LiOH(aq) + NH$_3$(g); lithium hydroxide, ammonia **5.** Lime is CaO(s), prepared by heating limestone: CaCO$_3$(s) ⟶ CaO(s) + CO$_2$(g); CaO(s) + H$_2$O(l) ⟶ Ca(OH)$_2$(s). Industrial uses include treating acid soil, the manufacture of glass and cement, and to remove silica as slag in metallurgical processes. **7.** 2.13 L. **9.** (b) NaH(s) + H$_2$O(l) ⟶ Na$^+$(aq) + OH$^-$(aq) + H$_2$(g) (d) Na$_2$O(s) + H$_2$O(l) ⟶ 2Na$^+$(aq) + 2OH$^-$(aq) (f) Na$_2$CO$_3$(aq) + H$_2$O(l) ⇌ 2Na$^+$(aq) + HCO$_3$$^-$(aq) + OH$^-$(aq) **11.** Solubilities of the hydroxides increase from Be(OH)$_2$ and Mg(OH)$_2$ (insoluble) to Sr(OH)$_2$; this is related to the decreasing strength of the ionic bonds with increasing size of the cation in descending the group. The reverse trend is found for the sulfates. Because the SO$_4$$^{2-}$ is large the ionic bonds are relatively weak and the dominant factor is the decreasing tendency for the cations to be hydrated as they increase in size in descending the group. **13.** (a) CaCO$_3$(s) ⟶ CaO(s) + CO$_2$(g); calcium oxide, carbon dioxide (b) Ca(OH)$_2$(s) ⟶ CaO(s) + H$_2$O(l); calcium oxide, water

(c) 2NaHCO$_3$(s) ⟶ Na$_2$CO$_3$(s) + H$_2$O(g) + CO$_2$(g); sodium carbonate, water, carbon dioxide (d) MgCl$_2$·6H$_2$O(s) ⟶ MgO(s) + 2HCl(g) + 5H$_2$O(g); magnesium oxide, hydrogen chloride, water **15.** 1.2 × 10^6 L. **17.** (a) O^{2-} + H$_2$O ⟶ 2OH$^-$; (b) H$^-$ + H$_2$O ⟶ OH$^-$ + H$_2$; (c) N^{3-} + 3H$_2$O ⟶ NH$_3$ + 3OH$^-$; (d) CO$_3$$^{2-}$ + H$_2$O ⇌ HCO$_3$$^-$ + OH$^-$. **19.** The small Mg^{2+} ion is strongly hydrated as Mg(H$_2$O)$_6$$^{2+}$, which behaves as an acid. When it is heated, two protons are transferred from the hydrated cation to the Cl$^-$ ions, Mg(H$_2$O)$_6$$^{2+}$ + 2Cl$^-$ ⇌ Mg(H$_2$O)$_4$(OH)$_2$ + 2HCl, and as the HCl(g) is driven off the equilibrium shifts to the right and eventually the Mg(H$_2$O)$_4$(OH)$_2$ is dehydrated to give MgO(s). In contrast, the relatively large Ba^{2+} is only weakly hydrated and has no acidic properties. When it is heated the water is simply driven off to leave anhydrous BaCl$_2$.

21. 1.05 × 10^6 kg CaCO$_3$(s). **23.** (a) The alkali metals readily donate their outer ns^1 electrons to form M$^+$ ions and are therefore reducing agents. Reactivity follows first ionization energy and electronegativity which decrease down the group. (b) The low core charge and the ease with which the ns^1 valence electron is removed mean that almost all alkali metal compounds are ionic and contain M$^+$ ions. (c) The oxides, M$_2$O(s), have the Lewis structure [M$^+$]$_2$:O:$^{2-}$; the oxide ion is a strong base in water; O^{2-} + 2H$_2$O ⟶ 2OH$^-$. (d) None of the alkali metals cations are sufficiently strongly hydrated to give hydrated cations M(H$_2$O)$_n$$^+$ that have acidic properties, so in combination with the anions of strong acid, which have no basic properties, alkali metal salts give neutral solutions in water. **25.** Be^{2+} is a small ion and has a large charge: radius ratio, so that it can strongly attract unshared electron pairs of other species, to a maximum of four pairs to complete its valence shell. Compounds such as BeCl$_2$ are covalent rather than ionic and attract, for example, two Cl$^-$ ions, to form complex ions such as BeCl$_4$$^{2-}$. The salt BeCl$_4$·4H$_2$O has the structure Be(H$_2$O)$_4$$^{2+}$(Cl$^-$)$_2$. The ion Be(H$_2$O)$_4$$^{2+}$ is also found in aqueous solution and behaves as a weak acid. Be(OH)$_2$ dissolves both in acid and in base and is amphoteric. In acid, Be(OH)$_2$ + 2H$_3$O$^+$ ⟶ Be(H$_2$O)$_4$$^{2+}$, and in base, Be(OH)$_2$ + 2OH$^-$ ⟶ BeOH$_4$$^{2-}$. **27.** Mg^{2+} and CO$_3$$^{2-}$ ions strongly attract each other and MgCO$_3$(s) is very insoluble in water. When CO$_2$(g) is bubbled through an aqueous suspension of MgCO$_3$(s), the reaction MgCO$_3$(s) + CO$_2$(aq) + H$_2$O(l) ⟶ Mg(HCO$_3$)$_2$(aq) occurs give a solution of soluble Mg(HCO$_3$)$_2$. Mg(HCO$_3$)$_2$(s) is soluble because Mg^{2+} and HCO$_3$$^-$ ions do not attract each other nearly as strongly as do Mg^{2+} and the doubly charged CO$_3$$^{2-}$ ion. **29.** (a) K_{sp} = [Fe^{3+}][OH$^-$]^3 (mol^4 L^{-4}); (b) K_{sp} = [Ca^{2+}]^3[PO$_4$$^{3-}$]^2 (mol^5 L^{-5}). **31.** (a) K_{sp} = [Ag$^+$][F$^-$], K_{sp} = [Ca^{2+}][F$^-$]^2, K_{sp} = [Pb^{2+}][F$^-$]^2; (b) K_{sp} = [Ag$^+$][OH$^-$], K_{sp} = [Mg^{2+}][OH$^-$]^2, K_{sp} = [Al^{3+}][OH$^-$]^3; (c) K_{sp} = [Ag$^+$]^2[SO$_4$$^{2-}$], K_{sp} = [Sr^{2+}][SO$_4$$^{2-}$]. **33.** 6.9 × 10^{-9} mol^3 L^{-3}. **35.** (a) 1.9 × 10^{-4} M, (b) 1.3 × 10^{-5} M, (c) 2 × 10^{-9} M, (d) 1.3 × 10^{-3} M. **37.** 5.3 × 10^{-6} M. **39.** (a) Yes; (b) No;

(c) $2.2 \times 10^{-2} M$, $1.0 \times 10^{6} M$. **41.** (a) No, (b) no.
43. (a) Yes; (b) PbI_2, 0.443 g; (c) $[Pb^{2+}] = 0.999 M$,
$[I^-] = 8 \times 10^{-5} M$, $[Na^+] = 2.0 \times 10^{-3} M$,
$[NO_3^-] = 2.00 M$. **45.** 0.25 g $BaF_2(s)$.
47. (a) Increase, (b) decrease. **49.** $MgCO_3(s)$.
51. The solubility of $PbCl_2(s)$ is reduced to $1.8 \times 10^{-6} M$ in
$3M$ HCl(aq), which ensures a more efficient separation of
lead. **53.** 10.38. **55.** 6×10^{-4} mol L^{-1}.
57. $Cu(OH)_2(s)$, $4.8 \times 10^{-3} M$; $Pb(OH)_2(s)$, very soluble; yes.
59. (a) $10^{-24} M$, (b) $8 \times 10^{-7} M$, (c) $3 \times 10^{-16} M$,
(d) $9 \times 10^{-20} M$. **61.** (a) $MgCl_2 \cdot 6H_2O$. (b) $MgCl_2(s)$.
(c) MgO(s). (d) Because of the acidity of the $Mg(H_2O)_6^{2+}$
ion, the reaction in air is $MgCl_2 \cdot 6H_2O \rightleftharpoons MgO(s) +$
$2HCl(g) + 5H_2O(g)$ and the equilibrium shifts to the *right*
as the HCl(g) is swept away. In an atmosphere of HCl(g)
the above equilibrium shifts to the *left* but $H_2O(g)$ in equilibrium with the hydrate is swept away in the stream of dry
HCl(g) and the reaction is $MgCl_2 \cdot 6H_2O \longrightarrow$
$MgCl_2(s) + 6H_2O(g)$.
63. (a) $MgCO_3(s)$ is dissolved in the requisite amount of
$H_2SO_4(aq)$, $MgCO_3(s) + H_2SO_4(aq) \longrightarrow$
$MgSO_4(aq) + CO_2(g) + H_2O(g)$, and the crystals of
$MgSO_4 \cdot 7H_2O(s)$ that are obtained on evaporating the resulting solution can be recrystallized from water.
(b) $BaCO_3(s)$ is dissolved in excess $H_2SO_4(aq)$ to give a precipitate of $BaSO_4(s)$, $BaCO_3(s) + H_2SO_4(aq) \longrightarrow$
$BaSO_4(s) + CO_2(g) + H_2O(l)$. **65.** The extent of hydration depends on the strength of attraction of the cations for
polar water molecules, which is proportional to the ionic
charge and inversely proportional to the ionic radius
(Coulomb's law). An approximate guide to this strength of
interaction is the charge: radius ratio, which decreases in
descending each group and is greater for all the group 2
cations than it is for the group 1 cations. All the M^{2+} ions
and Li^+ and Na^+ attract lone pairs of electrons on water
molecules sufficiently strongly to form strongly hydrated
ions, so that their salts separate from solution with water
of hydration. Although K^+, Rb^+, and Cs^+ are hydrated to
some extent in solution, their attraction for water molecules
is sufficiently weak that their salts crystallize as anhydrous
compounds (unless the anion of the salt is strongly hydrated).
67. $BaCO_3(s)$ is the only insoluble salt and $Cu(NO_3)_2(s)$
is the only colored compound. Flame tests distinguish
$Ca(NO_3)_2$ (orange-red), $BaCO_3$ (green), and the sodium salts
(c), (d), and (h) (yellow). The latter may be distinguished from
the pH of their aqueous solutions: $NaHSO_4$, (c), acidic;
Na_2CO_3, (d), basic; NaCl, (h), neutral. Of the remaining
salts, the ammonium compounds are distinguished from
$AgNO_3(s)$ by heating them, when $NH_3(g)$ is evolved (identified by its odor and its action on red litmus paper, which
turns blue). Thus, the $AgNO_3$ sample is identified by elimination. Addition of $AgNO_3(aq)$ to (a) and (e) gives a white
precipitate of AgCl(s) with $NH_4Cl(aq)$, and a brown precipitate of $Ag_2O(s)$ with $(NH_4)_2CO_3(aq)$. **69.** (a) $BaSO_4(s)$,
(b) $CaSO_4(s)$. **70.** Li (red), Na (yellow), K (lilac), Rb
(purple), Ca (orange-red), Sr (deep red), Ba (pale green).
72. (a) $0.10M$ NaCl(aq). (b) The decrease in solubility of
a salt in the presence of a common ion. (c) The presence

of a common ion shifts the position of equilibrium to minimize as far as is possible the concentration of the common
ion, which results in a decrease in solubility of the salt.
73. 1.45×10^{-3} g.
75. (a) $CaO(s) + H_2O(l) \longrightarrow Ca(OH)_2(s)$; calcium
hydroxide
(b) $CaO(s) + CO_2(g) \longrightarrow CaCO_3(s)$; calcium carbonate
(c) $CaO(s) + SO_2(g) \longrightarrow CaSO_3(s)$; calcium sulfite
(d) $CaO(s) + SiO_2(s) \longrightarrow CaSiO_3(s)$; calcium silicate
(e) $6CaO(s) + P_4O_{10}(s) \longrightarrow 2Ca_3(PO_4)_2$; calcium
phosphate
77. (a) $PbCl_2(s)$, $2.1 \times 10^{-3} M$ (water), $4.3 \times 10^{-6} M$ ($2M$
HCl(aq)); AgCl(s), $1.3 \times 10^{-5} M$ (water), $9.0 \times 10^{-11} M$ ($2M$
HCl(aq)); $Hg_2Cl_2(s)$, $5.5 \times 10^{-10} M$ (water), $3.0 \times 10^{-19} M$
($2M$ HCl(aq)). (b) The solubilities are reduced to very
small values in $2M$ HCl(aq), due to the common ion effect,
which ensures the very efficient removal of these cations
from solution under these conditions.

CHAPTER 16

1. (a) Water, (b) shuffled cards, (c) the collection, (d)
the solution. **3.** (a) $\Delta S > 0$, (b) $\Delta S < 0$, (c) $\Delta S < 0$.
5. (a) When $\Delta S_{system} > 0$, or $\Delta S_{system} < \Delta S_{surroundings}$ if
$\Delta S_{system} < 0$. (b) $\Delta H_{system} < 0$, or $\Delta S_{surroundings} < \Delta S_{system}$ if
$\Delta H_{system} > 0$. **7.** (a) $\Delta S > 0$, (b) $\Delta S < 0$, (c) $\Delta S < 0$,
(d) $\Delta S < 0$, (e) $\Delta S > 0$. **9.** (a) 2.9 J K^{-1} mol^{-1},
(b) -138.3 J K^{-1} mol^{-1}, (c) 289.8 J K^{-1} mol^{-1},
(d) 166.3 J K^{-1} mol^{-1}. **11.** (a) 11.1 J K^{-1} mol^{-1},
(b) 24.7 J K^{-1} mol^{-1}, (c) -835.1 J K^{-1} mol^{-1},
(d) -636.8 J K^{-1} mol^{-1}, (e) -326.2 J K^{-1} mol^{-1}.
13. $\Delta H° = 493$ kJ, $\Delta G° = 330.6$ kJ, $\Delta S° = 542.6$ J K^{-1}
mol^{-1}. The reverse reaction is spontaneous. For the forward
reaction $\Delta S° > 0$, which favors its spontaneity, but $\Delta H° > 0$,
which works against the spontaneity of the forward reaction
and dominates, and so it is the reverse reaction that is spontaneous. **15.** (a) -1302.6 kJ, spontaneous;
(b) 4.8 kJ, not spontaneous; (c) -18.5 kJ, spontaneous.
16. $NO_2(g)$. **18.** (a) Positive, (b) negative,
(c) negative, (d) positive. **20.** (a) 324 kJ, (b) -70.4
kJ, (c) -432.6 kJ, (d) 839.8 kJ. The data show that, in
order of oxidizing power, (a) $F_2(g) > Cl_2(g)$, and (b) $Cl_2(g) >$
$Br_2(l)$, and in order of reducing power, (c) Zn(s) > Pb(s), and
(d) Al(s) > Fe(s). **22.** 7.7×10^{24} atm^{-1}; spontaneous.
24. 3.1×10^{-7} atm. **26.** (a) $\Delta G° = -800.8$ kJ, spontaneous; (b) 10^{140}; (c) the reaction is very slow at room
temperature and proceeds to equilibrium at a negligible rate
unless initiated by a spark, whereupon combustion occurs
rapidly at high temperature. **28.** (a) $\Delta G > 0$, so K_p will
be very small, so that only a small concentration of $C_2H_2(g)$
will be present at equilibrium; no; (b) yes, T > 3891 K;
(c) 1.47×10^{-7}. **30.** (a) -52.0 kJ, spontaneous, $K_p =$
1.3×10^9 atm^{-8}; (b) 159.1 kJ, not spontaneous,
$K_p = 1.6 \times 10^{-28}$. **32.** (a) $\Delta G° = 1404.8$ kJ,
(b) $\Delta G° = -788.8$ kJ, and, for the coupled reaction,
$\Delta G° = 616.0$ kJ, not spontaneous. **35.** $Al_2O_3(s)$ cannot
be reduced to Al(s) by C(s); FeO(s) is reduced in the reaction
in which $CO_2(g)$ is formed; PbO(s) and CuO(s) are reduced
by carbon under all circumstances. **36.** (a) $\Delta G°$ is

negative only for the second reaction, which should be spontaneous at room temperature, although very slow. (b) A catalyst speeds up the reaction but does not affect the position of equilibrium. **38.** (a) 104 kJ, -316 kJ; (b) in each reaction 3 mol of gases are replaced by 1 mol of gases, so that for each $\Delta S° < 0$; (c) yes. The first reaction is not spontaneous under any conditions, but the second is spontaneous under most conditions. **40.** (a) $K_p = 1.3 \times 10^5$ atm^{-3}, (b) $K_p = 1.6 \times 10^5$ atm^{-3}, (c) yes. **42.** (a) $K_p = p_{H_2O(g)}$; (b) 23.6 mm Hg (cf. 23.8 mm Hg). **44.** (a) -300.1 kJ mol^{-1}, -70 kJ mol^{-1}, -371.1 kJ mol^{-1}; (b) yes; (c) yes; (d) SO_3 should predominate; (e) SO_3 is an important precursor of acid rain, due to the reaction $SO_3(g) + H_2O(l) \longrightarrow H_2SO_4(l)$.

CHAPTER 17

1. (a) (i) A process in which electrical energy is used to bring about a chemical change. (ii) A substance which when melted or dissolved in a solvent can *conduct* electricity. (iii) A substance which when melted or dissolved in a solvent is a *nonconductor* of electricity. (b) In *ionic conduction* the current is carried by ions; in *electronic conduction* the current is carried by electrons.

3. (a)

$2[Al^{3+} + 3e^- \longrightarrow Al(l)]$ *cathode*
$3[2Cl^- \longrightarrow Cl_2(g) + 2e^-]$ *anode*
$2AlCl_3(l) \longrightarrow 2Al(l) + 3Cl_2(g)$ *overall*

(b) In an aqueous solution of $AlCl_3(aq)$, water is more easily reduced than $Al^{3+}(aq)$, and in relatively concentrated solutions $Cl^-(aq)$ is more readily oxidized than $H_2O(l)$.

$2H_2O(l) + 2e^- \longrightarrow H_2(g) + 2OH^-(aq)$ *cathode*
$2Cl^-(aq) \longrightarrow Cl_2(g) + 2e^-$ *anode*
$2Cl^-(aq) + 2H_2O \longrightarrow Cl_2(g) + 2OH^-(g)$ *overall*

5. In the diagram, electrons move from the anode to the cathode in the external circuit. In the HBr(aq) solution, $Br^-(aq)$ ions move to the *anode* and $H_3O^+(aq)$ ions move to the *cathode*, where the following reactions occur.

$2H_3O^+(aq) + 2e^- \longrightarrow H_2(g) + 2H_2O(l)$ *cathode*
$2Br^-(aq) \longrightarrow Br_2(l) + 2e^-$ *anode*
$2HBr(aq) \longrightarrow H_2(g) + Br_2(l)$ *overall*

7. (a) 1.93×10^5 C, (b) 9.65×10^4 C, (c) 4.83×10^5 C, (d) 5.79×10^5 C. **9.** (a) 861 C, (b) 5.37×10^8 C, (c) 9.63×10^4 C, (d) 4.47×10^3 C. **11.** 0.896 g NaOH, 0.794 g Cl_2. **13.** 39.7 min. **15.** 3.05 days. **17.** 9.648×10^4 C. **19.** 726.9 g Al min^{-1}, 1.047×10^5 kg Al day^{-1}, 3.495×10^4 kg C day^{-1}. **21.** 1.758 g Cu. **23.** pH $= 13.15$. **25.** (a) 89.4 mA, (b) 0.360 g Ag(s), (c) 62.5 mL. **27.** (c), (d), (e), (g), and (h). **29.** Yes, but the rate of the reaction is very slow. **31.** Mg, Al, Zn, Fe, Cu. **33.** All are oxidized under standard conditions in acid solution. **35.** (a) and (c), no reaction; (b) $O_2 + 4Br^- + 4H^+ \longrightarrow 2H_2O + Br_2$, $E°_{cell} = +0.14$ V. **37.** (a) and (d). **39.** (b) $Mg + Fe^{2+} \longrightarrow Mg^{2+} + Fe$, $E°_{cell} = +1.92$ V; (d) $2Fe^{3+} + Fe \longrightarrow 3Fe^{2+}$, $E°_{cell} = +1.21$ V. **41.** (a) $Pb^{2+}(aq)$, (b) Fe(s). **43.** The ease of oxidation is Ca > Mg > Zn > Cu, which is confirmed, for example, by the following: (1) only Ca reacts with water at an observable

A-38

rate at room temperature, while Ca and Mg react with steam; (2) Ca, Mg, and Zn are oxidized by $H_3O^+(aq)$ in, for example, HCl(aq); (3) $Cu^{2+}(aq)$ is displaced from aqueous solution as Cu(s) by each of the other metals; $Zn^{2+}(aq)$ is displaced by Ca and Mg, and $Mg^{2+}(aq)$ is displaced only by Ca. **44.** (a) $Pb(s) + Cu^{2+}(aq) \longrightarrow Cu(s) + Pb^{2+}(aq)$, $E°_{cell} = +0.47$ V, and the bronze-like finish is due to the deposition of Cu(s). (b) $2Ag^+(aq) + Cu(s) \longrightarrow 2Ag(s) + Cu^{2+}(aq)$, $E°_{cell} = +0.46$ V, and the bronze-like finish is replaced by silver.

46. $Zn(s)|Zn^{2+}(aq)||Ag^+(aq)|Ag(s)$, in which Zn is the anode and Ag the cathode. Electrons move externally from Zn to Ag, Ag^+ ions move to the cathode, and anions (unspecified) move to the anode.

Anode $Zn \longrightarrow Zn^{2+} + 2e^-$ $E°_{ox} = +0.76$ V
Cathode $2[Ag^+ + e^- \longrightarrow Ag]$ $E°_{red} = +0.80$ V
    $Zn + 2Ag^+ \longrightarrow Zn^{2+} + 2Ag$ $E°_{cell} = +1.56$ V

48. (a)
$2[Al \longrightarrow Al^{3+} + 3e^-]$ $E°_{ox} = +1.66$ V
$3[Cu^{2+} + 2e^- \longrightarrow Cu]$ $E°_{red} = +0.34$ V
$2Al + 3Cu^{2+} \longrightarrow 2Al^{3+} + 3Cu$ $E°_{cell} = +2.00$ V

(b) $Pb \longrightarrow Pb^{2+} + 2e^-$ $E°_{ox} = +0.13$ V
$2[Ag^+ + e^- \longrightarrow Ag]$ $E°_{red} = +0.80$ V
$Pb + 2Ag^+ \longrightarrow Pb^{2+} + 2Ag$ $E°_{cell} = +0.93$ V

(c) $2[Ag \longrightarrow Ag^+ + e^-]$ $E°_{ox} = -0.80$ V
$Cl_2 + 2e^- \longrightarrow 2Cl^-$ $E°_{red} = +1.36$ V
$2Ag + Cl_2 \longrightarrow 2Ag^+ + 2Cl^-$ $E°_{cell} = +0.56$ V

50. Oxidation occurs at the anode on the left and reduction at the cathode on the right, and in each case the electron flow is from anode to cathode in the external circuit:
(a) $Zn(s)|Zn^{2+}(aq)||Br^-(aq)|Br_2(g), Pt(s)$
(b) $Pb(s)|Pb^{2+}(aq)||Ag^+(aq)|Ag(s)$
(c) $Pt(s)|Cu^+(aq), Cu^{2+}(aq)||Fe^{3+}(aq), Fe^{2+}(aq)|Pt(s)$
52. (a) Cathode, $MnO_4^- + 5e^- + 8H^+ \longrightarrow Mn^{2+} + 4H_2O$, reduction; anode, $Fe^{2+} \longrightarrow Fe^{3+} + e^-$, oxidation. (b) From anode (where oxidation occurs) to cathode (where reduction occurs). (c) $Fe^{2+}(aq)$ moves to the cathode and $MnO_4^-(aq)$ to the anode. (d) 0.72 V. **54.** (a) 0.46 V, (b) 0.43 V. **56.** 2.75×10^{-5} mol L^{-1}. **58.** (a) Fe(s) dissolves and Cu(s) increases in mass; (b) 0.78 V; (c) $FeSO_4(aq)$ increases in concentration; (d) decrease. **60.** (a) 0.47 V, (b) 0.32 V, (c) 0.50 V, (d) 0.54 V. **62.** (a) 0.38 V, (b) $K_c = [Fe^{2+}]^2[Cu^{2+}]/[Fe^{3+}]^2$, (c) 3.5×10^{14} mol L^{-1}. **64.** pH $= 9.10$, $K_b(CH_3CO_2^-) = x^2/(C_s - x)$ where x is $[OH^-]$ and C_s is the concentration of sodium acetate at the equivalence point, which can be calculated from the initial volume and concentration of acid and the volume of NaOH(aq) added to achieve the equivalence point. $[OH^-]$ can be found from the pH and $K_a(CH_3CO_2H)$ from the expression $K_a(CH_3CO_2H) \times K_b(CH_3CO_2^-) = K_w$. **66.** 5.94×10^7 mol^{-2} L^2. **68.** (a) E_{cell} increases; (b) E_{cell} decreases; (c) no change; (d) no change; (e) E_{cell} decreases; (f) E_{cell} decreases. **70.** (a) -145 kJ, (b) -143 kJ, (c) -284 kJ, (d) -79.1 kJ. **72.** (a) Anode, $Zn(s) \longrightarrow Zn^{2+} + 2e^-$; *cathode*, $O_2(g) + 4H^+(aq) + 4e^- \longrightarrow 2H_2O(l)$, (b) 1.99 V. (c) 12 years.

1. (a) HNO_3, (b) HNO_2, (c) KNO_2, (d) NO, (e) N_2O_5, (f) N_2H_4, (g) NaN_3. **3.** (a) Sodium nitrate, (b) potassium nitrite, (c) dinitrogen tetraoxide, (d) hydrazoic acid, (e) barium peroxide, (f) potassium superoxide, (g) hydrazine, (h) lithium azide, (i) lithium nitride. **5.** (a) NH_3, -3; (b) N_2H_4, -2; (c) N_2, 0; (d) HNO_3, $+5$; (e) N_2O, $+1$; (f) NO, $+2$; (g) HNO_2, $+3$; (h) N_2O_4, $+4$.

7. (a) $2KNO_3 \longrightarrow 2KNO_2 + O_2$; $+3$
(b) $NH_4NO_3 \longrightarrow N_2O + 2H_2O$; $+1$
(c) $2Pb(NO_3)_2 \longrightarrow 2PbO + 4NO_2 + O_2$; $+4$

9. (a) $3Cu(s) + 8HNO_3(aq) \longrightarrow$
$3Cu(NO_3)_2(aq) + 2NO(g) + 4H_2O(l)$
$HNO_3(aq) + NH_3(g) \longrightarrow NH_4NO_3(aq)$
$2HNO_3(aq) + Mg(OH)_2(s) \longrightarrow Mg(NO_3)_2(aq) + 2H_2O(l)$
(b) $Cu(s) + 4HNO_3(aq) \longrightarrow Cu(NO_3)_2(aq) +$
$2NO_2(g) + 2H_2O(l)$
$Mg(OH)_2(s) + 2HNO_3(aq) \longrightarrow Mg(NO_3)_2(aq) + 2H_2O(l)$
(c) $P_4O_{10}(s) + 4HNO_3(l) \longrightarrow 4HPO_3(s) + 2N_2O_5(g)$
$2H_2SO_4(l) + HNO_3(l) \longrightarrow NO_2^+(soln) + H_3O^+(soln) + 2HSO_4^-(soln)$

11. The white precipitate is $BaSO_4(s)$ formed from $Ba^{2+}(aq)$ ions and $SO_4^{2-}(aq)$ ions that result from the oxidation of $SO_2(g)$ by $NO_3^-(aq)$:

$$3[SO_2 + 2H_2O \longrightarrow SO_4^{2-} + 2e^- + 4H^+] \quad \text{oxidation}$$
$$\underline{2[NO_3^- + 3e^- + 4H^+ \longrightarrow NO + 2H_2O] \quad \text{reduction}}$$
$$2SO_2 + 2NO_3^- + 2H_2O \longrightarrow 3SO_4^{2-} + 2NO + 4H^+$$

13. (a) $3Cu(s) + 8HNO_3(aq) \longrightarrow 3Cu(NO_3)_2(aq) + 2NO(g) + 4H_2O(l)$; (b) $Cu(s) + 4HNO_3(aq) \longrightarrow Cu(NO_3)_2(aq) + 2NO_2(g) + 2H_2O(l)$. The first reaction uses dilute aqueous acid and the second uses concentrated aqueous acid.

15. (a) $NH_4NO_3(l) \xrightarrow{\text{Gentle heat}} N_2O(g) + 2H_2O(g)$
(b) $3Cu(s) + 8HNO_3(\text{dil, aq}) \longrightarrow$
$3Cu(NO_3)_2(aq) + 2NO(g) + 4H_2O(l)$
(c) $2Pb(NO_3)_2(s) \xrightarrow{\text{Heat}} 2PbO(s) + 4NO_2(g) + O_2(g)$
and $N_2O_4(s)$ results when the $NO_2(g)$ is condensed in a trap immersed in liquid nitrogen. **17.** $3N_2O_4(l) + 2H_2O(l) \longrightarrow 4HNO_3(aq) + 2NO(g); N_2O_4(l) + H_2O(l) \longrightarrow HNO_2(aq) + HNO_3(aq)$.

19. (a) $2HgO(s) \longrightarrow 2Hg(l) + O_2(g)$;
(b) $2PbO_2(s) \longrightarrow 2PbO(s) + O_2(g)$;
(c) $2BaO_2(s) \longrightarrow 2BaO(s) + O_2(g)$;
(d) $2KNO_3(s) \longrightarrow 2KNO_2(s) + O_2(g)$.

21. (a) $8H_2S + 8NO_2 \longrightarrow 8NO + 8H_2O + S_8$, oxidizing agent; (b) $NO_2 + 3I^- + H_2O \longrightarrow NO + I_3^- + 2OH^-$, oxidizing agent; (c) $MnO_4^- + 5NO_2 + 3H_2O \longrightarrow Mn^{2+} + 2H_3O^+ + 5NO_3^-$, reducing agent; (d) $2NO_2 + H_2O \longrightarrow HNO_2 + HNO_3$, both as an oxidizing agent and a reducing agent. **23.** 0.0727 mol L^{-1}.

25. (a)

(two resonance structures)

(b)

(three resonance structures)

(c) $H-\overset{..}{\underset{..}{O}}-N=\overset{..}{\underset{..}{O}}:$

(d) $:\overset{..}{\underset{..}{O}}-\overset{..}{N}=\overset{..}{\underset{..}{O}}:$ (two resonance structures)

(e) $^{1/2+}:\overset{..}{O}=\overset{..}{N}:^{1/2-}$

(f)

(four resonance structures)

(g)

(four resonance structures)

26. (a) NO_2, AX_2E, angular; (b) NO_2^+, AX_2, linear; (c) NO_3^-, AX_3, trigonal planar; (d) $HONO_2$, AX_3, trigonal planar; (e) N_2O, AX_2, linear; (f) $ClNO$, AX_2E, angular; (g) NF_3, AX_3E, trigonal pyramid; (h) $HONO$, AX_2E, angular; (i) CH_3NO_2, AX_3 trigonal planar.

28. (a), (b), (d), (e), (g), (h), (i), (j). **29.** N_2O, NO_2^+, N_3^-. **32.** (a) 2.46, (b) 7.72, (c) 3.14 (buffer solution). **34.** 303 kJ. **35.** -225.9 kJ. **37.** (a) -1.4 kJ, (b) yes, (c) 1.8 atm, (d) 15 °C. **39.** (a) -3, (b) -2, (c) $+3$, (d) $+4$, (e) -1, (f) $+1$, (g) -3 and $+5$, (h) $-\frac{1}{3}$, (i) -3.

41. (a) $NO_3^- + e^- + 2H^+ \longrightarrow NO_2 + H_2O$;
(b) $NO_3^- + 3e^- + 4H^+ \longrightarrow NO + 2H_2O$;
(c) $2NO_3^- + 10e^- + 12H^+ \longrightarrow N_2 + 6H_2O$;
(d) $NO_3^- + 8e^- + 10H^+ \longrightarrow NH_4^+ + 3H_2O$.

42. Oxygen in H_2O_2 is in the -1 oxidation state and can be oxidized to $O_2(g)$ or reduced to H_2O (oxidation states 0 and -2, respectively).
(a) $SO_2(g) + H_2O_2(l) + 2H_2O(l) \longrightarrow SO_4^{2-}(aq) + 2H_3O^+(aq)$
(b) $O_3(g) + 3H_2O_2(aq) \longrightarrow 3O_2(g) + 3H_2O(l)$
(c) $2I^-(aq) + H_2O_2(aq) + 2H_3O^+(aq) \longrightarrow I_2(s) + 4H_2O(l)$
(d) $2Cr(OH)_3(s) + 3H_2O_2(aq) + 4OH^-(aq) \longrightarrow 2CrO_4^{2-}(aq) + 8H_2O(l)$
(e) $NO_2^-(aq) + H_2O_2(aq) \longrightarrow NO_3^-(aq) + H_2O(l)$
44. (a), (b), (c), and (d).

46. (a) $BaO_2(s) + H_2SO_4(aq) \xrightarrow{\text{Cold}} BaSO_4(s) + H_2O_2(aq)$
(b) (i) $PbS(s) + 4H_2O_2(aq) \longrightarrow PbSO_4(s) + 4H_2O$, oxidizing agent
(ii) $2Fe^{2+}(aq) + H_2O_2(aq) + 2H_3O^+(aq) \longrightarrow 2Fe^{3+}(aq) + 4H_2O(l)$, oxidizing agent
(iii) $3I^-(aq) + H_2O_2(aq) + 2H_3O^+(aq) \longrightarrow I_3^-(aq) + 4H_2O(l)$, oxidizing agent
(iv) $HOCl(aq) + H_2O_2(aq) \longrightarrow HCl(aq) + O_2(g) + H_2O(l)$, reducing agent
(v) $2MnO_4^-(aq) + 5H_2O_2(aq) + 6H_3O^+(aq) \longrightarrow 2Mn^{2+}(aq) + 5O_2(g) + 14H_2O(l)$, reducing agent
48. $4NH_3(g) + 5O_2(g) \longrightarrow 4NO(g) + 6H_2O(g)$;
$2NO(g) + O_2(g) \longrightarrow 2NO_2(g)$ (brown fumes)

50.

Species	Valence Electrons	Lewis Structure	Bond Order	Bond Length
O_2	12	$:\ddot{O}=\ddot{O}:$	2.0	121
O_3	18	$:\overset{+}{\ddot{O}}=\overset{\cdot\cdot}{O}-\overset{\cdot\cdot}{\ddot{O}}:^-$	1.5	128
O_2^-	13	$^{1/2-}:\overset{\cdot\cdot}{O}-\overset{\cdot\cdot}{O}:^{1/2-}$	1.5	128
O_2^{2-}	14	$^-:\overset{\cdot\cdot}{O}-\overset{\cdot\cdot}{O}:^-$	1.0	149

As expected, bond length increases with decreasing bond order.

CHAPTER 19

1. Average rate (mol L^{-1} min^{-1}): 2.4×10^{-3}, 1.6×10^{-3}, 1.4×10^{-3}, 0.92×10^{-3}.

3. (a) $\text{Rate} = -\dfrac{\Delta[HI]}{\Delta t} = \dfrac{2\Delta[H_2]}{\Delta t} = \dfrac{2\Delta[I_2]}{t}$

(b) $\text{Rate} = -\dfrac{\Delta[NOCl]}{\Delta t} = \dfrac{\Delta[NO]}{\Delta t} = \dfrac{2\Delta[Cl_2]}{\Delta t}$

(c) $\text{Rate} = -\dfrac{\Delta[N_2O_5]}{\Delta t} = \dfrac{1}{2}\dfrac{\Delta[NO_2]}{\Delta t} = \dfrac{2\Delta[O_2]}{\Delta t}$

5. (a) 0.81 mol L^{-1} s^{-1}, (b) -0.54 mol L^{-1} s^{-1}, (c) -0.41 mol L^{-1} s^{-1}. **7.** 2.92×10^{-2} mol L^{-1} s^{-1}, 3.51×10^{-2} mol L^{-1} s^{-1}. **9.** $\text{Rate} = k[A][B][C]^2$ mol^{-3} L^3 time^{-1}. **11.** (a) 4.0×10^{-9} mol L^{-1} s^{-1}, (b) 5.0×10^{-7} mol L^{-1} s^{-1}. **13.** First order. **15.** 1.5, rate $= k[CH_3CHO]^{3/2}$. **17.** 1.0 in H_2 and 0.5 in Br_2, rate $= k[H_2][Br_2]^{1/2}$, no. **19.** 4.0 min, 20 min. **21.** $t_{1/2} = 13.9$ min; $k = 0.050$ min^{-1}. **23.** 200 years. **25.** Second order in NO, first order in O_2; $k = 7.0 \times 10^3$ mol^{-2} L^2 s^{-1}. **26.** (a) $BrO_3^- + 5Br^- + 6H_3O^+ \longrightarrow 3Br_2 + 9H_2O$. (b) $\text{Rate} = k[BrO_3^-][Br^-][H_3O^+]^2$. **28.** $E_a = 186$ kJ mol^{-1}, $k_{400\,°C} = 3.41 \times 10^{-4}$ mol^{-1} L s^{-1}. **29.** $\Delta H° = +9.6$ kJ mol^{-1}, $E_a = 176$ kJ mol^{-1}. **31.** 19 °C. **33.** 35.2 kJ mol^{-1}. **35.** 108 kJ mol^{-1}. **36.** 5.2×10^{-3} mol^{-1} L s^{-1}, 7.9×10^{-3} mol^{-1} L s^{-1}. **38.** 133 kJ mol^{-1}. **41.** $\text{Rate} = k[Cl_2]^{3/2}[CO]$; COCl and Cl are reaction intermediates. **43.** $\text{Rate} = k[O_3]^2[O_2]^{-1}$; O_2 is produced in the first step and any decrease in $[O_2]$ shifts the equilibrium to the left, thus decreasing $[O]$, which is a reactant in the rate-determining step; a decrease in its concentration decreases the reaction rate. **45.** Visible light contains photons with energies in the range 171–299 kJ mol^{-1}, insufficient to break the bonds in H_2 molecules. **47.** (a) First order in H_2O_2, first order in I^-, and zero order in H_3O^+; (b) no effect. **48.** (a) An increase; (b) decreases E_a; (c) none; (d) lowers that temperature; (e) no effect. **50.** Yes; since it lowers the activation energy of the slow step, it must be a reactant in that step. **53.** Each step must be thermodynamically favorable, with $E° > 0$; for example $2Fe^{3+} + 2I^- \longrightarrow 2Fe^{2+} + I_2$, $E° = +0.23$ V $2Fe^{2+} + S_2O_8^{2-} \longrightarrow 2Fe^{3+} + 2SO_4^{2-}$, $E° = +1.24$ V but the $E° > 0$ condition is not fulfilled for Cr^{3+}.

55. 480 nm light has an energy of 249 kJ mol^{-1}, sufficient to break the Cl—Cl bond in Cl_2 but not the C—H bond in CH_4, suggesting the mechanism

(a) $Cl_2 \xrightarrow{h\nu} 2Cl$ chain initiation

(b) $Cl + CH_4 \longrightarrow CH_3 + HCl$ ⎫
$CH_3 + Cl_2 \longrightarrow CH_3Cl + Cl$ ⎬ chain propagation

(c) $2Cl \longrightarrow Cl_2$ chain termination
$CH_3 + CH_3 \longrightarrow C_2H_6$ chain termination

57. (a) Not all molecular collisions have a favorable *steric orientation* or *sufficient energy* to overcome the activation energy barrier. (b) By establishing the experimental rate law. (c) *Increase in temperature* increases the proportion of molecular collisions with sufficient energy to exceed the activation energy E_a; a *catalyst* provides a different pathway a lower E_a. **59.** (i) Yes, (ii) no.

CHAPTER 20

2. (a) AX_5, trigonal bipyramidal; AX_6, octahedral. (b) For example, AX_5, PCl_5, PF_5; AX_6, SF_6, PCl_6^-, or SeF_6. (c) AX_4E, disphenoidal; AX_3E_2, T-shaped; AX_4E_2, square planar; AX_2E_3, linear. **4.** (a) Third period elements have d orbitals in addition to s and p. (b) (i) PF_3, AX_3E, trigonal pyramidal, and PF_5, AX_5, trigonal bipyramidal; (ii) SF_2, AX_2E_2, angular; SF_4, AX_4E, disphenoidal; SF_6, AX_6, octahedral.

6.

AX_3E, trigonal pyramidal AX_4, tetrahedral

8. (b), (c), (d). **10.** (a) AX_5E, square pyramidal; (b) AX_4E, disphenoidal; (c) AX_5, trigonal pyramidal. Oxidation states: $+6$, $+6$, $+8$. **11.** ClF and BrF, AXE_3, linear; ClF_3, BrF_3, and IF_3, AX_3E_2, T-shaped; ClF_5, BrF_5, and IF_5, AX_5E, square pyramidal; IF_7, AX_6E, distorted octahedral; all with dipole moments. **13.** $HClO_4 >$ $H_2SO_4 > H_3PO_4 > HOCl$, explained in terms of their $(HO)_nXO_m$ structures.

15. $XeF_2 + BrO_3^- + 2OH^- \longrightarrow BrO_4^- + Xe + 2F^- + H_2O$; BrO_4^-, AX_4, tetrahedral. **17.** $4KClO_3 \xrightarrow{\text{Heat}} 3KClO_4 + KCl$; $2KClO_3 \xrightarrow[\text{MnO}_2]{\text{Heat}} 2KCl + 3O_2$.

19. $Cl_2(g) + 2H_2O(l) \rightleftharpoons$ $HOCl(aq) + H_3O^+(aq) + Cl^-(aq)$. The equilibrium can be shifted to the right by precipitating the $Cl^-(aq)$ as $AgCl(s)$, $[Ag_2O(aq) + 2Cl^-(aq) + 2H_3O^+(aq) \longrightarrow$ $2AgCl(s) + 3H_2O(l)]$. Attempts to concentrate the solution by evaporation lead to the decomposition of the $HOCl(aq)$, $[2HOCl(aq) + 2H_2O(l) \longrightarrow$ $2H_3O^+(aq) + 2Cl^-(aq) + O_2(g)]$. **21.** (a) Hypobromous acid, (b) calcium hypochlorite, (c) potassium bromate, (d) potassium chlorite, (e) magnesium perchlorate, (f) iodic acid, (g) perbromic acid, (h) paraperiodic acid. **23.** $ClO_3^- + 6H_3O^+ + 6e^- \longrightarrow Cl^- + 9H_2O$; $ClO_2^- + 4H_3O^+ + 4e^- \longrightarrow Cl^- + 6H_2O$; $ClO^- + 2H_3O^+ + 2e^- \longrightarrow Cl^- + 3H_2O$.

26. ICl_3, I_2Cl_6,

$$\text{:} \overset{..}{\underset{..}{Cl}} \quad \overset{+}{\underset{..}{\overset{..}{Cl}}} \quad \overset{..}{\underset{..}{Cl}} \text{:}$$

(structure showing I_2Cl_6 bridged)

27. (a) $+5$, (b) $+3$, (c) $+7$, (d) $+3$, (e) $+4$, (f) $+7$. **29.** A disproportionation reaction is a redox reaction in which a single species behaves both as the oxidizing agent and the reducing agent. For example:

$Cl_2 + 2H_2O \longrightarrow HOCl + H_3O^+ + Cl^-$;

$4KClO_3 \longrightarrow KCl + 3KClO_4$;

$6XeF_4 + 12H_2O \longrightarrow 2XeO_3 + 4Xe + 3O_2 + 24HF$.

32. $2KIO_3(s) \xrightarrow{\text{Heat}} 2KI(s) + 3O_2(g)$.

33. (a) 6.25 g NaOCl, (b) 5.61 g $KClO_3$.

34. (a) H_2SO_4(conc) oxidizes Br^- to Br_2(l) which is soluble in tetrachloroethane to give a brown solution:

$2HBr + H_2SO_4 \longrightarrow Br_2 + SO_2 + 2H_2O$. (b) H_2S is oxidized by potassium bromate to give S(s) and potassium bromide: $KBrO_3 + 3H_2S \longrightarrow 3S + Br^- + K^+ + 3H_2O$. The acidified filtrate contains Br^-(aq) which reacts with Ag^+(aq) to give insoluble yellow AgBr(s), which dissolves in NH_3(aq) due to the formation of the complex ion $Ag(NH_3)_2^+$(aq): $Br^- + Ag^+ \longrightarrow AgBr(s)$;

$AgBr(s) + 2NH_3(aq) \longrightarrow Ag(NH_3)_2^+(aq) + Br^-(aq)$.

(c) Potassium bromate decomposes to give potassium bromide and oxygen, which reignites a glowing splint:

$2KBrO_3(s) \xrightarrow{\text{Heat}} 2KBr(s) + 3O_2(g)$. (d) Bromate ion oxidizes bromide ion to bromine, which is absorbed in the organic layer to give a brown solution:

$KBrO_3 + 5KBr + 6H_3O^+ \longrightarrow 3Br_2 + 6K^+ + 9H_2O$.

35. (a) 1.33×10^9, (b) 5.17×10^{27}, (c) 3.88×10^{18}.

37. (a) Acidity decreases, ClOH > BrOH > IOH, with decreasing electronegativity of the halogen and decreasing polarity of the O—H bond. (b) 4.20, 4.81, 5.82.

39. (a) $3ClO_3^- + 3H_2SO_4 \longrightarrow$

$ClO_4^- + 2ClO_2 + H_3O^+ + 3HSO_4^-$

(b) $2ClO_3^- + 2H_2C_2O_4 \longrightarrow$

$2ClO_2 + 2H_2O + 2CO_2 + C_2O_4^{2-}$

(c) $2AgClO_3 + Cl_2 \longrightarrow 2AgCl + 2ClO_2 + O_2$

41. Bond energies (kJ mol^{-1}): ClF, 252.9; BrF, 249.3; BrF_5, 187.6; IF_7, 230.9; ClF_3, 173.6; BrF_3, 202.0; IF_5, 268.4.

43. $^{\delta+}H—X^{\delta-}$, $\delta+ = \delta-$; HF, 0.43e; HCl, 0.17e; HBr, 0.12e; HI, 0.05e. **45.** (a) Enthalpy of vaporization increases with increasing size, from He to Rn, with increasing polarizability and increasing strength of the induced dipole–induced dipole (London) forces. (b) London forces between polar water molecules and noble gas atoms increase with increasing size of the noble gas atoms (from He to Rn).

47. Formation of neon compounds requires the expansion of the valence shell beyond the octet, which is not possible for this second-period element which has no low lying d orbitals, unlike Kr and Xe.

49. 142 kJ mol^{-1}. **51.** (a) 9.87%, (b) 2.24×10^{-3} L.

CHAPTER 21

$[Ar]3d^7$; Au^{3+}; $[Xe]4f^{14}5d^8$; Fe^{3+}; $[Ar]3d^5$.

3. Cu^{2+}, 9; Fe^{3+}, 5; Mn^{2+}, 5; Ag^+, 10; V^{3+}, 2; Ti^{2+}, 2.

5. (a) $+3$, (b) $+3$, (c) $+3$, (d) $+1$, (e) $+2$, (f) $+3$. **7.** (a) $+6$, (b) $+6$, (c) $+7$, (d) $+5$, (e) $+4$,

7. (a) $+6$, (b) $+6$, (c) $+7$, (d) $+5$, (e) $+4$, (f) $+6$. **9.** (a) NH_3, 0; H_2O, 0; Cl^-, -1. (b) $+2$. (c) -1. (d) Octahedral. **11.** (a) Tetraamminezinc(II) chloride; (b) trichlorotriamminecobalt(III); (c) chloropentamminecobalt(III) chloride; (d) potassium tetrachloroferrate(II).

13. (a) trans-Dichlorotetramminechromium(III) cation; (b) trioxalatocobaltate(III) anion; (c) Tetrabromooxalatochromate(III) anion; (d) cis-dicyanobisethylenediamineplatinate(0) anion.

15. Red $[Co(NH_3)_5SO_4]Br$; violet $[Co(NH_3)_5Br]SO_4$.

17. EDTA is a chelating ligand that forms a very stable soluble complex ion with Ca^{2+}(aq).

19. A, $Co(NH_3)_4Cl_3$; B, $Co(NH_3)_4Cl_3$; C, $Co(NH_3)_5Cl_3$; D, $Co(NH_3)_6Cl_3$. A, $[Co(NH_3)_4Cl_2^+]Cl^-$, cis or trans; B, the same as A but the other isomer; C, $[Co(NH_3)_5Cl^{2+}](Cl^-)_2$; D, $[Co(NH_3)_6^{3+}](Cl^-)_3$.

21. (a) $Na_2[Zn(OH)_4]$, (b) $Na[Al(OH)_4]$, (c) $[PtBr(H_2O)_3]Cl$, (d) $FeCl_4^{2-}$, (e) $[Ni(en)_3]Br_2$, (f) $K_2[Co(CN)_4]$. **23.** No, colorless because the Zn in $Zn(NH_3)_4^{2+}$ has a filled d shell.

27. (a) A, $[Cr(H_2O)_4Cl_2]Cl \cdot 2H_2O$; B, $[Cr(H_2O)_5Cl]Cl_2 \cdot H_2O$; C, $[Cr(H_2O)_6]Cl_3$. (b) A, 0.538 g AgCl; B, 1.08 g AgCl; C, 1.61 g AgCl.

29. Because of the interaction between the ligand electron pairs and the electrons in the d orbitals all the d orbitals increase in energy but there is greater repulsion with the electrons in the $d_{x^2-y^2}$ and d_{z^2} orbitals than with the electrons in the d_{xy}, d_{xz}, and d_{yz} orbitals, because the latter point in directions between the ligands, while the former point directly at the ligands. Thus, the $d_{x^2-y^2}$ and d_{z^2} orbitals have a higher energy than the d_{xy}, d_{xz}, and d_{yz} orbitals.

31. (a) The complexes contain Ni^{2+} with a d^8 valence-shell configuration. In tetrahedral $Ni(Cl)_4^{2-}$ the interaction of the ligands with the d orbitals on Ni is weak and the shape is not distorted from AX_4 tetrahedral. In planar $Ni(CN)_4^{2-}$ the interaction of the d orbitals with the ligands is strong and distorting the tetrahedral shape to a square planar shape. (b) $NiCl_4^{2-}$ (c) 2 **33.** Acidity increases with increased oxidation state. Lower oxidation state oxides such as Cr(II)O and Cr(III)$_2$O$_3$ are ionic and are basic and amphoteric, respectively; higher oxidation state oxides such as Cr(VI)O$_3$ are covalent and acidic.

35. In going along the first transition metal series from Sc to Zn, the number of d electrons increases and core charge is expected also to increase. From both effects it would be predicted that the strength of the metallic bonding, and therefore properties such as the melting point, should increase along the series. However, as core charge increases, the 3d orbitals contract and are less able to participate in bonding. The balance of these two factors results in the strongest metallic bonding, and melting point, at Cr. The densities show no similar maximum but the sharp increase from Sc to Cr, which is followed by a much more gradual increase, is due to the same effect. **37.** Sc^{3+} has no d electrons, whereas Ti^{3+} and V^{3+} have 1 and 2, respectively, that can be promoted from lower to higher energy d orbitals by the absorption of visible light.

39. (a) $Cu(H_2O)_4^{2+}(aq)$,(b) $CuCl_4^{2-}(aq)$,
(c) $Cu(NH_3)_4^{2+}(aq)$, (d) $Cu(CN)_4^{2-}(aq)$.
41. $MgO(s) + 2H_3O^+(aq) \longrightarrow$
$Mg^{2+}(aq) + 3H_2O(l)$; *basic*
$SO_3(g) + H_2O(l) \longrightarrow H_2SO_4(aq)$; *acidic*
$2CrO_3(s) + H_2O(l) \longrightarrow H_2Cr_2O_7(aq)$; *acidic*
$ZnO(s) + 2H_3O^+(aq) \longrightarrow Zn^{2+}(aq) + 3H_2O(l)$; *basic*
$Cr_2O_3(s) + 6H_3O^+(aq) \longrightarrow$
$2Cr^{3+}(aq) + 9H_2O(l)$ }
$Cr^{3+}(aq) + 4OH^-(aq) \longrightarrow Cr(OH)_4^-(aq)$ } *amphoteric*
$TiO_2(s) + 2OH^-(aq) \longrightarrow TiO_3^{2-}(aq) + H_2O(l)$; *basic*
$V_2O_5(s) + 6OH^-(aq) \longrightarrow$
$2VO_4^{3-}(aq) + 3H_2O(l)$ }
$V_2O_5(s) + 2H_3O^+(aq) \longrightarrow$ } *amphoteric*
$2VO_2^+(aq) + 3H_2O(l)$ }
43. pH = 2.5.   **45.** (a) (i) $Zn(s) + Ni_2^+(aq) \longrightarrow$
$Zn_2^+(aq) + Ni(s)$
(ii) $Zn(s) + 2H_3O^+(aq) \longrightarrow Zn^{2+}(aq) + 2H_2O(l) + H_2(g)$
(b) (i) No reaction
(ii) $Ni(s) + 2H_3O^+(aq) \longrightarrow Ni^{2+}(aq) + 2H_2O(l) + H_2(g)$
47. Fe^{3+} forms a very stable complex with oxalate ion,
$Fe_2O_3(s) + 6H_3O^+(aq) + 6C_2C_4^{2-}(aq) \rightleftharpoons$
$2Fe(C_2O_4)_3^{3-}(aq) + 9H_2O(l)$.
49. CrO_2Cl_2 is similar to SO_2Cl_2:

$$:\overset{\displaystyle \ddot{O}:}{\underset{\displaystyle :\ddot{Cl}:}{:\ddot{O}-\overset{|}{\underset{|}{Cr}}-\ddot{Cl}:}} \quad AX_4, \text{ tetrahedral}$$

51. 0.617 L, 4.17 L.   **53.** 278 kg.
55. $K_{sp} = 2.3 \times 10^{-4}$ mol^4 L^{-4}.
57. (a) 1.6×10^{-5} mol L^{-1}, (b) 1.8×10^{-9} mol L^{-1}.
59. 54.7 mass % $FeSO_4$, x = 7.   **61.** 2.1×10^{-6} mol L^{-1}.
63. $(NH_4)_2TiCl_6 \cdot 2H_2O$; Lewis structure:

$$\left[\begin{matrix} H \\ H-\overset{+}{N}-H \\ H \end{matrix} \right]_2 \left[\begin{matrix} Cl \quad Cl \\ Cl-Ti-Cl \\ Cl \quad Cl \end{matrix} \right]^{2-} \left[\begin{matrix} H-O \\ H \end{matrix} \right]_2$$

65. V, Cr, Mṅ, Fe, Co, Ni, Zn.   **67.** (a) Cobalt(II) chloride hexahydrate containing the octahedral $[Co(H_2O)_6^{2+}]$ion; (b) hexamminecobalt(III) chloride, containing the octahedral $[Co(NH_3)_6^{3+}]$ion;
(c) chloropentamminecobalt(III) chloride, containing the octahedral $[Co(NH_3)_5Cl^{2+}]$ion; (d) *cis-* or *trans-*dichlorotetramminecobalt(III) chloride containing the octahedral $[Co(NH_3)_4Cl_2^+]$ion, with the Cl ligands either in *cis* or *trans* positions.

CHAPTER 22

1. (a) The SiO_4^{4-} tetrahedron; (b) Si is partially replaced by Al^-.   **3.** An amphibole contains two chains of SiO_4 tetrahedra joined by sharing oxygen atoms on alternate tetrahedra. Its empirical formula is $Si_4O_{11}^{4-}$.
5. See page 993 for a diagram of $Si_6O_{18}^{12-}$ in beryl.
7. (a) Network, (b) sheet, (c) sheet, (d) network.
9. Sodium oxide, Na_2O; calcium oxide, CaO; and silica, SiO_2.   **11.** (a) Like a crystalline solid, glass is hard and does not flow to any perceptible extent at room temperature,

but (b) like a liquid its viscosity increases as the temperature decreases.   **13.** See page 991.

15.

$$:\ddot{Cl}::\ddot{Cl}: \quad\quad :\ddot{Cl}: \quad :\ddot{Cl}:$$
$$:\ddot{Cl}-\overset{|}{\underset{|}{Si}}-\overset{|}{\underset{|}{Si}}-\ddot{Cl}: \quad :\ddot{Cl}-\overset{|}{\underset{|}{Si}}-\ddot{O}-\overset{|}{\underset{|}{Si}}-\ddot{Cl}:$$
$$:\ddot{Cl}::\ddot{Cl}: \quad\quad :\ddot{Cl}: \quad :\ddot{Cl}:$$

Si_2Cl_6O is angular at the O atom and has a dipole moment.
17. Ketones are the analogs of the silicones. Whereas silicones are polymers linked by Si—O—Si bonds, ketones contain C=O groups. C in period 2 has a greater tendency to form double bonds than does Si in period 3.
19. Si_nH_{2n+2}; silanes are more reactive than alkanes because Si in period 3 has empty valence-shell d orbitals to use in bond making, while C in period 2 has no valence-shell d orbitals.   **21.** A is trimethylchlorosilane, while B is hexamethyldisiloxane:

A   B

23. (a) $2.85 \times 10^{-3}M$; (b) 5.59×10^{-2} g, $2.45 \times 10^{-2}M$, pH 1.3.   **25.** BE(B $\cdots$ H) = 209 kJ mol^{-1} $\ll$ BE(B—H) = 389 kJ mol^{-1}; while the B $\cdots$ H bridge bond has a bond order of $\frac{1}{2}$ the B—H terminal bond has a bond order of 1.
27. A Lewis acid is an electron pair acceptor; a Lewis base is an electron pair donor: $BF_3 + NH_3 \longrightarrow$
$F_3\overset{+}{B}-\overset{-}{N}H_3$; $BF_3 + F^- \longrightarrow BF_4^-$;
$B_2H_6 + 2H^- \longrightarrow 2BH_4^-$.

29.
$$H-\overset{\displaystyle H}{\underset{\displaystyle H}{\overset{|}{\underset{|}{B^-}}}}-H \quad AX_4, \text{ tetrahedral}; \quad ^-:\ddot{O}-\overset{:\ddot{O}:^-}{\underset{}{B}}-\ddot{O}-\overset{:\ddot{O}:^-}{\underset{}{B}}-\ddot{O}:^- \text{ each}$$

B atom has AX_3, trigonal planar geometry, while the central O atom is AX_2E_2, angular, so that the ion is overall planar.

$$\begin{matrix} & :\ddot{O}:^- & \\ & \overset{|}{B} & \\ :\ddot{O} & & \ddot{O}: \\ ^-:\ddot{O}-B & & B-\ddot{O}:^- \\ & \ddot{O}. & \end{matrix} \quad \text{each B is } AX_3, \text{ trigonal planar, and}$$

the ring O atoms are AX_2E_2, planar, so that the ion is overall planar.   **31.** BN is isoelectronic with C_2. In the graphite-like form there are linked B_3N_3 hexagons like the C_6 hexagons in graphite, so that each B atom is formally bonded by two single B—N bonds and a double B=N bond, for an AX_3 planar arrangement of bonds, and a BN bond order of $1\frac{1}{3}$, and the linked hexagons form a sheet structure similar to that of graphite. The high temperature and high pressure form resembles diamond in that each B atom, and each N

A-42

atom, has AX_4 tetrahedral geometry and it contains all B—N single bonds. In each form, there is a formal -1 charge on each B atom and a formal $+1$ charge on each N atom.

33. The empirical and molecular formulas are both $BO_3C_6H_{15}$. Since the reaction is that between an acid and an alcohol, the product is an *ester*, $B(OC_2H_5)_3$, with the

Lewis structure

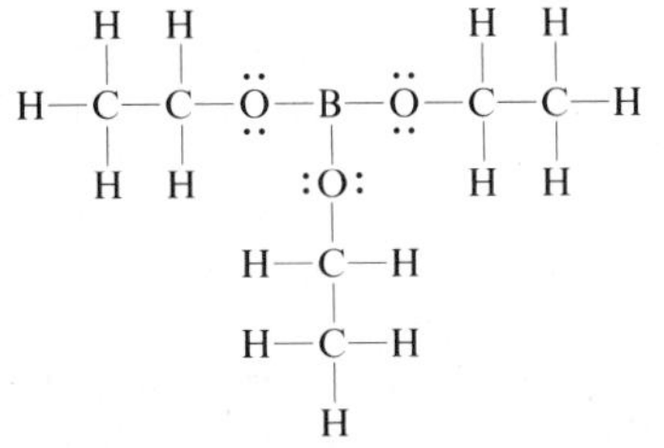

35. (a) $CaCO_3(s)$, (b) $CaSO_4 \cdot 2H_2O(s)$, (c) $SiO_2(s)$, (d) $Al_2O_3 \cdot xH_2O(s)$, (e) $Mg_3(Si_2O_5)_2(OH)_2(s)$, (f) $Be_3Al_2Si_6O_{18}$.   **37.** A three-center bond is one in which three nuclei are bonded together by one electron pair. The bridging BHB bonds in boranes such as diborane, B_2H_6, are examples of three-center bonds.

39. $B_2H_6(g) + 6H_2O(l) \longrightarrow 2B(OH)_3(aq) + 6H_2(g)$; pH = 5.09.

CHAPTER 23

1. (a) 4, (b) 3, (c) 8, (d) 5, (e) 6.

3. (a) Alkane, CH_4,

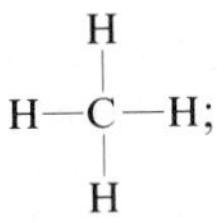

(b) alkene, C_2H_4,

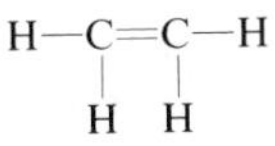

(c) alkyne, C_3H_4,

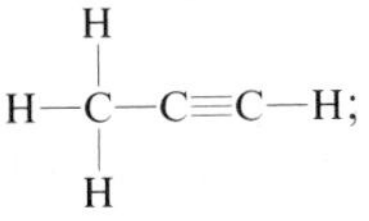

(d) alkane, C_4H_8,

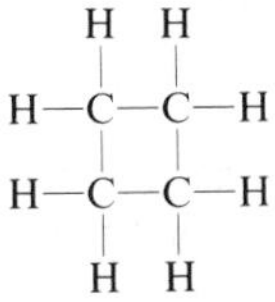

(e) alkene, C_3H_4,

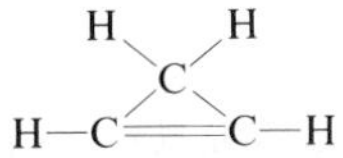

5.

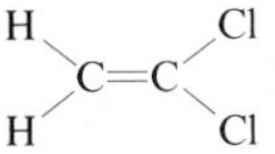

1,1-Dichloroethene;

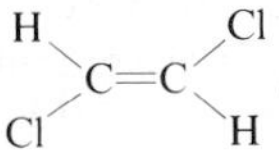

trans-1,2-Dichloroethene

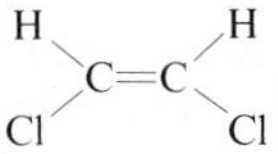

cis-1,2-Dichloroethene

7. $CH_3-CH_2-CH_2-CH_2-CH_2-CH_2-CH_3$
Heptane

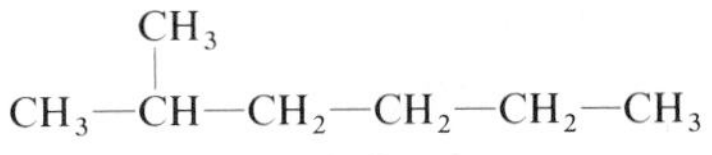

2-Methylhexane

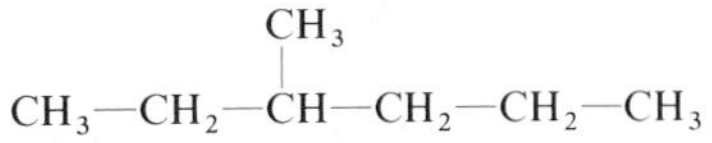

3-methylhexane

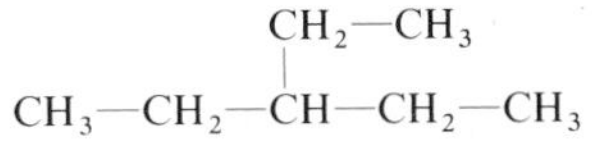

3-Ethylpentane

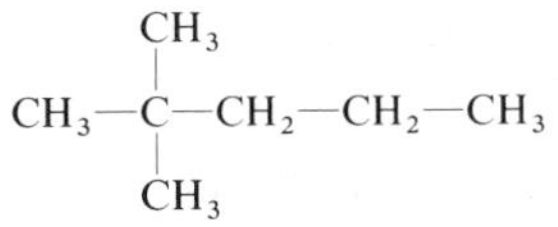

2,2-Dimethylpentane 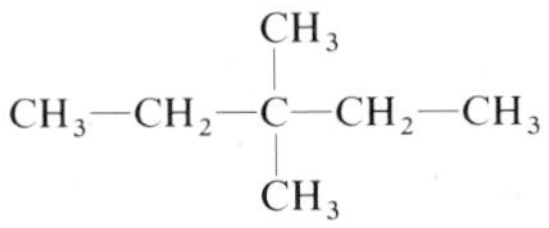   3,3-Dimethylpentane

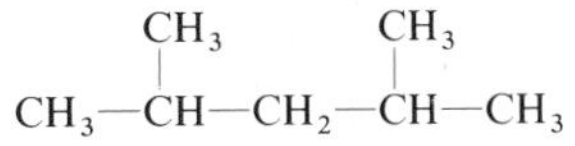

2,4-Dimethylpentane

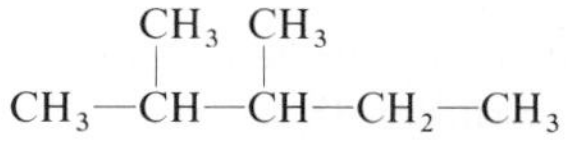

2,3-Dimethylpentane 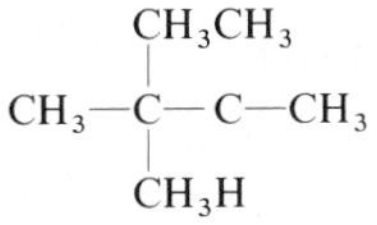   2,2,3-Trimethylbutane

9. (a)

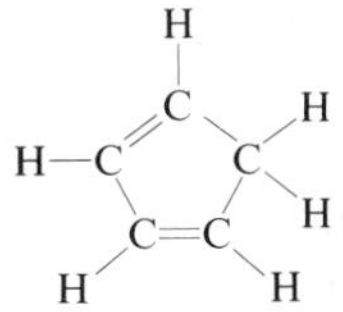

1,3-Cyclopentadiene   1,3-Cyclohexadiene

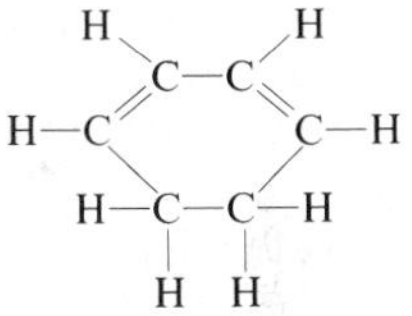

(c)

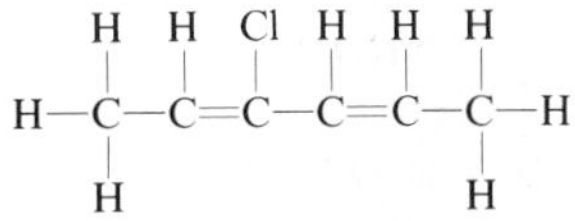

3-Chloro-3,4-hexadiene

11. (a) 2-Methylpropane, (b) 4(2-methylethyl)octane, (c) propene, (d) 2-methyl-2-butene, (e) 2-heptene, (f) 1,1,2,3-tetramethylcyclobutane, (g) 1,4-hexadiene.

13. 2,3-dimethylbutane.

15. (a)

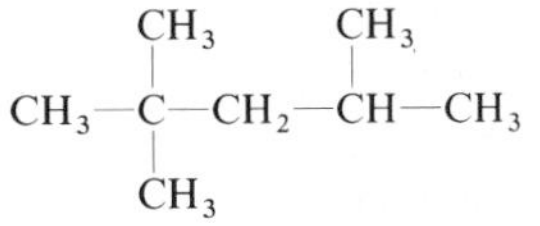

(b)

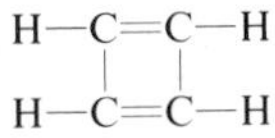

(c) 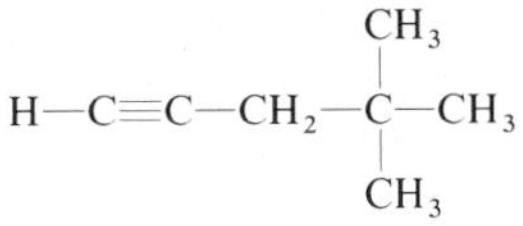 (d)

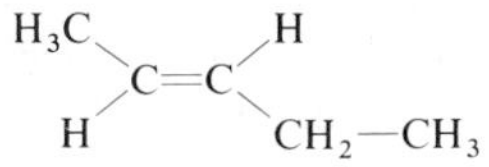

(e) $CH_2=CH-CH=CH_2$ (f)

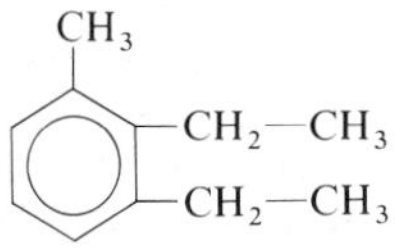

(g) [structure: benzene ring with Cl at positions 1,3,5] (h) [structure: alkene]

$$H_3C \diagup C=C \diagdown \begin{matrix} H \\ Cl \\ CH-CH_2-CH_2-CH_3 \\ | \\ Cl \end{matrix}$$

17. Empirical formula C_2H_5; molecular formula C_4H_{10}: $CH_3CH_2CH_2CH_3$, butane; $CH_3-CH-CH_3$, 2-methylpropane
with CH_3 branch

19. (a) $-C\diagup{}^O_H$, aldehyde; (b) $-C\diagup{}^O_{CH_3}$, ketone;

(c) $-C\diagup{}^O_{OH}$, carboxylic acid; (d) $-C\diagup{}^O_{NH_2}$, amide;

(e) $-NH_2$, primary amine; (f) $-OH$, secondary alcohol;

(g) $-C\diagup{}^O_{OH}$, carboxylic acid, and $-NH_2$,

amine(α-amino acid); (h) $-OH$, phenol.

21. (a) Methanol, (b) ethanol, (c) methanoic (formic) acid, (d) 2-butanol, (e) propanal, (f) 1,2-ethanediol, (g) propanone, (h) 2-methylpropanal, (i) 1-chloro-2-bromopropane.
23. (a) $CH_3CH_2CH_2CHO$, (b) $CH_3COCH_2CH_2CH_3$, (c) $CH_3COCH(CH_3)_2$, (d) $CH_3CH_2CH_2C(CH_3)_2CH_2CHO$, (e) CH_3COCH_3, (f) H_2CO.   **25.** (a) Tertiary, 2-methyl-2-butanol; (b) secondary, 2-butanol; (c) primary, 2,2-dimethylpropanol; (d) secondary, 2-propanol; (e) secondary, 3-pentanol.   **27.** (a) The products of complete combustion are CO_2 and water:
$CH_4(g) + 2O_2(g) \longrightarrow CO_2(g) + 2H_2O(g)$
(b) $C_2H_6(g) \longrightarrow C_2H_4(g) + H_2(g)$
(c) $C_2H_5OH(l) \longrightarrow C_2H_4(g) + H_2O(l)$
(d) $C_2H_4(g) + Br_2(g) \longrightarrow C_2H_4Br_2$
29. (a) $C_2H_5OH + OH^- \longrightarrow$
$C_2H_5O^- + H_2O$; ethoxide ion
(b) $C_2H_5OH + H_3O^+ \longrightarrow$
$C_2H_5OH_2^+ + H_2O$; ethanonium ion
(c) $2C_2H_5OH + 2Na \longrightarrow$
$2C_2H_5O^-Na^+ + H_2$; sodium ethoxide
(d) $C_2H_5OH + H_2SO_4 \longrightarrow$
$C_2H_4 + H_3O^+ + HSO_4^-$; ethene
31. (a) $C_6H_{10} + Br_2 \longrightarrow$
$C_6H_{10}Br_2$: 1,2-dibromocyclohexane
(b) $C_3H_8 + 5O_2 \xrightarrow{\text{Heat}} 3CO_2 + 4H_2O$;
carbon dioxide and water
(c) $C_2H_5OH \xrightarrow[\text{Catalyst}]{\text{Heat}} CH_3CHO + H_2$;
ethanal and hydrogen
(d) $CH_3OH + NH_3 \xrightarrow{\text{Heat}} CH_3NH_2 + H_2O$;
methylamine and water
33. (a) Ethanol, CH_3CH_2OH; (b) 2-propanol, $CH_3CH(OH)CH_3$; (c) 2-methyl-1-propanol,

$(CH_3)_2CHCH_2OH$; (d) 2-pentanol, $CH_3CH(OH)CH_2CH_2CH_3$.

35. (a) $CH_4(g) + H_2O(g) \xrightarrow[\text{Ni catalyst}]{700-800\ ^\circ C} CO(g) + 3H_2(g)$

(b) $CO(g) + 2H_2(g) \xrightarrow[\text{Al}_2\text{O}_3\ \text{catalyst}]{240-260\ ^\circ C,\ 50-100\ \text{atm}} CH_3OH(g)$

(c) $CH_3OH(g) \xrightarrow[\text{Cu catalyst}]{550-600\ ^\circ C} H_2CO(g) + H_2(g)$

(d) $CH_3OH(g) + NH_3(g) \xrightarrow[\text{Al}_2\text{O}_3\ \text{catalyst}]{400\ ^\circ C}$
$CH_3NH_2(g) + H_2O(g)$
37. (a) $3CH_3CH_2CHO + K_2Cr_2O_7 + 8H_3O^+ \xrightarrow{\text{Heat}}$
$3CH_3CH_2CO_2H + 2Cr^{3+} + 2K^+ + 12H_2O$
(b) $CH_3CH_2CO_2H + (CH_3)_2CHOH \longrightarrow$
$CH_3CH_2CO_2CH(CH_3)_2 + H_2O$
(c) $CH_3OH + NH_3 \xrightarrow{\text{Heat}} CH_3NH_2 + H_2O$
(d) $CH_3NH_2 + H_2O \rightleftharpoons CH_3NH_3^+ + OH^-$
(e) $CH_3NH_2 + HCl \longrightarrow CH_3NH_3^+Cl^-$
(f) $CH_3CO_2H + C_2H_5NH_2 \longrightarrow$
$CH_3CON(H)CH_2CH_3 + H_2O$
39. $CH_3OH + HCO_2H \longrightarrow CH_3OC(O)H + H_2O$
41. $HC\equiv CH + HCl \longrightarrow H_2C=CHCl$
43. (a) $CH_3CH=CH_2$, propene, and H_2O.
(b) $CH_3C\equiv CCH_3$, 2-butyne, and Br_2.
(c) $CH_3C\equiv CCH_3$, 2-butyne, and H_2. (d) $CH_3C\equiv CH$, propyne, and HBr.

45. (a) [structure AX_4 Tetrahedral] (b) [structure AX_2E_2 Angular] (c) [structure AX_3 Trigonal planar]

AX_4 — Tetrahedral AX_2E_2 — Angular AX_3 — Trigonal planar

(d) [structure AX_3 Trigonal planar] (e) [structure AX_3 Trigonal planar] (f) [structure AX_3 Trigonal planar]

AX_3 Trigonal planar AX_3 Trigonal planar AX_3 Trigonal planar

(g) [structure AX_3E Trigonal pyramidal] (h) [structure AX_4 Tetrahedral] (i) [structure AX_3E Trigonal pyramidal]

AX_3E Trigonal pyramidal AX_4 Tetrahedral AX_3E Trigonal pyramidal

47. Conformations are the different arrangements of the atoms or groups in a molecule that are possible as a consequence of rotation about single bonds. (a) Staggered and eclipsed (see Figure 23.4). (b) Boat and chair (see Figure 23.6).
49. (a) $^-{:}C\equiv O{:}^+$

(b) [resonance structures of carbonate ion]

(c) $H-C\overset{\displaystyle \overset{..}{\overset{..}{O}}:}{\underset{:\overset{..}{O}:^-}{}}$ $\longleftrightarrow$ $H-C\overset{:\overset{..}{O}:}{\underset{\overset{..}{O}:}{}}$ (d) $H-C\overset{\displaystyle \overset{..}{O}:}{\underset{:\overset{..}{O}-H}{}}$

(e) $H-C\overset{\displaystyle \overset{..}{O}:}{\underset{\overset{\displaystyle |}{\underset{H}{\overset{..}{N}-H}}}{}}$ (f) [benzene resonance structures]

51. Ethanol, CH_3CH_2OH, but not dimethyl ether, CH_3OCH_3, is capable of forming intermolecular hydrogen bonds. On reacting with sodium metal, dimethyl ether is inert while ethanol reacts to give $H_2(g)$: $2C_2H_5OH(l) + 2Na(s) \longrightarrow 2C_2H_5ONa + H_2(g)$. **53.** (a) Nonane, 34.4 kJ mL^{-1}; ethanol, 23.4 kJ mL^{-1}. (b) Nonane, 47.7 kJ g^{-1}; ethanol, 29.7 kJ g^{-1}. **55.** C_2H_5, C_4H_{10}, and there are two possible isomers, $CH_3CH_2CH_2CH_3$, butane, and $(CH_3)_2CHCH_3$, 2-methylpropane. **57.** C_4H_8, with the following possible structures: $CH_2{=}CH{-}CH_2{-}CH_3$, butene; $CH_3{-}\underset{H}{\overset{}{C}}{=}\underset{H}{\overset{}{C}}{-}CH_3$,

2-butene (*cis* or *trans*); $CH_2{=}C(CH_3)_2$, methylpropene;

$\underset{H_2C-CH_2}{\overset{H_2C-CH_2}{\big|\quad\big|}}$ cyclobutane; $\underset{H_2C\diagup\ \diagdown CH_2}{\overset{H_3C\diagdown\ \diagup H}{C}}$ methylcyclopropane.

59. The alkene is C_2H_4, ethene; mass % $CH_4 = 36.4$, mass % $C_2H_4 = 63.6$%. **61.** C_2H_4O, ethanal, with the structure CH_3CHO; to confirm that it is an aldehyde, see Experiment 23.5, for example. **63.** Molar mass 113 g mol^{-1}, $C_3H_6Cl_2$, with the possible structures $CH_2ClCHClCH_3$, $CH_3CCl_2CH_3$, and $CHCl_2CH_2CH_3$.
65. (a) $CH_3CH_2CH_2CO_2H$, (b) $CH_3CH_2CH_2CHO$, $CH_3CH_2CO_2H$, (c) $CH_3CH_2CH[CH_3]CHO$; $CH_3CH_2CH[CH_3]CO_2H$ (d) $(CH_3)_2CHCOCH_3$, (e) $(CH_3)_3CCH_2CO_2H$.

CHAPTER 24

2. (a) Polyvinyl chloride, PVC; (b) styrene; (c) teflon.
4. (a) $\{CH_2{-}CH(CH_3)\}_n$;

(b) $\{CH_2{-}CH{=}CH{-}CH_2\}_n$;

(c) $\{\underset{H}{\overset{}{N}}{-}\underset{O}{\overset{}{C}}{-}(CH_2)_4{-}\underset{O}{\overset{}{C}}{-}\underset{H}{\overset{}{N}}{-}(CH_2)_6\}_n$

(d) $\{O{-}CH_2{-}CH_2{-}O{-}\underset{O}{\overset{}{C}}{-}\langle\text{benzene}\rangle{-}\underset{O}{\overset{}{C}}\}_n$

6. $CH_2{=}CH(CN)$, acrylonitrile. **8.** H_3O^+ from the acid protonates the NH groups of nylon and then water attacks the C—N bonds, according to the equation

$-\underset{O}{\overset{H}{\overset{|}{C}}}{-}\underset{H}{\overset{H}{\overset{|}{N^+}}}{-}\ +\ H_2O \longrightarrow -\underset{O}{\overset{}{C}}{-}OH + H_3\overset{+}{N}-$

10. (a) London forces; (b) London forces, dipole–induced dipole, and dipole–dipole forces; (c) all the intermolecular forces described in Chapter 12. **12.** The two forms have different orientations of the —OH groups on C_1 of the ring.

14. (a) Their empirical formulas can be written as $C_x(H_2O)_y$ and they were originally thought to be hydrates of carbon. However, their molecular structures contain no water moleules, so the name is misleading.
(b) (i) Glucose or fructose, (ii) sucrose, (iii) starch, cellulose. (c) Glucose contains a six-membered ring of 5 carbon atoms and 1 oxygen atom; fructose contains a five-membered ring of 4 carbon atoms and 1 oxygen atom.
16. 1580 residues. **18.** The hydrogen-bonded distance in adenine–thymine is 1.1 nm, which is exactly the same distance as the hydrogen-bonded distance in guanine–cytosine. None of the four other possible combinations of these nitrogen bases gives this hydrogen-bonded distance, which is crucial to the separation of the strands in the DNA double helix. **21.** 22.6 g. **23.** (a) Methyl ester, (b) amide, (c) hydroxo groups, (d) carboxylate, (e) hydroxy groups and an ether link in the ring, (f) amino group and carboxylate group. **25.** Hydrogen bonds—specifically two hydrogen bonds between an adenine base of one strand and a thymine of the other and three hydrogen bonds between a guanine of one strand and a cytosine of the other.
27. (a) A protein molecule which catalyzes a specific metabolic reaction. (b) In their absence, most metabolic reactions, which are usually one step of a complex reaction, would proceed very slowly—at a rate which would not allow life to exist. (c) It lowers the activation energy. (d) See text. **29.** (a) 6.7×10^{-4} mol, (b) 9.38×10^{-3} g, (c) 18.8 mass % N (theoretical value 18.7%).

CHAPTER 25

1. (a) 3 protons, 3 neutrons; (b) 6 protons, 7 neutrons; (c) 40 protons, 54 neutrons; (d) 56 protons, 81 neutrons.
3. (a) Mass 4 u, charge $+2$; (b) mass 0 u, charge -1; (c) mass 0 u, charge $+1$. **5.** (a) The product isotope is located one group to the *left* in the periodic table; (b) the product isotope is located one group to the *left* in the periodic table. **7.** $Z = 92$. **9.** (a) $^{207}_{84}Po + ^{0}_{-1}e \longrightarrow ^{207}_{83}Bi$;
(b) $^{207}_{84}Po \longrightarrow ^{0}_{1}e + ^{207}_{83}Bi$; (c) $^{207}_{84}Po \longrightarrow$ $^{4}_{2}He + ^{203}_{82}Pb$. **11.** $^{235}_{92}U \longrightarrow ^{4}_{2}He + ^{231}_{90}Th$; $^{231}_{90}Th \longrightarrow ^{0}_{-1}e + ^{231}_{91}Pa$; $^{231}_{91}Pa \longrightarrow ^{4}_{2}He + ^{227}_{89}Ac$; $^{227}_{89}Ac \longrightarrow ^{0}_{-1}e + ^{227}_{90}Th$.
13. The missing products are (a) $^{32}_{15}P$, (b) $^{7}_{3}Li$, (c) $^{81}_{36}Kr$, (d) $^{4}_{2}He$, and (e) $^{99}_{38}Sr$. **15.** (a) $^{66}_{32}Ge \longrightarrow$ $^{0}_{1}e + ^{66}_{31}Ga$; (b) 1.56 g. **17.** (a) Decomposition of ^{131}I and ^{90}Sr to 10% in 26.6 days and 66.2 years, and (b) to 1% in 53.2 days and 132.3 years, respectively.
19. $t_{1/2} = 14.2$ days. **21.** 460 A.D.
23. 1.03×10^4 years. **25.** Binding energy is 4.78×10^{-11} J, or 1.36×10^{-12} J nucleon^{-1}. **27.** $^{2}_{1}H$, 1.78×10^{-13} J nucleon^{-1}; $^{16}_{8}O$, 1.28×10^{-12} J nucleon^{-1}; $^{56}_{26}Fe$, 1.41×10^{-12} J nucleon^{-1}; $^{125}_{52}Te$, 1.35×10^{-12} nucleon^{-1}; $^{235}_{92}U$, 1.21×10^{-12} J nucleon^{-1}.
29. ^{14}N, 1.20×10^{-12} J nucleon^{-1}; ^{15}N, 1.23×10^{-12} J nucleon^{-1}; ^{15}N is more stable; ^{14}N is more abundant.
31. Energy released in nuclear reaction is 5.84×10^8 kJ mol^{-1} D; energy released in combustion reaction is 1.43×10^2 kJ mol^{-1} D. **33.** For example, (a) $^{12}_{6}C$, (b) $^{11}_{5}B$, (c) $^{13}_{6}C$, (d) $^{14}_{7}N$.
37. $^{232}_{90}Th + ^{12}_{6}C \longrightarrow ^{240}_{96}Cm + 4^{1}_{0}n$.

INDEX

Boldface page numbers indicate end-of-chapter definitions (important terms); *t* indicates a table; *f* indicates a figure; *E* indicates an experiment; and *B* indicates a box.

The International System of Units (SI)

SI Base Units

PHYSICAL QUANTITY	NAME	SYMBOL
Length	meter	m
Mass	kilogram	kg
Time	second	s
Electric current	ampere	A
Thermodynamic temperature	kelvin	K
Amount of substance	mole	mol
Luminous intensity	candela	cd

SI Derived Units

PHYSICAL QUANTITY	NAME	SYMBOL	DEFINITION
Frequency	hertz	Hz	s^{-1}
Energy	joule	J	$kg\,m^2\,s^{-2}$
Force	newton	N	$kg\,m\,s^{-2} = J\,m^{-1}$
Power	watt	W	$kg\,m^2\,s^{-3} = J\,s^{-1}$
Pressure	pascal	Pa	$kg\,m^{-1}\,s^{-2} = N\,m^{-2} = J\,m^{-3}$
Electric charge	coulomb	C	$A\,s$
Electric potential difference	volt	V	$kg\,m^2\,s^{-3}\,A^{-1} = J\,A^{-1}\,s^{-1}$
Electric resistance	ohm	Ω	$kg\,m^2\,s^{-3}\,A^{-2} = V\,A^{-1}$

Physical Constants

CONSTANT	SYMBOL	VALUE
Atomic mass unit	u	$1.660\ 56 \times 10^{-27}\,kg$
Avogadro constant	N	$6.022\ 05 \times 10^{23}\,mol^{-1}$
Boltzmann constant	$k = R/N$	$1.380\ 66 \times 10^{-23}\,J\,K^{-1}$
Elementary charge	e	$1.602\ 19 \times 10^{-19}\,C$
Faraday constant	$F = Ne$	$9.648\ 46 \times 10^{4}\,C\,mol^{-1}$
Gas constant	R	$8.314\ 41\ J\,K^{-1}\,mol^{-1}$
		$0.082\ 06\ L\,atm\,K^{-1}\,mol^{-1}$
Mass of an electron	m_e	$9.109\ 53 \times 10^{-31}\,kg$
		$5.485\ 80 \times 10^{-4}\,u$
Mass of a neutron	m_n	$1.674\ 95 \times 10^{-27}\,kg$
		$1.008\ 66\,u$
Mass of a proton	m_p	$1.672\ 65 \times 10^{-27}\,kg$
		$1.007\ 28\,u$
The Planck constant	h	$6.626\ 18 \times 10^{-34}\,J\,s$
Speed of light	c	$2.997\ 924\ 6 \times 10^{8}\,m\,s^{-1}$